MW01329004

ELECTRONICS
A Complete Course

ELECTRONICS
A Complete Course

Second Edition

Nigel P. Cook

Upper Saddle River, New Jersey
Columbus, Ohio

Library of Congress Cataloging-in-Publication Data

Cook, Nigel P.
Electronics : a complete course / Nigel P. Cook—2nd ed.
p. cm.
Includes index.
ISBN 0-13-111066-7
1. Electronics—Textbooks. I. Title.

TK7816.C64 2004
621.381—dc21 2003051263

Editor in Chief: Stephen Helba
Acquisitions Editor: Dennis Williams
Development Editor: Kate Linsner
Production Editor: Rex Davidson
Production Coordinator: Holly Henjum, Carlisle Publishers Services
Design Coordinator: Diane Ernsberger
Cover Designer: Ali Mohrman
Production Manager: Pat Tonneman
Marketing Manager: Ben Leonard

This book was set in Times by Carlisle Communications, Ltd. It was printed and bound by R. R. Donnelley & Sons Company. The cover was printed by Phoenix Color Corp.

Pearson Education Ltd.
Pearson Education Singapore Pte. Ltd.
Pearson Education Canada, Ltd.
Pearson Education—Japan
Pearson Education Australia Pty. Limited
Pearson Education North Asia Ltd.
Pearson Educación de Mexico, S. A. de C.V.
Pearson Education Malaysia Pte. Ltd.

10 9 8 7 6 5 4
ISBN: 0-13-111066-7

ELECTRONICS
A Complete Course

Second Edition

Nigel P. Cook

Upper Saddle River, New Jersey
Columbus, Ohio

Library of Congress Cataloging-in-Publication Data

Cook, Nigel P.
Electronics : a complete course / Nigel P. Cook—2nd ed.
p. cm.
Includes index.
ISBN 0-13-111066-7
1. Electronics—Textbooks. I. Title.

TK7816.C64 2004
621.381—dc21 2003051263

Editor in Chief: Stephen Helba
Acquisitions Editor: Dennis Williams
Development Editor: Kate Linsner
Production Editor: Rex Davidson
Production Coordinator: Holly Henjum, Carlisle Publishers Services
Design Coordinator: Diane Ernsberger
Cover Designer: Ali Mohrman
Production Manager: Pat Tonneman
Marketing Manager: Ben Leonard

This book was set in Times by Carlisle Communications, Ltd. It was printed and bound by R. R. Donnelley & Sons Company. The cover was printed by Phoenix Color Corp.

Pearson Education Ltd.
Pearson Education Singapore Pte. Ltd.
Pearson Education Canada, Ltd.
Pearson Education—Japan
Pearson Education Australia Pty. Limited
Pearson Education North Asia Ltd.
Pearson Educación de Mexico, S. A. de C.V.
Pearson Education Malaysia Pte. Ltd.

10 9 8 7 6 5 4
ISBN: 0-13-111066-7

To Dawn, Candy, and Jon

Books by Nigel P. Cook

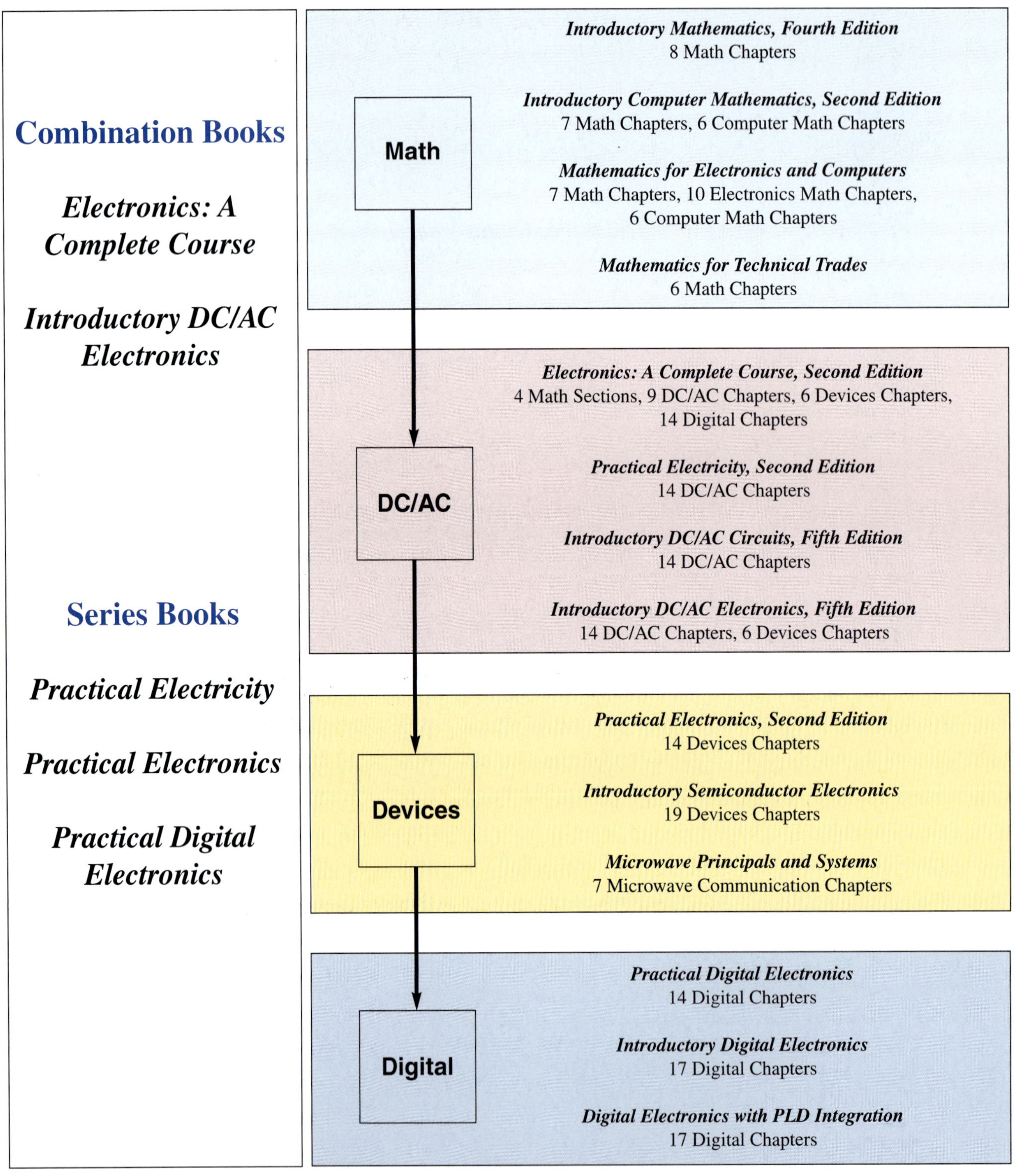

Combination Books
Electronics: A Complete Course
Introductory DC/AC Electronics
Series Books
Practical Electricity
Practical Electronics
Practical Digital Electronics
Math
Introductory Mathematics, Fourth Edition
8 Math Chapters
Introductory Computer Mathematics, Second Edition
7 Math Chapters, 6 Computer Math Chapters
Mathematics for Electronics and Computers
7 Math Chapters, 10 Electronics Math Chapters, 6 Computer Math Chapters
Mathematics for Technical Trades
6 Math Chapters
DC/AC
Electronics: A Complete Course, Second Edition
4 Math Sections, 9 DC/AC Chapters, 6 Devices Chapters, 14 Digital Chapters
Practical Electricity, Second Edition
14 DC/AC Chapters
Introductory DC/AC Circuits, Fifth Edition
14 DC/AC Chapters
Introductory DC/AC Electronics, Fifth Edition
14 DC/AC Chapters, 6 Devices Chapters
Devices
Practical Electronics, Second Edition
14 Devices Chapters
Introductory Semiconductor Electronics
19 Devices Chapters
Microwave Principals and Systems
7 Microwave Communication Chapters
Digital
Practical Digital Electronics
14 Digital Chapters
Introductory Digital Electronics
17 Digital Chapters
Digital Electronics with PLD Integration
17 Digital Chapters

For more information and a desk copy of any of these textbooks by Nigel P. Cook, call 1-800-228-7854, visit the Prentice Hall website at www.prenhall.com, or ask your local Prentice Hall representative.

Preface

INTRODUCTION

Since World War II, no branch of science has contributed more to the development of the modern world than *electronics.* It has stimulated dramatic advances in the fields of communication, computing, consumer products, industrial automation, test and measurement, and healthcare. It has now become the largest single industry in the world, exceeding the automobile and oil industries, with annual sales of electronic systems greater than $2 trillion. One of the most important trends in this huge industry has been a gradual shift from *analog electronics* to *digital electronics.* This movement began in the 1960s and is almost complete today. In fact, a recent statistic stated that, on average, 90% of the circuitry within electronic systems is now digital and only 10% is analog. This digitization of the electronics industry is merging sectors that were once separate. For example, two of the largest sectors or branches of electronics are *computing* and *communications.* Being able to communicate with each other using the common language of digital has enabled computers and communications to interlink, so that computers can now function within communication-based networks, and communications networks can now function through computer-based systems. Industry experts call this merging *convergence,* and predict that digital electronics will continue to unite the industry and stimulate progress in practically every field of human endeavor.

CONTENTS

With many programs compressing their courses in order to devote more time to the electronics system applications, such as communication, computing, industrial, and consumer, a strong need arises for a single electronics fundamentals text. Employing his now renowned student-friendly practical approach, *Electronics: A Complete Course* by Nigel Cook provides comprehensive coverage of the three electronics fundamental courses:

Part A: DC/AC Electronics
Part B: Semiconductor Devices
Part C: Digital Circuits

Part A **DC/AC Electronics**

Chapter 1 Current and Voltage
Chapter 2 Resistance and Power
Chapter 3 Series Circuits
Chapter 4 Parallel Circuits
Chapter 5 Series–Parallel Circuits
Chapter 6 Direct Current (DC) versus Alternating Current (AC)
Chapter 7 Capacitors

Chapter 8	Electromagnets, Inductors, and Transformers
Chapter 9	Resistive, Inductive, and Capacitive *(RLC)* Circuits
Part B	**Semiconductor Devices**
Chapter 10	Semiconductor Principles
Chapter 11	Diodes and Power Supply Circuits
Chapter 12	Bipolar Junction Transistors (BJTs)
Chapter 13	Field Effect Transistors (FETs)
Chapter 14	Thyristors and Transducers
Chapter 15	Op-Amps and Other Linear ICs
Part C	**Digital Circuits**
Chapter 16	Analog to Digital
Chapter 17	Number Systems and Codes
Chapter 18	Logic Gates
Chapter 19	Logic Circuit Simplification
Chapter 20	Standard Logic Devices (SLDs)
Chapter 21	Programmable Logic Devices (PLDs)
Chapter 22	Testing and Troubleshooting
Chapter 23	Combinational Logic Circuits
Chapter 24	Flip-Flops and Timers
Chapter 25	Sequential Logic Circuits
Chapter 26	Arithmetic Operations and Circuits
Chapter 27	Semiconductor Memory
Chapter 28	Analog and Digital Signal Converters
Chapter 29	Computer Hardware and Software

SUPPLEMENTS

The following ancillaries accompanying this text provide extensive opportunity for further study and support:

- **Electronics Workbench MultiSim® CD.** Packaged with each text, this CD includes the Enhanced Textbook Edition of Multisim 2001® by Electronics Workbench, along with more than 100 circuit files in Multisim simulated from circuits in the text. The Enhanced Textbook Edition of Multisim is all that is required to open and work with all of the circuit files on the CD. However, anyone who wishes to order the full version of Multisim may do so through Prentice Hall by visiting http://www.prenhall.com, phoning 1-800-282-0693, or sending a fax request to 1-800-835-5327.

  Technical questions relating specifically to Multisim can be referred to Electronics Workbench by writing to support@electronicsworkbench.com.
- **Laboratory Manual.** This important supplement is co-authored by Nigel Cook and Gary Lancaster and offers numerous experiments designed to translate all of the theory from the text into practical experimentation.
- **Instructor's Solutions Manual.** Solutions for text problems and the lab experiments are provided.
- **PowerPoint™ Transparencies.** All of the figures and tables appearing in the text are available as PowerPoint slides on CD.
- **TestGen,** a computerized test bank, is available to instructors who wish to create customized tests.
- **Companion Website.** Located at http://www.prenhall.com/cook, this site is an online study guide enabling students to review text material and test their understanding.

Contents

12
Bipolar Junction Transistors (BJTs) 422

13
Field Effect Transistors (FETs) 476

14
Thyristors and Transducers 518

15
Op-Amps and Other Linear ICs 540

PART C
Digital Circuits 592

16
Analog to Digital 592

17
Number Systems and Codes 600

18
Logic Gates 622

19
Logic Circuit Simplification 652

20
Standard Logic Devices (SLDs) 684

28
Analog and Digital Signal Converters 954

29
Computer Hardware and Software 980

Appendixes

Index 1033

Accessible Writing Style

3-1 COMPONENTS IN SERIES

Series Circuit
Circuit in which the components are connected end to end so that current has only one path to follow throughout the circuit.

Figure 3-1 illustrates five examples of **series** resistive **circuits.** In all five examples, you will notice that the resistors are connected "in-line" with one another so that the current through the first resistor must pass through the second resistor, and the current through the second resistor must pass through the third, and so on.

EXAMPLE:

In Figure 3-2(a), seven resistors are laid out on a table top. Using a protoboard, connect all the resistors in series, starting at R_1, and proceeding in numerical order through the resistors until reaching R_7. After completing the circuit, connect the series circuit to a dc power supply.

Solution:

In Figure 3-2(b), you can see that all the resistors are now connected in series (end-to-end), and the current has only one path to follow from negative to positive.

EXAMPLE:

Figure 3-3(a) shows four 1.5 V cells and three lamps. Using wires, connect all of the cells in series to create a 6 V battery source. Then connect all of the three lamps in series with one another, and finally, connect the 6 V battery source across the three-series-connected-lamp load.

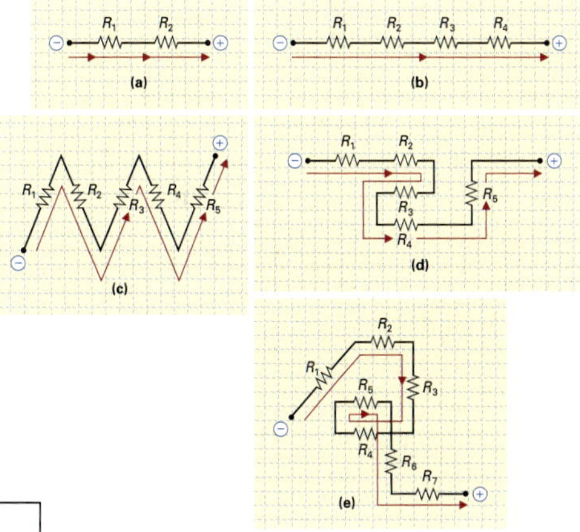

3-1 Five Series Resistive Circuits.

Descriptive Illustrations

FIGURE 28-2 Converting Information Signals.

For example, on the active edge of strobe pulse 1, the digital code 010 (decimal 2) is being applied to the DAC, so it will generate 2 V at its output. On the active edge of strobe pulse 2, the digital code 100 (decimal 4) is being applied to the DAC, so it will generate 4 V at its output. On the active edge of strobe pulse 3, the digital code 110 (decimal 6) is being applied to the DAC, so it will generate 6 V at its output. On the active edge of strobe pulse 4, the digital code 111 (decimal 7) is being applied to the DAC, so it will generate 7 V at its output. On the active edge of strobe pulse 5, the digital code 110 (decimal 6) is being applied to the DAC, so it will generate 6 V at its output, and so on. If the output of a DAC is then applied to a lowpass filter, the discrete voltage steps can be blended into a smooth wave that closely approximates the original analog wave, as shown by the dashed line in Figure 28-2(b).

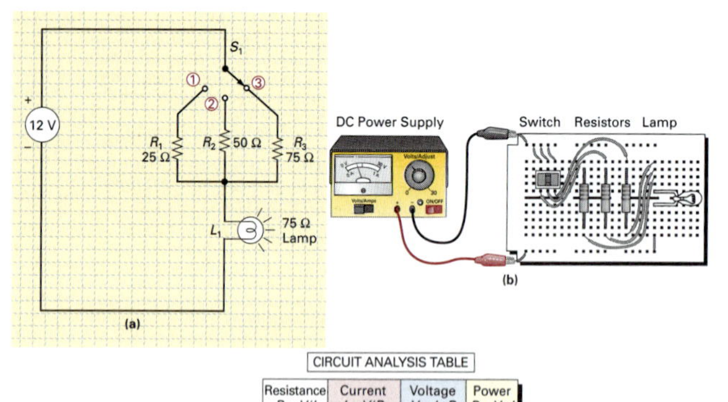

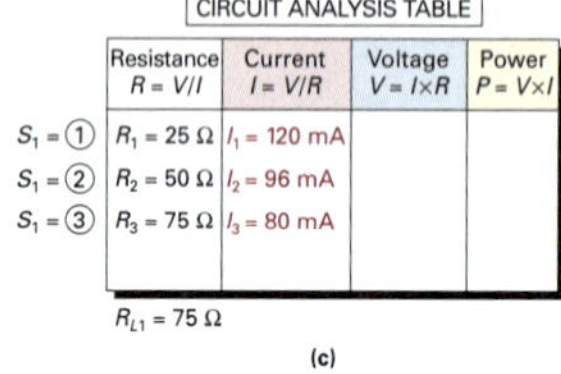

CIRCUIT ANALYSIS TABLE

	Resistance $R = V/I$	Current $I = V/R$	Voltage $V = I \times R$	Power $P = V \times I$
S_1 = ①	$R_1 = 25\ \Omega$	$I_1 = 120$ mA		
S_1 = ②	$R_2 = 50\ \Omega$	$I_2 = 96$ mA		
S_1 = ③	$R_3 = 75\ \Omega$	$I_3 = 80$ mA		

$R_{L1} = 75\ \Omega$

(c)

FIGURE 3-8 **Three-Position Switch Controlling Lamp Brightness. (a) Schematic. (b) Protoboard Circuit. (c) Circuit Analysis Table.**

EWB

The details of this circuit are summarized in the analysis table shown in Figure 3-8(c).

SELF-TEST EVALUATION POINT FOR SECTION 3-3

Use the following questions to test your understanding of Section 3-3.

1. State the total resistance formula for a series circuit.
2. Calculate R_T if $R_1 = 2\ \text{k}\Omega$, $R_2 = 3\ \text{k}\Omega$, and $R_3 = 4700\ \Omega$.

3-4 VOLTAGE IN A SERIES CIRCUIT

A potential difference or voltage drop will occur across each resistor in a series circuit when current is flowing. The amount of voltage drop is dependent on the value of the resistor and the amount of current flow. This idea of potential difference or voltage drop is best ex by returning to the water analogy. In Figure 3-9(a), you can see that the high pressure pump's outlet is present on the left side of the valve. On the right side of the valve, h

SECTION 3-4 / VOLTAGE IN A SERIES

Companion Website and Electronics Workbench

Introduction to Electronics

Introduction to Electronics

Your Course in Electronics

Your future in the electronics industry begins with this text. To give you an idea of where you are going and what we will be covering, Figure I-1 acts as a sort of road map, breaking up your study of electronics into four basic steps.

Step 1: Basics of Electricity
Step 2: Electronic Components
Step 3: Electronic Circuits
Step 4: Electronic Systems

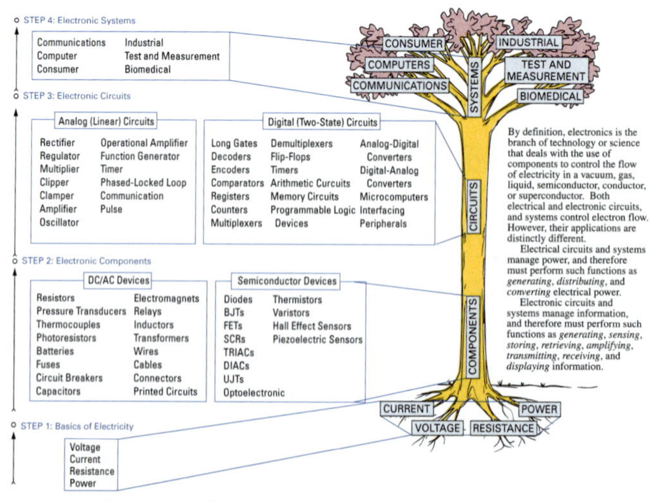

By definition, electronics is the branch of technology or science that deals with the use of components to control the flow of electricity in a vacuum, gas, liquid, semiconductor, conductor, or superconductor. Both electrical and electronic circuits, and systems control electron flow. However, their applications are distinctly different.

Electrical circuits and systems manage power, and therefore must perform such functions as *generating, distributing,* and *converting* electrical power.

Electronic circuits and systems manage information, and therefore must perform such functions as *generating, sensing, storing, retrieving, amplifying, transmitting, receiving,* and *displaying* information.

FIGURE I-1 ***The Steps Involved in Studying Electronics.***

xxiii

History Vignettes and Margin Timeline Illustrations

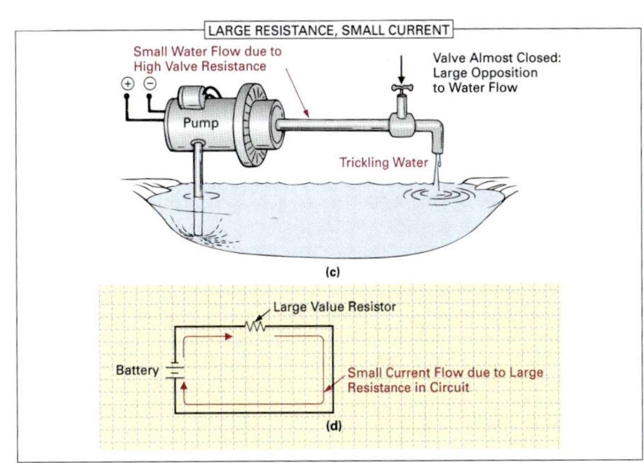

FIGURE 2-1 (continued) **The Opposite Relationship between Resistance and Current.**

TIME LINE

Ohm's law, the best known law in electrical circuits, was formulated by Georg S. Ohm (1787–1854), a German physicist. His law was so coldly received that his feelings were hurt and he resigned his teaching post. When his law was finally recognized, he was reinstated. In honor of his accomplishments, the unit of resistance is called the ohm.

low resistance or small opposition will allow a large amount of current to flow through the conductor, as illustrated by the thick current line.

In the fluid circuit in Figure 2-1(c), the valve is almost completely closed, resulting in a high resistance or opposition to water flow, so only a trickle of water passes through the pipe and out from the outlet. In the electrical circuit in Figure 2-1(d), a large value resistor has been placed in the circuit, causing a large resistance to the passage of current. This high resistance allows only a small amount of current to flow through the connecting wire, as illustrated by the thin current line.

In the previous examples of low resistance and high resistance shown in Figure 2-1, you may have noticed that resistance and current are inversely proportional ($1/\propto$) to one another. This means that if resistance is high the current is low, and if resistance is low, the current is high.

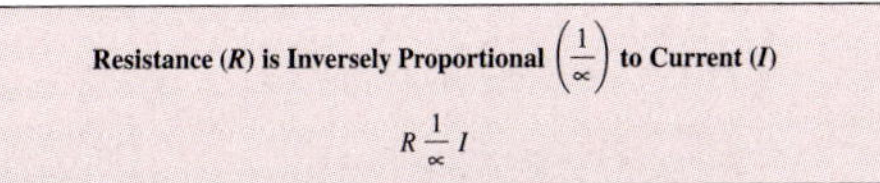

Resistance (R) is Inversely Proportional (1/∝) to Current (I)
R 1/∝ I

This relationship between resistance and current means that if resistance is increased by some value, current will be decreased by the same value (assuming a constant electrical pressure or voltage). For example, if resistance is doubled current is halved and similarly, if resistance is halved current is doubled.

$$R\uparrow \frac{1}{\propto} I\downarrow \qquad R\downarrow \frac{1}{\propto} I\uparrow$$

In Figure 2-1, the fluid analogy was used alongside the electrical circuit to help you under-
d the idea of low resistance and high resistance. In the following section, we will examine re-
nce in more detail, and define clearly exactly how much resistance exists within a circuit.

PTER 2 / RESISTANCE AND POWER

Self-Test Evaluation Points

FIGURE 15-24 **Clock Oscillators. (a) *RC* Controlled. (b) Crystal Controlled.**

Figure 15-24(b) shows how the same basic circuit configuration, but in this instance with a crystal acting as the frequency determining component, can be connected to generate a highly accurate 2 MHz clock signal.

SELF-TEST EVALUATION POINT FOR SECTION 15-3

Use the following questions to test your understanding of Section 15-3.

1. The astable multivibrator is also called the ____________ multivibrator.
2. Which of the multivibrator circuits has only one stable state?
3. Which multivibrator is also called a set-reset flip-flop?
4. Why is the set-reset flip-flop also called a latch?
5. The 555 timer derived its number identification from ____________.
6. List two applications of the 555 timer.
7. What is the key advantage of the crystal oscillator?

15-4 FUNCTION GENERATORS

A function generator is by definition a signal generator circuit that delivers a variety of different output waveforms whose frequency can be varied. In this section, we will examine some of the different linear function generator ICs available.

15-4-1 *Voltage Controlled Oscillators (VCOs)*

Voltage Controlled Oscillator (VCO) An oscillator whose frequency can be controlled by an input control voltage.

Control Voltage A voltage signal that starts, stops, or adjusts the operation of a device, circuit, or system.

In many communication circuit applications, we need an oscillator whose frequency can be controlled by an input "control voltage." The circuit that achieves this function is the **voltage controlled oscillator (VCO)** or, as it is sometimes called, a "voltage-to-frequency converter." The NE/SE 566 shown in Figure 15-25 is a good example of a linear IC VCO which can be connected to generate both a square wave and triangular wave output. The frequency of oscillation is determined by external resistors R_1 and R_2 and capacitor C_1, which determine the **control voltage** applied to pin 5. The triangular wave is generated by linearly charging and discharging the external capacitor C_1, using the 566's internal current source. The charge

582 CHAPTER 15 / OP-AMPS AND OTHER LINEAR ICS

$$P_7 = \frac{V_{R7}^2}{R_7} = \frac{(36\text{ V})^2}{40\ \Omega} = 32.4\text{ W}$$

$$\begin{aligned} P_T &= P_1 + P_2 + P_3 + P_4 + P_5 + P_6 + P_7 \\ &= 90 + 27 + 9 + 18 + 21.6 + 54 + 32.4 \\ &= 252\text{ W} \end{aligned}$$

or

$$\begin{aligned} P_T &= \frac{V_T^2}{R_T} = \frac{(84\text{ V})^2}{28\ \Omega} \\ &= 252\text{ W} \end{aligned}$$

The total power dissipated in this example circuit is 252 W. All the information can now be inserted in a final diagram for the example, as shown in Figure 5-18.

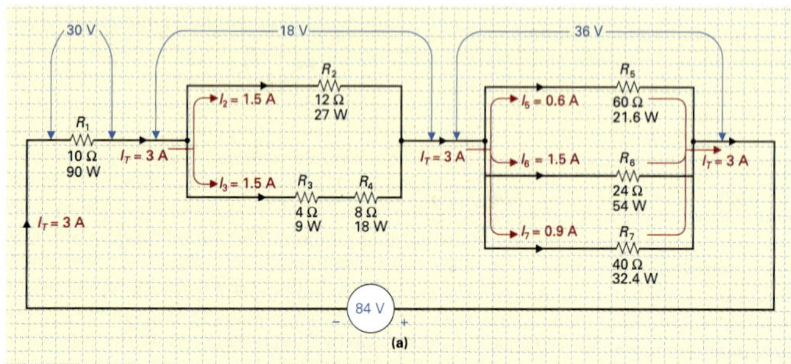

CIRCUIT ANALYSIS TABLE

Resistance $R = V/I$	Voltage $V = I \times R$	Current $I = V/R$	Power $P = V \times I$
R_1 = 10 Ω	V_{R1} = 30 V	I_1 = 3 A	P_1 = 90 W
R_2 = 12 Ω	V_{R2} = 18 V	I_2 = 1.5 A	P_2 = 27 W
R_3 = 4 Ω	V_{R3} = 6 V	I_3 = 1.5 A	P_3 = 9 W
R_4 = 8 Ω	V_{R4} = 12 V	I_4 = 1.5 A	P_4 = 18 W
R_5 = 60 Ω	V_{R5} = 36 V	I_5 = 0.6 A	P_5 = 21.6 W
R_6 = 24 Ω	V_{R6} = 36 V	I_6 = 1.5 A	P_6 = 54 W
R_7 = 40 Ω	V_{R7} = 36 V	I_7 = 0.9 A	P_7 = 32.4 W
R_T = 28 Ω	$V_T = V_S$ = 84 V	I_T = 3 A	P_T = 252 W

(b)

FIGURE 5-18 Series–Parallel Circuit Example with All Information Inserted. (a) Sc (b) Circuit Analysis Table.

SECTION 5-5 / POWER IN A SERIES–PARALLE

Circuit Analysis Tables

Protoboard Pictorials

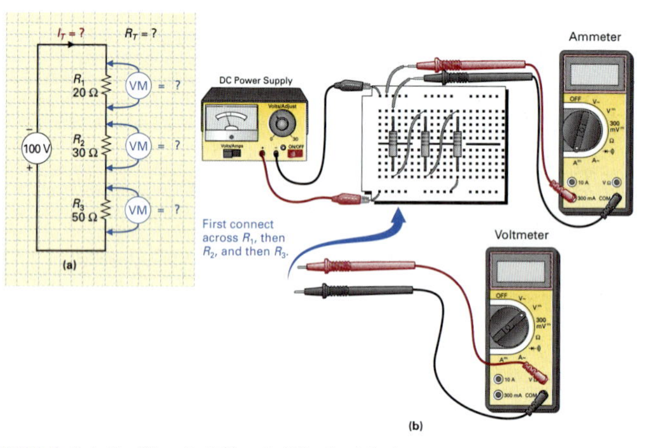

FIGURE 3-10 Series Circuit Example. (a) Schematic. (b) Protoboard Circuit.

The same current will flow through the complete series circuit, so the current through R_1 will equal 1 A, the current through R_2 will equal 1 A, and the current through R_3 will equal 1 A.

c. Voltage across R_1 (V_{R1}) $= I_1 \times R_1$
$= 1\text{ A} \times 20\ \Omega$
$= 20\text{ V}$

d. Voltage across R_2 (V_{R2}) $= I_2 \times R_2$
$= 1\text{ A} \times 30\ \Omega$
$= 30\text{ V}$

e. Voltage across R_3 (V_{R3}) $= I_3 \times R_3$
$= 1\text{ A} \times 50\ \Omega$
$= 50\text{ V}$

Figure 3-11(a) shows the schematic and Figure 3-11(b) shows the analysis table for this example, with all of the calculated data inserted. As you can see, the 20 Ω resistor drops 20 V, the 30 Ω resistor has 30 V across it, and the 50 Ω resistor has dropped 50 V. From this example, you will notice that the larger the resistor value, the larger the voltage drop. Resistance and voltage drops are consequently proportional to one another.

SECTION 3-4 / VOLTAGE IN A SERIES CIRCUIT 93

Practical Circuit Example

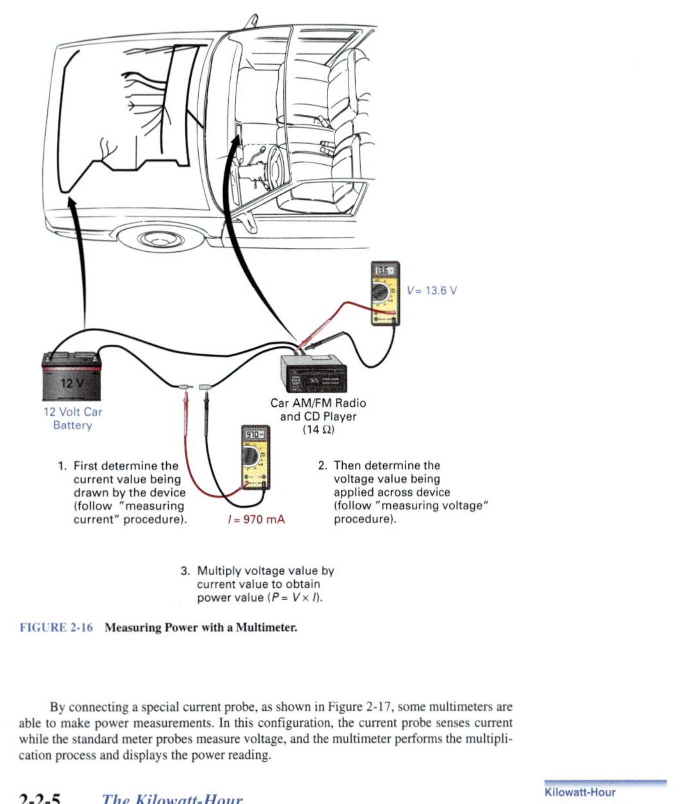

FIGURE 2-16 **Measuring Power with a Multimeter.**

By connecting a special current probe, as shown in Figure 2-17, some multimeters are able to make power measurements. In this configuration, the current probe senses current while the standard meter probes measure voltage, and the multimeter performs the multiplication process and displays the power reading.

2-2-5 *The Kilowatt-Hour*

You and I pay for our electric energy in a unit called the **kilowatt-hour** (kWh). The **kilowatt-hour meter**, shown in Figure 2-18, measures how many kilowatt-hours are consumed, and … accordingly. … Well, since power is the rate at which energy is used, if we … calculate how much energy has been consumed.

Kilowatt-Hour
1000 watts for 1 hour.

Kilowatt-Hour Meter
A meter used by electric companies to measure a customer's electric power use in kilowatt-hours.

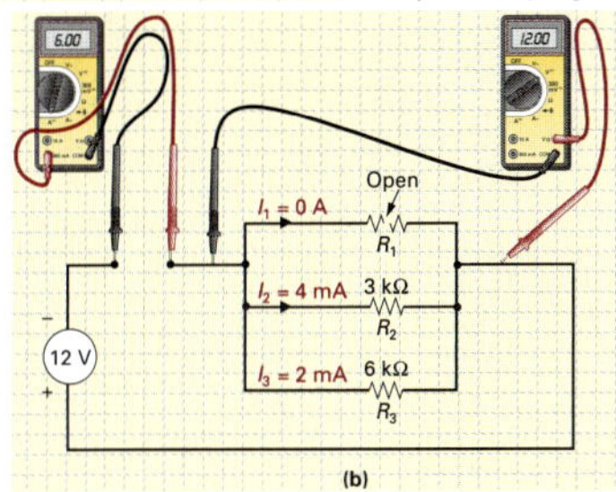

CIRCUIT ANALYSIS TABLE

Resistance $R = V/I$	Voltage $V = I \times R$	Current $I = V/R$	Power $P = V \times I$
R_1 = 12 kΩ	V_{R1} = 12 V	I_1 = 1 mA	P_1 = 12 mW
R_2 = 3 kΩ	V_{R2} = 12 V	I_2 = 4 mA	P_2 = 48 mW
R_3 = 6 kΩ	V_{R3} = 12 V	I_3 = 2 mA	P_3 = 24 mW
R_T = 1.7 kΩ	$V_T = V_S$ = 12 V	I_T = 7 mA	P_T = 84 mW

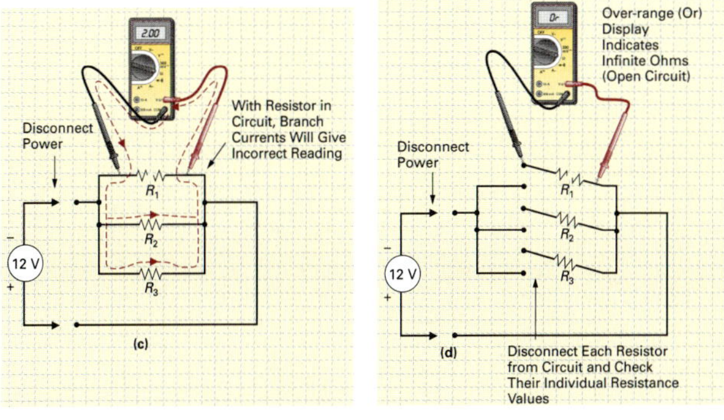

Disconnect Power

With Resistor in Circuit, Branch Currents Will Give Incorrect Reading

R_1 R_2 R_3 12 V

(c)

Over-range (Or) Display Indicates Infinite Ohms (Open Circuit)

Disconnect Power

R_1 R_2 R_3 12 V

(d)

Disconnect Each Resistor from Circuit and Check Their Individual Resistance Values

FIGURE 4-25 **Troubleshooting an Open in a Parallel Circuit.**

Component Testing, Circuit Troubleshooting, and Test Equipment

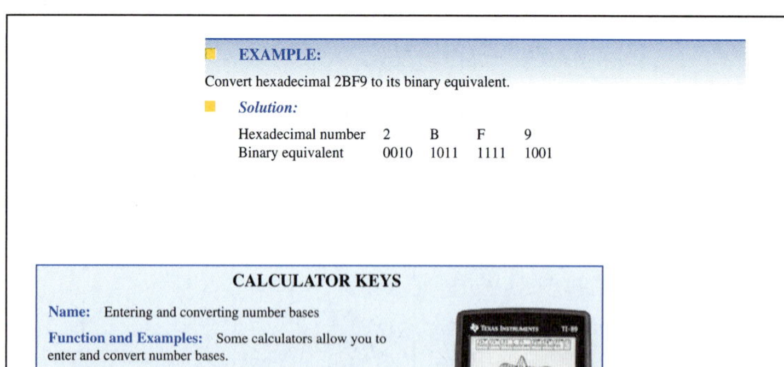

EXAMPLE:

Convert hexadecimal 2BF9 to its binary equivalent.

Solution:

Hexadecimal number	2	B	F	9
Binary equivalent	0010	1011	1111	1001

CALCULATOR KEYS

Name: Entering and converting number bases

Function and Examples: Some calculators allow you to enter and convert number bases.

To enter a binary number, use the form:

0b*binaryNumber* (for example: 0b11100110)
- Binary number with up to 32 digits
- Zero, not the letter *O*, and the letter *b*

To enter a hexadecimal number, use the form:

0h*hexadecimalNumber* (for example: 0h89F2C)
- Hexadecimal number with up to 8 digits
- Zero, not the letter *O*, and letter *h*

If you enter without the 0b or 0h prefix, such as 11, it is always treated as a decimal number. If you omit the 0h prefix on a hexadecimal number containing A–F, all or part of the entry is treated as a variable.

Use the ▶ conversion operator.

integerExpression ▶ Bin
integerExpression ▶ Dec
integerExpression ▶ Hex

For ▶, press 2nd [▶]. Also, you can select base conversions from the MATH/Base menu.

For example, to convert 256 from decimal to binary:

256 ▶ Bin

For a binary or hex entry, you must use the 0b or 0h prefix.

n 256 ▶ Bin	0b100000000
n 0b101110 ▶ Hex	0h2E
0b101110 ▶ hex	
MATH RAD AUTO FUNC	2738

To convert 101110 from binary to hexidecimal:

0b101110 ▶ Hex

Result use the 0b or 0h prefix to identify the base.

614 CHAPTER 17 / NUMBER SYSTEMS AND CODES

Calculator Sequences

End-of-Chapter Test Bank

A	B	LED
0 V	0 V	
0 V	+5 V	
+5 V	0 V	

FIGURE 11-45 An OR Gate Circuit with an LED Output.

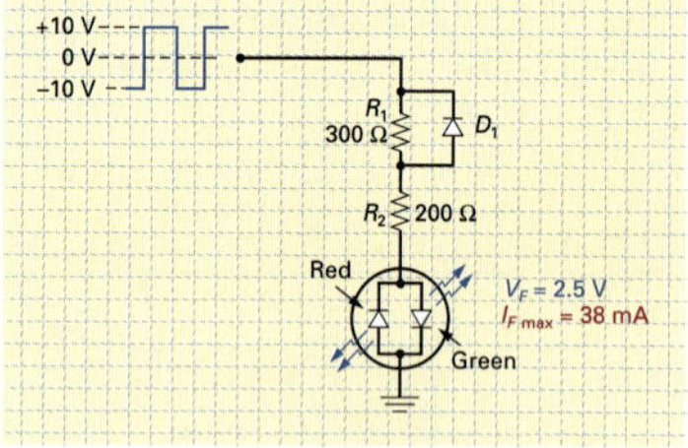

FIGURE 11-46 A Bi-Color Output Display Circuit.

63. Figure 11-45 shows how two basic P-N junction diodes and a resistor can be used to construct an OR gate circuit. In this circuit, an LED has been connected to the output so that any HIGH or positive 5 V output will turn ON the LED. Considering the *A* and *B* input combinations shown in the table, indicate whether the LED will be ON or OFF for each of these input conditions.

64. Assuming a 0.7 V voltage drop across the P-N junction diode and a 2 V voltage drop across the LED, what would be the value of current through the LED if one of the inputs in Figure 11-45 was HIGH (+5 V)?

65. Which of the bi-color LEDs will be ON in Figure 11-46 when the input voltage is +10 V, and which will be ON when the input voltage is −10 V?

66. In Figure 11-46, a basic P-N junction diode (D_1) has been included across R_1 so that when the input voltage goes negative this diode will bypass the additional current-limiting resistor R_1. When the input voltage is positive, D_1 is reversed biased and therefore both R_1 and R_2 will limit the value of series current. Which of the colors in the bi-color LED will be brighter?

67. Calculate the value of green and red LED current for the circuit in Figure 11-46.

68. In Figure 11-47 a comparator is used to compare 0 V (at the negative input) to a sine wave which varies between +5 V and −5 V (at the positive input). When the comparator's inputs are "true" (positive input is positive with respect to negative input), the comparator's output will be switched HIGH (+12 V). When the comparator's inputs are "not true" (positive input is negative with respect to the negative input), the comparator's output will be switched LOW (−12 V). What value of series current-limiting resistor should be used to ensure that the LEDs are bright, but do not burn out, for the +12 V and −12 V source voltages? Aim for a forward current that is about 90% of the maximum rating.

69. Would the 16-LED display in Figure 11-48 (D_{11}–D_{14}) be classified as a common-anode or common-cathode display?

70. In regard to Figure 11-48, construct a table to show the HIGH and LOW encoder outputs on *A, B, C,* and *D* for each of the rotary switch positions. Also show in the table which of the LEDs are ON or OFF for each of these codes.

71. In Figure 11-49, seven switches are used to turn ON and OFF the seven LEDs in a seven-segment display. Construct a table to show which switches will be CLOSED and which will be OPEN to display the digits 0 through 9 on the seven-segment display.

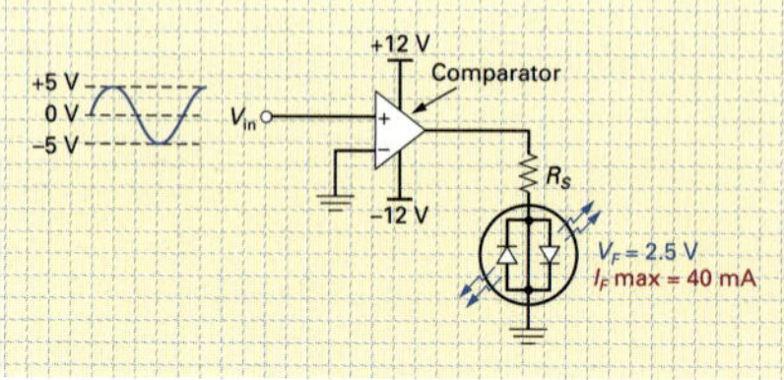

FIGURE 11-47 A Comparator Circuit with a Bi-Color Output Display.

418 CHAPTER 11 / DIODES AND POWER SUPPLY CIRCUITS

Introduction to Electronics

Your Course in Electronics

Your future in the electronics industry begins with this text. To give you an idea of where you are going and what we will be covering, Figure I-1 acts as a sort of road map, breaking up your study of electronics into four basic steps.

Step 1: Basics of Electricity
Step 2: Electronic Components
Step 3: Electronic Circuits
Step 4: Electronic Systems

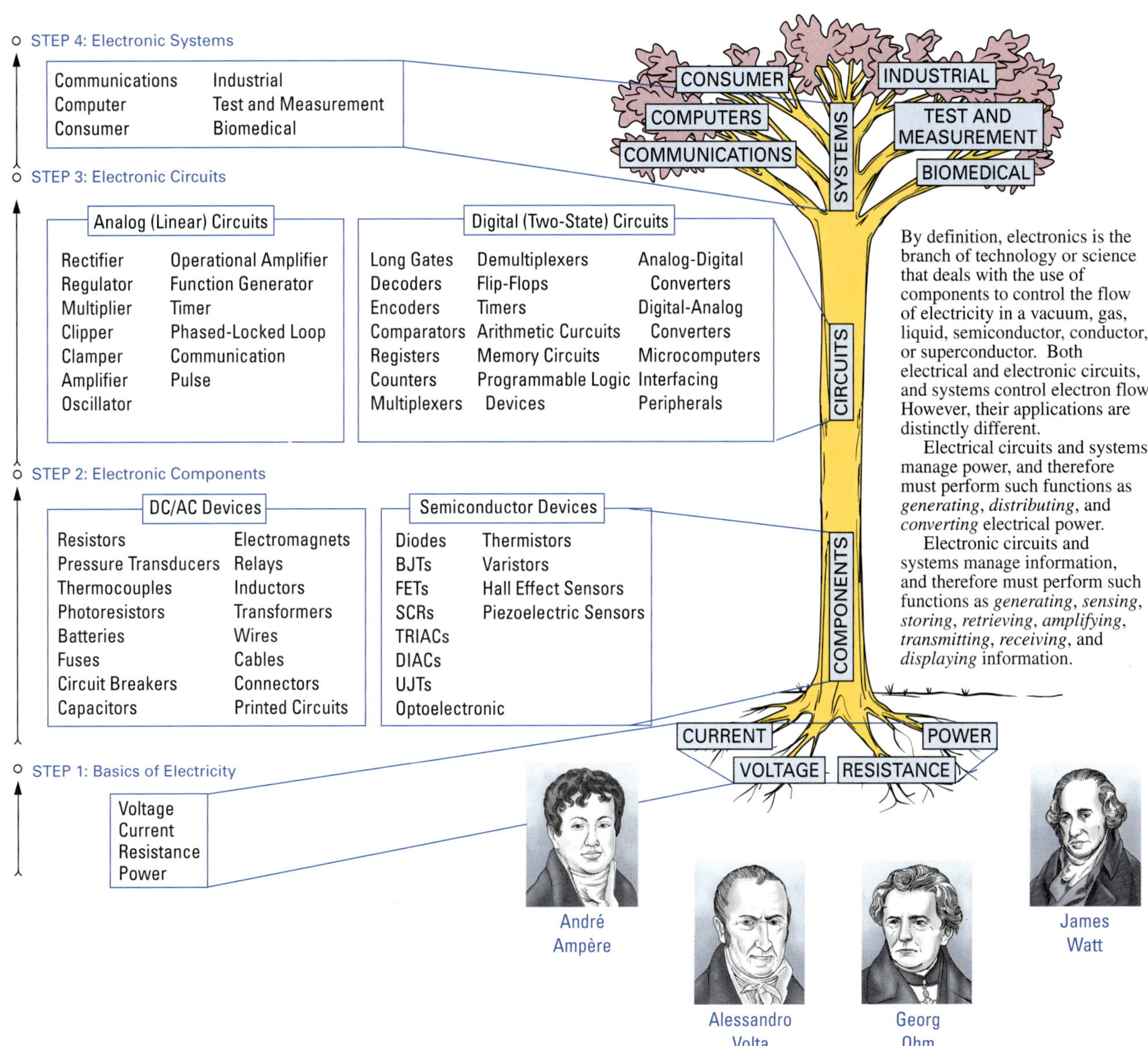

By definition, electronics is the branch of technology or science that deals with the use of components to control the flow of electricity in a vacuum, gas, liquid, semiconductor, conductor, or superconductor. Both electrical and electronic circuits, and systems control electron flow. However, their applications are distinctly different.

Electrical circuits and systems manage power, and therefore must perform such functions as *generating*, *distributing*, and *converting* electrical power.

Electronic circuits and systems manage information, and therefore must perform such functions as *generating*, *sensing*, *storing*, *retrieving*, *amplifying*, *transmitting*, *receiving*, and *displaying* information.

FIGURE I-1 **The Steps Involved in Studying Electronics.**

The main purpose of this introduction is not only to introduce you to the terms of the industry, but also to show you why the first two chapters in this text begin at the very beginning with "current and voltage," and then "resistance and power." **Components,** which are the basic electronic building blocks, were developed to control these four roots or properties, and when these devices are combined they form **circuits.** Moving up the tree to the six different branches of electronics, you will notice that just as components are the building blocks for circuits, circuits are in turn the building blocks for **systems.**

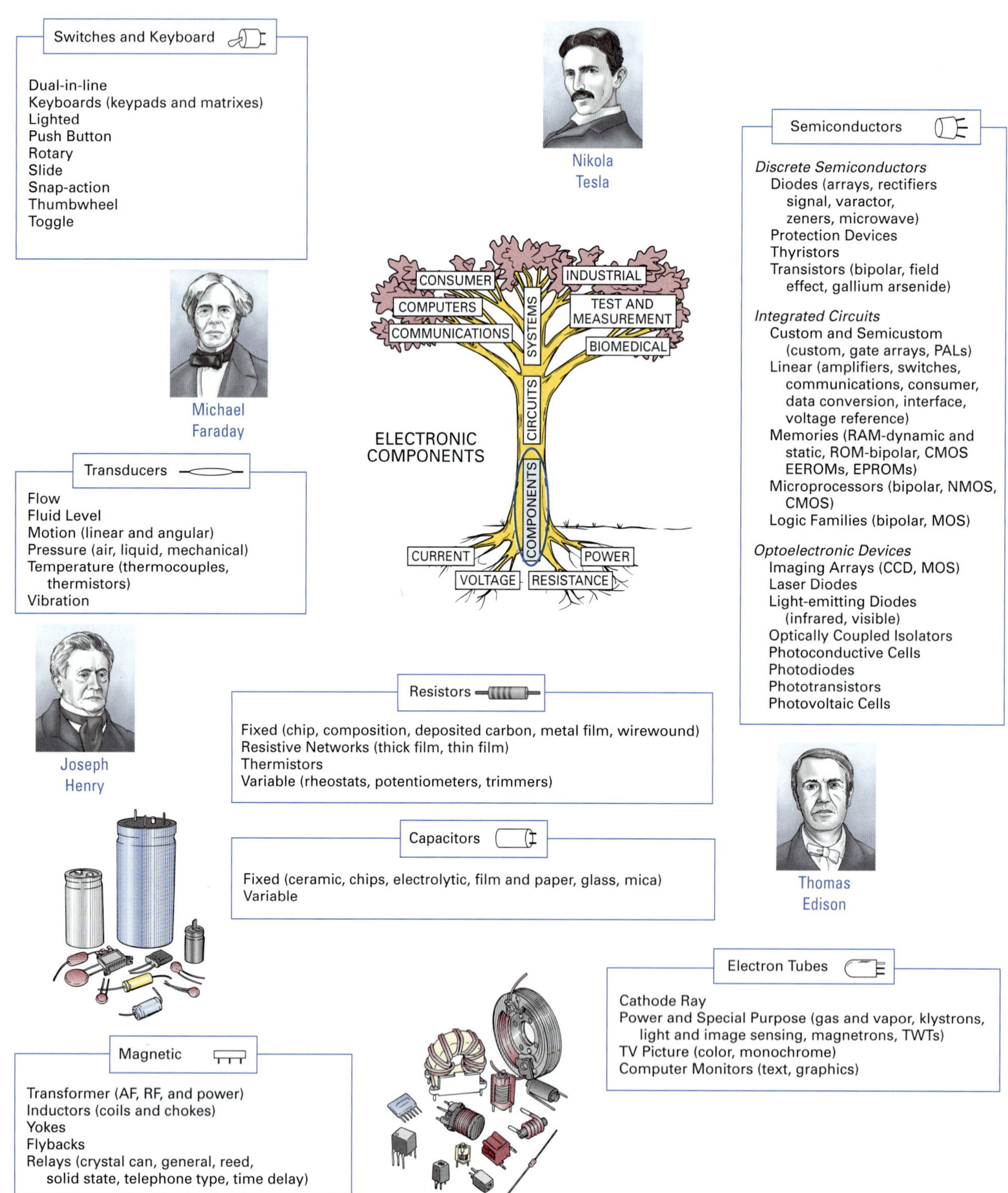

Charles Wheatstone

Signal Generators and Timers

Sine-wave Generators (oscillators)
Square- and Pulse-wave Generators
Ramp- and Triangular-wave Generators

Digital Circuits

Oscillators and Generators
Gates and Flip-flops
Display Drivers
Counters and Dividers
Encoders and Decoders
Memories
Input/Output
Microprocessors
Latches
Registers
Multiplexers
Demultiplexers
Gate Arrays

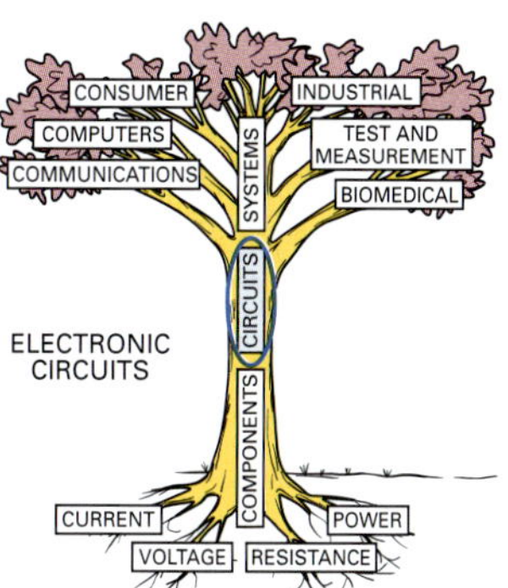

René Descartes

Amplifiers

Bipolar Transistor
Field-effect Transistor
Operational Amplifier

Power Supplies

Switching
Linear
Uninterruptible

Robert Noyce

Miscellaneous

Detectors and Mixers
Filters
Phase-locked Loops
Converters
Data Acquisition
Synthesizers

Gustav Kirchoff

Radio

Amateur (mobile and base stations)
Aviation Mobile and Ground Support Stations
Broadcast Equipment
Land Mobile (mobile and base stations)
Marine Mobile (ship and shore stations)
Microwave Systems
Satellite Systems
Radar and Sonar Systems

Guglielmo Marconi

Lee DeForest

Telecommunications

Switching Systems
Data and Voice Switching
Voice-only Switching
Cellular Systems
Telephones
Corded
Cordless
Telephone/Video Equipment
Facsimile Terminals
Fiber Optic Communication Systems

CONSUMER
INDUSTRIAL
COMPUTERS
TEST AND MEASUREMENT
COMMUNICATIONS
BIOMEDICAL
SYSTEMS
CIRCUITS
COMPONENTS
CURRENT
POWER
VOLTAGE
RESISTANCE

COMMUNICATION SYSTEMS

Electronic communications allow the transmission and reception of information between two points. Radio and television are obvious communication devices, broadcasting data or entertainment between two points.

John Baird

Heinrich Hertz

Television

Broadcast Equipment
CATV Equipment
CCTV Equipment
Satellite TV Equipment
HDTV (High Definition TV) Equipment

Data Communications

Concentrators
Front-end Communications Processors
Message-switching Systems
Modems
Multiplexers
Network Controllers
Mixed Service (combining voice, data, video, imaging)

COMPUTER SYSTEMS

The computer is proving to be one of the most useful of all systems. Its ability to process, store, and manipulate large groups of information at an extremely fast rate makes it ideal for almost any and every application. Systems vary in complexity and capability, ranging from the Cray supercomputer to the home personal computer. The applications of word processing, record keeping, inventory, analysis, and accounting are but a few examples of data processing systems.

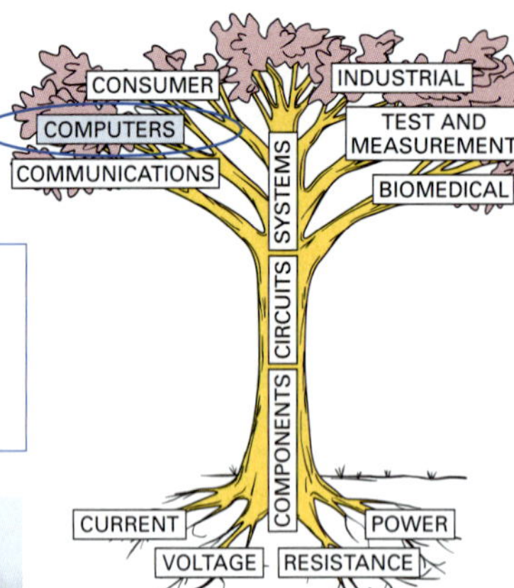

Data Terminals

CRT Terminals
- ASCII Terminals
- Graphics Terminals (color, monochrome)

Remote Batch Job Entry Terminals

Computer Systems

Microcomputers and Supermicrocomputers
Minicomputers (personal computers) and Superminicomputers (technical workstations, multiuser)
Mainframe Computers
Supercomputers

John von Neuman

Alan Turing

Charles Babbage

George Boole

Data Storage Devices

Fixed Disk (14, 8, $5\frac{1}{4}$, and $3\frac{1}{2}$ in.)
Flexible Disk (8, $5\frac{1}{4}$, and $3\frac{1}{2}$ in.)
Optical Disk Drives (read-only, write once, erasable)
Cassette
Cartridge Magnetic Tape ($\frac{1}{4}$ in.)
Cartridge Tape Drives ($\frac{1}{2}$ in.)
Reel-type Magnetic Tape Drives

I/O Peripherals

Computer Microfilm
Digitizers
Graphics Tablets
Light Pens
Trackball and Mice
Optical Scanning Devices
Plotters
Printers
- Impact
- Nonimpact (laser, thermal, electrostatic, inkjet)

CONSUMER SYSTEMS

From the smart computer-controlled automobiles, which provide navigational information and monitor engine functions and braking, to the compact disc players, video camcorders, satellite TV receivers, and wide-screen stereo TVs, this branch of electronics provides us with entertainment, information, safety, and, in the case of the pacemaker, life.

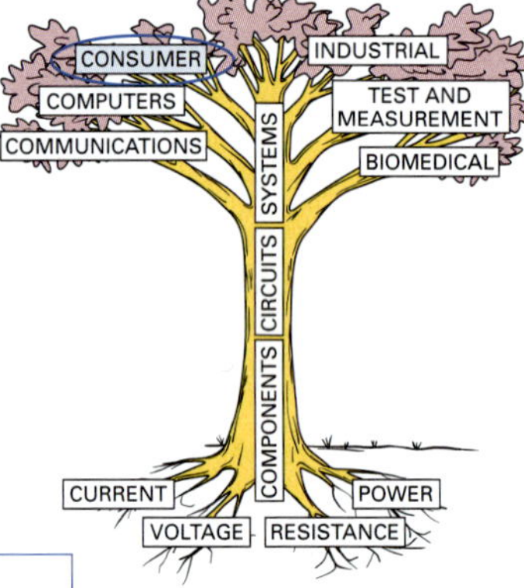

Jack Kilby

William Shockley

Automobile Electronics

Dashboard
Engine Monitoring and Analysis
Computer Navigation Systems
Alarms
Telephones

Video Equipment

TV Receiver (color, monochrome)
Projection TV Receivers
Video Cassette Recorders (VCRs)
Video Disk Players
Camcorders (8 mm, $\frac{1}{2}$ in.)
Home Satellite Receivers

Personal

Calculators, Cameras, Watches
Telephone Answering Equipment
Personal Computers
Microwave Ovens
Musical Equipment and Instruments
Pacemakers and Hearing Aids
Alarms and Smoke Detectors

Nolan Bushnell

Audio Equipment

Car
Stereo Equipment
- Compact Systems (miniature components)
- Components (speakers, amps, turntables, tuners, tape decks)

Phonographs and Radio Phonographs
Radios (table, clock, portable)
Tape Players/Recorders
Compact Disk Players
Digital Tape Players

Chester Carlson

Manufacturing Equipment

Energy Management Equipment
Inspection Systems
Motor Controllers (speed, torque)
Numerical-control Systems
Process Control Equipment (data-acquisition systems, process instrumentation, programmable controllers)
Robot Systems
Vision Systems

Computer-aided Design and Engineering CAD/CAE

Hardware Equipment
- Design Work Stations (PC based, 32 microprocessor based platform, host based)
- Application Specific Hardware

Design Software
- Design Capture (schematic capture, logic fault and timing simulators, model libraries)
- IC Design (design rule checkers, logic synthesizers, floor planners–place and route, layout editors)
- Printed Circuit Board Design Software
- Project Management Software
- Test Equipment

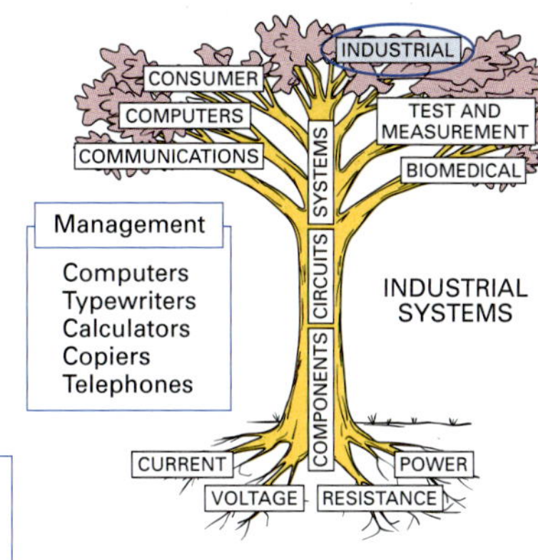

Management

Computers
Typewriters
Calculators
Copiers
Telephones

Almost any industrial company can be divided into three basic sections, all of which utilize electronic equipment to perform their functions. The manufacturing section will typically use power, motor, and process control equipment, along with automatic insertion, inspection, and vision systems, for the fabrication of a product. The engineering section uses computers and test equipment for the design and testing of a product, while the management section uses electronic equipment such as computers, copiers, telephones, and so on.

James Joule

Carl Gauss

Charles Steinmetz

Werner Von Siemens

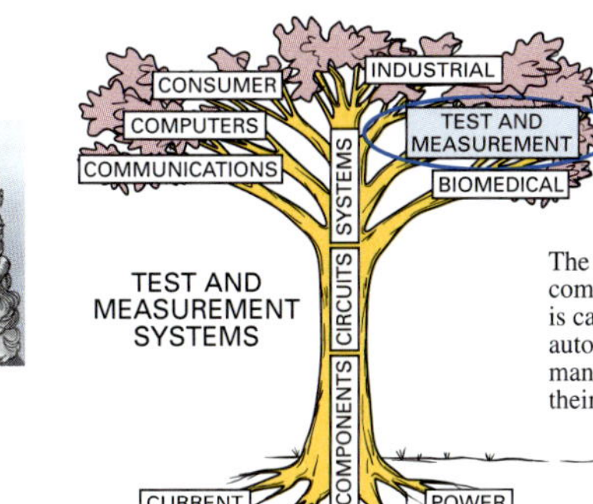

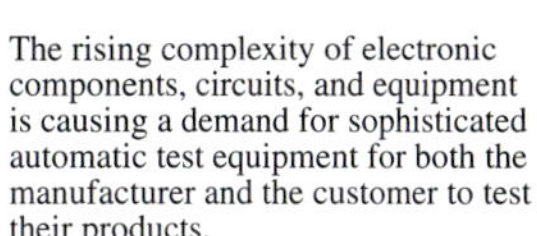

The rising complexity of electronic components, circuits, and equipment is causing a demand for sophisticated automatic test equipment for both the manufacturer and the customer to test their products.

Charles Napier

Isaac Newton

Benjamin Franklin

Grace Hopper

General Test and Measurement Equipment

Amplifiers (lab)
Arbitrary Waveform Generators
Analog Voltmeters, Ammeters, and Multimeters
Audio Oscillators
Audio Waveform Analyzers and Distortion Meters
Calibrators and Standards
Dedicated IEEE–488 Bus Controllers
Digital Multimeters
Electronic Counters (RF, Microwave, Universal)
Frequency Synthesizers
Function Generators
Pulse/Timing Generators
Signal Generators (RF, Microwave)
Logic Analyzers
Microprocessor Development Systems
Modulation Analyzers
Noise-measuring Equipment
Oscilloscopes (Analog, Digital)
Panel Meters
Personal Computer (PC) Based Instruments
Recorders and Plotters
RF/Microwave Network Analyzers
RF/Microwave Power-measuring Equipment
Spectrum Analyzers
Stand-alone In-circuit Emulators
Temperature-measuring Instruments

Automated Test Systems

Active and Discrete Component Test Systems
Automated Field Service Testers
IC Testers (benchtop, general purpose, specialized)
Interconnect and Bare Printed Circuit-Board Testers
Loaded Printed Circuit-Board Testers (in-circuit, functional, combined)

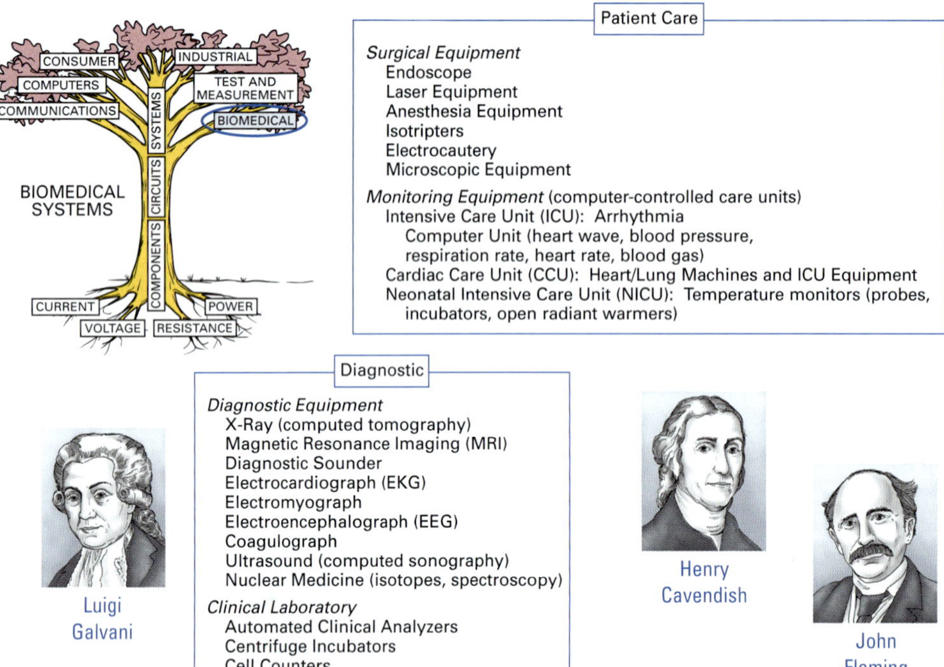

Electronic equipment is used more and more within the biological and medical fields, which can be categorized simply as being either patient care or diagnostic equipment. In the operating room, the endoscope, which is an instrument used to examine the interior of a canal or hollow organ, and the laser, which is used to coagulate, cut, or vaporize tissue with extremely intense light, both reduce the use of invasive surgery. A large amount of monitoring equipment is used both in and out of operating rooms, and the equipment consists of generally large computer-controlled systems that can have a variety of modules inserted (based on the application) to monitor, on a continuous basis, body temperature, blood pressure, pulse rate, and so on. In the diagnostic group of equipment, the clinical laboratory test results are used as diagnostic tools. With the advances in automation and computerized information systems, multiple tests can be carried out at increased speeds. Diagnostic imaging, in which a computer constructs an image of a cross-sectional plane of the body, is probably one of the most interesting equipment areas.

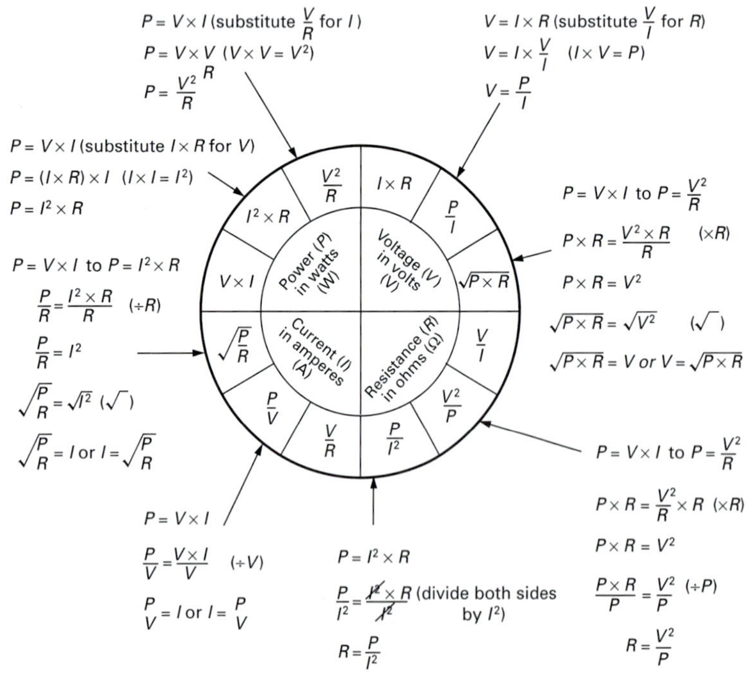

The formula circle shows how the four basic properties—voltage (V), current (I), resistance (R), and power (P)—are all related. This text will start at the very beginning with these basic roots of electronics. Voltage and current will be discussed in Ch. 1, resistance and power in Ch. 2, and then subsequent chapters will proceed to components, circuits, and systems.

Development of an Electronic Product

The flow chart below shows the order, from top to bottom, in which an electronic product is developed from conception to shipping.

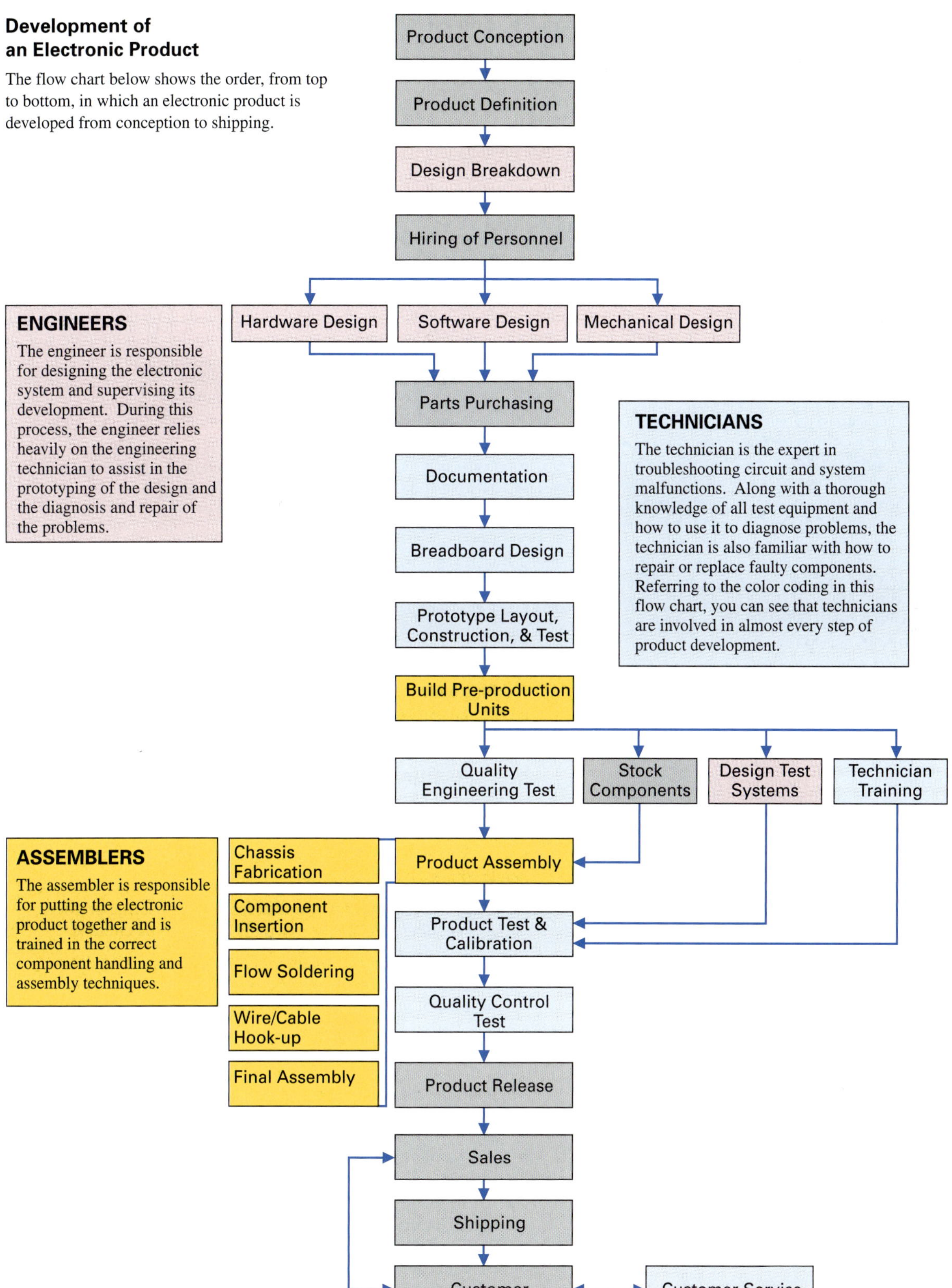

ENGINEERS

The engineer is responsible for designing the electronic system and supervising its development. During this process, the engineer relies heavily on the engineering technician to assist in the prototyping of the design and the diagnosis and repair of the problems.

TECHNICIANS

The technician is the expert in troubleshooting circuit and system malfunctions. Along with a thorough knowledge of all test equipment and how to use it to diagnose problems, the technician is also familiar with how to repair or replace faulty components. Referring to the color coding in this flow chart, you can see that technicians are involved in almost every step of product development.

ASSEMBLERS

The assembler is responsible for putting the electronic product together and is trained in the correct component handling and assembly techniques.

Electronic Technicians in the System Development Process

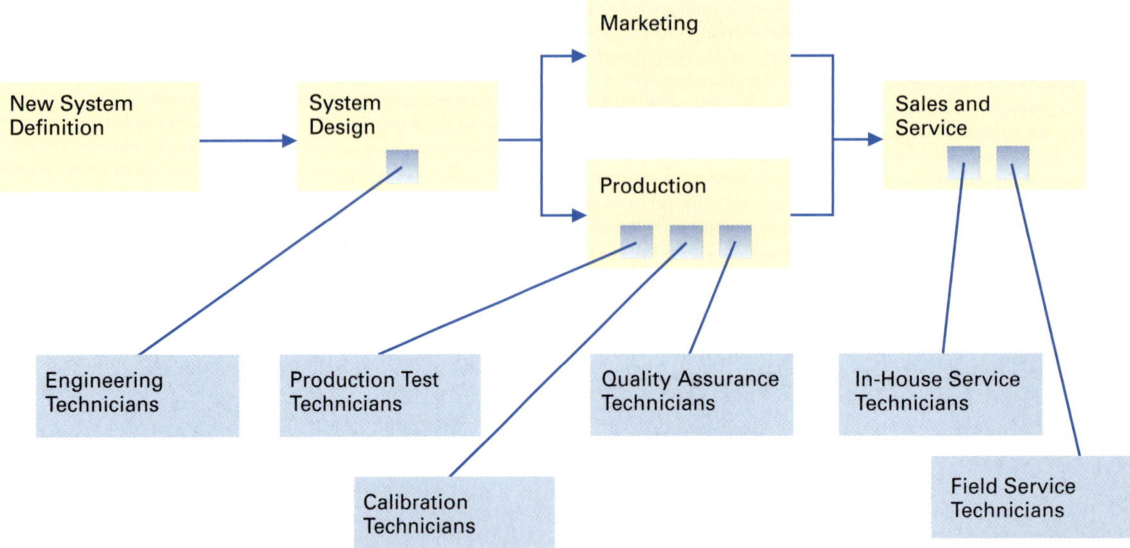

ENGINEERING TECHNICIAN

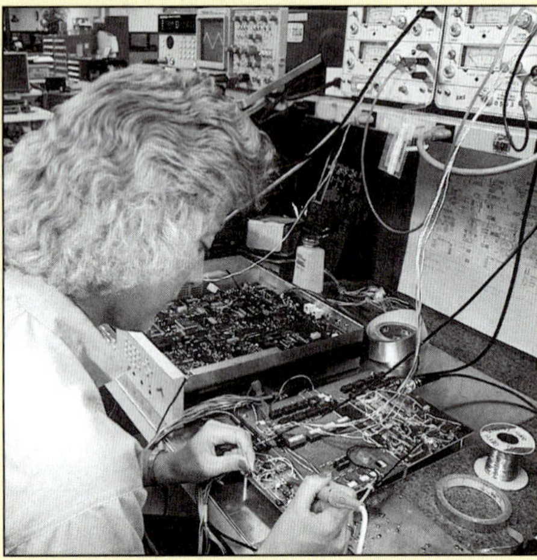

Here you can see an engineering technician breadboarding the design. From sketches supplied by the engineers, a breadboard model of the design is constructed. The breadboard model is an experimental arrangement of a circuit in which the components are temporarily attached to a flat board. In this arrangement, the components can be tested to prove the feasability of the circuit. A breadboard facilitates making easy changes when they are necessary.

Working under close supervision, the engineering technician performs all work assignments as given by all levels of engineers.

RESPONSIBILITIES:

- Breadboard electronic circuits from schematics.
- Test, evaluate, and document circuits and system performance under the engineer's direction.
- Check out, evaluate, and take data for the engineering prototypes including mechanical assembly of prototype circuits.
- Help to generate and maintain preliminary engineering documentations. Assist engineers and senior Engineering Technicians in ERN documentation.
- Maintain the working station equipment and tools in orderly fashion.
- Support the engineers in all aspects of the development of new products.

REQUIREMENTS:

AS/AAS Degree in Electronics or equivalent plus 1–2 years technician experience. Ability to read color code, to solder properly, and to bond wires where the skill is required to complete breadboards and prototypes. Ability to use common machinery required to build prototype circuits. Working knowledge of common electronic components: TTL logic circuits, op-amps, capacitors, resistors, inductors, semiconductor devices.

IN-HOUSE SERVICE TECHNICIAN

This photograph shows some in-house service technicians troubleshooting problems on returned units. Once the customer has received the electronic equipment, customer service provides assistance in maintenance and repair of the unit through direct in-house service or at service centers throughout the world.

Working under moderate supervision, the in-house service technician performs all work assignments given by lead tech or direct supervisor. Work necessary overtime as assigned by supervisors.

RESPONSIBILITIES:

- Utilizing all appropriate tools, troubleshoot and repair customer systems in a timely, quality manner, to the general component level.
- Working with basic test equipment, perform timely quality calibration of customer systems to specifications.
- Timely repair of QA rejects.
- Solder and desolder components where appropriate, meeting company standards.
- Aid marketing in solving customer problems via the telephone.
- When appropriate, instruct customers in the proper methods of calibration and repair of products.

REQUIREMENTS:

AS/AAS Degree in Electronics or equivalent plus 2–3 years experience troubleshooting analog and/or digital systems, at least 6–12 months of which should be in a service environment. Must be able to read and understand flow charts, block diagrams, schematics, and truth tables. Must be able to operate and utilize test equipment such as oscilloscopes, counters, voltmeters, and analyzers.

Ability to effectively communicate and work with customers.

FIELD SERVICE TECHNICIAN

The field service technician seen here has been requested by the customer to make a service call on a malfunctioning unit that currently is under test.

Working under moderate supervision, the field service technician performs all work assignments given by the lead tech or direct supervisor. Work necessary overtime as assigned by supervisors.

RESPONSIBILITIES:

- Utilizing all appropriate tools, troubleshoot and repair customer systems in a timely, quality manner, to the general component level.
- Working with basic test equipment, perform timely quality calibration of customer systems to specifications.
- When appropriate, instruct others in proper soldering techniques meeting company standards.
- Timely repair of QA rejects.
- Aid marketing in solving customer problems via the telephone or at the customers' facility at marketings' discretion.
- When appropriate, aid marketing with sales applications.
- Help QA, Production, and Engineering in solving field problems.
- Evaluate manuals and other customer documents for errors or omissions.

REQUIREMENTS:

AS/AAS Degree in Electronics or equivalent plus 4–5 years experience troubleshooting analog and/or digital systems, at least 6–12 months of which should be in a service environment. Must be able to read and understand flow charts, block diagrams, schematics, and truth tables. A demonstrated ability to effectively communicate and work with customers, and suggest alternative applications for product utilization is also required.

CALIBRATION TECHNICIAN

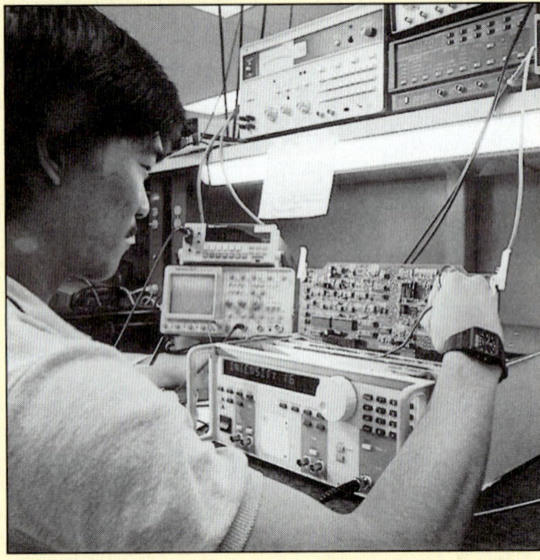

The calibration technician shown here is undertaking a complete evaluation of the newly constructed breadboard's mechanical and electrical form, design, and performance.

Working under general supervision, the calibration technician interfaces with test equipment in a system environment requiring limited decision-making.

RESPONSIBILITIES:

- Work in an interactive mode with test station. Test, align, and calibrate products to defined specifications.
- May set up own test stations and those of other operators.
- Perform multiple alignments to get products to meet specifications.
- May perform other manufacturing-related tasks as required.

REQUIREMENTS:

1–2 years experience with test and measurement equipment, experience with multiple alignment and calibration of assemblies including test station setups. Able to follow written instructions and write clearly.

QUALITY ASSURANCE TECHNICIANS

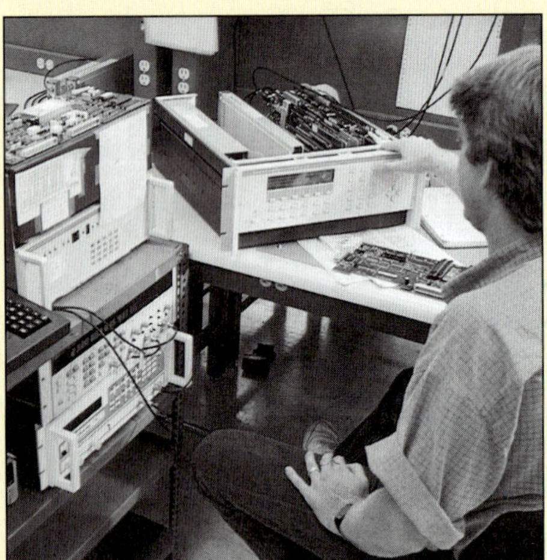

The quality assurance (QA) technician takes one of the pre-production units through an extensive series of tests to determine whether it meets the standards listed. This technician is evaluating the new product as it is put through an extensive series of tests.

Working under direct supervision, the quality assurance technician performs functional tests on completed instruments.

RESPONSIBILITIES:

- Using established Acceptance Test Procedures, perform operational tests of all completed systems to ensure that all functional and electrical parameters are within specified limits.
- Perform visual inspection of all completed systems for cleanliness and absence of cosmetic defects.
- Reject all systems that do not meet specifications and/or established parameters of function and appearance.
- Make appropriate notations on the system history sheet.
- Maintain the QA Acceptance Log in accordance with current instructions.
- Refer questionable characteristics to supervisor.

REQUIREMENTS:

AS/AAS Degree in Electronics or equivalent including the use of test equipment. Must know color code and be able to distinguish between colors. Must have a working knowledge of related test equipment. Must know how to read and interpret drawings.

PRODUCTION TEST TECHNICIAN

Once the system is fully operational, it is calibrated by a calibration technician. The more complex problems are handled by the production test technicians seen in this photograph.

TECHNICIAN I

Working under close supervision, the production test technician performs all work assignments, and exercises limited decision-making.

RESPONSIBILITIES:

- Perform routine, simple operational tests and fault isolation on simple components, circuits, and systems for verification of product performance to well-defined specifications.
- May perform standard assembly operations and simple alignment of electronic components and assemblies.
- May set up simple test equipment to test performance of products to specifications.

REQUIREMENTS:

AS/AAS Degree in Electronic Technology or equivalent work experience.

TECHNICIAN II

Working under moderate supervision, the production test technician exercises general decision-making involving simple cause and effect relationships to identify trends and common problems.

RESPONSIBILITIES:

- Perform moderately complex operational tests and fault isolation on components, circuits, and systems for verification of product performance to well-defined specifications.
- Perform simple mathematical calculations to verify test measurements and product performance to well-defined specifications.
- Set up general test stations, utilizing varied test equipment, including some sophisticated equipment.
- Perform standard assembly operations.

REQUIREMENTS:

AS/AAS Degree in Electronic Technology or equivalent plus 2–4 years directly related experience. Demonstrated experience working mathematical formulas and equations. Working knowledge of counters, scopes, spectrum analyzers, and related industry standard test equipment.

ELECTRONICS
A Complete Course

Current and Voltage

Problem-Solver

Charles Proteus Steinmetz (1865–1923) was an outstanding electrical genius who specialized in mathematics, electrical engineering, and chemistry. His three greatest electrical contributions were his investigation and discovery of the law of hysteresis, his investigations in lightning, which resulted in his theory on traveling waves, and his discovery that complex numbers could be used to solve ac circuit problems. Solving problems was in fact his specialty, and on one occasion he was commissioned to troubleshoot a failure on a large company system that no one else had been able to repair. After studying the symptoms and schematics for a short time, he chalked an X on one of the metal cabinets, saying that this was where they would find the problem, and left. He was right, and the problem was remedied to the relief of the company executives; however, they were not pleased when they received a bill for \$1000. When they demanded that Steinmetz itemize the charges, he replied—\$1 for making the mark and \$999 for knowing where to make the mark.

The strong message this vignette conveys is that you will get \$1 for physical labor and \$999 for mental labor, and this is a good example as to why you should continue in your pursuit of education.

1

Introduction

Before we begin, let me try to put you in the right frame of mind. As you proceed through this chapter and the succeeding chapters, it is imperative that you study every section, example, self-test evaluation point, and end-of-chapter question. If you cannot understand a particular section or example, go back and review the material that led up to the problem and make sure that you fully understand all the basics before you continue. Since each chapter builds on previous chapters, you may find that you need to return to an earlier chapter to refresh your understanding before moving on with the current chapter. This process of moving forward and then backtracking to refresh your understanding is very necessary and helps to engrave the material in your mind. Try never to skip a section or chapter because you feel that you already have a good understanding of the subject matter. If it is a basic topic that you have no problem with, read it anyway to refresh your understanding about the steps involved and the terminology because these may be used in a more complex operation in a later chapter.

In this chapter we will examine the smallest and most significant part of electricity and electronics—the **electron.** By having a good understanding of the electron and the atom, we will be able to obtain a clearer understanding of three basic electrical quantities: **voltage, current,** and **resistance.** In this chapter we will be discussing the basic building blocks of matter, the electrical quantities of voltage and current, and the difference between a conductor and insulator.

Electron
Smallest subatomic particle of negative charge that orbits the nucleus of the atom.

Element
There are 107 different natural chemical substances or elements that exist on the earth and can be categorized as either being a gas, solid, or liquid.

Atom
Smallest particle of an element.

Subatomic
Particles such as electrons, protons, and neutrons that are smaller than atoms.

Atomic Number
Number of positive charges or protons in the nucleus of an atom.

Atomic Weight
The relative weight of a neutral atom of an element, based on a neutral oxygen atom having an atomic weight of 16.

TIME LINE
The electron was first discovered by Jean Baptiste Perrin (1870–1942), a French physicist who was awarded the Nobel prize for physics.

1-1 THE STRUCTURE OF MATTER

All of the matter on the earth and in the air surrounding the earth can be classified as being either a solid, liquid, or gas. A total of approximately 107 different natural elements exist in, on, and around the earth. An **element,** by definition, is a substance consisting of only one type of atom; in other words, every element has its own distinctive atom, which makes it different from all the other elements. This **atom** is the smallest particle into which an element can be divided without losing its identity, and a group of identical atoms is called an element, shown in Figure 1-1(a).

1-1-1 *The Atom*

The word "atom" is a Greek word meaning a particle that is too small to be subdivided. At present, we cannot clearly see the atom; however, physicists and researchers do have the ability to record a picture as small as 12 billionths of an inch (about the diameter of one atom), and this image displays the atom as a white fuzzy ball.

In 1913, a Danish physicist, Neils Bohr, put forward a theory about the atom, and his basic model outlining the **subatomic** particles that make up the atom is still in use today and is illustrated in Figure 1-1(b). Bohr actually combined the ideas of Lord Rutherford's (1871–1937) nuclear atom with Max Planck's (1858–1947) and Albert Einstein's (1879–1955) quantum theory of radiation.

The three important particles of the atom are the *proton,* which has a positive charge, the *neutron,* which is neutral or has no charge, and the *electron,* which has a negative charge. Referring to Figure 1-1(b), you can see that the atom consists of a positively charged central mass called the *nucleus,* which is made up of protons and neutrons surrounded by a quantity of negatively charged orbiting electrons.

Table 1-1 lists the periodic table of the elements, in order of their atomic number. The **atomic number** of an atom describes the number of protons that exist within the nucleus.

The proton and the neutron are almost 2000 times heavier than the very small electron, so if we ignore the weight of the electron, we can use the fourth column in Table 1-1 (weight of an atom) to give us a clearer picture of the protons and neutrons within the atom's nucleus. For example, a hydrogen atom, shown in Figure 1-2(a), is the smallest of all atoms and has an atomic number of 1, which means that hydrogen has a one-proton nucleus. Helium, however [Figure 1-2(b)], is second on the table and has an atomic number of 2, indicating that two protons are within the nucleus. The **atomic weight** of helium, however, is 4, meaning that two protons and two neutrons make up the atom's nucleus.

The number of neutrons within an atom's nucleus can therefore be calculated by subtracting the atomic number (protons) from the atomic weight (protons and neutrons). For

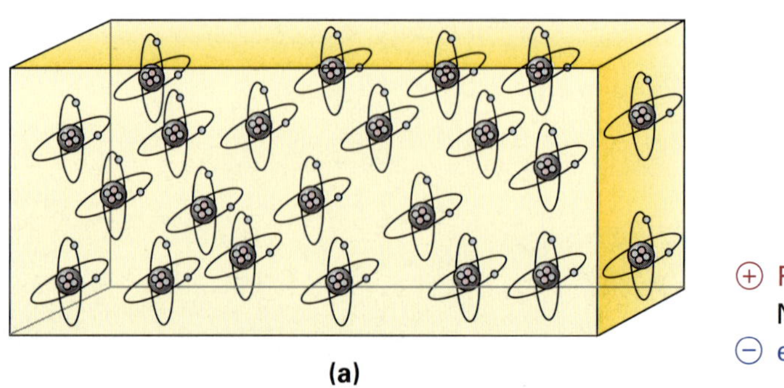

FIGURE 1-1 **(a) Element: Many Similar Atoms. (b) Atom: Smallest Unit.**

TABLE 1-1 **Periodic Table of the Elements**

ATOMIC NUMBER	ELEMENT NAME	SYMBOL	ATOMIC WEIGHT	ELECTRONS/SHELL K	L	M	N	O	P	Q	DISCOVERED	COMMENT
1	Hydrogen	H	1.007	1							1766	Active gas
2	Helium	He	4.002	2							1895	Inert gas
3	Lithium	Li	6.941	2	1						1817	Solid
4	Beryllium	Be	9.01218	2	2						1798	Solid
5	Boron	B	10.81	2	3						1808	Solid
6	Carbon	C	12.011	2	4						Ancient	Semiconductor
7	Nitrogen	N	14.0067	2	5						1772	Gas
8	Oxygen	O	15.9994	2	6						1774	Gas
9	Fluorine	F	18.998403	2	7						1771	Active gas
10	Neon	Ne	20.179	2	8						1898	Inert gas
11	Sodium	Na	22.98977	2	8	1					1807	Solid
12	Magnesium	Mg	24.305	2	8	2					1755	Solid
13	Aluminum	Al	26.98154	2	8	3					1825	Metal conductor
14	Silicon	Si	28.0855	2	8	4					1823	Semiconductor
15	Phosphorus	P	30.97376	2	8	5					1669	Solid
16	Sulfur	S	32.06	2	8	6					Ancient	Solid
17	Chlorine	Cl	35.453	2	8	7					1774	Active gas
18	Argon	Ar	39.948	2	8	8					1894	Inert gas
19	Potassium	K	39.0983	2	8	8	1				1807	Solid
20	Calcium	Ca	40.08	2	8	8	2				1808	Solid
21	Scandium	Sc	44.9559	2	8	9	2				1879	Solid
22	Titanium	Ti	47.90	2	8	10	2				1791	Solid
23	Vanadium	V	50.9415	2	8	11	2				1831	Solid
24	Chromium	Cr	51.996	2	8	13	1				1798	Solid
25	Manganese	Mn	54.9380	2	8	13	2				1774	Solid
26	Iron	Fe	55.847	2	8	14	2				Ancient	Solid (magnetic)
27	Cobalt	Co	58.9332	2	8	15	2				1735	Solid
28	Nickel	Ni	58.70	2	8	16	2				1751	Solid
29	Copper	Cu	63.546	2	8	18	1				Ancient	Metal conductor
30	Zinc	Zn	65.38	2	8	18	2				1746	Solid
31	Gallium	Ga	69.72	2	8	18	3				1875	Liquid
32	Germanium	Ge	72.59	2	8	18	4				1886	Semiconductor
33	Arsenic	As	74.9216	2	8	18	5				1649	Solid
34	Selenium	Se	78.96	2	8	18	6				1818	Photosensitive
35	Bromine	Br	79.904	2	8	18	7				1898	Liquid
36	Krypton	Kr	83.80	2	8	18	8				1898	Inert gas
37	Rubidium	Rb	85.4678	2	8	18	8	1			1861	Solid
38	Strontium	Sr	87.62	2	8	18	8	2			1790	Solid
39	Yttrium	Y	88.9059	2	8	18	9	2			1843	Solid
40	Zirconium	Zr	91.22	2	8	18	10	2			1789	Solid
41	Niobium	Nb	92.9064	2	8	18	12	1			1801	Solid
42	Molybdenum	Mo	95.94	2	8	18	13	1			1781	Solid
43	Technetium	Tc	98.0	2	8	18	14	1			1937	Solid
44	Ruthenium	Ru	101.07	2	8	18	15	1			1844	Solid
45	Rhodium	Rh	102.9055	2	8	18	16	1			1803	Solid
46	Palladium	Pd	106.4	2	8	18	18				1803	Solid
47	Silver	Ag	107.868	2	8	18	18	1			Ancient	Metal conductor
48	Cadmium	Cd	112.41	2	8	18	18	2			1803	Solid
49	Indium	In	114.82	2	8	18	18	3			1863	Solid
50	Tin	Sn	118.69	2	8	18	18	4			Ancient	Solid
51	Antimony	Sb	121.75	2	8	18	18	5			Ancient	Solid

(continued)

TABLE 1-1 Periodic Table of the Elements (continued)

ATOMIC NUMBER[a]	ELEMENT NAME	SYMBOL	ATOMIC WEIGHT	K	L	M	N	O	P	Q	DISCOVERED	COMMENT
52	Tellurium	Te	127.60	2	8	18	18	6			1783	Solid
53	Iodine	I	126.9045	2	8	18	18	7			1811	Solid
54	Xenon	Xe	131.30	2	8	18	18	8			1898	Inert gas
55	Cesium	Cs	132.9054	2	8	18	18	8	1		1803	Liquid
56	Barium	Ba	137.33	2	8	18	18	8	2		1808	Solid
57	Lanthanum	La	138.9055	2	8	18	18	9	2		1839	Solid
72	Hafnium	Hf	178.49	2	8	18	32	10	2		1923	Solid
73	Tantalum	Ta	180.9479	2	8	18	32	11	2		1802	Solid
74	Tungsten	W	183.85	2	8	18	32	12	2		1783	Solid
75	Rhenium	Re	186.207	2	8	18	32	13	2		1925	Solid
76	Osmium	Os	190.2	2	8	18	32	14	2		1804	Solid
77	Iridium	Ir	192.22	2	8	18	32	15	2		1804	Solid
78	Platinum	Pt	195.09	2	8	18	32	16	2		1735	Solid
79	Gold	Au	196.9665	2	8	18	32	18	1		Ancient	Solid
80	Mercury	Hg	200.59	2	8	18	32	18	2		Ancient	Liquid
81	Thallium	Tl	204.37	2	8	18	32	18	3		1861	Solid
82	Lead	Pb	207.2	2	8	18	32	18	4		Ancient	Solid
83	Bismuth	Bi	208.9804	2	8	18	32	18	5		1753	Solid
84	Polonium	Po	209.0	2	8	18	32	18	6		1898	Solid
85	Astatine	At	210.0	2	8	18	32	18	7		1945	Solid
86	Radon	Rn	222.0	2	8	18	32	18	8		1900	Inert gas
87	Francium	Fr	223.0	2	8	18	32	18	8	1	1945	Liquid
88	Radium	Ra	226.0254	2	8	18	32	18	8	2	1898	Solid
89	Actinium	Ac	227.0278	2	8	18	32	18	9	2	1899	Solid

ELECTRONS/SHELL: K L M N O P Q

[a]Rare earth series 58–71 and 90–107 have been omitted.

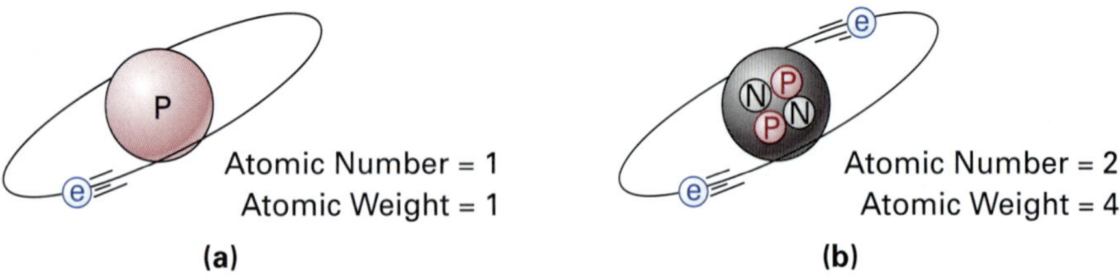

FIGURE 1-2 (a) Hydrogen Atom. (b) Helium Atom.

example, Figure 1-3(a) illustrates a beryllium atom which has the following atomic number and weight.

> **Beryllium**
> Atomic number: 4 (protons)
> Atomic weight: 9 (protons and neutrons)

Neutral Atom
An atom in which the number of positive charges in the nucleus (protons) is equal to the number of negative charges (electrons) that surround the nucleus.

Subtracting the beryllium atom's weight from the beryllium atomic number, we can determine the number of neutrons in the beryllium atom's nucleus, as shown in Figure 1-3(a).

In most instances, atoms like the beryllium atom will not be drawn in the three-dimensional way shown in Figure 1-3(a). Figure 1-3(b) shows how a beryllium atom could be more easily drawn as a two-dimensional figure.

A **neutral atom** or *balanced atom* is one that has an equal number of protons and orbiting electrons, so the net positive proton charge is equal but opposite to the net negative

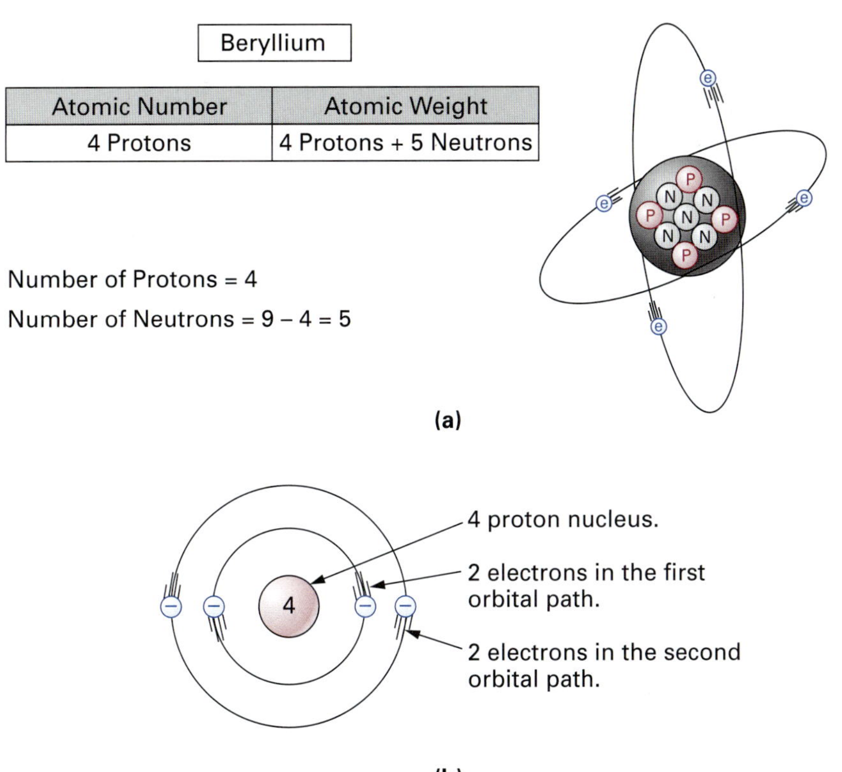

FIGURE 1-3 **Beryllium Atom.**

electron charge, resulting in a balanced or neutral state. For example, Figure 1-4 illustrates a copper atom, which is the most commonly used metal in the field of electronics. It has an atomic number of 29, meaning that 29 protons and 29 electrons exist within the atom when it is in its neutral state.

Orbiting electrons travel around the nucleus at varying distances from the nucleus, and these orbital paths are known as **shells** or **bands.** The orbital shell nearest the nucleus is referred to as the first or K shell. The second is known as the L, the third is M, the fourth is N, the fifth is O, the sixth is P, and the seventh is referred to as the Q shell. There are seven shells available for electrons (K, L, M, N, O, P, and Q) around the nucleus, and each of these seven shells can only hold a certain number of electrons, as shown in Figure 1-5.

Shells or Bands
An orbital path containing a group of electrons that have a common energy level.

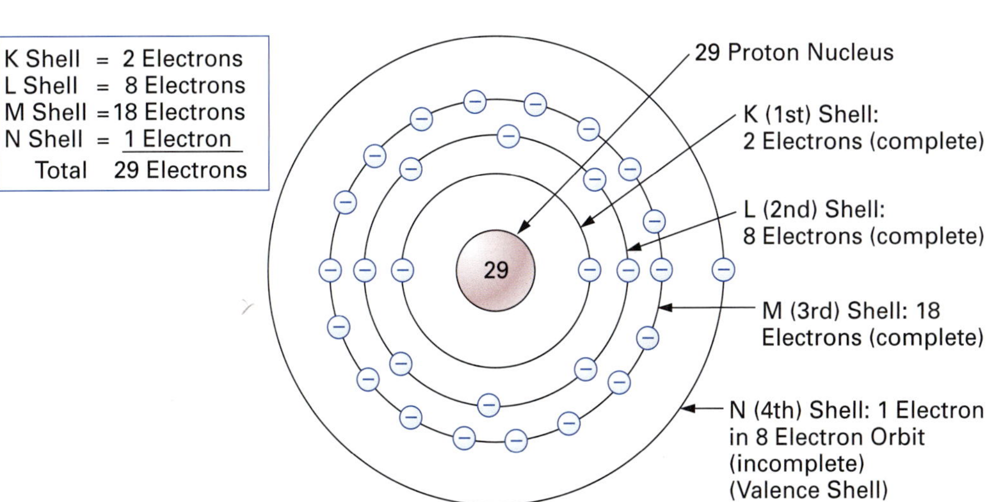

FIGURE 1-4 **Copper Atom.**

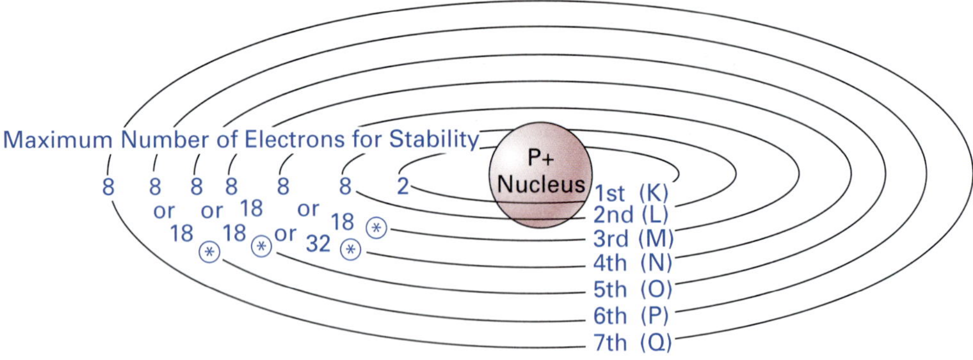

*The maximum number of electrons in these shells is dependent on the element's place in the periodic table.

FIGURE 1-5 Electrons and Shells.

Valence Shell or Ring
Outermost shell formed by electrons.

An atom's outermost electron-occupied shell is referred to as the **valence shell** or **ring,** and electrons in this shell are termed *valence electrons.* In the case of the copper atom, a single valence electron exists in the valence N shell.

All matter exists in one of three states: solid, liquid, or gas. The atoms of a solid are fixed in relation to one another but vibrate in a back-and-forth motion, unlike liquid atoms, which can flow over each other. The atoms of a gas move rapidly in all directions and collide with one another. The far-right column of Table 1-1 indicates whether the element is a gas, a solid, or a liquid.

1-1-2 *Laws of Attraction and Repulsion*

For the sake of discussion and understanding, let us theoretically imagine that we are able to separate some positive and negative subatomic particles. Using these separated protons and electrons, let us carry out a few experiments, the results of which are illustrated in Figure 1–6. Studying Figure 1-6, you will notice that:

1. *Like charges* (positive and positive or negative and negative) repel one another.
2. *Unlike charges* (positive and negative or negative and positive) attract one another.

Orbiting negative electrons are therefore attracted toward the positive nucleus, which leads us to the question of why the electrons do not fly into the atom's nucleus. The answer is that the orbiting electrons remain in their stable orbit due to two equal but opposite forces. The centrifugal outward force exerted on the electrons due to the orbit counteracts the attractive inward force trying to pull the electrons toward the nucleus due to the unlike charges.

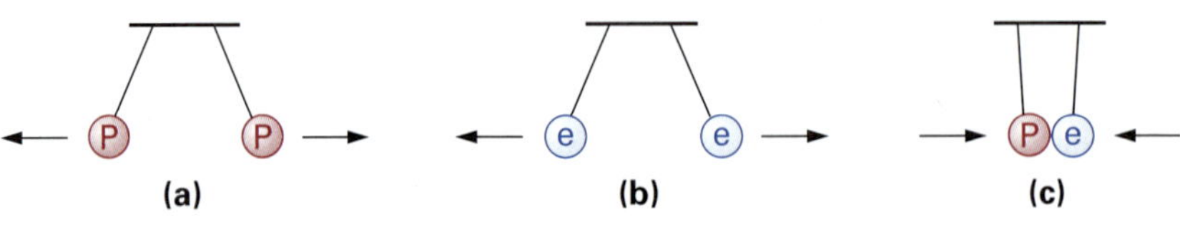

FIGURE 1-6 Attraction and Repulsion. (a) Positive Repels Positive. (b) Negative Repels Negative. (c) Unlike Charges Attract.

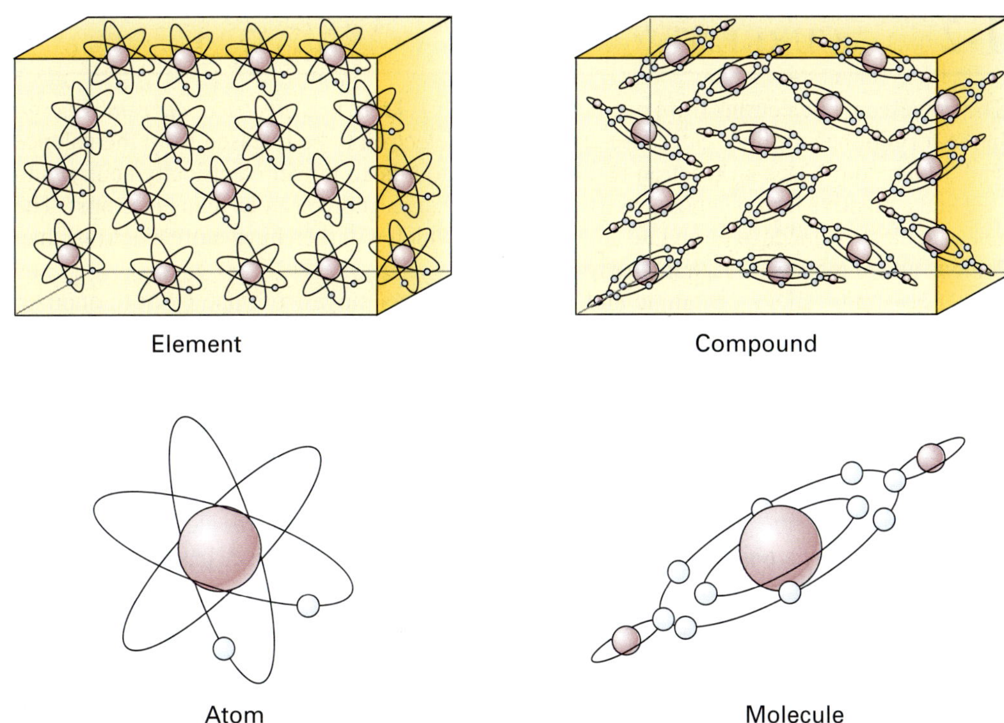

FIGURE 1-7 (a) An Element Is Made Up of Many Atoms. (b) A Compound Is Made Up of Many Molecules.

Due to their distance from the nucleus, valence electrons are described as being loosely bound to the atom. These electrons can easily be dislodged from their outer orbital shell by any external force, to become a **free electron.**

Free Electron
An electron that is not in any orbit around a nucleus.

1-1-3 *The Molecule*

An atom is the smallest unit of a natural element, or an element is a substance consisting of a large number of the same atom. Combinations of elements are known as **compounds,** and the smallest unit of a compound is called a **molecule,** just as the smallest unit of an element is an atom. Figure 1-7 summarizes how elements are made up of atoms and compounds are made up of molecules.

Compound
A material composed of united separate elements.

Molecule
Smallest particle of a compound that still retains its chemical characteristics.

Water is an example of a liquid compound in which the molecule (H_2O) is a combination of an explosive gas (hydrogen) and a very vital gas (oxygen). Table salt is another example of a compound; here the molecule is made up of a highly poisonous gas atom (chlorine) and a potentially explosive solid atom (sodium). These examples of compounds each contain atoms that, when alone, are both poisonous and explosive, yet when combined the resulting substance is as ordinary and basic as water and salt.

SELF-TEST EVALUATION POINT FOR SECTION 1-1

Use the following questions to test your understanding of Section 1-1:

1. Define the difference between an element and a compound.
2. Name the three subatomic particles that make up an atom.
3. What is the most commonly used metal in the field of electronics?
4. State the laws of attraction and repulsion.

1-2 CURRENT

Current (*I*)
Measured in amperes or amps, it is the flow of electrons through a conductor.

Positive Ion
Atom that has lost one or more of its electrons and therefore has more protons than electrons, resulting in a net positive charge.

Negative Ion
Atom that has more than the normal neutral amount of electrons.

Positive Charge
The charge that exists in a body which has fewer electrons than normal.

Negative Charge
An electric charge that has more electrons than protons.

Electric Current (*I*)
Measured in amperes or amps, it is the flow of electrons through a conductor.

The movement of electrons from one point to another is known as *electrical* ***current.*** Energy in the form of heat or light can cause an outer shell electron to be released from the valence shell of an atom. Once an electron is released, the atom is no longer electrically neutral and is called a **positive ion,** as it now has a net positive charge (more protons than electrons). The released electron tends to jump into a nearby atom, which will then have more electrons than protons and is referred to as a **negative ion.**

Let us now take an example and see how electrons move from one point to another. Figure 1-8 shows a broken metal conductor between two charged objects. The metal conductor could be either gold, silver, or copper, but whichever it is, one common trait can be noted: The valence electrons in the outermost shell are very loosely bound and can easily be pulled from their parent atom.

In Figure 1-9, the conductor between the two charges has been joined so that a path now exists for current flow. The negative ions on the right in Figure 1-9 have more electrons than protons, while the positive ions on the left in Figure 1-9 have fewer electrons than protons and so display a **positive charge.** The metal joining the two charges has its own atoms, which begin in the neutral condition.

Let us now concentrate on one of the negative ions. In Figure 1-9(a), the extra electrons in the outer shells of the negative ions on the right side will feel the attraction of the positive ions on the left side and the repulsion of the outer negative ions, or **negative charge.** This will cause an electron in a negative ion to jump away from its parent atom's orbit and land in an adjacent atom to the left within the metal wire conductor, as shown in Figure 1-9(b). This adjacent atom now has an extra electron and is called a negative ion, while the initial parent negative ion becomes a neutral atom, which will now receive an electron from one of the other negative ions, because their electrons are also feeling the attraction of the positive ions on the left side and the repulsion of the surrounding negative ions.

The electrons of the negative ion within the metal conductor feel the attraction of the positive ions, and eventually one of its electrons jumps to the left and into the adjacent atom, as shown in Figure 1-9(c). This continual movement to the left will produce a stream of electrons flowing from right to left. Millions upon millions of atoms within the conductor pass a continuous movement of billions upon billions of electrons from right to left. This electron flow is known as **electric current.**

To summarize, we could say that as long as a force or pressure, produced by the positive charge and negative charge, exists it will cause electrons to flow from the negative to the positive terminal. The positive side has a deficiency of electrons and the negative side has an abundance, and so a continuous flow or migration of electrons takes place between the negative and positive terminal through our metal conducting wire. This electric current or electron flow is a measurable quantity, as will now be explained.

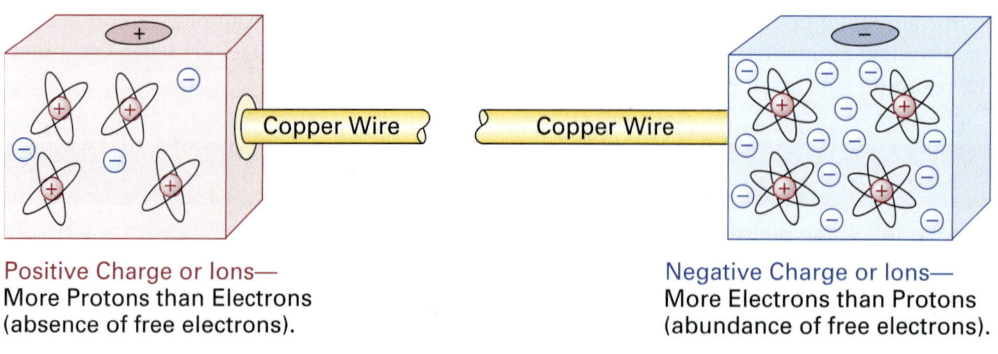

FIGURE 1-8 Positive and Negative Charges.

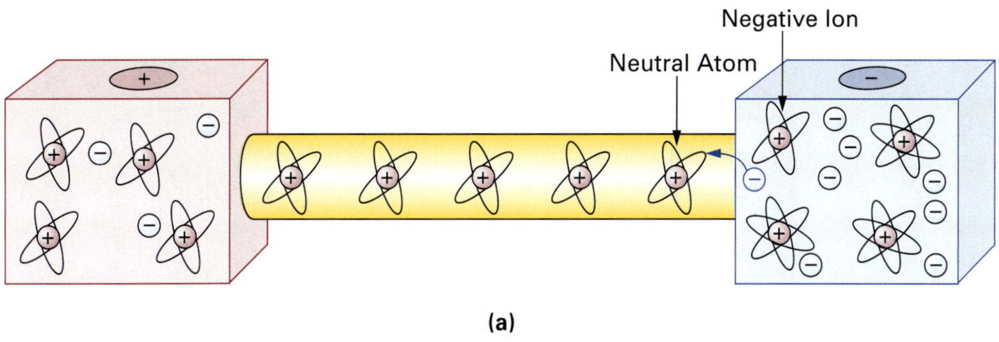

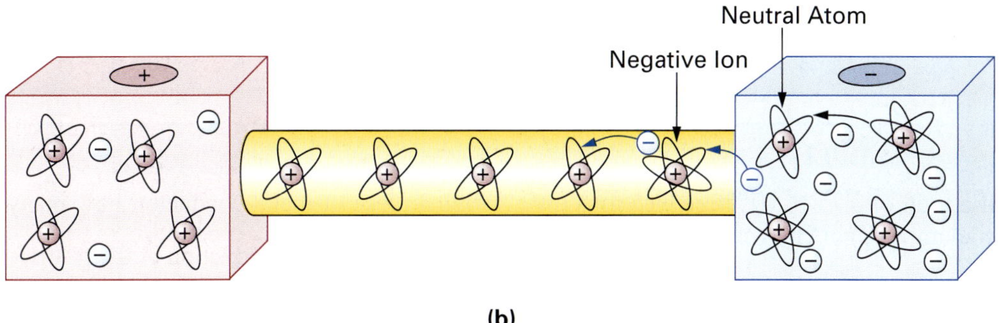

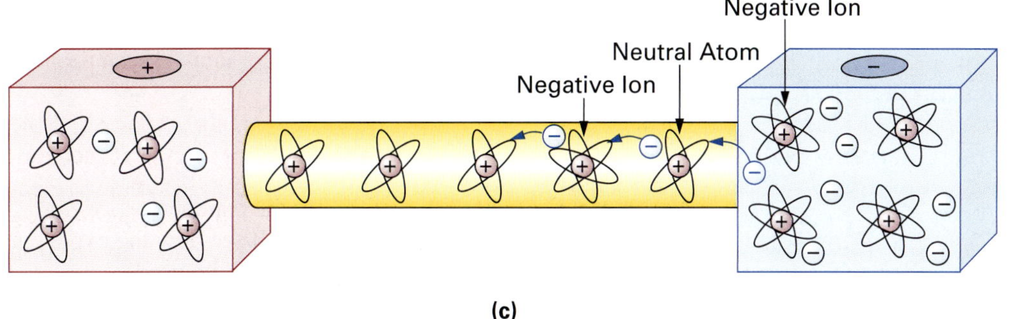

FIGURE 1-9 Electron Migration Due to Forces of Positive Attraction and Negative Repulsion on Electrons.

TIME LINE

Ironically, Coulomb's law was not first discovered by Charles A. de Coulomb (1736-1806), but by Henry Cavendish, a wealthy scientist and philosopher. Cavendish did not publish his discovery, which he made several years before Coulomb discovered the law independently. James Clerk Maxwell published the scientific notebooks of Cavendish in 1879, describing his experiments and conclusions. However, about 100 years had passed, and Coulomb's name was firmly associated with the law. Many scientists demanded that the law be called Cavendish's law, while other scientists refused to change, stating that Coulomb was the discoverer because he made the law known promptly to the scientific community.

1-2-1 *Coulombs per Second*

There are 6.24×10^{18} electrons in 1 **coulomb of charge,** as illustrated in Figure 1-10. To calculate coulombs of charge (designated Q), we can use the formula

Coulomb of Charge
Unit of electric charge. One coulomb equals 6.24×10^{18} electrons.

$$\text{charge, } Q = \frac{\text{total number of electrons } (n)}{6.24 \times 10^{18}}$$

where Q is the electric charge in coulombs.

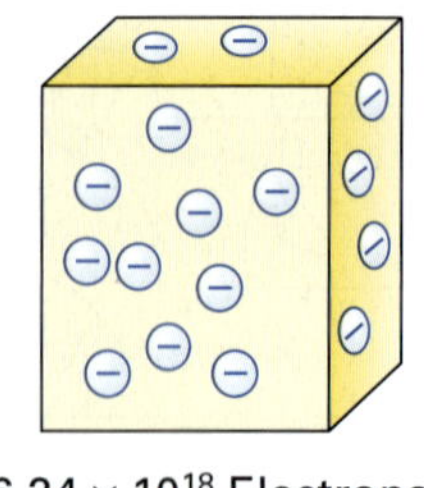
(a) 6.24×10^{18} Electrons = 1 Coulomb of Charge

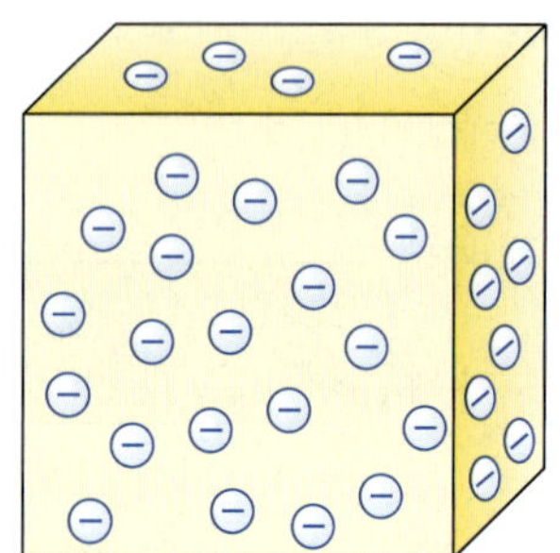
(b) 12.48×10^{18} Electrons = 2 Coulombs of Charge

FIGURE 1-10 **(a) 1 C of Charge. (b) 2 C of Charge.**

EXAMPLE:

If a total of 3.75×10^{19} free electrons exist within a piece of metal conductor, how many coulombs (C) of charge would be within this conductor?

Solution:

By using the charge formula (Q) we can calculate the number of coulombs (C) in the conductor.

CALCULATOR SEQUENCE

Step	Keypad Entry	Display Response
1.	3 . 7 5 E (Exponent) 1 9	3.75E19
2.	÷	
3.	6 . 2 4 E 1 8	6.24E18
4.	=	6.0096

$$Q = \frac{n}{6.24 \times 10^{18}}$$

$$= \frac{3.75 \times 10^{19}}{6.24 \times 10^{18}}$$

$$= 6 \text{ C}$$

A total of 6 C of charge exists within the conductor.

In the calculator sequence you will see how the exponent key (E, EE, or EXP) on your calculator can be used.

1-2-2 *The Ampere*

Ampere (A)
Unit of electric current.

A coulomb is a static amount of electric charge. In electronics, we are more interested in electrons in motion. Coulombs and time are therefore combined to describe the number of electrons and the rate at which they flow. This relationship is called *current (I)* flow and has the unit of **amperes (A).** By definition, 1 ampere of current is said to be flowing if 6.24×10^{18} electrons (1 C) are drifting past a specific point on a conductor in 1 second of time. Stated as a formula:

$$\text{current } (I) = \frac{\text{coulombs } (Q)}{\text{time } (t)}$$

$$1 \text{ ampere} = 1 \text{ coulomb per 1 second}$$

$$1 \text{ A} = \frac{1 \text{ C}}{1 \text{ s}}$$

In summary, 1 ampere equals a flow rate of 1 coulomb per second, and current is measured in amperes.

TABLE 1-2 Current Units

NAME	SYMBOL	VALUE
Picoampere	pA	$10^{-12} = \frac{1}{1{,}000{,}000{,}000{,}000}$
Nanoampere	nA	$10^{-9} = \frac{1}{1{,}000{,}000{,}000}$
Microampere	µA	$10^{-6} = \frac{1}{1{,}000{,}000}$
Milliampere	mA	$10^{-3} = \frac{1}{1000}$
Ampere	A	$10^{0} = 1$
Kiloampere	kA	$10^{3} = 1000$
Megaampere	MA	$10^{6} = 1{,}000{,}000$
Gigaampere	GA	$10^{9} = 1{,}000{,}000{,}000$
Teraampere	TA	$10^{12} = 1{,}000{,}000{,}000{,}000$

TIME LINE

The unit of electrical current is the ampere, named in honor of André Ampère (1775-1835), a French physicist who pioneered in the study of electromagnetism. After hearing of Hans Oerstad's discoveries, he conducted further experiments and discovered that two current-carrying conductors would attract and repel one another, just like two magnets.

EXAMPLE:

If 5×10^{19} electrons pass a point in a conductor in 4 s, what is the amount of current flow in amperes?

Solution:

Current (I) is equal to Q/t. We must first convert electrons to coulombs.

$$Q = \frac{n}{6.24 \times 10^{18}}$$

$$= \frac{5 \times 10^{19}}{6.24 \times 10^{18}}$$

$$= 8 \text{ C}$$

Now, to calculate the amount of current, we use the formula

$$I = \frac{Q}{t}$$

$$= \frac{8 \text{ C}}{4 \text{ s}}$$

$$= 2 \text{ A}$$

This means that 2 A or 1.248×10^{19} electrons (2 C) are passing a specific point in the conductor every second.

CALCULATOR SEQUENCE

Step	Keypad Entry	Display Response
1.	5 E (exponent) 1 9	5E19
2.	÷	
3.	6 . 2 4 E 1 8	6.24E18
4.	=	8.012
5.	÷	
6.	4	2.003
7.	=	2.003

1-2-3 *Units of Current*

Current within electronic equipment is normally a value in milliamperes or microamperes and very rarely exceeds 1 ampere. Table 1-2 lists all the prefixes related to current. For example, 1 milliampere is one-thousandth of an ampere, which means that if 1 ampere were divided into 1000 parts, 1 part of the 1000 parts would be flowing through the circuit.

EXAMPLE:

Convert the following:

a. 0.003 A = ____________ mA (milliamperes)

b. 0.07 mA = ____________ µA (microamperes)

c. 7333 mA = ____________ A (amperes)
d. 1275 μA = ____________ mA (milliamperes)

Solution:

a. 0.003A = ____________ mA. In this example, 0.003 A has to be converted so that it is represented in milliamperes (10^{-3} or 1/1000 of an ampere). The basic algebraic rule to be remembered is that both expressions on either side of the equals must be equal.

LEFT			RIGHT	
Base	Multiplier		Base	Multiplier
0.003	$\times 10^0$	=	______	$\times 10^{-3}$

The multiplier on the right in this example is going to be decreased 1000 times (10^0 to 10^{-3}), so for the statement to balance the number on the right will have to be increased 1000 times; that is, the decimal point will have to be moved to the right three places (0.003 or 3). Therefore,

$$0.003 \times 10^0 = 3 \times 10^{-3}$$

or

$$0.003 \text{ A} = 3 \times 10^{-3} \text{ A or } 3 \text{ mA}$$

b. 0.07 mA = ____________ μA. In this example the unit is going from milliamperes to microamperes (10^{-3} to 10^{-6}) or 1000 times smaller, so the number must be made 1000 times greater.

0.070 or 70.0

Therefore, 0.07 mA = 70 μA.

c. 7333 mA = ____________ A. The unit is going from milliamperes to amperes, increasing 1000 times, so the number must decrease 1000 times.

7333 or 7.333

Therefore, 7333 mA = 7.333 A.

d. 1275 μA = ____________ mA. The unit is changing from microamperes to milliamperes, an increase of 1000 times, so the number must decrease by the same factor.

1275.0 or 1.275

Therefore, 1275 μA = 1.275 mA.

TIME LINE

The unit of voltage, the volt, was named in honor of Alessandro Volta (1745–1827), an Italian physicist who is famous for his invention of the electric battery. In 1801, he was called to Paris by Napoleon to show his experiment on the generation of electric current.

1-2-4 *Conventional versus Electron Flow*

Electrons drift from a negative to a positive charge, as illustrated in Figure 1-11. As already stated, this current is known as **electron flow.**

In the eighteenth and nineteenth centuries, when very little was known about the atom, researchers believed that current was a flow of positive charges. Although this has now been proved incorrect, many texts still use **conventional current flow,** which is shown in Figure 1-12.

Whether conventional flow or electron flow is used, the same answers to problems, measurements, and designs are obtained. The key point to remember is that direction is not important, but the amount of current flow is.

Throughout this book we will be using electron flow so that we can relate back to the atom when necessary. If you wish to use conventional flow, just reverse the direction of the arrows. To avoid confusion, be consistent with your choice of flow.

Electron Flow
A current produced by the movement of free electrons toward a positive terminal.

Conventional Current Flow
A current produced by the movement of positive charges toward a negative terminal.

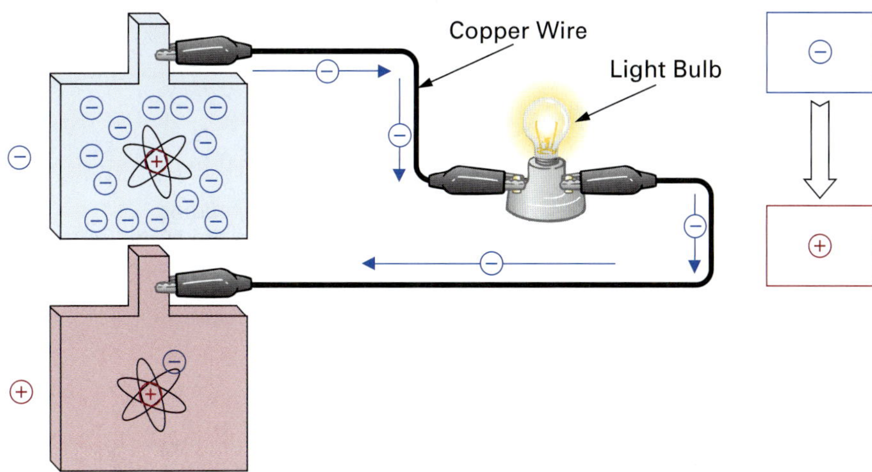

FIGURE 1-11 **Electron Current Flow.**

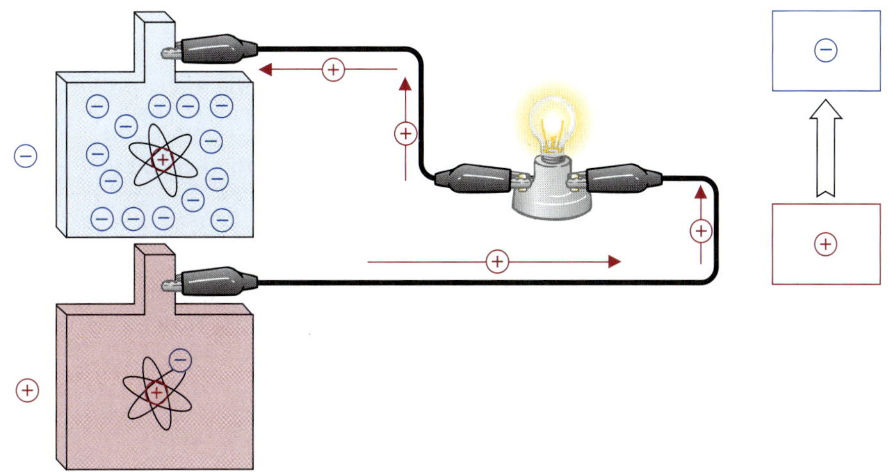

FIGURE 1-12 **Conventional Current Flow.**

TIME LINE
The galvanometer, which is used to measure electrical current, was named after Luigi Galvani (1737–1798), who conducted many experiments with electrical current, or, as it was known at the time, "galvanism."

1-2-5 *How Is Current Measured?*

Ammeters (ampere meters) are used to measure the current flow within a circuit. Stepping through the sequence detailed in Figure 1-13, you will see how an ammeter is used to measure the value of current within a circuit.

In the simple circuit shown in Figure 1-13, an ON/OFF switch is being used to turn ON or OFF a small light bulb. One of the key points to remember with ammeters is that if you wish to measure the value of current flowing within a wire, the current path must be opened and the ammeter placed in the path so that it can sense and display the value.

Ammeter
Meter placed in the path of current flow to measure the amount.

SELF-TEST EVALUATION POINT FOR SECTION 1-2

Use the following questions to test your understanding of Section 1-2:

1. What is the unit of current?
2. Define current in relation to coulombs and time.
3. What is the difference between conventional and electron current flow?
4. What test instrument is used to measure current?

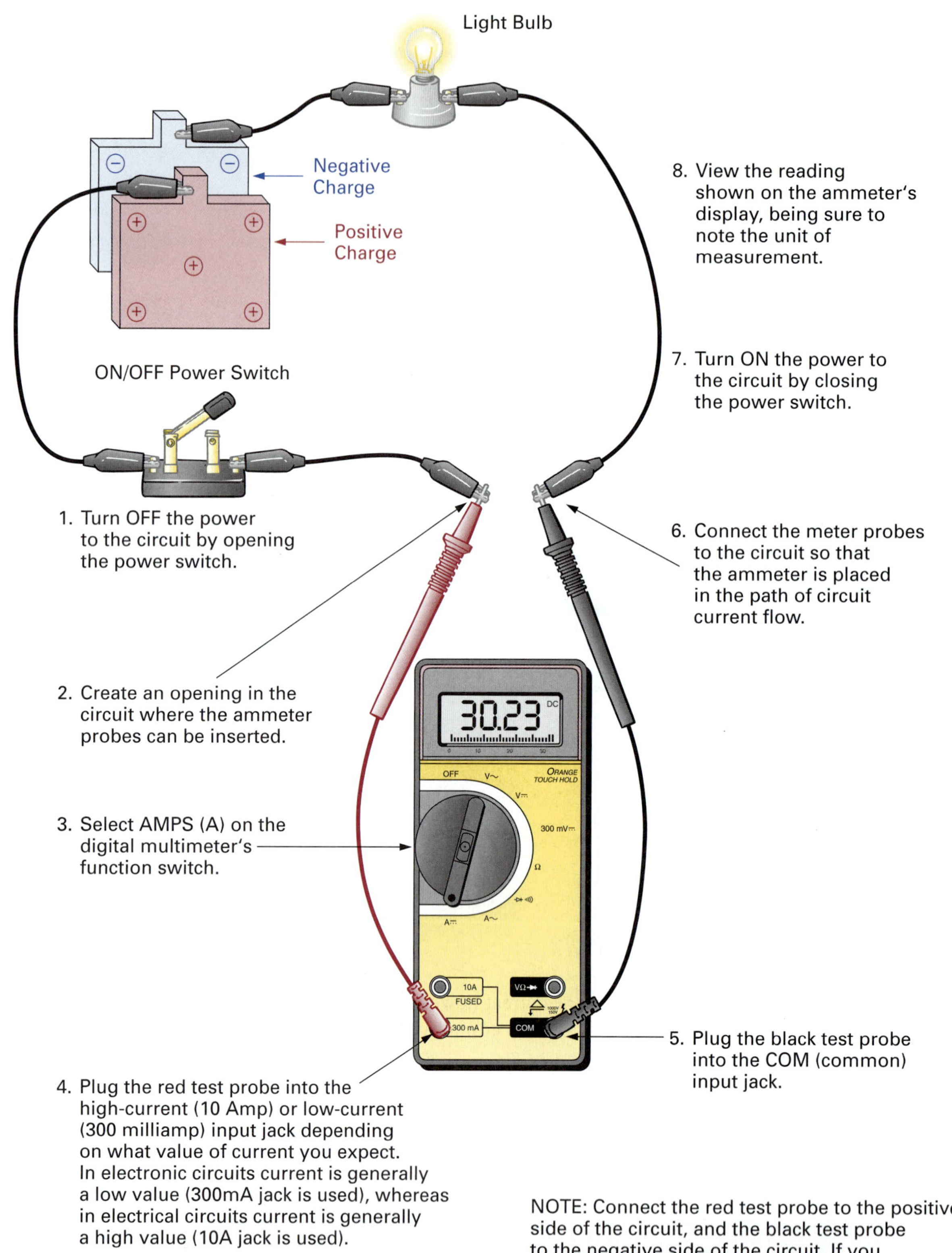

FIGURE 1-13 **Measuring Current with the Ammeter.**

1-3 VOLTAGE

Voltage is the force or pressure exerted on electrons. Referring to Figure 1-14(a) and (b), you will notice two situations. Figure 1-14(a) shows highly concentrated positive and negative charges or potentials connected to one another by a copper wire. In this situation, a large potential difference or voltage is being applied across the copper atom's electrons. This force or voltage causes a large amount of copper atom electrons to move from right to left. On the other hand, Figure 1-14(b) illustrates a low concentration of positive and negative potentials, so a small voltage or pressure is being applied across the conductor, causing a small amount of force, and therefore current, to move from right to left.

In summary, we could say that a highly concentrated charge produces a high voltage, whereas a low concentrated charge produces a low voltage. Voltage is also appropriately known as the "electron moving force" or **electromotive force (emf),** and since two opposite potentials exist (one negative and one positive), the strength of the voltage can also be

Voltage (*V* or *E*)
Term used to designate electrical pressure or the force that causes current to flow.

Electromotive Force (emf)
Force that causes the motion of electrons due to a potential difference between two points.

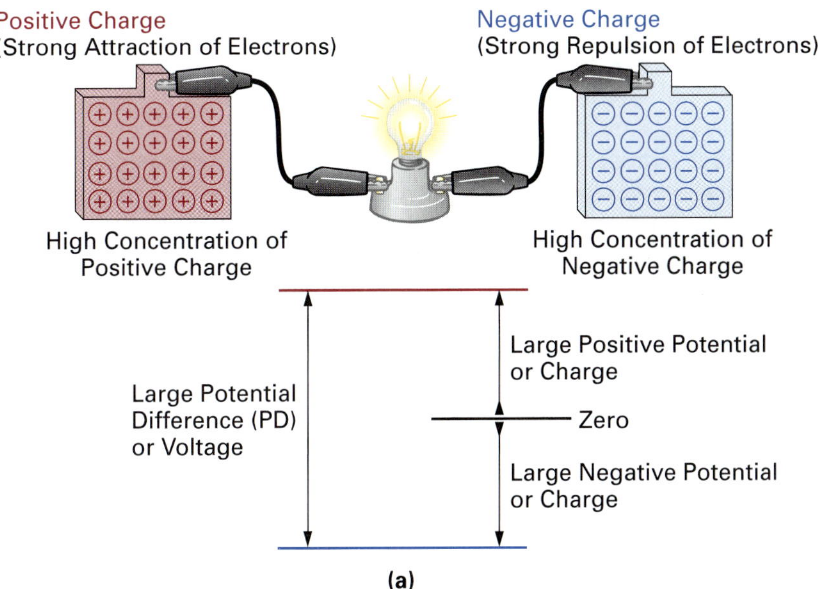

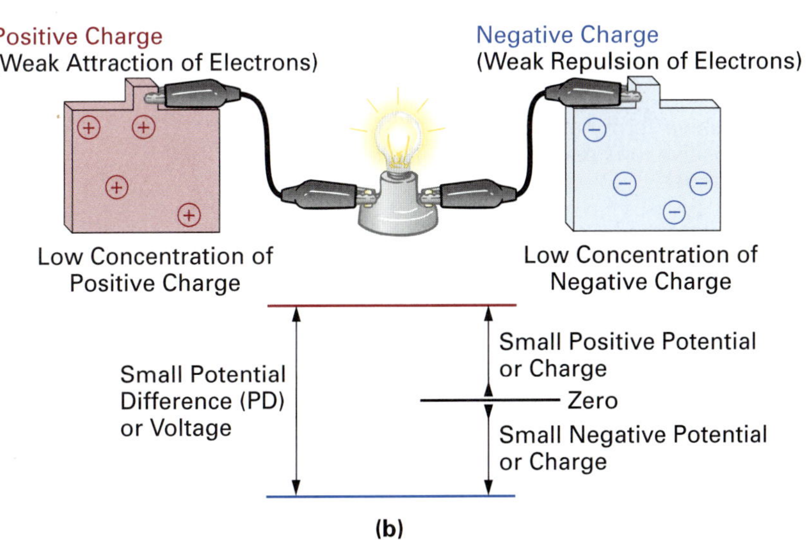

FIGURE 1-14 (a) Large Potential Difference or Voltage. (b) Small Potential Difference or Voltage.

Potential Difference (PD)

Voltage difference between two points, which will cause current to flow in a closed circuit.

Battery

DC voltage source containing two or more cells that converts chemical energy into electrical energy.

TIME LINE

One of the best known and most admired men in the latter half of the eighteenth century was the American Benjamin Franklin (1706–1790). He is renowned for his kite-in-a-storm experiment, which proved that lightning is electricity. Franklin also discovered that there was only one type of electricity, and that the two previously believed types were simply two characteristics of electricity. Vitreous was renamed *positive charge* and resinous was called *negative charge,* terms that were invented by Franklin, along with the terms *battery* and *conductor,* which are still used today.

referred to as the amount of **potential difference (PD)** applied across the circuit. To compare, we can say that a large voltage, electromotive force, or potential difference exists across the copper conductor in Figure 1-14(a), while a small voltage, potential difference, or electromotive force is exerted across the conductor in Figure 1-14(b).

Voltage is the force, pressure, potential difference (PD), or electromotive force (emf) that causes electron flow or current and is symbolized by italic uppercase V. The unit for voltage is the volt, symbolized by roman uppercase V. This can become a bit confusing. For example, when the voltage applied to a circuit equals 5 volts, the circuit notation would appear as

$$V = 5\ \text{V}$$

You know the first V represents "voltage," not "volt," because 1 volt cannot equal 5 volts. To avoid confusion, some texts and circuits use *E,* symbolizing electromotive force, to represent voltage; for example,

$$E = 5\ \text{V}$$

In this text, we will maintain the original designation for voltage (V).

1-3-1 *A Simple Voltage Source*

A **battery,** like the one shown in Figure 1-15(a), converts chemical energy into electrical energy. At the positive terminal of the battery, positive charges or ions (atoms with more protons than electrons) are present, and at the negative terminal, negative charges or ions (atoms with more electrons than protons) are available to supply electrons for current flow within a circuit. A battery, therefore, chemically generates negative and positive ions at its respective terminals. The symbol for the battery is shown in Figure 1-15(b).

1-3-2 *Schematic Symbols*

In Figure 1-16(a) you can see the schematic symbols for many of the devices discussed so far, along with their physical appearance.

In the circuit shown in Figure 1-16(b), a 9 V battery chemically generates positive and negative ions. The negative ions at the negative terminal force away the negative electrons, which are attracted by the positive charge or absence of electrons at the positive terminal. As the electrons proceed through the copper conductor wire, jumping from one atom to the next, they eventually reach the bulb. As they pass through the bulb, they cause it to glow. When emerging from the light bulb, the electrons travel through another connector cable and finally reach the positive terminal of the battery.

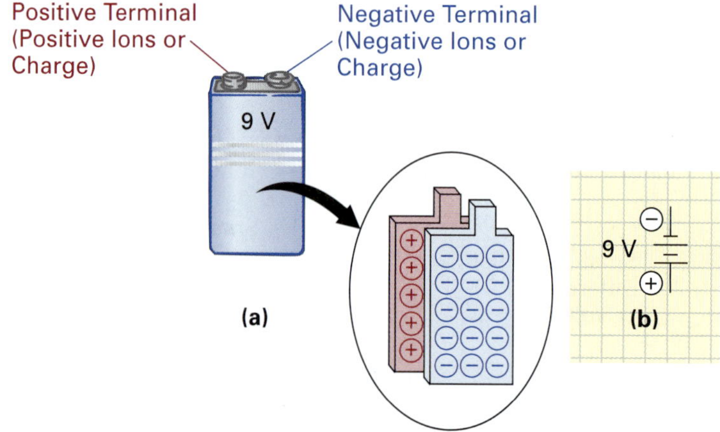

FIGURE 1-15 **The Battery—A Source of Voltage. (a) Physical Appearance. (b) Schematic Symbol.**

Studying Figure 1-16(b), you will notice two reasons why the circuit is drawn using symbols rather than illustrating the physical appearance:

1. A circuit with symbols can be drawn faster and more easily.
2. A circuit with symbols has less detail and clutter, and is therefore more easily comprehended since it has fewer distracting elements.

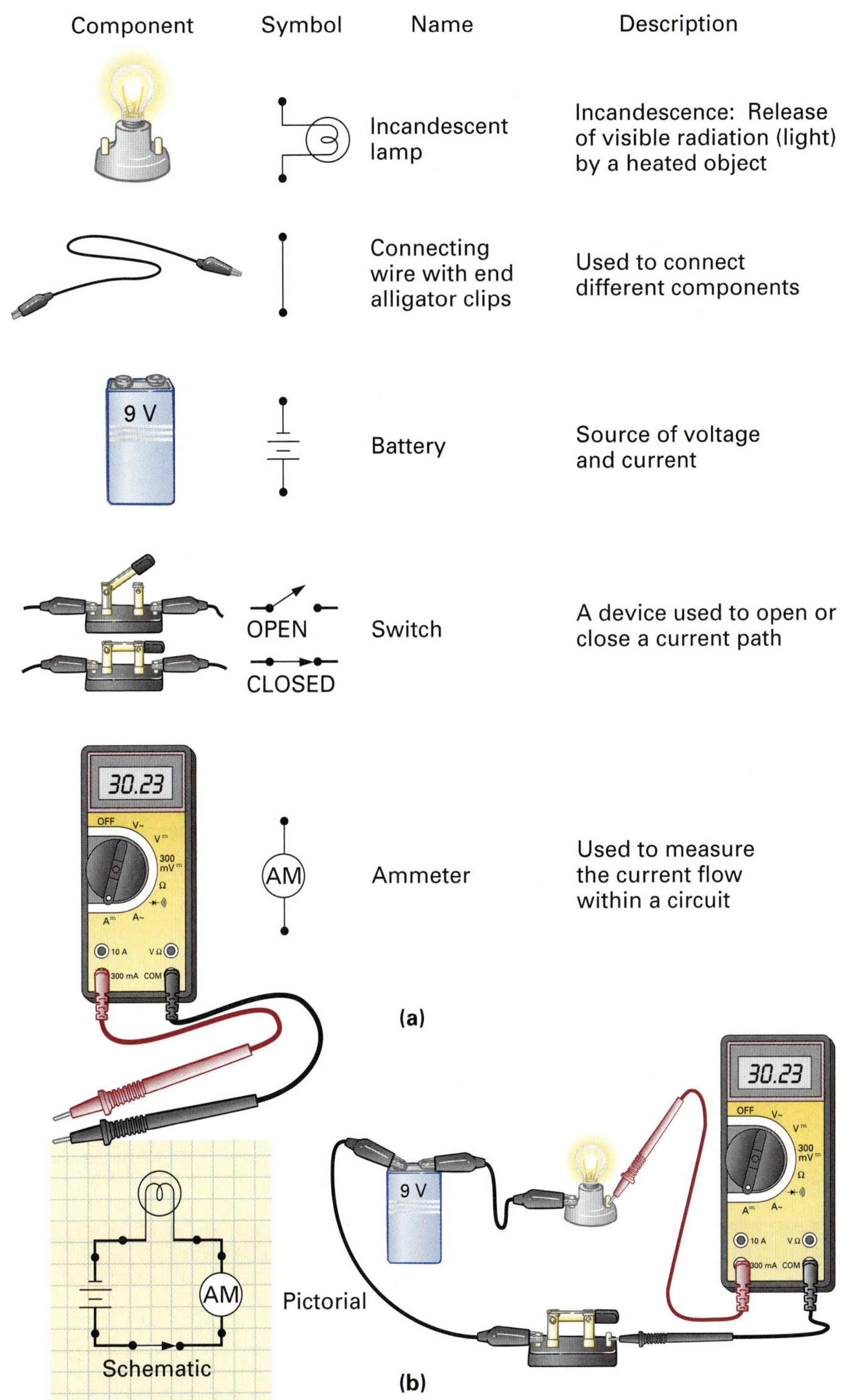

FIGURE 1-16 **(a) Components. (b) Example Circuit.**

TABLE 1-3 Voltage Units

NAME	SYMBOL	VALUE
Picovolts	pV	$10^{-12} = \frac{1}{1,000,000,000,000}$
Nanovolts	nV	$10^{-9} = \frac{1}{1,000,000,000}$
Microvolts	μV	$10^{-6} = \frac{1}{1,000,000}$
Millivolts	mV	$10^{-3} = \frac{1}{1000}$
Volts	V	$10^{0} = 1$
Kilovolts	kV	$10^{3} = 1000$
Megavolts	MV	$10^{6} = 1,000,000$
Gigavolts	GV	$10^{9} = 1,000,000,000$
Teravolts	TV	$10^{12} = 1,000,000,000,000$

1-3-3 *Units of Voltage*

The unit for voltage is the volt (V). Voltage within electronic equipment is normally measured in volts, whereas heavy-duty industrial equipment normally requires high voltages that are generally measured in kilovolts (kV). Table 1-3 lists all the prefixes and values related to volts.

EXAMPLE:

Convert the following:

a. 3000 V = ____________ kV (kilovolts)
b. 0.14 V = ____________ mV (millivolts)
c. 1500 kV = ____________ MV (megavolts)

Solution:

a. 3000 V = 3 kV or 3×10^{3} volts (multiplier ↑ 1000, number ↓ 1000)
b. 0.14 V = 140 mV or 140×10^{-3} volt (multiplier ↓ 1000, number ↑ 1000)
c. 1500 kV = 1.5 MV or 1.5×10^{6} volts (multiplier ↑ 1000, number ↓ 1000)

1-3-4 *How Is Voltage Measured?*

Voltmeter
Instrument designed to measure the voltage or potential difference. Its scale can be graduated in kilovolts, volts, or millivolts.

Voltmeters (voltage meters) are used to measure electrical pressure or voltage. Stepping through the sequence detailed in Figure 1-17, you will see how a voltmeter can be used to measure a battery's voltage.

In many instances, the voltmeter is used to measure the potential difference or voltage drop across a device, as shown in Figure 1-18(a) and (b). Figure 1-18(a) shows how to measure the voltage across light bulb 1 (L1), and Figure 1-18(b) shows how to measure the voltage across light bulb 2 (L2).

SELF-TEST EVALUATION POINT FOR SECTION 1-3

Use the following questions to test your understanding of Section 1-3:

1. What is the unit of voltage?
2. Convert 3 MV to kilovolts.
3. Which meter is used to measure voltage?

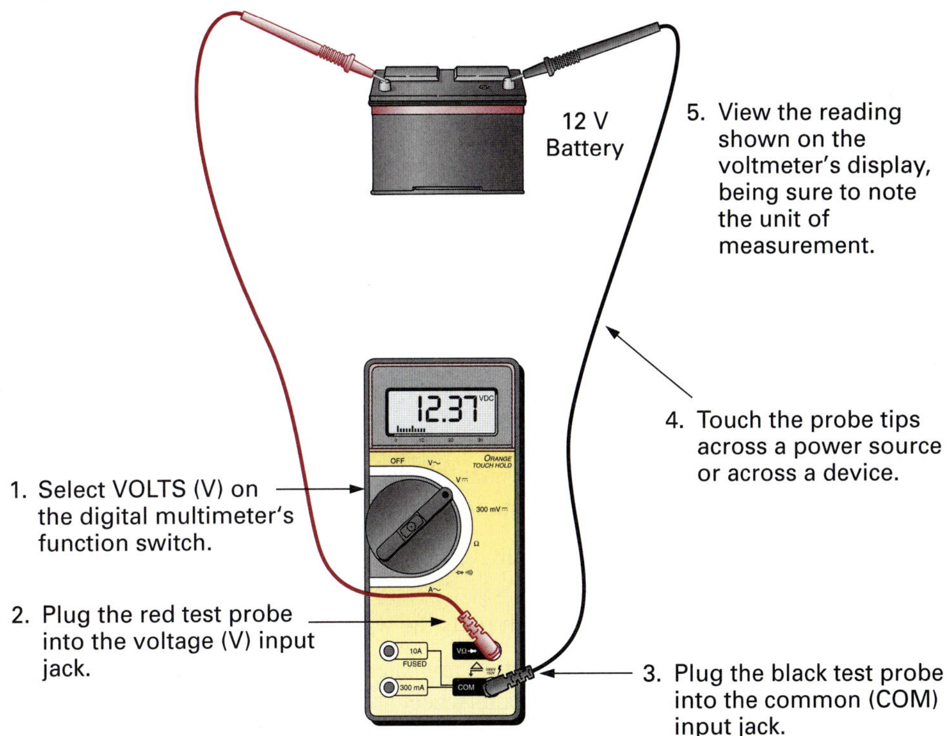

FIGURE 1-17 Using the Voltmeter to Measure Voltage.

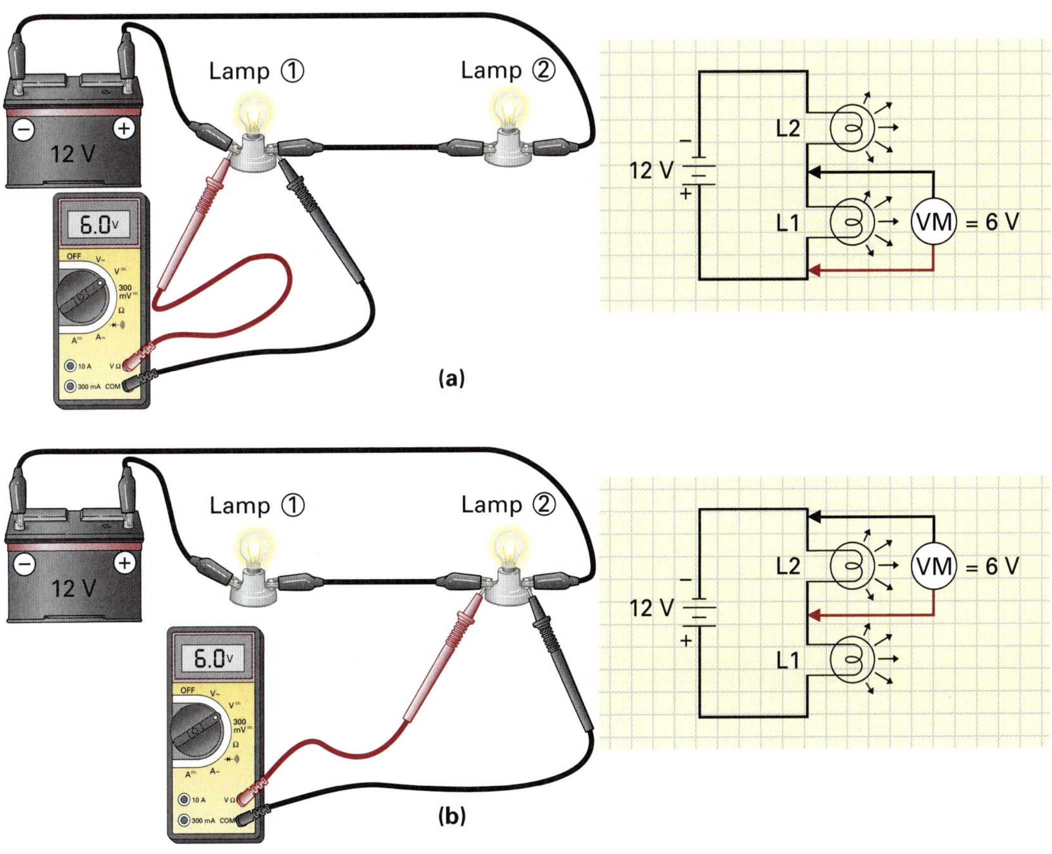

FIGURE 1-18 Measuring the Voltage Drop across Components (a) Lamp 1 and (b) Lamp 2.

1-4 COMPONENTS AND CIRCUITS

Just as a complete path exists around a track's circuit for an athlete, a complete path also exists around an electric circuit for current. Components are individual devices such as wires, batteries, lamps, and switches that serve as the building blocks for these circuits.

In this section, we will begin with a fluid analogy in order to reinforce your understanding of current and voltage and then examine some of the components introduced so far in more detail.

1-4-1 *Fluid Analogy of Current and Voltage*

In Figure 1-19(a), a system using a pump, pipes, and a waterwheel is being used to convert electrical energy into mechanical energy. When electrical energy is applied to the pump, it will operate and cause water to flow. The pump generates:

1. A high pressure at the outlet port, which pushes the water molecules out and into the system
2. A low pressure at the inlet port, which pulls the water molecules into the pump

The water current flow is in the direction indicated, and the high pressure or potential within the piping will be used to drive the waterwheel around, producing mechanical energy. The remaining water is attracted into the pump due to the suction or low pressure existing at the inlet port. In fact, the amount of water entering the inlet port is the same as the amount of water leaving the outlet port. It can therefore be said that the water flow rate is the same throughout the circuit. The only changing element is the pressure felt at different points throughout the system.

In Figure 1-19(b), an electric circuit containing a battery, conductors, and a bulb is being used to convert electrical energy into light energy. The battery generates a voltage just as the pump generates pressure. This voltage causes electrons to move through conductors, just as pressure causes water molecules to move through the piping. The amount of water flow is dependent on the pump's pressure, and the amount of current or electron flow is dependent on the battery's voltage. Water flow through the wheel can be compared to current flow through the bulb. The high pressure is lost in turning the wheel and producing mechanical energy, just as voltage is lost in producing light energy out of the bulb. We cannot say that pressure or voltage flows: Pressure and voltage are applied and cause water or current to flow, and it is this flow that is converted to mechanical energy in our fluid system and light energy in our electrical system. Voltage is the force of repulsion and attraction needed to cause current to flow through a circuit, and without this potential difference or pressure there cannot be current.

Current (I) is therefore said to be *directly proportional* ($\propto$) *to voltage (V),* as a voltage decrease ($V\downarrow$) results in a current decrease ($I\downarrow$), and similarly, a voltage increase ($V\uparrow$) causes a current increase ($I\uparrow$), as in Figure 1-20.

Current is directly proportional to voltage ($I \propto V$)

$V\downarrow$ causes a $I\downarrow$

$V\uparrow$ causes a $I\uparrow$

Proportional
A term used to describe the relationship between two quantities that have the same ratio.

As you can see from this example, directly **proportional** ($\propto$) is a phrase which means that one term will change in proportion, or in size, relative to another term.

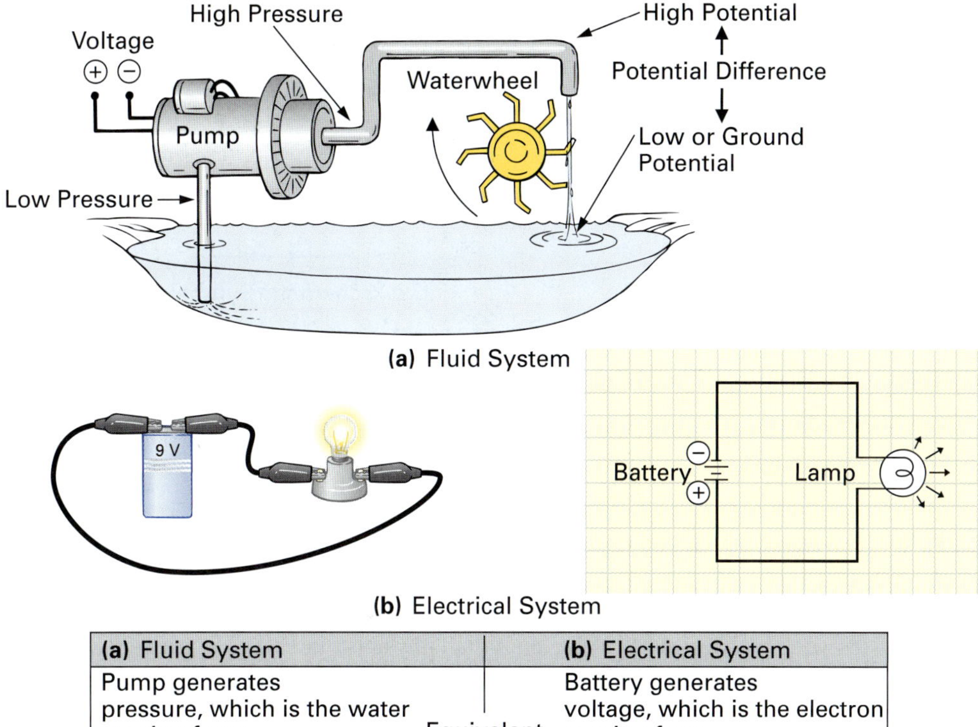

(a) Fluid System		(b) Electrical System
Pump generates pressure, which is the water moving force.	Equivalent to (≅)	Battery generates voltage, which is the electron moving force.
Water current flow.		Electron current flow.
High pressure or potential.		High voltage or potential.
Low pressure or potential.		Low voltage or potential.

FIGURE 1-19 Comparison between a Fluid System (a) and an Electrical System (b).

1-4-2 *Switches*

As we have seen previously, a *switch* is a device that completes (closes) or breaks (opens) the path of current, as seen in Figure 1-21. All mechanical switches can be classified into one of the eight categories shown in Figure 1-22. Many different variations of these eight different classifications exist, and throughout this text you will see many of them incorporated into a variety of circuit applications.

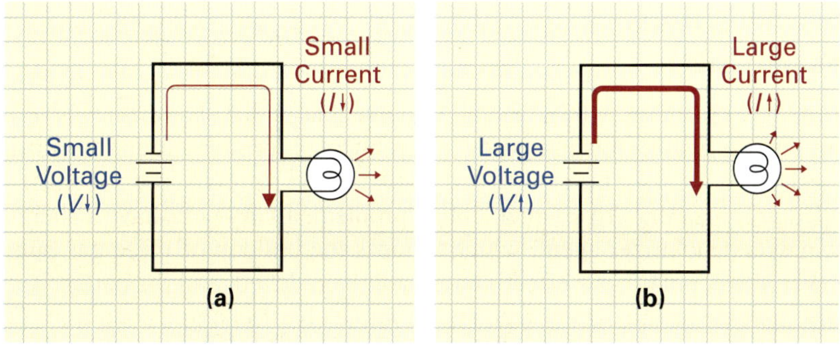

FIGURE 1-20 Current Is Directly Proportional to Voltage. (a) Small Voltage Produces a Small Current. (b) Large Voltage Produces a Large Current.

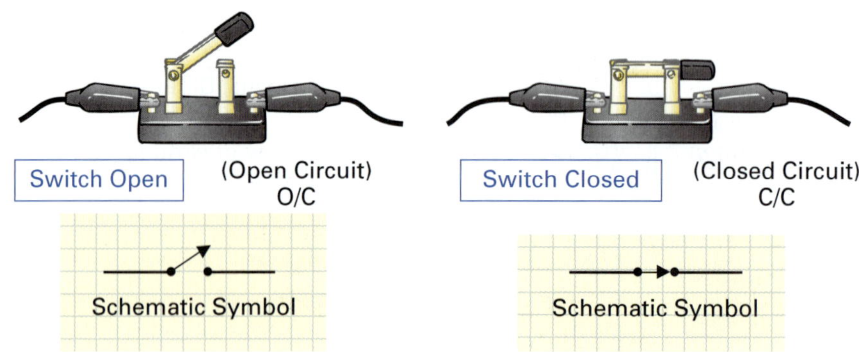

FIGURE 1-21 **Open and Closed Switches.**

1-4-3 *Batteries*

Voltaic Cell
A battery cell having two unlike metal electrodes immersed in a solution that chemically interacts with the plates to produce an emf.

The battery is a chemical voltage source that contains one or more voltaic cells. The **voltaic cell** was discovered by Alessandro Volta, an Italian physicist, in 1800. Figure 1-23 shows the three basic components within a battery's cell, which are a negative plate, a positive plate, and an electrolyte.

Cell Operation

Electrolyte
Electrically conducting liquid (wet) or paste (dry).

Two dissimilar, separated metal plates such as copper and zinc are placed within a container that is filled with a substance known as the **electrolyte,** usually an acid. A chemical reaction causes electrons to pass through the electrolyte as they are repelled from one plate and attracted to the other. This action causes a large number of electrons to collect on one plate (negative plate), and an absence or deficiency of electrons to exist on the opposite plate (positive plate). The electrolyte therefore acts on the two plates and transforms chemical energy into electrical energy, which can be taken from the cell at its two output terminals as an electrical current flow.

If nothing is connected across the battery, as shown in Figure 1-24(a), a chemical reaction between the electrolyte and the negative electrode produces free electrons that travel from atom to atom but are held in the negative electrode. On the other hand, if a light bulb is connected between the negative and positive electrodes, as shown in Figure 1-24(b), the mutual repulsion of the free electrons at the negative electrode combined with the attraction of the positive electrode will cause a migration of free electrons (current flow) through the light bulb, causing its filament to produce light.

Primary Cells

Primary Cell
Cell that produces electrical energy through an internal electrochemical action; once discharged, it cannot be reused.

Dry Cell
DC voltage-generating chemical cell using a nonliquid (paste) type of electrolyte.

The chemical reaction within the cell will eventually dissolve the negative plate. This discharging process also results in hydrogen gas bubbles forming around the positive plate, causing a resistance between the two plates, known as the battery's internal resistance, to increase (called *polarization*). To counteract this problem, all dry cells have a chemical within them known as a depolarizer agent, which reduces the buildup of these gas bubbles. With time, however, the depolarizer's effectiveness will be reduced, and the battery's internal resistance will increase as the battery reaches a completely discharged condition. These types of batteries are known as **primary cells** (first time, last time) because they discharge once and must then be discarded. Almost all primary cells have their electrolyte in paste form, which is why they are also referred to as **dry cells.** Wet cells have an electrolyte that is in liquid form.

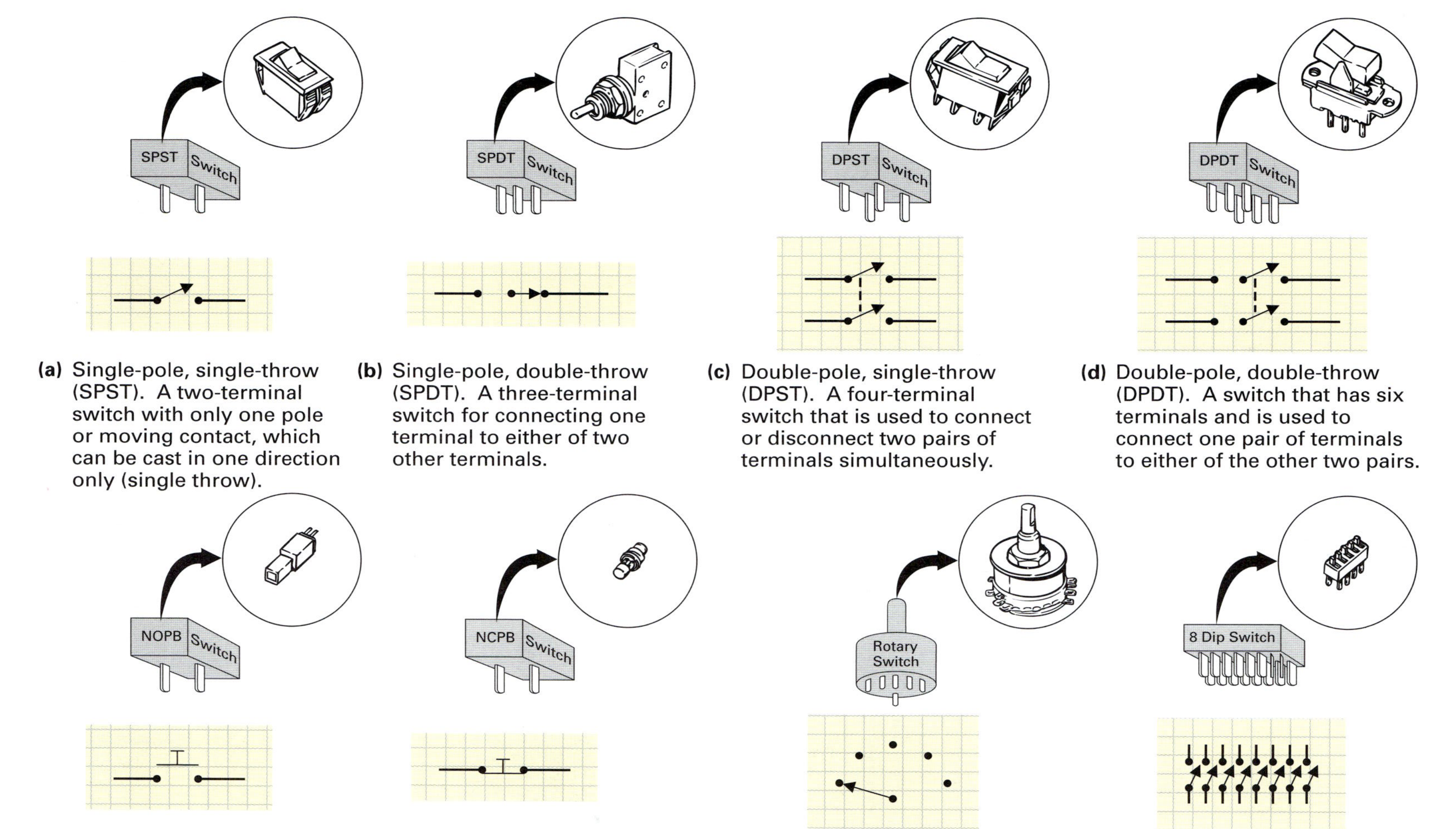

(a) Single-pole, single-throw (SPST). A two-terminal switch with only one pole or moving contact, which can be cast in one direction only (single throw).

(b) Single-pole, double-throw (SPDT). A three-terminal switch for connecting one terminal to either of two other terminals.

(c) Double-pole, single-throw (DPST). A four-terminal switch that is used to connect or disconnect two pairs of terminals simultaneously.

(d) Double-pole, double-throw (DPDT). A switch that has six terminals and is used to connect one pair of terminals to either of the other two pairs.

(e) Normally open push button (NOPB). A switch that will make contact and pass current when it is pressed.

(f) Normally closed push button (NCPB). A switch that normally makes contact and passes current, but disconnects or opens when pressed.

(g) Rotary. An electromechanical device that is capable of making (closing contacts) or breaking (opening contacts) in a circuit.

(h) Dual-in-line package (DIP) switch. A DIP switch is a group of separate miniature switches within two (dual) rows of external connecting pins or terminals.

Single pole: one moving contact.
Double pole: two moving contacts.

Single throw: pole can be thrown or cast in only one direction.
Double throw: pole can be thrown or cast in two directions.

FIGURE 1-22 Eight Basic Types of Switches.

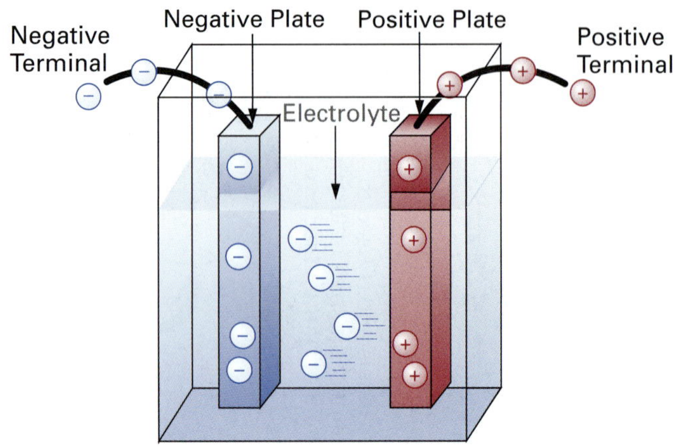

FIGURE 1-23 **A Battery's Cell.**

Primary cell types. The shelf life of a battery is the length of time a battery can remain on the shelf (in storage) and still retain its usability. Most primary cells will deteriorate in a three-year period to approximately 80% of their original capacity. Of all the types of primary cells, five dominate the market as far as applications and sales are concerned. These are the carbon–zinc, alkaline–manganese, mercury, silver oxide, and lithium types, all of which are described and illustrated in Table 1-4. The cell voltage in Table 1-4 describes the voltage produced by one *cell* (one set of plates). If two or more sets of plates are installed in one package, the component is called a *battery.*

Secondary Cells

Secondary Cells
An electrolytic cell for generating electric energy. Once discharged the cell may be restored or recharged by sending an electric current through the cell in the opposite direction to that of the discharge current.

Secondary cells operate on the same basic principle as that of primary cells. In this case, however, the plates are not eaten away or dissolved, they only undergo a chemical change during discharge. Once discharged, the secondary cell can have the chemical change that occurred during discharge reversed by recharging, resulting once more in a fully charged cell. Restoring a secondary cell to the charged condition is achieved by sending a current through the cell in the direction opposite to that of the discharge current, as illustrated in Figure 1-25.

In Figure 1-25(a), we have first connected a light bulb across the battery, and free electrons are being supplied by the negative electrode and moving through the light bulb or load and back to the positive electrode. After a certain amount of time, any secondary cell will

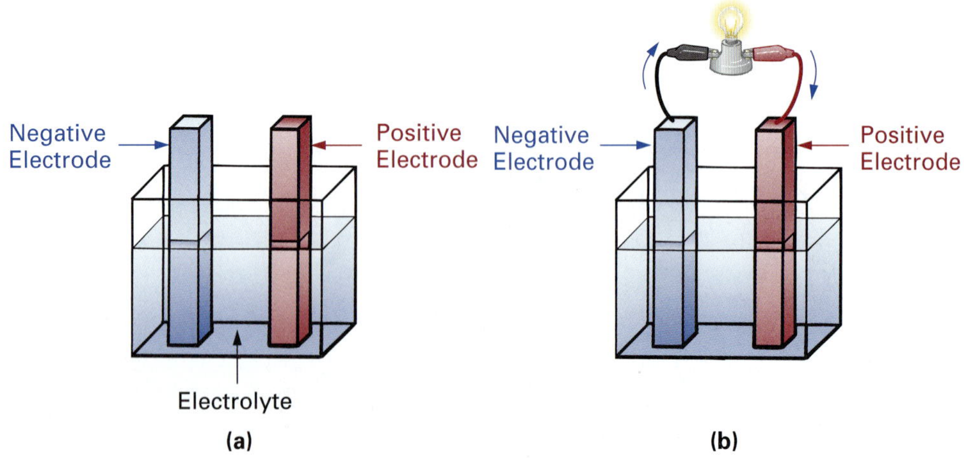

FIGURE 1-24 **Cell Operation.**

TABLE 1-4 Primary Cell Types

Battery type	Features	Cell Voltage (V)	Applications	Construction
Carbon–zinc (also called Leclanché cell in honor of its inventor)	Most popular due to its low cost. Cylindrical D and C cells are most commonly used. Flat cells are stacked in series to obtain voltages greater than 1.5 V, as in the case of the 9 V battery	1.5	Portable radios, tape players, televisions, toys.	Has a zinc case (negative), a carbon center rod (positive), with ammonium chloride as the electrolyte and manganese dioxide as the depolarizing agent. 9 V Flat cell has 6 stacked 1.5 V cells. Metal Cap (+ Terminal); Carbon Rod (+ Electrode); Electrolyte (Paste); Zinc Can (– Electrode); 9 V. (a) Carbon–Zinc
Alkaline–maganese (also called alkaline cell)	Has three times the shelf life and capacity of carbon zinc. Cylindrical or miniature cell sizes are available	1.4	Portable radios, tape players, televisions, toys; large capacity is worth investment	The negative plate is granular zinc mixed with an electrolyte (alkaline); the positive plate is a polarizer in contact with the outer metal can. Alkaline Miniature Cell. Anode; Cathode. (b) Alkaline–Manganese
Mercury (also called mercuric–oxide)	Higher energy density than preceding two with good shelf life and small size; used in low-power applications. Flat or cylindrical cells are available	1.35 and 1.4	Watches, hearing aids, pacemakers, cameras, test equipment	A zinc anode and a mercuric oxide cathode in a potassium hydroxide electrolyte. Amalgamated Zinc Anode; Mercuric Oxide Depolarizing Cathode; Potassium Hydroxide Electrolyte. (c) Mercury
Silver–oxide	High capacity; however, the material used (silver) makes it the most costly; can supply high currents for short periods of time; used in low-power applications	1.5	Watches, hearing aids, pacemakers, cameras	Contains a cathode of silver oxide, an alkaline electrolyte, and zinc anode. Anode; Cathode. (d) Silver-Oxide
Lithium	High discharge rate and long shelf life; light weight and higher output voltage; the familiar cylindrical and flat cell are available	1.9	Liquid crystal watches, semi-conductor memories, hand-held calculators, sensor circuits	Contains a lithium anode, porous carbon cathode, and sulfur dioxide electrolyte. Cathode; Solid Electrolyte; Anode

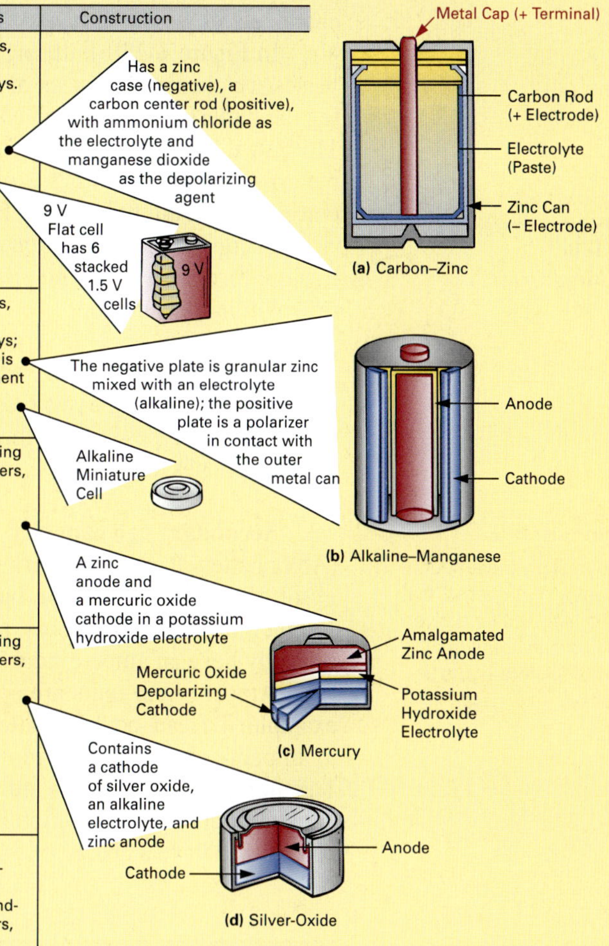

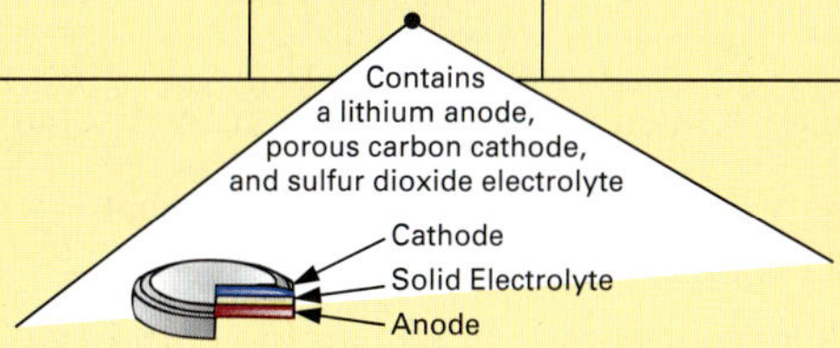

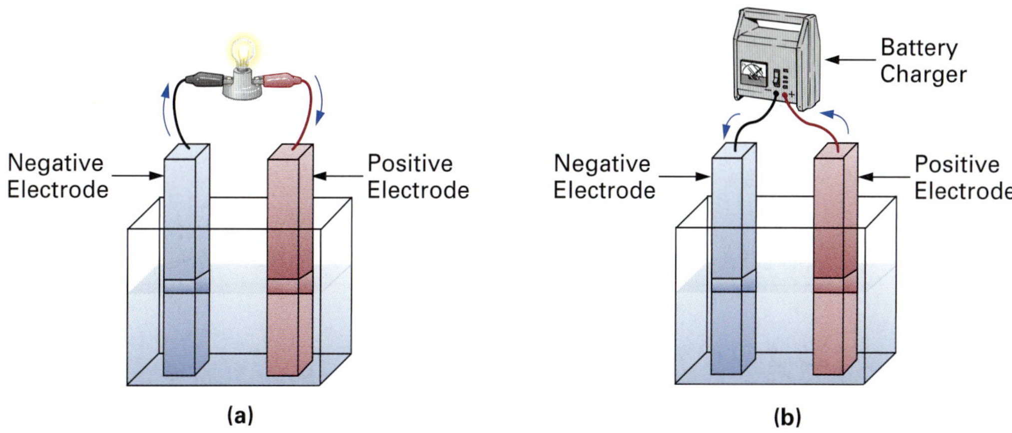

FIGURE 1-25 Secondary Cell Operation. (a) Discharge. (b) Charge.

run down or will have discharged to such an extent that no usable value of current can be produced. The surfaces of the plates are changed, and the electrolyte lacks the necessary chemicals.

In Figure 1-25(b), during the recharging process, we use a battery charger, which reverses the chemical process of discharge by forcing electrons back into the cell and restoring the battery to its charged condition. The battery charger voltage is normally set to about 115% of the battery voltage. To use an example, a battery charger connected across a 12 V battery would be set to 115% of 12 V, or 13.8 V. If this voltage is set too high, an excessive current can result, causing the battery to overheat.

Wet Cell
Cell using a liquid electrolyte.

Secondary cells are often referred to as **wet cells** because the electrolyte is not normally in paste form (dry) as in the primary cell, but is in liquid (wet) form and is free to move and flow within the container.

Secondary cell types. Of all the secondary cells on the market today, the lead–acid and nickel–cadmium are the two most popular, and are illustrated and described in Table 1-5. Other secondary cell types include the nickel–metal hydride and lithium ion, which are used in cellular telephones, camcorders, and other portable electronic systems.

Ampere-Hours
If you multiply the amount of current, in amperes, that can be supplied for a given time frame in hours, a value of ampere-hours is obtained.

Secondary cell capacity. Capacity (C) is measured by the amount of **ampere-hours** (Ah) a battery can supply during discharge. For example, if a battery has a discharge capacity of 10 ampere-hours, the battery could supply 1 ampere for 10 hours, 10 amperes for 1 hour, 5 amperes for 2 hours, and so on, during discharge. Automobile batteries (12 V) will typically have an ampere-hour rating of between 100 and 300 Ah.

Ampere-hour units are actually specifying the coulombs of charge in the battery. For example, if a lead–acid car battery was rated at 150 Ah, this value would have to be converted to ampere-seconds to determine coulombs of charge, as 1 ampere-second (As) is equal to 1 coulomb. Ampere-hours are easily converted to ampere-seconds simply by multiplying ampere-hours by 3600, which is the number of seconds in 1 hour. In our example, a 150 Ah battery will have a charge of 54×10^4 coulombs.

EXAMPLE:

A Ni–Cd battery is rated at 300 Ah and is fully discharged.

a. How many coulombs of charge must the battery charger put into the battery to restore it to full charge?
b. If the charger is supplying a charging current of 3 A, how long will it take the battery to fully charge?
c. Once fully charged, the battery is connected across a load that is pulling a current of 30 A. How long will it take until the battery is fully discharged?

Solution:

a. The same amount that was taken out: $300 \text{ Ah} \times 3600 = 1080 \times 10^3 \text{ C}$

b. $\dfrac{300 \text{ Ah}}{3 \text{ A}} = 100$ hours until charged

c. $\dfrac{300 \text{ Ah}}{30 \text{ A}} = 10$ hours until discharged

1-4-4 *Power Supply Unit*

A battery could be used to apply a voltage across a component or circuit, but batteries have a fixed output voltage and will, in time, run down and lose their potential.

TABLE 1-5 Secondary Cell Types

Battery type	Features	Applications	Construction
Lead–acid (lead cell)	High cycle (discharge–charge) life and high current capacity; 2.1 V per cell; must be operated upright if a wet lead–acid. Lead–acid gel cells can be used in any position	*Gel Cell Batteries* televisions, recorders, robots, alarms, and tools *Wet Cell Batteries* Starting source for automobiles and power for robotics equipment	The electrodes are lead oxide immersed in an electrolyte of dilute sulfuric acid Wet-type lead–acid battery with 6 internal 2.1 V cells making 12.6 V car battery
Nickel–cadmium (Ni–Cd)	High capacity and high cost; three times discharge current of lead–acid for same amp-hour rating; can be sealed and operated in any position, so ideal for drills and other such portable equipment; 1.2 V per cell. Nickel cadmium batteries suffer from "memory effect." Short discharge and charge cycles cause the batteries' capacity to drop, because the cell appears to "remember" the lower capacity.	*Ni–Cd Gel Cell* televisions, radios, toys, recorders, shavers, toothbrushes *Ni–Cd Wet Cell Batteries* Starter source for jet engines (resembles lead–acid car battery in appearance). Robotics equipment	The positive plate is made of cadmium, the negative plate of nickel hydroxide, and the electrolyte used is potassium hydroxide
Nickel–metal Hydride (Ni–MH) and Lithium Ion	Discharge characteristics are very similar to those of the nickel–cadmium cell. The cell's nominal voltage is 1.2 V, and it does not suffer from memory effect.	Biggest applications include cellular phones, portable computers, camcorders, and other portable electronic systems	

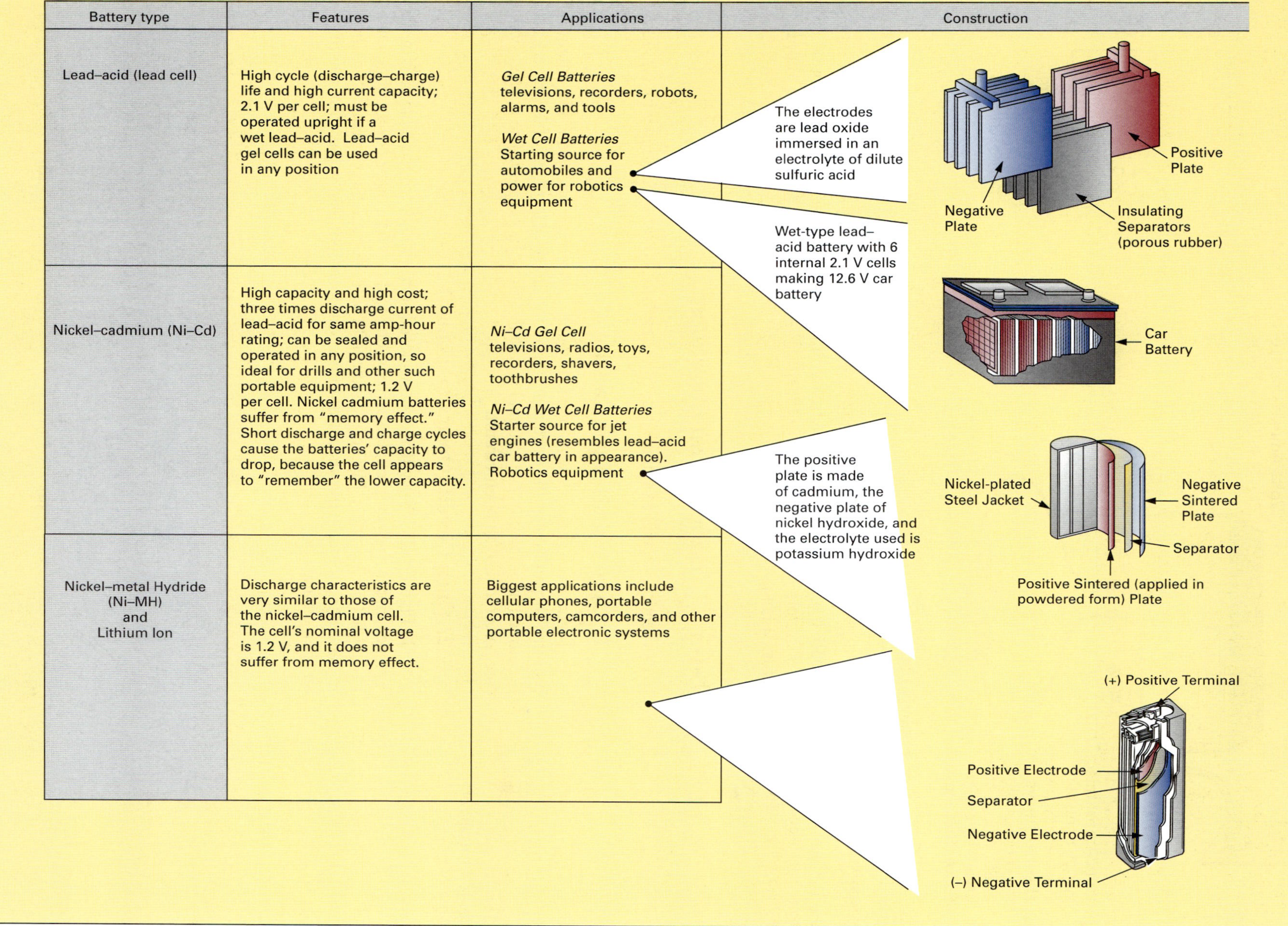

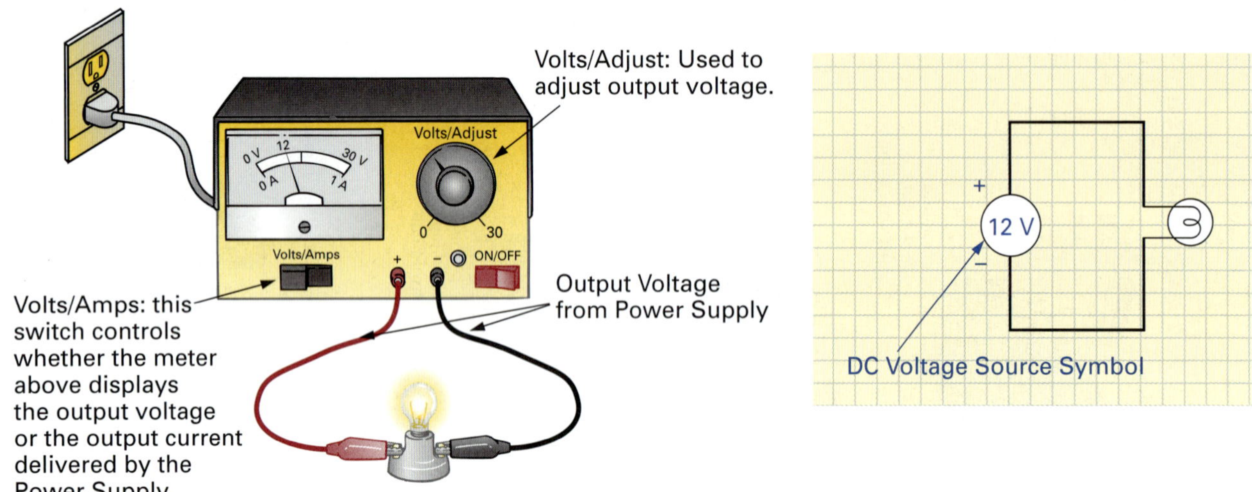

FIGURE 1-26 Power Supply Unit.

Power Supply Unit
Piece of electrical equipment used to deliver either ac or dc voltage.

The **power supply unit,** shown in Figure 1-26, can provide an accurate voltage at its output terminal that can be increased or decreased using the "volts/adjust" control. In the example shown in Figure 1-26, the power supply is being used to supply 12 volts across a lamp.

1-4-5 *Basic Circuit Conditions*

In this section we will examine some of the terms used in association with circuits.

1. Open Circuit (Open Switch)

Figure 1-27 shows how an opened switch can produce an "open" in a circuit. The open prevents current flow as the extra electrons in the negative terminal of the battery cannot feel the attraction of the positive terminal due to the break in the path. The opened switch therefore has produced an **open circuit.**

Open Circuit
Break in the path of current flow.

2. Open Circuit (Open Component)

In Figure 1-28, the switch is now closed so as to make a complete path in the circuit. An open circuit, however, could still exist due to the failure of one of the components in the circuit. In the example in Figure 1-28, the light bulb filament has burned out creating an open in the circuit, which prevents current flow. An open component therefore can also produce an open circuit.

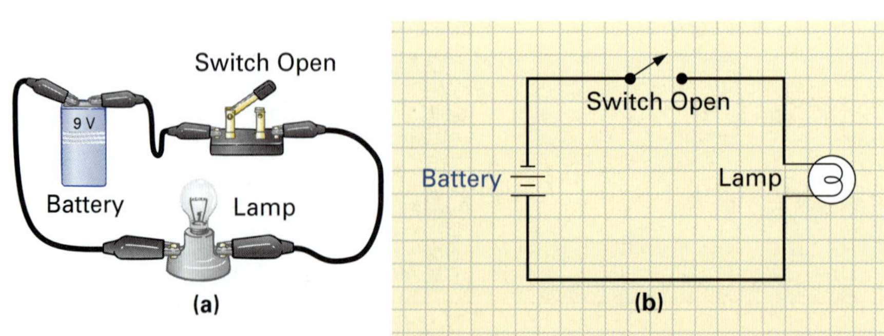

FIGURE 1-27 An Open Switch Causing an Open Circuit. (a) Pictorial. (b) Schematic.

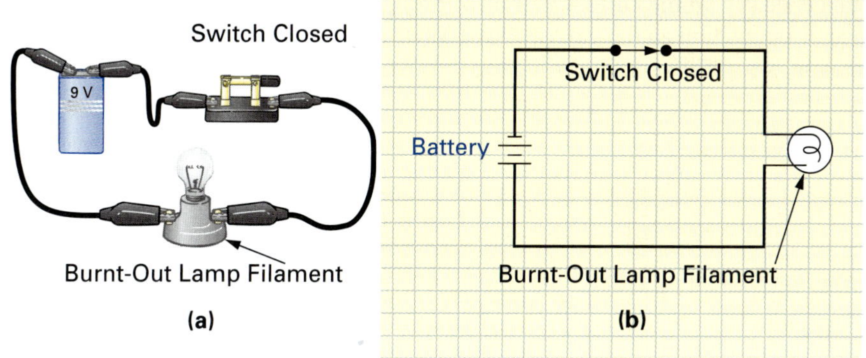

FIGURE 1-28 **An Open Lamp Filament Causing an Open Circuit. (a) Pictorial. (b) Schematic.**

3. Closed Circuit (Closed Switch)

If all of the devices in a circuit are operating properly and connected correctly, a closed switch produces a **closed circuit,** as shown in Figure 1-29. A closed circuit provides a complete path from the negative terminal of the battery to the positive, and so current will flow through the circuit.

Closed Circuit
Circuit having a complete path for current to flow.

4. Short Circuit (Shorted Component)

A **short circuit** normally occurs when one point in a circuit is accidentally connected to another. Figure 1-30(a) illustrates how a short circuit could occur. In this example, a set of pliers was accidentally laid across the two contacts of the light bulb. Figure 1-30(b) shows the circuit's schematic diagram, with the effect of the short across the light bulb drawn in. In

Short Circuit
Also called a short; it is a low-resistance connection between two points in a circuit, typically causing a large amount of current flow.

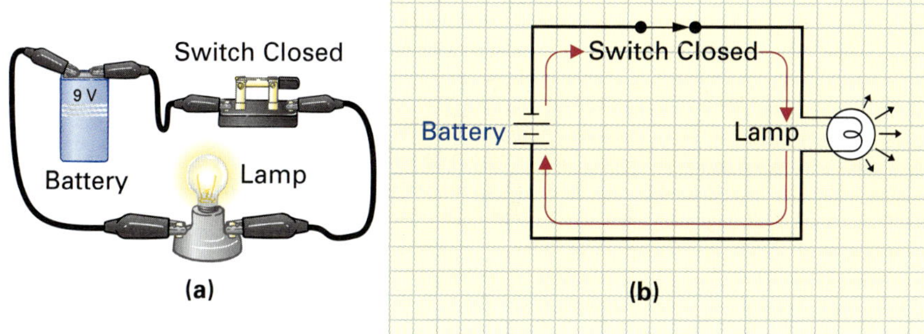

FIGURE 1-29 **Closed Switch Causing a Closed Circuit. (a) Pictorial. (b) Schematic.**

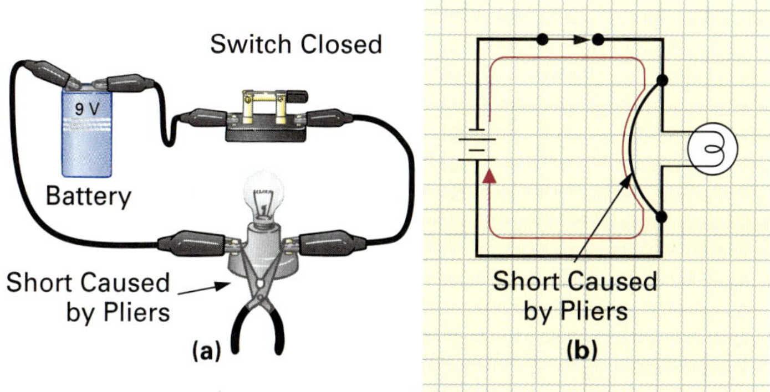

FIGURE 1-30 **Short Circuit Due to Pliers across a Lamp. (a) Pictorial. (b) Schematic.**

this instance, nearly all the current will flow through the metal of the pliers, which offers no resistance or opposition to the current flow. Since the light bulb does have some resistance or opposition, very little current will flow through the bulb, and therefore no light will be produced.

SELF-TEST EVALUATION POINT FOR SECTION 1-4

Use the following questions to test your understanding of Section 1-4.

1. Does current flow through an open circuit?
2. Does current flow through a short circuit?
3. Describe the difference between a closed circuit and a short circuit.
4. Describe the difference between an open switch and a closed switch.
5. What are the three parts within a battery?
6. What is the difference between a primary cell and a secondary cell?
7. What advantages does a power supply unit have over a battery?

REVIEW QUESTIONS

Multiple-Choice Questions

1. The most commonly used metal in the field of electronics is:
a. Silver **c.** Mica
b. Copper **d.** Gold

2. The smallest unit of an element is:
a. A compound **c.** A molecule
b. An atom **d.** A proton

3. The smallest unit of a compound is:
a. An element **c.** An electron
b. A neutron **d.** A molecule

4. A negative ion has:
a. More protons than electrons
b. More electrons than protons
c. More neutrons than protons
d. More neutrons than electrons

5. A positive ion has:
a. Lost some of its electrons **c.** Lost neutrons
b. Gained extra protons **d.** Gained more electrons

6. One coulomb of charge is equal to:
a. 6.24×10^{18} electrons **c.** 6.24×10^{8} electrons
b. 1018×10^{12} electrons **d.** 6.24×10^{81} electrons

7. If 14 C of charge passes by a point in 7 seconds, the current flow is said to be:
a. 2 A **c.** 21 A
b. 98 A **d.** 7 A

8. How many electrons are there within 16 C of charge?
a. 9.98×10^{19} **c.** 16
b. 14 **d.** 10.73×10^{19}

9. Current is measured in:
a. Volts **c.** Ohms
b. Coulombs/second **d.** Siemens

10. Voltage is measured in units of:
a. Amperes **c.** Siemens
b. Ohms **d.** Volts

11. Another word used to describe voltage is:
a. Potential difference
b. Pressure
c. Electromotive force (emf)
d. All of the above

12. A short circuit will cause:
a. No current flow
b. Maximum current flow
c. A break in the circuit
d. Both (a) and (c)

13. An open switch will cause:
a. No current flow
b. Maximum current flow
c. A break in the circuit
d. Both (a) and (c)

14. An open component will cause:
a. No current flow
b. Maximum current flow
c. A break in the circuit
d. Both (a) and (c)

15. An open circuit will cause:
a. No current flow
b. Maximum current flow
c. A break in the circuit
d. Both (a) and (c)

Practice Problems

16. What is the value of conductance in siemens for a 100-ohm resistor?

17. Calculate the total number of electrons in 6.5 C of charge.

18. Calculate the amount of current in amperes passing through a conductor if 3 C of charge passes by a point in 4s.

19. Convert the following:
    a. 0.014 A = ____________ Ma
    b. 1374 A = ____________ kA
    c. 0.776 μA = ____________ μA
    d. 0.91 mA = ____________ A

20. Convert the following:
    a. 1473 mV = ____________ V
    b. 7143 V = ____________ kV
    c. 0.139 kV = ____________ V
    d. 0.390 MV = ____________ kV

Web Site Questions

Go to the web site http://www.prenhall.com/cook, select the textbook *Electronics: A Complete Course,* select this chapter, and then follow the instructions when answering the multiple choice practice problems.

Resistance and Power

Genius of Chippewa Falls

In 1960, Seymour R. Cray, a young vice-president of engineering for Control Data Corporation, informed president William Norris that in order to build the world's most powerful computer he would need a small research lab built near his home. Norris would have shown any other employee the door, but Cray was his greatest asset, so in 1962 Cray moved into his lab, staffed by 34 people and nestled in the woods near his home overlooking the Chippewa River in Wisconsin. Eighteen months later the press was invited to view the 14- by 6-foot 6600 supercomputer that could execute 3 million instructions per second and contained 80 miles of circuitry and 350,000 transistors, which were so densely packed that a refrigeration cooling unit was needed due to the lack of airflow.

Cray left Control Data in 1972 and founded his own company, Cray Research. Four years later the $8.8 million Cray-1 scientific supercomputer outstripped the competition. It included some revolutionary design features, one of which is that since electronic signals cannot travel faster than the speed of light (1 foot per billionth of a second), the wire length should be kept as short as possible, because the longer the wire the longer it takes for a message to travel from one end to the other. With this in mind, Cray made sure that none of the supercomputer's conducting wires exceeded 4 feet in length.

In the summer of 1985, the Cray-2 was installed at Lawrence Livermore Laboratory. The Cray-2 was 12 times faster than the Cray-1, and its densely packed circuits are encased in clear Plexiglas and submerged in a bath of liquid coolant. The 60-year-old genius moved on from his latest triumph, nicknamed "Bubbles," and began working on another revolution in the supercomputer field, because for Seymour Cray a triumph was merely a point of departure.

Introduction

Voltage, current, resistance, and power are the four basic properties of prime importance in our study of electronics. In Chapter 1, voltage and current were introduced, and in this chapter we will examine resistance and power.

2

2-1 RESISTANCE

By definition, *resistance* is the opposition to current flow accompanied by the dissipation of heat. To help explain the concept of resistance, Figure 2-1 compares fluid opposition to electrical opposition.

In the fluid circuit in Figure 2-1(a), a valve has been opened almost completely, so a very small opposition to the water flow exists within the pipe. This small or low resistance within the pipe will not offer much opposition to water flow, so a large amount of water will flow through the pipe and gush from the outlet. In the electrical circuit in Figure 2-1(b), a small value **resistor** (which is symbolized by the zigzag line) has been placed in the circuit. The resistor is a device that is included within electrical and electronic circuits to oppose current flow by introducing a certain value of circuit **resistance.** In this circuit, a small value resistor has been included to provide very little resistance to the passage of current flow. This

Resistor
Device constructed of a material that opposes the flow of current by having some value of resistance.

Resistance
Symbolized R and measured in ohms (Ω), it is the opposition to current flow with the dissipation of energy in the form of heat.

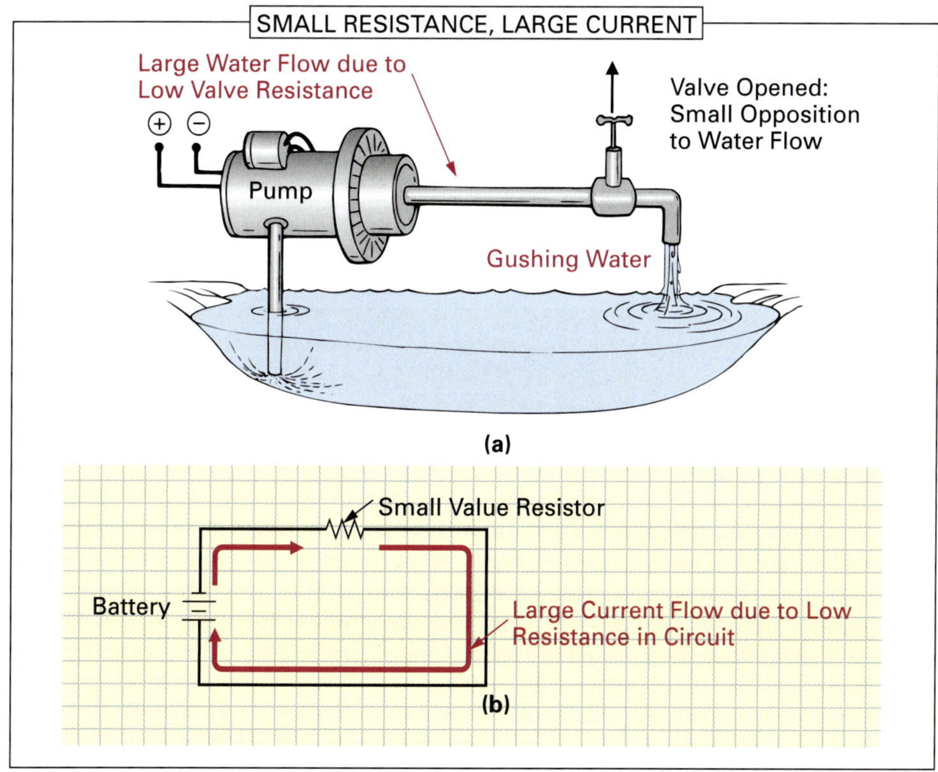

FIGURE 2-1 **The Opposite Relationship between Resistance and Current.**

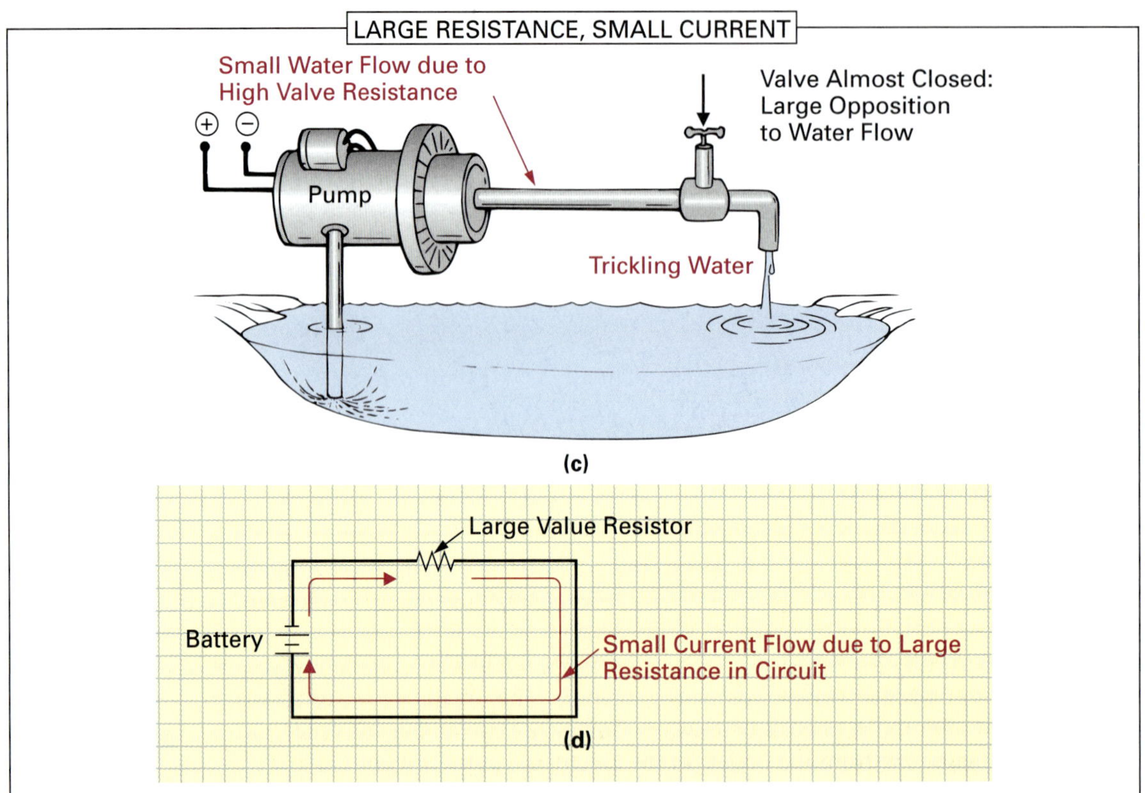

FIGURE 2-1 (continued) **The Opposite Relationship between Resistance and Current.**

TIME LINE

Ohm's law, the best known law in electrical circuits, was formulated by Georg S. Ohm (1787–1854), a German physicist. His law was so coldly received that his feelings were hurt and he resigned his teaching post. When his law was finally recognized, he was reinstated. In honor of his accomplishments, the unit of resistance is called the ohm.

low resistance or small opposition will allow a large amount of current to flow through the conductor, as illustrated by the thick current line.

In the fluid circuit in Figure 2-1(c), the valve is almost completely closed, resulting in a high resistance or opposition to water flow, so only a trickle of water passes through the pipe and out from the outlet. In the electrical circuit in Figure 2-1(d), a large value resistor has been placed in the circuit, causing a large resistance to the passage of current. This high resistance allows only a small amount of current to flow through the connecting wire, as illustrated by the thin current line.

In the previous examples of low resistance and high resistance shown in Figure 2-1, you may have noticed that resistance and current are inversely proportional ($1/\propto$) to one another. This means that if resistance is high the current is low, and if resistance is low, the current is high.

Resistance (R) is Inversely Proportional $\left(\frac{1}{\propto}\right)$ to Current (I)

$$R \frac{1}{\propto} I$$

This relationship between resistance and current means that if resistance is increased by some value, current will be decreased by the same value (assuming a constant electrical pressure or voltage). For example, if resistance is doubled current is halved and similarly, if resistance is halved current is doubled.

$$R\uparrow \frac{1}{\propto} I\downarrow \qquad R\downarrow \frac{1}{\propto} I\uparrow$$

In Figure 2-1, the fluid analogy was used alongside the electrical circuit to help you understand the idea of low resistance and high resistance. In the following section, we will examine resistance in more detail, and define clearly exactly how much resistance exists within a circuit.

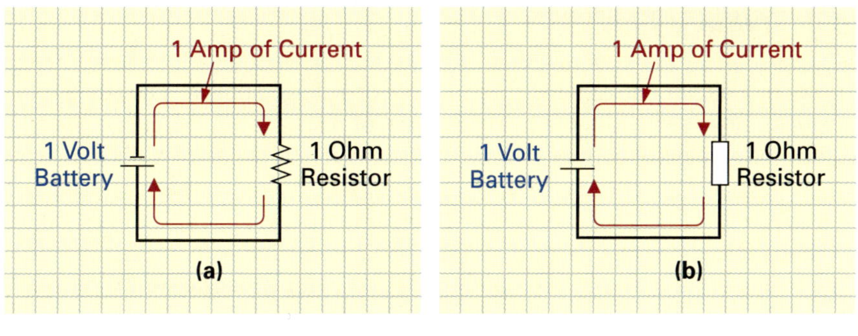

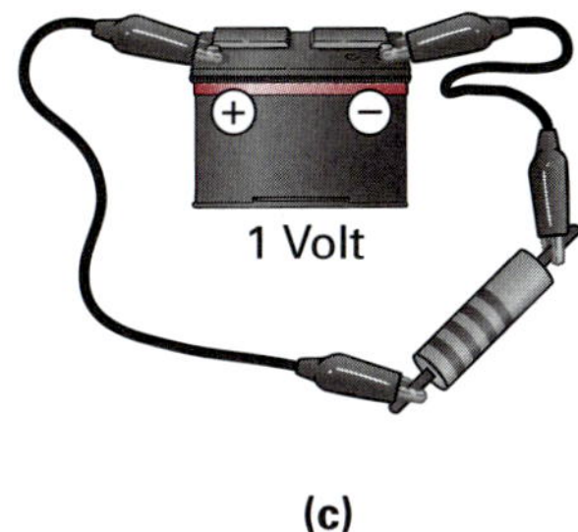

(c)

FIGURE 2-2 **One Ohm. (a) New and (b) Old Resistor Symbols. (c) Pictorial.**

2-1-1 *The Ohm*

As discussed in the previous chapter, current is measured in amperes and voltage is measured in volts. Resistance is measured in **ohms,** in honor of Georg Simon Ohm, who was the first to formulate the relationship between current, voltage, and resistance.

Ohm
Unit of resistance, symbolized by the Greek capital letter omega (Ω).

The larger the resistance, the larger the value of ohms and the more the resistor will oppose current flow. The ohm is given the symbol Ω, which is the Greek capital letter omega. By definition, 1 ohm is the value of resistance that will allow 1 ampere of current to flow through a circuit when a voltage of 1 volt is applied, as shown in Figure 2-2(a), where the resistor is drawn as a zigzag. In some schematics or circuit diagrams the resistor is drawn as a rectangular block, as shown in Figure 2-2(b). The pictorial of this circuit is shown in Figure 2-2(c).

The circuit in Figure 2-3 reinforces our understanding of the ohm. In this circuit, a 1 V battery is connected across a resistor, whose resistance can be either increased or decreased. As the resistance in the circuit is increased the current will decrease and, conversely, as the resistance of the resistor is decreased the circuit current will increase. So how much is 1 ohm of resistance? The answer to this can be explained by adjusting the resistance of the variable resistor. If the resistor is adjusted until exactly 1 amp of current is flowing around the circuit, the value of resistance offered by the resistor is referred to as 1 ohm (Ω).

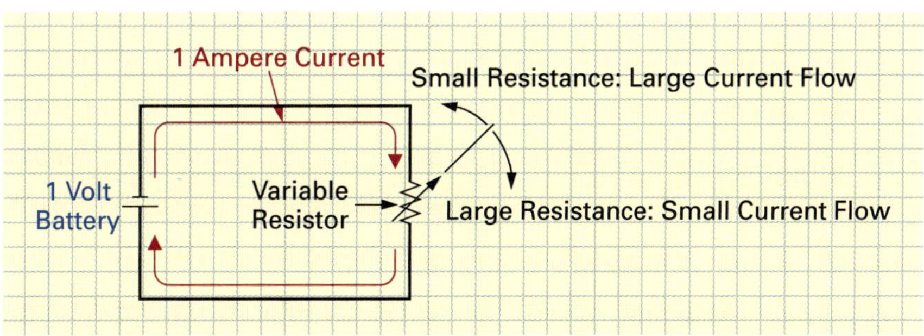

FIGURE 2-3 **1 Ohm Allows 1 Amp to Flow When 1 Volt Is Applied.**

2-1-2 *Ohm's Law*

Current flows in a circuit due to the electrical force or voltage applied. The amount of current flow in a circuit, however, is limited by the amount of resistance in the circuit. It can be said, therefore, that the amount of current flow around a circuit is dependent on both voltage and resistance. This relationship between the three electrical properties of current, voltage, and resistance was discovered by Georg Simon Ohm, a German physicist, in 1827. Published originally in 1826, **Ohm's law** states that the current flow in a circuit is directly proportional ($\propto$) to the source voltage applied and inversely proportional ($1/\propto$) to the resistance of the circuit.

Ohm's Law
Relationship between the three electrical properties of voltage, current, and resistance. Ohm's law states that the current flow within a circuit is directly proportional to the voltage applied across the circuit and inversely proportional to its resistance.

Stated in mathematical form, Ohm arrived at this formula:

$$\text{current } (I) = \frac{\text{voltage } (V)}{\text{resistance } (R)}$$

$$\text{current } (I) \propto \text{voltage } (V)$$

$$\text{current } (I) \frac{1}{\propto} \text{resistance } (R)$$

1. Current Is Proportional to Voltage (I ∝ V)

An increase in voltage (electron moving force), therefore, will exert a greater pressure on the circuit electrons and cause an increase in current flow. This point is reinforced with the Ohm's law formula:

$$I\uparrow = \frac{V\uparrow}{R}$$ Both properties are above the line and are therefore proportional.

In the two following examples, you will see how an increase in the voltage applied to a circuit results in a proportional increase in the circuit current.

EXAMPLE:

If the resistance in the circuit in Figure 2-4(a) remains constant at 1 Ω and the applied voltage equals 2 V, what is the value of current flowing through the circuit?

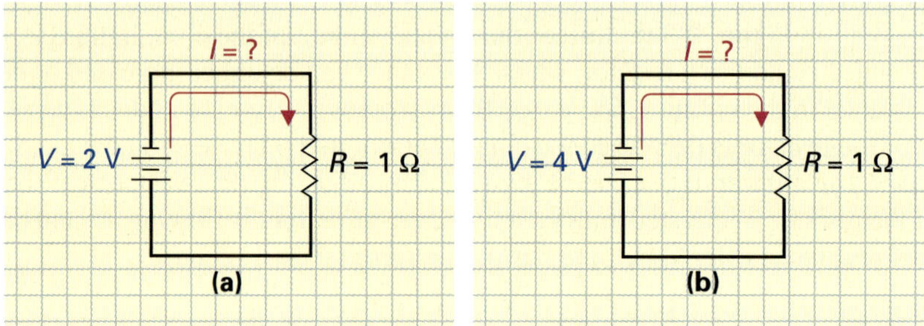

FIGURE 2-4

Solution:

CALCULATOR SEQUENCE

Step	Keypad Entry	Display Response
1.	2	2
2.	÷	
3.	1	1
4.	=	2

$$\text{current } (I) = \frac{\text{voltage } (V)}{\text{resistance } (R)} = \frac{2\text{ V}}{1\ \Omega} = 2\text{ A}$$

EXAMPLE:

If the voltage from the previous example is now doubled to 4 V as shown in Figure 2-4(b), what would be the change in current?

Solution:

$$\text{current } (I) = \frac{\text{voltage } (V)}{\text{resistance } (R)}$$
$$= \frac{4\text{ V}}{1\ \Omega}$$
$$= 4\text{ A}$$

On the other hand, a decrease in the circuit's applied voltage will cause a decrease in circuit current. This is again reinforced by Ohm's law.

$$I\downarrow = \frac{V\downarrow}{R}$$

In the following two examples, you will see how a decrease in the applied voltage will result in a proportional decrease in circuit current.

EXAMPLE:

If a circuit has a resistance of 2 Ω and an applied voltage of 8 V, what would be the circuit current?

Solution:

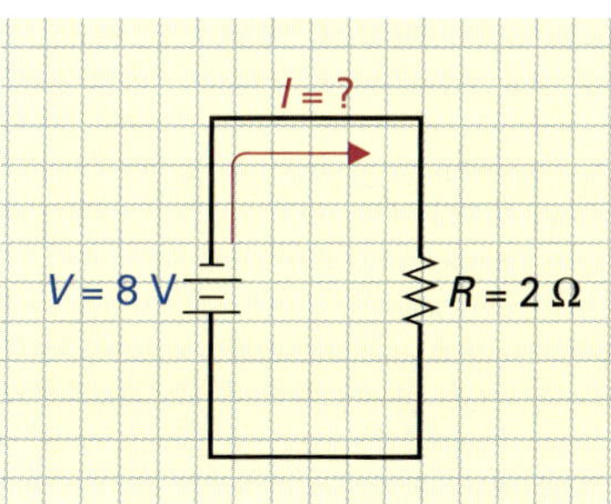

$$\text{current } (I) = \frac{\text{voltage } (V)}{\text{resistance } (R)}$$
$$= \frac{8\text{ V}}{2\ \Omega}$$
$$= 4\text{ A}$$

EXAMPLE:

If the circuit described in the previous example were to have its applied voltage halved to 4 V, what would be the change in circuit current?

Solution:

$$\text{current } (I) = \frac{\text{voltage } (V)}{\text{resistance } (R)}$$
$$= \frac{4\text{ V}}{2\ \Omega}$$
$$= 2\text{ A}$$

To summarize the relationship between circuit voltage and current, we can say that if the voltage were to double, the current within the circuit would also double (assuming the circuit's resistance were to remain at the same value). Similarly, if the voltage were halved, the current would also halve, making the two properties directly proportional to one another.

2. Current Is Inversely Proportional to Resistance $\left(I \frac{1}{\propto} R\right)$

The second relationship in Ohm's law states that the circuit current is inversely proportional to the circuit's resistance.

Therefore, a small resistance ($R\downarrow$) allows a large current ($I\uparrow$) flow around the circuit. This point is reinforced with Ohm's law:

$$I\uparrow = \frac{V}{R\downarrow}$$ One property is above the line, the other is below the line, and therefore the two are inversely proportional.

Conversely, a large resistance ($R\uparrow$) will always result in a small current ($I\downarrow$).

$$I\downarrow = \frac{V}{R\uparrow}$$

To help reinforce this relationship, let us look at a couple of examples.

EXAMPLE:

Figure 2-5 shows a circuit that has a constant 8 volts applied. If the resistance in this circuit were doubled from 2 ohms to 4 ohms, what would happen to the circuit current?

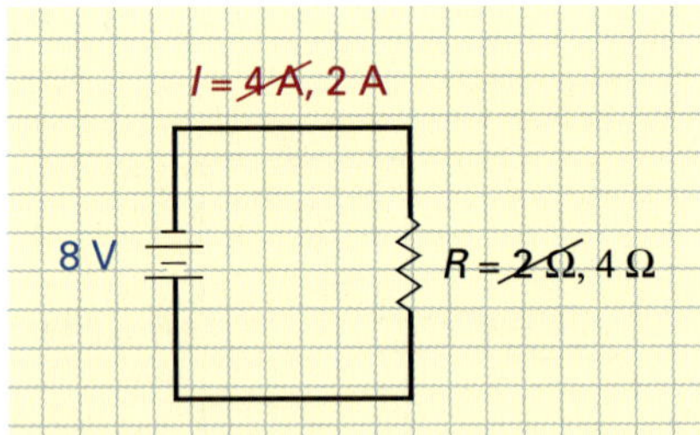

FIGURE 2-5

Solution:

If the resistance in Figure 2-5 were doubled from 2 Ω to 4 Ω, the circuit current would halve from 4 A to:

$$\text{current } (I) = \frac{\text{voltage } (V)}{\text{resistance } (R)} = \frac{8 \text{ V}}{4\ \Omega} = 2 \text{ A}$$

EXAMPLE:

If the resistance in Figure 2-5 were returned to its original value of 2 Ω, what would happen to the circuit current?

■ ***Solution:***

If the resistance in Figure 2-5 were halved from 4 Ω to 2 Ω, the circuit current would double from 2 A to:

$$\text{current } (I) = \frac{\text{voltage } (V)}{\text{resistance } (R)} = \frac{8 \text{ V}}{2 \ \Omega} = 4 \text{ A}$$

To summarize the relationship between circuit current and resistance, we can say that if the resistance were to double, the current within the circuit would be halved (assuming the circuit's voltage were to remain at the same value). Similarly, if the circuit resistance were halved, the circuit current would double, confirming that current is inversely proportional to resistance.

3. The Three Ohm's Law Formulas

By transposing Ohm's Law, we can obtain the following:

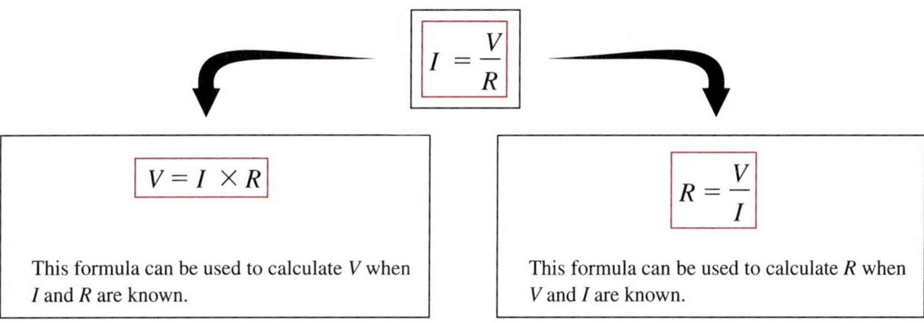

Figure 2-6 shows how these three Ohm's law formulas can be placed within a triangle so that their relative proximity to one another can be used for easy memory recall.

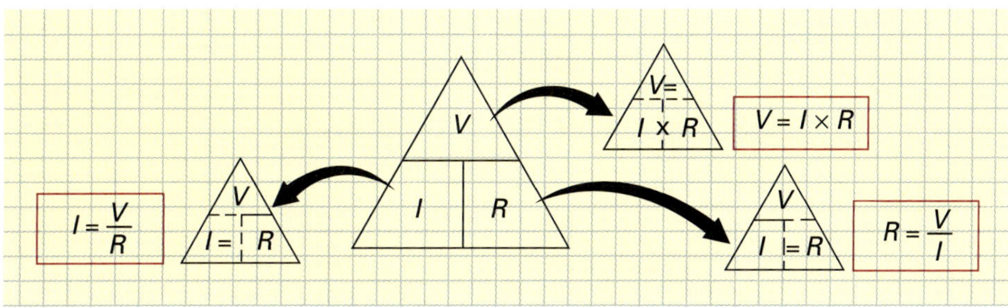

FIGURE 2-6 Ohm's Law Triangle.

■ **EXAMPLE:**

List the formula to use, and determine the unknown for the three examples shown in Figure 2-7.

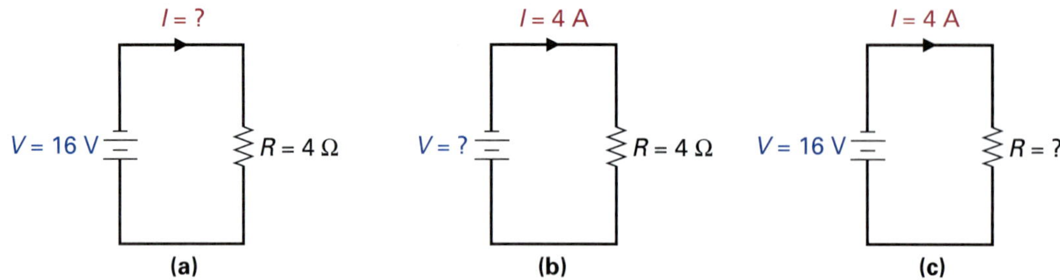

FIGURE 2-7 **Using Ohm's Law to Determine the Unknown.**

Solution:

a. In Figure 2-7(a), current (I) has to be determined when voltage (V) and resistance (R) are known. The Ohm's Law formula for current should therefore be used as follows:

$$I = \frac{V}{R} = \frac{16 \text{ volts}}{4 \text{ ohms}} = 4 \text{ amps.}$$

b. In Figure 2-7(b), voltage (V) has to be determined when current (I) and resistance (R) are known. The Ohm's Law formula for voltage should therefore be used as follows:

$$V = I \times R = 4 \text{ amps} \times 4 \text{ ohms} = 16 \text{ volts.}$$

c. In Figure 2-7(c), resistance (R) has to be determined when current (I) and voltage (V) are known. The Ohm's Law formula for resistance should therefore be used as follows:

$$R = \frac{V}{I} = \frac{16 \text{ volts}}{4 \text{ amps}} = 4 \text{ ohms.}$$

4. Source and Load

Source
Device that supplies the signal power or electric energy to a load.

Figure 2-8(a) illustrates a circuit with a battery connected across a light bulb. In this circuit, the potential energy of the battery (voltage) produces kinetic energy (current) that is used to produce light energy from the bulb. The battery is the **source** in this circuit and the bulb is

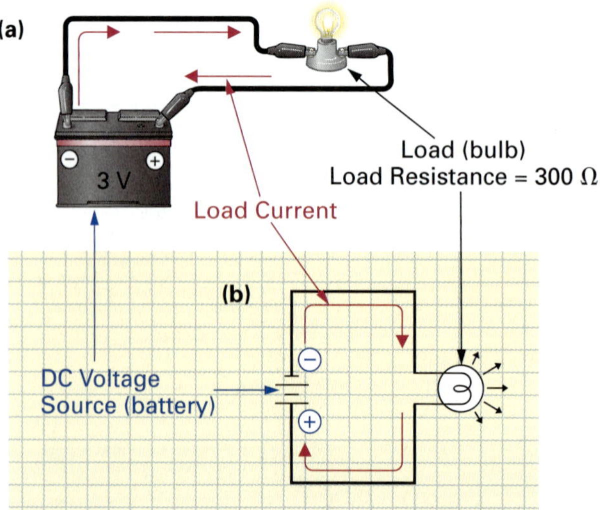

FIGURE 2-8 **Load on a Voltage Source. (a) Pictorial. (b) Schematic.**

the **load.** By definition, a load is a device that absorbs the energy being supplied and converts it into the desired form. The **load resistance** will determine how hard the voltage source has to work. For example, the bulb filament has a resistance of 300 Ω. The connecting wires have a combined resistance of 0.02 Ω (20 mΩ), and since this value is so small compared to the filament's resistance, it will be ignored, as it will have no noticeable effect. Consequently, a 300 Ω load resistance will permit 10 mA of **load current** to flow when a 3 V source is connected across the dc circuit ($I = V/R = 3\ \text{V}/300\ \Omega = 10\ \text{mA}$). If you were to change the filament resistance by using a different bulb, the circuit current would be changed. To be specific, if the load resistance is increased, the load current decreases ($I\downarrow = V/R\uparrow$). Conversely, if the load resistance is decreased, the load current increases ($I\uparrow = V/R\downarrow$) assuming of course that the voltage source remains constant.

Load
A component, circuit, or piece of equipment connected to the source. The load resistance will determine the load current.

Load Resistance
The resistance of the load.

Load Current
The current that is present in the load.

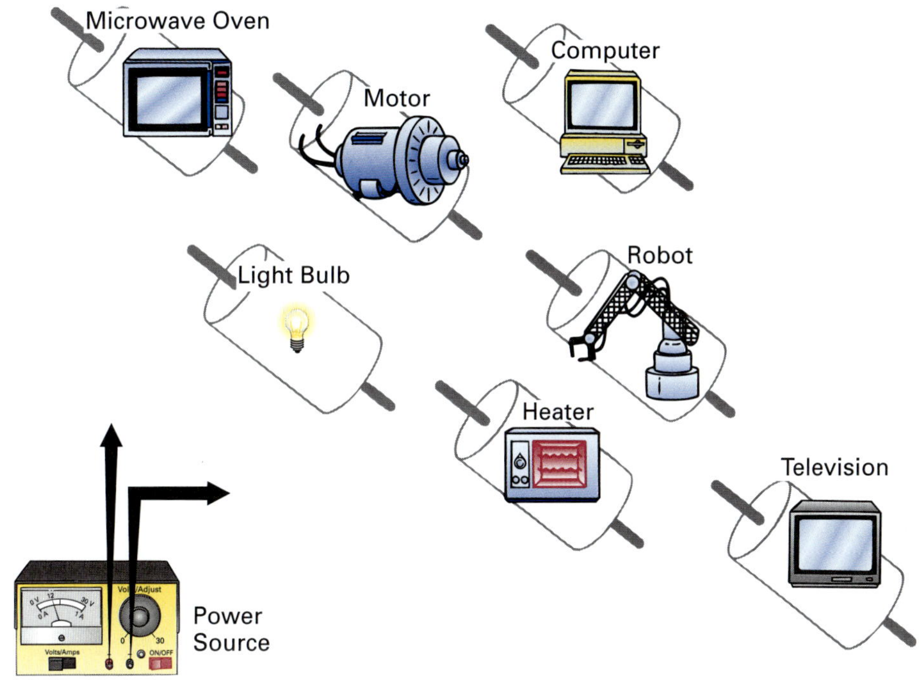

Figure 2-9 illustrates a heater, light bulb, motor, computer, robot, television, and microwave oven. All these and all other devices or pieces of equipment are connected to some source, such as a battery or dc power supply. This supply or source does not see all the circuitry and internal workings of the equipment; it simply sees the whole device or piece of equipment as having some value of resistance. The resistance of the equipment is referred to as the device's *load resistance,* and it is this value of load resistance and the value of *source voltage* that determines the value of *load current.*

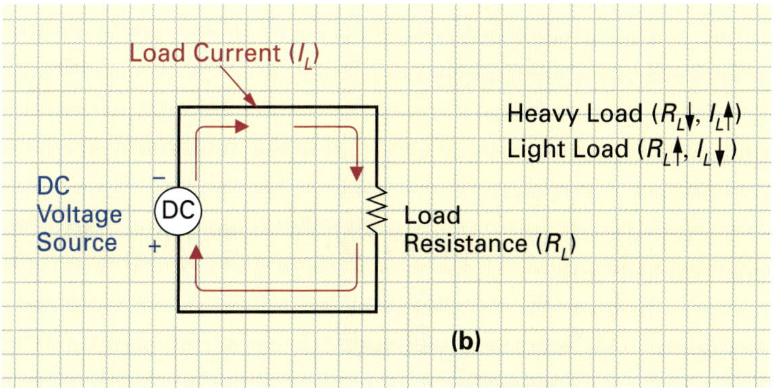

FIGURE 2-9 (a) Load Resistance of Equipment. (b) Schematic Symbols and Terms.

EXAMPLE:

A small portable radio offers a 390 Ω load resistance to a 9 V battery source. What current will flow in this circuit?

Solution:

$$\text{current } (I) = \frac{\text{voltage } (V)}{\text{resistance } (R)} = \frac{9\text{ V}}{390\ \Omega} = 23\text{ mA}$$

In summary, the phrase *load resistance* describes the device or equipment's circuit resistance, whereas the phrase *load current* describes the amount of current drawn by the device or equipment. A device that causes a large load current to flow (due to its small load resistance) is called a large or heavy load because it is heavily loading down or working the supply or source. On the other hand, a device that causes a small load current to flow (due to its large load resistance) is referred to as a small or light load because it is only lightly loading down or working the supply or source. It can be said, therefore, that load current and load resistance are inversely proportional to one another.

Ohmmeter
Measurement device used to measure electric resistance.

2-1-3 *How Is Resistance Measured?*

Resistance is measured with an **ohmmeter.** Stepping through the procedure described in Figure 2-10, you will see how the ohmmeter is used to measure the resistance of a resistor.

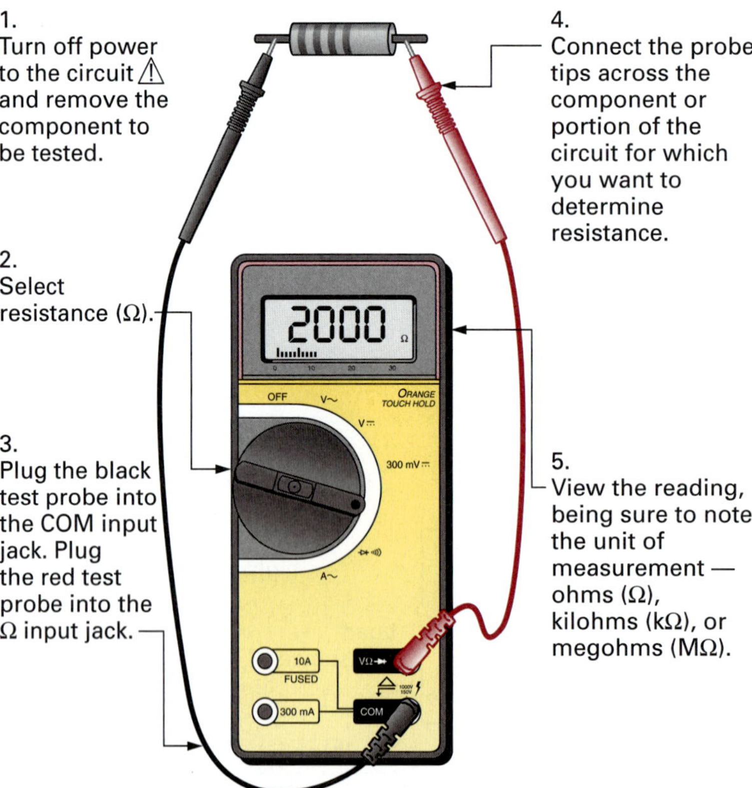

FIGURE 2-10 Measuring Resistance with the Ohmmeter.

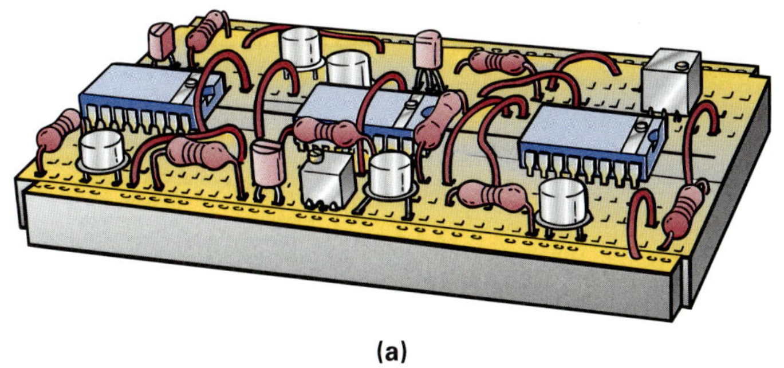

(a)

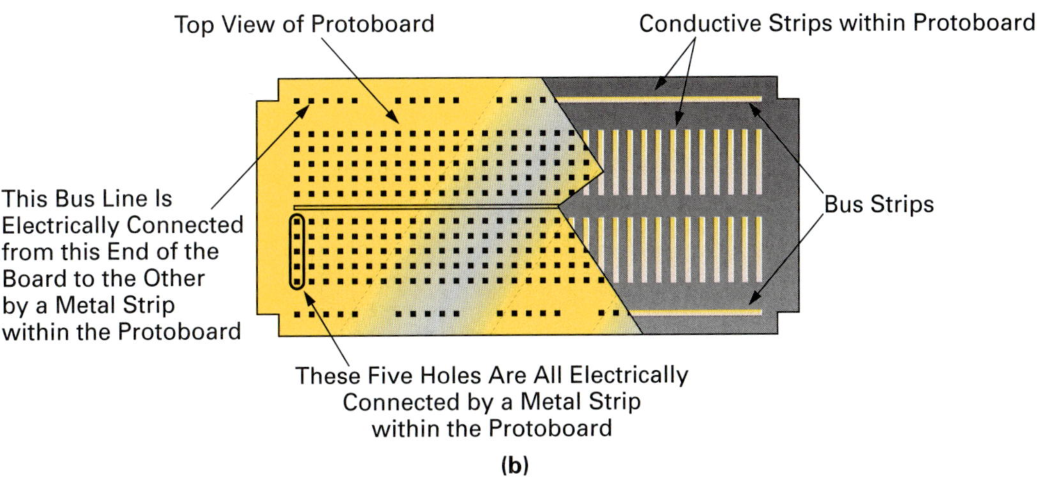

(b)

FIGURE 2-11 Experimenting with the Protoboard.

2-1-4 *Protoboards*

Protoboard
An experimental arrangement of a circuit on a board. Also called breadboard.

The solderless prototyping board (**protoboard**) or breadboard is designed to accommodate experiments. This protoboard will hold and interconnect resistors, lamps, switches and many other components, as well as provide a connecting point for electrical power. Figure 2-11(a) shows an experimental circuit wired up on a protoboard.

Figure 2-11(b) shows the top view of a basic protoboard. As you can see by the cross section on the right side, electrical connector strips are within the protoboard. These conductive strips make a connection between the five hole groups. The bus strips have an electrical connector strip running from end to end, as shown in the cross section. They are usually connected to a power supply as seen in the example circuit in Figure 2-11(a). In this circuit, you can see that the positive supply voltage is connected to the upper bus strip, while the negative supply (ground) is connected to the lower bus strip. These power supply "rails" can then be connected to a circuit formed on the protoboard with hookup wire, as shown in Figure 2-12(a). In this example, three resistors are connected end-to-end, as shown in the schematic diagram in Figure 2-12(b).

Figure 2-13 illustrates how the multimeter could be used to make voltage, current, and resistance measurements of a circuit constructed on the protoboard.

Other protoboards may vary slightly as far as layout, but you should be able to determine the pattern of conductive strips by making a few resistance checks with an ohmmeter.

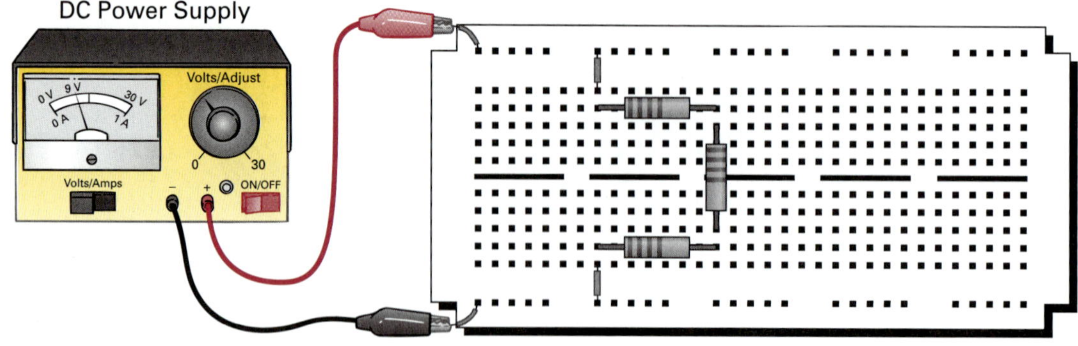

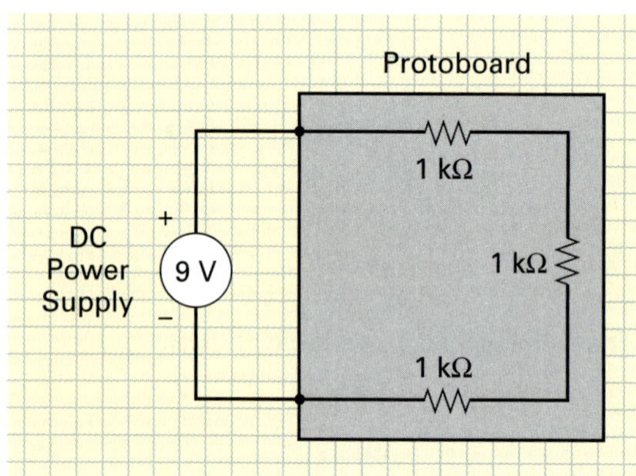

FIGURE 2-12 **Constructing Circuits on the Protoboard.**

SELF-TEST EVALUATION POINT FOR SECTION 2-1

Use the following questions to test your understanding of Section 2-1:

1. Define 1 ohm in relation to current and voltage.
2. Calculate I if $V = 24$ V and $R = 6\ \Omega$.
3. What is the Ohm's law triangle?
4. What is the relationship between: (a) current and voltage; (b) current and resistance?
5. Calculate V if $I = 25$ mA and $R = 1\ \text{k}\Omega$.
6. Calculate R if $V = 12$ V and $I = 100\ \mu$A.
7. Name the instrument used to measure resistance.
8. Describe the five-step procedure that should be used to measure the resistance.

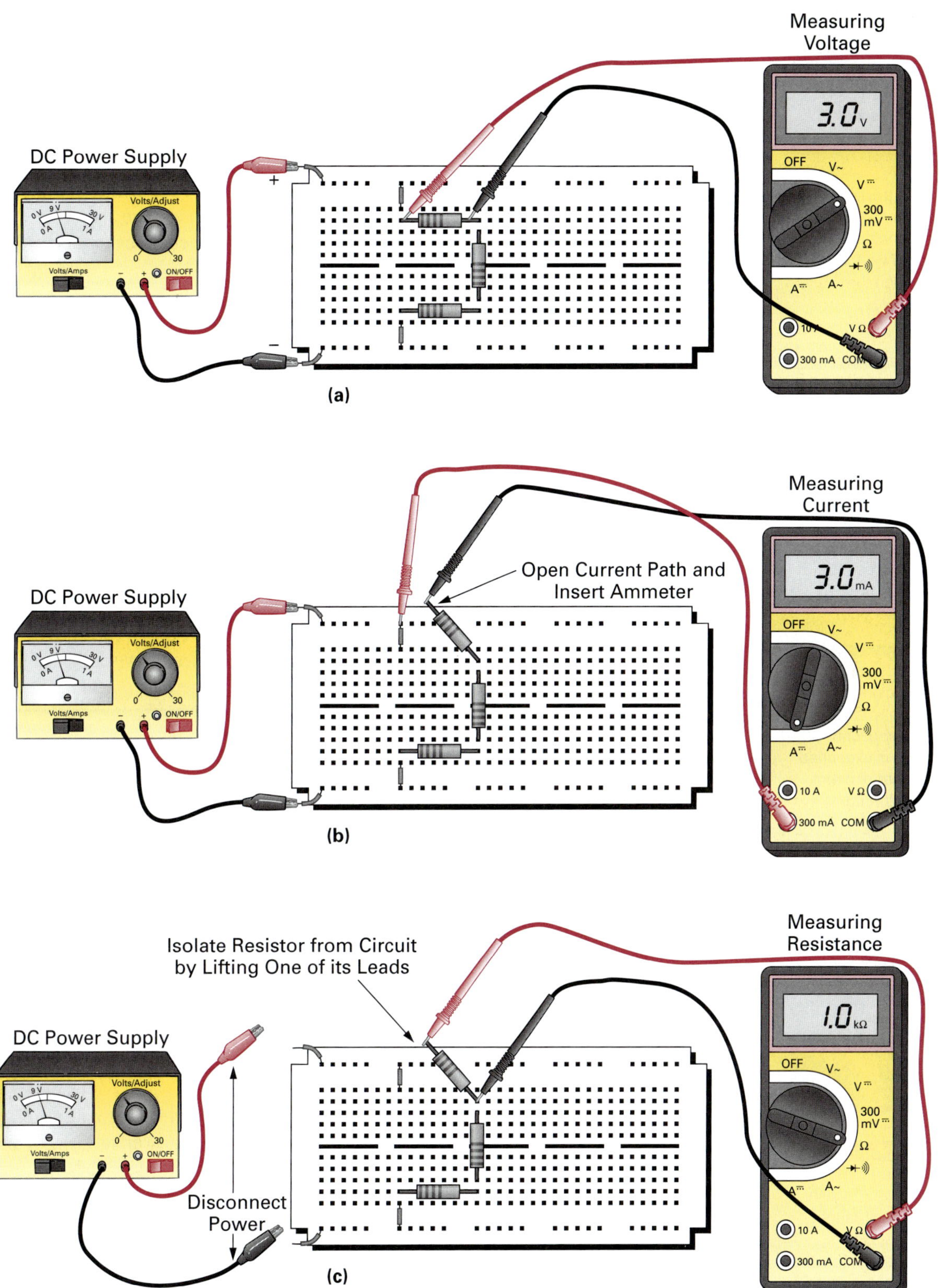

FIGURE 2-13 **Making Measurements of a Circuit on the Protoboard.**

2-2 POWER

Energy
Capacity to do work.

Work
Work is done anytime energy is transformed from one type to another, and the amount of work done is dependent on the amount of energy transformed.

Joule
The unit of work and energy.

Power
Amount of energy converted by a component or circuit in a unit of time, normally seconds. It is measured in units of watts (joules/second).

Watt (W)
Unit of electric power required to do work at a rate of 1 joule/second. One watt of power is expended when 1 ampere of direct current flows through a resistance of 1 ohm.

The sun provides us with a consistent supply of energy in the form of light. Coal and oil are fossilized vegetation that grew, among other things, due to the sun, and are examples of **energy** that the earth has stored for millions of years. It can be said, then, that all energy begins from the sun. On the earth, energy is not created or destroyed; it is merely transformed from one form to another. The transforming of energy from one form to another is called **work.** The greater the energy transformed, the more work that is done.

The six basic forms of energy are light, heat, magnetic, chemical, electrical, and mechanical energy. The unit for energy is the **joule** (J).

EXAMPLE:

One person walks around a track and takes 5 minutes, while another person runs around the track and takes 50 seconds. Both were full of energy before they walked or ran around the track, and during their travels around the track they converted the chemical energy within their bodies into the mechanical energy of movement.

a. Who exerted the most energy?
b. Who did the most work?

Solution:

Both exerted the same amount of energy. The runner exerted all his energy (for example, 100 J) in the short time of 50 seconds, while the walker spaced his energy (100 J) over 5 minutes. Since they both did the same amount of work, the only difference between the runner and the walker is time, or the rate at which their energy was transformed.

2-2-1 *What Is Power?*

Power (P) is the rate at which work is performed and is given the unit of **watt** (W), which is Joules per second (J/s).

Returning to the example involving the two persons walking and running around the track, we could say that the number of joules of energy exerted in 1 second by the runner was far greater than the number of joules of energy exerted in 1 second by the walker, although the total energy exerted by both persons around the entire track was equal and therefore the same amount of work was done. Even though the same amount of energy was used, and therefore the same amount of work was done by the runner and the walker, the output power of each was different. The runner exerted a large value of joules/second or watts (high power output) in a short space of time, while the walker exerted only a small value of joules/second or watts (low power output) over a longer period of time.

Whether discussing a runner, walker, electric motor, heater, refrigerator, light bulb, or compact disk player—power is power. The output power, or power ratings, of electrical, electronic, or mechanical devices can be expressed in watts and describes the number of joules of energy converted every second. The output power of rotating machines is given in the unit *horsepower* (hp), the output power of heaters is given in the unit British thermal unit per hour (Btu/h), and the output power of cooling units is given in the unit *ton of refrigeration.* Despite the different names, they can all be expressed quite simply in the unit of watts. The conversions are as follows:

$$\begin{aligned} 1 \text{ horsepower (hp)} &= 746 \text{ W} \\ 1 \text{ British thermal unit per hour (Btu/h)} &= 0.293 \text{ W} \\ 1 \text{ ton of refrigeration} &= 3.52 \text{ kW (3520 W)} \end{aligned}$$

Now we have an understanding of power, work, and energy. Let's reinforce our knowledge by introducing the energy formula and try some examples relating to electronics.

TIME LINE
James P. Joule (1818–1889), an English physicist and self-taught scientist, conducted extensive research into the relationships between electrical, chemical, and mechanical effects, which led him to the discovery that one energy form can be converted into another. For this achievement, his name was given to the unit of energy, the joule.

2-2-2 *Calculating Energy*

The amount of energy stored (W) is dependent on the coulombs of charge stored (Q) and the voltage (V).

$$W = Q \times V$$

where W = energy stored, in joules (J)
Q = coulombs of charge (1 coulomb = 6.24×10^{18} electrons)
V = voltage, in volts (V)

If you consider a battery as an example, you can probably better understand this formula. The battery's energy stored is dependent on how many coulombs of electrons it holds (current) and how much electrical pressure it is able to apply to these electrons (voltage).

TIME LINE
The unit of power, the watt, was named after Scottish engineer and inventor James Watt (1736–1819) in honor of his advances in the field of science.

EXAMPLE:

How many coulombs of electrons would a 9 V battery have to store to have 63 J of energy?

Solution:

If $W = Q \times V$, then by transposition:

$$Q = \frac{W}{V}$$

$$\text{coulombs of electrons } (Q) = \frac{\text{energy in joules } (W)}{\text{battery voltage } (V)}$$

$$= \frac{63 \text{ J}}{9 \text{ V}}$$

$$= 7 \text{ C of electrons}$$

or

$$7 \times 6.24 \times 10^{18} = 4.36 \times 10^{19} \text{ electrons}$$

CALCULATOR SEQUENCE

Step	Keypad Entry	Display Response
1.	6 3	63
2.	÷	
3.	9	9
4.	=	7
5.	×	7
6.	6 . 2 4 E 1 8	6.24E18
7.	=	4.36E19

2-2-3 *Calculating Power*

Power, in connection with electricity and electronics, is the rate at which electric energy (W) is converted into some other form.

$$P = I \times V$$

where P = power, in watts (W)
I = current, in amperes (A)
V = voltage, in volts (V)

This formula states that the amount of power delivered to a device is dependent on the electrical pressure (or voltage applied across the device) and the current flowing through the device.

EXAMPLE:

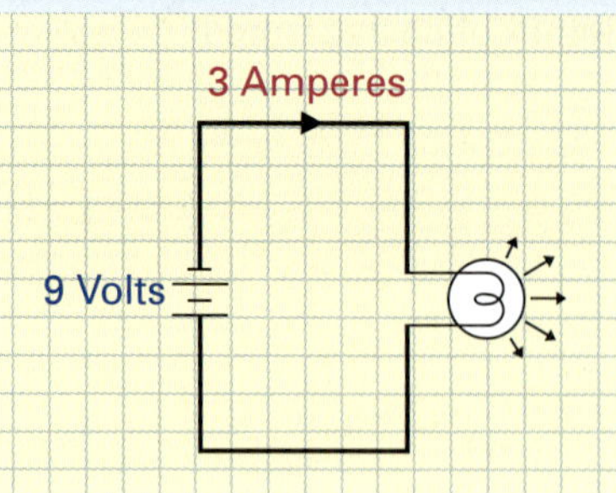

In regards to electrical and electronic circuits, power is the rate at which electric energy is converted into some other form. In the example in Figure 2-14, electric energy will be transformed into light and heat energy by the light bulb. Power has the unit of watts, which is the number of joules of energy transformed per second (J/s). If 27 J of electric energy is being transformed into light and heat per second, how many watts of power does the light bulb convert?

FIGURE 2-14 Calculating Power.

Solution:

$$\text{power} = \frac{\text{joules}}{\text{second}} = \frac{27\text{ J}}{1\text{ s}} = 27\text{ W}$$

The power output in the previous example could have easily been calculated by merely multiplying current by voltage to arrive at the same result.

$$\text{power} = I \times V = 3\text{ A} \times 9\text{ V} = 27\text{ W}$$

We could say, therefore, that the light bulb dissipates 27 watts of power, or 27 joules of energy per second.

Studying the power formula $P = I \times V$, we can also say that 1 watt of power is expended when 1 ampere of current flows through a circuit that has 1 volt applied.

Like Ohm's law, we can transpose the power formula as follows:

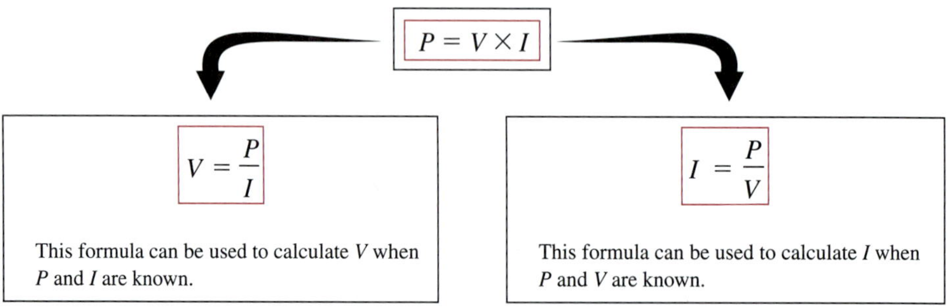

$P = V \times I$
$V = \frac{P}{I}$
This formula can be used to calculate V when P and I are known.
$I = \frac{P}{V}$
This formula can be used to calculate I when P and V are known.

EXAMPLE:

A coffee cup warming plate has a power rating of 120 V, 23 W. How much current is flowing through the heating element, and what is the heating element's resistance?

Solution:

$$P = V \times I \qquad I = \frac{P}{V}$$

$$I = \frac{23\text{ W}}{120\text{ V}} = 191.7\text{ mA}$$

$$R = \frac{V}{I} = \frac{120\text{ V}}{191.7\text{ mA}} = 626\ \Omega$$

EXAMPLE:

To illustrate a point, let us work through a hypothetical example. Calculate the current drawn by a 100 W light bulb if it is first used in the home (and therefore connected to a 120 V source) and then used in the car (and therefore connected across a 12 V source).

Solution:

$$P = V \times I \quad \text{therefore,} \quad I = \frac{P}{V}$$

In the home,

$$I = \frac{P}{V} = \frac{100 \text{ W}}{120 \text{ V}} = 0.83 \text{ A}$$

In the car,

$$I = \frac{P}{V} = \frac{100 \text{ W}}{12 \text{ V}} = 8.33 \text{ A}$$

This example brings out a very important point in relation to voltage, current, and power. In the home, only a small current needs to be supplied when the applied voltage is large ($P = V\uparrow \times I\downarrow$). However, when the source voltage is small, as in the case of the car, a large current must be supplied in order to deliver the same amount of power ($P = V\downarrow \times I\uparrow$).

When trying to calculate power we may not always have the values of V and I available. For example, we may know only I and R, or V and R. The following shows how we can substitute terms in the $P = V \times I$ formula to arrive at alternative power formulas for wattage calculations.

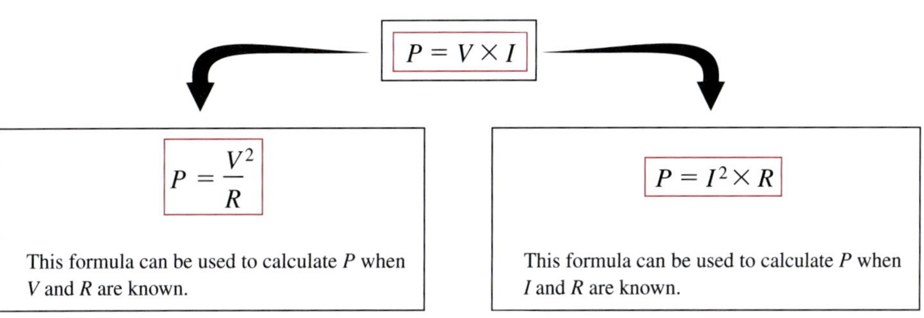

$P = V \times I$
$P = \frac{V^2}{R}$
This formula can be used to calculate P when V and R are known.
$P = I^2 \times R$
This formula can be used to calculate P when I and R are known.

EXAMPLE:

In Figure 2-15 a 12 V battery is connected across a 36 Ω resistor. How much power does the resistor dissipate?

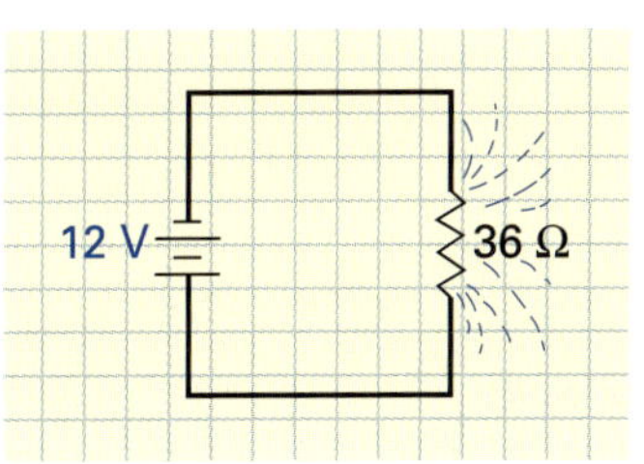

FIGURE 2-15

■ ***Solution:***

Since V and R are known, the V^2/R power formula should be used.

CALCULATOR SEQUENCE

Step	Keypad Entry	Display Response
1.	[1] [2]	12
2.	[X^2] (square key)	144
3.	[÷]	
4.	[3] [6]	36
5.	[=]	4

$$\text{power} = \frac{\text{voltage}^2}{\text{resistance}}$$

$$= \frac{(12\ V)^2}{36\ \Omega} = \frac{144}{36}$$

$$= 4\ \text{W}$$

Four joules of heat energy are being dissipated every second.

The atoms of a resistor obstruct the flow of moving electrons (current), and this friction between the two causes heat to be generated. Current flow through any resistance therefore will always be accompanied by heat. In some instances, such as a heater or oven, the device has been designed specifically to generate heat. In other applications it may be necessary to include a resistor to reduce current flow, and the heat generated will be an unwanted side effect.

Any one of the three power formulas can be used to calculate the power dissipated by a resistance. However, since power is determined by the friction between current (I) and the circuit's resistance (R), the $P = I^2 \times R$ formula is more commonly used to calculate the heat generated.

■ **EXAMPLE:**

If 100 mA of current is passing through a 33 Ω resistor, how much heat will it dissipate?

■ ***Solution:***

$$P = I^2 \times R$$

$$= (100\ \text{mA})^2 \times 33\ \Omega$$

$$= 0.33\ \text{W}$$

2-2-4 *Measuring Power*

The multimeter can be used to measure power by following the procedure described in Figure 2-16. In this example, the power consumed by a car's music system is being determined. After the current measurement and voltage measurement have been taken, the product of the two values will have to be calculated, since power = current × voltage ($P = I \times V$).

■ **EXAMPLE:**

Calculate the power consumed by the car music system shown in Figure 2-16.

■ ***Solution:***

Current drawn from battery by music system = 970 mA. Voltage applied to music system power input = 13.6 V.

$$P = I \times V = 970\ \text{mA} \times 13.6\ \text{V} = 13.2\ \text{watts}$$

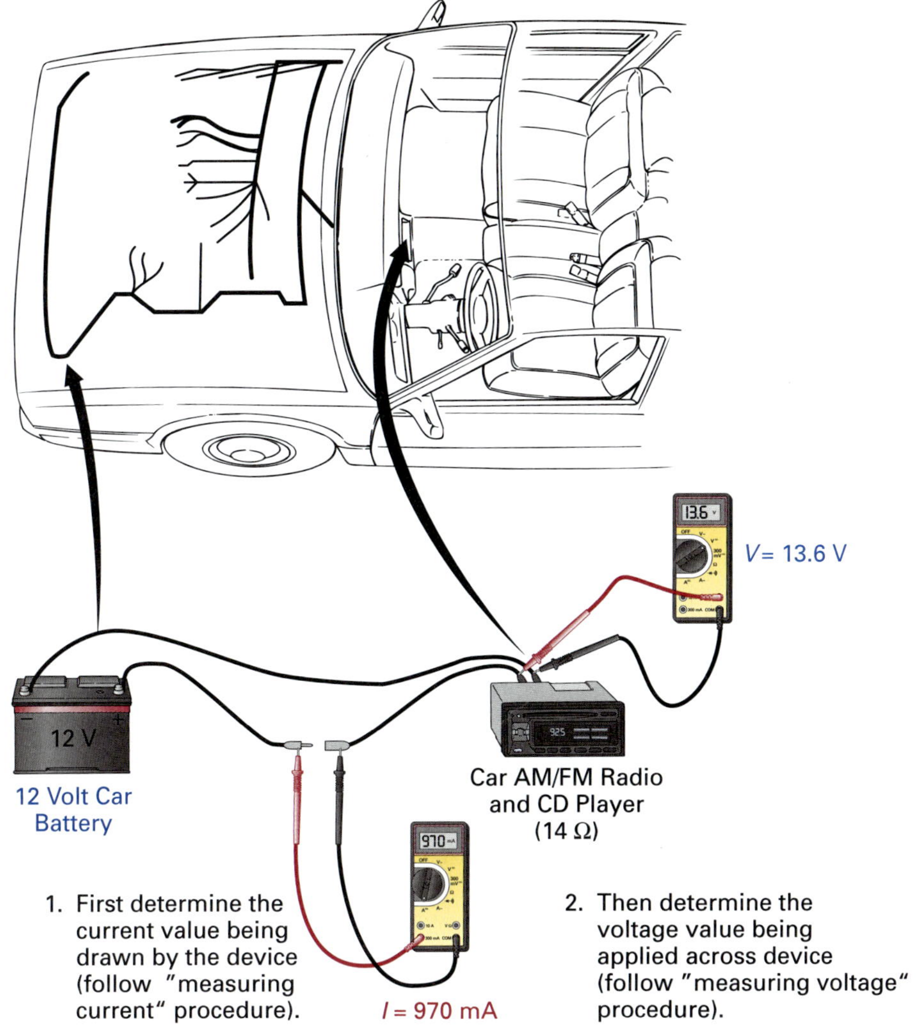

FIGURE 2-16 **Measuring Power with a Multimeter.**

By connecting a special current probe, as shown in Figure 2-17, some multimeters are able to make power measurements. In this configuration, the current probe senses current while the standard meter probes measure voltage, and the multimeter performs the multiplication process and displays the power reading.

2-2-5 *The Kilowatt-Hour*

You and I pay for our electric energy in a unit called the **kilowatt-hour** (kWh). The **kilowatt-hour meter,** shown in Figure 2-18, measures how many kilowatt-hours are consumed, and the electric company then charges accordingly.

So what is a kilowatt-hour? Well, since power is the rate at which energy is used, if we multiply power and time, we can calculate how much energy has been consumed.

Kilowatt-Hour
1000 watts for 1 hour.

Kilowatt-Hour Meter
A meter used by electric companies to measure a customer's electric power use in kilowatt-hours.

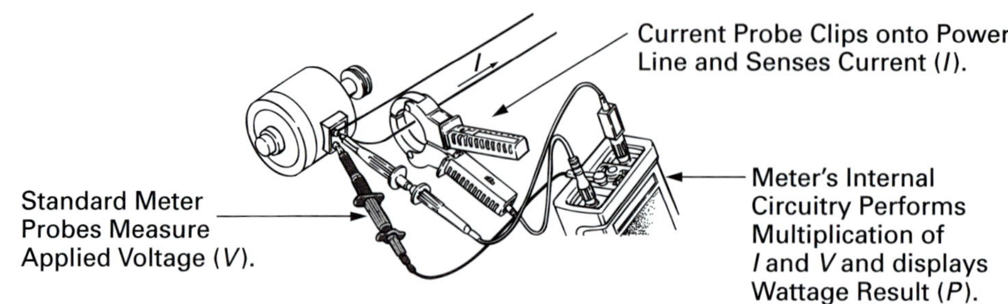

FIGURE 2-17 **Configuring the Multimeter to Measure Power.**

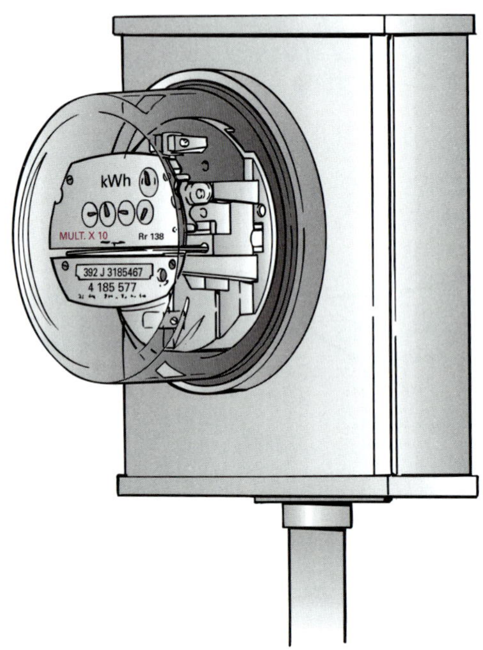

FIGURE 2-18 **Kilowatt-Hour Meter.**

$$\text{energy consumed } (W) = \text{power } (P) \times \text{time } (t)$$

This formula uses the product of power (in watts) and time (in seconds or hours) and so we can use one of three units: the watt-second (Ws), watt-hour (Wh), or kilowatt-hour (kWh). The kilowatt-hour is most commonly used by electric companies, and by definition, a kilowatt-hour of energy is consumed when you use 1000 watts of power in 1 hour (1 kW in 1 h).

$$\text{energy consumed (kWh)} = \text{power (kW)} \times \text{time (h)}$$

To see how this would apply, let us look at a couple of examples.

EXAMPLE:

If a 100 W light bulb is left on for 10 hours, how many kilowatt-hours will we be charged for?

■ ***Solution:***

$$\text{power consumed (kWh)} = \text{power (kW)} \times \text{time (hours)}$$
$$= 0.1 \text{ kW} \times 10 \text{ hours}$$
$$(100 \text{ W} = 0.1 \text{ kW})$$
$$= 1 \text{ kWh}$$

CALCULATOR SEQUENCE

Step	Keypad Entry	Display Response
1.	[0] [.] [1] [E] [3]	0.1E3
2.	[×]	
3.	[1] [0]	10
4.	[=]	1E3

■ **EXAMPLE:**

Figure 2-19 illustrates a typical household electric heater and an equivalent electrical circuit. The heater has a resistance of 7 Ω and the electric company is charging 12 cents/kWh. Calculate:

a. The energy consumed by the heater
b. The cost of running the heater for 7 hours

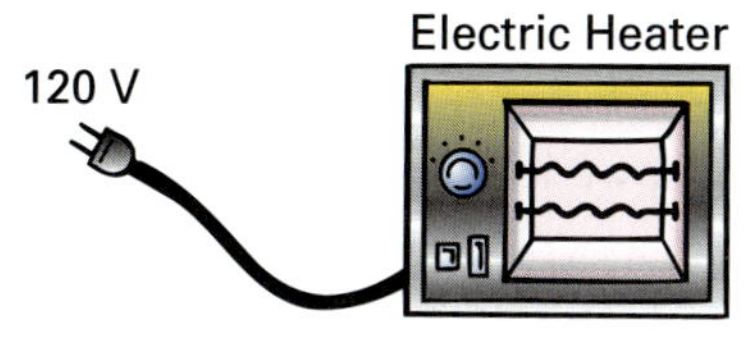

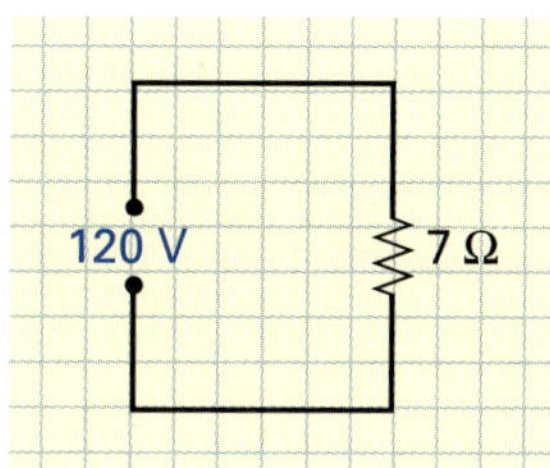

FIGURE 2-19 **An Electric Heater.**

■ ***Solution:***

a. $$\text{Power } (P) = \frac{V^2}{R} = \frac{120^2}{7}$$
$$= \frac{14400 \text{ V}}{7} = \frac{14.4 \text{ kV}}{7}$$
$$= 2057 \text{ W (approximately 2 kilowatts or 2 kW)}$$

b. $$\text{Energy consumed} = \text{power (kW)} \times \text{time (hours)}$$
$$= 2.057 \times 7$$
$$= 14.4 \text{ kWh}$$
$$\text{Cost} = \text{kWh} \times \text{rate} = 14.4 \times 12 \text{ cents}$$
$$= \$1.73$$

SELF-TEST EVALUATION POINT FOR SECTION 2-2

Use the following questions to test your understanding of Section 2-2:

1. List the six basic forms of energy.

2. What is the difference between energy, work, and power?

3. List the formulas for calculating energy and power.

4. What is 1 kilowatt-hour of energy?

2-3 CONDUCTORS

A lightning bolt that sets fire to a tree and the operation of your calculator are both electrical results achieved by the flow of electrons. The only difference is that your calculator's circuits control the flow of electrons, while the lightning bolt is the uncontrolled flow of electrons. In electronics, a **conductor** is used to channel or control the path in which electrons flow.

Conductor
Length of wire whose properties are such that it will carry an electric current.

Any material that passes current easily is called a conductor. These materials are said to have a "low resistance," which means that they will offer very little opposition to current flow. This characteristic can be explained by examining conductor atoms. As mentioned previously, the atom has a maximum of seven orbital paths known as shells, which are named K, L, M, N, O, P, and Q, stepping out toward the outermost or valence shell. Conductors are materials or natural elements whose valence electrons can easily be removed from their parent atoms. They are therefore said to be sources of free electrons, and these free electrons provide us with circuit current. The precious metals of silver and gold are the best conductors.

More specifically, a better conductor has:

1. Electrons in shells farthest away from the nucleus; these electrons feel very little nucleus attraction and can be broken away from their atom quite easily.
2. More electrons per atom.
3. An incomplete valence shell. This means that the valence shell does not have in it the maximum possible number of electrons. If the atom had its valence ring complete (full), there would be no holes (absence of an electron) in that shell, so no encouragement for adjacent atom electrons to jump from their parent atom into the next atom would exist, preventing the chain reaction known as current.

TIME LINE
Stephen Gray (1693–1736), an Englishman, discovered that certain substances would conduct electricity.

Economy must be considered when choosing a conductor. Large quantities of conductors using precious metals are obviously going to send the cost of equipment beyond reach. The conductor must also satisfy some physical requirements, in that we must be able to shape it into wires of different sizes and easily bend it to allow us to connect one circuit to the next.

Copper is the most commonly used conductor, as it meets the following three requirements:

1. It is a good source of electrons.
2. It is inexpensive.
3. It is physically pliable.

Aluminum is also a very popular conductor, and although it does not possess as many free electrons as copper, it has the two advantages of being less expensive and lighter than copper.

2-3-1 *Conductance*

Conductance (*G*)
Measure of how well a circuit or path conducts or passes current. It is equal to the reciprocal of resistance.

Conductance is the measure of how good a conductor is at carrying current. Conductance (symbolized *G*) is equal to the reciprocal of resistance (or opposition) and is measured in the unit siemens (S):

$$\text{conductance } (G) = \frac{1}{\text{resistance } (R)}$$

Conductance (G) values are measured in siemens (S) and resistance (R) in ohms (Ω).

This means that conductance is inversely proportional to resistance. For example, if the opposition to current flow (resistance) is low, the conductance is high and the material is said to have a good conductance.

$$\text{high conductance } G\uparrow = \frac{1}{R\downarrow \text{ (low resistance)}}$$

On the other hand, if the resistance of a conducting wire is high ($R\uparrow$), its conductance value is low ($G\downarrow$) and it is called a poor conductor. A good conductor therefore has a high conductance value and a very small resistance to current flow.

TIME LINE

Stephen Gray's lead was picked up in 1730 by Charles du Fay, a French experimenter, who believed that there were two types of electricity, which he called *vitreous* and *resinous* electricity.

EXAMPLE:

A household electric blanket offers 25 ohms of resistance to current flow. Calculate the conductance of the electric blanket's heating element.

Solution:

$$\text{conductance} = \frac{1}{\text{resistance}}$$

$$G = \frac{1}{R} = \frac{1}{25 \text{ ohms}} = 0.04 \text{ siemens}$$

$$\text{or } 40 \text{ mS (millisiemens)}$$

To express how good or poor a conductor is, we must have some sort of reference point. The reference point we use is the best conductor, silver, which is assigned a relative conductivity value of 1.0. Table 2-1 lists other conductors and their relative conductivity values with respect to the best, silver. The formula for calculating relative conductivity is

$$\text{Relative Conductivity} = \frac{\text{Conductor's Relative Conductivity}}{\text{Reference Conductor's Relative Conductivity}}$$

TABLE 2-1 Relative Conductivity of Conductors

MATERIAL	CONDUCTIVITY (RELATIVE)
Silver	1.000
Copper	0.945
Gold	0.652
Aluminum	0.575
Tungsten	0.297
Nichrome	0.015

EXAMPLE:

What is the relative conductivity of tungsten if copper is used as the reference conductor?

Solution:

$$\text{tungsten} = 0.297$$

$$\text{copper} = 0.945$$

$$\text{relative conductivity} = \frac{\text{conductivity of conductor}}{\text{conductivity of reference}} = \frac{0.297}{0.945} = 0.314$$

CALCULATOR SEQUENCE

Step	Keypad Entry	Display Response
1.	0 . 2 9 7	0.297
2.	÷	
3.	0 . 9 4 5	0.945
4.	=	0.314

EXAMPLE:

What is the relative conductivity of silver if copper is used as the reference?

Solution:

$$\text{relative conductivity} = \frac{\text{silver}}{\text{copper}}$$

$$= \frac{1.000}{0.945} = 1.058$$

2-3-2 *Conductors and Their Resistance*

If the current within a circuit needs to be reduced, a resistor is placed in the current path. Resistors are constructed using materials that are known to oppose current flow. Conductors, on the other hand, are not meant to offer any resistance or opposition to current flow. This, however, is not always the case since some are good conductors having a low resistance, while others are poor conductors having a high resistance. Conductance is the measure of how good a conductor is, and even the best conductors have some value of resistance. Up until this time, we have determined a conductor's resistance based on the circuit's electrical characteristics,

$$R = \frac{V}{I}$$

With this formula we could determine the conductor's resistance based on the voltage applied and circuit current. Another way to determine the resistance of a conductor is to examine the physical factors of the conductor.

The total resistance of a typical conductor, like the one shown in Figure 2-20, is determined by four main physical factors:

1. The type of conducting material used
2. The conductor's cross-sectional area
3. The total length of the conductor
4. The temperature of the conductor

By combining all of the physical factors of a conductor, we can arrive at a formula for its resistance.

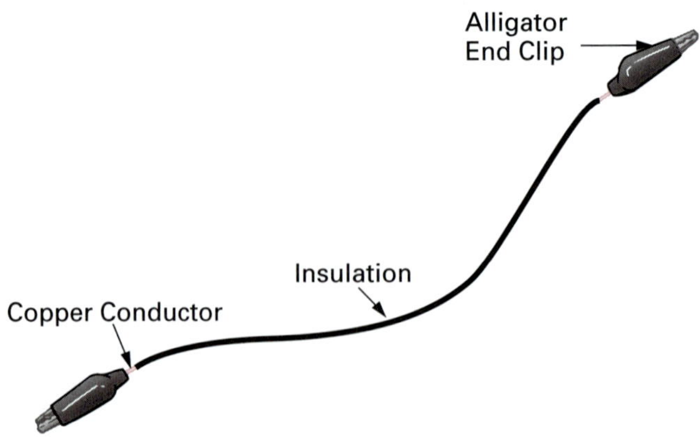

FIGURE 2-20 Copper Conductor.

TABLE 2-2 Material Resistivity

MATERIAL	RESISTIVITY (cmil/ft) IN OHMS
Silver	9.9
Copper	10.7
Gold	16.7
Aluminum	17.0
Tungsten	33.2
Zinc	37.4
Brass	42.0
Nickel	47.0
Platinum	60.2
Iron	70.0

$$R = \frac{\rho \times l}{a}$$

where R = resistance of conductor, in ohms
p = resistivity of conducting material
l = length of conductor, in feet
a = area of conductor, in circular mils

Resistivity
Measure of a material's resistance to current flow.

Resistivity, by definition, is the resistance (in ohms) that a certain length of conductive material (in feet) will offer to the flow of current. Table 2-2 lists the resistivity of the more commonly used conductors.

The resistance of a conductor, therefore, can be calculated based either on its physical factors or on electrical performance.

$$\frac{\rho \times l}{a} = R = \frac{V}{I}$$

Physical ⇔ Electrical

EXAMPLE:

Calculate the resistance of 333 ft of copper conductor with a conductor area of 3257 cmil.

Solution:

$$R = \frac{\rho \times l}{a}$$
$$= \frac{10.7 \times 333}{3257}$$
$$= 1.09\ \Omega$$

CALCULATOR SEQUENCE

Step	Keypad Entry	Display Response
1.	[1] [0] [.] [7]	10.7
2.	[×]	
3.	[3] [3] [3]	333
4.	[÷]	3563.1
5.	[3] [2] [5] [7]	3257
6.	[=]	1.09

EXAMPLE:

Calculate the resistance of 1274 ft of aluminum conductor with a diameter of 86.3 mils.

Solution:

$$R = \frac{\rho \times l}{a}$$

If the diameter is equal to 86.3 mils, the circular mil area equals d^2:

$$86.3^2 = 7447.7 \text{ cmil}$$

Referring to Table 2-1 you can see that the resistivity of aluminum is 17.0, and therefore

$$R = \frac{17 \times 1274}{7447.7}$$
$$= 2.9\ \Omega$$

2-3-3 *Temperature Effects on Conductors*

When heat is applied to a conductor, the atoms within the conductor convert this thermal energy into mechanical energy or movement. These random moving atoms cause collisions between the directed electrons (current flow) and the adjacent atoms, resulting in an opposition to the current flow (resistance).

Positive Temperature Coefficient of Resistance
The rate of increase in resistance relative to an increase in temperature.

Metallic conductors are said to have a **positive temperature coefficient of resistance** (+ Temp. Coe. of R). This means that the greater the heat applied to the conductor, the greater the atom movement, causing more collisions of atoms to occur, and consequently the greater the conductor's resistance.

heat ↑ resistance ↑

1. The Filament Resistor

Filament Resistor
The resistor in a light bulb or electron tube.

Incandescent Lamp
An electric lamp that generates light when an electric current is passed through its filament of resistance, causing it to heat to incandescence.

Figure 2-21(a) illustrates the **filament resistor** within a glass bulb. This component is more commonly known as the household light bulb. The filament resistor is just a coil of wire that glows white hot when current is passed through it and, in so doing, dissipates both heat and light energy.

This **incandescent lamp** is an electric lamp in which electric current flowing through a filament of resistive material heats the filament until it glows and emits light. Figure 2-21(b) shows a test circuit for varying current through a small incandescent lamp. Potentiometer R_1 is used to vary the current flow through the lamp. The lamp is rated at 6 V, 60 mA. Using Ohm's law we can calculate the lamp's filament resistance at this rated voltage and current:

$$\text{filament resistance, } R = \frac{V}{I} = \frac{6\text{ V}}{60\text{ mA}} = 100\ \Omega$$

Cold Resistance
The resistance of a device when cold.

Hot Resistance
The resistance of a device when hot due to the generation of heat by electric current.

The filament material (tungsten, for example) is like all other conductors in that it has a positive temperature coefficient of resistance. Therefore, as the current through the filament increases, so does the temperature and so does the filament's resistance (I↑, temperature ↑, R↑). Consequently, when variable resistor R_1's wiper is moved to the right so that it produces a high resistance (R_1↑), the circuit current will be small (I↓) and the lamp will glow dimly. Since the circuit current is small, the filament temperature will be small and so will the lamp's resistance (I↓ temperature ↓, lamp resistance ↓). This small value of resistance is called the lamp's **cold resistance.** On the other hand, when R_1's wiper is moved to the left so that it produces a small resistance (R↓), the circuit current will be large (I↑), and the lamp will glow brightly. With the circuit current high, the filament temperature will be high and so will the lamp's resistance (I↑, temperature ↑, lamp resistance ↑). This large value of resistance is called the lamp's **hot resistance.**

Figure 2-21(c) plots the filament voltage, which is being measured by the voltmeter, against the filament current, which is being measured by the ammeter. As you can see, an increase in current will cause a corresponding increase in filament resistance. Studying this graph, you may have also noticed that the lamp has been operated beyond its rated value of 6 V, 60 mA. Although the lamp can be operated beyond its rated value (for example, 10 V, 80 mA), its life expectancy will be decreased dramatically from several hundred hours to only a few hours.

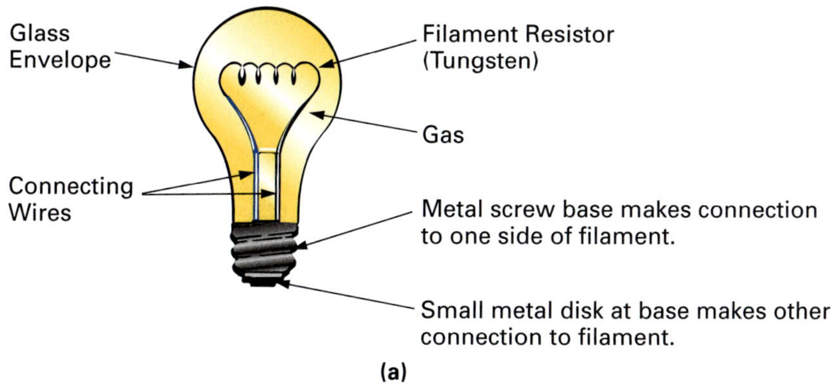

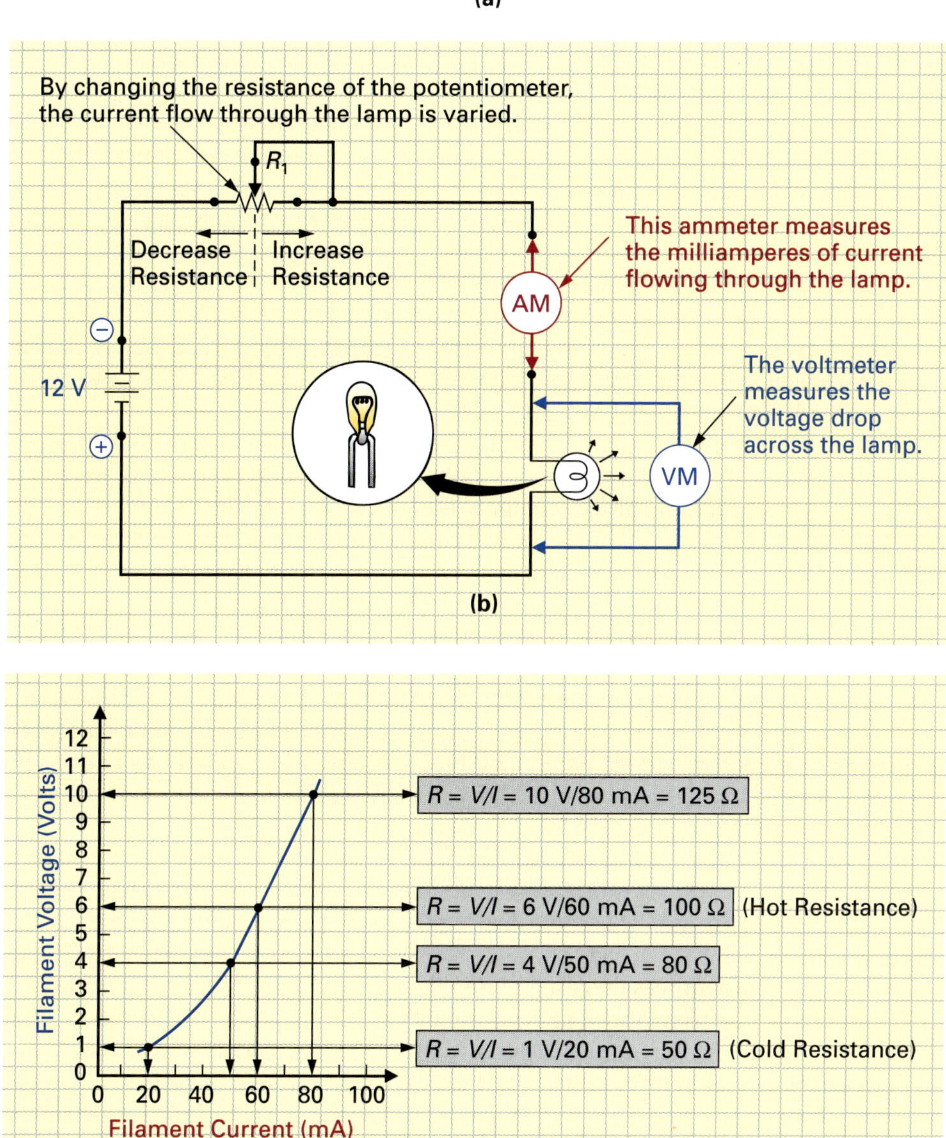

FIGURE 2-21 Filament Resistor.

Ballast Resistor

A resistor that increases in resistance when voltage increases. It can therefore maintain a constant current despite variations in line voltage.

A coil of wire similar to the one found in a light bulb, called a **ballast resistor,** can be used to maintain a constant current despite a variation in voltage. Since the coil of wire is a conductor and conductors have a positive temperature coefficient, an increase in voltage will cause a corresponding increase in current and therefore in the heat generated, which will result in an increase in the wire's resistance. This increase in resistance will decrease the initial current rise. Similarly, a decrease in voltage and therefore current ($I \propto V$) will result in a decrease in heat and wire resistance. This decrease in resistance will permit an increase in current to counteract the original decrease. Current is therefore regulated or maintained constant by the ballast resistor, despite variations in voltage.

2. Fuses

Fuse

This circuit- or equipment-protecting device consists of a short, thin piece of wire that melts and breaks the current path if the current exceeds a rated damaging level.

Current must be monitored and not be allowed to exceed a safe level so as to protect users from shock, to protect the equipment from damage, and to prevent fire hazards. A **fuse** is an equipment protection device that consists of a thin wire link within a casing. Figure 2-22(a) shows the physical appearance of a fuse and Figure 2-22(b) shows the fuse's schematic symbol.

To protect a system, a fuse needs to disconnect power from the unit the moment current exceeds a safe value. To explain how this happens, Figure 2-22(c) shows how a fuse is mounted in the rear of a dc power supply. All fuses have a current rating, and this rating indicates what value of current will generate enough heat to melt the thin wire link within the fuse. In the example in Figure 2-22(c), if the current is less than the fuse rating of 2 amps, the metal link of the fuse will remain intact. If, on the other hand, the current through the thin metal element exceeds the current rating of 2 A, the excessive current will create enough heat to melt the element and "open" or "blow" the fuse, thus disconnecting the dc power supply from the 120 volt source and protecting it from damage.

One important point to remember is that if the current increases to a damaging level, the fuse will open and protect the dc power supply. This implies that something went wrong with the 120 V source and it started to supply too much current. This is almost never the case. What actually happens is that, as with all equipment, eventually something internally breaks down. This can cause the overall load resistance of the piece of equipment to increase or decrease. An increase in load resistance ($R_L\uparrow$) means that the source sees a higher resistance in the current path and therefore a small current would flow from the source ($I\downarrow$), and the user would be aware of the problem due to the nonoperation of the equipment. If an internal equipment breakdown causes the equipment's load resistance to decrease ($R_L\downarrow$), the source will see a smaller circuit resistance and will supply a heavier circuit current ($I\uparrow$), which could severely damage the equipment before the user had time to turn it off. Fuse protection is needed to disconnect the current automatically in this situation, to protect the equipment from damage.

Fuse elements come in various shapes and sizes so as to produce either quick heating and then melting (fast blow) or delayed heating and then melting (slow blow), as illustrated in Figure 2-22(d). The reason for the variety is based in the application differences. When turned on, some pieces of equipment will be of such low resistance that a short momentary current surge, sometimes in the region of four times the fuse's current rating, will result for the first couple of seconds. If a fast-blow fuse were placed in the circuit, it would blow at the instant the equipment was turned on, even though the equipment did not need to be protected. The system merely needs a large amount of current initially to start up. A slow-blow fuse would be ideal in this application, as it would permit the initial heavy current and would begin to heat up and yet not blow due to the delay. Once the surge had ended, the current would decrease, and the fuse would still be intact (some slow-blow fuses will allow a 400% overload current to flow for a few seconds).

On the other hand, some equipment cannot take any increase in current without being damaged. The slow blow would not be good in this application, as it would allow an increase of current to pass to the equipment for too long a period of time. The fast blow would be ideal in this application, since it would disconnect the current instantly if the current rating of the fuse were exceeded, and so prevent equipment damage. The automobile was the first application for a fuse in a glass holder, and so a size standard named AG (automobile glass) was established. This physical size standard is still used today, and is listed in Figure 2-22(d).

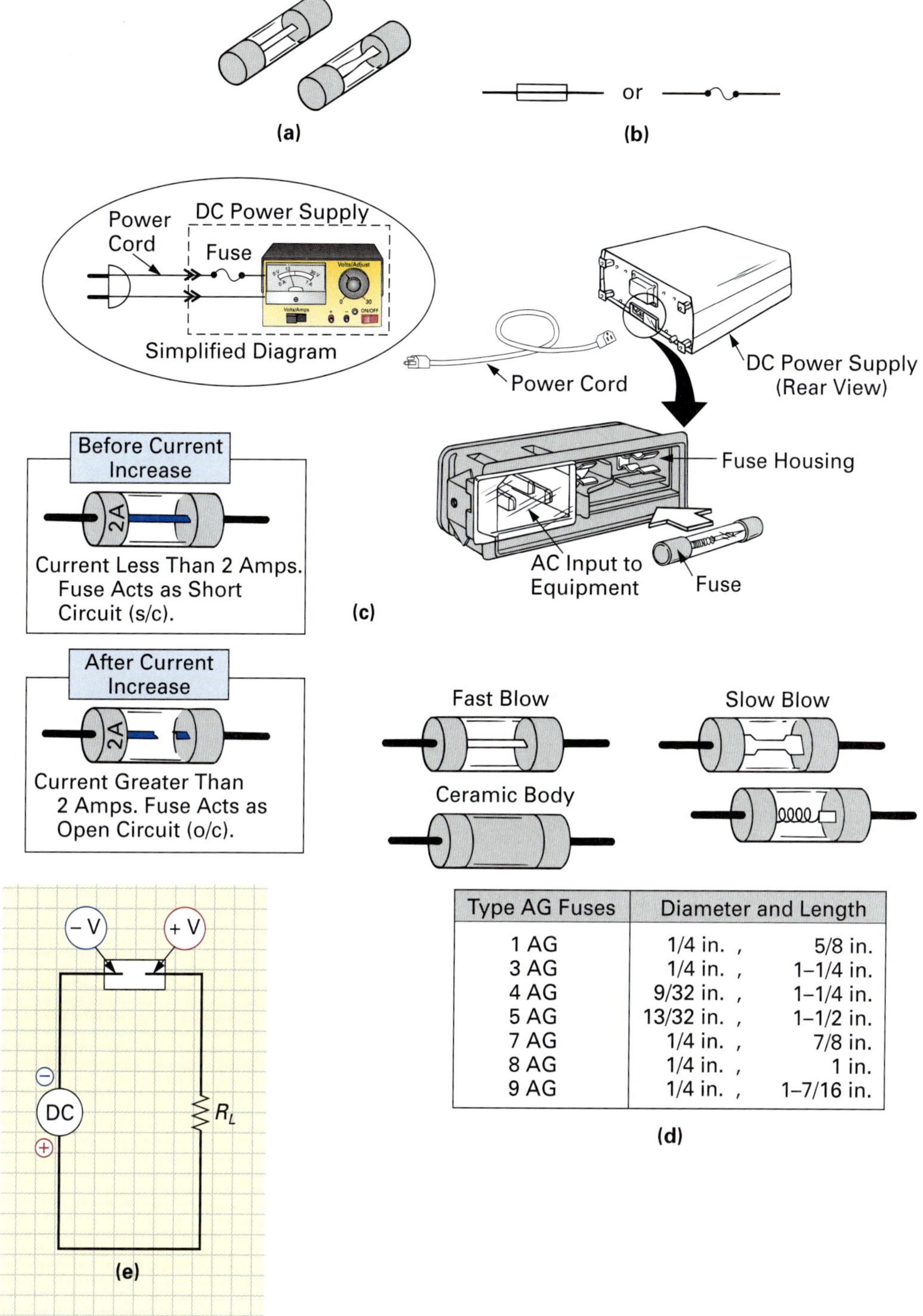

Type AG Fuses	Diameter and Length	
1 AG	1/4 in. ,	5/8 in.
3 AG	1/4 in. ,	1–1/4 in.
4 AG	9/32 in. ,	1–1/4 in.
5 AG	13/32 in. ,	1–1/2 in.
7 AG	1/4 in. ,	7/8 in.
8 AG	1/4 in. ,	1 in.
9 AG	1/4 in. ,	1–7/16 in.

FIGURE 2-22 **Fuses.**

Fuses also have a voltage rating that indicates the maximum circuit voltage that can be applied across the fuse by the circuit in which the fuse resides. This rating, which is important after the fuse has blown, prevents arcing across the blown fuse contacts, as illustrated in Figure 2-22(e). Once the fuse has blown, the circuit's positive and negative voltages are now connected across the fuse contacts. If the voltage is too great, an arc can jump across the gap, causing a sudden surge of current, damaging the equipment connected.

Fuses are mounted within fuse holders and normally placed at the back of the equipment for easy access, as was shown in Figure 2-22(c). When replacing blown fuses, you must

TABLE 2-3 **The American Wire Gauge (AWG) for Copper Conductor**

AWG NUMBER[a]	DIAMETER (mils)	MAXIMUM CURRENT (A)	Ω/1000 ft
0000	460.0	230	0.0490
000	409.6	200	0.0618
00	364.8	175	0.0780
0	324.9	150	0.0983
1	289.3	130	0.1240
2	257.6	115	0.1563
3	229.4	100	0.1970
4	204.3	85	0.2485
5	181.9	75	0.3133
6	162.0	65	0.3951
7	144.3	55	0.4982
8	128.5	45	0.6282
9	114.4	40	0.7921
10	101.9	30	0.9981
11	90.74	25	1.260
12	80.81	20	1.588
13	71.96	17	2.003
14	64.08	15	2.525
15	57.07		3.184
16	50.82	6	4.016
17	45.26		5.064
18	40.30	3	6.385
19	35.89	Wires of	8.051
20	31.96	this size	10.15
22	25.35	have	16.14
26	15.94	current	40.81
30	10.03	measured	103.21
40	3.145	in mA	1049.0

[a] The larger the AWG number, the smaller the size of the conductor.

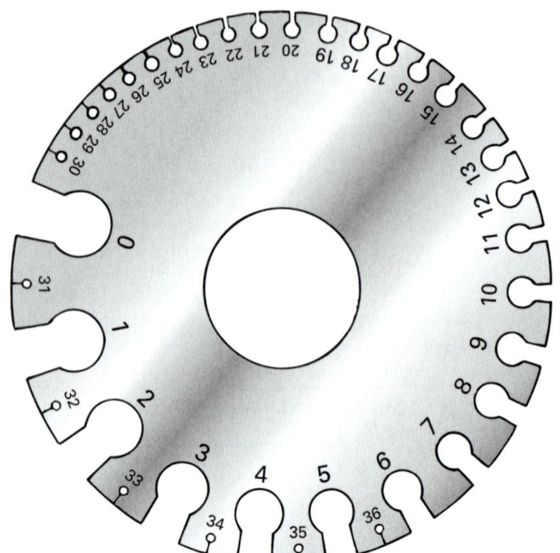

FIGURE 2-23 **Wire Gauge Size.**

be sure that the equipment is turned OFF and disconnected from the source. If you do not remove power you could easily get a shock, since the entire source voltage appears across the fuse as was shown in Figure 2-22(e). Also, make sure that a fuse with the correct current and voltage rating is used.

A good fuse should have a resistance of 0 Ω when it is checked with an ohmmeter. In circuit and with power ON, the fuse should have no voltage drop across it because it is just a piece of wire. A blown or burned-out fuse will, when removed, read infinite ohms, and when in its holder and power ON, will have the full applied voltage across its two terminals.

2-3-4 *Maximum Conductor Current*

American Wire Gauge

American wire gauge (AWG) is a system of numerical designations of wire sizes, with the first being 0000 (the largest size) and then going to 000, 00, 0, 1, 2, 3, and so on up to the smallest sizes of 40 and above.

Any time that current flows through any conductor, a certain resistance or opposition is inherent in that conductor. This resistance will convert current to heat, and the heat further increases the conductor's resistance (+ Temp. Coe. of R), causing more heat to be generated due to the opposition. As a result, a conductor must be chosen carefully for each application so that it can carry the current without developing excessive heat. This is achieved by selecting a conductor with a greater cross-sectional area to decrease its resistance. The National Fire Protection Association has developed a set of standards known as the **American Wire Gauge** for all copper conductors, which lists their diameter, resistance, and maximum safe current in amperes. This table is given in Table 2-3. A rough guide for measuring wire size is shown in Figure 2-23.

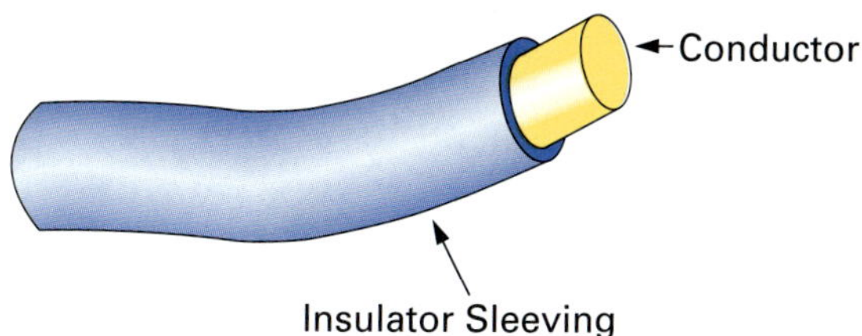

FIGURE 2-24 **Conductor with Insulator.**

Conducting wires are normally covered with a plastic or rubber type of material, known as insulation, as shown in Figure 2-24. This insulation is used to protect the users and technicians from electrical shock and also to keep the conductor from physically contacting other conductors within the equipment. If the current through the conductor is too high, this insulation will burn due to the heat and may cause a fire hazard. This is why a conductor's maximum safe current value should not be exceeded.

2-3-5 *Conductor and Connector Types*

A cable is made up of two or more wires. Figure 2-25 illustrates some of the different types of **wires** and **cables.** In Figure 2-25(a), (b), and (c), only one conductor exists within the insulation, so they are classified as wires, whereas the cables seen in Figure 2-25(e) and (f) have two conductors. The coaxial and twin-lead cables are most commonly used to connect TV signals into television sets. Figure 2-25(d) illustrates how conducting copper, silver, or gold strips are printed on a plastic insulating board and are used to connect components such as resistors that are mounted on the other side of the board.

Wire
Single solid or stranded group of conductors having a low resistance to current flow.

Cable
Group of two or more insulated wires.

Wires and cables have to connect from one point to another. Some are soldered directly, whereas others are attached to plugs that plug into sockets. A sample of connectors is shown in Figure 2-26.

SELF-TEST EVALUATION POINT FOR SECTION 2-3

Use the following questions to test your understanding of Section 2-3:

1. List the four factors that determine the total resistance of a conductor.
2. What is the relationship between the resistance of a conductor and its cross-sectional area, length, and resistivity?
3. True or false: Conductors are said to have a positive temperature coefficient.
4. True or false: The smaller the AWG number, the larger the size of the conductors.
5. What is a fuse, and what is its main purpose?

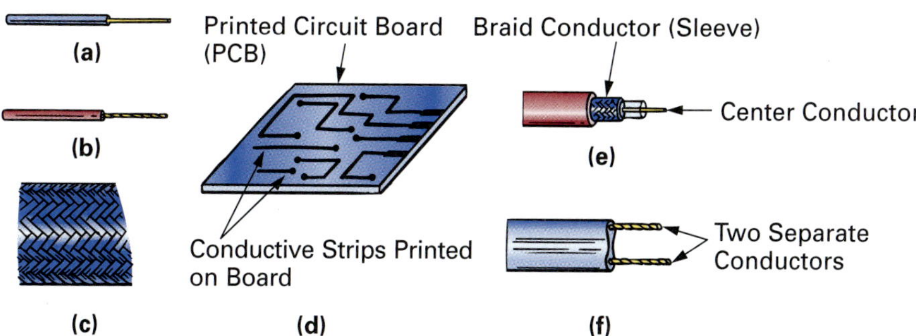

FIGURE 2-25 **Wires and Cables. (a) Solid Wire. (b) Stranded Wire. (c) Braided Wire. (d) Printed Wire. (e) Coaxial Cable. (f) Twin Lead.**

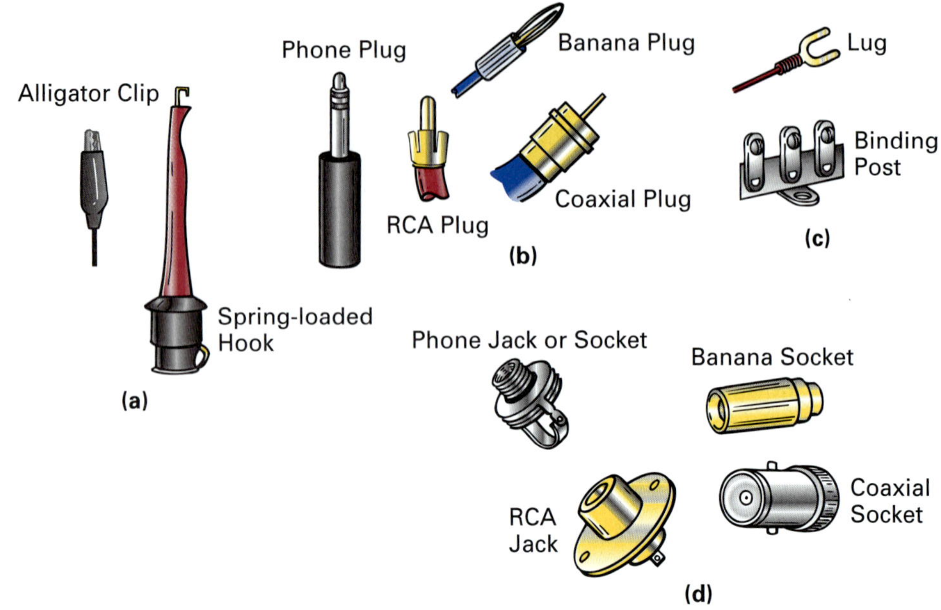

FIGURE 2-26 **Connectors. (a) Temporary Connectors. (b) Plugs. (c) Lug and Binding Post. (d) Sockets.**

2-4 INSULATORS

Insulator

A material that has few electrons per atom and those electrons are close to the nucleus and cannot be easily removed.

Breakdown Voltage

The voltage at which breakdown of an insulator occurs.

Dielectric Strength

The maximum potential a material can withstand without rupture.

Any material that offers a high resistance or opposition to current flow is called an **insulator.** Conductors permit the easy flow of current and so have good conductivity, while insulators allow small to almost no amount of free electrons to flow. Insulators can, with sufficient pressure or voltage applied across them, "break down" and conduct current. This **breakdown voltage** must be great enough to dislodge the electrons from their close orbital shells (K, L shells) and release them as free electrons.

Insulators are also called "dielectrics," and the best insulator or dielectric should have the maximum possible resistance and conduct no current at all. The **dielectric strength** of an insulator indicates how good or bad an insulator is by indicating the voltage that will cause the insulating material to break down and conduct a large current. Table 2-4 lists some of the more popular insulators and the value of kilovolts that will cause a

TABLE 2-4 **Breakdown Voltages of Certain Insulators**

MATERIAL	BREAKDOWN STRENGTH (kV/cm)
Mica	2000
Glass	900
Teflon	600
Paper	500
Rubber	275
Bakelite	151
Oil	145
Porcelain	70
Air	30

centimeter of insulator to break down. As an example, if 1 centimeter of paper is connected to a variable voltage source, a voltage of 500 kilovolts is needed to break down the paper and cause current to flow.

The following formula can be used to calculate the dielectric thickness needed to withstand a certain voltage.

$$\text{dielectric thickness} = \frac{\text{voltage to insulate}}{\text{insulator's breakdown voltage}}$$

EXAMPLE:

What thickness of mica would be needed to withstand 16,000 V?

Solution:

$$\text{mica strength} = 2000 \text{ kV/cm}$$

$$\text{dielectric thickness} = \frac{16{,}000 \text{ V}}{2000 \text{ kV/cm}} = 0.008 \text{ cm}$$

CALCULATOR SEQUENCE

Step	Keypad Entry	Display Response
1.	[1] [6] [E](exponent) [3]	16E3
2.	[÷]	
3.	[2] [E] [6](or 2000 E3)	2E6
4.	[=]	0.008

EXAMPLE:

What maximum voltage could 1 mm of air withstand?

Solution:

There are 10 mm in 1 cm. If air can withstand 30,000 V/cm, it can withstand 3000 V/mm.

SELF-TEST EVALUATION POINT FOR SECTION 2-4

Use the following questions to test your understanding of Section 2-4.

1. True or false: An insulator is a material used to block the flow of current.
2. What is considered to be the best insulator material?
3. Define *breakdown voltage.*
4. Would the conductance figure of a good insulator be large or small?

2-5 RESISTORS

Conductors are used to connect one device to another, and although they offer a small amount of resistance, this resistance is not normally enough. In electronic circuits, additional resistance is normally needed to control the amount of current flow, and the component used to supply this additional resistance is called a **resistor.** Resistors come in a variety of shapes and sizes. Some have a fixed value of resistance, while others can be adjusted manually to change their resistance. To begin, let us examine some of the basic resistor characteristics.

Resistor
Component made of a material that opposes the flow of current and therefore has some value of resistance.

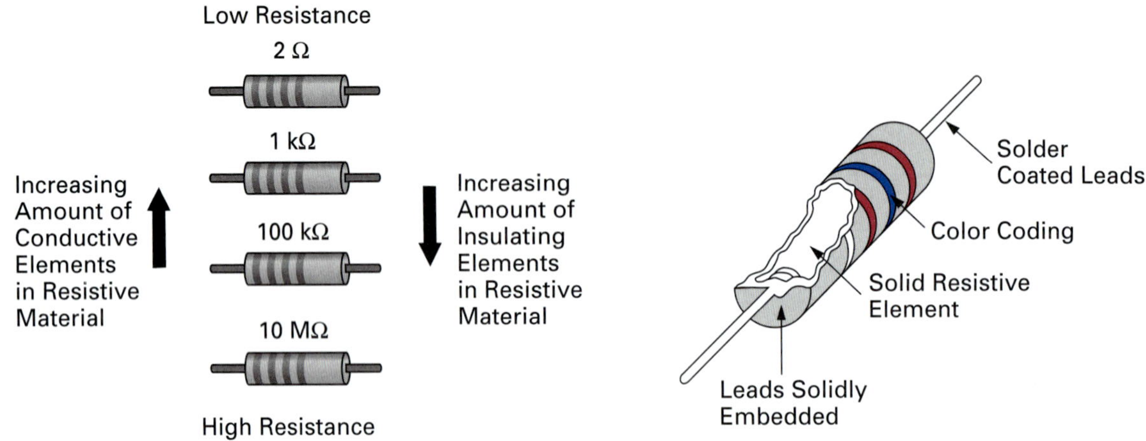

FIGURE 2-27 **Changing Resistor Resistance.**

FIGURE 2-28 **Resistor Construction.**

2-5-1 *Resistor Wattage, Tolerance, and Color Coding*

Figure 2-27 illustrates four resistors that range in value from 2 Ω to 10 MΩ (10 million Ω). Resistors are constructed by placing a piece of resistive material with embedded connecting leads at each end within an insulating cylindrical, molded case, as seen in Figure 2-28. The resistance of the resistor is changed by varying the ratio of conductor to insulator within the resistive material. The 2 Ω resistor in Figure 2-27, therefore, has more conductive elements than insulating elements in its resistive material, whereas the 10 MΩ resistor has fewer conducting elements than insulating elements in its resistive material.

1. Resistor Wattage Rating

Dissipation
Release of electrical energy in the form of heat.

Wattage Rating
Maximum power a device can safely handle continuously.

The physical size of the resistors lets the user know how much power in the form of heat can be **dissipated,** as shown in Figure 2-29. As you already know, resistance is the opposition to current flow, and this opposition causes heat to be generated whenever current is passing through a resistor. The amount of heat dissipated each second is measured in watts and each resistor has its own **wattage rating.** For example, a 2 watt size resistor can dissipate up to 2 joules of heat per second, whereas a 1/8 W size resistor can only dissipate up to 1/8 joule of heat per second. The key point to remember is that resistors in high current circuits should have a surface area that is large enough to dissipate away the heat faster than it is being generated. If the current passing through a resistor generates heat faster than the resistor can dissipate it, the resistor will burn up and no longer perform its function.

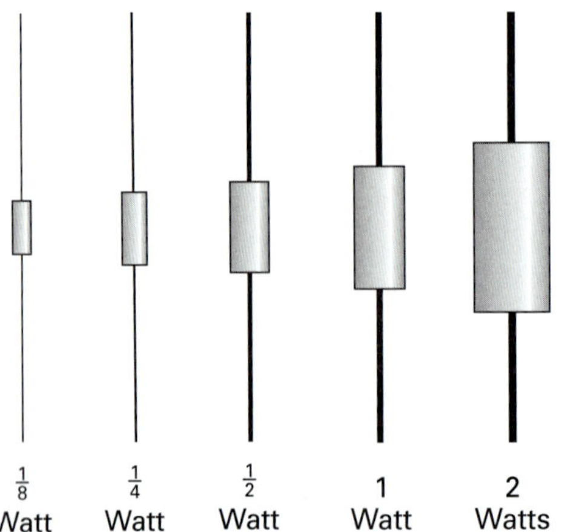

FIGURE 2-29 **Resistor Wattage Rating Guide. All Resistor Silhouettes Are Drawn to Scale.**

2. Resistor Tolerance

Another factor to consider when discussing resistors is their **tolerance.** Tolerance is the amount of deviation or error from the specified value. For example, a 1000 Ω (1 kΩ) resistor with a ± 10% (plus and minus 10%) tolerance when manufactured could have a resistance anywhere between 900 and 1100 Ω.

Tolerance
Permissible deviation from a specified value, normally expressed as a percentage.

$$\pm 10\% \text{ of } 1000 = 100$$
$$10\% - 1000 + 10\%$$
$$\downarrow \qquad\qquad \downarrow$$
$$900 \qquad\qquad 1100$$

This means that two identically marked resistors when measured could be from 900 to 1100 Ω, a difference of 200 Ω. In some applications, this may be acceptable. In other applications, where high precision is required, this deviation could be too large and so a more expensive, smaller tolerance resistor would have to be used.

EXAMPLE:

Calculate the amount of deviation of the following resistors:

a. 2.2 kΩ ± 10%
b. 5 MΩ ± 2%
c. 3 Ω ± 1%

Solution:

a. 10% of 2.2 kΩ = 220 Ω. For +10%, the value is

$$2200 + 220\ \Omega = 2420\ \Omega$$
$$= 2.42\ \text{k}\Omega$$

For − 10%, the value is

$$2200 - 220\ \Omega = 1980\ \Omega$$
$$= 1.98\ \text{k}\Omega$$

The resistor will measure anywhere from 1.98 kΩ to 2.42 kΩ.

b. 2% of 5 MΩ = 100 kΩ

$$5\ \text{M}\Omega + 100\ \text{k}\Omega = 5.1\ \text{M}\Omega$$
$$5\ \text{M}\Omega - 100\ \text{k}\Omega = 4.9\ \text{M}\Omega$$

Deviation = 4.9 MΩ to 5.1 MΩ

c. 1% of 3 Ω = 0.03 Ω or 30 milliohms (mΩ)

$$3\ \Omega + 0.03\ \Omega = 3.03\ \Omega$$
$$3\ \Omega - 0.03\ \Omega = 2.97\ \Omega$$

Deviation = 2.97 Ω to 3.03 Ω

CALCULATOR SEQUENCE

Step	Keypad Entry	Display Response
1.	[1] [0]	10
2.	[%]	10
3.	[×]	0.10
4.	[2] [.] [2] [E] [3]	2.2E3
5.	[=]	220

3. Resistor Color Code

Manufacturers indicate the value and tolerance of resistors on the body of the component using either a color code (colored rings or bands) or printed alphanumerics (alphabet and numerals).

Resistors that are smaller in size tend to have the value and tolerance of the resistor encoded using colored rings, as shown in Figure 2-30(a). To determine the specifications of a color-coded resistor, therefore, you will have to decode the rings.

Referring to Figure 2-30(b), you can see that when the value, tolerance, and wattage of a resistor are printed on the body, no further explanation is needed.

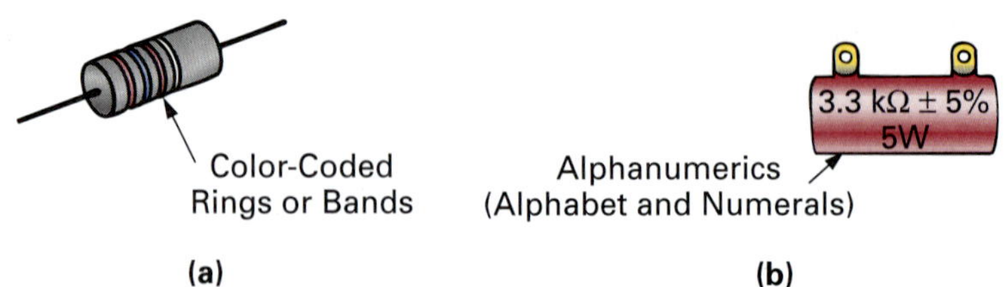

FIGURE 2-30 **Resistor Value and Tolerance Markings.**

There are basically two different types of fixed-value resistors: general purpose and precision. Resistors with tolerances of ±2% or less are classified as *precision resistors* and have five bands. Resistors with tolerances of ±5% or greater have four bands and are referred to as *general-purpose resistors.* The color code and differences between precision and general-purpose resistors are explained in Figure 2-31.

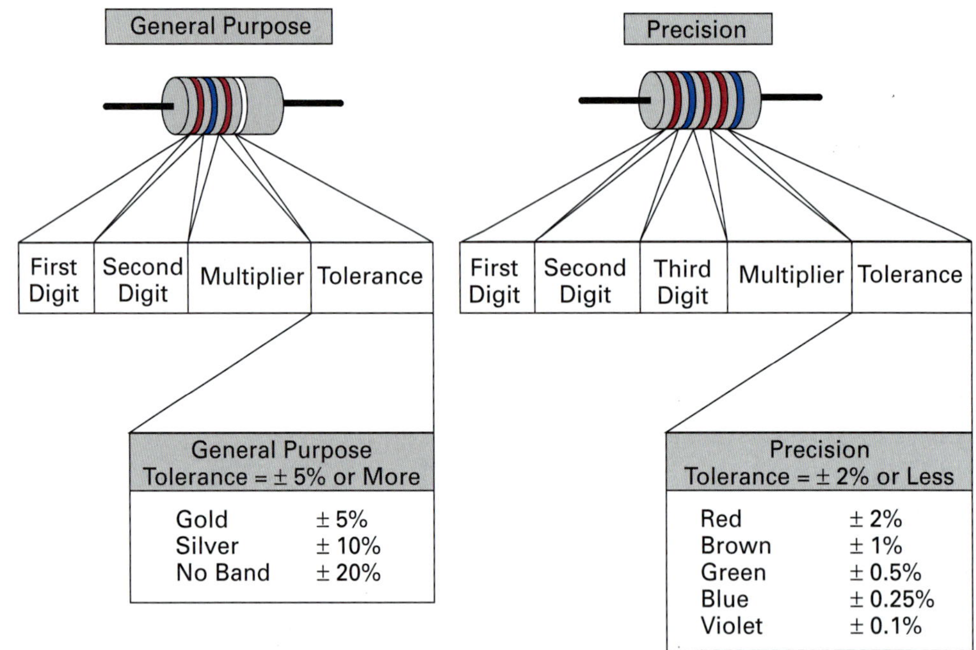

	Color	Digit Value	Multiplier		
Big	Black	0	1	One	1
Beautiful	Brown	1	10	One Zero	10
Roses	Red	2	100	Two Zeros	100
Occupy	Orange	3	1000	Three Zeros	1 k
Your	Yellow	4	10000	Four Zeros	10 k
Garden	Green	5	100000	Five Zeros	100 k
But	Blue	6	1000000	Six Zeros	1 M
Violets	Violet	7	10000000	Seven Zeros	10 M
Grow	Gray	8	–		
Wild	White	9	–		
So	Silver	–	10^{-2} or 0.01		1/100
Get some	Gold	–	10^{-1} or 0.1		1/10
Now	None	–			

FIGURE 2-31 **General-Purpose and Precision Resistor Color Code.**

When you pick up a resistor, look for the bands that are nearer to one end. This end should be held in your left hand. If there are four bands on the resistor, follow the general-purpose resistor code. If five bands are present, follow the precision resistor code.

General-Purpose Resistor Code

1. The first band on either a general-purpose or precision resistor can never be black, and it is the first digit of the number.
2. The second band indicates the second digit of the number.
3. The third band specifies the multiplier to be applied to the number, which ranges from $\times \frac{1}{100}$ to 10,000,000.
4. The fourth band describes the tolerance or deviation from the specified resistance, which is $\pm 5\%$ or greater.

Precision Resistor Code

1. The first band, like the general-purpose resistor, is never black and is the first digit of the three-digit number.
2. The second band provides the second digit.
3. The third band indicates the third and final digit of the number.
4. The fourth band specifies the multiplier to be applied to the number.
5. The fifth and final band indicates the tolerance figure of the precision resistor, which is always less than $\pm 2\%$, which is why precision resistors are more expensive than general-purpose resistors.

EXAMPLE:

Figure 2-32 illustrates a ½ W resistor.

a. Is this a general-purpose or a precision resistor?
b. What is the resistor's value of resistance?
c. What tolerance does this resistor have, and what deviation plus and minus could occur to this value?

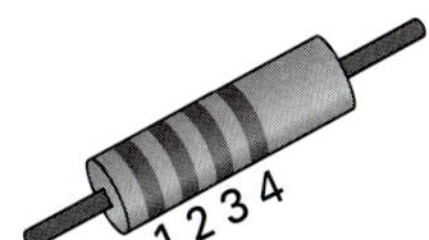

Band 1 = Green
Band 2 = Blue
Band 3 = Brown
Band 4 = Gold

FIGURE 2-32

Solution:

a. General purpose (four bands)
b. green blue × brown

   5 6 × 10 = 560 Ω
c. Tolerance band is gold, which is ± 5%.

$$\text{deviation} = 5\% \text{ of } 560 = 28$$
$$560 + 28 = 588\ \Omega$$
$$560 - 28 = 532\ \Omega$$

The resistor could, when measured, be anywhere from 532 to 588 Ω.

EXAMPLE:

State the resistor's value and tolerance and whether it is general purpose or precision for the examples shown in Figure 2-33.

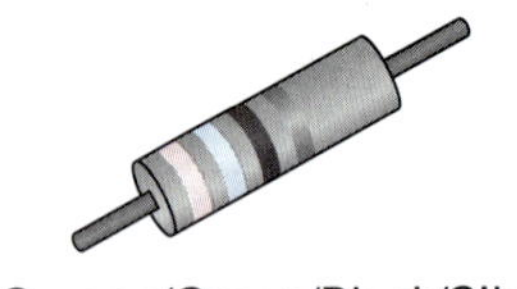

Orange/Green/Black/Silver

(a)

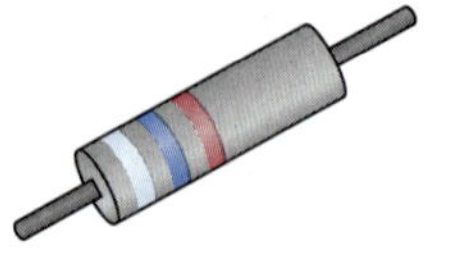

Green/Blue/Red/None

(b)

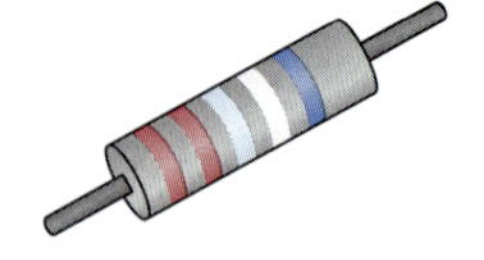

Red/Red/Green/Gold/Blue

(c)

FIGURE 2-33

Solution:

a. orange green black silver (four bands = general purpose)

3 5 $\times$ 1 10% $= 35\ \Omega \pm 10\%$

b. green blue red none (four bands = general purpose)

5 6 $\times$ 100 20% $= 5.6\ \text{k}\Omega \pm 20\%$

c. red red green gold blue (five bands = precision)

2 2 5 $\times$ 0.1 0.25% $= 22.5\ \Omega \pm 0.25\%$

Thick-Film Resistors
Fixed-value resistor consisting of a thick-film resistive element made from metal particles and glass powder.

Single In-Line Package (SIP)
Package containing several electronic components (generally resistors) with a single row of external connecting pins.

Dual In-Line Package (DIP)
Package that has two (dual) sets or lines of connecting pins.

Surface Mount Technology
A method of installing tiny electronic components on the same side of a circuit board as the printed wiring pattern that interconnects them.

Variable Resistor
A resistor whose value can be changed.

2-5-2 *Fixed-Value Resistor Types.*

Figure2-34(a) lists many of the fixed-value resistor types, showing their relative cost, resistive material used, and characteristics.

Figure 2-34(b) shows the two types of **thick-film resistor** networks, called SIPs and DIPs. The **single in-line package** is so called because all its lead connections are in a single line, whereas the **dual in-line package** has two lines of connecting pins. The chip resistors shown in Figure 2-34(c) are small thick-film resistors that are approximately the size of a pencil lead.

The SIP and DIP resistor networks, once constructed, are trimmed by lasers to obtain close tolerances of typically ±2%. Resistance values ranging from 22 Ω to 2.2 MΩ are available, with a power rating of ½ W

The chip resistor shown in Figure 2-34(c) is commercially available with resistance values from 10 Ω to 3.3 MΩ, a ±2% tolerance, and a ⅛ W heat dissipation capability. They are ideally suited for applications requiring physically small sized resistors, as explained in the inset in Figure 2-34(c). The chip resistor is called a **surface mount technology** (SMT) device. The key advantage of SMT devices over "through-hole" devices is that a through-hole device needs both a hole in the printed circuit board (PCB) and a connecting pad around the hole. With the SMT device, no holes are needed since the package is soldered directly onto the surface of the PCB. Pads can therefore be placed closer together. This results in a considerable space saving, as you can see in the inset in Figure 2-34(c) by comparing the through-hole resistor to the surface mount chip resistor.

2-5-3 *Variable-Value Resistors*

A **variable resistor** can have its resistance varied or changed while it is connected in a circuit. In certain applications, the ability to adjust the resistance of a resistor is needed. For example, the volume control on your television set makes use of a variable resistor to vary the amount of current passing to the speakers, and so change the volume of the sound.

Type	Cost	Resistive Material	Advantages/Disadvantages
Carbon Composition	$	Powdered carbon and carbon insulator.	Low cost but inherently noisy.
Carbon Film	$$	A thin film of carbon and insulator.	Smaller tolerances (±5% to ±2%) and better temperature stability than carbon composition.
Metal Film	$$$	Thin metal film spiral on substrate.	Best tolerances (±1% to ±0.1%) and temperature stability; however, high cost.
Wirewound	$$$$	Length of wire wrapped around ceramic core.	High power rating and good tolerance (±1%); however, very high cost.
Metal Oxide	$$$	Oxide of a metal on an insulating substrate.	Has the best temperature stability; however, high cost.
Thick Film Networks Chips	$$$$ $$	Thick film of resistive paste on insulating substrate.	Small size makes them ideal for high density circuits. Resistor networks (SIPs and DIPs) and chip resistors for SMT.

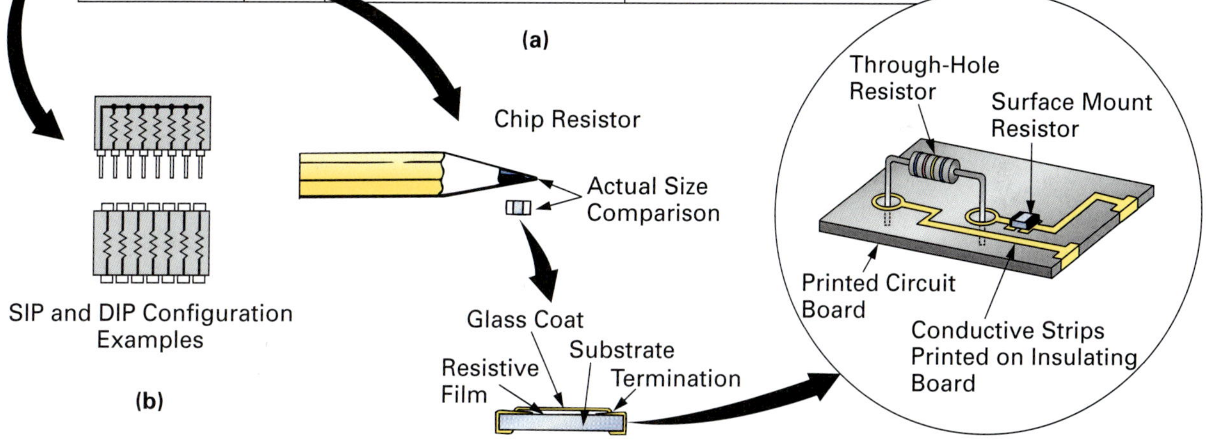

FIGURE 2-34 **Fixed-Value Resistor Types.**

Mechanically (User) Adjustable Variable Resistors

In this section we will discuss two variable resistor types, which will cause a change in resistance when a shaft is physically rotated.

Rheostat (two terminals: A and B). Figure 2-35(a) shows the physical appearance of different rheostats, while Figure 2-35(b) shows the rheostat's schematic symbols. As can be seen in the construction of a circular rheostat in Figure 2-35(c), one terminal is connected

Rheostat
Two-terminal variable resistor that, through mechanical turning of a shaft, can be used to vary its resistance and therefore its value of terminal to terminal current.

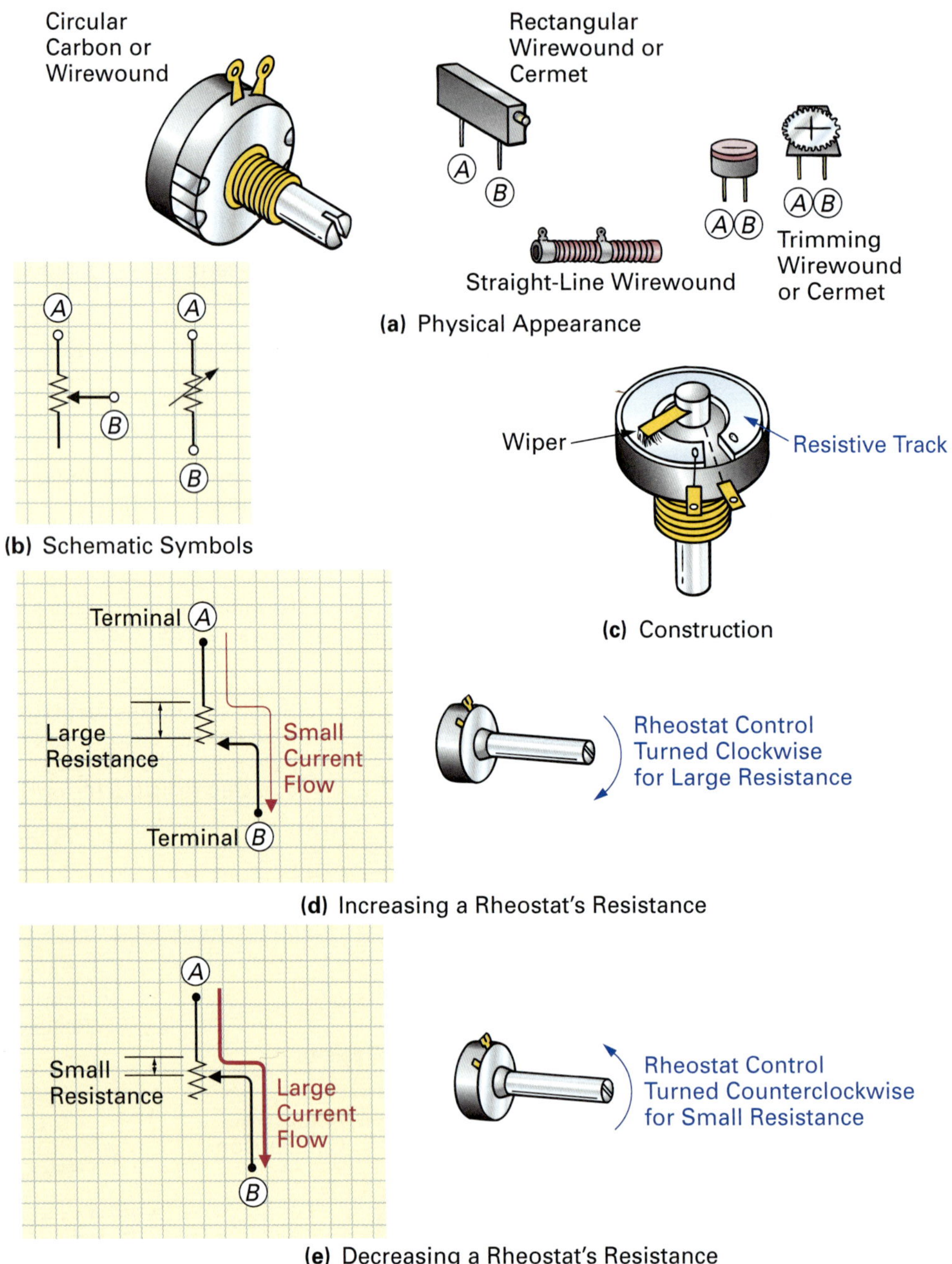

FIGURE 2-35 Rheostat. (a) Physical Appearance. (b) Schematic Symbols. (c) Construction. (d) Increasing a Rheostat's Resistance. (e) Decreasing a Rheostat's Resistance.

to one side of a resistive track and the other terminal of this two-terminal device is connected to a movable wiper. As the wiper is moved away from the end of the track with the terminal, the resistance between the stationary end terminal and the mobile wiper terminal increases. This is summarized in Figure 2-35(d), where the wiper has been moved down by a clockwise rotation of the shaft. Current would have to flow through a large resistance as it travels from one terminal to the other. On the other hand, as the wiper is moved closer to the end of the track connected to the terminal, the resistance decreases. This is summarized in Figure 2-35(e), which shows that as the wiper is moved up, as a result of turning the shaft counterclockwise, current will see only a small resistance between the two terminals.

Rheostats come in many shapes and sizes, as can be seen in Figure 2-35(a). Some employ a straight-line motion to vary resistance, while others are classified as circular-motion

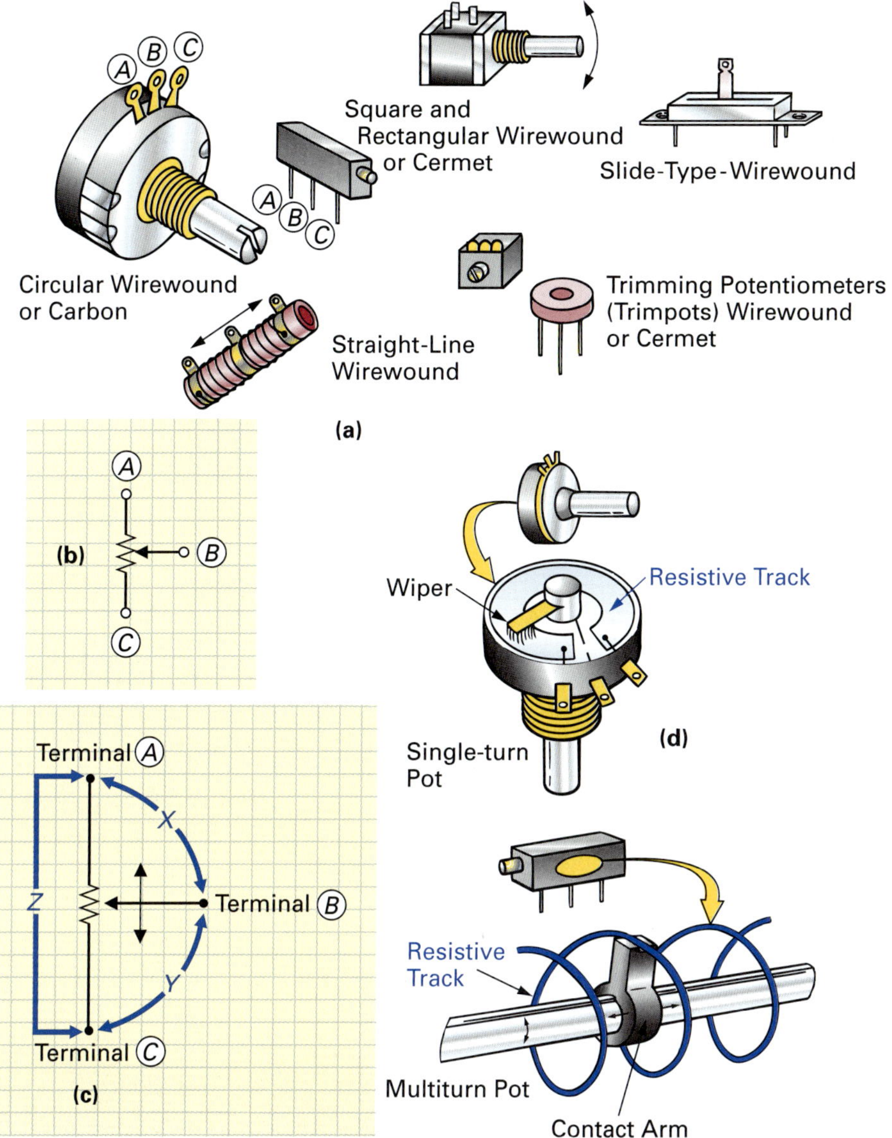

FIGURE 2-36 **Potentiometer. (a) Physical Appearance. (b) Schematic Symbol. (c) Operation. (d) Construction.**

rheostats. The resistive elements also vary; wirewound and carbon tracks are very popular. Cermet rheostats mix the ratio of ceramic (insulator) and metal (conductor) to produce different values of resistive tracks. A trimming rheostat is a miniature device used to change resistance by a small amount. Other circular-motion rheostats are available that require between two to ten turns to cover the full resistance range.

Potentiometer (three terminals: A, B, and C). Figure 2-36(a) illustrates the physical appearance of a variety of potentiometers, also called pots (slang), while Figure 2-36(b) shows the potentiometer's schematic symbol. You will probably notice that the difference between a rheostat and potentiometer is the number of terminals; the rheostat has two terminals while the potentiometer has three. With the rheostat, there were only two terminals and the resistance between the wiper and terminal varied as the wiper was adjusted. Referring to the potentiometer shown in Figure 2-36(c), you can see that resistance can actually be measured across three separate combinations: between *A* and *B* (*X*), between *B* and *C*(*Y*), and between *C* and *A* (*Z*).

Potentiometer
Three-lead variable resistor that through mechanical turning of a shaft can be used to produce a variable voltage or potential.

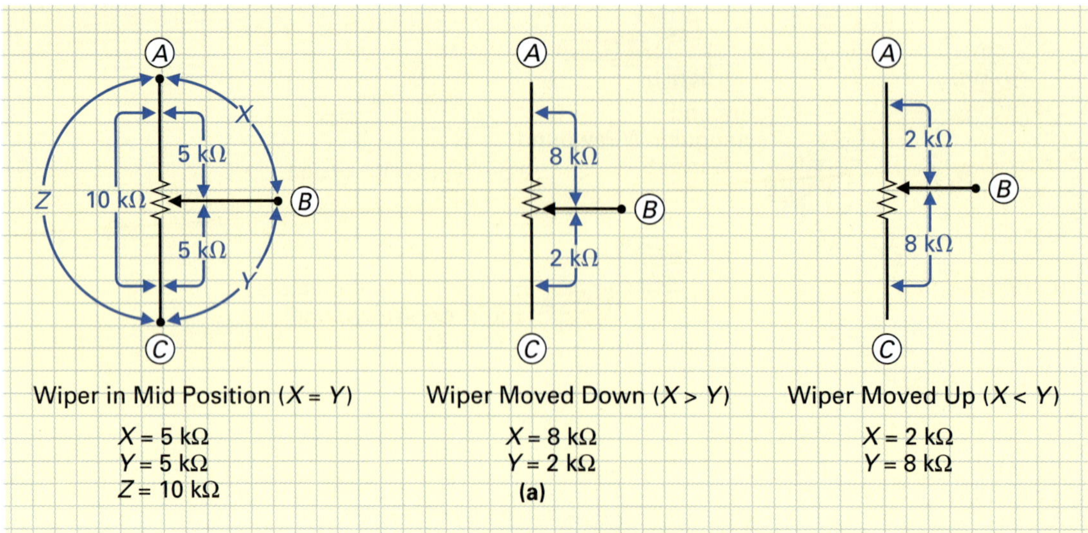

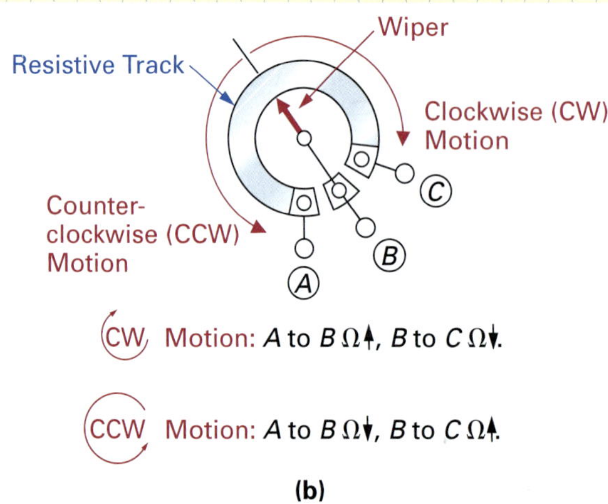

FIGURE 2-37 **The Resistance Changes between the Terminals of a Potentiometer.**

The only difference between the rheostat and the potentiometer in construction is the connection of a third terminal to the other end of the resistive track, as can be seen in Figure 2-36(d), which shows the single-turn potentiometer. Also illustrated in this section of the figure is the construction of a multiturn potentiometer in which a contact arm slides along a shaft and the resistive track is formed into a helix of 2 to 10 coils.

As shown in Figure 2-37, the resistance between *A* and *B* (*X*) and *B* and *C* (*Y*) will vary as the wiper's position is moved, as illustrated in Figure 2-37(a). If the user physically turns the shaft in a clockwise direction, the resistance between *A* and *B* increases, while the resistance between *B* and *C* decreases. Similarly, if the user mechanically turns the shaft counterclockwise, a decrease occurs between *A* and *B* and there is a resulting increase in resistance between *B* and *C*. This point is summarized in Figure 2-37(b).

Linear

Relationship between input and output in which the output varies in direct proportion to the input.

Tapered

Nonuniform distribution of resistance per unit length throughout the element.

Whether rheostat or potentiometer, the resistive track can be classified as having either a **linear** or a **tapered** (nonlinear) resistance. In Figure 2-38(a) we have taken a 1 kΩ rheostat and illustrated the resistance value changes between *A* and *B* for a linear and a tapered one-turn rheostat. The definition of linear is having an output that varies in direct proportion to the input. The input in this case is the user turning the shaft, and the output is the linearly increasing resistance between *A* and *B*.

With a tapered rheostat or potentiometer, the resistance varies nonuniformly along its resistor element, sometimes being greater or less for equal shaft movement at various points along the resistance element, as shown in the table in Figure 2-38(a). Figure 2-38(b) plots the position of the variable resistor's wiper against the resistance between

	Fully CCW	$\frac{1}{4}$ CW	$\frac{1}{2}$ CW	$\frac{3}{4}$ CW	Fully CW
1 kΩ Rheostat (A, B)	A, B	A, B	A, B	A, B	A, B
Linear	0 Ω	250 Ω ($\frac{1}{4}$ of 1000 Ω)	500 Ω ($\frac{1}{2}$ of 1000 Ω)	750 Ω ($\frac{3}{4}$ of 1000 Ω)	1000 Ω
Tapered	0 Ω	350 Ω	625 Ω	900 Ω	1000 Ω
		These values have been arbitrarily chosen to illustrate a nonlinear change.			

(a)

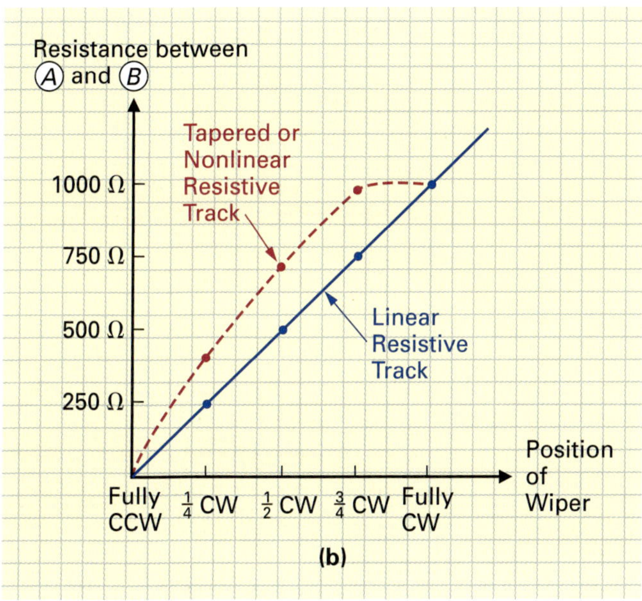

(b)

FIGURE 2-38 Linear versus Tapered Resistive Track.

the two output terminals, showing the difference between a linear increase and a nonlinear or tapered increase.

2-5-4 *Testing Resistors*

It is virtually impossible for fixed-value resistors to internally short-circuit (no resistance or zero ohms). Generally, the resistor's internal elements will begin to develop a higher resistance than its specified value (due to a partial internal open) or in some cases go completely open circuit (maximum resistance or infinite ohms).

Variable-value resistors have problems with the wiper making contact with the resistive track at all points. Faulty variable-value resistors in sound systems can normally be detected because they generate a scratchy noise whenever you adjust the volume control.

The ohmmeter is the ideal instrument for verifying whether or not a resistor is functioning correctly. When checking a resistor's resistance, here are some points that should be remembered.

1. The ohmmeter has its own internal power source (a battery), so always turn off the circuit power and remove from the circuit the resistor to be measured. If this is not done, you will not only obtain inaccurate readings, but you can damage the ohmmeter.
2. With an autoranging digital multimeter, the suitable ohms range is automatically selected by the ohmmeter. The range scales on a non-autoranging digital multimeter have to be selected, and often confuse people. For example, if a 100 kΩ range is chosen the highest reading that can be measured is 100 kΩ. If a 1 MΩ resistor is measured, the meter will indicate an infinite-ohms reading and you may be misled into believing that you have found the problem (an open resistor). To overcome this problem, always start on the highest range and then work down to a lower range for a more accurate reading.
3. Another point to keep in mind is tolerance. For example, if a suspected faulty 1 kΩ (1000 Ω) resistor is measured with an ohmmeter and reads 1.2 kΩ (1200 Ω), it could be within tolerance if no tolerance band is present on the resistor's body (±20%). A 1 kΩ resistor with ±20% tolerance could measure anywhere between 800 and 1200 Ω.
4. The ohmmeter's internal battery voltage is really too small to deliver an electrical shock. You should, however, avoid touching the bare metal parts of the probes or resistor leads, as your body resistance of approximately 50 kΩ will affect your meter reading.

SELF-TEST EVALUATION POINT FOR SECTION 2-5

Use the following questions to test your understanding of Section 2-5:

1. List the six types of fixed-value resistors.
2. What is the difference between SIPs and DIPs?
3. Name the two types of mechanically adjustable variable resistors and state the difference between the two.
4. Describe a linear and a tapered potentiometer.

REVIEW QUESTIONS

Multiple-Choice Questions

1. Resistance is measured in:
   a. Ohms **c.** Amperes
   b. Volts **d.** Siemens
2. Current is proportional to:
   a. Resistance **c.** Both (a) and (b)
   b. Voltage **d.** None of the above
3. Current is inversely proportional to:
   a. Resistance **c.** Both (a) and (b)
   b. Voltage **d.** None of the above
4. If the applied voltage is equal to 15 V and the circuit resistance equals 5 Ω, the total circuit current would be equal to:
   a. 4 A **b.** 5 A **c.** 3 A **d.** 75 A
5. Calculate the applied voltage if 3 mA flows through a circuit resistance of 25 kΩ.
   a. 63 mV **b.** 25 V **c.** 77 μV **d.** 75 V
6. Energy is measured in:
   a. Volts **b.** Joules **c.** Amperes **d.** Watts
7. Work is measured in:
   a. Joules **b.** Volts **c.** Amperes **d.** Watts
8. Power is the rate at which energy is transformed and is measured in:
   a. Joules **c.** Volts
   b. Watts **d.** Amperes
9. AWG is an abbreviation for:
   a. Alternate Wire Gauge **c.** American Wave Guide
   b. Alternating Wire Gauge **d.** American Wire Gauge

10. What would be the power dissipated by a 2 kΩ carbon composition resistor when a current of 20 mA is flowing through it?
   a. 40 W
   b. 0.8 W
   c. 1.25 W
   d. None of the above
11. A rheostat is a __________ terminal device, while the potentiometer is a __________ terminal device.
   a. 2, 3
   b. 1, 2
   c. 2, 4
   d. 3, 2
12. Which of the following is *not* true?
   a. Precision resistors have a tolerance of > 5%.
   b. General-purpose resistors can be recognized because they have either three or four bands.
   c. The fifth band indicates the tolerance of a precision resistor.
   d. The third band of a general-purpose resistor specifies the multiplier.
13. The first band on either a general-purpose or a precision resistor can never be:
   a. Brown
   b. Red or black stripe
   c. Black
   d. Red
14. The term *infinite ohms* describes:
   a. A small finite resistance
   b. A resistance so large that a value cannot be placed on it
   c. A resistance between maximum and minimum
   d. None of the above
15. A 10% tolerance, 2.7 MΩ carbon composition resistor measures 2.99 MΩ when checked with an ohmmeter. It is:
   a. Within tolerance
   b. Outside tolerance
   c. Faulty
   d. Both (a) and (c)
   e. Both (a) and (b)

Practice Problems

16. An electric heater with a resistance of 6 Ω is connected across a 120 V wall outlet.
   a. Calculate the current flow.
   b. Draw the schematic diagram.
17. What source voltage would be needed to produce a current flow of 8 mA through a 16 kΩ resistor?
18. If an electric toaster draws 10 A when connected to a power outlet of 120 V, what is its resistance?
19. Calculate the power used in Problems 16, 17, and 18.
20. Calculate the current flowing through the following light bulbs when they are connected across 120 V:
   a. 300 W b. 100 W c. 60 W d. 25 W
21. If an electric company charges 9 cents/kWh, calculate the cost for each light bulb in Problem 20 if on for 10 hours.
22. Indicate which of the following unit pairs is larger:
   a. Millivolts or volts
   b. Microamperes or milliamperes
   c. Kilowatts or watts
   d. Kilohms or megohms
23. Calculate the resistance of 200 ft of copper having a diameter of 80 mils.
24. What AWG size wire should be used to safely carry just over 15 A?
25. Calculate the voltage dropped across 1000 ft of No. 4 copper conductor when a current of 7.5 A is flowing through it.
26. Calculate the unknown resistance in a circuit when an ammeter indicates that a current of 12 mA is flowing and a voltmeter indicates 12 V.
27. What battery voltage would use 1000 J of energy to move 40 C of charge through a circuit?
28. Calculate the resistance of a light bulb that passes 500 mA of current when 120 V is applied. What is the bulb's wattage?
29. Which of the following circuits has the largest resistance and which has the smallest?
   a. $V = 120$ V, $I = 20$ mA
   b. $V = 12$ V, $I = 2$ A
   c. $V = 9$ V, $I = 100$ μA
   d. $V = 1.5$ V, $I = 4$ mA
30. Calculate the power dissipated in each circuit in Problem 29.
31. How many watts are dissipated if 5000 J of energy are consumed in 25 s?
32. Convert the following:
   a. 1000 W = __________ kW
   b. 0.345 W = __________ mW
   c. 1250×10^3 W = __________ MW
   d. 0.00125 W = __________ μW
33. What is the value of the resistor when a current of 4 A is causing 100 W to be dissipated?
34. How many kilowatt-hours of energy are consumed in each of the following:
   a. 7500 W in 1 hour
   b. 25 W for 6 hours
   c. 127,000 W for half an hour
35. What is the maximum output power of a 12 V, 300 mA power supply?
36. If a 5.6 kΩ resistor has a tolerance of ±10%, what would be the allowable deviation in resistance above and below 5.6 kΩ?
37. (a) If a current of 50 mA is flowing through a 10 kΩ, 25 W resistor, how much power is the resistor dissipating? (b) Can the current be increased, and if so by how much?
38. What minimum wattage size could be used if a 12 Ω wire-wound resistor were connected across a 12 V supply?
39. Calculate the resistance deviation for the tolerances of the resistors listed below.
   a. 1.2 MΩ, ±10%
   b. 10Ω, ±5%
   c. 27 kΩ, ±20%
   d. 273 kΩ, ±0.5%
   e. Orange, orange, black
   f. Red, red, green, red, red
   g. Brown, black, orange, gold
   h. White, brown, brown, silver
40. If a 1 kΩ rheostat has 50 V across it, calculate the different values of current if the rheostat is varied in 100 Ω steps. Calculate and insert the values into the table in Figure 2-40(a), and then plot these values in the graph in Figure 2-40(b).

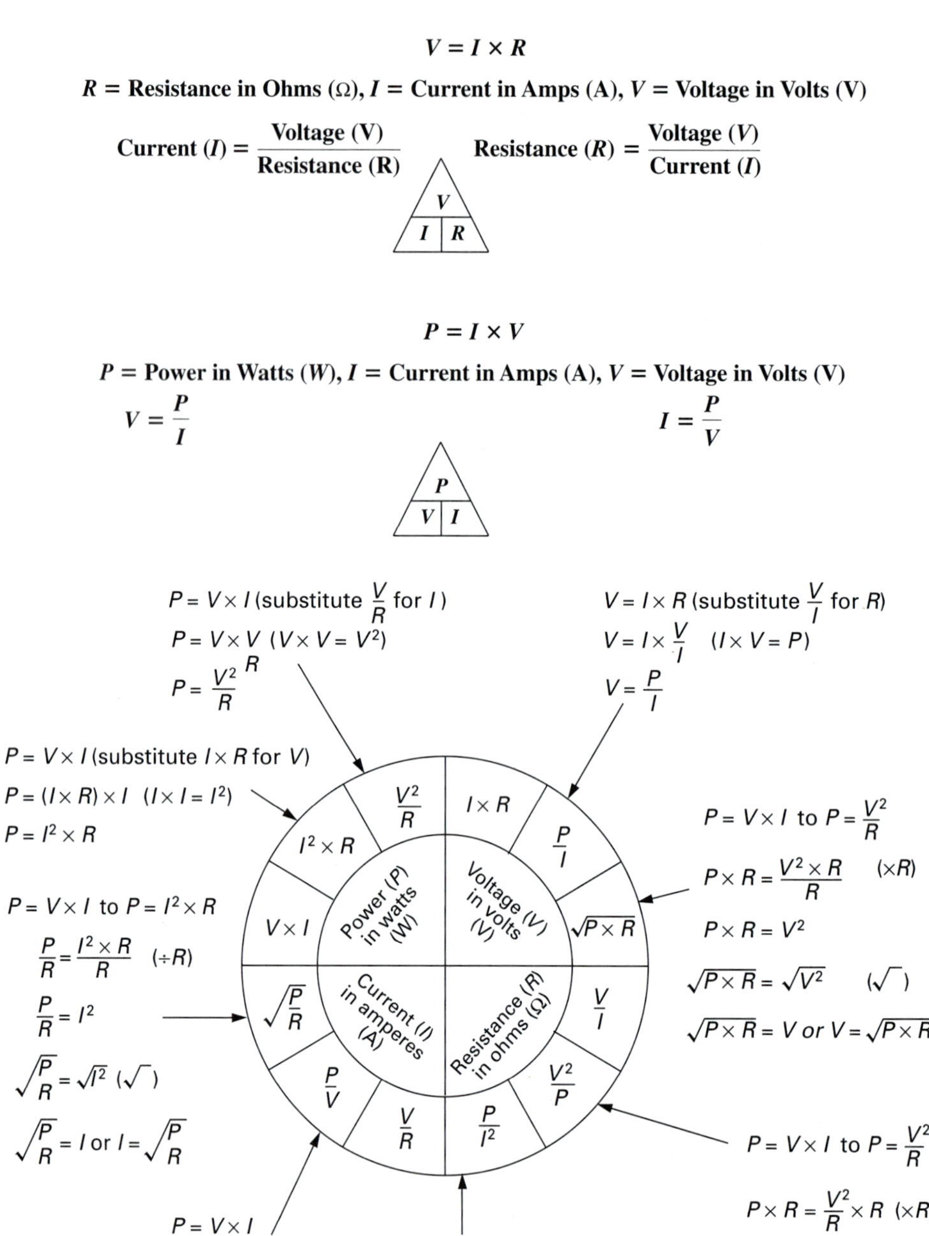

FIGURE 2-39 **The Ohm's Law and Power Formula Circle.**

Figure 2-39 shows how transposition and substitution can be used to develop alternative formulas.

V = 50 V

R (Ω)	*I* (mA)
100	
200	
300	
400	
500	
600	
700	
800	
900	
1000	

(a)

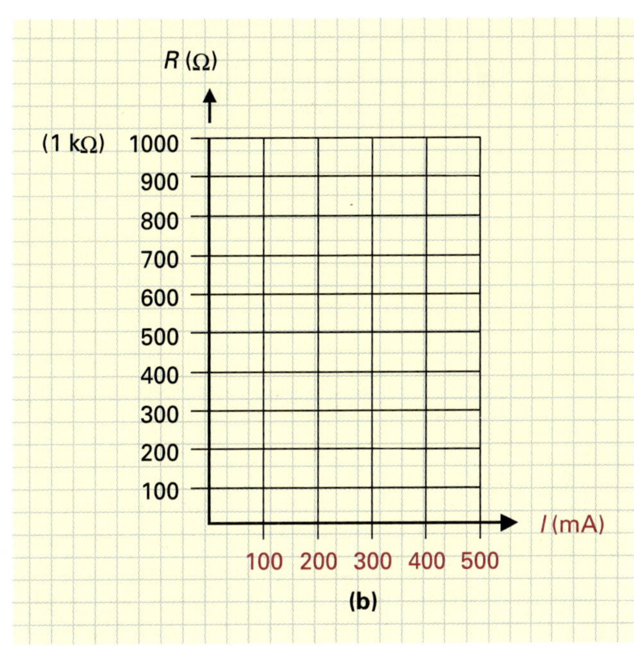

FIGURE 2-40 **Plotting a Rheostat's Current against Resistance.**

Troubleshooting Questions

41. Briefly describe some of the problems that can occur with fixed and variable resistors.

42. Describe why the tolerance of a resistor can make you think that it has a problem when in fact it does not.

43. If a 20 kΩ, 1-W, ±10% tolerance resistor measures 20.39 kΩ, is it in or out of tolerance?

44. If 33 μA is flowing through a carbon film 4.7 kΩ, ¼ W, ±5% resistor, will it burn up?

45. What points should be remembered when using the ohmmeter to verify a resistor's value?

Web Site Questions

Go to the web site http://www.prenhall.com/cook, select the textbook *Electronics: A Complete Course,* select this chapter, and then follow the instructions when answering the multiple choice practice problems.

Series Circuits

The First Pocket Calculator

During the seventeenth century, European thinkers were obsessed with any device that could help them with mathematical calculation. Scottish mathematician John Napier decided to meet this need, and in 1614 he published his new discovery of logarithms. In this book, consisting mostly of tediously computed tables, Napier stated that a logarithm is the exponent of a base number. For example:

The common logarithm (base 10) of 100 is 2 ($100 = 10^2$).

The common logarithm of 10 is 1 ($10 = 10^1$).

The common logarithm of 27 is 1.43136 ($27 = 10^{1.43136}$).

The common logarithm of 6 is 0.77815 ($6 = 10^{0.77815}$).

Any number, no matter how large or small, can be represented by or converted to a logarithm. Napier also outlined how the multiplication of two numbers could be achieved by simply adding the numbers' logarithms. For example, if the logarithm of 2 (which is 0.30103) is added to the logarithm of 4 (which is 0.60206), the result will be 0.90309, which is the logarithm of the number 8 ($0.30103 + 0.60206 = 0.90309$, $2 \times 4 = 8$). Therefore, the multiplication of two large numbers can be achieved by looking up the logarithms of the two numbers in a log table, adding them together, and then finding the number that corresponds to the sum in an antilog (reverse log) table. In this example, the antilog of 0.90309 is 8.

Napier's table of logarithms was used by William Oughtred, who developed, just 10 years after Napier's death in 1617, a handy mechanical device that could be used for rapid calculation. This device, considered the first pocket calculator, was the slide rule.

3

Introduction

A series circuit, by definition, is the connecting of components end to end in a circuit to provide a single path for the current. This is true not only for resistors, but also for other components that can be connected in series. In all cases, however, the components are connected in succession or strung together one after another so that only one path for current exists between the negative (−) and positive (+) terminals of the supply.

3-1 COMPONENTS IN SERIES

Series Circuit

Circuit in which the components are connected end to end so that current has only one path to follow throughout the circuit.

Figure 3-1 illustrates five examples of **series** resistive **circuits.** In all five examples, you will notice that the resistors are connected "in-line" with one another so that the current through the first resistor must pass through the second resistor, and the current through the second resistor must pass through the third, and so on.

EXAMPLE:

In Figure 3-2(a), seven resistors are laid out on a table top. Using a protoboard, connect all the resistors in series, starting at R_1, and proceeding in numerical order through the resistors until reaching R_7. After completing the circuit, connect the series circuit to a dc power supply.

Solution:

In Figure 3-2(b), you can see that all the resistors are now connected in series (end-to-end), and the current has only one path to follow from negative to positive.

EXAMPLE:

Figure 3-3(a) shows four 1.5 V cells and three lamps. Using wires, connect all of the cells in series to create a 6 V battery source. Then connect all of the three lamps in series with one another, and finally, connect the 6 V battery source across the three-series-connected-lamp load.

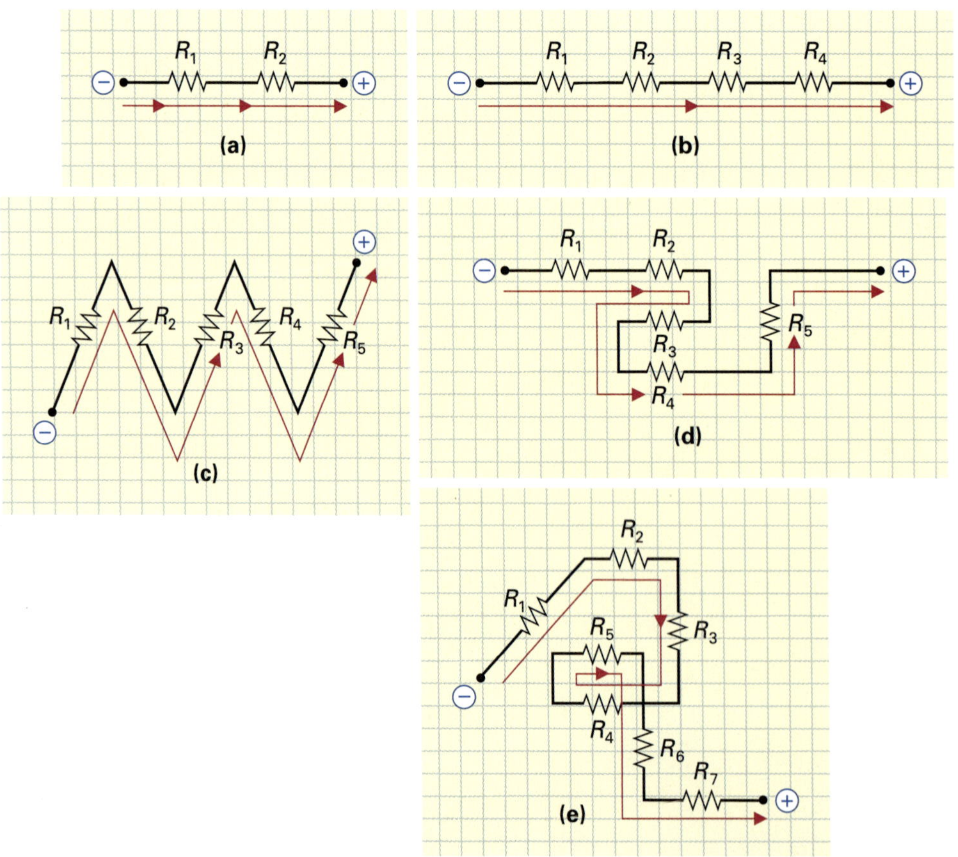

FIGURE 3-1 **Five Series Resistive Circuits.**

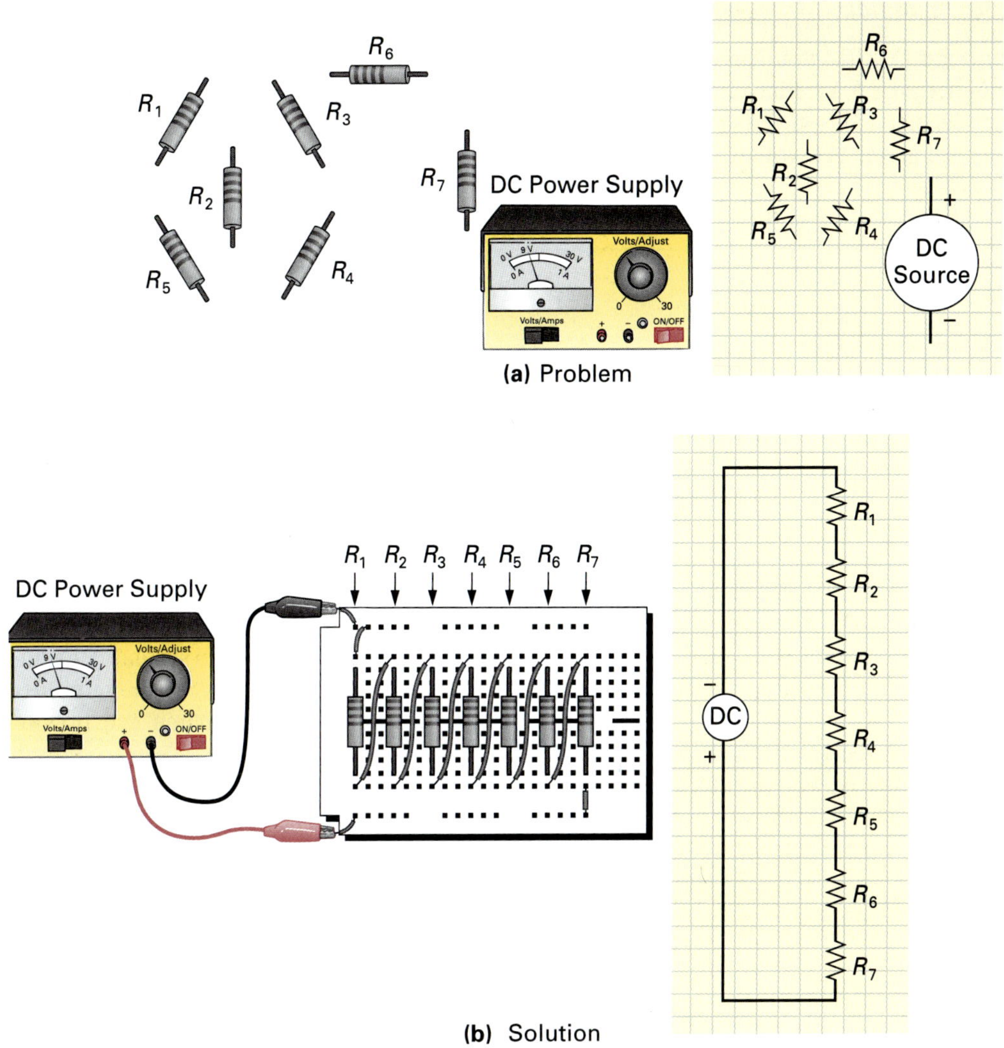

FIGURE 3-2 **Connecting Resistors in Series. (a) Problem. (b) Solution.**

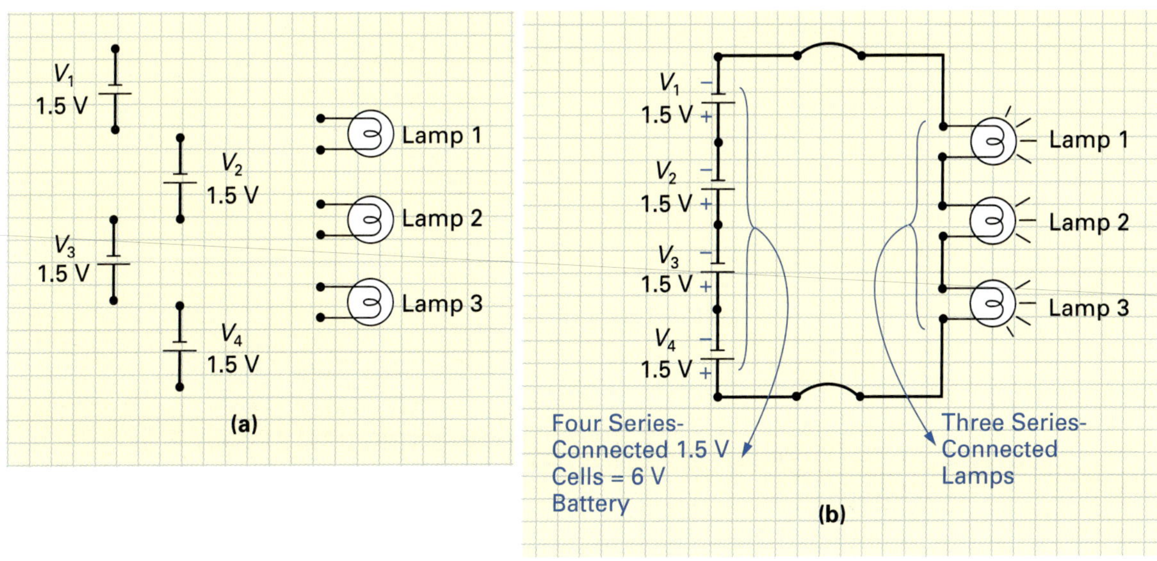

FIGURE 3-3 **Series-Connected Cells and Lamps.**

■ *Solution:*

In Figure 3-3(b) you can see the final circuit containing a source, made up of four series-connected 1.5 V cells, and a load, consisting of three series-connected lamps. As explained in Chapter 2, when cells are connected in series, the total voltage (V_T) will be equal to the sum of all the cell voltages:

$$V_T = V_1 + V_2 + V_3 + V_4 = 1.5\text{ V} + 1.5\text{ V} + 1.5\text{ V} + 1.5\text{ V} = 6\text{ V}$$

3-2 CURRENT IN A SERIES CIRCUIT

The current in a series circuit has only one path to follow and cannot divert in any other direction. The current through a series circuit, therefore, is the same throughout that circuit.

Returning once again to the water analogy, you can see in Figure 3-4(a) that if 2 gallons of water per second are being supplied by the pump, 2 gallons per second must be pulled into the pump. If the rate at which water is leaving and arriving at the pump is the same, 2 gallons of water per second must be flowing throughout the circuit. It can be said, therefore, that the same value of water flow exists throughout a series-connected fluid system. This rule will always remain true, for if the valves were adjusted to double the opposition to flow, then half the flow, or 1 gallon of water per second, would be leaving the pump and flowing throughout the system.

Similarly, with the electronic series circuit shown in Figure 3-4(b), there is a total of 2 A leaving and 2 A arriving at the battery, and so the same value of current exists throughout the series-connected electronic circuit. If the circuit's resistance were changed, a new value

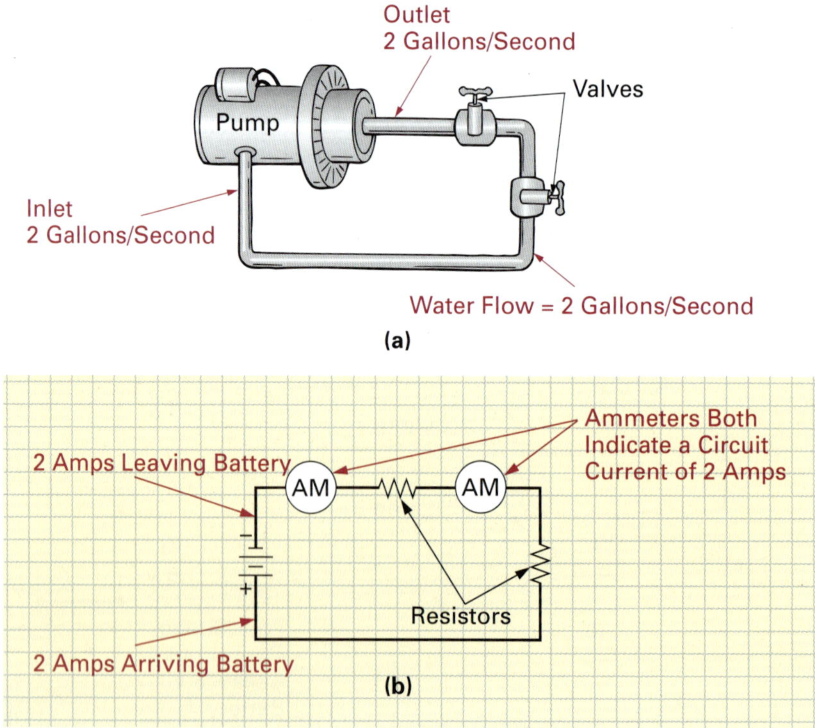

FIGURE 3-4 Series Circuit Current. (a) Fluid System. (b) Electric System.

of series circuit current would be present throughout the circuit. For example, if the resistance of the circuit was doubled, then half the current, or 1 A, will leave the battery, but that same value of 1 A will flow throughout the entire circuit. This series circuit current characteristic can be stated mathematically as

$$I_T = I_1 = I_2 = I_3 = \cdots$$

Total current = current through R_1 = current through R_2 = current through R_3, and so on.

EXAMPLE:

In Figure 3-5(a) and (b), a total current (I_T) of 1 A is flowing out of the negative terminal of the dc power supply, through two end-to-end resistors R_1 and R_2, and returning to the positive terminal of the dc power supply. Calculate:

a. The current through R_1 (I_1)

b. The current through R_2 (I_2)

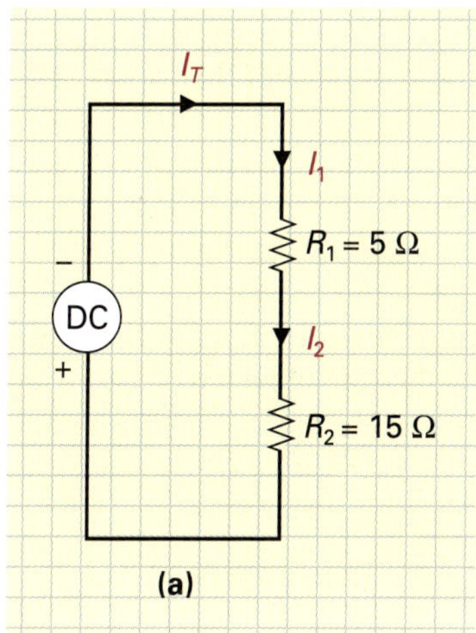

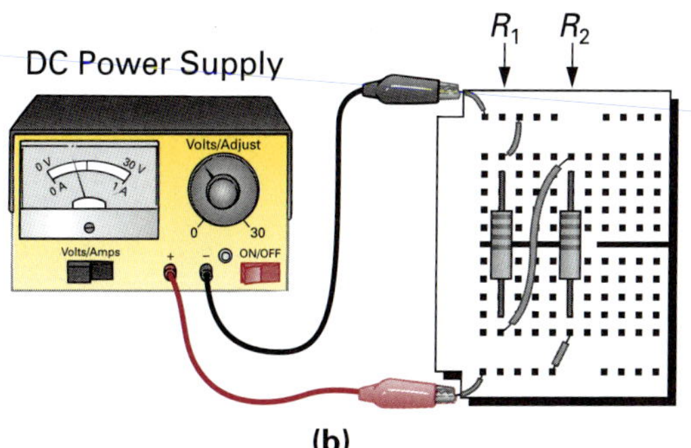

CIRCUIT ANALYSIS TABLE

Resistance $R = V/I$	Current $I = V/R$	Voltage $V = I \times R$	Power $P = V \times I$
$R_1 = 5\ \Omega$	$I_1 = 1$ A		
$R_2 = 15\ \Omega$	$I_2 = 1$ A		
	$I_T = 1$ A		

(c)

FIGURE 3-5 Total Current Example. (a) Schematic. (b) Protoboard Circuit. (c) Circuit Analysis Table.

■ *Solution:*

Since R_1 and R_2 are connected in series, the current through both will be the same as the circuit current, which is equal to 1 A.

$$I_T = I_1 = I_2$$
$$1\text{ A} = 1\text{ A} = 1\text{ A}$$

The details of this circuit are summarized in the analysis table shown in Figure 3-5(c).

SELF-TEST EVALUATION POINT FOR SECTIONS 3-1 AND 3-2

Use the following questions to test your understanding of Sections 3-1 and 3-2.

1. What is a series circuit?
2. What is the current flow through each of eight series-connected 8 Ω resistors if 8 A total current is flowing out of a battery?

3-3 RESISTANCE IN A SERIES CIRCUIT

Resistance is the opposition to current flow, and in a series circuit every resistor in series offers opposition to the current flow. In the water analogy of Figure 3-4, the total resistance or opposition to water flow is the sum of the two individual valve opposition values. Like the battery, the pump senses the total opposition in the circuit offered by all the valves or resistors, and the amount of current that flows is dependent on this resistance or opposition.

The total resistance in a series-connected electronic resistive circuit is thus equal to the sum of all the individual resistances, as shown in Figure 3-6(a) through (d). An equivalent circuit can be drawn for each of the circuits in Figure 3-6(b), (c), and (d) with one resistor of a value equal to the sum of all the series resistance values.

No matter how many resistors are connected in series, the total resistance or opposition to current flow is always equal to the sum of all the resistor values. This formula can be stated mathematically as

$$R_T = R_1 + R_2 + R_3 + \cdots$$

Total resistance = value of R_1 + value of R_2 + value of R_3, and so on.

Equivalent Resistance (R_{eq})
Total resistance of all the individual resistances in a circuit.

Total resistance (R_T) is the only opposition a source can sense. It does not see the individual separate resistors, but one **equivalent resistance.** Based on the source's voltage and the circuit's total resistance, a value of current will be produced to flow through the circuit (Ohm's law, $I = V/R$).

■ **EXAMPLE:**

Referring to Figure 3-7(a) and (b), calculate

a. The circuit's total resistance
b. The current flowing through R_2

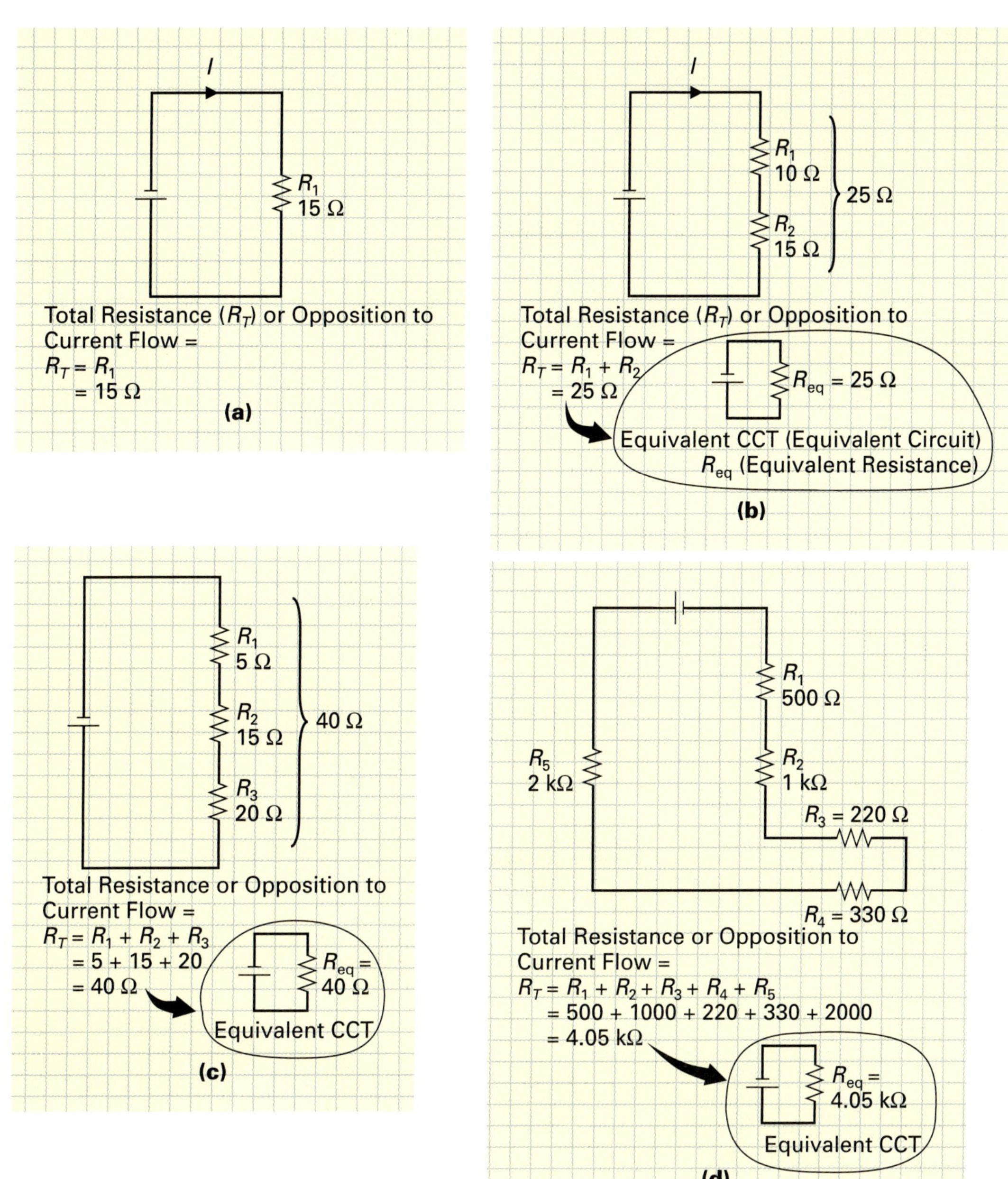

FIGURE 3-6 **Total or Equivalent Resistance.**

Solution:

a. $R_T = R_1 + R_2 + R_3 + R_4$
$= 25\ \Omega + 20\ \Omega + 33\ \Omega + 10\ \Omega$
$= 88\ \Omega$

b. $I_T = I_1 = I_2 = I_3 = I_4$. Therefore, $I_2 = I_T = 3$ A.

The details of this circuit are summarized in the analysis table shown in Figure 3-7(c).

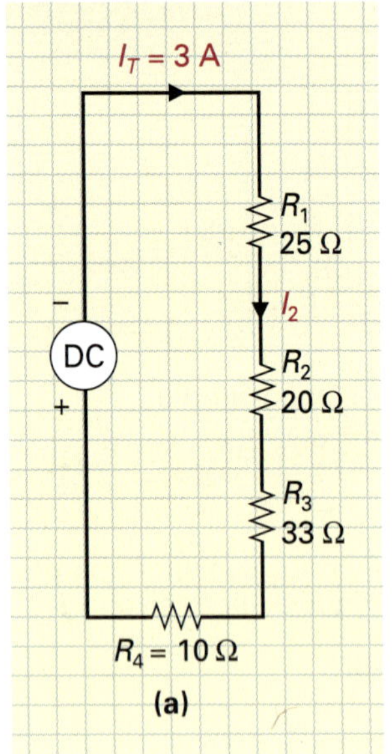

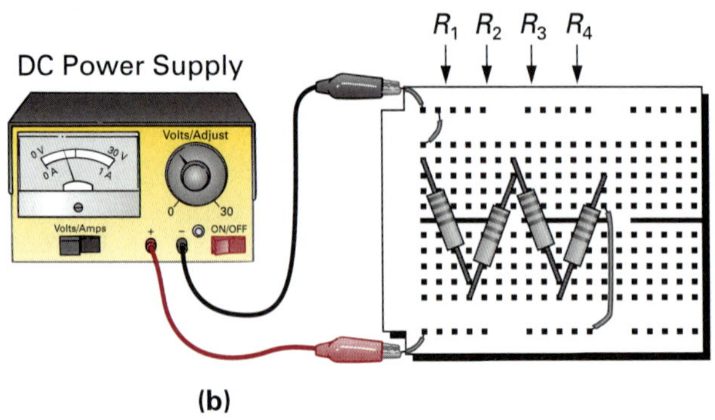

CIRCUIT ANALYSIS TABLE

Resistance $R = V/I$	Current $I = V/R$	Voltage $V = I \times R$	Power $P = V \times I$
$R_1 = 25\ \Omega$	$I_1 = 3$ A		
$R_2 = 20\ \Omega$	$I_2 = 3$ A		
$R_3 = 33\ \Omega$	$I_3 = 3$ A		
$R_4 = 10\ \Omega$	$I_4 = 3$ A		
$R_T = 88\ \Omega$	$I_T = 3$ A		

(c)

FIGURE 3-7 Total Resistance Example. (a) Schematic. (b) Protoboard Circuit. (c) Circuit Analysis Table.

EXAMPLE:

Figure 3-8(a) and (b) shows how a single-pole three-position switch is being used to provide three different lamp brightness levels. In position ① R_1 is placed in series with the lamp, in position ② R_2 is placed in series with the lamp, and in position ③ R_3 is placed in series with the lamp. If the lamp has a resistance of 75 Ω, calculate the three values of current for each switch position.

Solution:

Position 1:

$$R_T = R_1 + R_{\text{lamp}}$$
$$= 25\ \Omega + 75\ \Omega = 100\ \Omega$$
$$I_T = \frac{V_T}{R_T} = \frac{12\ \text{V}}{100\ \Omega} = 120\ \text{mA}$$

Position 2:

$$R_T = R_2 + R_{\text{lamp}} = 50\ \Omega + 75\ \Omega = 125\ \Omega$$
$$I_T = \frac{V_T}{R_T} = \frac{12\ \text{V}}{125\ \Omega} = 96\ \text{mA}$$

Position 3:

$$R_T = R_3 + R_{\text{lamp}} = 75\ \Omega + 75\ \Omega = 150\ \Omega$$
$$I_T = \frac{V_T}{R_T} = \frac{12\ \text{V}}{150\ \Omega} = 80\ \text{mA}$$

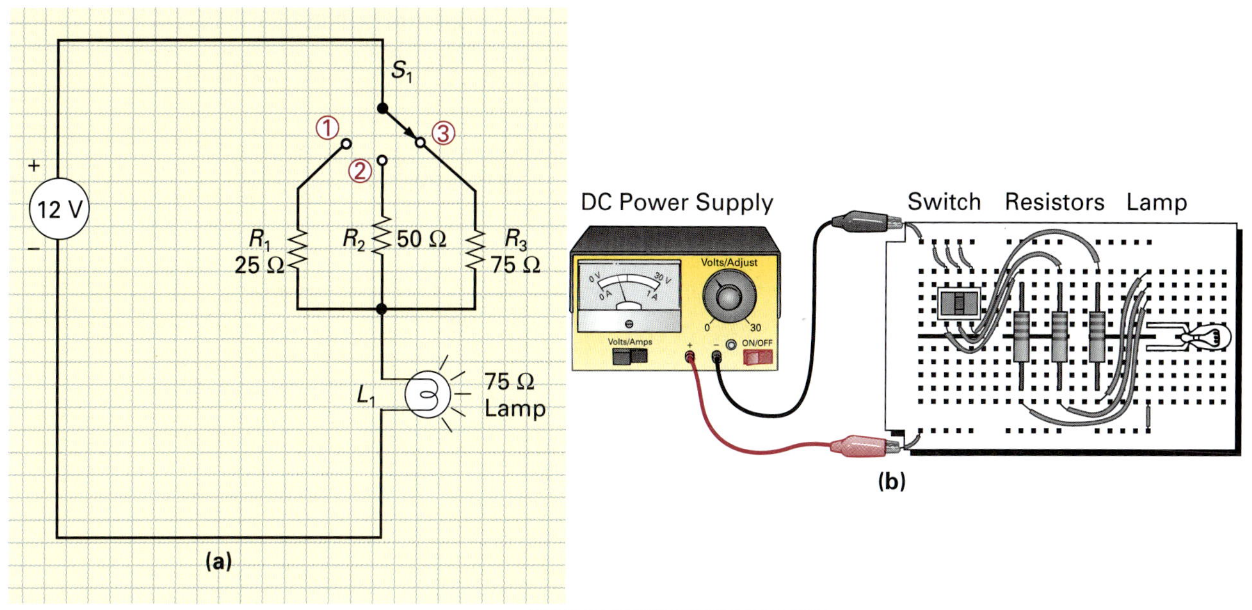

CIRCUIT ANALYSIS TABLE

	Resistance $R = V/I$	Current $I = V/R$	Voltage $V = I \times R$	Power $P = V \times I$
S_1 = ①	$R_1 = 25\ \Omega$	$I_1 = 120$ mA		
S_1 = ②	$R_2 = 50\ \Omega$	$I_2 = 96$ mA		
S_1 = ③	$R_3 = 75\ \Omega$	$I_3 = 80$ mA		

$R_{L1} = 75\ \Omega$

(c)

FIGURE 3-8 **Three-Position Switch Controlling Lamp Brightness. (a) Schematic. (b) Protoboard Circuit. (c) Circuit Analysis Table.**

The details of this circuit are summarized in the analysis table shown in Figure 3-8(c).

SELF-TEST EVALUATION POINT FOR SECTION 3-3

Use the following questions to test your understanding of Section 3-3.

1. State the total resistance formula for a series circuit.
2. Calculate R_T if $R_1 = 2\ \text{k}\Omega$, $R_2 = 3\ \text{k}\Omega$, and $R_3 = 4700\ \Omega$.

3-4 VOLTAGE IN A SERIES CIRCUIT

A potential difference or voltage drop will occur across each resistor in a series circuit when current is flowing. The amount of voltage drop is dependent on the value of the resistor and the amount of current flow. This idea of potential difference or voltage drop is best explained by returning to the water analogy. In Figure 3-9(a), you can see that the high pressure from the pump's outlet is present on the left side of the valve. On the right side of the valve, however,

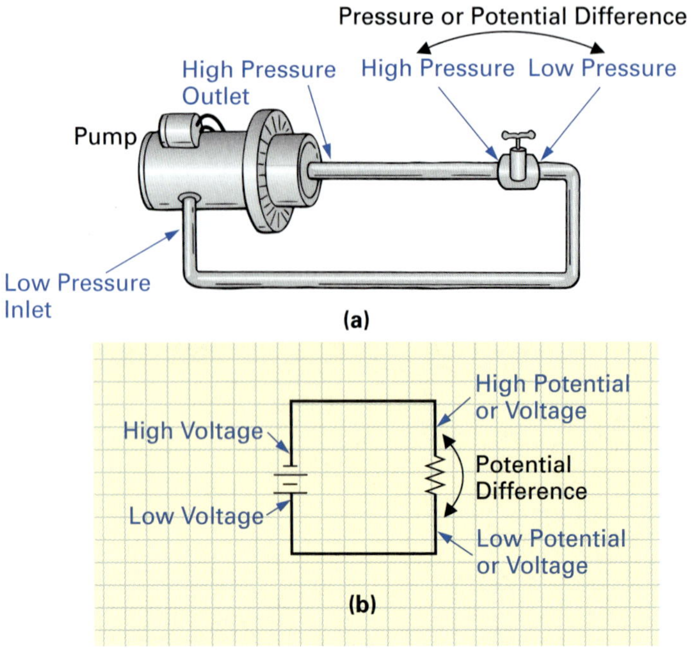

FIGURE 3-9 **Series Circuit Voltage. (a) Fluid Analogy of Potential Difference. (b) Electrical Potential Difference.**

the high pressure is no longer present. The high potential that exists on the left of the valve is not present on the right, so a potential or pressure difference is said to exist across the valve.

Similarly, with the electronic circuit shown in Figure 3-9(b), the battery produces a high voltage or potential that is present at the top of the resistor. The high voltage that exists at the top of the resistor, however, is not present at the bottom. Therefore, a potential difference or voltage drop is said to occur across the resistor. This voltage drop that exists across resistors can be found by utilizing Ohm's law: $V = I \times R$.

EXAMPLE:

Referring to Figure 3-10, calculate:

a. Total resistance (R_T)
b. Amount of series current flowing throughout the circuit (I_T)
c. Voltage drop across R_1
d. Voltage drop across R_2
e. Voltage drop across R_3

Solution:

a. Total resistance $(R_T) = R_1 + R_2 + R_3$

$$= 20\ \Omega + 30\ \Omega + 50\ \Omega$$
$$= 100\ \Omega$$

b. $$\text{Total current } (I_T) = \frac{V_{\text{source}}}{R_T} = \frac{100\ \text{V}}{100\ \Omega} = 1\ \text{A}$$

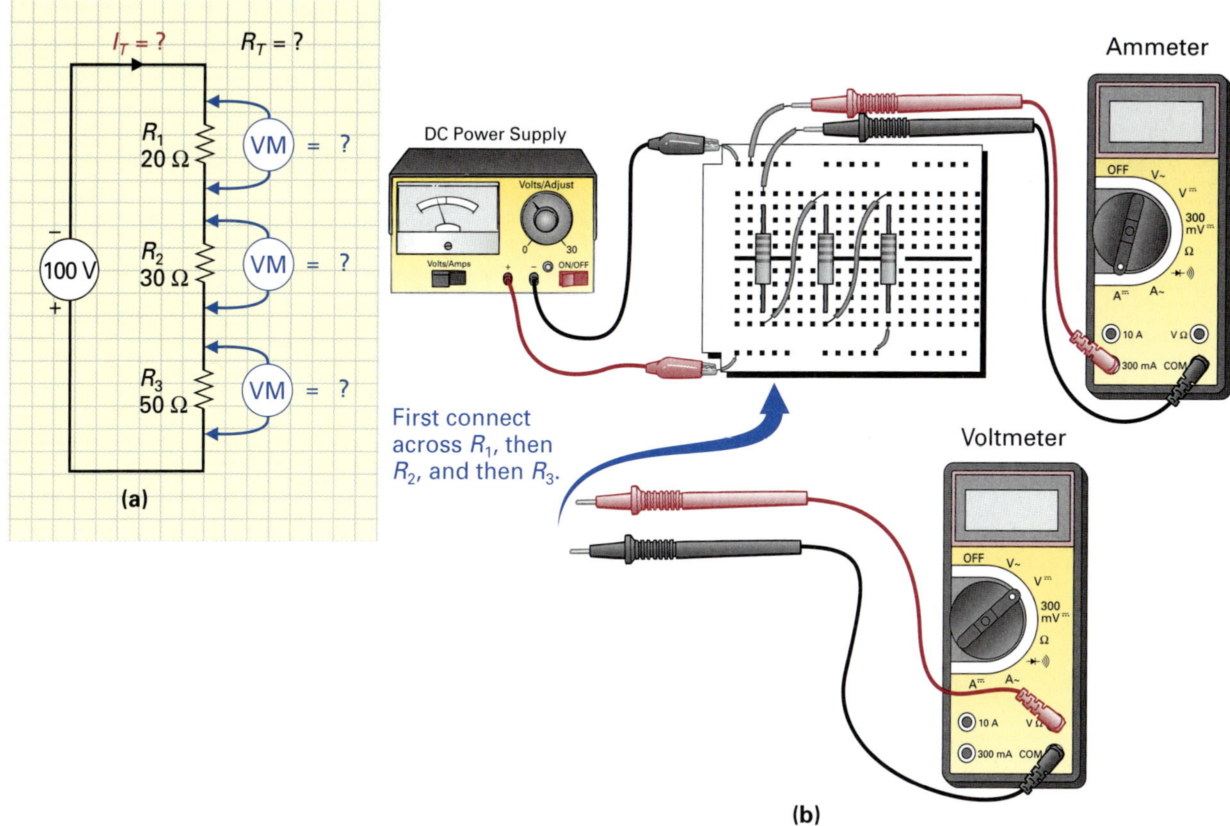

FIGURE 3-10 **Series Circuit Example. (a) Schematic. (b) Protoboard Circuit.**

The same current will flow through the complete series circuit, so the current through R_1 will equal 1 A, the current through R_2 will equal 1 A, and the current through R_3 will equal 1 A.

c. Voltage across R_1 (V_{R1}) $= I_1 \times R_1$
$= 1\text{ A} \times 20\ \Omega$
$= 20\text{ V}$

d. Voltage across R_2 (V_{R2}) $= I_2 \times R_2$
$= 1\text{ A} \times 30\ \Omega$
$= 30\text{ V}$

e. Voltage across R_3 (V_{R3}) $= I_3 \times R_3$
$= 1\text{ A} \times 50\ \Omega$
$= 50\text{ V}$

Figure 3-11(a) shows the schematic and Figure 3-11(b) shows the analysis table for this example, with all of the calculated data inserted. As you can see, the 20 Ω resistor drops 20 V, the 30 Ω resistor has 30 V across it, and the 50 Ω resistor has dropped 50 V. From this example, you will notice that the larger the resistor value, the larger the voltage drop. Resistance and voltage drops are consequently proportional to one another.

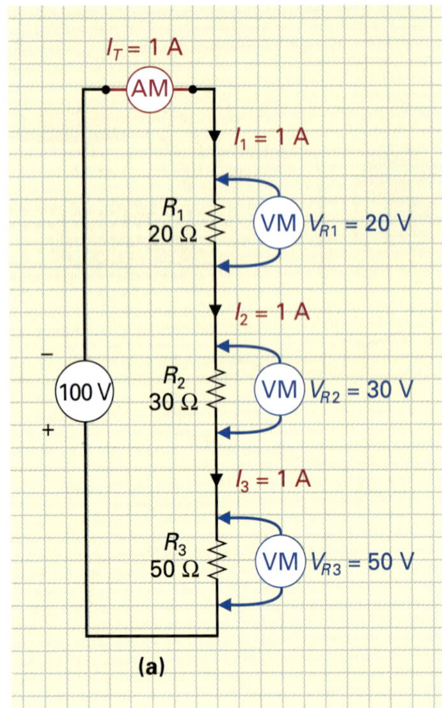

CIRCUIT ANALYSIS TABLE

Resistance $R = V/I$	Current $I = V/R$	Voltage $V = I \times R$	Power $P = V \times I$
$R_1 = 20\ \Omega$	$I_1 = 1$ A	$V_{R1} = 20$ V	
$R_2 = 30\ \Omega$	$I_2 = 1$ A	$V_{R2} = 30$ V	
$R_3 = 50\ \Omega$	$I_3 = 1$ A	$V_{R3} = 50$ V	
$R_T = 100\ \Omega$	$I_T = 1$ A	$V_T = V_S = 100$ V	

(b)

FIGURE 3-11 **Voltage Drop and Resistance. (a) Schematic. (b) Circuit Analysis Table.**

TIME LINE

German physicist Gustav Robert Kirchhoff (1824–1879) extended Ohm's law and developed two important laws of his own, known as Kirchhoff's voltage law and Kirchhoff's current law.

Kirchhoff's Voltage Law

The algebraic sum of the voltage drops in a closed path circuit is equal to the algebraic sum of the source voltage applied.

The Voltage Drop across a Device Is ∝ to Its Resistance

$$V_{\text{drop}} \uparrow = I(\text{constant}) \times R \uparrow$$
$$V_{\text{drop}} \downarrow = I(\text{constant}) \times R \downarrow$$

Another interesting point you may have noticed from Figure 3-11 is that, if you were to add up all the voltage drops around a series circuit, they would equal the source (V_S) applied:

$$\text{total voltage applied } (V_S \text{ or } V_T) = V_{R1} + V_{R2} + V_{R3} + \cdots$$

In the example in Figure 3-11, you can see that this is true, since

$$100\text{ V} = 20\text{ V} + 30\text{ V} + 50\text{ V}$$
$$100\text{ V} = 100\text{ V}$$

The series circuit has in fact divided up the applied voltage, and it appears proportionally across all the individual resistors. This characteristic was first observed by Gustav Kirchhoff in 1847. In honor of his discovery, this effect is known as **Kirchhoff's voltage law,** which states: The sum of the voltage drops in a series circuit is equal to the total voltage applied.

To summarize the effects of current, resistance, and voltage in a series circuit so far, we can say that:

1. The current in a series circuit has only one path to follow.
2. The value of current in a series circuit is the same throughout the entire circuit.
3. The total resistance in a series circuit is equal to the sum of all the resistances.

4. Resistance and voltage drops in a series circuit are proportional to one another, so a large resistance will have a large voltage drop and a small resistance will have a small voltage drop.
5. The sum of the voltage drops in a series circuit is equal to the total voltage applied.

EXAMPLE:

First, calculate the voltage drop across the resistor R_1 in the circuit in Figure 3-12(a) for a resistance of 4 Ω. Then, change the 4 Ω resistor to a 2 Ω resistor and recalculate the voltage drop across the new resistance value. Use a constant source of 4 V.

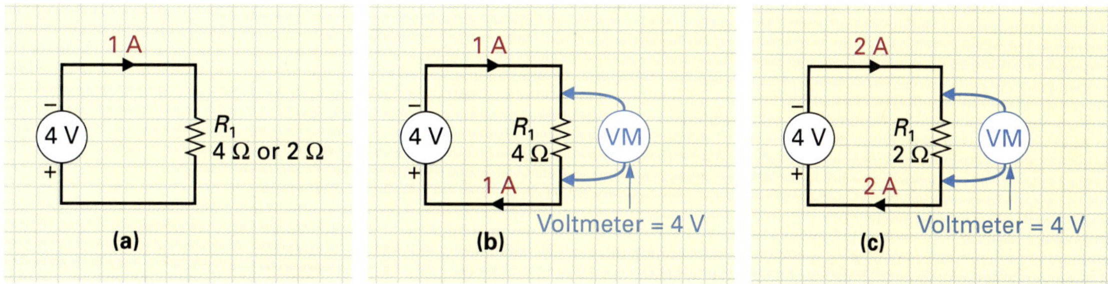

FIGURE 3-12 **Single Resistor Circuit.**

Solution:

Referring to Figure 3-12(b), you can see that when $R_1 = 4\ \Omega$, the voltage across R_1 can be calculated by using Ohm's law and is equal to

$$\begin{aligned} V_{R1} &= I_1 \times R_1 \\ V_{R1} &= 1\text{ A} \times 4\ \Omega \\ &= 4\text{ V} \end{aligned}$$

If the resistance is now changed to 2 Ω as shown in Figure 3-12(c), the current flow within the circuit will be equal to

$$I = \frac{V_S}{R} = \frac{4\text{ V}}{2\ \Omega} = 2\text{ A}$$

The voltage dropped across the 2 Ω resistor will still be equal to

$$\begin{aligned} V_{R1} &= I_1 \times R_1 \\ &= 2\text{ A} \times 2\ \Omega \\ &= 4\text{ V} \end{aligned}$$

As you can see from this example, if only one resistor is connected in a series circuit, the entire applied voltage appears across this resistor. The value of this single resistor will determine the amount of current flow through the circuit and this value of circuit current will remain the same throughout.

EXAMPLE:

Referring to Figure 3-13(a), calculate the following, and then draw the circuit schematic again with all of the new values inserted.

a. Total circuit resistance
b. Value of circuit current (I_T)
c. Voltage drop across each resistor

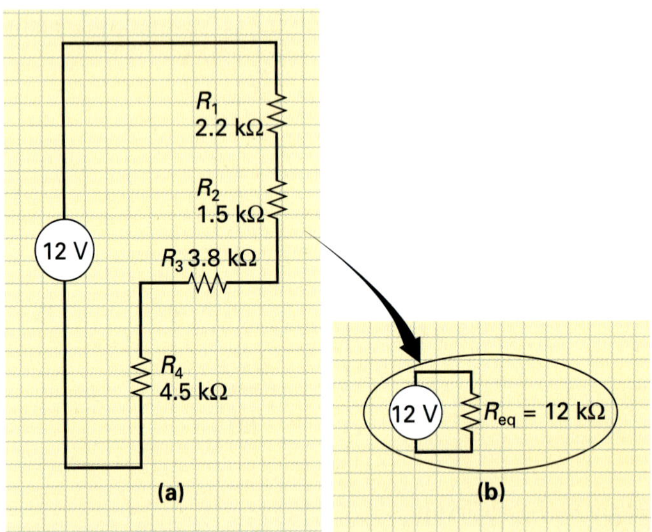

FIGURE 3-13 **Series Circuit Example.**

Solution:

a. $R_T = R_1 + R_2 + R_3 + R_4$

$$= 2.2\ \text{k}\Omega + 1.5\ \text{k}\Omega + 3.8\ \text{k}\Omega + 4.5\ \text{k}\Omega$$
$$= (2.2 \times 10^3) + (1.5 \times 10^3) + (3.8 \times 10^3) + (4.5 \times 10^3)$$
$$= 12\ \text{k}\Omega \qquad \text{[Figure 3–13(b)]}$$

b. $I_T = \dfrac{V_S}{R_T} = \dfrac{12\ \text{V}}{12\ \text{k}\Omega} = 1\ \text{mA}$

c. Voltage drop across each resistor:

$$V_{R1} = I_T \times R_1 = 1\ \text{mA} \times 2.2\ \text{k}\Omega = 2.2\ \text{V}$$

$$V_{R2} = I_T \times R_2 = 1\ \text{mA} \times 1.5\ \text{k}\Omega = 1.5\ \text{V}$$

$$V_{R3} = I_T \times R_3 = 1\ \text{mA} \times 3.8\ \text{k}\Omega = 3.8\ \text{V}$$

$$V_{R4} = I_T \times R_4 = 1\ \text{mA} \times 4.5\ \text{k}\Omega = 4.5\ \text{V}$$

Figure 3-14(a) shows the schematic diagram for this example with all of the values inserted, and Figure 3-14(b) shows this circuit's analysis table.

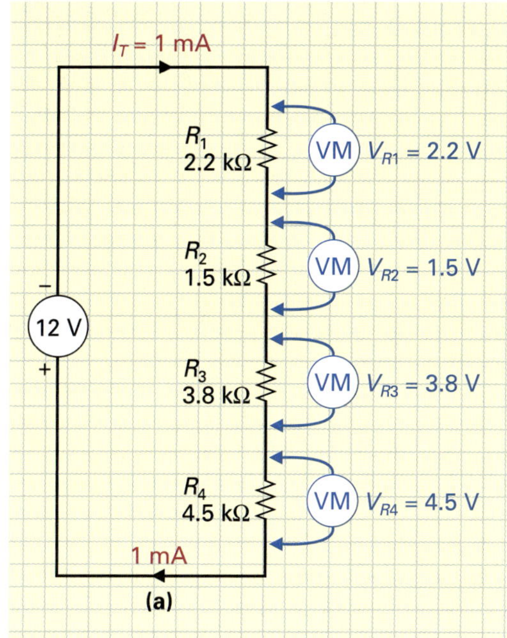

CIRCUIT ANALYSIS TABLE

Resistance $R = V/I$	Current $I = V/R$	Voltage $V = I \times R$	Power $P = V \times I$
R_1 = 2.2 kΩ	I_1 = 1 mA	V_{R1} = 2.2 V	
R_2 = 1.5 kΩ	I_2 = 1 mA	V_{R2} = 1.5 V	
R_3 = 3.8 kΩ	I_3 = 1 mA	V_{R3} = 3.8 V	
R_4 = 4.5 kΩ	I_4 = 1 mA	V_{R4} = 4.5 V	
R_T = 12 kΩ	I_T = 1 mA	$V_T = V_S$ = 12 V	

(b)

FIGURE 3-14 Series Circuit Example with All Values Inserted. (a) Schematic. (b) Circuit Analysis Table.

3-4-1 *Fixed Voltage Divider*

A series-connected circuit is often referred to as a *voltage-divider circuit,* because the total voltage applied (V_T) or source voltage (V_S) is divided and dropped proportionally across all the resistors in a series circuit. The amount of voltage dropped across a resistor is proportional to the value of resistance, and so a larger resistance will have a larger voltage drop, while a smaller resistance will have a smaller voltage drop.

The voltage dropped across a resistor is normally a factor that needs to be calculated. The voltage-divider formula allows you to calculate the voltage drop across any resistor without having to first calculate the value of circuit current. This formula is stated as

$$V_X = \frac{R_X}{R_T} \times V_S$$

where V_X = voltage dropped across selected resistor
R_X = selected resistor's value
R_T = total series circuit resistance
V_S = source or applied voltage

Figure 3-15 illustrates a circuit from a previous example. To calculate the voltage drop across each resistor, we would normally have to:

1. Calculate the total resistance by adding up all the resistance values.
2. Once we have the total resistance and source voltage (V_S), we could then calculate current.
3. Having calculated the current flowing through each resistor, we could then use the current value to calculate the voltage dropped across any one of the four resistors merely by multiplying current by the individual resistance value.

The voltage-divider formula allows us to bypass the last two steps in this procedure. If we know total resistance, supply voltage, and the individual resistance values, we can calculate the

FIGURE 3-15 **Series Circuit Example.**

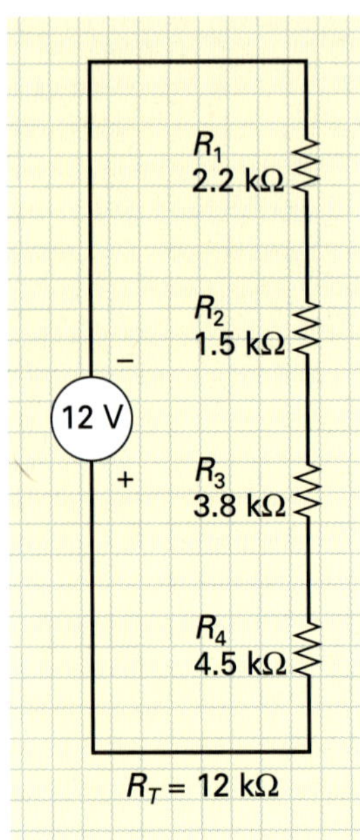

CALCULATOR SEQUENCE

Step	Keypad Entry	Display Response
1.	1 . 5 E 3	1.5E3
2.	+	
3.	1 2 E 3	12E3
4.	×	0.125
5.	1 2	12
6.	=	1.5

voltage drop across the resistor without having to calculate steps 2 and 3. For example, what would be the voltage dropped across R_2 and R_4 for the circuit shown in Figure 3-15? The voltage dropped across R_2 is

$$V_{R2} = \frac{R_2}{R_T} \times V_S$$

$$= \frac{1.5\text{ k}\Omega}{12\text{ k}\Omega} \times 12\text{ V}$$

$$= 1.5\text{ V}$$

The voltage dropped across R_4 is

$$V_{R4} = \frac{R_4}{R_T} \times V_S$$

$$= \frac{4.5\text{ k}\Omega}{12\text{ k}\Omega} \times 12\text{ V}$$

$$= 4.5\text{ V}$$

The voltage-divider formula could also be used to find the voltage drop across two or more series-connected resistors. For example, referring again to the example circuit shown in Figure 3-15, what would be the voltage dropped across R_2 and R_3 combined? The voltage across $R_2 + R_3$ can be calculated using the voltage-divider formula as follows:

$$V_{R2}\text{ and }V_{R3} = \frac{R_2 + R_3}{R_T} \times V_S$$

$$= \frac{5.3\text{ k}\Omega}{12\text{ k}\Omega} \times 12\text{ V}$$

$$= 5.3\text{ V}$$

As can be seen in Figure 3-16, the voltage drop across R_2 and R_3 is 5.3 V.

FIGURE 3-16 Series Circuit Example.

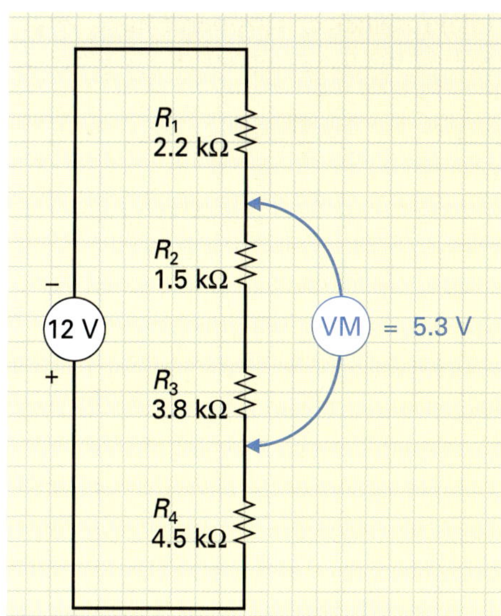

EXAMPLE:

Referring to Figure 3-17(a), calculate the voltage drop across:

a. R_1, R_2, and R_3 separately
b. R_2 and R_3 combined
c. R_1, R_2, and R_3 combined

Solution:

a. The voltage drop across a resistor is proportional to the resistance value. The total resistance (R_T) in this circuit is 100 Ω or 100% of R_T. R_1 is 20% of the total resistance, so 20% of the source voltage will appear across R_1. R_2 is 30% of the total resistance, so 30% of the source voltage will appear across R_2. R_3 is 50% of the total resistance, so 50% of the source voltage will appear across R_3. This was a very simple problem in which the figures worked out very neatly. The voltage-divider formula achieves the very same thing by calculating the ratio of the resistance value to

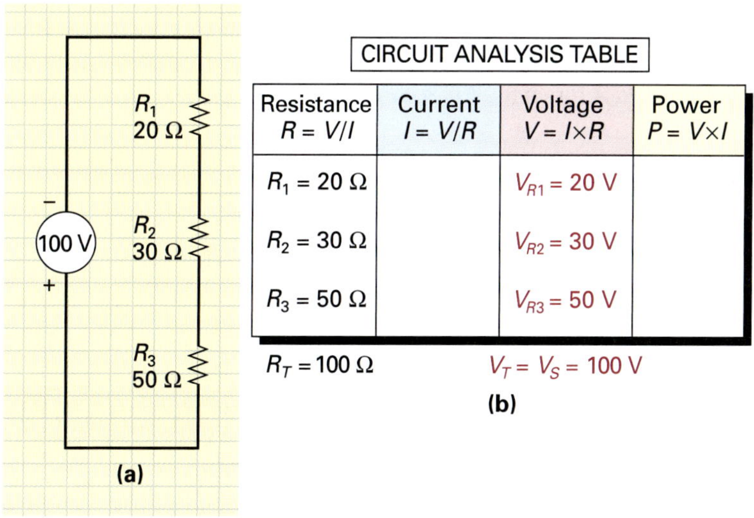

CIRCUIT ANALYSIS TABLE

Resistance $R = V/I$	Current $I = V/R$	Voltage $V = I \times R$	Power $P = V \times I$
$R_1 = 20\ \Omega$		$V_{R1} = 20$ V	
$R_2 = 30\ \Omega$		$V_{R2} = 30$ V	
$R_3 = 50\ \Omega$		$V_{R3} = 50$ V	
$R_T = 100\ \Omega$		$V_T = V_S = 100$ V	

(b)

FIGURE 3-17 Series Circuit Example. (a) Schematic. (b) Circuit Analysis Table.

the total resistance. This percentage is then multiplied by the source voltage in order to find the desired resistor's voltage drop:

$$V_{R1} = \frac{R_1}{R_T} \times V_S$$

$$= \frac{20\ \Omega}{100\ \Omega} \times V_S$$

$$= 0.2 \times 100\ \text{V} \qquad (20\%\ \text{of}\ 100\ \text{V})$$

$$= 20\ \text{V}$$

$$V_{R2} = \frac{R_2}{R_T} \times V_S$$

$$= \frac{30\ \Omega}{100\ \Omega} \times V_S$$

$$= 0.3 \times 100\ \text{V} \qquad (30\%\ \text{of}\ 100\ \text{V})$$

$$= 30\ \text{V}$$

$$V_{R3} = \frac{R_3}{R_T} \times V_S$$

$$= \frac{50\ \Omega}{100\ \Omega} \times V_S$$

$$= 0.5 \times 100\ \text{V} \qquad (50\%\ \text{of}\ 100\ \text{V})$$

$$= 50\ \text{V}$$

b. Voltage dropped across R_2 and $R_3 = 30 + 50 = 80$ V.

c. Voltage dropped across R_1, R_2, and $R_3 = 20 + 30 + 50 = 100$ V.

The details of this circuit are summarized in the analysis table shown in Figure 3-17(b).

To summarize the voltage-divider formula, we can say: The voltage drop across a resistor or group of resistors in a series circuit is equal to the ratio of that resistance (R_X) to the total resistance (R_T), multiplied by the source voltage (V_S).

To show an application of a voltage divider, let us imagine that three voltages of 50, 80, and 100 V were required by an electronic system in order to make it operate. To meet this need, we could use three individual power sources, which would be very expensive, or use one 100 V voltage source connected across three resistors, as shown in Figure 3-18, to divide up the 100 V.

FIGURE 3-18 **Series Circuit Example.**

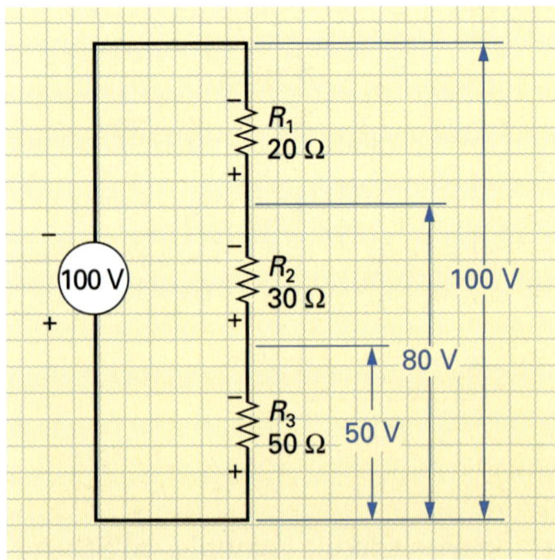

EXAMPLE:

Figure 3-19(a) shows a 24 V voltage source driving a 10 Ω resistor that is located 1000 ft from the battery. If two 1000 ft lengths of AWG No. 13 wire are used to connect the source to the load, what will be the voltage applied across the load?

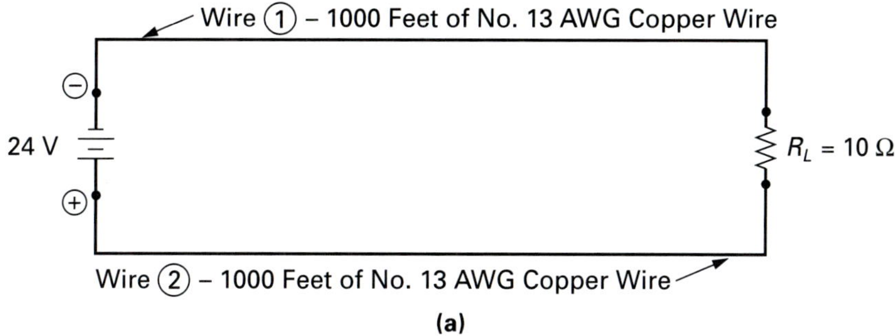

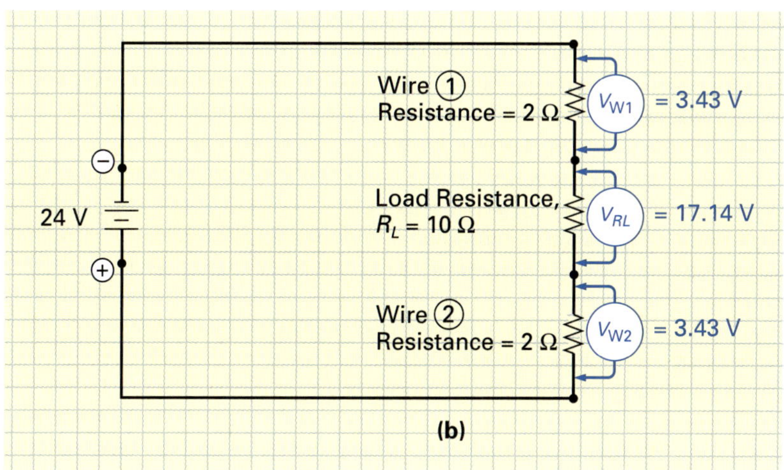

FIGURE 3-19 Series Wire Resistance.

Solution:

Referring back to Table 2-3, you can see that AWG No. 13 copper cable has a resistance of 2.003 Ω for every 1000 ft. To be more accurate, this means that our circuit should be redrawn as shown in Figure 3-19(b) to show the series resistances of wire 1 and wire 2. Using the voltage-divider formula, we can calculate the voltage drop across wire 1 and wire 2.

$$V_{W1} = \frac{R_{W1}}{R_T} \times V_T$$

$$= \frac{2\ \Omega}{14\ \Omega} \times 24\ \text{V} = 3.43\ \text{V}$$

Since the voltage drop across wire 2 will also be 3.43 V, the total voltage drop across both wires will be 6.86 V. The remainder, 17.14 V (24 V − 6.86 V = 17.14 V), will appear across the load resistor, R_L.

3-4-2 *Variable Voltage Divider*

When discussing variable-value resistors in Chapter 2, we talked about a potentiometer, or variable voltage divider, which consists of a fixed value of resistance between two terminals and a wiper that can be adjusted to vary resistance between its terminal and one of the other two.

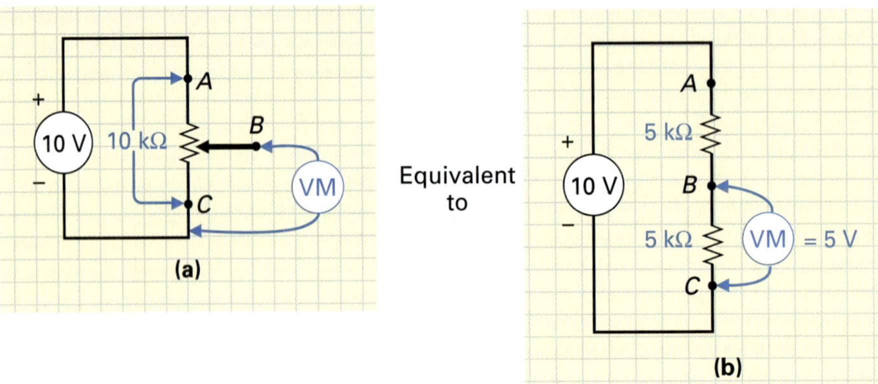

FIGURE 3-20 Potentiometer Wiper in Mid-Position.

As an example, Figure 3-20(a) illustrates a 10 kΩ potentiometer that has been hooked up across a 10 V dc source with a voltmeter between terminals *B* and *C*. If the wiper terminal is positioned midway between *A* and *C*, the voltmeter should read 5 V, and the potentiometer will be equivalent to two 5 kΩ resistors in series, as shown in Figure 3-20(b). Kirchhoff's voltage law states that the entire source voltage will be dropped across the resistances in the circuit, and since the resistance values are equal, each will drop half of the source voltage, that is, 5 V.

In Figure 3-21(a), the wiper has been moved down so that the resistance between *A* and *B* is equal to 8 kΩ, and the resistance between *B* and *C* equals 2 kΩ. This will produce 2 V on the voltmeter, as shown in Figure 3–21(b). The amount of voltage drop is proportional to the resistance, so a larger voltage will be dropped across the larger resistance. Using the voltage-divider formula, you can calculate that the 8 kΩ resistance is 80% of the total resistance and therefore will drop 80% of the voltage:

$$V_{AB} = \frac{R_{AB}}{R_{\text{total}}} \times V_S = \frac{8\text{ k}\Omega}{10\text{ k}\Omega} \times 10\text{ V} = 8\text{ V}$$

The 2 kΩ resistance between *B* and *C* is 20% of the total resistance and consequently will drop 20% of the total voltage:

$$V_{BC} = \frac{R_{BC}}{R_{\text{total}}} \times V_S = \frac{2\text{ k}\Omega}{10\text{ k}\Omega} \times 10\text{ V} = 2\text{ V}$$

In Figure 3-22(a), the wiper has been moved up and now 2 kΩ exists between *A* and *B*, and 8 kΩ is present between *B* and *C*. In this situation, 2 V will be dropped across the 2 kΩ between *A* and *B*, and 8 V will be dropped across the 8 kΩ between *B* and *C*, as shown in Figure 3-22(b).

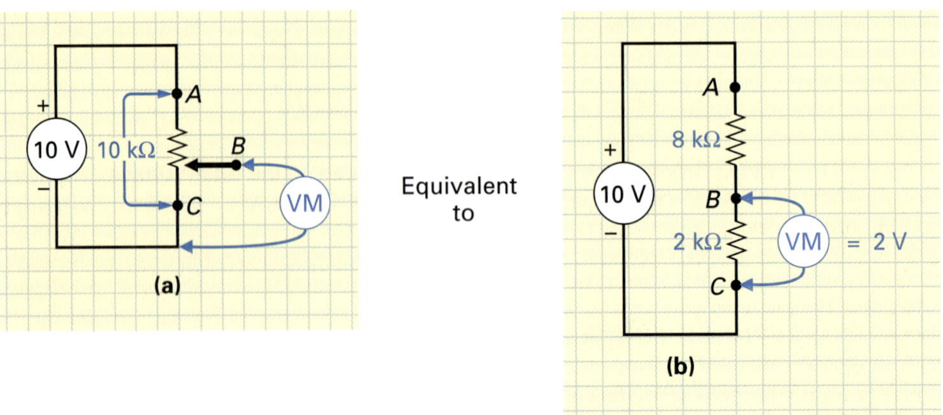

FIGURE 3-21 Potentiometer Wiper in Lower Position.

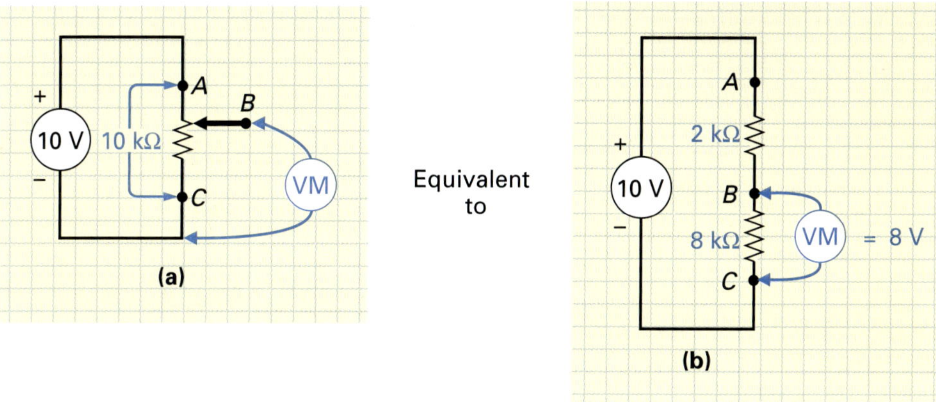

FIGURE 3-22 **Potentiometer Wiper in Upper Position.**

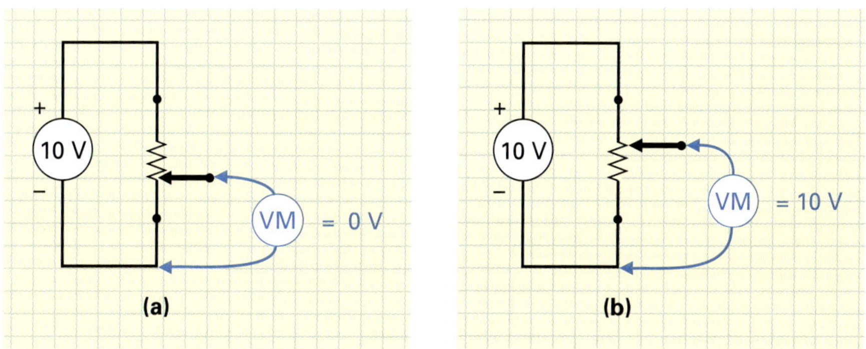

FIGURE 3-23 **Minimum and Maximum Settings of a Potentiometer.**

From this discussion, you can see that the potentiometer can be adjusted to supply different voltages on the wiper. This voltage can be decreased by moving the wiper down to supply a minimum of 0 V as shown in Figure 3-23(a), or the wiper can be moved up to supply a maximum of 10 V as shown in Figure 3-23(b). By adjusting the wiper position, the potentiometer can be made to deliver any voltage between its maximum and minimum value, which is why the potentiometer is known as a variable voltage divider.

EXAMPLE:

Figure 3-24 illustrates how a potentiometer can be used to control the output volume of an amplifier that is being driven by a compact disk (CD) player. The preamplifier is producing an output of 2 V, which is developed across a 50 kΩ potentiometer. If the wiper of the potentiometer is in its upper position, the full 2 V from the preamp will be applied into the input of the power amplifier. The power amplifier has a fixed voltage gain (A_V) of 12, and therefore the power amplifier's output is always 12 times larger than the input voltage. An input of 2 V (V_{in}) will therefore produce an output voltage (V_{out}) of 24 V ($V_{out} = V_{in} \times A_V = 2\text{ V} \times 12 = 24\text{ V}$).

As the wiper is moved down, less of the 2 V from the preamp will be applied to the power amplifier, and therefore the output of the power amplifier and volume of the music heard will decrease. If the wiper of the potentiometer is adjusted so that a resistance of 20 kΩ exists between the wiper and the lower end of the potentiometer, what will be the input voltage to the power amplifier and output voltage to the speaker?

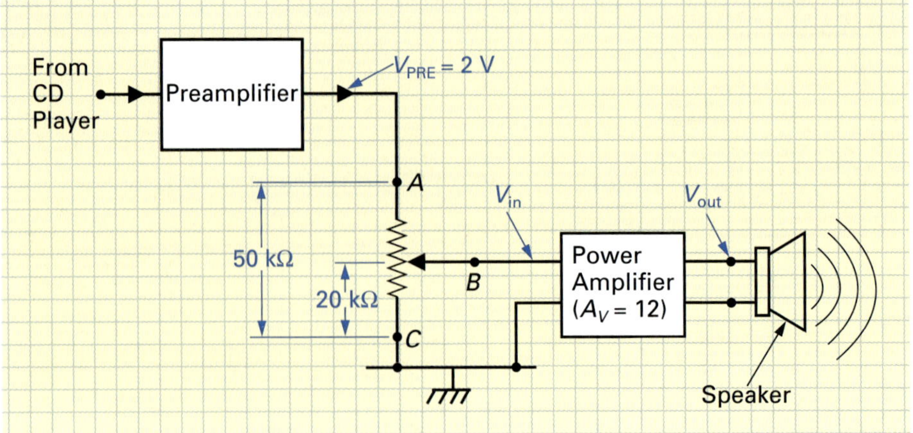

FIGURE 3-24 The Potentiometer as a Volume Control.

■ ***Solution:***

By using the voltage-divider formula, we can determine the voltage developed across the potentiometer with 20 kΩ of resistance between the wiper (B) and lower end (C).

$$V_{\text{in}} = \frac{R_{AB}}{R_{AC}} \times V_{\text{Pre}}$$

$$= \frac{20 \text{ k}\Omega}{50 \text{ k}\Omega} \times 2 \text{ V} = 0.8 \text{ V (800 mV)}$$

The voltage at the input of the power amplifier (V_{in}) is applied to the input of the power amplifier, which has an amplification factor or gain of 12, and therefore the output voltage will be

$$V_{\text{out}} = V_{\text{in}} \times A_V$$

$$= 0.8 \text{ V} \times 12 = 9.6 \text{ V}$$

SELF-TEST EVALUATION POINT FOR SECTION 3-4

Use the following questions to test your understanding of Section 3-4.

1. True or false: A series circuit is also known as a voltage-divider circuit.
2. True or false: The voltage drop across a series resistor is proportional to the value of the resistor.
3. If 6 Ω and 12 Ω resistors are connected across a dc 18 V supply, calculate I_T and the voltage drop across each.
4. State the voltage-divider formula.
5. Which component can be used as a variable voltage divider?
6. Could a rheostat be used in place of a potentiometer?

3-5 POWER IN A SERIES CIRCUIT

As discussed in Chapter 2, power is the rate at which work is done, and work is said to have been done when energy is converted from one energy form to another. Resistors convert electrical energy into heat energy, and the rate at which they dissipate energy is called *power* and is measured in *watts* (joules per second). Resistors all have a resistive value, a tolerance, and a wattage rating. The wattage of a resistor is the amount of heat energy a resistor can safely dissipate per second, and this wattage is directly proportional to the resistor's size.

EXAMPLE:

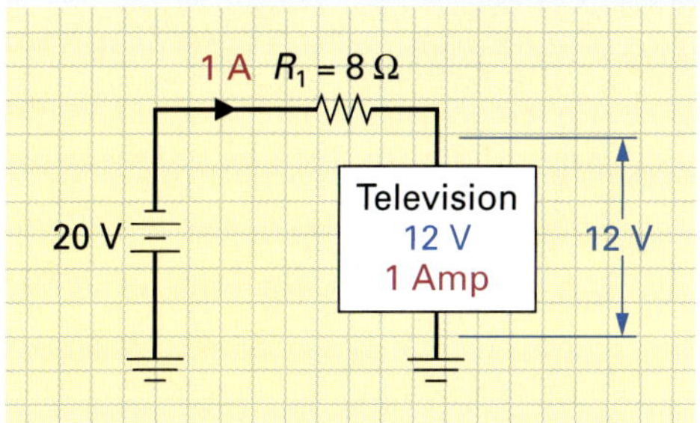

FIGURE 3-25 Series Circuit Example.

Figure 3-25 shows a 20 V battery driving a 12 V/1 A television set. R_1 is in series with the TV set and is being used to drop 8 V of the 20 V supply, so 12 V will be developed across the television.

a. What is the wattage rating for R_1?

b. What is the series load resistance of the TV set?

c. What is the amount of power being consumed by the TV set?

d. Compile a circuit analysis table for this circuit.

Solution:

a. Everything is known about R_1. Its resistance is 8Ω, it has 1 A of current flowing through it, and 8 V is being dropped across it. As a result, any one of the three power formulas can be used to calculate the wattage rating of R_1.

$$\text{power } (P) = I \times V = 1 \text{ A} \times 8 \text{ V} = 8 \text{ W}$$

or

$$P = I^2 \times R = 1^2 \times 8 = 8 \text{ W}$$

or

$$P = \frac{V^2}{R} = \frac{8^2}{8} = 8 \text{ W}$$

The nearest commercially available device would be a 10 W resistor. If size is not a consideration, it is ideal to double the wattage needed and use a 16 W resistor.

b. You may recall that any piece of equipment is equivalent to a load resistance. The TV set has 12 V across it and is pulling 1 A of current. Its load resistance can be calculated simply by using Ohm's law and deriving an equivalent circuit, as shown in Figure 3-26(a).

$$R_L \text{ (load resistance)} = \frac{V}{I} = \frac{12 \text{ V}}{1 \text{ A}} = 12 \ \Omega$$

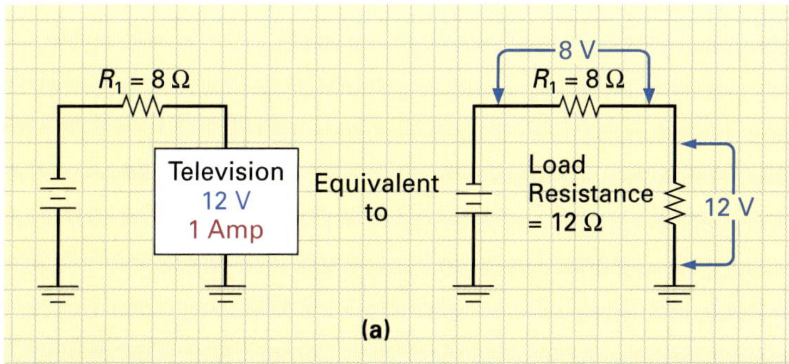

CIRCUIT ANALYSIS TABLE

Resistance $R = V/I$	Current $I = V/R$	Voltage $V = I \times R$	Power $P = V \times I$
$R_1 = 8\ \Omega$	$I_1 = 1$ A	$V_{R1} = 8$ V	$P_1 = 8$ W
$R_{TV} = 12\ \Omega$	$I_2 = 1$ A	$V_{RTV} = 12$ V	$P_{TV} = 12$ W
$R_T = 20\ \Omega$	$I_T = 1$ A	$V_T = V_S = 20$ V	$P_T = 20$ W

(b)

FIGURE 3-26 Series Circuit Example with Values Inserted. (a) Schematic. (b) Circuit Analysis Table.

c. The amount of power being consumed by the TV set is

$$P = V \times I = 12\text{ V} \times 1\text{ A} = 12\text{ W}$$

d. Figure 3–26(b) shows the circuit analysis table for this example.

EXAMPLE:

Calculate the total amount of power dissipated in the series circuit in Figure 3-27, and insert any calculated values in a circuit analysis table.

Solution:

The total power dissipated in a series circuit is equal to the sum of all the power dissipated by all the resistors. The easiest way to calculate the total power is to simplify the circuit to one resistance, as shown in Figure 3-27.

$$\begin{aligned} R_T &= R_1 + R_2 + R_3 + R_4 \\ &= 5\ \Omega + 33\ \Omega + 45\ \Omega + 75\ \Omega \\ &= 158\ \Omega \end{aligned}$$

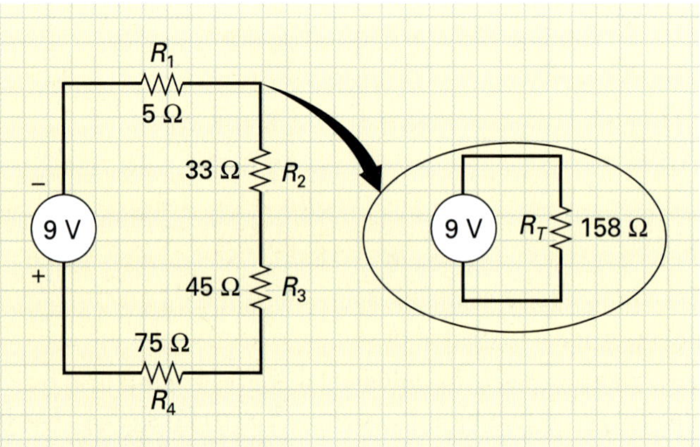

FIGURE 3-27 **Series Circuit Example.**

We now have total resistance and total voltage, so we can calculate the total power:

$$P_T = \frac{V_s^2}{R_T} = \frac{(9\ V)^2}{158\ \Omega} = \frac{81}{158} = 512.7 \text{ milliwatts (mW)}$$

The longer method would have been to first calculate the current through the series circuit:

$$I = \frac{V_s}{R_T} = \frac{9\text{ V}}{158\ \Omega} = 56.96\text{ mA} \quad \text{or} \quad 57\text{ mA}$$

We could then calculate the power dissipated by each separate resistor and add up all the individual values to gain a total power figure. This is illustrated in Figure 3-28(a).

$$P_T = P_1 + P_2 + P_3 + P_4 + \cdots$$

total power = addition of all the individual power losses

$$\begin{aligned} P_T &= 16\text{ mW} + 107\text{ mW} + 146\text{ mW} + 243\text{ mW} \\ &= 512\text{ mW} \end{aligned}$$

The calculated values for this example are shown in the circuit analysis table in Figure 3–28(b).

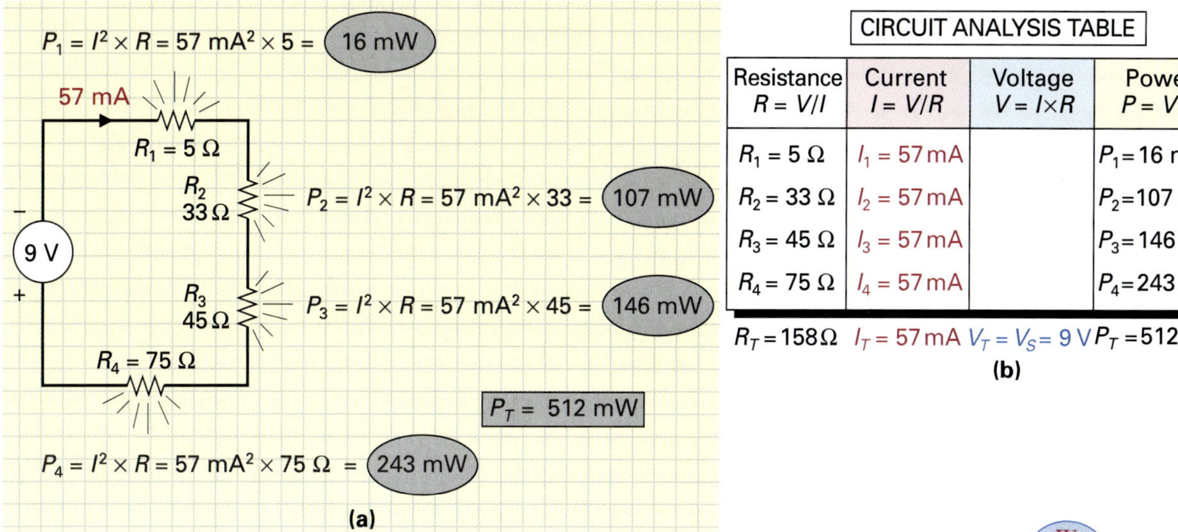

CIRCUIT ANALYSIS TABLE

Resistance $R = V/I$	Current $I = V/R$	Voltage $V = I \times R$	Power $P = V \times I$
$R_1 = 5\ \Omega$	$I_1 = 57$ mA		$P_1 = 16$ mW
$R_2 = 33\ \Omega$	$I_2 = 57$ mA		$P_2 = 107$ mW
$R_3 = 45\ \Omega$	$I_3 = 57$ mA		$P_3 = 146$ mW
$R_4 = 75\ \Omega$	$I_4 = 57$ mA		$P_4 = 243$ mW
$R_T = 158\Omega$	$I_T = 57$ mA	$V_T = V_S = 9$ V	$P_T = 512$ mW

(b)

FIGURE 3-28 Series Circuit Example with Values Inserted. (a) Schematic. (b) Circuit Analysis Table.

3-5-1 *Maximum Power Transfer*

The **maximum power transfer theorem** states that maximum power will be delivered to a load when the resistance of the load (R_L) is equal to the resistance of the source (R_S). The best way to see if this theorem is correct is to apply it to a series of examples and then make a comparison.

Maximum Power Transfer Theorem
A theorem that states maximum power will be transferred from source to load when the source resistance is equal to the load resistance.

EXAMPLE:

Figure 3-29(a) illustrates a 10 V battery with a 5 Ω internal resistance connected across a load. The load in this case is a light bulb, which has a load resistance of 1 Ω. Calculate the power delivered to this light bulb or load when $R_L = 1\ \Omega$.

Solution:

$$\text{circuit current, } I = \frac{V}{R}$$
$$= \frac{10\text{ V}}{R_S + R_L}$$
$$= \frac{10\text{ V}}{6\ \Omega}$$
$$= 1.66\text{ A}$$

The power supplied to the load is $P = I^2 \times R = 1.66^2 \times 1\ \Omega = 2.8$ W.

EXAMPLE:

Figure 3-29(b) illustrates the same battery and R_S, but in this case connected across a 5 Ω light bulb. Calculate the power delivered to this light bulb now that $R_L = 5\ \Omega$.

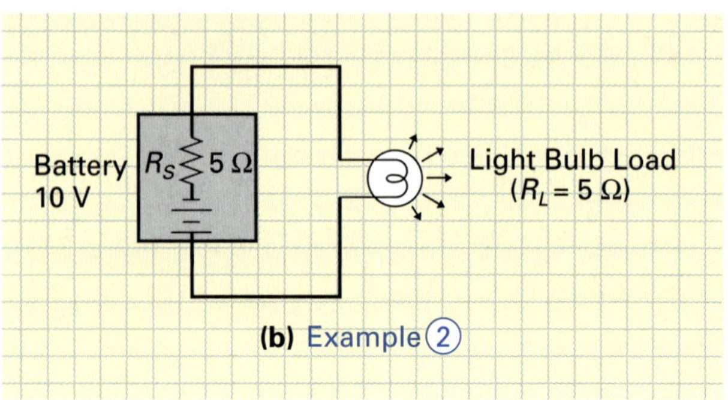

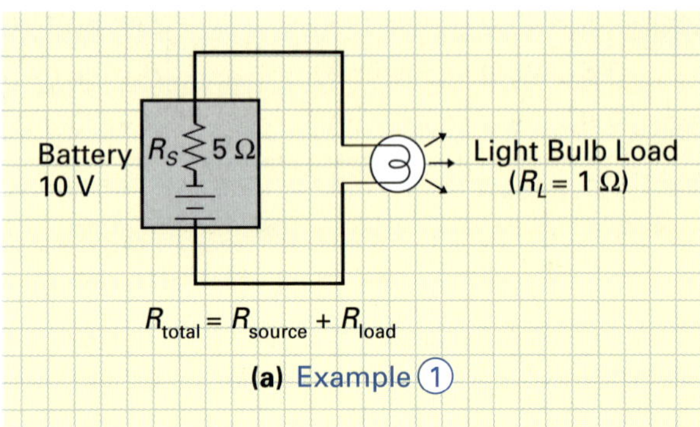

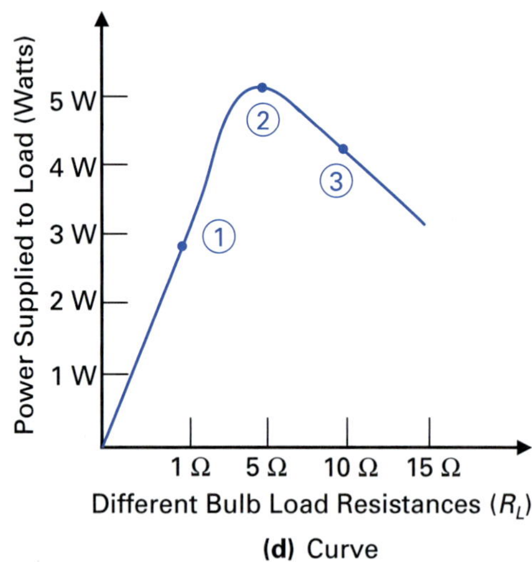

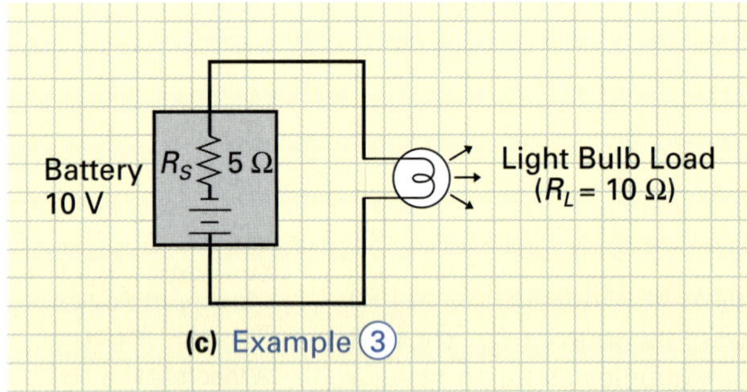

FIGURE 3-29 Maximum Power Transfer.

■ ***Solution:***

$$
\begin{aligned}
I &= \frac{V}{R} \\
&= \frac{10\text{ V}}{R_S + R_L} \\
&= \frac{10\text{ V}}{10\ \Omega} \\
&= 1\text{ A}
\end{aligned}
$$

Power supplied is $P = I^2 \times R = 1^2 \times 5\ \Omega = 5$ W.

EXAMPLE:

Figure 3-29(c) illustrates the same battery again, but in this case a 10 Ω light bulb is connected in the circuit. Calculate the power delivered to this light bulb now that $R_L = 10\ \Omega$.

■ ***Solution:***

$$
\begin{aligned}
I &= \frac{V}{R} \\
&= \frac{10\text{ V}}{R_S + R_L}
\end{aligned}
$$

$$= \frac{10\text{ V}}{15\ \Omega}$$

$$= 0.67\text{ A}$$

Thus $P = I^2 \times R = 0.67^2 \times 10\ \Omega = 4.5$ W.

As can be seen by the graph in Figure 3-29(d), which plots the power supplied to the load against load resistance, maximum power is delivered to the load (5 W) when the load resistance is equal to the source resistance.

The maximum power transfer condition is only used in special cases such as the automobile starter, where the load resistance remains constant and maximum power is needed. In most other cases, where load resistance can vary over a range of values, circuits are designed for a load resistance that will cause the best amount of power to be delivered. This is known as **optimum power transfer.**

Optimum Power Transfer
Since the ideal maximum power transfer conditions cannot always be achieved, most designers try to achieve optimum power transfer and have the source resistance and load resistance as close in value as possible.

SELF-TEST EVALUATION POINT FOR SECTION 3-5

Use the following questions to test your understanding of Section 3-5.

1. State the power formula.
2. Calculate the power dissipated by a 12 Ω resistor connected across a 12 V supply.
3. What fixed resistor type should probably be used for Question 2, and what would be a safe wattage rating?
4. What would be the total power dissipated if R_1 dissipates 25 W and R_2 dissipates 3800 mW?

3-6 TROUBLESHOOTING A SERIES CIRCUIT

A resistor will usually burn out and cause an open between its two leads when an excessive current flow occurs. This can normally, but not always, be noticed by a visual check of the resistor, which will appear charred due to the excessive heat. In some cases you will need to use your multimeter (combined ammeter, voltmeter, and ohmmeter) to check the circuit components to determine where a problem exists.

The two basic problems that normally exist in a series circuit are opens and shorts. In most instances a problem is not always as drastic as a short or an open, but may be a variation in a component's value over a long period of time, which will eventually cause a problem.

To summarize, then, we can say that one of three problems can occur to components in a series circuit:

1. A component will open (infinite resistance).
2. A component's value will change over a period of time.
3. A component will short (zero resistance).

The voltmeter is the most useful tool for checking series circuits as it can be used to measure voltage drops by connecting the meter leads across the component or resistor. Let's now analyze a circuit problem and see if we can solve it by logically **troubleshooting** the circuit and isolating the faulty component. To begin with, let us take a look at the effects of an open component.

Troubleshooting
The process of locating and diagnosing malfunctions or breakdowns in equipment by means of systematic checking or analysis.

3-6-1 *Open Component in a Series Circuit*

A component is open when its resistance is the maximum possible (infinity).

■ **EXAMPLE:**

Figure 3-30(a) illustrates a TV set with a load resistance of 3 Ω. The TV set is off because R_2 has burned out and become an open circuit. How would you determine that the problem is R_2?

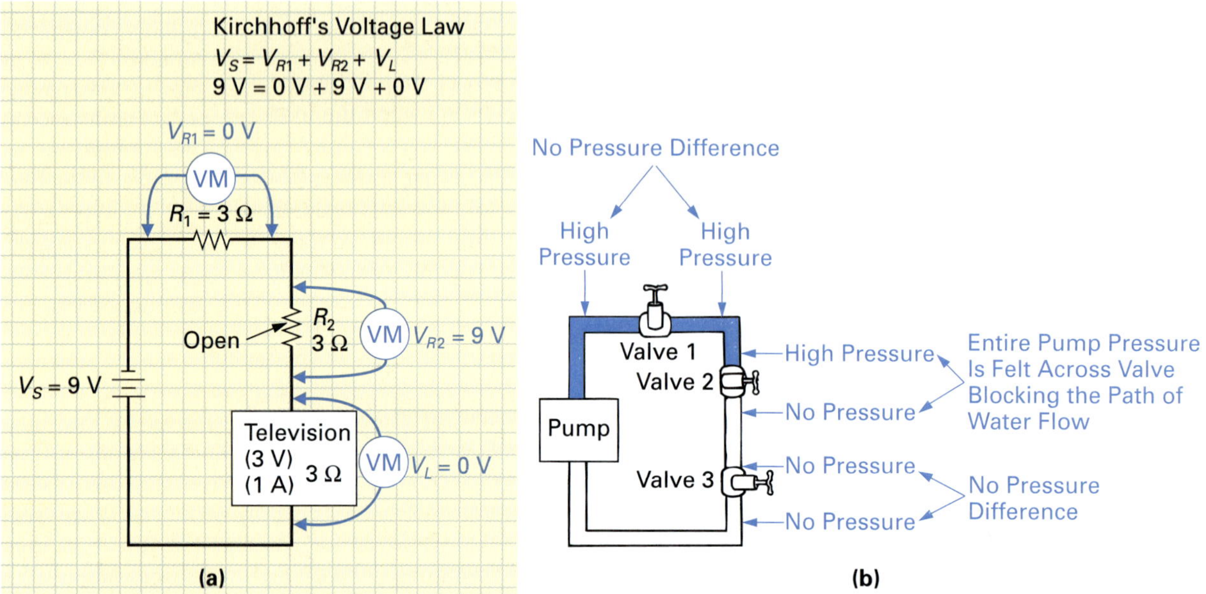

FIGURE 3-30 **Troubleshooting an Open in a Series Circuit.**

■ ***Solution:***

If an open circuit ever occurs in a series circuit, due in this case to R_2 having burned out, there can be no current flow, because series circuits have only one path for current to flow and that path has been broken ($I = 0$ A). Using the voltmeter to check the amount of voltage drop across each resistor, two results will be obtained:

1. The voltage drop across a good resistor will be zero volts.
2. The voltage drop across an open resistor will be equal to the source voltage V_S.

No voltage will be dropped across a good resistor because current is zero, and if $I = 0$, the voltage drop, which is the product of I and R, must be zero ($V = I \times R = 0 \times R = 0$ V). If no voltage is being dropped across the good resistor R_1 and the TV set resistance of 3 Ω, the entire source voltage will appear across the open resistor, R_2, in order that this series circuit comply with Kirchhoff's voltage law: V_S (9 V) = V_{R1} (0 V) + V_{R2} (9 V) + V_L (0 V).

To explain this point further, refer to the fluid analogy in Figure 3–30(b). Like R_2, valve 2 has completely blocked any form of flow (water flow = 0). Looking at the pressure differences across all three valves, you can see that no pressure difference occurs across valves 1 and 3, but the entire pump pressure is appearing across valve 2, which is the component that has opened the circuit.

3-6-2 *Component Value Variation in a Series Circuit*

Resistors will rarely go completely open unless severely stressed because of excessive current flow. With age, resistors will normally change their resistance value. This occurs slowly and will generally cause a decrease in the resistor's resistance and eventually cause a circuit problem. This lowering of resistance will cause an increase of current, which will cause an increase in the power dissipated. If the wattage of a resistor is exceeded, it can burn out. If they do not burn out but merely blow the circuit fuse due to an increase in current, the problem can be found by measuring the resistance values of each resistor or by measuring how

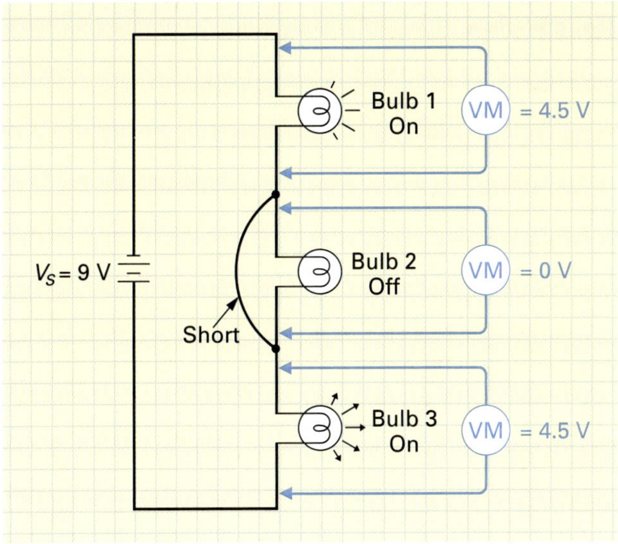

FIGURE 3-31 **Troubleshooting a Short in a Series Circuit.**

much voltage is dropped across each resistor and comparing these to the calculated expected voltage, based on the parts list supplied by the manufacturer.

3-6-3 *Shorted Component in a Series Circuit*

A component has gone short when its resistance is 0 W. Let's work out a problem using the same example of the three bulbs across the 9 V battery, as shown in Figure 3-31.

EXAMPLE:

Bulbs 1 and 3 in Figure 3-31 are on and bulb 2 is off. A piece of wire exists between the two terminals of bulb 2, and the current is taking the lowest resistance path through the wire rather than through the filament resistor of the bulb. Bulb 2 therefore has no current flow through it, and light cannot be generated without current. How would you determine what the problem is?

Solution:

If bulb 2 was open (burned out), there could be no current flow in the circuit, so bulbs 1 and 2 would be off. Bulb 2 must therefore be shorted. Using the voltmeter to investigate this further, you find that bulb 2 drops 0 V across it because it has no resistance except the very small wire resistance of the bypass, which means that 4.5 V must be dropped across each working bulb (1 and 3) in order to comply with Kirchhoff's voltage law. The loss of bulb 2's resistance causes the overall circuit resistance offered by bulbs 1 and 3 to decrease, which will cause an increase in current and so bulbs 1 and 3 should glow more brightly.

To summarize opens and shorts in series circuits, we can say that:

1. The supply voltage appears across an open component.
2. Zero volts appears across a shorted component.

SELF-TEST EVALUATION POINT FOR SECTION 3-6

Use the following questions to test your understanding of Section 3-6.

1. List the three basic problems that can occur with components in a series circuit.
2. How can an open component be detected in a series circuit?
3. True or false: If a series-connected resistor's value were to decrease, the voltage drop across that same resistor would increase.
4. How could a shorted component be detected in a series circuit?

REVIEW QUESTIONS

Multiple-Choice Questions

1. A series circuit:
   a. Is the connecting of components end to end
   b. Provides a single path for current
   c. Functions as a voltage divider
   d. All of the above

2. The total current in a series circuit is equal to:
   a. $I_1 + I_2 + I_3 + \cdots$ **c.** $I_1 = I_2 = I_3 = \cdots$
   b. $I_1 - I_2$ **d.** All of the above

3. If R_1 and R_2 are connected in series with a total current of 2 A, what will be the current flowing through R_1 and R_2, respectively?
   a. 1 A, 1 A
   b. 2 A, 1 A
   c. 2 A, 2 A
   d. All of the above could be true on some occasions.

4. The total resistance in a series circuit is equal to:
   a. The total voltage divided by the total current
   b. The sum of all the individual resistor values
   c. $R_1 + R_2 + R_3 + \cdots$
   d. All of the above
   e. None of the above are even remotely true.

5. Which of Kirchhoff's laws applies to series circuits:
   a. His voltage law
   b. His current law
   c. His power law
   d. None of them apply to series circuits, only parallel.

6. The amount of voltage dropped across a resistor is proportional to:
   a. The value of the resistor
   b. The current flow in the circuit
   c. Both (a) and (b)
   d. None of the above

7. If three resistors of 6 kΩ, 4.7 kΩ, and 330 Ω are connected in series with one another, what total resistance will the battery sense?
   a. 11.03 MΩ **c.** 6 kΩ
   b. 11.03 Ω **d.** 11.03 kΩ

8. The voltage-divider formula states that the voltage drop across a resistor or multiple resistors in a series circuit is equal to the ratio of that ____________ to the ____________ multiplied by the ____________.
   a. Resistance, source voltage, total resistance
   b. Resistance, total resistance, source voltage
   c. Total current, resistance, total voltage
   d. Total voltage, total current, resistance

9. The ____________ can be used as a variable voltage divider.
   a. Potentiometer **c.** SPDT switch
   b. Fixed resistor **d.** None of the above

10. A resistor of larger physical size will be able to dissipate ____________ heat than a small resistor.
    a. More **c.** About the same
    b. Less **d.** None of the above

11. The ____________ is the most useful tool when checking series circuits.
    a. Ammeter **c.** Voltmeter
    b. Wattmeter **d.** Both (a) and (b)

12. When an open component occurs in a series circuit, it can be noticed because:
    a. Zero volts appears across it
    b. The supply voltage appears across it
    c. 1.3 V appears across it
    d. None of the above

13. Power can be calculated by:
    a. The addition of all the individual power figures
    b. The product of the total current and the total voltage
    c. The square of the total voltage divided by the total resistance
    d. All of the above

14. A series circuit is known as a:
    a. Current divider **c.** Current subtractor
    b. Voltage divider **d.** All of the above

15. In a series circuit only ____________ path(s) exist(s) for current flow, while the voltage applied is distributed across all the individual resistors.
    a. Three **c.** Four
    b. Several **d.** One

Practice Problems

16. If three resistors of 1.7 kΩ, 3.3 kΩ, and 14.4 kΩ are connected in series with one another across a 24 V source as shown in Figure 3-32, calculate:
    a. Total resistance (R_T)
    b. Circuit current
    c. Individual voltage drops
    d. Individual and total power dissipated

17. If 40 Ω and 35 Ω resistors are connected across a 24 V source, what would be the current flow through the resistors, and what resistance would cause half the current to flow?

18. Calculate the total resistance (R_T) of the following series-connected resistors: 2.7 kΩ, 3.4 MΩ, 370 Ω, and 4.6 MΩ.

19. Calculate the value of resistors needed to divide up a 90 V source to produce 45 V and 60 V outputs, with a divider circuit current of 1 A.

20. If R_1 = 4.7 kΩ and R_2 = 6.4 kΩ and both are connected across a 9 V source, how much voltage will be dropped across R_2?

21. What current would flow through R_1 if it were one-third the ohmic value of R_2 and R_3, and all were connected in series with a total current of 6.5 mA flowing out of V_S?

22. Draw a circuit showing R_1 = 2.7 kΩ, R_2 = 3.3 kΩ, and R_3 = 0.027 MΩ in series with one another across a 20 V source. Calculate:
    a. I_T **d.** P_2 **g.** V_{R2}
    b. P_T **e.** P_3 **h.** V_{R3}
    c. P_1 **f.** V_{R1} **i.** I_{R1}

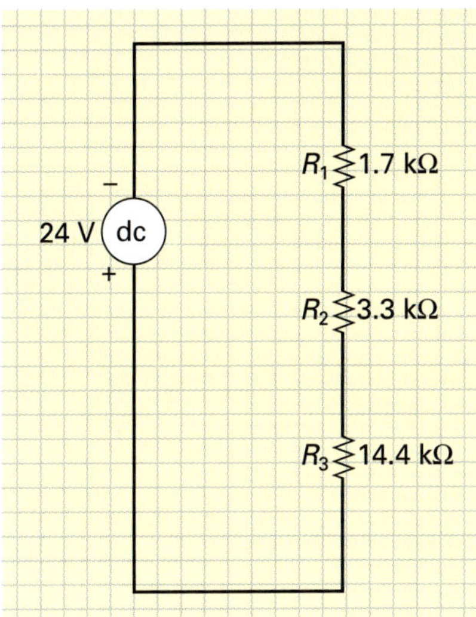

FIGURE 3-32

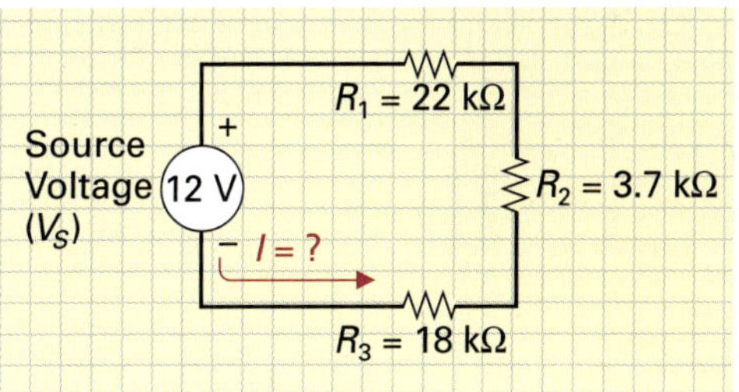

FIGURE 3-33

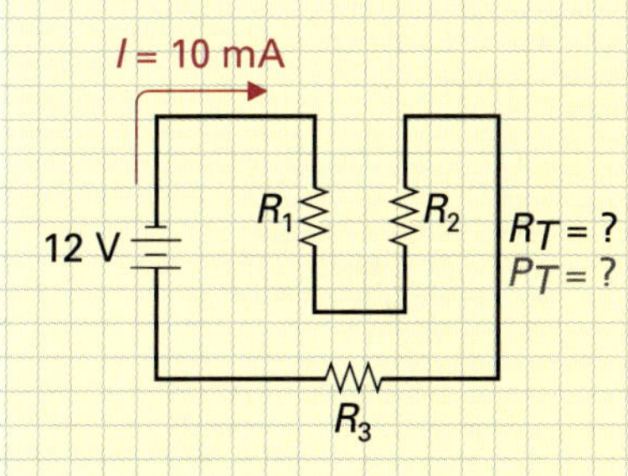

FIGURE 3-34

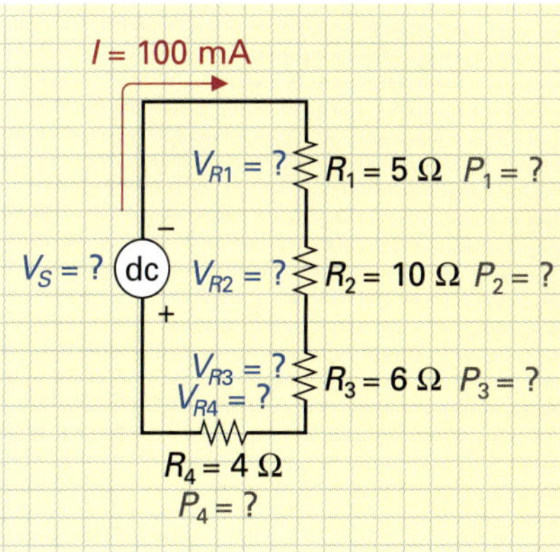

FIGURE 3-35

23. Calculate the current flowing through three light bulbs that are dissipating 120 W, 60 W, and 200 W when they are connected in series across a 120 V source. How is the voltage divided around the series circuit?

24. If three equal-value resistors are series connected across a dc power supply adjusted to supply 10 V, what percentage of the source voltage will appear across R_1?

25. Refer to the following figures and calculate:

a. I (Figure 3-33)

b. R_T and P_T (Figure 3-34)

c. V_S, V_{R1}, V_{R2}, V_{R3}, V_{R4}, P_1, P_2, P_3, and P_4 (Figure 3-35)

d. P_T, I, R_1, R_2, R_3, and R_4 (Figure 3-36)

Troubleshooting Questions

26. If three bulbs are connected across a 9 V battery in series, and the filament in one of the bulbs burned out, causing an open in the bulb, would the other lamps be on? Explain why.

27. Using a voltmeter, how would a short be recognized in a series circuit?

28. If one of three series-connected bulbs is shorted, will the other two bulbs be on? Explain why.

29. When one resistor in a series string is open, explain what would happen to the circuit's:

a. Current

b. Resistance

c. Voltage across the open component

d. Voltage across the other components

30. When one resistor in a series string is shorted, explain what would happen to the circuit's:

a. Current

b. Resistance

c. Voltage across the shorted component

d. Voltage across the other components

Web Site Questions

Go to the web site http://www.prenhall.com/cook, select the textbook *Electronics: A Complete Course,* select this chapter, and then follow the instructions when answering the multiple choice practice problems.

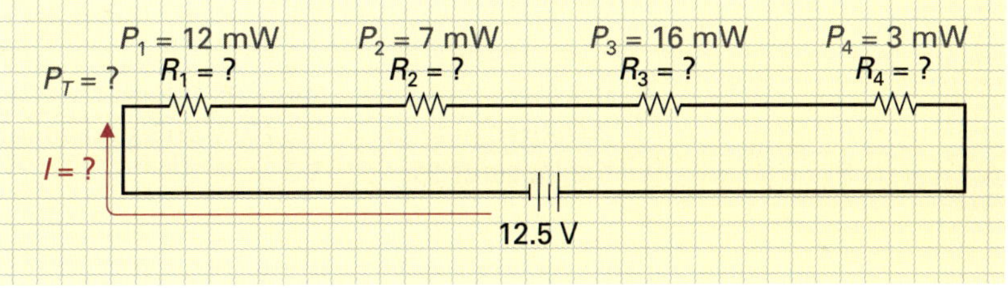

FIGURE 3-36

Parallel Circuits

An Apple a Day

One of the early pioneers who laid the foundations of many branches of science was Isaac Newton. He was born in a small farmhouse near Woolsthorpe in Lincolnshire, England, on Christmas Day in 1642. He was an extremely small, premature baby, which worried the midwives who went off to get medicine and didn't expect to find him alive when they came back. Luckily for science, however, he did survive.

Newton's father was an illiterate farmer who died three months before he was born. His mother married the local vicar soon after Newton's birth and left him in the care of his grandmother. This parental absence while he was growing up had a traumatic effect on him and throughout his life affected his relationships with people.

At school in the nearby town of Grantham, Newton showed no interest in classical studies but rather in making working models and studying the world around him. When he was in his early teens, his stepfather died and Newton had to return to the farm to help his mother. Newton proved to be a hopeless farmer; in fact, on one occasion when he was tending sheep he became so engrossed with a stream that he followed it for miles and was missing for hours. Luckily, a schoolteacher recognized Newton's single-minded powers of concentration and convinced his mother to let him return to school, where he performed better and later went off to Cambridge University.

In 1665, in his graduation year, Newton left Cambridge to return home to escape an epidemic of bubonic plague that had spread throughout London. During this time, Newton reflected on his years of seclusion at his mother's cottage and called them the most significant time in his life. It was here on a warm summer's day that Newton saw the apple fall to the ground, leading him to develop his laws of motion and gravitation. It was here that he wondered about the nature of light and later built a prism and proved that white light contains all the colors in a rainbow.

Later, when Newton returned to Cambridge, he demonstrated many of his discoveries but was reluctant to publish the details and did so finally only at the insistence of others. Newton went on to build the first working reflecting astronomical telescope and wrote a paper on optics that was fiercely challenged by the physicist Robert Hooke. Hooke quarreled bitterly with Newton over the years, and there were also heated debates about whether Newton or the German mathematician Gottfried Leibniz invented calculus.

The truth is that many of Newton's discoveries roamed around with him in the English countryside, and even though many of these would, before long, have been put forward by others, it was Newton's genius and skill (and long walks) that tied together all the loose ends.

4

Introduction

By tracing the path of current, we can determine whether a circuit has series-connected or parallel-connected components. In a series circuit, there is only one path for current, whereas in parallel circuits the current has two or more paths. These paths are known as *branches.* A parallel circuit, by definition, is when two or more components are connected to the same voltage source so that the current can branch out over two or more paths. In a parallel resistive circuit, two or more resistors are connected to the same source, so current splits to travel through each separate branch resistance.

4-1 COMPONENTS IN PARALLEL

Parallel Circuit
Also called shunt; circuit having two or more paths for current flow.

Many components, other than resistors, can be connected in parallel, and a **parallel circuit** can easily be identified because current is split into two or more paths. Being able to identify a parallel connection requires some practice, because they can come in many different shapes and sizes. The means for recognizing series circuits is that if you can place your pencil at the negative terminal of the voltage source (battery) and follow the wire connections through components to the positive side of the battery and only have one path to follow, the circuit is connected in series. If, however, you can place your pencil at the negative terminal of the voltage source and follow the wire and at some point have a choice of two or more routes, the circuit is connected with two or more parallel branches. The number of routes determines the number of parallel branches. Figure 4-1 illustrates five examples of parallel resistive circuits.

EXAMPLE:

Figure 4-2(a) illustrates four resistors laid out on a table top.

a. With wire leads, connect all four resistors in parallel on a protoboard, and then connect the circuit to a dc power supply.
b. Draw the schematic diagram of the parallel-connected circuit.

Solution:

Figure 4-2(b) shows how to connect the resistors in parallel and the circuit's schematic.

EXAMPLE:

Figure 4-3(a) shows four 1.5 V cells and three lamps. Using wires, connect all of the cells in parallel to create a 1.5 V source. Then connect all of the three lamps in parallel with one another and finally connect the 1.5 V source across the three parallel-connected lamp load.

Solution:

In Figure 4-3(b) you can see the final circuit containing a source consisting of four parallel-connected 1.5 V cells, and a load consisting of three parallel-connected lamps. When cells are connected in parallel, the total voltage remains the same as for one cell, but the current demand can now be shared.

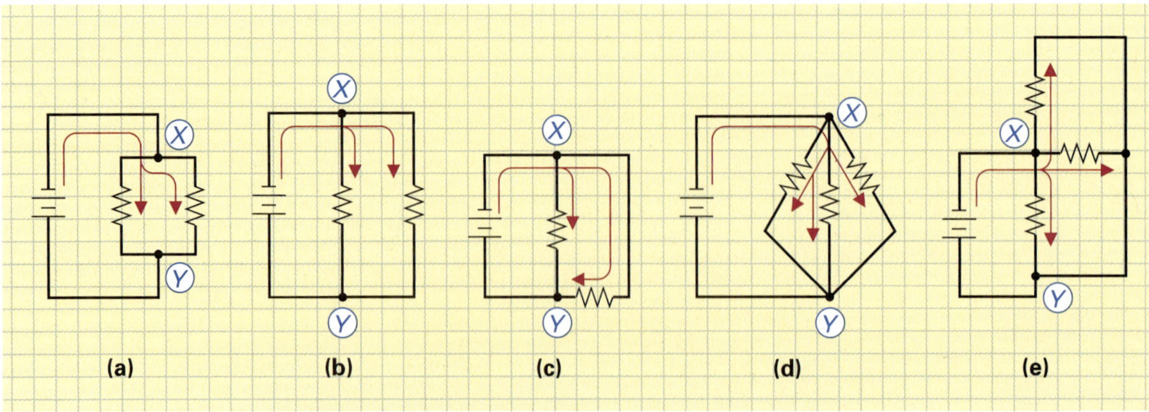

FIGURE 4-1 **Parallel Circuits.**

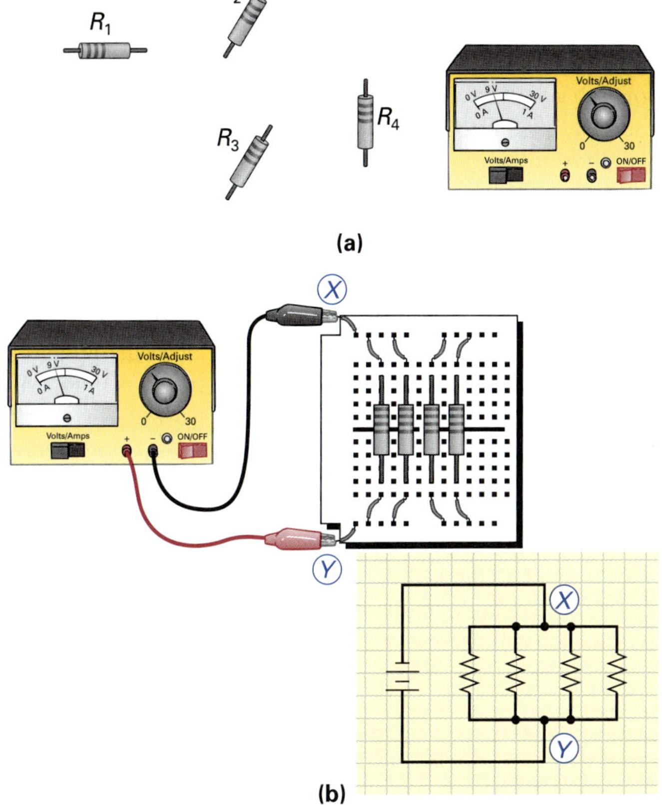

FIGURE 4-2 Connecting Resistors in Parallel. (a) Problem. (b) Solution.

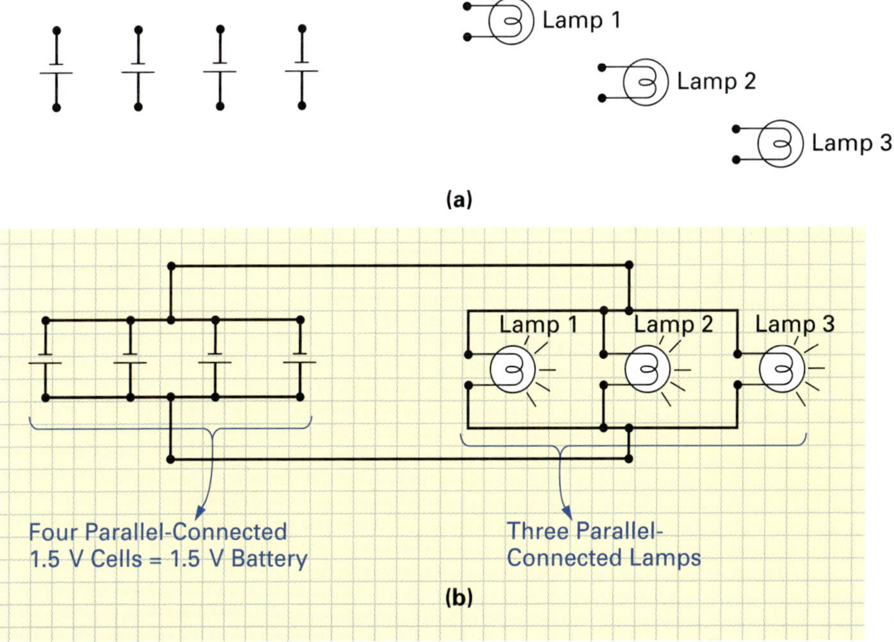

FIGURE 4-3 Parallel-Connected Cells and Lamps.

4-2 VOLTAGE IN A PARALLEL CIRCUIT

Figure 4-4(a) shows a simple circuit with four resistors connected in parallel across the voltage source of a 9 V battery. The current from the negative side of the battery will split between the four different paths or branches, yet the voltage drop across each branch of a parallel circuit is equal to the voltage drop across all the other branches in parallel. This means that if the voltmeter were to measure the voltage across *A* and *B* or *C* and *D* or *E* and *F* or *G* and *H,* they would all be the same or, in this example, would all drop 9 V.

It is quite easy to imagine why there will be the same voltage drop across all the resistors, seeing that points *A, C, E,* and *G* are all one connection and points *B, D, F,* and *H* are all one connection. Measuring the voltage drop with the voltmeter across any of the resistors is the same as measuring the voltage across the battery, as shown in Figure 4-4(b). As long as the voltage source remains constant, the voltage drop will always be common (9 V) across the parallel resistors, no matter what value or how many resistors are connected in parallel. The voltmeter is therefore measuring the voltage between two common points that are directly connected to the battery, and the voltage dropped across all these parallel resistors will be equal to the source voltage.

In Figure 4-5(a) and (b), the same circuit is shown in two different ways, so you can see how the same circuit can look completely different. In both examples, the voltage drop across any of the resistors will always be the same and, as long as the voltage source is not heavily loaded, equal to the source voltage. Just as you can trace the positive side of the battery to all four resistors, you can also trace the negative side to all four resistors.

Mathematically stated, we can say that in a parallel circuit

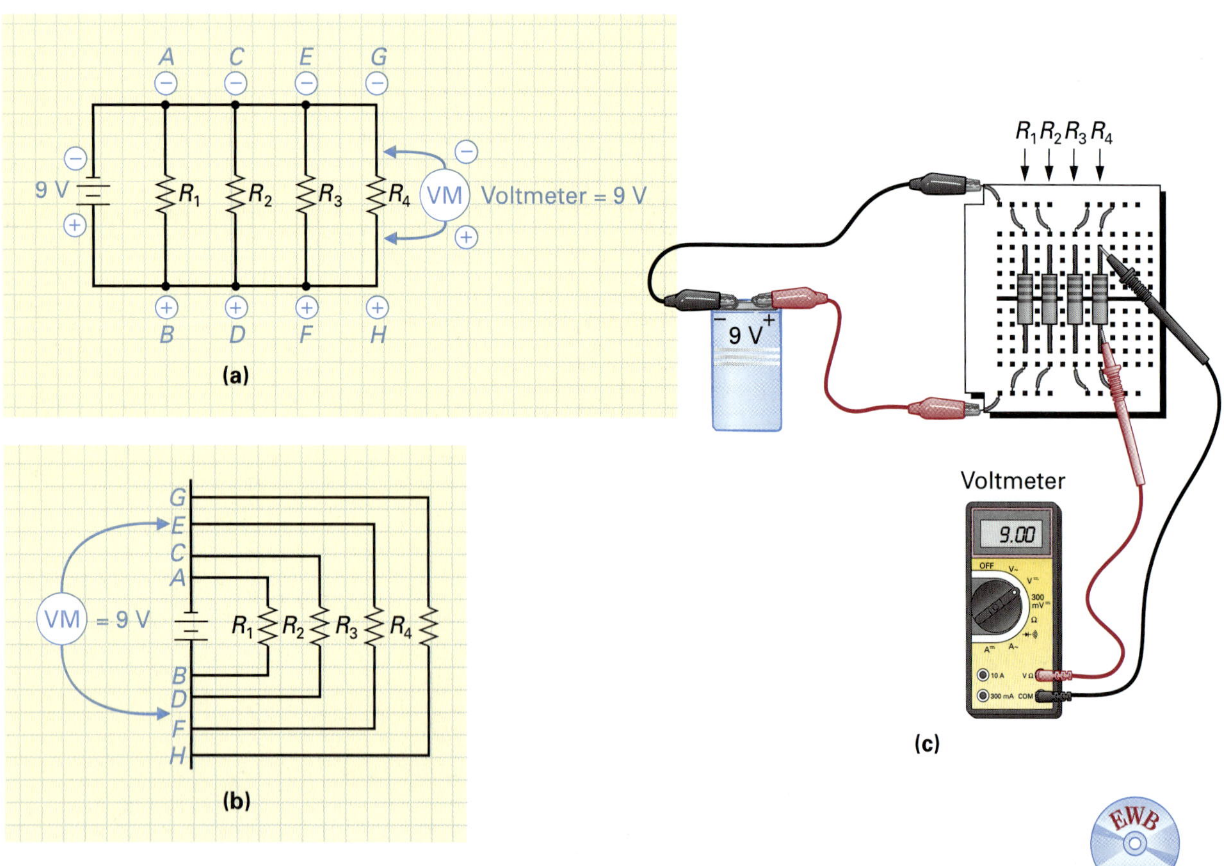

FIGURE 4-4 **Voltage in a Parallel Circuit.**

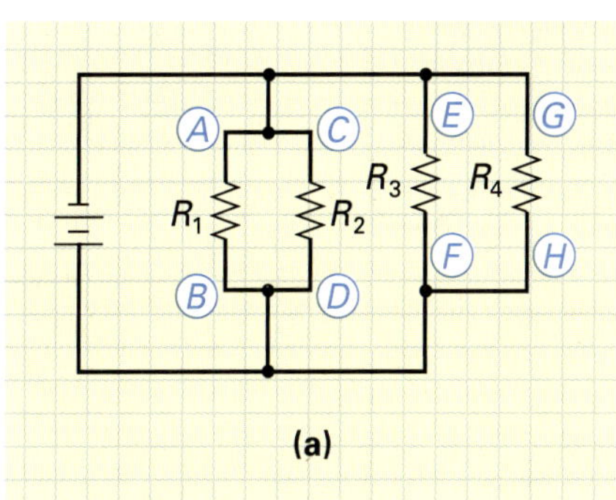

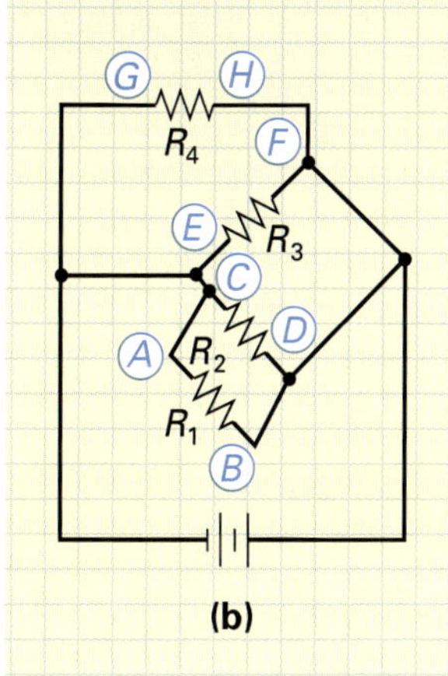

FIGURE 4-5 Parallel-Circuit Voltage Drop.

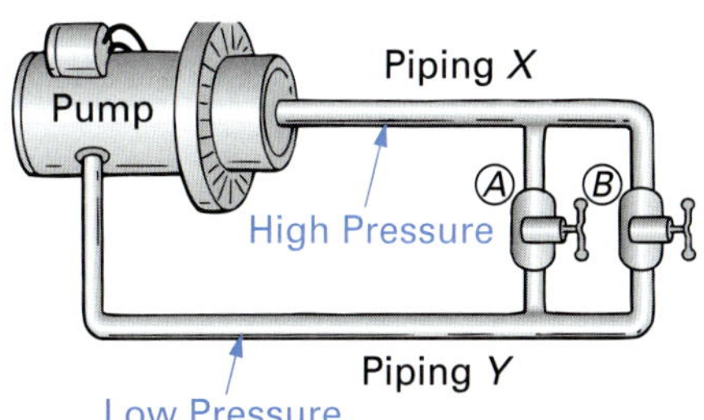

FIGURE 4-6 Fluid Analogy of Parallel-Circuit Pressure.

$$V_{B1} = V_{B2} = V_{B3} = V_{B4} = V_S$$

voltage drop across branch 1 = voltage drop across branch 2 = voltage drop across branch 3 (etc.) = source voltage

To reinforce the concept, let us compare this parallel circuit characteristic to a water analogy, as seen in Figure 4-6. The pressure across valves *A* and *B* will always be the same, even if one offers more opposition than the other. This is because the pressure measured across either valve will be the same as checking the pressure difference between piping *X* and *Y*. Since the piping at points *X* and *Y* runs directly back to the pump, the pressure across *A* and *B* is the same as the pressure difference across the pump.

EXAMPLE:

Refer to Figure 4-7 and calculate:

a. Voltage drop across R_1
b. Voltage drop across R_2
c. Voltage drop across R_3

Solution:

Since all these resistors are connected in parallel, the voltage across every branch will be the same and equal to the source voltage applied. Therefore,

$$V_{R1} = V_{R2} = V_{R3} = V_S$$
$$7.5 \text{ V} = 7.5 \text{ V} = 7.5 \text{ V} = 7.5 \text{ V}$$

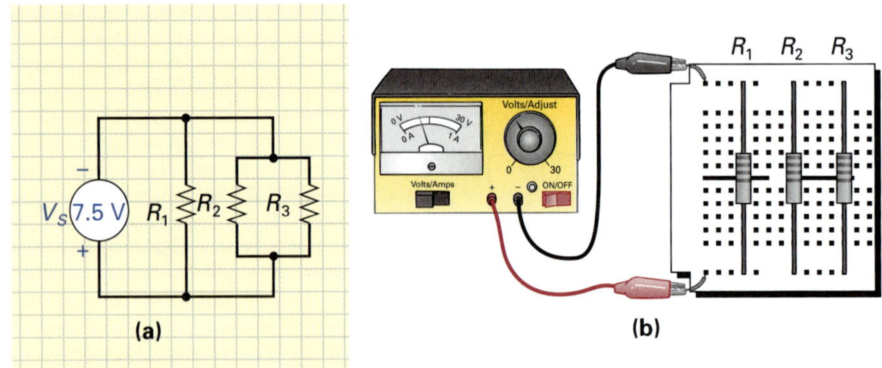

FIGURE 4-7 **A Parallel Circuit Example (a) Schematic. (b) Protoboard Circuit.**

SELF-TEST EVALUATION POINT FOR SECTIONS 4-1 AND 4-2

Use the following questions to test your understanding of Sections 4-1 and 4-2.

1. Describe a parallel circuit.
2. True or false: A parallel circuit is also known as a voltage-divider circuit.
3. What would be the voltage drop across R_1 if $V_S = 12$ V and R_1 and R_2 are both in parallel with one another and equal to 24 Ω each?
4. Can Kirchhoff's voltage law be applied to parallel circuits?

4-3 CURRENT IN A PARALLEL CIRCUIT

Branch Current
A portion of the total current that is present in one path of a parallel circuit.

Kirchhoff's Current Law
The sum of the currents flowing into a point in a circuit is equal to the sum of the currents flowing out of that same point.

In addition to providing the voltage law for series circuits, Gustav Kirchhoff (in 1847) was the first to observe and prove that the sum of all the **branch currents** in a parallel circuit ($I_1 + I_2 + I_3$, etc.) was equal to the total current (I_T). In honor of his second discovery, this phenomenon is known as **Kirchhoff's current law,** which states that the sum of all the currents entering a junction is equal to the sum of all the currents leaving that same junction.

Figure 4-8(a) and (b) illustrate two examples of how this law applies. In both examples, the sum of the currents entering a junction is equal to the sum of the currents leaving that same junction. In Figure 4-8(a) the total current arrives at a junction X and splits to pro-

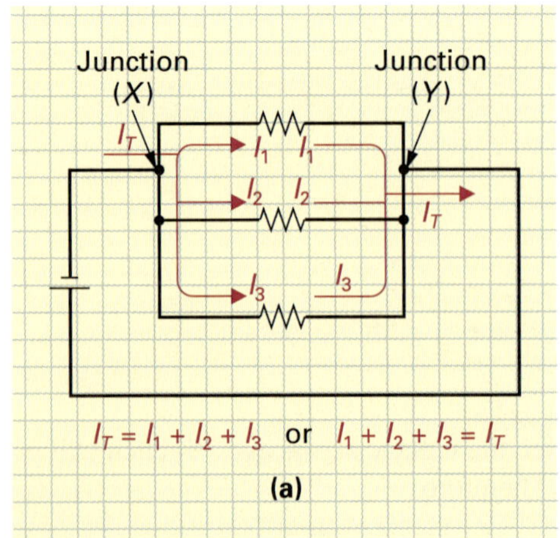

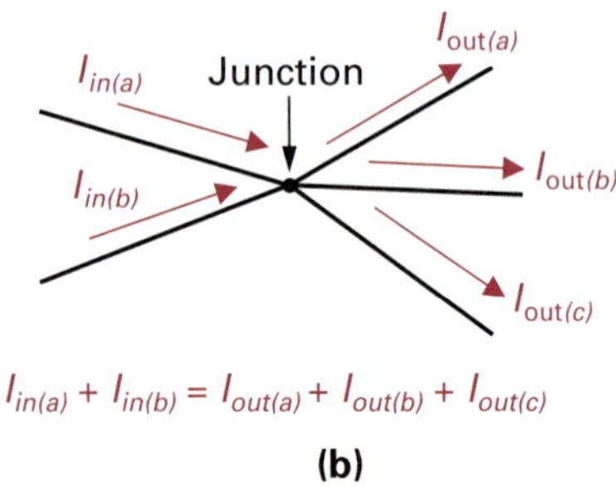

FIGURE 4-8 **Kirchhoff's Current Law.**

duce three branch currents, I_1, I_2, and I_3, which cumulatively equal the total current (I_T) that arrived at the junction *X*. The same three branch currents combine at junction *Y*, and the total current (I_T) leaving that junction is equal to the sum of the three branch currents arriving at junction *Y*. Stated mathematically:

$$I_T = I_1 + I_2 + I_3 + I_4 \cdots$$

As another example, in Figure 4-8(b) you can see that there are two branch currents entering a junction [$I_{in(a)}$ and $I_{in(b)}$] and three branch currents leaving that same junction [$I_{out(a)}$, $I_{out(b)}$, and $I_{out(c)}$]. As stated below the illustration, the sum of the input currents will equal the sum of the output currents:

$$I_{in(a)} + I_{in(b)} = I_{out(a)} + I_{out(b)} + I_{out(c)}.$$

EXAMPLE:

Refer to Figure 4-9 and calculate the value of I_T.

Solution:

By Kirchhoff's current law,

$$I_T = I_1 + I_2 + I_3 + I_4$$
$$? = 2 \text{ mA} + 17 \text{ mA} + 7 \text{ mA} + 37 \text{ mA}$$
$$I_T = 63 \text{ mA}$$

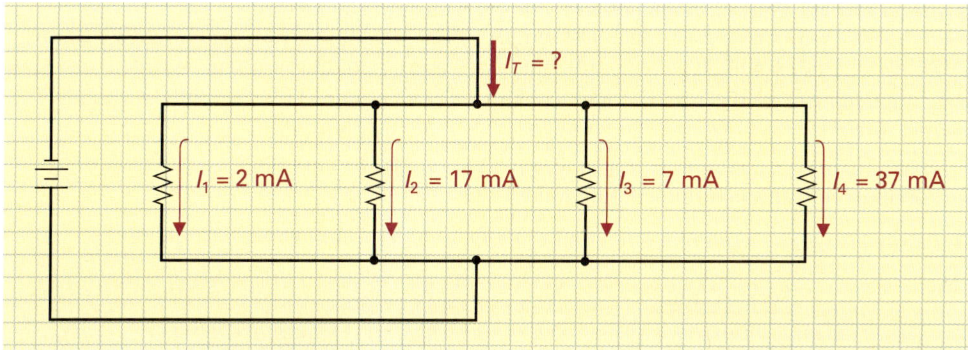

FIGURE 4-9 Calculating Total Current.

EXAMPLE:

Refer to Figure 4-10 and calculate the value of I_1.

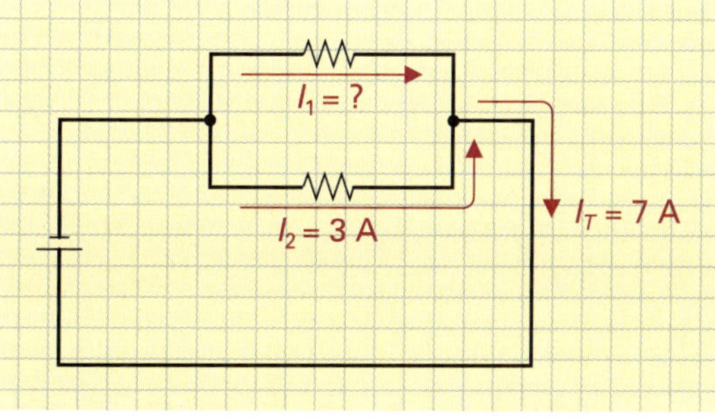

FIGURE 4-10 Calculating Branch Current.

■ *Solution:*

By transposing Kirchhoff's current law, we can determine the unknown value (I_1):

$$I_T = I_1 + I_2 \qquad I_T = I_1 + I_2 \qquad (-I_2)$$

$$7\text{ A} = ? + 3\text{ A} \quad \text{or} \quad I_T - I_2 = I_1$$

$$I_1 = 4\text{ A} \qquad I_1 = I_T - I_2 = 7\text{ A} - 3\text{ A} = 4\text{ A}$$

As with series circuits, to find out how much current will flow through a parallel circuit, we need to find out how much opposition or resistance is being connected across the voltage source.

$$I_T = \frac{V_S}{R_T}$$

Total current equals source voltage divided by total resistance.

When we connect resistors in parallel, the total resistance in the circuit will actually decrease. In fact, the total resistance in a parallel circuit will always be less than the value of the smallest resistor in the circuit.

To prove this point, Figure 4-11 shows how two sets of identical resistors (R_1, R_2, and R_3) were used to build both a series and a parallel circuit. The total current flow in the parallel circuit would be larger than the total current in the series circuit, because the parallel circuit has two or more paths for current to flow, while the series circuit only has one.

To explain why the total current will be larger in a parallel circuit, let us take the analogy of a freeway with only one path for traffic to flow. A single-lane freeway is equivalent to a series circuit, and only a small amount of traffic is allowed to flow along this freeway. If the freeway is expanded to accommodate two lanes, a greater amount of traffic can flow along the freeway in the same amount of time. Having more lanes permits a greater total amount of traffic flow. In parallel circuits, more branches will allow a greater total amount of current flow because there is less resistance in more paths than there is with only one path. This concept is summarized in Figure 4-11.

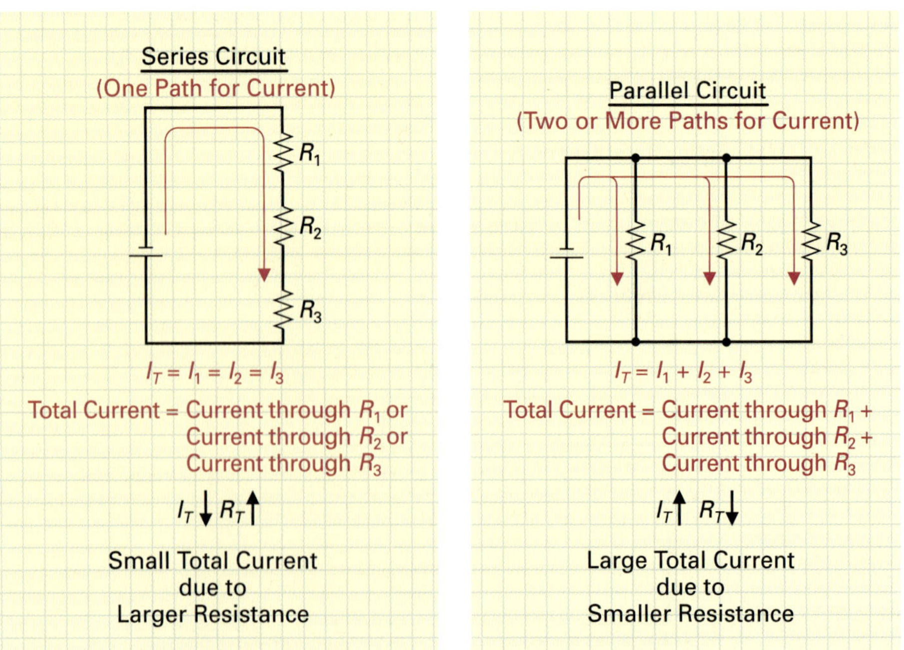

FIGURE 4-11 **Series Circuit and Parallel Circuit Current Comparison.**

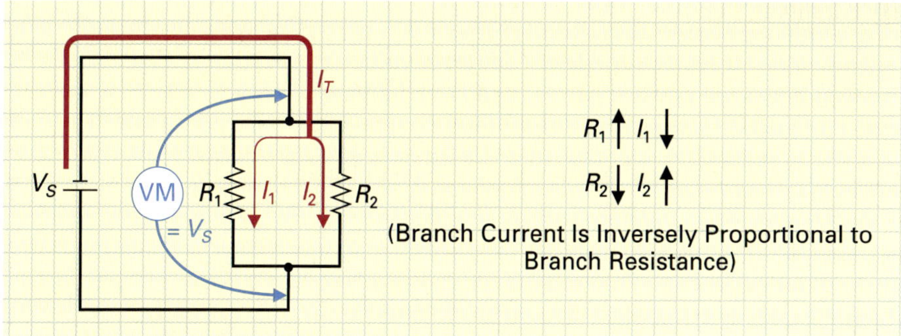

FIGURE 4-12 **The Parallel Circuit Current Divider.**

Just as a series circuit is often referred to as a voltage-divider circuit, a parallel circuit is often referred to as a **current-divider** circuit, because the total current arriving at a junction will divide or split into branch currents (Kirchhoff's law), as shown in Figure 4-12.

The current division is inversely proportional to the resistance in the branch, assuming that the voltage across both resistors is constant and equal to the source voltage (V_S). This means that a large branch resistance will cause a small branch current ($I\downarrow = V/R\uparrow$), and a small branch resistance will cause a large branch current ($I\uparrow = V/R\downarrow$).

Current Divider
A parallel network designed to proportionally divide the circuit's total current.

EXAMPLE:

Calculate the following for Figure 4-13(a), and then insert the values in a circuit analysis table.

a. I_1
b. I_2
c. I_T

Solution:

Since R_1 and R_2 are connected in parallel across the 10 V source, the voltage across both resistors will be 10 V.

a. $I_1 = \dfrac{V_{R1}}{R_1} = \dfrac{10\text{ V}}{6\text{ k}\Omega} = 1.6\text{ mA}$ (smaller branch current through larger branch resistance)

b. $I_2 = \dfrac{V_{R2}}{R_2} = \dfrac{10\text{ V}}{3\text{ k}\Omega} = 3.3\text{ mA}$ (larger branch current through smaller branch resistance)

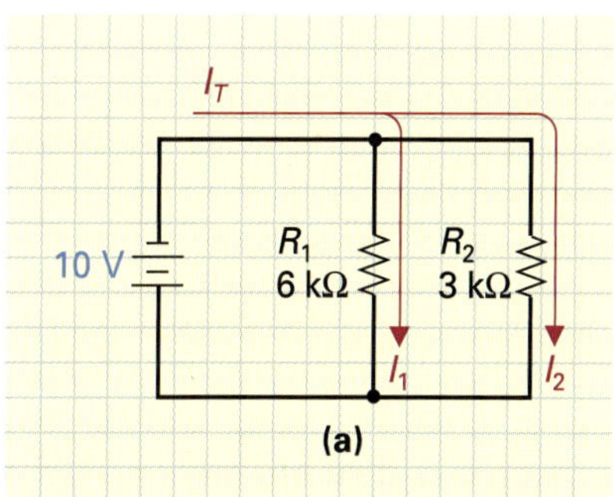

CIRCUIT ANALYSIS TABLE

Resistance $R = V/I$	Voltage $V = I \times R$	Current $I = V/R$	Power $P = V \times I$
$R_1 = 6\text{ k}\Omega$	$V_{R1} = 10\text{ V}$	$I_1 = 1.6\text{ mA}$	
$R_2 = 3\text{ k}\Omega$	$V_{R2} = 10\text{ V}$	$I_2 = 3.3\text{ mA}$	

$V_T = V_S = 10\text{ V}$ $I_T = 4.9\text{ mA}$

(b)

FIGURE 4-13 **Parallel Circuit Example. (a) Schematic. (b) Circuit Analysis Table.**

c. By Kirchhoff's current law,

$$I_T = I_1 + I_2$$
$$= 1.6 \text{ mA} + 3.3 \text{ mA}$$
$$= 4.9 \text{ mA}$$

The circuit analysis table for this example is shown in Figure 4-13(b).

By rearranging Ohm's law, we can arrive at another formula, which is called the *current-divider formula* and can be used to calculate the current through any branch of a multiple-branch parallel circuit.

$$I_x = \frac{R_T}{R_x} \times I_T$$

where I_x = branch current desired
R_T = total resistance
R_x = resistance in branch
I_T = total current

EXAMPLE:

Refer to Figure 4-14 and calculate the following if the total circuit resistance (R_T) is equal to 1 kΩ:

a. $I_1 =$
b. $I_2 =$
c. $I_3 =$

Solution:

Since the source and therefore the voltage across each branch resistor are not known, we will use the current-divider formula to calculate I_1, I_2, and I_3.

a. $I_1 = \frac{R_T}{R_1} \times I_T = \frac{1 \text{ k}\Omega}{2 \text{ k}\Omega} \times 10 \text{ mA} = 5 \text{ mA}$

(smallest branch resistance has largest branch current)

b. $I_2 = \frac{R_T}{R_2} \times I_T = \frac{1 \text{ k}\Omega}{3 \text{ k}\Omega} \times 10 \text{ mA}$
$= 3.33 \text{ mA}$

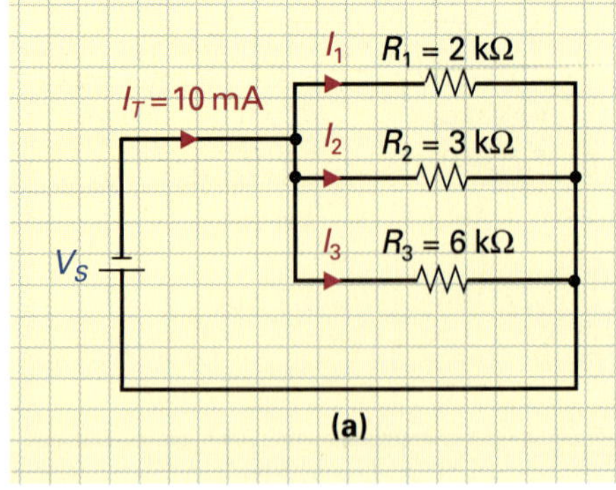

CIRCUIT ANALYSIS TABLE

Resistance $R = V/I$	Voltage $V = I \times R$	Current $I = V/R$	Power $P = V \times I$
R_1 = 2 kΩ		I_1 = 5 mA	
R_2 = 3 kΩ		I_2 = 3.33 mA	
R_3 = 6 kΩ		I_3 = 1.67 mA	
R_T = 1 kΩ		I_T = 10 mA	

(b)

FIGURE 4-14 Parallel Circuit Example. (a) Schematic. (b) Circuit Analysis Table.

c. $I_3 = \frac{R_T}{R_3} \times I_T = \frac{1\ \text{k}\Omega}{6\ \text{k}\Omega} \times 10\ \text{mA}$
$= 1.67\ \text{mA}$

(largest branch resistance has smallest branch current)

To double-check that the values for I_1, I_2, and I_3 are correct, you can apply Kirchhoff's current law, which is

$$I_T = I_1 + I_2 + I_3$$
$$10\ \text{mA} = 5\ \text{mA} + 3.33\ \text{mA} + 1.67\ \text{mA}$$
$$= 10\ \text{mA}$$

CALCULATOR SEQUENCE

Step	Keypad Entry	Display Response
1.	1 E 3	1. 3
2.	÷	
3.	2 E 3	2. 3
4.	×	0.5
5.	1 0 E 3 +/−	10 -3
6.	=	5.-03

EXAMPLE:

A common use of parallel circuits is in the residential electrical system. All of the household lights and appliances are wired in parallel, as seen in the typical room wiring circuit in Figure 4-15(a). If it is a cold winter morning, and lamps 1 and 2 are switched ON, together with the space heater and hair dryer, what will the individual branch currents be, and what will be the total current drawn from the source?

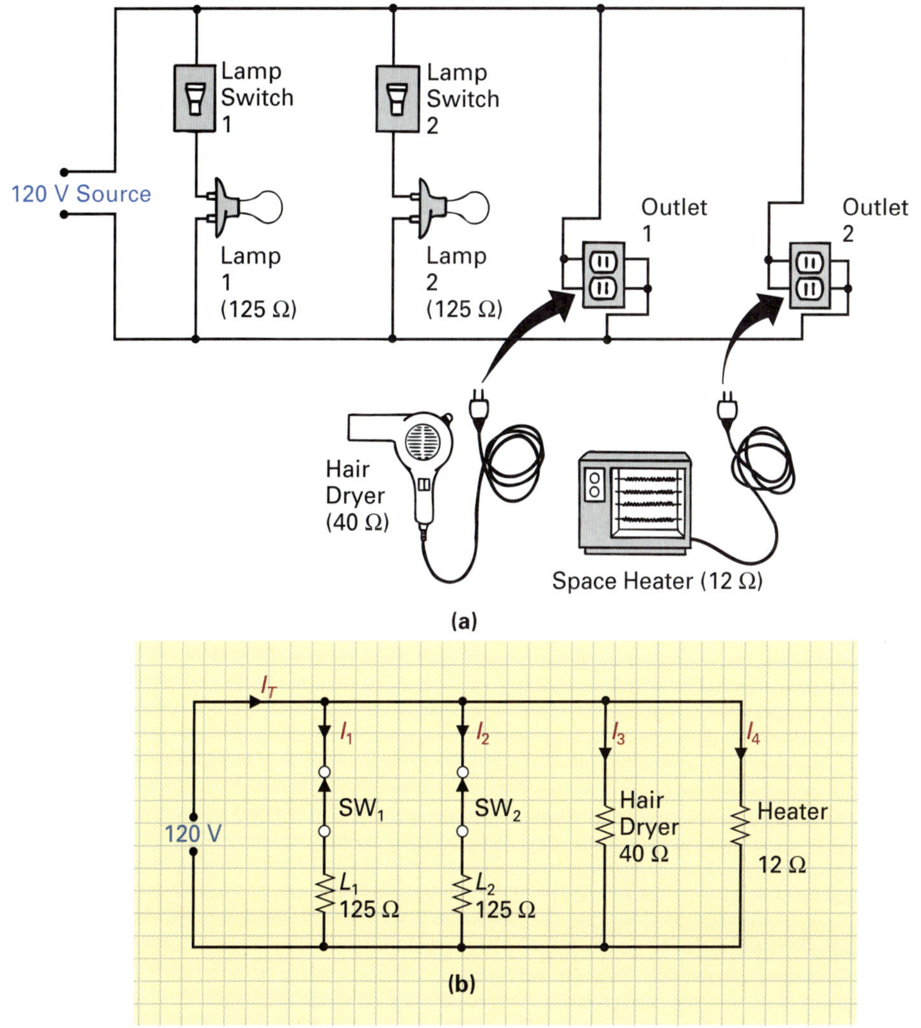

FIGURE 4-15 **Parallel Home Electrical System.**

■ *Solution:*

Figure 4-15(b) shows the schematic of the pictorial in Figure 4-15(a). Since all resistances are connected in parallel across a 120 V source, the voltage across all devices will be 120 V. Using Ohm's law we can calculate the four branch currents:

$$I_1 = \frac{V_{\text{lamp1}}}{R_{\text{lamp1}}} = \frac{120\text{ V}}{125\ \Omega} = 960\text{ mA}$$

$$I_2 = \frac{V_{\text{lamp2}}}{R_{\text{lamp2}}} = \frac{120\text{ V}}{125\ \Omega} = 960\text{ mA}$$

$$I_3 = \frac{V_{\text{hairdryer}}}{R_{\text{hairdryer}}} = \frac{120\text{ V}}{40\ \Omega} = 3\text{ A}$$

$$I_4 = \frac{V_{\text{heater}}}{R_{\text{heater}}} = \frac{120\text{ V}}{12\ \Omega} = 10\text{ A}$$

By Kirchhoff's current law,

$$\begin{aligned} I_T &= I_1 + I_2 + I_3 + I_4 \\ &= 960\text{ mA} + 960\text{ mA} + 3\text{ A} + 10\text{ A} \\ &= 14.92\text{ A} \end{aligned}$$

SELF-TEST EVALUATION POINT FOR SECTION 4-3

Use the following questions to test your understanding of Section 4-3.

1. State Kirchhoff's current law.
2. If $I_T = 4$ A and $I_1 = 2.7$ A in a two-resistor parallel circuit, what would be the value of I_2?
3. State the current-divider formula.
4. Calculate I_1 if $R_T = 1$ kΩ, $R_1 = 2$ kΩ, and $V_T = 12$ V.

4-4 RESISTANCE IN A PARALLEL CIRCUIT

We now know that parallel circuits will have a larger current flow than a series circuit containing the same resistors due to the smaller total resistance. To calculate exactly how much total current will flow, we need to be able to calculate the total resistance that the parallel circuit presents to the source.

The ability of a circuit to conduct current is a measure of that circuit's conductance, and you will remember from Chapter 1 that conductance (G) is equal to the reciprocal of resistance and is measured in siemens.

$$G = \frac{1}{R} \quad \text{(siemens)}$$

Every resistor in a parallel circuit will have a conductance figure that is equal to the reciprocal of its resistance, and the total conductance (G_T) of the circuit will be equal to the sum of all the individual resistor conductances. Therefore,

$$G_T = G_{R1} + G_{R2} + G_{R3} + \cdots$$

Total conductance is equal to the conductance of R_1 + the conductance of R_2 + the conductance of R_3 + · · ·

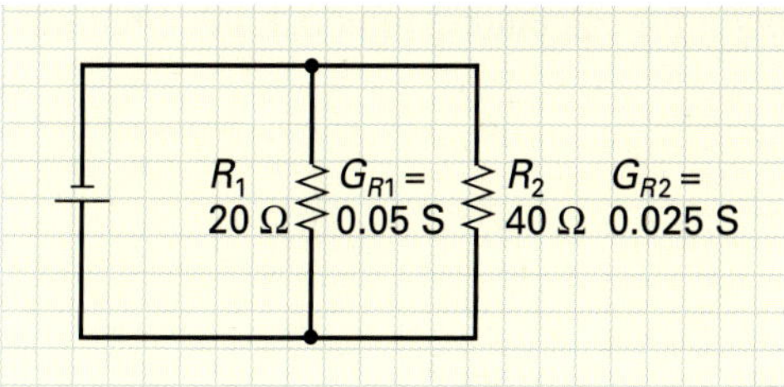

FIGURE 4-16 Parallel Circuit Conductance and Resistance.

Once you have calculated total conductance, the reciprocal of this figure will give you total resistance. If, for example, we have two resistors in parallel, as shown in Figure 4-16, the conductance for R_1 will equal

$$G_{R1} = \frac{1}{R_1} = \frac{1}{20\ \Omega} = 0.05\ \text{S}$$

The conductance for R_2 will equal

$$G_{R2} = \frac{1}{R_2} = \frac{1}{40\ \Omega} = 0.025\ \text{S}$$

The total conductance will therefore equal

$$\begin{aligned} G_{\text{total}} &= G_{R1} + G_{R2} \\ &= 0.05 + 0.025 \\ &= 0.075\ \text{S} \end{aligned}$$

Since total resistance is equal to the reciprocal of total conductance, total resistance for the parallel circuit in Figure 4-16 will be

$$R_{\text{total}} = \frac{1}{G_{\text{total}}} = \frac{1}{0.075\ \text{S}} = 13.3\ \Omega$$

Combining these three steps (first calculate individual conductances, total conductance and then total resistance) we can arrive at the following *reciprocal formula:*

$$R_{\text{total}} = \frac{1}{(1/R_1) + (1/R_2)}$$

This formula states that the conductance of R_1 (G_{R1})
+ conductance of R_2 (G_{R2})
= total conductance (G_T),
and the reciprocal of total conductance is equal to total resistance

In the example for Figure 4-16, this combined general formula for total resistance can be verified by plugging in the example values.

$$\begin{aligned} R_T &= \frac{1}{(1/R_1) + (1/R_2)} \\ &= \frac{1}{(1/20) + (1/40)} \\ &= \frac{1}{0.05 + 0.025} \\ &= \frac{1}{0.075} \\ &= 13.3\ \Omega \end{aligned}$$

CALCULATOR SEQUENCE

Step	Keypad Entry	Display Response
1.	(Clear Memory)	
2.	[2] [0]	20.
3.	[1/x]	5.E-2
4.	[+]	5.E-2
5.	[4] [0]	40
6.	[1/x]	2.5E-2
7.	[=]	7.5E-2
8.	[1/x]	13.33333

The *reciprocal formula* for calculating total parallel circuit resistance for any number of resistors is

$$R_T = \frac{1}{(1/R_1) + (1/R_2) + (1/R_3) + (1/R_4) + \cdots}$$

EXAMPLE:

Referring to Figure 4-17(a), calculate:

a. Total resistance
b. Voltage drop across R_2
c. Voltage drop across R_3

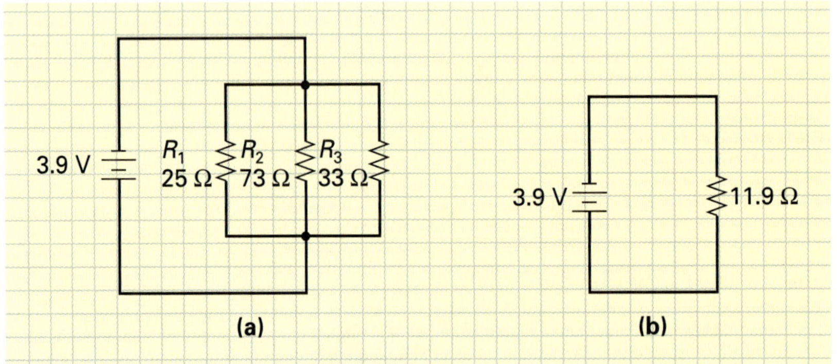

FIGURE 4-17 Parallel Circuit Example. (a) Schematic. (b) Equivalent Circuit.

Solution:

Total resistance can be calculated using the reciprocal formula.

$$\begin{aligned} R_T &= \frac{1}{(1/R_1) + (1/R_2) + (1/R_3)} \\ &= \frac{1}{(1/25\ \Omega) + (1/73\ \Omega) + (1/33\ \Omega)} \\ &= \frac{1}{0.04 + 0.014 + 0.03} \\ &= 11.9\ \Omega \end{aligned}$$

With parallel resistance circuits, the total resistance is always smaller than the smallest branch resistance. In this example the total opposition of this circuit is equivalent to 11.9 Ω, as shown in Figure 4-17(b).

With parallel resistive circuits, the voltage drop across any branch is equal to the voltage drop across each of the other branches and is equal to the source voltage, in this example 3.9 V.

4-4-1 *Two Resistors in Parallel*

If only two resistors are connected in parallel, a quick and easy formula called the *product-over-sum formula* can be used to calculate total resistance.

$$R_T = \frac{R_1 \times R_2}{R_1 + R_2}$$

$$\text{total resistance} = \frac{\text{product of both resistance values}}{\text{sum of both resistance values}}$$

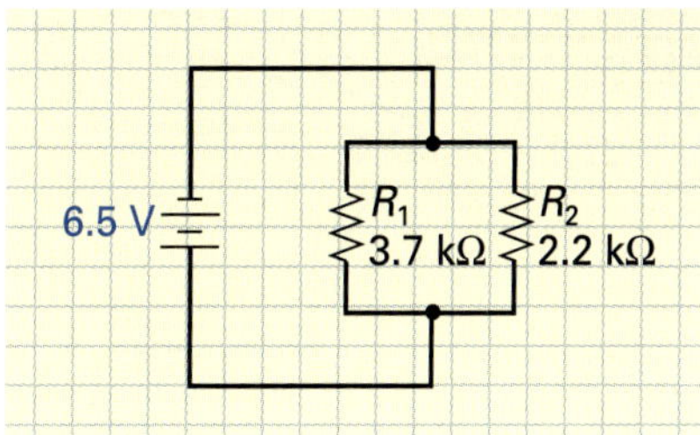

FIGURE 4-18 Two Resistors in Parallel.

Using the example shown in Figure 4-18, let us compare the *product-over-sum* formula with the *reciprocal* formula.

(A) PRODUCT-OVER-SUM FORMULA

$$R_T = \frac{R_1 \times R_2}{R_1 + R_2}$$

$$= \frac{3.7\ \text{k}\Omega \times 2.2\ \text{k}\Omega}{3.7\ \text{k}\Omega + 2.2\ \text{k}\Omega}$$

$$= \frac{8.14\ \text{k}\Omega^2}{5.9\ \text{k}\Omega}$$

$$= 1.38\ \text{k}\Omega$$

(B) RECIPROCAL FORMULA

$$R_T = \frac{1}{(1/R_1) + (1/R_2)}$$

$$= \frac{1}{(1/3.7\ \text{k}\Omega) + (1/2.2\ \text{k}\Omega)}$$

$$= \frac{1}{(270.2 \times 10^{-6}) + (454.5 \times 10^{-6})}$$

$$= \frac{1}{724.7 \times 10^{-6}}$$

$$= 1.38\ \text{k}\Omega$$

As you can see from this example, the advantage of the product-over-sum parallel resistance formula (a) is its ease of use. Its disadvantage is that it can only be used for two resistors in parallel. The rule to adopt, therefore, is that if a circuit has two resistors in parallel use the *product-over-sum* formula, and in circuits containing more than two resistors, use the *reciprocal* formula.

4-4-2 *Equal-Value Resistors in Parallel*

If resistors of equal value are connected in parallel, a special case *equal-value formula* can be used to calculate the total resistance.

$$R_T = \frac{\text{value of one resistor } (R)}{\text{number of parallel resistors } (n)}$$

EXAMPLE:

Figure 4-19(a) shows how a stereo music amplifier is connected to drive two 8 Ω speakers, which are connected in parallel with one another. What is the total resistance connected across the amplifier's output terminals?

Solution:

Referring to the schematic in Figure 4-19(b), you can see that since both parallel-connected speakers have the same resistance, the total resistance is most easily calculated by using the equal-value formula:

$$R_T = \frac{R}{n} = \frac{8\ \Omega}{2} = 4\ \Omega$$

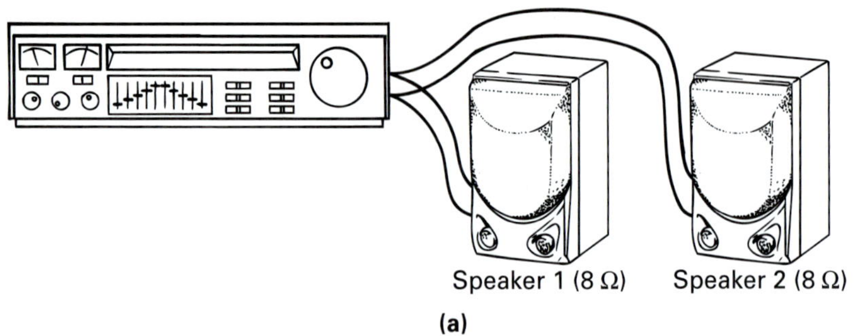

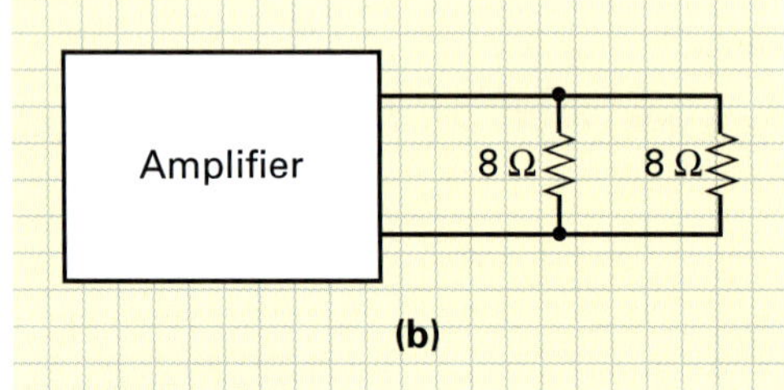

FIGURE 4-19 **Parallel-Connected Speakers.**

EXAMPLE:

Refer to Figure 4-20 and calculate:

a. Total resistance in part (a)
b. Total resistance in part (b)

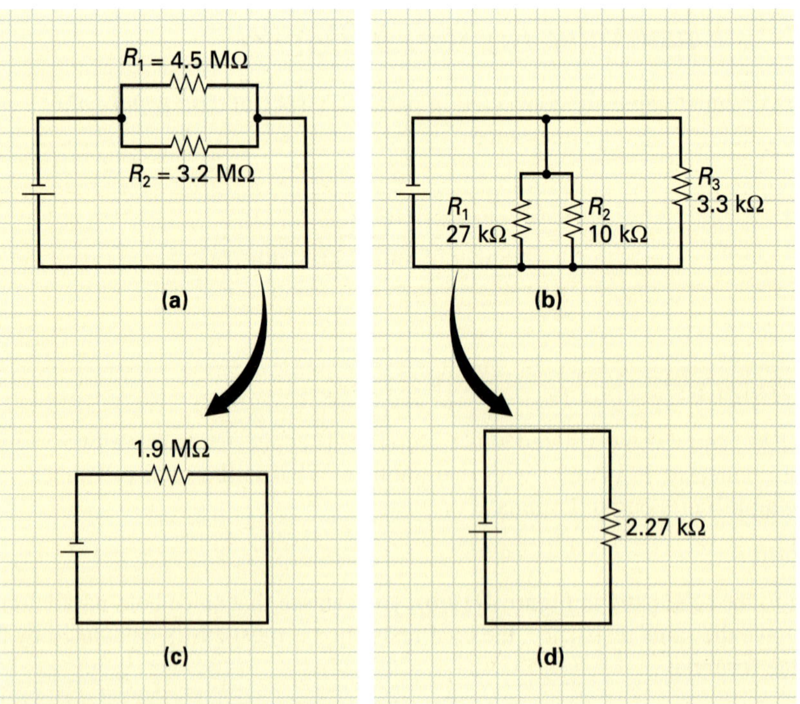

FIGURE 4-20 **Parallel Circuit Examples.**

Solution:

a. Figure 4-20(a) has only two resistors in parallel and therefore the product-over-sum resistor formula can be used.

$$\begin{aligned} R_T &= \frac{R_1 \times R_2}{R_1 + R_2} \\ &= \frac{4.5\ \text{M}\Omega \times 3.2\ \text{M}\Omega}{4.5\ \text{M}\Omega + 3.2\ \text{M}\Omega} \\ &= \frac{14.4\ \text{M}\Omega^2}{7.7\ \text{M}\Omega} \\ &= 1.9\ \text{M}\Omega \end{aligned}$$

The equivalent circuit is seen in Figure 4-20(c).

b. Figure 4-20(b) has more than two resistors in parallel, and therefore the reciprocal formula must be used.

$$\begin{aligned} R_T &= \frac{1}{(1/R_1) + (1/R_2) + (1/R_3)} \\ &= \frac{1}{(1/27\ \text{k}\Omega) + (1/10\ \text{k}\Omega) + (1/3.3\ \text{k}\Omega)} \\ &= \frac{1}{440.0 \times 10^{-6}} \\ &= 2.27\ \text{k}\Omega \end{aligned}$$

The equivalent circuit can be seen in Figure 4-20(d).

CALCULATOR SEQUENCE FOR (A)

Step	Keypad Entry	Display Response
1.	4 . 5 E 6	4.5E6
2.	+	
3.	3 . 2 E 6	3.2E6
4.	=	7.7E6
5.	STO (store in memory)	
6.	C/CE	0.
7.	4 . 5 E 6	4.5E6
8.	×	
9.	3 . 2 E 6	3.2E6
10.	=	1.44E13
11.	÷	
12.	RM (Recall memory)	7.7E6
13.	=	1.87E6

EXAMPLE:

Find the total resistance of the parallel circuit in Figure 4-21.

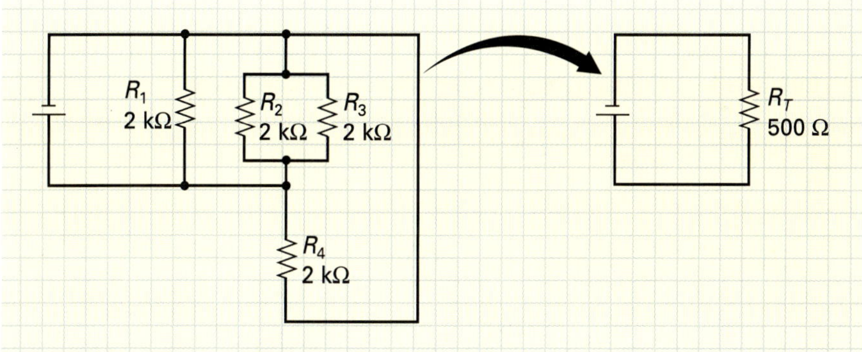

FIGURE 4-21 **Parallel Circuit Example.**

Solution:

Since all four resistors are connected in parallel and are all of the same value, the equal-value resistors in parallel formula can be used to calculate total resistance.

$$R_T = \frac{R}{n} = \frac{2\ \text{k}\Omega}{4} = 500\ \Omega$$

To summarize what we have learned so far about parallel circuits, we can say that:

1. Components are said to be connected in parallel when the current has to travel two or more paths between the negative and positive sides of the voltage source.

2. The voltage across all the parallel branches is always the same.
3. The total current from the source is equal to the sum of all the branch currents (Kirchhoff's current law).
4. The amount of current flowing through each branch is inversely proportional to the resistance value in that branch.
5. The total resistance of a parallel circuit is always less than the value of the smallest branch resistance.

SELF-TEST EVALUATION POINT FOR SECTION 4–4

Use the following questions to test your understanding of Section 4–4.

State what parallel resistance formula should be used in questions 1 through 3 for calculating R_T.

1. Two resistors in parallel.
2. More than two resistors in parallel.
3. Equal-value resistors in parallel.
4. Calculate the total parallel resistance when $R_1 = 2.7\ \text{k}\Omega$, $R_2 = 24\ \text{k}\Omega$, and $R_3 = 1\ \text{M}\Omega$.

4-5 POWER IN A PARALLEL CIRCUIT

As with series circuits, the total power in a parallel resistive circuit is equal to the sum of all the power losses for each of the resistors in parallel.

$$P_T = P_1 + P_2 + P_3 + P_4 + \cdots$$

total power = addition of all the power losses

The formulas for calculating the amount of power dissipated are

$$P = I \times V$$

$$P = \frac{V^2}{R}$$

$$P = I^2 \times R$$

EXAMPLE:

Calculate the total amount of power dissipated in Figure 4–22.

Solution:

The total power dissipated in a parallel circuit is equal to the sum of all the power dissipated by all the resistors. With P_1, we only know voltage and resistance and therefore we can use the formula

$$P_1 = \frac{V_{R1}^{\ 2}}{R_1} = \frac{20^2}{2\ \text{k}\Omega} = 2.0\ \text{W} \text{ or } 200\ \text{mW}$$

With P_2, we only know current and voltage, and therefore we can use the formula

$$P_2 = I_2 \times V_{R2} = 2\ \text{mA} \times 20\ \text{V} = 40\ \text{mW}$$

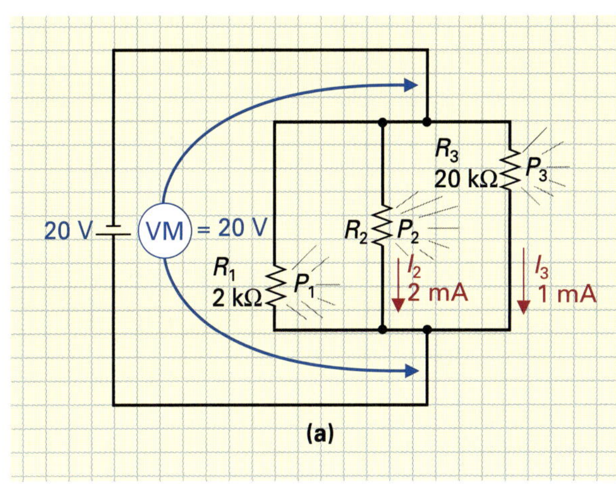

CIRCUIT ANALYSIS TABLE

Resistance $R = V/I$	Voltage $V = I \times R$	Current $I = V/R$	Power $P = V \times I$
$R_1 = 2$ kΩ	$V_{R1} = 20$ V	$I_1 =$	$P_1 = 200$ mW
$R_2 =$	$V_{R2} = 20$ V	$I_2 = 2$ mA	$P_2 = 40$ mW
$R_3 = 20$ kΩ	$V_{R3} = 20$ V	$I_3 = 1$ mA	$P_3 = 20$ mW

$V_T = V_S = 20$ V $\quad\quad P_T = 260$ mW

(b)

FIGURE 4-22 **Parallel Circuit Example. (a) Schematic. (b) Circuit Analysis Table.**

With P_3, we know V, I, and R; however, we will use the third power formula:

$$P_3 = I_3^2 \times \mathrm{R}_3 = 1\ \mathrm{mA}^2 \times 20\ \mathrm{k\Omega} = 20\ \mathrm{mW}$$

Total power (P_T) equals the sum of all the power or wattage losses for each resistor:

$$\begin{aligned} P_T &= P_1 + P_2 + P_3 \\ &= 200\ \mathrm{mW} + 40\ \mathrm{mW} + 20\ \mathrm{mW} \\ &= 260\ \mathrm{mW} \end{aligned}$$

Calculator Sequence

Step	Keypad Entry	Display Response
1.	[2] [0]	20.
2.	[x²]	400.
3.	[÷]	
4.	[2] [E] [3]	2.E3
5.	[=]	0.2

Calculator Sequence

Step	Keypad Entry	Display Response
1.	[2] [E] [3] [+/−]	2.E-3
2.	[×]	
3.	[2] [0]	20.
4.	[=]	40E-3

Calculator Sequence

Step	Keypad Entry	Display Response
1.	[1] [E] [3] [+/−] [x²]	1.E-6
2.	[×]	1.E-6
3.	[2] [0] [E] [3]	20E3
4.	[=]	20.E-3

EXAMPLE:

In Figure 4-23, there are two ½W (0.5 W) resistors connected in parallel. Should the wattage rating for each of these resistors be increased or decreased, or can they remain the same?

Solution:

Since current and voltage are known for both branches, the power formula used in both cases can be $P = I \times V$.

$$\begin{aligned} P_1 &= I \times V \\ &= 20\ \mathrm{mA} \times 5\ \mathrm{V} \\ &= 0.1\ \mathrm{W} \end{aligned} \qquad \begin{aligned} P_2 &= I \times V \\ &= 200\ \mathrm{mA} \times 5\ \mathrm{V} \\ &= 1\ \mathrm{W} \end{aligned}$$

R_1 is dissipating 0.1 W and so a 0.5 W resistor will be fine in this application. On the other hand, R_2 is dissipating 1 W and is only designed to dissipate 0.5 W. R_2 will therefore overheat unless it is replaced with a resistor of the same ohmic value, but with a 1 W or greater rating.

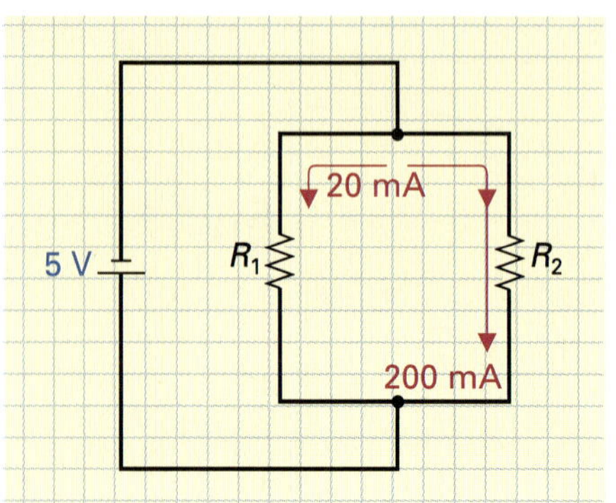

FIGURE 4-23 **Parallel Circuit Example.**

EXAMPLE:

Figure 4–24(a) shows a simplified diagram of an automobile external light system. A 12 V lead–acid battery is used as a source and is connected across eight parallel-connected lamps. The left and right brake lights are controlled by the brake switch, which is attached to the brake pedal. When the light switch is turned on, both the rear taillights and the low-beam headlights are brought into circuit and turned on. The high-beam set of headlights is activated

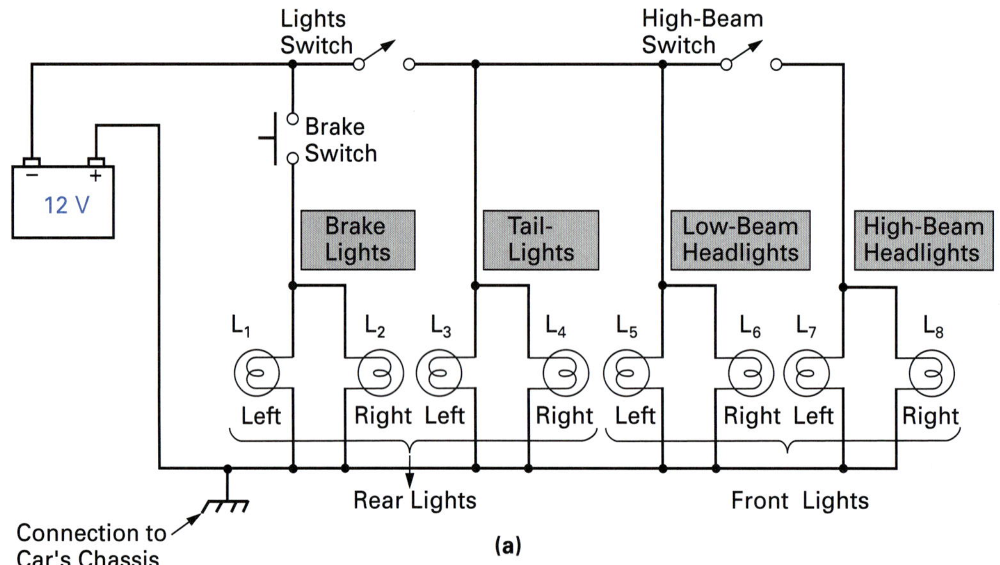

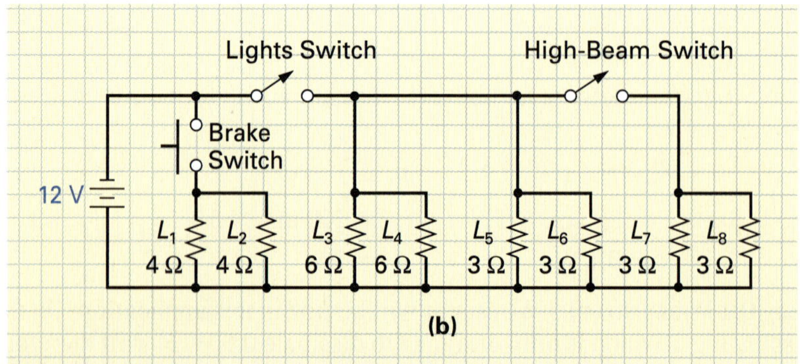

FIGURE 4-24 **Parallel Automobile External Light System.**

only if the high-beam switch is closed. For the lamp resistances given, calculate the output power of each lamp, when in use.

■ *Solution:*

Figure 4-24(b) shows the schematic diagram of the pictorial in Figure 4-24(a). Since both V and R are known, we can use the V^2/R power formula.

Brake lights: Left lamp wattage $(P) = \frac{V^2}{R} = \frac{(12\ V)^2}{4\Omega} = 36\ \text{W}$

Right lamp wattage is the same as left.

Taillights: Left lamp wattage $(P) = \frac{V^2}{R} = \frac{(12\ V)^2}{6\ \Omega} = 24\ \text{W}$

Right lamp wattage is the same as left.

Low-beam headlights: $(P) = \frac{V^2}{R} = \frac{(12V)^2}{3\ \Omega} = 48\ \text{W}$

Each low-beam headlight is a 48 W lamp.

High-beam headlights: $(P) = \frac{V^2}{R} = \frac{(12\ V)^2}{3\ \Omega} = 48\ \text{W}$

Each high-beam headlight is a 48 W lamp.

SELF-TEST EVALUATION POINT FOR SECTION 4-5

Use the following questions to test your understanding of Section 4-5.

1. True or false: Total power in a parallel circuit can be obtained by using the same total power formula as for series circuits.
2. If $I_1 = 2$ mA and $V = 24$ V, calculate P_1.
3. If $P_1 = 22$ mW and $P_2 = 6400\ \mu$W, $P_T = ?$
4. Is it important to observe the correct wattage ratings of resistors when they are connected in parallel?

4-6 TROUBLESHOOTING A PARALLEL CIRCUIT

In the troubleshooting discussion on series circuits, we mentioned that one of three problems can occur, and these three also apply to parallel circuits:

1. A component will open.
2. A component will short.
3. A component's value will change over a period of time.

As a technician, it is important that you know how to recognize these circuit malfunctions and know how to isolate the cause and make the repair. Since you could be encountering opens, shorts, and component value changes in the parallel circuits you construct in lab, let us step through the troubleshooting procedure you should use.

As an example, Figure 4-25(a) indicates the normal readings that should be obtained from a typical parallel circuit if it is operating correctly. Keep these normal readings in mind, since they will change as circuit problems are introduced in the following section.

4-6-1 *Open Component in a Parallel Circuit*

In Figure 4-25(b), R_1 has opened, so there can be no current flow through the R_1 branch. With one fewer branches for current, the parallel circuit's resistance will increase, causing the

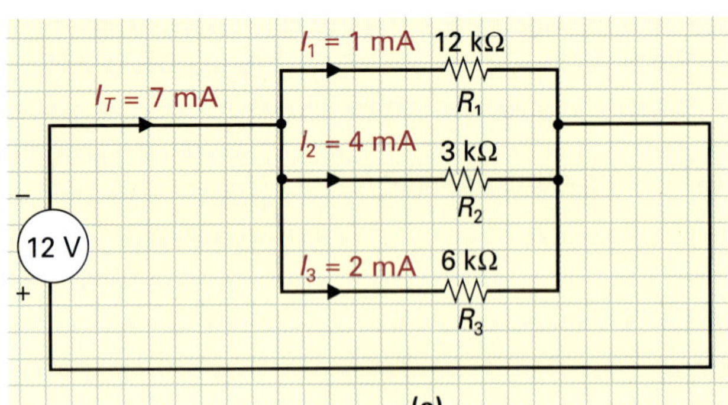

CIRCUIT ANALYSIS TABLE

Resistance $R = V/I$	Voltage $V = I \times R$	Current $I = V/R$	Power $P = V \times I$
R_1 = 12 kΩ	V_{R1} = 12 V	I_1 = 1 mA	P_1 = 12 mW
R_2 = 3 kΩ	V_{R2} = 12 V	I_2 = 4 mA	P_2 = 48 mW
R_3 = 6 kΩ	V_{R3} = 12 V	I_3 = 2 mA	P_3 = 24 mW
R_T = 1.7 kΩ	$V_T = V_S$ = 12 V	I_T = 7 mA	P_T = 84 mW

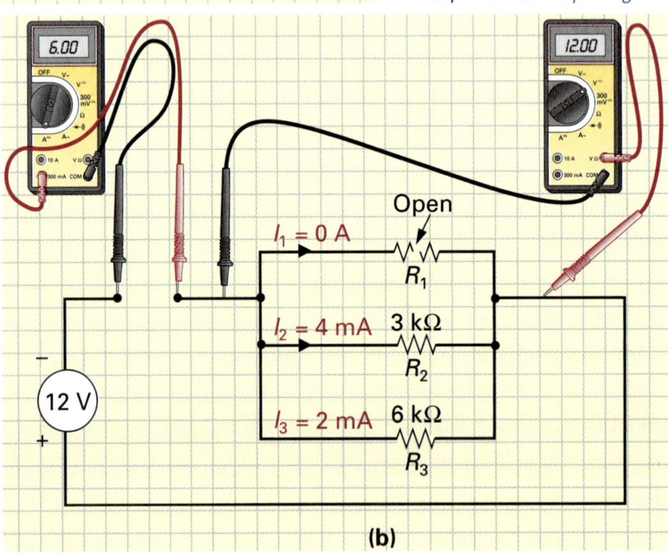

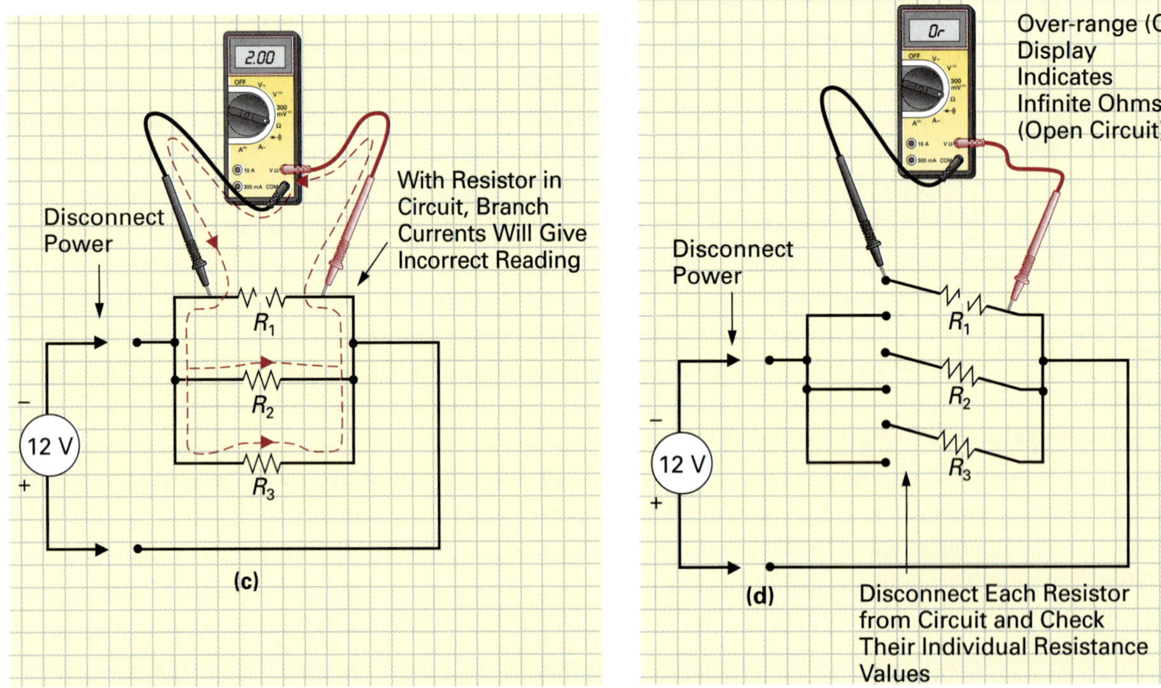

FIGURE 4-25 Troubleshooting an Open in a Parallel Circuit.

total current flow to decrease. The total current flow, which was 7 mA, will actually decrease by the amount that was flowing through the now open branch (1 mA) to a total current flow of 6 mA. If you had constructed this circuit in lab, you would probably not have noticed that there was anything wrong with the circuit until you started to make some multimeter measurements. A voltmeter measurement across the parallel circuit, as seen in Figure 4–25(b), would not have revealed any problem since the voltage measured across each resistor will always be the same and equal to the source voltage. A total current measurement, on the other hand, would indicate that there is something wrong since the measured reading does not equal the expected calculated value. Using your ammeter, you could isolate the branch fault by one of the following two methods:

1. If you measure each branch current, you will isolate the problem of R_1, because there will be no current flow through R_1. However, this could take three checks.
2. If you just measure total current (one check), you will notice that total current has decreased by 1 mA, and after making a few calculations you could determine that the only path that had a branch current of 1 mA was through R_1, so R_1 must have gone open.

We can summarize how method 2 could be used to isolate an open in this example circuit:

If R_1 opens, total current decreases by 1 mA to 6 mA ($I_T = I_1 + I_2 + I_3 = 0 \text{ mA} + 4 \text{ mA} + 2 \text{ mA} = 6 \text{ mA}$).
If R_2 opens, total current decreases by 4 mA to 3 mA ($I_T = I_1 + I_2 + I_3 = 1 \text{ mA} + 0 \text{ mA} + 2 \text{ mA} = 3 \text{ mA}$).
If R_3 opens, total current decreases by 2 mA to 5 mA ($I_T = I_1 + I_2 + I_3 = 1 \text{ mA} + 4 \text{ mA} + 0 \text{ mA} = 5 \text{ mA}$).

Method 2 can be used as long as the resistors are unequal. If they are equal, method 1 may be used to locate the 0 current branch.

In most cases, the ohmmeter is ideal for locating open circuits. To make a resistance measurement, you should disconnect power from the circuit, isolate the device from the circuit, and then connect the ohmmeter across the device. Figure 4–25(c) shows why it is necessary for you to isolate the device to be tested from the circuit. With R_1 in circuit, the ohmmeter will be measuring the parallel resistance of R_2 and R_3, even though the ohmmeter is connected across R_1. This will lead to a false reading and confuse the troubleshooting process. Disconnecting the device to be tested from the circuit, as shown in Figure 4–25(d), will give an accurate reading of R_1, then R_2, and then R_3's resistance and lead you to the cause of the problem.

4-6-2 *Shorted Component in a Parallel Circuit*

In Figure 4–26(a), R_2, which has a resistance of 3 kΩ, has been shorted out. The only resistance in this center branch is the resistance of the wire, which would typically be a fraction of an ohm. In this example, let us assume that the resistance in the branch is 1 Ω, and therefore the current in this branch will attempt to increase to:

$$I_2 = \frac{12 \text{ V}}{1 \, \Omega} = 12 \text{ A}$$

Before the current even gets close to 12 A, the fuse in the dc power supply will blow, as shown in the inset in Figure 4–26(a), and disconnect power from the circuit. This is typically the effect you will see from a circuit short. Replacing the fuse will be a waste of time and money since the fuse will simply blow again once power is connected across the circuit. As before, the tool to use to isolate the cause is the ohmmeter, as shown in Figure 4–26(b). To check the resistance of each branch, disconnect power from the circuit, disconnect the devices to be tested from the circuit to prevent false readings, and then test each branch until the shorted branch is located. Once you have corrected the fault, you will need to change the power supply fuse or reset its circuit breaker.

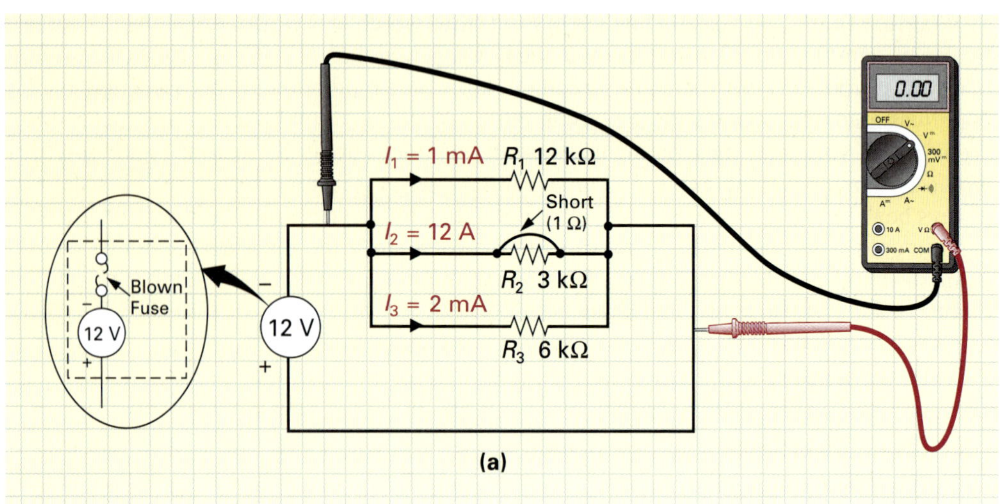

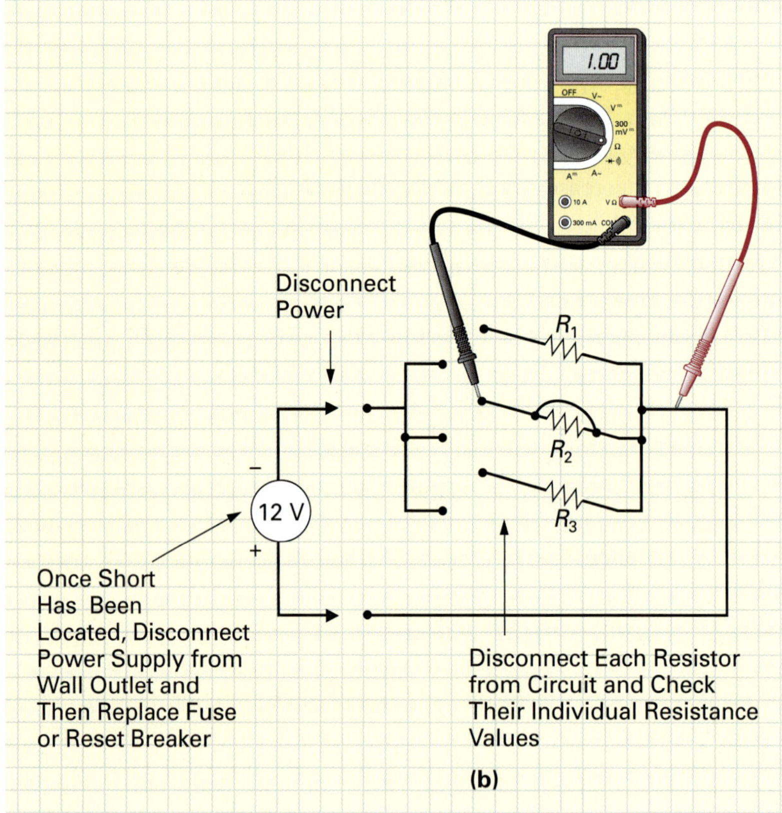

FIGURE 4-26 **Troubleshooting a Short in a Parallel Circuit.**

4-6-3 *Component Value Variation in a Parallel Circuit*

The resistive values of resistors will change with age. This increase or decrease in resistance will cause a corresponding decrease or increase in branch current and therefore total current. This deviation from the desired value can also be checked with the ohmmeter. Be careful to take into account the resistor's tolerance because this can make you believe that the resistor's value has changed, when it is, in fact, within the tolerance rating.

4-6-4 *Summary of Parallel Circuit Troubleshooting*

1. An open component will cause no current flow within that branch. The total current will decrease, and the voltage across the component will be the same as the source voltage.
2. A shorted component will cause maximum current through that branch. The total current will increase and the voltage source fuse will normally blow.
3. A change in resistance value will cause a corresponding opposite change in branch current and total current.

SELF-TEST EVALUATION POINT FOR SECTION 4-6

Use the following questions to test your understanding of Section 4-6.

How would problems 1 through 3 be recognized in a parallel circuit?

1. An open component
2. A shorted component
3. A component's value variation
4. True or false: An ammeter is typically used to troubleshoot series circuits, whereas a voltmeter is typically used to troubleshoot parallel circuits.

REVIEW QUESTIONS

Multiple-Choice Questions

1. A parallel circuit has ____________ path(s) for current to flow.
   a. One **c.** Only three
   b. Two or more **d.** None of the above
2. If a source voltage of 12 V is applied across four resistors of equal value in parallel, the voltage drops across each resistor would be equal to:
   a. 12 V **c.** 4 V
   b. 3 V **d.** 48 V
3. What would be the voltage drop across two 25 Ω resistors in parallel if the source voltage were equal to 9 V?
   a. 50 V **c.** 12 V
   b. 25 V **d.** None of the above
4. If a four-branch parallel circuit has 15 mA flowing through each branch, the total current into the parallel circuit will be equal to:
   a. 15 mA **c.** 30 mA
   b. 60 mA **d.** 45 mA
5. If the total three-branch parallel circuit current is equal to 500 mA, and 207 mA is flowing through one branch and 153 mA through another, what would be the current flow through the third branch?
   a. 707 mA **c.** 140 mA
   b. 653 mA **d.** None of the above
6. A large branch resistance will cause a ____________ branch current.
   a. Large **c.** Medium
   b. Small **d.** None of the above are true
7. What would be the conductance of a 1 kΩ resistor?
   a. 10 mS
   b. 1 mS
   c. 2 kΩ
   d. All of the above
8. If only two resistors are connected in parallel, the total resistance equals:
   a. The sum of the resistance values
   b. Three times the value of one resistor
   c. The product over the sum
   d. All of the above
9. If resistors of equal value are connected in parallel, the total resistance can be calculated by:
   a. One resistor value divided by the number of parallel resistors
   b. The sum of the resistor values
   c. The number of parallel resistors divided by one resistor value
   d. All of the above could be true.
10. The total power in a parallel circuit is equal to the:
    a. Product of total current and total voltage
    b. Reciprocal of the individual power losses
    c. Sum of the individual power losses
    d. Both (a) and (b)
    e. Both (a) and (c)

Practice Problems

11. Calculate the total resistance of four 30 kΩ resistors in parallel.

12. Find the total resistance for each of the following parallel circuits:
  a. 330 Ω and 560 Ω
  b. 47 kΩ, 33 kΩ, and 22 kΩ
  c. 2.2 MΩ, 3 kΩ, and 220 Ω
13. If 10 V is connected across three 25 Ω resistors in parallel, what will be the total and individual branch currents?
14. If a four-branch parallel circuit has branch currents equal to 25 mA, 37 mA, 220 mA, and 0.2 A, what is the total circuit current?
15. If three resistors of equal value are connected across a 14 V supply and the total resistance is equal to 700 Ω, what is the value of each branch current?
16. If three 75 W light bulbs are connected in parallel across a 110 V supply, what is the value of each branch current? What is the branch current through the other two light bulbs if one burns out?
17. If 33 kΩ and 22 kΩ resistors are connected in parallel across a 20 V source, calculate:
  a. Total resistance
  b. Total current
  c. Branch currents
  d. Total power dissipated
  e. Individual resistor power dissipated
18. If four parallel-connected resistors are each dissipating 75 mW, what is the total power being dissipated?
19. Calculate the branch currents through the following parallel resistor circuits when they are connected across a 10 V supply:
  a. 22 kΩ and 33 kΩ
  b. 220 Ω, 330 Ω, and 470 Ω
20. If 30 Ω and 40 Ω resistors are connected in parallel, which resistor will generate the greatest amount of heat?
21. Calculate the total conductance and resistance of the following parallel circuits:
  a. Three 5 Ω resistors
  b. Two 200 Ω resistors
  c. 1 MΩ, 500 MΩ, 3.3 MΩ
  d. 5 Ω, 3 Ω, 2 Ω
22. Connect the three resistors in Figure 4-27 in parallel across a 12 V battery and then calculate the following:
  a. V_{R1}, V_{R2}, V_{R3}
  b. I_1, I_2, I_3
  c. I_T
  d. P_T
  e. P_1, P_2, P_3
  f. G_{R1}, G_{R2}, G_{R3}
23. Calculate R_T in Figure 4-28 (a), (b), (c), and (d).
24. Calculate the branch currents through four 60 W bulbs connected in parallel across 110 V. How much is the total current, and what would happen to the total current if one of the bulbs were to burn out? What change would occur in the remaining branch currents?
25. Calculate the following in Figure 4-29:
  a. I_2
  b. I_T
  c. V_S, I_1, I_2
  d. R_2, I_1, I_2, P_T

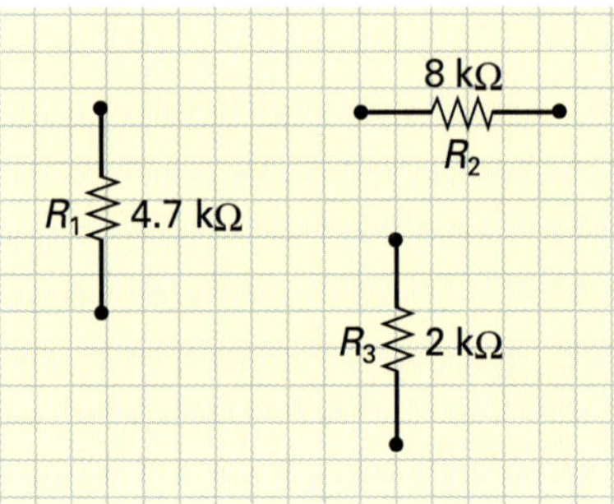

FIGURE 4-27 **Connect in Parallel across a 12-volt Source.**

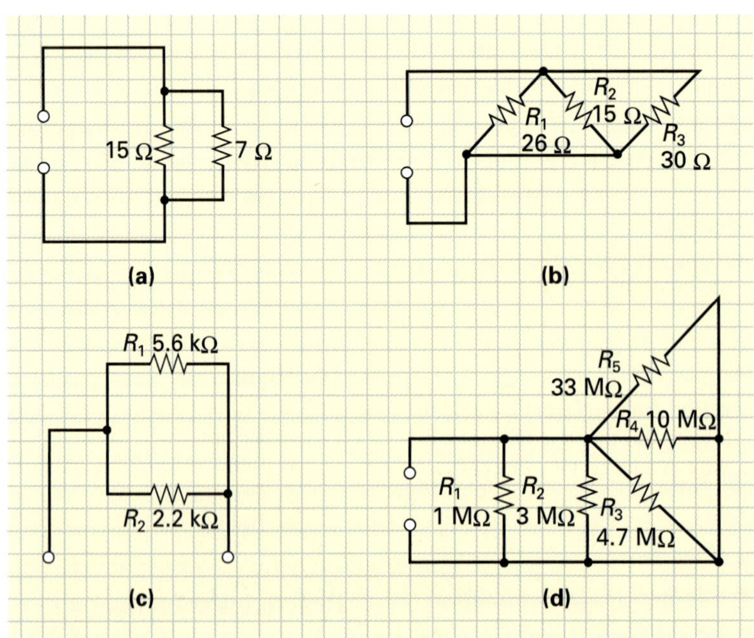

FIGURE 4-28 **Calculate Total Resistance.**

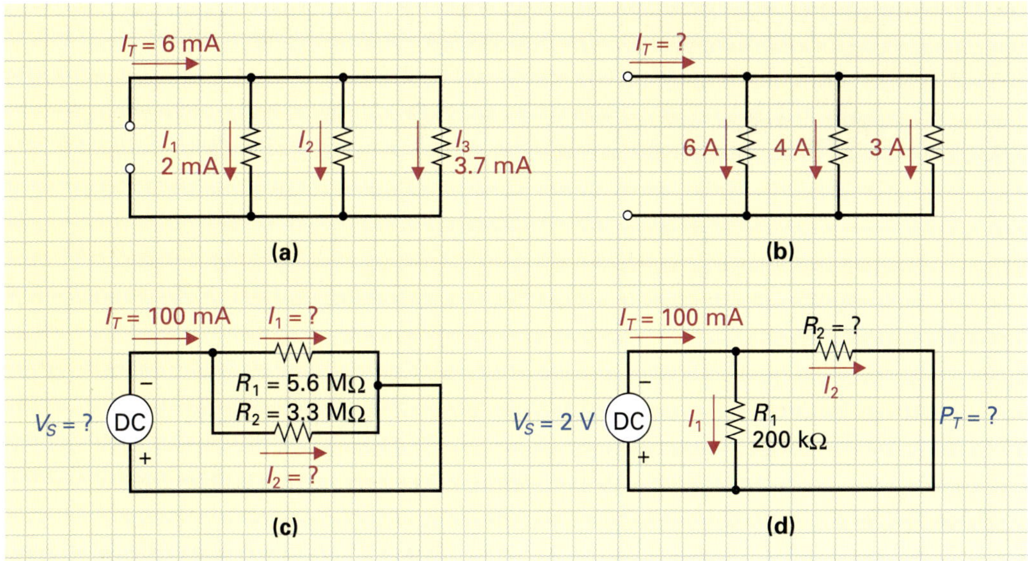

FIGURE 4-29 **Calculate the Unknown.**

Troubleshooting Questions

26. An open component in a parallel circuit will cause ____________ current flow within that branch, which will cause the total current to ____________.

a. Maximum, increase **c.** Maximum, decrease
b. Zero, decrease **d.** Zero, increase

27. A shorted component in a parallel circuit will cause ____________ current through a branch, and consequently the total current will ____________.

a. Maximum, increase **c.** Maximum, decrease
b. Zero, decrease **d.** Zero, increase

28. If a 10 kΩ and two 20 kΩ resistors are connected in parallel across a 20 V supply, and the total current measured is 2 mA, determine whether a problem exists in the circuit and, if it does, isolate the problem.

29. What situation would occur and how would we recognize the problem if one of the 20 kΩ resistors in Question 28 were to short?

30. With age, the resistance of a resistor will ____________, resulting in a corresponding but opposite change in ____________ ____________.

a. Increase, branch current
b. Change, source voltage
c. Decrease, source resistance
d. Change, branch current

Web Site Questions

Go to the web site http://www.prenhall.com/cook, select the textbook *Electronics: A Complete Course,* select this chapter, and then follow the instructions when answering the multiple choice practice problems.

Series–Parallel Circuits

The Christie Bridge Circuit

Charles Wheatstone

Who invented the Wheatstone bridge circuit? It was obviously Sir Charles Wheatstone. Or was it?

The Wheatstone bridge was actually invented by S.H. Christie of the Royal Military Academy at Woolwich, England. He described the circuit in detail in the *Philosophical Transactions* paper dated February 28, 1833. Christie's name, however, was unknown and his invention was ignored.

Ten years later, Sir Charles Wheatstone called attention to Christie's circuit. Sir Charles was very well known, and from that point on, and even to this day, the circuit is known as a Wheatstone bridge. Later, Werner Siemens would modify Christie's circuit and invent the variable-resistance arm bridge circuit, which would also be called a Wheatstone bridge.

No one has given full credit to the real inventors of these bridge circuits, until now!

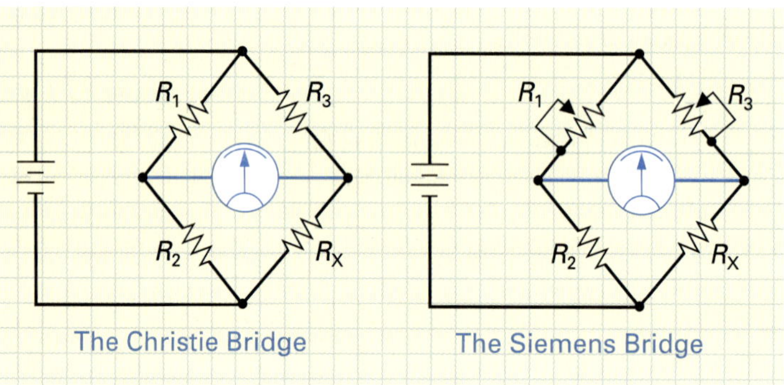

5

Introduction

Very rarely are we lucky enough to run across straightforward series or parallel circuits. In general, all electronic equipment is composed of many components that are interconnected to form a combination of series and parallel circuits. In this chapter, we will be combining our knowledge of the series and parallel circuits discussed in the previous two chapters.

5-1 SERIES- AND PARALLEL-CONNECTED COMPONENTS

Series–Parallel Circuit
Network or circuit that contains components that are connected in both series and parallel.

Figure 5-1(a) through (f) shows six examples of **series–parallel resistive circuits.** The most important point to learn is how to distinguish between the resistors that are connected in series and the resistors that are connected in parallel, which will take a little practice.

One thing that you may not have noticed when examining Figure 5-1 is that:

Circuit 5-1(a) is equivalent to 5-1 (b)
Circuit 5-1(c) is equivalent to 5-1 (d)
Circuit 5-1(e) is equivalent to 5-1 (f)

When analyzing these series–parallel circuits, always remember that current flow determines whether the resistor is connected in series or parallel. Begin at the negative side of the battery and apply these two rules:

1. If the total current has only one path to follow through a component, that component is connected in series.
2. If the total current has two or more paths to follow through two or more components, those components are connected in parallel.

Referring again to Figure 5-1, you can see that series or parallel resistor networks are easier to identify in parts (a), (c), and (e) than in parts (b), (d), and (f). Redrawing the circuit so that the components are arranged from left to right or from top to bottom is your first line of attack in your quest to identify series- and parallel-connected components.

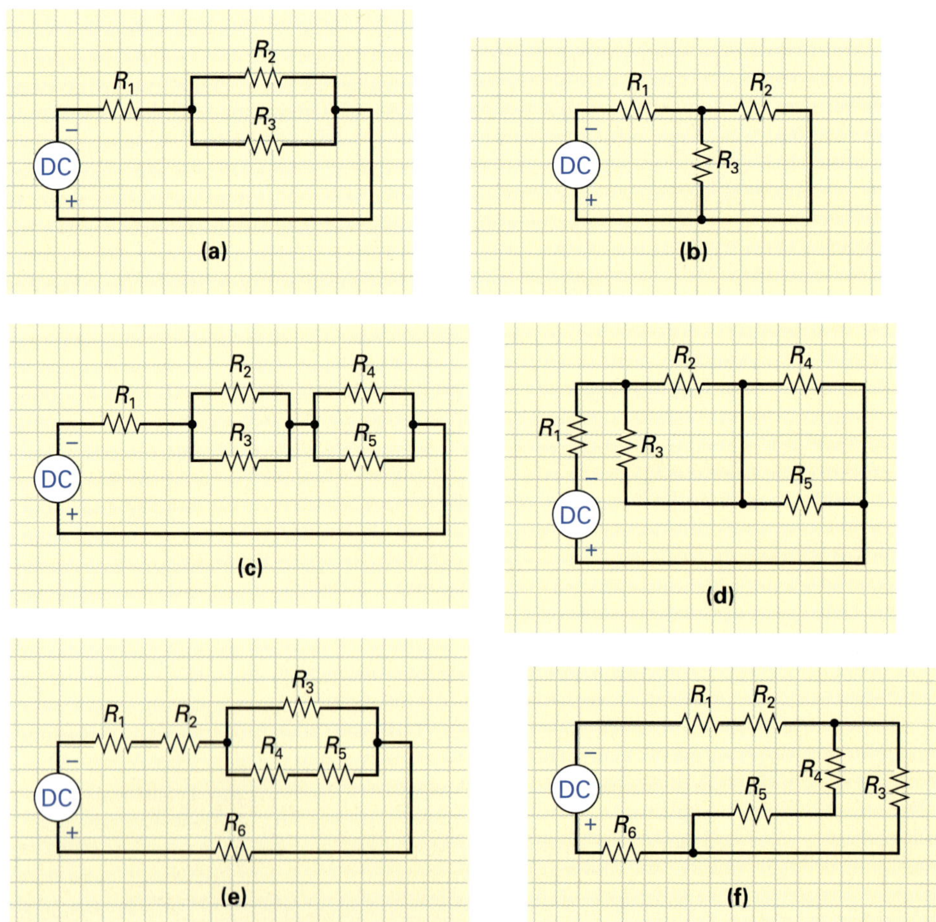

FIGURE 5-1 Series–Parallel Resistive Circuits.

EXAMPLE:

Refer to Figure 5-2 and identify which resistors are connected in series and which are in parallel.

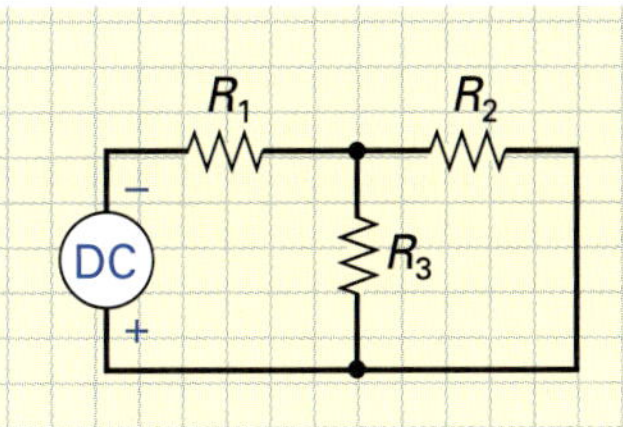

FIGURE 5-2 Series–Parallel Circuit Example.

Solution:

First, let's redraw the circuit so that the components are aligned either from left to right as shown in Figure 5-3(a), or from top to bottom as shown in Figure 5-3(b). Placing your pencil at the negative terminal of the battery on whichever figure you prefer, either Figure 5-3(a) or (b), trace the current paths through the circuit toward the positive side of the battery, as illustrated in Figure 5-4.

The total current arrives first at R_1. There is only one path for current to flow, which is through R_1, and therefore R_1 is connected in series. The total current proceeds on past R_1 and arrives at a junction where current divides and travels through two branches, R_2 and R_3. Since

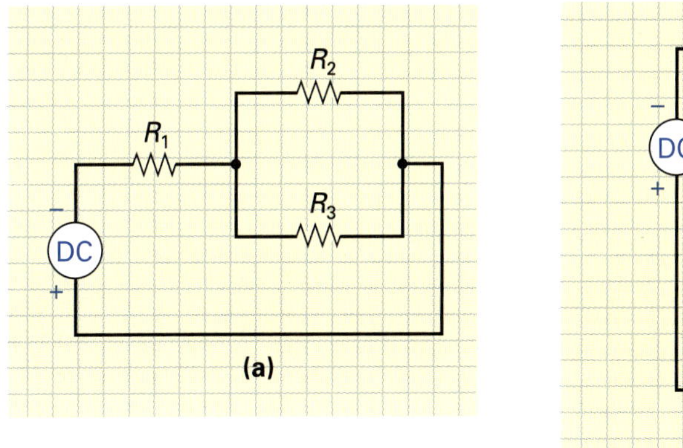

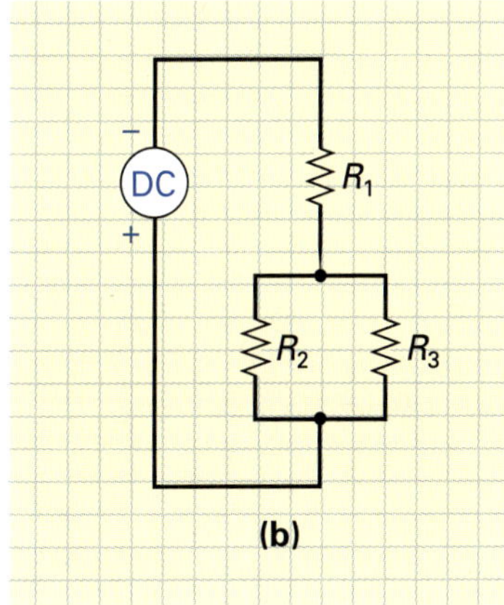

FIGURE 5-3 Redrawn Series–Parallel Circuit. (a) Left to Right. (b) Top to Bottom.

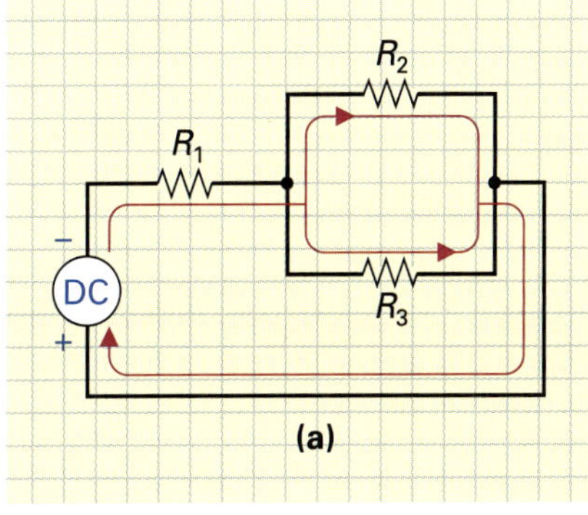

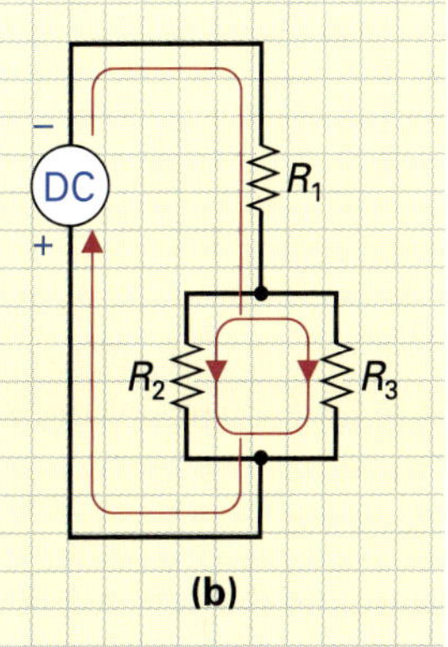

FIGURE 5-4 Tracing Current through a Series–Parallel Circuit.

current had to split into two paths, R_2 and R_3 are therefore connected in parallel. After the parallel connection of R_2 and R_3, total current combines and travels to the positive side of the battery.

In this example, therefore, R_1 is in series with the parallel combination of R_2 and R_3.

EXAMPLE:

Refer to Figure 5-5 and identify which resistors are connected in series and which are connected in parallel.

Solution:

Figure 5-6 illustrates the simplified, redrawn schematic of Figure 5-5. Total current leaves the negative terminal of the battery, and all of this current has to travel through R_1, which is therefore a series-connected resistor. Total current will split at junction A, and consequently, R_3 and R_4 with R_2 make up a parallel combination. The current that flows through R_3 (I_2) will also flow through R_4 and therefore R_3 is in series with R_4. I_1 and I_2 branch currents combine at junction *B* to produce total current, which has only one path to follow through the series resistor R_5, and finally, to the positive side of the battery.

In this example, therefore, R_3 and R_4 are in series with one another and both are in parallel with R_2, and this combination is in series with R_1 and R_5.

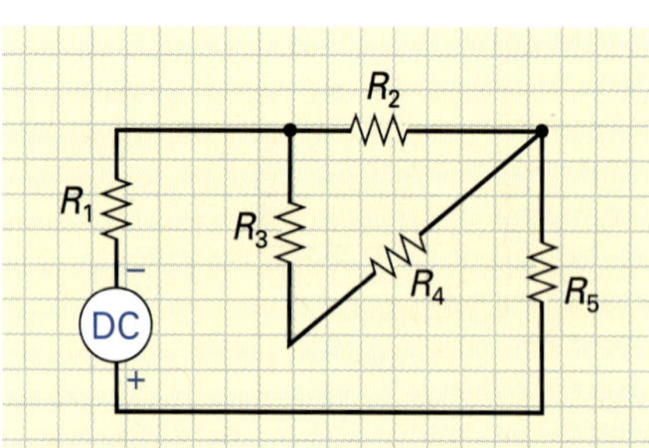

FIGURE 5-5 Series–Parallel Circuit Example.

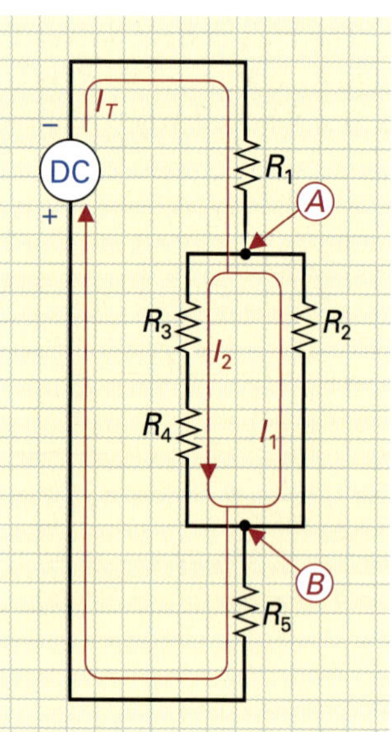

FIGURE 5-6 Redrawn Series–Parallel Circuit Example.

5-2 TOTAL RESISTANCE IN A SERIES–PARALLEL CIRCUIT

No matter how complex or involved the series–parallel circuit, there is a simple three-step method to simplify the circuit to a single equivalent total resistance. Figure 5-7 illustrates an example of a series–parallel circuit. Once you have analyzed and determined the series–parallel relationship, we can proceed to solve for total resistance.

The three-step method is:

Step A: Determine the equivalent resistances of all branch series-connected resistors.

Step B: Determine the equivalent resistances of all parallel-connected combinations.

Step C: Determine the equivalent resistance of the remaining series-connected resistances.

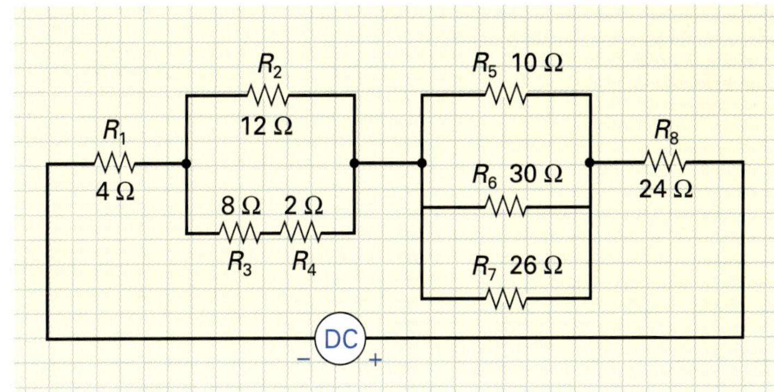

FIGURE 5-7 **Total Series–Parallel Circuit Resistance.**

EWB

Let's now put this procedure to work with the example circuit in Figure 5-7.

STEP A Solve for all branch series-connected resistors. In our example, this applies only to R_3 and R_4, and since this is a series connection, we have to use the series resistance formula.

$$R_{3,4} = R_3 + R_4 = 8 + 2 = 10\ \Omega \qquad \text{(series resistance formula)}$$

With R_3 and R_4 solved, the circuit now appears as indicated in Figure 5-8.

STEP B Solve for all parallel combinations. In this example, they are the two parallel combinations of (a) R_2 and $R_{3,4}$ and (b) R_5 and R_6 and R_7. Since these are parallel connections, use the parallel resistance formulas.

$$R_{2,3,4} = \frac{R_2 \times R_{3,4}}{R_2 + R_{3,4}} = \frac{12 \times 10}{12 + 10} = 5.5\ \Omega \qquad \text{(product-over-sum formula)}$$

$$R_{5,6,7} = \frac{1}{(1/R_5) + (1/R_6) + (1/R_7)} = 5.8\ \Omega \qquad \text{(reciprocal formula)}$$

With $R_{2,3,4}$ and $R_{5,6,7}$ solved, the circuit now appears as illustrated in Figure 5-9.

STEP C Solve for the remaining series resistances. There are now four remaining series resistances, which can be reduced to one equivalent resistance (R_{eq}) or total resistance (R_T). As seen in Figure 5-10, by using the series resistance formula, the total equivalent resistance for this example circuit will be

$$\begin{aligned} R_{eq} &= R_1 + R_{2,3,4} + R_{5,6,7} + R_8 \\ &= 4\ \Omega + 5.5\ \Omega + 5.8\ \Omega + 24\ \Omega \\ &= 39.3\ \Omega \end{aligned}$$

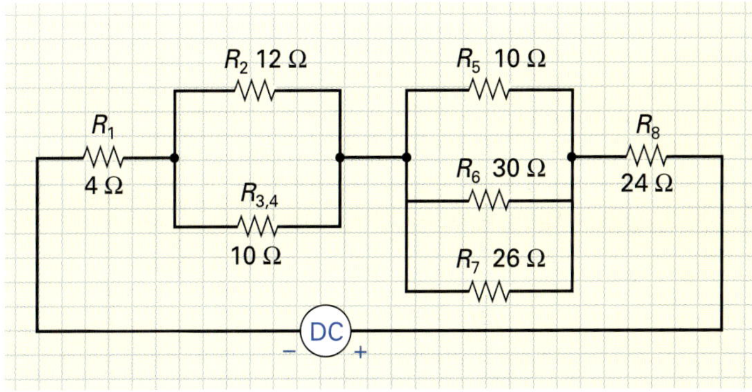

FIGURE 5-8 **After Completing Step A.**

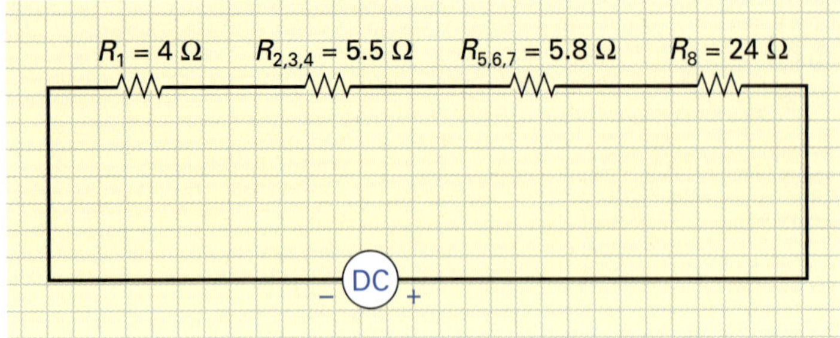

FIGURE 5-9 **After Completing Step B.**

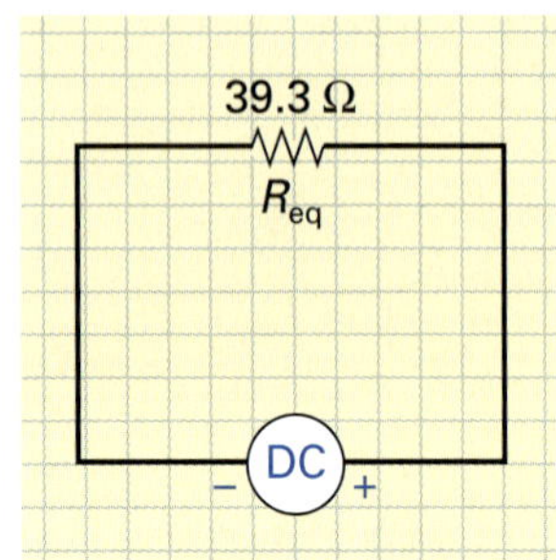

FIGURE 5-10 **After Completing Step C.**

EXAMPLE:

Find the total resistance of the circuit in Figure 5-11.

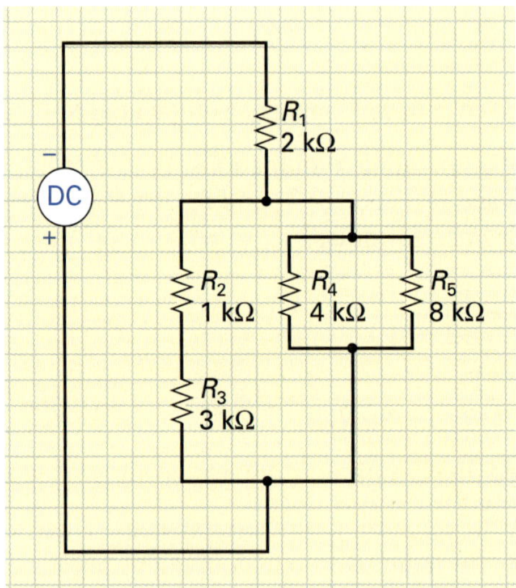

FIGURE 5-11 **Calculate Total Resistance.**

Solution:

STEP A Solve for all branch series-connected resistors. This applies to R_2 and R_3 (series connection):

$$R_{2,3} = R_2 + R_3 = 1\text{ k}\Omega + 3\text{ k}\Omega = 4\text{ k}\Omega$$

The resulting circuit, after completing step A, is illustrated in Figure 5-12(a).

STEP B Solve for all parallel combinations. Looking at Figure 5-12(a), which shows the circuit resulting from step A, you can see that current branches into three paths, so the parallel reciprocal formula must be used for this step.

$$R_{2,3,4,5} = \frac{1}{(1/R_{2,3}) + (1/R_4) + (1/R_5)}$$

$$= \frac{1}{(1/4\text{ k}\Omega) + (1/4\text{ k}\Omega) + (1/8\text{ k}\Omega)} = 1.6\text{ k}\Omega$$

The resulting circuit, after completing step B, is illustrated in Figure 5-12(b).

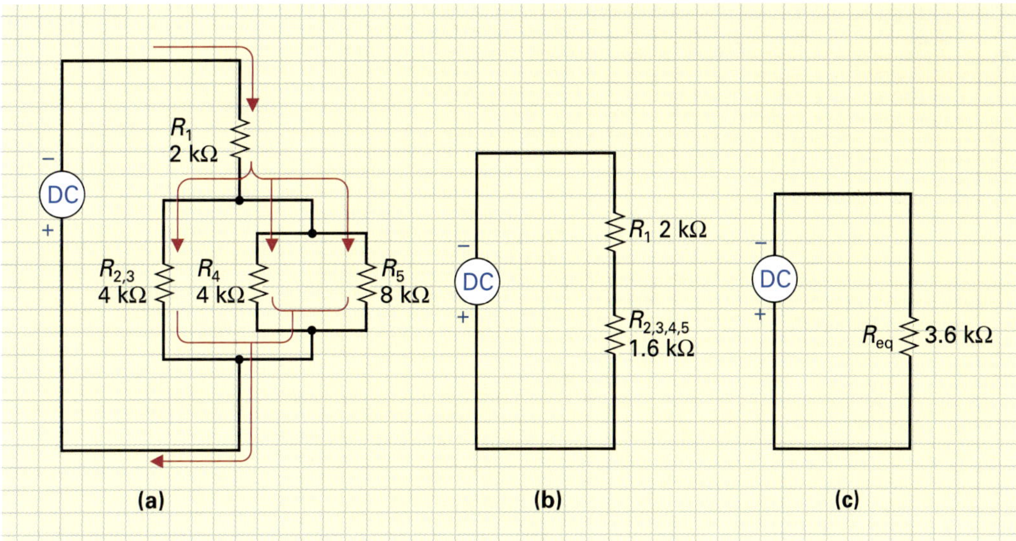

FIGURE 5-12 Calculating Total Resistance. (a) Step A. (b) Step B. (c) Step C.

STEP C Solve for the remaining series resistances. Looking at Figure 5-12(b), which shows the circuit resulting from step B, you can see that there are two remaining series resistances. The equivalent resistance (R_{eq}) is therefore equal to

$$R_{eq} = R_1 + R_{2,3,4,5} = 2\text{ k}\Omega + 1.6\text{ k}\Omega = 3.6\text{ k}\Omega$$

The total equivalent resistance, after completing all three steps, is illustrated in Figure 5-12(c).

SELF-TEST EVALUATION POINT FOR SECTIONS 5-1 AND 5-2

Use the following questions to test your understanding of Sections 5-1 and 5-2.

1. How can we determine which resistors are connected in series and which are connected in parallel in a series–parallel circuit?
2. Calculate the total resistance if two series-connected 12 kΩ resistors are connected in parallel with a 6 kΩ resistor.
3. State the three-step procedure used to determine total resistance in a circuit made up of both series and parallel resistors.
4. Sketch the following series–parallel resistor network made up of three resistors. R_1 and R_2 are in series with each other and are connected in parallel with R_3. If $R_1 = 470\ \Omega$, $R_2 = 330\ \Omega$, and $R_3 = 270\ \Omega$, what is R_T?

5-3 VOLTAGE DIVISION IN A SERIES–PARALLEL CIRCUIT

There is a simple three-step procedure for finding the voltage drop across each part of the series–parallel circuit. Figure 5-13 illustrates an example of a series–parallel circuit to which we will apply the three-step method for determining voltage drop.

STEP 1 Determine the circuit's total resistance. This is achieved by following the three-step method used previously for calculating total resistance.

Step A: $R_{3,4} = 4 + 8 = 12\ \Omega$

Step B:

$$R_{2,3,4} = \frac{1}{(1/R_2) + (1/R_{3,4})} = 6\ \Omega$$

$$R_{5,6,7} = \frac{1}{(1/R_5) + (1/R_6) + (1/R_7)} = 12\ \Omega$$

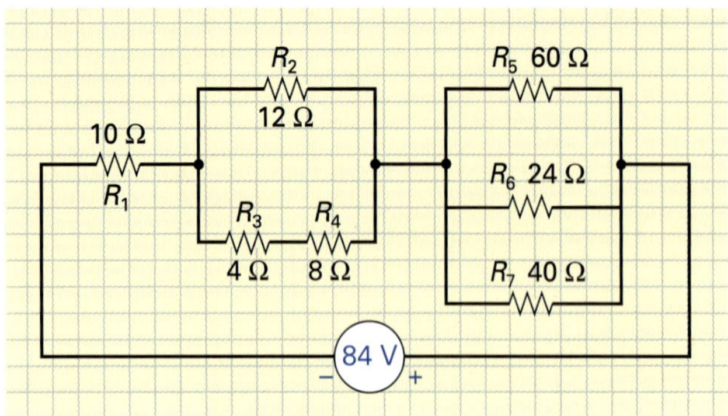

FIGURE 5-13 **Series–Parallel Circuit Example.**

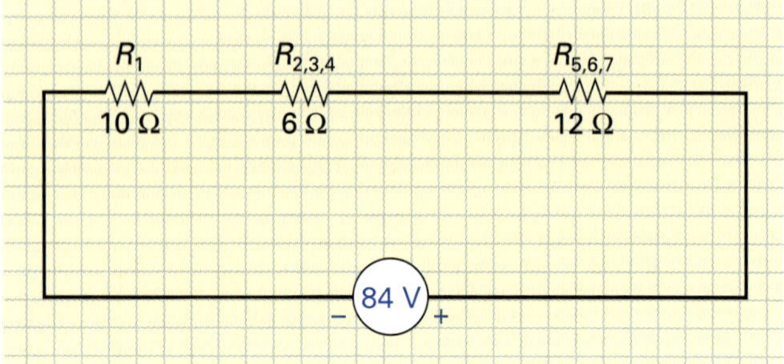

FIGURE 5-14 **After Completing Step 1.**

Figure 5-14 illustrates the equivalent circuit up to this point. We end up with one series resistor (R_1) and two series equivalent resistors ($R_{2,3,4}$ and $R_{5,6,7}$). R_T is therefore equal to 28 Ω.

STEP 2 Determine the circuit's total current. This step is achieved simply by utilizing Ohm's law.

$$I_T = \frac{V_T}{R_T} = \frac{84\text{ V}}{28\ \Omega} = 3\text{ A}$$

STEP 3 Determine the voltage across each series resistor and each parallel combination (series equivalent resistor) in Figure 5-14. Since these are all in series, the same current (I_T) will flow through all three.

$$V_{R1} = I_T \times R_1 = 3\text{ A} \times 10\ \Omega = 30\text{ V}$$
$$V_{R2,3,4} = I_T \times R_{2,3,4} = 3\text{ A} \times 6\ \Omega = 18\text{ V}$$
$$V_{R5,6,7} = I_T \times R_{5,6,7} = 3\text{ A} \times 12\ \Omega = 36\text{ V}$$

The voltage drops across the series resistor (R_1) and series equivalent resistors ($R_{2,3,4}$ and $R_{5,6,7}$) are illustrated in Figure 5-15.

Kirchhoff's voltage law states that the sum of all the voltage drops is equal to the source voltage applied. This law can be used to confirm that our calculations are all correct:

$$\begin{aligned} V_T &= V_{R1} + V_{R2,3,4} + V_{R5,6,7} \\ &= 30\text{ V} + 18\text{ V} + 36\text{ V} \\ &= 84\text{ V} \end{aligned}$$

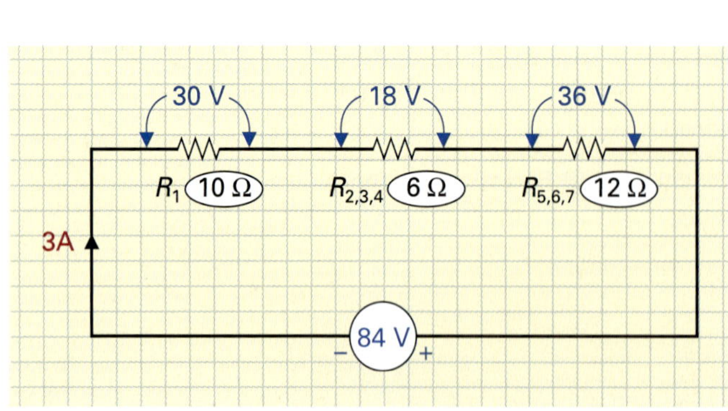

FIGURE 5-15 **After Completing Steps 2 and 3.**

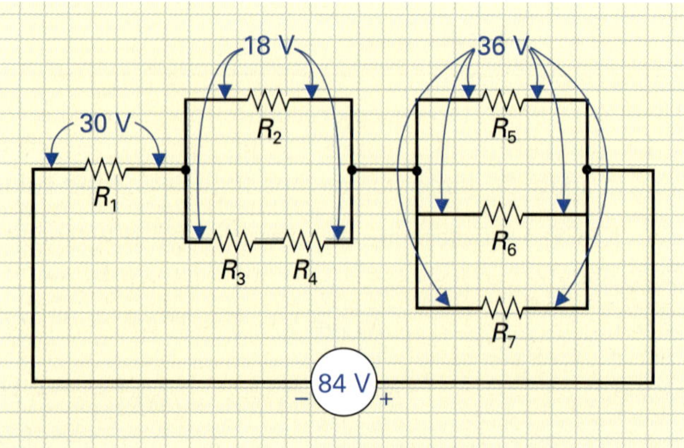

FIGURE 5-16 **Detail of Step 3.**

To summarize, refer to Figure 5-16, which shows these voltage drops inserted into our original circuit. As you can see from this illustration:

30 V is dropped across R_1.

18 V is dropped across R_2.

18 V is dropped across both R_3 and R_4.

36 V is dropped across R_5.

36 V is dropped across R_6.

36 V is dropped across R_7.

SELF-TEST EVALUATION POINT FOR SECTION 5-3

Use the following questions to test your understanding of Section 5-3.

1. State the three-step procedure used to calculate the voltage drop across each part of a series–parallel circuit.
2. Referring to Figure 5-13, double the values of all the resistors. Would the voltage drops calculated previously change, and if so, what would they be?

5-4 BRANCH CURRENTS IN A SERIES–PARALLEL CIRCUIT

In the preceding example, step 2 calculated the total current flowing in a series–parallel circuit. The next step is to find out exactly how much current is flowing through each parallel branch. This will be called step 4. Figure 5-17 shows the previously calculated data inserted in the appropriate places in our example circuit.

STEP 4 Total current (I_T) will exist at points *A, B, C,* and *D*. Between *A* and *B*, current has only one path in which to flow, which is through R_1. R_1 is therefore a series resistor, so, $I_1 = I_T = 3$ A. Between points *B* and *C*, current has two paths: through R_2 (12 Ω) and through R_3 and R_4 (12 Ω).

$$I_2 = \frac{V_{R2}}{R_2} = \frac{18 \text{ V}}{12 \ \Omega} = 1.5 \text{ A}$$

$$I_{3,4} = \frac{V_{R3,4}}{R_{3,4}} = \frac{18 \text{ V}}{12 \ \Omega} = 1.5 \text{ A}$$

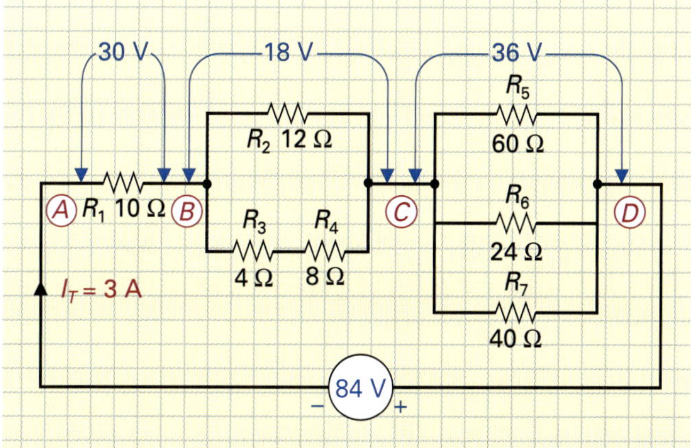

FIGURE 5-17 Series–Parallel Circuit Example with Previously Calculated Data.

Not surprisingly, the total current of 3 A is split equally due to both branches having equal resistance.

The two 1.5 A branch currents will combine at point C to produce once again the total current of 3 A. Between points C and D, current has three paths to flow through, R_5, R_6, and R_7.

$$I_5 = \frac{V_{R5}}{R_5} = \frac{36\text{ V}}{60\ \Omega} = 0.6\text{ A}$$

$$I_6 = \frac{V_{R6}}{R_6} = \frac{36\text{ V}}{24\ \Omega} = 1.5\text{ A}$$

$$I_7 = \frac{V_{R7}}{R_7} = \frac{36\text{ V}}{40\ \Omega} = 0.9\text{ A}$$

All three branch currents will combine at point D to produce the total current of 3 A ($I_T = I_5 + I_6 + I_7 = 0.6 + 1.5 + 0.9 = 3$ A), proving Kirchhoff's current law.

5-5 POWER IN A SERIES–PARALLEL CIRCUIT

Whether resistors are connected in series or in parallel, the total power in a series–parallel circuit is equal to the sum of all the individual power losses.

$$P_T = P_1 + P_2 + P_3 + P_4 + \cdots$$

total power = addition of all power losses

The formulas for calculating the amount of power lost by each resistor are

$$P = \frac{V^2}{R} \qquad P = I \times V \qquad P = I^2 \times R$$

Let us calculate the power dissipated by each resistor. This final calculation will be called step 5.

STEP 5 Since resistance, voltage, and current are known, either of the three formulas for power can be used to determine power.

$$P_1 = \frac{V_{R1}^2}{R_1} = \frac{(30\text{ V})^2}{10\ \Omega} = 90\text{ W}$$

$$P_2 = \frac{V_{R2}^2}{R_2} = \frac{(18\text{ V})^2}{12\ \Omega} = 27\text{ W}$$

$$P_3 = I_{R3}^2 \times R_3 = (1.5\text{ A})^2 \times 4\ \Omega = 9\text{ W}$$

$$P_4 = I_{R4}^2 \times R_4 = (1.5\text{ A})^2 \times 8\ \Omega = 18\text{ W}$$

$$P_5 = \frac{V_{R5}^2}{R_5} = \frac{(36\text{ V})^2}{60\ \Omega} = 21.6\text{ W}$$

$$P_6 = \frac{V_{R6}^2}{R_6} = \frac{(36\text{ V})^2}{24\ \Omega} = 54\text{ W}$$

$$P_7 = \frac{V_{R7}{}^2}{R_7} = \frac{(36\text{ V})^2}{40\ \Omega} = 32.4\text{ W}$$

$$\begin{aligned} P_T &= P_1 + P_2 + P_3 + P_4 + P_5 + P_6 + P_7 \\ &= 90 + 27 + 9 + 18 + 21.6 + 54 + 32.4 \\ &= 252\text{ W} \end{aligned}$$

or

$$\begin{aligned} P_T &= \frac{V_T{}^2}{R_T} = \frac{(84\text{ V})^2}{28\ \Omega} \\ &= 252\text{ W} \end{aligned}$$

The total power dissipated in this example circuit is 252 W. All the information can now be inserted in a final diagram for the example, as shown in Figure 5-18.

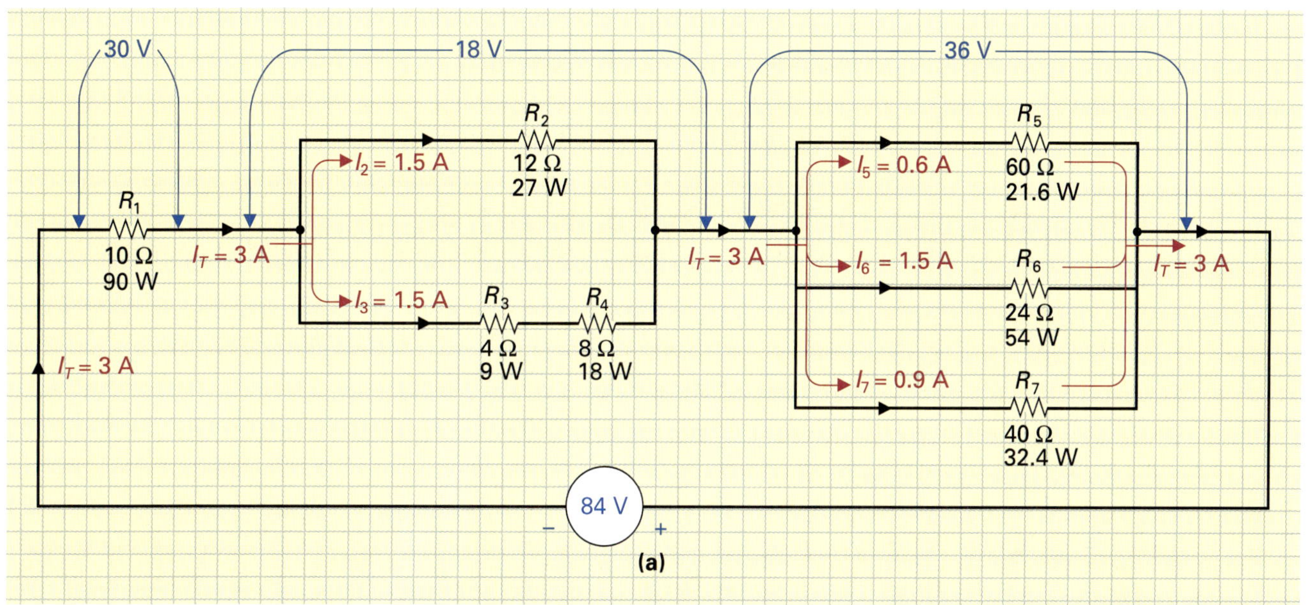

CIRCUIT ANALYSIS TABLE

Resistance $R = V/I$	Voltage $V = I \times R$	Current $I = V/R$	Power $P = V \times I$
$R_1 = 10\ \Omega$	$V_{R1} = 30$ V	$I_1 = 3$ A	$P_1 = 90$ W
$R_2 = 12\ \Omega$	$V_{R2} = 18$ V	$I_2 = 1.5$ A	$P_2 = 27$ W
$R_3 = 4\ \Omega$	$V_{R3} = 6$ V	$I_3 = 1.5$ A	$P_3 = 9$ W
$R_4 = 8\ \Omega$	$V_{R4} = 12$ V	$I_4 = 1.5$ A	$P_4 = 18$ W
$R_5 = 60\ \Omega$	$V_{R5} = 36$ V	$I_5 = 0.6$ A	$P_5 = 21.6$ W
$R_6 = 24\ \Omega$	$V_{R6} = 36$ V	$I_6 = 1.5$ A	$P_6 = 54$ W
$R_7 = 40\ \Omega$	$V_{R7} = 36$ V	$I_7 = 0.9$ A	$P_7 = 32.4$ W
$R_T = 28\ \Omega$	$V_T = V_S = 84$ V	$I_T = 3$ A	$P_T = 252$ W

(b)

FIGURE 5-18 Series–Parallel Circuit Example with All Information Inserted. (a) Schematic. (b) Circuit Analysis Table.

5-6 FIVE-STEP METHOD FOR SERIES–PARALLEL CIRCUIT ANALYSIS

Let's now combine and summarize all the steps for calculating resistance, voltage, current, and power in a series–parallel circuit by solving another problem. Before we begin, however, let us review the five-step procedure.

SOLVING FOR RESISTANCE, VOLTAGE, CURRENT, AND POWER IN A SERIES–PARALLEL CIRCUIT

STEP 1 Determine the circuit's total resistance.

Step A Solve for series-connected resistors in all parallel combinations.

Step B Solve for all parallel combinations.

Step C Solve for remaining series resistances.

STEP 2 Determine the circuit's total current.

STEP 3 Determine the voltage across each series resistor and each parallel combination (series equivalent resistor).

STEP 4 Determine the value of current through each parallel resistor in every parallel combination.

STEP 5 Determine the total and individual power dissipated by the circuit.

EXAMPLE:

Referring to Figure 5-19, calculate:

a. Total resistance
b. Total current
c. Voltage drop across all resistors
d. Current through each resistor
e. Total power dissipated by the circuit

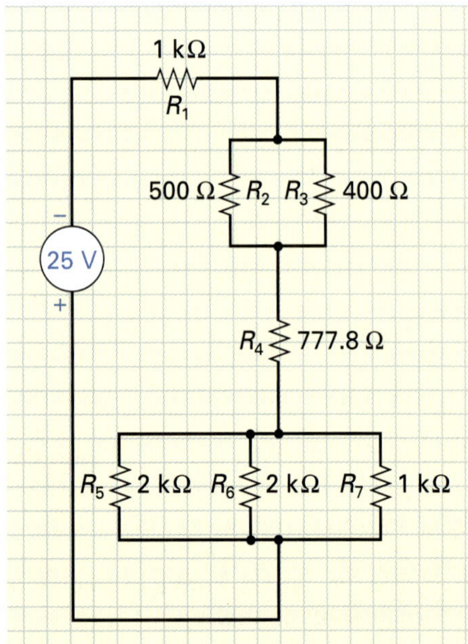

FIGURE 5-19 Apply the Five-Step Procedure to This Series–Parallel Circuit Example.

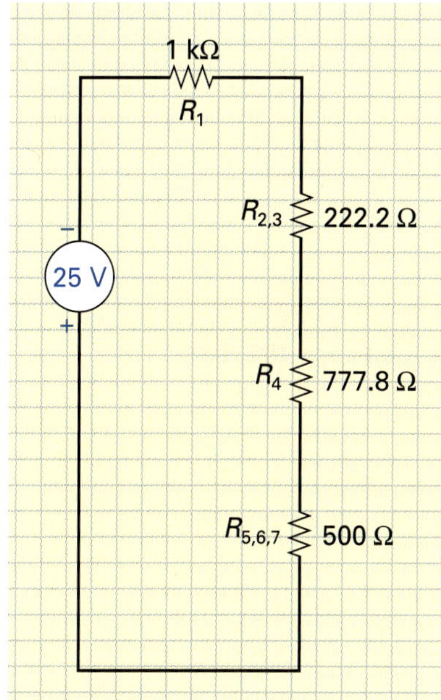

FIGURE 5-20 Circuit Resulting after Step 1B.

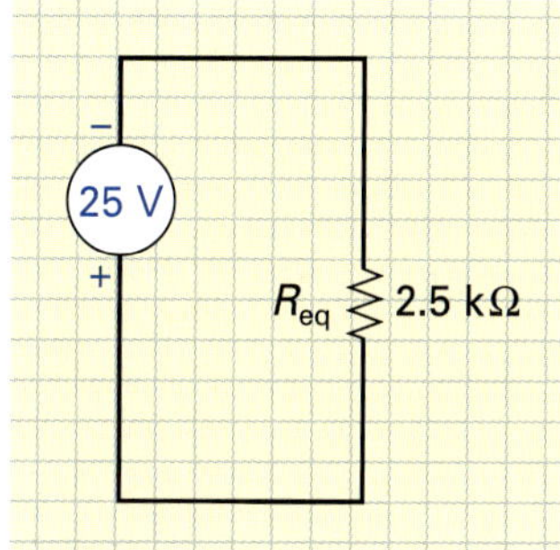

FIGURE 5-21 Circuit Resulting after Step 1C.

■ ***Solution:***

This problem has asked us to calculate everything about the series–parallel circuit shown in Figure 5-19, and is an ideal application for our five-step series–parallel circuit analysis procedure.

STEP 1 Determine the circuit's total resistance.

Step A: There are no series resistors within parallel combinations.

Step B: There are two-resistor (R_2, R_3) and three-resistor (R_5, R_6, R_7) parallel combinations in this circuit.

$$R_{2,3} = \frac{1}{(1/R_2) + (1/R_3)} = 222.2\ \Omega$$

$$R_{5,6,7} = \frac{1}{(1/R_5) + (1/R_6) + (1/R_7)} = 500\ \Omega$$

Figure 5-20 illustrates the circuit resulting after step B.

Step C: Solve for the remaining four resistances to gain the circuit's total resistance (R_T) or equivalent resistance (R_{eq}).

$$\begin{aligned} R_{eq} &= R_1 + R_{2,3} + R_4 + R_{5,6,7} \\ &= 1000\ \Omega + 222.2\ \Omega + 777.8\ \Omega + 500\ \Omega \\ &= 2500\ \Omega \quad \text{or} \quad 2.5\ \text{k}\Omega \end{aligned}$$

Figure 5-21 illustrates the circuit resulting after step C.

STEP 2 Determine the circuit's total current.

$$I_T = \frac{V_S}{R_T} = \frac{25\ \text{V}}{2.5\ \text{k}\Omega} = 10\ \text{mA}$$

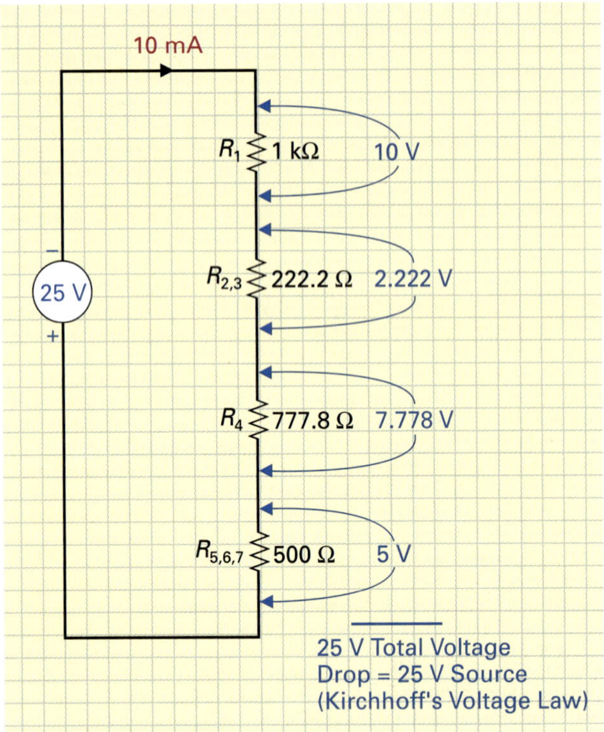

FIGURE 5-22 Circuit Resulting after Step 3.

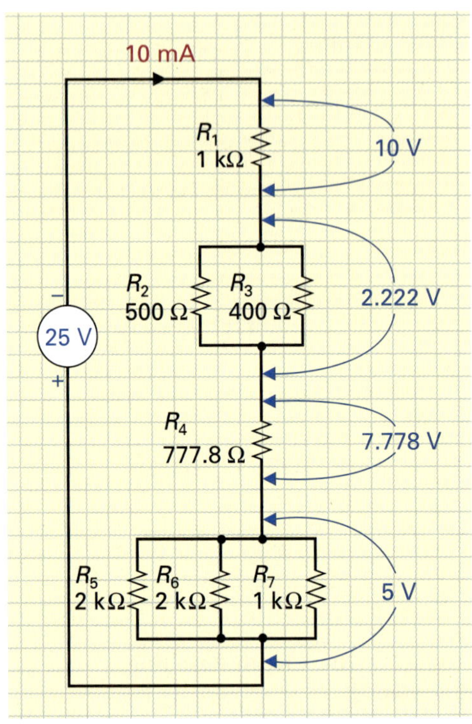

FIGURE 5-23 Series–Parallel Circuit Example with Step 1, 2, and 3 Data Inserted.

STEP 3 Determine the voltage across each series resistor and each series equivalent resistor. To achieve this, we utilize the diagram obtained after completing step B (Figure 5-20):

$$V_{R1} = I_T \times R_1 = 10 \text{ mA} \times 1 \text{ k}\Omega = 10 \text{ V}$$
$$V_{R2,3} = I_T \times R_{2,3} = 10 \text{ mA} \times 222.2 \ \Omega = 2.222 \text{ V}$$
$$V_{R4} = I_T \times R_4 = 10 \text{ mA} \times 777.8 \ \Omega = 7.778 \text{ V}$$
$$V_{R5,6,7} = I_T \times R_{5,6,7} = 10 \text{ mA} \times 500 \ \Omega = 5 \text{ V}$$

Figure 5-22 illustrates the results after step 3.

STEP 4 Determine the value of current through each parallel resistor (Figure 5-23). R_1 and R_4 are series-connected resistors, and therefore their current will equal 10 mA.

$$I_1 = 10 \text{ mA}$$
$$I_4 = 10 \text{ mA}$$

The current through the parallel resistors is calculated by Ohm's law.

$$I_2 = \frac{V_{R2}}{R_2} = \frac{2.222 \text{ V}}{500 \ \Omega} = 4.4 \text{ mA}$$

$$I_3 = \frac{V_{R3}}{R_3} = \frac{2.222 \text{ V}}{400 \ \Omega} = 5.6 \text{ mA}$$

$$\left.\begin{aligned} I_T &= I_2 + I_3 \\ 10 \text{ mA} &= 4.4 \text{ mA} + 5.6 \text{ mA} \end{aligned}\right\} \text{Kirchhoff's current law}$$

$$I_5 = \frac{V_{R5}}{R_5} = \frac{5 \text{ V}}{2 \text{ k}\Omega} = 2.5 \text{ mA}$$

$$I_6 = \frac{V_{R6}}{R_6} = \frac{5 \text{ V}}{2 \text{ k}\Omega} = 2.5 \text{ mA}$$

$$I_7 = \frac{V_{R7}}{R_7} = \frac{5 \text{ V}}{1 \text{ k}\Omega} = 5 \text{ mA}$$

$$\left.\begin{aligned} I_T &= I_5 + I_6 + I_7 \\ 10\text{ mA} &= 2.5\text{ mA} + 2.5\text{ mA} + 5\text{ mA} \end{aligned}\right\}\text{Kirchhoff's current law}$$

STEP 5 Determine the total power dissipated by the circuit.

$$P_T = P_1 + P_2 + P_3 + P_4 + P_5 + P_6 + P_7$$

or

$$P_T = \frac{V_T^2}{R_T}$$

Each resistor's power figure can be calculated and the sum would be the total power dissipated by the circuit. Since the problem does not ask for the power dissipated by each individual resistor, but for the total power dissipated, it will be easier to use the formula:

$$\begin{aligned} P_T &= \frac{V_T^2}{R_T} \\ &= \frac{(25\text{ V})^2}{2.5\text{ k}\Omega} \\ &= 0.25\text{ W} \end{aligned}$$

SELF-TEST EVALUATION POINT FOR 5-4, 5-5, AND 5-6

Use the following questions to test your understanding of Sections 5-4, 5-5, and 5-6.

1. State the five-step method used for series–parallel circuit analysis.
2. Design your own five-resistor series–parallel circuit, assign resistor values and a source voltage, and then apply the five-step analysis method.

5-7 SERIES–PARALLEL CIRCUITS

5-7-1 *Loading of Voltage-Divider Circuits*

Loading
The adding of a load to a source.

The straightforward voltage divider was discussed in Chapter 2, but at that point we did not explore some changes that will occur if a load resistance is connected to the voltage divider's output. Figure 5-24 shows a voltage divider, and as you can see, the advantage of a voltage-divider circuit is that it can be used to produce several different voltages from one main voltage source by the use of a few chosen resistor values.

In our discussion on load resistance, we discussed how every circuit or piece of equipment offers a certain amount of resistance, and this resistance represents how much a circuit or piece of equipment will load down the source supply.

Figure 5-25 shows an example voltage-divider circuit that is being used to develop a 10 V source from a 20 V dc supply. Figure 5-25(a) illustrates this circuit in the unloaded condition, and by making a few calculations you can analyze this circuit condition.

Bleeder Current
Current drawn continuously from a voltage source. A bleeder resistor is generally added to lessen the effect of load changes or provide a voltage drop across a resistor.

STEP 1 $R_T = R_1 + R_2 = 1\text{ k}\Omega + 1\text{ k}\Omega = 2\text{ k}\Omega$

STEP 2 $I_T = \dfrac{V_T}{R_T} = \dfrac{20\text{ V}}{2\text{ k}\Omega} = 10\text{ mA}$

The current that flows through a voltage divider, without a load connected, is called the **bleeder current.** In this example the bleeder current is equal to 10 mA. It is called the bleeder current because it is continually drawing or bleeding this current from the voltage source.

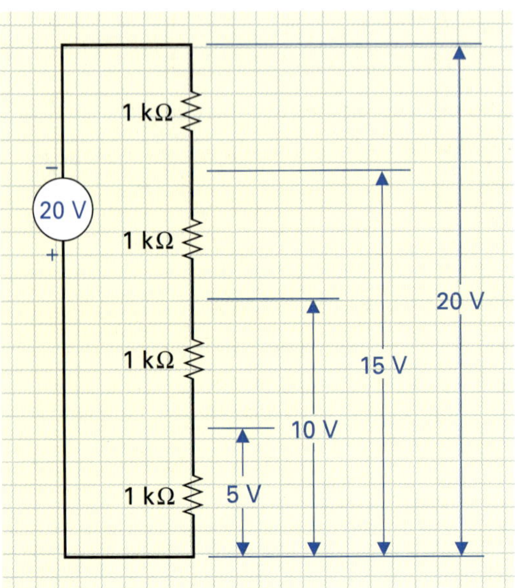

FIGURE 5-24 Voltage-Divider Circuit.

STEP 3 $V_{R1} = V_{R2}$ (as resistors are the same value)

$V_{R1} = 10 \text{ V}$

$V_{R2} = 10 \text{ V}$

In Figure 5-25(b) we have connected a piece of equipment represented as a resistance (R_3) across the 10 V supply. This automatically turns the previous series circuit of R_1 and R_2 into a series–parallel circuit made up of R_1, R_2, and the 100 kΩ load resistance. By making a few more calculations, we can discover the changes that have occurred by connecting this load resistance.

STEP 1 Total resistance (R_T)

Step B:

$$R_{2,3} = \frac{R_2 \times R_3}{R_2 + R_3} = \frac{1 \text{ k}\Omega \times 100 \text{ k}\Omega}{1 \text{ k}\Omega + 100 \text{ k}\Omega} = 990.1 \ \Omega$$

Step C:

$$\begin{aligned} R_{1,2,3} &= R_1 + R_{2,3} \\ &= 1 \text{ k}\Omega + 990.1 \ \Omega \\ &= 1.99 \text{ k}\Omega \end{aligned}$$

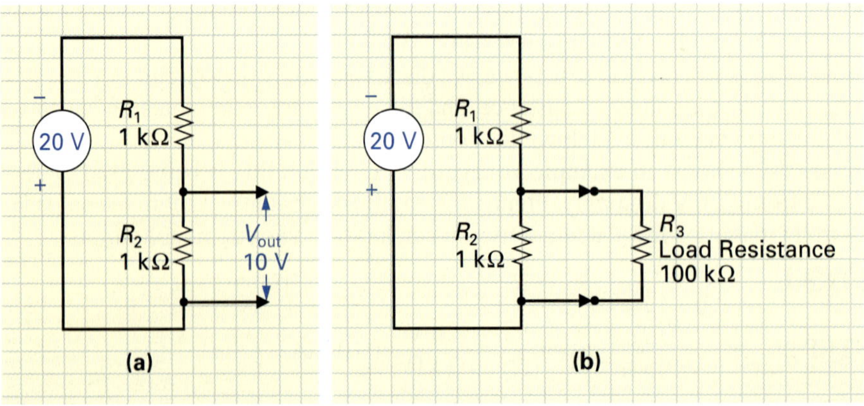

FIGURE 5-25 Voltage-Divider Circuit. (a) Unloaded Output Voltage. (b) Loaded Output Voltage.

STEP 2 Total current (I_T)

$$I_T = \frac{20 \text{ V}}{1.99 \text{ k}\Omega} = 10.05 \text{ mA}$$

STEP 3

$$V_{R1} = I_T \times R_1 = 10.05 \text{ mA} \times 1 \text{ k}\Omega = 10.05 \text{ V}$$

$$V_{R2,3} = I_T \times R_{2,3} = 10.05 \text{ mA} \times 990.1 \ \Omega = 9.95 \text{ V}$$

STEP 4

$$I_1 = I_T = 10.05 \text{ mA}$$

$$I_2 = \frac{V_{R2}}{R_2} = \frac{9.95 \text{ V}}{1 \text{ k}\Omega} = 9.95 \text{ mA}$$

$$I_3 = \frac{V_{R3}}{R_3} = \frac{9.95 \text{ V}}{100 \text{ k}\Omega} = 99.5 \ \mu\text{A}$$

$$\left.\begin{aligned} I_2 + I_3 &= I_T \\ 9.95 \text{ mA} + 99.5 \ \mu\text{A} &= 10.05 \text{ mA} \end{aligned}\right\} \text{Kirchhoff's current law}$$

As you can see, the load resistance is pulling 99.5 μA from the source, and this pulls the voltage down to 9.95 V from the required 10 V that was desired and is normally present in the unloaded condition.

When designing a voltage divider, design engineers need to calculate how much current a particular load will pull and then alter the voltage-divider resistor values to offset the loading effect when the load is connected.

5-7-2 *The Wheatstone Bridge*

In 1850, Charles Wheatstone developed a circuit to measure resistance. This circuit, which is still widely used today, is called the **Wheatstone bridge** and is illustrated in Figure 5-26(a). In Figure 5-26(b), the same circuit has been redrawn so that the series and parallel resistor connections are easier to see.

Wheatstone Bridge
A four-arm, generally resistive, bridge that is used to measure resistance.

Balanced Bridge

Figure 5-27 illustrates an example circuit in which four resistors are connected together to form a series–parallel arrangement. Let us now use the five-step procedure to find out exactly what resistance, current, voltage, and power values exist throughout the circuit.

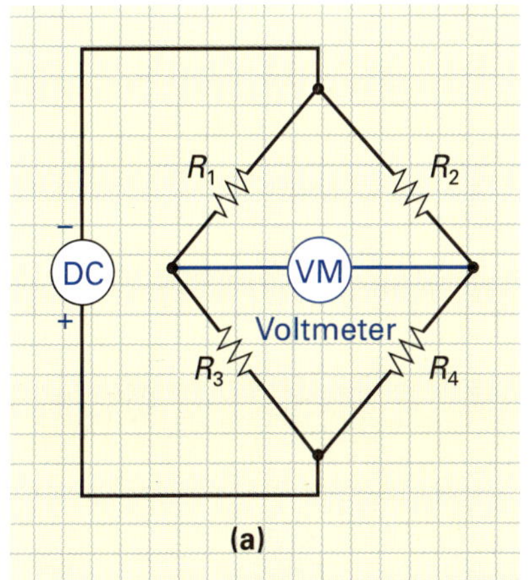

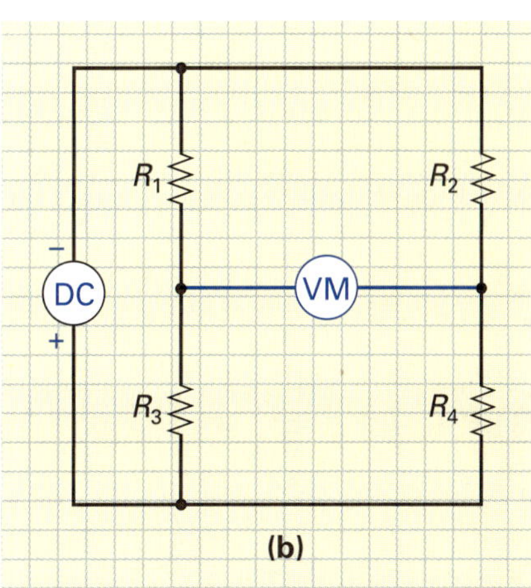

FIGURE 5-26 **Wheatstone Bridge. (a) Actual Circuit. (b) Redrawn Simplified Circuit.**

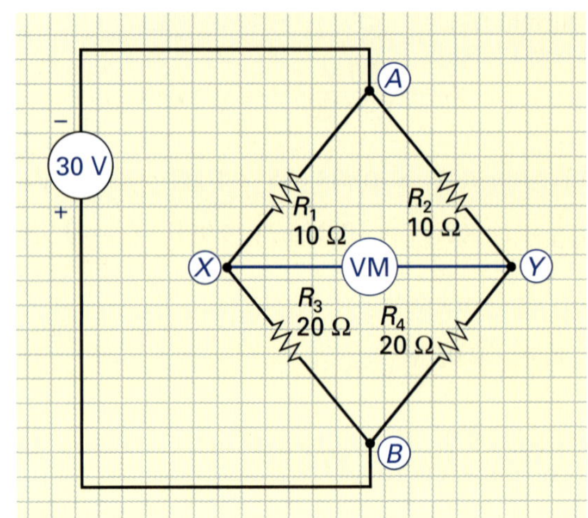

FIGURE 5-27 **Wheatstone Bridge Circuit Example.**

STEP 1 Total resistance (R_T)

Step A:

$$R_{1,3} = R_1 + R_3 = 10 + 20 = 30\ \Omega$$
$$R_{2,4} = R_2 + R_4 = 10 + 20 = 30\ \Omega$$

Step B:

$$R_T\text{: } (R_{1,2,3,4}) = \frac{R_{1,3} \times R_{2,4}}{R_{1,3} + R_{2,4}}$$
$$= \frac{30 \times 30}{30 + 30} = 15\ \Omega$$

Total resistance = 15 Ω

STEP 2 Total current (I_T)

$$I_T = \frac{V_T}{R_T} = \frac{30\text{ V}}{15\ \Omega} = 2\text{ A}$$

STEP 3 Since $R_{1,3}$ is in parallel with $R_{2,4}$, 30 V will appear across both $R_{1,3}$ and $R_{2,4}$.

$$V_T = V_{R1,3} = V_{R2,4} = 30\text{ V}$$

voltage-divider formula $\quad V_{R1} = \frac{R_1}{R_{1,3}} \times V_T$

$$= \frac{10}{30} \times 30 = 10\text{ V}$$
$$V_{R3} = \frac{R_3}{R_{1,3}} \times V_T = 20\text{ V}$$
$$V_{R2} = \frac{R_2}{R_{2,4}} \times V_T = 10\text{ V}$$
$$V_{R4} = \frac{R_4}{R_{2,4}} \times V_T = 20\text{ V}$$

STEP 4 $I_{1,3} = \frac{V_{R1,3}}{R_{1,3}} = \frac{30\text{ V}}{30\ \Omega} = 1\text{ A}$

$$I_{2,4} = \frac{V_{R2,4}}{R_{2,4}} = \frac{30\text{ V}}{30\ \Omega} = 1\text{ A}$$

$$\left.\begin{aligned} I_{1,3} + I_{2,4} &= I_T \\ 1\text{ A} + 1\text{ A} &= 2\text{ A} \end{aligned}\right\} \text{Kirchhoff's current law}$$

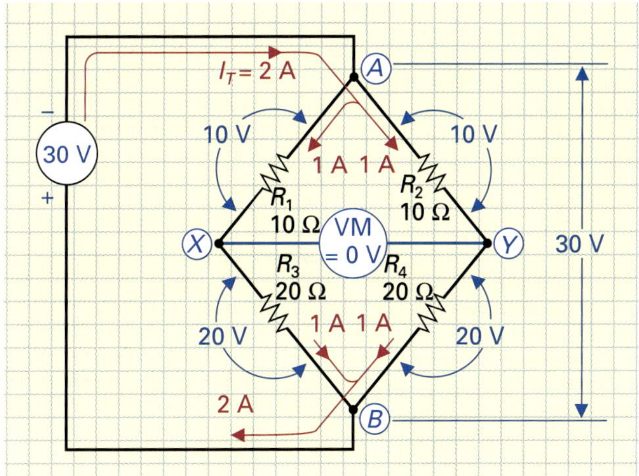

FIGURE 5-28 **Balanced Wheatstone Bridge.**

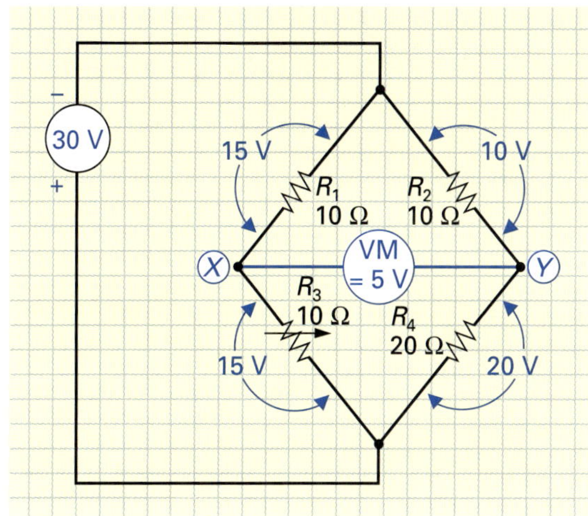

FIGURE 5-29 **Unbalanced Wheatstone Bridge.**

STEP 5 Total power dissipated (P_T)

$$
\begin{aligned}
P_T &= I_T^2 \times R_T \\
&= (2\text{ A})^2 \times 15\ \Omega \\
&= 4\text{ A}^2 \times 15\ \Omega \\
&= 60\text{ W}
\end{aligned}
$$

Figure 5-28 shows all of the step results inserted in the Wheatstone bridge example schematic. The Wheatstone bridge is said to be in the balanced condition when the voltage at point X equals the voltage at point Y ($V_{R3} = V_{R4}$, 20 V = 20 V). This same voltage exists across R_3 and R_4, so the voltmeter, which is measuring the voltage difference between X and Y, will indicate 0 V potential difference, and the circuit is said to be a *balanced bridge.*

Unbalanced Bridge

In Figure 5-29 we have replaced R_3 with a variable resistor and set it to 10 Ω. The R_2 and R_4 resistor combination will not change its voltage drop. However, R_1 and R_3, which are now equal, will each split the 30 V supply, producing 15 V across R_3. The voltmeter will indicate the difference in potential (5 V) from the voltage across R_3 at point X (15 V) and across R_4 at point Y (20 V). The voltmeter is actually measuring the imbalance in the circuit, which is why this circuit in this condition is known as an *unbalanced bridge.*

Determining Unknown Resistance

Figure 5-30 shows how a Wheatstone bridge circuit can be used to find the value of an unknown resistor (R_{un}). The variable-value resistor (R_{va}) is a calibrated resistor, which means that its resistance has been checked against a known, accurate resistance and its value can be adjusted and read from a calibrated dial.

The procedure to follow to find the value of the unknown resistor is as follows:

1. Adjust the variable-value resistor until the voltmeter indicates that the Wheatstone bridge is balanced (voltmeter indicates 0 V).
2. Read the value of the variable-value resistor. As long as $R_1 = R_2$, the variable resistance value will be the same as the unknown resistance value.

$$R_{va} = R_{un}$$

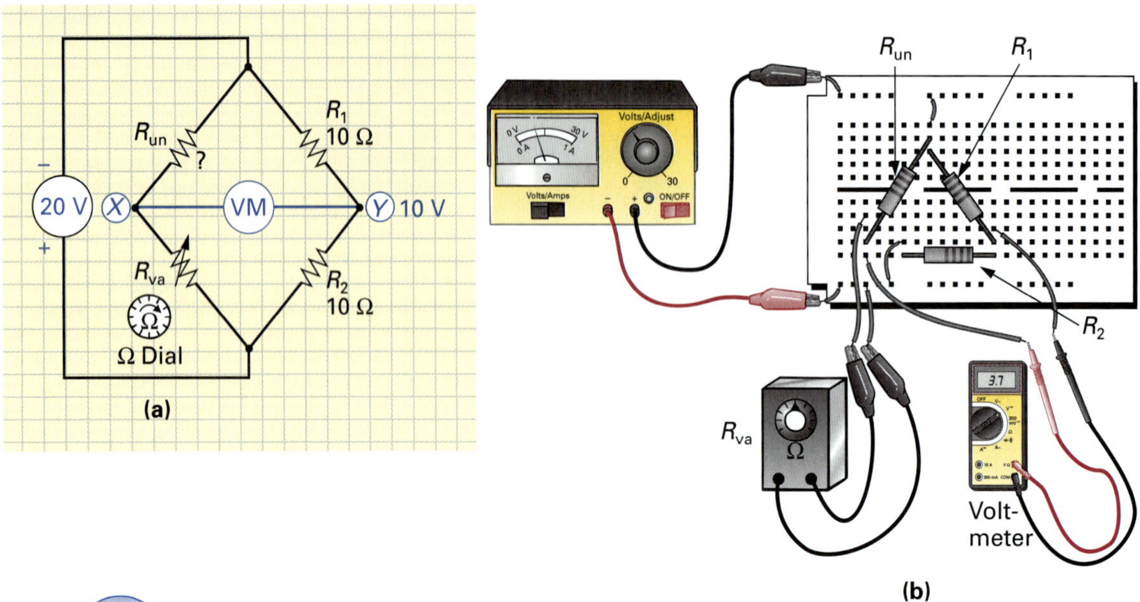

FIGURE 5-30 Using a Wheatstone Bridge to Determine an Unknown Resistance. (a) Schematic. (b) Pictorial.

Since R_1 and R_2 are equal to one another, the voltage will be split across the two resistors, producing 10 V at point *Y*. The variable-value resistor must therefore be adjusted so that it equals the unknown resistance, and therefore the same situation will occur, in that the 20 V source will be split, producing 10 V at point *X*, indicating a balanced zero-volt condition on the voltmeter. For example, if the unknown resistance is equal to 5 Ω, then only when the variable-value resistor is adjusted and equal to 5 Ω would 10 V appear at point *X* and allow the circuit meter to read zero volts, indicating a balance. The variable-value resistor resistance could be read (5 Ω) and the unknown resistor resistance would be known (5 Ω).

EXAMPLE:

What is the unknown resistance in Figure 5-31?

FIGURE 5-31 Wheatstone Bridge Circuit Example.

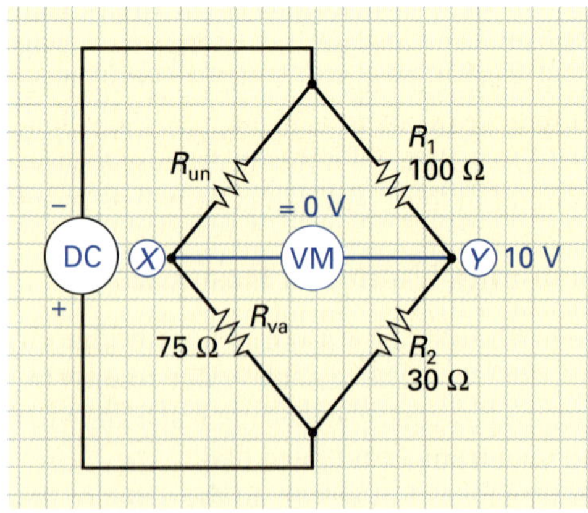

Solution:

The bridge is in a balanced condition as the voltmeter is reading a 0 V difference between points *X* and *Y*. In the previous section we discovered that if $R_1 = R_2$, then:

$$R_{va} = R_{un}$$

In this case, R_1 does not equal R_2, so a variation in the formula must be applied to take into account the ratio of R_1 and R_2.

$$R_{un} = R_{va} \times \frac{R_1}{R_2}$$

$$= 75\ \Omega \times \frac{100}{30}$$

$$= 75\ \Omega \times 3.33 = 250\ \Omega$$

Since R_1 is 3.33 times greater than R_2, then R_{un} must be 3.33 times greater than R_{va} if the Wheatstone bridge is in the balanced condition.

CALCULATOR SEQUENCE

Step	Keypad Entry	Display Response
1.	[1] [0] [0]	100
2.	[÷]	
3.	[3] [0]	30
4.	[=]	3.333
5.	[×]	
6.	[7] [5]	75
7.	[=]	250

SELF-TEST EVALUATION POINT FOR SECTION 5-7

Use the following questions to test your understanding of Section 5-7.

1. What is meant by loading of a voltage-divider circuit?
2. Sketch a Wheatstone bridge circuit and list an application of this circuit.

5-8 TROUBLESHOOTING SERIES–PARALLEL CIRCUITS

Troubleshooting is defined as the process of locating and diagnosing malfunctions or breakdowns in equipment by means of systematic checking or analysis. As discussed in previous resistive-circuit troubleshooting procedures, there are basically only three problems that can occur:

1. A component will open. This usually occurs if a resistor burns out or a wire or switch contact breaks.
2. A component will short. This usually occurs if a conductor, such as solder, wire, or some other conducting material, is dropped or left in the equipment, making or connecting two points that should not be connected.
3. There is a variation in a component's value. This occurs with age in resistors over a long period of time and can eventually cause a malfunction of the equipment.

Using the example circuit in Figure 5-32, we will step through a few problems, beginning with an open component. Throughout the troubleshooting, we will use the voltmeter whenever possible, as it can measure voltage by just connecting the leads across the component, rather than the ammeter, which has to be placed in the circuit, in which case the circuit path has to be opened. In some situations, using an ammeter can be difficult.

To begin, let's calculate the voltage drops and branch current obtained when the circuit is operating normally.

STEP 1 (A) $R_{3,4} = R_3 + R_4 = 3\ \Omega + 9\ \Omega = 12\ \Omega$

(B) $R_{2,3,4} = \dfrac{R_2 \times R_{3,4}}{R_2 + R_{3,4}} = \dfrac{6\ \Omega \times 12\ \Omega}{6\ \Omega + 12\ \Omega} = 4\ \Omega$

(C) $R_{1,2,3,4} = R_T = R_1 + R_{2,3,4}$
$= 2\ \Omega + 4\ \Omega = 6\ \Omega$

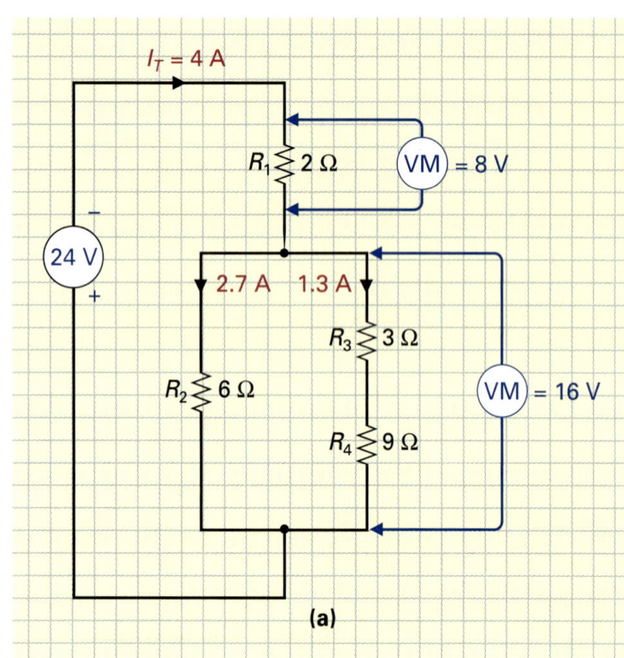

CIRCUIT ANALYSIS TABLE

Resistance $R = V/I$	Voltage $V = I \times R$	Current $I = V/R$	Power $P = V \times I$
$R_1 = 2\ \Omega$	$V_{R1} = 8$ V	$I_1 = 4$ A	$P_1 = 32$ W
$R_2 = 6\ \Omega$	$V_{R2} = 16$ V	$I_2 = 2.7$ A	$P_2 = 43.2$ W
$R_3 = 3\ \Omega$	$V_{R3} = 4$ V	$I_3 = 1.3$ A	$P_3 = 5.2$ W
$R_4 = 9\ \Omega$	$V_{R4} = 12$ V	$I_4 = 1.3$ A	$P_4 = 15.6$ W
$R_T = 6\ \Omega$	$V_T = V_S = 24$V	$I_T = 4$ A	$P_T = 96$ W

(b)

FIGURE 5-32 Series–Parallel Circuit Example. (a) Schematic. (b) Circuit Analysis Table.

STEP 2 $I_T = \dfrac{V_T}{R_T} = \dfrac{24\text{ V}}{6\ \Omega} = 4\text{ A}$

STEP 3 $V_{R1} = I_{R1} \times R_1 = 4\text{ A} \times 2\ \Omega = 8\text{ V}$

$V_{R2,3,4} = I_{R2,3,4} \times R_{2,3,4} = 4\text{ A} \times 4\ \Omega = 16\text{ V}$

(Kirchhoff's voltage law)

STEP 4 $I_1 = 4\text{ A}$ (series resistor)

$$I_2 = \frac{V_{R2}}{R_2} = \frac{16\text{ V}}{6\ \Omega} = 2.7\text{ A}$$

$$I_{3,4} = \frac{V_{R3,4}}{R_{3,4}} = \frac{16\text{ V}}{12\ \Omega} = 1.3\text{ A}$$

All these results have been inserted in the schematic in Figure 5-32(a) and in the circuit analysis table in Figure 5-32(b).

5-8-1 *Open Component*

R_1 Open (Figure 5-33)

With R_1 open, there cannot be any current flow through the circuit as there is not a path from one side of the power supply to the other. This fault can be recognized easily because approximately all of the applied 24 V will be measured across the open resistor (R_1), and 0 V will appear across all the other resistors.

R_3 Open (Figure 5-34)

With R_3 open, there will be no current through the branch made up of R_3 and R_4. The current path will be through R_1 and R_2, and therefore the total resistance will now increase ($R_T\uparrow$) from 6 Ω to

$$R_T = R_1 + R_2$$
$$= 2\ \Omega + 6\ \Omega = 8\ \Omega$$

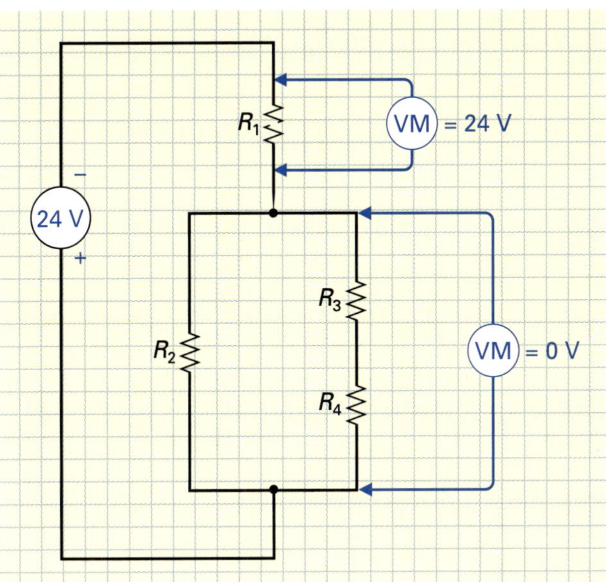

FIGURE 5-33 **Open Series-Connected Resistor in a Series–Parallel Circuit.**

This 8 Ω is an increase in circuit resistance from the normal resistance, which was 6 Ω, which implies that an open has occurred to increase resistance. The total current will decrease ($I_T\downarrow$) from 4 A to

$$I_T = \frac{V_T}{R_T} = \frac{24 \text{ V}}{8 \,\Omega} = 3 \text{ A}$$

The voltage drop across the resistors will be

$$V_{R1} = I_T \times R_1 = 3 \text{ A} \times 2 \,\Omega = 6 \text{ V}$$
$$V_{R2} = I_T \times R_2 = 3 \text{ A} \times 6 \,\Omega = 18 \text{ V}$$

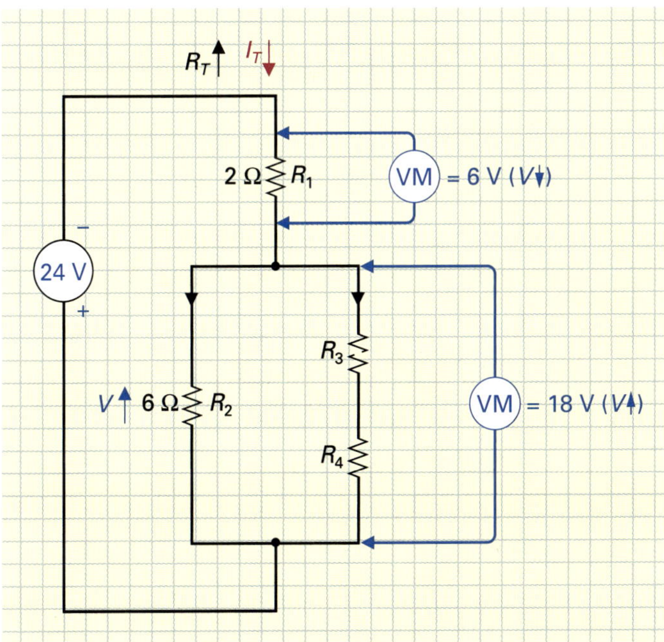

FIGURE 5-34 **Open Parallel-Connected Resistor in a Series–Parallel Circuit.**

If one of the parallel branches is opened, the overall circuit resistance will always increase. This increase in the total resistance will cause an increase in the voltage dropped across the parallel branch (the greater the resistance, the greater the voltage drop), which enables the technician to localize the fault area and also to determine that the fault is an open.

The voltage measured with a voltmeter will be

$$\begin{aligned} V_{R1} &= 6\text{ V} \\ V_{R2} &= 18\text{ V} \\ V_{R3} &= 18\text{ V (open)} \\ V_{R4} &= 0\text{ V} \end{aligned}$$

This identifies the problem as R_3 being open, as it drops the entire parallel circuit voltage (18 V) across itself, whereas normally the voltage would be dropped proportionally across R_3 and R_4, which are in series with one another.

5-8-2 *Shorted Component*

R_1 Shorted (Figure 5-35)

With R_1 shorted, the total circuit resistance will decrease ($R_T\downarrow$), causing an increase in circuit current ($I_T\uparrow$). This increase in current will cause an increase in the voltage dropped across the parallel branch. However, the fault can be located once you measure the voltage across R_1, which will read 0 V, indicating that this resistor has almost no resistance, as it has no voltage drop across it.

R_3 Shorted (Figure 5-36)

With R_3 shorted, there will be a decrease in the circuit's total resistance from 6 Ω to

$$\begin{aligned} R_{2,3,4} &= \frac{R_2 \times R_{3,4}}{R_2 + R_{3,4}} = \frac{6 \times 9}{6 + 9} = 3.6\ \Omega \\ R_{1,2,3,4} &= R_1 + R_{2,3,4} \\ &= 2\ \Omega + 3.6\ \Omega \\ &= 5.6\ \Omega \end{aligned}$$

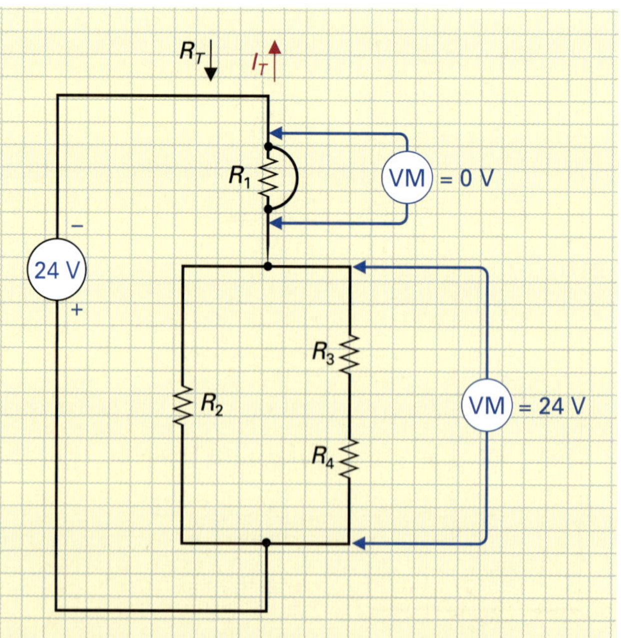

FIGURE 5-35 Shorted Series-Connected Resistor in a Series–Parallel Circuit.

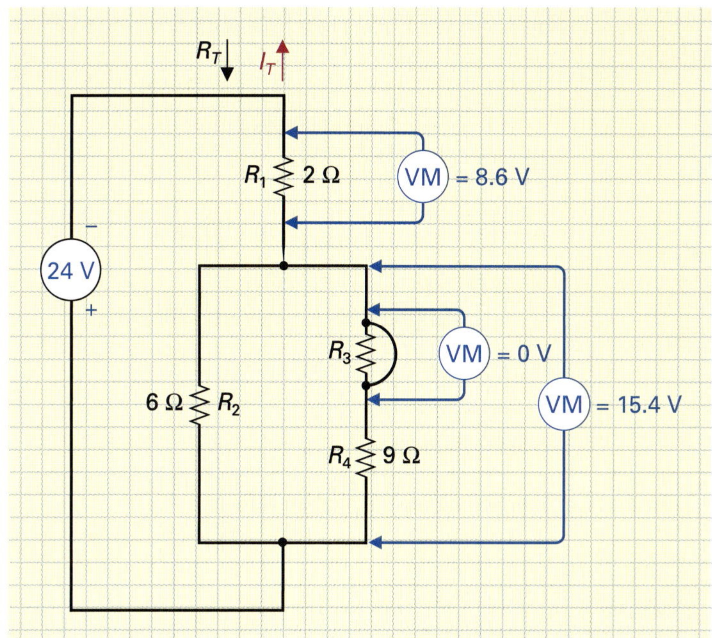

FIGURE 5-36 **Shorted Parallel-Connected Resistor in a Series–Parallel Circuit.**

This decrease in total resistance ($R_T\downarrow$) will cause an increase in total current ($I_T\uparrow$), which implies that a short has occurred to decrease resistance. The total current will now increase from 4 A to

$$I_T = \frac{V_T}{R_T} = \frac{24\text{ V}}{5.6\ \Omega} = 4.3\text{ A}$$

The voltage drops across the resistors will be

$$V_{R1} = I_T \times R_1 = 4.3\text{ A} \times 2\ \Omega = 8.6\text{ V}$$
$$V_{R2,3,4} = I_T \times R_{2,3,4} = 4.3\text{ A} \times 3.6\ \Omega = 15.4\text{ V}$$

When one of the parallel branch resistors is shorted, the overall circuit resistance will always decrease. This decrease in total resistance will cause a decrease in the voltage dropped across the parallel branch (the smaller the resistance, the smaller the voltage drop), and this enables the technician to localize the faulty area and also to determine that the fault is a short.

The voltage measured with the voltmeter (VM) will be

$$V_{R1} = 8.6\text{ V}$$
$$V_{R2} = 15.4\text{ V}$$
$$V_{R3} = 0\text{ V (short)}$$
$$V_{R4} = 15.4\text{ V}$$

This identifies the problem as R_3 being a short, as 0 V is being dropped across it.

In summary, you may have noticed that (Figure 5-37):

1. An open component causes total resistance to increase ($R_T\uparrow$) and, therefore, total current to decrease ($I_T\downarrow$), and the open component, if in series, has the supply voltage across it, and if in a parallel branch, has the parallel branch voltage across it.
2. A shorted component causes total resistance to decrease ($R_T\downarrow$) and, therefore, total current to increase ($I_T\uparrow$), and the shorted component, if in series or parallel branches, will have 0 V across it.

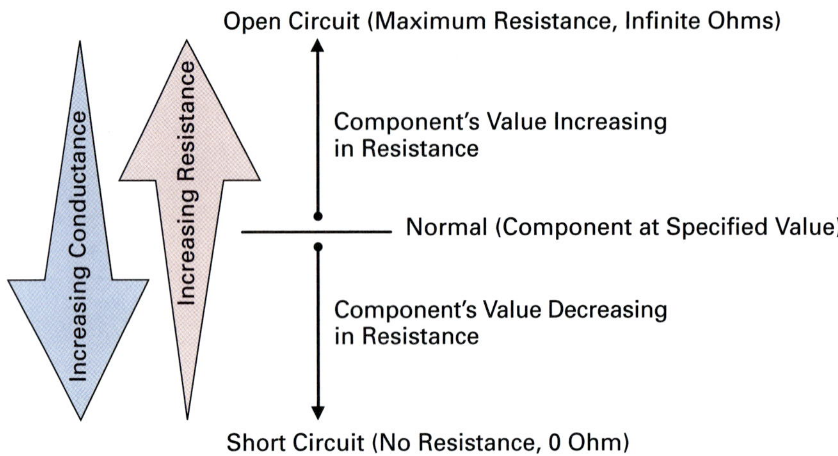

FIGURE 5-37 Summary of Symptoms for Opens and Shorts.

5-8-3 *Resistor Value Variation*

R_2 Resistance Decreases (Figure 5-38)

If the resistance of R_2 decreases, the total circuit resistance (R_T) will decrease and the total circuit current (I_T) will increase. The result of this problem and the way in which the fault can be located is that when the voltage drop across R_1 and the parallel branch is tested, there will be an increased voltage drop across R_1 due to the increased current flow, and a decrease in the voltage drop across the parallel branch due to a decrease in the parallel branch resistance.

With open and shorted components, the large voltage (open) or small voltage (short) drop across a component enables the technician to identify the faulty component. A variation in a component's value, however, will vary the circuit's behavior. With this example, the symptoms could have been caused by a combination of variations. So once the

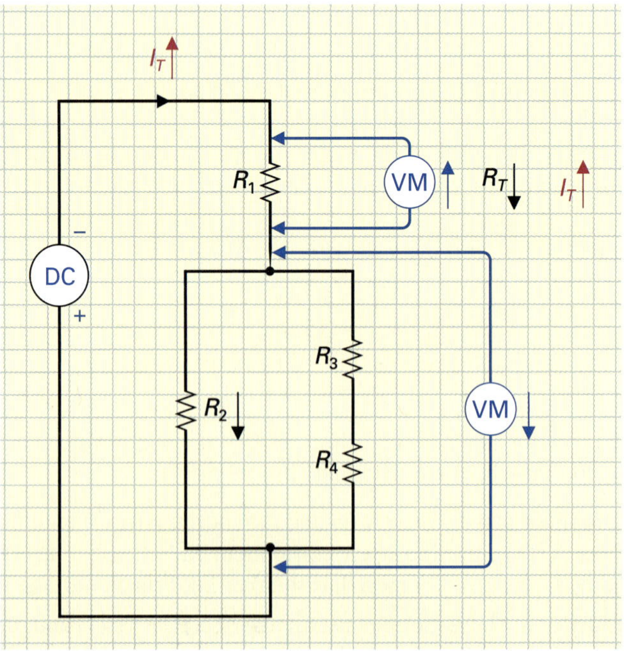

FIGURE 5-38 Parallel-Connected Resistor Value Decrease in Series–Parallel Circuit.

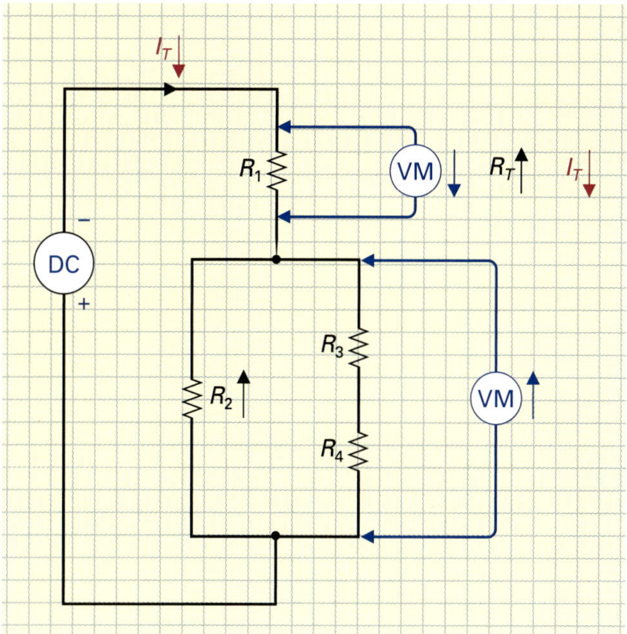

FIGURE 5-39 Parallel-Connected Resistor Value Increase in Series–Parallel Circuit.

area of the problem has been localized, the next troubleshooting step is to disconnect power and then remove each of the resistors in the suspected faulty area and verify that their resistance values are correct by measuring their resistances with an ohmmeter. In this problem we had an increase in voltage across R_1 and a decrease in voltage across the parallel branch when measuring with a voltmeter, which was caused by R_2 decreasing. The same swing in voltage readings could also have been obtained by an increase in the resistance of R_1.

R_2 Resistance Increases (Figure 5-39)

If the resistance of R_2 increases, the total circuit resistance (R_T) will increase and the total circuit current (I_T) will decrease. The voltage drop across R_1 will decrease and the voltage across the branch will increase in value. Once again, these measured voltage changes could be caused by the resistance of R_2 increasing or the resistance of R_1 decreasing.

To reinforce your understanding let's work out a few examples of troubleshooting other series–parallel circuits.

EXAMPLE:

If bulb 1 in Figure 5-40 goes open, what effect will it have, and how will the fault be recognized?

Solution:

A visual inspection of the circuit shows all bulbs off, as there is no current path from one side of the battery to the other, since bulb 1 is connected in series and has opened the only path. Since all bulbs are off, the faulty bulb cannot be visually isolated. However, you can easily localize the faulty bulb by using one of two methods:

1. Use the voltmeter and check the voltage across each bulb. Bulb 1 would have 12 V across it, while 2 and 3 would have 0 V across them. This isolates the faulty component to bulb 1, since all the supply voltage is being measured across it.

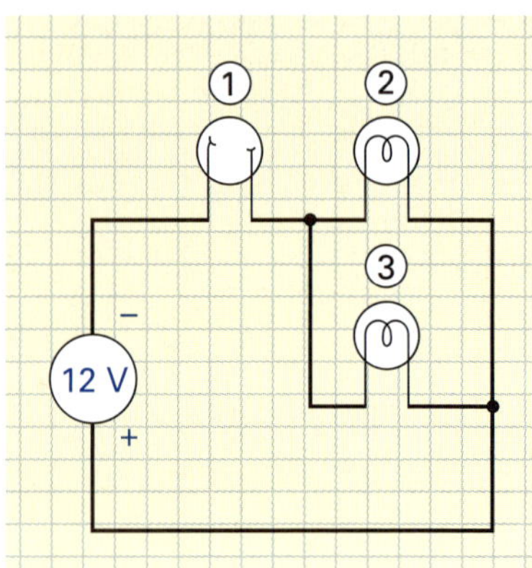

FIGURE 5-40 **Series-Connected Open Bulb in Series–Parallel Circuit.**

2. By analyzing the circuit diagram, you can see that only one bulb can open and cause all the bulbs to go out, and that is bulb 1. If bulb 2 opens, 1 and 3 would still be on, and if bulb 3 opens, 1 and 2 would still remain on. With power off, the ohmmeter could verify this open.

EXAMPLE:

One resistor in Figure 5-41 has shorted. From the voltmeter reading shown, determine which one.

Solution:

If the supply voltage is being measured across the parallel branch of R_2 and R_3, there cannot be any other resistance in circuit, so R_1 must have shorted. The next step would be to locate the component, R_1, and determine what has caused it to short. If we were not told that a resistor had shorted, the same symptom could have been caused if R_2 and R_3 were both open, and therefore the open parallel branch would allow no current to flow and maximum supply voltage would appear across it. The individual component resistance, when checked, will isolate the problem.

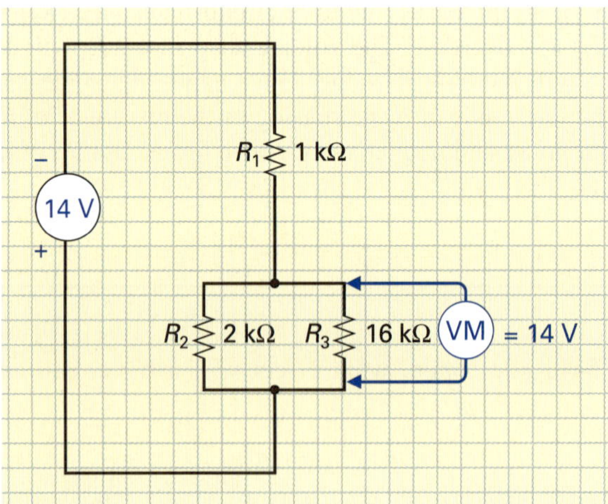

FIGURE 5-41 **Find the Shorted Resistor.**

EXAMPLE:

Determine if there is an open or short in Figure 5-42. If so, isolate it by the two voltage readings that are shown in the circuit diagram.

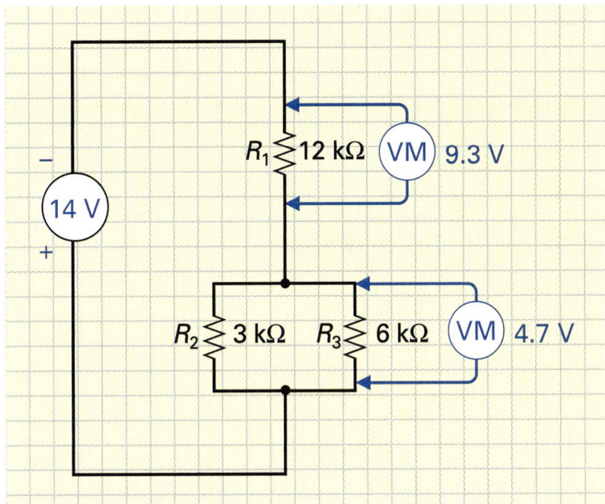

FIGURE 5-42 Does a Problem Exist?

Solution:

Performing a few calculations, you should come up with a normal total circuit resistance of 14 kΩ and a total circuit current of 1 mA. This should cause 12 V across R_1 and 2 V across the parallel branch under no-fault conditions. The decrease in the voltage drop across the series resistor R_1 leads you to believe that there has been a decrease in total circuit current, which must have been caused by a total resistance increase, which points to an open component (assuming that only an open or short can occur and not a component value variation).

If R_1 was open, all the 14 V would have been measured across R_1, which did not occur.

If R_3 opened,

$$\text{total resistance} = 15\ \text{k}\Omega$$

$$\text{total current} = 0.93\ \text{mA}\left(\frac{V_T}{R_T} = \frac{14\ \text{V}}{15\ \text{k}\Omega}\right)$$

$$V_{R1} = I_T \times R_1 = 11.16\ \text{V}$$

Since the voltage dropped across R_1 was 9.3 V, R_3 is not the open.

If R_2 opened,

$$\text{total resistance} = 18\ \text{k}\Omega$$

$$\text{total current} = 0.78\ \text{mA}\left(\frac{V_T}{R_T} = \frac{14\ \text{V}}{18\ \text{k}\Omega}\right)$$

$$V_{R1} = I_T \times R_1 = 9.36\ \text{V}$$

This circuit's problem is resistor R_2, which has opened.

SELF-TEST EVALUATION POINT FOR SECTION 5-8

Use the following questions to test your understanding of Section 5-8.

Describe how to isolate the following problems in a series–parallel circuit.

1. An open component
2. A shorted component
3. A resistor value variation

REVIEW QUESTIONS

Multiple-Choice Questions

1. A series–parallel circuit is a combination of:
   a. Components connected end to end
   b. Series (one current path) circuits
   c. Both series and parallel circuits
   d. Parallel (two or more current path) circuits
2. Total resistance in a series–parallel circuit is calculated by applying the __________ resistance formula to series-connected resistors and the __________ resistance formula to resistors connected in parallel.
   a. Series, parallel **c.** Series, series
   b. Parallel, series **d.** Parallel, parallel
3. Total current in a series–parallel circuit is determined by dividing the total __________ by the total __________.
   a. Power, current **c.** Current, resistance
   b. Voltage, resistance **d.** Voltage, power
4. Branch current within series–parallel circuits can be calculated by:
   a. Ohm's law
   b. The current-divider formula
   c. Kirchhoff's current law
   d. All of the above
5. A __________ ground has the negative side of the source voltage connected to ground, while a __________ ground has the positive side of the source voltage connected to ground.
   a. Positive, negative **c.** Positive, earth
   b. Chassis, earth **d.** Negative, positive
6. The output voltage will always __________ when a load or voltmeter is connected across a voltage divider.
   a. Decrease **c.** Increase
   b. Remain the same **d.** All of the above could be considered true.
7. A Wheatstone bridge was originally designed to measure:
   a. An unknown voltage
   b. An unknown current
   c. An unknown power
   d. An unknown resistance
8. A balanced bridge has an output voltage:
   a. Equal to the supply voltage
   b. Equal to half the supply voltage
   c. Of 0 V
   d. Of 5 V
9. In a series–parallel resistive circuit, an open series-connected resistor will cause __________ current, whereas an open parallel-connected resistor will result in a total current __________.
   a. An increase in, decrease **c.** Zero, decrease
   b. A decrease in, increase **d.** None of the above
10. In a series–parallel resistive circuit, a shorted series-connected resistor will cause __________ current, whereas a shorted parallel-connected resistor will result in a total current __________.
   a. An increase in, increase **c.** An increase in, decrease
   b. A decrease in, decrease **d.** A decrease in, increase

Practice Problems

11. R_3 and R_4 are in series with one another and are both in parallel with R_5. This parallel combination is in series with two series-connected resistors, R_1 and R_2. R_1 = 2.5 kΩ, R_2 = 10 kΩ, R_3 = 7.5 kΩ, R_4 = 2.5 kΩ, R_5 = 2.5 MΩ, and V_S = 100 V. For these values, calculate:
   a. Total resistance
   b. Total current
   c. Voltage across series resistors and parallel combinations
   d. Current through each resistor
   e. Total and individual power figures
12. Referring to the example in Question 11, calculate the voltage at every point of the circuit with respect to ground.
13. A 10 V source is connected across a series–parallel circuit made up of R_1 in parallel with a branch made up of R_2 in series with a parallel combination of R_3 and R_4. R_1 = 100 Ω, R_2 = 100 Ω, R_3 = 200 Ω, and R_4 = 300 Ω. For these values, apply the five-step procedure, and also determine the voltage at every point of the circuit with respect to ground.

Troubleshooting Questions

14. Referring to the example circuit in Figure 5-21, describe the effects you would get if a resistor were to short, and if a resistor were to open.

15. Design a simple five-resistor series–parallel circuit and insert a source voltage and resistance values. Apply the five-step series–parallel circuit procedure, and then theoretically open and short all the resistors and calculate what effect would occur and how you would recognize the problem.

16. Carbon composition resistors tend to increase in resistance with age, while most other types generally decrease in resistance. What effects would resistance changes have on their respective voltage drops?

Web Site Questions

Go to the web site http://www.prenhall.com/cook, select the textbook *Electronics: A Complete Course,* select this chapter, and then follow the instructions when answering the multiple choice practice problems.

Direct Current (DC) versus Alternating Current (AC)

The Laser

Charles Townes

In 1898, H. G. Wells's famous book, *The War of the Worlds,* had Martian invaders with laserlike death rays blasting bricks, incinerating trees, and piercing iron as if it were paper. In 1917, Albert Einstein stated that, under certain conditions, atoms or molecules could absorb light and then be stimulated to release this borrowed energy. In 1954, Charles H. Townes, a professor at Columbia University, conceived and constructed with his students the first "maser" (acronym for "microwave amplification by stimulated emission of radiation"). In 1958, Townes and Arthur L. Shawlow wrote a paper showing how stimulated emission could be used to amplify light waves as well as microwaves, and the race was on to develop the first "laser." In 1960, Theodore H. Maiman, a scientist at Hughes Aircraft Company, directed a beam of light from a flash lamp into a rod of synthetic crystal, which responded with a burst of crimson light so bright that it outshone the sun.

An avalanche of new lasers emerged, some as large as football fields, while others were no bigger than a pinhead. They can be made to produce invisible infrared or ultraviolet light or any visible color in the rainbow, and the high-power lasers can vaporize any material a million times faster and more intensely than a nuclear blast, while the low-power lasers are safe to use in children's toys.

At present, the laser is being used by the FBI to detect fingerprints that are 40 years old, in defense programs, in compact disk players, in underground fiber optic communication to transmit hundreds of telephone conversations, to weld car bodies, to drill holes in baby-bottle nipples, to create three-dimensional images called holograms, and as a surgeon's scalpel in the operating room. Not a bad beginning for a device that when first developed was called "a solution looking for a problem."

6

Introduction

In this chapter, we will begin by describing the difference between direct current (dc) and alternating current (ac), and then examine where and why ac is used. Following this we will discuss all the characteristics of ac waveform shapes, the differences between electricity and electronics, and finally, ac test equipment.

6-1 THE DIFFERENCE BETWEEN DC AND AC

Direct Current
Current flow in only one direction.

Alternating Current
Electric current that rises from zero to a maximum in one direction, falls to zero, and then rises to a maximum in the opposite direction, and then repeats another cycle, the positive and negative alternations being equal.

One of the best ways to describe anything new is to begin by redescribing something known and then discuss the unknown. The known topic in this case is **direct current.** Direct current (dc) is the flow of electrons in one DIRECTion and one direction only. DC voltage is nonvarying and normally obtained from a battery or power supply unit, as seen in Figure 6–1(a). The only variation in voltage from a battery occurs due to the battery's discharge, but even then, the current will still flow in only one direction, as seen in Figure 6-1(b). A dc voltage of 9 or 6 V could be illustrated graphically as shown in Figure 6-1(c). Whether 9 or 6 V, the voltage can be seen to be constant or the same at any time.

Alternating current (ac) flows first in one direction and then in the opposite direction. This reversing current is produced by an alternating voltage source, as shown in Figure 6–2(a), which reaches a maximum in one direction (positive), decreases to zero, and then reverses itself and reaches a maximum in the opposite direction (negative). This is graphically illustrated in Figure 6-2(b). During the time of the positive voltage alternation, the polarity of the voltage will be as shown in Figure 6-2(c), so current will flow from neg-

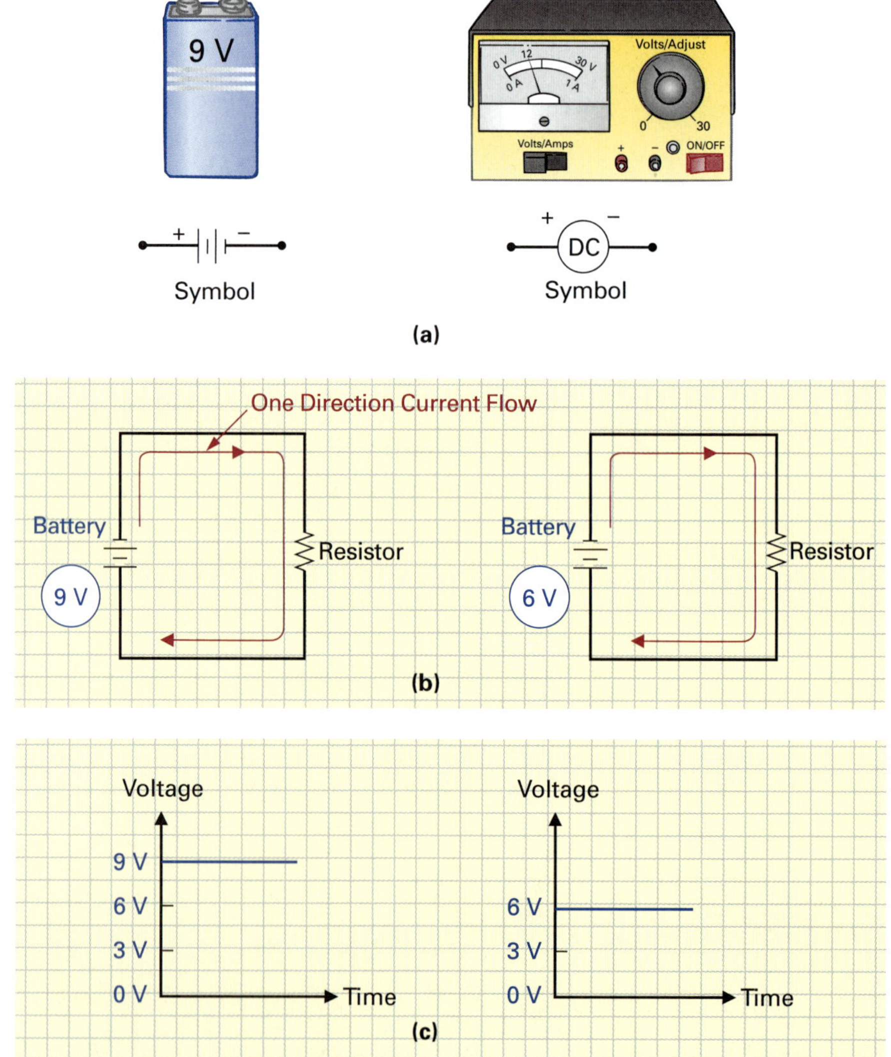

FIGURE 6-1 Direct Current. (a) DC Sources. (b) DC Flow. (c) Graphic Representation of DC.

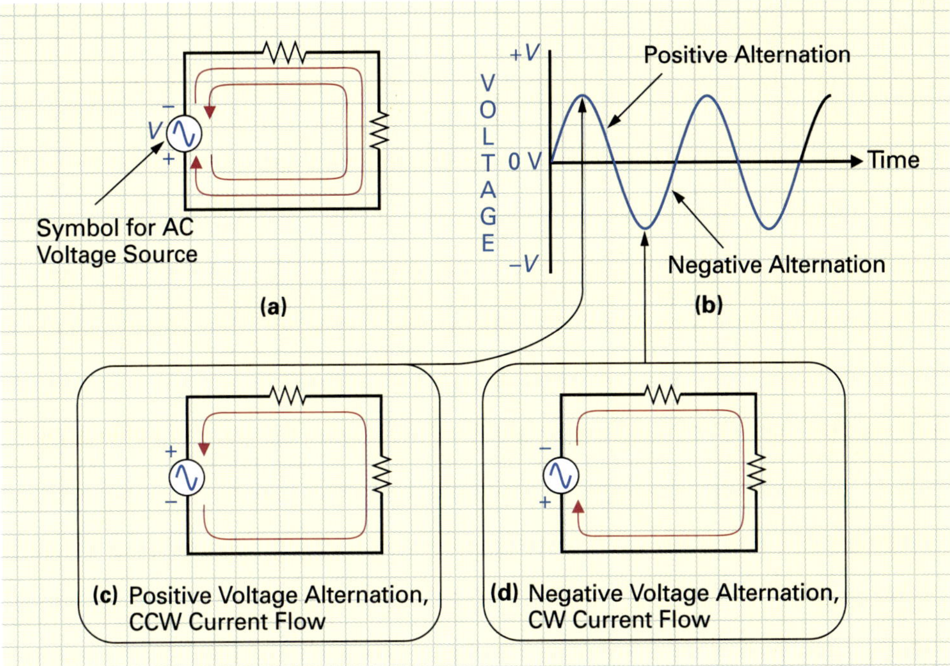

FIGURE 6-2 **Alternating Current.**

ative to positive in a counterclockwise direction. During the time of the negative voltage alternation, the polarity of the voltage will reverse, as shown in Figure 6-2(d), causing current to flow once again from negative to positive, but in this case, in the opposite clockwise direction.

SELF-TEST EVALUATION POINT FOR SECTION 6-1

Use the following questions to test your understanding of Section 6-1.

1. Give the full names of the following abbreviations: (a) ac; (b) dc.
2. The polarity of a/an ___________ voltage source will continually reverse, and therefore so will the circuit current.
3. The polarity of a/an ___________ voltage source will remain constant, and therefore current will flow in only one direction.
4. State briefly the difference between ac and dc.

6-2 WHY ALTERNATING CURRENT?

The question that may be troubling you at this point is: If we have been managing fine for the past chapters with dc, why do we need ac?

There are two main applications for ac:

1. *Power transfer:* to supply electrical power for lighting, heating, cooling, appliances, and machinery in both home and industry
2. *Information transfer:* to communicate or carry information, such as radio music and television pictures, between two points

To begin with, let us discuss the first of these applications, power transfer.

6-2-1 *Power Transfer*

There are three advantages that ac has over dc from a power point of view, and these are:

1. Flashlights, radios, and portable televisions all use batteries (dc) as a source of power. In these applications where a small current is required, batteries will last a good length of time before there is a need to recharge or replace them. Many appliances and most industrial equipment need a large supply of current, and in this situation a generator would have to be used to generate this large amount of current. Generators operate in the opposite way to motors, in that a **generator** converts a mechanical rotation input into an electrical output. Generators can be used to generate either dc or ac, but ac generators can be larger, less complex internally, and cheaper to operate, and this is the first reason why we use ac instead of dc for supplying power.

Generator
Device used to convert a mechanical energy input into an electrical energy output.

2. From a power point of view, ac is always used by electric companies when transporting power over long distances to supply both the home and industry with electrical energy. Recalling the power formula, you will remember that power is proportional to either current or voltage squared ($P \propto I^2$ or $P \propto V^2$), which means that to supply power to the home or industry, we would supply either a large current or voltage. As you can see in Figure 6-3, between the electric power plant and home or industry are power lines carrying the power. The amount of power lost (heat) in these power lines can be calculated by using the formula $P = I^2 \times R$, where I is the current flowing through the

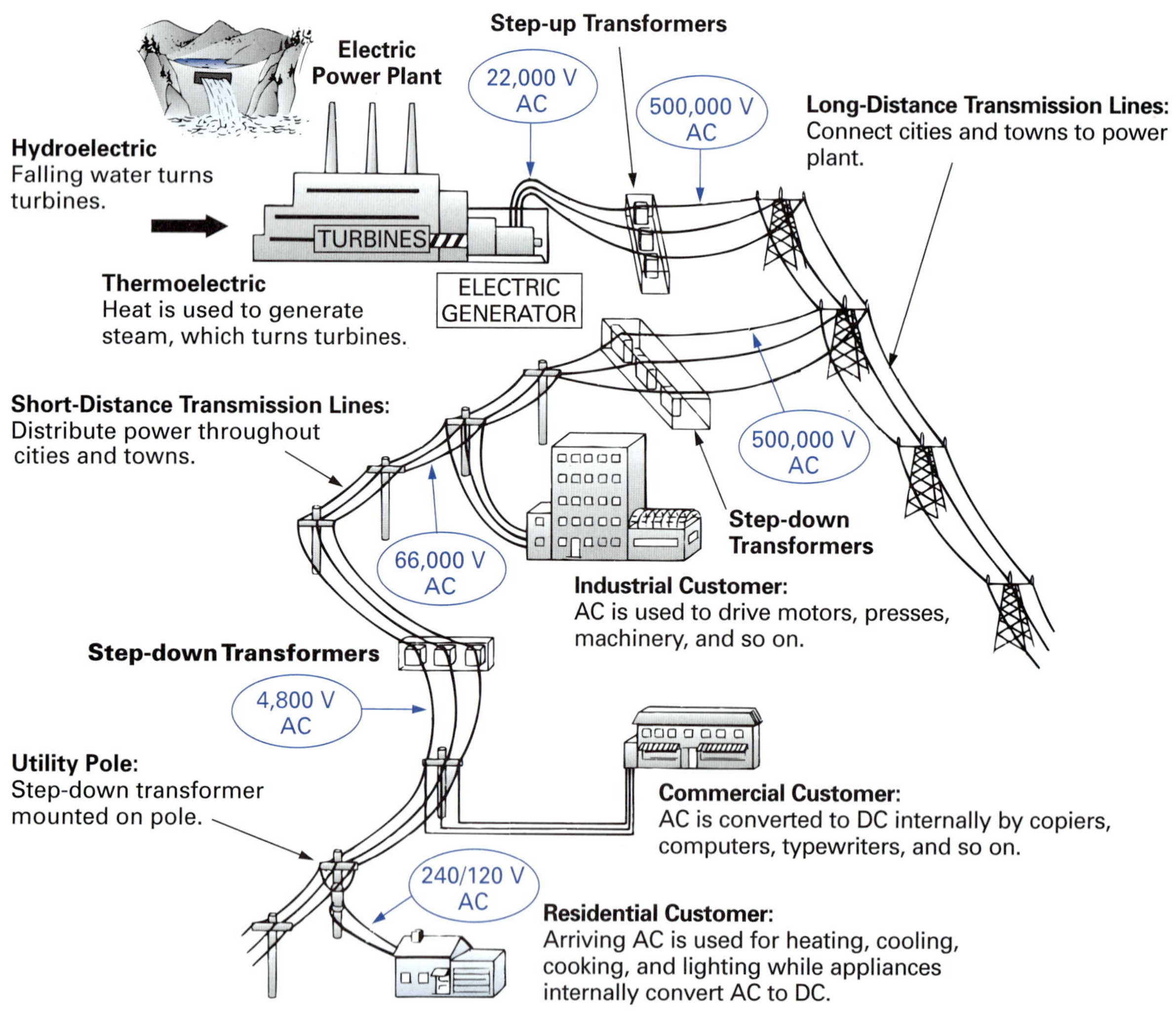

FIGURE 6-3 AC Power Distribution.

line and R is the resistance of the power lines. This means that the larger the current, the greater the amount of power lost in the lines in the form of heat and therefore the less the amount of power supplied to the home or industry. For this reason, power companies transport electric energy at a very high voltage between 200,000 and 600,000 V. Since the voltage is high, the current can be low and provide the same amount of power to the consumer ($P = V\uparrow \times I\downarrow$). Yet, by keeping the current low, the amount of heat loss generated in the power lines is minimal.

Now that we have discovered why it is more efficient over a long distance to transport high voltages than high current, what does this have to do with ac? An ac voltage can easily and efficiently be transformed up or down to a higher or lower voltage by utilizing a device known as a **transformer,** and even though dc voltages can be stepped up and down, the method is inefficient and more complex.

3. Nearly all electronic circuits and equipment are powered by dc voltages, which means that once the ac power arrives at the home or industry, in most cases it will have to be converted into dc power to operate electronic equipment. It is a relatively simple process to convert ac to dc, but conversion from dc to ac is a complex and comparatively inefficient process.

Transformer
Device consisting of two or more coils that are used to couple electric energy from one circuit to another, yet maintain electrical isolation between the two.

Figure 6-3 illustrates ac power distribution from the electric power plant to the home and industry. The ac power distribution system begins at the electric power plant, which has the powerful large generators driven by turbines to generate large ac voltages. The turbines can be driven by either falling water (hydroelectric), or from steam, which is produced with intense heat by burning either coal, gas, or oil or from a nuclear reactor (thermoelectric). The turbine supplies the mechanical energy to the generator, to be transformed into ac electrical energy.

The generator generates an ac voltage of approximately 22,000 V, which is stepped up by transformers to approximately 500,000 V. This voltage is applied to the long-distance transmission lines, which connect the power plant to the city or town. At each city or town, the voltage is tapped off the long-distance transmission lines and stepped down to approximately 66,000 V, and is distributed to large-scale industrial customers. The 66,000 V is stepped down again to approximately 4800 V and distributed throughout the city or town by short-distance transmission lines. This 4800 V is used by small-scale industrial customers and residential customers who receive the ac power via step-down transformers on utility poles, which step down the 4800 V to 240 V and 120 V.

Most equipment and devices within industry and the home will run directly from the ac power, such as heating, lighting, and cooling. Some equipment that runs on dc, such as televisions and computers, has an internal dc power supply that will accept the 120 V ac and convert it to the dc voltages required to power the system.

SELF-TEST EVALUATION POINT FOR SECTION 6-2-1

Use the following questions to test your understanding of Section 6-2-1.

1. In relation to power transfer, what three advantages does ac have over dc?
2. True or false: A generator converts an electrical input into a mechanical output.
3. What formula is used to calculate the amount of power lost in a transmission line?
4. What is a transformer?
5. What voltage is provided to the wall outlet in the home?
6. Most appliances internally convert the __________ input voltage into a __________ voltage.

6-2-2 *Information Transfer*

Communication
Transmission of information between two points.

Information, by definition, is the property of a signal or message that conveys something meaningful to the recipient. **Communication,** which is the transfer of information between

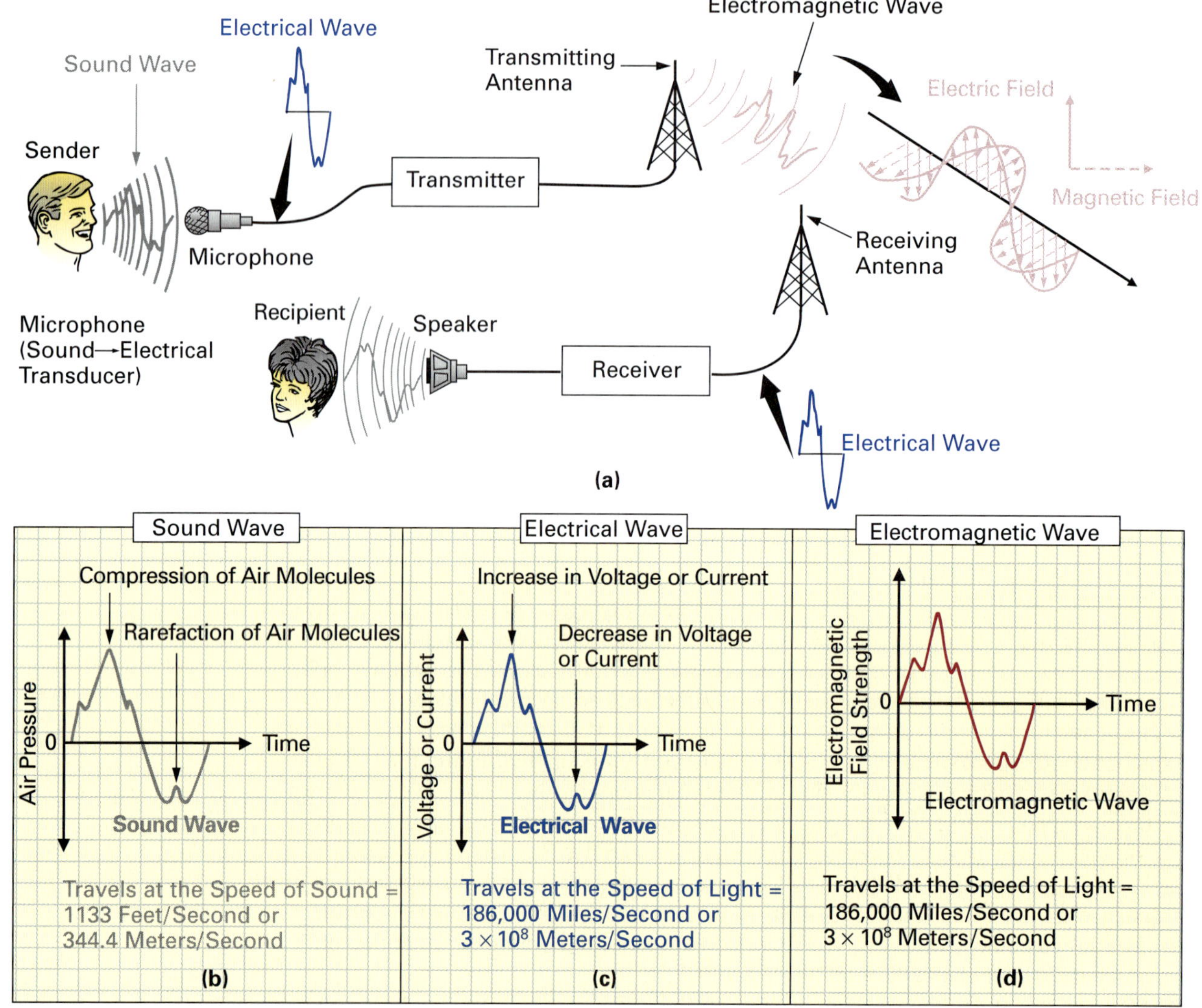

FIGURE 6-4 **Information Transfer.**

two points, began with speech and progressed to handwritten words in letters and printed words in newspapers and books. To achieve greater distances of communication, face-to-face communications evolved into telephone and radio communications.

A simple communication system can be seen in Figure 6-4(a).

The voice information or sound wave produced by the sender is a variation in air pressure, and travels at the speed of sound, as detailed in Figure 6-4. Sound waves or sounds are normally generated by a vibrating reed or plucked string in the case of musical instruments. In this example the sender's vocal cords vibrate backward and forward, producing a rarefaction or decreased air pressure, where few air molecules exist, and a compression or increased air pressure, where many air molecules exist. Like the ripples produced by a stone falling in a pond, the sound waves produced by the sender are constantly expanding and traveling outward.

Transducer
Any device that converts energy from one form to another.

Electrical Wave
Traveling wave propagated in a conductive medium that is a variation in voltage or current and travels at slightly less than the speed of light.

The microphone is in fact a **transducer** (energy converter), because it converts the sound wave (which is a form of mechanical energy) into electrical energy in the form of voltage and current, which varies in the same manner as the sound wave and therefore contains the sender's message or information.

The **electrical wave,** shown in Figure 6-4(c), is a variation in voltage or current and can only exist in a wire conductor or circuit. This electrical signal travels at the speed of light.

The speaker, like the microphone, is also an electroacoustical transducer that converts the electrical energy input into a mechanical sound-wave output. These sound waves strike the outer eardrum, causing the ear diaphragm to vibrate, and these mechanical vibrations actuate nerve endings in the ear, which convert the mechanical vibrations into electrochemical impulses that are sent to the brain. The brain decodes this information by comparing these impulses with a library of previous sounds and so provides the sensation of hearing.

To communicate between two distant points, a wire must be connected between the microphone and speaker. However, if an electrical wave is applied to an antenna, the electrical wave is converted into a radio or electromagnetic wave, as shown in the inset in Figure 6-4(a), and communication is established without the need of a connecting wire—hence the term **wireless communication.** Antennas are designed to radiate and receive electromagnetic waves, which vary in field strength, as shown in Figure 6-4(d), and can exist in either air or space. These radio waves, as they are also known, travel at the speed of light and allow us to achieve great distances of communication.

More specifically, radio waves are composed of two basic components. The electrical voltage applied to the antenna is converted into an electric field and the electrical current into a magnetic field. This **electromagnetic** (electric–magnetic) **wave** is used to carry a variety of information, such as speech, radio broadcasts, television signals, and so on.

In summary, the sound wave is a variation in air pressure, the electrical wave is a variation of voltage or current, and the electromagnetic wave is a variation of electric and magnetic field strength.

Wireless Communication
Term describing radio communication that requires no wires between the two communicating points.

Electromagnetic (Radio) Wave
Wave that consists of both an electric and magnetic variation, and travels at the speed of light.

EXAMPLE:

How long will it take the sound wave produced by a rifle shot to travel 9630.5 feet?

Solution:

This problem makes use of the following formula:

$$\text{Distance} = \text{velocity} \times \text{time}$$

or

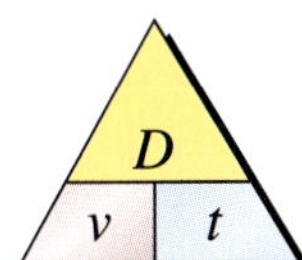

D = distance
v = velocity
t = time

If someone travels at 20 mph for 2 hours, the person will travel 40 miles ($D = v \times t = 20$ mph $\times$ 2 hours = 40 miles). In this problem, the distance (9630.5) and the sound wave's velocity (1130 ft/sec) are known, and so by rearranging the formula, we can find time:

$$\text{time} = \frac{\text{distance}}{\text{velocity}} = \frac{9630.5\ \text{ft}}{1130\ \text{ft/s}} = 8.5\ \text{s}$$

EXAMPLE:

How long will it take an electromagnetic (radio) wave to reach a receiving antenna that is 2000 miles away from the transmitting antenna?

Solution:

In this problem, both distance (2000 miles) and velocity (186,000 miles/s) are known, and time has to be calculated:

$$\text{time} = \frac{\text{distance}}{\text{velocity}} = \frac{2000\ \text{miles}}{186{,}000\ \text{miles/s}}$$
$$= 1.075 \times 10^{-2} \text{ or } 10.8\ \text{ms}$$

TIME LINE
Studying the experiments of Maxwell and Heinrich Hertz, Guglielmo Marconi (1874–1937) invented a practical system of telegraphy communication. In an evolutionary process, Marconi extended his distance of communication from 1 1/2 miles in 1896 to 6000 miles in 1902. In September 1899, Marconi equipped two U.S. ships with equipment and used them in the Atlantic Ocean to transmit to America the progress of the America's Cup yacht race.

6-2-3 *Electrical Equipment or Electronic Equipment?*

In the beginning of this chapter, it was stated that ac is basically used in two applications: (1) power transfer and (2) information transfer. These two uses for ac help define the difference between electricity and electronics. Electronic equipment manages the flow of information, while electrical equipment manages the flow of power. In summary:

EQUIPMENT	MANAGES
Electrical	Power (large values of V and I)
Electronic	Information (small values of V and I)

To use an example, we can say that a dc power supply is a piece of electrical equipment since it is designed to manage the flow of power. A TV set, however, is an electronic system since its electronic circuits are designed to manage the flow of audio (sound) and video (picture) information.

Since most electronic systems include a dc power supply, we can also say that the electrical circuits manage the flow of power, and this power supply enables the electronic circuits to manage the flow of information.

SELF-TEST EVALUATION POINT FOR SECTIONS 6-2-2 AND 6-2-3

Use the following questions to test your understanding of Sections 6-2-2 and 6-2-3.

1. The __________ wave is a variation in air pressure, the __________ wave is a variation in field strength, and the __________ wave is a variation of voltage or current.
2. Sound waves travel at the speed of sound, which is __________, while electrical and electromagnetic waves travel at the speed of light, which is __________.
3. A human ear is designed to receive __________ waves, an antenna is designed to transmit or receive __________ waves, and an electronic circuit is designed to pass only __________ waves.
4. __________ equipment manages the flow of information, and these ac waveforms normally have small values of current and voltage.
5. __________ equipment manages the flow of power, and these ac waveforms normally have large values of current and voltage.

6-3 AC WAVE SHAPES

In all fields of electronics, whether medical, industrial, consumer, or data processing, different types of information are being conveyed between two points, and electronic equipment is managing the flow of this information.

Let's now discuss the basic types of ac wave shapes. The way in which a wave varies in magnitude with respect to time describes its wave shape. All ac waves can be classified into one of six groups, and these are illustrated in Figure 6-5.

Sine Wave
Wave whose amplitude is the sine of a linear function of time. It is drawn on a graph that plots amplitude against time or radial degrees relative to the angular rotation of an alternator.

6-3-1 *The Sine Wave*

The **sine wave** is the most common type of waveform. It is the natural output of a generator that converts a mechanical input, in the form of a rotating shaft, into an electrical output in the form of a sine wave. In fact, for one cycle of the input shaft, the generator will produce one sinusoidal ac voltage waveform, as shown in Figure 6-6. When the input shaft of the generator is at 0°, the ac output is 0 V. As the shaft is rotated through 360°, the ac output voltage

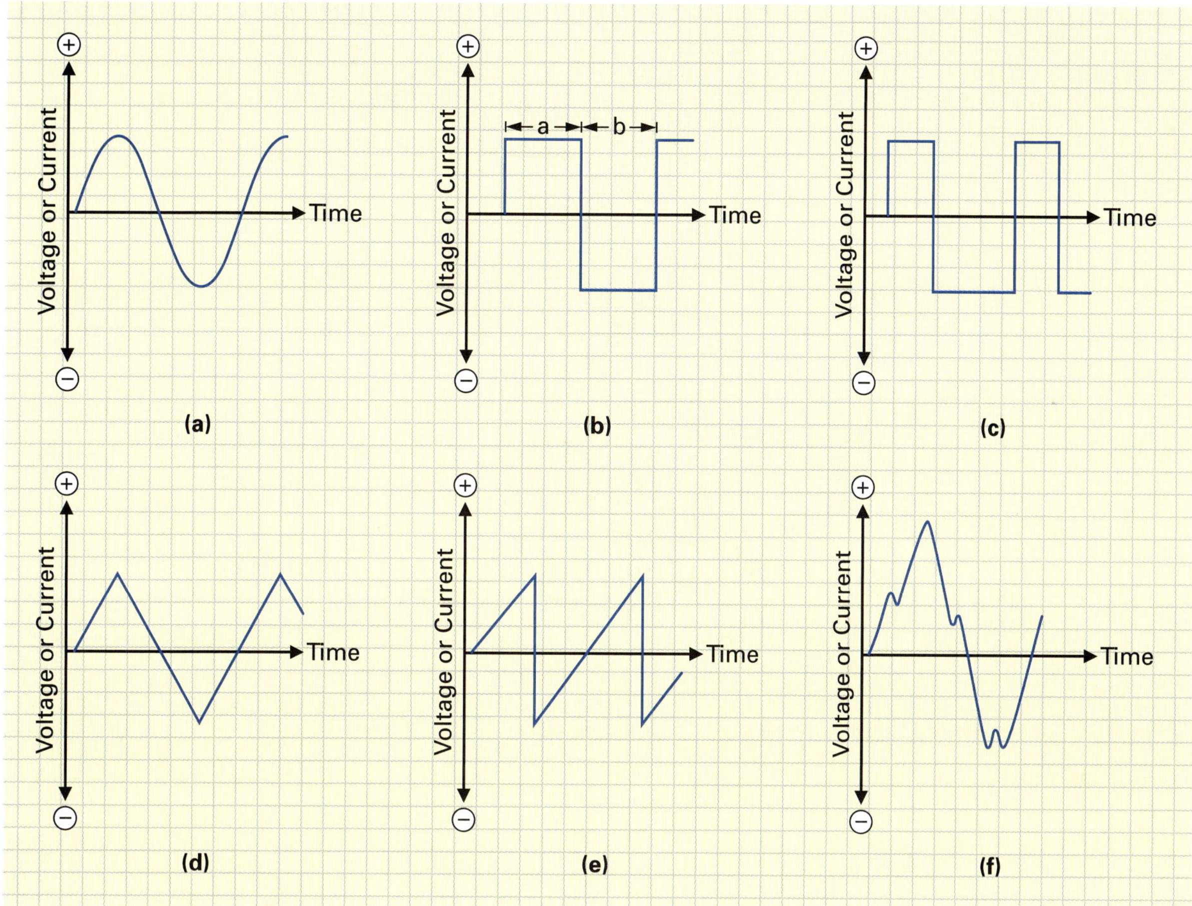

FIGURE 6-5 **AC Wave Shapes. (a) Sine Wave. (b) Square Wave. (c) Pulse Wave. (d) Triangular Wave. (e) Sawtooth Wave. (f) Irregular Wave.**

will rise to a maximum positive voltage at 90°, fall back to 0 V at 180°, and then reach a maximum negative voltage at 270°, and finally return to 0 V at 360°. If this ac voltage is applied across a closed circuit, it produces a current that continually reverses or alternates in each direction.

Figure 6-7 illustrates the sine wave, with all the characteristic information inserted, which at first glance looks a bit ominous. Let's analyze and discuss each piece of information individually, beginning with the sine wave's amplitude.

Amplitude

Figure 6-8 plots direction and amplitude against time. The **amplitude** or magnitude of a wave is often represented by a **vector** arrow, also illustrated in Figure 6-8. The vector's length indicates the magnitude of the current or voltage, while the arrow's point is used to show the direction, or polarity.

Amplitude
Magnitude or size an alternation varies from zero.

Vector
Quantity that has both magnitude and direction. They are normally represented as a line, the length of which indicates magnitude and the orientation of which, due to the arrowhead on one end, indicates direction.

Peak Value

The peak of an ac wave occurs on both the positive and negative alternation, but is only at the peak (maximum) for an instant. Figure 6-9(a) illustrates an ac current waveform rising to a positive peak of 10 A, falling to zero, and then reaching a negative peak of 10 A in the reverse direction. Figure 6-9(b) shows an ac voltage waveform reaching positive and negative peaks of 9 V.

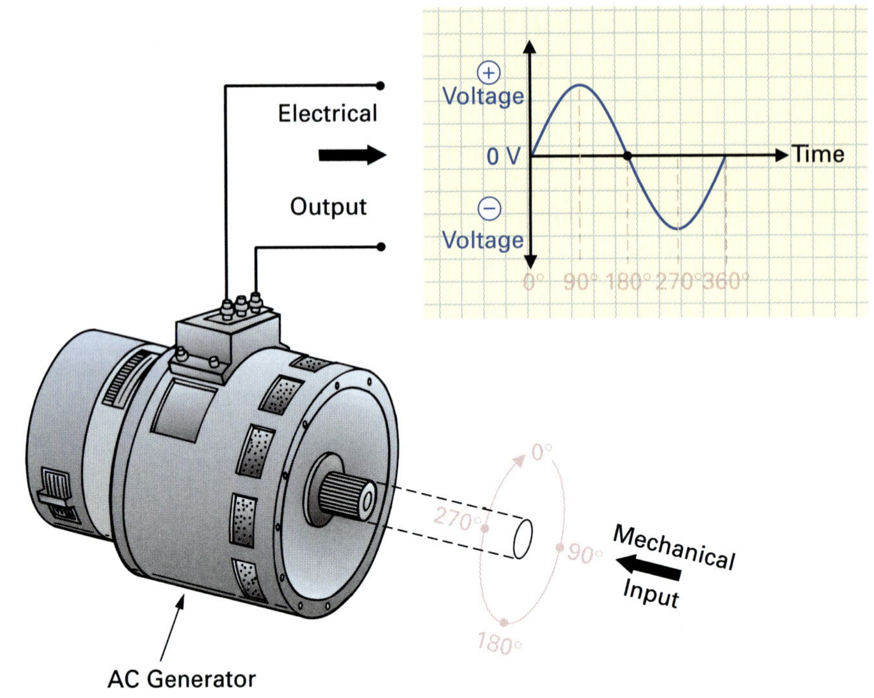

FIGURE 6-6 **Degrees of a Sine Wave.**

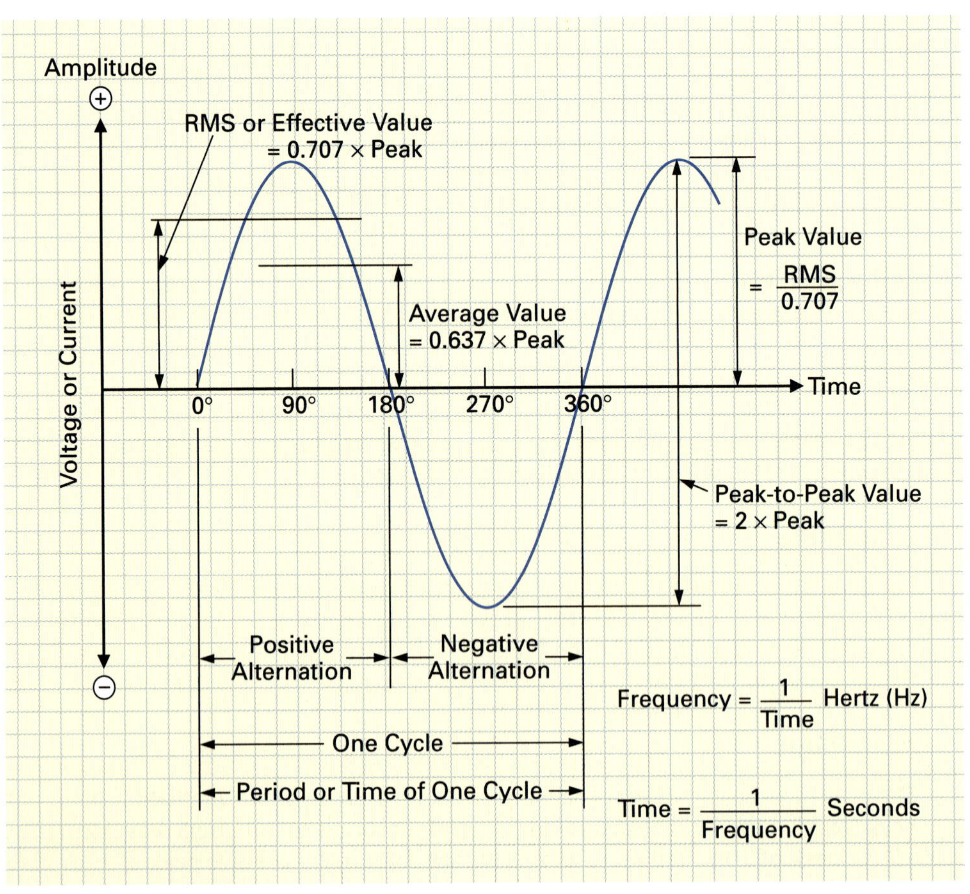

FIGURE 6-7 **Sine Wave.**

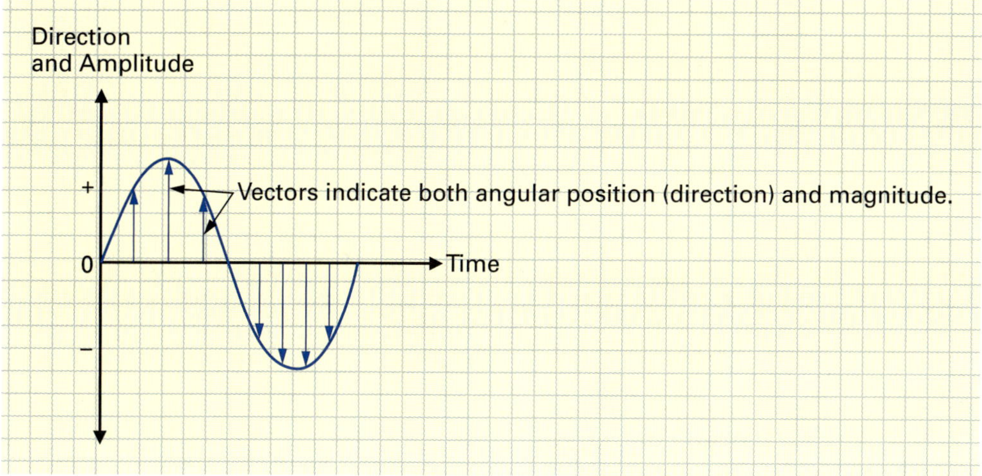

FIGURE 6-8 **Sine-Wave Amplitude.**

Peak-to-Peak Value

The **peak-to-peak value** of a sine wave is the value of voltage or current between the positive and negative maximum values of a waveform. For example, the peak-to-peak value of the current waveform in Figure 6-9(a) is equal to $I_{P\text{-}P} = 2 \times I_p = 20$ A. In Figure 6-9(b), it would be equal to $V_{P\text{-}P} = 2 \times V_p = 18$ V.

Peak Value
Maximum or highest-amplitude level.

Peak-to-Peak Value
Difference between the maximum positive and maximum negative values.

$$\text{p-p} = 2 \times \text{peak}$$

RMS or Effective Value

Both the positive and negative alternation of a sine wave can accomplish the same amount of work, but the ac waveform is only at its maximum value for an instant in time, spending most of its time between peak currents. Our examples in Figure 6-9(a) and (b), therefore, cannot supply the same amount of power as a dc value of 10 A or 9 V.

The effective value of a sine wave is equal to 0.707 of the peak value.

$$\text{rms} = 0.707 \times \text{peak}$$

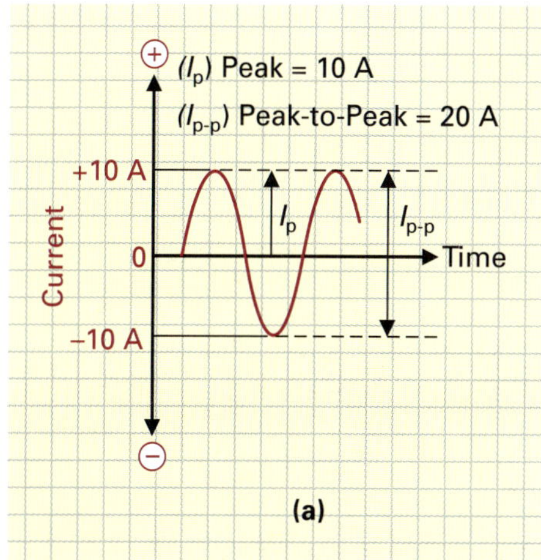

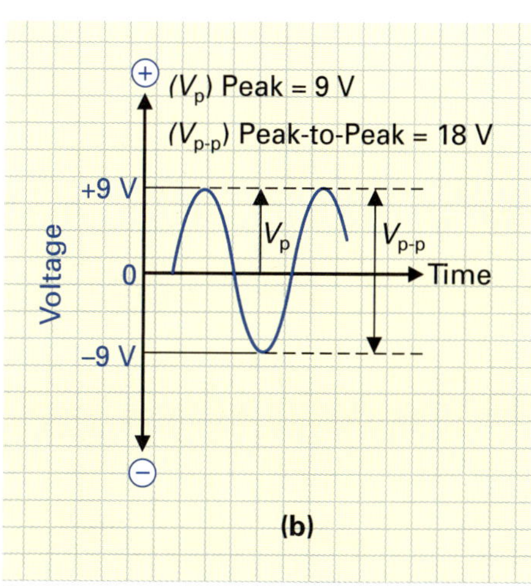

FIGURE 6-9 **Peak and Peak-to-Peak of a Sine Wave.**

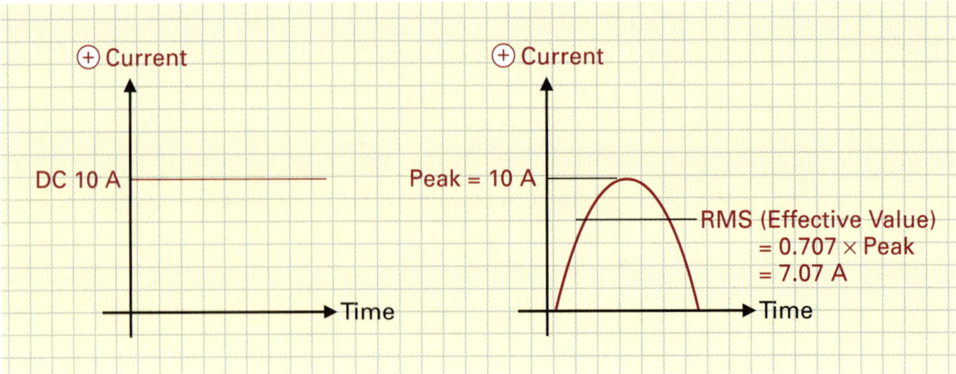

FIGURE 6-10 **Effective Equivalent.**

RMS Value
RMS value of an ac voltage, current, or power waveform is equal to 0.707 times the peak value. The rms value is the effective or dc value equivalent of the ac wave.

This **root-mean-square (rms) value** of 0.707 can always be used to tell us how effective an ac sine wave will be. For example, a 10 A dc source would be 10 A effective because it is continually at its peak value and always delivering power to the circuit to which it is connected, while a 10 A ac source would only be 7.07 A effective, as seen in Figure 6-10, because it is at 10 A for only a short period of time. As another example, a 10 V ac sine-wave alternation would be as effective or supply the same amount of power to a circuit as a 7.07 V dc source.

Unless otherwise stated, ac values of voltage or current are always given in rms. The peak value can be calculated by transposing the original rms formula of rms = peak × 0.707, and ending up with

$$\text{peak} = \frac{\text{rms}}{0.707}$$

Since 1/0.707 = 1.414, the peak can also be calculated by

$$\text{peak} = \text{rms} \times 1.414$$

Average Value

Average Value
Mean value found when the area of a wave above a line is equal to the area of the wave below the line.

The **average value** of the positive or negative alternation is found by taking either the positive or negative alternation, and listing the amplitude or vector length of current or voltages at 1° intervals, as shown in Figure 6-11(a). The sum of all these values is then divided by the total number of values (averaging), which for all sine waves will calculate out to be 0.637 of the peak voltage or current. For example, the average value of a sine-wave alternation with a peak of 10 V, as seen in Figure 6-11(b), is equal to

$$\text{average} = 0.637 \times \text{peak}$$

$$= 0.637 \times 10 \text{ V}$$

$$= 6.37 \text{ V}$$

EXAMPLE:

Calculate V_p, $V_{p\text{-}p}$, V_{rms}, and V_{avg} of a 16 V peak sine wave.

Solution:

$$V_p = 16 \text{ V}$$
$$V_{p\text{-}p} = 2 \times V_p = 2 \times 16 \text{ V} = 32 \text{ V}$$
$$V_{rms} = 0.707 \times V_p = 0.707 \times 16 \text{ V} = 11.3 \text{ V}$$
$$V_{avg} = 0.637 \times V_p = 0.637 \times 16 \text{ V} = 10.2 \text{ V}$$

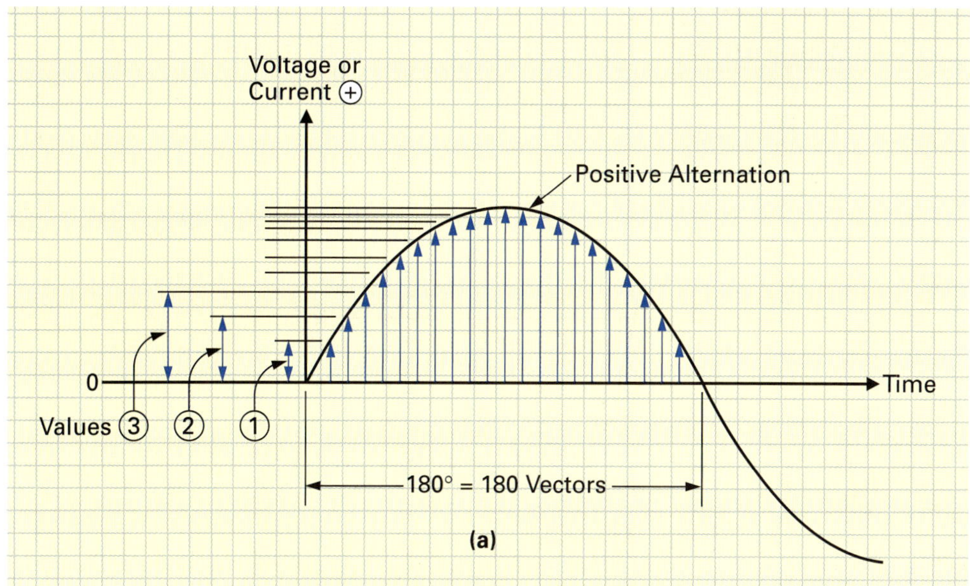

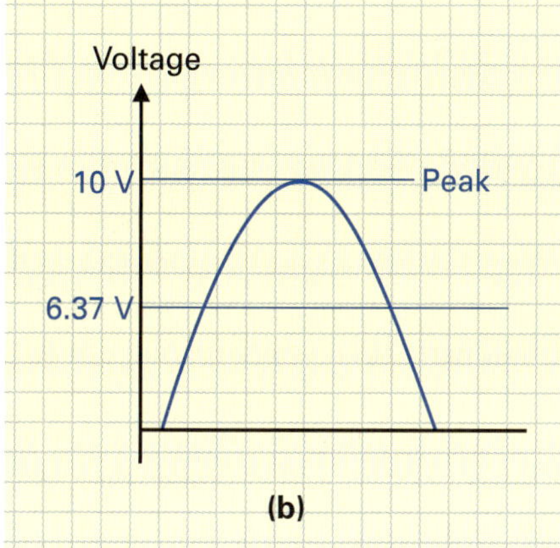

FIGURE 6-11 **Average Value of a Sine-Wave Alternation = 0.637 × Peak.**

EXAMPLE:

Calculate V_p, $V_{p\text{-}p}$, and V_{avg} of a 120 V (rms) ac main supply.

Solution:

$$V_p = \text{rms} \times 1.414 = 120\text{ V} \times 1.414 = 169.68\text{ V}$$
$$V_{p\text{-}p} = 2 \times V_p = 2 \times 169.68\text{ V} = 339.36\text{ V}$$
$$V_{avg} = 0.637 \times V_p = 0.637 \times 169.68\text{ V} = 108.09\text{ V}$$

The 120 V (rms) that is delivered to every home and business has a peak of 169.68 V. This ac value will deliver the same power as 120 V dc.

Frequency and Period

As shown in Figure 6-12, the **period** (t) is the time required for one complete cycle (positive and negative alternation) of the sinusoidal current or voltage waveform. A *cycle,* by definition, is the change of an alternating wave from zero to a positive peak, to zero, then to a negative peak, and finally, back to zero.

Period
Time taken to complete one complete cycle of a periodic or repeating waveform.

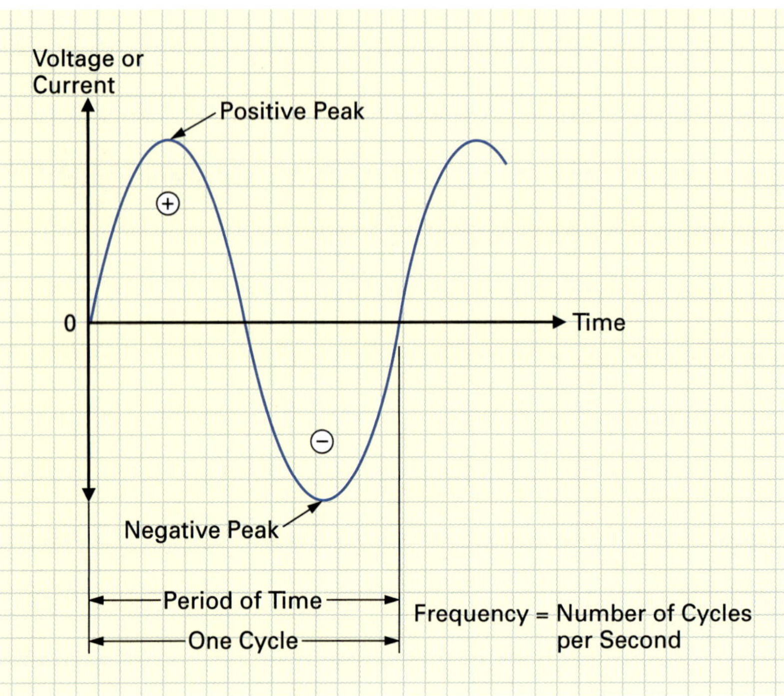

FIGURE 6-12 **Frequency and Period.**

Frequency
Rate of recurrences of a periodic wave normally within a unit of one second, measured in hertz (cycles/second).

Frequency is the number of repetitions of a periodic wave in a unit of time. It is symbolized by f and is given the unit hertz (cycles per second), in honor of a German physicist, Heinrich Hertz.

Sinusoidal waves can take a long or a short amount of time to complete one cycle. This time is related to frequency in that period and is equal to the reciprocal of frequency, and vice versa.

$$f\,(\text{hertz}) = \frac{1}{t}$$

$$t\,(\text{seconds}) = \frac{1}{f}$$

where t = period, f = frequency

TIME LINE
Heinrich R. Hertz (1857–1894), a German physicist, was the first to demonstrate the production and reception of electromagnetic (radio) waves. In honor of his work in this field, the unit of frequency is called the hertz.

For example, the ac voltage of 120 V (rms) arrives at the household electrical outlet alternating at a frequency of 60 hertz (Hz). This means that 60 cycles arrive at the household electrical outlet in 1 second. If 60 cycles occur in 1 second, as seen in Figure 6-13(a), it is actually taking $^1/_{60}$ of a second for one of the 60 cycles to complete its cycle, which calculates out to be

$$^1/_{60} \text{ of 1 second} = \frac{1}{60 \text{ cycles}} \times 1 \text{ second} = 16.67 \text{ milliseconds (ms)}$$

So the time or period of one cycle can be calculated by using the formula period $(t) = 1/f = 1/60$ Hz = 16.67 ms, as shown in Figure 6-13(b).

If the period or time of a cycle is known, the frequency can be calculated. For example:

$$\text{frequency } (f) = \frac{1}{\text{period}} = 1/16.67 \text{ ms} = 60 \text{ Hz}$$

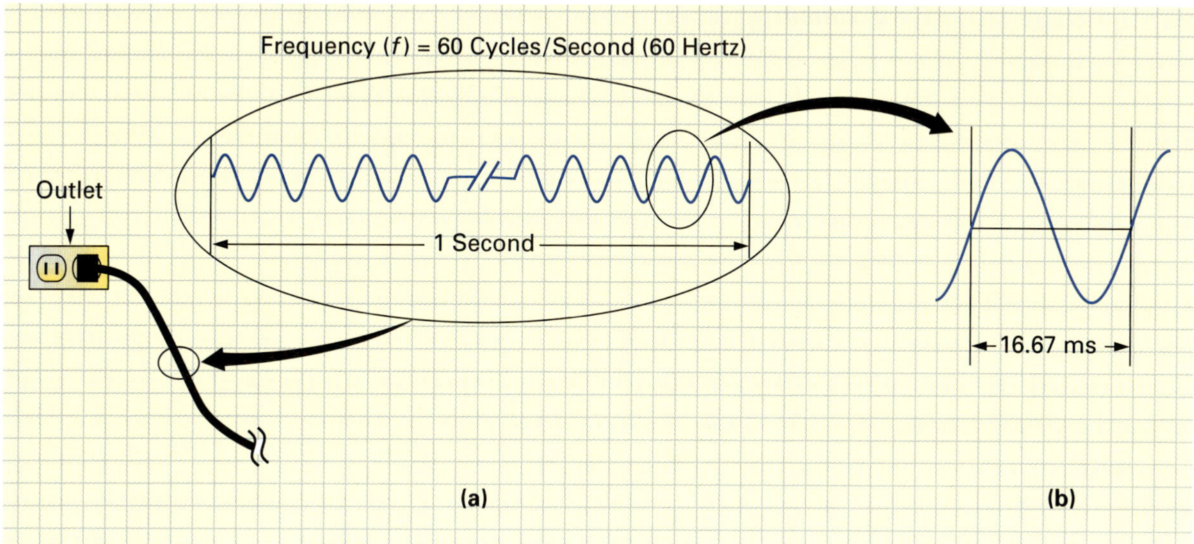

FIGURE 6-13 **120 V, 60 Hz AC Supply.**

As illustrated in Figure 6-13, all homes in the United States receive at their wall outlets an ac voltage of 120 V rms at a frequency of 60 Hz. This frequency was chosen for convenience, as a lower frequency would require larger transformers, and if the frequency were too low, the slow switching (alternating) current through light bulbs would cause the lights to flicker. A higher frequency than 60 Hz was found to cause an increase in the amount of heat generated in the core of all power distribution transformers due to eddy currents and hysteresis losses. A frequency of 60 Hz was chosen in the United States; however, other countries, such as England and most of Europe, use an ac power line frequency of 50 Hz (240 V).

EXAMPLE:

If a sine wave has a period of 400 μs, what is its frequency?

Solution:

$$\text{frequency } (f) = \frac{1}{\text{time } (t)} = \frac{1}{400\ \mu\text{s}} = 2.5 \text{ kHz or 2500 cycles per second}$$

EXAMPLE:

If it takes a sine wave 25 ms to complete two cycles, how many of the cycles will be received in 1 s?

Solution:

If the period of two cycles is 25 ms, one cycle period will equal 12.5 ms. The number of cycles per second or frequency will equal

$$f = \frac{1}{t} = \frac{1}{12.5 \text{ ms}} = 80 \text{ Hz} \ \text{ or } \ 80 \text{ cycles/second}$$

EXAMPLE:

Calculate the period of the following:

a. 100 MHz

b. 40 cycles every 5 seconds

c. 4.2 kilocycles/second

d. 500 kHz

■ *Solution:*

$$f = \frac{1}{t} \qquad \text{therefore, } t = \frac{1}{f}$$

a. $t = \dfrac{1}{100 \text{ MHz}} = 10$ nanoseconds (ns)

b. 40 cycles/5 s = 8 cycles/second (8 Hz)

$t = \dfrac{1}{8 \text{ Hz}} = 125$ ms

c. $t = \dfrac{1}{4.2 \text{ kHz}} = 238\ \mu\text{s}$

d. $t = \dfrac{1}{500 \text{ kHz}} = 2\ \mu\text{s}$

Wavelength

Wavelength, as its name states, is the physical length of one complete cycle and is generally measured in meters. The wavelength (λ, lambda) of a complete cycle is dependent on the frequency and velocity of the transmission:

Wavelength
Distance between two points of corresponding phase and equal to waveform velocity or speed divided by frequency.

$$\lambda = \frac{\text{velocity}}{\text{frequency}}$$

Electromagnetic waves. Radio waves travel at the speed of light in air or a vacuum, which is 3×10^8 meters/second or 3×10^{10} cm/second.

$$\lambda\ (\text{m}) = \frac{3 \times 10^8 \text{ m/s}}{f(\text{Hz})} \qquad \text{or} \qquad \lambda\ (\text{cm}) = \frac{3 \times 10^{10} \text{ cm/s}}{\text{frequency (Hz)}}$$

(There are 100 centimeters [cm] in 1 meter [m], therefore cm $= 10^{-2}$, and m $= 10^0$ or 1.) Subsequently, the higher the frequency, the shorter the wavelength, which is why a short-wave radio receiver is designed to receive high frequencies ($\lambda\downarrow = 3 \times 10^8/f\uparrow$).

■ **EXAMPLE:**

Calculate the wavelength of the electromagnetic waves illustrated in Figure 6-14.

■ *Solution:*

a. $\lambda = \dfrac{3 \times 10^8}{f(\text{Hz})} \text{ m/s} = \dfrac{3 \times 10^8}{10 \text{ kHz}} = 30{,}000$ m or 30 km

b. $\lambda = \dfrac{3 \times 10^{10}}{f(\text{Hz})} \text{ cm/s} = \dfrac{3 \times 10^{10}}{2182 \text{ kHz}} = 13{,}748.9$ cm or 137.489 m

c. $\lambda = \dfrac{3 \times 10^{10}}{f(\text{Hz})} \text{ cm/s} = \dfrac{3 \times 10^{10}}{4.0 \text{ GHz}} = \dfrac{3 \times 10^{10}}{4 \times 10^9} = 7.5$ cm or 0.075 m

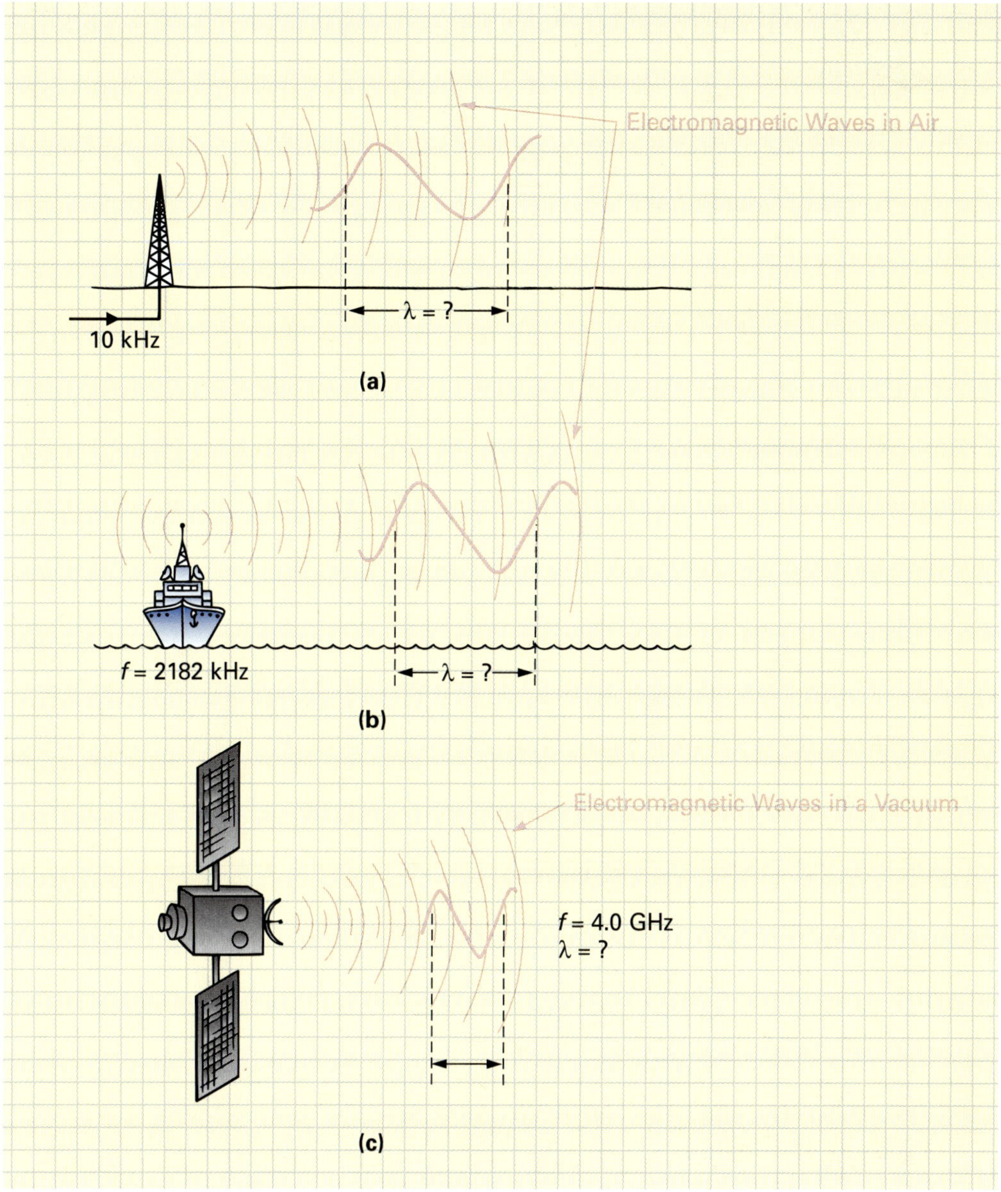

FIGURE 6-14 **Electromagnetic Wavelength Examples.**

TIME LINE

During World War II, there was a need for microwave-frequency vacuum tubes. British inventor Henry Boot developed the magnatron in 1939.

TIME LINE

In 1939, an American brother duo, Russel and Sigurd Varian, invented the klystron. In 1943, the travelling wave tube amplifier was invented by Rudolf Komphner, and up to this day these three microwave tubes are still used extensively.

Sound waves. Sound waves travel at a slower speed than electromagnetic waves, as their mechanical vibrations depend on air molecules, which offer resistance to the traveling wave. For sound waves, the wavelength formula will be equal to

$$\lambda\,(\text{m}) = \frac{344.4\ \text{m/s}}{f(\text{Hz})}$$

EXAMPLE:

Calculate the wavelength of the sound waves illustrated in Figure 6-15.

Solution:

a. $\lambda\,(\text{m}) = \dfrac{344.4\ \text{m/s}}{f(\text{Hz})} = \dfrac{344.4}{35\ \text{kHz}} = 9.8 \times 10^{-3}\ \text{m} = 9.8\ \text{mm}$

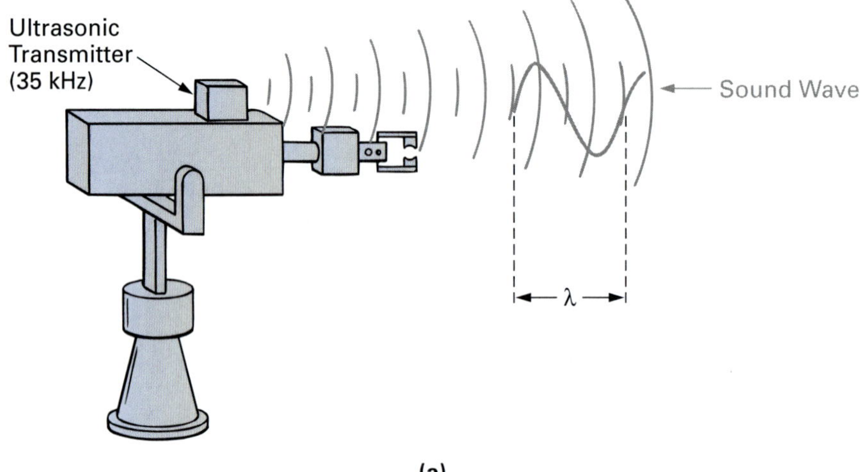

FIGURE 6-15 Sound Wavelength Examples.

b. $300 \text{ Hz: } \lambda \text{ (m)} = \dfrac{344.4 \text{ m/s}}{300 \text{ Hz}} = 1.15 \text{ m}$

$3000 \text{ Hz: } \lambda \text{ (m)} = \dfrac{344.4 \text{ m/s}}{3000 \text{ Hz}} = 0.115 \text{ m or } 11.5 \text{ cm}$

Phase
Angular relationship between two waves, normally between current and voltage in an ac circuit.

Phase Relationships

The **phase** of a sine wave is always relative to another sine wave of the same frequency. Figure 6-16(a) illustrates two sine waves that are in phase with one another, while

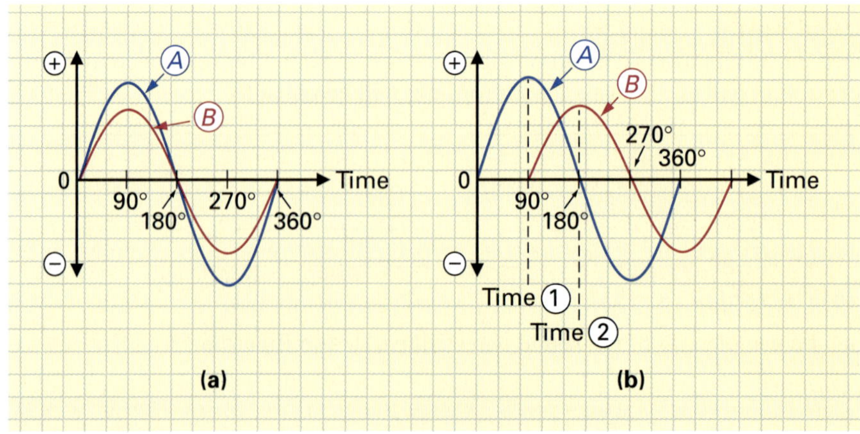

FIGURE 6-16 Phase Relationship. (a) In Phase. (b) Out of Phase.

Figure 6-16(b) shows two sine waves that are out of phase with one another. Sine wave *A* is our reference, since the positive-going zero crossing is at 0°, its positive peak is at 90°, its negative-going zero crossing is at 180°, its negative peak is at 270°, and the cycle completes at 360°. In Figure 6-16(a), sine wave *B* is in phase with *A* since its peaks and zero crossings occur at the same time as sine wave *A*'s.

In Figure 6-16(b), sine wave *B* has been shifted to the right by 90° with respect to the reference sine wave *A*. This **phase shift** or **phase angle** of 90° means that sine wave *A leads B* by 90°, or *B lags A* by 90°. Sine wave *A* is said to lead *B* as its positive peak, for example, occurs first at time 1, while the positive peak of *B* occurs later at time 2.

Phase Shift or Angle
Change in phase of a waveform between two points, given in degrees of lead or lag. Phase difference between two waves, normally expressed in degrees.

EXAMPLE:

What are the phase relationships between the two waveforms illustrated in Figure 6-17(a) and (b)?

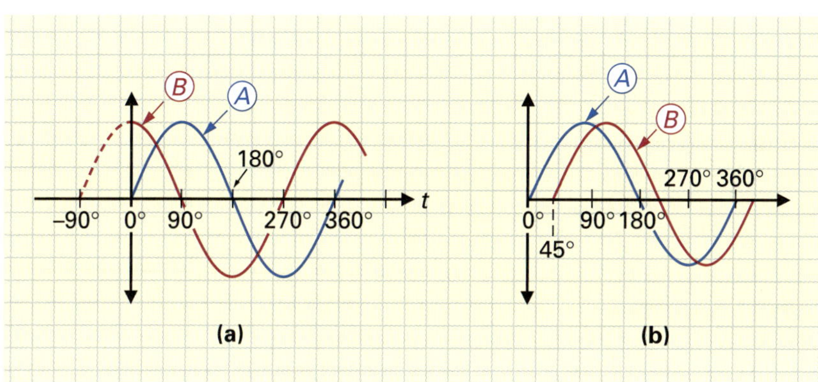

FIGURE 6-17 **Phase Relationship Examples.**

Solution:

a. The phase shift or angle is 90°. Sine wave *B* leads sine wave *A* by 90°, or *A* lags *B* by 90°.
b. The phase shift or angle is 45°. Sine wave *A* leads sine wave *B* by 45°, or *B* lags *A* by 45°.

6-3-2 *The Square Wave*

The **square wave** is a periodic (repeating) wave that alternates from a positive peak value to a negative peak value, and vice versa, for equal lengths of time.

Square Wave
Wave that alternates between two fixed values for an equal amount of time.

In Figure 6-18 you can see an example of a square wave that is at a frequency of 1 kHz and has a peak of 10 V. If the frequency of a wave is known, its period or time of one cycle can be calculated by using the formula $t = 1/f = 1/1$ kHz $= 1$ ms or $^{1}/_{1000}$ of a second. One complete cycle will take 1 ms to complete, so the positive and negative alternations will each last for 0.5 ms.

If the peak of the square wave is equal to 10 V, the peak-to-peak value of this square wave will equal $V_{\text{p-p}} = 2 \times V_{\text{p}} = 20$ V.

To summarize this example, the square wave alternates from a positive peak value of +10 V to a negative peak value of −10 V for equal time lengths (half-cycles) of 0.5 ms.

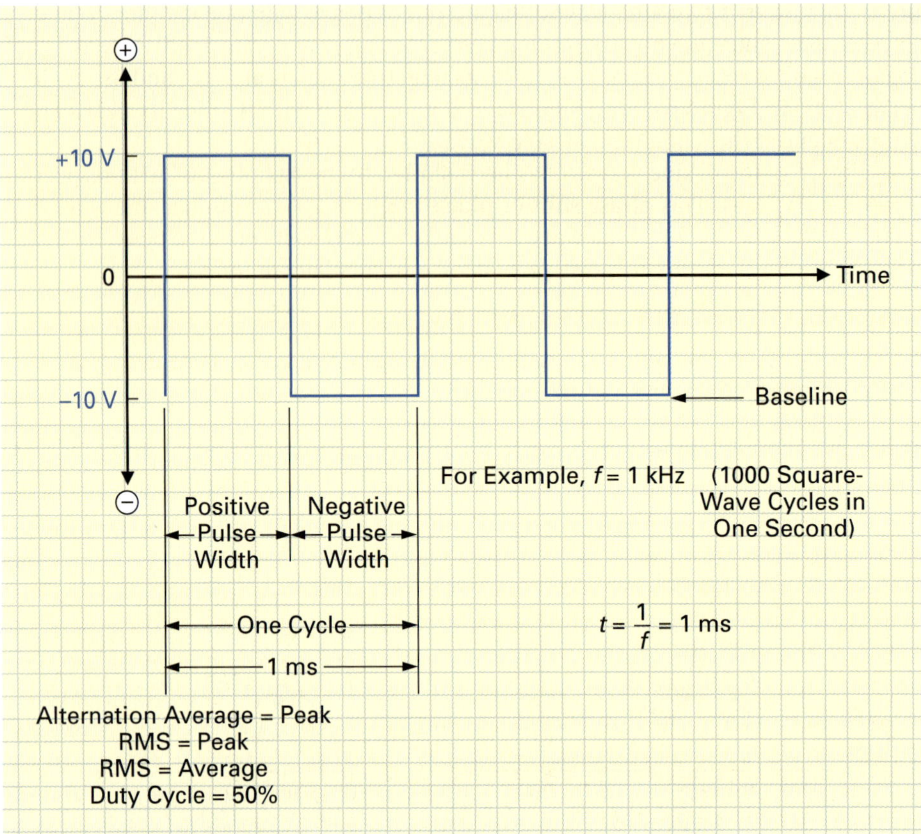

FIGURE 6-18 **Square Wave.**

Duty Cycle

Duty Cycle
A term used to describe the amount of ON time versus OFF time. ON time is usually expressed as a percentage.

Duty cycle is an important relationship, which has to be considered when discussing square waveforms. The **duty cycle** is the ratio of a pulse width (positive or negative pulse or cycle) to the overall period or time of the wave and is normally given as a percentage.

$$\text{duty cycle } (\%) = \frac{\text{pulse width } (P_w)}{\text{period } (t)} \times 100\%$$

The duty cycle of the example square wave in Figure 6-18 will equal

$$\begin{aligned}\text{duty cycle } (\%) &= \frac{\text{pulse width } (P_w)}{\text{period } (t)} \times 100\% \\ &= \frac{0.5 \text{ ms}}{1 \text{ ms}} \times 100\% \\ &= 50\%\end{aligned}$$

Since a square wave always has a positive and a negative alternation that are equal in time, the duty cycle of all square waves is equal to 50%, which actually means that the positive cycle lasts for 50% of the time of one cycle.

Average

The average or mean value of a square wave can be calculated by using the formula

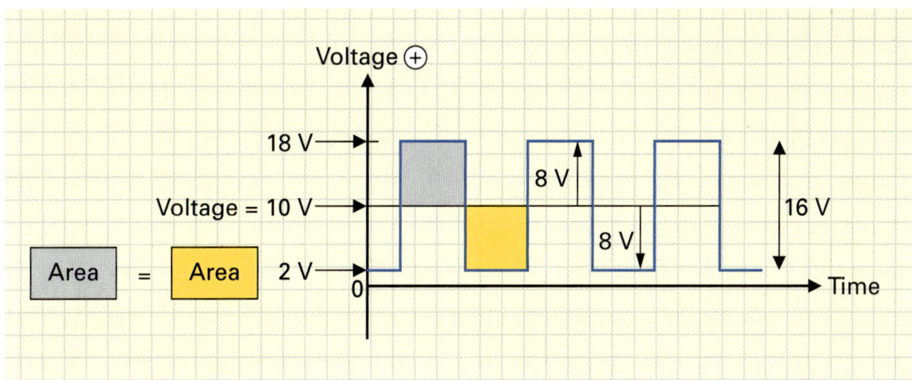

FIGURE 6-19 **2 to 18 V Square Wave.**

V or I average = baseline + (duty cycle × peak to peak)

The average of the complete square-wave cycle in Figure 6-18 should calculate out to be zero, as the amount above the line equals the amount below. If we apply the formula to this example, you can see that

$$\begin{aligned} V_{\text{avg}} &= \text{baseline} + (\text{duty cycle} \times \text{peak to peak}) \\ &= -10\text{ V} + (0.5 \times 20\text{ V}) \\ &= -10\text{ V} + 10\text{ V} \\ &= 0\text{ V} \end{aligned}$$

However, a square wave does not always alternate about 0. For example, Figure 6-19 illustrates a 16 $V_{\text{p-p}}$ square wave that rests on a baseline of 2 V. The average value of this square wave is equal to

$$\begin{aligned} V_{\text{avg}} &= \text{baseline} + (\text{duty cycle} \times \text{peak to peak}) \\ &= 2\text{ V} \times (0.5 \times 16\text{ V}) \\ &= (+2) + (+8\text{ V}) \\ &= 10\text{ V} \end{aligned}$$

EXAMPLE:

Calculate the duty cycle and V_{avg} of a square wave of 0 to 5 V.

Solution:

The duty cycle of a square wave is always 0.5 or 50%. The average is:

$$\begin{aligned} V_{\text{avg}} &= \text{baseline} + (\text{duty cycle} \times V_{\text{p-p}}) \\ &= 0\text{ V} + (0.5 \times 5\text{ V}) \\ &= 0\text{ V} + 2.5\text{ V} = 2.5\text{ V} \end{aligned}$$

Up to this point, we have seen the ideal square wave, which has instantaneous transition from the negative to the positive values, and vice versa, as shown in Figure 6-20(a). In fact, the transitions from negative to positive (positive or leading edge) and from positive to negative (negative or trailing edge) are not as ideal as shown here. It takes a small amount of time for the wave to increase to its positive value (the **rise time**), and an equal amount of time for a wave to decrease to its negative value (the **fall time**). Rise time (T_R), by definition, is the time it takes for an edge to rise from 10% to 90% of its full amplitude, while fall time (T_F) is the time it takes for an edge to fall from 90% to 10% of its full amplitude, as shown in Figure 6-20(b).

Rise Time
Time it takes a positive edge of a pulse to rise from 10% to 90% of its peak value.

Fall Time
Time it takes a negative edge of a pulse to fall from 90% to 10% of its peak value.

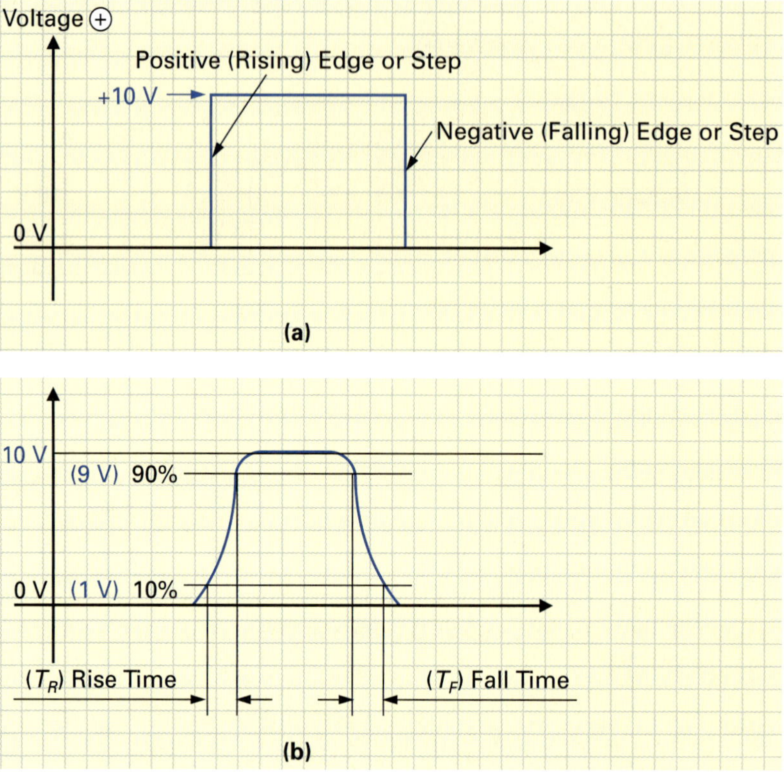

FIGURE 6-20 Square Wave's Rise and Fall Times. (a) Ideal. (b) Actual.

With a waveform such as that in Figure 6-20(b), it is difficult, unless a standard is used, to know exactly what points to use when measuring the width of either the positive or negative alternation. The standard width is always measured between the two 50% amplitude points, as shown in Figure 6-21.

Rectangular (Pulse) Wave

Also known as a pulse wave; it is a repeating wave that only alternates between two levels or values and remains at one of these values for a small amount of time relative to the other.

6-3-3 *The Rectangular or Pulse Wave*

The **rectangular wave** is similar to the square wave in many respects, in that it is a periodic wave that alternately switches between one of two fixed values. The difference in the rectangular wave is that it does not remain at the two peak values for equal lengths of time, as shown in the examples in Figure 6-22(a) and (b).

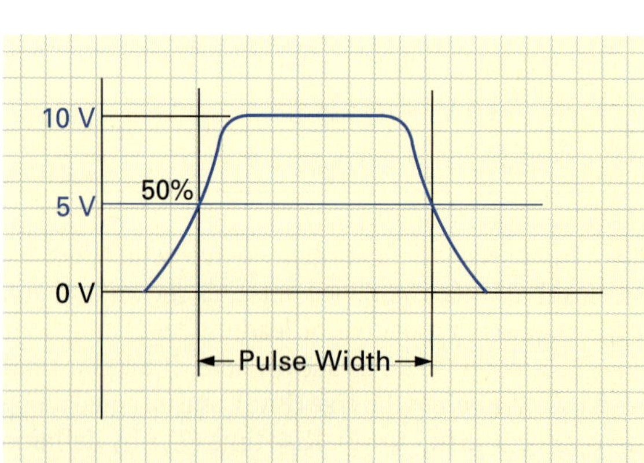

FIGURE 6-21 Pulse Width of a Square Wave.

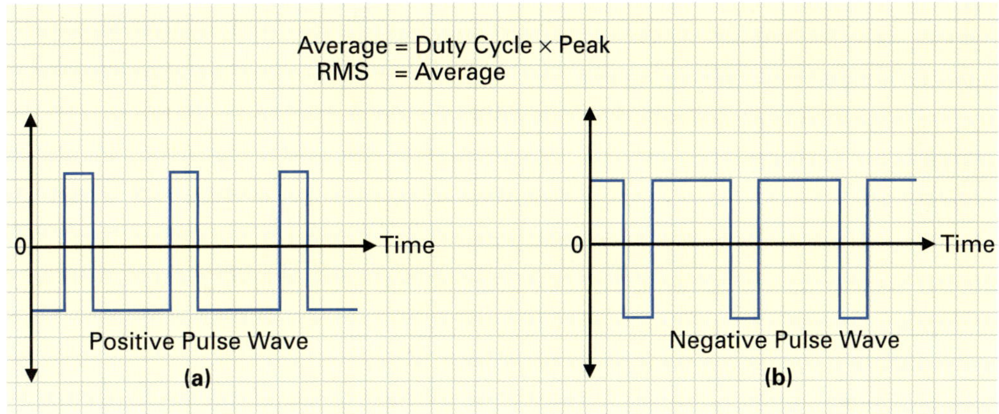

FIGURE 6-22 **Rectangular or Pulse Wave.**

In Figure 6-22(a), the rectangular wave remains at its negative level for a longer period than its positive, while the rectangular wave in Figure 6-22(b) stays at its positive value for the longer period of time and is only momentarily at its negative value.

Pulse Repetition Frequency
The number of times per second that a pulse is transmitted.

PRF, PRT, and Pulse Length

When discussing a rectangular wave, a few terms change. Instead of stating the cycles per second as frequency, it is called **pulse repetition frequency** (PRF), which is far more descriptive. The reciprocal of frequency is time, and with rectangular pulse waveforms the reciprocal of the PRF is **pulse repetition time** (PRT). With rectangular or pulse waves, therefore, frequency is equivalent to PRF, time to PRT, and the only difference is the name.

Pulse Repetition Time
The time interval between the start of two consecutive pulses.

Let us look at the example in Figure 6-23 of a 5 V rectangular wave at a frequency of 1 kHz and a pulse width of 1 μs, and practice with these new terms. With a pulse repetition frequency of 1 kHz, the time between the leading edges of pulses (PRT) will be 1/1 kHz = 1 ms. **Pulse width** (P_w), **pulse duration** (P_d), or **pulse length** (P_l) are all terms that describe the length of time for which the pulse lasts, and in this example it is equal to 1 μs, which means that 999 μs exists between the end of one pulse and the beginning of the next.

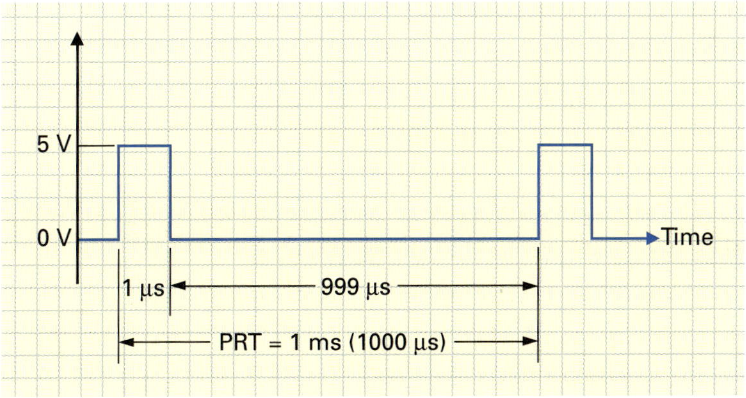

Pulse Width, Pulse Length, or Pulse Duration
The time interval between the leading edge and trailing edge of a pulse at which the amplitude reaches 50% of the peak pulse amplitude.

FIGURE 6-23 **PRF and PRT of a Pulse Wave.**

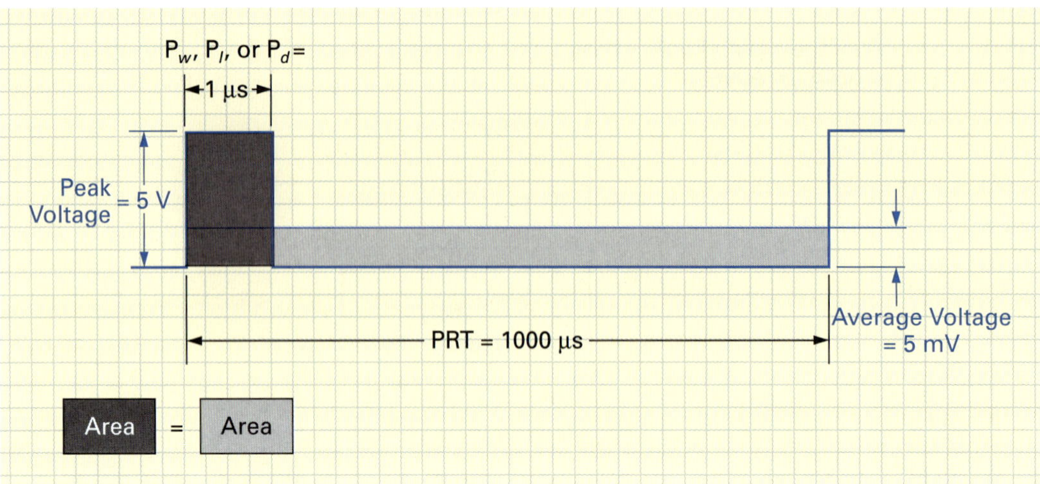

FIGURE 6-24 Average of a Pulse Wave.

Duty Cycle

The duty cycle is calculated in exactly the same way as for the square wave and is a ratio of the pulse width to the overall time (PRT). In our example in Figure 6-23 the duty cycle will be equal to

$$\begin{aligned}\text{duty cycle (\%)} &= \frac{\text{pulse width } (P_w)}{\text{PRT}} \times 100\% \\ &= \frac{1\ \mu\text{s}}{1000\ \mu\text{s}} \times 100\% \\ &= \text{duty cycle figure of } 0.001 \times 100\% \\ &= 0.001 \times 100\% \\ &= 0.1\%\end{aligned}$$

The result tells us that the positive pulse lasts for 0.1% of the total time (PRT).

Average

The average or mean value of this waveform is calculated by using the same square-wave formula. The average of the pulse wave in Figure 6-23 will be

$$\begin{aligned}V \text{ or } I \text{ average} &= \text{baseline} + (\text{duty cycle} \times \text{peak to peak}) \\ V_{\text{avg}} &= 0\text{ V} + (0.001 \times 5\text{ V}) \\ &= 0\text{ V} + (5\text{ mV}) \\ &= 5\text{ mV}\end{aligned}$$

Figure 6-24 illustrates the average value of this rectangular waveform. If the voltage and width of the positive pulse are taken and spread out over the entire PRT, they will have a mean level equal, in this example, to 5 mV.

EXAMPLE:

Calculate the duty cycle and average voltage of the following radar pulse waveform:

$$\begin{aligned}\text{peak voltage, } V_p &= 20\text{ kV} \\ \text{pulse length, } P_l &= 1\ \mu\text{s} \\ \text{baseline voltage} &= 0\text{ V} \\ \text{PRF} &= 3300 \text{ pulses per second (pps)}\end{aligned}$$

■ ***Solution:***

$$\text{duty cycle} = \frac{\text{pulse length } (P_l)}{\text{PRT}} \times 100\%$$
$$= \frac{1\ \mu\text{s}}{303\ \mu\text{s}} \times 100\% \left(\text{PRT} = \frac{1}{\text{PRF}} = \frac{1}{3300} = 303\ \mu\text{s}\right)$$
$$= (3.3 \times 10^{-3}) \times 100\%$$
$$= 0.33\%$$
$$V_{avg} = \text{baseline} + (\text{duty cycle} \times V_{\text{p-p}})$$
$$= 0\text{ V} + [(3.3 \times 10^{-3}) \times 20 \times 10^3]$$
$$= 66\text{ V}$$

6-3-4 *The Triangular Wave*

A **triangular wave** consists of a positive and negative ramp of equal values, as shown in Figure 6-25. Both the positive and negative ramps have a linear increase and decrease, respectively. Linear, by definition, is the relationship between two quantities that exists when a change in a second quantity is directly proportional to a change in the first quantity. The two quantities in this case are voltage or current and time. As shown in Figure 6-25, if the increment of change of voltage ΔV (pronounced "delta vee") is changing at the same rate as the increment of time, Δt ("delta tee"), the ramp is said to be **linear.**

Triangular Wave
A repeating wave that has equal positive and negative ramps that have linear rates of change with time.

Linear
Relationship between input and output in which the output varies in direct proportion to the input.

6-3-5 *The Sawtooth Wave*

On an oscilloscope display (time-domain presentation), the **sawtooth wave** is very similar to a triangular wave, in that a sawtooth wave has a linear ramp. However, unlike the triangular wave, which reverses and has an equal but opposite ramp back to its starting level, the sawtooth "flies" back to its starting point immediately and then repeats the previous ramp, as seen in Figure 6-26, which shows both a positive and a negative ramp sawtooth.

Sawtooth Wave
Repeating waveform that rises from zero to a maximum value linearly and then falls to zero and repeats.

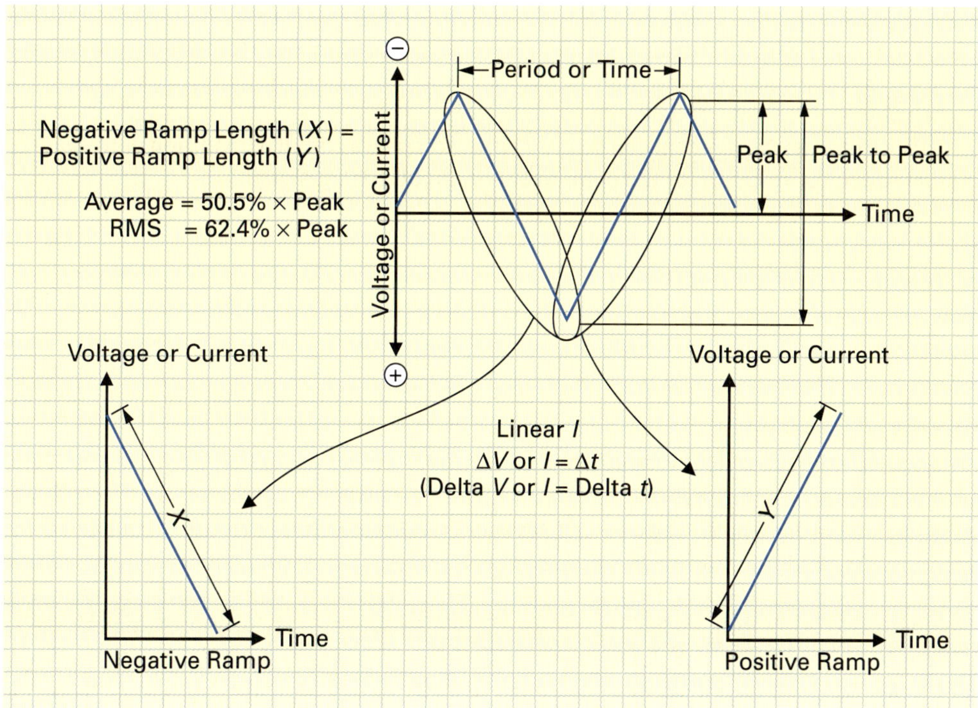

FIGURE 6-25 Triangular Wave.

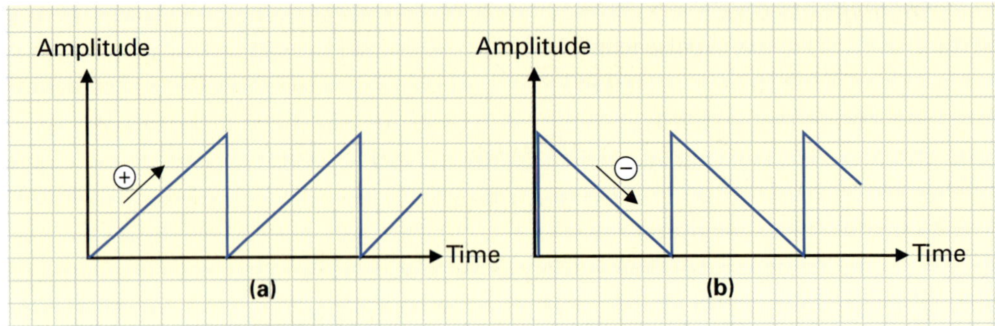

FIGURE 6-26 **Sawtooth Wave. (a) Positive Ramp. (b) Negative Ramp.**

SELF-TEST EVALUATION POINT FOR SECTION 6-3

Use the following questions to test your understanding of Section 6-3.

1. Sketch the following waveforms:
  a. Sine wave
  b. Square wave
  c. Rectangular wave
  d. Triangular wave
  e. Sawtooth wave

2. What are the wavelength formulas for sound waves and electromagnetic waves, and why are they different?

6-4 MEASURING AND GENERATING AC SIGNALS

As a technician or engineer, you are going to be required to diagnose and repair a problem in the shortest amount of time possible. To aid in the efficiency of this fault finding and repair process, you can make use of certain pieces of test equipment. Humans have five kinds of sensory systems: touch, taste, sight, sound, and smell. Four can be used for electronic troubleshooting: sight, sound, touch, and smell.

Electronic test equipment can be used either to *sense* a circuit's condition or to *generate* a signal to see the response of the component or circuit to that signal.

6-4-1 *The AC Meter*

Up to this point, we have seen how a multimeter can be used to measure direct current and voltage. Most multimeters can be used to measure either dc or ac. When the technician or engineer wishes to measure ac, the ac current or voltage is converted to dc internally by a circuit known as a **rectifier.**

Rectifier
Device that achieves rectification.

The dc produced by the rectification process is in fact pulsating, as shown in Figure 6-27, so the current through the meter is a series of pulses rising from zero to maximum (peak) and from maximum back to zero. Frequencies below 10 Hz (lower frequency limit) will cause the digits on a digital multimeter to continually increase in value and then decrease in value as the meter follows the pulsating dc. This makes it difficult to read the meter. This effect will also occur with an analog multimeter which, from 10 Hz to approximately 2 kHz, will not be able to follow the fluctuation, causing the needle to remain in a position equal to the average value of the pulsating dc from the rectifier (0.637 of peak). Most meters are normally calibrated internally to indicate rms values (0.707 of peak) rather than average values, because this effective value is most commonly used when expressing ac voltage or current.

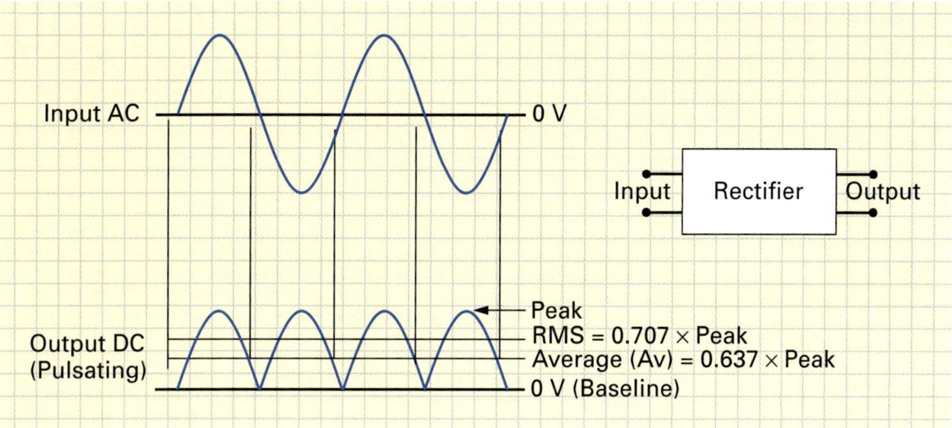

FIGURE 6-27 **Rectifier.**

The upper frequency limit of the ac meter is approximately 2 to 8 kHz. Beyond this limit the meter becomes progressively inaccurate due to the reactance of the capacitance in the rectifier. This reactance, which will be discussed later, will result in inaccurate indications due to the change in opposition at different ac input frequencies.

SELF-TEST EVALUATION POINT FOR SECTION 6-4-1

Use the following questions to test your understanding of Section 6-4-1.

1. Can a multimeter be used to measure an AC voltage?
2. What is a rectifier?

6-4-2 *The Oscilloscope*

Figure 6-28 illustrates a typical **oscilloscope** (sometimes abbreviated to *scope*), which is used primarily to display the shape and spacing of electrical signals. The oscilloscope displays the actual sine, square, rectangular, triangular, or sawtooth wave shape that is occurring at any point in a circuit. This display is made on a cathode-ray tube (CRT), which is also used in television sets and computers to display video information. From the display on the CRT, we can measure or calculate time, frequency, and amplitude characteristics such as rms, average, peak, and peak-to-peak.

Oscilloscope
Instrument used to view signal amplitude, frequency, and shape at different points throughout a circuit.

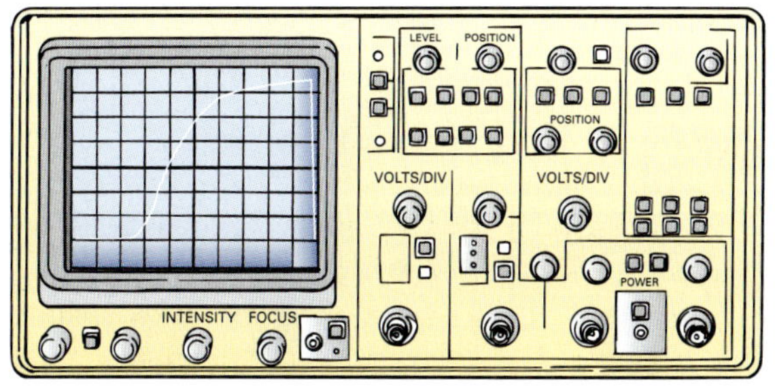

(a)

(b)

FIGURE 6-28 **Typical Oscilloscope. (a) Oscilloscope. (b) Oscilloscope Probe.**

Controls

Oscilloscopes come with a wide variety of features and functions, but the basic operational features are almost identical. Figure 6-29 illustrates the front panel of a typical oscilloscope. Some of these controls are difficult to understand without practice and experience, so practical experimentation is essential if you hope to gain a clear understanding of how to operate an oscilloscope.

General controls (see Figure 6-29)
Intensity control: Controls the brightness of the trace, which is the pattern produced on the screen of a CRT.
Focus control: Used to focus the trace.
Power OFF/ON: Switch will turn on oscilloscope while indicator shows when oscilloscope is turned on.

Some oscilloscopes have the ability to display more than one pattern or trace on the CRT screen, as seen by the examples in Figure 6-30. A dual-trace oscilloscope can produce two traces or patterns on the CRT screen at the same time, whereas a single-trace oscilloscope can trace out only one pattern on the screen. The dual-trace oscilloscope is very useful, as it allows us to make comparisons between the phase, amplitude, shape, and timing of two signals from two separate test points. One signal or waveform is applied to the channel *A* input of the oscilloscope, while the other waveform is applied to the channel *B* input.

Channel selection (see Figure 6-29)
Mode switch: This switch allows us to select which channel input should be displayed on the CRT screen.
CHA: The input arriving at channel *A*'s jack is displayed on the screen as a single trace.
CHB: The input arriving at channel *B*'s jack is displayed on the screen as a single trace.
Dual: The inputs arriving at jacks *A* and *B* are both displayed on the screen, as a dual trace.

Calibration
To determine, by measurement or comparison with a standard, the correct value of each scale reading.

(a) **Calibration** Output This output connection provides a point where a fixed 1 V peak-to-peak square-wave signal can be obtained at a frequency of 1 kHz. This signal is normally fed into either channel *A* or *B*'s input to test probes and the oscilloscope operation.
(b) Channel *A* and *B* Horizontal Controls

↔ *Position control:* This control will move the position of the one (single-trace) or two (dual-trace) waveforms horizontally (left or right) on the CRT screen.

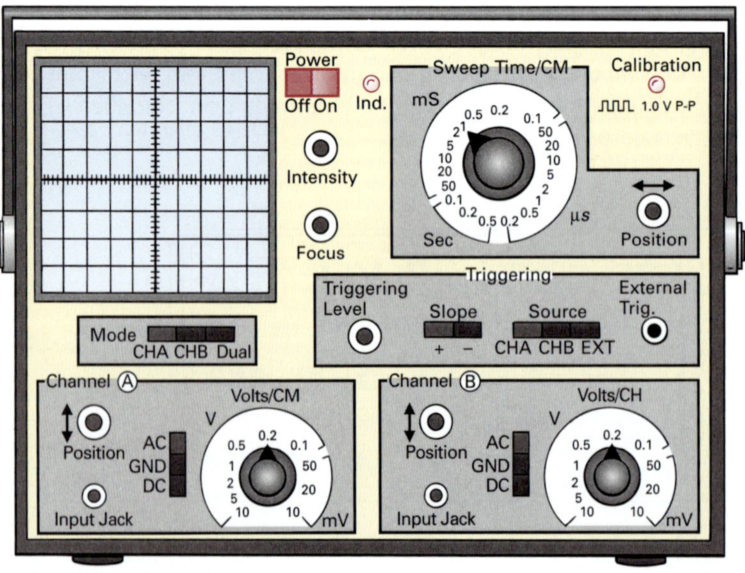

FIGURE 6-29 Oscilloscope Controls.

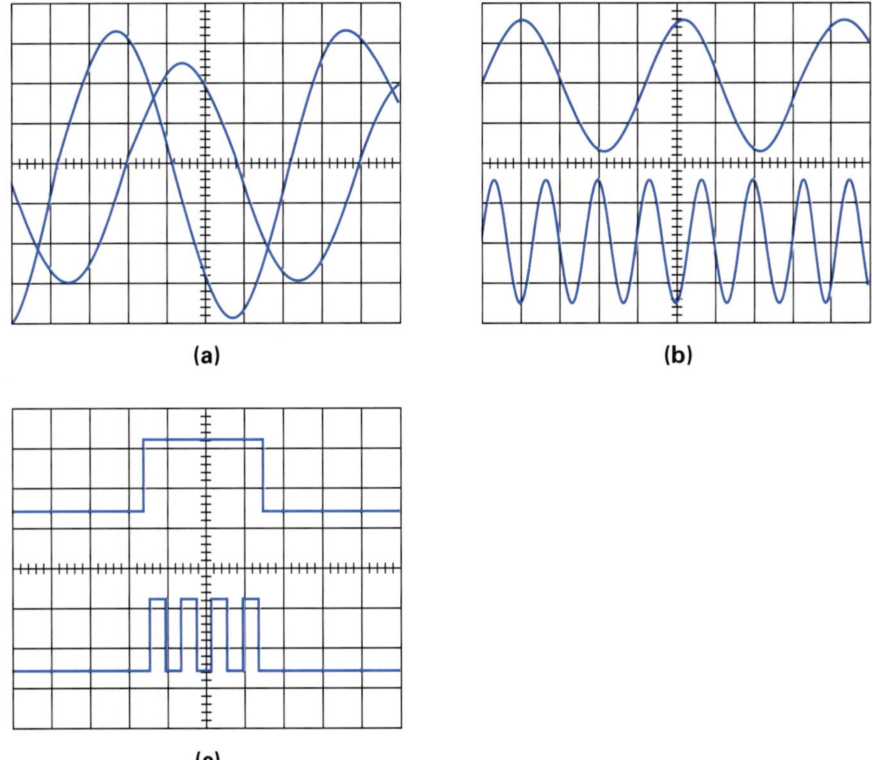

FIGURE 6-30 **Sample of Dual-Trace Oscilloscope Displays for Comparison.**

Sweep time/cm switch: The oscilloscope contains circuits that produce a beam of light that is swept continually from the left to the right of the CRT screen. When no input signal is applied, this sweep will produce a straight horizontal line in the center of the screen. When an input signal is present, this horizontal sweep is influenced by the input signal, which moves it up and down to produce a pattern on the CRT screen the same as the input pattern (sine, square, sawtooth, and so on). This sweep time/cm switch selects the speed of the sweep from left to right, and it can be either fast (0.2 microseconds per centimeter; 0.2 μs/cm) or slow (0.5 second per centimeter; 0.5 s/cm). A low-frequency input signal (long cycle time or period) will require a long time setting (0.5 s/cm) so that the sweep can capture and display one or more cycles of the input. A number of settings are available, with lower time settings displaying fewer cycles and higher time settings showing more cycles of an input.

(c) **Triggering** Controls These provide the internal timing control between the sweep across the screen and the input waveform.

Triggering
Initiation of an action in a circuit which then functions for a predetermined time, for example, the duration of one sweep in a cathode-ray tube.

Triggering level control: This determines the point where the sweep starts.

Slope switch (+): Sweep is triggered on positive-going slope.
(−): Sweep is triggered on negative-going slope.

Source switch, CHA*:* The input arriving into channel *A* jack triggers the sweep.
CHB*:* The input arriving into channel *B* jack triggers the sweep.
EXT: The signal arriving at the external trigger jack is used to trigger the sweep.

(d) Channel *A* and *B* Vertical Controls The *A* and *B* channel controls are identical.

Volts/cm switch: This switch sets the number of volts to be displayed by each major division on the vertical scale of the screen.

↕ *Position control:* Moves the trace up or down for easy measurement or viewing.

AC-DC-GND switch: In the AC position, a capacitor on the input will pass the ac component entering the input jack, but block any dc components.

In the GND position, the input is grounded (0 V) so that the operator can establish a reference.

In the DC position, both ac and dc components are allowed to pass on to and be displayed on the screen.

Measurements

The oscilloscope is probably the most versatile of test equipment, as it can be used to test:

DC voltage
AC voltage
Waveform duration
Waveform frequency
Waveform shape

(a) Voltage Measurement The screen is divided into eight vertical and ten horizontal divisions, as shown in Figure 6-31. This 8 × 10 cm grid is called the *graticule.* Every vertical division has a value depending on the setting of the volts/cm control. For example, if the volts/cm control is set to 5 V/div or 5 V/cm, the waveform shown in Figure 6-32 (a), which rises up four major divisions, will have a peak positive alternation value of 20 V (4 div × 5 V/div = 20 V).

As another example, look at the positive alternation in Figure 6-32(b). The positive alternation rises up three major divisions and then extends another three subdivisions, which are each equal to 1 V because five subdivisions exist within one major division, and one major division is, in this example, equal to 5 V. The positive alternation shown in Figure 6-32(b) therefore has a peak of three major divisions (3 cm × 5 V/cm = 15 V), plus three subdivisions (3 × 1 V = 3 V), which equals 18 volts peak.

In Figure 6-32(c), we have selected the 10 volt/cm position, which means that each major division is equal to 10 V and each subdivision is equal to 2 V. In this example, the waveform peak will be equal to two major divisions (2 × 10 *V* = 20 V), plus four subdivisions (4 × 2 V = 8 V), which is equal to 28 V. Once the peak value of a sine wave is known, the peak to peak, average, and rms can be calculated mathematically.

When measuring a dc voltage with the oscilloscope, the volts/cm is applied in the same way, as shown in Figure 6-33. A positive dc voltage in this situation will cause deflection

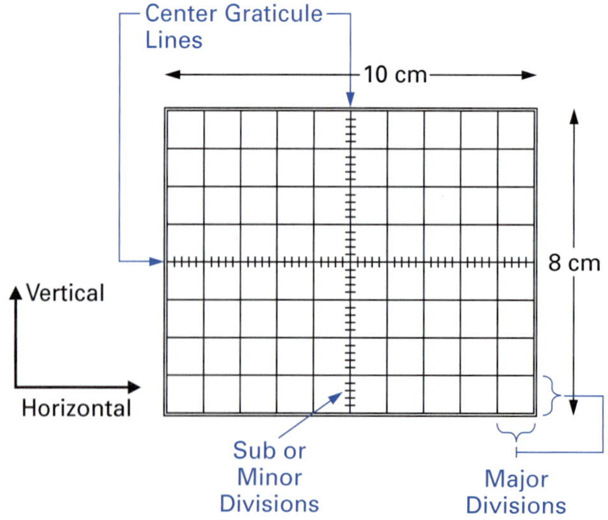

FIGURE 6-31 Oscilloscope Grid.

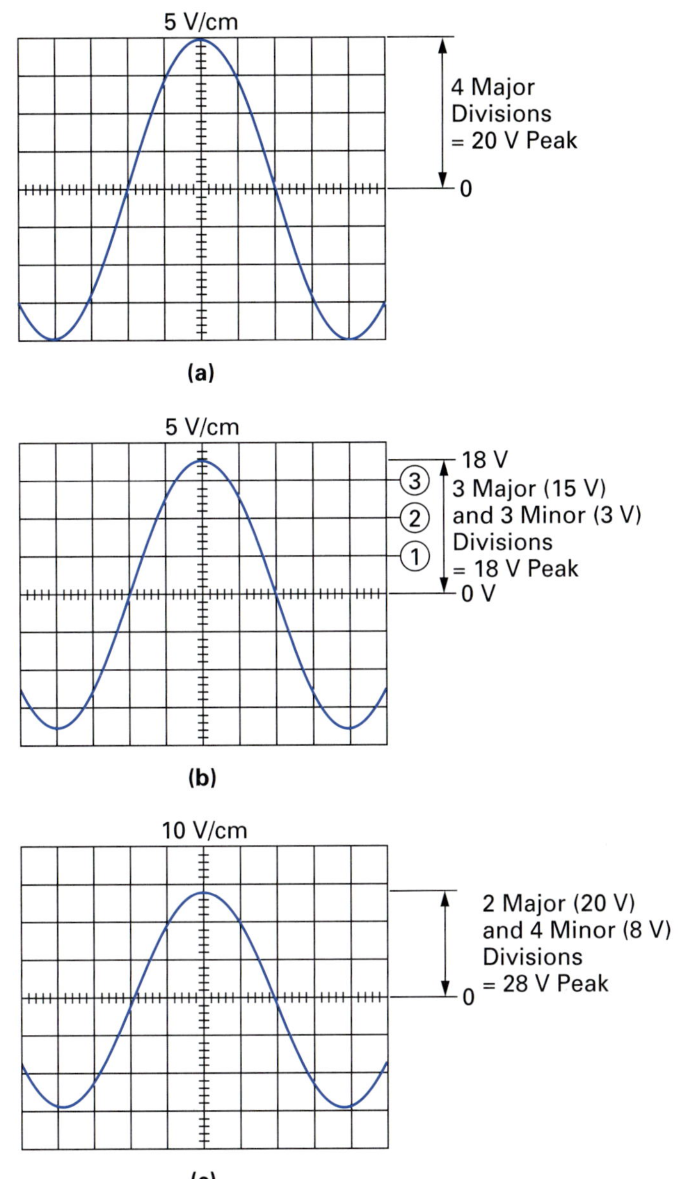

FIGURE 6-32 **Measuring AC Peak Voltage.**

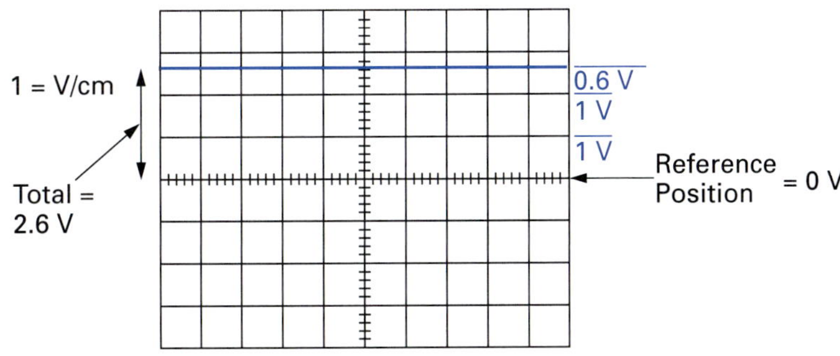

FIGURE 6-33 **Measuring DC Voltage.**

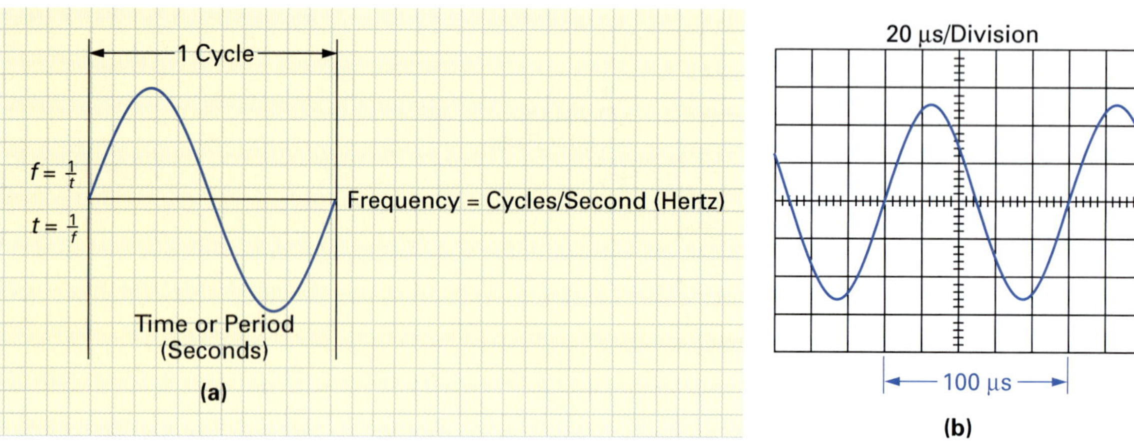

FIGURE 6-34 **Time and Frequency Measurement.**

toward the top of the screen, whereas a negative voltage will cause deflection toward the bottom of the screen.

To determine the dc voltage, count the number of major divisions and to this add the number of minor divisions. In Figure 6-33, a major division equals 1 V/cm and, therefore, a minor division equals 0.2 V/cm, so the dc voltage being measured is interpreted as +2.6 V.

(b) Time and Frequency Measurement The frequency of an alternating wave, such as that seen in Figure 6-34(a), is inversely proportional to the amount of time it takes to complete one cycle ($f = 1/t$). Consequently, if time can be measured, frequency can be determined.

The time/cm control relates to the horizontal line on the oscilloscope graticule and is used to determine the period of a cycle so that frequency can be calculated. For example, in Figure 6-34(b), a cycle lasts five major horizontal divisions, and since the 20 μs/division setting has been selected, the period of the cycle will equal 5 × 20 μs/division = 100 μs. If the period is equal to 100 μs, the frequency of the waveform will be equal to $f = 1/t$ = 1/100 μs = 10 kHz.

EXAMPLE:

A complete sine-wave cycle occupies four horizontal divisions and four vertical divisions from peak to peak. If the oscilloscope is set on 20 ms/cm and 500 mV/cm, calculate:

a. $V_{\text{p-p}}$
b. t
c. f
d. V_{p}
e. V_{avg} of peak
f. V_{rms}

Solution:

$$4 \text{ horizontal divisions} \times 20 \text{ ms/div} = 80 \text{ ms}$$
$$4 \text{ vertical divisions} \times 500 \text{ mV/div} = 2 \text{ V}$$

a. $V_{\text{p-p}} = 2\text{ V}$
b. $t = 80\text{ ms}$
c. $f = \dfrac{1}{t} = 12.5\text{ Hz}$

d. $V_p = 0.5 \times V_{p\text{-}p} = 1$ V
e. $V_{avg} = 0.637 \times V_p = 0.637$ V
f. $V_{rms} = 0.707 \times V_p = 0.707$ V

SELF-TEST EVALUATION POINT FOR SECTION 6-4-2

Use the following questions to test your understanding of Section 6-4-2.

1. Name the device within the oscilloscope on which the waveforms can be seen.
2. On the 20 μs/div time setting, a full cycle occupies four major divisions. What is the waveform's frequency and period?
3. On the 2 V/cm voltage setting, the waveform swings up and down two major divisions (a total of 4 cm of vertical swing). Calculate the waveform's peak and peak-to-peak voltage.
4. What is/are the advantage(s) of the dual-trace oscilloscope?

6-4-3 *The Function Generator and Frequency Counter*

In many cases you will wish to generate a waveform of a certain shape and frequency, and then apply this waveform to your newly constructed circuit. One of the most versatile waveform generators is the **function generator,** so called because it can function as a sine wave, square wave, rectangular wave, triangular wave, or sawtooth wave generator. Figure 6-35 shows a photograph of a typical function generator.

It is very important that anyone involved in the design, manufacture, and servicing of electronic equipment be able to accurately measure the frequency of a periodic wave. Without the ability to measure frequency accurately, there could be no communications, home entertainment, or a great number of other systems. A **frequency counter,** like the one shown in Figure 6-36, can be used to analyze the frequency of any periodic wave applied to its input jack and provide a readout on the display of its frequency.

Function Generator
Signal generator that can function as a sine, square, rectangular, triangular, or sawtooth waveform generator.

Frequency Counter
Meter used to measure the frequency or cycles per second of a periodic wave.

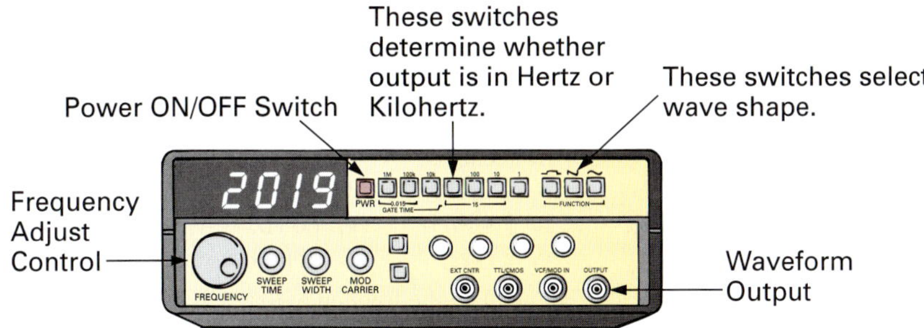

FIGURE 6-35 **Function Generator.**

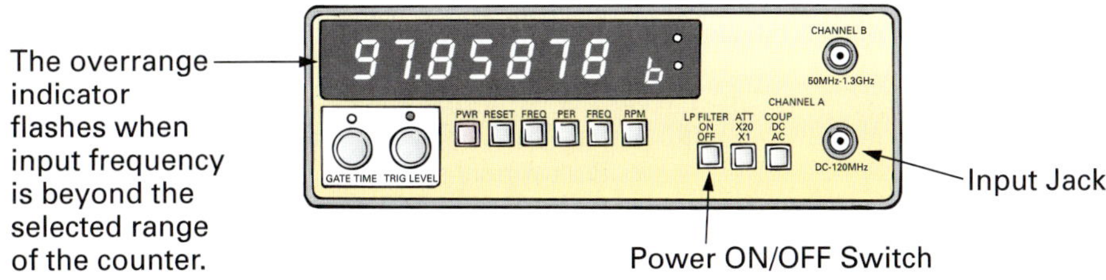

FIGURE 6-36 **Frequency Counter.**

SELF-TEST EVALUATION POINT FOR SECTION 6-4-3

Use the following questions to test your understanding of Section 6-4-3.

1. List some of the waveform shapes typically generated by a function generator.
2. What is the function of a frequency counter?

REVIEW QUESTIONS

Multiple-Choice Questions

1. A current that rises from zero to maximum positive, decreases to zero, and then reverses to reach a maximum in the opposite direction (negative) is known as:
   a. Alternating current
   b. Pulsating direct current
   c. Steady direct current
   d. All of the above
2. The advantage(s) of ac over dc from a power distribution point of view is/are:
   a. Generators can supply more power than batteries
   b. AC can be transformed to a high or low voltage easily, minimizing power loss
   c. AC can easily be converted into dc
   d. All of the above
   e. Only (a) and (c)
3. The most common type of alternating wave shape is the:
   a. Square wave c. Rectangular wave
   b. Sine wave d. Triangular wave
4. The peak-to-peak value of a sine wave is equal to:
   a. Twice the rms value
   b. 0.707 times the rms value
   c. Twice the peak value
   d. 1.14 × the average value
5. The rms value of a sine wave is also known as the:
   a. Effective value c. Peak value
   b. Average value d. All of the above
6. The peak value of a 115 V (rms) sine wave is:
   a. 115 V c. 162.7 V
   b. 230 V d. Two of the above could be true
7. The mathematical average value of a sine wave cycle is:
   a. 0.637 × peak c. 1.414 × rms
   b. 0.707 × peak d. Zero
8. The frequency of a sine wave is equal to the reciprocal of __________.
   a. The period d. Both (a) and (b)
   b. One cycle time e. None of the above
   c. One alternation
9. What is the period of a 1 MHz sine wave?
   a. 1 ms
   b. One millionth of a second
   c. 10 ms
   d. 100 μs
10. What is the frequency of a sine wave that has a cycle time of 1 ms?
    a. 1 MHz c. 200 m
    b. 1 kHz d. 10 kHz
11. The pulse width (P_w) is the time between the __________ points on the positive and negative edges of a pulse.
    a. 10% c. 50%
    b. 90% d. All of the above
12. The duty cycle is the ratio of __________ to period.
    a. Peak c. Pulse length
    b. Average power d. Both (a) and (c)
13. With a pulse waveform, PRF can be calculated by taking the reciprocal of:
    a. The duty cycle c. P_d
    b. PRT d. P_l
14. The sound wave exists in __________ and travels at approximately __________.
    a. Space, 1130 ft/s c. Air, 3×10^6 m/s
    b. Wires, 186,282.397 miles/s d. None of the above
15. The electrical and electromagnetic waves travel at a speed of:
    a. 186,000 miles/s c. 162,000 nautical miles/s
    b. 3×10^8 meters/s d. All of the above

Practice Problems

16. Calculate the periods of the following sine-wave frequencies:
    a. 27 kHz d. 365 Hz
    b. 3.4 MHz e. 60 Hz
    c. 25 Hz f. 200 kHz
17. Calculate the frequency for each of the following values of time:
    a. 16 ms d. 0.05 s
    b. 1 s e. 200 μs
    c. 15 μs f. 350 ms
18. A 22 V peak sine wave will have the following values:
    a. Rms voltage = c. Peak-to-peak voltage =
    b. Average voltage =
19. A 40 mA rms sine wave will have the following values:
    a. Peak current = c. Average current =
    b. Peak-to-peak current =
20. What is the duty cycle of a 10 V peak square wave at a frequency of 1 kHz?

21. If one cycle of a sine wave occupies 4 cm on the oscilloscope horizontal grid and 5 cm from peak to peak on the vertical grid, calculate frequency, period, rms, average, and peak for the following control settings:
  a. 0.5 V/cm, 20 μs/cm
  b. 10 V/cm, 10 ms/cm
  c. 50 mV/cm, 0.2 μs/cm
22. Assuming the same graticule and switch settings of the oscilloscope in Figure 6-29, what would be the lowest setting of the volts/cm and time/division switches to fully view a 6 V rms, 350 kHz sine wave?
23. If the volts/cm switch is positioned to 10 V/cm and the waveform extends 3.5 divisions from peak to peak, what is the peak-to-peak value of this wave?
24. If a square wave occupies 5.5 horizontal cm on the 1 μs/cm position, what is its frequency?
25. Which settings would you use on an autoset scopemeter to fully view an 8 V rms, 20 kHz triangular wave?

Web Site Questions

Go to the web site http://www.prenhall.com/cook, select the textbook *Electronics: A Complete Course,* select this chapter, and then follow the instructions when answering the multiple choice practice problems.

Capacitors

Back to the Future

Born in England in 1791, Charles Babbage became very well known for both his mathematical genius and eccentric personality. Babbage's ultimate pursuit was that of mathematical accuracy. He delighted in spotting errors in everything from log tables (used by astronomers, mathematicians, and navigators) to poetry. In fact, he once wrote to poet Alfred Lord Tennyson, pointing out an inaccuracy in his line "Every moment dies a man—every moment one is born." Babbage explained to Tennyson that since the world population was actually increasing and not, as he indicated, remaining constant, the line should be rewritten to read "Every moment dies a man—every moment one and one-sixteenth is born."

In 1822, Babbage described in a paper and built a model of what he called "a difference engine," which could be used to calculate mathematical tables. The Royal Society of Scientists described his machine as "highly deserving of public encouragement," and a year later the government awarded Babbage £1500 for his project. Babbage originally estimated that the project should take 3 years; however, the design had its complications, and after 10 years of frustrating labor, in which the government grants increased to £17,000, Babbage was still no closer to completion. Finally, the money stopped and Babbage reluctantly decided to let his brainchild go.

In 1833, Babbage developed an idea for a much more practical machine, which he named "the analytical engine." It was to be a more general machine that could be used to solve a variety of problems, depending on instructions supplied by the operator. It would include two units, called a "mill" and a "store," both of which would be made of cogs and wheels. The store, which was equivalent to a modern-day computer memory, could hold up to 100 forty-digit numbers. The mill, which was equivalent to a modern computer's arithmetic and logic unit (ALU), could perform both arithmetic and logic operations on variables or numbers retrieved from the store, and the result could be stored in the store and then acted upon again or printed out. The program of instructions directing these operations would be fed into the analytical engine in the form of punched cards.

The analytical engine was never built. All that remains are the volumes of descriptions and drawings, and a section of the mill and printer built by Babbage's son, who also had to concede defeat. It was, unfortunately for Charles Babbage, a lifetime of frustration to have conceived the basic building blocks of the modern computer a century before the technology existed to build it.

7

Introduction

Up to this point we have concentrated on circuits containing only resistance, which opposes the flow of current and then converts or dissipates power in the form of heat. Capacitance and inductance are two circuit properties that act differently from resistance in that they will charge or store the supplied energy and then return almost all the stored energy back to the circuit, rather than lose it in wasted heat. Inductance will be discussed in a following chapter.

Capacitance is the ability of a circuit or device to store electrical charge. A device or component specifically designed to have this capacity or capacitance is called a *capacitor.* A capacitor stores an electrical charge similar to a bucket holding water. Using the analogy throughout this chapter that a capacitor is similar to a bucket will help you gain a clear understanding of the capacitor's operation. For example, the capacitor holds charge in the same way that a bucket holds water. A larger capacitor will hold more charge and will take longer to charge, just as a larger bucket will hold more water and take longer to fill. A larger circuit resistance means a smaller circuit current, and therefore a longer capacitor charge time. Similarly, a smaller hose will have a greater water resistance producing a smaller water flow, and therefore the bucket will take a longer time to fill. Capacitors store electrons, and basically, the amount of electrons stored is a measure of the capacitor's capacitance.

7-1 CAPACITOR CHARACTERISTICS

Capacitor
Device that stores electric energy in the form of an electric field that exists within a dielectric (insulator) between two conducting plates each of which is connected to a lead. This device was originally called a condenser.

Several years ago, capacitors were referred to as condensers, but that term is very rarely used today. Figure 7-1 illustrates the main parts and schematic symbol of the **capacitor.** Two leads are connected to two parallel metal conductive plates, which are separated by an insulating material known as a *dielectric.* It is called a dielectric because it exists between two (*di*) plates and when the capacitor is charged an electric field exists within it (*di-electric*). The conductive plates are normally made of metal foil, while the dielectric can be paper, air, glass, ceramic, mica, or some other form of insulator.

1. The Unit of Capacitance

Farad
Unit of capacitance.

Capacitance is the ability of a capacitor to store an electrical charge, and the unit of capacitance is the **farad** (F), named in honor of Michael Faraday's work in 1831 in the

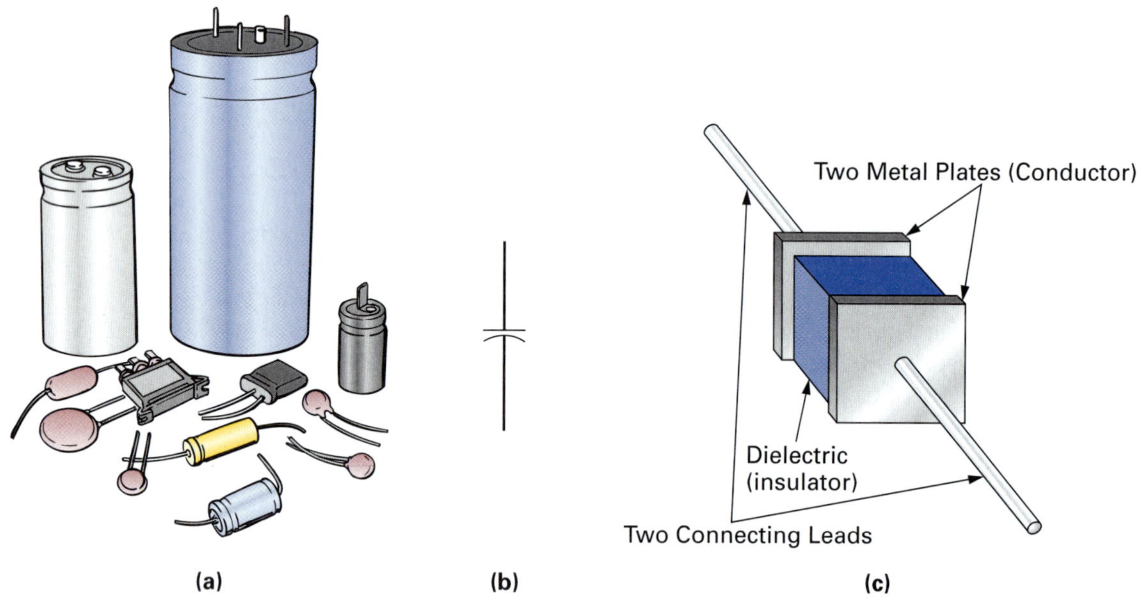

FIGURE 7-1 **Capacitor. (a) Physical Appearance. (b) Schematic Symbol. (c) Basic Construction.**

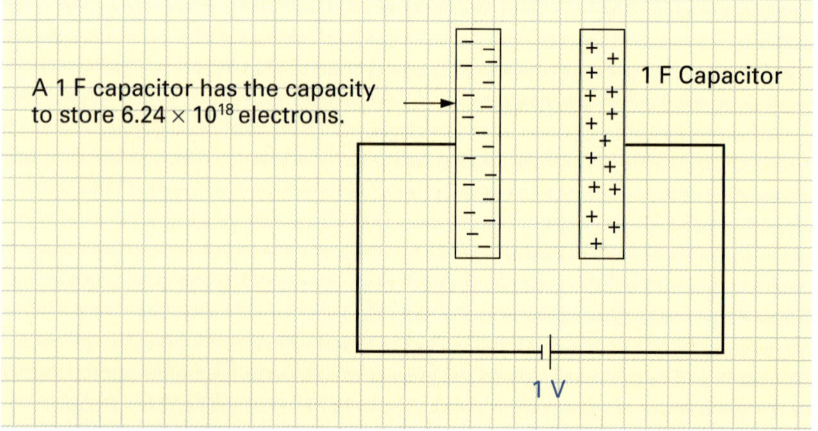

FIGURE 7-2 **One Farad of Capacitance.**

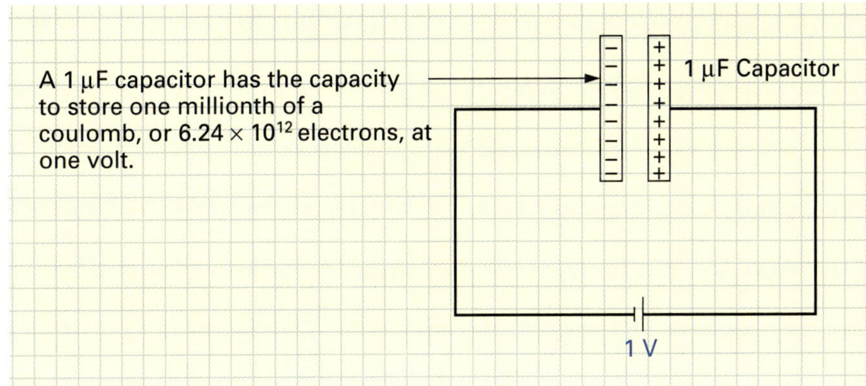

FIGURE 7-3 One-Millionth of a Farad.

field of capacitance. A capacitor with the capacity of 1 farad (1 F) can store 1 coulomb of electrical charge (6.24×10^{18} electrons) if 1 volt is applied across the capacitor's plates, as seen in Figure 7-2.

A 1 F capacitor is a very large value and not frequently found in electronic equipment. Most values of capacitance found in electronic equipment are in the units between the microfarad ($\mu F = 10^{-6}$ F) and picofarad ($pF = 10^{-12}$ F). A microfarad is 1 millionth of a farad (10^{-6}). So if a 1 F capacitor can store 6.24×10^{18} electrons with 1 V applied, a 1 μF capacitor, which has 1 millionth the capacity of a 1 F capacitor, can store only 1 millionth of a coulomb, or $(6.24 \times 10^{18}) \times (1 \times 10^{-6}) = 6.24 \times 10^{12}$ electrons when 1 V is applied, as shown in Figure 7-3.

EXAMPLE:

Convert the following to either microfarads or picofarads (whichever is more appropriate):

a. 0.00002 F
b. 0.00000076 F
c. 0.00047×10^{-7} F

Solution:

a. 20 μF
b. 0.76 μF
c. 47 pF

Since there is a direct relationship between capacitance, charge, and voltage, there must be a way of expressing this relationship in a formula.

$$\text{capacitance, } C \text{ (farads)} = \frac{\text{charge, } Q \text{ (coulombs)}}{\text{voltage, } V \text{(volts)}}$$

where C = capacitance, in farads
Q = charge, in coulombs
V = voltage, in volts

By transposition of the formula, we arrive at the following combinations for the same formula:

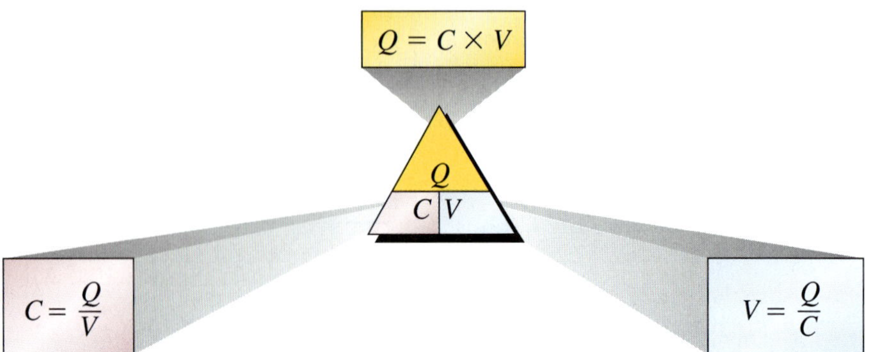

EXAMPLE:

If a capacitor has the capacity to hold 36 C ($36 \times 6.24 \times 10^{18} = 2.25 \times 10^{20}$ electrons) when 12 V is applied across its plates, what is the capacitance of the capacitor?

Solution:

$$C = \frac{Q}{V} = \frac{36\text{ C}}{12\text{ V}} = 3\text{ F}$$

EXAMPLE:

How many electrons could a 3 μF capacitor store when 5 V is applied across it?

Solution:

$$Q = C \times V = 3\ \mu\text{F} \times 5\text{ V} = 15\ \mu\text{C}$$

(15 microcoulombs is 15 millionths of a coulomb.) Since $1\text{ C} = 6.24 \times 10^{18}$ electrons, $15\ \mu\text{C} = (15 \times 10^{-6}) \times 6.24 \times 10^{18} = 9.36 \times 10^{13}$ electrons.

EXAMPLE:

If a capacitor of 2 F has stored 42 C of charge (2.63×10^{20} electrons), what is the voltage across the capacitor?

Solution:

$$V = \frac{Q}{C} = \frac{42\text{ C}}{2\text{ F}} = 21\text{ V}$$

2. Capacitive Time Constant

When a capacitor is connected across a dc voltage source, it will charge to a value equal to the voltage applied. If the charged capacitor is then connected across a load, the capacitor will then discharge through the load. The time it takes a capacitor to charge or discharge can be calculated if the circuit's resistance and capacitance are known. Let us now see how we can calculate a capacitor's charge time and discharge time.

7-1-1 *DC Charging*

When a capacitor is connected across a dc voltage source, such as a battery or power supply, current will flow and the capacitor will charge up to a value equal to the dc source voltage, as shown in Figure 7-4. When the charge switch is first closed, as seen in Figure 7-4(a), there is no voltage across the capacitor at that instant and therefore a potential difference exists between the battery and capacitor. This causes current to flow and begin charging the capacitor.

Once the capacitor begins to charge, the voltage across the capacitor does not instantaneously rise to 100 V. It takes a certain amount of time before the capacitor voltage is equal to the battery voltage. When the capacitor is fully charged no potential difference exists between the voltage source and the capacitor. Consequently, no more current flows in the circuit as the capacitor has reached its full charge, as seen in Figure 7-4(b). The amount of time it takes for a capacitor to charge to the supplied voltage (in this example, 100 V) is dependent on the circuit's resistance and capacitance value. If the circuit's resistance is increased, the opposition to current flow will be increased, and it will take the capacitor a longer period of time to obtain the same amount of charge because the circuit current available to charge the capacitor is less.

If the value of capacitance is increased, it again takes a longer time to charge to 100 V because a greater amount of charge is required to build up the voltage across the capacitor to 100 V.

The circuit's resistance (R) and capacitance (C) are the two factors that determine the charge time (τ). Mathematically, this can be stated as

Time Constant

Time needed for either a voltage or current to rise to 63.2% of the maximum or fall to 36.8% of the initial value. The time constant of an *RC* circuit is equal to the product of *R* and *C*.

$$\tau = R \times C$$

where τ = **time constant** (s)
R = resistance (Ω)
C = capacitance (F)

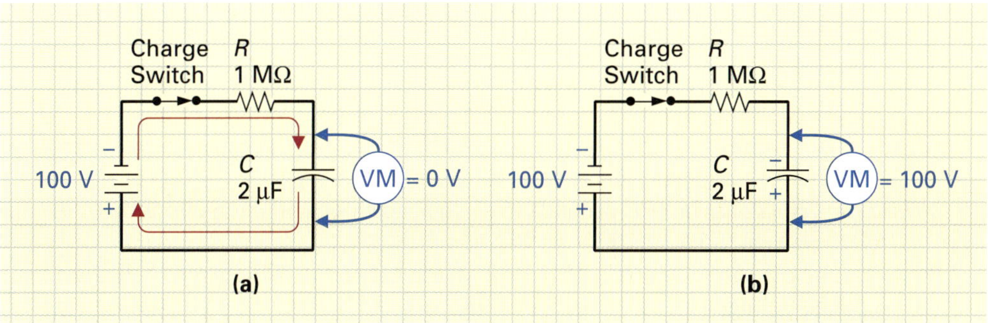

FIGURE 7-4 Capacitor Charging. (a) Switch Is Closed and Capacitor Begins to Charge. (b) Capacitor Charged.

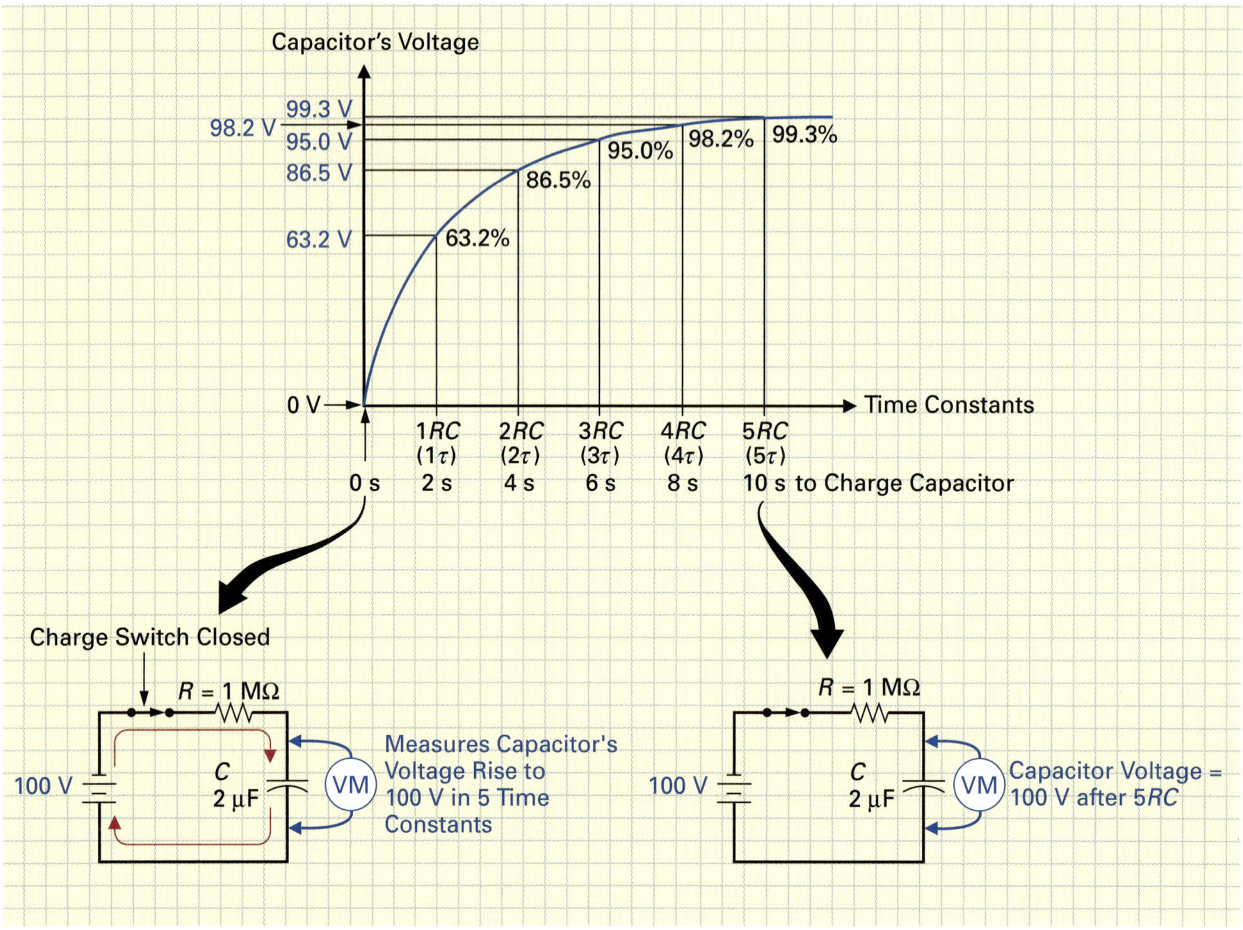

FIGURE 7-5 Charging Capacitor.

In this example, we are using a resistance of 1 MΩ and a capacitance of 2 μF, which means that the time constant is equal to

$$
\begin{aligned}
\tau &= R \times C \\
&= 2\ \mu\text{F} \times 1\ \text{M}\Omega \\
&= (2 \times 10^{-6}) \times (1 \times 10^{6}) \\
&= 2\ \text{s}
\end{aligned}
$$

Two seconds is the time, so what is the constant? The constant value that should be remembered throughout this discussion is "**63.2**."

Figure 7-5 illustrates the rise in voltage across the capacitor from 0 to a maximum of 100 V in five time constants (5×2 s = 10 s). So where does 63.2 come into all this?

First time constant: In 1*RC* seconds ($1 \times R \times C = 2$ s), the capacitor will charge to 63.2% of the applied voltage (63.2% × 100 V = 63.2 V).

Second time constant: In 2*RC* seconds ($2 \times R \times C = 4$ s), the capacitor will charge to 86.5% of the applied voltage (86.5% × 100 V = 86.5 V).

Third time constant: In 3*RC* seconds (6 s), the capacitor will charge to 95% of the applied voltage (95% × 100 V = 95 V).

Fourth time constant: In 4*RC* seconds (8 s), the capacitor will charge to 98.2% of the applied voltage (98.2% × 100 V = 98.2 V).

Fifth time constant: In 5*RC* seconds (10 s), the capacitor is considered to be fully charged since the capacitor will have reached 99.3% of the applied voltage (99.3%× 100 V = 99.3 V, approximately 100 V).

The voltage waveform produced by the capacitor acquiring a charge is known as an *exponential* waveform, and the voltage across the capacitor is said to rise exponentially. An exponential rise is also referred to as a *natural increase*. There are many factors that exponentially rise and fall. For example, we grow exponentially, in that there is quite a dramatic change in our height in the early years and then this increase levels off and reaches a maximum.

Before the switch is closed and even at the instant the switch is closed, the capacitor is not charged, which means that there is no capacitor voltage to oppose the supply voltage and, therefore, a maximum current of V/R, 100 V/1 MΩ = 100 μA flows. This current begins to charge the capacitor, a potential difference begins to build up across the plates of the capacitor, and this voltage opposes the supply voltage, causing a decrease in charging current. As the capacitor begins to charge, less of a potential difference exists between the supply voltage and capacitor voltage and so the current begins to decrease.

To calculate the current at any time, we can use the formula

$$i = \frac{V_S - V_C}{R}$$

where i = instantaneous current
V_S = source voltage
V_C = capacitor voltage
R = resistance

For example, the current flowing in the circuit after one time constant will equal the source voltage, 100 V, minus the capacitor's voltage, which in one time constant will be 63.2% of the source voltage or 63.2 V, divided by the resistance.

$$\begin{aligned} i &= \frac{V_S - V_C}{R} \\ &= \frac{100\text{ V} - 63.2\text{ V}}{1\text{ M}\Omega} \\ &= 36.8\ \mu\text{A} \end{aligned}$$

As the charging continues, the potential difference across the plates exponentially rises to equal the supply voltage, as seen in Figure 7-6(a), while the current exponentially falls to zero, as shown in Figure 7-6(b). The constant of 63.2 can be applied to the exponential fall of current from 100 μA to 0 μA in 5*RC* seconds.

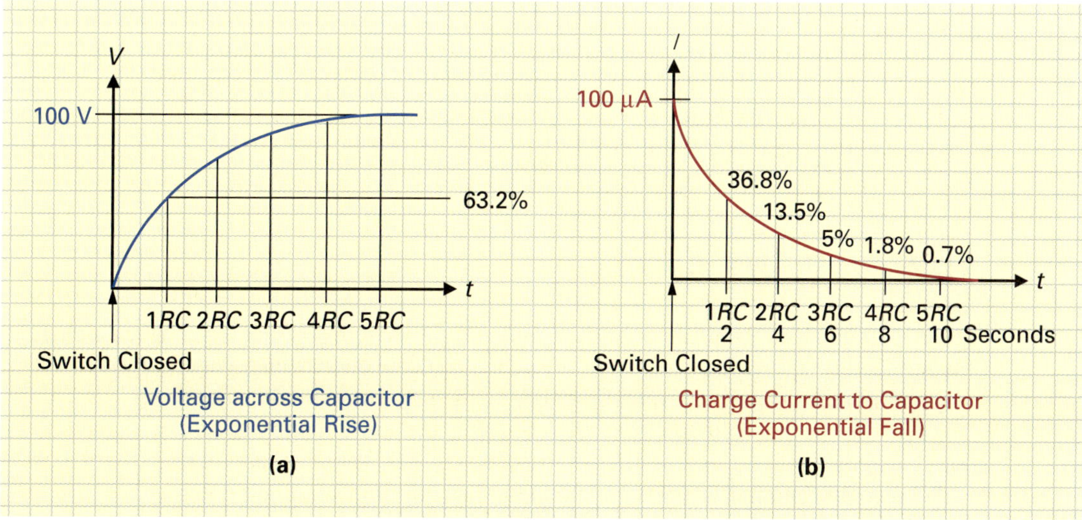

FIGURE 7-6 Exponential Rise in Voltage and Fall in Current in a Charging Capacitive Circuit.

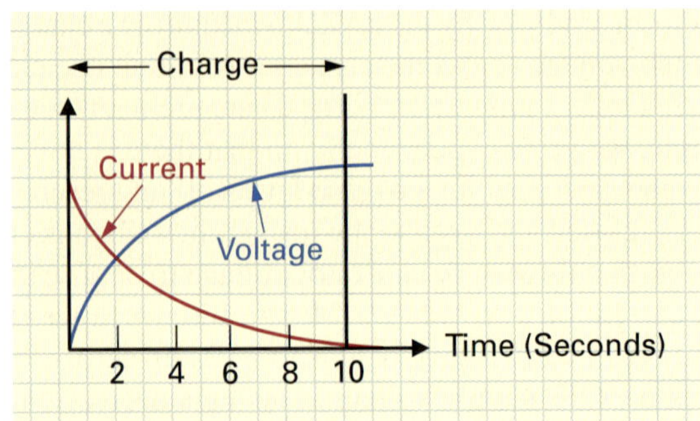

FIGURE 7-7 Phase Difference between Voltage and Current in a Charging Capacitive Circuit.

When the switch was closed to start the charging of the capacitor, there was no charge on the capacitor; therefore, a maximum potential difference existed between the battery and capacitor, causing a maximum current flow of 100 μA ($I = V/R$).

First time constant: In 1*RC* seconds, the current will have exponentially decreased 63.2% (63.2% of 100 μA = 63.2 μA) to a value of 36.8 μA (100 μA − 63.2 μA). In the example of 2 μF and 1 MΩ, this occurs in 2 s.

Second time constant: In 2*RC* seconds ($2 \times R \times C = 4$ s), the current will decrease to 13.5 μA or 13.5%.

Third time constant: In 3*RC* seconds (6 s), the capacitor's charge current will decrease to 5 μA or 5%.

Fourth time constant: In 4*RC* seconds (8 s), the current will have decreased to 1.8 μA or 1.8%.

Fifth time constant: In 5*RC* seconds (10 s), the charge current is now 0.7 μA or 0.7%. At this time, the charge current is assumed to be zero and the capacitor is now charged to a voltage equal to the applied voltage.

Studying the exponential rise of the voltage and the exponential decay of current in a capacitive circuit, you will notice an interesting relationship. In a pure resistive circuit, the current flow through a resistor will be in step with the voltage across that same resistor, in that an increased current will cause a corresponding increase in voltage drop across the resistor. Voltage and current are consequently said to be *in step* or *in phase* with one another. With the capacitive circuit, the current flow in the circuit and voltage across the capacitor are not in step or in phase with one another. When the switch is closed to charge the capacitor, the current is maximum (100 μA), while the voltage across the capacitor is zero. After five time constants (10 s), the capacitor's voltage is now maximum (100 V) and the circuit current is zero, as seen in Figure 7-7. The circuit current flow is out of phase with the capacitor voltage, and this difference is referred to as a *phase shift.* In any circuit containing capacitance, current will lead voltage.

7-1-2 *DC Discharging*

Figure 7-8 illustrates the circuit, voltage, and current waveforms that occur when a charged capacitor is discharged from 100 V to 0 V. The 2 μF capacitor, which was charged to 100 V in 10 s (5*RC*), is discharged from 100 to 0 V in the same amount of time.

Looking at the voltage curve, you can see that the voltage across the capacitor decreases exponentially, dropping 63.2% to 36.8 V in 1*RC* seconds, another 63.2% to 13.5 V in 2*RC* seconds, another 63.2% to 5 V in 3*RC* seconds, and so on, until zero.

The current flow within the circuit is dependent on the voltage in the circuit, which is across the 2 μF capacitor. As the voltage decreases, the current will also decrease by the same amount ($I\downarrow = V\downarrow/R$).

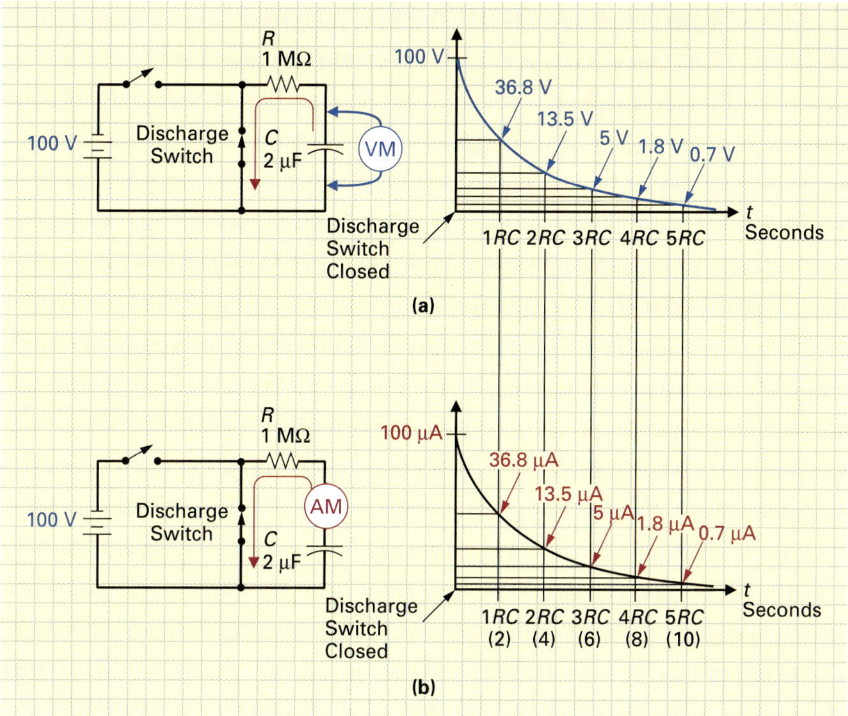

FIGURE 7-8 **Discharging Capacitor. (a) Voltage Waveform. (b) Current Waveform.**

$$\text{discharge switch closed: } I = \frac{V}{R} = \frac{100 \text{ V}}{1 \text{ M}\Omega} = 100\ \mu\text{A} \quad \text{maximum}$$

$$1RC\ (2) \text{ seconds: } I = \frac{V}{R} = \frac{36.8 \text{ V}}{1 \text{ M}\Omega} = 36.8\ \mu\text{A}$$

$$2RC\ (4) \text{ seconds: } I = \frac{V}{R} = \frac{13.5 \text{ V}}{1 \text{ M}\Omega} = 13.5\ \mu\text{A}$$

$$3RC\ (6) \text{ seconds: } I = \frac{V}{R} = \frac{5 \text{ V}}{1 \text{ M}\Omega} = 5.0\ \mu\text{A}$$

$$4RC\ (8) \text{ seconds: } I = \frac{V}{R} = \frac{1.8 \text{ V}}{1 \text{ M}\Omega} = 1.8\ \mu\text{A}$$

$$5RC\ (10) \text{ seconds: } I = \frac{V}{R} = \frac{0.7 \text{ V}}{1 \text{ M}\Omega} = 0.7\ \mu\text{A} \quad \text{zero}$$

3. Factors Determining Capacitance

The capacitance of a capacitor is determined by three factors:

1. The plate area of the capacitor
2. The distance between the plates
3. The type of dielectric used

The formula that combines these three factors is

$$C = \frac{(8.85 \times 10^{-12}) \times K \times A}{d}$$

where C = capacitance, in farads (F)
8.85×10^{-12} is a constant
K = dielectric constant (see Table 7-1)
A = plate area, in square meters (m^2)
d = distance between the plates, in meters (m)

TABLE 7-1 Dielectric Constants

MATERIAL	DIELECTRIC CONSTANT (K)[a]
Vacuum	1.0
Air	1.0006
Teflon	2.0
Wax	2.25
Paper	2.5
Amber	2.65
Rubber	3.0
Oil	4.0
Mica	5.0
Ceramic (low)	6.0
Bakelite	7.0
Glass	7.5
Water	78.0
Ceramic (high)	8000.

[a]The different material compositions can cause different values of K.

This formula states that the capacitance of a capacitor is directly proportional to the dielectric constant (K) and the plates' area (A) and is inversely proportional to the dielectric thickness or distance between the plates (d).

EXAMPLE:

What is the capacitance of a ceramic capacitor with a 0.3 m^2 plate area and a dielectric thickness of 0.0003 m?

Solution:

$$C = \frac{(8.85 \times 10^{-12}) \times K \times A}{d}$$

$$= \frac{(8.85 \times 10^{-12}) \times 6 \times 0.3 \text{ m}^2}{0.0003 \text{ m}}$$

$$= 5.31 \times 10^{-8} \text{F}$$

$$= 0.0531 \ \mu\text{F}$$

CALCULATOR SEQUENCE

Step	Keypad Entry	Display Response
1.	[8] [.] [8] [5] [E] [1] [2] [+/−]	8.85E-12
2.	[×]	
3.	[6]	6
4.	[×]	5.31E-11
5.	[0] [.] [3]	
6.	[÷]	1.59E-11
7.	[0] [.] [0] [0] [0] [3]	
8.	[=]	5.31E-8

SELF-TEST EVALUATION POINT FOR SECTION 7-1

Use the following questions to test your understanding of Section 7-1.

1. List the three variable factors that determine the capacitance of a capacitor.
2. State the formula for capacitance.

3. If a capacitor's plate area is doubled, the capacitance will __________.
4. If a capacitor's dielectric thickness is halved, the capacitance will __________.
5. In one time constant, a capacitor will have charged to what percentage of the applied voltage?
6. In one time constant, a capacitor will have discharged to what percentage of its full charge?

7-2 CAPACITORS IN COMBINATION

Like resistors, capacitors can be connected in either series or parallel. As you will see in this section, the rules for determining total capacitance for parallel- and series-connected capacitors are opposite to series- and parallel-connected resistors.

7-2-1 *Capacitors in Parallel*

In Figure 7-9(a), you can see a 2 μF and 4 μF capacitor connected in parallel with one another. As the top plate of capacitor *A* is connected to the top plate of capacitor *B* with a wire, and a similar situation occurs with the bottom plates, you can see that this is the same as if the top and bottom plates were touching one another, as shown in Figure 7-9(b). When drawn so that the respective plates are touching, the dielectric constant and plate separation is the same as shown in Figure 7-9(a), but now we can easily see that the plate area is actually increased. Consequently, if capacitors are connected in parallel, the effective plate area is increased; and since capacitance is proportional to plate area $[C\uparrow = (8.85 \times 10^{-12}) \times K \times A\uparrow/d]$, the capacitance will also increase. Total capacitance is actually calculated by adding the plate areas, so total capacitance is equal to the sum of all the individual capacitances in parallel.

$$C_T = C_1 + C_2 + C_3 + C_4 + \cdots$$

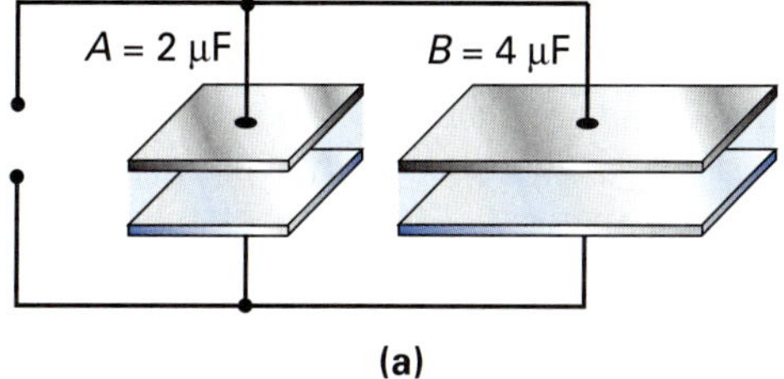

(a)

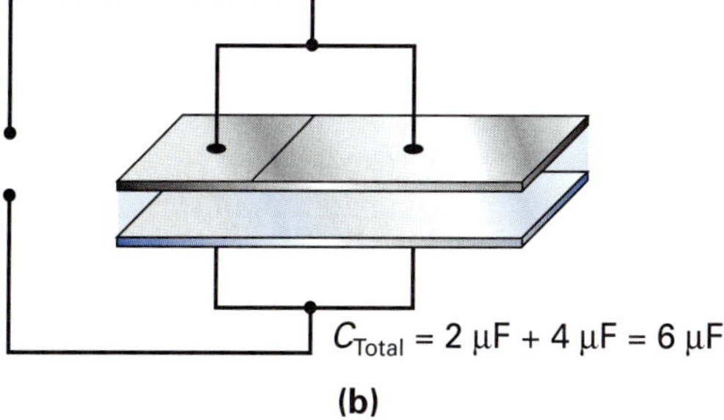

(b)

FIGURE 7-9 Capacitors in Parallel.

EXAMPLE:

Determine the total capacitance of the circuit in Figure 7-10(a). What will be the voltage drop across each capacitor?

Solution:

$$\begin{aligned} C_T &= C_1 + C_2 + C_3 \\ &= 1\ \mu\text{F} + 0.5\ \mu\text{F} + 0.75\ \mu\text{F} \\ &= 2.25\ \mu\text{F} \end{aligned}$$

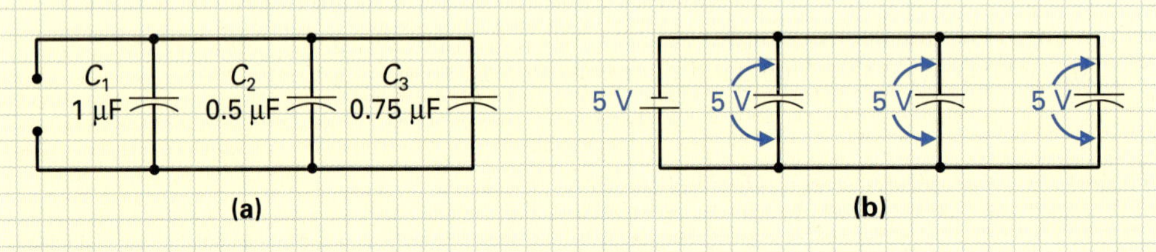

FIGURE 7-10 Example of Parallel-Connected Capacitors.

As with any parallel-connected circuit, the source voltage appears across all the components. If, for example, 5 V is connected to the circuit of Figure 7-10(b), all the capacitors will charge to the same voltage of 5 V because the same voltage always exists across each section of a parallel circuit.

7-2-2 *Capacitors in Series*

In Figure 7-11(a), we have taken the two capacitors of 2 μF and 4 μF and connected them in series. Since the bottom plate of the *A* capacitor is connected to the top plate of the *B* capacitor, they can be redrawn so that they are touching, as shown in Figure 7-11(b).

The top plate of the *A* capacitor is connected to a wire into the circuit, and the bottom plate of *B* is connected to a wire into the circuit. This connection creates two center plates that are isolated from the circuit and can therefore be disregarded, as shown in Figure 7-11(c). The first thing you will notice in this illustration is that the dielectric thickness ($d\uparrow$) has increased, causing a greater separation between the plates. The effective plate area of this capacitor has decreased, as it is just the area of the top plate only. Even though the bottom plate extends out further, the electric field can only exist between the two plates, so the surplus metal of the bottom plate has no metal plate opposite for the electric field to exist.

Consequently, when capacitors are connected in series the effective plate area is decreased ($A\downarrow$) and the dielectric thickness increased ($d\uparrow$), and both of these effects result in an overall capacitance decrease ($C\downarrow\downarrow = (8.85 \times 10^{-12}) \times K \times A\downarrow/d\uparrow$).

The plate area is actually decreased to the smallest individual capacitance connected in series, which in this example is the plate area of *A*. If the plate area were the only factor, then capacitance would always equal the smallest capacitor value. However, the dielectric thickness is always equal to the sum of all the capacitor dielectrics, and this factor always causes the total capacitance (C_T) to be less than the smallest individual capacitance when capacitors are connected in series.

The total capacitance of two or more capacitors in series therefore is calculated by using the following formulas: For two capacitors in series,

$$C_T = \frac{C_1 \times C_2}{C_1 + C_2}$$

(product-over-sum formula)

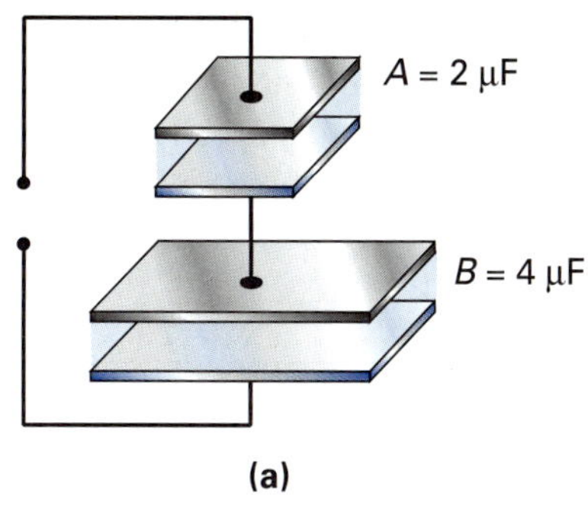

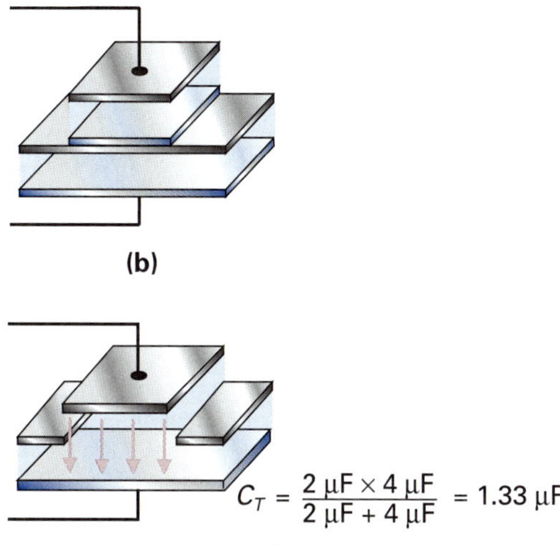

FIGURE 7-11 Capacitors in Series.

For more than two capacitors in series,

$$C_T = \frac{1}{(1/C_1) + (1/C_2) + (1/C_3) + \cdots}$$

(reciprocal formula)

EXAMPLE:

Determine the total capacitance of the circuit in Figure 7-12.

Solution:

$$\begin{aligned} C_T &= \frac{1}{(1/C_1) + (1/C_2) + (1/C_3)} \\ &= \frac{1}{(1/4\ \mu\text{F}) + (1/2\ \mu\text{F}) + (1/1\ \mu\text{F})} \\ &= \frac{1}{1.75 \times 10^6} = 5.7143 \times 10^{-7} \\ &= 0.5714\ \mu\text{F} \text{ or } 0.6\ \mu\text{F} \end{aligned}$$

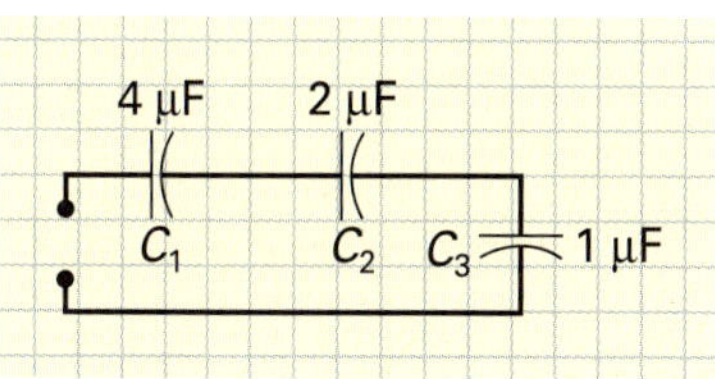

FIGURE 7-12 Example of Series-Connected Capacitors.

The total capacitance for capacitors in series is calculated in the same way as total resistance when resistors are in parallel.

As with series-connected resistors, the sum of all of the voltage drops across the series-connected capacitors will equal the voltage applied (Kirchhoff's voltage law). With capacitors connected in series, the charged capacitors act as a voltage divider, and therefore the voltage-divider formula can be applied to capacitors in series.

$$V_{cx} = \frac{C_T}{C_x} \times V_T$$

where V_{cx} = voltage across desired capacitor
C_T = total capacitance
C_x = desired capacitor's value
V_T = total supplied voltage

EXAMPLE:

Using the voltage-divider formula, calculate the voltage dropped across each of the capacitors in Figure 7-12 if $V_T = 24$ V.

Solution:

$$V_{C1} = \frac{C_T}{C_1} \times V_T = \frac{0.5714\ \mu\text{F}}{4\ \mu\text{F}} \times 24\text{ V} = 3.4\text{ V}$$

$$V_{C2} = \frac{C_T}{C_2} \times V_T = \frac{0.5714\ \mu\text{F}}{2\ \mu\text{F}} \times 24\text{ V} = 6.9\text{ V}$$

$$V_{C3} = \frac{C_T}{C_3} \times V_T = \frac{0.5714\ \mu\text{F}}{1\ \mu\text{F}} \times 24\text{ V} = 13.7\text{ V}$$

$$V_T = V_{C1} + V_{C2} + V_{C3} = 3.4 + 6.9 + 13.7 = 24\text{ V}$$

(Kirchhoff's voltage law)

If the capacitor values are the same, as seen in Figure 7-13(a), the voltage is divided equally across each capacitor, as each capacitor has an equal amount of charge and therefore has half of the applied voltage (in this example, 3 V across each capacitor).

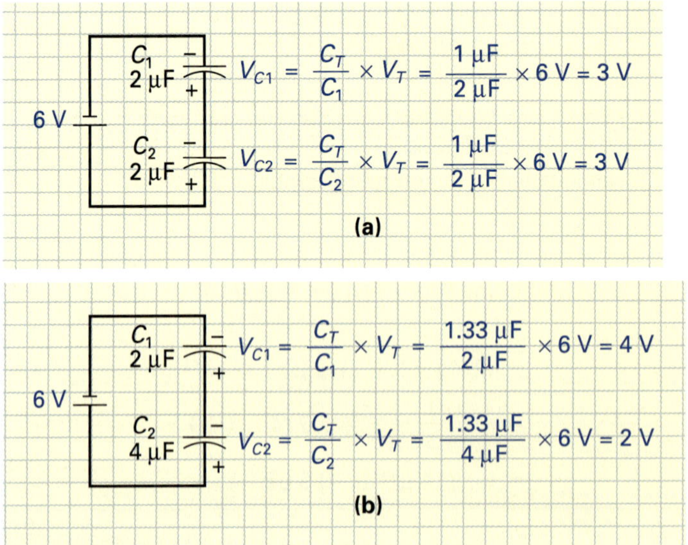

FIGURE 7-13 Voltage Drops across Series-Connected Capacitors.

When the capacitor values are different, the smaller value of capacitor will actually charge to a higher voltage than the larger capacitor. In the example in Figure 7-13(b), the smaller capacitor is actually half the size of the other capacitor, and it has charged to twice the voltage. Since Kirchhoff's voltage law has to apply to this and every series circuit, you can easily calculate that the voltage across C_1 will equal 4 V and is twice that of C_2, which is 2 V. To understand this fully, we must first understand that although the capacitance is different, both capacitors have an equal value of coulomb charge held within them, which in this example is 8 μC.

$$\begin{aligned} Q_1 &= C_1 \times V_1 \\ &= 2\ \mu\text{F} \times 4\ \text{V} = 8\ \mu\text{C} \\ Q_2 &= C_2 \times V_2 \\ &= 4\ \mu\text{F} \times 2\ \text{V} = 8\ \mu\text{C} \end{aligned}$$

This equal charge occurs because the same amount of current flow exists throughout a series circuit, so both capacitors are being supplied with the same number or quantity of electrons. The charge held by C_1 is large with respect to its small capacitance, whereas the same charge held by C_2 is small with respect to its larger capacitance.

If the charge remains the same (Q is constant) and the capacitance is small, the voltage drop across the capacitor will be large, because the charge is large with respect to the capacitance:

$$V\uparrow = \frac{Q}{C\downarrow}$$

On the other hand, for a constant charge, a large capacitance will have a small charge voltage because the charge is small with respect to the capacitance:

$$V\downarrow = \frac{Q}{C\uparrow}$$

We can apply the water analogy once more and imagine two series-connected buckets, one of which is twice the size of the other. Both are being supplied by the same series pipe, which has an equal flow of water throughout, and are consequently each holding an equal amount of water, for example, 1 gallon. The 1 gallon of water in the small bucket is large with respect to the size of the bucket, and a large amount of pressure exists within that bucket. The 1 gallon of water in the large bucket is small with respect to the size of the bucket, so a small amount of pressure exists within this bucket. The pressure within a bucket is similar to the voltage across a capacitor, and therefore a small bucket or capacitor will have a greater pressure or voltage associated with it, while a large bucket or capacitor will develop a small pressure or voltage.

To summarize capacitors in series, all the series-connected components will have the same charging current throughout the circuit, and because of this, two or more capacitors in series will always have equal amounts of coulomb charge. If the charge (Q) is equal, the voltage across the capacitor is determined by the value of the capacitor. A small capacitance will charge to a larger voltage ($V\uparrow = Q/C\downarrow$), whereas a large value of capacitance will charge to a smaller voltage ($V\downarrow = Q/C\uparrow$).

SELF-TEST EVALUATION POINT FOR SECTION 7-2

Use the following questions to test your understanding of Section 7-2.

1. If 2 μF, 3 μF, and 5 μF capacitors are connected in series, what will be the total circuit capacitance?
2. If 7 pF, 2 pF, and 14 pF capacitors are connected in parallel, what will be the total circuit capacitance?
3. State the voltage-divider formula as it applies to capacitance.
4. True or false: With resistors, the large value of resistor will drop a larger voltage, whereas with capacitors the smaller value of capacitor will actually charge to a higher voltage.

7-3 TYPES OF CAPACITORS AND CODING

Capacitors come in a variety of shapes and sizes and can be either fixed or variable in their values of capacitance. Within these groups, capacitors are generally classified by the dielectric used between the plates.

Fixed-Value Capacitor
A capacitor whose value is fixed and cannot be varied.

7-3-1 *Capacitor Types*

A **fixed-value capacitor** is a capacitor whose capacitance value remains constant and cannot be altered. Fixed capacitors normally come in a disk or a tubular package and consist of

Name	Construction	Approximate Range of Values and Tolerances	Characteristics	
Mica	Foil Mica Foil Mica Foil Mica Foil	1 pF–0.1 μF ±1% to ±5%	Lower voltage rating than other capacitors of the same size	Small Capacitor Values
Ceramic	Ceramic Dielectric Lead Wire	*Low Dielectric K*: 1 pF–0.01 μF ±0.5% to ±10% *High Dielectric K*: 1 pF–0.1 μF ±10% to ±80%	Most popular small value capacitor due to lower cost than mica, and its ruggedness	
Paper	Lead to Inner Foil Sheet Inner Foil Lead to Outer Foil Sheet Outer Foil	1 pF–1 μF ± 10%	Has a large plate area and therefore large capacitance for a small size	Large Capacitor Values
Plastic	Outside Foil Inside Foil	1 pF–10 μF ± 5% to ±10%	Has almost completely replaced paper capacitors; has large capacitance values for small size and high voltage ratings	
Electrolytic (Aluminum and Tantalum)	(Tantalum has tantalum rather than aluminum foil plates) Aluminum Foil Aluminum Oxide Aluminum Foil Gauze (Saturated with Electrolyte)	1 μF–1F ± 10% to ±50%	Most popular large value capacitor: large capacitance into small area, wide range of values. Disadvantages are: cannot be used in AC circuits as they are polarized; poor tolerances; low leakage resistance and so high leakage current. *Tantalum* advantages over aluminum include smaller size, longer life than aluminum, which has an approximate lifespan of 12 years. Disadvantages: 4 to 5 times the price.	

FIGURE 7-14 Summary of Fixed-Value Capacitors.

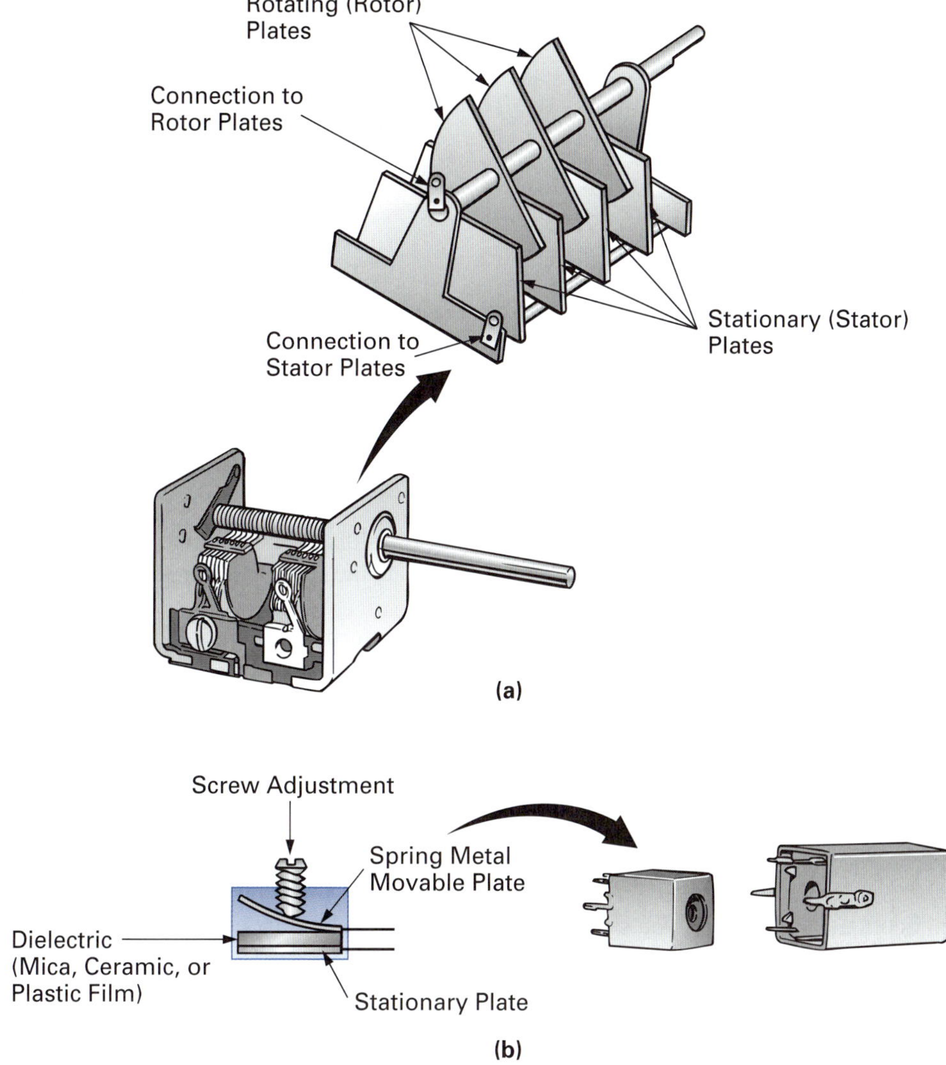

FIGURE 7-15 Variable Value Capacitors. (a) Larger Air-Variable Capacitor. (b) Smaller Adjustable Capacitor.

metal foil plates separated by one of the following types of insulators (dielectric), which is the means by which we classify them, as shown in Figure 7-14.

Variable-value capacitors are the second basic type, and like the fixed-value type of capacitors, they are classified by dielectric.

Variable-Value Capacitor
A capacitor whose value can be varied.

Figure 7-15(a) illustrates the construction of a typical air dielectric variable capacitor. With this type of capacitor, the effective plate area is adjusted to vary capacitance by causing a set of rotating plates (rotor) to mesh with a set of stationary plates (stator). When the rotor plates are fully out, the capacitance is minimum, and when the plates are fully in, the capacitance is maximum, because the maximum amount of rotor plate area is now opposite the stator plate, creating the maximum value of capacitance.

Figure 7-15(b) illustrates some of the typical packages for mica, ceramic, or plastic film types of adjustable capacitors, which are also referred to as *trimmers*. The adjustable capacitor generally has one stationary plate and one spring metal moving plate. The screw forces the spring metal plate closer or farther away from the stationary plate, varying the distance between the plates and so changing capacitance. The two plates are insulated from one another by either mica, ceramic, or plastic film, and the advantages of each are the same as for fixed-value capacitors.

FIGURE 7-16 **Alphanumeric Coding of Capacitors. 0.47 μF, +80%, −20% Tolerance, 10 V.**

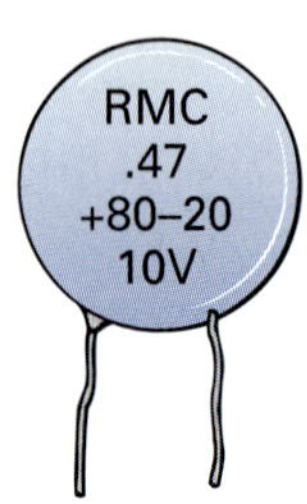

These types of capacitors should only be adjusted with a plastic or nonmetallic alignment tool, because a metal screwdriver may affect the capacitance of the capacitor when nearby, making it very difficult to adjust for a specific value of capacitance.

7-3-2 *Capacitor Coding*

Manufacturers today most commonly use letters of the alphabet and numbers (alphanumerics) printed on either the disk or tubular body to indicate the capacitor's specifications, as illustrated in Figure 7-16.

The tubular type is the easier of the two since the information is basically uncoded. The value of capacitance and unit, typically the microfarad (μF or MF), tolerance figure (preceded by ± or followed by %), and voltage rating (followed by a V for voltage) are printed on all sizes of tubular cases. The remaining letters or numbers are merely manufacturers' codes for case size, series, and the like.

With disk capacitors (dipped or molded), certain rules have to be applied when decoding the notations. Many capacitors of this type do not define the unit of capacitance; in this situation, try to locate a decimal point. If a decimal point exists, for example 0.01 or 0.001, the value is in microfarads (10^{-6}). If no decimal point exists, for example 50 or 220, the value is in picofarads (10^{-12}) and you must analyze the number in a little more detail.

EXAMPLE:

What is the value of a capacitor if it is labeled 50, 50 V, ±5?

Solution:

Since no decimal point is present, the unit is in picofarads:

50 pF, 50 V, ±5%

If no decimal point is present and three digits exist and the last digit is a zero, the value is as stands and in picofarads. If the third digit is a number other than 0 (1 to 9), it is a multiplier and describes the number of zeros to be added to the picofarad value.

EXAMPLE:

If two capacitors are labeled with the following coded values, how should they be interpreted?

a. 220
b. 104

Solution:

Since no decimal point is present, they are both in picofarads.

a. If the last of the three digits is a zero, the value is as it stands.

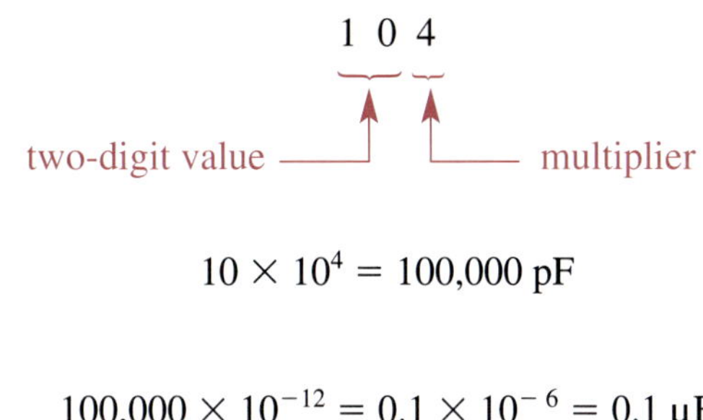

b. If the third digit is a number from 1 to 9, it is a multiplier.

1 0 4

two-digit value — multiplier

So

$$10 \times 10^4 = 100{,}000 \text{ pF}$$

or

$$100{,}000 \times 10^{-12} = 0.1 \times 10^{-6} = 0.1\ \mu\text{F}$$

The tolerance of the capacitor is sometimes clearly indicated, for example, ±5 or 10%; in other cases, a letter designation is used, such as

F = ±1%
G = ±2%
J = ±5%
K = ±10%
M = ±20%
Z = −20%, +80%

Unfortunately, there does not seem to be a standard among capacitor manufacturers, which can cause confusion when trying to determine the value of capacitance. Therefore, if you are not completely sure, you should always measure the value or consult technical data or information sheets from the manufacturer.

SELF-TEST EVALUATION POINT FOR SECTION 7-3

Use the following questions to test your understanding of Section 7-3.

What are the following values of capacitance?

1. 470 ± 2

2. 0.47 ± 5

3. Which fixed-value capacitor is the most popular in applications where:
  a. Large values are required?
  b. Small values are required?

7-4 CAPACITIVE REACTANCE

Resistance (R), by definition, is the opposition to current flow with the dissipation of energy and is measured in ohms. Capacitors oppose current flow like a resistor, but a resistor dissipates energy, whereas a capacitor stores energy (when it charges) and then gives back its energy into the circuit (when it discharges). Because of this difference, a new term had to be used to describe the opposition offered by a capacitor. **Capacitive reactance (X_C),** by definition, is the opposition to current without the dissipation of energy and is also measured in ohms.

Capacitive Reactance (X_C)
Measured in ohms, it is the ability of a capacitor to oppose current flow without the dissipation of energy.

If capacitive reactance is basically opposition, it is inversely proportional to the amount of current flow. If a large current is within a circuit, the opposition or reactance must be low ($I\uparrow$, $X_C\downarrow$). Conversely, a small circuit current will be the result of a large opposition or reactance ($I\downarrow$, $X_C\uparrow$).

When a dc source is connected across a capacitor, current will flow only for a short period of time ($5RC$ seconds) to charge the capacitor. After this time, there is no further current flow. Consequently, the capacitive reactance or opposition offered by a capacitor to dc is infinite (maximum).

Alternating current is continuously reversing in polarity, resulting in the capacitor continuously charging and discharging. This means that charge and discharge currents are always flowing around the circuit, and if we have a certain value of current, we must also have a certain value of reactance or opposition.

Initially, when the capacitor's plates are uncharged, they will not oppose or react against the charging current and therefore maximum current will flow ($I\uparrow$) and the reactance will be very low ($X_C\downarrow$). As the capacitor charges, it will oppose or react against the charge current, which will decrease ($I\downarrow$), so the reactance will increase ($X_C\uparrow$). The discharge current is also highest at the start of discharge ($I\uparrow$, $X_C\downarrow$) as the voltage of the charged capacitor is also high; but as the capacitor discharges, its voltage decreases and the discharge current will also decrease ($I\downarrow$, $X_C\uparrow$).

To summarize, at the start of a capacitor charge or discharge, the current is maximum, so the reactance is low. This value of current then begins to fall to zero, so the reactance increases.

If the applied alternating current is at a high frequency, as shown in Figure 7-17(a), it is switching polarity more rapidly than a lower frequency and there is very little time between the start of charge and discharge. As the charge and discharge currents are largest at the beginning of the charge and discharge of the capacitor, the reactance has very little time to build up and oppose the current, which is why the current is a high value and the capacitive reactance is small at higher frequencies. With lower frequencies, as shown in Figure 10-1(b), the applied alternating current is switching at a slower rate, and therefore the reactance, which is low at the beginning, has more time to build up and oppose the current.

Capacitive reactance is therefore inversely proportional to frequency:

$$\text{capacitive reactance } (X_C) \propto \frac{1}{f\,(\text{frequency})}$$

Frequency, however, is not the only factor that determines capacitive reactance. Capacitive reactance is also inversely proportional to the value of capacitance. If a larger capacitor value is used a longer time is required to charge the capacitor ($t\uparrow = C\uparrow R$), which means that current will be present for a longer period of time, so the overall current will be large ($I\uparrow$); consequently, the reactance must be small ($X_C\downarrow$). On the other hand, a small capacitance value will charge in a small amount of time ($t\downarrow = C\downarrow R$) and the current is present for only a short period of time. The overall current will therefore be small ($I\downarrow$), indicating a large reactance ($X_C\uparrow$).

$$\text{capacitive reactance } (X_C) \propto \frac{1}{C\,(\text{capacitance})}$$

Capacitive reactance (X_C) is therefore inversely proportional to both frequency and capacitance and can be calculated by using the formula

$$X_C = \frac{1}{2\pi fC}$$

where X_C = capacitive reactance, in ohms
2π = constant
f = frequency, in hertz
C = capacitance, in farads

FIGURE 7-17 Capacitive Reactance Is Inversely Proportional to Frequency.

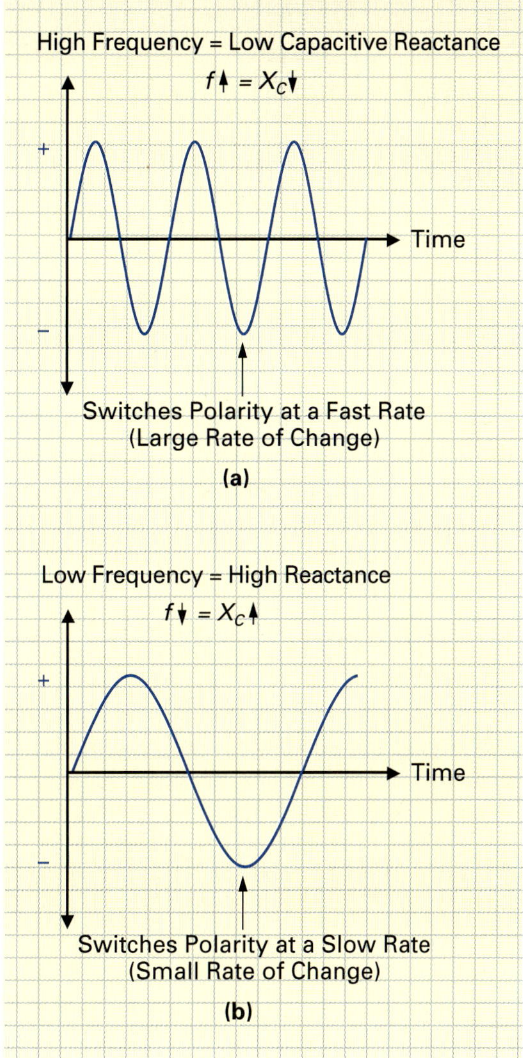

EXAMPLE:

Calculate the reactance of a 2 μF capacitor when a 10 kHz sine wave is applied.

Solution:

$$X_C = \frac{1}{2\pi fC}$$

$$= \frac{1}{2 \times \pi \times 10\ \text{kHz} \times 2\ \mu\text{F}} = 8\ \Omega$$

CALCULATOR SEQUENCE

Step	Keypad Entry	Display Response
1.	2	2.0
2.	×	
3.	π	3.1415927
4.	×	6.283185
5.	1 0 EE 3	10E3
6.	×	62831.85
7.	2 EE 6 +/−	2.-06
8.	=	0.1256637
9.	1/x	7.9577

SELF-TEST EVALUATION POINT FOR SECTION 7-4

Use the following questions to test your understanding of Section 7-4.

1. Define *capacitive reactance.*
2. State the formula for capacitive reactance.
3. Why is capacitive reactance inversely proportional to frequency and capacitance?
4. If $C = 4\ \mu F$ and $f = 4$ kHz, calculate X_C.

7-5 SERIES *RC* CIRCUIT

In a purely resistive circuit, as shown in Figure 7-18(a), the current flowing within the circuit and the voltage across the resistor are in phase with one another. In a purely capacitive circuit, as shown in Figure 7-18(b), the current flowing in the circuit leads the voltage across the capacitor by 90°.

Purely resistive: 0° phase shift (*I* is in phase with *V*)
Purely capacitive: 90° phase shift (*I* leads *V* by 90°)

If we connect a resistor and capacitor in series, as shown in Figure 7-19(a), we have probably the most commonly used electronic circuit, which has many applications. The voltage across the resistor (V_R) is always in phase with the circuit current (I), as can be seen in Figure 7-19(b), because maximum points and zero crossover points occur at the same time. The voltage across the capacitor (V_C) lags the circuit current by 90°.

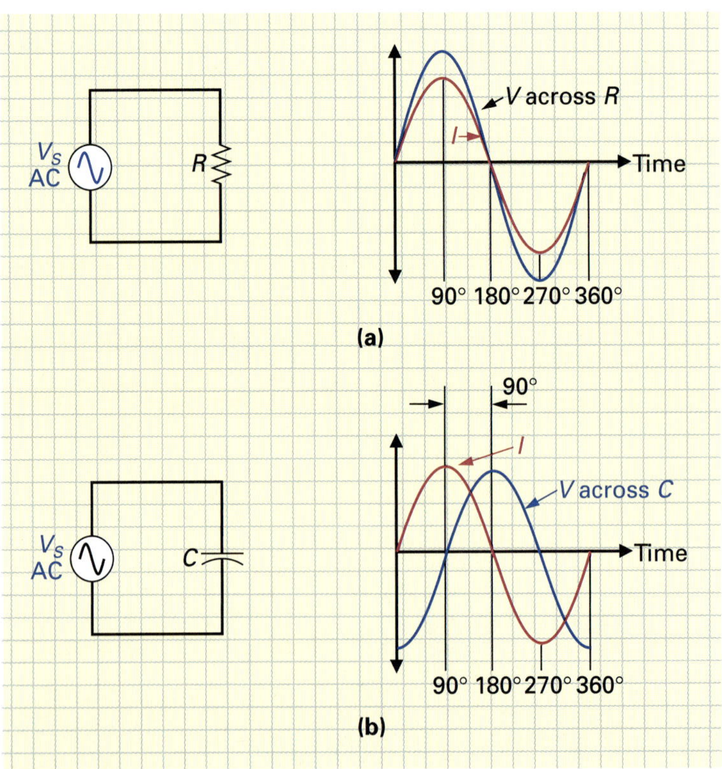

FIGURE 7-18 Phase Relationships between *V* and *I*. (a) Resistive Circuit: Current and Voltage Are in Phase. (b) Capacitive Circuit: Current Leads Voltage by 90°.

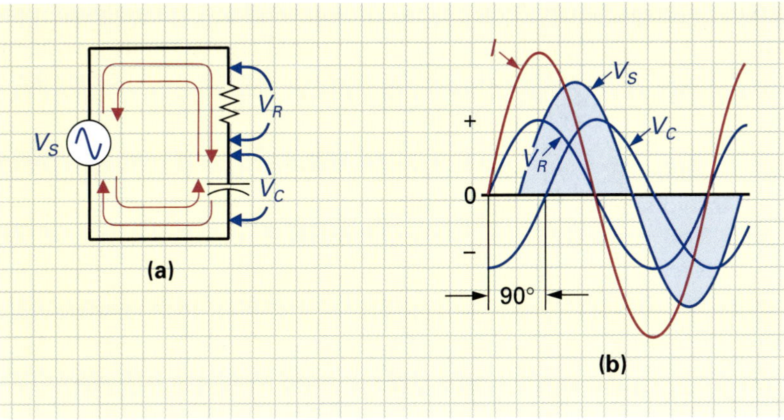

FIGURE 7-19 ***RC*** **Series Circuit. (a) Circuit. (b) Waveforms.**

Since the capacitor and resistor are in series, the same current is supplied to both components; Kirchhoff's voltage law can be applied, which states that the sum of the voltage drops around a series circuit is equal to the voltage applied (V_S). The voltage drop across the resistor (V_R) and the voltage drop across the capacitor (V_C) are out of phase with one another, which means that their peaks occur at different times. The signal for the applied voltage (V_S) is therefore obtained by adding the values of V_C and V_R at each instant in time, plotting the results, and then connecting the points with a line; this is represented in Figure 7-19(b) by the shaded waveform.

7-5-1 *Voltage*

Figure 7-20(a), (b), and (c) repeat our previous *RC* series circuit with waveforms and a vector diagram. In Figure 7-20(b), the current peak flowing in the series *RC* circuit occurs at 0° and will be used as a reference. The voltage across the resistor (V_R) is in phase or coincident with the current (I).

The voltage across the capacitor (V_C) is, as shown in Figure 7-20(b), 90° out of phase (lagging) with the circuit's current. Since the ohmic values of the resistor (R) and the capacitor (X_C) are equal, the voltage drop across both components is the same. The V_R and V_C waveforms are subsequently equal in size.

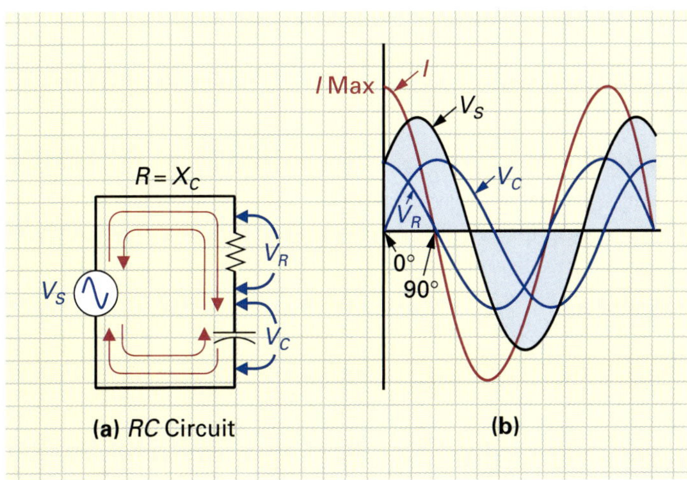

FIGURE 7-20 ***RC*** **Series Circuit Analysis.**

The source voltage (V_S) is, by Kirchhoff's voltage law, equal to the sum of the series voltage drops (V_C and V_R). However, since these voltages are not in phase with one another, we cannot simply add the two together. The source voltage (V_S) will be the sum of both V_C and V_R at a particular time. Studying the waveforms in Figure 7-20(b), you will notice that peak source voltage will occur at 45°. The angle theta (θ) formed between circuit current (I) and source voltage (V_S) will always be less than 90°, and in this example is equal to $-45°$ because the voltage drops across R and C are equal due to R and X_C being of the same ohmic value.

The Pythagorean theorem for right-angle triangles states that if you take the square of a (V_R) and add it to the square of b (V_C), the square root of the result will equal c (the source voltage, V_S).

$$V_S = \sqrt{V_R^2 + V_C^2}$$

By transposing the formula according to the rules of algebra we can calculate any unknown if two variables are known.

$$V_C = \sqrt{V_S^2 - V_R^2}$$
$$V_R = \sqrt{V_S^2 - V_C^2}$$

EXAMPLE:

Calculate the source voltage applied across an *RC* series circuit if $V_R = 12$ V and $V_C = 8$ V.

Solution:

$$V_S = \sqrt{V_R^2 + V_C^2}$$
$$= \sqrt{(12\text{ V})^2 + (8\text{ V})^2}$$
$$= \sqrt{144 + 64}$$
$$= 14.42\text{ V}$$

CALCULATOR SEQUENCE

Step	Keypad Entry	Display Response
1.	[1] [2]	12.0
2.	[x^2]	144.0
3.	[+]	
4.	[8]	8.0
5.	[x^2]	64.0
6.	[=]	208.0
7.	[$\sqrt{x}$]	14.42220

7-5-2 *Impedance*

Impedance (Z)
Measured in ohms, it is the total opposition a circuit offers to current flow (reactive and resistive).

Since resistance is the opposition to current with the dissipation of heat, and reactance is the opposition to current without the dissipation of heat, a new term is needed to describe the total resistive and reactive opposition to current. **Impedance** (designated Z) is also measured in ohms and is the total circuit opposition to current flow. It is a combination of resistance (R) and reactance (X_C); however, in our capacitive and resistive circuit, a phase shift or difference exists, and just as V_C and V_R cannot be merely added together to obtain V_S, R and X_C cannot be simply summed to obtain Z.

If the current within a series circuit is constant (the same throughout the circuit), the resistance of a resistor (R) or reactance of a capacitor (X_C) will be directly proportional to the voltage across the resistor (V_R) or the capacitor (V_C).

$$V_R\updownarrow = I \times R\updownarrow, \quad V_C\updownarrow = I \times X_C\updownarrow$$

The Pythagorean theorem can be used to calculate the total opposition or impedance (Z) to current flow, taking into account both R and X_C.

$$Z = \sqrt{R^2 + X_C^2}$$

$$R = \sqrt{Z^2 - X_C^2}$$
$$X_C = \sqrt{Z^2 - R^2}$$

EXAMPLE:

Calculate the total impedance of a series RC circuit if $R = 27\ \Omega$, $C = 0.005\ \mu\text{F}$, and the source frequency = 1 kHz.

Solution:

The total opposition (Z) or impedance is equal to

$$Z = \sqrt{R^2 + X_C^2}$$

R is known, but X_C will need to be calculated.

$$\begin{aligned} X_C &= \frac{1}{2\pi fC} \\ &= \frac{1}{2 \times \pi \times 1\ \text{kHz} \times 0.005\ \mu\text{F}} \\ &= 31.8\ \text{k}\Omega \end{aligned}$$

Since $R = 27\ \Omega$ and $X_C = 31.8\ \text{k}\Omega$, then

$$\begin{aligned} Z &= \sqrt{R^2 + X_C^2} \\ &= \sqrt{(27\ \Omega)^2 + (31.8\ \text{k}\Omega)^2} \\ &= \sqrt{729 + 1 \times 10^9} \\ &= 31.8\ \text{k}\Omega \end{aligned}$$

As you can see in this example, the small resistance of 27 Ω has very little effect on the circuit's total opposition or impedance, due to the relatively large capacitive reactance of 31,800 Ω.

EXAMPLE:

Calculate the total impedance of a series RC circuit if $R = 45\ \text{k}\Omega$ and $X_C = 45\ \Omega$.

Solution:

$$\begin{aligned} Z &= \sqrt{R^2 + X_C^2} \\ &= \sqrt{(45\ \text{k}\Omega)^2 + (45\ \Omega)^2} \\ &= 45\ \text{k}\Omega \end{aligned}$$

In this example, the relatively small value of X_C had very little effect on the circuit's opposition or impedance, due to the large circuit resistance.

EXAMPLE:

Calculate the total impedance of a series *RC* circuit if $X_C = 100\ \Omega$ and $R = 100\ \Omega$.

Solution:

$$
\begin{aligned}
Z &= \sqrt{R^2 + X_C{}^2} \\
&= \sqrt{100^2 + 100^2} \\
&= 141.4\ \Omega
\end{aligned}
$$

In this example R was equal to X_C.

We can define the total opposition or impedance in terms of Ohm's law, in the same way as we defined resistance.

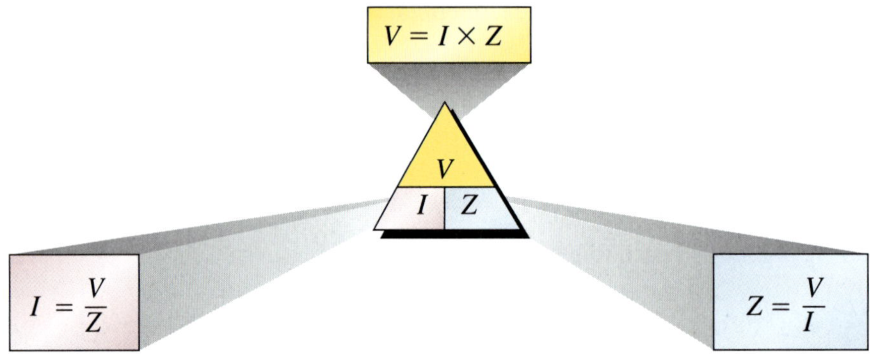

By transposition, we can arrive at the usual combinations of Ohm's law.

7-5-3 *Phase Angle or Shift* (θ)

In a purely resistive circuit, the total opposition (Z) is equal to the resistance of the resistor, so the phase shift (θ) is equal to 0°.

In a purely capacitive circuit, the total opposition (Z) is equal to the capacitive reactance (X_C) of the capacitor, so the phase shift (θ) is equal to −90°. When a circuit contains both resistance and capacitive reactance, the total opposition or impedance has a phase shift that is between 0 and 90°.

The phase angle, θ, is equal to

$$\theta = \text{invtan}\,\frac{X_C}{R}$$

This formula will determine by what angle Z leads R. Since X_C/R is equal to V_C/V_R, the phase angle can also be calculated if V_R and V_C are known.

$$\theta = \text{invtan}\,\frac{V_C}{V_R}$$

This formula will determine by what angle V_R leads V_S.

EXAMPLE:

Calculate the phase shift or angle in two different series *RC* circuits if:

a. $V_R = 12\text{ V}, V_C = 8\text{ V}$
b. $R = 27\ \Omega, X_C = 31.8\text{ k}\Omega$

Solution:

a. $\theta = \text{invtan}\,\dfrac{V_C}{V_R}$

$= \text{invtan}\,\dfrac{8\text{ V}}{12\text{ V}}$

$= 33.7°$ (V_R leads V_S by 33.7°)

b. $\theta = \text{invtan}\,\dfrac{X_C}{R}$

$= \text{invtan}\,\dfrac{31.8\text{ k}\Omega}{27\ \Omega}$

$= 89.95°$ (R leads Z by 89.95°)

7-5-4 *Power*

In this section we examine power in a series ac circuit. Let us begin with a simple resistive circuit and review the power formulas used previously.

Purely Resistive Circuit

In Figure 7-21 you can see the current, voltage, and power waveforms generated by applying an ac voltage across a purely resistive circuit. The applied voltage causes current to

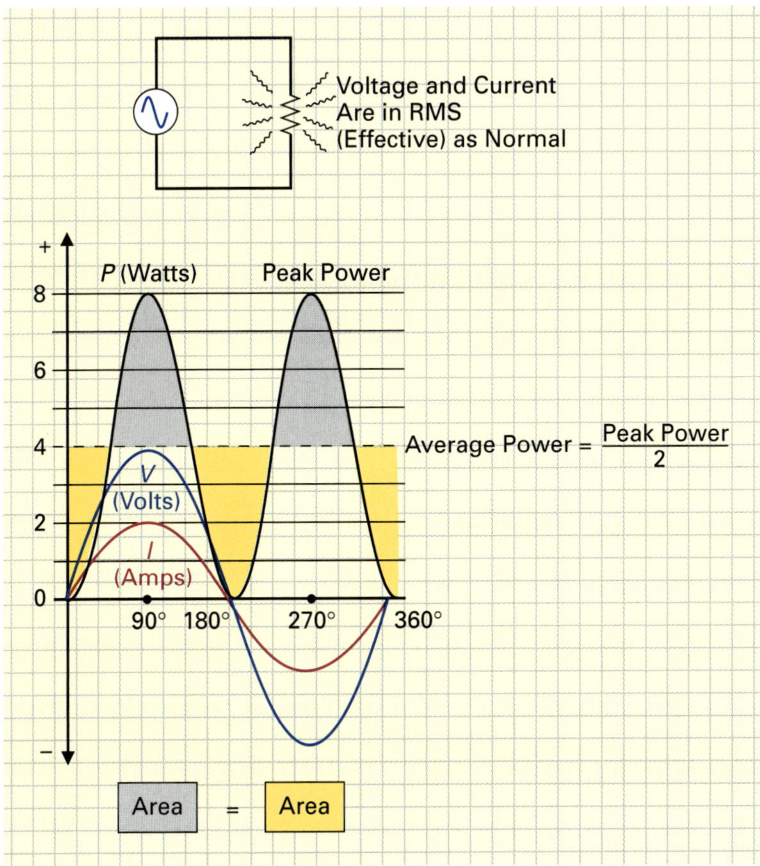

FIGURE 7-21 Power in a Purely Resistive Circuit.

flow around the circuit, and the electrical energy is converted into heat energy. This heat or power is dissipated and lost and can be calculated by using the power formula

$$P = V \times I$$
$$P = I^2 \times R$$
$$P = \frac{V^2}{R}$$

Voltage and current are in phase with one another in a resistive circuit, and instantaneous power is calculated by multiplying voltage by current at every instant through 360° ($P = V \times I$). The sinusoidal power waveform is totally positive, because a positive voltage multiplied by a positive current gives a positive value of power, and a negative voltage multiplied by a negative current will also produce a positive value of power. For these reasons, a resistor is said to generate a positive power waveform, which you may have noticed is twice the frequency of the voltage and current waveforms; two power cycles occur in the same time as one voltage and current cycle.

The power waveform has been split in half, and this line that exists between the maximum point (8 W) and zero point (0 W) is the average value of power (4 W) that is being dissipated by the resistor.

Purely Capacitive Circuit

In Figure 7-22 you can see the current, voltage, and power waveforms generated by applying an ac voltage source across a purely capacitive circuit. As expected, the current leads

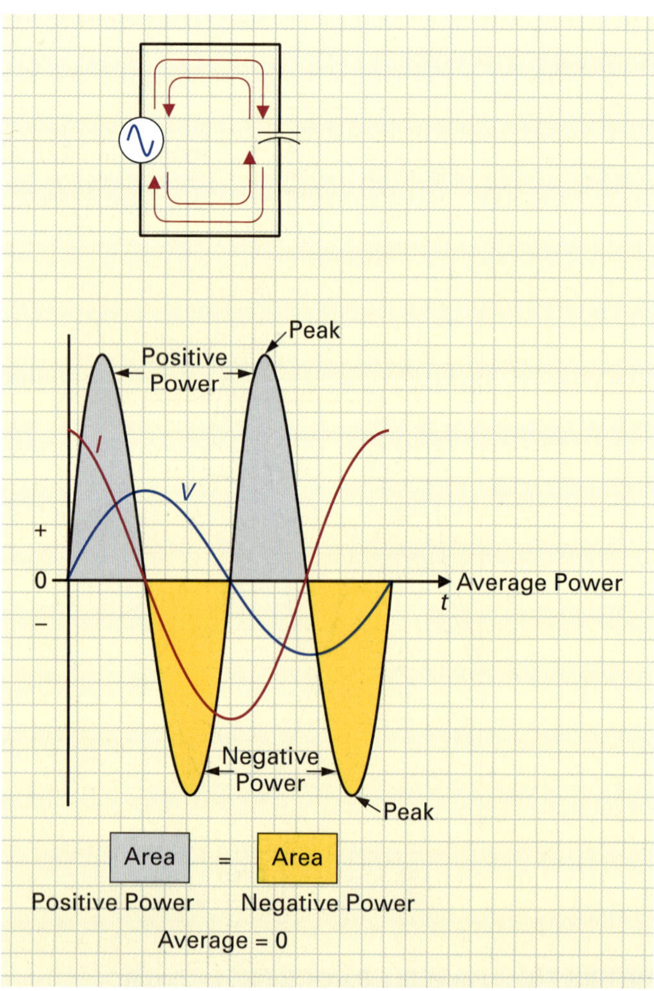

FIGURE 7-22 Power in a Purely Capacitive Circuit.

the voltage by 90°, and the power wave is calculated by multiplying voltage by current, as before, at every instant through 360°. The resulting power curve is both positive and negative. During the positive alternation of the power curve, the capacitor is taking power as the capacitor charges. When the power alternation is negative, the capacitor is giving back the power it took as it discharges back into the circuit.

The average power dissipated is once again the value that exists between the maximum positive and maximum negative points and causes the area above this line to equal the area below. This average power level calculates out to be zero, which means that no power is dissipated in a purely capacitive circuit.

Resistive and Capacitive Circuit

In Figure 7-23 you can see the current, voltage, and power waveforms generated by applying an ac voltage source across a series-connected *RC* circuit. The current leads the voltage by some phase angle less than 90°, and the power waveform is once again determined by the product of voltage and current. The negative alternation of the power cycle indicates that the capacitor is discharging and giving back the power that it consumed during the charge.

The positive alternation of the power cycle is much larger than the negative alternation because it is the combination of both the capacitor taking power during charge and the resistor consuming and dissipating power in the form of heat. The average power being dissipated will be some positive value, due to the heat being generated by the resistor.

Power Factor

Resistive Power or True Power
The average power consumed by a circuit during one complete cycle of alternating current.

In a purely resistive circuit, all the energy supplied to the resistor from the source is dissipated in the form of heat. This form of power is referred to as **resistive power** (P_R) or **true power,** and is calculated with the formula

$$P_R = I^2 \times R$$

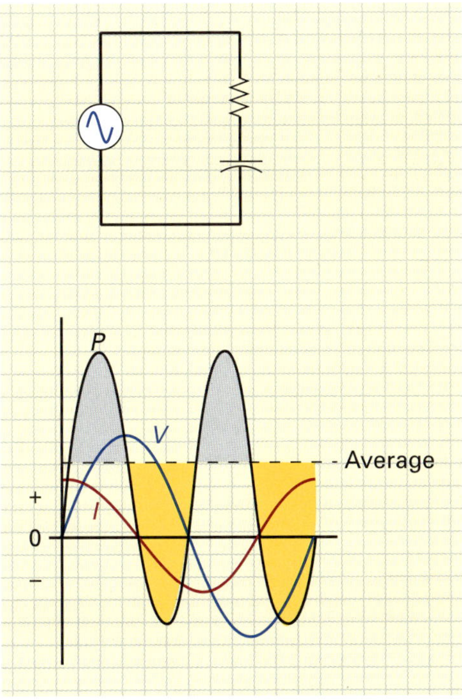

FIGURE 7-23 **Power in a Resistive and Capacitive Circuit.**

Reactive Power or Imaginary Power

Also called wattless power, it is the power value obtained by multiplying the effective value of current by the effective value of voltage and the sine of the angular phase difference between current and voltage.

In a purely capacitive circuit, all the energy supplied to the capacitor is stored from the source and then returned to the source, without energy loss. This form of power is referred to as **reactive power** (P_X) or **imaginary power.**

$$P_X = I^2 \times X_C$$

When a circuit contains both capacitance and resistance, some of the supply is stored and returned by the capacitor and some of the energy is dissipated and lost by the resistor.

Applying the Pythagorean theorem once again, we can calculate apparent power by

$$P_A = \sqrt{{P_R}^2 + {P_X}^2}$$

where P_A = apparent power, in volt-amperes (VA)
P_R = true power, in watts (W)
P_X = reactive power, in volt-amperes reactive (VAR)

Power Factor

Ratio of actual power to apparent power. A pure resistor has a power factor of 1 or 100% while a capacitor has a power factor of 0 or 0%.

The **power factor** is a ratio of true power to apparent power and is a measure of the loss in a circuit. It can be calculated by using the formula

$$\text{PF} = \frac{\text{true power } (P_R)}{\text{apparent power } (P_A)}$$

A power factor of 1 indicates a maximum power loss (circuit is resistive), while a power factor of 0 indicates no power loss (circuit is capacitive). With circuits that contain both resistance and reactance, the power factor will be somewhere between zero (0 = reactive) and one (1 = resistive).

Since true power is determined by resistance and apparent power is dependent on impedance, the power factor can also be calculated by using the formula

$$\text{PF} = \frac{R}{Z}$$

The power factor can also be determined by the cosine of angle θ.

$$\text{PF} = \cos\theta$$

EXAMPLE

Calculate the following for a series *RC* circuit if $R = 2.2\ \text{k}\Omega$, $X_C = 3.3\ \text{k}\Omega$, and $V_S = 5$ V.

a. Z
b. I
c. θ
d. P_R
e. P_X
f. P_A
g. PF

Solution:

a. $Z = \sqrt{R^2 + {X_C}^2}$
$= \sqrt{(2.2\ \text{k}\Omega)^2 + (3.3\ \text{k}\Omega)^2}$
$= 3.96\ \text{k}\Omega$

b. $I = \dfrac{V_S}{Z} = \dfrac{5\ \text{V}}{3.96\ \text{k}\Omega} = 1.26\ \text{mA}$

c. $\theta = \text{invtan}\,\dfrac{X_C}{R} = \text{invtan}\,\dfrac{3.3\ \text{k}\Omega}{2.2\ \text{k}\Omega}$
$= \text{invtan}\ 1.5 = 56.3°$

d. True power $= I^2 \times R$
$= (1.26 \text{ mA})^2 \times 2.2 \text{ k}\Omega$
$= 3.49 \text{ mW}$

e. Reactive power $= I^2 \times X_C$
$= (1.26 \text{ mA})^2 \times 3.3 \text{ k}\Omega$
$= 5.24 \times 10^{-3}$ or 5.24 mVAR

f. Apparent power $= \sqrt{P_R^2 + P_X^2}$
$= \sqrt{(3.49 \text{ mW})^2 + (5.24 \text{ mW})^2}$
$= 6.29 \times 10^{-3}$ or 6.29 mVA

g. Power factor $= \frac{R}{Z} = \frac{2.2 \text{ k}\Omega}{3.96 \text{ k}\Omega} = 0.55$

or

$= \frac{P_R}{P_A} = \frac{3.49 \text{ mW}}{6.29 \text{ mW}} = 0.55$

or

$= \cos \theta = \cos 56.3° = 0.55$

CALCULATOR SEQUENCE

Step	Keypad Entry	Display Response
1.	[3] [.] [3] [E] [3]	3.3E3
2.	[÷]	
3.	[2] [.] [2] [E] [3]	2.2E3
4.	[=]	1.5
5.	[inv] [tan]	56.309932

SELF-TEST EVALUATION POINT FOR SECTION 7-5

Use the following questions to test your understanding of Section 7-5.

1. What is the phase relationship between current and voltage in a series *RC* circuit?
2. Define and state the formula for impedance.

7-6 PARALLEL *RC* CIRCUIT

Now that we have analyzed the characteristics of a series *RC* circuit, let us connect a resistor and capacitor in parallel.

7-6-1 *Voltage*

As with any parallel circuit, the voltage across all components in parallel is equal to the source voltage; therefore,

$$V_R = V_C = V_S$$

7-6-2 *Current*

In Figure 7-24(a), you will see a parallel circuit containing a resistor and a capacitor. The current through the resistor and capacitor is simply calculated by applying Ohm's law:

$$\text{(resistor current) } I_R = \frac{V_S}{R}$$

$$\text{(capacitor current) } I_C = \frac{V_S}{X_C}$$

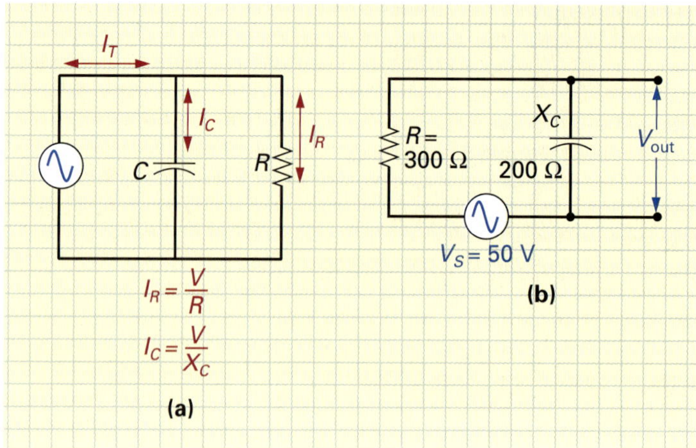

FIGURE 7-24 **Parallel *RC* Circuit.**

Total current (I_T), however, is not as simply calculated. As expected, resistor current (I_R) is in phase with the applied voltage (V_S), as shown in the waveforms in Figure 7-24(b). Capacitor current will always lead the applied voltage by 90°.

Total current is therefore the vector sum of both the resistor and capacitor currents. Using the Pythagorean theorem, total current can be calculated by

$$I_T = \sqrt{I_R^2 + I_C^2}$$

7-6-3 *Phase Angle*

The angle by which the total current (I_T) leads the source voltage (V_S) can be determined with either of the following formulas:

$$\theta = \text{invtan}\,\frac{I_C\,(\text{opposite})}{I_R\,(\text{adjacent})} \qquad \theta = \text{invtan}\,\frac{R}{X_C}$$

7-6-4 *Impedance*

Since the circuit is both capacitive and resistive, the total opposition or impedance of the parallel *RC* circuit can be calculated by

$$Z = \frac{V_S}{I_T}$$

The impedance of a parallel *RC* circuit is equal to the total voltage divided by the total current. Using basic algebra, this basic formula can be rearranged to express impedance in terms of reactance and resistance.

$$Z = \frac{R \times X_C}{\sqrt{R^2 + X_C^2}}$$

7-6-5 *Power*

With respect to power, there is no difference between a series circuit and a parallel circuit. The true power or resistive power (P_R) dissipated by an *RC* circuit is calculated with the formula

$$P_R = I_R^2 \times R$$

The imaginary power or reactive power (P_X) of the circuit can be calculated with the formula

$$P_X = I_C^2 \times X_C$$

The apparent power is equal to the vector sum of the true power and the reactive power.

$$P_A = \sqrt{P_R^2 + P_X^2}$$

As with series *RC* circuits the power factor is calculated as

$$\text{PF} = \frac{P_R \text{ (resistive power)}}{P_A \text{ (apparent power)}}$$

A power factor of 1 indicates a purely resistive circuit, while a power factor of 0 indicates a purely reactive circuit.

EXAMPLE:

Calculate the following for a parallel *RC* circuit in which $R = 24\ \Omega$, $X_C = 14\ \Omega$, and $V_S = 10$ V.

a. I_R
b. I_C
c. I_T
d. Z
e. θ

Solution:

a. $I_R = \dfrac{V_S}{R} = \dfrac{10V}{24\ \Omega} = 416.66 \text{ mA}$

b. $I_C = \dfrac{V_S}{X_C} = \dfrac{10 \text{ V}}{14\ \Omega} = 714.28 \text{ mA}$

c. $I_T = \sqrt{I_R^2 + I_C^2}$
$= \sqrt{(416.66 \text{ mA})^2 + (714.28 \text{ mA})^2}$
$= \sqrt{0.173 + 0.510}$
$= 826.5 \text{ mA}$

d. $Z = \dfrac{V_S}{I_T} = \dfrac{10 \text{ V}}{826.5 \text{ mA}} = 12\ \Omega$

or

$= \dfrac{R \times X_C}{\sqrt{R^2 + X_C^2}} = \dfrac{24 \times 14}{\sqrt{24^2 + 14^2}}$
$= 12\ \Omega$

e. $\theta = \arctan \dfrac{I_C}{I_R} = \arctan \dfrac{714.28 \text{ mA}}{416.66 \text{ mA}}$
$= \arctan 1.714 = 59.7°$

or

$\theta = \arctan \dfrac{R}{X_C} = \arctan \dfrac{24\ \Omega}{14\ \Omega}$
$= \arctan 1.714 = 59.7°$

CALCULATOR SEQUENCE

Step	Keypad Entry	Display Response
1.	7 1 4 . 2 8 E 3 +/−	714.28E-3
2.	÷	
3.	4 1 6 . 6 6 E 3 +/−	416.66E-3
4.	=	1.7142994
5.	inv tan	59.743762

SELF-TEST EVALUATION POINT FOR SECTION 7-6

Use the following questions to test your understanding of Section 7-6.

1. What is the phase relationship between current and voltage in a parallel *RC* circuit?
2. Could a parallel *RC* circuit be called a voltage divider?

7-7 TESTING CAPACITORS

Now that you have a good understanding as to how a capacitor should function, let us investigate how to diagnose a capacitor malfunction.

7-7-1 *The Ohmmeter*

A faulty capacitor may have one of three basic problems:

1. A short, which is easy to detect and is caused by a contact from plate to plate.
2. An open, which is again quite easy to detect and is normally caused by one of the leads becoming disconnected from its respective plate.
3. A leaky dielectric or capacitor breakdown, which is quite difficult to detect, as it may only short at a certain voltage. This problem is usually caused by the deterioration of the dielectric, which starts displaying a much lower dielectric resistance than it was designed for. The capacitor with this type of problem is referred to as a *leaky capacitor.*

Capacitors of 0.5 μF and larger can be checked by using an analog ohmmeter or a digital multimeter with a bar graph display by using the procedure shown in Figure 7-25.

Step 1: Ensure that the capacitor is discharged by shorting the leads together.

Step 2: Set the ohmmeter to the highest ohms range scale.

Step 3: Connect the meter to the capacitor, observing the correct polarity if an electrolytic is being tested, and observe the bar graph display. The capacitor will initially be discharged, and therefore maximum current will flow from the meter battery to the capacitor. Maximum current means low resistance, which is why the meter's display indicates 0 Ω.

Step 4: As the capacitor charges, it will cause current flow from the meter's battery to decrease, and consequently, the bar graph display will increase.

A good capacitor will cause the meter to react as just explained. A larger capacitance will cause the meter to increase slowly to infinity (∞), as it will take a longer time to charge, while a smaller value of capacitance will charge at a much faster rate, causing the meter to increase rapidly toward ∞. For this reason, the ohmmeter cannot be reliably used to check capacitors with values of less than 0.5 μF, because the capacitor charges up too quickly and the meter does not have enough time to respond.

A shorted capacitor will cause the meter to show zero ohms and remain in that position. An open capacitor will cause a maximum reading (infinite resistance) because there is no path for current to flow.

A leaky capacitor will cause the bar graph to reduce its bars to the left, and then return almost all the way back to ∞ if only a small current is still flowing and the capacitor has a small dielectric leak. If the meter bars only come back to halfway or a large distance away from infinity, a large amount of current is still flowing and the capacitor has a large dielectric leak (defect).

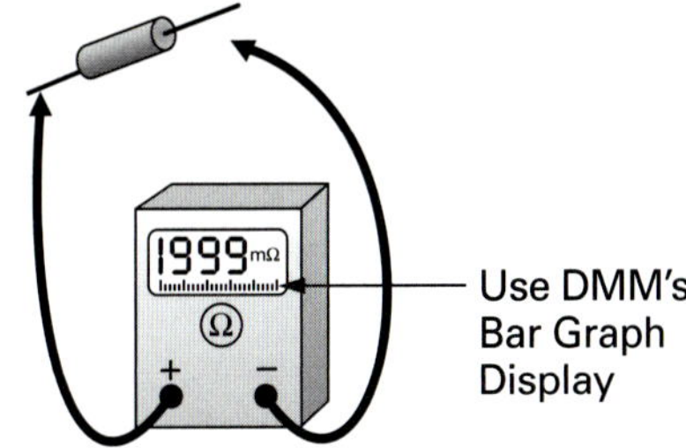

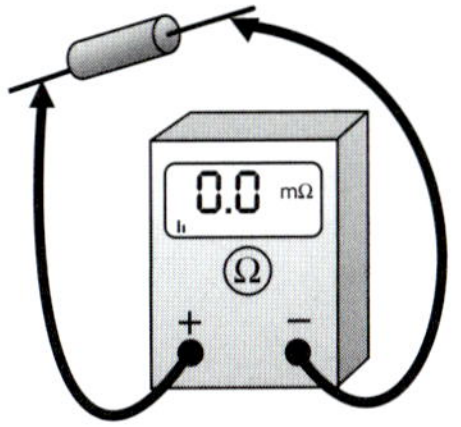

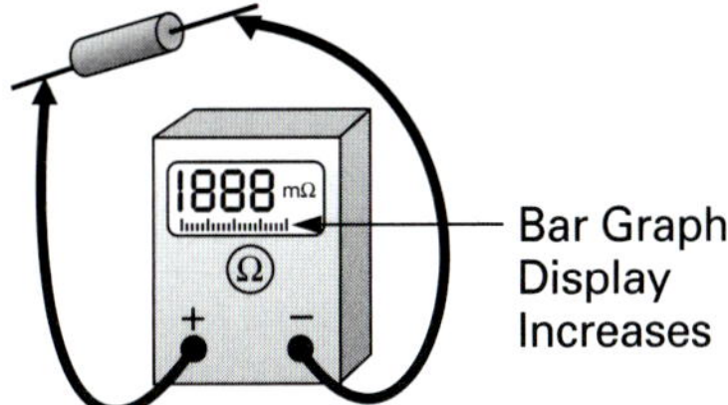

FIGURE 7-25 **Testing a Capacitor of More Than 0.5 μF.**

When using the ohmmeter to test capacitors, there are some other points that you should be aware of:

1. Electrolytics are noted for having a small yet noticeable amount of inherent leakage, and so do not expect the meter's bar display to move all the way to the right (∞ ohms). Most electrolytic capacitors that are still functioning normally will show a resistance of 200 kΩ or more.
2. Some ohmmeters utilize internal batteries of up to 15 V, so be careful not to exceed the voltage rating of the capacitor.

7-7-2 *The Capacitance Meter or Analyzer*

The ohmmeter check tests the capacitor under a low-voltage condition. This may be adequate for some capacitor malfunctions; however, a problem that often occurs with capacitors is that they short or leak at a high voltage. The ohmmeter test is also adequate for capacitors of 0.5 μF or greater; however, a smaller capacitor cannot be tested because its charge time is too fast for the meter to respond. The ohmmeter cannot check for high-voltage failure, for small-value capacitance, or if the value of capacitance has changed through age or extreme thermal exposure.

A **capacitance meter** or analyzer, which is illustrated in Figure 7-26, can totally check all aspects of a capacitor in a range of values from approximately 1 pF to 20 F.

Capacitance Meter
Instrument used to measure the capacitance of a capacitor or circuit.

FIGURE 7-26 **Capacitor Analyzer.**

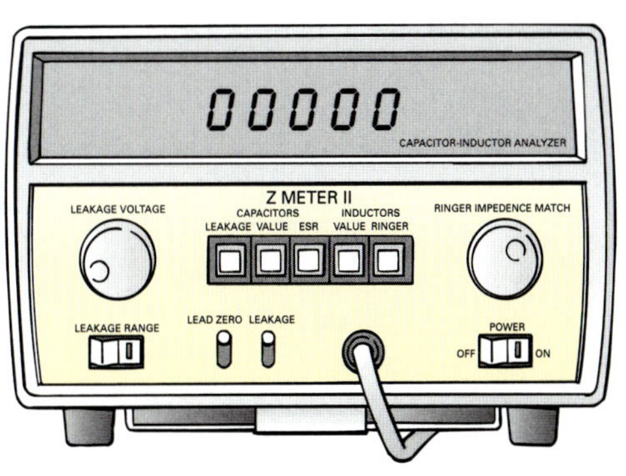

SELF-TEST EVALUATION POINT FOR SECTION 7-7

Use the following questions to test your understanding of Section 7-7.

1. The analog ohmmeter or digital multimeter with bar graph can only be used to test capacitors that are __________ or more in value.
2. A __________ __________ tests a capacitor for value change, leakage, dielectric absorption, and equivalent series resistance.

7-8 APPLICATIONS OF CAPACITORS

There are many applications of capacitors, some of which will be discussed now; others will be presented later and in your course of electronic studies.

7-8-1 *Combining AC and DC*

Figure 7-27(a) and (b) show how the capacitor can be used to combine ac and dc. The capacitor is large in value (electrolytic typically) and can be thought of as a very large bucket that will fill or charge to a dc voltage level. The ac voltage will charge and discharge the capacitor, which is similar to pouring in and pulling out (alternating) more and less water. The resulting waveform is a combination of ac and dc that varies above and below an average dc level. In this instance, the ac is said to be superimposed on a dc level.

7-8-2 *The Capacitive Voltage Divider*

Earlier we illustrated how capacitors could be connected in series across a dc voltage source to form a voltage divider. This circuit is repeated in Figure 7-28(a). Figure 7-28(b) shows how two capacitors can be used to divide up an ac voltage source of 30 V rms. Like a resistive voltage divider, voltage drop is proportional to current opposition, which with a capacitor is called *capacitive reactance* (X_C). The larger the capacitive reactance ($X_C\uparrow$), the larger the voltage drop ($V\uparrow$). Since capacitive reactance is inversely proportional to the capacitor value and the frequency of the ac source, a change in input frequency will cause a change in the capacitor's ca-

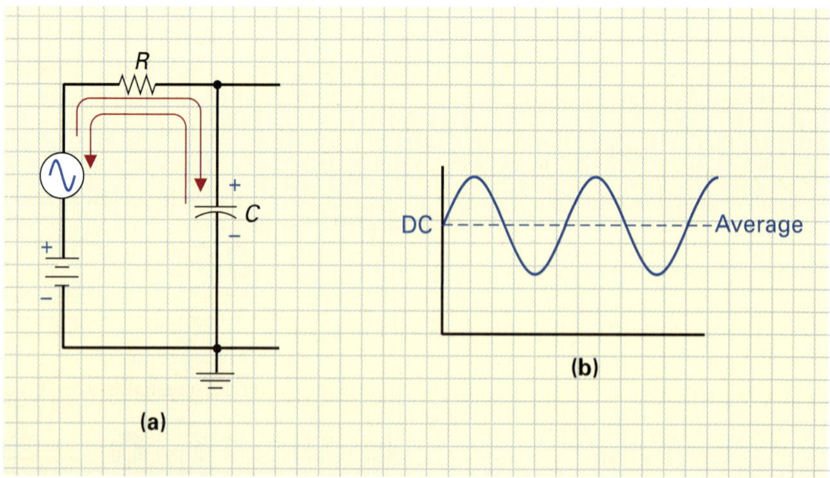

FIGURE 7-27 Superimposing AC on a DC Level.

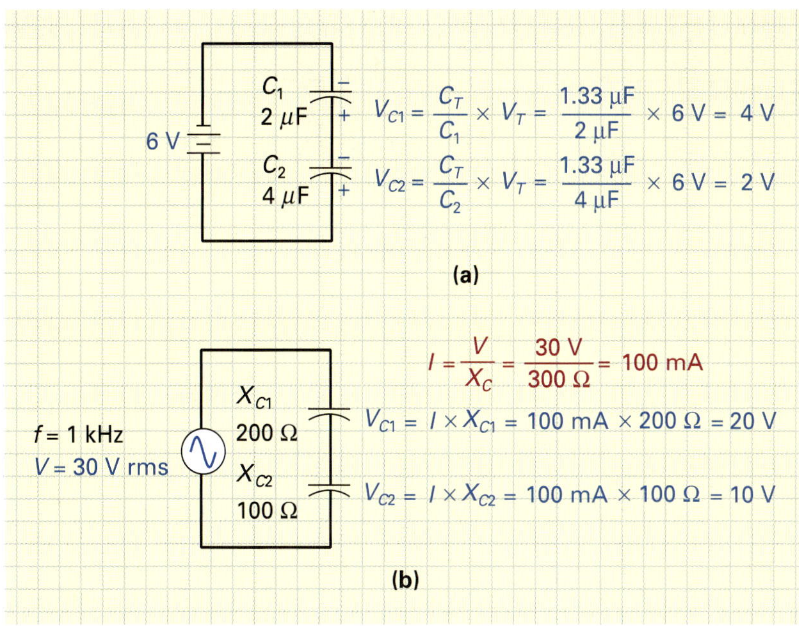

FIGURE 7-28 Capacitive Voltage Divider. (a) DC Circuit. (b) AC Circuit.

pacitive reactance, and this will change the voltage drop across the capacitor. Although a 90° phase shift is present between voltage and current in a purely capacitive circuit (I leads V by 90°), there is no phase difference between the input voltage and the output voltages.

7-8-3 *RC Filters*

A **filter** is a circuit that allows certain frequencies to pass but blocks other frequencies. In other words, it filters out the unwanted frequencies but passes the wanted or selected ones.

There are two basic RC filters:

1. The **low-pass filter,** which can be seen in Figure 7-29(a); as its name implies, it passes the low frequencies but heavily **attenuates** the higher frequencies.
2. The **high-pass filter,** which can be seen in Figure 7-29(b); as its name implies, it allows the high frequencies to pass but heavily attenuates the lower frequencies.

With either the low- or high-pass filter, it is important to remember that capacitive reactance is inversely proportional to frequency ($X_C \propto 1/f$) as stated in the center of Figure 7-29.

With the low-pass filter shown in Figure 7-29(a), the output is connected across the capacitor. As the frequency of the input increases, the amplitude of the output decreases. At dc (0 Hz) and low frequencies, the capacitive reactance is very large ($X_C\uparrow = 1/f\downarrow$) with respect to the resistor. All the input will appear across the capacitor, because the capacitor and resistor form a voltage divider, as shown in Figure 7-30(a). As with any voltage divider, the larger opposition to current flow will drop the larger voltage. Since the output voltage is determined by the voltage drop across the capacitor, almost all the input will appear across the capacitor and therefore be present at the output.

If the frequency of the input increases, the reactance of the capacitor will decrease ($X_C\downarrow = 1/f\uparrow$), and a larger amount of the signal will be dropped across the resistor. As frequency increases, the capacitor becomes more of a short circuit (lower reactance), and the output, which is across the capacitor, decreases, as shown in Figure 7-30(b).

Below the circuit of the low-pass filter in Figure 7-29(a), you will see a graph known as the **frequency response curve** for the low-pass filter. This curve illustrates that as the frequency of the input increases the voltage at the output will decrease.

Filter
Network composed of resistor, capacitor, and inductors used to pass certain frequencies yet block others through heavy attenuation.

Low-Pass Filter
Network or circuit designed to pass any frequencies below a critical or cutoff frequency and reject or heavily attenuate all frequencies above.

Attenuate
To reduce in amplitude an action or signal.

High-Pass Filter
Network or circuit designed to pass any frequencies above a critical or cutoff frequency and reject or heavily attenuate all frequencies below.

Frequency Response Curve
A graph indicating how effectively a circuit or device responds to the frequency spectrum.

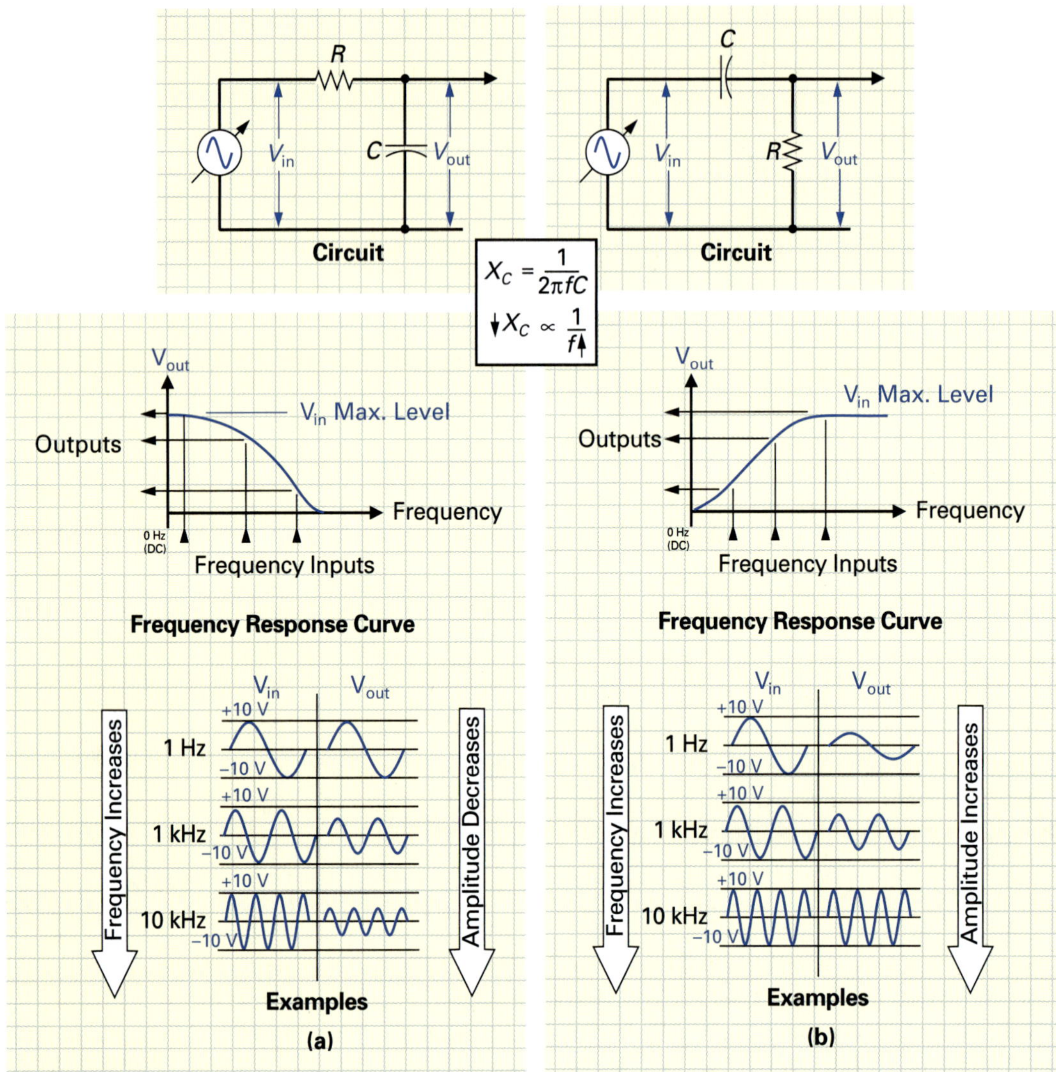

FIGURE 7-29 ***RC*** **Filters. (a) Low-Pass Filter. (b) High-Pass Filter.**

With the high-pass filter, seen in Figure 7-29(b), the capacitor and resistor have traded positions to show how the opposite effect for the low-pass filter occurs.

At low frequencies, the reactance will be high and almost all of the signal will be dropped across the capacitor. Very little signal appears across the resistor, and, consequently, the output, as shown in Figure 7-31(a). As the frequency of the input increases, the reactance of the capacitor decreases, allowing more of the input signal to appear across the resistor and therefore appear at the output, as shown in Figure 7-31(b).

Below the circuit of the high-pass filter in Figure 7-29(b), you will see the frequency response curve for the high-pass filter. This curve illustrates that as the frequency of the input increases the voltage at the output increases.

7-8-4 *The* RC *Integrator*

Integrator
Device that approximates and whose output is proportional to an integral of the input signal.

Up until now we have analyzed the behavior of an *RC* circuit only when a sine-wave input signal was applied. In both this and the following sections we demonstrate how an *RC* circuit will react to a square-wave input, and show two other important applications of capacitors. The term **integrator** is derived from a mathematical function in calculus. This

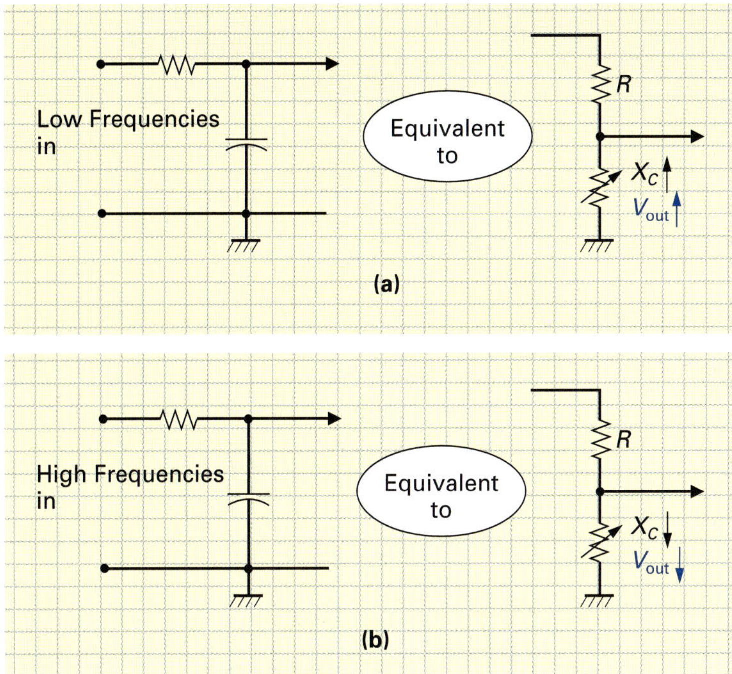

FIGURE 7-30 **Low-Pass Filter. (a) Low-Pass Filter, High X_C. (b) Low-Pass Filter, Low X_C.**

combination of *R* and *C*, in some situations, displays this mathematical function. Figure 7-32(a) illustrates an integrator circuit that can be recognized by the series connection of *R* and *C*, but mainly from the fact that the output is taken across the capacitor.

If a 10 V square wave is applied across the circuit, as shown in the waveforms in Figure 7-32(b), and the time constant of the *RC* combination calculates out to be 1 second,

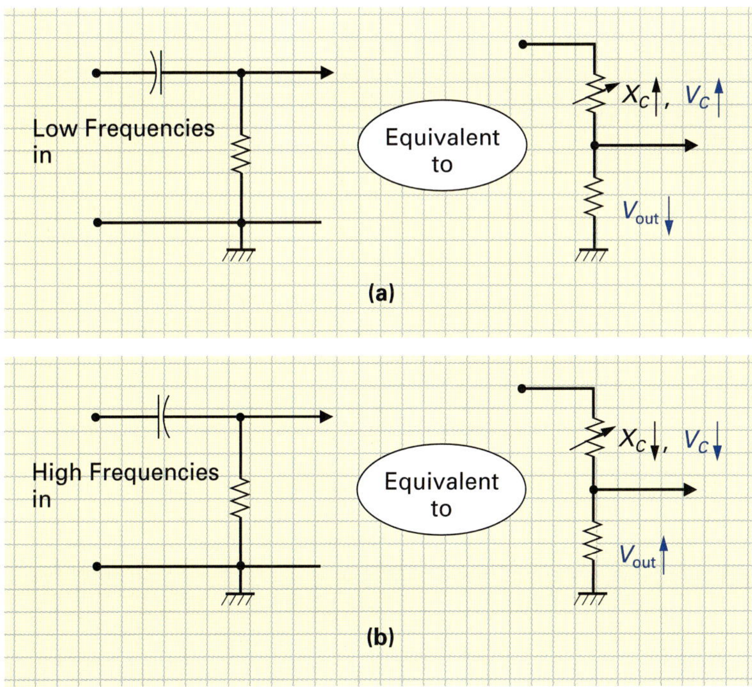

FIGURE 7-31 **High-Pass Filter. (a) High X_C. (b) Low X_C.**

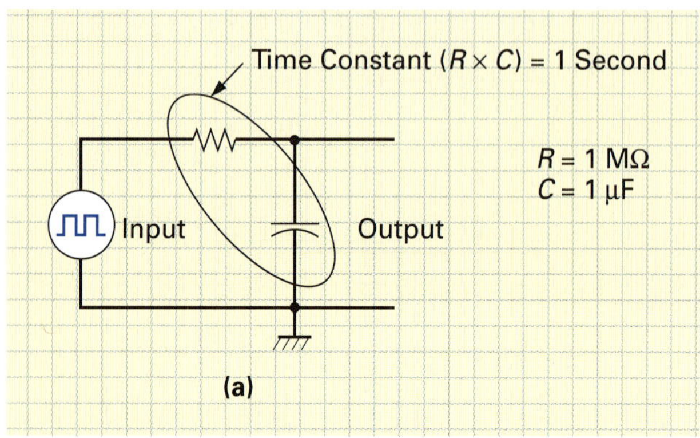

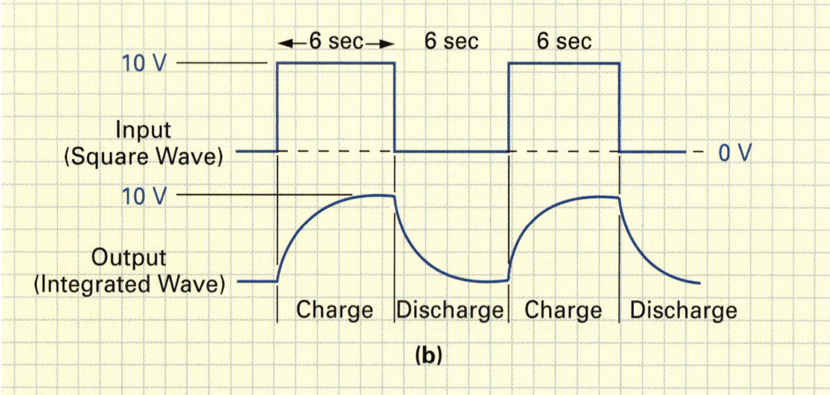

FIGURE 7-32 ***RC* Integrator. (a) Circuit. (b) Waveforms.**

the capacitor will charge when the square wave input is positive toward the applied voltage (10 V) and reaches it in five time constants (5 seconds). Since the positive alternation of the square wave lasts for 6 seconds, the capacitor will be fully charged 1 second before the positive alternation ends, as shown in Figure 7-33.

When the positive half-cycle of the square-wave input ends after 6 seconds, the input falls to 0 V and the circuit is equivalent to that shown in Figure 7-34(a). The 10 V charged capacitor now has a path to discharge and in five time constants (5 seconds) is fully discharged, as shown in Figure 7-34(b).

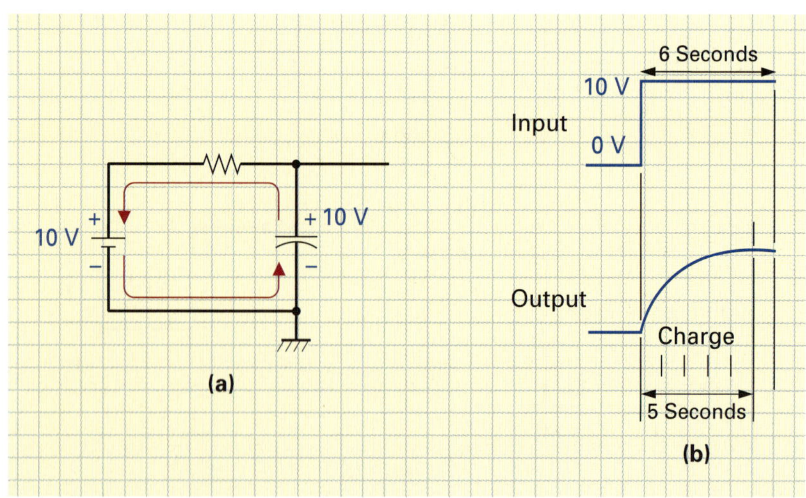

FIGURE 7-33 Integrator Response to Positive Step. (a) Equivalent Circuit. (b) Waveforms for Charge.

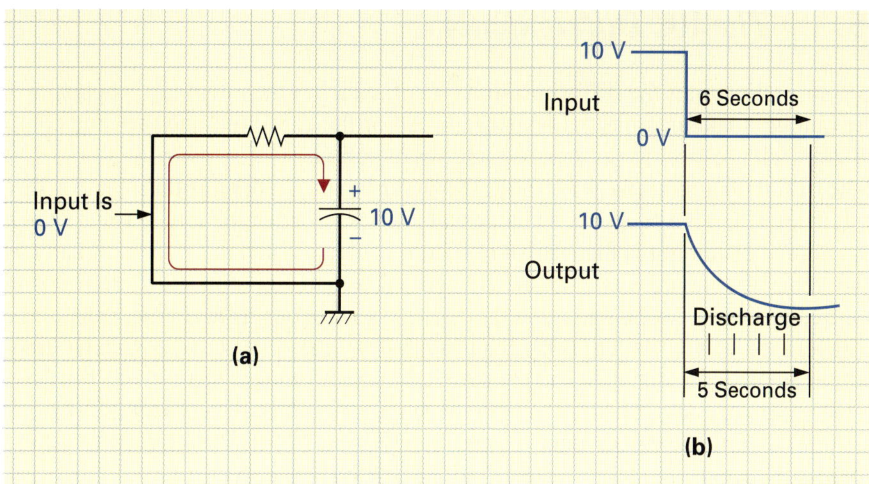

FIGURE 7-34 Integrator Response to Negative Step. (a) Equivalent Circuit. (b) Waveforms for Discharge.

If the same *RC* integrator circuit was connected to a square wave that has a 1-second positive half-cycle, the capacitor will not be able to charge fully toward 10 V. In fact, during the positive alternation of 1 second (one time constant), the capacitor will reach 63.2% of the applied voltage (6.32 V), and then during the 0 V half-cycle it will discharge to 63.2% of 6.32 V, to 2.33 V, as shown by the waveform in Figure 7-35. The voltage across the capacitor, and therefore the output voltage, will gradually build up and eventually level off to an average value of 5 V in about five time constants (5 seconds).

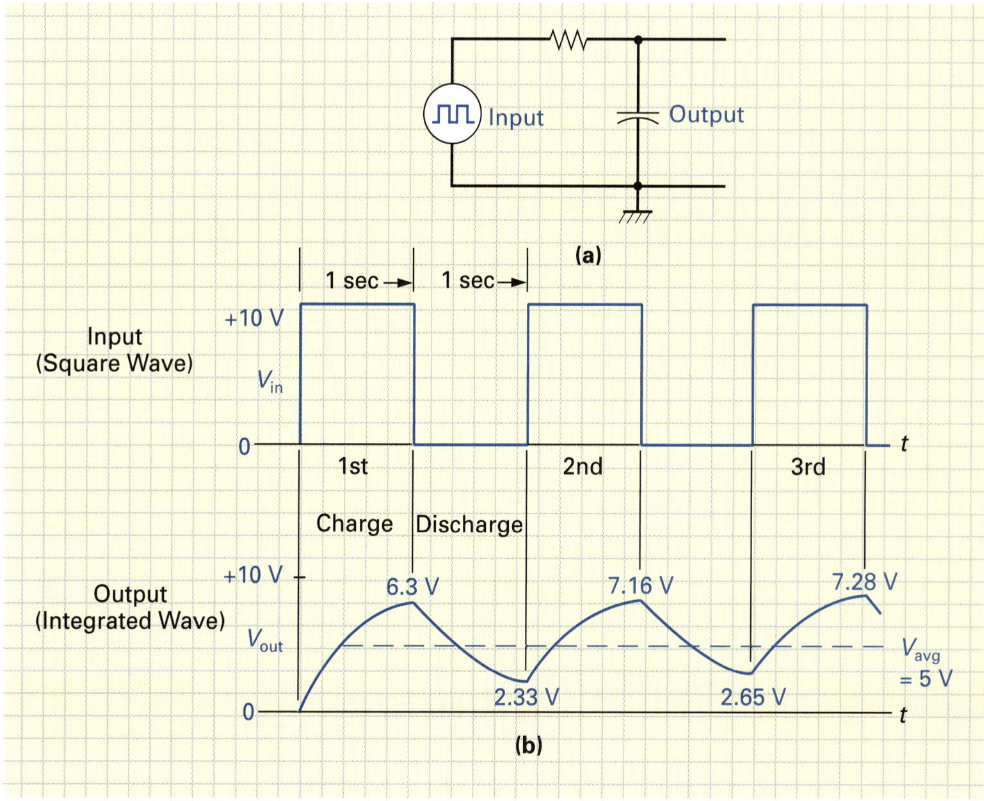

FIGURE 7-35 Integrator Response to Square Wave. (a) Circuit. (b) Waveforms.

7-8-5 *The* RC *Differentiator*

Differentiator

A circuit whose output voltage is proportional to the rate of change of the input voltage. The output waveform is then the time derivative of the input waveform, and the phase of the output waveform leads that of the input by 90°.

Figure 7-36(a) illustrates the **differentiator** circuit, which is the integrator's opposite. In this case the output is taken across the resistor instead of the capacitor, and the time constant is always short with respect to the input square wave period.

The differentiator output waveform, shown in Figure 7-36(b), is taken across the resistor and is the result of the capacitor's charge and discharge. When the square wave swings positive, the equivalent circuit is that shown in Figure 7-37(a). When the 10 V is initially applied (positive step of the square wave), all the voltage is across the resistor, and therefore at the output, as the capacitor cannot charge instantly. As the capacitor begins to charge, more of the voltage is developed across the capacitor and less across the resistor. The voltage across the capacitor exponentially increases and reaches 10 V in five time constants (5×1 ms = 5 ms), while the voltage across the resistor, and therefore the output, exponentially falls from its initial 10 V to 0 V in five time constants as shown in Figure 7-37(b), at which time all the voltage is across the capacitor and no voltage will be across the resistor.

When the positive half-cycle of the square wave ends and the input falls to zero, the circuit is equivalent to that shown in Figure 7-38(a). The negative plate of the capacitor is now applied directly to the output. Since the capacitor cannot instantly discharge, the output drops suddenly down to -10 V as shown in Figure 7-38(b). This is the voltage across the resistor, and therefore the output. The capacitor is now in series with the resistor and therefore has a path through the resistor to discharge, which it does in five time constants to 0 V.

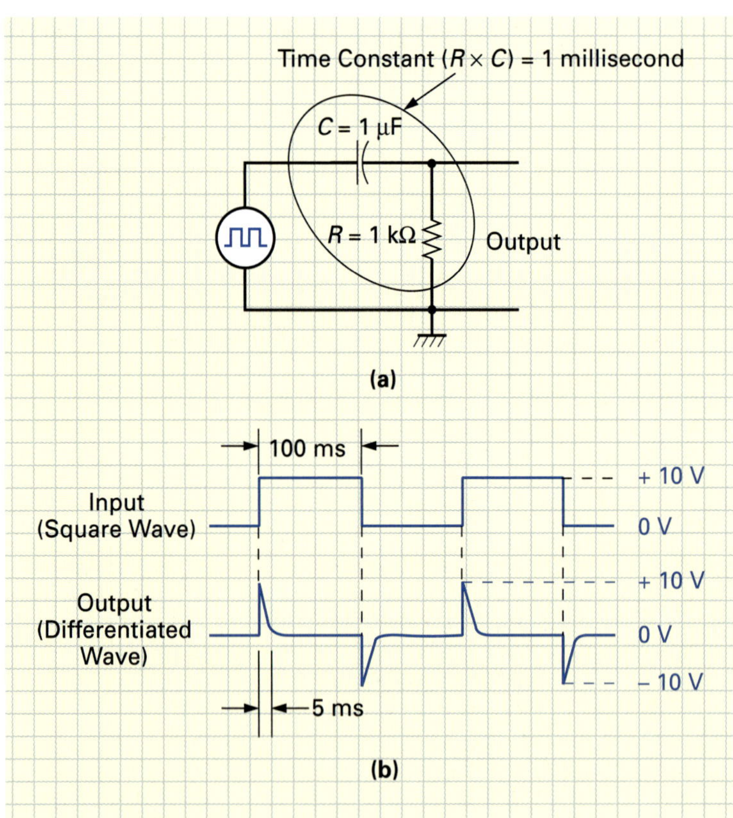

FIGURE 7-36 ***RC* Differentiator. (a) Circuit. (b) Waveforms.**

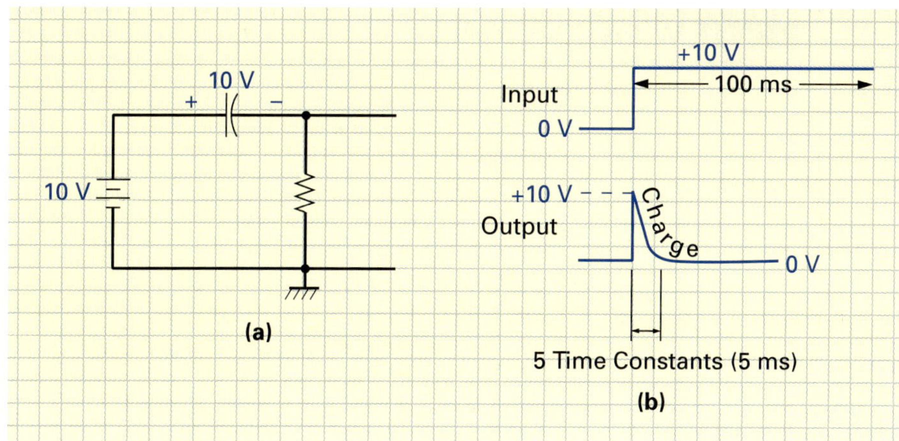

FIGURE 7-37 **Differentiator's Response to Positive Step. (a) Equivalent Circuit. (b) Waveforms.**

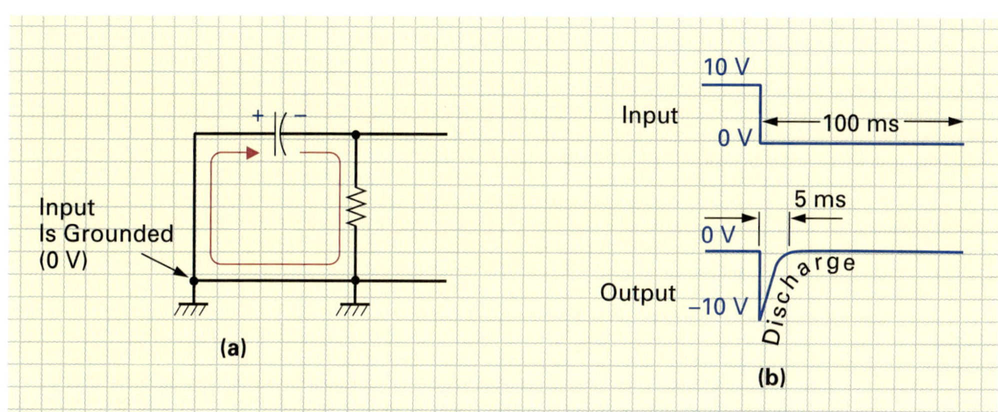

FIGURE 7-38 **Differentiator's Response to a Negative Step. (a) Equivalent Circuit. (b) Waveforms.**

SELF-TEST EVALUATION POINT FOR SECTION 7-8

Use the following questions to test your understanding of Section 7-8.

1. Give three circuit applications for a capacitor.
2. With an *RC* low-pass filter, the ___________ is connected across the output. Whereas with an *RC* high-pass filter, the ___________ is connected across the output.
3. An integrator has a ___________ time constant compared to the period of the input square wave.
4. A ___________ circuit produces positive and negative spikes at the output when a square wave is applied at the input.

REVIEW QUESTIONS

Multiple-Choice Questions

1. When a capacitor charges:
- **a.** The voltage across the plates rises exponentially
- **b.** The circuit current falls exponentially
- **c.** The capacitor charges to the source voltage in $5RC$ seconds
- **d.** All of the above

2. What is the capacitance of a capacitor if it can store 24 C of charge when 6 V is applied across the plates?
a. 2 μF **b.** 3 μF **c.** 4.7 μF **d.** None of the above

3. The capacitance of a capacitor is directly proportional to:
- **a.** The plate area
- **b.** The distance between the plates
- **c.** The constant of the dielectric used
- **d.** Both (a) and (c)
- **e.** Both (a) and (b)

4. The capacitance of a capacitor is inversely proportional to:
- **a.** The plate area
- **b.** The distance between the plates
- **c.** The dielectric used
- **d.** Both (a) and (c)
- **e.** Both (a) and (b)

5. Total series capacitance of two capacitors is calculated by:
- **a.** Using the product-over-sum formula
- **b.** Using the voltage-divider formula
- **c.** Using the series resistance formula on the capacitors
- **d.** Adding all the individual values

6. Total parallel capacitance is calculated by:
- **a.** Using the product-over-sum formula
- **b.** Using the voltage-divider formula
- **c.** Using the parallel resistance formula on capacitors
- **d.** Adding all the individual values

7. The mica and ceramic fixed capacitors have:
- **a.** An arrangement of stacked plates
- **b.** An electrolyte substance between the plates
- **c.** An adjustable range
- **d.** All of the above

8. An electrolytic capacitor:
- **a.** Is the most popular large-value capacitor
- **b.** Is polarized
- **c.** Can have either aluminum or tantalum plates
- **d.** All of the above

9. Variable capacitors normally achieve a large variation in capacitance by varying ____________, while adjustable trimmer capacitors only achieve a small capacitance range by varying ____________.
- **a.** Dielectric constant, plate area
- **b.** Plate area, plate separation
- **c.** Plate separation, dielectric constant
- **d.** Plate separation, plate area

10. In one time constant, a capacitor will charge to ____________ of the source voltage.
a. 86.5% **b.** 63.2% **c.** 99.3% **d.** 98.2%

11. Capacitive reactance is inversely proportional to:
- **a.** Capacitance and resistance
- **b.** Frequency and capacitance
- **c.** Capacitance and impedance
- **d.** Both (a) and (c)

12. The impedance of an RC series circuit is equal to:
- **a.** The sum of R and X_C
- **b.** The square root of the sum of R^2 and X_C^2
- **c.** The square of the sum of R and X_C
- **d.** The sum of the square root of R and X_C

13. In a purely resistive circuit:
- **a.** The current flowing in the circuit leads the voltage across the capacitor by 90°.
- **b.** The circuit current and resistor voltage are in phase with one another.
- **c.** The current leads the voltage by 45°.
- **d.** The current leads the voltage by a phase angle between 0 and 90°.

14. In a purely capacitive circuit:
- **a.** The current flowing in the circuit leads the voltage across the capacitor by 90°.
- **b.** The circuit current and resistor voltage are in phase with one another.
- **c.** The current leads the voltage by 45°.
- **d.** The current leads the voltage by a phase angle between 0 and 90°.

15. In a series circuit containing both capacitance and resistance:
- **a.** The current flowing in the circuit leads the voltage across the capacitor by 90°.
- **b.** The circuit current and resistor voltage are in phase with one another.
- **c.** The current leads the voltage by 45°.
- **d.** Both (a) and (b).

16. In a series RC circuit, the source voltage is equal to:
- **a.** The sum of V_R and V_C
- **b.** The difference between V_R and V_C
- **c.** The vectoral sum of V_R and V_C
- **d.** The sum of V_R and V_C squared

17. As the source frequency is increased, the capacitive reactance will:
a. Increase **c.** Be unaffected
b. Decrease **d.** Increase, depending on harmonic content

18. The phase angle of a series RC circuit indicates by what angle V_S ____________ V_R.
a. Lags **c.** Leads or lags
b. Leads **d.** None of the above

19. In a series RC circuit, the vector combination of R and X_C is the circuit's ____________.
a. Phase angle **c.** Source voltage
b. Apparent power **d.** Impedance

20. In a parallel *RC* circuit, the total current is equal to:
   a. The sum of I_R and I_C
   b. The difference between I_R and I_C
   c. The vectoral sum of I_R and I_C
   d. The sum of I_R and I_C squared

21. ____________ is the opposition offered by a capacitor to current flow without the dissipation of energy.
   a. Capacitive reactance
   b. Resistance
   c. Impedance
   d. Phase angle
   e. The power factor

22. ____________ is the total reactive and resistive circuit opposition to current flow.
   a. Capacitive reactance
   b. Resistance
   c. Impedance
   d. Phase angle
   e. The power factor

23. ____________ is the ratio of true (resistive) power to apparent power and is therefore a measure of the loss in a circuit.
   a. Capacitive reactance
   b. Resistance
   c. Impedance
   d. Phase angle
   e. The power factor

24. In a series *RC* circuit, the leading voltage will be measured across the:
   a. Resistor c. Source
   b. Capacitor d. Any of the choices are true.

25. In a series *RC* circuit, the lagging voltage will be measured across the:
   a. Resistor c. Source
   b. Capacitor d. Any of the choices are true.

Practice Problems

26. If a 10 μF capacitor is charged to 10 V, how many coulombs of charge has it stored?

27. If a 0.006 μF capacitor has stored 125×10^{-6} C of charge, what potential difference would appear across the plates?

28. Calculate the capacitance of the capacitor that has the following parameter values: $A = 0.008\ \text{m}^2$; $d = 0.00095$ m; the dielectric used is paper.

29. Calculate the total capacitance if the following are connected in:
   a. Parallel: 1.7 μF, 2.6 μF, 0.03 μF, 1200 pF
   b. Series: 1.6 μF, 1.4 μF, 4 μF

30. If three capacitors of 0.025 μF, 0.04 μF, and 0.037 μF are connected in series across a 12 V source, as shown in Figure 7-39, what would be the voltage drop across each?

31. Give the value of the following alphanumeric capacitor value codes:
   a. 104 b. 125 c. 0.01 d. 220

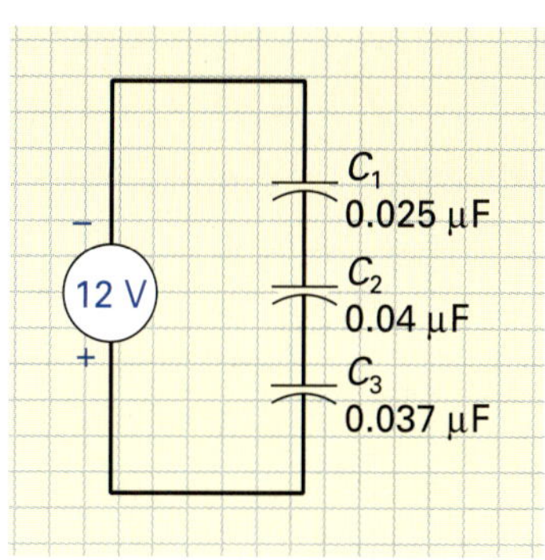

FIGURE 7-39

32. What would be the time constant of the following *RC* circuits?
   a. $R = 6\ \text{k}\Omega$, $C = 14\ \mu\text{F}$
   b. $R = 12\ \text{M}\Omega$, $C = 1400\ \text{pF}$
   c. $R = 170\ \Omega$, $C = 24\ \mu\text{F}$
   d. $R = 140\ \text{k}\Omega$, $C = 0.007\ \mu\text{F}$

33. If 10 V were applied across all the *RC* circuits in Question 32, what would be the voltage across each capacitor after one time constant, and how much time would it take each capacitor to fully charge?

34. Calculate the capacitive reactance of the capacitor circuits with the following parameters:
   a. $f = 1\ \text{kHz}$, $C = 2\ \mu\text{F}$
   b. $f = 100\ \text{Hz}$, $C = 0.01\ \mu\text{F}$
   c. $f = 17.3\ \text{MHz}$, $C = 47\ \mu\text{F}$

35. In a series *RC* circuit, the voltage across the capacitor is 12 V and the voltage across the resistor is 6 V. Calculate the source voltage.

36. Calculate the impedance for the following series *RC* circuits:
   a. 2.7 MΩ, 3.7 μF, 20 kHz
   b. 350 Ω, 0.005 μF, 3 MHz
   c. $R = 8.6\ \text{k}\Omega$, $X_C = 2.4\ \Omega$
   d. $R = 4700\ \Omega$, $X_C = 2\ \text{k}\Omega$

37. In a parallel *RC* circuit with parameters of $V_S = 12$ V, $R = 4\ \text{M}\Omega$, and $X_C = 1.3\ \text{k}\Omega$, calculate:
   a. I_R d. Z
   b. I_C e. θ
   c. I_T

38. Calculate the total reactance in:
   a. A series circuit where $X_{C1} = 200\ \Omega$, $X_{C2} = 300\ \Omega$, $X_{C3} = 400\ \Omega$
   b. A parallel circuit where $X_{C1} = 3.3\ \text{k}\Omega$, $X_{C2} = 2.7\ \text{k}\Omega$

39. Calculate the capacitance needed to produce 10 kΩ of reactance at 20 kHz.

40. At what frequency will a 4.7 μF capacitor have a reactance of 2000 Ω?

41. A series *RC* circuit contains a resistance of 40 Ω and a capacitive reactance of 33 Ω across a 24 V source.
   a. Sketch the schematic diagram.
   b. Calculate Z, I, V_R, V_C, I_R, I_C, and θ.

FIGURE 7-40

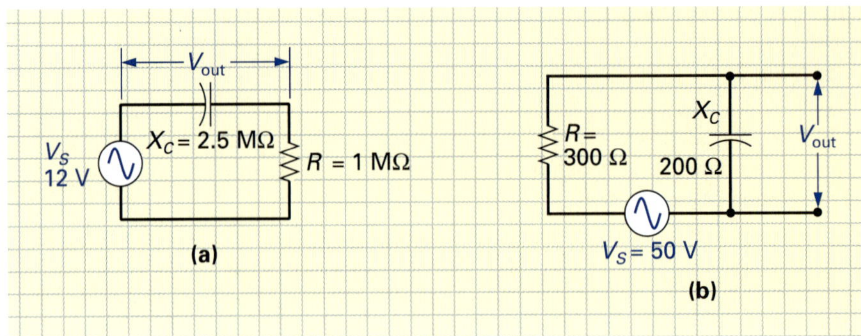

42. A parallel *RC* circuit contains a resistance of 10 kΩ and a capacitive reactance of 5 kΩ across a 100 V source.
    a. Sketch the schematic diagram.
    b. Calculate I_R, I_C, I_T, Z, V_R, V_C, and θ.
43. Calculate V_R and V_C for the circuits seen in Figure 7-40(a) and (b).
44. Calculate the impedance of the four circuits shown in Figure 7-41.
45. In Figure 7-42, the output voltage, since it is taken across the capacitor, will ___________ the voltage across the resistor by ___________ degrees.
46. If the positions of the capacitor and resistor in Figure 7-42 are reversed, the output voltage, since it is now taken across the resistor, will ___________ the voltage across the capacitor by ___________ degrees.
47. Calculate the resistive power, reactive power, apparent power, and power factor for the circuit seen in Figure 7-43; V_{in} = 24 V and f = 35 kHz.
48. Refer to Figure 7-43 and calculate the following:
    a. [Figure 7-43(a)] X_C, I, Z, I_R, θ, V_R, V_C
    b. [Figure 7-43(b)] V_R, V_C, I_R, I_C, I_T, Z, θ

Web Site Questions

Go to the web site http://www.prenhall.com/cook, select the textbook *Electronics: A Complete Course,* select this chapter, and then follow the instructions when answering the multiple choice practice problems.

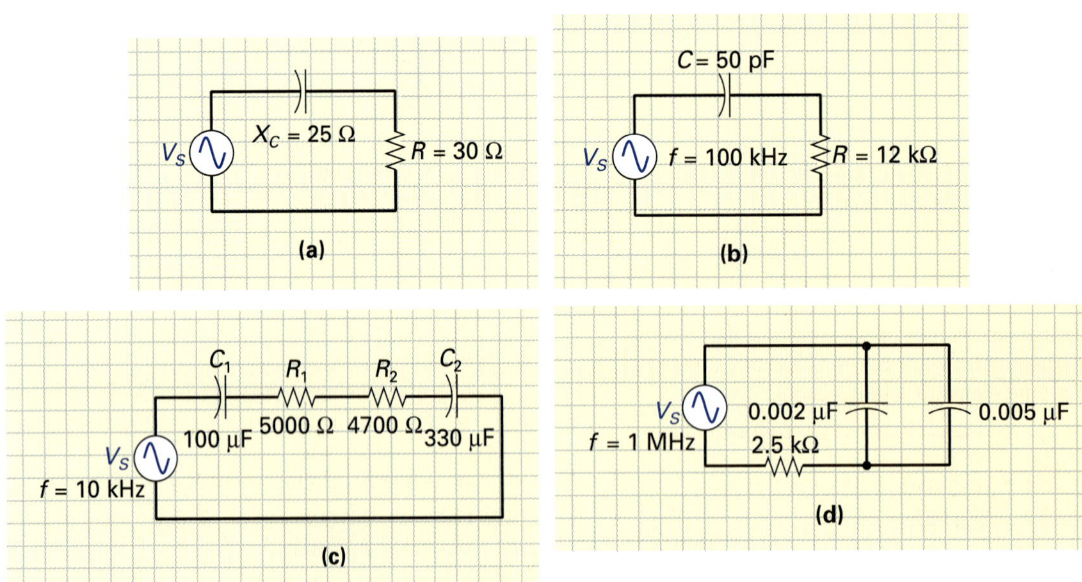

FIGURE 7-41

FIGURE 7-42

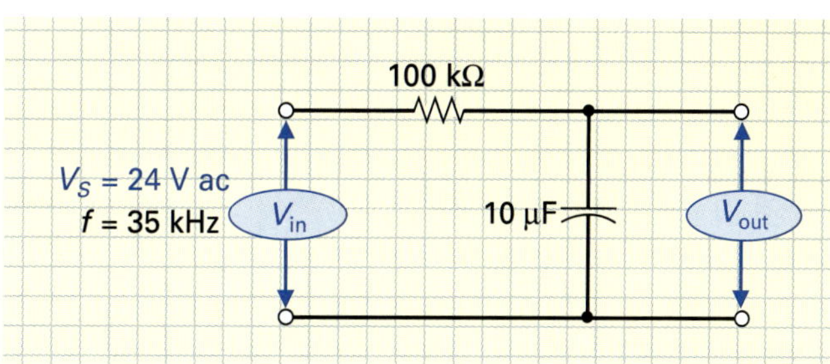

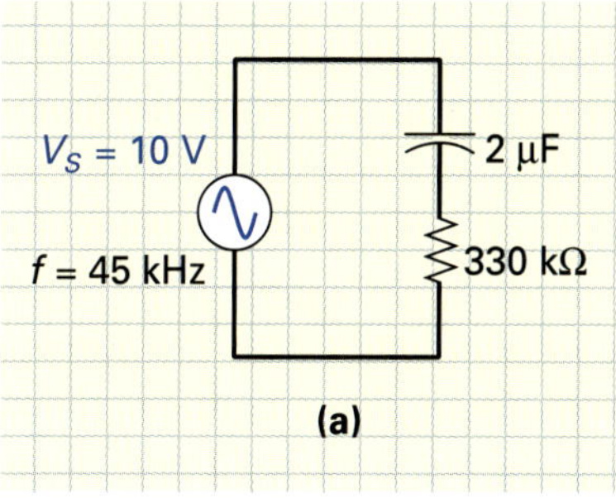

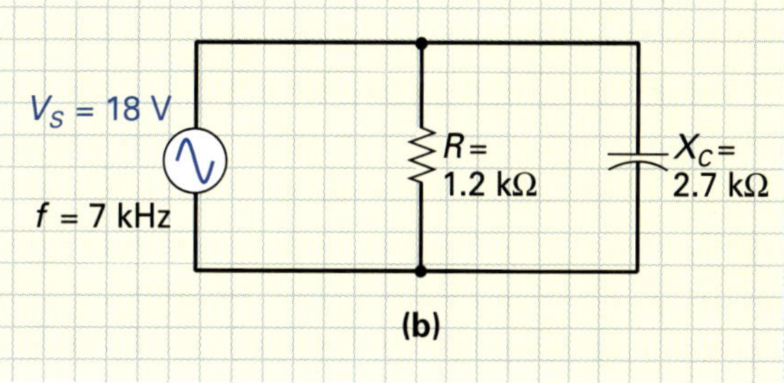

FIGURE 7-43

Electromagnets, Inductors, and Transformers

The Great Experimenter

Michael Faraday was born to James and Margaret Faraday on September 22, 1791. At twenty-two, the gifted and engaging Faraday was at the right place at the right time, and with the right talents. He impressed the brilliant professor Humphry Davvy, who made him his research assistant at Count Rumford's Royal Institution. After only two years, Faraday was given a promotion and an apartment at the Royal Institution, and in 1821 he married. He lived at the Royal Institution for the rest of his active life, working in his laboratory, giving very dynamic lectures, and publishing over 450 scientific papers. Unlike many scientific papers, Faraday's papers would never use a calculus formula to explain a concept or idea. Instead, Faraday would explain all of his findings using logic and reasoning, so that a person trying to understand science did not need to be a scientist. It was this gift of being able to make even the most complex areas of science easily accessible to the student, coupled with his motivational teaching style, that made him so popular.

In 1855 he had written three volumes of papers on electromagnetism, the first dynamo, the first transformer, and the foundations of electrochemistry, a large amount on dielectrics and even some papers on plasma. The unit of capacitance is measured in *farads,* in honor of his work in these areas of science.

Faraday began two series of lectures at the Royal Institution, which have continued to this day. Michael and Sarah Faraday were childless, but they both loved children, and in 1860 Faraday began a series of Christmas lectures expressly for children, the most popular of which was called "The Chemical History of a Candle." The other lecture series was the "Friday Evening Discourses," of which he himself delivered over a hundred. These very dynamic, enlightening, and entertaining lectures covered areas of science or technology for the lay person, and were filled with demonstrations. On one evening in 1846, an extremely nervous and shy speaker ran off just moments before he was scheduled to give the Friday Evening Discourse. Faraday had to fill in, and to this day a tradition is still enforced whereby the lecturer for the Friday Evening Discourse is locked up for half an hour before the presentation with a single glass of whiskey.

Faraday was often referred to as "the great experimenter," and it was this consistent experimentation that led to many of his findings. He was fascinated by science and technology, and was always exploring new and sometimes dangerous horizons. In fact, in one of his reports he states, "I have escaped, not quite unhurt, from four explosions." When asked to comment on experimentation, his advice was to "Let your imagination go, guiding it by judgment and principle, but holding it in and directing it by experiment. Nothing is so good as an experiment which, while it sets an error right, gives you as a reward for your humility an absolute advance in knowledge."

8

Introduction

It was during a classroom lecture in 1820 that Danish physicist Hans Christian Oersted accidentally stumbled on an interesting reaction. As he laid a compass down on a bench he noticed that the compass needle pointed to an adjacent conductor that was carrying a current, instead of pointing to the earth's north pole. It was this discovery that first proved that magnetism and electricity were very closely related to one another. This phenomenon is now called *electromagnetism* since it is now known that any conductor carrying an electrical current will produce a magnetic field.

In 1831, the English physicist Michael Faraday explored further Oersted's discovery of electromagnetism and found that the process could be reversed. Faraday observed that if a conductor was passed through a magnetic field, a voltage would be induced in the conductor and cause a current to flow. This phenomenon is referred to as *electromagnetic induction.*

In this chapter we will first examine the characteristics and applications of **electromagnetism** (electricity to magnetism) and electromagnetic induction (magnetism to electricity). These two rules form the basis for three key components: the electromagnet, the inductor and the transformer. Inductance, by definition, is the ability of a device to oppose a change in current flow, and the device designed specifically to achieve this function is called the *inductor.* There is, in fact, no physical difference between an inductor and an electromagnet, since they are both coils. The two devices are given different names because they are used in different applications even though their construction and principle of operation are the same. An electromagnet is used to generate a magnetic field in response to current, while an inductor is used to oppose any change in current.

Electromagnetism
Relates to the magnetic field generated around a conductor when current is passed through it.

The transformer is an electrical device that makes use of electromagnetic induction to transfer alternating current from one circuit to another. The transformer consists of two inductors that are placed in very close proximity to one another. When an alternating current flows through the first coil or **primary winding** the inductor sets up a magnetic field. The expanding and contracting magnetic field produced by the primary cuts across the windings of the second inductor or **secondary winding** and induces a voltage in this coil.

Primary Winding
First winding of a transformer that is connected to the source.

Secondary Winding
Output winding of a transformer that is connected across the load.

By changing the ratio between the number of turns in the secondary winding to the number of turns in the primary winding, some characteristics of the ac signal can be changed or transformed as it passes from primary to secondary. For example, a low ac voltage could be stepped up to a higher ac voltage, or a high ac voltage could be stepped down to a lower ac voltage.

8-1 ELECTROMAGNETISM

Coil
Number of turns of wire wound around a core to produce magnetic flux (an electromagnet) or to react to a changing magnetic flux (an inductor).

Electromagnet
A magnet consisting of a coil wound on a soft iron or steel core. When current is passed through the coil a magnetic field is generated and the core is strongly magnetized to concentrate the magnetic field.

A magnetic field results whenever a current flows through any piece of conductor or wire, as shown in Figure 8-1(a). If a conductor is wound to form a spiral, as illustrated in Figure 8-1(b), the conductor, which is now referred to as a **coil,** will, as a result of current flow, develop a magnetic field, which will sum or intensify within the coil. If many coils are wound in the same direction, an **electromagnet** is formed, as shown in Figure 8-1(c), which will produce a concentrated magnetic field whenever a current is passed through its coils.

Up until now, we have only been discussing current flow through a coil in one direction (dc). A dc voltage produces a fixed current in one direction and therefore generates a magnetic field in a coil of fixed polarity, as shown in Figure 8-2(a).

Alternating current (ac) is continually varying, and as the polarity of the magnetic field is dependent on the direction of current flow, the magnetic field will also be alternating in polarity, as shown in Figure 8-2(b). Let us look at times 1 through 4 in Figure 8-2 in more detail.

Time 1: The alternating voltage has risen to a maximum positive level and causes current to flow in the direction seen in the circuit. This will cause a magnetic field with a south pole above and a north pole below.

Time 2: Between positions 1 and 2 the voltage, and therefore current, will decrease from a maximum positive value to zero. This will cause a corresponding collapse of the magnetic field from maximum (time 1) to zero (time 2).

Time 3: Voltage and current increase from zero to maximum negative between positions 2 and 3. The increase in current flow causes a similar increase or buildup of magnetic flux, producing a north pole above and south pole below.

Time 4: From time 3 to time 4, the current within the circuit diminishes to zero, and the magnetic field once again collapses. The cycle then repeats.

In regard to current flow, therefore, we can say that:

1. A direct current (dc) produces a constant magnetic field of a fixed polarity, for example, north–south.
2. An alternating current (ac) produces an alternating magnetic field, which continuously switches polarity, for example, north–south, south–north, north–south, and so on.

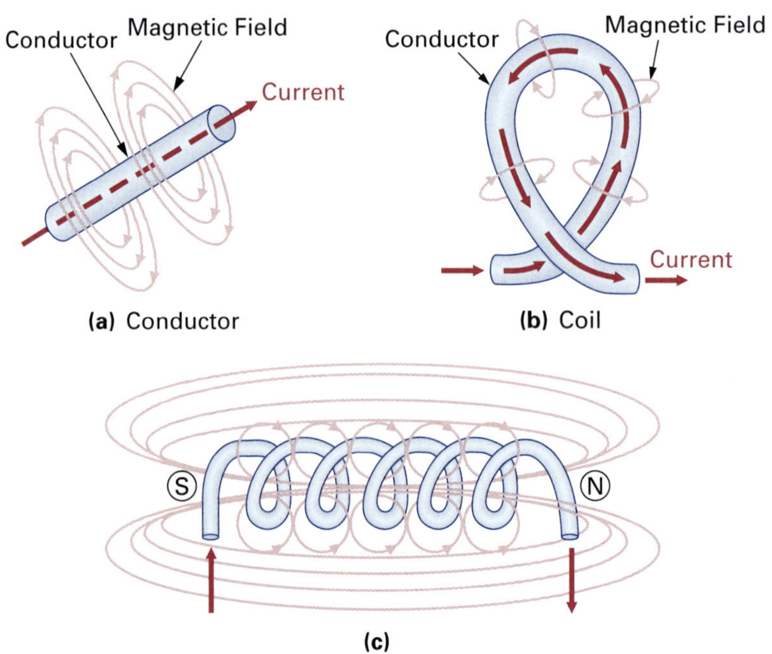

FIGURE 8-1 Conductor and Coil Electromagnetism.

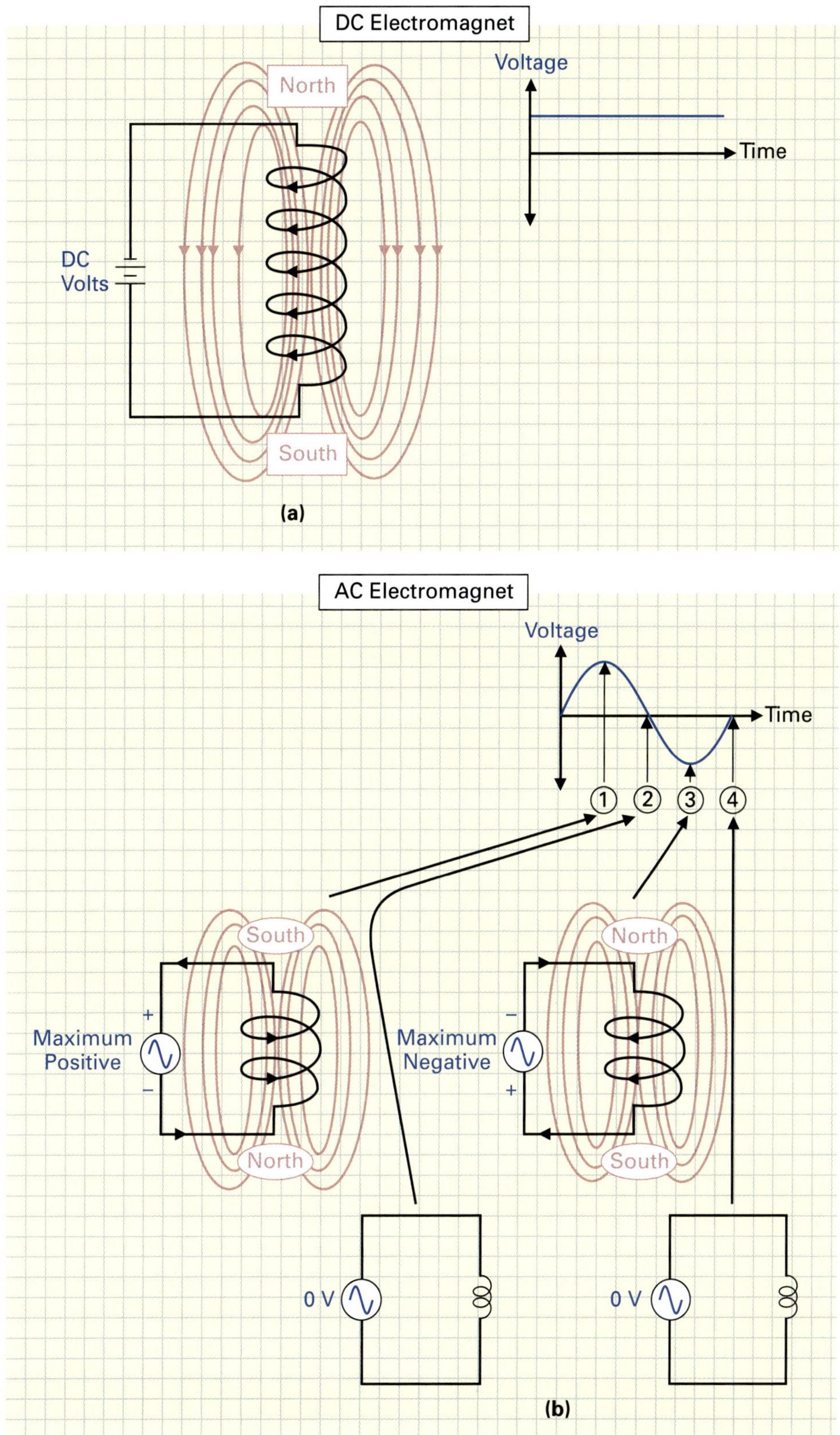

FIGURE 8-2 **Electromagnets. (a) DC Electromagnet. (b) AC Electromagnet.**

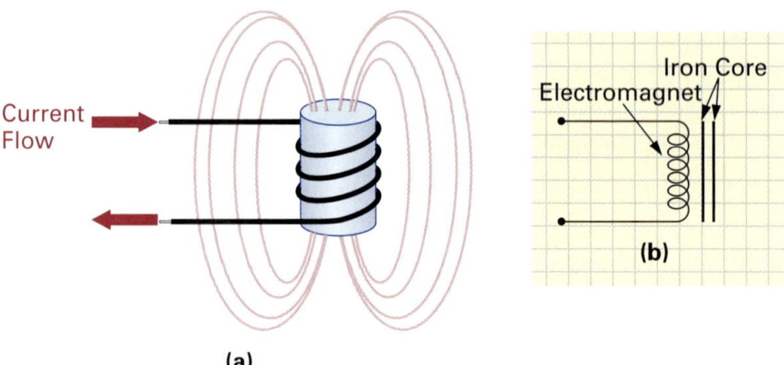

FIGURE 8-3 **Iron-Core Electromagnet. (a) Operation. (b) Symbol.**

8-1-1 *Magnetic Terms*

***B–H* Curve**
Curve plotted on a graph to show successive states during magnetization of a ferromagnetic material.

The magnetic field strength of an electromagnet can be increased by increasing the number of turns in its coil, increasing the current through the electromagnet, or decreasing the length of the coil ($H = I \times N/l$). The field strength can be increased further if an iron core is placed within the electromagnet, as illustrated in Figure 8-3(a). This is because an iron core has less *reluctance* or opposition to the magnetic lines of force than air, so the *flux density (B)* is increased. Another way of saying this is that the *permeability* or conductance of magnetic lines of force within iron is greater than that of air, and if the permeability of iron is large, its reluctance must be small. The symbol for an iron-core electromagnet is seen in Figure 8-3(b).

The ***B–H* curve** in Figure 8-4 illustrates the relationship between the two most important magnetic properties: flux density (B) and *magnetizing force* (H). Figure 8-4(a) illustrates

TIME LINE
In 1820, a Danish physicist, Hans C. Oersted (1777-1851), accidentally discovered an interesting phenomenon. Placing a compass near a current-carrying conductor, he noticed that the needle of the compass pointed to the conductor rather than north. He was quick to realize that electricity and magnetism were related, and in honor of his work, the unit oersted was adopted for the unit of magnetic field strength.

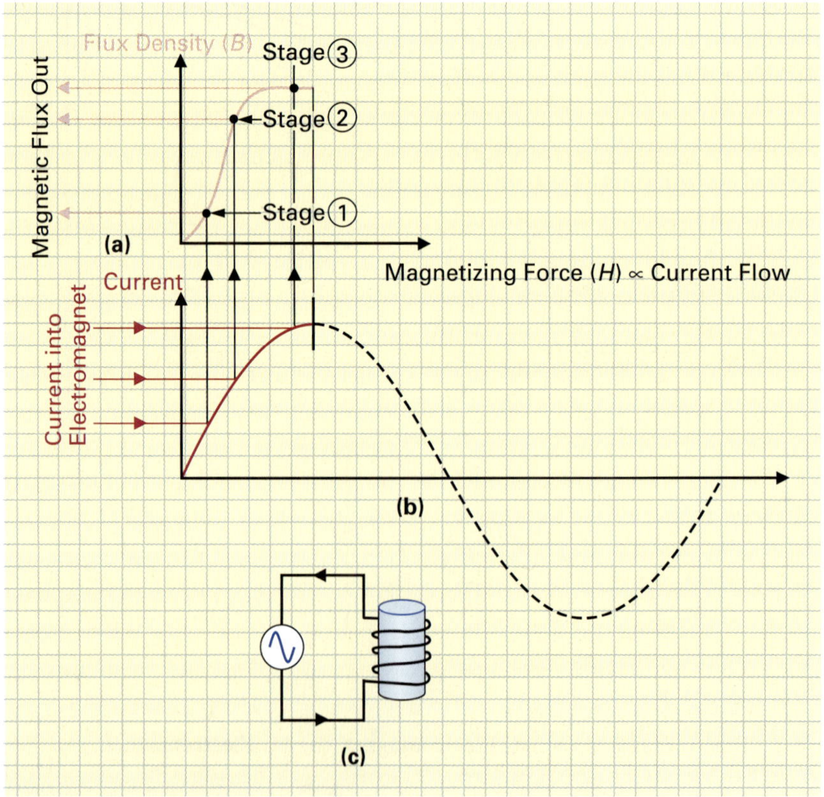

FIGURE 8-4 **(a) Flux Density versus Magnetizing Force Curve (*B–H* Curve). (b) Current Applied to Circuit. (c) Circuit.**

the *B–H* curve, while Figure 8-4(b) illustrates the positive rising portion of the ac current that is being applied to the iron-core electromagnetic circuit in Figure 8-4(c).

The magnetizing force is actually equal to $H = I$ (current) $\times N$ (number of turns) $/l$ (length of coil); but since the number of turns and length of coil are fixed for the coil being used, the magnetizing force (H) is proportional to the current (I) applied, which is shown in Figure 8-4(b). The positive rise of the current from zero to maximum positive is applied through the electromagnet and will produce a corresponding bloom or buildup in magnetic flux at a rate indicated by the *B–H* curve shape in Figure 8-4(a).

It is important to note that as the magnetizing force (current) is increased, there are three distinct stages in the change of flux density or magnetic flux out.

Stage 1: Up to this point, the increase in flux density is slow, as a large amount of force is required to begin alignment of the molecule magnets.

Stage 2: Increase in flux density is now rapid and almost linear as the molecule magnets are aligning easily.

Stage 3: In this state the molecule magnets cannot be magnetized any further because they are all fully aligned and no more flux density can be easily obtained. This is called the **saturation point,** and it is the state of magnetism beyond which an electromagnet is incapable of further magnetic strength. It is the point beyond which the *B–H* curve is a straight, horizontal line, indicating no change.

Saturation Point
The point beyond which an increase in one of two quantities produces no increase in the other.

Saturation can easily be described by a simple analogy of a sponge. A dry sponge can only soak up a certain amount of water. As it continues to absorb water, a point will be reached where it will have soaked up the maximum amount of water possible. At this point, the sponge is said to be saturated with water, and no matter how much extra water you supply, it cannot hold any more.

The electromagnet is saturated at stage 3 and cannot produce any more magnetic flux, even though more magnetizing force is supplied as the sine wave continues on to its maximum positive level.

Looking at these three stages and the *B–H* curve that is produced, you can see that, in fact, the magnetization (setting up of the magnetic field, B) lags the magnetizing force (H) because of molecular friction. This lag or time difference between magnetizing force (H) and flux density (B) is known as **hysteresis.**

Hysteresis
A lag between cause and effect. With magnetism it is the amount that the magnetization of a material lags the magnetizing force due to molecular friction.

Figure 8-5(a) illustrates what is known as a *hysteresis loop,* which is formed when you plot magnetizing force (H) against flux density (B) through a complete cycle of alternating current, as seen in Figure 8-5(b). Initially, when the electric circuit switch is open, the iron core is unmagnetized. Therefore, both H and B are zero at point a. When the switch is closed, as seen in Figure 8-5(c), the current [Figure 8-5(b)] is increased and flux density [Figure 8-5(a)] increases until saturation point b is reached. This part of the waveform (from point a to b) is exactly the same as the *B–H* curve, discussed previously. The current continues on beyond saturation point b, but the flux density cannot increase beyond saturation.

At point c, the magnetizing force (current) is zero and B (flux density) falls to a value c that is the positive magnetic flux remaining after the removal of the magnetizing force (H). This particular value of flux density is termed **remanence** or **retentivity.**

Remanence or Retentivity
Amount a material remains magnetized after the magnetizing force has been removed.

The current or magnetizing force now reverses, and the amount of current in the reverse direction that causes flux density to be brought down from c to zero (point d) after the core has been saturated is termed the **coercive force.** The current, and therefore magnetizing force, continues on toward a maximum negative until saturation in the opposite magnetic polarity occurs at point e.

Coercive Force
Magnetizing force needed to reduce the residual magnetism within a material to zero.

At point f, the magnetizing force (current) is zero and B falls to remanence, which is the negative magnetic flux remaining after the removal of the magnetizing force H. The value of current between f and g is the coercive force needed in the reverse direction to bring flux density down to zero (point g).

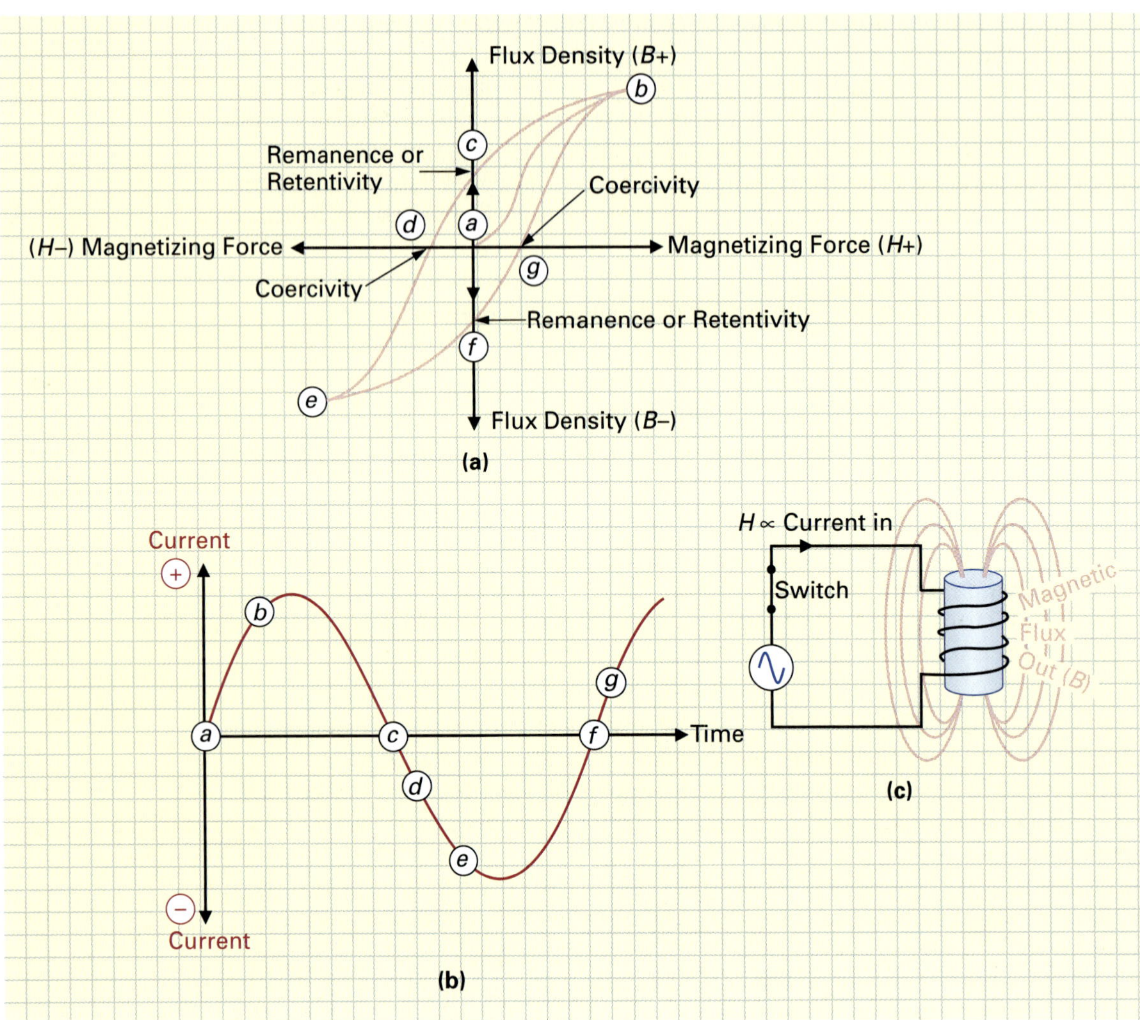

FIGURE 8-5 ***B–H*** **Hysteresis Loop.**

8-1-2 *Applications of Electromagnetism*

Relay

Relay
Electromechanical device that opens or closes contacts when a current is passed through a coil.

Energized
Being electrically connected to a voltage source so that the device is activated.

A **relay** is an electromechanical device that either makes (closes) or breaks (opens) a circuit by moving contacts together or apart. Figure 8–6(a) shows the normally open (NO) relay and Figure 8-6(b) shows the normally closed (NC) relay.

Operation. In both cases, the relay consists basically of an electromagnet connected to lines *x* and *y*, a movable iron arm known as the armature, and a set of contacts. When current passes from *x* to *y*, the electromagnet generates a magnetic field, or is said to be **energized,** which attracts the armature toward the electromagnet. When this occurs, it closes or makes the normally open relay contacts and opens or breaks the normally closed relay contacts.

If the electromagnet is deenergized by discontinuing the current through the coil, the spring will pull back the armature to open the NO relay contacts or close the NC relay contacts between *A* and *B*.

The "normal" condition for the contacts between *A* and *B* is when the electromagnet is deenergized. In the deenergized condition, the normally open relay contacts are open and the normally closed relay contacts are closed.

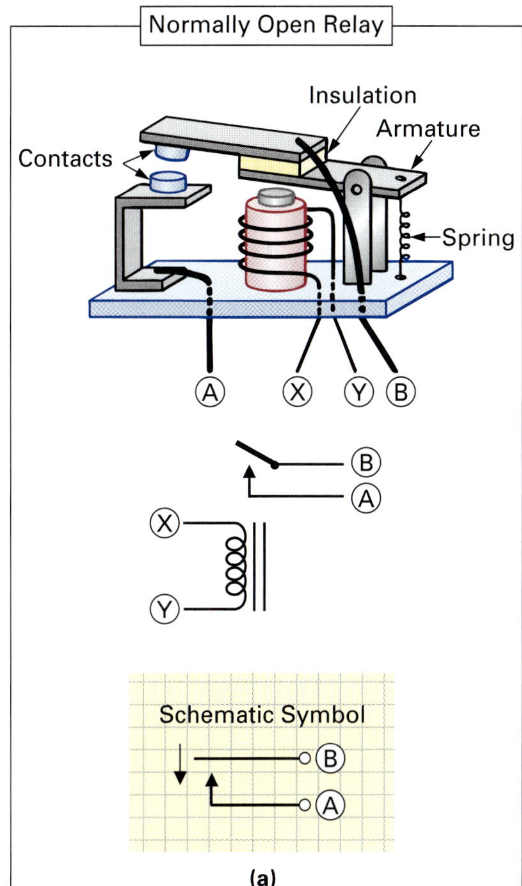

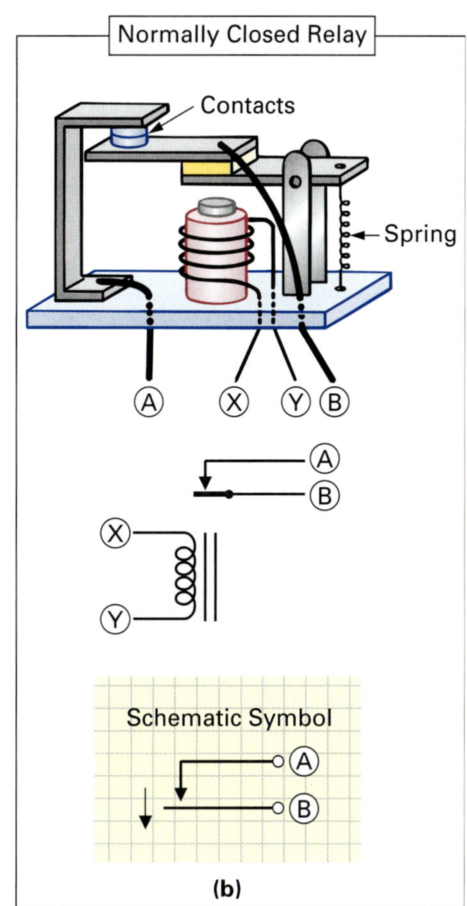

FIGURE 8-6 Relays. (a) Single-Pole, Single-Throw (SPST), Normally Open (NO) Relay (Contacts Are Open until Activated). (b) Single-Pole, Single-Throw (SPST), Normally Closed (NC) Relay (Contacts Are Closed until Activated).

The two relays discussed so far are actually single-pole, single-throw relays, as they have one movable contact (single pole) and one stationary contact that the pole can be thrown to, as shown in Figure 8-6. There are actually four basic configurations for relays and all are illustrated in Table 8-1(a). Variations on these basic four can come in all shapes and sizes, with one relay controlling sometimes several sets of contacts. Table 8-1(b) shows several different styles and packages of relays that are available.

Applications of a relay. The relay is generally used in two basic applications:

1. To enable one master switch to operate several remote or difficultly placed contact switches, as illustrated in Figure 8-7. When the master switch is closed, the relay is energized, closing all its contacts and turning on all the lights. The advantage of this is twofold in that, first, the master switch can turn on three lights at one time, which saves time for the operator, and second, only one set of wires need be taken from the master switch to the lights, rather than three sets for all three lights.
2. The second basic application of the relay is to enable a switch in a low-voltage circuit to operate relay contacts in a high-voltage circuit, as shown in Figure 8-8. The operator activates the switch in the safer low-voltage circuit, which will energize the relay, closing its contacts and connecting the more dangerous high voltage to the motor.

TABLE 8-1 Relay Types

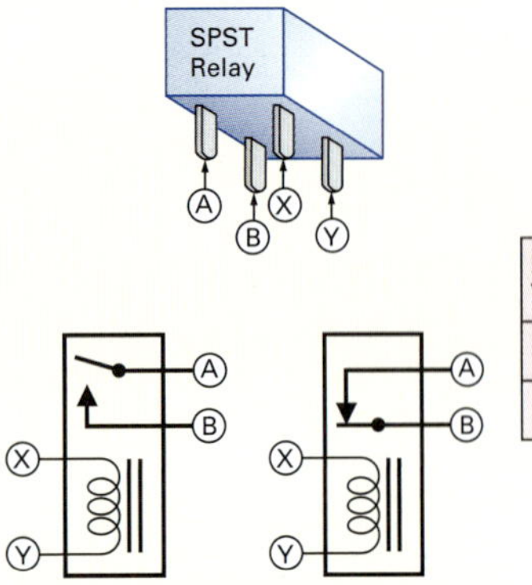

SPST Relay

Relay Type	De-energized	Energized
NO	A to B Open	A to B Closed
NC	A to B Closed	A to B Open

Normally Open (NO) Normally Closed (NC)

Single-Pole, Single-Throw (SPST)

A relay with only one moving and one stationary contact. Available as either normally open (NO) or normally closed (NC).

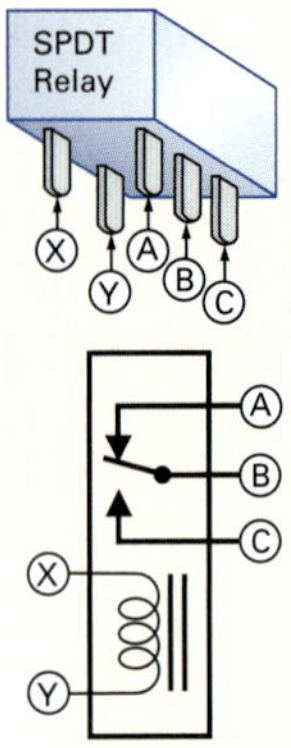

SPDT Relay

Contacts	De-energized	Energized
A to B	Closed	Open
B to C	Open	Closed

Single-Pole, Double-Throw (SPDT)

A relay with one moving and two stationary contacts. One set of contacts is normally closed (A and B), while the other set is normally open (B and C).

Single pole: one moving contact. *Double pole:* two moving contacts. *Single throw:* pole can be thrown or cast in only one direction. *Double throw:* pole can be thrown or cast in one of two directions.

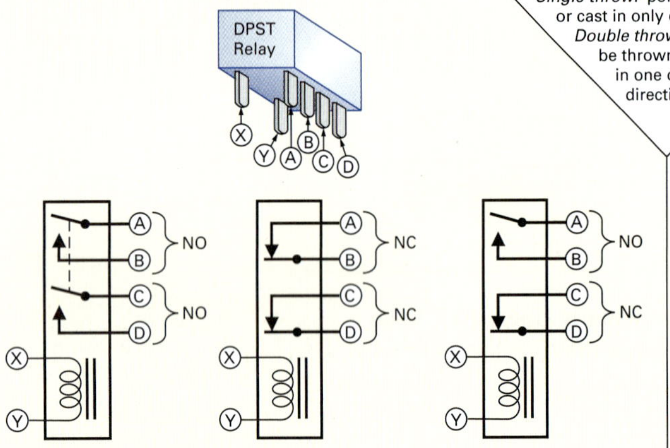

2 Normally Open   2 Normally Closed   1 Normally Open/1 Normally Closed

Double-Pole, Single-Throw (DPST)

A relay with two moving poles or armatures and two stationary contacts. Available as either both sets normally closed, both normally open, or one set normally open and the other set normally closed.

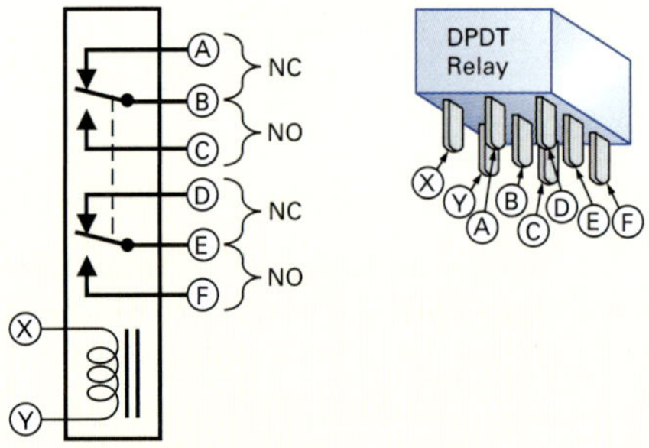

Double-Pole, Double-Throw (DPDT)

A relay with two moving contacts and four stationary contacts. Two sets of contacts are normally closed (A and B/D and E), while the other two sets are normally open (B and C/E and F).

(a)

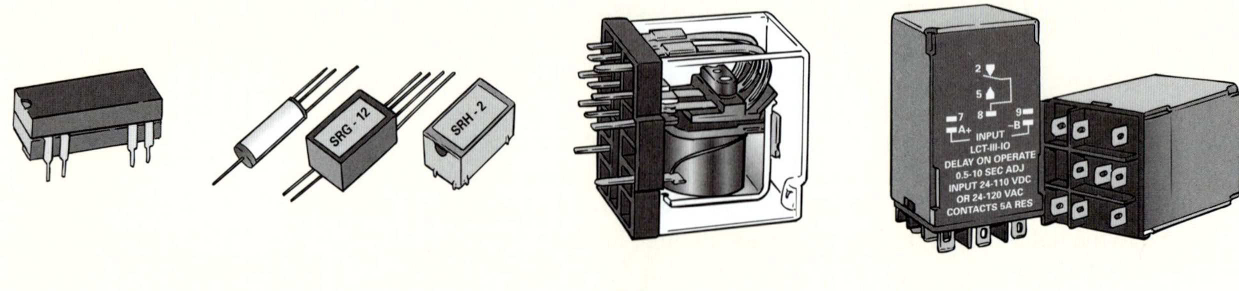

(b)

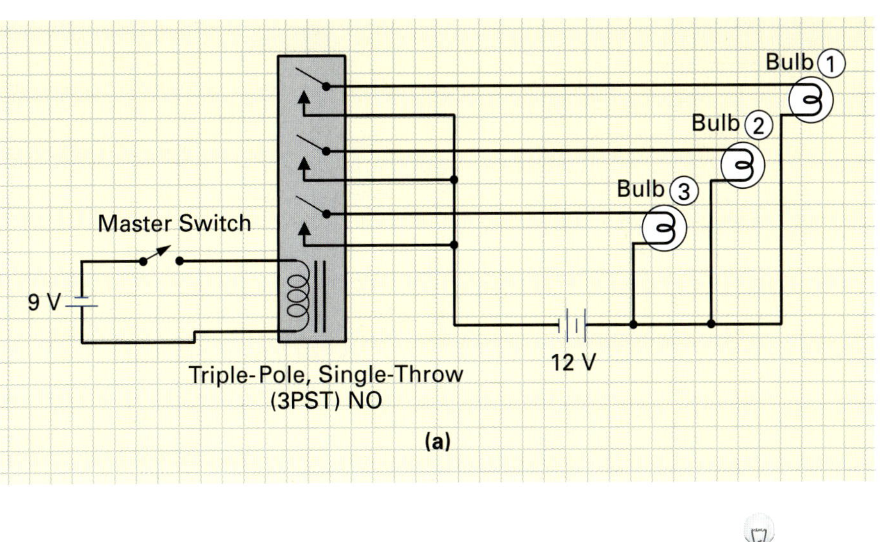

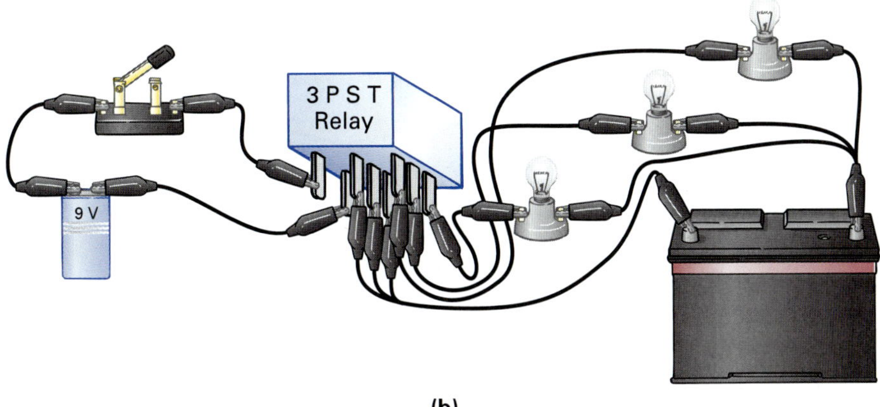

FIGURE 8-7 One Master Switch Operating Several Remotes. (a) Schematic. (b) Pictorial.

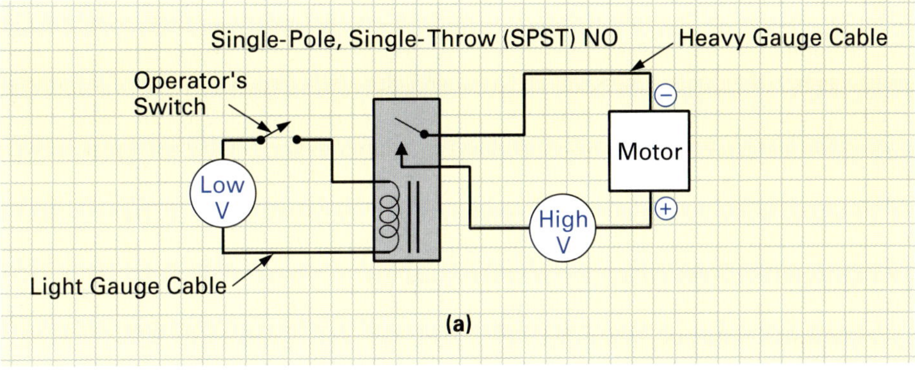

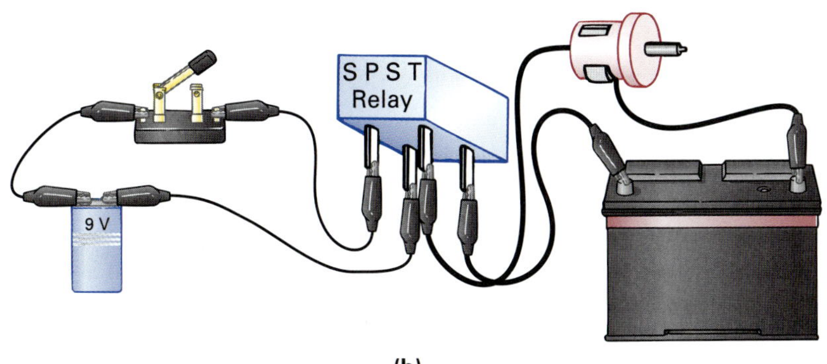

FIGURE 8-8 Low-Current Switch Enabling a High-Current Circuit. (a) Schematic. (b) Pictorial.

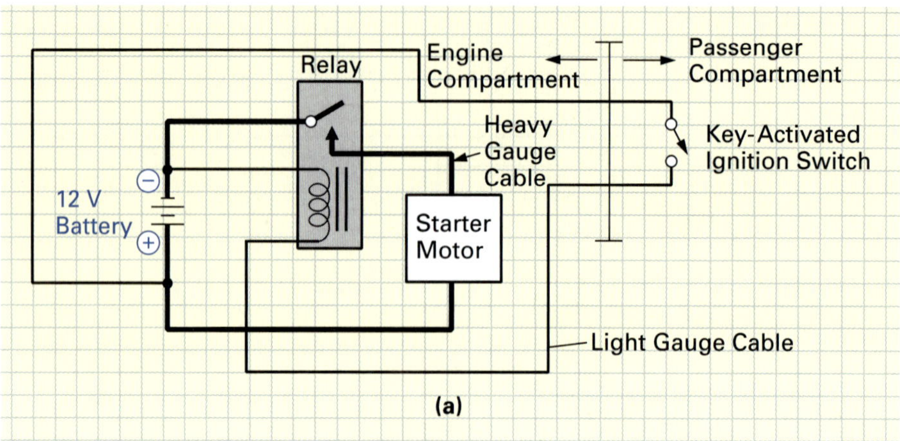

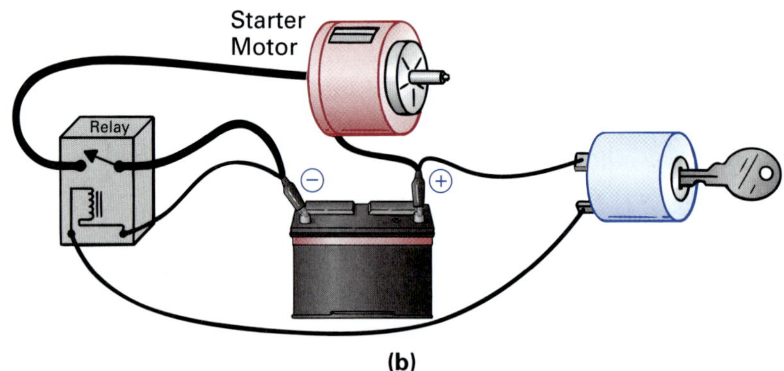

FIGURE 8-9 Automobile Starter Circuit.

As an application, Figure 8-9 shows an automobile starter circuit. In this circuit a relay is being used to supply the large dc current needed to activate a starter motor in an automobile. When the ignition switch is engaged in the passenger compartment by the driver, current flows through a light-gauge wire from the negative side of the battery, through the relay's electromagnet, through the ignition switch, and back to the positive side of the battery. This current flow through the electromagnet of the relay energizes the relay and closes the relay's contacts. Closing the relay's contacts makes a path for the current to flow through the heavy-gauge cable from the negative side of the battery, through the relay contacts and starter motor, and back to the positive side of the battery. The starter motor's output shaft spins the engine, causing it to start.

This application is a perfect example of how a relay can be used to close contacts in a heavy-current (heavy-gauge cable) circuit, while the driver has only to close contacts in a small-current (light-gauge cable) circuit.

If the relay were omitted, the driver's ignition switch would have to be used to connect the 12 V and large current to the starter motor. This would mean that:

1. Heavy-gauge, expensive cable would need to be connected in a longer path between starter motor and passenger compartment.
2. The ignition switch would need to be larger to handle the heavier current.
3. The driver would be in closer proximity to a more dangerous, high-current circuit.

Solenoid-Type Electromagnet

Solenoid
Coil and movable iron core that when energized by an alternating or direct current will pull the core into a central position.

Up to now, electromagnets have been used to close or open a set, or sets, of contacts to either make or break a current path. These electromagnets have used a stationary soft-iron core. Some electromagnets are constructed with movable iron cores, as shown in Figure 8-10, which can be used to open or block the passage of a gas or liquid through a valve. These are known as **solenoid**-type electromagnets.

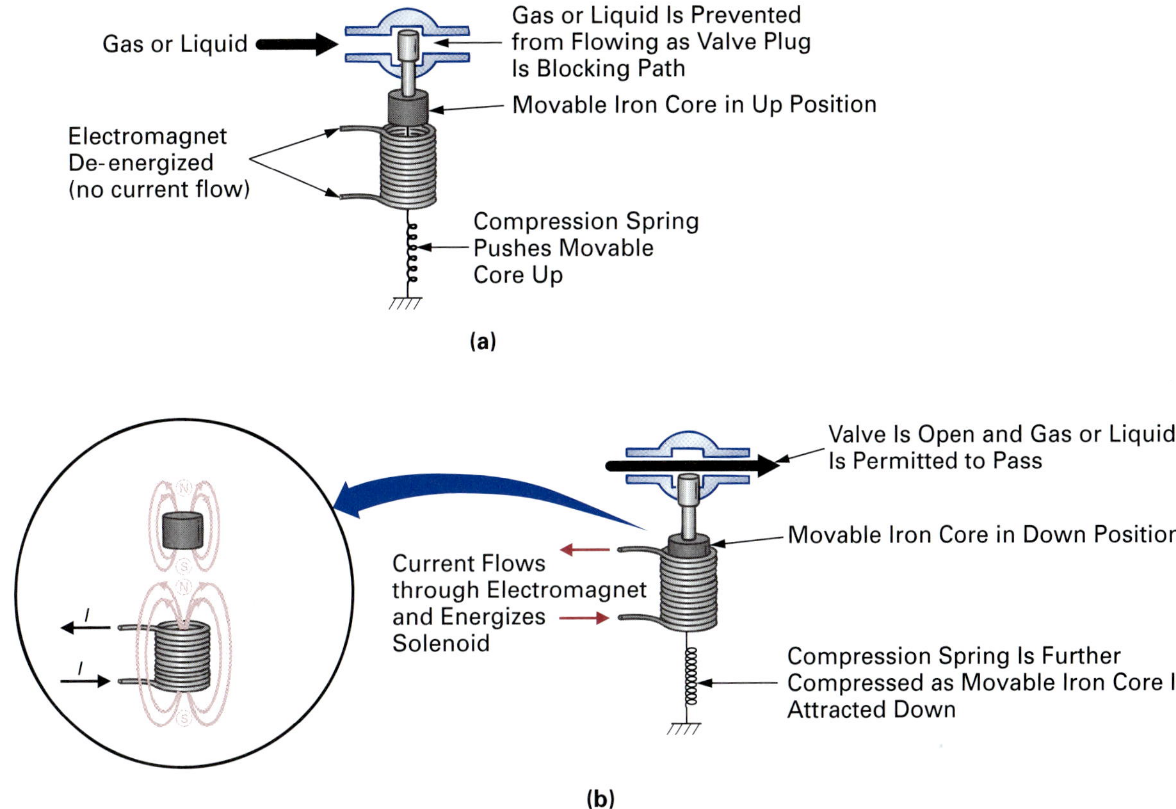

FIGURE 8-10 Solenoid-Type Electromagnet. (a) Deenergized. (b) Energized.

When no current is flowing through the solenoid coil, no magnetic field is generated, so no magnetic force is exerted on the movable iron core, as shown in Figure 8-10(a), and therefore the compression spring maintains it in the up position, with the valve plug on the end of the core preventing the passage of either a liquid or gas through the valve (valve closed).

When a current flows through the electromagnet, the solenoid coil is energized, creating a magnetic field, as shown in Figure 8-10(b). Due to the influence of the coil's magnetic field, the movable soft-iron core will itself generate a magnetic field, as seen in the inset in Figure 8-10(b). This condition will create a north pole at the top of the solenoid coil and a south pole at the bottom of the movable core, and the resulting attraction will pull down the core (which is free to slide up and down), pulling with it the valve plug and opening the valve.

Solenoid-type electromagnets are actually constructed with the core partially in the coil and are used in washing machines to control water and in furnaces to control gas.

Circuit Breaker

In many appliances and in a home electrical system, **circuit breakers** are used in place of fuses to prevent damaging current. A circuit breaker can open a current-carrying circuit without damaging itself, then be manually reset and used repeatedly, unlike a fuse, which must be replaced when it blows. We can say, therefore, that a circuit breaker is a reusable fuse. Figure 8–11 illustrates a circuit breaker's schematic symbol, physical appearance, and a typical application in which it is used to protect a television.

Circuit Breaker
Reusable fuse. This device will open a current-carrying path without damaging itself once the current value exceeds its maximum current rating.

The construction of a magnetic circuit breaker is shown in Figure 8-12, and it will operate as follows. A small level of current flowing through the coil of the electromagnet will provide only a small amount of magnetic pull to the left on the iron arm. This magnetic force cannot overcome the pull to the right being generated by spring *A*. This safe value of current will therefore be allowed to pass from the *A* terminal, through the coil to the top contact, out of the bottom contact, and then exit the breaker at terminal *B*.

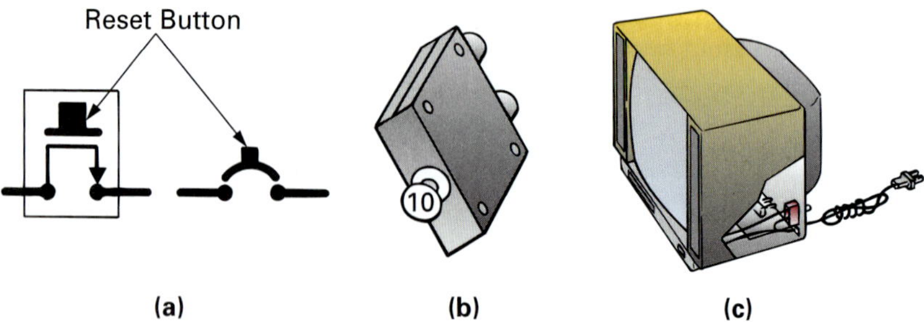

FIGURE 8-11 **Circuit Breaker. (a) Symbol. (b) Appearance. (c) Typical Application.**

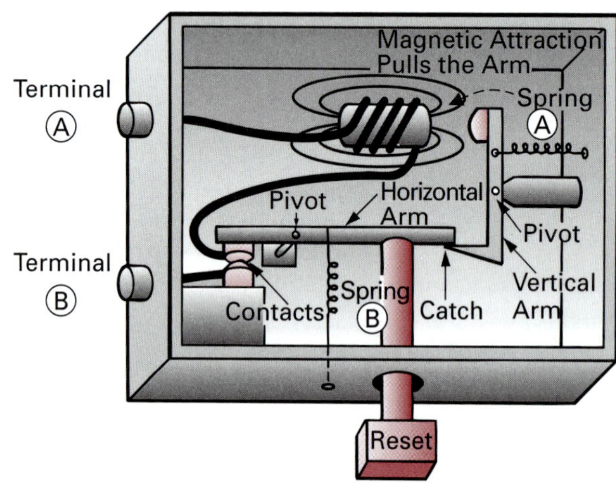

FIGURE 8-12 **Magnetic-Type Circuit Breaker.**

If the current exceeds the current rating of the circuit breaker, an increase in current through the coil of the electromagnet generates a greater magnetic force on the vertical arm, which pulls the top half of the vertical arm to the left and the lower half below the pivot to the right. This action releases the catch holding the horizontal arm, allowing spring *B* to pull the right side of the lower arm down, opening the contacts and disconnecting or tripping the breaker. The reset button must now be pressed to close the contacts. If, however, the problem still exists, the breaker will continue to trip, as the excessive circuit current still exists.

SELF-TEST EVALUATION POINT FOR SECTION 8-1

Use the following questions to test your understanding of Section 8-1.

1. True or false: The greater the number of loops in an electromagnet, the greater or stronger the magnetic field.
2. What form of current flow produces a magnetic field that maintains a fixed magnetic polarity?
3. What form of current flow produces a magnetic field that continuously switches in polarity?
4. List two applications of the electromagnet.
5. What is the difference between an NO and an NC relay?

8-2 ELECTROMAGNETIC INDUCTION

Electromagnetic Induction
The voltage produced in a coil due to relative motion between the coil and magnetic lines of force.

Electromagnetic induction is the name given to the action that causes electrons to flow within a conductor when that conductor is moved through a magnetic field. Stated another way, electromagnetic induction is the voltage or emf induced or produced in a coil as the magnetic lines of force link with the turns of a coil.

8-2-1 *Faraday's Law*

In 1831, Michael Faraday carried out an experiment in which he used a coil, a zero center ammeter (galvanometer), and a bar permanent magnet. Faraday's discoveries, collectively known as **Faraday's Law,** illustrated in Figure 8-13(a) through (f), are that:

Faraday's Law
When a magnetic field cuts a conductor, or when a conductor cuts a magnetic field, an electric current will flow in the conductor if a closed path is provided over which the current can circulate.

a. When the magnet is moved into a coil the magnetic lines of flux cut the turns of the coil. This action that occurs when the magnetic lines of flux link with a conductor is known as *flux linkage.* Whenever flux linkage occurs, an emf is induced in the coil known as an induced voltage, which causes current to flow within the circuit and the meter to deflect in one direction, for example, to the right. Faraday discovered, in fact, that if the magnet was moved into the coil, or if the coil was moved over the magnet, an emf or voltage was induced within the coil. What actually occurs is that the electrons within the coil are pushed to one end of the coil by the magnetic field, creating an abundance of electrons at one end of the coil (negative charge) and an absence of electrons at the other end of the coil (positive charge). This potential difference across the coil will produce a current flow if a complete path for current (closed circuit) exists.

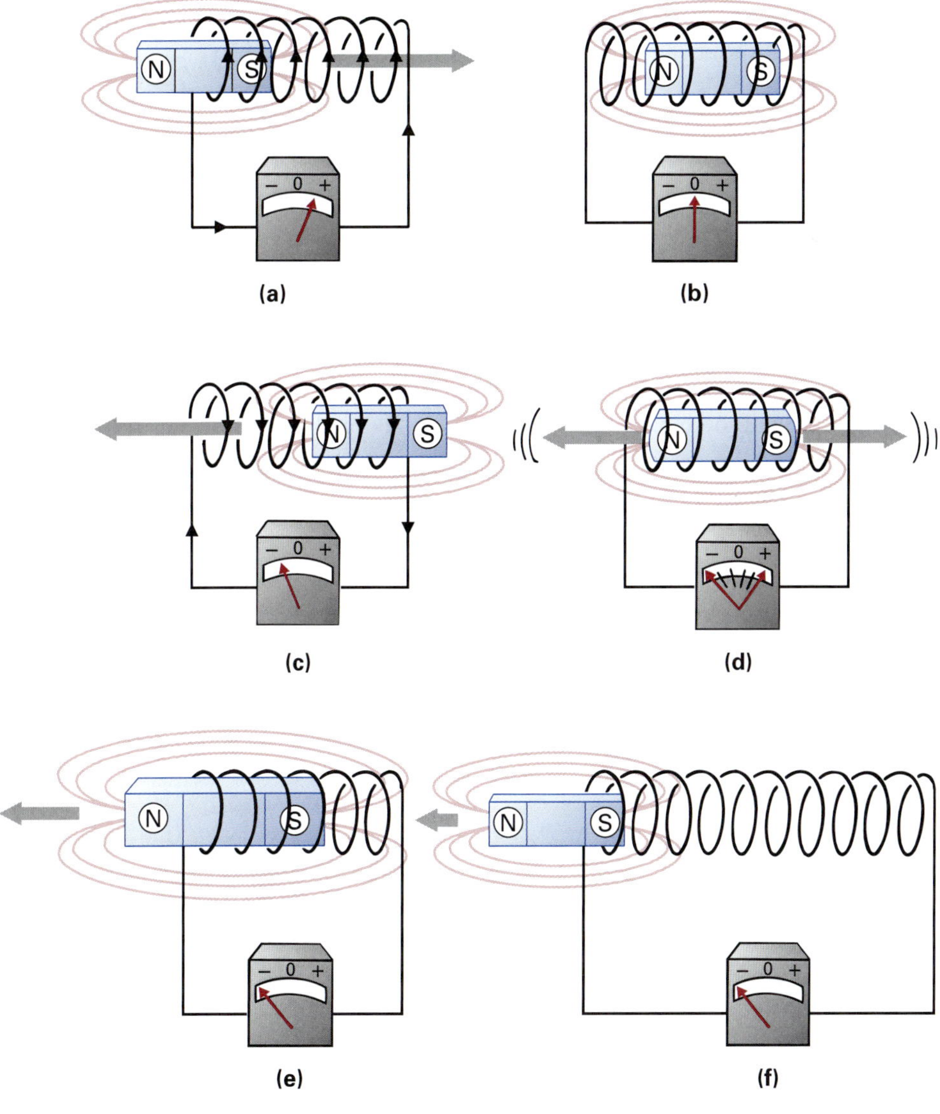

FIGURE 8-13 **Faraday's Electromagnetic Induction Discoveries.**

b. When the magnet is stationary within the coil, the magnetic lines are no longer cutting the turns of the coil, and so there is no induced voltage and the meter returns to zero.
c. When the magnet is pulled back out of the coil, a voltage is induced that causes current to flow in the opposite direction to that of (a) and the meter deflects in the opposite direction, for example, to the left.
d. If the magnet is moved into or out of the coil at a greater speed, the voltage induced also increases, and therefore so does current.
e. If the size of the magnet and therefore the magnetic flux strength are increased, the induced voltage also increases.
f. If the number of turns in the coil is increased, the induced voltage also increases.

TIME LINE

Eduard W. Weber (1804–1891), a German physicist, made enduring contributions to the modern system of electric units, and magnetic flux is measured in webers in honor of his work.

In summary, whenever there is relative motion or movement between the coil of a conductor and the magnetic lines of flux, a potential difference will be induced and this action is called *electromagnetic induction.* The magnitude of the induced voltage depends on the number of turns in the coil, rate of change of flux linkage, and flux density.

8-2-2 *The Weber*

Weber (Wb)

Unit of magnetic flux. One weber is the amount of flux that when linked with a single turn of wire for an interval of 1 second, will induce an electromotive force of 1 V.

Consideration of Faraday's law enables us to take a closer look at the unit of flux (ϕ). The **weber** is equal to 10^8 magnetic lines of force, and from the electromagnetic induction point of view, if 1 weber of magnetic flux cuts a conductor for a period of 1 second, a voltage of 1 volt will be induced.

8-2-3 *Application of Electromagnetic Induction*

Generator

Device used to convert a mechanical energy input into an electrical energy output.

Armature

Rotating or moving component of a magnetic circuit.

The ac **generator** or alternator is an example of a system that uses electromagnetic induction to generate electricity. From the large 700,000 kW power plant generators to the mobile units shown in Figure 8-14(b), the basic principle of operation is that the mechanical drive energy input will produce ac electrical energy out by means of electromagnetic induction.

A loop of conductor, known as an **armature,** is rotated continually through 360° by a mechanical drive. This armature resides within a magnetic field produced by a dc electromagnet. Voltage will be induced into the armature and will appear on slip rings, which are

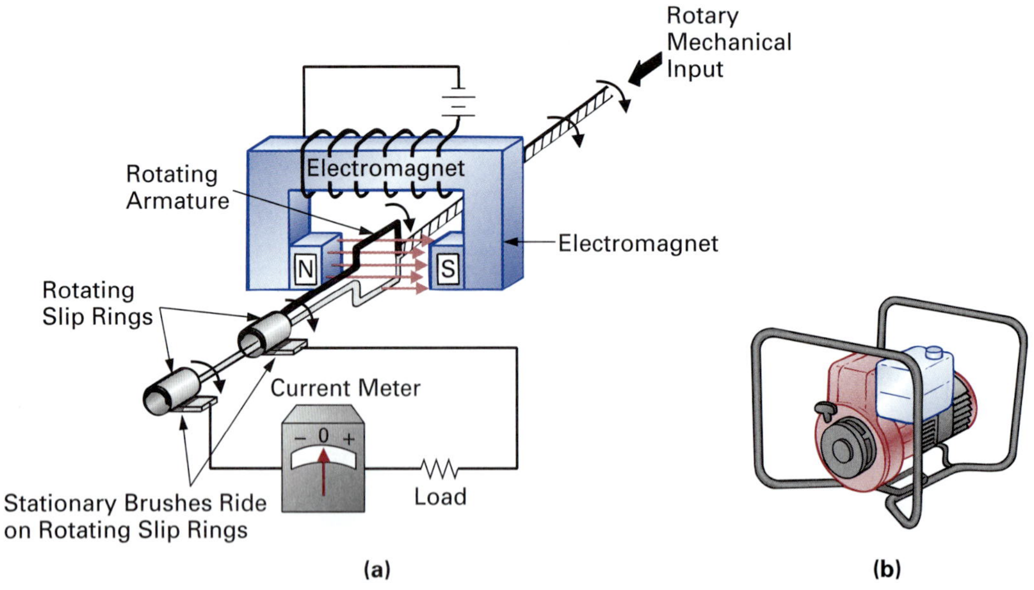

FIGURE 8-14 **Basic Generator Construction. (a) Construction. (b) Physical Appearance.**

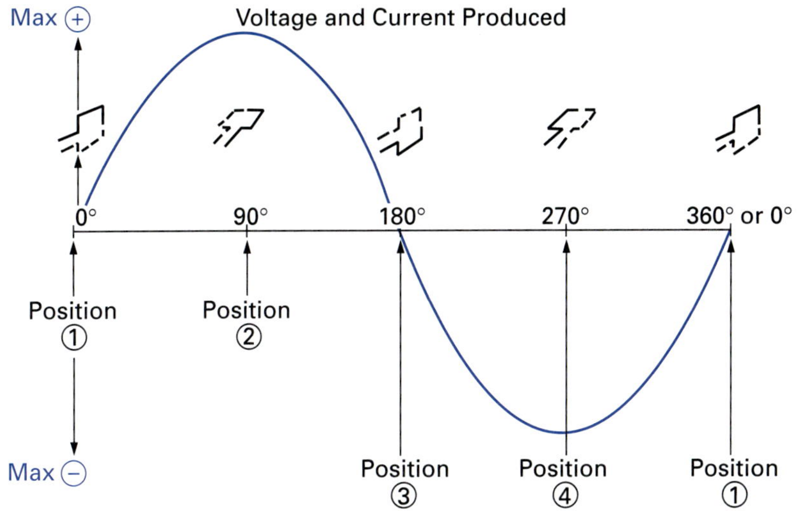

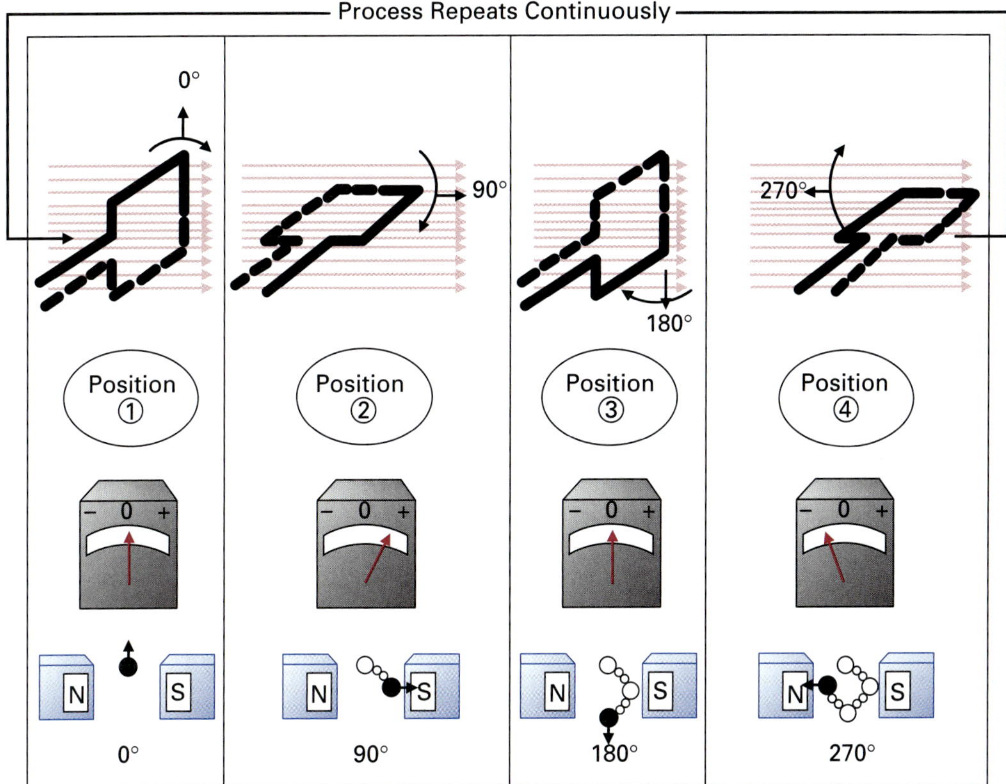

FIGURE 8-15 **360° Generator Operation.**

also being rotated. A set of stationary brushes rides on the rotating slip rings and picks off the generated voltage and applies this voltage across the load. This voltage will cause current to flow within the circuit and be indicated by the zero center ammeter.

Let's now take a closer look at the armature as it sweeps through 360°, or one complete revolution. Figure 8-15 illustrates four positions of the armature as it rotates through 360° in the clockwise direction.

Position 1: At this instant, the armature is in a position such that it does not cut any magnetic lines of force. The induced voltage in the armature conductor is equal to 0 V and there is no current flow through the circuit.

Position 2: As the conducting armature moves from position 1 to position 2, you can see that more and more magnetic lines of flux will be cut, and the induced emf in the armature (being coupled off by the brushes from the slip rings) will also increase to a maximum value. The current flow throughout the circuit will rise to a maximum as the voltage increases, and this can be seen by the zero center ammeter deflection to the right. From the ac waveform in Figure 8-15, you can see the sinusoidal increase from zero to a maximum positive as the armature is rotated from 0 to 90°.

Position 3: The armature continues its rotation from 90 to 180°, cutting through a maximum quantity and then fewer magnetic lines of force. The induced voltage decreases from the maximum positive at 90° to 0 V at 180°. At the 180° position, as with the 0° position, the armature is once again perpendicular to the magnetic field, and so no lines are cut and the induced voltage is equal to 0 V.

Position 4: From 180 to 270°, the armature is still moving in a clockwise direction, and as it travels toward 270°, it cuts more and more magnetic lines of force. The direction of the cutting action between 0 and 90° causes a positive induced voltage; and since the cutting position between 180 and 270° is the reverse, a negative induced voltage will result in the armature causing current flow in the opposite direction, as indicated by the deflection of the zero center ammeter to the left. The voltage induced when the armature is at 270° will be equal but opposite in polarity to the voltage generated when the armature was at the 90° position. The current will therefore also be equal in value but opposite in its direction of flow.

From position 4 (270°), the armature turns to the 360° or 0° position, which is equivalent to position 1, and the induced voltage decreases from maximum negative to zero. The cycle then repeats.

To summarize, in Figure 8-15, one complete revolution of the mechanical energy input causes one complete cycle of the ac electrical energy output, with a sine wave being generated as a result of circular motion.

SELF-TEST EVALUATION POINT FOR SECTION 8-2

Use the following questions to test your understanding of Section 8-2.

1. Define *electromagnetic induction.*
2. Briefly describe Faraday's law in relation to electromagnetic induction.
3. What waveform shape does the ac generator produce?

8-3 SELF-INDUCTION

In Figure 8-16 an external voltage source has been connected across a coil, forcing a current through the coil. This current will generate a magnetic field that will increase in field strength in a short time from zero to maximum, expanding from the center of the conductor (electromagnetism). The expanding magnetic lines of force have relative motion with respect to the stationary conductor, so an induced voltage results (electromagnetic induction). The blooming magnetic field generated by the conductor is actually causing a voltage to be induced in the conductor that is generating the magnetic field. This effect of a current-carrying coil of conductor inducing a voltage within itself is known as **self-inductance.** This phenomenon was first discovered by Heinrich Lenz, who observed that the induced voltage causes an induced (bucking) current to flow in the coil, which opposes the source current producing it.

Self-Inductance
The property that causes a counter-electromotive force to be produced in a conductor when the magnetic field expands or collapses with a change in current.

Figure 8-17(a) shows an inductor connected across a dc source. When the switch is closed, a circuit current will exist through the inductor and the resistor. As the current rises toward its maximum value, the magnetic field expands, and throughout this time of relative

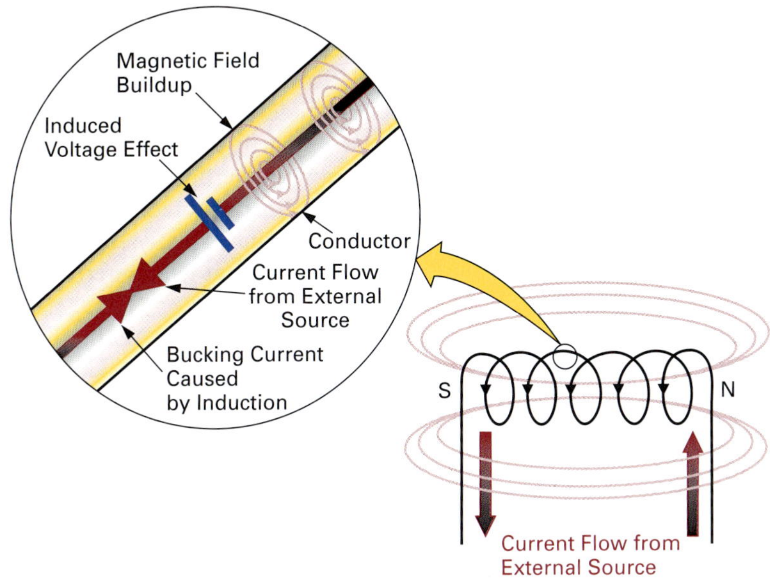

FIGURE 8-16 **Self-Induction of a Coil.**

motion between field and conductor, an induced voltage will be present. This induced voltage will produce an induced current to oppose the change in the circuit current.

When the current reaches its maximum, the magnetic field, which is dependent on current, will also reach a maximum value and then no longer expand but remain stationary. When the current remains constant, no change will occur in the magnetic field and therefore no relative motion will exist between the conductor and magnetic field, resulting in no induced voltage or current to oppose circuit current, as shown in Figure 8-17(b). The coil has accepted electrical energy and is storing it in the form of a magnetic energy field, just as the capacitor stored electrical energy in the form of an electric field.

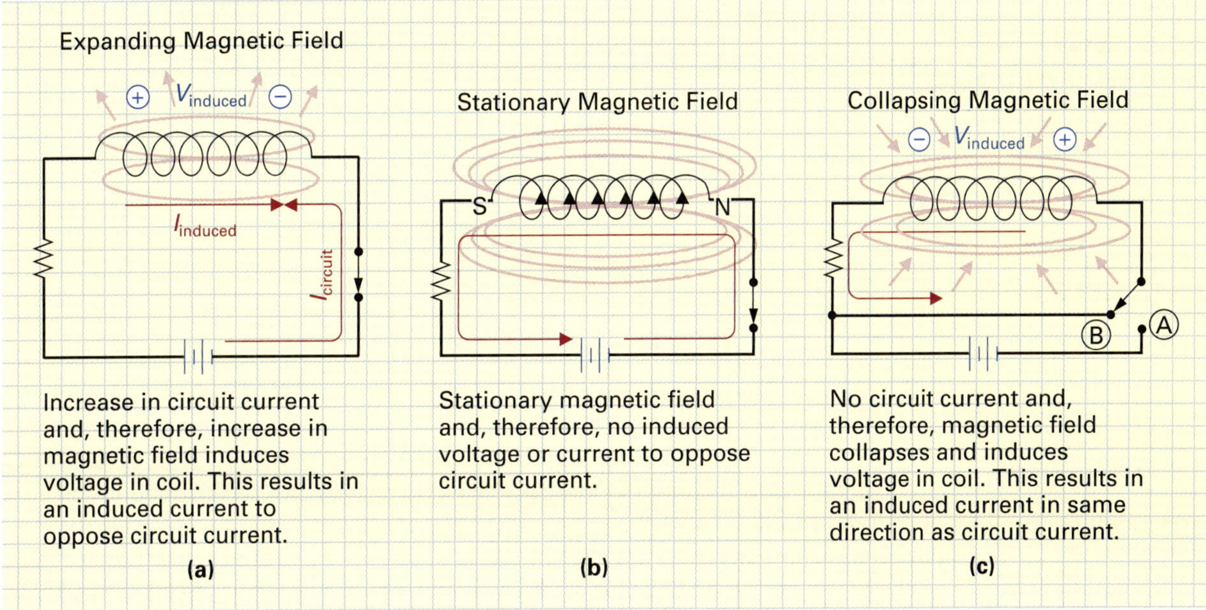

FIGURE 8-17 **Self-Inductance. (a) Switch Closed. (b) Constant Circuit Current. (c) Switch Opened.**

Counter emf (Counter Electromotive Force)

Abbreviated "counter emf," or "back emf," it is the voltage generated in an inductor due to an alternating or pulsating current and is always of opposite polarity to that of the applied voltage.

Inductance

Property of a circuit or component to oppose any change in current as the magnetic field produced by the change in current causes an induced countercurrent to oppose the original change.

Henry

Unit of inductance.

TIME LINE

Joseph Henry (1797–1878), an American physicist, conducted extensive studies into electromagnetism. Henry was the first to insulate the magnetic coil of wire and developed coils for telegraphy and motors. In recognition of his discovery of self-induction in 1832, the unit of inductance is called the henry.

If the switch is put in position *B,* as shown in Figure 8-17(c), the current from the battery will be zero, and the magnetic field will collapse as it no longer has circuit current to support it. As the magnetic lines of force collapse, they cut the conducting coils, causing relative motion between the conductor and magnetic field. A voltage is induced in the coil, which will produce an induced current to flow in the same direction as the circuit current was flowing before the switch was opened. The coil is now converting the magnetic field energy into electrical energy and returning the original energy that it stored.

After a short period of time, the magnetic field will have totally collapsed, the induced voltage will be zero, and the induced current within the circuit will therefore also no longer be present.

This induced voltage is called a **counter emf** or *back emf.* It opposes the applied emf (or battery voltage). The ability of a coil or conductor to induce or produce a counter emf within itself as a result of a change in current is called *self-inductance,* or more commonly **inductance** (symbolized L). The unit of inductance is the **henry** (H), named in honor of Joseph Henry, an American physicist, for his experimentation within this area of science. The inductance of an inductor is 1 henry when a current change of 1 ampere per second causes an induced voltage of 1 volt. Inductance is therefore a measure of how much counter emf (induced voltage) can be generated by an inductor for a given amount of current change through that same inductor.

This counter emf or induced voltage can be calculated by the formula

$$V_{ind} = L \times \frac{\Delta I}{\Delta t}$$

where L = inductance, in henrys (H)
ΔI = increment of change of current (I)
Δt = increment of change with respect to time (t)

A larger inductance ($L\uparrow$) will create a larger induced voltage ($V_{ind}\uparrow$), and if the rate of change of current with respect to time is increased ($\Delta I/\Delta t\uparrow$), the induced voltage or counter emf will also increase ($V_{ind}\uparrow$).

EXAMPLE:

What voltage is induced across an inductor of 4 H when the current is changing at a rate of:

a. 1 A/s?
b. 4 A/s?

Solution:

a. $V_{ind} = L \times \frac{\Delta I}{\Delta t} = 4\text{ H} \times 1\text{ A/s} = 4\text{ V}$

b. $V_{ind} = L \times \frac{\Delta I}{\Delta t} = 4\text{ H} \times 4\text{ A/s} = 16\text{ V}$

The faster the coil current changes, the larger the induced voltage.

Inductor

Coil of conductor used to introduce inductance into a circuit.

An **inductor,** therefore, is basically an electromagnet, as its construction and principle of operation are the same. We use the two different names because they have different applications. The purpose of the electromagnet or solenoid is to generate a magnetic field, while the purpose of an inductor or coil is to oppose any change of circuit current.

In Figure 8-18 a steady value of direct current is present within the circuit and the inductor is creating a steady or stationary magnetic field. If the current in the circuit is suddenly increased (by lowering the circuit resistance), the change in the expanding magnetic field

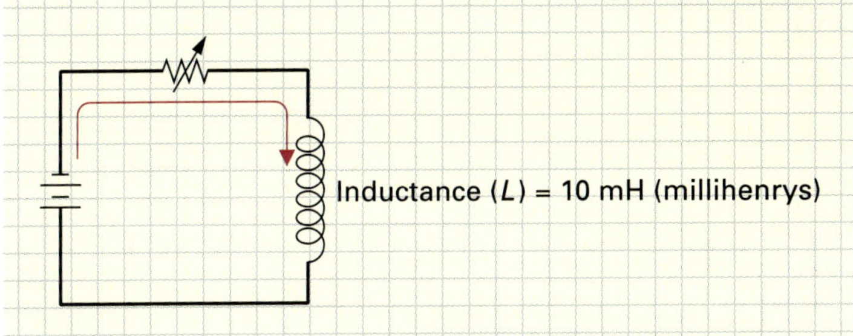

FIGURE 8-18 Inductor's Ability to Oppose Current Change.

will induce a counter emf within the inductor. This induced voltage will oppose the source voltage from the battery and attempt to hold current at its previous low level.

The counter emf cannot completely oppose the current increase, for if it did, the lack of current change would reduce the counter emf to zero. Current therefore incrementally increases up to a new maximum, which is determined by the applied voltage and the circuit resistance ($I = V/R$). Once the new higher level of current has been reached and remains constant, there will no longer be a change. This lack of relative motion between field and conductor will no longer generate a counter emf, so the current will remain at its new higher constant value.

This effect also happens in the opposite respect. If current decreases (by increasing circuit resistance), the magnetic lines of force will collapse because of the reduction of current and induce a voltage in the inductor, which will produce an induced current in the same direction as the circuit current. These two combine and tend to maintain the current at the higher previous constant level. Circuit current will fall, however, as the induced voltage and current are only present during the change (in this case the decrease from the higher current level to the lower); and once the new lower level of current has been reached and remains constant, the lack of change will no longer induce a voltage or current. So the current will then remain at its new lower constant value.

The inductor is therefore an electronic component that will oppose any changes in circuit current, and this ability or behavior is referred to as *inductance.* Since the current change is opposed by a counter emf, inductance may also be defined as the ability of a device to induce a counter emf within itself for a change in current.

8-3-1 *Factors Determining Inductance*

The inductance of an inductor is determined by four factors:

1. Number of turns
2. Area of the coil
3. Length of the coil
4. Core material used within the coil

All the four factors listed can be placed in a formula to calculate inductance.

$$L = \frac{N^2 \times A \times \mu}{l}$$

where L = inductance, in henrys (H)
N = number of turns
A = cross-sectional area, in square meters (m^2)
μ = permeability
l = length of core, in meters (m)

EXAMPLE:

Refer to Figure 8-19(a) and (b) and calculate the inductance of each.

Solution:

$$\text{a. } L = \frac{5^2 \times 0.01 \times (6.28 \times 10^{-5})}{0.001} = 15.7 \text{ mH}$$

$$\text{b. } L = \frac{10^2 \times 0.1 \times (1.1 \times 10^{-4})}{0.1} = 11 \text{ mH}$$

TABLE 8-2 Permeabilities of Various Materials

MATERIAL	PERMEABILITY (μ)
Air or vacuum	1.26×10^{-6}
Nickel	6.28×10^{-5}
Cobalt	7.56×10^{-5}
Cast iron	1.1×10^{-4}
Machine steel	5.65×10^{-4}
Transformer iron	6.9×10^{-3}
Silicon iron	8.8×10^{-3}
Permalloy	0.126
Superalloy	1.26

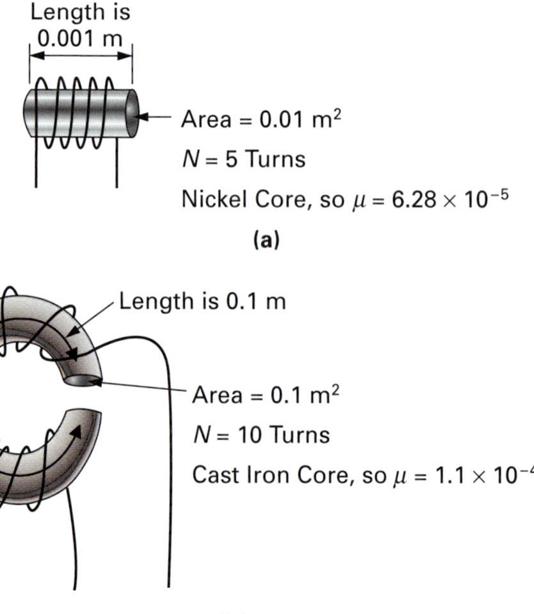

FIGURE 8-19 Inductor Examples.

8-3-2 *Inductors in Combination*

Inductors oppose the change of current in a circuit and so are treated in a manner similar to resistors connected in combination. Two or more inductors in series merely extend the coil length and increase inductance. Inductors in parallel are treated in a manner similar to resistors, with the total inductance being less than that of the smallest inductor's value.

1. Inductors in Series

When inductors are connected in series with one another, the total inductance is calculated by summing all the individual inductances.

$$L_T = L_1 + L_2 + L_3 + \cdots$$

EXAMPLE:

Calculate the total inductance of the circuit shown in Figure 8-20.

Solution:

$$\begin{aligned} L_T &= L_1 + L_2 + L_3 \\ &= 5 \text{ mH} + 7 \text{ mH} + 10 \text{ mH} \\ &= 22 \text{ mH} \end{aligned}$$

FIGURE 8-20 Inductors in Series.

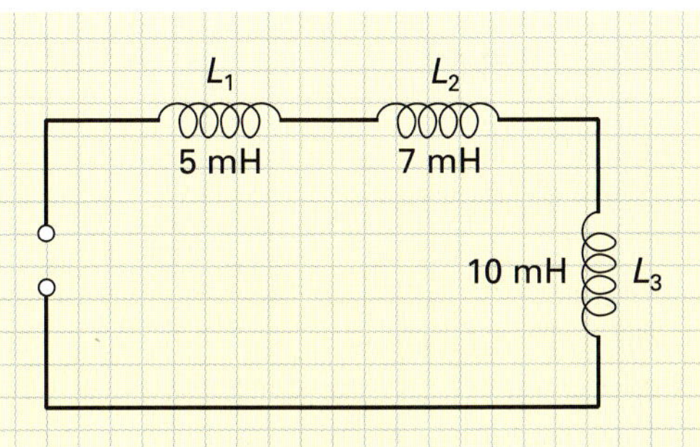

2. Inductors in Parallel

When inductors are connected in parallel with one another, the reciprocal (two or more inductors) or product-over-sum (two inductors) formula can be used to find total inductance, which will always be less than the smallest inductor's value.

$$L_T = \frac{1}{(1/L_1) + (1/L_2) + (1/L_3) + \cdots}$$

$$L_T = \frac{L_1 \times L_2}{L_1 + L_2}$$

EXAMPLE:

Determine L_T for the circuits in Figure 8-21(a) and (b).

Solution:

a. Reciprocal formula:

$$\begin{aligned} L_T &= \frac{1}{(1/L_1) + (1/L_2) + (1/L_3)} \\ &= \frac{1}{(1/10 \text{ mH}) + (1/5 \text{ mH}) + (1/20 \text{ mH})} \\ &= 2.9 \text{ mH} \end{aligned}$$

b. Product over sum:

$$\begin{aligned} L_T &= \frac{L_1 \times L_2}{L_1 + L_2} \\ &= \frac{10\ \mu\text{H} \times 2\ \mu\text{H}}{10\ \mu\text{H} + 2\ \mu\text{H}} \\ &= \frac{20 \times 10^{-12}\ \text{H}^2}{12\ \mu\text{H}} = 1.67\ \mu\text{H} \end{aligned}$$

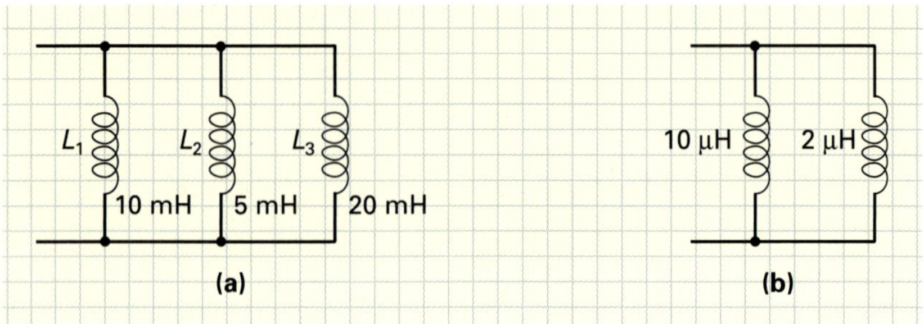

FIGURE 8-21 **Inductors in Parallel.**

8-3-3 *Types of Inductors*

As with resistors and capacitors, inductors are basically divided into the two categories of fixed-value and variable-value inductors, as shown by the symbols in Figure 8-22(a) and (b). Within these two categories, inductors are generally classified by the type of core material used. Figure 8-22(c) shows a variety of different types.

8-3-4 *Inductive Time Constant*

Inductors will not have any effect on a steady value of direct current (dc) from a dc voltage source. If, however, the dc is changing (pulsating), the inductor will oppose the change

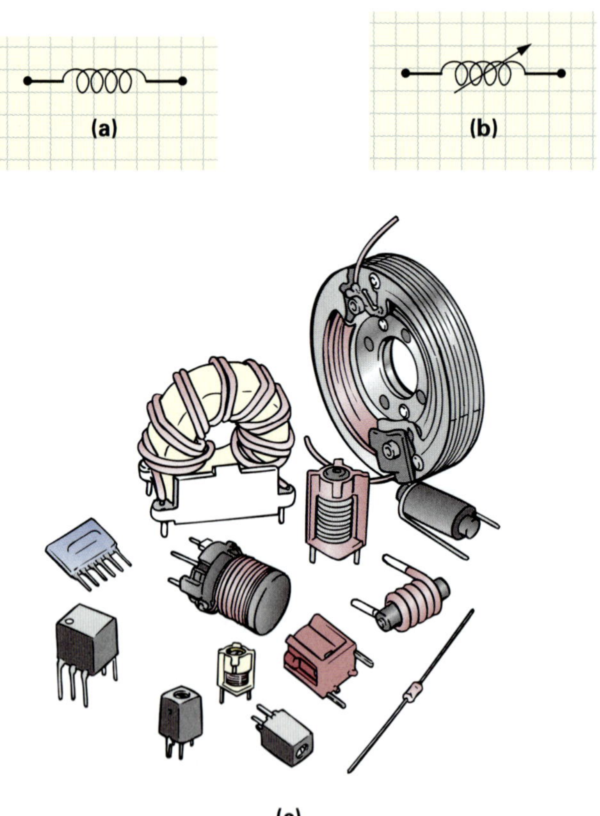

FIGURE 8-22 **Inductor Types. (a) Fixed-Value Inductor Symbol. (b) Variable-Value Inductor Symbol. (c) Physical Appearance of Inductors.**

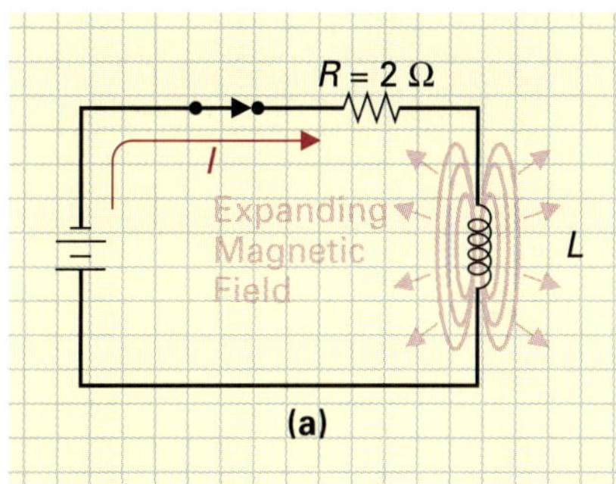

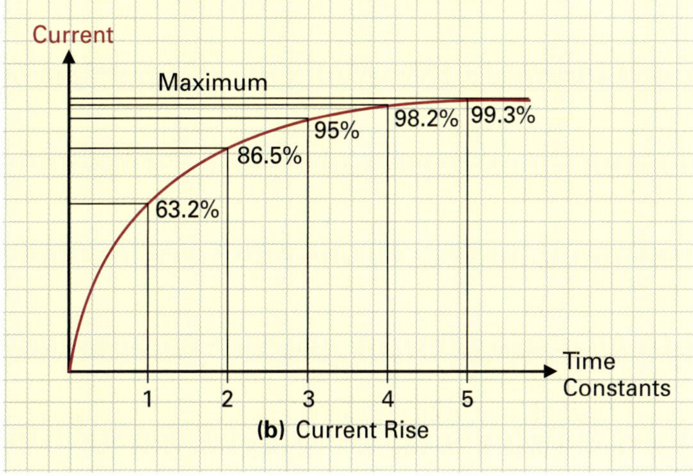

FIGURE 8-23 **DC Inductor Current Rise.**

whether it is an increase or decrease in direct current, because a change in current causes the magnetic field to expand or contract, and in so doing it will cut the coil of the inductor and induce a voltage that will counter the applied emf.

DC Current Rise

Figure 8-23(a) illustrates an inductor (L) connected across a dc source (battery) through a switch and series-connected resistor. When the switch is closed, current will flow and the magnetic field will begin to expand around the inductor. This field cuts the coils of the inductor and induces a counter emf to oppose the rise in current. Current in an inductive circuit, therefore, cannot rise instantly to its maximum value, which is determined by Ohm's law ($I = V/R$). Current will in fact take a time to rise to maximum, as graphed in Figure 8-23(b), due to the inductor's ability to oppose change.

It will actually take five time constants (5τ) for the current in an inductive circuit to reach maximum value. This time can be calculated by using the formula

$$\tau = \frac{L}{R} \text{ seconds}$$

The constant to remember is the same as before: 63.2%. In one time constant ($1 \times L/R$) the current in the RL circuit will have reached 63.2% of its maximum value. In two time constants ($2 \times L/R$), the current will have increased 63.2% of the remaining current, and so on, through five time constants.

For example, if the maximum possible circuit current is 100 mA and an inductor of 4 H is connected in series with a resistor of 2 Ω, the current will increase as shown in Figure 8-24.

Referring back to the $\tau = L/R$ formula, you will notice that how quickly an inductor will allow the current to rise to its maximum value is proportional to the inductance and

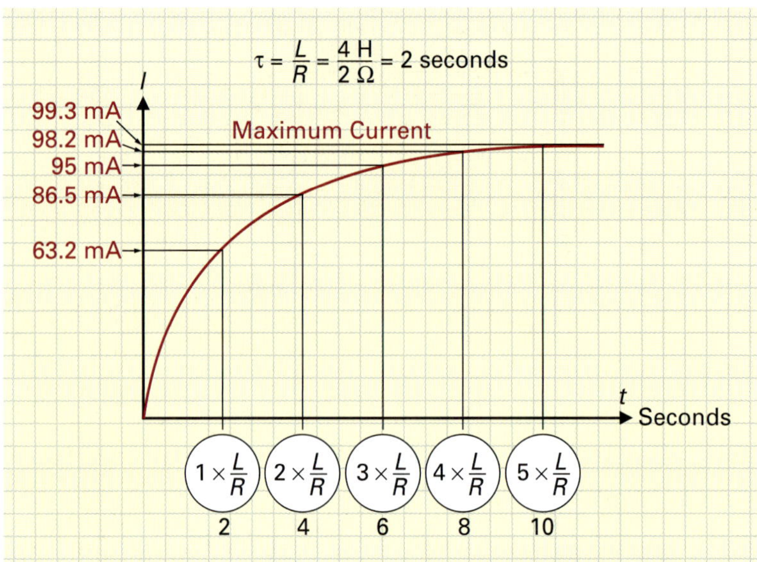

FIGURE 8-24 Exponential Current Rise.

inversely proportional to the resistance. A larger inductance increases the strength of the magnetic field, so the opposition or counter emf increases, and the longer it takes for current to rise to a maximum ($\tau\uparrow = L\uparrow/R$). If the circuit resistance is increased, the maximum current will be smaller, and a smaller maximum is reached more quickly than a higher ($\tau\downarrow = L/R\uparrow$).

DC Current Fall

When the inductor's dc source of current is removed, as shown in Figure 8-25(a) by placing the switch in position *B,* the magnetic field will collapse and cut the coils of the inductor, inducing a voltage and causing a current to flow in the same direction as the original source current. This current will exponentially decay, or fall from the maximum to zero level, in five time constants ($5 \times L/R = 5 \times 4/2 = 10$ seconds), as shown in Figure 8-25(b).

EXAMPLE:

Calculate the circuit current at each of the five time constants if a 12 V dc source is connected across a series *RL* circuit, and $R = 60\ \Omega$ and $L = 24$ mH. Plot the results on a graph showing current against time.

Solution:

$$\text{maximum current, } I_{\max} = \frac{V_S}{R} = \frac{12\text{ V}}{60\ \Omega} = 200\text{ mA}$$

$$\text{time constant, } \tau = \frac{L}{R} = \frac{24\text{ mH}}{60\ \Omega} = 400\ \mu\text{s}$$

At one time constant (400 μs after source voltage is applied), the current will be

$$I = 63.2\%\text{ of } I_{\max}$$
$$= 0.632 \times 200\text{ mA} = 126.4\text{ mA}$$

At two time constants (800 μs after source voltage is applied):

$$I = 86.5\%\text{ of } I_{\max}$$
$$= 0.865 \times 200\text{ mA} = 173\text{ mA}$$

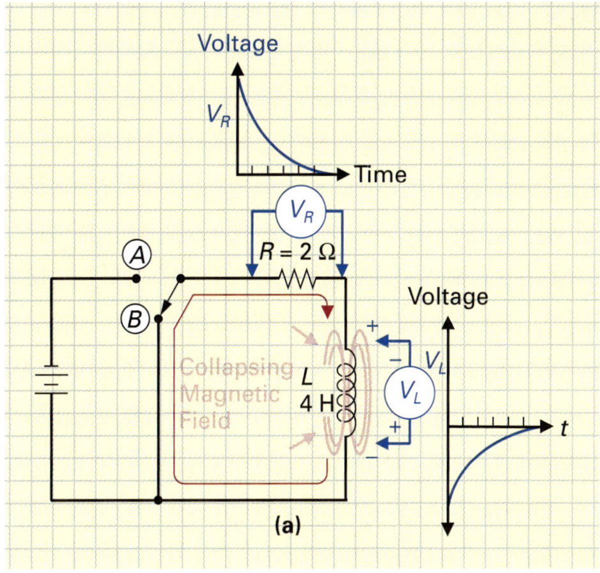

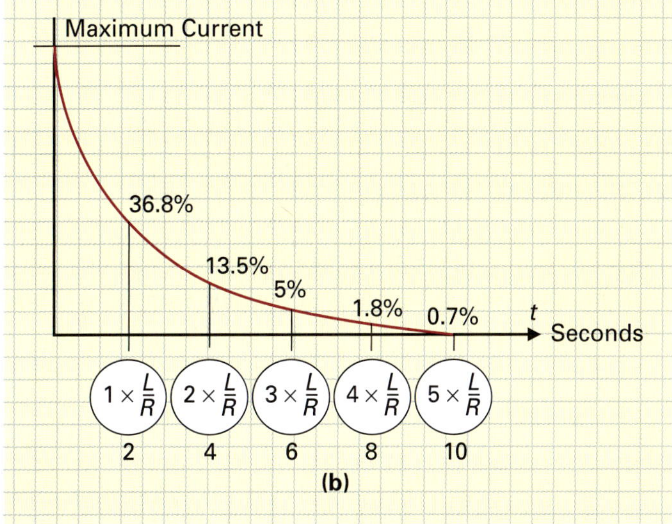

FIGURE 8-25 Exponential Current Fall.

At three time constants (1200 μs or 1.2 ms):

$$\begin{aligned} I &= 95\% \text{ of } I_{max} \\ &= 0.95 \times 200 \text{ mA} = 190 \text{ mA} \end{aligned}$$

At four time constants (1.6 ms):

$$\begin{aligned} I &= 98.2\% \text{ of } I_{max} \\ &= 0.982 \times 200 \text{ mA} = 196.4 \text{ mA} \end{aligned}$$

At five time constants (2 ms):

$$\begin{aligned} I &= 99.3\% \text{ of } I_{max} \\ &= 0.993 \times 200 \text{ mA} = 198.6 \text{ mA, approximately maximum (200 mA)} \end{aligned}$$

See Figure 8-26.

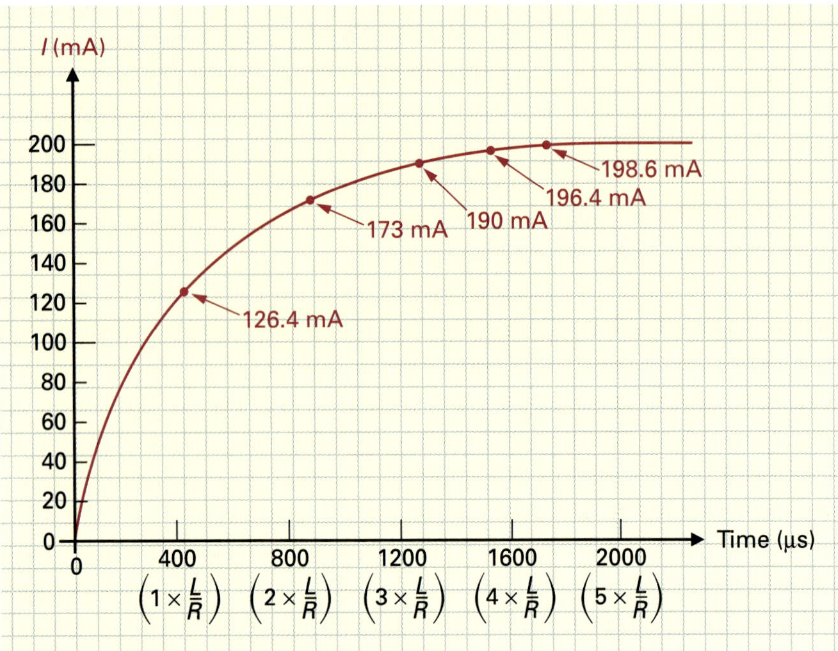

FIGURE 8-26 Exponential Current Rise Example.

8-3-5 *Inductive Reactance*

Inductive Reactance
Measured in ohms, it is the opposition to alternating or pulsating current flow without the dissipation of energy.

Reactance is the opposition to current flow without the dissipation of energy, as opposed to resistance, which is the opposition to current flow with the dissipation of energy.

Inductive reactance (X_L) is the opposition to current flow offered by an inductor without the dissipation of energy. It is measured in ohms and can be calculated by using the formula:

$$X_L = 2\pi \times f \times L$$

where X_L = inductive reactance, in ohms (Ω)
2π = 2π radians, 360° or 1 cycle
f = frequency, in hertz (Hz)
L = inductance, in henrys (H)

Inductive reactance is proportional to frequency ($X_L \propto f$) because a higher frequency (fast-switching current) will cause a greater amount of current change, and a greater change will generate a larger counter emf, which is an opposition or reactance against current flow. When 0 Hz is applied to a coil (dc), there exists no change, so the inductive reactance of an inductor to dc is zero ($X_L = 2\,\pi \times 0 \times L = 0$).

Inductive reactance is also proportional to inductance because a larger inductance will generate a greater magnetic field and subsequent counter emf, which is the opposition to current flow.

Ohm's law can be applied to inductive circuits just as it can be applied to resistive and capacitive circuits. The current flow in an inductive circuit (I) is proportional to the voltage applied (V), and inversely proportional to the inductive reactance (X_L). Expressed mathematically,

$$I = \frac{V}{X_L}$$

EXAMPLE:

Calculate the current flowing in the circuit illustrated in Figure 8-27.

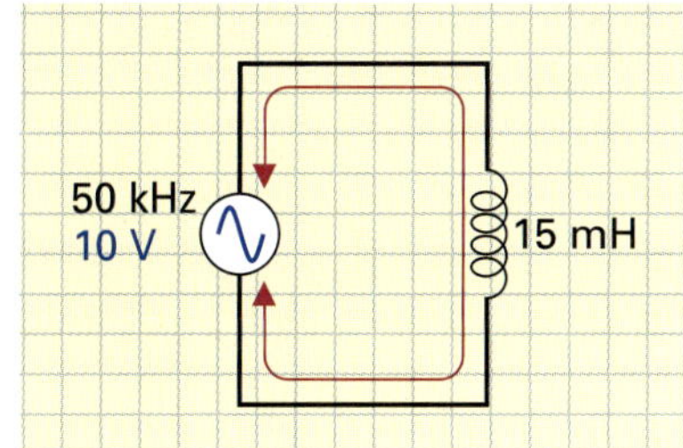

FIGURE 8-27

Solution:

The current can be calculated by Ohm's law and is a function of the voltage and opposition, which in this case is inductive reactance.

$$I = \frac{V}{X_L}$$

However, we must first calculate X_L:

$$\begin{aligned} X_L &= 2\pi \times f \times L \\ &= 6.28 \times 50 \text{ kHz} \times 15 \text{ mH} \\ &= 4710\ \Omega \text{ or } 4.71\ \text{k}\Omega \end{aligned}$$

Current is therefore equal to

$$I = \frac{V}{X_L} = \frac{10\text{ V}}{4.71\ \text{k}\Omega} = 2.12 \text{ mA}$$

EXAMPLE:

What opposition or inductive reactance will a motor winding or coil offer if $V = 12$ V and $I = 4.5$ A?

Solution:

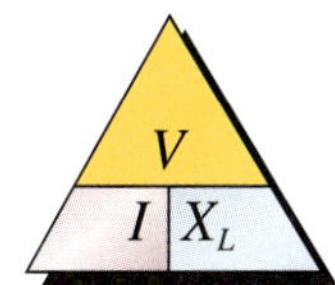

$$X_L = \frac{V}{I} = \frac{12\text{ V}}{4.5\text{ A}} = 2.66\ \Omega$$

8-3-6 *Series* RL *Circuit*

In a purely resistive circuit, as seen in Figure 8-28, the current flowing within the circuit and the voltage across the resistor are in phase with one another. In a purely inductive circuit, as shown in Figure 8-29, the current will lag the applied voltage by 90°. If we connect a resistor and inductor in series, as shown in Figure 8-30(a), we will have the most common combination of *R* and *L* used in electronic equipment.

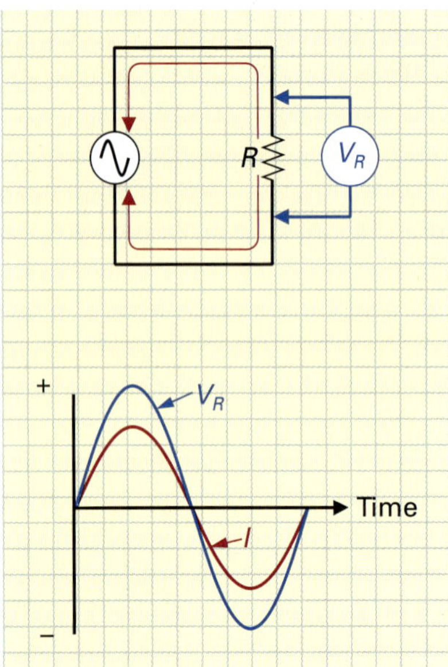

FIGURE 8-28 Purely Resistive Circuit: Current and Voltage Are in Phase.

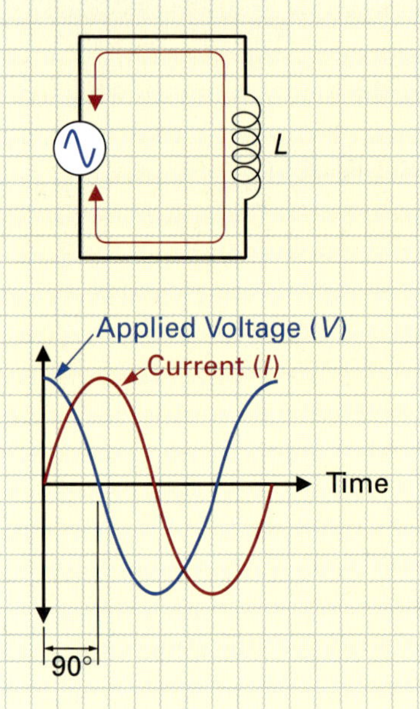

FIGURE 8-29 Purely Inductive Circuit: Current Lags Applied Voltage by 90°.

Voltage

The voltage across the resistor and inductor shown in Figure 8-30(a) can be calculated by using Ohm's law:

$$V_R = I \times R$$

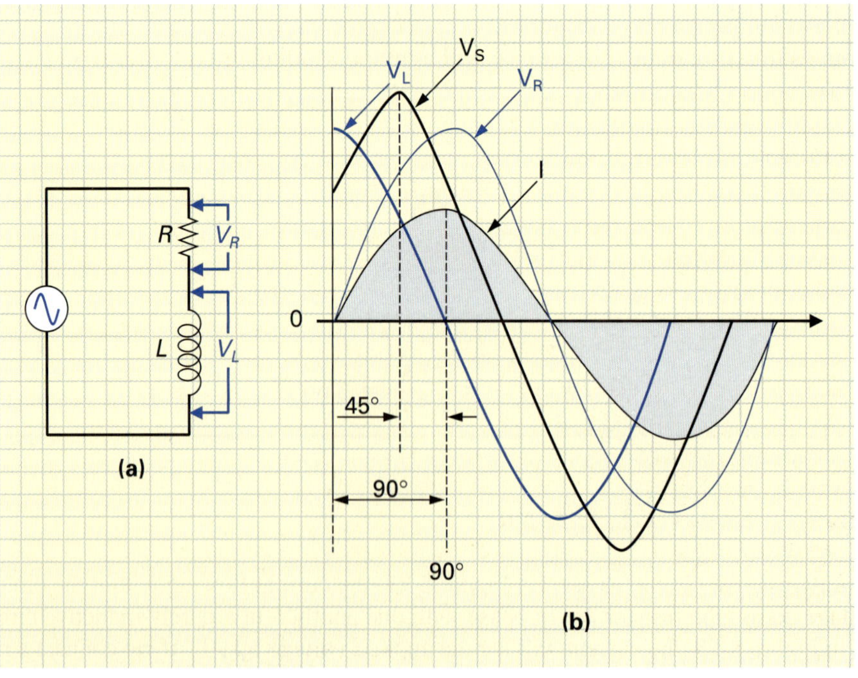

FIGURE 8-30 Series *RL* Circuit.

$$V_L = I \times X_L$$

As with any series circuit, we have to apply Kirchhoff's voltage law when calculating the value of applied or source voltage (V_S), which, due to the phase difference between V_R and V_L, is the vector sum of all the voltage drops. Applying the Pythagorean theorem, we arrive at a formula for source voltage.

$$V_S = \sqrt{V_R^2 + V_L^2}$$

As with any formula with three quantities, if two are known, the other can be calculated by simply rearranging the formula to

$$V_S = \sqrt{V_R^2 + V_L^2}$$

$$V_R = \sqrt{V_S^2 - V_L^2}$$

$$V_L = \sqrt{V_S^2 - V_R^2}$$

EXAMPLE:

Calculate V_R, V_L, and V_S for the circuit shown in Figure 8-31.

Solution:

$$\begin{aligned} V_R &= I \times R \\ &= 100 \text{ mA} \times 55\ \Omega \\ &= 5.5 \text{ V} \\ V_L &= I \times X_L \\ &= 100 \text{ mA} \times 26\ \Omega \\ &= 2.6 \text{ V} \\ V_S &= \sqrt{V_R^2 + V_L^2} \\ &= \sqrt{(5.5 \text{ V})^2 + (2.6 \text{ V})^2} \\ &= 6 \text{ V} \end{aligned}$$

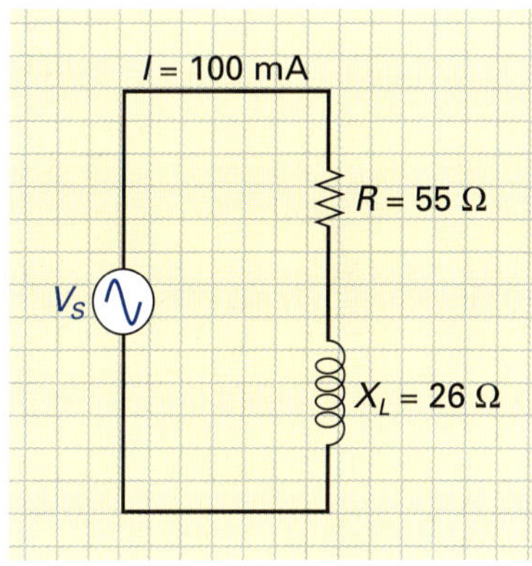

FIGURE 8-31 Voltage in a Series *RL* Circuit.

Impedance (Z)

Impedance is the total opposition to current flow offered by a circuit with both resistance and reactance. It is measured in ohms and can be calculated by using Ohm's law:

$$Z = \frac{V}{I}$$

The impedance of a series RL circuit is equal to the square root of the sum of the squares of resistance and reactance, and by rearrangement, X_L and R can also be calculated if the other two values are known:

$$Z = \sqrt{R^2 + X_L^2}$$

$$R = \sqrt{Z^2 - X_L^2}$$

$$X_L = \sqrt{Z^2 - R^2}$$

EXAMPLE:

Referring back to Figure 8-31, calculate Z.

Solution:

$$Z = \sqrt{R^2 + X_L^2}$$
$$= \sqrt{55^2 + 26^2}$$
$$= 60.8\ \Omega$$

Phase Shift

If a circuit is purely resistive, the phase shift (θ) between the source voltage and circuit current is zero. If a circuit is purely inductive, voltage leads current by 90°; therefore, the phase shift is +90°. If the resistance and inductive reactance are equal, the phase shift will equal +45°.

The phase shift in an inductive and resistive circuit is the degrees of lead between the source voltage (V_S) and current (I). Mathematically, it can be expressed as

$$\theta = \arctan\frac{X_L}{R}$$

As the current is the same in both the inductor and resistor in a series circuit, the voltage drops across the inductor and resistor are directly proportional to reactance and resistance:

$$V_R\updownarrow = I(\text{constant}) \times R\updownarrow,\ V_L\updownarrow = I(\text{constant}) \times X_L\updownarrow$$

The phase shift can also be calculated by using the voltage drop across the inductor and resistor:

$$\theta = \arctan\frac{V_L}{V_R}$$

EXAMPLE:

Referring back to Figure 8-31, calculate the phase shift between source voltage and circuit current.

■ ***Solution:***

Since the ratio of X_L/R and V_L/V_R are known for this example circuit, either of the phase shift formulas can be used.

$$\begin{aligned}\theta &= \arctan\frac{X_L}{R} \\ &= \arctan\frac{26\ \Omega}{55\ \Omega} \\ &= \arctan 0.4727 \\ &= 25.3^\circ\end{aligned}$$

or

$$\begin{aligned}\theta &= \arctan\frac{V_L}{V_R} \\ &= \arctan\frac{2.6\ \text{V}}{5.5\ \text{V}} \\ &= \arctan 0.4727 \\ &= 25.3^\circ\end{aligned}$$

The source voltage in this example circuit leads the circuit current by 25.3°.

Power

Purely Resistive Circuit. Figure 8-32 illustrates the current, voltage, and power waveforms produced when applying an ac voltage across a purely resistive circuit. Voltage and current are in phase, and true power (P_R) in watts can be calculated by multiplying current by voltage ($P = V \times I$).

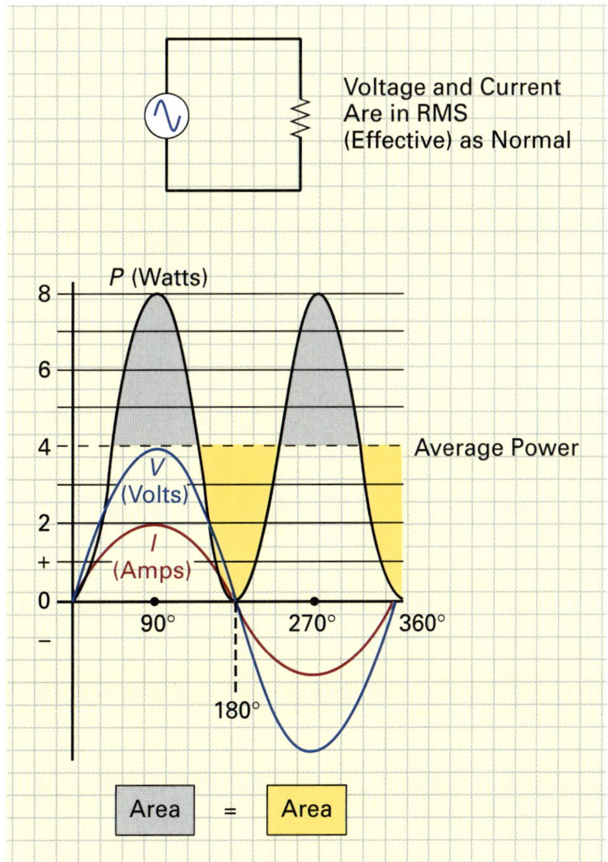

FIGURE 8-32 **Purely Resistive Circuit.**

The sinusoidal power waveform is totally positive, as a positive voltage multiplied by a positive current produces a positive value of power, and a negative voltage multiplied by a negative current also produces a positive value of power. For this reason, the resistor is said to develop a positive power waveform that is twice the frequency of the voltage or current waveform.

The average value of power dissipated by a purely resistive circuit is the halfway value between maximum and zero, in this example 4 watts.

Purely Inductive Circuit. The pure inductor, like the capacitor, is a reactive component, which means that it will consume power without the dissipation of energy. The capacitor holds its energy in an electric field, while the inductor consumes and holds its energy in a magnetic field and then releases it back into the circuit.

The power curve alternates equally above and below the zero line, as seen in Figure 8-33. During the first positive power half-cycle, the circuit current is on the increase, to maximum (point *A*), the magnetic field is building up, and the inductor is storing electrical energy. When the circuit current is on the decline between *A* and *B*, the magnetic field begins to collapse and self-induction occurs and returns electrical energy back into the circuit. The power alternation is both positive when the inductor is consuming power and negative when the inductor is returning the power back into the circuit. As the positive and negative power alternations are equal but opposite, the average power dissipated is zero.

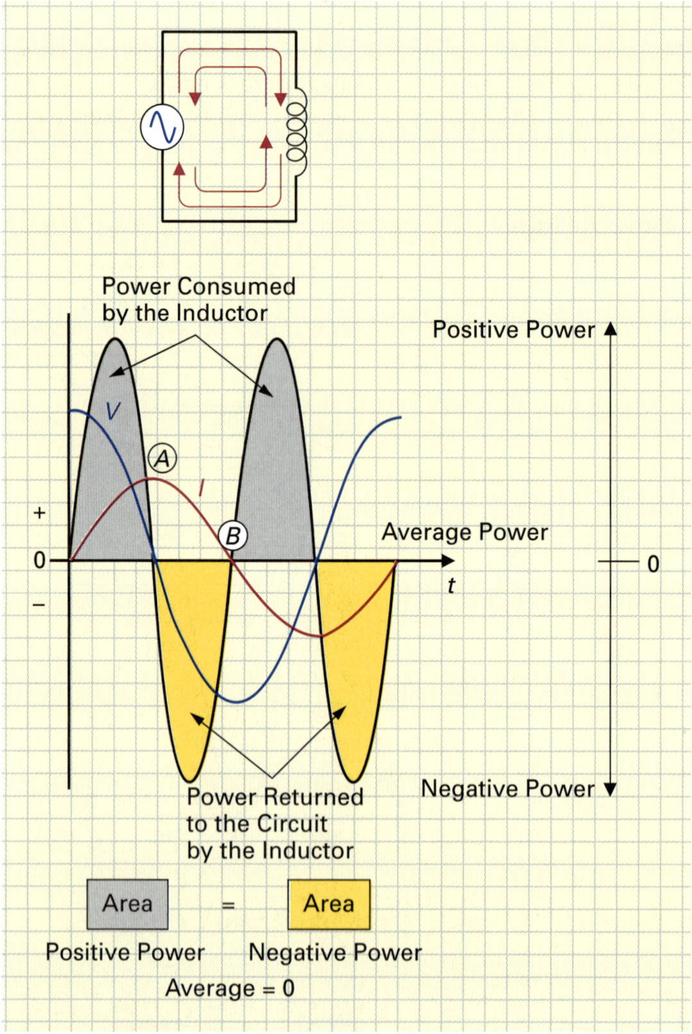

FIGURE 8-33 Purely Inductive Circuit.

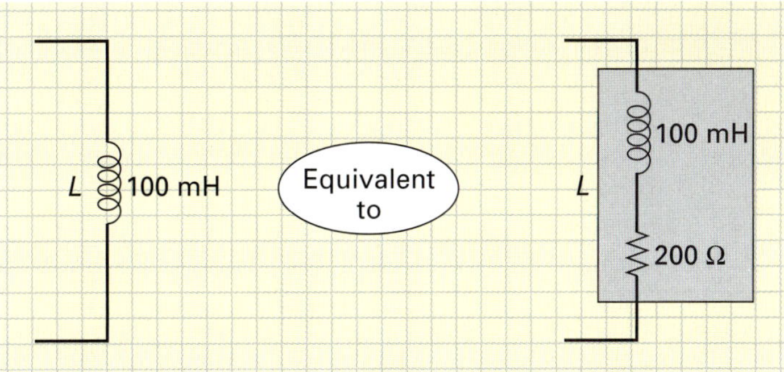

FIGURE 8-34 Resistance within an Inductor.

Resistive and Inductive Circuit. An inductor is different from a capacitor in that it has a small amount of resistance no matter how pure the inductor. For this reason, inductors will never have an average power figure of zero, because even the best inductor will have some value of inductance and resistance within it, as seen in Figure 8-34. The reason an inductor has resistance is that it is simply a piece of wire, and any piece of wire has a certain value of resistance, as well as inductance. This coil resistance should be, and normally is, very small and can usually be ignored; however, in some applications even this small resistance can prevent the correct operation of a circuit, so a value or term had to be created to specify the differences in the quality of inductor available.

The **quality factor** (Q) of an inductor is the ratio of the energy stored in the coil by its inductance to the energy dissipated in the coil by the resistance; therefore, the higher the Q, the better the coil is at storing energy rather than dissipating it:

Quality Factor
Quality factor of an inductor or capacitor; it is the ratio of a component's reactance (energy stored) to its effective series resistance (energy dissipated).

$$\text{quality } (Q) = \frac{\text{energy stored}}{\text{energy dissipated}}$$

The energy stored is dependent on the inductive reactance (X_L) of the coil, and the energy dissipated is dependent on the resistance (R) of the coil. The quality factor of a coil or inductor can therefore also be calculated by using the formula

$$Q = \frac{X_L}{R}$$

EXAMPLE:

Calculate the quality factor Q of a 22 mH coil connected across a 2 kHz, 10 V source if its internal coil resistance is 27 Ω.

Solution:

$$Q = \frac{X_L}{R}$$

The reactance of the coil is not known but can be calculated by the formula

$$\begin{aligned} X_L &= 2\pi fL \\ &= 2\pi \times 2 \text{ kHz} \times 22 \text{ mH} = 276.5 \ \Omega \end{aligned}$$

Therefore,

$$Q = \frac{X_L}{R} = \frac{276.5 \ \Omega}{27 \ \Omega} = 10.24$$

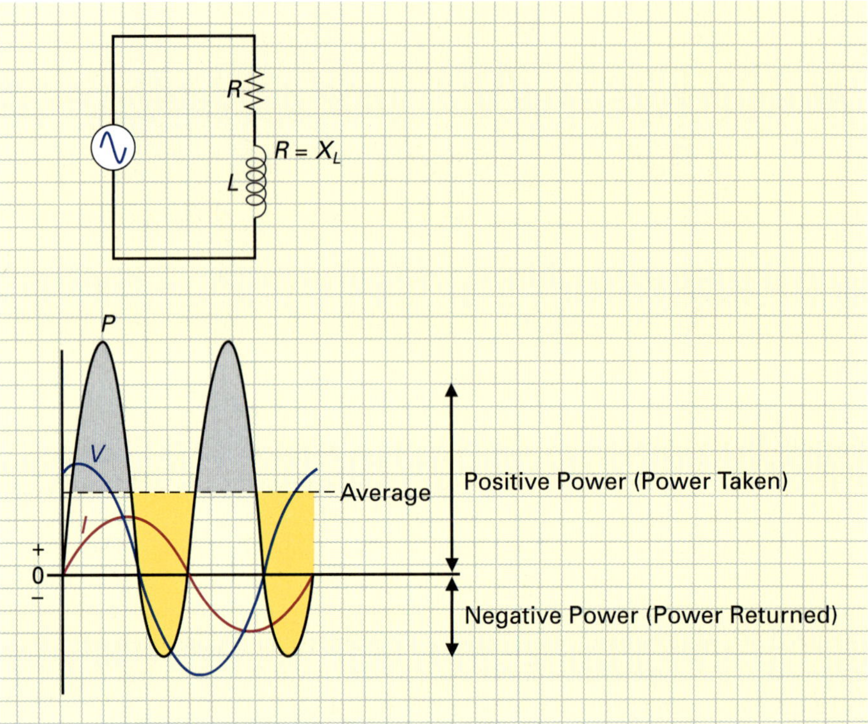

FIGURE 8-35 **Power in a Resistive and Inductive Circuit.**

Inductors, therefore, will never appear as pure inductance, but rather as an inductive and resistive (*RL*) circuit, and the resistance within the inductor will dissipate *true power.*

Figure 8-35 illustrates a circuit containing *R* and *L* and the power waveforms produced when $R = X_L$; the phase shift (θ) is equal to 45°.

The positive power alternation, which is above the zero line, is the combination of the power dissipated by the resistor and the power consumed by the inductor while circuit current was on the rise. The negative power alternation is the power that was given back to the circuit by the inductor while the inductor's magnetic field was collapsing and returning the energy that was consumed.

Power Factor. When a circuit contains both inductance and resistance, some of the energy is consumed and then returned by the inductor (reactive or imaginary power), and some of the energy is dissipated and lost by the resistor (resistive or true power).

Apparent power is the power that appears to be supplied to the load and is the vector sum of both the reactive and true power; it can be calculated by using the formula

$$P_A = \sqrt{P_R^2 + P_X^2}$$

where P_A = apparent power, in volt-amperes (VA)
P_R = true power, in watts (W)
P_X = reactive power, in volt-amperes reactive (VAR)

The power factor is a ratio of the true power to the apparent power and is therefore a measure of the loss in a circuit.

$$\text{PF} = \frac{\text{true power } (P_R)}{\text{apparent power } (P_A)}$$

or

$$PF = \frac{R}{Z}$$

or

$$PF = \cos\theta$$

EXAMPLE:

Calculate the following for a series *RL* circuit if $R = 40\ \text{k}\Omega$, $L = 450\ \text{mH}$, $f = 20\ \text{kHz}$, and $V_S = 6\ \text{V}$:

a. X_L
b. Z
c. I
d. θ
e. Apparent power
f. PF

Solution:

a. $X_L = 2\pi fL = 2\pi \times 20\ \text{kHz} \times 450\ \text{mH}$
$= 56.5\ \text{k}\Omega$

b. $Z = \sqrt{R^2 + X_L^{\,2}}$
$= \sqrt{(40\ \text{k}\Omega)^2 + (56.5\ \text{k}\Omega)^2} = 69.23\ \text{k}\Omega$

c. $I = \frac{V_S}{Z} = \frac{6\ \text{V}}{69.23\ \text{k}\Omega} = 86.6\ \mu\text{A}$

d. $\theta = \arctan\frac{X_L}{R} = \arctan\frac{56.5\ \text{k}\Omega}{40\ \text{k}\Omega} = 54.7°$

e. Apparent power $= \sqrt{(\text{true power})^2 + (\text{reactive power})^2}$
$P_R = I^2 \times R = (86.6\ \mu\text{A})^2 \times 40\ \text{k}\Omega = 300\ \mu\text{W}$
$P_X = I^2 \times X_L = (86.6\ \mu\text{A})^2 \times 56.5\ \text{k}\Omega = 423.7\ \mu\text{VAR}$
$P_A = \sqrt{P_R^{\,2} + P_X^{\,2}}$
$= \sqrt{(300\ \mu\text{W})^2 + (423.7\ \mu\text{W})^2} = 519.2\ \mu\text{VA}$

f. $PF = \frac{P_R}{P_A} = \frac{300\ \mu\text{W}}{519.2\ \mu\text{W}} = 0.57$
$= \frac{R}{Z} = \frac{40\ \text{k}\Omega}{69.23\ \text{k}\Omega} = 0.57$
$= \cos\theta = \cos 54.7° = 0.57$

8-3-7 *Parallel* RL *Circuit*

Now that we have seen the behavior of resistors and inductors in series, let us analyze the parallel *RL* circuit.

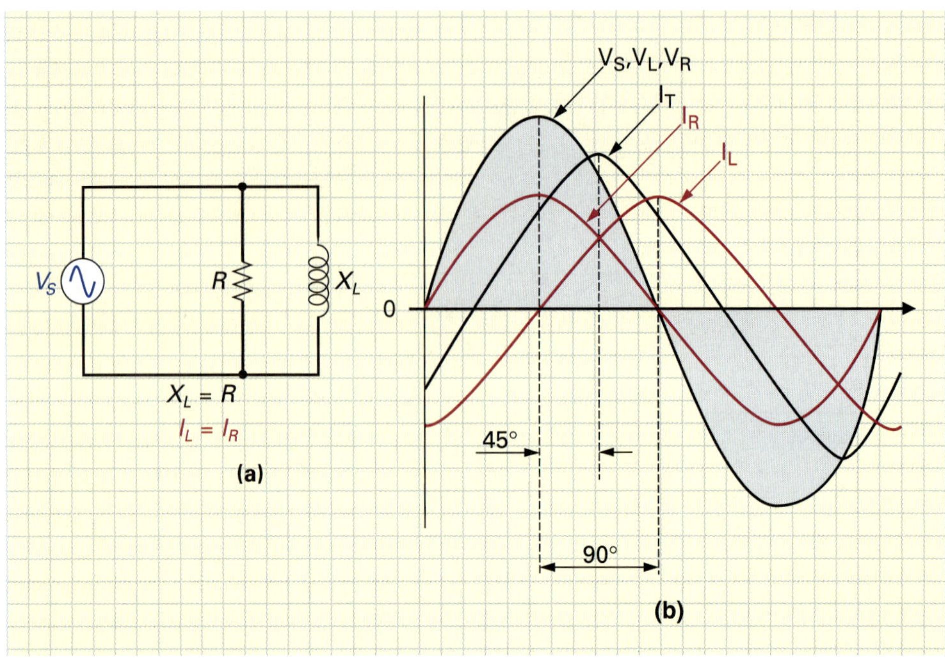

FIGURE 8-36 **Parallel *RL* Circuit.**

Current

In Figure 8-36(a) you will see a parallel combination of a resistor and inductor. The voltage across both components is equal because of the parallel connection, and the current through each branch is calculated by applying Ohm's law.

$$\text{(resistor current) } I_R = \frac{V_S}{R}$$

$$\text{(inductor current) } I_L = \frac{V_S}{X_L}$$

Total current (I_T) is equal to the vector combination of the resistor current and inductor current, as shown in Figure 8-36(b).

$$I_T = \sqrt{I_R^{\,2} + I_L^{\,2}}$$

Phase Angle

The angle by which the total current (I_T) leads the source voltage (V_S) can be determined with either of the following formulas:

$$\theta = \arctan\frac{I_L}{I_R}$$

or

$$\theta = \arctan\frac{R}{X_L}$$

Impedance

The total opposition or impedance of a parallel *RL* circuit can be calculated by

$$Z = \frac{V_S}{I_T}$$

Using basic algebra, this formula can be rearranged to express impedance in terms of reactance and resistance.

$$Z = \frac{R \times X_L}{\sqrt{R^2 + X_L^2}}$$

Power

As with series *RL* circuits, resistive power and reactive power can be calculated by

$$P_R = I_R^2 \times R$$

$$P_X = I_L^2 \times X_L$$

The apparent power of the circuit is calculated by

$$P_A = \sqrt{P_R^2 + P_X^2}$$

and, finally, the power factor is equal to

$$\text{PF} = \cos\theta = \frac{P_R}{P_A}$$

EXAMPLE:

Calculate the following for a parallel *RL* circuit if $R = 45\ \Omega$, $X_L = 1100\ \Omega$, and $V_S = 24$ V:

a. I_R b. I_L c. I_T d. Z e. θ

Solution:

a. $I_R = \frac{V_S}{R} = \frac{24\text{ V}}{45\ \Omega} = 533.3\text{ mA}$

b. $I_L = \frac{V_S}{X_L} = \frac{24\text{ V}}{1100\ \Omega} = 21.8\text{ mA}$

c. $I_T = \sqrt{I_R^2 + I_L^2}$

$= \sqrt{(533.3\text{ mA})^2 + (21.8\text{ mA})^2} = 533.7\text{ mA}$

d. $Z = \frac{R \times X_L}{\sqrt{R^2 + X_L^2}} = \frac{45\ \Omega \times 1100\ \Omega}{\sqrt{(45\ \Omega)^2 + (1100\ \Omega)^2}}$

$= \frac{49.5\text{ k}\Omega^2}{1100.9\ \Omega} = 44.96\ \Omega$

e. $\theta = \arctan\frac{R}{X_L} = \arctan\frac{45\ \Omega}{1100\ \Omega} = 2.34°$

Therefore, I_T lags V_S by 2.34°.

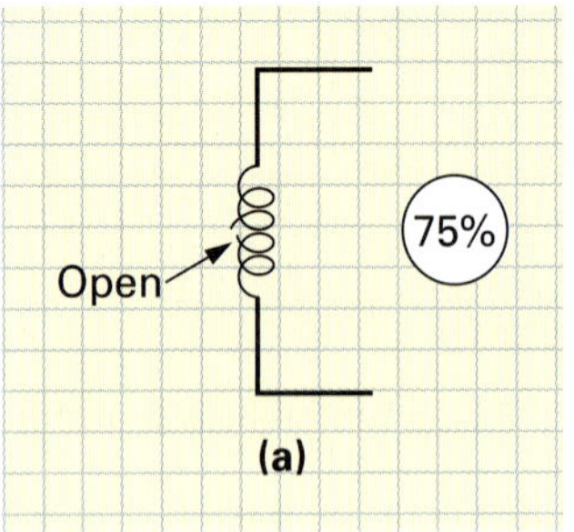

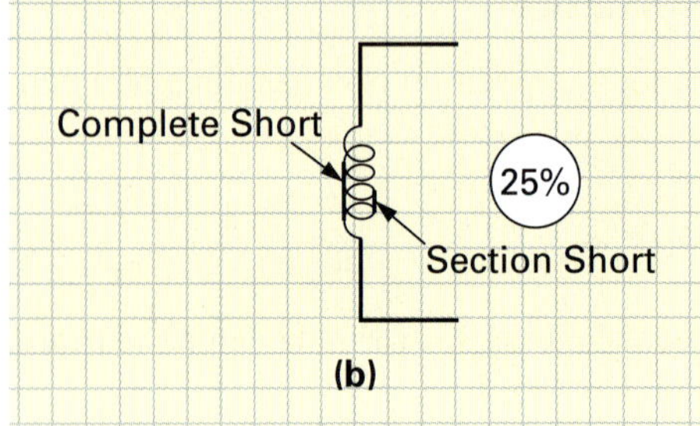

FIGURE 8-37 **Defective Inductors.**

8-3-8 *Testing Inductors*

Basically, only three problems can occur with inductors:

1. An open
2. A complete short
3. A section short (value change)

Open [Figure 8-37(a)]

This problem can be isolated with an ohmmeter. Depending on the winding's resistance, the coil should be in the range of zero to a few hundred ohms. An open accounts for 75% of all defective inductors.

Complete or Section Short [Figure 8-37(b)]

Inductor Analyzer
A test instrument designed to test inductors.

A coil with one or more shorted turns or a complete short can be checked with an ohmmeter and thought to be perfectly good because of the normally low resistance of a coil, as it is just a piece of wire. But if it is placed in a circuit with a complete or section short present, it will not function effectively as an inductor, if at all. For these checks, an **inductor analyzer** needs to be used like the one seen in Figure 8-38, which can be used to check capacitance and inductance. Complete or section shorts account for 25% of all defective inductors.

8-3-9 *Applications of Inductors*

RL Filter
A selective circuit of resistors and inductors which offers little or no opposition to certain frequencies while blocking or attenuating other frequencies.

RL Filters

The ***RL* filter** will achieve results similar to the *RC* filter in that it will pass some frequencies and block others, as seen in Figure 8-39. The inductive reactance of the coil and the resistance of the resistor form a voltage divider. Since inductive reactance is proportional to frequency ($X_L \propto f$), the inductor will drop less voltage at lower frequencies ($f\downarrow, X_L\downarrow, V_L\downarrow$) and more voltage at higher frequencies ($f\uparrow, X_L\uparrow, V_L\uparrow$).

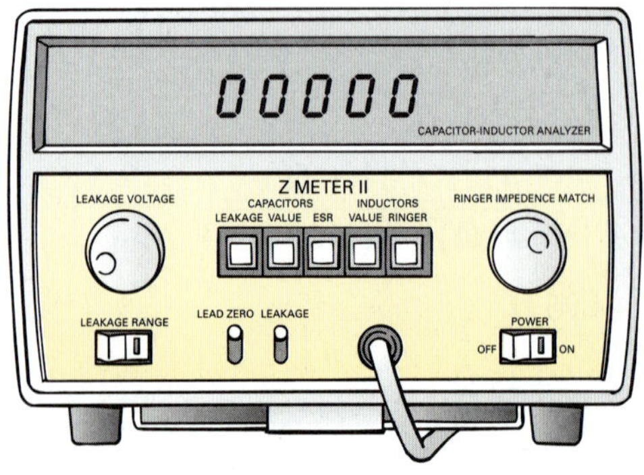

FIGURE 8-38 **Capacitor and Inductor Analyzer.**

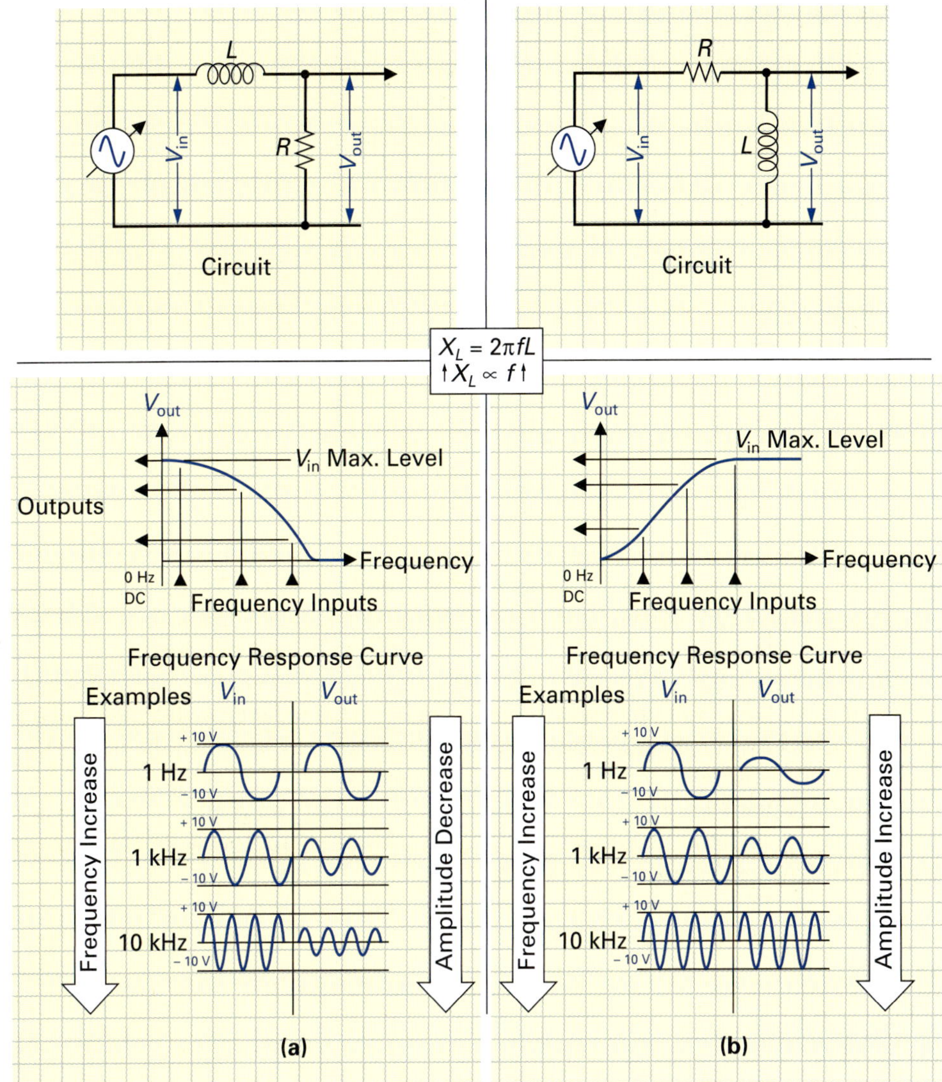

FIGURE 8-39 ***RL*** **Filter. (a) Low-Pass Filter. (b) High-Pass Filter.**

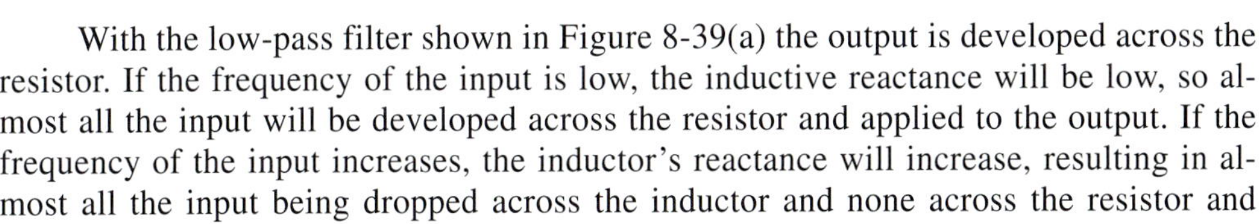

With the low-pass filter shown in Figure 8-39(a) the output is developed across the resistor. If the frequency of the input is low, the inductive reactance will be low, so almost all the input will be developed across the resistor and applied to the output. If the frequency of the input increases, the inductor's reactance will increase, resulting in almost all the input being dropped across the inductor and none across the resistor and therefore the output.

With the high-pass filter, shown in Figure 8-39(b), the inductor and resistor have been placed in opposite positions. If the frequency of the input is low, the inductive reactance will be low, so almost all the input will be developed across the resistor and very little will appear across the inductor and therefore at the output. If the frequency of the input is high, the inductive reactance will be high, resulting in almost all of the input being developed across the inductor and therefore appearing at the output.

RL Integrator

In the preceding section on *RL* filters we saw how an *RL* circuit reacted to a sine-wave input of different frequencies. In this section and the following we will see how an *RL* circuit will react to a square-wave input, and show two other important applications of inductors. In the

FIGURE 8-40 *RL* Integrator.

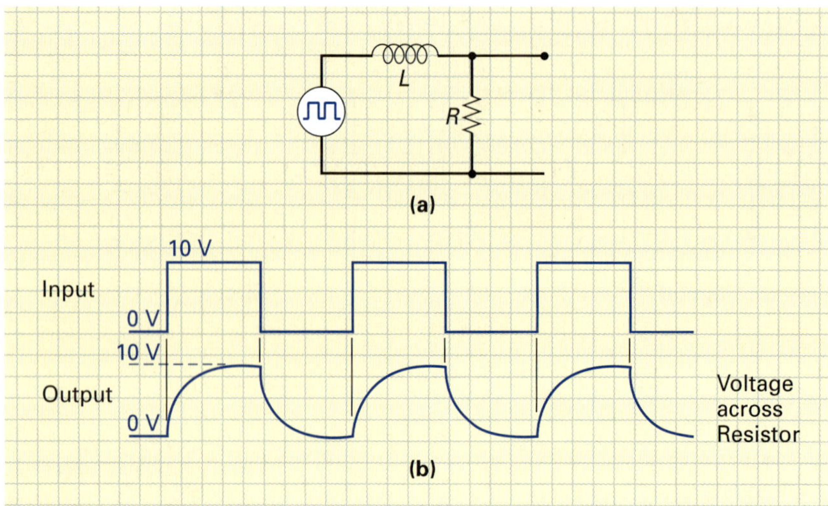

***RL* Integrator**

An *RL* circuit with an output proportionate to the integral of the input signal.

***RL* integrator,** the output is taken across the resistor, as seen in the circuit in Figure 8-40(a). The output shown in Figure 8-40(b) is the same as the previously described *RC* integrator's output.

The inductor's 10 V counter emf opposes the sudden input change from 0 to 10 V, and if 10 V is across the inductor, 0 V must be across the resistor (Kirchhoff's voltage law), and therefore appearing at the output. After five time constants ($5 \times L/R$), the inductor's current in the circuit will have built up to maximum (V_{in}/R), and the inductor will be an equivalent short circuit, because no change and consequently no back emf exist. All the input voltage will now be across the resistor and therefore at the output.

When the square-wave input drops to zero, the collapsing magnetic field will cause an induced voltage within the conductor, which will cause current to flow within the circuit for five time constants, whereupon it will reach 0.

As with the *RC* integrator, if the period of the square wave is decreased or if the time constant is increased, the output will reach an average value of half the input square wave's amplitude.

***RL* Differentiator**

An *RL* circuit whose output voltage is proportional to the rate of change of the input voltage.

RL *Differentiator*

With the ***RL* differentiator,** the output is taken across the inductor, as shown in Figure 8-41, and is the same as the *RC* differentiator's output. When the square-wave in-

FIGURE 8-41 *RL* Differentiator.

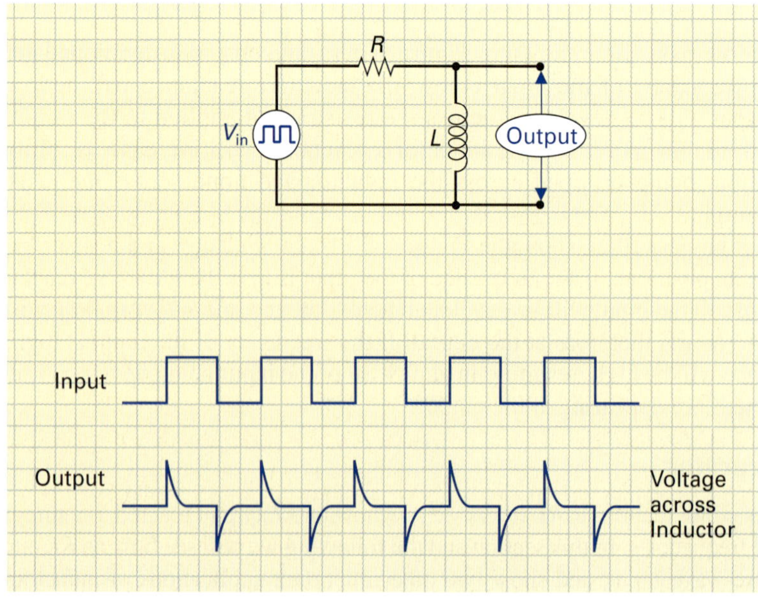

put rises from 0 to 10 V, the inductor will generate a 10 V counter emf across it, and this 10 V will appear across the output. As the circuit current exponentially rises, the voltage across the inductor, and therefore at the output, will fall exponentially to 0 V after five time constants.

When the square wave input falls from 10 to 0 V, the collapsing magnetic field induces a counter emf. The sudden −10 V impulse at the output will decrease to 0 V as the circuit current decreases in five time constants.

SELF-TEST EVALUATION POINT FOR SECTION 8-3

Use the following questions to test your understanding of Section 8-3.

1. List the four factors that determine the inductance of an inductor.
2. State the formula for inductance.
3. How does the inductive time constant relate to the capacitive time constant?
4. True or false: The greater the value of the inductor, the longer it would take for current to rise to a maximum.
5. True or false: A constant dc level is opposed continuously by an inductor.
6. Calculate the total circuit inductance if 4 mH and 2 mH are connected:
   a. In series
   b. In parallel
7. True or false: In a purely inductive circuit, the current will lead the applied voltage by 90°.
8. Calculate the applied source voltage V_S in an R_L circuit where $V_R = 4$ V and $V_L = 2$ V.
9. Define and state the formula for impedance when R and X_L are known.

8-4 MUTUAL INDUCTANCE

As was discussed in the previous section, self-inductance is the process by which a coil induces a voltage within itself. The principle on which a transformer is based is an inductive effect known as **mutual inductance,** which is the process by which an inductor induces a voltage in another inductor.

Mutual Inductance
Ability of one inductor's magnetic lines of force to link with another inductor.

Figure 8-42 illustrates two inductors that are magnetically linked, yet electrically isolated from one another. As the alternating current continually rises, falls, and then rises in the opposite direction, a magnetic field will build up, collapse, and then build up in the opposite direction.

If a second inductor or secondary coil (L_2) is in close proximity with the first inductor or primary coil (L_1), which is producing the alternating magnetic field, a voltage will be induced into the nearby inductor, which causes current to flow in the secondary

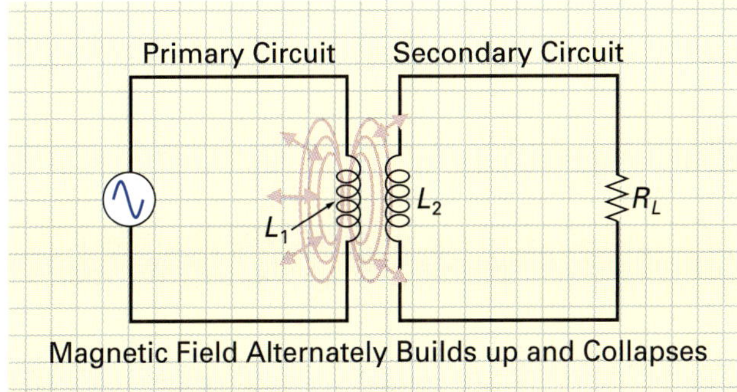

FIGURE 8-42 Mutual Inductance.

circuit through the load resistor. This phenomenon is known as mutual inductance or transformer action.

As with self-inductance, mutual inductance is dependent on change. Direct current (dc) is a constant current and produces a constant or stationary magnetic field that does not change. Alternating current, however, is continually varying, and as the polarity of the magnetic field is dependent on the direction of current flow, the magnetic field will also be alternating in polarity, and it is this continual building up and collapsing of the magnetic field that cuts the adjacent inductor's conducting coils and induces a voltage in the secondary circuit. Mutual induction is possible only with ac and cannot be achieved with dc due to the lack of change.

In summary, therefore, self-induction is a measure of how much voltage an inductor can induce within itself. Mutual inductance is a measure of how much voltage is induced in the secondary coil due to the change in current in the primary coil.

8-4-1 *Basic Transformer*

Transformer

Device consisting of two or more coils that are used to couple electric energy from one circuit to another, yet maintain electrical isolation between the two.

Figure 8-43(a) illustrates a basic **transformer,** which consists of two coils within close proximity to one another, to ensure that the second coil will be cut by the magnetic flux lines produced by the first coil, and thereby ensure mutual inductance. The ac voltage source is electrically connected (through wires) to the primary coil or winding, and the load (R_L) is electrically connected to the secondary coil or winding.

In Figure 8-43(b), the ac voltage source has produced current flow in the primary circuit, as illustrated. This current flow produces a north pole at the top of the primary winding and, as the ac voltage input swings more negative, the current increase causes the magnetic field being developed by the primary winding to increase. This expanding magnetic field cuts the coils of the secondary winding and induces a voltage, and a subsequent current flows in the secondary circuit, which travels up through the load resistor. The ac voltage follows a sinusoidal pattern and moves from a maximum negative to zero and then begins to build up toward a maximum positive.

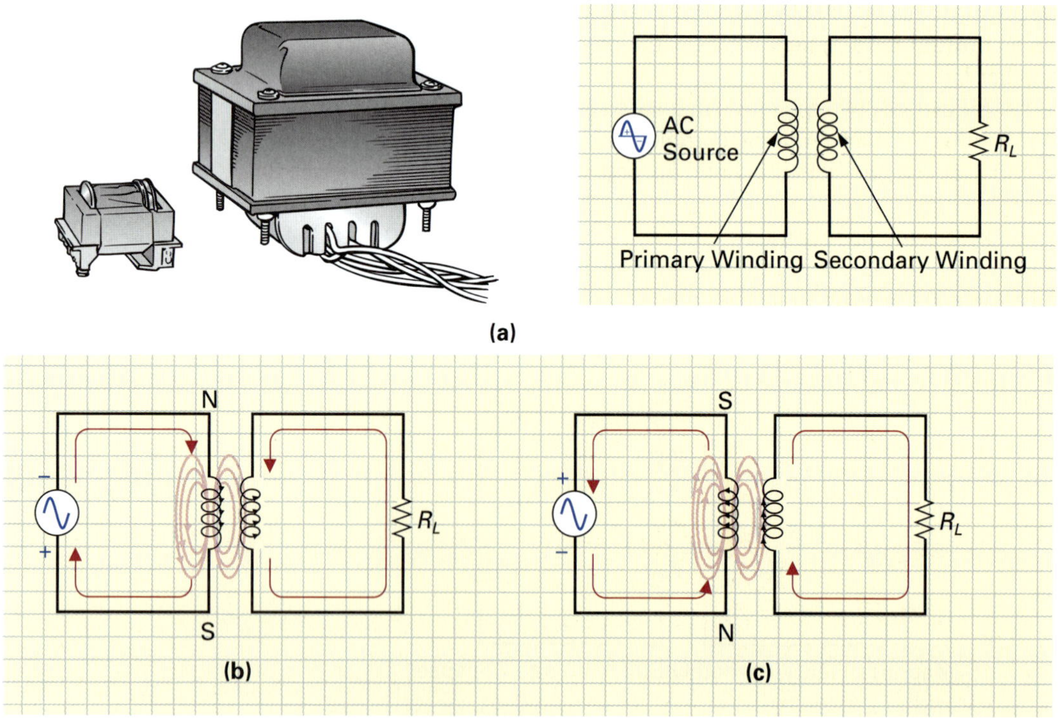

FIGURE 8-43 **Transformer Action.**

In Figure 8-43(c), the current flow in the primary circuit is in the opposite direction due to the ac voltage increase in the positive direction. As voltage increases, current increases and the magnetic field expands and cuts the secondary winding, inducing a voltage and causing current to flow in the reverse direction down through the load resistor.

You may have noticed a few interesting points about the basic transformer just discussed.

1. As primary current increases, secondary current increases, and as primary current decreases, secondary current also decreases. It can therefore be said that the frequency of the alternating current in the secondary is the same as the frequency of the alternating current in the primary.
2. Although the two coils are electrically isolated from one another, energy can be transferred from primary to secondary, because the primary converts electrical energy into magnetic energy, and the secondary converts magnetic energy back into electrical energy.

8-4-2 *Transformer Loading*

Let's now carry our discussion of the basic transformer a little further and see what occurs when the transformer is not connected to a load, as shown in Figure 8-44. Primary circuit current is determined by $I = V/Z$, where Z is the impedance of the primary coil (both its inductive reactance and resistance) and V is the applied voltage. Since no current can flow in the secondary, because an open in the circuit exists, the primary acts as a simple inductor, and the primary current is small due to the inductance of the primary winding. This small primary current lags the applied voltage due to the counter emf by approximately 90° because the coil is mainly inductive and has very little resistance.

When a load is connected across the secondary, as shown in Figure 8-45, a change in conditions occurs and the transformer acts differently. The important point that will be observed is that as we go from a no-load to a load condition the primary current will increase due to mutual inductance. Let's follow the steps one by one.

1. The ac applied voltage sets up an alternating magnetic field in the primary winding.
2. The continually changing flux of this primary field induces and produces a counter emf into the primary to oppose the applied voltage.
3. The primary's magnetic field also induces a voltage in the secondary winding, which causes current to flow in the secondary circuit through the load.
4. The current in the secondary winding produces another magnetic field that is opposite to the field being produced by the primary.
5. This secondary magnetic field feeds back to the primary and induces a voltage that tends to cancel or weaken the counter emf that was set up in the primary by the primary current.
6. The primary's counter emf is therefore reduced, so primary current can now increase.

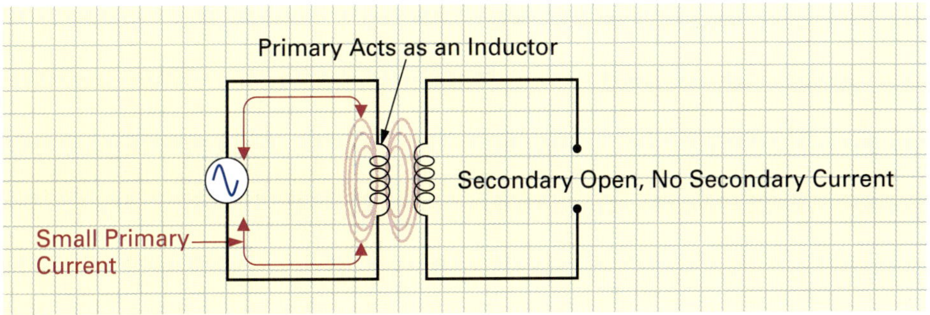

FIGURE 8-44 Unloaded Transformer.

FIGURE 8-45 **Loaded Transformer.**

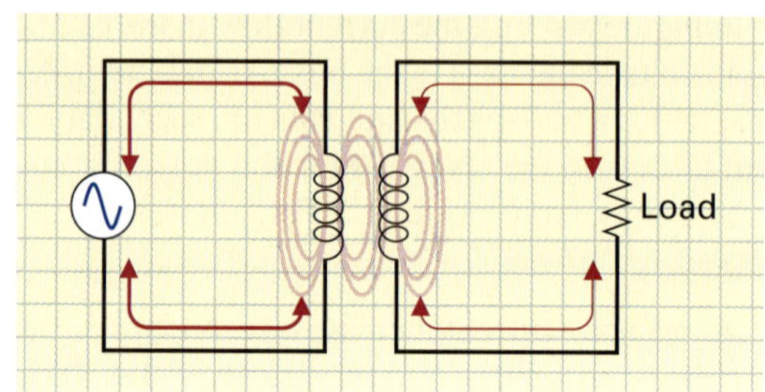

7. This increase in primary current is caused by the secondary's magnetic field; consequently, the greater the secondary current, the stronger the secondary magnetic field, which causes a reduction in the primary's counter emf, and therefore a primary current increase.

In summary, an increase in secondary current ($I_s\uparrow$) causes an increase in primary current ($I_p\uparrow$), and this effect in which the primary induces a voltage in the secondary (V_s) and the secondary induces a voltage into the primary (V_p) is known as *mutual inductance.*

8-4-3 *Transformer Ratios and Applications*

Basically, transformers are used for one of three applications:

1. To step up (increase) or step down (decrease) voltage
2. To step up (increase) or step down (decrease) current
3. To match impedances

In all three cases, any of the applications can be achieved by changing the ratio of the number of turns in the primary winding compared to the number of turns in the secondary winding. This ratio is appropriately called the turns ratio.

1. Turns Ratio

Turns Ratio
Ratio of the number of turns in the secondary winding to the number of turns in the primary winding of a transformer.

The **turns ratio** is the ratio between the number of turns in the secondary winding (N_s) and the number of turns in the primary winding (N_p).

$$\text{turns ratio} = \frac{N_s}{N_p}$$

Let us use a few examples to see how the turns ratio can be calculated.

EXAMPLE:

If the primary has 200 turns and the secondary has 600, what is the turns ratio (Figure 8-46)?

FIGURE 8-46 **Step-Up Transformer Example.**

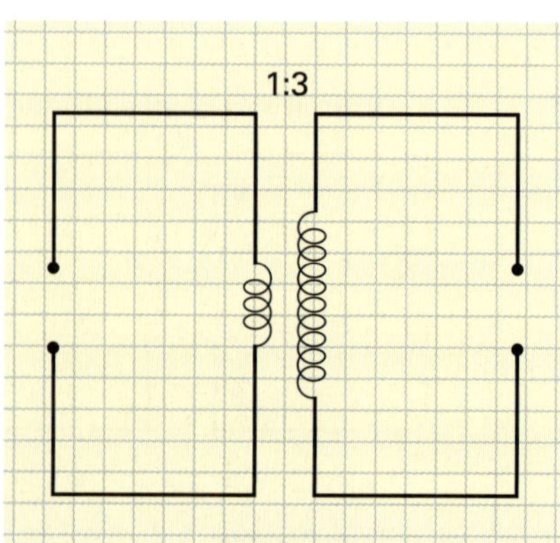

Solution:

$$\text{turns ratio} = \frac{N_s}{N_p} = \frac{600}{200} = \frac{3(\text{secondary})}{1(\text{primary})} = 3$$

This simply means that there are three windings in the secondary to every one winding in the primary. Moving from a small number (1) to a larger number (3) means that we *stepped up* in value. Stepping up always results in a turns ratio figure greater than 1, in this case, 3.

EXAMPLE:

If the primary has 120 turns and the secondary has 30 turns, what is the turns ratio (Figure 8-47)?

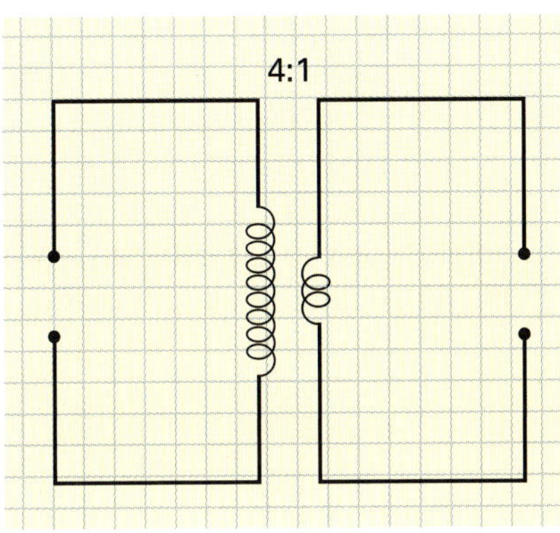

FIGURE 8-47 Step-Down Transformer Example.

Solution:

$$\text{turns ratio} = \frac{N_s}{N_p} = \frac{30}{120} = \frac{1\ (\text{secondary})}{4\ (\text{primary})} = 0.25$$

Said simply, there are four primary windings to every one secondary winding. Moving from a larger number (4) to a smaller number (1) means that we *stepped down* in value. Stepping down always results in a turns ratio figure of less than 1, in this case 0.25.

2. Voltage Ratio

Transformers are used within the power supply unit of almost every piece of electronic equipment to step up or step down the 115 V ac from the outlet. Some electronic circuits require lower-power supply voltages, while other devices may require higher-power supply voltages. The transformer is used in both instances to convert the 115 V ac to the required value of voltage.

Step Up. If the secondary voltage (V_s) is greater than the primary voltage (V_p), the transformer is called a **step-up transformer** ($V_s > V_p$), as shown in Figure 8-48. The voltage

Step-up Transformer
Transformer in which the ac voltage induced in the secondary is greater (due to more secondary windings) than the ac voltage applied to the primary.

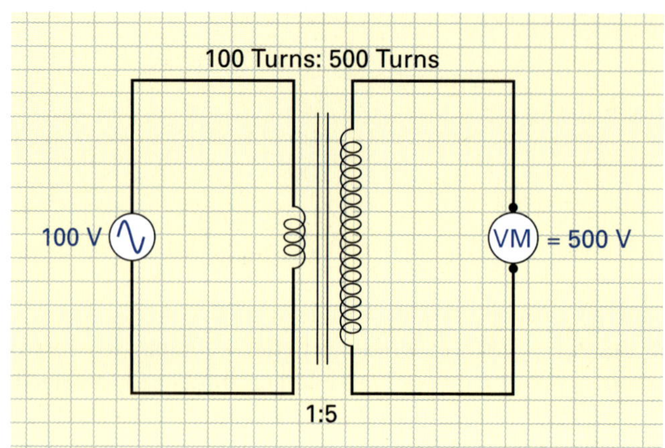

FIGURE 8-48 **Step-Up Transformer.**

is stepped up or increased in much the same way as a generator voltage can be increased by increasing the number of turns.

If the ac primary voltage is 100 V and the turns ratio is a 1:5 step up, the secondary voltage will be five times that of the primary voltage, or 500 V, because the magnetic flux established by the primary cuts more turns in the secondary and therefore induces a larger voltage.

In this example, you can see that the ratio of the secondary voltage to the primary voltage is equal to the turns ratio; in other words,

$$\frac{V_s}{V_p} = \frac{N_s}{N_p}$$

or

$$\frac{500}{100} = \frac{500}{100}$$

To calculate V_s, therefore, we can rearrange the formula and arrive at

$$V_s = \frac{N_s}{N_p} \times V_p$$

In our example, this is

$$\begin{aligned} V_s &= \frac{500}{100} \times 100\ \text{V} \\ &= 500\ \text{V} \end{aligned}$$

Step-down Transformer

Transformer in which the ac voltage induced in the secondary is less (due to fewer secondary windings) than the ac voltage applied to the primary.

Step Down. If the secondary voltage (V_s) is smaller than the primary voltage (V_p), the transformer is called a **step-down transformer** ($V_s < V_p$), as shown in Figure 8-49. The secondary voltage will be equal to

$$\begin{aligned} V_s &= \frac{N_s}{N_p} \times V_p \\ &= \frac{10}{100} \times 1000\ \text{V} = 100\ \text{V} \end{aligned}$$

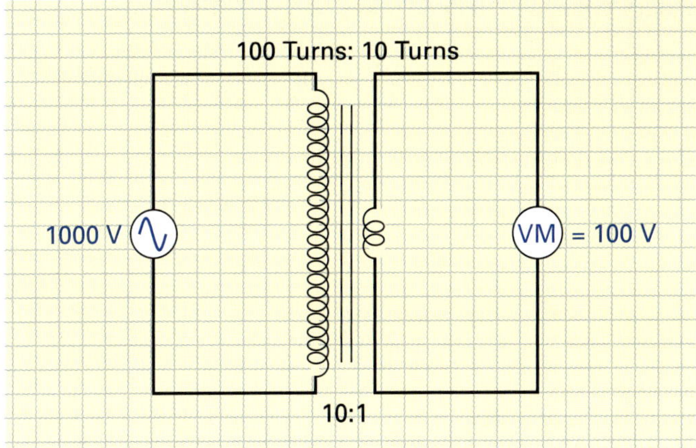

FIGURE 8-49 **Step-Down Transformer.**

EXAMPLE:

Calculate the secondary voltage (V_s) if a 1:6 step-up transformer has 24 V ac applied to the primary.

Solution:

$$V_s = \frac{N_s}{N_p} \times V_p$$

$$= \frac{6}{1} \times 24 \text{ V} = 144 \text{ V}$$

The coupling coefficient (k) in this formula is always assumed to be 1, which for most iron-core transformers is almost always the case. This means that all the primary magnetic flux is linking the secondary, and the secondary voltage is dependent on the number of secondary turns that are being cut by the primary magnetic flux.

The transformer can be used to transform the primary ac voltage into any other voltage, either up or down, merely by changing the transformer's turns ratio.

3. Power and Current Ratio

The power in the secondary of the transformer is equal to the power in the primary ($P_p = P_s$). Power, as we know, is equal to $P = V \times I$, and if voltage is stepped up or down, the current automatically is stepped down or up, respectively, in the opposite direction to voltage to maintain the power constant.

For example, if the secondary voltage is stepped up ($V_s \uparrow$), the secondary current is stepped down ($I_s \downarrow$), so the output power is the same as the input power.

$$P_s = V_s \uparrow \times I_s \downarrow$$

This is an equal but opposite change. Therefore, $P_s = P_p$; and you cannot get more power out than you put in. The current ratio is therefore inversely proportional to the voltage ratio:

$$\frac{V_s}{V_p} = \frac{I_p}{I_s}$$

If the secondary voltage is stepped up, the secondary current goes down:

$$\frac{V_s\uparrow}{V_p} = \frac{I_p}{I_s\downarrow}$$

If the secondary voltage is stepped down, the secondary current goes up:

$$\frac{V_s\downarrow}{V_p} = \frac{I_p}{I_s\uparrow}$$

If the current ratio is inversely proportional to the voltage ratio, it is also inversely proportional to the turns ratio:

$$\frac{I_p}{I_s} = \frac{V_s}{V_p} = \frac{N_s}{N_p}$$

By rearranging the current and turns ratio, we can arrive at a formula for secondary current, which is

$$I_s = \frac{N_p}{N_s} \times I_p$$

EXAMPLE:

The step-up transformer in Figure 8-50 has a turns ratio of 1 to 5. Calculate:

a. Secondary voltage (V_s)
b. Secondary current (I_s)
c. Primary power (P_p)
d. Secondary power (P_s)

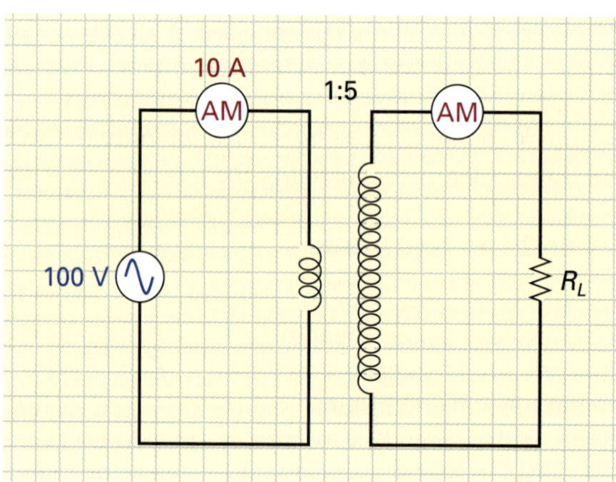

FIGURE 8-50 Step-Up Transformer Example.

Solution:

The secondary has five times as many windings as the primary, and, consequently, the voltage will be stepped up by a factor of 5 between primary and secondary. If the secondary voltage is going to be five times that of the primary, the secondary current is going to decrease to one-fifth of the primary current.

a. $V_s = \frac{N_s}{N_p} \times V_p$
$= \frac{5}{1} \times 100\text{ V}$
$= 500\text{ V}$

b. $I_s = \frac{N_p}{N_s} \times I_p$
$= \frac{1}{5} \times 10\text{ A}$
$= 2\text{ A}$

c. $P_p = V_p \times I_p = 100\text{ V} \times 10\text{ A} = 1000\text{ VA}$

d. $P_s = V_s \times I_s = 500\text{ V} \times 2\text{ A} = 1000\text{ VA}$

Therefore, $P_p = P_s$.

4. Impedance Ratio

The **maximum power transfer theorem,** which has been discussed previously and is summarized in Figure 8-51, states that maximum power is transferred from source (ac generator) to load (equipment) when the impedance of the load is equal to the internal impedance of the source. If these impedances are different, a large amount of power could be wasted.

Maximum Power Transfer Theorem

A theorem that states maximum power will be transferred from source to load when the source resistance is equal to the load resistance.

In most cases it is required to transfer maximum power from a source that has an internal impedance (Z_s) that is not equal to the load impedance (Z_L). In this situation, a transformer can be inserted between the source and the load to make the load impedance appear to equal the source's internal impedance.

For example, let's imagine that your car stereo system (source) has an internal impedance of 100 Ω and is driving a speaker (load) of 4 Ω impedance, as seen in Figure 8-52.

By choosing the correct turns ratio, the 4 Ω speaker can be made to appear as a 100 Ω load impedance, which will match the 100 Ω internal source impedance of the stereo system, resulting in maximum power transfer.

The turns ratio can be calculated by using the formula

$$\text{turns ratio} = \sqrt{\frac{Z_L}{Z_s}}$$

Z_L = Load impedance in ohms
Z_s = Source impedance in ohms

In our example, this will calculate out to be

$$\begin{aligned}\text{turns ratio} &= \sqrt{\frac{Z_L}{Z_s}} \\ &= \sqrt{\frac{4}{100}} = \frac{\sqrt{4}}{\sqrt{100}} \\ &= \frac{2}{10} = \frac{1}{5} \\ &= 0.2\end{aligned}$$

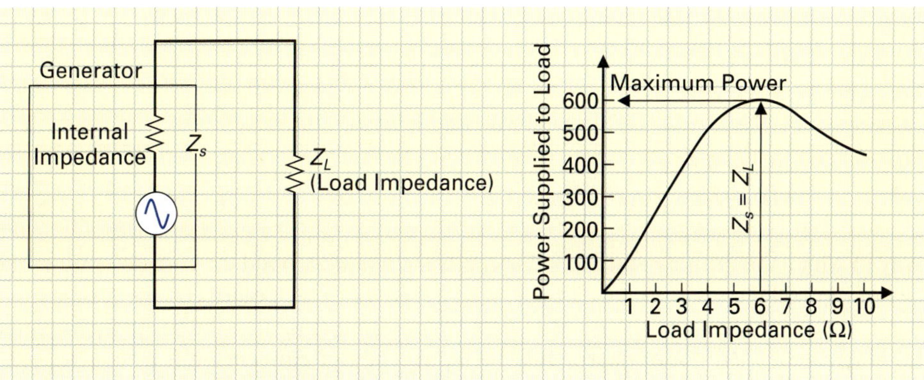

FIGURE 8-51 Maximum Power Transfer Theorem.

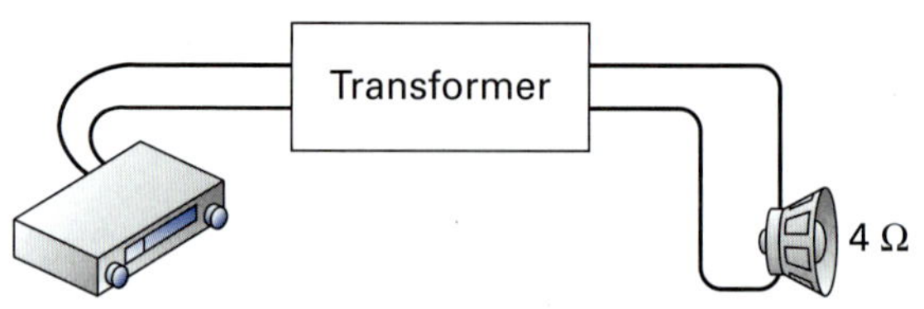

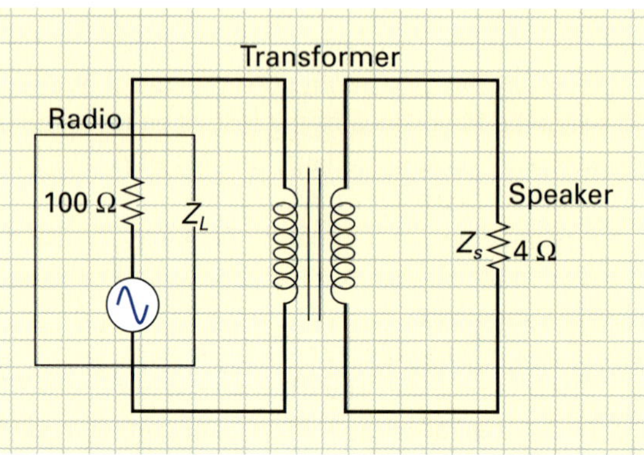

FIGURE 8-52 Impedance Matching.

If the turns ratio is less than 1, a step-down transformer is required. A turns ratio of 0.2 means a step-down transformer is needed with a turns ratio of 5:1 (⅕ = 0.2).

EXAMPLE:

Calculate the turns ratio needed to match the 22.2 Ω output impedance of an amplifier to two 16 Ω speakers connected in parallel.

Solution:

The total load impedance of two 16 Ω speakers in parallel will be

$$Z_L = \frac{\text{product}}{\text{sum}} = \frac{16 \times 16}{16 + 16} = 8\ \Omega$$

The turns ratio will be

$$\begin{aligned} \text{turns ratio} &= \sqrt{\frac{Z_L}{Z_s}} \\ &= \sqrt{\frac{8\ \Omega}{22.2\ \Omega}} \\ &= \sqrt{0.36} = 0.6 \end{aligned}$$

Therefore, a step-down transformer is needed with a turns ratio of 1.67:1.

8-4-4 *Windings and Phase*

Dot Convention
A standard used with transformer symbols to indicate whether the secondary voltage will be in phase or out of phase with the primary voltage.

The way in which the primary and secondary coils are wound around the core determines the polarity of the voltage induced into the secondary relative to the polarity of the primary.

In a schematic diagram, there has to be a way of indicating that the secondary voltage will be in phase or 180° out of phase with the input. The **dot convention** is a standard used with transformers and is illustrated in Figure 8-53(a) and (b). A positive on the primary dot causes a positive on the secondary dot, and similarly, since ac is applied, a negative on the

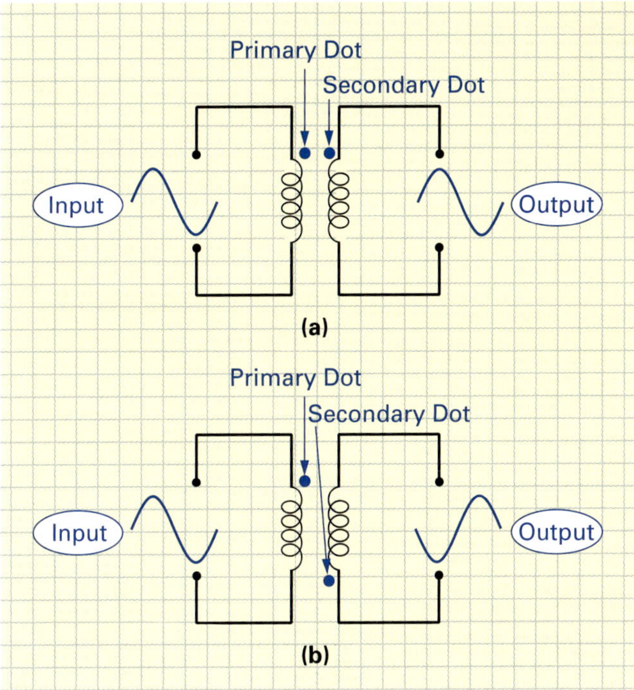

FIGURE 8-53 **Dot Convention. (a) In-Phase Dots. (b) Out-of-Phase Dots.**

primary dot will cause a negative on the secondary dot. In Figure 8-53(a), you can see that when the top of the primary swings positive or negative, the top of the secondary will also follow suit and swing positive or negative, respectively, and you will now be able to determine that the transformer secondary voltage is in phase with the primary voltage.

In Figure 8-53(b), however, the dots are on top and bottom, which means that as the top of the primary winding swings positive, the bottom of the secondary will go positive, which means the top of the secondary will actually go negative, resulting in the secondary ac voltage being out of phase with the primary ac voltage.

8-4-5 *Transformer Types*

Fixed turns ratio transformers have a turns ratio that cannot be varied. They are generally wound on a common core to ensure efficient magnetic coupling between primary and secondary and can be classified by the type of core material used.

The air-core transformers typically have a nonmagnetic core, such as ceramic or a cardboard hollow shell, and are used in high-frequency applications. The electronic circuit symbol just shows the primary and secondary coils. The more common iron- or ferrite-core transformers concentrate the magnetic lines of force, resulting in improved transformer performance; they are symbolized by two lines running between the primary and secondary. The iron-core transformer's lines are solid, while the ferrite-core transformer's lines are dashed.

Variable turns ratio transformers have a turns ratio that can be varied. Figure 8-54 illustrates the first two of these, the center-tapped and the multiple-tapped secondary types. Transformers can often have tapped secondaries. A tapped winding will have a lead connected to one of the loops other than the two end connections.

If the tapped lead is in the exact center of the secondary, the transformer is said to have a center-tapped secondary, as shown in Figure 8-55. With the **center-tapped transformer,** the two secondary voltages are each half of the total secondary voltage. If we assume a 1:1 turns ratio (primary turns = secondary turns) and a 20 V ac primary voltage and therefore secondary voltage, each of the output voltages between either end of the secondary and the center tap will be 10 V waveforms, as shown in Figure 8-55. The two secondary outputs will

Center-Tapped Transformer
A transformer that has a connection at the electrical center of the secondary winding.

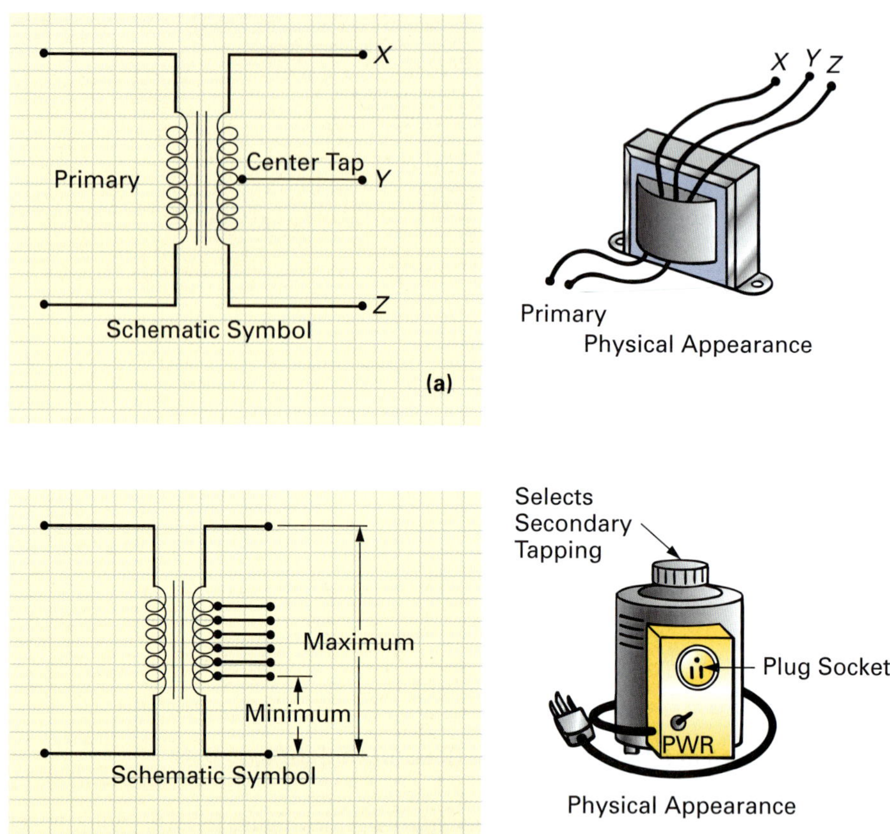

FIGURE 8-54 Secondary Tapped Transformers. (a) Center Tapped. (b) Multiple Tapped.

be 180° out of phase with one another from the center tap. When the top of the secondary swings positive (+), the bottom of the secondary will be negative, and vice versa.

Figure 8-56 illustrates a typical power company utility pole, which as you can see makes use of a center-tapped secondary transformer. These transformers are designed to step down the high 4.8 kV from the power line to 120 V/240 V for commercial and residential customers. Within the United States, the actual secondary voltage may be anywhere between 225 and 245 V, depending on the demand. The demand is actually higher during the day and in the winter months (source voltage is pulled down by smaller customer load resistance). The multiple taps on the primary are used to change the turns ratio and compensate for power line voltage differences.

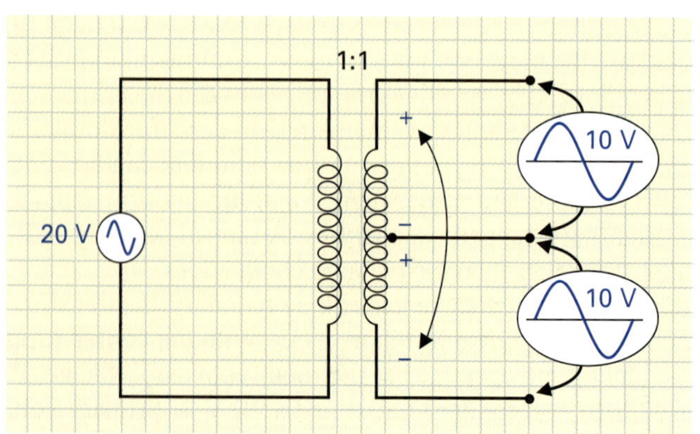

FIGURE 8-55 Center-Tapped Secondary Transformer.

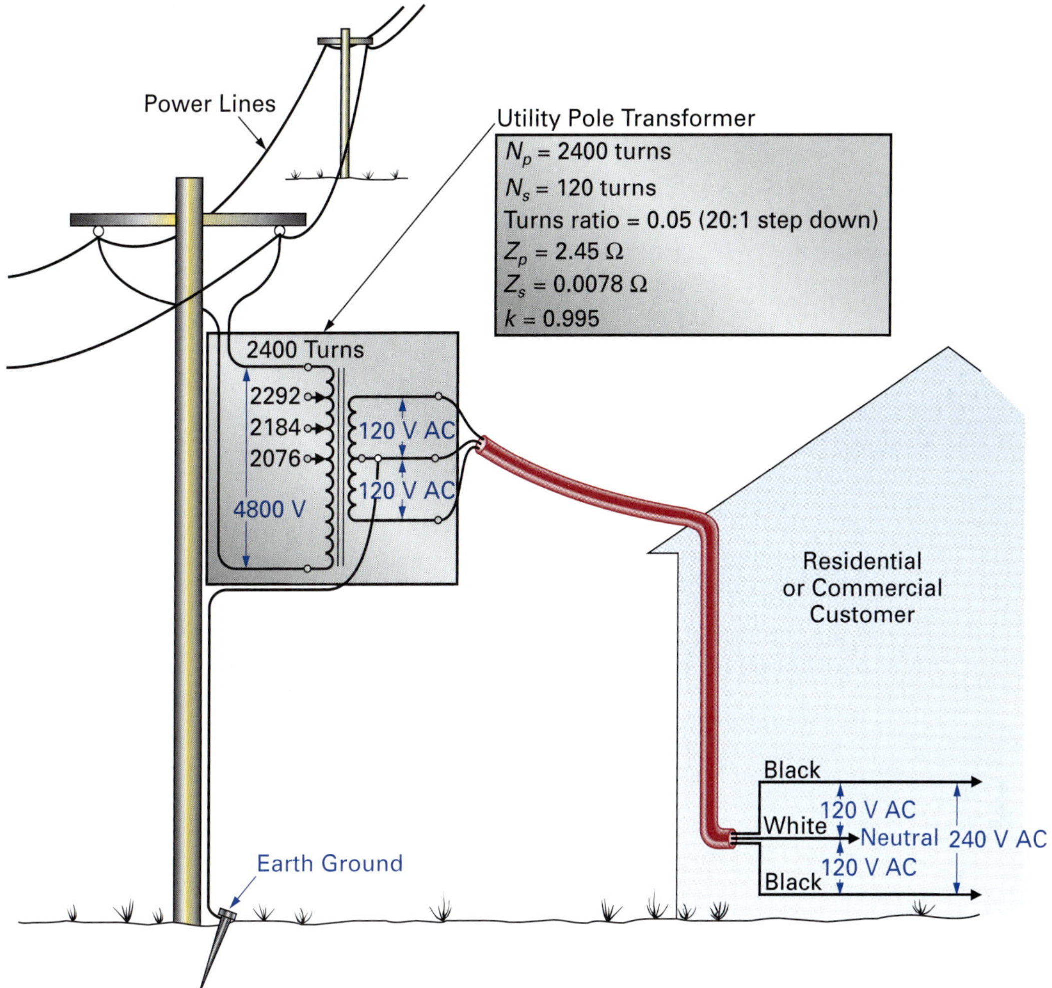

FIGURE 8-56 **Center-Tapped Utility Pole Transformer.**

The center tap of the secondary winding is connected to a copper earth grounding rod. This center tap wire is called the **neutral wire,** and within the building it is color coded with a white insulation. Residential or commercial appliances or loads that are designed to operate at 240 V are connected between the two black wires, while loads that are designed to operate at 120 V are connected between a black wire and the white wire (neutral).

Neutral Wire

The conductor of a polyphase circuit or of a single-phase three-wire circuit that is intended to have a ground potential. The potential differences between the neutral and each of the other conductors are approximately equal in magnitude and are also equally spaced in phase.

8-4-6 *Transformer Ratings*

A typical transformer rating could read 1 kVA, 500/100, 60 Hz. The 500 normally specifies the maximum primary voltage, the 100 normally specifies the maximum secondary voltage, and the 1 kVA is the apparent power rating. In this example, the maximum load current will equal

$$I_s = \frac{\text{apparent power } (P_A)}{\text{secondary voltage } (V_s)} \qquad \left(P = V \times I; \text{ therefore, } I = \frac{P}{V}\right)$$

$$= \frac{1 \text{ kVA}}{100 \text{ V}} = 10 \text{ A}$$

With this secondary voltage at 100 V and a maximum current of 10 A, the smallest load resistor that can be connected across the output of the secondary is

$$R_L = \frac{V_s}{I_s}$$

$$= \frac{100 \text{ V}}{10 \text{ A}}$$
$$= 10 \ \Omega$$

Exceeding the rating of the transformer will cause overheating and even burning out of the windings.

EXAMPLE:

Calculate the smallest value of load resistance that can be connected across a 3 kVA, 600/200, 60 Hz step-down transformer.

Solution:

$$I = \frac{\text{apparent power } (P_A)}{\text{secondary voltage } (V_s)}$$

$$\frac{3 \text{ kVA}}{200 \text{ V}} = 15 \text{ A}$$

$$R_L = \frac{V_s}{I_s} = \frac{200 \text{ V}}{15 \text{ A}} = 13.3 \ \Omega$$

8-4-7 *Testing Transformers*

An open in the primary winding will prevent any primary current, and therefore there will be no induced voltage in the secondary and therefore no voltage will be present across the load. An open secondary winding will prevent the flow of secondary current, and once again, no voltage will be present across the load.

An open in the primary or secondary winding is easily detected by disconnecting the transformer from the circuit and testing the resistance of the windings with an ohmmeter. Like an inductor, a transformer winding should have a low resistance, in the tens to hundreds range. An open will easily be recognized because of its infinite (maximum) resistance.

A partial or complete short in the primary winding of a transformer will result in an excessive source current that will probably blow the circuit's fuse or trip the circuit's breaker. A partial or complete short in the secondary winding will cause an excessive secondary current, which in turn will result in an excessive primary current.

In both instances, a short in the primary or secondary will generally burn out the primary winding unless the fuse or circuit breaker opens the excessive current path. An ohmmeter can be used to test for partial or complete shorts in transformer windings; however, all coils have a naturally low resistance which can be mistaken for a short. An inductive or reactive analyzer can accurately test transformer windings, checking the inductance value of each coil.

It is virtually impossible to repair an open, partially shorted, or completely shorted transformer winding, and therefore defective transformers are always replaced with a transformer that has an identical rating.

SELF-TEST EVALUATION POINT FOR SECTION 8-4

Use the following questions to test your understanding of Section 8-4.

1. True or false: An increase in secondary current causes an increase in primary current.
2. True or false: The greater the secondary current, the greater the primary's counter emf.
3. What is the turns ratio of a 402 turn primary and 1608 turn secondary, and is this transformer step up or step down?
4. State the formula for calculating the secondary voltage (V_s).
5. True or false: A transformer can, by adjusting the turns ratio, be made to step up both current and voltage between primary and secondary.
6. State the formula for calculating the secondary current (I_s).
7. What turns ratio is needed to match a 25 Ω source to a 75 Ω load?
8. Calculate V_s if $N_s = 200$, $N_p = 112$, and $V_p = 115$ V.
9. What do each of the values mean when a transformer is rated as a 10 kVA, 200/100, 60 Hz?
10. If a 100 Ω resistor is connected across the secondary of a transformer that is to supply 1 kV and is rated at a maximum current of 8 A, will the transformer overheat and possibly burn out?

REVIEW QUESTIONS

Multiple-Choice Questions

1. Relays can be used to:
   a. Allow one master switch to enable several others
   b. Allow several switches to enable one master
   c. Allow a switch in a low-current circuit to close contacts in a high-current circuit
   d. Both (a) and (b)
   e. Both (a) and (c)
2. The starter relay in an automobile is used to:
   a. Allow one master switch to enable several others
   b. Allow several switches to enable one master
   c. Allow a switch in a low-current circuit to close contacts in a high-current circuit
   d. Both (a) and (b)
3. An electromagnet is also known as a:
   a. Coil
   b. Solenoid
   c. Resistor
   d. Both (a) and (c)
   e. Both (a) and (b)
4. A normally open relay (NO) will have:
   a. Contacts closed until activated
   b. Contacts open until activated
   c. All contacts permanently open
   d. All contacts permanently closed
5. Direct current produces a/an ____________ magnetic field of ____________ polarity.
   a. Alternating, unchanging
   b. Constant, alternating
   c. Constant, unchanging
   d. Both (a) and (c)
6. Electromagnetism:
   a. Is the magnetism resulting from electrical current flow
   b. Is the electrical voltage resulting in a coil from the relative motion of a magnetic field
   c. Both (a) and (b)
   d. None of the above
7. Electromagnetic induction:
   a. Is the magnetism resulting from electrical current flow
   b. Is the electrical voltage resulting in a coil from the relative motion of a magnetic field
   c. Both (a) and (b)
   d. None of the above
8. Self-inductance is a process by which a coil will induce a voltage within ____________.
   a. Another inductor
   b. Two or more close proximity inductors
   c. Itself
   d. Both (a) and (b)
9. The inductor stores electrical energy in the form of a (an) ____________ field, just as a capacitor stores electrical energy in the form of a (an) ____________ field.
   a. Electric, magnetic
   b. Magnetic, electric
10. The inductor is basically:
    a. An electromagnet
    b. A coil of wire
    c. A coil of conductor formed around a core material
    d. All of the above
    e. None of the above
11. The total inductance of a series circuit is:
    a. Less than the value of the smallest inductor
    b. Equal to the sum of all the inductance values
    c. Equal to the product over sum
    d. All of the above

12. The total inductance of a parallel circuit can be calculated by:
    a. Using the product-over-sum formula
    b. Using L divided by N for equal-value inductors
    c. Using the reciprocal resistance formula
    d. All of the above
13. The time constant for a series inductive/resistive circuit is equal to:
    a. $L \times R$
    b. L/R
    c. V/R
    d. $2\pi \times f \times L$
14. Inductive reactance (X_L) is proportional to:
    a. Time or period of the ac applied
    b. Frequency of the ac applied
    c. The stray capacitance that occurs due to the air acting as a dielectric between two turns of a coil
    d. The value of inductance
    e. Two of the above are true.
15. In a series RL circuit, the source voltage (V_S) is equal to:
    a. The square root of the sum of V_R^2 and V_L^2
    b. The vector sum of V_R and V_L
    c. $I \times Z$
    d. Two of the above are partially true.
    e. Answers (a), (b), and (c) are correct.
16. With an RL integrator, the output is taken across the:
    a. Inductor
    b. Capacitor
    c. Resistor
    d. Transformer's secondary
17. With an RL differentiator, the output is taken across the:
    a. Inductor
    b. Capacitor
    c. Resistor
    d. Transformer's secondary
18. An inductor or choke between the input and output forms a:
    a. High-pass filter
    b. Low-pass filter
19. An inductor or choke connected to ground or in shunt forms a:
    a. Low-pass filter
    b. High-pass filter
20. When tested with an ohmmeter, an open coil would show:
    a. Zero resistance
    b. An infinite resistance
    c. A 100 Ω to 200 Ω resistance
    d. Both (b) and (c)
21. Transformer action is based on:
    a. Self-inductance
    b. Air between the coils
    c. Mutual capacitance
    d. Mutual inductance
22. An increase in transformer secondary current will cause a(an) ____________ in primary current.
    a. Decrease
    b. Increase
23. A step-up transformer will always have a turns ratio ____________, while a step-down transformer has a turns ratio ____________.
    a. $< 1, > 1$
    b. $> 1, > 1$
    c. $> 1, < 1$
    d. $< 1, < 1$
24. With an 80-V ac secondary voltage center-tapped transformer, what would be the voltage at each output, and what would be the phase relationship between the two secondary voltages?
    a. 20 V, in phase with one another
    b. 30 V, 180° out of phase
    c. 40 V, in phase
    d. 40 V, 180° out of phase
25. Alternating current can be used only with transformers because:
    a. It produces an alternating magnetic field.
    b. It produces a fixed magnetic field.
    c. Its magnetic field is greater than that of dc.
    d. Its rms is 0.707 of the peak.

Practice Problems

26. Convert the following:
    a. 0.037 H to mH
    b. 1760 μH to mH
    c. 862 mH to H
    d. 0.256 mH to μH
27. Calculate the impedance (Z) of the following series RL combinations:
    a. 22 MΩ, 25 μH, f = 1 MHz
    b. 4 kΩ, 125 mH, f = 100 kHz
    c. 60 Ω, 0.05 H, f = 1 MHz
28. Calculate the total inductance of the following series circuits:
    a. 75 μH, 61 μH, 50 mH
    b. 8 mH, 4 mH, 22 mH
29. Calculate the total inductance of the following parallel circuits:
    a. 12 mH, 8 mH
    b. 75 μH, 34 μH, 27 μH
30. Calculate the total inductance of the following series–parallel circuits:
    a. 12 mH in series with 4 mH, and both in parallel with 6 mH
    b. A two-branch parallel arrangement made up of 6 μH and 2 μH in series with one another, and 8 μH and 4 μH in series with one another
    c. Two parallel arrangements in series with one another, made up of 1 μH and 2 μH in parallel and 4 μH and 15 μH in parallel

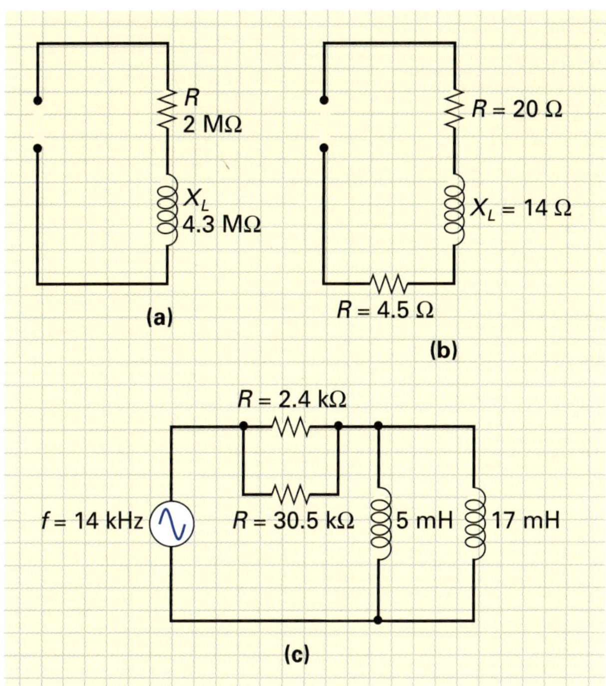

FIGURE 8-57 Calculating Impedance.

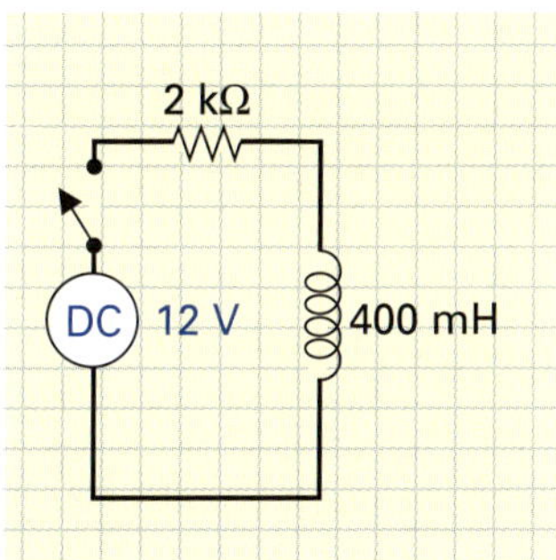

FIGURE 8-58 Inductive Time Constant Example.

31. In a series *RL* circuit, if $V_L = 12$ V and $V_R = 6$ V, calculate:

- **a.** V_S
- **b.** I if $Z = 14$ kΩ
- **c.** Phase angle
- **d.** Q
- **e.** Power factor

32. Calculate the impedance of the circuits seen in Figure 8-57.

33. Referring to Figure 8-58, calculate the voltage across the inductor for all five time constants after the switch has been closed.

34. Referring to Figure 8-59, calculate:

- **a.** L_T
- **b.** X_L
- **c.** Z
- **d.** I_{R1}, I_{L1}, and I_{L2}
- **e.** θ
- **f.** True power, reactive power, and apparent power
- **g.** Power factor

35. Referring to Figure 8-60, calculate:

- **a.** R_T
- **b.** L_T
- **c.** X_L
- **d.** Z
- **e.** V_{RT}, V_{LT}
- **f.** I_{RT}, I_{LT}
- **g.** I_T
- **h.** θ
- **i.** Apparent power
- **j.** PF

36. Calculate the turns ratio of the following transformers and state whether they are step up or step down:

- **a.** P = 12 T, S = 24 T
- **b.** P = 3 T, S = 250 T
- **c.** P = 24 T, S = 5 T
- **d.** P = 240 T, S = 120 T

37. What turns ratio would be needed to match a source impedance of 24 Ω to a load impedance of 8 Ω?

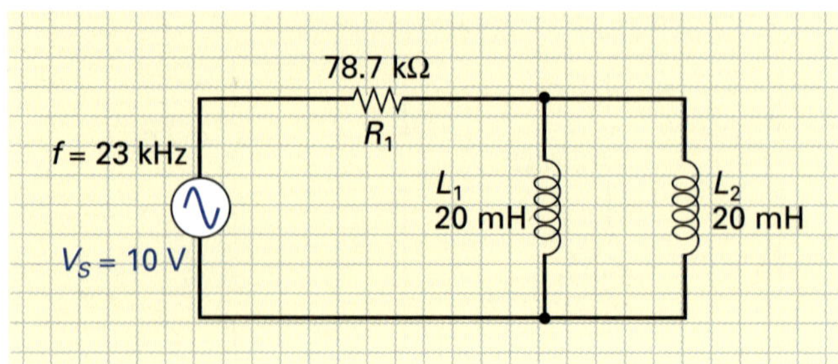

FIGURE 8-59 **Series *RL* Circuit.**

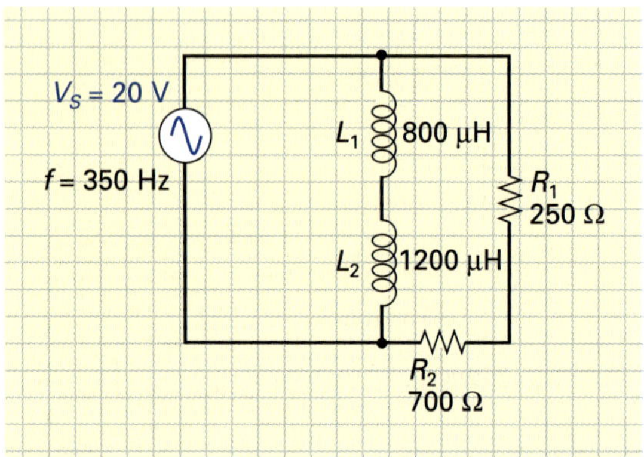

FIGURE 8-60 **Parallel *RL* Circuit.**

38. What turns ratio would be needed to step:
a. 120 V to 240 V **c.** 30 V to 14 V
b. 240 V to 720 V **d.** 24 V to 6 V

39. Referring to Figure 8-61(a) and (b), sketch the outputs, showing polarity and amplitude with respect to the inputs.

40. If a transformer is rated at 500 VA, 60 Hz, the primary voltage is 240 V ac, and the secondary voltage is 600 V ac, calculate:
a. Maximum load current
b. Smallest value of R_L

Web Site Questions

Go to the web site http://www.prenhall.com/cook, select the textbook *Electronics: A Complete Course,* select this chapter, and then follow the instructions when answering the multiple choice practice problems.

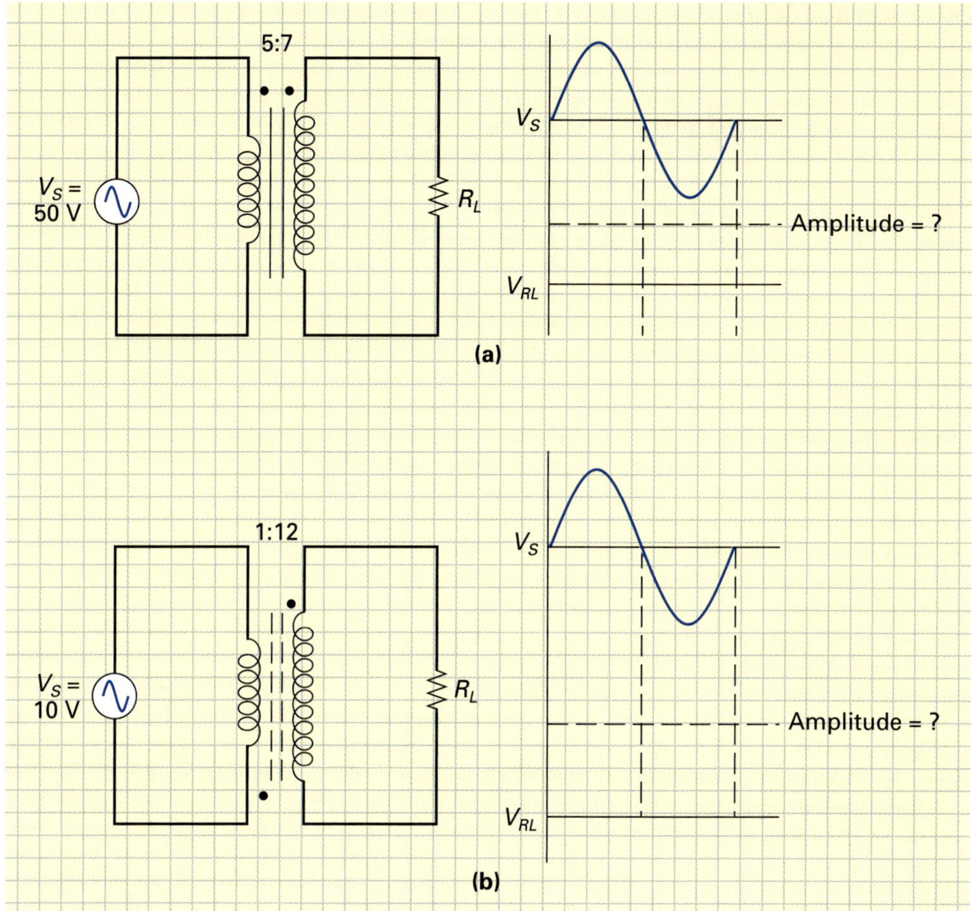

FIGURE 8-61 **Input/Output Polarity Examples.**

Resistive, Inductive, and Capacitive (*RLC*) Circuits

The First Computer Bug

Mathematician Grace Murray Hopper, an extremely independent U.S. naval officer, was assigned to the Bureau of Ordnance Computation Project at Harvard during World War II. As Hopper recalled, "We were not programmers in those days, the word had not yet come over from England. We were 'coders,' " and with her colleagues she was assigned to compute ballistic firing tables on the Harvard Mark 1 computer. In carrying out this task, Hopper developed programming method fundamentals that are still in use.

Hopper is also credited, on a less important note, with creating a term frequently used today falling under the category of computer jargon. During the hot summer of 1945, the computer developed a mysterious problem. Upon investigation, Hopper discovered that a moth had somehow strayed into the computer and prevented the operation of one of the thousands of electromechanical relay switches. In her usual meticulous manner, Hopper removed the remains and taped and entered it into the logbook. In her own words, "From then on, when an officer came in to ask if we were accomplishing anything, we told him we were 'debugging' the computer," a term that is still used to describe the process of finding problems in a computer program.

9

Introduction

In this chapter we will combine resistors (R), inductors (L), and capacitors (C) into series and parallel ac circuits. Resistors, as we have discovered, operate and react to voltage and current in a very straightforward way, in that the voltage across a resistor is in phase with the resistor current.

Inductors and capacitors operate in essentially the same way, in that they both store energy and then return it back to the circuit. However, they have completely opposite reactions to voltage and current. To help you remember the phase relationships between voltage and current for capacitors and inductors, you may wish to use the following memory phrase:

ELI the ICE man

This phrase states that voltage (symbolized E) leads current (I) in an inductive (L) circuit (abbreviated by the word "ELI"), while current (I) leads voltage (E) in a capacitive (C) circuit (abbreviated by the word "ICE").

In this chapter we will study the relationships between voltage, current, impedance, and power in both series and parallel *RLC* circuits. We will also examine the important *RLC* circuit characteristic called resonance, and see how *RLC* circuits can be made to operate as filters. In the final section we will discuss how complex numbers can be used to analyze series and parallel ac circuits containing resistors, inductors, and capacitors.

9-1 SERIES *RLC* CIRCUIT

Figure 9-1 begins our analysis of series *RLC* circuits by illustrating the current and voltage relationships. The circuit current is always the same throughout a series circuit and can therefore be used as a reference. Studying the waveforms and vector diagrams shown alongside the components, you can see that the voltage across a resistor is always in phase with the current, while the voltage across the inductor leads the current by 90°, and the voltage across the capacitor lags the current by 90°.

Now let's analyze the impedance, current, voltage, and power distribution of this circuit in a little more detail.

9-1-1 *Impedance*

Impedance is the total opposition to current flow and is a combination of both reactance (X_L, X_C) and resistance (R).

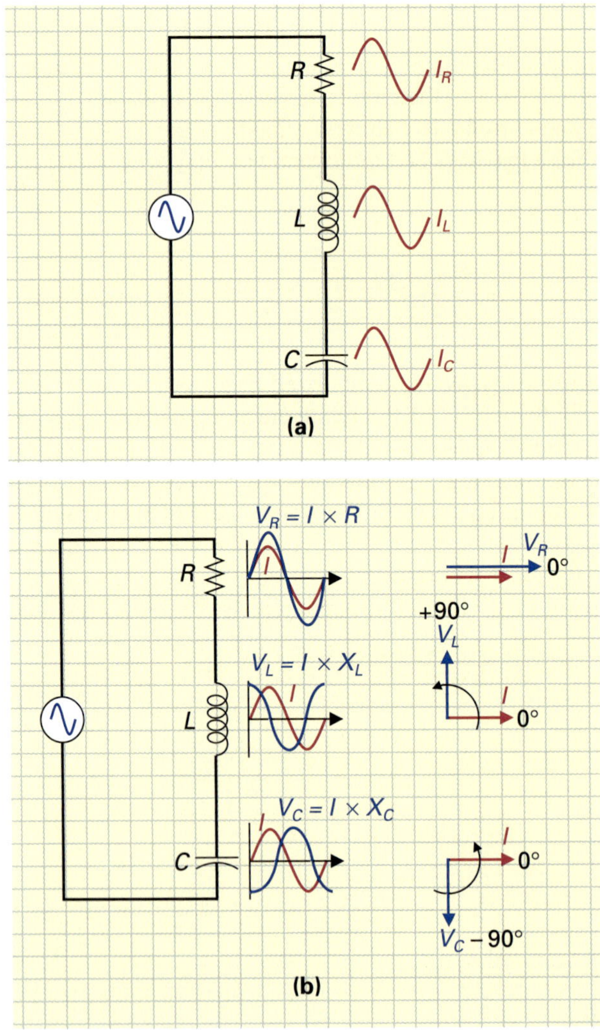

FIGURE 9-1 **Series *RLC* Circuit. (a) *RLC* Series Circuit Current: Current Flow Is Always the Same in All Parts of a Series Circuit. (b) *RLC* Series Circuit Voltages: *I* Is in Phase with V_R, *I* Lags V_L by 90°, and *I* Leads V_C by 90°.**

Capacitive reactance can be calculated by using the formula

$$X_C = \frac{1}{2\pi fC}$$

In the example,

$$X_C = \frac{1}{2\pi \times 60 \times 10\ \mu\text{F}} = 265.3\ \Omega$$

Inductive reactance is calculated by using the formula

$$X_L = 2\pi fL$$

In the example,

$$X_L = 2\pi \times 60 \times 20\ \text{mH} = 7.5\ \Omega$$

The difference between X_L and X_C is equal to 257.8, and since X_C is greater than X_L, the circuit is said to act capacitively. Reactance, however, is not in phase with resistance, and impedance is the vector sum of the reactive (X) and resistive (R) vectors. The formula, based on the Pythagorean theorem, is

$$Z = \sqrt{R^2 + X^2}$$

In this example, therefore, the circuit impedance will be equal to

$$Z = \sqrt{R^2 + X^2} = \sqrt{33^2 + 257.8^2} = 260\ \Omega$$

Since reactance (X) is equal to the difference between (symbolized~) X_L and X_C ($X_L \sim X_C$), the impedance formula can be modified slightly to incorporate the calculation to determine the difference between X_L and X_C.

$$Z = \sqrt{R^2 + (X_L \sim X_C)^2}$$

Using our example with this new formula, we arrive at the same value of impedance, and since the difference between X_L and X_C resulted in a capacitive vector, the circuit is said to act capacitively.

$$\begin{aligned} Z &= \sqrt{R^2 + (X_L \sim X_C)^2} \\ &= \sqrt{33^2 + (7.5 \sim 265.3)^2} \\ &= \sqrt{33^2 + 257.8^2} \\ &= 260\ \Omega \end{aligned}$$

If, on the other hand, the component values were such that the difference was inductive, then the circuit would be said to act inductively.

9-1-2 *Current*

The current in a series circuit is the same at all points throughout the circuit, and therefore

$$I = I_R = I_L = I_C$$

Once the total impedance of the circuit is known, Ohm's law can be applied to calculate the circuit current:

$$I = \frac{V_S}{Z}$$

In the example, circuit current is equal to

$$I = \frac{V_S}{Z}$$
$$= \frac{115\text{ V}}{260\ \Omega}$$
$$= 0.44\text{ A} \quad \text{or} \quad 440\text{ mA}$$

9-1-3 *Voltage*

Now that you know the value of current flowing in the series circuit, you can calculate the voltage drops across each component, as shown in Figure 9-2.

$$V_R = I \times R$$

$$V_L = I \times X_L$$

$$V_C = I \times X_C$$

Since none of these voltages are in phase with one another, they must be added vectorially to obtain the applied voltage.

$$V_S = \sqrt{V_R^2 + (V_L \sim V_C)^2}$$

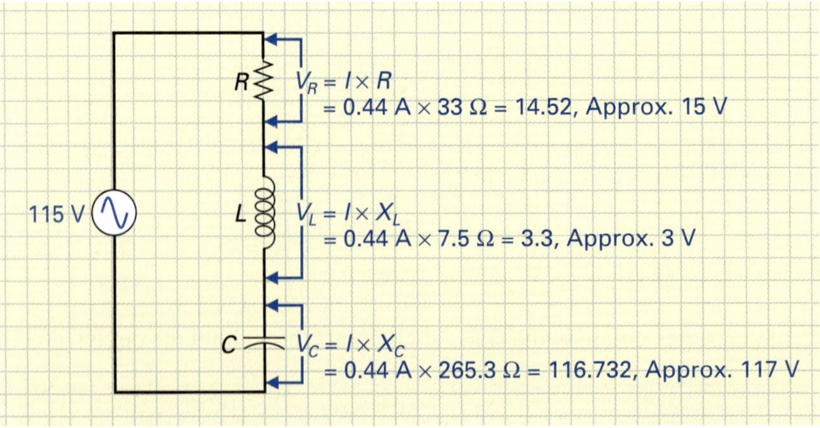

FIGURE 9-2 **Series Voltage Drops.**

In the example circuit, the applied voltage is, as we already know, 115 V.

$$\begin{aligned} V_S &= \sqrt{15^2 + (3 \sim 117)^2} \\ &= \sqrt{225 + 12{,}996} \\ &= \sqrt{13{,}221} \\ &= 115 \text{ V} \end{aligned}$$

9-1-4 *Phase Angle*

As can be seen in Figure 9-1(b), there is a phase difference between the source voltage (V_S) and the circuit current (I). This phase difference can be calculated with either of the following formulas.

$$\theta = \arctan\frac{V_L \sim V_C}{V_R}$$

$$\theta = \arctan\frac{X_L \sim X_C}{R}$$

In the example circuit, θ is

$$\begin{aligned} \theta &= \arctan\frac{3.3 \text{ V} \sim 116.732 \text{ V}}{14.52 \text{ V}} \\ &= \arctan 7.812 \\ &= 82.7° \end{aligned}$$

Since the example circuit is capacitive (ICE), the phase angle will be $-82.7°$ since V_S lags I in a circuit that acts capacitively.

9-1-5 *Power*

The true power or resistive power (P_R) dissipated by a circuit can be calculated using the formula

$$P_R = I^2 \times R$$

which in our example will be

$$P_R = (0.44 \text{ A})^2 \times 33\ \Omega = 6.4 \text{ W}$$

The apparent power (P_A) consumed by the circuit is calculated by

$$P_A = V_S \times I$$

which in our example will be

$$P_A = 115 \text{ V} \times 0.44 \text{ A} = 50.6 \text{ volt-amperes (VA)}$$

The true or actual power dissipated by the resistor is, as expected, smaller than the apparent power that appears to be being used.

The power factor can be calculated, as usual, by

$$\text{PF} = \cos\theta = \frac{R}{Z} = \frac{P_R}{P_A}$$

PF of 0 = reactive circuit
PF of 1 = resistive circuit

In the example circuit, PF = 0.126, indicating that the circuit is mainly reactive.

EXAMPLE:

For a series circuit where $R = 10\ \Omega$, $L = 5$ mH, $C = 0.05\ \mu$F, and $V_S = 100$ V/2 kHz, calculate:

a. X_C
b. X_L
c. Z
d. I
e. V_R, V_C, and V_L
f. Apparent power
g. True power
h. Power factor
i. Phase angle

Solution:

a. $X_C = \dfrac{1}{2\pi fC} = 1.6\ \text{k}\Omega$

b. $X_L = 2\pi fL = 62.8\ \Omega$

c. $Z = \sqrt{R^2 + (X_L \sim X_C)^2}$
$= \sqrt{(10\ \Omega)^2 + (1.6\ \text{k}\Omega \sim 62.8\ \Omega)^2}$
$= \sqrt{(10\ \Omega)^2 + (1.54\ \text{k}\Omega)^2}$
$= 1.54\ \text{k}\Omega$ (capacitive circuit due to high X_C)

d. $I = \dfrac{V_S}{Z} = \dfrac{100\ \text{V}}{1.54\ \text{k}\Omega} = 64.9\ \text{mA}$

e. $V_R = I \times R = 64.9\ \text{mA} \times 10\ \Omega = 0.65\ \text{V}$
$V_C = I \times X_C = 64.9\ \text{mA} \times 1.6\ \text{k}\Omega = 103.9\ \text{V}$
$V_L = I \times X_L = 64.9\ \text{mA} \times 62.8\ \Omega = 4.1\ \text{V}$

f. Apparent power $= V_S \times I = 100\ \text{V} \times 64.9\ \text{mA} = 6.49\ \text{VA}$

g. True power $= I^2 \times R = (64.9\ \text{A})^2 \times 10\ \Omega = 42.17\ \text{mW}$

h. $\text{PF} = \dfrac{R}{Z} = \dfrac{10\ \Omega}{1.5\ \text{k}\Omega} = 0.006$ (reactive circuit)

i. $\theta = \arctan \dfrac{V_L \sim V_C}{V_R}$
$= \arctan \dfrac{4.1\ \text{V} \sim 103.9\ \text{V}}{0.65\ \text{V}}$
$= \arctan - 153.54$
$= 89.63°$

Capacitive circuit (ICE); therefore, V_S lags I by $-89.63°$.

SELF-TEST EVALUATION POINT FOR SECTION 9-1

Use the following questions to test your understanding of Section 9-1.

1. List in order the procedure that should be followed to fully analyze a series *RLC* circuit.

2. State the formulas for calculating the following in relation to a series *RLC* circuit.

a. Impedance
b. Current
c. Apparent power
d. V_S
e. True power
f. V_R
g. V_L
h. V_C
i. Phase angle (θ)
j. Power factor (PF)

9-2 PARALLEL *RLC* CIRCUIT

Now that the characteristics of a series circuit are understood, let us connect a resistor, inductor, and capacitor in parallel with one another. Figure 9-3(a) and (b) show the current and voltage relationships of a parallel *RLC* circuit.

9-2-1 *Voltage*

As can be seen in Figure 9-3(a), the voltage across any parallel circuit will be equal and in phase. Therefore,

$$V_R = V_L = V_C = V_S$$

9-2-2 *Current*

With current, we must first calculate the individual branch currents (I_R, I_L, and I_C) and then calculate the total circuit current (I_T). An example circuit is illustrated in Figure 9-4, and the branch currents can be calculated by using the formulas

$$I_R = \frac{V}{R}$$

$$I_L = \frac{V}{X_L}$$

$$I_C = \frac{V}{X_C}$$

The 180° phase difference between I_C and I_L results in a cancellation.

The total current (I_T) can be calculated by using the Pythagorean theorem on the right triangle.

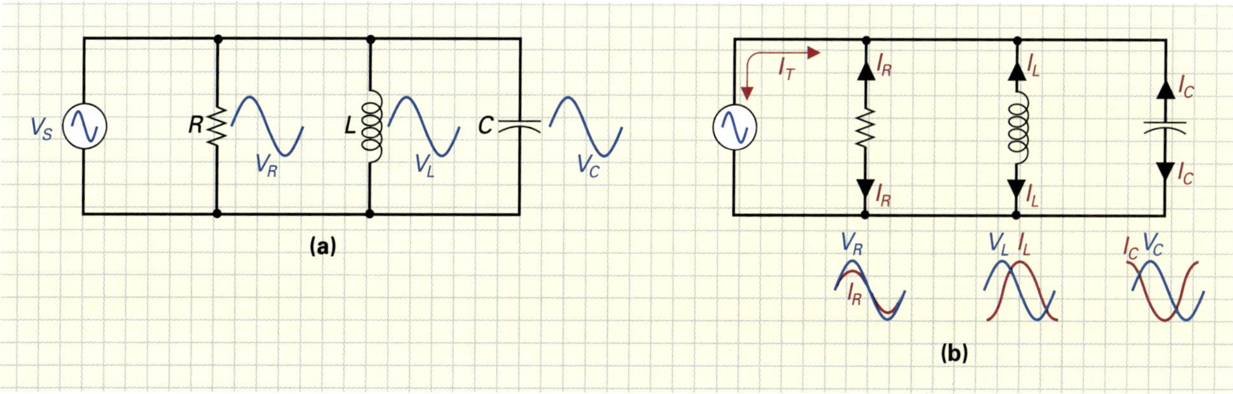

FIGURE 9-3 **Parallel *RLC* Circuit. (a) *RLC* Parallel Circuit Voltage: Voltages across Each Component Are All Equal and in Phase with One Another in a Parallel Circuit. (b) *RLC* Parallel Circuit Currents: I_R Is in Phase with V_R, I_L Lags V_L by 90°, and I_C Leads V_C by 90°.**

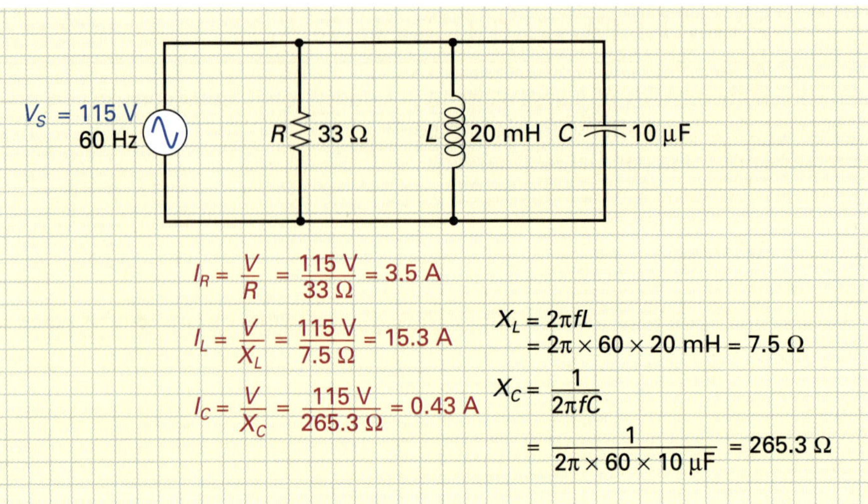

FIGURE 9-4 **Example *RLC* Parallel Circuit.**

$$I_T = \sqrt{I_R^2 + I_X^2}$$

$$(I_X = I_L \sim I_C)$$

$$= \sqrt{3.5^2 + 14.9^2}$$
$$= 15.3\text{ A}$$

9-2-3 *Phase Angle*

There is a phase difference between the source voltage (V_S) and the circuit current (I_T). This phase difference can be calculated using the formula

$$\theta = \arctan \frac{I_L \sim I_C}{I_R}$$

$$= \arctan \frac{15.3\text{ A} \sim 0.43\text{ A}}{3.5\text{ A}}$$
$$= \arctan 4.25$$
$$= 76.7°$$

Since this is an inductive circuit (ELI), the total current (I_T) will lag the source voltage (V_S) by $-76.7°$.

9-2-4 *Impedance*

With the total current (I_T) known, the impedance of all three components in parallel can be calculated by the formula

$$Z = \frac{V}{I_T}$$

$$= \frac{115\text{ V}}{15.3\text{ A}}$$
$$= 7.5\ \Omega$$

9-2-5 *Power*

The true power dissipated can be calculated using

$$P_R = I_R^2 \times R$$

$$= (3.5\text{ A})^2 \times 33\ \Omega$$
$$= 404.3\text{ W}$$

The apparent power consumed by the circuit is calculated by

$$P_A = V_S \times I_T$$

$$= 115\text{ V} \times 15.3\text{ A}$$
$$= 1759.5\text{ volt-amperes (VA)}$$

Finally, the power factor can be calculated, as usual, with

$$\text{PF} = \cos\theta = \frac{P_R}{P_A}$$

In the example circuit, PF = 0.23.

SELF-TEST EVALUATION POINT FOR SECTION 9-2

Use the following questions to test your understanding of Section 9-2.

1. State the formulas for calculating the following in relation to a parallel *RLC* circuit:
   - **a.** I_R
   - **b.** I_T
   - **c.** I_C
   - **d.** I_L
   - **e.** θ
   - **f.** Z
2. State the formulas for:
   - **a.** P_R
   - **b.** P_X
   - **c.** P_A
   - **d.** PF

9-3 RESONANCE

Resonance is a circuit condition that occurs when the inductive reactance (X_L) and the capacitive reactance (X_C) have been balanced. Figure 9-5 illustrates a parallel- and a series-connected *LC* circuit. If a dc voltage is applied to the input of either circuit, the capacitor will act as an open (X_C = infinite Ω) and the inductor will act as a short ($X_L = 0\ \Omega$).

Resonance
Circuit condition that occurs when the inductive reactance (X_L) is equal to the capacitive reactance (X_C).

If a low-frequency ac is now applied to the input, X_C will decrease from maximum, and X_L will increase from zero. As the ac frequency is increased further, the capacitive reactance will continue to fall ($X_C \downarrow \propto 1/f\uparrow$) and the inductive reactance to rise ($X_L\uparrow \propto f\uparrow$), as shown in Figure 9-6.

As the input ac frequency is increased further, a point will be reached where X_L will equal X_C, and this condition is known as *resonance*. The frequency at which $X_L = X_C$ in either a parallel or a series *LC* circuit is known as the *resonant frequency* (f_0) and can be

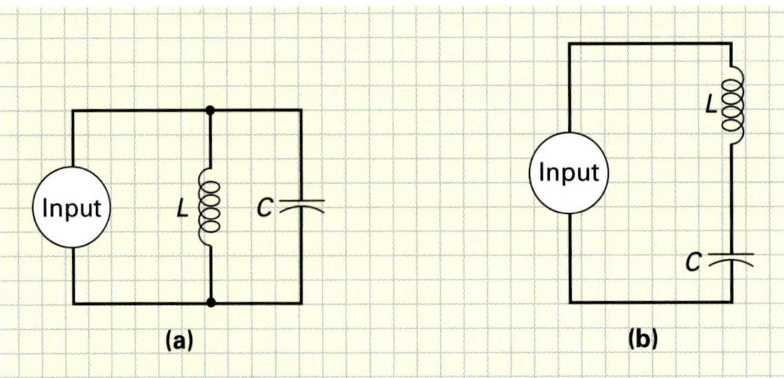

FIGURE 9-5 **Resonance. (a) Parallel *LC* Circuit. (b) Series *LC* Circuit.**

calculated by the following formula, which has been derived from the capacitive and inductive reactance formulas:

$$f_0 = \frac{1}{2\pi\sqrt{LC}}$$

where f_0 = resonant frequency, in hertz (Hz)
L = inductance, in henrys (H)
C = capacitance, in farads (F)

EXAMPLE:

Calculate the resonant frequency (f_0) of a series LC circuit if L = 750 mH and C = 47 μF.

Solution:

$$\begin{aligned} f_0 &= \frac{1}{2\pi\sqrt{L \times C}} \\ &= \frac{1}{2\pi\sqrt{(750 \times 10^{-3}) \times (47 \times 10^{-6})}} \\ &= 26.8 \text{ Hz} \end{aligned}$$

FIGURE 9-6 **Frequency versus Reactance.**

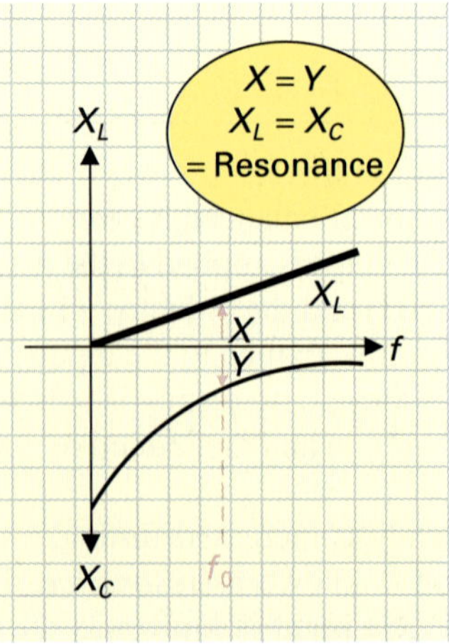

9-3-1 *Series Resonance*

Figure 9-7(a) illustrates a series *RLC* circuit at resonance ($X_L = X_C$), or **series resonant circuit.** The ac input voltage causes current to flow around the circuit, and since all the components are connected in series, the same value of current (I_S) will flow through all the components. Since *R*, X_L, and X_C are all equal to 100 Ω and the current flow is the same throughout, the voltage dropped across each component will be equal.

Series Resonant Circuit
A resonant circuit in which the capacitor and coil are in series with the applied ac voltage.

The voltage across the resistor is in phase with the series circuit current (I_S); however, since the voltage across the inductor (V_L) is 180° out of phase with the voltage across the capacitor (V_C), and both are equal to one another, V_L cancels V_C, when both are measured in series.

Three unusual characteristics occur when a circuit is at resonance, which do not occur at any other frequency.

(1) The first is that if V_L and V_C cancel, the voltage across *L* and *C* will measure 0 V on a voltmeter. Since there is effectively no voltage being dropped across these two components, all the voltage must be across the resistor (V_R = 12 V). This is true; however, since the same current flows throughout the series circuit, a voltmeter will measure 12 V across *C*, 12 V across *L*, and 12 V across *R*, as shown in Figure 9-7(b). It now appears that the voltage drops around the series circuit (36 V) do not equal the voltage applied (12 V). This is not true, as V_L and V_C cancel, because they are out of phase with one another, so Kirchhoff's voltage law is still valid.

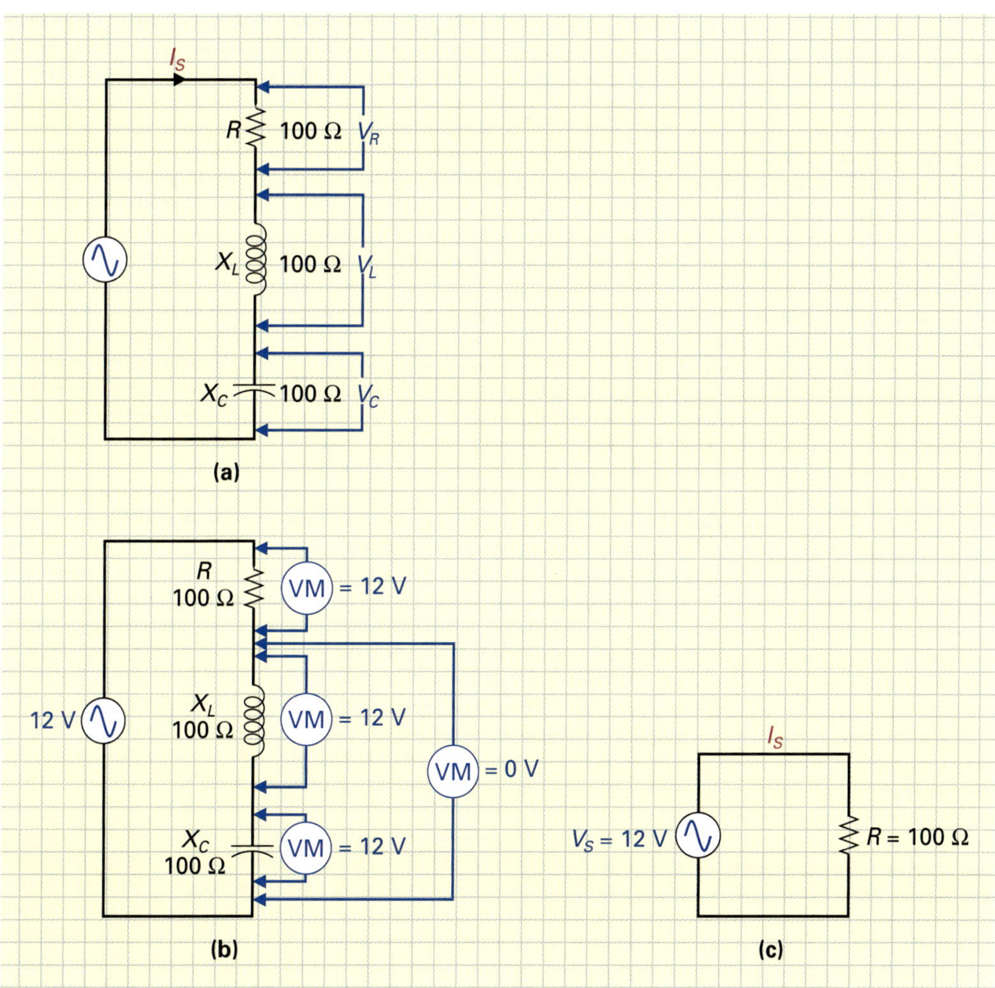

FIGURE 9-7 Series Resonant Circuit.

(2) The second unusual characteristic of resonance is that because the total opposition or impedance (Z) is equal to

$$Z = \sqrt{R^2 + (X_L \sim X_C)^2}$$

and the difference between X_L and X_C is 0 ($Z = \sqrt{R^2 + 0}$), the impedance of a series circuit at resonance is equal to the resistance value R ($Z = \sqrt{R^2} = R$). As a result, the applied ac voltage of 12 V is forcing current to flow through this series *RLC* circuit. Since current is equal to $I_S = V/Z$ and $Z = R$, the circuit current at resonance is dependent only on the value of resistance. The capacitor and inductor are invisible and are seen by the source as simply a piece of conducting wire with no resistance, as illustrated in Figure 9-7(c). Since only resistance exists in the circuit, current (I_S) and voltage (V_S) are in phase with one another, and as expected for a purely resistive circuit, the power factor will be equal to 1.

(3) To emphasize the third strange characteristic of series resonance, we will take another example, shown in Figure 9-8. In this example, R is made smaller (10 Ω) than X_L and X_C (100 Ω each). The circuit current in this example is equal to $I = V/R = 12\text{ V}/10\ \Omega = 1.2\text{ A}$, as $Z = R$ at resonance. Since the same current flows throughout a series circuit, the voltage across each component can be calculated.

$$V_R = I \times R = 1.2\text{ A} \times 10\ \Omega = 12\text{ V}$$
$$V_L = I \times X_L = 1.2\text{ A} \times 100\ \Omega = 120\text{ V}$$
$$V_C = I \times X_C = 1.2\text{ A} \times 100\ \Omega = 120\text{ V}$$

As V_L is 180° out of phase with V_C, the 120 V across the capacitor cancels with the 120 V across the inductor, resulting in 0 V across L and C combined, as shown in Figure 9-8(b).

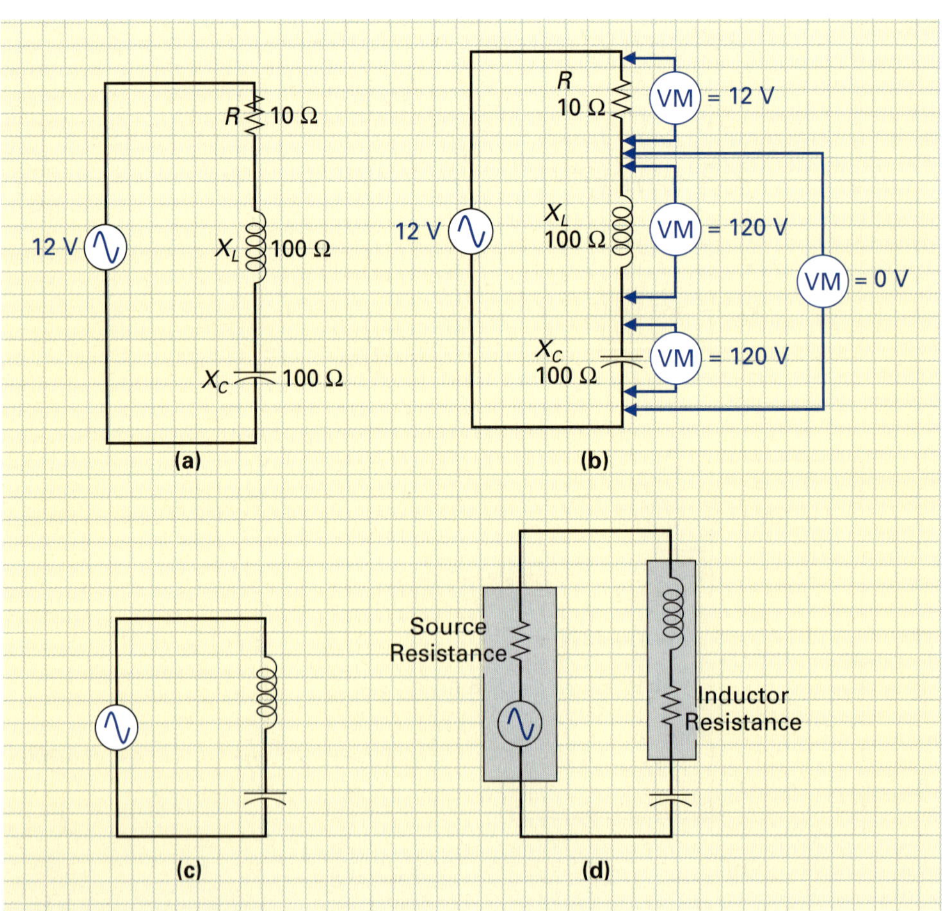

FIGURE 9-8 Circuit Effects at Resonance.

Since L and C have the ability to store energy, the voltage across them individually will appear larger than the applied voltage.

If the resistance in the circuit is removed completely, as shown in Figure 9-8(c), the circuit current, which is determined by the resistance only, will increase to a maximum ($I\uparrow = V/R\downarrow$) and, consequently, cause an infinitely high voltage across the inductor and capacitor ($V\uparrow = I\uparrow \times R$). In reality, the ac source will have some value of internal resistance, and the inductor, which is a long length of thin wire ($R\uparrow$), will have some value of resistance, as shown in Figure 9-8(d), which limits the series resonant circuit current.

In summary, we can say that in a series resonant circuit:

1. The inductor and capacitor electrically disappear due to their equal but opposite effect, resulting in a 0 V drop across the series combination, and the circuit seems purely resistive.
2. The current flow is large because the impedance of the circuit is low and equal to the series resistance (R), which has the source voltage developed across it.
3. The individual voltage drops across the inductor or capacitor can be larger than the source voltage if R is smaller than X_L and X_C.

Quality Factor

As discussed previously in the inductance chapter, the Q factor is a ratio of inductive reactance to resistance and is used to express how efficiently an inductor will store rather than dissipate energy. In a series resonant circuit, the Q factor indicates the quality of the series resonant circuit, or is the ratio of the reactance to the resistance.

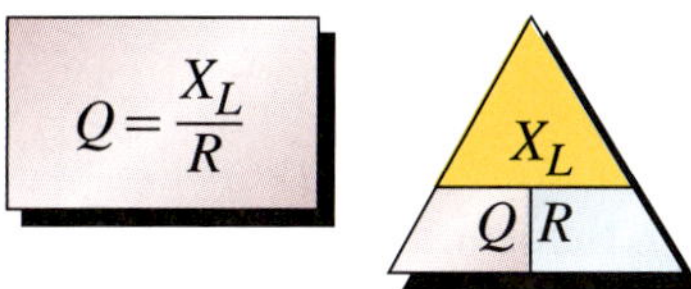

$$Q = \frac{X_L}{R}$$

or, since $X_L = X_C$,

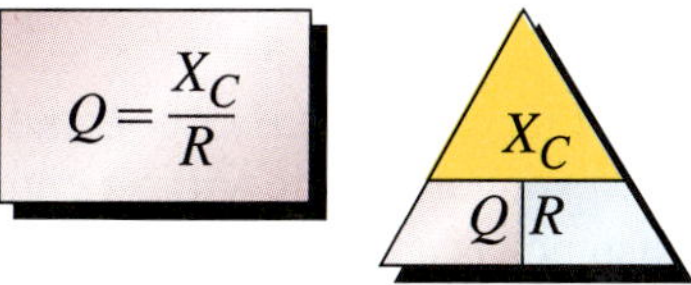

$$Q = \frac{X_C}{R}$$

Another way to calculate the Q of a series resonant circuit is by using the formula

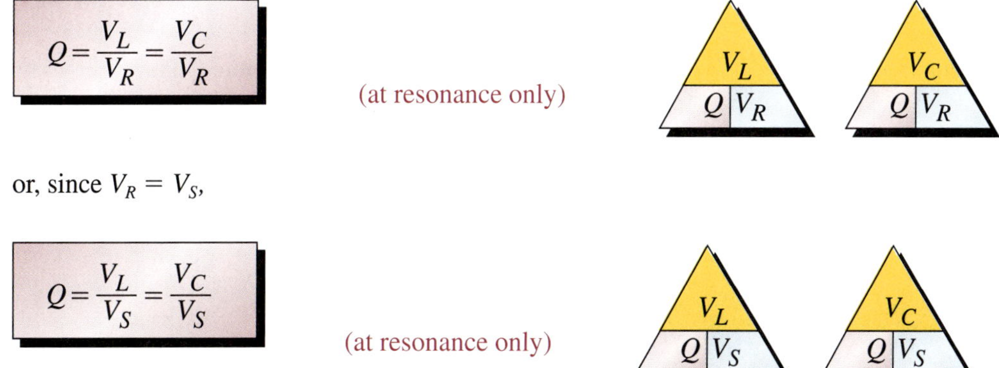

$$Q = \frac{V_L}{V_R} = \frac{V_C}{V_R}$$

(at resonance only)

or, since $V_R = V_S$,

$$Q = \frac{V_L}{V_S} = \frac{V_C}{V_S}$$

(at resonance only)

If the Q and source voltage are known, the voltage across the inductor or capacitor can be found by transposition of the formula, as can be seen in the example in Figure 9-9.

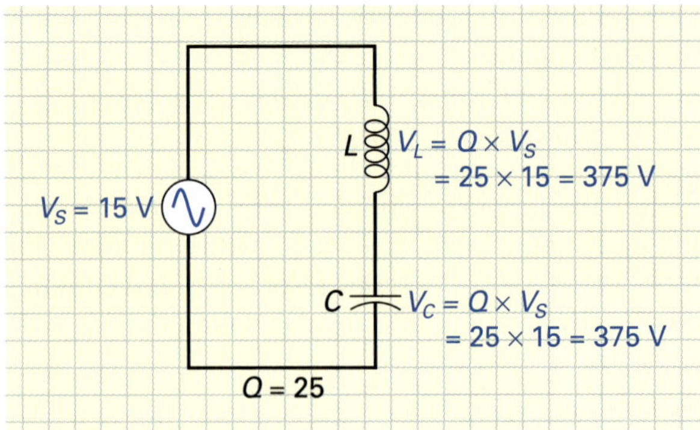

FIGURE 9-9 **Quality Factor at Resonance.**

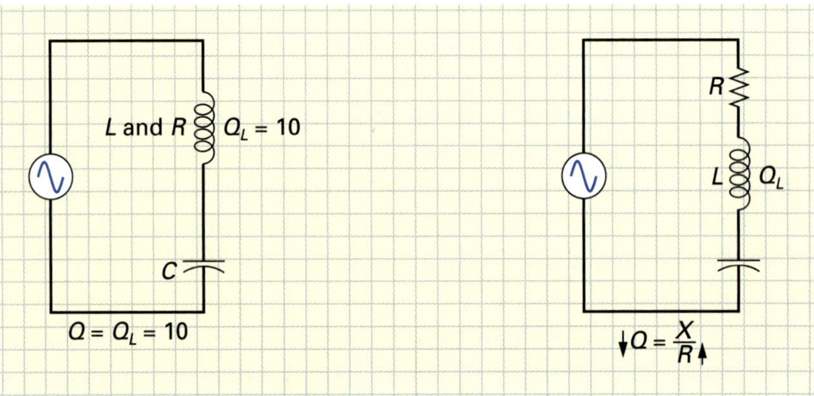

FIGURE 9-10 **Resistance within Inductor.**

The Q of a resonant circuit is almost entirely dependent on the inductor's coil resistance, because capacitors tend to have almost no resistance at all, only reactance, which makes them very efficient.

The inductor has a Q value of its own, and if only L and C are connected in series with one another, the Q of the series resonant circuit will be equal to the Q of the inductor, as shown in Figure 9-10. If the resistance is added in with L and C, the Q of the series resonant circuit will be less than that of the inductor's Q.

EXAMPLE:

Calculate the resistance of the series resonant circuit illustrated in Figure 9-11.

FIGURE 9-11 **Series Resonant Circuit Example.**

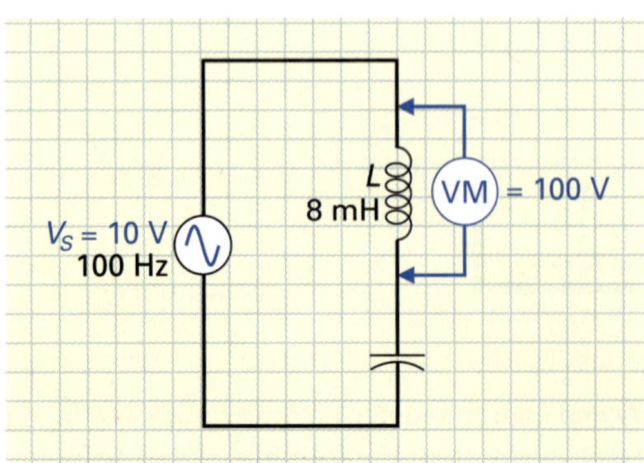

■ ***Solution:***

$$Q = \frac{V_L}{V_S} = \frac{100\text{ V}}{10\text{ V}} = 10$$

Since $Q = X_L/R$ and $R = X_L/Q$, if the inductive reactance can be found, then R can be determined.

$$\begin{aligned} X_L &= 2\,\pi \times f \times L \\ &= 2\,\pi \times 100 \times 8\text{ mH} \\ &= 5\ \Omega \end{aligned}$$

R will therefore equal

$$\begin{aligned} R &= \frac{X_L}{Q} \\ &= \frac{5}{10} \\ &= 0.5\ \Omega \end{aligned}$$

Bandwidth

A series resonant circuit is selective in that frequencies at resonance or slightly above or below will cause a larger current than frequencies well above or below the circuit's resonant frequency. The group or band of frequencies that causes the larger current is called the circuit's **bandwidth.**

Bandwidth
Width of the group or band of frequencies between the half-power points.

Figure 9-12 illustrates a series resonant circuit and its bandwidth. The × marks on the curve illustrate where different frequencies were applied to the circuit and the resulting value of current measured in the circuit. The resulting curve produced is called a **frequency response curve,** as it illustrates the circuit's response to different frequencies. At resonance, $X_L = X_C$ and the two cancel, which is why maximum current was present in the circuit (100 mA) when the resonant frequency (100 Hz) was applied.

Frequency Response Curve
A graph indicating a circuit's response to different frequencies.

The bandwidth includes the group or band of frequencies that cause 70.7% or more of the maximum current to flow within the series resonant circuit. In this example, frequencies from 90 to 110 Hz cause 70.7 mA or more, which is 70.7% of maximum (100 mA), to flow. The bandwidth in this example is equal to

$$\text{BW} = 110 - 90 = 20\text{ Hz}$$

(110 Hz and 90 Hz are known as **cutoff frequencies.**)

Cutoff Frequency
Frequency at which the gain of the circuit falls below 0.707 of the maximum current or half-power (−3 dB).

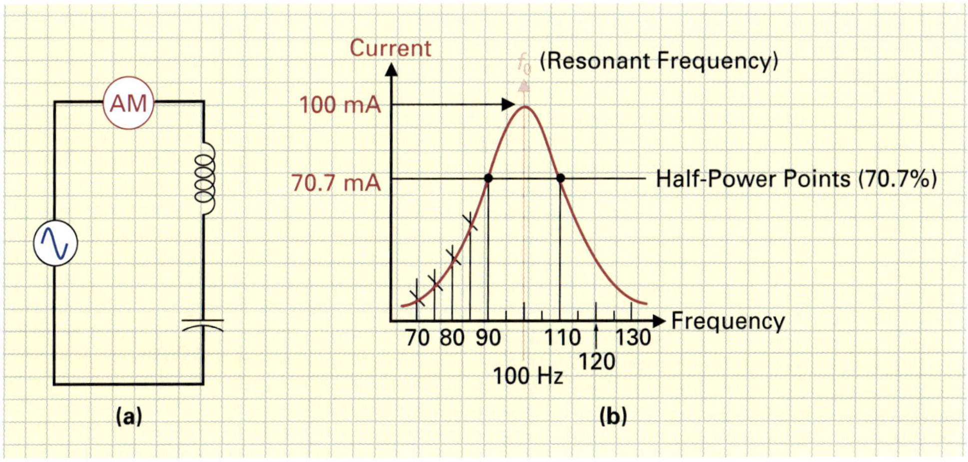

FIGURE 9-12 **Series Resonant Circuit Bandwidth. (a) Circuit. (b) Frequency Response Curve.**

Half-Power Point

A point at which power is 50%. This half-power point corresponds to 70.7% of the total current.

Referring to the bandwidth curve in Figure 9-12(b), you may notice that 70.7% is also called the **half-power point,** although it does not exist halfway between 0 and maximum. This value of 70.7% is not the half-current point but the half-power point, as we can prove with a simple example.

EXAMPLE:

$R = 2\ \text{k}\Omega$ and $I = 100\ \text{mA}$; therefore, power $= I^2 \times R = (100\ \text{mA})^2 \times 2\ \text{k}\Omega = 20\ \text{W}$. If the current is now reduced so that it is 70.7% of its original value, calculate the power dissipated.

Solution:

$$P = I^2 \times R = (70.7\ \text{mA})^2 \times 2\ \text{k}\Omega = 10\ \text{W}$$

In summary, the 70.7% current points are equal to the 50% or half-power points. A circuit's bandwidth is the band of frequencies that exists between the 70.7% current points or half-power points.

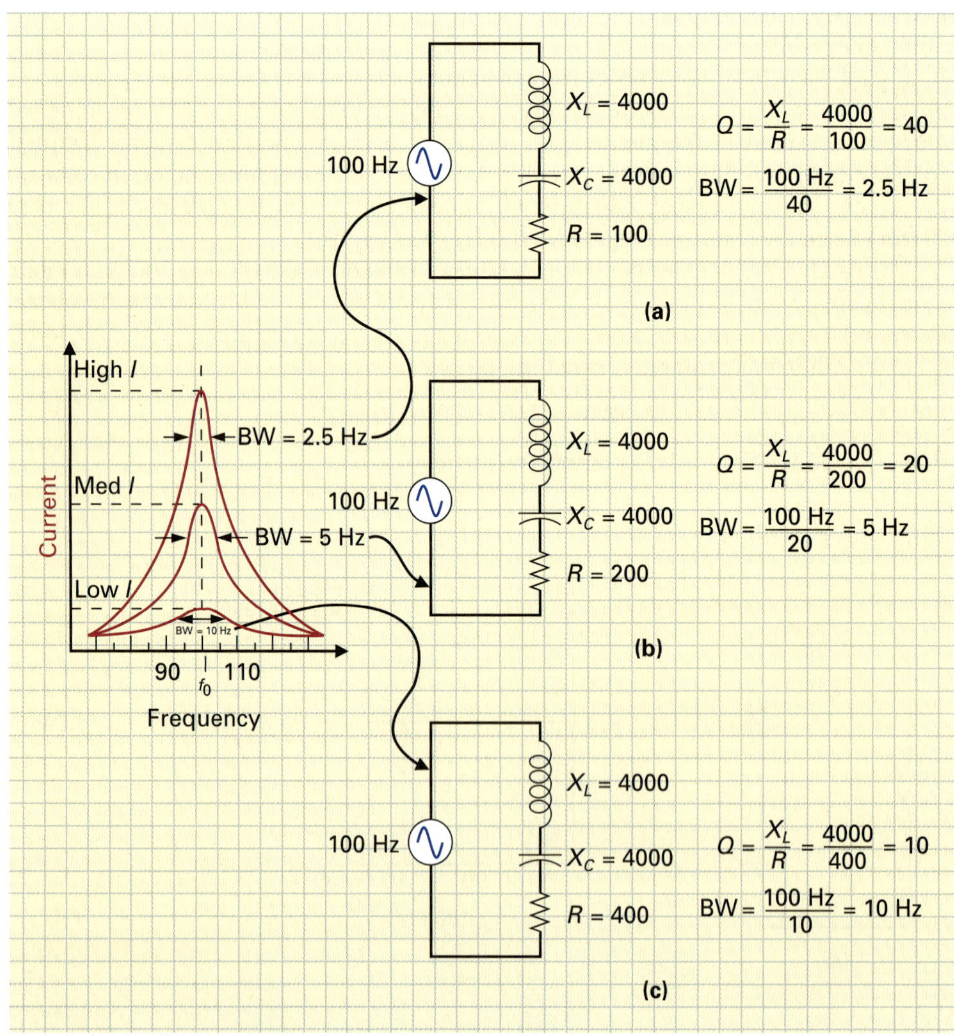

FIGURE 9-13 Bandwidth of a Series Resonant Circuit.

The bandwidth of a series resonant circuit can also be calculated by use of the formula

$$BW = \frac{f_0}{Q_{f_0}}$$

f_0 = Resonant frequency
Q_{f_0} = Quality factor at resonance

This formula states that the BW is proportional to the resonant frequency of the circuit and inversely proportional to the Q of the circuit.

Figure 9-13 illustrates three example response curves. In these three examples, the value of R is changed from 100 Ω to 200 Ω to 400 Ω. This does not vary the resonant frequency, but simply alters the Q and therefore the BW. The resistance value will determine the Q of the circuit, and since Q is inversely proportional to resistance, Q is proportional to current; consequently, a high value of Q will cause a high value of current.

In summary, the bandwidth of a series resonant circuit will increase as the Q of the circuit decreases (BW$\uparrow = f_0/Q\downarrow$), and vice versa.

SELF-TEST EVALUATION POINT FOR SECTION 9-3-1

Use the following questions to test your understanding of Section 9-3-1.

1. Define *resonance.*
2. What is series resonance?
3. In a series resonant circuit, what are the three rather unusual circuit phenomena that take place?
4. How does Q relate to series resonance?
5. Define *bandwidth.*
6. Calculate BW if $f_0 = 12$ kHz and $Q = 1000$.

9-3-2 *Parallel Resonance*

The **parallel resonant circuit** acts differently from the series resonant circuit, and these different characteristics need to be analyzed and discussed. Figure 9-14 illustrates a parallel resonant circuit. The inductive current could be calculated by using the formula

Parallel Resonant Circuit
Circuit having an inductor and capacitor in parallel with one another, offering a high impedance at the frequency of resonance.

$$I_L = \frac{V_L}{X_L} = \frac{10\text{ V}}{1\text{ k}\Omega} = 10\text{ mA}$$

The capacitive current could be calculated by using the formula

$$I_C = \frac{V_C}{X_C} = \frac{10\text{ V}}{1\text{ k}\Omega} = 10\text{ mA}$$

Looking at the vector diagram in Figure 9-14(b), you can see that I_C leads the source voltage by 90° (ICE) and I_L lags the source voltage by 90° (ELI), creating a 180° phase difference between

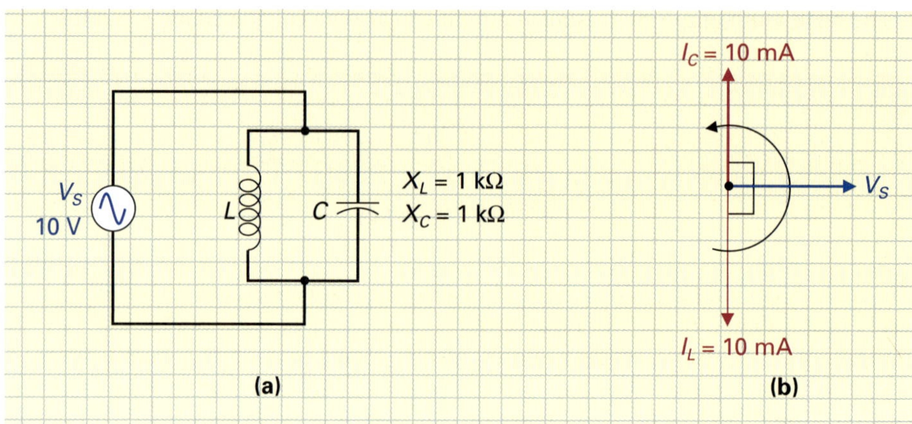

FIGURE 9-14 **Parallel Resonant Circuit. (a) Circuit. (b) Vector Diagram.**

I_C and I_L. This means that when 10 mA of current flows up through the inductor, 10 mA of current will flow in the opposite direction down through the capacitor, as shown in Figure 9-15(a). During the opposite alternation, 10 mA will flow down through the inductor and 10 mA will travel up through the capacitor, as shown in Figure 9-15(b).

If 10 mA arrives into point *X* and 10 mA of current leaves point *X*, no current can be flowing from the source (V_S) to the parallel *LC* circuit; the current is simply swinging or oscillating back and forth between the capacitor and inductor.

The source voltage (V_S) is needed initially to supply power to the *LC* circuit and start the oscillations; but once the oscillating process is in progress (assuming the ideal case), current is only flowing back and forth between inductor and capacitor, and no current is flowing from the source. So the *LC* circuit appears as an infinite impedance and the source can be disconnected, as shown in Figure 9-15(c).

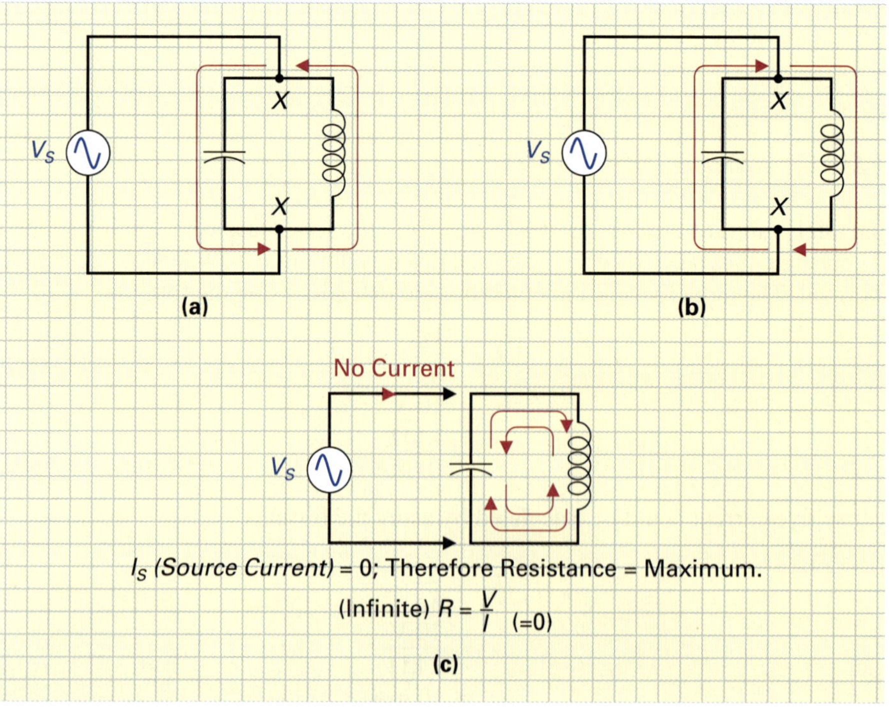

FIGURE 9-15 **Current in a Parallel Resonant Circuit.**

Flywheel Action

Let's discuss this oscillating effect, called **flywheel action,** in a little more detail. The name is derived from the fact that it resembles a mechanical flywheel, which, once started, will keep going continually until friction reduces the magnitude of the rotations to zero.

The electronic equivalent of the mechanical flywheel is a resonant parallel-connected *LC* circuit. Figure 9-16(a) through (h) illustrate the continual energy transfer between capacitor and inductor, and vice versa. The direction of the circulating current reverses each half-cycle at the frequency of resonance. Energy is stored in the capacitor in the form of an electric field between the plates on one half-cycle, and then the capacitor discharges, supplying current to build up a magnetic field on the other half-cycle. The inductor stores its energy in the form of a magnetic field, which will collapse, supplying a current to charge the capacitor, which will then discharge, supplying a current back to the inductor, and so on. Due to the "storing action" of this circuit, it is sometimes related to the fluid analogy and referred to as a **tank circuit.**

Flywheel Action
Sustaining effect of oscillation in an *LC* circuit due to the charging and discharging of the capacitor and the expansion and contraction of the magnetic field around the inductor.

Tank Circuit
Circuit made up of a coil and capacitor that is capable of storing electric energy.

The Reality of Tanks

Under ideal conditions a tank circuit should oscillate indefinitely if no losses occur within the circuit. In reality, the resistance of the coil reduces that 100% efficiency, as does friction with the mechanical flywheel. This coil resistance is illustrated in

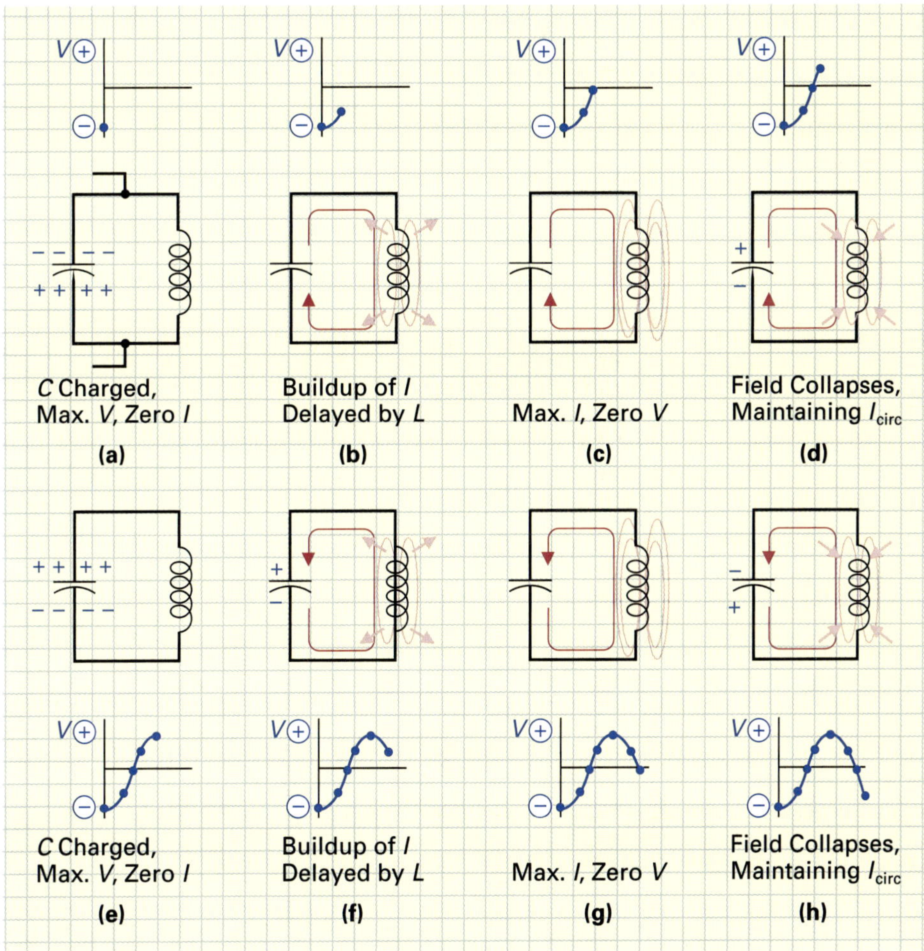

FIGURE 9-16 **Energy and Current in an *LC* Parallel Circuit at Resonance.**

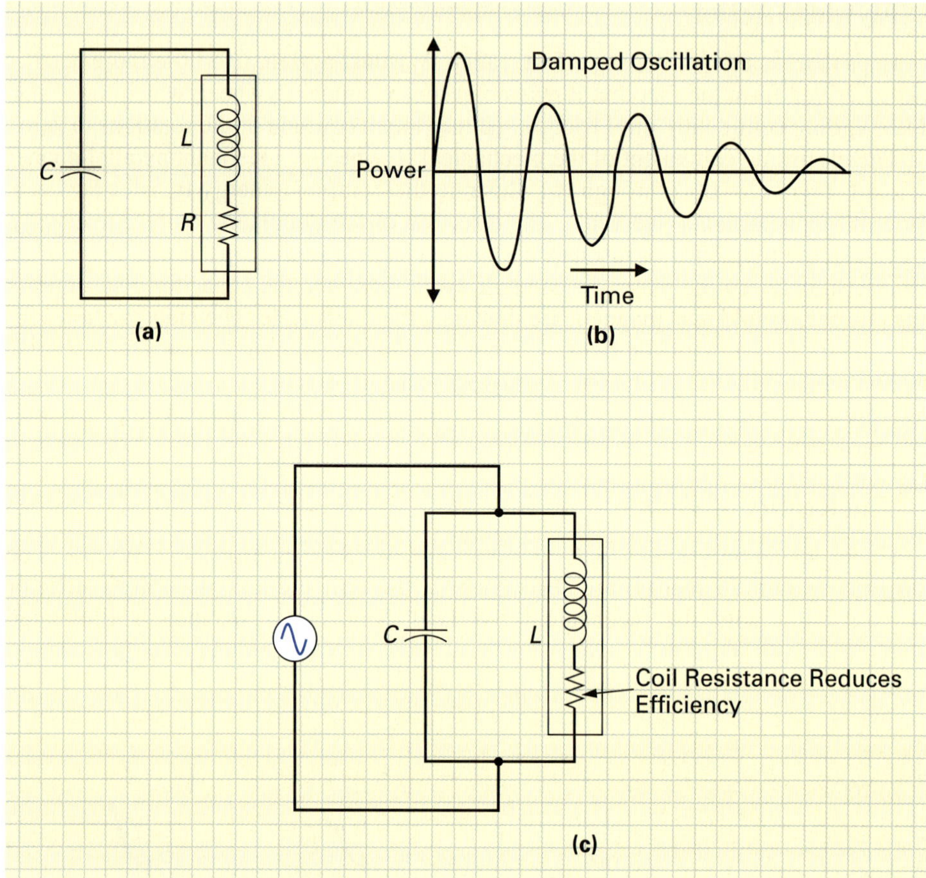

FIGURE 9-17 **Losses in Tanks.**

Figure 9-17(a), and, unlike reactance, resistance is the opposition to current flow, with the dissipation of energy in the form of heat. As a small part of the energy is dissipated with each cycle, the oscillations will be reduced in size and eventually fall to zero, as shown in Figure 9-17(b).

If the ac source is reconnected to the tank, as shown in Figure 9-17(c), a small amount of current will flow from the source to the tank to top up the tank or replace the dissipated power. The higher the coil resistance is, the higher the loss and the larger the current flow from source to tank to replace the loss.

Quality Factor

In the series resonant circuit, we were concerned with voltage drops since current remains the same throughout a series circuit, so

$$Q = \frac{V_C \text{ or } V_L}{V_S} \quad \text{(at resonance only)}$$

In a parallel resonant circuit, we are concerned with circuit currents rather than voltage, so

$$Q = \frac{I_{\text{tank}}}{I_S}$$

(at resonance only)

The quality factor, Q, can also be expressed as the ratio between reactance and resistance:

$$Q = \frac{X_L}{R} \quad \text{(at any frequency)}$$

Another formula, which is the most frequently used when discussing and using parallel resonant circuits, is

$$Q = \frac{Z_{\text{tank}}}{X_L}$$

(at resonance only)

This formula states that the Q of the tank is proportional to the tank impedance. A higher tank impedance results in a smaller current flow from source to tank. This assures less power is dissipated, and that means a higher-quality tank.

Of all the three Q formulas for parallel resonant circuits, $Q = I_{\text{tank}}/I_S$, $Q = X_L/R$, and $Q = Z_{\text{tank}}/X_L$, the latter is the easiest to use as both X_L and the tank impedance can easily be determined in most cases where C, L, and R internal for the inductor are known.

Bandwidth

Figure 9-18 illustrates a parallel resonant circuit and two typical response curves. These response curves summarize what we have described previously, in that a parallel resonant circuit has maximum impedance [Figure 9-18(b)] and minimum current [Figure 9-18(c)] at resonance. The current versus frequency response curve shown in Figure 9-18(c) is the complete opposite to the series resonant response curve. At frequencies below resonance (< 10 kHz), X_L is low and X_C is high, and the inductor offers a low reactance, producing a high current path and low impedance. On the other hand, at frequencies above resonance (> 10 kHz), the capacitor displays a low reactance, producing a high current path and low impedance. The parallel resonant circuit is like the series resonant circuit in that it responds to a band of frequencies close to its resonant frequency.

The bandwidth (BW) can be calculated by use of the formula

$$\text{BW} = \frac{f_0}{Q_{f_0}}$$

f_0 = Resonant frequency
Q_{f_0} = Quality factor at resonance

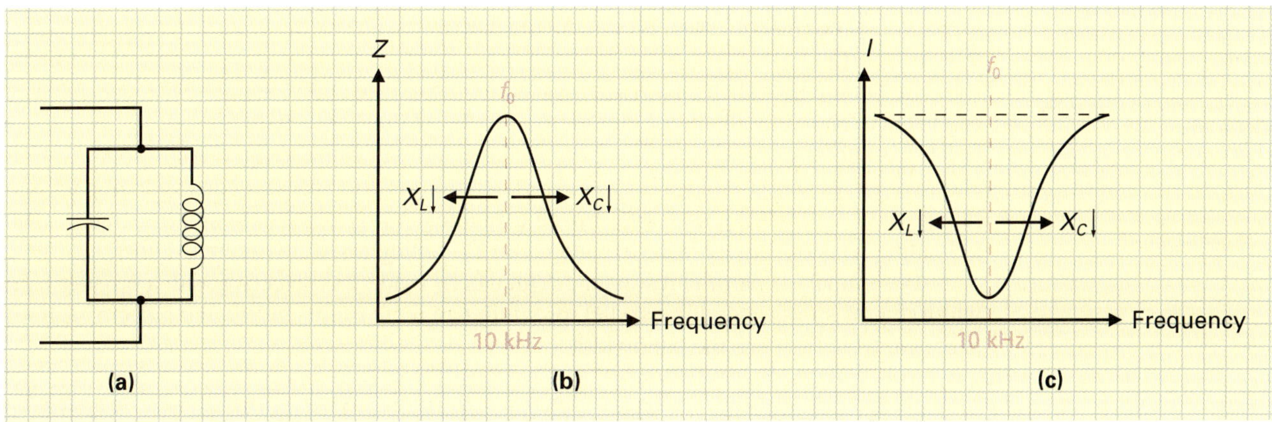

FIGURE 9-18 Parallel Resonant Circuit Bandwidth.

EXAMPLE:

Calculate the bandwidth of the circuit illustrated in Figure 9-19.

FIGURE 9-19 **Bandwidth Example.**

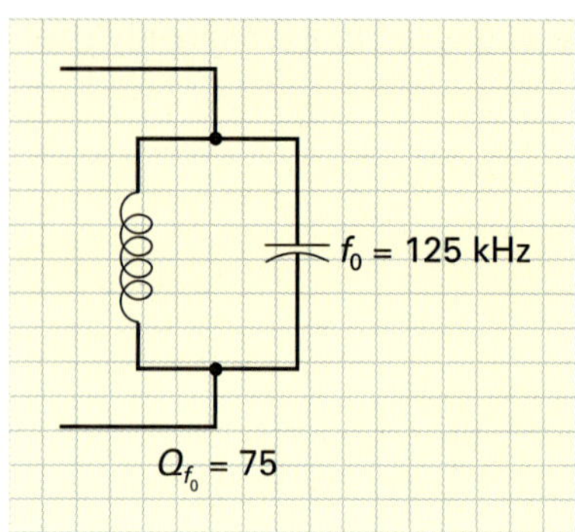

Solution:

$$\text{BW} = \frac{f_0}{Q_{f_0}}$$
$$= \frac{125 \text{ kHz}}{75}$$
$$= 1.7 \text{ kHz}$$
$$\frac{1.7 \text{ kHz}}{2} = 0.85 \text{ kHz}$$

therefore the bandwidth extends from

$$f_0 + 0.85 \text{ kHz} = 125.85 \text{ kHz}$$
$$f_0 - 0.85 \text{ kHz} = 124.15 \text{ kHz}$$
$$\text{BW} = 124.15 \text{ kHz to } 125.85 \text{ kHz}$$

Tuned Circuit
Circuit that can have its components' values varied so that the circuit responds to one selected frequency yet heavily attenuates all other frequencies.

Selectivity
Characteristic of a circuit to discriminate between the wanted signal and the unwanted signal.

Selectivity

Circuits containing inductance and capacitance are often referred to as **tuned circuits** since they can be adjusted to make the circuit responsive to a particular frequency (the resonant frequency). **Selectivity,** by definition, is the ability of a tuned circuit to respond to a desired frequency and ignore all others. Parallel resonant *LC* circuits are sometimes too selective, as the *Q* is too large, producing too narrow a bandwidth, as shown in Figure 9-20(a) ($\text{BW}\downarrow = f_0/Q\uparrow$).

In this situation, because of the very narrow response curve, a high resistance value can be placed in parallel with the *LC* circuit to provide an alternative path for source current. This process is known as *loading* or *damping* the tank and will cause an increase in source current and decrease in *Q* ($Q\downarrow = I_{\text{tank}}/I_{\text{source}}\uparrow$). The decrease in *Q* will cause a corresponding increase in BW ($\text{BW}\uparrow = f_0/Q\downarrow$), as shown by the examples in Figure 9-20, which illustrates a 1000 Ω loading resistor [Figure 9-20(b)] and a 100 Ω loading resistor [Figure 9-20(c)].

In summary, a parallel resonant circuit can be made less selective with a broader bandwidth if a resistor is added in parallel, providing an increase in current and a decrease in impedance, which widens the bandwidth.

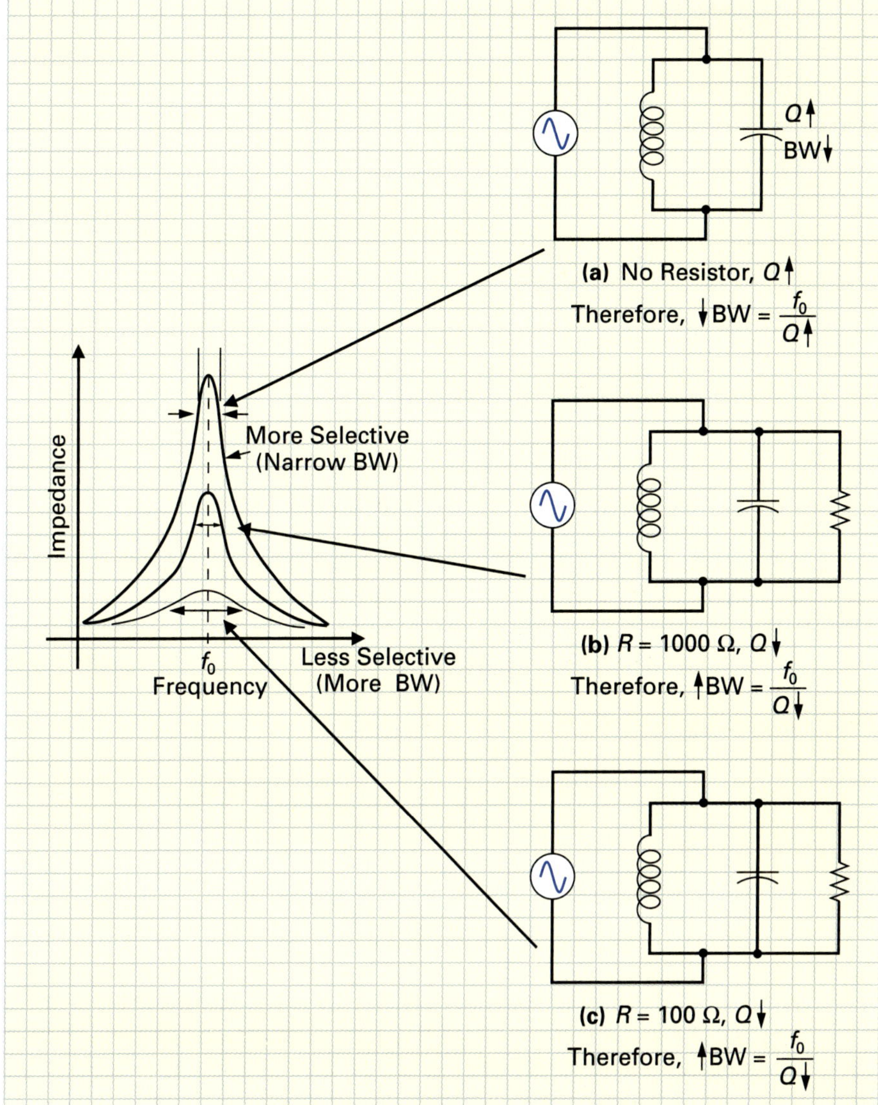

FIGURE 9-20 Varying Bandwidth by Loading the Tank Circuit.

SELF-TEST EVALUATION POINT FOR SECTION 9-3-2

Use the following questions to test your understanding of Section 9-3-2.

1. What are the differences between a series and a parallel resonant circuit?
2. Describe flywheel action.
3. Calculate the value of Q of a tank if $X_L = 50\ \Omega$ and $R = 25\ \Omega$.
4. When calculating bandwidth for a parallel resonant circuit, can the series resonant bandwidth formula be used?
5. What is selectivity?

9-4 APPLICATIONS OF *RLC* CIRCUITS

In the previous chapters, you saw how *RC* and *RL* filter circuits are used as low- or high-pass filters to pass some frequencies and block others. There are basically four types of filters:

1. Low-pass filter, which passes frequencies below a cutoff frequency
2. High-pass filter, which passes frequencies above a cutoff frequency
3. Bandpass filter, which passes a band of frequencies
4. Band-stop filter, which stops a band of frequencies

9-4-1 *Low-Pass Filter*

Figure 9-21(a) illustrates how an inductor and capacitor can be connected to act as a low-pass filter. At low frequencies, X_L has a small value compared to the load resistor (R_L), so nearly all the low-frequency input is developed and appears at the output across R_L. Since X_C is high at low frequencies, nearly all the current passes through R_L rather than *C*.

At high frequencies, X_L increases and drops more of the applied input across the inductor rather than the load. The capacitive reactance, X_C, aids this low-output-at-high-frequency effect by decreasing its reactance and providing an alternative path for current to flow.

Since the inductor basically blocks alternating current and the capacitor shunts alternating current, the net result is to prevent high-frequency signals from reaching the load. The way in which this low-pass filter responds to frequencies is graphically illustrated in Figure 9-21(b).

9-4-2 *High-Pass Filter*

Figure 9-22(a) illustrates how an inductor and capacitor can be connected to act as a high-pass filter. At high frequencies, the reactance of the capacitor (X_C) is low while the reactance of the inductor (X_L) is high, so all the high frequencies are easily passed by the capacitor and blocked by the inductor, so they all are routed through to the output and load.

At low frequencies, the reverse condition exists, resulting in a low X_L and a high X_C. The capacitor drops nearly all the input, and the inductor shunts the signal current away from the output load.

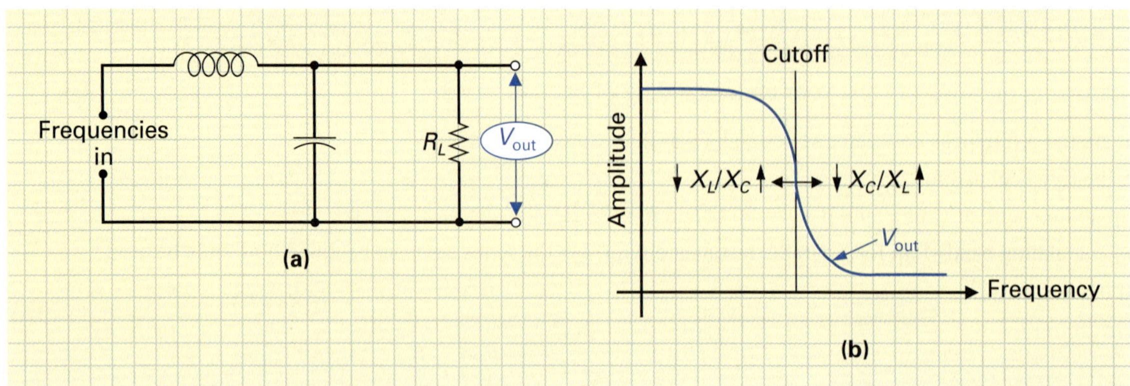

FIGURE 9-21 Low-Pass Filter. (a) Circuit. (b) Frequency Response.

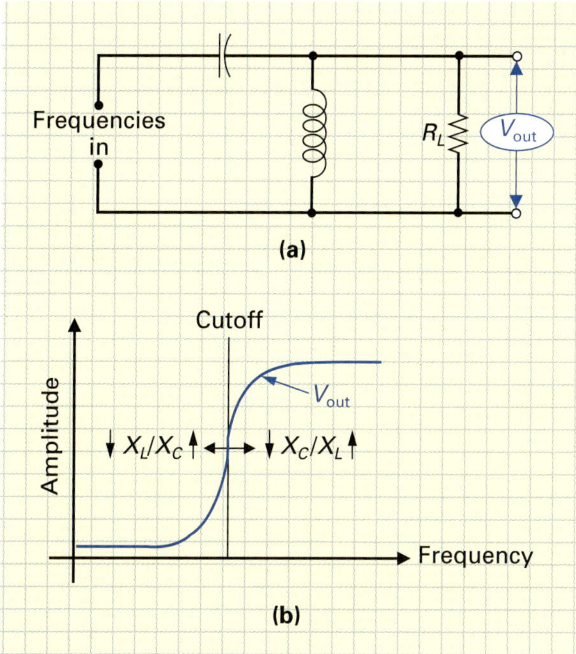

FIGURE 9-22 High-Pass Filter. (a) Circuit. (b) Frequency Response.

9-4-3 *Bandpass Filter*

Figure 9-23(a) illustrates a series resonant **bandpass filter,** and Figure 9-23(b) shows a parallel resonant bandpass filter. Figure 9-23(c) shows the frequency response curve produced by the bandpass filter. At resonance, the series resonant LC circuit has a very low impedance and will consequently pass the resonant frequency to the load with very little drop across the L and C components.

Bandpass Filter
Filter circuit that passes a group or band of frequencies between a lower and an upper cutoff frequency, while heavily attenuating any other frequency outside this band.

Below resonance, X_C is high, and the capacitor drops a large amount of the input signal; above resonance, X_L is high and the inductor drops most of the input frequency voltage. This circuit will therefore pass a band of frequencies centered around the resonant frequency of the series LC circuit and block all other frequencies above and below this resonant frequency.

Figure 9-23(b) illustrates how a parallel resonant LC circuit can be used to provide a bandpass response. The series resonant circuit was placed in series with the output, whereas the parallel resonant circuit will have to be placed in parallel with the output to provide the same results. At resonance, the parallel resonant circuit or tank has a high impedance, so very little current will be shunted away from the output; it will be passed on to the output, and almost all the input will appear at the output across the load.

Above resonance, X_C is small, so most of the input is shunted away from the output by the capacitor; below resonance, X_L is small, and the shunting action occurs again, but this time through the inductor.

Figure 9-24 illustrates how a transformer can be used to replace the inductor to produce a bandpass filter. At resonance, maximum flywheel current flows within the parallel circuit made up of the capacitor and the primary of the transformer (L), which is known as a *tuned transformer.* With maximum flywheel current, there will be a maximum magnetic field, which means that there will be maximum power transfer between primary and secondary. So nearly all the input will be coupled to the output (coupling coefficient $k = 1$) and appear across the load at and around a small band of frequencies centered on resonance.

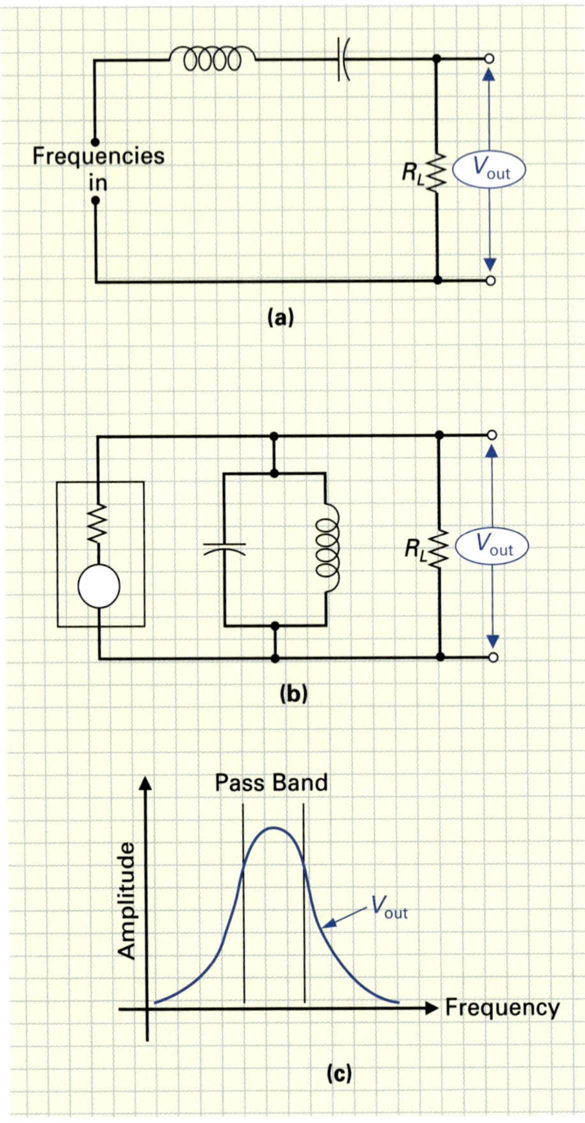

FIGURE 9-23 **Bandpass Filter. (a) Series Resonant Bandpass Filter. (b) Parallel Resonant Bandpass Filter. (c) Frequency Response.**

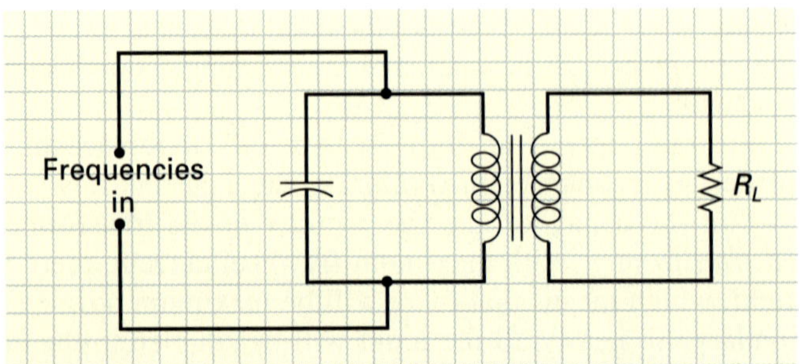

FIGURE 9-24 **Parallel Resonant Bandpass Circuit Using a Transformer.**

Above and below resonance, current within the parallel resonant circuit will be smaller, so the power transfer ability will be less, effectively keeping the frequencies outside the passband from appearing at the output.

9-4-4 *Band-Stop Filter*

Figure 9-25(a) illustrates a series resonant and Figure 9-25(b) a parallel resonant **band-stop filter.** Figure 9-25(c) shows the frequency response curve produced by a band-stop filter. The band-stop filter operates exactly the opposite to a bandpass filter in that it blocks or attenuates a band of frequencies centered on the resonant frequency of the *LC* circuit.

Band-stop Filter
A filter that attenuates alternating currents whose frequencies are between given upper and lower cutoff values while passing frequencies above and below this band.

In the series resonant circuit in Figure 9-25(a), the *LC* impedance is very low at and around resonance, so these frequencies are rejected or shunted away from the output. Above and below resonance, the series circuit has a very high impedance, which results in almost no shunting of the signal away from the output.

In the parallel resonant circuit in Figure 9-25(b), the *LC* circuit is in series with the load and output. At resonance, the impedance of a parallel resonant circuit will be very high, and the band of frequencies centered around resonance will be blocked. Above and below resonance, the impedance of the tank is very low, so nearly all the input is developed across the output.

Filters are necessary in applications such as television or radio, where we need to tune in (select or pass) one frequency that contains the information we desire, yet block all the millions of other frequencies that are also carrying information, as shown in Figure 9-26.

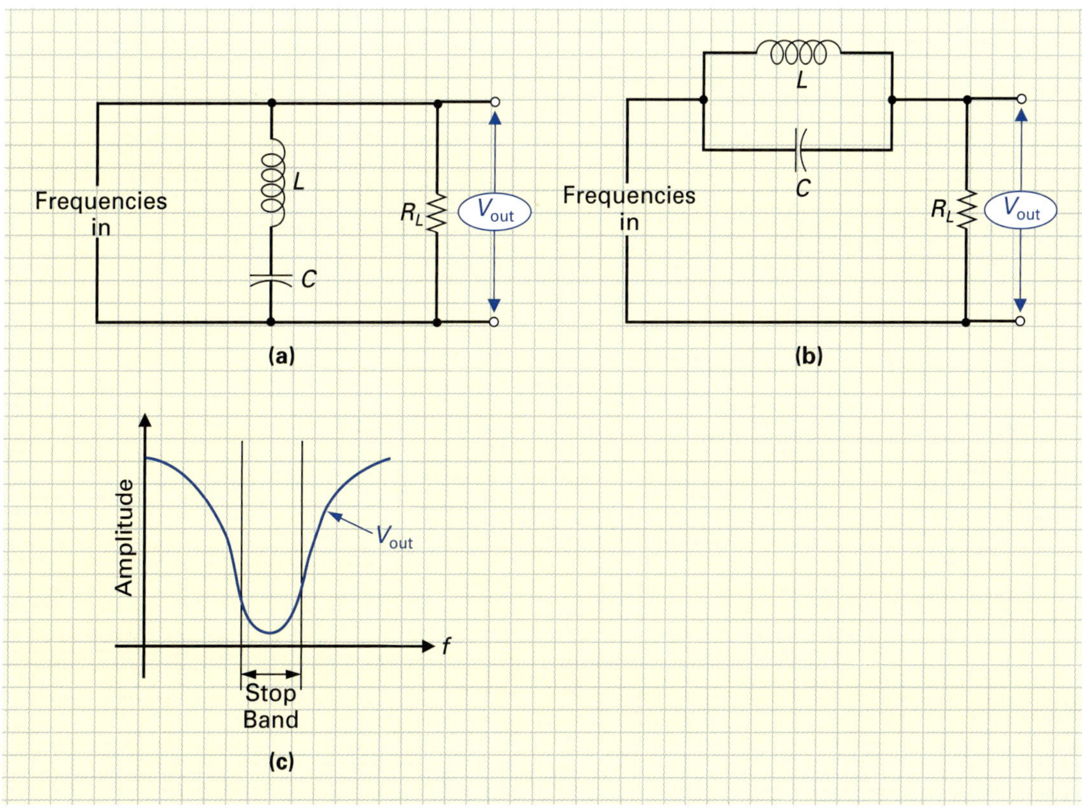

FIGURE 9-25 **Band-Stop Filter. (a) Series Resonant Band-Stop Filter. (b) Parallel Resonant Band-Stop Filter. (c) Frequency Response.**

FIGURE 9-26 **Tuning in of Station by Use of a Bandpass Filter.**

SELF-TEST EVALUATION POINT FOR SECTION 9-4

Use the following questions to test your understanding of Section 9-4.

1. Of the four types of filters, which:
   a. Would utilize the inductor as a shunt?
   b. Would utilize the capacitor as a shunt?
   c. Would use a series resonant circuit as a shunt?
   d. Would use a parallel resonant circuit as a shunt?

2. In what applications can filters be found?

REVIEW QUESTIONS

Multiple-Choice Questions

1. Capacitive reactance is ____________ to frequency and capacitance, while inductive reactance is ____________ to frequency and inductance.
   a. Proportional, inversely proportional
   b. Inversely proportional, proportional
   c. Proportional, proportional
   d. Inversely proportional, inversely proportional

2. Resonance is a circuit condition that occurs when:
   a. V_L equals V_C
   b. X_L equals X_C
   c. L equals C
   d. Both (a) and (c)
   e. Both (a) and (b)

3. As frequency is increased, X_L will ____________ while X_C will ____________.
   a. Decrease, increase
   b. Increase, decrease
   c. Remain the same, decrease
   d. Increase, remain the same

4. In an *RLC* series resonant circuit, with $R = 500\ \Omega$ and $X_L = 250\ \Omega$, what would be the value of X_C?
   a. 2 Ω
   b. 125 Ω
   c. 250 Ω
   d. 500 Ω

5. At resonance, the voltage drop across both a series-connected inductor and a capacitor will equal:
   a. 70.7 V
   b. 50% of the source
   c. 10 V
   d. Zero

6. In a series resonant circuit the current flow is ____________, as the impedance is ____________ and equal to ____________.
   a. Large, small, *R*
   b. Small, large, *X*
   c. Large, small, *X*
   d. Small, large, *R*

7. A circuit's bandwidth includes a group or band of frequencies that cause ____________ or more of the maximum current, or more than ____________ of the maximum power to appear at the output.
   a. 110, 90
   b. 50%, 70.7%
   c. 70.7%, 50%
   d. Both (a) and (c)

8. The bandwidth of a circuit is proportional to the:
   a. Frequency of resonance
   b. *Q* of the tank
   c. Tank current
   d. Two of the above

9. Series or parallel resonant circuits can be used to create:
   a. Low-pass filters
   b. Low-pass and high-pass filters
   c. Bandpass and band-stop filters
   d. All of the above

10. Flywheel action occurs in:
   a. A tank circuit
   b. A parallel *LC* circuit
   c. A series *LC* circuit
   d. Two of the above
   e. None of the above

Practice Problems

11. Calculate the values of capacitive and inductive reactance for the following when connected across a 60 Hz source:
   a. 0.02 μF
   b. 18 μF
   c. 360 pF
   d. 2700 nF
   e. 4 mH
   f. 8.18 H
   g. 150 mH
   h. 2 H

12. If a 1.2 k Ω resistor, a 4 mH inductor, and an 8 μF capacitor are connected in series across a 120 V/60 Hz source, calculate:
   a. X_C
   b. X_L
   c. Z
   d. I
   e. V_R
   f. V_L
   g. V_C
   h. Apparent power
   i. True power
   j. Resonant frequency
   k. Circuit quality factor
   l. Bandwidth

13. If a 270 Ω resistor, a 150 mH inductor, and a 20 μF capacitor are all connected in parallel with one another across a 120 V/60 Hz source, calculate:
   a. X_L
   b. X_C
   c. I_R
   d. I_L
   e. I_C
   f. I_T
   g. Z
   h. Resonant frequency
   i. Q factor
   j. Bandwidth

14. Calculate the impedance of a series circuit if $R = 750\ \Omega$, $X_L = 25\ \Omega$, and $X_C = 160\ \Omega$.

15. Calculate the impedance of a parallel circuit with the same values as those of Question 14 when a 1 V source voltage is applied.

Web Site Questions

Go to the web site http://www.prenhall.com/cook, select the textbook *Electronics: A Complete Course,* select this chapter, and then follow the instructions when answering the multiple choice practice problems.

Semiconductor Principles

The Turing Enigma

During the Second World War, the Germans developed a cipher-generating apparatus called "Enigma." This electromechanical teleprinter would scramble messages with several randomly spinning rotors that could be set to a predetermined pattern by the sender. This key and plug pattern was changed three times a day by the Germans and cracking the secrets of Enigma became of the utmost importance to British Intelligence. With this objective in mind, every brilliant professor and eccentric researcher was gathered at a Victorian estate near London called Bletchley Park. They specialized in everything from engineering to literature and were collectively called the Backroom Boys.

By far the strangest and definitely most gifted of the group was an unconventional theoretician from Cambridge University named Alan Turing. He wore rumpled clothes and had a shrill stammer and crowing laugh that aggravated even his closest friends. He had other legendary idiosyncrasies that included setting his watch by sighting on a certain star from a specific spot and then mentally calculating the time of day. He also insisted on wearing his gas mask whenever he was out, not for fear of a gas attack, but simply because it helped his hay fever.

Turing's eccentricities may have been strange but his genius was indisputable. At the age of twenty-six he wrote a paper outlining his "universal machine" that could solve any mathematical or logical problem. The data or, in this case, the intercepted enemy messages could be entered into the machine on paper tape and then compared with known Enigma codes until a match was found.

In 1943 Turing's ideas took shape as the Backroom Boys began developing a machine that used 2,000 vacuum tubes and incorporated five photoelectric readers that could process 25,000 characters per second. It was named "Colossus," and it incorporated the stored program and other ideas from Turing's paper written seven years earlier.

Turing could have gone on to accomplish much more. However, his idiosyncrasies kept getting in his way. He became totally preoccupied with abstract questions concerning machine intelligence. His unconventional personal lifestyle led to his arrest in 1952 and, after a sentence of psychoanalysis, his suicide two years later.

Before joining the Backroom Boys at Bletchley Park, Turing's genius was clearly apparent at Cambridge. How much of a role he played in the development of Colossus is still unknown and remains a secret guarded by the British Official Secrets Act. Turing was never fully recognized for his important role in the development of this innovative machine, except by one of his Bletchley Park colleagues at his funeral who said, "I won't say what Turing did made us win the war, but I daresay we might have lost it without him."

10

Introduction

Materials can be divided into three main types according to the way they react to current when a voltage is applied across them. **Insulators** (nonconductors), for example, are materials that have a very high resistance and therefore oppose current, whereas **conductors** are materials that have a very low resistance and therefore pass current easily. The third type of material is the **semiconductor** which, as its name suggests, has properties that lie between the insulator and the conductor. Semiconductor materials are not good conductors or insulators and so the next question is: What characteristic do they possess that makes them so useful in electronics? The answer is that they can be controlled to either increase their resistance and behave more like an insulator or decrease their resistance and behave more like a conductor. *It is this ability of a semiconductor material to vary its resistive properties that makes it so useful in electrical and electronic applications.*

In this chapter we will examine the characteristics of semiconductor materials so that we can better understand the operation and characteristics of semiconductor devices.

Insulators
Materials that have a very high resistance and oppose current.

Conductors
Materials that have a very low resistance and pass current easily.

Semiconductors
Materials that have properties that lie between insulators and conductors.

TIME LINE

In the early days of World War II, German scientist Konrad Zuse (1910–1995) who designed and built the first general-purpose computer, proposed constructing a computer that would operate 1000 times faster than anything else at that time. This proposal was rejected by Hitler, who was not interested in this long-term, two-year project, as he was sure that the war was going to be, for him, a certain, quick victory. Due to Hitler's shortsightedness, this powerful computer, which could have been used to break British communication codes, was never developed. However, unknown to both Hitler and Zuse, the British code-breaking computer project, called Ultra, had highest priority and was moving rapidly toward completion.

Solid State Device
Uses a solid semiconductor material, such as silicon, between the input and output whereas a vacuum tube has vacuum between input and output.

Discrete Components
Separate active and passive devices that were manufactured before being used in a circuit.

10-1 SEMICONDUCTOR DEVICES

Semiconductor materials such as *germanium* and *silicon* are used to construct semiconductor devices like the *diodes, transistors,* and *integrated circuits* (*ICs*) shown in Figure 10-1. These devices are used in electrical and electronic circuits to control current and voltage, so as to produce a desired result. For example, a diode could be used as the controlling element in a rectifier circuit that would convert ac to pulsating dc. A transistor, on the other hand, could be made to act like a variable resistance so it could amplify a radio signal. Conversely, an integrated circuit could be used to generate an oscillating signal or be made to perform arithmetic operations.

The most significant development in electronics since World War II has been a small semiconductor device called the transistor. It was first introduced in 1948 by its inventors William Schockley, Walter Bratten, and John Bardeen in the Bell Telephone Laboratories and was described as a **solid state device.** This term was used because the transistor contained a solid semiconductor material between its input and output pins, unlike its predecessor the vacuum tube, which had a vacuum between its input and output pins.

The first *point-contact transistor* unveiled in 1948 was extremely unreliable, and it took its inventors another twelve years to develop the superior *bipolar junction transistor* (*BJT*) and make it available in commercial quantities.

In 1960, many electronic system manufacturers began to use the bipolar junction transistor instead of the vacuum tube in low-power and low-frequency applications. Research and development into semiconductor or solid state devices mushroomed and a variety of semiconductor devices began to appear. A different type of transistor emerged called the *field effect transistor* (*FET*), which had characteristics similar to those of the vacuum tube. Once it was discovered that semiconductor materials could also generate and sense light, a new line of optoelectronic devices became available. Later it was discovered that semiconductor materials could sense magnetism, temperature, and pressure and, as a result, a variety of sensor devices or transducers (energy converters) appeared on the market. Along with all these different types of semiconductor devices, a wide variety of semiconductor diodes emerged that could rectify, regulate, and oscillate at high frequencies. Even to this day it is clear that we have not yet seen all the potential value of semiconductors. Figure 10-2 illustrates many of these semiconductor or solid state devices.

Although semiconductor diodes and transistors are still widely used as individual or **discrete components,** in 1959 Robert Noyce discovered that more than one transistor could be constructed on a single piece of semiconductor material. Soon other components such as resistors, capacitors, and diodes were added with transistors and then interconnected to form a complete circuit on a single chip or piece of semiconductor material. This integrating of various components on a single chip of semiconductor was called an *integrated circuit* (*IC*)

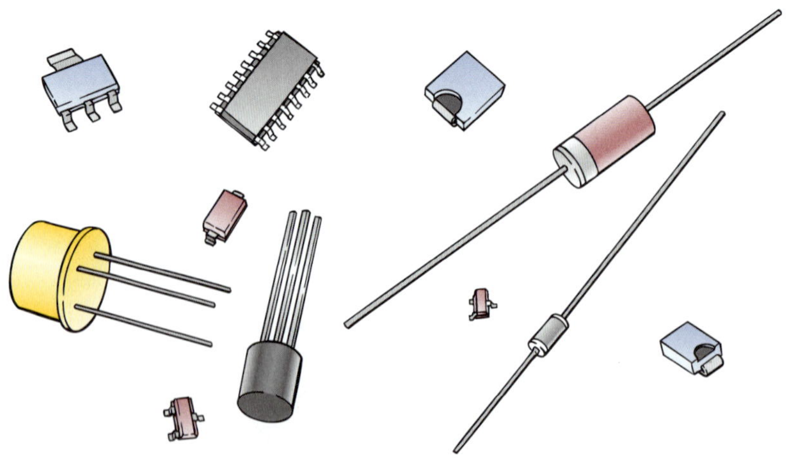

FIGURE 10-1 Semiconductor Devices.

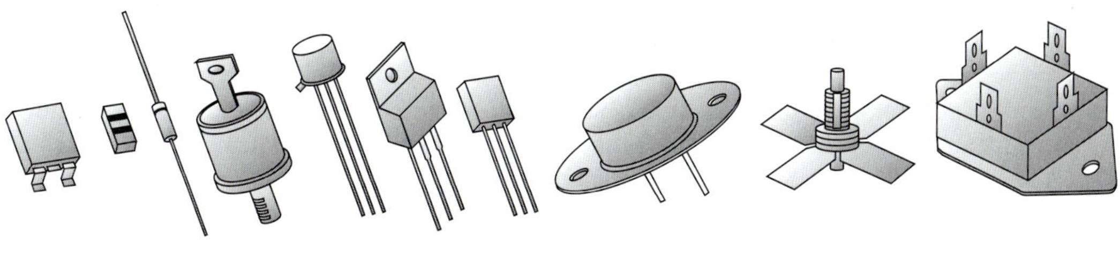

Diodes, Transistors, and Thyristor Devices

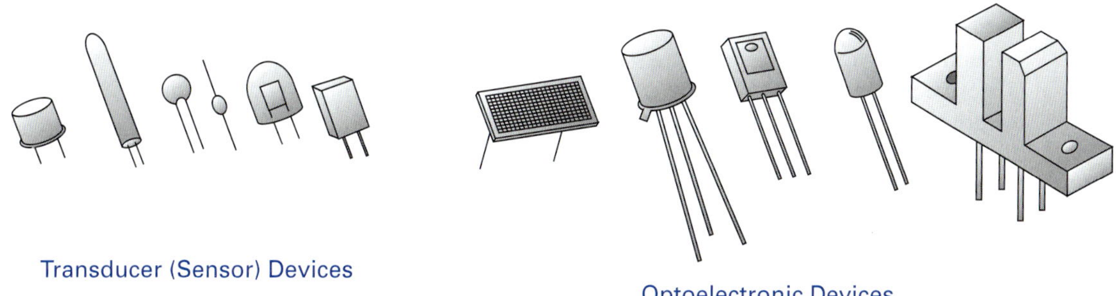

Transducer (Sensor) Devices

Optoelectronic Devices

FIGURE 10-2 **Discrete Semiconductor (Solid State) Devices.**

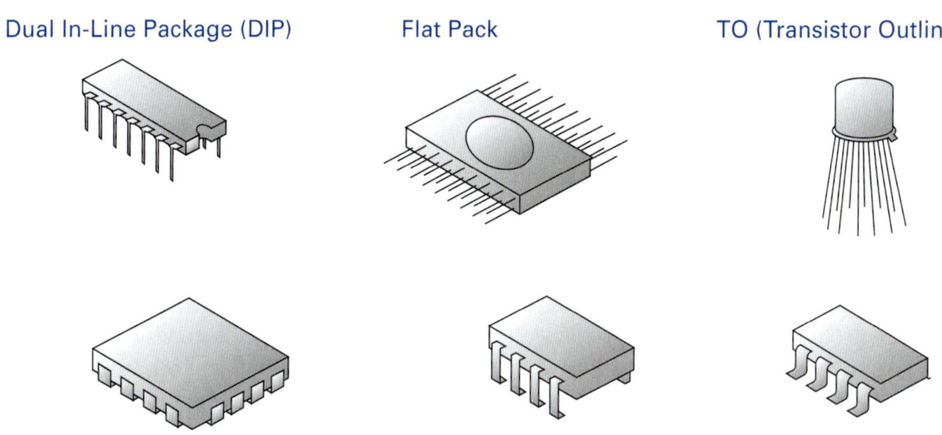

Dual In-Line Package (DIP)

Flat Pack

TO (Transistor Outline) Can

Surface Mount Technology (SMT) Packages

FIGURE 10-3 **Semiconductor Integrated Circuits (ICs).**

or *IC chip.* Today the IC is used extensively in every branch of electronics with hundreds of thousands of transistors and other components being placed on a chip of semiconductor no bigger than this ■. Figure 10-3 illustrates some of the different types of integrated circuits.

Like an evolving species, semiconductors have come to dominate the products of which they used to be only a part. For example, there used to be 400 components in a typical cellular telephone. Now there are 40, and soon only 3 or 4 IC chips will make up the entire phone circuitry. Today the semiconductor business—once regarded as a technical sideshow—occupies center stage and is key to the development of new products for all industries.

TIME LINE

In 1904 John A. Fleming, a British scientist, saw the value of an effect that was discovered by Thomas Edison but for which he saw no practical purpose. The "Edison effect" permitted Fleming to develop the "Fleming value," which passes current in only one direction. Its operation made it the first device able to convert alternating current into direct current and to detect radio waves.

SELF-TEST EVALUATION POINT FOR SECTION 10-1

Use the following questions to test your understanding of Section 10-1.

1. Name the three most frequently used semiconductor devices in electrical and electronic equipment.
2. The main function of a semiconductor device is to control the ____________ or ____________ in an electrical or electronic circuit.

10-2 SEMICONDUCTOR MATERIALS

A semiconductor material is one that is neither a conductor nor a nonconductor (insulator). This means simply that it will not conduct current as well as a conductor or block current as well as an insulator. Some semiconductor materials are pure or natural elements such as carbon (C), germanium (Ge), and silicon (Si), while other semiconductor materials are compounds.

Silicon and germanium are used most frequently in the construction of semiconductor devices for electrical and electronic applications. Germanium is a brittle grayish-white element that may be recovered from the ash of certain types of coals. Silicon, the most popular semiconductor material due to its superior temperature stability, is a white element normally derived from sand. Let us now examine the silicon, germanium, and carbon semiconductor atoms in more detail.

10-2-1 *Semiconductor Atoms*

TIME LINE

Jean Perrin discovered that cathode rays consisted of negatively charged particles, and these particles, which later became known as electrons, were measured by an English physicist, Joseph Thomson (1846–1914).

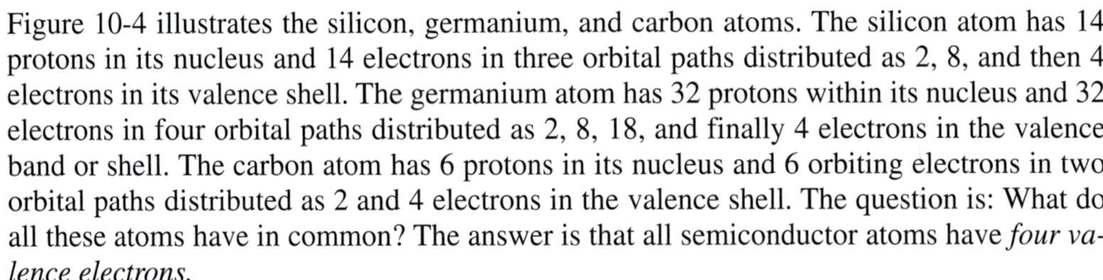

Figure 10-4 illustrates the silicon, germanium, and carbon atoms. The silicon atom has 14 protons in its nucleus and 14 electrons in three orbital paths distributed as 2, 8, and then 4 electrons in its valence shell. The germanium atom has 32 protons within its nucleus and 32 electrons in four orbital paths distributed as 2, 8, 18, and finally 4 electrons in the valence band or shell. The carbon atom has 6 protons in its nucleus and 6 orbiting electrons in two orbital paths distributed as 2 and 4 electrons in the valence shell. The question is: What do all these atoms have in common? The answer is that all semiconductor atoms have *four valence electrons.*

The valence shell of an atom can contain up to 8 electrons, and it is the number of electrons in this valence shell that determines the conductivity of the atom. For example, an atom with only 1 valence electron would be classed as a good conductor whereas an atom having 8 valence electrons, and therefore a complete valence shell, would be classed as an insulator.

To summarize, Figure 10-4 shows three semiconductor atoms, all of which contain four valence electrons. Since the number of valence electrons determines the conductivity of the element, semiconductor atoms are midway between conductors (which have 1 valence electron) and insulators (which have 8 valence electrons). Silicon and germanium are used to manufacture semiconductor devices, whereas carbon is combined with other elements to construct resistors.

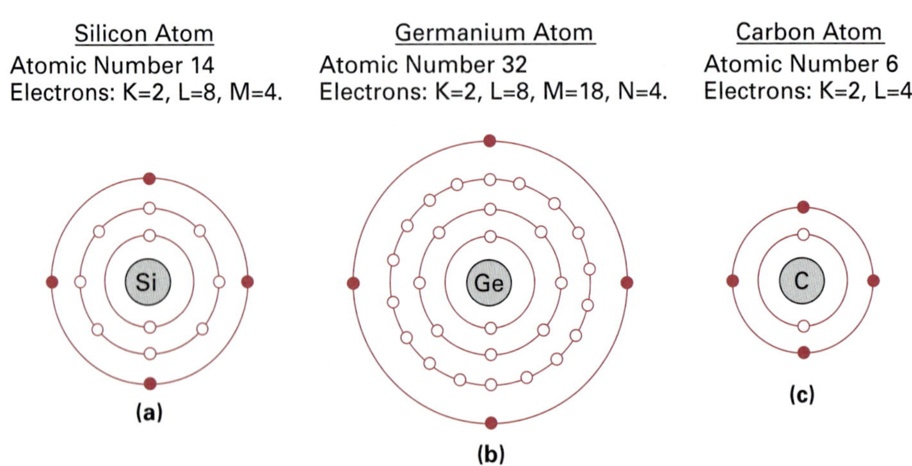

FIGURE 10-4 Semiconductor Atoms with Their Four Valence Shell Electrons.

10-2-2 *Crystals and Covalent Bonding*

So far we have discussed only isolated atoms. When two or more similar semiconductor atoms are combined to form a solid element, they automatically arrange themselves into an orderly lattice-like structure or pattern known as a **crystal,** as shown in Figure 10-5(a). This pattern is formed because each atom shares its four valence electrons with its four neighboring atoms. Since each atom shares one electron with a neighboring atom, two atoms will share two, or a pair, of electrons between the two cores. These two atom cores are pulling the two electrons with equal but opposite force and it is this pulling action that holds the atoms together in this solid crystal-lattice structure. The joining together of two semiconductor atoms is called an **electron-pair bond** or **covalent bond.** When many atoms combine, or bond, in this way the result is a crystal (smooth, glassy, solid) lattice structure. To illustrate this bonding process, each atom in

Crystal

A solid element with an orderly lattice-like structure.

Electron-Pair Bond or Covalent Bond

A pair of electrons shared by two neighboring atoms.

TIME LINE

Although the Fleming valve was an advance, it could not amplify or boost a signal. The "audion," developed by U.S. inventor Lee de Forest (1873-1961), sparked an era known as "vacuum-tube electronics" that brought about transcontinental telephony in 1915, radio broadcasting in 1920, radar in 1936, and television between 1927 and 1946, because of this triode vacuum tube's ability to amplify small signals.

TIME LINE

In 1947, J. Presper Eckert (right) and John Mauchly (left) unveiled ENIAC, which used over 300,000 vacuum tubes. ENIAC, which is an acronym for "electronic numerical integrator and computer," was the first large-scale electronic digital computer.

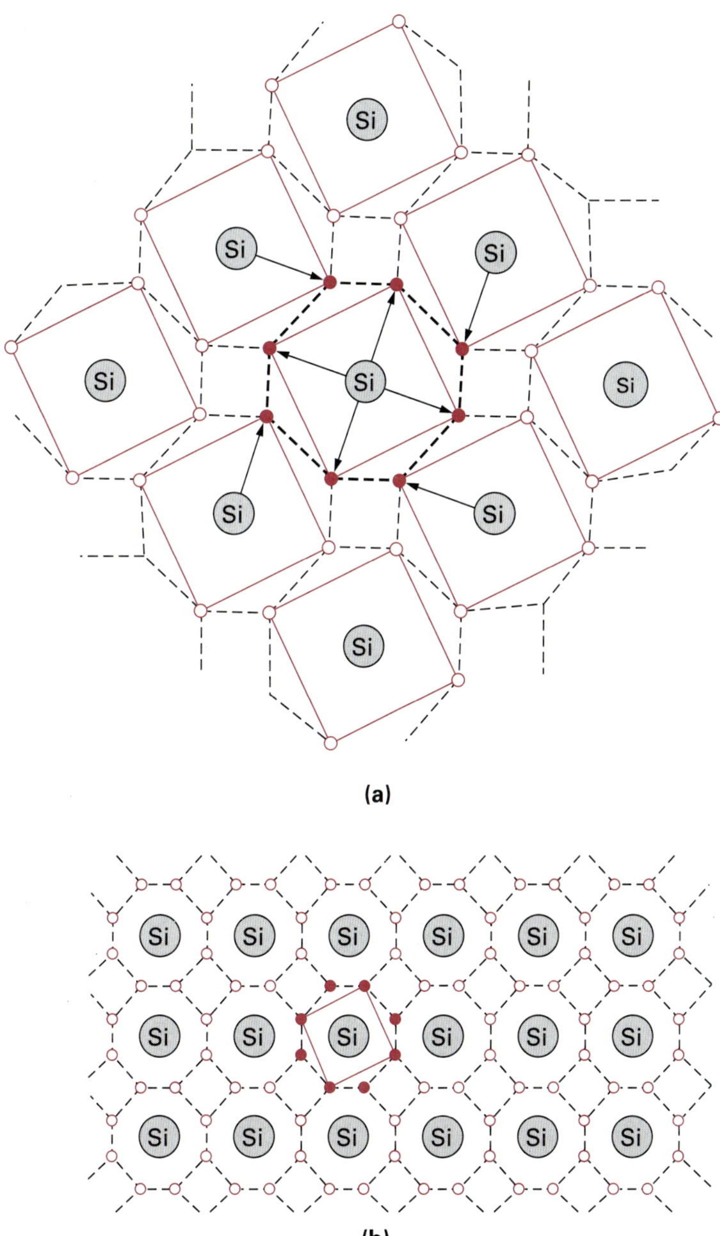

FIGURE 10-5 Covalent Bonding. (a) Silicon Atoms Sharing Valence Electrons. (b) Silicon Crystal Lattice Structure.

Figure 10-5(a) has been drawn as a square and each valence shell has been drawn as an octagon (eight-sided figure) so that we can easily see which electrons belong to which atom. As you can see, the atom in the center of the diagram has 4 valence electrons (shown at the corners of the Si square), and shares one electron from each of its four neighbors.

Figure 10-5(b) shows a larger view of a silicon crystal structure. All of the atoms in this structure are electrically stable because all of their valence shells are complete (they all contain eight electrons). These completed valence shells cause the pure semiconductor crystal structure to act as an insulator since it will not easily give up or accept electrons. Pure semiconductor materials, which are often called **intrinsic** materials, are therefore very poor conductors. Once this pure material is available, it must then be modified by a *doping* process to give it the qualities necessary to construct semiconductor devices. Silicon is most frequently used to construct solid state or semiconductor devices such as diodes and transistors because germanium has poor temperature stability and carbon crystals (diamonds) are too expensive to use.

Intrinsic Semiconductor Materials
Pure semiconductor materials.

10-2-3 *Energy Gaps and Energy Levels*

Let us now examine the relationships between electrons and orbital shells in a little more detail so that we can better understand charge and conduction within a semiconductor material.

As mentioned previously, there are seven shells available for electrons (K, L, M, N, O, P, and Q) around the nucleus. Electrons must travel or orbit in one of these orbital paths because they cannot exist in any of the spaces between orbital shells. Each orbital shell has its own specific energy level. Therefore, electrons traveling in a specific orbital shell will contain the shell's energy level. Figure 10-6 shows an example of an atom's orbital shell energy

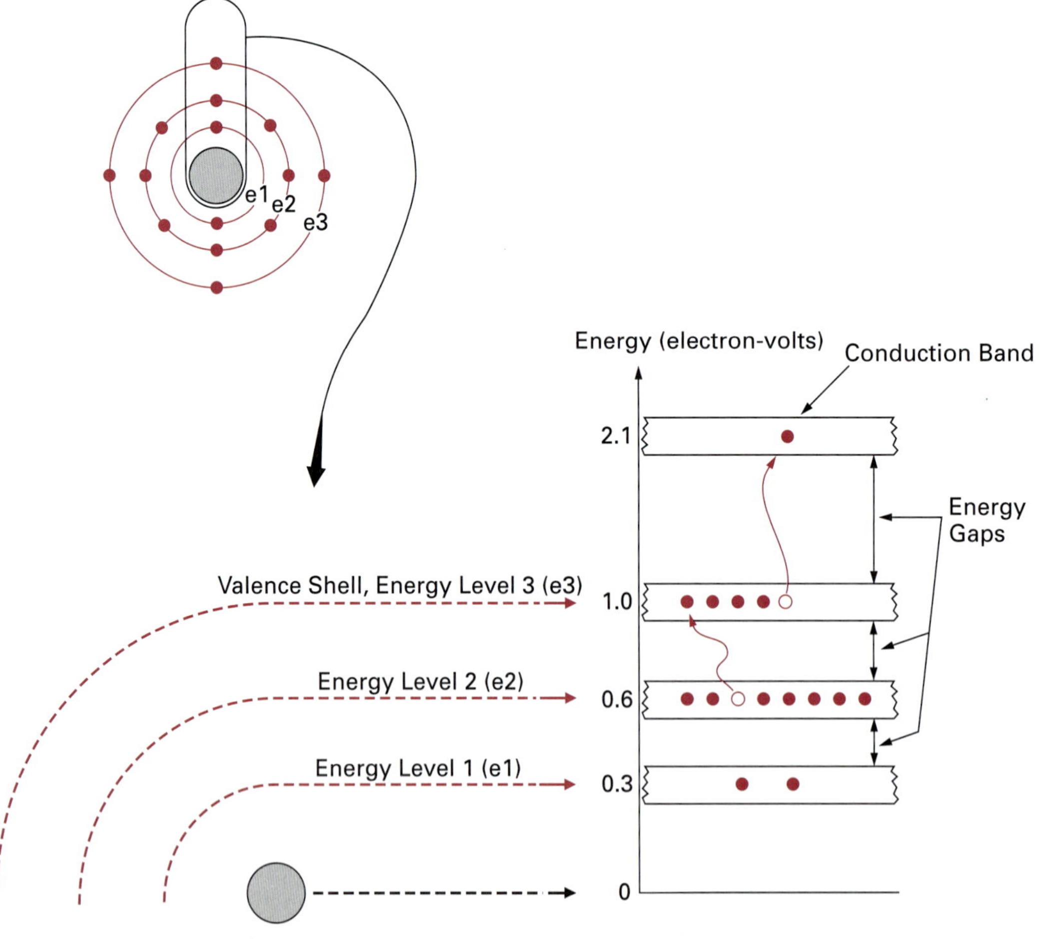

FIGURE 10-6 An Atom's Orbital Shell Energy Levels.

levels. The energy levels for each shell increase as you move away from the nucleus of the atom. The valence shell and the valence electrons will always have the highest energy level for a given atom. The space between any two orbital shells is called the **energy gap.** Electrons can jump from one shell to another if they absorb enough energy to make up the difference between their initial energy level and the energy level of the shell that they are jumping to. For example, in Figure 10-6 the valence shell has an energy level of 1.0 **electron-volts (eV).** Because this atom has three orbital shells, the valence shell will be energy level 3 (e3). The second energy level or orbital shell (e2) has an energy level of 0.6 eV. Therefore, for an electron to jump from energy level 2 (shell 2) to energy level 3 (e3 or valence shell), it will have to absorb a value of energy equal to the difference between e2 and e3. This will equal:

$$1.0 \text{ eV} - 0.6 \text{ eV} = 0.4 \text{ eV}$$

In this example, when either heat, light, or electrical energy was applied, one of the electrons in shell 2 (e2) absorbed 0.4 electron-volts of energy and jumped to valence shell (e3).

If a valence (e3) electron absorbs enough energy it can jump from the valence shell into the **conduction band.** The conduction band is an energy band in which electrons can move freely or wander within a solid. When an electron jumps from the valence shell into the conduction band, it is released from the atom and no longer travels in one of its orbital paths. The electron is now free to move within the semiconductor material and is said to be in the **excited state.** An excited electron in the conduction band will eventually give up the energy it absorbed in the form of light or heat and return to its original energy level in the atom's valence shell.

When an electron jumps from the valence shell or band to the conduction band, it leaves a gap in the covalent bond called a **hole.** This action is shown in Figure 10-7(a). A hole

Energy Gap
The space between two orbital shells.

Electron-Volt (eV)
A unit of energy equal to the energy acquired by an electron when it passes through a potential difference of 1 V in a vacuum.

Conduction Band
An energy band in which electrons can move freely within a solid.

Excited State
An energy level in which an electron may exist if given sufficient energy to reach this state from a lower state.

Hole
The gap in the covalent bond left when an electron jumps from the valence shell or band to the conduction band.

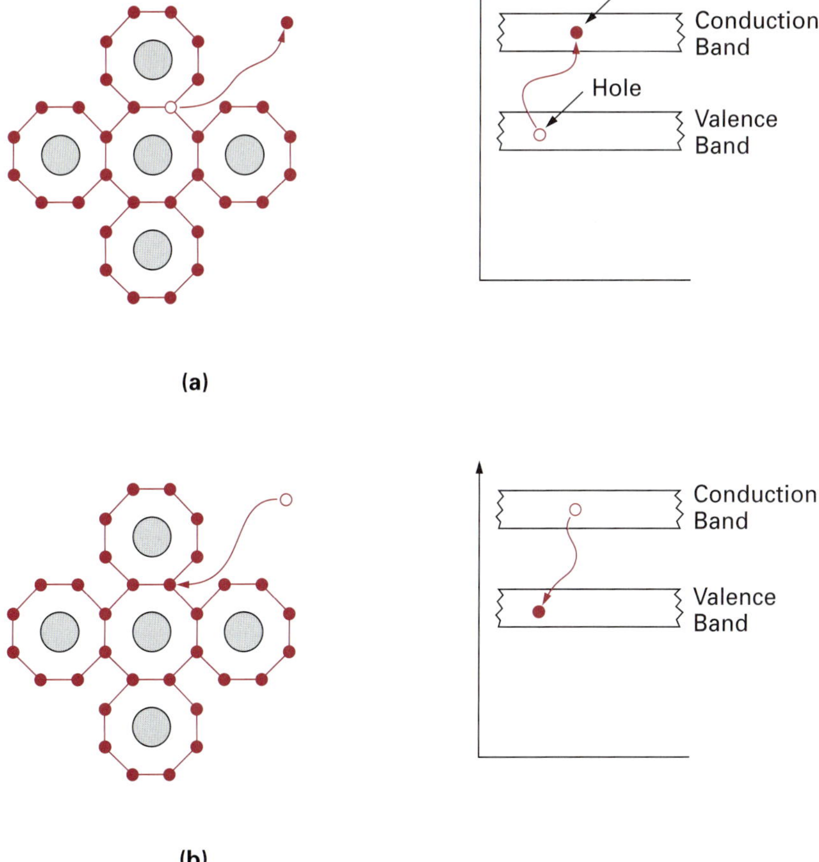

FIGURE 10-7 Valence Band and Conduction Band Actions. (a) Generating an Electron-Hole Pair. (b) Recombination.

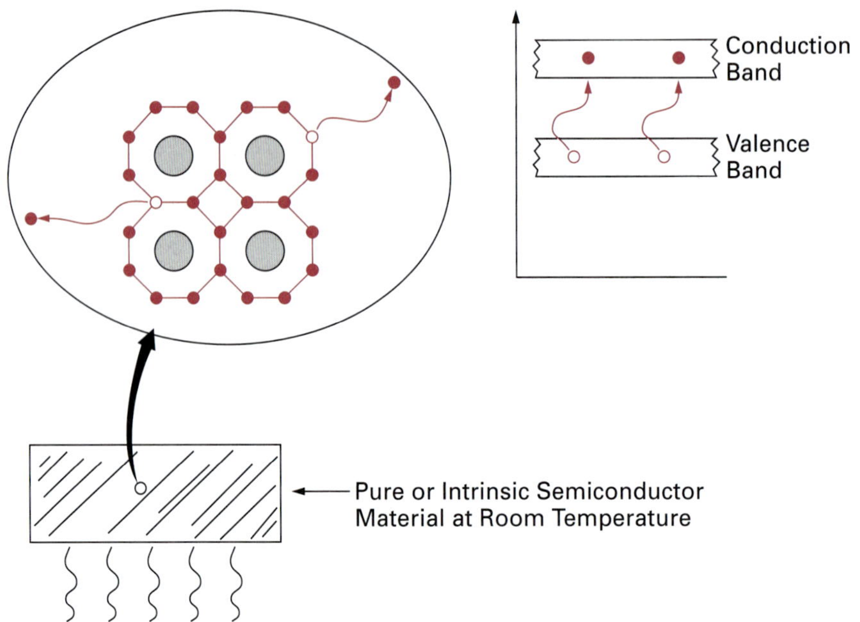

FIGURE 10-8 **Temperature Effects on Semiconductor Materials.**

Electron-Hole Pair
When an electron jumps from the valence shell or band to the conduction band, it leaves a gap in the covalent bond called a hole. This action creates an electron-hole pair.

Recombination
Action occurring in microseconds as an electron in the conduction band gives up its energy and falls into one of the valence shell holes in the covalent bond.

Lifetime
The time difference between an electron jumping into the conduction band and then falling back into a hole.

is created every time an electron enters the conduction band. This action creates an **electron-hole pair.**

It only takes a few microseconds before a free electron in the conduction band will give up its energy and fall into one of the valence shell holes in the covalent bond. This action is called **recombination** and is shown in Figure 10-7(b). The time difference between an electron jumping into the conduction band (becoming a free electron) and then falling back into a hole (recombination) is called the **lifetime** of the electron-hole pair.

10-2-4 *Temperature Effects on Semiconductor Materials*

At extremely low temperatures the valence electrons are tightly bound to their parent atoms, preventing valence electrons from drifting between atoms. Therefore, pure or intrinsic semiconductor materials function as insulators at temperatures close to absolute zero (−273.16°C or − 459.69°F).

At room temperature, however, the valence electrons absorb enough heat energy to break free of their covalent bonds creating electron-hole pairs, as shown in Figure 10-8. Therefore, *the conductivity of a semiconductor material is directly proportional to temperature, in that an increase in temperature will cause an increase in the semiconductor material's conductance.* This means: *an increase in temperature* ($T\uparrow$) *will cause an increase in a semiconductor's conductivity* ($G\uparrow$) *and current* ($I\uparrow$). This is why all circuits containing a semiconductor device tend to consume more current once they have warmed up.

Stated another way, *semiconductor materials, and therefore semiconductor devices, have a negative temperature coefficient of resistance, which means as temperature increases* ($T\uparrow$), *their resistance decreases* ($R\downarrow$).

10-2-5 *Applying a Voltage across a Semiconductor*

Free Electron
An electron that is able to move freely when an external force is applied.

If a voltage was applied across a room-temperature section of intrinsic semiconductor material, **free electrons** in the conduction band would make up a small electrical current as shown in Figure 10-9. In this illustration you can see how the negatively charged free electrons are attracted to the positive terminal of the voltage source. For every free electron that leaves the

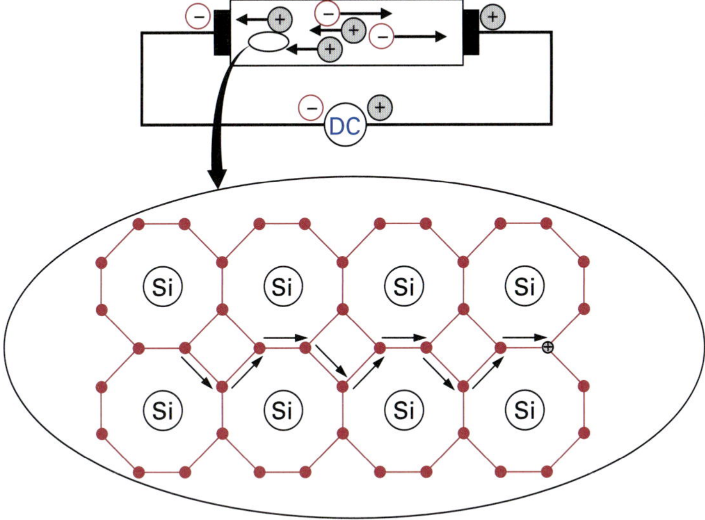

FIGURE 10-9 **Electron Flow and Hole Flow in an Intrinsic Semiconductor.**

semiconductor material on the right side and travels to the positive terminal of the source, another electron is generated at the negative terminal of the voltage source and is injected into the left side of the semiconductor material. These injected electrons are captured by holes in the semiconductor material (recombination). As you can see from this illustration, current in a semiconductor material is made up of both electrons and holes. The holes act like positively charged particles while the electrons act like negatively charged particles. As electrons jump between atoms in a migration to the positive terminal of the source voltage, they leave behind them holes, which are then filled by other advancing electrons. These advancing electrons leave behind them other holes, making it appear as though these holes are traveling toward the negative terminal of the source voltage. This **hole flow** is a new phenomenon to us, and it is one of the key differences between a semiconductor and a conductor. With conductors we were only interested in free-electron flow, but with semiconductors we must consider the movement of free electrons (negative charge carriers) and the apparent movement of holes (positive charge carriers).

Hole Flow
Conduction in a semiconductor when electrons move into holes when a voltage is applied.

In summary, therefore, *when a potential difference is applied across a semiconductor, the electrons move toward the positive potential and the holes travel toward the negative potential. The total current flow is equal to the sum of the electron flow and the hole flow currents.*

SELF-TEST EVALUATION POINT FOR SECTION 10-2

Use the following questions to test your understanding of Section 10-2.

1. What do all semiconductor atoms have in common?
2. The electrical conductivity of an element is determined by the number of electrons in the valence shell. Semiconductor atoms are midway between conductors, which have ____________ valence electron(s), and insulators, which have ____________ valence electron(s).
3. Each of the semiconductor atoms in a crystal lattice shares its electrons with four neighboring atoms. This joining of atoms is called a ____________.
4. Semiconductor materials have a ____________ temperature coefficient of resistance, which means as temperature increases, resistance ____________.
5. The number of electron-hole pairs within a semiconductor will increase as temperature ____________.
6. When a pure or ____________ semiconductor is connected across a voltage, free electrons travel toward the ____________ terminal of the applied voltage, whereas holes appear to travel toward the ____________ terminal of the applied voltage.

10-3 DOPING SEMICONDUCTOR MATERIALS

At room temperature pure or intrinsic semiconductors will not permit a large enough value of current. Therefore, some modification has to be applied in order to increase the semiconductor's current-carrying capability or conductivity. **Doping** is a process wherein impurities are added to the intrinsic semiconductor material either to increase the number of free electrons (negative doping) or to increase the number of holes (positive doping).

Doping
The process wherein impurities are added to the intrinsic semiconductor material either to increase the number of free electrons or to increase the number of holes.

Basically, there are two types of impurities that can be added to semiconductor crystals. One type of impurity is called a *pentavalent material* because its atom has five (*penta*) valence electrons. The second type of impurity is called a *trivalent material* because its atoms have three (*tri*) valence electrons. A doped semiconductor material is referred to as an **extrinsic semiconductor** material because it is no longer pure.

Extrinsic Semiconductor
A semiconductor whose electrical properties are dependent on impurities added to the semiconductor crystal.

10-3-1 n-*Type Semiconductor*

Figure 10-10(a) shows how a semiconductor material's atoms will appear after pentavalent atom impurities have been added. The pentavalent atoms, which are listed in Figure 10-10(b), can be added to molten silicon to create, when cooled, a crystalline structure that has an extra electron due to the pentavalent (5 valence-electron impurity) atoms. The fifth pentavalent electron is not part of the covalent bonding and requires little energy to break free and enter the conduction band, as shown in Figure 10-10(c). Because millions of pentavalent atoms are added to the pure semiconductor, there will be millions of free electrons available for flow through the material.

Even though the doped semiconductor material has millions of free electrons, the material is still electrically neutral. This is because each arsenic atom has the same number of protons as electrons and so do the silicon atoms. Therefore, the overall number of protons and electrons in the semiconductor is still equal and the result is a net charge of zero. However, because we now have more electrons than valence-band holes, the material is called an ***n*-type semiconductor.** *n*-Type semiconductors have more conduction-band electrons than valence-band holes. The electrons are therefore called the **majority carriers** and the valence-band holes are called the **minority carriers.** In Figure 10-10(c) you can see the abundance of conduction-band electrons. The holes in the valence band are few and are generated by thermal energy because the semiconductor is at room temperature.

***n*-Type Semiconductor**
A material that has more conduction-band electrons than valence-band holes.

Majority Carriers
The type of carrier that makes up more than half the total number of carriers in a semiconductor device.

Minority Carriers
The type of carrier that makes up less than half of the total number of carriers in a semiconductor device.

When a voltage is applied across an *n*-type semiconductor, as shown in Figure 10-10(d), the additional free conduction-band electrons travel toward the positive terminal of the dc source. The applied voltage will cause extra electrons to break away from their covalent bonds to create holes, resulting in an increase in current and conductivity. Although the total current flow in this *n*-type semiconductor is the sum of the electron and hole currents, the conduction-band electrons make up the majority of the flow.

10-3-2 p-*Type Semiconductor*

Figure 10-11(a) shows how a semiconductor material's atoms will appear after trivalent atom impurities have been added. The trivalent atoms, which are listed in Figure 10-11(b), can be added to molten silicon to create, when cooled, a crystalline structure that has a hole in the valence band of every trivalent (3 valence-electron impurity) atom. Instead of an excess of electrons, we now have an excess of holes. Because millions of trivalent atoms are added to the pure semiconductor, there will be millions of holes available for flow through the material.

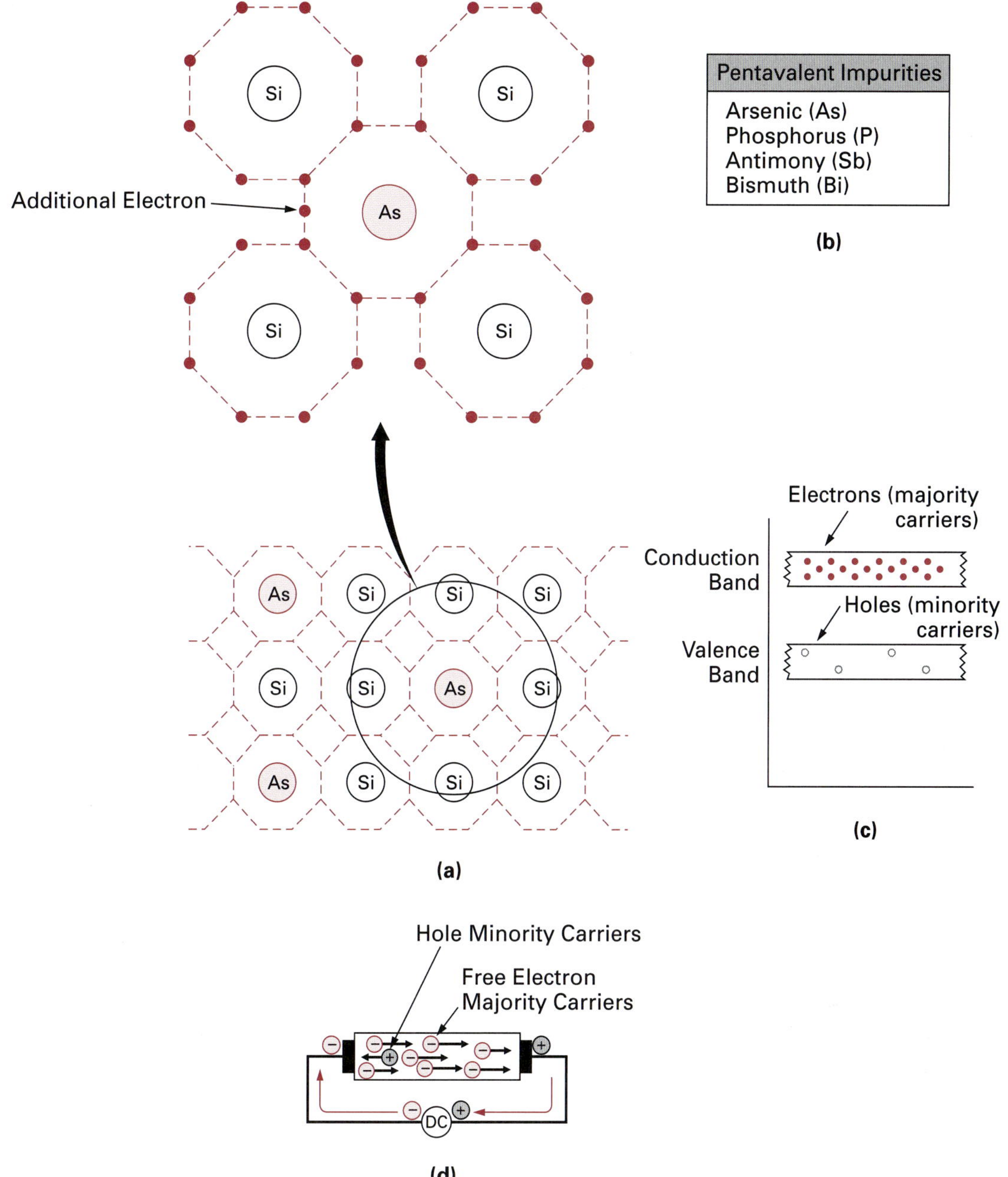

FIGURE 10-10 **Adding Pentavalent Impurities to Create an *n*-Type Semiconductor Material.**

Even though the doped semiconductor material has millions of holes, the material is still electrically neutral. This is because each aluminum atom has the same number of protons as electrons and so do the silicon atoms. Therefore the overall number of protons and electrons in the semiconductor is still equal and the result is a net charge of zero. However, because we now have more valence band holes than electrons the material is called a ***p*-type semiconductor.** *p*-Type semiconductors have more valence-band holes than conduction-band electrons. The holes are called the *majority carriers* and the electrons are called the *minority carriers.* In Figure 10-11(c) you can see the abundance of valence-band holes. The few electrons in the conduction band are generated by thermal energy because the semiconductor is at room temperature.

***p*-Type Semiconductor**
A material that has more valence-band holes than conduction-band electrons.

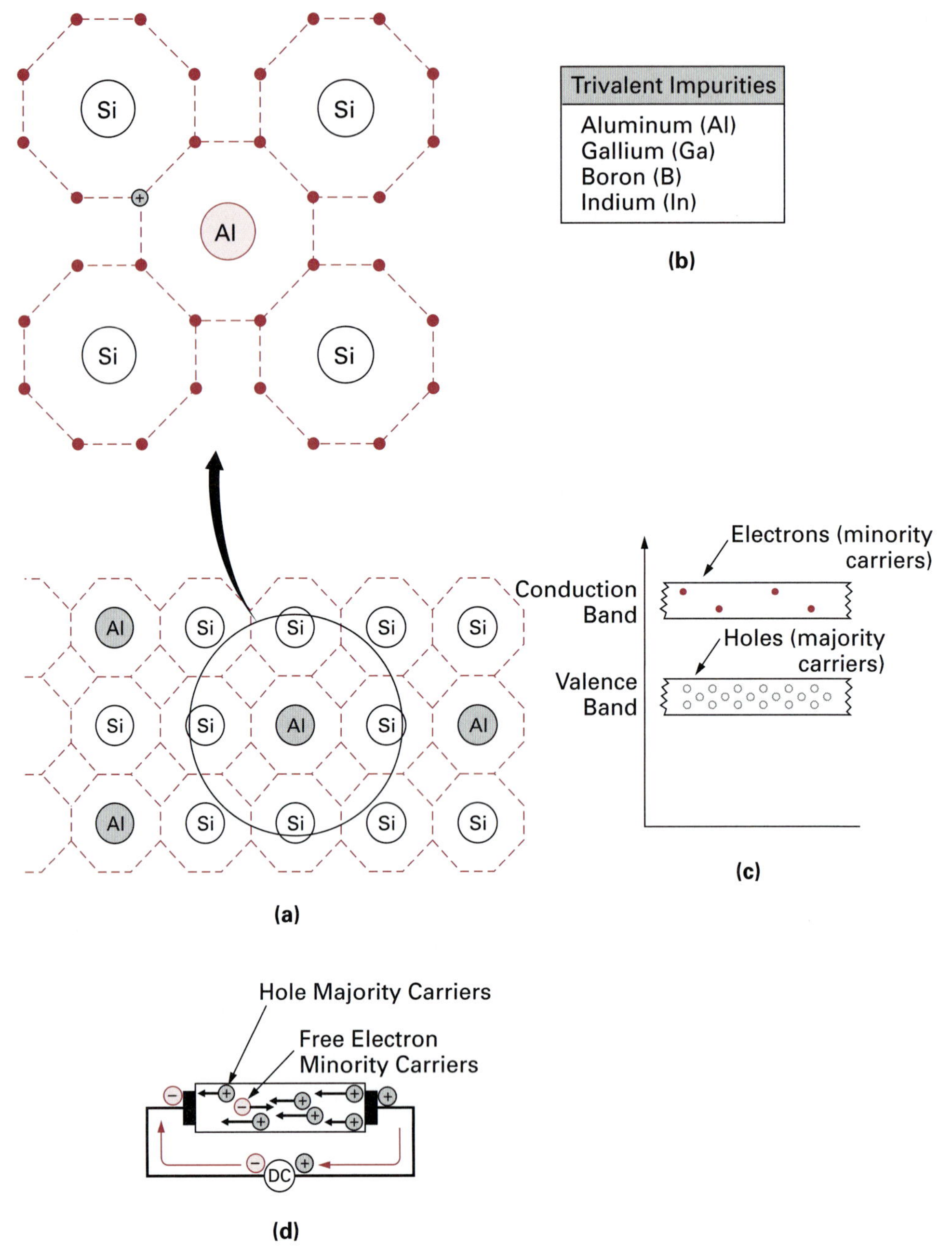

FIGURE 10-11 Adding Trivalent Impurities to Create a *p*-Type Semiconductor Material.

When a voltage is applied across a *p*-type semiconductor, as illustrated in Figure 10-11(d), the large number of holes within the material will attract electrons from the negative terminal of the dc source into the *p*-type semiconductor. These holes appear to move because each time an electron moves into a hole it creates a hole behind it, and the holes appear to move in the opposite direction to the electrons (toward the negative terminal of the dc source). The applied voltage will cause some electrons to break away from the covalent bond resulting in an increased current and conductivity. Although the total current flow in this *p*-type semiconductor is the sum of the hole and electron currents, the valence-band holes make up the majority of the flow.

SELF-TEST EVALUATION POINT FOR SECTION 10-3

Use the following questions to test your understanding of Section 10-3.

1. Why are impurities added to pure semiconductor materials?
2. Pentavalent atoms add ____________ to semiconductor crystals, to create ____________-type semiconductors.
3. Trivalent atoms add ____________ to semiconductor crystals, to create ____________-type semiconductors.
4. In an *n*-type semiconductor the majority carriers are ____________, whereas in a *p*-type semiconductor the majority carriers are ____________.

10-4 THE P-N JUNCTION

On their own, *n*-type semiconductor materials and *p*-type semiconductor materials are of little use. Together, however, these two form a **P-N semiconductor junction.** Semiconductor devices such as diodes and transistors are constructed using these P-N junctions, which give specific current flow characteristics. In this section we will examine the characteristics of the P-N junction in detail.

P-N Junction
The point at which two oppositely doped materials come in contact with one another.

10-4-1 *The Depletion Region*

Figure 10-12(a) shows the individual *n*-type and *p*-type materials. The *n*-type material is represented as a block containing an excess of electrons (solid circles), while the *p*-type material is represented as a block containing an excess of holes (open circles). The energy diagrams below the two semiconductor sections show the differences between the two materials. Because different impurity atoms were added to the pure semiconductor material, the atomic make-up of the *n*-type and *p*-type materials is slightly different, which is why the valence bands and conduction bands are at slightly different energy levels.

Figure 10-12(b) shows the two *n*-type and *p*-type semiconductor sections joined together. A manufacturer of semiconductor devices would not join two individual pieces in this way to create a P-N junction. Instead, a single piece of pure semiconductor material would have each of its halves doped to create a *p*-type and *n*-type section.

The point at which the two oppositely doped materials come in contact with one another is called the *junction.* This junction of the two materials now permits the free electrons in the *n*-type material to combine with the holes in the *p*-type material as shown in Figure 10-12(c). As free electrons in the *n* material cross the junction and combine with holes in the *p* material, they create negative ions (atoms with more electrons than protons) in the *p* material, and leave behind positive ions (atoms with less electrons than protons) in the *n* material, as shown in Figure 10-12(d). An area or region on either side of the junction becomes emptied or depleted of free electrons and holes. This small layer containing positive and negative ions is called the **depletion region.**

Depletion Region
A small layer on either side of the junction that becomes empty, or depleted, of free electrons or holes.

As the ion layer on either side of the junction builds up, it has the effect of diminishing and eventually preventing any further recombination of free electrons and holes across the junction. In other words, the negative ions in the *p* region near the junction repel and prevent free electrons in the *n* region from recombination. This action prevents the depletion region from becoming larger and larger.

These positive ions or charges and negative ions or charges accumulate a certain potential. Since these charges are opposite in polarity, a potential difference or voltage called the **barrier potential** or **barrier voltage** exists across the junction as shown in Figure 10-12(e). At room temperature, the barrier voltage of a silicon P-N junction is approximately 0.7 V, and a germanium P-N junction is approximately 0.3 V.

Barrier Potential or Barrier Voltage
The potential difference, or voltage, that exists across the junction.

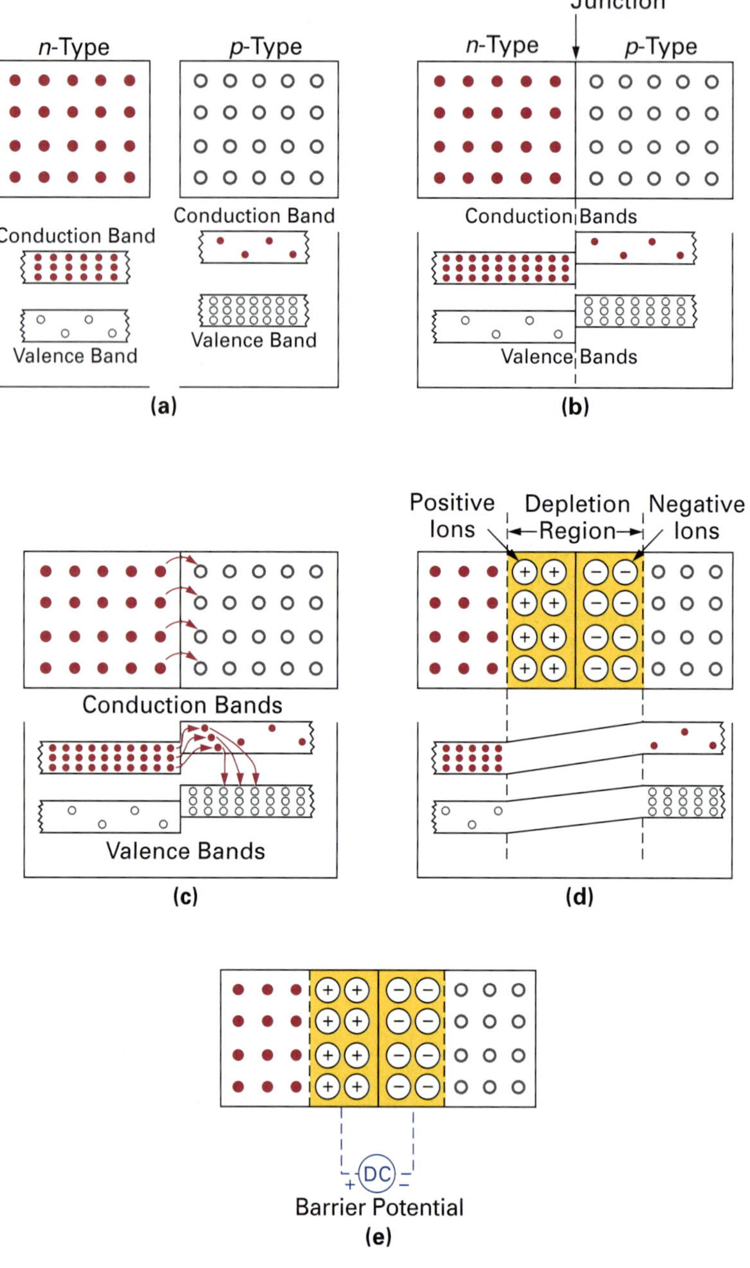

FIGURE 10-12 **The Depletion Region.**

10-4-2 *Biasing a P-N Junction*

Bias Voltages

The dc voltages applied to control a device's operation.

Forward Biased

A small depletion region at the junction will offer a small resistance and permit a large current. Such a junction is forward biased.

Semiconductor devices are constructed using P-N junctions. These P-N junctions need voltages of a certain amplitude and polarity to control their operation. These voltages, which incline or cause the device to operate in a certain manner, are known as **bias voltages.** Bias voltages control the width of the depletion region, which in turn controls the resistance of the P-N junction and, therefore, the amount of current that can pass through the P-N junction or semiconductor device.

To be specific, a small depletion region ($dr\downarrow$) will offer a small P-N junction resistance ($R\downarrow$) and therefore permit a large P-N junction current ($I\uparrow$). In this instance, the P-N semiconductor junction is said to be **forward biased** and acts like a conductor.

On the other hand, a large depletion region ($dr\uparrow$) will offer a large P-N junction resistance ($R\uparrow$) and therefore permit only a small P-N junction current ($I\downarrow$). In this instance the P-N semiconductor junction is said to be **reverse biased** and acts like an insulator.

Reverse Biased

A large depletion region at the junction will offer a large resistance and permit only a small current. Such a junction is reverse biased.

Forward Biasing a P-N Junction

Figure 10-13 shows in detail why a forward biased P-N junction will pass current with almost no opposition (act like a conductor). To begin, Figure 10-13(a) shows a P-N junction

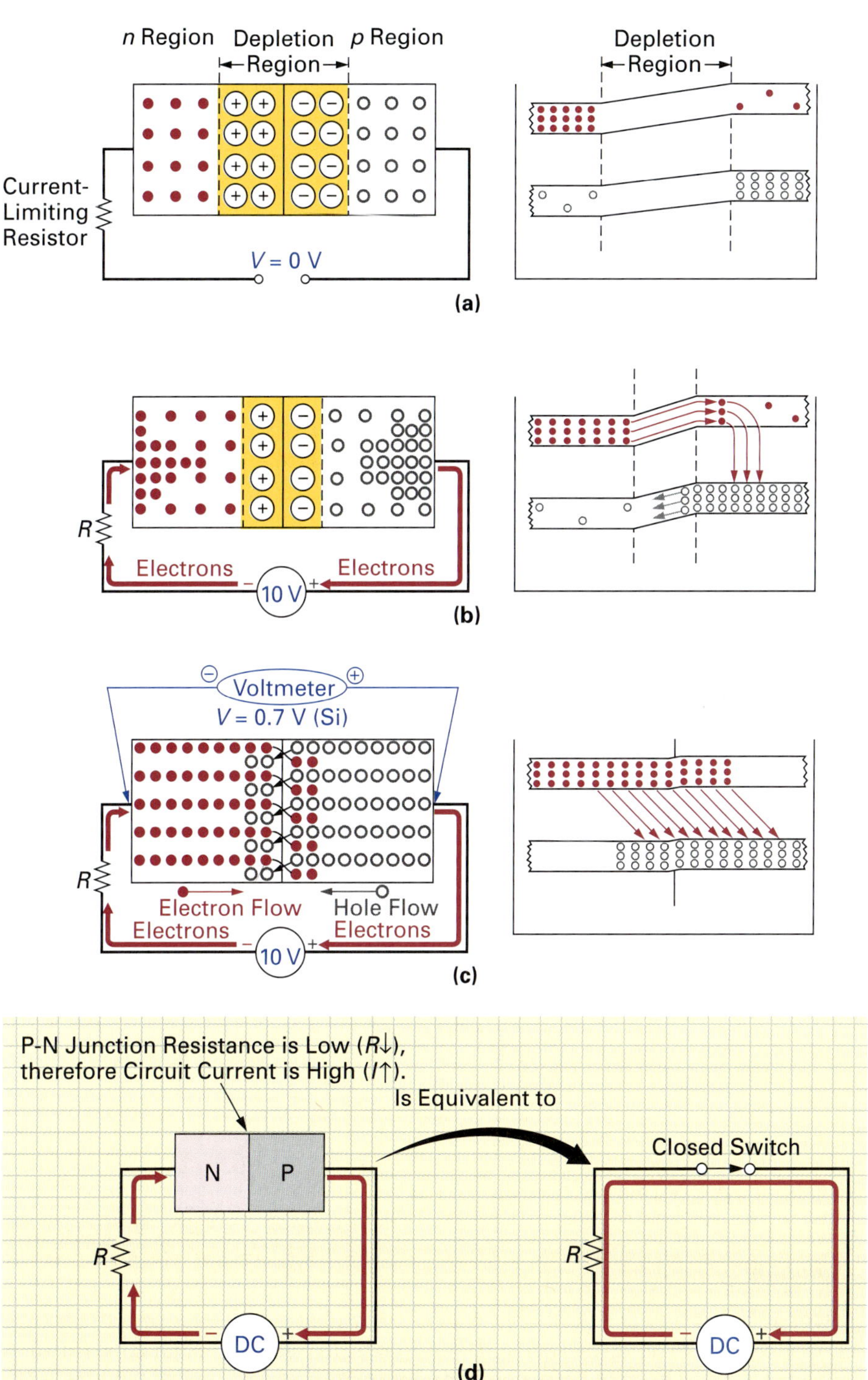

FIGURE 10-13 Forward Biasing a P-N Semiconductor Junction.

with wires attached. A resistor has been included to limit the amount of current passing through the P-N junction to a safe level. Energy diagrams have also been included on the right side of each part of Figure 10-13 to show the relationship between the conduction band and valence band of the p and n regions.

Let us now connect a dc voltage across the P-N junction to see how it reacts. This is shown in Figure 10-13(b). The negative potential of the dc source has been applied to the n region and the positive potential of the dc source has been applied to the p region. Referring to the energy diagram in Figure 10-13(b), you can see that the conduction band electrons in the n region are repelled by the negative voltage source towards the junction. On the opposite side, valence band holes in the p region are repelled by the positive voltage source towards the junction. A forward-conducting current will begin to flow if the external source voltage is large enough to overcome the internal barrier voltage of the P-N junction. In this example we will assume that the dc source voltage is 10 volts and therefore this will be more than enough to overcome the silicon P-N junction's barrier potential of 0.7 volts.

Forward Voltage Drop (V_F)
The forward voltage drop is equal to the junction's barrier voltage.

Conduction through the P-N junction is shown in Figure 10-13(c). When forward biased, a P-N junction will act as a conductor and have a low but finite resistance value that will cause a corresponding voltage drop across its terminals. This **forward voltage drop** **(V_F)** is approximately equal to the P-N junction's barrier voltage:

$$\text{Forward Voltage Drop } (V_F) \text{ for Silicon} = 0.7 \text{ V}$$
$$\text{Forward Voltage Drop } (V_F) \text{ for Germanium} = 0.3 \text{ V}$$

Figure 10-13(c) shows how a voltmeter can be used to measure the forward voltage drop of 0.7 V across a silicon P-N junction when it is forward biased.

To summarize, Figure 10-13(d) shows that when a P-N junction is forward biased ($+V \rightarrow p$ region, $-V \rightarrow n$ region) the P-N junction resistance is low ($R\downarrow$), and therefore the circuit current is high ($I\uparrow$). When forward biased, therefore, the P-N junction acts like a conductor and is equivalent to a closed switch.

EXAMPLE:

Calculate the current for the circuit in Figure 10-14.

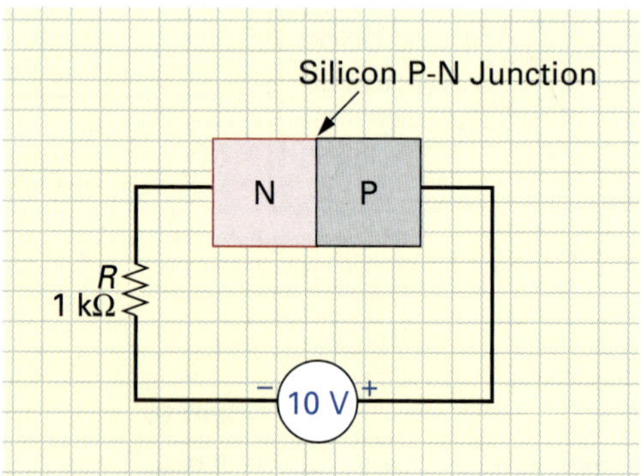

FIGURE 10-14 A P-N Junction Circuit.

Solution:

The silicon P-N junction is forward biased ($+V \rightarrow p$ region, $-V \rightarrow n$ region). The applied voltage of 10 V will be more than enough to overcome the silicon P-N junction forward voltage drop of 0.7 volts ($V_F = 0.7$ V for silicon). Since 10 volts are applied, and the P-N junction is dropping 0.7 V, the remaining voltage of 9.3 V is being dropped across the 1 kΩ resistor. Consequently, the forward-biased current (I_F) will equal:

$$I_F = \frac{V_S - V_{P\text{-}N}}{R}$$

$$= \frac{10\text{ V} - 0.7\text{ V}}{1\text{ k}\Omega}$$

$$= 9.3\text{ mA}$$

Reverse Biasing a P-N Junction

Figure 10-15 shows in detail why a reverse biased P-N junction will reduce current to almost zero (act like an insulator). To begin, Figure 10-15(a) shows a P-N junction

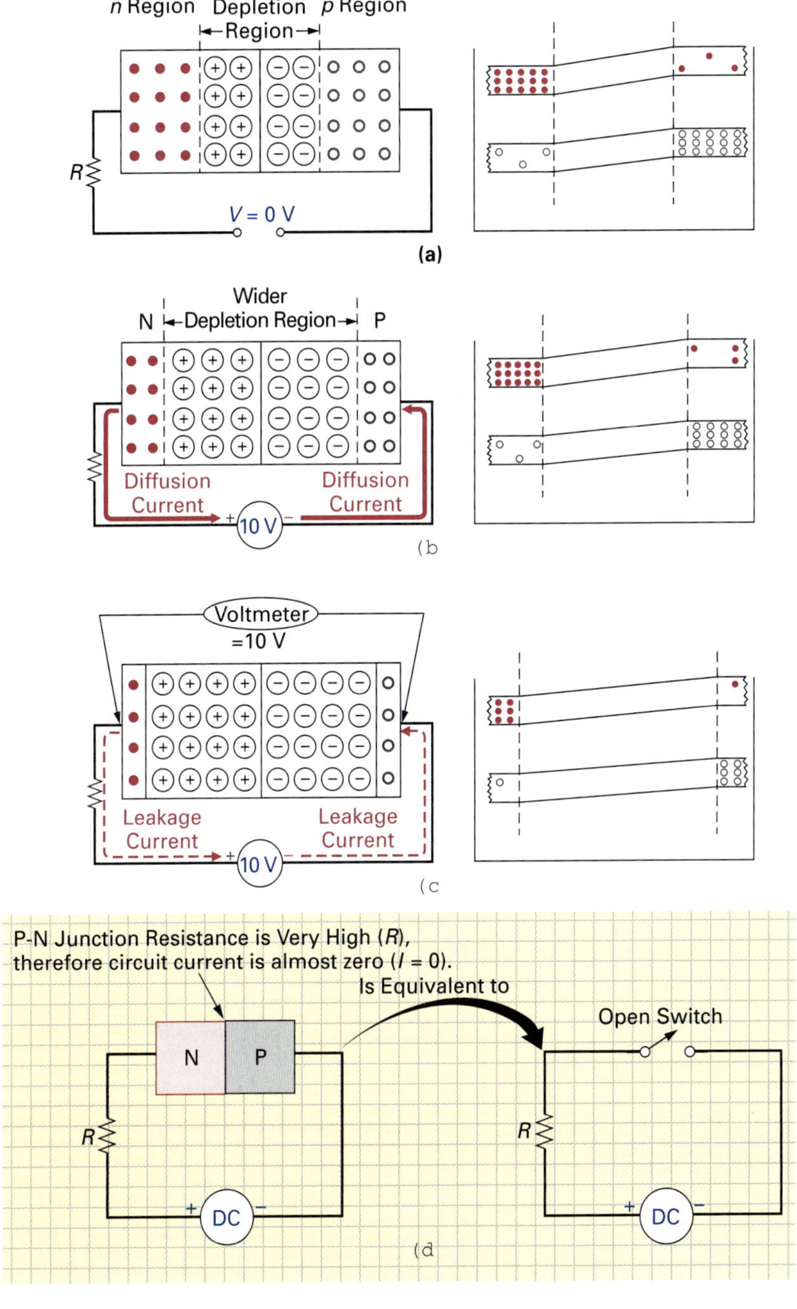

FIGURE 10-15 **Reverse Biasing a P-N Semiconductor Junction.**

with wires attached and no voltage being applied. Energy diagrams have again been included to show the relationship between the conduction band and valence band of the *p* and *n* regions.

Let us now connect a dc voltage across the P-N junction to see how it reacts. This is shown in Figure 10-15(b). The positive potential of the dc source is now being applied to the *n* region, and the negative potential of the dc source has been applied to the *p* region.

A forward biased P-N junction is able to conduct current because the external bias voltage forces the majority carriers in the *n* and *p* regions to combine at the junction. In this instance, however, the dc bias voltage polarity has been reversed, causing free electrons in the *n* region to travel to the positive terminal of the voltage source leaving behind a large number of positive ions at the junction. This increases the width of the depletion region. At the same time, electrons from the negative terminal of the source are attracted to the holes in the *p* region of the P-N junction. These electrons fill the holes in the *p* region near the junction creating a large number of negative ions. This further increases the width of the depletion region. The current that is present at the time the depletion layer is expanding is called the **diffusion current.** Referring to Figure 10-15(b), you can see that the depletion region is now wider than the unbiased P-N junction shown in Figure 10-15(a).

Diffusion Current
The current that is present when the depletion layer is expanding.

The ions on either side of the junction build up until the P-N junction's internal-barrier voltage is equal to the external-source voltage, as shown by the voltmeter in Figure 10-15(c). When reverse biased, therefore, the **reverse voltage drop (V_R)** across a P-N junction is equal to the source or applied voltage. At this time the resistance of the junction has been increased to a point that current drops to zero.

Reverse Voltage Drop (V_R)
The reverse voltage drop is equal to the source voltage (applied voltage).

Actually, an extremely small current called the **leakage current** or **reverse current (I_R)** will pass through the P-N junction, as shown in Figure 10-15(c). It is present because the minority carriers (holes in the *n* region, electrons in the *p* region) are forced toward the junction where they combine, producing a constant small current. The current in the P-N junction is still considered to be at zero because the leakage or reverse current is so small (nanoamps in silicon diodes).

Leakage Current or Reverse Current (I_R)
The extremely small current present at the junction.

To summarize, Figure 10-15(d) shows that when a P-N junction is reverse biased ($+V \rightarrow n$ region, $-V \rightarrow p$ region), the P-N junction resistance is extremely high ($R \uparrow\uparrow$), and the circuit current is effectively zero ($I = 0$ amps). When reverse biased, therefore, the P-N junction acts like an insulator and is equivalent to an open switch.

EXAMPLE:

Referring to Figure 10-16(a) and (b), calculate each circuit's:

a. current value
b. P-N junction voltage drop

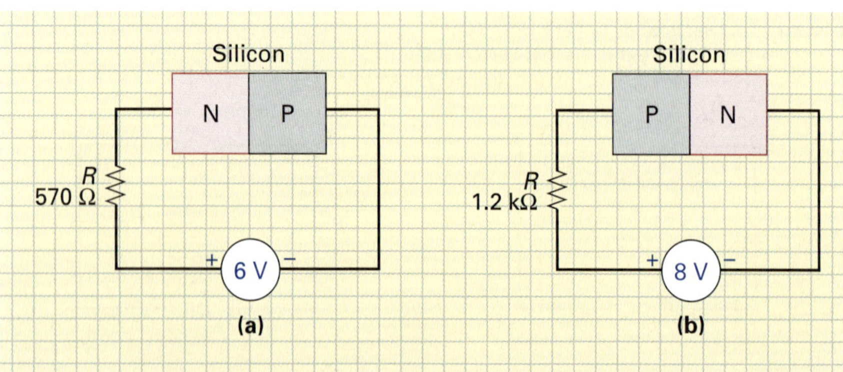

FIGURE 10-16 **P-N Junction Circuit Examples.**

■ ***Solution:***

The P-N junction in Figure 10-16(a) is reverse biased ($+V \rightarrow n$ region, $-V \rightarrow p$ region); therefore, the P-N junction resistance is extremely high ($R \uparrow\uparrow$), and the circuit current is effectively zero ($I = 0$). When reverse biased, the P-N junction acts like an insulator and is equivalent to an open switch, and the voltage developed across the open P-N junction will equal the source voltage applied. For Figure 10-16(a):

$$\text{Circuit Current} = 0$$

$$\text{P-N Junction Voltage Drop} = V_S = 6\text{ V}$$

The P-N junction in Figure 10-16(b) is forward biased ($+V \rightarrow p$ region, $-V \rightarrow n$ region). Therefore, the P-N junction resistance is low ($R\downarrow$), and the circuit current is high ($I\uparrow$). When forward biased, the P-N junction acts like a conductor and is equivalent to a closed switch, and the P-N junction's voltage drop will equal the forward voltage drop (V_F) for a silicon P-N junction. For Figure 10-16(b):

$$\text{Circuit Current} = I_F = \frac{V_S - V_{P\text{-}N}}{R}$$

$$= \frac{8\text{ V} - 0.7\text{ V}}{1.2\text{ k}\Omega}$$

$$= 6.08\text{ mA}$$

$$\text{P-N Junction Voltage Drop} = V_F = 0.7\text{ V}$$

SELF-TEST EVALUATION POINT FOR SECTION 10-4

Use the following questions to test your understanding of Section 10-4.

1. When a P-N junction is formed, a ____________ region is created on either side of the junction.
2. The barrier voltage within a silicon diode is:
   a. 700 mV
   b. 7.0 V
   c. 0.3 V
   d. None of the above
3. True or false: A P-N junction is forward biased when its P terminal is made positive relative to its N terminal.
4. A reverse biased P-N junction acts like a/an ____________ switch, whereas a forward biased P-N junction acts like a/an ____________ switch.

REVIEW QUESTIONS

Multiple-Choice Questions

1. What is the atomic number of silicon?
   a. 14
   b. 16
   c. 10
   d. 32
2. How many valence electrons are normally present in the valence shell of a semiconductor material?
   a. 2
   b. 4
   c. 6
   d. 8
3. Adding trivalent impurities to an intrinsic semiconductor will produce a/an __________ material.
   a. Extrinsic
   b. *n*-type
   c. *p*-type
   d. Both (a) and (c) are true
4. What is the majority carrier in an *n*-type material?
   a. Holes
   b. Electrons
   c. Neutrons
   d. Protons

5. Adding pentavalent impurities to an intrinsic semiconductor will produce a/an ____________ material.
   a. Extrinsic
   b. *n*-type
   c. *p*-type
   d. Both (a) and (b) are true
6. What are the majority carriers in a *p*-type semiconductor?
   a. Holes
   b. Electrons
   c. Neutrons
   d. Protons
7. A semiconductor material has a ____________ temperature coefficient of resistance, which means that as temperature increases its resistance ____________.
   a. Positive, increases
   b. Positive, decreases
   c. Negative, increases
   d. Negative, decreases
8. A hole is considered to be ____________.
   a. Negative
   b. Positive
   c. Neutral
   d. Both (b) and (c) are true
9. Intrinsic semiconductors are doped to increase their ____________.
   a. Resistance
   b. Conductance
   c. Inductance
   d. Reactance
10. As temperature increases, a semiconductor acts more like a/an ____________.
   a. Conductor
   b. Insulator
11. A negative ion has more:
   a. Protons than electrons
   b. Electrons than protons
   c. Neutrons than protons
   d. Neutrons than electrons
12. A positive ion has:
   a. Lost some of its electrons
   b. Gained extra protons
   c. Lost neutrons
   d. Gained more electrons
13. The resistance of a semiconductor material is more than the resistance of:
   a. Glass
   b. Copper
   c. Ceramic
   d. Both (a) and (c) are true
14. The basic function of a semiconductor device in an electrical or electronic circuit is to:
   a. Control current
   b. Control voltage
   c. Increase the price of the equipment
   d. Both (a) and (b) are true
15. For a silicon P-N junction, V_F = ?
   a. The value of the applied voltage
   b. 300 mV
   c. 0.7 V
   d. 10 V

Practice Problems

16. Which of the silicon P-N junctions in Figure 10-17 are forward biased, and which are reverse biased?
17. Determine the current for the circuits shown in Figure 10-18.
18. What would be the voltage drop (V_F) across each of the P-N junctions shown in Figure 10-18?
19. What would be the voltage drop across each of the resistors in Figure 10-18?
20. Which of the P-N junctions in Figure 10-18 are equivalent to open switches, and which are equivalent to closed switches?

Web Site Questions

Go to the web site http://www.prenhall.com/cook, select the textbook *Electronics: A Complete Course,* select this chapter, and then follow the instructions when answering the multiple-choice practice problems.

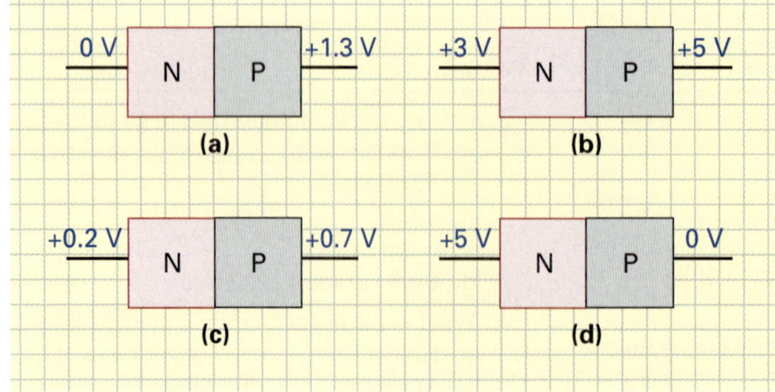

FIGURE 10-17 **Biased P-N Junctions.**

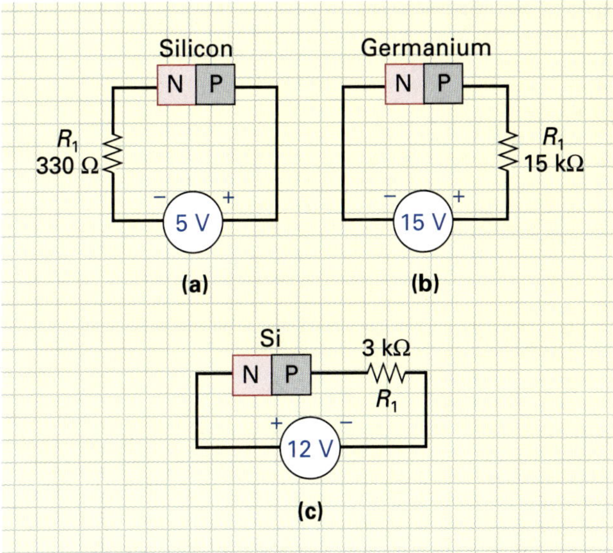

FIGURE 10-18 P-N Junction Circuit.

Diodes and Power Supply Circuits

A Problem with Early Mornings

René Descartes was born in Brittany, France, in 1596. At the age of eight he had surpassed most of his teachers at school and was sent on to the Jesuit College in La Flèche, one of the best in Europe. It was here that his genius in mathematics became apparent; however, due to his extremely delicate health, his professors allowed him to study in bed until midday.

In 1616 he had an urge to see the world and so he joined the army, which made use of Descartes' mathematical genius in military engineering. While traveling, Descartes met Dutch philosopher Isaac Beekman who convinced him to leave the army and, in his words, "turn his mind back to science and more worthier occupations."

After leaving the army Descartes traveled looking for some purpose, and then on November 10th, 1619, he found it. Descartes was in Neuberg, Germany, where he had shut himself in a well-heated room for the winter. It was on the eve of St. Martin's that a freezing blizzard forced Descartes to retire early. That night he described having an extremely vivid dream that clarified his purpose and showed him that physics and all sciences could be reduced to geometry and were therefore all interconnected like a chain.

In his time, and to this day, he is heralded as an analytical genius. In fact, Descartes' procedure can still be used as a guide to solving any problem.

Descartes' four-step procedure for solving a problem:

1. Never accept anything as true unless it is clear and distinct enough to exclude doubt from your mind.
2. Divide the problem into as many parts as necessary to reach a solution.
3. Start with the simplest things and proceed step by step toward the complex.
4. Review the solution so completely and generally that you are sure nothing was omitted.

For me, this four-step procedure has been especially helpful as a troubleshooting guide for system and circuit malfunctions.

Descartes' fame was so renowned that he was asked in 1649 to tutor Queen Christina of Sweden. The Queen demanded that her lessons begin at 5 o'clock in the morning, which conflicted with Descartes' lifetime practice of remaining in bed until midday. After several unsuccessful attempts to change her majesty's mind, and with pressure being applied by the French ambassador, Descartes agreed to the early morning lessons. A short time later on his way to the palace one cold winter morning, Descartes caught a severe chill and died within two weeks.

11

Introduction

The first diode was accidentally created by Edison in 1883 when he was experimenting with his light bulb. At this time he did not place any importance on the device and its effect, as he could not see any practical application for it. The word *diode* is derived from the fact that the device has two (*di*) electrodes (*ode*).

Once the importance of diodes was realized, construction of the device began. The first diodes were vacuum-tube devices having a hot-filament negative cathode, which released free electrons that were collected by a positive plate called the anode. Today's diode is made of a P-N semiconductor junction but still operates on the same principle. The *n*-type region (cathode) is used to supply free electrons, which are then collected by the *p*-type region (anode). The operation of both the vacuum tube and semiconductor diode is identical in that the device will only pass current in one direction. That is, it will act as a conductor and pass current easily in one direction when the bias voltage across it is of one polarity, yet it will block current and imitate an insulator when the bias voltage applied is of the opposite polarity.

11-1 THE JUNCTION DIODE

The two electrodes or terminals of the diode are called the anode and cathode, as seen in Figure 11-1(a) which shows the schematic symbol of a diode. To help you remember which terminal is the anode and which is the cathode, and which terminal is positive and which is negative, Figure 11-1(b) shows how a line drawn through the triangle section of the symbol will make the letter "A" and indicate the "anode" terminal. Similarly, if the vertical flat side of the diode symbol is aligned horizontally "—", as in Figure 11-1(b), it becomes the "negative" symbol. This memory system helps us to remember that the anode terminal of the diode is next to the triangle part of the symbol and is positive, while the cathode terminal of a diode is next to the vertical line of the symbol and is negative.

The diode is generally mounted in one of the three basic packages shown in Figure 11-2. These packages are designed to protect the diode from mechanical stresses and the environment. The difference in the size of the packages is due to the different current rating of the diode. A black band or stripe is generally placed on the package closest to the cathode terminal for identification purposes, as seen in Figure 11-2(a) and (b). Larger diode packages, like the one seen in Figure 11-2(c), usually have the diode symbol stamped on the package to indicate anode/cathode terminals.

11-1-1 *Diode Operation*

As far as operation is concerned, the diode operates like a switch. If you give the diode what it wants, that is make the anode terminal positive with respect to the cathode terminal as seen in Figure 11-3(a), the device is equivalent to a closed switch as seen in Figure 11-3(b). In this condition, the diode is said to be ON or *forward biased.*

On the other hand, if you do not give the diode what it wants, that is if you make the anode terminal negative with respect to the cathode as seen in Figure 11-3(c), the device is equivalent to an open switch as seen in Figure 11-3(d). In this condition the diode is said to be OFF or *reverse biased.*

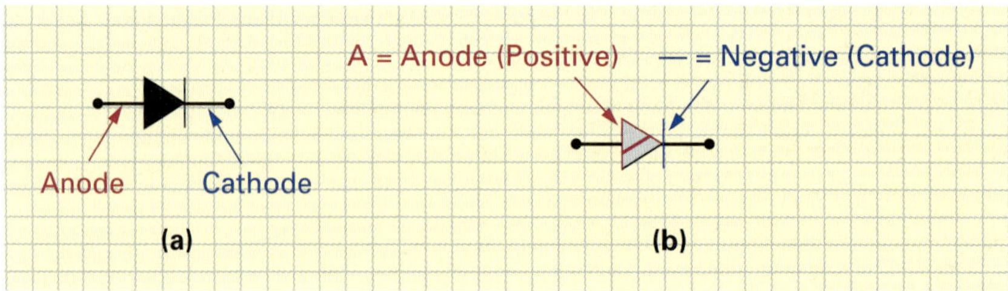

FIGURE 11-1 Schematic Symbol of a Diode.

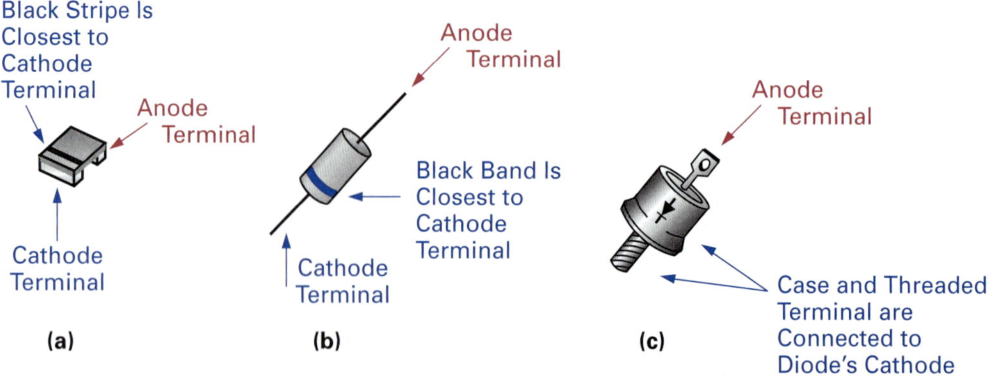

FIGURE 11-2 Diode Packaging. (a) Chip Package—1/4 A. (b) Small Current Package—Less than 3 A. (c) Large Current Package—Greater than 3 A.

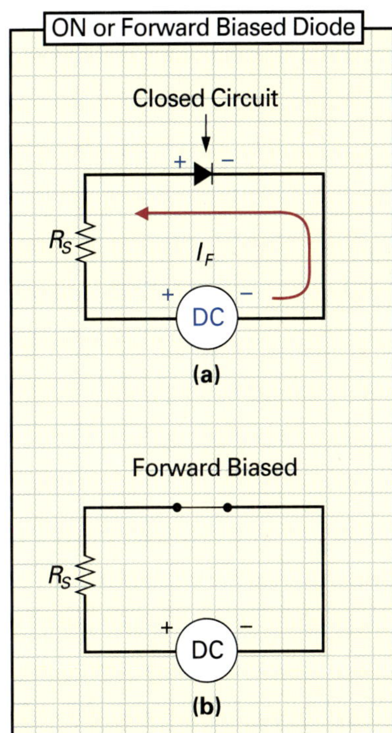

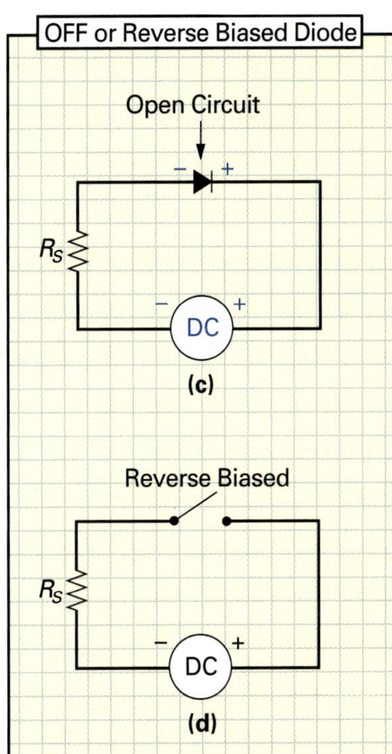

FIGURE 11-3 **Diode Operation. (a)(b) Forward Biased (ON) Diode. (c)(d) Reverse Biased (OFF) Diode.**

11-1-2 *Basic Diode Application*

As an application, Figure 11-4 shows how the diode can be used as a switch within an **encoder circuit.** The pull-up resistors R_1, R_2, and R_3 ensure that lines *A, B,* and *C* are normally all at +5 V. This is the output voltage on each line when the rotary switch is in position 2, as seen in the table in Figure 11-4.

Encoder Circuit
A circuit that produces different output voltage codes, depending on the position of a rotary switch.

When the rotary switch is turned to position 1, D_1 is connected in-circuit and, because its anode is made positive via R_2 and its cathode is at 0 V, the diode D_1 will turn ON and be

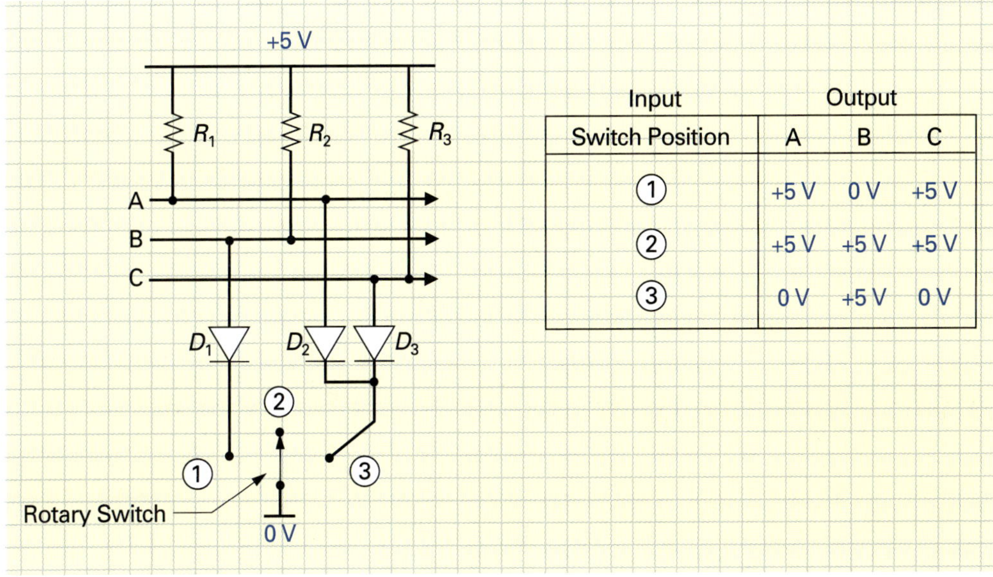

Input	Output		
Switch Position	A	B	C
①	+5 V	0 V	+5 V
②	+5 V	+5 V	+5 V
③	0 V	+5 V	0 V

FIGURE 11-4 **Diode Application: A Switch Encoder Circuit.**

equivalent to a closed switch. The 0 V on the cathode of D_1 will be switched through to line B (all of the five volts will be dropped across R_2) producing an output voltage code of $A = +5$ V, $B = 0$ V, $C = +5$ V as seen in the table in Figure 11-4.

When the rotary switch is turned to position 3, D_2 and D_3 are connected in circuit and because both anodes are made positive via R_1 and R_3, and both diode cathodes are at 0 V, D_2 and D_3 will turn ON. These forward biased diodes will switch 0 V through to lines A and C, producing an output voltage code of $A = 0$ V, $B = +5$ V, $C = 0$ V, as seen in the table in Figure 11-4.

This *code generator* or *encoder circuit* will produce three different output voltage codes for each of the three positions of the rotary switch. These codes could then be used to initiate one of three different operations based on the operator setting of the rotary control switch.

11-1-3 *A Junction Diode's Characteristic Curve*

Junction Diode

A semiconductor diode whose ON/OFF characteristics occur at a junction between the *n*-type and *p*-type semiconductor materials.

Bias Voltage

Voltage that inclines or causes the diode to operate in a certain manner.

Semiconductor devices such as diodes and transistors are constructed using P-N junctions. A diode, for example, has only one P-N junction and is created by doping a single piece of pure semiconductor to produce an *n*-type and *p*-type region. A bipolar junction transistor, on the other hand, has two P-N junctions and is created by doping a single piece of pure semiconductor with three alternate regions (NPN or PNP). The point at which these two opposite-doped materials come in contact with each other is called a junction, which is why these devices are called **junction diodes** and *bipolar junction transistors.*

These P-N junctions need voltages of a certain amplitude and polarity to control their operation. These voltages, which incline or cause the diode to operate in a certain manner, are known as **bias voltages.** Bias voltages control the resistance of the junction and, therefore, the amount of current that can pass through the P-N junction diode.

The upper right quadrant of the four sections in Figure 11-5 shows what forward current will pass through the diode when a forward bias voltage is applied. As you can see from the inset, the diode is forward biased by applying a positive potential to its anode and a negative potential to its cathode. In this instance the diode is said to be ON, and is equivalent to a closed switch. Beginning at the graph origin and following the curve into the forward quadrant, you can see that the forward current through a diode is extremely small until the forward bias voltage exceeds the diode's internal barrier voltage, which for silicon is 0.7 V and for germanium is 0.3 V.

Referring to the linearly increasing current portion of the forward curve in Figure 11-5, you will notice that although there is a large change in forward current, the forward voltage drop across the diode remains almost constant between 0.7 V and 0.75 V.

The amount of heat produced in the diode is proportional to the value of current through the diode ($P\uparrow = I^2\uparrow \times R$). For example, an IN4001 diode, which is a commonly used low-power silicon diode, has a manufacturer's maximum forward (I_F max.) rating of 1 A. If this value of current is exceeded, the diode will begin generating more heat than it can dissipate and burn out. A series current limiting resistor (R_S) is generally always included to limit the forward current, as shown in the inset in Figure 11-5. Although the series resistor will limit forward current, it cannot prevent a damaging forward current if enough pressure or forward voltage is applied ($V\uparrow = I\uparrow \times R$). The value of forward current is equal to

$$I_F = \frac{V_S - V_{\text{diode}}}{R_S}$$

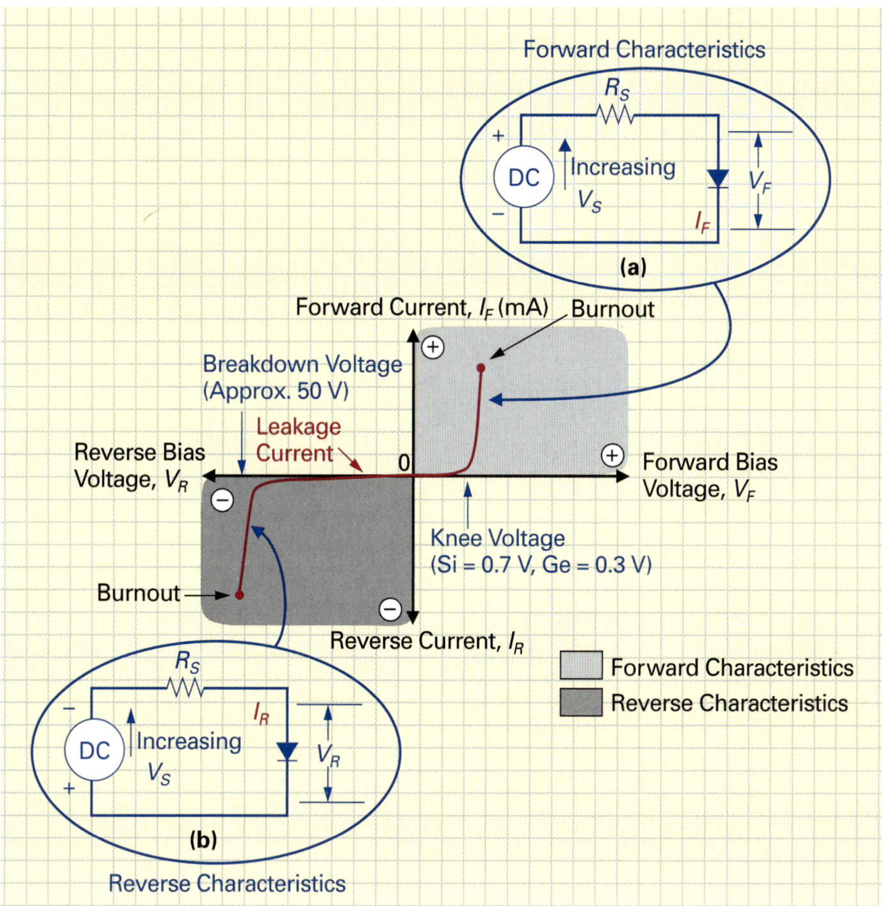

FIGURE 11-5 **The Junction Diode Voltage-Current Characteristic Curve.**

EXAMPLE:

Calculate the value of current for the circuit shown in Figure 11-6.

Solution:

The diode is forward biased because the applied voltage is connected so that its positive terminal is applied to the anode and the negative terminal is applied to the cathode. Because a silicon diode is being used, the forward voltage drop will be 0.7 V. With an applied voltage of 8.5 V and a circuit resistance of 1.2 kΩ, the circuit current will equal

$$I_F = \frac{V_S - V_{\text{diode}}}{R_S}$$

$$I_F = \frac{8.5\text{ V} - 0.7\text{ V}}{1.2\text{ k}\Omega}$$

$$I_F = \frac{7.8\text{ V}}{1.2\text{ k}\Omega}$$

$$I_F = 6.5\text{ mA}$$

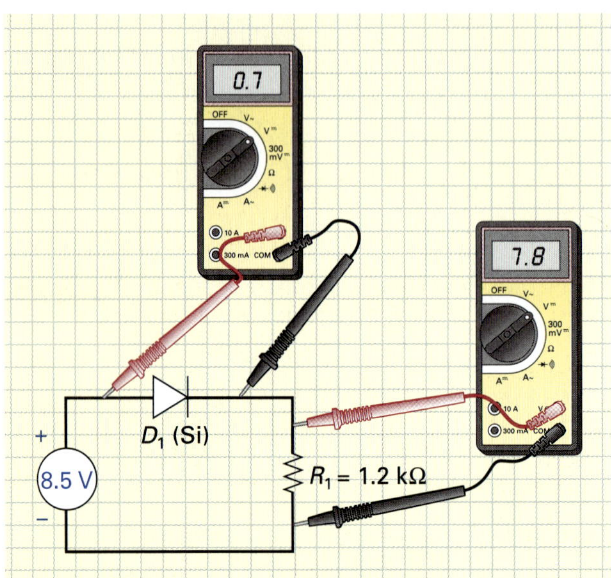

FIGURE 11-6 A P-N Junction Diode Circuit.

The lower left quadrant of the four sections in Figure 11-5 shows what reverse current will pass through the diode when a reverse bias voltage is applied. As you can see from the inset, a diode is reverse biased by applying a negative potential to its anode and a positive potential to its cathode. In this instance, current is effectively reduced to zero and the diode is said to be OFF and equivalent to an open switch.

These characteristics can be seen in the reverse curve in Figure 11-5. Beginning at the graph origin and following the curve into the reverse quadrant, you can see that the reverse current through the diode increases only slightly (approximately 100 μA). Throughout this part of the curve the diode is said to be blocking current because the leakage current is generally so small it is ignored for most practical applications. If the reverse voltage (V_R) is further increased, a point will be reached where the diode will break down, resulting in a sudden increase in current. The point on the reverse voltage scale at which the diode breaks down and there is a sudden increase in reverse current is called the **breakdown voltage.** Referring to the reverse curve in Figure 11-5, you can see that most silicon diodes break down as the reverse bias voltage approaches 50 V. For example, the IN4001 low-power silicon diode has a reverse breakdown voltage (which is sometimes referred to as the **Peak Inverse Voltage** or **PIV**) of 50 V listed on its manufacturer's data sheet. If this reverse bias voltage is exceeded, an avalanche of continuously rising current will eventually generate more heat than can be dissipated, resulting in the destruction of the diode.

Breakdown Voltage or Peak Inverse Voltage (PIV)
The point on the reverse voltage scale at which the diode breaks down and there is a sudden increase in the reverse current.

Semiconductor materials, and therefore diodes, have a negative temperature coefficient of resistance. This means as temperature increases ($T\uparrow$), their resistance decreases ($R\downarrow$).

11-1-4 *Testing Junction Diodes*

Figure 11-7 shows how an ohmmeter can be used to check whether a diode has malfunctioned or is operating correctly. A good diode should display a very low resistance when it is biased ON, and a very high resistance when it is biased OFF. Figure 11-7(a) shows how a diode can be forward biased by an ohmmeter's internal battery (+ lead to anode, − lead to cathode), and if good, should display a low value of resistance (typically less than 10 Ω). Figure 11-7(b) shows how the diode is then flipped over and reverse biased by the ohmmeter's internal battery (+ lead to cathode, − lead to anode). If the diode is good, the ohmmeter should display a very high resistance (typically greater than 1000 MΩ). Since this value is generally off the ohmmeter's scale, you will probably have the display showing OL or OR.

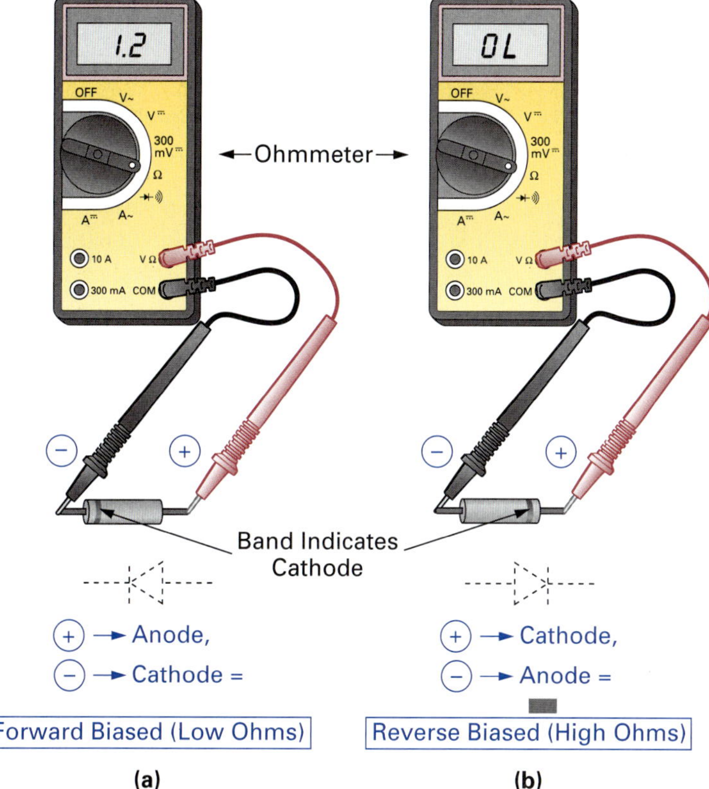

FIGURE 11-7 **Testing Diodes with an Ohmmeter.**

This is what you should expect since it means that the reverse biased diode's resistance is so high that it is over the range, or off the scale, selected.

SELF-TEST EVALUATION POINT FOR SECTION 11-1

Use the following questions to test your understanding of Section 11-1.

1. What is the typical forward drop across a silicon diode?
2. What value of barrier voltage has to be overcome in order to forward bias a silicon diode?
3. How many P-N junctions are within a junction diode?

11-2 THE ZENER DIODE

Figure 11-8(a) shows the two schematic symbols used to represent the **zener diode.** As you can see, the zener diode symbol resembles the basic P-N junction diode symbol in appearance; however, the zener diode symbol has a zig-zag bar instead of the straight bar. This zig-zag bar at the cathode terminal is included as a memory aid since it is "**Z**" shaped and will always remind us of zener.

Zener Diode
Diodes constructed to operate at voltages that are equal to or greater than the reverse breakdown voltage rating.

Figure 11-8(b) shows two typical low-power zener diode packages, and one high-power zener diode package. The surface mount low-power zener package has two metal pads for direct mounting to the surface of a circuit board, while the axial lead low-power zener package has the zener mounted in a glass or epoxy case. The high-power zener package is generally stud mounted and contained in a metal case. These packages are

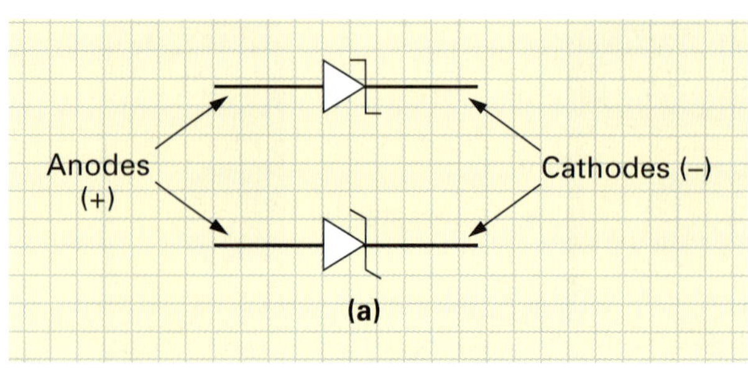

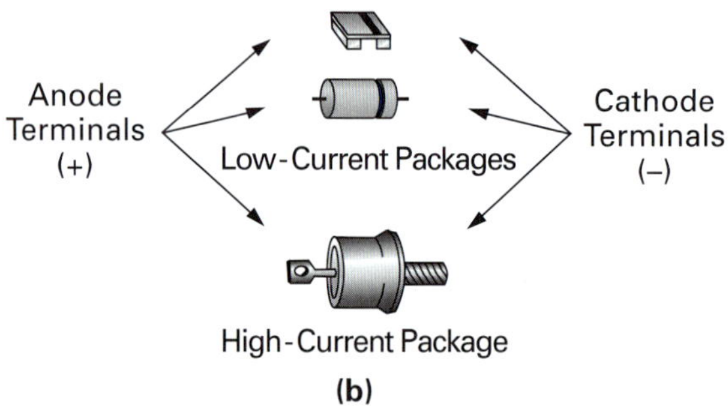

FIGURE 11-8 The Zener Diode. (a) Zener Diode Schematic Symbols. (b) Packages.

identical to the basic P-N junction diode low-power and high-power packages. Once again, a band or stripe is used to identify the cathode end of the zener diode in the low-power packages, whereas the threaded terminal of a high-power package is generally always the cathode.

11-2-1 *Zener Diode Voltage-Current* (V-I) *Characteristics*

Figure 11-9 shows the *V-I* (voltage-current) characteristic curve of a typical zener diode. This characteristic curve is almost identical to the basic P-N junction diode's characteristic curve. For example, when forward biased at or beyond 0.7 V, the zener diode will turn ON and be equivalent to a closed switch; whereas, when reverse biased, the zener diode will turn OFF and be equivalent to an open switch. The main difference, however, is that the zener diode has been specifically designed to operate in the reverse breakdown region of the curve. This is achieved, as can be seen in the inset in Figure 11-9, by making sure that the external bias voltage applied to a zener diode will not only reverse bias the zener diode (+ → cathode, − → anode) but also be large enough to drive the zener diode into its reverse breakdown region.

Reverse Leakage Current (I_R)
The undesirable flow of current through a device in the reverse direction.

As the reverse voltage across the zener diode is increased from the graph origin (which represents 0 volts), the value of **reverse leakage current (I_R)** begins to increase. Comparing the voltage developed across the zener (V_Z) to the value of current through the zener (I_Z), you may have noticed that *the voltage drop across a zener diode (V_Z) remains almost constant when it is operated in the reverse zener breakdown region, even though current through the zener (I_Z) can vary considerably. This ability of the zener diode to maintain a relatively constant voltage regardless of variations in zener current is the key characteristic of the zener diode.*

Zener Voltage (V_Z)
The voltage drop across the zener when it is being operated in the reverse zener breakdown region.

Generally, manufacturers rate zener diodes based on their **zener voltage (V_Z)** rather than their breakdown voltage (V_{Br}). A wide variety of zener diode voltage ratings are available ranging from 1.8 V to several hundred volts. For example, many of the frequently used low-voltage zener diodes have ratings of 3.3 V, 4.7 V, 5.1 V, 5.6 V, 6.2 V, and 9.1 V.

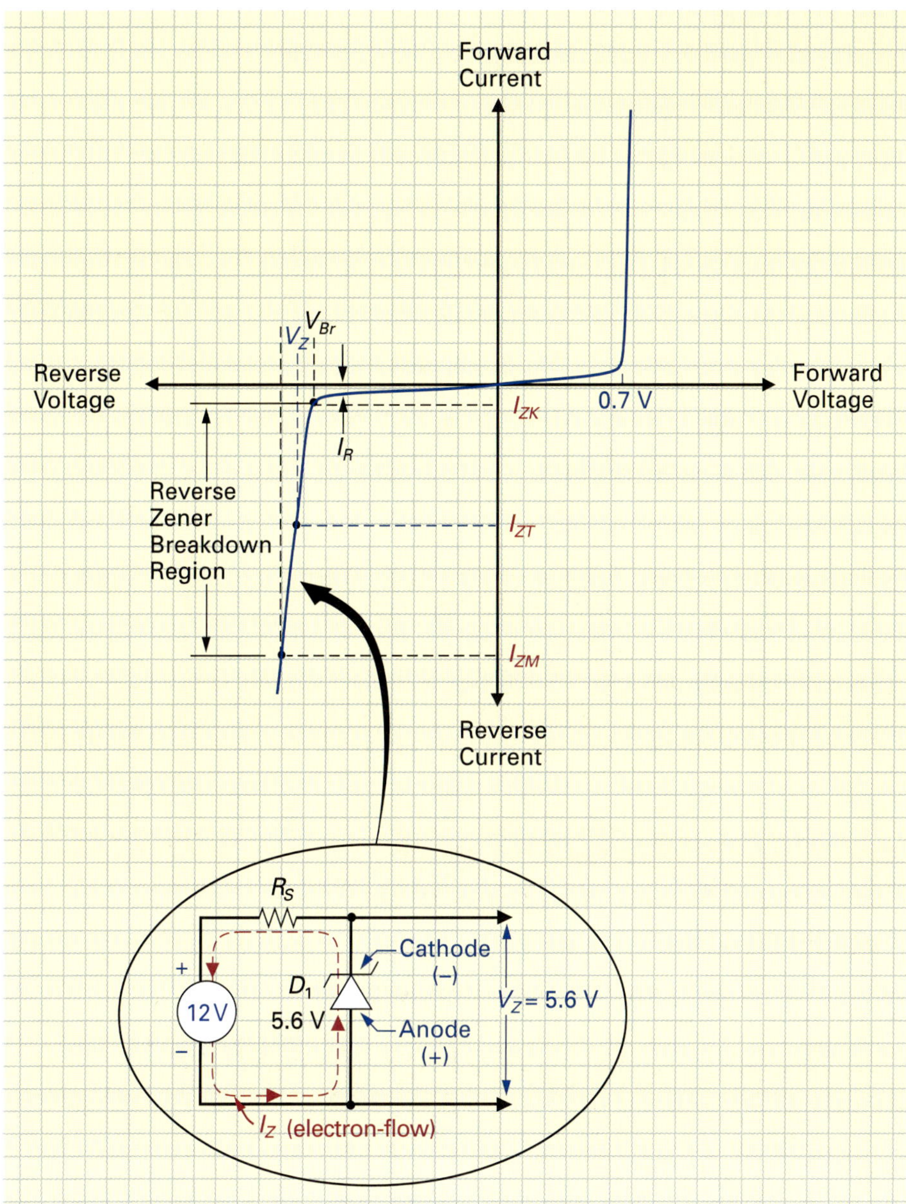

FIGURE 11-9 The Zener Diode Voltage-Current Characteristic Curve.

11-2-2 *Testing Zener Diodes*

Because a zener diode is designed to conduct in both directions, we cannot test it with the ohmmeter as we did the basic P-N junction diode. The best way to test a zener diode is to connect the voltmeter across the zener while it is in circuit and power is applied, as seen in Figure 11-10. If the voltage across the zener is at its specified voltage, then the zener is functioning properly. If the voltage across the zener is not at the nominal value, then the following checks should be made:

1. Check the source input voltage. If this voltage (V_{in}) does not exceed the zener voltage (V_Z), the zener diode will not be at fault because the source voltage is not large enough to send the zener into its reverse breakdown region.
2. Check the series resistor (R_S) to determine that it has not opened or shorted. An open series resistor will have all of the input voltage developed across it and there will be no

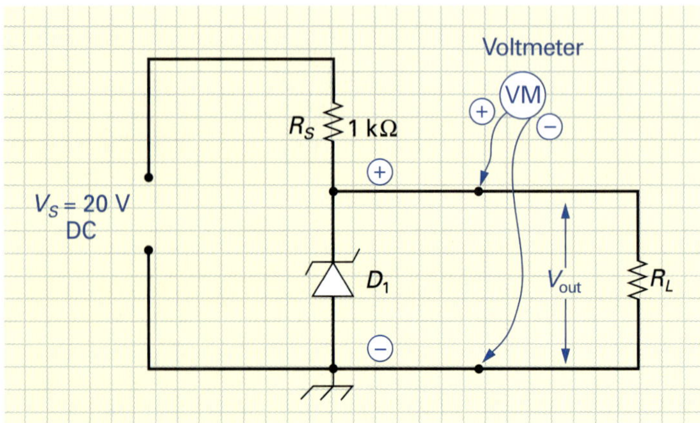

FIGURE 11-10 Zener Diode Testing.

voltage across the zener. A shorted series resistor will not provide any current-limiting capability and the zener could possibly burn out.

3. Check that there is not a short across the load because this would show up as 0 V across the zener and make the zener look faulty. To isolate this problem, disconnect the load and see if the zener functions normally.

If these three tests check out okay, the zener diode is probably at fault and should be replaced.

SELF-TEST EVALUATION POINT FOR SECTION 11-2

Use the following questions to test your understanding of Section 11-2.

1. True or false: The zener diode is designed specifically to operate at voltages exceeding breakdown.
2. In most applications, a zener diode is ____________ biased.
3. What is the difference between a zener diode's schematic symbol and a basic P-N junction's symbol?

11-3 THE LIGHT-EMITTING DIODE

Light-Emitting Diode (LED)
A semiconductor device that produces light when an electrical current or voltage is applied to its terminals.

The **light-emitting diode (LED)** is a semiconductor device that produces light when an electrical current or voltage is applied to its terminals. Figure 11-11(a) shows the two schematic symbols used most frequently to represent an LED. The two arrows leaving the diode symbol represent light.

A typical LED package is shown in Figure 11-11(b). The package contains the two terminals for connection to the anode and cathode and a semi-clear case which contains the light-emitting diode and a lens. Looking at this illustration, you can see that the LED chip is directly connected to the anode lead, while the cathode lead is connected to the LED chip by a thin wire. The dome-shaped top of the plastic (epoxy) case serves as the lens and acts as a magnifier to conduct light away from the LED chip. By adjusting the lens material, lens shape, and the distance the LED chip is from the lens, manufacturers can obtain a variety of radiation patterns.

There are three methods used to identify the anode and cathode leads of the LED, and these are shown in Figure 11-11(c). In all three cases, the cathode lead is distinguished from the anode lead by having its lead shorter or its lead flattened, or being nearest to the flat side of the case.

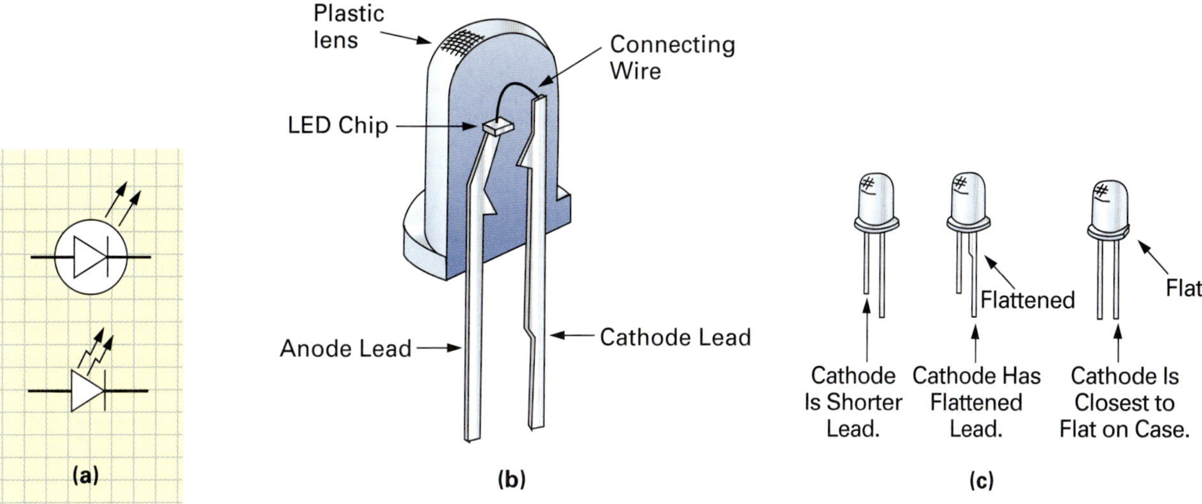

FIGURE 11-11 **The Light-Emitting Diode (LED). (a) Schematic Symbol. (b) Construction. (c) Lead Identification.**

11-3-1 *LED Characteristics*

Light-emitting diodes have *V-I* characteristic curves that are almost identical to the basic P-N junction diode as shown in Figure 11-12.

Studying the forward characteristics, you can see that LEDs have a high forward voltage rating (V_F is normally between +1 V and +3 V) and a low maximum forward current rating (I_F is normally between 20 mA and 50 mA). In most circuit applications, the LED will have a forward voltage drop of 2 V, and a forward current of 20 mA. The LED will usually always need to be protected from excessive forward current damage by a series current-limiting resistor, which will limit the forward current so that it does not exceed the LED's *maximum forward current* (I_{FM} or I_{FMAX}) rating, as seen in the inset in Figure 11-12. The circuit in the inset shows how to forward bias an LED. In this example circuit, the source voltage of 5 V is being applied to a 130 Ω series resistor (R_S) and an LED with an I_{FM} rating of 50 mA and a V_F rating of 1.8 V. To calculate the value of current in this circuit, we would first deduct the 1.8 V developed across the LED (V_{LED}) from the source voltage (V_S) to determine the voltage drop across the resistor (V_{RS}). Then using Ohm's law, divide V_{RS} by R_S to determine current.

$$I_S = \frac{V_S - V_{LED}}{R_S}$$

$$= \frac{5\text{ V} - 1.8\text{ V}}{130\ \Omega} = \frac{3.2\text{ V}}{130\ \Omega} = 24.6\text{ mA}$$

Studying the reverse characteristics, you can see that LEDs have lower reverse breakdown voltage values than junction diodes (V_{Br} is typically −3 V to −10 V), which means that even a low reverse voltage will cause the LED to break down and become damaged.

11-3-2 *Testing LEDs*

Light-emitting diodes have a very long life expectancy and are much more rugged than their predecessor, the small incandescent light bulb. They do, however, break down and have to be replaced with either exactly the same device or a similar device with the same characteristics.

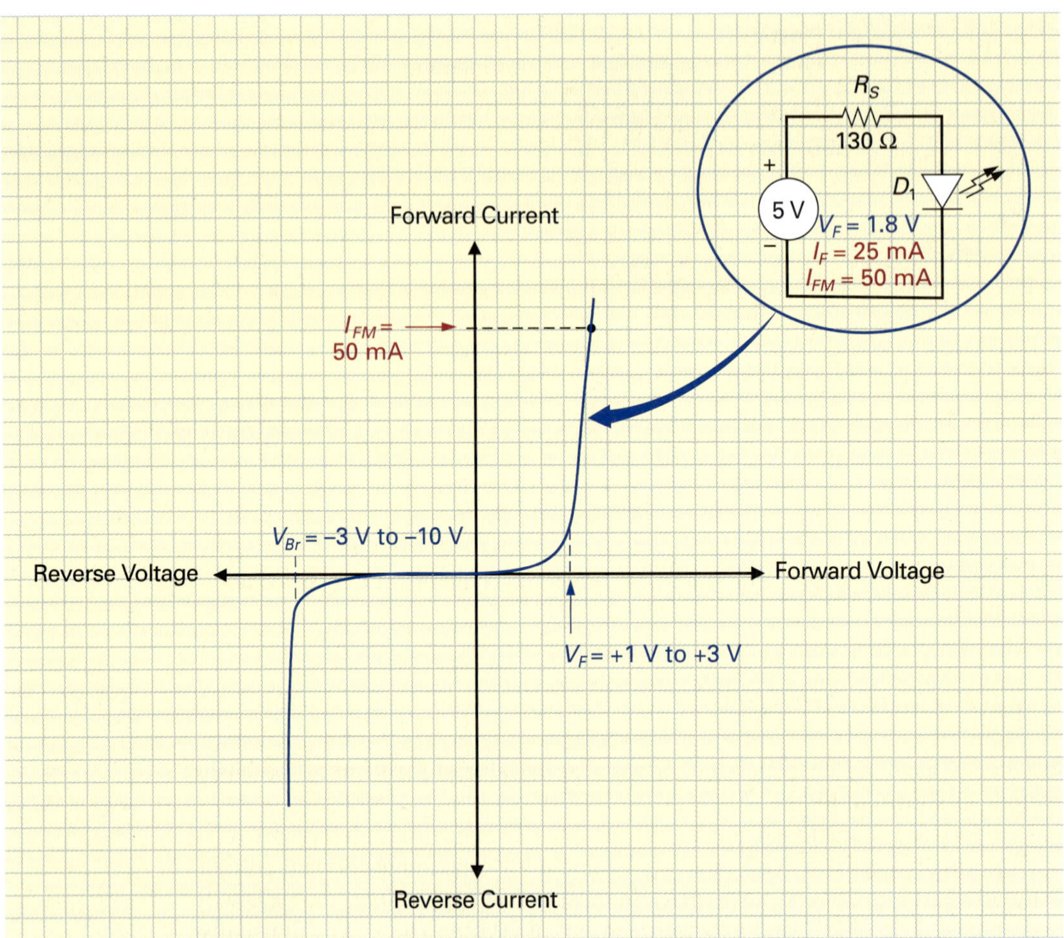

FIGURE 11-12 **The LED Voltage-Current Characteristic Curve.**

Figure 11-13 shows how to construct a simple LED test circuit using a dc power supply and a series current-limiting resistor. If a circuit problem is isolated to the LED, the new LED should be tested with a similar circuit to ensure it is operating correctly before it is inserted into the circuit.

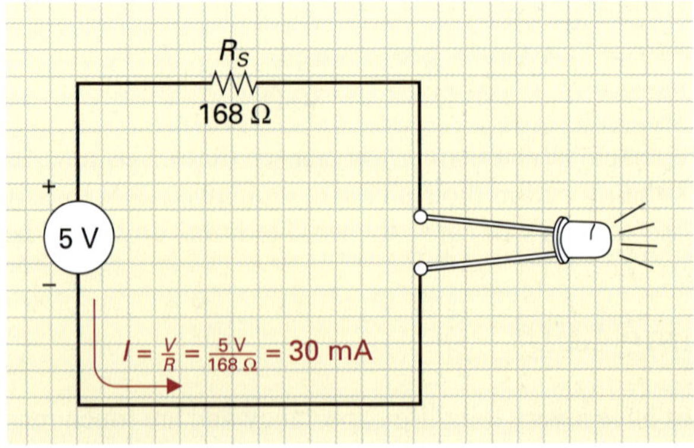

FIGURE 11-13 **Testing LEDs.**

SELF-TEST EVALUATION POINT FOR SECTION 11-3

Use the following questions to test your understanding of Section 11-3.

1. What would be the typical voltage drop across a forward biased LED?
   a. 0.7 V **b.** 2 V **c.** 0.3 V **d.** 5.6 V
2. A ____________ is normally always included to limit I_F to just below its maximum.

11-4 THE TRANSIENT SUPPRESSOR DIODE

Lightning, power line faults, and the switching on and off of motors, air conditioners, and heaters can cause the normal 120 V rms ac line voltage at the wall outlet to contain under-voltage dips and over-voltage spikes. Although these *transients* only last for a few microseconds, the over-voltage spikes can cause the input line voltage to momentarily increase by 1000 V or more. In sensitive equipment, such as televisions and computers, shunt filtering devices are connected between the ac line input and the primary of the dc power supply's transformer to eliminate these transients before they get into, and possibly damage, the system.

One such device that can be used to filter the ac line voltage is the **transient suppressor diode.** Referring to Figure 11-14(a), you can see that this diode contains two zener diodes that are connected back-to-back. The schematic symbol for this diode is shown in Figure 11-14(b).

Transient suppressor diodes are also called **transorbs** because they "absorb transients." Figure 11-14(c) shows how a transorb would be connected across the ac power line input to a dc power supply. Because the zeners within the transient suppressor diode are connected

Transient Suppressor Diode
A device used to protect voltage-sensitive electronic devices in danger of destruction by high-energy voltage transients.

Transorb
Absorb transients. Another name for transient suppressor diode.

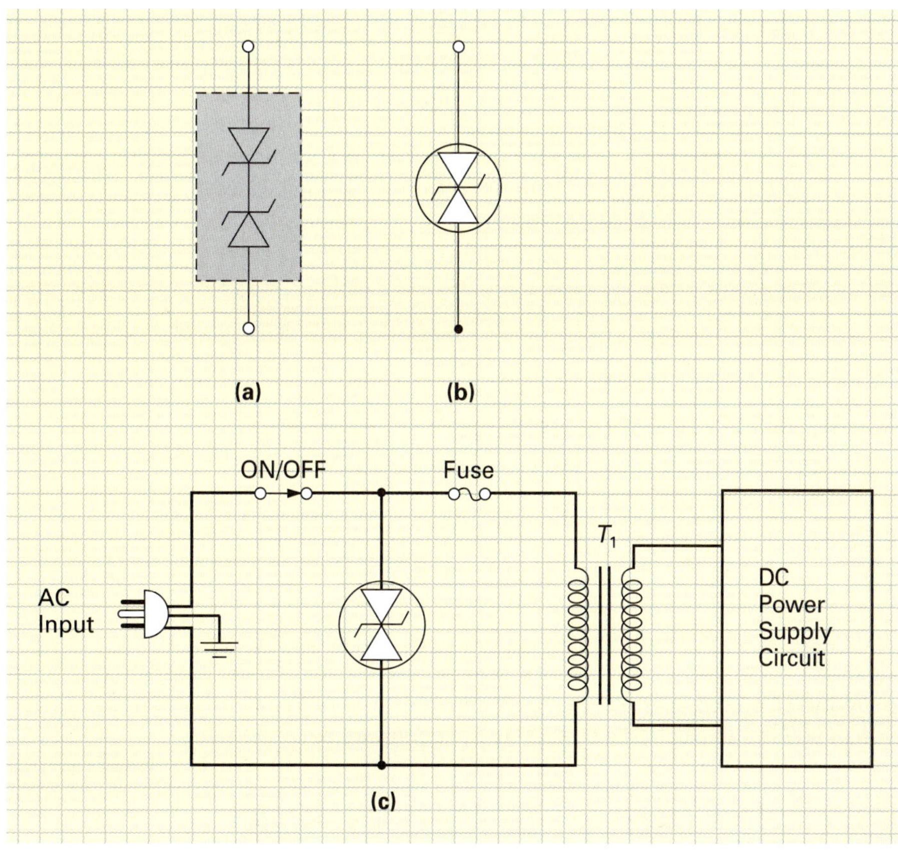

FIGURE 11-14 Transient Suppressor Diodes. (a) Construction. (b) Schematic Symbol. (c) Bidirectional Circuit Application (AC Line Voltage).

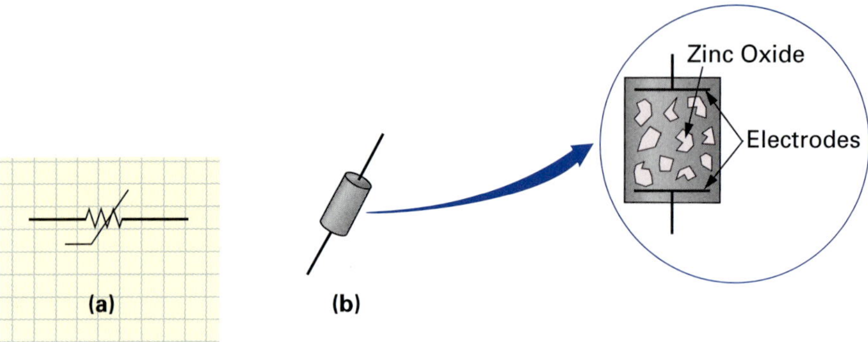

FIGURE 11-15 **Metal Oxide Varistors (MOVs). (a) Schematic Symbol. (b) Physical Appearance and Construction.**

back-to-back, they will operate in either direction (the device is "bi-directional") and monitor both alternations of the ac input. If a voltage surge occurs that exceeds the V_Z (zener voltage) of the diodes, they will break down and shunt the surge away from the power supply.

Most manufacturers' transorbs have a high power dissipation rating because they may have to handle momentary power line surges in the hundreds of watts. For example, the Motorola 1N5908-1N6389 series of transorbs can dissipate 1.5 kW for a period of approximately 10 ms (most surges last for a few milliseconds). The devices must also have a fast turn-on time so that they can limit or clamp any voltage spikes. For example, the Motorola P6KE6.8 series has a response time of less than 1 ns.

Metal Oxide Varistors (MOVs)
Devices that are replacing zener-diode and transient-diode suppressors because they are able to shunt a much higher current surge and are cheaper.

Varistor
Voltage dependent resistor.

In dc applications, a single unidirectional (one-direction) transient suppressor can be used instead of a bidirectional (two-direction) transient suppressor. These single transorbs have the same schematic symbol as a zener.

Metal oxide varistors (MOVs) are currently replacing zener-diode and transient-diode suppressors because they are able to shunt a much higher current surge and are cheaper. These are not semiconductor devices; in fact, they contain a zinc-oxide and bismuth-oxide compound in a ceramic body but are connected in the same way as a transient suppressor diode. They are called **varistors** because they operate as a "voltage dependent resistor" that will have a very low resistance at a certain breakdown voltage. The MOV's schematic symbol, typical appearance, and construction are shown in Figure 11-15.

SELF-TEST EVALUATION POINT FOR SECTION 11-4

Use the following questions to test your understanding of Section 11-4.

1. True or false: A bidirectional transient suppressor diode would be used to suppress ac power surges.
2. True or false: A unidirectional transient suppressor diode would be used to suppress dc power surges.

11-5 DIODE POWER SUPPLY CIRCUITS

In the previous sections we discussed the P-N junction diode, the zener diode, and the light-emitting diode. In this section we will see how we can use all of these diodes to construct a dc power supply. This section is also of great importance to our study of troubleshooting, since greater than 90% of all electronic system malfunctions are power related.

11-5-1 *Block Diagram of a DC Power Supply*

Figure 11-16(a) shows the four basic blocks of a dc power supply. Almost every piece of electronic equipment that makes use of 120 V ac as a source of power will have a built-in dc

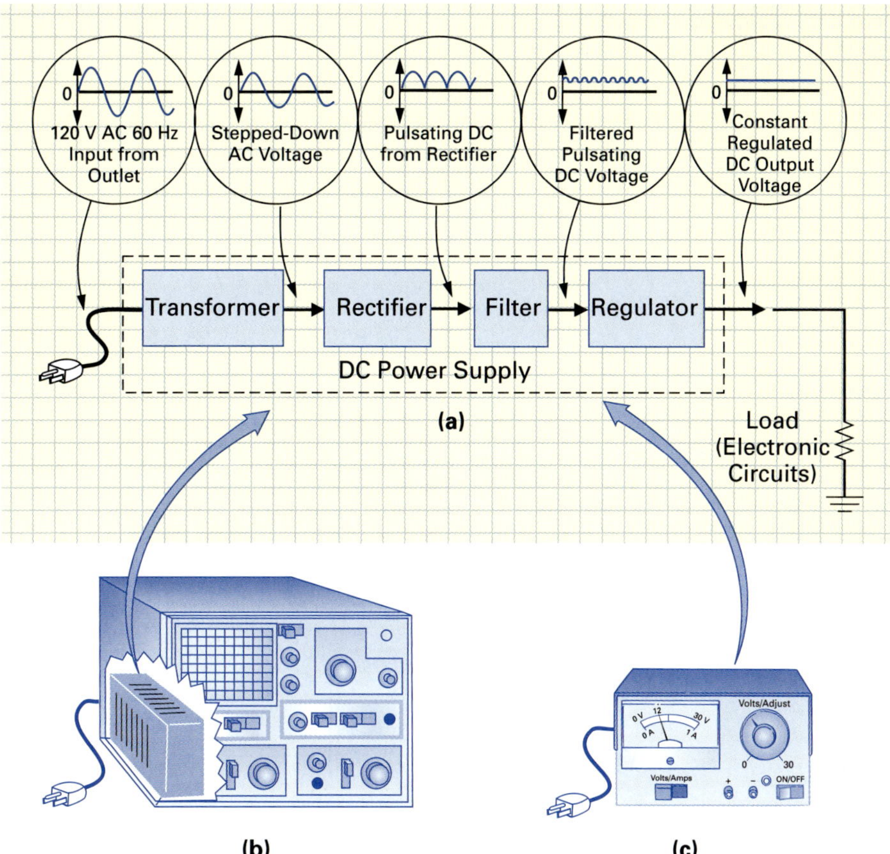

FIGURE 11-16 **DC Power Supply. (a) Block Diagram. (b) Built-In Subsystem. (c) Stand-Alone Unit.**

power supply, as shown in Figure 11-16(b). Stand-alone dc power supplies, such as the one shown in Figure 11-16(c), are also available for use in laboratory experimentation.

Let us now refer to the block diagram of the dc power supply in Figure 11-16(a) and describe the function of each block.

Since the final voltage desired is generally not 120 V, a transformer is usually included to step the ac line voltage up or down to a desired value. Electronic circuits generally require low-voltage supply values such as 12 and 5 V dc, and a step-down transformer would be used. For example, a step-down transformer with a turns ratio of 10:1 would reduce the 120 V ac input to a 12 V ac output. The output current capability of this transformer will be 1:10, which will be ideal because most electronic circuits require low supply voltages with high current capacity.

As shown by the waveforms in Figure 11-16(a), the rectifier converts the stepped-down ac input from the transformer to a pulsating dc output. This pulsating dc output could not be used to power an electronic circuit because of the continuous changes between zero volts and a peak voltage. The filter smooths out the pulsating dc ripples into an almost constant dc level, as seen in the waveform after the filter.

The final block is called a regulator, and although there appears to be no difference between the regulator's input and output waveforms, it provides a very important function. The regulator maintains the dc output voltage from the power supply constant, or stable, despite variations in the ac input voltage or variations in the output load resistance.

Many dc power supplies have several rectifier, filter, and regulator stages, depending on how many dc output voltages are desired for the electronic system. In the following sections we will examine these electrical circuits in more detail and then combine all of them in a working dc power supply circuit.

11-5-2 *Transformers*

The turns ratio of a transformer in a dc power supply can be selected to either increase or decrease the 120 V ac input. With most electronic equipment, a supply voltage of less than 120 V is required, and therefore a step-down transformer is used. The secondary output voltage V_s from the transformer can be calculated with the following formula, which was introduced previously:

$$V_s = \frac{N_s}{N_p} \times V_p$$

To apply this formula to an example, refer to the power supply transformer shown in Figure 11-17. This transformer can be connected so that it delivers the same secondary volt-

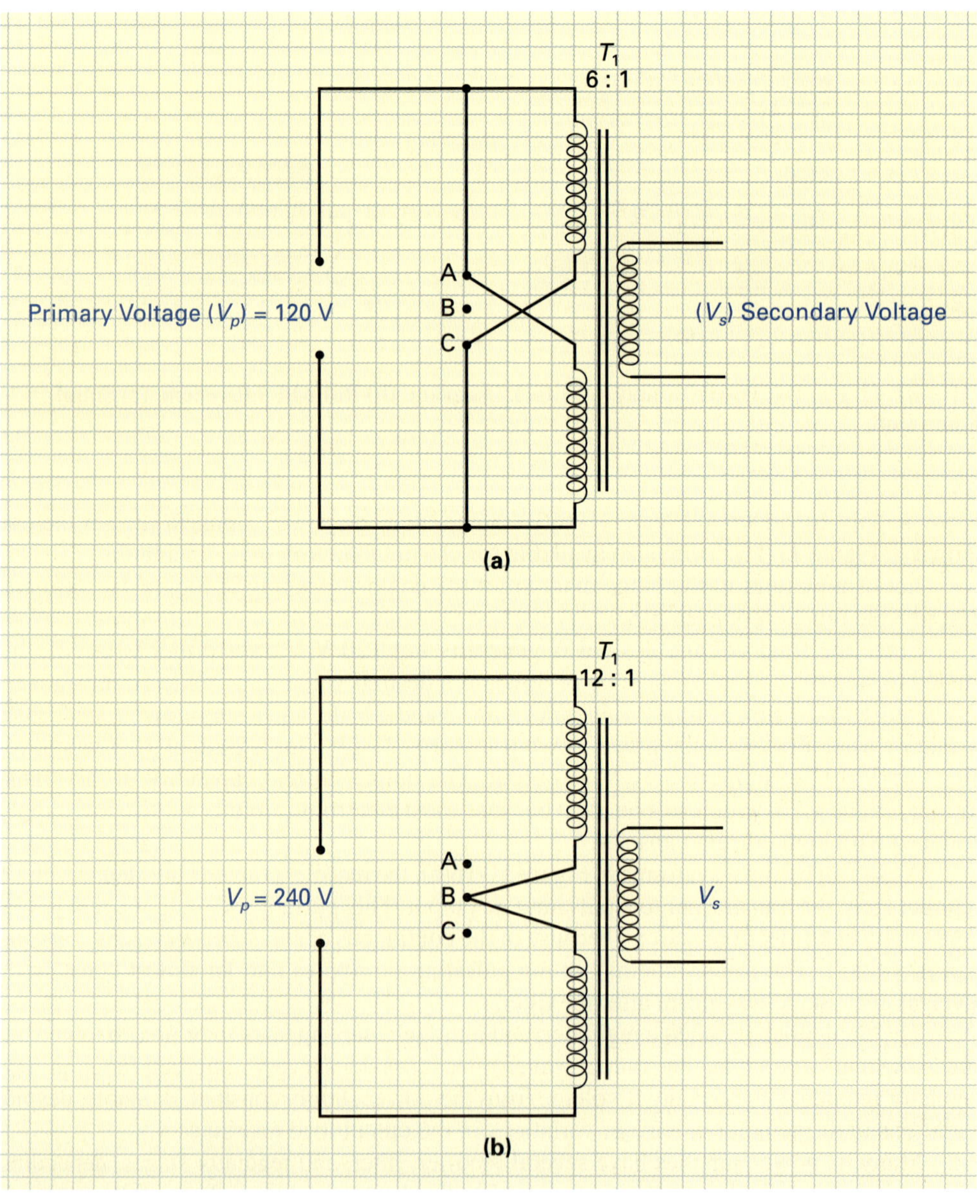

FIGURE 11-17 A 120 V/240 V Power Supply Transformer. (a) 120 V Connection. (b) 240 V Connection.

age for either a 120 V or a 240 V rms ac input. For example, when the two primary windings are connected in parallel, as shown in Figure 11-17(a), the transformer turns ratio is 6:1 step-down, and therefore the secondary voltage will be:

$$V_s = \frac{N_s}{N_p} \times V_p$$

$$= \frac{1}{6} \times 120 \text{ V rms} = 20 \text{ V rms}$$

When the two primary windings are connected in series, as shown in Figure 11-17(b), the transformer turns ratio is 12:1 step-down, and therefore the secondary voltage will be

$$V_s = \frac{N_s}{N_p} \times V_p$$

$$= \frac{1}{12} \times 240 \text{ V rms} = 20 \text{ V rms}$$

By doubling the number of primary turns, we can accept twice the input voltage and deliver the same output voltage.

11-5-3 *Rectifiers*

The junction diode's ability to switch current in only one direction makes it ideal for converting two-direction alternating current into one-direction direct current. In this section we will discuss the three basic diode rectifier circuits: the half-wave rectifier, the full-wave center-tapped rectifier, and the full-wave bridge rectifier.

Half-Wave Rectifiers

The **half-wave rectifier circuit** is constructed simply by connecting a diode between the power supply transformer and the load, as shown in Figure 11-18(a). When the secondary ac voltage swings positive, as shown in Figure 11-18(b), the anode of the diode is made positive, causing the diode to turn ON and connect the positive half-cycle of the secondary ac voltage across the load (R_L). When the secondary ac voltage swings negative, as shown in Figure 11-18(c), the anode of the diode is made negative, and therefore the diode will turn OFF. This will prevent any circuit current, and no voltage will be developed across the load (R_L).

Half-Wave Rectifier Circuit

A circuit that converts ac to dc by allowing current to flow during only one-half of the ac input cycle.

Output voltage. Figure 11-18(d) illustrates the input and output waveforms for the half-wave rectifier circuit. The 120 V ac rms input, or 169.7 V ac peak input, is applied to the 17:1 step-down transformer, which produces an output of:

$$V_s = \frac{N_s}{N_p} \times V_p$$

$$= \frac{1}{17} \times 169.7 \text{ V peak} = 10 \text{ V peak}$$

Because the diode will only connect the positive half-cycle of this ac input across the load (R_L), the output voltage (V_{RL}) is a positive pulsating dc waveform of 10 V peak. In this final waveform in Figure 11-18(d), you can see that the circuit is called a half-wave rectifier because only half of the input wave is connected across the output.

The average value of two half-cycles is equal to 0.637 V peak. Therefore, the average value of one half-cycle is equal to 0.318 V peak (0.637/2 = 0.318):

$$V_{\text{avg}} = 0.318 \times V_{s\,\text{peak}}$$

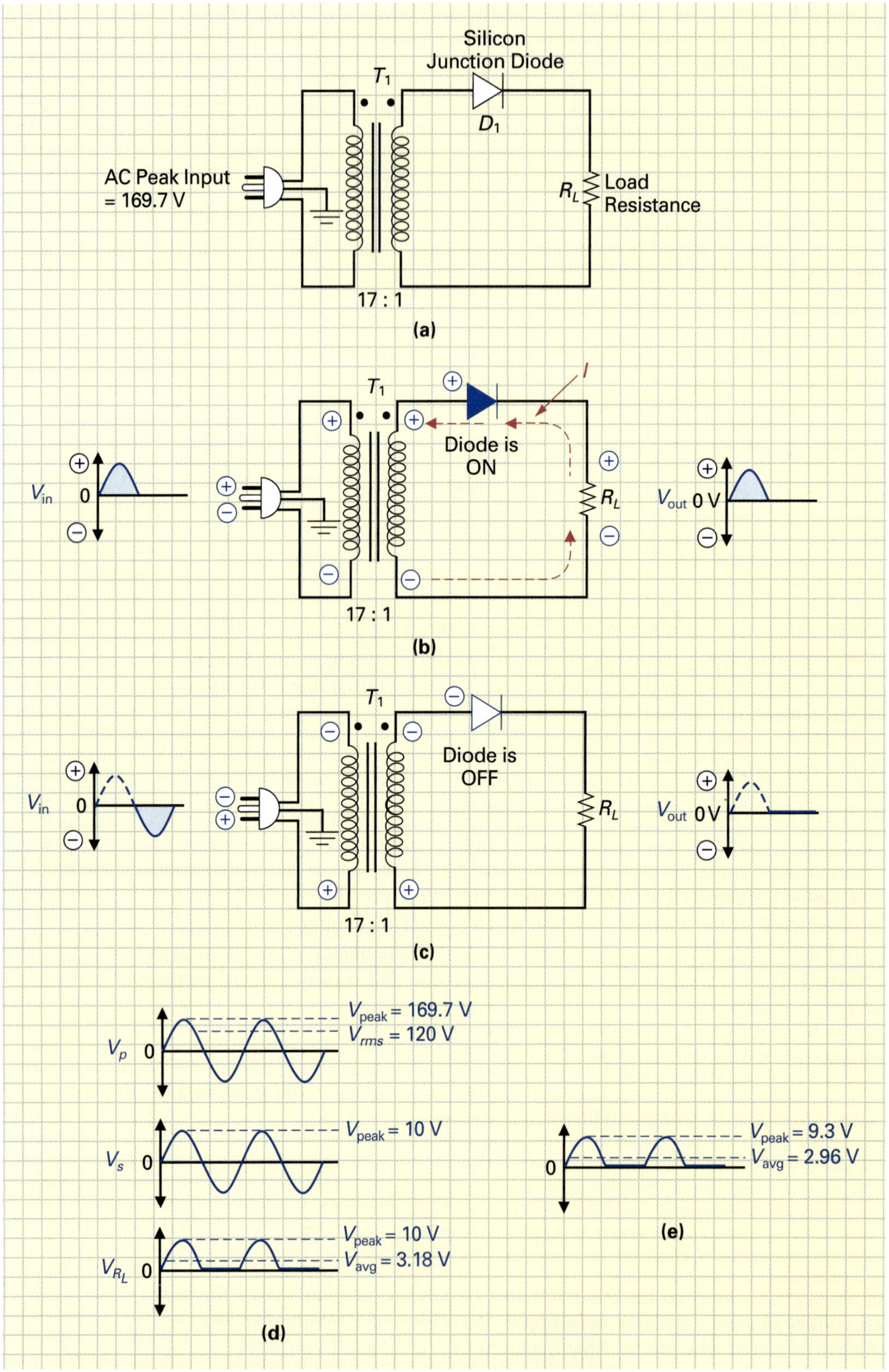

FIGURE 11-18 **Half-Wave Rectifier. (a) Basic Circuit. (b) Positive Input Half-Cycle Operation. (c) Negative Input Half-Cycle Operation. (d) Input/Output Waveforms. (e) Half-Wave Output Minus Diode Barrier Voltage.**

In the example in Figure 11-18, the average voltage of the half-wave output will be:

$$V_{avg} = 0.318 \times V_{s\ peak}$$
$$V_{avg} = 0.318 \times 10\ V$$
$$= 3.18\ V$$

To be more accurate, there will, of course, be a small voltage drop across the diode due to its barrier voltage of 0.7 V for silicon and 0.3 V for germanium. The output from the circuit in Figure 11-18 would actually have a peak of 9.3 V (10 V − 0.7 V), and therefore an average of 2.96 V (0.318 × 9.3 V), as shown in Figure 11-18(e).

$$V_{out} = V_s - V_{diode}$$

Another point to consider is the reverse breakdown voltage of the junction or rectifier diode. When the input swings negative, as illustrated in Figure 11-18(c), the entire negative supply voltage will appear across the open or OFF diode. The maximum reverse breakdown voltage, or peak inverse voltage (PIV) rating of the rectifier diode, must therefore be larger than the peak of the ac voltage at the diode's input.

Output polarity. The half-wave rectifier circuit can be arranged to produce either a positive pulsating dc output, as shown in Figure 11-19(a), or a negative pulsating dc output, as shown in Figure 11-19(b). Studying the difference between these circuits you can see that in Figure 11-19(a) the rectifier diode is connected to conduct the positive half-cycles of the ac input, while in Figure 11-19(b) the rectifier diode is reversed so that it will conduct the negative half-cycles of the ac input. By changing the direction of the diode in this manner, the rectifier can be made to produce either a positive or a negative dc output.

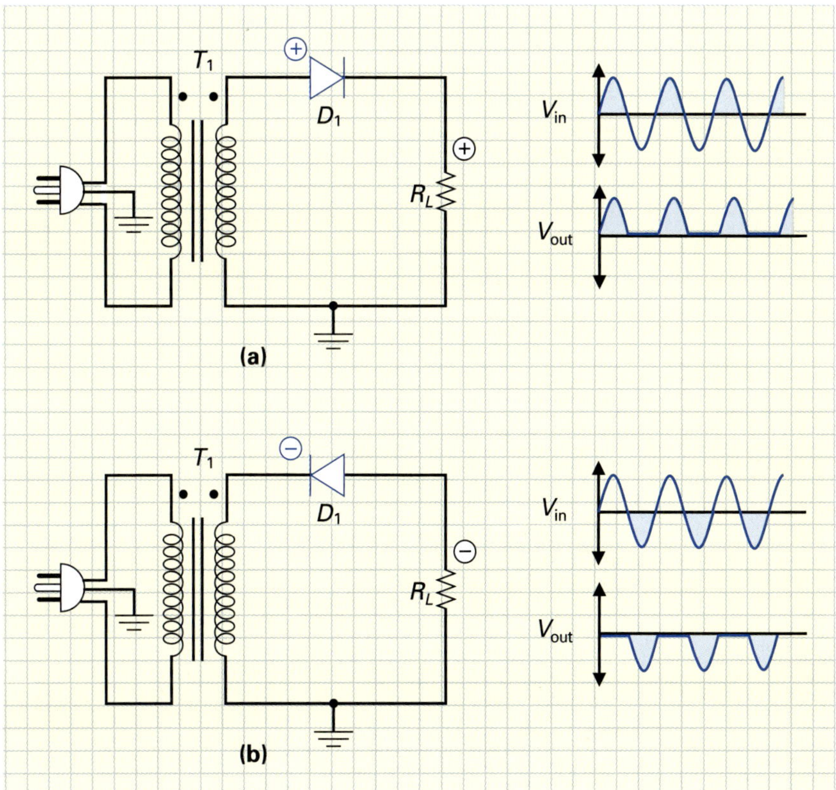

FIGURE 11-19 Changing the Output Polarity of a Half-Wave Rectifier. (a) Positive Pulsating DC. (b) Negative Pulsating DC.

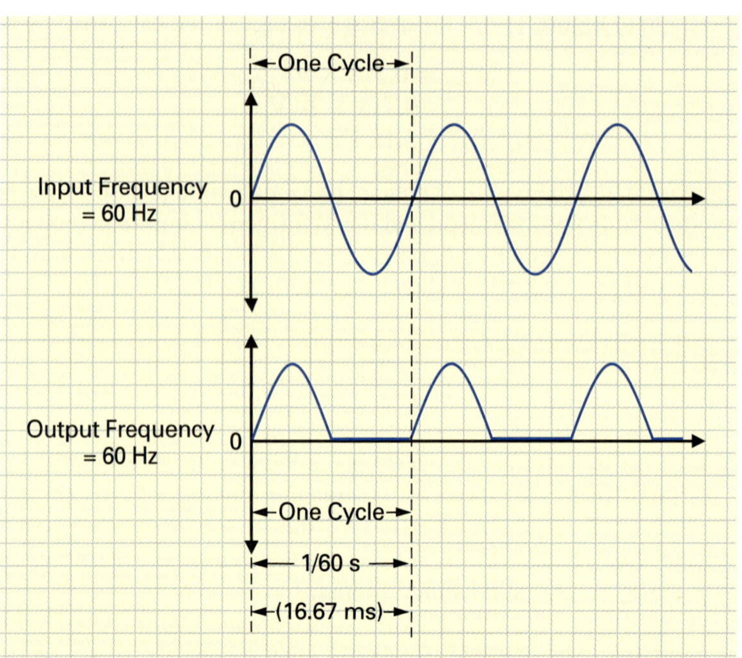

FIGURE 11-20 Ripple Frequency of a Half-Wave Rectifier.

Ripple frequency. Referring to the input/output waveforms of the half-wave rectifier circuit shown in Figure 11-20, you can see that one output ripple is produced for every complete cycle of the ac input. Consequently, if a half-wave rectifier is driven by the 120 V ac 60 Hz line voltage, each complete cycle of the ac input and each complete cycle of the output will last for one-sixtieth or 16.67 ms (1/60 = 1 ÷ 60 = 16.67 ms). The frequency of the pulsating dc output from a rectifier is called the ripple frequency, and for half-wave rectifier circuits, the

output pulsating dc ripple frequency = input ac frequency

EXAMPLE:

Calculate the following for the rectifier circuit shown in Figure 11-21:

a. Output polarity
b. Peak and average output voltage, taking into account the diode's barrier potential
c. Output ripple frequency

Solution:

a. Negative pulsating dc

b.
$$V_{\text{in peak}} = 169.7\text{ V}$$
$$V_s = \frac{N_s}{N_p} \times V_p$$
$$= \frac{1}{8} \times 169.7\text{ V peak} = -21.2\text{ V peak}$$
$$V_{\text{out}} = V_s - V_{\text{diode}}$$
$$= 21.2\text{ V} - 0.7\text{ V} = -20.5\text{ V}$$
$$V_{\text{avg}} = 0.318 \times V_{\text{out}}$$
$$= 0.318 \times 20.5\text{ V} = -6.5\text{ V}$$

c. Output ripple frequency = input frequency = 60 Hz

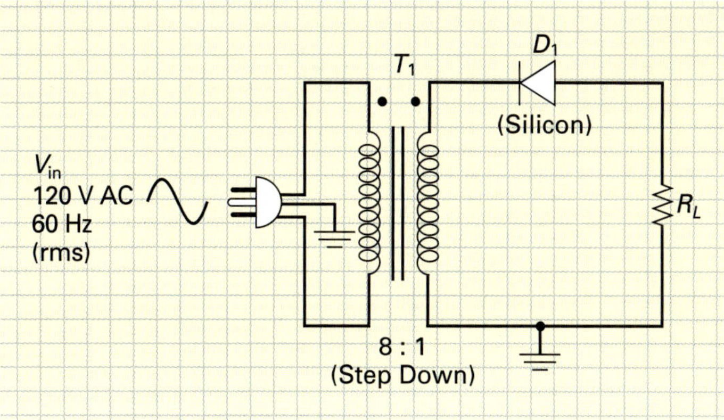

FIGURE 11-21 Half-Wave Rectifier.

Full-Wave Rectifiers

The half-wave rectifier's output is difficult to filter to a smooth dc level because the output voltage and current are applied to the load for only half of each input cycle. In this section we will examine two full-wave rectifier circuits which, as their name implies, switch both half-cycles (or the full ac input wave) of the input through the load in only one direction.

Full-Wave Center-Tapped Rectifier
A rectifier circuit that makes use of a center-tapped transformer to cause an output current to flow in the same direction during both half-cycles of the ac input.

Center-tapped rectifier. The basic **full-wave center-tapped rectifier** circuit, which is shown in Figure 11-22(a), contains a center-tapped transformer and two diodes. The center tap of the transformer secondary is grounded (0 V) to create a 180° phase difference between the top and bottom of the secondary winding.

When the ac input (V_{in}) swings positive, the circuit operates in the manner illustrated in Figure 11-22(b). A positive voltage is developed on the top of T_1 secondary turning ON D_1, while a negative voltage is developed on the bottom of T_1 secondary turning OFF D_2. This will permit a flow of electrons up through the load as indicated by the dashed current (I) line, developing a positive output half-cycle across the load (V_{out}).

When the ac input (V_{in}) swings negative, the circuit operates in the manner illustrated in Figure 11-22(c). A negative voltage is developed on the top of T_1 secondary turning OFF D_1, while a positive voltage is developed on the bottom of T_1 secondary turning ON D_2. This will permit a flow of electrons up through the load as indicated by the dashed current (I) line, again developing a positive output half-cycle across the load (V_{out}).

The two diodes and center-tapped transformer in this circuit switch the two-direction ac current developed across the secondary through the load in only one direction and develop an output voltage across the load of the same polarity. The input/output voltage details of a full-wave center-tapped rectifier are shown in Figure 11-23 (p. 394). A 1:1 turns ratio has been selected in this example to emphasize how the center tap in the secondary of the transformer causes the secondary voltage (V_s) to be divided into two equal halves (V_{s1} and V_{s2}). Referring to the waveforms, you can see that the ac line voltage of 120 V rms, or 169.7 V peak (V_{in}), is applied to the primary of the transformer T_1 (V_p). A 1:1 turns ratio ensures that this same voltage will be developed across the secondary winding ($V_s = V_p = 169.7$ V peak). This secondary voltage will be split in two as indicated by the waveforms V_{s1} and V_{s2}, which will each have a peak voltage of 1/2 V_s ($V_{s1\ peak} = V_{s2\ peak} = 1/2\ V_{s\ peak} = 1/2 \times 169.7\text{ V} = 84.9$ V peak). Because V_{s1} will be connected across the output during one half-cycle, and V_{s2} will be connected across the output during the other half-cycle, the output voltage (V_{out}) will also have a peak voltage equal to 1/2 V_s or 84.9 V. Compared to a half-wave rectifier, therefore, whose peak output voltage is equal to the peak secondary voltage, the full-wave center-tapped rectifier

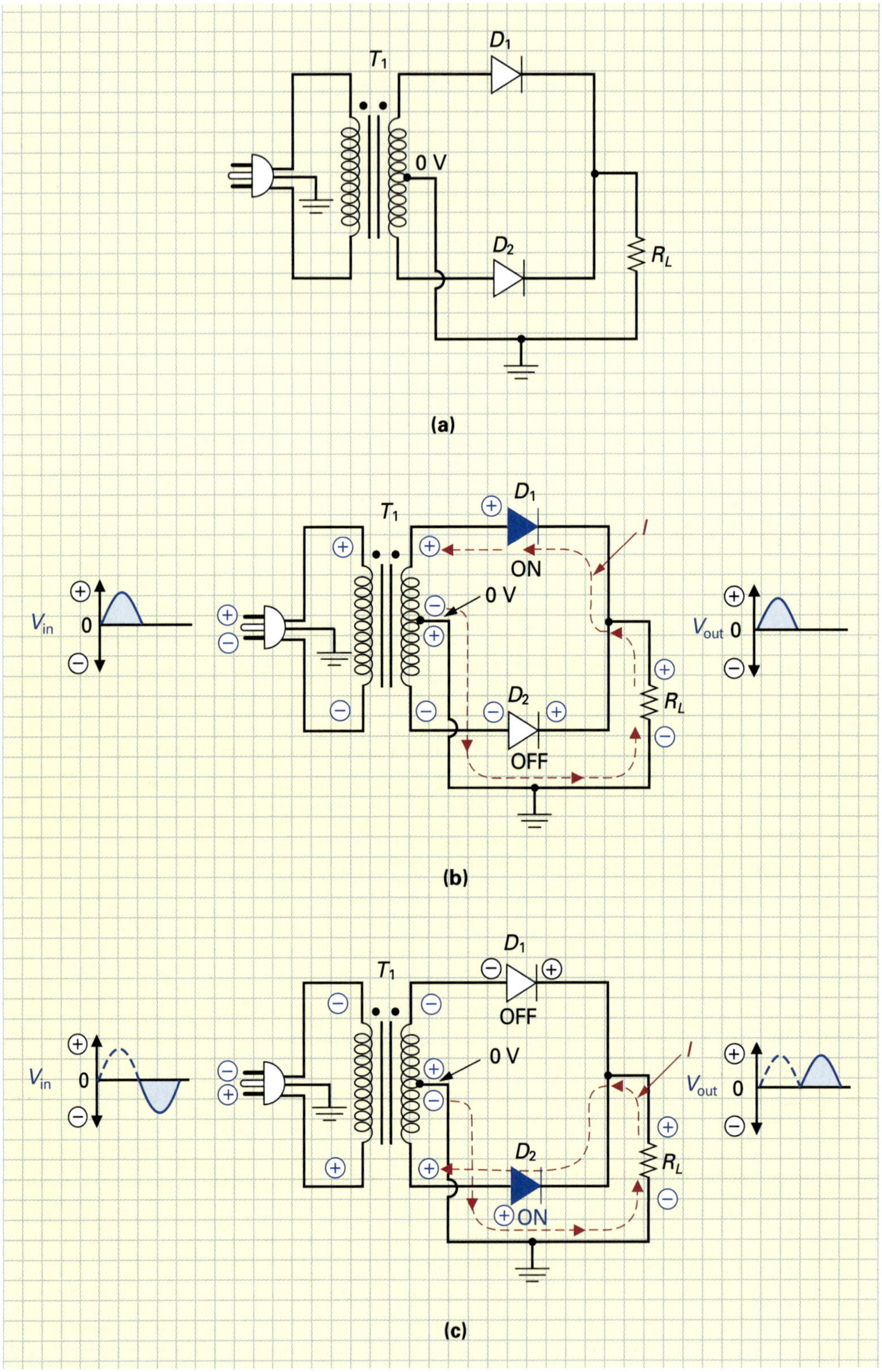

FIGURE 11-22 Full-Wave Center-Tapped Rectifier. (a) Basic Circuit. (b) Positive Input Half-Cycle Operation. (c) Negative Input Half-Cycle Operation.

seems to do worse, delivering a peak output of 1/2 V_s. However, the full-wave center-tapped rectifier compensates for the halving of the peak output voltage by doubling the number of half-cycles at the output, compared to a half-wave rectifier. The average output voltage (V_{avg}) can be calculated for a full-wave rectifier with the following formula:

$$V_{avg} = 0.636 \times \frac{1}{2} V_{s\,peak}$$

EXAMPLE:

Calculate the average output voltage from a half-wave and full-wave center-tapped rectifier if $V_{s\,peak} = 169.7$ V.

Solution:

Half-wave rectifier:

$$V_{avg} = 0.318 \times V_{s\,peak}$$
$$= 0.318 \times 169.7 \text{ V} = 54 \text{ V}$$

Full-wave center-tapped rectifier:

$$\text{If } V_{s\,peak} = 169.7 \text{ V, then } V_{s1\,peak} \text{ and } V_{s2\,peak} = 84.9 \text{ V}$$
$$V_{avg} = 0.636 \times \frac{1}{2} V_{s\,peak}$$
$$= 0.636 \times 84.9 \text{ V} = 54 \text{ V}$$

As you can see from this example, even though the peak output of the full-wave center-tapped rectifier was half that of the half-wave rectifier, the average output was the same because the full-wave center-tapped rectifier doubles the number of half-cycles at the output, compared to a half-wave rectifier.

Because two half-cycles appear at the output for every one cycle at the input, as shown in Figure 11-23, the ripple frequency will be twice that of the input frequency, and this higher frequency will be easier to filter or smooth of fluctuations.

$$\text{output pulsating dc ripple frequency} = 2 \times \text{input ac frequency}$$

To be completely accurate, we should take into account the barrier voltage drop across the diodes. Since only one diode is on for each half-cycle, the peak output voltage will only be less 0.7 V (silicon diode) as shown by the last waveform in Figure 11-23. Therefore

$$V_{out} = \frac{1}{2} V_{s\,peak} - 0.7 \text{ V}$$

With regard to the peak inverse voltage, if you refer back to Figure 11-22(b), you will see that the full secondary peak voltage ($V_{s\,peak}$) appears across the OFF diode D_2 for one half-cycle of the input. Similarly, the full $V_{s\,peak}$ voltage appears across the OFF D_1 during the other half-cycle of the input, as shown in Figure 11-22(c). Both diodes must therefore have a peak inverse voltage (PIV) rating that is larger than the peak secondary voltage ($V_{s\,peak}$). For the circuit in Figure 11-23, the diode's maximum reverse voltage rating (PIV) must be greater than 169.7 V.

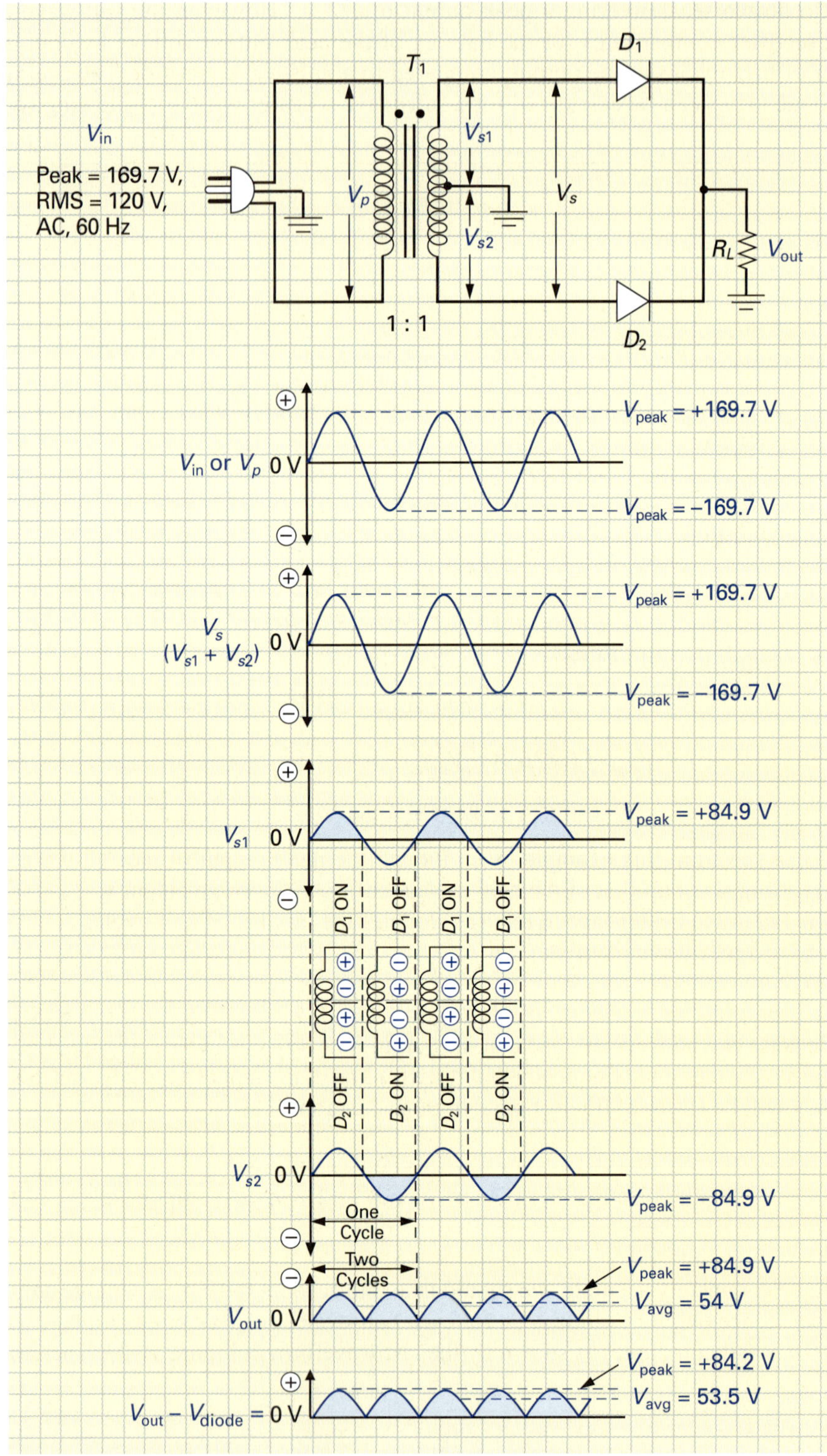

FIGURE 11-23 Input/Output Waveforms of a Full-Wave Center-Tapped Rectifier.

Bridge rectifier. By center tapping the secondary of the transformer and having two diodes instead of one, we were able to double the ripple frequency and ease the filtering process. However, that seems to be the only advantage the full-wave center-tapped rectifier has over the half-wave rectifier, and the price we pay is having a more expensive center-tapped transformer and an extra diode.

With the **bridge rectifier circuit** shown in Figure 11-24(a), we can have the peak secondary output voltage of the half-wave circuit and the full-wave ripple frequency of the center-tapped circuit. This circuit was originally called a "bridge" rectifier because its shape resembled the framework of a suspension bridge. Figure 11-24(b) and (c) illustrate how the bridge rectifier circuit will behave when an ac input cycle is applied.

Bridge Rectifier Circuit
A full-wave rectifier circuit using four diodes that will convert an alternating voltage input into a direct voltage output.

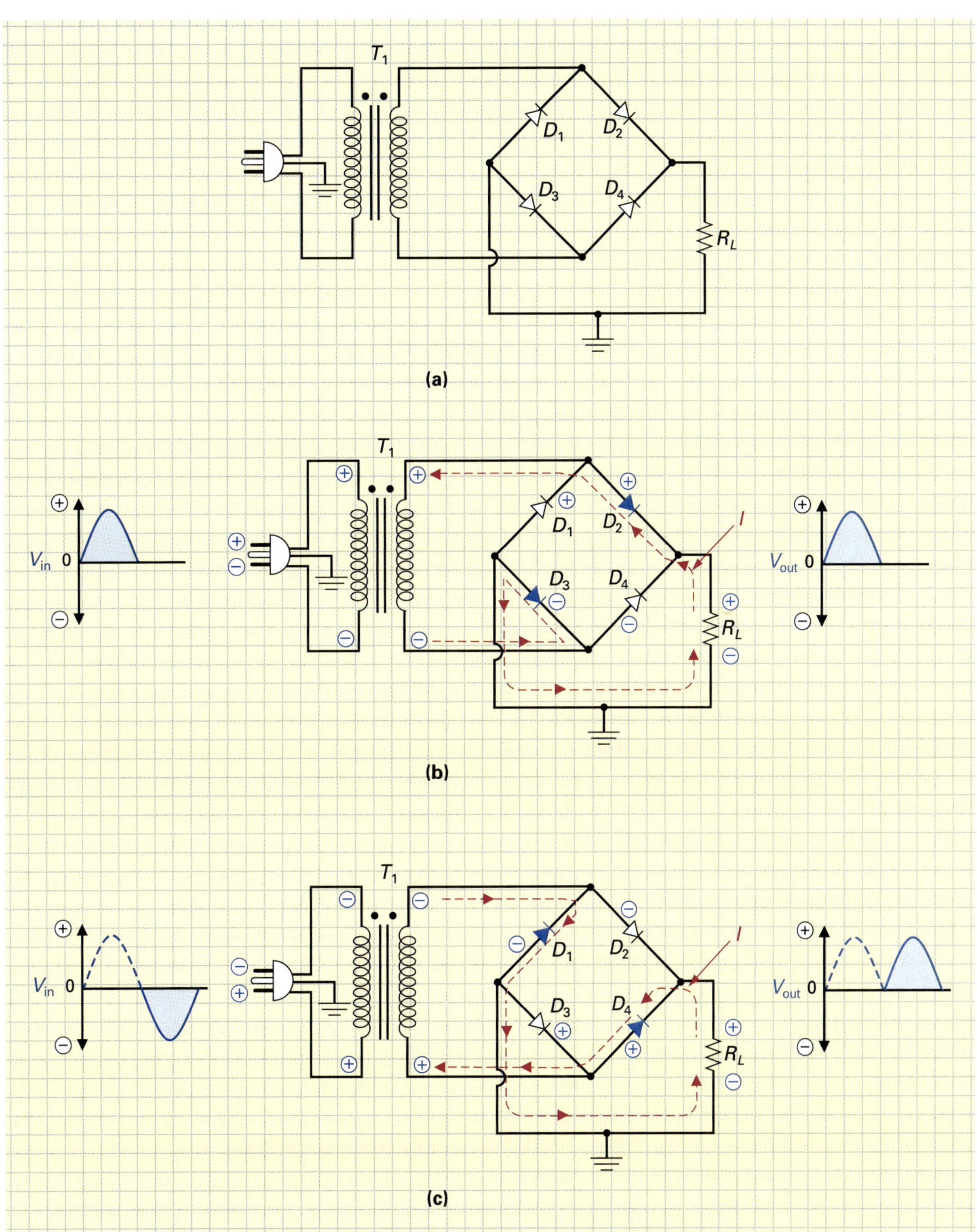

FIGURE 11-24 **Full-Wave Bridge Rectifier. (a) Basic Circuit. (b) Positive Half-Cycle Operation. (c) Negative Half-Cycle Operation.**

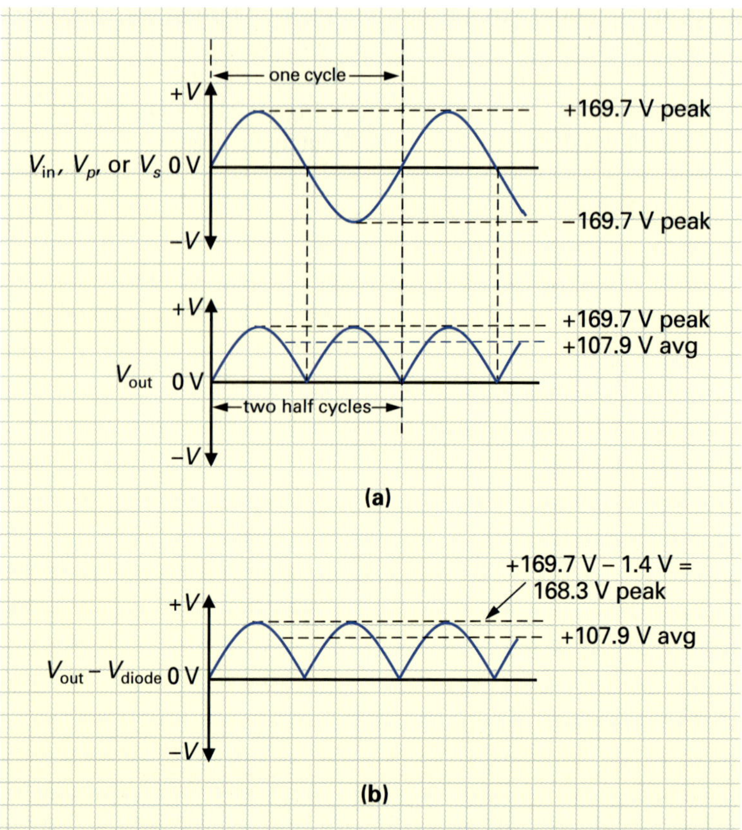

FIGURE 11-25 Input/Output Waveforms of a Full-Wave Bridge Rectifier.

When the ac input (V_{in}) swings positive, the circuit operates in the manner illustrated in Figure 11-24(b). A positive potential is applied to the top of the bridge, causing D_2 to turn ON, while a negative potential is applied to the bottom of the bridge, causing D_3 to turn ON. With D_2 and D_3 ON, and D_1 and D_4 OFF, electrons will flow up through the load as indicated by the dashed current line (I), developing a positive output half-cycle across the load (V_{out}).

When the ac input (V_{in}) swings negative, the circuit operates in the manner illustrated in Figure 11-24(c). A negative potential is applied to the top of the bridge, causing D_1 to turn ON, while a positive potential is applied to the bottom of the bridge, causing D_4 to turn ON. With D_1 and D_4 ON and D_2 and D_3 OFF, electrons will flow up through the load as indicated by the dashed current line (I), again developing a positive output half-cycle across the load (V_{out}).

Like the center-tapped rectifier, the bridge rectifier switches two half-cycles of the same polarity through to the load. However, unlike the center-tapped rectifier, the bridge rectifier connects the total peak secondary voltage across the load. A secondary peak voltage of 169.7 V ac would produce a pulsating dc peak across the load of 169.7 V, or the same voltage, as shown in Figure 11-25(a). The average voltage in this case would be larger than that of a center-tapped rectifier, and equal to

$$V_{avg} = 0.636 \times V_{s\,peak}$$

$$= 0.636 \times 169.7\text{ V} = 107.9\text{ V}$$

EXAMPLE:

Calculate the average output voltage from a center-tapped rectifier and a bridge rectifier if $V_{s\,peak} = 169.7$ V.

■ ***Solution:***

Center-tapped:

$$\begin{aligned} V_{avg} &= 0.636 \times 1/2\ V_s \\ &= 0.636 \times 84.9\text{ V} \\ &= 54\text{ V} \end{aligned}$$

Bridge:

$$\begin{aligned} V_{avg} &= 0.636 \times V_s \\ &= 0.636 \times 169.7\text{ V} \\ &= 107.9\text{ V} \end{aligned}$$

As you can see from this example, unlike the center-tapped rectifier that only connects half of the peak secondary voltage across the load, the bridge rectifier connects the total peak secondary voltage across the load.

Because two half-cycles appear at the output for every one cycle at the input, as shown in Figure 11-25(a) on the previous page, the ripple frequency will be twice that of the input frequency, and this higher frequency will be easier to filter or smooth of fluctuations.

$$\text{output pulsating dc ripple frequency} = 2 \times \text{input ac frequency}$$

To be completely accurate, we should take into account the barrier voltage drop across the diodes. Because two diodes are ON for each half-cycle and both diodes are connected in series with the load, a 1.4 V (2 × 0.7 V) drop will occur between the peak secondary voltage (V_s) and the peak output voltage, as shown in Figure 11-25(b).

$$V_{out} = V_{s\text{ peak}} - 1.4\text{ V}$$

With regard to the peak inverse, if you refer to Figure 11-26, you can see that the full $V_{s\text{ peak}}$ voltage appears across the two OFF diodes. Therefore, each diode must have a peak inverse voltage (PIV) rating that is greater than the peak secondary voltage (V_s). For the circuit and values given in Figures 11-24 and 11-25, the diode's maximum reverse voltage rating (PIV) must be greater than 169.7 V.

11-5-4 *Filters*

The filter in a dc power supply converts the pulsating dc output from the half-wave or full-wave rectifier into an unvarying dc voltage, as was shown in the waveforms in the basic

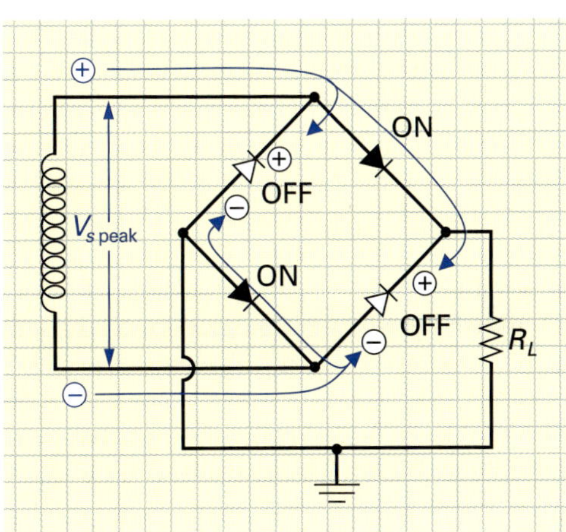

FIGURE 11-26 **Peak Inverse Voltage.**

Capacitive Filter
A capacitor used in a power supply filter system to suppress ripple currents while not affecting direct currents.

***RC* Filter**
A selective circuit which makes use of a resistance-capacitance network.

***LC* Filter**
A selective circuit which makes use of an inductance-capacitance network.

block diagram in Figure 11-16. In this section, we will examine the three basic types of filters: the **capacitive filter,** the ***RC* filter,** and the ***LC* filter.**

The Capacitive Filter

Figure 11-27(a) shows a positive output half-wave rectifier and capacitive filter circuit. The waveforms in Figure 11-27(b), (c), and (d) show how the capacitive filter will respond to the half-wave pulsating dc output from the rectifier. Let us examine how the filter operates by referring to the dashed lines in the circuit in Figure 11-27(a), the output wave-

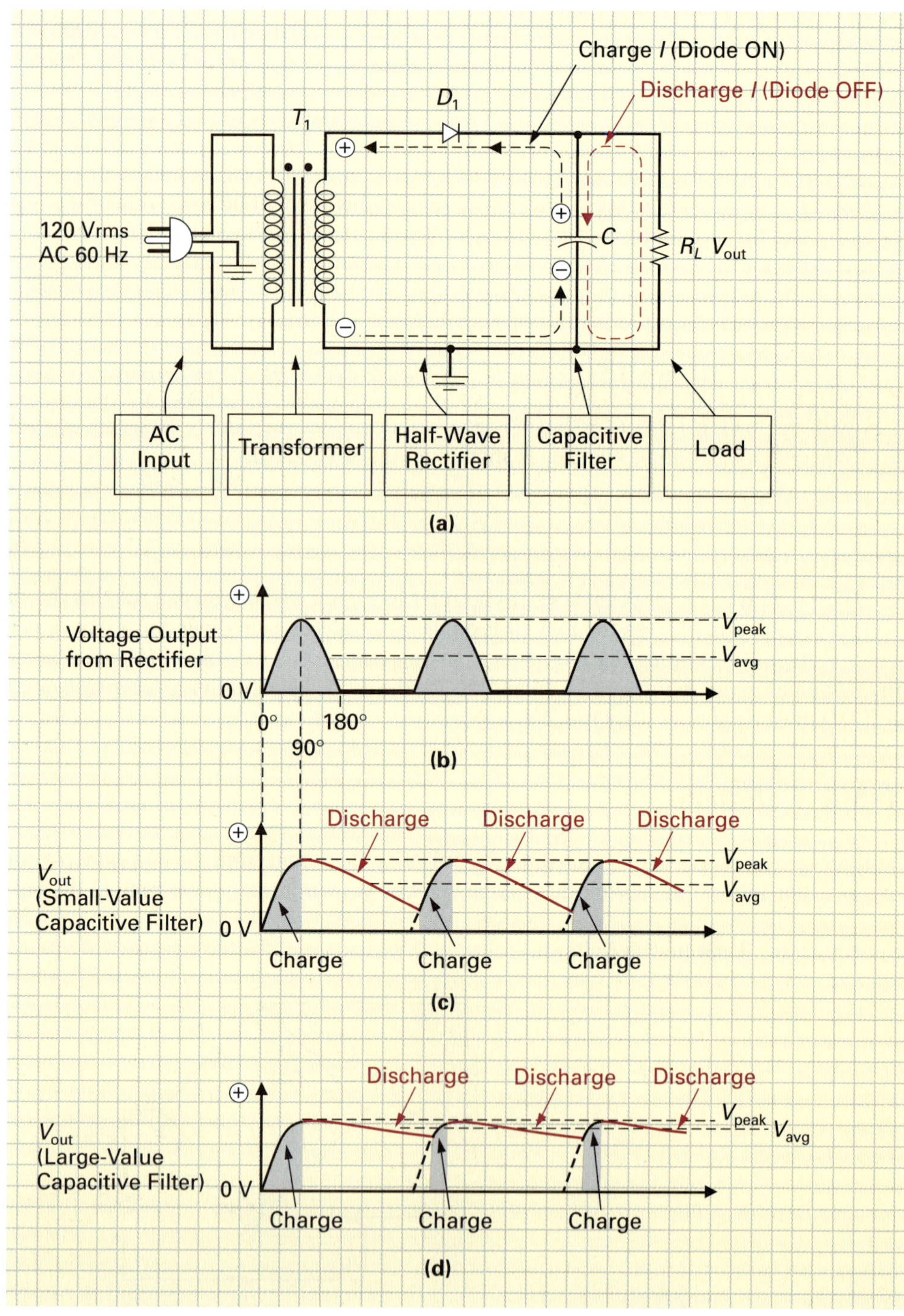

FIGURE 11-27 **Capacitive Filtering of a Half-Wave Rectifier Output.**

form from the rectifier shown in Figure 11-27(b), and the small-value capacitive filter waveform shown in Figure 11-27(c). When the ac input swings positive, the diode is turned ON and the capacitor charges as indicated by the black dashed charge current line in Figure 11-27(a). The charge time constant will be small because no resistance exists in the charge path except for that of the resistance of the connecting wires (charge $\tau\downarrow = R\downarrow \times C$). This charge time is indicated by the gray shaded section between 0 and 90° in Figure 11-27(c). When the ac input begins to fall from its positive peak (at 90°), the diode is turned OFF by the large positive potential on the diode's cathode being supplied by the charged capacitor, and the decreasing positive potential on the diode's anode being supplied by the input. With the diode OFF, the capacitor begins to discharge as indicated by the colored dashed discharge current line in Figure 11-27(a). The discharge time constant is a lot longer than the charge time because of the load resistance (discharge $\tau\uparrow = R\uparrow \times C$). The decreasing slope in Figure 11-27(c) illustrates the decreasing voltage across the capacitor, and therefore across the load at the output (V_{out}), as the capacitor discharges. As the ac input and the rectifier's pulsating dc cycle repeat, the output (V_{out}) is, as shown in Figure 11-27(c), an almost constant dc output with a slight variation or ripple above and below the average value.

Figure 11-27(d) shows how a larger-value capacitor will make the charge and discharge time constants longer ($\tau\uparrow = R \times C\uparrow$), therefore decreasing the amount of ripple and increasing the average output voltage.

Figure 11-28(a) shows a positive output voltage full-wave bridge rectifier and capacitive filter circuit. Figure 11-28(b), (c), and (d) show the output waveforms from the rectifier and the filtered output from a small-value and a large-value capacitor filter. The ripple frequency from a full-wave rectifier is twice that of a half-wave rectifier, and therefore the capacitor does not have too much time to discharge before another positive half-cycle reoccurs. This results in a higher average voltage than in a half-wave circuit, and if a larger-value capacitor is used, the ripple will be even less because of the greater time constant ($\tau\uparrow = R \times C\uparrow$), causing an increased average voltage, as shown in Figure 11-28(d).

Percent of Ripple

In the previous two capacitive filter circuits, you could see that even though the output from a filter should be a constant dc level, there is a slight fluctuation or ripple. This fluctuation is called the percent ripple and its value is used to rate the action of the filter. It can be calculated with the following formula:

$$\text{Percent ripple} = \frac{V_{rms} \text{ of ripple}}{V_{avg} \text{ of ripple}} \times 100$$

To help show how this is calculated, let us apply this formula to the example in Figure 11-29. In this example, the filter's output is fluctuating between +20 and +30 V, and therefore the ripple's peak-to-peak value is

$$\begin{aligned} \text{Peak-to-peak of ripple} &= 20 \text{ to } 30 \text{ V} \\ &= 10 \text{ V pk-pk} \end{aligned}$$

A peak-to-peak value of 10 V means that the ripple has a peak value of

$$\begin{aligned} \text{Peak of ripple} &= 1/2 \text{ of pk-pk value} \\ &= 1/2 \text{ of } 10 \text{ V} \\ &= 5 \text{ V peak} \end{aligned}$$

Once the peak of the ripple is known, the ripple rms value can be calculated:

$$\begin{aligned} \text{RMS of ripple} &= 0.707 \text{ of peak} \\ &= 0.707 \text{ of } 5 \text{ V} \\ &= 3.54 \text{ V} \end{aligned}$$

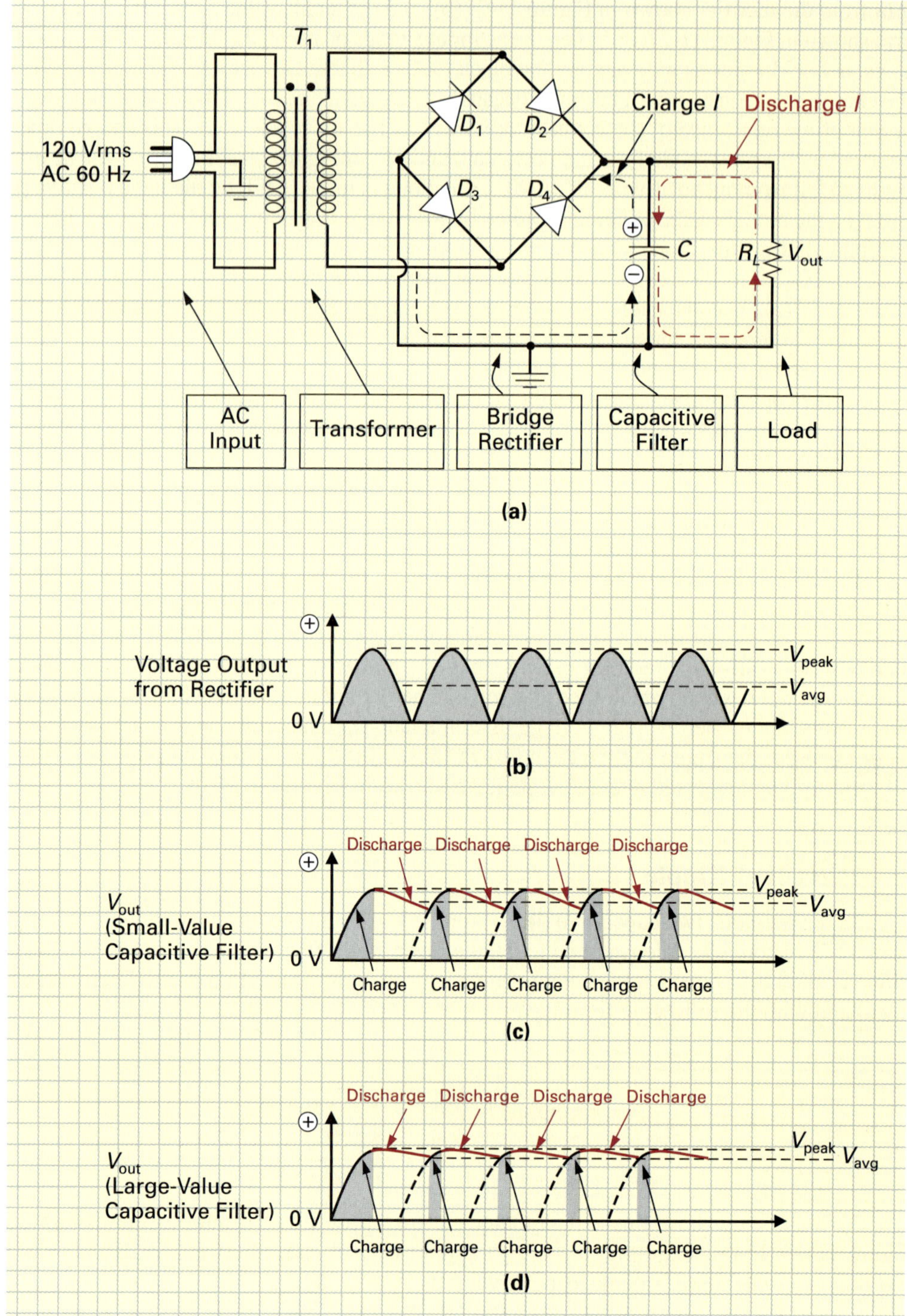

FIGURE 11-28 **Capacitive Filtering of a Full-Wave Rectifier Output.**

The average of the ripple is approximately midway between peaks, which in the example in Figure 11-29 is +25 V. Inserting these values in the formula, we can calculate the percent of ripple:

$$\% \text{ ripple} = \frac{V_{rms} \text{ of ripple}}{V_{\text{avg}} \text{ of ripple}} \times 100$$

$$= \frac{3.54 \text{ V}}{25 \text{ V}} \times 100$$

$$= 14\%$$

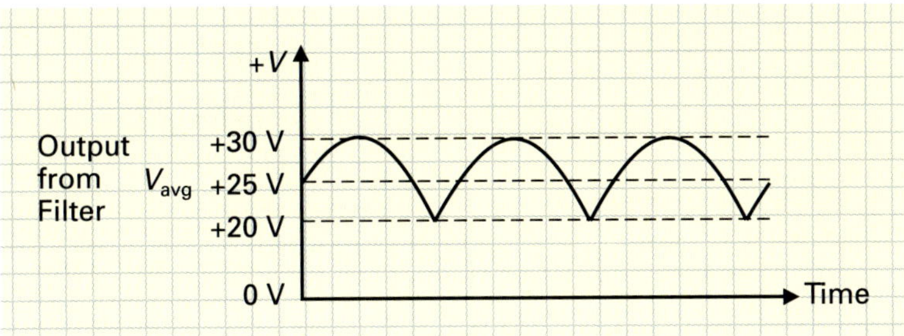

FIGURE 11-29 **Percent of Ripple Out of a Filter.**

This means that the average voltage out of the rectifier (+25 V) will fluctuate 14%.

LC Filters

Adding an inductor in series (L_1), as shown in Figure 11-30, will increase the efficiency of the filter because the inductor will oppose current without generating heat. The inductor, L_1, will offer a very low opposition or reactance to dc ($X_L\downarrow \propto f\downarrow$) and a very high reactance to the ac ripple ($X_L\uparrow \propto f\uparrow$).

11-5-5 *Regulators*

The **voltage regulator** in a dc power supply maintains the dc output voltage constant despite variations in the ac input voltage and the output load resistance. To help explain why these variations occur, let us first examine the relationship between a source and its load.

Voltage Regulator
A device or circuit that will maintain the output voltage of a voltage source constant despite variations in the input voltage and load resistance.

Source and Load

Ideally, a dc power supply should convert all of its *ac electrical energy input* into a *dc electrical energy output.* However, like all devices, circuits, and systems, a dc power supply is not 100% efficient and, along with the electrical energy output, the dc power supply generates wasted heat. This is why a dc power supply can be represented as a voltage source with an internal resistance (R_{int}), as shown in Figure 11-31(a). The internal resistance, which is normally very small (in this example, 1 Ω), represents the heat energy loss or inefficiency of the dc power supply. The load resistance (R_L) represents the resistance of the electronic circuits in the electronic system. Since R_{int} and R_L are connected in series with one another,

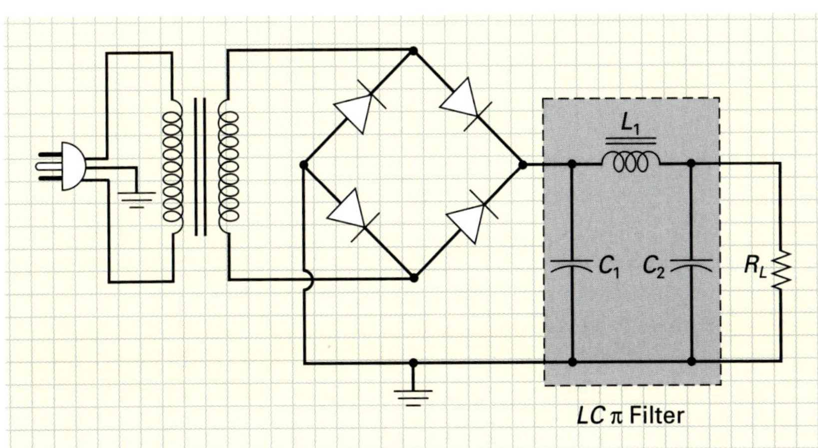

FIGURE 11-30 ***LC* π Filter.**

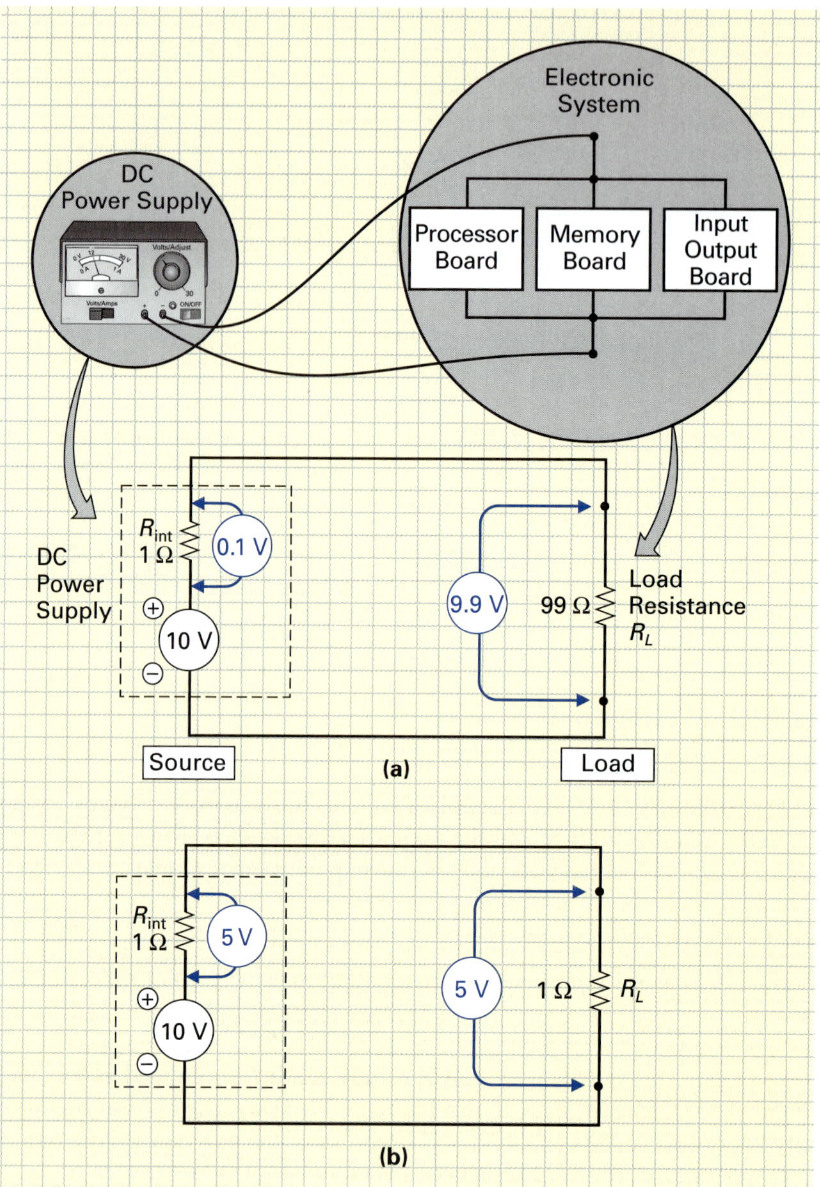

FIGURE 11-31 Source and Load.

the 10 V supply in this example will be developed proportionally across these resistors, resulting in

$$V_{R_{\text{int}}} = 0.1\text{ V}$$
$$V_{R_L} = \frac{9.9\text{ V}}{10\text{ V total}}$$

If the load resistance (R_L) were to decrease dramatically to 1 Ω, as shown in Figure 11-31(b), the decrease in load resistance ($R_L\downarrow$) would cause an increase in load current ($I_L\uparrow$). This current increase would cause an increase in the heat generated by the power supply ($P_{R_{\text{int}}}\uparrow = I^2\uparrow \times R$). This loss is seen at the output of the dc power supply, which is now delivering only a 5 V output to the load because the 10 V source voltage is being divided equally across R_{int} and R_L.

$$V_{R_{\text{int}}} = 5\text{ V}$$
$$V_{R_L} = \frac{5\text{ V}}{10\text{ V total}}$$

If the load resistance decreases, the load current will increase, and if too much current is drawn from the source, it will pull down the source voltage. Although a load resistance will generally not change as dramatically as shown in Figure 11-31(b), it will change slightly as different control settings are selected, and this will *load the source.* For example, the load resistance of the electronic circuits in your music system will change as the volume, bass, treble, and other controls are adjusted. This load resistance change would affect the output of a dc power supply if a regulator were not included to maintain the output voltage constant despite variations in load resistance.

To the electric company, each user appliance is a small part of its load, and it is the source. The ac voltage from the electric company can be anywhere between 105 V and 125 V ac rms (148 to 177 V ac peak), depending on consumer use, which depends on the time of day. When many appliances are in use, the load current will be high and the overall load resistance low, causing the source voltage to be pulled down. If a dc power supply did not have a regulator, these different input ac voltages from the electric company would produce different dc output voltages, when we really want the dc supply voltage to the electronic circuits to always be the same value no matter what ac input is present. This is the other reason that a regulator is included in a dc power supply; it maintains the dc output voltage constant despite variations in the ac input voltage.

In summary, a regulator is included in a dc power supply to maintain the dc output voltage constant despite variations in the output load resistance and the ac input voltage, as shown in Figure 11-32.

Percent of Regulation

The percent of regulation is a measure of the regulator's ability to regulate—or maintain constant—the output dc voltage. It is calculated with the formula

$$\text{percent regulation} = \frac{V_{nl} - V_{fl}}{V_{nl}} \times 100$$

V_{nl} = no-load voltage
V_{fl} = full-load voltage

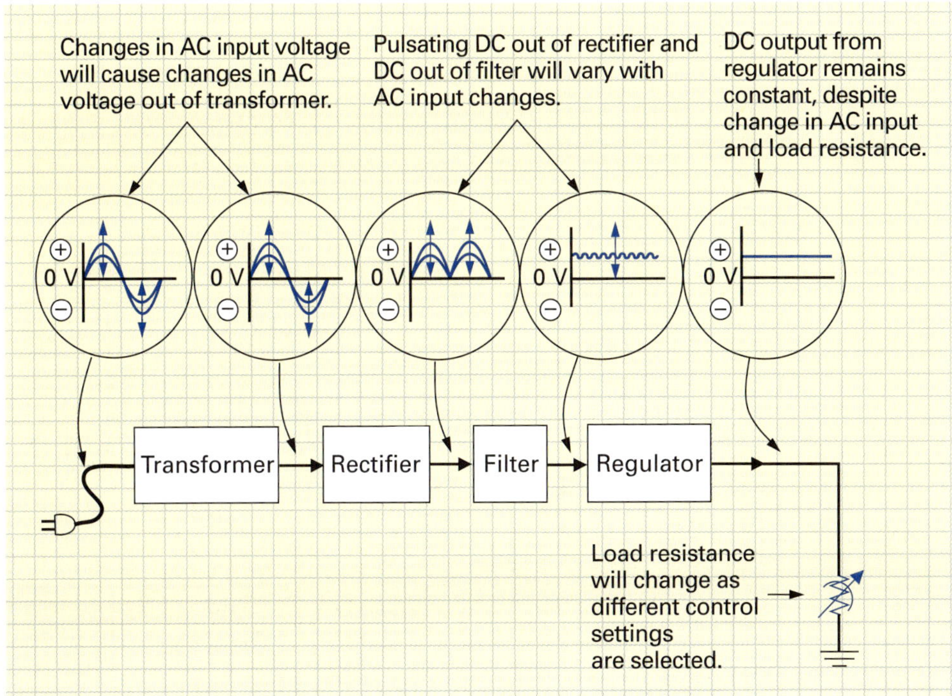

FIGURE 11-32 Function of a Regulator.

If we apply this formula to the example in Figure 11-31, we get

$$\% \text{ regulation} = \frac{V_{nl} - V_{fl}}{V_{nl}} \times 100$$

$$= \frac{10 \text{ V} - 5 \text{ V}}{10 \text{ V}} \times 100 = 50\%$$

In the ideal situation, a regulator would be included and would maintain the output voltage constant between no-load and full-load, resulting in a percent regulation figure of

$$\% \text{ regulation} = \frac{V_{nl} - V_{fl}}{V_{nl}} \times 100$$

$$= \frac{10 \text{ V} - 10 \text{ V}}{10 \text{ V}} \times 100$$

$$= 0\%$$

Most regulators achieve a percent regulation figure that is not perfect (0%) but is generally in single digits (2% to 8%).

Zener Regulator

Figure 11-33 shows how a zener diode (D_5) and a series resistor (R_1) would be connected in a dc power supply circuit to provide regulation. An increase in the ac input voltage would cause an increase in the dc output from the filter, which would cause an increase in the current through the series resistor (I_S), the zener diode (I_Z), and the load (I_L). The reverse biased zener diode will, however, maintain a constant voltage across its anode and cathode (in this example, 12 V) despite these input voltage and current variations, with the additional voltage being dropped across the series resistor. For example, if the output from the filter could be between +15 and +20 V, the zener diode would always drop 12 V, while the series resistor would drop between 3 V (when input is 15 V) and 8 V (when input is 20 V). Consequently, the zener regulator will maintain a constant output voltage despite variation in the ac input voltage.

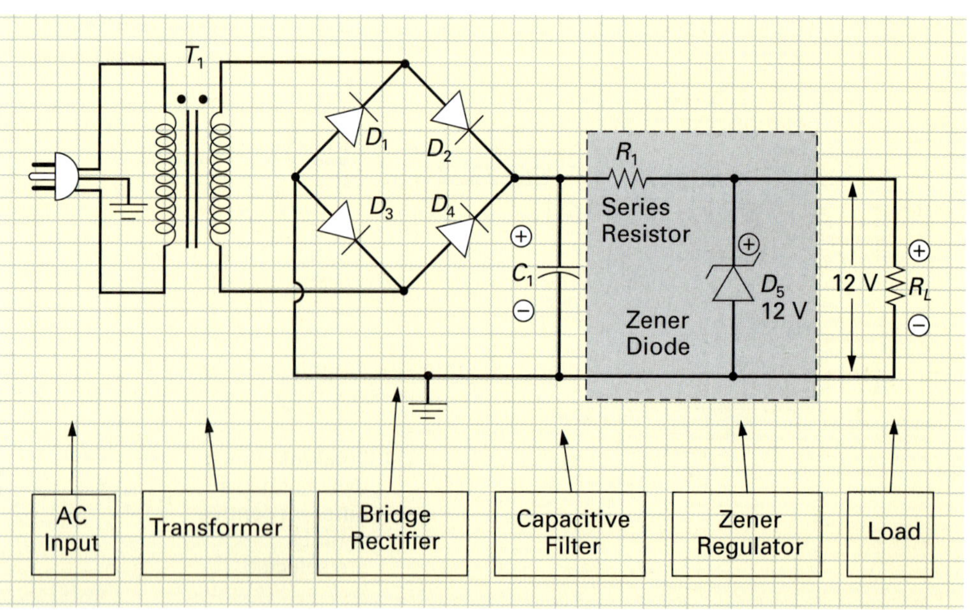

FIGURE 11-33 Zener Diode Regulator.

From the other standpoint, the zener regulator will also maintain a constant output voltage despite variations in load resistance. The zener diode achieves this by increasing and decreasing its current (I_Z) in response to load resistance changes. Despite these changes in zener and load current however, the zener voltage (V_Z), and therefore the output voltage (V_{out} or V_{R_L}), always remains constant. The disadvantage with this regulator is that the series-connected resistor will limit load current and, in addition, generate unwanted heat.

The IC Regulator

Most dc power supply circuits today make use of integrated circuit (IC) regulators, such as the one shown in Figure 11-34. These IC regulators contain about 50 individual or discrete components all integrated on one silicon semiconductor chip and then encapsulated in a three-pin package. The inset in Figure 11-34 shows how all of the IC regulator's internal components form circuits which are shown as blocks. For example, a short-circuit protection and thermal shutdown circuit will protect the IC regulator by turning it OFF if the

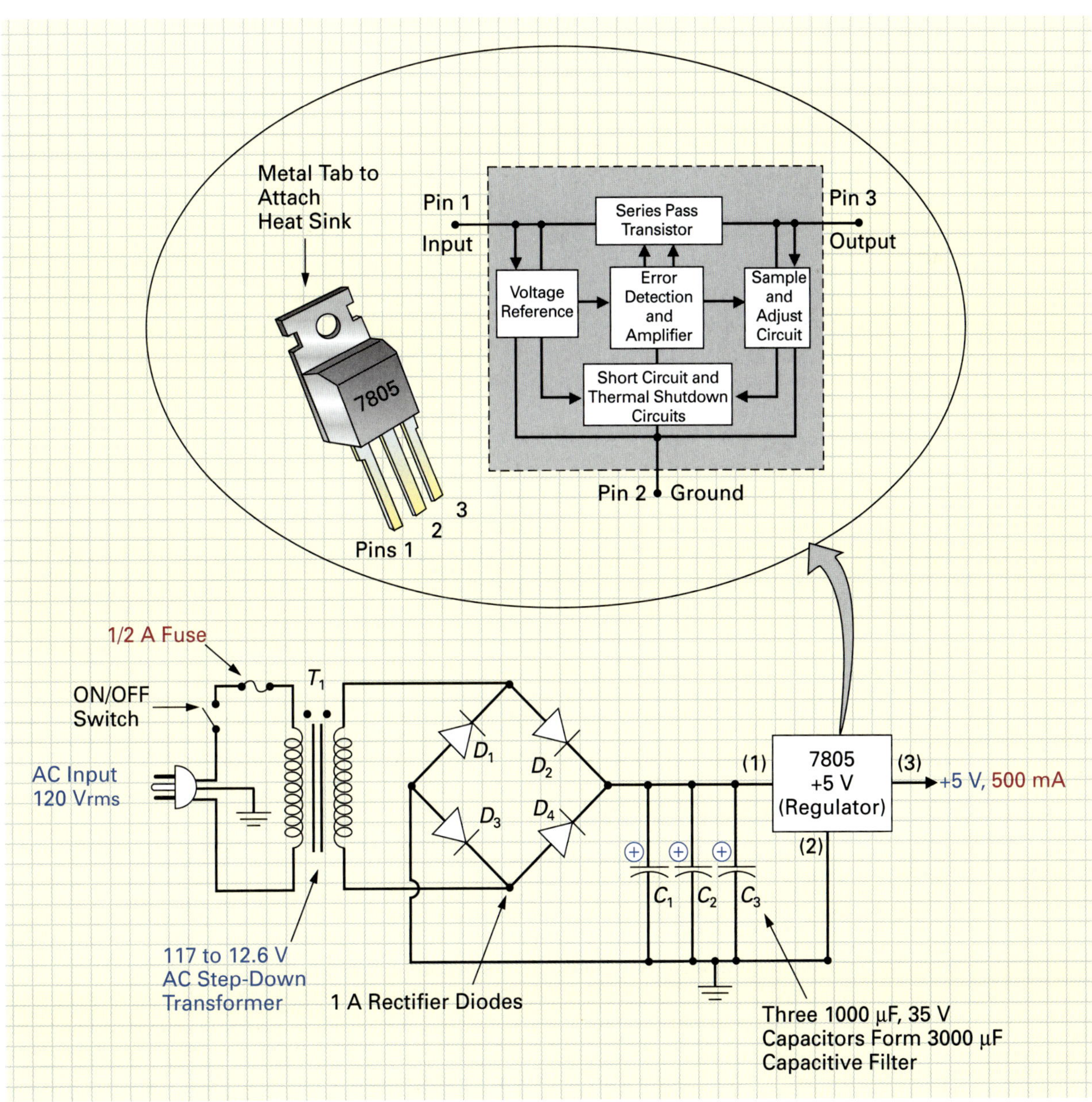

FIGURE 11-34 Integrated Circuit Regulator within a Power Supply Circuit.

load current drawn exceeds the regulator's rated current, or if the heat sink is too small and the IC regulator is generating heat faster than it can dissipate it.

11-5-6 *Troubleshooting a DC Power Supply*

In this section we will discuss how to troubleshoot a typical dc power supply circuit. The three-step troubleshooting procedure for fixing a failure can be broken down into three basic steps:

Step 1: DIAGNOSE
The first step is to determine whether a problem really exists. To carry out this step, a technician must collect as much information as possible about the system, circuit, and components used, and then diagnose the problem.

Step 2: ISOLATE
The second step is to apply a logical and sequential reasoning process to isolate the problem. In this step, a technician will operate, observe, test, and apply troubleshooting techniques in order to isolate the malfunction.

Step 3: REPAIR
The third and final step is to make the actual repair and then test the circuit.

Let us examine this three-step troubleshooting process in more detail, and apply it to a typical dc power supply circuit.

Typical Power Supply Circuit

As an example, Figure 11-35 shows the schematic diagram for a dc power supply circuit that can generate a +5 V, +12 V, and −12 V dc output from a 120 V or 240 V ac input. You should recognize most of the pieces of this picture because this circuit contains nearly all of the devices discussed in the previous sections of this chapter.

The power supply ON/OFF switch (S_1) switches a 120 V ac input to the 120 V primary-winding connection, or a 240 V ac input to the 240 V primary-winding connection of the transformer (T_1). By doubling the number of primary turns, we can accept twice the input voltage and deliver the same output voltage. The upper secondary winding of T_1 supplies a bridge rectifier module (BR_1), that generates a +12 V peak full-wave pulsating dc output, which is filtered by C_1 and regulated by the 7805 (U1) to produce a +5 V dc 750 mA output. The +5 V output will turn on a "power ON" LED (D_3), which is mounted on the power supply printed circuit board (PCB) along with the filter and regulator. R_1 is a current-limiting resistor for D_3.

The lower secondary winding of T_1 supplies the two half-wave rectifier diodes D_1 and D_2, which generate −50 V and +50 V peak outputs, respectively. These inputs are filtered by C_2 and C_3 and then regulated by a 7912 (U_2), which generates a −12 V dc 500 mA output, and by a 7812 (U_3), which generates a +12 V dc 500 mA output. The devices C_2, C_3, U_2, and U_3 are also all mounted on the power supply printed circuit board (PCB).

Step 1: Diagnose

It is extremely important that you first understand how a system, circuit, and all of its components are supposed to work so that you can determine whether or not a problem exists. If you were preparing to troubleshoot the dc power supply circuit in Figure 11-35, your first step should be to read through the circuit description and review the operation of each device used in the circuit until you feel completely confident with the correct operation of the circuit.

For other circuits and systems, technicians usually refer to service or technical manuals which generally contain circuit descriptions and troubleshooting guides. As far as each of the devices is concerned, technicians refer to manufacturer data books that contain a full description of the device.

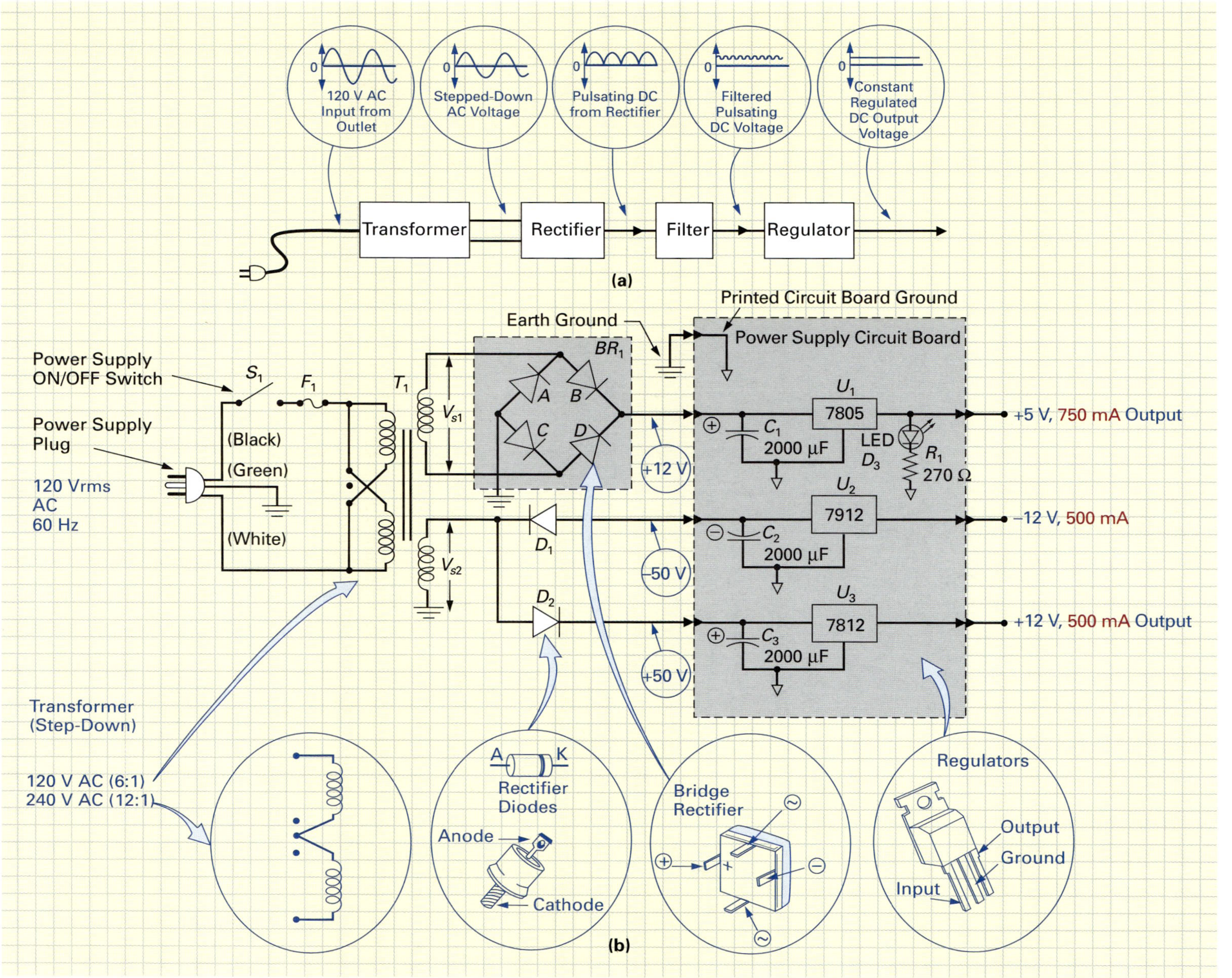

FIGURE 11-35 **DC Power Supply Circuit. (a) Block Diagram. (b) Circuit Diagram.**

Referring to all of this documentation before you begin troubleshooting will generally speed up and simplify the isolation process. Once you are fully familiar with the operation of the circuit, you will be ready to *diagnose the problem* as either an *operator error* or as a *circuit malfunction.*

Here are some examples of operator errors that can be mistaken for circuit malfunctions if a technician is not fully familiar with the operation of the dc power supply circuit in Figure 11-35.

SYMPTOMS	DIAGNOSIS
1. No +5 V, −12 V, or +12 V output.	Check first that the power supply is plugged in and turned ON because either of these operator errors will give the appearance of a circuit malfunction.
2. The +5 V LED is ON, but the −12 V and +12 V LEDs are not lit.	There are no LED indicators for the −12 V and +12 V outputs, therefore this is a normal condition.

Once you have determined that the problem is not an operator error but, in fact, a circuit malfunction, proceed to Step 2 and isolate the circuit problem.

Step 2: Isolate

A technician will generally spend most of the time in this phase of the process isolating a circuit problem. Steps 1 (diagnose) and 3 (repair) may only take a few minutes to complete compared to Step 2, which could take a few hours to complete. However, with practice and a good logical and sequential reasoning process you can quickly isolate even the most obscure of problems.

The directions you take as you troubleshoot a circuit malfunction may be different for each problem. However, certain troubleshooting techniques apply to all problems. Let us now review some of these techniques and apply them to our dc power supply circuit in Figure 11-35.

1. Check first for *obvious errors:*
   a. Power fuse blown
   b. Wiring errors if circuit is newly constructed
   c. Devices are incorrectly oriented. For example, if the rectifier diodes, bridge rectifier module, IC regulators, or the LED were placed in the circuit in the opposite direction to what is shown in Figure 11-35, that device would not operate as it should and therefore neither would the circuit.
2. Use your *senses* to check for broken wires, loose connections, overheating or smoking components, leads or pins not making contact, and so on.
3. and 4. Use a *cause-and-effect troubleshooting process* which involves studying the effects you are getting from the faulty circuit and then trying to reason out what could be the cause. Also, apply the *half-split method* of troubleshooting first to the entire system, then to a circuit within the system, and then to a section within the circuit to help speed up the isolation process. To explain how to use these two methods with the circuit in Figure 11-35, let us examine a few examples.

EXAMPLE:

The dc power supply in Figure 11-35 was being used to supply power to several circuits in an electronic system. If a power problem were to occur, how would we apply the half-split method?

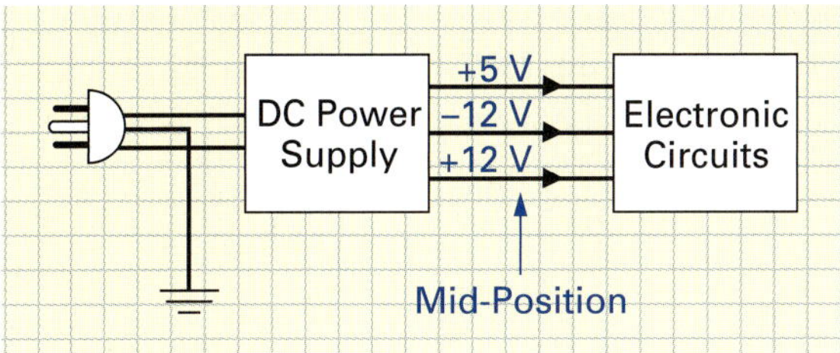

FIGURE 11-36 **Using the Half-Split Method to Isolate a System Problem.**

Solution:

The half-split process involves first choosing a point roughly in the middle of the problem area. By making a test at this midpoint you can determine whether the problem exists either before or after this point. For example, Figure 11-36 shows how this half-split method can be applied to a power supply problem. By testing the output of the dc power supply (+5 V, −12 V, +12 V) you can determine whether the power problem is in the power supply or within the circuits that are being supplied power.

a. If the dc supply voltages are present and of the correct value, then the dc power supply is functioning normally.
b. If these dc supply voltages are not present or are not of the correct value, then we will need to isolate the problem.

In some instances you will have to be careful not to assume the obvious when applying the half-split method. For example, let us imagine that we have tested the output of the dc power supply and its output voltage is incorrect. Your first instinct would be to assume that the problem lies within the dc power supply. However, as we discovered in this chapter, *a short in the load can pull down a source, making it appear to be supplying a faulty output.* Therefore, most problems need to be further isolated because the problem could be in the source or it could be in the load. Let us look at another example in which the source appears to be the problem when, in fact, the fault is in the load.

EXAMPLE:

The +5 V output from a dc power supply in Figure 11-37 measures +2 V on the multimeter. How would we proceed to troubleshoot this problem?

Solution:

Figure 11-37 shows how a +5 V output from a dc power supply could be pulled down to +2 V due to a short or low resistance path in one of the electronic circuits that are being supplied power. These problems can cause the power supply input fuse to blow or the IC regulator to disconnect its output due to the excessive current being drawn by the short. Because the output of a power supply generally goes to several boards, and within each board it is connected across many devices, it becomes very difficult to isolate a short. If the fuse does not blow and the regulator IC does not switch OFF, a short will definitely load or pull down the dc supply voltage. At first, it appears that the power supply itself is malfunctioning because of the decreased output voltage, but once you disconnect the output of the dc power supply (at point *A*) from the external circuit boards, the +5 V should go back up to its correct voltage indicating that the problem lies in the load and not in the source. To isolate the faulty circuit board, you will have to reconnect the supply voltage (at point *A*) and then disconnect each circuit board one at a time in the following way to isolate the faulty board:

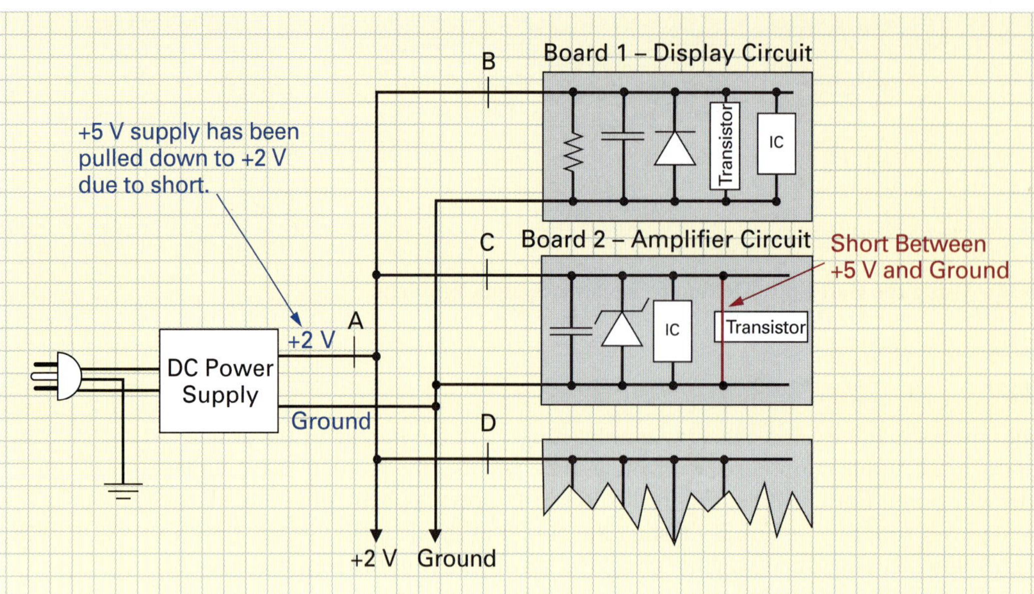

FIGURE 11-37 **Locating Power Line Shorts.**

a. First disconnect board 1 at point *B* and then check the +5 V output from the dc power supply to see if it has returned to its normal voltage. If the dc supply voltage at point *A* remains LOW, then the short is not within board 1 and you should reconnect this board.
b. Disconnect board 2 at point *C* and again check the +5 V output from the dc power supply at point *A* to see if it has returned to its normal voltage. If its output voltage remains LOW, then the short is not within board 2, and you should reconnect this board. In the example in Figure 11-37, however, the short does exist within board 2 and, once this board is disconnected at point *C*, the output from the dc power supply at point *A* will return to +5 V.
c. If the fault was not in board 2, you would have continued to disconnect each board until you isolated the short.

Once the faulty board is located, you will have to troubleshoot that circuit until the short is isolated. For example, if this board were a seven-segment encoder and display circuit, we would follow the troubleshooting procedure for that board. A dramatic approach to localizing a difficult-to-find short on a complex printed circuit board is to freeze spray the entire board using an aerosol freeze can. Once the board is reconnected to the power supply, the short will become quickly visible since its path of excess current will generate heat and therefore defrost faster than the rest of the board.

What would happen if an open developed in one of the lines connecting power to a circuit? Let us answer this question by again looking at another example.

EXAMPLE:

How would you isolate the open shown in Figure 11-38?

Solution:

An open in any of the dc supply lines connecting power to a circuit will cause that circuit not to operate. These opens can be easily traced with a voltmeter to determine the point at which voltage is no longer present.

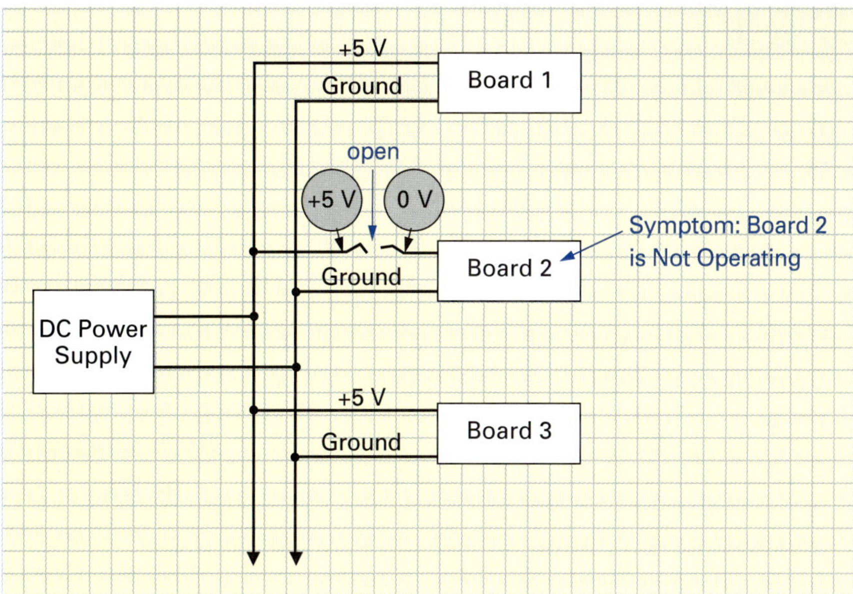

FIGURE 11-38 Locating Power Line Opens.

In summary, once you have disconnected the circuit load from a dc power supply's output, you can isolate the problem as being either a source fault or load fault, because, *if the power supply's output voltage goes back to its rated voltage after it has been disconnected from the circuit load, the problem is with the load. Whereas, if the output voltage remains LOW or at zero after the source has been disconnected from the load, the problem is with the source.*

Assuming that we have isolated the problem to the power supply, let us now examine several types of common failures. Once again, we will apply these problems to the example circuit in Figure 11-35.

EXAMPLE:

What would happen to the circuit in Figure 11-35 if the power supply's fuse were to blow?

Solution:

A short in the power supply or the load will cause an excessive current to be drawn from the secondary of the transformer. The primary of the transformer will try to respond and deliver the power needed by increasing its current. Since the power supply's fuse (F_1) is connected in series with the transformer's primary, as shown in Figure 11-35, the increase in primary current will blow or open the fuse. When a fuse needs to be replaced, always make sure that the replacement has the same voltage and current rating. In most cases, power supplies use a **slow-blow fuse,** which allows a surge of current to pass through the power supply when it is first turned on. This surge current is momentary, usually lasting for less than a second, while all of the system circuits are warming up. If the surge or high current persists for longer than a second, however, the fuse will blow and protect the power supply. Under no circumstances should you ever use a fuse that has a higher current rating because this will permit a higher value of current to pass through to the power supply and load circuits. Never be tempted to "defeat" or bypass the fuse if you do not have a replacement by placing a piece of wire across the fuse terminals. This could introduce other faults, start a fire, or possibly electrocute somebody, since there is now nothing to prevent a damaging value of current.

Slow-Blow Fuse
Allows a surge of current to pass through the power supply when it is first turned on.

EXAMPLE:

What would happen to the circuit in Figure 11-35 if the power supply transformer (T_1) developed the following problems:

a. An open primary or secondary winding
b. A shorted primary or secondary winding
c. A short between the primary winding and its grounded case, or between the secondary and its grounded case

Solution:

a. An open in the primary or secondary winding will cause the power supply's output voltage to drop to zero. Making a few checks with the voltmeter, you will find that you have primary ac voltage but no secondary ac voltage. If you unplug the power supply and disconnect the transformer and check the resistance of its windings, you will find the open winding using the ohmmeter (it will have an infinite resistance).

b. A shorted primary or secondary will provide a low resistance path, resulting in an excessive primary current and a blown fuse. If the fuse does not blow because the current is not quite high enough, the shorted transformer will produce almost no output voltage and be generating a lot of heat. Once again, by unplugging and disconnecting the transformer and then checking its primary and secondary winding resistance, you should find the short (the transformer in Figure 11-35 should typically have a primary resistance of 50 Ω, and a secondary resistance of about 10 Ω).

c. If either the primary winding or secondary winding were to short to the transformer's casing (which is generally grounded), the result will be a low-resistance path to ground, resulting in an excessive primary current and a blown fuse. Once again, by unplugging and disconnecting the transformer and then checking the winding-to-case resistance for both the primary and secondary, you should be able to find out if a winding has shorted to the case. This winding-to-case resistance should normally be infinite ohms.

Once you have determined that the transformer has developed an internal problem, you will have to get a replacement because these problems cannot be repaired. Be sure that the replacement has the same transformer rating.

EXAMPLE:

What would happen to the circuit in Figure 11-35 if one of the rectifier diodes were to open in the

a. Half-wave rectifiers
b. Full-wave rectifier

Solution:

a. An open rectifier diode is quite a common power supply problem. If a half-wave rectifier diode such as D_1 or D_2 in Figure 11-35 were to open, the symptoms would be easy to diagnose because no pulsating dc would be applied to the filters and regulators, and there would not be a −12 or +12 V output from the power supply. Like all opens, all of the supply voltage (V_{s2}) will appear across the defective device's terminals when measured with the multimeter.

b. If one of the diodes in the bridge rectifier in Figure 11-35 were to open, it would prevent current flow for one-half of the input cycle. This would result in a half-wave output that would be difficult to filter, and if the average voltage into U_1 is less than +5 V, the regulator would not function, producing no +5 V output.

EXAMPLE:

What would happen to the circuit in Figure 11-35 if one of the rectifier diodes were to short in the

a. Half-wave rectifiers
b. Full-wave rectifier

Solution:

a. A shorted half-wave rectifier diode, such as D_1 or D_2 in Figure 11-35, would connect ac to the filter capacitor, causing the filter capacitor to charge and discharge. Because a constant dc voltage is not being applied to the regulator, there would be no dc output voltage.
b. If a diode, such as *B,* in the bridge rectifier module BR_1 in Figure 11-35 were to short, it would act like a forward biased diode when diodes *B* and *C* are turned ON. However, when diodes *A* and *D* are forward biased, the ON *D* and shorted *B* will short the secondary winding, causing an excessive current. It is likely that this high current will burn open the shorted diode *B* and the ON diode *D.*

EXAMPLE:

What would happen to the circuit in Figure 11-35 if a filter capacitor were leaky or were to short?

Solution:

A shorted or leaky filter capacitor will generally cause an excessive current that will usually burn open the rectifier diodes, producing no dc output. If the leakage is not too large, the problem will be an increase in the ripple voltage into the regulator.

EXAMPLE:

What would happen to the circuit in Figure 11-35 if a regulator's output current rating was exceeded?

Solution:

Referring to Figure 11-35, you can see that each output of the regulator has a rated voltage and maximum load current value. If this current is exceeded due to a short in the load, the regulator will shut down to protect itself. These built-in short-circuit protection and thermal shutdown circuits normally prevent the regulator from damage. In most cases, if the regulator's input is fine but there is no output voltage, the regulator has probably shut itself down, due to a short in the load or an insufficient heat sink. By disconnecting the output of the regulator from the load, you should be able to determine if the problem is the source or the load.

Step 3: Repair

The final step is to repair the circuit, which could involve simply removing a wire clipping or some excess solder, resoldering a broken connection, reconnecting a connector, or some other easy repair. In most instances, however, the repair will involve the replacement of a faulty component. For a circuit that has been constructed on a breadboard or prototyping board, the removal and replacement of the component is simple. However, when a printed circuit board is involved you should make a note of the component's orientation (in the case of rectifier diodes, rectifier modules, electrolytic capacitors, IC regulators, and other similar devices), and observe good desoldering and soldering practices.

When the circuit has been repaired, always perform a final test to see that the circuit is now fully operational.

SELF-TEST EVALUATION POINT FOR SECTION 11-5

Use the following questions to test your understanding of Section 11-5.

1. What are the four blocks that make up a dc power supply?
2. What is the function of
   a. A rectifier **b.** A regulator
3. Why is a filter needed in a dc power supply?
4. What is the three-step troubleshooting procedure?

REVIEW QUESTIONS

Multiple-Choice Questions

1. What is the barrier voltage for a silicon junction diode?
   a. 0.3 V **b.** 0.4 V **c.** 0.7 V **d.** 2.0 V
2. Which of the following junction diodes are forward biased?
   a. Anode = +7 V, cathode = +10 V
   b. Anode = +5 V, cathode = +3 V
   c. Anode = +0.3 V, cathode = +5 V
   d. Anode = −9.6 V, cathode = −10 V
3. The junction diode ____________ current when it is forward biased, and ____________ current when it is reverse biased.
   a. Blocks, conducts
   b. Conducts, passes
   c. Blocks, prevents
   d. Conducts, blocks
4. The *n*-type region of a junction diode is connected to the ____________ terminal and the *p*-type region is connected to the ____________.
   a. Cathode, anode **b.** Anode, cathode
5. Semiconductor devices need voltages of a certain amplitude and polarity to control their operation. These voltages are called:
   a. Barrier potentials **c.** Knee voltages
   b. Depletion voltages **d.** Bias voltages
6. When reverse biased, a junction diode has a leakage current passing through it which is typically measured in:
   a. Amps **b.** Milliamps **c.** Microamps **d.** Kiloamps
7. What happens to the forward voltage drop across the diode (V_F) if temperature increases? V_F will:
   a. Decrease **c.** Remain the same
   b. Increase **d.** Be unpredictable
8. When forward biased, a junction diode is equivalent to a/an ____________ switch, whereas when it is reverse biased it is equivalent to a/an ____________ switch.
   a. Open, closed **c.** Open, open
   b. Closed, closed **d.** Closed, open
9. The black band on a diode's package is always closest to the ____________.
   a. Anode **c.** *p*-type material
   b. Cathode **d.** Both (a) and (c) are true
10. When current dramatically increases, the voltage point on the diode's forward *V-I* characteristic curve is called the:
    a. Breakdown voltage **c.** Barrier voltage
    b. Knee voltage **d.** Both (b) and (c) are true
11. When current dramatically increases, the voltage point on the diode's reverse *V-I* characteristic curve is called the:
    a. Breakdown voltage **c.** Barrier voltage
    b. Knee voltage **d.** Both (b) and (c) are true
12. A rectifier is:
    a. A circuit that converts ac to dc
    b. An analog circuit
    c. A two-state decision-making circuit
    d. A circuit that converts dc to ac
13. What resistance should a good diode have when it is reverse biased?
    a. Less than 10 Ω
    b. More than 1000 MΩ
    c. Between 120 Ω and 1.2 kΩ
    d. Both (b) and (c) are true
14. What resistance should a good diode have when it is forward biased?
    a. Less than 10 Ω
    b. More than 1000 MΩ
    c. Between 120 Ω and 1.2 kΩ
    d. Both (b) and (c) are true
15. The ____________ diode is designed to withstand high reverse currents that result when the diode is operated in the reverse breakdown region.
    a. Basic P-N junction **c.** Zener
    b. Light emitting **d.** Both (a) and (c) are true
16. When a zener diode's breakdown voltage is exceeded, the reverse current through the diode increases from a small leakage value to a high reverse current value.
    a. True
    b. False

17. When operating in the reverse breakdown region, the ____________ the zener will vary over a wide range, while the ____________ the zener will vary by only a small amount.
   a. Voltage drop across, forward current through
   b. Forward current through, voltage drop across
   c. Reverse current through, forward current through
   d. Reverse current through, forward drop across
18. The zener diode's symbol is different from all other diode symbols due to its:
   a. Z shaped cathode bar c. Straight bar cathode
   b. Two exiting arrows d. None of the above
19. The ability of a zener diode to maintain a relatively constant ____________ regardless of variations in zener ____________ is the key characteristic of a zener diode.
   a. Current, voltage
   b. Impedance, voltage
   c. Current, impedance
   d. Voltage, current
20. A 12 V, ± 5% zener diode will have a voltage drop in the ____________ range.
   a. 10.8 V to 13.2 V c. 5.04 V to 18.96 V
   b. 11.4 V to 12.6 V d. 11.88 V to 12.12 V
21. A 9.1 V zener diode has a power rating of 10 W. What is the diode's value of maximum zener current?
   a. 1.1 mA b. 91 mA c. 1.1 A d. None of the above
22. A/an ____________ circuit maintains the output voltage of a voltage source constant despite variations in the input voltage and the load resistance.
   a. Encoder c. Comparator
   b. Logic gate d. Voltage regulator
23. The zener diode is able to maintain the voltage drop across its terminals constant by continually changing its ____________ in response to a change in input voltage.
   a. Impedance c. Power rating
   b. Voltage d. Both (a) and (c) are true
24. In a voltage regulator circuit, the voltage developed across the zener diode remains constant and therefore any changes in the input voltage must appear across the:
   a. Load c. Source terminals
   b. Series resistor d. Zener diode
25. The light emitting diode is a semiconductor device that converts ____________ energy into ____________ energy.
   a. Chemical, electrical c. Electrical, light
   b. Light, electrical d. Heat, electrical
26. The ____________ lead of an LED is distinguished from the other terminal by its longer lead, flattened lead, or its close proximity to the flat side of the case.
   a. Anode b. Cathode
27. The forward voltage drop across a typical LED is usually:
   a. 0.7 V b. 0.3 V c. 5 V d. 2.0 V
28. Since the forward voltage rating of an LED remains almost constant, the output power of an LED is directly proportional to its:
   a. Impedance c. Reverse voltage
   b. Forward current d. Both (a) and (c) are true
29. If all seven cathodes in a seven-segment display are connected to 0 V, the display is referred to as a common ____________ configuration and requires a ____________ voltage input to turn on an LED segment.
   a. Anode, LOW c. Anode, HIGH
   b. Cathode, LOW d. Cathode, HIGH
30. Electrical systems manage the flow of ____________, while electronic systems manage the flow of ____________.
   a. Information, power b. Power, information
31. The four main circuit blocks of a dc power supply listed in order from input to output are:
   a. Transformer, rectifier, filter, regulator
   b. Filter, regulator, rectifier, transformer
   c. Transformer, rectifier, regulator, filter
   d. Rectifier, filter, regulator, transformer
32. Which of the four circuit blocks of a dc power supply converts an ac input into a pulsating dc output?
   a. Transformer c. Filter
   b. Regulator d. Rectifier
33. Which of the four circuit blocks of a dc power supply converts a high ac voltage into a low ac voltage?
   a. Transformer c. Filter
   b. Regulator d. Rectifier
34. Which of the four circuit blocks of a dc power supply maintains the output voltage constant despite variations in the ac input and output load?
   a. Transformer c. Filter
   b. Regulator d. Rectifier
35. Which of the four circuit blocks of a dc power supply converts a pulsating dc input into a steady dc output?
   a. Transformer c. Filter
   b. Regulator d. Rectifier
36. Which rectifier circuit can be used to generate a negative pulsating dc output?
   a. Half-wave c. Bridge
   b. Center-tapped d. All of the above
37. Which rectifier uses four diodes?
   a. Half-wave c. Bridge
   b. Center-tapped d. All of the above
38. A typical dc power supply will supply a ____________ -voltage ____________-current output.
   a. Low, high c. Low, low
   b. High, low d. High, high
39. Which is typically the most commonly used filter for a dc power supply?
   a. Pi b. *RC* c. Capacitive d. *LC*
40. What rating is used to measure the action of a filter?
   a. Percent rectification
   b. Percent ripple
   c. Percent regulation
   d. Percent filtration
41. A small load resistance will cause a ____________ load current and possibly pull ____________ the source voltage.
   a. Small, down c. Small, up
   b. Large, up d. Large, down

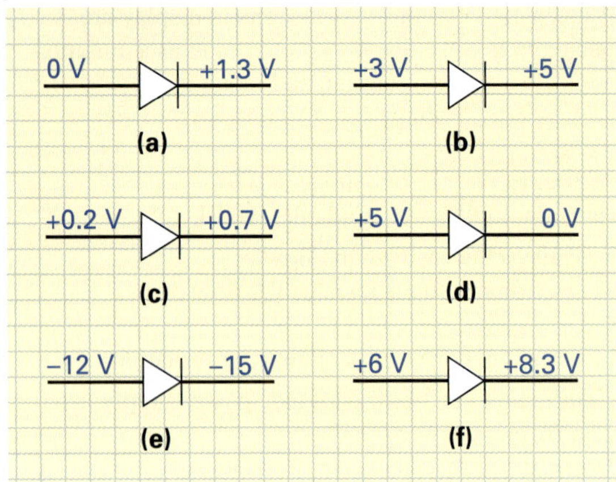

FIGURE 11-39 **Biased Junction Diodes.**

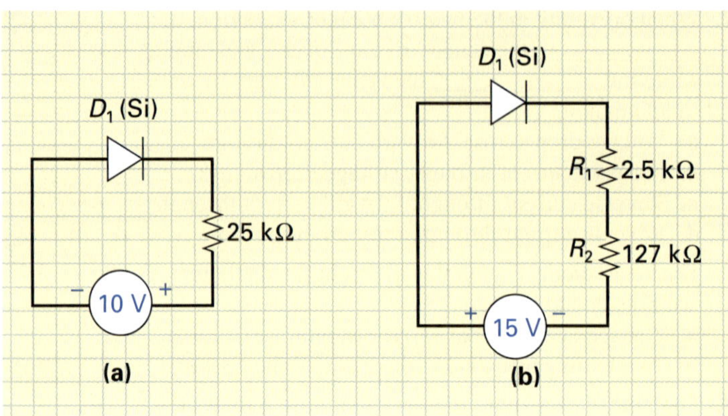

FIGURE 11-40 **Forward Current Examples.**

42. What would be the output voltage of a 7915 IC regulator?
  a. +5 V **c.** −5 V
  b. +15 V **d.** −15 V

43. A _____________ diode would typically be used in a power supply as a power indicator.
  a. Rectifier **c.** Zener
  b. Light-emitting **d.** Junction

44. A _____________ diode would typically be used as a regulator.
  a. Rectifier **c.** Zener
  b. Light-emitting **d.** Junction

Practice Problems

45. Which of the silicon diodes in Figure 11-39 are forward biased and which are reverse biased?

46. Calculate I_F for the circuits in Figure 11-40.

47. What would be the voltage drop across each of the diodes in Figure 11-40?

48. What would be the voltage drop across each of the resistors in Figure 11-40?

49. In reference to the polarity of the applied voltage, which of the circuits in Figure 11-41 are correctly biased for normal zener operation?

50. In reference to the magnitude of the applied voltage, which of the circuits in Figure 11-41 are correctly biased for normal zener operation?

51. Calculate the value of circuit current for each of the zener diode circuits in Figure 11-41.

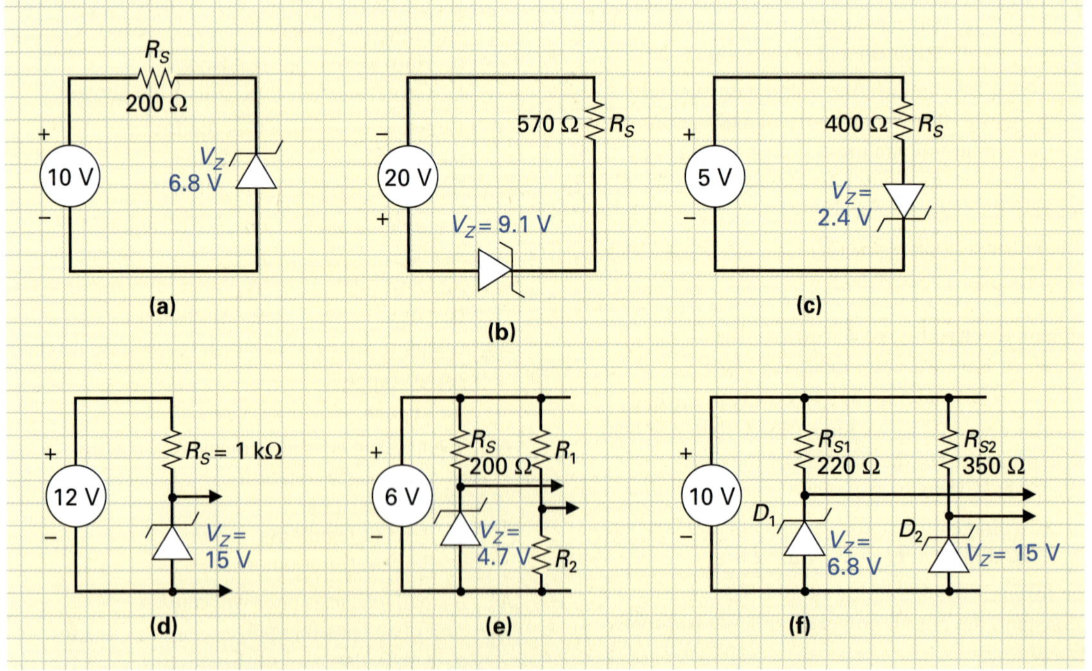

FIGURE 11-41 **Biasing Voltage Polarity and Magnitude.**

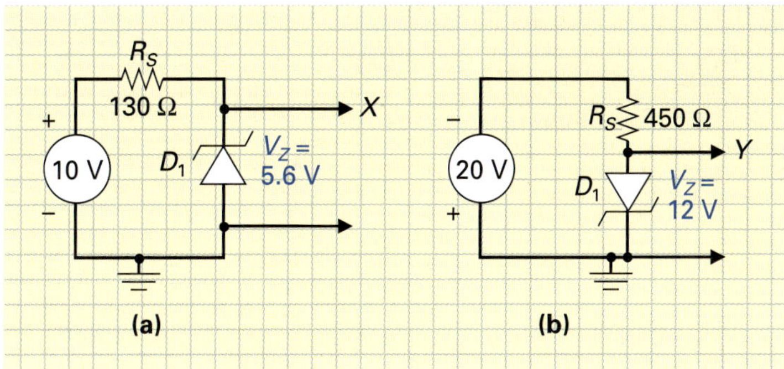

FIGURE 11-42 **Voltage Regulator Circuits (Unloaded).**

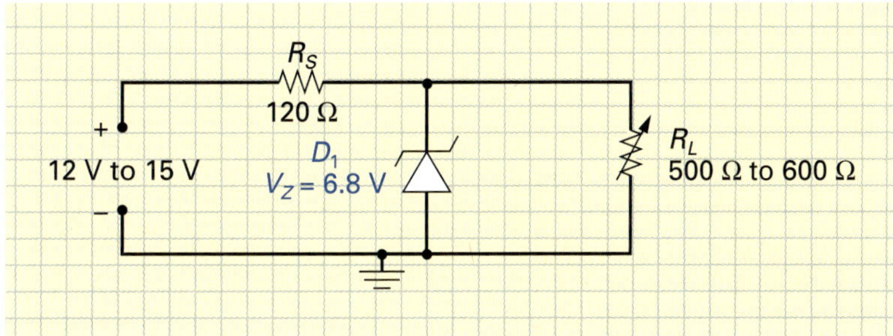

FIGURE 11-43 **A Voltage Regulator Circuit (Loaded).**

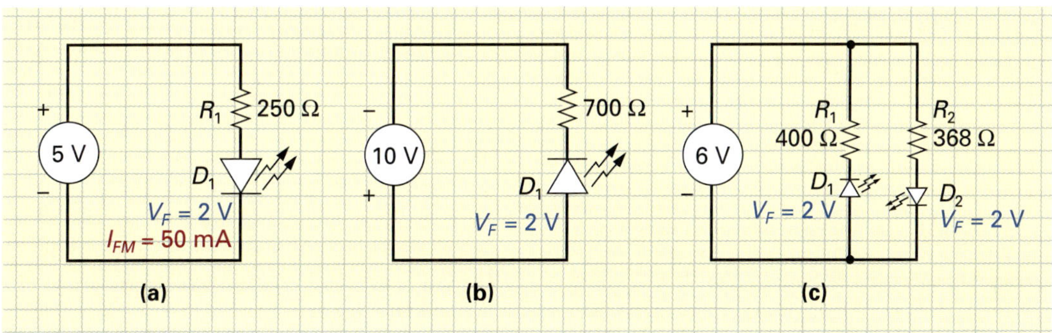

FIGURE 11-44 **Biasing Light Emitting Diodes.**

52. Would a 1 watt zener diode have a suitable power dissipation rating for the circuit in Figure 11-41(a)?

53. What wattage or maximum power dissipation rating would you choose for the zener diode in Figure 11-41(e)?

54. What would be the regulated output voltage and polarity at points *X* and *Y* for the circuits shown in Figure 11-42(a) and (b)?

55. Knowing that the zener diode must always be reverse biased, how could we change the circuit connection in Figure 11-42(a) and (b) to obtain opposite polarity supply voltages?

56. Calculate the value of circuit current for both the circuits shown in Figure 11-43.

57. Calculate the value of I_{RS}, I_Z, I_{RL}, and V_{RL} for the two extreme low and high voltages shown in Figure 11-43. Assume R_L remains constant at 500 Ω.

58. Calculate the values of I_{RS}, I_{RL}, and I_Z for all values of maximum and minimum input voltage and load resistance for the circuit in Figure 11-43.

59. Which of the light-emitting diodes in Figure 11-44 are biased correctly?

60. Calculate the value of circuit current for each of the LEDs in Figure 11-44.

61. Would a maximum forward current rating of 18 mA be adequate for the LED in Figure 11-44(a)?

62. Would a maximum reverse voltage rating of 5 V be adequate for the LEDs in Figure 11-44(c)?

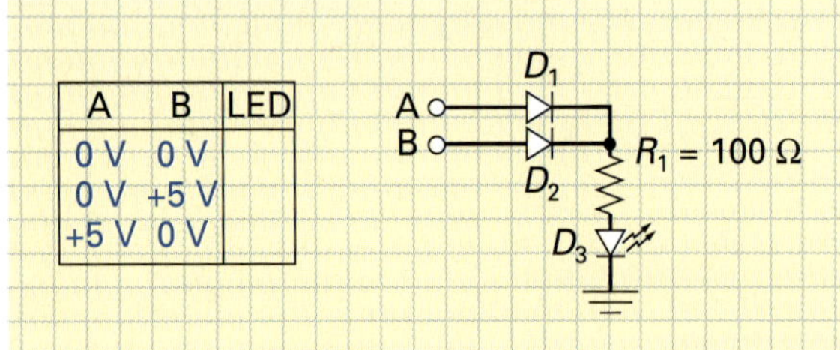

FIGURE 11-45 **An OR Gate Circuit with an LED Output.**

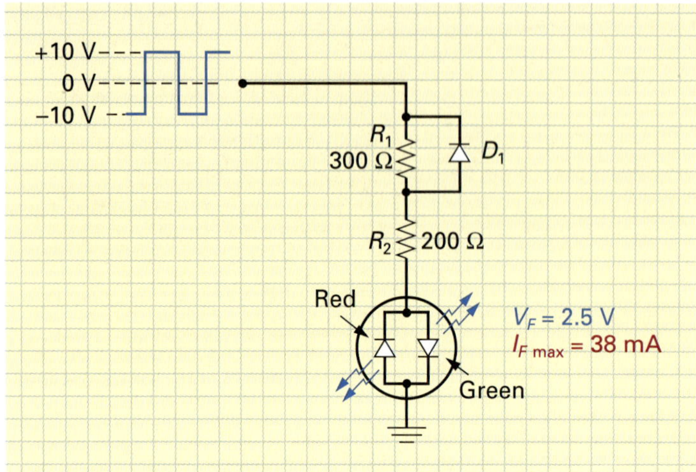

FIGURE 11-46 **A Bi-Color Output Display Circuit.**

63. Figure 11-45 shows how two basic P-N junction diodes and a resistor can be used to construct an OR gate circuit. In this circuit, an LED has been connected to the output so that any HIGH or positive 5 V output will turn ON the LED. Considering the *A* and *B* input combinations shown in the table, indicate whether the LED will be ON or OFF for each of these input conditions.

64. Assuming a 0.7 V voltage drop across the P-N junction diode and a 2 V voltage drop across the LED, what would be the value of current through the LED if one of the inputs in Figure 11-45 was HIGH (+5 V)?

65. Which of the bi-color LEDs will be ON in Figure 11-46 when the input voltage is +10 V, and which will be ON when the input voltage is −10 V?

66. In Figure 11-46, a basic P-N junction diode (D_1) has been included across R_1 so that when the input voltage goes negative this diode will bypass the additional current-limiting resistor R_1. When the input voltage is positive, D_1 is reversed biased and therefore both R_1 and R_2 will limit the value of series current. Which of the colors in the bi-color LED will be brighter?

67. Calculate the value of green and red LED current for the circuit in Figure 11-46.

68. In Figure 11-47 a comparator is used to compare 0 V (at the negative input) to a sine wave which varies between +5 V and −5 V (at the positive input). When the comparator's inputs are "true" (positive input is positive with respect to negative input), the comparator's output will be switched HIGH (+12 V). When the comparator's inputs are "not true" (positive input is negative with respect to the negative input), the comparator's output will be switched LOW (−12 V). What value of series current-limiting resistor should be used to ensure that the LEDs are bright, but do not burn out, for the +12 V and −12 V source voltages? Aim for a forward current that is about 90% of the maximum rating.

69. Would the 16-LED display in Figure 11-48 (D_{11}–D_{14}) be classified as a common-anode or common-cathode display?

70. In regard to Figure 11-48, construct a table to show the HIGH and LOW encoder outputs on *A, B, C,* and *D* for each of the rotary switch positions. Also show in the table which of the LEDs are ON or OFF for each of these codes.

71. In Figure 11-49, seven switches are used to turn ON and OFF the seven LEDs in a seven-segment display. Construct a table to show which switches will be CLOSED and which will be OPEN to display the digits 0 through 9 on the seven-segment display.

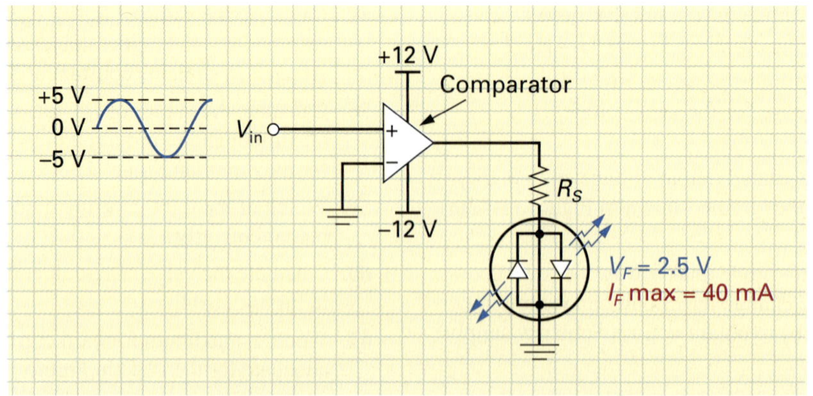

FIGURE 11-47 **A Comparator Circuit with a Bi-Color Output Display.**

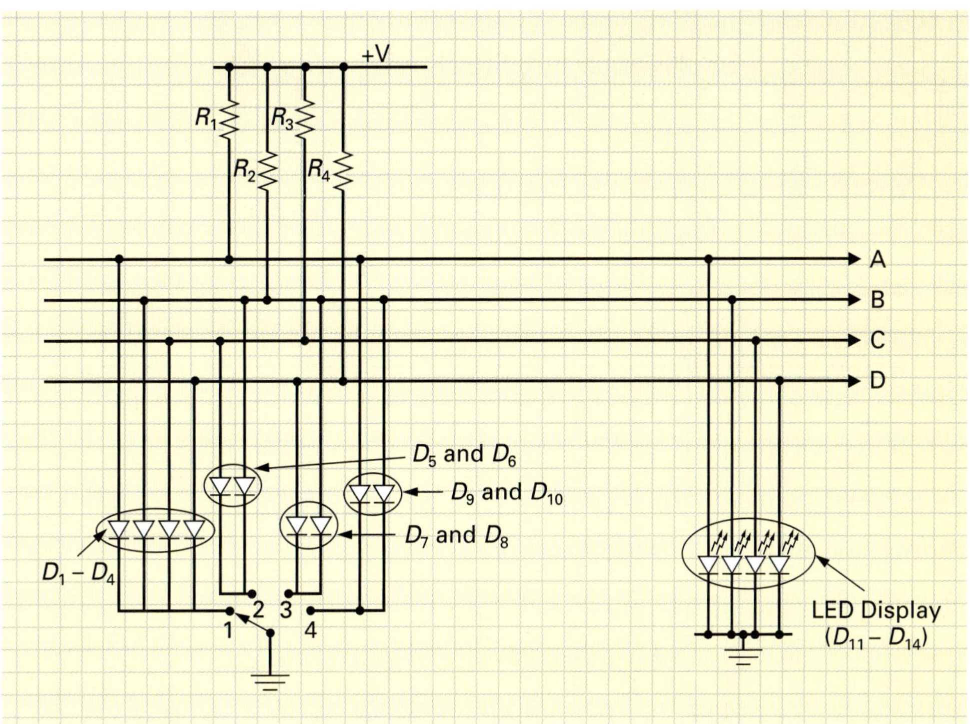

FIGURE 11-48 **A Switch Encoder Circuit with an LED Output Display.**

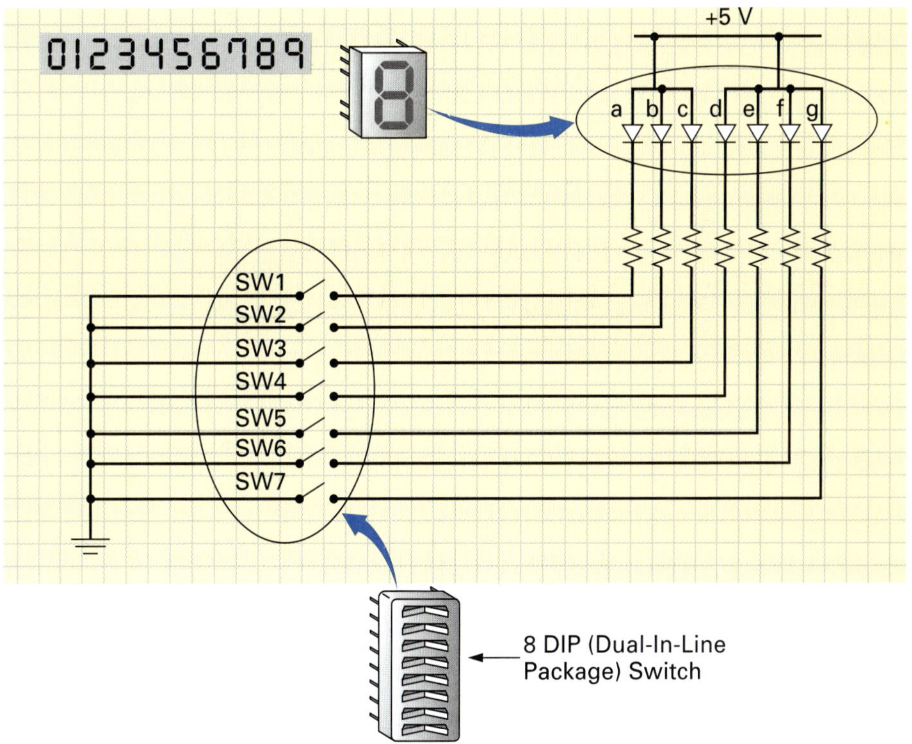

FIGURE 11-49 **A Seven-Segment Display Circuit.**

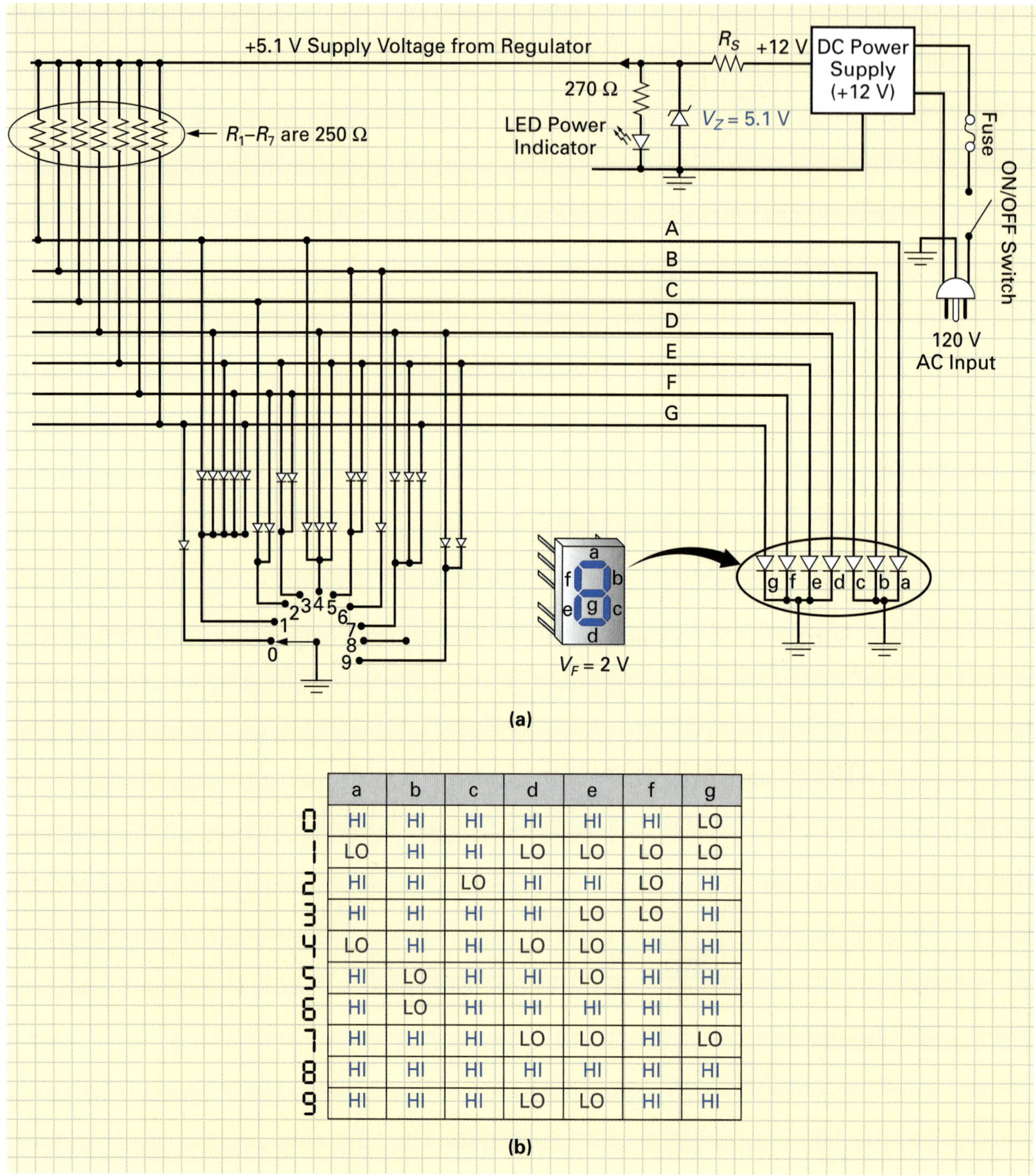

	a	b	c	d	e	f	g
0	HI	HI	HI	HI	HI	HI	LO
1	LO	HI	HI	LO	LO	LO	LO
2	HI	HI	LO	HI	HI	LO	HI
3	HI	HI	HI	HI	LO	LO	HI
4	LO	HI	HI	LO	LO	HI	HI
5	HI	LO	HI	HI	LO	HI	HI
6	HI	LO	HI	HI	HI	HI	HI
7	HI	HI	HI	LO	LO	HI	LO
8	HI	HI	HI	HI	HI	HI	HI
9	HI	HI	HI	LO	LO	HI	HI

FIGURE 11-50 A Seven-Segment Encoder and Display Circuit.

72. Are the seven-segment displays in Figure 11-49 and Figure 11-50 common-anode or common-cathode types?

73. Figure 11-50(a) shows how a 5.1 V zener voltage regulator circuit can be used to supply power to a seven-segment encoder and display circuit. Figure 11-50(b) shows how the LOW and HIGH codes generated by the encoder circuit will turn ON and OFF the necessary LEDs in the seven-segment display to produce the digits 0 through 9. For example, when the rotary switch is in position 0, a single P-N junction diode makes line *G* LOW. This will turn segment "g" of the seven-segment display OFF, while the pull up resistors R_1 through R_6 will make lines *A* through *F* HIGH and therefore turn ON segments "a" through "f." This condition is shown in the first row of the table in Figure 11-50(b).

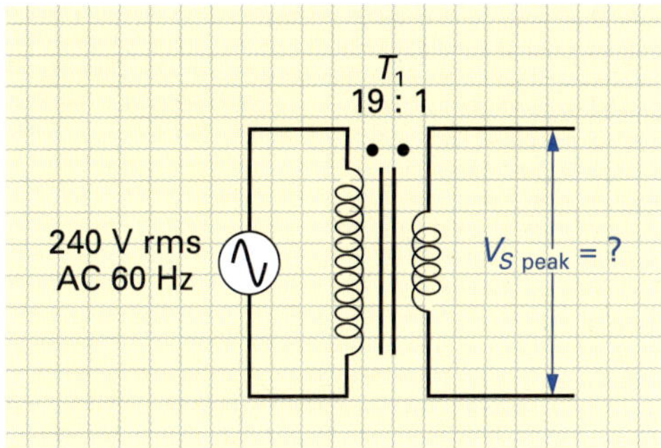

FIGURE 11-51

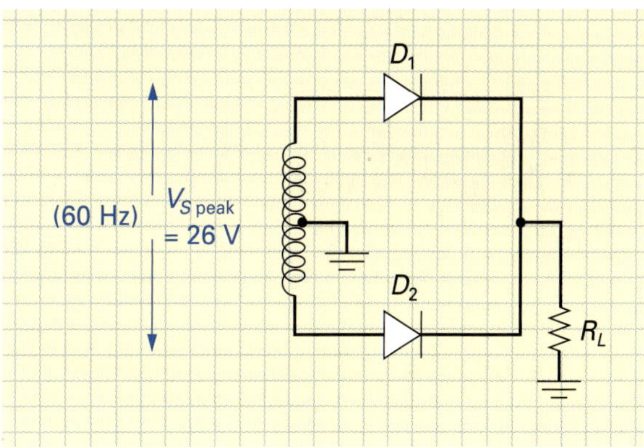

FIGURE 11-52

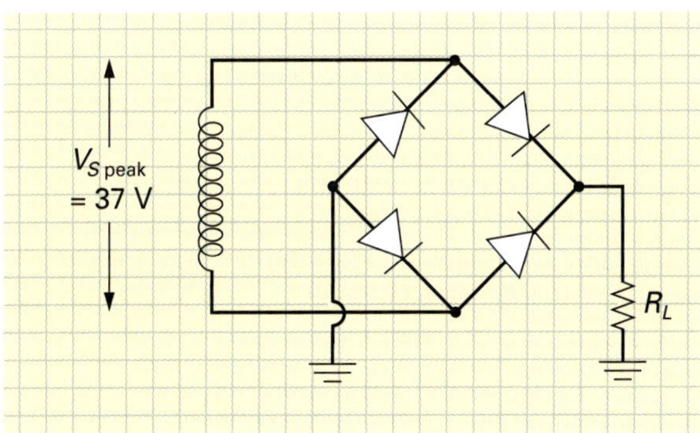

FIGURE 11-53

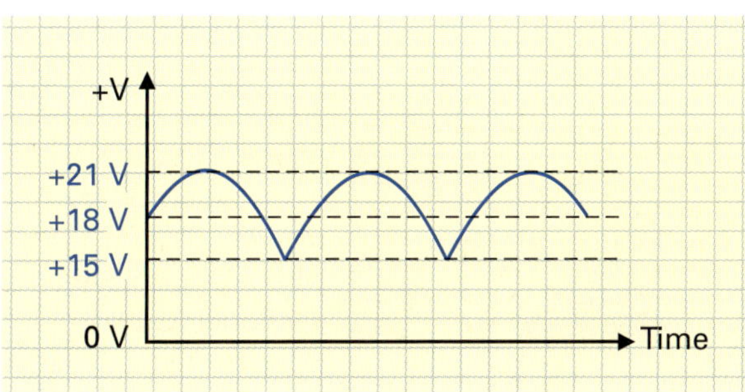

FIGURE 11-54

Calculate the following encoder circuit details:

a. The amount of current that will be drawn by each of the basic P-N junction diodes in the encoder circuit.

b. What rotary switch position will cause the maximum value of current, and what will this current value be?

Calculate the following display circuit details:

c. The amount of current that will be drawn by each of the LEDs in the seven-segment display.

d. What digit on the seven-segment display will cause the maximum value of current, and what will this current value be?

Calculate the following encoder-display circuit details:

e. Which path draws more current: an encoder diode or a display diode?

f. When the digit 8 is being displayed, a total of 99.2 mA of current is being drawn from the 5.1 V source. When the digit 1 is being displayed, a total of 112.8 mA is being drawn from the 5.1 V source. Because the source only sees the encoder-display circuit as a single load resistance, what will this resistance be equal to?

74. If a 240 V rms ac 60 Hz input is applied to the 19:1 step-down transformer shown in Figure 11-51, what would be the peak secondary output voltage?

75. Calculate the peak output voltage (taking into account V_{diode}) for the circuit shown in Figure 11-52.

76. What would be the output ripple frequency from the circuit in Figure 11-52?

77. Calculate the average output voltage of the bridge rectifier circuit shown in Figure 11-53, taking into account the diode voltage drop.

78. A capacitive filter produces the output shown in Figure 11-54. What is the filter's percent of ripple?

79. If a regulator delivers +12 V when no load is connected, and +10.6 V when a full load is connected, what would be the regulator's percent of regulation?

Web Site Questions

Go to the web site http://www.prenhall.com/cook, select the textbook *Electronics: A Complete Course,* select this chapter, and then follow the instructions when answering the multiple choice practice problems.

Bipolar Junction Transistors (BJTs)

There's No Sleeping When He's Around!

Carl Friedrich Gauss was born April 30, 1777, to poor, uneducated parents in Brunswick, Germany. He was a child of precocious abilities, particularly in mental computation. In elementary school he soon impressed his teachers, who said that mathematical ability came easier to Gauss than speech.

In secondary school he rapidly distinguished himself in ancient languages and mathematics. At 14, Gauss was presented to the court of the duke of Brunswick, where he displayed his computing skill. Until his death in 1806, the duke generously supported Gauss and his family, encouraging the boy with textbooks and a laboratory.

In the early years of the nineteenth century, Gauss's interest was in astronomy, and his accumulated work on celestial mechanics was published in 1809. In 1828, at a conference in Berlin, Gauss met physicist Wilhelm Weber, who would eventually become famous for his work on electricity. They worked together for many years and became close friends, investigating electromagnetism and the use of a magnetic needle for current measurement. In 1833 they constructed an electric telegraph system that could communicate across Göttingen from Gauss's observatory to Weber's physics laboratory. (This telegraph system of communication was later developed independently by U.S. inventor Samuel Morse.)

Gauss conceived almost all of his fundamental mathematical discoveries between the ages of 14 and 17. There are many stories of his genius in his early years, one of which involved a sarcastic teacher who liked giving his students long-winded problems and then resting, or on some occasions sleeping, in class. On his first day with Gauss, who was 8 years old, the teacher began, as usual, by telling the students to find the sum of all the numbers from 1 to 100. The teacher barely had a chance to sit down before Gauss raised his hand and said "5050." The dumbfounded teacher, who believed Gauss must have heard the problem before and memorized the answer, asked Gauss to explain how he had solved the problem. He replied: "The numbers 1, 2, 3, 4, 5, and so on to 100 can be paired as 1 and 100, 2 and 99, 3 and 98, and so on. Since each pair has a sum of 101, and there are 50 pairs, the total is 5050."

12

Introduction

In 1948, a component known as a transistor sparked a whole new era in electronics, the effects of which have not been fully realized even to this day. A transistor is a three-element device made of semiconductor materials used to control electron flow, the amount of which can be controlled by varying the voltages applied to its three elements. Having the ability to control the amount of current through the transistor allows us to achieve two very important applications: switching and amplification.

Like the diode, transistors are formed by *p* and *n* regions and, as we are already aware, the point at which a *p* and an *n* region join is known as a junction. Transistors in general are classified as being either the *bipolar* or *unipolar* type. The bipolar type has two P-N junctions, while unipolar transistors have only one P-N junction. In this chapter we will study all of the details relating to the *bipolar* transistor, or as it is also known, the *bipolar junction transistor* or *BJT.*

12-1 INTRODUCTION TO THE TRANSISTOR

In most cases it is easier to build a jigsaw puzzle when you can refer to the completed picture on the box. The same is true whenever anyone is trying to learn anything new, especially a science that contains many small pieces. This introduction to the transistor is a means for you to quickly see the complete picture without having to wait until you connect all of the pieces, and will cover the transistor's basic construction, schematic symbol, physical appearance, basic operation, and main applications.

12-1-1 *Transistor Types (NPN and PNP)*

Like the diode, a bipolar transistor is constructed from a semiconductor material. However, unlike the diode, which has two oppositely doped regions and one P-N junction, the transistor has three alternately doped semiconductor regions and two P-N junctions. These three alternately doped regions are arranged in one of two different ways, as shown in Figure 12-1.

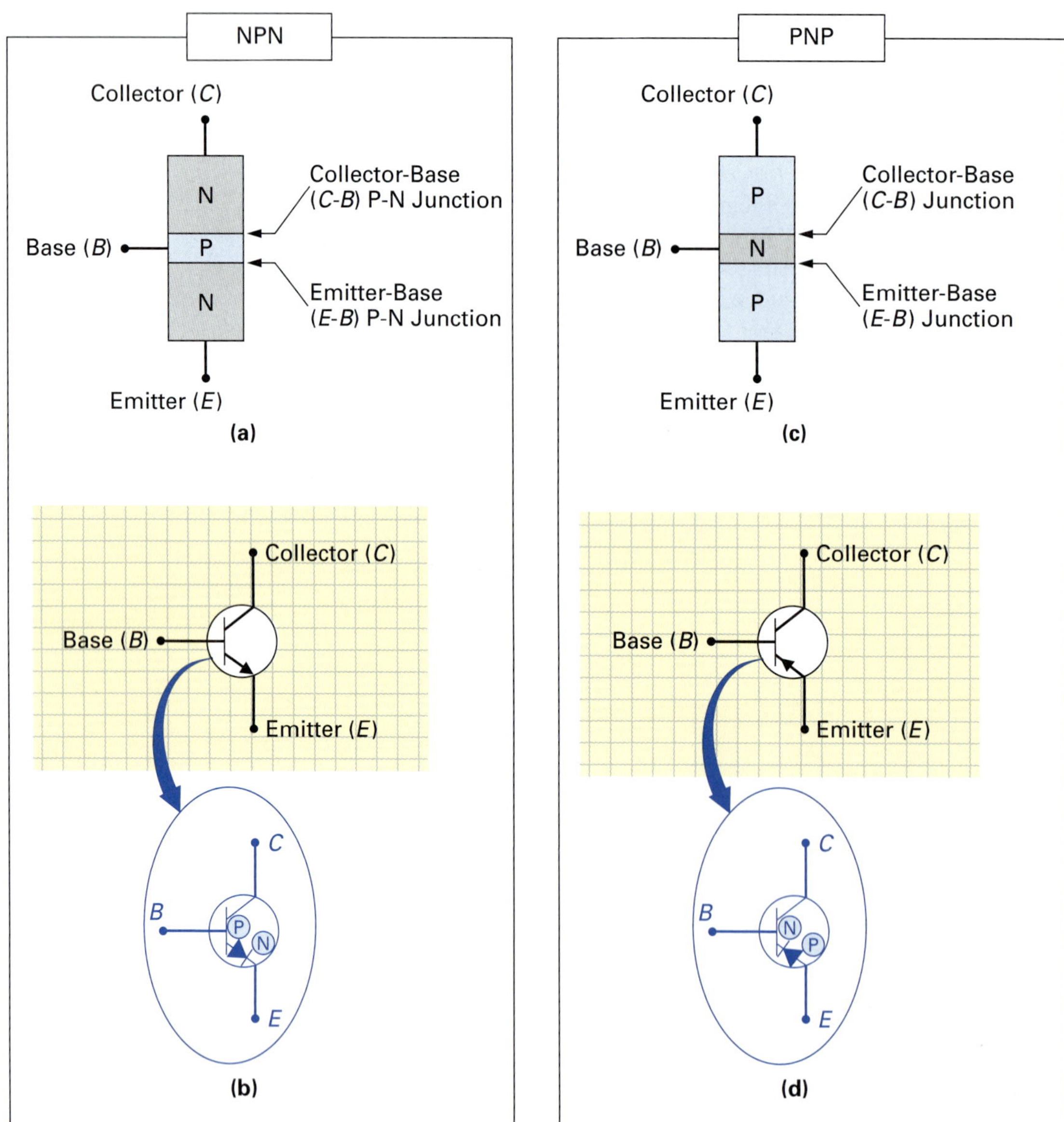

FIGURE 12-1 Bipolar Junction Transistor (BJT) Types.

With the **NPN transistor** shown in Figure 12-1(a), a thin, lightly doped *p*-type region known as the **base** (symbolized *B*) is sandwiched between two *n*-type regions called the **emitter** (symbolized *E*) and the **collector** (symbolized *C*). Looking at the NPN transistor's schematic symbol in Figure 12-1(b), you can see that an arrow is used to indicate the emitter lead. As a memory aid for the NPN transistor's schematic symbol, you may want to remember that when the emitter arrow is "**N**ot **P**ointing i**N**" to the base, the transistor is an **"NPN."** An easier method is to think of the arrow as a diode, with the tip of the arrow or cathode pointing to an *n* terminal and the back of the arrow or anode pointing to a *p* terminal, as seen in the inset in Figure 12-1(b).

The **PNP transistor** can be seen in Figure 12-1(c). With this transistor type, a thin, lightly doped *n*-type region (base) is placed between two *p*-type regions (emitter and collector). Figure 12-1(d) illustrates the PNP transistor's schematic symbol. Once again, if you think of the emitter arrow as a diode, as shown in the inset in Figure 12-1(d), the tip of the arrow or cathode is pointing to an *n* terminal and the back of the arrow or anode is pointing to a *p* terminal.

NPN Transistor
A thin, lightly doped *p*-type region (base) is sandwiched between two *n*-type regions (emitter and collector).

Base
The region that lies between an emitter and a collector of a transistor and into which minority carriers are injected.

Emitter
A transistor region from which charge carriers are injected into the base.

Collector
A semiconductor region through which a flow of charge carriers leaves the base of the transistor.

PNP Transistor
A thin, lightly doped *n*-type region (base) is placed between two *p*-type regions (emitter and collector).

12-1-2 *Transistor Construction and Packaging*

Like the diode, the three layers of an NPN or PNP transistor are not formed by joining three alternately doped regions. These three layers are formed by a "diffusion process," which first melts the base region into the collector region, and then melts the emitter region into the base region. For example, with the NPN transistor shown in Figure 12-2(a), the construction process would begin by diffusing or melting a *p*-type base region into the *n*-type collector region. Once this *p*-type base region is formed, an *n*-type emitter region is diffused or melted into the newly diffused *p*-type base region to form an NPN transistor. Keep in mind that manufacturers will generally construct thousands of these transistors simultaneously on a thin semiconductor wafer or disc, as shown in Figure 12-2(a). Once tested, these discs, which are about 3 inches in diameter, are cut to separate the individual transistors. Each transistor is placed in a package, as shown in Figure 12-2(b). The package will protect the transistor from humidity and dust, provide a means for electrical connection between the three semiconductor regions and the three transistor terminals, and serve as a heat sink to conduct away any heat generated by the transistor.

Figure 12-3 illustrates some of the typical low-power and high-power transistor packages. Most low-power, small-signal transistors are hermetically sealed in a metal, plastic, or

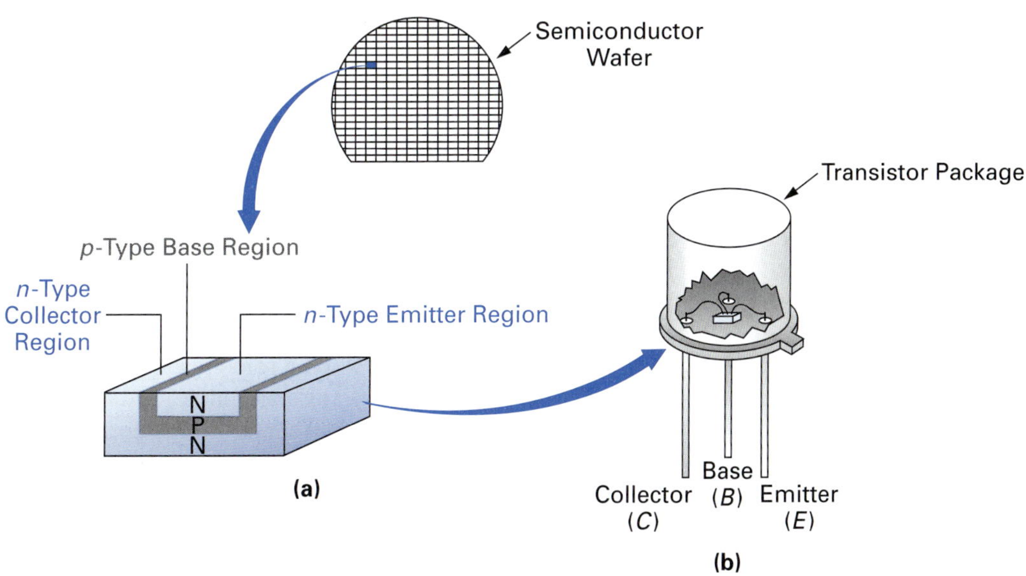

FIGURE 12-2 Bipolar Junction Transistor Construction and Packaging.

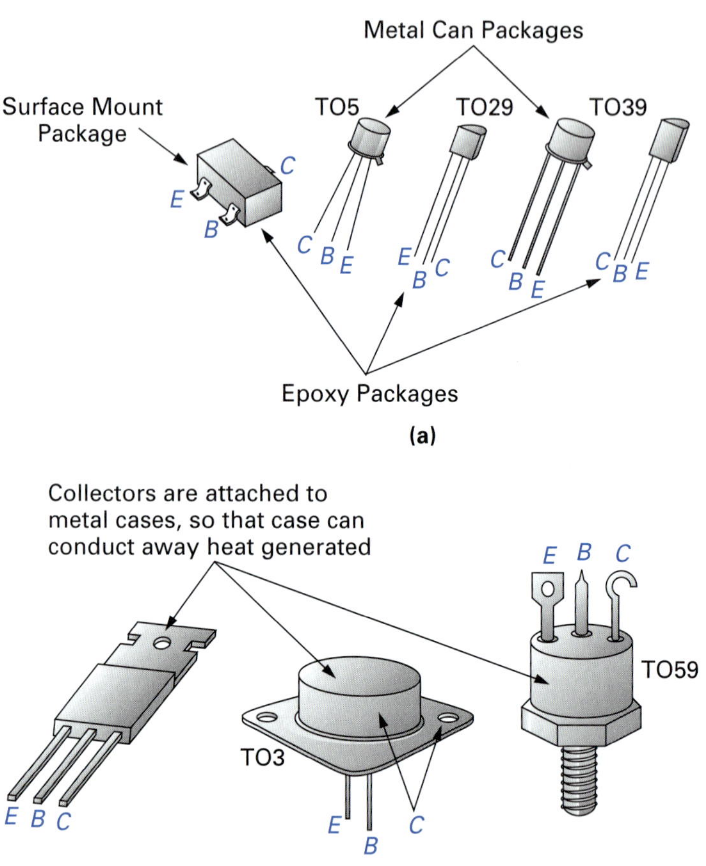

FIGURE 12-3 Bipolar Junction Transistor Package Types. (a) Low Power. (b) High Power.

epoxy package. Four of the low-power packages shown in Figure 12-3(a) have their three leads protruding from the bottom of the package because these package types are usually inserted and soldered into holes in printed circuit boards (PCBs). The surface mount technology (SMT) low-power transistor package, on the other hand, has flat metal legs that mount directly onto the surface of the PCB. These transistor packages are generally used in high component density PCBs because they use less space than a "through-hole" package. To explain this in more detail, a through-hole transistor package needs a hole through the PCB and a connecting pad around the hole to make a connection to the circuit. With an SMT package, however, no holes are needed, only a small connecting pad. Without the need for holes, pads on printed circuit boards can be smaller and placed closer together, resulting in considerable space savings.

The high-power packages, shown in Figure 12-3(b), are designed to be mounted onto the equipment's metal frame or chassis so that the additional metal will act as a heat sink and conduct the heat away from the transistor. With these high-power transistor packages, two or three leads may protrude from the package. If only two leads are present, the metal case will serve as a collector connection, and the two pins will be the base and emitter.

Transistor package types are normally given a reference number. These designations begin with the letters "TO," which stands for transistor outline, and are followed by a number. Figure 12-3 includes some examples of TO reference designators.

12-1-3 *Transistor Operation*

Figure 12-4 shows an NPN bipolar transistor, and the inset shows how a transistor can be thought of as containing two diodes: a *base-to-collector diode* and a *base-to-emitter diode.*

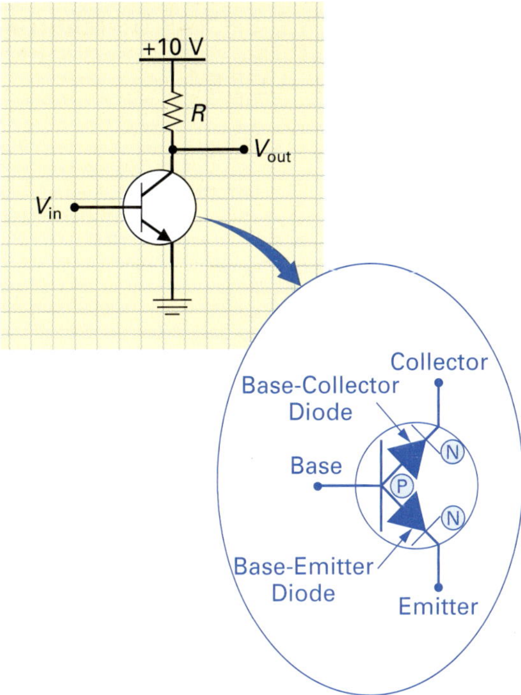

FIGURE 12-4 **The Base-Collector and Base-Emitter Diodes Within a Bipolar Transistor.**

With an NPN transistor, both diodes will be back-to-back and "**N**ot be **P**ointing i**N**" (NPN) to the base, as shown in the inset in Figure 12-4. For a PNP transistor, the base-collector and base-emitter diodes will both be pointing into the base.

Transistors are basically controlled to operate as a switch, or they are controlled to operate as a variable resistor. Let us now examine each of these operating modes.

The Transistor's ON/OFF Switching Action

Figure 12-5 illustrates how the transistor can be made to operate as a switch. This ON/OFF switching action of the transistor is controlled by the transistor's base-to-emitter (*B-E*) diode. If the *B-E* diode of the transistor is forward biased, the transistor will turn ON; if the *B-E* diode of the transistor is reverse biased, the transistor will turn OFF.

To begin with, let us see how the transistor can be switched ON. In Figure 12-5(a), the *B-E* diode of the transistor is forward biased (anode at base is +5 V, cathode at emitter is 0 V), and the transistor will turn ON. Its collector and emitter output terminals will be equivalent to a closed switch, as shown in Figure 12-5(b). This low resistance between the transistor's collector and emitter will cause a current (*I*), as shown in Figure 12-5(b). The output voltage in this condition will be zero volts because all of the +10 V supply voltage will be dropped across *R*. Another way to describe this would be to say that the low resistance path between the transistor's emitter and collector connects the zero volt emitter potential through to the output.

Now let us see how the transistor can be switched OFF. In Figure 12-5(c), the transistor has 0 V being applied to its base input. In this condition, the *B-E* diode of the transistor is reverse biased (anode at base is 0 V, cathode at emitter is 0 V), and so the transistor will turn OFF and its collector and emitter output terminals will be equivalent to an open switch, as shown in Figure 12-5(d). This high resistance between the transistor's collector and emitter will prevent any current and any voltage drop, resulting in the full +10 V supply voltage being applied to the output, as shown in Figure 12-5(d).

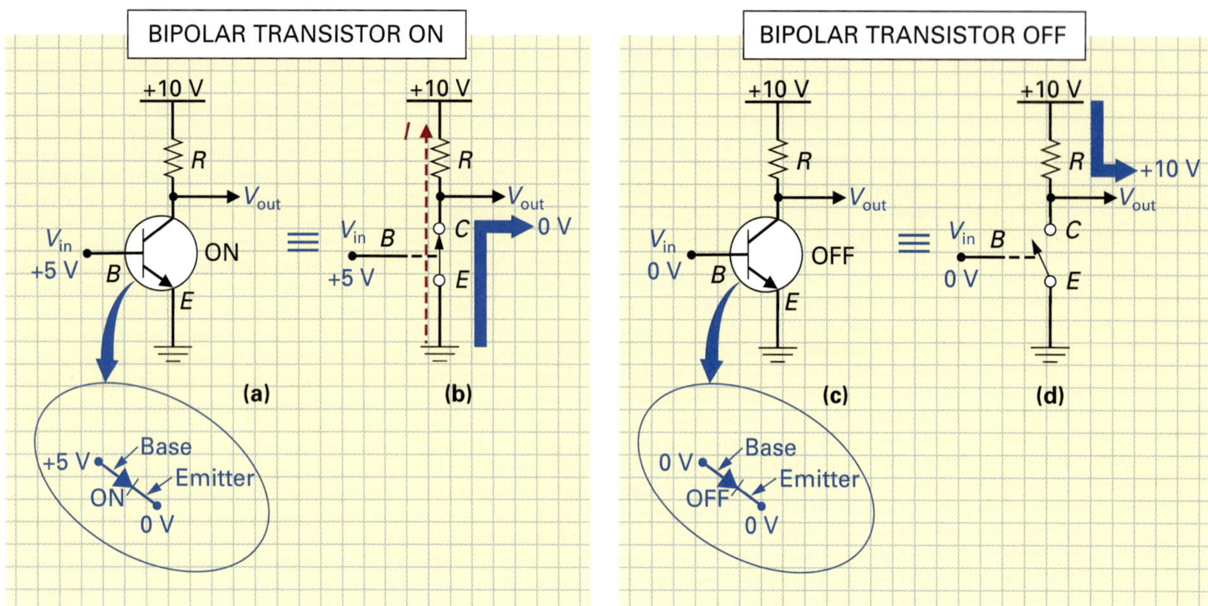

FIGURE 12-5 **The Bipolar Transistor's ON/OFF Switching Action.**

The Transistor's Variable-Resistor Action

In the previous section we saw how the transistor can be biased to operate in one of two states: ON or OFF. When operated in this two-state way, the transistor is being switched ON and OFF in almost the same way as a junction diode. The transistor, however, has another ability that the diode does not have—it can also function as a variable resistor, as shown in the equivalent circuit in Figure 12-6(a). In Figure 12-5 we saw how +5 V base input bias voltage would result in a low resistance between emitter and collector (closed switch) and how a 0 V base input bias would result in a high resistance between emitter and collector (open switch). The table in Figure 12-6(b) shows an example of the relationship between base input bias voltage (V_B) and emitter-to-collector resistance (R_{CE}). In this table, you can see that the transistor is not only going to be driven between the two extremes of fully ON and fully OFF. When the base input voltage is at some voltage level between +5 V and 0 V, the transistor is partially ON; therefore, the transistor's emitter-to-collector resistance is somewhere between 0 Ω and maximum Ω. For example, when $V_B = +4$ V, the transistor is not fully ON, and its emitter-to-collector resistance will be slightly higher, at 100 Ω. If the base input bias voltage is further reduced to +3 V, for example, you can see in the table that the emitter-to-collector resistance will further increase to 10 kΩ. Further decreases in base input voltage ($V_B\downarrow$) will cause further increases in emitter-to-collector resistance ($R_{CE}\uparrow$) until $V_B = 0$ V and $R_{CE} =$ maximum Ω.

As a matter of interest, the name transistor was derived from the fact that through base control we can "transfer" different values of "resistance" between the emitter and collector. This effect of "transferring resistance" is known as **transistance** and the component that functions in this manner is called the transistor.

Transistance
The effect of transferring resistance.

Now that we have seen how the transistor can be made to operate as either a switch or a variable resistor, let us see how these characteristics can be made use of in circuit applications.

12-1-4 *Transistor Applications*

The transistor's impact on electronics has been phenomenal. It initiated the multibillion dollar semiconductor industry and was the key element behind many other inventions, such as integrated circuits (ICs), optoelectronic devices, and digital computer electronics. In all of

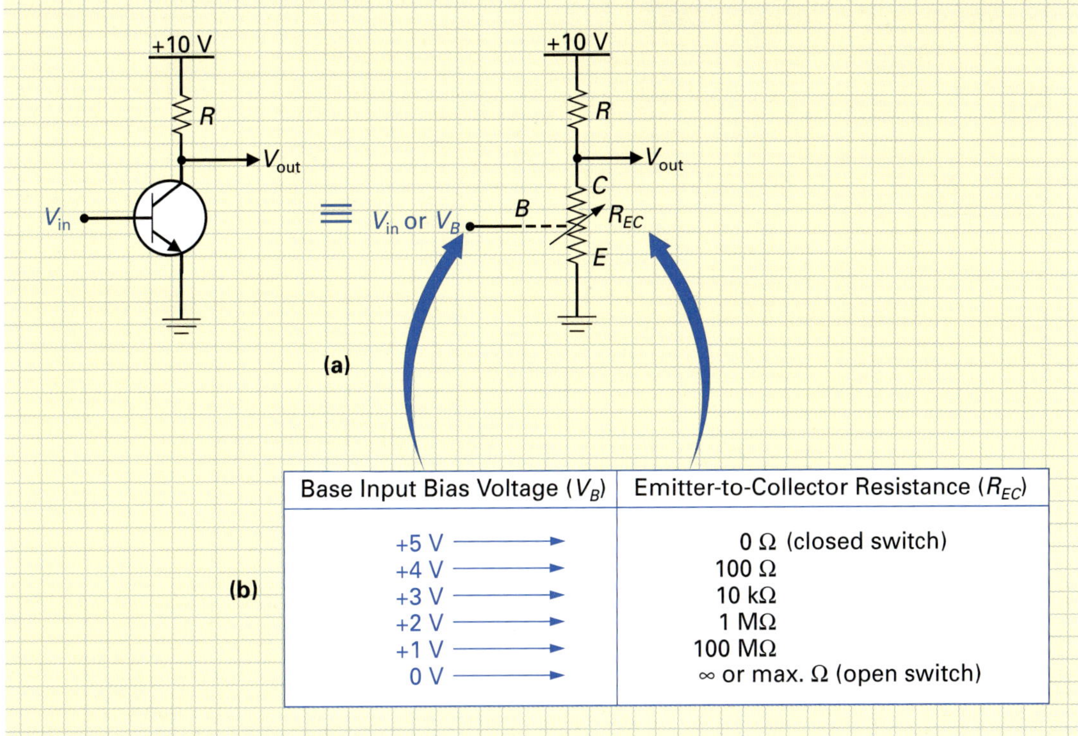

FIGURE 12-6 The Bipolar Transistor's Variable-Resistor Action.

these applications, however, the transistor is basically made to operate in one of two ways: as a switch or as a variable resistor. Let us now briefly examine an example of each.

Digital Logic Gate Circuit

A digital logic gate circuit makes use of the transistor's ON/OFF switching action. Digital circuits are often referred to as "switching" or "two-state" circuits because their main control device (the transistor) is switched between the two states of ON and OFF. The transistor is at the very heart of all digital electronic circuits. For example, transistors are used to construct logic gate circuits, gates are used to construct flip-flop circuits, flip-flops are used to construct register and counter circuits, and these circuits are used to construct microprocessor, memory, and input/output circuits—the three basic blocks of a digital computer.

Figure 12-7(a) shows how the transistor can be used to construct a NOT gate or INVERTER gate. The basic NOT gate circuit is constructed using one NPN transistor and two resistors. This logic gate has only one input (*A*) and one output (*Y*), and its schematic symbol is shown in Figure 12-7(b). Figure 12-7(c) shows how this logic gate will react to the two different input possibilities. When the input is 0 V (logic 0), the transistor's base-emitter P-N diode will be reverse biased and so the transistor will turn OFF. Referring to the inset for this circuit condition in Figure 12-7(c), you can see that the OFF transistor is equivalent to an open switch between emitter and collector, and therefore the +5 V supply voltage will be connected to the output. In summary, a logic 0 input (0 V) will be converted to a logic 1 output (+5 V). On the other hand, when the input is +5 V (logic 1), the transistor's base-emitter P-N diode will be forward biased and so the transistor will turn ON. Referring to the inset for this circuit condition in Figure 12-7(c), you can see that the ON transistor is equivalent to a closed switch between emitter and collector, and therefore 0 V will be connected to the output. In summary, a logic 1 input (+5 V) will be converted to a logic 0 output (0 V).

Referring to the function table in Figure 12-7(c), you can see that the output logic level is "NOT" the same as the input logic level—hence the name NOT gate.

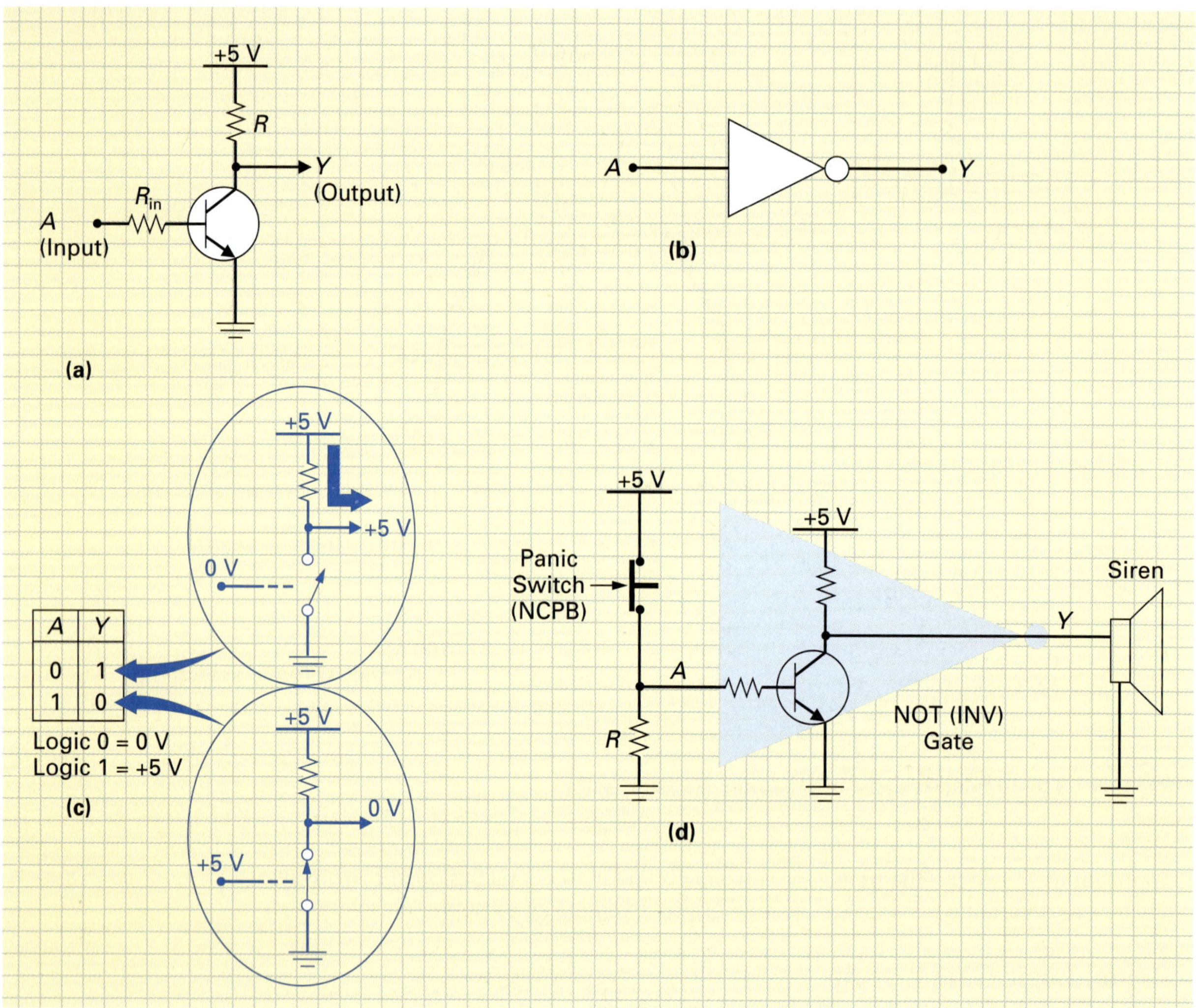

FIGURE 12-7 The Transistor Being Used in a Digital Electronic Circuit. (a) Basic NOT or INVERTER Gate Circuit. (b) NOT Gate Schematic Symbol. (c) NOT Gate Function Table. (d) NOT Gate Security System Application.

As an application, Figure 12-7(d) shows how a NOT or INV gate can be used to invert an input control signal. In this circuit, you can see that a normally closed push button (NCPB) switch is used as a panic switch to activate a siren in a security system. Because the push button is normally closed, it will produce +5 V at *A* when it is not in alarm. If this voltage were connected directly to the siren, the siren would be activated incorrectly. By including the NOT gate between the switch circuit and the siren, the normally HIGH output of the NCPB will be inverted to a LOW, and not activate the siren when we are not in alarm. When the panic switch is pressed, however, the NCPB contacts will open producing a LOW input voltage to the NOT gate. This LOW input will be inverted to a HIGH output and activate the siren.

Analog Amplifier Circuit

Amplification
Boosting in strength, or increasing amplitude, of electronic signals.

When used as a variable resistor, the transistor is the controlling element in many analog or linear circuit applications such as amplifiers, oscillators, modulators, detectors, regulators, and so on. The most important of these applications is **amplification,** which is the boosting in strength or increasing in amplitude of electronic signals.

Figure 12-8(a) shows a simplified transistor amplifier circuit, while Figure 12-8(b) shows the voltage waveforms present at different points in the circuit. As you can see, the transistor is labeled Q_1 because the letter "Q" is the standard letter designation used for transistors.

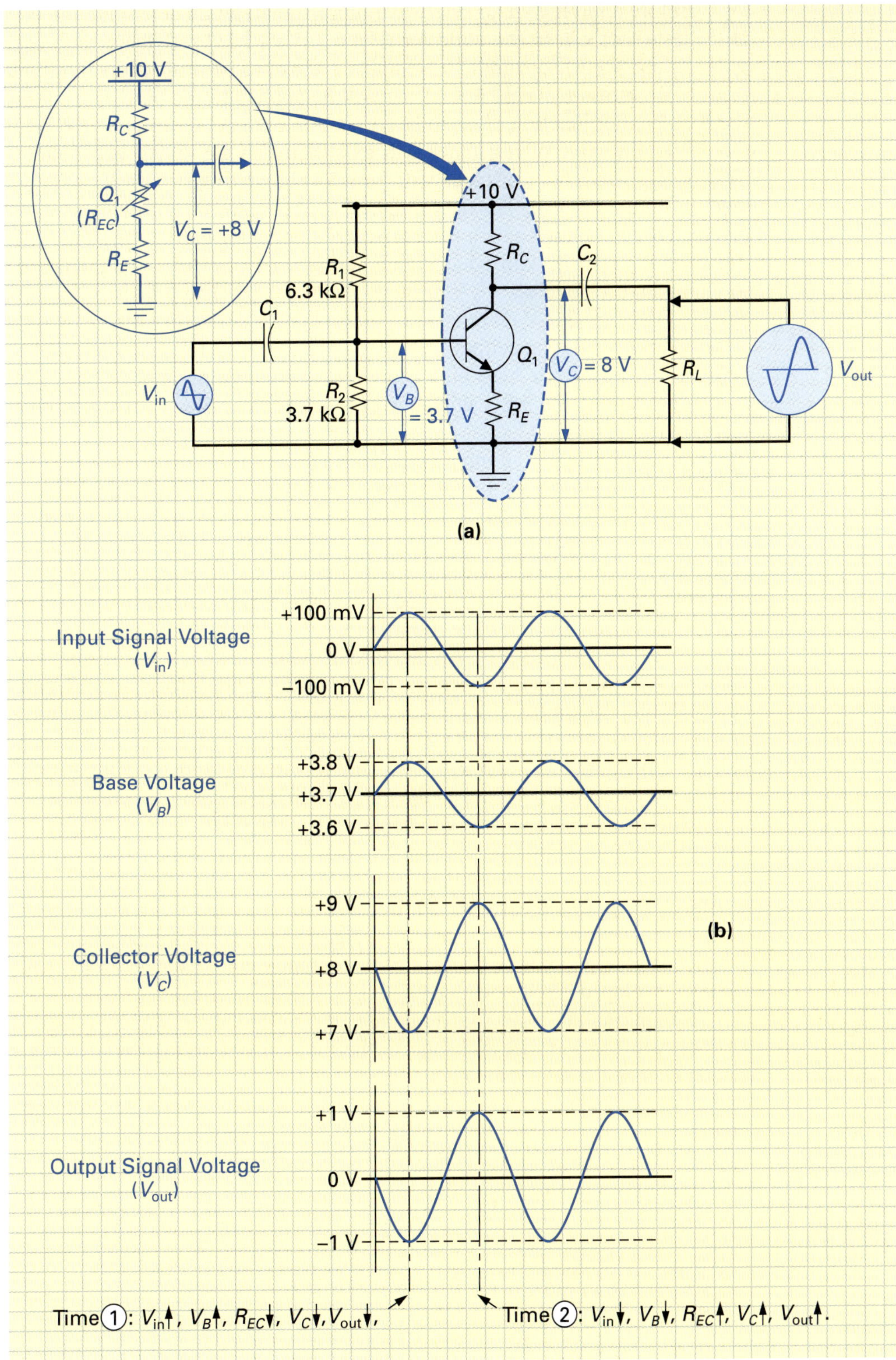

FIGURE 12-8 **The Transistor Being Used in an Analog Electronic Circuit. (a) Basic Amplifier Circuit. (b) Input/Output Voltage Waveforms.**

Before applying an ac sine-wave input signal, let us determine the dc voltage levels at the transistor's base and collector. The 6.3 kΩ/3.7 kΩ resistance ratio of the voltage divider R_1 and R_2 causes the +10 V supply voltage to be proportionally divided, producing +3.7 V dc across R_2. This +3.7 V dc will be applied to the base of the transistor, causing the base-emitter junction of Q_1 to be forward biased and Q_1 to turn ON. With transistor Q_1 ON, a certain value of resistance will exist between the transistor's collector and emitter (R_{CE} or R_{EC}), and this resistance will form a voltage divider with R_E and R_C, as seen in the inset in Figure 12-8(a). The dc voltage at the collector of Q_1 relative to ground (V_C) will be equal to the voltage developed across Q_1's collector-emitter resistance (R_{CE}) and R_E. In this example circuit, with no ac signal applied, a V_B of +3.7 V dc will cause R_{CE} and R_E to cumulatively develop +8 V at the collector of Q_1. The transistor has a base bias voltage (V_B) that is +3.7 V dc relative to ground and a collector voltage (V_C) that is +8 V dc relative to ground. Capacitors C_1 and C_2 are included to act as dc blocks, with C_1 preventing the +3.7 V dc base bias voltage (V_B) from being applied back to the input (V_{in}) and C_2 preventing the +8 V dc collector reference voltage (V_C) from being applied across the output (V_{R_L} or V_{out}).

Let us now apply an input signal and see how it is amplified by the amplifier circuit in Figure 12-8(a). The alternating input sine-wave signal (V_{in}) is applied to the base of Q_1 via C_1, which, like most capacitors, offers no opposition to this ac signal. This input signal, which has a peak-to-peak voltage change of 200 mV, is shown in the first waveform in Figure 12-8(b). The alternating input signal will be superimposed on the +3.7 V dc base bias voltage and cause the +3.7 V dc at the base of Q_1 to increase by 100 mv (3.7 V + 100 mV = 3.8 V), and decrease by 100 mV (3.7 V − 100 mV = 3.6 V), as seen in the second wave-form in Figure 12-8(b). An increase in the input signal ($V_{in}\uparrow$), and therefore the base voltage ($V_B\uparrow$), will cause an increase in the emitter diode's forward bias, causing Q_1 to turn more ON and the emitter-to-collector resistance of Q_1 to decrease ($R_{EC}\downarrow$). Because voltage drop is always proportional to resistance, a decrease in $R_{EC}\downarrow$ will cause a decrease in the voltage drop across R_{CE} and R_E ($V_C\downarrow$), and this decrease in V_C will be coupled to the output via C_2, causing a decrease in the output voltage developed across the load ($V_{out}\downarrow$).

Now let us examine what will happen when the sine-wave input signal decreases. A decrease in the input signal ($V_{in}\downarrow$), and therefore the base voltage ($V_B\downarrow$), will cause a decrease in the emitter diode's forward bias, causing Q_1 to turn less ON and the emitter-to-collector resistance of Q_1 to increase ($R_{EC}\uparrow$). Because voltage drop is always proportional to resistance, an increase in $R_{EC}\uparrow$ will cause an increase in the voltage drop across R_{EC} and R_E ($V_C\uparrow$). This increase in V_C will be coupled to the output via C_2, causing an increase in the output voltage developed across the load ($V_{out}\uparrow$).

Comparing the input signal voltage (V_{in}) to the output signal voltage (V_{out}) in Figure 12-8(b), you can see that a change in the input signal voltage produces a corresponding greater change in the output signal voltage. The ratio (comparison) of output signal voltage change to input signal voltage change is a measure of this circuit's **voltage gain (A_V)**. In this example, the **output signal voltage change** (ΔV_{out}) is between +1 V and −1 V, and the **input signal voltage change** (ΔV_{in}) is between +100 mV and −100 mV. The circuit's voltage gain between input and output will therefore be:

Voltage Gain (A_V)
The ratio of the output signal voltage change to input signal voltage change.

Output Signal Voltage Change
Change in output signal voltage in response to a change in the input signal voltage.

Input Signal Voltage Change
The input voltage change that causes a corresponding change in the output voltage.

$$\text{Voltage Gain } (A_V) = \frac{\text{Output Voltage Change } (\Delta V_{out})}{\text{Input Voltage Change } (\Delta V_{in})}$$

$$A_V = \frac{+1 \text{ to } -1 \text{ V}}{+100 \text{ mV to } -100 \text{ mV}} = \frac{2 \text{ V}}{200 \text{ mV}} = 10$$

A voltage gain of 10 means that the output voltage is ten times larger than the input voltage. The transistor does not produce this gain magically within its NPN semiconductor structure. The gain or amplification is achieved by the input signal controlling the conduction of the transistor, which takes energy from the collector supply voltage and develops this energy across the load resistor. Amplification is achieved by having a small input voltage control a transistor and its large collector supply voltage, so that a small input voltage change results in a similar but larger output voltage change.

Comparing the input signal to the output signal at time 1 and time 2 in Figure 12-8(b), you can see that this circuit will invert the input signal voltage in the same way that the NOT gate inverts its input voltage (positive input voltage swing produces a negative output voltage swing, and vice versa). This inversion always occurs with this particular transistor circuit arrangement; however, it is not a problem since the shape of the input signal is still preserved at the output (both input and output signals are sinusoidal).

A Switching Regulator Circuit

In the previous chapter, it was shown how a series resistor and zener diode could be used in a dc power supply to function as a voltage regulator. These regulator types maintain a constant output voltage because variations in input voltage or load current are dissipated as heat. These **series dissipative regulators** generally have a low "conversion efficiency" of typically 60% to 70% and should be used only in low- to medium-load current applications.

Series Dissipative Regulators
Voltage regulators that maintain a constant output voltage by causing variations in input voltage or load current to be dissipated as heat.

Series switching regulators, on the other hand, have a conversion efficiency of typically 90%. To explain the operation of these regulator types, refer to the simplified circuit in Figure 12-9(a). To improve efficiency, a series-pass transistor (Q_1) is operated as a switch, rather than as a variable resistor. This means that Q_1 is switched ON and OFF, and therefore either switches the +12 V input at its collector through to its emitter, or blocks the +12 V from passing through to the emitter. These +12 V pulses at the emitter of Q_1 charge capacitor C_1 to an average voltage (which in this example is +5 V) and this voltage is applied to the load (R_L). To explain this in more detail, when Q_1 is turned ON by a HIGH base voltage from the switching regulator IC, the unregulated +12 V at Q_1's collector is switched through

Series Switching Regulators
A regulator circuit containing a power transistor in series with the load, that is switched ON and OFF to regulate the dc output voltage delivered to the load.

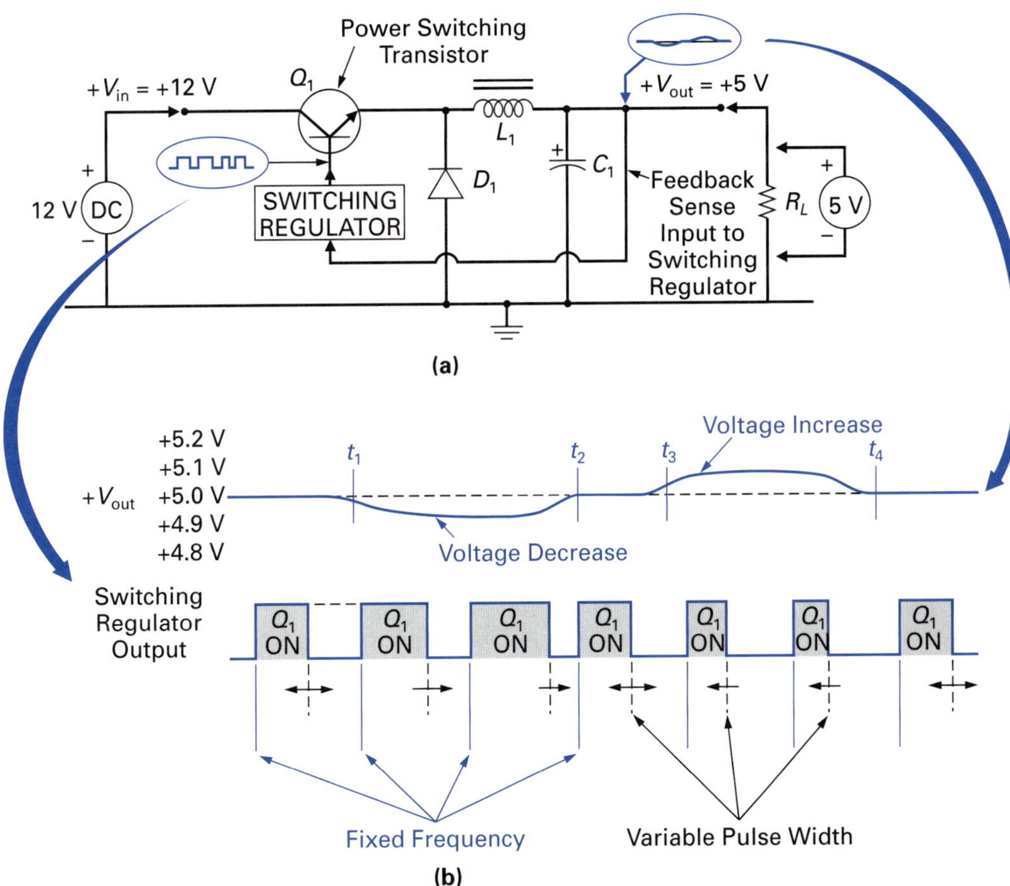

FIGURE 12-9 Basic Switching Regulator Action.

to Q_1's emitter, where it reverse biases D_1, and is applied to the series-connected inductor L_1 and parallel capacitor C_1. Inductor L_1 and capacitor C_1 act as a low-pass filter because series-connected L_1 opposes the ON/OFF changes in current and passes a relatively constant current to the load: shunt- or parallel-connected C_1 opposes the ON/OFF changes in voltage and holds the output voltage relatively constant at +5 V. When Q_1 is turned OFF by a LOW base voltage from the switching regulator IC, the unregulated +12 V input is disconnected from the *LC* filter, the inductor's magnetic field will collapse and produce a currrent through the load, and the +5 V charge held by C_1 will still be applied across the load. Inductor L_1, therefore, smoothes out the current changes, while capacitor C_1 smoothes out the voltage changes caused by the ON/OFF switching of transistor Q_1. The next, and most important, question is: how does this circuit regulate, or maintain constant, the output voltage? The answer is: throught a closed-loop "sense and adjust" system controlled by a switching regulator IC. The switching regulator IC operates by comparing an internal fixed reference voltage to a sense input, which is taken from the +5 V output, as is shown in Figure 12-9(a). Referring to the waveforms shown in Figure 12-9(b), you can see that whenever the output voltage falls below +5 V (from time t_1 to t_2), the switching regulator responds by increasing the width of the positive output pulse applied to the base of Q_1. This increases the ON time of Q_1, which raises the average output voltage, bringing the output back up to +5 V. On the other hand, whenever the output voltage rises above +5 V (between time t_3 and t_4), the switching regulator responds by decreasing the width of the positive output pulse applied to the base of Q_1. This decreases the ON time of Q_1, which lowers the average output voltage, bringing the output back down to +5 V. The net result is that the output voltage will remain locked at +5 V despite variations in the input voltage and variations in the load.

Switching Power Supply
A dc power supply that makes use of a series switching regulator controlled by a pulse-width-modulator to regulate the output voltage.

A switching regulator, therefore, is a voltage regulator that chops up, or switches ON and OFF (at typically a 20 kHz rate), a dc input voltage to efficiently produce a regulated dc output voltage. A **switching power supply** uses switching regulators and is generally small in size and very efficient. The only disadvantage is that the circuitry is generally a little more complex, and therefore a little more costly.

SELF-TEST EVALUATION POINT FOR SECTION 12-1

Use the following questions to test your understanding of Section 12-1.

1. What are the two basic types of bipolar transistor?
2. Name the three terminals of a bipolar transistor.
3. What are the two basic ways in which a transistor is made to operate?
4. Which of the modes of operation mentioned in question 3 is made use of in digital circuits and which is made use of in analog circuits?

12-2 DETAILED DESCRIPTION OF A BIPOLAR TRANSISTOR

Now that we have a good understanding of the bipolar junction transistor's (BJT's) general characteristics, operation, and applications, let us examine all of these aspects in a little more detail.

12-2-1 *Basic Bipolar Transistor Action*

When describing diodes previously, we saw how the P-N junction of a diode could be either forward or reverse biased to either permit or block the flow of current through the device. The transistor must also be biased correctly; however, in this case, two P-N junctions rather than one must have the correct external supply voltages applied.

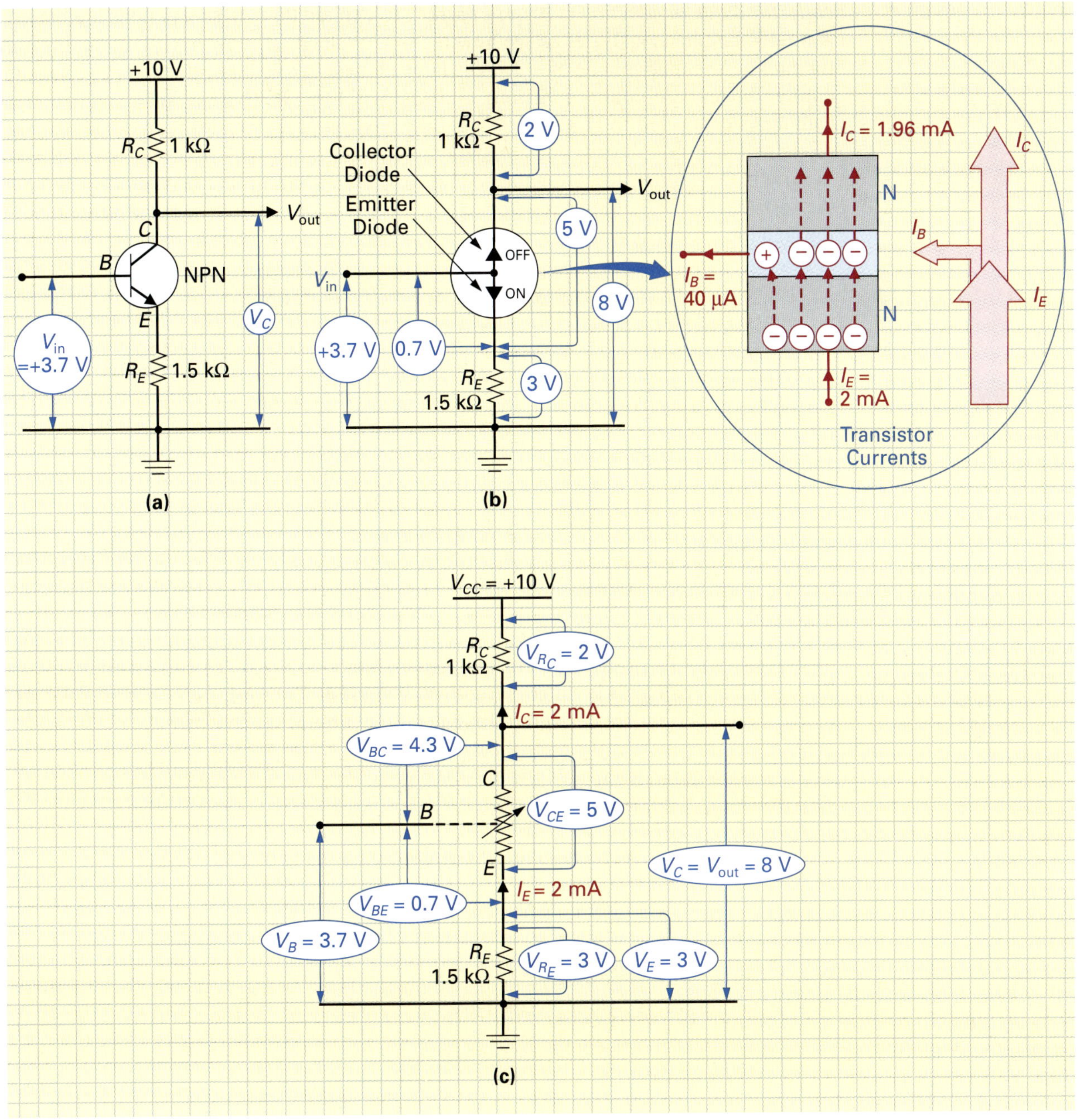

FIGURE 12-10 **A Correctly Biased NPN Transistor Circuit.**

A Correctly Biased NPN Transistor Circuit

Figure 12-10(a) shows how an NPN transistor should be biased for normal operation. In this circuit, a +10 V supply voltage is connected to the transistor's collector (*C*) via a 1 kΩ collector resistor (R_C). The emitter (*E*) of the transistor is connected to ground via a 1.5 kΩ emitter resistor (R_E), and, as an example, an input voltage of +3.7 V is being applied to the base (*B*). The output voltage (V_{out}) is taken from the collector, and this collector voltage (V_C) will be equal to the voltage developed across the transistor's collector-to-emitter and the emitter resistor R_E.

As previously mentioned in the first approximation description of the transistor, the transistor can be thought of as containing two diodes, as shown in Figure 12-9(b). In normal operation, *the transistor's emitter diode or junction is forward biased, while the transistor's collector diode or junction is reverse biased.* To explain how these junctions are biased ON and OFF simultaneously, let us see how the input voltage of +3.7 V will affect this transistor circuit. An input voltage of +3.7 V is large enough to overcome the barrier voltage of the emitter diode (base-emitter junction), and so it will turn ON (base or anode is +, emitter or cathode is connected to ground or 0 V). Like any forward biased silicon diode, the emitter diode will drop 0.7 V between base and emitter, and so the +3.7 V at the base will produce +3.0 V at the emitter. Knowing the voltage drop across the emitter resistor ($V_{R_E} = 3$ V) and the resistance of the emitter resistor ($R_E = 1.5$ kΩ), we can calculate the value of current through the emitter resistor.

$$I_{R_E} = \frac{V_{R_E}}{R_E} = \frac{3\text{ V}}{1.5\text{ k}\Omega} = 2\text{ mA}$$

Emitter Current (I_E)
The current at the transistor's emitter terminal.

Base Current (I_B)
The relatively small current at the transistor's base terminal.

Collector Current (I_C)
The current emerging out of the transistor's collector.

This emitter resistor current of 2 mA will leave ground, travel through R_E, and then enter the transistor's *n*-type emitter region. This current at the transistor's emitter terminal is called the **emitter current (I_E).** The forward biased emitter diode will cause the steady stream of electrons entering the emitter to head toward the base region, as shown in the inset in Figure 12-10(b). The base is a very thin, lightly doped region with very few holes in relation to the number of electrons entering the transistor from the emitter. Consequently, only a few electrons combine with the holes in the base region and flow out of the base region. This relatively small current at the transistor's base terminal is called the **base current (I_B).** Because only a few electrons combine with holes in the base region, there is an accumulation of electrons in the base's *p* layer. These free electrons, feeling the attraction of the large positive collector supply voltage (+10 V), will travel through the *n*-type collector junction and out of the transistor to the positive external collector supply voltage. The current emerging out of the transistor's collector is called the **collector current (I_C).** Because both the collector current and base current are derived from the emitter current, we can state that:

$$I_E = I_B + I_C$$

In the example in the inset in Figure 12-10(b), you can see that this is true because

$$I_E = I_B + I_C$$
$$I_E = 40\ \mu\text{A} + 1.96\text{ mA} = 2\text{ mA} \quad (40\ \mu\text{A} = 0.04\text{ mA})$$

Stated another way, we can say that the collector current is equal to the emitter current minus the current that is lost out of the base.

$$I_C = I_E - I_B$$
$$I_C = 2\text{ mA} - 40\ \mu\text{A} = 1.96\text{ mA}$$

Approximately 98% of the electrons entering the emitter of a transistor will arrive at the collector. Because of the very small percentage of current flowing out of the base (I_B equals about 2% of I_E), we can approximate and assume that I_C is equal to I_E.

$$I_C \cong I_E$$

(I_C approximately equals I_E)

The Current-Controlled Transistor

In the previous section, we discovered that because the collector and base currents (I_C and I_B) are derived from the emitter current (I_E), an increase in the emitter current ($I_E\uparrow$), for example, will cause a corresponding increase in collector and base current ($I_C\uparrow$, $I_B\uparrow$). Looking at this from a different angle, an increase in the applied base voltage (base input increases to +3.8 V) will increase the forward bias applied to the emitter diode of the transistor, which will draw more electrons up from the emitter and cause an increase in I_E, I_B, and I_C. Similarly, a decrease in the applied base voltage (base input decreases to +3.6 V) will decrease the forward bias applied to the emitter diode of the transistor, which will decrease the number of electrons being drawn up from the emitter and cause a decrease in I_E, I_B, and I_C. The applied input base voltage will control the amount of base current, which will in turn control the amount of emitter and collector current, and therefore the conduction of the transistor. This is why *the bipolar transistor is known as a current-controlled device.*

Continuing our calculations for the example circuit in Figure 12-10(b), let us apply this current relationship and assume that I_C is equal to I_E, which, as we previously calculated, is equal to 2 mA. Knowing the value of current for the collector resistor ($I_{RC} = 2$ mA) and the resistance of the collector resistor ($R_C = 1$ kΩ), we can calculate the voltage drop across the collector resistor.

$$V_{RC} = I_{RC} \times R_C = 2 \text{ mA} \times 1 \text{ k}\Omega = 2 \text{ V}$$

With 2 V being dropped across R_C, the voltage at the transistor's collector (V_C) will be:

$$V_C = +10 \text{ V} - V_{RC} = 10 \text{ V} - 2 \text{ V} = 8 \text{ V}$$

Because the voltage at the transistor's collector relative to ground is applied to the output, the output voltage will also be equal to 8 V.

$$V_C = V_{\text{out}} = 8 \text{ V}$$

At this stage, we can determine a very important point about any correctly biased NPN transistor circuit. *A properly biased transistor will have a forward biased base-emitter junction (emitter diode is ON), and a reverse biased base-collector junction (collector diode is OFF).* We can confirm this with our example circuit in Figure 12-10(b), because we now know the voltages at each of the transistor's terminals.

Emitter diode (base-emitter junction) is forward biased (ON) because
Anode (base) is connected to +3.7 V (V_{in})
Cathode (emitter) is connected to 0 V via R_E

Collector diode (base-collector junction) is reverse biased (OFF) because
Anode (base) is connected to +3.7 V (V_{in})
Cathode (collector) is at +8 V (due to 2 V drop across R_C)

Keep in mind that even though the collector diode (base-collector junction) is reverse biased, current will still flow through the collector region. This is because most of the electrons traveling from emitter-to-base (through the forward biased emitter diode) do not find many holes in the thin, lightly doped base region, and therefore the base current is always very small. Almost 98% of the electrons accumulating in the base region feel the strong attraction of the positive collector supply voltage and flow up into the collector region and then out of the collector as collector current.

With the example circuit in Figure 12-10(a) and (b), the emitter diode is ON and the collector diode is OFF, and the transistor is said to be operating in its normal, or *active, region.*

Operating a Transistor in the Active Region

A transistor is said to be in **active operation,** or in the **active region,** when its base-emitter junction is forward biased (emitter diode is ON), and the base-collector junction is reverse biased (collector diode is OFF). In this mode, the transistor is equivalent to a variable resistor between collector and emitter.

Active Operation or in the Active Region
When the base-emitter junction is forward biased and the base-collector junction is reverse biased. In this mode, the transistor is equivalent to a variable resistor between collector and emitter.

In Figure 12-10(c), our transistor circuit example has been redrawn with the transistor this time being shown as a variable resistor between collector and emitter and with all of our calculated voltage and current values inserted. Before we go any further with this circuit, let us discuss some of the letter abbreviations used in transistor circuits. To begin with, the term V_{CC} is used to denote the "stable collector voltage" and this dc supply voltage will typically be positive for an NPN transistor. Two *C*s are used in this abbreviation ($+V_{CC}$) because V_C (*V* sub single *C*) is used to describe the voltage at the transistor's collector relative to ground. The doubling up of letters such as V_{CC}, V_{EE}, or V_{BB} is used to denote a constant dc bias voltage for the collector (V_{CC}), emitter (V_{EE}), and base (V_{BB}). A single sub letter abbreviation such as V_C, V_E, or V_B is used to denote a transistor terminal voltage relative to ground. The other voltage abbreviations, V_{CE}, V_{BE}, and V_{CB}, are used for the voltage difference between two terminals of the transistor. For example, V_{CE} is used to denote the potential difference between the transistor's collector and emitter terminals. Finally, I_E, I_B, and I_C are, as previously stated, used to denote the transistor's emitter current (I_E), base current (I_B), and collector current (I_C).

Because the transistor's resistance between emitter and collector in Figure 12-10(c) is in series with R_C and R_E, we can calculate the voltage drop between collector and emitter (V_{CE}) because V_{R_C} and V_{R_E} are known.

$$V_{CE} = V_{CC} - (V_{R_E} + V_{R_C})$$
$$V_{CE} = 10\text{ V} - (3\text{ V} + 2\text{ V}) = 10\text{ V} - 5\text{ V} = 5\text{ V}$$

Now that we know the voltage drop between the transistor's collector and emitter (V_{CE}), we can calculate the transistor's equivalent resistance between collector and emitter (R_{CE}) because we know that the current through the transistor is 2 mA.

$$R_{CE} = \frac{V_{CE}}{I_C} = \frac{5\text{ V}}{2\text{ mA}} = 2.5\text{ k}\Omega$$

Operating the Transistor in Cutoff and Saturation

Figure 12-11 shows the three basic ways in which a transistor can be operated. As we have already discovered, the bias voltages applied to a transistor control the transistor's operation by controlling the two P-N junctions (or diodes) in a bipolar transistor. For example, the center column reviews how a transistor will operate in the active region. As you can see, our previous circuit example with all of its values has been used. To summarize: *when a transistor is operated in the active region, its emitter diode is biased ON, its collector diode is biased OFF, and the transistor is equivalent to a variable resistor between the collector and the emitter.*

Cutoff

A transistor is in cutoff when the bias voltage is reduced to a point that it stops current in the transistor.

Saturation

A transistor is in saturation when the bias voltage is increased to such a point that further increase will not cause any increase in current through the transistor.

The left column in Figure 12-11 shows how the same transistor circuit can be driven into **cutoff.** A transistor is in cutoff when the bias voltage is reduced to a point that it stops current in the transistor. In this example circuit, you can see that when the base input bias voltage (V_B) is reduced to 0 V, the transistor is cut off. *In cutoff, both the emitter and the collector diode of the transistor will be biased OFF, the transistor is equivalent to an open switch between the collector and the emitter, and the transistor current is zero.*

The right column in Figure 12-11(c) shows how the same transistor circuit can be driven into **saturation.** A transistor is in saturation when the bias voltage is increased to such a point that any further increase in bias voltage will not cause any further increase in current through the transistor. In the equivalent circuit in Figure 12-11(c), you can see that when the base input bias voltage (V_B) is increased to +6.7 V, the emitter diode of the transistor will be heavily forward biased and the emitter current will be large.

$$I_E = \frac{V_B - V_{BE}}{R_E} = \frac{6.7\text{ V} - 0.7\text{ V}}{1.5\text{ k}\Omega} = 4\text{ mA}$$

Because I_B and I_C are both derived from I_E, an increase in I_E will cause a corresponding increase in both I_B and I_C. These high values of current through the transistor account for why a transistor operating in saturation is said to be equivalent to a closed switch (high conductance, low resistance). Although the transistor's resistance between the collector and emitter (R_{CE}) is assumed to be 0 Ω, there is still some small value of R_{CE}. Typically, a saturated transistor will have a 0.3 V drop between the collector and emitter ($V_{CE} =$ 0.3 V), as shown in the equivalent circuit in Figure 12-11(c). If $V_{BE} = 0.7$ V and $V_{CE} =$ 0.3 V, then the voltage drop across the collector diode of a saturated transistor (V_{BC}) will be 0.4 V. This means that the base of the transistor (anode) is now +0.4 V relative to the collector (cathode), and there is not enough reverse bias voltage to turn OFF the collec-

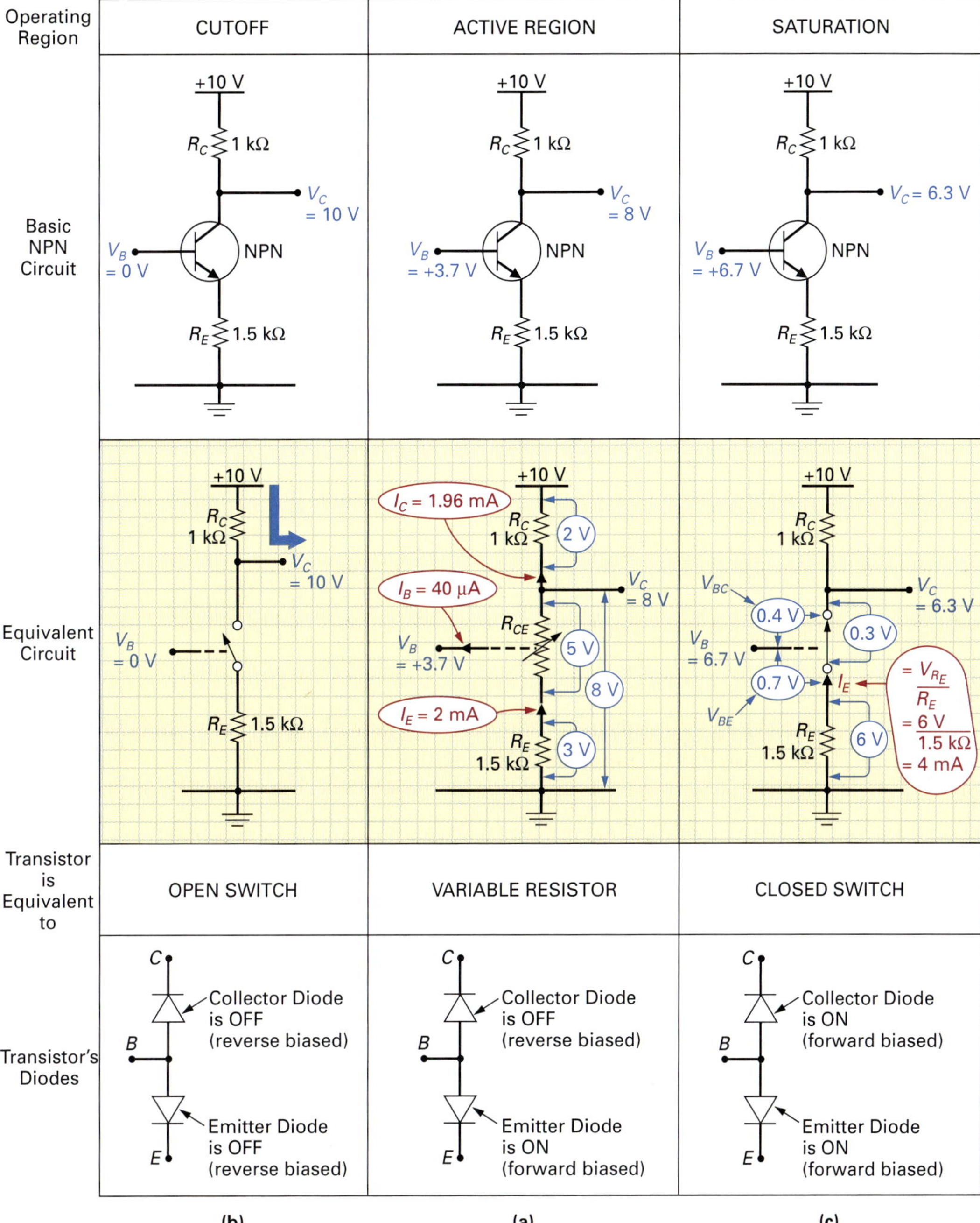

FIGURE 12-11 The Three Bipolar Transistor Operating Regions.

tor diode. *In saturation, both the emitter and collector diodes are said to be forward biased, the transistor is equivalent to a closed switch, and any further increase in bias voltage will not cause any further increase in current through the transistor.*

Biasing PNP Bipolar Transistors

Generally the PNP transistor is not employed as much as the NPN transistor in most circuit applications. The only difference that occurs with PNP transistor circuits is that the polarity of V_{CC} and the base bias voltage (V_B) need to be reversed to a negative voltage, as shown in Figure 12-12. The PNP transistor has the same basic operating characteristics as the NPN transistor, and all of the previously discussed equations still apply.

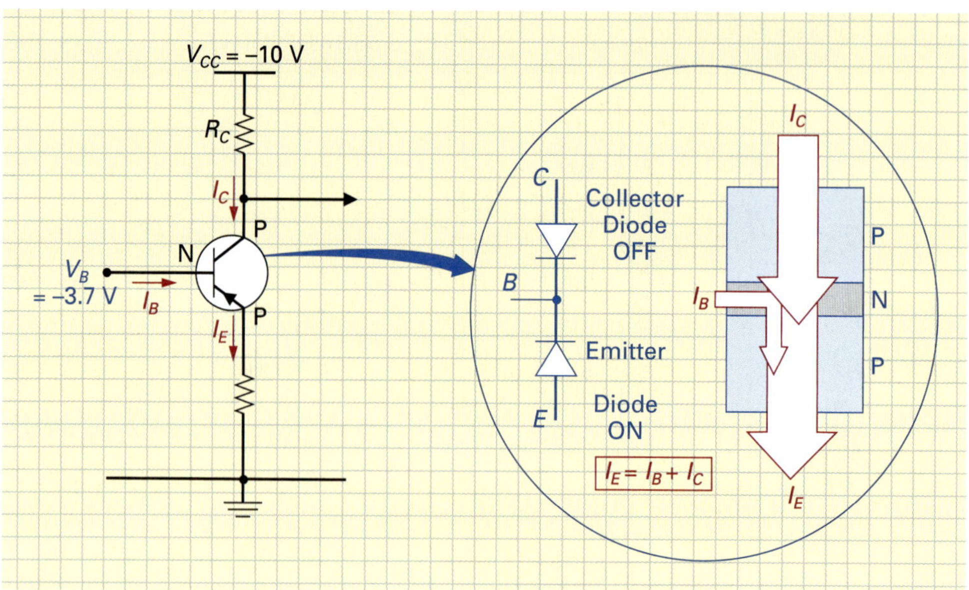

FIGURE 12-12 A Correctly Biased PNP Transistor Circuit.

Referring to the inset in Figure 12-12, you will see that the −3.7 V base bias voltage will forward bias the emitter diode, and the −10 V V_{CC} will reverse bias the collector diode, so that the transistor is operating in the active region. Also, the electron transistor currents are in the opposite direction. This, however, makes no difference because the sum of the collector current entering the collector and base current entering the base is equal to the value of emitter current leaving the emitter, so $I_E = I_B + I_C$ still applies.

12-2-2 *Bipolar Transistor Circuit Configurations and Characteristics*

Configurations
Different circuit interconnections.

Common
Shared by two or more services, circuits, or devices. Although the term "common ground" is frequently used to describe two or more connections sharing a common ground, the term "common" alone does not indicate a ground connection, only a shared connection.

In the previous sections, we have seen how the bipolar junction transistor can be used in digital two-state switching circuits and analog or linear circuits such as the amplifier. In all of these different circuit interconnections or **configurations,** the bipolar transistor was used as the main controlling element, with one of its three leads being used as a common reference and the other two leads being used as an input and an output. Although there are many thousands of different bipolar transistor circuit applications, all of these circuits can be classified in one of three groups based on which of the transistor's leads is used as the **common** reference. These three different circuit configurations are shown in Figure 12-13. With the *common-emitter (C-E)* bipolar transistor circuit configuration, shown in Figure 12-13(a), the input signal is applied between the base and emitter, while the output signal appears between the transistor's collector and emitter. With this circuit arrangement, the input signal controls the transistor's base current, which in turn controls the transistor's output collector current, and the emitter lead is common to both the input and output. Similarly, with the *common-base (C-B)* circuit configuration shown in Figure 12-13(b), the input signal is applied between the transistor's emitter and base, the output signal is developed across the transistor's collector and base, and the base is common to both input and output. Finally, with the *common-collector (C-C)* circuit configuration shown in Figure 12-13(c), the input is applied between the base and collector, the output is developed across the emitter and collector, and the collector is common to both the input and output.

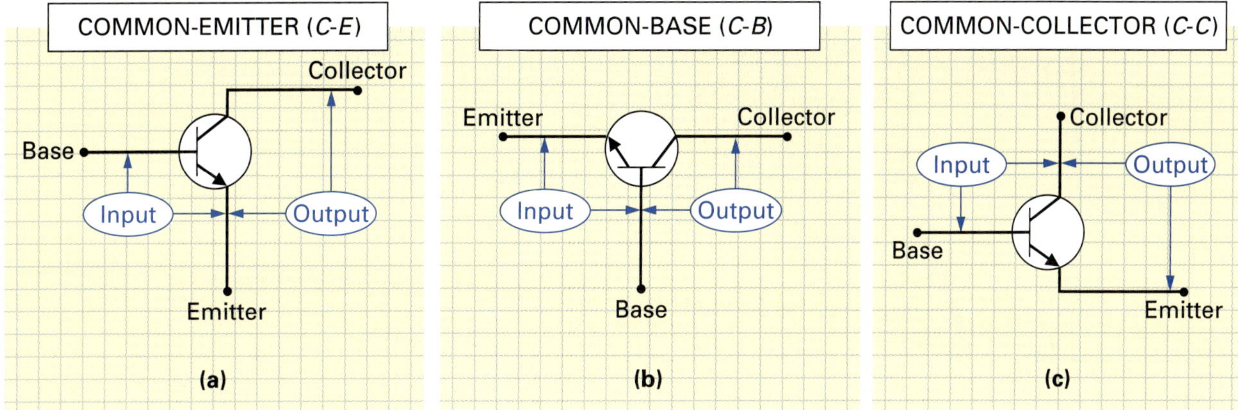

FIGURE 12-13 **Bipolar Transistor Circuit Configurations.**

To begin with, we will discuss the common-emitter circuit configuration characteristics because it has been this circuit arrangement that we have been using in all of the circuit examples in this chapter.

Common-Emitter Circuits

With the **common-emitter circuit,** the transistor's emitter lead is common to both the input and output signals. In this circuit configuration, the base serves as the input lead, and the collector serves as the output lead. Figure 12-14 contains a basic common-emitter (*C-E*) circuit, its associated input/output voltage and current waveforms, characteristic curves, and table of typical characteristics. Using this illustration, we will examine the operation and characteristics of the *C-E* circuit configuration.

Common-Emitter (*C-E*) Circuit
Configuration in which the input signal is applied between the base and the emitter, while the output signal appears between the transistor's collector and emitter.

DC Current Gain. Referring to the *C-E* circuit in Figure 12-14(a), and its associated waveforms in Figure 12-14(b), let us now examine this circuit's basic operation.

Before applying the ac sine-wave input signal (V_{in}), let us assume that $V_{in} = 0$ V and examine the transistor's dc operating characteristics, or the "no input signal" condition. The voltage divider R_1 and R_2 will divide the V_{CC} supply voltage, producing a positive dc base bias voltage across R_2. This base bias voltage will be applied to the base of Q_1, and because it is generally greater than 0.7 V, it will forward bias the transistor's base-emitter junction, turning Q_1 ON (in the example in Figure 12-14, $V_B = 3$ V dc). Capacitor C_1 is included to act as a dc block, preventing the base bias voltage (V_B) from being applied back to the input (V_{in}). The value of dc base bias voltage will determine the value of base current (I_B) flowing out of the transistor's base, and this value of I_B will in turn determine the value of collector current (I_C) flowing out of the transistor's collector and through R_C. Because the transistor's output current (I_C) is so much larger than the transistor's very small input current (I_B), the circuit produces an increase in current, or a **current gain.** The current gain in a common-emitter configuration is called the transistor's **beta** (symbolized β). A transistor's **dc beta** (β_{DC}) indicates a common-emitter transistor's "dc current gain," and it is the ratio of its output current (I_C) to its input current (I_B). This ratio can be expressed mathematically as:

Current Gain
The increase in current produced by the transistor circuit.

Beta (β)
The transistor's current gain in a common-emitter configuration.

DC Beta (β_{DC})
The ratio of a transistor's dc output current to its input current.

$$\beta_{DC} = \frac{I_C}{I_B}$$

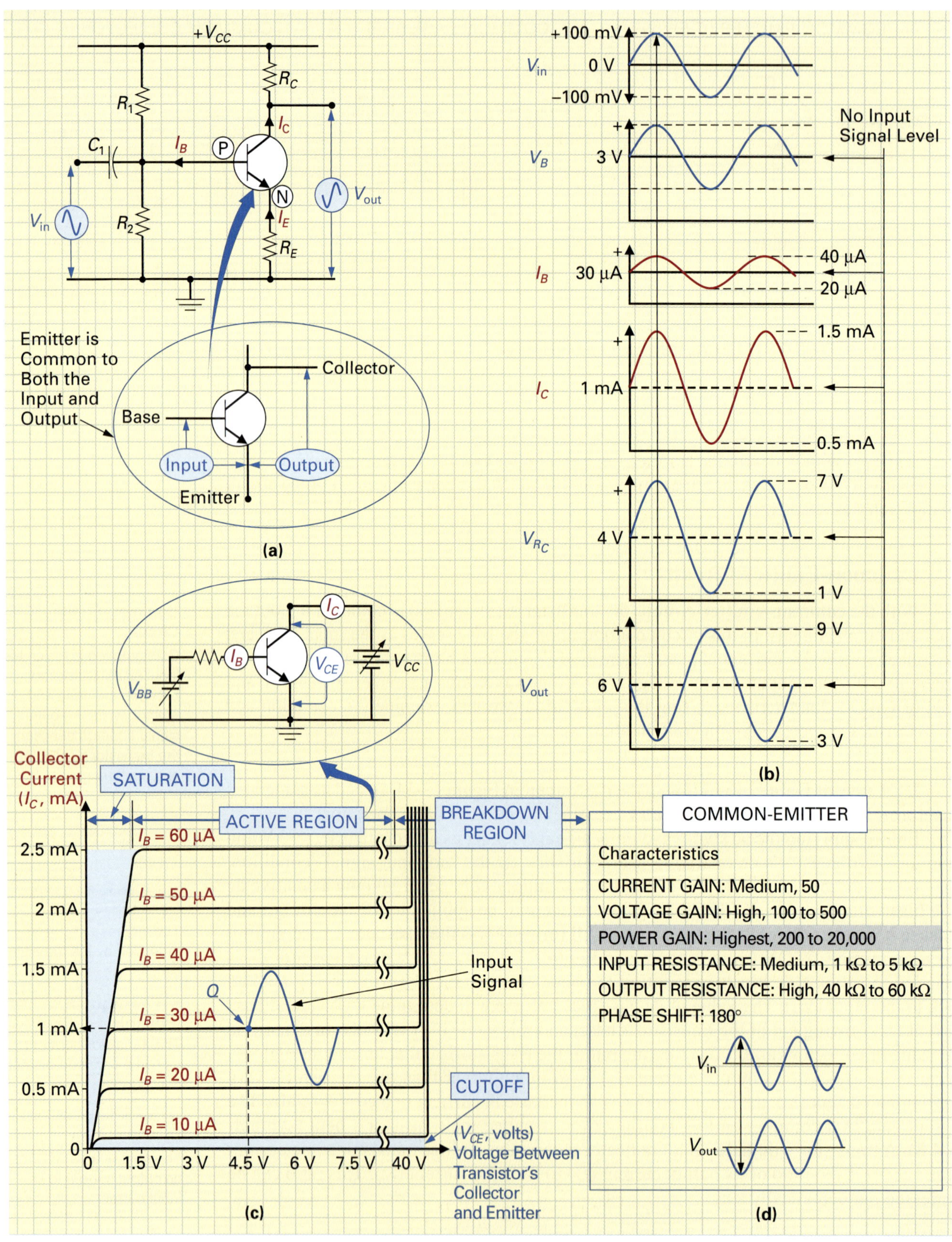

FIGURE 12-14 Common-Emitter *(C-E)* Circuit Configuration Characteristics.

EXAMPLE:

As can be seen in Figure 12-14(b), the "no input signal" level of $I_C = 1$ mA, and $I_B = 30$ μA. What is the transistor's dc beta?

Solution:

$$\beta_{DC} = \frac{I_C}{I_B}$$

$$= \frac{1 \text{ mA}}{30 \text{ μA}}$$

$$= 33.3$$

This value indicates that I_C is 33.3 times greater than I_B, therefore the dc current gain between input and output is 33.3.

Another form of system analysis, known as "hybrid parameters," uses the term *hfe* instead of β_{DC} to indicate a transistor's dc current gain.

AC current gain. When amplifying an ac waveform, a transistor has applied to it both the dc voltage to make it operational and the ac signal voltage to be amplified that varies the base bias voltage and the base current. Referring to the first four waveforms in Figure 12-14(b), let us now examine what will happen when we apply a sine-wave input signal. As V_{in} increases above 0 V to a peak positive voltage ($V_{in}\uparrow$), it will cause the forward base bias voltage applied to the transistor to increase ($V_B\uparrow$) above its dc reference or "no input signal level." This increase in V_B will increase the forward conduction of the transistor's emitter diode, resulting in an increase in the input base current ($I_B\uparrow$) and a corresponding larger increase in the output collector current ($I_C\uparrow$). The ratio of output collector current change (ΔI_C) to input base current change (ΔI_B) is the transistor's ac current gain or **ac beta** (β_{AC}). The formula for calculating a transistor's ac current gain is:

AC Beta (β_{AC})
The ratio of a transistor's ac output current to input current.

$$\beta_{AC} = \frac{\Delta I_C}{\Delta I_B}$$

EXAMPLE:

Calculate the ac current gain of the example in Figure 12-14(b).

Solution:

$$\beta_{AC} = \frac{\Delta I_C}{\Delta I_B}$$

$$= \frac{1.5 \text{ to } 0.5 \text{ mA}}{40 \text{ to } 20 \text{ μA}}$$

$$= \frac{1 \text{ mA}}{20 \text{ μA}}$$

$$= 50$$

This means that the alternating collector current at the output is 50 times greater than the alternating base current at the input.

Common-emitter transistors will typically have beta, or current gain values, of 50.

Voltage Gain. The common-emitter circuit is not only used to increase the level of current between input and output. It can also be used to increase the amplitude of the input signal voltage, or produce a voltage gain. This action can be seen by examining the *C-E* circuit in Figure 12-14(a) and by following the changes in the associated waveforms in Figure 12-14(b). An input signal voltage increase from 0 V to a positive peak ($V_{in}\uparrow$) causes an increase in the dc base bias voltage ($V_B\uparrow$), causing the emitter diode of Q_1 to turn more ON and result in an increase in both I_B and I_C. Because the I_C flows through R_C, and because voltage drop is proportional to current, an increase in I_C will cause an increase in the voltage drop across R_C ($V_{R_C}\uparrow$). The output voltage is equal to the voltage developed across Q_1's collector-to-emitter resistance (R_{CE}) and R_E. Because R_C is in series with Q_1's collector-to-emitter resistance (R_{CE}) and R_E, an increase in $V_{R_C}\uparrow$ will cause a decrease in the voltage developed across R_{CE} and R_E, which is $V_{out}\downarrow$. This action is summarized with Kirchhoff's voltage law, which states that the sum of the voltages in a series circuit is equal to the voltage applied ($V_{R_C} + V_{out} = V_{CC}$). Using the example in Figure 12-14(b), you can see that when:

$$V_{R_C}\uparrow \text{ to } 7\text{ V}, V_{out}\downarrow \text{ to } 3\text{ V}. \qquad (V_{R_C} + V_{out} = V_{CC}, 7\text{ V} + 3\text{ V} = 10\text{ V})$$
$$V_{R_C}\downarrow \text{ to } 1\text{ V}, V_{out}\uparrow \text{ to } 9\text{ V}. \qquad (V_{R_C} + V_{out} = V_{CC}, 1\text{ V} + 9\text{ V} = 10\text{ V})$$

Although the input voltage (V_{in}) and output voltage (V_{out}) are out of phase with one another, you can see from the example values in Figure 12-14(b) that there is an increase in the signal voltage between input and output. This voltage gain between input and output is possible because the output current (I_C) is so much larger than the input current (I_B). The amount of voltage gain (which is symbolized A_V) can be calculated by comparing the output voltage change (ΔV_{out}) to the input voltage change (ΔV_{in}).

$$A_V = \frac{\Delta V_{out}}{\Delta V_{in}}$$

EXAMPLE:

Calculate the voltage gain of the circuit and its associated waveforms in Figure 12-14(a) and (b).

Solution:

$$A_V = \frac{\Delta V_{out}}{\Delta V_{in}} = \frac{+9\text{ V to }+3\text{ V}}{+100\text{ mV to }-100\text{ mV}} = \frac{6\text{ V}}{200\text{ mV}} = 30$$

This value indicates that the output ac signal voltage is 30 times larger than the input ac signal voltage.

Most common-emitter transistor circuits have high voltage gains between 100 to 500.

Power Gain (A_P)
The ratio of the output signal power to the input signal power.

Power Gain. As we have seen so far, the common-emitter circuit provides both current gain and voltage gain. Because power is equal to the product of current and voltage ($P = V \times I$), it is not surprising that the *C-E* circuit configuration also provides **power gain (A_P)**. The power gain of a circuit can be calculated by dividing the output signal power (P_{out}) by the input signal power (P_{in}).

$$A_P = \frac{P_{out}}{P_{in}}$$

To calculate the amount of input power (P_{in}) applied to the *C-E* circuit, we will have to multiply the change in input signal voltage (ΔV_{in}) by the accompanying change in input signal current (ΔI_{in} or ΔI_B).

$$P_{in} = \Delta V_{in} \times \Delta I_{in}$$

To calculate the amount of output power (P_{out}) delivered by the *C-E* circuit, we will have to multiply the change in output signal voltage (ΔV_{out}) produced by the change in output signal current (ΔI_{out} or ΔI_C).

$$P_{out} = \Delta V_{out} \times \Delta I_{out}$$

The power gain of a common-emitter circuit is therefore calculated with the formula:

$$A_P = \frac{P_{out}}{P_{in}} = \frac{\Delta V_{out} \times \Delta I_{out}}{\Delta V_{in} \times \Delta I_{in}}$$

EXAMPLE:

Calculate the power gain of the example circuit in Figure 12-14.

Solution:

$$\begin{aligned} A_P &= \frac{P_{out}}{P_{in}} = \frac{\Delta V_{out} \times \Delta I_{out}}{\Delta V_{in} \times \Delta I_{in}} \\ &= \frac{(9\text{ V to }3\text{ V}) \times (1.5\text{ mA to }0.5\text{ mA})}{(+100\text{ mV to }-100\text{ mV}) \times (40\ \mu\text{A to }20\ \mu\text{A})} \\ &= \frac{6\text{ V} \times 1\text{ mA}}{200\text{ mV} \times 20\ \mu\text{A}} = \frac{6\text{ mW}}{4\ \mu\text{W}} = 1500 \end{aligned}$$

In this example, the common-emitter circuit has increased the input signal power from 4 μW to 6 mW—a power gain of 1500.

The power gain of the circuit in Figure 12-14 can also be calculated by multiplying the previously calculated *C-E* circuit voltage gain (A_V) by the previously calculated *C-E* circuit current gain (β_{AC}).

$$\text{Since } A_V = \frac{\Delta V_{out}}{\Delta V_{in}}, \text{ and } \beta_{AC} = \frac{\Delta I_{out}\ (\text{or } \Delta I_C)}{\Delta I_{in}\ (\text{or } \Delta I_B)}$$

$$A_P = V \times I = \frac{\Delta V_{out}}{\Delta V_{in}} \times \frac{\Delta I_{out}\ (\text{or } \Delta I_C)}{\Delta I_{in}\ (\text{or } \Delta I_B)} \text{ or } A_P = A_V \times \beta_{AC}$$

$$A_P = A_V \times \beta_{AC}$$

For the example circuit in Figure 12-14, this will be:

$$A_P = A_V \times \beta_{AC} = 30 \times 50 = 1500$$

indicating that the common-emitter circuit's output power in Figure 12-14 is 1500 times larger than the input power.

The power gain of common-emitter transistor circuits can be as high as 20,000, making this characteristic the circuit's key advantage.

Collector characteristic curves. One of the easiest ways to compare several variables is to combine all of the values in a graph. Figure 12-14(c) shows a special graph, called the **collector characteristic curves,** for a typical common-emitter transistor circuit. The data

Collector Characteristic Curves
Graph for a typical common-emitter transistor circuit.

for this graph are obtained by using the transistor test circuit, shown in the inset in Figure 12-14(c), which will apply different values of base bias voltage (V_{BB}) and collector bias voltage (V_{CC}) to an NPN transistor. The two ammeters and one voltmeter in this test circuit are used to measure the circuit's I_B, I_C, and V_{CE} response to each different circuit condition. The values obtained from this test circuit are then used to plot the transistor's collector current (I_C in mA) on the vertical axis against the transistor's collector-emitter voltage (V_{CE}) drop on the horizontal axis for various values of base current (I_B in μA). This graph shows the relationship between a transistor's input base current, output collector current, and collector-to-emitter voltage drop.

Let us now examine the typical set of collector characteristic curves shown in Figure 12-14(c). When V_{CE} is increased from zero, by increasing V_{CC}, the collector current rises very rapidly, as indicated by the rapid vertical rise in any of the curves. When the collector diode of the transistor is reverse biased by the voltage V_{CE}, the collector current levels off. At this point, any one of the curves can be followed based on the amount of base current, which is determined by the value of base bias voltage (V_{BB}) applied. This flat part of the curve is known as the transistor's **active region.** The transistor is normally operated in this region, where it is equivalent to a variable resistor between the collector and emitter.

Active Region

Flat part of the collector characteristic curve. A transistor is normally operated in this region, where it is equivalent to a variable resistor between the collector and emitter.

As an example, let us use these curves in Figure 12-14(c) to calculate the value of output current (I_C) for a given value of input current (I_B) and collector-to-emitter voltage (V_{CE}). If V_{BB} is adjusted to produce a base current of 30 μA, and V_{CC} is adjusted until the voltage between the transistor's collector and emitter (V_{CE}) is 4.5 V, the output collector current (I_C) will be equal to approximately 1 mA. This is determined by first locating 4.5 V on the horizontal axis (V_{CE} = 4.5 V), following this point directly up to the I_B = 30 μA curve, and then moving directly to the left to determine the value of output current on the vertical axis (I_C = 1 mA). When these values of V_{CE} and I_B are present, the transistor is said to be operating at point "*Q*." This dc operating point is often referred to as a **quiescent operating point (*Q* point),** which means a dc steady-state or no input signal operating point. The *Q* point of a transistor is set by the circuit's dc bias components and supply voltages. For instance, in the circuit in Figure 12-14(a), R_1 and R_2 were used to set the dc base bias voltage (V_B, and therefore I_B), and the values of V_{CC}, R_C, and R_E were chosen to set the transistor's dc collector-emitter voltage (V_{CE}). At this dc operating point, we can calculate the transistor's dc current gain (β_{DC}), since both I_B and I_C are known:

Quiescent Operating Point (*Q* Point)

The voltage or current value that sets up the no input signal or operating point bias voltage.

$$\beta_{DC} = \frac{I_C}{I_B} = \frac{1 \text{ mA}}{30 \text{ μA}} = 33.3$$

If a sine-wave signal was applied to the circuit, as shown in the waveforms in Figure 12-14(b) and the characteristic curves in Figure 12-14(c), it would cause the transistor's input base current, and therefore output collector current, to alternate above and below the transistor's *Q* point (dc operating point). This ac input signal voltage will cause the input base current (I_B) to increase between 20 μA and 40 μA, and this input base current change will generate an output collector current change of 0.5 mA to 1.5 mA. The transistor's ac current gain (β_{AC}) will be:

$$\begin{aligned}\beta_{AC} &= \frac{\Delta I_C}{\Delta I_B} \\ &= \frac{1.5 \text{ to } 0.5 \text{ mA}}{40 \text{ to } 20 \text{ μA}} \\ &= \frac{1 \text{ mA}}{20 \text{ μA}} \\ &= 50\end{aligned}$$

Returning to the collector characteristic curves in Figure 12-14(c), you can see that if the collector supply voltage (V_{CC}) is increased to an extreme, a point will be reached where the V_{CE} voltage across the transistor will cause the transistor to break down, as indicated by the rapid rise in I_C. This section of the curve is called the **breakdown region** of the graph, and the damaging value of current through the transistor will generally burn out and destroy

Breakdown Region

The point at which the collector supply voltage will cause a damaging value of current through the transistor.

the device. As an example, for the 2N3904 bipolar transistor, breakdown will occur at a V_{CE} voltage of 40 V.

There are two shaded sections shown in the set of collector characteristic curves in Figure 12-14(c). These two shaded sections represent the other two operating regions of the transistor. To begin with, let us examine the vertically shaded **saturation region.** If the base bias voltage (V_{BB}) is increased to a large positive value, the emitter diode of the transistor will turn ON heavily, I_B will be a large value, and the transistor will be operating in saturation. In this operating region, the transistor is equivalent to a closed switch between its collector and emitter (both the emitter and collector diode are forward biased), and therefore the voltage drop between collector and emitter will be almost zero (V_{CE} = typically 0.3 V, when transistor is saturated), and I_C will be a large value that is limited only by the externally connected components. The horizontally shaded section represents the **cutoff region** of the transistor. If the base bias voltage (V_{BB}) is decreased to zero, the emitter diode of the transistor will turn OFF, I_B will be zero, and the transistor will be operating in cutoff. In this operating region, the transistor is equivalent to an open switch between its collector and emitter (both the emitter and collector diode are reverse biased), and the voltage drop between collector and emitter will be equal to V_{CC} and I_C will be zero.

Saturation Region
The point at which the collector supply voltage has the transistor operating at saturation.

Cutoff Region
The point at which the collector supply voltage has the transistor operating in cutoff.

A set of collector characteristic curves are therefore generally included in a manufacturer's device data sheet, and can be used to determine the values of I_B, I_C, and V_{CE} at any operating point.

Input Resistance. The **input resistance (R_{in})** of a common-emitter transistor is the amount of opposition offered to an input signal by the input base-emitter junction (emitter diode). Because the base-emitter junction is normally forward biased when the transistor is operating in the active region, the opposition to input current is relatively small. However, the extremely small base region will only support a very small input base current. On average, if no additional components are connected in series with the transistor's base-emitter junction, the input resistance of a *C-E* transistor circuit is typically a medium value between 1 kΩ and 5 kΩ. This typical value is an average because the transistor's input resistance is a "dynamic or changing quantity" that will vary slightly as the input signal changes the conduction of the *C-E* transistor's emitter diode, and this changes I_B ($R \updownarrow\updownarrow = V/I \updownarrow\updownarrow$).

Input Resistance (R_{in})
The amount of opposition offered to an input signal by the input base-emitter junction (emitter diode).

Because the transistor has a small value of input P-N junction capacitance and input terminal inductance, the opposition to the input signal is not only resistive but, to a small extent, reactive. For this reason, the total opposition offered by the transistor to an input signal is often referred to as the **input impedance (Z_{in})** because impedance is the total combined resistive and reactive input opposition.

Input Impedance (Z_{in})
The total opposition offered by the transistor to an input signal.

Output Resistance. The **output resistance (R_{out})** of a common-emitter transistor is the amount of opposition offered to an output signal by the output base-collector junction (collector diode). This junction is normally reverse biased when the transistor is operating in the active region, and therefore the *C-E* transistor's output resistance is relatively high. However, because a unique action occurs within the transistor and allows current to flow through this reverse biased junction (electron accumulation at the base and then conduction through collector diode due to attraction of $+V_{CC}$), the output current (I_C) is normally large, and so the output resistance is not an extremely large value. On average, if no load resistor is connected in series with the transistor's collector diode, the output resistance of a *C-E* transistor circuit is typically a high value between 40 kΩ and 60 kΩ.

Output Resistance (R_{out})
The amount of opposition offered to an output signal by the output base-collector junction (collector diode).

Because the transistor has a small value of output P-N junction capacitance and output terminal inductance, the opposition to the output signal is not only resistive but also reactive. For this reason, the total opposition offered by the transistor to the output signal is often referred to as the **output impedance (Z_{out}).**

Output Impedance (Z_{out})
The total opposition offered by the transistor to the output signal.

Common-Base Circuits

With the **common-base circuit,** the transistor's base lead is common to both the input and output signal. In this circuit configuration, the emitter serves as the input lead, and the collector serves as the output lead. Figure 12-15 contains a basic common-base (*C-B*) circuit,

Common-Base (*C-B*) Circuit
Configuration in which the input signal is applied between the transistor's emitter and base, while the output is developed across the transistor's collector and base.

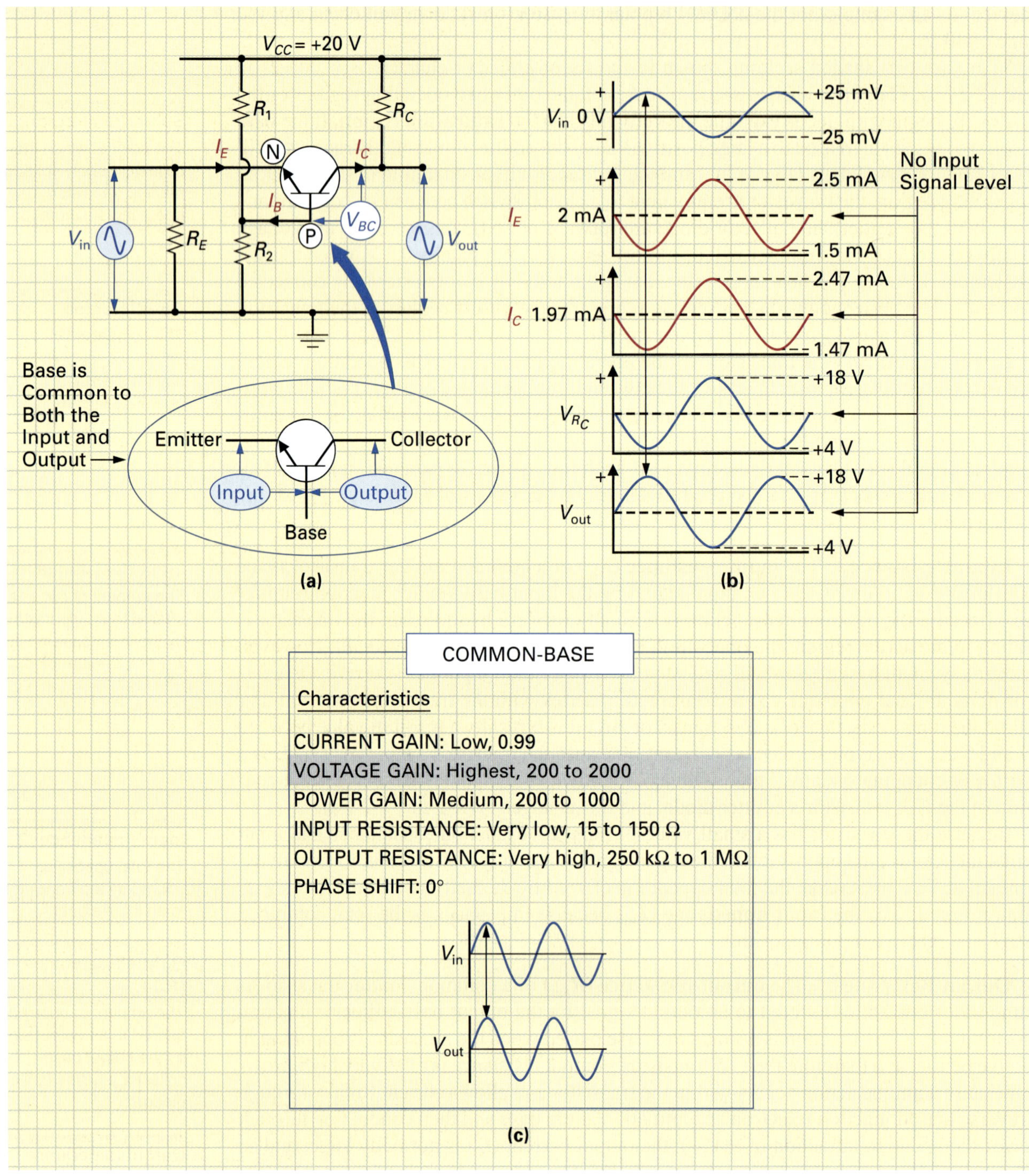

FIGURE 12-15 **Common-Base *(C-B)* Circuit Configuration Characteristics.**

its associated input/output voltage and current waveforms, and table of typical characteristics. Using this illustration, we will examine the operation and characteristics of the *C-B* circuit configuration.

DC current gain. Referring to the *C-B* circuit in Figure 12-15(a), and its associated waveforms in Figure 12-15(b), let us now examine this circuit's basic operation.

Before applying the ac sine-wave input signal (V_{in}), let us assume that V_{in} = 0 V and examine the transistor's dc operating characteristics, or the "no input signal" condition. The voltage divider R_1 and R_2 will divide the V_{CC} supply voltage, producing a positive dc base bias voltage across R_2. This base bias voltage will be applied to the base of Q_1. Because it is generally greater than 0.7 V, it will forward bias the transistor's base-emitter junction, turning Q_1 ON. Because the common-base circuit's input current is I_E and its output current is I_C, the current gain between input and output will be determined by the ratio of I_C to I_E. This ratio for calculating a *C-B* transistor's dc current gain is called the transistor's **dc alpha (α_{DC})** and is equal to

DC Alpha (α_{DC})
The ratio for calculating a *C-B* transistor's dc current gain.

$$\alpha_{DC} = \frac{I_C}{I_E}$$

The "no input signal" or "steady state" dc levels of I_E and I_C are determined by the value of voltage developed across R_2, which is controlling the conduction of the transistor's forward biased base-emitter junction (emitter diode). Because the output current I_C is always slightly lower than the input current I_E (due to the small I_B current flow out of the base), the *C-B* transistor circuit does not increase current between input and output. In fact, there is a slight loss in current between input and output, which is why the *C-B* circuit configuration is said to have a current gain that is less than 1.

EXAMPLE:

Calculate the dc alpha of the circuit in Figure 12-15(a) if I_C = 1.97 mA and I_E = 2 mA.

Solution:

$$\alpha_{DC} = \frac{I_C}{I_E} = \frac{1.97 \text{ mA}}{2 \text{ mA}} = 0.985$$

This value of 0.985 indicates that I_C is 98.5% of I_E (0.985 × 100 = 98.5).

As you can see from this example, the difference between I_C and I_E is generally so small that we always assume that the dc alpha is 1, which means that $I_C = I_E$.

AC current gain. When amplifying an ac waveform, a transistor has applied to it both the dc voltage to make it operational and the ac signal voltage to be amplified that varies the base-emitter bias and the input emitter current. Referring to the waveforms in Figure 12-15(b), let us now examine what will happen when we apply a sine-wave input signal. As mentioned previously, the positive voltage developed across R_2 will make the NPN transistor's base positive with respect to the emitter and forward bias the P-N base-emitter junction.

As the input voltage swings positive ($V_{in}\uparrow$), it will reduce the forward bias across the transistor's P-N base-emitter junction. For example, if V_B = +5 V and the transistor's *n*-type emitter is made positive, the P-N base-emitter diode will be turned more OFF. Turning the transistor's emitter diode less ON will cause a decrease in emitter current ($I_E\downarrow$), a decrease in collector current ($I_C\downarrow$), and a decrease in the voltage developed across R_C ($V_{R_C}\downarrow$). Because V_{R_C} and V_{out} are connected in series across V_{CC}, a decrease in $V_{R_C}\downarrow$ must be accompanied by an increase in $V_{out}\uparrow$. To explain this another way, the decrease in I_E and the subsequent decrease in both I_C and I_B means that the conduction of the transistor has decreased. This decrease in conduction means that the normally forward biased base-emitter junction has turned less ON, and the normally reverse biased base-collector junction has turned more OFF. Because the transistor's base-collector junction is in series with R_C and R_2 across V_{CC}, an increase in the transistor's base-collector resistance ($R_{BC}\uparrow$) will cause an increase in the voltage developed across the transistor's base-collector junction ($V_{BC}\uparrow$), which will cause an increase in $V_{out}\uparrow$.

Similarly, as the input voltage swings negative ($V_{in}\downarrow$), it will increase the forward bias across the transistor's P-N base-emitter junction. For example, if $V_B = -5$ V and the transistor's *n*-type emitter is made negative, the P-N base-emitter diode will be turned more ON. Turning the transistor's emitter diode more ON will cause an increase in emitter current ($I_E\uparrow$), an increase in collector current ($I_C\uparrow$), and an increase in the voltage developed across R_C ($V_{R_C}\uparrow$). Because V_{R_C} and V_{out} are connected in series across V_{CC}, an increase in $V_{R_C}\uparrow$ must be accompanied by a decrease in $V_{out}\downarrow$.

Now that we have seen how the input voltage causes a change in input current (I_E) and output current (I_C), let us examine the *C-B* circuit's ac current gain. The ratio of input emitter current change (ΔI_E) to output collector current change (ΔI_C) is the *C-B* transistor's ac current gain or **AC alpha (α_{AC}).** The formula for calculating a *C-B* transistor's ac current gain is:

AC Alpha (α_{AC})
The ratio of input emitter current change to output collector current change.

$$\alpha_{AC} = \frac{\Delta I_C}{\Delta I_E}$$

EXAMPLE:

Calculate the ac current gain of the example in Figure 12-15(b).

Solution:

$$\alpha_{AC} = \frac{\Delta I_C}{\Delta I_E}$$

$$= \frac{2.47 \text{ to } 1.47 \text{ mA}}{2.5 \text{ to } 1.5 \text{ mA}}$$

$$= \frac{1 \text{ mA}}{1 \text{ mA}}$$

$$= 1$$

This means that the change in output collector current is equal to the change in input emitter current, and therefore the ac current gain is 1. (The output is 1 times larger than the input, 1 mA × 1 = 1 mA.)

Common-base transistors will typically have an ac alpha, or ac current gain, of 0.99.

Voltage Gain. Although the common-base circuit does not achieve any current gain, it does make up for this disadvantage by achieving a very large voltage gain between input and output. Returning to the *C-B* circuit in Figure 12-15(a) and its waveforms in Figure 12-15(b), let us see how this very high voltage gain is obtained. Only a small input voltage (V_{in}) is needed to control the conduction of the transistor's emitter diode, and therefore the input emitter current (I_E) and output collector current (I_C). Even though I_C is slightly lower than I_E, it is still a relatively large value of current and will develop a large voltage change across R_C for a very small change in V_{in}. Because V_{R_C} and V_{out} are in series and connected across V_{CC}, a large change in voltage across R_C will cause a large change in the voltage developed across the transistor output (V_{BC}) and V_{out}. As before, the amount of voltage gain (which is symbolized A_V) can be calculated by comparing the output voltage change (ΔV_{out}) to the input voltage change (ΔV_{in}).

$$A_V = \frac{\Delta V_{out}}{\Delta V_{in}}$$

EXAMPLE:

Calculate the voltage gain of the circuit and its associated waveforms in Figure 12-15(a) and (b).

Solution:

$$A_V = \frac{\Delta V_{out}}{\Delta V_{in}} = \frac{+18\text{ V to }+4\text{ V}}{+25\text{ mV to }-25\text{ mV}} = \frac{14\text{ V}}{50\text{ mV}} = 280$$

This value indicates that the output ac signal voltage is 280 times larger than the ac input signal voltage.

Most common-base transistor circuits have very high voltage gains between 200 and 2000. While on the topic of comparing the input voltage to output voltage, you can see by looking at Figure 12-15(b) that, unlike the *C-E* circuit, the common-base circuit has no phase shift between input and output (V_{out} is in phase with V_{in}).

Power Gain. Although the common-base circuit achieves no current gain, it does have a very high voltage gain and therefore can provide a medium amount of power gain ($P\uparrow = V\uparrow \times I$). The power gain of a circuit can be calculated by dividing the output signal power (P_{out}) by the input signal power (P_{in}).

$$A_P = \frac{P_{out}}{P_{in}}$$

To calculate the amount of input power (P_{in}) applied to the *C-B* circuit, we will have to multiply the change in input signal voltage (ΔV_{in}) by the accompanying change in input signal current (ΔI_{in} or ΔI_E).

$$P_{in} = \Delta V_{in} \times \Delta I_{in}$$
$$P_{in} = 50\text{ mV} \times 1\text{ mA} = 50\ \mu\text{W}$$

To calculate the amount of output power (P_{out}) delivered by the *C-B* circuit, we will have to multiply the change in output signal voltage (ΔV_{out}) produced by the change in output signal current (ΔI_{out} or ΔI_C).

$$P_{out} = \Delta V_{out} \times \Delta I_{out}$$
$$P_{out} = 14\text{ V} \times 1\text{ mA} = 14\text{ mW}$$

The power gain of a common-base circuit is calculated with the formula:

$$A_P = \frac{P_{out}}{P_{in}} = \frac{\Delta V_{out} \times \Delta I_{out}}{\Delta V_{in} \times \Delta I_{in}}$$

EXAMPLE:

Calculate the power gain of the example circuit in Figure 12-15.

Solution:

$$A_P = \frac{P_{out}}{P_{in}} = \frac{\Delta V_{out} \times \Delta I_{out}}{\Delta V_{in} \times \Delta I_{in}}$$
$$= \frac{14\text{ V} \times 1\text{ mA}}{50\text{ mV} \times 1\text{ mA}} = \frac{14\text{ mW}}{50\ \mu\text{W}} = 280$$

In this example, the common-base circuit has increased the input signal power from 50 μW to 14 mW—a power gain of 280.

The power gain of the circuit in Figure 12-15 can also be calculated by multiplying the previously calculated *C-B* circuit voltage gain (A_V) by the previously calculated *C-B* circuit current gain (α_{AC}).

$$A_P = A_V \times \alpha_{AC}$$

For the example circuit in Figure 12-15, this will be

$$A_P = A_V \times \alpha_{AC} = 280 \times 1 = 280$$

indicating that the common-base circuit's output power in Figure 12-15 is 280 times larger than the input power.

Typical common-base circuits will have power gains from 200 to 1000.

Input Resistance. The input resistance (R_{in}) of a common-base transistor is the amount of opposition offered to an input signal by the input base-emitter junction (emitter diode). Because the base-emitter junction is normally forward biased and the input emitter current (I_E) is relatively large, the input signal sees a very low input resistance. On average, if no additional components are connected in series with the transistor's base-emitter junction, the input resistance of a *C-B* transistor circuit is typically a low value between 15 Ω and 150 Ω. This typical value is an average because the transistor's input resistance is a dynamic or changing quantity that will vary slightly as the input signal changes the conduction of the *C-B* transistor's emitter diode, and this changes I_E ($R \updownarrow = V/I \updownarrow$).

Output Resistance. The output resistance (R_{out}) of a common-base transistor is the amount of opposition offered to an output signal by the output base-collector junction (collector diode). This junction is normally reverse biased when the transistor is operating in the active region, and therefore the *C-B* transistor's output resistance is relatively high. On average, if no load resistor is connected in series with the transistor's collector diode, the output resistance of a *C-B* transistor circuit is typically a very high value between 250 kΩ and 1 MΩ.

Common-Collector Circuits

Common-Collector (*C-C*) Circuit

Configuration in which the input signal is applied between the transistor's base and collector, while the output is developed across the transistor's collector and emitter.

With the **common-collector circuit,** the transistor's collector lead is common to both the input and output signal. In this circuit configuration therefore, the base serves as the input lead, and the emitter serves as the output lead. Figure 12-15 contains a basic common-collector (*C-C*) circuit, its associated input/output voltage and current waveforms, and table of typical characteristics. Using this illustration, we will examine the operation and characteristics of the *C-C* circuit configuration.

DC Current Gain. Referring to the *C-C* circuit in Figure 12-16(a) and its associated waveforms in Figure 12-16(b), let us now examine this circuit's basic operation.

Before applying the ac sine-wave input signal (V_{in}), let us assume that V_{in} = 0 V and examine the transistor's dc operating characteristics, or the "no input signal" condition. The voltage divider R_1 and R_2 will divide the V_{CC} supply voltage, producing a positive dc base bias voltage across R_2. This base bias voltage will be applied to the base of Q_1. Because it is generally greater than 0.7 V, it will forward bias the transistor's base-emitter junction, turning Q_1 ON. Because the common-collector (*C-C*) circuit's input current I_B is much smaller than the output current I_E, the circuit provides a high current gain. In fact, the common-collector circuit provides a slightly higher gain than the *C-E* circuit because the common-collector's output current (I_E) is slightly higher than the *C-E*'s output current (I_C).

Like any circuit configuration, the dc current gain is equal to the ratio of output current to input current. For the *C-C* circuit, this is equal to the ratio of I_E to I_B:

$$\text{DC Current Gain} = \frac{I_E}{I_B}$$

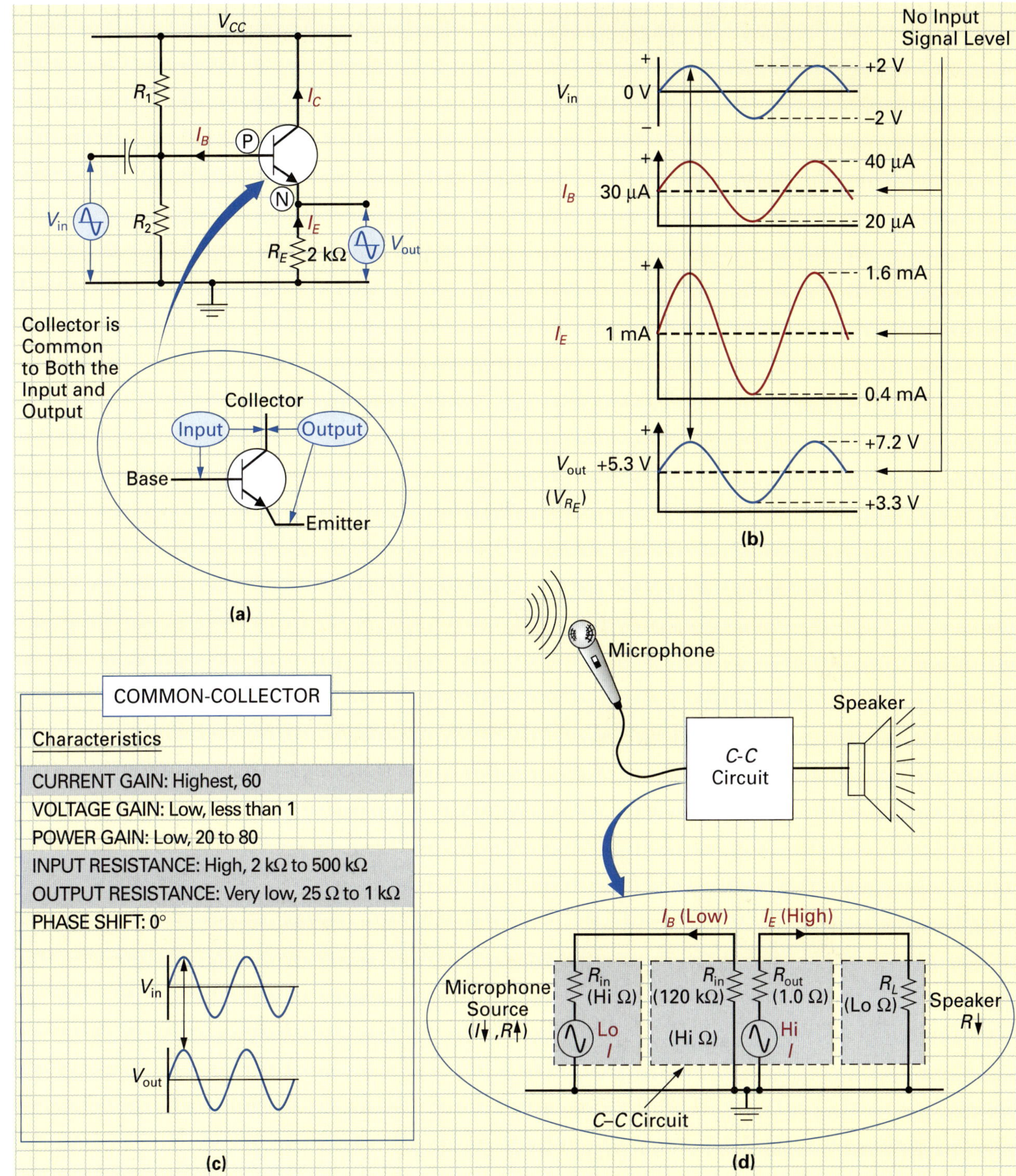

FIGURE 12-16 **Common-Collector *(C-C)* Circuit Configuration Characteristics.**

Transistor manufacturers will generally not provide specifications for all three circuit configurations. In most cases, because the common-emitter (*C-E*) circuit configuration is most frequently used, manufacturers will give the transistor's characteristics for only the *C-E* circuit configuration. In these instances, we will have to convert this *C-E* circuit data to equivalent specifications for other configurations. For example, in most data sheets the transistor's dc current gain will be listed as β_{DC}. As we know, dc beta is the measure of a *C-E* circuit's current gain because it compares input current I_B to output current I_C. How, then, can we convert this value so that it indicates the dc current gain of a common-collector circuit? The answer is as follows:

$$\text{Common-Collector DC Current Gain} = \frac{\text{Output Current}}{\text{Input Current}} = \frac{I_E}{I_B}$$

Since $I_E = I_B + I_C$,

$$\text{DC Current Gain} = \frac{I_E}{I_B} = \frac{(I_B + I_C)}{I_B}$$

Since $I_B \div I_B = 1$,

$$\text{DC Current Gain} = 1 + \frac{I_C}{I_B}$$

Since $\frac{I_C}{I_B} = \beta_{DC}$,

$$\text{DC Current Gain} = 1 + \frac{I_C}{I_B} = 1 + \beta_{DC}$$

$$\text{DC Current Gain} = 1 + \beta_{DC}$$

EXAMPLE:

Calculate the dc current gain of the *C-C* circuit in Figure 12-16 if the transistor's $\beta_{DC} = 32.33$.

Solution:

$$\text{DC Current Gain} = 1 + \frac{I_C}{I_B} = 1 + \frac{(I_E - I_B)}{I_B} = 1 + \frac{1\text{ mA} - 30\ \mu\text{A}}{30\ \mu\text{A}}$$

$$= 1 + \frac{970\ \mu\text{A}}{30\ \mu\text{A}} = 1 + 32.33 = 33.33$$

or,

$$\text{DC Current Gain} = 1 + \beta_{DC} = 1 + 32.33 = 33.33$$

As you can see in this example, the dc current gain of a common-collector circuit ($\beta_{DC} + 1$) is slightly higher than the dc current gain of a *C-E* circuit (β_{DC}). In most instances, the extra 1 makes so little difference when the transistor's dc current gain is a large value of about 30, as in this example, that we assume that the current gain of a common-collector circuit is equal to the current gain of a *C-E* circuit.

$$C\text{-}C \text{ DC Current Gain} \cong C\text{-}E \text{ DC Current Gain } (\beta_{DC})$$

AC Current Gain. When amplifying an ac waveform, a transistor has applied to it both the dc voltage to make it operational and the ac signal voltage that varies the base-emitter bias and the input base current. Referring to the waveforms in Figure 12-15(b), let us now examine what will happen when we apply a sine-wave input signal. As mentioned previously, the positive voltage developed across R_2 will make the NPN transistor's base positive with respect to the emitter, and therefore forward bias the P-N base-emitter junction.

As the input voltage swings positive ($V_{in}\uparrow$), it will add to the forward bias applied across the transistor's P-N base-emitter junction. This means that the transistor's emitter diode will turn more ON, cause an increase in the $I_B\uparrow$, and therefore a proportional but much larger increase in the output current $I_E\uparrow$.

Similarly, as the input voltage swings negative ($V_{in}\downarrow$), it will subtract from the forward bias applied across the transistor's P-N base-emitter junction. This means that the transistor's emitter diode will turn less ON, cause a decrease in the $I_B\downarrow$, and therefore a proportional but larger decrease in the output current $I_E\downarrow$.

The ac current gain of a common-collector transistor is calculated using the same formula as dc current gain. However, with an ac current, we will compare the output current change (ΔI_E) to the input current change (ΔI_B).

$$\text{AC Current Gain} = \frac{\Delta I_E}{\Delta I_B}$$

EXAMPLE:

Calculate the ac current gain of the circuit in Figure 12-16(a), using the values in Figure 12-16(b).

Solution:

$$\text{AC Current Gain} = \frac{\Delta I_E}{\Delta I_B} = \frac{1.6\text{ mA} - 0.4\text{ mA}}{40\ \mu\text{A} - 20\ \mu\text{A}} = 60$$

Like the common-collector's dc current gain, because there is so little difference between I_E and I_C, we can assume that the ac current gain is equivalent to β_{AC}.

$$C\text{-}C\text{ AC Current Gain} \cong C\text{-}E\text{ AC Current Gain } (\beta_{AC})$$

Common-collector transistor circuit configurations can have current gains as high as 60, indicating that I_E is sixty times larger than I_B.

Voltage Gain. Although the common-collector circuit has a very high current gain rating, it cannot increase voltage between input and output. Returning to the *C-C* circuit in Figure 12-16(a) and its waveforms in Figure 12-16(b), let us see why this circuit has a very low voltage gain.

As the input voltage swings positive ($V_{in}\uparrow$), it will add to the forward bias applied across the transistor's P-N base-emitter junction ($V_{BE}\uparrow$). As the transistor's emitter diode turns more ON, it will cause an increase in $I_B\uparrow$, a proportional but larger increase in $I_E\uparrow$, and therefore an increase in the voltage developed across R_E (V_{R_E}, V_{out}, or $V_E\uparrow$). This increase in the voltage developed across R_E has a **degenerative effect** because an increase in the emitter voltage ($V_E\uparrow$) will counter the initial increase in base voltage ($V_B\uparrow$), and therefore the voltage difference between the transistor's base and emitter will remain almost constant (V_{BE} is almost constant). In other words, if the base goes positive and then the emitter goes positive, there is almost no increase in the potential difference between the base and the emitter and so the change in forward bias is almost zero. There is, in fact, a very small change in forward bias between base and emitter, and this will cause a small change in I_B and I_E, and therefore a small output voltage will be developed across R_E. Comparing the input and output

Degenerative Effect
An effect that causes a reduction in amplification due to negative feedback.

Emitter-Follower or Voltage-Follower
The common-collector circuit in which the emitter output voltage seems to track or follow the phase and amplitude of the input voltage.

voltage signals in Figure 12-15(b), you can see that both are about 4 V pk-pk, and both are in phase with one another. The common-collector circuit is often referred to as an **emitter-follower** or **voltage-follower** because the emitter output voltage seems to track or follow the phase and amplitude of the input voltage.

As with all circuit configurations, the amount of voltage gain (A_V) can be calculated by comparing the output voltage change (ΔV_{out}) to the input voltage change (ΔV_{in}).

$$A_V = \frac{\Delta V_{out}}{\Delta V_{in}}$$

EXAMPLE:

Calculate the voltage gain of the circuit and its associated waveforms in Figure 12-16(a) and (b).

Solution:

$$A_V = \frac{\Delta V_{out}}{\Delta V_{in}} = \frac{+7.2 \text{ V to } +3.3 \text{ V}}{+2 \text{ V to } -2 \text{ V}} = \frac{3.9 \text{ V}}{4 \text{ V}} = 0.975$$

This value indicates that the output ac signal voltage is 0.975 or 97.5% of the ac input signal voltage ($0.975 \times 4 \text{ V} = 3.9 \text{ V}$).

Most common-collector transistor circuits have a voltage gain that is less than 1. However, in most circuit examples it is assumed that output voltage change equals input voltage change.

Power gain. Although the common-collector circuit achieves no voltage gain, it does have a very high current gain, and therefore can provide a small amount of power gain ($P\uparrow = V \times I\uparrow$). As before, the power gain of a circuit can be calculated by dividing the output signal power (P_{out}) by the input signal power (P_{in}).

$$A_P = \frac{P_{out}}{P_{in}}$$

To calculate the amount of input power (P_{in}) applied to the *C-C* circuit, we will have to multiply the change in input signal voltage (ΔV_{in}) by the accompanying change in input signal current (ΔI_{in} or ΔI_B).

$$P_{in} = \Delta V_{in} \times \Delta I_{in}$$
$$P_{in} = 4 \text{ V} \times 20 \text{ µA} = 80 \text{ µW}$$

To calculate the amount of output power (P_{out}) delivered by the *C-C* circuit, we will have to multiply the change in output signal voltage (ΔV_{out}) produced by the change in output signal current (ΔI_{out} or ΔI_E).

$$P_{out} = \Delta V_{out} \times \Delta I_{out}$$
$$P_{out} = 3.9 \text{ V} \times 1.2 \text{ mA} = 4.68 \text{ mW}$$

The power gain of a common-collector circuit is therefore calculated with the formula:

$$A_P = \frac{P_{out}}{P_{in}} = \frac{\Delta V_{out} \times \Delta I_{out}}{\Delta V_{in} \times \Delta I_{in}}$$

EXAMPLE:

Calculate the power gain of the example circuit in Figure 12-16.

Solution:

$$A_P = \frac{P_{\text{out}}}{P_{\text{in}}} = \frac{\Delta V_{\text{out}} \times \Delta I_{\text{out}}}{\Delta V_{\text{in}} \times \Delta I_{\text{in}}}$$

$$= \frac{3.9 \text{ V} \times 1.2 \text{ mA}}{4 \text{ V} \times 20\ \mu\text{A}} = \frac{4.68 \text{ mW}}{80\ \mu\text{W}} = 58.5$$

In this example, the common-collector circuit has increased the input signal power from 80 μW to 4.68 mW—a power gain of 58.5.

The power gain of the circuit in Figure 12-16 can also be calculated by multiplying the previously calculated *C-C* circuit voltage gain (A_V) by the previously calculated *C-C* circuit current gain.

$$A_P = A_V \times \text{AC Current Gain}$$

For the example circuit in Figure 12-15, this will be

$$A_P = A_V \times \text{AC Current Gain} = 0.975 \times 60 = 58.5$$

indicating that the common-collector circuit's output power in Figure 12-16 is 58.5 times larger than the input power.

Typical common-collector circuits will have power gains from 20 to 80.

Input Resistance. An input signal voltage will see a very large input resistance when it is applied to a common-collector circuit. This is because the input signal sees the very large emitter-connected resistor ($R_E\uparrow\uparrow$) and, to a smaller extent, the resistance of the forward biased base-emitter junction ($R_{\text{in}}\uparrow = V_{\text{in}}/I_B\downarrow$: R_{in} is large because I_{in} or I_B is small). Using these two elements, we can derive a formula for calculating the input resistance of a *C-C* transistor circuit.

$$R_{\text{in}} = R_E \times \text{AC Current Gain}$$

Since

$$C\text{-}C \text{ AC Current Gain} \cong C\text{-}E \text{ AC Current Gain } (\beta_{AC})$$

the input resistance can also be calculated with the formula:

$$R_{\text{in}} = R_E \times \beta_{AC}$$

EXAMPLE:

Calculate the input resistance of the circuit in Figure 12-16, assuming $\beta_{AC} = 60$.

Solution:

$$R_{\text{in}} = R_E \times \beta_{AC} = 2 \text{ k}\Omega \times 60 = 120 \text{ k}\Omega$$

This means that an input voltage signal will see this *C-C* circuit as a resistance of 120 kΩ.

The input resistance of a *C-C* transistor circuit is typically a very large value between 2 kΩ and 500 kΩ.

Output Resistance. The output signal from a common-collector circuit sees a very low output resistance, as proved by this circuit's very high output current gain. Like this circuit's input resistance, the output resistance is largely dependent on the value of the emitter resistor R_E.

The output resistance of a typical *C-C* transistor circuit is a very low value between 25 Ω and 1 kΩ.

Impedance or Resistance Matching. Do not be misled into thinking that the very high input resistance and low output resistance of the common-collector transistor circuit are disadvantages. On the contrary, the very high input resistance and low output resistance of this configuration are made use of in many circuit applications, along with the *C-C* circuit's other advantage of high current gain.

To explain why a high input resistance and a low output resistance are good circuit characteristics, refer to the application circuit in Figure 12-16(d). In this example, a microphone is connected to the input of a *C-C* amplifier, and the output of this circuit is applied to a speaker. As we know, the sound wave input to the microphone will physically move a magnet within the microphone, which will in turn interact to induce a signal voltage into a stationary coil. This voice signal voltage from the microphone, which is our source, is then applied across the input resistance of our example *C-C* circuit, which is our load.

In the inset in Figure 12-16(d), you can see that the microphone has been represented as a low-current ac source with a high internal resistance, and the input resistance of the *C-C* circuit is shown as a high value (in the previous example, 120 kΩ) resistor. Remembering our previous discussion on sources and loads, we know that a small load resistance will cause a large current to be drawn from the source, and this large current will drain or pull down the source voltage. Many signal sources, such as microphones, can only generate a small signal source voltage because they have a high internal resistance. If this small signal source voltage is applied across an amplifier with a small input resistance, a large current will be drawn from the source. This heavy load will pull the signal voltage down to such a small value that it will not be large enough to control the amplifier. A large amplifier input resistance ($R_{in}\uparrow$), on the other hand, will not load the source. Therefore the input voltage applied to the amplifier will be large enough to control the amplifier circuit, to vary its transistor currents, and achieve the gain between amplifier input and output. In summary, the high input resistance of the *C-C* circuit can be connected to a high resistance source because it will not draw an excessive current and pull down the source voltage.

Referring again to the inset in Figure 12-16(d), you can see that at the output end, the *C-C*'s output circuit has been represented as a high current source with a low value internal output resistor, and the speaker has been represented as a low resistance load. The low output resistance of the *C-C* circuit means that this circuit can deliver the high current output that is needed to drive the low resistance load.

As you will see later in application circuits, most *C-C* circuits are used as a resistance or **impedance matching circuit** that can match, or isolate, a high resistance (low current) source, such as a microphone, to a low resistance (high current) load, such as the speaker. By acting as a **buffer current amplifier,** the *C-C* circuit can ensure that power is efficiently transferred from source to load.

Impedance Matching Circuit

A circuit that can match, or isolate, a high resistance (low current) source.

Buffer Current Amplifier

The *C-C* circuit that can ensure that power is efficiently transferred from source to load.

12-2-3 *Testing Bipolar Junction Transistors*

Although transistors are exceptionally more reliable than their counterpart, the vacuum tube, they still will malfunction. These failures are normally the result of excessive temperature, current, or mechanical abuse and generally result in one of three problems:

1. An open between two or three of the transistor's leads
2. A short between two or three of the transistor's leads
3. A change in the transistor's characteristics

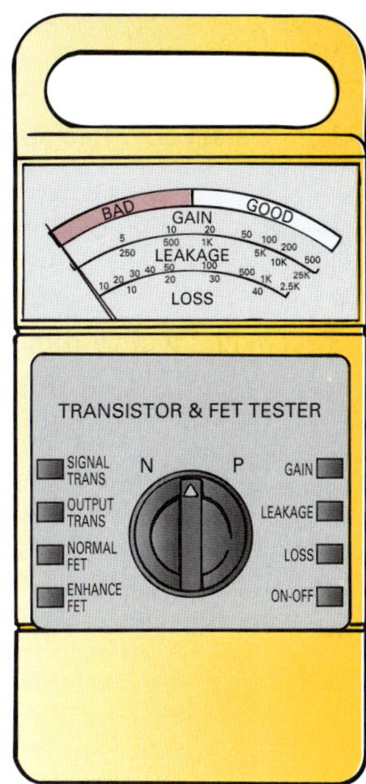

FIGURE 12-17 **Transistor Tester.**

Transistor Tester

The **transistor tester** shown in Figure 12-17 is a special test instrument that can be used to test both NPN and PNP bipolar transistors. This special meter can be used to determine whether an open or short exists between any of the transistor's three terminals, the transistor's dc current gain (β_{DC}), and whether an undesirable value of leakage current is present through one of the transistor's junctions.

Transistor Tester
Special test instrument that can be used to test both NPN and PNP bipolar transistors.

Ohmmeter Transistor Test

If the transistor tester is not available, the ohmmeter can be used to detect open and shorted junctions, which are the most common transistor failures. Figure 12-18 shows the step-by-step procedure for testing an NPN transistor. Following through this test procedure, we begin by reverse biasing the collector diode and then the emitter diode, and then forward biasing the collector diode and then the emitter diode. The table in Figure 12-18 shows the order and action to be performed for each step and the reading that should result if the NPN transistor junction is operating correctly.

12-2-4 *Bipolar Transistor Biasing Circuits*

As we discovered in the previous discussion on transistor circuit configurations, the ac operation of a transistor is determined by the "dc bias level," or "no input signal level." This steady-state or dc operating level is set by the value of the circuit's dc supply voltage (V_{CC}) and the value of the circuit's biasing resistors. This single supply voltage and the one or more biasing resistors set up the initial dc values of transistor current (I_B, I_E and I_C) and transistor voltage (V_{BE}, V_{CE} and V_{BC}).

In this section, we will examine some of the more commonly used methods for setting the "initial dc operating point" of a bipolar transistor circuit. As you encounter different circuit applications, you will see that many of these circuits include combinations of these basic biasing techniques and additional special-purpose components for specific functions. Because the common-emitter (*C-E*) circuit configuration is used more extensively than the *C-B* and the *C-C*, we will use this configuration in all of the following basic biasing circuit examples.

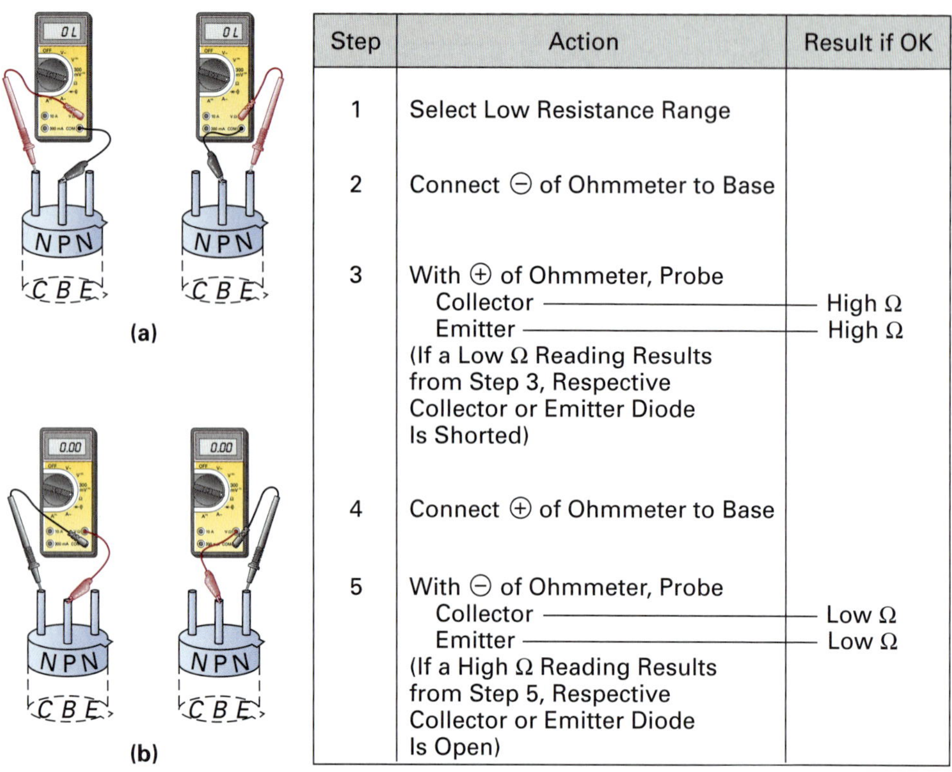

Step	Action	Result if OK
1	Select Low Resistance Range	
2	Connect ⊖ of Ohmmeter to Base	
3	With ⊕ of Ohmmeter, Probe Collector Emitter (If a Low Ω Reading Results from Step 3, Respective Collector or Emitter Diode Is Shorted)	 High Ω High Ω
4	Connect ⊕ of Ohmmeter to Base	
5	With ⊖ of Ohmmeter, Probe Collector Emitter (If a High Ω Reading Results from Step 5, Respective Collector or Emitter Diode Is Open)	 Low Ω Low Ω

FIGURE 12-18 NPN Ohmmeter Test Procedure. (a) Reverse-Biasing Collector, Then Emitter, Diode. (b) Forward-Biasing Collector, Then Emitter, Diode.

Base Biasing

Base Biasing
A transistor biasing method in which the dc supply voltage is applied to the base of the transistor via a base bias resistor.

Figure 12-19(a) shows how a common-emitter transistor circuit could be base biased. With **base biasing,** the emitter diode of the transistor is forward biased by applying a positive base bias voltage ($+V_{BB}$) via a current-limiting resistor (R_B) to the base of Q_1. In Figure 12-19(b), the transistor circuit from Figure 12-19(a) has been redrawn so as to simplify the analysis of the circuit. The transistor is now represented as a diode between base and emitter (emitter diode), and the transistor's emitter to collector has been represented as a variable resistor. Assuming Q_1 is a silicon bipolar transistor, the forward biased emitter diode will have a standard base-emitter voltage drop of 0.7 V (emitter diode drop = 0.7 V).

$$V_{BE} = 0.7\ \text{V}$$

The base bias resistor (R_B) and the transistor's emitter diode form a series circuit across V_{BB}, as seen in Figure 12-18(b). Therefore, the voltage drop across R_B (V_{R_B}) will be equal to the difference between V_{BB} and V_{BE}.

$$\begin{aligned} V_{R_B} &= V_{BB} - V_{BF} \\ &= V_{BB} - 0.7\ \text{V} \end{aligned}$$

$$V_{R_B} = V_{BB} - V_{BE} = 10\ \text{V} - 0.7\ \text{V} = 9.3\ \text{V}$$

Now that the resistance and voltage drop across R_B are known, we can calculate the current through R_B (I_{R_B}). Because a series circuit is involved, the current through R_B (I_{R_B}) will also be equal to the transistor base current I_B.

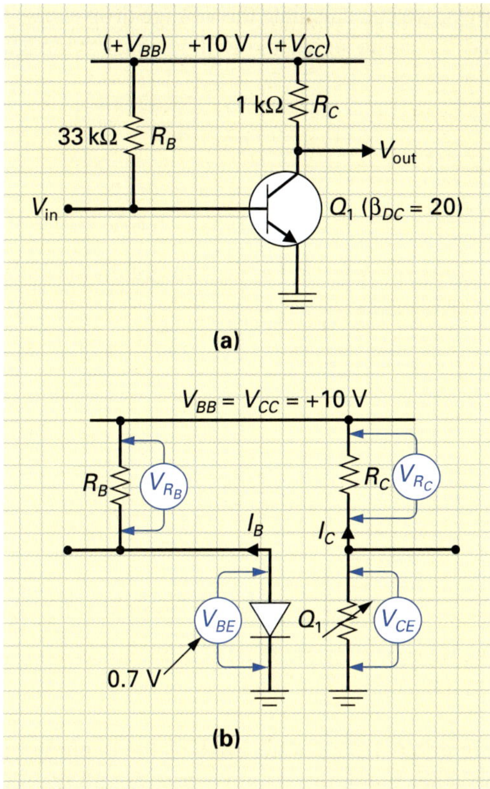

FIGURE 12-19 A Base Biased Common Emitter Circuit. (a) Basic Circuit. (b) Simplified Equivalent Circuit.

$$I_B = \frac{V_{R_B}}{R_B}$$

$$I_B = \frac{V_{R_B}}{R_B} = \frac{9.3 \text{ V}}{33 \text{ k}\Omega} = 282 \text{ }\mu\text{A}$$

Because the transistor's dc current gain (β_{DC}) is given in Figure 17-19(a), we can calculate I_C because β_{DC} tells us how much greater the output current I_C is compared to the input current I_B.

$$I_C = I_B \times \beta_{DC}$$

$$I_C = I_B \times \beta_{DC} = 282 \text{ }\mu\text{A} \times 20 = 5.6 \text{ mA}$$

Because the current through R_C is I_C, we can now calculate the voltage drop across R_C (V_{R_C}).

$$V_{R_C} = I_C \times R_C$$

$$V_{R_C} = I_C \times R_C = 5.6 \text{ mA} \times 1 \text{ k}\Omega = 5.6 \text{ V}$$

Now that V_{R_C} is known, we can calculate the voltage drop across the transistor's collector-to-emitter because V_{CE} and V_{R_C} are in series and will be equal to the applied voltage V_{CC}.

$$V_{CE} = V_{CC} - V_{R_C}$$

$$V_{CE} = V_{CC} - V_{R_C} = 10 \text{ V} - 5.6 \text{ V} = 4.4 \text{ V}$$

Combining the previous two equations, we can obtain the following V_{CE} formula:

$$V_{CE} = V_{CC} - V_{R_C}$$

Since

$$V_{R_C} = I_C \times R_C$$

$$V_{CE} = V_{CC} - (I_C \times R_C)$$

$$V_{CE} = V_{CC} - (I_C \times R_C) = 10 \text{ V} - (5.6 \text{ mA} \times 1 \text{ k}\Omega) = 4.4 \text{ V}$$

Using the above formulas, which are all basically Ohm's law, you can calculate the current and voltage values in a base biased circuit.

DC Load Line. In a transistor circuit, such as the example in Figure 12-19, V_{CC} and V_{R_C} are constants. On the other hand, the input current I_B and the output current I_C are variables. Using the example circuit in Figure 12-19 let us calculate what collector-to-emitter voltage drops (V_{CE}) will result for different values of I_C.

a. When Q_1 is OFF, $I_C = 0$ mA, and therefore V_{CE} equals:

$$V_{CE} = V_{CC} - (I_C \times R_C) = 10 \text{ V} - (0 \text{ mA} \times 1 \text{ k}\Omega) = 10 \text{ V} - 0 \text{ V} = 10 \text{ V}$$

This would make sense because Q_1 would be equivalent to an open switch between collector and emitter when it is OFF, and therefore all of the 10 V V_{CC} supply voltage would appear across the open. Figure 12-20 shows how this point would be plotted on a graph (point *A*).

b. When $I_C = 1$ mA,

$$V_{CE} = 10 \text{ V} - (1 \text{ mA} \times 1 \text{ k}\Omega) = 10 \text{ V} - 1 \text{ V} = 9 \text{ V (point } B)$$

c. When $I_C = 2$ mA,

$$V_{CE} = 10 \text{ V} - (2 \text{ mA} \times 1 \text{ k}\Omega) = 10 \text{ V} - 2 \text{ V} = 8 \text{ V (point } C)$$

d. When $I_C = 3$ mA,

$$V_{CE} = 10 \text{ V} - (3 \text{ mA} \times 1 \text{ k}\Omega) = 10 \text{ V} - 3 \text{ V} = 7 \text{ V (point } D)$$

e. When $I_C = 4$ mA, $V_{CE} = 6$ V (point *E*)
f. When $I_C = 5$ mA, $V_{CE} = 5$ V (point *F*)
g. When $I_C = 6$ mA, $V_{CE} = 4$ V (point *G*)
h. When $I_C = 7$ mA, $V_{CE} = 3$ V (point *H*)
i. When $I_C = 8$ mA, $V_{CE} = 2$ V (point *I*)
j. When $I_C = 9$ mA, $V_{CE} = 1$ V (point *J*)
k. When $I_C = 10$ mA, the only resistance is that of R_C because Q_1 is fully ON and is equivalent to a closed switch between collector and emitter. It is not a surprise that the voltage drop across Q_1's collector-to-emitter is almost 0 V.

$$V_{CE} = 10 \text{ V} - (10 \text{ mA} \times 1 \text{ k}\Omega) = 10 \text{ V} - 10 \text{ V} = 0 \text{ V (point } K)$$

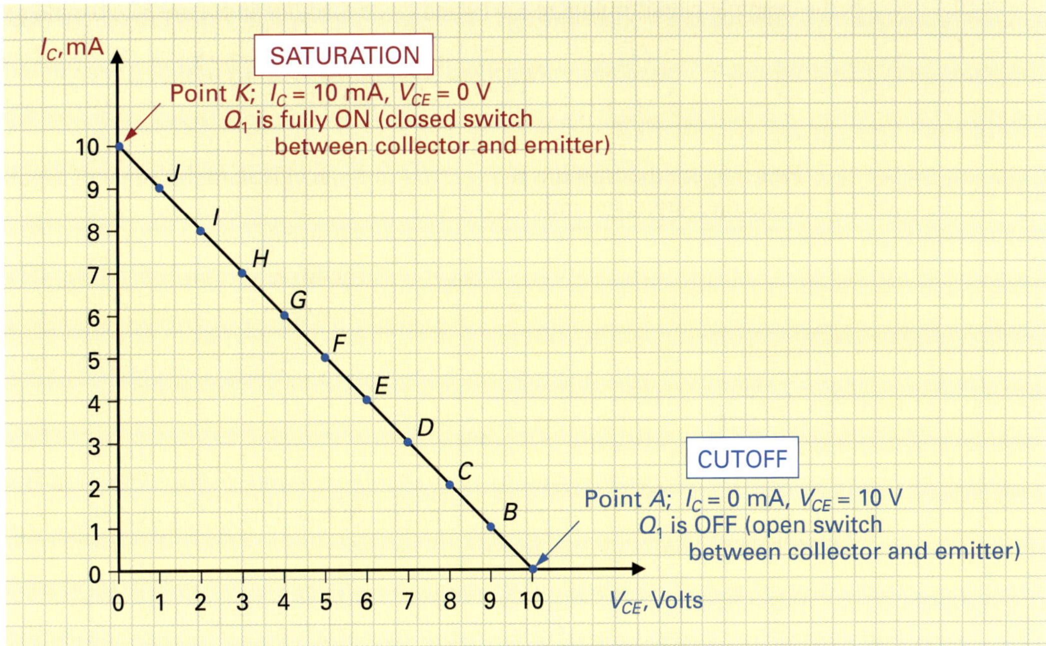

FIGURE 12-20 **A Transistor DC Load Line with Cutoff and Saturation Points.**

DC Load Line

A line representing all the dc operating points of the transistor for a given load resistance.

The line drawn in the graph in Figure 12-20 is called the **dc load line** because it is a line representing all the dc operating points of the transistor for a given load resistance. In this example, the transistor's load was the 1 kΩ collector-connected resistor R_C in Figure 12-19.

Cutoff and Saturation Points. Let us now examine the two extreme points in a transistor's dc load line, which in the example in Figure 12-20 were points *A* and *K*. If a transistor's base input bias voltage is reduced to zero, its input current I_B will be zero, Q_1 will turn OFF and be equivalent to an open switch between the collector and emitter, the output current I_C will be 0 mA, and a V_{CE} will be 10 V. This point in the transistor dc load line is called *cutoff* (point *A* in Figure 12-19) because the output collector current is reduced to zero, or cut off. In summary, at cutoff:

$$I_{C(\text{Cutoff})} = 0 \text{ mA}$$
$$V_{CE(\text{Cutoff})} = V_{CC}$$

In the example circuit in Figure 12-19 and its dc load line in Figure 12-20, with Q_1 cut OFF:

$$I_{C(\text{Cutoff})} = 0 \text{ mA}, \; V_{CE(\text{Cutoff})} = V_{CC} = 10 \text{ V}$$

If the base input bias voltage is increased to a large positive value, the transistor's collector diode (which is normally reverse biased) will be forward biased. In this condition, I_B will be at its maximum, Q_1 will be fully ON and equivalent to a closed switch between the collector and emitter, I_C will be at its maximum of 10 mA, and V_{CE} will be 0 V. This point in the transistor's dc load line is called *saturation* (point *K* in Figure 12-20) because, just as a point is reached where a wet sponge is saturated and cannot hold any more water, the transistor at saturation cannot increase I_C beyond this point. In summary, at saturation:

$$I_{C\,(\text{Sat.})} = \frac{V_{CC}}{R_C}$$
$$V_{CE\,(\text{Sat.})} = 0 \text{ V}$$

In the example circuit in Figure 12-19 and its dc load line in Figure 12-20, with Q_1 saturated:

$$I_{C\,(\text{Sat.})} = \frac{V_{CC}}{R_C} = \frac{10\text{ V}}{1\text{ k}\Omega} = 10\text{ mA}$$
$$V_{CE\,(\text{Sat.})} = 0\text{ V}$$

Rearranging the formula $\beta_{DC} = I_C/I_B$, we can calculate the value of input base current that causes the output saturation current:

$$\beta_{DC} = \frac{I_C}{I_B}, \text{ therefore,}$$

$$I_{B\,(\text{Sat.})} = \frac{I_{C(\text{Sat.})}}{\beta_{DC}}$$

In the example circuit in Figure 12-19 and its dc load line in Figure 12-20, the input current that will cause saturation will be

$$I_{B\,(\text{Sat.})} = \frac{I_{C\,(\text{Sat.})}}{\beta_{DC}} = \frac{10\text{ mA}}{20} = 500\ \mu\text{A}$$

Figure 12-21 summarizes all of our base bias circuit calculations so far by including the dc load line from Figure 12-20 in a set of collector characteristic curves for the transistor circuit example in Figure 12-19. As you can see in the graph in Figure 12-21, at cutoff: $I_B = 0\ \mu\text{A}$, $I_C = 0$ mA, and $V_{CE} = V_{CC}$ which is 10 V. On the other hand, at saturation: $I_B = 500\ \mu\text{A}$, $I_C = 10$ mA, and $V_{CE} = 0$ V.

Quiescent Point. Generally, the value of the base bias resistor (R_B) is chosen so that the value of base current (I_B) is near the middle of the dc load line. For example, if a base bias resistance of 37.2 kΩ was used in the example circuit in Figure 12-20 ($R_B = 37.2$ kΩ), it would produce a base current of 250 mA ($I_B = 9.3\text{ V}/37.2\text{ k}\Omega = 250\ \mu\text{A}$). Referring to the

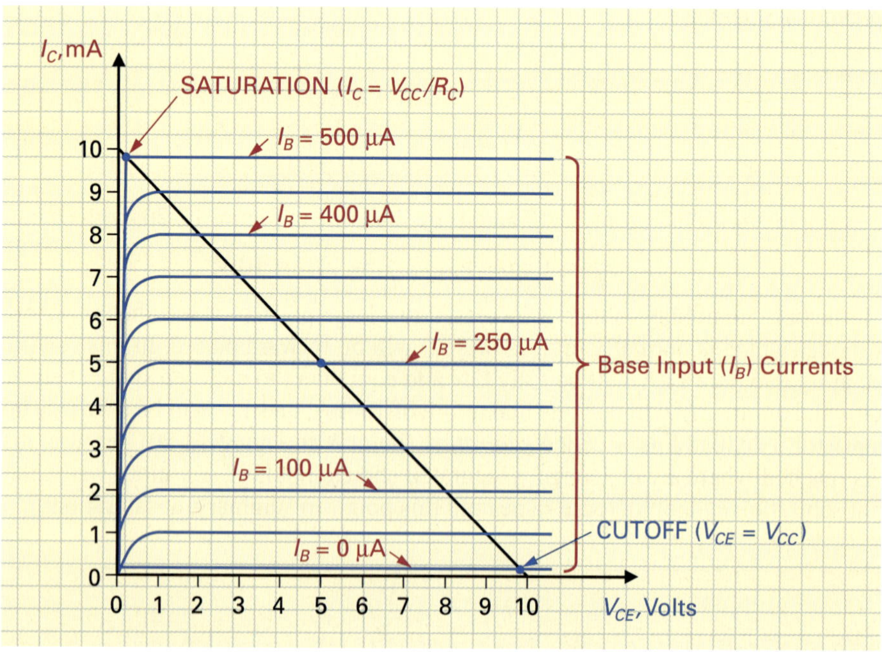

FIGURE 12-21 **Transistor Input/Output Characteristic Graph.**

dc load line in Figure 12-21, you can see that this value of base current is halfway between cutoff at 0 μA, and saturation at 500 μA. This point is called the *quiescent* (at rest) or *Q point* and is defined as *the dc bias point at which the circuit rests when no ac input signal is applied.* An ac input signal voltage will vary I_B above and below this Q point, resulting in a corresponding but larger change in I_C.

EXAMPLE:

Complete the following for the circuit shown in Figure 12-22.

a. Calculate I_B.
b. Calculate I_C.
c. Calculate V_{CE}.
d. Sketch the circuit's dc load line with saturation and cutoff points.
e. Indicate where the Q point is on the circuit's dc load line.

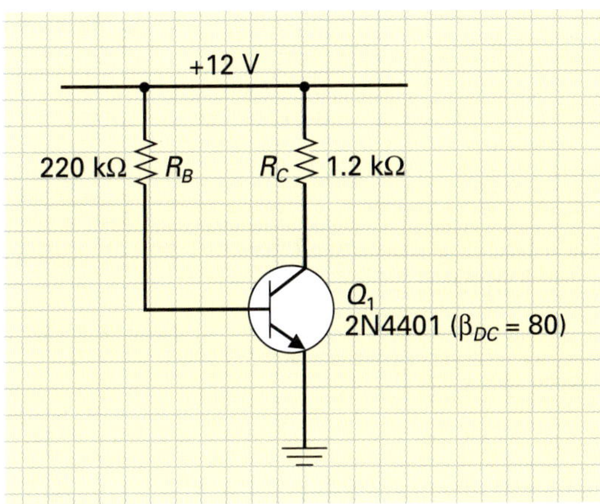

FIGURE 12-22 **Bipolar Transistor Example.**

Solution:

a. Since $V_{BE} = 0.7$ V and $V_{BB} = 12$ V,

$$V_{R_B} = 12\text{ V} - 0.7\text{ V} = 11.3\text{ V}$$

$$I_B = \frac{V_{R_B}}{R_B} = \frac{11.3\text{ V}}{220\text{ k}\Omega} = 51.4\ \mu\text{A}$$

b.

$$I_C = I_B \times \beta_{DC} = 51.4\ \mu\text{A} \times 80 = 4.1\text{ mA}$$

c.

$$V_{R_C} = I_C \times R_C = 4.1\text{ mA} \times 1.2\text{ k}\Omega = 4.92\text{ V}$$

$$V_{CE} = V_{CC} - V_{R_C} = 12\text{ V} - 4.92\text{ V} = 7.08\text{ V}$$

d. At cutoff, the transistor is OFF and therefore equivalent to an open switch between collector and emitter. All of the V_{CC} supply voltage will therefore be across Q_1.

At cutoff, $V_{CE} = V_{CC} = 12$ V (see cutoff in the dc load line in Figure 12-23).

At saturation, the transistor is fully ON and therefore equivalent to a closed switch between the collector and emitter. The only resistance is that of R_C, and so:

At saturation,

$$I_{C\,(\text{Sat.})} = \frac{V_{CC}}{R_C} = \frac{12\text{ V}}{1.2\text{ k}\Omega} = 10\text{ mA}$$

(see saturation in the dc load line in Figure 12-22).

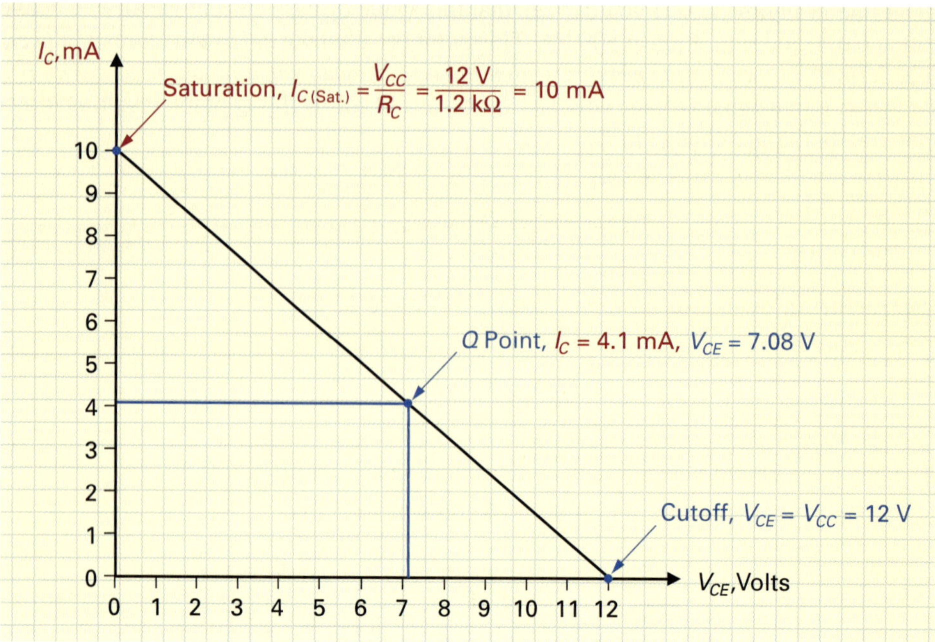

FIGURE 12-23 The DC Load Line for the Circuit in Figure 12-22.

e. The operating point or Q point of this circuit is set by the base bias resistor R_B. This Q point will be at

$$I_C = 4.1\text{ mA}$$

which produces a

$$V_{CE} = 7.08\text{ V}$$

This quiescent (Q) point is also shown on Figure 12-23.

EXAMPLE:

Calculate the current through the lamp in Figure 12-24.

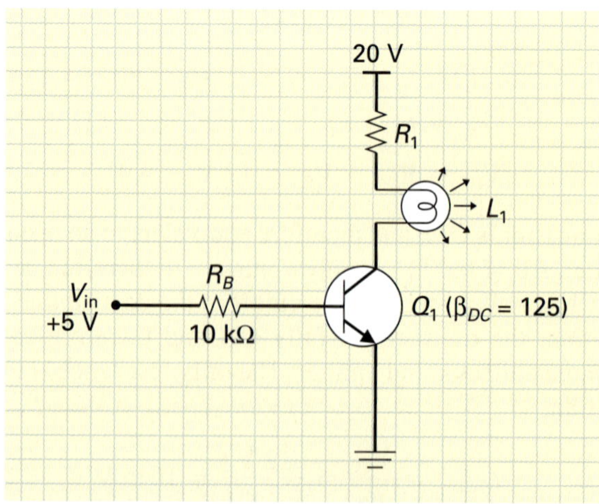

FIGURE 12-24 Two-State Lamp Circuit.

■ ***Solution:***

$$\begin{aligned} V_{BE} &= 0.7\text{ V},\ V_{in} = +5\text{ V, therefore,} \\ V_{R_B} &= V_{in} - 0.7\text{ V} \\ &= 5\text{ V} - 0.7\text{ V} = 4.3\text{ V} \\ I_B &= \frac{V_{R_B}}{R_B} = \frac{4.3\text{ V}}{10\text{ k}\Omega} = 430\ \mu\text{A} \\ I_C &= I_B \times \beta_{DC} \\ &= 430\ \mu\text{A} \times 125 = 53.75\text{ mA} \end{aligned}$$

An input of zero volts (V_{in} = 0 V) will turn OFF Q_1, and therefore lamp L_1. On the other hand, an input of +5 V will turn ON Q_1 and permit a collector current, and therefore lamp current, of 53.75 mA.

Base Biasing Applications. Base bias circuits are used in switching circuit applications like the two-state ON/OFF lamp circuit discussed in the previous example. In these circuits, the bipolar transistor is equivalent to a switch and is controlled by a HIGH/LOW input voltage that drives the transistor between the two extremes of cutoff and saturation.

The advantage of this biasing technique is circuit simplicity because only one resistor is needed to set the base bias voltage. The disadvantage of the base biased circuit is that it cannot compensate for changes in its dc bias current due to changes in temperature. To explain this in more detail, a change in temperature will result in a change in the internal resistance of the transistor (all semiconductor devices have a negative temperature coefficient of resistance—temperature ↑ causes internal resistance ↓). This change in the transistor's internal resistance will change the transistor's dc bias currents (I_B and I_C), which will change or shift the transistor's dc operating point or Q point away from the desired midpoint.

Voltage-Divider Biasing

Figure 12-25(a) shows how a common-emitter transistor circuit could be **voltage-divider biased.** The name of this biasing method comes from the two- resistor series voltage divider (R_1 and R_2) connected to the transistor's base. In this most widely used biasing method, the emitter diode of Q_1 is forward biased by the voltage developed across R_2 (V_{R_2}), as seen in the simplified equivalent circuit in Figure 12-25(b). To calculate the voltage developed across R_2, and therefore the voltage applied to Q_1's base, we can use the voltage-divider formula.

Voltage-Divider Biasing
A biasing method used with amplifiers in which a series arrangement of two fixed-value resistors is connected across the voltage source. The result is that a desired fraction of the total voltage is obtained at the center of the two resistors and is used to bias the amplifier.

$$V_{R_2} \text{ or } V_B = \frac{R_2}{R_1 + R_2} \times V_{CC}$$

$$V_{R_2} \text{ or } V_B = \frac{R_2}{R_1 + R_2} \times V_{CC} = \frac{10\text{ k}\Omega}{20\text{ k}\Omega + 10\text{ k}\Omega} \times 20\text{ V} = 0.333 \times 20\text{ V} = 6.7\text{ V}$$

Because the current through R_1 and R_2 (from ground to $+V_{CC}$) is generally more than 10 times greater than the base current of Q_1 (I_B), it is normally assumed that I_B will have no effect on the voltage-divider current through R_1 and R_2. The R_1 and R_2 voltage divider can be assumed to be independent of Q_1, and the previous voltage-divider formula can be used to calculate V_{R_2} or V_B.

Because V_B = 6.7 V, the emitter diode of Q_1 will be forward biased. Assuming a 0.7 V drop across the transistor's base-emitter junction (V_{BE} = 0.7 V), the voltage at the emitter terminal of Q_1 (V_E) will be:

$$V_{R_E} \text{ or } V_E = V_B - 0.7\text{ V}$$

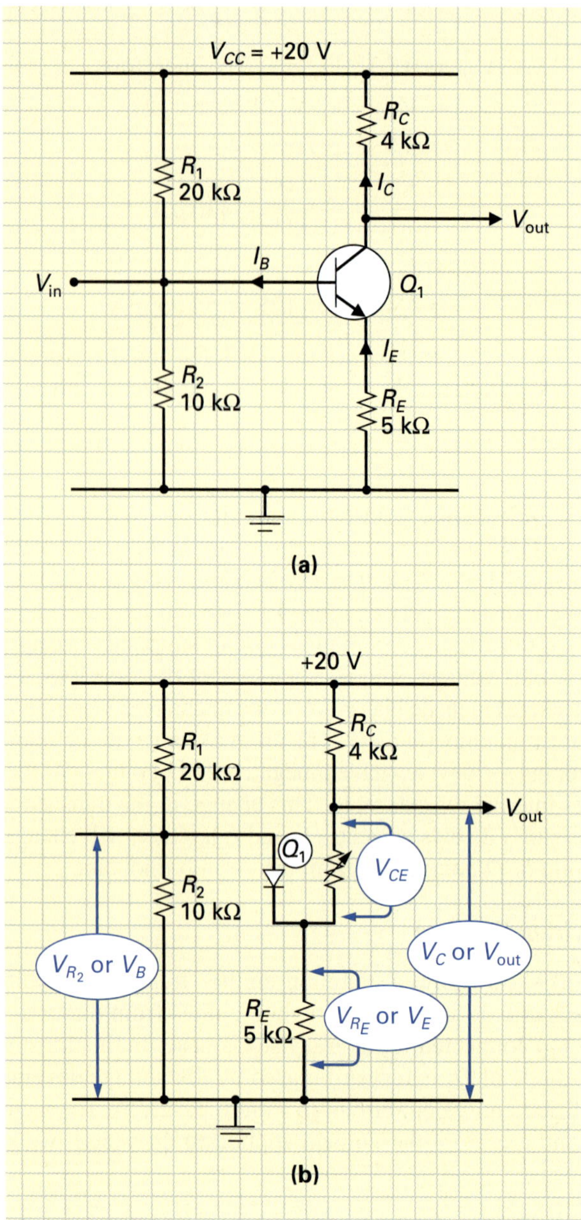

FIGURE 12-25 A Voltage-Divider Biased Common-Emitter Circuit. (a) Basic Circuit. (b) Simplified Equivalent Circuit.

$$V_{R_E} \text{ or } V_E = V_B - 0.7\text{ V} = 6.7\text{ V} - 0.7\text{ V} = 6\text{ V}$$

Now that the voltage drop across R_E (V_{R_E}) is known, along with its resistance, we can calculate the current through R_E and the value of current being injected into the transistor's emitter.

$$I_{R_E} = I_E = \frac{V_{R_E}}{R_E}$$

$$I_{R_E} = I_E = \frac{V_{R_E}}{R_E} = \frac{6\text{ V}}{5\text{ k}\Omega} = 1.2\text{ mA}$$

Because we know that a transistor collector current (I_C) is approximately equal to the emitter current (I_E), we can state that

$$I_E \cong I_C$$

$$I_E \cong I_C = 1.2 \text{ mA}$$

Now that I_C is known, we can calculate the voltage drop across R_C (V_{R_C}) because both its resistance and current are known.

$$V_{R_C} = I_C \times R_C$$

$$V_{R_C} = I_C \times R_C = 1.2 \text{ mA} \times 4 \text{ k}\Omega = 4.8 \text{ V}$$

The dc quiescent voltage at the collector of Q_1 with respect to ground (V_C), which is also V_{out}, will be equal to the dc supply voltage (V_{CC}) minus the voltage drop across R_C.

$$V_C \text{ or } V_{out} = V_{CC} - V_{R_C}$$

$$V_C \text{ or } V_{out} = V_{CC} - V_{R_C} = 20 \text{ V} - 4.8 \text{ V} = 15.2 \text{ V}$$

Because V_{CC} is connected across the series voltage divider formed by R_C, Q_1's collector-to-emitter resistance (R_{CE}), and R_E, we can calculate V_{CE} if both V_{R_C} and V_E are known:

$$V_{CE} = V_{CC} - (V_{R_C} + V_E)$$

$$V_{CE} = V_{CC} - (V_{R_C} + V_E) = 20 \text{ V} - (4.8 \text{ V} + 6 \text{ V}) = 20 \text{ V} - 10.8 \text{ V} = 9.2 \text{ V}$$

DC Load Line. Figure 12-26 shows the dc load line for the example circuit in Figure 12-25. Referring to the dc load line's two end points, let us examine this circuit's saturation and cutoff points.

When transistor Q_1 is fully ON or saturated, it will have approximately 0 Ω of resistance between its collector and emitter. As a result, R_C and R_E determine the value of I_C when Q_1 is saturated.

$$I_{C\,(\text{Sat.})} = \frac{V_{CC}}{R_C + R_E}$$

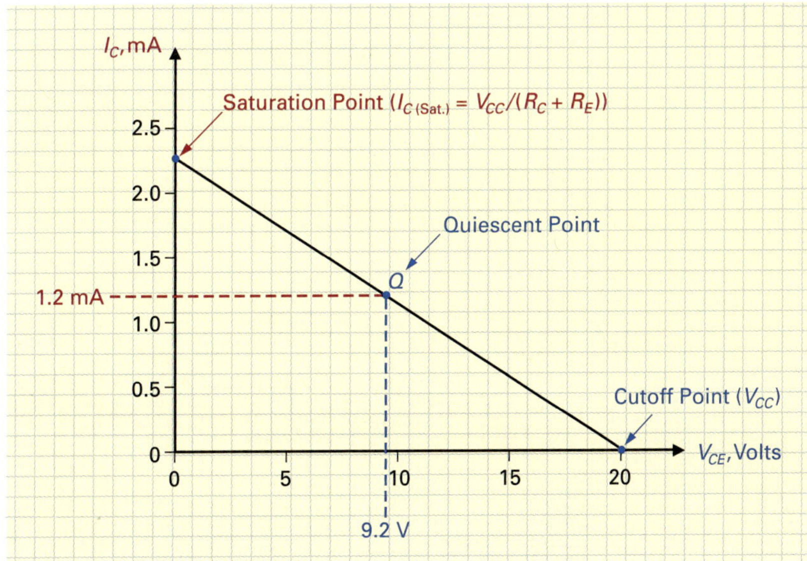

FIGURE 12-26 **The DC Load Line for the Circuit in Figure 12-24.**

$$I_{C\,(\text{Sat.})} = \frac{V_{CC}}{R_C + R_E} = \frac{20\text{ V}}{4\text{k}\Omega + 5\text{ k}\Omega} = \frac{20\text{ V}}{9\text{ k}\Omega} = 2.2\text{ mA}$$

As you can see in Figure 12-26, at saturation, I_C is maximum at 2.2 mA, and V_{CE} is 0 V because Q_1 is equivalent to a closed switch (0 Ω) between Q_1's collector and emitter.

$$V_{CE(\text{Sat.})} = 0\text{ V}$$

At the other end of the dc load line in Figure 12-26, we can see how the transistor's characteristics are plotted when it is cut off. When Q_1 is cut OFF, it is equivalent to an open switch between collector-to-emitter. Therefore all of the V_{CC} supply voltage will appear across the series circuit open.

$$V_{CE(\text{Cutoff})} = V_{CC}$$

$$V_{CE(Cutoff)} = V_{CC} = 20\text{ V}$$

As you can see in Figure 12-26, when Q_1 is cut OFF, all of the V_{CC} supply voltage will appear across Q_1's collector-to-emitter terminals, and I_C will be blocked and equal to zero.

$$I_{C(\text{Cutoff})} = 0\text{ mA}$$

Generally, the value of the voltage-divider resistors R_1 and R_2 are chosen so that the value of base current (I_B) is near the middle of the dc load line. Referring to Figure 12-26, you can see that by plotting our previously calculated values of I_C (which at rest was 1.2 mA) and V_{CE} (V_{CE} = 9.2 V), we obtain a Q point that is near the middle of the dc load line.

EXAMPLE:

Calculate the following for the circuit shown in Figure 12-27.

a. V_B and V_E
b. Determine whether C_E will have any effect on the dc operating voltages
c. I_C
d. V_C and V_{CE}
e. Sketch the circuit's dc load line and include the saturation, cutoff, and Q points

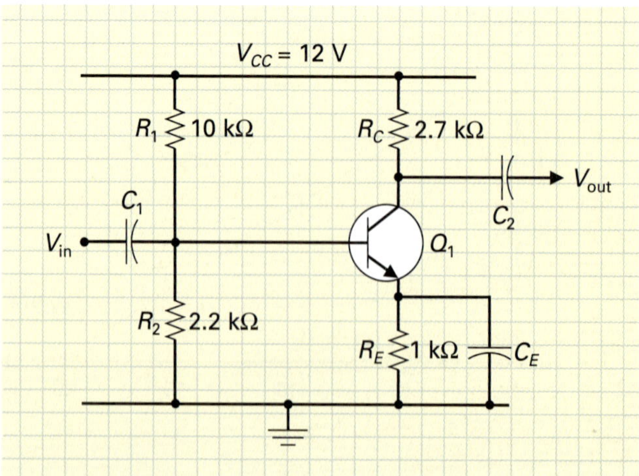

FIGURE 12-27 A Common-Emitter Amplifier Circuit Example.

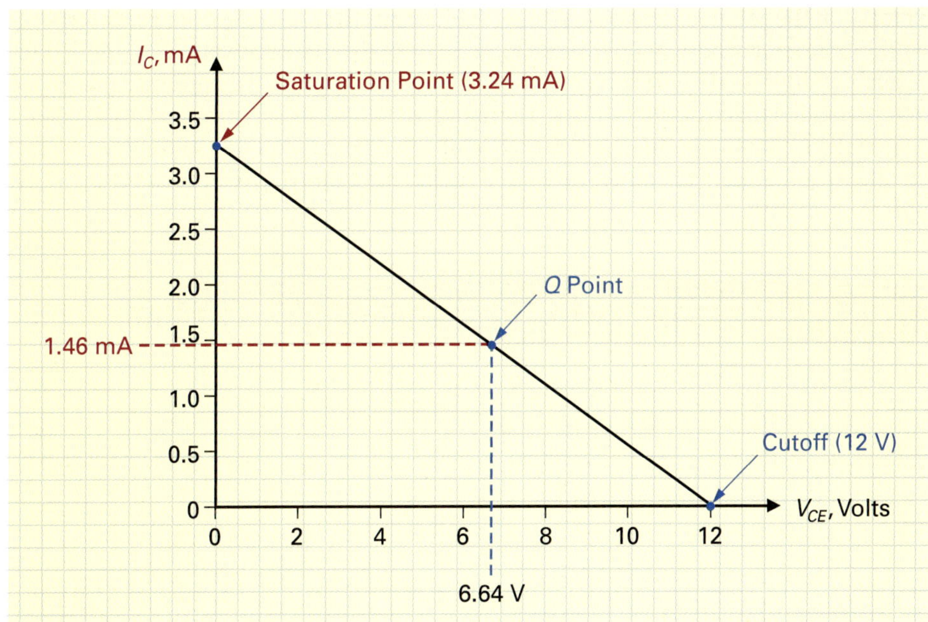

FIGURE 12-28 **The DC Load Line for the Circuit in Figure 12-27.**

■ ***Solution:***

a.
$$V_B = \frac{R_2}{R_1 + R_2} \times V_{CC} = \frac{2.2\ \text{k}\Omega}{10\ \text{k}\Omega + 2.2\ \text{k}\Omega} \times 12\ \text{V} = 2.16\ \text{V}$$
$$V_E = V_B - 0.7\ \text{V} = 2.16\ \text{V} - 0.7\ \text{V} = 1.46\ \text{V}$$

b. Since all capacitors can be thought of as a dc block, C_E will have no effect on the circuit's dc operating voltages.

c.
$$I_E = \frac{V_E}{R_E} = \frac{1.46\ \text{V}}{1\ \text{k}\Omega} = 1.46\ \text{mA}$$
$$I_C \cong I_E = 1.46\ \text{mA}$$

d. $V_{R_C} = I_C \times R_C = 1.46\ \text{mA} \times 2.7\ \text{k}\Omega = 3.9\ \text{V}$
V_{out} or $V_C = V_{CC} - V_{R_C} = 12\ \text{V} - 3.9\ \text{V} = 8.1\ \text{V}$
$V_{CE} = V_{CC} - (V_{R_C} + V_E) = 12\ \text{V} - (3.9\ \text{V} + 1.46\ \text{V}) = 12\ \text{V} - 5.36\ \text{V} = 6.64\ \text{V}$

e. $I_{C(\text{Sat.})} = \dfrac{V_{CC}}{R_C + R_E} = \dfrac{12\ \text{V}}{2.7\ \text{k}\Omega + 1\ \text{k}\Omega} = \dfrac{12\ \text{V}}{3.7\ \text{k}\Omega} = 3.24\ \text{mA}$
$V_{CE(\text{Cutoff})} = V_{CC} = 12\ \text{V}$
Q point, $I_C = 1.46$ mA and $V_{CE} = 6.64$ V
(This information is plotted on the graph in Figure 12-28.)

Voltage-Divider Bias Applications. Voltage-divider biased circuits are used in analog or linear circuit applications such as the amplifier circuit discussed in the previous example. In these circuits, the bipolar transistor is equivalent to a variable resistor and is controlled by an alternating input signal voltage.

Unlike the base biased circuit, the voltage-divider biased circuit has very good temperature stability due to the emitter resistor R_E. To explain this in more detail, let us assume that there is an increase in the temperature surrounding a voltage-divider circuit, such as the example circuit in Figure 12-27. As temperature increases, it causes an increase in the transistor's internal currents ($I_B\uparrow$, $I_E\uparrow$, $I_C\uparrow$) because all semiconductor devices have a negative temperature coefficient of resistance (temperature $\uparrow$, $R\downarrow$, $I\uparrow$). An increase in $I_E\uparrow$ will cause an increase in the voltage drop across $R_E\uparrow$, which will decrease the voltage difference between the transistor's base and emitter ($V_{BE}\downarrow$). Decreasing the forward bias applied to the transistor's emitter diode will decrease all of the transistor's internal currents ($I_B\downarrow$, $I_E\downarrow$, $I_C\downarrow$) and return them to their original values. Therefore, a change in output current (I_C) due to temperature will effectively be fed back to the input and change the input current (I_B), which is why a circuit containing an emitter resistor is said to have **emitter feedback** for temperature stability.

Emitter Feedback
The coupling from the emitter output to the base input in a transistor amplifier.

SELF-TEST EVALUATION POINT FOR SECTION 12-2

Use the following questions to test your understanding of Section 12-2.

1. The bipolar transistor is a ____________ (voltage/current) controlled device.
2. When a bipolar transistor is being operated in the active region, its emitter diode is ____________ biased and its collector diode is ____________ biased.
3. Which of the following is correct:
   a. $I_E = I_C - I_B$
   b. $I_C = I_E - I_B$
   c. $I_B = I_C - I_E$
4. When a transistor is in cutoff, it is equivalent to a/an ____________ between its collector and emitter.
5. When a transistor is in saturation, it is equivalent to a/an ____________ between its collector and emitter.
6. Which of the bipolar transistor circuit configurations has the best
   a. Voltage gain
   b. Current gain
   c. Power gain
7. Which biasing method makes use of two series-connected resistors across the V_{CC} supply voltage?
8. Which biasing technique has a single resistor connected in series with the base of the transistor?

REVIEW QUESTIONS

Multiple-Choice Questions

1. The bipolar junction transistor has three terminals called the:
   a. Drain, source, gate
   b. Anode, cathode, gate
   c. Main terminal 1, main terminal 2, gate
   d. Emitter, base, collector
2. The term bipolar junction transistor was given to the device because it has:
   a. Two P-N junctions
   b. Two magnetic poles
   c. One *p* region and one *n* region
   d. Two magnetic junctions
3. An NPN transistor is normally biased so that its base is ____________.
   a. Positive
   b. Negative
4. Which is considered the most common bipolar junction transistor configuration?
   a. Common-base
   b. Common-collector
   c. Common-emitter
   d. None of the above
5. A common-collector circuit is often called a/an ____________.
   a. Base-follower
   b. Emitter-follower
   c. Collector-follower
   d. None of the above
6. With the NPN transistor schematic symbol, the emitter arrow will point ____________ the base, whereas with the PNP transistor schematic symbol, the emitter arrow will point ____________ the base.
   a. Toward, away from
   b. Away from, toward
7. The transistor's ON/OFF switching action is made use of in ____________ circuits.
   a. Analog
   b. Digital
   c. Linear
   d. Both (a) and (c) are true
8. The transistor's variable resistor action is made use of in ____________ circuits.
   a. Analog
   b. Digital
   c. Linear
   d. Both (a) and (c) are true

9. Approximately 98 percent of the electrons entering the _____________ of a bipolar transistor will arrive at the _____________, and the remainder will flow out of the _____________.
   a. Emitter, collector, base
   b. Base, collector, emitter
   c. Collector, emitter, base
   d. Emitter, base, collector

10. The common-base circuit configuration achieves the highest _____________ gain, the common-emitter achieves the highest _____________ gain, and the common-collector achieves the highest _____________ gain.
   a. Voltage, current, power
   b. Current, power, voltage
   c. Voltage, power, current
   d. Power, voltage, current

11. Which of the following abbreviations is used to denote the voltage drop between a transistor's base and emitter?
   a. I_{BE}
   b. V_{CC}
   c. V_{CE}
   d. V_{BE}

12. Which of the following abbreviations is used to denote the voltage drop between a transistor's collector and emitter?
   a. V_C
   b. V_{CE}
   c. V_E
   d. V_{CC}

13. A transistor's _____________ specification indicates the gain in dc current between the input and output of a common-emitter circuit.
   a. α_{AC}
   b. α_{DC}
   c. β_{AC}
   d. β_{DC}

14. Consider the following for a base biased bipolar transistor circuit: $R_B = 33\ \text{k}\Omega$, $R_C = 560\ \Omega$, $Q_1\ (\beta_{DC}) = 25$, $V_{CC} = +10$ V. What is V_{BE}?
   a. 1.43 mV
   b. $25 \times 33\ \text{k}\Omega$
   c. 0.7 V
   d. Not enough information given to calculate

15. Which point on the dc load line results in an $I_C = V_{CC}/R_C$ and a $V_{CE} = 0$ V?
   a. Saturation point
   b. Cutoff point
   c. *Q* point
   d. None of the above

16. Which point on the dc load line results in a $V_{CE} = V_{CC}$, and an $I_C = 0$?
   a. Saturation point
   b. Cutoff point
   c. *Q* point
   d. None of the above

17. The midway point on the dc load line at which a transistor is biased with dc voltages when no input signal is applied is called the:
   a. Saturation point
   b. Cutoff point
   c. *Q* point
   d. None of the above

18. Which transistor biasing method makes use of one current limiting resistor in the base circuit?
   a. Base bias
   b. Voltage-divider bias
   c. Emitter-follower bias
   d. Current-divider bias

19. A forward biased transistor emitter or collector diode should have a _____________ resistance, while a reverse biased emitter or collector diode should have a _____________ resistance.
   a. Low, low
   b. High, low
   c. High, high
   d. Low, high

20. A transistor tester will check a transistor's:
   a. Opens or shorts between any of the terminals
   b. Gain
   c. Reverse leakage current value
   d. All of the above

Practice Problems

21. Identify the type and terminals of the transistors shown in Figure 12-29.

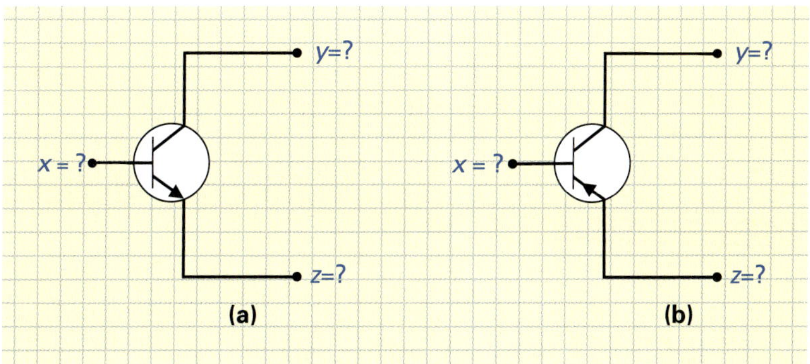

FIGURE 12-29 **Identify the Transistor Type and Terminals.**

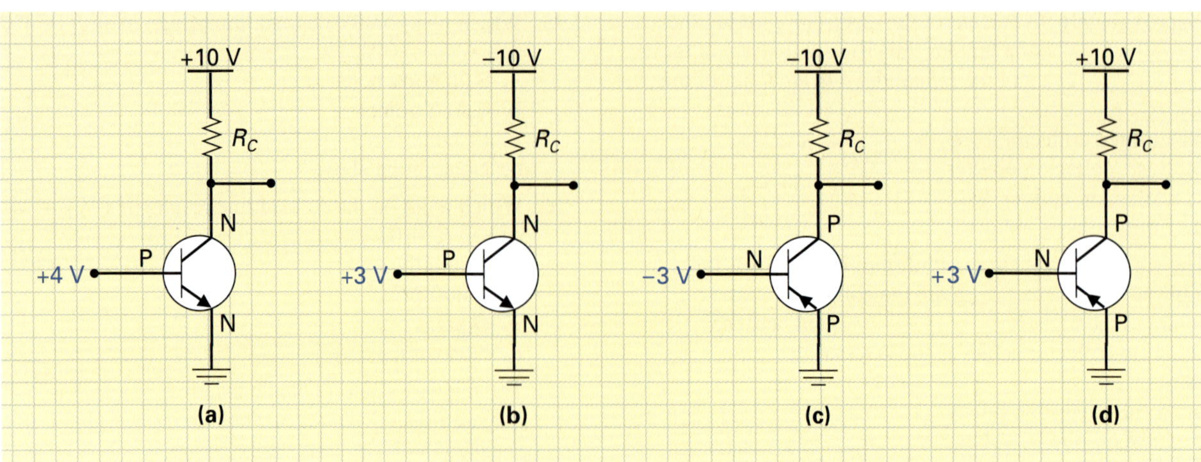

FIGURE 12-30 Identifying the Correctly Biased (Active Region) Bipolar Transistors.

22. A bipolar transistor is correctly biased for operation in the active region when its emitter diode is forward biased and its collector diode is reverse biased. Referring to Figure 12-30, which of the bipolar transistor circuits is correctly biased?

23. Calculate the value of the missing current in the following examples:

a. $I_E = 25$ mA, $I_C = 24.6$ mA, $I_B = ?$

b. $I_B = 600$ μA, $I_C = 14$ mA, $I_E = ?$

c. $I_E = 4.1$ mA, $I_B = 56.7$ μA, $I_C = ?$

24. Calculate the voltage gain (A_V) of the transistor amplifier whose input/output waveforms are shown in Figure 12-31.

25. Identify the configuration of the actual bipolar transistor electronic system circuits shown in Figure 12-32.

26. Identify the bipolar transistor type and the biasing technique used in each of the circuits in Figure 12-32.

27. Calculate the following for the base biased transistor circuit shown in Figure 12-33:

a. I_B

b. I_C

c. V_{CE}

28. Sketch the dc load line for the circuit in Figure 12-33, showing the saturation, cutoff, and Q points.

29. Calculate the following for the voltage-divider biased transistor circuit shown in Figure 12-34:

a. V_B and V_E

b. I_C

c. V_C

d. V_{CE}

30. Sketch the dc load line for the circuit in Figure 12-34, showing the saturation, cutoff, and Q points.

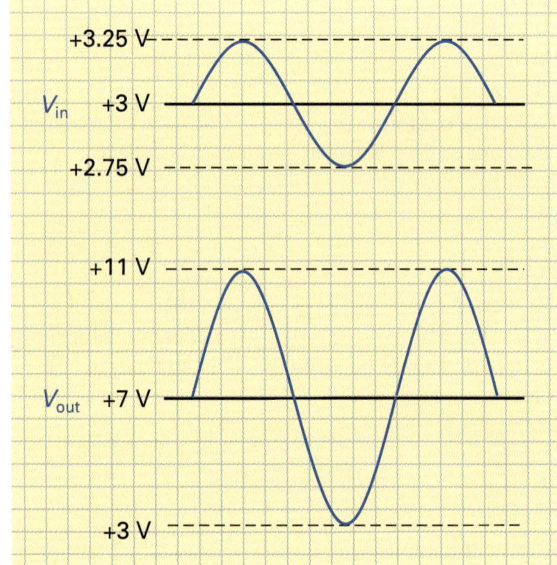

FIGURE 12-31 Transistor Amplifier Input/Output Waveforms.

31. In Figure 12-34, does Kirchhoff's voltage law apply to the voltage divider made up of R_C, R_{CE}, and R_E?

Web Site Questions

Go to the web site http://www.prenhall.com/cook, select the textbook *Electronics: A Complete Course,* select this chapter, and then follow the instructions when answering the multiple-choice practice problems.

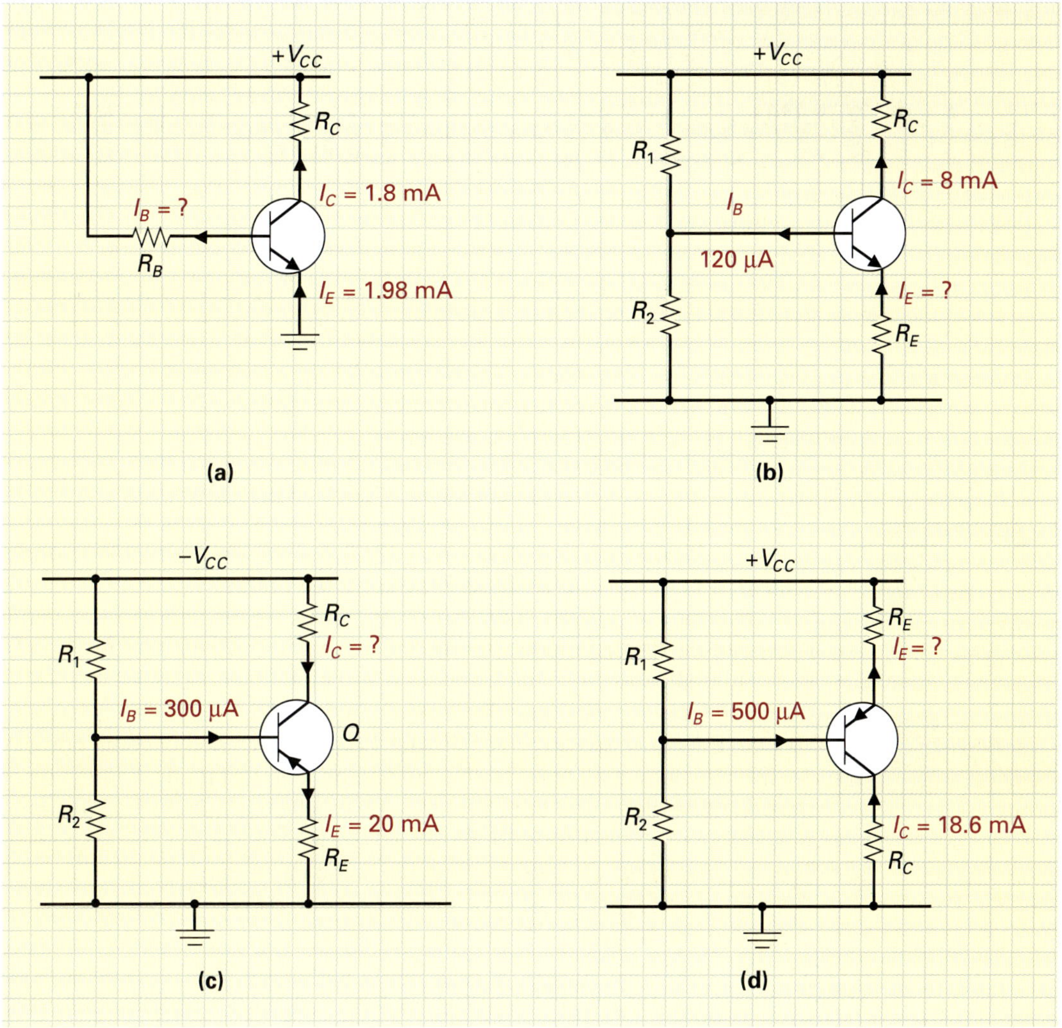

FIGURE 12-32 **Identifying the Configuration of Actual BJT Circuits.**

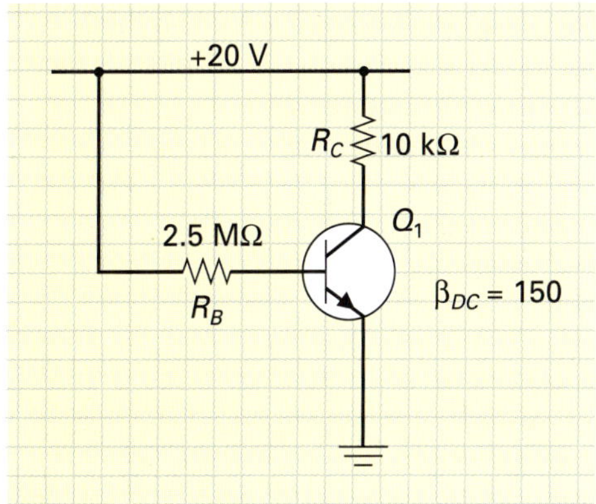

FIGURE 12-33 **Base Biased Transistor Circuit.**

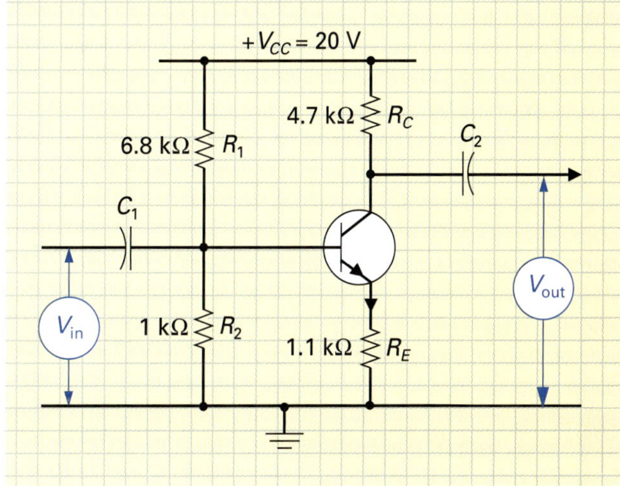

FIGURE 12-34 **Voltage-Divider Biased Transistor Circuit.**

Field Effect Transistors (FETs)

Spitting Lightning Bolts

Nikola Tesla was born in Yugoslavia in 1856. He studied mathematics and physics in Prague, and in 1884 he emigrated to the United States.

In New York he met Thomas Edison, the self-educated inventor who is best known for his development of the phonograph and the incandescent light bulb. Both men were gifted and eccentric and, due to their common interest in "invention," they got along famously. Because Tesla was unemployed, Edison offered him a job. In his lifetime, Edison would go on to take out 1,033 patents and become one of the most prolific inventors of all time. Tesla would go on to invent many different types of motors, generators, and transformers, one of which is named the "Tesla coil" and produces five-foot lightning bolts. With this coil, Tesla investigated "wireless power transmission," the only one of his theories that has not come into being.

Both Tesla and Edison had very strong views on different aspects of electricity and, as time passed, the two men began to engage in very long, loud, and angry arguments. One such discussion concerned whether power should be distributed as alternating current or direct current. Eventually, the world would side with Tesla and choose ac. At the time, this topic, like many others, would cause a hatred to develop between the two men. Eventually Tesla left Edison and started his own company. However, the anger remained, and on one occasion when they were both asked to attend a party for their friend Mark Twain, both refused to come because the other had been invited.

In 1912, Tesla and Edison were both nominated for the Nobel prize in physics, but because neither one would have anything to do with the other, the prize went to a third party—proving that bitterness really will cause a person to cut off his nose to spite his face.

When angry count up to four; when very angry, swear.

Mark Twain

13

Introduction

In the previous chapters, we have concentrated on all aspects of the bipolar junction transistor or BJT. We have examined its construction, operation, characteristics, testing, and basic circuit applications. In this chapter we will examine another type of transistor called the *field effect transistor,* which is more commonly called an FET (pronounced "eff-ee-tee"). Like the BJT, the FET has three terminals and can operate as a switch and can be used in digital circuit applications. It can also operate as a variable resistor and be used in analog or linear circuit applications. In fact, as we step through this chapter, you will see many similarities between the BJT and FET. You will also notice a few distinct differences between these two transistor types, and these differences are what make the BJT ideal in some applications and the FET ideal in other applications.

There are two types of field effect transistors or FETs. One type is the *junction field effect transistor,* which is more typically called a JFET (pronounced "jay-fet"). The other type is the *metal oxide semiconductor field effect transistor,* which is more commonly called a MOSFET (pronounced "moss-fet"). In this chapter we will examine the operation, characteristics, applications, and testing of these two types of field effect transistors.

13-1 JUNCTION FIELD EFFECT TRANSISTOR (JFET)

Junction Field Effect Transistor (JFET)
A field-effect transistor made up of a gate region diffused into a channel region. When a control voltage is applied to the gate, the channel is depleted or enhanced, and the current between source and drain is thereby controlled.

***n*-Channel JFET**
A junction field effect transistor having an *n*-type channel between source and drain.

***p*-Channel JFET**
A junction field effect transistor having a *p*-type channel between source and drain.

Gate
One of the field effect transistor's electrodes (also used for thyristor devices).

Source
One of the field effect transistor's electrodes.

Drain
One of the field effect transistor's electrodes.

Channel
A path for a signal.

Drain Supply Voltage ($+V_{DD}$)
The bias voltage connected between the drain and source of the JFET, which causes current to flow.

Gate-Source Bias Voltage (V_{GS})
The bias voltage applied between the gate and source of a field effect transistor.

Drain Current (I_D)
A JFET's source-to-drain current.

Like the bipolar junction transistor, the **junction field effect transistor** or **JFET** is constructed from *n*-type and *p*-type semiconductor materials. However, the JFET's construction is very different from the BJT's construction, and therefore we will need to first see how the JFET device is built before we can understand how it operates.

13-1-1 *JFET Construction*

Just as the bipolar junction transistor has two basic types (NPN BJT or PNP BJT), there are two types of junction field effect transistor called the ***n*-channel JFET** and ***p*-channel JFET.** The construction and schematic symbols for these two JFET types are shown in Figure 13-1.

To begin with, let us examine the construction of the more frequently used *n*-channel JFET, shown in Figure 13-1(a). This type of JFET basically consists of an *n*-type block of semiconductor material on top of a *p*-type substrate, with a "U" shaped *p*-type section attached to the surface of a *p*-type substrate. Like the BJT, the JFET has three terminals called the **gate, source,** and **drain.** The gate lead is attached to the *p*-type substrate, and the source and drain leads are attached to either end of an *n*-type channel that runs through the middle of the "U" shaped *p*-type section. In the simplified two-dimensional view in the inset in Figure 13-1(a), you can see the *n*-type **channel** that exists between the *n*-channel JFET's source and drain. The schematic symbol for the *n*-channel JFET is shown in Figure 13-1(b), and, as you can see, the gate lead's arrowhead points into the device. To aid your memory, you can imagine this arrowhead as a P-N junction diode as shown in the inset in Figure 13-1(b). The gate lead is connected to the diode's anode, which is a *p*-type material, and the source and drain leads are connected to either end of the diode's cathode, which is an *n*-type material. Because the source-to-drain channel is made from an *n*-type material, this is an *n*-channel JFET. The *p*-type gate and *n*-type source and drain makes this *n*-channel JFET equivalent to an NPN BJT, which has a *p*-type base and *n*-type emitter and collector, as shown in the inset in Figure 13-1(b).

Figure 13-1(c) shows the construction of the *p*-channel JFET. This JFET type is constructed in exactly the same way as the *n*-channel JFET except that the gate lead is attached to an *n*-type substrate, and the source and drain leads are attached to either end of a *p*-type channel. Looking at the schematic symbol for the *p*-channel JFET in Figure 13-1(d), you can see that the gate's arrowhead points out of the device. To help you distinguish the symbols used for the *n*-channel JFET from the *p*-channel JFET, once again imagine the arrowhead as a P-N junction diode as shown in the inset in Figure 13-1(d). Because the gate lead is connected to the diode's cathode, the gate must be an *n*-type material. Because the source and drain leads are connected to either end of the diode's anode, the source-to-drain channel is therefore made from a *p*-type material, and this is a *p*-channel JFET. The *n*-type gate and *p*-type source and drain makes this *p*-channel JFET equivalent to a PNP BJT, which has an *n*-type base and *p*-type emitter and collector, as shown in the inset in Figure 13-1(d).

13-1-2 *JFET Operation*

As you know, an NPN bipolar transistor needs both a collector supply voltage ($+V_{CC}$) and a base-emitter bias voltage (V_{BE}) in order to operate correctly. The same is true for the JFET, which requires both a **drain supply voltage ($+V_{DD}$)** and a **gate-source bias voltage (V_{GS}),** as shown in Figure 13-2(a). The $+V_{DD}$ bias voltage is connected between the drain and source of the *n*-channel JFET and will cause a current to flow through the *n*-channel. This source-to-drain current—which is made up of electrons because they are the majority carriers within an *n*-type material—is called the JFET's **drain current (I_D).** The value of drain current passing through a JFET's channel is dependent on two elements: the value of $+V_{DD}$ applied between the drain and source, and the value of V_{GS} applied between gate and source.

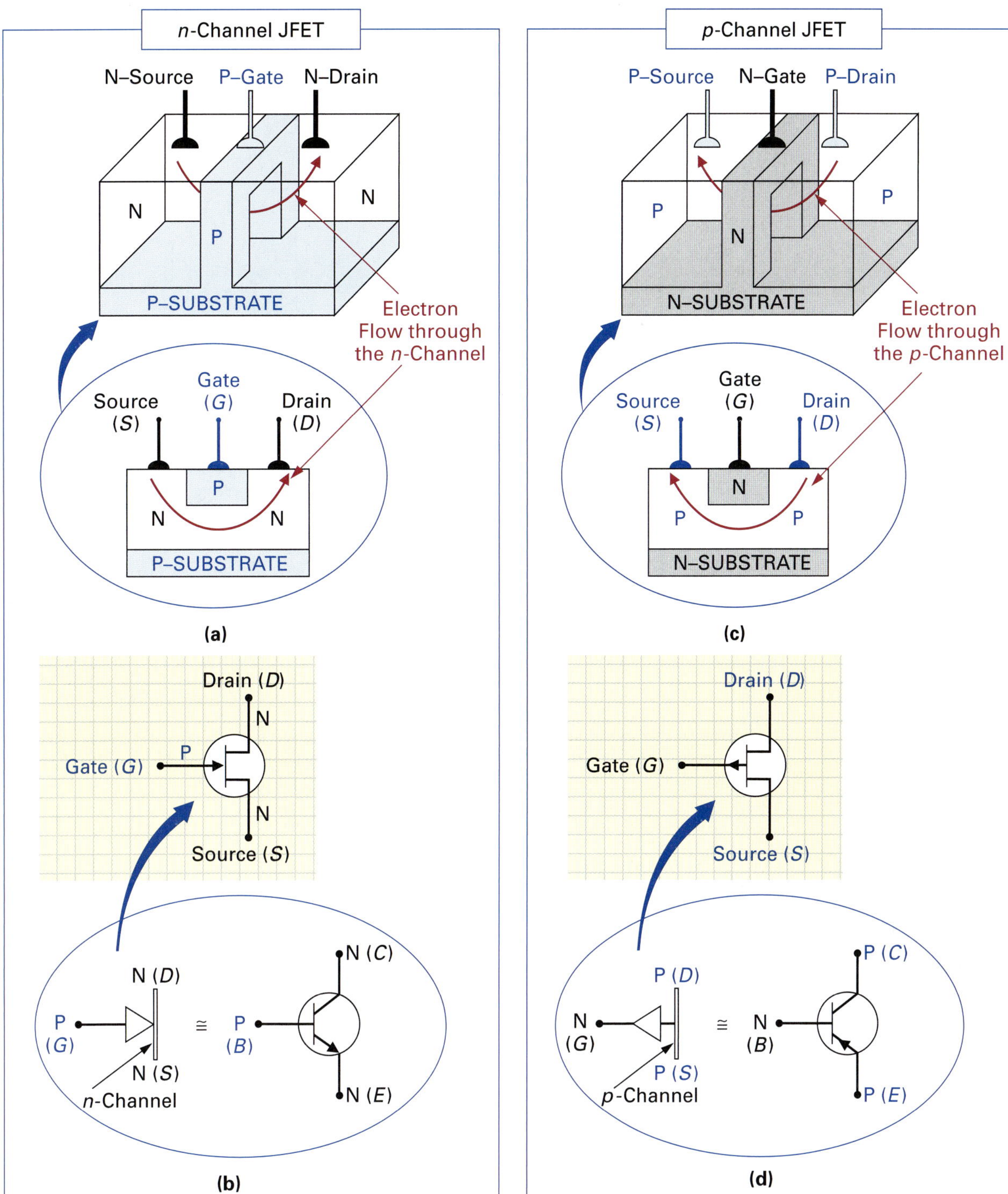

FIGURE 13-1 **The Junction Field Effect Transistor (JFET) Types.**

Let us examine in more detail why these applied voltages control the value of drain current passing through the JFET's channel.

The Relationship between $+V_{DD}$ and I_D

The value of $+V_{DD}$ controls the amount of drain current between source and drain because it is this supply voltage that controls the potential difference applied across the channel, as seen in Figure 13-2(a). Therefore, an increase in the voltage applied across the JFET's

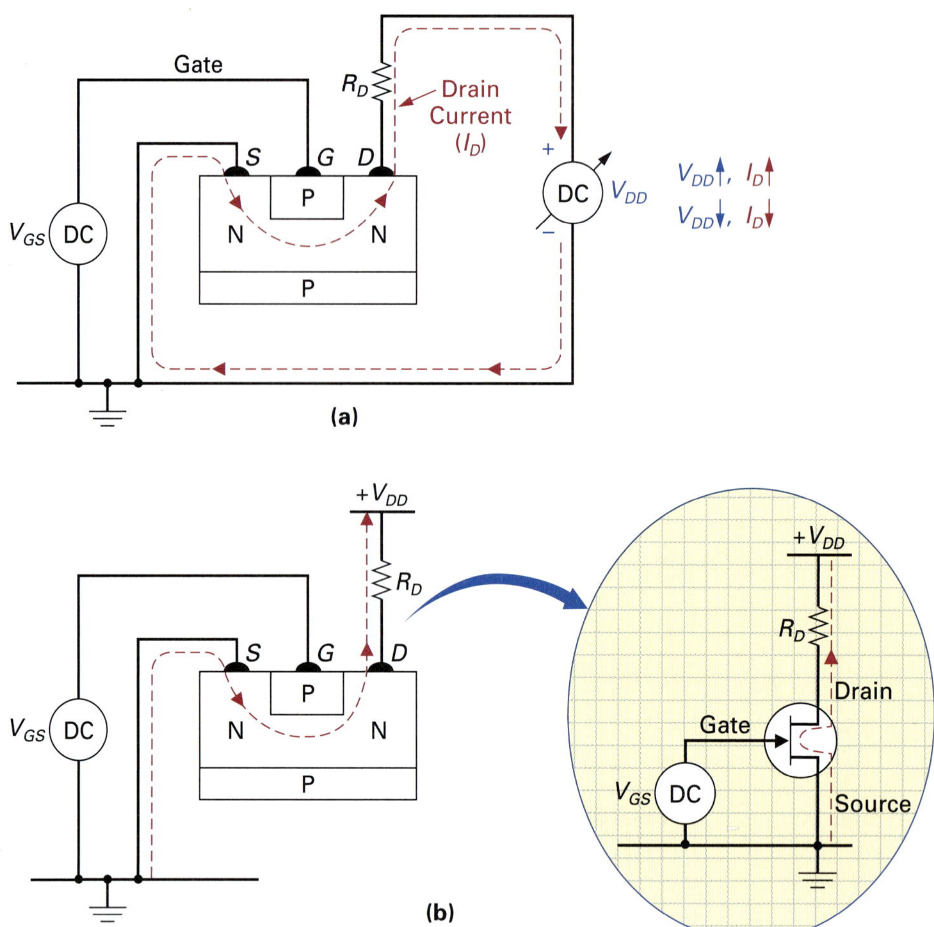

FIGURE 13-2 The Relationship between a JFET's DC Supply Voltage ($+V_{DD}$) and Output Current (I_D).

drain and source ($+V_{DD}\uparrow$) will increase the amount of drain current ($I_D\uparrow$) passing through the channel. Similarly, a decrease in the voltage applied across the JFET's drain and source ($+ V_{DD}\downarrow$) will decrease the amount of drain current ($I_D\downarrow$) passing through the channel. In most cases, a schematic diagram will show the V_{DD} supply voltage connection to the JFET as a source connection to ground and a drain connection up to $+V_{DD}$, as shown in Figure 13-2(b).

The Relationship between V_{GS} and I_D

TIME LINE

In 1961, Steven Hofstein devised the field-effect transistor used in MOS (metal oxide semiconductor) integrated circuits.

The value of V_{GS} controls the amount of drain current between source and drain because it is this voltage that controls the resistance of the channel. Figure 13-3(a) shows that when the V_{GS} bias voltage is 0 volts (which is the same as connecting the gate to ground), there is no potential difference between the gate and source. With no bias voltage applied, a small depletion layer will form and spread into the channel. Although it appears as though two depletion regions exist, in fact they are both part of the same depletion region that extends around the wall of the *n*-channel. This extremely small depletion region will offer very little opposition to I_D, and so drain current will be large. In Figure 13-3(b), V_{GS} is increased to -2 V (made more negative), and therefore the gate-to-source P-N junction will be further reverse biased. This causes an increase in the depletion region, decrease in the channel's width, and a decrease in drain current. In Figure 13-3(c), V_{GS} is increased to -4 V, and therefore the gate-to-source P-N junction will be further reverse biased, causing an increase in the depletion region, decrease in the channel's width, and a further decrease in drain current.

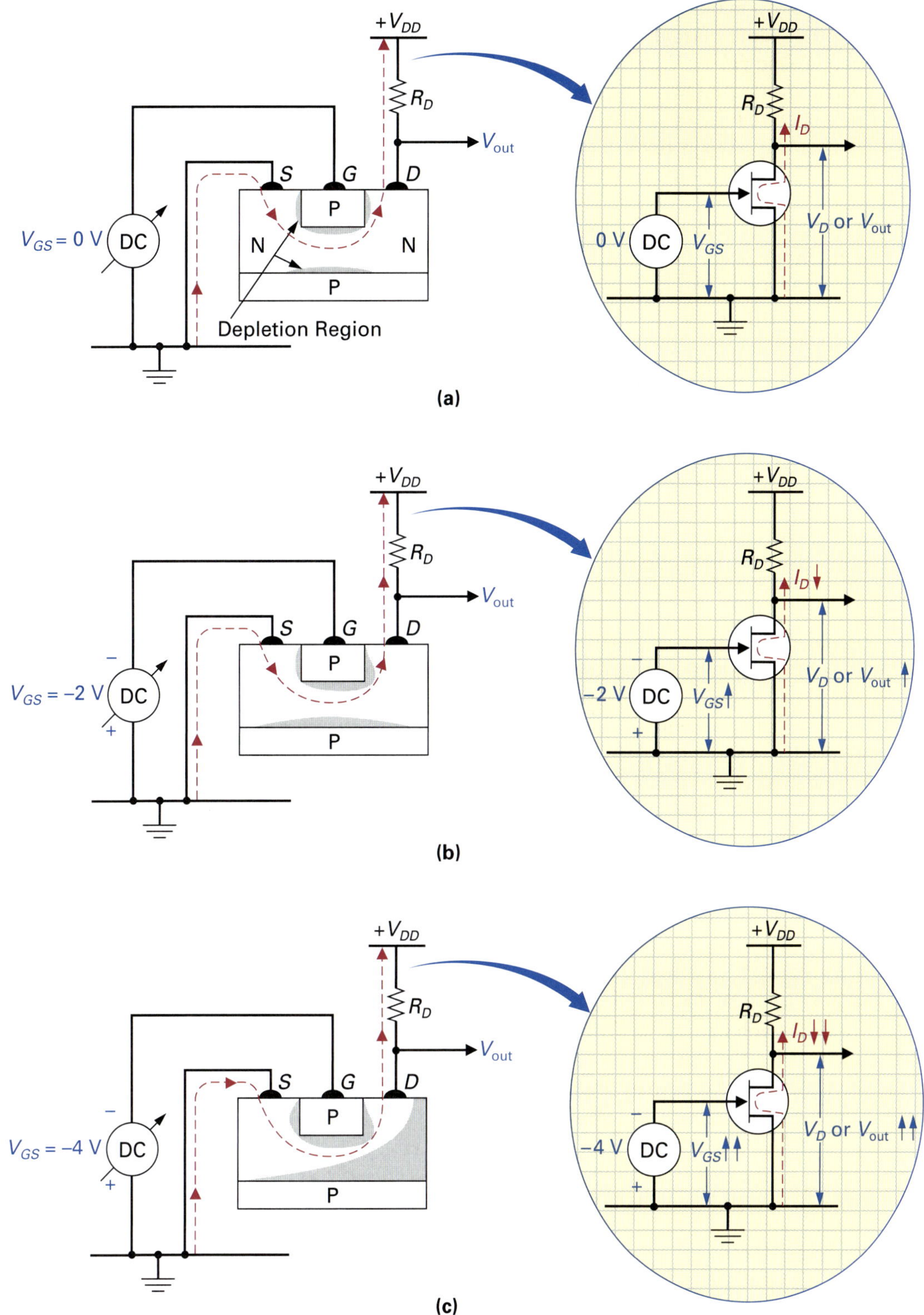

FIGURE 13-3 **The Relationship between a JFET's Input Voltage (V_{GS}) and Output Current (I_D).**

In most circuit applications, the $+V_{DD}$ supply voltage is maintained constant and the V_{GS} input voltage is used to control the resistance of the channel and the value of the output current, I_D. This can be seen more clearly in the insets in Figure 13-3. Because the output voltage (V_D or V_{out}) is dependent on the resistance between the JFET's source and drain, by controlling the value of I_D, we can control the output voltage. For instance, if the input

Current-Controlled Device
A device in which the input junction is normally forward biased and the input current controls the output current.

Voltage-Controlled Device
A device in which the input junction is normally reverse biased and the input voltage controls the output current.

Unipolar Device
A device in which only one type of semiconductor material exists between the output terminals and therefore the charge carriers have only one polarity (unipolar).

Bipolar Device
A device in which there is a change in semiconductor material between the output terminals (NPN or PNP between emitter and collector), so the charge carriers can be one of two polarities (bipolar).

Drain Characteristic Curve
A plot of the drain current (I_D) versus the drain-to-source voltage (V_{DS}).

Pinch-Off Voltage (V_P)
The value of V_{DS} at which further increases in V_{DS} will cause no further increase in I_D.

Constant-Current Region
The flat portion of the drain characteristic curve. In this region I_D remains constant despite changes in V_{DS}.

Breakdown Voltage (V_{BR})
The voltage at which a damaging value of I_D will pass through the JFET.

voltage V_{GS} is made more negative, the resistance of the channel will be increased, causing the output current I_D to decrease, and therefore the voltage developed between drain and ground (V_D or V_{out}) to increase. The gate-to-source junction of an FET is normally always reverse biased by the input voltage (V_{GS}), and it is this input voltage that controls the output current (I_D) and output voltage (V_{out}). Because the gate-to-source junction of an FET is normally always reverse biased by the input voltage (V_{GS}), there will be no input current. This characteristic accounts for the FET's naturally high input impedance. The operation of an FET is very different from the BJT, which normally uses an input voltage to forward bias the base-to-emitter junction and vary the input current (which varies the output current), and therefore the output voltage. This is the distinct difference between an FET and a BJT. An FET's input junction is normally reverse biased, and therefore the input voltage controls the output current. A BJT's input junction is normally forward biased and therefore the input current controls the output current. This difference is why BJTs are known as **current-controlled devices** and FETs are known as **voltage-controlled devices.** In fact, the name "field effect transistor" is derived from this voltage-control action because the applied input voltage will generate an electric field. It is this electric field that varies the size of the depletion region, and therefore the resistance of the channel between the FET's drain and source output terminals. In other words, the "effect" of the electric "field" causes "transistance," which is the transferring of different values of resistance between the output terminals.

The term "junction" is attached to this type of FET because of the single P-N junction formed between the gate and the source-to-drain channel. Therefore, an *n*-channel JFET has a single P-N junction between gate to channel, and a *p*-channel JFET has a single N-P junction between gate to channel.

The field effect transistor is also often referred to as a **unipolar device** because only one type of semiconductor material exists between the output terminals (*n*-type or *p*-type channel between source and drain), and therefore the charge carriers have only one polarity (unipolar). Compare this to a BJT, which is a **bipolar device** because there is a change in semiconductor material between the output terminals (NPN or PNP between emitter and collector), and the charge carriers can be one of two polarities (bipolar, because both majority and minority carriers are used).

13-1-3 *JFET Characteristics*

Like the BJT, the JFET's response to certain variables is best described by using a graph. Figure 13-4(a) shows a graph plotting drain current (I_D) against drain-to-source voltage (V_{DS}). As you have probably already observed, this **drain characteristic curve** is very similar to a bipolar transistor's collector characteristic curve. Starting at 0 V and moving right along the horizontal axis, you can see that an increase in the drain supply voltage ($+V_{DD}$), and therefore an increase in V_{DS}, will result in a continual increase in I_D. At a certain V_{DS} voltage (in this example 5 V), further increases in V_{DS} will cause no further increase in I_D. This value of V_{DS} is called the **pinch-off voltage (V_P)** because it is the point at which the bias voltage has caused the depletion region to pinch off or restrict drain current. From this point on, further increases in V_{DS} are counteracted by increases in the resistance of the channel, and therefore I_D remains constant. This is shown by the flat portion of the graph in Figure 13-4(a) and is called the **constant-current region** because I_D remains constant despite changes in V_{DS}. If V_{DS} is further increased (by increasing $+V_{DD}$), the JFET will eventually reach its **breakdown voltage (V_{BR}),** at which time a damaging value of I_D will pass through the JFET.

In the example graph in Figure 13-4(a), we plotted what would happen to I_D as V_{DS} increased with V_{GS} at 0 V. In Figure 13-4(b) we will examine what will happen to an *n*-channel JFET when the gate-source junction is reverse biased by several negative voltages. As previously described in the JFET operation section, a negative voltage is normally applied to reverse bias the gate and set up a depletion region. As V_{GS} is made more negative, the gate will be further reverse biased, and the corresponding I_D value will be smaller. Therefore, when V_{GS} is at 0 V, a maximum value of drain current is passing through the JFET's

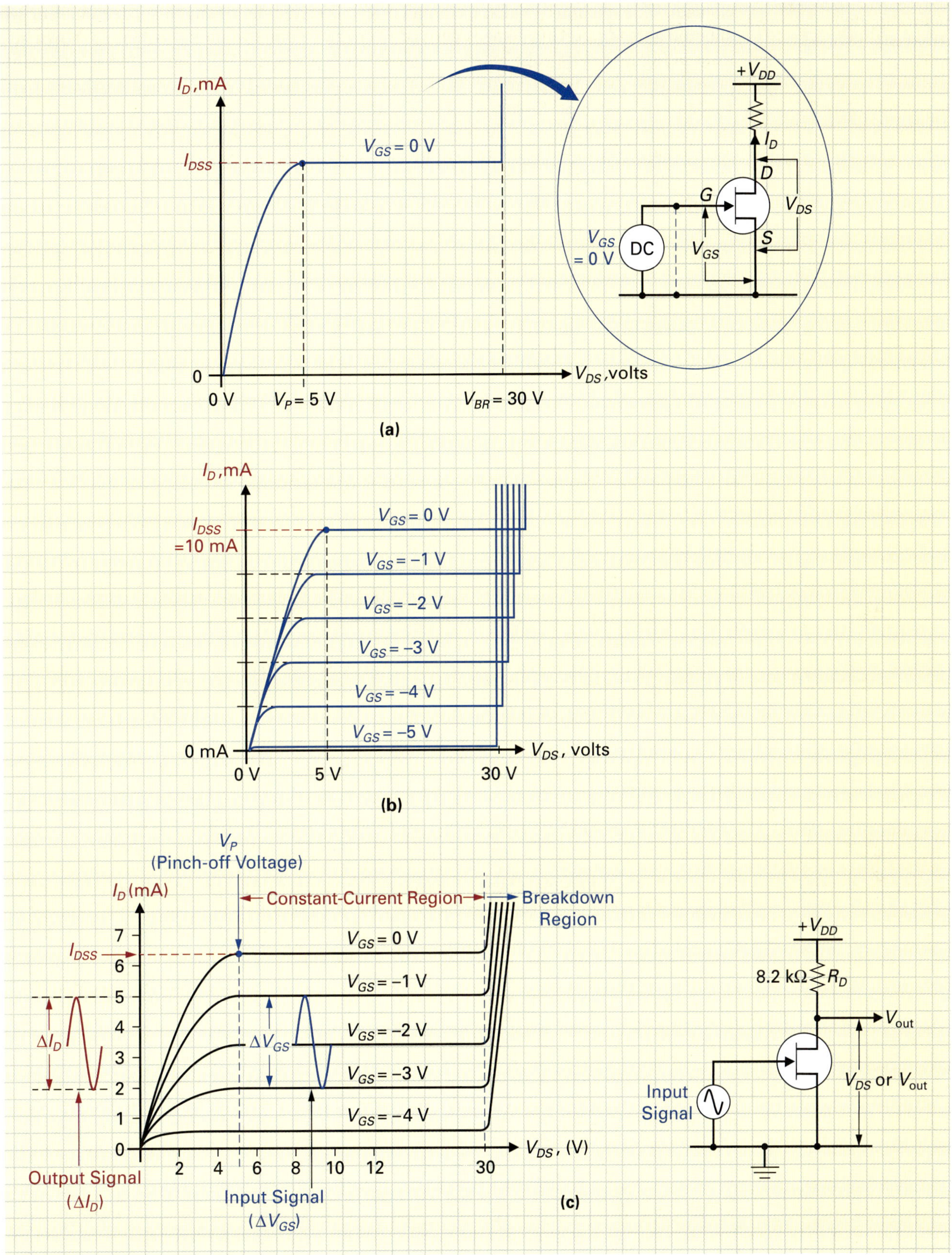

FIGURE 13-4 **JFET Characteristics.**

Drain-to-Source Current with Shorted Gate (I_{DSS})
The maximum value of drain current, achieved by holding V_{GS} at 0 V.

Gate-to-Source Cutoff Voltage or $V_{GS(OFF)}$
The negative V_{GS} bias voltage that causes I_D to drop to approximately zero.

Transconductance
Also called mutual conductance, it is the ratio of a change in output current to the initiating change in input voltage.

channel. This maximum value of drain current is called the **drain-to-source current with shorted gate (I_{DSS}).** This name is derived from the fact that when $V_{GS} = 0$ V, as shown in the inset in Figure 13-4(a), the gate and source terminals of the JFET are at the same potential of zero volts, and therefore the gate is effectively shorted to the source as shown by the dashed line. The drain-to-source current with shorted gate (I_{DSS}) rating is therefore the maximum current that can pass through the channel of a given JFET. When given on a specification sheet, this rating is equivalent to a bipolar transistor's $I_{C(\text{Sat.})}$ rating.

Returning to Figure 13-4(b), you can see that if V_{GS} is made more negative, the depletion regions within the JFET will get closer and closer and eventually touch, cutting off drain current. This negative V_{GS} bias voltage that causes I_D to drop to approximately zero is called the **gate-to-source cutoff voltage** or $V_{GS(OFF)}$. In the example in Figure 13-4(b), when $V_{GS} = -5$ V, I_D is almost zero and therefore $V_{GS(OFF)} = -5$ V. When cut OFF, the JFET will be equivalent to an open circuit between drain and source, and subsequently all of the drain supply voltage (V_{DD}) will appear across the open JFET ($V_{DS} = V_{DD}$).

To summarize the specifications in Figure 13-4(b),

When $V_{GS} = 0$ V, $I_D = I_{DSS} = 10$ mA
$V_P = 5$ V
$V_{BR} = 30$ V
$V_{GS(OFF)} = -5$ V
Constant-Current Region $= V_P$ to $V_{BR} = 5$ V to 30 V

13-1-4 *Transconductance*

Figure 13-4(c) illustrates a JFET test circuit and its associated characteristic graph. Before we see how a JFET can be made to amplify, let's first summarize the details given in this graph. If we first consider the curve when $V_{GS} = 0$ V, you can see that up to V_P, I_D increases in almost direct proportion to V_{DS}. This is because the depletion region is not sufficiently large enough to affect I_D, so the channel is simply behaving as a semiconductor with a fixed resistance value between source and drain.

When V_{DS} is equal to V_P, the drain current (I_D) will be pinched into an extremely narrow channel between the wedge-shaped depletion region. Any further increase in V_{DS} will have two effects:

1. increase the pinching effect on the channel, which will resist current flow, and
2. increase the potential between the drain and source, which will encourage current flow.

The net result is that channel resistance increases in direct proportion with V_{DS} and consequently I_D remains constant, as shown by the flat portion of the characteristic curve.

Assuming a fixed value of V_{DD}, any increase in the negative voltage of V_{GS} will cause a corresponding decrease in I_D. Therefore, beyond V_P, I_D is controlled by small-signal changes (such as the input signal) in V_{GS} and is independent of changes in V_{DS}. This section of the curve between V_P and V_{BR} is called the constant-current region.

Like the bipolar transistor, an FET can be used to amplify a signal, as shown in Figure 13-4(c). As before, the amount of amplification achieved is a ratio between output and input. For a bipolar transistor, the amount of gain is equal to the ratio of input current to output current (beta). For an FET, there is no input current, and therefore an FET's gain is equal to the ratio of output current change (ΔI_D) to input voltage change (ΔV_{GS}). This ratio is called the FET's **transconductance** (symbolized δ_m).

$$\delta_m = \frac{\Delta I_D}{\Delta V_{GS}}$$

δ_m = transconductance in siemens (S)
ΔI_D = change in drain current
ΔV_{GS} = change in gate-source voltage

EXAMPLE:

Calculate the transconductance of the FET for the example shown in Figure 13-4(c).

Solution:

$$g_m = \frac{\Delta I_D}{\Delta V_{GS}} = \frac{5\text{ mA} - 2\text{ mA}}{-1\text{ V} - (-3\text{ V})} = \frac{3\text{ mA}}{2\text{ V}} = 1.5\text{ millisiemens}$$

A high-gain FET will produce a large change in I_D for a small change in V_{GS}, resulting in a high transconductance figure ($g_m\uparrow$).

13-1-5 *Voltage Gain*

Because transconductance is the ratio of output current change (ΔI_D) to input voltage change (ΔV_{GS}), it is no surprise that this ratio is used to determine a JFET's voltage gain. The voltage gain formula is as follows

$$A_V = g_m \times R_D$$

EXAMPLE:

Calculate the voltage gain for the circuit example shown in Figure 13-4(c).

Solution:

$$A_V = g_m \times R_D = 1.5\text{ mS} \times 8.2\text{ k}\Omega = 12.3$$

This means that the output voltage will be 12.3 times greater than the input voltage.

13-1-6 *JFET Biasing*

The biasing methods used in FET circuits are very similar to those employed in BJT circuits. In this section we will examine the circuit calculations for the three most frequently used JFET biasing methods: gate biasing, self biasing and voltage-divider biasing.

Gate Biasing

Figure 13-5(a) shows a gate biased JFET circuit. The gate supply voltage ($-V_{GG}$) is used to reverse bias the gate-source junction of the JFET. With no gate current, there can be no voltage drop across R_G and therefore the voltage at the gate of the JFET will equal the dc gate supply voltage.

$$V_{GS} = V_{GG}$$

In the example in Figure 13-5,

$$V_{GS} = V_{GG} = -1.5\text{ V}$$

Knowing V_{GS}, we can calculate I_D if the JFET's current (I_{DSS}) and voltage ($V_{GS(\text{OFF})}$) specification limits are known by using the following formula.

$$I_D = I_{DSS}\left(1 - \frac{V_{GS}}{V_{GS(\text{OFF})}}\right)^2$$

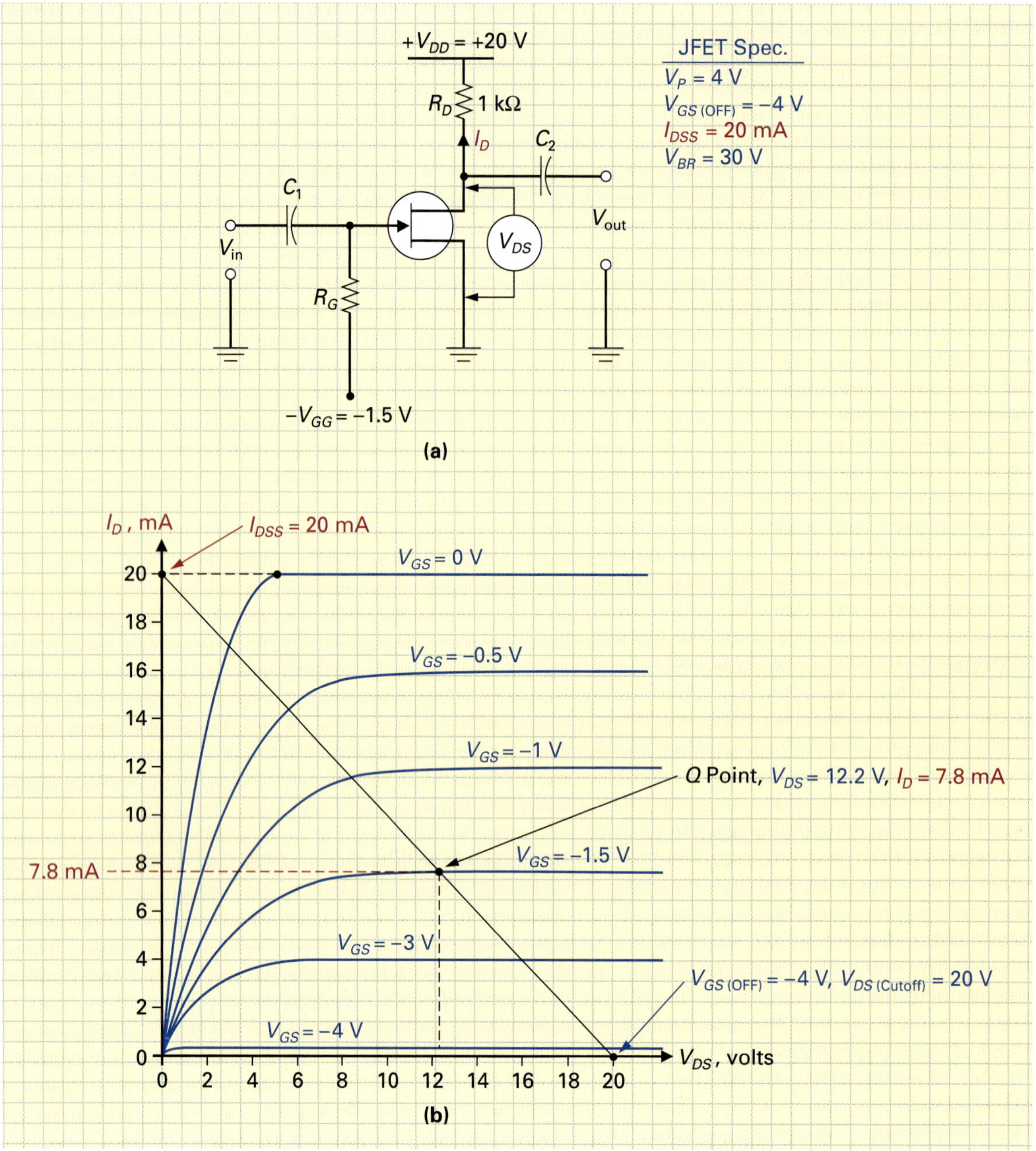

FIGURE 13-5 A Gate-Biased JFET Circuit. (a) Basic Circuit. (b) Drain Characteristic Curves and DC Load Line.

In the example in Figure 13-5,

$$I_D = 20\text{ mA}\left(1 - \frac{-1.5\text{ V}}{-4\text{ V}}\right)^2$$
$$= 20\text{ mA }(1 - 0.375)^2$$
$$= 20\text{ mA} \times 0.625^2$$
$$= 20\text{ mA} \times 0.39 = 7.8\text{ mA}$$

Now that I_D is known, we can calculate the voltage drop across R_D using Ohm's law.

$$V_{R_D} = I_D \times R_D$$

In the example in Figure 13-5,

$$V_{R_D} = I_D \times R_D = 7.8 \text{ mA} \times 1 \text{ k}\Omega = 7.8 \text{ V}$$

Because V_{R_D} plus V_{DS} will equal V_{DD}, we calculate V_{DS} once V_{R_D} is known with the following formula.

$$V_{DS} = V_{DD} - V_{R_D}$$

In the example in Figure 13-5,

$$V_{DS} = V_{DD} - V_{R_D} = 20 \text{ V} - 7.8 \text{ V} = 12.2 \text{ V}$$

Figure 13-5(b) shows the drain characteristic curves and dc load line for the example JFET circuit in Figure 13-5(a). Like the bipolar transistor, the JFET's dc load line extends between the maximum output current point, or saturation point (when the JFET is fully ON, $I_{DSS} = 20$ mA), to the maximum output voltage point (when the JFET is cut OFF, $V_{DS} = 20$ V). The dc operating point, or Q point, which was determined with the previous calculations, is also plotted on the dc load line in Figure 13-5(b).

EXAMPLE:

Calculate the following for the circuit shown in Figure 13-6.

a. V_{GS}
b. I_D
c. V_{DS}
d. Maximum value of I_D
e. V_{DS} when $V_{GS} = V_{GS(\text{OFF})}$
f. Q point

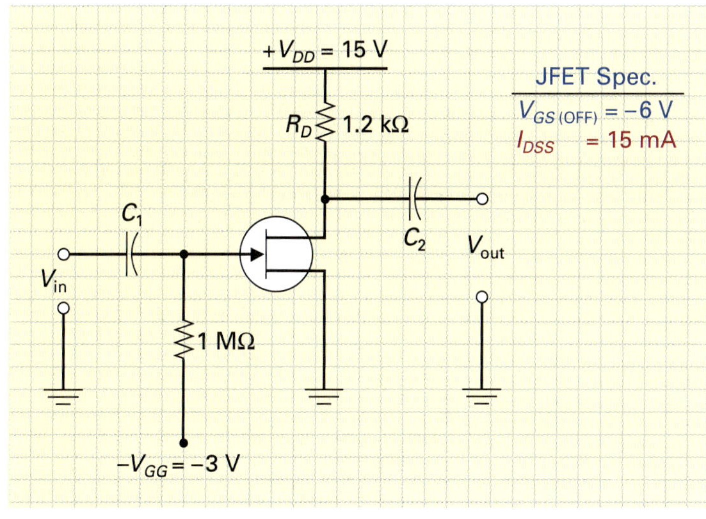

FIGURE 13-6 A Gate Biased JFET Circuit Example.

Solution:

a. $V_{GS} = V_{GG} = -3$ V

b. $I_D = I_{DSS}\left(1 - \frac{V_{GS}}{V_{GS(\text{OFF})}}\right)^2 = 15 \text{ mA}\left(1 - \frac{-3 \text{ V}}{-6 \text{ V}}\right)^2 = 15 \text{ mA}(1 - 0.5)^2 = 3.75 \text{ mA}$

c. $V_{DS} = V_{DD} - V_{R_D}$ (since $V_{RD} = I_D \times R_D$, we can substitute)

$V_{DS} = V_{DD} - (I_D \times R_D) = 15\text{ V} - (3.75\text{ mA} \times 1.2\text{ k}\Omega) = 15\text{ V} - 4.5\text{ V} = 10.5\text{ V}$

d. Maximum value of $I_D = I_{DSS} = 15$ mA

e. When $V_{GS} = V_{GS(\text{OFF})}$, the JFET is cut off and equivalent to an open switch between drain and source. In this condition all of the drain supply voltage will appear across the open JFET.

$$V_{DS(\text{Cutoff})} = V_{DD} = 15\text{ V}$$

f. The dc operating or Q point is

$$V_{GS} = -3\text{ V}$$
$$I_D = 3.75\text{ mA}$$
$$V_{DS} = 10.5\text{ V}$$

Self Biasing

Figure 13-7(a) shows how to self bias a JFET circuit. One advantage of this biasing method over gate biasing is that only a single drain supply voltage is needed (V_{DD}) instead of both V_{DD} and a negative gate supply voltage ($-V_{GG}$). The other difference you may have noticed is that a source resistor (R_S) has been included, and R_G has been connected to ground. Although this arrangement seems completely different to the gate bias circuit, the inclusion of R_S and the grounding of R_G will achieve the same result, which is to reverse bias the JFET's gate-source junction. Figure 13-7(b) illustrates how this is achieved. Since there is no gate current in a JFET circuit ($I_G = 0$), all of the current flowing into the source will travel through the channel and flow out of the drain. Therefore

$$I_S = I_D$$

For the example in Figure 13-7,

$$I_S = I_D = 7\text{ mA}$$

Now that I_S is known, we can calculate the voltage drop across the source resistor (V_{R_S}), and therefore the voltage at the JFET's source (V_S).

$$V_{R_S} = V_S = I_S \times R_S$$

For the example in Figure 13-7,

$$V_{RS} = V_S = I_S \times R_S = 7\text{ mA} \times 500\ \Omega = +3.5\text{ V}$$

Because $I_G = 0$ A, there will be no voltage drop across R_G, and so the voltage at the gate of the JFET will be 0 V.

$$V_G = 0\text{ V}$$

Now that we know that $V_S = +3.5$ V and $V_G = 0$ V, we can see how the JFET's gate-source junction is reverse biased. To reverse bias a gate biased JFET, we simply made the gate voltage negative with respect to the source that is at 0 V. With a self biased JFET, we achieve the same result by making the source voltage positive with respect to the gate that is

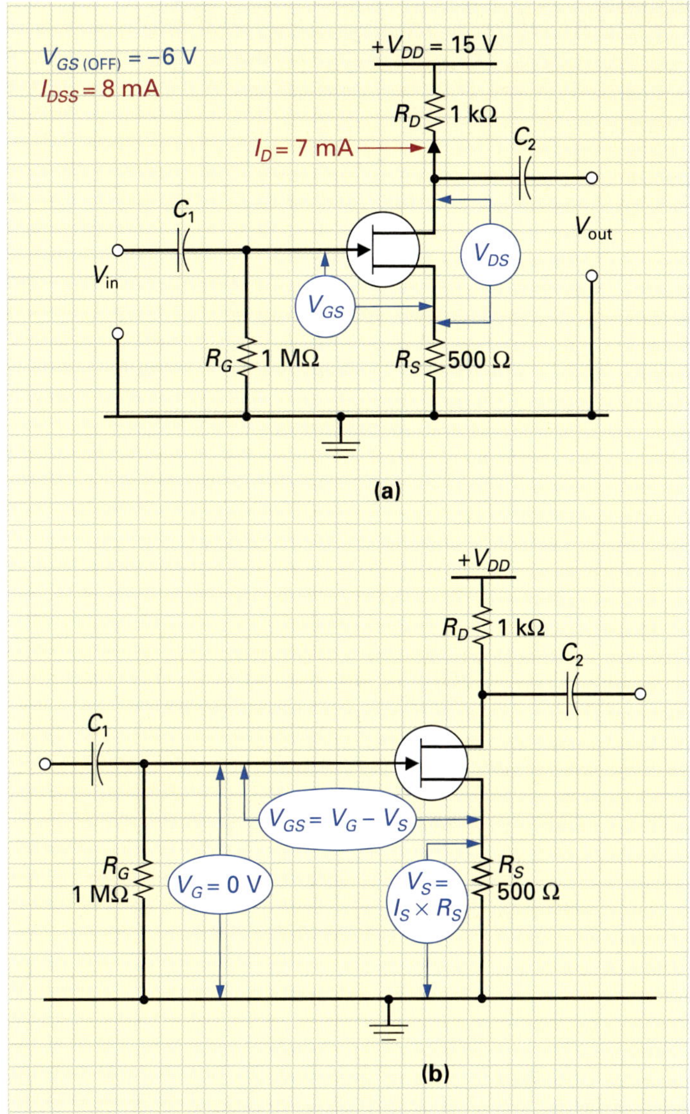

FIGURE 13-7 **A Self Biased JFET Circuit. (a) Basic Circuit. (b) How R_S Develops a $-V_{GS}$.**

EWB

at 0 V. This makes the gate of the JFET negative with respect to the source. This potential difference from gate-to-source (V_{GS}) is therefore equal to

$$V_{GS} = V_G - V_S$$

Because $V_S = I_S \times R_S$ and $V_G = 0$ V, we can substitute the previous formula to obtain

$$V_{GS} = 0\text{ V} - (I_S \times R_S)$$

or

$$V_{GS} = -(I_S \times R_S)$$

or because $I_S = I_D$

$$V_{GS} = -(I_D \times R_S)$$

In the example in Figure 13-7,

$$\begin{aligned} V_{GS} &= -(I_S \text{ or } I_D \times R_S) \\ &= -(7 \text{ mA} \times 500\ \Omega) \\ &= -3.5 \text{ V} \end{aligned}$$

If $-V_{GS}$ and R_S are known, we could transpose the above equation to calculate I_D.

$$I_D = \frac{V_{GS}}{R_S}$$

In the example in Figure 13-7,

$$I_D = \frac{V_{GS}}{R_S} = \frac{3.5 \text{ V}}{500\ \Omega} = 7 \text{ mA}$$

The final calculation is to determine the voltage at the JFET's drain with respect to ground (V_D) and the drain-to-source voltage drop across the JFET.

$$V_D = V_{DD} - V_{R_D}$$

Because $V_{R_D} = I_D \times R_D$,

$$V_D = V_{DD} - (I_D \times R_D)$$

In the example in Figure 13-7,

$$V_D = V_{DD} - (I_D \times R_D) = 15 \text{ V} - (7 \text{ mA} \times 1 \text{ k}\Omega) = 15 \text{ V} - 7 \text{ V} = 8 \text{ V}$$

Now that the voltage drops across R_D (V_{R_D}) and R_S (V_{R_S}) are known, we can calculate the voltage drop across the JFET's drain to source (V_{DS}).

$$V_{DS} = V_{DD} - (V_{R_D} + V_{R_S})$$

In the example in Figure 13-7,

$$V_{DS} = V_{DD} - (V_{R_D} + V_{R_S}) = 15 \text{ V} - (7 \text{ V} + 3.5 \text{ V}) = 15 \text{ V} - 10.5 \text{ V} = 4.5 \text{ V}$$

EXAMPLE:

Calculate the following for the circuit shown in Figure 13-8.

a. V_S
b. V_{GS}
c. V_{DS}
d. I_D maximum
e. V_{DS} when the JFET is OFF
f. V_D

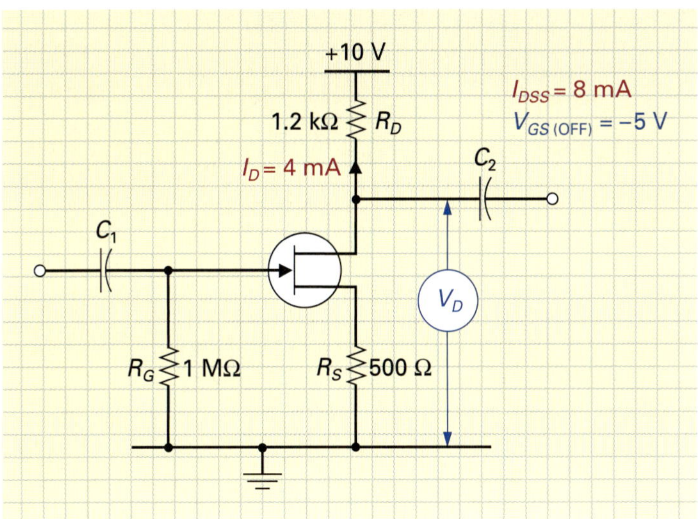

FIGURE 13-8 **A Self Biased JFET Circuit Example.**

Solution:

a. Because $I_S = I_D$, $V_S = I_S \times R_S = 4\text{ mA} \times 500\ \Omega = 2\text{ V}$

b. $V_{GS} = V_G - V_S = 0\text{ V} - 2\text{ V} = -2\text{ V}$

c. $V_{DS} = V_{DD} - (V_{R_D} + V_{R_S}) = 10\text{ V} - [(I_D R_D) + 2\text{ V}]$
$= 10\text{ V} - [(4\text{ mA} \times 1.2\text{ k}\Omega) + 2\text{ V}] = 10\text{ V} - (4.8\text{ V} + 2\text{ V})$
$= 10\text{ V} - 6.8\text{ V} = 3.2\text{ V}$

d. I_D maximum $= I_{DSS} = 8\text{ mA}$

e. $V_{DS(\text{Cutoff})} = V_{DD} = 10\text{ V}$

f. $V_D = V_{DS} + V_{R_S} = 3.2\text{ V} + 2\text{ V} = 5.2\text{ V}$

As previously mentioned, one advantage of this self biased JFET method is that only a drain supply voltage is needed (V_{DD}). The gate supply voltage (V_{GG}) is not needed due to the inclusion of a source resistor that reverse biases the JFET's gate-source junction by applying a positive voltage to the source with respect to the 0 V on the gate. This method of effectively sending back a negative voltage from the source to the gate is known as "negative feedback." It not only enables us to bias a JFET with one supply voltage, it also provides temperature stability. Any change in the ambient temperature will cause a change in the semiconductor JFET's conduction, which would move the JFET's *Q* point away from its desired setting. The inclusion of R_S will prevent the *Q* point from shifting due to temperature in the same way as a BJT's emitter resistor. If temperature were to increase, for instance ($T\uparrow$), the resistance of the semiconductor would decrease ($R\downarrow$) because all semiconductor materials have a negative temperature coefficient of resistance ($T\uparrow$, $R\downarrow$), and this will cause the channel current to increase. If the drain current increases ($I_D\uparrow$), the voltage drop across R_S will increase (V_{R_S} or $V_S\uparrow = I_D\uparrow \times R_S$). This increase in V_S will increase the gate-source reverse voltage ($-V_{GS}\uparrow$), causing the JFET's channel to get narrower and the drain current to decrease ($I_D\downarrow$) and counteract the original increase. Similarly, a decrease in temperature will cause a decrease in I_D, which will decrease the gate-source reverse bias, resulting in an increase in I_D. The *Q* point will remain relatively stable despite changes in temperature when a JFET circuit has a source resistor included.

Voltage-Divider Biasing

Referring to the voltage-divider biased JFET circuit shown in Figure 13-9, you will probably notice that it is very similar to the voltage-divider biased BJT circuit discussed

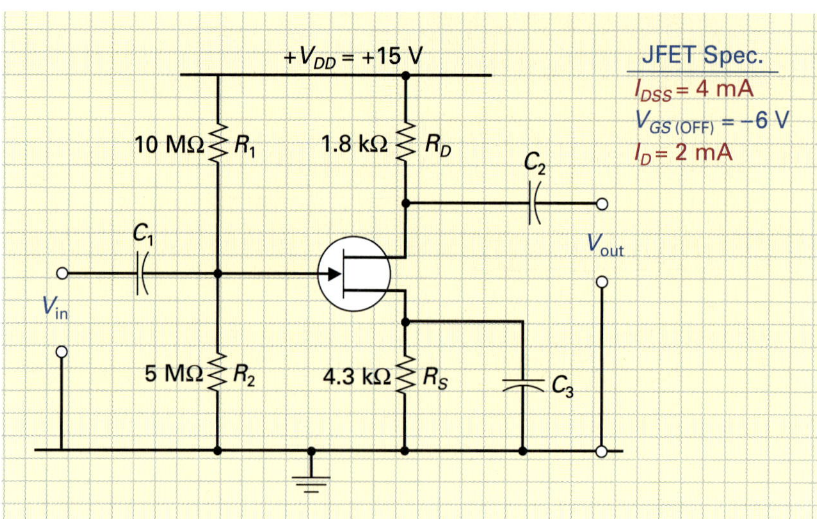

FIGURE 13-9 A Voltage-Divider Biased JFET Circuit.

previously. Like the self biased circuit, the inclusion of a source resistor stabilizes the Q point despite ambient temperature changes. In addition, using a voltage divider to determine the gate-source bias voltage ensures that V_{GS}, and therefore the circuit, has increased stability.

The gate voltage (V_G) is calculated using the following voltage-divider formula:

$$V_{R_2} \text{ or } V_G = \frac{R_2}{R_1 + R_2} \times V_{DD}$$

For the example in Figure 13-9,

$$V_{R_2} \text{ or } V_G = \frac{R_2}{R_1 + R_2} \times V_{DD} = \frac{5\ \text{M}\Omega}{10\ \text{M}\Omega + 5\ \text{M}\Omega} \times 15\ \text{V} = 5\ \text{V}$$

Because $I_D = I_S$ ($I_G = 0$) and the drain resistance and current are known, we can next calculate the voltage drop across the source resistor (V_{R_S}), drain resistor (V_{R_D}), and the JFET's source-drain junction (V_{DS}).

$$I_S = I_D$$

$$I_S = I_D = 2\ \text{mA}$$

$$V_{R_S} = I_S \times R_S$$

$$V_{R_S} = 2\ \text{mA} \times 4.3\ \text{k}\Omega = 8.6\ \text{V}$$

$$V_{R_D} = I_D \times R_D$$

$$V_{R_D} = 2\ \text{mA} \times 1.8\ \text{k}\Omega = 3.6\ \text{V}$$

$$V_{DS} = V_{DD} - (V_{R_S} + V_{R_D})$$

$$V_{DS} = 15 \text{ V} - (8.6 \text{ V} + 3.6 \text{ V}) = 2.8 \text{ V}$$

Now that the JFET's gate and source voltages are known (V_G and V_S), we can calculate the value of gate-source reverse bias ($- V_{GS}$).

$$V_{GS} = V_G - V_S$$

$$(V_G = V_{R_2}, V_S = V_{R_S})$$

For the example in Figure 13-9,

$$V_{GS} = 5 \text{ V} - 8.6 \text{ V} = -3.6 \text{ V}$$

EXAMPLE:

Calculate the following for the voltage-divider biased JFET circuit shown in Figure 13-10:

a. V_G
b. I_S
c. V_S
d. V_{DS}
e. V_{GS}
f. V_D, when $V_{GS} = V_{GS(OFF)}$
g. I_D, when $V_{GS} = 0$ V

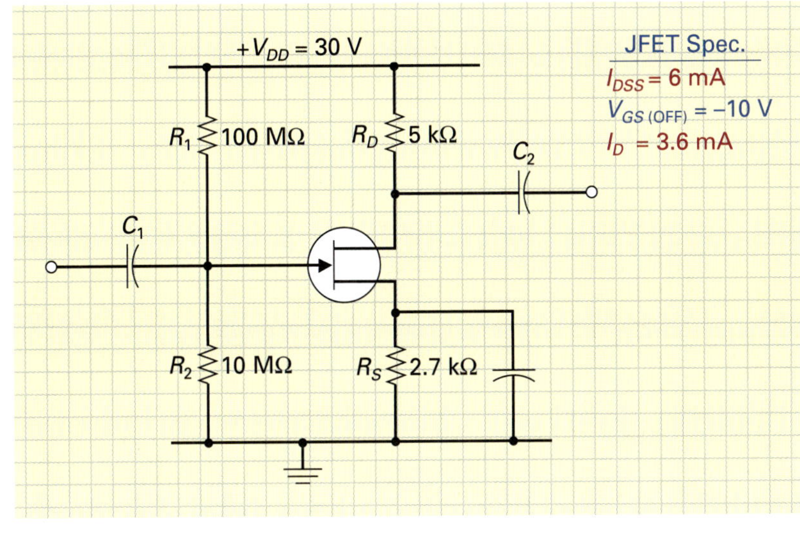

FIGURE 13-10 A Voltage-Divider Biased Circuit Example.

Solution:

a. $V_G = \dfrac{R_2}{R_1 + R_2} \times V_{DD} = \dfrac{10 \text{ M}\Omega}{100 \text{ M}\Omega + 10 \text{ M}\Omega} \times 30 \text{ V} = 2.7 \text{ V}$

b. $I_S = I_D = 3.6$ mA

c. $V_S = V_{R_S} = I_S \times R_S = 3.6 \text{ mA} \times 2.7 \text{ k}\Omega = 9.7 \text{ V}$

d. $V_{DS} = V_{DD} - (V_{R_S} + V_{R_D}) = 30 \text{ V} - [9.7 \text{ V} + (I_D \times R_D)]$
$= 30 \text{ V} - [9.7 \text{ V} + (3.6 \text{ mA} \times 5 \text{ k}\Omega)] = 30 \text{ V} - (9.7 \text{ V} + 18 \text{ V}) = 2.3 \text{ V}$

e. $V_{GS} = V_G - V_S = 2.7\text{ V} - 9.7\text{ V} = -7\text{ V}$
f. When $V_{GS} = V_{GS(\text{OFF})}$, JFET is OFF and $V_D = V_{DD} = 30\text{ V}$
g. When $V_{GS} = 0\text{ V}$, $I_D = \text{maximum} = I_{DSS} = 6\text{ mA}$

13-1-7 *JFET Circuit Configurations*

The three JFET circuit configurations are illustrated in Figure 13-11 along with their typical circuit characteristics. Like the bipolar transistor configurations, the term "common" is used to indicate which of the JFET's leads is common to both the input and output. In this section we will examine the characteristics of these three configurations: common-source, common-gate, and common-drain.

	Voltage Gain	Input Impedance	Output Impedance	Circuit Appearance and Application	Waveforms
Common-Source (a)	5–10 (Voltage Amp)	Very High 1–15 MΩ	Low 2–10 kΩ	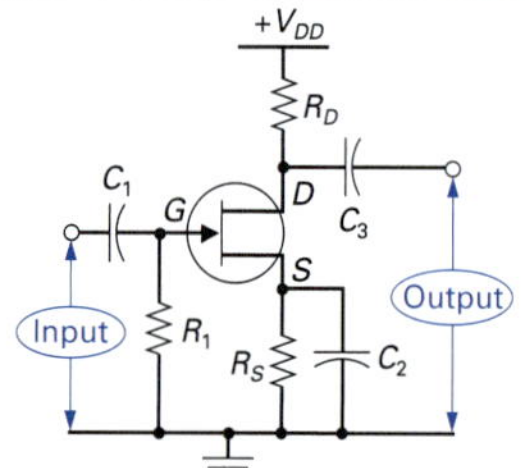 $+V_{DD}$, R_D, C_1, G, D, C_3, S, Input, R_1, R_S, C_2, Output. Most widely used FET configuration. It is mainly used as a voltage amplifier, however it is also used as an impedance matching device and can handle the high radio frequency signals.	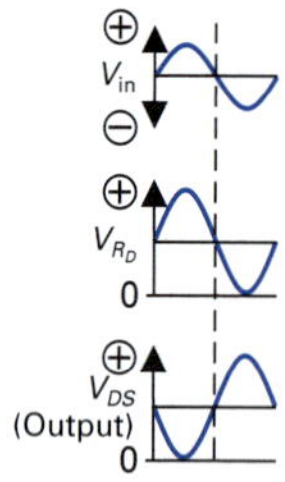 V_{in}, V_{R_D}, V_{DS} (Output). V_{in} and V_{out} are out of phase (180° phase shift)
Common-Gate (b)	2–5	Very Low 200–1500 Ω	Medium 5–15 kΩ	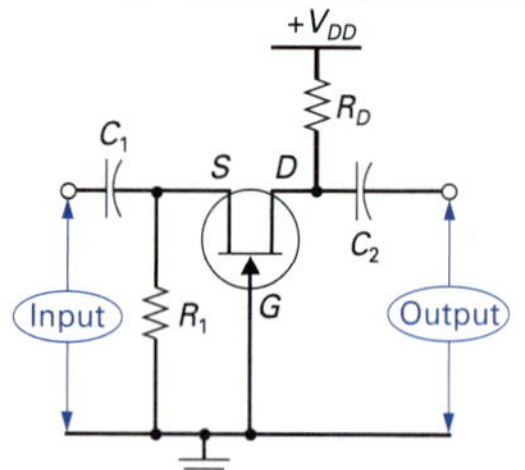 $+V_{DD}$, R_D, C_1, S, D, C_2, Input, R_1, G, Output. This configuration is used to amplify radio frequency signals due to its very stable nature at high frequencies. It is also used as a buffer to match a low impedance source to a high impedance load.	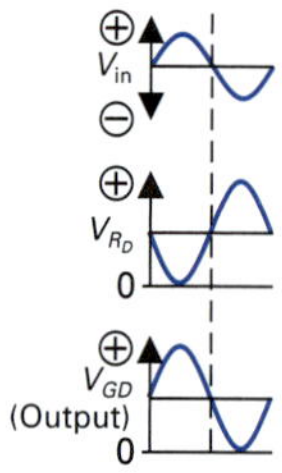 V_{in}, V_{R_D}, V_{GD} (Output). V_{in} and V_{out} are in phase (0° phase shift)
Common-Drain (c)	0.98	Very High 1500 MΩ	Low 10 kΩ	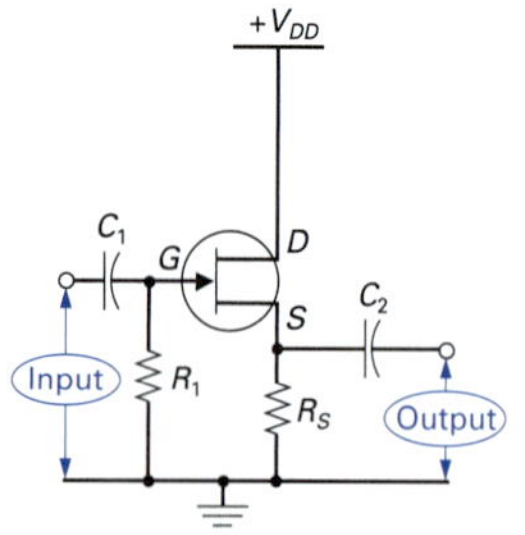 $+V_{DD}$, C_1, G, D, S, C_2, Input, R_1, R_S, Output. This amplifier is commonly called a source-follower as the source follows whatever is applied to the gate. Its very high input impedance will not load down (and therefore not distort) signals from high-impedance signal sources, such as a microphone, and its low output impedance is ideal to drive a low-impedance load such as an audio amplifier.	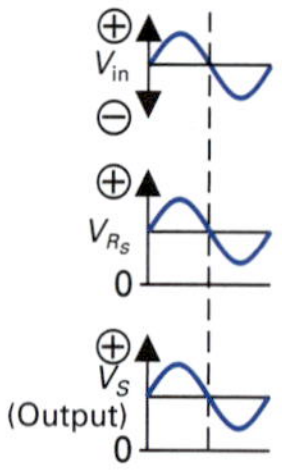 V_{in}, V_{R_S}, V_S (Output). V_{in} and V_{out} are in phase (0° phase shift)

FIGURE 13-11 JFET Circuit Configurations.

Common-Source (*C-S*) Circuits

Similar to its bipolar counterpart, the common-emitter configuration, the **common-source configuration** is the most widely used JFET circuit and is detailed in Figure 13-11(a). The input is applied between the gate and source and the output is taken between the drain and source, with the source being common to both input and output. The ac input will pass through the coupling capacitor C_1 and be superimposed on the dc gate-source bias voltage provided by resistor R_1, which sets up the dc operating or Q point. As the signal input changes, it will cause a change in gate voltage, which will cause a corresponding change in the output drain current. The output voltage developed between the FET's drain and ground is 180° out of phase with the input because an increase in $V_{in}\uparrow$, and therefore $V_{GS}\uparrow$, will cause an increase in $I_D\uparrow$, a decrease in the voltage drop across the FET ($V_{DS}\downarrow$), and a decrease in the output voltage $V_{out}\downarrow$. Resistor R_S is included to provide temperature stability and, as with the bipolar transistor, the source decoupling capacitor C_2 is included to prevent degenerative feedback.

Common-Source Configuration
An FET configuration in which the source is grounded and common to the input and output signals.

When a small ac input signal is applied to the gate of a common-source amplifier, the variations in voltage at the gate control the JFET, which effectively acts as a variable resistor, varying the output drain current. These changes in drain current will vary the voltage drop across R_D and the drain-to-source voltage drop, which, with R_S, determines the output voltage. Referring to the characteristics listed in Figure 13-11(a), you can see that the output voltage (V_{out}) of the common-source JFET configuration can be five to ten times larger than the gate control input voltage (V_{in}). If a high amount of voltage gain is desired, R_D is made relatively large (typically greater than 20 kΩ) and the JFET is biased so that its drain-to-source resistance is also high. A larger resistance will develop a larger voltage.

Also listed in the common-source characteristics in Figure 13-11(a) is the very high input resistance and the relatively low output resistance of this circuit. The high input resistance is due to the JFET's reverse biased gate-source junction, which permits no gate input current, and therefore has a very large resistance. This key characteristic means that the common-source JFET circuit is ideal in applications where we need to provide voltage amplification but do not want to load down a source that can only generate a small input signal. Such applications include the following:

1. Digital circuits in which the outputs of many circuits are connected to one another, and therefore the output resistances of all the circuits load one another. As a result, the signals generated by these circuits are small, and a circuit is needed that will not load the signal source but will still provide voltage gain.
2. Analog circuits in which it can amplify both dc and low- and high-frequency ac input signal voltages. The *C-S* circuit's high input impedance makes it ideal at the front end of systems such as the first RF amplifier stage following the antenna and the first stage in a voltmeter, in which it will not load the source yet will amplify a wide range of input signal voltages.

Common-Gate (*C-G*) Circuits

The **common-gate circuit configuration** shown in Figure 13-11(b) is very similar to its bipolar counterpart, the common-base circuit. The input is applied between the source and gate, while the output appears across the drain and gate. Self bias resistor R_1 sets up the static Q point, and the input is applied through the coupling capacitor C_1 and will cause a change in the JFET's source voltage. An increase in source voltage will cause a decrease in the V_{GS} forward bias (n-type source is driven positive), a decrease in I_D, a decrease in the voltage drop across R_D, and therefore an increase in the voltage dropped between the FET's drain and gate. Because the voltage developed across the JFET's drain and gate is applied to the output, an increase in the input produces an increase in the output, and so the input and output voltage are in phase with one another. Similarly, as the input voltage decreases, the gate-source forward bias will increase. Therefore, I_D will increase, and there will be more voltage developed across R_D and less voltage developed at the output.

Common-Gate Configuration
An FET configuration in which the gate is grounded and common to the input and output signal.

Referring to the common-gate characteristics listed in Figure 13-11(b), you can see that this circuit can be used to provide a small voltage gain. Because the input is applied to the

JFET's high-current source terminal, the input resistance is very low. This low input resistance and relatively high output resistance makes the circuit ideal in applications where we need to efficiently transfer power between a low-resistance source and a high-resistance load.

Common-Drain (*C-D*) Circuits

Common-Drain Configuration
An FET configuration in which the drain is grounded and common to the input and output signal.

Source-Follower
Another name used for a common-drain circuit configuration.

Comparable to the bipolar transistor's common-collector or emitter-follower, the **common-drain configuration** shown in Figure 13-11(c) is sometimes called a **source-follower** because the source output voltage follows in polarity and amplitude the input voltage at the gate. Once again, self bias resistor R_1 sets up the quiescent operating point, and an ac gate input voltage will cause a variation in I_D. When the input voltage at the gate swings positive, the FET will conduct more current, less voltage will be developed across the FET drain to source, and therefore more voltage will be developed across R_S and the output. Similarly, a decrease in the input voltage will cause the resistance of the JFET's drain-source junction to increase. Therefore, V_{DS} will increase and V_{R_L}, or V_{out}, will decrease.

Referring to the common-drain characteristics listed in Figure 13-11(c), you can see that the output voltage is slightly less than the input voltage (circuit does not provide any voltage gain). The input resistance of the common-drain circuit configuration is extremely high due to the JFET's reverse biased gate and R_S connection, and the output resistance is relatively very low. Inserting a common-drain circuit between a high-resistance source and a low-resistance load will ensure that the two opposite resistances are matched and power is efficiently transferred.

13-1-8 *JFET Applications*

It is the high input impedance of the JFET, and therefore its ability not to load a source, and the voltage amplification ability that are mainly made use of in circuit applications. Like the bipolar junction transistor, the JFET can be made to function as a switch or as a variable resistor. Let us begin by examining how the JFET's switching ability can be made use of in digital or two-state circuits.

Digital (Two-State) JFET Circuits

As a switch, the JFET makes use of only two points on the load line: saturation (in which it is equivalent to a closed switch between source and drain) and cutoff (in which it is equivalent to an open switch between source and drain). Figure 13-12(a) shows an ON/OFF JFET switch circuit and its associated load line in Figure 13-12(b). Figure 13-12(c) shows the input/output voltages for each of the circuit's two operating states. When $V_{GS} = V_{GS(OFF)}$ (-4 V), the JFET is cut OFF (lower end of the load line) and is equivalent to an open switch between source and drain. With the JFET's drain-source open, $I_D = 0$ mA, and the drain supply voltage will be applied to the output ($V_{DS} = V_{out} = +V_{DD}$). On the other hand, when $V_{GS} = 0$ V, the JFET is saturated (upper end of the load line) and is equivalent to a closed switch between source and drain. With the JFET's drain-source closed, I_D = max. = I_{DSS}, and the 0 V at the source will be applied to the output ($V_{DS} = V_{out} = 0$ V).

A typical FET application in digital circuits would be a buffer circuit, which is used to isolate one device from another. The high input impedance of the FET does not load the input circuit or circuits, while the low output impedance of the FET provides a high output current to the output circuit. The high output current and buffering or isolating characteristics of these circuits account for why they are also called buffer-drivers. The schematic symbol of the buffer-driver is shown in Figure 13-12(d).

The high input impedance ($Z_{in}\uparrow$) of the FET is also made use of in other FET integrated circuits (ICs). When Z_{in} is high, circuit current is low ($I\downarrow$), and therefore power dissipation is low ($P_D\downarrow$). This condition is ideal for digital integrated circuits (ICs), which contain thousands of transistors all formed onto one small piece of silicon. The low power dissipation of the JFET enables us to densely pack many more components into a very small area.

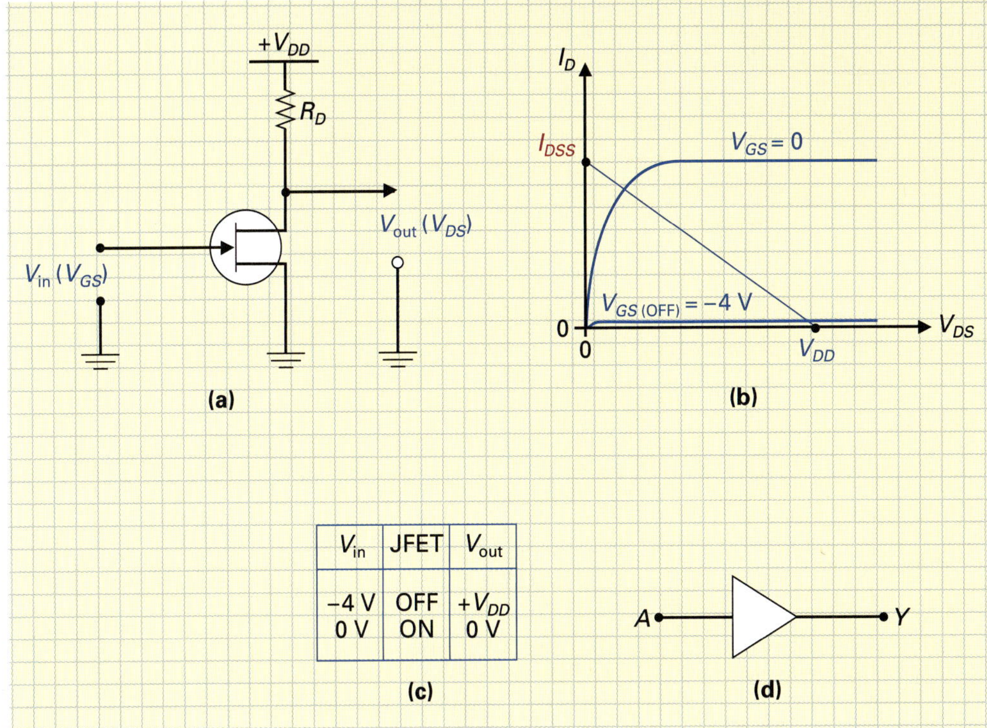

V_{in}	JFET	V_{out}
−4 V	OFF	$+V_{DD}$
0 V	ON	0 V

(c)

FIGURE 13-12 The JFET's Switching Action—Digital Circuit Applications. (a) Basic Switching Circuit. (b) Load Line. (c) Input/Output Voltages. (d) Digital Buffer-Driver Schematic Symbol.

Analog (Linear) JFET Circuits

In two-state applications, the JFET is made to operate between the two extreme points of saturation (0 Ω) and cutoff (max. Ω). By controlling the gate-source bias voltage (V_{GS}), the resistance between the JFET's drain and source (R_{DS}) can be changed to be any value between 0 Ω and maximum Ω. The JFET can therefore be made to act as a variable resistor, with an increase in negative V_{GS} causing a larger R_{DS}. In contrast, a decrease in negative V_{GS} causes a smaller R_{DS}. This is illustrated in Figure 13-13(a).

It is the reverse biased gate-source junction of a JFET that gives the JFET its key advantage: *an extremely high input impedance* (typically in the high-MΩ range). In addition, the JFET can provide a small voltage gain and has been found to be a very low noise component. All these characteristics make it an ideal choice as an amplifier.

In previous chapters, we have seen how a light load (large resistance $R_L\uparrow$ and small $I\downarrow$) does not pull down the source voltage by any large amount, whereas a heavy load (small resistance $R_L\downarrow$ and large $I\uparrow$) will pull down the source voltage. A heavy load results in less output voltage ($V_{R_L}\downarrow$) and an increased current and heat loss at the source. The overall effect is that a small load resistance or impedance causes less power to be delivered to the load.

The circuit in Figure 13-13(b) shows how a common-source JFET has been connected to function as a preamplifier, which is a circuit that provides gain for a very weak input signal. In this example, the JFET preamplifier matches the high-impedance (small signal) crystal microphone to a low-impedance power amplifier. The reverse biased gate-source junction of a JFET preamplifier will offer a large load impedance to the source or microphone. This light load ($R_L\uparrow$) input resistance of the JFET will therefore permit most of the signal voltage being generated by the microphone to be applied to the JFET's gate and then be amplified. In other words, the high input impedance of the JFET amplifier circuit will not pull down the voltage signal being generated by the microphone. Therefore, maximum power will be transferred from source to load.

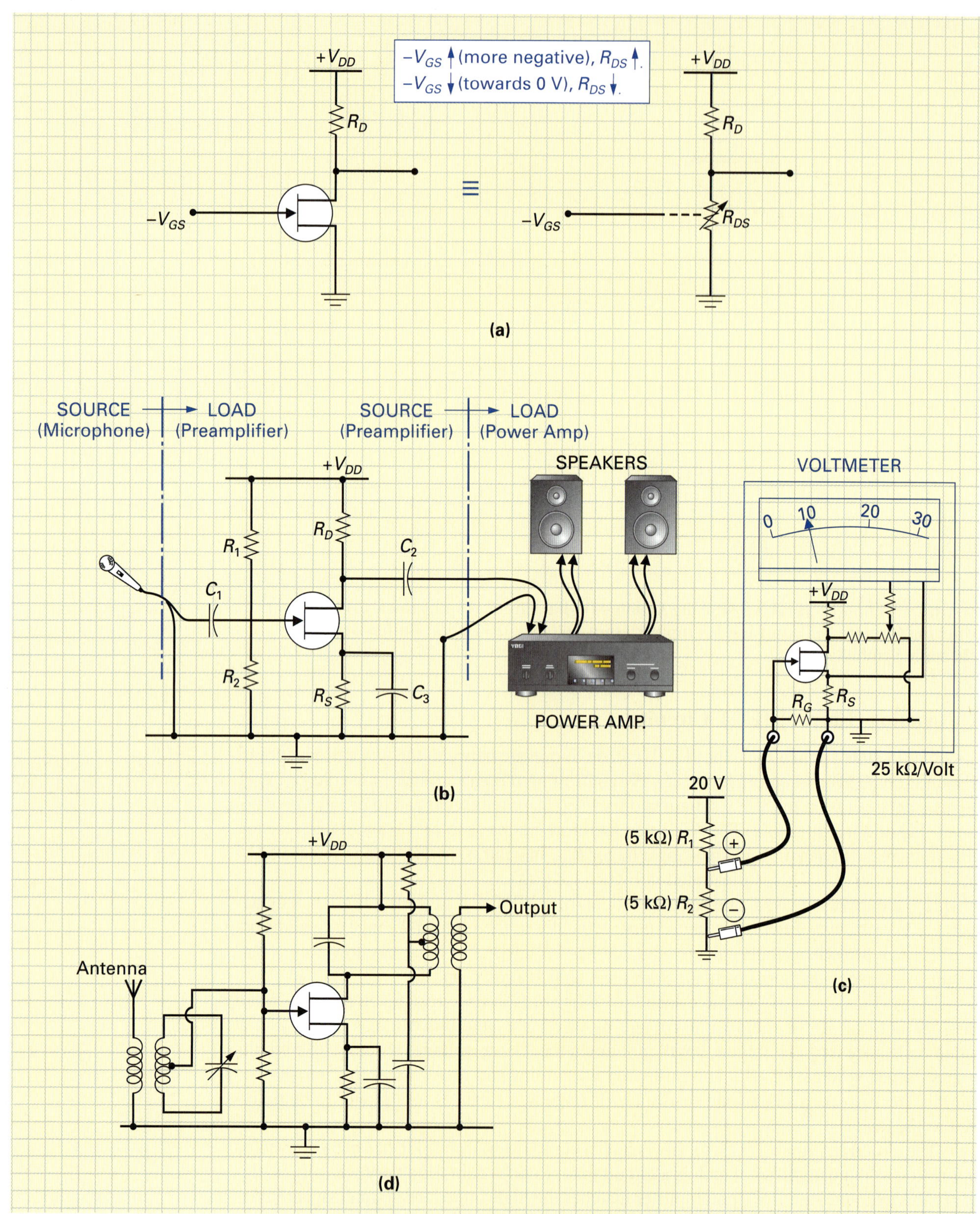

FIGURE 13-13 The JFET's Variable-Resistor Action—Analog Circuit Applications. (a) Equivalent Circuit. (b) Application 1: An Audio Preamplifier Circuit. (c) Application 2: A Voltmeter High Input Impedance Circuit. (d) Application 3: An RF Amplifier Circuit.

Figure 13-13(c) shows how a JFET at the front end of a voltmeter or oscilloscope will provide a very high input impedance and therefore not load the circuit under test. In this example, the meter will measure 10 V across R_2, and because the ohms per volt (Ω/V) rating of the voltmeter is 125 kΩ/V, the meter input impedance is

$$Z_{in} = \Omega/\text{V} \times V_{measured}$$
$$= 125 \text{ k}\Omega/\text{V} \times 10 \text{ V} = 1.25 \text{ M}\Omega$$

A 1.25 MΩ meter resistance in parallel with the 5 kΩ resistance of R_2 will have very little effect (1.25 MΩ in parallel with 5 kΩ = about 5 kΩ), so an accurate reading will be obtained.

Figure 13-13(d) shows how the JFET can be used as a radio frequency (RF) amplifier. Studying this circuit, you can see that both the gate and drain contain tuned circuits in the same way as the previously discussed BJT RF amplifier circuits. However, there are two advantages that the JFET has over the bipolar transistor as a front end RF amp.

1. The very weak signals injected into the antenna will have a very small value of current. Because the JFET is a voltage-controlled device, it requires no input current, and it will respond well to the small voltage signal variations picked up by the antenna.
2. The JFET is a very low noise component. Because any noise generated at the front end will be amplified along with the signal at each of the following amplifier stages, this JFET characteristic is ideal in this application.

13-1-9 Testing JFETs

The transistor tester shown in Figure 13-14(a) can be used to test both BJTs and FETs. This tester can be used to determine

1. whether an open or short exists between any of the terminals,
2. the FET's transconductance/gain, and
3. the FET's value of I_{DSS} and leakage current.

If a transistor tester is not available, the ohmmeter can be used to detect the most common failures: opens and shorts. Figure 13-14(b) indicates what resistance values should be

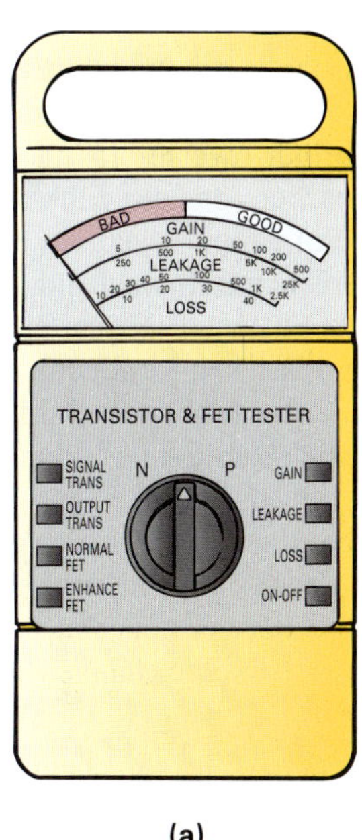

(a)

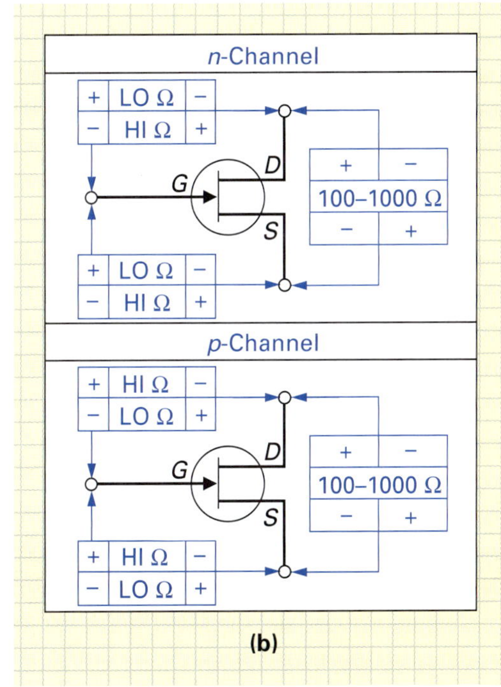

(b)

FIGURE 13-14 Testing JFETs.

obtained between the terminals of a good *n*-channel and *p*-channel JFET. Looking at these ohmmeter readings, you can see that because the JFET has only one P-N junction (gate-to-channel), it is relatively simple to test with an ohmmeter for an open or shorted junction.

SELF-TEST EVALUATION POINT FOR SECTION 13-1

Use the following questions to test your understanding of Section 13-1.

1. The BJT is a ____________ controlled device while the FET is a ____________ controlled device.
2. The gate-source junction of a JFET is always ____________ biased.
3. True or false: When $V_{GS} = 0$ V, $I_D = I_{DSS}$.
4. True or false: When $V_{GS} = V_{GS(\text{OFF})}$, $I_D = \text{max}$.
5. ____________ is a ratio of an FET's output current change to input voltage change.
6. A JFET has a ____________ input impedance due to its ____________ biased gate-source junction.
7. What component in a JFET circuit provides temperature stability?
8. Like self bias, ____________ bias has negative feedback and therefore maintains the Q point stable.
9. Which FET circuit configuration is most widely used like its BJT common-emitter counterpart?
10. Which JFET circuit configuration could provide a high input impedance and a good value of voltage gain?
11. Which JFET circuit configuration is best suited for providing a very high input impedance and low output impedance?
12. Which JFET characteristic is made use of in most circuit applications?
13. Why is the JFET ideal as an RF preamplifier?

13-2 THE METAL OXIDE SEMICONDUCTOR FIELD EFFECT TRANSISTOR (MOSFET)

With the JFET, an input voltage of zero volts would reverse bias the P-N junction, resulting in a maximum channel size and a maximum value of source-to-drain current. To decrease the size of the channel, the input voltage was made negative to further reverse bias the gate-source junction. This action would deplete the channel of free carriers, reducing the size of the channel and therefore the source-to-drain current. This type of action is actually called "depletion-mode operation," since an input voltage is used to deplete the channel and, therefore, reduce the channel's size and current. The MOSFET does not have a P-N gate-channel junction like the JFET. It has a "metal gate" that is insulated from the "semiconductor channel" by a layer of "silicon dioxide," hence the name "metal oxide semiconductor." Like all "field effect transistors" (FETs), the input voltage will generate an "electric field" which will have the "effect" of changing the channel's size.

The key difference between the JFET and MOSFET is that the JFET's input voltage would always have to be zero or a negative voltage in order to reverse bias the gate-source junction. With the MOSFET, the input voltage can be either a positive or negative voltage since gate current will always be zero because the gate is insulated from the channel. To examine each of these input voltage possibilities:

a. If the input voltage is negative, the resulting electric field depletes the channel, reducing its size, and the MOSFET is said to be operating in the depletion mode.
b. If, on the other hand, the input voltage is positive, the resulting electric field enhances the channel, increasing its size, and the MOSFET is said to be operating in the enhancement mode.

The MOSFET can therefore be operated in either the depletion or enhancement mode due to its insulated gate. The two different types of MOSFETs are given names based on their normal mode of operation. For instance, the *depletion-type MOSFET (D-type MOSFET* or *D-MOSFET),* should actually be called a DE-MOSFET because it can be operated in both the depletion mode and the enhancement mode, whereas the *enhancement-type MOSFET (E-type MOSFET* or *E-MOSFET)* is correctly named since it can only be operated in the enhancement mode. In this chapter, we will examine the construction, operation, characteristics, circuit biasing, applications, and testing of these two MOSFET types.

13-2-1 *The Depletion-Type (D-Type) MOSFET*

The **depletion-type MOSFET** construction is slightly different from the JFET, and therefore we will need to first see how the D-MOSFET device is built before we can understand how it operates.

Depletion-Type MOSFET
A field effect transistor with an insulated gate (MOSFET) that can be operated in either the depletion or enhancement mode.

***n*-Channel D-Type MOSFET**
A depletion type MOSFET having an *n*-type channel between its source and drain terminals.

***p*-Channel D-Type MOSFET**
A depletion type MOSFET having a *p*-type channel between its source and drain terminal.

D-MOSFET Construction

Like the JFET and BJT, the D-MOSFET has two basic transistor types called the ***n*-channel D-type MOSFET** and ***p*-channel D-type MOSFET.** The construction and schematic symbols for these two D-MOSFET types are shown in Figure 13-15.

To begin with, let us examine the construction of the more frequently used *n*-channel D-type MOSFET, shown in Figure 13-15(a). This type of MOSFET basically consists of an *n*-type channel formed on a *p*-type substrate. A source and drain lead are connected to either end of the *n*-channel, and an additional lead is attached to the substrate. In addition, a thin insulating (silicon dioxide) layer is placed on top of the *n*-channel, and a metal plated area with a gate lead attached is formed on top of this insulating layer. Figure 13-15(b) shows the schematic symbol for an *n*-channel D-MOSFET, and, as you can see, the arrow on the substrate (*SS*) or base (*B*) lead points into the device. As a memory aid, imagine this arrowhead as a P-N junction diode as shown in the inset. The source and drain leads are connected to either end of the diode's *n*-type cathode. Therefore, this device must be an *n*-channel D-MOSFET. The basic difference in the construction and schematic symbol of the *p*-channel D-MOSFET can be seen in Figure 13-15(c) and (d).

Figure 13-15(e) and (f) show how the MOSFET is available as a four-terminal or three-terminal device. In some applications, a separate bias voltage will be applied to the substrate terminal for added control of drain current and the four-terminal device will be used. In most circuit applications, however, the three-terminal device, which has its source and substrate lead internally connected, is all that is needed.

D-MOSFET Operation

Figure 13-16(a) shows the typical drain characteristic curves for an *n*-channel depletion-type MOSFET. As you can see, this set of curves has the same general shape as the JFET's set of drain curves and the BJT's set of collector curves. The key difference is that V_{GS} is plotted for both positive and negative values. This is because the D-MOSFET should actually be called a DE-MOSFET because it can be operated in both the depletion mode (in which V_{GS} is a negative value) and the enhancement mode (in which V_{GS} is a positive value). To best understand the operation of the D-MOSFET, let us examine the three operation diagrams shown in Figure 13-16(b), (c), and (d).

Zero-Volt Operation: The center operation diagram, Figure 13-16(b), shows how the *n*-channel D-MOSFET will respond to a V_{GS} input of zero volts. When $V_{GS} = 0$ V, the gate and source terminals are at the same zero volt potential, and therefore the gate is effectively shorted to the source. The value of drain current passing through the channel is called the I_{DSS} value (I_{DSS} is the drain-to-source current passing through the channel when the gate is shorted to the source). Therefore, when

$$V_{GS} = 0 \text{ V}, I_D = I_{DSS}$$

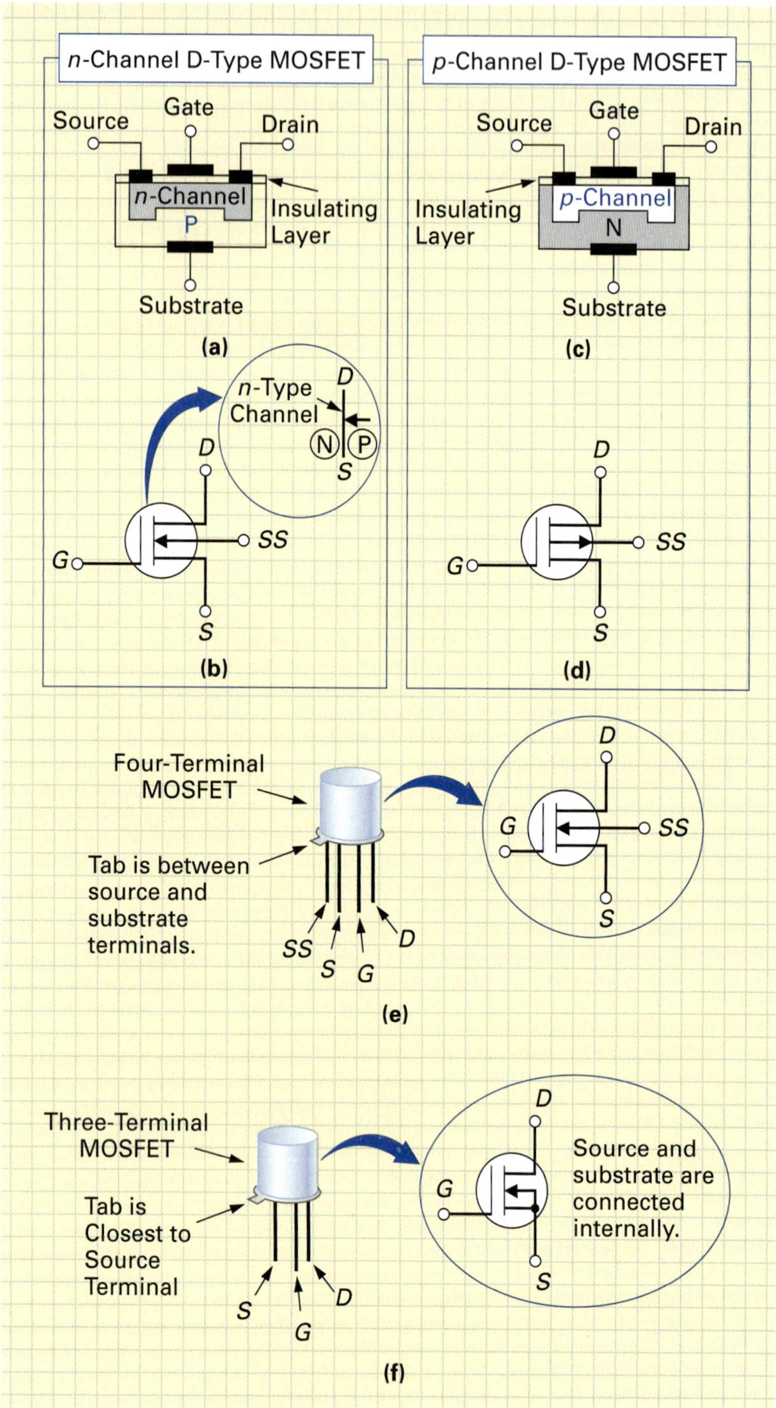

FIGURE 13-15 **D-Type MOSFET Construction and Types.**

When zero volts is applied to the input of a D-type MOSFET, therefore, it will conduct a value of drain current. With no input, this device is ON, which is why the D-type MOSFET is known as a "normally ON" device.

Enhancement Mode: The upper operation diagram, Figure 13-16(c), shows how the n-channel D-MOSFET will respond when V_{GS} is made positive. In this condition, the channel is enhanced or widened, and the value of I_D is increased above I_{DSS}. Therefore, when

$$V_{GS} = +V, I_D > I_{DSS}$$

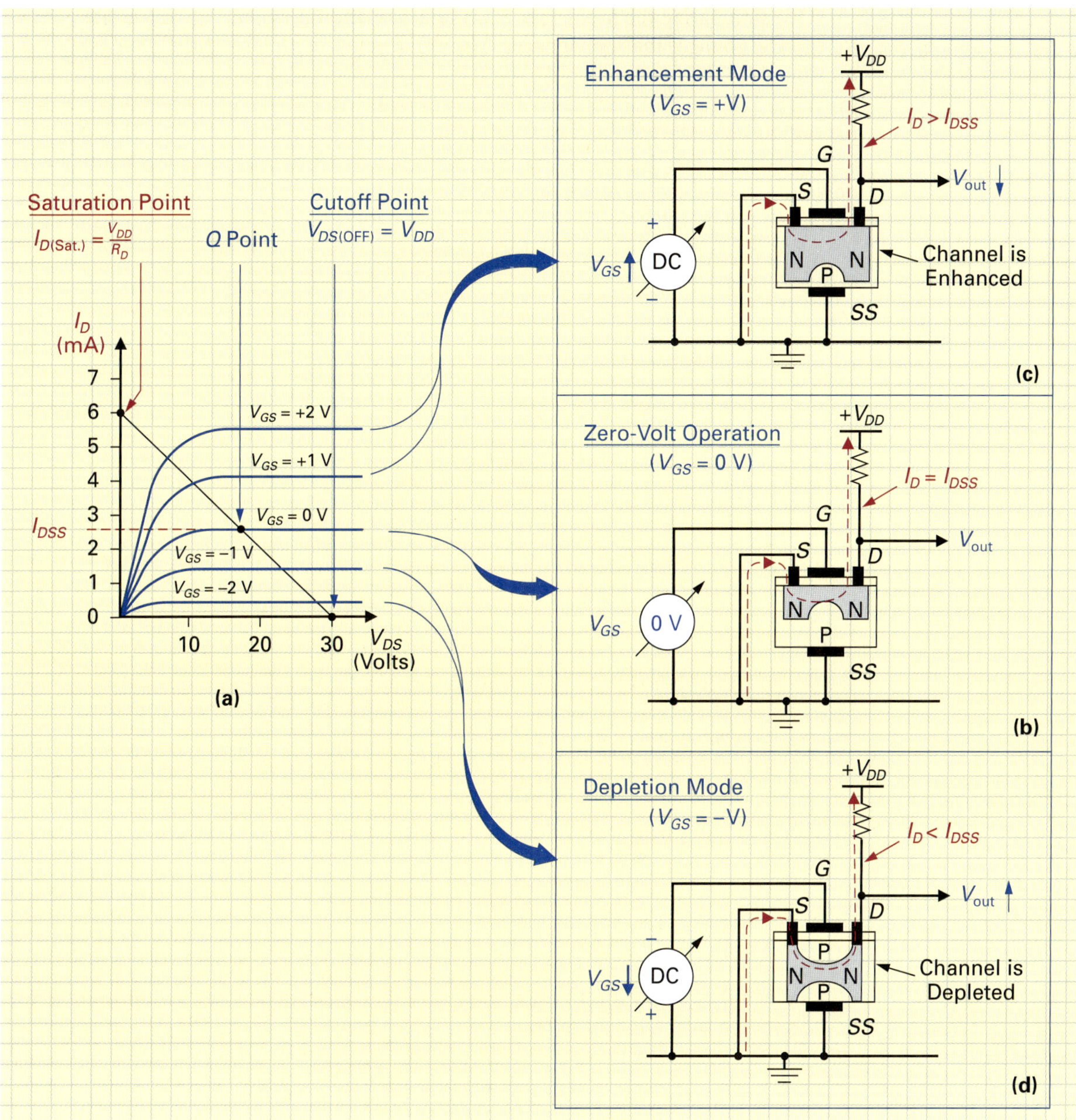

FIGURE 13-16 **D-Type MOSFET Operation and Characteristics.**

Let us examine in more detail why the channel is widened by a positive gate voltage. Because the valence-band holes in the *p*-type material (majority carriers) will be repelled by a positive gate voltage, and the conduction band electrons in the *p*-type material (minority carriers) will be attracted to the channel by the positive gate voltage, there will be a build-up of electrons in the *p*-type material near the channel. This build-up of electrons in the *p*-type material below the channel will effectively widen the size of the channel, reducing its resistance, and therefore increasing I_D to a value greater than I_{DSS}.

Depletion Mode: The lower operation diagram, Figure 13-16(d), shows how the *n*-channel D-MOSFET will respond when V_{GS} is made negative. In this condition, the channel is depleted of free carriers, and therefore the value of I_D is decreased below I_{DSS}. Therefore, when

$$V_{GS} = -V, I_D < I_{DSS}$$

To summarize the n-channel MOSFET's operation, when V_{GS} was either zero volts or a negative voltage, the n-channel D-MOSFET acted in almost exactly the same way as an n-channel JFET. However, unlike the JFET, the D-MOSFET can have a forward biased gate-to-source P-N junction because the silicon dioxide insulating layer prevents any current from passing through the gate and will still maintain a high input resistance. This dual operating ability is why the depletion-type MOSFET or D-MOSFET should actually be called a depletion-enhancement or DE-MOSFET.

The drain characteristic curves of the D-MOSFET can be used to plot the device's dc load line, as shown in Figure 13-16(a), with

$$I_{D(\text{Sat.})} = \frac{V_{DD}}{R_D} \text{ at saturation, and}$$

$$V_{DS(\text{OFF})} = V_{DD} \text{ at cutoff}$$

As with the JFET, the D-MOSFET's transconductance is equal to the ratio of output current change ($\Delta\, I_D$) to input voltage change ($\Delta\, V_{GS}$),

$$g_m = \frac{\Delta I_D}{\Delta V_{GS}}$$

and the D-MOSFET's voltage gain is equal to

$$A_V = g_m \times R_D$$

EXAMPLE:

A D-MOSFET circuit has the following specifications:

$$I_{DSS} = 2 \text{ mA},\ V_{GS(\text{OFF})} = -6 \text{ V},\ R_D = 3 \text{ k}\Omega,\ V_{DD} = 12 \text{ V}$$

Calculate the following two extremes on the D-MOSFET's load line:

a. $I_{D(\text{Sat.})}$
b. $V_{DS(\text{OFF})}$

Solution:

a. When the D-MOSFET is saturated, it is equivalent to a closed switch and therefore the only resistance is that of R_D.

$$I_{D(Sat.)} = \frac{V_{DD}}{R_D} = \frac{12 \text{ V}}{3 \text{ k}\Omega} = 4 \text{ mA}$$

b. When the D-MOSFET is cut off, it is equivalent to an open switch, and therefore the full drain supply voltage will appear across the open between drain and source.

$$V_{DS(\text{OFF})} = V_{DD} = 12 \text{ V}$$

D-MOSFET Biasing

Like the JFET, the D-MOSFET can be configured in the same way as a common-drain, common-gate, or common-source circuit, with all of the dc and ac configuration characteristics being the same. As far as biasing, the D-MOSFET is easier to bias than the JFET because of its ability to operate in either the depletion mode ($-V_{GS}$) or the enhancement mode ($+V_{GS}$). In fact, one of the most frequently used D-MOSFET biasing methods is to simply have no biasing at all. This biasing method is called **zero biasing** because the Q point is set at zero volts ($V_{GS} = 0$ V), as seen in Figure 13-17. This makes biasing the D-MOSFET very simple because no gate or source bias voltages are needed. The ac input signal developed across R_G is therefore applied to the extremely high input impedance of the D-MOSFET,

Zero Biasing
A configuration in which no bias voltage is applied at all.

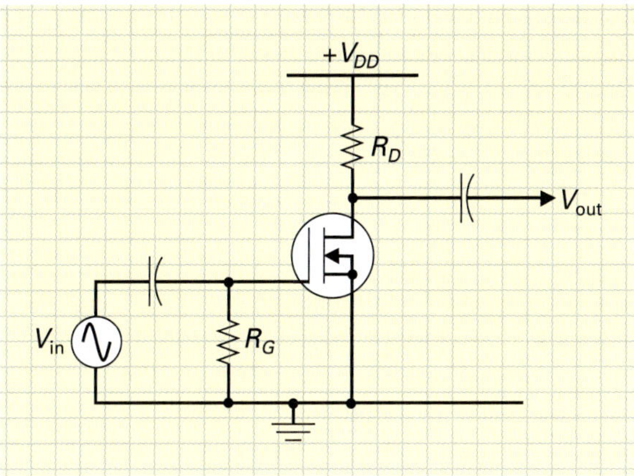

FIGURE 13-17 **Zero Biasing a D-MOSFET.**

causing an increase and decrease in the conduction of the MOSFET above and below the $V_{GS} = 0$ V, Q point.

D-MOSFET Applications

The D-MOSFET is most frequently used in analog or linear circuit applications. This is because the D-MOSFET can be very simply biased at a midpoint in the load line and then have its output current varied above and below this natural Q point in a linear fashion. This, coupled with the D-MOSFET's almost infinite input impedance and low noise properties, makes it ideal as a preamplifier at the front end of a system. Figure 13-18 shows how two D-MOSFETs can be used to construct a typical front end **cascode amplifier circuit.** This cascode amplifier circuit consists of a self biased common-source amplifier (Q_1) in series with a voltage-divider biased common-gate amplifier (Q_2). The input signal (V_{in}) is applied to Q_1's gate, and the amplified output at Q_1's drain is then passed to Q_2's source, where it is further amplified by Q_2 before appearing at Q_2's drain, and therefore at the output (V_{out}).

Cascode Amplifier Circuit

An amplifier circuit consisting of a self-biased common-source amplifier in series with a voltage-divider biased common-gate amplifier.

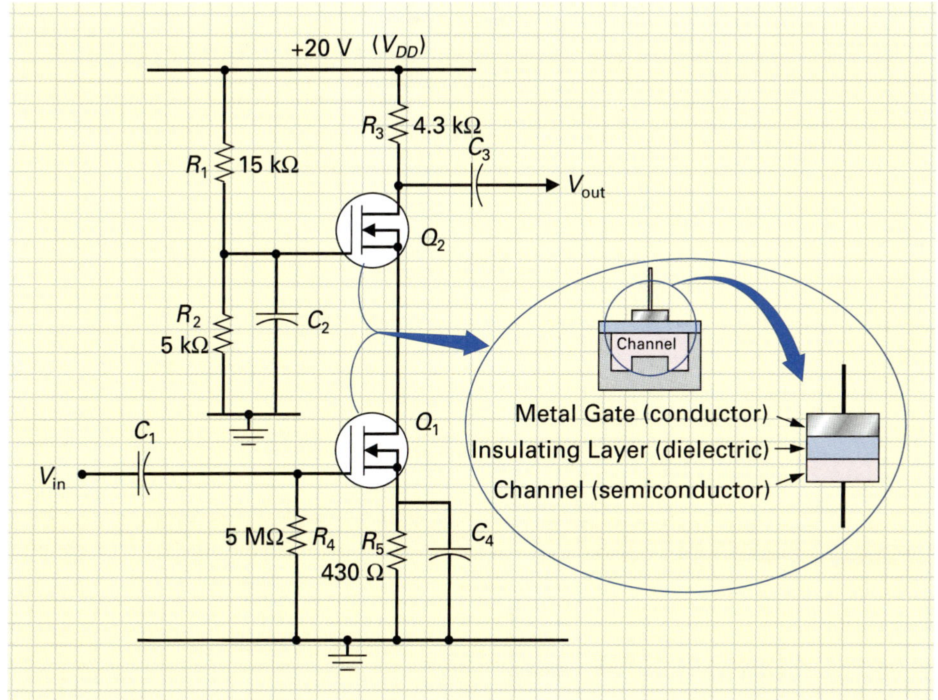

FIGURE 13-18 **A D-MOSFET Analog Circuit Application—Cascode Amplifier.**

The FET's only limiting factor is that its high input impedance starts to decrease as the input signal's frequency increases. Refer to the inset in Figure 13-18, which shows how the gate, insulator, and channel of a D-MOSFET form a capacitor. This input capacitance of typically 5 pF has very little effect at low input signal frequencies ($X_C\uparrow = 1/2\pi f\downarrow C$) because the input impedance is high ($X_C\uparrow$ therefore $Z_{in}\uparrow$) and the loading effect is negligible. At higher radio frequency ($X_C\downarrow = 1/2\pi f\uparrow C$), however, the input impedance is lowered ($X_C\downarrow$ therefore $Z_{in}\downarrow$) and the D-MOSFET loses its high input impedance advantage. To compensate for this disadvantage, FETs are often connected in series, as in Figure 13-18, so that their input capacitances are also in series. Recall that series-connected capacitors have a lower total capacitance than any of the individual capacitance values. Therefore, the overall input capacitance of two series-connected D-MOSFETs will be less than that of a single D-MOSFET, making this cascode amplifier ideal as a high radio frequency (RF) amplifier: a low input capacitance ($C_{in}\downarrow$) means a high input reactance ($X_{C(in)}\uparrow$), and therefore a high input impedance ($Z_{in}\uparrow$) at high frequencies.

Dual-Gate D-MOSFET

Dual-Gate D-MOSFET
A metal oxide semiconductor FET having two separate gate electrodes.

To compensate for the D-MOSFET's input capacitance problem, the **dual-gate D-MOSFET** was developed. The construction and schematic symbol for the dual-gate D-MOSFET is shown in Figure 13-19(a). In most applications, the dual-gate D-MOSFET is connected so it acts as two series-connected D-MOSFETs, as shown in the cascode amplifier circuit in Figure 13-19(b). With this amplifier, the ac input signal drives the lower gate, which acts like a common-source amplifier. The output of the common-source lower section

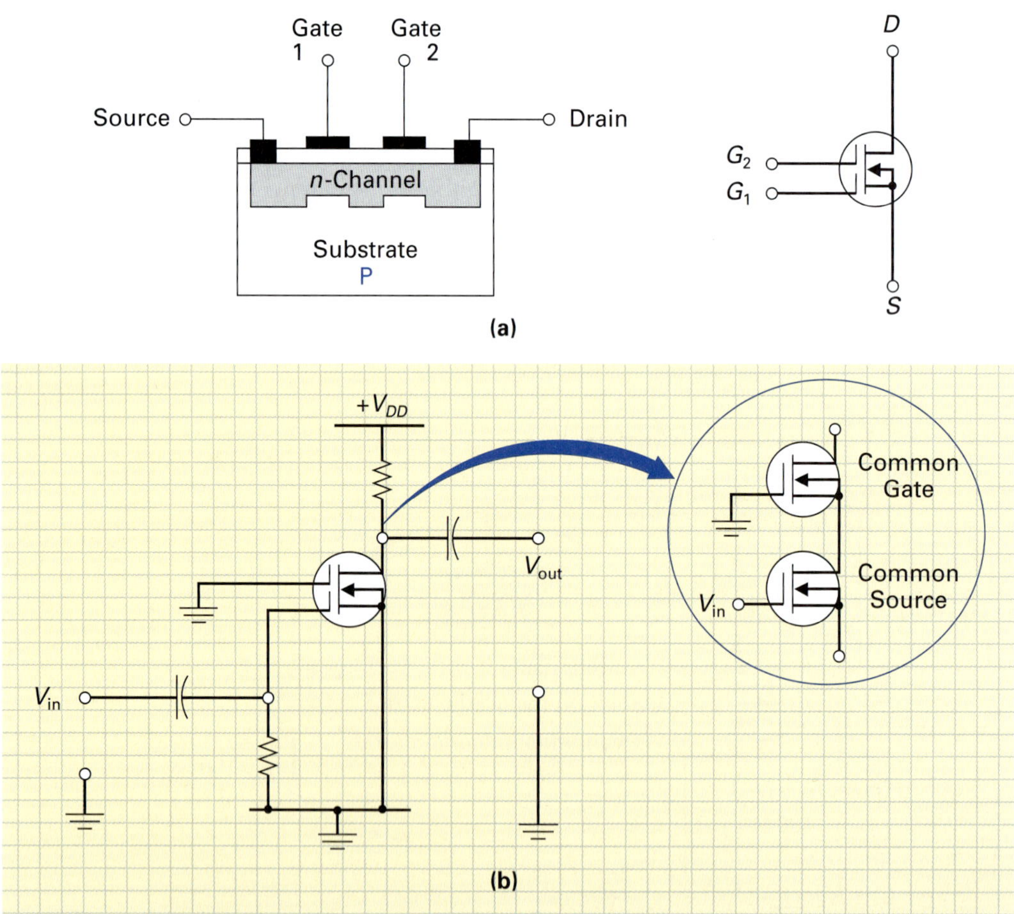

FIGURE 13-19 The Dual-Gate D-MOSFET. (a) Dual-Gate D-MOSFET Construction and Schematic Symbol. (b) Dual-Gate D-MOSFET Application—Cascode Amplifier.

of the dual-gate D-MOSFET drives the upper half, which acts like a common-gate amplifier. The inset in Figure 13-19(b) shows how the dual-gate D-MOSFET is equivalent to two series-connected D-MOSFETs. As with the previous cascode amplifier, the overall input capacitance of a dual-gate D-MOSFET is less than that of a standard D-MOSFET, and if capacitance is low, X_C, and therefore Z_{in}, are high.

SELF-TEST EVALUATION POINT FOR SECTION 13-2-1

Use the following questions to test your understanding of Section 13-2-1.

1. The two different types of MOSFETs are called the ____________-type MOSFET and ____________-type MOSFET.
2. True or false: The D-type MOSFET can be operated in both the depletion and enhancement mode.
3. The D-type MOSFET is a normally ____________ (ON/OFF) device.
4. Which FET has a higher input impedance: JFET or MOSFET?
5. Why is the D-MOSFET ideal as a preamplifier?
   a. It can be mid-load-line biased when 0 V is applied.
   b. It has a high input impedance.
   c. It has low noise properties.
   d. All of the above
6. The ____________ MOSFET was developed to lower input capacitance so that it can handle high-frequency signals.

13-2-2 *The Enhancement-Type (E-Type) MOSFET*

With an input of zero volts, a D-MOSFET will be ON, and a certain value of current will pass through the channel between source and drain. If the input to the D-MOSFET is made positive, the channel is enhanced, causing the source-to-drain current to increase. If the input is made negative, the channel is depleted, causing the source-to-drain current to decrease. The D-MOSFET can therefore operate in either the enhancement or depletion mode and is called a "normally ON" device because it is ON when nothing (0 V) is applied.

The **enhancement-type MOSFET** or **E-MOSFET** can only operate in the enhancement mode. In other words, when the input is either zero volts or a negative voltage, the transistor is OFF and there is no source-to-drain current. However, when the input is made positive, the E-MOSFET will turn ON, resulting in a source-to-drain channel current. The E-MOSFET is therefore a "normally OFF" device because it is OFF when nothing (0 V) is applied.

Enhancement-Type MOSFET or E-MOSFET
A field effect transistor with an insulated gate (MOSFET) that can only be turned on if the channel is enhanced.

E-MOSFET Construction

As with all of the other transistor types, it is easier to understand the operation and characteristics of a device once we have seen how the component is constructed. The E-MOSFET has two basic transistor types called the ***n*-channel E-type MOSFET** and ***p*-channel E-type MOSFET.** The construction and schematic symbols for these two E-MOSFET types are shown in Figure 13-20.

_n_-Channel E-Type MOSFET
An enhancement type MOSFET having an *n*-type channel between its source and drain terminals.

To begin with, let us examine the construction of the more frequently used *n*-channel E-type MOSFET, shown in Figure 13-20(a). Studying the construction of this E-MOSFET, notice that no channel exists between the source and drain. Consequently, with no gate bias voltage, the device will be OFF. When the gate is made positive, however, electrons will be attracted from the substrate, causing a channel to be induced between the source and drain. This enhanced channel will permit drain current to flow, and any further increase in gate voltage will cause a corresponding increase in the size of the channel and therefore the value of I_D.

_p_-Channel E-Type MOSFET
An enhancement type MOSFET having a *p*-type channel between its source and drain terminals.

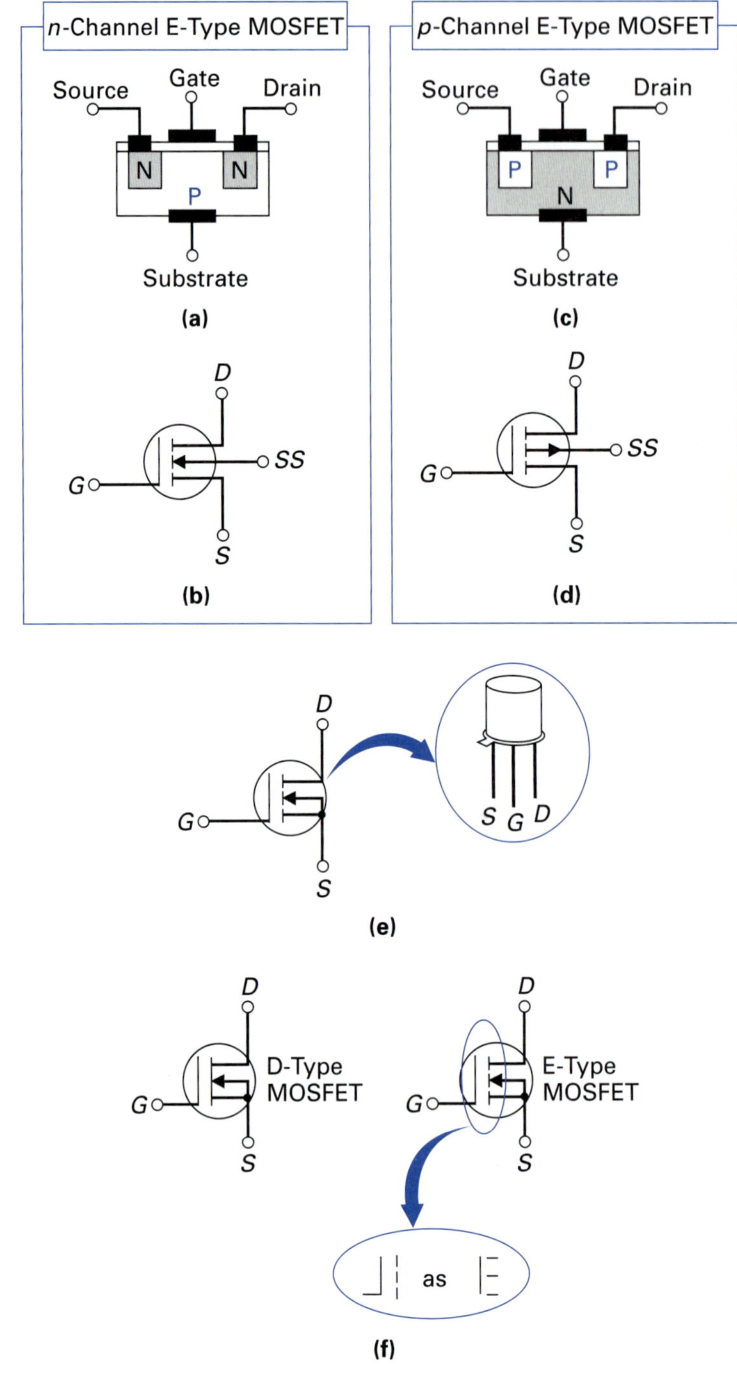

FIGURE 13-20 E-Type MOSFET Construction and Type.

The schematic symbol for the *n*-channel E-type MOSFET can be seen in Figure 13-20(b). The construction and schematic symbol for the *p*-channel E-MOSFET can be seen in Figure 13-20(c) and (d). As with the D-MOSFET, both four-terminal and three-terminal devices are available, with the three-terminal device having a common connection between source and substrate as shown in Figure 13-20(e).

The only difference between the E-type and D-type MOSFET symbols is the three dashed lines representing the drain, substrate, and source regions. The dashed line is used instead of the solid line to indicate that an E-MOSFET has a normally broken path, or channel, between drain, substrate, and source. To aid memory, the three dashed lines used in the "E"-MOSFET symbol could be thought of as the three horizontal prongs in the capital letter "E," as shown in Figure 13-20(f).

E-MOSFET Operation

Figure 13-21(a) shows the typical drain characteristic curves for an *n*-channel enhancement-type MOSFET. As you can see, this set of drain curves is very similar to the D-MOSFET's set of drain curves; however, in this case only positive values of V_{GS} are plotted. Looking at the relationship between V_{GS} and I_D, you may have noticed that any increase in V_{GS} will cause a corresponding increase in I_D.

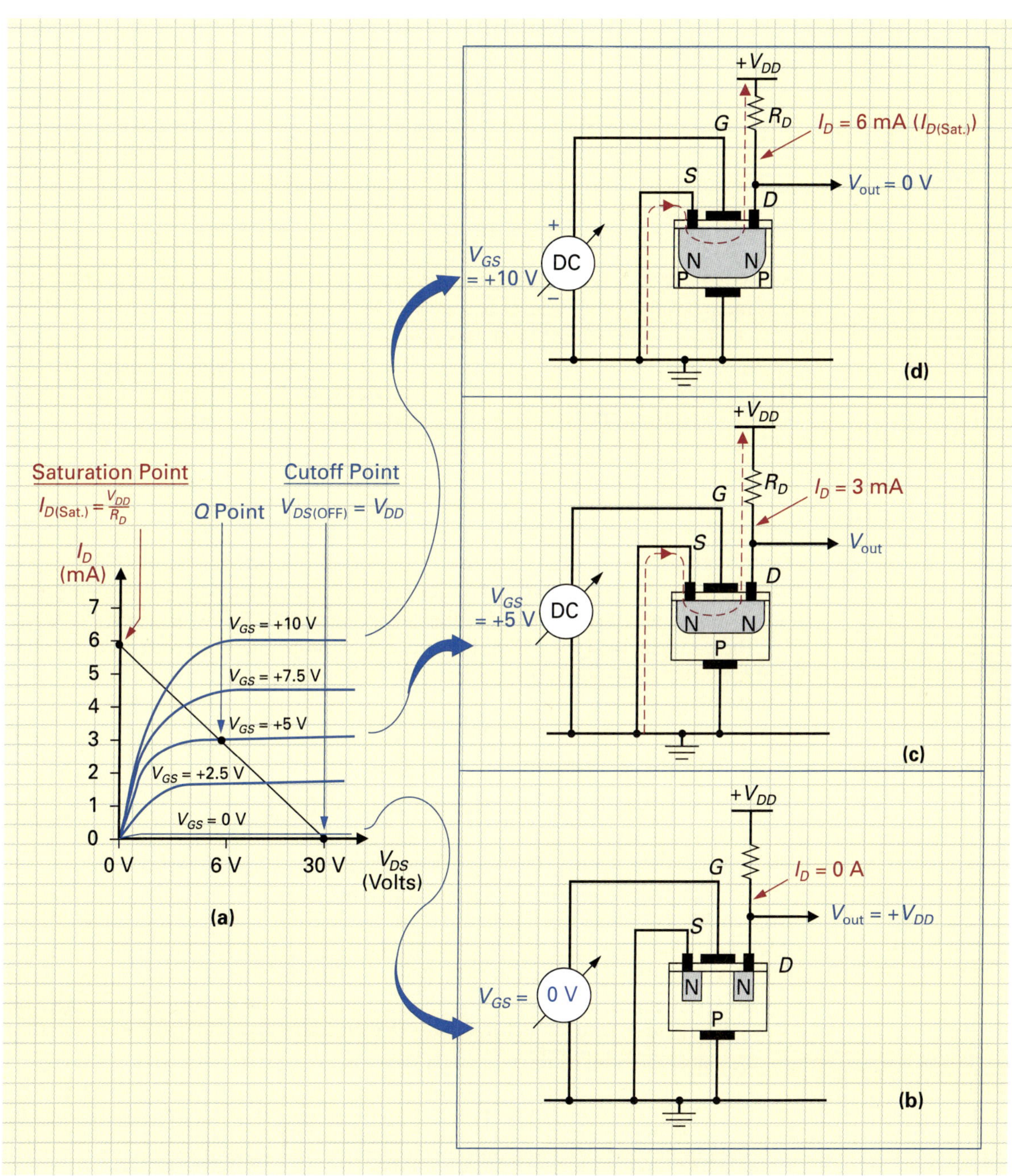

FIGURE 13-21 E-Type MOSFET Operation and Characteristics.

To best understand the operation of the E-MOSFET, let us examine the three operation diagrams shown in Figure 13-21(b), (c), and (d).

$V_{GS} = 0$ **V Curve:** The lower operation diagram, Figure 13-21(b), shows how the *n*-channel E-MOSFET will respond to a V_{GS} input of zero volts. When $V_{GS} = 0$ V, there is no channel connecting the source and drain, and therefore the drain current will be zero. As a result, $+V_{DD}$ will be present at the output because the E-MOSFET is equivalent to an open switch between drain and source.

$V_{GS} = +5$ **V Curve:** The center operation diagram, Figure 13-21(c), shows how the *n*-channel E-MOSFET will respond to a V_{GS} input of +5 volts. When $V_{GS} = +5$ V, the E-MOSFET will act in almost exactly the same way as an enhanced D-MOSFET. The positive gate voltage will repel the *p*-type material's majority carriers (holes) away from the gate, while attracting the *p*-type material's minority carriers (electrons) toward the gate. This action will form an *n*-type bridge between the source and drain, and therefore a value of I_D will flow between source and drain, as shown.

$V_{GS} = +10$ **V Curve:** The upper operation diagram, Figure 13-21(d), shows how the *n*-channel E-MOSFET will respond if the V_{GS} input is further increased to +10 volts. When $V_{GS} = +10$ V, the attraction of electrons and repulsion of holes within the *p*-type material is increased, causing the channel's width to increase and I_D to also increase. As a result, 0 V will be present at the output because the E-MOSFET is equivalent to a closed switch between drain and source.

In summary, when V_{GS} is zero volts, drain current is also zero. As the value of V_{GS} is increased (made more positive), the channel becomes wider, causing I_D to increase. On the other hand, as the value of V_{GS} is decreased (made less positive), the channel becomes narrower, causing I_D to decrease. In other words, when the input is either zero volts or a negative voltage, the transistor is OFF, and there is no source-to-drain current. However, when the input is made positive, the E-MOSFET will turn ON, resulting in a source-to-drain channel current. The E-MOSFET is therefore a "normally OFF" device because it is OFF when nothing (0 V) is applied.

Although it cannot be seen in the set of drain curves in Figure 13-21(a), V_{GS} will have to increase to a positive threshold voltage of about +1 V before a channel will be induced and a small value of drain current will flow. This "threshold level" is a highly desirable characteristic because it prevents noise or any low-level input signal voltage from turning ON the device. This advantage makes the E-type MOSFET ideally suited as a switch because it can be turned ON by an input voltage and turned OFF once the input voltage falls below the threshold level.

The *p*-channel E-type MOSFET operates in much the same way as the *n*-channel device except that holes are attracted from the substrate to form a *p*-channel and the V_{GS} and V_{DS} bias voltages are reversed.

E-MOSFET Biasing

Drain-Feedback Biased
A configuration in which the gate receives a bias voltage fed back from the drain.

Like the D-MOSFET, the E-MOSFET can be configured as a common-drain, common-gate, or common-source circuit. Unlike the D-MOSFET and JFET, the E-MOSFET cannot be biased using self bias or zero bias because V_{GS} must be a positive voltage. As a result, gate bias and voltage-divider bias can be used; however, more frequently E-MOSFETs are **drain-feedback biased,** as shown in Figure 13-22(a). In this example, R_D equals 8 kΩ, and R_G (which feeds back a positive voltage from the drain, hence the name drain-feedback bias) equals 100 MΩ. Because an E-MOSFET has an extremely high input impedance (due to the insulated gate), no current will flow in the gate circuit. With no gate current, there will be no voltage drop across the gate resistor ($V_{R_G} = 0$ V), and therefore the voltage at the gate will be at the same potential as the voltage at the drain.

$$V_{GS} = V_{DS}$$

To help set up the Q point, most manufacturer's data sheets specify a load line midpoint drain current $I_{D(\text{ON})}$ and drain voltage $V_{DS(\text{ON})}$. In the example in Figure 13-22(a), when the E-MOSFET is ON, or conducting, and $I_D = 3$ mA, the E-MOSFET's drain-to-source volt-

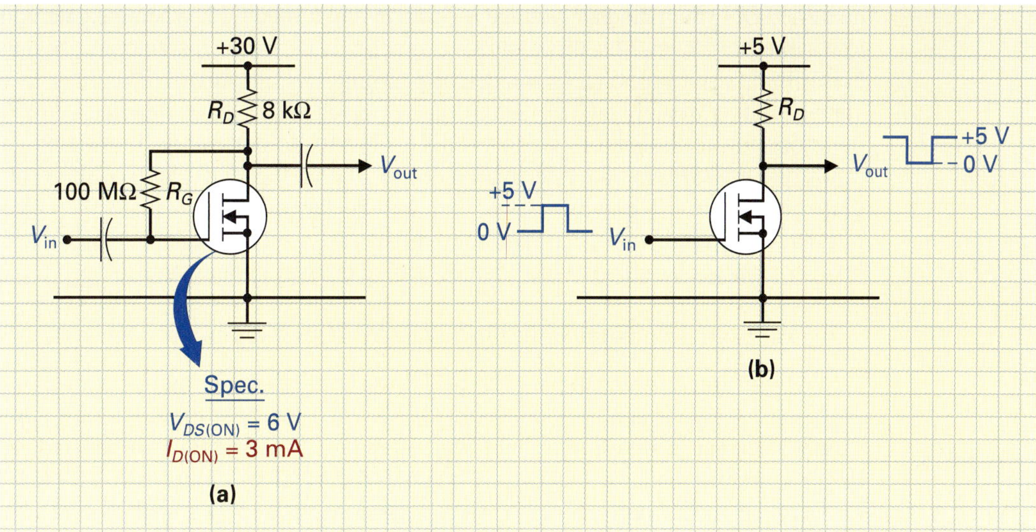

FIGURE 13-22 **Biasing E-Type MOSFETs. (a) Drain-Feedback Biasing. (b) Voltage Controlled Switching.**

age drop ($V_{DS(ON)}$) is 6 V. The value of R_D in Figure 13-22(a) has been chosen so that this E-MOSFET circuit will be biased at its specified Q point. To check, we can use the formula

$$V_{DS} = V_{DD} - (I_{D(ON)} \times R_D)$$

In the example in Figure 13-22(a)

$$V_{DS} = V_{DD} - (I_{D(ON)} \times R_D) = 30\text{ V} - (3\text{ mA} \times 8\text{ k}\Omega)$$
$$= 30\text{ V} - 24\text{ V} = 6\text{ V}$$

EXAMPLE:

A drain-feedback biased E-MOSFET circuit has the following specifications: $R_D = 1\text{ k}\Omega$, $I_{D(ON)} = 10\text{ mA}$, $V_{DD} = 20\text{ V}$. Calculate V_{DS} and V_{R_D}.

Solution:

$$V_{DS} = V_{DD} - (I_{D(ON)} \times R_D)$$
$$= 20\text{ V} - (10\text{ mA} \times 1\text{ k}\Omega)$$
$$= 20\text{ V} - 10\text{ V} = 10\text{ V}$$

The constant drain supply voltage (V_{DD}) will be evenly divided across R_D and the E-MOSFET's drain-to-source ($V_{DS} = 10$ V, $V_{R_D} = 10$ V).

In most instances, the E-MOSFET will be used in digital two-state switching circuit applications. In this case, there will be no need for a gate resistor (R_G) because the input voltage (V_{in}) will either turn ON or OFF the E-MOSFET, as shown in Figure 13-22(b). For example, when $V_{in} = 0$ V, the E-MOSFET is OFF, therefore $V_{out} = +V_{DD} = +5$ V, whereas when $V_{in} = +5$ V, the E-MOSFET is ON and therefore $V_{out} = 0$ V.

E-MOSFET Applications

The E-MOSFET is more frequently used in digital or two-state circuit applications. One reason is that it naturally operates as a "normally OFF voltage-controlled switch" because it can be turned ON when the gate voltage is positive and turned OFF when the gate voltage falls below a threshold level. This threshold level is a highly desirable characteristic because it prevents noise from false triggering, or accidentally turning ON, the device. The other E-MOSFET advantage is its extremely high input impedance, which means that the device's circuit current, and therefore power dissipation, are low. This enables us to densely pack or integrate many thousands of E-MOSFETs onto one small piece of silicon, forming a high component density integrated circuit (IC). These low-power and high-density advantages make the E-MOSFET ideal in battery-powered small-size (portable) applications such as calculators, wristwatches, notebook computers, hand-held video games, digital cellular phones, and so on.

Vertical-Channel E-MOSFET (VMOS FET)

Vertical-Channel E-MOSFET

An enhancement type MOSFET that, when turned on, forms a vertical channel between source and drain.

As just mentioned, the E-MOSFET is generally used in two-state or digital circuit applications, where it acts as a normally OFF switch. In most digital circuit applications, the channel current is small, and therefore a standard E-MOSFET can be used. However, if a larger current-carrying capability is needed, the **vertical-channel E-MOSFET** or VMOS FET can be used. Figure 13-23(a) shows the construction of a VMOS FET. The gate at the top of the device is insulated from the source (which is also at the top), and as with all E-MOSFETs, no channel exists between the source terminal and the drain terminal with no bias voltage applied. The VMOS FET's semiconductor materials are labeled P, N+, and N− and indicate different levels of doping.

When this *n*-channel VMOS FET is biased ON (gate is made positive with respect to source), as seen in Figure 13-23(b), a vertical *n*-type channel is formed between source and drain. This channel is much wider than a standard E-MOSFET's horizontal channel, which is why VMOS FETs can handle a much higher drain current.

Figure 13-23(c) shows how the high-current capability of a VMOS FET can be made use of in an interfacing circuit application. In this circuit, a VMOS FET is used to interface a low-power source input signal to a high-power load. Referring to the current specifications of the VMOS FET, relay, and motor, you can see that each device is used to step up current. The standard E-MOSFET supplies a low-power input signal to the VMOS FET, which can handle enough current to actuate a relay whose contacts can handle enough current to switch power to the dc motor. To be more specific, if V_{in} is LOW, the standard E-MOSFET will turn OFF, producing a HIGH output to the gate of the VMOS FET, turning it ON. When the VMOS FET turns ON, it effectively switches ground through to the lower end of the relay coil, energizing the relay and closing its normally open contacts. The closed relay contacts switch ground through to the lower end of the motor, which turns ON because it now has the full +12 V supply across its terminals. The motor will stay ON as long as V_{in} stays LOW. If V_{in} were to go HIGH, the standard E-MOSFET inverter would produce a LOW output, which would turn OFF the VMOS FET, relay, and motor.

Other than its high-current capability, the VMOS FET also has a positive temperature coefficient of resistance, which means an increase in temperature will cause a decrease in drain current ($T\uparrow$, $R\uparrow$, $I\downarrow$), and this will prevent thermal runaway. This gives the VMOS FET a distinct advantage over the BJT power amplifier, which has a negative temperature coefficient of resistance ($T\uparrow$, $R\downarrow$, $I\uparrow$). This means that if maximum ratings are exceeded, an increase in temperature will cause an increase in current, which will generate a further increase in heat (temperature) and current, and so on.

MOSFET Handling Precautions

Certain precautions must be taken when handling any MOSFET devices. The very thin insulating layer between the gate and the substrate of a MOSFET can easily be punctured if an excessive voltage is applied. Your body can build up extremely large electrostatic charges due to friction. If this charge came in contact with the pins of a MOSFET device, an electrostatic-discharge (ESD) would occur, resulting in a possible arc across the thin insulating layer causing permanent damage. Most MOSFETs presently manufactured have zeners internally connected between gate and source to bypass high-voltage static or in-circuit potentials and protect the MOSFET. However, it is important to remember the following.

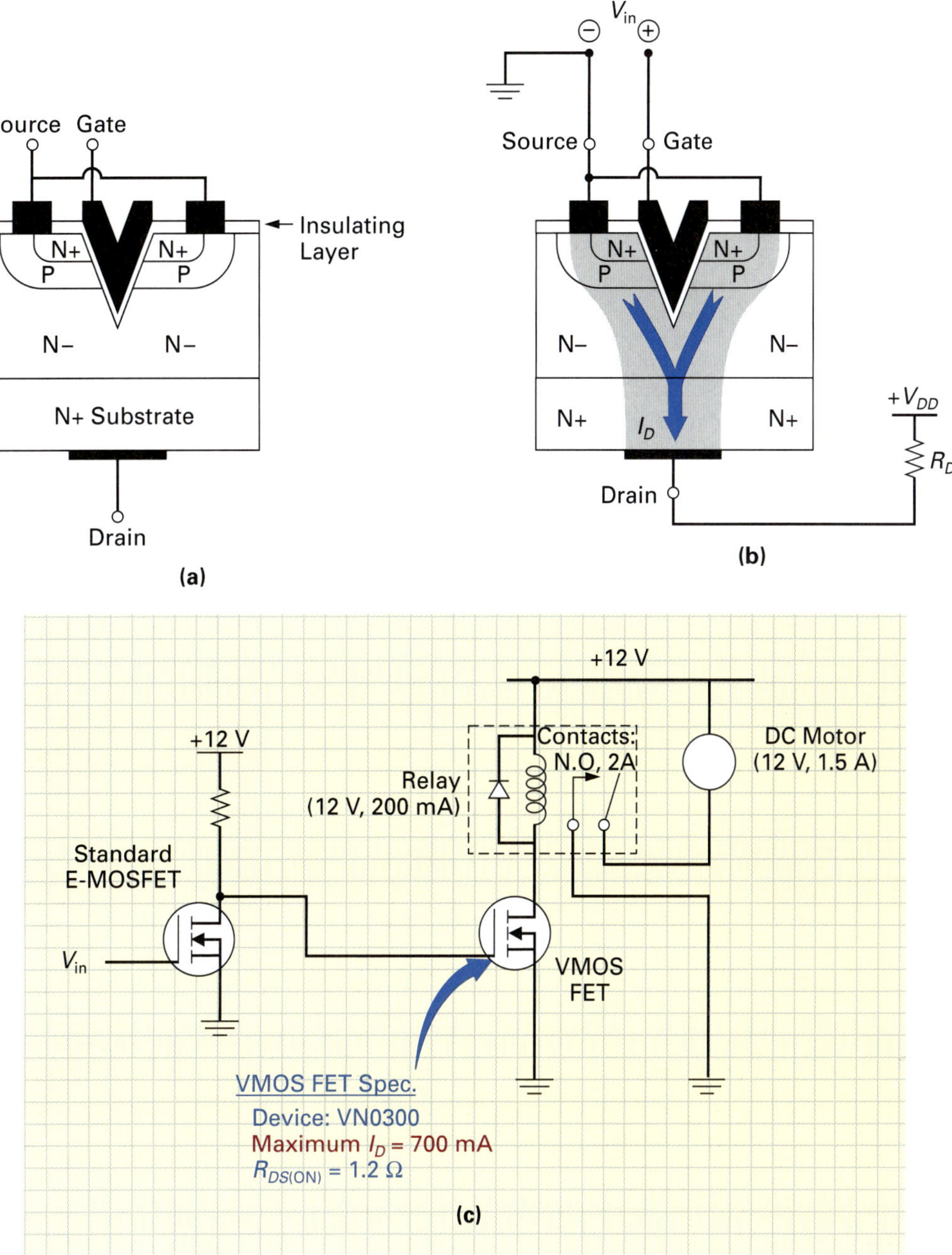

FIGURE 13-23 The Vertical Channel E-MOSFET (VMOS FET). (a) VMOS FET OFF. (b) VMOS FET ON. (c) VMOS FET Application Circuit—Low-Power Signal to High-Power Load Circuit.

1. All MOS devices are shipped and stored in a "conductive foam" or "protective foil" so that all of the IC pins are kept at the same potential, and therefore electrostatic voltages cannot build up between terminals.
2. When MOS devices are removed from the conductive foam, be sure not to touch the pins because your body may have built up an electrostatic charge.
3. When MOS devices are removed from the conductive foam, always place them on a grounded surface such as a metal tray.
4. When continually working with MOS devices, use a "wrist grounding strap," which is a length of cable with a 1 MΩ resistor in series. This prevents electrical shock if you come in contact with a voltage source.

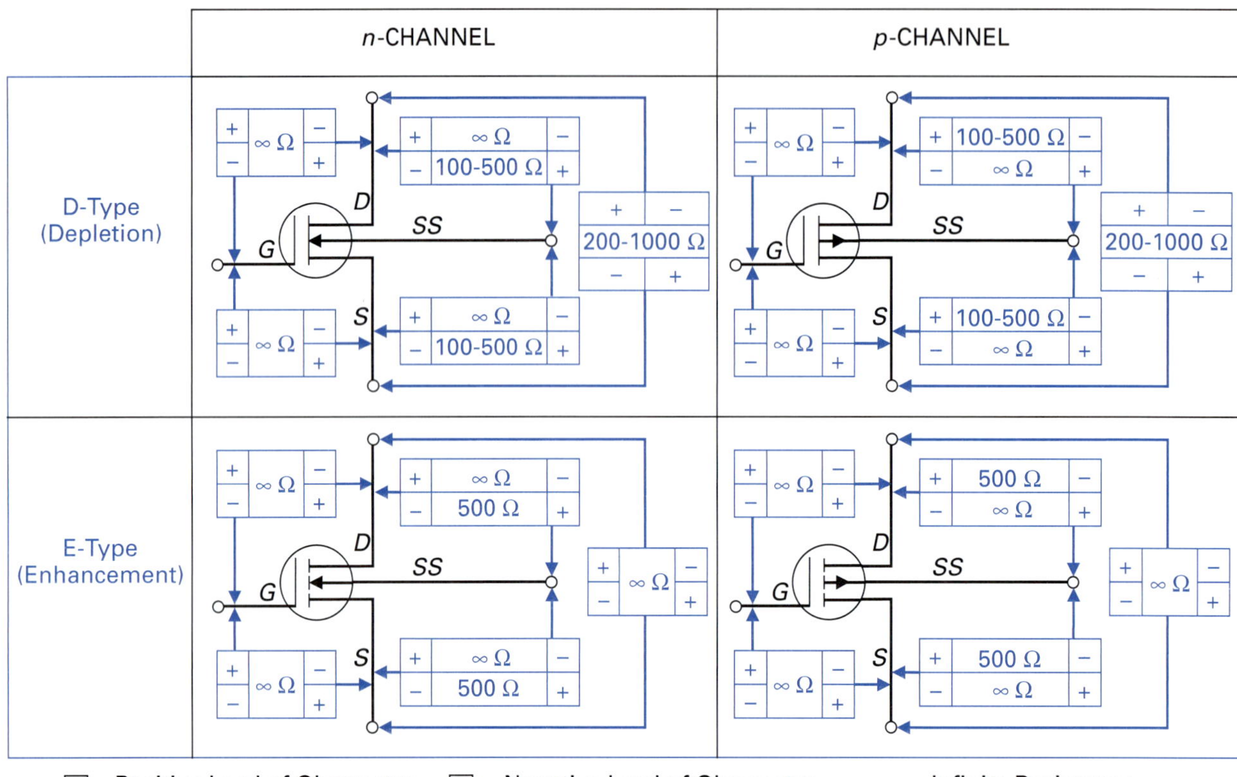

FIGURE 13-24 Testing MOSFETs with the Ohmmeter.

5. All test equipment, soldering irons, and work benches should be properly grounded.
6. All power in equipment should be off before MOS devices are removed or inserted into printed circuit boards.
7. Any unused MOSFET terminals must be connected because an unused input left open can build up an electrostatic charge and float to high voltage levels.
8. Any boards containing MOS devices should be shipped or stored with the connection side of the board in conductive foam.

MOSFET Testing

Like the BJT and JFET, MOSFETs can be tested with the transistor tester to determine: first, whether an open or short exists between any of the terminals; second, the transistor's transconductance/gain; and third, the value of I_{DSS} and leakage current.

If a transistor tester is not available, the ohmmeter can be used to determine the most common failures: opens and shorts. Figure 13-24 shows what resistance values should be obtained between the terminals of a good *n*-channel or *p*-channel D-MOSFET or E-MOSFET when testing with an ohmmeter. Remember that the gate-to-channel resistance of a MOSFET is always an open due to the insulated gate.

SELF-TEST EVALUATION POINT FOR SECTION 13-2-2

Use the following questions to test your understanding of Section 13-2-2.

1. True or false: The E-type MOSFET can be operated in both the depletion and enhancement mode.
2. The E-type MOSFET is a normally ____________ (ON/OFF) device.

3. The E-type MOSFET is generally
   a. Zero biased
   b. Base biased
   c. Drain-feedback biased
   d. None of the above
4. True or false: The E-MOSFET naturally operates as a voltage-controlled switch.

REVIEW QUESTIONS

Multiple-Choice Questions

1. The BJT is a ____________ controlled device whereas the FET is a ____________ controlled device.
   a. voltage, current
   b. current, voltage
2. The gate-source junction of a JFET is always ____________ biased since the input voltage is normally ____________ or some ____________ voltage.
   a. reverse, 0 V, negative
   b. forward, 0 V, negative
   c. reverse, −4 V, positive
   d. forward, −4 V, negative
3. When $V_{GS} = V_{GS(OFF)}$, $I_D = ?$
   a. I_{DSS} c. Maximum
   b. Zero d. V_P
4. Which JFET circuit configuration is also known as a source-follower?
   a. Common-drain c. Common-gate
   b. Common-source d. Both (a) and (b) are true
5. Which JFET circuit configuration provides a high input impedance and a good voltage gain?
   a. Common-drain c. Common-gate
   b. Common-source d. Both (a) and (c) are true
6. Which biasing method makes use of a $-V_{GG}$ supply voltage?
   a. Self biasing c. Base biasing
   b. Gate biasing d. Voltage-divider biasing
7. Which biasing method uses a source resistor?
   a. Self-biasing c. Voltage-divider biasing
   b. Gate biasing d. Both (a) and (c)
8. Transconductance is a ratio of
   a. ΔI_D to ΔV_{DS} c. ΔI_D to ΔV_{GS}
   b. ΔV_{GD} to ΔI_D d. ΔV_{GS} to ΔV_{DS}
9. Which JFET circuit configuration has phase inversion between input and output?
   a. Common-drain c. Common-gate
   b. Common-source d. Both (b) and (c)
10. The current between ____________ leads of a JFET is controlled by varying the reverse bias voltage applied to the ____________ leads.
    a. gate and source, source and drain
    b. source and drain, drain and gate
    c. source and drain, gate and source
    d. drain and gate, gate and source
11. The ____________ has an insulated gate that allows the use of either a positive or negative gate input voltage.
    a. JFET c. BJT
    b. MOSFET d. VJT
12. The ____________ has drain current when the gate voltage is zero and is therefore known as a normally ____________ device.
    a. D-MOSFET, OFF c. E-MOSFET, OFF
    b. D-MOSFET, ON d. E-MOSFET, ON
13. Which type of MOSFET can use zero biasing to bias it to a mid-load-line point?
    a. D-MOSFET c. VMOS FET
    b. E-MOSFET d. Both (b) and (c)
14. With the ____________, the gate voltage must be greater than the device's threshold voltage to produce drain current.
    a. JFET c. E-MOSFET
    b. D-MOSFET d. All of the above
15. The vertical channel MOSFET is ideal for interfacing low-voltage devices to high-power loads.
    a. True b. False
16. Which of the following transistors has the highest input impedance?
    a. BJT c. MOSFET
    b. JFET d. Both (a) and (c) are true
17. D-MOSFETs are more frequently used in ____________ circuit applications whereas E-MOSFETs are more often used in ____________ circuit applications.
    a. analog, digital c. digital, linear
    b. linear, analog d. two-state, digital
18. A digital code consists of a series of HIGH and LOW voltages called ____________.
    a. NOT gates c. Logic levels
    b. CMOS d. Both (a) and (b)
19. The ____________ has made its mark in two-state circuits in which it uses only two points on the load line and acts like a voltage-controlled ____________.
    a. E-MOSFET, variable resistor
    b. E-MOSFET, switch
    c. D-MOSFET, variable resistor
    d. D-MOSFET, switch

20. To avoid possible damage to MOS devices while handling, testing, or in operation, the following precaution should be taken.
   a. All leads should be connected except when being tested or in actual operation.
   b. Pick up devices by plastic case instead of leads.
   c. Do not insert or remove devices when power is applied.
   d. All of the above

Practice Problems

21. Calculate the following for the amplifier circuit in Figure 13-25:
   a. Transconductance b. Voltage gain
22. Calculate the following for the circuit in Figure 13-26:
   a. V_{GS} d. $I_{D(\text{maximum})}$
   b. I_D e. Q point
   c. V_{DS}
23. Calculate the following for the circuit in Figure 13-27:
   a. V_S b. V_{GS} c. V_{DS}
24. Calculate the following for the circuit in Figure 13-28:
   a. V_G b. V_S c. V_{GS} d. V_{DS}
25. Calculate the following for the circuit shown in Figure 13-29:
   a. $I_{D(\text{Sat.})}$ b. $V_{DS(\text{OFF})}$
26. Calculate the following for the circuit shown in Figure 13-30:
   a. I_D b. V_{DS}
27. What biasing method was used in Figures 13-29 and 13-30?

Web Site Questions

Go to the web site http://www.prenhall.com/cook, select the textbook *Electronics: A Complete Course,* select this chapter, and then follow the instructions when answering the multiple-choice practice problems.

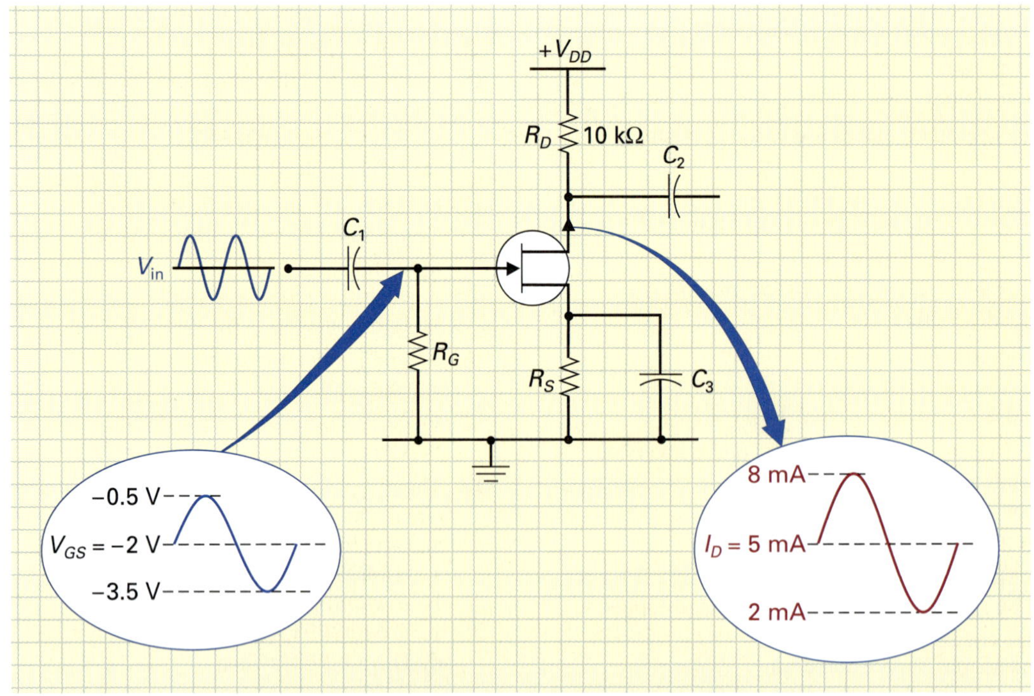

FIGURE 13-25 A Common-Source Amplifier.

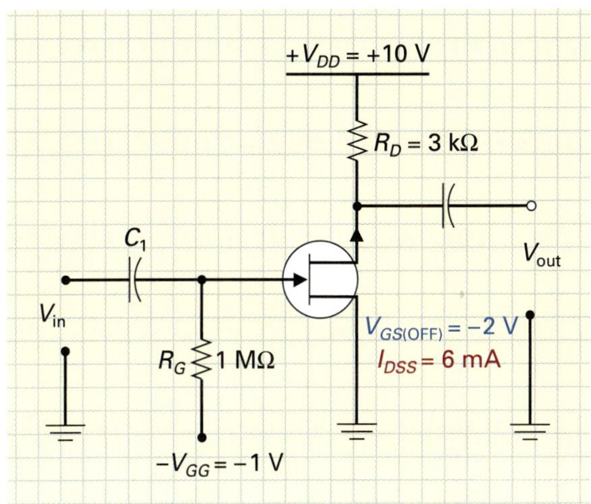

FIGURE 13-26 **A Gate-Biased JFET Circuit.**

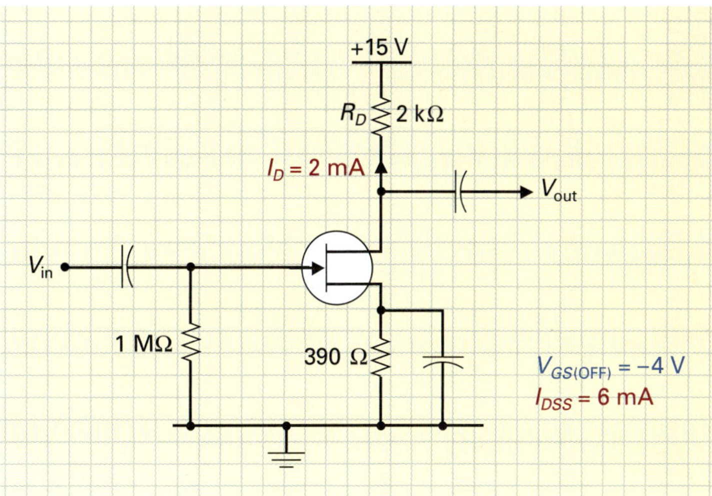

FIGURE 13-27 **A Self Biased JFET Circuit.**

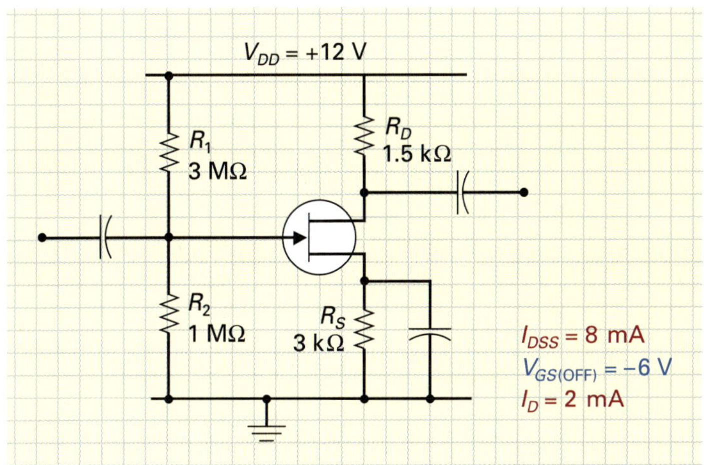

FIGURE 13-28 **A Voltage-Divider Biased JFET Circuit.**

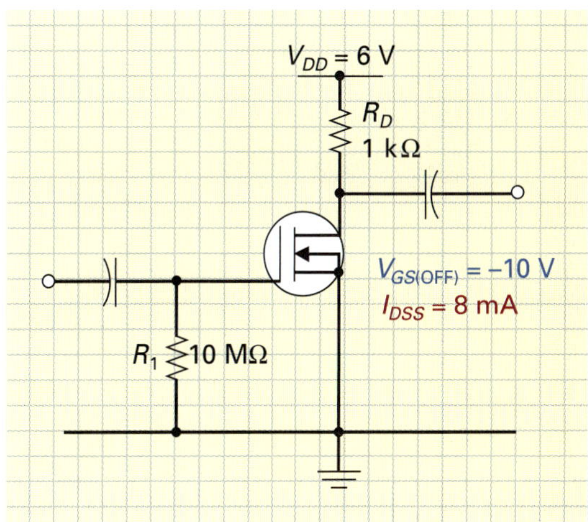

FIGURE 13-29 **A D-MOSFET Amplifier Circuit.**

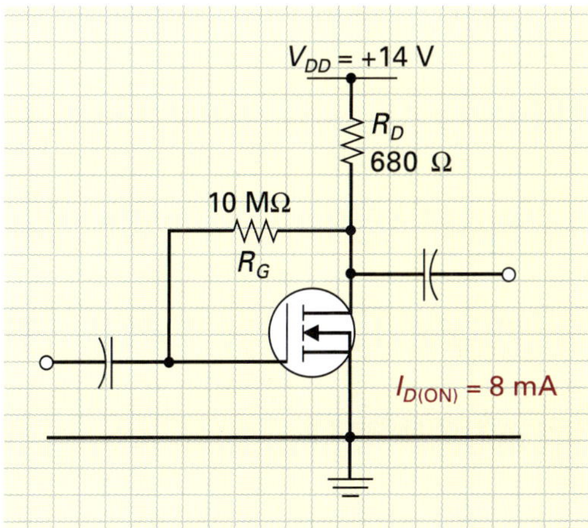

FIGURE 13-30 **An E-MOSFET Amplifier Circuit.**

Thyristors and Transducers

Leibniz's Language of Logic

Gottfried Wilhelm Leibniz was born in Leipzig, Germany, in 1646. His father was a professor of moral philosophy and spent much of his time discussing thoughts and ideas with his son. Tragically, his father died when he was only six, and from that time on Leibniz spent hour upon hour in his late father's library reading through all of his books.

By the age of twelve he had taught himself history, Latin, and Greek. At fifteen he entered the University of Leipzig, and it was here that he came across the works of scholars such as Johannes Kepler and Galileo. The new frontiers of science fascinated him, so he added mathematics to his curriculum.

In 1666, while finishing his university studies, the twenty-year-old Leibniz wrote what he called modestly a schoolboy essay, "De Arte Combinatoria," which means "On the Art of Combination." In this work, he described how all thinking of any sort on any subject could be reduced to exact mathematical statements. Logic or, as he called it, the laws of thought, could be converted from the verbal realm—which is full of ambiguities—into precise mathematical statements. In order to achieve this, however, Leibniz stated that a "universal language" would be needed. Most of his professors found the paper either baffling or outrageous and this caused Leibniz not to pursue the idea any further.

After graduating from university, Leibniz was offered a professorship at the University of Nuremberg, which he turned down for a position as an international diplomat. This career proved not to be as glamorous as he imagined because most of his time was spent in uncomfortable horse-drawn coaches traveling between the European capitals.

In 1672, his duties took him to Paris where he met Dutch mathematician and astronomer Christian Huygens. After seeing the hours that Huygens spent on endless computations, Leibniz set out to develop a mechanical calculator. A year later Leibniz unveiled the first machine that could add, subtract, multiply, and divide decimal numbers.

In 1676 Leibniz began to concentrate more on mathematics. It was at this time that he invented calculus, which was also independently discovered by Isaac Newton in England. Leibniz's focus, however, was on the binary number system, which occupied him for years. He worked tirelessly to document the long combinations of ones and zeros that make up the modern binary number system and to perfect binary arithmetic. What is ironic is that for all his genius, Leibniz failed to make the connection between his 1666 essay and binary, which was the universal language of logic that he was seeking. It would be a century and a quarter after Leibniz's death (in 1716) when another self-taught mathematician named George Boole would discover it.

14

Introduction

In this chapter, we will examine thyristors and transducers. Thyristors are generally used as dc and ac power control devices, while transducers are generally used as sensing, displaying, and actuating devices.

The semiconductor thyristor acts as an electronically controlled switch, switching power ON and OFF to adjust the average amount of power delivered to a load or connecting or disconnecting power from a load. Some thyristors are "unidirectional," which means that they will only conduct current in one direction (dc), while others are "bi-directional," which means that they can conduct current in either direction (ac). Thyristors are generally used as electronically controlled switches instead of the transistors because they have better power handling capabilities and are more efficient.

A semiconductor transducer is an electronic device that converts one form of energy to another. Electronic transducers can be classified as either input transducers (such as the photodiode) that generate input control signals, or output transducers (such as the LED) that convert output electrical signals to some other energy form. Input transducers are sensors that convert thermal, optical, mechanical, and magnetic energy variations into equivalent voltage and current variations. Output transducers, on the other hand, perform the exact opposite, converting voltage and current variations into optical or mechanical energy variations.

14-1 THYRISTORS

The most frequently used thyristors are the "silicon controlled rectifier (SCR)," "triode ac semiconductor switch (TRIAC)," "diode ac semiconductor switch (DIAC)," "unijunction transistor (UJT)," and "programmable unijunction transistor (PUT)." In this section, we will examine the operation, characteristics, applications, and testing of all these electronic devices.

14-1-1 *The Silicon Controlled Rectifier (SCR)*

Silicon Controlled Rectifier or SCR

A three-junction, three-terminal, unidirectional P-N-P-N thyristor that is normally an open circuit. When triggered with the proper gate signal it switches to a conducting state and allows current to flow in one direction.

The **silicon controlled rectifier** or **SCR** is the most frequently used of the thyristor family. Figure 14-1(a) shows how this device is a four-layered, alternately-doped component, with three terminals labeled anode (A), cathode (K) and gate (G).

SCR Operation

Complementary Latch Circuit

A circuit containing an NPN and PNP transistor that once triggered ON will remain latched ON.

To simplify the operation of the SCR, Figure 14-1(b) shows how the four layers of the SCR can be thought of, when split, as being a PNP and NPN transistor, and Figure 14-1(c) shows how this interconnection forms a **complementary latch circuit.** To operate as an ON/OFF switch, the SCR must be biased like a diode, with the anode of the SCR made positive relative to the cathode. The gate of the SCR is an active-HIGH input and must be triggered by a voltage that is positive relative to the cathode. In the example in Figure 14-1(c), you can see that the anode of the SCR is connected to +50 V via the load (R_L), the cathode is connected to 0 V, and the gate has a control input that is either 0 V or +5 V. When the gate input is at 0 V, the NPN and PNP transistors are OFF, and the SCR is equivalent to an open switch, as shown in the inset in Figure 14-1(c). On the other hand, when the gate input is +5 V, the NPN transistor's base is made positive with respect to the emitter and so the NPN transistor will turn ON. Turning the NPN transistor ON will connect the 0 V at the emitter of the NPN transistor through to the PNP's base, causing it to also turn ON. Turning the PNP transistor ON will connect the large positive voltage on the PNP's emitter through to the collector and the base of the NPN transistor, keeping it ON even after the +5 V input trigger is removed. As a result, both transistors will be latched ON and held ON by one another, allowing a continuous flow of current from cathode to anode. A momentary positive input trigger, therefore, will cause the SCR to be latched ON and be equivalent to a closed switch, as seen in the inset in Figure 14-1(c).

SCR Characteristics

Forward Breakover Voltage

Voltage needed to turn ON an SCR.

Figure 14-1(d) shows a correctly biased SCR: the left inset shows a typical low-power and high-power package, and the right inset shows a typical SCR characteristic curve. This graph plots the forward and reverse voltage applied across the SCR's anode-to-cathode against the current through the SCR between cathode and anode. Looking at the forward conduction quadrant, you can see that the voltage needed to turn ON an SCR is called the **forward breakover voltage.** An SCR's forward breakover voltage, or turn ON voltage, is inversely proportional to the value of gate current. For example, a larger gate current will cause the SCR to turn ON when only a small forward voltage is applied between the SCR's anode-to-cathode. If no gate current is applied, the forward voltage applied across the SCR's anode to cathode will have to be very large to make the SCR turn ON and conduct current between cathode and anode. Once the SCR is turned ON, a holding current latches the SCR's complementary latch ON, independent of the gate current. If the forward current between the SCR's cathode and anode falls below this minimum holding current value, the SCR will turn OFF because the NPN/PNP latch will not have enough current to keep each other ON. A gate trigger is therefore used to turn ON the SCR; however, once the SCR is ON, it can only be turned OFF by decreasing the anode-to-cathode voltage so that the cathode-to-anode current passing through the SCR drops below the holding current value.

In the reverse direction with the gate switch open, the SCR acts in almost the same way as a diode because a large reverse voltage is needed between anode and cathode to cause the SCR to break down.

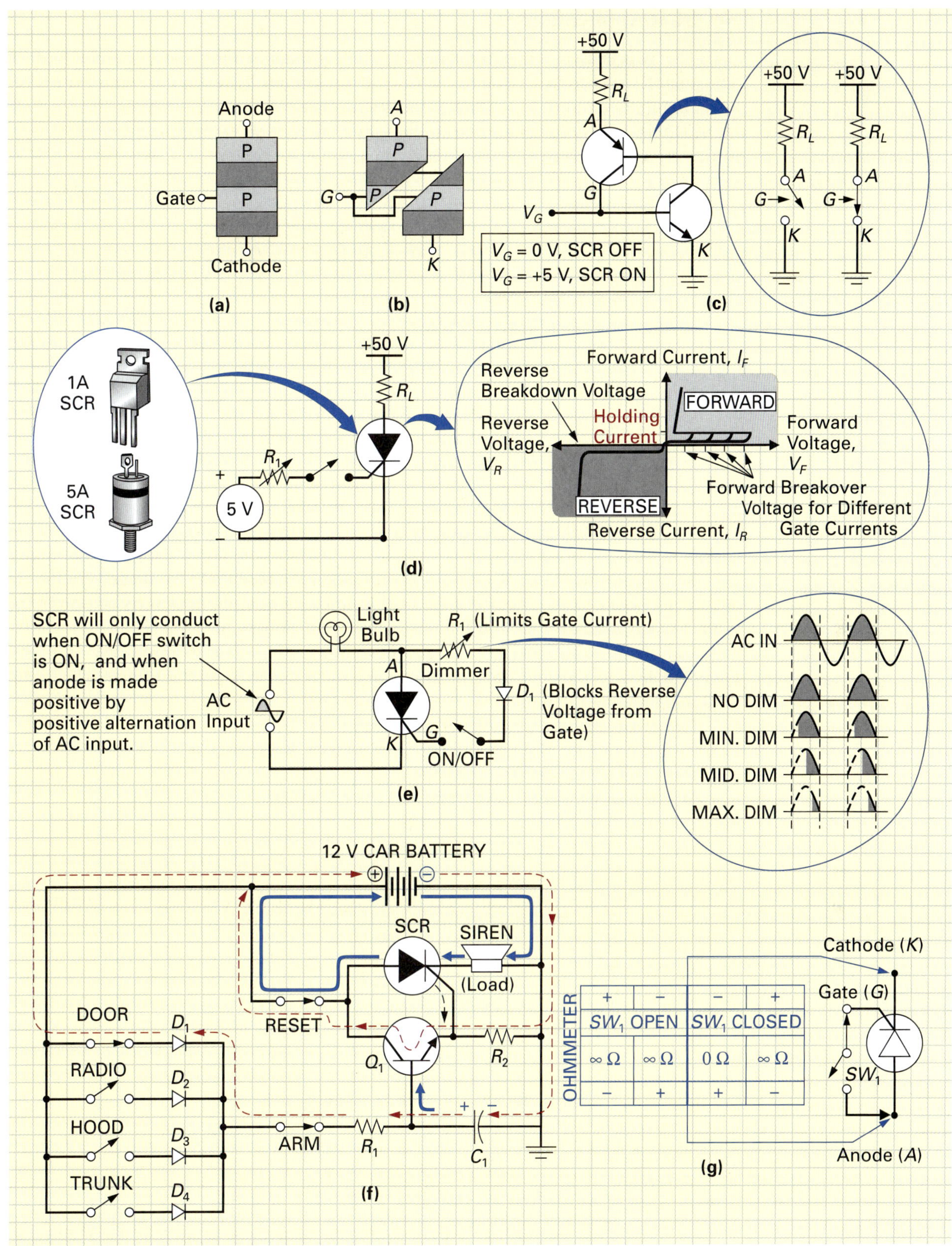

+	−	−	+
SW_1 OPEN		SW_1 CLOSED	
∞ Ω	∞ Ω	0 Ω	∞ Ω
−	+	+	−

FIGURE 14-1 Silicon-Controlled Rectifier. (a) Construction. (b) Complementary Latch. (c) Closed, Open Latch Action. (d) Correctly Biased. (e) SCR Application—Basic Light Dimmer. (f) Application—Alarm System. (g) Testing SCRs.

SCR Applications

Unidirectional Device

A device that will conduct current in only one direction.

The SCR is used in a variety of electrical power applications, such as light dimmer circuits, motor-speed control circuits, battery charger circuits, temperature control systems, and power regulator circuits. From the characteristic curve in Figure 14-1(d), we discovered that the SCR will only conduct current in the forward direction, which is why it is classed as a **unidirectional device.** This means that if an ac signal is applied across the SCR, it will only respond to a gate trigger during the time that the ac alternation makes the anode positive with respect to the cathode. For example, Figure 14-1(e) shows how the SCR could be connected to form a simple light dimmer circuit. When the ON/OFF switch is open, the SCR will be OFF because its gate current is zero, and the light will be OFF. When the ON/OFF switch is closed, diode D_1 will connect a positive voltage to the gate of the SCR whenever the ac input is positive. The value of gate current applied to the SCR is controlled by the variable dimmer resistor (R_1), and therefore this resistor value will determine the SCR's forward breakover (turn ON) voltage. Referring to the waveforms in the inset in Figure 14-1(e), you can see that when the resistance of R_1 is zero (no dim), gate current will be maximum, the SCR will turn ON for the full positive half cycle of the ac input, and the average power delivered to the light bulb will be HIGH. As the resistance of R_1 is increased, the SCR gate current is decreased, causing the SCR to turn ON for less of the positive alternation, and therefore the average power delivered to the light bulb to decrease. You may ask: why don't we simply connect a variable resistor in series with the light bulb to vary the light bulb's current and therefore brightness? The problem with this arrangement is identical to the disadvantages of the previously discussed "series dissipative regulator." The wattage rating, size, and cost of the variable resistor would be very large, and the circuit would be very inefficient because power is taken away from the light bulb by dissipating it away as heat from the resistor. Varying the SCR's ON/OFF time is a much more efficient system because we will only switch through to the light bulb the power that is desired, and therefore vary the average power applied to the load in almost exactly the same way as the previously discussed "switching regulator circuit."

Now that we have seen the SCR in an ac circuit application, let us now see how it could be used in a dc circuit application. Figure 14-1(f) shows a basic car-alarm system. When both the ARM switch and RESET switch are closed, the alarm system is active, and any of the four sensor switches (door, radio, hood, or trunk) will activate the alarm. For example, Figure 14-1(f) shows what will happen if the car door is opened when the alarm system is armed. Opening the door will cause capacitor C_1 to charge via diode D_1 and resistor R_1. After a short delay, the charge on C_1 is large enough to turn ON Q_1, which will switch the positive potential on its collector from the battery through to its emitter and to the gate of the SCR. This positive gate trigger will turn ON the SCR and activate the siren because the SCR is equivalent to a closed switch when ON and when triggered will connect the full 12-V battery across the siren. Once the SCR is turned ON, it will remain latched ON independent of the ARM switch and the sensor switches. Only by opening the RESET switch, which is hidden within the vehicle, can the siren be shut OFF. The values of R_1 and C_1 are chosen so that a small delay occurs before Q_1 and the SCR are triggered. This delay is included so that the vehicle owner has enough time to enter the car and disarm the alarm system by opening the ARM switch.

SCR Testing

Using the oscilloscope, you can monitor the gate trigger input and ON/OFF switching of an SCR while it is operating in circuit. If you suspect that the SCR is the cause of a circuit malfunction, you should remove the SCR and use the ohmmeter test circuit shown in Figure 14-1(g) to check for terminal-to-terminal opens and shorts. With this test, we will be using the ohmmeter's internal battery to apply different polarities to the different terminals of the SCR and SW_1 to either apply or disconnect gate current. To explain the ohmmeter response table in this illustration, you can see that if SW_1 is open (no gate current), the resistance between anode and cathode should be almost infinite ohms (actually about 250 kΩ), no matter what polarity is applied between anode and cathode. On the other hand, if SW_1 is closed and the anode is made positive with respect to the cathode, the gate will also be made positive due to SW_1, and so the SCR should turn ON and have a very low resistance between anode and cathode. If SW_1 is closed, and a reverse polarity is applied across the SCR (anode is made negative, cathode is made positive), the ohmmeter should once again read infinite ohms.

14-1-2 *The Triode AC Semiconductor Switch (TRIAC)*

The disadvantage with the SCR is that it is unidirectional, which means that it can only be activated when the applied anode-to-cathode voltage makes the anode positive with respect to the cathode, and it will conduct current in one direction. As a result, the SCR can only control a dc supply voltage, or one-half cycle of the ac supply voltage. To gain control of the complete ac input cycle, we would need to connect two SCRs in parallel, facing in opposite directions, as shown in Figure 14-2(a). This is exactly what was done to construct the **triode ac semiconductor switch** or **TRIAC,** which has three terminals called main terminal 1 (MT_1), main terminal 2 (MT_2), and gate (G). The *P-N* doping for this **bidirectional device** is shown in Figure 14-2(b), and its schematic symbol is shown in Figure 14-2(c).

Triode AC Semiconductor Switch or TRIAC
A bidirectional gate-controlled thyristor that provides full-wave control of ac power.

Bidirectional Device
A device that will conduct current in either direction.

TRIAC Operation and Characteristics

Since the TRIAC is basically two SCRs connected in parallel, back-to-back, it comes as no surprise that its operation and characteristics are very similar to the SCR. The characteristic curve for a typical TRIAC is shown in Figure 14-2(d). Looking at the identical forward and reverse curves, you can see that the key difference with the TRIAC is that it can be triggered or activated by either a positive or negative input gate trigger. This means that the TRIAC can be used to control both the positive and negative alternation of an ac supply voltage. To explain this in more detail, Figure 14-2(e) shows how a TRIAC could be connected across an ac input. When the ON/OFF gate switch is open, the TRIAC will not receive a gate trigger and so it will remain OFF. When the ON/OFF gate switch is closed, the TRIAC will be triggered by the positive and negative cycles of the ac input, via R_1. By adjusting the resistance of R_1, we can control the TRIAC's value of gate current. By controlling gate current, we can control the TRIAC's turn-ON voltage (positive and negative breakover voltage), so that the ac input voltage can be chopped up to adjust the average value of voltage applied to the load.

TRIAC Applications

Figure 14-2(f) shows how a TRIAC could be connected as an automatic night light for home or business security and safety. Later in this chapter, we will discuss the photocell in more detail. However, for now just think of it as a variable resistor that changes its resistance based on the amount of light present. During the day when the photocell is exposed to light, its resistance is less than a few ohms, and since it is in parallel with the energizing coil of a reed relay, most of the current will pass through the photocell keeping the reed relay de-energized. As the sun goes down and the photocell is deprived of light, its resistance increases to a few mega ohms, and this high resistance will cause the current through the reed relay coil to increase, and therefore the reed relay to energize. With the reed-relay switch closed, the TRIAC will be triggered by each half cycle of the ac supply voltage, causing it to turn ON and connect power to the night light.

TRIAC Testing

Using the oscilloscope, you can monitor the gate trigger input, and ON/OFF switching of a TRIAC, while it is operating in-circuit. If you suspect that the TRIAC is the cause of a circuit malfunction, you should remove the TRIAC and use the ohmmeter test circuit shown in Figure 14-2(g), to check for terminal-to-terminal opens and shorts. The ohmmeter response table in this illustration shows that if SW_1 is open (no gate current), the resistance between MT_2 and MT_1 should be almost infinite ohms (actually about 250 kΩ), no matter what polarity is

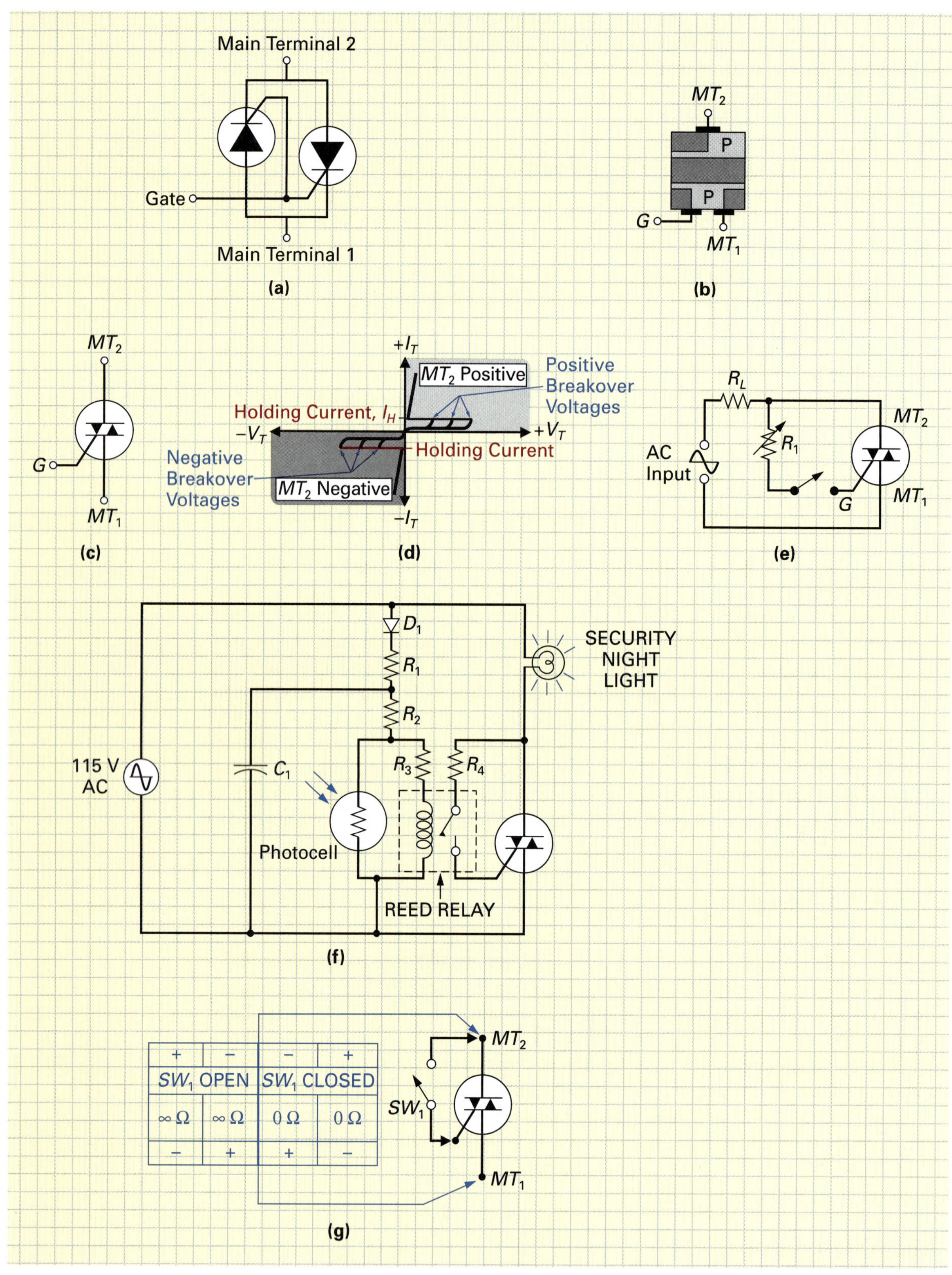

FIGURE 14-2 TRIAC. (a) TRIAC Equivalent Circuit. (b) Construction. (c) Schematic Symbol. (d) *V-I* Characteristics. (e) Application—AC TRIAC Switch. (f) Application—Automatic Night Light. (g) Testing TRIACs.

applied. When switch 1 is closed, the gate of the TRIAC will receive a trigger. Since the TRIAC operates on either a positive or negative trigger, it should turn ON no matter what polarity is applied, and therefore have a very low resistance between MT_2 and MT_1.

14-1-3 *The Diode AC Semiconductor Switch (DIAC)*

One disadvantage with the TRIAC is that its positive breakover voltage is usually slightly different from its negative breakover voltage. This nonsymmetrical trigger characteristic can be compensated for by using a **diode ac semiconductor switch** or **DIAC** to trigger a TRIAC. The DIAC's construction is shown in Figure 14-3(a), its schematic symbol in Figure 14-3(b), and its equivalent circuit in Figure 14-3(c). Equivalent to two back-to-back, series-connected junction diodes, the DIAC has two terminals.

Diode AC Semiconductor Switch or DIAC
A bidirectional diode that has a symmetrical switching mode.

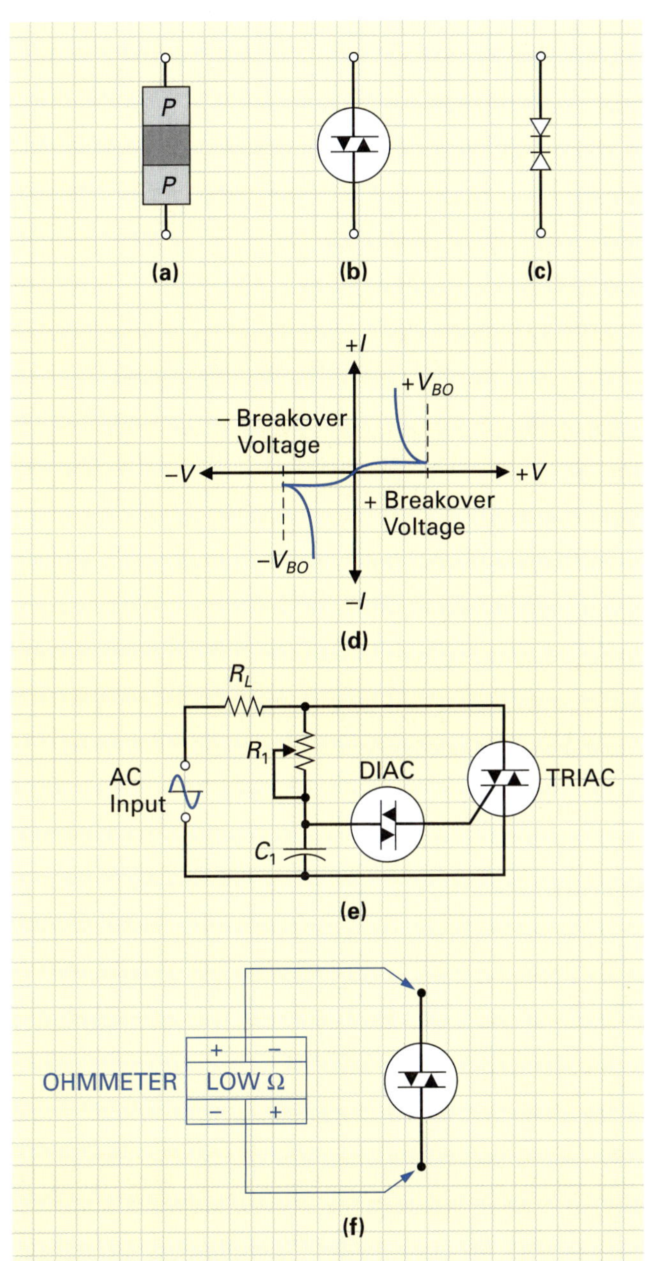

FIGURE 14-3 DIAC. (a) Construction. (b) Schematic Symbol. (c) Equivalent Circuit. (d) *V-I* Characteristics. (e) Application—TRIAC Control. (f) Testing DIACs.

DIAC Operation and Characteristics

Since the PNP regions of a DIAC are all equally doped, the DIAC will have the same forward and reverse characteristics, as shown in Figure 14-3(d). As a result, the DIAC is classed as a **symmetrical bidirectional switch,** which means that it will have the same value of breakover voltage in both the forward and reverse direction.

Symmetrical Bidirectional Switch
A device that has the same value of breakover voltage in both the forward and reverse direction.

DIAC Applications

Figure 14-3(e) shows how a DIAC could be connected as a pulse-triggering device in a TRIAC ac power control circuit. The DIAC will turn ON when the capacitor has charged to either the positive or negative breakover voltage ($+V_{BO}$ or $-V_{BO}$). Once this voltage is reached the DIAC turns ON, and the capacitor discharges through the DIAC, triggering the TRIAC into conduction, which then connects the ac supply voltage across the load. The variable resistor R_1 is used to adjust the RC charge time constant, so that the DIAC turn-ON time, and therefore TRIAC turn-ON time, can be changed.

DIAC Testing

Using the oscilloscope, you can monitor the ON/OFF switching of a DIAC while it is operating in-circuit. If you suspect that the DIAC is the cause of a circuit malfunction, you should remove the DIAC and check it with the ohmmeter, as seen in Figure 14-3(f). Since the DIAC is basically two diodes connected back-to-back in series, the ohmmeter should show a low resistance reading between its terminals no matter what polarity is applied.

14-1-4 *The Unijunction Transistor (UJT)*

Unijunction Transistor or UJT
A P-N device that has an emitter connected to the P-N junction on one side of the bar and two bases at either end of the bar. Used primarily as a switching device.

The **unijunction transistor** or **UJT** operates in a very different way to the SCR, TRIAC, and DIAC. Although it is given the name transistor, it is never used as an amplifying device like the BJT and FET: it is only ever used as a voltage-controlled switch. Figure 14-4(a) shows the construction of the UJT and illustrates how the uni (one) junction transistor derives its name from the fact that it has only one P-N junction. Looking at this illustration you can see that the UJT is a three-terminal device with an emitter lead (E) attached to a small p-type pellet, that is fused into a bar of n-type silicon with contacts at either end [labeled base 1 (B_1) and base 2 (B_2)].

The schematic symbol for the UJT is shown in Figure 14-4(b). To remember whether the junction is a P-N or N-P type, think of the arrow as a junction diode: the emitter (diode-anode) is positive and the bar (diode-cathode) is negative.

UJT Operation and Characteristics

Figure 14-4(c) illustrates the UJT's equivalent circuit. The emitter-to-bar P-N junction is equivalent to a junction diode, and the bar is equivalent to a two-resistor voltage divider (R_{B1} and R_{B2}). Referring once again to the UJT construction in Figure 14-4(a), you can see that the emitter pellet is closer to terminal B_2 than B_1, and this is why the resistance of R_{B2} is smaller than the resistance of R_{B1}. Figure 14-4(d) shows how a UJT should be correctly biased. The voltage source V_{BB} is connected to make B_2 positive relative to B_1 and the input voltage V_S is connected to make the emitter positive with respect to B_1. The resistor R_E is used to limit emitter current. When V_S is zero, the UJT's emitter diode is OFF, and the resistance between B_2 and B_1 allows only a very small amount of current between ground and $+V_{BB}$. If the emitter supply voltage is increased so that V_E exceeds the voltage at B_1, the emitter diode will turn ON and inject holes into the p region. Flooding the lower half of the UJT with holes increases the amount of current flow through the UJT, dramatically reducing the resistance of R_{B1}. A lower R_{B1} resistance results in a lower V_{B1} voltage drop, and this further increases the E to B_1 P-N forward bias permitting more holes to be injected into the n-type bar between E and B_1. Figure 14-4(e) graphically illustrates this action by plotting the UJT's emitter voltage (between E and B_1) against the UJT's emitter current. An increase in V_S, and therefore

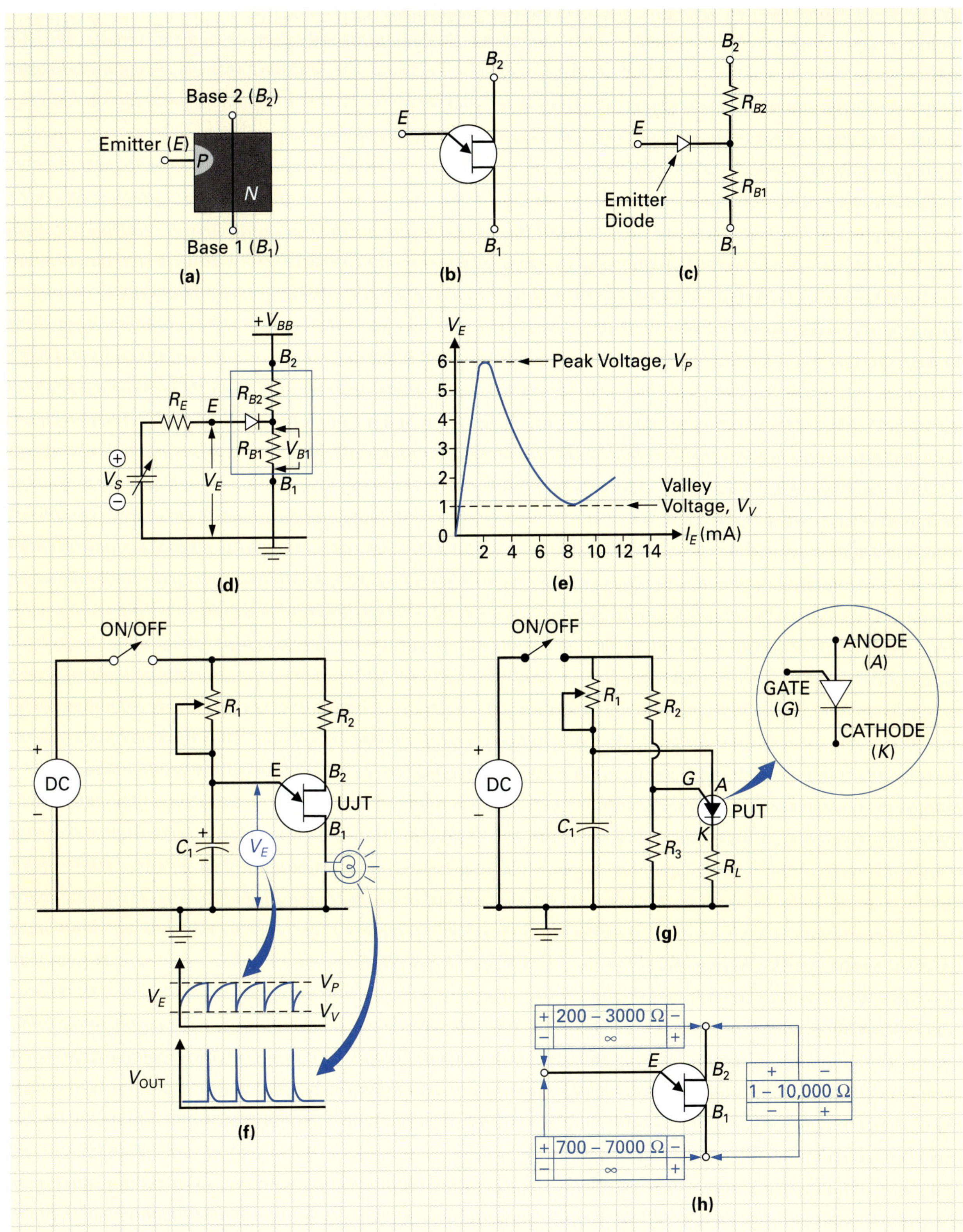

FIGURE 14-4 Unijunction Transistors. (a) Construction. (b) Schematic Symbol. (c) Equivalent Circuit. (d) Correctly Biased. (e) *V-I* Characteristic Curve. (f) Application—Relaxation Oscillator. (g) Programmable Unijunction Transistor (PUT). (h) Ohmmeter Resistances.

Peak Voltage (V_P)
The maximum value of voltage.

Valley Voltage (V_V)
The voltage at the dip or valley in the characteristic curve.

V_E, produces very little emitter current until the **peak voltage (V_P)** is reached. Beyond V_P, V_E has exceeded V_{B1} and the emitter diode is forward biased, causing IE to increase and V_E to decrease due to the lower resistance of R_{B1}. This negative resistance region reaches a low point known as the **valley voltage (V_V),** which is a point at which V_E begins to increase and the UJT no longer exhibits a negative resistance. The point at which a UJT turns ON and increases the current between B_1 and B_2 can be controlled, and used in switching applications.

UJT Applications

The UJT's negative resistance characteristic is useful in switching and timing applications. Figure 14-4(f) shows how a UJT could be connected to form a relaxation oscillator in an emergency flasher circuit. When the ON/OFF switch is closed, capacitor C_1 charges by resistor R_1. When the voltage across C_1 reaches the UJT's V_P value, the UJT will turn ON and its resistance between E and B_1 will drop LOW. This low resistance will allow C_1 to discharge through the UJT's E-to-B_1 junction and into the flasher light bulb, causing it to momentarily flash. As C_1 discharges, its voltage decreases and this causes the UJT to turn OFF. The cycle then repeats since the off UJT will allow capacitor C_1 to begin charging towards V_P, at which time it will trigger the UJT and repeat the process. The circuit's repetition rate, or frequency, is determined by the UJT's V_P rating, the supply voltage, and the RC time constant. To change the flashing rate, the value of R_1 can be changed to vary the rate at which C_1 is charged and therefore how soon the UJT is triggered.

The Programmable UJT (PUT)

Programmable Unijunction Transistor or PUT
A unijunction transistor that can have its peak voltage controlled.

The **programmable unijunction transistor** or **PUT** is a variation on the basic UJT thyristor. This four-layer thyristor has three terminals labeled cathode (K), anode (A), and gate (G). The key difference between the basic UJT and the PUT is that the PUT's peak voltage (V_P) can be controlled. Figure 14-4(g) shows how a PUT could also be connected to form a relaxation oscillator circuit. To differentiate the PUT's schematic symbol from the SCR, the gate input is connected into the anode side of the diode symbol instead of the cathode side. This circuit will produce exactly the same output waveform as the circuit shown in Figure 14-4(f). The gate-to-cathode voltage is derived from R_3, which is connected with R_2 to form a voltage divider. This circuit will operate in exactly the same way as the previous relaxation oscillator, in that C_1 will charge via R_1 until the charge across C reaches the V_P value. In this circuit, however, the V_P trigger voltage is set by R_3. When the PUT's anode-to-cathode voltage exceeds the gate voltage by 0.7 V (single diode voltage drop), the PUT will turn ON, and C_1 will discharge through the PUT and develop an output pulse across R_L. To vary the frequency of this circuit, we can change the resistance of R_1 as before, or change the ratio of R_2 to R_3, which controls the V_P value of the PUT. For example, if R_3 is made larger than R_2, the gate voltage and therefore V_P voltage will be larger. A high V_P value will mean that C_1 will have to charge to a larger voltage before the PUT will turn ON. Increasing the time needed for C_1 to charge will decrease the triggering rate of the PUT and therefore decrease the circuit's frequency of operation.

UJT Testing

Using the oscilloscope, you can monitor the ON/OFF switching of a UJT while it is operating in-circuit. If you suspect that the UJT is the cause of a circuit malfunction, you should remove it and check it with the ohmmeter, as seen in Figure 14-4(h).

SELF-TEST EVALUATION POINT FOR SECTION 14-1

Use the following questions to test your understanding of Section 14-1.

1. The SCR's forward breakover voltage will ____________ as gate current is increased.
2. The SCR can only be used to control dc power. (True/False)
3. The TRIAC is a ____________ directional device that can be either positive or negative triggered.

4. Which device can be used to trigger the TRIAC to compensate for the TRIAC's non-symmetrical triggering characteristic?
5. TRIACs should always be used instead of SCRs when we want to control the complete ac power cycle. (True/False)
6. The unijunction transistor has ____________ P-N semiconductor junction(s).
7. The UJT can, like the bipolar and field effect transistors, function as either a switch or an amplifier. (True/False)
8. The PUT's peak voltage can be varied by changing the voltage between ____________ and cathode.

14-2 TRANSDUCERS

A **semiconductor transducer** is an electronic device that converts one form of energy to another. Electronic transducers can be classified as either **input transducers** (such as the photodiode) that generate input control signals or **output transducers** (such as the LED) that convert output electrical signals to some other energy form. Input transducers are sensors that convert thermal, optical, mechanical, and magnetic energy variations into equivalent voltage and current variations. Output transducers, on the other hand, perform the exact opposite, converting voltage and current variations into optical or mechanical energy variations.

Semiconductor Transducer
An electronic device that converts one form of energy to another.

Input Transducer
A transducer that generates input control signals.

Output Transducers
Transducers that convert output electrical signals to some other energy form.

14-2-1 *Optoelectronic Transducers*

The LED and photodiode are the most frequently used optoelectronic devices; however, they are not the only semiconductor devices in the optoelectronic family. In this section, we will review the operation of the LED and photodiode, along with the operation of other types of semiconductor light transducers.

Light-Sensitive Devices

Figure 14-5 illustrates some of the different types of **light-sensitive semiconductor devices.**

Light-Sensitive Semiconductor Devices
Semiconductor devices that change their characteristics in response to light.

The **photoconductive cell** or **light-dependent resistor (LDR)** shown in Figure 14-5(a) is a two-terminal device that changes its resistance (conductance) when light (photo) is applied. The photoconductive cell is normally mounted in a metal or plastic case with a glass window that allows the sensed light to strike the S-shaped light-sensitive material (typically cadmium sulfide). When light strikes the photoconductive atoms, electrons are released into the conduction band and the resistance between the device's terminals is reduced. When light is not present, the electrons and holes recombine, and the resistance is increased. A photoconductive cell will typically have a "dark resistance" of several hundred mega ohms and a "light resistance" of a few hundred ohms. The photoconductive cell's key advantage is that it can withstand a high operating voltage (typically a few hundred volts). Its disadvantages are that it responds slowly to changes in light level, and that its power rating is generally low (typically a few hundred milliwatts). The schematic symbol for the photoconductive cell is shown in Figure 14-5(b). Figure 14-5(c) shows how the photoconductive cell could be connected to control a street light. During the day, the LDR's resistance is LOW due to the high light levels, and the current through D_1, R_1, R_2, the LDR, and the control relay coil is HIGH. This HIGH value of current will energize the coil and cause the relay's normally closed (*NC*) contacts to open, and therefore the street light to be OFF. When dark, the resistance of the LDR will be HIGH, and the relay will be de-energized, its contacts will return to their normal condition, which is closed, and the street light will be ON.

Photoconductive Cell or Light-Dependent Resistor (LDR)
A two-terminal device that changes its resistance when light is applied.

The **photovoltaic cell** or **solar cell,** shown in Figure 14-5(d), generates a voltage across its terminals that will increase as the light level increases. The solar cell is usually made from silicon, and its schematic symbol is shown in Figure 14-5(e). The solar cell is available as either a discrete device or as a solar panel in which many solar cells are interconnected to form a series-aiding power source, as shown in the calculator application in Figure 14-5(f). The output of a solar cell is normally rated in volts and milliamps. For example, a typical photovoltaic cell could generate 0.5 V and 40 mA. To increase the output voltage simply connect solar cells in series; to increase the output current simply connect solar cells in parallel.

Photovoltaic Cell or Solar Cell
A device that generates a voltage across its terminal that will increase as the light level increases.

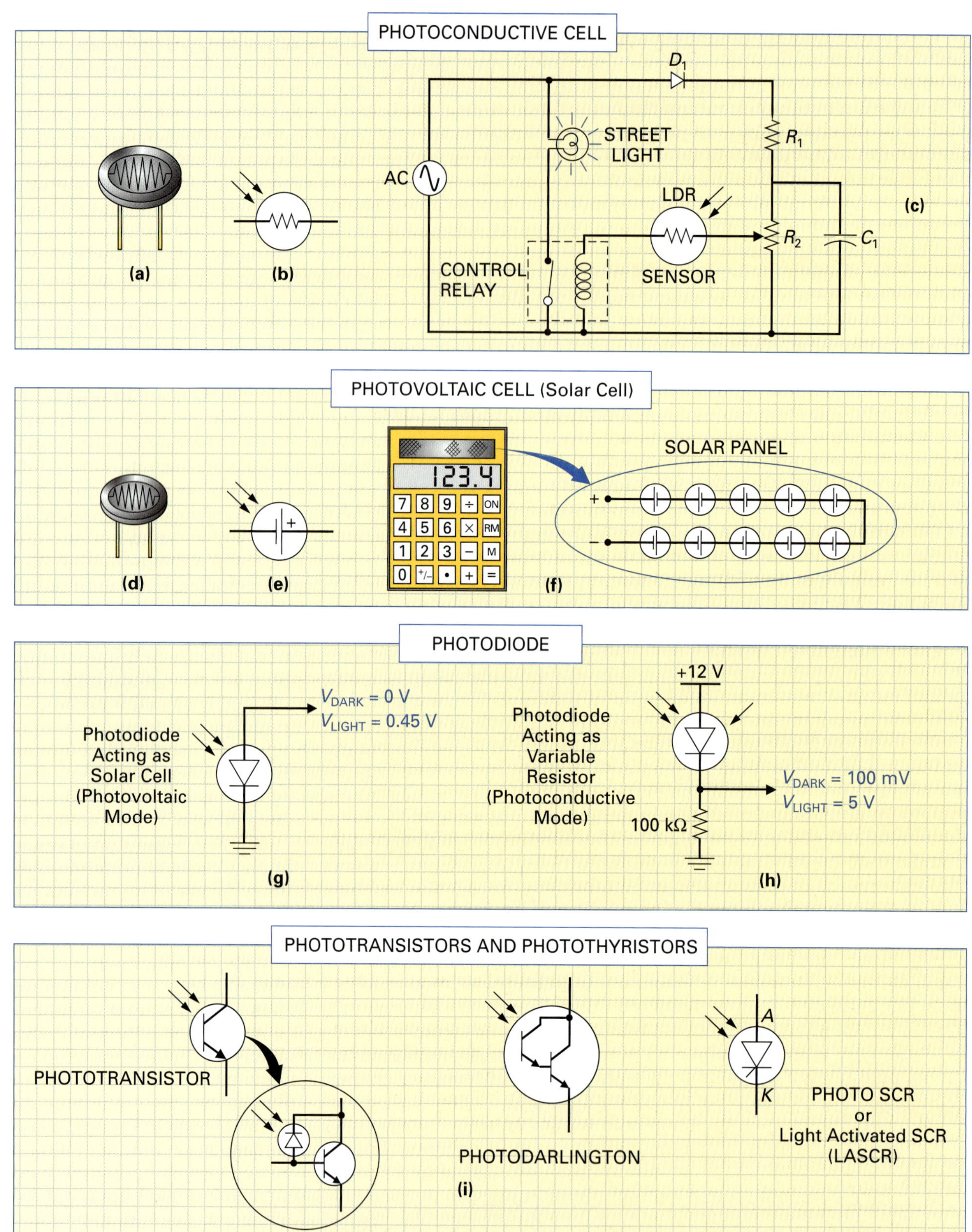

FIGURE 14-5 Light-Sensitive Devices.

A **photodiode** is a photo-detecting or light-receiving device that contains a semiconductor P-N junction. When used in the "photovoltaic mode" as shown in Figure 14-5(g), the photodiode will generate an output voltage (voltaic) in response to a light (photo) input (will operate like a solar cell). Photodiodes are most widely used in the "photoconductive mode," in which they will change their conductance (conductive) when light (photo) is applied (will operate like an LDR). In this mode, the photodiode is reverse biased (*n*-type region is made positive, *p*-type region is made negative).

The **photo-transistor, photo-darlington,** and **photo-SCR** (or light-activated SCR, LASCR), are all examples of light-reactive devices. As seen in the phototransistor inset, all three devices basically include a photodiode to activate the device whenever light is present. The advantage of the phototransistor is that it can produce a higher output current than the photodiode; however, the photodiode has a faster response time. The advantage of the photodarlington is its high current gain, and therefore it is ideal in low-light applications. When larger current switching is needed (typically a few amps), the LASCR can be used.

Photodiode

A photo-detecting or light-receiving device that contains a semiconductor P-N junction.

Photo-Transistor, Photo-Darlington, and Photo-SCR

Examples of light-reactive devices.

Light-Emitting Devices

Figure 14-6 illustrates some of the different types of **light-emitting semiconductor devices.**

Light-Emitting Semiconductor Devices

Semiconductor devices that will emit light when an electrical signal is applied.

Figure 14-6(a) shows the schematic symbols used to represent a **light-emitting diode** or **LED,** Figure 14-6(b) shows its typical physical appearance, and Figure 14-6(c) shows how the LED can be used as an ON/OFF indicator. The LED is basically a P-N junction diode and like all semiconductor diodes it can be either forward biased or reverse biased. When forward biased it will emit energy in response to a forward current. This emission of energy may be in the form of heat energy, light energy, or both heat and light energy depending on the type of semiconductor material used. The type of material also determines the color, and therefore frequency, of the light emitted. For example, different compounds are available that will cause the LED to emit red, yellow, green, blue, white, orange, or infrared light when it is forward biased.

Light-Emitting Diode or LED

A semiconductor diode that converts electric energy into electromagnetic radiation at visible and near infrared frequencies when its P-N junction is forward biased.

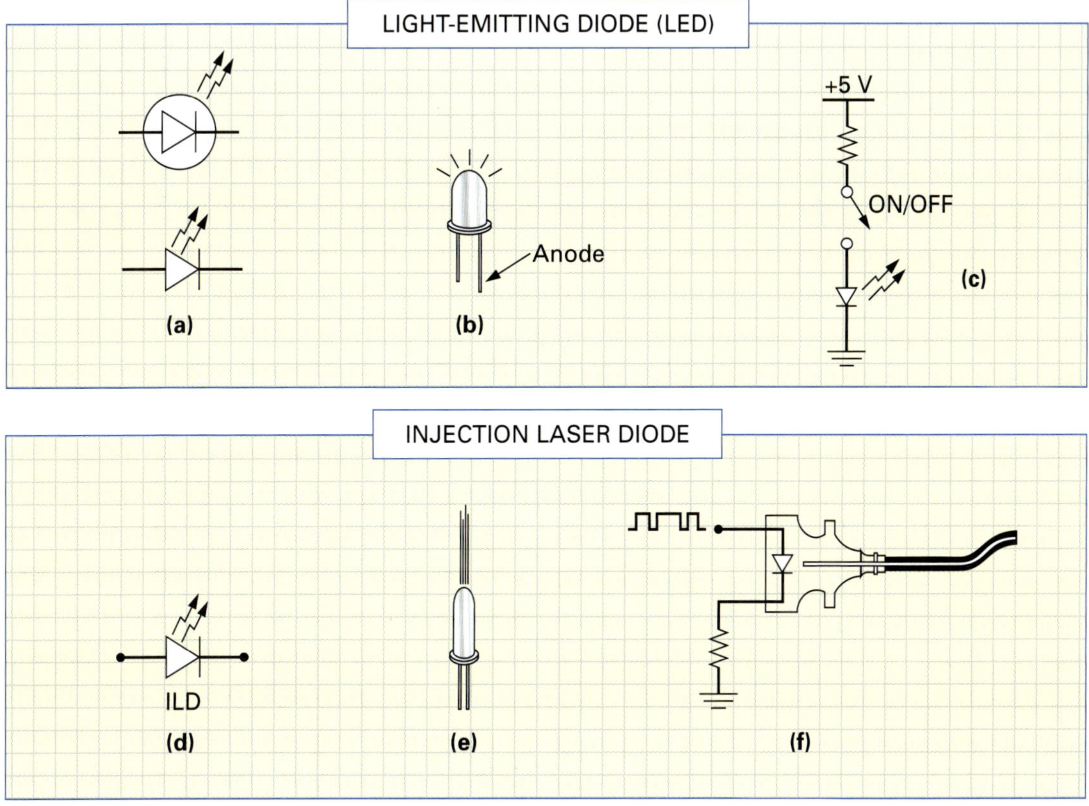

FIGURE 14-6 Light-Emitting Devices.

Injection Laser Diode or ILD

A semiconductor P-N junction diode that uses a lasing action to increase and concentrate the light output.

Figure 14-6(d) shows the schematic symbol for the **injection laser diode** or **ILD,** Figure 14-6(e) shows its typical physical appearance, and Figure 14-6(f) shows how it is usually used in fiber optic communication. The ILD differs from the LED in that it generates monochromatic (one color, or one frequency) light. Although it appears as though the LED is only generating one color light, it is in fact emitting several wavelengths or different frequencies that combined make up a color. Generating only one frequency is ideal in applications such as fiber optics, where we want to keep the light beam tightly focused so that the beam can travel long distances down a very thin piece of glass or plastic fiber. To explain this point in more detail, let us compare the light from a light bulb to the light produced by an ILD. An ILD generates light of only one frequency (monochromatic), and since all of the small packets of light being generated are of the same frequency, they act like an organized army, marching together in the same direction as a tight concentrated beam. The light bulb, on the other hand, generates white light that is composed of every color or frequency (panchromatic), and since these frequencies are all different, they act completely disorganized, and therefore radiate in all directions.

Optoelectronic Application—Character Displays

One of the biggest applications of LEDs is in the multisegment display. Figure 14-7(a) reviews the variety of LED display types available. Figure 14-7(b) shows the construction of

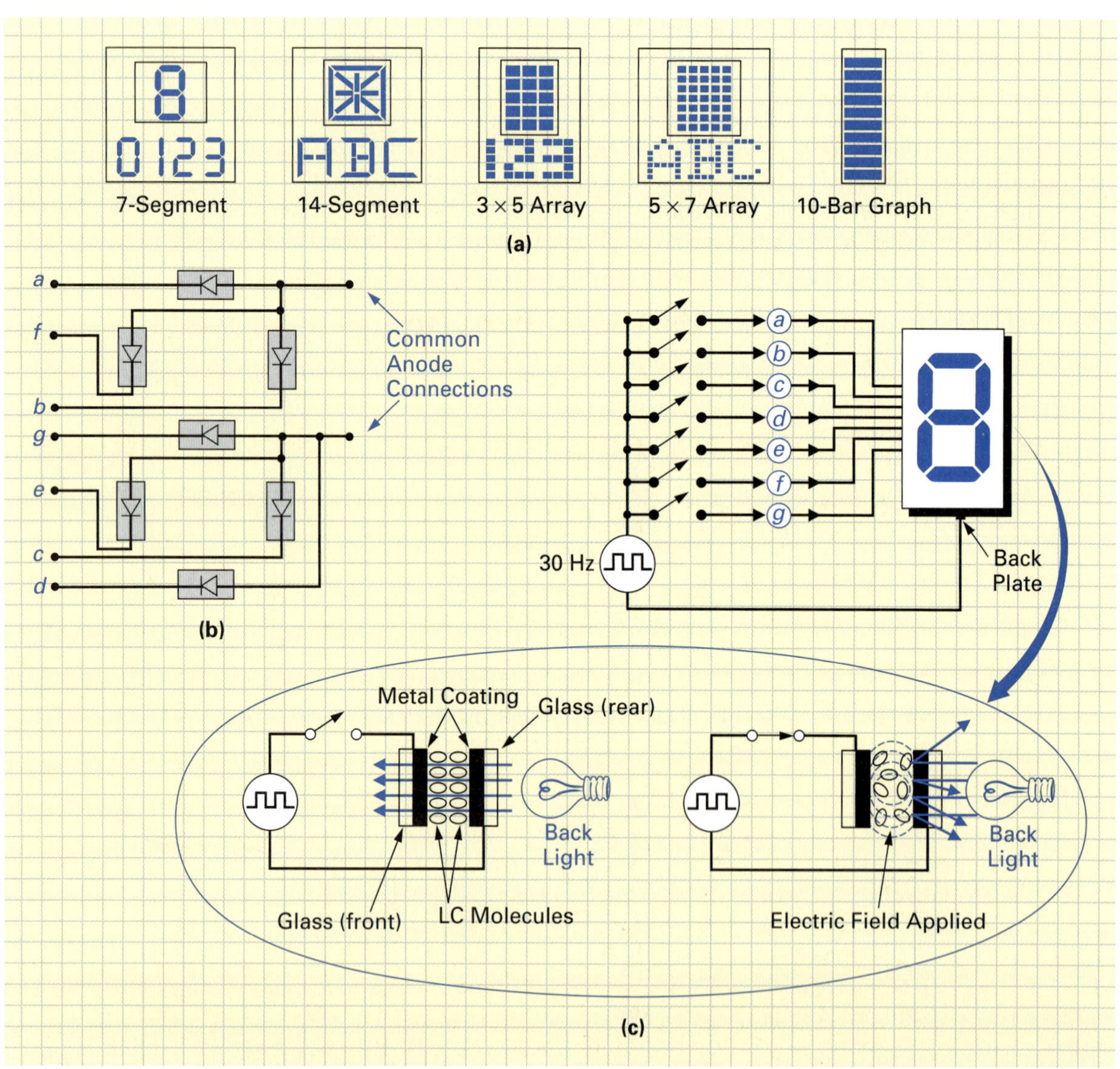

FIGURE 14-7 **Optoelectronic Application—Character Displays.**

a common-anode seven segment display. The need to display letters, in addition to numerals, resulted in a display with more segments such as the 14-segment display. For greater character definition, dot matrix displays are used, the most popular of which contains 35 small LEDs arranged in a grid of five vertical columns and seven horizontal rows, forming a 5 × 7 matrix. The bar display is rapidly replacing analog meter movements for displaying a quantity. For example, a quantity of fuel can be represented by the number of activated segments: no bars lit being zero and all bars lit being maximum.

The variety of character displays shown in Figure 14-7(a) are also available as **liquid crystal displays (LCDs),** as shown in Figure 14-7(c). The two key differences between an LED and LCD display are that an LED display generates light while the LCD display controls light, and the LCD display consumes a lot less power than an LED display. The low power consumption feature of the LCD display makes it ideal in portable battery-operated systems such as wristwatches, calculators, video games, portable test equipment, and so on. The LCD's only disadvantage is that the display is hard to see, but this can be compensated for by including back-lighting to highlight the characters. The liquid crystal display contains two pieces of glass that act as a sandwich for a "nematic liquid" or liquid crystal material, as seen in the inset in Figure 14-7(c). The rear piece of glass is completely coated with a very thin layer of transparent metal, while the front piece of glass is coated with the same transparent metal segments in the shape of the desired display. The operation of the liquid crystal display is explained in the two illustrations in the inset in Figure 14-7(c). When a segment switch is open, no electric field is generated between the two LCD metal plates, the nematic liquid molecules remain in their normal state which is parallel to the plane of the glass, and so all of the back-lighting passes through to the front display making the segment invisible. When a segment switch is closed, an ac voltage is applied between the two metal plates, generating an electric field between the two LCD metal plates. This electric field will cause the nematic liquid molecules to turn by 90°, and so all of the back-lighting for that segment is blocked making the segment visible.

Liquid Crystal Displays (LCDs)

A digital display having two sheets of glass separated by a sealed quantity of liquid crystal material. When a voltage is applied across the front and back electrodes, the liquid crystal's molecules become disorganized, causing the liquid to darken.

Optoelectronic Application—Optically Coupled Isolators

Up until this point, we have considered the optoelectronic emitter and detector as discrete or individual devices. There are, however, some devices available that include both a light-emitting and light-sensing device in one package. These devices are called **optically coupled isolators,** a name that describes their basic function: they are used to *optically couple two electrically isolated points.* To examine these devices in more detail, refer to the three basic types shown in Figure 14-8.

Figure 14-8(a) shows an **optically coupled isolator DIP module.** This optocoupler contains an infrared-emitting diode (IRED) and a silicon phototransistor and is generally used to transfer switching information between two electrically isolated points. Figure 14-8(b) shows how DIP optically coupled isolator ICs can be used in a motor control circuit. Since both circuits are identical, let us explain only the upper circuit's operation. Digital logic circuits generate the +5 V peak multiphase ON/OFF switching signal that is applied to IRED in the optocoupler. A HIGH input, for example, will turn ON the IRED, which will emit light to the phototransistor, turning it ON and switching the +12 V on its collector through to the emitter and the base of Q_1. This +12 V input to the base of Q_1 will cause it to turn ON and act as a closed switch, connecting ground to the top of the *A* winding. Since +12 V is connected to the center of the *A*/*B* winding, the *A* winding will be energized and the motor will move a step. The optocoupler, therefore, couples the ON/OFF switching information from the digital logic circuits and at the same time provides the necessary electrical isolation between the +5 V digital logic circuits and the +12 V motor supply circuits. In the past, isolation was provided by relays or isolation transformers that were larger in size, consumed more power, and were more expensive.

Figure 14-8(c) shows a different type of optocoupler or optoisolator called the **optically coupled isolator interrupter module.** This device consists of a matched and aligned emitter and detector and is used to detect opaque or nontransparent targets. Figure 14-8(d) shows how the interrupter optocoupler module could be used as an optical tachometer. The IRED is permanently ON, as seen in the inset in Figure 14-8(d), and emits a constant light beam towards the phototransistor. A transparent disc mounted to a shaft has opaque targets evenly spaced around

Optically Coupled Isolators

Devices that contain a light-emitting and light-sensing device in one package. They are used to optically couple two electrically isolated points.

Optically Coupled Isolator DIP Module

An optocoupler that contains an infrared-emitting diode and a silicon photo-transistor and is generally used to transfer switching information between two electrically isolated points.

Optically Coupled Isolator Interrupter Module

A device that consists of a matched and aligned emitter and detector and that is used to detect opaque or nontransparent targets.

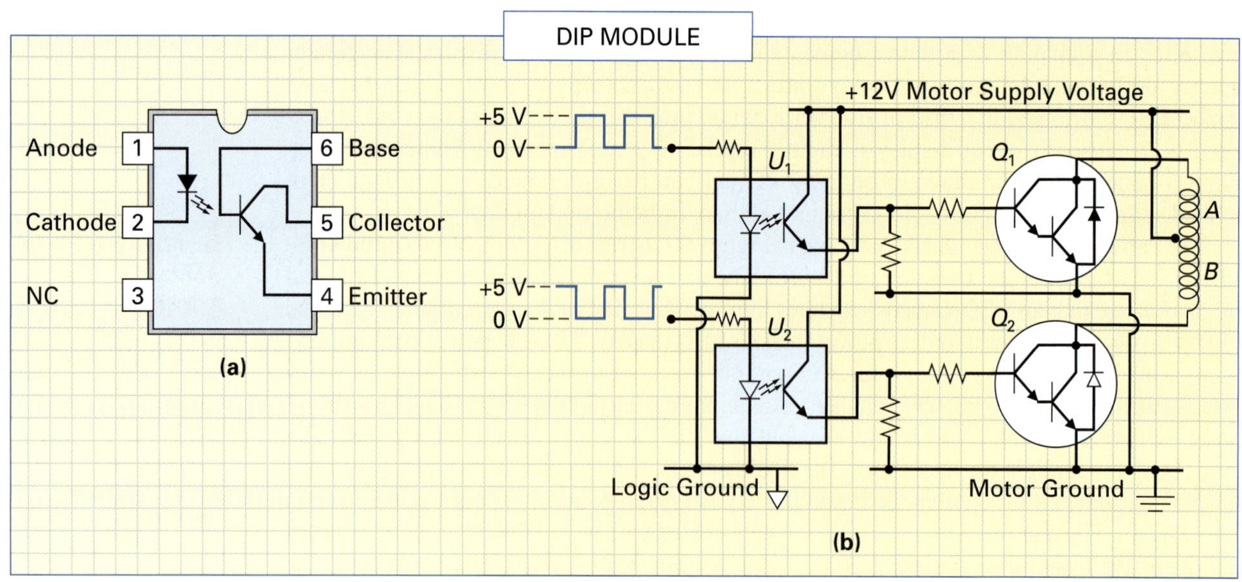

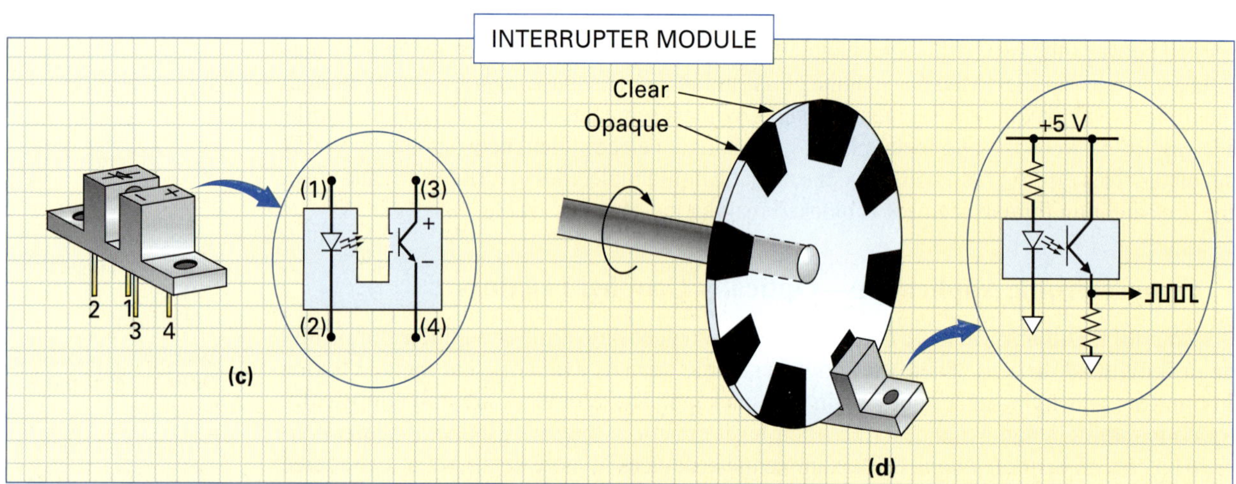

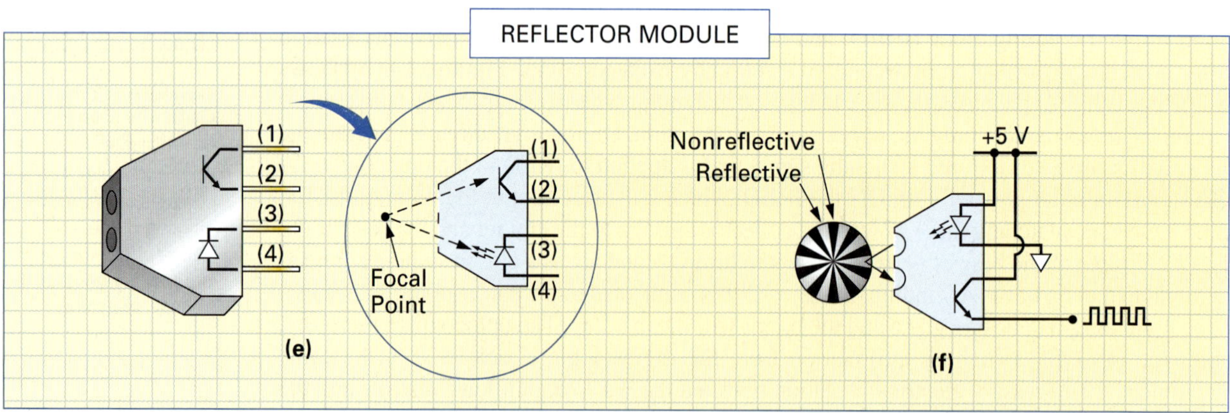

FIGURE 14-8 Optoelectronic Application—Optically Coupled Isolators.

the disc. As these opaque targets pass through the infrared beam, they will cause the phototransistor to momentarily turn OFF. On the other hand, a transparent section of the disc will not block the infrared light, and so the phototransistor will remain ON. As a result the phototransistor will turn ON and OFF, generating pulses, the number of which is an indication of the shaft's speed of rotation.

Figure 14-8(e) shows another type of optocoupler called the **optically coupled isolator reflector module.** Like the interrupter module, this device consists of a matched and aligned emitter and detector, and it is also used to detect targets. Figure 14-8(f) shows how the reflector module could also be used as an optical tachometer. The disadvantage with the interrupter module is that the disc has to be exactly aligned between the emitter or detector sections. The reflector module does not have to be positioned so close to the target disc, and therefore the alignment is not so crucial. The target disc used for reflector modules is different in that it is composed of reflective and nonreflective target areas. As the disc rotates, almost no light will be reflected by the dark areas when they are present at the focal point of the IRED and phototransistor. However, when a reflective target is present at the focal point, light is reflected directly back to the phototransistor, turning it ON and therefore generating an output pulse.

Optically Coupled Isolator Reflector Module
A device that consists of a matched and aligned emitter and detector and that is used to detect targets.

14-2-2 *Temperature Transducers*

A **thermistor** is a semiconductor device that acts as a temperature sensitive resistor. Figure 14-9(a) shows a few different types of thermistors. There are basically two different types: **positive temperature coefficient (PTC) thermistors** and the more frequently used **negative temperature coefficient (NTC) thermistors.** To explain the difference, Figure 14-9(b) shows how a temperature increase causes the resistance of an NTC thermistor to decrease. On the other hand, Figure 14-9(c) shows how a temperature increase causes the resistance of a PTC thermistor to increase.

Thermistors are used in a variety of applications. For example, an NTC could be used in a fire alarm circuit. When the ambient temperatures are LOW, the resistance of the NTC is HIGH, and therefore current cannot energize the alarm. If the ambient temperature increases to a HIGH level, however, the resistance of the thermistor drops LOW, and the alarm circuit is energized.

The PTC, on the other hand, could be used as a sort of circuit protection device (similar to a circuit breaker). With the PTC, the "switch temperature" (which is the temperature at which the resistance rapidly increases) can be varied by different construction techniques from below 0° C to above 160° C. The PTC therefore could be used as a current-limiting circuit protection device because currents lower than a limiting value will not generate enough heat to cause the PTC thermistor to switch to its high resistance state. On the other hand, circuit currents that go above the limiting value will generate enough heat to cause the PTC thermistor to switch to its high-resistance state. Current surges therefore are limited to a safe value until the surge is over.

Thermistor
A semiconductor device that acts as a temperature sensitive resistor.

Positive Temperature Coefficient (PTC) Thermistor
A thermistor in which a temperature increase causes the resistance to increase.

Negative Temperature Coefficient (NTC) Thermistor
A thermistor in which a temperature increase causes the resistance to decrease.

14-2-3 *Pressure Transducers*

A semiconductor pressure transducer will change its resistance in accordance with changes in pressure. Figure 14-9(d) shows a **piezoresistive diaphragm pressure sensor.** Piezoresistance of a semiconductor is described as a change in resistance due to a change in the applied pressure. This device has a dc excitation voltage applied, as seen in the inset in Figure 14-9(d), and will typically generate a 0 to 100 mV output voltage based on the pressure sensed. Figure 14-9(e) shows an application for these devices in which they are used in automobiles to sense the cooling-system pressure, hydraulic-transmission pressure, and fuel-injection pressure.

Piezoresistive Diaphragm Pressure Sensor
A sensor that changes its resistance in response to pressure.

14-2-4 *Magnetic Transducers*

The **hall effect sensor** was discovered by Edward Hall in 1879 and is used in computers, automobiles, sewing machines, aircraft, machine tools, and medical equipment. Figure 14-9(f) shows a few different types of hall effect sensors. To explain the hall effect principle, Figure 14-9(g) shows that when a magnetic field whose polarity is north is applied to the sensor, it causes a separation of charges, generating in this example a negative hall voltage output. On the other hand, Figure 14-9(h) shows that if the magnetic field polarity is reversed so that a south pole is applied

Hall Effect Sensor
A sensor that generates a voltage in response to a magnetic field.

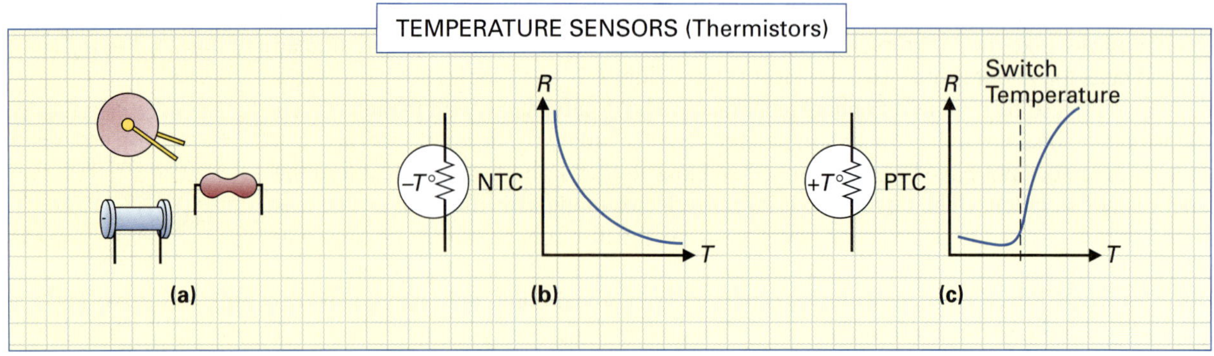

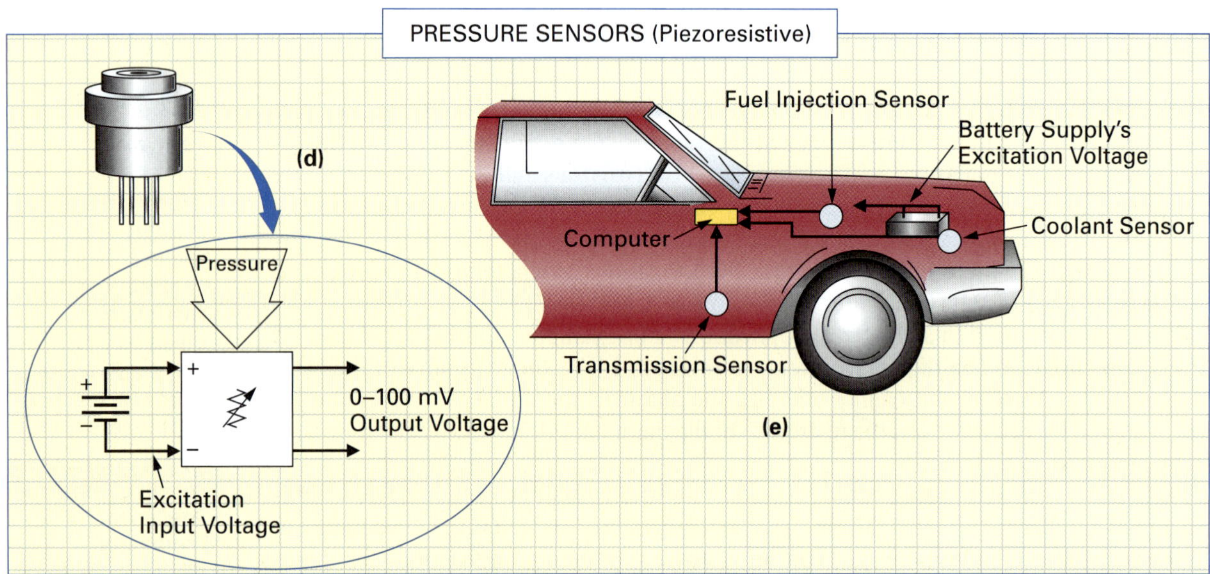

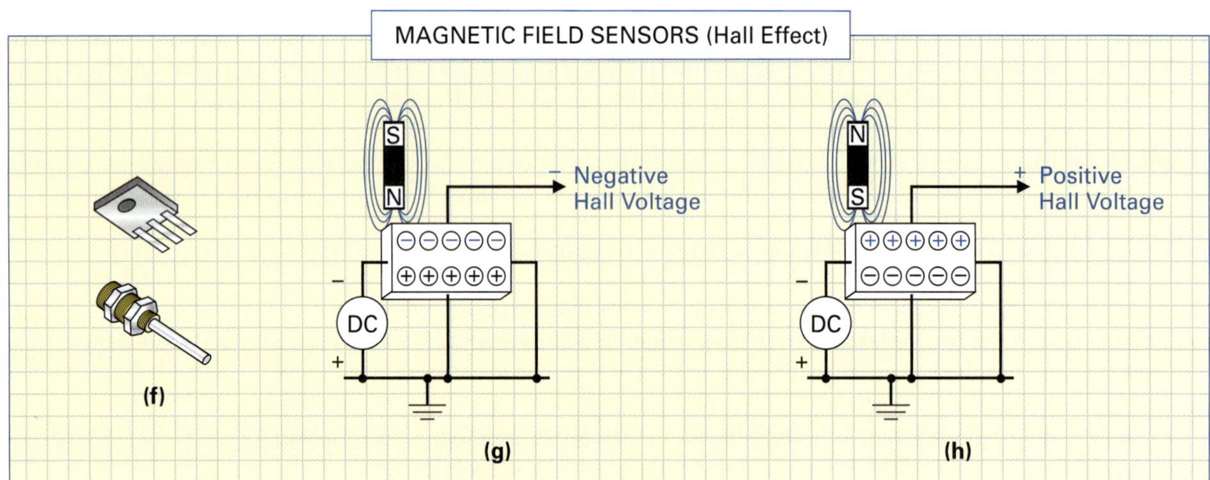

FIGURE 14-9 **Semiconductor Transducers.**

to the sensor, it will cause a polarity separation of charges within the sensor that will result in a positive hall voltage output. The amplitude of the generated positive or negative output voltage from the hall effect sensor is directly dependent on the strength of the magnetic field.

Their small size, light weight, and ruggedness make the hall effect sensors ideal in a variety of commercial and industrial applications. For example, hall effect sensors are embedded in the human heart to serve as timing elements. They are also used to sense shaft rotation, camera shutter positioning, rotary position, flow rate, and so on.

SELF-TEST EVALUATION POINT FOR SECTION 14-2

Use the following questions to test your understanding of Section 14-2.

1. The photoconductive cell basically operates as a light-sensitive __________.
2. The photovoltaic cell converts light energy into __________ energy, and is often referred to as a __________ cell.
3. What are the two modes of operation for the photodiode?
4. With the photoconductive cell, photodiode, phototransistor, photo-darlington, photothyristor, and other light-sensitive devices, a light increase always causes a __________ in conductance.
5. LEDs convert __________ energy into __________ energy, whereas an ILD will convert __________ energy into __________ energy.
6. LED displays emit light while LCD displays control light. (True/False)
7. An optically coupled isolator DIP module will electrically connect two points, while isolating information from the two points. (True/False)
8. Would a PTC thermistor or NTC thermistor be ideal as a series current limiter?

REVIEW QUESTIONS

Multiple-Choice Questions

1. Which of the following thyristors would normally be used to control dc power?
   a. TRIAC
   b. SCR
   c. UJT
   d. DIAC
   e. Both (b) and (c)
2. Which of the following thyristors is a bidirectional device?
   a. TRIAC
   b. SCR
   c. UJT
   d. ENIAC
3. __________ an SCR's gate current will __________ its forward breakover voltage.
   a. Increasing, decrease
   b. Increasing, increase
   c. Decreasing, increase
   d. Decreasing, decrease
4. Which of the following thyristors has a symmetrical switching characteristic?
   a. TRIAC
   b. SCR
   c. UJT
   d. DIAC
   e. Both (b) and (c)
5. Which of the following light sensitive devices will generate a voltage when light is applied?
   a. Photoconductive cell
   b. Photodiode
   c. Photovoltaic cell
   d. Phototransistor
   e. Both (b) and (c)
6. Which of the following is a monochromatic light source?
   a. LED
   b. Light bulb
   c. ILD
   d. LCD
   e. Both (b) and (c)
7. Which of the following devices would be best suited to detect whenever a coin has been inserted in a slot?
   a. Piezoresistive sensor
   b. Hall effect sensor
   c. PTC thermistor
   d. Interrupter optoisolator
   e. Both (a) and (c)
8. What advantage does a liquid crystal display have over an LED display?
   a. Consumes less power
   b. Is easier to read
   c. Is ideal in portable applications
   d. Both (a) and (c)
   e. Both (a) and (b)
9. What device could be used as a series connected current limiter?
   a. Hall effect sensor
   b. Piezoresistive sensor
   c. PTC thermistor
   d. NTC thermistor
10. Which of the following sensors could be used to control the switching of current through a motor's stator windings by sensing the position of the motor's permanent magnet rotor?
    a. Hall effect sensor
    b. Piezoresistive sensor
    c. PTC thermistor
    d. NTC thermistor

Practice Problems

11. Identify the schematic symbols shown in Figure 14-10.

To practice your circuit recognition and operation ability, refer to the circuit in Figure 14-11, and answer the following questions.

12. Describe the operation of the circuit when the input is normally HIGH.
13. Describe the operation of the circuit when the input is taken LOW.
14. How is the SCR turned ON, and how is the SCR turned OFF?

To practice your circuit recognition and operation ability, refer to the circuit in Figure 14-12, and answer the following questions.

15. Describe the basic operation of this circuit.
16. Why do you think a TRIAC is being used in this circuit instead of an SCR?

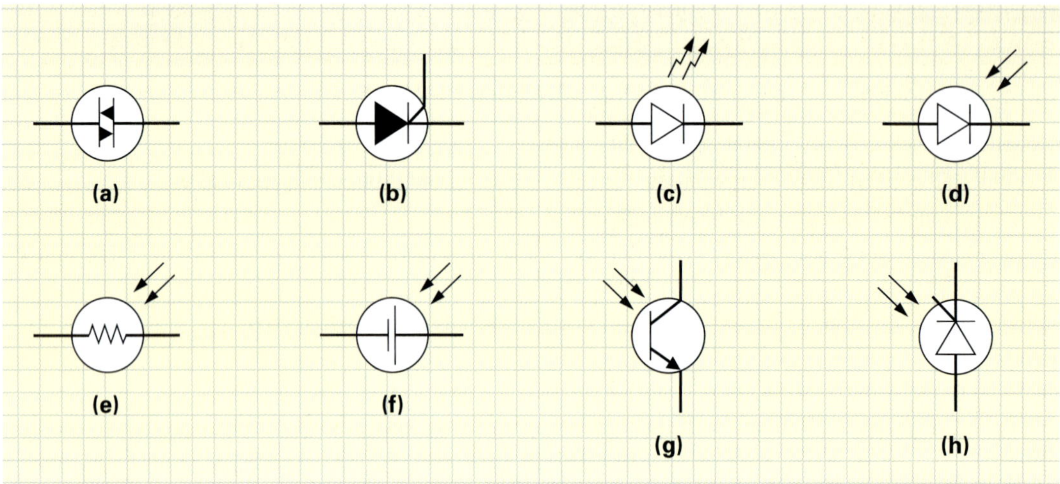

FIGURE 14-10 Schematic Symbols.

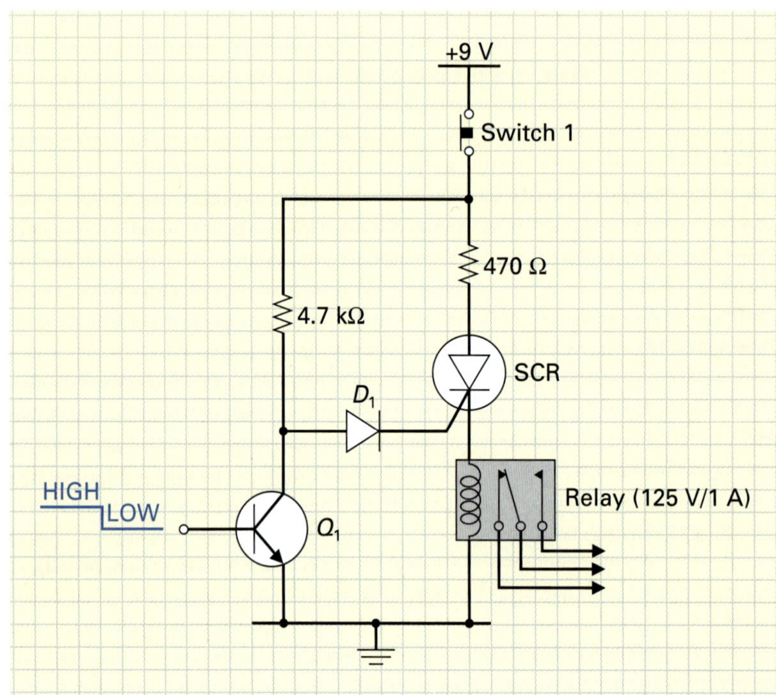

FIGURE 14-11 Low-Power to High-Power ON/OFF Switching.

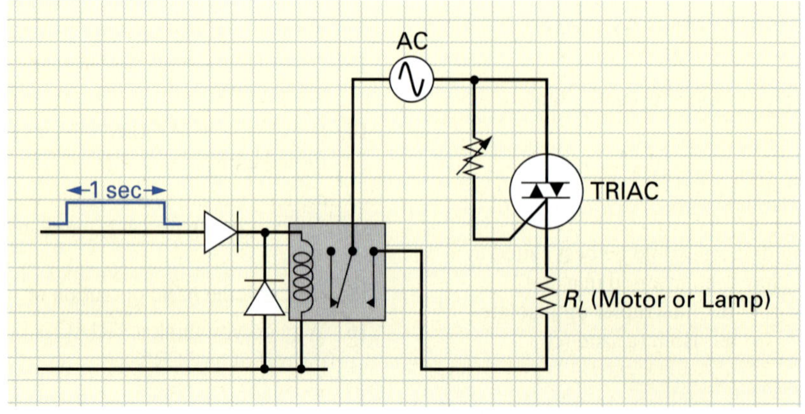

FIGURE 14-12 One-Shot Timer Control of TRIAC.

To practice your circuit recognition and operation ability, refer to the circuit in Figure 14-13(a) and (b), and answer the following questions.

17. Identify the light-emitting and light-sensitive devices used in these circuits.
18. Why was a different light-reactive device used for each of these circuits?
19. Why are these optically coupled isolators needed for computer output device control?

To practice your circuit recognition and operation ability, refer to the circuit in Figure 14-14, and answer the following questions.

20. Identify the light-sensitive device used in this circuit.
21. What is the function of this circuit?
22. Describe the operation of the circuit.

Web Site Questions

Go to the web site http://www.prenhall.com/cook, select the textbook *Electronics: A Complete Course,* select this chapter, and then follow the instructions when answering the multiple-choice practice problems.

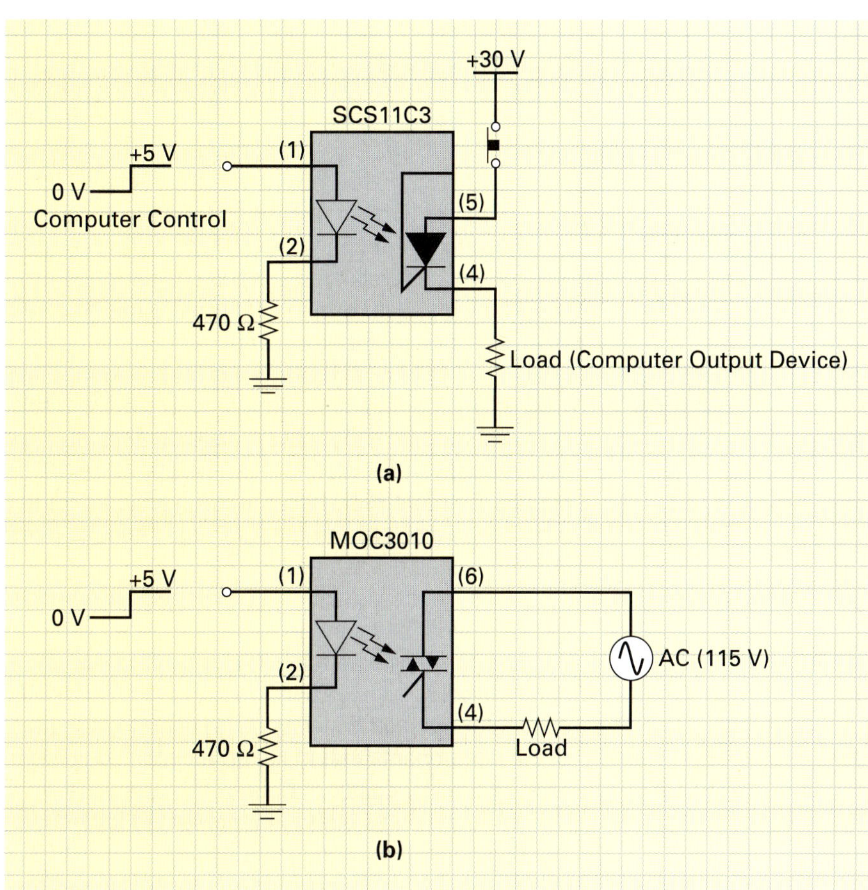

FIGURE 14-13 **Computer Output Device Control.**

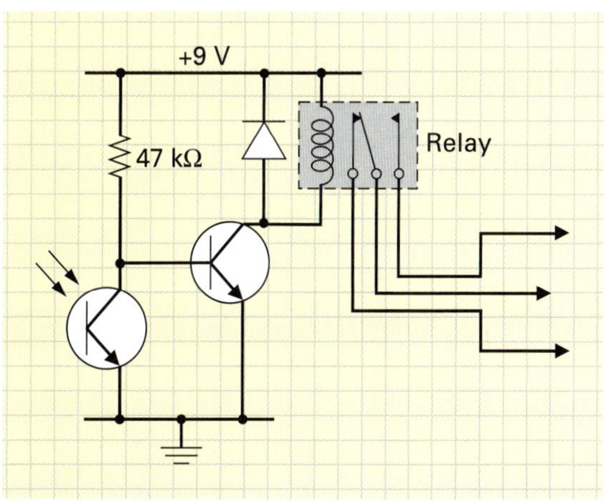

FIGURE 14-14 **Optical ON/OFF Relay Control Circuit.**

Op–Amps and Other Linear ICs

I See

Standing six feet six inches tall, Jack St. Clair Kilby was a quiet, introverted man from Kansas. Excited about the prospect of joining the very well-respected Massachusetts Institute of Technology (MIT) to further his education, he was thoroughly disappointed when he failed the mathematics entrance exam by three points. For the next ten years he worked for a manufacturer of radio and television parts, paying particularly close attention to the new component on the block, the transistor.

In May of 1958, Texas Instruments, a new, fast-growing company who developed the first commercially available silicon transistor just four years earlier, offered him a job in their development lab. A project was under way to print electronic components on ceramic wafers and then wire them and stack them together to make a circuit. The more Kilby became involved in the project, the more he realized how complicated and ridiculous the method was. The idea to miniaturize was a good one, but a different solution was needed.

Two months later in July, the company shut down for summer vacation, but Kilby was forced to work because he had not accrued vacation time. This proved to be a blessing in disguise for Kilby who found himself in the lab with a lot of time and resources available to him to develop his idea. His idea was to build resistors and capacitors from the same semiconductor material that was being used to manufacture transistors. This would mean that all of the components that make up a circuit could be manufactured simultaneously on a single slice of semiconductor material.

A few months later Kilby presented his prototype to a very skeptical boss. It contained five components all connected by tiny wires with the complete assembly held together by large blobs of wax. Kilby suddenly found himself the owner of a patent and the richly deserved acclaim that always goes along with being first. The first integrated circuit or IC was born on a thin wafer of germanium just two-fifths of an inch long. Texas Instruments demonstrated its miniaturization advantage by building a computer for the Air Force using their newly developed technique in a 587 IC. Its rewards were immediately apparent when a 78 cubic foot monster computer was replaced with a more powerful unit measuring only 6.5 cubic inches.

15

Introduction

In 1963, Fairchild Semiconductor introduced the µA702, which was the first commercially available IC operational amplifier, or op-amp. Realizing the op-amp's ability to adapt to a wide variety of applications, National Semiconductor followed suit and introduced the LM101 op-amp in 1965. In 1967, Fairchild once again made history by unveiling the very popular µA741 op-amp IC, which is still widely used today. From that point on, improvements in technology have refined the op-amp and led to a variety of other spin-off ICs that internally include several op-amps along with their other circuitry. These ICs include *voltage regulators, timers, function generators, phase-locked loops, analog-to-digital (A/D) converters, digital-to-analog (D/A) converters, sample-and-hold amplifiers,* and a variety of other specialized ICs. These ICs, along with the op-amp, are collectively called "linear integrated circuits (linear ICs)" because they provide linear signal amplification, in contrast to "digital integrated circuits (digital ICs)" that are primarily used for pulse signal processing.

As you continue your course in electronics, you will see how these ICs can be made use of in a variety of circuit applications. To begin with in this chapter, we will examine the most popular of the linear ICs, the op-amp, and see how it can be used as a comparator, inverting or noninverting amplifier, signal processor, signal generator, and active filter. Following this, we will continue our coverage of linear circuits by discussing the operation and application of four additional frequently used linear IC types: *Voltage regulators, timers, function generators* and *phase-locked loops.* Other linear ICs such as *digital-to-analog converters, analog-to-digital converters,* and *sample-and-hold amplifiers,* will be covered in the latter part of the digital section of this book.

15-1 OPERATIONAL AMPLIFIERS

The operational amplifier was initially a vacuum tube circuit used in the early 1940s in analog computers. The name "operational amplifier" or "op-amp" was chosen because the circuit was used as a high-gain dc "amplifier" performing mathematical "operations." These early circuits were expensive and bulky, and they found very little application until the semiconductor integrated circuit was developed in 1958 by Jack Kilby at Texas Instruments. Circuits that once needed hundreds of discrete or individual components can now be integrated into a single IC, making equipment smaller, more energy efficient, cheaper, and easier to design and troubleshoot.

Today's IC op-amp is a very high-gain dc amplifier that can have its operating characteristics changed by connecting different external components. This makes the op-amp very versatile, and it is this versatility that has made the op-amp the most widely used linear IC.

15-1-1 *Operational Amplifier Basics*

Operational Amplifier (Op-Amp)
Special type of high-gain amplifier.

To begin with, Figure 15-1 introduces the **operational amplifier,** or **op-amp,** by showing its schematic symbol in Figure 15-1(a) and internal circuit in Figure 15-1(b). It would be safe to say that you will not really be learning anything that has not already been covered because the op-amp's internal circuit is simply a combination of three previously covered amplifier circuits. These three circuits are all interconnected and contained within a single IC, and together they function as a "high-gain, high input impedance, low output impedance amplifier."

1. Op-Amp Symbol

Referring again to Figure 15-1(a), you can see that the triangle-shaped amplifier symbol is used to represent the op-amp in an electronic schematic diagram. Comparing the two symbols, you may have noticed that in some cases the two power supply connections are not shown, even though power is obviously applied.

Inverting Input
The inverting or negative input of an op-amp.

Noninverting Input
The noninverting or positive input of an op-amp.

Let us now examine the op-amp's input and output terminals shown in Figure 15-1. The two op-amp inputs are labeled "−" and "+." The "−" or negative input is called the **inverting input** because any signal applied to this input will be amplified and inverted between input and output (output is 180° out of phase with input). On the other hand, the "+" or positive input is called the **noninverting input** because any signal applied to this input will be amplified but not inverted between input and output (output is in phase with input). An input signal will normally be applied to only one of these inputs, while the other input is used to control the op-amp's operating characteristics.

The two power supply connections to the op-amp are labeled "$+V$" and "$-V$." Figure 15-1(c) shows how power to the op-amp can be supplied by dual supply voltages or by a single supply voltage. When two supply voltages are used (dual supply voltages), the voltage values are of the same value but of opposite polarity (for example, +12 V and −12 V). On the other hand, when only one supply voltage is used (single supply voltage), a positive or negative voltage is applied to its respective terminal while the other terminal is grounded (for example, +5 V and ground or −5 V and ground). Having both a positive and negative power supply voltage will allow the output signal to swing positive and negative, above and below zero. As with all high gain amplifiers, however, the output voltage can never exceed the value of the $+V$ and $-V$ supply voltages.

2. Op-Amp Packages

The entire op-amp circuit is placed within one of two basic packages, shown in Figure 15-2(a) and (b). The TO-5 metal can package is available with 8, 10, or 12 leads, while the dual in-line through-hole and surface-mount packages typically have 8 or 14 pins.

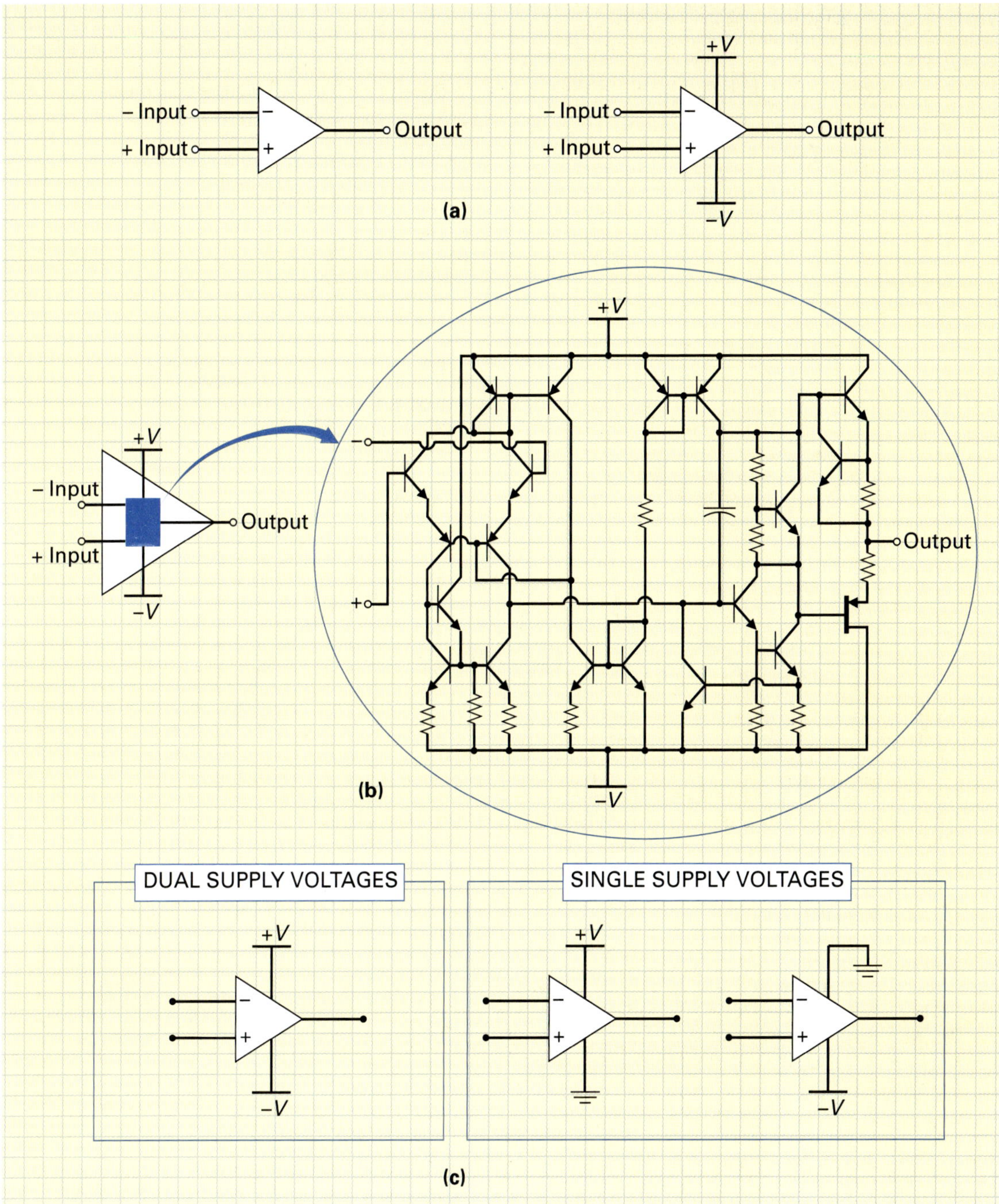

FIGURE 15-1 **The Operational Amplifier. (a) Schematic Symbols. (b) Internal Circuit. (c) Power Supply Connections.**

Like all ICs, an identification code is used to indicate the device manufacturer, device type, and key characteristics. Figure 15-2(c) lists some of the more common manufacturer prefix codes, operating temperature codes, and package codes. In this example, the "MC 741C N" code indicates that the 741 op-amp is made by Motorola, it is designed for commercial application in which the temperature range is between 0 to 70°C, and the package is a through-hole DIP with longer leads.

Referring back to the IC packages in Figure 15-2(a) and (b), you can see that in addition to the two inputs, single output, and two power supply terminals, there are two additional leads labeled offset. These two inputs will normally be connected to a potentiometer that can

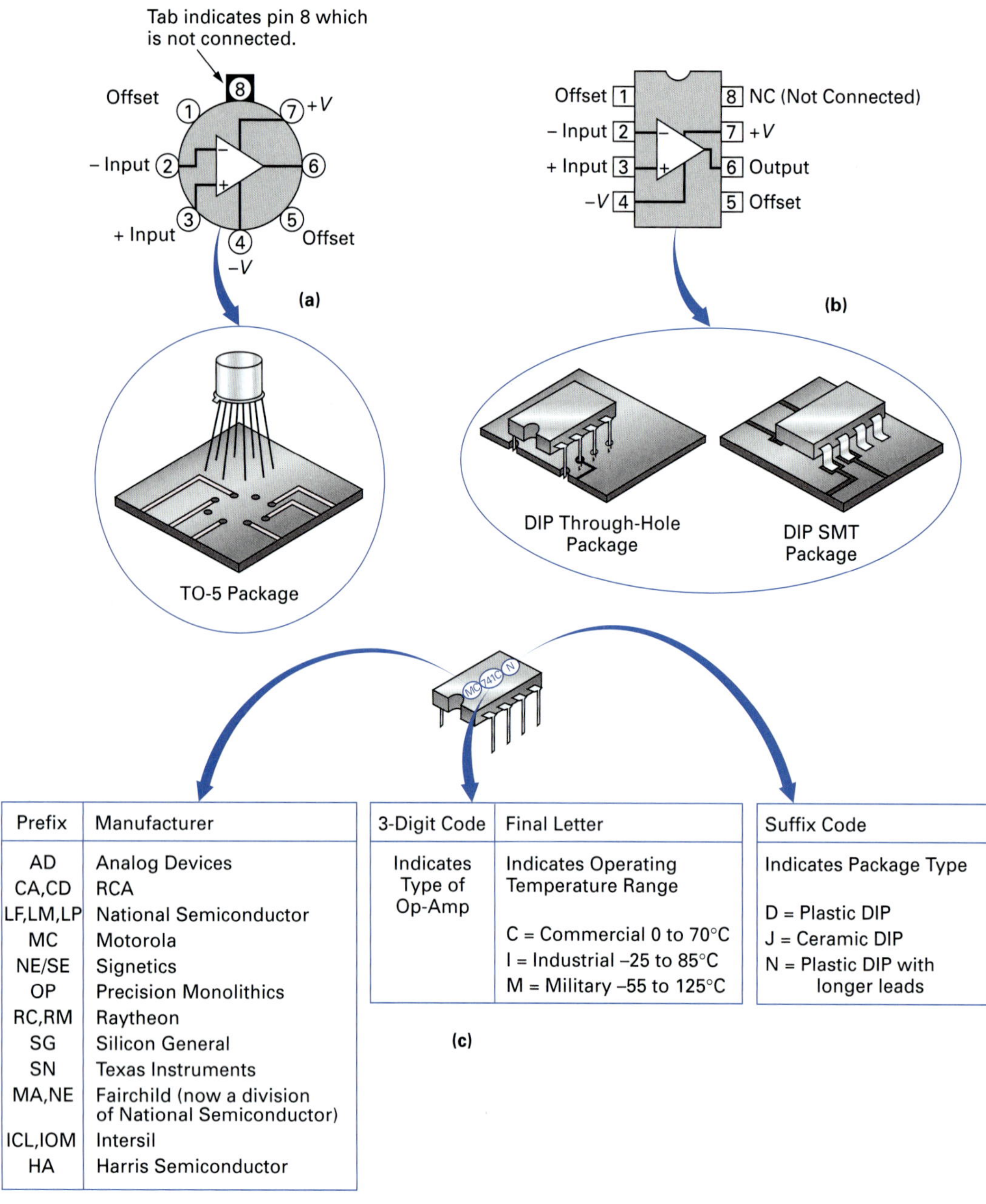

Prefix	Manufacturer
AD	Analog Devices
CA,CD	RCA
LF,LM,LP	National Semiconductor
MC	Motorola
NE/SE	Signetics
OP	Precision Monolithics
RC,RM	Raytheon
SG	Silicon General
SN	Texas Instruments
MA,NE	Fairchild (now a division of National Semiconductor)
ICL,IOM	Intersil
HA	Harris Semiconductor

3-Digit Code	Final Letter
Indicates Type of Op-Amp	Indicates Operating Temperature Range C = Commercial 0 to 70°C I = Industrial –25 to 85°C M = Military –55 to 125°C

Suffix Code
Indicates Package Type D = Plastic DIP J = Ceramic DIP N = Plastic DIP with longer leads

FIGURE 15-2 Op-Amp Package Types and Identification Codes.

Balancing
Setting the output of an op-amp to zero volts when both inverting and noninverting inputs are at zero volts.

be adjusted to set the output at zero volts when both inverting and noninverting inputs are at zero volts. **Balancing** the op-amp in this way is generally needed due to imbalances within the op-amp's internal circuit.

15-1-2 *Op-Amp Operation and Characteristics*

As mentioned previously, the operational amplifier contains three amplifier circuits, and these three circuits are all interconnected and contained within a single IC. Referring to the block diagram of the op-amp in Figure 15-3(a) you can see that these three circuits are *a differential amplifier, a voltage amplifier,* and *an output amplifier.* Combined, these three circuits give the op-amp its key characteristics, which are *high gain, high input impedance,* and *low output impedance.* We will briefly review their characteristics because combined they determine the characteristics of the op-amp.

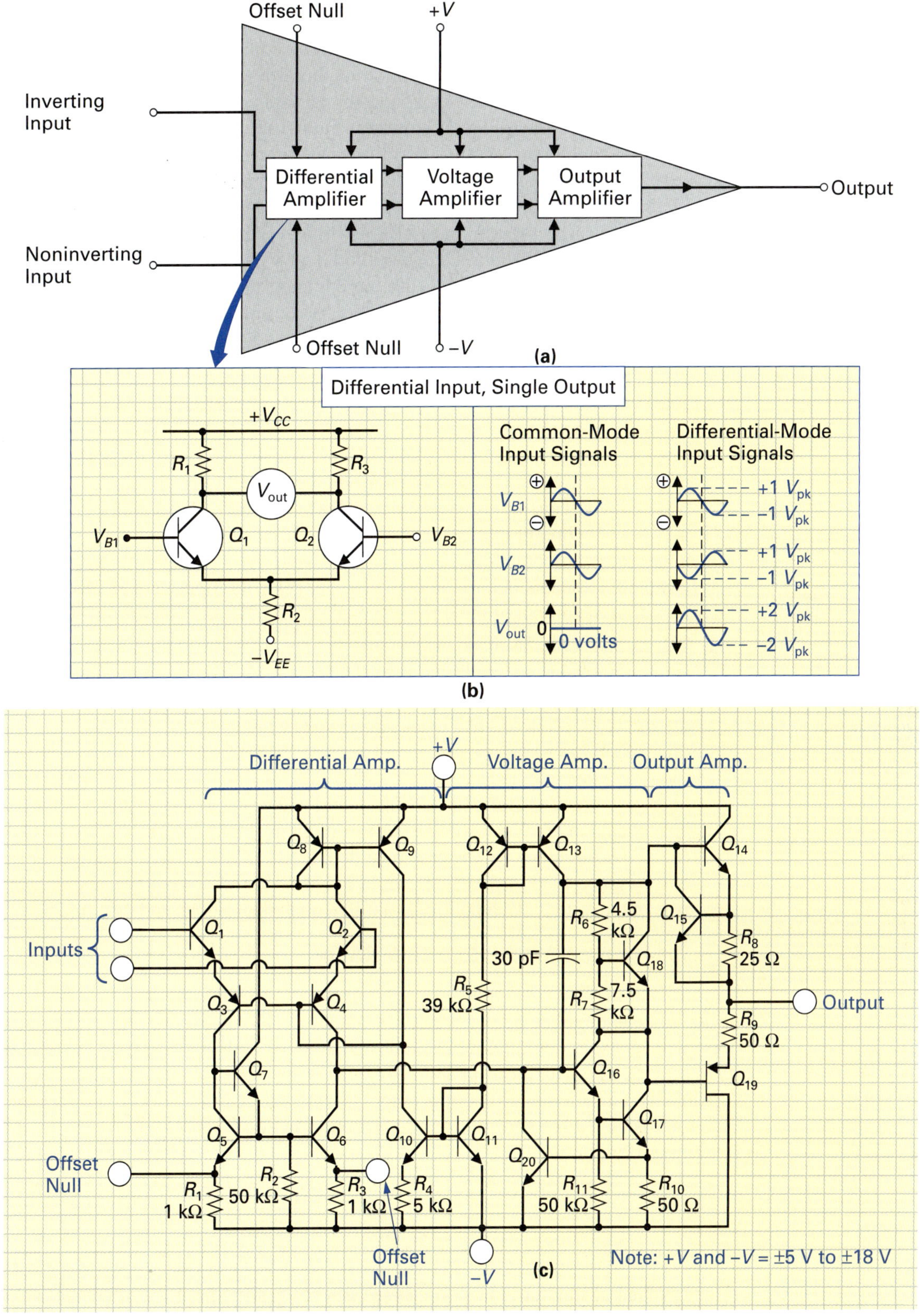

FIGURE 15-3 **The Operational Amplifier's Internal Circuit. (a) Block Diagram. (b) The Differential Amplifier's Operation. (c) Circuit Diagram.**

1. The Differential Amplifier within the Op-Amp

The differential amplifier within the op-amp is connected to operate in its "differential-input, single-output mode." The operation of the differential amplifier in this mode is reviewed in Figure 15-3(b). When both input signals are equal in amplitude and in phase with one another, they are referred to as **common-mode input signals,** as seen in the waveforms in Figure 15-3(b). On the other hand, if the input signals are out of phase with one another, they are referred to as **differential-mode input signals,** as seen in the other set of waveforms in Figure 15-3(b). The differential amplifier will amplify differential input signals while rejecting common-mode input signals. The questions you may be asking at this stage are what are common-mode input signals and why should we want to reject them? The answers are as follows: temperature changes and noise are common-mode input signals, and they are unwanted signals. Let us examine these common-mode signals in more detail.

Common-Mode Input Signals
Input signals to an op-amp that are in phase with one another.

Differential-Mode Input Signals
Input signals to an op-amp that are out of phase with one another.

1. Temperature variations within electronic equipment affect the operation of semiconductor materials and, therefore, the operation of semiconductor devices. These temperature variations can cause the dc output voltage of the first stage to drift away from its normal Q point. The second-stage amplifier will amplify this voltage change in the same way as it would amplify any dc input signal and so will all of the following amplifier stages. The increase or decrease in the normal Q-point bias for all of the amplifier stages will get progressively worse due to this thermal instability, and the final stage may have a Q point that is so far off of its mid position that an input signal may drive it into saturation or cutoff, causing signal distortion. With the differential amplifier, any change that occurs due to temperature changes will affect both stages and so will not appear at the output of the differential amplifier, due to its **common-mode rejection.**
2. The second common-mode input signal that the differential amplifier removes is noise. It is often necessary to amplify low-level signals from low-sensitivity sources such as microphones, light detectors, and other transducers. High-gain amplifiers are used to increase the amplitude of these small input signals up to a more usable level that is large enough to drive or control a load, such as a loudspeaker. The 60 Hz ac power line, or any other electrical variation, can induce a noise signal along with the input signal at the input of this high-gain amplifier. Since these noise signals will be induced at all points in the circuit, and be identical in amplitude and phase, the differential amplifier will block these unwanted signals because they will be present at both inputs of the differential amplifier (noise will be a common-mode input). A true input signal, on the other hand, will appear at the two inputs of the differential amplifier as a differential input signal, and therefore be amplified.

Common-Mode Rejection
The ability of a device to reject a voltage signal applied simultaneously to both input terminals.

2. Common-Mode Rejection Ratio

To summarize, a differential input will be amplified by the op-amp's differential amplifier and passed to the output, while unwanted signals caused by temperature variations or noise will appear as common-mode input signals and therefore be rejected. An op-amp's ability to provide a high **differential gain (A_{VD})** and a low **common-mode gain (A_{CM})** is directly dependent on its internal differential amplifier and is a measure of an op-amp's performance. This ratio is called the **common-mode rejection ratio (CMRR)** and is calculated with the following formula:

Differential Gain (A_{VD})
The amplification of differential-mode input.

Common-Mode Gain (A_{CM})
The amplification of common-mode input.

$$\text{CMRR} = \frac{A_{VD}}{A_{CM}}$$

Looking at this formula, you can see that the higher the A_{VD} (differential gain), or the smaller the A_{CM} (common-mode gain), the higher the CMRR value, and therefore the better the operational amplifier. This ratio can also be expressed in dBs by using the following formula:

Common-Mode Rejection Ratio (CMRR)
The ratio of an operational amplifier's differential gain to the common-mode gain.

$$\text{CMRR} = 20 \times \log \frac{A_{VD}}{A_{CM}}$$

EXAMPLE:

If an op-amp's differential amplifier has a differential gain of 5000 and a common-mode gain of 0.5, what is the operational amplifier's CMRR? Express the answer in standard gain and dBs.

Solution:

$$\text{CMRR} = \frac{A_{VD}}{A_{CM}} = \frac{5000}{0.5} = 10{,}000$$

$$\text{CMRR} = 20 \times \log \frac{A_{VD}}{A_{CM}} = 20 \times \log 10{,}000 = 20 \times 4 = 80 \text{ dB}$$

A CMRR of 10,000 or 80 dB means that the op-amp's desired input signals will be amplified 10,000 times more than the unwanted common-mode input signals.

3. The Op-Amp Block Diagram

Now that we have reviewed the differential amplifier's circuit characteristics, let us return to the op-amp block diagram in Figure 15-3(a). It is the op-amp's differential-amplifier stage that provides the good common-mode rejection and high differential gain. Because the op-amp's "−" and "+" inputs are applied to either base of the diff-amp, we know that input current will be very small. It is this circuit characteristic that provides the op-amp with another key feature, which is a high input impedance ($I\downarrow$, $Z\uparrow$). The voltage-amplifier stage following the diff-amp usually consists of several darlington-pair stages that provide an overall op-amp voltage gain of typically 50,000 to 200,000. The final-output stage consists of a complementary emitter-follower stage to provide a low output impedance and high current gain, so that the op-amp can deliver up to several milliamps, depending on the value of the load.

4. The Op-Amp Circuit Diagram

The complete internal circuit of a typical op-amp can be seen in Figure 15-3(c). With integrated circuits, it is better to have transistors function as resistors wherever possible because they occupy less chip space than actual resistors. This accounts for why the circuit seems to contain many transistors that have their base and collector leads connected. You may also have noticed that no coupling capacitors have been used so that the op-amp can amplify both ac and dc input signals. As with most schematics, the inputs are shown on the left, output on the right, and power is above and below. As discussed previously, the two balancing, or **offset null, inputs** will normally be connected to an external potentiometer that can be adjusted to set the output at zero volts when both the inverting and noninverting inputs are at zero volts. Balancing the op-amp to find the zero-volt output point, or null, in this way is generally needed due to slight imbalances within the op-amp's internal circuit.

Offset Null Inputs
The two balancing inputs used to balance an op-amp.

An important point to realize at this time is that the op-amp is a single component, and up until this time we have concentrated on an understanding of the op-amp's internal circuitry because this helps us to better understand the circuit's normal input/output relationships and characteristics. These operational characteristics are important if we are going to be able to isolate whether a circuit malfunction is internal or external to the op-amp. However, because it is impossible to repair any internal op-amp failures, we will not concentrate on every detail of the op-amp's internal circuit.

5. Basic Op-Amp Circuit Applications

Now that we have an understanding of the op-amp's characteristics, let us put it to use in some basic circuit applications. To begin with, we will examine the comparator circuit.

The Open-Loop Comparator Circuit. Figure 15-4 shows how the op-amp can be used to function as a **comparator,** which is a circuit that is used to detect changes in voltage level. In Figure 15-4(a), the inverting input (−) of the op-amp is grounded and the input signal is applied to the op-amp's noninverting input (+). Referring to the associated waveforms,

Comparator
An op-amp used without feedback to detect changes in voltage level.

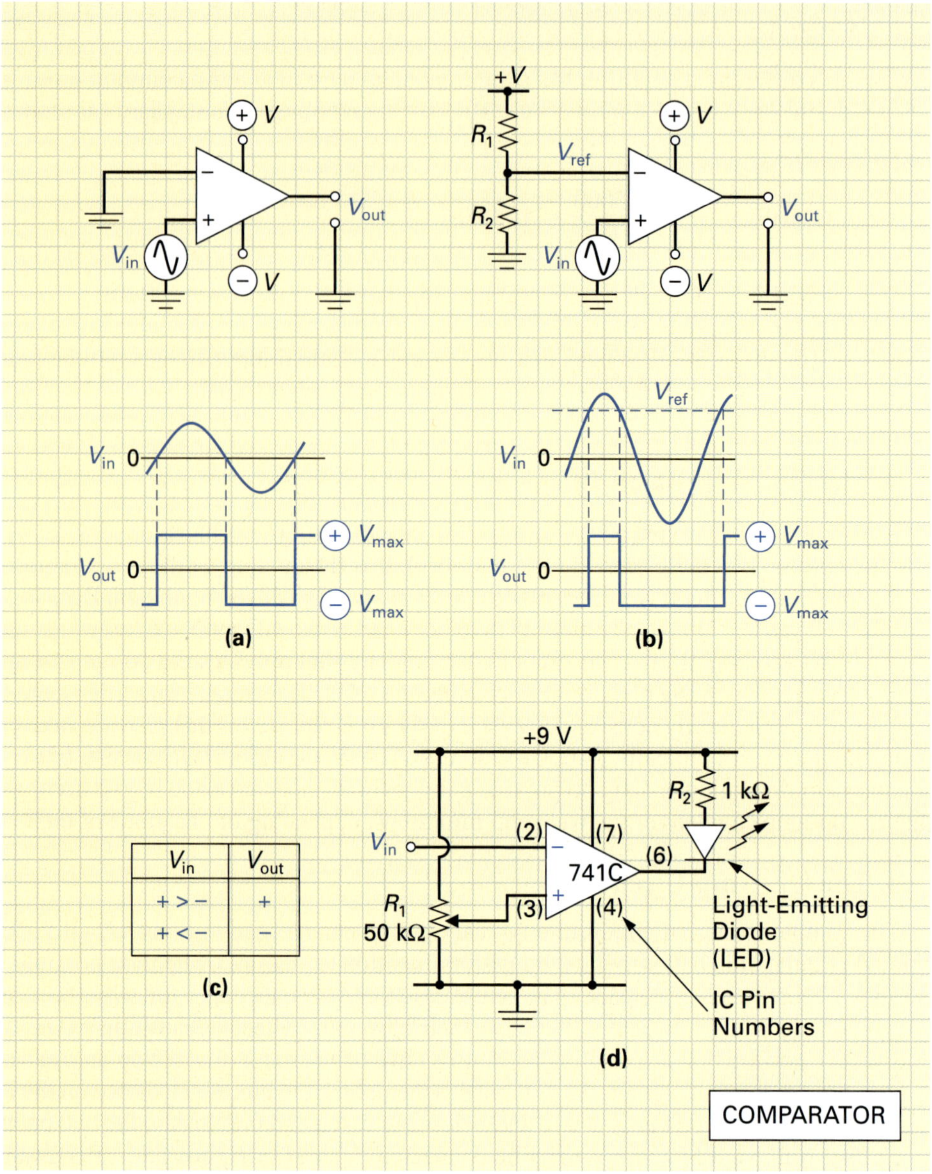

FIGURE 15-4 The Open-Loop Comparator Circuit.

you can see that when the input swings positive relative to the negative input (which is 0), the output of the amplifier goes into immediate saturation due to the very large gain of the op-amp. For example, if the op-amp had a voltage gain of 25,000 ($A_V = 25{,}000$), even a small input of +25 mV would cause the op-amp to try and drive its output to 625 V ($V_{out} = V_{in} \times A_V = 25 \text{ mV} \times 25{,}000 = 625 \text{ V}$). Since the maximum possible positive output voltage cannot exceed the positive supply voltage ($+V$), the output goes to its maximum positive limit, which is equal to the $+V$ supply voltage. When the input swings negative, the amplifier is driven immediately into its opposite state (cutoff), and the output goes to its maximum negative limit, which is equal to the $-V$ supply voltage.

Figure 15-4(b) shows how a voltage divider made up of R_1 and R_2 can be used to supply the inverting input (−) of the op-amp with a reference voltage (V_{ref}) that can be determined by using the voltage-divider formula.

$$V_{ref} = \frac{R_2}{R_1 + R_2} \times (+V)$$

Referring to the associated waveforms in Figure 15-4(b), you can see that whenever the ac input signal is more positive than the reference voltage, the output is positive. On the other hand, whenever the ac input signal is less than the reference voltage, the output is negative.

The table in Figure 15-4(c) summarizes the operation of the comparator circuit. In the first line of the table you can see that when the negative input is negative relative to the positive input, the output will go to its positive limit ($V_{out} = +V$). In the second line of the table you can see that when the opposite occurs (the negative input is positive, or the positive input is negative), the output will go to its negative limit ($V_{out} = -V$).

EXAMPLE:

Briefly describe the operation of the 741C op-amp comparator circuit shown in Figure 15-4(d).

Solution:

Potentiometer R_1 sets up the reference voltage (V_{ref}), which can be anywhere between 0 V and +9 V. When the input voltage (V_{in}) to the negative input of the op-amp is greater than the positive reference voltage, the op-amp's output will be equal to the $-V$ voltage supply (which is ground), and so the LED will turn ON.

The closed-loop inverting amplifier circuit. The op-amp is usually operated in either the **open-loop mode** or **closed-loop mode.** With the previously discussed comparator circuit, the op-amp was operating in its open-loop mode because there was no signal feedback from output to input. In most instances, the op-amp is operated in the closed-loop mode, in which there is signal feedback from output back to input. This feedback signal will always be out of phase with the input signal and therefore oppose the original signal, which is why it is called "degenerative or negative feedback." Negative feedback, however, is necessary in nearly all op-amp circuits for the following reasons:

Open-Loop Mode
A control system that has no means of comparing the output with the input for control purposes.

Closed-Loop Mode
A control system containing one or more feedback control loops in which functions of the controlled signals are combined with functions of the commands that tend to maintain prescribed relationships between the commands and the controlled signals.

1. Because the op-amp has such an extremely high gain, even a very small input signal will be amplified to a very large signal, which will drive the op-amp out of its linear region and into saturation and cutoff. Negative feedback will lower the op-amp's gain, and therefore control the op-amp to prevent output waveform distortion.
2. Having such a high gain can cause the amplifier to go into oscillation due to positive feedback. Negative feedback prevents an amplifier from going into oscillation by reducing the op-amp's gain.
3. The open-loop gain of an op-amp can have a very large range of value for the same device. For example, the 741's open-loop gain can be anywhere from a minimum of 25,000 to 200,000. Including negative feedback in the op-amp circuit will reduce the gain to a consistent value so that the same part can be relied on to provide the same response.

Figure 15-5(a) shows how an op-amp can be connected as an **inverting amplifier circuit,** which produces an amplified output signal that is 180° out of phase with the input signal. Looking at the output voltage label ($-V_{out}$), notice that the negative symbol preceding V_{out} is being used to indicate the 180° phase inversion between input and output. In this circuit arrangement, the input signal (V_{in}) is applied through an input resistor (R_{in}) to the inverting input (−) of the op-amp, while the noninverting input (+) is connected to ground. A feed-back loop is connected from the output back to the inverting input via the feed-back resistor R_F.

Inverting Amplifier Circuit
An op-amp circuit that produces an amplified output signal that is 180° out of phase with the input signal.

Let us now take a closer look at the closed-loop feedback system that occurs within this amplifier circuit. If the applied input voltage was zero volts ($V_{in} = 0$ V), the differential input signal (which is the difference between the op-amp's "+" and "−" inputs) will be 0 V, because both the inverting and noninverting inputs will now be at 0 V. A differential input of zero volts therefore will generate an output of zero volts. If the input signal were now to swing positive toward +5 V, the output (V_{out}) would swing negative due to the internal op-amp circuit phase inversions. This negative output voltage swing would be applied back to the inverting input via R_F to counteract the original positive input change. The feedback path

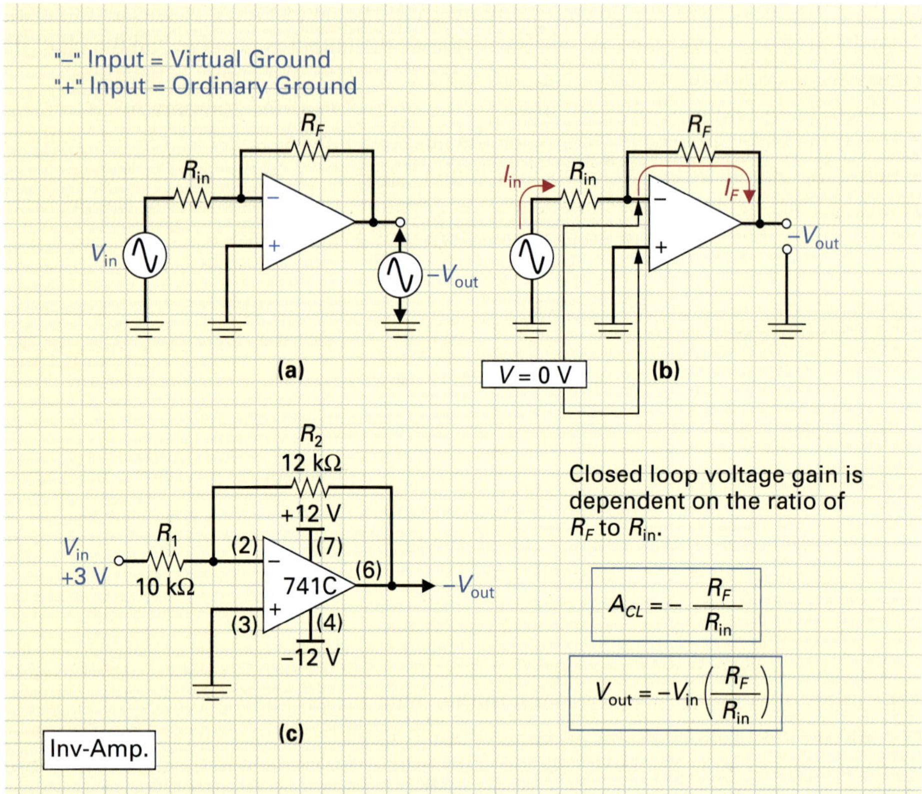

FIGURE 15-5 **The Closed-Loop Inverting Amplifier Circuit.**

is designed so that it cannot completely cancel the input signal, for if it did, there would be no input, and therefore no output or feedback. In most instances, the feedback voltage (V_F) will greatly restrain the input voltage change to the point that a +5 V input change at V_{in} will only be felt as a +5 microvolt change at the op-amp's inverting input. Therefore, even though V_{in} seems to change in values measured in volts, the inverting input of the op-amp will only change in values measured in microvolts. In fact, if the voltage at the inverting input of the op-amp is measured with a voltmeter as shown in Figure 15-5(b), the "−" input appears to remain at 0 V due to the very minute change at the "−" input. The inverting input of the op-amp in this case would be defined as a **virtual ground,** which is different from an **ordinary ground.** A virtual ground is a voltage ground because this point is at zero volts; however, it is not a current ground because it cannot sink or conduct away any current. An ordinary ground, on the other hand, is at zero volts and can sink any amount of current.

Virtual Ground
A ground for voltage but not for current.

Ordinary Ground
A connection in the circuit that is said to be at ground potential or zero volts. Because of its connection to earth it has the ability to conduct electrical current to and from earth.

Returning to the inverting amplifier circuit in Figure 15-5(b), we can now analyze this circuit in a little more detail now that we know its basic operation. To begin with, if the "−" input of the op-amp is at 0 V, then all of the input voltage (V_{in}) will be dropped across the input resistor (R_{in}). Therefore, the input current can be calculated if we know the value of V_{in} and R_{in}, with the following formula:

$$I_{in} = \frac{V_{in}}{R_{in}}$$

Knowing that the left side of R_F is at 0 V means that all of the output voltage will be developed across R_F, and therefore the value of feedback current (I_F) can be calculated with the formula:

$$I_F = \frac{-V_{out}}{R_F}$$

The extremely high input impedance of the op-amp means that only a very small fraction of the input current will enter the inverting input. In fact, nearly all of the current flowing through R_{in} and reaching the "−" op-amp input will leave this virtual ground point via the easiest path, which is through R_F. Therefore, it can be said that the feedback current is equal to the input current, or

$$I_{in} = I_F$$

If I_{in} (which equals V_{in}/R_{in}) and I_F (which equals $-V_{out}/R_F$) are equal, then

$$\frac{V_{in}}{R_{in}} = \frac{-V_{out}}{R_F}$$

If this is rearranged, we arrive at the following:

$$\frac{V_{out}}{V_{in}} = -\frac{R_F}{R_{in}}$$

Because the voltage gain of an amplifier is equal to

$$A_V = \frac{V_{out}}{V_{in}}$$

and because $V_{out}/V_{in} = -R_F/R_{in}$, the **closed-loop voltage gain (A_{CL})** of the inverting operational amplifier is equal to the ratio of R_F to R_{in}.

Closed-Loop Voltage Gain (A_{CL})
The voltage gain of an amplifier when it is operated in the closed-loop mode.

$$A_{CL} = -\frac{R_F}{R_{in}}$$

(Negative symbol preceding R_F/R_{in} indicates signal inversion)

By rearranging the previous equation $V_{out}/V_{in} = -R_F/R_{in}$, we can arrive at the following formula for calculating the output voltage of this circuit.

Since

$$\frac{V_{out}}{V_{in}} = -\frac{R_F}{R_{in}}$$

$$V_{out} = -V_{in}\left(\frac{R_F}{R_{in}}\right)$$

The input impedance of this inverting op-amp is equal to the value of R_{in} because the input voltage (V_{in}) is developed across the input resistor (R_{in}).

EXAMPLE:

Calculate the output voltage of the inverting amplifier shown in Figure 15-5(c).

Solution:

$$V_{out} = -V_{in}\left(\frac{R_F}{R_{in}}\right) = -3\text{ V} \times \frac{12\text{ k}\Omega}{10\text{ k}\Omega} = -3\text{ V} \times 1.2 = -3.6\text{ V}$$

The Closed-Loop Noninverting Amplifier Circuit. Figure 15-6(a) shows how an op-amp can be configured to operate as a **noninverting amplifier circuit.** The input voltage

Noninverting Amplifier Circuit
An op-amp circuit that produces an amplified output signal that is in phase with the input signal.

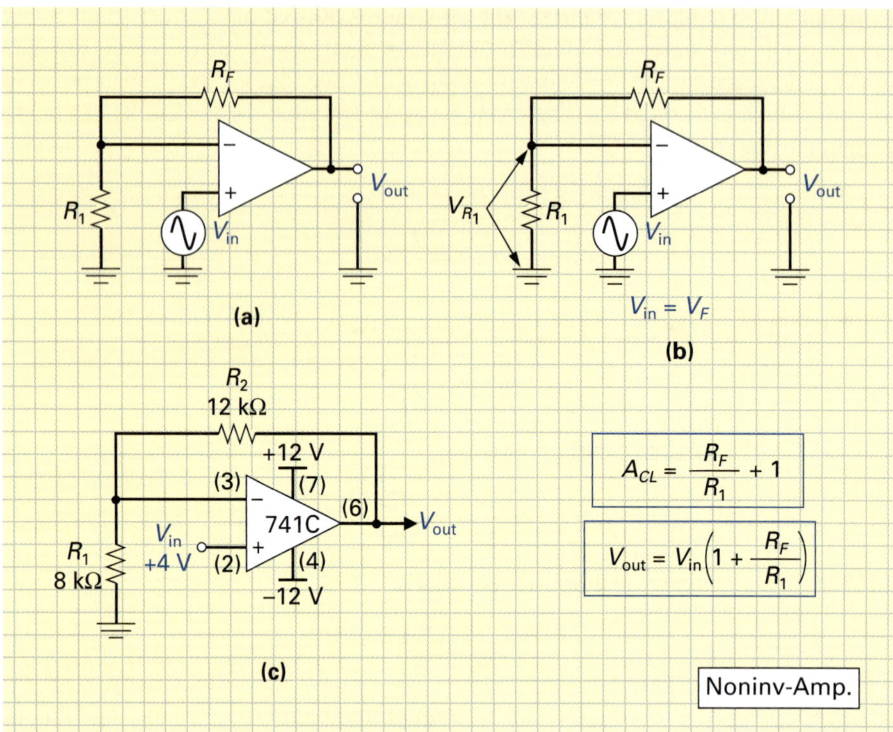

FIGURE 15-6 The Closed-Loop Noninverting Amplifier Circuit.

(V_{in}) is applied to the op-amp's noninverting input (+), and therefore the output voltage (V_{out}) will be in phase with the input. To achieve negative feedback, the output is applied back to the inverting input (−) of the op-amp via the feedback network formed by R_F and R_1.

Let us now take a closer look at the closed-loop negative feedback system that occurs within this amplifier circuit. The output voltage (V_{out}) is proportionally divided across R_F and R_1, with the feedback voltage (V_F) developed across R_1 being applied to the inverting input (−) of the op-amp, as shown in Figure 15-6(b). Because V_{out} is in phase with V_{in}, the feedback voltage (V_F) will also be in phase with V_{in}, and therefore these two in-phase inputs to the op-amp will be common-mode input signals. As a result, feedback will be degenerative. However, because V_{out} is slightly larger than V_{in}, there will be a small difference between V_{in} and V_F, and this differential input will be amplified. To summarize, the noninverting op-amp provides negative feedback by feeding back an in-phase common-mode signal, and this degenerative feedback will lower the op-amp's gain to

1. prevent output waveform distortion,
2. prevent the amplifier from going into oscillation, and
3. reduce the gain of the op-amp to a consistent value.

The very small microvolt difference between V_{in} and V_F will be amplified; however, since there is such a very small difference between these two input signals, it can be said that

$$V_{in} = V_F$$

Because the closed-loop voltage gain of any op-amp is equal to

$$A_{CL} = \frac{V_{out}}{V_{in}}$$

the gain of the noninverting amplifier could also be calculated with

$$A_{CL} = \frac{V_{out}}{V_F}$$

By using the voltage-divider formula, we can develop a formula for calculating V_F.

$$V_F = \frac{R_1}{R_1 + R_2} \times V_{out}$$

By rearranging the above formula as follows:

$$\frac{V_{out}}{V_F} = \frac{R_1 + R_F}{R_1} \quad \text{or} \quad \frac{V_{out}}{V_F} = \frac{R_1}{R_1} + \frac{R_F}{R_1} = \frac{R_F}{R_1} + 1$$

and because $V_{out}/V_F = A_{CL}$, the noninverting op-amp's gain can also be calculated with the formula

$$A_{CL} = \frac{R_F}{R_1} + 1$$

Because the output voltage of an amplifier is equal to the product of input voltage and gain ($V_{out} = V_{in} \times A_{CL}$), we can add V_{in} to the previous closed-loop gain formula in order to calculate output voltage.

$$V_{out} = V_{in}\left(1 + \frac{R_F}{R_1}\right)$$

With the inverting op-amp, the input impedance is determined by the input resistor (R_{in}). With the noninverting amplifier, there is no resistor connected because V_{in} is applied directly into the very high input impedance of the op-amp. As a result, the noninverting amplifier circuit has an extremely high input impedance.

EXAMPLE:

Calculate the output voltage from the noninverting amplifier shown in Figure 15-6(c).

Solution:

$$V_{out} = V_{in}\left(1 + \frac{R_F}{R_1}\right) = 4\text{ V}\left(1 + \frac{12\text{ k}\Omega}{8\text{ k}\Omega}\right) = 4\text{ V} \times (1 + 1.5) = 4\text{ V} \times 2.5 = +10\text{ V}$$

The open-loop comparator and closed-loop inverting and noninverting amplifier circuits are just three of many op-amp application circuits. In the final section of this chapter, we will examine many other typical op-amp circuit applications.

Figure 15-7 compares the characteristics and cost of several different op-amp types to the 741, which is very popular due to its good performance and low price. Referring to the cost-factor column, you can see, for example, that the LF351 is twice the price of the 741. Studying the key characteristics in this figure, you can see that the 741, for example, will typically have an input impedance of 2 MΩ, an output impedance of 75 Ω, an open-loop gain of 25,000, and a common-mode rejection ratio of 70 dB for any input signal from 0 Hz up to

OP-AMP TYPE	INPUT IMPEDANCE	OUTPUT IMPEDANCE	OPEN-LOOP GAIN (Min.)	CMRR (dB)	COST FACTOR	SLEW RATE (V/μs)	GAIN BANDWIDTH	FEATURES
741 C	2 MΩ	75Ω	25,000	70	1	0.5	1 MHz	Low cost
101	800 kΩ	Low	25,000	70	1	0.5	1 MHz	Low cost
108 A	70 MΩ	Low	80,000	96	—	0.3	1 MHz	Precision low drift
351	High	Low	25,000	70	2	13	4 MHz	Low bias current
318	High	Low	25,000	70	5	70	15 MHz	High slew rate
357	High	Low	50,000	80	4	30	20 MHz	High CMRR
363	High	Low	1,000,000	94	45	—	2 MHz	Low noise; high rejection
356	High	Low	25,000	80	3	10	5 MHz	Improved 741

FIGURE 15-7 **Comparing the Characteristics of Several Op-Amps.**

1 MHz. The gain bandwidth column lists only the upper frequency limit because the op-amp has no internal coupling capacitors, and therefore the lower frequency limit extends down to dc signals, or 0 Hz.

15-1-3 *Additional Op-Amp Circuit Applications*

The operational amplifier's flexibility and characteristics make it the ideal choice for a wide variety of circuit applications. In fact, because the op-amp is the most frequently used linear IC, it is safe to say that you will find several op-amps in almost every electronic system. Although it is impossible to cover all of these applications, many circuits predominate, and others are merely variations on the same basic theme. In this section we will concentrate on the operation and characteristics of all of the most frequently used op-amp circuit applications.

1. The Voltage-Follower Circuit

Voltage-Follower Circuit
An op-amp circuit that has a direct feedback to give unity gain so the output voltage follows the input voltage. Used in applications where a very high input impedance and very low output impedance are desired.

Figure 15-8 shows how an op-amp can be connected to form a noninverting **voltage-follower circuit.** Using the noninverting closed-loop gain formula discussed previously, we can calculate the voltage gain of this circuit.

$$A_{CL} = \frac{R_F}{R_1} + 1 = \frac{0\ \Omega}{0\ \Omega} + 1 = 0 + 1 = 1$$

With a gain of 1, the output voltage will be equal to the input voltage—so what is the advantage of this circuit? The answer is the op-amp characteristics of a high input impedance and a low output impedance. Similar to the BJT's emitter-follower and the FET's source-follower, the op-amp voltage-follower circuit derives its name from the fact that the output voltage follows the input voltage in both polarity and amplitude. This circuit is therefore ideal as a buffer, interfacing a high-impedance source to a low-impedance load.

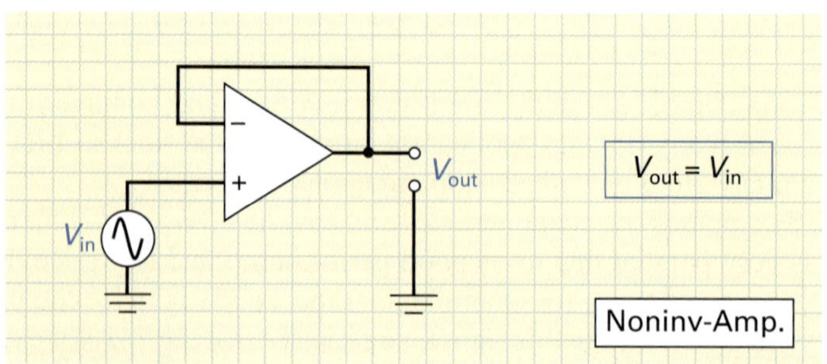

FIGURE 15-8 **Voltage-Follower Circuit.**

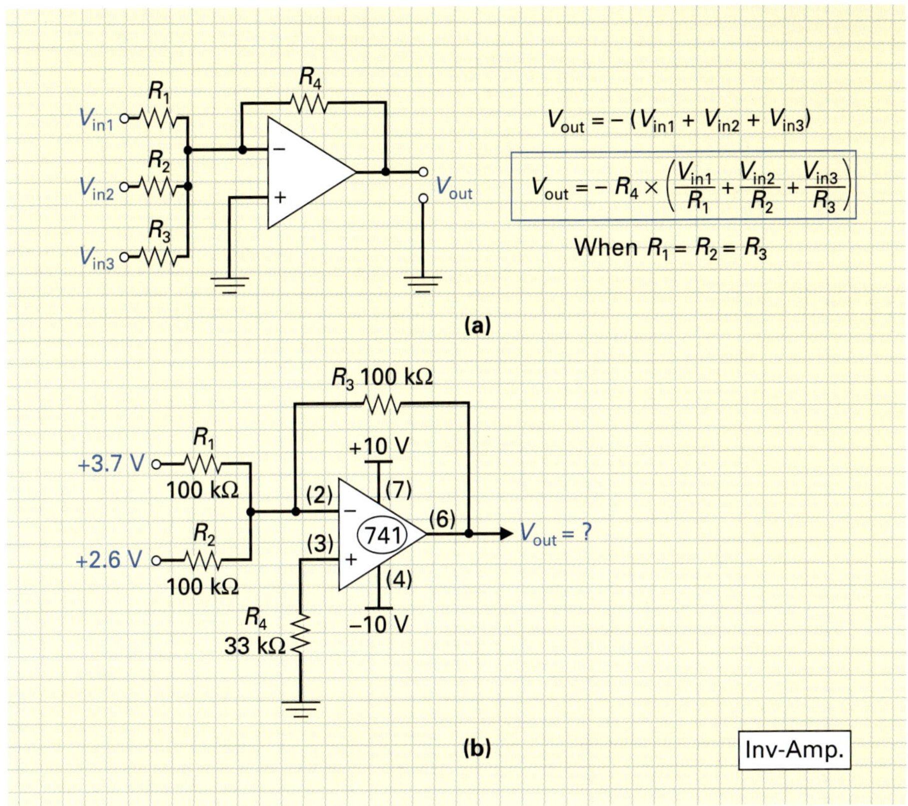

FIGURE 15-9 **Summing Amplifier Circuit.**

2. The Summing Amplifier Circuit

Summing Amplifier Circuit (or Adder Circuit)
An op-amp circuit that will sum or add all of the input voltages.

The **summing amplifier circuit,** or adder amplifier, consists of two or more input resistors connected to the inverting input of an op-amp as shown in Figure 15-9(a). This circuit will sum or add all of the input voltages, and therefore the output voltage will be

$$V_{out} = -(V_{in1} + V_{in2} + V_{in3})$$

Using Ohm's law ($V = R \times I$), we can also calculate the output voltage with the formula

$$V_{out} = -R_4 \times \left(\frac{V_{in1}}{R_1} + \frac{V_{in2}}{R_2} + \frac{V_{in3}}{R_3}\right)$$

Once again, the negative sign preceding the formula indicates that the output signal will be opposite in polarity to the two or three input signals.

EXAMPLE:

Calculate the output voltage of the summing amplifier circuit shown in Figure 15-9(b).

Solution:

$$V_{out} = -(V_{in1} + V_{in2}) = -(3.7\text{ V} + 2.6\text{ V}) = -6.3\text{ V}$$

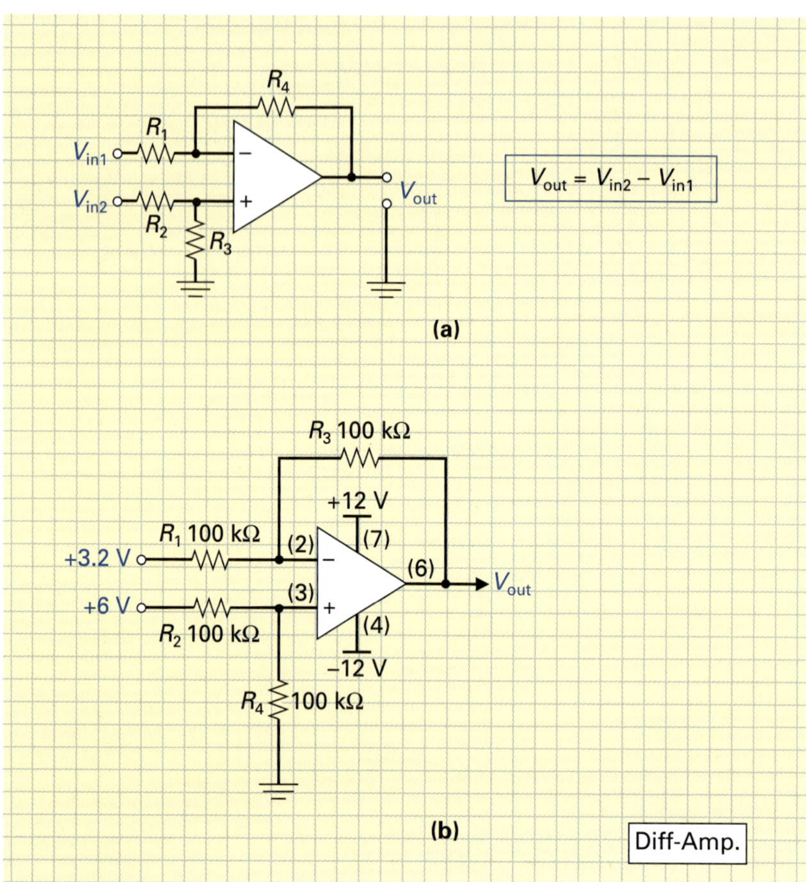

FIGURE 15-10 **The Difference Amplifier Circuit.**

Differential Amplifier Circuit

An op-amp circuit in which the output voltage is equal to the difference between the two input voltages.

3. The Difference Amplifier Circuit

Figure 15-10(a) shows how the op-amp can be connected to operate as a difference or **differential amplifier circuit.** In this circuit application, the op-amp will simply be making use of its first internal amplifier stage, which (as mentioned earlier) is a diff-amp. In this circuit, all four resistors are normally of the same value, and the output voltage is equal to the difference between the two input voltages.

$$V_{out} = V_{in2} - V_{in1}$$

EXAMPLE:

Calculate the output voltage of the difference amplifier circuit shown in Figure 15-10(b).

Solution:

$$V_{out} = V_{in2} - V_{in1} = 6\text{ V} - 3.2\text{ V} = 2.8\text{ V}$$

Differentiator Circuit

A circuit whose output voltage is proportional to the rate of change of the input voltage. The output waveform is the time derivative of the input waveform.

4. The Differentiator Circuit

The op-amp **differentiator circuit** (not to be confused with the previous differential circuit) is similar to the basic inverting amplifier except that R_{in} is replaced by a capacitor, as shown in Figure 15-11(a). Including a capacitor in any circuit means that we will develop

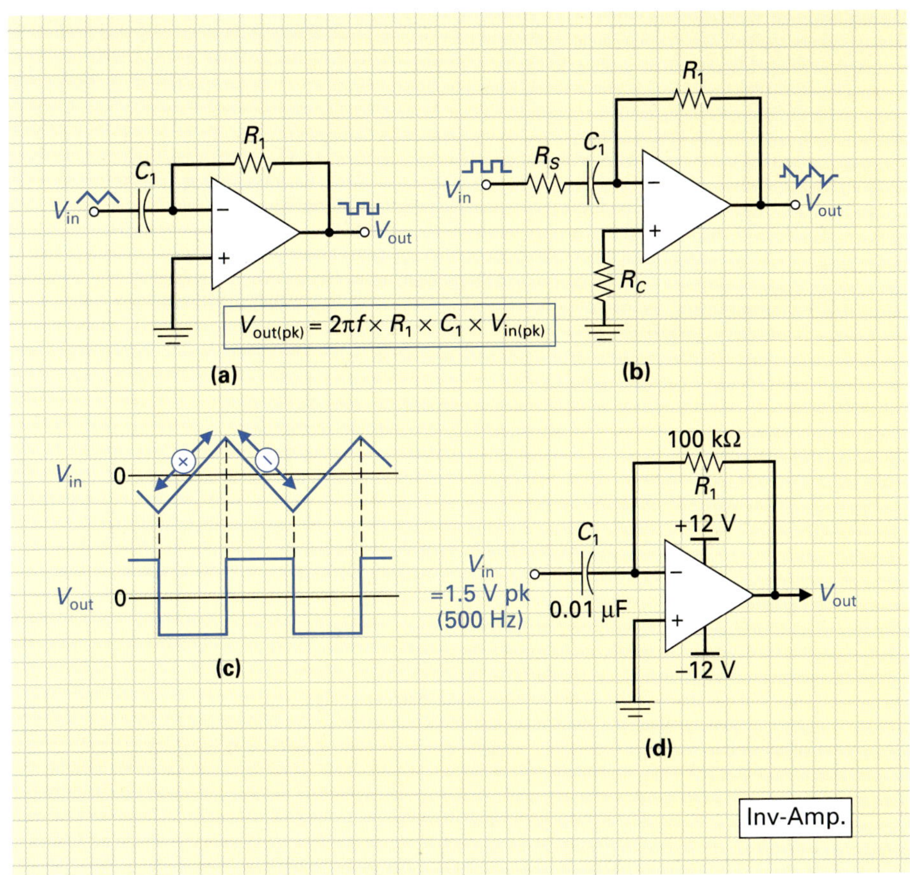

FIGURE 15-11 **Differentiator Circuit.**

problems as the frequency of the input signal increases because capacitive reactance is inversely proportional to frequency. This means that the reactance of the input capacitor will decrease for input signals that are higher in frequency, and therefore the input voltage applied to the op-amp and output voltage from the op-amp will increase with frequency. Including an additional resistor (R_S) in series with the input capacitor, as shown in Figure 15-11(b), will decrease the high-frequency gain because gain will now be a ratio of R_F/R_S.

Figure 15-11(c) shows the differentiator's input/output waveforms, with the peak of the output square wave being equal to

$$V_{\text{out(pk)}} = 2\pi f \times R_1 \times C_1 \times V_{\text{in(pk)}}$$

EXAMPLE:

Calculate the peak output voltage of the differentiator circuit shown in Figure 15-11(d).

Solution:

$$V_{\text{out(pk)}} = 2\,\pi f \times R_1 \times C_1 \times V_{\text{in(pk)}}$$
$$= (2 \times \pi \times 500\text{ Hz}) \times 100\text{ k}\Omega \times 0.01\ \mu\text{F} \times (1.5\text{ V pk}) = 4.7\text{ V}$$

5. The Integrator Circuit

Integrator Circuit
A circuit with an output which is the integral of its input with respect to time.

Figure 15-12(a) shows how the position of the resistor and capacitor in the differentiator circuit can be reversed to construct an op-amp **integrator circuit.** As in the differentiator circuit, the capacitor will alter the gain of the op-amp because its capacitive reactance

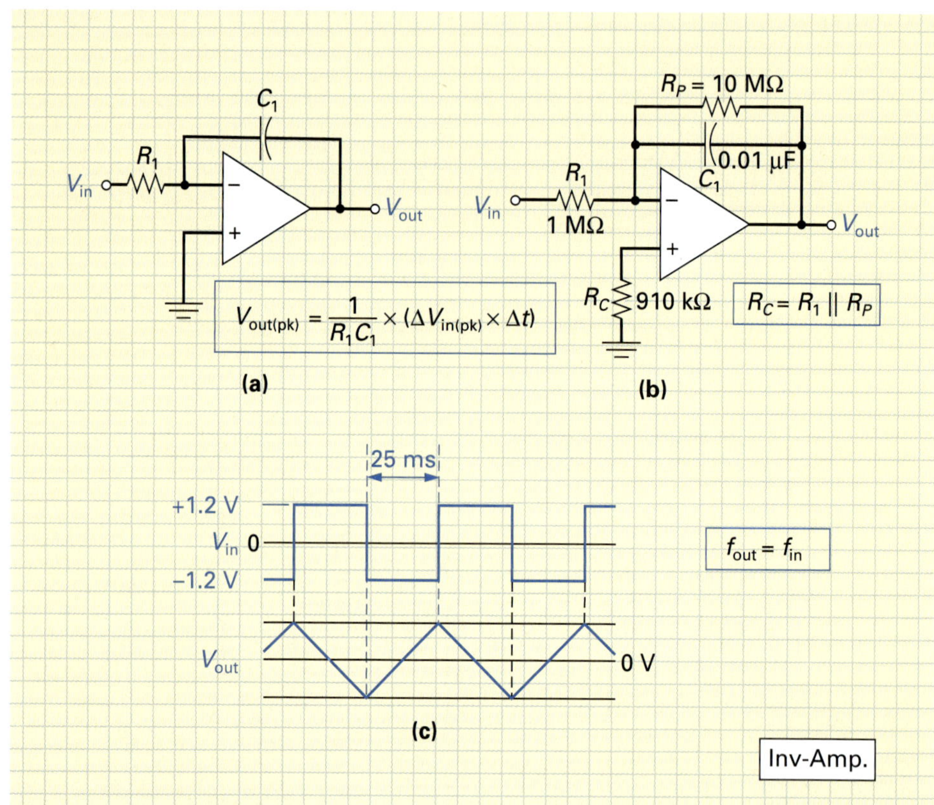

FIGURE 15-12 **Integrator Circuit.**

changes with frequency. To compensate for this effect, a parallel resistor (R_P) is included in shunt with the capacitor, as shown in Figure 15-12(b), to decrease the low-frequency gain: gain will now be a ratio of R_P/R_1.

Figure 15-12(c) shows the integrator's input/output waveforms, with the peak of the output triangular wave being equal to

$$V_{\text{out(pk)}} = \frac{1}{R_1 C_1} \times (\Delta V_{\text{in(pk)}} \times \Delta t)$$

EXAMPLE:

Calculate the peak output voltage of the integrator circuit shown in Figure 15-12(b) considering the input/output waveforms given in Figure 15-12(c).

Solution:

$$V_{\text{out(pk)}} = \frac{1}{R_1 C_1} \times (\Delta V_{\text{in(pk)}} \times \Delta t) = \frac{1}{1\ \text{M}\Omega \times 0.01\ \mu\text{F}} \times (0\ \text{V to } 1.2\ \text{V pk} \times 25\ \text{ms})$$

$$= 100 \times (1.2\ \text{V pk} \times 25\ \text{ms}) = 100 \times 0.03 = 3\ \text{V pk}$$

6. Signal Generator Circuits

Figure 15-13 shows how the op-amp can be connected to act as a signal generator, which is a circuit that will convert a dc supply voltage into a repeating output signal.

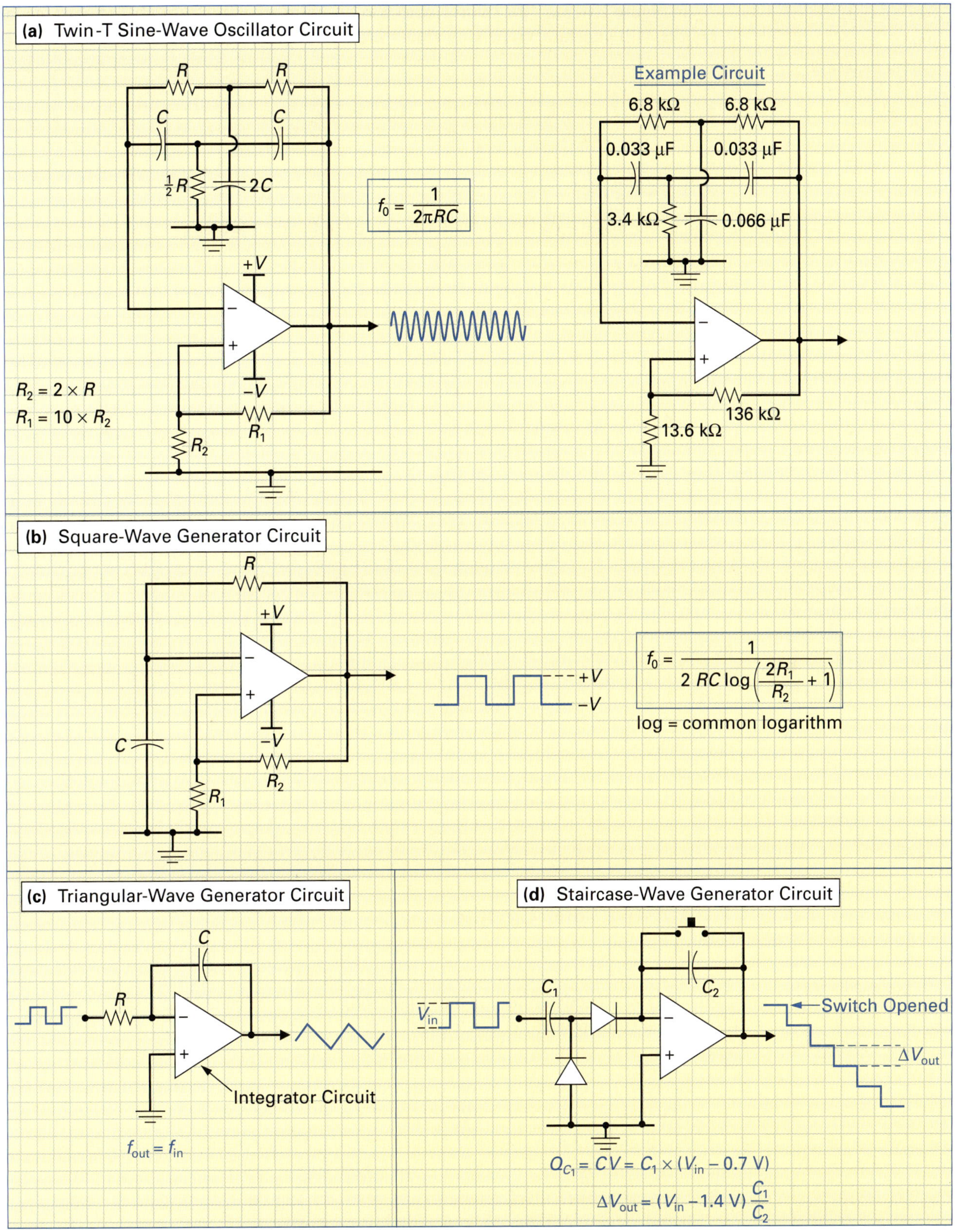

(a) Twin-T Sine-Wave Oscillator Circuit

R R C C $\frac{1}{2}R$ $2C$ $+V$ $-V$ R_1 R_2

$$f_0 = \frac{1}{2\pi RC}$$

$R_2 = 2 \times R$

$R_1 = 10 \times R_2$

Example Circuit

6.8 kΩ 6.8 kΩ 0.033 μF 0.033 μF 3.4 kΩ 0.066 μF 136 kΩ 13.6 kΩ

(b) Square-Wave Generator Circuit

R $+V$ $-V$ C R_1 R_2

$+V$ $-V$

$$f_0 = \frac{1}{2\,RC \log\left(\frac{2R_1}{R_2} + 1\right)}$$

log = common logarithm

(c) Triangular-Wave Generator Circuit

C R

Integrator Circuit

$f_{out} = f_{in}$

(d) Staircase-Wave Generator Circuit

V_{in} C_1 C_2

Switch Opened

ΔV_{out}

$Q_{C_1} = CV = C_1 \times (V_{in} - 0.7\text{ V})$

$\Delta V_{out} = (V_{in} - 1.4\text{ V})\,\frac{C_1}{C_2}$

FIGURE 15-13 **Signal Generator Circuits.**

Twin-T Sine-Wave Oscillator

An oscillator circuit that makes use of two T-shaped feedback networks.

In Figure 15-13(a), the **twin-T sine-wave oscillator,** which has two T-shaped feedback networks, will generate a repeating sine-wave output at a frequency equal to

$$f_0 = \frac{1}{2\pi RC}$$

EXAMPLE:

Calculate the frequency of the oscillator shown in Figure 15-13(a).

Solution:

$$f_0 = \frac{1}{2\pi RC} = \frac{1}{2 \times \pi \times 6.8\ \text{k}\Omega \times 0.033\ \mu\text{F}} = 709\ \text{Hz}$$

Square-Wave Generator

A circuit that generates a continuously repeating square wave.

Relaxation Oscillator

An oscillator circuit whose frequency is determined by an *RL* or *RC* network, producing a rectangular or sawtooth output waveform.

In Figure 15-13(b), the op-amp has been connected to act as a **square-wave generator,** or as it is more frequently called a **relaxation oscillator.** The output signal is fed back to both the inverting and noninverting inputs of the op-amp. The capacitor will charge and discharge through R controlling the frequency of the output square wave, which is equal to

$$f_0 = \frac{1}{2\,RC\log\left(\dfrac{2R_1}{R_2} + 1\right)}$$

Triangular-Wave Generator

A signal generator circuit that produces a continuously repeating triangular wave output.

Staircase-Wave Generator

A signal generator circuit that produces an output signal voltage that increases in steps.

Connecting the output of the square-wave generator in Figure 15-13(b) into the inputs of the circuits in Figure 15-13(c) and (d), we can generate a triangular or staircase output waveform. The **triangular-wave generator** in Figure 15-13(c) is simply the integrator circuit discussed previously, with its output frequency equal to the square-wave input frequency. When the switch is closed in the **staircase-wave generator** in Figure 15-13(d) the capacitor is bypassed and will therefore not charge. On the other hand, when the switch is open, C_2 will be charged by each input cycle, producing equal output steps that have the following voltage change

$$\Delta V_{out} = (V_{in} - 1.4\ \text{V})\frac{C_1}{C_2}$$

7. Active Filter Circuits

Active Filter

A circuit that uses an amplifier with passive filter elements to provide frequency paths with rejection characteristics.

Passive filters are circuits that contain passive or nonamplifying components (resistors, capacitors, and inductors) connected in such a way that they will pass certain frequencies while rejecting others. An **active filter,** on the other hand, is a circuit that uses an amplifier with passive filter elements to provide frequency paths with rejection characteristics. Active filters, like the op-amp circuits seen in Figure 15-14, have several advantages over passive filters.

1. Because the op-amp provides gain, the input signal passed to the output will not be attenuated, and therefore better response curves can be obtained.
2. The high input impedance and low output impedance of the op-amp means that the filter circuit does not interfere with the signal source or load.
3. Because active filters provide gain, resistors can be used instead of inductors, and therefore active filters are generally less expensive.

Figure 15-14 illustrates how the op-amp can be connected to form the four basic active filter types.

Active High-Pass Filter

A circuit that uses an amplifier with passive filter elements to pass all frequencies above a cutoff frequency.

Active High-Pass Filter. Figure 15-14(a) illustrates the simple op-amp circuit, frequency response, and relevant formulas for an **active high-pass filter.** As before, the gain of

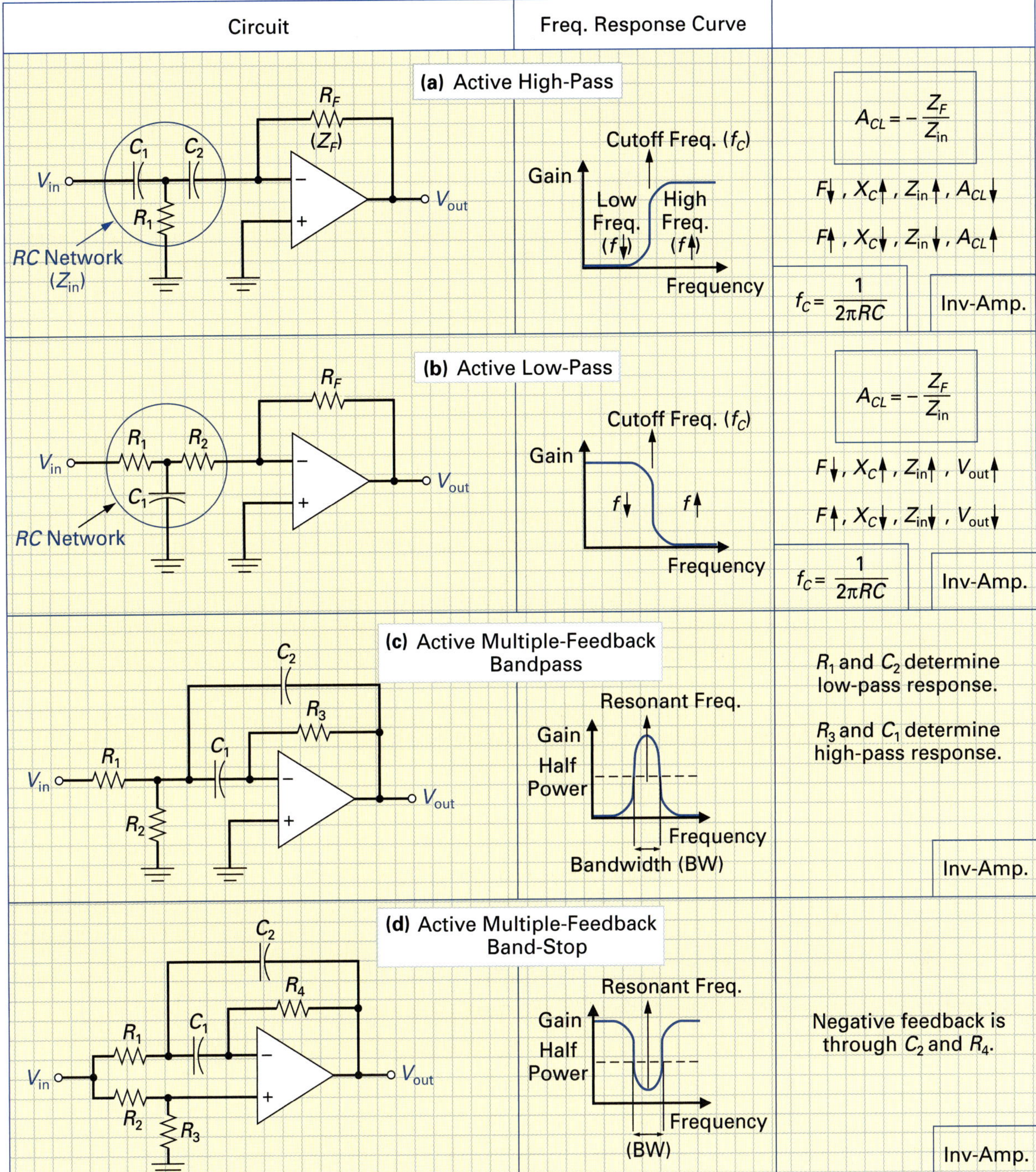

FIGURE 15-14 Active Filter Circuits.

this inverting amplifier is dependent on the ratio of R_F to R_{in}. When capacitors are included in any circuit, impedance (Z) must be considered instead of simply resistance, and gain is now equal to the ratio of feedback impedance to input impedance.

$$A_{CL} = -\frac{Z_F}{Z_{in}}$$

The input *RC* network will offer a high impedance to low frequencies, resulting in a low voltage gain. At high frequencies, the *RC* network will have a low impedance, causing a high voltage gain. The cutoff frequency for this circuit can be calculated with the following formula when $C_1 = C_2$.

$$fc = \frac{1}{2\pi RC}$$

Active Low-Pass Filter
Amplifier circuit with passive filter elements to pass all frequencies below a cutoff frequency.

Active Low-Pass Filter. Figure 15-14(b) illustrates the op-amp circuit, frequency response curve, and relevant formulas for an **active low-pass filter.** At low frequencies, the capacitor's reactance is high, and low-frequency signals will be passed to the op-amp's input to be amplified and passed to the output. As frequency increases, the capacitive reactance of C_1 will decrease; more of the signal will be shunted away from the op-amp and will not appear at the output. The cutoff frequency for this circuit can be calculated with the following formula when $R_1 = R_2$.

$$fc = \frac{1}{2\pi RC}$$

Active Bandpass Filter
A circuit that uses an amplifier with passive filter elements to pass only a band of input frequencies.

Active Bandpass Filter. Figure 15-14(c) illustrates how the op-amp can be connected to form an **active bandpass filter.** At frequencies outside of the band, V_{out} is fed back to the input without being attenuated, and therefore the input signal amplitude is almost equal to the feedback signal amplitude. This results in almost complete cancellation of the signal and therefore a very small output voltage. On the other hand, for the narrow band of frequencies within the band, the feedback network will increase its amount of attenuation. This increase of attenuation means that a very small feedback signal will appear back at the negative input of the op-amp and will have a very small degenerative effect. As a result, the change at the input of the op-amp will be larger when the input signal frequencies are within this band, and the voltage out will also be larger.

Active Band-Stop Filter
A circuit that uses an amplifier with passive filter elements to block a band of input frequencies.

Active Band-Stop Filter. Figure 15-14(d) illustrates how the op-amp can be connected to form an **active band-stop filter,** also known as a band-reject or notch filter. The basic operation of this circuit is opposite to that of the previously discussed bandpass filter. At frequencies outside of the band, the feedback signal will be heavily attenuated, and therefore the degenerative effect will be small and the output voltage large. On the other hand, at frequencies within the band, the feedback signal will not be heavily attenuated, and therefore the degenerative effect will be large and the output voltage small.

SELF-TEST EVALUATION POINT FOR SECTION 15-1

Use the following questions to test your understanding of Section 15-1.

1. Can the op-amp be used to amplify dc as well as ac signal inputs?
2. What is the difference between an open-loop and closed-loop op-amp circuit?
3. List the key characteristics of an op-amp.
4. What are the three basic amplifier types within an op-amp?
5. Which of the following circuits is connected in an open-loop mode?
   a. Comparator **b.** Inverting amplifier **c.** Noninverting amplifier
6. Why is it important for an op-amp circuit to have negative feedback?
7. Which op-amp circuit provides a voltage gain of 1 and is used as a buffer?
8. Which op-amp circuit will sum all of the input voltages?
9. What is the basic circuit difference and input/output waveform difference between the integrator and differentiator circuit?
10. Which op-amp circuit will generate an output that is equal to the difference between the two inputs?
11. Sketch a circuit showing how the op-amp can be connected to generate a repeating square-wave output.
12. What is the difference between an active filter and a passive filter?

15-2 VOLTAGE REGULATORS

In Chapter 11 you were introduced to fixed-output IC voltage regulators, and in Chapter 12 to the switching voltage regulator. These and other IC regulators are shown and reviewed in Figure 15-15. Linear voltage regulator ICs can be categorized into one of three basic groups:

1. Fixed Output Voltage Regulators (Chapter 11)
2. Variable Output Voltage Regulators
3. Switching Voltage Regulators (Chapter 12)

Let us now examine each of these three types, beginning with the fixed output voltage regulator ICs.

15-2-1 *Fixed Output Voltage Regulators [Figure 15-15(a)]*

As the name implies, a **fixed-output voltage regulator** will produce a regulated output voltage that is not variable. These fixed output voltage regulators can be broken down into two basic groups: those producing a positive voltage (7800—seventy-eight hundred series) and those producing a negative output voltage (7900—seventy-nine hundred series).

Fixed Output Voltage Regulator
A voltage regulator that produces a regulated voltage that is not variable.

These regulators will deliver a constant regulated output voltage, as long as the input voltage is greater than the regulator's rated output voltage. They can also deliver a maximum output current of up to 1.5 A if properly heat sunk. The three terminals of the IC regulator are labeled input, output, and ground. The package is generally given the identification code of either 78XX or 79XX. The 78XX series of regulators are used to supply a positive output voltage, with the last two digits specifying the output voltage (for example, 7805 = +5 V). On the other hand, the 79XX series of regulators are used to supply a negative output voltage, with the last two digits, once again, specifying the output voltage (for example, 7912 = −12 V).

15-2-2 *Variable Output Voltage Regulators [Figure 15-15(b)]*

As the name implies, a **variable output voltage regulator** will produce a regulated output voltage that can be adjusted. Like the fixed output voltage regulators, there are positive variable output voltage regulators and negative variable output voltage regulators.

Variable Output Voltage Regulator
A voltage regulator that produces a regulated voltage that can be adjusted.

The LM317 series, the most commonly used general-purpose positive variable-output voltage regulators, have three terminals called V_{IN}, V_{OUT}, and ADJ or adjustment, and are normally connected in circuit. Resistors R_1 and R_2 form a voltage divider across the output, with R_2 connected as a rheostat. The output voltage of the LM317 is adjusted by changing the resistance of R_2, and therefore the voltage applied to the ADJ input of the regulator.

The LM317's counterpart, the LM337 series, are a group of general-purpose negative variable-output voltage regulators. These regulators are available with the same voltage and current options as the LM317 devices.

15-2-3 *Switching Voltage Regulators [Figure 15-15(c)]*

The fixed and variable output voltage regulators just discussed are all called **series dissipative regulators** because these regulators control the resistance of an internal transistor connected in series between the input voltage and the load. These regulator types maintain a constant output voltage by constantly changing the resistance of the series-connected internal transistor so that variations in input voltage or load current are dissipated as heat. Series dissipative regulators generally have a low "conversion efficiency" of typically 60% to 70% and should only be used in low- to medium-load current applications.

Series Dissipative Regulators
Voltage regulators that maintain a constant output voltage by constantly changing the resistance of the series-connected internal transistor so that variations in input voltage or load current are dissipated as heat.

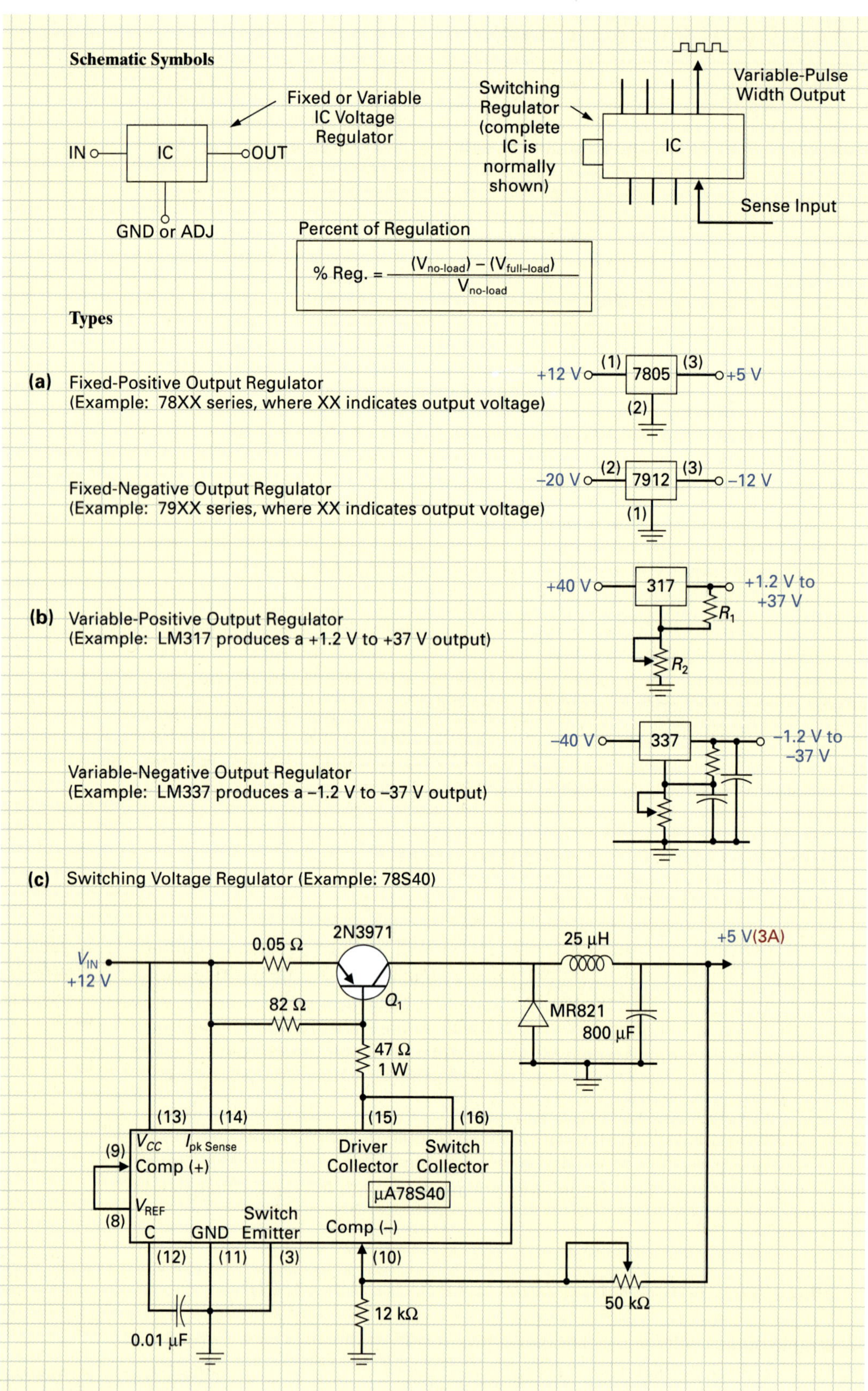

FIGURE 15-15 Linear IC Voltage Regulators.

Series switching regulators, or a **switching power supply** on the other hand, have a conversion efficiency of typically 90% and are generally small in size. The only disadvantage is that the circuitry is generally a little more complex, and therefore a little more costly.

As an example, Figure 15-15(c) shows how a 78S40 switching regulator IC could be connected to regulate a +5 V output when a +12 V input is applied. The switching frequency is set by the timing capacitor connected between pin 12 and the ground pin 11. The noninverting input (+) of the 78S40's internal comparator (pin 9) is connected to the internal 1.25 V reference voltage at pin 8, while the inverting input (−) of the comparator (pin 10) senses the output voltage that is developed across a voltage divider. The 78S40's internal comparator generates an output signal that varies the OFF time of the PNP switching transistor Q_1 instead of varying the ON time of an NPN transistor. To explain in more detail, whenever the output voltage falls below +5 V, the comparator responds by increasing the width of the OFF or LOW output pulse applied to the base of Q_1. Because Q_1 is a PNP transistor, this LOW input increases the ON time of Q_1, which raises the average output voltage, bringing the output back up to +5 V. On the other hand, whenever the output voltage rises above +5 V, the comparator responds by decreasing the width of the LOW output pulse applied to the base of Q_1. This decreases the ON time of Q_1, which lowers the average output voltage, bringing the output back down to +5 V.

Series Switching Regulator

A regulator circuit containing a power transistor in series with the load, that is switched ON and OFF to regulate the dc output voltage delivered to the load.

Switching Power Supply

A dc power supply that makes use of a series switching regulator controlled by a pulse-width-modulator to regulate the output voltage.

SELF-TEST EVALUATION POINT FOR SECTION 15-2

Use the following questions to test your understanding of Section 15–2.

1. List four examples of linear integrated circuits.
2. A voltage regulator will maintain its output voltage constant despite variations in ____________ and ____________.
3. Series dissipative regulators, such as the 7800, 7900, 317, and 377 series, have an internal resistor connected in series between the input voltage and load. This internal resistor is operated as a ____________. (switch, variable resistor, capacitor)
4. Series switching regulators have a transistor connected in series between the input voltage and the load. This transistor is operated as a ____________. (switch, variable resistor, capacitor)

15-3 TIMERS AND CLOCK GENERATORS

Timing is everything in electronic circuits, and to control the timing, a clock signal is distributed throughout the system. This square wave **clock signal** is generated by a **clock oscillator,** and its sharp positive (leading) and negative (trailing) edges are used to control the sequence of operations in a circuit. In this section we will discuss the **astable multivibrator,** which is commonly used as a clock oscillator, and the **monostable multivibrator,** which when triggered will generate a rectangular pulse of a fixed duration. To complete our discussion on multivibrator circuits, we will also discuss the **bistable multivibrator,** which is a digital control device that can be either set or reset, and the pulse-shaping Schmitt trigger circuit.

To help explain the operation and function of the astable, monostable, bistable, and Schmitt trigger circuits, we will first step through a detailed circuit description of each. Following this, you will be introduced to the extremely versatile linear "555 Times" IC, which can be configured to function in a wide variety of applications.

Clock Signal

Generally a square wave used for the synchronization and timing of several circuits.

Clock Oscillator

A device for generating a clock signal.

Astable Multivibrator

A device commonly used as a clock oscillator.

Monostable Multivibrator

A device that when triggered will generate a rectangular pulse of fixed duration.

Bistable Multivibrator

A digital control device that can be either set or reset.

15-3-1 *The Astable Multivibrator Circuit*

The astable multivibrator circuit, seen in Figure 15-16(a), is used to produce an alternating two-state square or rectangular output waveform. This circuit is often called a **free-running multivibrator** because the circuit requires no input signal to start its operation. It will simply begin oscillating the moment the dc supply voltage is applied.

Free-Running Multivibrator

A circuit that requires no input signal to start its operation, but simply begins to oscillate the moment the dc supply voltage is applied.

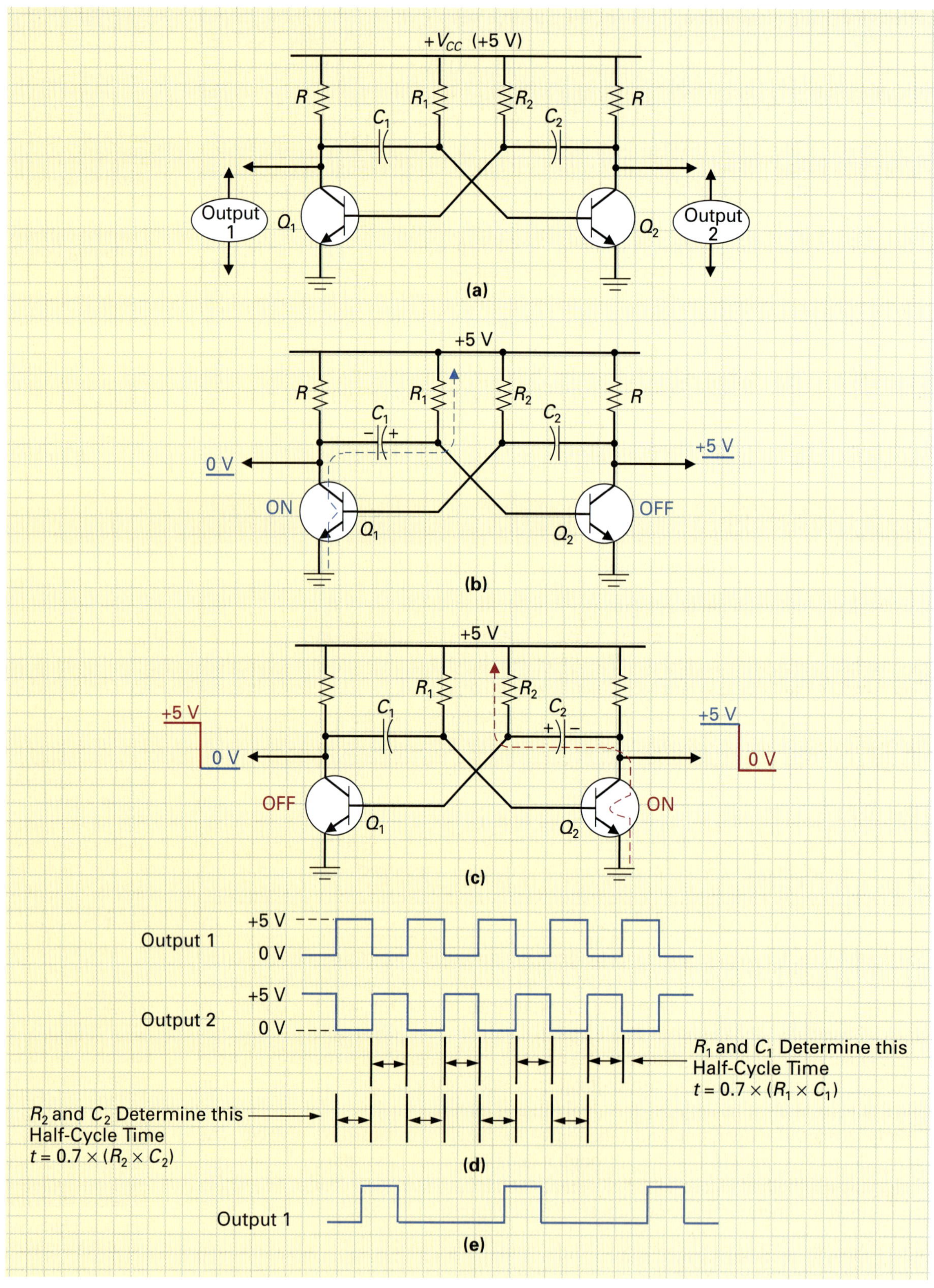

FIGURE 15-16 The Astable Multivibrator. (a) Basic Circuit. (b) Q_1 ON Condition. (c) Q_2 ON Condition. (d) Square Wave Mode. (e) Rectangular Wave Mode.

The circuit consists of two *cross-coupled bipolar transistors,* which means that there is a cross connection between the base and the collector of the two transistors Q_1 and Q_2. This circuit also contains two *RC* timing networks: R_1/C_1 and R_2/C_2.

Let us now examine the operation of this astable multivibrator circuit. When no dc supply voltage is present ($V_{CC} = 0$ V), both transistors are OFF, and therefore there is no output. When a V_{CC} supply voltage is applied to the circuit (for example, +5 V), both transistors will receive a positive bias base voltage via R_1 and R_2. Although both Q_1 and Q_2 are matched bipolar transistors, which means that their manufacturer ratings are identical, no two transistors are ever the same. This difference, and the differences in R_1 and R_2 due to resistor tolerances, means that one transistor will turn ON faster than the other. Let us assume that Q_1 turns on first, as seen in Figure 15-16(b). As Q_1 conducts, its collector voltage decreases because it is like a closed switch between the collector and emitter. This decrease in collector voltage is coupled through C_1 to the base of Q_2, causing it to conduct less and eventually turn OFF. With Q_2 OFF, its collector voltage will be high (+5 V) because Q_2 is equivalent to an open switch between the collector and emitter. This increase in collector voltage is coupled through C_2 to the base of Q_1 causing it to conduct more and eventually turn fully ON. The cross coupling between these two bipolar transistors will reinforce this condition with the LOW Q_1 collector voltage keeping Q_2 OFF and the HIGH Q_2 collector voltage keeping Q_1 ON. With Q_1 equivalent to a closed switch, a current path now exists for C_1 to charge as seen in Figure 15-16(b). As soon as the charge on C_1 reaches about 0.7 V, Q_2 will conduct because its base-emitter junction will be forward biased. This condition is shown in Figure 15-16(c). When Q_2 conducts, its collector voltage will drop, cutting OFF Q_1 and creating a charge path for C_2. As soon as the charge on C_2 reaches 0.7 V, Q_1 will conduct again and the cycle will repeat.

TIME LINE

John L. Baird (1888-1946) was a British inventor and television pioneer. He was the first to transmit television over a distance. He reproduced objects in 1924, transmitted recognizable human faces in 1926, demonstrated the first true television in 1926, and in 1939 he developed television in natural color.

The output waveforms switch between the supply voltage ($+V_{CC}$) when a transistor is cut off, and zero volts when a transistor is saturated (ON). The result is two square-wave outputs that are out of phase with one another, as seen in the waveforms in Figure 15-16(d). Referring to the output waveforms in Figure 15-16(d), you can see that the time constant of R_1 and C_1, and R_2 and C_2, determine the complete cycle time. If the R_1/C_1 time constant is equal to the R_2/C_2 time constant, both halves of the cycle will be equal (50% duty cycle) and the result will be a square wave. Referring to Figure 15-16(d), you can see that the R_1/C_1 time constant will determine the time of one half-cycle, while the R_2/C_2 time constant will determine the time of the other half-cycle. The formula for calculating the time of one half-cycle is equal to:

$$t = 0.7 \times (R_1 \times C_1) \quad \text{or} \quad T = 0.7 \times (R_2 \times C_2)$$

The frequency of this square wave can be calculated by taking the reciprocal of both half-cycles, which will be

$$f = \frac{1}{1.4 \times RC}$$

EXAMPLE:

Calculate the positive and negative cycle time and circuit frequency of the astable multivibrator circuit in Figure 15-16, if

$$R_1 \text{ and } R_2 = 100\text{ k}\Omega \qquad C_1 \text{ and } C_2 = 1\ \mu\text{F}$$

Solution:

Because the time constant of R_1/C_1 and R_2/C_2 are the same, each half-cycle time will be the same and equal to

$$\begin{aligned} t &= 0.7 \times (R \times C) \\ &= 0.7 \times (100\text{ k}\Omega \times 1\ \mu\text{F}) = 0.07\text{ s} \quad \text{or} \quad 70\text{ ms} \end{aligned}$$

TIME LINE

During his years as a student at the University of Utah, Nolan Bushnell (1943–) enjoyed playing the ancient Chinese game called GO. In this game, a word is used frequently throughout as a warning to your opponents that they are in jeopardy of losing with your next move.

With an initial investment of $500, Bushnell launched the video game industry in 1972, developing the first coin-operated Ping-Pong game, Pong. After almost an overnight success, Bushnell sold his company four years later for $15 million; the company was named after the polite Chinese term he learned in college, and as a warning to the competition, ATARI.

Following ATARI, Busnell stepped up his ambition, taking high-tech to the masses through his Chuck E. Cheese's Pizza Time Theatres.

One-Shot Multivibrator
Produces one output pulse or shot for each input trigger.

Trigger Input
Pulse used to initiate a circuit action.

The frequency of the astable circuit will be equal to the reciprocal of the complete cycle, or the reciprocal of twice the half-alternation time.

$$f = \frac{1}{1.4 \times RC} = \frac{1}{1.4 \times (100\ \text{k}\Omega \times 1\ \mu\text{F})} = 7.14\ \text{Hz}$$

$$\text{or}\ f = \frac{1}{2 \times t} = \frac{1}{2 \times 70\ \text{ms}} = \frac{1}{0.14} = 7.14\ \text{Hz}$$

If the time constants of the two *RC* timing networks in the astable circuit are different, however, the result will be a rectangular or pulse waveform, as seen in Figure 15-16(e). In this instance, the same formula can be used to calculate the time for each alternation, and the frequency will be equal to the reciprocal of the time for both alternations.

15-3-2 *The Monostable Multivibrator Circuit*

The astable multivibrator is often referred to as an "unstable multivibrator" because it is continually alternating or switching back and forth, and therefore it has no stable condition or state. The monostable multivibrator has, as its name implies, one (mono) stable state. The circuit will remain in this stable state indefinitely until a trigger is applied and forces the monostable multivibrator into its unstable state. It will remain in its unstable state for a small period of time and then switch back to its stable state and await another trigger. The monostable multivibrator is often compared to a gun and is called a **one-shot multivibrator** because it will produce one output pulse or shot for each input trigger.

Referring to the monostable multivibrator circuit in Figure 15-17(a), you can see that the monostable is similar to the astable except for the trigger input circuit and for the fact that it has only one *RC* timing network. To begin with, let us consider the stable state of the monostable. Components R_2, D_1, and R_5 form a voltage divider, the values of which are chosen to produce a large positive Q_2 base voltage. This large positive base bias voltage will cause Q_2 to saturate (turn heavily ON), which in turn will produce a LOW Q_2 collector voltage, which will be coupled via R_4 to the base of Q_1, cutting it OFF. The circuit remains in this stable state (Q_2 ON, Q_1 OFF) until a **trigger input** is received.

Referring to the timing waveforms in Figure 15-17(b), you can see how the circuit reacts when a positive input trigger is applied. The pulse is first applied to a differentiator circuit (C_2 and R_5) that converts the pulse into a positive and a negative spike. These spikes are then applied to the positive clipper diode D_1, which only allows the negative spike to pass to the base of Q_2. This negative spike will reverse bias Q_2's base-emitter junction, turning Q_2 OFF and causing its collector voltage to rise to $+V_{CC}$, as seen in the waveforms. This increased Q_2 collector voltage will be coupled to the base of Q_1, turning it ON. The monostable multivibrator is now in its unstable state, which is indicated in the second color in Figure 15-17(a). In this condition, C_1 will charge as shown by the dashed current line. However, as soon as the voltage across C_1 reaches 0.7 V (which is dependent on the R_2/C_2 time constant), it will force Q_2 to conduct, which in turn will cause Q_1 to cut OFF and the monostable to return to its stable state. The output pulse width or pulse time (*t*) seen in Figure 15-17(b) can be calculated with the same formula used for the astable multivibrator:

$$t = 0.7 \times (R_2 \times C_1)$$

The one-shot multivibrator is sometimes used in *pulse-stretching* applications. For example, referring to the waveforms in Figure 15-17(b), imagine the input positive trigger pulse is 1 μs in width, and the *RC* time constant of R_2 and C_1 is such that the output pulse width (t) is 500 μs. In this example, the input pulse would be effectively stretched from 1 μs to 500 μs. The monostable multivibrator, or one-shot timer circuit, is also used to introduce a *time delay.* Referring again to the waveforms in Figure 15-17(b), imagine a differentiator circuit connected to the output of the monostable circuit. If the output pulse width was again set to 500 μs, there would be a 500 μs delay between the differentiated negative edge of the input pulse and the differentiated negative edge of the output pulse.

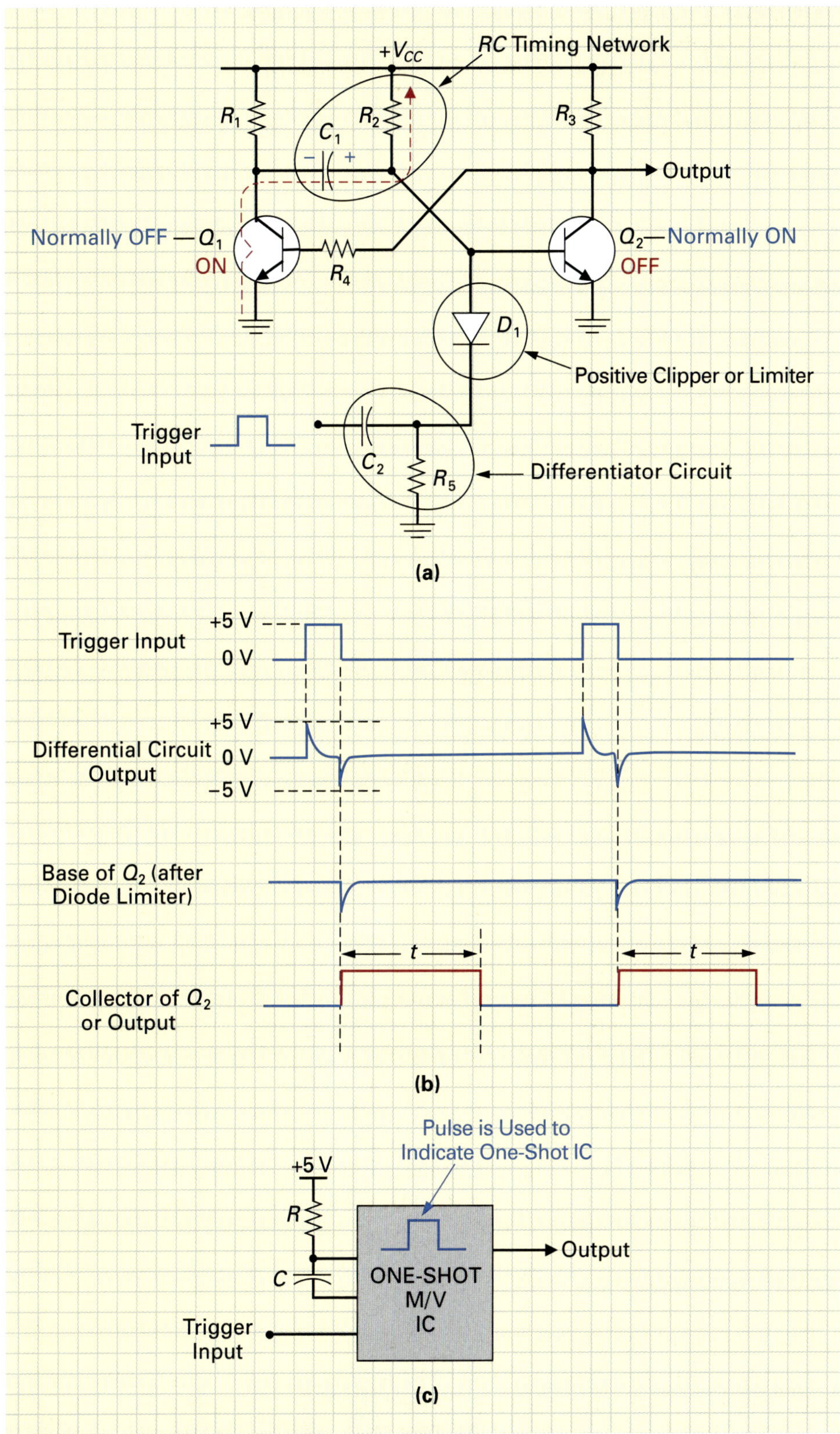

FIGURE 15-17 The Monostable Multivibrator Circuit. (a) Bipolar Transistor Circuit. (b) Input/Output Timing Waveforms. (c) One-Shot IC.

TIME LINE

In 1971, Ted Hoff of Intel Corporation designed a microprocessor, the 4004, that had all the basic parts of a central processor. Intel improved on the 4-bit 4004 microprocessor and unveiled an 8-bit microprocessor in 1974 that could add two numbers in 3.2 billionths of a second.

Figure 15-17(c) shows the logic symbol for a monostable (one-shot) multivibrator. Nearly all one-shot circuits in use today are in integrated form. These IC one-shots operate in exactly the same way as their discrete component counterparts. For example, the 74LS123 IC contains two fully independent monostable multivibrators. Like the symbol in Figure 15-17(c), the 74LS123 has pins for connecting external timing resistors and capacitors.

15-3-3 *The Bistable Multivibrator Circuit*

The bistable multivibrator has two (bi) stable states, and its bipolar transistor circuit is illustrated in Figure 15-18(a). The circuit has two inputs called the "SET" and "RESET" inputs, and these inputs drive the base of Q_1 and Q_2. The two outputs from this circuit are taken from the collectors of Q_1 and Q_2 and are called "Q" and "$\overline{Q}$" (pronounced "Q not"). The $\overline{Q}$ output derives its name from the fact that its voltage level is always the opposite of the Q output. For example, if Q is HIGH, $\overline{Q}$ will be LOW, and if Q is LOW, $\overline{Q}$ will be HIGH. The bistable multivibrator circuit is often called an ***S-R* (set-reset) flip-flop** because

> *A pulse on the SET input will "flip" the circuit into the set state (Q output is set HIGH), while*
>
> *A pulse on the RESET input will "flop" the circuit into its reset state (Q output is reset LOW).*

Figure 15-18(b) shows the logic symbol for a *S-R* or *R-S* flip-flop, or bistable multivibrator circuit.

To fully understand the operation of the circuit, refer to the waveforms in Figure 15-18(c). When power is first applied, one of the transistors will turn "ON" first and, because of the cross coupling, turn the other transistor "OFF." Let us assume that the circuit in Figure 15-18(a) starts with Q_1 ON and Q_2 OFF. The low voltage (approximately 0.3 V) on Q_1's collector will be coupled to Q_2's base, thus keeping it OFF. The high voltage on Q_2's collector (approximately +5 V) will be coupled to the base of Q_1, keeping it ON. This condition is called the **reset state** because the primary output (output 1, or Q) has been reset to binary 0, or 0 V. The cross-coupling action between the transistors will keep the transistors in the reset state (output 1 or $Q = 0$ V, and output 2 or $\overline{Q} = 5$ V) until an input appears.

Following the waveforms in Figure 15-17(c), you can see that the first input to go active is the SET input at time "t_1." This positive pulse will be applied to the base of Q_2 and forward bias its base-emitter junction. As a result, Q_2 will go ON and its LOW collector voltage will be applied to output 2 ($\overline{Q}$). This LOW on Q_2's collector will also be cross-coupled to the base of Q_1, turning it OFF and therefore making output 1 (Q) go HIGH. When a pulse appears on the SET input, the circuit will be put in its **set state** which means that the primary output (output 1 or Q) will be set HIGH. Studying the waveforms in Figure 15-17(c) once again, you will notice that after the positive SET input pulse has ended, the bistable will still remain **latched** or held in its last state, due to the cross-coupling between the transistors. This ability of the bistable multivibrator to remain in its last condition or state explains why the *S-R* flip-flop is also called an ***S-R* latch.**

Following the waveforms you can see that a RESET pulse occurs at time t_2, and resets the primary output (output 1 or Q) LOW. The flip-flop then remains latched in its reset state, until a SET pulse is applied to the set input at time t_3, setting Q HIGH. Finally, a positive RESET pulse is applied to the reset input at time t_4, and Q is reset LOW.

The operation of the *S-R* flip-flop is summarized in the truth table or function table shown in Figure 15-18(d). When only the R input is pulsed HIGH (reset condition), the Q output is reset to a binary 0 or reset LOW ($\overline{Q}$ will be the opposite, or HIGH). On the other hand, when only the S input is pulsed HIGH (set condition), the Q output is set to a binary 1 or set HIGH ($\overline{Q}$ will be the opposite, or LOW). When both the S and R inputs are LOW, the *S-R* flip-flop is said to be in the **no-change** or **latch condition** because there will be no change in the output Q. For example, if the output Q is SET, and then the S and R inputs are made LOW, the Q output will remain SET, or HIGH. On the other hand, if the output Q is RESET, and then the S and R inputs are made LOW, the Q output will remain RESET, or LOW.

_S-R_ (Set-Reset) Flip-Flop
A multivibrator circuit in which a pulse on the SET input will "flip" the circuit into the set state while a pulse on the RESET input will "flop" the circuit into its reset state.

Reset State
Primary output set LOW.

Set State
Primary output set HIGH.

Latched
Held in the last state.

_S-R_ Latch
Another name for *S-R* flip-flop, so called because the output remains latched in the set or reset state even though the input is removed.

No-Change or Latch Condition
When both inputs are LOW, the *S-R* flip-flop is said to be in the no-change or latch condition because there will be no change in the output.

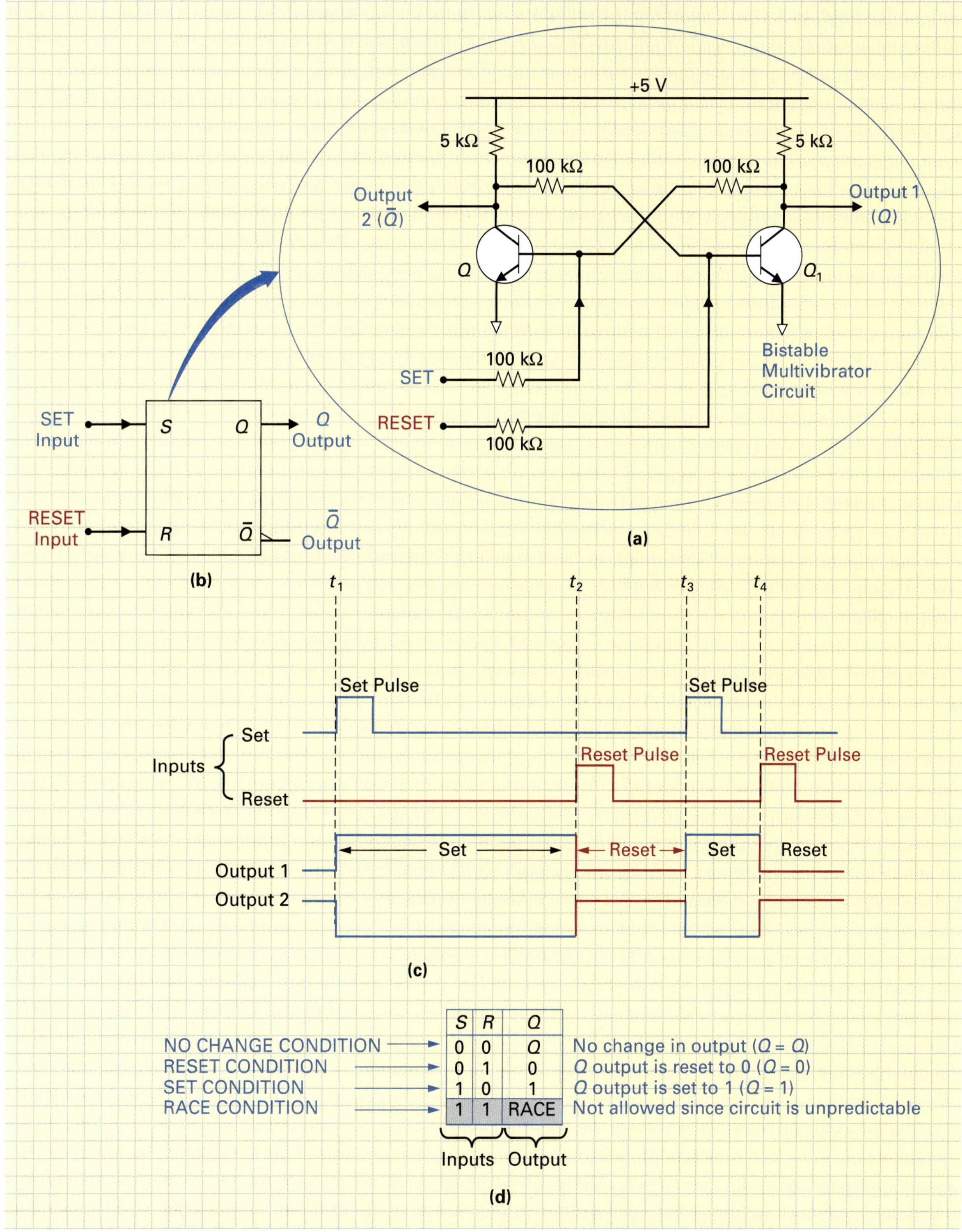

S	R	Q
0	0	Q
0	1	0
1	0	1
1	1	RACE

FIGURE 15-18 **The Bistable Multivibrator or Set-Reset (*S-R* or *R-S*) Flip-Flop. (a) Bipolar Transistor Circuit. (b) SR Symbol. (c) Timing Waveforms. (d) Truth Table.**

The external circuits driving the S and R inputs will be designed so that these inputs are never both HIGH, as shown in the last condition in the table in Figure 15-18(d). This is called the *race condition* because both bipolar transistors will have their bases made positive, and therefore they will race to turn ON, and then shut the other transistor OFF via the cross-coupling. This input condition is not normally applied because the output condition is unpredictable.

The Schmitt trigger circuit discussed previously is a bistable multivibrator circuit because its output voltage can be either one of two states. That is, its output can be either SET HIGH or RESET LOW, based on the voltage level of the input control voltage.

The bistable multivibrator S-R flip-flop or S-R latch has become one of the most important circuits in digital electronics. It is used in a variety of applications ranging from data storage to counting and frequency division. These applications will be covered later in the digital circuits chapter of this textbook. You will, however, see how this S-R flip-flop circuit is made use of in the following 555 timer circuit.

15-3-4 *Schmitt-Trigger Circuit*

Schmitt-Trigger Circuit
A level-sensitive input circuit that has a two-state output.

The **Schmitt-trigger circuit,** named after its inventor, is a two-state device that is used for pulse shaping. A typical Schmitt-trigger circuit can be seen in Figure 15-19(a); the schematic symbol for the Schmitt-trigger logic symbol is shown in Figure 15-19(b). The waveforms shown in Figure 15-19(c) and (d) illustrate the two basic applications of the Schmitt trigger, which are to convert a sine wave into a rectangular wave or to sharpen the rise and fall times of a rectangular wave.

Like all of the previously discussed multivibrators, the Schmitt-trigger circuit may be constructed with discrete or individual components or purchased as an IC containing several complete Schmitt-trigger circuits. Using the circuit shown in Figure 15-19(a) and the example input/output waveforms shown in Figure 15-19(c) and (d), let us see how this bipolar transistor circuit will operate. The input signal is applied to the base of Q_1, and if this signal is below the ON voltage (V_{ON}), Q_1 will be cut OFF and its collector voltage will be HIGH. The HIGH Q_1 collector voltage is coupled to the base of Q_2 causing it to saturate (to turn heavily ON). As a result, Q_2 conducts a large current (shown as electron flow) through R_6, Q_2 emitter-to-collector, and R_2 to $+V_{CC}$. The voltage developed across R_6 (V_{R6}) establishes the V_{ON} voltage of the Schmitt trigger, since the base of Q_1, and therefore the input, will have to be 0.7 V greater than the emitter of Q_1, or V_{R6}, if Q_1 is to turn ON.

When the input does reach a value that is 0.7 V above V_{R6}, Q_1 will turn ON and its collector voltage will fall. This decrease is coupled to the base of Q_2, turning it OFF and causing its collector voltage, and therefore the output, to rise. With Q_1 now ON and Q_2 OFF, the voltage across R_6 is now being established by the current path through R_6, Q_1 emitter-to-collector, and R_1 to $+V_{CC}$. This will set up a different V_{R6} voltage and therefore a different V_{OFF} voltage since the input must fall 0.7 V below V_{R6} for Q_1 to turn OFF. When the input falls below this voltage, Q_1 will turn OFF, its collector voltage will rise and turn ON Q_2, causing its collector voltage and the output voltage to fall. The circuit is now returned to its original condition, awaiting the input to once again exceed the V_{ON} threshold voltage.

The symbol for any Schmitt-trigger will contain the distinctive hysteresis symbol shown in Figure 15-19(b). Hysteresis can be generally defined as the lag between cause and effect. Referring to the waveforms in Figure 15-19(c) and (d), you can see that there is in fact a lag between the input (cause) and output (effect). The input has to rise to an "ON voltage" (V_{ON}) before the output will rise to its positive peak, and must drop below an "OFF voltage" (V_{OFF}) before the output will drop to its negative peak. This lag is desirable since it produces a sharp, well-defined, output signal. The difference between the V_{ON} and V_{OFF} voltage is called the **hysteresis voltage.**

Hysteresis Voltage
The difference between two voltages.

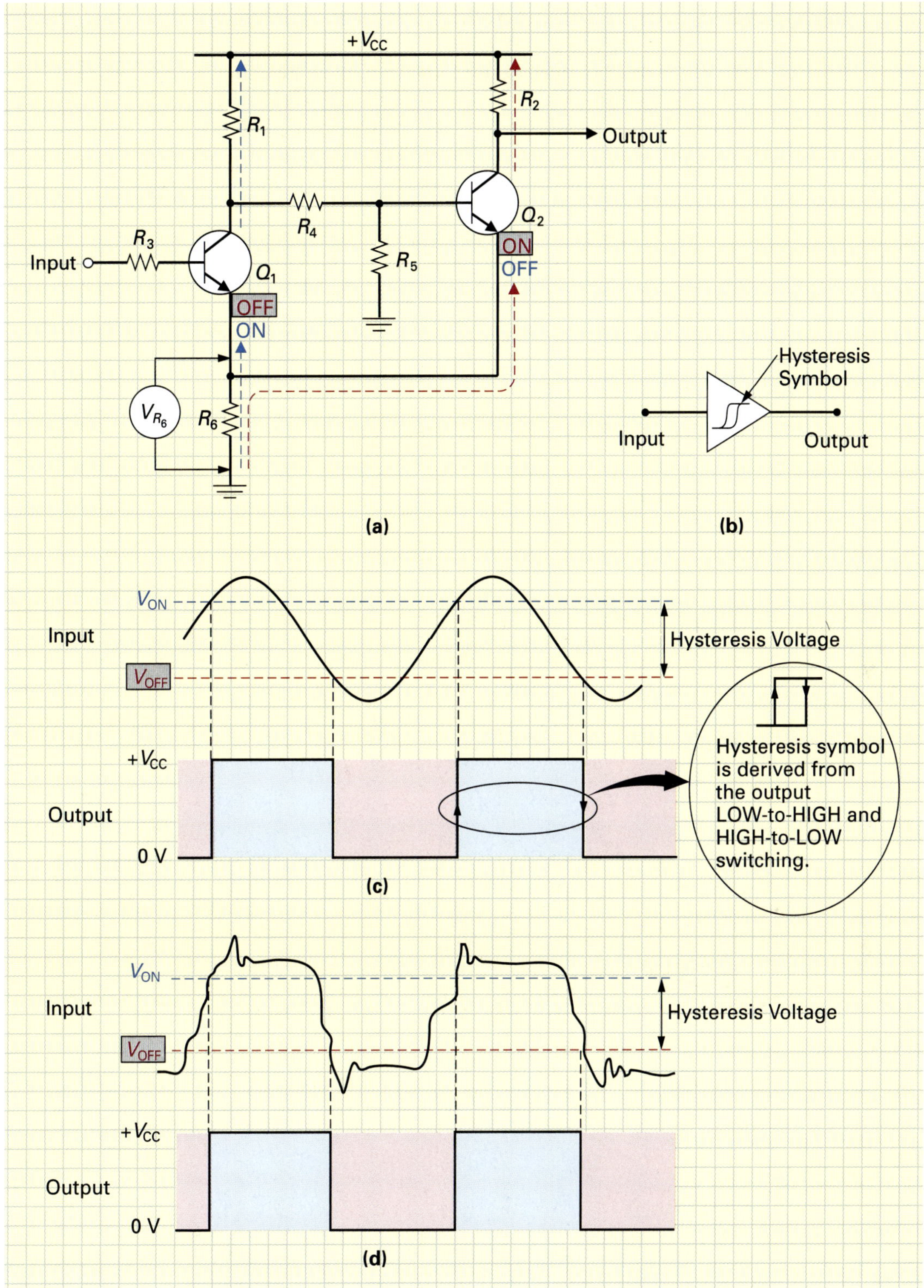

FIGURE 15-19 The Schmitt Trigger. (a) Bipolar Transistor Circuit. (b) Logic Gate Symbol. (c) Converting a Sine Wave into a Rectangular Wave. (d) Sharpening the Rise and Fall Times of a Rectangular Wave.

15-3-5 *The 555 Timer Circuit*

One of the most frequently used low-cost integrated circuit timers is the *555 timer.* Its IC package consists of 8 pins, as seen in Figure 15-20(a), and derives its number identification from the distinctive voltage divider circuit, seen in Figure 15-20(b), consisting of three 5 kΩ resistors. It is a highly versatile timer that can be made to function as an astable multivibrator, monostable multivibrator, frequency divider, or modulator depending on the connection of external components.

Nearly all the IC manufacturers produce a version of the 555 timer, which can be labeled in different ways, for example: SN72 555, MC14 555, SE 555, and so on. Two 555 timers are also available in a 16 pin dual IC package that is labeled with the numbers 556.

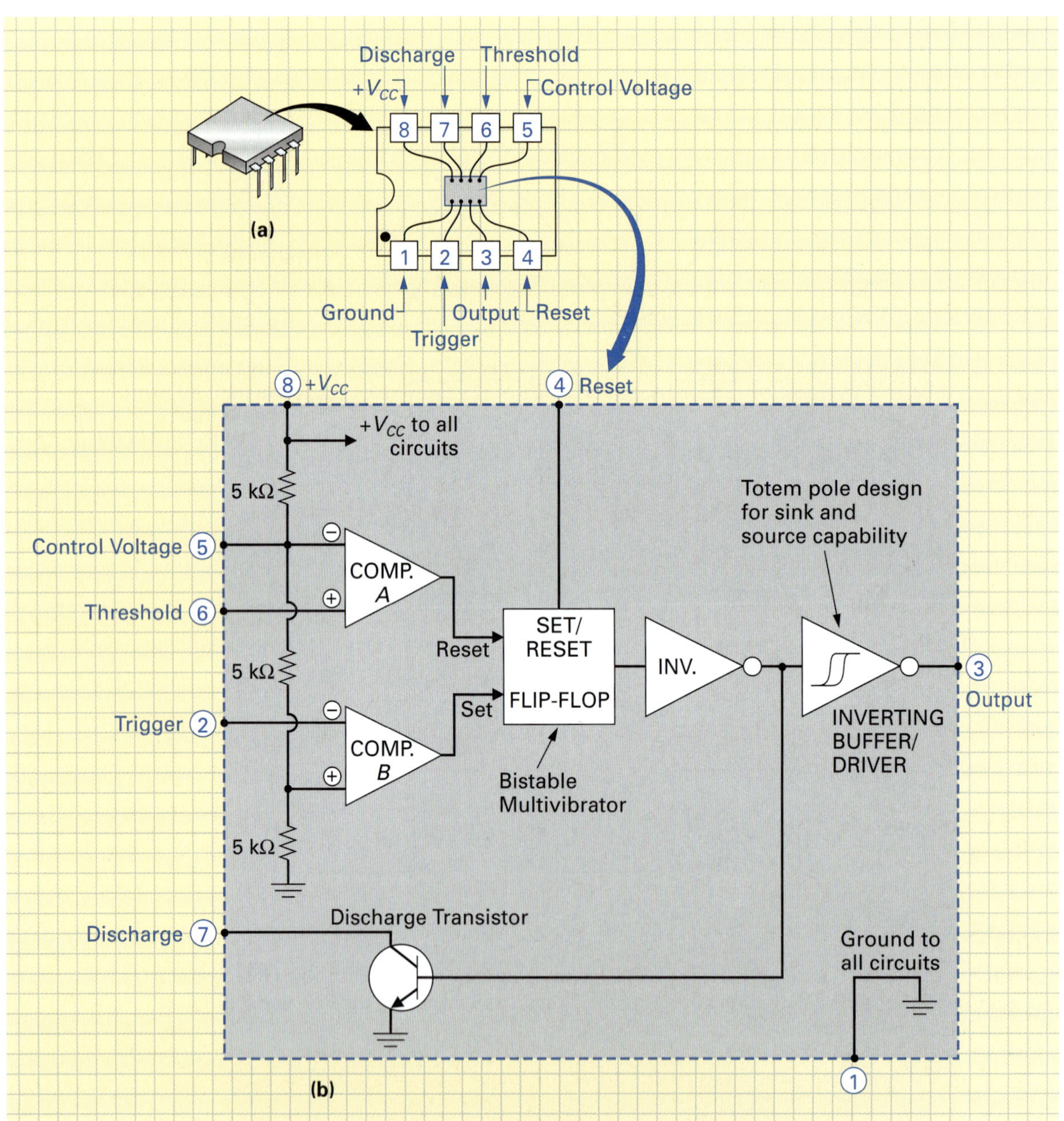

FIGURE 15-20 **The 555 Timer. (a) IC Pin Layout. (b) Basic Block Diagram.**

1. Basic 555 Timer Circuit Action

Referring to the basic block diagram in Figure 15-20(b), let us examine the basic action of all the devices in a 555 timer circuit. The three 5 kΩ resistors develop reference voltages at the inputs of the two comparators *A* and *B*. As previously mentioned, a comparator is a circuit that compares an input signal voltage to an input reference voltage and then produces a YES/NO or HIGH/LOW decision output. The negative input of comparator *A* will have a reference that is 2/3 of V_{CC}, and therefore the positive input (pin 6, threshold) will have to be more positive than 2/3 of V_{CC} for the output of comparator *A* to go HIGH. With comparator *B*, the positive input has a reference that is 1/3 of V_{CC}, and therefore the negative input (pin 2, trigger) will have to be more negative, or fall below, 1/3 of V_{CC} for the output of comparator *B* to go HIGH. If the output of comparator *A* were to go HIGH, the set/reset flip-flop output would be reset LOW to 0 V. This LOW output would be inverted by the INVERTER to a HIGH, and then inverted and buffered (boosted in current) by the final Schmitt INVERTER to appear as a LOW at the output pin 3. If the output of comparator *B* were to go high, the set/reset flip-flop would be set HIGH to +5 V. This HIGH output would be inverted by the INVERTER to a LOW, and then inverted and buffered by the final INVERTER to appear as a HIGH at the output pin 3. When the output of the set/reset flip-flop is LOW (reset), the input at the base of the discharge transistor will be HIGH. The transistor will therefore turn ON, and its low emitter-to-collector resistance will ground pin 7. When the output of the set/reset flip-flop is HIGH (set), the input at the base of the discharge transistor will be LOW. The transistor will therefore turn OFF and its high emitter-to-collector resistance will cause pin 7 to ground.

2. The 555 Timer as an Astable Multivibrator

Figure 15-21(a) shows how the 555 timer can be connected to operate as an astable or free-running multivibrator. The waveforms in Figure 15-21(b) show how the externally connected capacitor *C* will charge and discharge and how the output will continually switch between its positive ($+V_{CC}$) and negative (0 V) peaks.

To begin with, let us assume that the output of the *S-R* flip-flop is set HIGH, and therefore the output will be HIGH (blue condition in Figure 15-21(a), time T_1 in Figure 15-21(b)). The HIGH output of the *S-R* flip-flop will be inverted to a LOW and turn OFF the 555's internal discharge transistor. With this transistor OFF, the external capacitor *C* can begin to charge towards $+V_{CC}$ via R_A and R_B.

At time T_2, the capacitor's charge has increased beyond 2/3 of V_{CC}, and therefore the output of comparator *A* will go HIGH and RESET the *S-R* flip-flop's output LOW. This will cause the output (pin 3) of the 555 to go LOW. However, the discharge transistor's base will be HIGH, and so it will turn ON. With the discharge transistor ON, the capacitor can begin to discharge (black condition in Figure 15-21(a), time T_2 in Figure 15-21(b)). At time T_3, the capacitor's charge has fallen below 1/3 of V_{CC}, or the trigger level of comparator *B*. As a result, comparator *B*'s output will go HIGH and SET the output of the *S-R* flip-flop HIGH, or back to its original state. The discharge transistor will once again be cut OFF, allowing the capacitor to charge and the cycle to repeat.

As you can see in Figure 15-21(a), the capacitor charges through R_A and R_B to 2/3 of V_{CC}, and then discharges through R_B to 1/3 of V_{CC}. As a result, the positive half-cycle time (t_p) can be calculated with the formula:

$$t_p = 0.7 \times C \times (R_A + R_B)$$

The negative half-cycle time (t_n) can be calculated with the formula:

$$t_n = 0.7 \times C \times R_B$$

The total cycle time will equal the sum of both half-cycles ($t = t_p + t_n$), and the frequency will equal the reciprocal of time ($f = 1/t$).

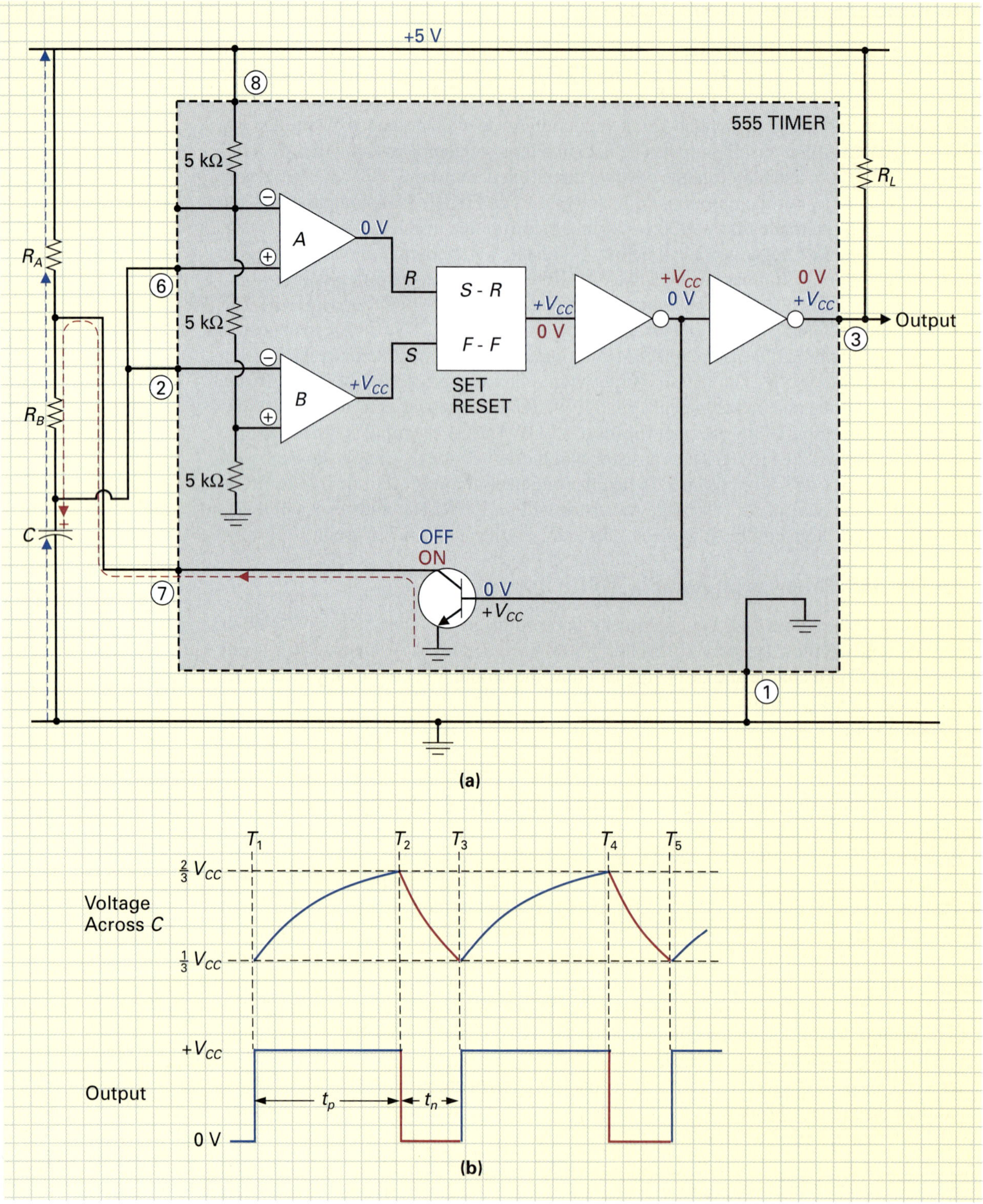

FIGURE 15-21 The 555 Timer as an Astable Multivibrator. (a) Circuit. (b) Waveforms.

EXAMPLE:

Calculate the positive half-cycle time, negative half-cycle time, complete cycle time, and frequency of the 555 astable multivibrator circuit in Figure 15-21, if

$$R_A = 1\ \text{k}\Omega \qquad R_B = 2\ \text{k}\Omega \qquad C = 1\ \mu\text{F}$$

■ ***Solution:***

The positive half-cycle will last for

$$t_p = 0.7 \times C \times (R_A + R_B)$$
$$= 0.7 \times 1\ \mu F \times (1\ k\Omega + 2\ k\Omega) = 2.1\ ms$$

The negative half-cycle will last for

$$t_n = 0.7 \times C \times R_B$$
$$= 0.7 \times 1\ \mu F \times 2\ k\Omega = 1.4\ ms$$

The complete cycle time will be

$$t = t_p + t_n$$
$$= 2.1\ ms + 1.4\ ms = 3.5\ ms$$

The frequency of this 555 astable multivibrator will be

$$f = \frac{1}{t}$$
$$= \frac{1}{3.5\ ms} = 285.7\ Hz$$

---

3. The 555 Timer as a Monostable Multivibrator

Figure 15-22(a) shows how the 555 timer can be connected to operate as a monostable or one-shot multivibrator. The waveforms in Figure 15-22(b) show the time relationships between the input trigger, the charge and discharge of the capacitor, and the output pulse. The width of the output pulse (P_W) is dependent on the values of the external timing components R_A and C.

At time T_1 in Figure 15-22(b), the set-reset flip-flop (*S-R F-F*) is in the reset condition and is therefore producing a LOW output. This LOW from the *S-R F-F* is inverted by the INVERTER, and then inverted and buffered by the final stage to produce a LOW (0 V) output from the 555 timer at pin 3. The LOW output from the *S-R F-F* will be inverted and appear as a HIGH at the base of the discharge transistor, turning it ON and providing a discharge path for the capacitor to ground.

At time T_2, a trigger is applied to pin 2 of the 555 monostable multivibrator. This negative trigger will cause negative input of comparator *B* to fall below 1/3 of V_{CC}, and so the output of comparator *B* will go HIGH and SET the output of the *S-R F-F* HIGH. This HIGH from the *S-R F-F* will send the output of the 555 timer (pin 3) HIGH, and turn OFF the discharge transistor. Once the path to ground through the discharge transistor has been removed from across the capacitor, the capacitor can begin to charge via R_A to $+V_{CC}$, as seen in the waveforms in Figure 15-22(b). The output of the 555 timer remains HIGH until the charge on the capacitor exceeds 2/3 of V_{CC}. At this time (T_3,) the output of comparator *A* will go HIGH, resetting the *S-R F-F* and causing the output of the 555 timer to go LOW and also turning ON the discharge transistor to discharge the capacitor. The circuit will then remain in this stable condition until a new trigger arrives to initiate the cycle once again.

The leading edge of the positive output pulse is initiated by the input trigger, while the trailing edge of the output pulse is determined by the R_A and C charge time, which is dependent on their values. Because the capacitor can charge to 2/3 of V_{CC} in a little more than one time constant (1 time constant = 0.632, 2/3 = 0.667), the following formula can be used to calculate the pulse width (P_W):

$$P_W = 1.1 \times (R_A \times C)$$

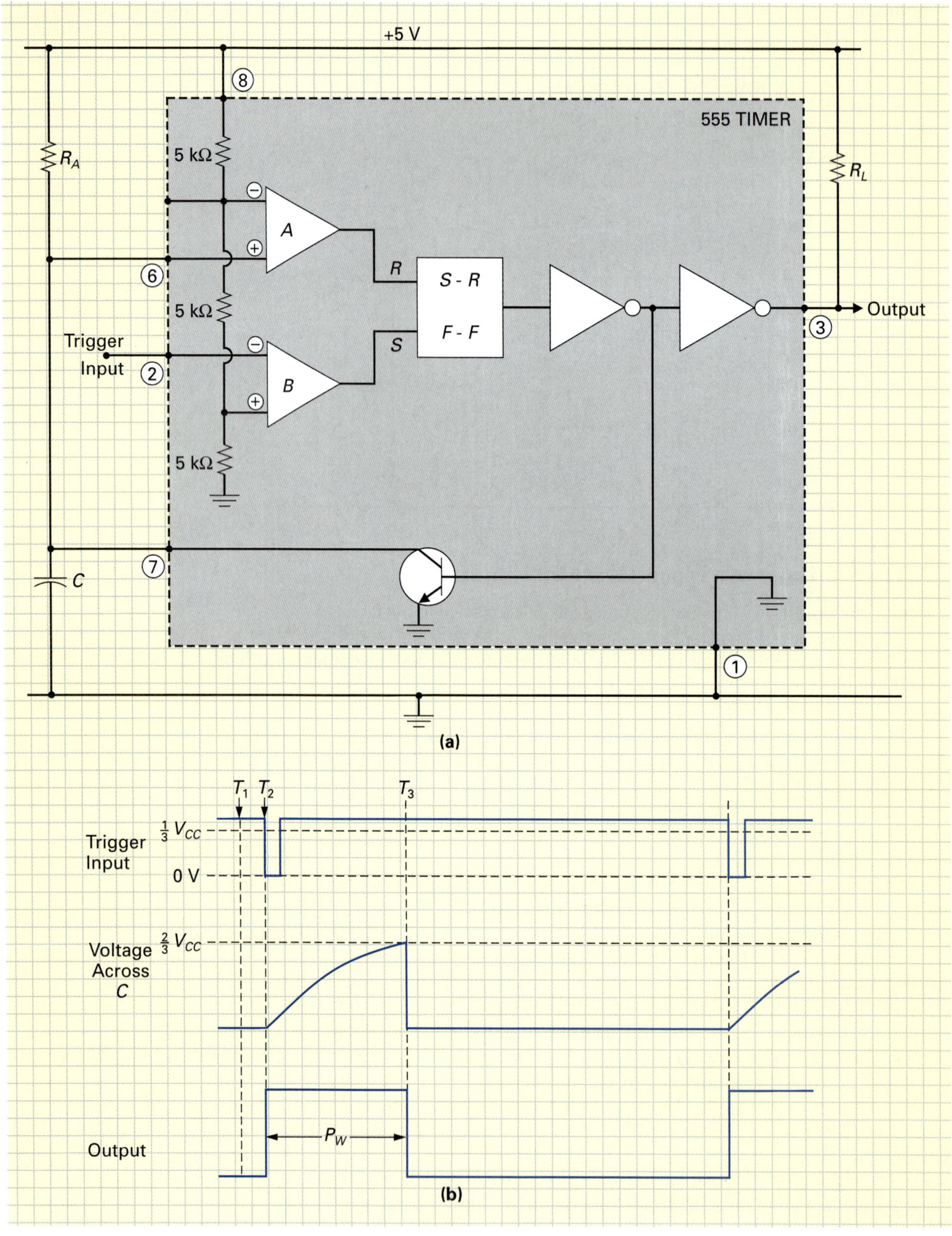

FIGURE 15-22 The 555 Timer as a Monostable Multivibrator. (a) Circuit. (b) Waveforms.

EXAMPLE:

Calculate the pulse width of a 555 monostable multivibrator, if

$$R_A = 2\ \text{M}\Omega, \text{ and } C = 1\ \mu\text{F}.$$

Solution:

The width of the output pulse will be

$$\begin{aligned} P_W &= 1.1 \times (R_A \times C) \\ &= 1.1 \times (2\ \text{M}\Omega \times 1\ \mu\text{F}) = 2.2 \end{aligned}$$

15-3-6 *Crystal Oscillator Circuits*

The clock generator circuits discussed in the previous section would typically have a 0.8% frequency drift. This frequency drift is due to circuit temperature changes, the aging of the components, and changes in the resistance of the load connected to the oscillator. If an *RC* oscillator were used to generate the timing signal for a wristwatch, a 0.8% frequency drift would mean that the watch may gain or lose approximately $11\frac{1}{2}$ minutes in a day.

$$60 \text{ minutes} \times 24 \text{ hours} = 1{,}440 \text{ minutes per day}$$
$$0.8\% \text{ of } 1{,}440 = 11.5 \text{ minutes per day}$$

If a high degree of oscillator stability is needed, crystal oscillators are used to generate the timing signal. Crystal-controlled oscillators will typically have a frequency drift of 0.0001%, and if used in a wristwatch, may gain or lose a maximum of half a minute a year.

$$60 \text{ minutes} \times 24 \text{ hours} \times 365 \text{ days} = 525{,}600 \text{ minutes per year}$$
$$0.0001\% \text{ of } 525{,}600 = 0.5 \text{ minutes per year}$$

Therefore, in applications where a high degree of frequency stability is needed, such as communication and computer systems, crystal-controlled oscillators are used.

1. The Characteristics of Crystals

The quartz crystal is made of silicon dioxide and is naturally a six-sided (hexagonal) compound with pyramids at either end, as seen in Figure 15-23(a). To construct an electronic component, a thin slice or slab of crystal is cut from the mother stone, mounted between two metal plates that make electrical contact, then placed in a protective holder, as seen in Figure 15-23(b). On a schematic diagram, a crystal is generally labeled either *XTAL* or *Y*, and has the symbol shown in Figure 15-23(c).

The crystal is basically operated as a transducer, or energy converter, transforming mechanical energy to electrical energy, as shown in Figure 15-23(d), (e), and (f), or electrical energy to mechanical energy, as shown in Figure 15-23(g).

2. Mechanical-to-Electrical Conversion

In Figure 15-23(d), you can see that a crystal normally has its internal charges evenly distributed throughout, and the potential difference between its two plates is zero. If the crystal is compressed by applying pressure to either side, as shown in Figure 15-23(e), opposite charges accumulate on either side of the crystal and a potential difference is generated. Similarly, if the crystal is expanded by applying pressure to the top and bottom, as shown in Figure 15-23(f), opposite charges accumulate on either side of the crystal, and a potential difference of the opposite polarity is generated. If a crystal were subjected to an alternating pressure that caused it to continually expand and compress, an alternating, or ac, voltage would be generated. Crystal microphones make use of this principle. Sound waves are applied to

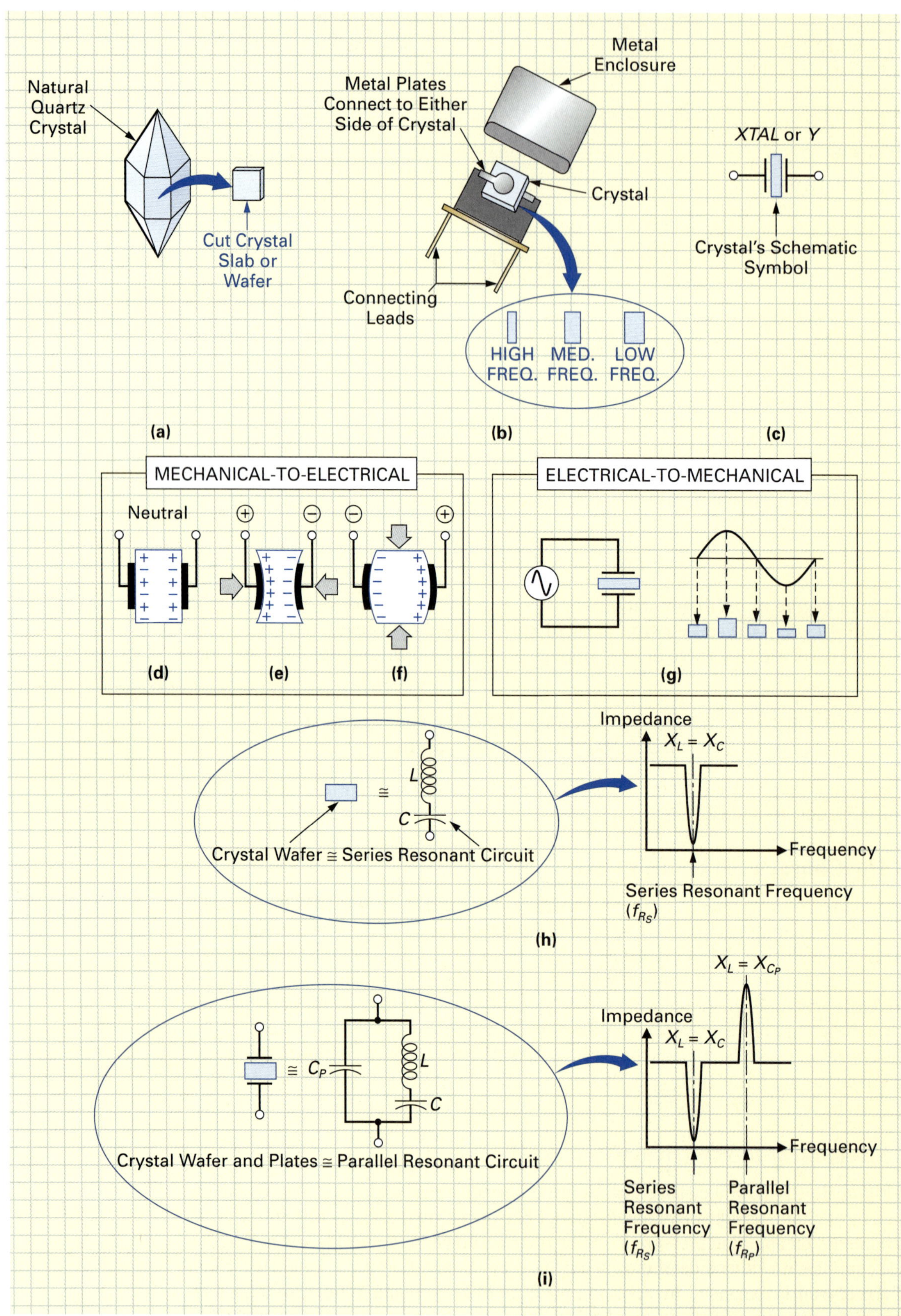

FIGURE 15-23 The Characteristics of Crystals.

the microphone, and these mechanical waves continually compress and expand the crystal, generating an electrical wave that is equivalent to the original sound wave.

3. Electrical-to-Mechanical Conversion

Due to their composition, crystals have a "natural frequency of vibration." This means that if the ac voltage applied across a crystal matches the crystal's natural frequency of vibration, as shown in Figure 15-23(g), the crystal will physically expand and contract by a relatively large amount. If, on the other hand, the frequency of the applied ac voltage is either above or below the crystal's natural frequency, the vibration is only slight. Crystals are, therefore, very frequency selective—a characteristic we can make use of in filter circuits and oscillator circuits. This action is called the **piezoelectric effect,** which by definition is the tendency of a crystal to vibrate at a constant rate when it is subjected to a changing electric field produced by an applied ac voltage. The crystal's natural frequency of vibration is dependent on the thickness of the crystal between the two plates, as seen in the inset in Figure 15-23(b). By cutting a crystal to the right size, we can obtain a crystal that will naturally vibrate at an exact frequency. This rating is usually printed on the crystal's package.

Piezoelectric Effect
The tendency of a crystal to vibrate at a constant rate when subjected to a changing electric field.

When the frequency of the ac voltage applied to a crystal matches the crystal's natural frequency of vibration, the impedance or current opposition of the crystal drops to a minimum, as seen in Figure 15-23(h). This makes the crystal wafer or slab equivalent to a series resonant circuit, and, as you can see in the frequency response curve in Figure 15-23(h), the very sharp response at the crystal's series resonant frequency (f_{R_S}) means that the crystal is extremely frequency selective.

When a crystal is mounted between two connecting plates, the device is equivalent to the crystal's original series resonant circuit in parallel with a small value of capacitance due to the connecting plates (C_P), as seen in Figure 15-23(i). Referring to the frequency response curve in Figure 15-23(i), you can see that the crystal component will still respond in exactly the same way when the applied ac frequency is equal to the crystal's natural series resonant frequency. As frequency is increased beyond the series resonant frequency of the crystal, however, the reactance of C will decrease, and the crystal will become inductive. At a higher frequency, there will be a point at which the inductive reactance of the crystal is equal to the capacitive reactance of the connecting plates ($X_L = X_{C_P}$). Because the crystal's inductance is in parallel with the mounting plate's capacitance, the two form a parallel resonant circuit. Therefore, the device's impedance will be maximum at this parallel resonant frequency (f_{R_P}), as shown in Figure 15-23(i). Circuit applications can make use of either the crystal's series resonant selectivity or parallel resonant selectivity.

Most crystal oscillator circuits will make use of the crystal's series resonant frequency response to feed back only the desired frequency to the input and tightly maintain the frequency stability of the oscillator. A crystal's series resonant tuned circuit will typically have a very high Q of 40,000. If you compare this to an LC tuned circuit that would typically have a Q of 200, it is easy to see why crystal-controlled oscillators are much more stable than LC-controlled oscillators.

4. Crystal Oscillator

Figure 15-24(a) shows how an astable multivibrator clock oscillator can be constructed using a Schmitt-trigger INVERTER, with a resistor and capacitor (RC) determining the output frequency. When power is first applied to this circuit, the capacitor will have no charge and this LOW input is inverted, giving a HIGH output (seen as red in the illustration). The capacitor (C) will begin to charge via the resistor (R), and the increasing positive charge across the capacitor will be felt at the input of the INVERTER. After a time (which is dependent on the values of R and C), the capacitor charge will be large enough to apply a valid HIGH to the INVERTER input. This HIGH input will cause the output of the INVERTER to go LOW, so the capacitor will begin to discharge (seen as blue in illustration). When the capacitor's charge falls to a valid LOW logic level, the INVERTER will generate a HIGH output, and the cycle will repeat.

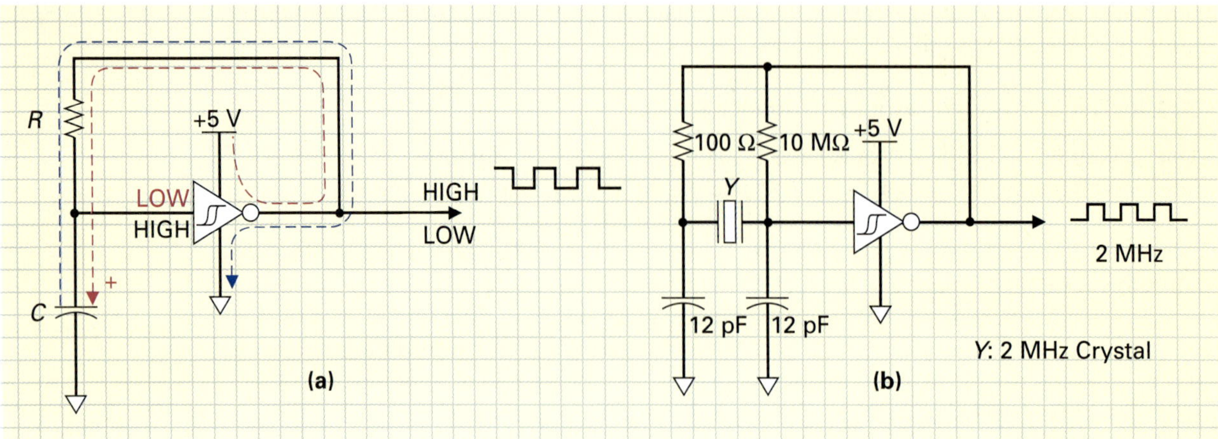

FIGURE 15-24 **Clock Oscillators. (a) *RC* Controlled. (b) Crystal Controlled.**

Figure 15-24(b) shows how the same basic circuit configuration, but in this instance with a crystal acting as the frequency determining component, can be connected to generate a highly accurate 2 MHz clock signal.

SELF-TEST EVALUATION POINT FOR SECTION 15-3

Use the following questions to test your understanding of Section 15-3.

1. The astable multivibrator is also called the ____________ multivibrator.
2. Which of the multivibrator circuits has only one stable state?
3. Which multivibrator is also called a set-reset flip-flop?
4. Why is the set-reset flip-flop also called a latch?
5. The 555 timer derived its number identification from ____________.
6. List two applications of the 555 timer.
7. What is the key advantage of the crystal oscillator?

15-4 FUNCTION GENERATORS

A function generator is by definition a signal generator circuit that delivers a variety of different output waveforms whose frequency can be varied. In this section, we will examine some of the different linear function generator ICs available.

Voltage Controlled Oscillator (VCO)
An oscillator whose frequency can be controlled by an input control voltage.

Control Voltage
A voltage signal that starts, stops, or adjusts the operation of a device, circuit, or system.

15-4-1 *Voltage Controlled Oscillators (VCOs)*

In many communication circuit applications, we need an oscillator whose frequency can be controlled by an input "control voltage." The circuit that achieves this function is the **voltage controlled oscillator (VCO)** or, as it is sometimes called, a "voltage-to-frequency converter." The NE/SE 566 shown in Figure 15-25 is a good example of a linear IC VCO which can be connected to generate both a square wave and triangular wave output. The frequency of oscillation is determined by external resistors R_1 and R_2 and capacitor C_1, which determine the **control voltage** applied to pin 5. The triangular wave is generated by linearly charging and discharging the external capacitor C_1, using the 566's internal current source. The charge

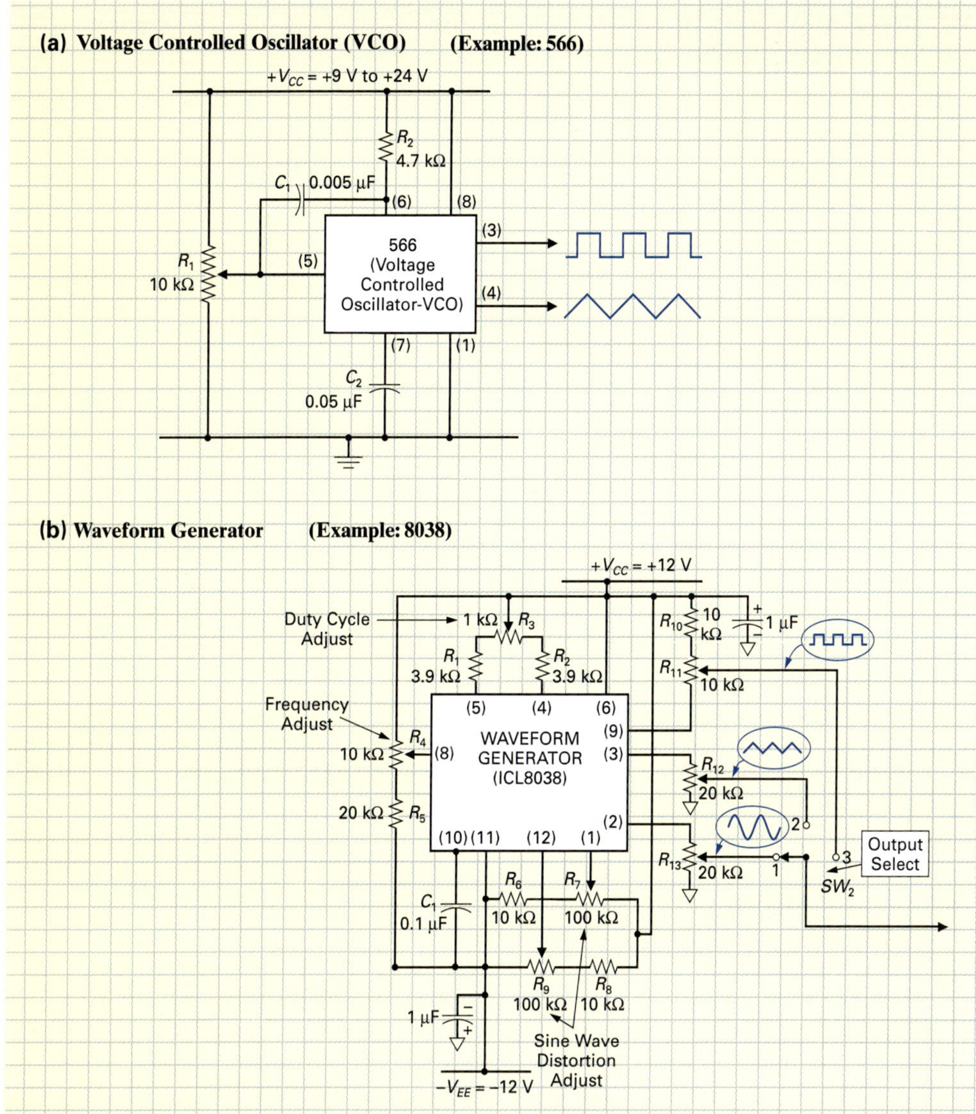

FIGURE 15-25 **Linear IC Function Generators.**

and discharge levels are determined by the 566's internal Schmitt trigger, which is also used to generate the square wave output. The triangular wave will typically have an amplitude of 2.4 V pk-pk, and the square wave will typically have an amplitude of 5.4 V pk-pk.

15-4-2 *Waveform Generators*

The 8038 function generator circuit in Figure 15-25(b) produces a 20 Hz to 100 kHz sine wave, square wave, or triangular wave output. Resistor R_4 is used to adjust the frequency output to the desired frequency. Switch SW_2 is used to select either the square wave, triangular wave, or sine wave output. The output amplitude of the square wave can be varied by R_{11}, and the duty cycle of the square wave can be adjusted by R_3. Similarly, the triangular wave's output amplitude can be adjusted by R_{12} and the sine wave's output amplitude can be adjusted by R_{13}. Variable resistors R_7 and R_9 are included to fine tune the sine wave output for minimum distortion.

SELF-TEST EVALUATION POINT FOR SECTION 15-4

Use the following questions to test your understanding of Section 15-4.

1. Give the full name of the abbreviation VCO.
2. A VCO IC generates a repeating output waveform based on a __________ at its input.
3. Which three waveforms are generated by the 8038 function generator IC?

15-5 PHASE-LOCKED LOOPS

Phase-Locked Loop (PLL) Circuit
A circuit consisting of a phase comparator that compares the output frequency of a voltage controlled oscillator with an input frequency. The error voltage out of the phase comparator is then coupled via an amplifier and low-pass filter to the control input of the voltage controlled oscillator to keep it in phase, and therefore at exactly the same frequency as the input.

Error Voltage
A voltage that is proportional to the error that exists between input and output.

Locked
To automatically follow a signal.

Capture
The act of gaining control of a signal.

Free-Running
Operating without any external control.

The operation of a **phase-locked loop** is best described by referring to the block diagram of the LM 565 PLL linear IC shown in Figure 15-26(a). As can be seen in this diagram, the PLL contains a phase comparator, amplifier and low-pass filter, and a voltage controlled oscillator. The input frequency (f_{IN}) is applied to one input of the phase comparator and compared with the output frequency (f_{OUT}) from the VCO. If a frequency difference exists between the two comparator inputs, the comparator will detect the difference in phase and generate an **error voltage,** which will be amplified and filtered before it is applied as the control input voltage to the VCO. If the input frequency (f_{IN}) and VCO output frequency (f_{OUT}) differ, the error voltage generated by the phase comparator will shift the VCO frequency until it matches the input frequency, and therefore $f_{IN} = f_{OUT}$. When f_{IN} matches the VCO frequency or f_{OUT}, the PLL is said to be in the **locked** condition. If the input frequency is varied, the VCO output frequency will follow or track f_{IN}, provided that the input frequency is within the capture range of the PLL. Referring to the inset in Figure 15-26(a), you can see how the error voltage from the phase comparator varies while the PLL is trying to **capture** the input frequency. When the input frequency is outside the capture range of the PLL, the VCO will generate a **free-running** or idle frequency.

To summarize: a phase-locked loop circuit consists of a phase comparator that compares the output frequency of a voltage controlled oscillator with an input frequency. The error voltage out of the phase comparator is then coupled via an amplifier and low-pass filter to the control input of the voltage controlled oscillator to keep it in phase—and therefore at exactly the same frequency—as the input frequency. These phase-locked loop circuits will operate in either the free-running state, capture state, or locked state.

To demonstrate an application, Figure 15-26(b) shows how a PLL could be connected to generate three stable output frequencies. When switch 1 is in the lower position, the output of the VCO is fed directly back to the phase comparator, and the circuit will operate in exactly the same way as the circuit in Figure 15-26(a). It will generate an output frequency that tracks the input frequency and is a very stable 1 kHz due to the locking action of the PLL. When switch 1 is in the mid position, a divide-by-five circuit is switched into circuit between the VCO output and the phase comparator input. Referring to the waveforms in Figure 15-26(b), you can see that the divide-by-five circuit will generate one output cycle for every five input cycles. This means that the VCO feedback signal applied to the phase comparator will be one-fifth of the output frequency. Because this signal will not be in phase with the 1 kHz signal from the crystal oscillator, the phase comparator will generate an error voltage instructing the VCO to increase its output frequency by five times. When the VCO generates an output frequency of 5 kHz, the divide-by-five circuit will produce a 1 kHz output (5 kHz $\div$ 5 = 1 kHz) to the phase comparator, which will match the phase comparator's other 1 kHz input from the crystal oscillator. As a result, the error voltage from the phase comparator will be zero and the VCO—and therefore the output—will remain locked at 5 kHz. If switch SW_1 of the circuit in Figure 15-26(b) is put in the upper position, a divide-by-ten circuit will be included between the VCO output and the phase comparator input. As a result, the PLL will generate an output frequency that is ten times the input frequency ($f_{OUT} = 10 \times f_{IN} = 10 \times 1\text{ kHz} = 10\text{ kHz}$).

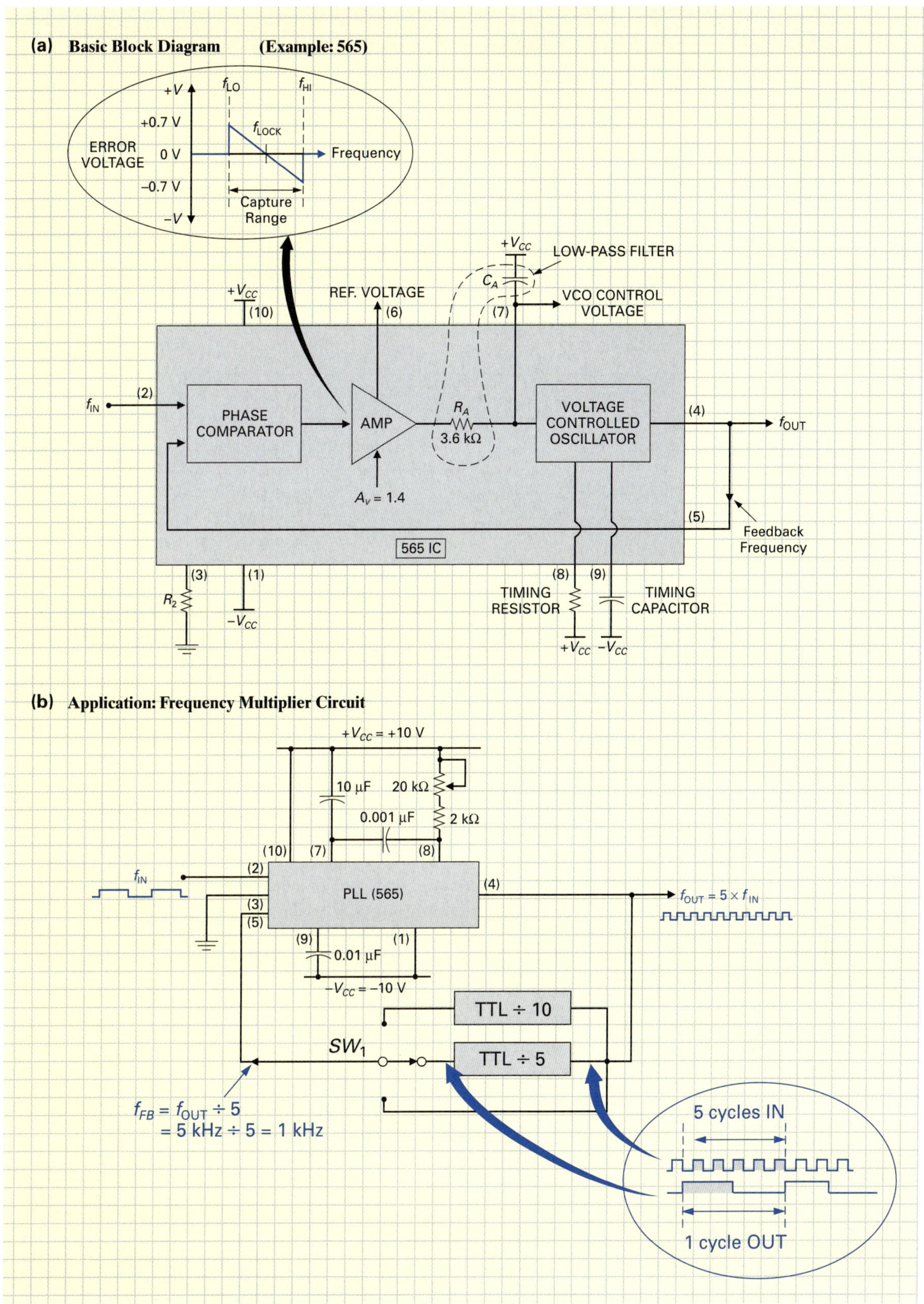

FIGURE 15-26 Linear IC Phase Locked Loops (PLLs).

SELF-TEST EVALUATION POINT FOR SECTION 15-5

Use the following questions to test your understanding of Section 15-5.

1. What are the three basic blocks in a phase-locked loop circuit?
2. Briefly describe the operation of the phase-locked loop.
3. What is included in the feedback path between the VCO and the phase comparator of a PLL when the PLL is used as a frequency multiplier?
4. In some applications, the PLL will be used to generate an output voltage that follows or tracks changes in frequency at the input. In such an application, which pin of the 565 PLL would be used to obtain this varying voltage?

REVIEW QUESTIONS

Multiple-Choice Questions

1. When a differential amplifier is used in the differential-input, single-output mode, it has a ____________ differential gain and a ____________ common-mode gain.
   a. high, high **c.** low, low
   b. high, low **d.** low, high
2. The op-amp's internal circuit contains a ____________, ____________, and ____________ amplifier stage.
   a. differentiator, current, power
   b. integrator, voltage, output
   c. darlington-pair, emitter-follower, summing
   d. differential, darlington-pair, emitter-follower
3. The op-amp's differential-amplifier stage provides the op-amp with a
   a. low common mode gain
   b. high differential gain
   c. high input impedance
   d. all of the above
4. Which transistor circuit is used in the op-amp's final output stage to provide a low output impedance and high current gain?
   a. Common-emitter **c.** Common-collector
   b. Common-base **d.** Both (a) and (c)
5. Could an op-amp circuit be constructed using discrete components?
   a. Yes **b.** No
6. The comparator is considered a/an ____________ loop op-amp circuit.
   a. common **c.** differential
   b. open **d.** closed
7. What is the lower frequency limit of an op-amp?
   a. 20 Hz **c.** dc
   b. 6 Hz **d.** 7.34 Hz
8. A virtual ground is a ground to ____________ but not to ____________.
   a. current, voltage
   b. voltage, current
9. The feedback loop in a closed-loop op-amp circuit provides
   a. positive feedback
   b. negative feedback
   c. degenerative feedback
   d. both (a) and (b)
   e. both (b) and (c)
10. The ____________ input(s) of an op-amp is used to compensate for slight differences in the transistors in the differential-amplifier stage.
    a. inverting **c.** $+V$
    b. dc offset **d.** V_{out}
11. Which semiconductor manufacturer was the first to develop an op-amp IC?
    a. National Semiconductor **c.** Fairchild
    b. Motorola **d.** Signetics
12. Which of the following is not a linear IC?
    a. Voltage regulator
    b. Operational amplifier
    c. VCO
    d. NAND gate
13. What would be the polarity and voltage output from a 7818 voltage regulator?
    a. −8 V
    b. +18 V
    c. +1.2 V to +37 V
    d. −18 V
14. What is the difference between a 78XX and 79XX voltage regulator?
    a. The 78XX has a variable output
    b. The 79XX has a + output
    c. The 78XX has a + output
    d. The 79XX has a variable output
15. Which of the following is an example of a series dissipative regulator?
    a. Zener regulator
    b. 317 regulator
    c. 7912 regulator
    d. All of the above

16. What advantage does the series dissipative have over the series switching regulator?
   a. Circuit simplicity
   b. Conversion efficiency
17. The frequency output from a VCO is determined by
   a. An *LC* network
   b. An *RC* network
   c. A control voltage
   d. A crystal
18. A PLL circuit contains a
   a. Crystal oscillator
   b. Bypass filter
   c. VCO
   d. VCR
19. What is the similarity between a switching voltage regulator and a phase-locked loop circuit?
   a. Both generate a constant output voltage
   b. Both employ a feedback mechanism to control operation
   c. Both generate a high-frequency output
   d. Both make use of high-pass filters
20. Which is the most frequently used linear IC?
   a. Voltage regulator
   b. Comparator
   c. Function generator
   d. Op-amp

Practice Problems

21. Explain the meaning of the following op-amp manufacturer codes:
   a. LM 318C N **b.** NE 101C D
22. Calculate the common-mode rejection ratio of an op-amp if it has a common-mode gain of 0.8 and a differential gain of 27,000. Also give the answer in dBs.
23. Identify the circuit shown in Figure 15-27. Why must the input to this circuit always be a negative voltage?
24. What would be the output voltage from the circuit in Figure 15-27 if the input voltage were −1.6 V?
25. Identify the circuits shown in Figure 15-28(a) and (b), and then sketch the shape of the output waveform if a square wave were applied to the inputs.

FIGURE 15-27 A 741C Op-Amp Circuit.

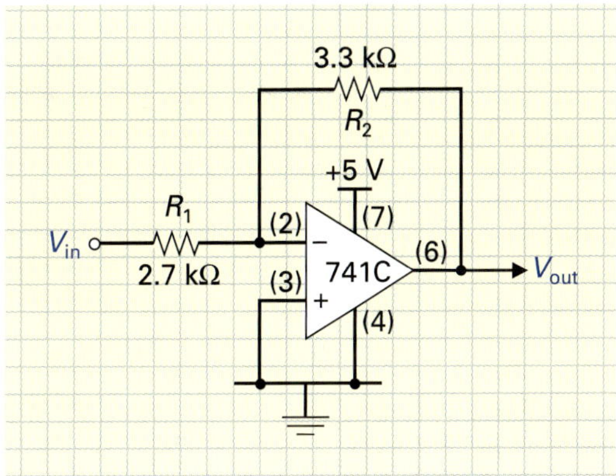

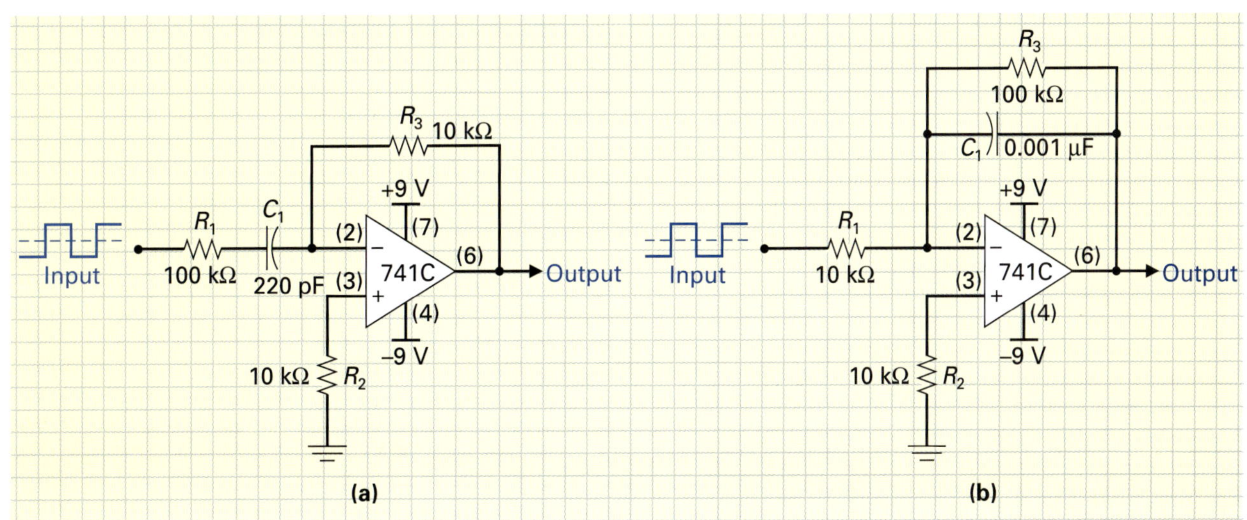

FIGURE 15-28 Two Applications for the 741C Op-Amp.

26. Referring to the circuit shown in Figure 15-29, what would be the voltage out if a +7.3-V input were applied?

27. Identify the circuits shown in Figure 15-30(a) and (b), and then calculate the output voltages for the given input voltages.

28. Which of the circuits shown in Figure 15-31 is an active high-pass filter and which is an active low-pass filter?

29. Calculate the cutoff frequency of the circuit in Figure 15-31(a) if $C_1 = C_2 = 0.1\ \mu F$, $R_1 = 10\ k\Omega$.

30. Calculate the cutoff frequency of the circuit in Figure 15-31(b), if $R_1 = R_2 = 33\ k\Omega$, $C_1 = 0.33\ \mu F$.

FIGURE 15-29 An Op-Amp Circuit.

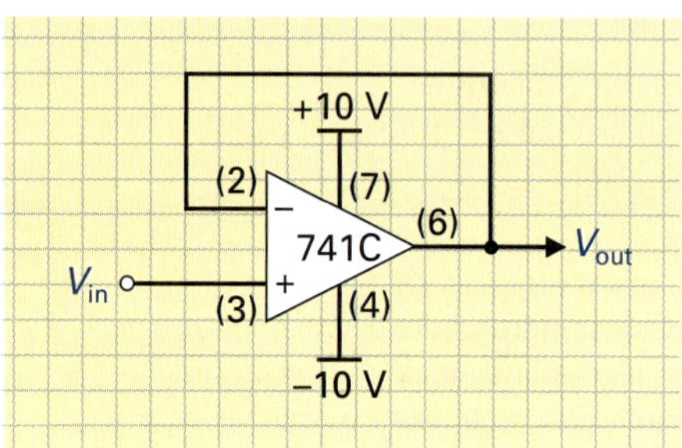

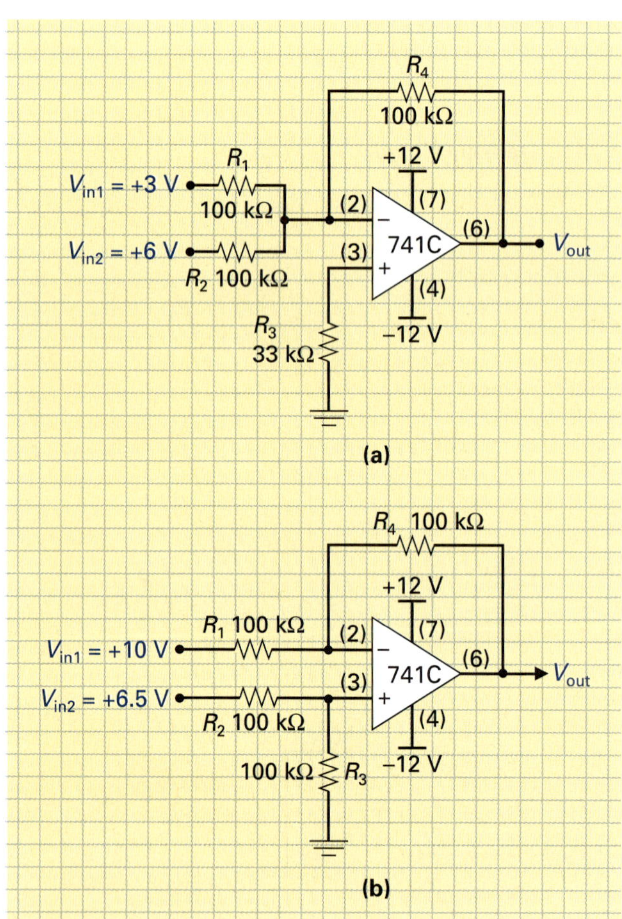

FIGURE 15-30 Op-Amp Circuit Examples.

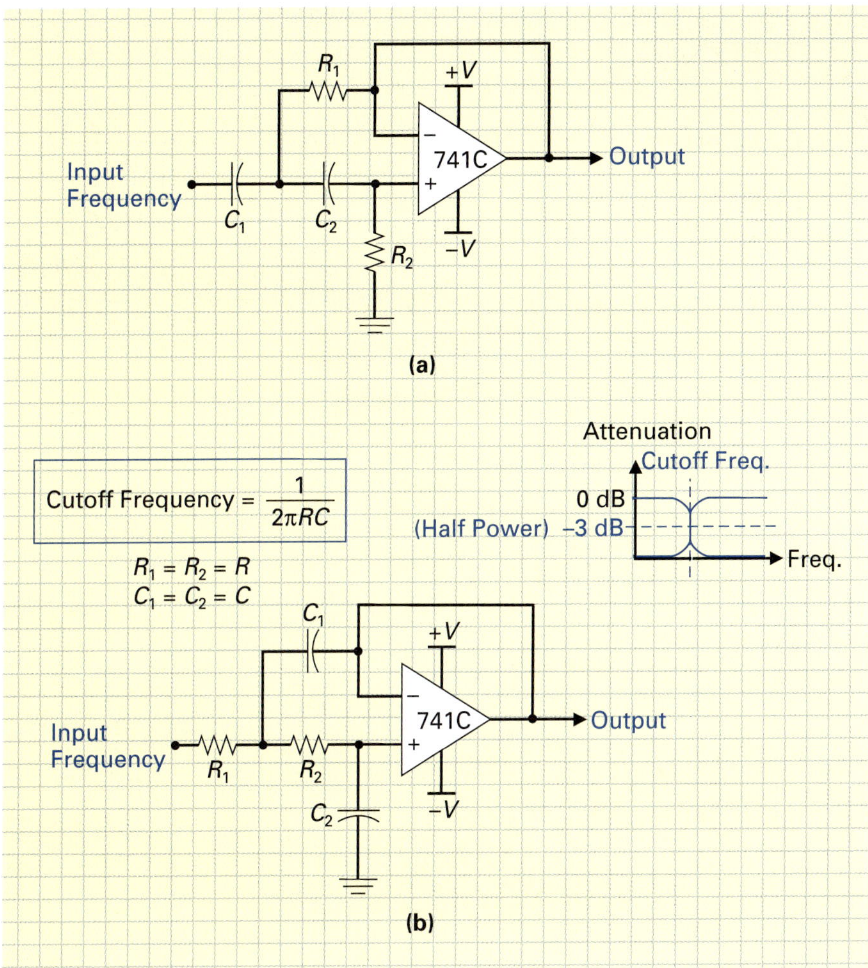

FIGURE 15-31 Op-Amp Active Filter Circuits.

31. To improve your circuit recognition and operation ability, study the circuit in Figure 15-32 and then answer the following questions:

a. Determine the output voltage and polarity at points *A*, *B*, *C* and *D*.

b. Describe whether the regulators used are fixed-positive, fixed-negative, variable-positive, variable-negative, or switching types.

c. What circuit is formed by diodes D_9 and D_{10}?

d. What is the purpose of the 555 timer?

32. To improve your circuit recognition and operation ability, study the circuit in Figure 15-33 and then answer the following questions:

a. What type of regulator is being used?

b. Why do you think this circuit is called a programmable dc power supply circuit?

c. Briefly describe the operation of this circuit.

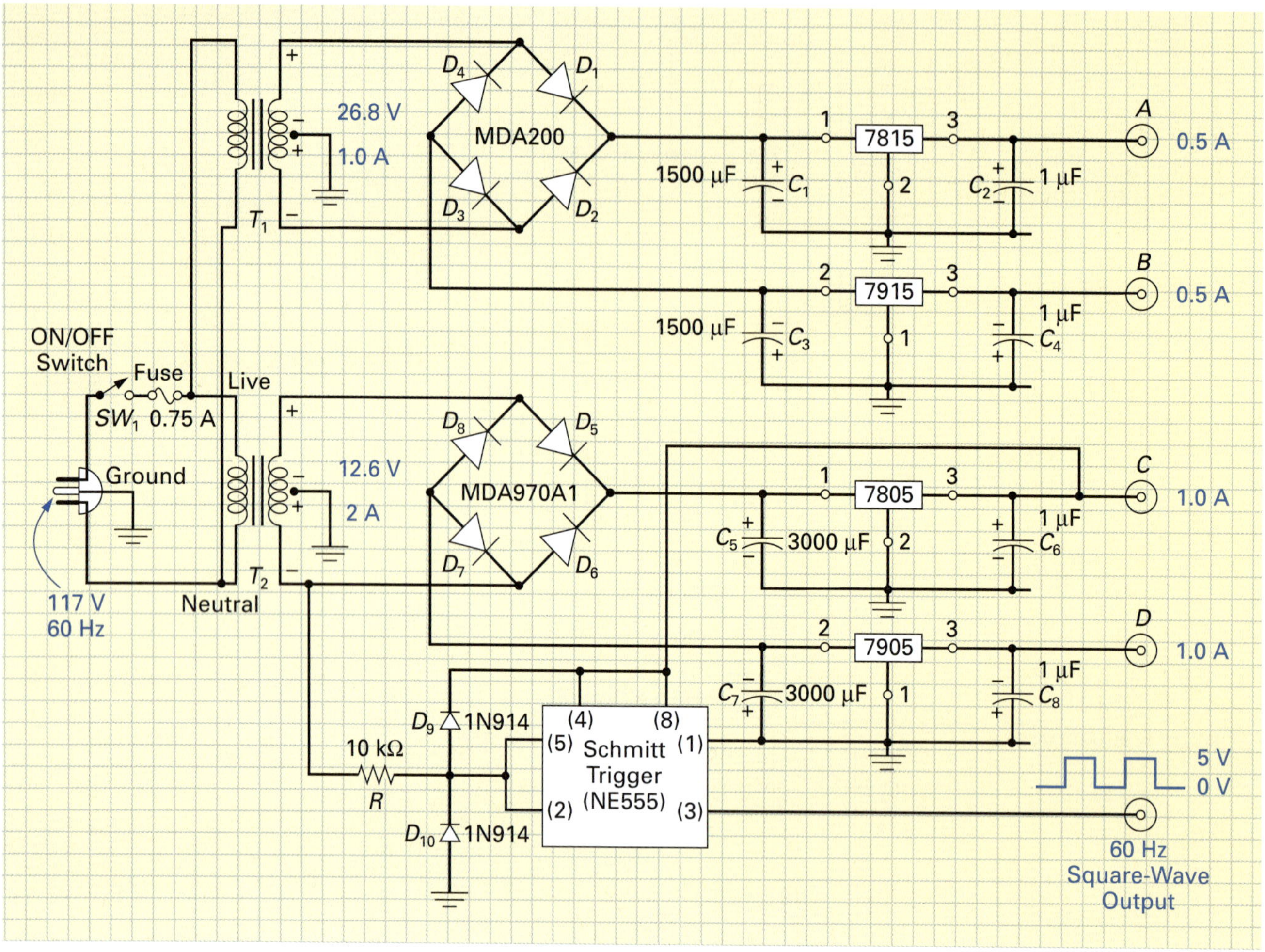

FIGURE 15-32 Typical DC Power Supply Circuit.

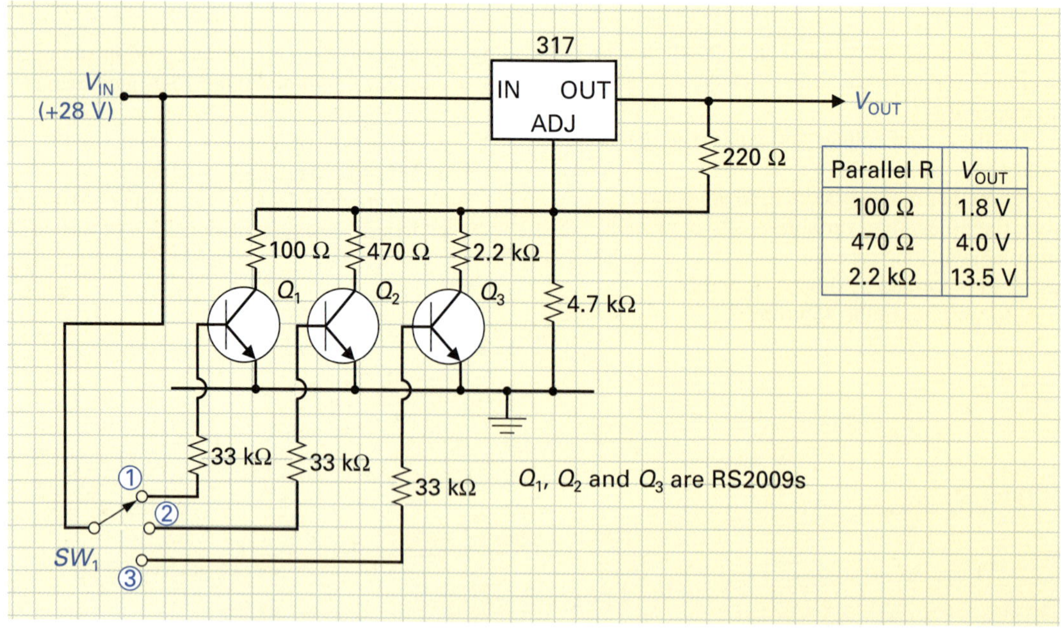

Parallel R	V_{OUT}
100 Ω	1.8 V
470 Ω	4.0 V
2.2 kΩ	13.5 V

FIGURE 15-33 A Programmable DC Power Supply Circuit.

33. Referring to Figure 15-34,
    a. What is the linear IC being used?
    b. What would be the waveform output at points A and B?
    c. What is the purpose of R_1?
34. Referring to Figure 15-35,
    a. What is the full name of the linear IC being used?
    b. What would be the output frequency for each of the switch positions?

Web Site Questions

Go to the web site http://www.prenhall.com/cook, select the textbook *Electronics: A Complete Course,* select this chapter, and then follow the instructions when answering the multiple choice practice problems.

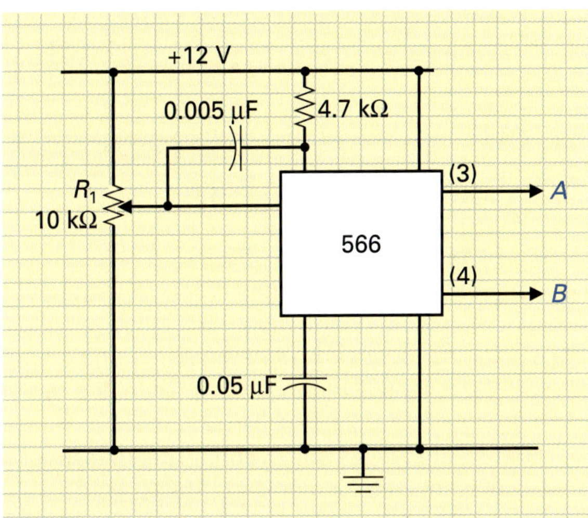

FIGURE 15-34 **An Oscillator Circuit.**

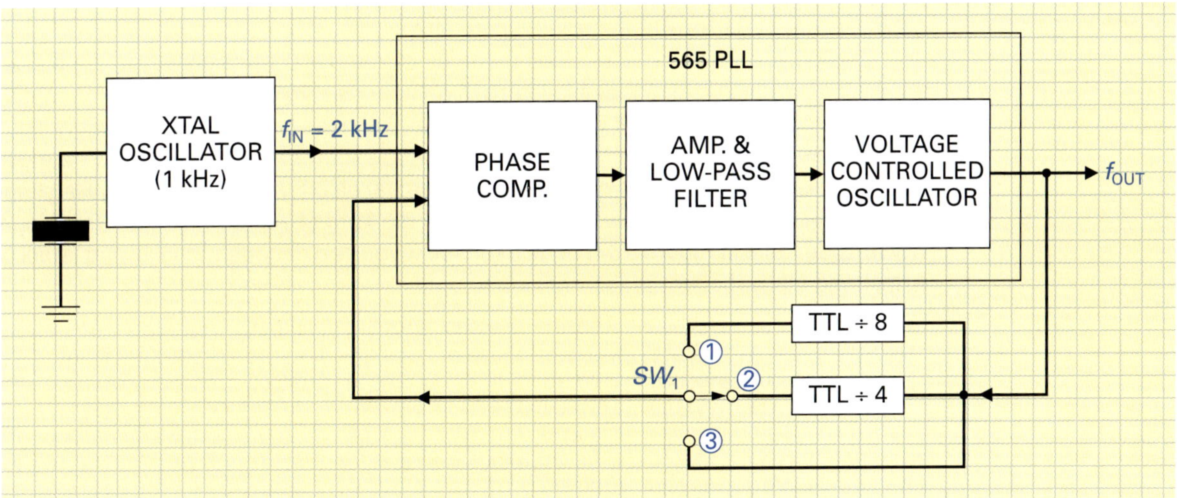

FIGURE 15-35 **A Frequency Multiplier Circuit.**

Analog to Digital

Moon Walk

In July of 1969 almost everyone throughout the world was caught at one stage or another gazing up to the stars and wondering. For U.S. astronaut Neil Armstrong, the age-old human dream of walking on the moon was close to becoming a reality. On earth, millions of people and thousands of newspapers and magazines waited anxiously to celebrate a successful moon landing. Then, the rather brief but eloquent message from Armstrong was transmitted 240,000 miles to Houston, Texas, where it was immediately retransmitted to a waiting world. This message was: "That's one small step for man, one giant leap for mankind." These words were received by many via television, but for magazines and newspapers this entire mission, including the speech from Armstrong, was converted into a special code made up of ON–OFF pulses that traveled from computer to computer. Every letter of every word was converted into a code that used the two symbols of the binary number system, 0 and 1. This code is still used extensively by all modern computers and is called *American Standard Code for Information Interchange,* or ASCII (pronounced "askey").

It is only fitting that these codes made up of zeros and ones conveyed the finale to this historic mission, since they played such an important role throughout. Commands encoded into zeros and ones were used to control almost everything, from triggering the takeoff to keeping the spacecraft at the proper angle for reentry into the earth's atmosphere.

No matter what its size or application, a digital electronic computer is quite simply a system that manages the flow of information in the form of zeros and ones. Referring to the ASCII code table in Figure 16-2 on p. 596, see if you can decode the following famous Armstrong message that reads from left to right:

0100010 1010100 1101000 1100001 1110100 0100111 1110011 0100000 1101111 1101110
1100101 0100000 1110011 1101101 1100001 1101100 1101100 0100000 1110011 1110100
1100101 1110000 0100000 1100110 1101111 1110010 0100000 1100001 0100000 1101101
1100001 1101110 0101100 0100000 1101111 1101110 1100101 0100000 1100111 1101101
1100001 1101110 1110100 0100000 1101100 1100101 1100001 1110000 0100000 1100110
1101111 1110010 0100000 1101101 1100001 1101110 1101011 1101001 1101110 1100100
0101110 0100010 0100000 0100000 0101101 1001110 1100101 1101001 1101100 0100000
1000001 1110010 1101101 1110011 1110100 1110010 1101111 1101110 1100111 0101100
0100000 1000001 1110000 1101111 1101100 1101100 1101111 0100000 0110001 0110001

16

Introduction

Since World War II, no branch of science has contributed more to the development of the modern world than electronics. It has stimulated dramatic advances in the fields of communication, computing, consumer products, industrial automation, test and measurement, and health care. It has now become the largest single industry in the world, exceeding the automobile and oil industries, with annual sales of electronic systems exceeding $2 trillion.

One of the most important trends in this huge industry has been a gradual shift from *analog electronics* to *digital electronics.* This movement began in the 1960s and is almost complete today. In fact, a recent statistic stated that, on average, 90% of the circuitry within electronic systems is now digital and only 10% is analog. This digitalization of the electronics industry is bringing sectors that were once separate closer together. For example, two of the largest sectors or branches of electronics are *computing* and *communications.* Being able to communicate with each other using the common language of digital has enabled computers and communications to interlink so that computers can now function within communication-based networks, and communications networks can now function through computer-based systems. Industry experts call this merging *convergence* and predict that digital electronics will continue to unite the industry and stimulate progress in practically every field of human endeavor.

Needless to say, this part of the book you are about to read involving digital electronic mathematics and concepts is an essential element in your study of electronics.

Analog
The representation of physical properties by a proportionally varying signal.

AC Analog Signal
An analog signal that alternates positive and negative.

AC Amplifier Circuit
An amplifier designed to increase the magnitude of an ac signal.

Analog Device or Component
A device or component that makes up a circuit designed to manage analog signals.

Analog Circuit
A circuit designed to manage analog signals.

DC Analog Signal
An analog signal that is always either positive or negative.

DC Amplifier Circuit
An amplifier designed to increase the magnitude of a dc signal.

Linear Circuit
A circuit in which the output varies in direct proportion to the input.

16-1 ANALOG AND DIGITAL DATA AND DEVICES

Before we begin, let's first review analog electronics and then compare it to digital, so that we can clearly see the differences between the two.

16-1-1 *Analog Data and Devices*

Figure 16-1(a) shows an electronic circuit designed to amplify speech information detected by a microphone. One of the easiest ways to represent data or information is to have a voltage change in direct proportion to the information it is representing. In the example in Figure 16-1(a), the *pitch and loudness* of the sound waves applied to the microphone should control the *frequency and amplitude* of the voltage signal from the microphone. The output voltage signal from the microphone is said to be an analog of the input speech signal. The word **analog** means "similar to," and in Figure 16-1(a) the electronic signal produced by the microphone is an analog (or similar) to the speech signal, since a change in speech "loudness or pitch" will cause a corresponding change in signal voltage "amplitude or frequency."

In Figure 16-1(b), a light detector or solar cell converts light energy into an electronic signal. This information signal represents the amount of light present, since changes in voltage amplitude result in a change in light-level intensity. Once again, the output electronic signal is an analog (or similar) to the sensed light level at the input.

Figure 16-1, therefore, indicates two analog circuits. The microphone in Figure 16-1(a) generates an **ac analog signal** that is amplified by an **ac amplifier circuit.** The microphone is considered an **analog device** or **component** and the amplifier an **analog circuit.** The light detector in Figure 16-1(b) would also be an analog component; however, in this example it generates a **dc analog signal** that is amplified by a **dc amplifier circuit.**

Both of the information signals in Figure 16–1 vary smoothly and continuously, in accordance with the natural quantities they represent (sound and light). Analog circuits are often called **linear circuits,** since linear by definition is *the variation of an output in direct*

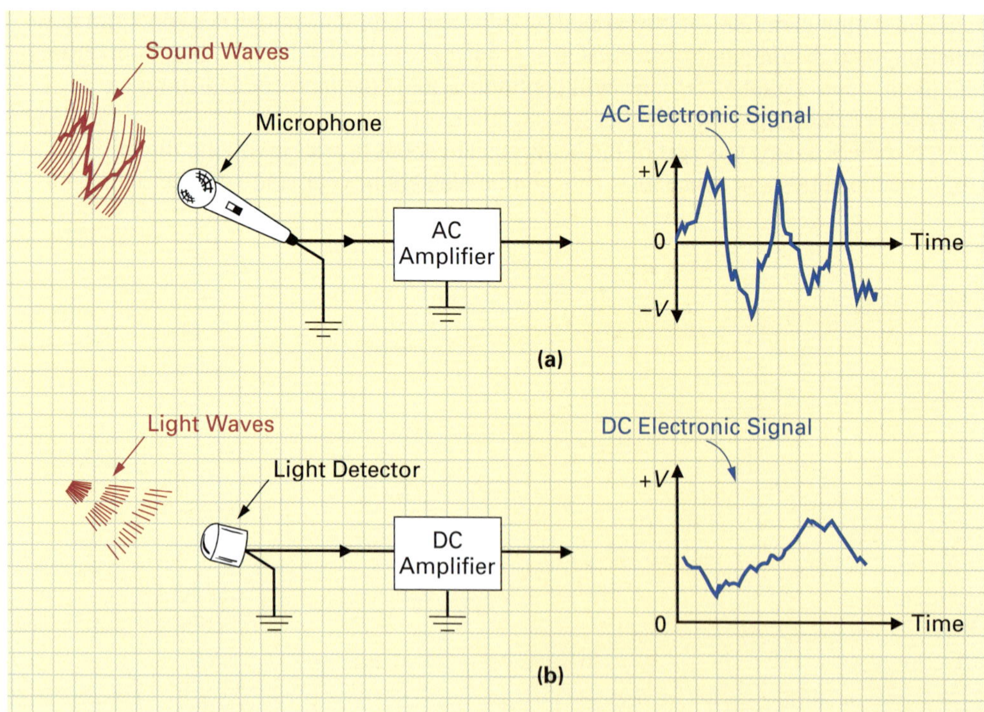

FIGURE 16-1 **Analog Data and Devices.**

proportion to the input. This linear circuit response is evident in Figure 16-1, where you can see that in both the dc and ac circuit, the output signal voltage varies in direct proportion to the sound or light signal input.

16-1-2 *Digital Data and Devices*

In **digital electronic circuits,** information is first converted into a coded group of pulses. To explain this concept, let's take a closer look at the example shown in Figure 16-2(a). This code consists of a series of HIGH and LOW voltages, in which the HIGH voltages are called "1s" (ones) and the LOW voltages are called "0s" (zeros). Figure 16-2(b) lists the ASCII code, which is one example of a digital code. Referring to Figure 16-2(a), you will notice that the "1101001" information or data stream code corresponds to lowercase *i* in the ASCII table shown highlighted in Figure 16-2(b). Computer keyboards are one of many devices that make use of the digital ASCII code. In Figure 16-2(c), you can see how the lowercase *i* ASCII code is generated whenever the *i* key is pressed, encoding the information *i* into a group of pulses (1101001).

Digital Electonic Circuit
A circuit designed to manage digital information signals.

The next question you may have is: Why do we go to all this trouble to encode our data or information into these two-state codes? The answer can best be explained by examining history. The early digital systems constructed in the 1950s made use of a decimal code that used 10 levels or voltages, with each of these voltages corresponding to one of the 10 digits in the decimal number system (0 = 0 V, 1 = 1 V, 2 = 2 V, 3 = 3 V, up to 9 = 9 V). The circuits that had to manage these decimal codes, however, were very complex since they had to generate one of 10 voltages and sense the difference among all 10 voltage levels. This complexity led to inaccuracy, since some circuits would periodically confuse one voltage level for a different voltage level. *The solution to the problem of circuit complexity and inaccuracy was solved by adopting a two-state system instead of a 10-state system.* Using a two-state or two-digit system, you can generate codes for any number, letter, or symbol, as we have seen in the ASCII table in Figure 16-2. The electronic circuits that manage these two-state codes are less complex, since they have to generate and sense only either a HIGH or LOW voltage. In addition, two-state circuits are much more accurate, since there is little room for error between the two extremes of ON and OFF, or HIGH voltage and LOW voltage.

Binary
Having only two alternatives, two-state.

Binary Digit
Abbreviated bit; it is either a 0 or 1 from the binary number system.

Abandoning the 10-state system and adopting the two-state system for the advantages of circuit simplicity and accuracy meant that we were no longer dealing with the decimal number system. As a result, having only two digits (0 and 1) means that we are now operating in the two-state number system, which is called **binary.** Figure 16-3(a) shows how the familiar decimal system has 10 digits or levels labeled 0 through 9, and Figure 16-3(b) shows how we could electronically represent each decimal digit. With the base 2, or binary number system, we have only two digits in the number scale, and therefore only the digits 0 (zero) and 1 (one) exist in binary, as shown in Figure 16-3(c). These two states are typically represented in an electronic circuit as two different values of voltage (binary 0 = LOW voltage; binary 1 = HIGH voltage), as shown in Figure 16-3(d).

Bit
Binary digit.

Digital Signal
An electronic signal made up of binary digits.

Using combinations of **binary digits** (abbreviated as **bits**) we can represent information as a binary code. This code is called a **digital signal** because it is an *information signal* that makes use of *binary digits.* Today, almost all information, from your telephone conversations to the music on your compact discs, is **digitized** or converted to binary data form.

Digitize
To convert an analog signal to a digital signal.

You will probably hear the terms *analog* and *digital* used to describe the difference between the two types of meters shown in Figure 16-4. On the **analog readout multimeter** shown in Figure 16-4(a), the amount of pointer deflection across the scale is an analog (or similar) to the magnitude of the electrical property being measured. On the other hand, with the **digital readout multimeter** shown in Figure 16-4(b), the magnitude of the electrical property being measured is displayed using digits, which in this case are decimal digits.

Analog Readout Multimeter
A multimeter that uses a movement on a calibrated scale to indicate a value.

Digital Readout Multimeter
A multimeter that uses digits to indicate a value.

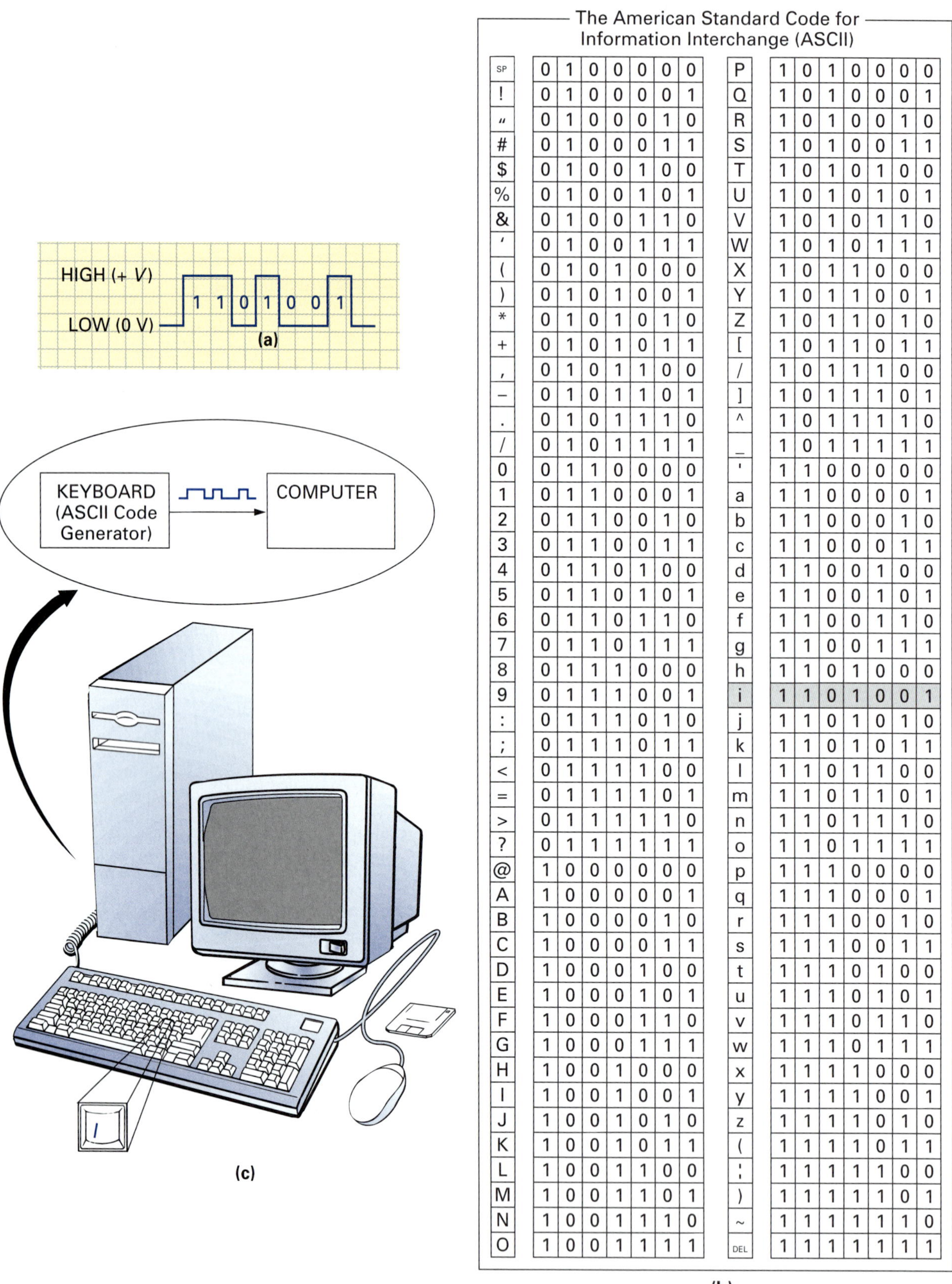

The American Standard Code for Information Interchange (ASCII)

Character	Code	Character	Code
SP	0100000	P	1010000
!	0100001	Q	1010001
"	0100010	R	1010010
#	0100011	S	1010011
$	0100100	T	1010100
%	0100101	U	1010101
&	0100110	V	1010110
'	0100111	W	1010111
(	0101000	X	1011000
)	0101001	Y	1011001
*	0101010	Z	1011010
+	0101011	[	1011011
,	0101100	/	1011100
–	0101101	]	1011101
.	0101110	^	1011110
/	0101111	_	1011111
0	0110000	'	1100000
1	0110001	a	1100001
2	0110010	b	1100010
3	0110011	c	1100011
4	0110100	d	1100100
5	0110101	e	1100101
6	0110110	f	1100110
7	0110111	g	1100111
8	0111000	h	1101000
9	0111001	i	1101001
:	0111010	j	1101010
;	0111011	k	1101011
<	0111100	l	1101100
=	0111101	m	1101101
>	0111110	n	1101110
?	0111111	o	1101111
@	1000000	p	1110000
A	1000001	q	1110001
B	1000010	r	1110010
C	1000011	s	1110011
D	1000100	t	1110100
E	1000101	u	1110101
F	1000110	v	1110110
G	1000111	w	1110111
H	1001000	x	1111000
I	1001001	y	1111001
J	1001010	z	1111010
K	1001011	(	1111011
L	1001100	¦	1111100
M	1001101	)	1111101
N	1001110	~	1111110
O	1001111	DEL	1111111

(b)

FIGURE 16-2 Two-State Digital Information. (a) Information or Data Code for *i*. (b) ASCII Code. (c) Keyboard ASCII Generator.

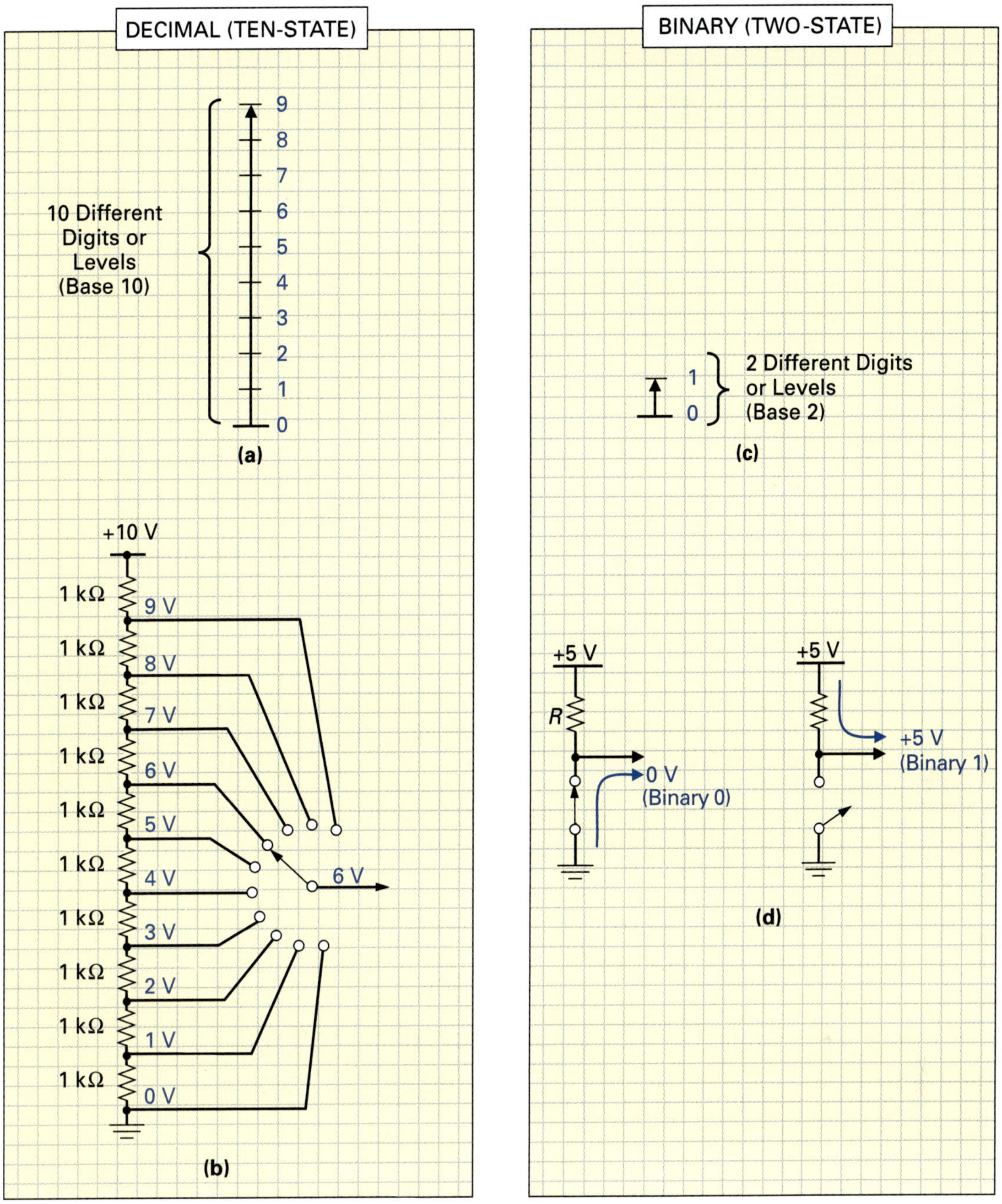

DECIMAL (TEN-STATE)

10 Different Digits or Levels (Base 10)

9
8
7
6
5
4
3
2
1
0

(a)

+10 V

1 kΩ 9 V
1 kΩ 8 V
1 kΩ 7 V
1 kΩ 6 V
1 kΩ 5 V
1 kΩ 4 V
1 kΩ 3 V
1 kΩ 2 V
1 kΩ 1 V
1 kΩ 0 V

6 V

(b)

BINARY (TWO-STATE)

1
0

2 Different Digits or Levels (Base 2)

(c)

+5 V
R
0 V (Binary 0)

+5 V
+5 V (Binary 1)

(d)

FIGURE 16-3 **Electronically Representing the Digits of a Number System.**

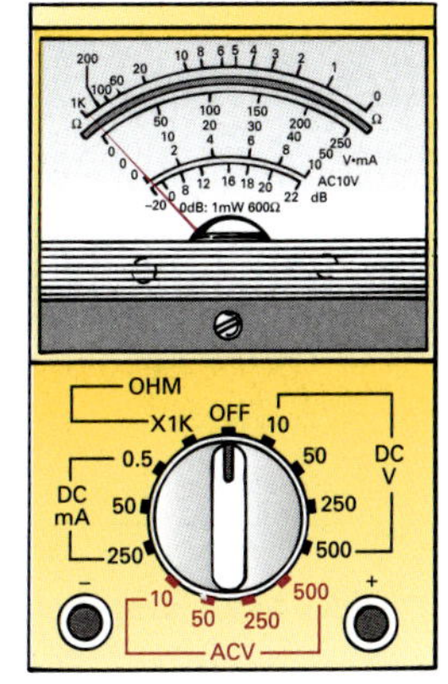

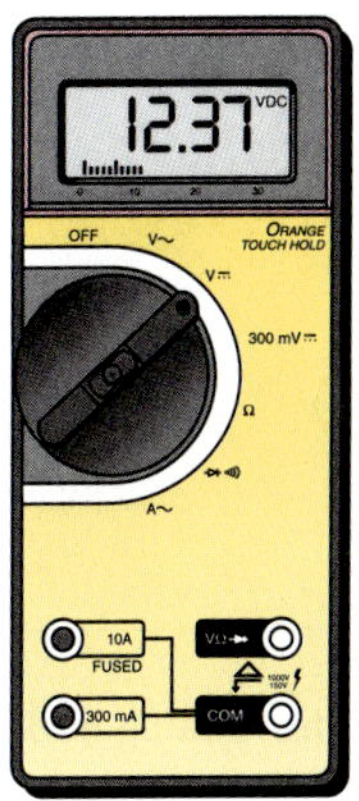

FIGURE 16-4 **Analog and Digital Readouts.**

SELF-TEST EVALUATION POINT FOR SECTION 16-1

Use the following questions to test your understanding of Section 16-1.

1. What does the word *analog* mean?
2. What is an analog signal?
3. What is the ASCII code?
4. What were the two key reasons for adopting the two-state binary system in place of the 10-state decimal system?
5. How many digits does the binary number system have?
6. Describe the difference between an analog readout watch and a digital readout watch.

16-2 ANALOG AND DIGITAL SIGNAL CONVERSION

Whenever you have two different forms, like analog and digital, there is always a need to convert back and forth between the two. Computers are used at the heart of almost every electronic system today because of their ability to quickly process and store large amounts of data, make systems more versatile, and perform many functions. Many of the input signals applied to a computer for storage or processing are analog in nature, and therefore a data conversion circuit is needed to interface these analog inputs with a digital system. Similarly, a data conversion circuit may also be needed to interface the digital computer system with an analog output device.

Analog-to-Digital Converter
A circuit that converts analog input signals into equivalent digital output signals.

Figure 16-5 shows a simplified block diagram of a computer. No matter what the application, a computer has three basic blocks: a microprocessor unit, a memory unit, and an input/output unit.

Getting information into the computer is the job of input devices, and as you can see in Figure 16-5, there are two input paths into a computer: one path from analog input devices,

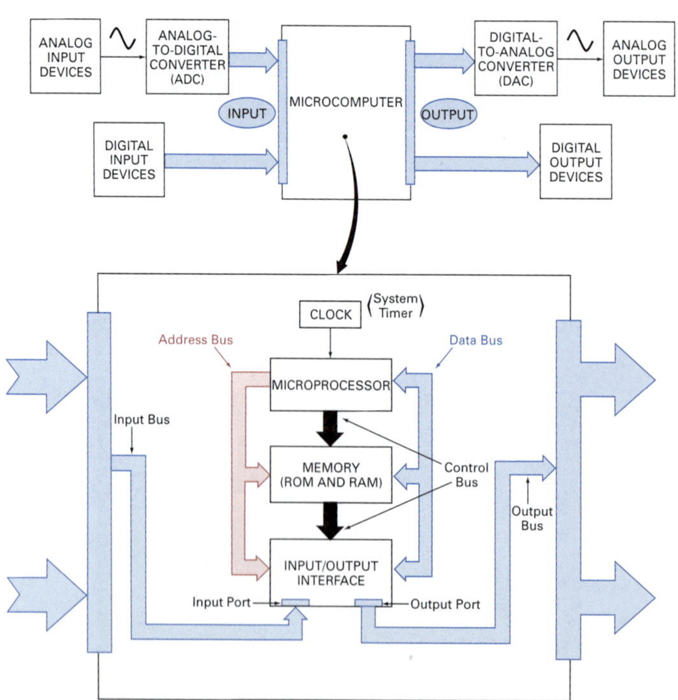

FIGURE 16-5 Connecting Analog and Digital Devices to a Computer.

and the other path from digital input devices. Information in the form of light, sound, heat, pressure, or any other real-world quantity is analog in nature and has an infinite number of input levels. Sensors or transducers of this type, such as photoelectric cells for light or microphones for sound, will generate analog signals. Since the computer operates on information in only digital or binary form, a translation or conversion is needed if the computer is going to be able to understand or interpret an analog signal. This signal processing is achieved by an **analog-to-digital converter (ADC),** which transforms the varying analog input voltage into equivalent digital codes that the computer can understand. Once the analog information has been encoded into digital form, the information can then enter the computer through an electronic doorway called the **input port.** Some input devices, such as a computer keyboard, automatically generate a digital code for each key that is pressed (ASCII codes). The codes generated by these digital input devices can be connected directly to the computer, without having to pass through a converter.

Input Port
An electronic doorway used for receiving information into a computer.

Output Port
An electronic doorway used for transmitting information out of a computer.

Digital-to-Analog Converter
A circuit that converts digital input signals into equivalent analog output signals.

In contrast, the digital output information from a computer's **output port** can be used to drive either an analog device or a digital device. If an analog output is desired, as in the case of a voice signal for a speaker, the digital computer's output will have to be converted into a corresponding equivalent analog signal by a **digital-to-analog converter (DAC).** If only a digital output is desired, as in the case of a printer, which translates the binary codes into printed characters, the digital computer output can be connected directly to the output device without the need for a converter.

SELF-TEST EVALUATION POINT FOR SECTION 16-2

Use the following questions to test your understanding of Section 16-2.

1. What are the three basic blocks of a computer?
2. Give the full names of the following acronyms:
   a. ADC   b. DAC
3. An ADC will convert a/an ____________ input into an equivalent ____________ output.
4. A DAC will convert a/an ____________ input into an equivalent____________ output.

REVIEW QUESTIONS

Multiple Choice Questions

1. Which of the following is an example of a digital electronic circuit?
   a. Amplifier   c. Power supply
   b. MPU   d. ON/OFF lighting circuit
2. Which of the following is an example of an analog electronic circuit?
   a. Amplifier   c. ASCII keyboard
   b. CPU   d. ON/OFF lighting circuit
3. Digital circuits based on the binary number system are often referred to as ____________ circuits. On the other hand, the output of analog circuits varies in direct proportion to the input, which is why analog circuits are often called ____________ circuits.
   a. Ten-state, linear   c. Digital, 10-state
   b. Linear, two-state   d. Two-state, linear
4. A digital multimeter displays the measured quantity using
   a. Binary digits   c. Decimal digits
   b. A pointer and scale   d. An abacus
5. Why do digital electronic systems use the two-state system instead of the 10-state system?
   a. Circuit simplicity   c. Both (a) and (b)
   b. Accuracy   d. None of the above
6. A ____________ converts an analog input voltage into a proportional digital output code.
   a. ROM   c. ADC
   b. DAC   d. RWM
7. A ____________ converts a digital input code into a proportional analog output voltage.
   a. ROM   b. DAC   c. ADC   d. RWM
8. What are the three basic blocks of a computer?
   a. ADC, I/O, memory   c. MPU, I/O, memory
   b. DAC, ADC, I/O   d. ROM, MPU, I/O

Web Site Questions

Go to the web site http://www.prenhall.com/cook, select the textbook *Electronics: A Complete Course,* select this chapter, and then follow the instructions when answering the multiple-choice practice problems.

Number Systems and Codes

Conducting Achievement

Werner Von Siemens

Werner Von Siemens (1816–1892) was born in Lenthe, Germany. He became familiar with the recently developed electric telegraph during military service. Later in 1847, together with skilled mechanic J. G. Halske, he founded the electrical firm of Siemens and Halske. This firm, under Siemens's guidance, became one of the most important electrical undertakings in the world. Siemens invented cable insulation, an armature for large generators, and the dynamo, or electric generator, which converts mechanical energy to electric energy.

Karl Wilhelm Siemens (1823–1883), Werner's brother, is also well known for his work in the fields of electricity and heat. At the age of nineteen he visited England expressly to patent his electroplating invention and never left, making England his home. From 1848 onward, Wilhelm represented his brother's firm in London and became an acknowledged authority. The company installed much of the overland telegraph cable then in existence, as well as an underwater telegraph cable, using submarines. One month before he died, Wilhelm was knighted as Sir William Siemens in acknowledgment of his achievements.

In the family tradition, Alexander Siemens (1847–1928), a nephew of William, went to England in 1867 and worked his way up, beginning in Siemens's workshops. In 1878 he became manager of the electric department and was responsible for the installation of electric light at Godalming, Surrey, the first English town to be lit with electricity. Like many other members of the family, he patented several inventions and, after the death of Sir William, he became company director.

In honor of the Siemens family's achievements, conductance (G) is measured in the unit siemens (S), with a resistance or impedance of 1 Ω being equal to a conductance or admittance of 1 S.

17

Introduction

One of the best ways to understand anything new is to compare it with something that you are already familiar with so that the differences are highlighted. In this chapter we will be examining the *binary number system,* which is the language used by digital computer electronic circuits. To best understand this new number system, we will compare it with the number system most familiar to you, the *decimal* (base 10) *number system.* Although this system is universally used to represent quantities, many people do not fully understand its weighted structure. This review of decimal has been included so that you can compare the base 10 number system with the system used internally by digital electronics, the binary (base 2) number system.

In addition to binary, two other number systems are widely used in conjunction with digital electronics: the octal (base 8) number system, and the hexadecimal (base 16) number system. These two number systems are used as a type of shorthand for large binary numbers and enable us to represent a large group of binary digits with only a few octal or hexadecimal digits.

The last section in this chapter covers many of the different binary codes, which use a combination of 1s and 0s to represent letters, numbers, symbols, and other information.

Keep in mind throughout this chapter that different number systems are just another way to count, and as with any new process, you will become more proficient and more at ease with practice as we proceed through the chapters that follow.

17-1 THE DECIMAL NUMBER SYSTEM

The decimal system of counting and keeping track of items was first created by Hindu mathematicians in India in A.D. 400. Since it involved the use of fingers and thumbs, it was natural that this system would have 10 digits. The system found its way to all the Arab countries by A.D. 800, where it was named the Arabic number system, and from there it was eventually adopted by nearly all the European countries by A.D. 1200, where it was called the **decimal number system.**

Decimal Number System
A base 10 number system.

Base
With number systems, it describes the number of digits used.

The key feature that distinguishes one number system from another is the number system's **base** or *radix*. This base indicates the number of digits that will be used. The decimal number system, for example, is a base 10 number system, which means that it uses 10 digits (0 through 9) to communicate information about an amount. A subscript is sometimes included after a number when different number systems are being used, to indicate the base of the number. For example, $12{,}567_{10}$ is a base 10 number, whereas 10110_2 is a base 2 number.

17-1-1 *Positional Weight*

The position of each digit of a decimal number determines the weight of that digit. A 1 by itself, for instance, is worth only 1, whereas a 1 to the left of three 0s makes the 1 worth 1000.

In decimal notation, each position to the left of the decimal point indicates an increased positive power of 10, as seen in Figure 17-1(a). The total quantity or amount of the number is therefore determined by the size and the weighted position of each digit. For example, the value shown in Figure 17-1(a) has six thousands, zero hundreds, one ten, and nine ones, which combined makes a total of 6019_{10}.

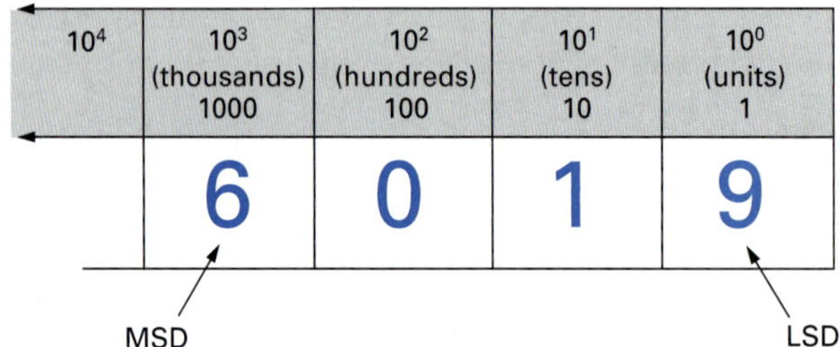

$$\begin{aligned}\text{Value of number} &= (6 \times 10^3) + (0 \times 10^2) + (1 \times 10^1) + (9 \times 10^0)\\ &= (6 \times 1000) + (0 \times 100) + (1 \times 10) + (9 \times 1)\\ &= 6000 + 0 + 10 + 9 = 6019_{10}\end{aligned}$$

(a)

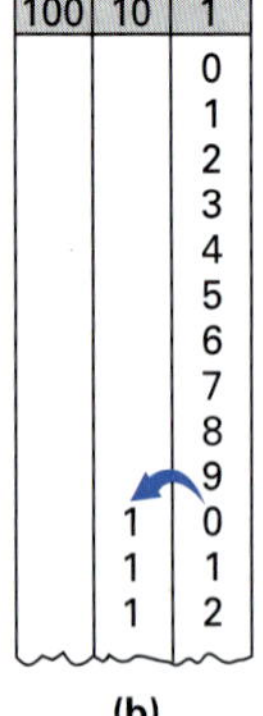

(b)

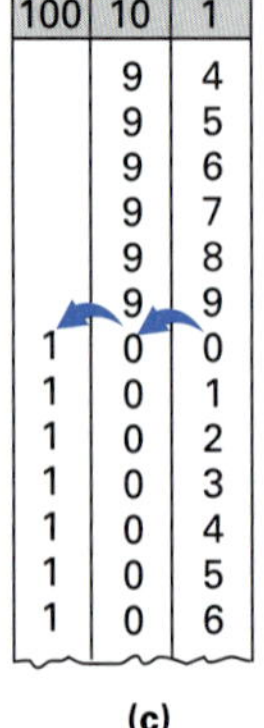

(c)

FIGURE 17-1 The Decimal (Base 10) Number System.

In the decimal number system, the leftmost digit is called the **most significant digit (MSD),** and the rightmost digit is called the **least significant digit (LSD).** Applying these definitions to the example in Figure 17-1(a), we see that the 6 is the MSD, since its position carries the most weight, and the 9 is the LSD, since its position carries the least weight.

Most Significant Digit (MSD)
The leftmost, largest-weight digit in a number.

17-1-2 *Reset and Carry*

Least Significant Digit (LSD)
The rightmost, smallest-weight digit in a number.

Before proceeding to the binary number system, let us review one other action that occurs when counting in decimal. This action, which is familiar to us all, is called **reset and carry.** Referring to Figure 17-1(b), you can see that a reset and carry operation occurs after a count of 9. The units column, which has reached its maximum count, resets to 0 and carries a 1 to the tens column, resulting in a final count of 10.

Reset and Carry
An action that occurs when a column has reached its maximum count.

This reset and carry action will occur in any column that reaches its maximum count. For example, Figure 17-1(c) shows how two reset and carry operations will take place after a count of 99. The units column resets and carries a 1 to the tens column, which in turn resets and carries a 1 to the hundreds column, resulting in a final count of 100.

SELF-TEST EVALUATION POINT FOR SECTION 17-1

Use the following questions to test your understanding of Section 17-1.

1. What is the difference between the Arabic number system and the decimal number system?
2. What is the base or radix of the decimal number system?
3. Describe the positional weight of each of the digits in the decimal number 2639.
4. What action occurs when a decimal column advances beyond its maximum count?

17-2 THE BINARY NUMBER SYSTEM

As in the decimal system, the value of a binary digit is determined by its position relative to the other digits. In the decimal system, each position to the left of the decimal point increases by a power of 10. Similarly, in the binary number system, since it is a base 2 (bi) number system, each place to the left of the **binary point** increases by a power of 2. Figure 17-2 shows how columns of the binary and decimal number systems have different weights. For example, with binary, the columns are weighted so that 2^0 is one, 2^1 is two, 2^2 is four, 2^3 is eight ($2 \times 2 \times 2 = 8$), and so on.

Binary Point
A symbol used to separate the whole from the fraction in a binary number.

As we know, the base or radix of a number system also indicates the maximum number of digits used by the number system. The base 2 binary number system uses only the

FIGURE 17-2 A Comparison of Decimal and Binary.

DECIMAL

10^2	10^1	10^0
100	10	1
		0
		1
		2
		3
		4
		5
		6
		7
		8
		9
	1	0
	1	1
	1	2

BINARY

2^3	2^2	2^1	2^0
8	4	2	1
			0
			1
		1	0
		1	1
	1	0	0
	1	0	1
	1	1	0
	1	1	1
1	0	0	0
1	0	0	1
1	0	1	0
1	0	1	1
1	1	0	0

first two digits on the number scale, 0 and 1. The 0s and 1s in binary are called binary digits, or *bits,* for short.

17-2-1 *Positional Weight*

As in the decimal system, each column in binary carries its own weight, as seen in Figure 17-3(a). With the decimal number system, each position to the left increases 10 times. With binary, the weight of each column to the left increases 2 times. The first column, therefore, has a weight of 1, the second column has a weight of 2, the third column 4, the fifth 8, and so on. The value or quantity of a binary number is determined by the digit in each column and the positional weight of the column. For example, in Figure 17-3(a), the binary number 101101_2 is equal in decimal to 45_{10}, since we have 1×32, 1×8, 1×4, and 1×1 ($32 + 8 + 4 + 1 = 45$). The 0s in this example are not multiplied by their weights (16 and 2), since they are "0" and of no value. The leftmost binary digit is called the most significant bit (MSB) since it carries the most weight, and the rightmost digit is called the least significant bit (LSB), since it carries the least weight. Applying these definitions to the example in Figure 17-3(a), we see that the 1 in the thirty-twos column is the MSB, and the 1 in the units column is the LSB.

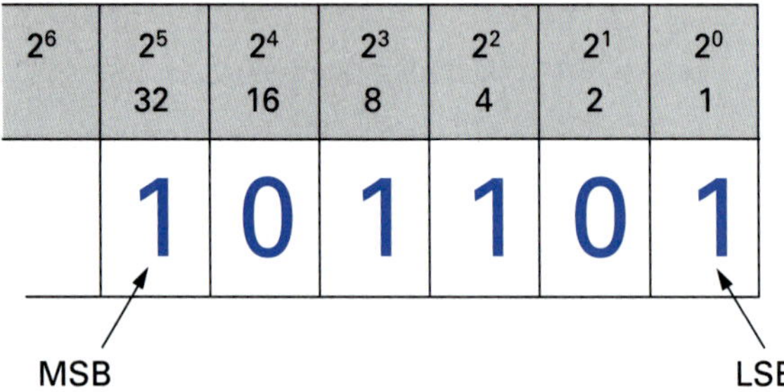

$$
\begin{aligned}
\text{Value of number} &= (1 \times 2^5) + (0 \times 2^4) + (1 \times 2^3) + (1 \times 2^2) + (1 \times 2^1) + (1 \times 2^0) \\
&= (1 \times 32) + (0 \times 16) + (1 \times 8) + (1 \times 4) + (0 \times 2) + (1 \times 1) \\
&= 32 + 0 + 8 + 4 + 0 + 1 \\
&= 45_{10}
\end{aligned}
$$

(a)

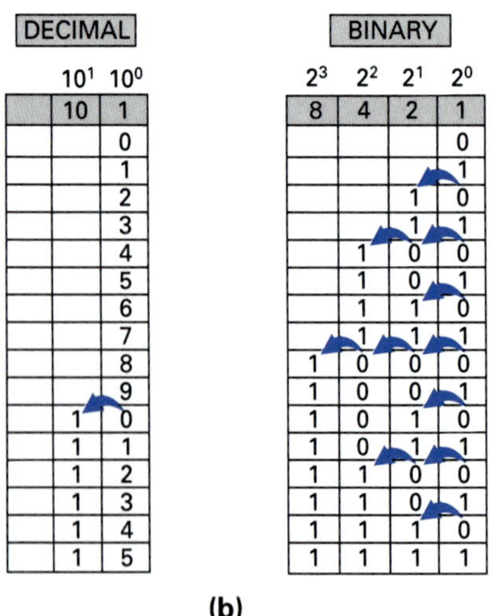

DECIMAL

10^1	10^0
10	1
	0
	1
	2
	3
	4
	5
	6
	7
	8
	9
1	0
1	1
1	2
1	3
1	4
1	5

BINARY

2^3	2^2	2^1	2^0
8	4	2	1
			0
			1
		1	0
		1	1
	1	0	0
	1	0	1
	1	1	0
	1	1	1
1	0	0	0
1	0	0	1
1	0	1	0
1	0	1	1
1	1	0	0
1	1	0	1
1	1	1	0
1	1	1	1

(b)

FIGURE 17-3 The Binary (Base 2) Number System.

17-2-2 *Reset and Carry*

The reset and carry action occurs in binary in exactly the same way as it does in decimal; however, since binary has only two digits, a column will reach its maximum digit much sooner, and therefore the reset and carry action in the binary number system will occur much more frequently.

Referring to Figure 17-3(b), you can see that the binary counter begins with 0 and advances to 1. At this stage, the units column has reached its maximum, and therefore the next count forces the units column to reset and carry into the next column, producing a count of 0010_2 (2_{10}). The units column then advances to a count of 0011_2 (3_{10}). At this stage, both the units column and twos column have reached their maximums. As the count advances by 1 it will cause the units column to reset and carry a 1 into the twos column, which will also have to reset and carry a 1 into the fours column. This will result in a final count of 0100_2 (4_{10}). The count will then continue to 0101_2 (5_{10}), 0110_2 (6_{10}), 0111_2 (7_{10}), and then 1000_2 (8_{10}), which, as you can see in Figure 17-3(b), is a result of three reset and carries.

Comparing the binary reset and carry with the decimal reset and carry in Figure 17-3(b), you can see that since the binary number system uses only two digits, binary numbers quickly turn into multidigit figures. For example, a decimal eight (8) uses only one digit, whereas a binary eight (1000) uses four digits.

17-2-3 *Converting Binary Numbers to Decimal Numbers*

Binary numbers can easily be converted to their decimal equivalent by simply adding together all the column weights that contain a binary 1, as we did previously in Figure 17-3(a).

EXAMPLE:

Convert the following binary numbers to their decimal equivalents.

a. 1010
b. 101101

Solution:

a. Binary column weights 32 16 8 4 2 1
Binary number 1 0 1 0
Decimal equivalent $= (1 \times 8) + (0 \times 4) + (1 \times 2) + (0 \times 1)$
$= 8 + 0 + 2 + 0 = 10_{10}$

b. Binary column weights 32 16 8 4 2 1
Binary number 1 0 1 1 0 1
Decimal equivalent $= (1 \times 32) + (0 \times 16) + (1 \times 8) + (1 \times 4)$
$+ (0 \times 2) + (1 \times 1)$
$= 32 + 0 + 8 + 4 + 0 + 1 = 45_{10}$

EXAMPLE:

The LEDs in Figure 17-4 are being used as a 4-bit (4-binary-digit) display. When the LED is OFF, it indicates a binary 0, and when the LED is ON, it indicates a binary 1. Determine the decimal equivalent of the binary displays shown in Figure 17-4(a), (b), and (c).

Solution:

a. $0101_2 = 5_{10}$
b. $1110_2 = 14_{10}$
c. $1001_2 = 9_{10}$

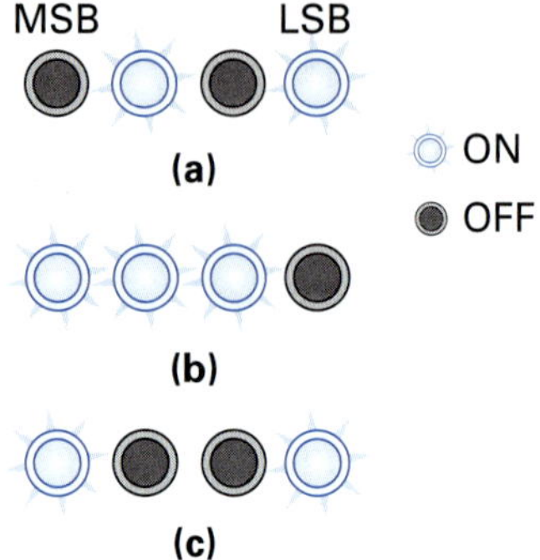

FIGURE 17-4 Using LEDs to Display Binary Numbers.

17-2-4 *Converting Decimal Numbers to Binary Numbers*

To convert a decimal number to its binary equivalent, continually subtract the largest possible power of two until the decimal number is reduced to zero, placing a binary 1 in columns that are used and a binary 0 in the columns that are not used. To see how simple this process is, refer to Figure 17-5, which shows how decimal 53 can be converted to its binary equivalent. As you can see in this example, the first largest power of two that can be subtracted from decimal 53 is 32, and therefore a 1 is placed in the thirty-twos column, and 21 remains. The largest power of two that can be subtracted from the remainder 21 is 16, and therefore a 1 is placed in the sixteens column, and 5 remains. The next largest power of two that can be sub-

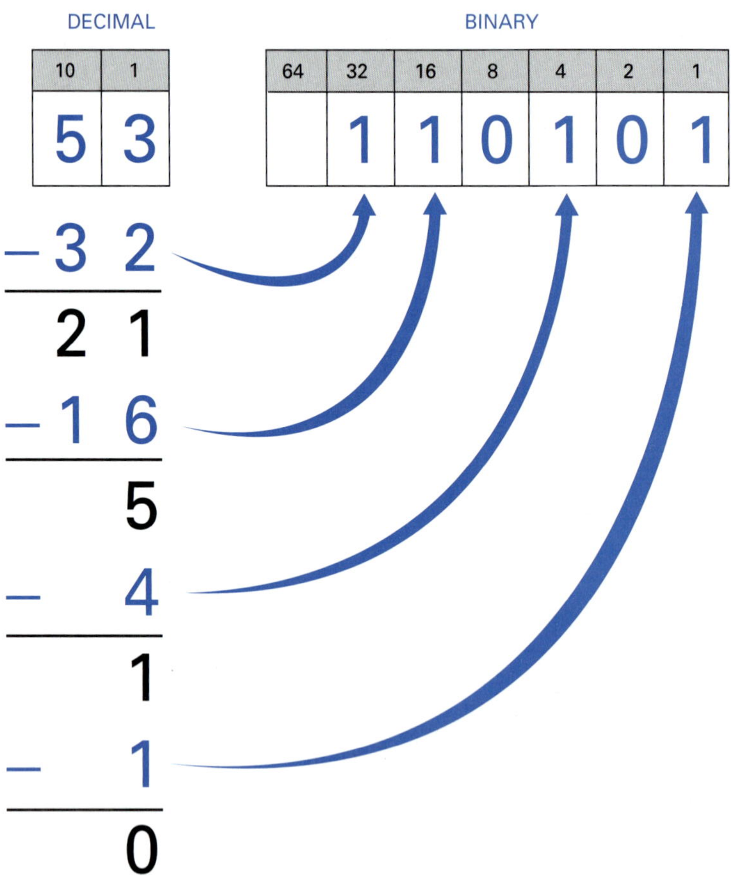

FIGURE 17-5 Decimal-to-Binary Conversion.

tracted from 5 is 4, and therefore a 1 is placed in the fours column, and 1 remains. The final 1 is placed in the units column, and therefore decimal 53 is represented in binary as 110101, which indicates that the value is the sum of 1×32, 1×16, 1×4, and 1×1.

EXAMPLE:

Convert the following decimal numbers to their binary equivalents.

a. 25
b. 55

Solution:

a.

	32	16	8	4	2	1
Binary column weights	32	16	8	4	2	1
Binary number		1	1	0	0	1

Decimal equivalent $25 - 16 = 9, 9 - 8 = 1, 1 - 1 = 0$

b.

	32	16	8	4	2	1
Binary column weights	32	16	8	4	2	1
Binary number	1	1	0	1	1	1

Decimal equivalent $55 - 32 = 23, 23 - 16 = 7, 7 - 4 = 3,$
$3 - 2 = 1, 1 - 1 = 0$

EXAMPLE:

Figure 17-6 shows a strip of magnetic tape containing binary data. The shaded dots indicate magnetized dots (which are equivalent to binary 1s), and the unshaded circles on the tape represent unmagnetized points (which are equivalent to binary 0s). What binary values should be stored in rows 4 and 5 so that decimal 19_{10} and 33_{10} are recorded?

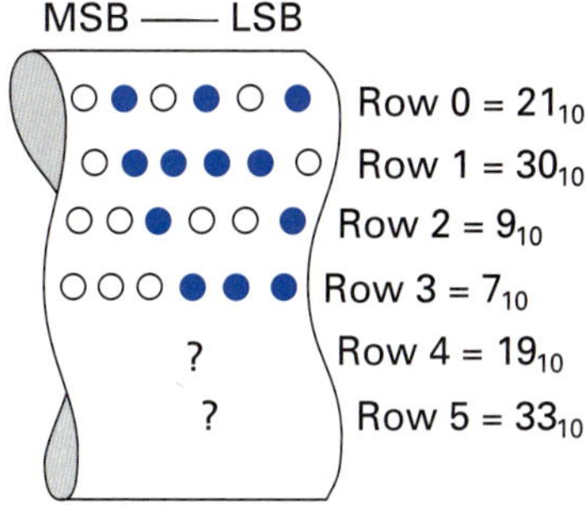

FIGURE 17-6 Binary Numbers on Magnetic Tape.

Solution:

a.

	32	16	8	4	2	1
Binary column weights	32	16	8	4	2	1
Binary number		1	0	0	1	1

Decimal equivalent $19 - 16 = 3, 3 - 2 = 1, 1 - 1 = 0$

Stored magnetic data ○●○○●●

b.

	32	16	8	4	2	1
Binary column weights	32	16	8	4	2	1
Binary number	1	0	0	0	0	1

Decimal equivalent $32 - 32 = 1, 1 - 1 = 0$

Stored magnetic data ●○○○○●

17-2-5 *Converting Information Signals*

Figure 17-7(a) shows how analog data can be converted into digital data, and Figure 17-7 (b) shows how digital data can be converted into analog data. To give this process purpose, let us consider how the music information stored on a compact disc (CD) is first recorded and then played back. Like all sound, music is made up of waves of compressed air. When these

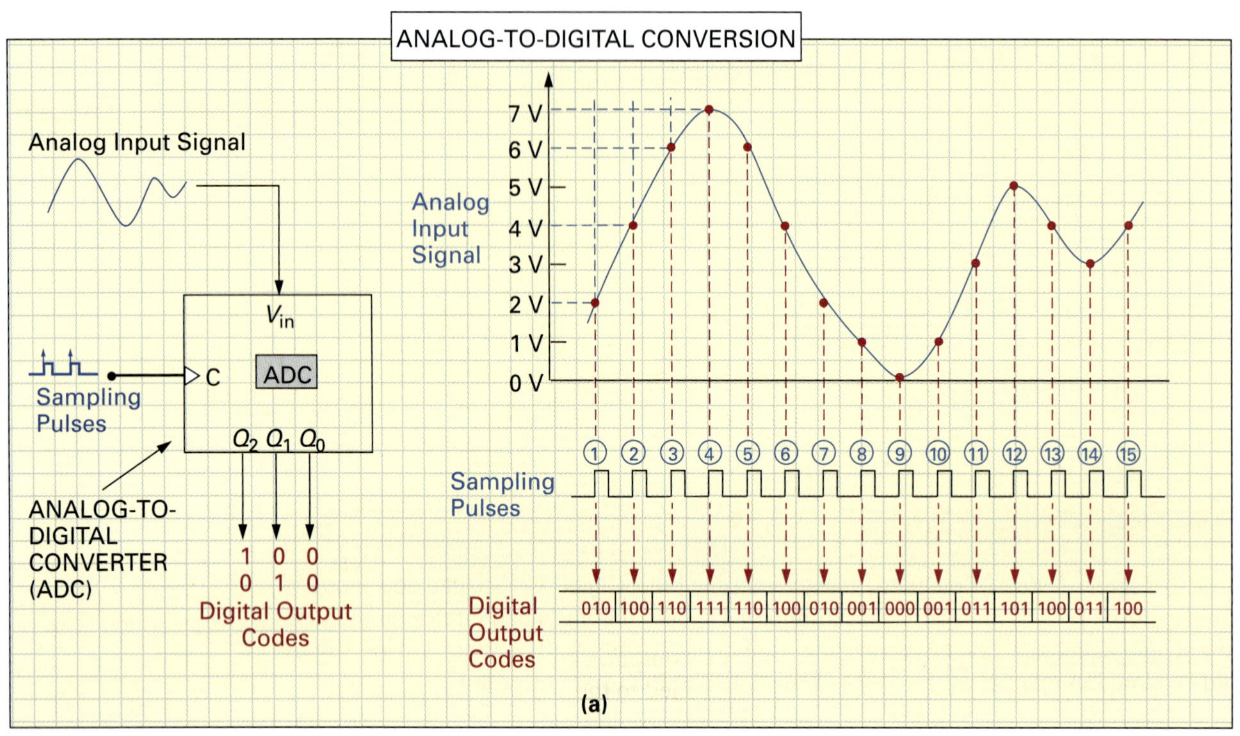

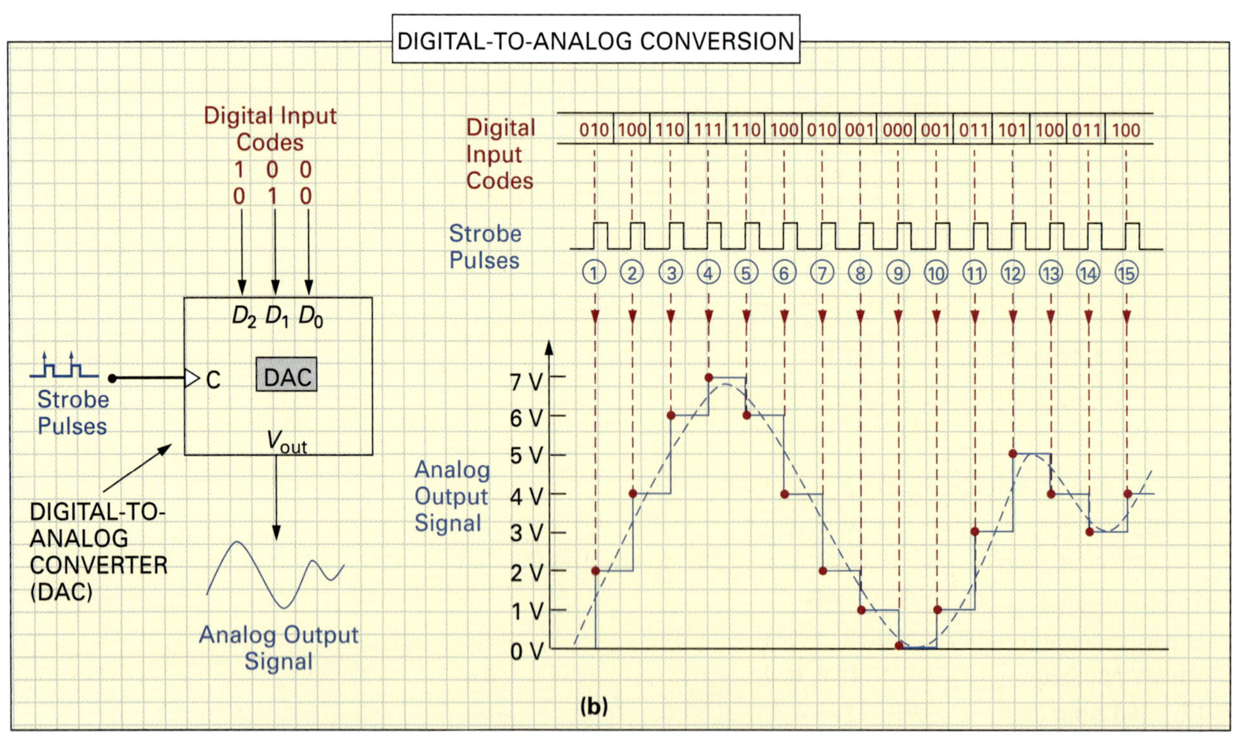

FIGURE 17-7 **Converting Information Signals.**

waves strike the diaphragm of a microphone, an analog voltage signal is generated. In the past, this analog data was recorded on magnetic tapes or as a grooved track on a record. These data storage devices were susceptible to wear and tear, temperature, noise, and age. Using digital recording techniques, binary codes are stored on a compact disc to achieve near-perfect fidelity to live sound.

Figure 17-7(a) shows how an analog-to-digital converter (ADC) is used during the recording process to convert the analog music input into a series of digital output codes. These digital codes are used to control the light beam of a recording laser so that it will engrave the binary 0s and 1s onto a compact disc (CD) in the form of pits or spaces. The ADC is triggered into operation by a sampling pulse that causes it to measure the input voltage of the analog signal at that particular time and generate an equivalent digital output code. For example, on the active edge of sampling pulse 1, the analog input voltage is at 2 V, so the binary code 010 (decimal 2) is generated at the output of the ADC. On the active edge of sampling pulse 2, the analog voltage has risen to 4 V, so the binary code 100 (decimal 4) is generated at the output of the ADC. On the active edge of sampling pulses 3, 4, and 5, the binary codes 110 (decimal 6), 111 (decimal 7), and 110 (decimal 6) are generated at the output, representing the analog voltages 6 V, 7 V, and 6 V, respectively, and so on.

Figure 17-7(b) shows how a digital-to-analog converter (DAC) is used during the music playback process to convert the digital codes stored on a CD to an analog output signal. Another laser is used during playback to read the pits and spaces on the compact disc as 0s and 1s. These codes are then applied to the DAC, which converts the digital input codes into discrete voltages. The DAC is triggered into operation by a strobe pulse that causes it to convert the code currently being applied at the input into an equivalent output voltage.

For example, on the active edge of strobe pulse 1, the digital code 010 (decimal 2) is being applied to the DAC, so it will generate 2 V at its output. On the active edge of strobe pulse 2, the digital code 100 (decimal 4) is being applied to the DAC, so it will generate 4 V at its output. On the active edge of strobe pulse 3, the digital code 110 (decimal 6) is being applied to the DAC, so it will generate 6 V at its output. On the active edge of strobe pulse 4, the digital code 111 (decimal 7) is being applied to the DAC, so it will generate 7 V at its output. On the active edge of strobe pulse 5, the digital code 110 (decimal 6) is being applied to the DAC, so it will generate 6 V at its output, and so on. If the output of a DAC is then applied to a low-pass filter, the discrete voltage steps can be blended into a smooth wave that closely approximates the original analog wave, as shown by the dashed line in Figure 17-7(b).

SELF-TEST EVALUATION POINT FOR SECTION 17-2

Use the following questions to test your understanding of Section 17-2.

1. What is the decimal equivalent of 11010_2?
2. Convert 23_{10} to its binary equivalent.
3. What are the full names for the acronyms LSB and MSB?
4. Convert 110_{10} to its binary equivalent.

17-3 THE HEXADECIMAL NUMBER SYSTEM

If digital electronic circuits operate using binary numbers and we operate using decimal numbers, why is there any need for us to have any other system? The hexadecimal, or hex, system is used as a sort of shorthand for large strings of binary numbers, as will be explained in this section. To begin, let us examine the basics of this number system and then look at its application.

Hexadecimal
A base 16 number system.

Hexadecimal means "sixteen," and this number system has 16 different digits, as shown in Figure 17-8(a), which shows a comparison between decimal, hexadecimal, and binary. Looking at the first 10 digits in the decimal and hexadecimal columns, you can see that there is no difference between the two columns; however, beyond 9, hexadecimal makes use of the letters A, B, C, D, E, and F. Digits 0 through 9 and letters A through F make up the 16 total digits of the hexadecimal number system. Comparing hexadecimal with decimal once again, you can see that $A_{16} = 10_{10}$, $B_{16} = 11_{10}$, $C_{16} = 12_{10}$, $D_{16} = 13_{10}$, $E_{16} = 14_{10}$ and $F_{16} = 15_{10}$. Having these extra digits means that a column reset and carry will not occur until the count has reached the last and largest digit, F. The hexadecimal column in Figure 17-8(a) shows how reset and carry will occur whenever a column reaches its maximum digit, F, and this action is further illustrated in the examples in Figure 17-8(b).

DECIMAL			HEXADECIMAL				BINARY					
10^2	10^1	10^0	16^3	16^2	16^1	16^0	2^5	2^4	2^3	2^2	2^1	2^0
100	10	1	4096	256	16	1	32	16	8	4	2	1
		0				0	0	0	0	0	0	0
		1				1	0	0	0	0	0	1
		2				2	0	0	0	0	1	0
		3				3	0	0	0	0	1	1
		4				4	0	0	0	1	0	0
		5				5	0	0	0	1	0	1
		6				6	0	0	0	1	1	0
		7				7	0	0	0	1	1	1
		8				8	0	0	1	0	0	0
		9				9	0	0	1	0	0	1
	1	0				A	0	0	1	0	1	0
	1	1				B	0	0	1	0	1	1
	1	2				C	0	0	1	1	0	0
	1	3				D	0	0	1	1	0	1
	1	4				E	0	0	1	1	1	0
	1	5				F	0	0	1	1	1	1
	1	6			1	0	0	1	0	0	0	0
	1	7			1	1	0	1	0	0	0	1
	1	8			1	2	0	1	0	0	1	0
	1	9			1	3	0	1	0	0	1	1
	2	0			1	4	0	1	0	1	0	0
	2	1			1	5	0	1	0	1	0	1
	2	2			1	6	0	1	0	1	1	0
	2	3			1	7	0	1	0	1	1	1
	2	4			1	8	0	1	1	0	0	0
	2	5			1	9	0	1	1	0	0	1
	2	6			1	A	0	1	1	0	1	0
	2	7			1	B	0	1	1	0	1	1
	2	8			1	C	0	1	1	1	0	0
	2	9			1	D	0	1	1	1	0	1
	3	0			1	E	0	1	1	1	1	0
	3	1			1	F	0	1	1	1	1	1
	3	2			2	0	1	0	0	0	0	0
	3	3			2	1	1	0	0	0	0	1
	3	4			2	2	1	0	0	0	1	0

(a)

A9	DE7	78	F9	10C
AA	DE8	79	FA	10D
AB	DE9	7A	FB	10E
AC	DEF	7B	FC	10F
AD	DF0	7C	FD	110
AE	DF1	7D	FE	111
AF	DF2	7E	FF	112
B0	DF3	7F	100	
B1		80	101	
		81		

(b)

FIGURE 17-8 Hexadecimal Rest and Carry. (a) Number System Comparison. (b) Hex Counting Examples.

16^4	16^3	16^2	16^1	16^0
65,536	4096	256	16	1
			4	C

$$\text{Value of Number} = (4 \times 16) + (C \times 1) = (4 \times 16) + (12 \times 1) = 64 + 12 = 76_{10}$$

FIGURE 17-9 Positional Weight of the Hexadecimal Number System.

17-3-1 *Converting Hexadecimal Numbers to Decimal Numbers*

To find the decimal equivalent of a hexadecimal number, simply multiply each hexadecimal digit by its positional weight. For example, referring to Figure 17-9, you can see that the positional weight of each of the hexadecimal columns, as expected, is progressively 16 times larger as you move from right to left. The hexadecimal number 4C, therefore, indicates that the value is 4×16 ($4 \times 16 = 64$) and $C \times 1$. Since C is equal to 12 in decimal (12×1), the result of $12 + 64$ is 76, so hexadecimal 4C is equivalent to decimal 76.

EXAMPLE:

Convert hexadecimal 8BF to its decimal equivalent.

Solution:

Hexadecimal column weights	4096	256	16	1
		8	B	F

$$\begin{aligned}\text{Decimal equivalent} &= (8 \times 256) + (B \times 16) + (F \times 1)\\ &= 2048 + (11 \times 16) + (15 \times 1)\\ &= 2048 + 176 + 15\\ &= 2239\end{aligned}$$

17-3-2 *Converting Decimal Numbers to Hexadecimal Numbers*

Decimal-to-hexadecimal conversion is achieved in the same way as decimal to binary. First, subtract the largest possible power of 16, and then keep subtracting the largest possible power of 16 from the remainder. Each time a subtraction takes place, add a 1 to the respective column until the decimal value has been reduced to zero. Figure 17-10 illustrates this procedure with an example showing how decimal 425 is converted to its hexadecimal equivalent. To begin, the largest possible power of sixteen (256) is subtracted once from 425 and therefore a 1 is placed in the 256s column, leaving a remainder of 169. The next largest power of 16 is 16, which can be subtracted 10 times from 169, and therefore the hexadecimal equivalent of 10, which is A, is placed in the sixteens column, leaving a remainder of 9. Since nine 1s can be subtracted from the remainder of 9, the units column is advanced nine times, giving us our final hexadecimal result, 1A9.

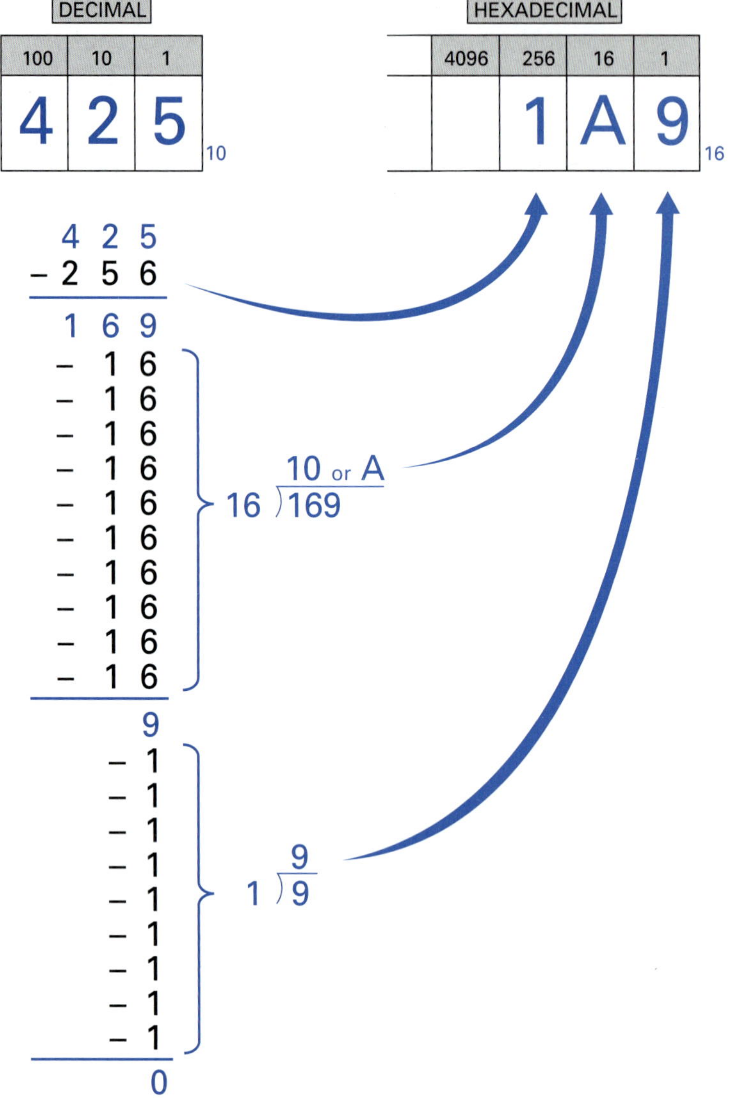

FIGURE 17-10 **Decimal-to-Hexadecimal Conversion.**

EXAMPLE:

Convert decimal 4525 to its hexadecimal equivalent.

Solution:

Hexadecimal column weights	4096	256	16	1
	1	1	A	D

Decimal equivalent $= 4525 - 4096 = 429, 429 - 256 = 173,$
$173 - 16 - 16 - 16 - 16 - 16 - 16 -$
$16 - 16 - 16 - 16 = 13, 13 - 1 -$
$1 - 1 - 1 - 1 - 1 - 1 - 1 - 1 -$
$1 - 1 - 1 - 1 = 0$

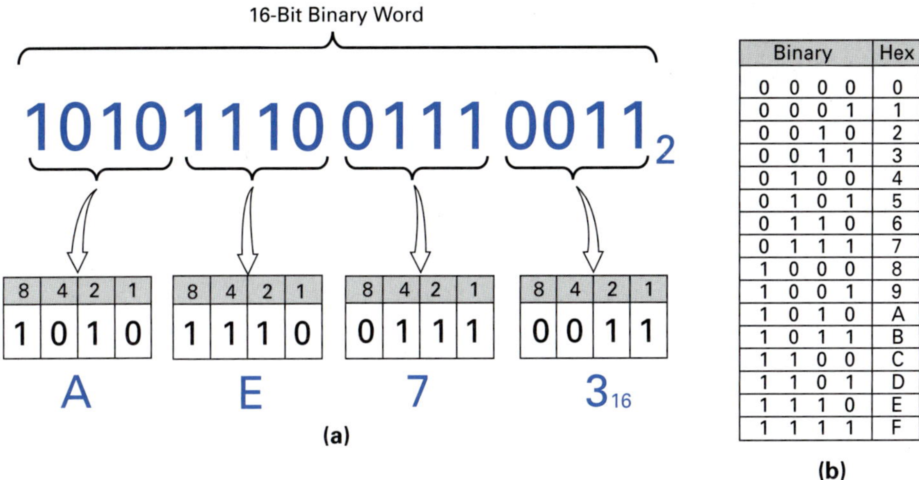

Binary	Hex
0 0 0 0	0
0 0 0 1	1
0 0 1 0	2
0 0 1 1	3
0 1 0 0	4
0 1 0 1	5
0 1 1 0	6
0 1 1 1	7
1 0 0 0	8
1 0 0 1	9
1 0 1 0	A
1 0 1 1	B
1 1 0 0	C
1 1 0 1	D
1 1 1 0	E
1 1 1 1	F

(b)

FIGURE 17-11 **Representing Binary Numbers in Hexadecimal.**

17-3-3 *Converting between Binary and Hexadecimal*

As mentioned at the beginning of this section, hexadecimal is used as a shorthand for representing large groups of binary digits. To illustrate this, Figure 17-11(a) shows how a 16-bit binary number, which is more commonly called a 16-bit binary **word,** can be represented by 4 hexadecimal digits. To explain this, Figure 17-11(b) shows how a 4-bit binary word can have any value from 0_{10} (0000_2) to 15_{10} (1111_2), and since hexadecimal has the same number of digits (0 through F), we can use one hexadecimal digit to represent 4 binary bits. As you can see in Figure 17-11(a), it is much easier to work with a number like $AE73_{16}$ than with 1010111001110011_2.

Word
An ordered set of characters that is treated as a unit.

To convert from hexadecimal to binary, we simply do the opposite, as shown in Figure 17-12. Since each hexadecimal digit represents 4 binary digits, a 4-digit hexadecimal number will convert to a 16-bit binary word.

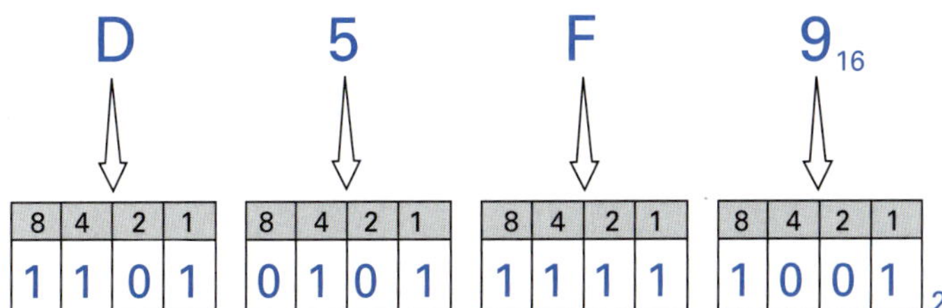

FIGURE 17-12 **Hexadecimal-to-Binary Conversion.**

EXAMPLE:

Convert hexadecimal 2BF9 to its binary equivalent.

Solution:

Hexadecimal number	2	B	F	9
Binary equivalent	0010	1011	1111	1001

CALCULATOR KEYS

Name: Entering and converting number bases

Function and Examples: Some calculators allow you to enter and convert number bases.

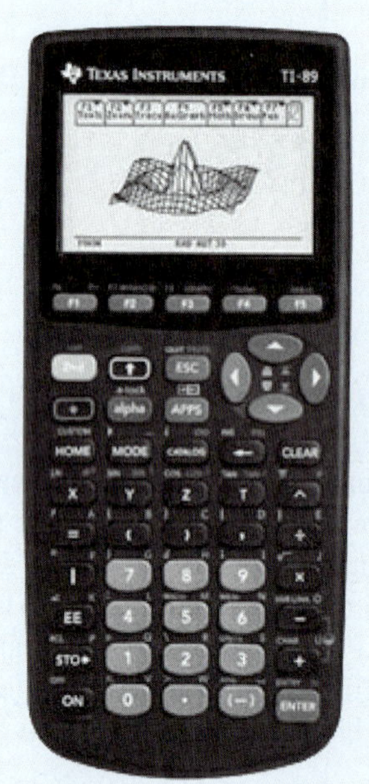

To enter a binary number, use the form:

0b*binaryNumber* (for example: 0b11100110)

- Binary number with up to 32 digits
- Zero, not the letter *O*, and the letter *b*

To enter a hexadecimal number, use the form:

0h*hexadecimalNumber* (for example: 0h89F2C)

- Hexadecimal number with up to 8 digits
- Zero, not the letter *O*, and letter *h*

If you enter without the 0b or 0h prefix, such as 11, it is always treated as a decimal number. If you omit the 0h prefix on a hexadecimal number containing A–F, all or part of the entry is treated as a variable.

Use the ▶ conversion operator.

integerExpression ▶ Bin
integerExpression ▶ Dec
integerExpression ▶ Hex

For ▶, press 2nd [▶]. Also, you can select base conversions from the MATH/Base menu.

For example, to convert 256 from decimal to binary:

256 ▶ Bin

For a binary or hex entry, you must use the 0b or 0h prefix.

n 256 ▶ Bin	0b100000000
n 0b101110 ▶ Hex	0h2E
0b101110 ▶ hex	
MATH RAD AUTO FUNC	2738

To convert 101110 from binary to hexidecimal:

0b101110 ▶ Hex

Result use the 0b or 0h prefix to identify the base.

EXAMPLE:

Convert binary 110011100001 to its hexadecimal equivalent.

Solution:

Binary number	1100	1110	0001
Hexadecimal equivalent	C	E	1

SELF-TEST EVALUATION POINT FOR SECTION 17-3

Use the following questions to test your understanding of Section 17-3.

1. What is the base of each?
   a. Decimal number system
   b. Binary number system
   c. Hexadecimal number system
2. What are the decimal and hexadecimal equivalents of 10101_2?
3. Convert 1011 1111 0111 1010_2 to its hexadecimal equivalent.
4. What are the binary and hexadecimal equivalents of 33_{10}?

17-4 BINARY CODES

The process of converting a decimal number to its binary equivalent is called *binary coding.* The result of the conversion is a binary number or code that is called **pure binary.** There are, however, binary codes used in digital circuits other than pure binary, and in this section we will examine some of the most frequently used.

Pure Binary
Uncoded binary.

17-4-1 *The Binary Coded Decimal (BCD) Code*

No matter how familiar you become with binary, it will always be less convenient to work with than the decimal number system. For example, it will always take a short time to convert 1111000_2 to 120_{10}. Designers realized this disadvantage early on and developed a binary code that had decimal characteristics and that was appropriately named **binary coded decimal (BCD).** Being a binary code, it has the advantages of a two-state system, and since it has a decimal format, it is also much easier for an operator to interface via a decimal keypad or decimal display with systems such as pocket calculators and wristwatches.

Binary-Coded Decimal (BCD)
A code in which each decimal digit is represented by a group of 4 binary bits.

The BCD code expresses each decimal digit as a 4-bit word, as shown in Figure 17-13(a). In this example, decimal 1753 converts to a BCD code of 0001 0111 0101 0011, with the first 4-bit code ($0001_2 = 1_{10}$) representing the 1 in the thousands column, the second 4-bit code ($0111_2 = 7_{10}$) representing the 7 in the hundreds column, the third 4-bit code ($0101_2 = 5_{10}$) representing the 5 in the tens column, and the fourth 4-bit code ($0011_2 = 3_{10}$) representing the 3 in the units column. As can be seen in Figure 17-13(a), the subscript BCD is often used after a BCD code to distinguish it from a pure binary number (1753_{10} = 0001 0111 0101 0011_{BCD}).

Figure 17-13(b) compares decimal, binary, and BCD. As you can see, the reset and carry action occurs in BCD at the same time it does in decimal. This is because BCD was designed to have only ten 4-bit binary codes, 0000, 0001, 0010, 0011, 0100, 0101, 0110, 0111, 1000, and 1001 (0 through 9), to make it easy to convert between this binary code and decimal. Binary codes 1010, 1011, 1100, 1101, 1110, and 1111 (A_{16} through F_{16}), are invalid codes that are not used in BCD because they are not used in decimal.

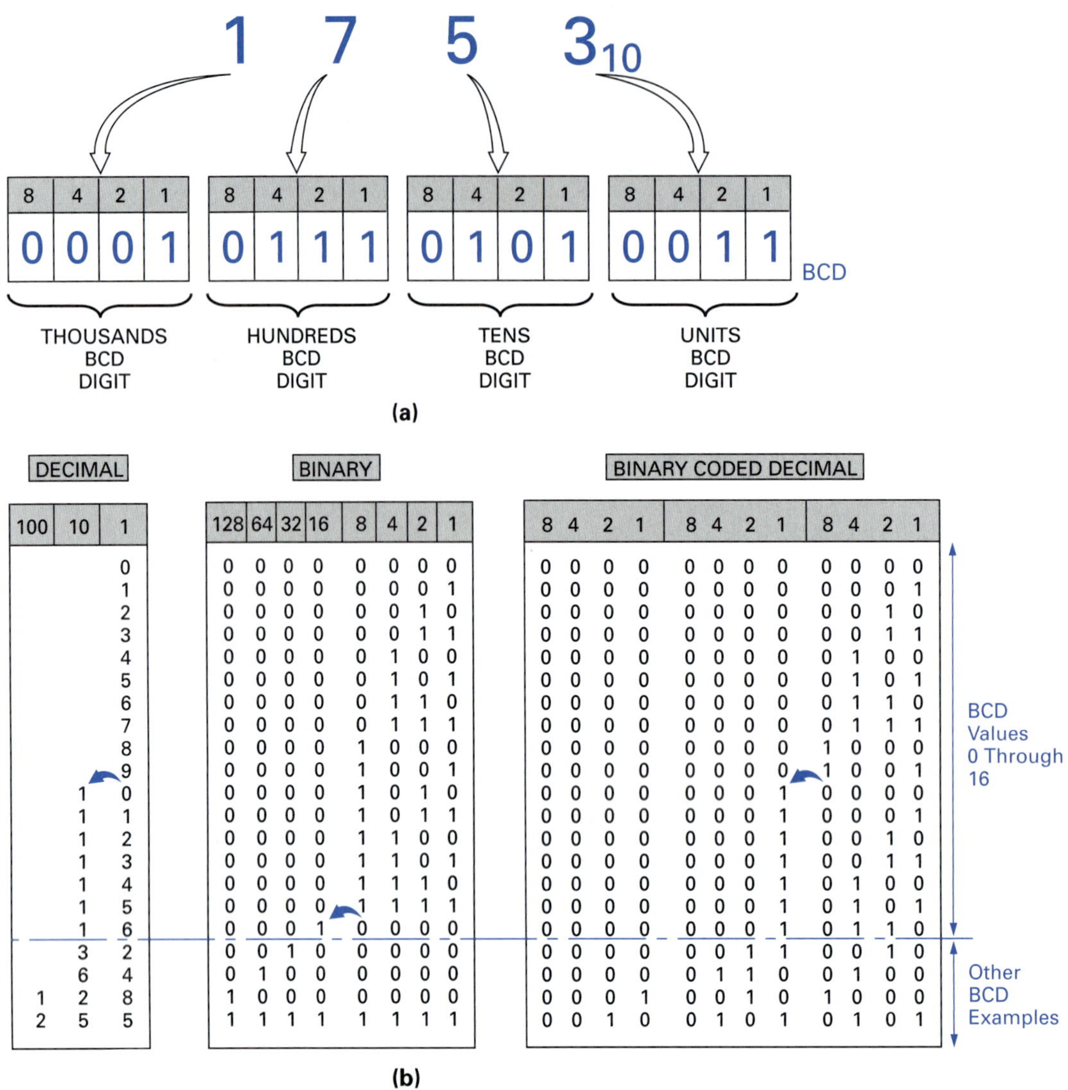

DECIMAL (100 10 1)	BINARY (128 64 32 16 8 4 2 1)	BINARY CODED DECIMAL (8 4 2 1 8 4 2 1 8 4 2 1)
0	0000 0000	0000 0000 0000
1	0000 0001	0000 0000 0001
2	0000 0010	0000 0000 0010
3	0000 0011	0000 0000 0011
4	0000 0100	0000 0000 0100
5	0000 0101	0000 0000 0101
6	0000 0110	0000 0000 0110
7	0000 0111	0000 0000 0111
8	0000 1000	0000 0000 1000
9	0000 1001	0000 0000 1001
10	0000 1010	0000 0001 0000
11	0000 1011	0000 0001 0001
12	0000 1100	0000 0001 0010
13	0000 1101	0000 0001 0011
14	0000 1110	0000 0001 0100
15	0000 1111	0000 0001 0101
16	0001 0000	0000 0001 0110
32	0010 0000	0000 0011 0010
64	0100 0000	0000 0110 0100
128	1000 0000	0001 0010 1000
255	1111 1111	0010 0101 0101

FIGURE 17-13 Binary Coded Decimal (BCD). (a) Decimal-to-BCD Conversion Example. (b) A Comparison of Decimal, Binary, and BCD.

EXAMPLE:

Convert the BCD code 0101 1000 0111 0000 to decimal, and the decimal number 369 to BCD.

Solution:

BCD code	0101	1000	0111	0000
Decimal equivalent	5	8	7	0
Decimal number	3	6	9	
BCD code	0011	0110	1001	

17-4-2 *The Excess-3 Code*

Excess-3 Code
A code in which the decimal digit n is represented by the 4-bit binary code $n + 3$.

The **excess-3 code** is similar to BCD and is often used in some applications because of certain arithmetic operation advantages. Referring to Figure 17-14(a), you can see that, like BCD, each decimal digit is converted into a 4-bit binary code. The difference with the excess-3 code, however, is that a value of 3 is added to each decimal digit before it is con-

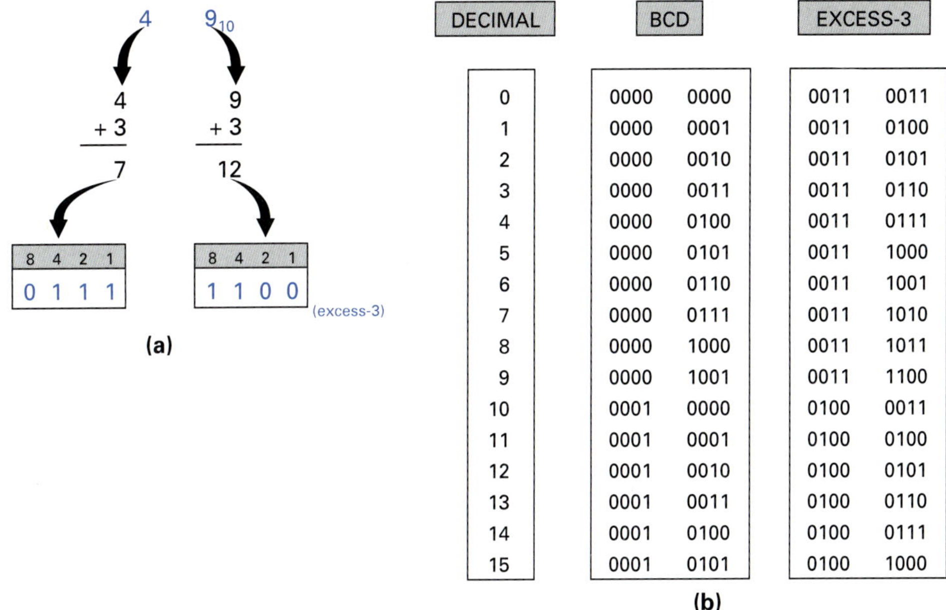

DECIMAL	BCD		EXCESS-3	
0	0000	0000	0011	0011
1	0000	0001	0011	0100
2	0000	0010	0011	0101
3	0000	0011	0011	0110
4	0000	0100	0011	0111
5	0000	0101	0011	1000
6	0000	0110	0011	1001
7	0000	0111	0011	1010
8	0000	1000	0011	1011
9	0000	1001	0011	1100
10	0001	0000	0100	0011
11	0001	0001	0100	0100
12	0001	0010	0100	0101
13	0001	0011	0100	0110
14	0001	0100	0100	0111
15	0001	0101	0100	1000

(b)

FIGURE 17-14 The Excess-3 Code. (a) Example. (b) Code Comparison.

verted into a 4-bit binary code. Although this code has an offset or excess of 3, it still uses only ten 4-bit binary codes (0011 through 1100) like BCD. The invalid 4-bit values in this code are therefore 0000, 0001, 0010 (decimal 0 through 2), and 1101, 1110, 1111 (decimal 13 through 15). Figure 17-14(b) shows a comparison between decimal, BCD, and excess-3.

EXAMPLE:

Convert decimal 408 to its excess-3 code equivalent.

Solution:

Decimal number	4	0	8
+3 =	7	3	11
Excess-3 binary	0111	0011	1011

17-4-3 *The Gray Code*

The **gray code,** shown in Figure 17-15, is a nonweighted binary code, which means that the position of each binary digit carries no specific weight or value. This code, named after its inventor, was developed so that only one of the binary digits changes as you step from one code group to the next code group in sequence. For example, in pure binary, a change from 3 (0011) to 4 (0100) results in a change in 3 bits, whereas in the gray code only 1 bit changes.

Gray Code
A nonweighted binary code in which sequential numbers are represented by codes that differ by one bit.

The minimum-change gray code is used to reduce errors in digital electronic circuitry. To explain this in more detail, when a binary count changes, it takes a very small amount of time for the bits to change from 0 to 1, or from 1 to 0. This transition time can produce an incorrect intermediate code. For example, if when changing from pure binary 0011 (3) to 0100 (4), the LSB were to switch slightly faster than the other bits, an incorrect code would be momentarily generated, as shown in the following example.

Binary	Decimal
0 0 1 1	3
0 0 1 0	Error
0 1 0 0	4

DECIMAL	BINARY	GRAY CODE
0	0000	0000
1	0001	0001
2	0010	0011
3	0011	0010
4	0100	0110
5	0101	0111
6	0110	0101
7	0111	0100
8	1000	1100
9	1001	1101
10	1010	1111
11	1011	1110
12	1100	1010
13	1101	1011
14	1110	1001
15	1111	1000

FIGURE 17-15 The Gray Code.

This momentary error could trigger an operation that should not occur, resulting in a system malfunction. Using the gray code eliminates these timing errors, since only 1 bit changes at a time. The disadvantage of the gray code is that the codes must be converted to pure binary if numbers need to be added, subtracted, or used in other computations.

17-4-4 *The American Standard Code for Information Interchange (ASCII)*

Alphanumeric Code

A code used to represent the letters of the alphabet and decimal numbers.

To this point we have discussed only how we can encode numbers into binary codes. Digital electronic computers must also be able to generate and recognize binary codes that represent letters of the alphabet and symbols. The ASCII code is the most widely used **alphanumeric code** (alphabet and numerical code). As I write this book the ASCII codes for each of these letters is being generated by my computer keyboard, and the computer is decoding these ASCII codes and then displaying the alphanumeric equivalent (letter, number, or symbol) on my computer screen.

Figure 17-16 lists all the 7-bit ASCII codes and the full names for the abbreviations used. This diagram also shows how each of the 7-bit ASCII codes is made up of a 4-bit group that indicates the row of the table and a 3-bit group that indicates the column of the table. For example, the uppercase letter K is in column 100 (4_{10}) and in row 1011 (11_{10}) and is therefore represented by the ASCII code 1001011.

EXAMPLE:

List the ASCII codes for the message "Digital."

Solution:

$$
\begin{aligned}
D &= 1000100 \\
i &= 1101001 \\
g &= 1100111 \\
i &= 1101001 \\
t &= 1110100 \\
a &= 1100001 \\
l &= 1101100
\end{aligned}
$$

Row	Bits 4321 ↓ / 765 →	0 (000)	1 (001)	2 (010)	3 (011)	**4 (100)**	5 (101)	6 (110)	7 (111)
0	0000	NUL	DLE	SP	0	@	P	\	p
1	0001	SOH	DC1	!	1	A	Q	a	q
2	0010	STX	DC2	"	2	B	R	b	r
3	0011	ETX	DC3	#	3	C	S	c	s
4	0100	EOT	DC4	$	4	D	T	d	t
5	0101	ENQ	NAK	%	5	E	U	e	u
6	0110	ACK	SYN	&	6	F	V	f	v
7	0111	BEL	ETB	'	7	G	W	g	w
8	1000	BS	CAN	(	8	H	X	h	x
9	1001	HT	EM	)	9	I	Y	i	y
10	1010	LF	SUB	*	:	J	Z	j	z
11	**1011**	VT	ESC	+	;	**K**	[	k	{
12	1100	FF	FS	,	<	L	\	l	¦
13	1101	CR	GS	-	=	M	]	m	}
14	1110	SO	RS	.	>	N	⌒	n	~
15	1111	SI	US	/	?	O	—	o	DEL

Example: Upper Case "K" = Column 4 (100), Row 11 (1011)

Bits	7	6	5	4	3	2	1
ASCII Code	1	0	0	1	0	1	1

NUL Null
SOH Start of Heading
STX Start of Text
ETX End of Text
EOT End of Transmission
ENQ Enquiry
ACK Acknowledge
BEL Bell (audible signal)
BS Backspace
HT Horizontal Tabulation (punched card skip)
LF Line Feed
VT Vertical Tabulation
FF Form Feed
CR Carriage Return
SO Shift Out
SI Shift In
SP Space (blank)
DLE Data Link Escape
DC1 Device Control 1
DC2 Device Control 2
DC3 Device Control 3
DC4 Device Control 4
NAK Negative Acknowledge
SYN Synchronous Idle
ETB End of Transmission Block
CAN Cancel
EM End of Medium
SUB Substitute
ESC Escape
FS File Separator
GS Group Separator
RS Record Separator
US Unit Separator
DEL Delete

(ASCII Abbreviations)

FIGURE 17-16 **The ASCII Code.**

EXAMPLE:

Figure 17-17 illustrates a 7-bit register. A register is a circuit that is used to store or hold a group of binary bits. If a switch is closed, the associated "Q output" is grounded and 0 V (binary 0) is applied to the output. On the other hand, if a switch is open, the associated "Q output" is pulled HIGH due to the connection to + 5 V via the 10 kΩ resistor, and therefore + 5 V (binary 1) is applied to the output. What would be the pure binary value, hexadecimal value, and ASCII code stored in this register?

Solution:

	Q_6	Q_5	Q_4	Q_3	Q_2	Q_1	Q_0
Pure binary	1	0	0	0	0	1	1_2
Hexadecimal				43_{16}			
ASCII			C	(uppercase C)			

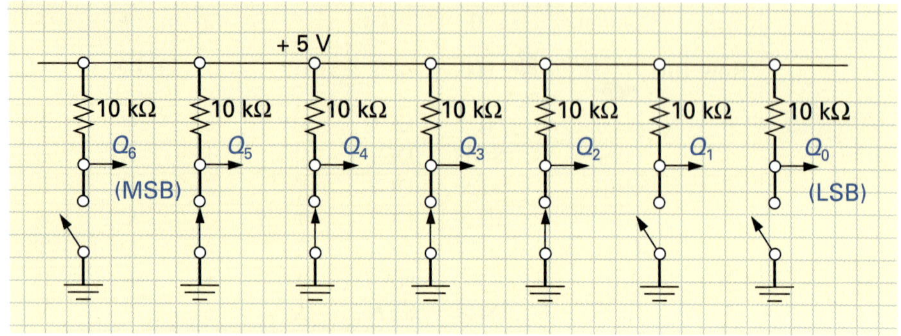

FIGURE 17-17 **A 7-Bit Switch Register.**

SELF-TEST EVALUATION POINT FOR SECTION 17-4

Use the following questions to test your understanding of Section 17-4.

1. Convert 7629_{10} to BCD.
2. What is the decimal equivalent of the excess-3 code 0100 0100?
3. What is a minimum-change code?
4. What does the following string of ASCII codes mean?
   1000010 1100001 1100010 1100010 1100001 1100111 1100101

REVIEW QUESTIONS

Multiple Choice Questions

1. Binary is a base __________ number system, decimal a base __________ number system, and hexadecimal a base __________ number system.
   a. 1, 2, 3 **c.** 10, 2, 3
   b. 2, 10, 16 **d.** 8, 4, 2
2. What is the binary equivalent of decimal 11?
   a. 1010 **c.** 1011
   b. 1100 **d.** 0111
3. Which of the following number systems is used as a shorthand for binary?
   a. Decimal **c.** Binary
   b. Hexadecimal **d.** Both a and b
4. What would be displayed on a hexadecimal counter if its count was advanced by 1 from $39FF_{16}$?
   a. 4000 **c.** 3900
   b. 3A00 **d.** 4A00
5. What is the base of 101?
   a. 10 **c.** 16
   b. 2 **d.** Could be any base
6. Which of the following is an invalid BCD code?
   a. 1010 **c.** 0111
   b. 1000 **d.** 0000
7. What is the decimal equivalent of $1001\ 0001\ 1000\ 0111_{BCD}$?
   a. 9187 **c.** 8659
   b. 9A56 **d.** 254,345
8. What is the pure binary equivalent of 15_{16}?
   a. 0000 1111 **c.** 0101 0101
   b. 0001 0101 **d.** 0000 1101
9. Which of the following is an invalid excess-3 code?
   a. 1010 **c.** 0111
   b. 1000 **d.** 0000
10. Since the lower 4 bits of the ASCII code for the numbers 0 through 9 are the same as their pure binary equivalent, what is the 7-bit ASCII code for the number 6?
    a. 0111001 **c.** 0110011
    b. 0110111 **d.** 0110110

Practice Problems

11. What is the pure binary output of the switch register shown in Figure 17-18 and what is its decimal equivalent?
12. Convert the following into their decimal equivalents:
    a. 110111_2
    b. $2F_{16}$
    c. 10110_{10}
13. What would be the decimal equivalent of the LED display in Figure 17-19 if it were displaying each of the following:
    a. Pure binary **c.** Excess-3
    b. BCD

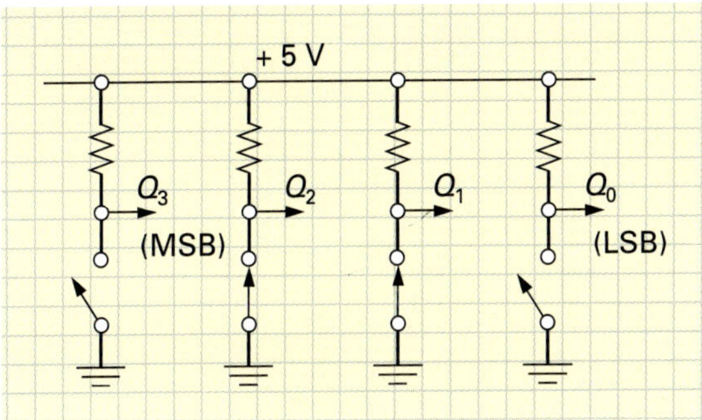

FIGURE 17-18 **Four-Bit Switch Register.**

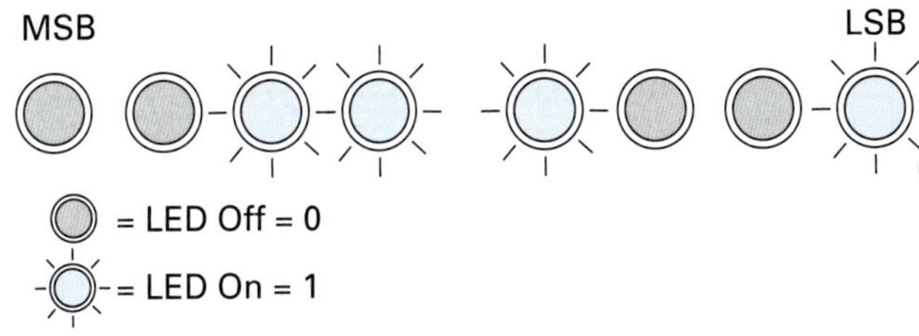

FIGURE 17-19 **8-Bit LED Display.**

14. Convert the following binary numbers to octal and hexadecimal:
a. 111101101001 **c.** 111000
b. 1011 **d.** 1111111

15. Convert the following decimal numbers to BCD:
a. 2,365 **b.** 24

16. Give the ASCII codes for the following:
a. ? **c.** 6
b. $

17. Identify the following codes.

DECIMAL	(A)		(B)		(C)		(D)
0	0000	0000	0000	0011	0011	011	0000
1	0000	0001	0001	0011	0100	011	0001
2	0000	0010	0011	0011	0101	011	0010
3	0000	0011	0010	0011	0110	011	0011
4	0000	0100	0110	0011	0111	011	0100
5	0000	0101	0111	0011	1000	011	0101

18. Convert the decimal number 23 into each of the following:
a. ASCII
b. Pure binary
c. BCD
d. Hexadecimal
e. Octal
f. Excess-3

19. If a flashlight is used to transmit ASCII codes, with a dot (short flash) representing a 0 and a dash (long flash) representing a 1, what is the following message?

0000010	1010000	1110010	1100001	1100011	1110100	1101001
1100011	1100101	0100000	1001101	1100001	1101011	1100101
1110011	0100000	1010000	1100101	1110010	1100110	1100101
1100011	1110100	0000011				

20. Convert the following:
a. $1110110_2 = \_\_\_\_\_\_\_\_\_\__{16}$
b. $175_{10} = \_\_\_\_\_\_\_\_\_\__{2}$
c. $ABC_{16} = \_\_\_\_\_\_\_\_\_\__{2}$
d. $00110110_2 = \_\_\_\_\_\_\_\_\_\__{8}$

Web Site Questions

Go to the web site http://prenhall.com/cook, select the textbook *Electronics: A Complete Course*, select this chapter, and then follow the instructions when answering the multiple-choice practice problems.

Logic Gates

Wireless

Lee De Forest was born in Council Bluffs, Iowa on August 26, 1873. He obtained his Ph.D. from Yale Sheffield Scientific School in 1899 with a thesis that is recognized as the first paper on radio communications in the United States.

De Forest was an astounding practical inventor; however, he wasn't a good businessman and many of his inventions were stolen and exploited by business partners.

In 1902, he invented an electrolytic radio detector and an alternating-current transmitter while working for the Western Electric Company. He also developed an optical sound track for movies which was rejected by film producers. Later another system was used that was based on De Forest's principles.

In 1907, De Forest patented his *audion* detector or triode (three-electrode) vacuum tube. This tube was a more versatile vacuum tube than the then available diode (two-electrode) vacuum tube, or *Fleming Valve,* invented by English electrical engineer, Sir John Fleming. By adding a third electrode, called a control grid, De Forest made amplification possible because the tube could now be controlled. The triode's ability to amplify as well as rectify led to the development of radio communications, or wireless as it was called, and later to television. De Forest also discovered that by cascading (connecting end-to-end) amplifiers, a higher gain could be attained. This ability led to long distance communications because weak long-range input signals could be increased in magnitude to a usable level.

In 1905, he began experimenting with speech and music broadcasts. In 1910 De Forest transmitted the singing voice of Enrico Caruso and, in so doing, became one of the pioneers of radio broadcasting.

Slowly the usefulness of the triode as a generator, amplifier, and detector of radio waves became apparent. However, it was not until World War I that the audion became an invaluable electronic device and was manufactured in large quantities.

As well as working on the technical development of radio, De Forest gave many practical public demonstrations of wireless communications. He became internationally renowned and was often referred to as "the father of radio." Before his fame and fortune, however—just before everyone became aware of the profound effect that his audion tube would have on the world—an incident occurred. While trying to sell stock in his company, De Forest was arrested on charges of fraud and his device, which was to launch a new era in electronics and make possible radio and television, was at this time called "a strange device like an incandescent lamp . . . which has proved worthless."

18

Introduction

Within any digital electronic system you will find that diodes and transistors are used to construct **logic gate circuits.** Logic gates are in turn used to construct flip-flop circuits, and flip-flops are used to construct register, counter, and a variety of other circuits. These logic gates, therefore, are used as the basic building blocks for all digital circuits, and their purpose is to control the movement of binary data and binary instructions. Logic gates, and all other digital electronic circuits, are often referred to as **hardware** circuits. By definition, the hardware of a digital electronic system includes all of the electronic, magnetic, and mechanical devices of a digital system. In contrast, the **software** of a digital electronic system includes the binary data (like pure binary and ASCII codes) and binary instructions that are processed by the digital electronic system hardware. To use an analogy, we could say that a compact disk player is hardware, and the music information stored on a compact disk and processed by the player is software. Just as a CD player is useless without CDs, a digital system's hardware is useless without software. In other words, the information on the CD determines what music is played, and similarly the digital software determines the actions of the digital electronic hardware.

Logic Gate Circuits
Circuits containing predictable gate functions that either open or close their outputs.

Hardware
A term used to describe the electronic, magnetic, and mechanical devices of a digital system.

Software
A term used to describe the binary instructions and data processed by digital electronic system hardware.

Every digital electronic circuit uses logic-gate circuits to manipulate the coded pulses of binary language. These logic-gate circuits are constructed using diodes and transistors, and they are the basic decision-making elements. Digital circuits are often referred to as switching circuits because their control devices (diodes and transistors) are switched between the two extremes of ON and OFF. They are also called **two-state circuits** because their control devices are driven into one of two states: either into the saturation state (fully ON), or cutoff state (fully OFF). These two modes of operation are used to represent the two binary digits of 1 and 0.

Two-state Circuits
Switching circuits that are controlled to be either ON or OFF.

In summary, a logic gate accepts inputs in the form of HIGH or LOW voltages, judges these input combinations based on a predetermined set of rules, and then produces a single output in the form of a HIGH or LOW voltage. The term *logic* is used because the output is predictable or logical, and the term *gate* is used because only certain input combinations will "unlock the gate." For example, any HIGH input to an OR gate will unlock the gate and allow the HIGH at the input to pass through to the output.

18-1 BASIC LOGIC GATES

The two basic types of logic gates used in digital circuits are the OR gate and the AND gate.

18-1-1 *The OR Gate*

OR Gate

A logic gate that will give a HIGH output if either of its inputs are HIGH.

Truth Table or Function Table

A table used to show the action of a device as it reacts to all possible input combinations.

The **OR gate** can have two or more inputs, but will always have a single output. Figure 18-1(a) shows how an OR gate can be constructed using two diodes and a resistor. Figure 18-1(b) shows a table listing all of the input possibilities for this two-input OR gate. This table is often referred to as a **truth table** or **function table** since it details the "truth" or the way in which this circuit will function. The insets in Figure 18-1(b) simplify how the circuit will basically produce one of two outputs. If both inputs are LOW, both diode switches are open and the output will be LOW. On the other hand, any HIGH input will cause the associated diode switch to close and connect the HIGH input through to the output. We could summarize the operation of this gate, therefore, by saying that *if either* A *OR* B *is HIGH, the output* Y *will be HIGH, and only when both inputs are LOW will the output be LOW.*

Figure 18-1(c) shows how an OR gate can be constructed using transistors as switches instead of diodes. In this OR gate circuit, three two-state transistor switches are included to construct a three-input OR gate. The truth table for this OR gate circuit is shown in Figure 18-1(d). Although another input has been added to this circuit, the action of the gate still remains the same since any HIGH input will still produce a HIGH output, and only when all of the inputs are LOW will the output be LOW. To explain the operation of this transistor logic gate in more detail, the insets in Figure 18-1(d) show how the transistor switches operate to produce one of two output voltage levels. When all inputs are LOW, the transistor's base-emitter junctions are reverse-biased, all transistors are OFF, and the pull-down resistor will take the Y output LOW. On the other hand, any HIGH input to the base of an NPN transistor will forward-bias the associated transistor and cause the transistor's collector-to-emitter junction to be equivalent to a closed switch, and therefore switch the HIGH ($+V_{CC}$) collector supply voltage through to the Y output.

Figure 18-1(e) shows the logic schematic symbols for a two- and three-input OR gate and the abbreviated operation expressions for the OR gate. Logic gates with a large number of inputs will have a large number of input combinations; however, the gate's operation will still be the same. To calculate the number of possible input combinations, you can use the following formula:

$$n = 2^x$$

n = number of input combinations
2 is included because we are dealing with a base-2 number system
x = number of inputs

EXAMPLE:

Construct a truth table for a four-input OR gate and show the output logic level for each and every input combination.

Solution:

Figure 18-2(a) shows the truth table for a four-input OR gate. The number of possible input combinations will be

$$n = 2^x$$
$$n = 2^4 = 16$$

(Calculator Sequence: press keys, 2 y^x 4 =)

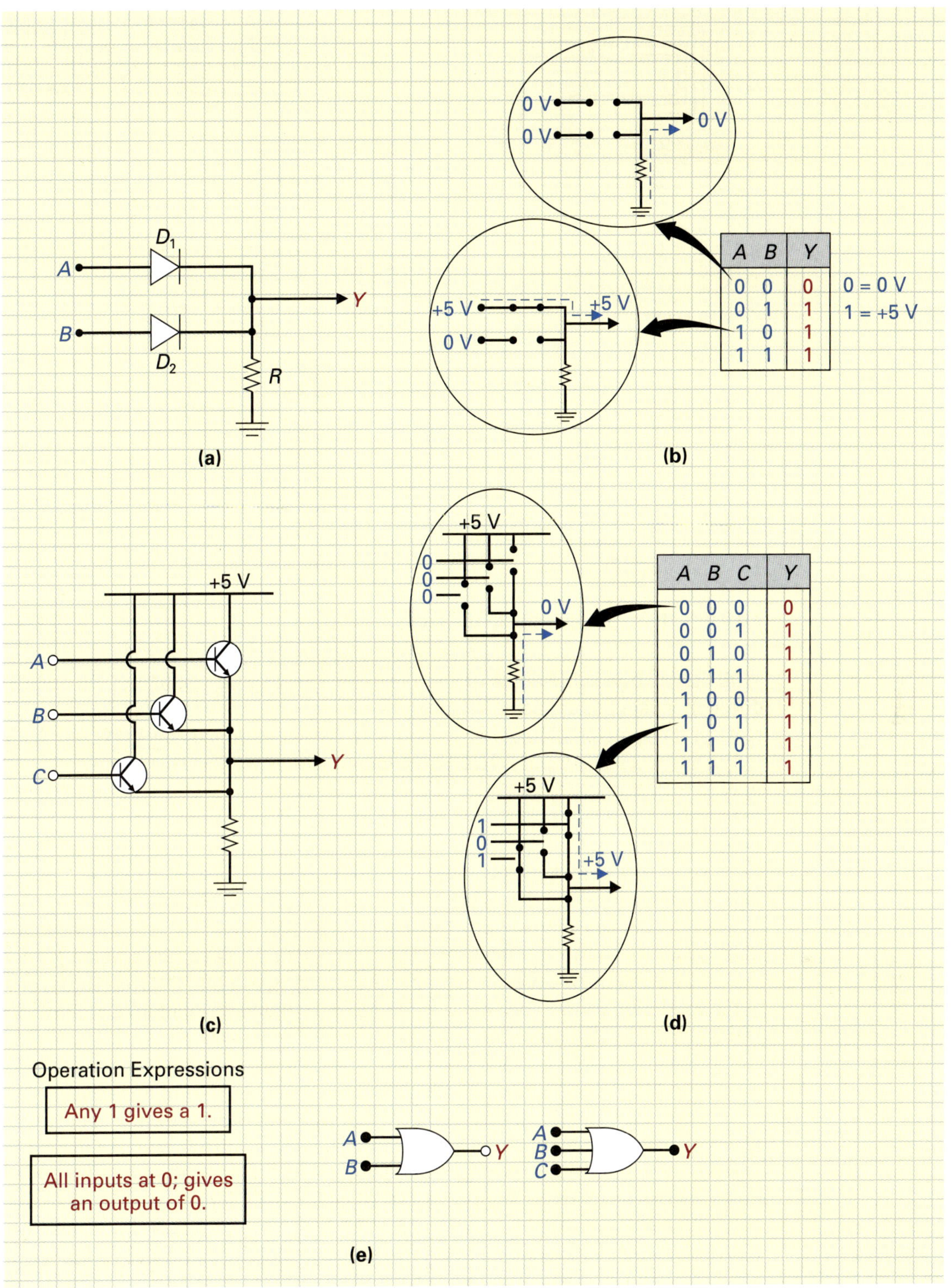

FIGURE 18-1 The OR Gate. (a) Two-Input Diode Circuit. (b) Two-Input Truth Table. (c) Three-Input Transistor Circuit. (d) Three-Input Truth Table. (e) OR Gate Symbols and Operation Expressions.

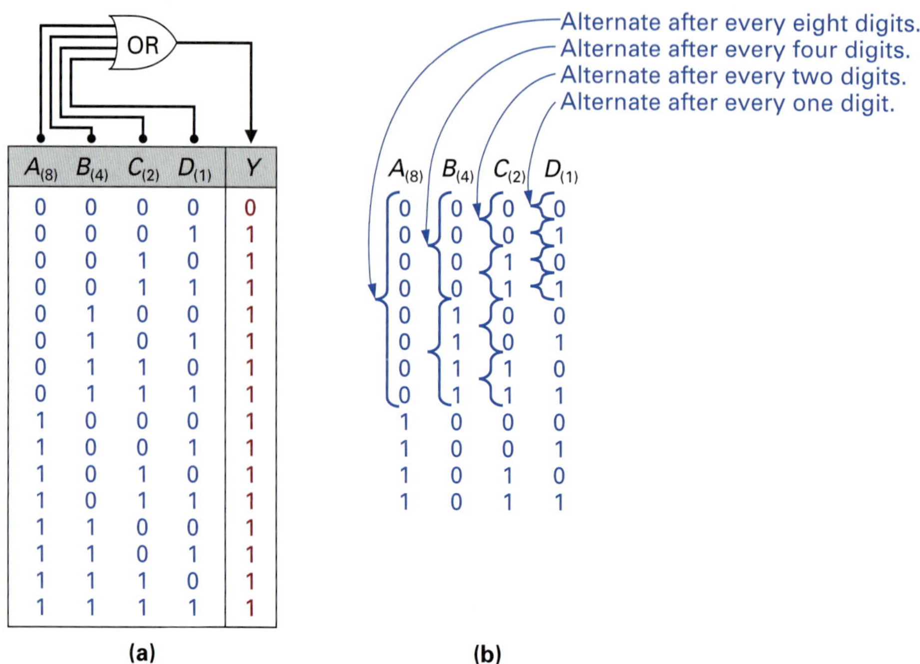

$A_{(8)}$	$B_{(4)}$	$C_{(2)}$	$D_{(1)}$	Y
0	0	0	0	0
0	0	0	1	1
0	0	1	0	1
0	0	1	1	1
0	1	0	0	1
0	1	0	1	1
0	1	1	0	1
0	1	1	1	1
1	0	0	0	1
1	0	0	1	1
1	0	1	0	1
1	0	1	1	1
1	1	0	0	1
1	1	0	1	1
1	1	1	0	1
1	1	1	1	1

FIGURE 18-2 A Four-Input OR Gate.

With only two possible digits (0 and 1) and four inputs, we can have a maximum of sixteen different input combinations. Whether the OR gate has 2, 3, 4, or 444 inputs, it will still operate in the same predictable or logical way: When all inputs are LOW, the output will be LOW, and when any input is HIGH, the output will be HIGH. This is shown in the *Y* column in Figure 18-2(a).

An easy way to construct these truth tables is to first calculate the maximum number of input combinations and then start in the units (1s) column (D column) and move down, alternating the binary digits after every single digit (0101010101, and so on) up to the maximum count, as shown in Figure 18-2(b). Then go to the twos (2s) column (C column) and move down, alternating the binary digits after every two digits (001100110011, and so on) up to the maximum count. Then go to the fours (4s) column (B column) and move down, alternating the binary digits after every four digits (000011110000, and so on) up to the maximum count. Finally, go to the eights (8s) column (A column) and move down, alternating the binary digits after every eight digits (0000000011111111, and so on) up to the maximum count.

These input combinations—or binary words—will start at the top of the truth table with binary 0 and then count up to a maximum value that is always one less than the maximum number of combinations. For example, with the four-input OR gate in Figure 18-2(a), the truth table begins with a count of 0000_2 (0_{10}) and then counts up to a maximum of 1111_2 (15_{10}). The maximum count within a truth table (1111_2 or 15_{10}) is always one less than the maximum number of combinations (16_{10}) because 0000_2 (0_{10}) is one of the input combinations (0000, 0001, 0010, 0011, 0100, 0101, 0110, 0111, 1000, 1001, 1010, 1011, 1100, 1101, 1110, 1111 is a total of sixteen different combinations, with fifteen being the maximum count). Stated with a formula:

$$Count_{max} = 2^x - 1$$

For example, in Figure 18-2: $Count_{max} = 2^4 - 1 = 16 - 1 = 15$ (Calculator Sequence: press keys, 2 y^x 4 − 1 =)

EXAMPLE:

Referring to the circuit in Figure 18-3(a), list the 4-bit binary words generated at the outputs of the four OR gates when each of the 0 through 9 push buttons is pressed.

Solution:

An encoder circuit is a code generator, and the circuit in Figure 18-3(a) is a decimal-to-binary code-generator circuit. Whenever a decimal push button is pressed, a 4-bit binary word equivalent to the decimal value is generated at the outputs of OR gates *D, C, B,* and *A*. The table in Figure 18-3(b) lists the pure binary codes generated for each of the decimal push buttons. For example, when push button 6 is pressed, a HIGH input is applied to OR gates *C* and *B,* and therefore their outputs will go HIGH, while the outputs of OR gates *C* and *A* will stay LOW because all of their inputs are LOW due to the pull-down resistors. In this instance, the 4-bit binary code 0110_2 (6_{10}) will be generated.

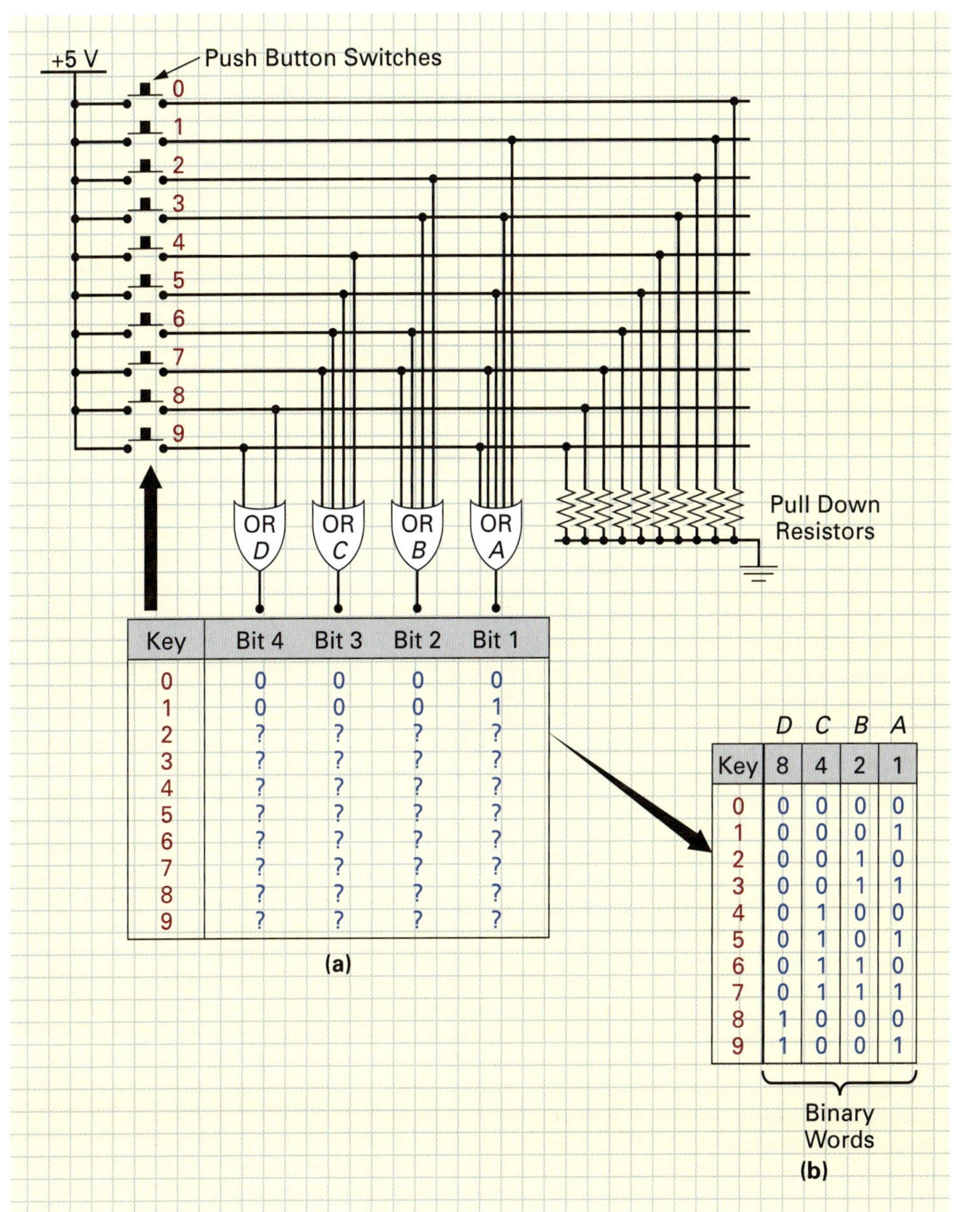

Key	Bit 4	Bit 3	Bit 2	Bit 1
0	0	0	0	0
1	0	0	0	1
2	?	?	?	?
3	?	?	?	?
4	?	?	?	?
5	?	?	?	?
6	?	?	?	?
7	?	?	?	?
8	?	?	?	?
9	?	?	?	?

Key	8 (D)	4 (C)	2 (B)	1 (A)
0	0	0	0	0
1	0	0	0	1
2	0	0	1	0
3	0	0	1	1
4	0	1	0	0
5	0	1	0	1
6	0	1	1	0
7	0	1	1	1
8	1	0	0	0
9	1	0	0	1

FIGURE 18-3 Application—A Decimal-to-Binary Encoder Circuit Using OR Gates.

Rather than construct OR gates using discrete components such as the diode and transistor circuits shown in Figure 18-1(a) and (c), you can experiment with some of the digital integrated circuits (ICs) available from semiconductor manufacturers. These digital ICs contain complete logic-gate circuits. Figure 18-4(a) introduces the "7432 digital IC," which contains four (quad) two-input OR gates within a single 14-pin package. All four OR gates within this IC will operate if power is connected to the IC by using a dc power supply to apply +5 V to pin 14 and ground to pin 7. To test the OR gate that has its inputs connected to pins 1 and 2 and its output at pin 3, we will apply a square wave from a function generator to the OR gate's pin 1 input and a HIGH/LOW control input to the OR gate's pin 2 input, and monitor the output of the OR gate at pin 3 with an oscilloscope.

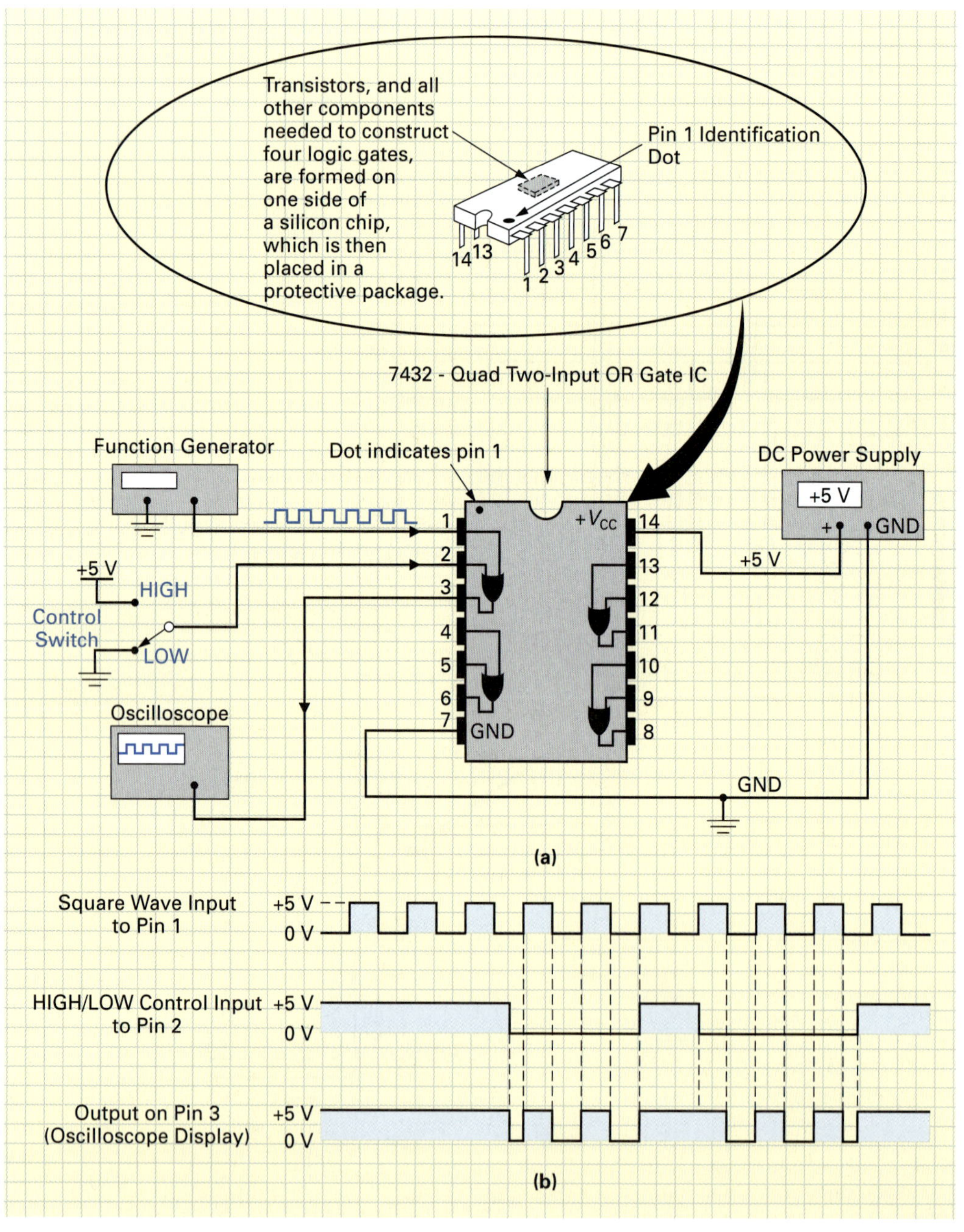

FIGURE 18-4 **Testing the Operation of Digital IC Containing Four Two-Input OR Gates.**

Referring to the timing diagram in Figure 18-4(b), you can see that whenever the square wave input (pin 1) or the control input (pin 2) are HIGH (shown shaded), the output will also be HIGH. Only when both inputs to the OR gate are LOW will the output be LOW. You may want to try constructing this test circuit in a lab and test each one of the four OR gates in a 7432 IC.

18-1-2 *The AND Gate*

Like the OR gate, an **AND gate** can have two or more inputs but will always have a single output. Figure 18-5(a) shows how an AND gate can be constructed using two diodes

AND Gate
A logic gate that will give a HIGH output only if all inputs are HIGH.

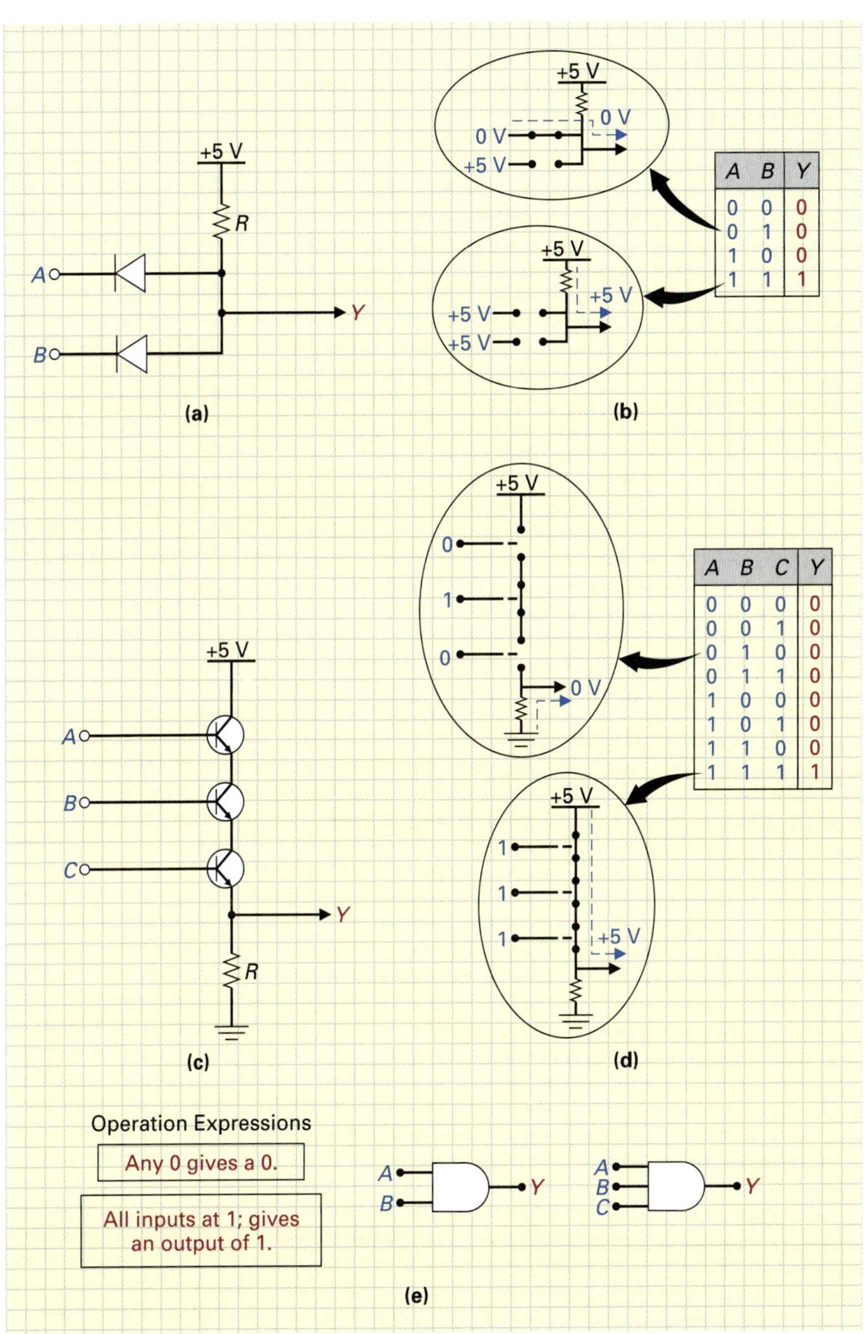

FIGURE 18-5 **The AND Gate. (a) Two-Input Diode Circuit. (b) Two-Input Truth Table. (c) Three-Input Transistor Circuit. (d) Three-Input Truth Table. (e) AND Gate Symbols.**

and a resistor. Figure 18-5(b) shows the truth table for this two-input AND gate. The insets in Figure 18-5(b) simplify how the circuit will basically produce one of two outputs. Since both of the diode anodes are connected to +5 V via *R,* any LOW input to the AND gate will turn ON the associated diode, as shown in the first inset in Figure 18-5(b). In this instance, the diode or diodes will be equivalent to a closed switch and therefore switch the LOW input through to the output. On the other hand, when both *A* AND *B* inputs are HIGH, both diodes will be reverse-biased and equivalent to open switches, and the output *Y* will be pulled HIGH, as shown in the second inset in Figure 18-5(b). We could summarize the operation of this gate, therefore, by saying that *any LOW input will cause a LOW output, and only when both* A *AND* B *inputs are HIGH will the output* Y *be HIGH.*

Figure 18-5(c) shows how a three-input AND gate can be constructed using three two-state transistor switches. The truth table for this AND gate circuit is shown in Figure 18-5(d). Although another input has been added to this circuit, the action of the gate still remains the same since any LOW input will still produce a LOW output, and only when all of the inputs are HIGH will the output be HIGH. To explain the operation of this transistor logic gate in more detail, the insets in Figure 18-5(d) show how the transistor switches operate to produce one of two output voltage levels. When any of the inputs are LOW, the transistor's base-emitter junction is reverse-biased, the OFF transistor is equivalent to an open switch between collector and emitter, and the pull-down resistor will take the *Y* output LOW. On the other hand, when all inputs are HIGH, all of the NPN transistor's base-emitter junctions will be forward-biased, causing all of the transistor's collector-to-emitter junctions to be equivalent to closed switches and therefore switching the HIGH ($+V_{CC}$) collector supply voltage through to the *Y* output.

Figure 18-5(e) shows the logic schematic symbols for a two- and three-input AND gate, and the abbreviated operation expressions for the AND gate.

EXAMPLE:

Develop a truth table for a five-input AND gate. Show the output logic level for each and every input combination, and indicate the range of values within the truth table.

Solution:

Figure 18-6 shows the truth table for a five-input AND gate. The number of possible input combinations will be

$$n = 2^x$$

$$n = 2^5 = 32$$

A five-input AND gate will still operate in the same predictable or logical way: When any input is LOW, the output will be LOW, and when all inputs are HIGH, the output will be HIGH. This is shown in the *Y* column in Figure 18-6.

The range of values within the truth table will be

$$Count_{\text{max}} = 2^5 - 1 = 32 - 1 = 31$$

With the five-input OR gate in Figure 18-6, the truth table begins with a count of 00000_2 (0_{10}) and then counts up to a maximum of 11111_2 (31_{10}).

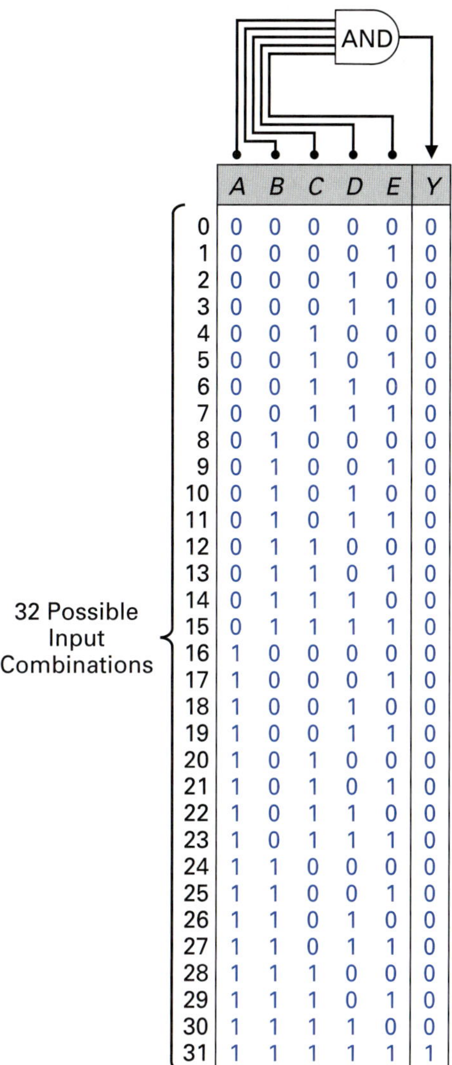

	A	B	C	D	E	Y
0	0	0	0	0	0	0
1	0	0	0	0	1	0
2	0	0	0	1	0	0
3	0	0	0	1	1	0
4	0	0	1	0	0	0
5	0	0	1	0	1	0
6	0	0	1	1	0	0
7	0	0	1	1	1	0
8	0	1	0	0	0	0
9	0	1	0	0	1	0
10	0	1	0	1	0	0
11	0	1	0	1	1	0
12	0	1	1	0	0	0
13	0	1	1	0	1	0
14	0	1	1	1	0	0
15	0	1	1	1	1	0
16	1	0	0	0	0	0
17	1	0	0	0	1	0
18	1	0	0	1	0	0
19	1	0	0	1	1	0
20	1	0	1	0	0	0
21	1	0	1	0	1	0
22	1	0	1	1	0	0
23	1	0	1	1	1	0
24	1	1	0	0	0	0
25	1	1	0	0	1	0
26	1	1	0	1	0	0
27	1	1	0	1	1	0
28	1	1	1	0	0	0
29	1	1	1	0	1	0
30	1	1	1	1	0	0
31	1	1	1	1	1	1

FIGURE 18-6 A Five-Input AND Gate.

EXAMPLE:

Referring to the circuit in Figure 18-7(a), describe what would be present at the output of the AND gate for each of the CONTROL switch positions.

Solution:

When the CONTROL switch is put in the ENABLE position, a HIGH is applied to the lower input of the AND gate, as shown in Figure 18-7(b). In this mode, the output Y will follow the square wave input, since when the square wave is HIGH, the AND gate inputs are both HIGH, and Y is HIGH. When the square wave is LOW, the AND gate will have a LOW input, and Y will be LOW. The Y output, therefore, follows the square wave input, and the AND gate is said to be equivalent to a closed switch, as shown in Figure 18-7(b). On the other hand, when the CONTROL switch is put in the DISABLE position, a

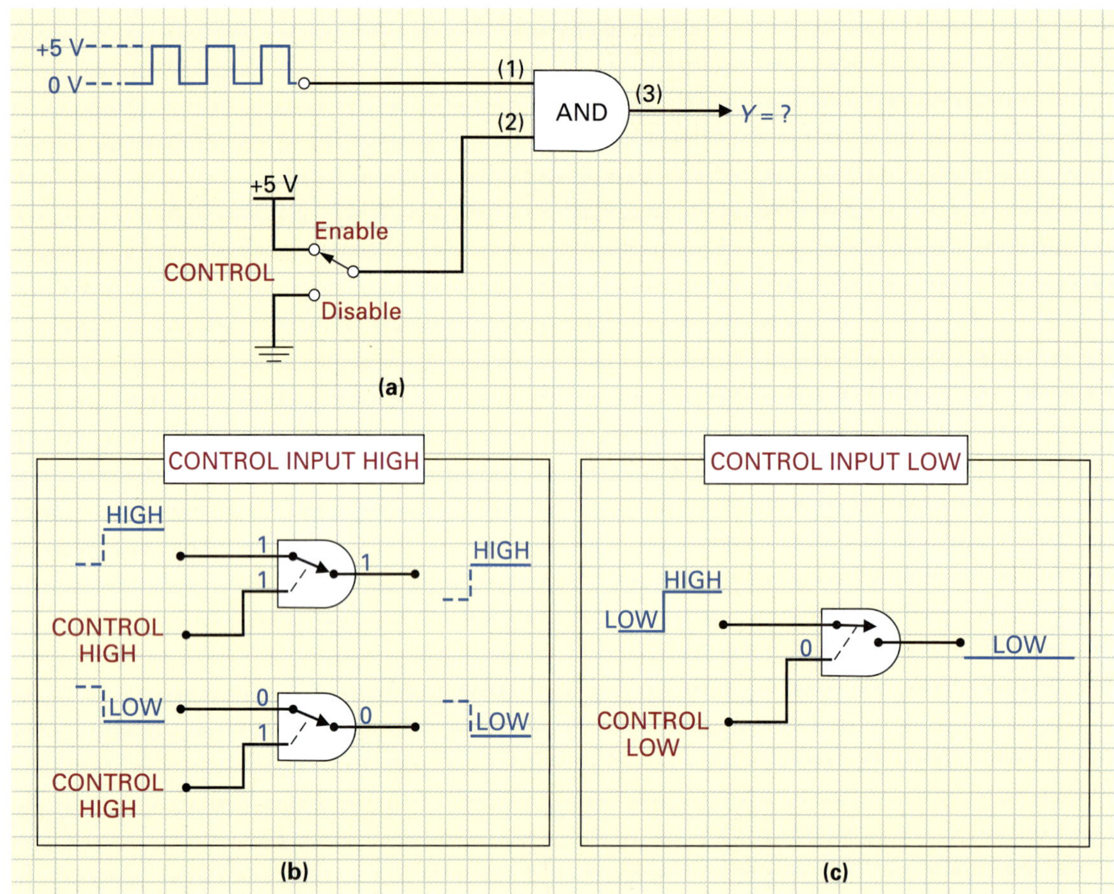

FIGURE 18-7 **Application—An Enable/Disable Control Circuit Using an AND Gate.**

LOW is applied to the lower input of the AND gate, as shown in Figure 18-7(c). In this mode, the output *Y* will always remain LOW, since any LOW input to an AND gate will always result in a LOW output. In this situation, the AND gate is said to be equivalent to an open switch, as shown in Figure 18-7(c).

Controlled Switch
An electronically controlled switch.

When used in applications such as this, the AND gate is said to be acting as a **controlled switch.**

Like the OR gate digital IC, there are several digital ICs available containing AND gates. Figure 18-8(a) introduces the "7408 digital IC," which contains four (quad) two-input AND gates and shows how to test the AND gates to see if they operate as controlled switches. As before, all four AND gates within this IC will operate if power is connected to the IC by using a dc power supply to apply +5 V to pin 14 and ground to pin 7. To test the AND gate that has its inputs connected to pins 1 and 2 and its output at pin 3, we will apply a square wave from a function generator to the AND gate's pin 1 input and a HIGH/LOW control input to the AND gate's pin 2 input and monitor the output of the AND gate at pin 3 with an oscilloscope. Referring to the timing diagram in Figure 18-8(b), you can see that only when both inputs are HIGH (shown shaded), will the output be HIGH. You may want to try constructing this test circuit in a lab and testing each one of the four AND gates in a 7408 IC.

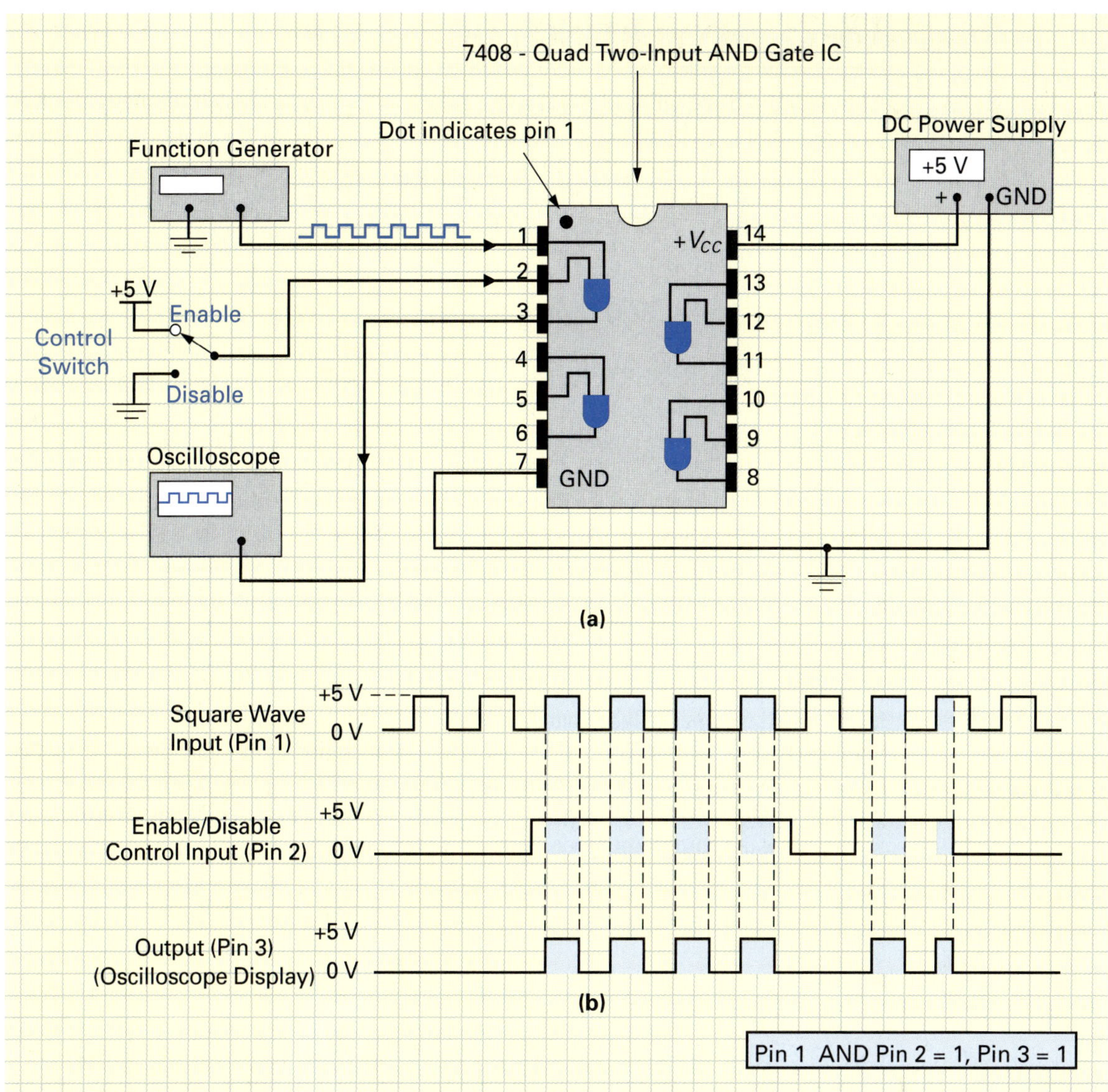

FIGURE 18-8 **Testing the Operation of a Digital IC Containing Four Two-Input AND Gates.**

SELF-TEST EVALUATION POINT FOR SECTION 18-1

Use the following questions to test your understanding of Section 18-1.

1. The two basic logic gates are the ____________ gate and the ____________ gate.
2. With an OR gate, any binary 1 input will give a binary ____________ output.
3. Which logic gate can be used as a controlled switch?
4. Only when all inputs are LOW will the output of an ____________ gate be LOW.
5. With an AND gate, any binary 0 input will give a binary ____________ output.
6. Only when all inputs are HIGH will the output of an ____________ gate be HIGH.

18-2 INVERTING LOGIC GATES

Although the basic OR and AND logic gates can be used to construct many digital circuits, in some circuit applications other logic operations are needed. For this reason, semiconductor manufacturers made other logic-gate types available. In this section, we will examine three inverting-type logic gates, called the NOT gate, the NOR gate, and the NAND gate. We will begin with the NOT or INVERTER gate, which was introduced earlier.

18-2-1 *The NOT Gate*

Complement
Opposite.

The NOT, or logic INVERTER gate, is the simplest of all the logic gates because it has only one input and one output. Figure 18-9(a) shows how a NOT gate can be constructed using two resistors and a transistor. Figure 18-9(b) shows the truth table for the NOT gate. The insets in Figure 18-9(b) simplify how the circuit will produce one of two outputs. Referring to the upper inset, you can see that when the input is LOW, the transistor's base-emitter diode will be reverse-biased and so the transistor will turn OFF. In this instance, the transistor is equivalent to an open switch between emitter and collector, and therefore the +5 V supply voltage will be connected to the output. A LOW input, therefore, will be converted or inverted to a HIGH output. On the other hand, referring to the lower inset in Figure 18-9(b), you can see that when the input is HIGH, the transistor's base-emitter diode will be forward-biased and the transistor will turn ON. In this circuit condition, the transistor is equivalent to a closed switch between emitter and collector, and the output will be switched LOW. A HIGH input, therefore, will be converted or inverted to a LOW output. We could summarize the operation of this gate by saying that *the output logic level is "NOT" the same as the input logic level* (a 1 input is NOT the same as the 0 output, or a 0 input is NOT the same as the 1 output), hence the name NOT gate. The output is, therefore, always the **complement,** or opposite, of the input.

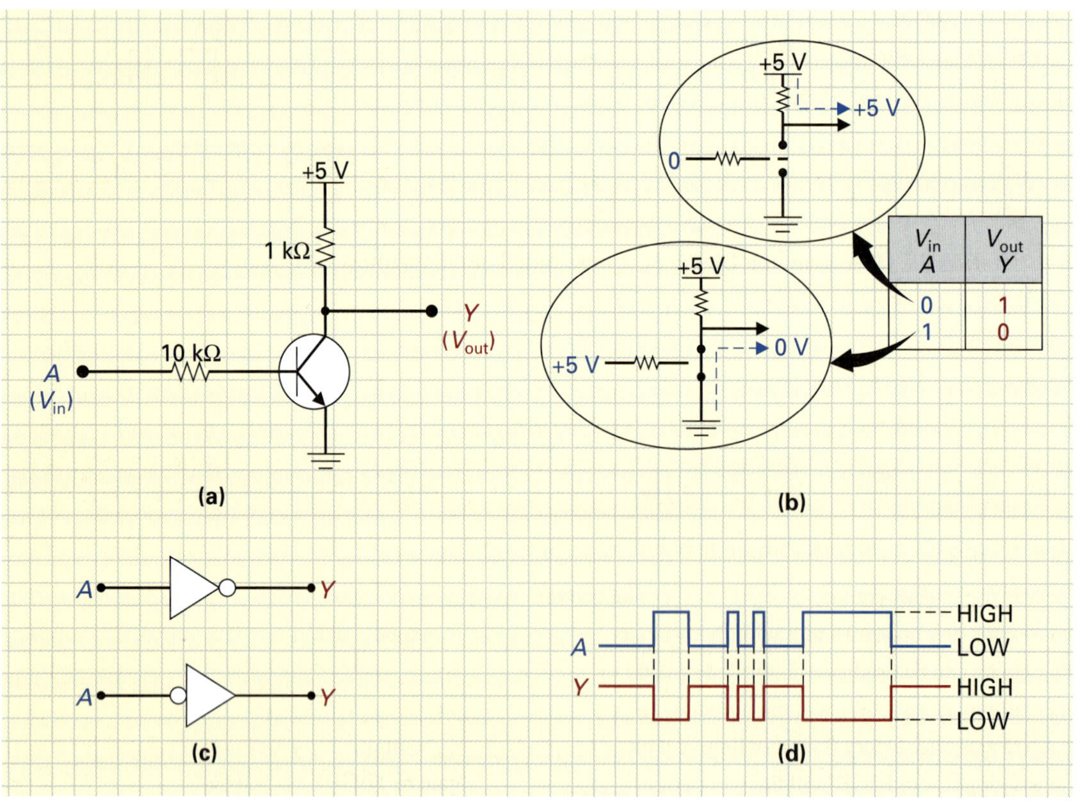

V_{in} A	V_{out} Y
0	1
1	0

FIGURE 18-9 **The NOT Gate or Logic Inverter. (a) Transistor Circuit. (b) Truth Table. (c) Logic Symbols for a NOT Gate. (d) Timing Diagram Showing Input-to-Output Inversion.**

Figure 18-9(c) shows the schematic symbols normally used to represent the NOT gate. The triangle is used to represent the logic circuit, while the **bubble,** or small circle either before or after the triangle, is used to indicate the complementary or inverting nature of this circuit. Figure 18-9(d) shows a typical example of a NOT gate's input/output waveforms. As you can see from these waveforms, the output waveform is an inverted version of the input waveform.

Bubble
A small circle symbol used to signify an invert function.

EXAMPLE:

What will be the outputs from the circuit in Figure 18-10(a), assuming the square wave input shown?

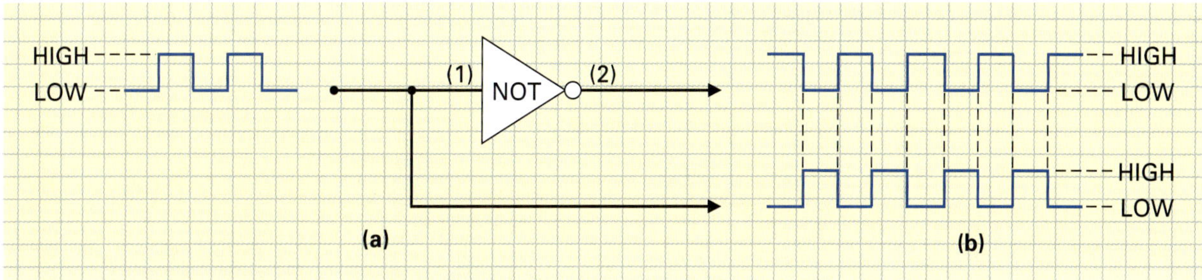

FIGURE 18-10 **Application—A Two-Phase Signal Generator Circuit Using a NOT Gate.**

Solution:

As can be seen in Figure 18-10(b), the two-phase signal generator circuit will supply two square wave output signals that are 180° out of phase due to the inverting action of the NOT gate.

EXAMPLE:

What will be the outputs from each of the six NOT gates within the 7404 digital IC shown in Figure 18-11?

Solution:

The NOT gate will always invert its input, and therefore:

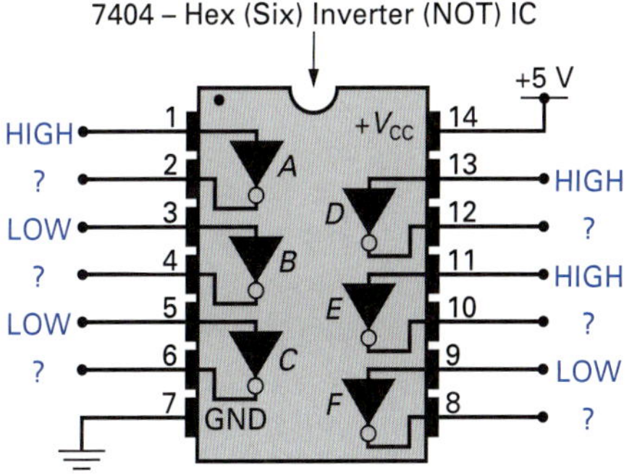

FIGURE 18-11 **A Digital IC Containing Six NOT Gates.**

	NOT GATE INPUT	NOT GATE OUTPUT
A	HIGH	LOW
B	LOW	HIGH
C	LOW	HIGH
D	HIGH	LOW
E	HIGH	LOW
F	LOW	HIGH

18-2-2 *The NOR Gate*

NOR Gate
A NOT-OR gate that will give a LOW output if any of its inputs are HIGH.

The name "NOR" is a combination of the two words "NOT" and "OR" and is used to describe a type of logic gate that contains an OR gate followed by a NOT gate, as shown in Figure 18-12(a). The **NOR gate** could be constructed using the previously discussed transistor OR gate circuit followed by a transistor NOT gate circuit, as shown in the inset in Figure 18-12(a). The NOR gate should not be thought of as a new type of logic gate because it isn't. It is simply a combination of two previously discussed gates—an OR gate and a NOT gate.

Figure 18-12(b) shows the standard symbol used to represent the NOR gate. Studying the NOR gate symbol, you can see that the OR symbol is used to show how the two inputs are first "ORed," and that the bubble is used to indicate that the result from the OR operation is inverted before being applied to the output. As a result, the NOR gate's output is simply the opposite, or complement, of the OR gate, as can be seen in the truth table for a two-input NOR gate shown in Figure 18-12(c). If you compare the NOR gate truth

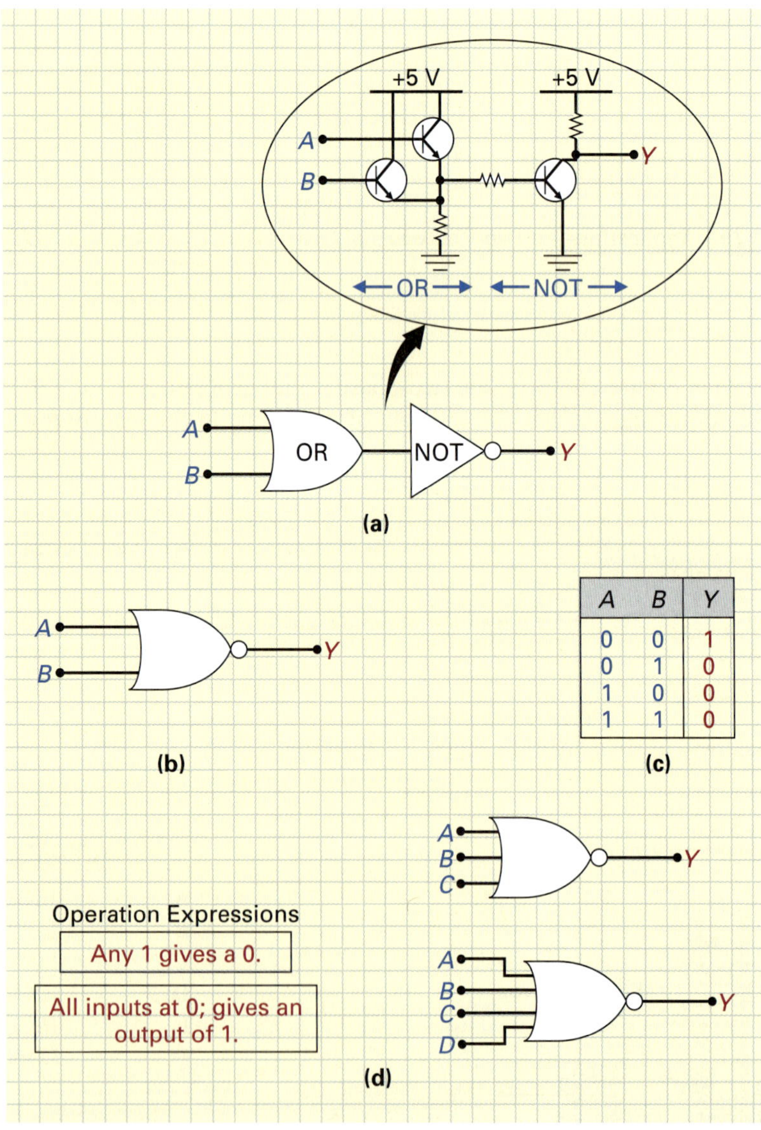

FIGURE 18-12 The NOR Gate. (a) Two-Input Circuit. (b) Two-Input Logic Symbol. (c) Two-Input Truth Table. (d) NOR Gate Operation Expressions and Three- and Four-Input Logic Symbols.

table to the OR gate truth table, you can see that the only difference is that the output is inverted:

OR Gate: When all inputs are 0, the output is **0,** and any 1 input gives a **1** output.
NOR Gate: When all inputs are 0, the output is **1,** and any 1 input gives a **0** output.

The operation expressions for the NOR gate are listed in Figure 18-12(d), along with the standard symbols for a three-input and four-input NOR gate.

EXAMPLE:

What will be the outputs from each of the four NOR gates within the 7402 digital IC shown in Figure 18-13?

Solution:

When all inputs are 0, the output is 1, and when any input is 1, the output is 0. Therefore:

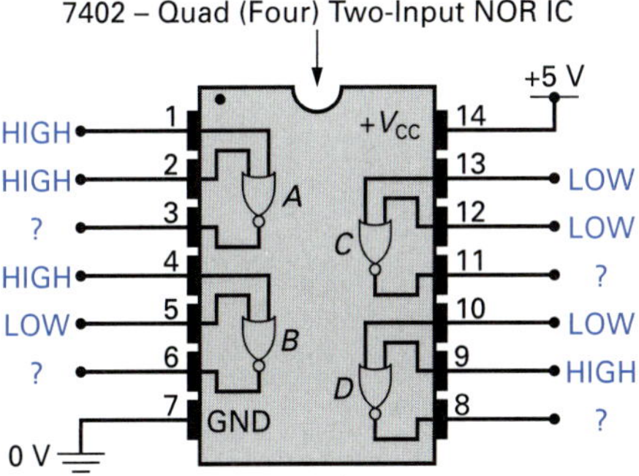

FIGURE 18-13 **A Digital IC Containing Four NOR gates.**

	NOR GATE INPUTS		NOR GATE OUTPUT
A	HIGH	HIGH	LOW
B	HIGH	LOW	LOW
C	LOW	LOW	HIGH
D	LOW	HIGH	LOW

EXAMPLE:

Referring to the circuit shown in Figure 18-14, determine whether the LED will flash at a rate of 1 Hz or 2 Hz for the inputs and switch position shown.

Solution:

The circuit shown in Figure 18-14 is an example of a multiplexer circuit, which is a circuit that switches only one of its inputs through to a single output. In this circuit, two AND gates

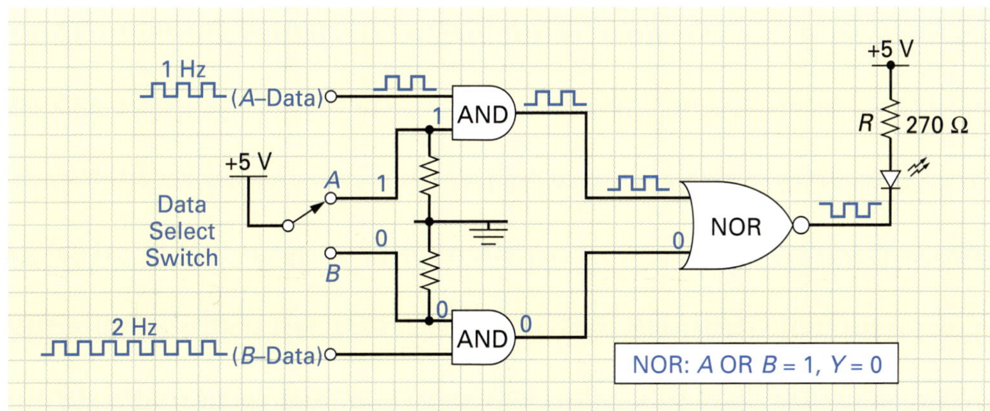

FIGURE 18-14 **Application—A Data Selector or Multiplexer Circuit Using a NOR Gate.**

are operated as controlled switches, with the data select switch determining which of the AND gates is enabled. In this example, the upper AND gate is enabled, and so the *A*-data input (1 Hz signal) will be switched through to the NOR gate's upper input. The lower input to the NOR gate will be LOW, because the lower AND gate has been disabled by the data select switch. (AND: Any 0 input gives a 0 output.) Since any HIGH input to a NOR gate will give a LOW output, the NOR gate will basically act as an INVERTER and invert the 1 Hz square wave input. The anode of the LED is connected to +5 V, so when the output of the NOR gate goes LOW, the LED will turn ON, and when the output of the NOR gate goes HIGH, the LED will turn OFF. This multiplexer circuit will therefore switch the 1 Hz *A*-data input through to the output, and the LED will flash ON and OFF at a 1 Hz rate.

If the data select switch were placed in the *B* position, the 2 Hz *B*-data input would be switched through to the output and the LED would flash ON and OFF at a 2 Hz rate.

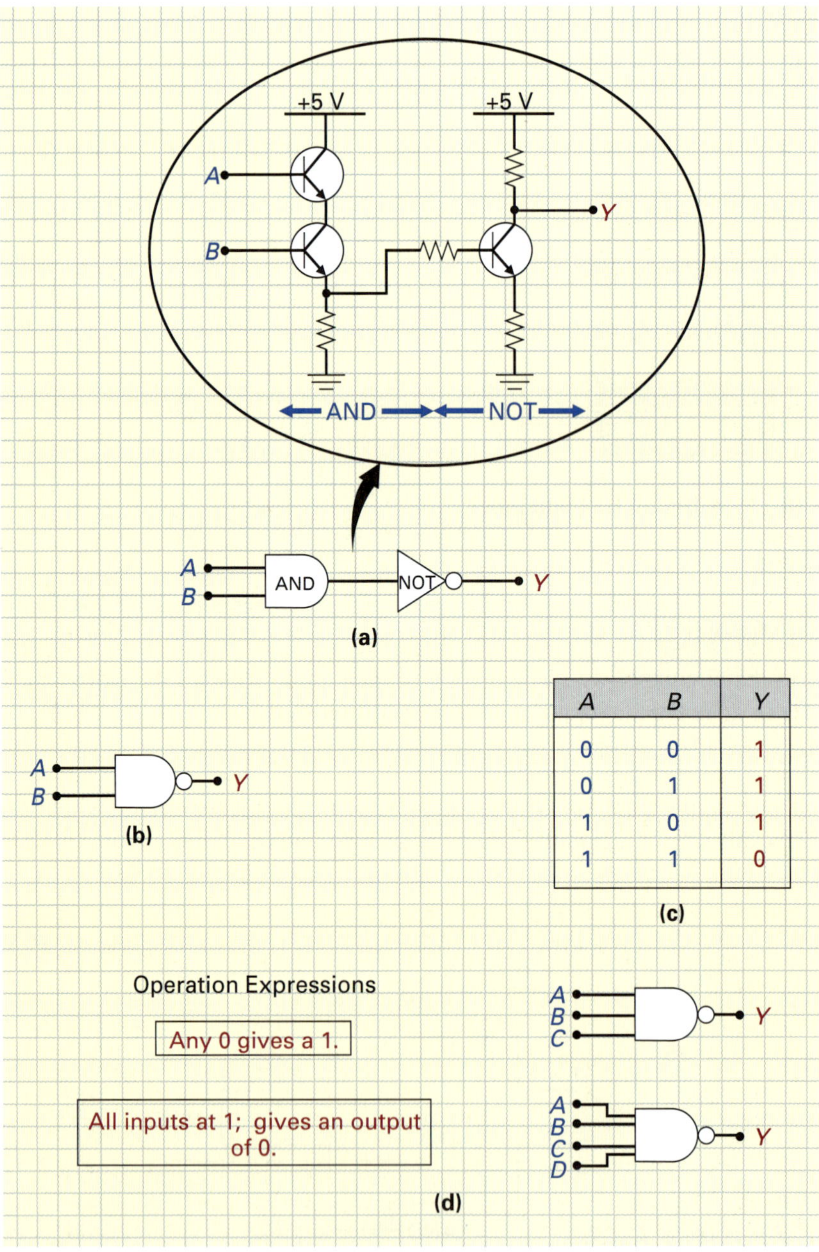

A	B	Y
0	0	1
0	1	1
1	0	1
1	1	0

FIGURE 18-15 The NAND Gate. (a) Two-Input Circuit. (b) Two-Input Logic Symbol. (c) Two-Input Truth Table. (d) NAND Gate Operation Expressions and Three- and Four-Input Logic Symbols.

18-2-3 *The NAND Gate*

Like NOR, the name "NAND" is a combination of the two words "NOT" and "AND," and it describes another type of logic gate that contains an AND gate followed by a NOT gate, as shown in Figure 18-15(a). The **NAND gate** could be constructed using the previously discussed transistor AND gate circuit followed by a transistor NOT gate circuit, as shown in the inset in Figure 18-15(a). The NAND gate should also not be thought of as a new type of logic gate because it is simply an AND gate and a NOT gate combined to perform the NAND logic function.

NAND Gate
A NOT-AND logic gate circuit that will give a HIGH output if any of its inputs are LOW.

Figure 18-15(b) shows the standard symbol used to represent the NAND gate. Studying the NAND gate symbol, you can see that the AND symbol is used to show how the two inputs are first "ANDed," and the bubble is used to indicate that the result from the AND operation is inverted before being applied to the output. As a result, the NAND gate's output is simply the opposite, or complement, of the AND gate, as can be seen in the truth table for a two-input NAND gate shown in Figure 18-15(c). If you compare the NAND gate truth table to the AND gate truth table, you can see that the only difference is that the output is inverted. To compare:

AND Gate: Any 0 input gives a **0** output, and only when both inputs are 1 will the output be **1.**

NAND Gate: Any 0 input gives a **1** output, and only when both inputs are 1 will the output be **0.**

The operation expressions for the NAND gate are listed in Figure 18-15(d), along with the standard symbols for a three-input and four-input NAND gate.

EXAMPLE:

What will be the outputs from each of the four NAND gates within the 7400 digital IC shown in Figure 18-16?

Solution:

With the NAND gate, any 0 input gives a 1 output, and only when both inputs are 1 will the output be 0. Therefore:

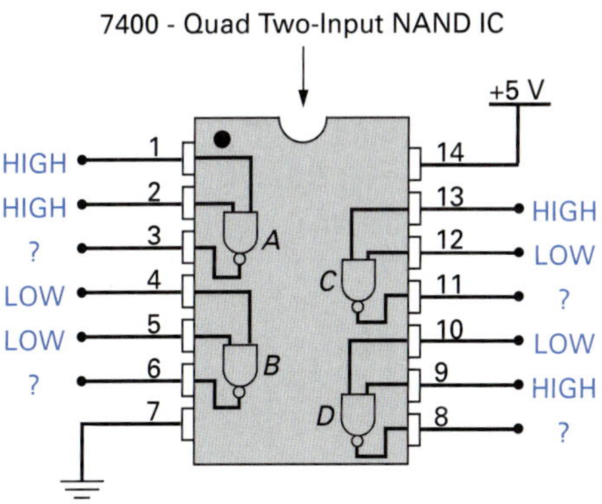

FIGURE 18-16 A Digital IC Containing Four NAND Gates.

	NAND GATE INPUTS		NAND GATE OUTPUT
A	HIGH	HIGH	LOW
B	LOW	LOW	HIGH
C	HIGH	LOW	HIGH
D	LOW	HIGH	HIGH

EXAMPLE:

Which of the LEDs in Figure 18-17 would turn ON if the binary input at *A* and *B* were both HIGH?

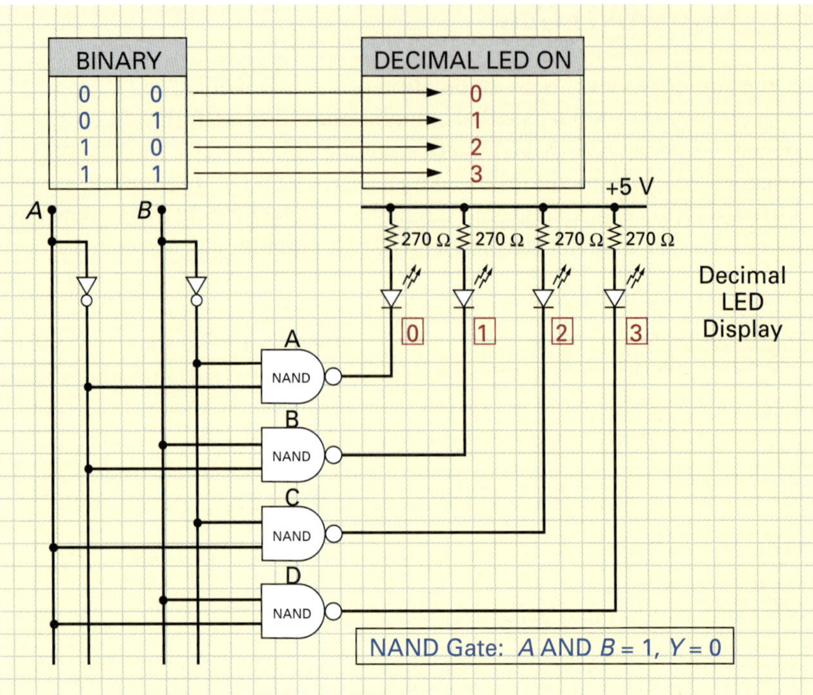

FIGURE 18-17 **Application—A Binary-to-Decimal Decoder Circuit Using NAND Gates.**

■ ***Solution:***

The circuit shown in Figure 18-17 is an example of a decoder circuit, which is a circuit that translates coded characters into a more understandable form. In this example, the circuit is used to decode a 2-bit binary code and then activate a decimal display. A set of four NAND gates is used to drive the decimal display, which contains four LEDs numbered 0 through 3. Since the NAND gate outputs are connected to the cathodes of the LEDs, and the anodes are connected to +5 V, a LOW output from any NAND gate will turn ON its associated LED. As we now know, a NAND gate will only give a LOW output when both of its inputs are HIGH; and the inputs in this circuit are controlled by the *A* and *B*—and inverted *A* and *B*—binary input lines. For example, when *A* and *B* inputs are both LOW, the inverted *A* and *B* lines will both be HIGH. Only NAND gate *D* will have both of its inputs HIGH, so its output will go LOW, causing the number 3 LED to turn ON. As a result, when binary 11_2 is applied to the input of the decoder, the output will display decimal 3_{10}. Therefore, each of the four 2-bit binary input codes will activate one of the equivalent decimal output LEDs.

SELF-TEST EVALUATION POINT FOR SECTION 18-2

Use the following questions to test your understanding of Section 18-2.

1. Which of the logic gates has only one input?
2. With a NOT gate, any binary 1 input will give a binary ____________ output.
3. Which logic gate is a combination of an AND gate followed by a logic INVERTER?
4. Only when all inputs are LOW will the output of a/an ____________ gate be HIGH.
5. With a NAND gate, any binary 0 input will give a binary ____________ output.
6. Only when all inputs are HIGH will the output of a/an ____________ gate be LOW.

18-3 EXCLUSIVE LOGIC GATES

In this section, we will discuss the final two logic gate types: the exclusive-OR (XOR) gate and exclusive-NOR (XNOR) gate. Although these two logic gates are not used as frequently as the five basic OR, AND, NOT, NOR, and NAND gates, their function is ideal in some circuit applications.

18-3-1 *The XOR Gate*

Exclusive-OR Gate
A logic gate circuit that will give a HIGH output if any odd number of binary 1s are applied to the input.

The **exclusive-OR (XOR) gate** logic symbol is shown in Figure 18-18(a), and its truth table is shown in Figure 18-18(b). Like the basic OR gate, the operation of the XOR is dependent

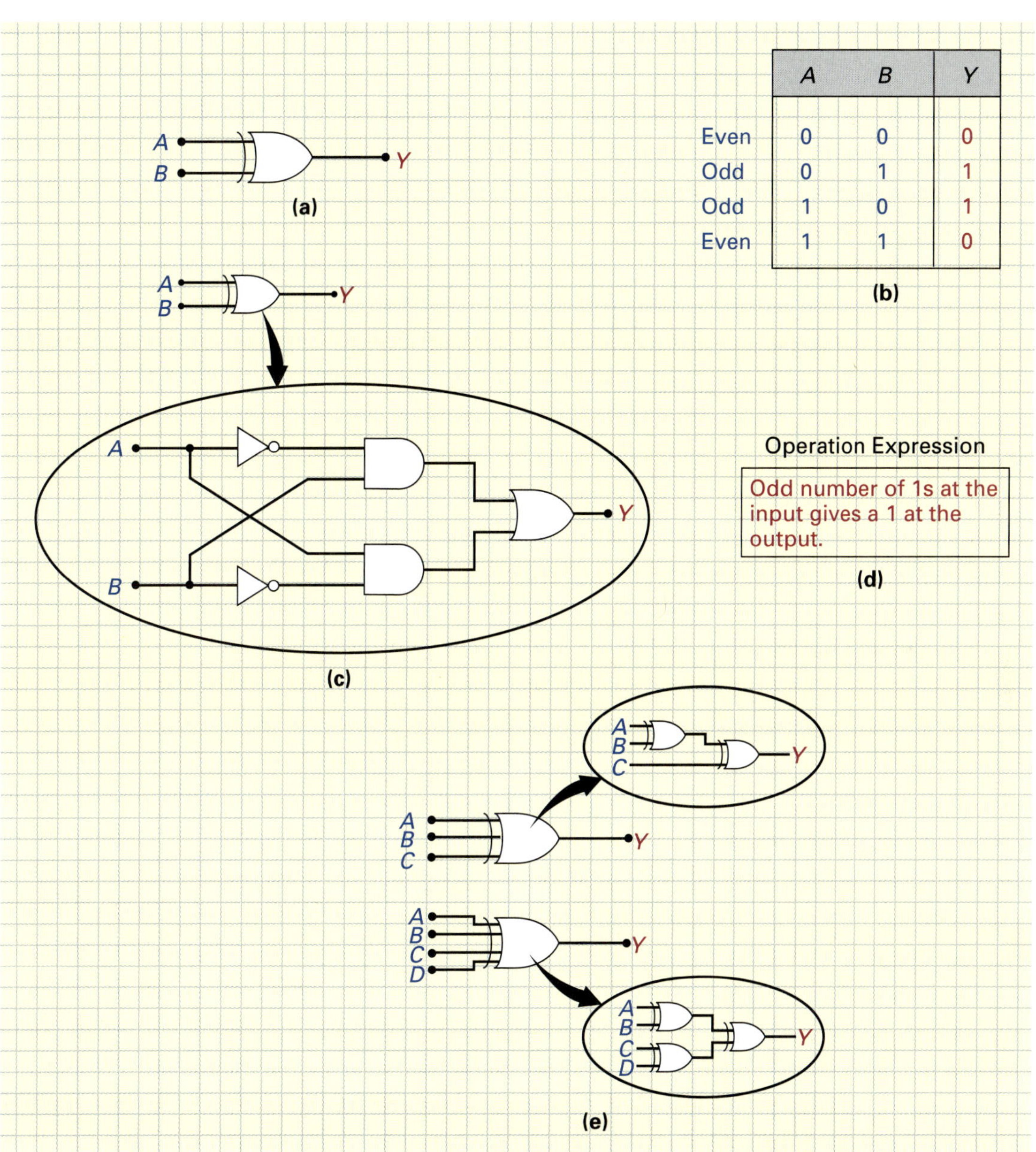

FIGURE 18-18 **The Exclusive-OR (XOR) Gate. (a) Two-Input Logic Symbol. (b) Two-Input Truth Table. (c) Basic Gate Circuit. (d) Operation Expression. (e) Three- and Four-Input XOR Gates.**

on the HIGH or binary 1 inputs. With the basic OR, any 1 input would cause a 1 output. With the XOR, any odd number of binary 1s at the input will cause a binary 1 at the output. Looking at the truth table in Figure 18-18(b), you can see that when the binary inputs are 01 or 10, there is one binary 1 at the input (odd) and the output is 1. When the binary inputs are 00 or 11, however, there are two binary 0s or two binary 1s at the input (even) and the output is 0. To distinguish the XOR gate symbol from the basic OR gate symbol, you may have noticed that an additional curved line is included across the input.

Exclusive-OR logic gates are constructed by combining some of our previously discussed basic logic gates. Figure 18-18(c) shows how two NOT gates, two AND gates, and an OR gate can be connected to form an XOR gate. If inputs A and B are both HIGH or both LOW (even input), the AND gates will both end up with a LOW at one of their inputs, so the OR gate will have both of its inputs LOW, and therefore the final output will be LOW. On the other hand, if A is HIGH and B is LOW or if B is HIGH and A is LOW (odd input), one of the AND gates will have both of its inputs HIGH, giving a HIGH to the OR gate and, therefore, a HIGH to the final output. We could summarize the operation of this gate therefore by saying that *the output is only HIGH when there is an odd number of binary 1s at the input.*

Figure 18-18(d) lists the operation expression for the XOR gate and Figure 18-18(e) shows how two-input XOR gates can be used to construct XOR gates with more than two inputs.

EXAMPLE:

List the outputs you would expect for all the possible inputs to a three-input XOR gate.

Solution:

$$n = 2^x$$

$$n = 2^3 = 8$$

A	B	C	Y	
0	0	0	0	Even
0	0	1	1	Odd
0	1	0	1	Odd
0	1	1	0	Even
1	0	0	1	Odd
1	0	1	0	Even
1	1	0	0	Even
1	1	1	1	Odd

The output is only 1 when there is an odd number of binary 1s at the input.

EXAMPLE:

What will be the outputs from each of the four XOR gates within the 7486 digital IC shown in Figure 18-19?

Solution:

With the XOR gate, the output is only HIGH when there is an odd number of binary 1s at the input. Therefore:

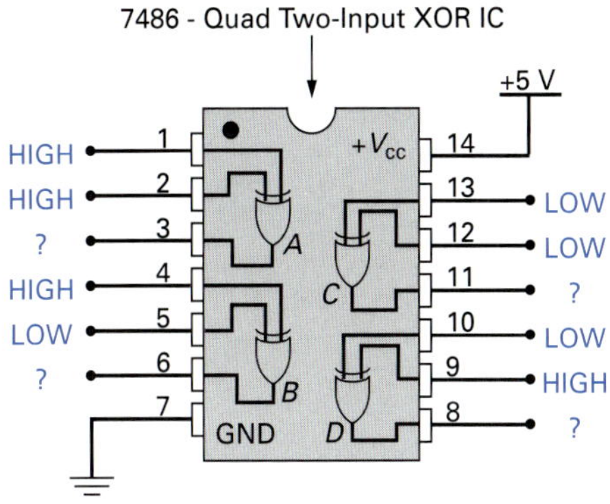

	XOR GATE INPUTS		XOR GATE OUTPUT
A	HIGH	HIGH	LOW
B	HIGH	LOW	HIGH
C	LOW	LOW	LOW
D	LOW	HIGH	HIGH

FIGURE 18-19 A Digital IC Containing Four XOR Gates.

18-3-2 *The XNOR Gate*

The **exclusive-NOR (XNOR) gate** is simply an XOR gate followed by a NOT gate, as shown in Figure 18-20(a). Its logic symbol, which is shown in Figure 18-20(b), is the same as the XOR symbol except for the bubble at the output, which is included to indicate that the result of the XOR operation is inverted before it appears at the output. Like the NAND and NOR gates, the XNOR is not a new gate, but simply a previously discussed gate with an inverted output.

Exclusive-NOR Gate
A NOT-exclusive OR gate that will give a HIGH output if any even number of binary 1s are applied to the input.

The truth table for the XNOR gate is shown in Figure 18-20(c). Comparing the XOR gate to the XNOR gate, the XOR gate's output is only HIGH when there is an odd number of binary 1s at the input, whereas with the XNOR gate, the output is only HIGH when there is an even number of binary 1s at the input. The operation expression for the XNOR gate is listed in Figure 18-20(d).

A
B
Y
(a)

A
B
Y
(b)

A	*B*	Number of 1s	*Y*
0	0	Even	1
0	1	Odd	0
1	0	Odd	0
1	1	Even	1

(c)

Operation Expression

Even number of 1s at the input gives a 1 at the output.

(d)

FIGURE 18-20 The Exclusive-NOR (XNOR) Gate. (a) Circuit. (b) Logic Symbol. (c) Truth Table. (d) Operation Expression.

EXAMPLE:

What will be the X and Y outputs from the circuit in Figure 18-21(a) if the A and B inputs are either out-of-phase or in-phase?

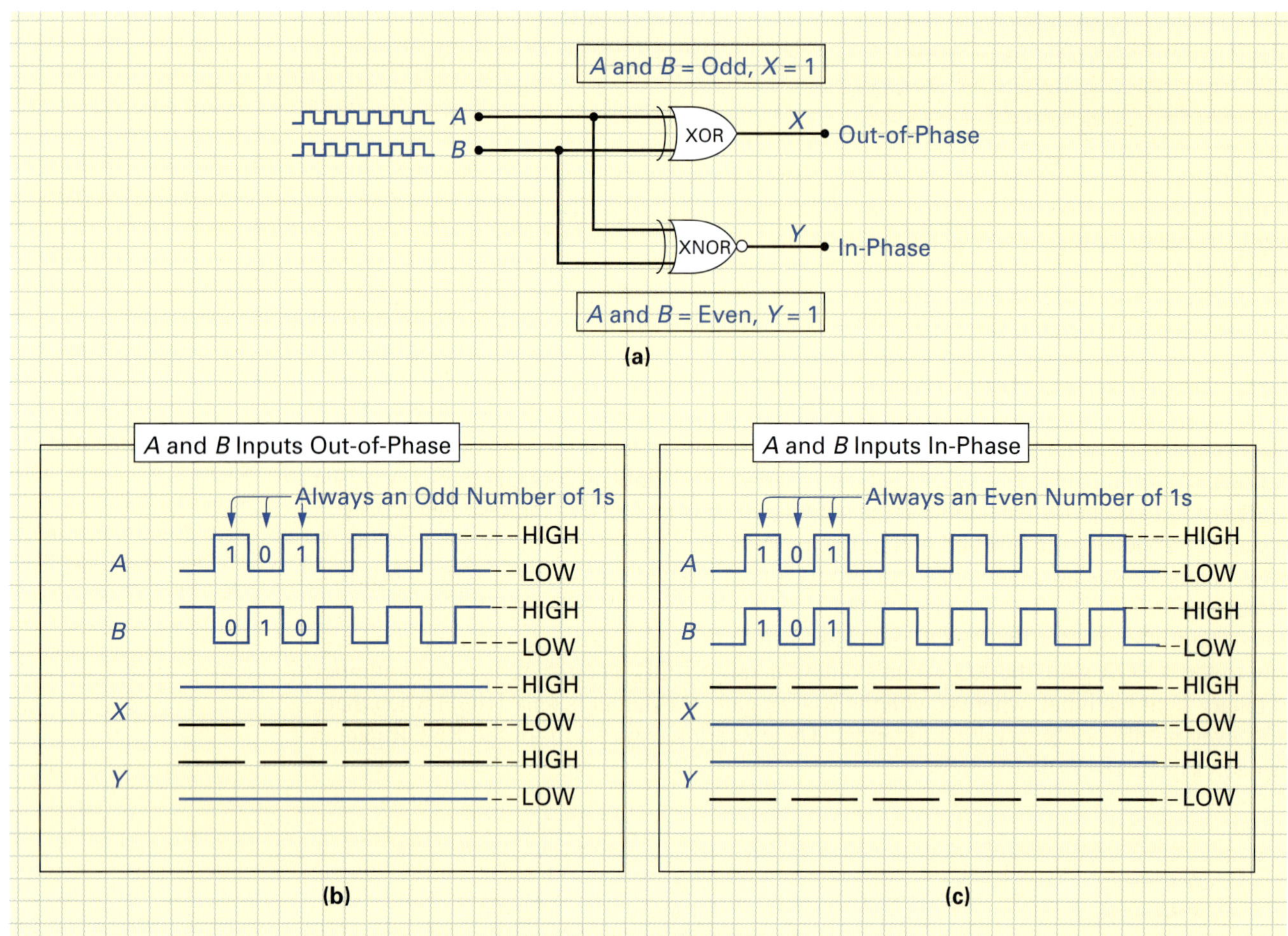

FIGURE 18-21 **Application—A Phase Detector Circuit Using an XOR and XNOR Gate.**

Solution:

The circuit in Figure 18-21(a) is a phase detector circuit. The two square wave inputs at A and B are applied to both the XOR and XNOR gates. If these two waveforms are out-of-phase, as shown in Figure 18-21(b), an odd number of binary 1s will be constantly applied to both gates, and the XOR output (X) will go HIGH and the XNOR output (Y) will go LOW. On the other hand, if the two waveforms are in-phase, as shown in Figure 18-21(c), an even number of binary 1s will constantly be applied to both gates, and the XOR output (X) will go LOW and the XNOR output (Y) will go HIGH.

EXAMPLE:

The four logic circuits within the 74135 digital IC, shown in Figure 18-22(a), can be made to operate as either XOR or XNOR logic gates. Their logic function is determined by the HIGH or LOW control input applied to pins 4 and 12.

Consider the logic circuit that has its inputs at pins 1 and 2, its output at pin 3, and a control input at pin 4. How will this logic circuit function if its control input is first made LOW and then made HIGH?

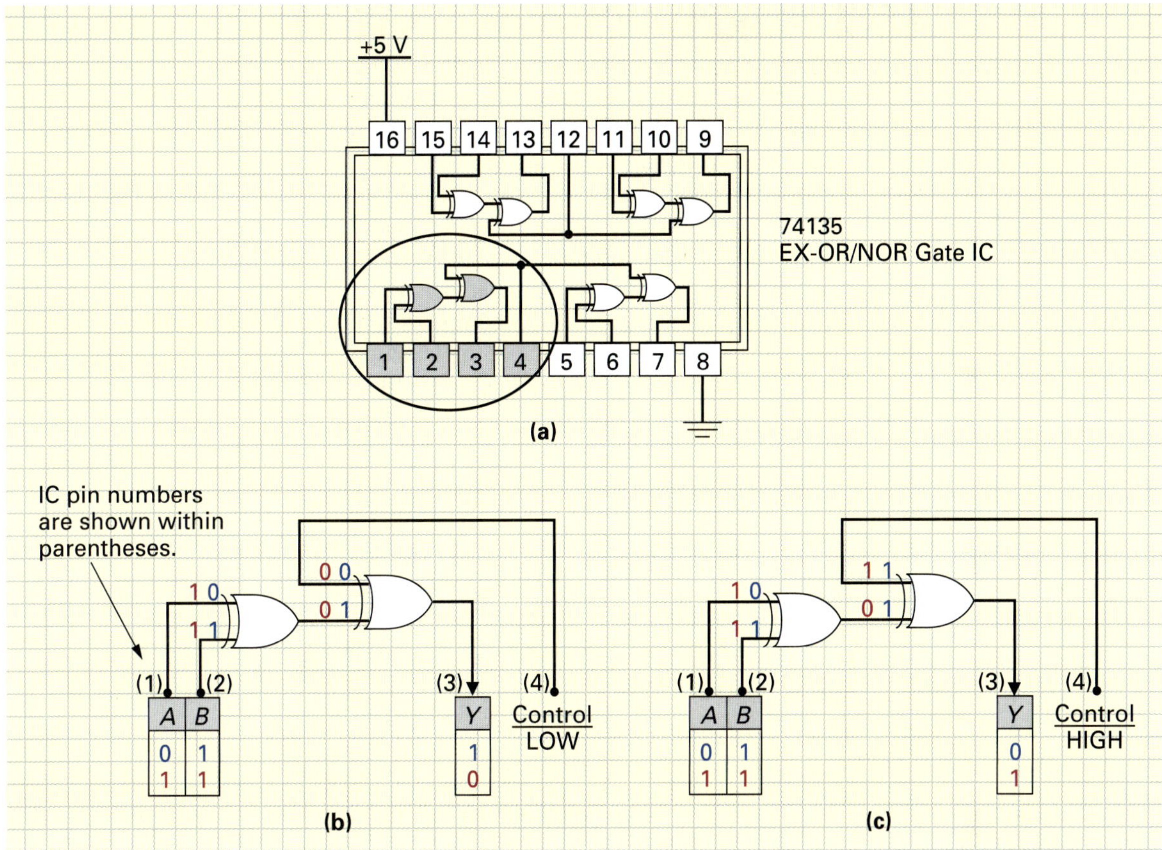

FIGURE 18-22 A Digital IC Containing Four Logic Gates That Can Be Controlled to Act as XOR Gates or XNOR Gates.

■ *Solution:*

Figure 18-22(b) follows the logic levels throughout the logic circuit when the control input is LOW, and an odd and even number of binary 1s are applied to the inputs *A* and *B*. Studying the *Y* output for these odd and even inputs, you can see that an odd input results in a HIGH output, and an even input results in a LOW output. When the control input is LOW, therefore, the logic circuits function as XOR gates.

Figure 18-22(c) follows the logic levels throughout the logic circuit when the control input is HIGH, and an odd and even number of binary 1s are applied to the inputs *A* and *B*. Studying the *Y* output for these odd and even inputs, you can see that an odd input results in a LOW output, and an even input results in a HIGH output. When the control input is HIGH, therefore, the logic circuits function as XNOR gates.

SELF-TEST EVALUATION POINT FOR SECTION 18-3

Use the following questions to test your understanding of Section 18-3.

1. Which type of logic gate will always give a HIGH output when any of its inputs are HIGH?
2. Only when an odd number of 1s is applied to the input, will the output of an ____________ gate be HIGH.
3. Only when an even number of 1s is applied to the input, will the output of an ____________ gate be HIGH.
4. Which type of logic gate will always give a LOW output when an odd number of 1s is applied at the input?

18-4 IEEE/ANSI SYMBOLS FOR LOGIC GATES

The logic symbols presented so far in this chapter have been used for many years in the digital electronics industry. In 1984 the *Institute of Electrical and Electronic Engineers (IEEE)* and the *American National Standards Institute (ANSI)* introduced a new standard for logic symbols, which is slowly being accepted by more and more electronics companies. The advantage of this new standard is that instead of using distinctive shapes to represent logic gates, it uses a special **dependency notation system.** Simply stated, the new standard uses a *notation* (or note) within a rectangular or square block to indicate how the output is *dependent* on the input. Figure 18-23 compares traditional logic symbols with the newer IEEE/ANSI logic symbols and describes the meaning of the dependency notations.

Dependency Notation
A coding system on schematic diagrams that uses notations to indicate how an output is dependent on inputs.

You should be familiar with both logic symbol types since you will come across both in industry schematics.

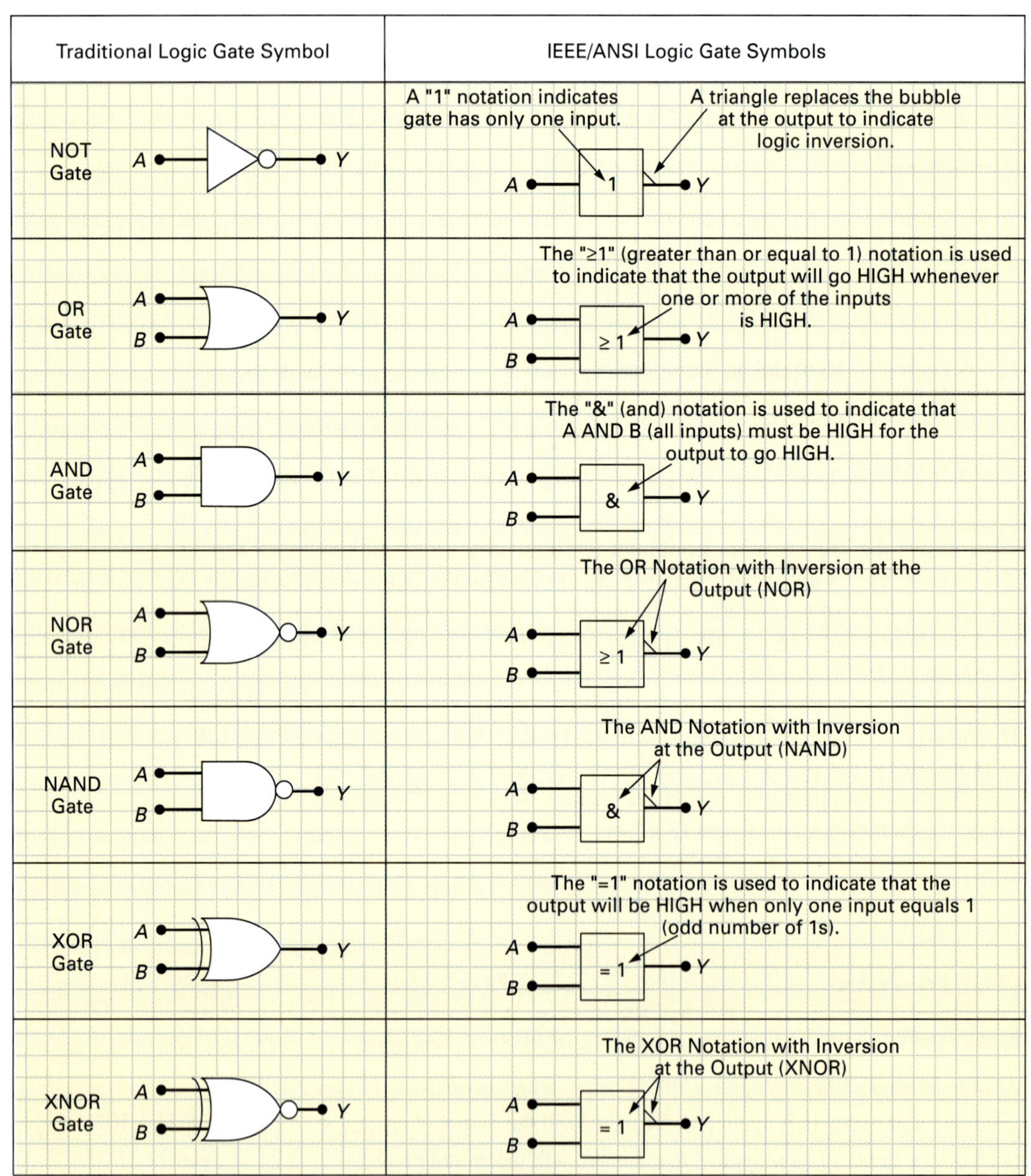

FIGURE 18-23 **Traditional and IEEE/ANSI Symbols for Logic Gates.**

EXAMPLE:

Identify which of the logic circuits shown in Figure 18-24 is traditional and which is using the IEEE/ANSI standard.

Solution:

Figure 18-24(a) shows the traditional logic symbol and Figure 18-24(b) shows the IEEE/ANSI logic symbol for a 7400 digital IC (quad two-input NAND).

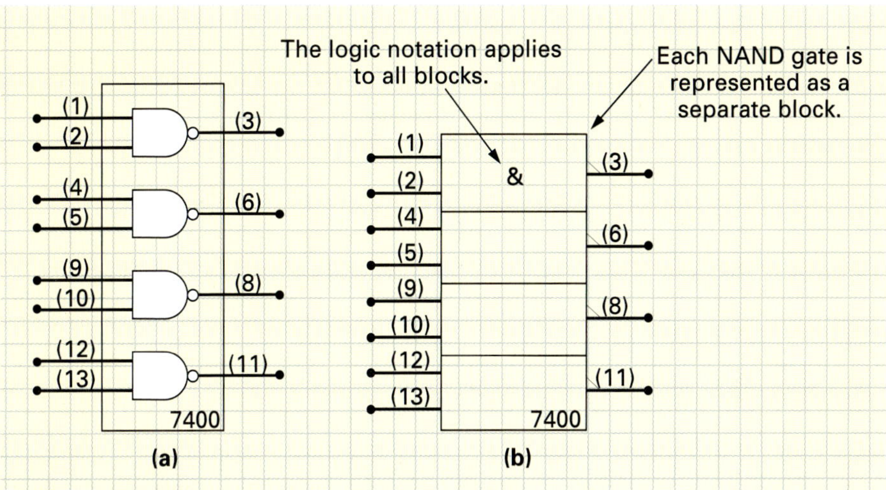

FIGURE 18-24 **Traditional and IEEE/ANSI Logic Symbols for a 7400 Digital IC (Quad Two-Input NAND).**

SELF-TEST EVALUATION POINT FOR SECTION 18-4

Use the following questions to test your understanding of Section 18-4.

1. Which type of logic-gate symbols makes use of rectangular blocks?
2. What advantage does the IEEE/ANSI standard have over the traditional symbol standard?

REVIEW QUESTIONS

Multiple-Choice Questions

1. Which device is currently being used extensively as a two-state switch in digital electronic circuits?
   a. Transistor
   b. Vacuum tube
   c. Relay
   d. Toggle switch
2. How many input combinations would a four-input logic gate have?
   a. Eight
   b. Sixteen
   c. Thirty-two
   d. Four
3. What would be the maximum count within the truth table for a four-input logic gate?
   a. 7
   b. 15
   c. 17
   d. 31
4. What would be the output from a three-input OR gate if its inputs were 101?
   a. 1
   b. 0
   c. Unknown
   d. None of the above

5. What would be the output from a three-input AND gate if its inputs were 101?
   a. 1
   b. 0
   c. Unknown
   d. None of the above
6. What would be the output from a three-input NAND gate if its inputs were 101?
   a. 1
   b. 0
   c. Unknown
   d. None of the above
7. What would be the output from a three-input NOR gate if its inputs were 101?
   a. 1
   b. 0
   c. Unknown
   d. None of the above
8. What would be the output from a three-input XOR gate if its inputs were 101?
   a. 1
   b. 0
   c. Unknown
   d. None of the above
9. What would be the output from a three-input XNOR gate if its inputs were 101?
   a. 1
   b. 0
   c. Unknown
   d. None of the above
10. What would be the output from a NOT gate if its input were 1?
    a. 1
    b. 0
    c. Unknown
    d. None of the above
11. Which of the following logic gates is ideal as a controlled switch?
    a. OR
    b. AND
    c. NOR
    d. NAND
12. Which of the following logic gates will always give a LOW output whenever a LOW input is applied?
    a. OR
    b. AND
    c. NOR
    d. NAND
13. Which of the following logic gates will always give a HIGH output whenever a LOW input is applied?
    a. OR
    b. AND
    c. NOR
    d. NAND
14. Which of the following logic gates will always give a HIGH output whenever a HIGH input is applied?
    a. OR
    b. AND
    c. NOR
    d. NAND
15. Which of the following logic gates will always give a LOW output whenever a HIGH input is applied?
    a. OR
    b. AND
    c. NOR
    d. NAND
16. An XOR gate will always give a HIGH output whenever an _____ number of 1s is applied to the input.
    a. odd
    b. even
    c. unknown
    d. none of the above
17. An XNOR gate will always give a HIGH output whenever an _____ number of 1s is applied to the input.
    a. odd
    b. even
    c. unknown
    d. none of the above
18. Which of the dependency notation logic symbols contains an "&" within the square block and no triangle at the output?
    a. OR
    b. AND
    c. NOR
    d. NAND
19. Which of the dependency notation logic symbols contains a "≥ 1" within the square block and no triangle at the output?
    a. OR
    b. AND
    c. NOR
    d. NAND
20. Which of the dependency notation logic symbols contains a "≥ 1" within the square block and a triangle at the output?
    a. OR
    b. AND
    c. NOR
    d. NAND

Practice Problems

21. In the home security system shown in Figure 18-25(a), a two-input logic gate is needed to actuate an alarm (output *Y*) if either the window is opened (input *A*) or the door is opened (input *B*). Which decision-making logic gate should be used in this application?
22. In the office temperature control system shown in Figure 18-25(b), a thermostat (input *A*) is used to turn ON and OFF a fan (output *Y*). A thermostat-enable switch is connected to the other input of the logic gate

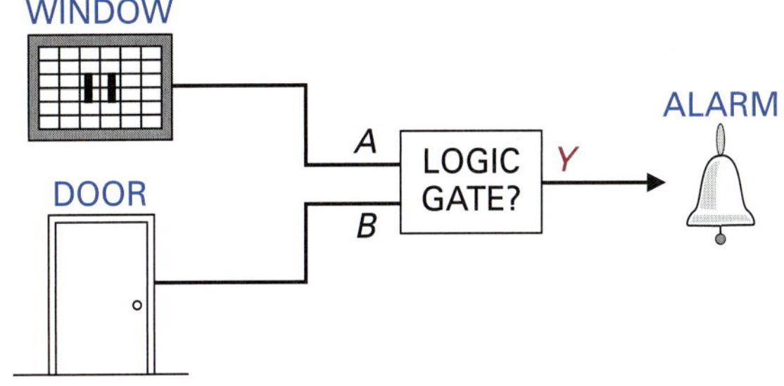

WINDOW	DOOR	ALARM
Closed (0)	Closed (0)	OFF (0)
Closed (0)	Open (1)	ON (1)
Open (1)	Closed (0)	ON (1)
Open (1)	Open (1)	ON (1)

(a)

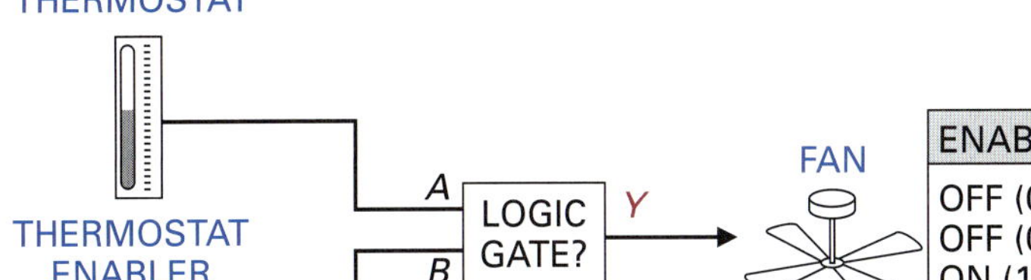

ENABLE	THERMOSTAT	FAN
OFF (0)	COLD (0)	OFF (0)
OFF (0)	HOT (1)	OFF (0)
ON (1)	COLD (0)	OFF (0)
ON (1)	HOT (1)	ON (1)

(b)

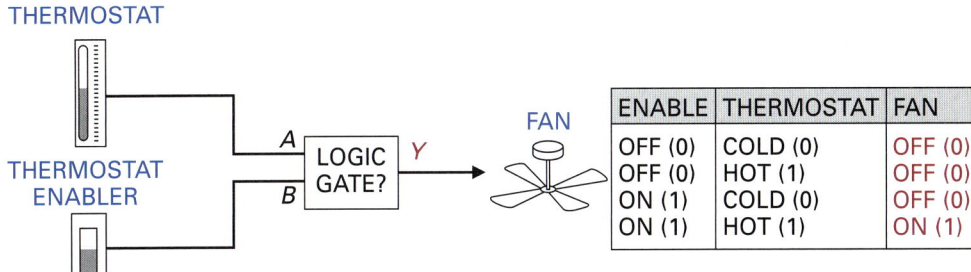

FIGURE 18-25 Application—Decision-Making Logic Gates.

(input *B*), and it will either enable thermostat control of the fan or disable thermostat control of the fan. Which decision-making logic gate should be used in this application?

23. Briefly describe the operation of the photodiode receiver circuit in Figure 18-26. What will be the output at *Y* when light is being applied to the photodiode and when light is not being applied?

24. Briefly describe the operation of the gated LED flasher circuit shown in Figure 18-27.

25. Which logic gate would produce the output waveform shown in Figure 18-28?

26. Choose one of the six logic gates shown in Figure 18-29(a) and then sketch the output waveform that would result for the input waveforms shown in Figure 18-29(b).

27. Develop truth tables for each of the logic circuits shown in Figure 18-30.

28. State the logic operation and sketch the IEEE/ANSI logic symbol for each of the logic circuits shown in Figure 18-30.

29. Referring to the circuit in Figure 18-31, what will be the outputs at Y_0, Y_1, Y_2, and Y_3 if the control line input is
a. HIGH?
b. LOW?

30. Referring to the circuit in Figure 18-31, what will the output logic levels be if the control line input is low?

Web Site Questions

Go to the web site http://www.prenhall.com/cook, select the textbook *Electronics: A Complete Course,* select this chapter, and then follow the instructions when answering the multiple choice practice problems.

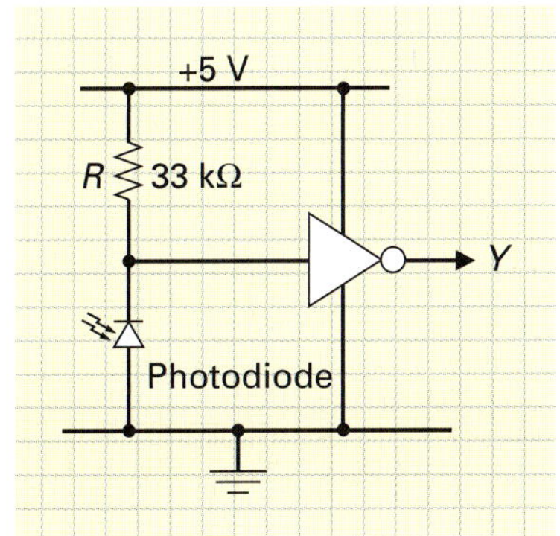

FIGURE 18-26 Application—Photodiode Receiver.

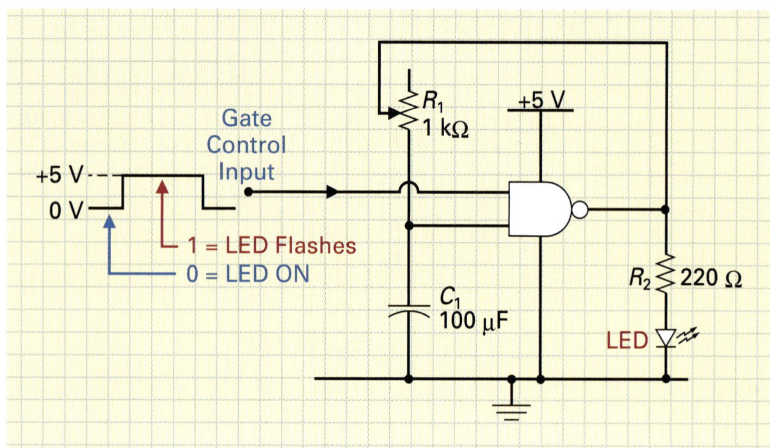

FIGURE 18-27 **Application—Gated LED Flasher Circuit.**

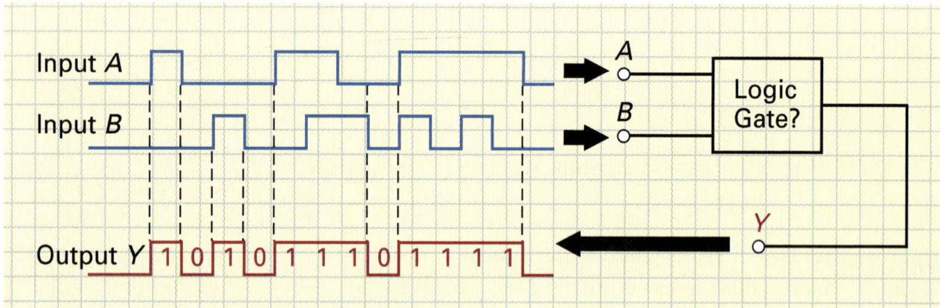

FIGURE 18-28 **Timing Analysis of a Logic Gate.**

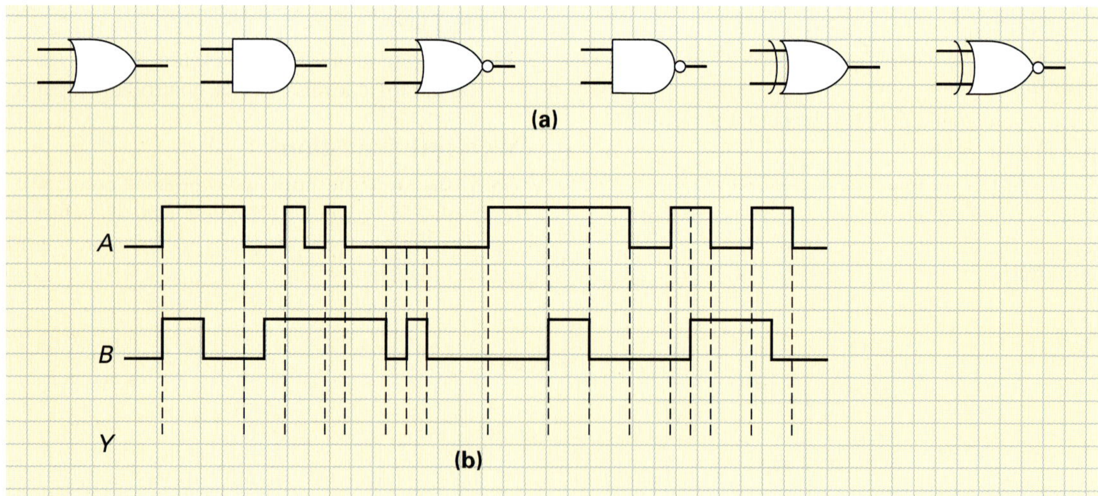

FIGURE 18-29 **Timing Waveforms.**

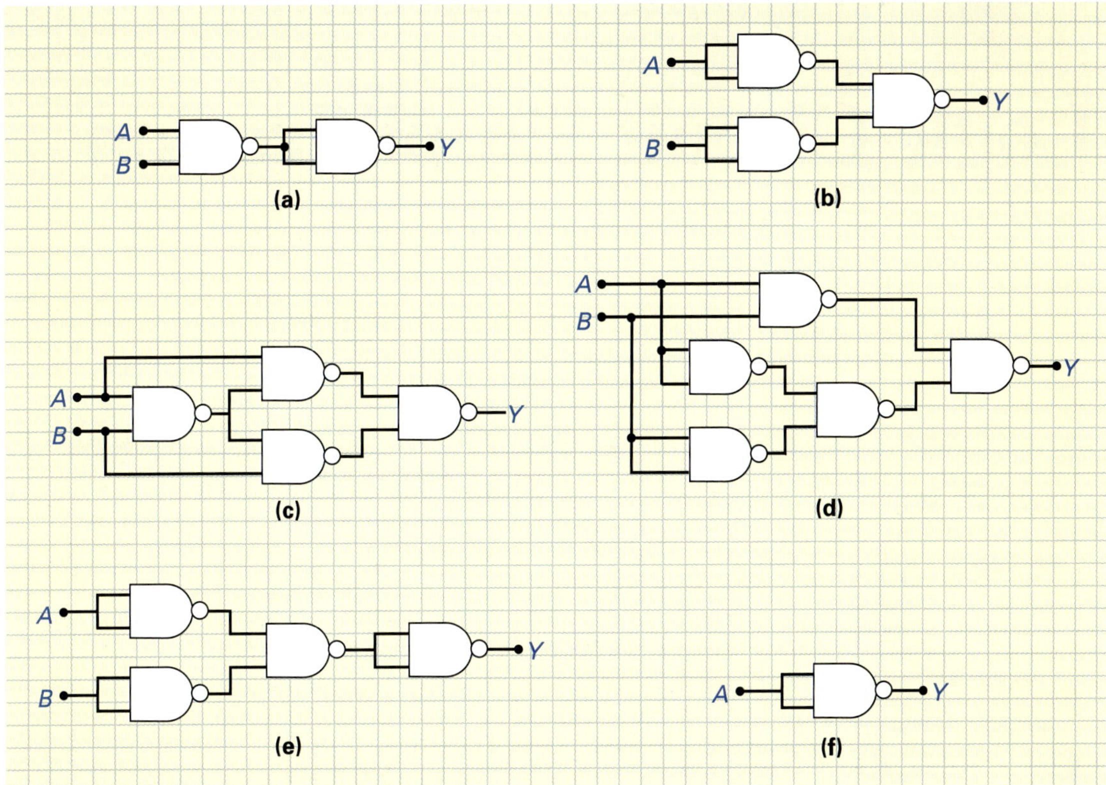

FIGURE 18-30 Using NAND Gates as Building Blocks for Other Gates.

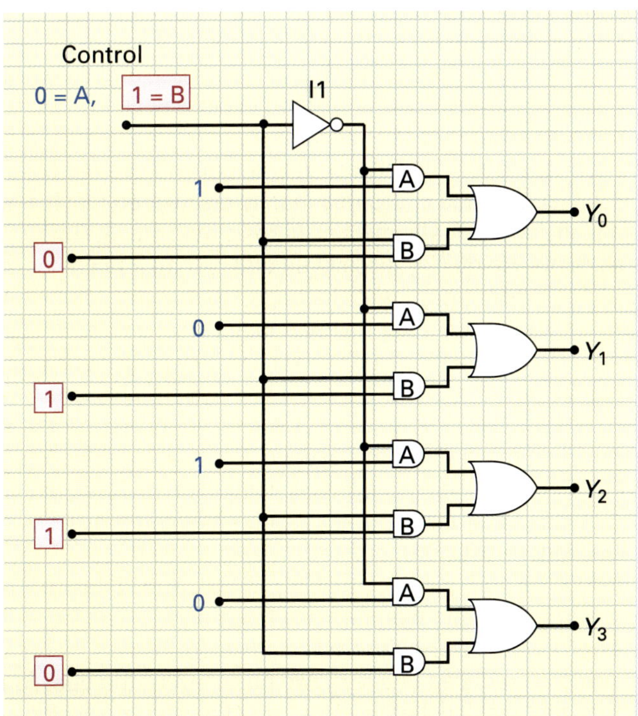

FIGURE 18-31 Application—Four-Bit Word Multiplexer.

Logic Circuit Simplification

From Folly to Foresight

George Boole was born in the industrial town of Lincoln in eastern England in 1815. His parents were poor tradespeople, and even though there was a school for boys in Lincoln, there is no record of him ever attending. In those hard times, children of the working class had no hope of receiving any form of education, and their lives generally followed the familiar pattern of their parents. George Boole, however, was to break the mold. He would rise up from these humble beginnings to become one of the most respected mathematicians of his day.

Boole's father had taught himself a small amount of mathematics, and since his son of six seemed to have a thirst for learning, he began to pass on all of his knowledge. At eight years old, George had surpassed his father's understanding and craved more. He quickly realized that his advancement was heavily dependent on his understanding Latin. Luckily, a family friend who owned a local book shop knew enough about the basics of Latin to get him on his way, and once he had taught Boole all he knew, Boole continued with the books at his disposal. By the age of twelve he had conquered Latin, and by fourteen he had added Greek, French, German, and Italian to his repertoire.

At the age of sixteen, however, poverty stood in his way. Since his parents could no longer support him, he was forced to take a job as a poorly paid teaching assistant. After studying the entire school system, he left four year laters and opened his own school in which he taught all subjects. It was in this role that he discovered that his mathematics was weak, so he began studying the mathematical journals at the local library in an attempt to stay ahead of his students. He quickly discovered that he had a talent for mathematics. As well as mastering all the present-day ideas, he began to develop some of his own, which were later accepted for publication. After a stream of articles, he became so highly regarded that he was asked to join the mathematics faculty at Queens College in 1849.

After accepting the position, Boole concentrated more on his ideas, one of which was to develop a system of symbolic logic. In this system he created a form of algebra that had its own set of symbols and rules. Using this system, Boole could encode any statement that had to be proven (a proposition) into his symbolic language and then manipulate it to determine whether it was true or false. Boole's algebra had three basic operations that are often called logic functions—AND, OR, and NOT. Using these three operations, Boole could perform such operations as add, subtract, multiply, divide, and compare. These logic functions were binary in nature and therefore dealt with only two entities—TRUE or FALSE, YES or NO, OPEN or CLOSED, ZERO or ONE, and so on. Boole's theory was that if all logical arguments could be reduced to one of two basic levels, the questionable middle ground would be removed, making it easier to arrive at a valid conclusion.

At the time, Boole's system, which was later called "Boolean algebra," was either ignored or criticized by colleagues who called it a folly with no practical purpose. Almost a century later, however, scientists would combine George Boole's Boolean algebra with binary numbers and make possible the digital electronic computer.

19

Introduction

In 1854 George Boole invented a symbolic logic that linked mathematics and logic. Boole's logical algebra, which today is known as **Boolean algebra,** states that each variable (input or output) can assume one of two values or states—true or false.

Up until 1938, Boolean algebra had no practical application until Claude Shannon used it to analyze telephone switching circuits in his MIT thesis. In his paper he described how the two variables of the Boolean algebra (true and false) could be used to represent the two states of the switching relay (open and closed).

Today, the mechanical relays have been replaced by semiconductor switches. However, Boolean algebra is still used to express both simple and complex two-state logic functions in a convenient mathematical format. These mathematical expressions of logic functions make it easier for technicians to analyze digital circuits and are a primary design tool for engineers. By using Boolean algebra, circuits can be made simpler, less expensive, and more efficient.

In this chapter we will discuss the basics of Boolean algebra and then see how it can be applied to logic gate circuit simplification.

Boolean Algebra
A form of algebra invented by George Boole that deals with classes, propositions, and ON/OFF circuit elements associated with operators such as NOT, OR, NOR, AND, NAND, XOR, and XNOR.

19-1 BOOLEAN EXPRESSIONS FOR LOGIC GATES

The operation of each of the seven basic logic gates can be described with a Boolean expression. To explain this in more detail, let us first consider the most basic of all the logic gates, the NOT gate.

19-1-1 *The NOT Expression*

In Figure 19-1 you can see the previously discussed NOT gate, and because of the inversion that occurs between input and output, Y is always equal to the opposite, or complement, of input A.

$$Y = \text{NOT } A$$

Therefore, if the input were 0, the output would be 1.

$$Y = \text{NOT } 0 = 1$$

If, on the other hand, the A input were 1, the output would be 0.

$$Y = \text{NOT } 1 = 0$$

In Boolean algebra, this inversion of the A input is indicated with a bar over the letter A as follows:

$Y = \overline{A}$ (pronounced "Y equals not A")

Using this Boolean expression, we can easily calculate the output Y for either of the two input A conditions. For example, if $A = 0$,

$$Y = \overline{A} = \overline{0} = 1$$

If, on the other hand, the A input is 1,

$$Y = \overline{A} = \overline{1} = 0$$

19-1-2 *The OR Expression*

The operation of the OR gate can also be described with a Boolean expression, as seen in Figure 19-2(a). In Boolean algebra, the "+" sign is used to indicate the OR operation.

$Y = A + B$ (pronounced "Y equals A or B")

The expression $Y = A + B$ (Boolean equation), is the same as $Y = A$ OR B (word equation). If you find yourself wanting to say "plus" instead of "OR," don't worry; it will require some

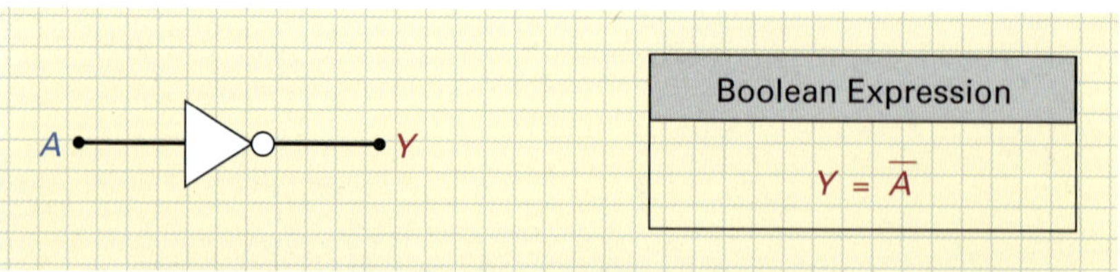

FIGURE 19-1 The Boolean Expression for a NOT Gate.

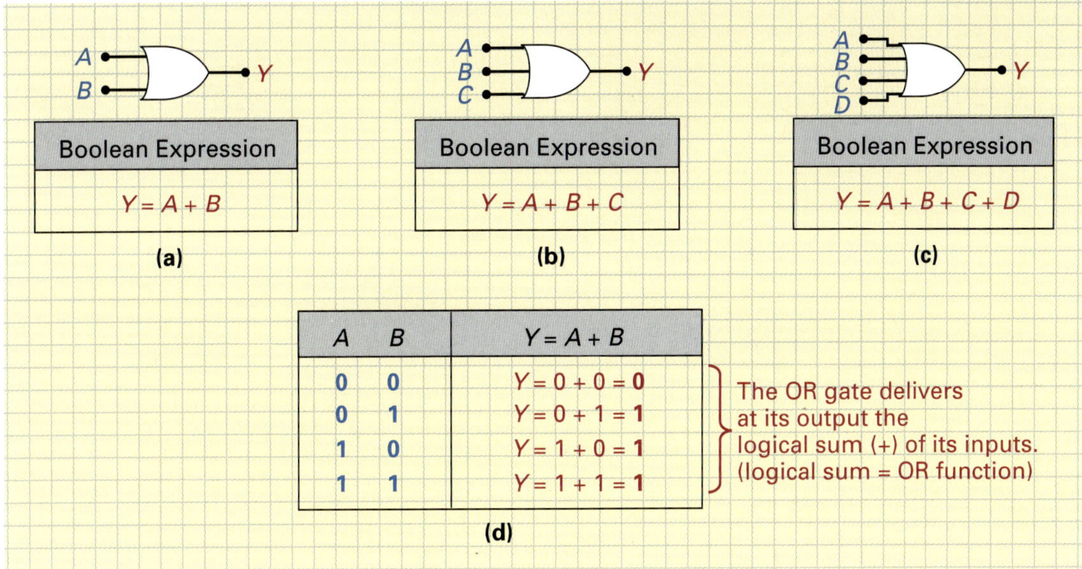

FIGURE 19-2 **The Boolean Expression for an OR Gate.**

practice to get used to this traditional symbol being used in a new application. Just as the word "wind" can be used in two different ways ("a gusty wind" or "to wind a clock"), then so can the "+" symbol be used to mean either addition or the OR operation. With practice you will get more comfortable with this symbol's dual role.

If OR gates have more than two inputs, as seen in Figure 19-2(b) or (c), the OR expression is simply extended, as follows:

For a three-input OR gate: $Y = A + B + C$
For a four-input OR gate: $Y = A + B + C + D$

The truth table for a two-input OR gate has been repeated in Figure 19-2(d). Using the Boolean expression for this gate, we can determine the output Y by substituting all of the possible 1 and 0 input combinations for the input variables A and B. Studying this truth table you can see that the output of the OR gate is a 1 when either A OR B is 1. George Boole described the OR function as *Boolean addition,* which is why the "plus" (addition) symbol was used. Boolean addition, or *logical addition,* is different from normal addition, and the OR gate is said to produce at its output the *logical sum* of the inputs.

19-1-3 *The AND Expression*

In Boolean algebra, the AND operation is indicated with a "·", and therefore the Boolean expression for the AND gate shown in Figure 19-3(a) would be

$Y = A \cdot B$ (pronounced "Y equals A and B")

This "·" is also used to indicate multiplication, and therefore like the "+" symbol, the "·" symbol has two meanings. In multiplication, the "·" is often dropped; however, the meaning is still known to be multiplication. For example:

$$V = I \times R \text{ is the same as } V = I \cdot R \text{ or } V = IR$$

In Boolean algebra the same situation occurs, in that

$$Y = A \cdot B \text{ is the same as } Y = AB$$

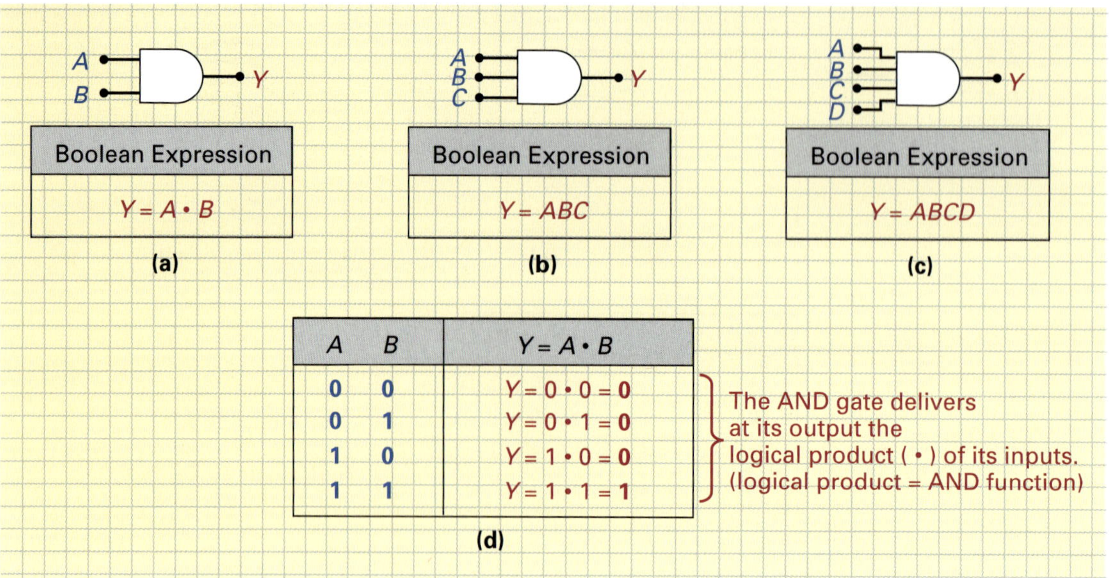

FIGURE 19-3 **The Boolean Expression for an AND Gate.**

If AND gates have more than two inputs, as seen in Figure 19-3(b) and (c), the AND expression is simply extended.

The truth table for a two-input AND gate is shown in Figure 19-3(d). Using the Boolean expression, we can calculate the output *Y* by substituting all of the possible 1 and 0 input combinations for the input variables *A* and *B*. Looking at the truth table, you can see that the output of the AND gate is a 1 only when both *A* and *B* are 1. George Boole described the AND function as *Boolean multiplication* or *logical multiplication,* which is why the period (multiplication) symbol was used. Boolean multiplication is different from standard multiplication as you have seen, and the AND gate is said to produce at its output the *logical product* of its inputs.

EXAMPLE:

What would be the Boolean expression for the circuit shown in Figure 19-4, and what would be the output if $A = 0$ and $B = 1$?

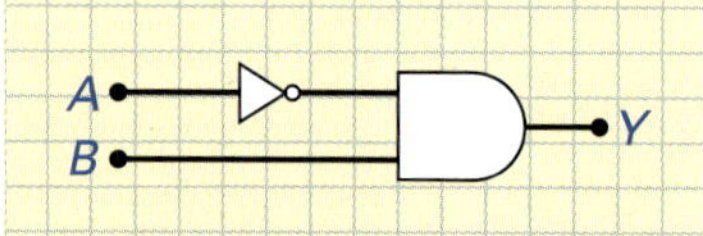

FIGURE 19-4 **Single, Bubble Input AND Gate**

Solution:

The *A* input in Figure 19-4 is first inverted before it is ANDed with the *B* input. The Boolean equation would be, therefore

$$Y = \overline{A} \cdot B$$

Using this expression, we can substitute the *A* and *B* for the 0 and 1 inputs, and then determine the output *Y.*

$$Y = \overline{A} \cdot B = \overline{0} \cdot 1 = 1 \cdot 1 = 1$$

Y equals NOT *A* and *B* = NOT 0 AND 1 = 1 AND 1 = 1)

EXAMPLE:

Give the Boolean expression and truth table for the logic circuit shown in Figure 19-5(a).

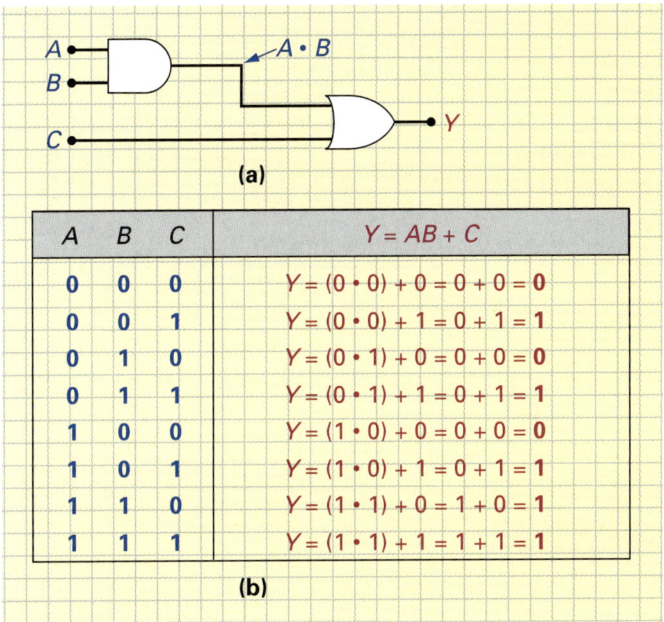

A	B	C	Y = AB + C
0	0	0	Y = (0 • 0) + 0 = 0 + 0 = 0
0	0	1	Y = (0 • 0) + 1 = 0 + 1 = 1
0	1	0	Y = (0 • 1) + 0 = 0 + 0 = 0
0	1	1	Y = (0 • 1) + 1 = 0 + 1 = 1
1	0	0	Y = (1 • 0) + 0 = 0 + 0 = 0
1	0	1	Y = (1 • 0) + 1 = 0 + 1 = 1
1	1	0	Y = (1 • 1) + 0 = 1 + 0 = 1
1	1	1	Y = (1 • 1) + 1 = 1 + 1 = 1

FIGURE 19-5 **(*A* AND *B*) OR (*C*) Logic Circuit.**

Solution:

To begin with, there are three inputs or variables to this circuit: *A, B,* and *C*. Inputs *A* and *B* are ANDed together ($A \cdot B$) and the result is then ORed with input *C* (+ *C*). The resulting equation will be

$$Y = (A \cdot B) + C \text{ or } Y = AB + C$$

As you can see in the truth table in Figure 19-5(b), the *Y* output is HIGH when *A* AND *B* are HIGH, OR if *C* is HIGH.

In the two previous examples, we have gone from a logic circuit to a Boolean equation. Now let us reverse the procedure and see how easy it is to go from a Boolean equation to a logic circuit.

EXAMPLE:

Sketch the logic gate circuit for the following Boolean equation.

$$Y = AB + CD$$

Solution:

Studying the equation, we can see that there are four variables: *A, B, C,* and *D*. Input *A* is ANDed with input *B*, giving $A \cdot B$ or *AB*, as seen in Figure 19-6(a). Input *C* is ANDed with input *D*, giving *CD*, as seen in Figure 19-6(b). The ANDed *AB* output and the ANDed *CD* output are then ORed to produce a final output *Y*, as seen in Figure 19-6(c).

$Y = \overline{A + B}$ (pronounced "Y equals not A or B")

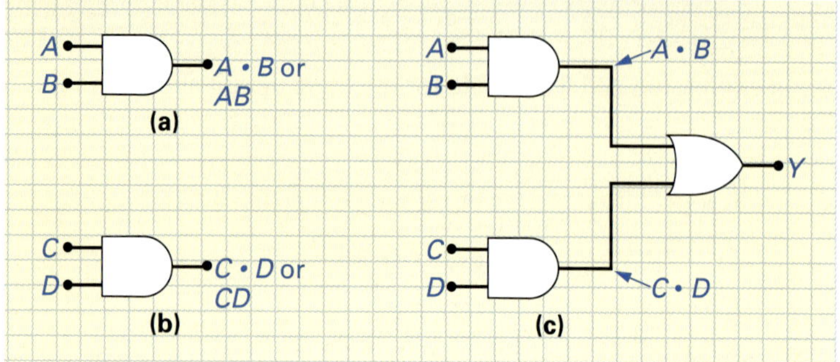

FIGURE 19-6 **(*A* AND *B*) OR (*C* AND *D*) Logic Circuit.**

19-1-4 *The NOR Expression*

The Boolean expression for the two-input NOR gate shown in Figure 19-7(a) will be

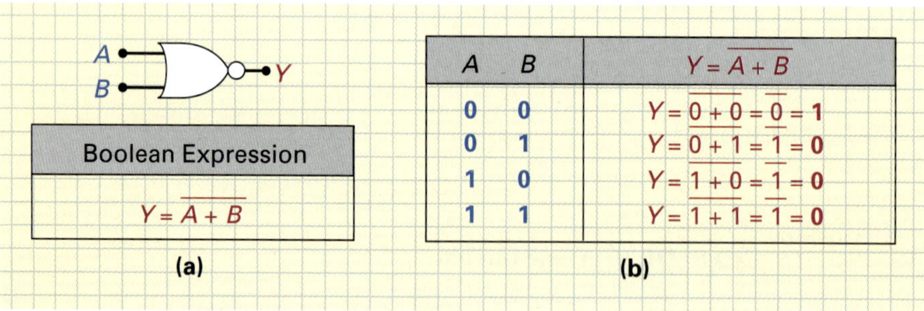

A	B	$Y = \overline{A + B}$
0	0	$Y = \overline{0 + 0} = \overline{0} = 1$
0	1	$Y = \overline{0 + 1} = \overline{1} = 0$
1	0	$Y = \overline{1 + 0} = \overline{1} = 0$
1	1	$Y = \overline{1 + 1} = \overline{1} = 0$

FIGURE 19-7 **The Boolean Expression for a NOR Gate.**

To explain why this expression describes a NOR gate, let us examine each part in detail. In this equation, the two input variables *A* and *B* are first ORed (indicated by the $A + B$ part of the equation), and then the result is complemented, or inverted (indicated by the bar over the whole OR output expression). The truth table in Figure 19-7(b) tests the Boolean expression for a NOR gate for all possible input combinations.

While studying Boolean algebra, Augustus DeMorgan discovered two important theorems. The first theorem stated that a bubbled input AND gate was logically equivalent to a NOR gate. These two logic gates are shown in Figure 19-8(a) and (b), respectively, with their associ-

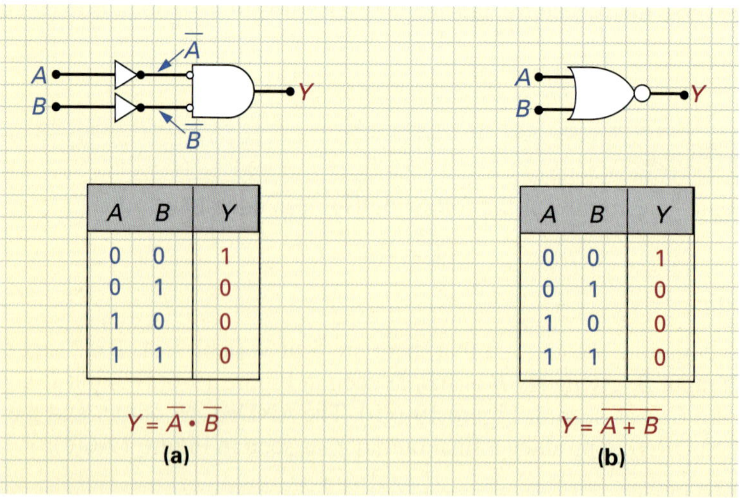

A	B	Y
0	0	1
0	1	0
1	0	0
1	1	0

A	B	Y
0	0	1
0	1	0
1	0	0
1	1	0

FIGURE 19-8 **DeMorgan's First Theorem.**

ated truth tables. Comparing the two truth tables you can see how any combination of inputs to either gate circuit will result in the same output, making the two circuits interchangeable.

EXAMPLE:

Give the two Boolean equations for the two logic gate circuits shown in Figure 19-8(a) and (b).

Solution:

With the bubbled input AND gate shown in Figure 19-8(a), you can see that input A is first inverted by a NOT gate, giving $\overline{A}$. Input B is also complemented by a NOT gate, giving $\overline{B}$. These two complemented inputs are then ANDed, giving the following final function:

$$Y = \overline{A} \cdot \overline{B}$$

With the NOR gate shown in Figure 19-8(b), inputs A and B are first ORed and then the result is complemented, giving

$$Y = \overline{A + B}$$

DeMorgan's first theorem, therefore, is as follows:

DeMorgan's First Theorem: $\overline{A} \cdot \overline{B} = \overline{A + B}$

Studying the differences between the left equation and the right equation, you can see that there are actually two basic changes: The line is either broken or solid, and the logical sign is different. Therefore, if we wanted to convert the left equation into the right equation, all we would have to do is follow this simple procedure:

$$Y = \overline{A} \cdot \overline{B} \rightarrow Y = \overline{A + B}$$

"Mend the line, and change the sign."

To convert the right equation into the left equation, all we would have to do is follow this simple procedure:

"Break the line, and change the sign."

$$Y = \overline{A} \cdot \overline{B} \leftarrow Y = \overline{A + B}$$

This ability to interchange an AND function with an OR function, and to interchange a bubbled input with a bubbled output, will come in handy when we are trying to simplify logic circuits, as you will see in this chapter. For now, remember the rules on how to "DeMorganize" an equation.

Mend the line, and change the sign.
Break the line, and change the sign.

19-1-5 *The NAND Expression*

The Boolean expression for the two-input NAND gate shown in Figure 19-9(a) will be

$Y = \overline{A \cdot B}$ or $Y = \overline{AB}$ (pronounced "Y equals not A and B")

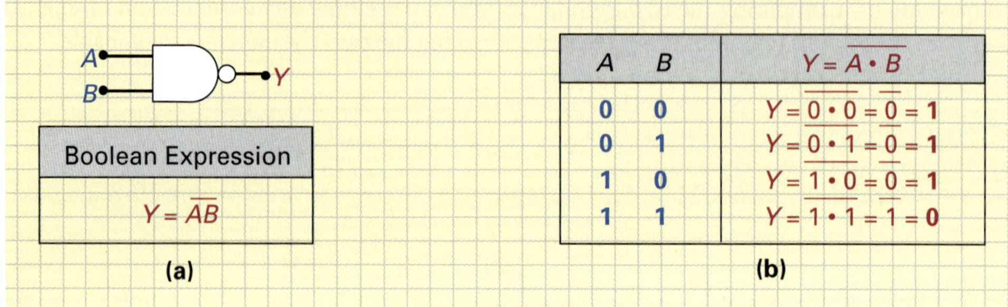

A
B
Y
Boolean Expression
$Y = \overline{AB}$
(a)
A B $Y = \overline{A \cdot B}$
0 0 $Y = \overline{0 \cdot 0} = \overline{0} = 1$
0 1 $Y = \overline{0 \cdot 1} = \overline{0} = 1$
1 0 $Y = \overline{1 \cdot 0} = \overline{0} = 1$
1 1 $Y = \overline{1 \cdot 1} = \overline{1} = 0$
(b)

FIGURE 19-9 The Boolean Expression for a NAND Gate.

This expression states that the two inputs A and B are first ANDed (indicated by the $A \cdot B$ part of the equation), and then the result is complemented (indicated by the bar over the AND expression).

The truth table shown in Figure 19-9(b) tests this Boolean equation for all possible input combinations.

EXAMPLE:

DeMorganize the following equation, and then sketch the logically equivalent circuits with their truth tables and Boolean expressions.

$$Y = \overline{A} + \overline{B}$$

Solution:

To apply DeMorgan's theorem to the equation $Y = \overline{A} + \overline{B}$, we simply apply the rule "mend the line, and change the sign," as follows:

$$\overline{A} + \overline{B} = \overline{A \cdot B}$$

The circuits for these two equations can be seen in Figure 19-10(a) and (b), and if you study the truth tables, you can see that the outputs of both logic gates are identical for all input combinations.

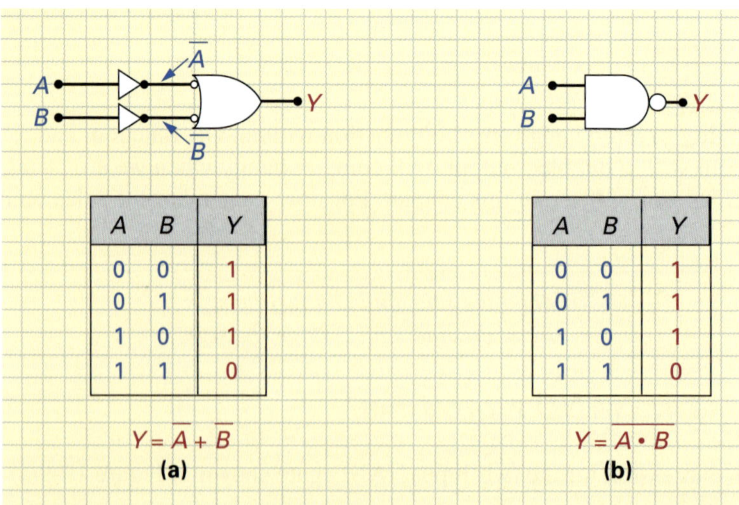

FIGURE 19-10 DeMorgan's Second Theorem.

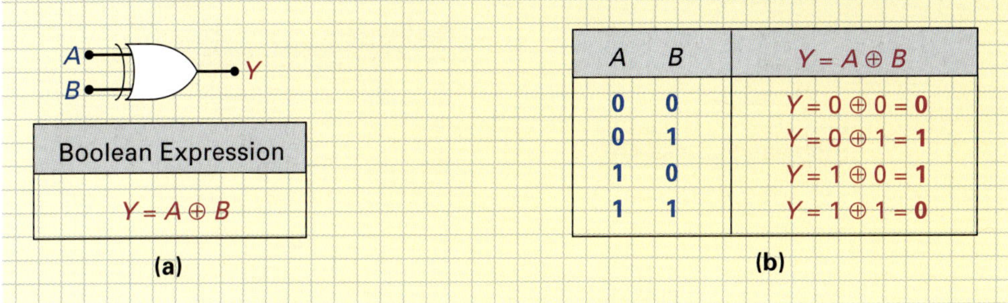

A	B	$Y = A \oplus B$
0	0	$Y = 0 \oplus 0 = 0$
0	1	$Y = 0 \oplus 1 = 1$
1	0	$Y = 1 \oplus 0 = 1$
1	1	$Y = 1 \oplus 1 = 0$

(b)

FIGURE 19-11 The Boolean Expression for an XOR Gate.

This observation—that a bubbled input OR gate is interchangeable with a bubbled output AND gate, or NAND gate—was first made by DeMorgan and is referred to as DeMorgan's second theorem.

DeMorgan's Second Theorem: $\overline{A} + \overline{B} = \overline{A \cdot B}$

19-1-6 *The XOR Expression*

In Boolean algebra, the "⊕" symbol is used to describe the exclusive-OR action. This means that the Boolean expression for the two-input XOR gate shown in Figure 19-11(a) will be

$Y = A \oplus B$ (pronounced "Y equals A exclusive-OR B")

The truth table in Figure 19-11(b) shows how this Boolean expression is applied to all possible input combinations.

EXAMPLE:

Give the Boolean expression and truth table for the circuit shown in Figure 19-12(a).

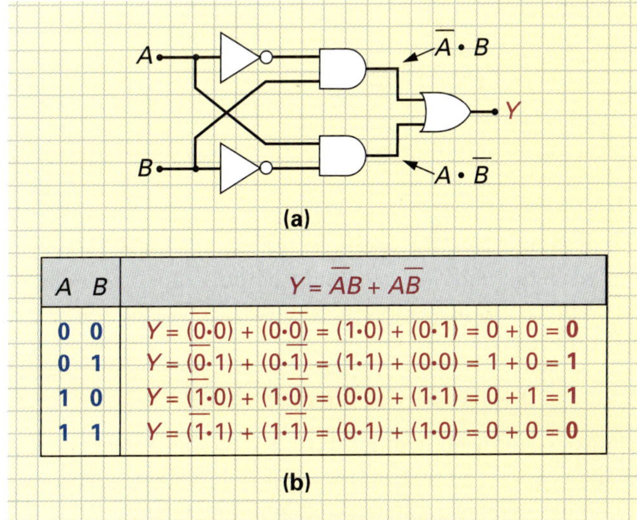

A	B	$Y = \overline{A}B + A\overline{B}$
0	0	$Y = (\overline{0}\cdot 0) + (0\cdot \overline{0}) = (1\cdot 0) + (0\cdot 1) = 0 + 0 = 0$
0	1	$Y = (\overline{0}\cdot 1) + (0\cdot \overline{1}) = (1\cdot 1) + (0\cdot 0) = 1 + 0 = 1$
1	0	$Y = (\overline{1}\cdot 0) + (1\cdot \overline{0}) = (0\cdot 0) + (1\cdot 1) = 0 + 1 = 1$
1	1	$Y = (\overline{1}\cdot 1) + (1\cdot \overline{1}) = (0\cdot 1) + (1\cdot 0) = 0 + 0 = 0$

(b)

FIGURE 19-12 (NOT *A* AND *B*) OR (*A* AND NOT *B*).

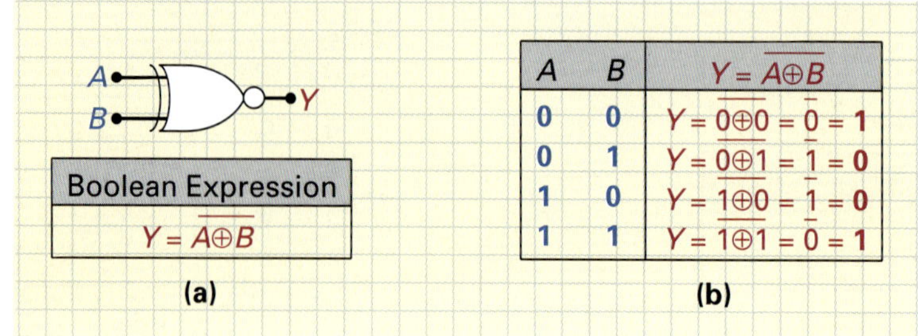

A	B	$Y = \overline{A \oplus B}$
0	0	$Y = \overline{0 \oplus 0} = \overline{0} = 1$
0	1	$Y = \overline{0 \oplus 1} = \overline{1} = 0$
1	0	$Y = \overline{1 \oplus 0} = \overline{1} = 0$
1	1	$Y = \overline{1 \oplus 1} = \overline{0} = 1$

(b)

FIGURE 19-13 The Boolean Expression for an XNOR Gate.

■ *Solution:*

Looking at the circuit in Figure 19-12(a), you can see that the upper AND gate has an inverted A input $(\overline{A})$ and a B input, giving a result of $\overline{A} \cdot B$. This $\overline{A} \cdot B$ output is then ORed with the input from the lower AND gate, which has an inverted B input and A input, giving a result of $A \cdot \overline{B}$. The final equation will therefore be

$$Y = (\overline{A} \cdot B) + (A \cdot \overline{B}) \text{ or } Y = \overline{A}B + A\overline{B}$$

The truth table for this logic circuit that combines five logic gates is shown in Figure 19-12(b). Looking at the input combinations and the output, you can see that the output is only 1 when an odd number of 1s appears at the input. This circuit is therefore acting as an XOR gate, and so

$$\overline{A}B + A\overline{B} = A \oplus B$$

19-1-7 *The XNOR Expression*

The Boolean expression for the two-input XNOR gate shown in Figure 19-13(a) will be

$Y = \overline{A \oplus B}$ (pronounced "Y equals not A exclusive-OR B")

From this equation, we can see that the two inputs A and B are first XORed and then the result is complemented. The truth table in Figure 19-13(b) tests this Boolean expression for all possible input combinations.

■ **EXAMPLE:**

Give the Boolean expression and truth table for the circuit shown in Figure 19-14(a).

■ *Solution:*

Studying the circuit in Figure 19-14(a), you can see that the upper AND gate has an inverted A input and an inverted B input, giving $\overline{A} \cdot \overline{B}$ at its output. The lower AND gate simply ANDs the A and B input, giving $A \cdot B$ at its output. The two AND gate outputs are then ORed, so the final equation for this circuit will be

$$Y = \overline{A} \cdot \overline{B} + A \cdot B \text{ or } Y = \overline{A}\,\overline{B} + AB$$

The truth table for this circuit is shown in Figure 19-14(b). Comparing this truth table to an XNOR gate's truth table, you can see that the two are equivalent since a 1 output is only present when an even number of 1s is applied to the input; therefore

$$\overline{A}\,\overline{B} + AB = \overline{A \oplus B}$$

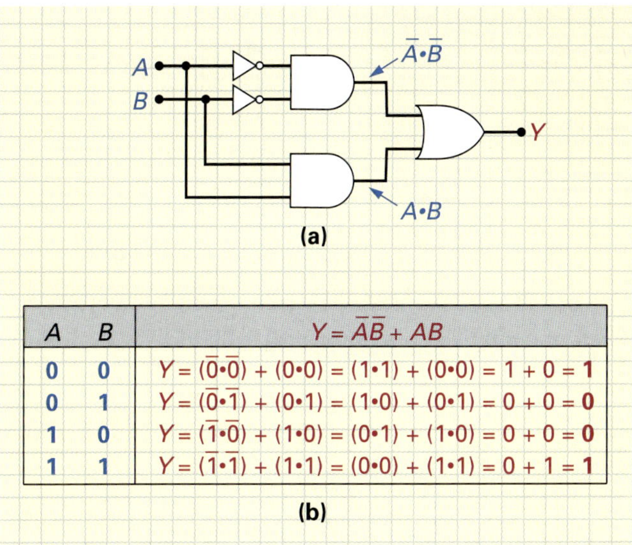

A	B	$Y = \overline{A}\overline{B} + AB$
0	0	$Y = (\overline{0} \cdot \overline{0}) + (0 \cdot 0) = (1 \cdot 1) + (0 \cdot 0) = 1 + 0 = \mathbf{1}$
0	1	$Y = (\overline{0} \cdot \overline{1}) + (0 \cdot 1) = (1 \cdot 0) + (0 \cdot 1) = 0 + 0 = \mathbf{0}$
1	0	$Y = (\overline{1} \cdot \overline{0}) + (1 \cdot 0) = (0 \cdot 1) + (1 \cdot 0) = 0 + 0 = \mathbf{0}$
1	1	$Y = (\overline{1} \cdot \overline{1}) + (1 \cdot 1) = (0 \cdot 0) + (1 \cdot 1) = 0 + 1 = \mathbf{1}$

FIGURE 19-14 **(NOT *A* AND NOT *B*) OR (*A* AND *B*).**

SELF-TEST EVALUATION POINT FOR SECTION 19-1

Use the following questions to test your understanding of Section 19-1.

1. Which logic gate would be needed to perform each of the following Boolean expressions?
   a. $A\,B$ **c.** $N + M$ **e.** $\overline{S + T}$
   b. $A \oplus B$ **d.** $\overline{X \cdot Y}$ **f.** $\overline{GH}$
2. State DeMorgan's first theorem using Boolean expressions.
3. Apply DeMorgan's theorem to the following:
   a. $\overline{A} + \overline{B}$ **b.** $Y = \overline{(A + B)} \cdot C$
4. Sketch the logic gates or circuits for the expressions:
   a. $A\,B\,C$ **c.** $\overline{A + B}$ **e.** $A \cdot \overline{(L \oplus D)}$
   b. $\overline{L \oplus D}$ **d.** $(\overline{A} \cdot \overline{B}) + \overline{C}$ **f.** $\overline{(A + B) \cdot C}$

19-2 BOOLEAN ALGEBRA LAWS AND RULES

In this section we will discuss some of the rules and laws that apply to Boolean algebra. Many of these rules and laws are the same as those of ordinary algebra and, as you will see in this section, are quite obvious.

19-2-1 *The Commutative Law*

Commutative Law
Combining elements in such a manner that the result is independent of the order in which the elements are taken.

The word *commutative* is defined as "combining elements in such a manner that the result is independent of the order in which the elements are taken." This means that the order in which the inputs to a logic gate are ORed or ANDed, for example, is not important since the result will be the same, as shown in Figure 19-15.

In Figure 19-15(a), you can see that ORing *A* and *B* will achieve the same result as reversing the order of the inputs and ORing *B* and *A*. As stated in the previous section, the OR

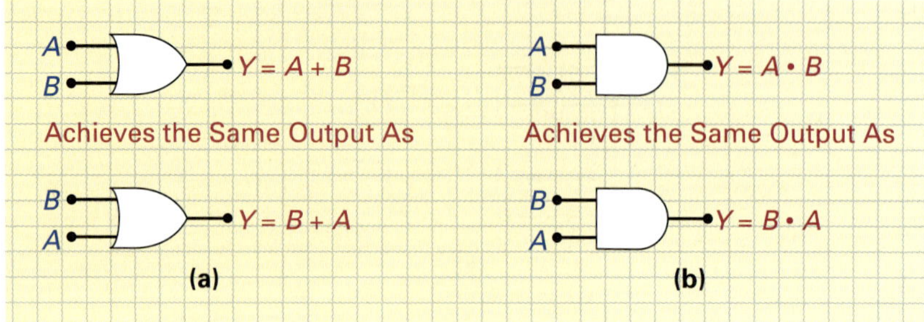

FIGURE 19-15 **The Commutative Law. (a) Logical Addition. (b) Logical Multiplication.**

function is described in Boolean algebra as logical addition, so the *commutative law of addition* can be algebraically written as

Commutative Law of Addition: $A + B = B + A$

Figure 19-15(b) shows how this law relates to an AND gate, which in Boolean algebra is described as logical multiplication. The *commutative law of multiplication* can be algebraically written as

Commutative Law of Multiplication: $A \cdot B = B \cdot A$

Associative Law
Combining elements such that when the order of the elements is preserved, the result is independent of the grouping.

19-2-2 *The Associative Law*

The word *associative* is defined as "combining elements such that when the order of the elements is preserved, the result is independent of the grouping." To explain this in simple terms, Figure 19-16 shows that how you group the inputs in an ORing process or ANDing process will have no effect on the output.

In Figure 19-16(a), you can see that ORing B and C and then ORing the result with A will achieve the same result as ORing A and B and then ORing the result with C. This *associative law of addition* can be algebraically written as

Associative Law of Addition: $A + (B + C) = (A + B) + C$

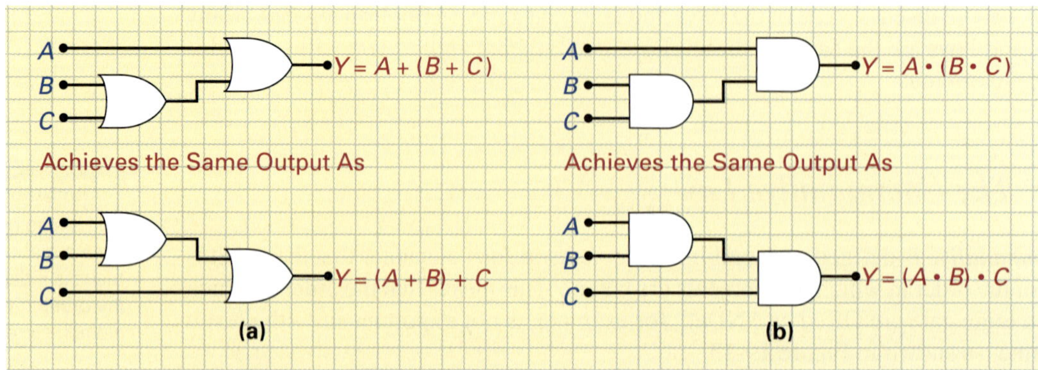

FIGURE 19-16 **The Associative Law. (a) Logical Addition. (b) Logical Multiplication.**

Figure 19-16(b) shows how this law relates to an AND gate, which in Boolean algebra is described as logical multiplication. The *associative law of multiplication* can be algebraically written as

Associative Law of Multiplication: $A \cdot (B \cdot C) = (A \cdot B) \cdot C$

19-2-3 *The Distributive Law*

Distributive Law
Producing the same element when operating on a whole as when operating on each part, and collecting the results.

By definition, the word *distributive* means "producing the same element when operating on a whole as when operating on each part, and collecting the results." It can be algebraically stated as

Distributive Law: $A \cdot (B + C) = (A \cdot B) + (A \cdot C)$

Figure 19-17 illustrates this law by showing that ORing two or more inputs and then ANDing the result, as shown in Figure 19-17(a), achieves the same output as ANDing the single variable (A) with each of the other inputs (B and C) and then ORing the results, as shown in Figure 19-17(b). To help reinforce this algebraic law, let us use an example involving actual values for A, B, and C.

EXAMPLE:

Prove the distributive law by inserting the values $A = 2$, $B = 3$ and $C = 4$.

Solution:

$$A \cdot (B + C) = (A \cdot B) + (A \cdot C)$$
$$2 \times (3 + 4) = (2 \times 3) + (2 \times 4)$$
$$2 \times 7 = 6 + 8$$
$$14 = 14$$

In some instances, we may wish to reverse the process performed by the distributive law to extract the common factor. For example, consider the following equation:

$$Y = \overline{A}\,B + AB$$

Since "$\cdot B$" (AND B) seems to be a common factor in this equation, by *factoring,* we could obtain

$$Y = \overline{A}B + AB \quad \text{Original Expression}$$
$$Y = (\overline{A} + A) \cdot B \quad \text{Factoring AND B}$$

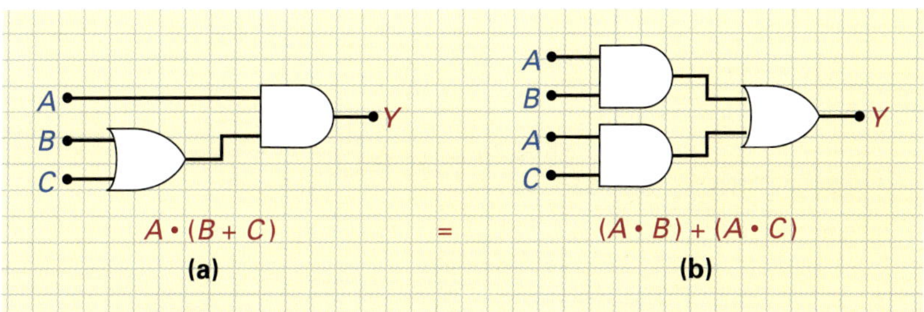

FIGURE 19-17 **The Distributive Law.**

19-2-4 *Boolean Algebra Rules*

Now that we have covered the three laws relating to Boolean algebra, let us concentrate on these Boolean algebra rules, beginning with those relating to OR operations.

OR Gate Rules

The first rule can be seen in Figure 19-18(a). This shows what happens when one input to an OR gate is always 0 and the other input (A) is a variable. If $A = 0$, the output equals 0, and if $A = 1$, the output equals 1. Therefore, a variable input ORed with 0 will always equal the variable input. Stated algebraically:

$$A + 0 = A$$

Another OR gate Boolean rule is shown in Figure 19-18(b), and states that

$$A + A = A$$

As you can see in Figure 19-18(b), when a variable is ORed with itself, the output will always equal the logic level of the variable input.

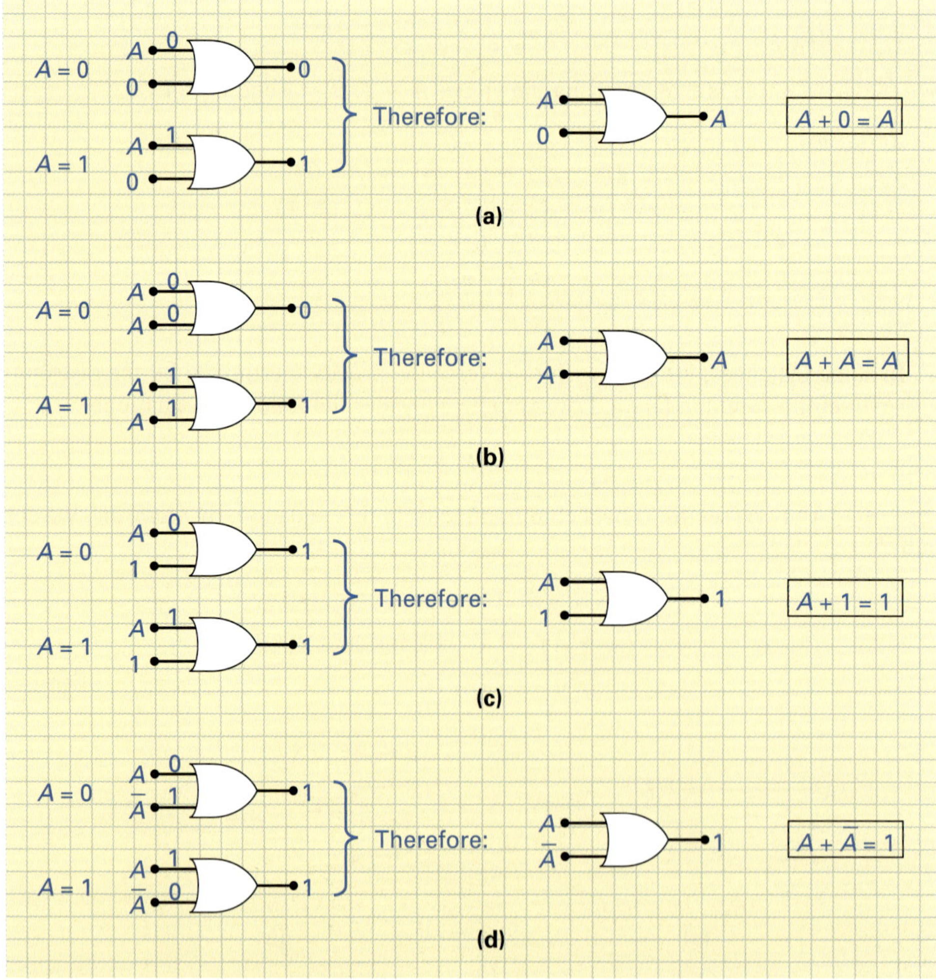

FIGURE 19-18 Boolean Rules for OR Gates.

Figure 19-18(c) shows the next OR gate Boolean rule. This rule states that any 1 input to an OR gate will result in a 1 output, regardless of the other input. Stated algebraically:

$$A + 1 = 1$$

The final Boolean rule for OR gates is shown in Figure 19-18(d). This rule states that when any variable (A) is ORed with its complement ($\overline{A}$) the result will be a 1.

$$A + \overline{A} = 1$$

AND Gate Rules

As with the OR gate, there are four Boolean rules relating to AND gates. The first is illustrated in Figure 19-19(a) and states that if a variable input A is ANDed with a 0, the output will be 0 regardless of the other input.

$$A \cdot 0 = 0$$

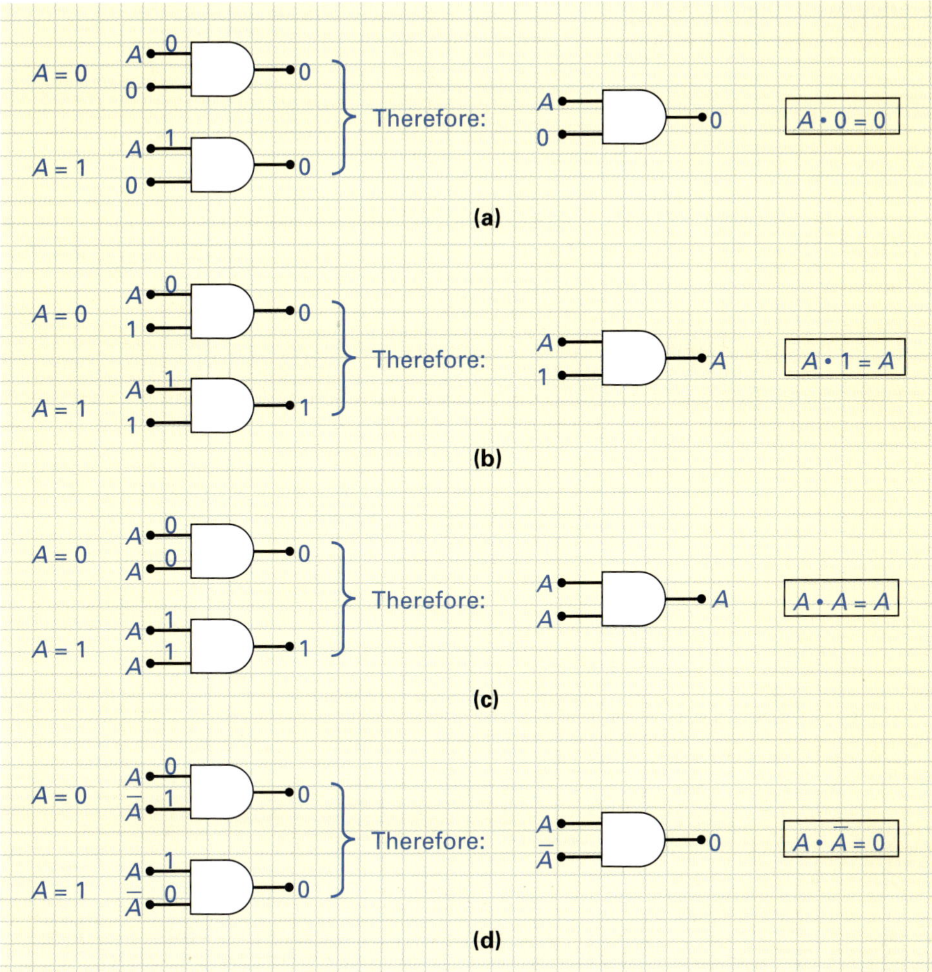

FIGURE 19-19 Boolean Rules for AND Gates.

The second Boolean rule for AND gates is shown in Figure 19-19(b). In this illustration you can see that if a variable is ANDed with a 1, the output will equal the variable.

$$A \cdot 1 = A$$

Another AND gate Boolean rule is shown in Figure 19-19(c). In this instance, any variable that is ANDed with itself will always give an output that is equal to the variable.

$$A \cdot A = A$$

The last of the AND gate Boolean rules is shown in Figure 19-19(d). In this illustration you can see that if a variable (A) is ANDed with its complement ($\overline{A}$), the output will always equal 0.

$$A \cdot \overline{A} = 0$$

Double-Inversion Rule

The *double-inversion rule* is illustrated in Figure 19-20(a) and states that if a variable is inverted twice, then the variable will be back to its original state. To state this algebraically, we use the double bar as follows:

$$\overline{\overline{A}} = A$$

In Figure 19-20(b), you can see that if the NOR gate were replaced with an OR gate, the INVERTER would not be needed since the double inversion returns the logic level to its original state.

DeMorgan's Theorems

DeMorgan's first and second theorems were discussed earlier in the previous section and are repeated here since they also apply as Boolean rules.

DeMorgan's First Theorem: $\overline{A} \cdot \overline{B} = \overline{A + B}$

DeMorgan's Second Theorem: $\overline{A} + \overline{B} = \overline{A \cdot B}$

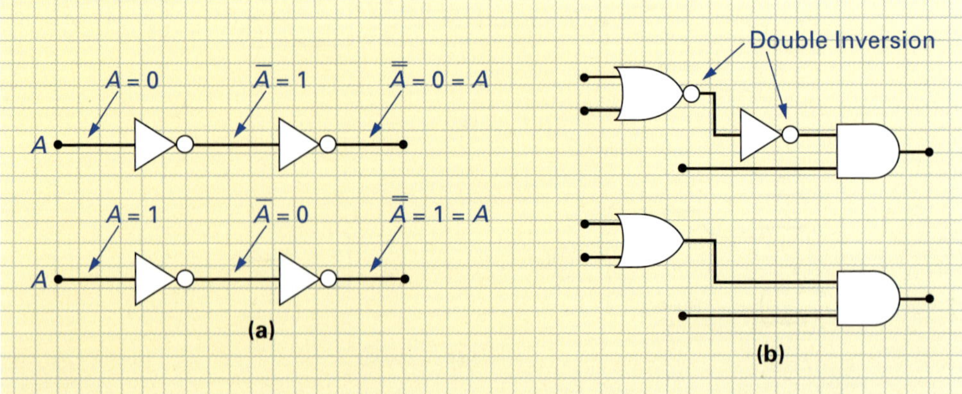

FIGURE 19-20 **Double-Inversion Rule.**

■ **EXAMPLE:**

Apply DeMorgan's theorem and the double-inversion rule to find equivalent logic circuits of the following:

$$\text{a. } \overline{\overline{A} + B} \qquad \text{b. } \overline{A \cdot \overline{B}}$$

■ ***Solution:***

With both equations we simply "break the line and change the sign" and then cancel any double inversions to find the equivalent logic circuits, as follows:

$$\text{a. } \overline{\overline{A} + B} = \overline{\overline{A}} \cdot \overline{B} = A \cdot \overline{B}$$

In the first example we found that a NOR gate with a bubbled A input is equivalent to an AND gate with a bubbled B input.

$$\text{b. } \overline{A \cdot \overline{B}} = \overline{A} + \overline{\overline{B}} = \overline{A} + B$$

In the second example we found that a NAND gate with a bubbled B input is equivalent to an OR gate with a bubbled A input.

The Duality Theorem

The *duality theorem* is very useful since it allows us to change Boolean equations into their *dual* and to produce new Boolean equations. The two steps to follow for making the change are very simple.

a. First, change each OR symbol to an AND symbol and each AND symbol to an OR symbol.

b. Second, change each 0 to a 1 and each 1 to a 0.

To see if this works, let us apply the duality theorem to a simple example.

■ **EXAMPLE:**

Apply the duality theorem to $A + 1 = 1$ and then state whether the resulting rule is true.

■ ***Solution:***

Changing the OR symbol to an AND symbol, and changing the 1s to 0s, we arrive at the following:

Original Rule:	$A + 1 = 1$
Dual Rule:	$A \cdot 0 = 0$

OR symbol is changed to AND symbol. ↑ ↑ ↑ 1s are changed to 0s.

The dual Boolean rule states that any variable ANDed with a 0 will always yield a 0 output, which is true.

■ **EXAMPLE:**

Determine the dual Boolean equation for the following:

$$A \cdot (B + C) = (A \cdot B) + (A \cdot C)$$

Once you have determined the new Boolean equation, sketch circuits and give the truth tables for both the left and right parts of the equation.

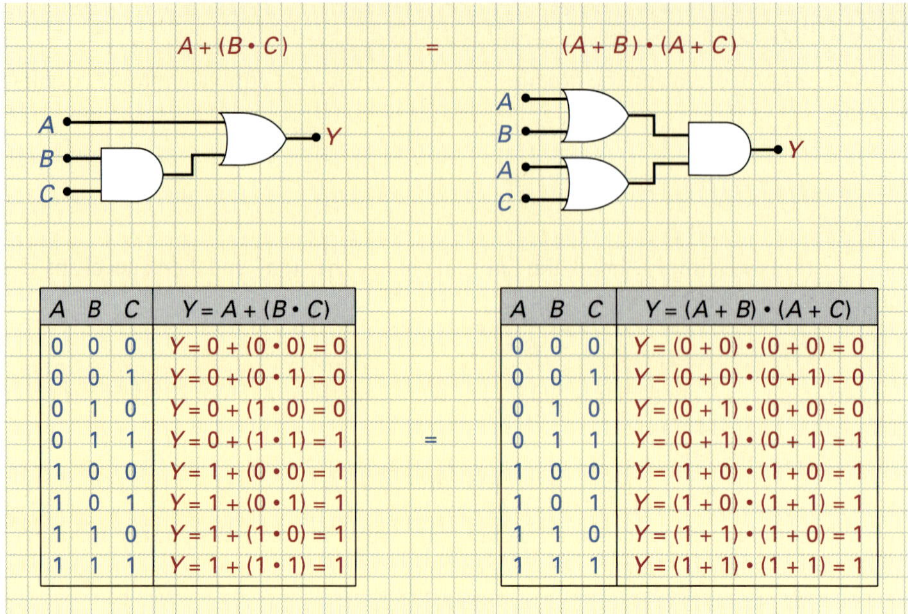

A	B	C	Y = A + (B • C)
0	0	0	Y = 0 + (0 • 0) = 0
0	0	1	Y = 0 + (0 • 1) = 0
0	1	0	Y = 0 + (1 • 0) = 0
0	1	1	Y = 0 + (1 • 1) = 1
1	0	0	Y = 1 + (0 • 0) = 1
1	0	1	Y = 1 + (0 • 1) = 1
1	1	0	Y = 1 + (1 • 0) = 1
1	1	1	Y = 1 + (1 • 1) = 1

A	B	C	Y = (A + B) • (A + C)
0	0	0	Y = (0 + 0) • (0 + 0) = 0
0	0	1	Y = (0 + 0) • (0 + 1) = 0
0	1	0	Y = (0 + 1) • (0 + 0) = 0
0	1	1	Y = (0 + 1) • (0 + 1) = 1
1	0	0	Y = (1 + 0) • (1 + 0) = 1
1	0	1	Y = (1 + 0) • (1 + 1) = 1
1	1	0	Y = (1 + 1) • (1 + 0) = 1
1	1	1	Y = (1 + 1) • (1 + 1) = 1

FIGURE 19-21 ***A* OR (*B* AND *C*) Equals (*A* OR *B*) AND (*A* OR *C*).**

Solution:

Applying the duality theorem to $A \cdot (B + C) = (A \cdot B) + (A \cdot C)$ means that we will have to change all the AND symbols to OR symbols, and vice versa.

Original Equation: $A \cdot (B + C) = (A \cdot B) + (A \cdot C)$

Dual Equation: $A + (B \cdot C) = (A + B) \cdot (A + C)$

Figure 19-21 shows the dual equation and its circuits and truth tables. As you can see by comparing the truth tables, the left and right parts of the equation are equivalent.

As a summary, here is a list of the Boolean laws and rules described in this section, with their dual equations.

	ORIGINAL EQUATION	DUAL EQUATION
1. Commutative Law	$A + B = B + A$	$A \cdot B = B \cdot A$
2. Associative Law	$A \cdot (B \cdot C) = (A \cdot B) \cdot C$	$A + (B + C) = (A + B) + C$
3. Distributive Law	$A + (B \cdot C) = (A + B) \cdot (A + C)$	$A \cdot (B + C) = (A \cdot B) + (A \cdot C)$
4. OR-AND Rules	$A + 0 = A$	$A \cdot 1 = A$
	$A + A = A$	$A \cdot A = A$
	$A + 1 = 1$	$A \cdot 0 = 0$
	$A + \overline{A} = 1$	$A \cdot \overline{A} = 0$
5. DeMorgan's Laws	$\overline{A} \cdot \overline{B} = \overline{A + B}$	$\overline{A} + \overline{B} = \overline{A \cdot B}$

SELF-TEST EVALUATION POINT FOR SECTION 19-2

Use the following questions to test your understanding of Section 19-2.

1. Give the answers to the following Boolean algebra rules.
   a. $A + 1 = ?$ **b.** $A + \overline{A} = ?$ **c.** $A \cdot 0 = ?$ **d.** $A \cdot A = ?$
2. What is the dual equation of $A + (B \cdot C) = (A + B) \cdot (A + C)$?
3. Apply the associative law to the equation $A \cdot (B \cdot C)$.
4. Apply the distributive law to the equation $(A \cdot B) + (A \cdot C) + (A \cdot D)$.

19-3 FROM TRUTH TABLE TO GATE CIRCUIT

Now that you understand the Boolean rules and laws for logic gates, let us put this knowledge to some practical use. If you need a logic gate circuit to perform a certain function, the best way to begin is with a truth table that details what input combinations should drive the output HIGH and what input combinations should drive the output LOW. As an example, let us assume that we need a logic circuit that will follow the truth table shown in Figure 19-22(a). This means that the output (Y) should only be 1 when

$$A = 0, B = 0, C = 1$$
$$A = 0, B = 1, C = 1$$
$$A = 1, B = 0, C = 1$$
$$A = 1, B = 1, C = 1$$

In the truth table in Figure 19-22(a), you can see the **fundamental products** listed for each of these HIGH outputs. A fundamental product is a Boolean expression that describes what the inputs will need to be in order to generate a HIGH output. For example, the first fundamental product ($Y = \overline{A} \cdot \overline{B} \cdot C$) states that A must be 0, B must be 0, and C must be 1 for the output Y to be 1 ($Y = \overline{A} \cdot \overline{B} \cdot C = \overline{0} \cdot \overline{0} \cdot 1 = 1 \cdot 1 \cdot 1 = 1$). The second fundamental product ($Y = \overline{A} \cdot B \cdot C$) states that A must be 0, B must be 1, and C must be 1 for the output Y to be 1 ($Y = \overline{A} \cdot B \cdot C = \overline{0} \cdot 1 \cdot 1 = 1 \cdot 1 \cdot 1 = 1$). The third fundamental product ($Y = A \cdot \overline{B} \cdot C$) states that A must be 1, B must be 0, and C must be 1 for Y to be 1 ($Y = A \cdot \overline{B} \cdot C = 1 \cdot \overline{0} \cdot 1 = 1 \cdot 1 \cdot 1 = 1$). The fourth and final fundamental product ($Y = A \cdot B \cdot C$) states that A must be 1, B must be 1, and C must be 1 for Y to be 1 ($Y = A \cdot B \cdot C = 1 \cdot 1 \cdot 1 = 1$). The fundamental products for each of the HIGH outputs are listed in the truth table in Figure 19-22(a). By ORing all of these fundamental products, a Boolean equation can be derived, as follows:

Fundamental Products
The truth table input combinations that are essential since they produce a HIGH output.

$$Y = \overline{A}\overline{B}C + \overline{A}BC + A\overline{B}C + ABC$$

From this Boolean equation we can create a logic network that is the circuit equivalent of the truth table. To complete this step, we will need to study in detail the different parts of the Boolean equation. First, the Boolean equation states that the outputs of four three-input AND gates are connected to the input of a four-input OR gate. The first AND gate is connected to inputs $\overline{A}\,\overline{B}$ and C, the second to inputs $\overline{A}\ B$ and C, the third to inputs $A\ \overline{B}$ and C, and finally the fourth to inputs $A\ B$ and C. The resulting logic circuit equivalent for the truth table in Figure 19-22(a) is shown in Figure 19-22(b).

Sum-of-Products (SOP) Form
A Boolean expression that describes the ORing of two or more AND functions.

The Boolean equation for the logic circuit in Figure 19-22 is in the **sum-of-products (SOP) form.** To explain what this term means, we must recall once again that in Boolean

A	B	C	Y	Fundamental Product
0	0	0	0	
0	0	1	1	$\overline{A}\overline{B}C$
0	1	0	0	
0	1	1	1	$\overline{A}BC$
1	0	0	0	
1	0	1	1	$A\overline{B}C$
1	1	0	0	
1	1	1	1	ABC

(a)

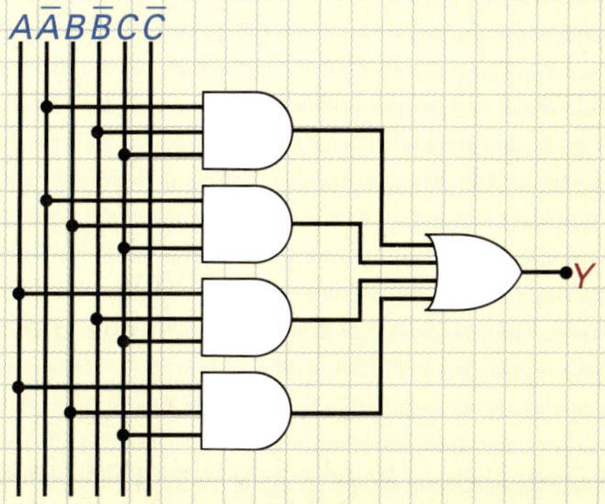

$A\ \overline{A}\ B\ \overline{B}\ C\ \overline{C}$

Y

$Y = \overline{A}\overline{B}C + \overline{A}BC + A\overline{B}C + ABC$

(b)

FIGURE 19-22 Sum-of-Products (SOP) Form.

algebra, an AND gate's output is the *logical product* of its inputs. On the other hand, the OR gate's output is the *logical sum* of its inputs. A sum-of-products expression, therefore, describes the ORing together of (sum of) two or more AND functions (products). Here are some examples of sum-of-product expressions.

$$Y = AB + \overline{A}B$$
$$Y = A\overline{B} + C\overline{D}E$$
$$Y = \overline{A}BCD + AB\overline{C}\overline{D} + A\overline{B}C\overline{D}$$

Product-of-Sums (POS) Form

A Boolean expression that describes the ANDing of two or more OR functions.

The other basic form for Boolean expressions is the **product-of-sums (POS) form.** This form describes the ANDing of (product of) two or more OR functions (sums). Here are some examples of product-of-sums expressions.

$$Y = (A + B + \overline{C}) \cdot (C + D)$$
$$Y = (\overline{A} + \overline{B}) \cdot (A + B)$$
$$Y = (A + B + C + \overline{D}) \cdot (\overline{A} + B + \overline{C} + D) \cdot (A + \overline{B} + \overline{C} + \overline{D})$$

EXAMPLE:

Is the circuit shown in Figure 19-23 an example of sum-of-products or product-of-sums, and what would this circuit's Boolean expression be?

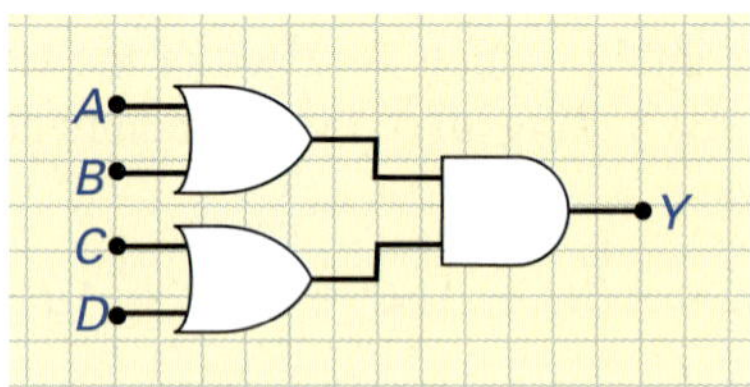

FIGURE 19-23 **SOP or POS?**

Solution:

The circuit in Figure 19-23 is a product-of-sums equivalent circuit, and its Boolean expression is

$$Y = (A + B) \cdot (C + D)$$

EXAMPLE:

Determine the logic circuit needed for the truth table shown in Figure 19-24(a).

Solution:

The first step is to write the fundamental product for each HIGH output in the truth table, as seen in Figure 19-24(a). By ORing (+) these fundamental products, we can obtain the following sum-of-products Boolean equation:

$$Y = \overline{A}\,\overline{B}\,\overline{C}D + \overline{A}B\overline{C}\,\overline{D} + A\overline{B}\,\overline{C}D + AB\overline{C}D$$

The next step is to develop the circuit equivalent of this equation. The equation states that the outputs of four four-input AND gates are connected to a four-input OR gate. The first AND gate had inputs $\overline{A}\,\overline{B}\,\overline{C}D$, the second $\overline{A}B\overline{C}\,\overline{D}$, the third has $A\overline{B}\,\overline{C}D$, and the fourth has $AB\overline{C}D$. The circuit equivalent for this Boolean expression can be seen in Figure 19-24(b). This circuit will operate in the manner detailed by the truth table in Figure 19-24(a), and generate a 1 output only when

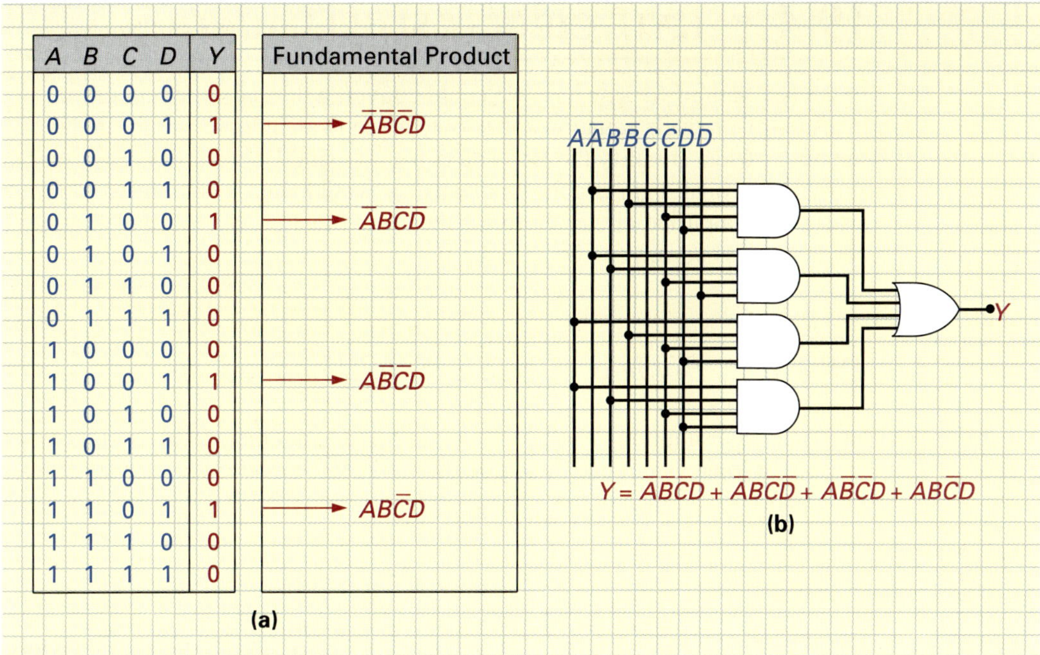

A	B	C	D	Y	Fundamental Product
0	0	0	0	0	
0	0	0	1	1	$\overline{A}\,\overline{B}\,\overline{C}D$
0	0	1	0	0	
0	0	1	1	0	
0	1	0	0	1	$\overline{A}B\overline{C}\,\overline{D}$
0	1	0	1	0	
0	1	1	0	0	
0	1	1	1	0	
1	0	0	0	0	
1	0	0	1	1	$A\overline{B}\,\overline{C}D$
1	0	1	0	0	
1	0	1	1	0	
1	1	0	0	0	
1	1	0	1	1	$AB\overline{C}D$
1	1	1	0	0	
1	1	1	1	0	

FIGURE 19-24 **Sum-of-Products Example.**

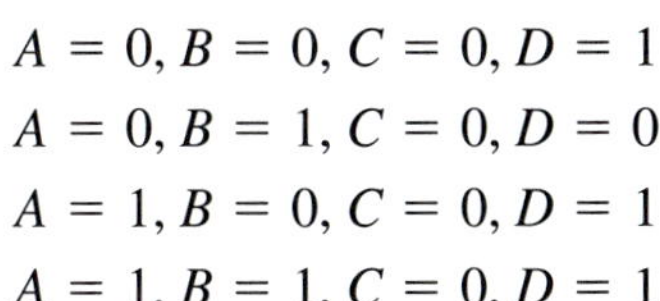

$$A = 0, B = 0, C = 0, D = 1$$
$$A = 0, B = 1, C = 0, D = 0$$
$$A = 1, B = 0, C = 0, D = 1$$
$$A = 1, B = 1, C = 0, D = 1$$

In all other instances, the output of this circuit will be 0.

EXAMPLE:

Apply the distributive law to the following Boolean expression in order to convert it into a sum-of-products form:

$$Y = A\overline{B} + C(D\overline{E} + F\overline{G})$$

Solution:

$Y = A\overline{B} + C(D\overline{E} + F\overline{G})$ Original Expression

$Y = A\overline{B} + CD\overline{E} + CF\overline{G}$ Distributive Law Applied

SELF-TEST EVALUATION POINT FOR SECTION 19-3

Use the following questions to test your understanding of Section 19-3.

1. A two-input AND gate will produce at its output the logical ____________ of *A* and *B*, while an OR gate will produce at its output the logical ____________ of *A* and *B*.
2. What are the fundamental products of each of the following input words?
   a. *A, B, C, D* = 1011 **b.** *A, B, C, D* = 0110
3. The Boolean expression $Y = AB + \overline{A}B$ is an example of a/an ____________ (SOP/POS) equation.
4. Describe the steps involved to develop a logic circuit from a truth table.

19-4 GATE CIRCUIT SIMPLIFICATION

In the last section we saw how we could convert a desired set of conditions in a truth table to a sum-of-products equation and then to an equivalent logic circuit. In this section we will see how we can simplify the Boolean equation using Boolean laws and rules and therefore simplify the final logic circuit. As to why we would want to simplify a logic circuit, the answer is: A simplified circuit performs the same function, but is smaller, easier to construct, consumes less power, and costs less.

19-4-1 *Boolean Algebra Simplification*

Generally, the simplicity of one circuit compared to another circuit is judged by counting the number of logic gate inputs. For example, the logic circuit in Figure 19-25(a) has a total of six inputs (two on each of the AND gates and two on the OR gate). The Boolean equation for this circuit would be

$$Y = \overline{A}B + AB$$

Since "$\cdot B$" (AND B) seems to be a common factor in this equation, by factoring we could obtain the following:

$$Y = \overline{A}B + AB \quad \text{Original Expression}$$
$$Y = (\overline{A} + A) \cdot B \quad \text{Factoring AND B}$$

This equivalent logic circuit is shown in Figure 19-25(b), and if you count the number of logic gate inputs in this circuit (two for the OR and two for the AND, a total of four), you can see that the circuit has been simplified from a six-input circuit to a four-input circuit.

This, however, is not the end of our simplification process, since another rule can be applied to further simplify the circuit. Remembering that any variable ORed with its complement results in a 1, we can replace $\overline{A} + A$ with a 1, as follows:

$$Y = (\overline{A} + A) \cdot B$$
$$Y = (1) \cdot B$$
$$Y = 1 \cdot B$$

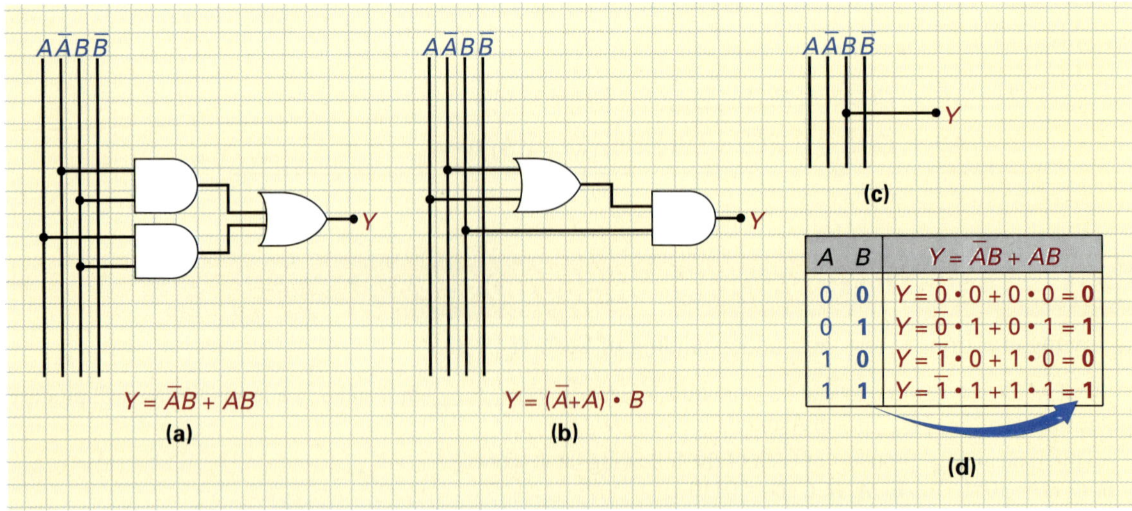

FIGURE 19-25 Simplifying Gate Circuits.

The resulting equation $Y = 1 \cdot B$ can be further simplified since any variable ANDed with a 1 will always cause the output to equal the variable. Therefore, $Y = 1 \cdot B$ can be replaced with B, as follows:

$$Y = 1 \cdot B$$
$$Y = B$$

This means that the entire circuit in Figure 19-25(a) can be replaced with a single wire connected from the B input to the output, as seen in Figure 19-25(c). This fact is confirmed in the truth table in Figure 19-25(d), which shows that output Y exactly follows the input B.

EXAMPLE:

Simplify the Boolean equation shown in Figure 19-26(a).

Solution:

The Boolean equation for this circuit is

$$Y = A\overline{B}\,\overline{C} + \overline{A}B\overline{C} + A\overline{B}C + \overline{A}\,\overline{B}\,\overline{C}$$

The first step in the simplification process is to try and spot common factors in each of the ANDed terms and then rearrange these terms as shown in Figure 19-26(b). The next step is to factor out the common expressions, which are $A\overline{B}$ and $\overline{A}\,\overline{C}$, resulting in

$Y = A\overline{B}\,\overline{C} + \overline{A}B\overline{C} + A\overline{B}C + \overline{A}\,\overline{B}\,\overline{C}$	Original Equation
$Y = A\overline{B}\,\overline{C} + A\overline{B}C + \overline{A}B\overline{C} + \overline{A}\,\overline{B}\,\overline{C}$	Rearranged Equation
$Y = A\overline{B}(\overline{C} + C) + \overline{A}\,\overline{C}(B + \overline{B})$	Factored Equation

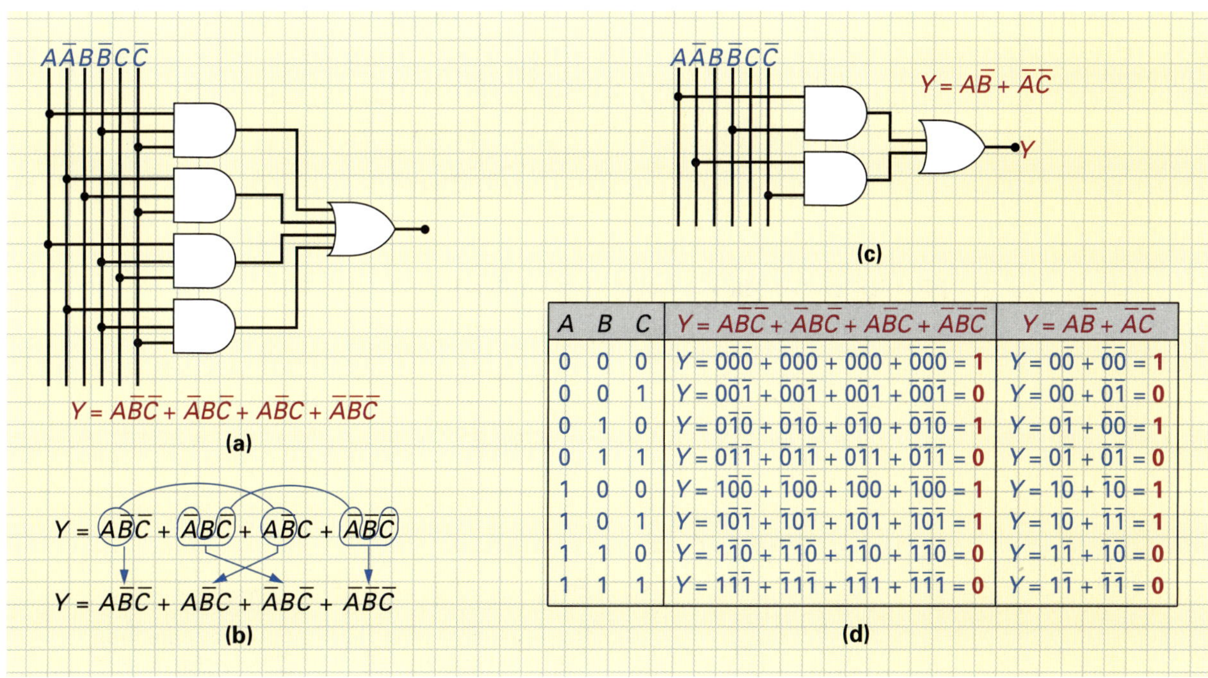

A	B	C	$Y = A\overline{B}\,\overline{C} + \overline{A}B\overline{C} + A\overline{B}C + \overline{A}\,\overline{B}\,\overline{C}$	$Y = A\overline{B} + \overline{A}\,\overline{C}$
0	0	0	$Y = 0\overline{0}\,\overline{0} + \overline{0}0\overline{0} + 0\overline{0}0 + \overline{0}\,\overline{0}\,\overline{0} = 1$	$Y = 0\overline{0} + \overline{0}\,\overline{0} = 1$
0	0	1	$Y = 0\overline{0}\,\overline{1} + \overline{0}0\overline{1} + 0\overline{0}1 + \overline{0}\,\overline{0}\,\overline{1} = 0$	$Y = 0\overline{0} + \overline{0}\,\overline{1} = 0$
0	1	0	$Y = 0\overline{1}\,\overline{0} + \overline{0}1\overline{0} + 0\overline{1}0 + \overline{0}\,\overline{1}\,\overline{0} = 1$	$Y = 0\overline{1} + \overline{0}\,\overline{0} = 1$
0	1	1	$Y = 0\overline{1}\,\overline{1} + \overline{0}1\overline{1} + 0\overline{1}1 + \overline{0}\,\overline{1}\,\overline{1} = 0$	$Y = 0\overline{1} + \overline{0}\,\overline{1} = 0$
1	0	0	$Y = 1\overline{0}\,\overline{0} + \overline{1}0\overline{0} + 1\overline{0}0 + \overline{1}\,\overline{0}\,\overline{0} = 1$	$Y = 1\overline{0} + \overline{1}\,\overline{0} = 1$
1	0	1	$Y = 1\overline{0}\,\overline{1} + \overline{1}0\overline{1} + 1\overline{0}1 + \overline{1}\,\overline{0}\,\overline{1} = 1$	$Y = 1\overline{0} + \overline{1}\,\overline{1} = 1$
1	1	0	$Y = 1\overline{1}\,\overline{0} + \overline{1}1\overline{0} + 1\overline{1}0 + \overline{1}\,\overline{1}\,\overline{0} = 0$	$Y = 1\overline{1} + \overline{1}\,\overline{0} = 0$
1	1	1	$Y = 1\overline{1}\,\overline{1} + \overline{1}1\overline{1} + 1\overline{1}1 + \overline{1}\,\overline{1}\,\overline{1} = 0$	$Y = 1\overline{1} + \overline{1}\,\overline{1} = 0$

FIGURE 19-26 Boolean Equation Simplification.

Studying the last equation, you can see that we can apply the OR rule $(\overline{A} + A) = 1$ and simplify the terms within the parentheses to obtain the following:

$$Y = A\overline{B}(\overline{C} + C) + \overline{A}\,\overline{C}(B + \overline{B})$$
$$Y = A\overline{B} \cdot (1) + \overline{A}\,\overline{C} \cdot (1)$$

Now we can apply the $A \cdot 1 = A$ rule to further simplify the terms, since $A\overline{B} \cdot 1 = A\overline{B}$ and $\overline{A}\,\overline{C} \cdot 1 = \overline{A}\,\overline{C}$; therefore:

$$Y = A\overline{B} + \overline{A}\,\overline{C}$$

This equivalent logic circuit is illustrated in Figure 19-26(c) and has only six logic gate inputs compared to the original circuit shown in Figure 19-26(a), which has sixteen logic gate inputs. The truth table in Figure 19-26(d) compares the original equation with the simplified equation and shows how the same result is obtained.

19-4-2 *Karnaugh Map Simplification*

Karnaugh Map
Also called K-map, it is a truth table that has been rearranged to show a geometrical pattern of functional relationships for gating configurations. Using this map, essential gating requirements can be more easily recognized and reduced to their simplest form.

In most instances, engineers and technicians will simplify logic circuits using a **Karnaugh map,** or **K-map.** The Karnaugh map, named after its inventor, is quite simply a rearranged truth table, as shown in Figure 19-27. With this map, the essential gating requirements can be more easily recognized and reduced to their simplest form.

The total number of boxes or cells in a K-map depends on the number of input variables. For example, in Figure 19-27, only two inputs (A and B) and their complements ($\overline{A}$ and $\overline{B}$) are present, and therefore the K-map contains (like a two-variable truth table) only four combinations (00, 01, 10, and 11).

Each cell of this two-variable K-map represents one of the four input combinations. In practice, the input labels are placed outside of the cells, as shown in Figure 19-28(a), and apply to either a column or row of cells. For example, the row label $\overline{A}$ applies to the two upper

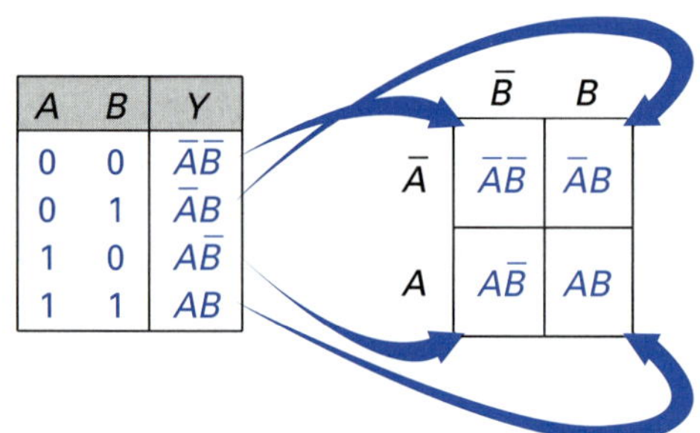

FIGURE 19-27 **The Karnaugh Map—A Rearranged Truth Table.**

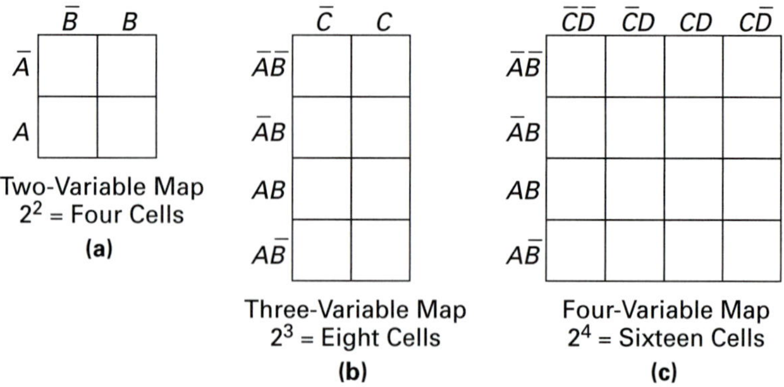

FIGURE 19-28 **Karnaugh Maps. (a) Two-Variable. (b) Three-Variable. (c) Four-Variable.**

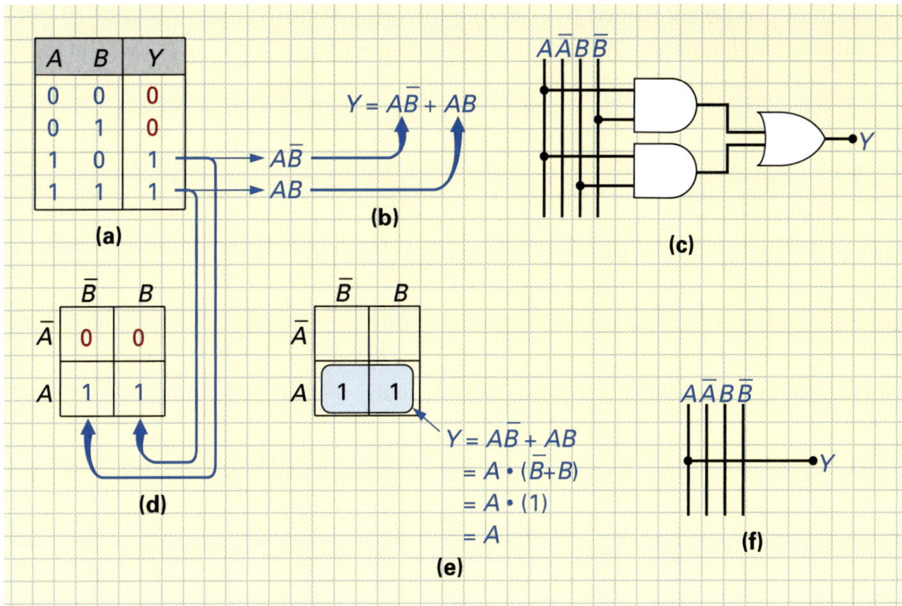

FIGURE 19-29 **Karnaugh Map Simplification.**

cells, while the row label A applies to the two lower cells. Running along the top of the K-map, the label $\overline{B}$ applies to the two left cells, while the label B applies to the two right cells. As an example, the upper right cell represents the input combination $\overline{A}B$. Figures 19-28(b) and (c) show the formats for a three-variable ($2^3 = 8$ cells) and four-variable ($2^4 = 16$ cells) K-map.

Now that we have an understanding of the K-map, let us see how it can be used to simplify a logic circuit. As an example, imagine that we need to create an equivalent logic circuit for the truth table given in Figure 19-29(a). The first step is to develop a sum-of-products Boolean expression. This is achieved by writing the fundamental product for each HIGH output in the truth table and then ORing all of the fundamental products, as shown in Figure 19-29(b). The equivalent logic circuit for this equation is shown in Figure 19-29(c). The next step is to plot this Boolean expression on a two-variable K-map, as seen in Figure 19-29(d). When plotting a sum-of-products expression on a K-map, remember that each cell corresponds to each of the input combinations in the truth table. A HIGH output in the truth table should appear as a 1 in its equivalent cell in the K-map, and a LOW output in the truth table should appear as a 0 in its equivalent cell. A 1, therefore, will appear in the lower left cell (corresponding to $A\overline{B}$) and in the lower right cell (corresponding to AB). The other input combinations ($\overline{A}\,\overline{B}$ and $\overline{A}B$) both yield a 0 output, and therefore a 0 should be placed in these two upper cells.

Reducing Boolean equations is largely achieved by applying the rule of complements, which states that $A + \overline{A} = 1$. Now that the SOP equation has been plotted on the K-map shown in Figure 19-29(d), the next step is to group terms and then factor out the common variables. If you study the K-map in Figure 19-29(d), you will see that adjacent cells differ by only one input variable. This means that if you move either horizontally or vertically from one cell to an adjacent cell, only one variable will change. By grouping adjacent cells containing a 1, as seen in Figure 19-29(e), cells can be compared and simplified (using the rule of complements) to create one-product terms. In this example, cells $A\overline{B}$ and AB contain B and $\overline{B}$, so these opposites or complements cancel, leaving A, as follows:

$Y = A\overline{B} + AB$ Grouped Pair
$Y = A \cdot (\overline{B} + B)$ Factoring A AND
$Y = A \cdot 1$
$Y = A$

This procedure can be confirmed by studying the original truth table in Figure 19-29(a) in which you can see that output *Y* exactly follows input *A*. The equivalent circuit, therefore, is shown in Figure 19-29(f).

EXAMPLE:

Determine the simplest logic circuit for the truth table shown in Figure 19-30(a). Illustrate each step in the process.

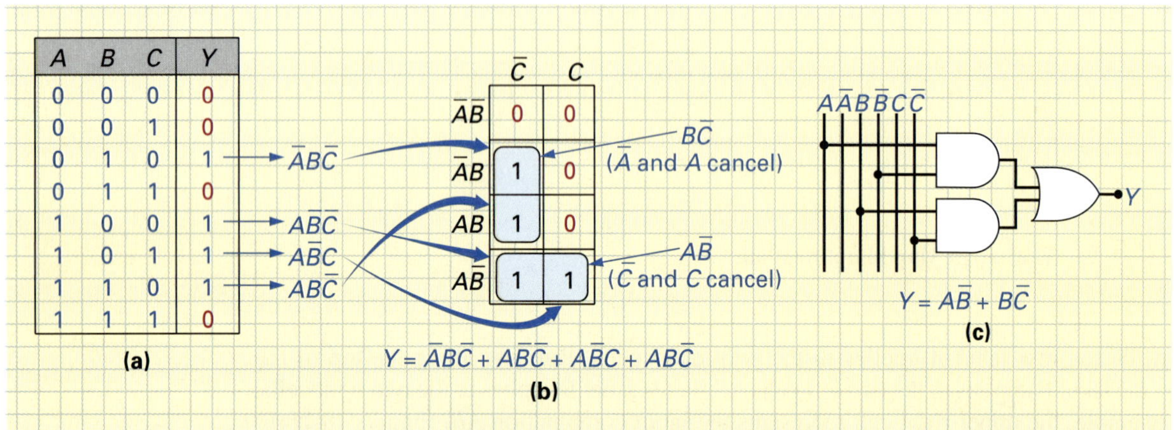

A	B	C	Y
0	0	0	0
0	0	1	0
0	1	0	1
0	1	1	0
1	0	0	1
1	0	1	1
1	1	0	1
1	1	1	0

FIGURE 19-30 **Three-Variable K-Map Simplification.**

Solution:

Since this example has three input variables, the first step is to draw a three-variable K-map, as shown in Figure 19-30(b). The next step is to look for the HIGH outputs in the truth table in Figure 19-30(a) and plot these 1s in their equivalent cells in the K-map, as shown in Figure 19-30(b). After inserting 0s in the remaining cells, group the 1s into pairs, as shown in Figure 19-30(b), and then study the row and column variable labels associated with that grouped pair to see which variable will drop out due to the rule of complements. In the upper group, $\bar{A}$ and A will cancel, leaving $B\bar{C}$, and in the lower group $\bar{C}$ and C will cancel, leaving $A\bar{B}$. These reduced products will form an equivalent Boolean equation and logic circuit, as seen in Figure 19-30(c). In this example, the original sixteen-input equation in Figure 19-30(b) has been simplified to the equivalent six-input logic gate circuit shown in Figure 19-30(c).

The 1s in a K-map can be grouped in pairs (groups of two), quads (groups of four), octets (groups of eight), and all higher powers of two. Figure 19-31 shows some examples of grouping and how the K-map has been used to reduce large Boolean equations. Notice that the larger groups will yield smaller terms and, therefore, gate circuits with fewer inputs. For this reason, you should begin by looking for the largest possible groups, and then step down in group size if none are found (for example, begin looking for octets, then quads, and finally pairs). Also notice that you can capture 1s on either side of a map by wrapping a group around behind the map.

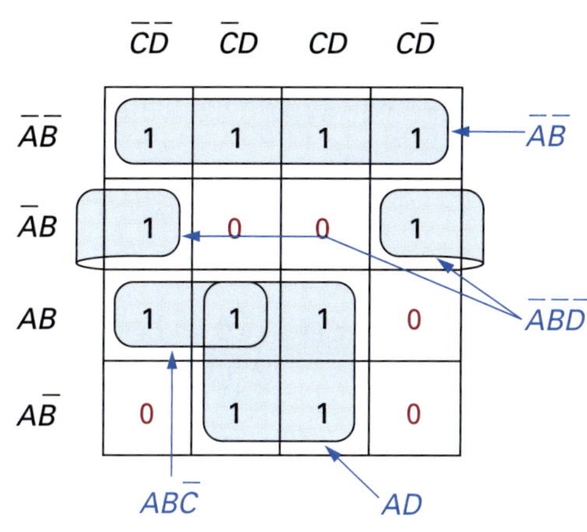

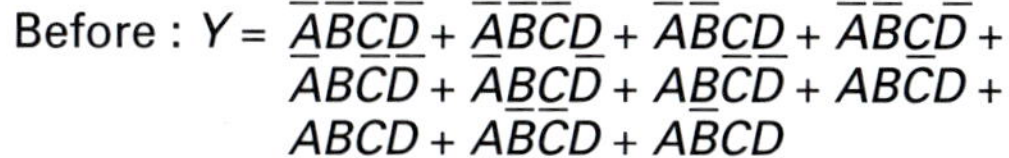

Before : $Y = \overline{A}\,\overline{B}\,\overline{C}\,\overline{D} + \overline{A}\,\overline{B}\,\overline{C}D + \overline{A}\,\overline{B}CD + \overline{A}\,\overline{B}C\overline{D} +$
$\overline{A}B\overline{C}\,\overline{D} + \overline{A}BC\overline{D} + AB\overline{C}\,\overline{D} + AB\overline{C}D +$
$ABCD + A\overline{B}\,\overline{C}D + A\overline{B}CD$

After : $Y = AB\overline{C} + AD + \overline{A}B\overline{D} + \overline{A}\,\overline{B}$

(a)

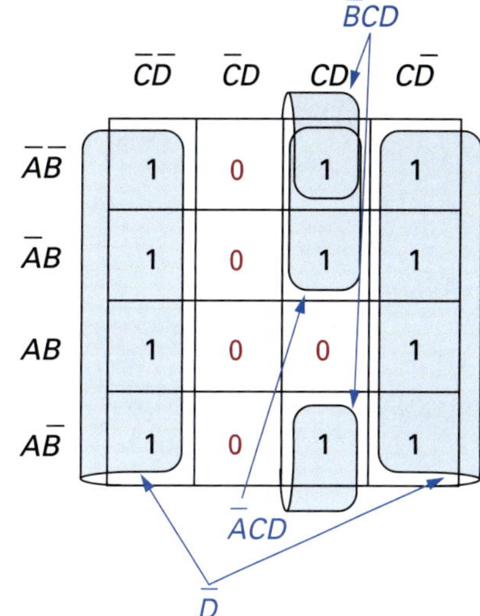

$\overline{B}CD$
$\overline{C}\,\overline{D}$ $\overline{C}D$ CD $C\overline{D}$
$\overline{A}\,\overline{B}$ 1 0 1 1
$\overline{A}B$ 1 0 1 1
AB 1 0 0 1
$A\overline{B}$ 1 0 1 1
$\overline{A}CD$
$\overline{D}$

Before : $Y = \overline{A}\,\overline{B}\,\overline{C}\,\overline{D} + \overline{A}\,\overline{B}CD + \overline{A}\,\overline{B}C\overline{D} +$
$\overline{A}B\overline{C}\,\overline{D} + \overline{A}BCD + \overline{A}BC\overline{D} +$
$A\overline{B}\,\overline{C}\,\overline{D} + A\overline{B}C\overline{D} + AB\overline{C}\,\overline{D} +$
$A\overline{B}CD + ABC\overline{D}$

After : $Y = \overline{A}CD + \overline{B}CD + \overline{D}$

(b)

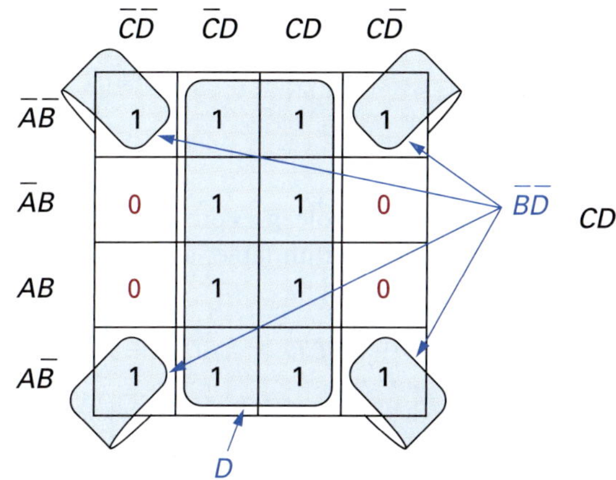

Before : $Y = \overline{A}\,\overline{B}\,\overline{C}\,\overline{D} + \overline{A}\,\overline{B}\,\overline{C}D + \overline{A}\,\overline{B}CD + \overline{A}\,\overline{B}C\overline{D} +$
$\overline{A}B\overline{C}D + \overline{A}BCD + AB\overline{C}D + A\overline{B}C\overline{D} +$
$A\overline{B}\,\overline{C}\,\overline{D} + A\overline{B}\,\overline{C}D + A\overline{B}CD + ABCD$

After : $Y = \overline{B}\,\overline{D} + D$

(c)

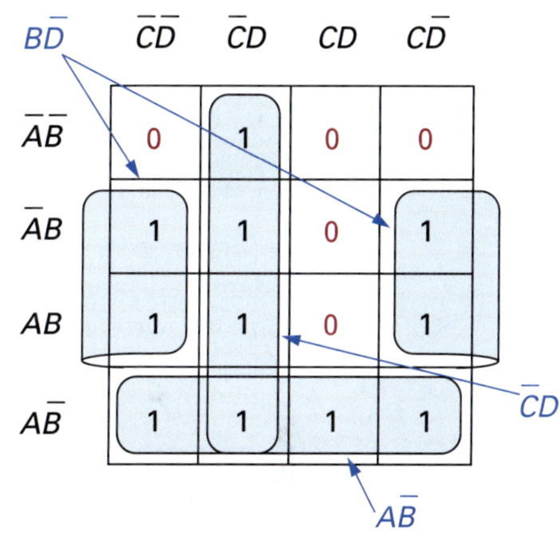

Before : $Y = \overline{A}\,\overline{B}\,\overline{C}D + \overline{A}B\overline{C}\,\overline{D} + \overline{A}B\overline{C}D + \overline{A}BC\overline{D} +$
$AB\overline{C}\,\overline{D} + AB\overline{C}D + ABC\overline{D} + A\overline{B}\,\overline{C}\,\overline{D} +$
$A\overline{B}\,\overline{C}D + A\overline{B}CD + A\overline{B}C\overline{D}$

After : $Y = A\overline{B} + B\overline{D} + \overline{C}D$

(d)

FIGURE 19-31 **K-Map Grouping Examples.**

EXAMPLE:

What would be the sum-of-products Boolean expression for the truth table in Figure 19-32(a)? After determining the Boolean expression, plot it on a K-map to see if it can be simplified.

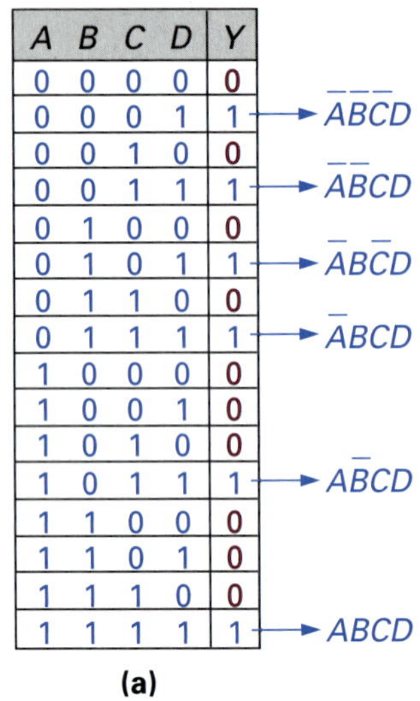

A	B	C	D	Y	
0	0	0	0	0	
0	0	0	1	1	→ $\overline{A}\,\overline{B}\,\overline{C}D$
0	0	1	0	0	
0	0	1	1	1	→ $\overline{A}\,\overline{B}CD$
0	1	0	0	0	
0	1	0	1	1	→ $\overline{A}B\overline{C}D$
0	1	1	0	0	
0	1	1	1	1	→ $\overline{A}BCD$
1	0	0	0	0	
1	0	0	1	0	
1	0	1	0	0	
1	0	1	1	1	→ $A\overline{B}CD$
1	1	0	0	0	
1	1	0	1	0	
1	1	1	0	0	
1	1	1	1	1	→ $ABCD$

(a)

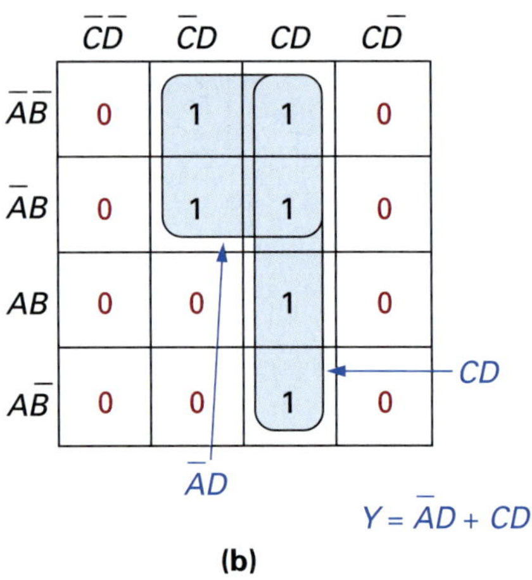

FIGURE 19-32 Example of Four-Variable K-Map.

Solution:

The first step in developing the sum-of-products Boolean expression is to write down the fundamental products for each HIGH output in the truth table, as seen in Figure 19-32(a). From this, we can derive an SOP equation, which will be

$$Y = \overline{A}\,\overline{B}\,\overline{C}D + \overline{A}\,\overline{B}CD + \overline{A}B\overline{C}D + \overline{A}BCD + A\overline{B}CD + ABCD$$

The next step is to draw a four-variable K-map, as seen in Figure 19-32(b), and then plot the HIGH outputs of the truth table in their equivalent cells in the K-map. Looking at Figure 19-32(b), you can see that the 1s can be grouped into two quads. With the square-shaped quad, row labels B and $\overline{B}$ will cancel and column labels C and $\overline{C}$ will cancel, leaving $\overline{A}D$. With the rectangular-shaped quad, all of its row labels will cancel, leaving CD. The simplified Boolean expression, therefore, will be

$$Y = \overline{A}D + CD$$

Referring to the truth table in Figure 19-33(a), you can see that input words 0000 through 1001 have either a 0 or 1 at the output Y. Input words 1010 through 1111 on the other hand, have an "X" written in the output column to indicate that the output Y for these input combinations is unimportant (output can be either a 0 or 1). These Xs in the output are appropriately called "don't-care conditions" and can be treated as either a 0 or 1. In K-maps, these Xs can be grouped with 1s to create larger groups, and therefore a simpler logic circuit, as we will see in the next example.

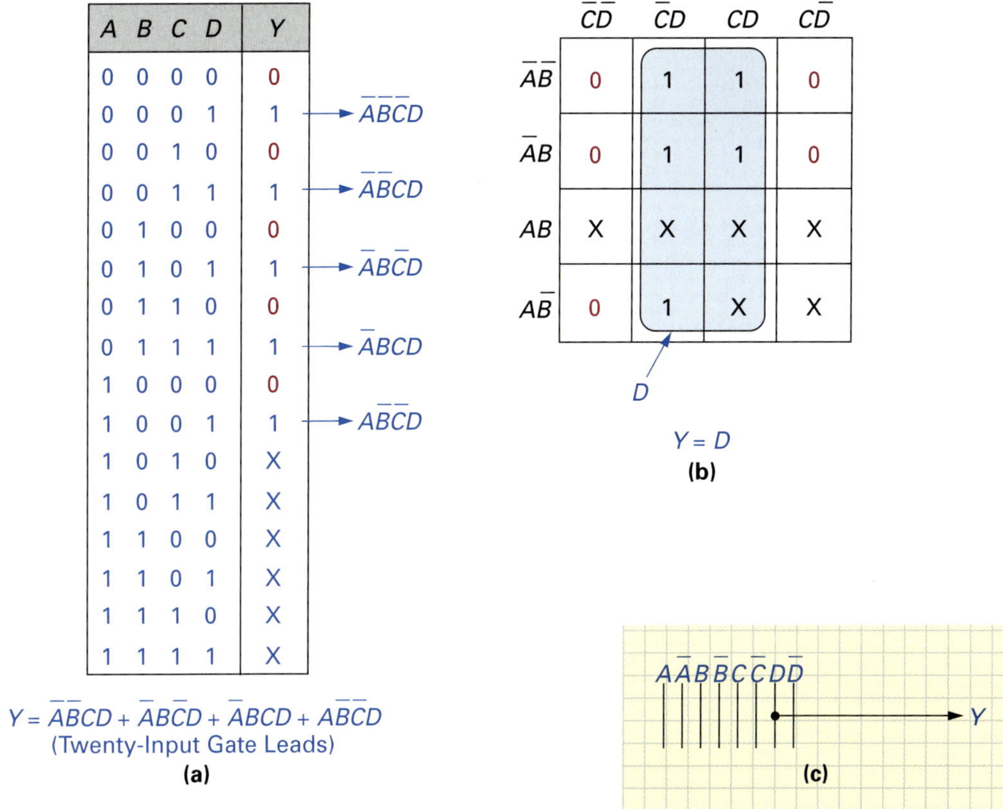

FIGURE 19-33 **Four-Variable K-Map Example with Don't-Care Conditions.**

EXAMPLE:

Sketch an equivalent logic circuit for the truth table given in Figure 19-33(a) using the least amount of inputs.

Solution:

Figure 19-33(b) shows a four-variable K-map with the 0s, 1s, and Xs (don't-care conditions) inserted into their appropriate cells. Since a larger group will always yield a smaller term, visualize the Xs as 1s and then try for as big a group as possible. In this example, the Xs have allowed us to group an octet, resulting in the single product *D*. As a result, the twenty-input Boolean equation in Figure 19-33(a) can be replaced with a single line connected to *D*, as shown in Figure 19-33(c).

SELF-TEST EVALUATION POINT FOR SECTION 19-4

Use the following questions to test your understanding of Section 19-4.

1. A two-variable K-map will have ____________ cells.
2. Simplify the equation $Y = \bar{A}B + AB$.
3. When grouping 1s in a K-map, the ____________ (smaller/larger) the groups, the smaller the terms.
4. Can the don't-care conditions in a K-map be grouped with 0s or with 1s, or doesn't it matter?

REVIEW QUESTIONS

Multiple-Choice Questions

1. In Boolean, the OR function is described as logical ______________ and the AND function is described as logical ______________.
   a. addition, division
   b. multiplication, addition
   c. addition, multiplication
   d. multiplication, subtraction

2. The OR gate is said to produce at its output the logical ______________ of its inputs, while the AND gate is said to produce at its output the logical ______________ of its inputs.
   a. sum, product **c.** sum, subtrahend
   b. product, quotient **d.** product, sum

3. If inputs A and B are ANDed together and then the result is ORed with C, the Boolean expression equation describing this operation would be
   a. $Y = AB + BC$ **c.** $Y = AB + C$
   b. $Y = (A + B)C$ **d.** $Y = AB + AC$

4. The Boolean expression $Y = (A \cdot B) + (A \cdot C)$ describes a logic circuit that has
   a. six inputs
   b. two OR and one AND gates
   c. two AND and one OR gates
   d. four inputs

5. The Boolean expression $Y = \overline{(A + B)}$ describes a two-input ______________.
   a. NAND gate **c.** AND gate
   b. NOR gate **d.** OR gate

6. DeMorgan's first theorem states that
   a. $\overline{A} \cdot \overline{B} = \overline{A + B}$ **c.** $\overline{A} + \overline{B} = \overline{A \cdot B}$
   b. $\overline{A} + \overline{\overline{B}} = \overline{A} + B$ **d.** $\overline{\overline{A}} \cdot \overline{B} = A \cdot \overline{B}$

7. DeMorgan's second theorem states that
   a. $\overline{A} \cdot \overline{B} = \overline{A + B}$ **c.** $\overline{A} + \overline{B} = \overline{A \cdot B}$
   b. $\overline{A} + \overline{B} = \overline{A} + B$ **d.** $\overline{\overline{A}} \cdot \overline{B} = A \cdot \overline{B}$

8. In Boolean algebra, which symbol is used to describe the exclusive OR action?
   a. "+" **c.** "×"
   b. "·" **d.** "⊕"

9. Which of the following algebraically states the commutative law?
   a. $A + B = B + A$
   b. $A \cdot (B \cdot C) = (A \cdot B) \cdot C$
   c. $A + (B \cdot C) = (A + B) \cdot (A + C)$
   d. None of the above

10. $A + (B \cdot C) = (A + B) \cdot (A + C)$ is the algebraic definition of the ______________.
    a. associative law **c.** commutative law
    b. distributive law **d.** complements law

11. The four Boolean rules for an OR gate are $A + 0 = ?, A + A = ?, A + 1 = ?, A + \overline{A} = ?.$
    a. A, 0, 1, 1 **c.** A, A, 1, 1
    b. A, A, 0, 0 **d.** A, A, 0, 1

12. The four Boolean rules for an AND gate are $A \cdot 1 = ?, A \cdot A = ?, A \cdot 0 = ?, A \cdot \overline{A} = ?.$
    a. A, 0, 1, 1 **c.** A, A, 1, 1
    b. A, A, 0, 0 **d.** A, A, 0, 1

13. If you applied DeMorgan's theorem and the double-inversion rule to the equation $Y = \overline{\overline{A} + B\overline{\overline{C}}}$, what would be the result?
    a. $A\overline{B} + C$ **c.** $\overline{A}B + C$
    b. $A + B\overline{C}$ **d.** $AB + \overline{C}$

14. What would be the dual equation of $A + (B \cdot C) = (A + B) \cdot (A + C)$?
    a. $A \cdot (B + C) = (A \cdot B) + (A \cdot C)$
    b. $\overline{\overline{A} + B} = AB + \overline{C}$
    c. $A \cdot (B \cdot C) = (A \cdot B) + (A \cdot C)$
    d. $A + (B \cdot C) = (A \cdot B) \cdot (A + C)$

15. $Y = AB + BC + AC$ is an example of a ______________.
    a. product-of-sums expression
    b. sum-of-products expression
    c. quotient-of-sums expression
    d. sum-of-quotients expression

16. What would be the fundamental product of the input word $ABCD = 1101$?
    a. $ABCD$ **c.** $AB\overline{C}D$
    b. $\overline{A}\overline{B}C\overline{D}$ **d.** $ABC\overline{D}$

17. How many cells would a three-variable Karnaugh map have?
    a. Four **c.** Eight
    b. Six **d.** Sixteen

18. What would be the Boolean expression for a four-variable logic circuit that NANDs inputs A and B and also NANDs inputs C and D and then NANDs the results?
    a. $\overline{\overline{AB} + \overline{CD}}$ **c.** $\overline{\overline{(AB)} \cdot \overline{(CD)}}$
    b. $\overline{(AB)} \cdot \overline{(CD)}$ **d.** $\overline{\overline{(A + B)} \cdot \overline{(C + D)}}$

19. If you applied DeMorgan's theorem and the double-inversion rule to the answer in question 18, you will obtain which of the following equations?
    a. $AB + CD$
    b. $A + B + C + D$
    c. $(A + B) \cdot (C + D)$
    d. $(A + B) + (CD)$

20. In reference to questions 18 and 19, you can see that the DeMorgan equivalent of a NAND-NAND circuit is a/an ______________.
    a. OR-AND **c.** AND-OR
    b. AND-NOR **d.** NOR-AND

Practice Problems

21. Determine the Boolean expressions for the logic circuits shown in Figure 19-34.

22. Which of the expressions in question 21 would be considered POS, and which SOP?

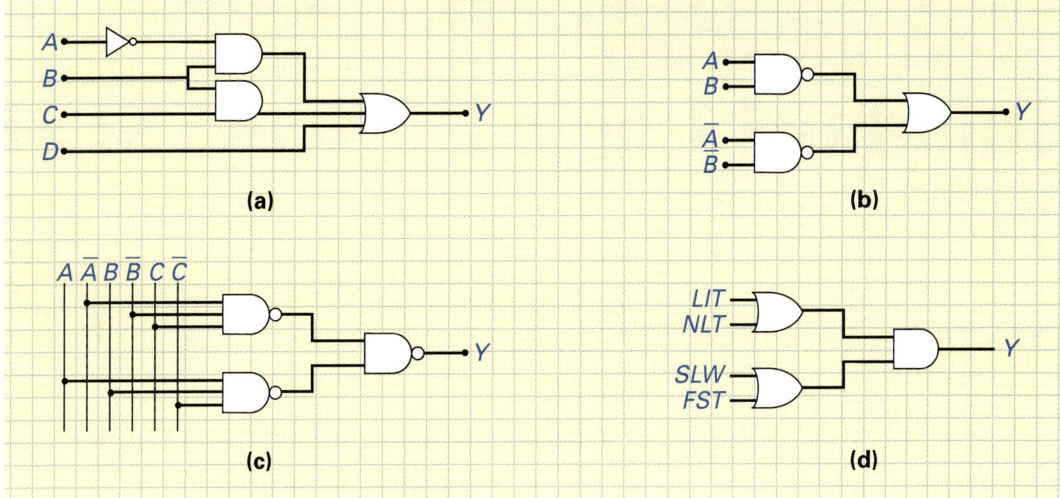

FIGURE 19-34 **From Logic Circuit to Boolean Equation.**

23. Sketch the equivalent logic circuit for the following Boolean equations:

a. $Y = (AB) + (\overline{AB}) + (\overline{A + B}) + (A \oplus B)$

b. $Y = (\overline{\overline{A} + B}) \oplus ABC$

24. Determine the Boolean expression for the Y_2 output in Table 19-1.

25. Use a K-map to simplify the Y_2 expression from question 24.

26. How much was the Y_2 logic circuit simplified in question 25?

27. Referring to Table 19-1, (a) determine the Boolean expression for the Y_1 output. (b) Simplify the expression using a K-map. (c) Describe how much the logic circuit was simplified.

28. Determine the Boolean expression for the Y_3 output in Table 19-1, simplify the expression using a K-map, and then describe how much the logic circuit was simplified.

29. Sketch the simplified logic circuits for the Y_1, Y_2, and Y_3 outputs listed in Table 19-1.

30. If the don't-care condition in the Y_3 output in Table 19-1 were 0s, what would the simplified expression be? Is the expression more or less simplified with don't-care conditions?

Web Site Questions

Go to the web site http://www.prenhall.com/cook, select the textbook *Electronics: A Complete Course,* select this chapter, and then follow the instructions when answering the multiple choice practice problems.

TABLE 19-1 **The Y_1, Y_2, and Y_3 Truth Table**

A	B	C	D	Y_1	Y_2	Y_3
0	0	0	0	1	1	1
0	0	0	1	1	1	0
0	0	1	0	1	0	1
0	0	1	1	1	0	0
0	1	0	0	0	0	1
0	1	0	1	0	0	1
0	1	1	0	0	1	1
0	1	1	1	0	1	1
1	0	0	0	0	1	0
1	0	0	1	1	1	0
1	0	1	0	0	0	0
1	0	1	1	1	0	0
1	1	0	0	1	0	1
1	1	0	1	1	0	x
1	1	1	0	1	0	x
1	1	1	1	1	0	x

Standard Logic Devices (SLDs)

A Noyce Invention

On June 3, 1990, Robert Noyce, who was best known as the inventor of the silicon integrated circuit, or IC, died of a heart attack at the age of 62.

Noyce was the son of a Congregational minister and became fascinated by computers and electronics while studying to earn his bachelor's degree at Grinnell College in Grinnell, Iowa. Luckily, he was the student of Grant Gale, who in 1948 was given one of the first transistors by one of its inventors, John Bardeen. Using this device in his lectures. Gale taught one of the first courses in solid-state physics to a class of 18 physics majors, one of whom was Noyce.

Wanting to pursue further interests in electronics, Noyce went on to obtain his doctorate in physics from the Massachusetts Institute of Technology in 1953. Later in that same year, Noyce took his first job as a research engineer at Philco Corp. in Philadelphia. He left in 1956 to join Shockley Semiconductor Laboratory in Mountain View, a company that was founded by another one of the inventors of the transistor, William Shockley.

A year later Noyce and seven other colleagues, whom Shockley called "the traitorous eight," resigned in what was to become a common pattern in Silicon Valley to join Fairchild Camera and Instruments Corp.'s semiconductor division. While at Fairchild, Noyce invented a process for interconnecting transistors on a single silicon chip, which was officially named an IC and was nicknamed "chip." This technological breakthrough, which brought the miniaturization of electronic circuits, was to be used in almost every electronic product, but probably the first to make use of the process was the digital electronic computer.

The honor for this discovery is actually shared by Jack Kilby, a staff scientist at Texas Instruments, who independently developed the same process and also holds patents for the invention.

In 1968 Noyce and More, another Fairchild scientist, founded Intel, a company which grew to become the nation's leading semiconductor company and a pioneer in the development of memory and microprocessor circuits.

In 1979 Noyce was awarded the National Medal of Science by President Jimmy Carter, and in 1987 he received the National Medal of Technology from President Ronald Reagan. He held more than a dozen patents and in 1983 was inducted into the National Inventors Hall of Fame.

While Noyce's name never became widely known, his invention, the IC, is used in every electronic product today, and he was an instrumental figure in creating the $50 billion semiconductor industry, which is at the heart of the $500 billion electronics industry. Sadly, he did not possess a feeling of success or real accomplishment, and this was made clear when he once described his career as "the result of a succession of dissatisfactions."

20

Introduction

In Chapter 18, we saw how diodes and transistors could be used to construct digital-logic gate circuits, which are the basic decision-making elements in all digital circuits. In reality, many more components are included in a typical logic gate circuit to obtain better input and output characteristics. In this chapter, we will see how the bipolar transistor and the MOS transistor are used to construct the two most frequently used digital IC logic families.

BIPOLAR LOGIC FAMILY	MOS LOGIC FAMILY
TTL (Transistor-Transistor Logic) Standard TTL Low-Power TTL High-Speed TTL Schottky TTL Low-Power Schottky TTL Advanced Low-Power Schottky TTL	PMOS (P-Channel MOSFET) NMOS (N-Channel MOSFET) CMOS (Complementary MOSFET) High-Speed CMOS High-Speed CMOS TTL Compatible Advanced CMOS Logic Advanced CMOS TTL Compatible
ECL (Emitter-Coupled Logic)	
IIL (Integrated-Injection Logic)	

A logic family is a group of digital circuits with nearly identical characteristics. Each of these two groups or families of digital ICs has its own characteristics and, therefore, advantages. For instance, as you will discover in this chapter, logic gates constructed using MOS transistors use less space due to their simpler construction, have a very high noise immunity, and consume less power than equivalent bipolar transistor logic gates; however, the high input impedance and input capacitance of the E-MOSFET transistor due to its insulated gate means that time constants are larger and, therefore, the transistor ON/OFF switching speeds are slower than equivalent bipolar gates. There are, therefore, trade-offs between the two logic families, and those are

a. Bipolar circuits are faster than MOS circuits, but their circuits are larger and consume more power.

b. MOS circuits are smaller and consume less power than bipolar circuits, but they are generally slower.

In this chapter, we will be examining the operation and characteristics of these bipolar and MOS circuits within digital ICs. This understanding is important if you are going to be able to determine the use of digital ICs within electronic circuits, interpret the characteristics of manufacturer data sheets, test digital ICs, and troubleshoot digital electronic circuits.

20-1 THE BIPOLAR FAMILY OF DIGITAL INTEGRATED CIRCUITS

In Chapter 18, diodes and bipolar transistors were used to construct each of the basic logic gate types. These simplified logic gate circuits were used to show how we could use the bipolar transistor to perform each logic gate function. In reality, many more components are included in a typical logic gate circuit to obtain better input and output characteristics. For example, Figure 20-1 shows a typical digital IC containing four logic gates. Referring to the inset in this figure, you can see that a logic gate is actually constructed using a number of bipolar transistors, diodes, and resistors. All of these components are formed and interconnected on one side of a silicon chip. This single piece of silicon actually contains four logic gate transistor circuits, with the inputs and output of each logic gate connected to the external pins of the IC package. For example, looking at the top view of the IC in Figure 20-1, you can see that one logic gate has its inputs on pins 1 and 2 and its output on pin 3. Since all four logic gate circuits will need a $+V_{CC}$ supply voltage (typically +5 V) and a ground (0 V), two pins are assigned for this purpose (pin 14 = $+V_{CC}$, pin 7 = ground). In the example in Figure 20-1, the IC contains four two-input NAND gates. This IC package type is called a **dual-in-line package (DIP)** because it contains two rows of connecting pins.

Dual-In-Line Package (DIP)
An electronic package having two rows of connecting pins.

Transistor-Transistor Logic (TTL)
A line of digital ICs that have logic gates containing several interconnected transistors.

In 1964 Texas Instruments introduced the first complete range of logic gate integrated circuits. They called this line of products **transistor-transistor logic (TTL) ICs** because the circuit used to construct the logic gates contained several interconnected transistors, as shown in the example in the inset in Figure 20-1. These ICs became the building blocks for all digital cir-

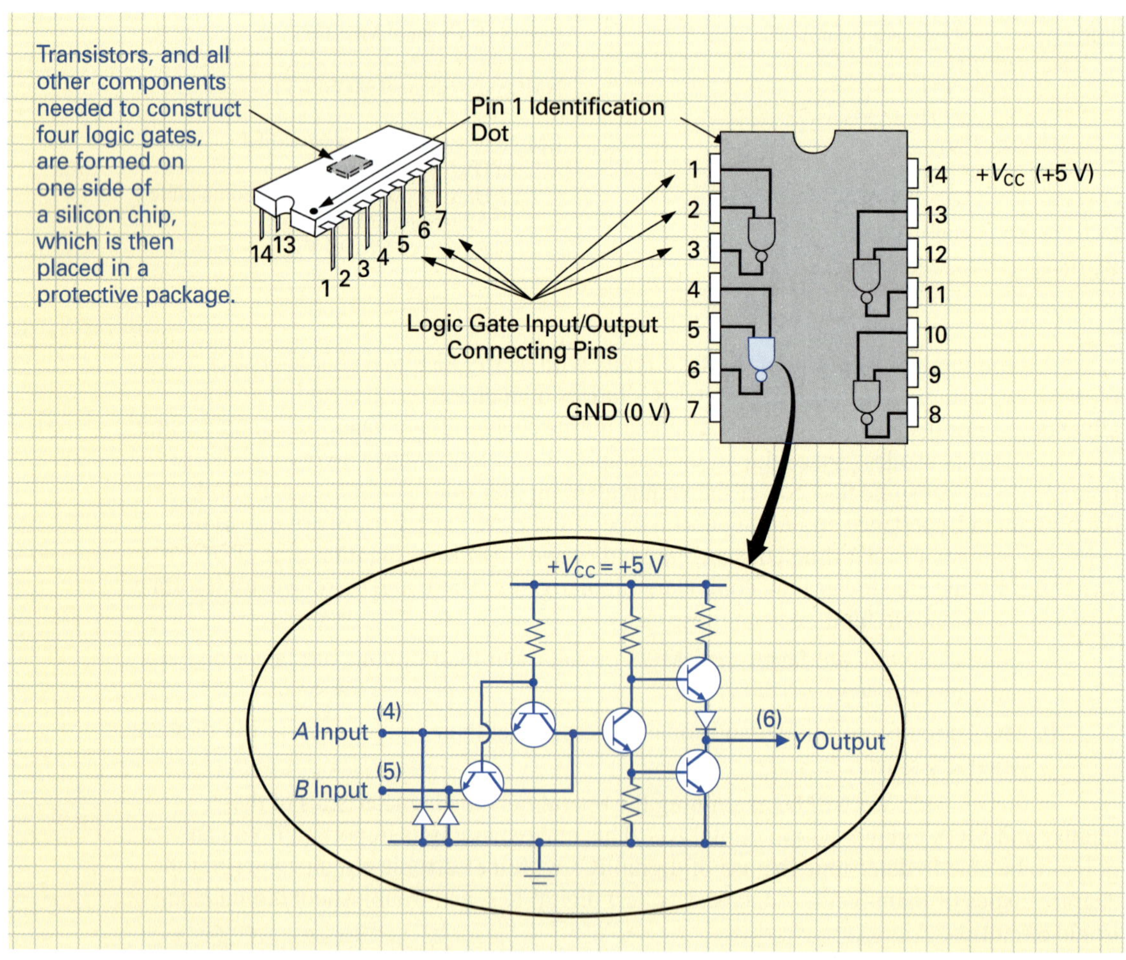

FIGURE 20-1 **Basic Logic Gate IC.**

TABLE 20-1 **A Few of the 7400 Standard TTL Series of Digital ICs**

DEVICE NUMBER	LOGIC GATE CONFIGURATION
7400	Quad (four) two-input NAND gates
7402	Quad two-input NOR gates
7404	Hex (six) NOT or INVERTER gates
7408	Quad two-input AND gates
7432	Quad two-input OR gates
7486	Quad two-input XOR gates
74135	Quad two-input XOR or XNOR gates

cuits. They were in fact, in the true sense of the word, building blocks—no knowledge of electronic circuit design was necessary because the electronic circuits had already been designed and fabricated on a chip. All that had to be done was connect all of the individual ICs in a combination that achieved the specific or the desired circuit operation. All of these TTL ICs were compatible, which meant they all responded and generated the same logic 1 and 0 voltage levels, and they all used the same value of supply voltage; therefore, the output of one TTL IC could be directly connected to the input of one or more other TTL ICs in any combination.

By modifying the design of the bipolar transistor circuit within the logic gate, TTL IC manufacturers can change the logic function performed. The circuit can also be modified to increase the number of inputs to the gate. The most common type of TTL circuits are the 7400 series, which were originally developed by Texas Instruments but are now available from almost every digital IC semiconductor manufacturer. As an example, Table 20-1 lists all of the 7400 series of TTL ICs discussed in Chapter 18.

The TTL IC example shown in Figure 20-1, therefore, would have a device part number of "7400" since it contains four (quad) two-input NAND gates.

20-1-1 *Standard TTL Logic Gate Circuits*

It is important that you understand the inner workings of a logic gate circuit, since you will be able to diagnose if a logic gate is working or not working only if you know how a logic gate is supposed to operate. In this section, we will examine the bipolar transistor circuits within two typical **standard TTL** logic gates.

Standard TTL
The original transistor-transistor logic circuit type.

Bipolar Transistor NOT Gate Circuit

Figure 20-2(a) shows the pin assignment for the six (hex) NOT gates within a 7404 TTL IC. All six NOT gates are supplied power by connecting the $+V_{CC}$ terminal (pin 14) to +5 V and the ground terminal (pin 7) to 0 V. Connecting power and ground to the 7404 will activate all six of the NOT gate circuits, so they are ready to function as inverters.

The inset in Figure 20-2(a) shows the circuit for a bipolar TTL NOT or logic INVERTER gate. As expected, the INVERTER has only one input, which is applied to the **coupling transistor** Q_1, and one output, which is developed by the **totem pole circuit** made up of Q_3 and Q_4. This totem pole circuit derives its name from the fact that Q_3 sits on top of Q_4, like the elements of a Native American totem pole. Transistor Q_2 acts as a **phase splitter** since its collector and emitter outputs will be out-of-phase with one another (base-to-emitter = 180° phase shift, base-to-collector = 0° phase shift), and it is these two opposite outputs that will drive the bases of Q_3 and Q_4.

Coupling Transistor
A transistor connected to couple or connect one part of a circuit to another.

Totem Pole Circuit
A circuit arrangement in which two devices are stacked so that the operation of one affects the operation of the other.

Phase Splitter
A circuit designed to generate two out-of-phase outputs.

Let us now see how this circuit will respond to a LOW input. A 0 V (binary 0) input to the NOT gate circuit will apply 0 V to the emitter of Q_1. This will forward-bias the PN base-emitter junction of Q_1 (base is connected to +5 V, emitter has an input of 0 V), causing electron flow from the 0 V input through Q_1's base emitter, through R_1 to + V_{CC}. Since the base of Q_1 is only 0.7 V above the emitter voltage, the collector diode of Q_1 will be reverse-biased,

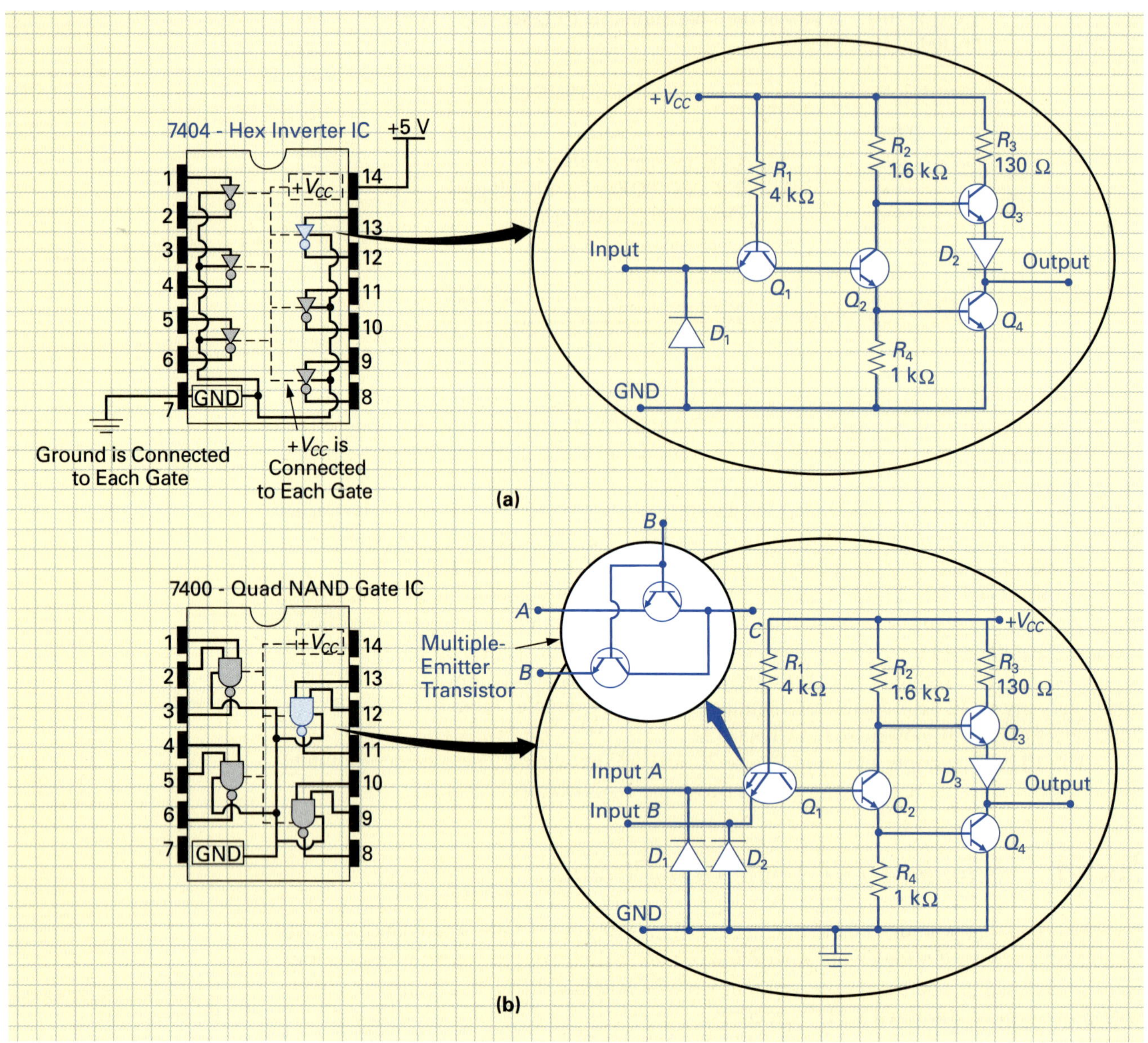

FIGURE 20-2 **Standard TTL Logic Gate Circuits. (a) NOT (Inverter) Gate. (b) NAND Gate.**

so the base of Q_2 will receive no input voltage and therefore turn OFF. With Q_2 cut OFF, all of the +5 V supply voltage will be applied via R_2 to the base of Q_3, sending Q_3 into saturation. With Q_3 saturated, its collector-to-emitter junction is equivalent to a closed switch between collector and emitter. As a result, the +5 V supply voltage (+V_{CC}) will be connected to the output via the low output resistance path of R_3, Q_3's collector-to-emitter, and D_2. A LOW input, therefore, will result in a HIGH output. In this condition, transistor Q_4 is cut OFF and equivalent to an open switch between collector and emitter because Q_2 is OFF and therefore the emitter of Q_2 and the base of Q_4 are at 0 V.

Let us now see how this circuit will respond to a logic 1 input. A HIGH input (+5 V) to the emitter of Q_1 will reverse-bias the PN base-emitter junction of Q_1 (base is connected to +5 V, emitter has an input of +5 V). The +5 V at the base of Q_1 will forward-bias Q_1's base-collector junction, applying +5 V to the base of Q_2, sending it into saturation. In the last condition, when Q_2 was cut OFF, all of the +V_{CC} supply voltage was applied to the base of Q_3. In this condition, Q_2 is heavily ON and therefore equivalent to a closed switch between collector and emitter, so the +V_{CC} supply voltage will be proportionally developed across R_2, Q_2's collector-to-emitter, and R_4. Transistor Q_4 will turn ON due to the positive voltage drop across R_4 and switch its emitter voltage of 0 V to the collector and, therefore, the output. A HIGH input will therefore result in a LOW output. In this condition, Q_3 is kept OFF by a sufficiently low voltage level on Q_2's collector and by diode D_2, which adds an additional diode to the emitter diode of Q_3, ensuring Q_3 stays OFF when Q_4 is ON.

The only other component that has not been discussed is the input diode D_1, which is included to prevent any negative input voltage spikes (negative transient) from damaging Q_1. D_1 will conduct these negative input voltages to ground.

Bipolar Transistor NAND Gate Circuit

Figure 20-2(b) shows the pin-out for the standard TTL 7400 IC and the bipolar transistor circuitry needed for each of the four NAND gates within this IC. The circuit is basically the same as the INVERTER or NOT gate circuit previously discussed, except for the **multiple-emitter input transistor,** Q_1. This transistor is used in all two-input logic gates, and if three inputs were needed, you would see a three-emitter input transistor. Q_1 can be thought of as having two transistors with separate emitters, but with the bases and collectors connected, as shown adjacent to the NAND circuit in Figure 20-2(b). Any LOW input to either *A* or *B* will cause the respective emitter diode to conduct current from the LOW input, through Q_1's base emitter and R_1, to $+V_{CC}$. The LOW voltage at the base of Q_1 will reverse-bias Q_1's collector diode, preventing any voltage from reaching Q_2's base, and so Q_2 will cut OFF. The remainder of the circuit will operate in exactly the same way as the NOT gate circuit. With Q_2 OFF, Q_3 will turn ON and connect $+V_{CC}$ through to the output. Any LOW input, therefore, will generate a HIGH output, which is in keeping with the NAND gate's truth table. Only when both inputs are HIGH will both of Q_1's emitter diodes be OFF and the HIGH at the base of Q_1 be applied via the forward-biased collector diode of Q_1 to the base of Q_2. With Q_2 ON, Q_4 will receive a positive base voltage and therefore switch ground, or a LOW voltage, through to the output.

Multiple-Emitter Input Transistor
A bipolar transistor having more than one emitter control terminal.

TTL Logic Gate Characteristics

Almost every digital IC semiconductor manufacturer has its own version of the standard TTL logic gate family or series of ICs. For instance, Fairchild has its 9300 series of TTL ICs, Signetics has the 8000 series, and so on. No matter what manufacturer is used, all of the TTL circuits are compatible in that they all have the same input and output characteristics. To examine the compatibility of these bipolar transistor logic gate circuits, we will have to connect two logic gates to see how they will operate when connected to form a circuit. Figure 20-3 shows a **TTL driver,** or logic source, connected to a **TTL load.** (In this example a NAND gate output is connected to the input of a NOT gate.)

TTL Driver
A transistor-transistor logic source device.

TTL Load
A transistor-transistor logic device that will encumber or burden the source.

Figure 20-4 lists all of the standard TTL logic gate characteristics. Let us now step through the items mentioned in this list of characteristics and explain their meaning.

a. For TTL circuits, the valid output logic levels from a gate and valid input logic levels to a gate are shown in the blocks in Figure 20-4. Looking at the block on the left you can see that a gate must generate an output voltage that is between the minimum and maximum points in order for that output to be recognized as a valid logical 1 or 0 (valid LOW output = 0 V to +0.4 V, valid HIGH output = +2.4 V to +5 V). Referring to the block on the right

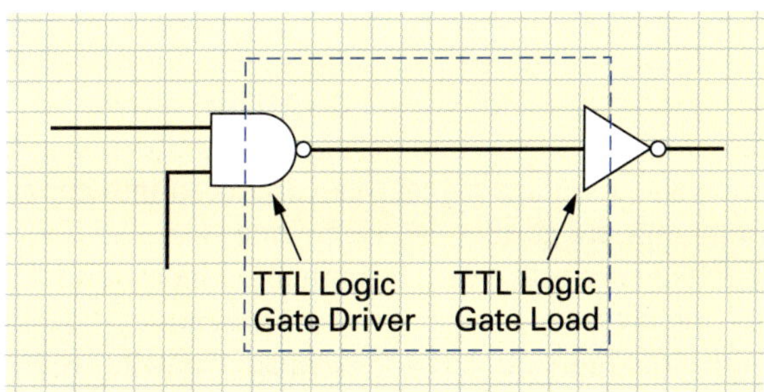

FIGURE 20-3 **Connecting Logic Gates.**

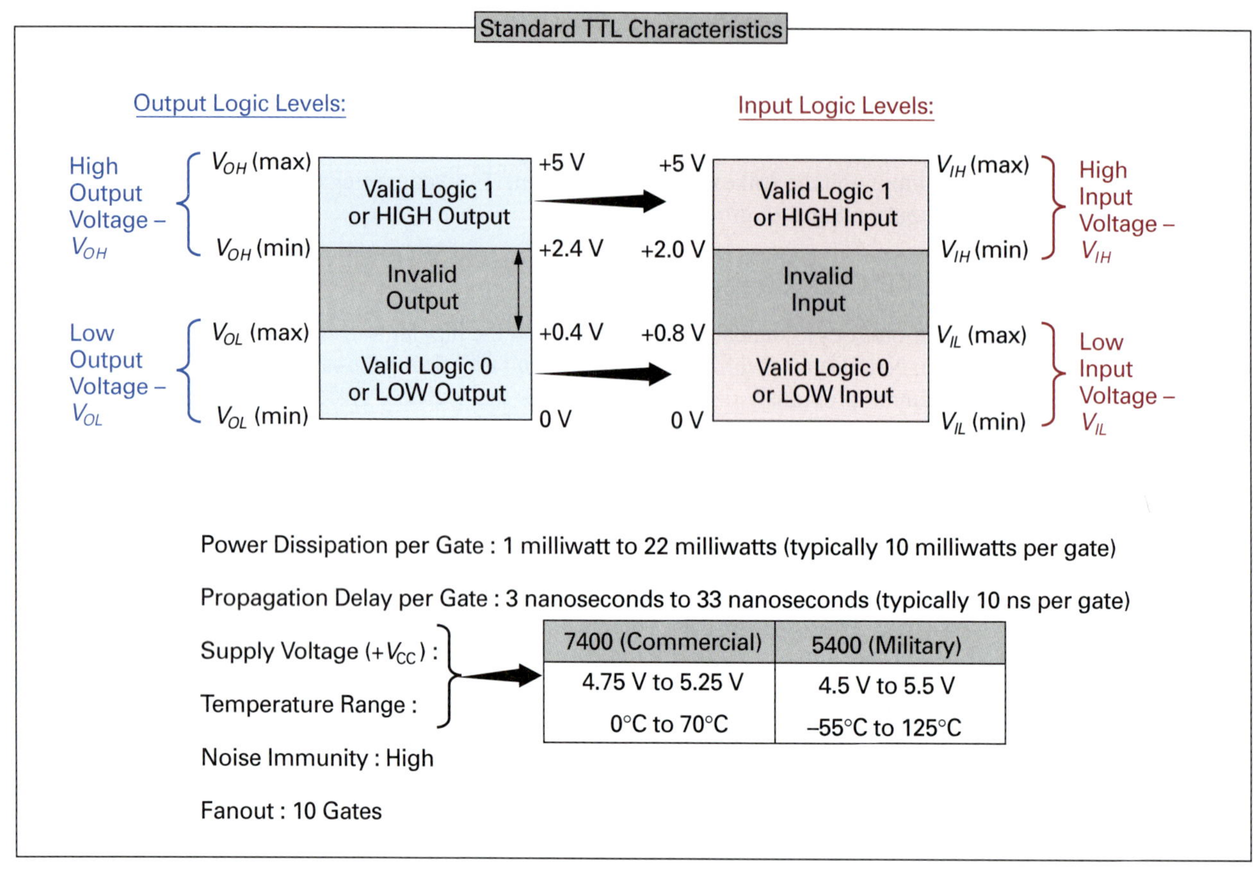

FIGURE 20-4 **Standard TTL Logic Gate Characteristics.**

of Figure 20-4, you can see that as long as the input voltages to a gate are between the minimum and maximum points, the gate will recognize the input as a valid logical 1 or 0 (valid LOW input = +0 V to +0.8 V, valid HIGH input = +2.0 V to +5 V).

Power Dissipation
The power dissipated by a device or circuit.

b. The second item on our list of TTL characteristics in Figure 20-4 is **power dissipation.** A standard TTL circuit will typically dissipate 10 mW per gate.

Propagation Delay Time
The time it takes for an output to change after a new input has been applied.

c. The third item on our list of TTL characteristics in Figure 20-4 is the **propagation delay time** of a gate, which is the time it takes for the output of a gate to change after the inputs have changed. This delay time is typically about 10 nanoseconds (ns) for a standard TTL gate. For example, it would normally take 10 ns for the output of an AND gate to go HIGH after all of its inputs go HIGH.

d. As you can see from the next two items mentioned in Figure 20-4, there are two standard TTL series of ICs available. The 7400 TTL series is used for all commercial applications, and it will operate reliably when the supply voltage is between 4.75 V and 5.25 V and when the ambient temperature is between 0°C and 70°C. The 5400 series is a more expensive line used for military applications because of its increased supply voltage tolerance (4.5 V to 5.5 V) and temperature tolerance (−55°C to 125°C). A 5400 series TTL IC will have the same pin configuration and perform almost exactly the same logic function as a 7400 series device. For example, a 5404 (hex INVERTER) has the same gate input and output characteristics, pin numbers, $+V_{CC}$, and ground pin numbers as a 7404.

Noise Immunity
The ability of a device to not be affected by noise.

Noise Margin
The range in which noise will not affect the operation of a circuit.

e. Noise is an unwanted voltage signal that is induced into conductors due to electromagnetic radiation from adjacent current-carrying conductors. All TTL logic circuits have a high **noise immunity,** which means they will be immune to most noise fluctuations at the gate's inputs. A standard TTL logic gate actually has a **noise margin** of 0.4 V between a TTL driver and TTL load, as is shown in Figure 20-4, if you compare the valid HIGH and LOW outputs to the HIGH and LOW inputs. This means that if a TTL driver were to output its worst-case LOW output (0.4 V) and induced noise increased this signal further away from 0.4 V to 0.8 V, the

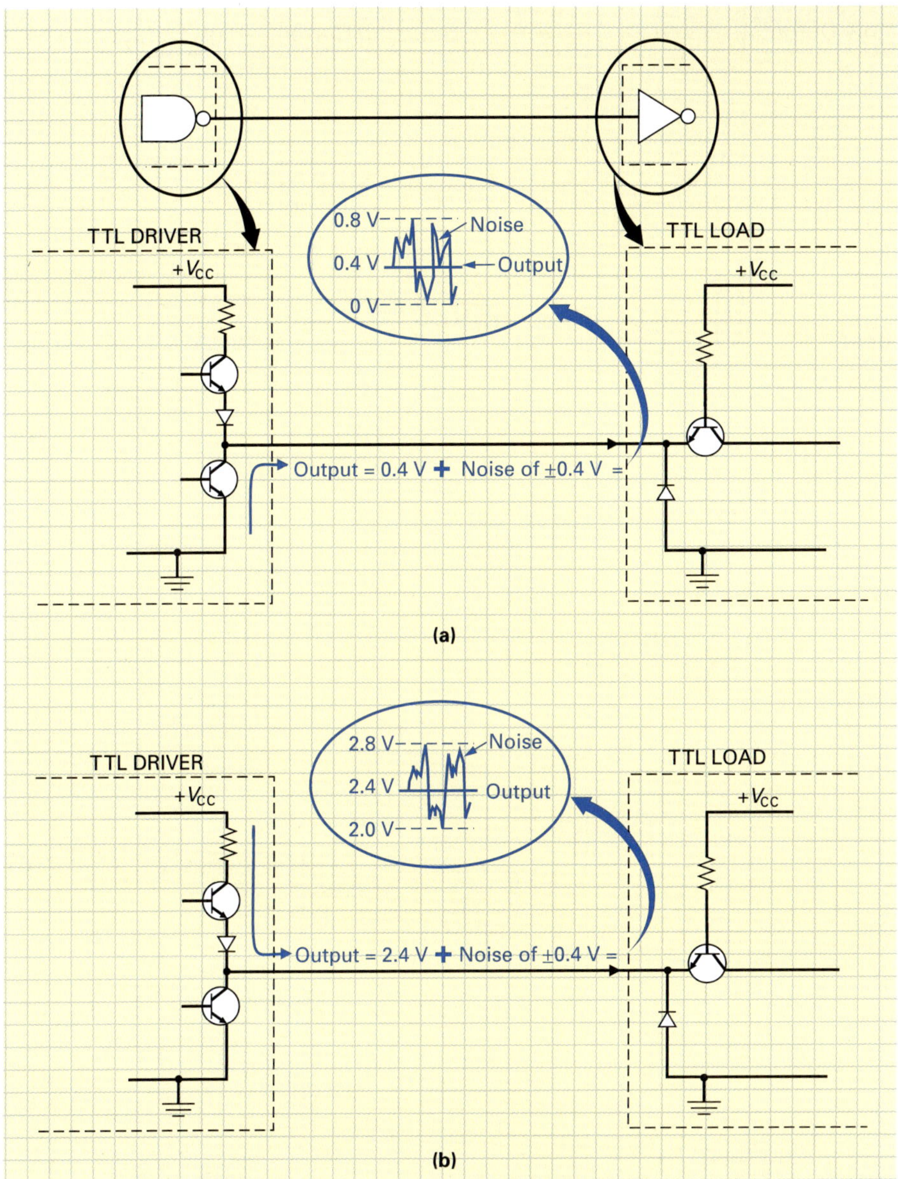

FIGURE 20-5 **TTL Noise Margin.**

TTL load would still accept this input as a valid LOW input since it is between 0 V and 0.8 V. This is shown in the simplified diagram in Figure 20-5(a). Similarly, if a TTL driver were to output its worst-case HIGH output (2.4 V) and induced noise decreased this signal further away from 2.4 V to 2.0 V, the TTL load would still accept this input as a valid logic 1 since it is between 2.0 V and 5 V. This is shown in the simplified diagram in Figure 20-5(b).

f. The last specification that has to be discussed in Figure 20-4 is the listing called **fanout.** To explain this term, Figure 20-6 shows how a TTL driver has to act as both a current **sink** and a current **source.** Looking at the direction of current in this illustration, you can see that this characteristic is described using conventional current flow (current travels from $+V \rightarrow -V$) instead of electron current flow (current travels from $-V \rightarrow +V$). Whether the current flow is from ground to $+V_{CC}$ (electron flow) or from $+V_{CC}$ to ground (conventional flow), the value of current is always the same if the same circuit exists between these two points. The only reason we have changed to conventional at this point is that most semiconductor manufacturers' specifications for the terms "sinking" and "sourcing" are detailed using conventional current flow.

Fanout
The number of parallel loads that can be driven simultaneously by one logic source, keeping logic levels within spec.

Sink
A power-consuming device or circuit.

Source
A circuit or device that supplies power.

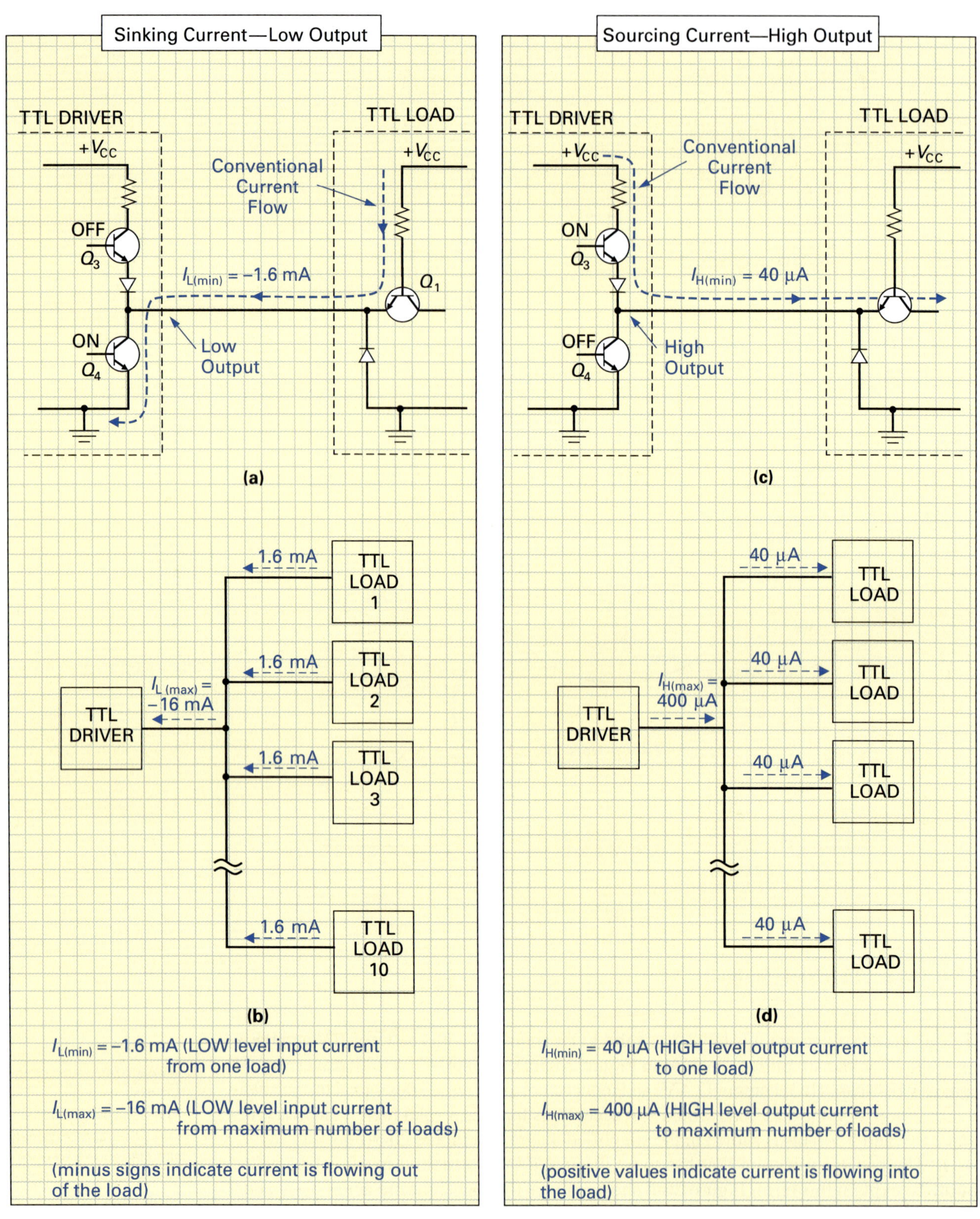

FIGURE 20-6 Sinking and Sourcing Current.

In Figure 20-6(a), a single logic gate load has been connected to a TTL logic gate driver, which is producing a LOW output (Q_4 in driver's totem pole is ON). In this instance, Q_4 will "sink" the single gate load current of 1.6 mA from Q_1 to ground. Manufacturer's data sheets state that Q_4 can sink a maximum of 16 mA and still produce a LOW output voltage of 0.4 V or less (the voltage drop across Q_4 will always be 0.4 V or less when it is sinking 16 mA). Since the maximum sinking current of 16 mA is ten times larger than the single gate load current of 1.6 mA, we can connect ten TTL loads to each TTL driver or source, as seen in Figure 20-6(b), and still stay "within spec" (within the parameters defined in the data sheet).

To examine both sides of the coin, Figure 20-6(c) shows a single TTL load connected to a TTL driver, which is producing a HIGH output (Q_3 in driver's totem pole is ON). In this instance, Q_3 will "source" the single gate load current of 40 μA. Manufacturers' data sheets state that Q_3 can source a maximum of 400 μA and still maintain a HIGH output voltage of 2.4 V or more. Since the maximum source current of 400 μA is ten times larger than the single gate load current of 40 μA, we can connect ten TTL loads to each TTL driver, as seen in Figure 20-6(d), and still stay within spec.

As we have seen in Figure 20-6, the maximum number of TTL loads that can be reliably driven by a single TTL driver is called the fanout. Referring to the standard TTL logic gate characteristics in Figure 20-4, you will now understand why a standard TTL gate has a fanout of 10.

g. Using Figure 20-6, we can describe another TTL circuit characteristic. When a TTL load receives a LOW input, as seen in Figure 20-6(a), a large current exists in the emitter of Q_1. On the other hand, when a TTL load receives a HIGH input, as seen in Figure 20-6(c), the emitter current in Q_1 is almost zero. This action explains why a TTL logic gate acts like it is receiving a HIGH input when its input lead is not connected. For example, if the input lead to a NOT gate was not connected to a circuit (an open input) and power was applied to the gate, the output of the NOT gate would be LOW. An unconnected input is called a **floating input.** In this condition, no emitter current will exist in Q_1 since the input is disconnected; so we can say *a floating TTL input is equivalent to a HIGH input.*

Floating Input
An input that is not connected.

20-1-2 *Low-Power and High-Speed TTL Logic Gates*

Low-Power TTL
A line of transistor-transistor logic ICs that have a low power-dissipation figure.

High-Speed TTL
A line of transistor-transistor logic ICs that have a high speed or low propagation-delay-speed figure.

A low-power TTL circuit is basically the same as the standard TTL circuit previously discussed except that all of the internal resistance values in the logic gate circuit are ten times larger. This means that the power dissipation of a low-power TTL circuit is one-tenth that of a standard TTL circuit (low-power TTL logic gates will typically have a power dissipation of about 1 mW per gate).

Increasing the internal resistances, however, will decrease the circuit's internal currents, and since the bipolar transistor is a current-operated device, the response or switching time of these logic gate types will decrease. (Low-power TTL logic gates will typically have a propagation delay time of 33 ns per gate.) With low-power TTL circuits, therefore, high speed is sacrificed for low power consumption. Low-power logic gate ICs can be identified because the device number will have the letter "L" following the series number 74 or 54; for example, 74L00, 74L01, 74L02, and so on.

By decreasing the resistances within a standard logic gate, manufacturers can lower the logic gate circuit's internal time constants and therefore decrease the gate's propagation delay time (typically 6 ns). The lower resistances, however, increase the gate's power dissipation (typically 22 mW), so the trade-off is therefore once again speed versus power. High-speed logic gate ICs can be identified because the device number will have the letter "H" following the series number 74 or 54; for example, 74H00, 74H01, 74H02, and so on.

20-1-3 *Schottky TTL Logic Gates*

Schottky TTL
A line of transistor-transistor logic ICs that use Schottky transistors.

With standard TTL and low-power TTL logic gates, transistors go into saturation when they are switched ON and are flooded with extra carriers. When a transistor is switched OFF, it

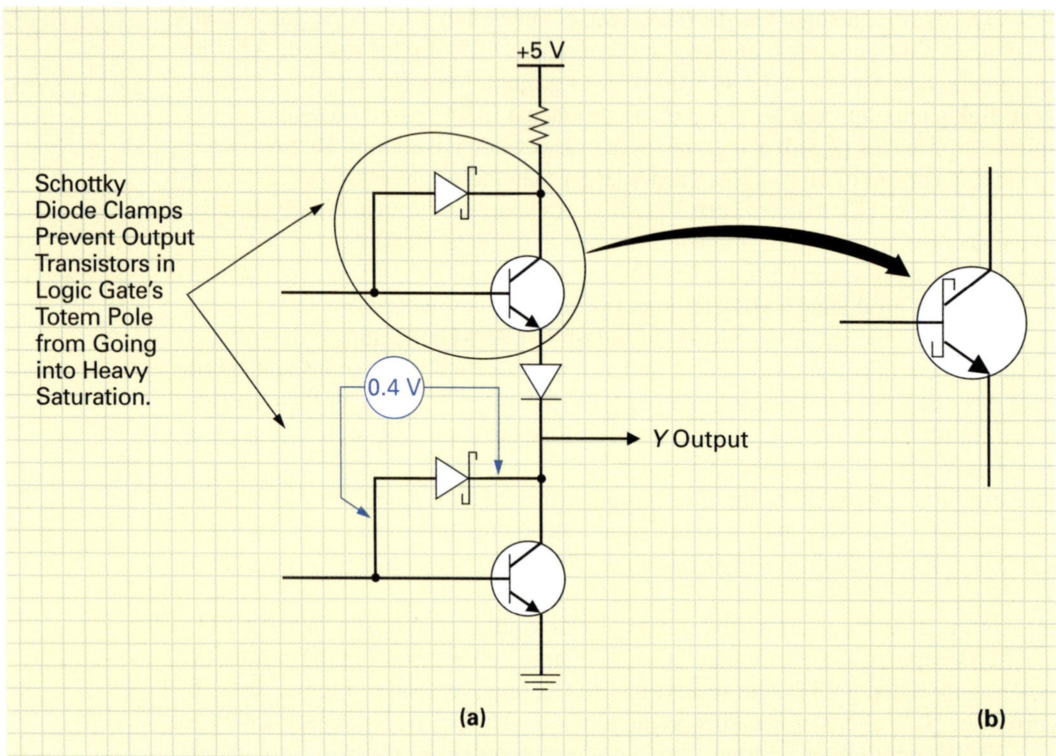

FIGURE 20-7 Schottky Clamped Bipolar Transistors. (a) Connecting a Schottky Diode across a Transistor Collector-Base Junction. (b) Schottky Transistor Symbol.

Saturation Delay Time
The additional time it takes a saturated transistor to turn off.

Schottky Transistor
A transistor having a Schottky diode connected between its collector and base.

takes a short time for these extra carriers to leave the transistor, and therefore it takes a short time for the transistor to cut OFF. This **saturation delay time** accounts for the slow propagation delay times, or switching times, of standard and low-power TTL logic gates.

Schottky TTL logic gates overcome this problem by fabricating a Schottky diode between the collector and base of each bipolar transistor, as seen in Figure 20-7(a). This transistor is called a **Schottky transistor** and uses the symbol shown in Figure 20-7(b). Let us now see how this Schottky diode will affect the operation of the transistor and decrease saturation delay time. When the transistor begins to turn ON, its collector voltage will fall. When the collector drops below 0.4 V of the base voltage, the Schottky diode will conduct and shunt current away from the collector junction. This action effectively clamps the collector to 0.4 V of the base voltage, keeping the collector diode of the transistor reverse-biased. Keeping the collector diode reverse-biased prevents the transistor from slipping into heavy saturation and therefore decreases the transistor's saturation delay time.

Schottky TTL devices are fast, with propagation delay times of typically 3 ns. This fast switching time means that the input signals can operate at extremely high frequencies of typically 100 MHz. Schottky TTL ICs are labeled with an "S" after the first two digits of the series number; for example, 74S04, 74S08, and so on.

By using Schottky transistors and increasing the internal resistances in a logic gate circuit, manufacturers compromised the speed of Schottky to obtain a lower dissipation rating. Schottky TTL devices will typically have a power dissipation of 19 mW per gate, whereas *low-power Schottky TTL devices* typically have a power dissipation of 2 mW per gate and propagation delay times of 9.5 ns. These low-power Schottky TTL devices include the letters "LS" in the device number; for example, 74LS32, 74LS123, and so on.

Special Schottky TTL devices are available that use a new integration process called oxide isolation, in which the transistor's collector diode is isolated by a thin oxide layer instead of a reverse-bias junction. These *advanced Schottky (AS) TTL devices* (labeled 74AS00) have a typical power dissipation of 8.5 mW per gate and a propagation delay time of 1.5 ns. By increasing internal resistances once again, manufacturers came up with an *advanced low-power*

DEVICE: SN7400, SN5400 (Quad Two-Input NAND)

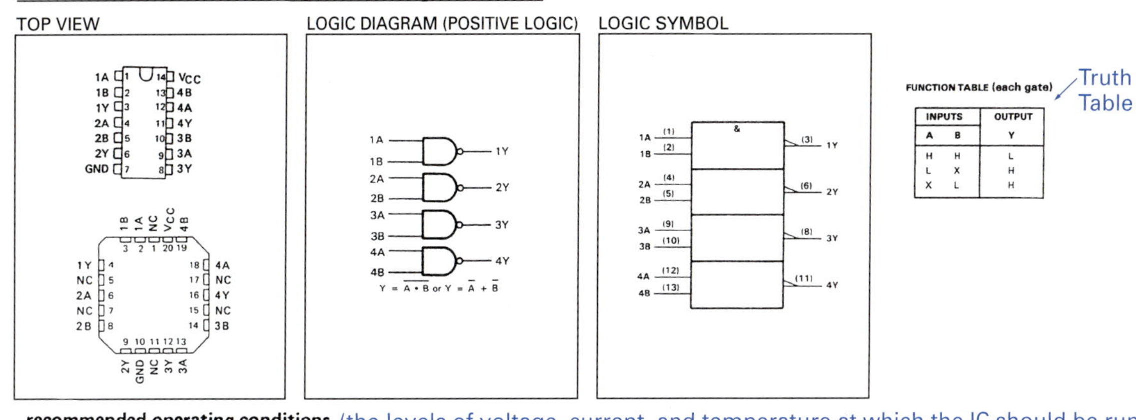

INPUTS		OUTPUT
A	B	Y
H	H	L
L	X	H
X	L	H

recommended operating conditions (the levels of voltage, current, and temperature at which the IC should be run under normal conditions)

	SN5400			SN7400			UNIT
	MIN	NOM	MAX	MIN	NOM	MAX	
V_{CC} Supply voltage (min. to max. supply voltage range)	4.5	5	5.5	4.75	5	5.25	V
V_{IH} High-level input voltage (the min. input voltage for a HIGH)	2			2			V
V_{IL} Low-level input voltage (the max. input voltage for a LOW)			0.8			0.8	V
I_{OH} High-level output current (the max. output current for a HIGH)			– 0.4			– 0.4	mA
I_{OL} Low-level output current (the max. output current for a LOW)			16			16	mA
T_A Operating free-air temperature (min. to max. operating temp.)	– 55		125	0		70	°C

DC Specifications (Electrical Characteristics)

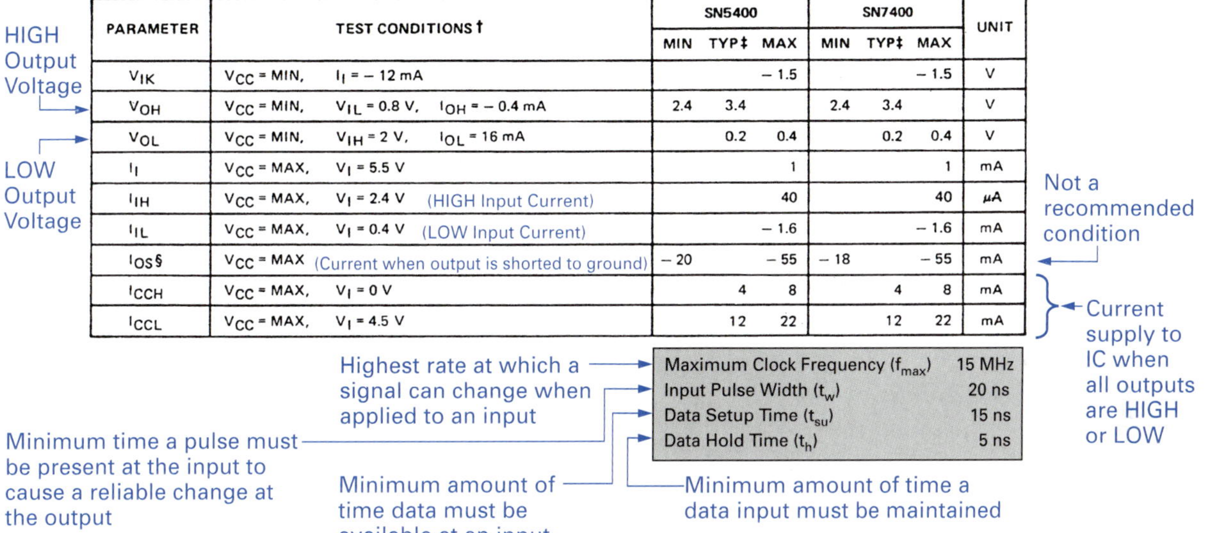

PARAMETER	TEST CONDITIONS†	SN5400			SN7400			UNIT
		MIN	TYP‡	MAX	MIN	TYP‡	MAX	
V_{IK}	V_{CC} = MIN, I_I = – 12 mA			– 1.5			– 1.5	V
V_{OH}	V_{CC} = MIN, V_{IL} = 0.8 V, I_{OH} = – 0.4 mA	2.4	3.4		2.4	3.4		V
V_{OL}	V_{CC} = MIN, V_{IH} = 2 V, I_{OL} = 16 mA		0.2	0.4		0.2	0.4	V
I_I	V_{CC} = MAX, V_I = 5.5 V			1			1	mA
I_{IH}	V_{CC} = MAX, V_I = 2.4 V (HIGH Input Current)			40			40	μA
I_{IL}	V_{CC} = MAX, V_I = 0.4 V (LOW Input Current)			– 1.6			– 1.6	mA
I_{OS}§	V_{CC} = MAX (Current when output is shorted to ground)	– 20		– 55	– 18		– 55	mA
I_{CCH}	V_{CC} = MAX, V_I = 0 V		4	8		4	8	mA
I_{CCL}	V_{CC} = MAX, V_I = 4.5 V		12	22		12	22	mA

Maximum Clock Frequency (f_{max})	15 MHz
Input Pulse Width (t_w)	20 ns
Data Setup Time (t_{su})	15 ns
Data Hold Time (t_h)	5 ns

FIGURE 20-8 A Standard TTL Data Sheet.

Schottky TTL line of products (labeled 74ALS00) which have power dissipation and propagation time figures of 1 mW and 4 ns. Some manufacturers, such as Fair-child, use the letter "F" for fast in their AS series of ICs; for example, 74F10, 74F14, and so on.

As you have seen, an IC contains quite a complex circuit, and this circuit has a detailed list of circuit characteristics. These characteristics or specifications are listed in a manufacturer's data sheet or spec sheet. As an example, Figure 20-8 shows a typical data sheet for a 7400 (quad two-input NAND) TTL IC.

Most data sheets can be broken down into five basic blocks:

a. pin-out and logic symbols

b. maximum ratings

c. recommended ratings

d. DC specifications or electrical characteristics

e. AC specifications or switching characteristics

Study all of the annotations included in this data sheet, because they explain the meaning of the terms used and the characteristics.

Open-Collector Output

A logic circuit in which the final output, which is a transistor's collector, is open or floating.

Pull-up Resistor

The name given to a resistor that is connected between a power line and a signal line. Its function is to pull up or make HIGH a signal line when it is not switched LOW.

20-1-4 *Open-Collector TTL Gates*

Some TTL devices have an **open-collector output** instead of the totem pole output circuit previously discussed. Figure 20-9(a) shows a standard TTL INVERTER circuit with an open-collector output, and Figure 20-9(b) shows the logic symbol used to represent an open-collector output gate.

In order to get a HIGH or LOW output from an open-collector gate, an external **pull-up resistor** (R_P) must be connected between $+V_{CC}$ and the collector of Q_3 (which is the output) as seen in Figure 20-9(c). In this circuit, two open-collector inverters are driving the same output, *Y*. If two standard TTL logic gates with totem pole outputs were wired together in this way and the *A* INVERTER produced a HIGH output and the *B* INVERTER a LOW

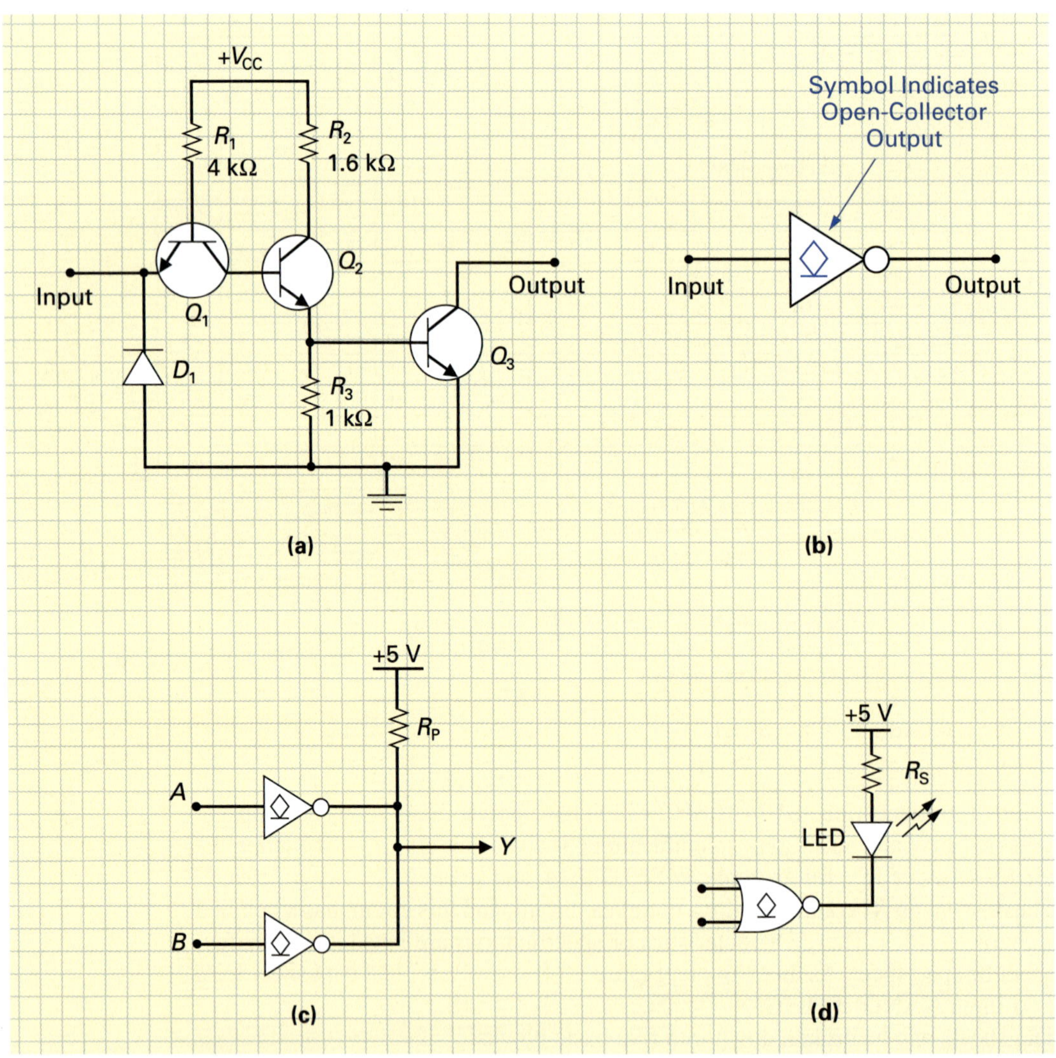

FIGURE 20-9 **Open Collector Gates. (a) Bipolar Transistor Open Collector INVERTER Circuit. (b) Open Collector Symbol. (c) Wired INVERTER Open Collector Gates. (d) Open Collector NOR Driver Gate.**

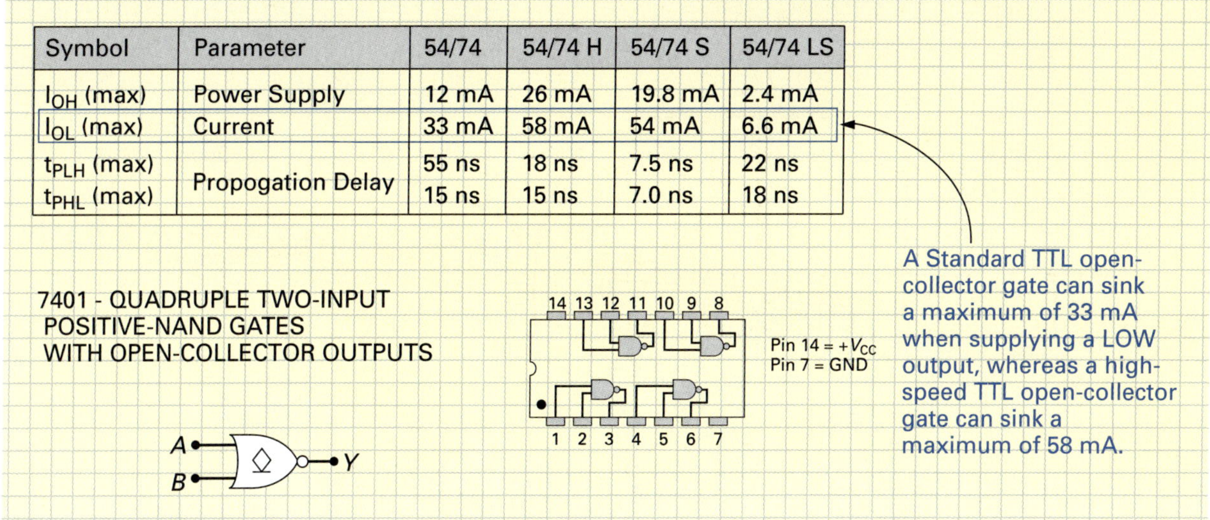

Symbol	Parameter	54/74	54/74 H	54/74 S	54/74 LS
I_{OH} (max)	Power Supply	12 mA	26 mA	19.8 mA	2.4 mA
I_{OL} (max)	Current	33 mA	58 mA	54 mA	6.6 mA
t_{PLH} (max)	Propogation Delay	55 ns	18 ns	7.5 ns	22 ns
t_{PHL} (max)		15 ns	15 ns	7.0 ns	18 ns

FIGURE 20-10 An Open-Collector Output Data Sheet.

output, there would be a direct short between the two, causing both gates to burn out. With open-collector gates, the outputs can be wired together or connected to a common output line without causing any HIGH-LOW output conflicts between gates. With the circuit in Figure 20-9(c), only when both inputs are LOW will the output transistors of both inverters be floating and the output be pulled up by R_P to +5 V ($A = 0$, $B = 0$, $Y = 1$). If either A input or B input is HIGH, one of the output transistors will turn ON and pull down or ground the output Y ($A = 1$ or $B = 1$, $Y = 0$). You may recognize this logic operation as that of a NOR gate.

Open-collector gates can sink around 40 mA, which is a vast improvement over standard TTL gates that only sink 16 mA (74ASXX gates can sink 20 mA).

In Figure 20-9(d), you can see how the high sinking current ability of an open-collector NOR gate can be used to drive the high current demand of a light emitting diode (LED) load. The current limiting resistor (R_S) is being used to limit the current through the LED to about 25 mA, which produces a good level of brightness from an LED. In many applications, open-collector gates are used to drive external loads that require increased current, such as LED displays, lamps, relays, and so on. As an example, Figure 20-10 shows typical open-collector current characteristics.

20-1-5 *Three-State (Tri-State) Output TTL Gates*

The **three-state output TTL gate** has, as its name implies, three output states or conditions. These three output states are LOW, HIGH, and FLOATING, or a "high-impedance output" state. Before we discuss why we need this third output state, let us see how this circuit operates.

Three-State Output TTL Gate
A line of transistor-transistor logic gates that can have three possible outputs: LOW, HIGH, or OPEN.

Figure 20-11(a) illustrates the internal bipolar transistor circuit for a three-state NOT gate. The circuit is basically the same as the previously discussed INVERTER gate, except for the additional control input applied to transistor Q_4. When the enable input is LOW, Q_4 is OFF and has no effect on the normal inverting operation of the circuit. However, when the enable input is HIGH, Q_4 is turned ON, and the ground on its emitter will be switched through to the second emitter of Q_1. The ON Q_1 will switch ground from its emitter through to the base of Q_2, turning Q_2 OFF and therefore Q_5 OFF. Transistor Q_4 will also ground the base of Q_3 via D_1, and so Q_3 will turn OFF. The result will be that both totem pole transistors are cut OFF, as seen in the equivalent circuit shown in the inset in Figure 20-11(a). This high-impedance output state, therefore, is created by turning OFF both totem pole output transistors so that the output is isolated from ground and the supply voltage. With the output of the gate completely

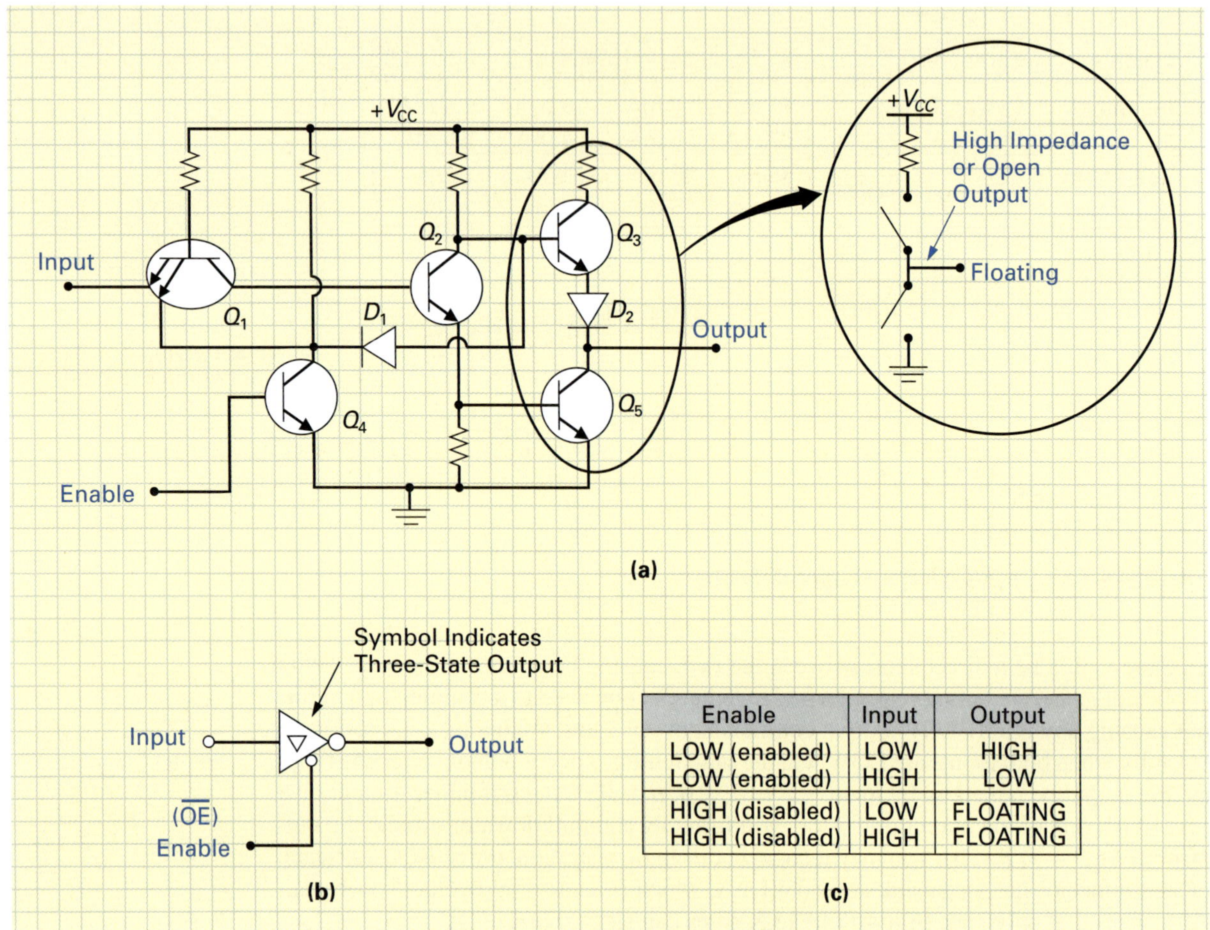

Enable	Input	Output
LOW (enabled)	LOW	HIGH
LOW (enabled)	HIGH	LOW
HIGH (disabled)	LOW	FLOATING
HIGH (disabled)	HIGH	FLOATING

FIGURE 20-11 **Three-State Output Gates. (a) Bipolar Transistor Circuits. (b) Symbol. (c) Truth Table.**

disconnected from the circuit, this lead is said to be a disconnected, or a FLOATING high-impedance line.

Active-LOW Control Line
A control line that goes LOW when activated.

Figure 20-11(b) shows the schematic symbol for this three-state output NOT gate. The bubble attached to the enable input is used to indicate an **active-LOW control line,** which means that if you want to "activate the gate," you must make this control line "LOW." To disable the gate, therefore, we would not make this line active, so the enable line would be HIGH. Studying the truth table or function table in Figure 20-11(c), you can see this gate's operation summarized. When the enable control line is LOW, the circuit is enabled for normal INVERTER operation. When the enable control line is HIGH, however, the output is always FLOATING, no matter what input logic level is present. The label for active-LOW control lines will usually have a bar over the label's letters, as shown with the output enable ($\overline{OE}$) in Figure 20-11(b). This bar over the label is used to distinguish an active-LOW control line label from an active-HIGH control line label.

Like the previously discussed open-collector gates, these devices are used in applications where two or more gate outputs are connected to a single output line. These common signal lines are called "signal buses," and they allow logic devices to individually take control of a common set of signal lines without causing signal interferences with other devices on the bus. To avoid HIGH/ LOW output line conflicts between gates, only one gate is enabled at one time to drive the output line, with all other gates being disabled and therefore their outputs FLOATING. These applications will be discussed in more detail in future chapters.

20-1-6 *Buffer/Driver TTL Gates*

A **buffer** is a device that isolates one circuit from another. In order to achieve this isolation, a buffer should have a high input impedance (low input current) and a low output impedance (high output current). A standard TTL gate provides a certain amount of isolation or buffering, since the output current is ten times that of the input current.

Buffer
A device that isolates one device from another. Its schematic symbol is a triangle with one input and one output.

	Input	Output	
Standard TTL	$I_{IL} = -1.6$ mA	$I_{OL} = -16$ mA	(output is ten times larger)
	$I_{IH} = 40$ μA	$I_{OH} = 400$ μA	(output is ten times larger)

A buffer or driver logic gate is a basic gate that has been slightly modified to increase the output current. For example, a 7428 is a "quad two-input NOR gate buffer" whose output current is thirty times greater than the input currents, enabling it to drive heavier loads.

	Input	Output	
Buffer/Driver	$I_{IL} = -1.6$ mA	$I_{OL} = -48$ mA	(output is thirty times larger)
TTL	$I_{IH} = 40$ μA	$I_{OH} = 1.2$ mA	(output is thirty times larger)

20-1-7 *Schmitt-Trigger TTL Gates*

The **Schmitt-trigger circuit,** named after its inventor, is a two-state device that is used for pulse shaping. Its schematic is shown in Figure 20-12.

Several Schmitt-trigger TTL ICs are available, including 7413 (dual four-input NAND Schmitt-trigger), 7414 (hex Schmitt-trigger inverters), 74132 (quad two-input NAND Schmitt-trigger), and so on. These logic circuits will have a Schmitt-trigger circuit connected to the output of each logic gate circuit so that the gate will perform its normal function and then sharpen the rise and fall times of pulses that have been corrupted by noise and attenuation.

Schmitt Trigger
A level-sensitive input circuit that has a two-state output.

20-1-8 *Emitter-Coupled Logic (ECL) Gate Circuits*

Schottky TTL circuits improved the switching speed of digital logic circuits by preventing the bipolar transistor from saturating. **Emitter-coupled logic,** or **ECL,** is another non-saturating bipolar transistor logic circuit arrangement; however, ECL is the fastest of all logic circuits. It is used in large digital computer systems that require fast propagation delay times (typically 0.8 ns per gate) and are able to sacrifice power dissipation (typically 40 mW per gate).

Figure 20-13(a) shows a typical ECL gate circuit. The two inputs to this circuit, *A* and *B,* are applied to the transistors Q_1 and Q_2, which are connected in parallel. If additional inputs are needed, more input transistors can be connected in parallel across Q_1 and Q_2, as shown in the dashed lines in Figure 20-13(a). The two collector outputs of Q_2 and Q_3 drive the two low output impedance emitter-followers Q_5 and Q_6. These emitter-follower outputs provide an OR output from Q_6 and its complement, or opposite, which is a NOR output from Q_5. This dual output is one of the advantages of ECL gates.

Emitter-Coupled Logic (ECL)
A bipolar digital logic circuit family that uses emitter input coupling.

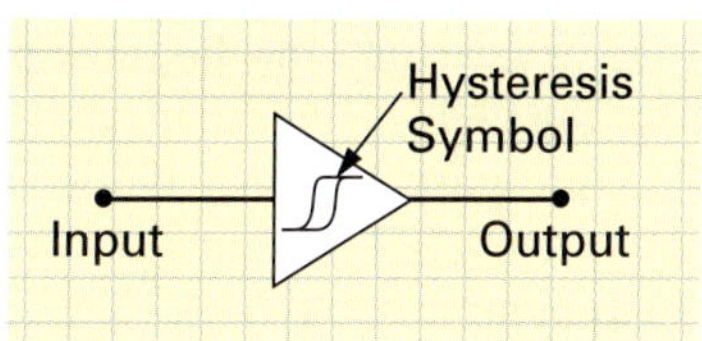

FIGURE 20-12 **Figure 20-12**

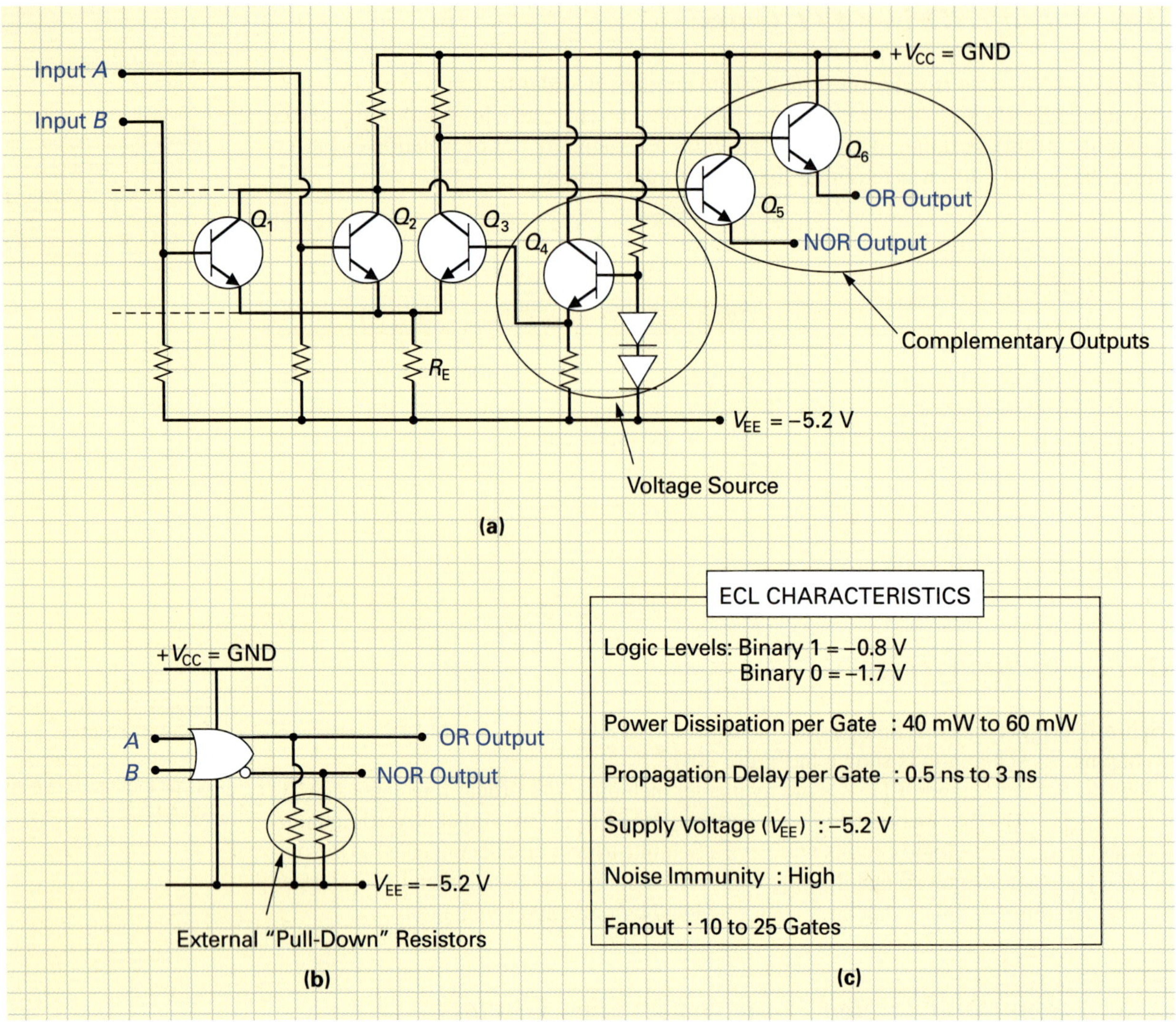

FIGURE 20-13 **Emitter-Coupled Logic (ECL) Gates. (a) Bipolar Transistor OR/NOR Circuit. (b) Logic Symbol. (c) Characteristics.**

Pull-Down Resistor

A name given to a resistor that is connected between a signal line and ground. Its function is to pull down or make LOW a signal line when it is not switched HIGH.

External **pull-down resistors** are required, as seen in Figure 20-13(b), which also shows the logic symbol for this OR/NOR ECL gate. The open outputs of ECL gates mean that two or more gate outputs can drive the same line without causing any HIGH/LOW voltage conflicts. The V_{EE} pin of an ECL IC is typically connected to −5.2 V, and the V_{CC} pin is connected to ground. Although these dc supply values seem to not properly bias the circuit, remember that the V_{CC} power line is still positive (since it is at ground) relative to the V_{EE} power line (which is at −5.2 V). The transistor Q_4 has a constant base bias voltage and an emitter resistor and will produce a temperature-stabilized constant voltage, or reference voltage, of approximately −1.3 V to the base of Q_3. The bipolar transistors in the circuit simply conduct more or conduct less depending on whether they represent a LOW or HIGH, ensuring that none of the transistors saturate.

Preventing transistors from going into saturation accounts for the fast speed of ECL logic gate circuits, and the fact that transistors, for the most part, are conducting and therefore drawing a current accounts for the ECL's increase in power dissipation. The typical characteristics of ECL gates are listed in Figure 20-13(c). Notice that the binary 0 and binary 1 voltage levels are slightly different from TTL circuits; however, the binary 1 voltage level (−0.8 V) is still positive relative to binary 0 (−1.7 V).

Let us now examine this ECL circuit's operation. When either one or both of the inputs go to a HIGH (−0.8 V), that transistor will conduct more (since its base is now positive with respect to the emitter, which is at −1.3 V). In so doing, the input transistor will make transistor Q_3 turn more OFF (since the base-emitter junction of the input transmitter will make Q_3's emitter positive with respect to its large negative base voltage). The conducting input transistor will switch its negative emitter voltage through to the collector and therefore to the base of Q_5. Q_5 will turn more OFF, and its output will be pulled LOW. For a NOR gate output, any binary 1 input will produce a binary 0 output.

On the other hand, with Q_3 more OFF, its collector will be more positive (closer to 0 V) and Q_6 will turn more ON and connect the more positive V_{CC} to the OR output. For an OR gate output, any binary 1 input will produce a binary 1 output. When all inputs are LOW (−1.7 V), all of the input transistors will be more OFF, producing a HIGH output to Q_5 (all inputs are 0, output from NOR is 1). With Q_1 and Q_2 more OFF, Q_3 will conduct more, producing a LOW output to Q_6 (all inputs are 0, output of OR is 0).

The high speed, high power consumption, and high-cost logic of these circuits account for why they are only used when absolutely necessary. The most widely used ECL family of ICs is Motorola's ECL (MECL) 10K and 100K series, in which devices are numbered MC10XXX or MC100XXX.

20-1-9 *Integrated-Injection Logic (I^2L) Gate Circuits*

Integrated-injection logic (IIL) or, as they are sometimes abbreviated, I^2L (pronounced "I squared L"), is a series of digital bipolar transistor circuits that have good speed, low power consumption, and very good circuit **packing density.** In fact, one I^2L gate is one-tenth the size of a standard TTL gate, making it ideal for compact or high-density circuit applications.

A typical I^2L logic NOT gate is shown in Figure 20-14(a). Transistor Q_1 acts as a current source and the multiple-collector transistor Q_2 acts as the INVERTER with two outputs. Studying the circuit, you can see that the base of Q_1 is connected to the emitter of Q_2, and the collector of Q_1 is connected to the base of Q_2. These common connections enable the entire I^2L gate to be constructed as one transistor with two emitters and two collectors and use the same space on a silicon chip as one standard TTL multiple-emitter transistor.

The circuit in Figure 20-14(a) operates in the following way. The emitter of Q_1 is connected to an external supply voltage ($+V_S$), which can be anywhere from 1 V to 15 V, depending on the injector current required. The constant base-emitter bias of Q_1 means that a constant current is available at the collector of Q_1 (called the **current injector transistor**), and this current will either be applied to the base of Q_2 or out of the input, depending on the input voltage logic level. To be specific, a LOW input will pull the injector current out of the input and away from the base of Q_2. This will turn Q_2 OFF, and its outputs will be open, which is equivalent to a HIGH output. On the other hand, a HIGH input will cause the injector current of Q_1 to drive the base of Q_2. This will turn Q_2 ON and switch the LOW on its emitter through to the collector outputs.

Figure 20-14(b) illustrates an I^2L NAND gate. As you can see, the only difference between this circuit and the INVERTER gate circuit in Figure 20-14(a) is the two inputs *A* and *B*. Transistors Q_3 and Q_4 are shown in this circuit; however, they are the multiple-collector transistors of the gates, driving inputs *A* and *B* of this NAND gate circuit. If either one of these transistors is ON (a low input to the NAND), the injector current of Q_1 will be pulled away from Q_2's base, turning OFF Q_2 and causing a HIGH output. Only when both inputs are HIGH (Q_3 and Q_4 are OFF) will the injector current from Q_1 turn ON Q_2 and produce a LOW output.

Figure 20-14(c) lists the characteristics of I^2L logic gate circuits. These circuits are used almost exclusively in digital watches, cameras, and almost all compact, battery-powered circuit applications. As mentioned earlier, its good speed, low power consumption, and small size make it ideal for any portable system. Also, the simplicity of I^2L circuits means that it is easy to combine analog or linear circuits (such as amplifiers and oscillators) with digital circuits to create complete electronic systems on one chip. The circuit complexity of standard TTL digital circuits means that it is normally too difficult and not cost-effective (due to fabrication difficulties and therefore high cost) to place digital and linear circuits on the same IC chip.

Integrated-Injection Logic (IIL)
A bipolar digital logic family that uses a current injector transistor.

Packing Density
A term used to describe the number of devices that can be packed into a given area.

Current Injector Transistor
A transistor connected to generate a constant current.

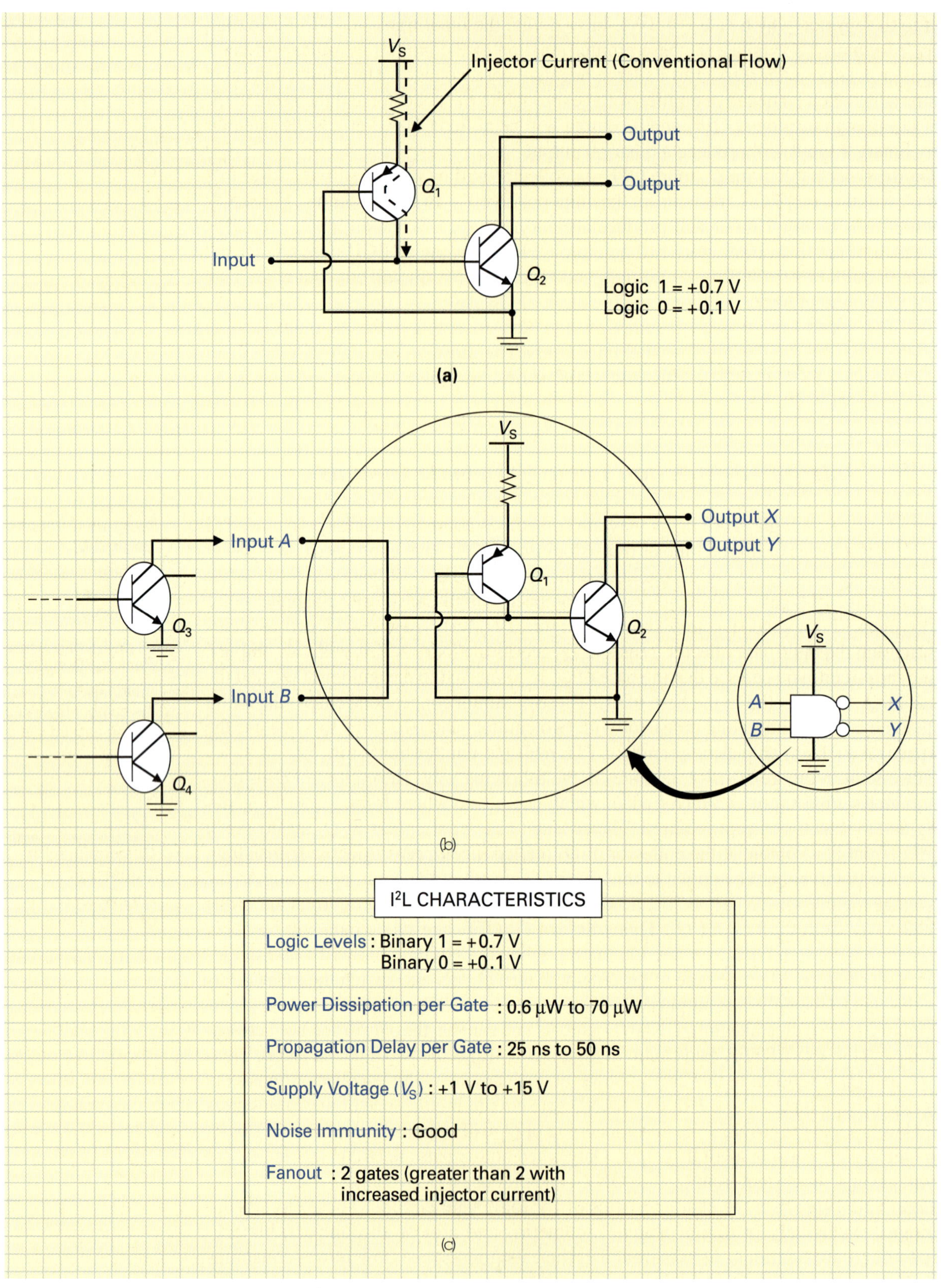

FIGURE 20-14 **Integrated-Injection Logic (IIL or I²L) Gates. (a) Bipolar Transistor INVERTER (NOT) Gate Circuit. (b) Bipolar Transistor NAND Gate Circuit. (c) Characteristics.**

SELF-TEST EVALUATION POINT FOR SECTION 20-1

Use the following questions to test your understanding of Section 20-1.

1. TTL is an abbreviation for ___________.
2. What is the fanout of a standard TTL gate?
3. A floating TTL input is equivalent to a ___________ (HIGH/LOW) input.
4. The three output states of a tri-state logic gate are ___________, ___________, and ___________.
5. In what application would ECL circuits be used?
6. What is the advantage of I^2L logic circuits?

20-2 THE MOS FAMILY OF DIGITAL INTEGRATED CIRCUITS

In this section we will examine the three basic types of MOS ICs, which are called PMOS (pronounced "pea-moss"), NMOS (pronounced "en-moss"), and CMOS (pronounced "sea-moss").

The E-MOSFET is ideally suited for digital or two-state circuit applications. One reason is that it naturally operates as a "normally-OFF voltage controlled switch," since it can be turned ON when the gate voltage is positive and turned OFF when the gate voltage falls below a threshold level. This threshold level is a highly desirable characteristic because it prevents noise from false triggering, or accidentally turning ON, the device. The other E-MOSFET advantage is its extremely high input impedance, which means that the device's circuit current, and therefore power dissipation, is low. This enables us to densely pack or integrate many thousands of E-MOSFETs onto one small piece of silicon, forming a high-component-density IC. These low-power and high-density advantages make the E-MOSFET ideal in battery-powered, small-sized (portable) applications such as calculators, wristwatches, notebook computers, hand-held video games, digital cellular phones, and so on.

In the following sections we will see how the E-MOSFET can be used to construct digital logic gate circuits and then compare these circuits and their characteristics to the previously discussed bipolar logic gate circuit characteristics.

20-2-1 *PMOS (P-Channel MOS) Logic Circuits*

The name **PMOS logic** is used to describe this logic circuitry because the logic gates are constructed using P-channel E-MOSFETs. Figure 20-15(a) shows how three P-channel E-MOSFETs can be used to construct a NOR logic gate. Figure 20-15(b) shows the logic levels used in PMOS circuits, and Figure 20-15(c) reviews the operation of a P-channel E-MOSFET.

PMOS Logic
A digital logic family of ICs that use P-channel MOSFETs.

The PMOS NOR gate circuit in Figure 20-15(a) will operate as follows. Transistor Q_3 can be thought of as a "current limiting resistor" between source and drain because its gate is constantly connected to $-V_{GG}$, and therefore Q_3 is always ON. Any HIGH input will turn OFF its respective E-MOSFET, resulting in a negative or LOW (-8 V) output due to the pull-down action of Q_3. Only when both inputs are LOW, will both Q_1 and Q_2 be ON, allowing $+V_{DD}$ (0 V, which is a HIGH) to be connected through Q_1 and Q_2 to the output.

20-2-2 *NMOS (N-Channel MOS) Logic Circuits*

The name **NMOS logic** is used to describe this logic circuitry because the logic gates are constructed using N-channel E-MOSFETs. Figure 20-16(a) shows how three N-channel E-MOSFETs can be used to construct a NAND gate. Figure 20-16(b) shows the logic levels used in NMOS circuits, and Figure 20-16(c) reviews the operation of an N-channel E-MOSFET.

NMOS Logic
A digital logic family of ICs that use N-channel MOSFETs.

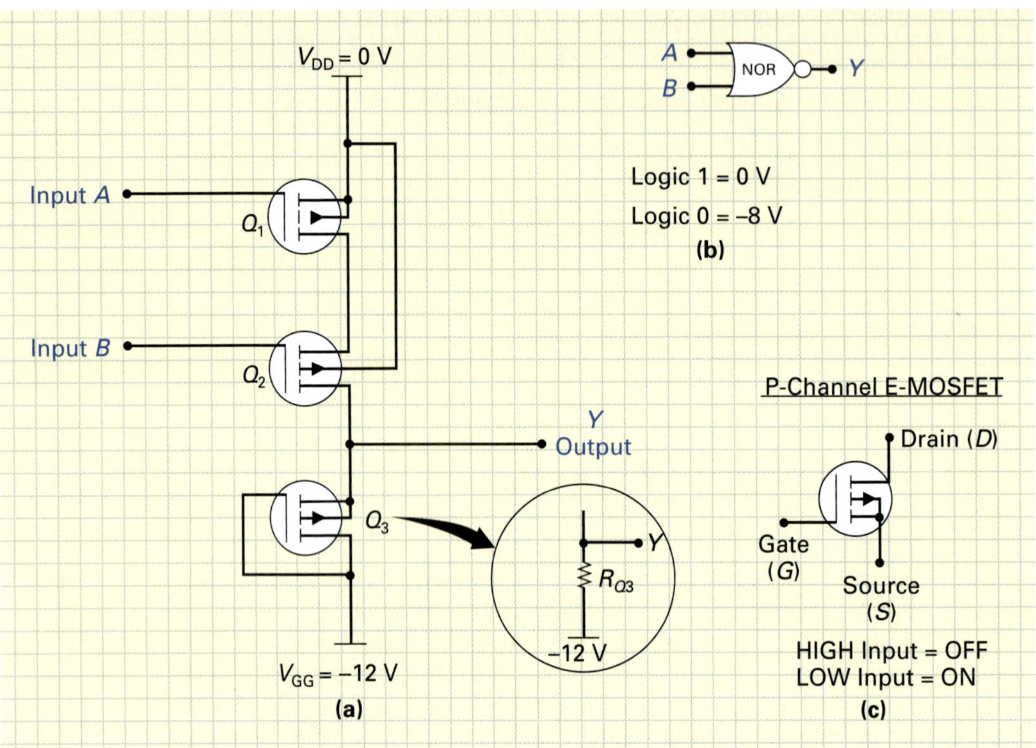

FIGURE 20-15 A P-Channel MOS (PMOS) NOR Gate.

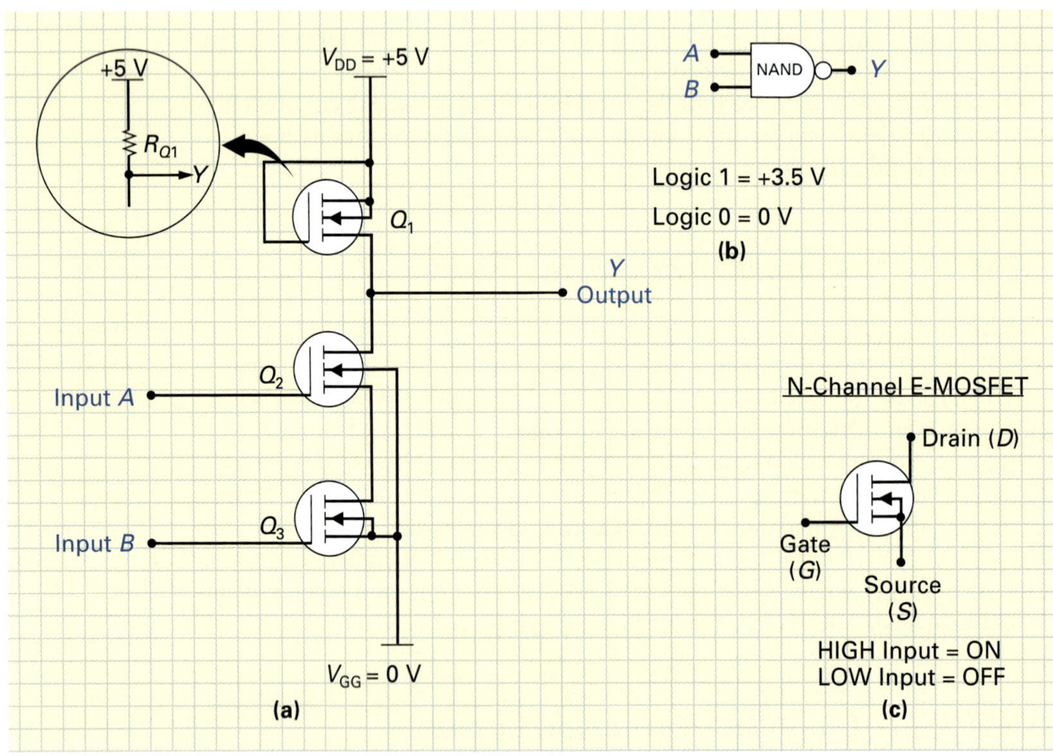

FIGURE 20-16 An N-Channel MOS (NMOS) NAND Gate.

The NMOS NAND gate circuit in Figure 20-16(a) will operate as follows. Transistor Q_1 functions as a "current limiting resistor" between source and drain because its gate is constantly connected to $+V_{DD}$, and therefore Q_1 is always ON. A LOW (0 V) on either or both of the inputs will turn OFF one or both of the transistors Q_2 and Q_3, causing the output to be pulled up to a HIGH (+3.5 V) by Q_1. Only when both inputs are HIGH, will Q_1 and Q_2 turn ON and switch the ground on Q_3's source through Q_3 and Q_2 to the output.

20-2-3 *CMOS (Complementary MOS) Logic Circuits*

The name **CMOS logic** is used to describe this logic circuitry because the logic gates are constructed using both a P-channel E-MOSFET (Q_1) and its complement, or opposite, an N-channel E-MOSFET (Q_2).

CMOS Logic
A digital logic family that makes use of both P-channel and N-channel MOSFETs.

A CMOS NOT Gate Circuit

Figure 20-17(a) shows how a CMOS NOT or INVERTER gate can be constructed. Notice the N-channel E-MOSFET (Q_2) has its source connected to 0 V, and therefore its gate needs to be positive (relative to the 0 V source) for it to turn ON. On the other hand, the P-channel E-MOSFET has its source connected to +10 V, and therefore its gate needs to be negative (relative to the +10 V source) for it to turn ON. The table in Figure 20-17(a) shows how this circuit will function. When V_{IN} or A is LOW (0 V), Q_1 will turn ON (N-P gate-source is forward-biased; $G = 0$ V, $S = +10$ V) and Q_2 will turn OFF (P-N gate-source is reverse-biased; $G = 0$ V, $S = 0$ V). A LOW input will therefore turn ON Q_1 creating a channel between source and drain, connecting the +10 V supply voltage to the output, or Y. On the other hand, when V_{IN} or A is HIGH (+10 V), Q_1 will turn OFF (N-P gate-source is reverse-biased; $G = +10$ V, $S = +10$ V) and Q_2 will turn ON (P-N gate-source is forward-biased; $G = +10$ V, $S = 0$ V). A HIGH input will therefore turn ON Q_2, creating a channel between source and drain connecting the 0 V at Q_2's source to the output, or Y.

A CMOS NAND Gate Circuit

Figure 20-17(b) shows how two complementary MOSFET pairs (two N-channel and two P-channel) could be connected to form a two-input CMOS NAND gate. If either input A or B is LOW, one or both of the P-channel MOSFETs (Q_1 and Q_2) will turn ON and switch $+V_{DD}$ or a HIGH through to the output. Only when both inputs A and B are HIGH will Q_1 and Q_2 be OFF, and Q_3 and Q_4 will be ON, switching ground or a LOW through to the output.

CMOS Logic Gate Characteristics

CMOS logic is probably the best all-round logic circuitry because of its low power consumption, very high noise immunity, large power supply voltage range, and high fanout. These characteristics are given in detail in Figure 20-18. The low power consumption achieved by CMOS circuits is due to the "complementary pairs" in each CMOS gate circuit. As was shown in Figure 20-17, during gate operation there is never a continuous path for current between the supply voltage and ground. When the P-channel device is OFF, $+V_{DD}$ is disconnected from the circuit, and when the N-channel device is OFF, ground is disconnected from the circuit. It is only when the output voltage switches from a HIGH to LOW, or LOW to HIGH, that the P-channel and N-channel E-MOSFET are momentarily ON at the same time. It is during this time of switching over from one state to the next that a small current will flow due to the complete path from $+V_{DD}$ to ground. This is why the power dissipation of CMOS logic gates increases as the operating frequency of the circuit increases.

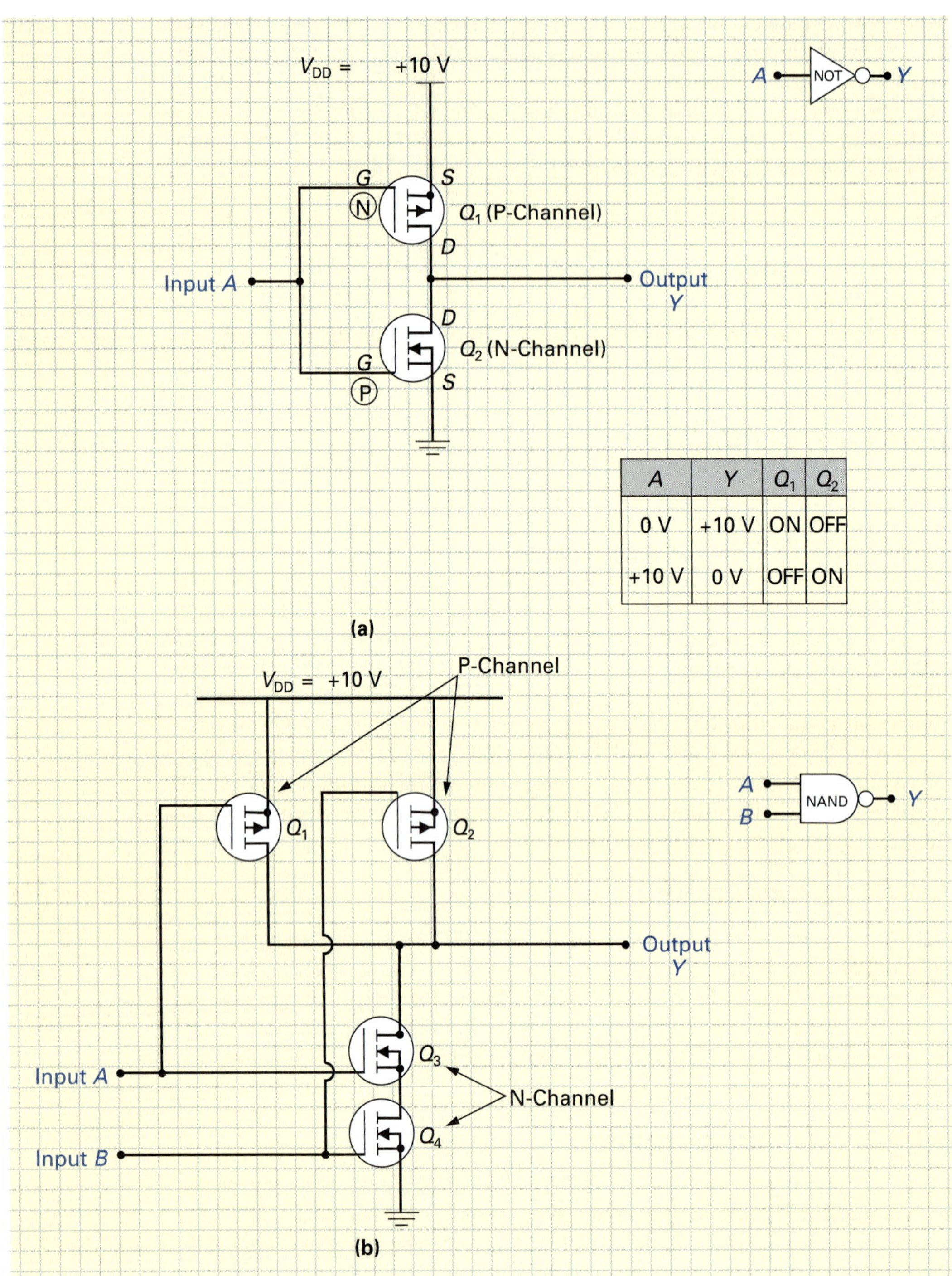

A	Y	Q_1	Q_2
0 V	+10 V	ON	OFF
+10 V	0 V	OFF	ON

FIGURE 20-17 Complementary MOS (CMOS) Logic Gates. (a) An Inverter. (b) A NAND Gate.

The CMOS Series of ICs

There is a full line of CMOS ICs available that can provide almost all the same functions as their TTL counterparts. These different types are given different series numbers to distinguish them from others. Some of the more frequently used CMOS ICs are described in the following paragraphs.

Output Logic Levels:

Input Logic Levels:

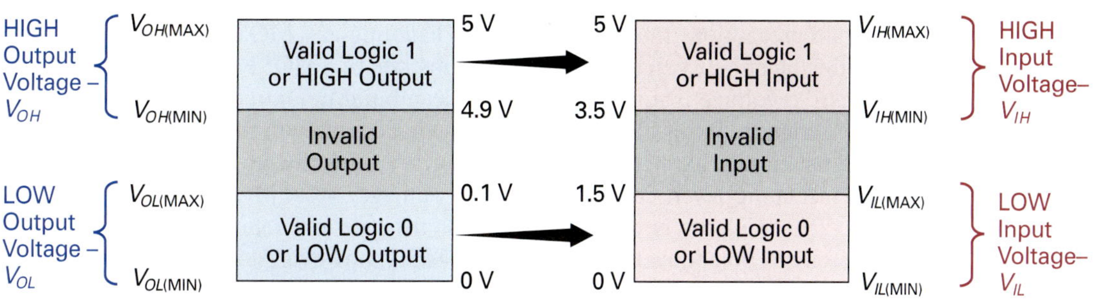

Power Dissipation per Gate: 2.7 nW (static) to 170 μW (at 100 kHz)
Propagation Delay per Gate: 10 ns
Supply Voltage: +3 V to +6 V
Noise Immunity: Very High
Fanout: 10

FIGURE 20-18 CMOS Logic Gate Characteristics (High-Speed CMOS—74HCXX).

a. *4000 Series* This first line of CMOS digital ICs was originally made by RCA and uses the numbers 40XX, with the number in the "XX" position indicating the circuit type. For example, the 4001 is a quad two-input NOR gate, while the 4011 is a quad two-input NAND gate. This series was eventually improved and called the 4000B series, and the original was labeled the 4000A series. Both use a V_{DD} supply of between +3 V and +15 V, with a logic 0 (LOW) = ⅓ of V_{DD} and logic 1 (HIGH) = ⅔ of V_{DD}. Propagation delay times for this series are typically 40 ns to 175 ns.

b. *40H00 Series* This "high-speed" CMOS series improved the propagation delay to approximately 20 ns; however, it was still not able to match the speed of the bipolar TTL 74LS00 series.

c. *74C00 Series* This series was designed to be "pin-compatible" with its TTL counterpart. This meant that all of the ICs used the same input and output pin numbers. Supply voltages can be anywhere from 3 V_{dc} to 15 V_{dc}, and propagation delays are long (90 ns, typically) and so applications are limited to signal frequency applications below 2 MHz. Low frequency and dc input signals result in almost no power dissipation; however, at high frequencies the gate power dissipation can exceed 10 mW to 15 mW.

d. *74HC00 and 74HCT00 Series* This high-speed CMOS series of ICs matches the propagation delay times of the 74LS00 bipolar ICs and still maintains the CMOS advantage of low power consumption. It was introduced to overcome some of the disadvantages that plagued older CMOS series. The HC and HCT series are both pin-compatible with TTL. The HCT (high-speed CMOS TTL-compatible) series is a truly compatible CMOS line of ICs since it uses the same input and output voltage levels for a HIGH and LOW. This advantage makes it easier to swap a TTL IC for a CMOS IC and not cause any interfacing problems. The 74HC series logic gates can operate from a supply voltage of 2 V_{dc} to 6 V_{dc} (typically 5 V_{dc} so that it is identical to TTL levels). They have a typical gate propagation delay of about 20 ns and a power dissipation of about 1 mW. At a supply voltage of 5 V_{dc}, a 74HCT gate will typically have a propagation delay of 40 ns and a power dissipation of 1 mW.

e. *74AC00 and 74ACT00 Series* This "advanced CMOS logic" series and "advanced CMOS, TTL-compatible" series has even better characteristics than the 74HC00 and 74HCT00 series. When operated from a supply voltage of 5 V_{dc}, a 74AC gate has a 3 ns propagation delay, and a 74ACT gate has a 5 ns propagation delay, enabling both families to operate beyond 100 MHz yet dissipate only 0.5 mW of power per gate. The 74AC series is not completely compatible with TTL; however, the 74ACT series is fully compatible.

20-2-4 *MOSFET Handling Precautions*

Certain precautions must be taken when handling any MOSFET devices. The very thin insulating layer between the gate and the substrate of a MOSFET can easily be punctured if an excessive voltage is applied. Your body can build up extremely large electrostatic charges due to friction. If this charge were to come in contact with the pins of a MOSFET device, an electrostatic-discharge (ESD) would occur, resulting in a possible arc across the MOSFET's thin insulating layer, causing permanent damage. Most MOSFETs presently manufactured have zeners internally connected between the gate and source to bypass high voltage static or in-circuit potentials to protect the MOSFET; however, it is important that

a. All MOS devices are shipped and stored in a "conductive foam" when not in use so that all of the IC pins are kept at the same potential, and therefore electrostatic voltages cannot build up between terminals.
b. When MOS devices are removed from the conductive foam, be sure not to touch the pins, since your body may have built up an electrostatic charge.
c. When MOS devices are removed from the conductive foam, always place them on a grounded surface such as a metal tray.
d. When continually working with MOS devices, use a wrist-grounding strap, which is a length of cable with a 1 MΩ resistor in series to prevent electrical shock if you were to come in contact with a voltage source.
e. All test equipment, soldering irons, and work benches should be properly grounded.
f. All power in equipment should be off before MOS devices are removed or inserted into printed circuit boards.
g. Any unused MOSFET terminals must be connected, since an unused input left open can build up an electrostatic charge and float to high voltage levels.
h. Any boards containing MOS devices should be shipped or stored with the connection side of the board in conductive foam.

SELF-TEST EVALUATION POINT FOR SECTION 20-2

Use the following questions to test your understanding of Section 20-2.

1. True or false: The E-MOSFET naturally operates as a voltage-controlled switch.
2. List three advantages that MOS logic gates have over bipolar logic gates.
3. Which type of E-MOSFETs are used in CMOS logic circuits?
4. What is the key advantage of CMOS logic?

20-3 DIGITAL IC PACKAGE TYPES AND COMPLEXITY CLASSIFICATION

There are four basic methods of packaging integrated circuits, and these are illustrated in Figure 20-19.

20-3-1 *Early Digital IC Package Types*

Transistor Outline
A package type used to house a transistor or circuit that resembles a small can.

Figure 20-19(a) shows the first type of package used for ICs, called the **transistor outline** or **TO can** package. This transistor package was modified from the basic three-lead transistor package to have the extra leads needed and is still used to house linear ICs due to its good

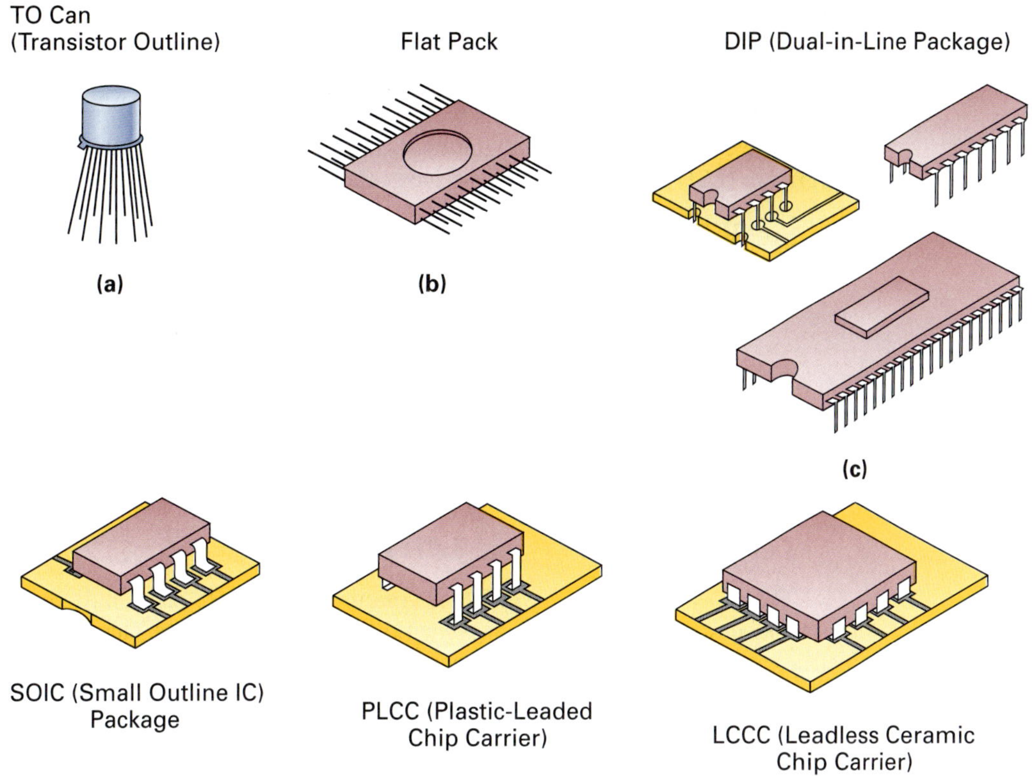

FIGURE 20-19 **IC Package Types.**

heat dissipation ability but is rarely used for digital ICs. Figure 20-19(b) illustrates another early IC package called the flat pack. The leads of the package were designed to be soldered directly to the tracks of the printed circuit board. This package allowed circuits to be placed close together for small-size applications, such as avionics and military systems. The package is generally made of ceramic to withstand the high temperatures due to the densely packaged circuits.

20-3-2 *Present-Day Digital IC Package Types*

Figure 20-19(c) shows the dual-in-line package, or DIP, which has been a standard for many years. Called a DIP because of its two (dual) lines of parallel connecting pins, its leads feed through holes punched in the printed circuit board. Connection pads circle the holes both on the top and the bottom of the printed circuit board (PCB) so that when the DIP package is soldered in place it makes a connection.

Figure 20-19(d) illustrates the three types of **surface-mount technology (SMT)** packages. The key advantage of SMT over DIP is that a DIP package needs both a hole and a connecting pad around the hole. With SMT, no holes are needed since the package is mounted directly onto the printed circuit board, and the space can be saved when a surface-mount resistor is used instead of a through-hole axial lead resistor. Without the need for holes, pads can be placed closer together, resulting in a considerable space savings. Figure 20-19(d) shows the three basic types of SMT packages. The "small outline IC" (SOIC) package uses L-shaped leads, the "plastic-leaded chip carrier" (PLCC) uses J-shaped leads, and the "leadless ceramic chip carrier" (LCCC) has metallic contacts molded into its ceramic body.

Surface-Mount Technology
A package type that has connections that connect to the surface of a printed circuit board.

20-3-3 *Digital IC Circuit Complexity Classification*

Although ICs are generally classified as being either linear (analog) or digital, another method of classification groups ICs based on their internal circuit complexity. The four categories are called small-scale integration (SSI), medium-scale integration (MSI), large-scale integration (LSI) and very-large-scale integration (VLSI).

Small-Scale Integration (SSI) ICs

These circuits are the simplest form of IC, such as single-function digital logic gate circuits. They contain less than 180 interconnected components on a single chip.

Medium-Scale Integration (MSI) ICs

These ICs contain between 180 and 1,500 interconnected components, and their advantage over an SSI IC is that the number of ICs needed in a system is less, resulting in a reduced cost and assembly time.

Large-Scale Integration (LSI) ICs

These circuits contain several MSI circuits all integrated on a single chip to form larger functional circuits or systems such as memories, microprocessors, calculators, and basic test instruments. These ICs contain between 1,500 and 15,000 interconnected components.

Very-Large-Scale Integration (VLSI) ICs

VLSI circuits contain extremely complex circuits such as large microprocessors, memories, and single-chip computers. These ICs contain more than 15,000 interconnected components.

SELF-TEST EVALUATION POINT FOR SECTION 20-3

Use the following questions to test your understanding of Section 20-3.

1. Give the full names for the following abbreviations:
   a. TO Can **b.** DIP **c.** SMT
2. Which package type saves the most printed circuit board space?
3. An IC containing four digital bipolar logic gates with about 44 components would be classified as a/an ___________ IC.
   a. SSI **b.** MSI **c.** LSI **d.** VLSI
4. The low circuit current and therefore low heat dissipation of ____________ circuits make it possible for small-size circuits, and this is why ____________ technology dominates the VLSI circuit market.

20-4 COMPARING AND INTERFACING BIPOLAR AND MOS LOGIC FAMILIES

Table 20-2(a) summarizes the differences between the two main digital IC logic families.

20-4-1 *The Bipolar Family*

The bipolar family of digital ICs has three basic types, called TTL (transistor-transistor logic), ECL (emitter-coupled logic), and IIL, or I^2L (integrated-injection logic). The characteristics of these three basic bipolar transistor logic gate types are summarized in

TABLE 20-2 Comparing Logic Types. (a) Logic Families (b) TTL and CMOS Series

	BIPOLAR FAMILY		
	LS TTL	ECL	IIL
Cost	Low	High	Medium
Fanout	20	10 to 25	2
Power Dissipation/Gate	2 mW	40 to 60 mW	0.06 to 70 μW
Propagation Delay/Gate	8 ns	0.5 to 3 ns	25 to 50 ns
Typical Input Signal Frequency	15 to 120 MHz	200 to 1000 MHz	1 to 10 MHz
External Noise Immunity	Good	Good	Fair–Good
Typical Supply Voltage	+5 V	−5.2 V	1 V to 15 V
Temperature Range	−55 to 125°C 0 to 70°C	−55 to 125°C	0 to 70°C
Internally Generated Noise	Medium–High	Low–Medium	Low

	MOS FAMILY		
	PMOS	NMOS	HCMOS
Cost	High	High	Medium
Fanout	20	20	10
Power Dissipation/Gate	0.2 to 10 mW	0.2 to 10 mW	2.7 nW to 170 μW
Propagation Delay/Gate	300 ns	50 ns	10 ns
Typical Input Signal Frequency	2 MHz	5 to 10 MHz	5 to 100 MHz
External Noise Immunity	Good	Good	Very Good
Typical Supply Voltage	−12 V	+5 V	+3 V to + 6 V
Temperature Range	−55 to 125°C 0 to 70°C	−55 to 125°C 0 to 70°C	−55 to 125°C −40 to 85°C
Internally Generated Noise	Medium	Medium	Low–Medium

(a)

Series	74	74L	74S	74LS	74AS	74ALS	74F	74C	4000B	74HC	74HCT	7AC	74ACT
Supply Voltage Range (V_{CC})	4.75 VDC to 5.25 VDC	4.75 VDC to 5.25 VDC	4.75 VDC to 5.25 VDC	4.75 VDC to 5.25 VDC	4.5 VDC to 5.5 VDC	4.5 VDC to 5.5 VDC	4.5 VDC to 5.5 VDC	3.0 VDC to 15.0 VDC	3.0 VDC to 18.0 VDC	2.0 VDC to 6.0 VDC	4.5 VDC to 5.5 VDC	3.0 VDC to 5.5 VDC	4.5 VDC to 5.5 VDC
Min. Logic-1 Input Voltage (V_{IN})	2.0 VDC	2.0 VDC	2.0 VDC	2.0 VDC	2.0 VDC	2.0 VDC	2.0 VDC	2.0 VDC	2.3 V_{CC}	3.15 VDC	2.0 VDC	3.15 VDC	2.0 VDC
Max. Logic-0 Input Voltage (V_{IL})	0.8 VDC	0.8 VDC	0.8 VDC	0.8 VDC	0.8 VDC	0.8 VDC	0.8 VDC	1.5 VDC	1.3 V_{CC}	0.9 VDC	0.8 VDC	1.35 VDC	0.8 VDC
Min. Logic-1 Output Voltage (V_{ON})	2.4 VDC	2.4 VDC	2.7 VDC	2.7 VDC	2.7 VDC	2.7 VDC	2.7 VDC	4.5 VDC	≈ V_{CC}	4.4 VDC	3.84 VDC	4.2 VDC	3.8 VDC
Max. Logic-0 Output Voltage (V_{OL})	0.4 VDC	0.4 VDC	0.5 VDC	0.4 VDC	0.4 VDC	0.4 VDC	0.5 VDC	0.5 VDC	≈0 VDC	0.1 VDC	0.33 VDC	0.5 VDC	0.5 VDC
Max. Logic-0 Input Current (I_G)	−1.6 mA	−0.18 mA	−2.0 mA	−0.36 mA	−0.5 mA	−0.1 mA	−0.6 mA	−0.5 nA	±1.0 μA	±1.0 μA	±1.0 μA	±1.0 μA	±1.0 μA
Max. Logic-1 Input Current (I_{IK})	40.0 μA	10.0 μA	50.0 μA	20.0 μA	20.0 μA	20.0 μA	20.0 μA	5.0 nA	±1.0 μA	±1.0 μA	±1.0 μA	±1.0 μA	±1.0 μA
Max. Logic-0 Output Current (I_{OS})	16.0 mA	3.6 mA	20.0 mA	4.0 mA	20.0 mA	8.0 mA	20.0 mA	0.4 mA	3.0 mA	20.0 μA	20.0 μA	24.0 mA	24.0 mA
Max. Logic-1 Output Current ($_{OH}$)	−400.0 μA	−200.0 μA	−1.0 mA	−400.0 μA	−2.0 mA	−400.0 μA	−1.0 mA	−0.36 mA	−3.0 mA	−20.0 μA	−20.0 μA	−24.0 mA	−24.0 mA
Propagation Delay (High to Low) (t_{PH})	8.0 ns	30.0 ns	5.0 ns	8.0 ns	1.5 ns	7 ns	3.7 ns	90.0 ns	50.0 ns	20.0 ns	40.0 ns	3.0 ns	5.0 ns
Propagation Delay (Low to High) (t_{PLH})	13.0 ns	60.0 ns	5.0 ns	8.0 ns	1.5 ns	5 ns	3.2 ns	90.0 ns	65.0 ns	20.0 ns	40.0 ns	3.0 ns	5.0 ns
Max. Operating Frequency (f_{max})	35 MHz	3 MHz	125 MHz	45 MHz	80 MHz	35 MHz	100 MHz	2 MHz	6 MHz	20 MHz	24 MHz	125 MHz	125 MHz
Power Dissipation Per Gate (mW)	10.0 mW	1.0 mW	20.0 mW	2.0 mW	4.0 mW	1.0 mW	4.0 mW	Note 1	Note 1	25.0 μW	80.0 μW	440 μW/ device	440 μW/ device

NOTE 1: Value is Frequency-Dependent

(b)

the upper table in Table 20-2(a). Generally, the power dissipation of TTL and ECL logic circuits limits their packing density and therefore they are only used for SSI and MSI circuit applications. On the other hand, I^2L digital circuits are easy to combine with linear circuits (such as op-amps and oscillators) to create LSI and VLSI systems on a single chip.

20-4-2 *The MOS Family*

The lower table in Table 20-2(a) shows the three basic E-MOSFET digital IC types, which are more frequently called the MOS family of digital ICs. The three basic types of MOS ICs are called PMOS, NMOS, and CMOS. Comparing the characteristics, you can see that MOS logic gates use less space due to their simpler construction, have a very high noise immunity, and consume less power than equivalent bipolar logic gates. However, the high input impedance and capacitance due to the E-MOSFET's insulated gate means that time constants are larger and therefore transistor ON/OFF switching speeds are slower than equivalent bipolar gates. Both PMOS and NMOS ICs can be used in LSI and VLSI applications such as large memory circuits, single-chip microcomputers, and test instruments. On the other hand, CMOS ICs are generally used in SSI and MSI circuit applications.

The trade-offs, therefore, between the two logic families are

a. Bipolar circuits are faster than MOS circuits, but their circuits are larger and consume more power.

b. MOS circuits are smaller and consume less power than bipolar circuits, but they are generally slower.

20-4-3 *Interfacing Logic Families*

Interface
A device or circuit that connects an output to an input.

Compatible
In relation to ICs, it describes a circuit family that has input/output logic levels and operating characteristics that are compatible with other logic IC families.

Table 20-2(b) lists the key differences between the TTL series and CMOS series of digital ICs. In some circuit applications, there is a need to connect or **interface** one logic gate family type to another logic gate family type. For example, in some circuit applications we may need to interface a TTL logic gate to a CMOS logic gate or connect a CMOS gate to a TTL gate. In these instances, it is important that both devices are **compatible,** meaning that the HIGH and LOW logic levels at the output of the source logic gate will be of the correct amplitude and polarity so that they will be interpreted correctly at the input of the load logic gate. If we were to use the 74HCT00 CMOS series, there would be no interfacing problems at all, since this series generates and recognizes the TTL HIGH and LOW voltage levels (it is TTL-compatible). Other CMOS series, however, are not voltage- and current-compatible and therefore an interfacing device will have to be included between the two different types of logic gates to compensate for the differences.

Interfacing Standard TTL and 4000B Series CMOS

The voltage and current differences for the key TTL and CMOS series of logic gates are listed in Figure 20-20(a). Figure 20-20(b) shows the incompatibility between the HIGH and LOW voltage levels of standard TTL series and 4000B CMOS series logic gates. A LOW logic level output voltage from the TTL logic gate will be recognized at the CMOS input as a LOW, since even a worst-case LOW output of 0.8 V is still less than the worst-case LOW level CMOS input of 1.67 V. The problem arises with a HIGH output from the TTL gate, which even if it is at 2.4 V, will still be less than the minimum acceptable CMOS input voltage of 3.33 V. To compensate for this problem, a 10 kΩ pull-up resistor is connected between the TTL and CMOS gate, as shown in Figure 20-20(c). This pull-up resistor will pull a HIGH output from the TTL gate up towards 5 V so that it will be recognized at the input as a HIGH by the CMOS gate.

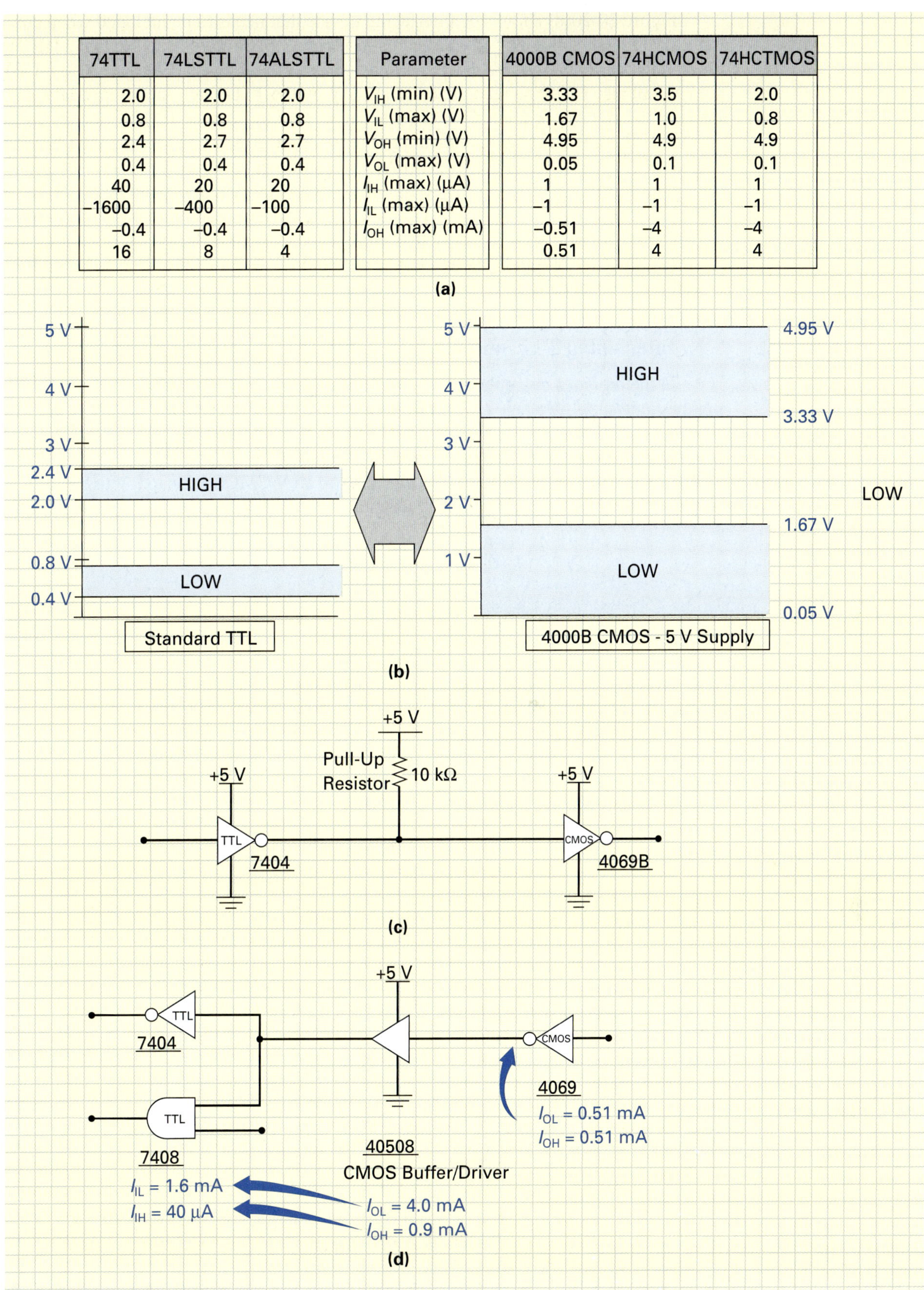

FIGURE 20-20 **Interfacing a TTL Gate and a 4000B CMOS Gate that Are Using the Same Supply Voltage Value.**

Figure 20-20(d) shows the interfacing device that needs to be included when connecting a CMOS gate to a TTL gate. Referring to Figure 20-20(b), you can see that a HIGH output voltage of 4.95 V will be easily recognized as a HIGH by the TTL gate, and a LOW of 0.05 V will also easily be recognized as a LOW by the TTL gate. The voltage levels in this instance are not the problem, but the current levels are because CMOS devices are low-current devices. For example, referring to Figure 20-20(d) you can see that the LOW output current of the 4069 CMOS gate (0.51 mA) is not sufficient to drive the TTL LOW input current requirement (1.6 mA). Similarly, the HIGH output current of the 4069 CMOS gate (0.51 mA) is not sufficient to drive the TTL HIGH input current requirement (40 μA). Including the 40508 CMOS buffer/driver will ensure current compatibility, as shown in Figure 20-20(d), since the output LOW and HIGH current values from this buffer are more than sufficient for two standard TTL loads.

Interfacing TTL and 4000B CMOS Using Level-Shifters

In the previous example, the 4000B CMOS gates and the standard TTL gates were using the same supply voltage (5 V). The 4000B CMOS series can use a supply voltage that is anywhere between +3 V and +15 V. To interface two different logic gate types with different supply voltages, we will have to include a **level-shifter** or **translator,** as shown in Figure 20-21. The 4050B level-shifter will convert 0 V to 15 V CMOS logic to 0 V to 5 V TTL logic, as shown in Figure 20-21(a), enabling us to interface CMOS to TTL. The 4504B level-shifter

Level-Shifter or Translator

A circuit that will shift or change logic levels.

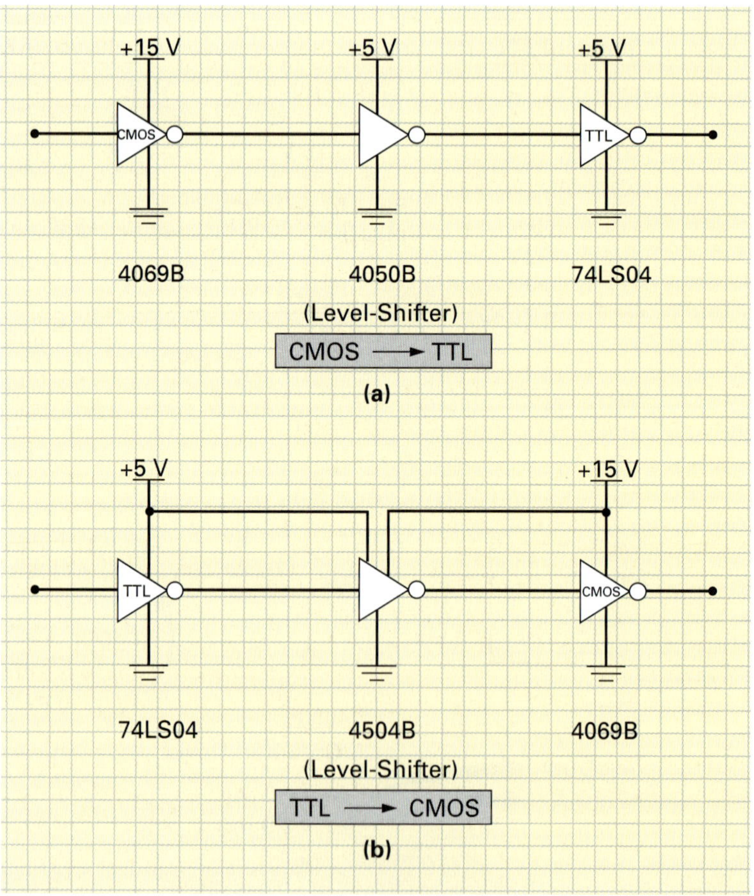

FIGURE 20-21 **Using Level-Shifters or Translators to Interface TTL and 4000B CMOS Gates with Different Supply Voltages.**

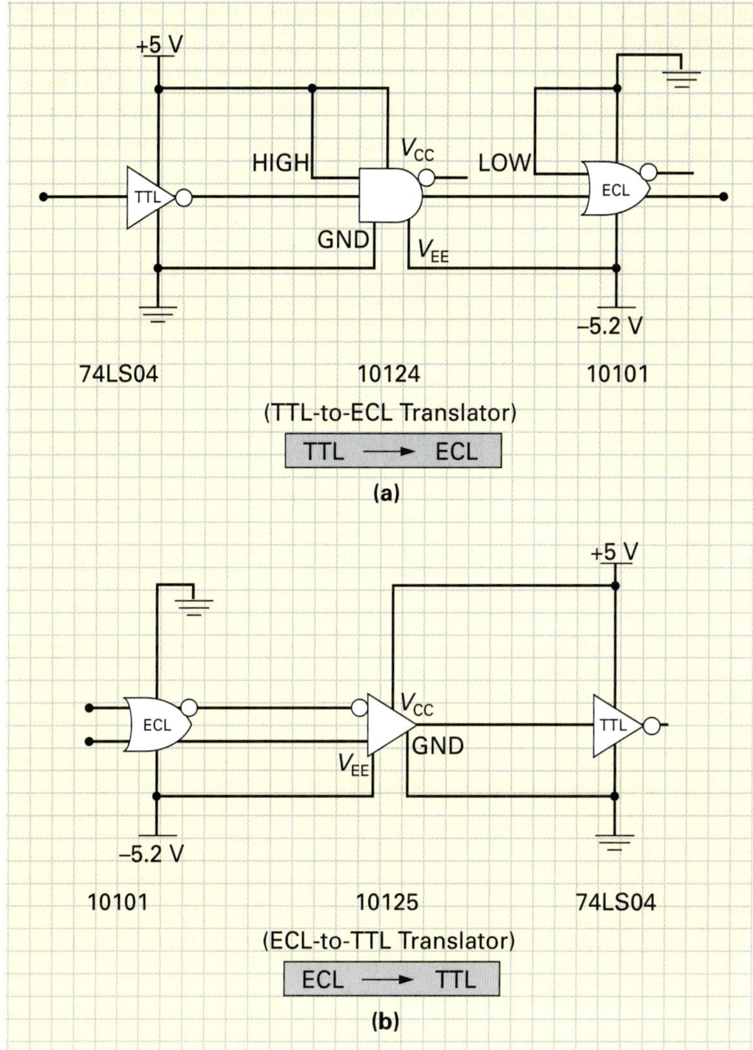

FIGURE 20-22 **Using Level-Shifters or Translators to Interface ECL and TTL.**

will convert 0 V to 5 V TTL logic to 0 V to 15 V CMOS logic, as shown in Figure 20-21(b), enabling us to interface TTL to CMOS.

Interfacing ECL and TTL Using Level-Shifters

The 0 V to +5 V logic levels of TTL are very much different than the −5.2 V to 0 V logic levels of ECL. Figure 20-22(a) shows how a 10124 level-shifter can be used to interface a TTL logic gate to an ECL logic gate, and Figure 20-22(b) shows how a 10125 level-shifter can be used to interface an ECL logic gate to a TTL logic gate.

20-4-4 *Other Logic Gate Families*

Semiconductor manufacturers are continually searching for new semiconductor circuits that have faster propagation delay times (since this enables them to operate at higher frequencies), lower power consumption, better noise immunity, and increased packing density. Some of the new hopefuls are gallium arsenide (GaAs) circuits, BiCMOS circuits (which is a combination of bipolar and CMOS on the same chip, an example of which was featured in Figure 20-12), silicon-on-sapphire (SOS) circuits, and Josephson junction circuits. It is therefore important in this ever-changing semiconductor industry that you read electronics magazines and manufacturers' technical publications to stay abreast of the latest technological advances. Most of the semiconductor manufacturers have their own home page on the Internet with access to all data sheets, circuit applications, technical support, and new product information.

SELF-TEST EVALUATION POINT FOR SECTION 20-4

Use the following questions to test your understanding of Section 20-4.

1. What is the speed/power trade-off with regard to the selection of TTL logic gates?
2. Which logic gate type is probably the best all-round logic because of its low power consumption, very high noise immunity, wide power supply voltage range, and high fanout?
3. What interfacing problems would occur between a 74HCT00 gate and a 74ALS00 gate?
4. What is a level-shifter gate?
5. Which logic family combines both bipolar and CMOS on a single chip?
6. Which logic gate series would be best suited if a circuit is processing input signal frequencies in the 110 MHz range?

REVIEW QUESTIONS

Multiple-Choice Questions

1. Which stage within a standard TTL gate provides a low output impedance for both a HIGH and LOW output?
   a. Multiple-emitter
   b. Phase splitter
   c. Totem pole
   d. Injection emitter
2. A floating TTL input has exactly the same effect as a/an
   a. LOW input.
   b. HIGH input.
   c. invalid input.
   d. both (a) and (c).
3. A standard TTL logic gate can sink a maximum of ____________ and source a maximum of ____________.
   a. 400 μA, 1.6 mA
   b. 40 μA, 16 mA
   c. 16 mA, 400 μA
   d. 1.6 mA, 40 μA
4. Which TTL logic gates require an external pull-up resistor?
   a. ECL
   b. Open-collector
   c. Schottky TTL
   d. Both (a) and (b)
5. The output current of a standard TTL gate is ____________ times greater than the input current, while the output current of a buffer/driver TTL gate is typically ____________ times greater than the input.
   a. 10, 30
   b. 10, 15
   c. 30, 10
   d. 15, 30
6. ____________ logic gates are generally incorporated in digital circuits to sharpen the rise and fall times of input pulses.
   a. Buffer/driver
   b. Schmitt trigger
   c. Open-collector
   d. Tri-state
7. The fanout of a standard TTL logic gate is ____________.
   a. 10
   b. 20
   c. 2
   d. 40
8. The output current of a standard TTL logic gate is ____________ times greater than the input current.
   a. 10
   b. 20
   c. 2
   d. 40

9. Which is the fastest of all the logic gate types?
   a. CMOS
   b. Open-collector
   c. Schottky TTL
   d. ECL

10. Which of the following CMOS series types is both pin- and voltage-compatible with TTL?
   a. 4000
   b. 74C00
   c. 74HCT00
   d. 74HC00

11. Which logic gate type contains both a P-channel and N-channel E-MOSFET?
   a. CMOS
   b. NMOS
   c. PMOS
   d. TREEMOS

12. A ____________ ____________ gate is used to convert a sine wave into a rectangular wave or to sharpen the rise and fall times of a rectangular wave.
   a. open-collector
   b. high-speed
   c. Schmitt trigger
   d. MOS logic

13. ____________ circuits have a combination of bipolar and CMOS logic circuits on the same chip.
   a. GaAs
   b. SOS
   c. Josephson junction
   d. BiCMOS

14. ____________ circuits can combine linear circuits (such as amplifiers and oscillators) with digital circuits to create complete electronic systems on one chip.
   a. CMOS
   b. IIL
   c. ECL
   d. NMOS

15. ____________ circuits are faster, but their circuits are larger and consume more power, whereas ____________ circuits are smaller and consume less power, but they are generally slower.
   a. Bipolar, MOS
   b. CMOS, ECL
   c. MOS, CMOS
   d. MOS, bipolar

Practice Problems

16. Referring to the TTL logic circuit shown in Figure 20-23, complete the truth table.

17. Sketch the logic symbol and describe the logic function being performed by the logic circuit in Figure 20-23.

18. What circuit arrangement is formed by Q_5 and Q_6 in Figure 20-23, and what is the circuit's purpose?

19. What does the symbol within the logic gate in Figure 20-24 indicate?

20. Referring to Figure 20-24(a), complete the truth table.

21. Referring to Figure 20-24(a), determine whether the 7401 can handle the LOW output sinking current for this circuit (assuming a 2 V LED drop).

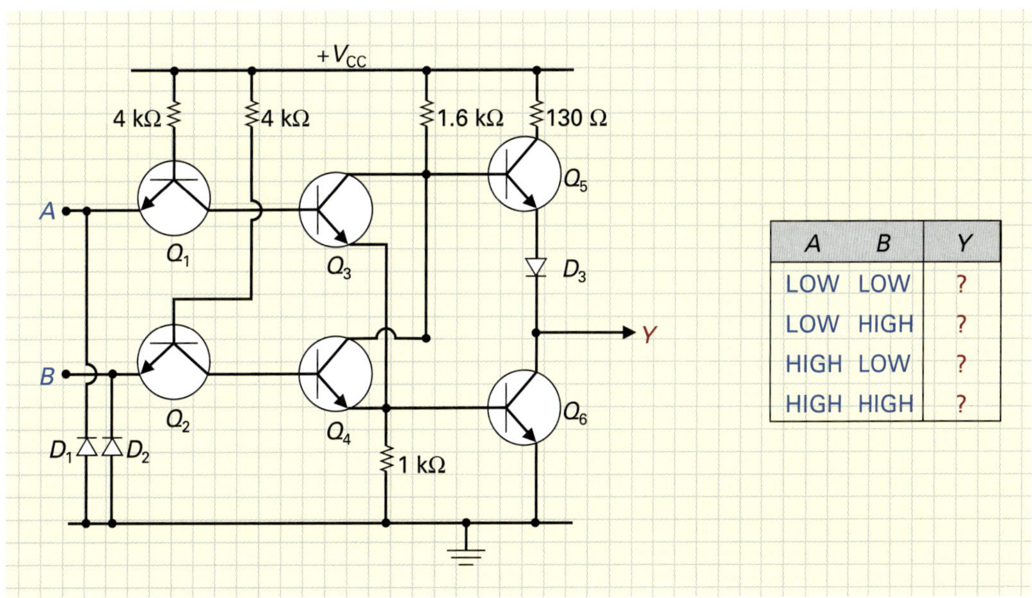

A	B	Y
LOW	LOW	?
LOW	HIGH	?
HIGH	LOW	?
HIGH	HIGH	?

FIGURE 20-23 A Standard TTL Logic Gate Circuit.

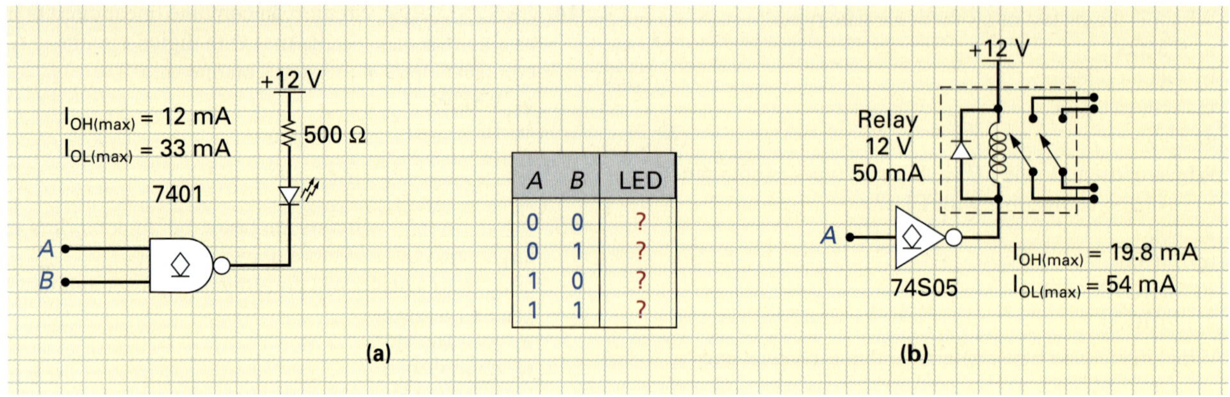

A	B	LED
0	0	?
0	1	?
1	0	?
1	1	?

FIGURE 20-24 Application—Controlling an LED with a Logic Gate.

22. Will the 74LS05 be able to handle the LOW output sinking current needed to energize the 12 V, 50 mA relay?
23. What open-collector output gate in Figure 20-10 would be ideal for a greater-than-50 mA load?
24. Identify the logic symbol used within the logic gate in Figure 20-25(a).
25. Referring to Figure 20-25(a), complete the truth table given.
26. Referring to Figure 20-25(b), complete the truth table given.
27. What is the difference between the 74LS125 shown in Figure 20-25(a) and the 74LS126 shown in Figure 20-25(b)?
28. Identify the logic symbol used within the logic gate in Figure 20-26(a).
29. Describe the operation of the complete circuit shown in Figure 20-26(a).
30. Comparing the inputs to the outputs in Figure 20-26(b), (c), (d), and (e), describe the circuit function being performed.
31. What type of E-MOSFET is being used in the logic circuit in Figure 20-27?
32. Complete the truth table for the logic gate shown in Figure 20-27.
33. Sketch the logic symbol and describe the logic function being performed by the logic circuit in Figure 20-27.

Web Site Questions

Go to the web site http://www.prenhall.com/cook, select the textbook *Electronics: A Complete Course,* select this chapter, and then follow the instructions when answering the multiple choice practice problems.

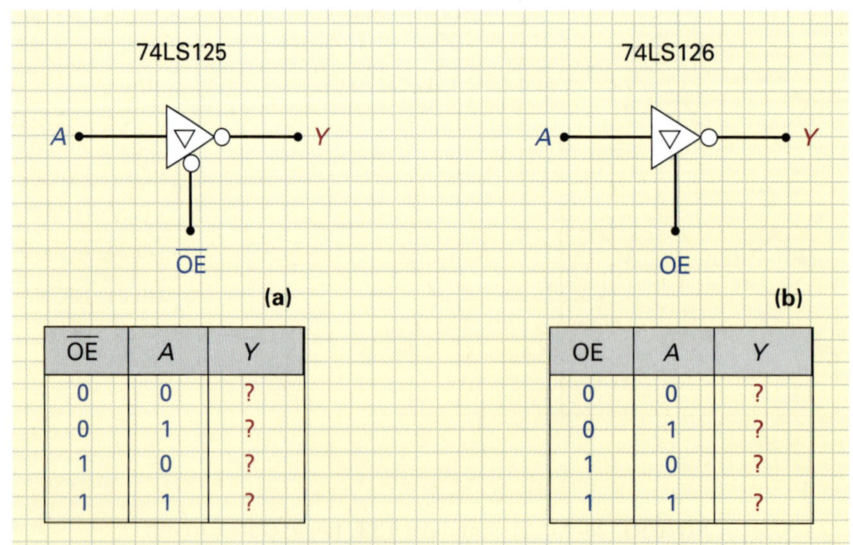

$\overline{OE}$	A	Y
0	0	?
0	1	?
1	0	?
1	1	?

OE	A	Y
0	0	?
0	1	?
1	0	?
1	1	?

FIGURE 20-25 Logic Gates With Control Inputs.

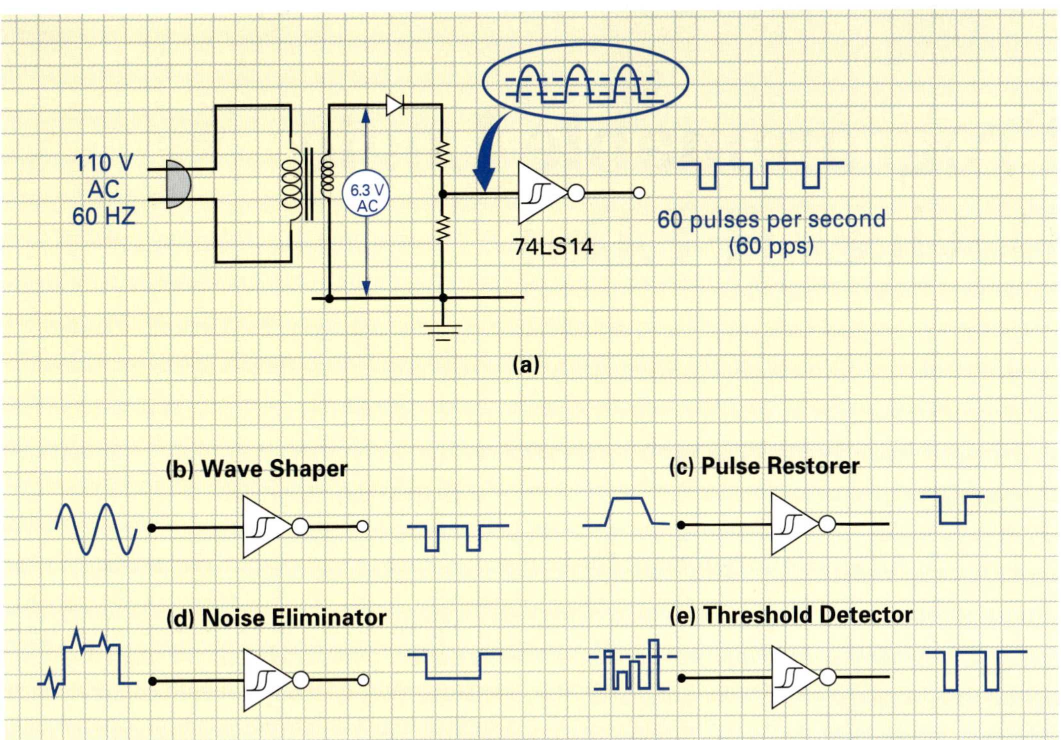

FIGURE 20-26 Applications—Logic Gate Circuits.

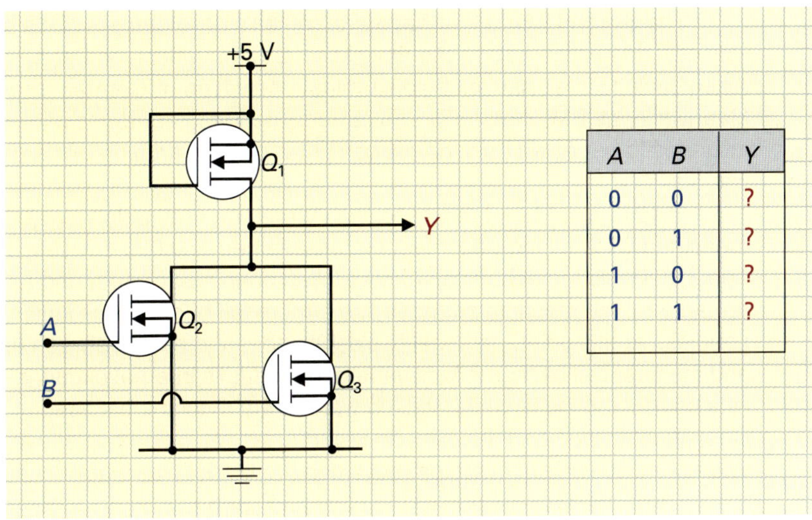

A	B	Y
0	0	?
0	1	?
1	0	?
1	1	?

FIGURE 20-27 A MOS Logic Gate.

Programmable Logic Devices (PLDs)

Don't Mention It!

Two years before the Japanese attacked Pearl Harbor, IBM president Thomas J. Watson decided to invest half a million dollars of the company's money in a project conceived by a young Harvard mathematician named Howard Aiken. Aiken's plan was to ignore the architecture of the sorters and calculators that were then available and concentrate on building a machine based on the nineteenth century concepts of Charles Babbage. Using Babbage's original description of his "analytical engine," Aiken set about to build a general-purpose programmable computer. In fact, Aiken stated at the time that "If Babbage had lived seventy-five years later, I would be out of a job."

The Second World War hampered Aiken's development of his machine because he was called to active duty in the navy soon after the attack on Pearl Harbor. In the navy, Aiken distinguished himself by disarming a new type of German torpedo single-handed. Meanwhile, at IBM, Watson did not appreciate the interference of the war and used his influence to get Aiken back on his computer project, arguing that it could greatly help the war effort.

In 1943, the Mark I computer was turned on and underwent its first successful test. It measured 51 feet long and 8 feet high, contained 750,000 devices, and 500 miles of wire. The 3,304 electromechanical relays clicked open and closed, and as one observer remarked, "The machine sounded like a room full of old ladies knitting away with steel needles." At this point, personality clashes began to erupt between the Mark I's inventor and the financial backer. Aiken wanted the Mark I's internal circuitry exposed so that scientists could examine it, while Watson wanted the machine encased in glass and stainless steel. Watson won on this point; however, Aiken retaliated by not even mentioning IBM or Watson's involvement in the project when the Mark I was introduced to the press in 1944.

The Mark I was leased to the navy and used to calculate the trajectory of cannon shells. It could add or subtract 23-digit numbers in a few tenths of a second, and multiply them in three seconds. This phenomenal speed—which was only slightly faster than Babbage had imagined 122 years earlier—enabled the navy, in a single day, to speed through calculations that previously took six months.

After the war Aiken stayed at Harvard to develop the second, third, and fourth generations of the Mark I, and had nothing further to do with Watson. Tom Watson was so angered by Aiken's failure to acknowledge him and IBM that he ordered all of his researchers to construct a better and faster system. Revenge launched IBM into the computer business and, when Watson died at the age of 82 in 1956, he had seen his wish fulfilled. IBM had overtaken all opposition, and the name had become so synonymous with computers that most people believed that IBM had invented them.

21

Introduction

Digital Technology has advanced, and continues to advance, almost meteorically each and every year. Proof of this is easily evident when we see the performance of personal computers (PCs) advance in leaps and bounds. To quote a recently posted statistic, microprocessor ICs are improving at a rate of 60% per year, while memory ICs are quadrupling their capacity every three years. These and other rapid changes in all the key building blocks of digital electronic systems mean that consumer products are generally obsolete in less than three years.

To keep up with this fast pace, electronic companies have to design and manufacture new products in a cycle of typically less than six months. To meet this accelerated schedule, engineers and technicians have looked for shortcuts that enable them to construct a digital prototype circuit and evaluate its performance in a much more timely manner.

In the previous chapter, you were introduced to most of the **standard logic devices (SLDs).** To construct a circuit using these ICs, you would first need to insert them into a protoboard and then connect them with a spaghetti-like maze of hookup wire. In this chapter, you will be introduced to **programmable logic devices (PLDs),** and as the industry has, you will see how much easier it is to construct your digital lab circuits using the highly versatile PLD.

Standard Logic Devices (SLDs)
Digital logic ICs that have a fixed design.

Programmable Logic Devices (PLDs)
Digital logic ICs that can have their function changed through programming.

21-1 WHY USE PROGRAMMABLE LOGIC DEVICES?

To clearly demonstrate the PLD advantage, let us compare the differences that occur when we construct a circuit using standard logic devices and then the same circuit using a programmable logic device.

21-1-1 *Constructing a Circuit Using Standard Logic Devices*

As an example, Figure 21-1(b) shows an application in which two NOT gates, four AND gates, and four OR gates are connected to achieve the logic function described in the truth table in Figure 21-1(a). The two NOT gates and four AND gates form a 1-of-4 decoder. This decoder will make only one of its four outputs HIGH based on the binary value applied to inputs A and B, as shown in the table in Figure 21-1(b). When $AB = 00$, the AND gate 0's output is HIGH and, since a connection exists to the inputs of OR gates 0 and 1, the output will be Q_3, Q_2, Q_1, $Q_0 = 0011$. Referring to the second line of the truth table in Figure 21-1(a), when $AB = 01$, AND gate 1's output is HIGH and, since a connection exists to the inputs of OR gates 2 and 3, the output is Q_3, Q_2, Q_1, $Q_0 = 1100$. When $AB = 10$, AND gate 2's output is HIGH, giving a HIGH to the input of OR gate 0 and an output of 0001. Finally, when $AB = 11$, AND gate 3's output is HIGH, giving a HIGH to OR gate 2 and an output of 0100.

Figure 21-1(c) shows how this circuit could be constructed on a protoboard. Once completed, switches should be connected to the inputs and LEDs to the outputs so that the circuit can be tested to see if it performs the desired logic function specified in the function table.

This standard logic prototyping method has the following disadvantages:

- Hookup wire cutting and stripping is time-consuming.
- Wires can easily be inserted incorrectly, causing possible device damage and lengthy delays while the errors are isolated.
- A large and costly inventory of all standard logic ICs must be maintained.
- If the desired standard logic IC is not available, further delays will result.
- To modify or add to a working circuit, the wires and ICs will normally all have to be removed from the protoboard and the new design built again from scratch.

Like industry, you have had a good taste of this frustrating practice when you have had to construct the hands-on lab experiments and are probably quite ready to welcome an alternative.

Personal Computer
Abbreviated PC, it is a data processing digital system for personal use.

Computer-Aided Design (CAD)
Application software used in conjunction with a PC for design.

Hardware Description Language (HDL)
A programming language used to specify hardware designs.

Text Editor
An element of a software program used to design a circuit using a hardware description language.

Schematic Editor
An element of a software program used to graphically design a circuit.

Compiled
A term used to describe a process in which a program is converted into a format to make it compatible with the device it is being sent to.

Circuit Simulation
A process in which a theoretical model of a circuit design is checked to see how its outputs will respond to all possible inputs.

21-1-2 *Constructing a Circuit Using Programmable Logic Devices*

By using an inexpensive **personal computer (PC),** a **computer-aided design (CAD)** software program, and a programmable logic device (PLD) IC, you can easily prototype a digital circuit.

Figure 21-2(a) and (b) shows the same truth table and application circuit from Figure 21-1, but in this instance a programmable logic device has been used to construct the circuit, as shown in Figure 21-2(c).

The single PLD contains a large inventory of logic gates, along with interconnecting devices, all within a single IC. Using the PC, you can write logic programs in a **hardware description language (HDL)** using a **text editor** or simply draw your logic circuit using a **schematic editor.** The HDL or schematic is then **compiled** by the CAD program to create a detailed logic circuit that will perform the actions you specified in your original HDL or schematic. The operation of this circuit can then be simulated to make sure that it is functioning as it should. If the **circuit simulation** is successful, the

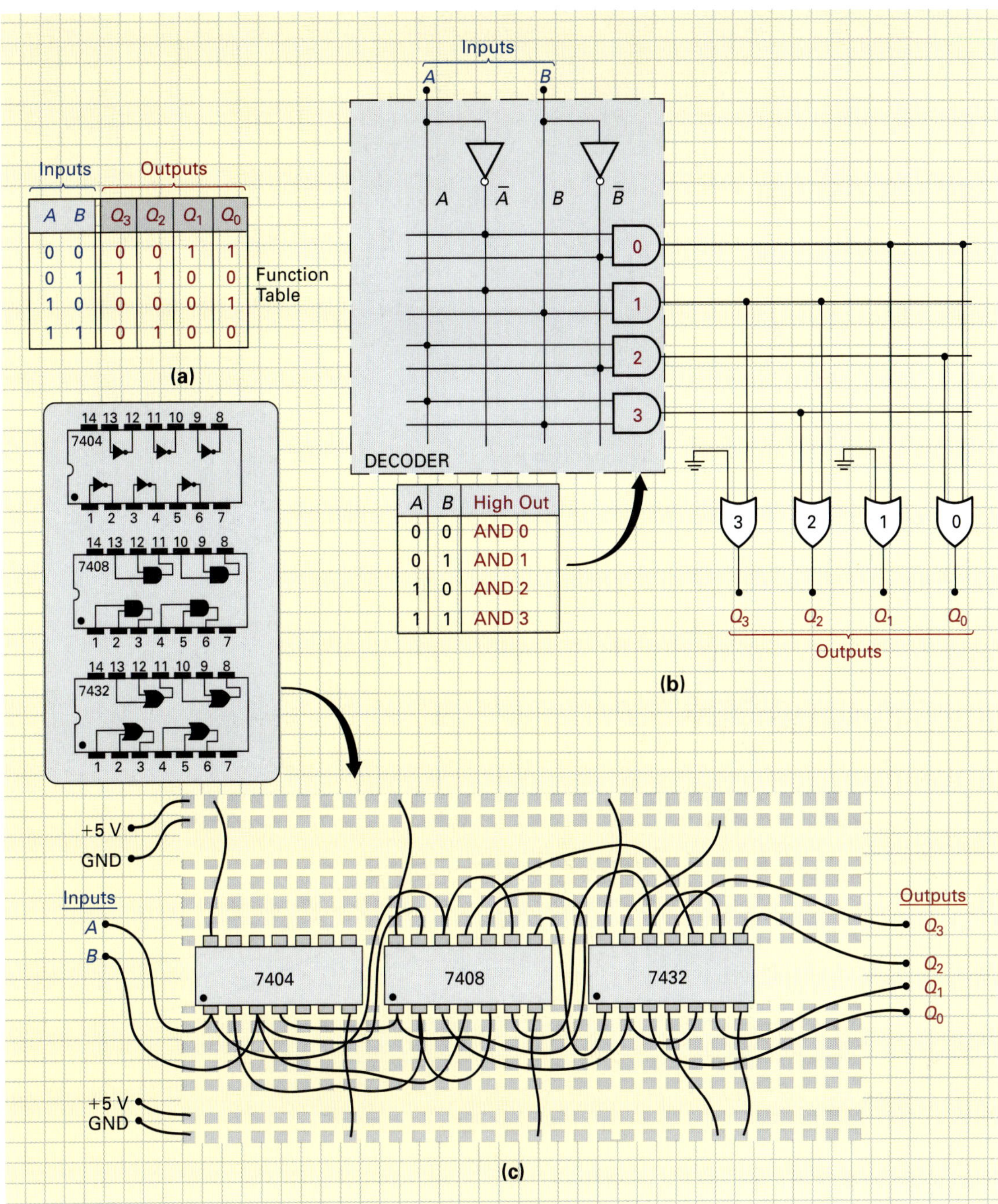

FIGURE 21-1 **Constructing a Circuit Using Standard Logic Devices (SLDs).**

design is then **downloaded** through the parallel port of the PC into the PLD. Switches are then connected to the inputs and LEDs are connected to the outputs to make a final test of the prototype.

Downloaded
A term used to describe a transfer of data from a computer to a destination.

In summary, the five-step process for creating a prototype using a PLD is

PLD Prototyping Process

Step 1: Create the new circuit using the software's schematic editor.

Step 2: Compile the circuit into a bitstream file that when loaded into the CPLD, will instruct it to act like the entered schematic.

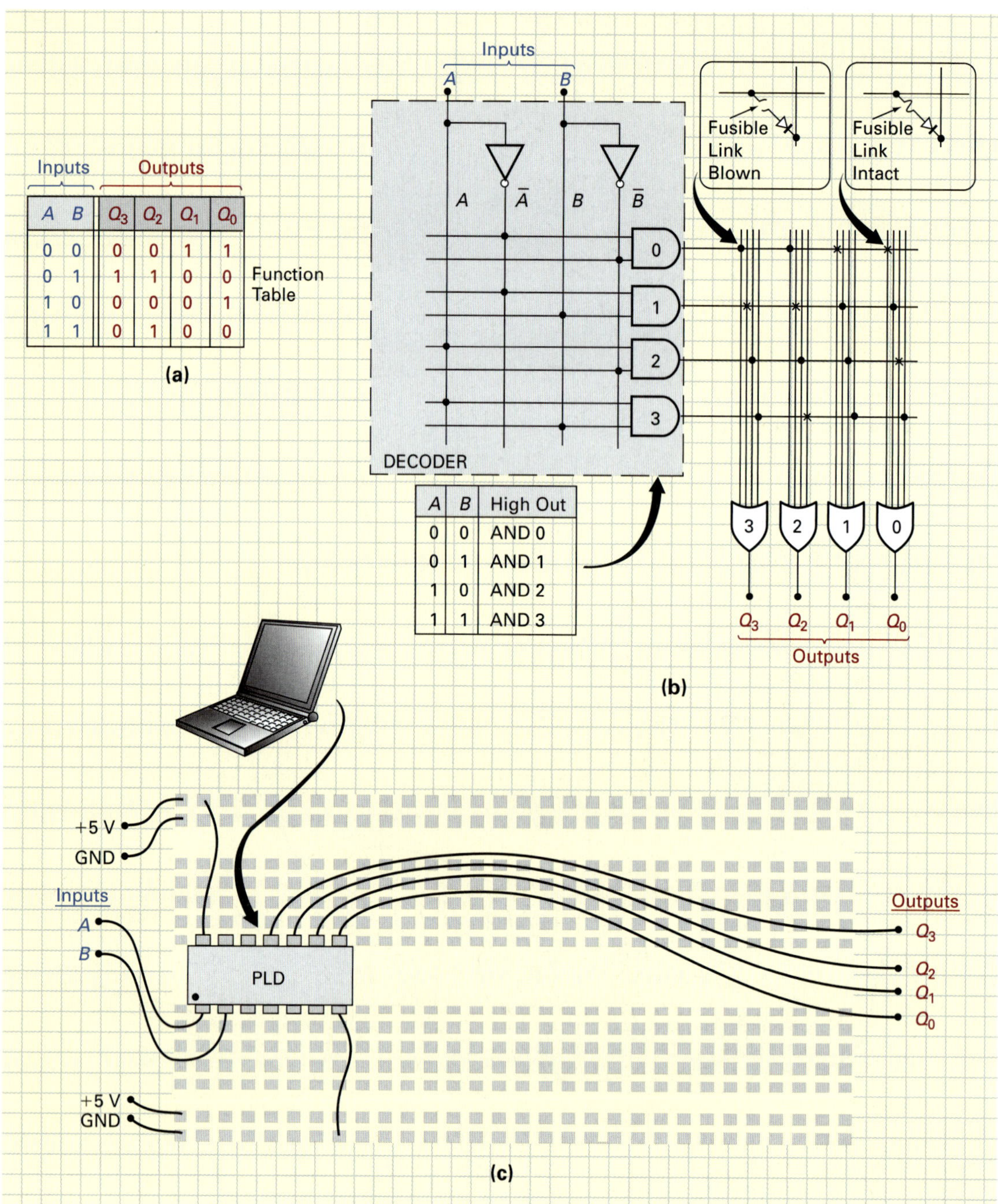

FIGURE 21-2 **Constructing a Circuit Using Programmable Logic Devices (PLDs).**

Step 3: Verify the operation of your circuit using the software's functional and timing simulator.

Step 4: Download the circuit file from the PC to the PLD.

Step 5: Physically test the PLD by activating its inputs and monitoring its outputs.

This programmable logic prototyping method has the following advantages:

- With manual wiring reduced to a minimum, prototypes can be constructed, tested, and modified at a much faster rate.
- Wiring errors can be avoided.

- You can experiment with many digital IC types without having to stock them in your supply cabinet.
- Circuit designs can be saved as electronic files within the PC and used again when needed.
- Since the PLD can be used over and over, modifications can easily be made by altering the circuit in the PC and then downloading the new design into the PLD.
- Larger and more complex projects can be undertaken now that the tedious manual procedures are automated.

SELF-TEST EVALUATION POINT FOR SECTION 21-1

Use the following questions to test your understanding of Section 21-1.

1. What are the full names and brief definitions of the following abbreviations:
   a. PLD
   b. SLD
   c. HDL
   d. CAD
   e. PC
2. What are five disadvantages of constructing a digital circuit prototype using standard logic devices?
3. What are five advantages of constructing a digital circuit prototype using programmable logic devices?
4. What are the basic steps you would follow to create and load a digital circuit into a PLD?

21-2 TYPES OF PROGRAMMABLE LOGIC DEVICES

The programmable logic devices of today evolved from the **programmable logic array (PLA),** which was first introduced in the early 1970s. To understand the advantages of today's powerful PLDs, let us first study the limitations of the first PLD devices.

Programmable Logic Array (PLA)
A PLD that has both a programmable AND array and a programmable OR array.

Programmable Array Logic (PAL)
A PLD that has a programmable AND array driving a hardwired OR array.

One-time Programmable (OTP)
A device that, after the initial programming, cannot be reprogrammed.

Complex Programmable Logic Devices (CPLDs)
A digital IC type that has several PLD blocks and a global interconnection matrix.

Field-Programmable Logic Arrays (FPGAs)
A digital IC type that has several programmable logic blocks that are interconnected by switches.

21-2-1 *Early Programmable Logic Devices*

The basic programmable logic array, or PLA, contains a set of NOT gates, AND gates, and OR gates, all interconnected with programmable switches, as shown in Figure 21-3(a). Although it was an extremely flexible arrangement, having programmable AND gate inputs and programmable OR gate inputs was unnecessary and costly, and so industry soon followed up with the simpler **programmable array logic (PAL)** structure, shown in Figure 21-3(b). With the PAL, the AND gate inputs were programmable but the OR gate inputs were not.

The PLA and PAL gave the technician and engineer a versatile device that could be customized to fit any application. However, their small size meant that they still had to be placed on a protoboard and wired to each other, and so they still had many of the same disadvantages of standard logic devices.

Their other disadvantage was that they were **one-time programmable (OTP).** This meant that they included specialized circuitry within the IC that used a high-voltage input to burn out the unwanted fuses at the wire cross points. Once the fuses were blown, they remained that way, and so they could only be used once.

21-2-2 *Today's Programmable Logic Devices*

To combat the size limitation of the early PLDs, IC manufacturers came up with two extremely large PLD types, called **complex programmable logic devices (CPLDs)** and **field-programmable gate arrays (FPGA).** Figure 21-4 shows the internal structure of

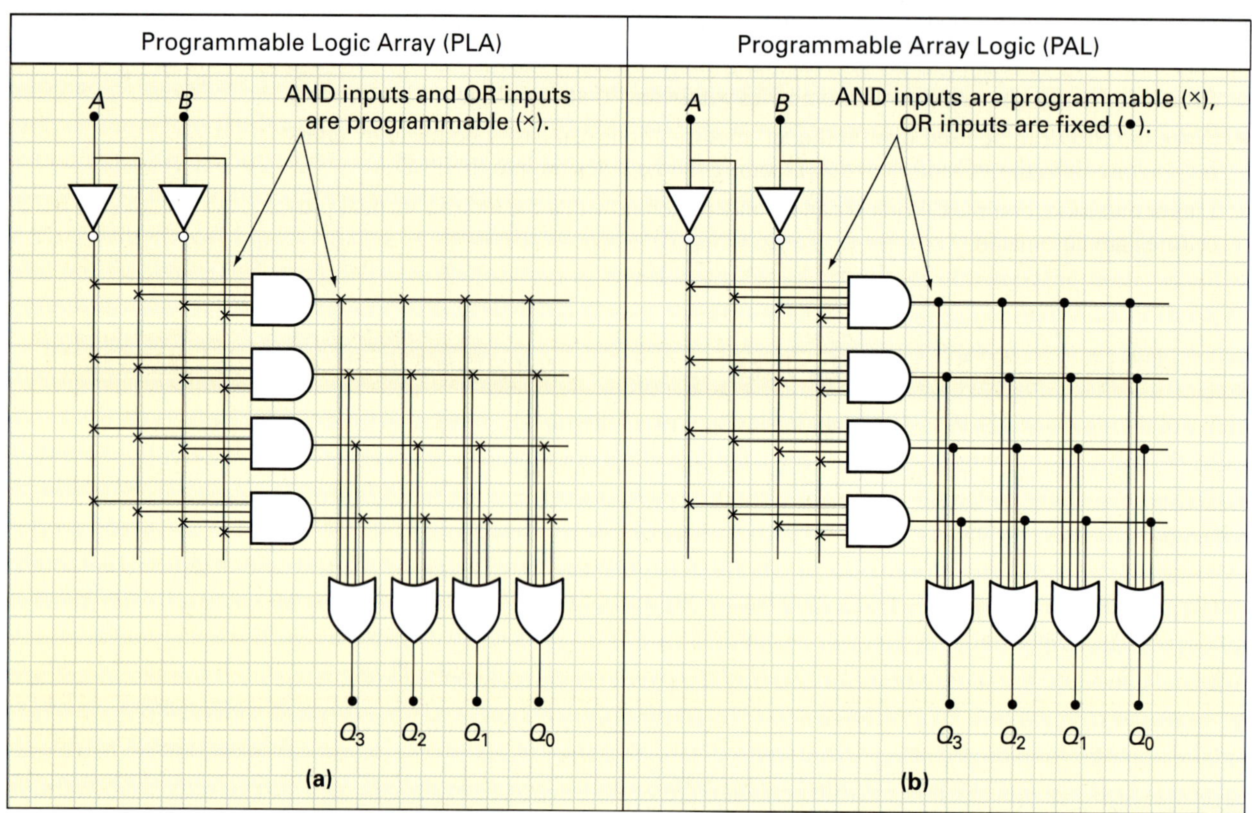

FIGURE 21-3 **First Types of Programmable Logic Devices. (a) PLA. (b) PAL.**

these two basic types, which can literally have a complete digital system programmed into a single IC.

The CPLD shown in Figure 21-4(a) contains several PLD blocks and a global interconnection matrix that determines the connection between the PLD blocks. To program a CPLD, you must first program each PLD block and then program the interconnections between the PLD blocks.

The FPGA shown in Figure 21-4(b) has a different internal structure. It has several programmable logic blocks that are interconnected by switches. To program an FPGA, you

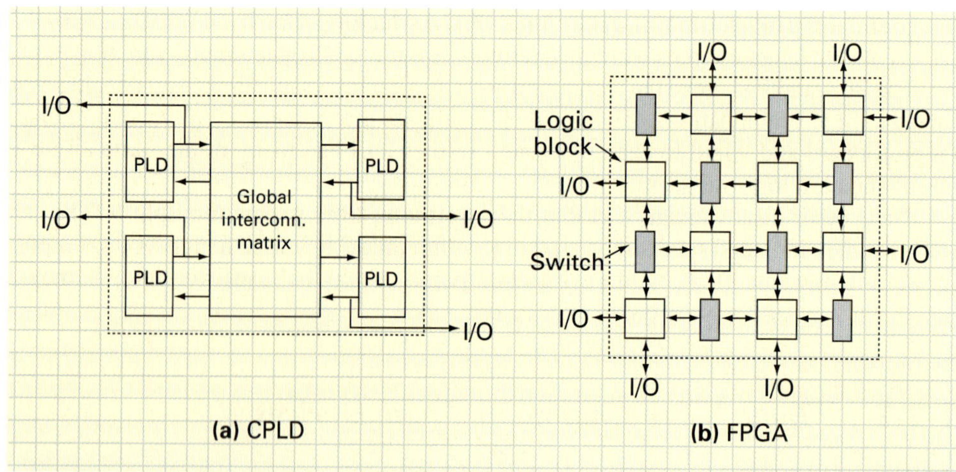

FIGURE 21-4 **Programmable Logic Device Types. (a) Complex Programmable Logic Device. (b) Field Programmable Gate Array.**

would first program the logic blocks to perform a logic function, such as AND, NOR, or XOR, and then program the switches to connect the logic blocks.

To overcome the one-time programmable disadvantage of the early PLDs, today's CPLDs and FPGAs contain reprogrammable switches at the cross points, which enable them to be used over and over.

SELF-TEST EVALUATION POINT FOR SECTION 21-2

Use the following questions to test your understanding of Section 21-2.

1. What are the full names and brief definitions of the following abbreviations:
   - **a.** PLA
   - **b.** PAL
   - **c.** OTP
   - **d.** CPLD
   - **e.** FPGA
2. Which inputs are fuse-programmable in the PLA, and which are fuse-programmable in the PAL?
3. What are the difference(s) between the internal architecture of the CPLD and the FPGA?
4. What two key advantages do the present-day CPLDs and FPGAs have over the early PLAs and PALs?

REVIEW QUESTIONS

Multiple-Choice Questions

1. The abbreviation PLD stands for
   - **a.** Program Logic Display
   - **b.** Programmable Language Decoder
   - **c.** Programmable Logic Device
   - **d.** Programming Language Device
2. Which of the following would be considered disadvantages of the standard logic prototyping method?
   - **a.** Wires can easily be inserted incorrectly, causing possible device damage and lengthy delays while the errors are isolated.
   - **b.** A large and costly inventory of all standard logic ICs must be maintained.
   - **c.** If the desired standard logic IC is not available, further delays will result.
   - **d.** To modify or add to a working circuit, the wires and ICs will normally all have to be removed from the protoboard and the new design built again from scratch.
   - **e.** All of the above.
3. The abbreviation CAD stands for
   - **a.** Computer-Assisted Designer
   - **b.** Code Activated Device
   - **c.** Computer-Aided Design
   - **d.** Computer-Activated Decoder
4. What three key devices are needed to construct a circuit using a programmable logic device?
   - **a.** CAD, PLD, PC
   - **b.** PLD, PC, SLDs
   - **c.** PC, SLDs, CAD
   - **d.** All of the above.
5. The abbreviation HDL stands for
   - **a.** Hardware Description Language
   - **b.** Hexadecimal Decoder Logic
   - **c.** Hard Description Language
   - **d.** Hardware Decoder Logic
6. Using the PC, you can write logic programs in a ____________ using a text editor or simply draw your logic circuit using a ____________.
   - **a.** schematic editor, HAL
   - **b.** PAL, HDL
   - **c.** HDL, schematic editor
   - **d.** HDL, PLA
7. Once the logic program or logic circuit has been created, it is then ____________ by the CAD program to create a detailed logic circuit that will perform the actions you specified.
   - **a.** compiled
   - **b.** simulated
   - **c.** downloaded
   - **d.** tested
8. What is the correct order for prototyping a digital circuit using a PLD?
   - **a.** Create, simulate, compile, download, test.
   - **b.** Create, compile, simulate, download, test.
   - **c.** Create, simulate, download, compile, test.
   - **d.** Compile, create, simulate, download, test.
9. Which of the following would be considered advantages of the programmable logic prototyping method?
   - **a.** Wiring errors can be avoided.
   - **b.** You can experiment with many digital IC types without having to stock them in your supply cabinet.
   - **c.** Circuit designs can be saved as electronic files within the PC and used again when needed.
   - **d.** Since the PLD can be used over and over again, modifications can easily be made by altering the circuit in the PC and then downloading the new design into the PLD.
   - **e.** All of the above.

10. The programmable logic devices of today evolved from the _____________, which was first introduced in the early 1970s.
    a. PAL c. PLA
    b. CPLD d. FPGA
11. With the _____________, the AND gate inputs were programmable but the OR gate inputs were not.
    a. PAL c. PLA
    b. CPLD d. FPGA
12. With _____________ devices, once the fuses were blown they remained that way, and so they could only be used once.
    a. OTP c. CAD
    b. HDL d. PLD
13. The _____________ contains several PLD blocks and a global interconnection matrix that determines the connection between the PLD blocks.
    a. PAL c. PLA
    b. CPLD d. FPGA
14. The _____________ has several programmable logic blocks that are interconnected by switches.
    a. PAL c. PLA
    b. CPLD d. FPGA
15. To program a(n) _____________ you would first program the logic blocks to perform a logic function, such as AND, NOR, or XOR, and then program the switches to connect the logic blocks.
    a. PAL c. PLA
    b. CPLD d. FPGA
16. To program a _____________ you must first program each PLD block and then program the interconnections between the PLD blocks.
    a. PAL c. PLA
    b. CPLD d. FPGA
17. To overcome the OTP disadvantage of the early PLDs, today's CPLDs and FPGAs contain _____________ at the cross points, which enable them to be used over and over again.
    a. programmable fuses
    b. reprogrammable switches
    c. reprogrammable circuit breakers
    d. Both a and c are true.

Practice Problems

To practice your device recognition and operation skills, refer to Figure 21-5 and answer the following questions.

18. How many inputs are being applied to the PLD, and how many outputs from the PLD are being applied to the 7-segment display?
19. Referring to the PLD's internal schematic diagram shown in part (b), would you say that this PLD is a PLA or a PAL?
20. What is the function of this PLD?
21. Describe the function table in part (a).
22. What do *a* through *f* refer to?
23. What would be the *a* through *f* outputs if the DCBA input is
    a. 0011 c. 0101
    b. LHLH d. HLLH
24. What digit will be shown on the 7-segment display if the following DCBA inputs were applied:
    a. 0011 c. 0101
    b. LHLH d. HLLH
25. Describe the similarity between the function table and the intact fuses in the PLD's schematic diagram.
26. What is the purpose of resistors R_1 through R_7?

Web Site Questions

Go to the web site http://www.prenhall.com/cook, select the textbook *Electronics: A Complete Course,* select this chapter, and then follow the instructions when answering the multiple choice practice problems.

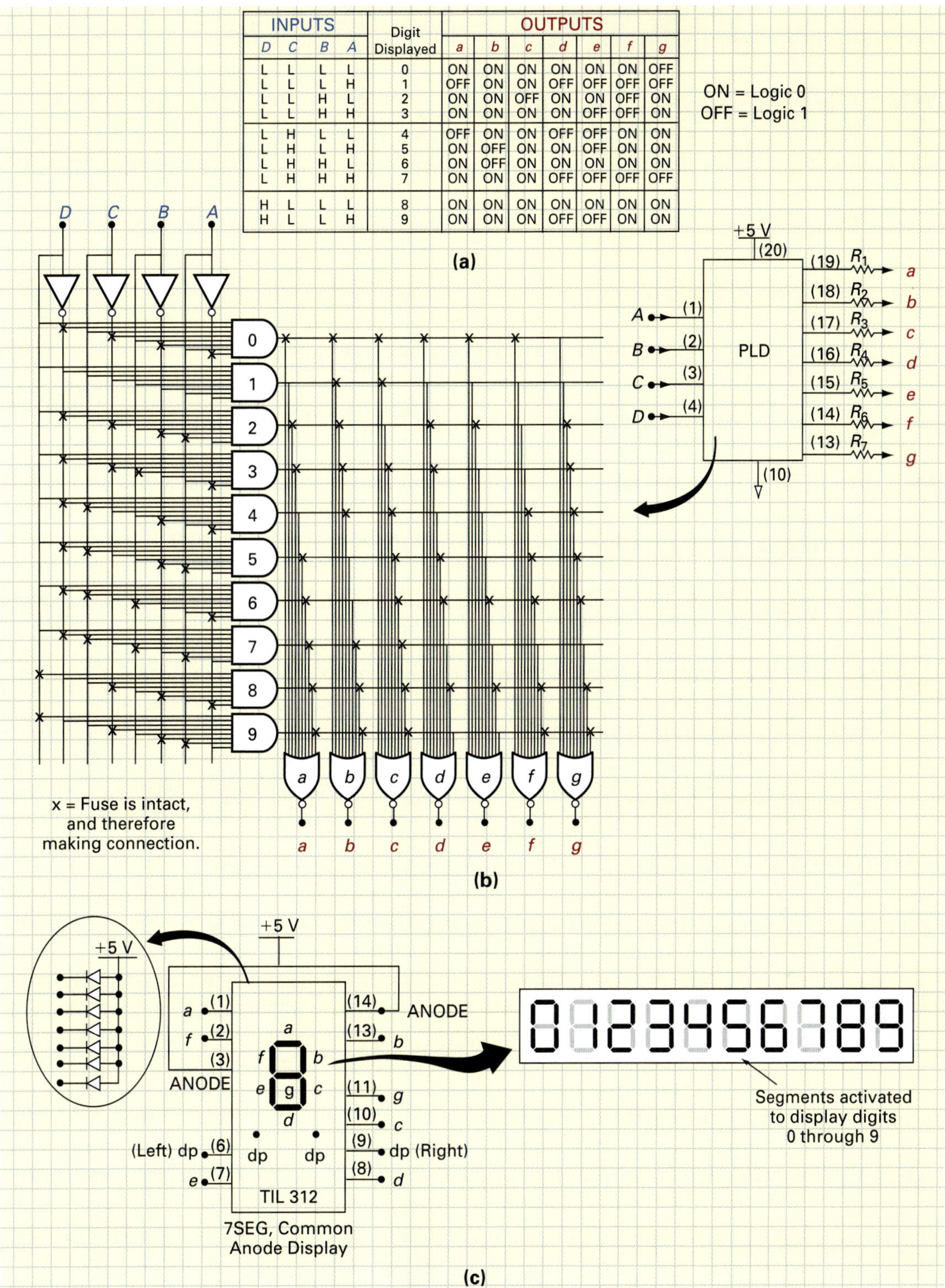

INPUTS				Digit Displayed	OUTPUTS						
D	*C*	*B*	*A*		*a*	*b*	*c*	*d*	*e*	*f*	*g*
L	L	L	L	0	ON	ON	ON	ON	ON	ON	OFF
L	L	L	H	1	OFF	ON	ON	OFF	OFF	OFF	OFF
L	L	H	L	2	ON	ON	OFF	ON	ON	OFF	ON
L	L	H	H	3	ON	ON	ON	ON	OFF	OFF	ON
L	H	L	L	4	OFF	ON	ON	OFF	OFF	ON	ON
L	H	L	H	5	ON	OFF	ON	ON	OFF	ON	ON
L	H	H	L	6	ON	OFF	ON	ON	ON	ON	ON
L	H	H	H	7	ON	ON	ON	OFF	OFF	OFF	OFF
H	L	L	L	8	ON	ON	ON	ON	ON	ON	ON
H	L	L	H	9	ON	ON	ON	OFF	OFF	ON	ON

FIGURE 21-5 Application—Using a PLD to Drive a 7-Segment LED Display.

Testing and Troubleshooting

Space, the Final Frontier

On May 24, 1962 a deep silence and tension filled the Mercury Control Center at Cape Canaveral, Florida. The only voice to be heard was that of Gus Grissom as he repeatedly tried to make contact with fellow astronaut Scott Carpenter aboard the *Aurora 7,* which had just completed three orbits around the earth and was now attempting a reentry. The normal procedure at this stage of the mission is to fire the thrusters to correctly orient the craft to the proper angle for returning to earth. The next step is to slow down the capsule by applying the brakes or retrorockets, which causes the spacecraft to fall into the atmosphere and to earth. The final step is to pop open the parachutes at about 25,000 feet and lower the capsule into the sea. All of these steps are crucial, and if any one of the steps is not followed exactly, it can spell disaster. For example, if the spacecraft is pitched at the wrong angle, or if it is traveling at too great a speed, the craft and astronaut will burn up due to the friction of the earth's atmosphere.

This was the second American orbital flight; however, unlike the first, this mission had not gone at all according to plan. The problems began when Carpenter was asked to position *Aurora 7* for the return and the capsule's automatic stabilization system had failed. Carpenter responded immediately by switching to manual control, but due to the malfunction, the craft was misaligned by 25 degrees. To make matters worse, the retrorockets fired three seconds too late, failing to produce the expected braking power.

Nine minutes had now passed and Gus Grissom at the Cape urgently repeated the same message into his microphone, not knowing whether Carpenter was dead or alive: "*Aurora 7, Aurora 7,* do you read me? Do you read me?"

Hundreds of miles away, a pair of mainframe computers at Goddard Space Flight Center in Maryland knew exactly where he was. These machines had continually digested the direction and altitude data from a radar ground station in California that had tracked the capsule from its time of reentry. Eleven minutes after the last radar contact, the mainframe displayed an incredibly accurate estimate of the capsule's splashdown position. Aircraft were immediately dispatched and homed in on the capsule's signal beacon, where Carpenter was found riding in an inflatable life raft alongside his spacecraft. Luckily, the combined effects of the errors had brought Carpenter safely through the atmosphere, but some 250 miles away from the planned splashdown point and well out of range of Grissom's radio transmission and the awaiting naval recovery task force.

At this time in history, the computer was not completely accepted as a valued partner in the space business. Astronauts, aerospace engineers, and ground control personnel were wary of putting too much trust in a machine that at that time would sometimes fail to operate for no apparent reason.

Today, the computer has proved itself indispensable, lending itself to almost every phase of our exploration of the final frontier. Computers are used by aerospace engineers to design space vehicles and to simulate the designs to determine their flight characteristics. These digital electronic computer systems can also simulate missions into deep space;

22

an astronaut can take a flight simulation to almost any place in the galaxy, developing good flight skills on the way. Computers are also used to continually monitor the thousands of tests preceding a launch and will stop the countdown for any abnormality. Once the mission is underway, the on-board computer supplies the astronaut quickly and accurately with all information needed, enabling mankind to explore new worlds and new civilizations and to boldly go where no one has gone before!

Introduction

This chapter will introduce you to digital test equipment and digital troubleshooting techniques. An effective electronics technician or troubleshooter must have a thorough knowledge of electronics, test equipment, troubleshooting techniques, and equipment repair. Like analog circuits, digital circuits occasionally fail, and in most cases a technician is required to quickly locate the problem within the system and then make the repair. The procedure for fixing a failure can be broken down into three basic steps.

Step 1: DIAGNOSE
The first step is to determine whether a problem really exists. To carry out this step, a technician must collect as much information about the system, the circuit, and the components used, and then diagnose the problem.

Step 2: ISOLATE
The second step is to apply a logical and sequential reasoning process to isolate the problem. In this step, a technician will operate, observe, test, and apply troubleshooting techniques in order to isolate the malfunction.

Step 3: REPAIR
The third and final step is to make the actual repair and final test the circuit.

Troubleshooting, by definition, is the process of locating and diagnosing breakdowns in equipment by means of systematic checking and analysis. To troubleshoot, you will need a thorough knowledge of troubleshooting techniques, a very good understanding of test equipment, documentation in the form of technical and service manuals, and experience. Troubleshooting experience can only be acquired with practice; therefore, in future chapters, complete troubleshooting sections will be included and applied to all new devices and circuits. As for practical experience, very few of your lab circuits will work perfectly the first time, so you will gain troubleshooting experience each and every time you experiment in lab. Although it seems very frustrating when a circuit is not operating correctly, remember that the more problems you have with a circuit, the better.

In this chapter, you will first be introduced to digital test equipment, and then in the following section you will see how these test instruments can be used to isolate internal and external digital IC failures. In the final section we will discuss how to make repairs to an electronics circuit once the problem has been isolated.

22-1 DIGITAL TEST EQUIPMENT

A variety of test equipment is available to help you troubleshoot digital circuits and systems. Some are standard, such as the multimeter and oscilloscope, which can be used for either analog or digital circuits. Other test instruments, such as the logic clip, logic probe, logic pulser, and current tracer, have been designed specifically to test digital logic circuits. In this section we will be examining each of these test instruments, beginning with the multimeter.

22-1-1 *Testing with the Multimeter*

Multimeter

Test instrument able to perform multiple tasks in that it can be used to measure voltage, current, or resistance.

Both analog and digital **multimeters** can be used to troubleshoot digital circuits. In most cases, you will only use the multimeter to test voltage and resistance. The current settings are rarely used since a path needs to be opened to measure current, and as most circuits are soldered onto a printed circuit board, making current measurements is impractical. Some of the more common digital circuit tests using the multimeter include the following:

a. You can use the multimeter to test power supply voltages. On the AC VOLTS setting, you can check the 120 V ac input into the system's dc power supply, and using the DC VOLTS setting you can check all of the dc power supply voltages out of the power supply circuit, as seen in Figure 22-1(a). The multimeter can also be used to check that these dc supply voltages are present at each printed circuit board within the system, as in the case of the 5 V supply to the display board shown in Figure 22-1(b).

b. You can also use the multimeter on the DC VOLTS setting to test the binary 0 and binary 1 voltages throughout a digital circuit, as shown in Figure 22-1(b).

c. The multimeter is also frequently used on the OHMS setting to test components such as resistors, fuses, diodes, transistors, wires, cables, printed circuit board tracks, and other disconnected devices, as shown in Figure 22-1(c).

22-1-2 *Testing with the Oscilloscope*

Pulse Wave

A repeating wave that only alternates between two levels or values and remains at one of these values for a small amount of time relative to the other.

Pulse Repetition Frequency

The number of times a second a pulse is generated.

Pulse Repetition Time

The time interval between the start of two consecutive pulses.

Since digital circuits manage two-state information, most of the signals within a digital system will be either square or rectangular. To review, the **pulse, or rectangular, wave** alternates between two peak values; however, unlike the square wave, the pulse wave does not remain at the two peak values for equal lengths of time. The *positive pulse waveform* seen in Figure 22-2(a), for instance, remains at its negative value for long periods of time and only momentarily pulses positive. On the other hand, the *negative pulse waveform* seen in Figure 22-2(b) remains at its positive value for long periods of time and momentarily pulses negative.

When referring to pulse waveforms, the term **pulse repetition frequency** (PRF) is used to describe the frequency or rate of the pulses, and the term **pulse repetition time** (PRT) is used to describe the period or time of one complete cycle, as seen in Figure 22-2(c). For example, if 1000 pulses are generated every second (PRF = 1000 pulses per second, pps or 1 kHz), each cycle will last 1/1000 of a second or 1 ms (PRT = 1/PRF = 1/1 kHz = 1 ms or 1000 μs).

The duty cycle of a pulse waveform indicates the ratio of pulse width time to the complete cycle time (it compares ON time to OFF time) and is calculated with the formula

$$\text{Duty Cycle} = \frac{\text{Pulse Width } (P_w)}{\text{Period } (t)} \times 100\%$$

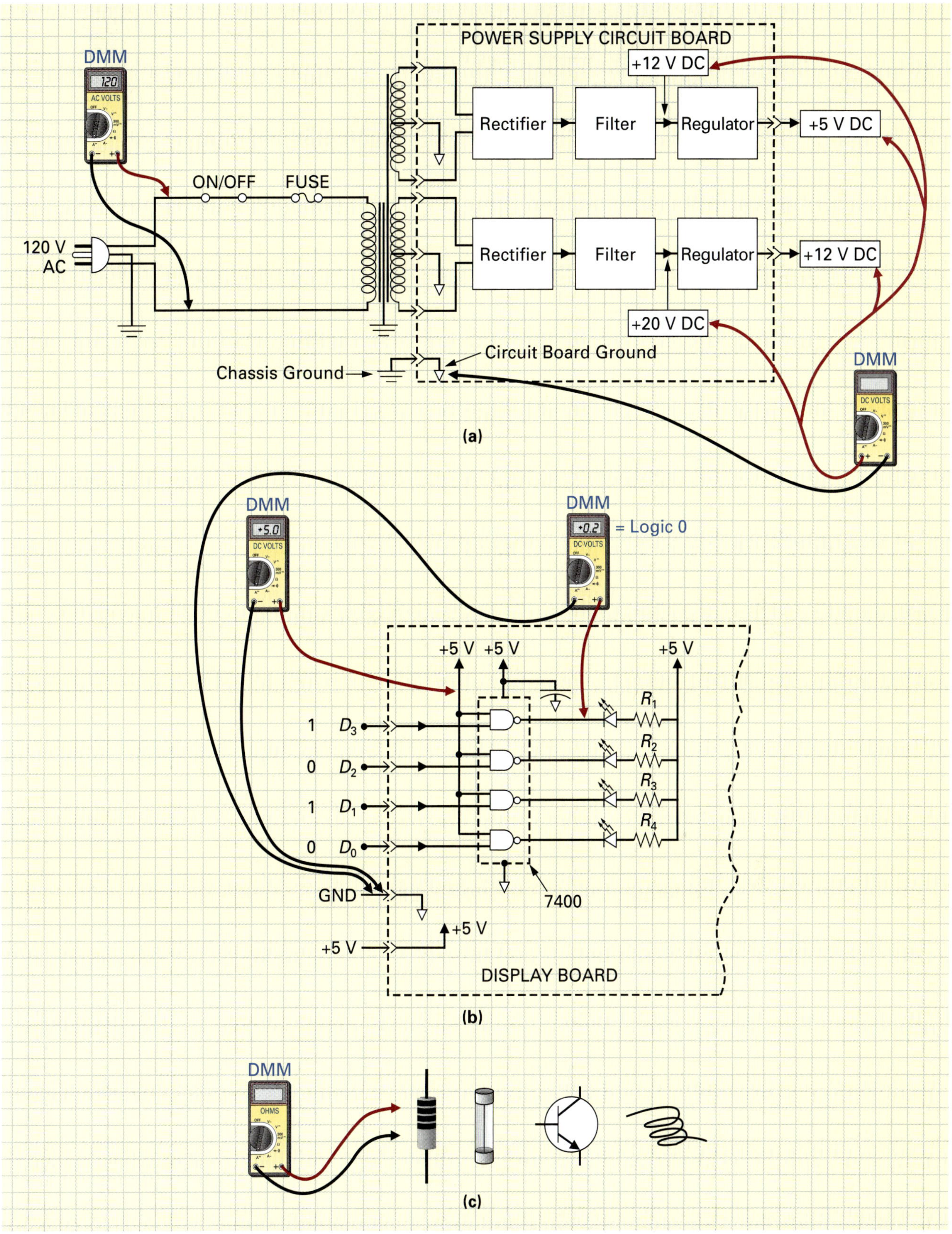

POWER SUPPLY CIRCUIT BOARD

DMM

120

AC VOLTS

ON/OFF

FUSE

120 V AC

+12 V DC

Rectifier

Filter

Regulator

+5 V DC

Rectifier

Filter

Regulator

+12 V DC

+20 V DC

Circuit Board Ground

Chassis Ground

DMM

DC VOLTS

(a)

DMM

+5.0

DC VOLTS

DMM

+0.2

DC VOLTS

= Logic 0

+5 V +5 V

+5 V

R_1

R_2

R_3

R_4

1 D_3

0 D_2

1 D_1

0 D_0

GND

+5 V

+5 V

7400

DISPLAY BOARD

(b)

DMM

OHMS

(c)

FIGURE 22-1 **Testing with the Multimeter.**

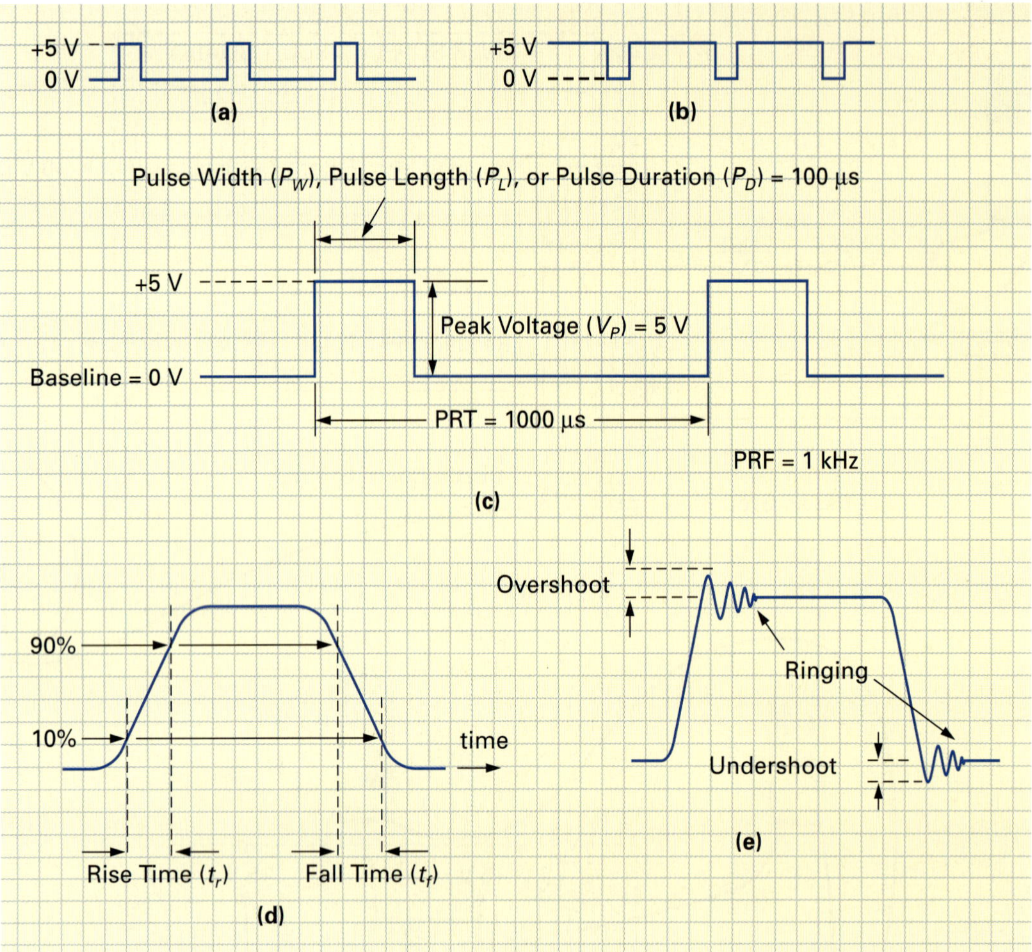

FIGURE 22-2 **The Pulse Waveform. (a) Positive Pulse Waveform. (b) Negative Pulse Waveform. (c) Peak Voltage, Pulse Width, PRF, and PRT. (d) Rise Time and Fall Time. (e) Overshoot, Undershoot, and Ringing.**

EXAMPLE:

Calculate the pulse width, PRT, PRF, and duty cycle for the waveform shown in Figure 22-2(c).

Solution:

In the example in Figure 22-2(c), a pulse width of 100 μs and a period of 1000 μs will produce a duty cycle of

$$\text{Duty Cycle} = \frac{\text{Pulse Width } (P_w)}{\text{Period } (t)} \times 100\%$$

$$= \frac{100\ \mu\text{s}}{1000\ \mu\text{s}} \times 100\% = 10\%$$

A duty cycle of 10% means that the positive pulse lasts for 10% of the complete cycle time, or PRT.

Since the pulse repetition time is 1000 μs, the pulse repetition frequency will equal the reciprocal of this value, or 1 kHz (PRF = 1/PRT = 1/1000 μs = 1 kHz).

Another area that needs to be reviewed with regard to digital signals is rise and fall time. The pulse waveforms seen in Figure 22-2(a), (b), and (c) all rise and fall instantly between their two states or peak values. In reality, there is a time lapse referred to as the *rise*

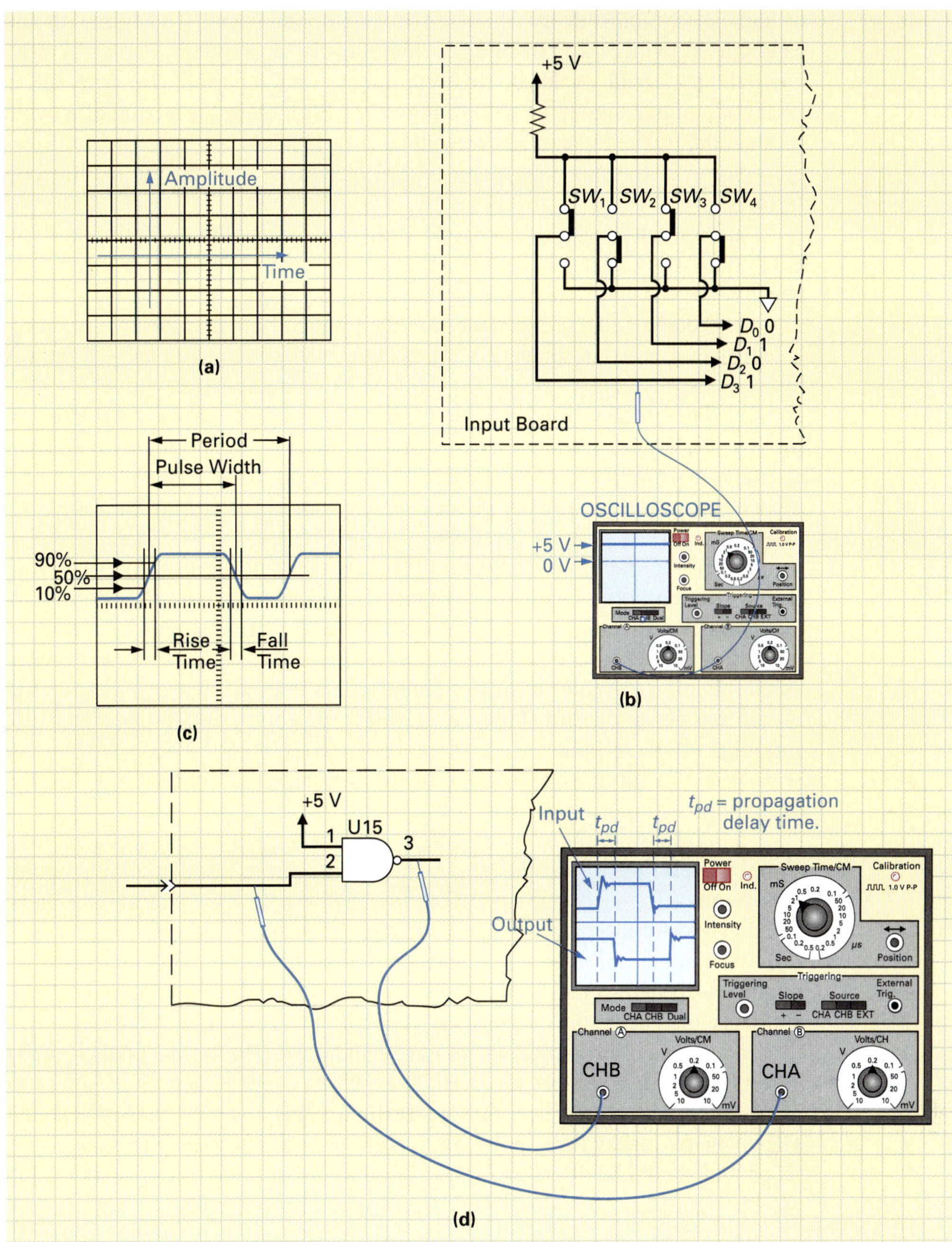

FIGURE 22-3 **Testing with the Oscilloscope.**

time and *fall time,* as seen in Figure 22-2(d). The rise time is defined as the time needed for the pulse to rise from 10% to 90% of the peak amplitude, while the fall time is the time needed for the pulse to fall from 90% to 10% of the peak amplitude.

When carefully studying pulse waveforms on the oscilloscope, you will be able to measure pulse width, PRT, rise time, and fall time. You will also see a few other conditions known as *overshoot, undershoot,* and *ringing.* These unwanted conditions tend to always accompany high-frequency pulse waveforms due to imperfections in the circuit, as seen in Figure 22-2(e). As you can see, the positive rising edge of the waveform "overshoots" the peak, and then a series of gradually decreasing sine wave oscillations occur, called ringing. At the end of the pulse this ringing action reoccurs, as the negative falling edge "undershoots" the negative peak value.

The graticules on an oscilloscope's display can be used to measure amplitude and time, as seen in Figure 22-3(a). This time/amplitude measuring ability of the oscilloscope enables us to measure almost every characteristic of a digital signal. To begin with, the vertical scale can be used to measure HIGH/LOW logic levels, as shown in Figure 22-3(b). On the other hand, the horizontal scale can be used to measure the period, rise time, fall time, and pulse width of a digital signal, as seen in Figure 22-3(c). Once the period of a cycle has been calculated, the frequency can be determined ($f = 1/t$).

Some of the more common digital circuit tests using the oscilloscope include the following:

a. You can use the oscilloscope to check the clock or master timing signal within a digital system to see that it is not only present at every point in the circuit but also that its frequency, wave shape, and amplitude are correct.

b. You will also use the oscilloscope frequently to monitor two or more signals simultaneously, as shown in Figure 22-3(d). In this instance, an input and output waveform are being compared to determine the NAND gate's propagation delay time.

c. The oscilloscope can also be used to measure waveform distortions, such as the ringing shown in Figure 22-3(d), to see if these undershoots or overshoots are causing any false operations, such as incorrectly triggering a logic gate. A noise problem in a digital circuit is often called a **glitch,** which stands for "gremlins loose in the computer housing."

Glitch

An acronym for "gremlins loose in the computer housing." The term is used to describe a problem within an electronic system.

From the previous discussion you can see that the oscilloscope is a very versatile test instrument, able to measure almost every characteristic of a digital signal. Its accuracy, however, does not match the multimeter for measuring logic level voltages, or the frequency counter for measuring signal frequency. The scope is a more visual instrument, able to display a picture image of signals at each point in a circuit and enabling its operator to see phase relationships, wave shape, pulse widths, rise and fall times, distortion, and other characteristics.

When using an oscilloscope to troubleshoot a digital circuit, try to remember the following:

a. Always use a *multitrace oscilloscope* when possible so you can compare input and output waveforms. Remember that a chopped or multiplexed display will not be accurate at high frequencies since one trace is trying to display two waveforms. A dual-beam oscilloscope is best for high-frequency digital circuit troubleshooting.

b. Trigger the horizontal sweep of the oscilloscope with the input signal or a frequency-related signal. Having a *triggered sweep* will ensure more accurate timing measurements and enable the oscilloscope to lock on to your input more easily.

c. Use an oscilloscope with a large *bandwidth.* The input signals are fed to vertical amplifiers within the oscilloscope, and these circuits need to have a high upper-frequency limit. Since all scopes can measure dc signals (0 Hz), this lower frequency is generally not listed. The upper-frequency limit, however, is normally printed somewhere on the front panel of the scope and should be 40 MHz for most digital measurements, 60 MHz for high-speed TTL, and between 100 and 1000 MHz for high-speed ECL circuits.

22-1-3 *Testing with the Logic Clip*

Logic Clip

A spring-loaded digital testing clip that can be clamped onto an IC package so that the signal present on any pin is available at the end of the clip. Some include LEDs to display logic levels present.

The **logic clip,** which is shown in Figure 22-4(a), was specially designed for troubleshooting digital circuits. It consists of a spring-loaded clip that clamps on to a standard IC package, where it makes contact with all the pins of the IC, as shown in Figure 22-4(b). These contacts are then available at the end of the clip, so a multimeter or oscilloscope probe can easily be connected.

Some logic clips simply have connections on the top of the clip, while others have light-emitting diodes (LEDs) and test points on the top of the clip, as seen in Figure 22-4(a). The LEDs give a quick indication of the binary level on the pins of the IC, with a logic 0 turning an LED OFF and a binary 1 turning an LED ON.

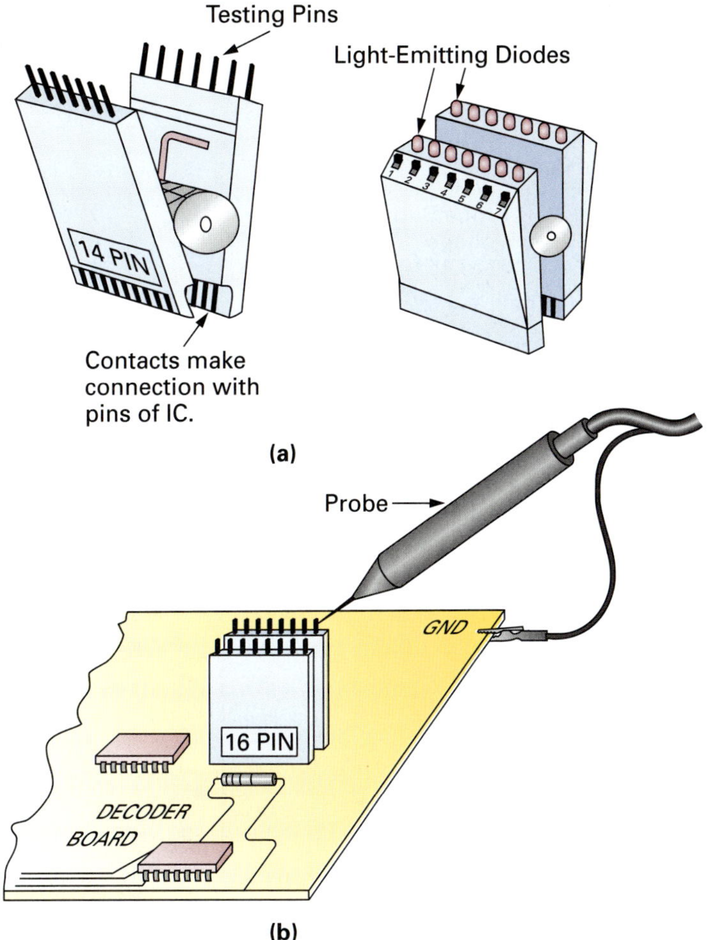

FIGURE 22-4 **Testing with the Logic Clip.**

22-1-4 *Testing with the Logic Probe*

Another test instrument designed specifically for digital circuit troubleshooting is the **logic probe,** shown in Figure 22-5(a). In order for the probe to operate, its red and black power leads must be connected to a power source. This power source can normally be obtained from the circuit being tested, as shown in Figure 22-5(b). Once the power leads have been connected, the logic probe is ready to sense the voltage at any point in a digital circuit and give an indication as to whether that voltage is a valid binary 1 or 0. The HI and LO LEDs on the logic probe are used to indicate the logic level, as seen in Figure 22-5(b). Since the voltage thresholds for a binary 0 and 1 can be different, most logic probes have a selector switch to select different internal threshold detection circuits for TTL, ECL, or CMOS.

Logic Probe
A digital test instrument that can indicate whether a HIGH, LOW, or PULSE is present at a circuit point.

Static Test
A test performed on a circuit that has been held constant in one condition.

Dynamic Test
A test performed on a circuit that is running at its normal operating speed.

In addition to being able to perform **static tests** (constant or nonchanging HIGH and LOW signals) like those of the multimeter, the logic probe can also perform **dynamic tests** (changing signals) like those of the oscilloscope, such as detecting a single momentary pulse (single pulse) or a pulse train (sequence of pulses), such as a clock signal. Nearly all logic probes have an internal low-frequency oscillator that will flash the "PULSE" LED ON and OFF if a pulse train is detected at a frequency of up to 100 MHz. The logic probe can also detect if a single pulse has occurred, even if the width of the pulse is so small that the operator was unable to see the PULSE LED blink. In this instance, a memory circuit can be

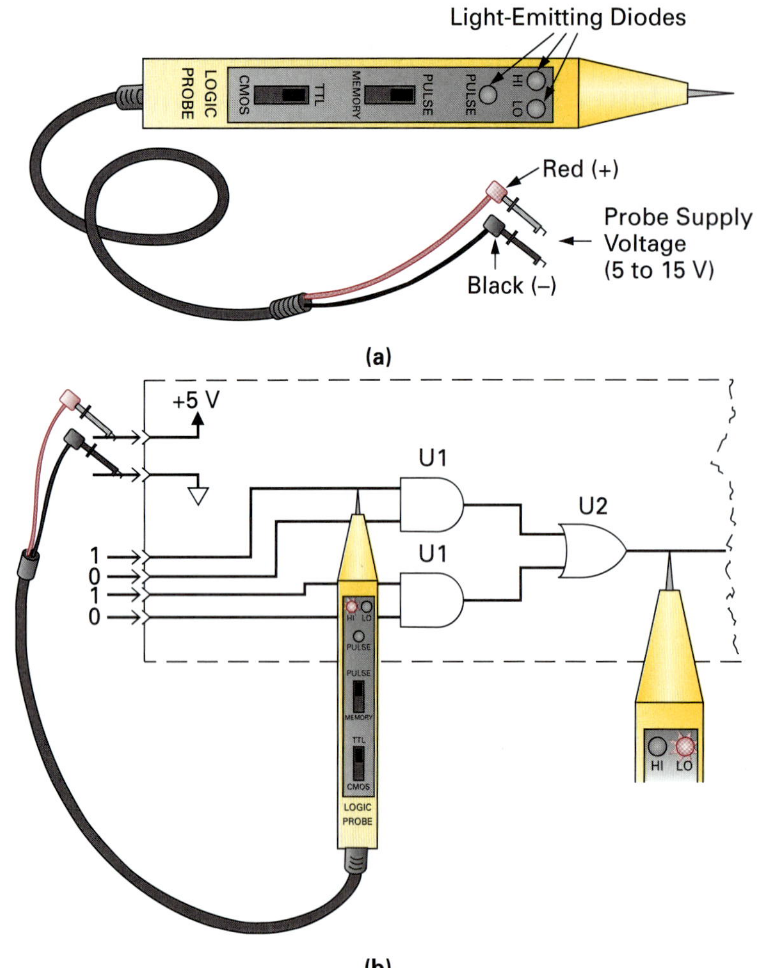

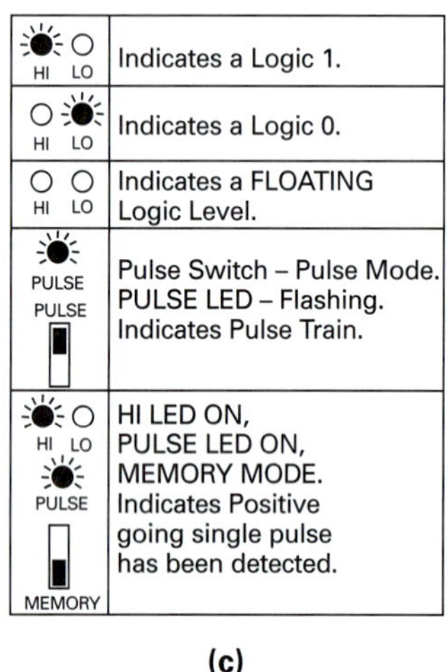

HI LO	Indicates a Logic 1.
HI LO	Indicates a Logic 0.
HI LO	Indicates a FLOATING Logic Level.
PULSE PULSE	Pulse Switch – Pulse Mode. PULSE LED – Flashing. Indicates Pulse Train.
HI LO PULSE MEMORY	HI LED ON, PULSE LED ON, MEMORY MODE. Indicates Positive going single pulse has been detected.

(c)

FIGURE 22-5 **Testing with the Logic Probe.**

enabled using the MEMORY switch, so that any pulse (even as narrow as 10 ns) will turn ON and keep ON the PULSE LED. This memory circuit within the logic probe senses the initial voltage level present at the test point (either HIGH or LOW) and then turns ON the PULSE LED if there is a logic transition or change. To reset this memory circuit, simply toggle (switch OFF and then ON) the memory switch. Figure 22-5(c) summarizes the operation and display interpretation of a typical logic probe.

Even though the logic probe is not as accurate as the multimeter, and cannot determine as much about a signal as the oscilloscope, it is ideal for digital circuit troubleshooting because of its low cost, small size, and versatility.

22-1-5 *Testing with the Logic Pulser*

Logic Pulser
A digital test instrument that can generate either single or repeating logic pulses.

Using the logic probe to test the static HIGH and LOW inputs and output of a gate does not fully test the operation of that logic gate. In most instances it is necessary to trigger the input of a circuit and then monitor its response to a dynamic test. A **logic pulser** is a signal generator designed to produce either a single pulse or a pulse train. Its appearance is similar to the logic probe, as shown in Figure 22-6(a), in that it requires a supply voltage to operate and has a probe which is used to apply pulses to the circuit under test, as shown in Figure 22-6(b).

The logic pulser is operated by simply touching the probe tip to any conducting point on a circuit and then pressing the SINGLE PULSE button for a single pulse or selecting the

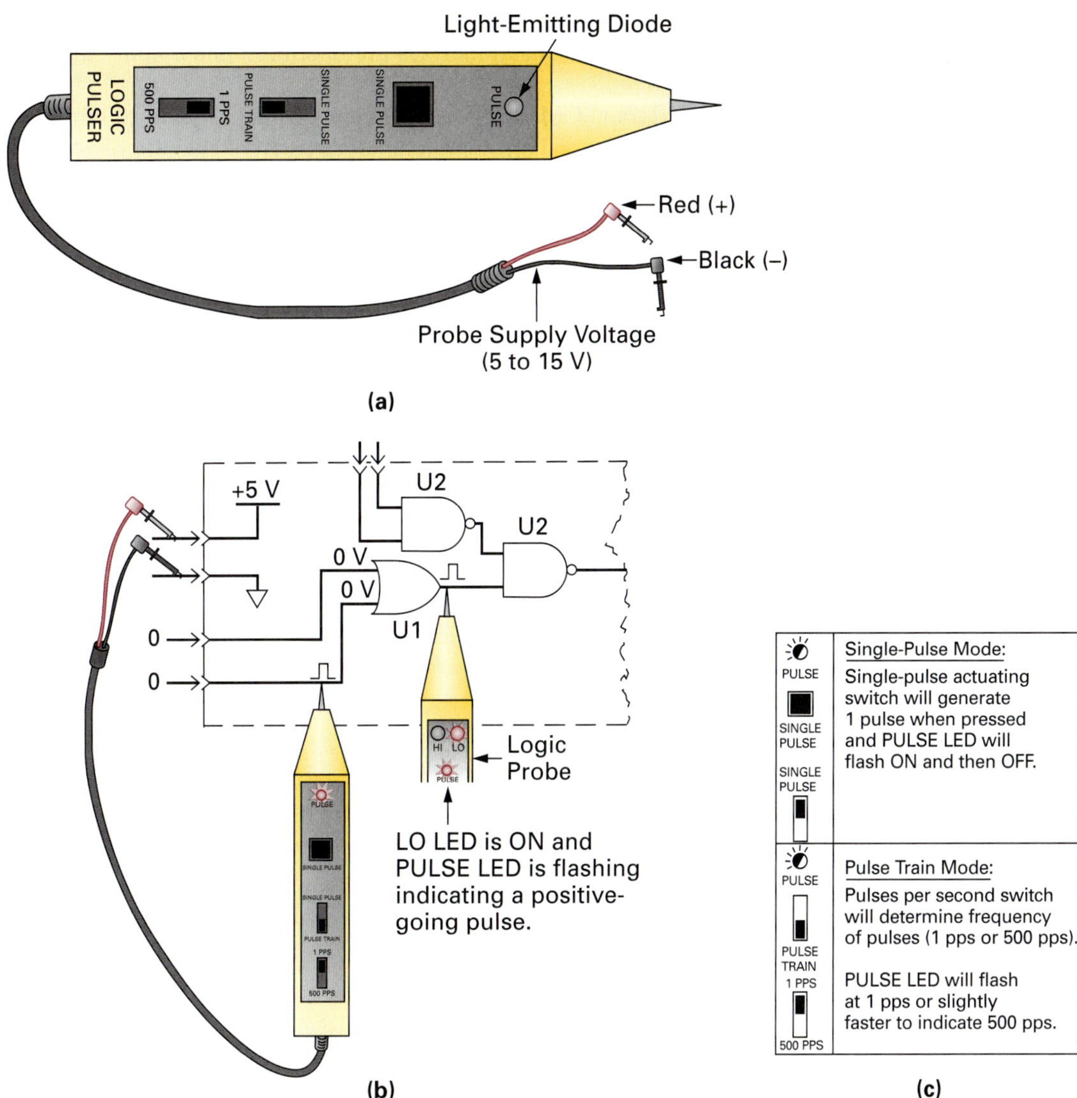

FIGURE 22-6 **Testing with the Logic Pulser.**

PULSE TRAIN switch position for a constant sequence of pulses. The frequency of the pulse train, with this example model, can be either 1 pulse per second (1 pps) or 500 pulses per second (500 pps), based on the pps selector switch.

The logic probe is generally used in conjunction with the logic pulser to sense the pulse or pulses generated by the logic pulser, as shown in Figure 22-6(b). In this example, the logic pulser is being used to inject a single pulse into the input of an OR gate, and a logic probe is being used to detect or sense this pulse at the output of the OR gate. If the input to the OR gate was HIGH, the logic pulser would have automatically sensed the binary 1 voltage level and then pulsed the input to the OR gate LOW when the pulse button was pressed. In this example, however, the input logic level to the OR gate was LOW, so the logic pulser sensed the binary 0 voltage level and then pulsed the OR gate input HIGH. This ability of the logic pulser to sense the logic level present at any point and then pulse the line to its opposite state means that the operator does not have to determine these logic levels before a point is tested, making the logic pulser a fast and easy instrument to use. The logic pulser achieves this feature with an internal circuit that will override the logic level in the circuit under test and either source current or sink current when a pulse needs to be generated. This complete in-circuit testing ability means that components do not need to be removed and input and output paths do not need to be opened in order to carry out a test. Figure 22-6(c) summarizes the operation and display interpretation of a typical logic pulser.

Since TTL, ECL, and CMOS circuits all have different input and output voltages and currents, logic pulsers will need different internal circuits for each family of logic ICs. Typical pulse widths for logic pulsers are between 500 ns and 1 μs for TTL logic pulsers, and 10 μs for CMOS logic pulsers.

22-1-6 *Testing with the Current Tracer*

Current Tracer

A digital test instrument used to sense relative values of current in a conductor using an inductive pick-up tip.

The logic pulser can also be used in conjunction with another very useful test instrument called a **current tracer,** which is shown in Figure 22-7(a). Like the logic probe, the current tracer is also a sensing test instrument, but unlike the logic probe, the current tracer senses the relative values of current in a conductor. It achieves this by using an insulated inductive pick-up tip, which senses the magnetic field generated by the current in a conductor. By adjusting the sensitivity control on the current tracer and observing the lamp's intensity when

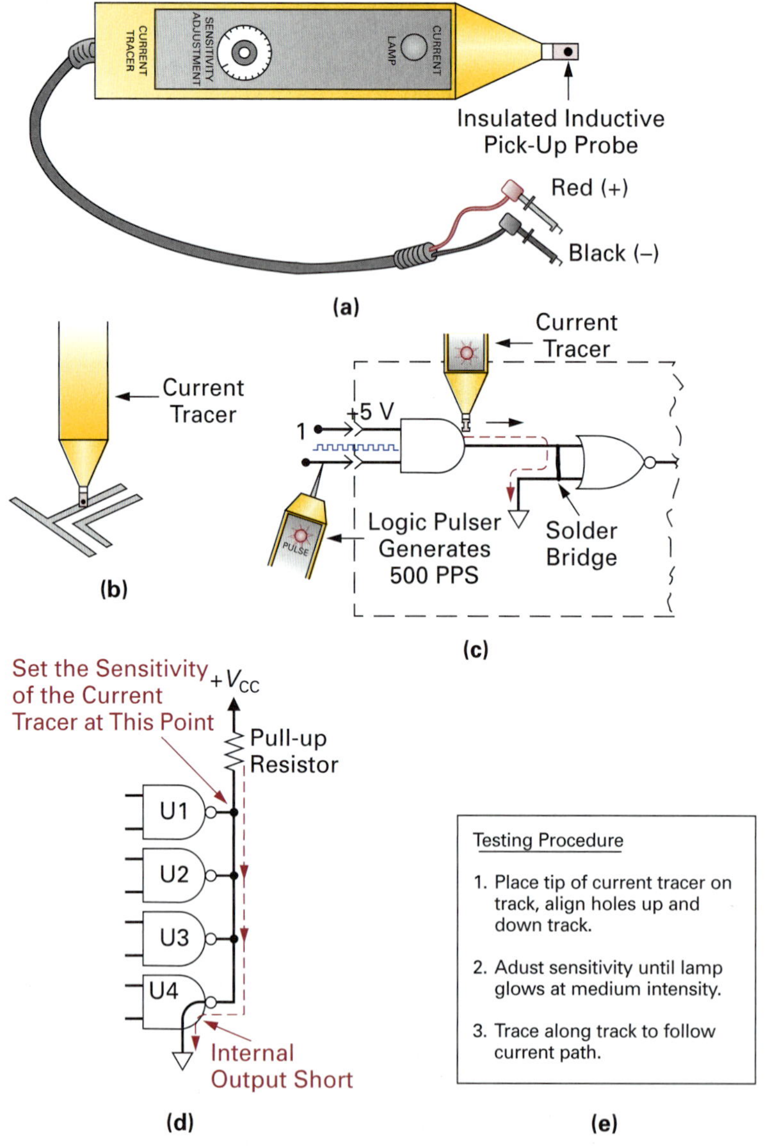

FIGURE 22-7 **Testing with the Current Tracer.**

the probe is placed on a pulsating logic signal line, a shorted path can be found by simply tracing the path of high current.

To troubleshoot a short in an IC's input or output circuit, a power supply, or a printed circuit board (PCB) track or cable, we would traditionally have to cut PCB tracks, snip pins, or open component leads. This would isolate the short, since the excessive current path or short would be broken when the path was opened. The current tracer allows the troubleshooter to isolate the shorted path without tampering with the circuit. Figure 22-7(b) illustrates how the current tracer should be correctly aligned on a conducting PCB track. Like the logic probe and logic pulser, the current tracer can use the same supply voltage as the circuit under test. Since the probe's tip is insulated, it can be placed directly on the track; however, be sure that the probe is always perpendicular to the board. The other important point to remember is to ensure that the small holes or dots on either side of the tip are aligned so they face up and down the track being traced at all times. This is so that the inductive coil in the insulated tip is oriented to pick up the maximum amount of magnetic flux.

Figure 22-7(c) shows an example of how a current tracer and logic pulser can be used to locate a short caused by a solder bridge. In this example, we will first connect the logic pulser so that it will generate a pulsating current, and therefore a changing magnetic field, which can be detected by the inductive tip of the current tracer. The next step is to touch the tip of the current tracer on the output of the AND gate and adjust the sensitivity level of the current tracer to light the current lamp (the sensitivity adjustment allows the current tracer to sense a wide range of current levels, from 1 mA to 1 A, typically). Once this reference level of brightness has been set, trace along the path between the output of the AND gate and the input of the OR gate. Since the current value should be the same at all points along this track, the lamp should glow at the same level of brightness. Once the current tracer moves past the solder bridge, the lamp will go out, since no current is present beyond this point. At this point you should, after a visual inspection, be able to locate the short.

The current tracer is also an ideal tool for troubleshooting a circuit in which many outputs are tied to a common point (such as open-collector gates), as seen in the example in Figure 22-7(d). If any of these gates were to develop an internal short, it would be difficult to isolate which of the parallel paths is causing the problem. In the past, you would have to remove each gate from the circuit (which is difficult with a soldered PCB) or clip the output IC pin of each gate until the short was removed and the faulty gate isolated. The current tracer can quickly isolate faults of this nature since it will highlight the current path provided by the short and therefore lead you directly to the faulty gate.

SELF-TEST EVALUATION POINT FOR SECTION 22-1

Use the following questions to test your understanding of Section 22-1.

1. What would be the most accurate instrument for measuring dc voltages?
2. Which test instrument would be best for measuring gate propagation delay time?
3. A logic ____________ is used to generate a single or train of test pulses.
4. A ____________ detects current pulses, while a ____________ detects voltage levels and pulses.

22-2 DIGITAL CIRCUIT PROBLEMS

Since ICs account for almost 85% of the components within digital systems, it is highly likely that most digital circuit problems will be caused by a faulty IC. Digital circuit failures can be basically divided into two categories: failures within ICs, or failures outside of ICs. When troubleshooting digital circuits, therefore, you will have to first isolate the faulty IC and then determine whether the problem is internal or external to the IC. An internal IC problem cannot be fixed and therefore the IC will have to be replaced, whereas an external IC

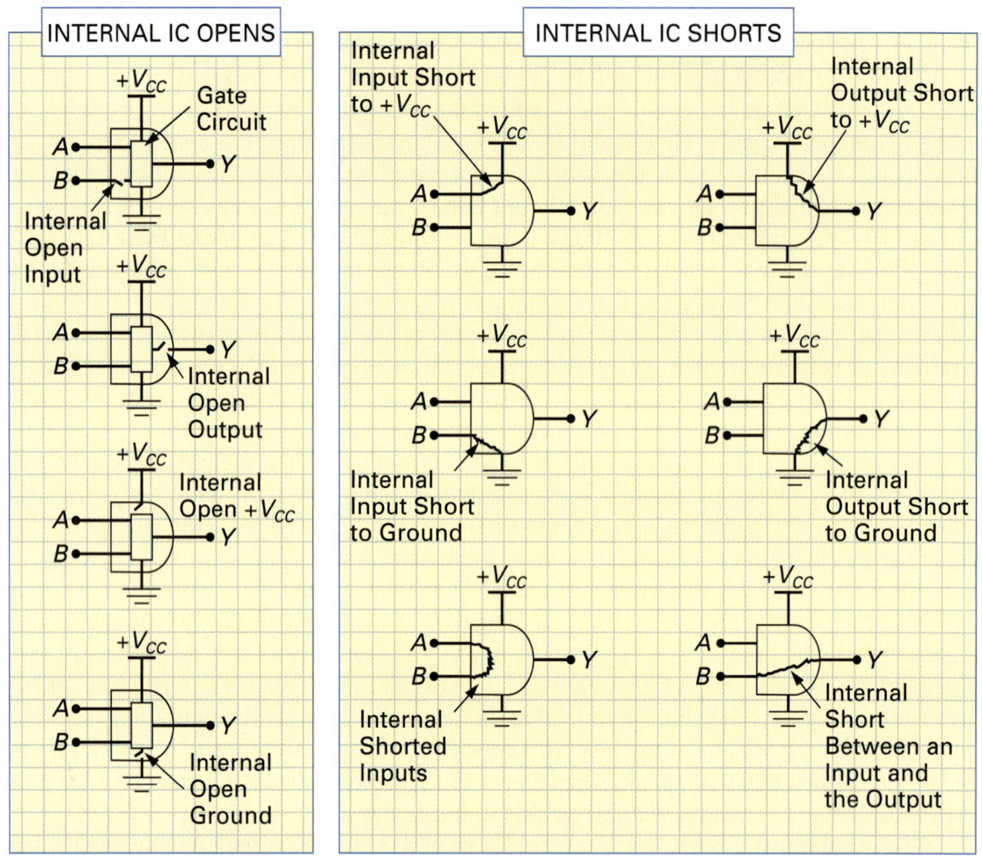

FIGURE 22-8 Internal IC Failures.

problem can be caused by electronic devices, connecting devices, mechanical or magnetic devices, power supply voltages, system clock signals, and so on. To begin with, let us discuss some of the typical internal IC failures.

22-2-1 *Digital IC Problems*

Digital IC logic gate failures can basically be classified as either opens or shorts in either the inputs or output, as summarized in Figure 22-8. In all cases, these failures will result in the IC having to be discarded and replaced. The following examples will help you to isolate an internal IC failure.

Internal IC Opens

An internal open gate input or an open gate output are very common internal IC failures. These failures are generally caused by the very thin wires connecting the IC to the pins of the package coming loose or burning out due to excessive values of current or voltage.

Figure 22-9 illustrates how a logic probe and logic pulser can be used to isolate an internal open NAND gate input. In Figure 22-9(a) the open input on pin 10 has been jumper connected to $+V_{CC}$, and the other gate input (pin 9) is being driven by the logic pulser. As you know, any LOW input to a NAND gate will produce a HIGH output, and only when both inputs are HIGH will the output be LOW, as shown in the inset in Figure 22-9(a). The logic probe, which is monitoring the gate on pin 8, indicates that a negative-going pulse waveform is present at the output, and therefore the gate seems to be functioning normally. When the jumper and pulser inputs are reversed, however, as seen in

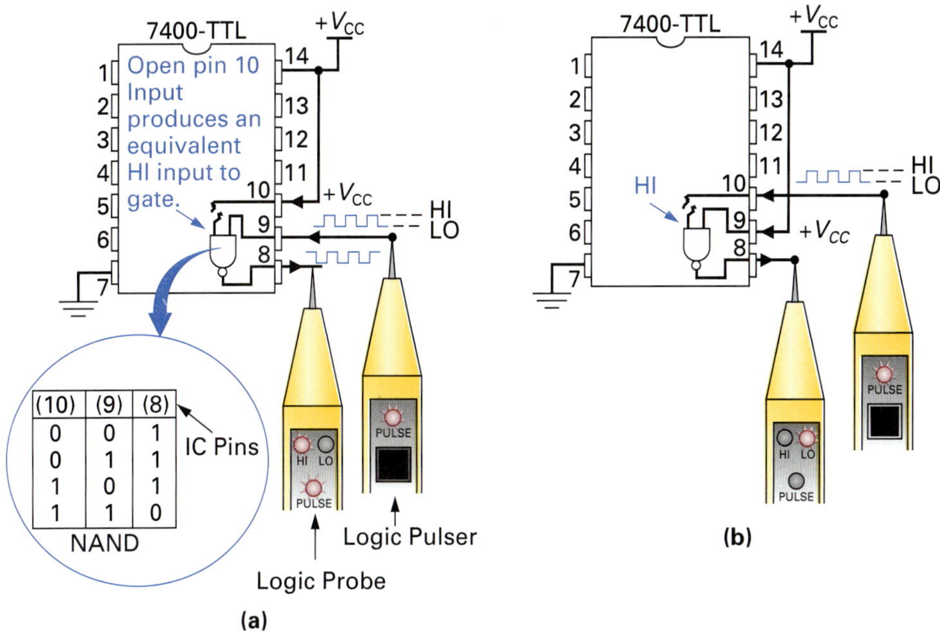

FIGURE 22-9 An Internal Open Gate Input.

Figure 22-9(b), the output remains permanently LOW. These conditions could only occur if the pin 10 input was permanently HIGH. As discussed previously, an open or floating input is equivalent to HIGH input, since both conditions have next-to-no input current. The cause, therefore, must be either an internal open at pin 10 (since this is equivalent to a HIGH input) or a $+V_{CC}$ short to the pin 10 input of this gate (which is less common than an open). In either case, however, the logic gate IC has an internal failure and will have to be replaced.

Figure 22-10 illustrates how a logic pulser and logic probe can be used to isolate an internal open gate output. In Figure 22-10(a), the logic gate's input at pin 10 has been jumper

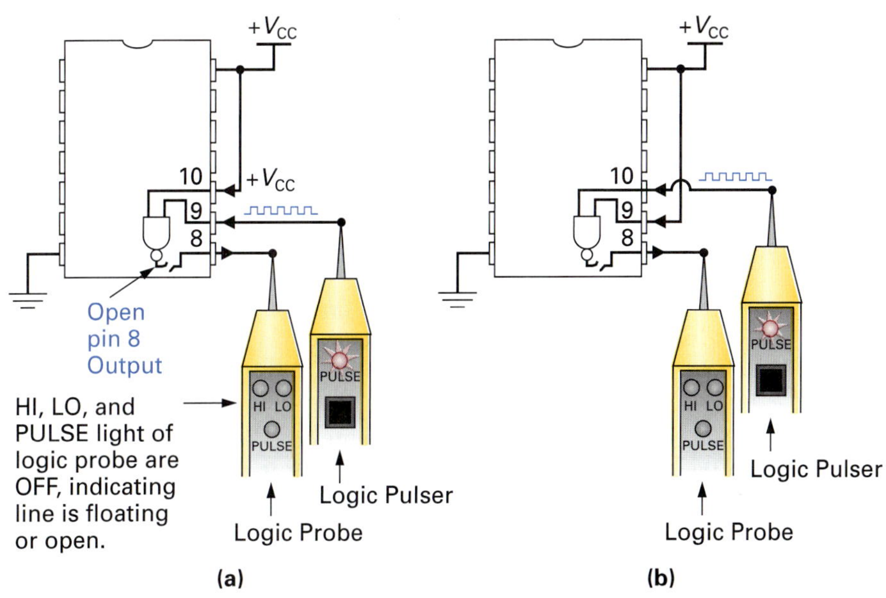

FIGURE 22-10 An Internal Open Gate Output.

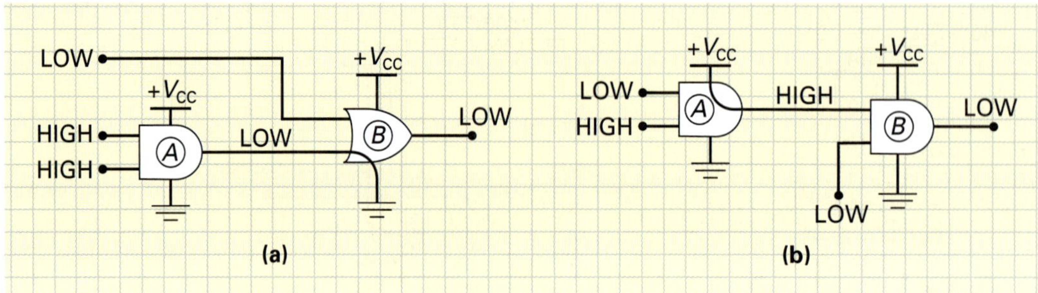

FIGURE 22-11 **Internal IC Shorts.**

connected to $+V_{CC}$, and the other input at pin 9 is being driven by the logic pulser. The logic probe, which is monitoring the output on pin 8, has none of its indicator LEDs ON, which indicates that the logic level is neither a valid logic 1 or a valid logic 0 (line is probably floating). In Figure 22-10(b), the inputs have been reversed; however, the logic probe is still indicating the same effect. This test highlights the failure, which is an open gate output. Once again, the logic gate IC has an internal failure and will have to be replaced.

Internal IC Shorts

Internal logic gate shorts to either $+V_{CC}$ or ground will cause the inputs or output to be stuck either HIGH or LOW. These internal shorts are sometimes difficult to isolate, so let us examine a couple of typical examples.

Figure 22-11(a) shows the effect an internal input lead short to ground will have on a digital logic gate circuit. The input to gate *B* is shorted to ground, so this input is being pulled LOW. This LOW between *A* output and *B* input makes it appear as though gate *A* is malfunctioning, since it has two HIGH inputs and therefore should be giving a LOW output. The problem is solved by disconnecting gate *A* from gate *B*. Since gate *A*'s output will go HIGH when it is no longer being pulled down by gate *B*, the fault will be isolated to a short at the input of gate *B*.

In Figure 22-11(b), the output of gate *A* has shorted to $+V_{CC}$. This circuit condition gives the impression that gate *A* is malfunctioning; however, if the *B* gate input had shorted to $+V_{CC}$ we would get the same symptoms. Once again, we will have to isolate, by disconnecting the *A* output from the *B* input, to determine which gate is faulty.

Troubleshooting Internal IC Failures

Let us now try a few examples to test our understanding of digital test equipment and troubleshooting techniques.

EXAMPLE:

Determine whether a problem exists in the circuit in Figure 22-12(a) based on the logic probe readings indicated. If a problem does exist, indicate what you think could be the cause.

Solution:

Gates *B* and *C* seem to be producing the correct outputs based on their input logic levels. Gate *A*'s output, on the other hand, should be LOW, and it is reading a HIGH. To determine whether the problem is the source gate *A* or the load gate *B*, the two gates have been isolated as shown in Figure 22-12(b). After isolating gate *A* from gate *B*, you can see that the pin 10 input to gate *C* seems to be stuck HIGH, probably due to an internal short to $+V_{CC}$.

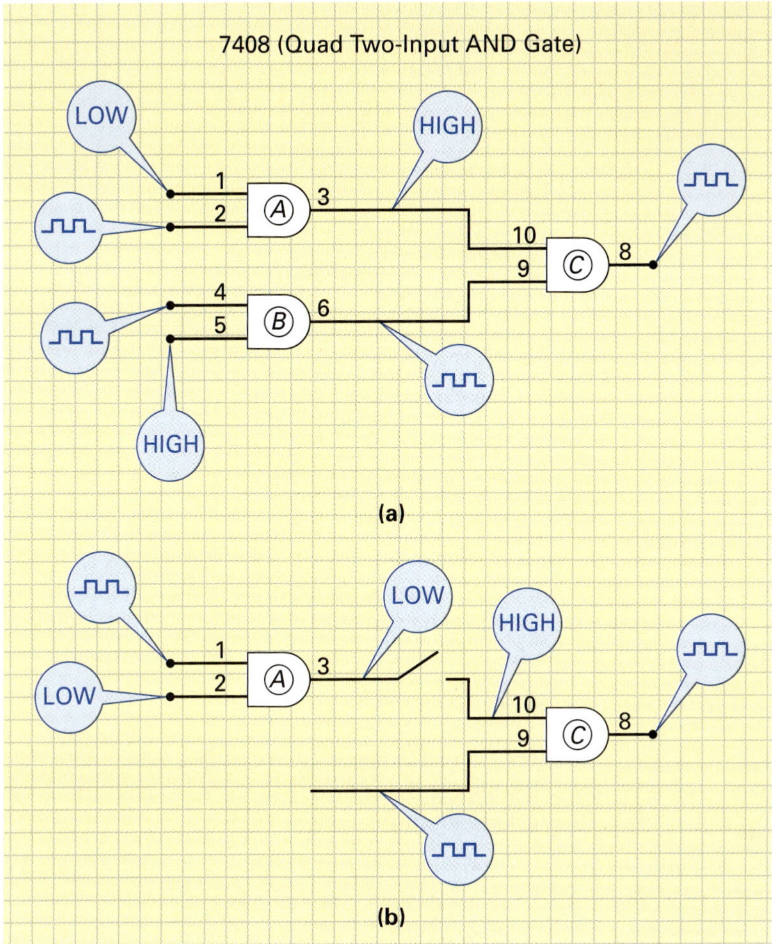

FIGURE 22-12 **Troubleshooting Gate Circuits.**

Even though gate *C* seems to be the only gate malfunctioning, the whole 7408 IC will have to be replaced. In situations like this, it is a waste of time isolating the problem to a specific gate when the problem has already been isolated down to a single component. Use your understanding of internal logic gate circuits to isolate the problem to a specific component and not to a device within that component.

EXAMPLE:

What could be the possible fault in the circuit shown in Figure 22-13?

Solution:

The output of NOT gate *A* should be HIGH; however, it is pulsating. This pulsating signal can only have come from pin 2 of the NAND gate. By disconnecting the pulsating input at pin 2 of the NAND gate and then single-pulsing this input using the logic pulser, you can use the logic probe to see if the pin 1 input of the NAND gate always follows what is on pin 2. Once an input short has been determined, visually inspect the NAND gate IC to be sure that there is no external short between pins 1 and 2.

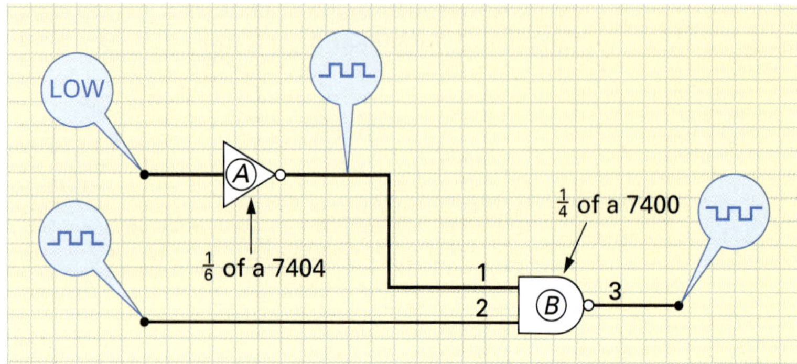

FIGURE 22-13 **Troubleshooting Gate Circuits.**

22-2-2 *Other Digital Circuit Device Problems*

Failures external to ICs can produce almost exactly the same symptoms as internal IC failures, so your task is to determine whether the problem is internal or external to the IC. The external IC circuit problems can be wide and varied; however, let us discuss some of the more typical failures.

External IC Shorts and Opens

Most digital signal line shorts can be isolated by injecting a signal onto the line using the logic pulser and then using the current tracer to find the shorted path. Here are some examples of typical external IC short circuit problems.

a. A short between the pins of the IC package due to a solder bridge, sloppy wiring, wire clippings, or an improperly etched printed circuit board.
b. A bending in or out of the IC pins as they are inserted into a printed circuit board.
c. The shorting of an externally connected component, such as a shorted capacitor, resistor, diode, transistor, LED, switch, and so on.
d. A shorted connector such as an IC socket, circuit board connector, or cable connector.

Most digital signal line opens are easily located since the digital signal does not arrive at its destination. The logic probe is ideal for tracing these open paths. Here are some examples of typical external IC open circuit problems.

a. An open in a signal line due to an improperly soldered pin, a deep scratch in the etched printed circuit board, a bent or broken IC pin, a broken wire, and so on.
b. An IC that is inserted into the printed circuit board backwards.
c. The failure of externally connected components such as resistors, diodes, transistors, and so on.
d. An open connector such as an IC socket, circuit board connector, or cable connector.

In summary, the logic pulser and current tracer are ideal for tracing shorts and the logic probe is ideal for tracing opens.

Power Supply Problems

Like analog circuits, digital circuits suffer from more than their fair share of power supply problems. When a digital circuit seems to have a power supply problem, remember to use the half-split method of troubleshooting first, as seen in Figure 22-14(a). First check the dc supply voltages from the dc power supply unit to determine whether the power problem is in the power supply (source) or in one of the digital circuits (load).

If the dc output voltages are not present at the output of the dc power supply, then the fault probably exists before this test point. In this instance, you should test the dc power supply circuit; however, a short in the digital circuits can pull down the dc supply voltage and

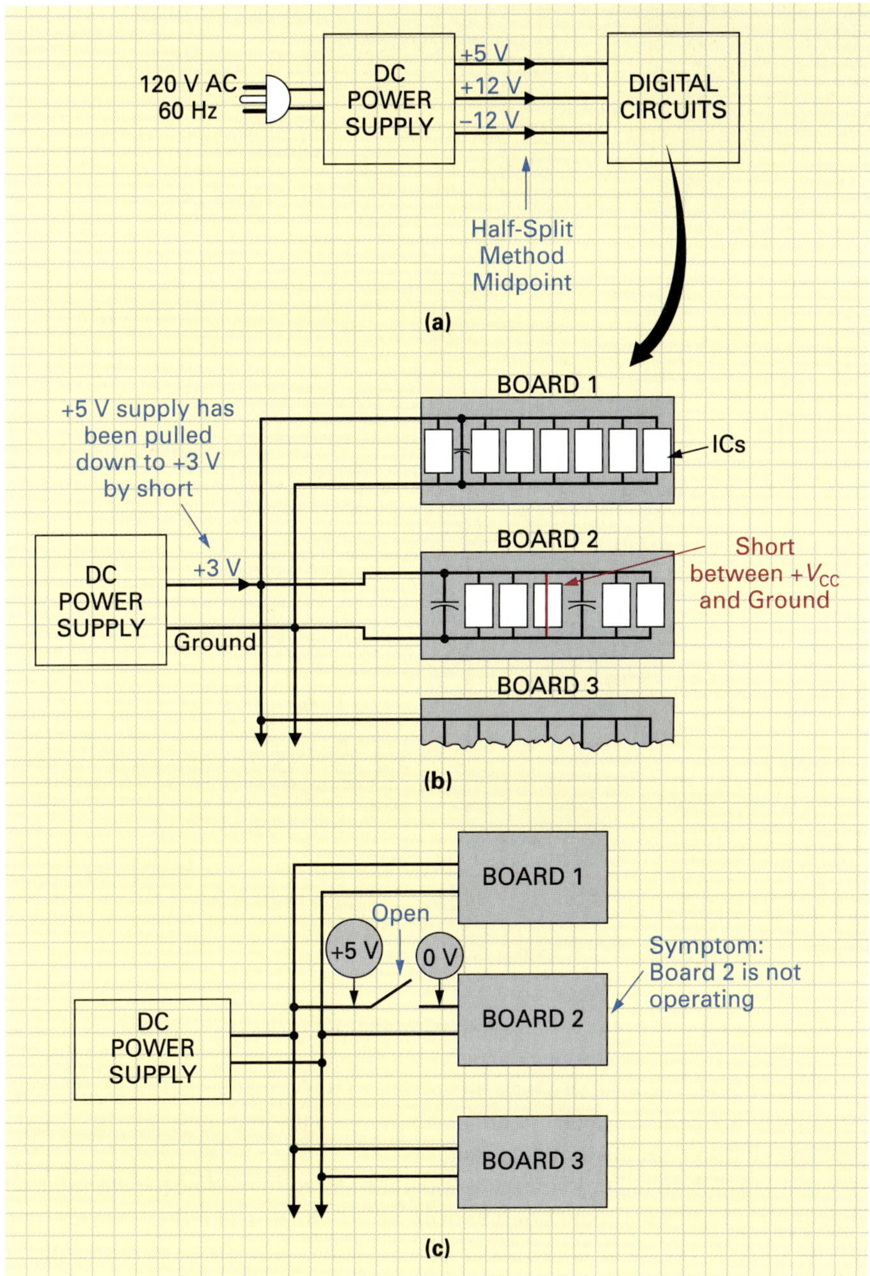

FIGURE 22-14 **Power Supply Problems in Digital Circuits. (a) Using the Half-Split Method. (b) Tracing Shorts. (c) Tracing Opens.**

make it appear as though the problem is in the dc power supply, as shown in Figure 22-14(b). Once again, the current tracer is ideal for tracing the faulty board and then the faulty device in that printed circuit board. If you cannot get power to the digital circuits because the short causes the dc power supply fuse to continually blow, use the logic pulser to inject a current into the supply line and then use the current tracer to follow the current path to the short.

Open power supply lines are easy to locate since power is prevented from reaching its destination. The logic probe is ideal for tracking down power line opens, as shown in Figure 22-14(c).

Clock
A two-state timing signal that is distributed throughout a digital electronic system to ensure that all circuits operate in sync or at the correct time.

Clock Signal Problems

All digital systems use a master timing signal called a **clock.** This two-state timing signal is generated by a digital oscillator circuit. This clock or timing signal is distributed

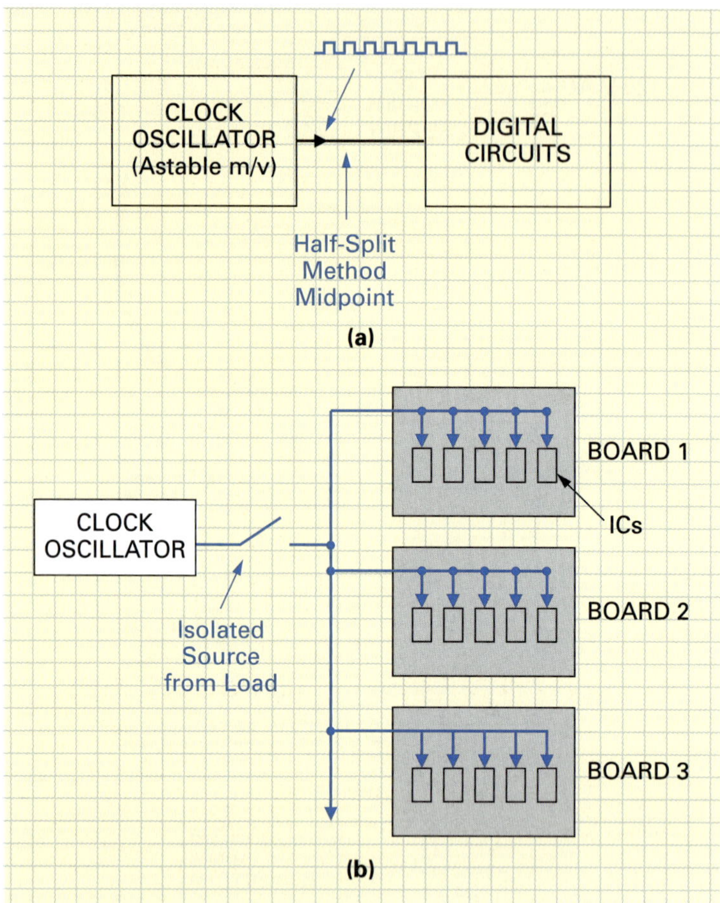

FIGURE 22-15 **Clock Signal Problems.**

throughout the digital electronic system to ensure that all circuits operate in sync or at their correct time. The presence or absence of this clock signal can be checked with a logic probe; however, in most cases it is best to check the quality of the signal with an oscilloscope. Even small variations in frequency, pulse width, amplitude, and other characteristics can cause many problems throughout the digital system.

To isolate a clock signal problem, first apply the half-split method to determine whether the fault is in the oscillator circuit (source) or in one of the digital circuits (load), as shown in Figure 22-15(a). If the clock signal from the oscillator is incorrect, or if no signal exists, the problem will generally be in the oscillator clock circuit; however, like the dc supply voltage, the clock signal is distributed throughout the digital system, and therefore you may have to disconnect the clock signal output from the oscillator to determine whether a short is pulling down the signal or whether the oscillator clock circuit is not generating the proper signal, as shown in Figure 22-15(b). If there is a short somewhere in one of the digital circuits, the current tracer can be used to find the shorted path, whereas an open in a clock signal line can easily be traced with a logic probe.

SELF-TEST EVALUATION POINT FOR SECTION 22-2

Use the following questions to test your understanding of Section 22-2.

1. How are most internal IC failures repaired?
2. An ____________ is easier to isolate and will generally not affect other components while a ____________ is harder to isolate and can often damage other components (short or open).
3. What is the half-split method?

4. What is a clock signal?
5. The ____________ and ____________ are ideal for tracing shorts, and the ____________ is ideal for tracing opens.

22-3 CIRCUIT REPAIR

Isolating the faulty component will always take a lot longer than repairing the circuit once the problem has been found. Some repairs are simple, such as replacing a cable, removing a solder bridge from a PCB, reseating a connector, or adjusting a variable resistor. In most instances, however, the repair will involve replacing a component such as an IC, transistor, diode, resistor, or capacitor. This will mean that you will need to use the soldering iron to remove the component from the PCB. When soldering and desoldering components to and from a PCB, follow the techniques discussed in your dc/ac electronics course and also remember the following:

a. Always make a note of the component's orientation in the PCB since certain components such as ICs, diodes, transistors, and electrolytic capacitors have to be correctly oriented.
b. When desoldering components, use either a vacuum bulb or solder wick to remove the solder.
c. Be extremely careful not to overheat the PCB or the component being soldered or desoldered.
d. Avoid using too much or too little solder because this may cause additional problems.
e. Use a grounded soldering iron and a grounding strap when working with MOS devices to prevent any damage to the IC from static discharge.

Once the equipment has been repaired, always "final test" the equipment to see if it is now fully operational. Check that the system operates correctly in all aspects, especially in the area that was previously malfunctioning, and then reassemble the system.

SELF-TEST EVALUATION POINT FOR SECTION 22-3

Use the following questions to test your understanding of Section 22-3.

1. List the steps you would follow to remove an IC from a PCB.
2. What is a "final test"?

REVIEW QUESTIONS

Multiple-Choice Questions

1. Which test instrument would most accurately measure logic voltage levels?
   a. Logic probe
   b. Multimeter
   c. Oscilloscope
   d. Logic pulser
2. Which test instrument would be best suited for testing a clock signal's duty cycle?
   a. Logic probe
   b. Multimeter
   c. Oscilloscope
   d. Logic pulser
3. Which test instrument would be best suited for testing the HIGH and LOW logic levels at the inputs and outputs of a series of several logic gates?
   a. Logic probe
   b. Multimeter
   c. Oscilloscope
   d. Logic pulser
4. Which test instrument would be best suited for generating a test signal?
   a. Logic probe
   b. Multimeter
   c. Oscilloscope
   d. Logic pulser

5. The voltmeter can be used to make _____________ tests, while the oscilloscope can be used to make _____________ tests.
   a. dynamic, static
   b. static, dynamic
6. The logic pulser
   a. has an insulated tip.
   b. senses current pulses at different points in a circuit.
   c. is a signal generator.
   d. is a signal-sensing instrument.
7. Which digital test instrument can be used to sense the logic levels at different points in a circuit?
   a. Multimeter
   b. Logic probe
   c. Oscilloscope
   d. Current tracer
8. The current tracer is ideal for locating what type of circuit faults?
   a. Shorts
   b. Opens
   c. Invalid logic levels
   d. Rise/fall signal problems
9. A scratch in a PCB track will more than likely cause a/an _____________.
   a. short
   b. open
   c. no problem
   d. any of the above
10. Which of the following is most likely to cause a short?
    a. Broken IC pin
    b. Low current
    c. Solder bridge
    d. IC pin not in an IC socket

Practice Problems

11. Identify the waveform shown in Figure 22-16.
12. What test instrument would you use to check the rise and fall time of the waveform shown in Figure 22-16?
13. Referring to Figure 22-16, calculate the following characteristics for the waveform:
    a. Pulse width
    b. Period
    c. Frequency
    d. Duty cycle
14. Calculate the PRF and PRT of the waveform shown in Figure 22-16.
15. Could a logic probe be used to identify whether the waveform in Figure 22-16 was a square or rectangular wave?

Troubleshooting Questions

16. Why is it important to be able to recognize different types of internal IC failures if the ICs themselves cannot be repaired?
17. In Figure 22-17(a), a logic probe has been used to check the circuit since it is not functioning as it should. In Figure 22-17(b), you can see what logic levels were obtained when the output of the NAND gate was isolated. Which IC should be replaced?
18. What do you suspect is wrong with the faulty IC in Figure 22-17?
19. In Figure 22-18, an oscilloscope has been used to test several points on the digital circuit. Which IC should be replaced, and what do you think is the problem?
20. Are the logic gates in Figure 22-19 operating as they should?

Web Site Questions

Go to the web site http://www.prenhall.com/cook, select the textbook *Electronics: A Complete Course,* select this chapter, and then follow the instructions when answering the multiple choice practice problems.

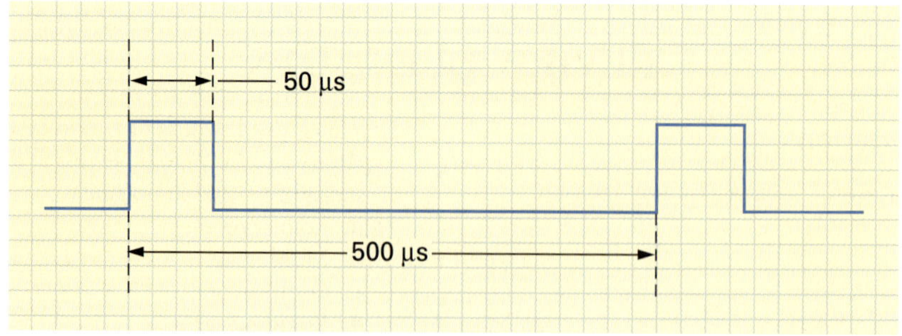

FIGURE 22-16 **A Repeating or Pulse Waveform.**

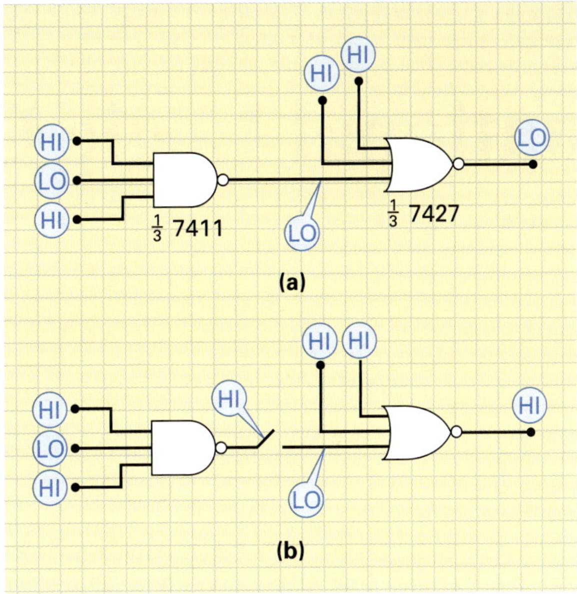

FIGURE 22-17 **Troubleshooting Exercise 1.**

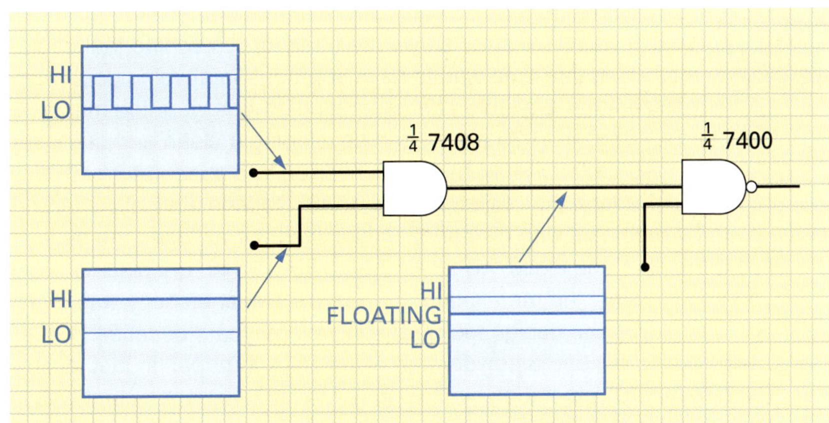

FIGURE 22-18 **Troubleshooting Exercise 2.**

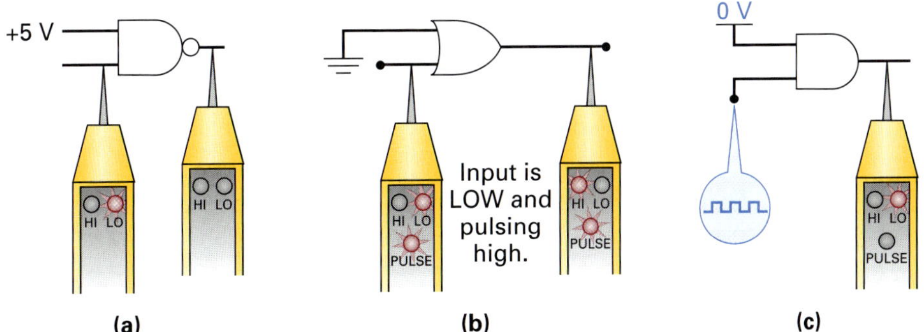

FIGURE 22-19 **Troubleshooting Exercise 3.**

Combinational Logic Circuits

Copy Master

Born in 1906 to two invalid Swedish immigrants, Chester Carlson worked part-time day and night to support his parents and still achieved excellent grades at high school. Sadly, when Carlson was seventeen both of his parents died, so Carlson left New York and went to the California Institute of Technology where he graduated in 1930 with a degree in physics.

Carlson's first job was as a researcher for Bell Telephone Laboratories in New York. It was here that he saw the need for a machine that could copy documents—the method at that time was to have someone in a typing pool retype the original. Carlson realized that available photographic methods were too messy and time consuming. He left Bell Labs in 1935 and began developing a clean and quick copying machine in the small rented bedroom that he had converted into a laboratory.

On the morning of October 23, 1938, Carlson statically charged a metal sulfur-coated plate by rubbing it with his handkerchief. He then exposed the plate to a glass slide that had on it the date and place "10-23-38 ASTORIA." When dry black powder on paper was pressed against the metal plate, the world's first photocopy was created.

Refining the design took Carlson several years because he had to develop more sensitive plates and a powder that would stick to paper. In fact, even after he had a good working prototype, it took him two years to find a company that was interested in manufacturing and selling his machine. However, in January, 1947, Haloid Company of Rochester, New York, which was a small photography firm, signed an agreement with Carlson for what he called his "dry printing machine." They were very dubious about whether it would really catch on and told Carlson that they would not pay him for his invention, but only give him a percentage of the profits. The company called the process "Xerography," which meant dry printing. After an almost overnight success, the company gave up all of their other products and renamed the company after their product, "Xerox Corporation."

The early Xerox machines in 1950 needed an operator to actuate the mechanism. However, by 1960 a fully automatic machine produced perfect copies by pressing one button.

Carlson's invention that nobody wanted transformed his lifestyle from a rented bedroom to royalties that paid him several million dollars a year.

23

Introduction

The first digital ICs contained only four logic gates, so if you wanted to decode a digital signal, you would have to construct a digital decoder circuit on a printed circuit board (PCB) using several logic gate ICs. As semiconductor manufacturers improved their miniaturization techniques, they were able to incorporate larger circuits within a single IC package. Studying the evolving digital electronics industry, semiconductor manufacturers noticed that several circuit functions are always needed in digital electronic systems, so they formed these complete functional circuits within a single IC. They called these digital ICs **combinational logic circuits** since logic gates were combined on a single chip of semiconductor to create functional logic circuits, such as decoders, encoders, multiplexers, demultiplexers, code converters, comparators, and parity generators and checkers. As time has passed, this practice of reducing a frequently needed digital PCB circuit to a single IC has continued, in step with the ability of semiconductor manufacturers to continuously improve their circuit miniaturization techniques.

In this chapter we will take a close look at two types of combinational logic circuits—decoders and encoders.

Combinational Logic Circuits
Digital logic circuits in which the outputs are dependent only on the present input states and the logic path delays.

23-1 DECODERS

Decoder Circuit
A circuit that responds to a particular coded signal while rejecting all others.

The term *decode* means to translate coded characters into a more understandable form, or to reverse a previous encoding process. A digital **decoder circuit** is a logic circuit that responds to one specific input word or code, while rejecting all others.

23-1-1 *Basic Decoder Circuits*

Figure 23-1(a) shows how an AND gate and two INVERTERS can be connected to act as a basic decoder. As you know, all inputs to an AND gate need to be 1 for the output to be 1. Referring to the function table for this circuit, you can see that only when the input code is 001 will the AND gate have all of its inputs HIGH and therefore produce a HIGH output.

While some decoders are designed to recognize only one input word combination, other decoders are designed to activate one output for every input word combination. For example, Figure 23-1(b) shows how four AND gates and two INVERTERS can be connected to activate one of four outputs based on the binary input word applied to inputs *A* and *B*. The two INVERTERS are included to invert the *A* and *B* inputs to create two additional input lines called "NOT *A*" and "NOT *B*" (symbolized $\overline{A}$ and $\overline{B}$). Looking at the function table for this circuit, you can see that when inputs *A* and *B* are 00_2 (0), only output 0 will be activated or made HIGH. This is because both of the inputs to AND gate 0 are connected to the $\overline{A}$ and $\overline{B}$ input lines, so when *A* and *B* are LOW, $\overline{A}$ and $\overline{B}$ are both HIGH, and therefore the output of AND gate 0 will be HIGH. The inputs of the other AND gates are connected to the input lines so that a 01_2 (1) input will make output 1 HIGH, a 10_2 (2) input will make output 2 HIGH, and a 11_2 (3) input will make output 3 HIGH. This

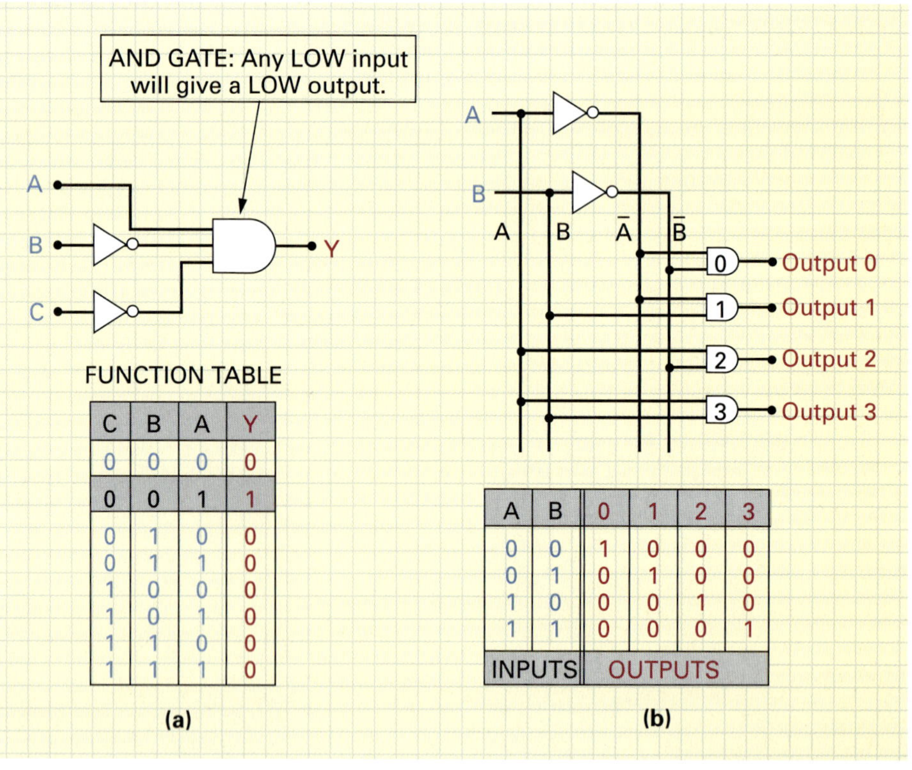

C	B	A	Y
0	0	0	0
0	0	1	1
0	1	0	0
0	1	1	0
1	0	0	0
1	0	1	0
1	1	0	0
1	1	1	0

A	B	0	1	2	3
0	0	1	0	0	0
0	1	0	1	0	0
1	0	0	0	1	0
1	1	0	0	0	1
INPUTS		OUTPUTS			

FIGURE 23-1 **Basic Decoders.**

circuit would be described as a *one-of-four decoder* since an input code is used to select or activate one of four outputs. If the four outputs were connected to light-emitting diodes (LEDs) that were labeled 0 through 3, the decoder would function as a **binary-to-decimal decoder,** since the circuit is translating a binary coded input into a decimal output display.

Binary-to-Decimal Decoder
A decoder circuit that selects one of ten outputs (decimal), depending on the 4-bit binary input applied.

23-1-2 *Decimal Decoders*

Binary-to-decimal conversion is one of the most common applications for digital decoder circuits.

Application—BCD-to-Decimal Counter Circuit

Figure 23-2 shows how a 7442 can be used in conjunction with a 7490 decade (ten) counter, to produce a BCD-to-decimal counter/decoder circuit. Counters will be discussed in more detail in Chapter 25; for now, however, simply think of the counter as a circuit that counts the number of clock pulses applied to its input (pin 14) and then generates a 4-bit binary equivalent value at its outputs (pins 12, 9, 8, 11). In this circuit, the input clock pulses are continuous, so the output binary count from the 7490 will continually advance or cycle through all ten BCD codes—0000, 0001, 0010, 0011, 0100, 0101, 0110, 0111, 1000, 1001—and then reset to 0000 and repeat the cycle. This incrementing 4-bit BCD count from the 7490 is applied to the 4-bit input of the 7442 decoder.

An input of 0000 to the 7442 will cause it to make its 0 output LOW, an input of 0001 to the 7442 will cause it to make its 1 output LOW, an input of 0010 to the 7442 will cause it to make its 2 output LOW, and so on. Since the LEDs are arranged as a common-anode display, a LOW at any of the 7442 outputs will turn ON the respective LED. As a result, the LED decimal display will turn ON each LED in sequence (0 through 9), and then repeat the cycle.

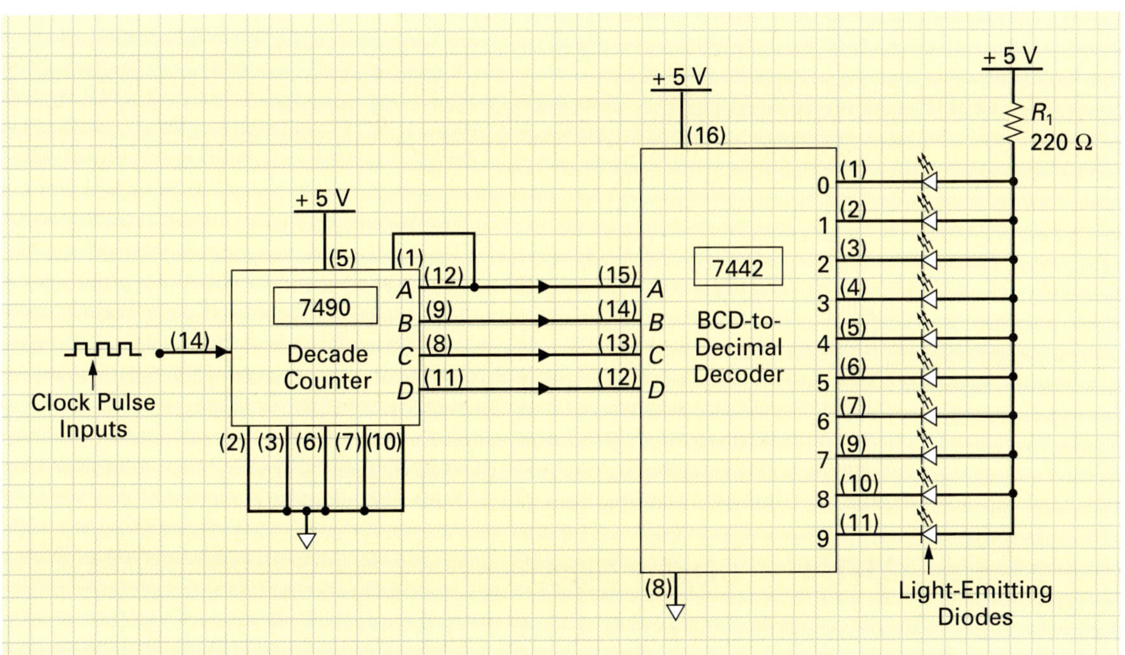

FIGURE 23-2 Application—BCD-to-Decimal Counter Circuit.

EXAMPLE:

Imagine you had constructed the circuit in Figure 23-2 in lab, and when you switched on the power the decimal display did not flash sequentially. What tests would you make to try and isolate the problem?

Solution:

The following tests should help you isolate a problem with the circuit shown in Figure 23-2.

a. Verify +5 V is applied to the 7490, 7442, and LED anodes using the logic probe.
b. Verify ground is applied to all necessary IC pins using the logic probe.
c. Check clock pulse input at pin 14 of the 7490 using the logic probe (dynamic test). You can also use the logic pulser to single pulse the counter input (static test).
d. Apply the half-split method to the circuit—isolate outputs from 7490 and then test to see if counter is operating as it should. This will determine whether the problem is in the counter half of the circuit or the decoder half of the circuit.
e. Further isolate to determine whether the problem is internal or external to the suspected IC.

23-1-3 *Hexadecimal Decoders*

Hexadecimal Decoder
A decoder circuit that selects one of sixteen outputs (hexadecimal) depending on the 4-bit binary input applied.

The 74154 is a good example of a **hexadecimal decoder** that will select one of sixteen (hexadecimal) outputs, depending on the 4-bit binary input value.

Application—Sixteen-LED Back-and-Forth Flasher Circuit

Figure 23-3 shows how the 74154 can be used in conjunction with a 74193 up/down counter to produce a sixteen-LED back-and-forth flasher circuit. The circuit in Figure 23-3 operates in the following way. Let us begin by assuming that *B* NAND gate has a HIGH on pin 5, and *A* NAND gate has a LOW on pin 1. These two points are always at opposite logic levels due to the cross-coupling between NAND gates *C* and *D*. With *B* NAND gate pin 5 HIGH, the clock pulses on pin 4 appear at the COUNT-UP input of the counter, and therefore the counter's count will increase in binary (0000, 0001, 0010, and so on toward 1111). This binary count appears at the *ABCD* outputs of the 74193, and is applied to the *ABCD* inputs of the 74154 decoder. The decoder will respond to the binary input and make LOW the equivalent output and consequently turn ON the associated LED (due to the LOW outputs). When the count reaches 1111 (15), the 15 output (pin 17) goes LOW, turning ON the left-most LED. This LOW output at pin 17 of the 74154 also appears at pin 10 of NAND gate *C,* which produces a HIGH output to drive pin 1 of NAND gate *A* and allows the clock pulses through to the COUNT-DOWN input of the 74193. The HIGH output of NAND gate *C* also drives pin 12 of NAND gate *D,* whose other input on pin 13 is also HIGH since the 0 output of the decoder is not active at this time. The two HIGH inputs into NAND gate *D* therefore produce a LOW output, which is applied to NAND gate *B* to block the clock pulses from driving the COUNT-UP input of the 74193. The LOW output of NAND gate *D* is also applied to pin 9 of NAND gate *C* to keep the COUNT-DOWN control line active (HIGH), even after the LOW is no longer present at the 74154's pin 17. The counter, therefore, counts down from 15 to 0, causing the LED's ON/OFF sequence to move from left to right. When the counter reaches a count of 0, the NAND gates will switch the clock pulses once again to the COUNT-UP input of the counter, and the cycle will repeat.

23-1-4 *Display Decoders*

Many systems such as digital watches, calculators, pagers, and cellular phones make use of a multisegment display. Decoders are needed in these systems to decode the binary data into the multisegment data needed to drive the display.

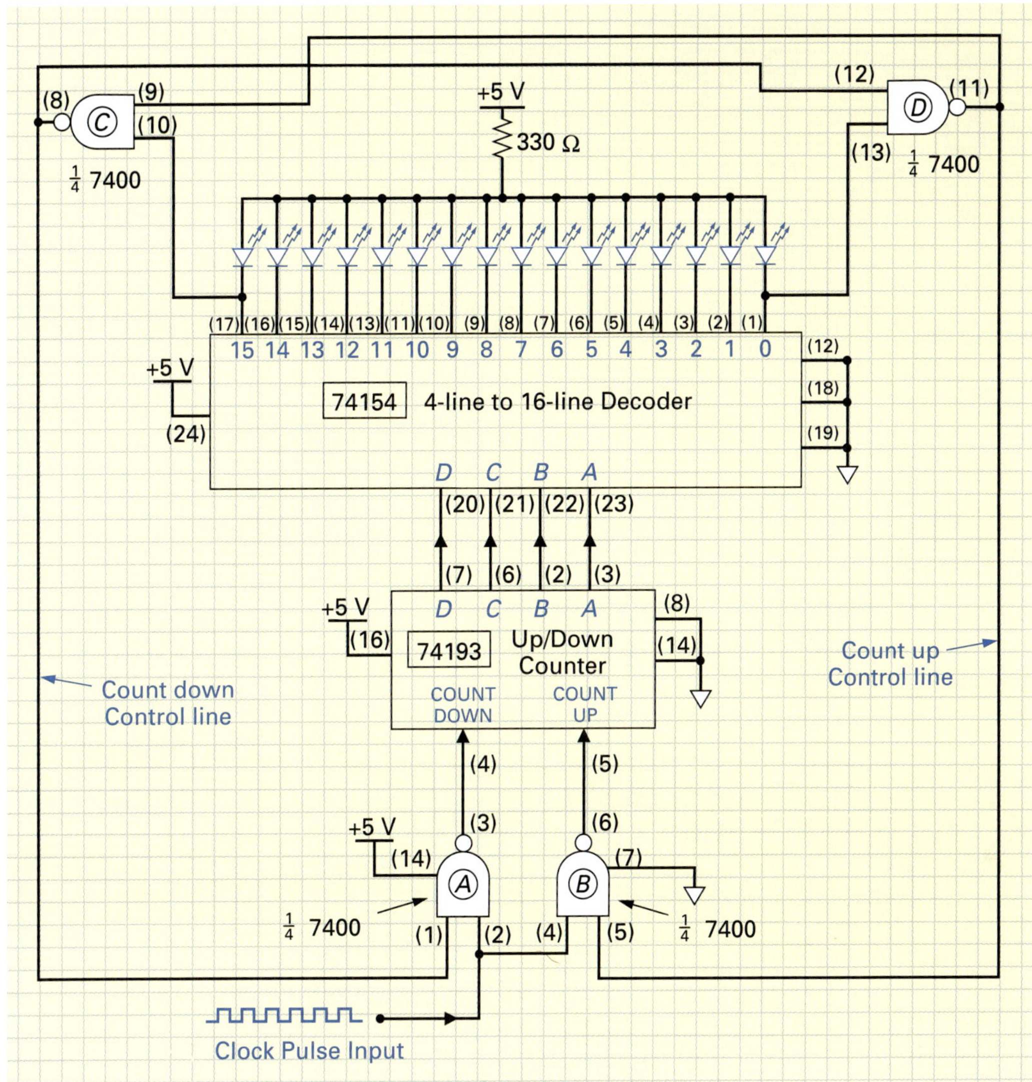

FIGURE 23-3 **Application—Sixteen-LED Back-and-Forth Flasher Circuit.**

Light-Emitting Diode (LED) Display Decoders

The 7447 IC is one example of a display decoder/driver. Figure 23-4(a) shows how the seven segments of an LED display are labeled—*a, b, c, d, e, f,* and *g*—and Figure 23-4(b) shows how certain ON/OFF segment combinations can be used to display the decimal digits numbers 0 through 9.

Application—Seven-Segment LED Display Decoder/Driver

Figure 23-4(c) shows how a 7447 **BCD-to-seven-segment decoder** IC can be connected to drive a common-anode seven-segment LED display. The function table for this digital IC has been repeated in the upper inset in Figure 23-4(c). Looking at the first line of this table, for example, you can see that if a binary input of 0000 (0) is applied to the *ABCD* inputs, the 7447 will make LOW outputs *abcdef* and make HIGH output *g*. These outputs are applied to the cathodes of the LEDs within the common-anode display (TIL 312), as seen in the lower inset, and since all of the anodes are connected to +5 V, a LOW output will turn ON a segment, and a HIGH output will turn OFF a segment. The 7447, therefore, generates active-LOW outputs, which means that a "LOW" output signal will "activate" its associated output LED. In this example, the active-LOW outputs from the 7447 will turn ON segments *abcdef,* while the HIGH output will turn OFF segment *g,* causing

BCD-to-Seven-Segment Decoder

A decoder circuit that converts a BCD input code into an equivalent seven-segment output code.

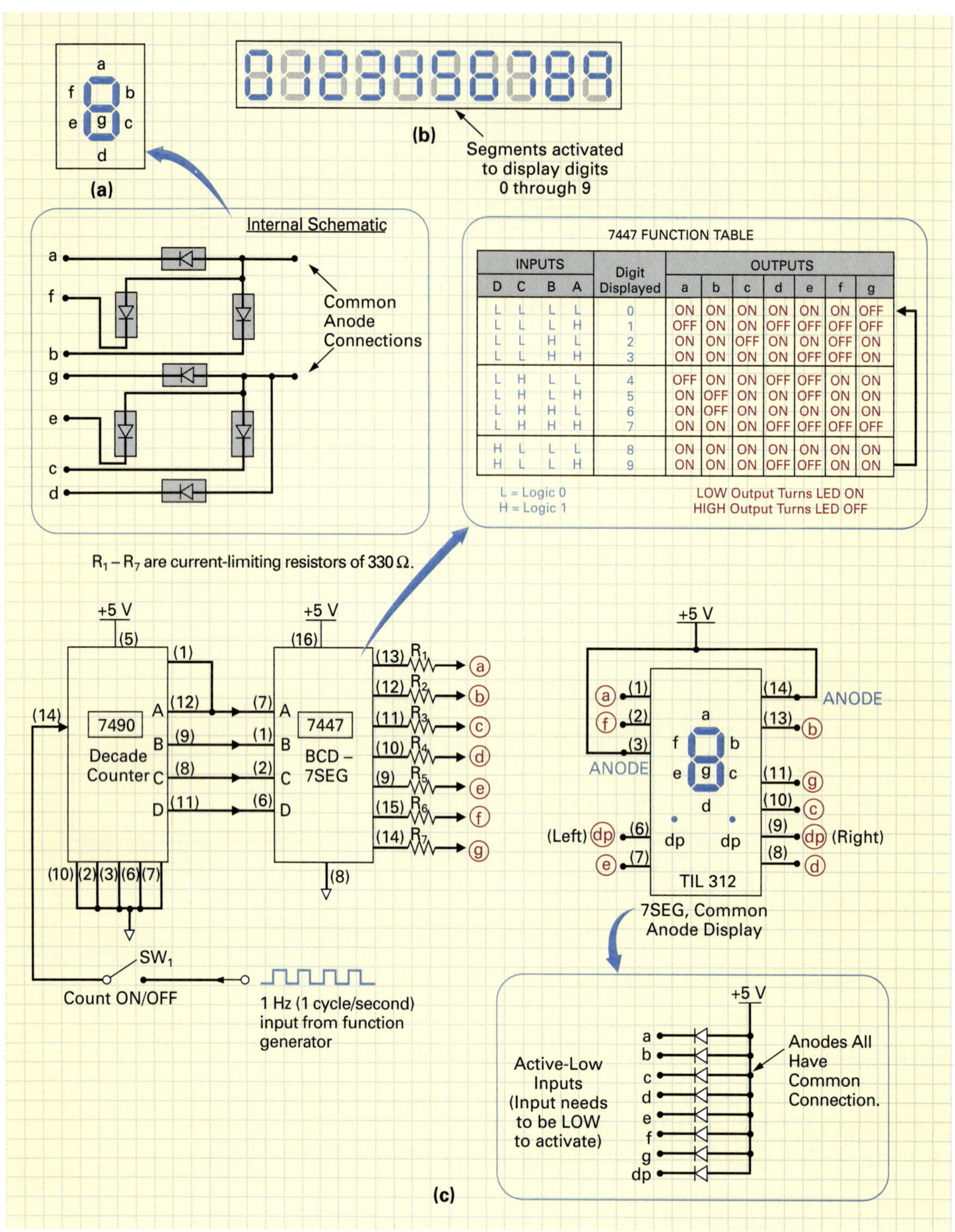

7447 FUNCTION TABLE

INPUTS D	C	B	A	Digit Displayed	OUTPUTS a	b	c	d	e	f	g
L	L	L	L	0	ON	ON	ON	ON	ON	ON	OFF
L	L	L	H	1	OFF	ON	ON	OFF	OFF	OFF	OFF
L	L	H	L	2	ON	ON	OFF	ON	ON	OFF	ON
L	L	H	H	3	ON	ON	ON	ON	OFF	OFF	ON
L	H	L	L	4	OFF	ON	ON	OFF	OFF	ON	ON
L	H	L	H	5	ON	OFF	ON	ON	OFF	ON	ON
L	H	H	L	6	ON	OFF	ON	ON	ON	ON	ON
L	H	H	H	7	ON	ON	ON	OFF	OFF	OFF	OFF
H	L	L	L	8	ON	ON	ON	ON	ON	ON	ON
H	L	L	H	9	ON	ON	ON	OFF	OFF	ON	ON

L = Logic 0
H = Logic 1

LOW Output Turns LED ON
HIGH Output Turns LED OFF

EWB

FIGURE 23-4 Application—Seven-Segment LED Display/Driver.

the decimal digit 0 to be displayed. The remaining lines in the function table indicate which segments are made active for the other binary count inputs 0001 (1) through 1001 (9).

EXAMPLE:

What would be the *abcdefg* HIGH/LOW code generated by a 7447 if 0101 were applied to the input?

Solution:

Referring to the 7447 function table in Figure 23-4, you can see that when 0101, or LHLH (decimal 5), is applied at the input, the 7447 will turn ON (by generating a LOW output) all of the segments except *b* and *e*. These two segment outputs will therefore be HIGH and the rest will be LOW, so the output code will be ON, OFF, ON, ON, OFF, ON, ON (LHLLHLL).

The circuit shown in Figure 23-4(c) operates in the following way. A 1 Hz clock input signal is applied via a count ON/OFF switch to the input of a 7490 digital counter IC. This counter will produce a 4-bit binary word at its outputs *ABCD* based on the number of input clock pulses received at its input (pin 14). This 4-bit binary word at the output of the 7490 is applied to the 7447 seven-segment decoder/driver, where it will be decoded into a seven-segment code that is applied to a seven-segment common-anode display. When the count ON/OFF switch is closed, the 7490 will count, the 7447 will decode the count, and the display will constantly cycle through the digits 0 through 9. When the count ON/OFF switch is opened, the counter will freeze at its present count, and, therefore, so will the number shown on the display.

The 7447 is an example of a BCD-to-seven-segment decoder/driver that has active-LOW outputs for a common-anode seven-segment display. Also available is a 7448 IC, which is a BCD-to-seven-segment decoder/driver that has active-HIGH outputs for a common-cathode seven-segment display. In most applications, the 7447 is more commonly used because a 7447 can sink more output current from a common-anode display than a 7448 can source current to a common-cathode display. The 7447 and 7448 are also referred to as **code converters,** because they convert a BCD code into a seven-segment code.

Code Converter
A decoder circuit that converts an input code into a different but equivalent output code.

The 7447 has three other inputs that should be explained. They are called lamp test input ($\overline{LT}$), blanking input ($\overline{BI}$), and ripple blanking input ($\overline{RBI}$). All three of these input control lines are active-LOW inputs, as indicated by the triangle on the input lines and the NOT bar over the abbreviated letter designations.

When the lamp test input ($\overline{LT}$) is made active (taken LOW), all of the 7447 will be pulled LOW, and therefore all of the seven-segment LEDs will go ON. This feature is good for quickly testing the decoder and all of the LEDs within the seven-segment display. The blanking input ($\overline{BI}$) can be used to blank the display (turn OFF all of the segments), overriding the present binary input being applied. This feature is used when several seven-segment displays are grouped together to form a multidigit display and the most significant displays need to be blanked when they are zero. For example, if we had to display the value "003456," there would be no need to display the first two zeros, so they could be blanked so that the display shows "3456." The third input, the ripple blanking input ($\overline{RBI}$), is an active-LOW control input that allows us to dim or brighten the display without having to adjust the display voltage applied to the LEDs. The ripple blanking input is usually driven by a pulse waveform, as shown in Figure 23-5. The duty cycle, or ratio of the pulse waveform's HIGH time to LOW time, is made variable under the control of a dimmer control. When the pulse waveform is LOW, the display is blanked, and when the pulse waveform is HIGH, the display is ON and controlled by the 4-bit input word applied to the 7447's *ABCD* inputs. By using a pulse frequency that is greater than 25 Hz so no display flicker is visible, the brightness of the display can be varied by changing the pulse waveform's duty cycle, and therefore the display ON-to-OFF ratio, as seen in the examples in the inset in Figure 23-5.

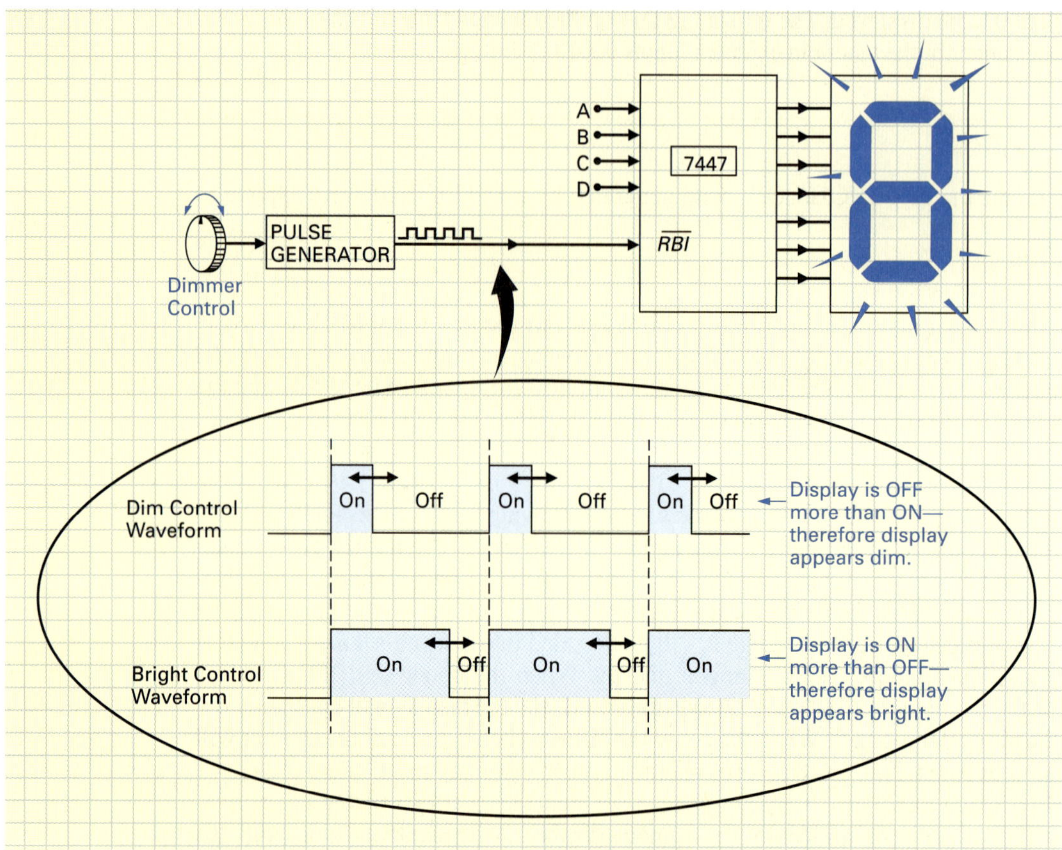

FIGURE 23-5 **Ripple Blanking of a Display.**

Liquid-Crystal Display (LCD) Decoders

Liquid-crystal displays are also frequently used in electronic systems to display information. Figure 23-6(a) shows a seven-segment LCD. The sealed liquid crystal segments are normally transparent, but when an ac voltage (typically 3 V to 15 V) is applied between a contact and the back electrode or plate, the crystal molecules become disorganized, resulting in a darkening of the segment. In some instances, edge- or back-lighting is used to improve the display's visibility. Displays using LCDs draw much less current than LED displays, which is why you will often see LCDs used in battery-operated devices such as watches and calculators.

Application—Seven-Segment LCD Display Decoder/Driver

Figure 23-6(b) shows how a CMOS 4511 (BCD-to-seven-segment decoder/driver) can be used in conjunction with two CMOS 4070 ICs (quad exclusive-OR gates) to drive a seven-segment liquid crystal display. The XOR gates act as controlled inverters, and the circuit operates as follows. To turn ON a segment, the 4511 generates a HIGH output, which is applied to an XOR gate input. This HIGH input to the XOR will cause the 30 Hz square wave applied to the other input to be inverted due to the action of the XOR, and therefore the 30 Hz being applied to the LCD's segment will be out-of-phase with the 30 Hz being applied to the LCD's backplate. This potential difference between the backplate and a specific segment will cause the respective segment to darken since an electric field will disorganize the segment molecules, as shown in the inset in Figure 23-6(b). On the other hand, when a 4511 output is LOW, there is no inversion of the 30 Hz square wave, so the 30 Hz being applied to the LCD's segment will be in-phase with the 30 Hz being applied to the LCD's backplate. In this instance, there is no potential difference between the backplate and segment, so the molecules remain in their normally organized condition and appear transparent, as shown in the inset in Figure 23-6(b).

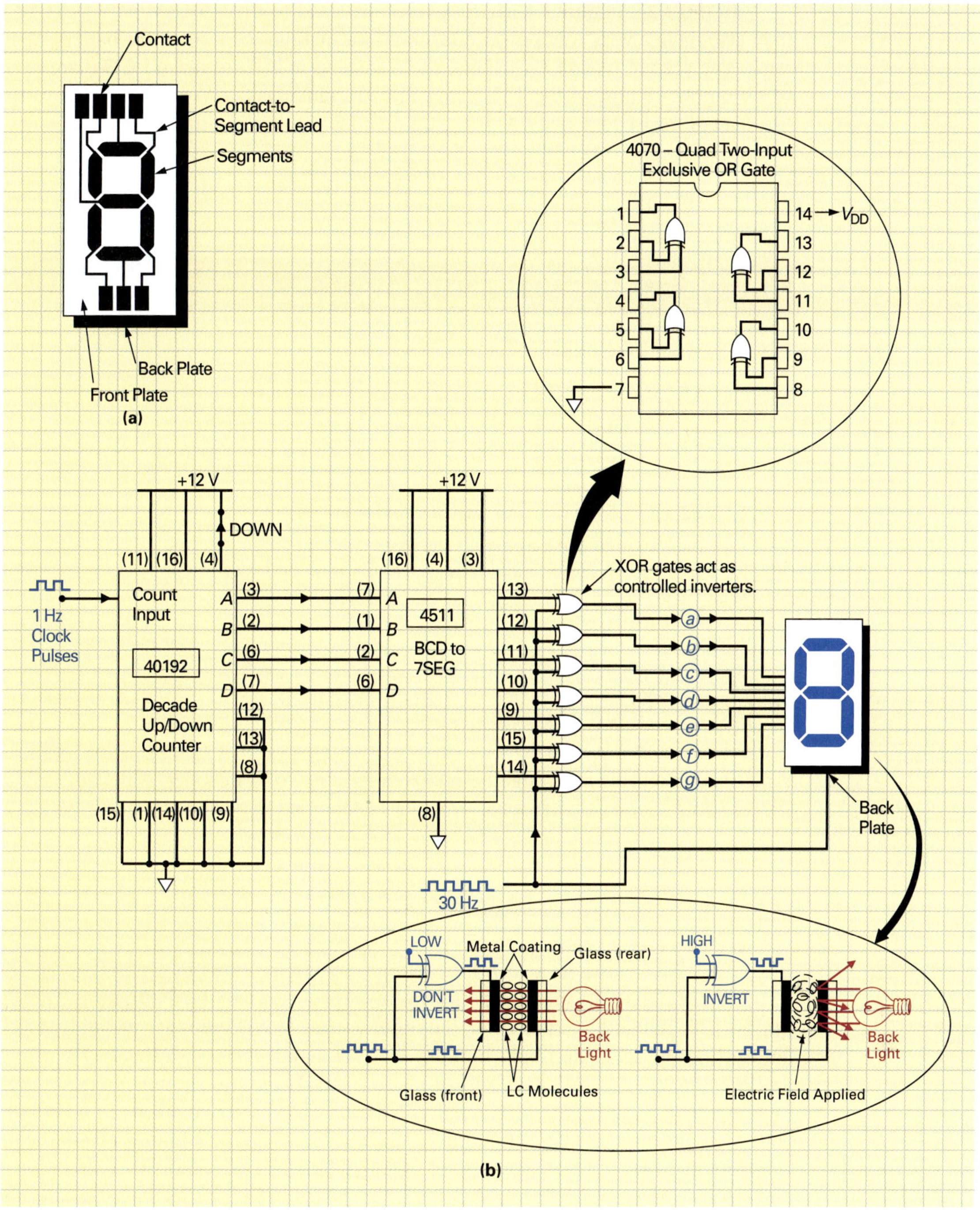

FIGURE 23-6 Application—A Liquid-Crystal Display (LCD) Decoder Circuit.

EXAMPLE:

Would the 4511 BCD-to-seven-segment decoder/driver in Figure 23-6(b) have active-HIGH or active-LOW outputs?

Solution:

A HIGH output from the 4511 in Figure 23-6(b) will cause the XOR gate to invert the square wave and turn ON the associated segment. A HIGH output therefore activates the segment, so the outputs of the 4511 are active-HIGH.

In most applications CMOS devices are used to drive LCDs because, like the LCD, they also have a very low power consumption, making them ideal for battery-operated systems. The second reason for using CMOS instead of TTL to drive LCD displays is that the LOW output voltage of the TTL gate can be as much as 0.4 V, and this dc level reduces the life span of the LCD display. This condition does not occur with CMOS logic since its worst-case LOW output is 0.1 V.

SELF-TEST EVALUATION POINT FOR SECTION 23-1

Use the following questions to test your understanding of Section 23-1.

1. What would be the 0 through 9 output from a 7442 BCD-to-decimal decoder if the input were 1001?
2. How many outputs would the following decoders have?
   a. BCD-to-decimal
   b. Binary-to-hexadecimal
3. What would be the *abcdefg* logic levels out of a 7447 if 0110 were applied at the input?
4. Why are CMOS decoder/drivers normally used to drive liquid-crystal displays?

23-2 ENCODERS

Encoder Circuit
A circuit that generates specific output codes in response to certain input conditions.

An **encoder circuit** performs the opposite function of a decoder. A decoder is designed to *detect* specific codes, whereas an encoder is designed to *generate* specific codes.

23-2-1 *Basic Encoder Circuits*

Figure 23-7(a) shows how a simple decimal-to-binary encoder circuit can be constructed using three push buttons, three pull-up resistors, and two NAND gates. The pull-up resistors are included to ensure that the input to the NAND gates are normally HIGH. When button 1 is pressed, the upper input of NAND gate *A* will be pulled LOW, and since any LOW input to a NAND gate will produce a HIGH output, A_0 will be driven HIGH. The two inputs to NAND gate *B* are unaffected by SW_1, so the two HIGH inputs to this NAND gate will produce a LOW output at A_1. Referring to the truth table in Figure 23-7(a), you can see that the 2 bit code generated when switch 1 is pressed is 01 (binary 1). Studying the circuit and the rest of the function table shown in Figure 23-7(a), you can see that the code 10 (binary 2) will be generated when switch 2 is pressed, and the code 11 (binary 3) will be generated when switch 3 is pressed. This basic decimal-to-binary encoder circuit, which has three inputs and two outputs, would be described as a three-line to two-line encoder.

Figure 23-7(b) expands on the basic encoder circuit discussed in Figure 23-7(a) by including extra push-button switches and NAND gates so we can generate a 4-bit BCD code that is equivalent to the decimal key pressed. This circuit, therefore, is a *decimal-to-BCD encoder* circuit, and it will operate in the same manner as the previously discussed basic encoder. For example, when the decimal 2 key is pressed, it grounds an input to NAND gate *B,* so only the A_1 output goes HIGH, giving a 0010 (binary 2) output. As another example, when the decimal 7 key is pressed, it grounds inputs to NAND gates *A, B,* and *C,* so the A_0, A_1, and A_2 outputs go HIGH, giving a 0111 (binary 7) output.

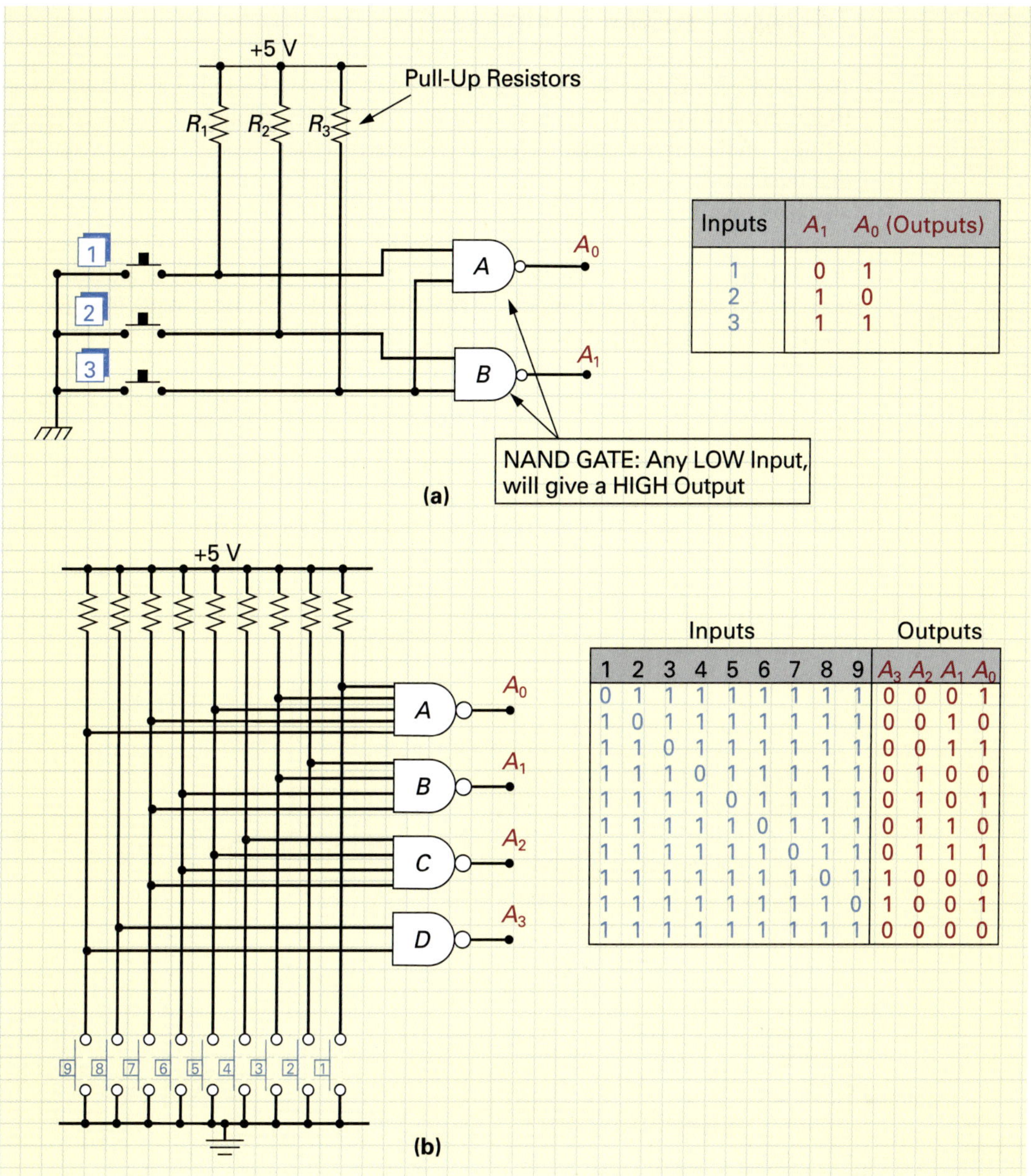

Inputs	A_1	A_0 (Outputs)
1	0	1
2	1	0
3	1	1

Inputs									Outputs			
1	2	3	4	5	6	7	8	9	A_3	A_2	A_1	A_0
0	1	1	1	1	1	1	1	1	0	0	0	1
1	0	1	1	1	1	1	1	1	0	0	1	0
1	1	0	1	1	1	1	1	1	0	0	1	1
1	1	1	0	1	1	1	1	1	0	1	0	0
1	1	1	1	0	1	1	1	1	0	1	0	1
1	1	1	1	1	0	1	1	1	0	1	1	0
1	1	1	1	1	1	0	1	1	0	1	1	1
1	1	1	1	1	1	1	0	1	1	0	0	0
1	1	1	1	1	1	1	1	0	1	0	0	1
1	1	1	1	1	1	1	1	1	0	0	0	0

FIGURE 23-7 Basic Encoder Circuits.

EXAMPLE:

Referring to the encoder circuit in Figure 23-7(b), determine what the output code would be if push buttons 2 and 4 were pressed simultaneously.

Solution:

If push buttons 2 and 4 in Figure 23-7(b) were pressed simultaneously, NAND gates *B* and *C* would both have a LOW input, so both would generate a HIGH output. The output codes for buttons 2 (0010) and 4 (0100) therefore would be combined, producing a 0110 output code.

Decimal-to-BCD Encoder

An encoder circuit that in response to decimal input conditions generates equivalent BCD output codes.

Negative-Logic

1. An active-LOW signal or code. 2. Digital logic in which the more negative logic level represents a binary 1.

Positive-Logic

1. An active-HIGH signal or code. 2. Digital logic in which the more positive logic level represents a binary 1.

23-2-2 Decimal-to-BCD Encoders

The 74147 IC, shown in Figure 23-8, is an example of a ten-line to four-line (decimal-to-BCD) priority encoder. The bars over the 74147's $\overline{A}_1$ through $\overline{A}_9$ input labels and the small triangle on the input lines are used to indicate that these are active-LOW inputs. This means that the encoder's inputs will be activated whenever these inputs are made LOW. There are also bars over the 74147's $\overline{Q}_0$ through $\overline{Q}_3$ outputs. This is again used to indicate that the output is an active-LOW code. This active-LOW output code is also known as **negative-logic,** and if you need to convert the output to an active-HIGH code, or **positive-logic,** you will need to invert the output from the 74147. For example, looking at the function table in Figure 23-8, you can see that if the $\overline{A}_1$ input is made LOW (active), the 74147 will generate the negative-logic output code HHHL (1110), which will have to be inverted, or reversed, to the positive-logic code LLLH (0001 or binary 1) if the BCD output code is to be equivalent to the activated input $\overline{A}_1$.

As another example, you can see that if the $\overline{A}_9$ input is made LOW (active), the 74147 will generate the negative-logic code LHHL (0110), which will have to be inverted to the positive-logic code HLLH (1001 or binary 9) in order to be equivalent to the activated input

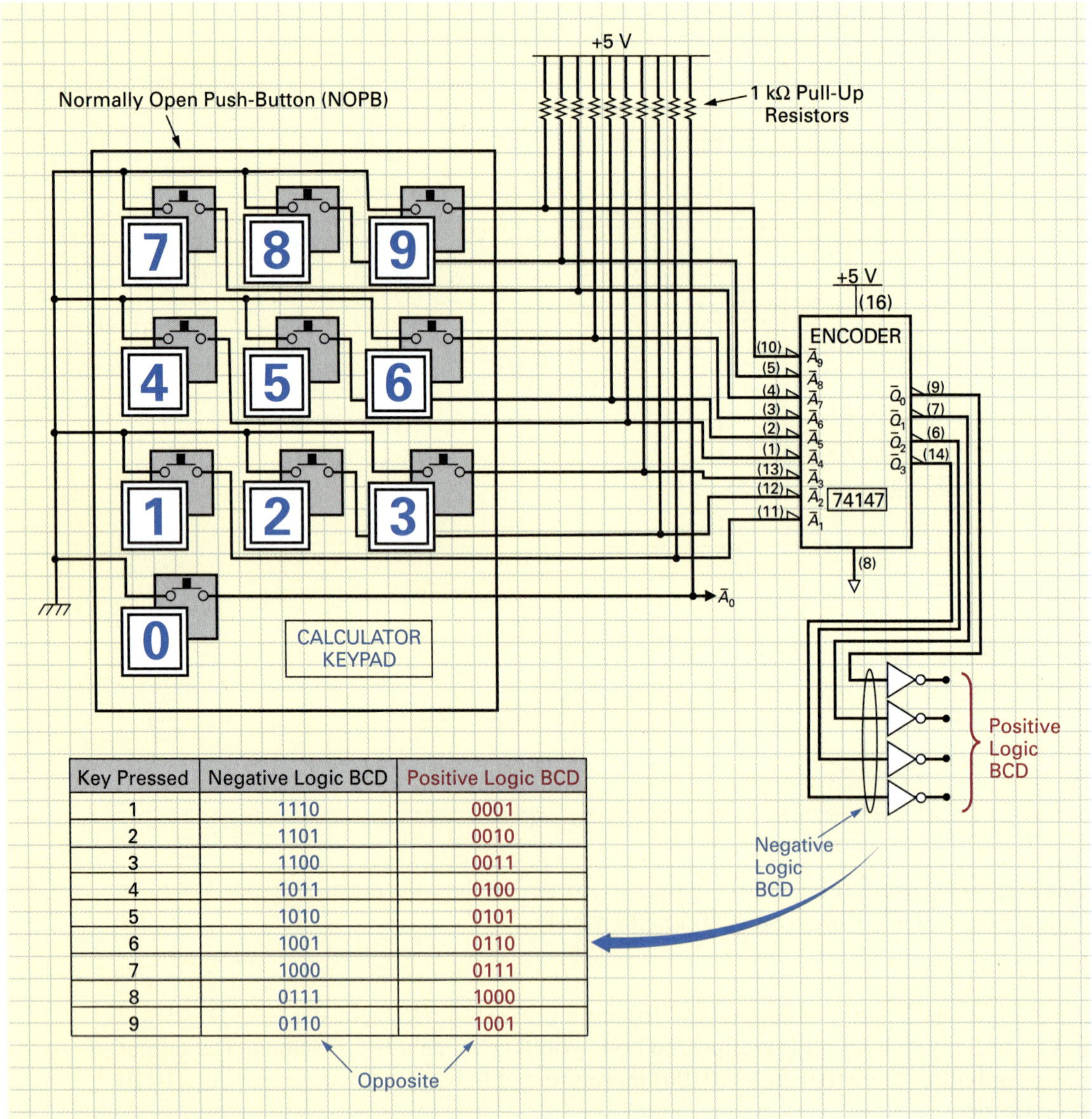

Key Pressed	Negative Logic BCD	Positive Logic BCD
1	1110	0001
2	1101	0010
3	1100	0011
4	1011	0100
5	1010	0101
6	1001	0110
7	1000	0111
8	0111	1000
9	0110	1001

FIGURE 23-8 **Application—A 0-to-9 Keypad Encoder Circuit.**

$\overline{A}_9$. The rest of the function table shows how a LOW on any of the $\overline{A}_1$ through $\overline{A}_9$ inputs will generate an equivalent negative-logic code. To explain one other detail, the 74147 is a **priority encoder,** or high-priority encoder (HPRI), which means that if more than one key is pressed at the same time, the 74147 will only generate the code for the larger digit, or higher priority. For example, if the 9 and 6 keys are pressed at the same time the 74147 will generate a nine output code, responding only to the LOW on the $\overline{A}_9$ input and ignoring the LOW on the $\overline{A}_6$ input. As another example, if input $\overline{A}_8$ is taken LOW, the output code will be LHHH, which is the inverse of 1000, the BCD code for 8. This BCD code of 8, however, will only be present at the output if the $\overline{A}_9$ input is inactive, since the $\overline{A}_9$ input would have a higher priority than the $\overline{A}_8$ input.

Priority Encoder
An encoder circuit that in response to several simultaneous inputs will generate an output code that is equivalent to the highest priority input.

Application—A 0-to-9 Keypad Encoder Circuit

As an application, Figure 23-8 shows how a 74147 can be used to encode a calculator's keypad. Looking first at the push-button switches, which are controlled by the 1 through 9 keypads, you can see that pull-up resistors are included to ensure that the 74147's inputs are normally HIGH. When a key is pressed, the respective input is taken LOW, making that input to the 74147 active. The table shown in the inset in Figure 23-8 shows how this encoder circuit will respond to different inputs. For example, when the 1 key is pressed, the 74147 will receive a LOW at the $\overline{A}_1$ input, so it will generate a 1110 negative-logic output code, which will be inverted by a set of four INVERTER logic gates to the positive-logic code 0001. Therefore, a binary 1 (0001) output code is generated whenever the 1 key is pressed. This circuit, therefore, will generate a positive-logic BCD code at the output of the INVERTER gates that is equivalent to the value of the keypad key pressed.

EXAMPLE:

What would be the negative-logic and positive-logic code generated by the circuit in Figure 23-8 if none of the keys were pressed?

Solution:

Referring to the first line of the function table for the 74147 data sheet shown in Figure 23-7, you can see that when none of the inputs is active (LOW), a 1111 negative-logic output code is generated at the output of the 74147, which, after the inversion, will be a positive-logic code 0000 (0).

SELF-TEST EVALUATION POINT FOR SECTION 23-2

Use the following questions to test your understanding of Section 23-2.

1. How many inputs and outputs would a decimal-to-BCD encoder have?
2. How can the negative BCD logic output of a 74147 be converted to a positive BCD logic output?
3. If the 3, 4, and 7 inputs of a priority encoder were all activated at the same time, what would be the binary output code?
4. Convert the following negative-logic codes to positive-logic codes:
   - **a.** 0010
   - **b.** 1110
   - **c.** LHLL
   - **d.** ON, OFF, ON, OFF

23-3 MULTIPLEXERS

Multiplexer (MUX)
Also called a data selector, it is a circuit that is controlled to switch one of several inputs through to one output.

Analog Multiplexer
An analog signal multiplexer circuit.

Digital Multiplexer
A digital signal multiplexer circuit.

A digital **multiplexer** (MUX, pronounced "mucks"), or selector, is a circuit that is controlled to switch one of several inputs through to one output. The most basic of all multiplexers is the rotary switch, shown in Figure 23-9(a), which will switch any of the six inputs through to the one output based on the switch position selected. In most high-speed applications, the multiplexer will need to be electronically rather than mechanically controlled. **Analog multiplexer** circuits use relays or transistors to electronically switch analog signals, while **digital multiplexer** circuits use logic gates to switch one of several digital input signals through to one output. Figure 23-9(b) illustrates a simple digital multiplexer designed to switch one of its two inputs through to the output (a one-of-two multiplexer). The logic level applied to the SELECT input determines whether AND gate A or B is enabled and therefore which of the data inputs (D_0 or D_1) is switched to the output. If SELECT = 0, AND gate A is enabled while AND gate B is disabled, and therefore output Y will follow input D_0 ($Y = D_0$). On the other hand, if SELECT = 1, AND gate B is enabled while AND gate A is disabled, and output Y will follow input D_1 ($Y = D_1$).

23-3-1 *One-of-Eight Data Multiplexer/Selector*

The 74151 IC shown in Figure 23-10 is an eight-input data multiplexer, or selector. The binary value applied to the three DATA SELECT control inputs (*ABC*) will determine which one of the eight inputs is switched through to the output. For example, when $ABC = 000$ (binary 0) data input D_0 is switched through to the Y output, or when $ABC = 110$ (binary 6) data input D_6 is switched through to the Y output. An eight-input MUX will therefore need three data select input control lines so that eight binary combinations can be obtained ($2^3 = 8$), and each of these combinations can be used to select one of the eight data inputs. The 74151 has a direct output Y (pin 5) and a complemented, or inverted, output W (pin 6). The active-LOW strobe enable or gate enable ($\overline{G}$) is used to either enable ($\overline{G} = 0$) or disable ($\overline{G} = 1$), as detailed in the function table.

Application—Parallel-to-Serial Data Conversion

Parallel-to-Serial Converter
A circuit that will convert a parallel data word input into a serial data stream output.

Figure 23-10(a) shows how a 74151 can be used as a **parallel-to-serial converter.** Binary data can be transferred either in parallel form or serial form between two digital circuits. The circuit in Figure 23-10 shows the differences between these two data formats by showing how we can convert an 8-bit parallel-data input into an 8-bit serial-data output. Two 4-bit storage registers hold the 8-bit parallel word that has to be converted to serial form. This 8-bit parallel binary word (0110 1001) is applied to the data inputs of the 8-bit multiplexer.

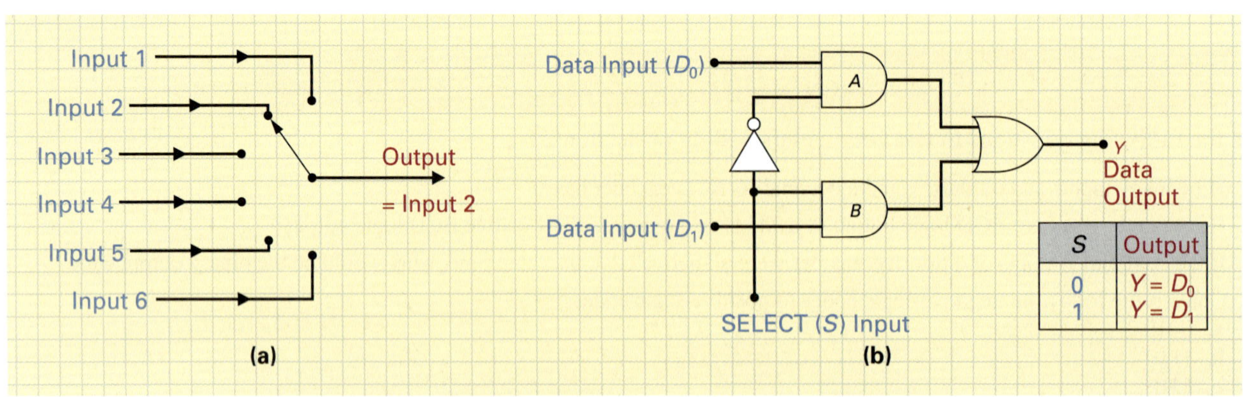

S	Output
0	$Y = D_0$
1	$Y = D_1$

FIGURE 23-9 **Basic Multiplexers.**

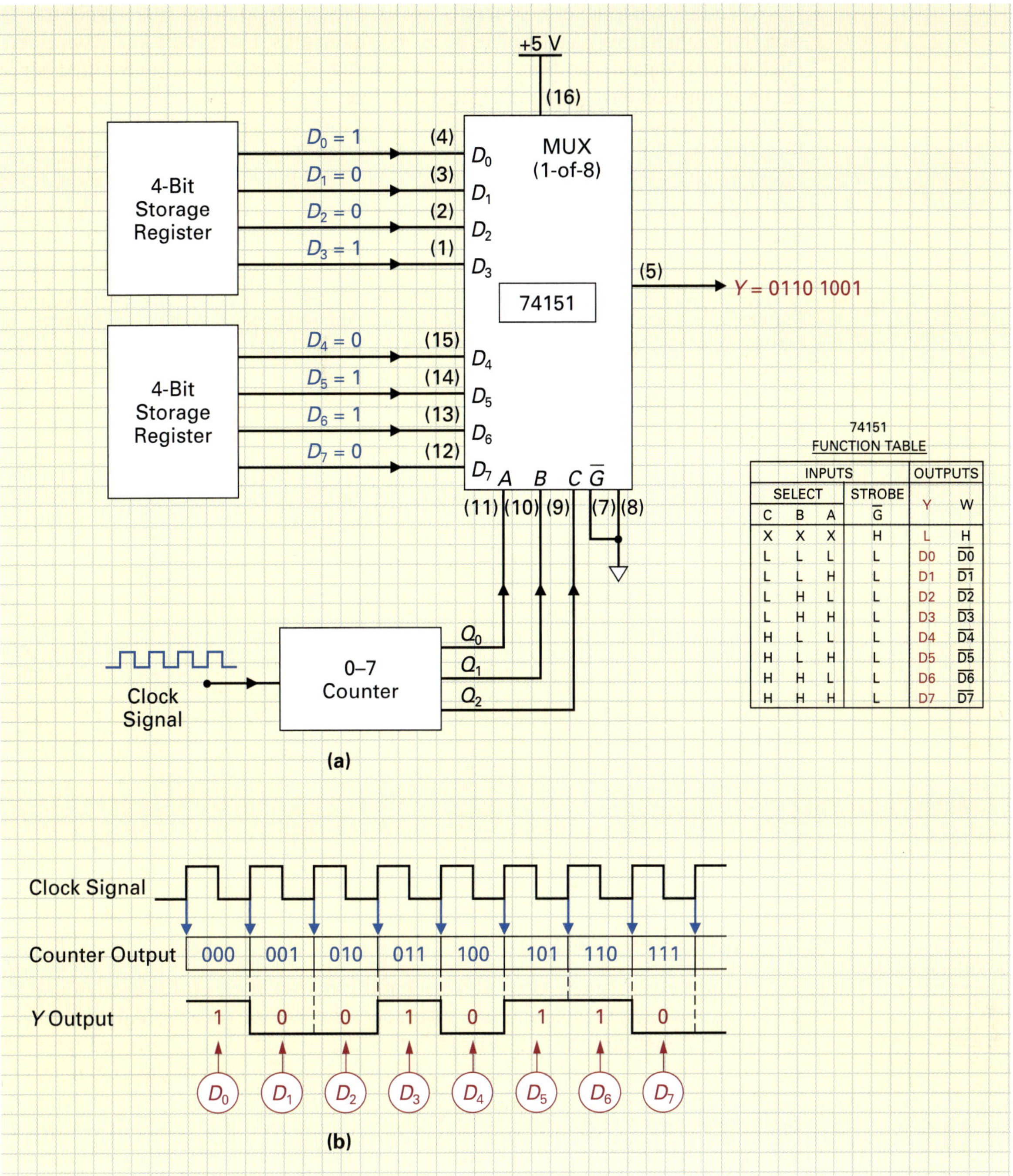

74151 FUNCTION TABLE

INPUTS				OUTPUTS	
SELECT			STROBE		
C	B	A	$\overline{G}$	Y	W
X	X	X	H	L	H
L	L	L	L	D0	$\overline{D0}$
L	L	H	L	D1	$\overline{D1}$
L	H	L	L	D2	$\overline{D2}$
L	H	H	L	D3	$\overline{D3}$
H	L	L	L	D4	$\overline{D4}$
H	L	H	L	D5	$\overline{D5}$
H	H	L	L	D6	$\overline{D6}$
H	H	H	L	D7	$\overline{D7}$

FIGURE 23-10 Application—Parallel-to-Serial Data Conversion.

A clock or timing signal is applied to the input of a number 0 through 7 counter whose three outputs (Q_0, Q_1, and Q_2) are applied to the three multiplexer data select inputs (*A, B,* and *C*). Referring to the waveforms in Figure 23-10(b), you can see that each positive-going edge of the clock signal causes the counter to advance by one count. As the 3-bit counter is incremented through each of its eight output states (000, 001, 010, 011, 100, 101, 110, and 111), the ABC data-select MUX inputs are sequenced in order, causing each of the eight data inputs (D_0 through D_7) to be selected one after the other. As a result, the 8-bit parallel binary word at the input of the 74151 is switched, one bit at a time, producing an 8-bit serial binary word at the output.

Parallel data transfer is generally not used for the long-distance transmission of binary data (for example, over a telephone line) since it requires several data lines connected in parallel. Parallel data is therefore converted to serial data before it is transmitted, since serial data transmission only requires a single data line. The serial data transfer of binary digits is, however, a lot slower than parallel data transmission. To explain this, let us assume that the time interval allotted to each bit in the example in Figure 23-10 is one microsecond. It will therefore take eight microseconds for the 8-bit word to be transmitted in serial form; however, all of the binary bits in a parallel word are transferred simultaneously in one microsecond over a group of parallel lines. The difference between these two data transfer methods therefore seems to come down to the standard speed-versus-cost trade-off. Serial data transfer needs only one transmitter circuit, one receiver circuit, and one data line to transfer any number of binary bits. Parallel data transfer, on the other hand, requires a separate transmitter, receiver, and line for each bit of the word. The transmission and reception of parallel data is therefore more expensive; however, when high-speed data transfer is required, the data handling time for a parallel transfer of data is extremely fast compared to the data handling time needed for a serial transfer of data.

EXAMPLE:

Which input would be switched through to the *Y* output if the following logic levels were applied to the 74151?

a. $A = 0, B = 1, C = 0, D_0 = 1, D_2 = 1, D_3 = 0, D_4 = 1, D_5 = 1, D_6 = 1, D_7 = 0, G = 1$

b. $A = 1, B = 1, C = 1, D_0 = 1, D_2 = 1, D_3 = 0, D_4 = 1, D_5 = 1, D_6 = 1, D_7 = 0, G = 0$

Solution:

a. $Y =$ LOW because the gate enable or strobe is not active.

b. $Y = D_7 = 0$ because select inputs $ABC = 111$, or decimal 7.

Application—A Logic-Function Generator Circuit

Logic-Function Generator
A circuit that will generate the desired logic level at the output for each input combination.

Multiplexers are also very useful as **logic-function generators.** As an example, Figure 23-11(a) shows a truth table and sum-of-products expression for a desired logic-function. By connecting the multiplexer's data inputs either HIGH or LOW, as shown in Figure 23-11(b), we can generate the desired HIGH or LOW output at *Y* for each of the *ABC* input combinations. Using a MUX instead of discrete logic gates will greatly reduce the number of ICs used and therefore the circuit cost, size, and power consumption. For example, to implement the sum-of-products expression in Figure 23-11 using discrete logic gates would require four three-input AND gates, one four-input OR gate, and three INVERTERS—a total of four ICs compared to one 74151 IC.

23-3-2 *Four-of-Eight Data Multiplexer/Selector*

Figure 23-12 shows the 74157 IC, which is an eight-input four-output multiplexer. This multiplexer will switch either the *A* 4-bit word (1A, 2A, 3A, and 4A) or the *B* 4-bit word (1B, 2B, 3B, and 4B) through to the 4-bit output (1Y, 2Y, 3Y, and 4Y). The active-LOW gate enable input ($\overline{G}$) is used to either enable ($\overline{G} = 0$) or disable ($\overline{G} = 1$) all of the logic gates within the IC. When the gate-enable line is active, the *Y* outputs will follow whichever inputs have been selected by the $\overline{A}/B$ select control input. For example, when $\overline{A}/B = 0$, the *Y* outputs will follow whatever is applied to the *A* inputs, whereas when $\overline{A}/B = 1$, the *Y* outputs will follow whatever is applied to the *B* inputs.

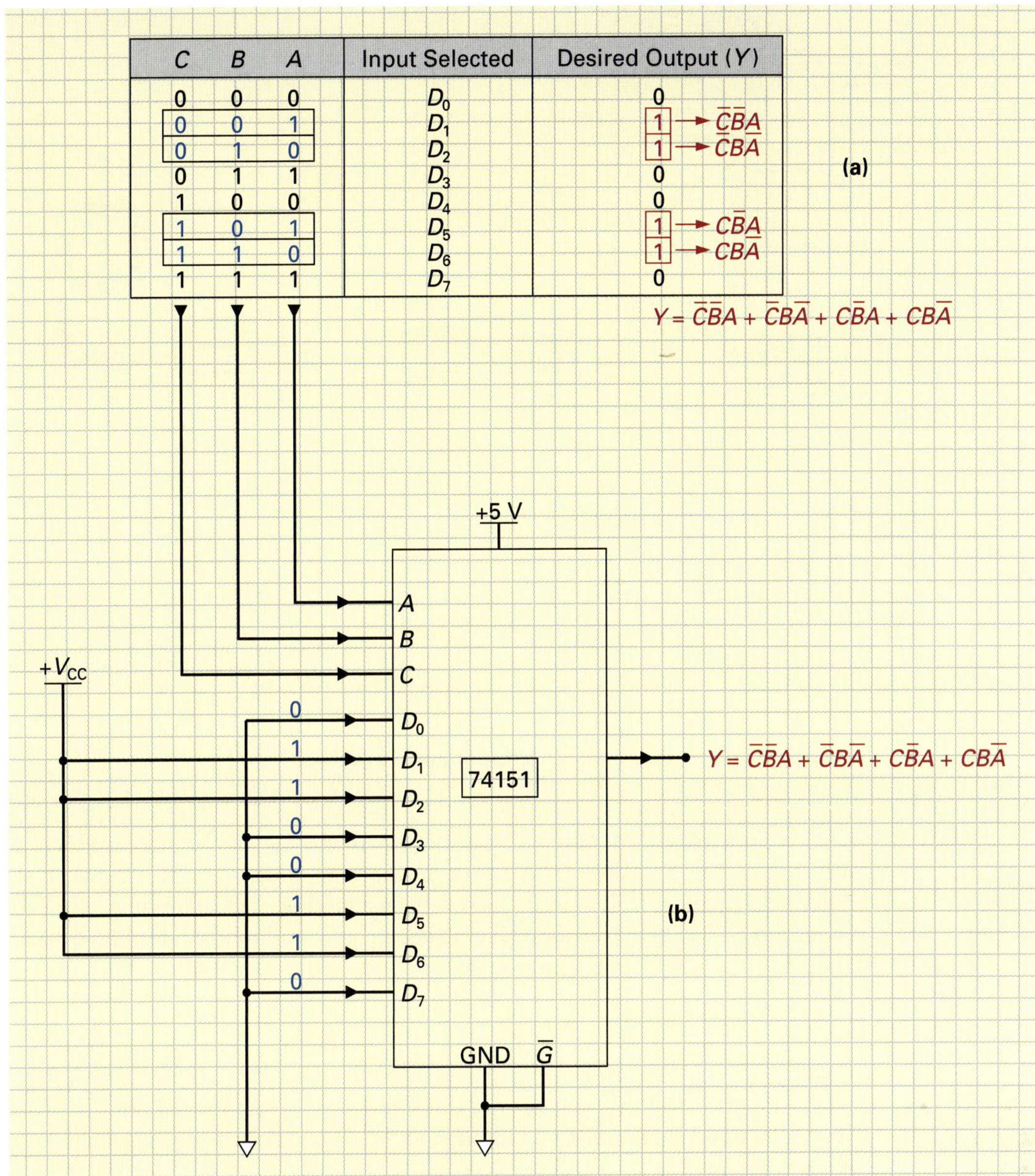

FIGURE 23-11 **Application—A Logic-Function Generator Circuit.**

Application—A Multiplexed Display

Most calculators have an $\boxed{X \leftrightarrow M}$ key that is used to select whether the calculator's seven-segment display shows the current value being operated on (*X*) or the value stored in memory (*M*). Each time the $\boxed{X \leftrightarrow M}$ key is pressed, the display switches between these two values. Figure 23-12 shows how this function could be implemented using a 74157, 7447, and a seven-segment display. The calculator's $\boxed{X \leftrightarrow M}$ key controls the $\bar{A}/B$ input of the 74157. When $\bar{A}/B = 0$, the contents of the *X* storage register (which is applied to the 74157's *A* inputs) is switched through to the *Y* outputs ($Y = 0110$), decoded by the 7447, and displayed on the seven-segment display (display will show decimal 6). On the other hand, when $\bar{A}/B = 1$, the contents of the *M* storage register (which is applied to the 74157's *B* inputs) is switched through to the *Y* outputs ($Y = 1001$), decoded by the 7447, and displayed on the seven-segment display (display will show decimal 9).

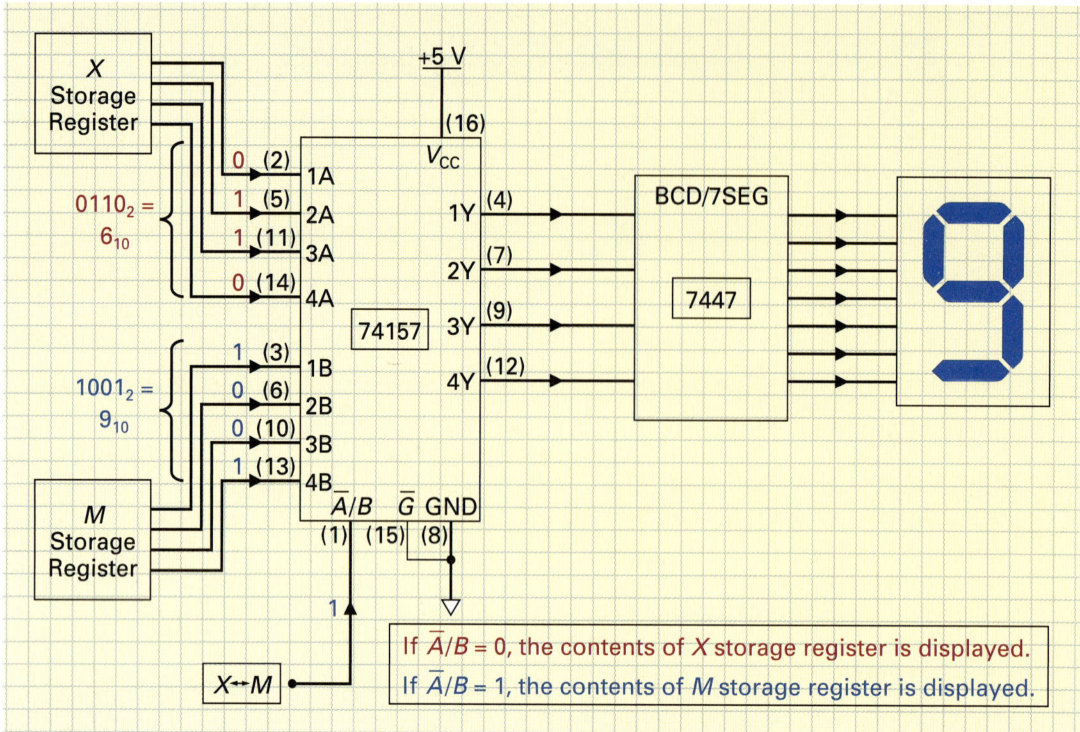

FIGURE 23-12 **Application—A Multiplexed Display.**

Multiplexed Display
A display circuit that timeshares the display or the display drive circuitry.

A **multiplexed display** uses one display to present two or more different pieces of information. This "timesharing" of the decoder driver, display, and interconnect wiring greatly reduces the number of ICs needed, power consumption, size, and printed circuit board interconnections. The big disadvantage of using a single multiplexed display instead of two separate displays is that you cannot see both values at the same time.

EXAMPLE:

Which input would be switched through to the output if the following logic levels were applied to the 74157?

a. $\overline{A}/B = 0, \overline{G} = 0, A = 1001, B = 0110.$

b. $\overline{A}/B = 1, \overline{G} = 0, A = 1001, B = 0110.$

Solution:

a. $Y = A = 1001$. Because the $\overline{G}$ input is LOW, the $\overline{A}/B$ input is LOW and so the A inputs are selected.

b. $Y = A = 0110$. Because the $\overline{G}$ input is LOW, $\overline{A}/B$ input is HIGH and so the B inputs are selected.

SELF-TEST EVALUATION POINT FOR SECTION 23-3

Use the following questions to test your understanding of Section 23-3.

1. A multiplexer is a circuit that through a control input will switch ____________ input(s) through to ____________ output(s).

a. one, one of several **c.** one, one

b. one of several, one **d.** one of several, one of several

2. How many data selects inputs would a one-of-sixteen multiplexer need?
3. True or false: multiplexers are often used for serial-to-parallel data conversion.
4. Which data transmission method is considered the fastest: serial or parallel?
5. Which data transmission method requires more hardware: serial or parallel?
6. Would the 74151 or 74157 best be described as a 4-bit word multiplexer?

23-4 DEMULTIPLEXERS

A digital multiplexer (MUX), or selector, is a circuit that is controlled to switch one of several inputs through to one output. On the other hand, a digital **demultiplexer** (DMUX, pronounced "demucks"), or data distributor, operates in exactly the opposite way since it is a circuit that is controlled to switch a single input through to one of several outputs. The demultiplexer is also known as a *data distributor* since the data applied to the single input is distributed to one of the many outputs.

Demultiplexer (DMUX)
Also called a data distributor, it is a circuit that is controlled to switch a single input through to one of several outputs.

Figure 23-13(a) shows how a rotary switch can be made to act as a mechanically controlled demultiplexer. In this circuit the switch is controlled to connect a single input through to one of several outputs. Like the multiplexer, mechanical devices are too slow for high-speed applications, and therefore most demultiplexers are also electronically rather than mechanically controlled.

Analog demultiplexers will switch a single analog signal input through to one of several outputs, while **digital demultiplexers** will switch a single digital signal input through to one of several outputs. Figure 23-13(b) shows a simple digital demultiplexer circuit designed to switch its single input through to one of two outputs. This one-line to two-line demultiplexer is controlled by the SELECT INPUT which will determine which AND gate is enabled and which AND gate is disabled, as described in the function table.

Analog Demultiplexer
An analog signal demultiplexer circuit.

Digital Demultiplexer
A digital signal demultiplexer circuit.

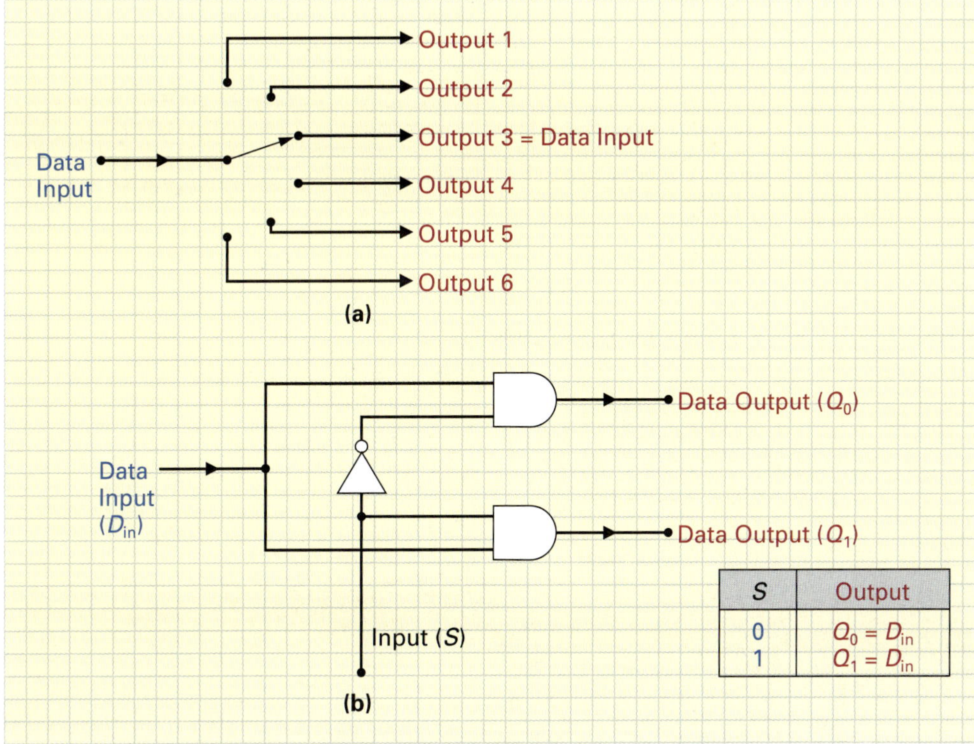

S	Output
0	$Q_0 = D_{in}$
1	$Q_1 = D_{in}$

FIGURE 23-13 Basic Demultiplexers/Data Distributors.

In the following sections you will be introduced to three typical demultiplexer ICs. Two of these ICs were discussed previously in the decoder section and they are reintroduced in this section since most decoder ICs can be made to function as either decoders or demultiplexers.

23-4-1 *A One-Line to Eight-Line Demultiplexer*

The 7442 BCD-to-decimal decoder can be used as a demultiplexer, as shown in Figure 23-14. The internal schematic of the 7442 IC is shown in Figure 23-14(a). As a decoder, any one of

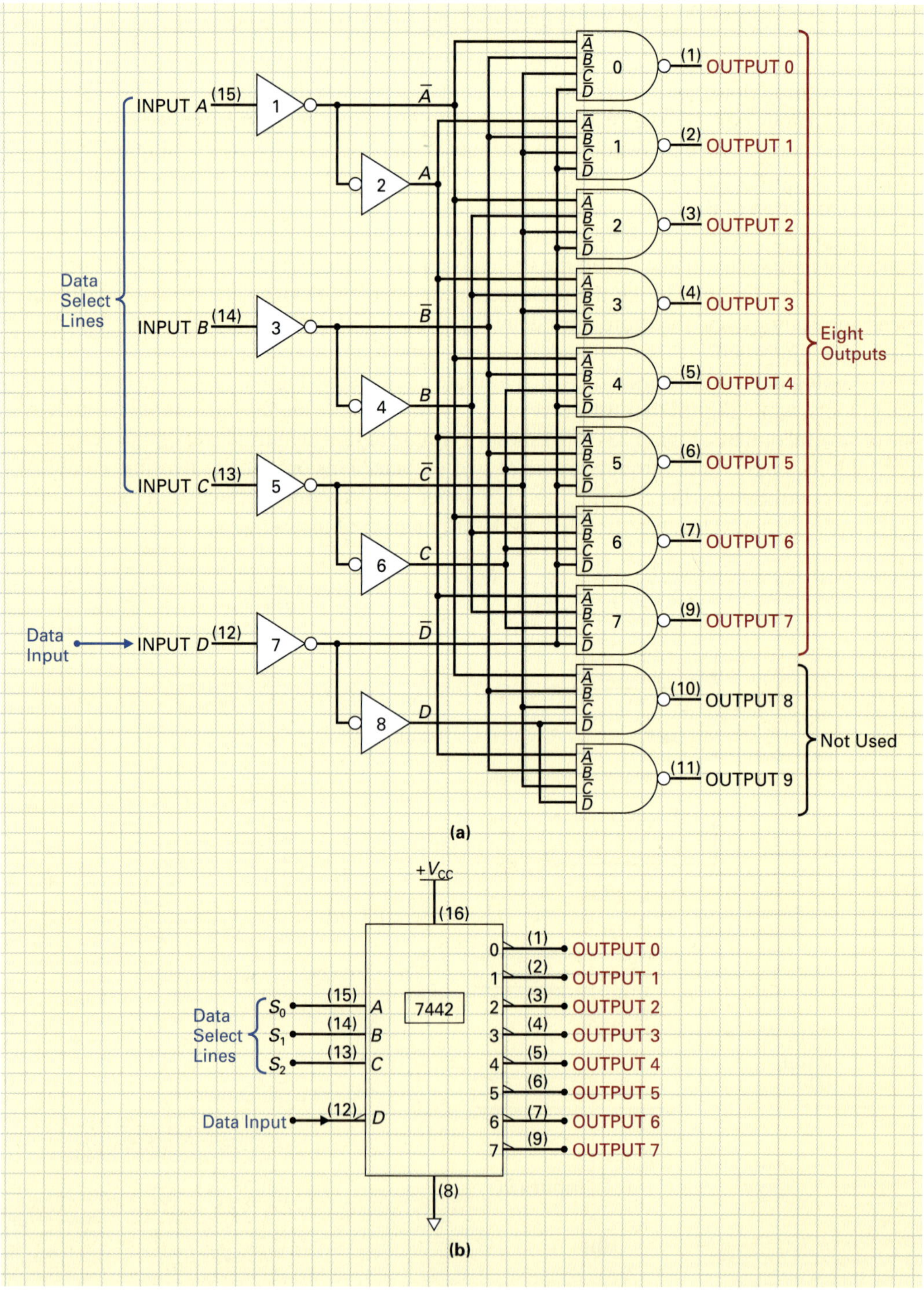

FIGURE 23-14 Application—The 7442 Decoder as a Demultiplexer.

the 7442's ten outputs could be driven LOW by applying an equivalent *BCD* input code to the *ABCD* inputs. As a demultiplexer, the 7442 can be made to function as a one-line to eight-line data distributor. To operate in this way, inputs *A, B,* and *C* are used to select one of the internal logic gates 0 through 7 (three select lines will give a total of eight combinations). For example, if 010 is applied to inputs ABC, internal gate 2 will be enabled. The serial data to be distributed is applied to input D, and will be applied to all of the lower inputs of NAND gates 0 through 7. INVERTER 7 will invert the input data, but this process will be reversed by the inversion of the NAND gates. The output of NAND gate 2 will therefore be under the control of the *D* input, so a HIGH input on *D* will appear as a HIGH on output 2 and a LOW input on *D* will appear as a LOW on output 2. When a 7442 IC is used as a demultiplexer, its logic symbol is drawn differently from that of a decoder logic symbol, as shown in Figure 23-14(b).

EXAMPLE:

The *D* data input would be switched through to which output if the following logic levels were applied to the 7442?

a. *CBA* = 011 b. *CBA* = 101

Solution:

a. Output 3 b. Output 5

23-4-2 *A One-Line to Sixteen-Line Demultiplexer*

Figure 23-15 shows how a 74154 decoder IC can be made to function as a one-line to sixteen-line demultiplexer. With this IC all four of the decoder OUTPUT SELECT inputs are used to select one of the sixteen available outputs ($2^4 = 16$), and the DATA input is applied to one

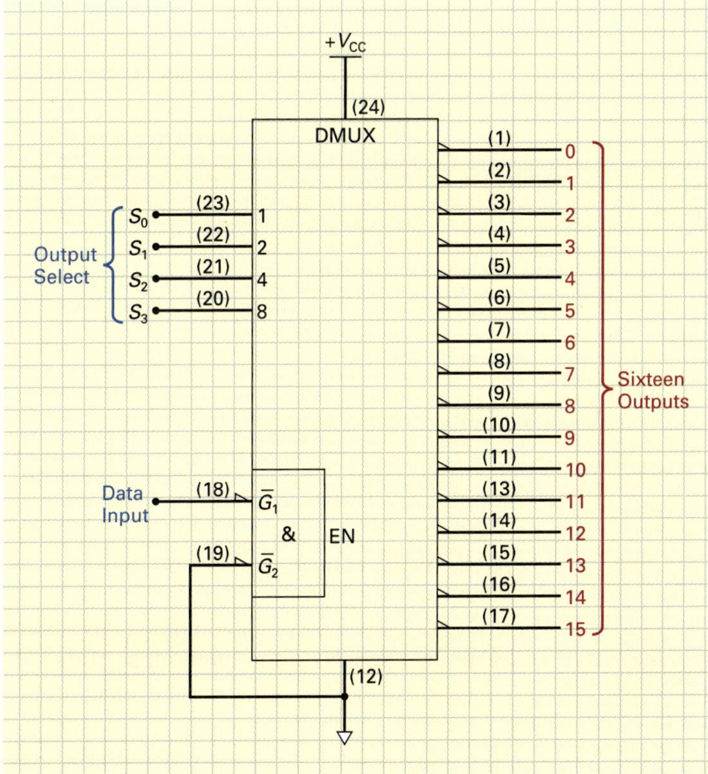

FIGURE 23-15 **Application—The 74154 as a Four-line to Sixteen-line Demultiplexer.**

of the gate-enable inputs. When the DATA input is LOW, all of the 74154's internal gates are enabled, so the selected output (based on the OUTPUT SELECT code applied) is driven LOW. On the other hand, when the DATA input is HIGH, all of the 74154's internal gates are disabled, so the selected output is HIGH.

EXAMPLE:

The D input would be switched through to which output if the following logic levels were applied to the 74154?

a. $S = 1011$ b. $S = 1110$

Solution:

a. Output 11 b. Output 14

23-4-3 *A Three-Line to Eight-Line Decoder/Demultiplexer*

Like the 7442 and 74154 ICs, the 74138 shown in Figure 23-16 is listed as a decoder/demultiplexer in the IC data books. The 74138 has three inputs. The G_1 input is active-HIGH, while the $\overline{G}_{2A}$ and the $\overline{G}_{2B}$ inputs are both active-LOW. All three of these gate-enable inputs must be active in order for the IC to be enabled. The three SELECT inputs (ABC) are used to select one of the eight ($2^3 = 8$) active-LOW outputs (Y_0 through Y_7). For example, when $CBA = 101$ (decimal 5), the Y_5 output is made active (LOW) while all the other outputs remain inactive (HIGH).

EXAMPLE:

What would be the outputs from a 74138 for the following input conditions?

a. $CBA = 011$, $G_1 =$ LOW, $\overline{G}_2 =$ LOW

b. $CBA = 101$, $G_1 =$ HIGH, $\overline{G}_2 =$ LOW

Solution:

a. All outputs will be HIGH because G_1 is not active.

b. All outputs will be HIGH, except for Y_5 (101), which will be LOW.

Application—A Basic Home Security System

Figure 23-16 shows how a 74151 eight-line to one-line multiplexer (MUX), and a 74138 one-line to eight-line demultiplexer (DMUX) could be connected to form a basic home security system. A binary 0 (000) to 7 (111) counter drives the ABC select or control inputs of both the MUX and DMUX. As the counter counts from 0 to 7, it "selects" in turn each of the 74151 inputs and switches the HIGH or LOW data through to the Y output, as seen in the function table in the inset in Figure 23-16. For example, when the counter is at a count of 011 (3), the HIGH or LOW data present at the three-input of the 74151 (pin 1) is switched through to the Y output. In this example circuit, the three-input of the 74151 is connected to a window switch. If the window is open, its associated switch will also be open and the input to the 74151 will be HIGH due to the pull-up resistors. On the other hand, if the window and its associated switch are closed, ground or a LOW logic level will be applied to the input of the 74151.

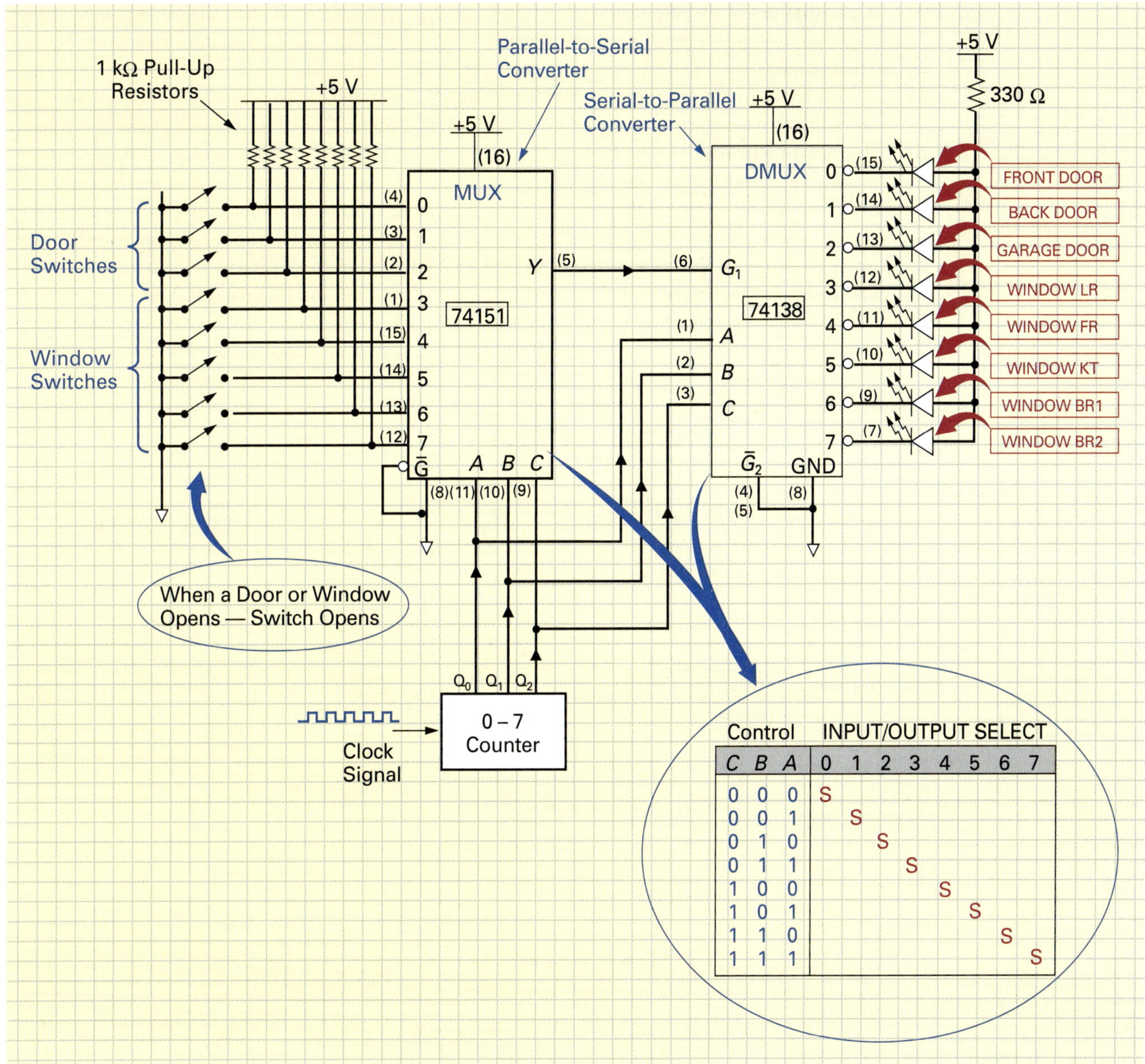

FIGURE 23-16 **Application—Basic Home Security System.**

The 74138 DMUX in this circuit application operates in exactly the opposite way of the 74151. For example, as we mentioned earlier, when the counter is at a count of 011 (3), the HIGH or LOW data present at the three-input of the 74151 (pin 1) is switched through to the *Y* output. This HIGH or LOW data is applied to the G_1 input of the 74138, which is also receiving a 011 (3) select input from the counter, so the data at G_1 is switched through to the three-output of the 74138 (pin 12). Since the 74138 has active-LOW outputs, there will be an inversion between input and output. This inversion between the 74138's input and outputs is needed since we are driving a common-anode LED display whose inputs, as we know, are also active-LOW (LEDs are activated by a LOW). To explain the operation from input to output, if the window connected to the three-input is opened, the switch will open and give a HIGH to the three-input of the 74151. The 74151 will switch this HIGH data through to the *Y* output, where it will be applied to the G_1 input of the 74138, which will invert the HIGH to a LOW and then switch it to the three-output, where it will turn ON the living room (LR) window LED. This circuit, therefore, first senses the condition of switch 0 and then transmits its data to LED

0, and then senses the condition of switch 1 and then transmits its data to LED 1, and then senses the condition of switch 2 and then transmits its data to LED 2, and so on until it reaches 7 (111). The counter then resets and the process is repeated.

Serial-to-Parallel Converter
A circuit that will convert serial data into a parallel format.

The 74151 functions as a parallel-to-serial converter since it converts the parallel 8-bit word applied to its inputs into an 8-bit serial data stream. Conversely, the 74138 functions as a **serial-to-parallel converter,** since it converts an 8-bit serial data stream applied to its input into a parallel 8-bit output word. One question you may have about this circuit is, why don't we simply connect the switches straight to the LEDs and bypass the 74151 and 74138? The answer lies in the basic difference between serial data transmission and parallel data transmission. With this circuit, the 74151 can be placed in close proximity to all the windows and doors, and then only four wires (three for the counter and one for data) need to be run to the 74138 and its display. If the switches were connected directly to the display, eight wires would be needed to connect the switch sensors to the LED display. The advantage of serial data transmission, therefore, is that only one data line is needed. On the other hand, the advantage of parallel data transmission is speed, since a parallel connection would mean that all of the data is transmitted at one time.

SELF-TEST EVALUATION POINT FOR SECTION 23-4

Use the following questions to test your understanding of Section 23-4.

1. A demultiplexer or data distributor is a circuit that through a control input will switch ________ input(s) through to ________ output(s).
   a. one, one of several
   b. one, one
   c. one of several, one
   d. one of several, one of several
2. A/an ________ can also be used as a demultiplexer. (encoder/decoder)
3. Which 3-bit code must be applied to the *CBA* inputs of a 74138 to select the Y_4 output?
4. The main difference between a decoder and a demultiplexer is that a decoder has only SELECT inputs, while a demultiplexer has both SELECT inputs and a DATA input. (true/false)

23-5 COMPARATORS

Comparator
A circuit that compares two inputs to determine whether or not the two inputs are equal.

A digital **comparator** compares two binary input words to see if they are equal. The exclusive NOR (XNOR) gate is a comparator logic gate since it compares the two binary input digits applied, as seen in Figure 23-17(a). When the two input bits are the same, the output of the XNOR gate is a 1; however when the input bits are different, the output of the XNOR gate is 0. Let us now see how we could use a group of XNOR gates to create a 4-bit binary comparator circuit.

23-5-1 *A 4-Bit Binary Comparator*

Figure 23-17(b) shows a 4-bit binary comparator circuit that will compare two 4-bit words being applied by two storage registers. Four XNOR gates are used to compare WORD *A* (A_0, A_1, A_2, and A_3) with WORD *B* (B_0, B_1, B_2, and B_3). Logic gate XNOR 0 compares A_0 and B_0, and if they are the same, Y_0 is a 1. Logic gate XNOR 1 compares A_1 and B_1, and if they are the same, Y_1 is a 1. In turn, XNOR 2 compares A_2 and B_2, and XNOR 3 compares A_3 and B_3, producing a 1 output if the bits are equal (=) and a 0 output if the bits are not equal (≠). If WORD *A* is identical to WORD *B,* all of the XNOR outputs will be HIGH, and therefore the output of the AND gate will also be HIGH. If WORD *A* differs from WORD *B* in any way, the AND gate will receive a LOW input and will produce a LOW output. Therefore, if WORD *A* equals WORD *B,* the output $Y = 1$, whereas if WORD *A* is not equal to WORD *B,* the output $Y = 0$.

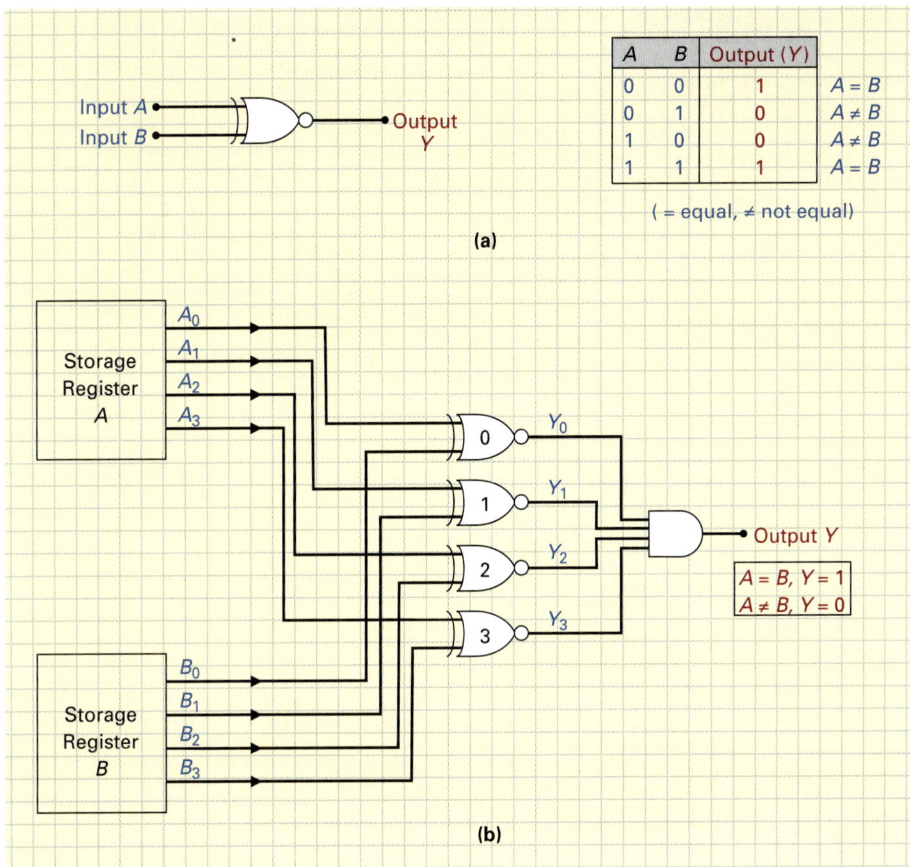

FIGURE 23-17 Basic Comparators.

23-5-2 *A 4-Bit Magnitude Comparator*

A wide variety of digital comparator ICs are available that contain a complete internal binary comparator circuit. Let us now examine the operation and application of a frequently used digital comparator IC, the TTL 7485, which is shown in Figure 23-18. The 7485 compares the 4-bit word inputs (WORD A and WORD B) and provides an active-HIGH output that indicates whether the two input words are equal ($A = B$, pin 6). In addition, the 7485 provides two other active-HIGH outputs indicating whether WORD A is greater than WORD B ($A > B$, pin 5), and whether WORD A is less than WORD B ($A < B$, pin 7). The greater-than and less-than magnitude outputs account for why this IC is called a magnitude comparator.

Application—A Photocopier Control Circuit

Figure 23-18 shows how a 7485 comparator could be used in a photocopier to control the number of copies made. A storage register (A) holds the 4-bit binary word 0010 (2), indicating the number of copies that have been made so far. The storage register (B) holds the 4-bit binary word 1000 (8), indicating the number of copies requested. The 7485 compares these two inputs and generates three output control signals. If the 4-bit input A is less than the 4-bit input B, the 7485 pin 7 output will be HIGH, signaling that the copier should continue copying. When input A is equal to input B, the 7485 pin 6 output will be HIGH, signaling that the requested number of copies have been made and the copier should stop copying.

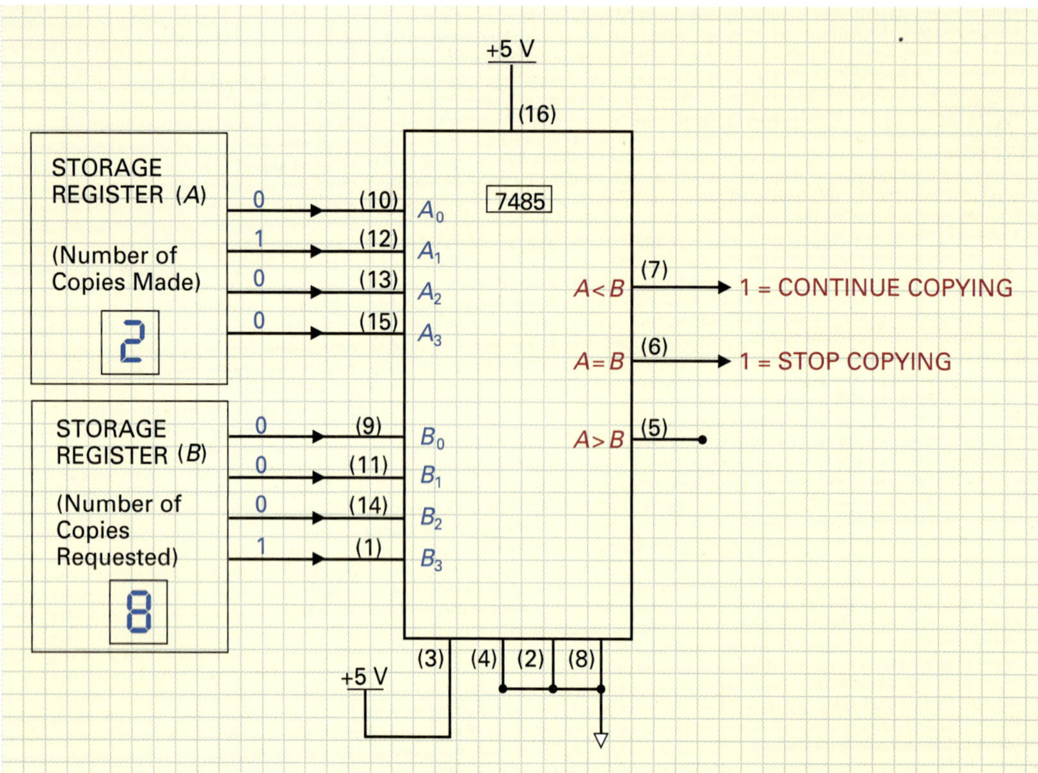

FIGURE 23-18 **Application—Comparator Control of a Photocopier.**

SELF-TEST EVALUATION POINT FOR SECTION 23-5

Use the following questions to test your understanding of Section 23-5.

1. Which logic gate type is considered a basic comparator?
2. What are the three outputs generally present at the output of a magnitude comparator IC, such as the 7485?
3. Why would a comparator IC have cascading inputs?
4. What would be the logic levels at the three outputs of a 7485 if WORD $A = 1010$ and WORD $B = 1011$?

23-6 PARITY GENERATORS AND CHECKERS

Parity Check
An odd-even check that makes the total number of 1s in a binary word either an odd or even number.

The main objective of a digital electronic system is to manage the flow of digital information or data. In all digital systems, therefore, it is no surprise that a large volume of binary data is continually being transferred from one point to another. The accuracy of this data is susceptible to electromagnetic noise, interference, and circuit fluctuations that can cause a binary 1 to be misinterpreted as a binary 0, or a binary 0 to be misinterpreted as a binary 1. Although the chance of a data transfer error in a two-state system is very remote, it can occur, and if it does, it is important that some checking system is in place so that a data error can be detected. The **parity check** system uses circuitry to check whether the odd or even number of binary 1s present in the transmitted code matches the odd or even number of binary 1s present in the received code.

23-6-1 *Even or Odd Parity*

Figure 23-19(a) shows a simplified 4-bit parallel data transmission system that has a **parity generator** circuit and a **parity checker** circuit included for data error detection. The parity generator circuit monitors the 4-bit code being transmitted from the keyboard storage register and generates a **parity bit** based on the total number of binary 1s present in the 4-bit parallel word. The table in Figure 23-19(b) shows what EVEN-PARITY BIT (P_E) or ODD-PARITY BIT (P_O) would be generated for each of the possible sixteen 4-bit codes from the keyboard. For example, let us assume that we wish to transmit an even-parity check bit. The P_E will be made a 1 or 0 so that the number of binary 1s in the final transmitted 5-bit code (D_0, D_1, D_2, D_3, and P_E) will always equal an even number (which means that there are zero, two, or four binary 1s in the transmitted 5-bit code). If, on the other hand, we wish to transmit an odd-parity check bit, the P_O will be made a 1 or 0 so that the number of binary 1s in the final transmitted 5-bit code (D_0, D_1, D_2, D_3, and P_O) will always equal an odd number (which means that there are one, three, or five binary 1s in the transmitted 5-bit code).

Parity Generator
A circuit that generates the parity bit that is added to a binary word so that a parity check can later be performed.

Parity Checker
A circuit that tests the parity bit that has been added to a binary word so that a parity check can be performed.

Parity Bit
A binary digit that is added to a binary word to make the sum of the bits either odd or even.

Referring to the circuit in Figure 23-19(a), you can see that XOR gates are used in both the parity generator and parity checker circuits. In this example circuit, an even-parity check system is being used, and therefore an even-parity bit is generated by XOR *A* and transmitted along with the 4-bit code from keyboard to computer. Within the computer, the 4-bit code is stored in a register, and at the same time it is monitored by XOR gates *B* and *C*, which will produce an active-HIGH PARITY ERROR output. This HIGH error signal could be used to turn on a DATA ERROR LED, signal the computer to display a keyboard data transmission error, or signal the computer to reject the transmitted code and signal the keyboard to retransmit the code for the key pressed.

23-6-2 *A 9-Bit Parity Generator/Checker*

Parity generator and checker ICs, such as the 74180 shown in Figure 23-20, are available for use in parity check applications. The 74180, like all parity generators and checkers, assumes that if an error were to occur, it would more than likely occur in only one of the bits within the transmitted word. If an error were to occur in two bits of the transmitted word, it is possible that an incorrect word will pass the parity test (if a 1 is degraded by noise to a 0, and at the same time a 0 is upgraded by noise to a 1). Multi-bit errors, however, are extremely rare due to the reliability and accuracy of two-state digital electronic systems.

The 74180 is able to function as either a parity generator or as a parity checker. The best way to fully understand the operation of this IC is to see it in an application.

Application—8-Bit Parallel Data Transmission System with Even-Parity Error Detection

Figure 23-20 shows how two 74180 ICs can be used in an 8-bit parallel data transmission system with even-parity error detection. The parity generator (74180 *A*) has its EVEN control input LOW and its ODD control input HIGH. To explain all of the possibilities, Figure 23-21 lists each condition.

a. In the first condition, the 8-bit code to be transmitted from the keyboard contains an even number of 1s, and therefore the parity generator (74180 *A*) generates a LOW, even-parity bit output. The parity checker IC (74180 *B*) receives an even number of 1s at its *A* through *H* inputs, and the same control inputs as the parity generator IC (EVEN input = LOW, ODD input = HIGH), so the even output is LOW, indicating no transmission error has occurred.

b. In the second condition, the 8-bit code to be transmitted from the keyboard contains an even number of 1s, and therefore the parity generator (74180 *A*) generates a LOW, even-parity bit output. In this condition, however, a bit error occurs due to electrical noise so an

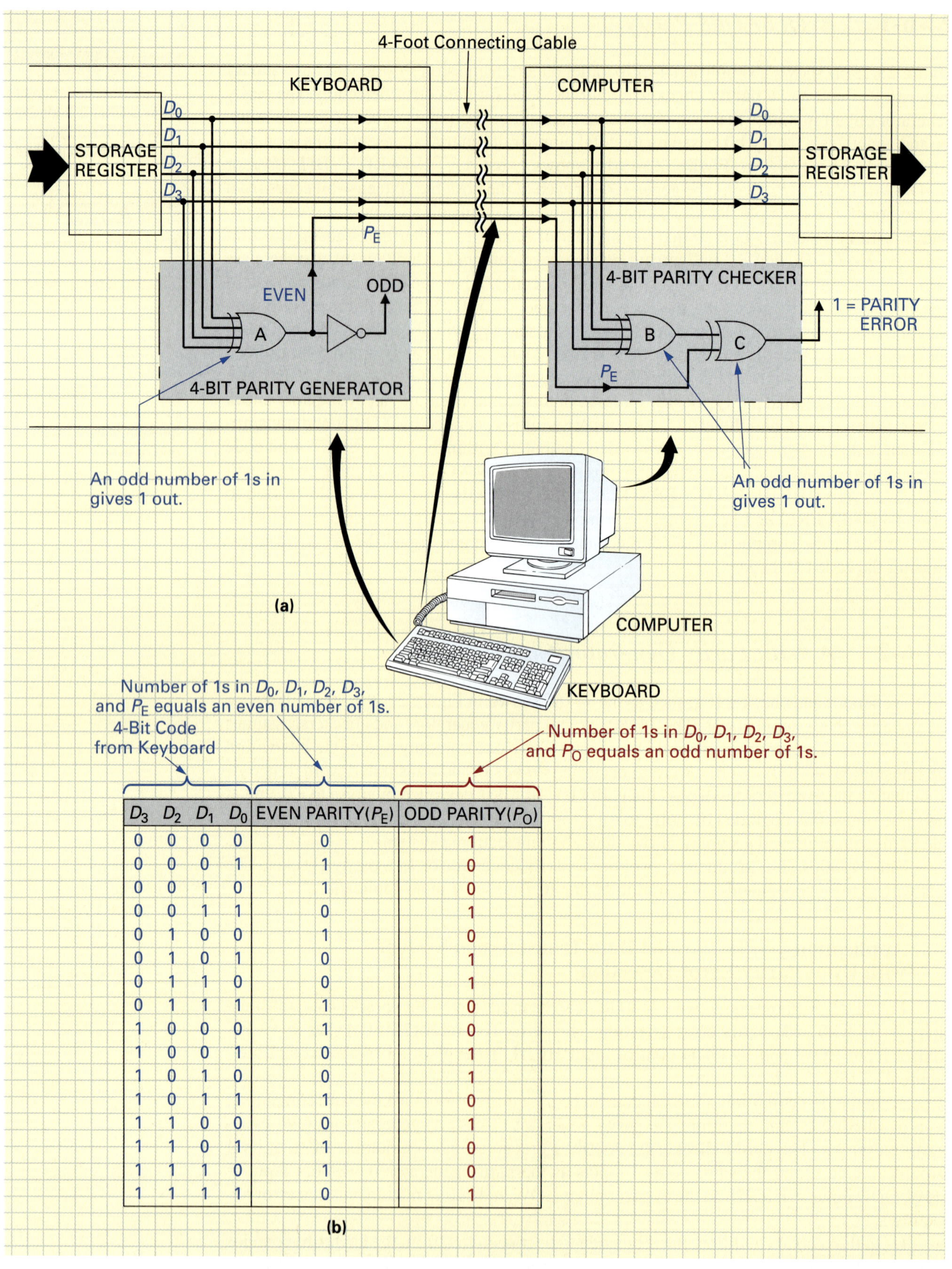

D_3	D_2	D_1	D_0	EVEN PARITY(P_E)	ODD PARITY(P_O)
0	0	0	0	0	1
0	0	0	1	1	0
0	0	1	0	1	0
0	0	1	1	0	1
0	1	0	0	1	0
0	1	0	1	0	1
0	1	1	0	0	1
0	1	1	1	1	0
1	0	0	0	1	0
1	0	0	1	0	1
1	0	1	0	0	1
1	0	1	1	1	0
1	1	0	0	0	1
1	1	0	1	1	0
1	1	1	0	1	0
1	1	1	1	0	1

FIGURE 23-19 Simplified 4-Bit Parallel Data Transmission System with Even-Parity Error Detection.

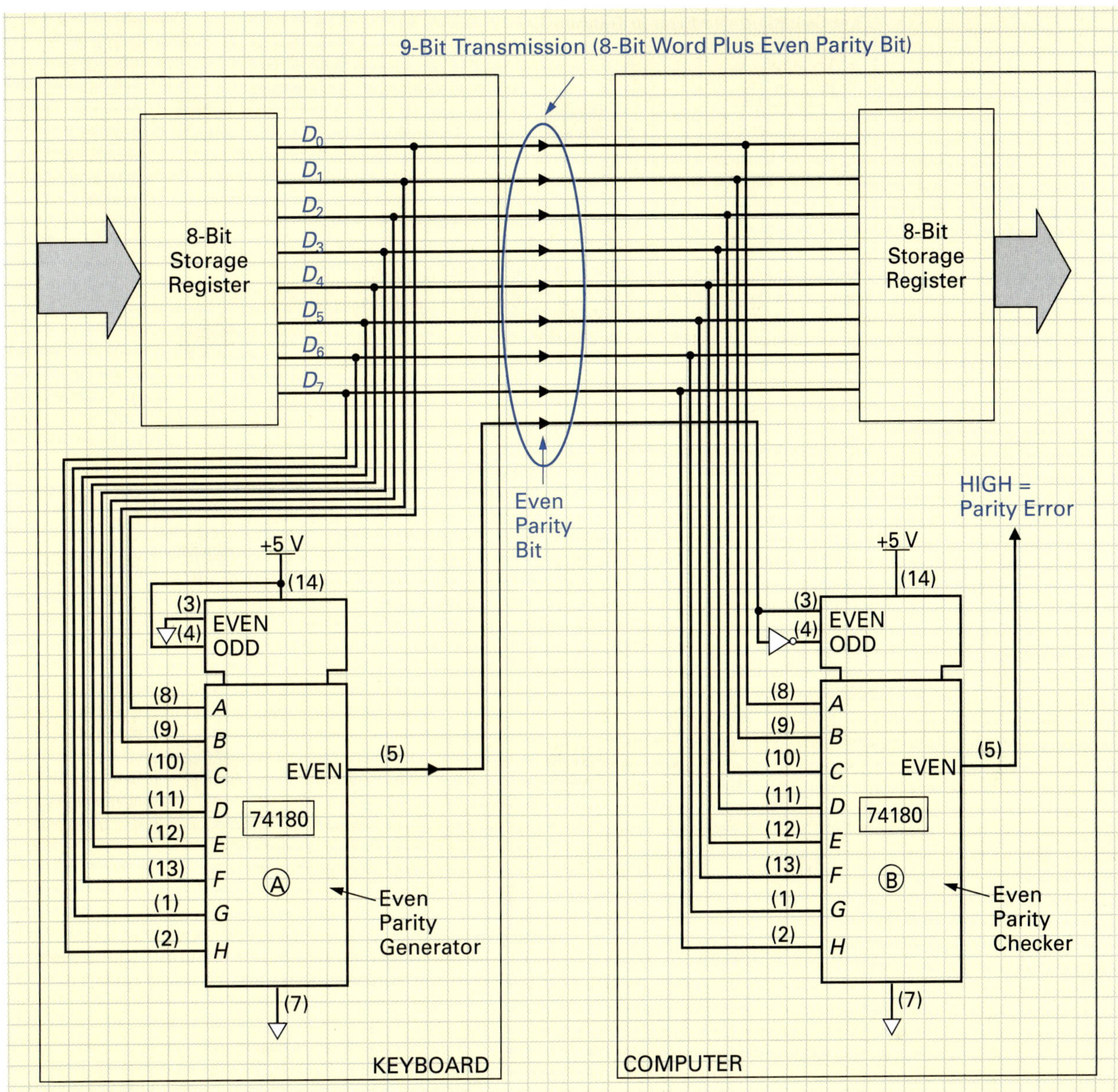

FIGURE 23-20 **Application—Simplified 8-Bit Parallel Data Transmission System with Even-Parity Error Detection.**

ODD number of 1s is received at the computer. The parity checker IC (74180 *B*) receives an ODD number of 1s at its *A* through *H* inputs, and the same control inputs as the parity generator IC (EVEN input = LOW, ODD input = HIGH), so the even output is HIGH, indicating a transmission error has occurred.

c. In the third condition, the 8-bit code to be transmitted from the keyboard contains an odd number of 1s, and therefore the parity generator (74180 *A*) generates a HIGH, even-parity bit output. The parity checker IC (74180 *B*) receives an odd number of 1s at its *A* through *H* inputs, and opposite control inputs to the parity generator IC (EVEN input = HIGH, ODD input = LOW), so the even output is LOW, indicating no transmission error has occurred.

d. In the fourth condition, the 8-bit code to be transmitted from the keyboard contains an odd number of 1s, and therefore the parity generator (74180 *A*) generates a HIGH, even-parity bit output. In this condition, however, a bit error occurs due to electrical noise so an even number of 1s is received at the computer. The parity checker IC (74180 *B*) receives an

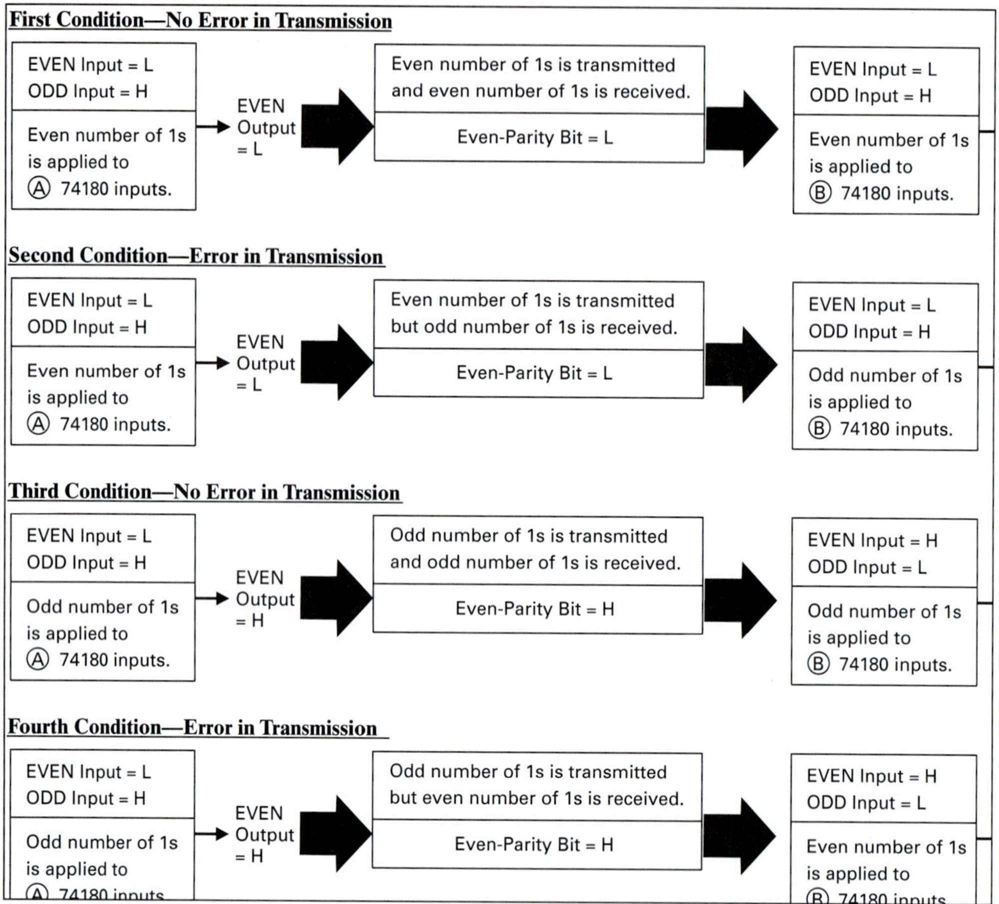

FIGURE 23-21 **Four Basic Conditions of 74180 Even-Parity Circuit.**

even number of 1s at its *A* through *H* inputs, and the opposite control inputs to the parity generator IC (EVEN input = HIGH, ODD input = LOW), so the even output is HIGH, indicating a transmission error has occurred.

SELF-TEST EVALUATION POINT FOR SECTION 23-6

Use the following questions to test your understanding of Section 23-6.

1. Include an odd-parity bit for the following 4-bit gray codes.

DECIMAL	GRAY CODE	ODD PARITY BIT
0	0000	?
1	0001	?
2	0011	?
3	0010	?
4	0110	?
5	0111	?

2. The parity system can only be used to detect single-bit errors. (true/false)
3. Is the 74180 used as a parity generator or as a parity checker?
4. If an even-parity bit were added to the ASCII codes being generated by a keyboard, what would be the size of the words being transmitted?

23-7 TROUBLESHOOTING COMBINATIONAL LOGIC CIRCUITS

To be an effective electronics technician or troubleshooter, you must have a thorough knowledge of electronics, test equipment, troubleshooting techniques, and equipment repair. In most cases a technician is required to quickly locate the problem within an electronic system and then make the repair. To review, the procedure for fixing a failure can be broken down into three steps.

Step 1: DIAGNOSE
The first step is to determine whether a circuit problem really exists, or if it is simply an operator error. To carry out this step, a technician must collect as much information as possible about the system, circuit, and components used, and then diagnose the problem.

Step 2: ISOLATE
The second step is to apply a logical and sequential reasoning process to isolate the problem. In this step, a technician will operate, observe, test, and apply troubleshooting techniques in order to isolate the malfunction.

Step 3: REPAIR
The third and final step is to make the actual repair, and then final test the circuit.

As far as troubleshooting is concerned, practice really does make perfect. Since this chapter has been devoted to additional combinational logic circuits, let us first examine the operation of a typical combinational logic circuit and then apply our three-step troubleshooting process to this circuit.

23-7-1 *A Combinational Logic Circuit*

Figure 23-22 shows a 2-digit counter with a multiplexed display circuit. The 555 timer A functions as a square-wave oscillator with its frequency controlled by R_2. The clock signal generated by this 555 timer serves as a multiplexing clock signal and is applied to both the 74157 multiplexer and the 1458 dual op-amp. The A op-amp of the 1458 is connected to operate as a comparator, while the B op-amp is connected to operate as an INVERTER. These two op-amps provide the two complementary, or opposite, waveforms shown at the top of Figure 23-22. Since both the units and 10s seven-segment displays are common-anode types, only when the anode pin (14) is made HIGH will the display be activated. Since these anode control signals are out-of-phase with one another, the displays will alternate between ON and OFF, with the units ON and the 10s OFF during one half-cycle, and then the units OFF and the 10s ON during the following half-cycle.

As mentioned previously, the multiplexer control signal from 555 timer A is also applied to the $\overline{A}/B$ switching control input of the 74157 multiplexer, making this multiplexer synchronized with the ON/OFF switching of the seven-segment displays. When the $\overline{A}/B$ multiplexer control line is HIGH, the 74157 switches the 4-bit B-input word from the units counter through to the 74157 Y-outputs. This unit count is decoded by the seven-segment decoder (7447) and the seven-segment code is applied to both of the displays; however, only the units display is enabled at this time and therefore the units count will be displayed on the units display. When the $\overline{A}/B$ multiplexer control line is LOW, the 74157 switches the 4-bit A-input word from the 10s counter through to the 74157 Y-outputs. This 10s count is decoded by the seven-segment decoder (7447), and the seven-segment code is applied to both of the displays; however, only the 10s display is enabled at this time, and therefore the 10s count will be displayed on the 10s display.

The 555 timer B provides a clock signal for the counters. By varying this timer's clock frequency (adjusting R_5), you can vary the rate at which the counters count. Switch 1 is used to clear both counters to 0 when the normally open push-button switch is pressed.

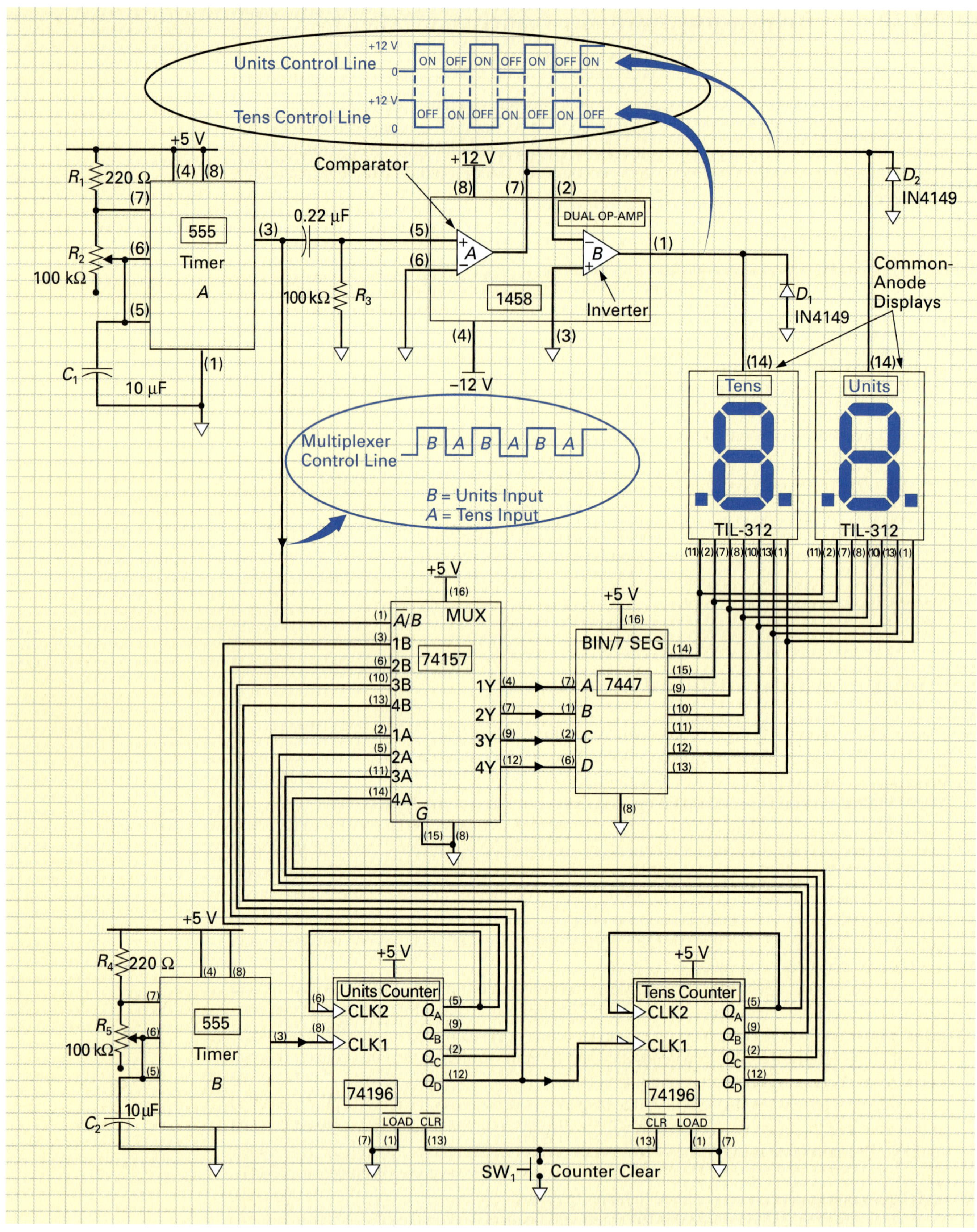

FIGURE 23-22 **Troubleshooting Combinational Logic Circuits—A 2-Digit Counter with a Multiplexed Display Circuit.**

Let us now apply our three-step troubleshooting procedure to this combinational logic circuit so that we can practice troubleshooting procedures and methods.

Step 1: Diagnose

It is extremely important that you first understand the operation of a circuit and how all of the devices within it are supposed to work so that you are able to determine whether or not a circuit malfunction really exists. If you were preparing to troubleshoot the circuit in Figure 23-22, your first step should be to read through the circuit description and review the operation of each integrated circuit until you feel completely confident with the correct operation of the entire circuit. The circuit description, or theory of operation, for an electronic circuit can generally be found in a service or technical manual, along with troubleshooting guides. As far as the circuit's ICs are concerned, manufacturer's digital data books contain a full description of the IC's operation, characteristics, and pin allocation. Referring to all of this documentation before you begin troubleshooting will generally speed up and simplify the isolation step. Many technicians bypass this data collection step and proceed directly to the isolation step. If you are completely familiar with the circuit's operation, this shortcut would not hurt your performance. However, if you are not completely familiar with the circuit, keep in mind the following expression and let it act as a brake to stop you from racing past the problem: *Less haste, more speed.*

Once you are fully familiar with the operation of the circuit, you will easily be able to diagnose the problem as either an *operator error* or a *circuit malfunction.* Distinguishing an operator error from an actual circuit malfunction is an important first step, and a wrong diagnosis can waste a lot of time and effort. For example, the following could be interpreted as circuit malfunctions, when in fact they are simply operator errors:

Example 1.
Symptom: If switch 1 is pressed, both displays go to 0.
Diagnosis: Operator error—This is normal since switch 1 is a display reset switch.

Example 2.
Symptom: Only one display is on at a time. The circuit seems to be switching between the units and tens.
Diagnosis: Operator error—The 555 timer *A* clock frequency is set too low. Adjust R_2 until the display switching is so fast that the eye sees both displays as being constantly ON.

Example 3.
Symptom: The circuit is malfunctioning because the display is counting up in half-seconds instead of seconds.
Diagnosis: Operator error—There is not a circuit malfunction; the 555 timer *B* clock frequency is set too high. Adjust R_5 until the display counts up in seconds.

Once you have determined that the problem is not an operator error, but is in fact a circuit malfunction, proceed to step 2 and isolate the circuit failure.

Step 2: Isolate

No matter what circuit or system failure has occurred, you should always follow a logical and sequential troubleshooting procedure. Let us review some of the isolating techniques and apply them to our example circuit in Figure 23-22.

a. Use a cause-and-effect troubleshooting process, which means study the effects you are getting from the faulty circuit and then logically reason out what could be the cause.

b. Check first for obvious errors before leaping into a detailed testing procedure. Is the power OFF or not connected to the circuit? Are there wiring errors? Are all of the ICs correctly oriented?

c. Using a logic probe or voltmeter, test that power and ground are connected to the circuit and are present at all points requiring power and ground. If the whole circuit, or a large section of the circuit, is not operating, the problem is normally power. Using a multimeter, check that all of the dc voltages for the circuit are present at all IC pins that should have power or a HIGH input and are within tolerance. Secondly, check that 0 V or ground is connected to all IC ground pins and all inputs that should be tied LOW.

d. Use your senses to check for broken wires, loose connectors, overheating or smoking components, pins not making contact, and so on.

e. Test the clock signals using the logic probe or the oscilloscope. Although clock signals can be easily checked with a logic probe, the oscilloscope is best for checking the signal's amplitude, frequency, pulse width, and so on. The oscilloscope will also display timing problems and noise spikes (glitches), which could be false-triggering a circuit into operation at the wrong time.

f. Perform a static test on the circuit. With our circuit example in Figure 23-22, you could static test the circuit by disconnecting the clock input to the units counter, and then use a logic pulser's SINGLE PULSE feature to clock the counter in single steps. A logic probe could then be used to test that valid logic levels are appearing at the outputs of the units and tens counters. For example, after resetting the counter and then applying nineteen clock-pulse inputs from the logic pulser, the logic probe should indicate the following logic levels at the counter outputs:

	TENS	UNITS
Counter	0001	1001_{BCD}
Display	1	9_{10}

By also disconnecting the clock output from 555 timer *A* and then switching the multiplexer control line first HIGH and then LOW, you can trace the 4-bit values through the 74157 multiplexer and 7447 decoder to the displays. Holding the circuit stationary at different stages in its operation (static testing) and using the logic pulser to single-step the circuit and the logic probe to detect logic levels at the inputs and outputs during each step will enable you to isolate any timing problems within the circuit.

g. With a dynamic test, the circuit is tested while it is operating at its normal clock frequency, and therefore all of the inputs and outputs are continually changing. Although a logic probe can detect a pulse waveform, the oscilloscope will display more signal detail, and since it can display two signals at a time, it is ideal for making signal comparisons and looking for timing errors. A *logic analyzer* is a type of oscilloscope that can display typically eight to sixteen digital signals at one time.

In some instances you will discover that some circuits will operate normally when undergoing a static test and yet fail a dynamic test. This effect usually points to a timing problem involving the clock signals and/or the propagation delay times of the ICs used within the circuit.

h. "Noise" due to electromagnetic interference (EMI) can false-trigger an IC, such as the counters in Figure 23-22. This problem can be overcome by not leaving any of the IC's inputs unconnected and therefore floating. Connect unneeded active-HIGH inputs to ground and active-LOW inputs to $+V_{CC}$.

i. Apply the half-split method of troubleshooting first to a circuit and then to a section of a circuit to help speed up the isolation process. With our circuit example in Figure 23-22, a good midpoint check would be to first static-test the inputs to the seven-segment displays. If the units control signal or the tens control signal to the seven-segment displays is not as it should be, then the problem is more than likely in the upper section of the circuit (555 timer *A* and the 1458). If, on the other hand, the seven-segment code from the 7447 is not as it should be, then the problem more than likely lies in the lower section of the circuit (555 timer *B,* the counters, the 74157, and the 7447).

Remember that a load problem can make a source appear at fault. If an output is incorrect, it would be safer to disconnect the source from the rest of the circuit and then recheck the output. If the output is still incorrect, the problem is definitely within the source; however, if the output is correct, the load is probably shorting the line either HIGH or LOW.

j. Substitution can be used to help speed up your troubleshooting process. Once the problem is localized to an area containing only a few ICs, substitute suspect ICs with known-working ICs (one at a time) to see if the problem can be quickly remedied.

Step 3: Repair

Once the fault has been found, the final step is to repair the circuit, which could involve simply removing an excess piece of wire, resoldering a broken connection, reconnecting a connector, or adjusting the power supply voltage or clock frequency. In most instances, however, the repair will involve the replacement of a faulty component. For a circuit that has been constructed on a prototyping board or bread board, the removal and replacement of a component is simple; however, when a printed circuit board is involved, you should make a note of the component's orientation and observe good soldering and desoldering techniques. Also be sure to handle any MOS ICs with care to prevent any damage due to static discharge.

When the circuit has been repaired, always perform a final test to see that the circuit and the system are now fully operational.

23-7-2 *Sample Problems*

Once you have constructed a circuit like the 2-digit counter with a multiplexed display shown in Figure 23-22, introduce a few errors to see what effect or symptoms they produce. Then logically reason out why a particular error or cause has a particular effect on the circuit. Never short any two points together unless you have carefully thought out the consequences. It is, however, generally safe to open a path and see the results. Here are some problems for our example circuit in Figure 23-22.

a. Disconnect the output of 555 timer *A* from the circuit.
b. Disconnect the output of 555 timer *A* from the 1458 input.
c. Open the pin 14 connection to the tens display.
d. Disconnect an input to the 7447.
e. Disconnect the output of 555 timer *B* from the counter.
f. Open the connection between 555 timer *A* and the 74157 $\overline{A}/B$ input.
g. Disconnect power to the 7447.
h. Disconnect one of the 7447 outputs from the 7447 and connect it to ground.

SELF-TEST EVALUATION POINT FOR SECTION 23-7

Use the following questions to test your understanding of Section 23-7.

1. What are the three basic troubleshooting steps?
2. What is a static test?
3. What is the half-split troubleshooting technique?
4. What symptom would you get from the circuit in Figure 23-22 if the units and tens control signals from the 1458 were in-phase with one another?

REVIEW QUESTIONS

Multiple-Choice Questions

1. A ____________ is controlled to switch a single input through to one of several outputs.
a. multiplexer **c.** demultiplexer
b. comparator **d.** encoder

2. A ____________ is controlled to switch one of several inputs through to one output.
a. multiplexer **c.** demultiplexer
b. comparator **d.** encoder

3. What code conversion is performed by a 7448 IC?
a. BCD to gray **c.** BCD to seven-segment
b. Hex to BCD **d.** Gray to seven-segment

4. ____________ are designed to detect specific codes, while ____________ are designed to generate specific codes.
a. Code converters, decoders
b. Decoders, encoders
c. Encoders, code converters
d. Encoders, decoders

5. A decimal-to-BCD encoder will have ____________ inputs and ____________ outputs.
   a. ten, four c. four, ten
   b. three, ten d. ten, three

6. A ____________ translates coded characters into a more understandable form.
   a. decoder c. multiplexer
   b. encoder d. code converter

7. How many inputs and outputs would a BCD-to-decimal decoder have?
   a. three inputs, sixteen outputs
   b. ten inputs, four outputs
   c. ten inputs, one output
   d. four inputs, ten outputs

8. A common-cathode seven-segment display would need active-____________ inputs, while a common-anode seven-segment display would need active-____________ inputs.
   a. HIGH, LOW c. LOW, HIGH
   b. LOW, LOW d. HIGH, HIGH

9. Multiplexers can be used for ____________ data conversion, while demultiplexers can be used for ____________ data conversion.
   a. parallel-to-serial, serial-to-parallel
   b. parallel-to-serial, parallel-to-serial
   c. serial-to-parallel, serial-to-parallel
   d. serial-to-parallel, parallel-to-serial

10. Demultiplexers are often referred to as ____________.
    a. encoders c. comparators
    b. data selectors d. data distributors

11. Multiplexers are often referred to as ____________.
    a. encoders c. comparators
    b. data selectors d. data distributors

12. ____________ ICs can also be used as ____________ ICs.
    a. Encoder, decoder c. Decoder, demultiplexer
    b. Decoder, multiplexer d. Demultiplexer, encoder

13. A digital ____________ is a circuit used to determine whether two parallel binary input words are equal or unequal.
    a. multiplexer c. decoder
    b. comparator d. parity generator

14. What would the even-parity bit be for the code 0011011?
    a. HIGH c. Either 1 or 0
    b. LOW d. None of the above

15. That would be the odd-parity bit for the code 0011011?
    a. HIGH c. Either 1 or 0
    b. LOW d. None of the above

16. The parity check system can be used to detect single- or multiple-bit errors.
    a. True b. False

17. Which basic logic gate functions as an even-parity generator?
    a. XNOR c. NOR
    b. NAND d. XOR

18. What are the three steps in the three-step troubleshooting procedure?
    a. Isolate, diagnose, repair c. Diagnose, isolate, repair
    b. Repair, diagnose, isolate d. Isolate, repair, diagnose

19. Disconnecting the clock signal to a circuit and then point-to-point testing all the logic levels throughout the circuit is an example of ____________ testing.
    a. cause-and-effect c. half-split
    b. dynamic d. static

20. Running a circuit at its normal operating frequency and testing different points throughout the circuit is an example of ____________ testing.
    a. cause-and-effect c. half-split
    b. dynamic d. static

Practice Problems

To practice your circuit recognition and operation, refer to Figure 23-23 and answer the following six questions.

21. What is the function of the 7442 in this circuit?
22. What is the purpose of R_1 through R_4?
23. Do the LEDs in this circuit form a common-anode or common-cathode display?
24. Are the outputs from the 7442 active-LOW or active-HIGH?
25. What would happen to the output display if the following input switches were pressed simultaneously?
    a. 3 and 1
    b. 0, 2, and 3
    c. 2 and 1
26. If the number 4 LED were ON, what binary logic levels would you expect to find on the *ABCD* inputs of the 7442, if you were to check using the logic probe?

To practice your circuit recognition and operation, refer to Figure 23-24 and answer the following six questions.

27. What is the function of the 7448 in this circuit?
28. Is the seven-segment display in this circuit a common-anode or common-cathode display?
29. Are the outputs of the 7448 in this circuit active-LOW or active-HIGH?
30. What is the purpose of the push-button switch in this circuit?
31. What would be displayed if the start/stop push button were held down permanently?
32. If decimal 8 were displayed on the seven-segment display, what binary logic levels would you expect to find on the 7448's input and output pins if you were to check using a logic probe?

To practice your circuit recognition and operation, refer to Figure 23-25 and answer the following three questions.

33. Is this circuit a decoder or encoder?
34. Briefly describe the operation of this circuit.
35. Determine the *A*, *B*, and *C* outputs for each of the rotary switch positions and insert the results in the function table alongside the circuit.

To practice your circuit recognition and operation, refer to Figure 23-26 and answer the following two questions.

36. Is the circuit operating as a serial-to-parallel or parallel-to-serial data transmission circuit?
37. What binary data will appear at the output as the counter cycles from 0 through 7?

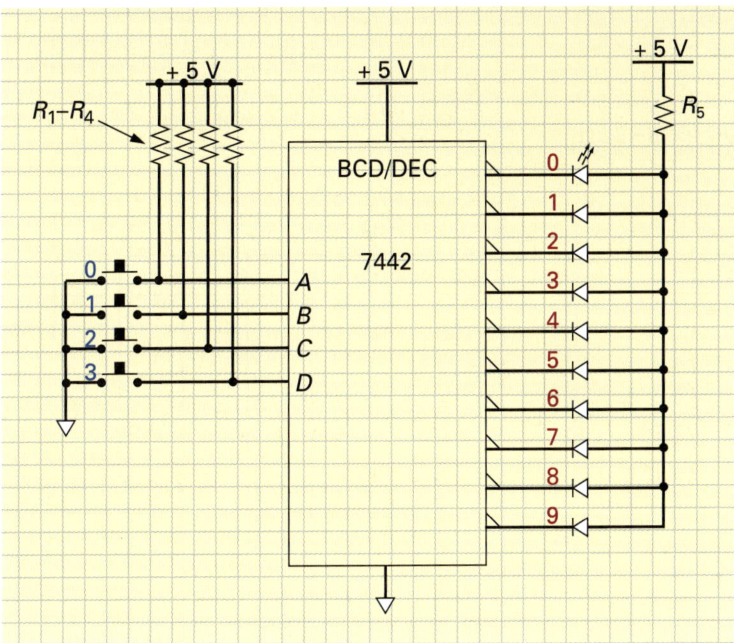

FIGURE 23-23 Application—A Decimal Decoder Circuit.

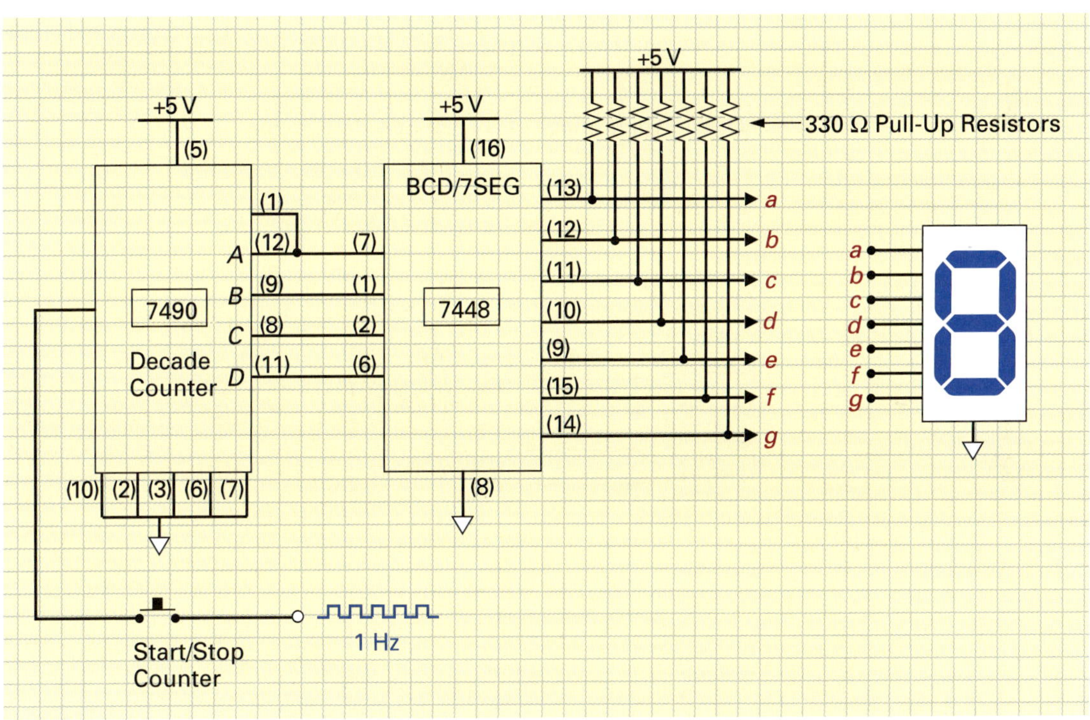

FIGURE 23-24 Application—A 0–9 Second Timer Circuit.

To practice your circuit recognition and operation, refer to Figure 23-27 and answer the following five questions.

38. Briefly describe the operation of the circuit, including how the display will show the binary value of whichever push-button key is pressed.

39. If push button 4 is pressed, the output of the 74151 will output a LOW when the counter has reached a count of ____________, and this LOW output from the 74151 will cause the NAND gate to ____________ (block/pass) the clock pulses to the 74193. The binary display will therefore display ____________.

40. The display in this circuit is a common-____________ display, and therefore a LED ON indicates a (LOW/HIGH) logic level.

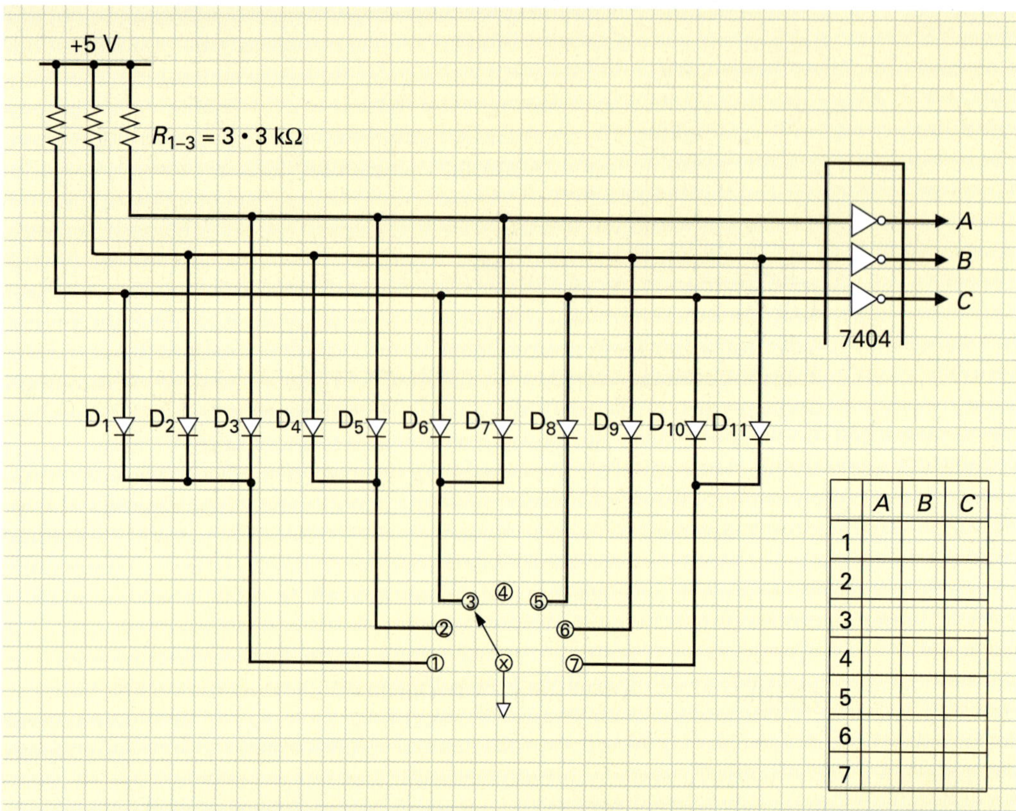

FIGURE 23-25 **Application—A Switch Encoder Circuit.**

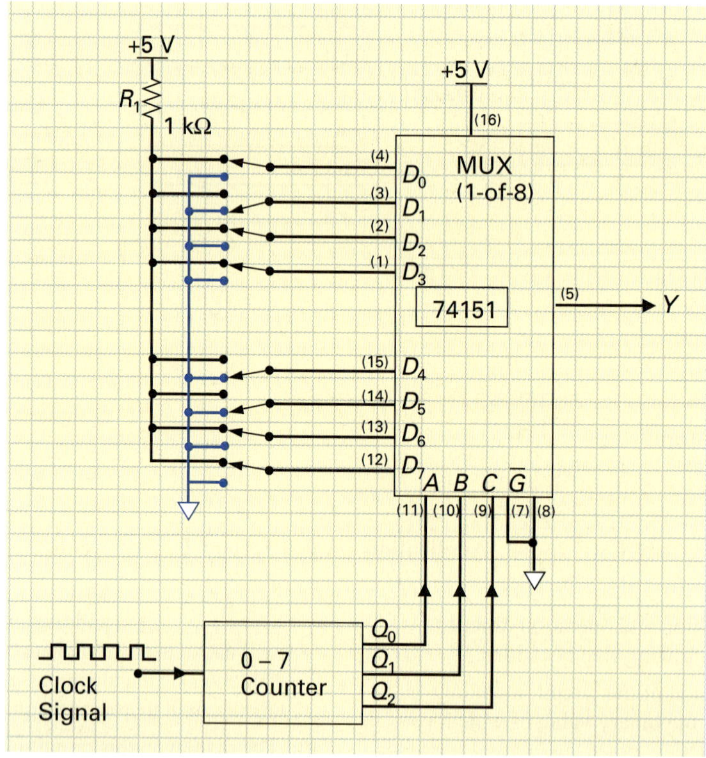

FIGURE 23-26 **Application—8-Bit Serial Word Generator.**

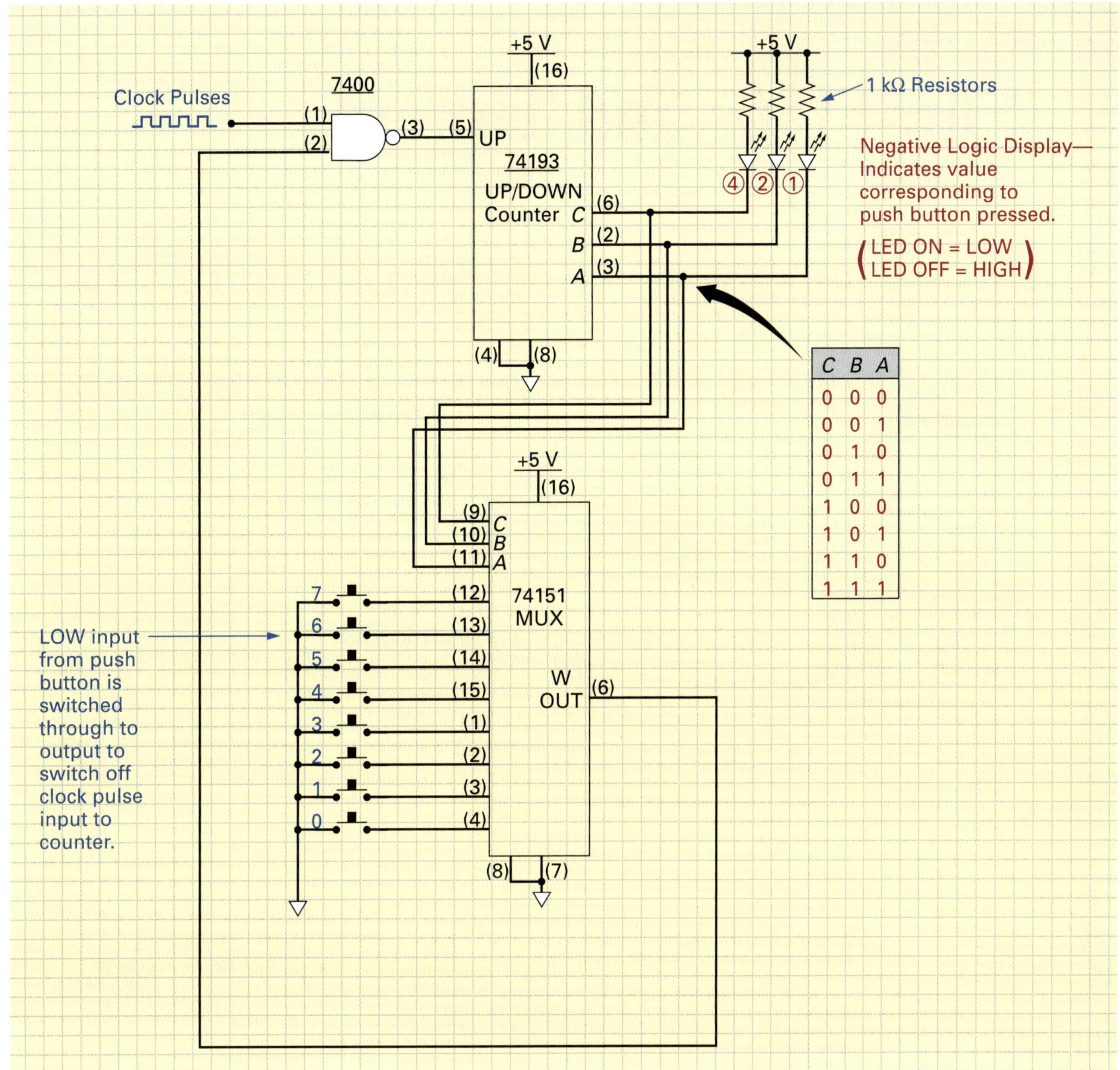

FIGURE 23-27 Application—0 through 7 Keyboard Encoder Circuit.

Web Site Questions

Go to the web site http://www.prenhall.com/cook, select the textbook *Electronics: A Complete Course,* select this chapter, and then follow the instructions when answering the multiple choice practice problems.

Flip-Flops and Timers

The Persistor

William Shockley

In the late 1930s, experimenters had successfully demonstrated that semiconductors could act as rectifiers and be used in place of the then-dominant vacuum tube. The only problem was that nobody could work out how to control semiconductors to make them predictable so they could function like the vacuum tube as an amplifier or a switch. It was not until the beginning of World War II that researchers and physicists began seriously investigating semiconductors in hopes that they could be used to create superior components for radar applications.

Convinced that semiconductor devices would be playing a major role in the future of electronics, AT&T's Bell Laboratories launched an extensive research program. Their goal was to develop a component to replace the vacuum-tube amplifier and mechanical-relay switch that were being used extensively throughout the nation's telephone system. The three men leading the assault were Walter Brattain, a sixteen-year Bell experimenter, John Bardeen, a young theoretician, and the team leader, William Shockley.

Shockley had been fascinated by semiconductors for the past ten years and had a talent for simplifying a research problem to its basic elements and then directing the project into its next avenue of investigation. He had a dominant, intensely serious, and competitive personality and on many occasions forged ahead without waiting to evaluate all of the details of the present situation. On one occasion, after several months of extensive research on a semiconductor amplifier, one of Shockley's ideas that worked perfectly in theory failed completely when tested. Shockley forged off in another direction, but Bardeen and Brattain became fascinated by the reasons for the failure.

On December 23, 1947, after three years of research and at a cost of $1 million, Brattain and Bardeen conducted an experiment on their newly constructed semiconductor prototype. Applying an audio signal to the input of the device, a replica was viewed on an oscilloscope at the output which was fifty times larger in size—they had achieved amplification. The input signal seemed to control the resistance between the output terminals, and this controlled the amount of current flowing through the resistive section of the device. Since the input controlled the amount of current transferring through the resistive section of the device, the component was called a *transresistor,* a name which was later shortened to *transistor.* Six months later, the device was released on the market, but its unpredictable behavior and cost of eight dollars, compared to its seventy-five-cent vacuum-tube counterpart, resulted in it not making much of an impression.

To perhaps make up for his lack of involvement in the final design, Shockley immediately began work on refining the design and removing what he called "the mysterious witchcraft" that the device seemed to possess. In a matter of days, Shockley had much of the theory outlined, but perfecting the design was going to be a long and arduous task. One colleague working with Shockley jokingly referred to the device as a "persistor" because of its stubbornness to work. Shockley's persistence, however, eventually won out, and in 1951

24

the first reliable commercial transistor was released. It could do everything that its vacuum-tube counterpart could do and it was smaller, had no fragile glass envelope, did not need a heating filament to warm it up, and only consumed a fraction of the power.

In 1956, Shockley, Brattain, and Bardeen were all recognized by the world's scientific community when they shared the Nobel Prize in physics. Today this point in history is recognized as the most significant milestone in the history of electronics. Every electronic system manufactured today has an abundance of transistors within its circuitry—not bad for a device that was originally ridiculed by many as "a flash in the pan."

Introduction

In the previous chapter we discussed how logic gates could be used in combination to decode, encode, select, distribute, and check binary data. In this chapter you will see how logic gates can be used in combination to form a memory circuit that is able to store binary information. This memory circuit is called a flip-flop because it can be flipped into its set condition in which it stores a binary 1 or flopped into its reset condition in which it stores a binary 0.

There are three basic types of flip-flop circuits available:

the Set-Reset (*S-R*) flip-flop
the Data-type (D-type) flip-flop
the *J-K* flip-flop

24-1 SET-RESET (*S-R*) FLIP-FLOPS

The flip-flop is a digital logic circuit that is capable of storing a single bit of data. It is able to store either a binary 1 or a binary 0 because of the circuit's two stable operating states—SET and RESET. Once the flip-flop has been flipped into its set condition (in which it stores a binary 1) or flopped into its reset condition (in which it stores a binary 0), the output of the circuit remains latched or locked in this state as long as power is applied to the circuit. This latching or holding action accounts for why the SET or RESET (*S-R*) flip-flop is also known as a SET or RESET latch. Let us now examine this circuit in detail.

24-1-1 *NOR* S-R *Latch and NAND* S-R *Latch*

Figure 24-1 shows how set-reset flip-flop circuits can be constructed using logic gates. Figure 24-1(a) shows how two NOR gates could be connected to form a set-reset NOR latch, or flip-flop. Looking at the truth table for the NOR latch in Figure 24-1(b), you can see that

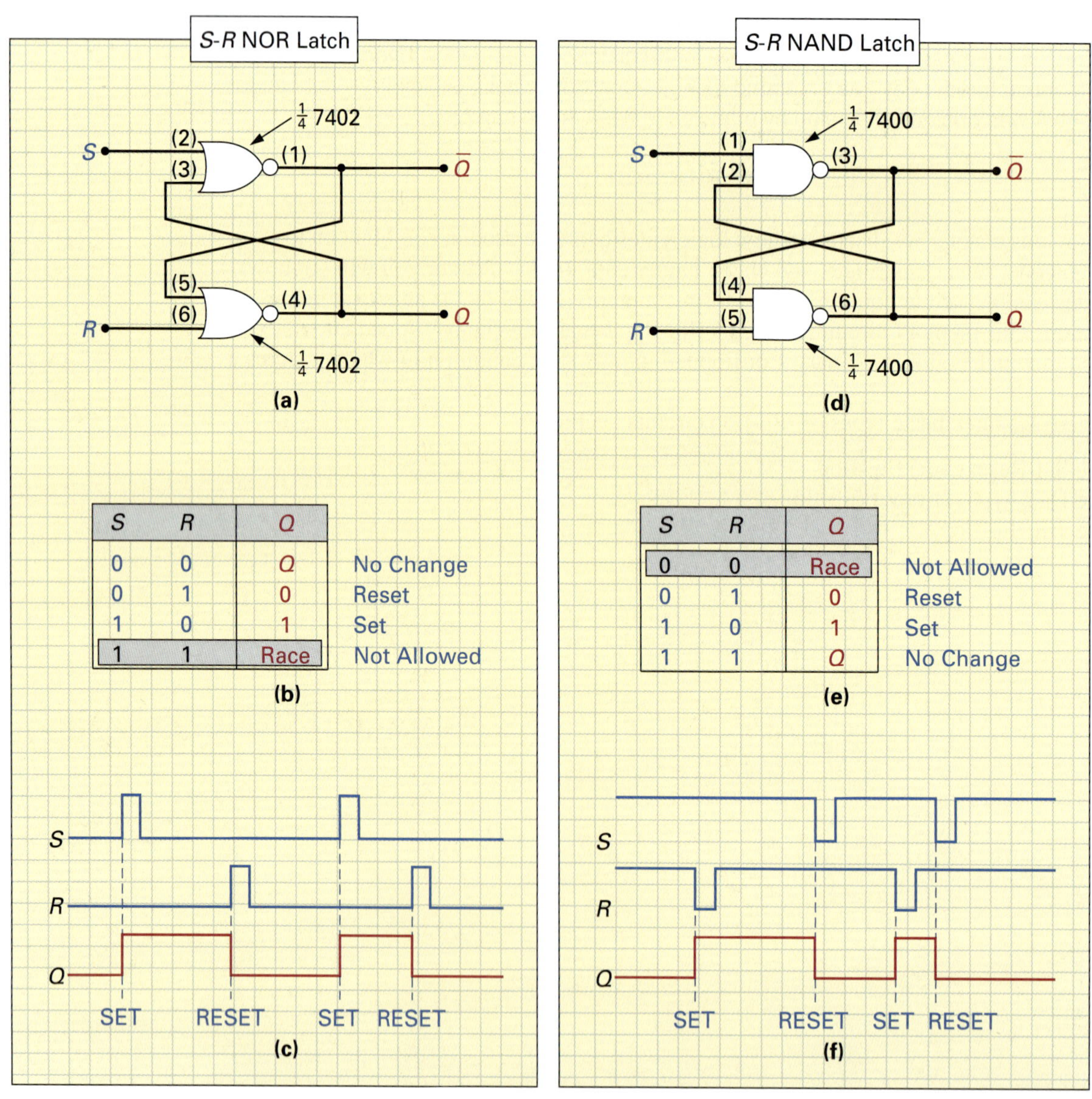

FIGURE 24-1 The *S-R* NOR Latch and the *S-R* NAND Latch.

when both the S and R inputs are LOW, there will be a **NO CHANGE condition** in the Q's output logic level. This means that if the Q output is HIGH it will stay HIGH, and similarly, if the Q output is LOW it will stay LOW. In the second line of the truth table, you can see that a HIGH on the R input will RESET the Q output to 0, while the third line of the truth table shows how a HIGH on the S input will SET the Q output to 1. The last line of the truth table shows that when the S and the R inputs are both HIGH, the circuit will race and the Q output will be unpredictable. The S-R latch is only operated in the NO CHANGE, RESET, and SET conditions. The *RACE* condition is not normally used because the output cannot be controlled or predicted.

NO CHANGE Condition
An input combination or condition that when applied to a flip-flop circuit will cause no change in the output logic level.

The timing diagram in Figure 24-1(c) shows how the Q output can be either SET HIGH by momentarily pulsing the S input HIGH or RESET LOW by momentarily pulsing the R input HIGH. Let us study a few examples to see how this circuit operates.

EXAMPLE:

Remembering that any HIGH into a NOR gate will give a LOW output, and that only when both inputs are LOW will the output be HIGH, study Figure 24-2, which shows a variety of input-output conditions for the set-reset NOR latch. Determine the Q output for each of the examples in Figure 24-2.

Solution:

By considering the S and R inputs to the NOR gates and the Q and $\overline{Q}$ outputs that are cross-coupled to the other NOR gate inputs, we can determine the final NOR gate outputs.

a. In Figure 24-2(a), NOR gate X inputs $= 0$ and 0, therefore $\overline{Q}$ will be 1. NOR gate Y inputs $= 0$ and 1, therefore Q will be 0. In this instance, the Q and $\overline{Q}$ outputs will not change ($Q = 0$, $\overline{Q} = 1$). This ties in with the truth table given in Figure 24-1(b), which states that when $S = 0$ and $R = 0$, there will be no change in Q ($Q = Q$).

b. Moving to the example in Figure 24-2(b), you can see that the only change that has occurred is that the S input is now HIGH. This HIGH input to NOR gate X will produce a LOW output at $\overline{Q}$, which will be coupled to the input of NOR gate Y. With both inputs LOW,

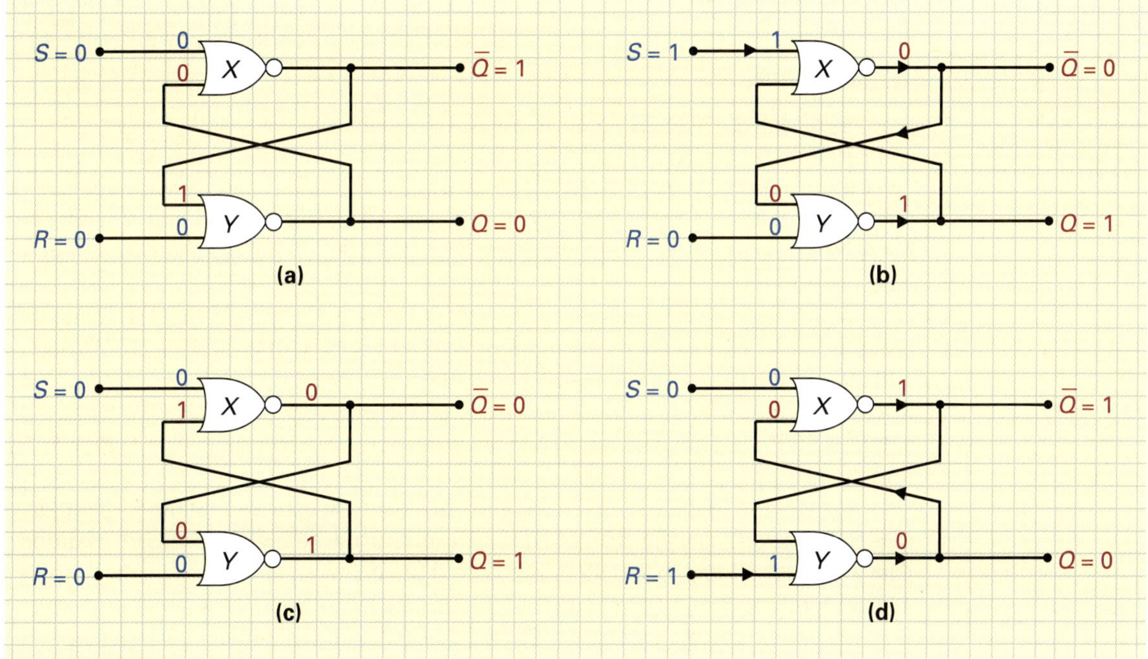

FIGURE 24-2 **NOR Latch Input Conditions.**

the output of NOR gate Y will go HIGH, and Q will therefore change to a 1 and $\overline{Q}$ to a 0. In this example, a SET condition was applied to the input to the NOR latch and the Q output was therefore SET to 1.

c. Moving to the example in Figure 24-2(c), you can see that the Q output is now SET HIGH. The S input, which was taken HIGH momentarily, has now gone LOW and therefore the NOR latch inputs are in the NO CHANGE condition. NOR gate X has inputs of 0 and 1, and therefore $\overline{Q} = 0$ (no change). NOR gate Y has inputs 0 and 0, and therefore $Q = 1$ (no change).

d. Moving to the example in Figure 24-2(d), you can see that the R input is now being pulsed HIGH. This HIGH input to NOR gate Y will produce a LOW output at Q, which will be cross-coupled back to the input of NOR gate X. With two LOW inputs to NOR gate X, its $\overline{Q}$ output will be HIGH. This HIGH level at $\overline{Q}$ will be coupled to the input of NOR gate Y and will keep Q at 0 even when R returns to a LOW. This RESET input condition ($S = 0$, $R = 1$) will therefore reset the Q output to 0.

From these examples you may have noticed that it is the cross-coupling that keeps the circuit outputs latched in either the SET or RESET condition.

An S-R latch can also be constructed using two NAND gates, as shown in Figure 24-1(d). Referring to the truth table in Figure 24-1(e), you can see that the NO CHANGE and RACE conditions for a NAND latch are opposite that of the NOR latch. With the NAND latch, a LOW on both inputs will cause the circuit to race and result in an unpredictable output, while a HIGH on both inputs will result in no change at the outputs. Since the NO CHANGE condition occurs when both inputs are HIGH, the S and R inputs will normally be HIGH, as seen in the timing diagram in Figure 24-1(f). To SET the Q output to a 1, therefore, the R input must be pulsed LOW so that $S = 1$ and $R = 0$ (SET condition). On the other hand, to reset the Q output to a 0, the S input must be pulsed LOW so that $S = 0$ and $R = 1$ (RESET condition). Let us study a few examples to see how this circuit operates.

EXAMPLE:

Remembering that any 0 into a NAND gate gives a 1 output, and only when both inputs are 1 will the output be 0, study Figure 24-3, which shows a variety of input-output conditions for the set-reset NAND latch. Determine the Q output for each of the examples in Figure 24-3.

Solution:

By considering the S and R inputs to the NAND gates and the Q and $\overline{Q}$ outputs that are cross-coupled to the other NAND gate inputs, we can determine the final NAND gate outputs.

a. In Figure 24-3(a), NAND gate X inputs $=$ 1 and 0 and $\overline{Q}$ will therefore be 1, and NAND Y inputs $=$ 1 and 1 and Q will therefore be 0. With both S and R inputs HIGH, the NAND latch is in the NO CHANGE condition, and the outputs Q and $\overline{Q}$ therefore remain in their last condition.

b. Moving to the example in Figure 24-3(b), you can see that the R input has been brought LOW. This LOW to the input of NAND gate Y will produce a HIGH output, which will change the Q to a 1. The HIGH Q will be cross-coupled back to the input of NAND gate X, which will now have both inputs HIGH and therefore produce a LOW output. In this example, a SET condition was applied to the input of the NAND latch ($S = 1, R = 0$), and therefore the Q output was set HIGH.

c. Moving to the example in Figure 24-3(c), you can see that the R input has returned to a HIGH level, causing the NAND latch to return to a NO CHANGE condition. NAND gate X inputs $=$ 1 and 1, and therefore $\overline{Q} = 0$ (no change), and NAND gate Y inputs $=$ 0 and 1, and therefore $Q = 1$ (no change).

d. Moving to the example in Figure 24-3(d), you will notice that the S input is now being momentarily pulsed LOW. This 0 applied to NAND gate X will produce a HIGH output, and therefore $\overline{Q} = 1$. This HIGH at $\overline{Q}$ will be cross-coupled to the input of NAND gate Y,

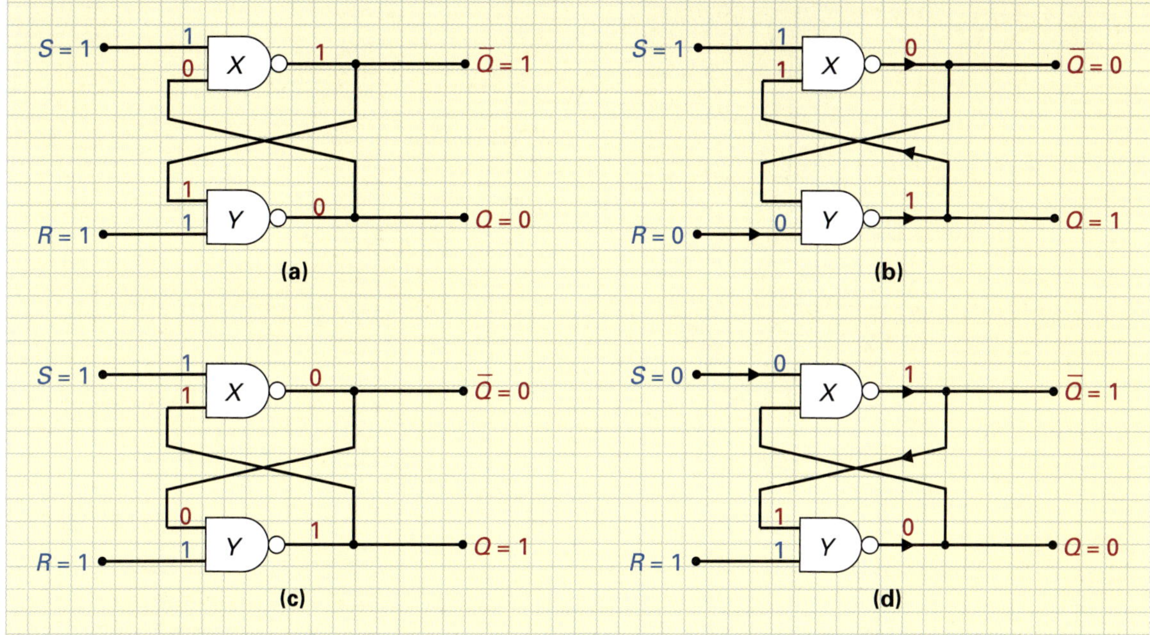

FIGURE 24-3 NAND Latch Input Conditions.

and since both of its inputs are now HIGH, it will produce a LOW output at Q. This RESET input condition ($S = 0, R = 1$) will therefore reset Q to a 0.

Like the NOR latch, it is the cross-coupling between the NAND gates that keeps the circuit latched into either the SET or RESET condition.

Application—Contact-Bounce Eliminator Circuit

Set-reset flip-flops or latches are often used as **switch debouncers,** as shown in Figure 24-4. Whenever a switch is thrown into a new position, the moving and stationary contacts of the switch tend to bounce, causing the switch to make and break for a few milliseconds before finally settling in the new position. This effect is illustrated in Figure 24-4(a) and (b), which show the before and after conditions. In Figure 24-4(a), the ON/OFF is in the OFF position, and due to the action of the pull-up resistor R_1, the input to the INVERTER gate is HIGH and the final output is therefore LOW. In Figure 24-4(b), the switch has been thrown from the OFF to the ON position, and due to the bouncing of the switch's moving contact (pole) on the stationary ON contact, the output vibrates HIGH and LOW several times before it settles down permanently in the ON position. If the output of this circuit were connected to the input of a counter, the counter would erroneously assume that several OFF-to-ON transitions had occurred instead of just one. It is therefore vital in any digital system that contact bounce be eliminated.

Switch Debouncer
A circuit designed to eliminate the contact bounce that occurs whenever a switch is thrown into a new position.

Figure 24-4(c) shows how a *S-R* NAND latch can be used as a *contact-bounce eliminator.* With the switch in the OFF position, the S input is LOW and the R input is HIGH. This RESET input condition ($S = 0, R = 1$) resets the Q output to 0, as seen in the timing diagram in Figure 24-4(c) at time t_0. When the switch is thrown to the ON position (time t_1), the S input is pulled permanently HIGH and the R input will bounce between LOW and HIGH for a few milliseconds. The first time that the switch's pole makes contact and the R input goes LOW, the latch will be SET since $S = 1$ and $R = 0$. Subsequent bounces will have no effect on the SET or HIGH Q output, since the alternating R input is only causing the latch to switch between its SET ($S = 1, R = 0$) and NO CHANGE ($S = 1, R = 1$) input conditions.

Similarly, when the switch is thrown into its OFF position, the R input will be pulled permanently HIGH and the S input will bounce between LOW and HIGH for a few milliseconds.

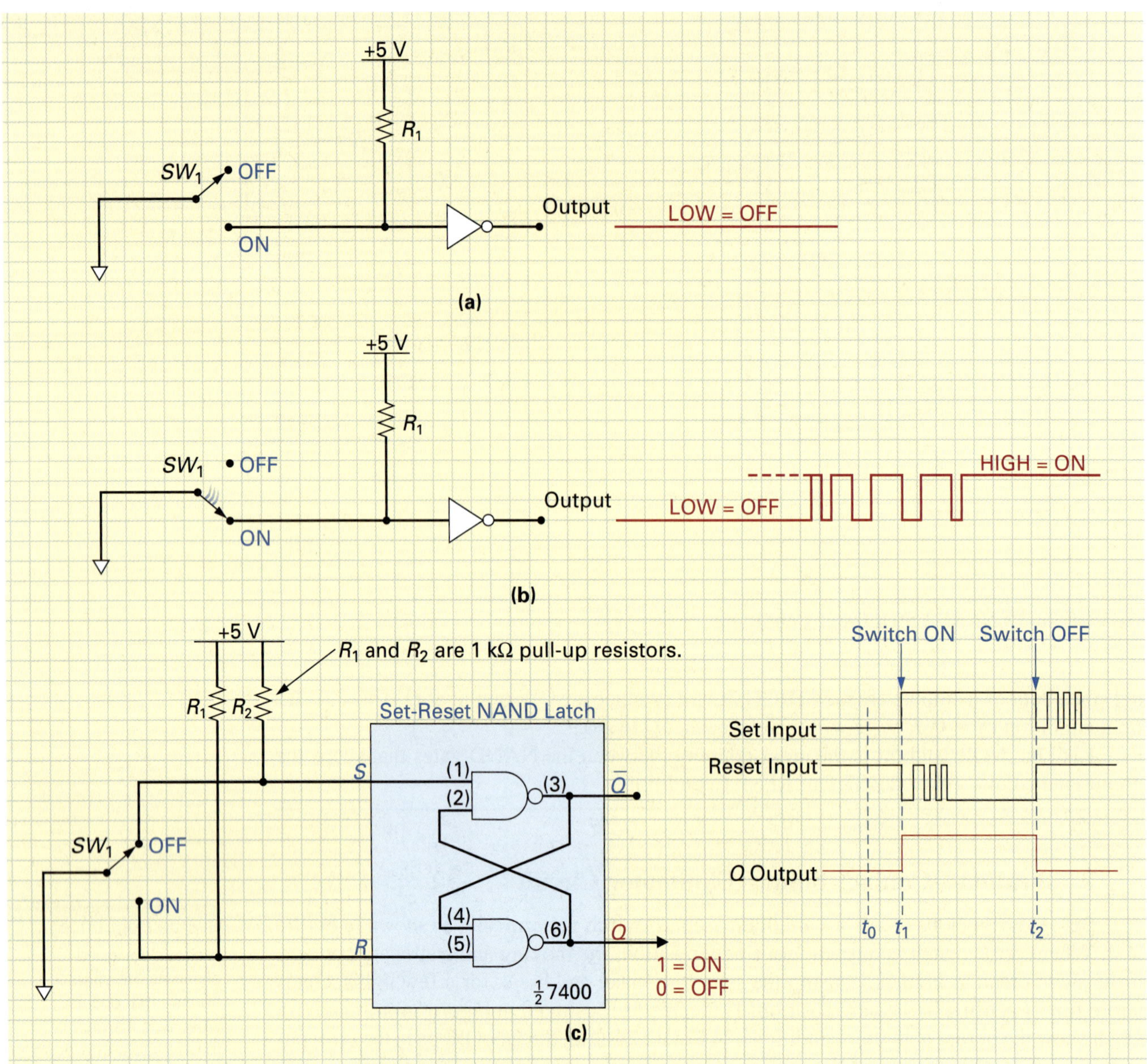

FIGURE 24-4 **Application—The *S-R* Latch as a Contact-Bounce Eliminator.**

The first time that the switch's pole makes contact and the S input goes LOW (t_2), the latch will be RESET ($S = 0$, $R = 1$). Subsequent bounces will have no effect on the RESET or LOW Q output, since the alternating S input is only causing the latch to switch between its RESET ($S = 0, R = 1$) and NO CHANGE ($S = 1, R = 1$) input conditions.

24-1-2 *Level-Triggered* S-R *Flip-Flops*

Triggered
A device or circuit that initiates an action in response to a trigger pulse.

A typical digital electronic system contains many thousands of flip-flops. To coordinate the overall operation of a digital system, a clock signal is applied to each flip-flop to ensure that each device is **triggered** into operation at the right time. A clock signal, therefore, controls when a flip-flop is enabled or disabled and when its outputs change state.

Figure 24-5(a) illustrates how a *S-R* NAND latch can be controlled by the HIGH or positive level of a clock signal. NAND gates *A* and *B* act as controlled switches, only allowing the *S* and *R* inputs through to the inputs of the latch when the clock signal is HIGH.

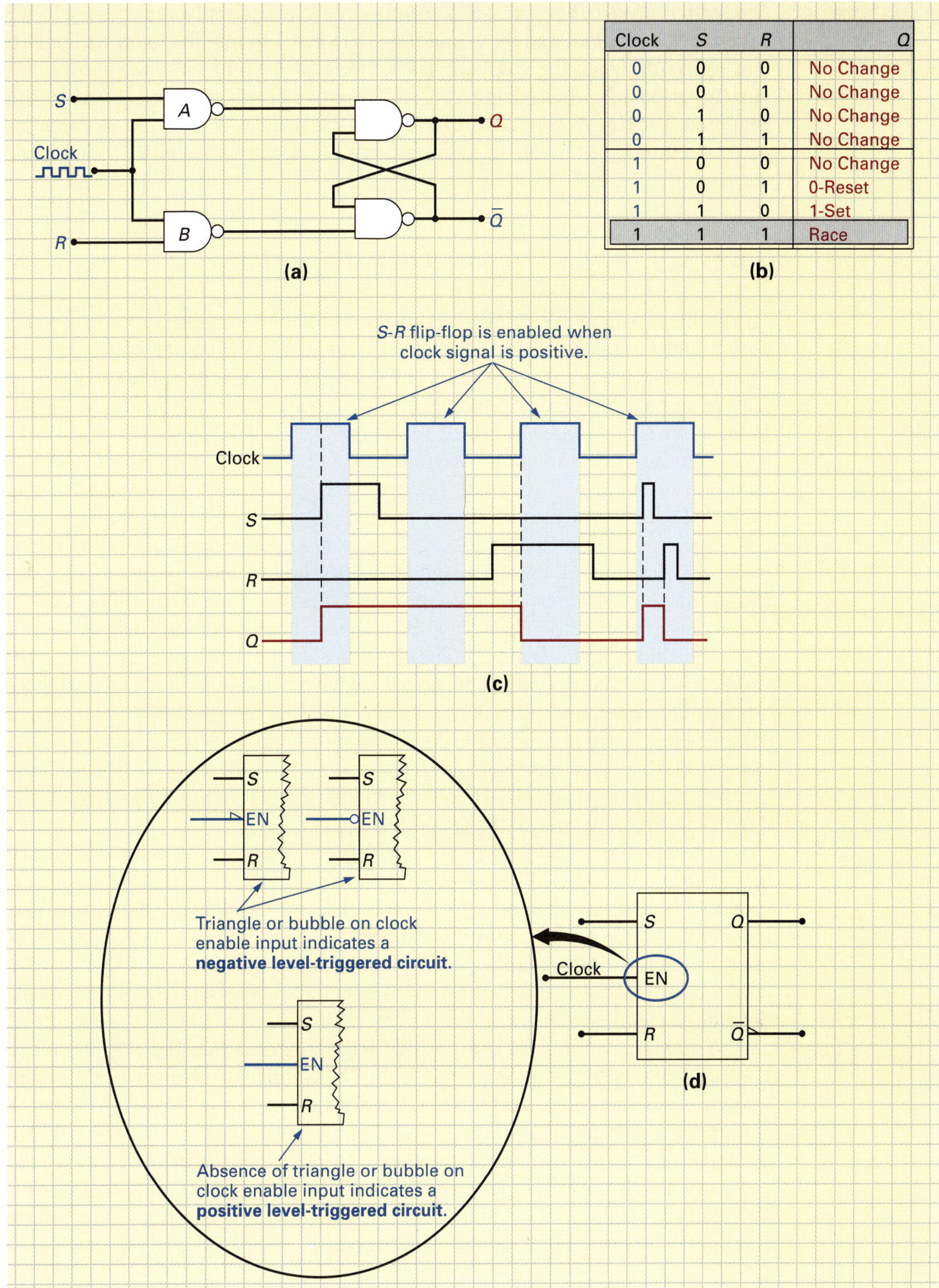

Clock	S	R	Q
0	0	0	No Change
0	0	1	No Change
0	1	0	No Change
0	1	1	No Change
1	0	0	No Change
1	0	1	0-Reset
1	1	0	1-Set
1	1	1	Race

FIGURE 24-5 **Level-Triggered or Gated Set-Reset Flip-Flop.**

This operation is summarized in the truth table in Figure 24-5(b). Whenever the clock signal is 0, the latch is in the NO CHANGE condition; however, when the clock signal is 1, the *S-R* latch functions.

Referring to the waveforms in Figure 24-5(c), you can see that this *S-R* latch is only enabled when the clock signal is positive. This circuit is normally referred to as a *positive* ***level-triggered*** *S-R flip-flop* since a positive clock level will trigger the circuit to operate.

Level-Triggered
A device or circuit that initiates an action in response to a trigger's logic level.

A *negative level-triggered S-R flip-flop* circuit will simply include an INVERTER between the clock input and the *A* and *B* NAND gates so that the latch will only be triggered into operation when the clock signal is 0 or LOW. Figure 24-5(d) shows the logic symbols for a positive level-triggered and negative level-triggered *S-R* flip-flop. A level-triggered *S-R* flip-flop is also referred to as a *gated S-R flip-flop or latch* since NAND gates *A* and *B* act as a gate to either pass or block the SET and RESET inputs through to the latch.

24-1-3 *Edge-Triggered* S-R *Flip-Flops*

Edge-Triggered

A device or circuit that initiates an action in response to a trigger's positive or negative edge.

The problem with a level-triggered flip-flop is that the *S* and *R* inputs have to be held in the desired input condition (SET, RESET, or NO CHANGE) for the entire time that the clock signal is enabling the flip-flop. With **edge-triggered** flip-flops, the device is only enabled during the positive edge of the clock signal *(positive edge-triggering)* or the negative edge of the clock signal *(negative edge-triggering).*

Figure 24-6(a) shows how the previously discussed level-triggered *S-R* flip-flop can be modified so that it will only be triggered into operation when the clock is changing from LOW to HIGH (positive transition or positive edge of the clock signal). The modification to this circuit involves including a *positive-transition pulse generator* in the internal circuitry of the *S-R* flip-flop. The inset in Figure 24-6(a) shows a basic positive edge pulse generator circuit. As you can see from the waveforms in the inset, one input of the AND gate receives the clock signal directly (input *Y*) while the other input receives a delayed and inverted version of the clock input (input *X*). The two inputs to the AND gate are both HIGH for only a short period of time, starting at the positive edge of the clock input and lasting as long as the propagation delay time introduced by the INVERTER gate. The final output from this positive-edge pulse-generator circuit is one narrow positive pulse for every LOW to HIGH transition of the clock input signal. This positive pulse will enable NAND gates *A* and *B* during the time this signal is HIGH and connect the *S* and *R* inputs through to the *S-R* latch section of the circuit to control the Q and $\overline{Q}$ outputs.

The operation of a positive edge-triggered *S-R* flip-flop is summarized by the truth table in Figure 24-6(b). The up and down arrows in the clock column represent the rising (positive transition) and falling (negative transition) edges of the clock. The first three rows of this table indicate that if the clock is at a LOW level (0), HIGH level (1), or on a negative edge (↓), the output Q will remain unchanged. The last four rows of this truth table indicate how the *S-R* flip-flop will operate as expected whenever a positive edge of the clock is applied (↑).

The logic symbol for the positive edge-triggered *S-R* flip-flop circuit discussed in Figure 24-6(a) is shown in Figure 24-6(c). The *C* input stands for clock, or control, and the triangle within the block next to the *C* is called the dynamic (changing) input indicator and is used to identify an edge-triggered flip-flop. As you can see in the inset in Figure 24-6(c), a bubble or triangle outside of the block on the clock input indicates a negative edge-triggered flip-flop. A negative edge-triggered flip-flop circuit would be exactly the same as the circuit in Figure 24-6(a) except that the clock input would have to be inverted before it was applied to the transition pulse generator.

24-1-4 *Pulse-Triggered* S-R *Flip-Flops*

Pulse-Triggered

A device or circuit that initiates an action in response to a complete trigger pulse.

Up to this point, we have seen how set-reset flip-flops can be controlled by the level or edge of a clock signal. A **pulse-triggered** *S-R* flip-flop is a level-clocked flip-flop; however, both the HIGH and LOW levels of the input clock are needed before the Q outputs will reflect the *S* and *R* input condition. With this circuit, therefore, a complete cycle of the input clock (a complete clock pulse) is needed to trigger the circuit into operation, which is why this circuit is called a pulse-triggered flip-flop.

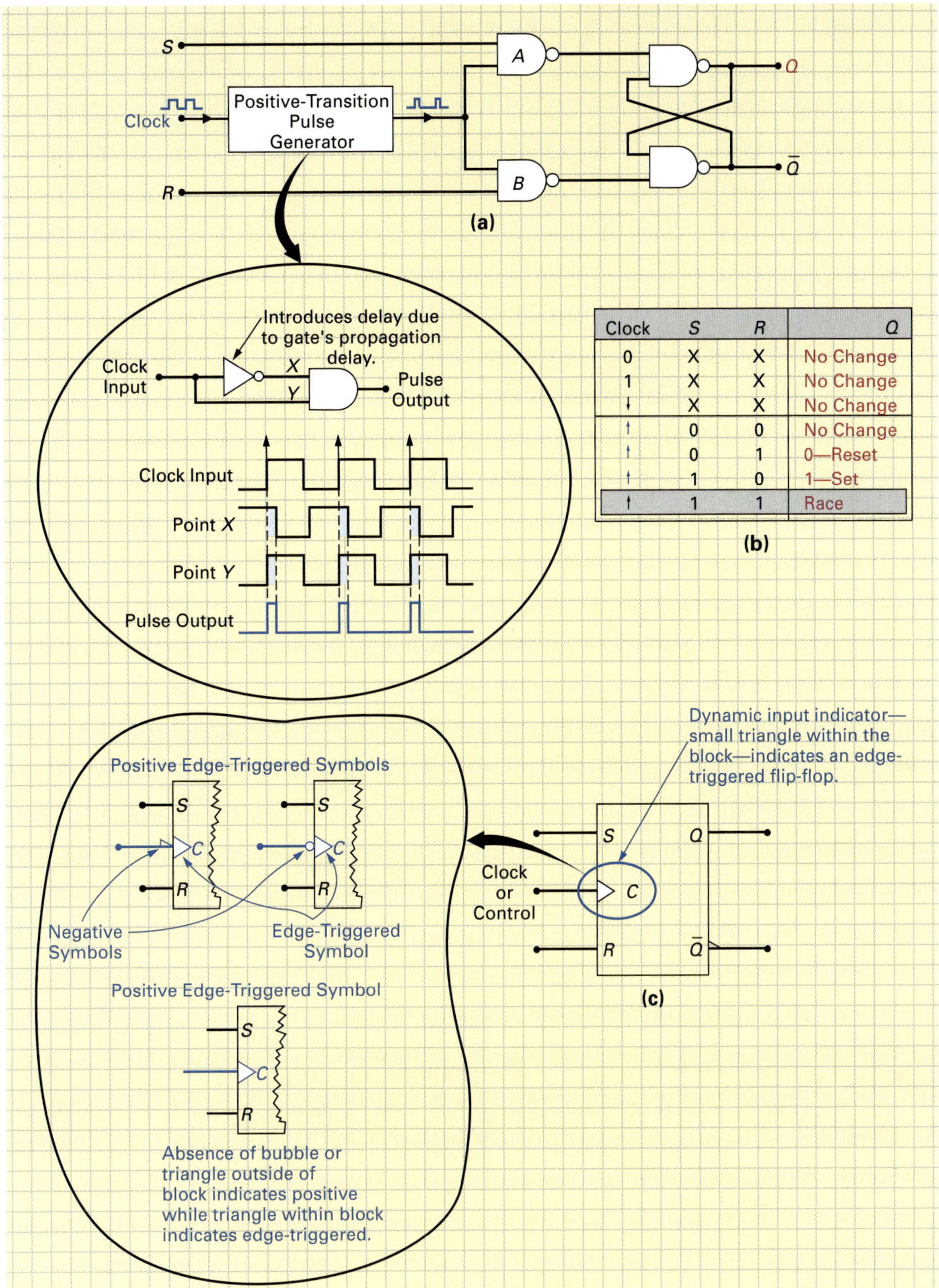

Clock	S	R	Q
0	X	X	No Change
1	X	X	No Change
↓	X	X	No Change
↑	0	0	No Change
↑	0	1	0—Reset
↑	1	0	1—Set
↑	1	1	Race

FIGURE 24-6 Edge-Triggered Set-Reset Flip-Flops.

Figure 24-7(a) shows a simplified circuit of a pulse-triggered *S-R* flip-flop. Pulse-triggered circuits are also referred to as **master-slave flip-flops** since they contain two clocked *S-R* latches called the master latch and the slave latch. The first-stage master latch receives the clock directly *(C),* while the second-stage slave latch receives an inverted version of the clock ($\overline{C}$). Referring to the C and $\overline{C}$ waveforms in Figure 24-7(c), you can see that the master section is enabled while the clock *(C)* is positive. The slave section, on the other

Master-Slave Flip-Flop

A two-latch flip-flop circuit in which the master latch is loaded on the leading edge of the clock pulse, and the slave latch is loaded on the trailing edge of the clock pulse.

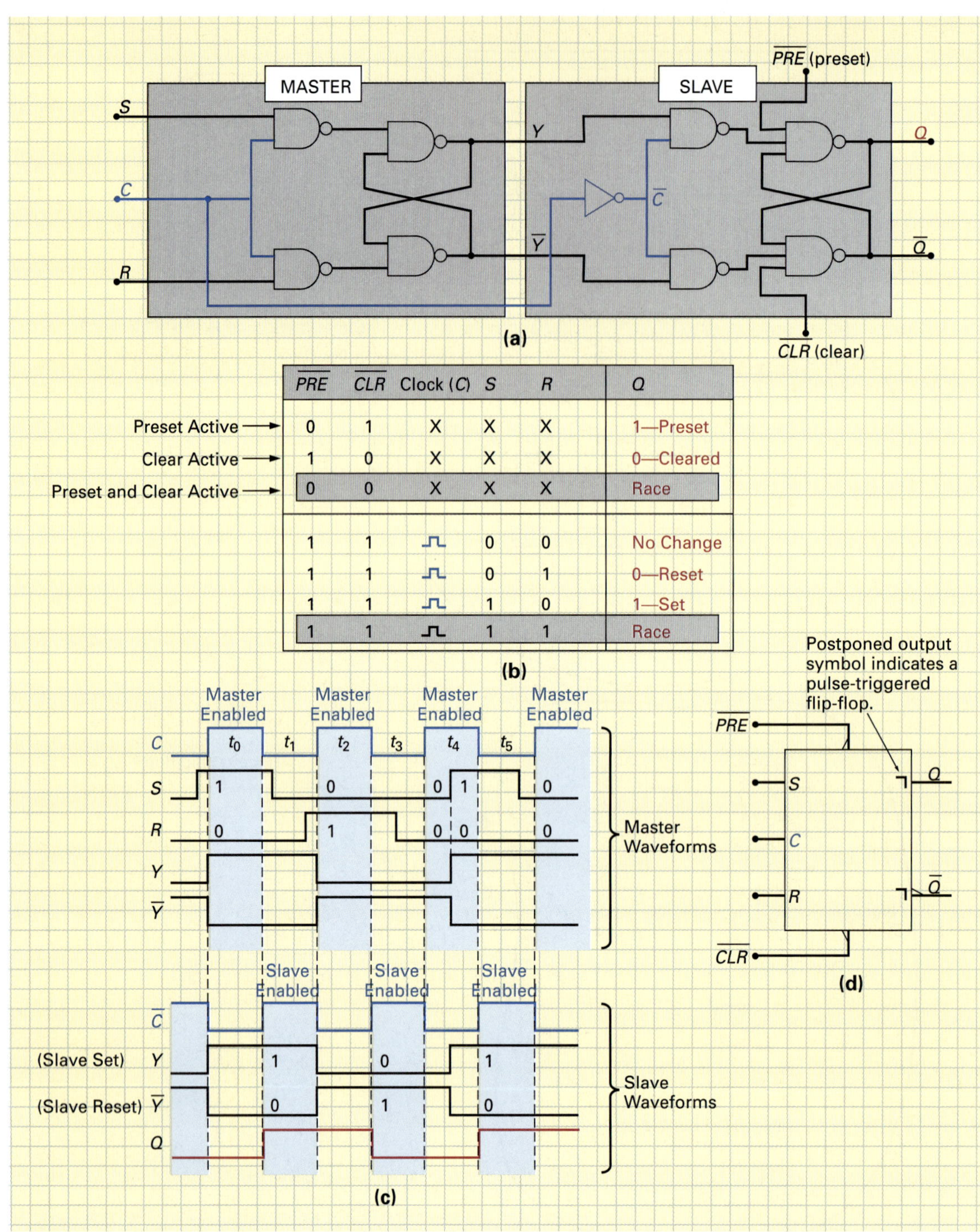

	$\overline{PRE}$	$\overline{CLR}$	Clock (C)	S	R	Q
Preset Active →	0	1	X	X	X	1—Preset
Clear Active →	1	0	X	X	X	0—Cleared
Preset and Clear Active →	0	0	X	X	X	Race
	1	1	⎍	0	0	No Change
	1	1	⎍	0	1	0—Reset
	1	1	⎍	1	0	1—Set
	1	1	⎍	1	1	Race

FIGURE 24-7 Pulse-Triggered (Master-Slave) Set-Reset Flip-Flop.

hand, is enabled when the clock is negative, because it is at this time that the inverted clock ($\overline{C}$) is positive. There are therefore two distinct steps involved before the Q and $\overline{Q}$ outputs reflect the input condition applied to the SET and RESET inputs.

Step 1: During the HIGH level of the clock input, the master SET-RESET latch is enabled and its output Y is either SET, RESET, or left unchanged (NO CHANGE).

Step 2: During the LOW level of the clock input, the slave SET-RESET latch is enabled and its output Q will follow whatever logic level is present on the Y input.

This two-step action is sometimes referred to as *cocking and triggering.* The master latch is cocked during the positive level of the clock and the slave latch is triggered during the negative level of the clock.

The truth table in Figure 24-7(b) summarizes the operation of the pulse-triggered (master-slave) *S-R* flip-flop. The last four lines of the truth table reflect the normal operating conditions (NO CHANGE, RESET, SET, and RACE) of a *S-R* flip-flop. A complete clock pulse is shown in the clock column for these conditions since both the HIGH and LOW levels of the clock input are needed to enable both sections of the flip-flop. The first three lines of the truth table describe how the active-LOW **PRESET** ($\overline{PRE}$) and active-LOW **CLEAR** ($\overline{CLR}$) input functions. When power is first applied to a digital system, flip-flops may start up in either the SET ($Q = 1$) or RESET ($Q = 0$) condition. This could be both dangerous and damaging if these Q outputs were controlling external devices. For this reason, the direct PRESET (synonymous with SET) and direct CLEAR (synonymous with RESET) inputs are generally always available with most commercially available flip-flop ICs. They are called direct inputs because they do not need the clock signal to be active in order to PRESET the Q to a 1 or CLEAR the Q to a 0. One application for these direct PRESET and CLEAR inputs would be to control the outputs of any flip-flops controlling machinery in an automated manufacturing plant so that when power is first applied, all devices start off in their OFF state.

Preset
A control input that is independent of the clock and is used to set the output HIGH.

Clear
A control input that is independent of the clock and is used to reset the output LOW.

Returning to the truth table in Figure 24-7(b), you can see that when the CLEAR is HIGH and the PRESET is LOW (active), the output is PRESET to 1 regardless of the *C, S,* and *R* inputs. In the second line of the table, the PRESET is HIGH and the CLEAR input is LOW (active), causing the Q to be cleared to 0, regardless of the *C, S,* and *R* inputs. These active-LOW inputs are therefore normally pulled HIGH to make them inactive. Referring to the circuit in Figure 24-7(a) you can see that the PRESET and CLEAR inputs control the final NAND gates of the slave latch. As with any NAND gate, any LOW input will produce a HIGH output, and therefore a LOW PRESET will produce a HIGH Q (LOW $\overline{Q}$), while a LOW CLEAR will produce a LOW Q (HIGH $\overline{Q}$). Bringing both the PRESET and CLEAR LOW at the same time will produce a RACE condition, as summarized in the truth table in Figure 24-7(b), since both Q and $\overline{Q}$ will try to go HIGH.

Figure 24-7(c) shows a timing diagram for a pulse-triggered (master-slave) set-reset flip-flop. Referring to the clock input (*C*), let us step through times t_0 through t_5 and see how this circuit responds to a variety of input conditions.

At time t_0, the master latch section is enabled by the positive level of *C*. Since the *S* input is HIGH and the *R* input is LOW at this time (SET condition), the *Y* output of the master latch will be SET to 1 ($\overline{Y} = 0$).

At time t_1, the master latch section is disabled by a LOW *C,* while the slave latch section is enabled by a HIGH $\overline{C}$. Since the *Y* and $\overline{Y}$ inputs to the slave latch section are in the SET condition ($Y = 1$, $\overline{Y} = 0$), the Q output is SET to 1. This action demonstrates how the slave section simply switches its inputs through to the outputs when enabled by the clock (if $Y = 1$ and $\overline{Y} = 0$, $Q = 1$ and $\overline{Q} = 0$ when $\overline{C}$ is HIGH). It can therefore be said that the second latch is a slave to the master latch.

At time t_2, the master latch section is enabled by a HIGH *C*. Since its SET input is LOW and its RESET input is HIGH at this time, the *Y* output will be reset to 0.

At time t_3, the master latch section is disabled and the slave latch section is enabled. The reset output of the master latch is clocked into the enabled slave section and transferred through to the output, resetting Q LOW (if $Y = 0$ and $\overline{Y} = 1$, $Q = 0$ and $\overline{Q} = 1$ when $\overline{C}$ is HIGH).

At time t_4, the master-slave *S-R* flip-flop is a level-clocked circuit, which means that the inputs have to remain constant for the entire time that the clock is enabling the latch. At the beginning of time t_4, the *S* and *R* inputs are both LOW and therefore *Y* remains in its last state, which was RESET ($Y = 0$). Half-way through time t_4, the *S* input goes HIGH and the *Y* output is therefore SET HIGH.

At time t_5, the master latch is disabled while the slave latch is enabled, connecting the HIGH *Y* input through to the Q output.

The logic symbol for a pulse-triggered master-slave flip-flop is shown in Figure 24-7(d). The active-LOW PRESET ($\overline{PRE}$)and active-LOW CLEAR ($\overline{CLR}$) inputs are included, with the traditional triangle being used to represent the fact that they are activated by a LOW logic level.

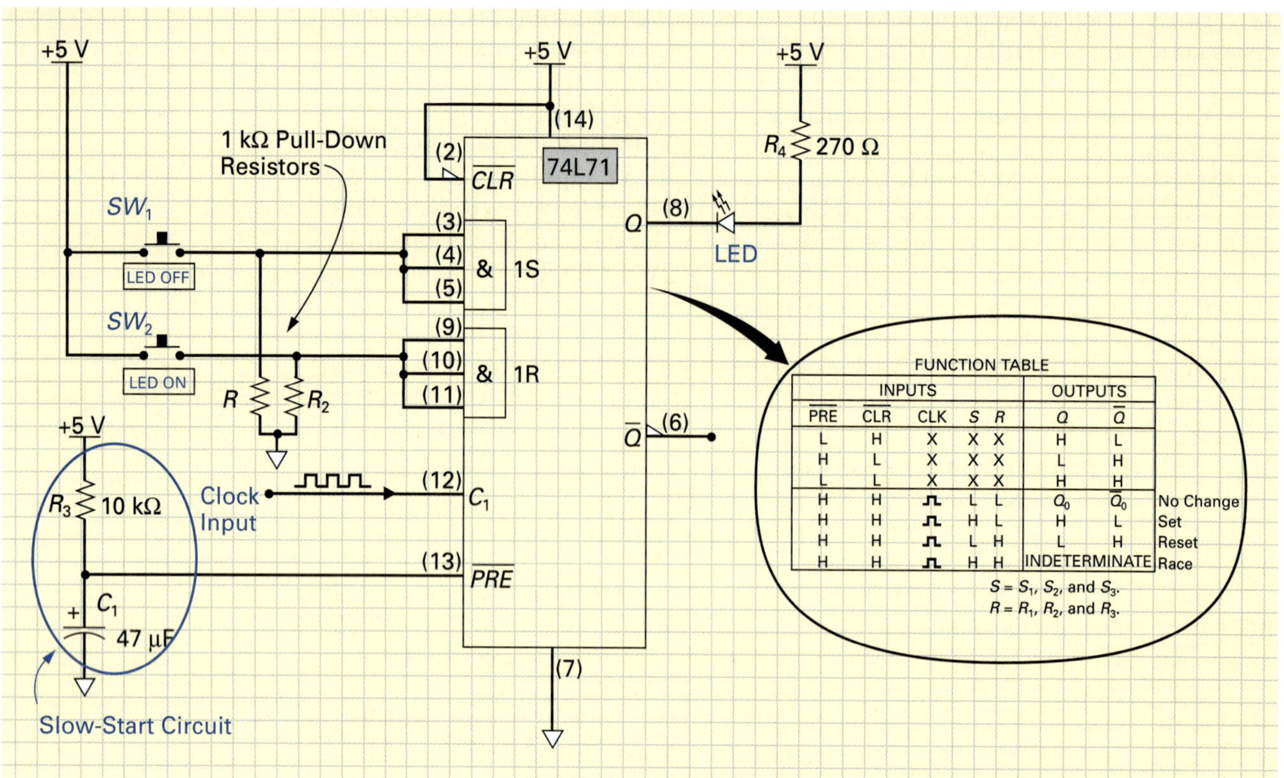

FIGURE 24-8 Application—A Memory Latch/Unlatch Circuit.

The *postponed output symbol* (⅂) at the Q and $\overline{Q}$ outputs indicates that these outputs do not change state until the clock input has fallen from a HIGH to a LOW level. This is due to the slave section, which, as you know, is only enabled—and the output updated—when the clock signal drops LOW. If the postponed output were turned the other way, it would indicate that the master latch would be enabled by a LOW clock input and the slave latch would be enabled by a HIGH clock input. The output with this type of *S-R* master-slave flip-flop would be updated whenever the clock rose to a HIGH level.

Application–A Memory Latch/Unlatch Circuit

The inset in Figure 24-8 shows the function table for a 74L71, which is a pulse-triggered set-reset master-slave flip-flop IC with three ANDed SET inputs and three ANDed RESET inputs. By ANDing the SET and RESET inputs in this way, the IC has more control versatility, since all inputs will have to be HIGH for the control line to be HIGH and any LOW input will pull the corresponding control line LOW. The first three lines of the function table show how the active-LOW PRESET and CLEAR inputs will override the outputs, and the last four lines of the function table show the NO CHANGE, SET, RESET, and RACE input conditions.

EXAMPLE:

What would be the Q output from the 7471 for the following input conditions?

a. $S_1 = 1$, $S_2 = 1$, $S_3 = 1$, $R_1 = 0$, $R_2 = 0$, $R_3 = 0$, CLK = complete pulse, PRE = 0, CLR = 1.

b. $S_1 = 1, S_2 = 1, S_3 = 0, R_1 = 0, R_2 = 0, R_3 = 0$, CLK = complete pulse, PRE = 1, CLR = 1.

Solution:

a. Q would be HIGH since PRESET is active.

b. Q would not change, since $S = 0$ (due to S_3) and $R = 0$.

To understand the operation of this IC better, let us see how it could be used in an application.

The circuit in Figure 24-8 shows how a pulse-triggered set-reset master-slave flip-flop can be used to latch or unlatch a control output. The output Q is connected to +5 V via an LED and current-limiting resistor (R_4), so a HIGH output (SET) will turn OFF the LED and a LOW output (RESET) will turn ON the LED.

All of the SET inputs are tied together and connected to the LED OFF push-button switch, SW_1. The RESET inputs are also all tied together and connected to the LED ON push-button switch, SW_2. The SET and RESET inputs are normally both LOW (NO CHANGE) due to the pull-down resistors R_1 and R_2; however, when a push-button switch is pressed, the corresponding input is switched HIGH. For example, if SW_1 were momentarily pressed, the Q output would be SET HIGH after one clock pulse and the LED would turn OFF. On the other hand, if SW_2 were momentarily pressed, the Q output would be RESET LOW after one clock pulse and the LED would turn ON. A momentary press of SW_2, therefore, will lock or ON the LED, while a momentary press of SW_1 will lock or OFF the LED.

As described earlier, a flip-flop's output may start out either HIGH or LOW when power is first applied to a circuit. To ensure that a flip-flop's output always starts off in the desired condition, a start circuit is normally included and connected to either the PRESET or CLEAR input. In Figure 24-8, R_3 and C_1 form a *slow-start circuit* that is connected to the flip-flop's PRESET input to ensure that the LED is initially OFF when the circuit is powered up. This circuit operates in the following way. When power is first applied to the circuit, there is no charge on C_1 and therefore the PRESET input to the 7471 will be active, so the Q output will be PRESET to a 1, causing the LED to turn OFF. After a small delay (equal to five time constants of R_3 and C_1), C_1 will have charged to 5 V and the PRESET input will be inactive (HIGH). This slow-start circuit therefore delays the PRESET input from going immediately HIGH to ensure that the Q output starts off SET or HIGH, causing the LED to be OFF when power is first applied. If you wished the Q output to be 0 at power-up, you would simply connect the PRESET input to +5 V and connect the CLEAR input to the slow-start circuit.

SELF-TEST EVALUATION POINT FOR SECTION 24-1

Use the following questions to test your understanding of Section 24-1.

1. Once triggered, the SET-RESET flip-flop's output will remain locked in either the SET condition or RESET condition, which is why a flip-flop is also known as a ____________.
2. List the three different ways a flip-flop can be triggered.
3. If $S = 0$ and $R = 0$, the NOR latch would be in the ____________ input condition and the NAND latch would be in the ____________ input conditions.
4. Which form of triggering makes use of a
   a. transition pulse generator? **b.** master latch and slave latch?
5. What is a contact-bounce eliminator circuit?
6. What is a slow-start circuit?

24-2 DATA-TYPE (D-TYPE) FLIP-FLOPS

D-Type Flip-Flop or Latch

A data-type flip-flop circuit that can latch or store a binary 1 or 0.

The **D-type flip-flop** or **D-type latch** is basically a SET-RESET flip-flop with a small circuit modification. This modification was introduced so that the data-type flip-flop could RACE like the *S-R* flip-flop.

The basic D-type flip-flop logic circuit is shown in Figure 24-9(a). The modification to the circuit is the inclusion of an INVERTER gate, which ensures that the *R* and *S* inputs to the NAND latch are never at the same logic level. The single data input bit (*D*) appears at the *S* input of the latch, and its complement $(\overline{D})$ appears at the *R* input of the latch. Therefore, when the *D* input is 1, the *Q* output is SET to 1, and when the *D* input is 0, the *Q* output is RESET to 0, as summarized in the truth table in Figure 24-9(b). The logic symbol for this unclocked D-type flip-flop is shown in Figure 24-9(c). As can be seen in the timing waveforms in Figure 24-9(d), the *Q* output is either SET or RESET as soon as the *D* input goes HIGH or LOW.

To coordinate the overall action of a digital system, devices are triggered into operation at specific times. This synchronization is controlled by a master timing signal called a clock. Like the *S-R* flip-flop, the D-type flip-flop can be either level-triggered, edge-triggered, or pulse-triggered. Let us now examine each of these three types in more detail.

24-2-1 *Level-Triggered D-Type Flip-Flops*

Figure 24-10(a) shows the basic logic circuit for a *level-triggered* or *gated D-type flip-flop*. A LOW clock level will disable the input gates *A* and *B* and prevent the latch from changing states, as indicated in the first line of the truth table in Figure 24-10(b). When the clock level is HIGH, however, gates *A* and *B* are enabled and the *D* input controls the *Q* output, as seen in the second and third lines of the truth table. In the timing diagram in Figure 24-10(c), the operation of the level-clocked or level-triggered D-type flip-flop is further reinforced. This circuit's operation can be summarized by saying that when the clock input is HIGH, the *Q* output equals the *D* input.

The logic symbol for both the positive level-triggered and negative level-triggered D-type flip-flop is shown in Figure 24-10(d).

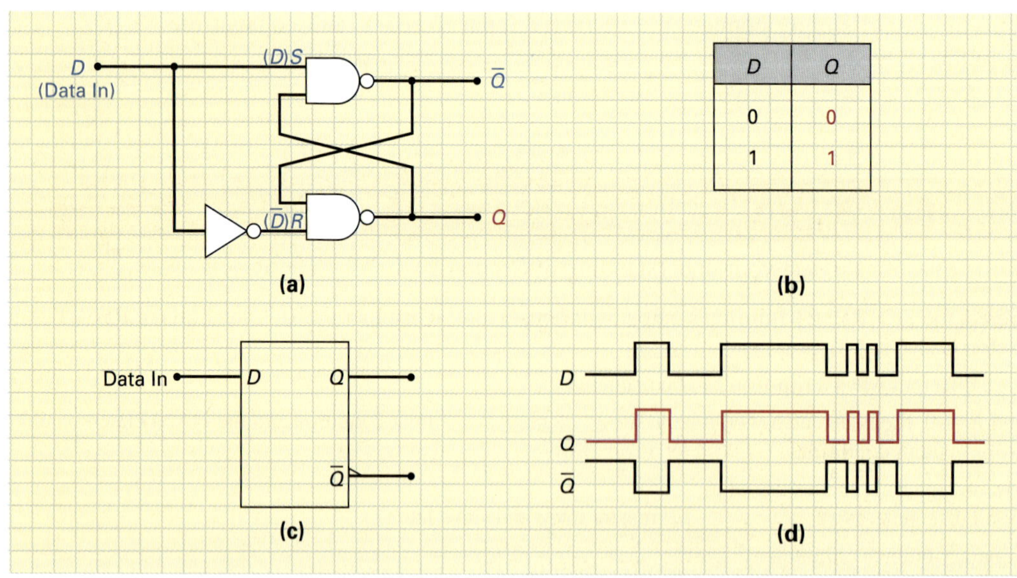

FIGURE 24-9 **The Basic Data-Type (D-type) Flip-Flop or Latch.**

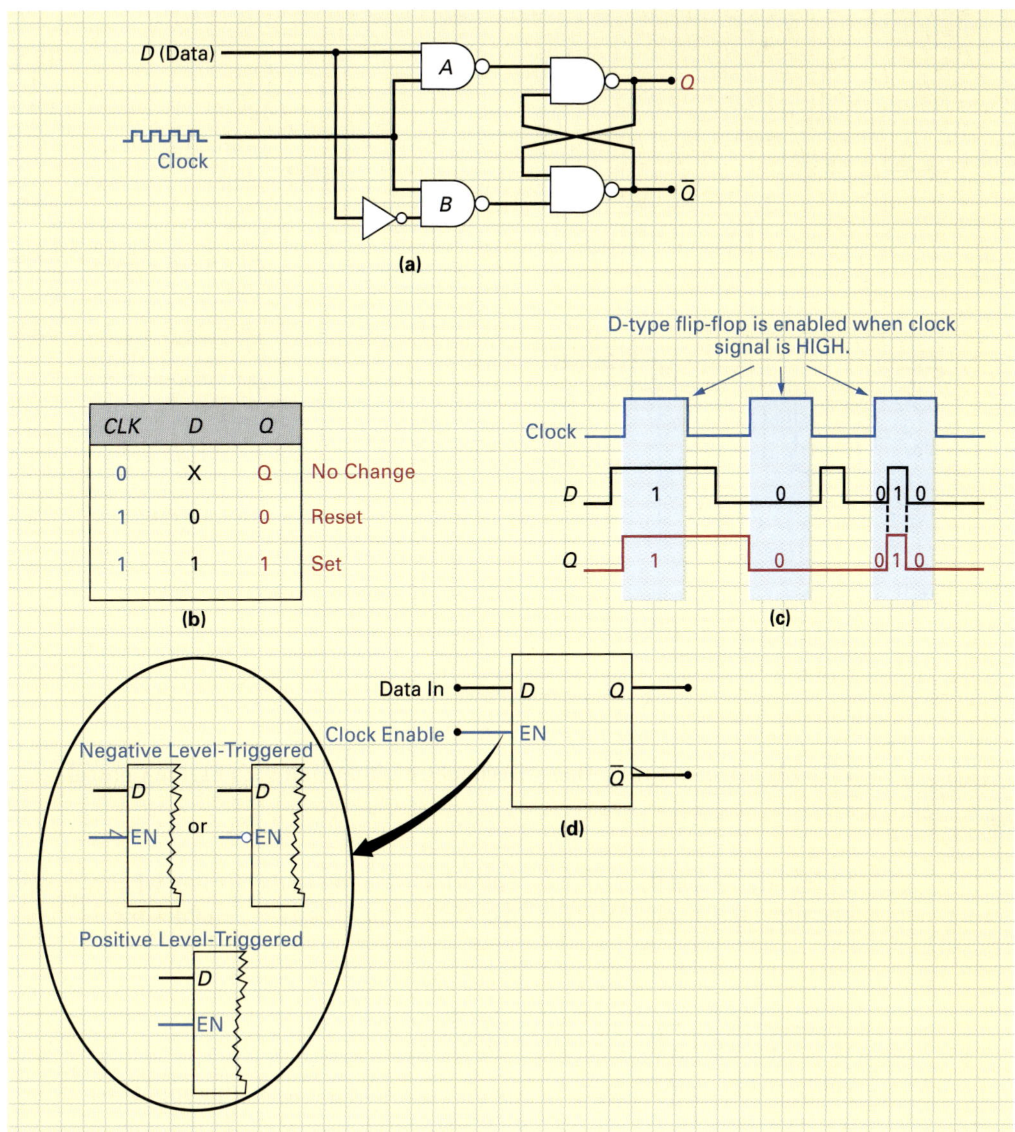

FIGURE 24-10 Level-Triggered or Gated D-Type Flip-Flop.

The 7475 IC is shown in Figure 24-11. This device contains four level-triggered D-type latches. Referring to the logic symbol, you can see that the four data inputs are labeled 1D, 2D, 3D, and 4D. There are only two enable or control inputs (*C*); pin 13 will enable the first and second latches, and pin 4 will enable the third and fourth latches.

Application—A Decimal Counting Unit with Freeze Control

The four latches of a 7475 are ideally suited for use as temporary storage of binary information between a processing unit and an input-output, or indicator, unit. Figure 24-11 shows how a 7475 can be connected to operate in such an application. The two enable inputs for the 7475 (pins 4 and 13) are tied together and connected to switch 2 (SW_2). When this switch is placed in the LATCH position, all four of the 7475 D-type latches are enabled and the BCD count from the 7490 will pass through the 7475 to the 7447 and finally be displayed on the common-anode seven-segment display. When SW_2 is placed in the DISABLE position, however, all four of the 7475 D-type latches are disabled, but the previously latched BCD code will remain at the output and, therefore, on the display, regardless of the BCD

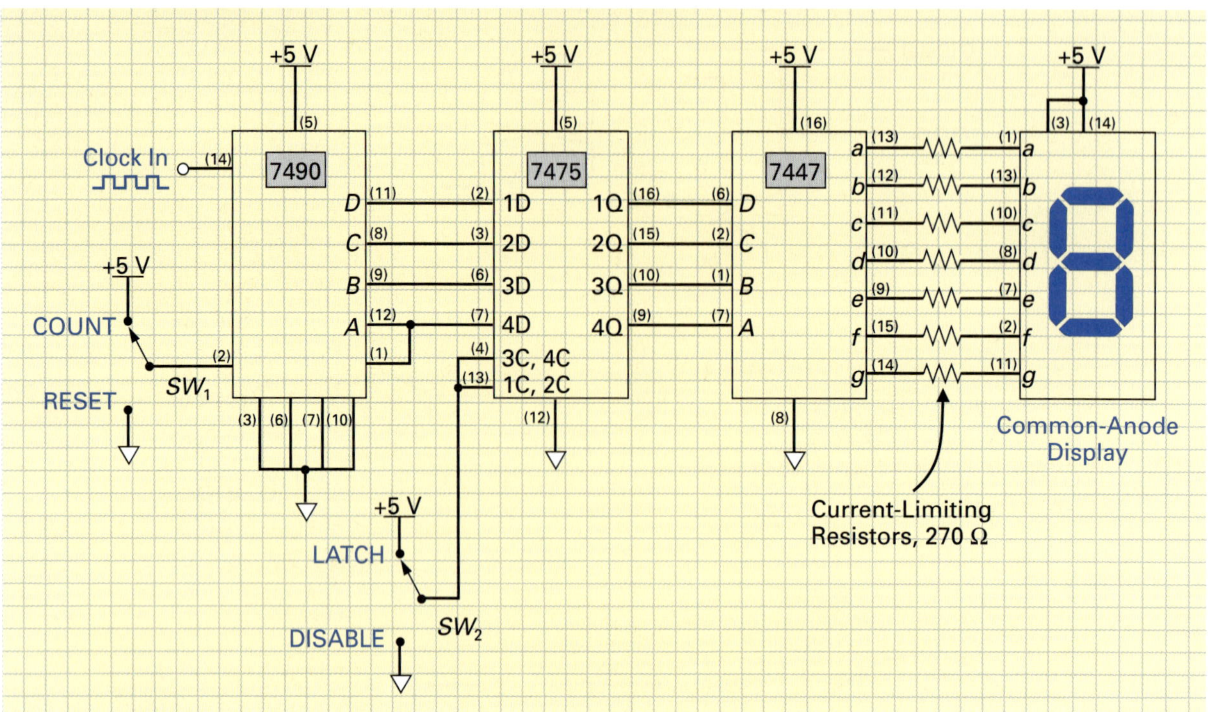

FIGURE 24-11 **Application—Decimal Counting Unit with Freeze Control.**

code variations at the *D* inputs. The effect is to freeze the display when SW_2 disables the 7475 but allow the display to reflect the 7490 count when SW_2 enables the 7475.

EXAMPLE:

What would be the *Q* output from the 7475 for the following input conditions?

$$1D = 1, 2D = 0, 1C, 2C = H, 3D = 1, 4D = 1, 3C, 4C = L$$

Solution:

Outputs 1Q = 1 and 2Q = 0, since the enables (1C and 2C = HIGH) for these two flip-flops are active. On the other hand, 3Q = NO CHANGE and 4Q = NO CHANGE, since these two flip-flops are disabled (3C and 4C = LOW).

24-2-2 *Edge-Triggered D-Type Flip-Flops*

To be useful in fast-acting digital circuits, a flip-flop needs to respond to the edge of a clock signal rather than the level of a clock. The advantage of edge-triggered devices is that the inputs do not need to be held stable for the entire time of the HIGH or LOW level of the clock.

Figure 24-12(a) shows how an edge-triggered D-type flip-flop incorporates the same positive transition pulse generator as the edge-triggered *S-R* flip-flop. Referring to the truth table in Figure 24-12(b), you can see that the active-LOW PRESET and CLEAR inputs are not dependent on the clock input or data input and can be used to PRESET or CLEAR the *Q* output. The last three lines of the truth table summarize how a positive edge-triggered D-type flip-flop will operate. If the clock input is at a HIGH level, LOW level, or making a negative transition from LOW to HIGH (↓), the D-type flip-flop will not be enabled and therefore NO

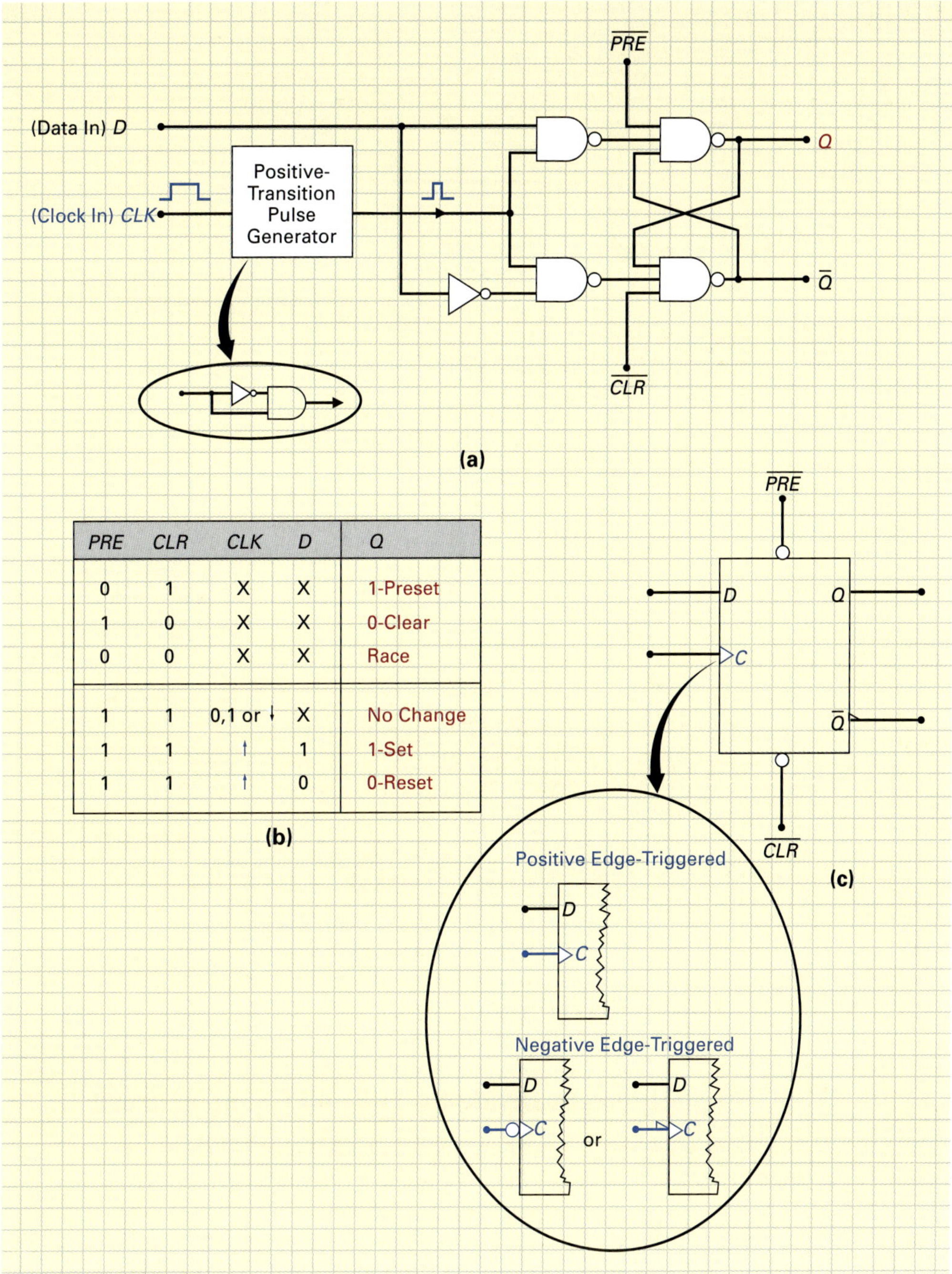

PRE	CLR	CLK	D	Q
0	1	X	X	1-Preset
1	0	X	X	0-Clear
0	0	X	X	Race
1	1	0,1 or ↓	X	No Change
1	1	↑	1	1-Set
1	1	↑	0	0-Reset

FIGURE 24-12 Edge-Triggered D-type Flip-Flop.

CHANGE will occur at the output. On a positive transition of the clock (↑), however, the Q output will follow the D input.

The logic symbol for a positive edge-triggered D-type flip-flop is shown in Figure 24-12(c). The inset illustrates how positive edge-triggered and negative edge-triggered symbols will differ.

Application—A Storage Register Circuit

Figure 24-13(a) shows how the two D-type flip-flops within a 7474 IC can be used as a 2-bit storage register. In this example, a combinational logic circuit produces a Y and Z data output, which is only present for a short space of time.

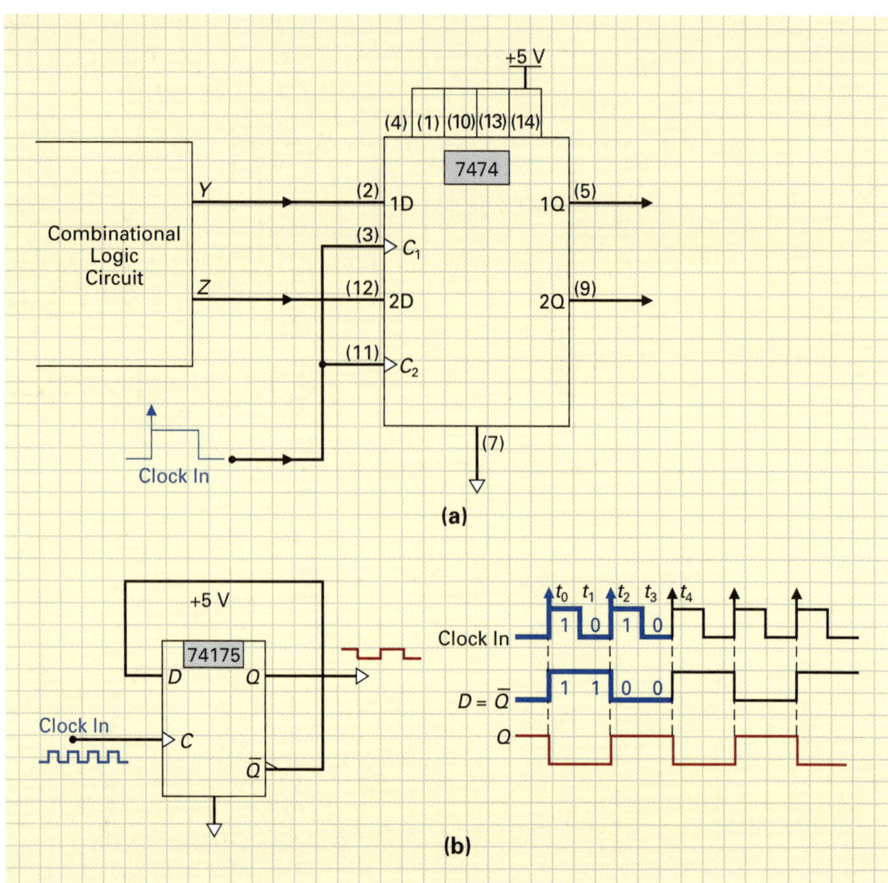

FIGURE 24-13 Application—D-type Flip-Flop. (a) Storage Register. (b) Divide-by-Two Counter.

The two D-type flip-flops are included to store the 2-bit output from the combinational logic circuit on the positive edge of the clock signal. Once this information is safely stored in the D-type latches, the combinational logic circuit can begin working on another task, since the 2-bit word is permanently stored and appears at the *Q* outputs of the D-type latches.

Application—A Divide-by-Two and Counter Circuit

Figure 24-13(b) shows how a D-type latch with a 74175 (quadruple D-type positive edge-triggered flip-flop) can function as a *divide-by-two* and *counter circuit.* To make a D-type latch divide the clock input by two, you need to connect the $\overline{Q}$ output back to the *D* input. Referring to the thick lines in the waveform in Figure 24-13(b), you can see that two cycles of the clock input are needed to produce one cycle at the *Q* output. The D-type latch has therefore divided the clock input by two by making use of the phase reversal between the *D* input and the $\overline{Q}$.

Referring to the binary values within the thick lines of the waveforms, you may have noticed that the divide-by-two D-type latch is also counting down in binary from 3 to 0. For example, at time t_0, the binary count present on the clock input and the $\overline{Q}$ output is 11_2, or 3_{10}. At time t_1, the count has dropped to 10_2, or 2_{10}, and then at time t_2 the count has decreased to 01_2, or 1_{10}. For each clock pulse following, the count decreases by 1, starting at the maximum 2-bit count of 3_{10} (11_2), then decreasing to 0 (00), and then repeating the cycle.

24-2-3 *Pulse-Triggered D-Type Flip-Flops*

The logic circuit, truth table, and logic symbol for a pulse-triggered (master-slave) D-type flip-flop is shown in Figure 24-14. Like the pulse-triggered *S-R* flip-flop, the pulse-triggered D-type flip-flop requires a complete clock pulse input before the *Q* output will reflect the *D* input.

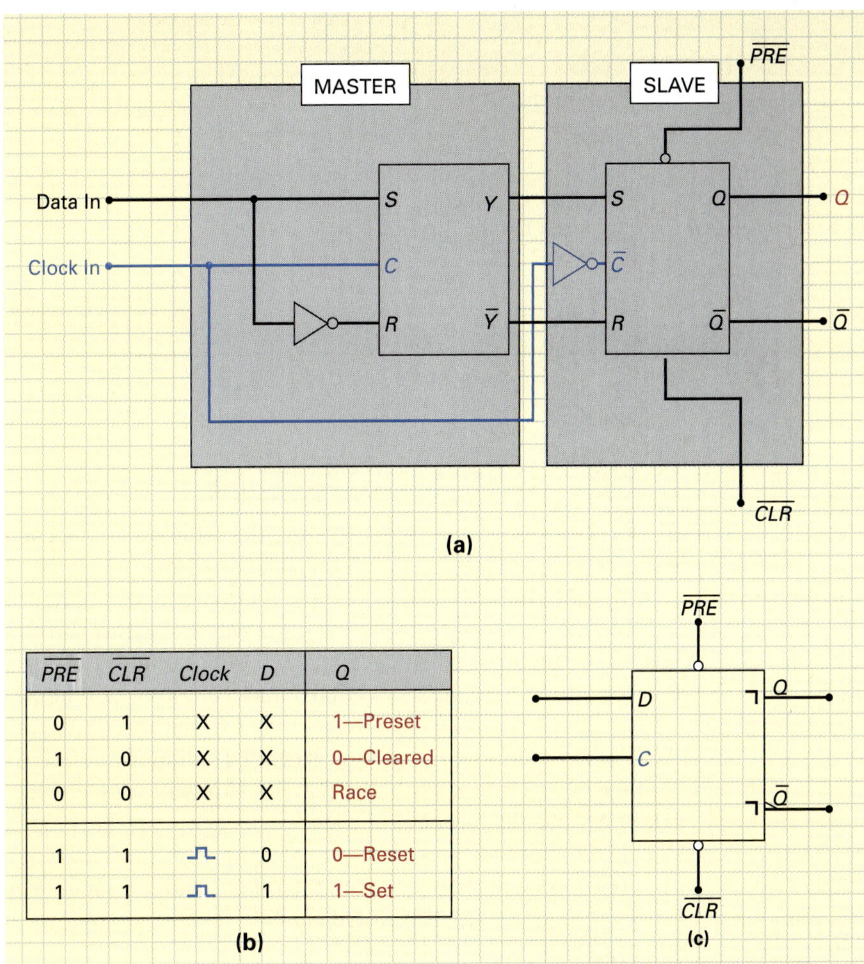

$\overline{PRE}$	$\overline{CLR}$	Clock	D	Q
0	1	X	X	1—Preset
1	0	X	X	0—Cleared
0	0	X	X	Race
1	1	⎍	0	0—Reset
1	1	⎍	1	1—Set

FIGURE 24-14 Pulse-Triggered (Master-Slave) D-Type Flip-Flop.

EXAMPLE:

What would be the Q output for the pulse-triggered D-type flip-flop shown in Figure 24-15, assuming the D and C inputs shown in the waveforms?

Solution:

All pulse-triggered flip-flops are level-clocked devices, which means that the master section is enabled for the entire time that the C input is HIGH and the slave section is enabled during the LOW input of C. The D input is therefore transferred through the master section when the clock is HIGH and then transferred through the slave section, and to the Q output, when the clock input is LOW, as shown in Figure 24-15.

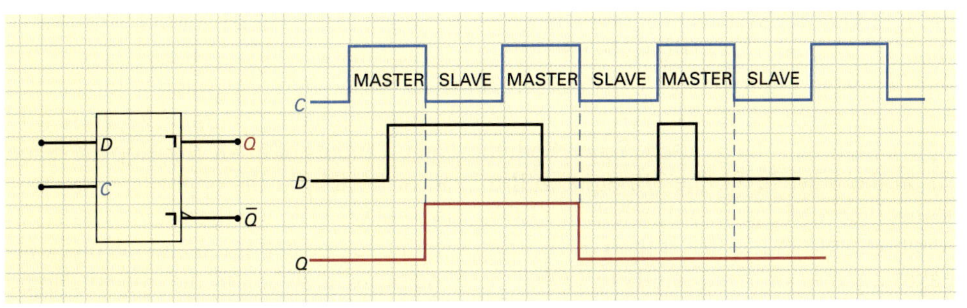

FIGURE 24-15 Input/Output Waveforms for a Pulse-Triggered D-Type Flip-Flop.

SELF-TEST EVALUATION POINT FOR SECTION 24-2

Use the following questions to test your understanding of Section 24-2.

1. The "D" in D-type stands for ____________.
2. Can the D-type flip-flop RACE?
3. What are the input and output differences between an *S-R* flip-flop and a D-type flip-flop?
4. A D-type flip-flop could be used as a/an ____________.
   a. binary counter
   b. frequency divider
   c. storage register
   d. all of the above

24-3 *J-K* FLIP-FLOPS

***J-K* Flip-Flop**
A circuit that will operate in the same way as a set-reset flip-flop, except that it will toggle its output instead of RACE when both inputs are HIGH.

Toggle
An input combination or condition that causes the output to toggle or switch its output to the opposite logic level.

With the previously discussed *S-R* flip-flop and D-type flip-flop, the letters "S," "R," and "D" were abbreviations for SET, RESET, and DATA. With the ***J-K* flip-flop,** the letters "J" and "K" are not abbreviations, they are arbitrarily chosen letters. As you will see in the following section, the *J-K* flip-flop operates in almost exactly the same way as the *S-R* flip-flop in that there will be NO CHANGE at the output when the *J* and *K* inputs are both LOW, the output will be SET HIGH when the *J* input (set input) is HIGH, and the output will be RESET LOW when the *K* input (reset input) is HIGH. The distinctive difference with the *J-K* flip-flop is that it will not RACE when both its SET (*J*) and RESET (*K*) inputs are HIGH. When this input condition is applied ($J = 1$ and $K = 1$), the *J-K* flip-flop will **toggle,** which means that it will simply switch or reverse the present logic level at its *Q* output. This toggle feature of the *J-K* flip-flop is achieved by modifying the basic *S-R* flip-flop's internal logic circuit to include two cross-coupled feedback lines between the output and the input. This circuit modification means, however, that the *J-K* flip-flop cannot be level-triggered; it can only be edge-triggered or pulse-triggered.

24-3-1 *Edge-Triggered* J-K *Flip-Flops*

Figure 24-16(a) shows the internal logic circuit for an edge-triggered *J-K* flip-flop circuit, and Figure 24-16(b) shows this circuit's function table. The first three lines of the function table show that if the clock input line is LOW, HIGH, or making a transition from LOW to HIGH (a positive edge), there will be no change at the *Q* output regardless of the *J* and *K* inputs. This is because this *J-K* flip-flop is a negative edge-triggered device, which means it will only "wake up" and perform the operation applied to the *J* and *K* inputs when the clock input is making a transition from HIGH to LOW (a negative edge). The negative-transition pulse generator will generate a positive output pulse every time the clock pulse input drops from a HIGH to a LOW level. In the last four lines of the function table, you can see that when a negative edge is applied to the clock input, the *J-K* flip-flop will react to its *J* (SET) and *K* (RESET) inputs. For example, when negative-edge-triggered, the *J-K* flip-flop will operate in almost the same way as a *S-R* flip-flop, since it will not change its *Q* output when *J* and *K* are LOW, RESET its *Q* output when the reset input (*K*) is HIGH, and SET its *Q* output when the set input (*J*) is HIGH. In the last line of the function table, you can see why the *J-K* flip-flop is an improvement over the *S-R* flip-flop. When both the *S* and *R* inputs of a *S-R* flip-flop are HIGH, it will race and generate an unpredictable *Q* output, which is why this input condition is never used. When both the *J* and *K* inputs of a *J-K* flip-flop are made HIGH, on the other hand, the *Q* output will toggle, or switch, to the opposite state. This means that if *Q* is HIGH, it will switch to a LOW, and if *Q* is LOW, it will switch to a HIGH. The function table shows that if the *J-K* flip-flop is in the toggle condition and the clock input makes a transition from HIGH to LOW, the *Q* output will switch to the opposite logic level, or the logic level of the $\overline{Q}$ output.

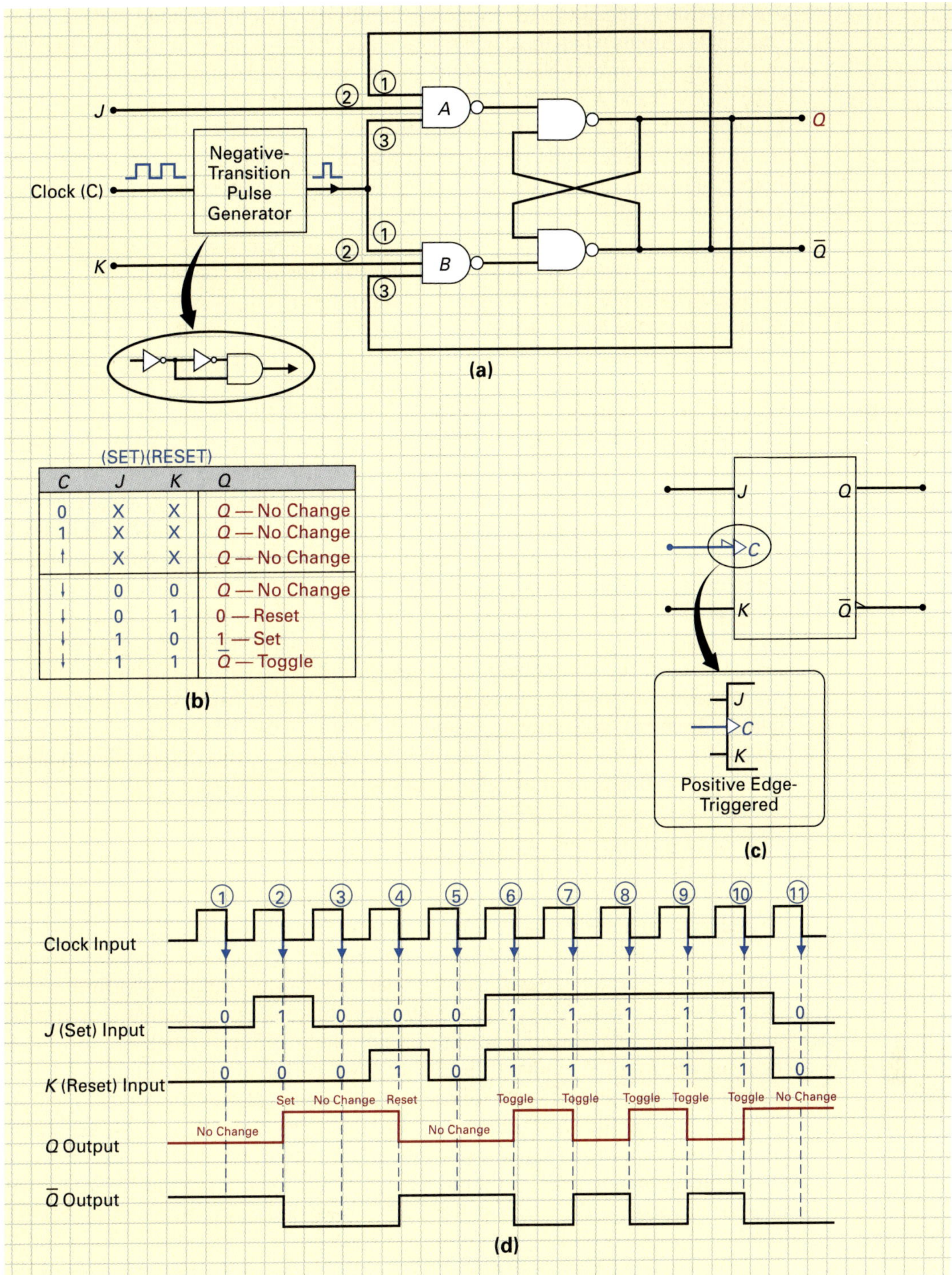

FIGURE 24-16 **Edge-Triggered *J-K* Flip-Flops.**

Figure 24-16(c) shows the logic symbol for a negative edge-triggered *J-K* flip-flop. A full triangle at the clock input, inside the rectangular block, is used to indicate that this device is edge-triggered, and the smaller right triangle (or bubble) outside the rectangular block is used to indicate that this input is active-LOW, or negative edge-triggered. The inset in Figure 24-16(c) shows the logic symbol for a positive edge-triggered *J-K* flip-flop.

To reinforce your understanding of the *J-K* flip-flop, Figure 24-16(d) shows how a negative edge-triggered *J-K* flip-flop will respond to a variety of input combinations. Since this

J-K flip-flop is negative edge-triggered, the device will only respond to the *J* and *K* inputs and change its outputs when the clock signal makes a transition from HIGH to LOW, as indicated by the negative arrows shown on the square-wave clock input waveform. At negative clock edge 1, $J = 0$ and $K = 0$, so there is no change in the Q output (which stays LOW), and the $\overline{Q}$ output (which stays HIGH). At negative clock edge 2, $J = 1$ and $K = 0$ (set condition), so the Q output is set HIGH ($\overline{Q}$ is switched to the opposite, LOW). At negative clock edge 3, both *J* and *K* are logic 0, so once again the output will not change (Q will remain latched in the set condition). At negative clock edge 4, $J = 0$ and $K = 1$ (reset condition), so the Q output will be reset LOW. At negative edge 5, both *J* and *K* are again LOW, so there will be no change in the Q output (Q will remain latched in the reset condition). At negative clock edge 6, both *J* and *K* are HIGH (toggle condition), so the output will toggle, or switch, to its opposite state (since Q is LOW it will be switched HIGH). At negative clock edges 7, 8, 9, and 10, the *J* and *K* inputs remain HIGH, and therefore the Q output will continually toggle or switch to its opposite logic level. At negative edge 11, both *J* and *K* return to 0, so the Q output remains in its last state, which in this example is HIGH.

Because of the toggle condition, the *J-K* flip-flop cannot be level-triggered. If the *J* and *K* inputs were both HIGH, and the clock were connected directly to the input gates without a transition pulse generator, the output would toggle continuously for as long as the clock input was HIGH. To explain this in a little more detail, when the clock is at its HIGH level, the HIGH *J* and *K* inputs would be passed through to the flip-flop outputs, and because of the cross-coupling, the outputs would toggle. These new outputs would then be fed back to the input gates, producing a new input condition, and the cycle would repeat. The result would be a continual change in the output or oscillations during the time the clock signal was active. For this reason, the *J-K* flip-flop is only either edge-triggered (so that the input gates are only enabled momentarily) or pulse-triggered (master is first enabled while slave—and therefore the output—is disabled). With edge-triggered *J-K* flip-flops, the pulse from the transition generator is always less than the propagation delay of the flip-flop, so that output oscillation will not occur when the flip-flop is in the toggle condition.

Application–Frequency Divider/Binary Counter Circuit

Figure 24-17(a) shows how a 555 timer and a 74LS76A *J-K* flip-flop can be connected to form a frequency divider or binary counter circuit. The 555 timer, which will be discussed in detail in the second section of this chapter, is connected to operate as an astable multivibrator and will generate the square-wave output waveform shown in Figure 24-17(b). The 74LS76A digital IC actually contains two complete *J-K* flip-flop circuits, and since only one is needed for this circuit, the flip-flop is labeled "½ of 74LS76A." Notice that the *J* and *K* inputs of the 74LS76A flip-flop are both connected HIGH, so the Q output will continually toggle for each negative edge of the clock signal input from the 555 timer, as seen in the waveforms in Figure 24-17(b). Since two negative edges are needed at the clock input to produce one cycle at the output, the *J-K* flip-flop is in fact dividing the input frequency by 2. An input frequency of 2 kHz from the 555 timer will therefore appear as 1 kHz at the *J-K* flip-flop's Q output, since two pulses in to the 74LS76A will produce one pulse out.

Now that we have seen how the *J-K* flip-flop can be used as a frequency divider, let us see how it can function as a binary counter. Figure 24-17(c) repeats the output waveforms from the 555 timer and the 74LS76A Q output. If these two outputs were connected to two LEDs, the LEDs would count up in binary, as shown in the table in Figure 24-17(c). To explain this in more detail, at time t_0, both the 555 and 74LS76A Q output are LOW, so our LED binary display shows a count of 00 (binary 0). At time t_1, the binary display will be driven by a HIGH from the 555 timer output and a LOW from the 74LS76A Q output and therefore will display 01 (binary 1). At time t_2, the display will show 10 (binary 2), and at time t_3 the display will show 11 (binary 3). Combined, the 555 timer and the divided-by-two output of the 74LS76A can be used to generate a 2-bit word that will count from binary 0 (00) to binary 3 (11) and then continuously repeat the count.

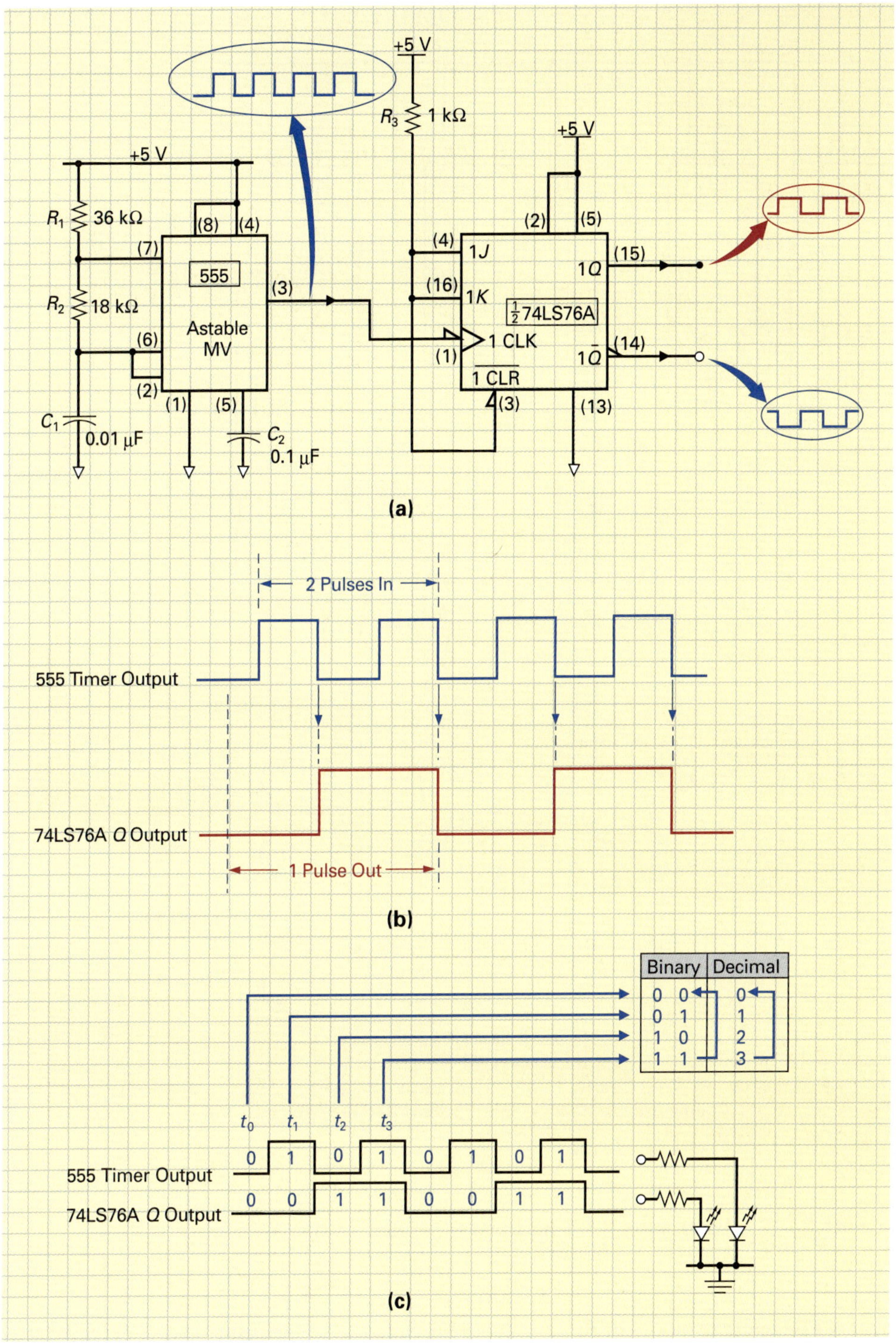

FIGURE 24-17 Application—The Edge-Triggered *J-K* Flip-Flop Acting as a Frequency Divider and Binary Counter.

Figure 24-18(a) shows how two 74LS76A ICs can be connected to produce a 4-bit binary up-counter. This and other counters will be discussed in more detail in Chapter 25; however, for now let us understand how this circuit operates. Since the *J* and *K* inputs of all the flip-flops in this circuit are tied HIGH, the flip-flops will permanently be in the toggle-input condition. The timing diagram for this circuit is shown in Figure 24-18(b). In this circuit, the clock input triggers flip-flop 1 (FF1), the output of FF1 (1Q) triggers FF2, the output of FF2

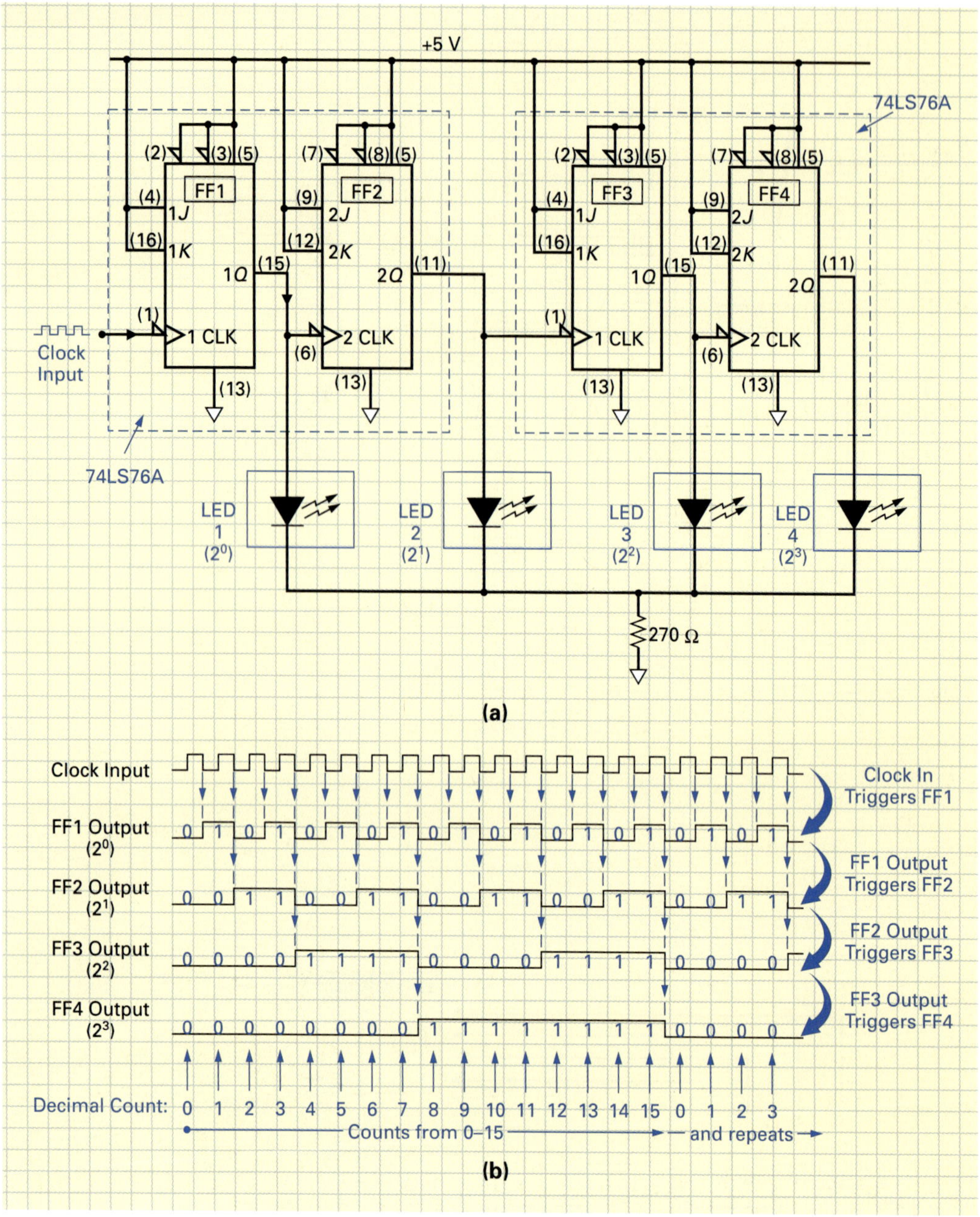

FIGURE 24-18 Application—A 4-Bit Binary Counter.

triggers FF3, and the output of FF3 triggers FF4. Since all of the flip-flops are negative edge-triggered, each negative edge of the clock input will toggle FF1, each negative edge of the FF1 output will toggle FF2, each negative edge of the FF2 output will toggle FF3, and each negative edge of the FF3 output will toggle FF4. Four active-HIGH light-emitting diodes are connected to display the logic level at each of the outputs. As far as the binary count is concerned, the display is in its reverse order, since LED1 is the display's LSB (2^0) and LED4 is the display's MSB (2^3). As a result, we will see a 1248 binary display instead of the customary 8421 binary display.

Referring to the decimal count line in Figure 24-18(b), you can see that at first all outputs are LOW and therefore all LEDs are OFF (a binary count of 0000 or decimal 0). The count then advances by 1 for every clock pulse input until it reaches its maximum 4-bit count with all LEDs ON (a binary count of 1111 or decimal 15).

24-3-2 *Pulse-Triggered* J-K *Flip-Flops*

The basic circuit for a pulse-triggered (master-slave) *J-K* flip-flop is shown in Figure 24-19(a). The circuit is almost identical to the master-slave *S-R* flip-flop except for the distinctive *J–K* cross-coupled feedback connections from the Q and $\overline{Q}$ slave outputs back to the master input gates. Figure 24-19(b) summarizes the operation of the pulse-triggered *J–K* flip-flop.

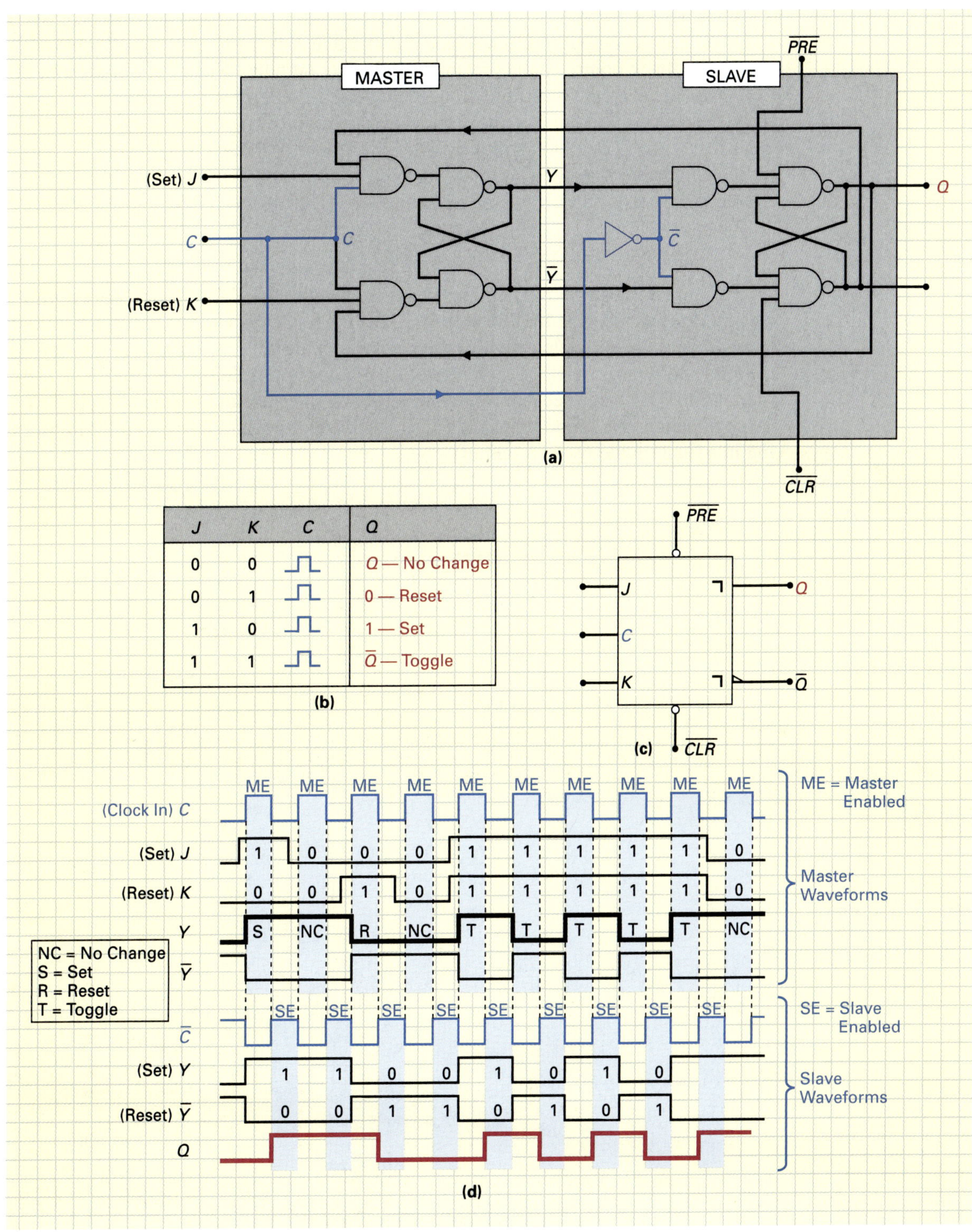

J	K	C	Q
0	0	⎍	Q — No Change
0	1	⎍	0 — Reset
1	0	⎍	1 — Set
1	1	⎍	$\overline{Q}$ — Toggle

FIGURE 24-19 **Pulse-Triggered (Master-Slave) *J-K* Flip-Flop.**

The operation of the PRESET and CLEAR inputs are not included in this table, since they will operate in exactly the same way as any other flip-flop. The logic symbol for this flip-flop is shown in Figure 24-19(c). The postponed output symbol (¬) is used to indicate that the Q and $\overline{Q}$ outputs will only change when the clock input (C) falls from a HIGH to a LOW level (which is when the slave is enabled).

The timing diagram in Figure 24-19(d) serves as a visual summary of the master-slave *J-K* flip-flop's operation. When the clock input (C) goes HIGH, the master section is enabled (ME). If $J = 1$ and $K = 0$, the Y output will be SET to 1 (S), whereas if $J = 0$ and $K = 1$, the Y output will be RESET to 0 (R). On the other hand, if $J = 0$ and $K = 0$, there will be NO CHANGE at the output (NC), whereas if $J = 1$ and $K = 1$, the output will toggle (T) based on the feedback control inputs from the Q and $\overline{Q}$ outputs. When the clock input (C) goes LOW and disables the master section, the inverted clock input ($\overline{C}$) goes HIGH and enables the slave section. Looking at the thick Y and Q output waveforms in Figure 24-19(d), you can see that the Q output is a replica of the Y output except that the Q lags Y by half a clock pulse. This is because the Q output will only follow the Y output when the slave section is enabled, and this occurs when the clock input (C) falls from a HIGH to a LOW (¬).

Application–Frequency Divider

Figure 24-20(a) shows how the two pulse-triggered *J-K* flip-flops within a 74107 could be connected as a divide-by-four circuit. Referring to the waveforms in Figure 24-20(b), you can see that two complete cycles of the clock input are needed to produce one complete cycle at the output of flip-flop 1 (FF_1 therefore divides by 2). Comparing the clock input to the output of FF_2, you can see that four complete cycles of the clock input are needed to produce one complete cycle at the output of FF_2 (FF_1 divides by 2 and then FF_2 divides by 2, resulting in a final divide-by-four output).

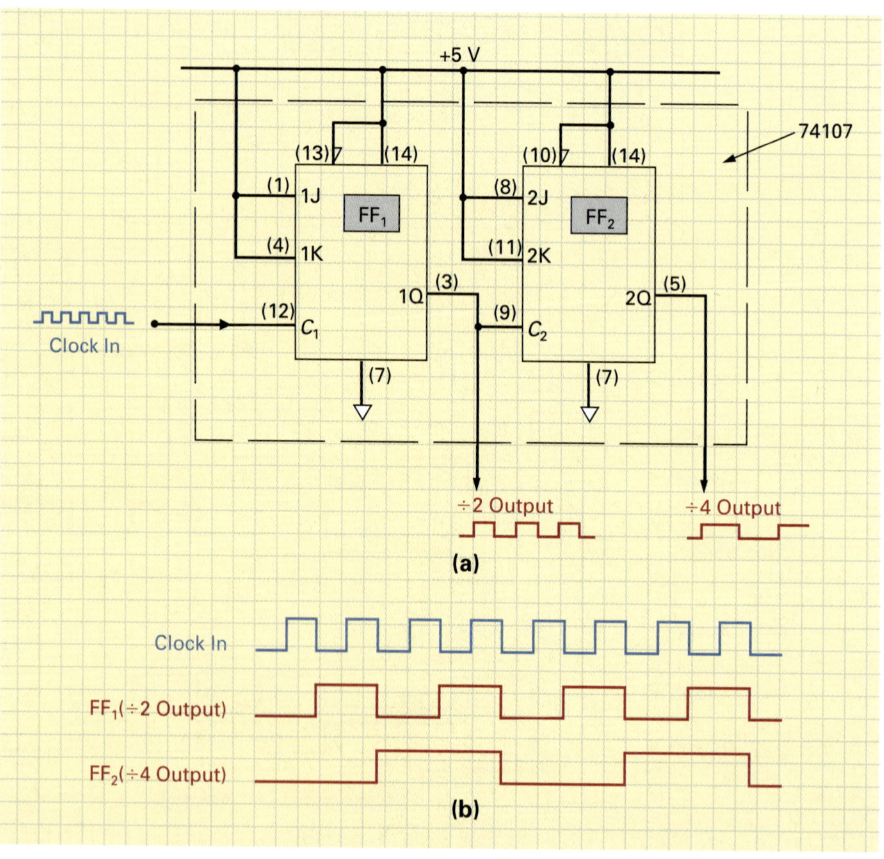

FIGURE 24-20 Application—Divide-by-Four Circuit.

SELF-TEST EVALUATION POINT FOR SECTION 24-3

Use the following questions to test your understanding of Section 24-3.

1. Which of the following input conditions will set HIGH the Q output of a negative edge-triggered J-K flip-flop?
   a. $J = 1, K = 1, C = \uparrow, \overline{PRE} = 1, \overline{CLR} = 1$.
   b. $J = 1, K = 0, C = \downarrow, \overline{PRE} = 1, \overline{CLR} = 0$.
   c. $J = 1, K = 1, C = \downarrow, \overline{PRE} = 1, \overline{CLR} = 1$.
   d. $J = 1, K = 0, C = \downarrow, \overline{PRE} = 1, \overline{CLR} = 1$.
2. When both the inputs of a J-K flip-flop are ____________, the Q output will ____________ or switch to the opposite state.
3. If the clock input to a J-K flip-flop were a 126 kHz square wave and $J = 1, K = 1, \overline{PRE} = 1, \overline{CLR} = 1$, what would the Q output be?
   a. A 126 kHz square wave.
   b. A 63 kHz square wave.
   c. A 31.5 kHz square wave.
   d. A 252 kHz square wave.
4. The ____________ flip-flop can perform all the functions of the ____________ flip-flop and ____________ flip-flop.
   a. *S-R, J-K,* D-type
   b. D-type, *J-K, S-R*
   c. *J-K,* D-type, *S-R*
   d. *S-R,* D-type, *J-K*

24-4 DIGITAL TIMER AND CONTROL CIRCUITS

Timing is everything in digital logic circuits. To control the timing of digital circuits, a clock signal is distributed throughout the digital system. This square-wave clock signal is generated by a clock oscillator, and its sharp positive (leading) and negative (trailing) edges are used to control the sequence of operations in a digital circuit.

The *S-R,* D-type, and *J-K* flip-flops are all examples of bistable multivibrators, since they have two (bi) stable states (SET and RESET). In this section we will discuss the astable, or unstable, multivibrator, which has no stable states and is commonly used as a clock oscillator. The third type of multivibrator is the monostable multivibrator which has only one (mono) stable state and when triggered will generate a rectangular pulse of a fixed duration.

Logic Gate Astable Multivibrator Circuits

The astable multivibrator can also be constructed using logic gates, as seen in Figure 24-21. In Figure 24-21(a), a Schmitt-trigger INVERTER is connected to operate as a clock oscillator. When power is first applied to this circuit, the capacitor will have no charge and this LOW input is inverted by the NOT gate, giving a HIGH output (seen as red in the illustration). The capacitor (C) will begin to charge via the resistor (R), and the increasing positive charge across the capacitor will be felt at the input of the INVERTER. After a time (which is dependent on the values of R and C), the capacitor charge will be large enough to apply a valid HIGH to the INVERTER input. This HIGH input will cause the output of the INVERTER to go LOW, so the capacitor will begin to discharge (seen as blue in illustration). When the capacitor's charge falls to a valid LOW logic level, the INVERTER will generate a HIGH output, and the cycle will repeat.

Figure 24-21(b) shows another basic astable multivibrator circuit using two INVERTERS. The operation of this circuit is similar to the one in Figure 24-21(a). A continual capacitor charge and discharge causes the astable or free-running multivibrator to switch back and forth between its two unstable states, producing a repeating rectangular wave at the output (two conditions are shown in the illustration as red and blue).

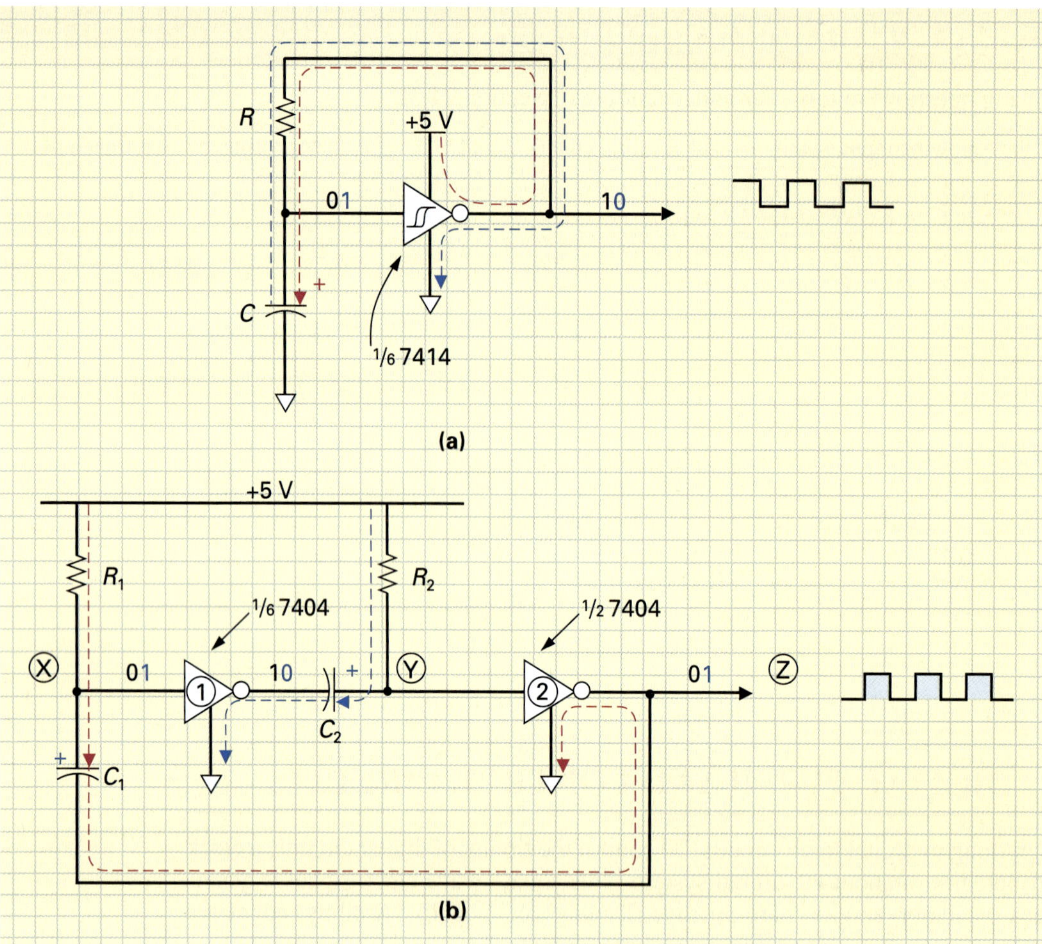

FIGURE 24-21 **Application—Logic-Gate Astable Multivibrators.**

Logic-Gate Monostable Multivibrator Circuits

Like the astable multivibrator, the monostable multivibrator can be constructed using logic gates, as seen in Figure 24-22(a). When the trigger input is LOW and the *Q* output is LOW, the output from the NOR gate is HIGH, and therefore the output from the INVERTER is LOW, keeping the circuit in its stable state. When the trigger input pulses HIGH, it causes the output of the NOR gate to go LOW. This HIGH-to-LOW transition is coupled through the capacitor to the INVERTER, which produces a HIGH *Q* output, as can be seen in the timing diagram below the circuit. This HIGH *Q* output is fed back to the NOR gate's other input and keeps the NOR gate output LOW even after the trigger pulse has ended. A LOW NOR gate output will produce a potential difference across the resistor-capacitor network, and therefore the capacitor will begin to charge. After a time, dependent on the values of *R* and *C,* the charge on *C* is large enough for the INVERTER to recognize it as a valid HIGH input, and therefore it generates a LOW *Q* output ending the pulse.

The typical block logic symbols for a monostable multivibrator, or one-shot, are shown in Figure 24-22(b). Some logic symbols group the entire circuit within a block, while others block only the logic gates and show the time-determining capacitor and resistor separately.

Today, one-shot circuits are very rarely constructed with discrete transistors or logic gates since integrated circuits are available. These one-shot ICs are classified as being either *nonretriggerable one-shots* or *retriggerable one-shots.* The difference between these two types is best described by comparing the waveforms in Figure 24-23.

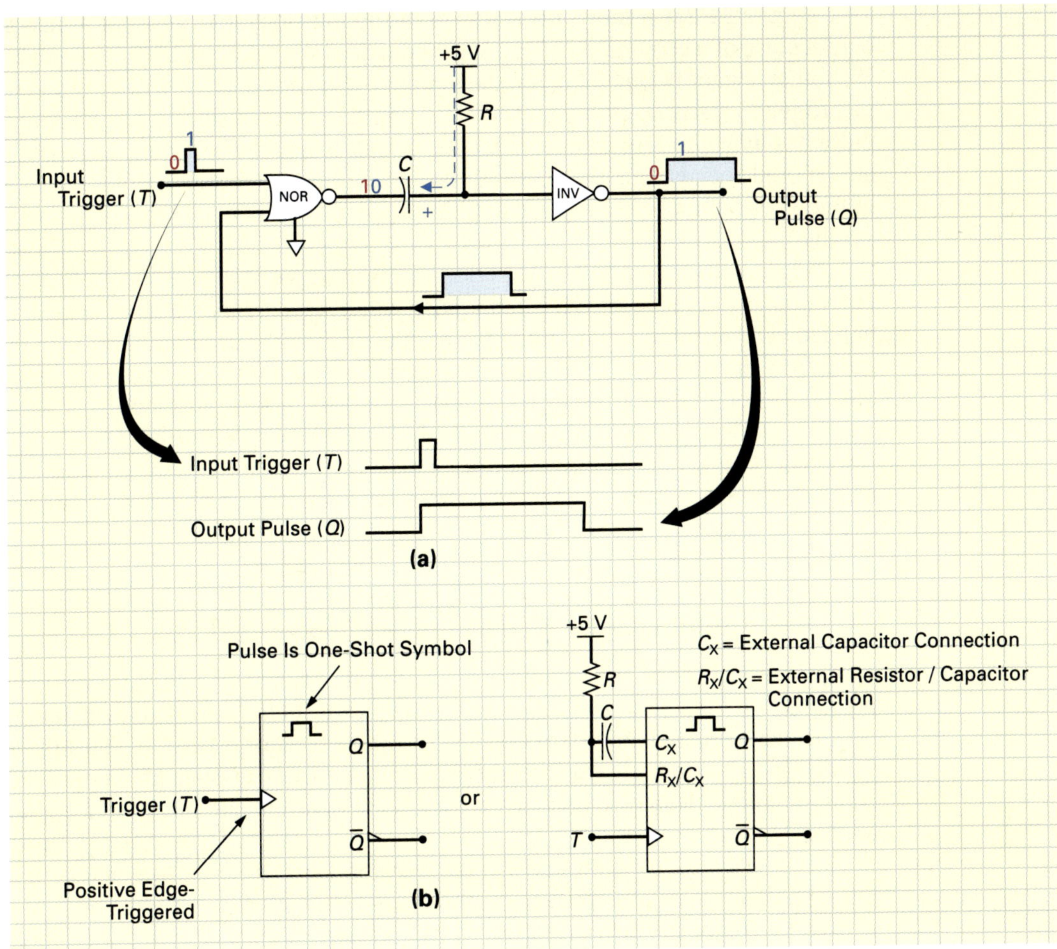

FIGURE 24-22 **Application—Logic-Gate Monostable (One-Shot) Multivibrator.**

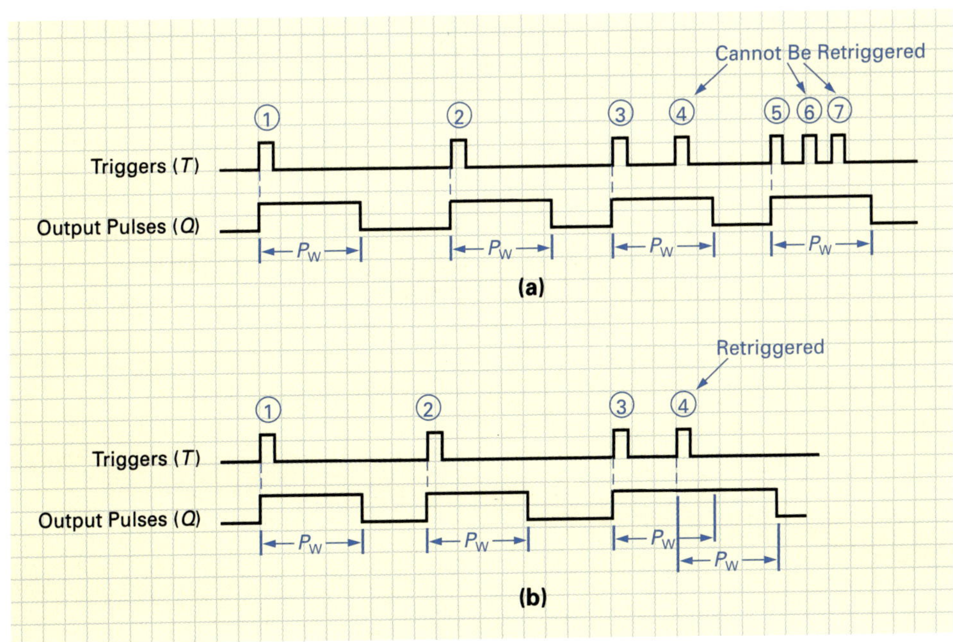

FIGURE 24-23 **One-Shot IC Types. (a) Nonretriggerable Action. (b) Retriggerable Action.**

Figure 24-23(a) shows the action of the nonretriggerable one-shot. When triggered, an output pulse of a certain width (P_w) is generated, as shown when triggers 1 and 2 are applied. If another trigger is applied before the output pulse has ended, it will be ignored. For example, trigger 3 starts an output pulse, but trigger 4 is ignored since it occurs before the output pulse has *timed out.* As another example, trigger 5 *fires* the one-shot, while triggers 6 and 7 are ignored because the one-shot is in its unstable state. Once triggered, therefore, this type of one-shot cannot be retriggered until its output pulse has ended (the one-shot is nonretriggerable).

Figure 24-23(b) shows the action of the retriggerable one-shot. When triggered, an output pulse is produced of a certain pulse width (P_w), as shown when triggers 1 and 2 are applied. The difference with this type of one-shot can be seen with triggers 3 and 4. Trigger 3 fires the one-shot, so it begins to produce an output pulse; however, before this pulse has timed out, trigger 4 retriggers the one-shot. The output pulse will now be extended for a time equal to a pulse width starting at the time trigger 4 occurred. Once triggered, therefore, this type of one-shot can be retriggered (the one-shot is retriggerable). The 74121 is an example of a nonretriggerable one-shot IC while the 74122 is an example of a retriggerable one-shot.

Applications–Pulse Generator, Sequencer and Delay Circuits

Figure 24-24(a) shows how a 74121 can be connected to function as a variable output pulse width generator. The normally open push-button (NOPB) switch is used to trigger the one-shot, and the external 100 kΩ variable resistor is used to adjust the output pulse width. The $\overline{Q}$ output is used to drive an output LED instead of the Q output, since a common-anode display is being used for a higher current and therefore a higher light-level display.

Figure 24-24(b) shows how three one-shots can be used to generate a sequence of timing pulses. Since all inputs are negative edge-triggered, the trailing edge of each output is used to trigger the next one-shot, as can be seen in the associated waveforms in Figure 24-24(b).

Figure 24-24(c) shows how two one-shots can be used to introduce a delay. An input trigger fires one-shot *Y,* which introduces a certain time delay, and its trailing edge triggers one-shot *Z,* which then produces an output pulse. The result is that the input pulse is delayed from appearing at the output for a time equal to the pulse width time of the *Y* one-shot. This time delay function is shown in the accompanying waveforms in Figure 24-24(c).

24-4-1 *The 555 Timer Circuit*

555 Timer

A versatile IC timer that can, for example, be configured to function as a modulator, multivibrator, or frequency divider.

One of the most frequently used low-cost integrated circuit timers is the **555 timer.** It is a highly versatile timer that can be made to function as an astable multivibrator, monostable multivibrator, frequency divider, or modulator, depending on the connection of external components.

Nearly all the IC manufacturers produce a version of the 555 timer, which can be labeled in different ways, such as SN72 555, MC14 555, SE 555, and so on.

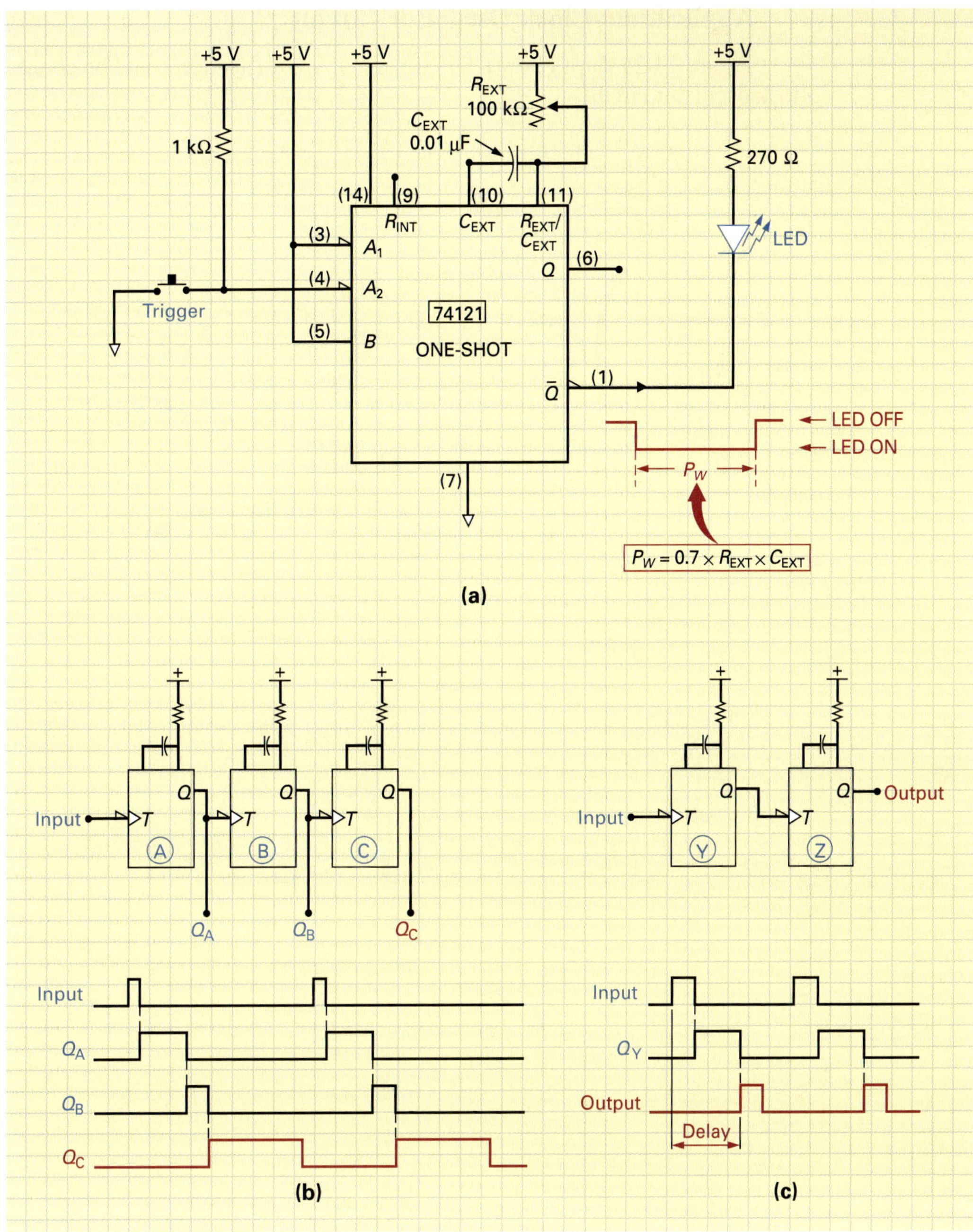

FIGURE 24-24 **Application—One-Shot. (a) Pulse Generation. (b) Timing and Sequencing. (c) Time Delay.**

Application–Clock Oscillator Circuit

Figure 24-25 shows how a 555 timer and a *J-K* flip-flop could be connected to form a clock oscillator circuit.

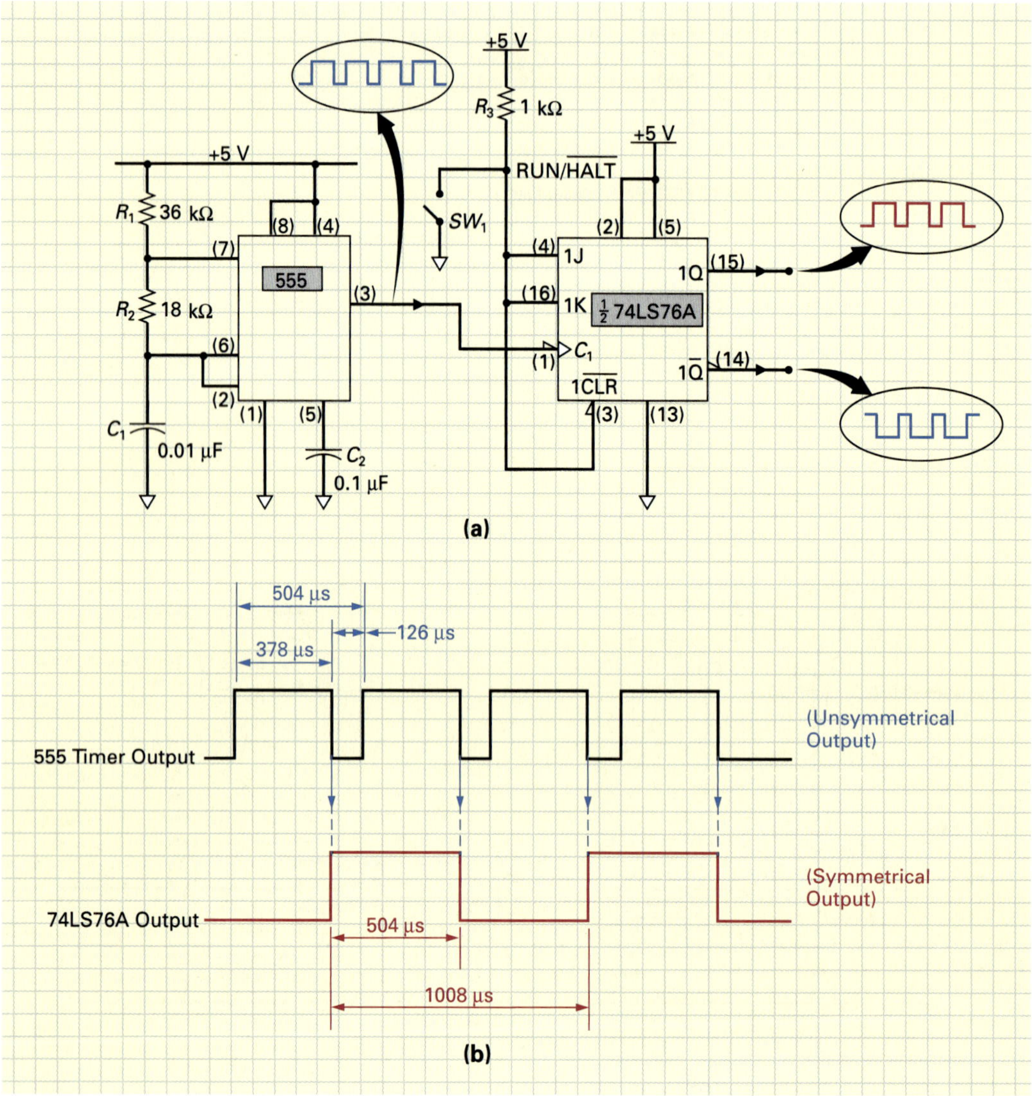

FIGURE 24-25 Application—1 kHz Clock Oscillator with RUN/$\overline{\text{HALT}}$ Control and Low-Level Start.

EXAMPLE:

Figure 24-25(a) shows a typical clock-oscillator circuit using a 555 timer and a *J-K* flip-flop. Perform the following:

a. Calculate the output frequency from the 555 timer.

b. Determine the HIGH time and LOW time of the 555 timer's output.

c. Determine the output frequency from the *J-K* flip-flop.

d. Sketch the 555 timer output and the 74LS76A's *Q* output in time relation.

e. Explain the purpose of the RUN/$\overline{\text{HALT}}$ switch.

■ ***Solution:***

a.
$$f = \frac{1.43}{(R_1 + 2R_2) \times C_1}$$
$$= \frac{1.43}{[36\ \text{k}\Omega + (2 \times 18\ \text{k}\Omega)] \times 0.01\ \mu\text{F}} = 2\ \text{kHz}$$

b. The positive half-cycle will last for

$$t_p = 0.7 \times C \times (R_A + R_B)$$
$$= 0.7 \times 0.01\ \mu\text{F} \times (36\ \text{k}\Omega + 18\ \text{k}\Omega) = 378\ \mu\text{s}$$

The negative half-cycle will last for

$$t_p = 0.7 \times C \times R_B$$
$$= 0.7 \times 0.01\ \mu\text{F} \times 18\ \text{k}\Omega = 126\ \mu\text{s}$$

c. Since the *J* and *K* inputs of the flip-flop are normally tied HIGH, the output will toggle and therefore divide the input frequency by 2. Since we know that the 555 timer's output is 2 kHz, the *J-K* flip-flop's output will be 1 kHz (2 kHz ÷ 2 = 1 kHz).

d. The 555 timer output and the 74LS76A's *Q* output in time relation are shown in Figure 24-25(b). The 555 timer output drives the clock input (*C*) of the negative edge-triggered 74LS76A *J-K* flip-flop. Each negative edge of the unsymmetrical 2 kHz clock input will cause the output to toggle, or change state, producing a symmetrical 1 kHz output.

e. With SW_1 open, the RUN/$\overline{\text{HALT}}$ line will be HIGH, so the *J* and *K* inputs of the flip-flop will also be in HIGH (toggle) condition. If SW_1 is closed, the RUN/$\overline{\text{HALT}}$ line will be LOW, forcing the *J-K* flip-flop into the NO CHANGE condition. This will freeze the flip-flop's outputs and therefore halt or stop the clock signal division. The RUN/$\overline{\text{HALT}}$ line is also connected to the active-LOW clear input of the *J-K* flip-flop. This will ensure that when we switch from HALT to the RUN mode, the clock output at *Q* will always start LOW (in the HALT mode *Q* is reset LOW, and therefore when we switch to the RUN mode, *Q* will start LOW and then switch HIGH). Having both a *Q* and $\overline{Q}$ output means that this circuit can be used to supply either a single-phase clock signal (*Q* only) or a two-phase clock signal (*Q* and $\overline{Q}$).

SELF-TEST EVALUATION POINT FOR SECTION 24-4

Use the following questions to test your understanding of Section 24-4.

1. Which of the multivibrator circuits has only one stable state?
2. Which multivibrator is also called a set-reset flip-flop?
3. In what applications are one-shots used?
4. The 555 timer derived its number identification from ____________.

REVIEW QUESTIONS

Multiple-Choice Questions

1. What is another name for a flip-flop circuit?
   a. Astable multivibrator
   b. Monostable multivibrator
   c. Bistable multivibrator
   d. Horstable multivibrator

2. Which two multivibrator types require trigger inputs?
   a. Astable and monostable
   b. Monostable and bistable
   c. Bistable and astable
   d. Both a. and c. are true

3. What input condition does the *J-K* flip-flop have that the *S-R* and D-type flip-flops don't have?
   a. SET c. NO CHANGE
   b. RESET d. TOGGLE
4. What would be the output from a negative edge-triggered *J-K* flip-flop if $J = 1$, $K = 0$, C = LOW to HIGH, and $Q = 0$?
   a. NO CHANGE c. $Q = 0$
   b. $Q = 1$ d. Both a. and c. are true
5. A four-stage *J-K* flip-flop can be made to function as a ____________ binary counter.
   a. 0-to-7 c. 0-to-15
   b. 0-to-3 d. 0-to-16
6. If $J = 0$ and $K = 1$, a *J-K* flip-flop when clocked is said to be in the ____________ condition.
   a. SET c. NO CHANGE
   b. RESET d. TOGGLE
7. If $J = 1$, $K = 1$, and the input clock frequency is 2 MHz, what would be the Q output frequency?
   a. 2 MHz c. 0.5 MHz
   b. 1 MHz d. 4 MHz
8. Which triggering system is used to control a master-slave latch?
   a. Level-triggering c. Pulse-triggering
   b. Edge-triggering d. Either a. or c.
9. The *J-K* flip-flop responds to only ____________.
   a. Level-triggering c. Pulse-triggering
   b. Edge-triggering d. Both b. and c.
10. Which multivibrator type is generally used as a clock oscillator?
   a. Astable multivibrator
   b. Monostable multivibrator
   c. Bistable multivibrator
   d. Tristable multivibrator
11. Which multivibrator type is used as a pulse generator?
   a. Astable multivibrator
   b. Monostable multivibrator
   c. Bistable multivibrator
   d. Tristable multivibrator
12. Monostable or one-shot multivibrators are generally used as a
   a. pulse generator. c. delay.
   b. timer and sequencer. d. all of the above.
13. Which multivibrator type has two stable states?
   a. Astable multivibrator
   b. Monostable multivibrator
   c. Bistable multivibrator
   d. Tristable multivibrator
14. Which multivibrator is also called a set-reset flip-flop?
   a. Astable c. Bistable
   b. Monostable d. Schmitt
15. How many *J-K* flip-flop stages will be needed to achieve a divide-by-eight frequency division?
   a. Two c. Four
   b. Three d. Five

Practice Problems

To practice your circuit recognition and operation, refer to Figure 24-26 and answer the following six questions.

16. What type of flip-flop circuit is being used in this application?
17. If the LOAD control line were LOW, what would be applied to the *D* inputs of the flip-flops?
18. If the LOAD control line were HIGH, what would be applied to the *D* inputs of the flip-flops?
19. What type of triggering is used in this circuit?
20. Describe the basic steps involved in loading a 4-bit word into this storage register circuit.
21. How is the $\overline{CLR}$ control line used in this circuit?

To practice your circuit recognition and operation, refer to Figure 24-27 and answer the following three questions.

22. The logic gates *A* and *B* form what type of flip-flop circuit?
23. What will be the output at point *X* when the switch is in the OFF position and in the ON position?
24. What will be the output at point *Y* when the switch is in the OFF position and in the ON position?

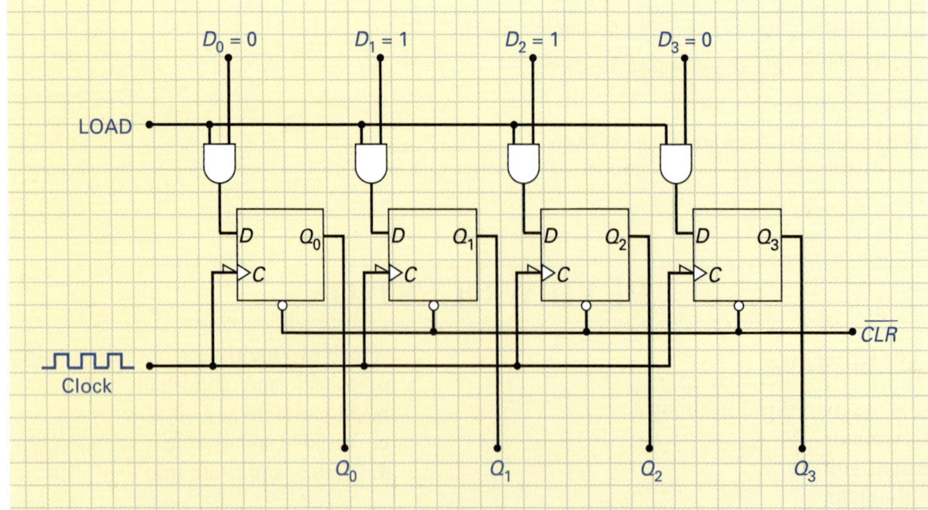

FIGURE 24-26 **Application—A 4-Bit Storage Register.**

To practice your circuit recognition and operation, refer to Figure 24-28 and answer the following three questions.

25. What type of flip-flop is being used in the application in this circuit?

26. What will be the output frequency at
  a. Q_A? **b.** Q_B?

27. Sketch the input/output waveforms for this circuit.

To practice your circuit recognition and operation, refer to Figure 24-29 and answer the following four questions.

28. What type of flip-flop is being used in this circuit?

29. What type of triggering is being used in this circuit?

30. Describe how this circuit will operate when inputs *A* and *B* are out-of-phase with one another.

31. Describe how this circuit will operate when inputs *A* and *B* are in-phase with one another.

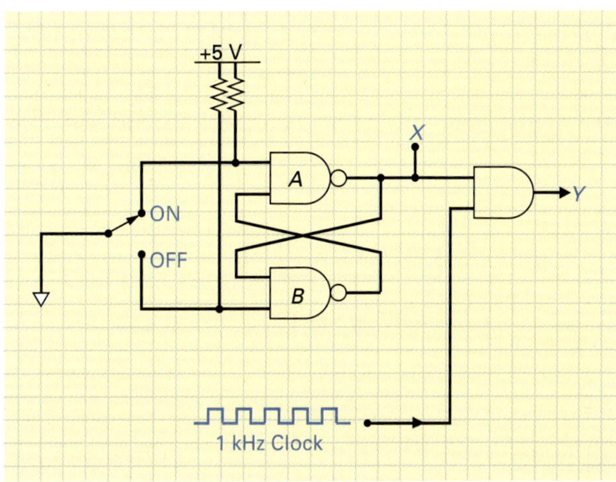

FIGURE 24-27 Application—ON/OFF Clock-Control Circuit.

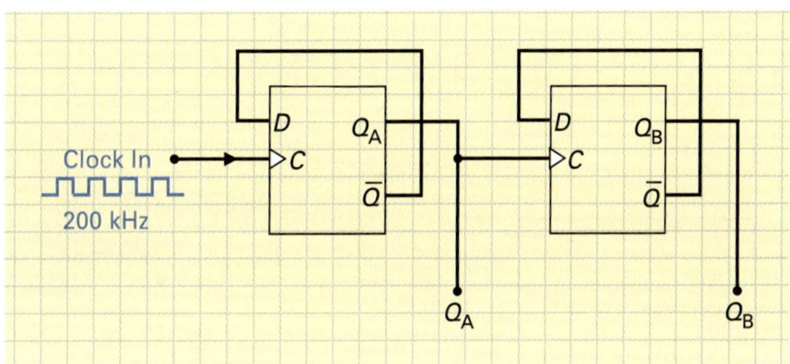

FIGURE 24-28 Application—Clock-Division Circuit.

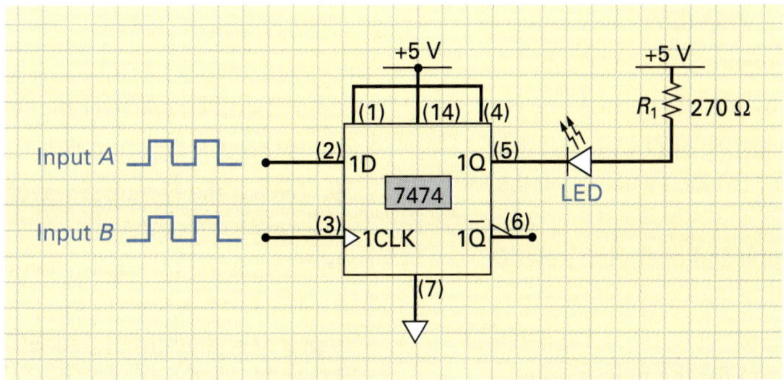

FIGURE 24-29 Application—A Phase-Detector Circuit.

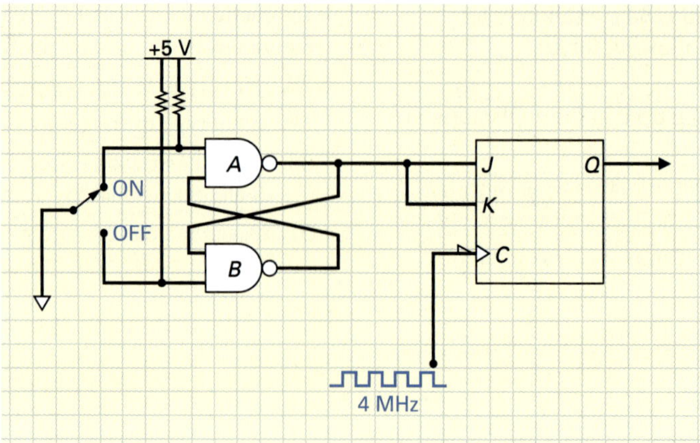

FIGURE 24-30 **Application—Divide-by-Two Clock Circuit with ON/OFF Control.**

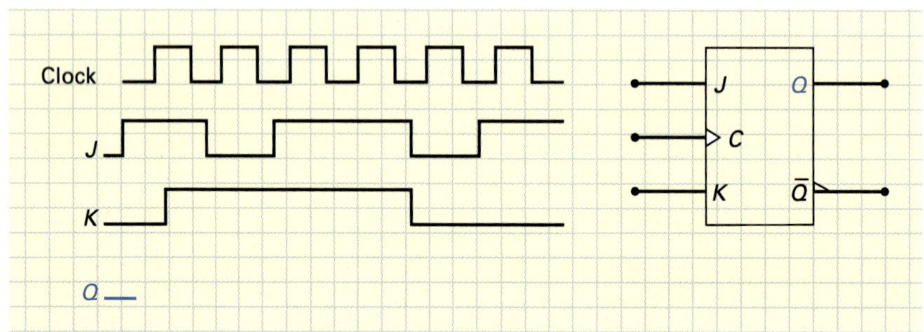

FIGURE 24-31 **A *J-K* Flip-Flop Timing Example.**

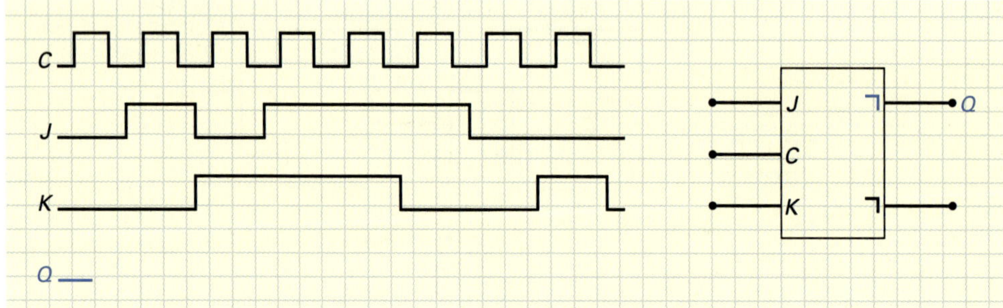

FIGURE 24-32 **A *J-K* Flip-Flop Timing Example.**

To practice your circuit recognition and operation, refer to Figure 24-30 and answer the following three questions.

32. How is the *J-K* flip-flop in Figure 24-30 triggered?

33. Identify the circuit formed by NAND gates *A* and *B*.

34. What will be the output frequency at *Q* when (a) the switch is in the ON position? (b) the switch is on the OFF position?

To practice your circuit recognition and operation, refer to Figure 24-31 and answer the following three questions.

35. Is this *J-K* flip-flop edge-triggered or pulse-triggered?

36. Sketch the *Q* output waveform for the input timing waveforms shown.

37. Sketch the $\overline{Q}$ output waveform for the input timing waveforms shown.

To practice your circuit recognition and operation, refer to Figure 24-32 and answer the following three questions.

38. Is this *J-K* flip-flop edge-triggered or pulse-triggered?

39. Sketch the *Q* output waveform for the input timing waveforms shown.

40. Sketch the $\overline{Q}$ output waveform for the input timing waveforms shown.

To practice your circuit recognition and operation, refer to Figure 24-33 and answer the following four questions.

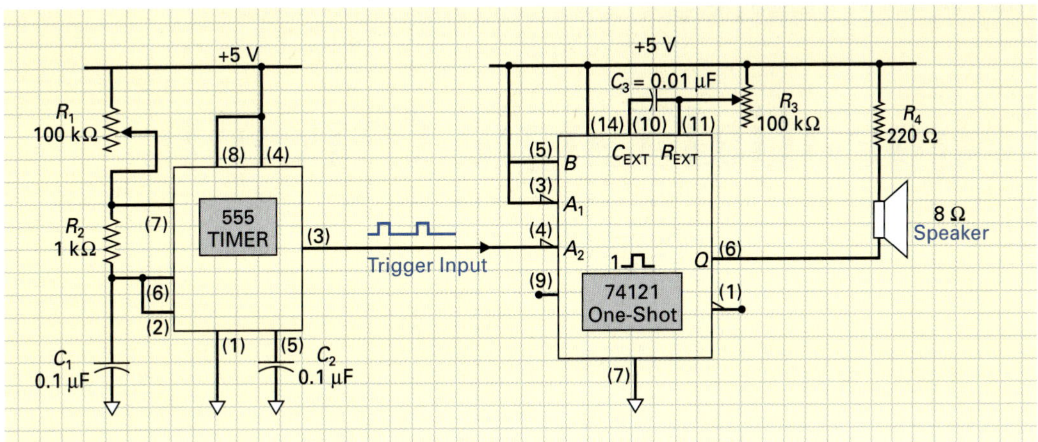

FIGURE 24-33 Application—Tone Generator.

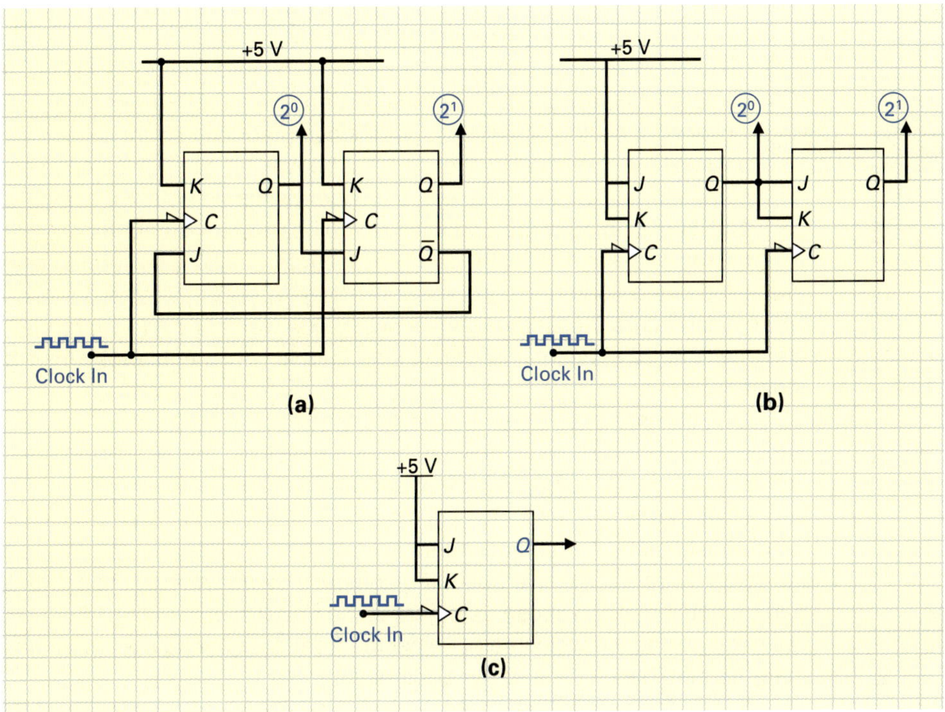

FIGURE 24-34 Application—Divider Circuits.

41. The 555 timer is connected to operate as a/an _____ multivibrator.
42. Calculate the frequency range of the 555 timer if R_1 were adjusted between the two extremes of 1 Ω and 100 kΩ.
43. Is the 74121 nonretriggerable or retriggerable?
44. Calculate the pulse length range of the 74121 one-shot if R_3 were adjusted between the two extremes of 1 Ω and 100 kΩ.

To practice your circuit recognition and operation, refer to Figure 24-34 and answer the following four questions.

45. Which type of *J-K* flip-flops are being used?
46. Which of the *J-K* flip-flop circuits will function as a divide-by-two circuit?
47. Which of the *J-K* flip-flop circuits will function as a divide-by-three circuit?
48. Which of the *J-K* flip-flop circuits will function as a divide-by-four circuit?

Web Site Questions

Go to the web site http://www.prenhall.com/cook, select the textbook *Electronics: A Complete Course,* select this chapter, and then follow the instructions when answering the multiple choice practice problems.

Sequential Logic Circuits

Trash

In 1976 a Texas-based chain of stores known as Tandy Radio Shack, which up to that time had only been involved with electronic kits and parts for citizen band (CB) radios and stereo equipment, sensed a sudden demand among its hobbyists for a personal computer. In July of that year the company recruited Steven Leininger, who had been working for National Semiconductor since graduating from college. He seemed to possess the right qualifications in that he was an electronics engineer and a computer buff. In a double feat, Leininger designed both the computer's architecture and its built-in software and used at the heart of his computer the brand-new Z80 microprocessor. This new microprocessor from the newly formed company, Zilog, resembled Intel's 8080 microprocessor but was exceedingly more powerful. The reason it resembled Intel's processor was because Zilog was founded by a group of disgruntled Intel engineers; a situation that would later lead to some legal problems—but then, that's another story.

On February 2, 1977, after Leininger had not left the plant for the last two weeks and had been working almost completely around the clock, the TRS-80-Model 1 was demonstrated faultlessly to company owner Charles Tandy. Blowing smoke from his ever-present large cigar, Tandy said that it was "quite a nice little gadget." An executive then asked, "How many should be built—one thousand, two thousand?" The company's financial controller answered by saying, "We've got 3,500 company-owned stores. I think we can build that many. If nothing else, we can use them in the back for accounting." For Leininger, his baby had not received the welcome he expected from the general management, who did not share his confidence in its ability or attraction.

Not even Leininger was prepared for the TRS-80 (nicknamed Trash 80) splash that was going to take place. Going on sale in September of 1977, 10,000 orders flooded in almost overnight and the demand remained so strong that the company's production plant took over a year to catch up with the orders. A year later, in 1978, Tandy Computers took an impressive lead in the personal computer race, leaving behind their chief competitor, Commodore, and its PET computer.

25

Introduction

Flip-flops are memory elements that are the basic building blocks we will use to construct sequential logic circuits, which can perform such functions as storing, sequencing, and counting. To help define the term **sequential logic circuits,** let us compare them to combinational logic circuits, which perform such functions as decoding, encoding, and comparing. The output from a combinational logic circuit is determined by the present state of the inputs, whereas the output of a sequential logic circuit, because of its memory ability, is determined by the previous state of the inputs.

Like combinational logic circuits, there is a large variety of sequential logic circuits; however, two types predominate in digital systems: registers and counters. Both of these circuit types have been briefly discussed in previous applications; however, in this chapter we will examine them in more detail.

Sequential Logic Circuits
Logic circuits in which outputs are dependent on the input states, delays encountered in the logic path, the presence of a discrete timing interval, and the previous state of the logic circuit.

25-1 BUFFER REGISTERS

Register

A group of flip-flops or memory elements that store and shift a group of bits or binary word.

Buffer Register

A temporary data storage circuit able to store a digital word.

A **register** is a group of flip-flops or memory elements that work together to store and shift a group of bits or a binary word. The most basic type of register is called the **buffer register,** and its function is simply to store a binary word. Other registers, such as the shift register, modify the stored word by shifting the stored bits to the left or to the right.

The buffer register simply stores a digital word. Figure 25-1(a) shows how a buffer register could be constructed using positive edge-triggered D-type flip-flops. This buffer register has actually been constructed using the four D-type flip-flops within a 74175 IC, which was covered in Chapter 24. The 4-bit data input word (1D, 2D, 3D, and 4D) will be stored in the register and appear at the Q output (Q_1, Q_2, Q_3, and Q_4) when the first positive clock edge occurs at the 74175 clock input (pin 9). The 4-bit Q output word drives a 4-bit LED display, which will display the contents of the 4-bit buffer register. Referring to the waveforms in

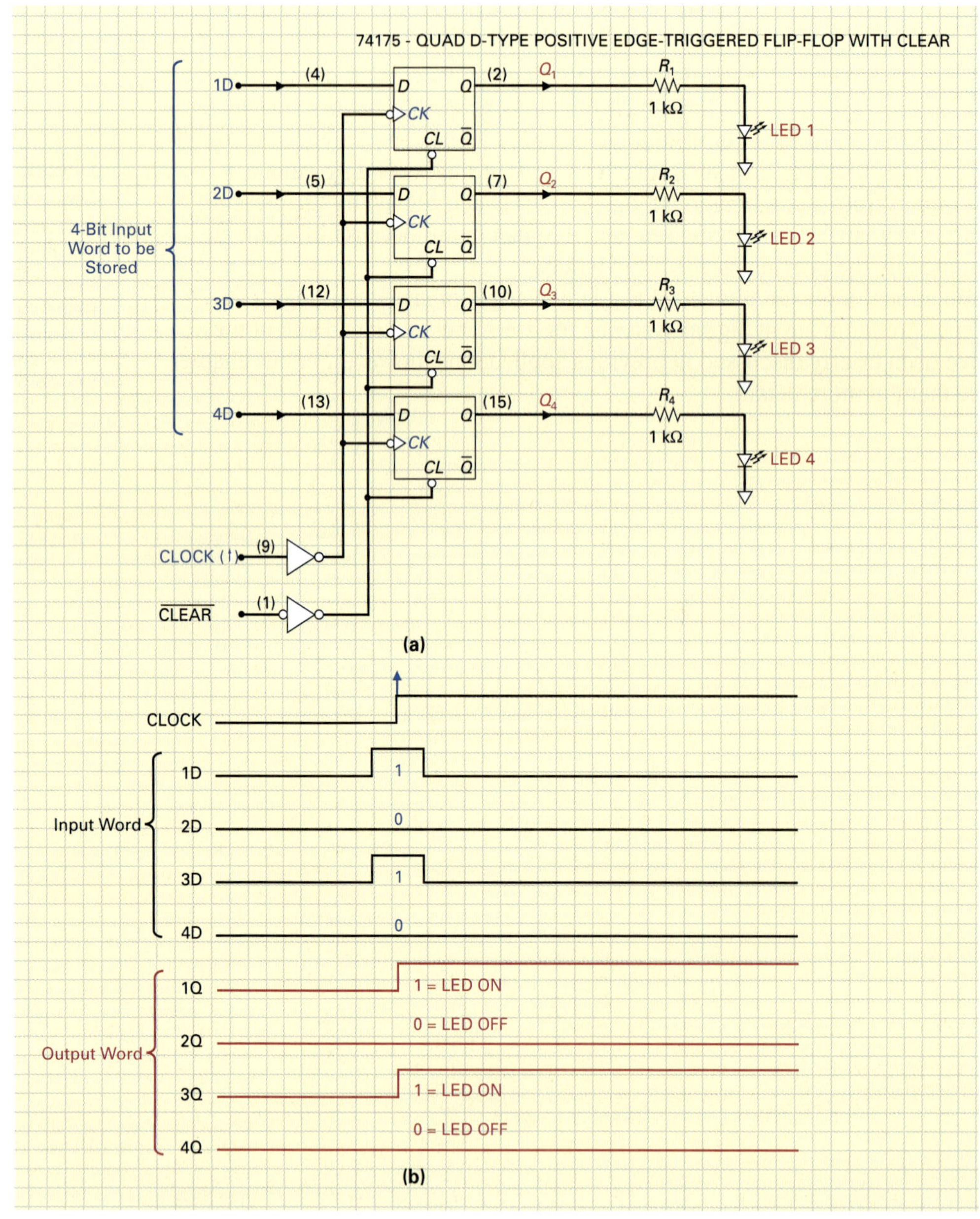

FIGURE 25-1 A Buffer Register. (a) 74175 Parallel-In, Parallel-Out Register. (b) Timing Waveforms.

Figure 25-1(b), you can see that the 4-bit data input, which is only momentarily present on the input data lines (1D, 2D, 3D, and 4D), is stored, or retained, in the buffer register when a rising clock edge occurs. This data is permanently present at the output, and therefore displayed on the LED display.

Since a 4-bit parallel word is delivered to the input of the register, and a 4-bit parallel word is present at the output, buffer registers are often referred to as **parallel-in, parallel-out (PIPO) registers.** The active-LOW clear input of the 74175 can be used to clear all of the flip-flops (erase the stored word) and therefore turn OFF all of the LEDs.

Parallel-In, Parallel-Out (PIPO) Register
A register that will parallel-load an input word and parallel-output its stored contents.

SELF-TEST EVALUATION POINT FOR SECTION 25-1

Use the following questions to test your understanding of Section 25-1.

1. Buffer registers are also referred to as __________ registers.
   a. SISO b. PISO c. PIPO d. SIPO
2. What is the main function of a buffer register?

25-2 SHIFT REGISTERS

A **shift register** is a storage register that will move or shift the bits of the stored word either to the left or to the right. The three basic types of shift registers are

a. serial-in, serial-out (SISO) shift registers
b. serial-in, parallel-out (SIPO) shift registers
c. parallel-in, serial-out (PISO) shift registers

Shift Register
A temporary data storage circuit able to shift or move the stored word either left or right.

Each of these shift register classifications is shown in Figure 25-2. The names given to these three types actually describe how data is entered, or inputted, into the shift register

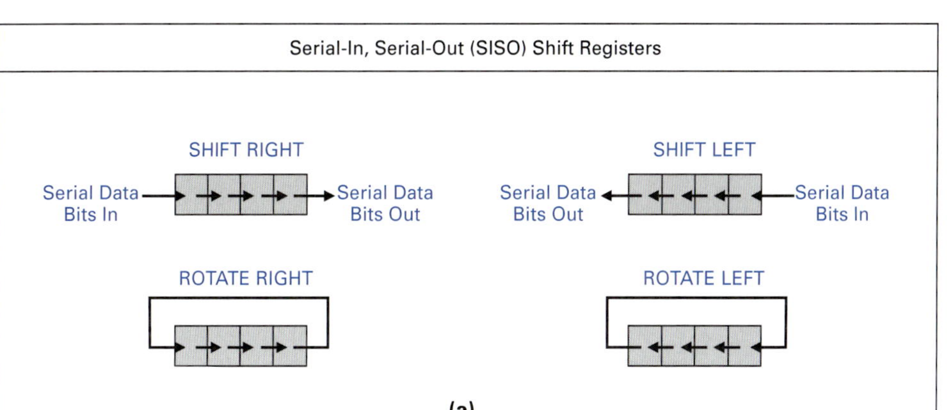

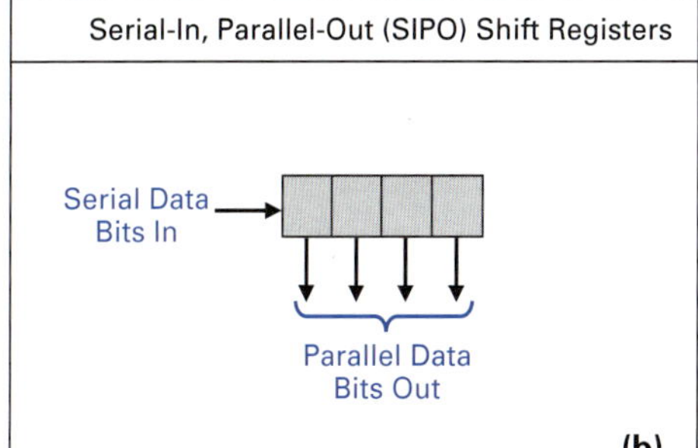

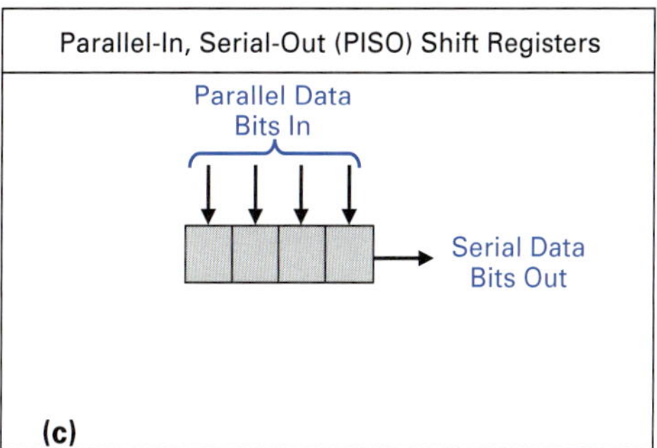

FIGURE 25-2 **Shift Register Classifications. (a) SISO. (b) SIPO. (c) PISO.**

for storage, and how data exits, or is outputted from, the shift register. To understand these operations in more detail, let us take a close look at each of these three shift registers in the following sections.

25-2-1 *Serial-In, Serial-Out (SISO) Shift Registers*

To begin with, Figure 25-3(a) shows how a shift register operates. In this example, the initial condition shows that the 4-bit word 0110 is stored within the 4-bit shift register,

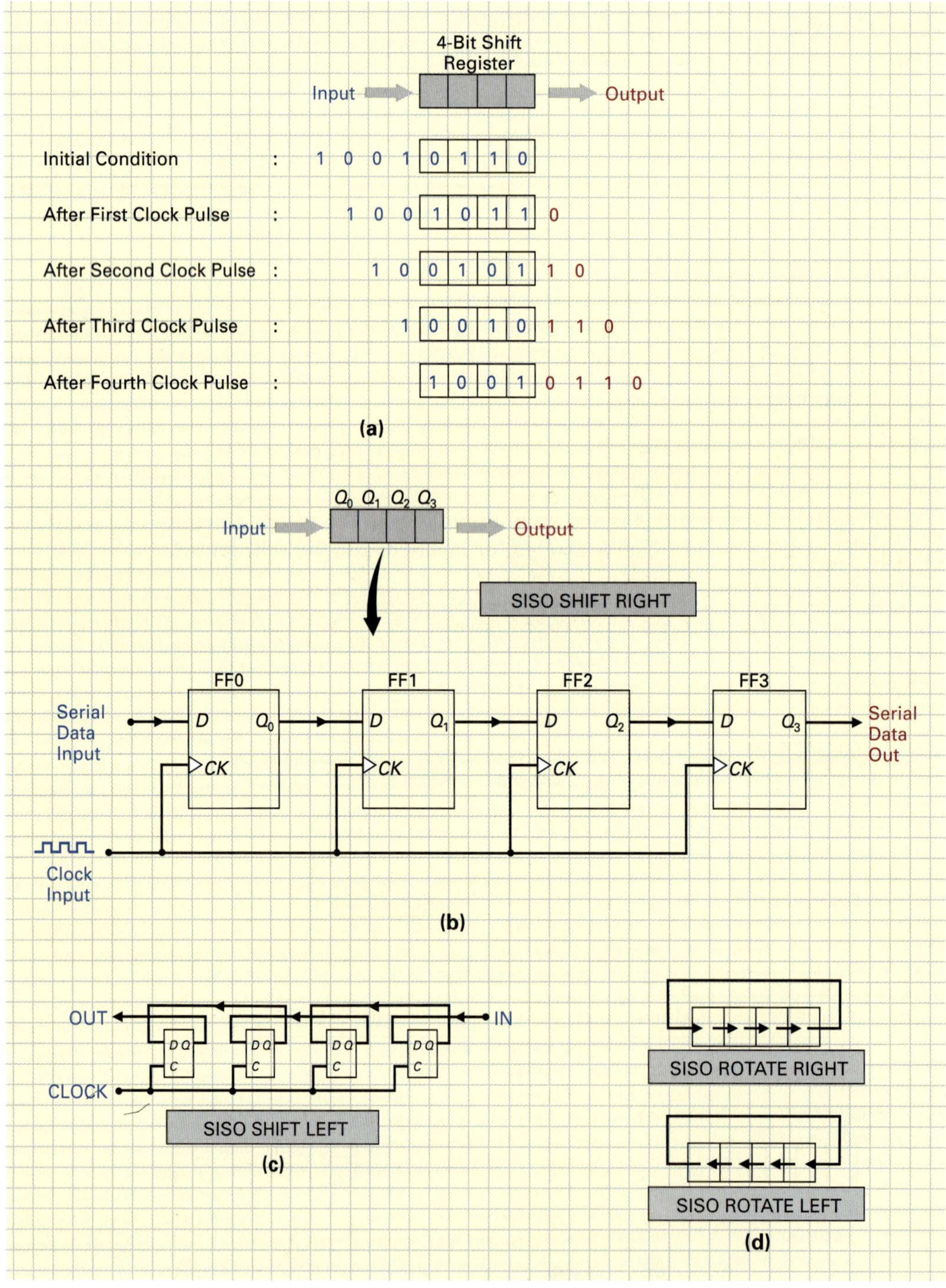

FIGURE 25-3 Serial-In, Serial-Out (SISO) Shift Registers.

while the external serial word 1001 is being applied to the shift register's input. After the first clock pulse, the data stored within the shift register is shifted one position to the right, and the first bit of the applied serial word is shifted into the first position of the shift register. After the second clock pulse, two bits of the stored 0110 word have been shifted out of the shift register, while two bits of the applied 1001 word have been shifted into the shift register. After the third clock pulse, three shift-right operations have occurred. After four clock pulses, the originally stored 0110 word has been completely shifted out of the register, while the applied 1001 input word has been completely shifted into the register and is now being stored.

Now that the basic operation of the shift register is understood, let us see how we can use flip-flops to construct a shift register circuit. Figure 25-3(b) shows how a 4-bit shift register circuit can be constructed using four D-type flip-flops. The serial data input is applied to the D input of flip-flop 0 (FF0). The output from FF0 (Q_0) is applied to the D input of FF1, the output of FF1 (Q_1) is applied to the D input of FF2, the output of FF2 (Q_2) is applied to the D input of FF3, and the output of FF3 (Q_3) is the final serial data output of the 4-bit shift register. The clock input is applied simultaneously to all of the D-type flip-flops, with each positive edge of the clock causing the stored 4-bit word to shift one position to the right. With this **serial-in, serial-out shift register,** data is serially fed into the register for storage under the control of the clock (1 bit is stored per clock pulse, so four clock pulses will be needed to shift in a serially applied 4-bit word). To extract a stored 4-bit word from the shift register, four clock pulses will have to be applied to serially shift the stored word out of the register.

Serial-In, Serial-Out (SISO) Register
A register that will serial-load an input word, and serial-output its stored contents.

In summary, the circuit in Figure 25-3(b) shows how four D-type flip-flops should be connected to form a *SISO shift-right shift register.* Figure 25-3(c) shows how the *D* and *Q* connections can be altered to form a *SISO shift-left shift register.* In some applications, the serial data outputs of the circuits shown in Figure 25-3(b) and (c) are connected directly back to the serial data inputs so that the outputted data is immediately inputted. These SISO operations are called *SISO rotate-right* and *SISO rotate-left* and are shown in Figure 25-3(d).

The 7491A shown in Figure 25-4 is an 8-bit SISO shift-right shift register IC. *S-R* flip-flops are used to form an 8-bit SISO shift register, which is positive edge-triggered and has a gated *A* and *B* data input. Since the *A* and *B* data inputs are ANDed, both inputs must be HIGH to shift in a binary 1, while either input can be LOW to shift in a binary 0.

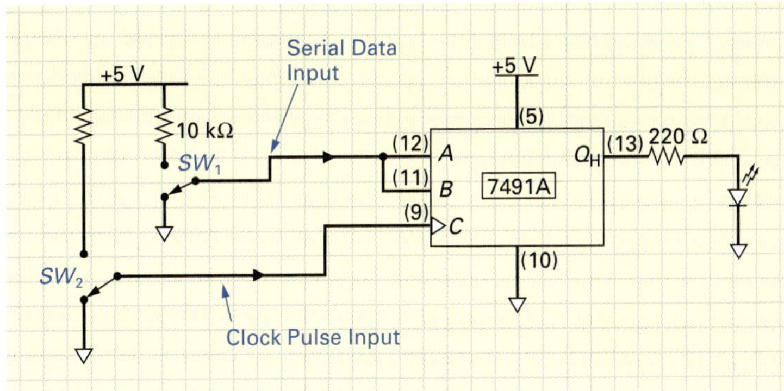

FIGURE 25-4 **Test Circuit for a 7491A Shift Register.**

EXAMPLE:

Referring to the circuit shown in Figure 25-4, describe how you would

a. input or store the word "11001010" within the 7491A SISO shift register.

b. output the stored 8-bit word so that it can be displayed on the LED.

■ *Solution:*

This test circuit operates as follows. SW_1 controls the data input to the IC, while SW_2 controls the dock input to the IC. If SW_1 is in the upper position, a HIGH will be applied to the IC's serial data input, whereas if SW_1 is in the lower position, a LOW will be applied to the IC's serial data input. To shift-right the data stored within the 7491A, and therefore load the input data applied by SW_1, SW_2 should be momentarily moved into its upper position and then placed back in its lower position. Switching SW_2 up and then down in this way will apply a positive edge to the clock input of the 7491A and therefore all of the *S-R* flip-flops within the 7491A.

a. To store the word "11001010" within the 7491A SISO shift register, first apply a LOW (LSB) from SW_1 and then clock the IC using SW_2. Then apply a HIGH and clock, then a LOW and clock, and then a HIGH and clock, and so on until all eight bits are inputted.

b. To display the previously stored 8-bit word on the LED, simply clock the 7491A eight times, and each bit of the word 11001010 will be shifted right and appear in turn at the 7491A output (pin 13). Since the 8-bit word's LSB was the first to be applied to the input of the shift register, it will be the first to appear at the output of the shift register.

25-2-2 *Serial-In, Parallel-Out (SIPO) Shift Registers*

Serial-In, Parallel-Out (SIPO) Register

A register that will serial-load an input word and parallel-output its stored contents.

Figure 25-5 shows the second shift register type, which is called a **serial-in, parallel-out (SIPO) shift register.** As can be seen in the simplified diagram in Figure 25-5(a), data is entered into this shift register type in serial form, and the stored data is available at the output as a parallel word.

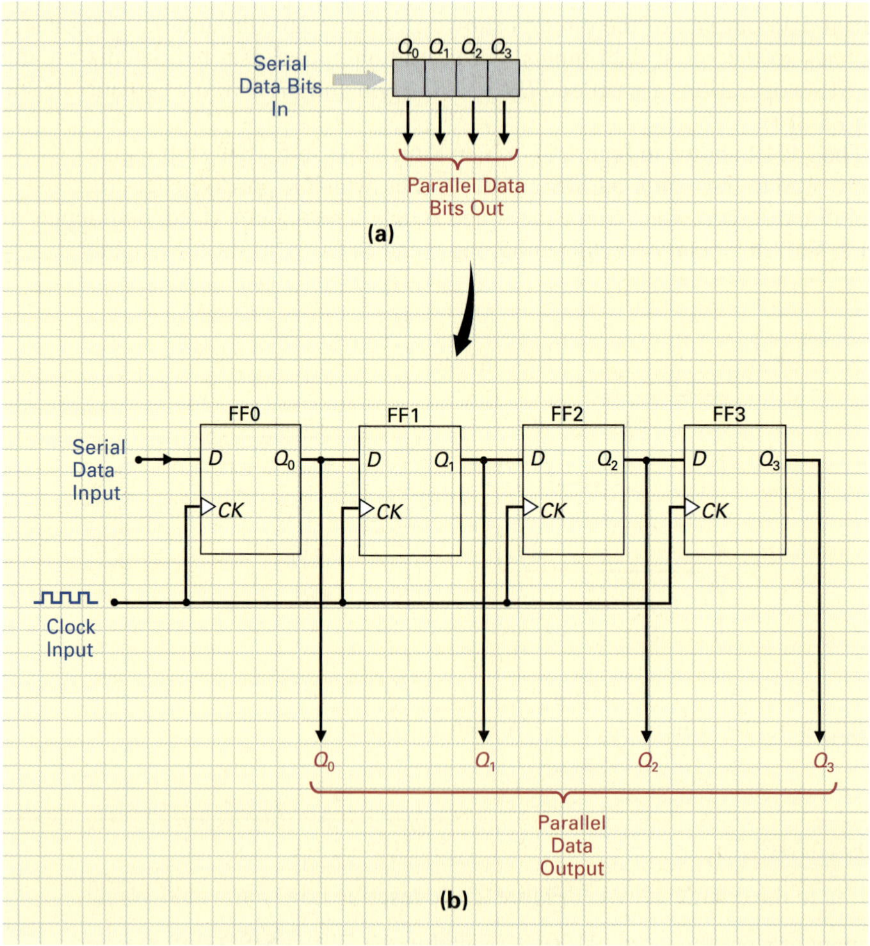

FIGURE 25-5 **Serial-In, Parallel-Out (SIPO) Shift Registers.**

Figure 25-5(b) shows how a 4-bit SIPO shift register can be constructed using D-type flip-flops. To input data into this shift register, a 4-bit serial word is applied at the serial data input and shifted in under the control of the clock input (one shift-right/clock pulse). To input or store a 4-bit serial word into this shift register, therefore, would require four clock pulses. The data stored within the shift register is available at the four Q outputs (Q_0, Q_1, Q_2, and Q_3) as a 4-bit parallel data output.

Figure 25-6 features the 74164, which is an 8-bit SIPO shift-right shift register IC.

Application—Serial-to-Parallel Data Converter

Figure 25-6(a) shows how a 74164 can be used to convert a serial data bitstream into a parallel data word output. Referring to the waveforms in Figure 25-6(b), you can see that one bit of serial data is loaded into the 74164 for each positive transition of the clock. The clock signal is also fed to a divide-by-eight circuit which divides the clock input by eight and loads the 8-bit data from the 74164 into an 8-bit buffer register (made up of two 74175s) at 8-bit intervals.

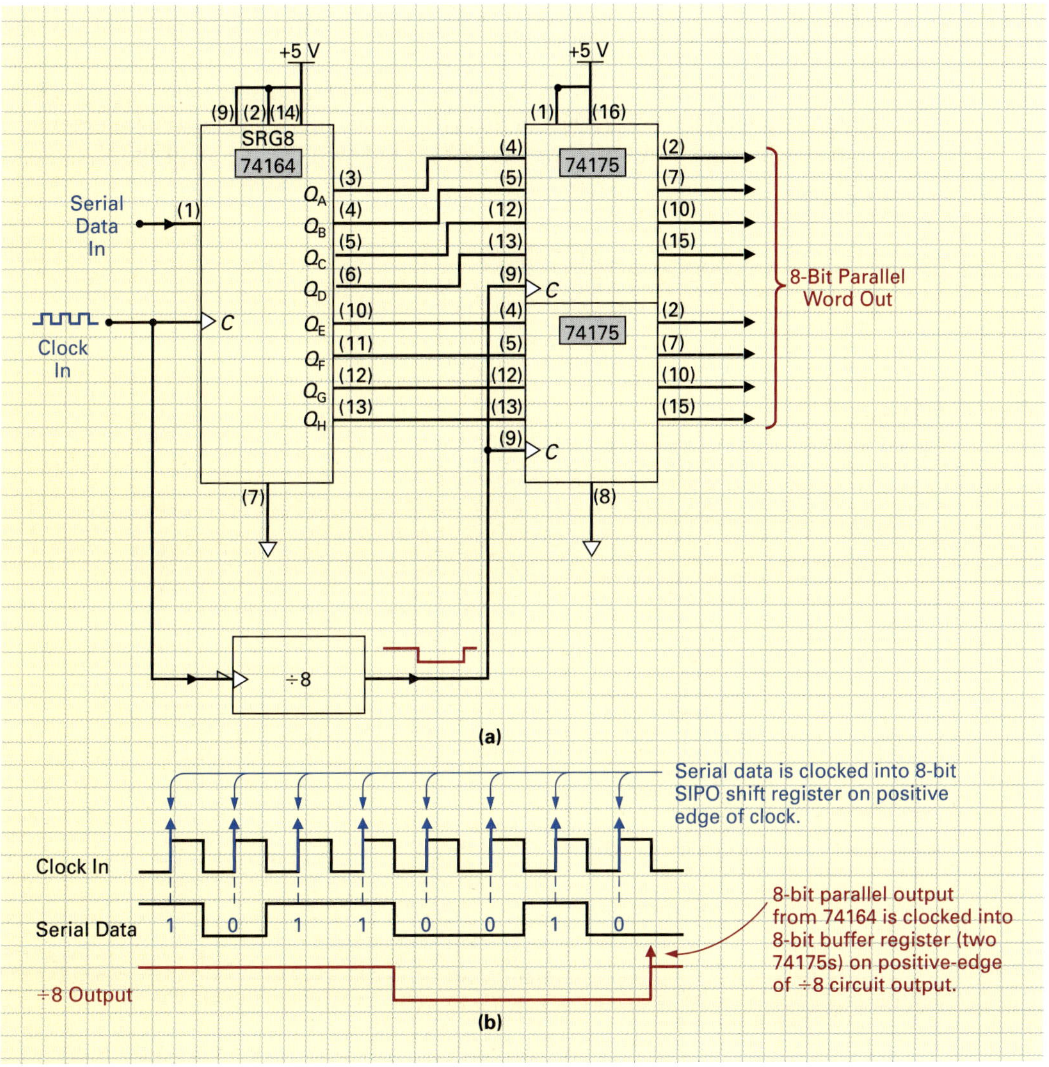

FIGURE 25-6 Application—Serial-to-Parallel Data Converter. (a) Circuit. (b) Waveforms.

25-2-3 *Parallel-In, Serial-Out (PISO) Shift Registers*

Parallel-In, Serial-Out (PISO) Register
A register that will parallel-load an input word and serial-output its stored contents.

The simplified diagram in Figure 25-7(a) shows the basic operation of a **parallel-in, serial-out shift register.** With this shift register type, the data bits are entered simultaneously, or in parallel, into the register, and then the binary word is shifted out of the register bit by bit as a serial data stream.

Figure 25-7(b) shows how a 4-bit PISO shift register can be constructed using D-type flip-flops. The circuit is controlled by the $\mathrm{SHIFT}/\overline{\mathrm{LOAD}}$ control input. When the $\mathrm{SHIFT}/\overline{\mathrm{LOAD}}$ control line is LOW, all of the shaded AND gates are enabled, due to the inversion of this control signal by the shaded INVERTER. These enabled AND gates connect the 4-bit parallel input word on the data input lines (D_0, D_1, D_2, and D_3) to the data inputs of the flip-flops. When a clock pulse occurs, this 4-bit parallel word will be latched simultaneously into the four flip-flops and appear at the Q outputs (Q_0, Q_1, Q_2, and Q_3). When the $\mathrm{SHIFT}/\overline{\mathrm{LOAD}}$ control line is HIGH, all of the unshaded AND gates are enabled. These enabled AND gates connect the Q_0 output to the D input of FF_1, the Q_1 output to the D input of FF_2, and the Q_2 output to the D input of FF_3. In this mode, therefore, the data stored in the shift register will be shifted to the right by one position for every clock pulse applied.

The 74165 IC can function as either an 8-bit parallel-in (inputs A-H), serial-out (output Q_H), or an 8-bit serial-in (input SER) shift register. A LOW on the $SH/\overline{LD}$ (pin 1) control line will enable all of the NAND gates and allow simultaneous loading of the eight S-R flip-flops. A LOW on any of the parallel A through H inputs will RESET the respective flip-flop, while a HIGH on any of the parallel A through H inputs will SET the respective flip-flop. The loading of the register is called an *asynchronous operation* since the loading operation will not be in step with the occurrence of an active clock signal (the loading operation and clock are not synchronized). A HIGH on the $SH/\overline{LD}$ line will disable all of the NAND gates and shift

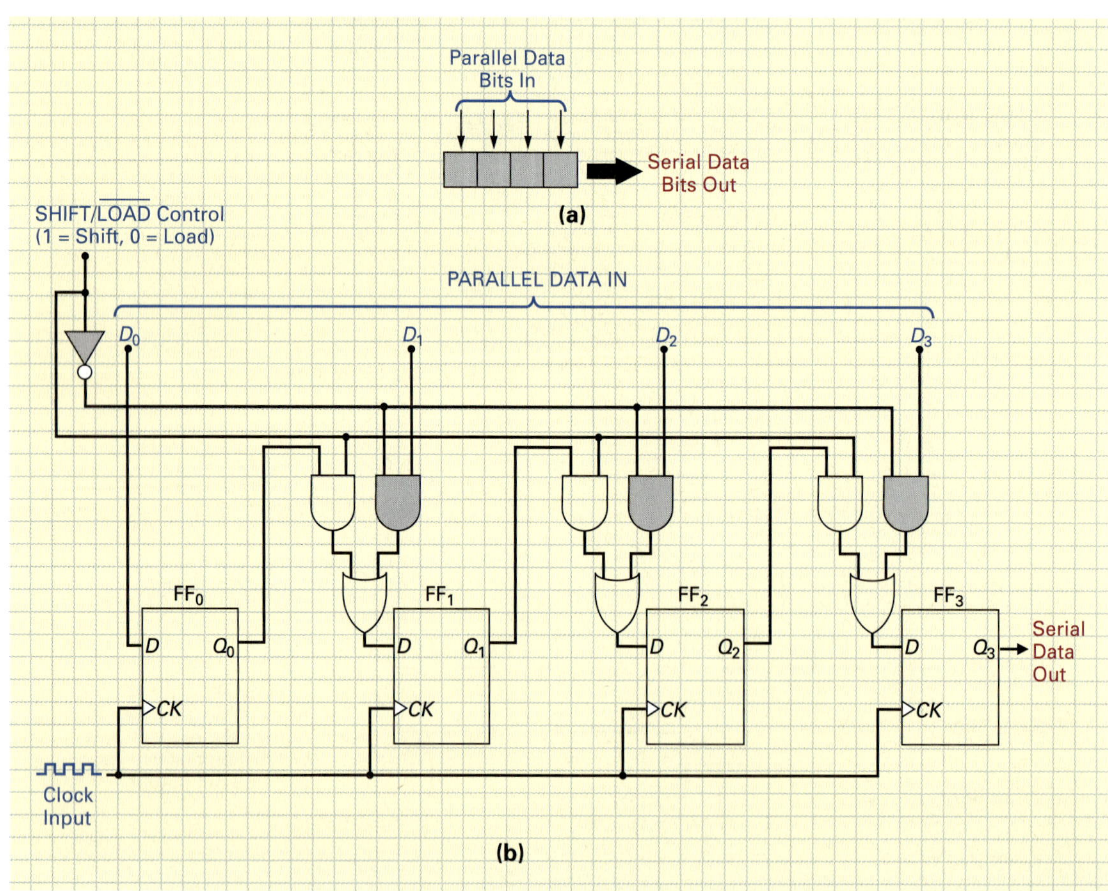

FIGURE 25-7 Parallel-In, Serial-Out (PISO) Shift Registers.

the stored 8-bit word to the right at a rate of one position for every one clock pulse. The shifting of data within the shift register is called a *synchronous operation* since the shifting of data will be in step or in sync with the occurrence of an active clock signal (the shifting operation and clock are synchronized). The clock input for this IC can be inhibited, if desired, by applying a HIGH to the *CLK INH* (pin 15) input.

Application—Parallel-to-Serial Data Converter

Figure 25-8 shows how a 74165 can be connected to convert an 8-bit parallel data input into an 8-bit serial data output. In this example, switches have been used to supply an 8-bit parallel input of 11010110 to the *A* through *H* inputs of the 74165. Moving SW_1 to the LOAD position will make pin 1 of the 74165 ($SH/\overline{LD}$) LOW, so the 8-bit word applied to the parallel inputs *A* through *H* will be loaded into the 74165's eight internal flip-flops. Since the LSB *H* input is LOW, the output LED will initially be OFF since Q_H is LOW. The 74165 is positive edge-triggered, and therefore as SW_2 is toggled, the stored bits will be shifted to the output (in the order *H, G, F, E, D, C, B,* and then *A*) at a rate of one bit per HIGH/LOW switching of SW_2.

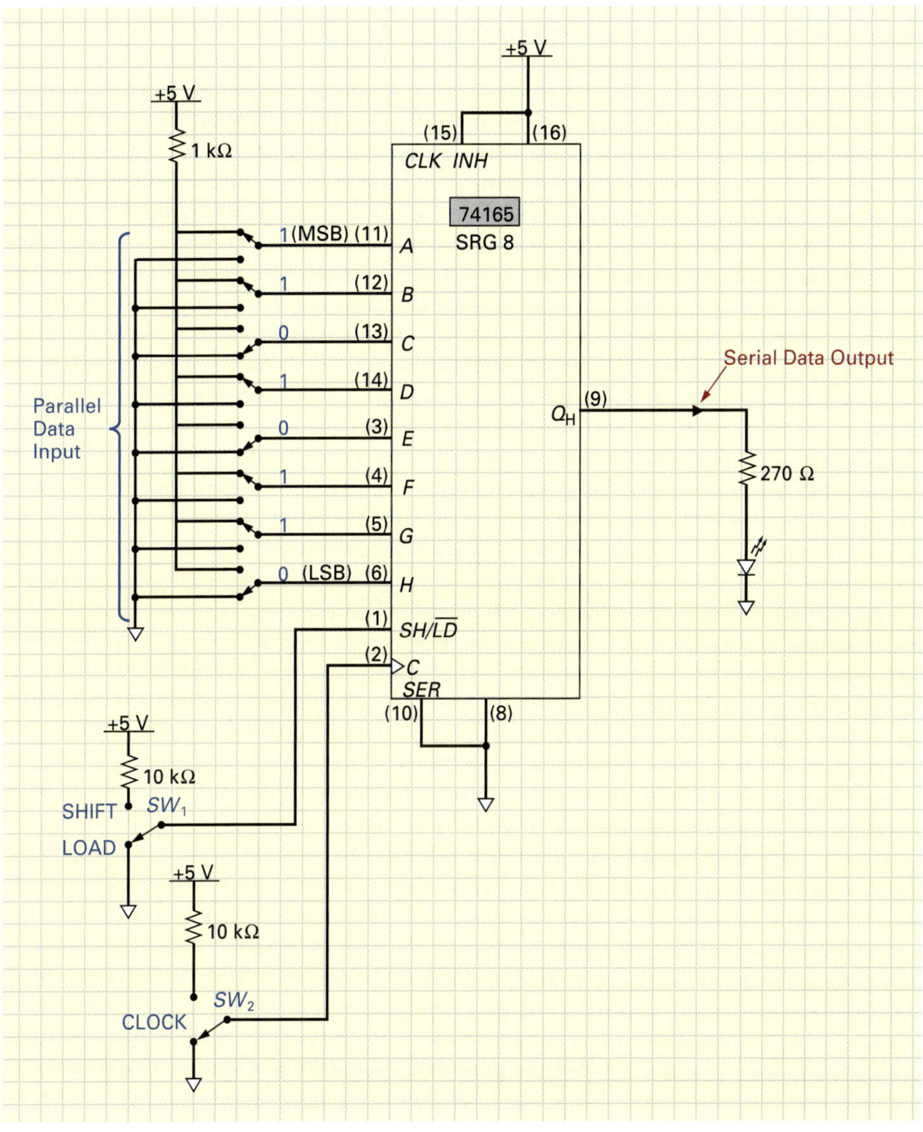

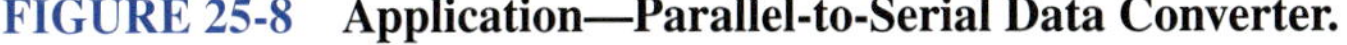

FIGURE 25-8 **Application—Parallel-to-Serial Data Converter.**

25-2-4 *Bidirectional Universal Shift Register*

A bidirectional shift register can shift stored data either left or right depending on the logic level applied to a control line. The 74194 IC shown in Figure 25-9 is frequently used in applications requiring bidirectional shifting of data. This IC contains forty-six gates and can be made to operate in one of four modes by applying different logic levels to the mode control inputs S_0 and S_1. These modes of operation are listed below.

To parallel-load a 4-bit word, therefore, the S_1 and S_0 mode control inputs must both be HIGH, and the 4-bit word to be loaded must be applied to inputs *A, B, C,* and *D*. Parallel loading is accomplished synchronously (in step or in sync) with an active clock signal, which for the 74194 is a positive clock transition. Once loaded, the stored 4-bit word will appear at the outputs Q_A, Q_B, Q_C, and Q_D. During a parallel-loading operation, serial data flow is inhibited.

S_1	S_0	MODE OF OPERATION
H	H	Parallel (broadside) load
L	H	Shift right (Q_A is shifted toward Q_D)
H	L	Shift left (Q_D is shifted toward Q_A)
L	L	Inhibit clock (do nothing)

To shift data to the right, the mode control input S_1 must be LOW and the mode control input S_0 must be HIGH. In this mode, data within the shift register will be shifted right synchronously from Q_A to Q_D, with new data being inputted from the shift-right serial input (pin 2).

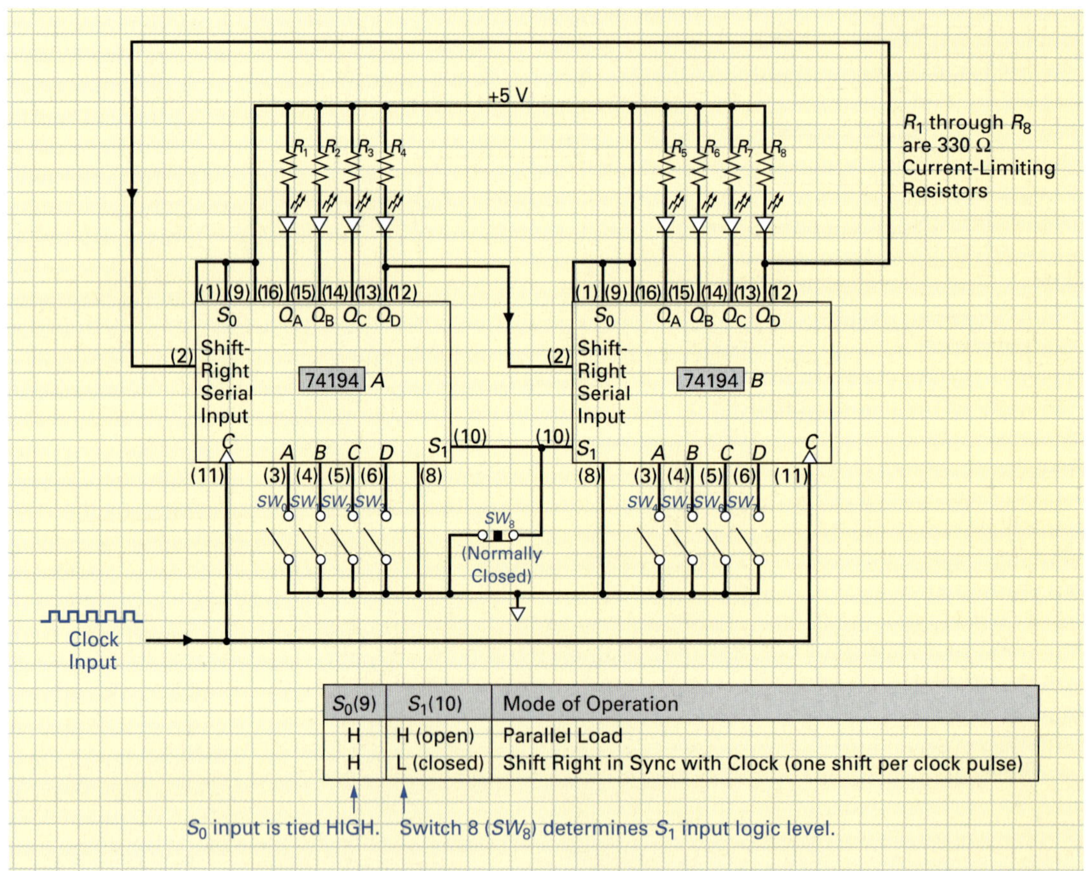

S_0(9)	S_1(10)	Mode of Operation
H	H (open)	Parallel Load
H	L (closed)	Shift Right in Sync with Clock (one shift per clock pulse)

FIGURE 25-9 Application—Shift-Right Sequence Generator.

To shift data to the left, the mode control input S_1 must be HIGH and the mode control input S_0 must be LOW. In this mode, data within the shift register will be shifted left synchronously from Q_D to Q_A, with new data being inputted from the shift-left serial input (pin 7).

When both the mode control inputs S_1 and S_0 are LOW, the clocking of the shift register is inhibited. The mode control inputs S_1 and S_0 should only be changed when the clock input is HIGH (inactive).

Application—Shift Register Sequence Generator

Figure 25-9 shows how two 74194 ICs could be connected to form an 8-bit shift-right sequence generator. Switches SW_0 through SW_7 can be set so that they apply any 8-bit pattern or word value (open switch = HIGH, closed switch = LOW). This 8-bit word can be loaded into the shift register by pressing the normally closed push-button (NCPB) SW_8. SW_8 controls the logic level applied to the mode control input S_1 (pin 10), as shown in the table in Figure 25-9. With S_0 connected permanently HIGH, an open SW_8 (switch pressed) will apply an equivalent HIGH input to S_1, which will instruct the two 74194s to parallel-load the 8-bit word, which, when loaded by a positive transition of the clock, will appear at the Q_A, Q_B, Q_C, and Q_D outputs. When SW_8 is closed (switch released), the S_1 mode control input will be pulled LOW, instructing both 74194s to shift right the data stored synchronously with each positive edge of the clock. In this shift-right mode, the Q_D output of 74194 *B* on the right of Figure 25-9 is connected back to the shift-right serial input of the 74194 *A*, and therefore the data is continually rotated right.

The LEDs in Figure 25-9 form a common-anode display, which means that a HIGH output will turn OFF an LED, and a LOW output will turn ON an LED.

SELF-TEST EVALUATION POINT FOR SECTION 25-2

Use the following questions to test your understanding of Section 25-2.

1. How many clock pulses would be needed to serially load an 8-bit SISO shift register with an 8-bit word?
2. The ____________ shift register could be used for serial-to-parallel conversion, while the ____________ shift register could be used for parallel-to-serial conversion. (SISO, SIPO, PISO).
3. How many clock pulses would be needed to parallel-load an 8-bit PISO shift register with an 8-bit word?
4. The ____________ shift register can parallel-load, shift right, shift left, input serially, and output serially.

25-3 THREE-STATE OUTPUT REGISTERS

Many registers have three-state or tri-state output gates connected between the flip-flop's *Q* outputs and the IC's output pins, as illustrated in Figure 25-10(a). The three-state output gate, which was discussed previously in Section 20-1-5, has as its name implies three output states, which are LOW (logic 0), HIGH (logic 1), and float (an open output or HIGH impedance state). With the active-HIGH three-state gates shown in Figure 25-10(a), a HIGH on the enable output control line (*E*) will connect the *Q* outputs from the D-type flip-flops directly through to the *Y* outputs of the IC. On the other hand, when the enable output control line is LOW, the three-state gates are disabled and the *Y* outputs of the IC will float, since they have effectively been disconnected from the *Q* outputs. Figure 25-10(b) shows how a small inverted triangle on each output is used to indicate that this 4-bit controlled buffer register (PIPO) has *three-state output control.*

Referring to Figure 25-10(a), let us now see how this PIPO buffer register with three-state output control operates. Other than the enable output control line (*E*), this register has a load control line (*L*) and a clock input (*C*). When the load control line is LOW (inactive), the contents of each of the D-type flip-flops within the register remain unchanged since the data is recirculated from each *Q* output back to its respective *D* input for each positive edge of the clock input via the enabled unshaded AND gates. When the load control input is HIGH,

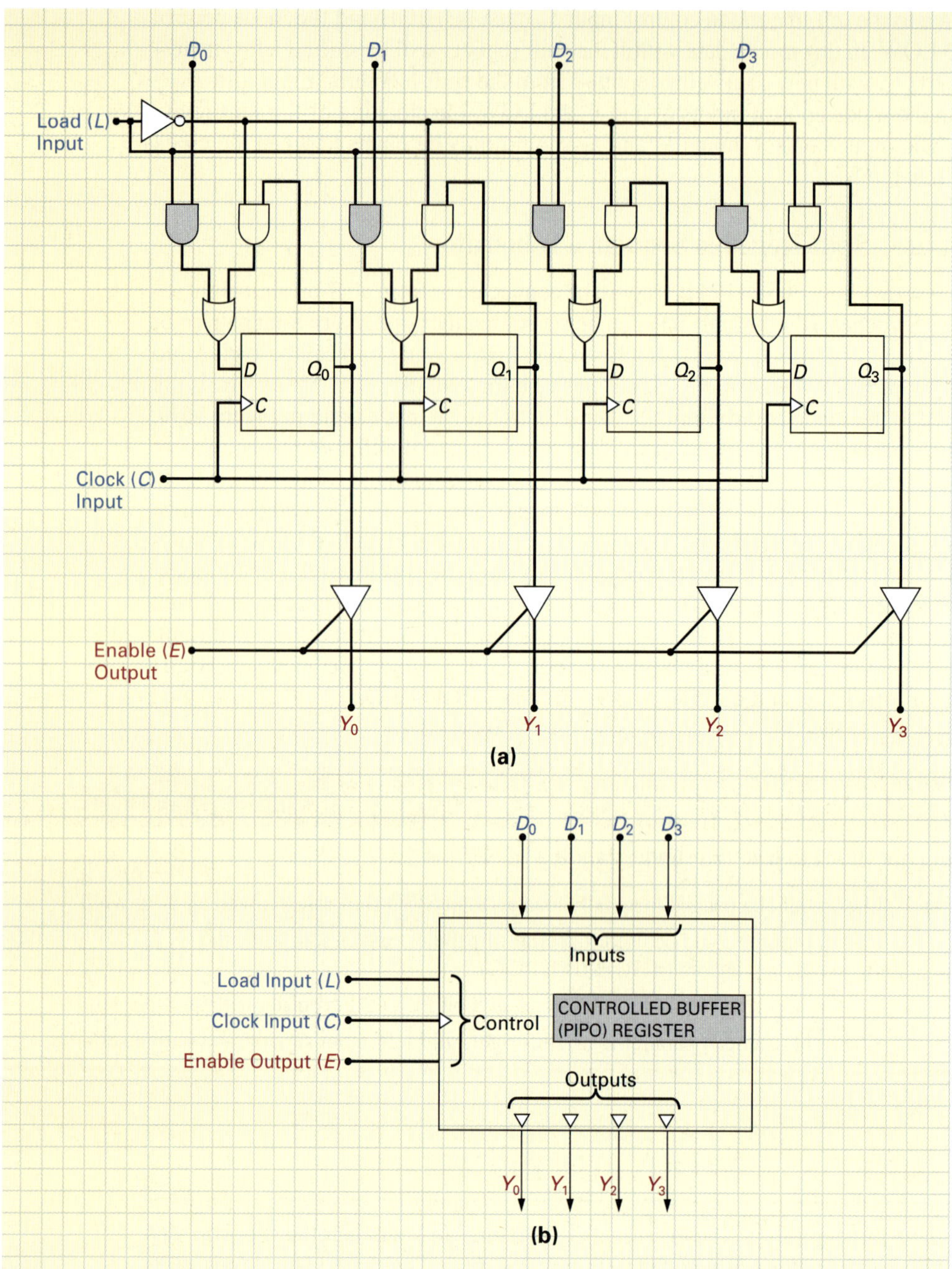

FIGURE 25-10 A Controlled Buffer Register with Three-State Output.

Bus-Organized Digital System

A digital system interconnection technique in which a common bus is time-shared by all the devices connected to the bus.

Bus

A group of conductors used as a path for transmitting digital information from one of several sources to one of several destinations.

however, all of the shaded AND gates are enabled, so the parallel input word on the D_0, D_1, D_2, and D_3 inputs are connected to the flip-flop data inputs and latched into the flip-flops on the next positive edge of the clock.

The three-state gate was originally developed by National Semiconductor in the early 1970s, and its purpose was to reduce the number of wires in a digital system. Three-state gates made possible a **bus-organized digital system.** To explain this concept, Figure 25-11 shows a simple circuit consisting of four input switches, three registers, and four output LEDs. A group of wires used to carry binary data is called a **bus,** and in Figure 25-11 a 4-bit bus is used as a common transmission path between the input device, output device, and the registers. Since the input device and the registers all have three-state outputs, all of these devices (unless enabled) are effectively isolated or disconnected from the bus. If they were not

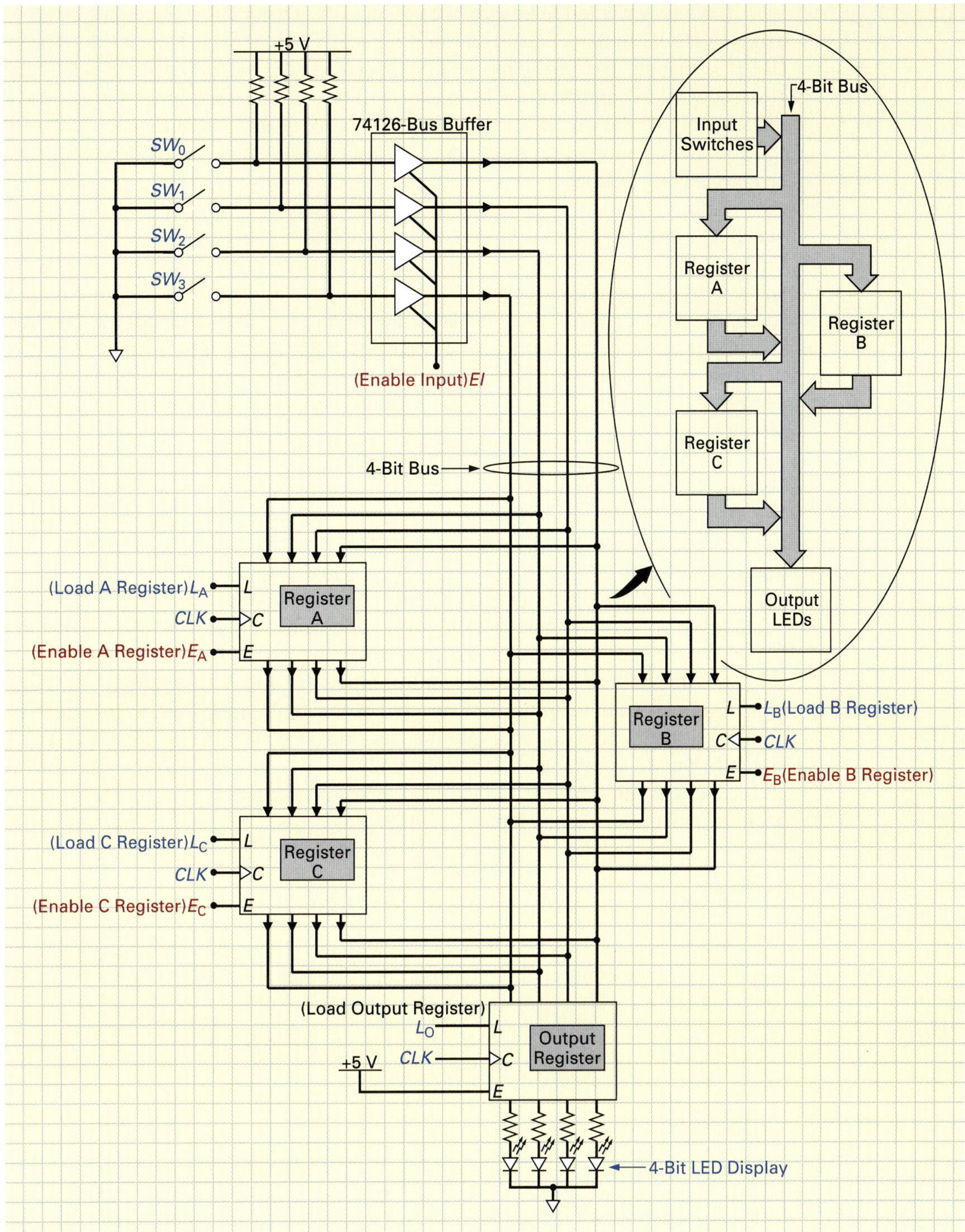

FIGURE 25-11 **A Bus-Organized Digital System.**

disconnected, a *bus conflict* would occur since one register could be driving a bus line HIGH while another register could be trying to drive that same bus line LOW.

To transfer a 4-bit word from register A to register B, we must connect the contents of register A on to the bus, making E_A HIGH, and then load this data into register B by making L_B HIGH. Since the clock (CLK) is connected to all registers, the next positive edge of the clock will load the 4-bit word from register A into register B.

If we now wanted to transfer the contents of register B to the output, we would simply make E_B HIGH so that contents of the B register would now be connected to the bus and L_O HIGH so that the data on the bus will be loaded into the output register. On the next positive clock edge, the contents of the B register will be stored in the output register and displayed immediately on the output LEDs, since the output of the output register has no three-state control. The output register does not connect to the common bus, so its output can therefore be permanently enabled.

Before three-state outputs were available, separate groups of wires were needed between all of the devices in a circuit. For example, without three-state control, the circuit in Figure 25-11 would need a set of wires connecting register A to register B, another set connecting A to C, another set connecting B to C, another set connecting the input to A, and so on. As you can easily imagine, even the smallest of digital electronic systems containing only several devices would need a mass of connecting wires. With a bus-organized digital system, only one set of connecting wires is needed, and these can be connected to all devices as long as those devices that transmit data onto the common bus have three-state output control.

In Figure 25-11, only a 4-bit bus was used. In many digital systems 8-bit (8 wires), 16-bit (16 wires), and 32-bit (32 wires) buses are common and are used to transfer 8-bit, 16-bit, or 32-bit parallel words from one device to another. In schematic diagrams, a solid bar with arrows is generally used to represent a bus of any size, as shown in the inset in Figure 25-11.

The 74173 IC shown in Figure 25-12 is a 4-bit (PIPO) buffer register with three-state output control. This IC contains four D-type flip-flops with an active-HIGH clear input (pin 15). The three-state output buffers are controlled by the active-LOW inputs *M* and *N* (pins 1 and 2). When both of these output control inputs (*M* and *N*) are LOW, the HIGH or LOW logic levels stored in the four flip-flops will be available at the four outputs (pins 3, 4, 5, and 6). The parallel loading of data into the flip-flops is controlled by the active-LOW ANDed inputs $\overline{G}_1$ and $\overline{G}_2$ (pins 9 and 10). When both of these data enable inputs are LOW, the data at the D-inputs (pins 14, 13, 12, and 11) are loaded into their respective flip-flops on the net positive transmission clock input.

Application—A Bus-Organized Digital System

Figure 25-12 shows how four 74173s and one 74126 (quad buffer gates with three-state outputs) could be connected to construct the bus-organized digital system discussed previously in Figure 25-11.

EXAMPLE:

Referring to the circuit in Figure 25-12, list which control lines need to be made active in order to perform the following operations:

a. loading the 4-bit word from the input bus buffer into register A
b. copying the contents of register B into register C
c. displaying the contents of register C

Solution:

a. E_I(HIGH), $\overline{L}_A$(LOW), $CLK\uparrow$
b. E_B(HIGH), $\overline{L}_C$(LOW), $CLK\uparrow$
c. E_C(HIGH), $\overline{L}_O$(LOW), $CLK\uparrow$

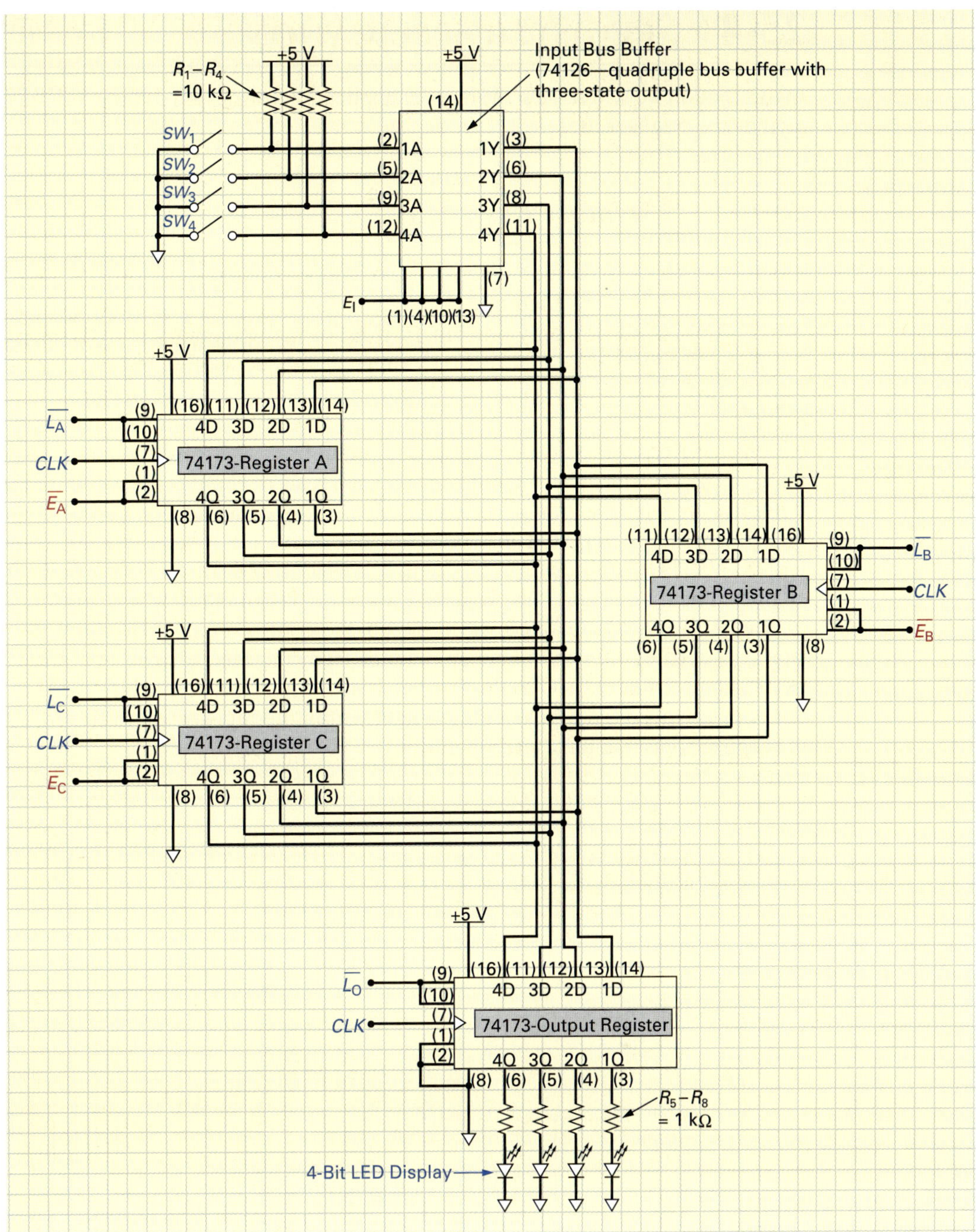

FIGURE 25-12 **Application—A 4-Bit Bus-Organized Digital System Containing Four Input Switches and an Input Buffer, Three PIPO Storage Registers, and an Output Register Driving a Four-LED Display.**

SELF-TEST EVALUATION POINT FOR SECTION 25-3

Use the following questions to test your understanding of Section 25-3.

1. A three-state output can be either HIGH, LOW, or ___________.
2. A group of parallel wires designed to carry binary data is called a ___________.
3. What is bus conflict, and how is it avoided?
4. What are the three main inputs to a controlled buffer register with three-state output control?

25-4 REGISTER APPLICATIONS

In this section we will review the register applications discussed so far in this chapter and examine other register uses.

25-4-1 *Memory Registers*

Memory Register
A register capable of storing several binary words.

Read
An operation in which data is acquired from some source.

Write
An operation in which data is transferred to some destination.

Since a register is basically a storage or memory element used to store a binary word, it is often referred to as a **memory register.** A memory unit may contain one memory register capable of storing only one binary word or many memory registers capable of storing many binary words. Figure 25-13 shows how a controlled buffer register PIPO with three-state output control can be configured to function as a memory register. In order to perform as a memory register, a register must be able to achieve two basic operations, which are referred to as **read** (which is an input operation, since you are reading the contents of the register in the same way that we read or input words in a book) and **write** (which is an output operation, since you are writing data into the register in the same way that you write or output words onto a pad). These two operations are controlled by a READ/$\overline{\text{WRITE}}$ ($R/\overline{W}$) control line, which is used to control the buffer register's load and enable control inputs, as shown in Figure 25-13. When the $R/\overline{W}$ control line is LOW, the load control input is enabled while the enable control input is disabled, so new data is stored or written into the memory register from the bus (a write operation). When the ($R/\overline{W}$) control line is HIGH, the enable control input is active while the load control input is inactive, so the 8 bits of stored data within the register are retrieved, or read, from the register onto the bus (read operation).

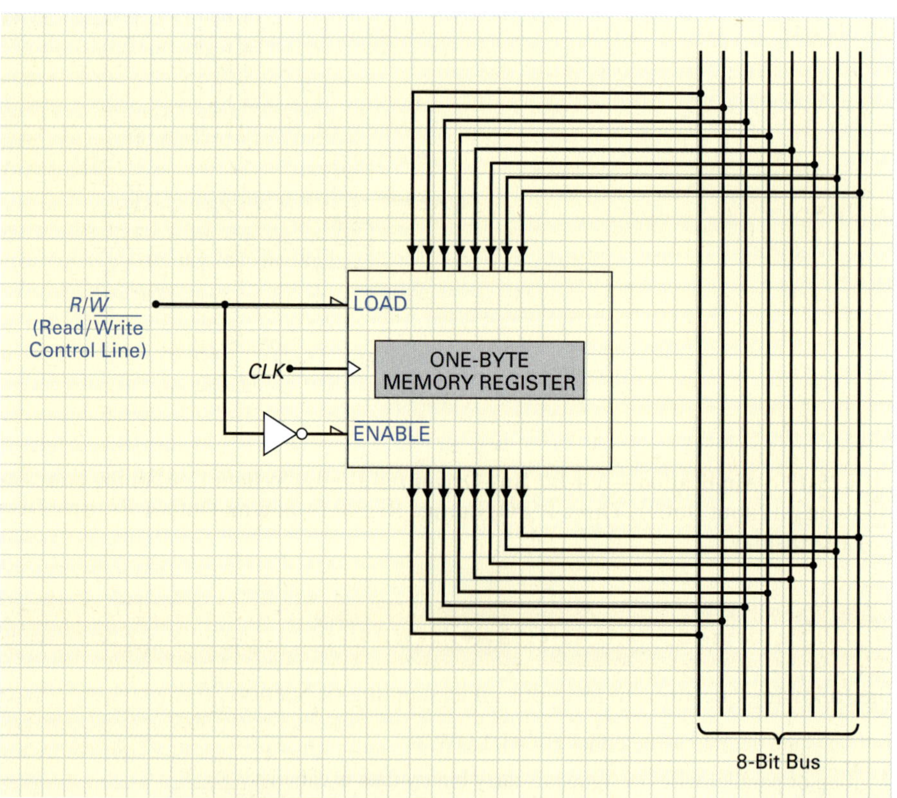

FIGURE 25-13 **Application—A One-Byte (8-Bit) Memory Register.**

25-4-2 *Serial-to-Parallel and Parallel-to-Serial Conversions*

Figure 25-14(a) shows how a SIPO shift register can be used for serial-to-parallel conversion. This shift register shifts in an 8-bit serial input at a rate of 1 bit per clock pulse, and after eight clock pulses all of the byte is stored and all bits are simultaneously available as an 8-bit parallel output word. Figure 25-14(b) shows how a PISO shift register can be used for

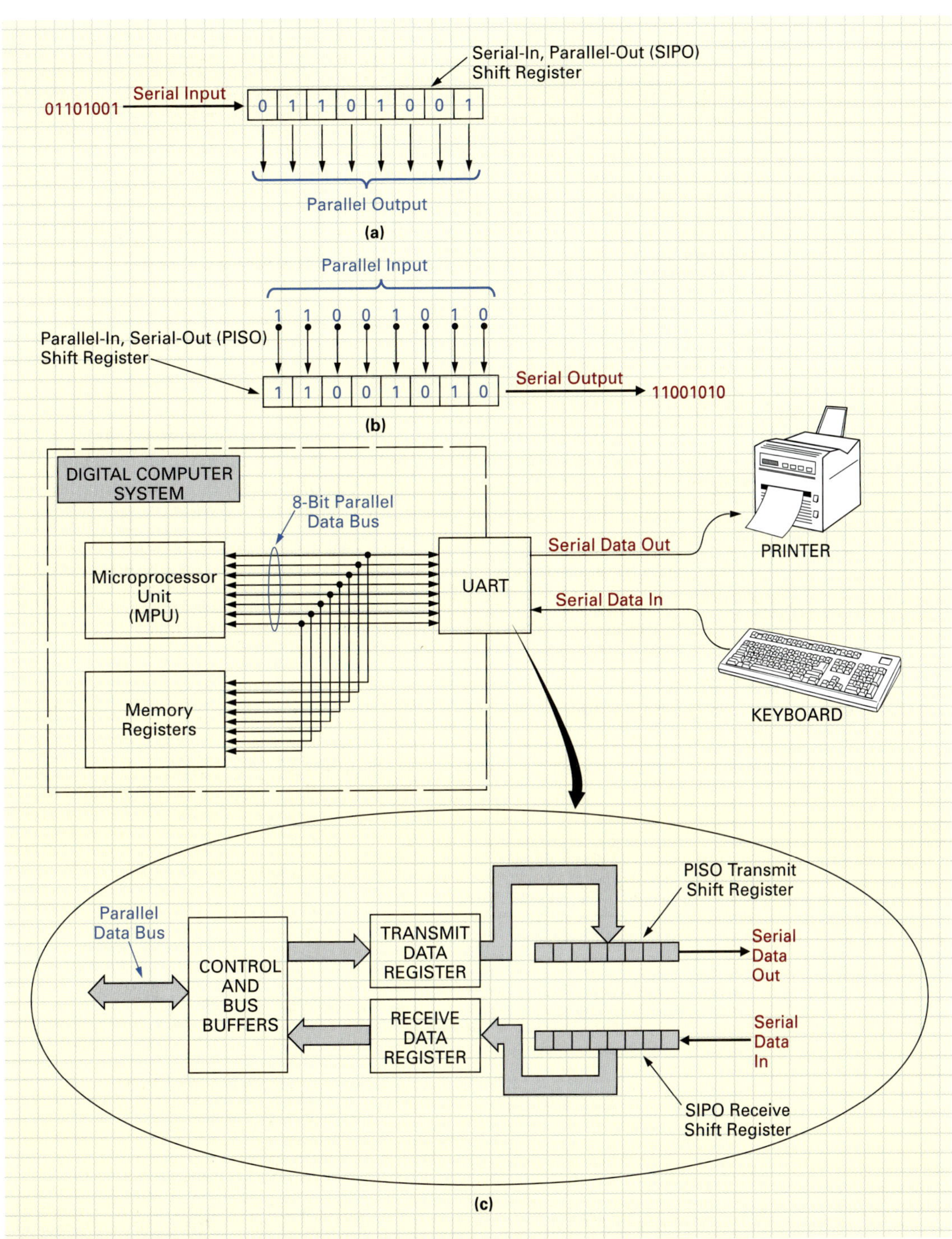

FIGURE 25-14 Application—Serial-to-Parallel and Parallel-to-Serial Conversions. (a) SIPO. (b) PISO. (c) UART.

parallel-to-serial conversion. This shift register simultaneously loads all 8 bits of a parallel input and then shifts the data out serially at a rate of 1 bit per clock pulse, and therefore after eight clock pulses the byte has been outputted.

UART
An abbreviation for universal asynchronous receiver/transmitter. A communication device that can interface a parallel-word system to a serial communication network.

The ability of a shift register to convert binary data from serial-to-parallel or parallel-to-serial is made use of in an interfacing (data transfer) device known as a **UART,** which is an abbreviation for **universal asynchronous receiver and transmitter.** Data within a digital system is generally transferred from one device to another in parallel format on a bus, as shown in Figure 25-14(c). In many cases these digital electronic systems must communicate with external equipment that handles data in serial format. Under these circumstances, a UART can be used to interface the parallel data format within a digital electronics system to the serial data format of external equipment. The inset in Figure 25-14(c) shows the basic block diagram of a UART's internal circuit. In this particular example, the UART will be receiving data in parallel format from the microprocessor in the digital system via an 8-bit data bus. This parallel data will be converted to serial data and transmitted to a printer. The serial data received by the UART from a keyboard is converted to parallel data and then transmitted to the digital system microprocessor via the parallel 8-bit data bus. To carry out these operations, you can see from the inset in Figure 25-14(c) that the UART makes use of a PISO shift register to transmit serial data and a SIPO shift register to receive serial data in.

25-4-3 *Arithmetic Operations*

The shift register can also be used to perform arithmetic operations such as multiplication and division. Since binary is a base-2 number system, binary numbers can be changed by some power of 2 by simply moving the position of the number. For example, the contents of the upper shift register in Figure 25-15(a) is 00000111 (decimal 7). By shifting the contents once to the left and shifting a zero into the least significant bit (LSB) position, we can multiply the initial number of 7 by 2 to obtain 14 ($7 \times 2 = 14$). If we continue to shift to the left and add zeros, the value within the shift register in Figure 25-15(a) is continually doubled. The multiplying factor is equal to 2^n, where n is the number of times the contents are shifted. In the example in Figure 25-15(a), the contents of the shift register are shifted three times, and therefore the original number is multiplied by $2^3=8$ ($7 \times 8 = 56$).

If shifting a binary value to the left causes it to double ($\times$ 2), then shifting a binary value to the right should cause it to halve ($\div$ 2). As an example, the contents of the upper shift register in Figure 25-15(b) is 00011000 (decimal 24). By shifting the contents once to the

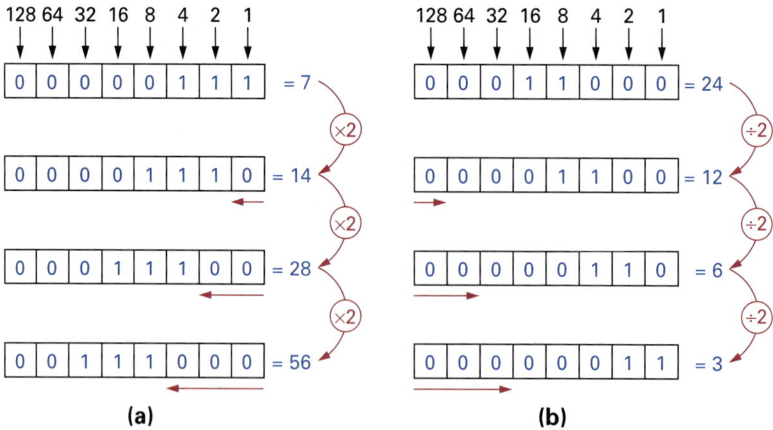

FIGURE 25-15 **Application—Shifting Binary Numbers. (a) Left to Multiply by 2. (b) Right to Divide by 2.**

right, and shifting a zero into the most significant bit (MSB) position, we can divide the initial number of 24 by 2 to obtain 12 (24 2 = 12). If we continue to shift to the right and add zeros, the value within the shift register in Figure 25-15(b) is continually halved. The divide ratio is also equal to 2^n, where n is the number of times the contents are shifted. In the example in Figure 25-15(b), the contents of the shift register are shifted three times, and therefore the original number is divided by $2^3 = 8$ ($24 \div 8 = 3$).

25-4-4 *Shift Register Counters/Sequencers*

One common application of shift registers is as a **sequencer,** in which the parallel outputs of the shift register are used to generate a sequence of equally spaced timing pulses. When connected to operate in this way, shift registers are often referred to as counters, since they generate a specific sequence of logic levels at their parallel outputs. Two of the most widely used shift register *counters/sequencers* are the ring counter and the johnson counter, both of which we will now discuss.

Sequencer
A circuit that generates a sequence of equally spaced timing pulses.

The Ring Counter

Figure 25-16(a) shows how four D-type flip-flops could be connected to form a **ring counter** or **shift register sequencer.** All shift register counters have a feedback connection from the output of the last flip-flop (Q_3) to the input of the first flip-flop (D_0), as shown in Figure 25-16(a). This characteristic circle or "ring" that causes the stored data to be constantly cycled through the shift register accounts for the name *ring counter.*

Ring Counter
A shift register sequencer circuit in which the position of a binary 1 moves through the counter in an ordered sequence.

The circuit shown in Figure 25-16(a) operates as follows. When the active-LOW START control input is brought LOW, Q_0 will be preset HIGH and Q_1, Q_2, and Q_3 will be cleared LOW, as shown in the waveforms in Figure 25-16(b). The stored word (1000) is then shifted right as clock pulses are applied, and since a feedback loop exists between Q_3 and D_0, the single HIGH "start bit" will be continually recirculated. This ring counter action is shown in the waveforms in Figure 25-16(b) and the logic state table in Figure 25-16(c).

Figure 25-17 shows how two 74194s (4-bit bidirectional universal shift registers) could be connected to form an 8-bit ring counter. The ring or feedback connection is made by connecting Q_D of 74194 B back to the shift-right serial input of 74194 A. This ring counter circuit is initiated by pressing SW_1, which will put a HIGH on the mode control S_1. The HIGH applied to the S_1 input will instruct both 74194s to parallel-load their *A, B, C,* and *D* inputs, and since only the *A* input of 74194 A is HIGH, the word 10000000 will be stored. When SW_1 is released, the normally closed switch will connect a LOW to the S_1 mode inputs, instructing them to shift right the stored data. The stored 8-bit value 10000000 will be shifted to the right in sync with the positive edges of the clock input, and the result is shown in the table below.

	74194 A				74194 B			
CLOCK PULSES	Q_A	Q_B	Q_C	Q_D	Q_A	Q_B	Q_C	Q_D
0	1	0	0	0	0	0	0	0
1	0	1	0	0	0	0	0	0
2	0	0	1	0	0	0	0	0
3	0	0	0	1	0	0	0	0
4	0	0	0	0	1	0	0	0
5	0	0	0	0	0	1	0	0
6	0	0	0	0	0	0	1	0
7	0	0	0	0	0	0	0	1

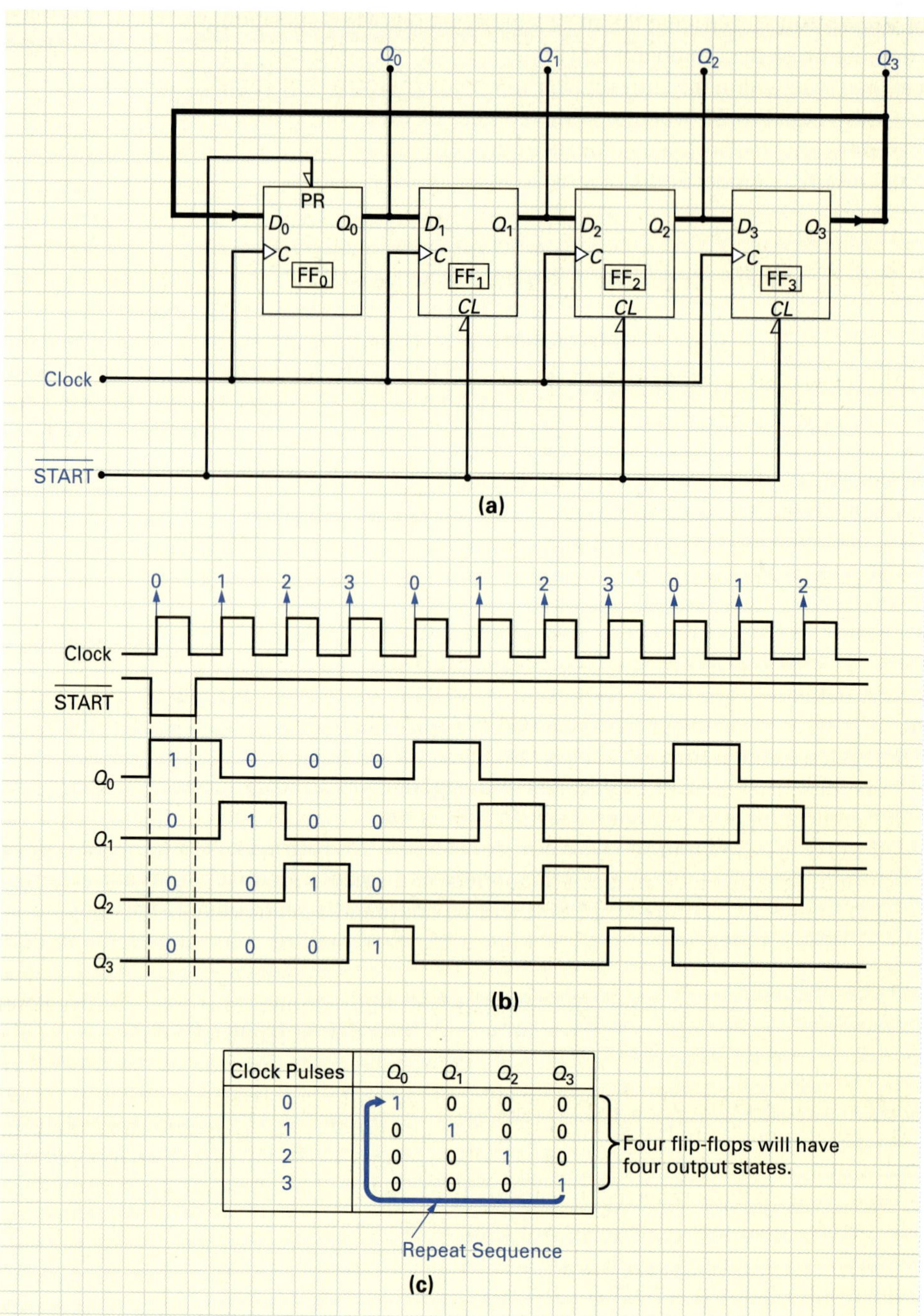

Clock Pulses	Q_0	Q_1	Q_2	Q_3
0	1	0	0	0
1	0	1	0	0
2	0	0	1	0
3	0	0	0	1

FIGURE 25-16 Application—the Ring Counter. (a) Basic Circuit. (b) Timing Waveforms. (c) Logic State Table.

In this example circuit, the outputs Q_A, Q_B, Q_C, and Q_D of both 74194s are used to sequence ON and OFF LEDs (common-anode display, HIGH = LED OFF, LOW = LED ON). In a real digital circuit application, the outputs would be connected to logic circuits, and the ON/OFF sequence generated by the ring counter would control when each of these logic circuits is sequenced ON and then OFF. Since the frequency of the clock pulses is fixed, a sequence of equally spaced timing pulses would be generated to control the external logic circuits.

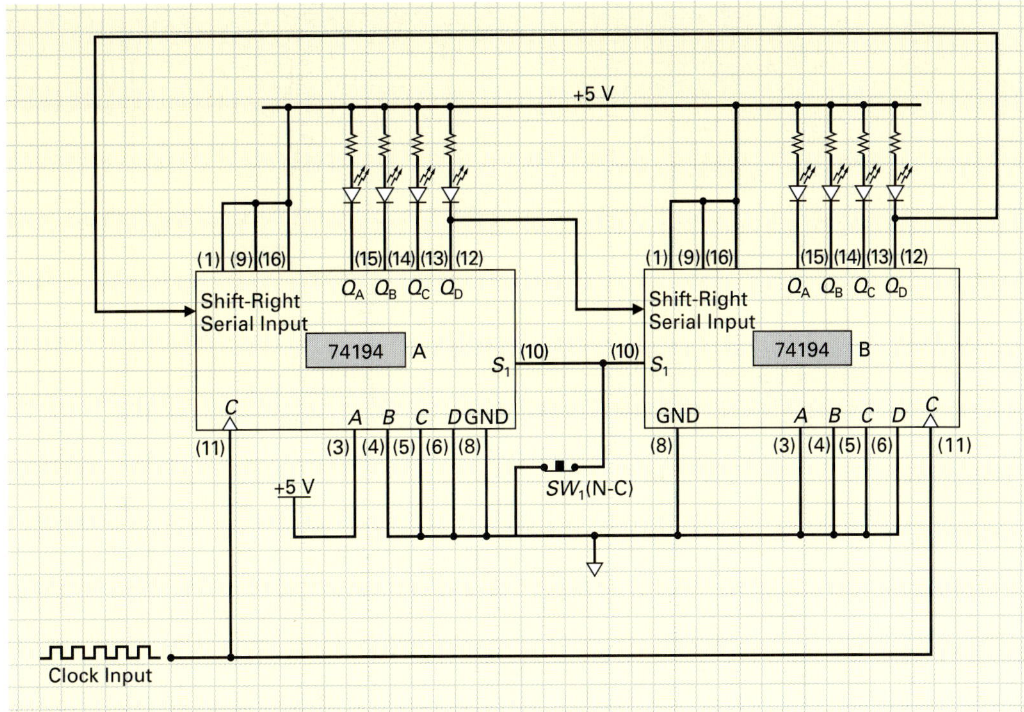

FIGURE 25-17 **Application—8-Bit Ring Counter Using Two 74194s.**

The Johnson Counter

Figure 25-18(a) shows how four D-type flip-flops could be connected to form a **johnson counter.** As you can see from this illustration, the johnson counter is constructed in almost exactly the same way as the ring counter except that the inverted output of the last flip-flop ($\overline{Q}_3$) is connected to the input of the first flip-flop (D_0). This physical difference in the feedback path of the johnson counter accounts for why this circuit is often referred to as a *twisted-ring counter* instead of being named after its inventor.

Johnson Counter
Also known as a twisted-ring counter because of its inverted feedback path, it is a circuit that first fills with 1s and then fills with 0s and then repeats.

Like the ring counter, the johnson counter needs a start circuit to initialize the first output sequence, which, as you can see in the timing waveforms in Figure 25-18(b) and the logic state diagram in Figure 25-18(c), is 1000. Since Q_3 is LOW at the start, $\overline{Q}_3$ will be HIGH, and this HIGH will be fed back to the D_0 input, so HIGH inputs will be clocked into the shift register in a left-to-right motion until the shift register is full of 1s. When Q_3 finally goes HIGH (after the third clock pulse), $\overline{Q}_3$ will go LOW and so will D_0. The shift register will now shift in LOW inputs in a left-to-right motion until it is fully loaded with 0s. When Q_3 finally goes LOW (after the seventh clock pulse), $\overline{Q}_3$ will go HIGH and therefore so will D_0, causing the entire cycle to repeat.

With the ring counter, the number of different output states is governed by the number of flip-flops in the register, and therefore a four-stage ring counter would have four different output states. Since the johnson counter first fills up with 1s (in four clock pulse inputs) and then fills up with 0s (in the next four clock pulses), the number of different output states is equal to twice the number of flip-flops. In the example circuit in Figure 25-18, the four-stage johnson counter will have eight different output states (2×4 flip-flops $= 8$), as listed in the logic state table in Figure 25-18(c).

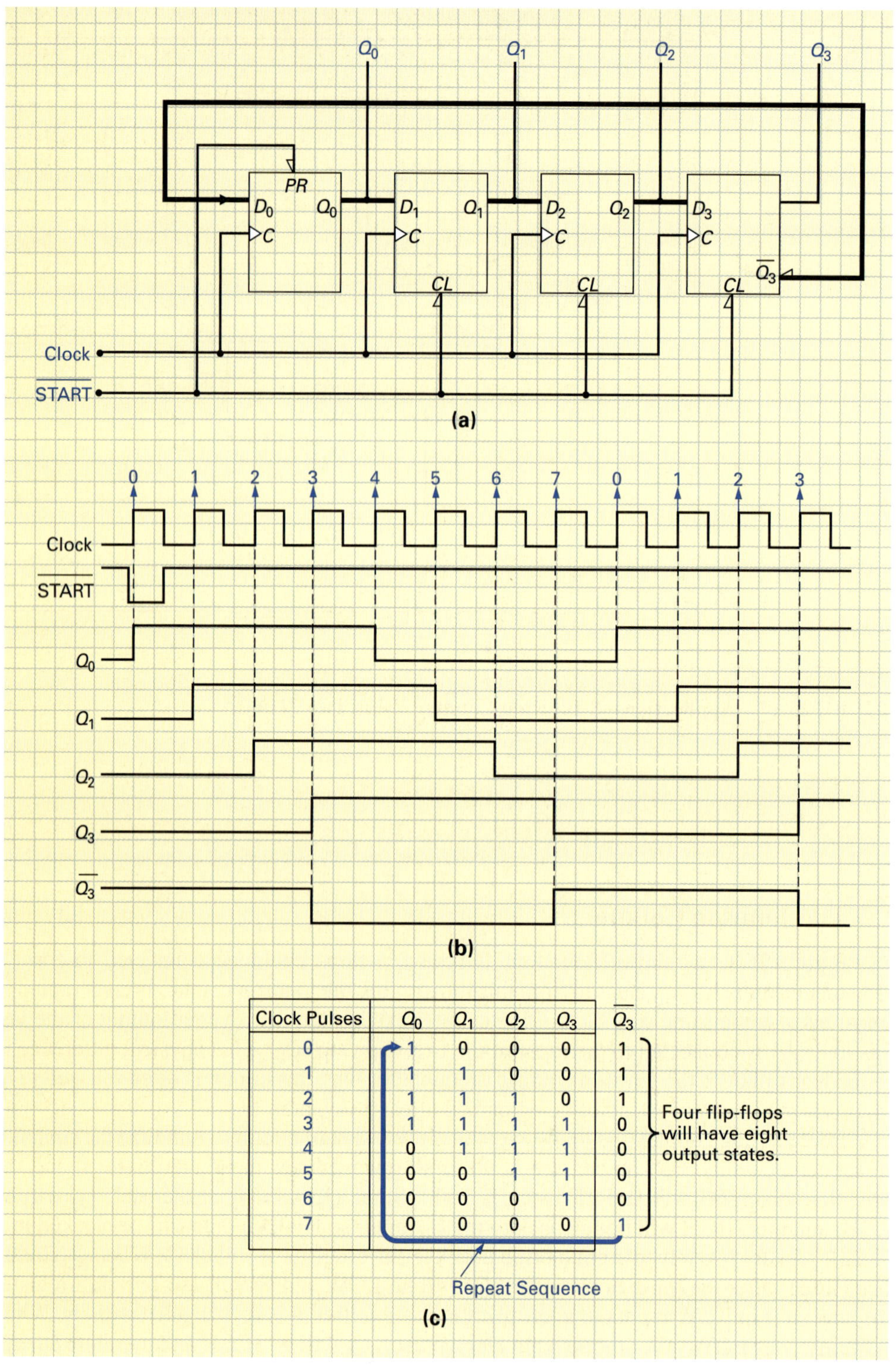

FIGURE 25-18 Application—the Johnson Counter. (a) Basic Circuit. (b) Timing Waveforms. (c) Logic State Table.

SELF-TEST EVALUATION POINT FOR SECTION 25-4

Use the following questions to test your understanding of Section 25-4.

1. What is the function of the load and enable inputs to a controlled buffer register with three-state outputs?
2. Define the following terms.
   a. Bus b. Bus-organized digital system c. Bus conflict d. 16-bit bus
3. Give the full name of the abbreviation UART and briefly describe the device's purpose.
4. If a binary number is shifted six positions to the left, the resulting number will be ____________ (16, 32, 64, 128) times ____________ (larger/smaller).
5. Describe the basic circuit difference between a ring counter and a johnson counter.

25-5 ASYNCHRONOUS COUNTERS

A *counter* is like a register in that it is a sequential logic circuit made up of several flip-flops. A register, however, is designed to store a binary word, whereas the binary word stored in a counter represents the number of clock pulses that have occurred at the clock input. The input clock pulses, therefore, cause the flip-flops of the counter to change state, and by monitoring the outputs of the flip-flops, you can determine how many clock pulses have been applied.

There are basically two types of counters—*asynchronous* and *synchronous*. The key difference between these two counter types is whether or not the operation of the counter is in synchronism with the clock signal. Most of the flip-flops of an asynchronous counter are not connected to the clock signal, and so the operation of this counter is not in sync with the master clock signal. On the other hand, all of the flip-flops of a synchronous counter are connected to the clock signal, and so the operation of this counter is in sync with the master clock signal.

Most of the flip-flops in an **asynchronous counter,** therefore, are not connected to the clock signal, and so the flip-flops do not change state in sync with the master clock signal. To better understand this distinction, let us examine some of the different types of asynchronous counters.

Asynchronous Counter
A counter in which an action starts in response to a signal generated by a previous operation, rather than in response to a master clock signal.

25-5-1 *Asynchronous Binary Up Counters*

Figure 25-19(a) shows how a 4-bit or 4-stage asynchronous binary up counter can be constructed with four *J-K* flip-flops. In this circuit you can see that the *J-K* flip-flops are cascaded, or strung together, in series so that the output of one flip-flop will drive the clock input of the next flip-flop. Since the *J* and *K* inputs of all four flip-flops are tied HIGH, the output of each flip-flop will toggle, or change state, each time a negative edge occurs at the flip-flop's clock input.

The clock input to this circuit and the Q output waveforms for each of the four flip-flops are shown in Figure 25-19(b). The Q_0, Q_1, Q_2, and Q_3 outputs make up a 4-bit word, which we will initially assume is 0000, as shown at the far left of the waveforms and on the first line of the truth table in Figure 25-19(c). The output of FF0 (Q_0) generates the least significant bit (LSB) of the 4-bit output word, while the output of FF3 (Q_3) supplies the most significant bit (MSB) of the 4-bit output word. Since FF0 is driven by the clock input, Q_0 will toggle once for every negative edge of this clock input, as shown in the Q_0 timing waveform. This means that clock input negative edge 1 will cause Q_0 to change from a 0 to a 1, clock input negative edge 2 will cause Q_0 to change from a 1 to a 0, clock input negative edge 3 will cause Q_0 to change from a 0 to a 1, and so on. Since Q_0 is connected to the clock input of FF1, each negative edge of Q_0 will cause the Q_1 output to toggle. Similarly, a negative edge at the Q_1 output will cause the output of FF2 to toggle, and a negative edge at the Q_2 output will cause the output of FF3 to toggle.

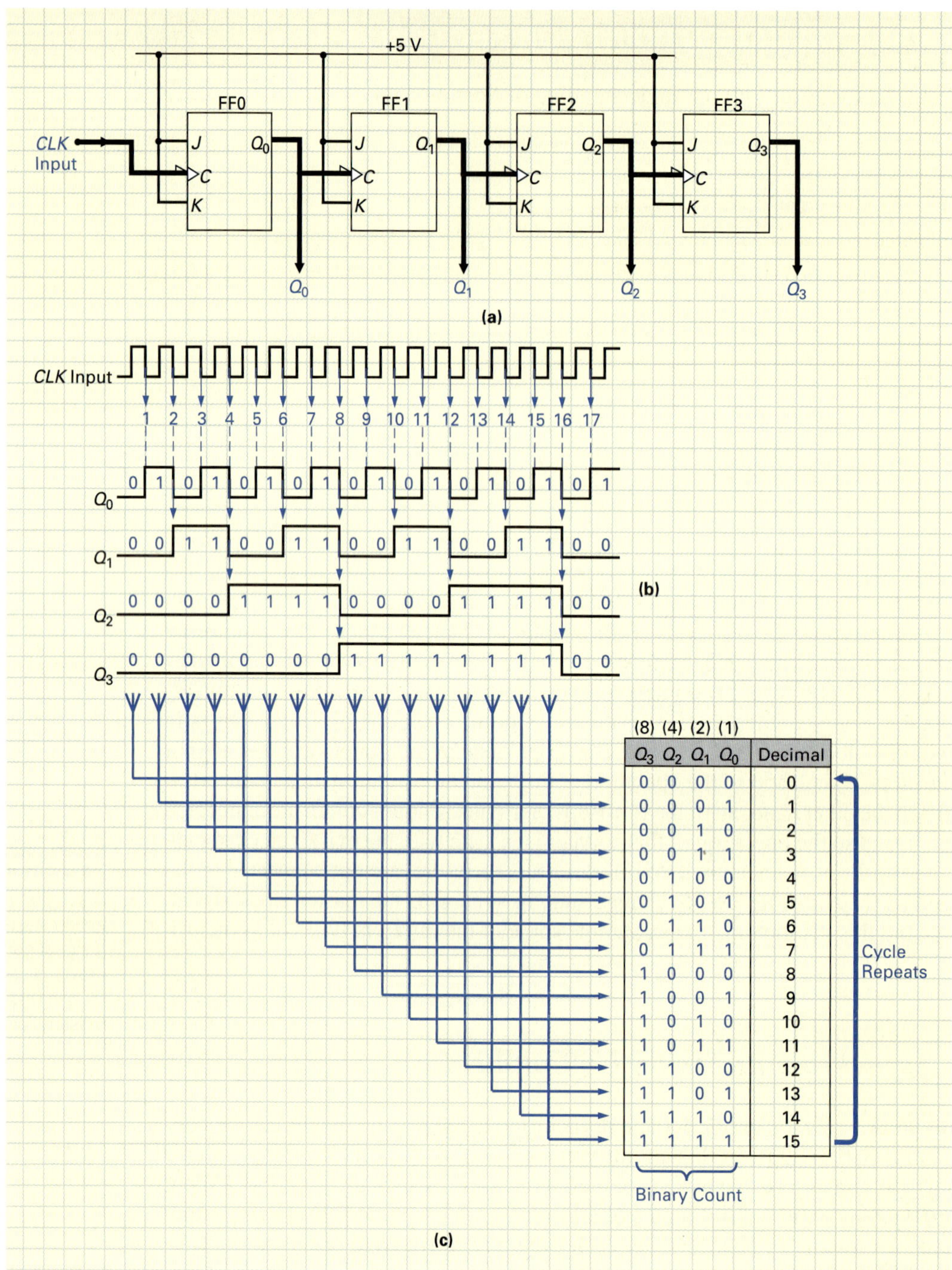

(8) Q_3	(4) Q_2	(2) Q_1	(1) Q_0	Decimal
0	0	0	0	0
0	0	0	1	1
0	0	1	0	2
0	0	1	1	3
0	1	0	0	4
0	1	0	1	5
0	1	1	0	6
0	1	1	1	7
1	0	0	0	8
1	0	0	1	9
1	0	1	0	10
1	0	1	1	11
1	1	0	0	12
1	1	0	1	13
1	1	1	0	14
1	1	1	1	15

FIGURE 25-19 An Asynchronous Binary Up Counter. (a) Logic Circuit. (b) Waveforms. (c) Truth Table.

The Maximum Count (N) of a Counter

Maximum Count

For a counter, the maximum count is $2^n - 1$ or one less than the counter's modulus.

Referring to the truth table in Figure 25-19(c), you may have noticed that the 4-bit binary value appearing at the Q_0, Q_1, Q_2, and Q_3 outputs indicated the number of clock pulses applied to this circuit. For example, before the clock pulse, the output was 0000 (decimal 0). After the first clock pulse, the count was 0001 (decimal 1); after the second clock pulse, the count was 0010 (decimal 2); after the third clock pulse, the count was 0011 (decimal 3); and so on. The **maximum count** (N) of a binary counter is governed by the number of flip-flops in the counter.

$$N = 2^n - 1$$

N = maximum count before cycle repeats

n = number of flip-flops in the counter circuit

With the counter circuit illustrated in Figure 25-19, the maximum count will be

$$\begin{aligned} N &= 2^n - 1 \\ &= 2^4 - 1 \\ &= 16 - 1 \\ &= 15_{10}(1111_2) \end{aligned}$$

This can be verified by referring to the waveforms in Figure 25-19(b) and the last line of the truth table in Figure 25-19(c), which shows the final maximum count of the 4-bit binary output is 1111 (decimal 15). You may have noticed that the maximum binary count that can be output from a binary counter before its cycle repeats is calculated in the same way that we determine the maximum count of a binary word containing a certain number of bits.

EXAMPLE:

Calculate the maximum count of a

a. 1-bit counter
b. 2-bit counter
c. 3-bit counter
d. 4-bit counter
e. 8-bit counter

Solution:

a. A 1-bit counter will have a maximum count of

$$N = 2^n - 1 = 2^1 - 1 = 2 - 1 = 1$$

b. A 2-bit counter will have a maximum count of

$$N = 2^n - 1 = 2^2 - 1 = 4 - 1 = 3$$

c. A 3-bit counter will have a maximum count of

$$N = 2^n - 1 = 2^3 - 1 = 8 - 1 = 7$$

d. A 4-bit counter will have a maximum count of

$$N = 2^n - 1 = 2^4 - 1 = 16 - 1 = 15$$

e. An 8-bit count will have a maximum count of

$$N = 2^n - 1 = 2^8 - 1 = 256 - 1 = 255$$

The Modulus (mod) of a Counter

The **modulus** of a counter describes the number of different output combinations generated by the counter. As an example, the 4-bit counter in Figure 25-19 has a modulus of 16, since the counter generates sixteen different output words or combinations. These sixteen different output combinations (0000 through 1111) are listed in the truth table in Figure 25-19(c). The

Modulus
The maximum number of output combinations generated by a counter, equal to 2^n.

modulus of a counter, therefore, can be calculated in the same way that we determine the number of different combinations generated by a binary word with a certain number of bits.

$$\text{mod} = 2^n$$

mod = modulus of the counter

n = number of flip-flops in the counter

With the counter circuit illustrated in Figure 25-19, the modulus of the counter will be

$$\begin{aligned} \text{mod} &= 2^n \\ &= 2^4 \\ &= 16 \end{aligned}$$

EXAMPLE:

Calculate the modulus of a

a. 1-bit counter
b. 2-bit counter
c. 3-bit counter
d. 4-bit counter
e. 8-bit counter

Solution:

a. A 1-bit counter will have a modulus of mod $= 2^n = 2^1 = 2$
b. A 2-bit counter will have a modulus of mod $= 2^n = 2^2 = 4$
c. A 3-bit counter will have a modulus of mod $= 2^n = 2^3 = 8$
d. A 4-bit counter will have a modulus of mod $= 2^n = 2^4 = 16$
e. An 8-bit counter will have a modulus of mod $= 2^n = 2^8 = 256$

The Frequency Division of a Counter

Frequency Divider
A counter circuit that will divide the input signal frequency.

Divide-by-Two Circuit
A counter circuit that will divide the input signal frequency by 2.

Referring again to the waveforms in Figure 25-19(b), you can see how a binary counter is also a **frequency divider.** Each of the flip-flops divides its input frequency by 2. For example, two clock pulse cycles (or two negative edges) are needed at the C input of FF0 to produce one cycle at the Q_0 output. Likewise, two complete cycles of the clock are needed at the input of FF1 to produce one cycle at the Q_1 output, and two clock cycles are needed at the input of FF2 to produce one cycle at the Q_2 output. Finally, two clock cycles have to be applied at the input of FF3 in order to produce one cycle at the Q_3 output. Each flip-flop, therefore, acts as a **divide-by-two circuit.** If two flip-flops were connected together, the input clock frequency would be divided by 2, and then divided by 2 again. This overall action would divide the input clock frequency by 4, as shown in Figure 25-19(b), which shows that four negative edges are needed at the clock input in order to produce one complete cycle at the Q_1 output. Therefore, one flip-flop divides by 2, two flip-flops will divide by 4, three flip-flops will divide by 8, four flip-flops will divide by 16, and so on. The frequency division performed by a counter circuit can be calculated with the following formula:

$$\textit{Division Factor} = 2^n$$

n = number of flip-flops in the counter circuit

EXAMPLE:

Calculate the following for a 3-bit asynchronous binary counter:

a. maximum count
b. modulus
c. division factor

Solution:

a.
$$N = 2^n - 1$$
$$= 2^3 - 1 = 8 - 1 = 7$$

b.
$$\text{mod} = 2^n$$
$$= 2^3 = 8$$

c.
$$\textit{Division Factor} = 2^n$$
$$= 2^3 = 8$$

A 3-bit counter, therefore, has eight different output combinations (a modulus of 8), a maximum count of 7 (111_2), and will divide the input frequency by 8.

The Propagation Delay Time (t_p) of a Counter

The asynchronous counter is also known as a **ripple counter.** This term is used because the clock pulses are applied to the input of only the first flip-flop, and as the count is advanced, the effect "ripples" through to the other flip-flops in the circuit. Since each flip-flop in the counter circuit is triggered by the preceding flip-flop, it can take a certain amount of time for a pulse to ripple through all of the flip-flops and update the outputs to their new value. For example, when the eighth negative clock pulse edge occurs, all four Q outputs need to be changed from 0111 to 1000. If each flip-flop has a propagation delay time (t_p) of 10 ns, it will take 40 ns (4 flip-flops $\times$ 10 ns) to update the counter's count from 0111 to 1000. The counting speed or clock pulse frequency is therefore limited by the propagation delay time of all the flip-flops in the counter circuit. The HIGH input clock frequency limit can be calculated with the following formula:

Ripple Counter
An asynchronous counter in which the first flip-flop signal change affects the second flip-flop, which in turn affects the third, and so on until the last flip-flop in the series is changed.

$$f = \frac{1 \times 10^9}{n \times t_p}$$

f = upper clock pulse frequency limit
n = number of flip-flops in the counter circuit
t_p = propagation delay time of each flip-flop in nanoseconds

EXAMPLE:

If a 4 MHz clock input were applied to a 4-bit asynchronous (ripple) binary counter, and each flip-flop had a propagation delay time of 32 ns,

a. would this input frequency be too fast for the counter?
b. what would be the frequency at the Q_0, Q_1, Q_2, and Q_3 outputs?

Solution:

a.
$$f = \frac{1 \times 10^9}{n \times t_p}$$
$$= \frac{1 \times 10^9}{4 \times 32} = 7.8 \text{ MHz (upper frequency limit)}$$

A 4 MHz clock input is less than the 7.8 MHz upper frequency limit, and so this input is not too fast.

b. Frequency at Q_0 = 4 MHz ÷ 2 = 2 MHz
Frequency at Q_1 = 4 MHz ÷ 4 = 1 MHz
Frequency at Q_2 = 4 MHz ÷ 8 = 500 kHz
Frequency at Q_3 = 4 MHz ÷ 16 = 250 kHz

25-5-2 *Asynchronous Binary Down Counters*

With the asynchronous binary up counter just described, each input clock pulse would cause the parallel output binary word to be increased by 1, or incremented. By making a small modification to the up counter circuit, it is possible to produce a down counter circuit that will cause the parallel output word to be decreased by 1, or decremented, for each input clock pulse.

Figure 25-20(a) shows how a 4-bit or four-stage asynchronous binary down counter circuit can be constructed with four *J-K* flip-flops. Once again, the *J-K* flip-flops are cascaded, or strung together. In this instance, however, the $\overline{Q}$ outputs of each stage are connected to the clock inputs of the next stage. This small circuit difference causes the count sequence to be the complete opposite of the previously described up counter.

The clock input and Q output waveforms for this 4-bit binary down counter are shown in Figure 25-20(b). Referring to the far left side of the waveforms, you can see that we will begin with all the flip-flops initially RESET, and so the Q_0, Q_1, Q_2, and Q_3 outputs are 0000. If all the Q outputs are LOW, then all of the $\overline{Q}$ outputs must be HIGH, and therefore so are all of the C inputs to FF1, FF2, and FF3. Since the J and K inputs of all four flip-flops are tied HIGH, the output of each flip-flop will toggle each time a negative edge occurs at the circuit's clock (*CLK*) input. When the first clock pulse negative edge is applied to FF0, the Q_0 output will toggle from a binary 0 to a 1. This will cause the $\overline{Q}_0$ output to change from a binary 1 to a 0, and this negative edge or transition will clock the input of FF2, causing it to toggle, sending Q_2 HIGH and $\overline{Q}_2$ LOW. This HIGH-to-LOW transition at $\overline{Q}_2$ will clock the input of FF3, which will toggle FF3, sending its Q_3 output HIGH and its $\overline{Q}_3$ LOW. After the first clock pulse, therefore, the Q_0, Q_1, Q_2, and Q_3 outputs from the down counter will be 1111 (decimal 15), as shown in the waveforms in Figure 25-20(b) and on the first line of the truth table in Figure 25-20(c).

The down counter circuit will then proceed to count down by 1 for every clock pulse input applied. Referring again to the waveforms in Figure 25-20(b), you can see that FF0 will toggle for each negative edge of the *CLK* input, producing a Q_0 output that is half the input frequency (since each flip-flop divides by 2). Referring to the Q_1, Q_2, and Q_3 waveforms, you can see that these outputs seem to toggle in response to each positive edge of the previous flip-flop's Q output. For instance, Q_1 will toggle for every positive edge of the Q_0; Q_2 will toggle for every positive edge of the Q_1; and Q_3 will toggle for every positive edge of the Q_2. This is because a positive edge at any of the Q outputs is a negative edge at the associated $\overline{Q}$ output, and it is the $\overline{Q}$ outputs that clock each following stage.

Referring again to the truth table in Figure 25-20(c), you can see that this mod-16 (a 4-bit counter has sixteen possible output combinations) down counter will count down from 1111 (decimal 15) to 0000 (decimal 0), and then repeat the down-count cycle.

25-5-3 *Asynchronous Binary Up/Down Counters*

Comparing the asynchronous binary up counter in Figure 25-19(a) to the asynchronous down counter in Figure 25-20(a), you can see that the only difference between the two circuits is whether we clock FF1, FF2, and FF3 with the Q or $\overline{Q}$ outputs from the previous flip-flops.

Figure 25-21 shows how three AND-OR arrangements can be controlled by an UP/$\overline{\text{DOWN}}$ control line to create an asynchronous up/down binary counter. If the UP/$\overline{\text{DOWN}}$ control line is made HIGH, all of the shaded AND gates will be enabled,

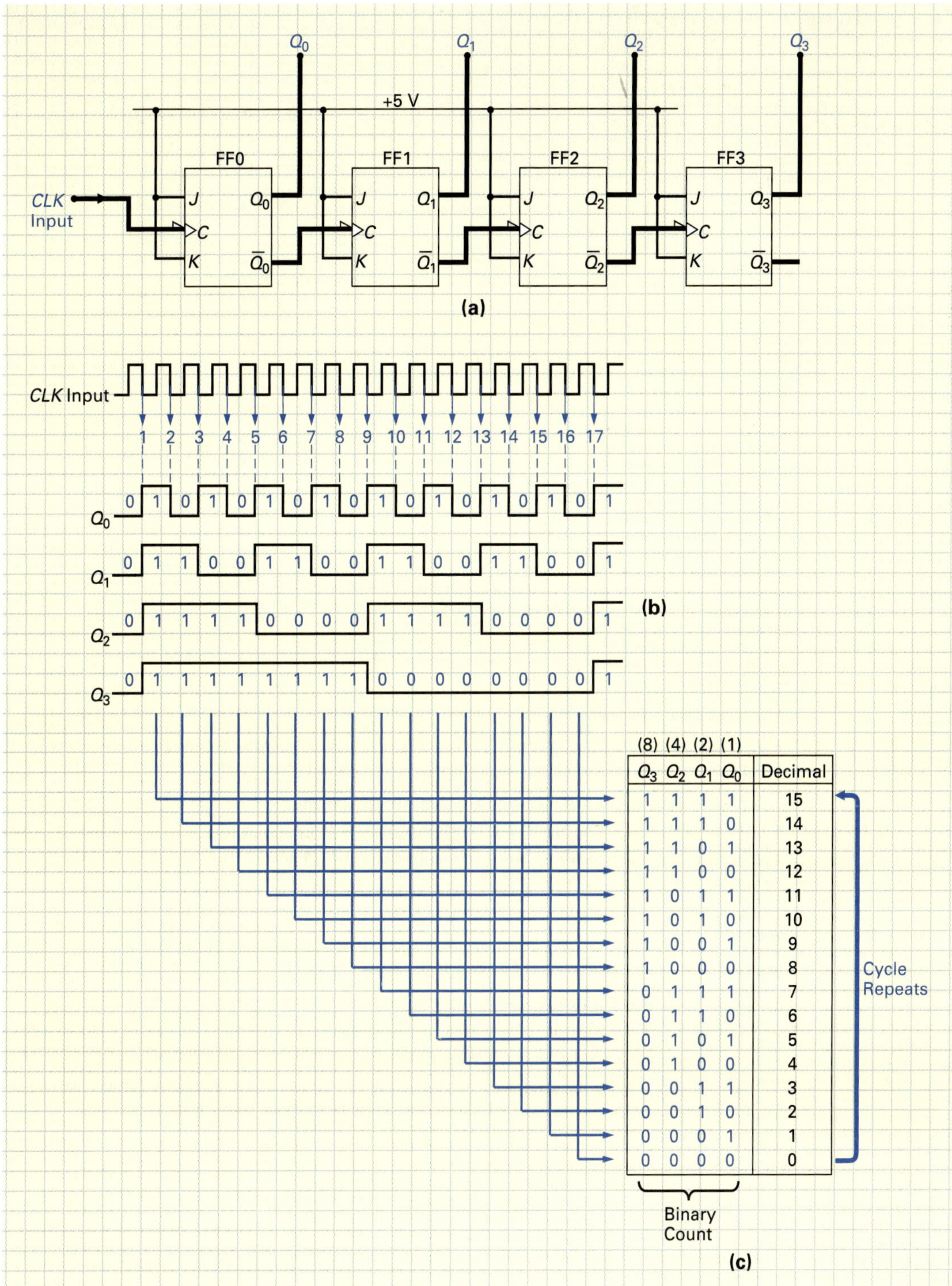

Q_3 (8)	Q_2 (4)	Q_1 (2)	Q_0 (1)	Decimal
1	1	1	1	15
1	1	1	0	14
1	1	0	1	13
1	1	0	0	12
1	0	1	1	11
1	0	1	0	10
1	0	0	1	9
1	0	0	0	8
0	1	1	1	7
0	1	1	0	6
0	1	0	1	5
0	1	0	0	4
0	0	1	1	3
0	0	1	0	2
0	0	0	1	1
0	0	0	0	0

FIGURE 25-20 An Asynchronous Binary Down Counter. (a) Logic Circuit. (b) Waveforms. (c) Truth Table.

connecting the Q outputs through to the flip-flop's C inputs, causing the counter circuit to count up. On the other hand, if the UP/$\overline{\text{DOWN}}$ control line is made LOW, all of the shaded AND gates will be disabled and all of the unshaded AND gates will be enabled, connecting the $\overline{Q}$ outputs through to the flip-flop's C inputs, causing the counter circuit to count down.

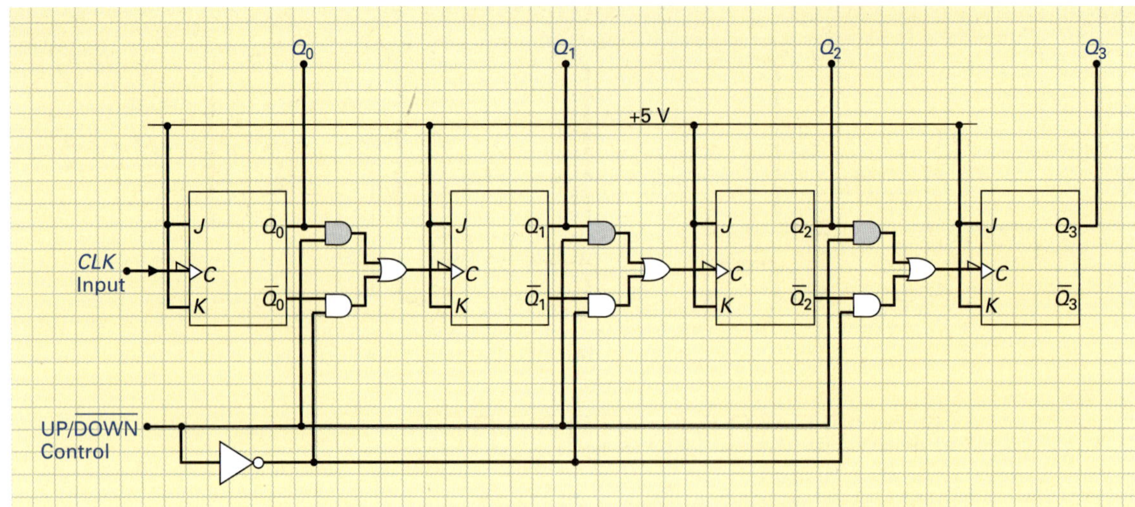

FIGURE 25-21 **An Asynchronous Binary Up/Down Counter.**

25-5-4 *Asynchronous Decade (mod-10) Counters*

Decade Counter
A counter that has a modulus of 10 and therefore counts from 0 to a maximum of 9.

Figure 25-22(a) shows how the previously discussed asynchronous mod-16 up counter circuit could be modified to construct a mod-10 or **decade counter** circuit. This counter will count from 0000 (decimal 0) to 1001 (decimal 9), and then the cycle will repeat, as shown in the timing waveforms in Figure 25-22(b) and the truth table in Figure 25-22(c). The reason this counter skips the counts 1010 through 1111 (decimal 10 through 15) is due to the action of the NAND gate, which controls the active-LOW clear inputs $(\overline{CLR})$ of all four flip-flops. Since the two inputs to this NAND gate are connected to the Q_1 and Q_3 outputs, when the counter reaches a count of 1010 (decimal 10) both Q_1 and Q_3 will go HIGH, and therefore the output of the NAND gate will go LOW and CLEAR the counter. Referring to the timing waveforms in Figure 25-22(b), you can see that the $(\overline{CLR})$ line is inactive for the counts 0000 through 1001. However, when the tenth clock pulse is applied, both Q_1 and Q_3 are HIGH. This condition of having both Q_1 and Q_3 HIGH is only temporary since the active $\overline{CLR}$ will almost immediately go LOW and therefore CLEAR or RESET all of the flip-flop outputs, producing a 0000 counter output. The truth table in Figure 25-22(c) summarizes the operation of this decade counter by showing how it counts from 0 to 9 and then repeats the cycle.

Figure 25-22(d) illustrates another way to show how a counter cycles through its different output states. In this *state transition diagram,* each circled binary number represents one of the output states, and the solid arrow indicates how one state changes to another as each clock pulse is received. For example, if we start at the count of 0000, after one clock pulse, the counter will go to 0001; after the next clock pulse, the counter will go to 0010; after the next clock pulse, the counter will go to 0011; and so on. The dotted line and shaded circle in this diagram indicate a temporary state that is only momentarily reached. For most purposes, this temporary state is ignored, and the counter is said to go directly from 1001 to 0000, as indicated by the solid arrow.

In summary, a decade counter counts from 0 up to a maximum count of 9, a total of ten different output states (mod-10). Since the counter requires ten input clock pulses before the output of the counter is reset, the Q_3 output frequency will be one-tenth that of the input clock frequency (*CLK* input). A decade counter, therefore, can also function as a divide-by-ten circuit. Remember that the only difference between a counter circuit and frequency divider circuit is how many of the outputs are used. For example, with a binary counter circuit, all of the flip-flops' outputs are used to deliver a parallel output word that represents the count, whereas with a frequency divider circuit, the clock input is applied to the first flip-flop, and the last flip-flop delivers serial output that is some fraction of the input clock frequency.

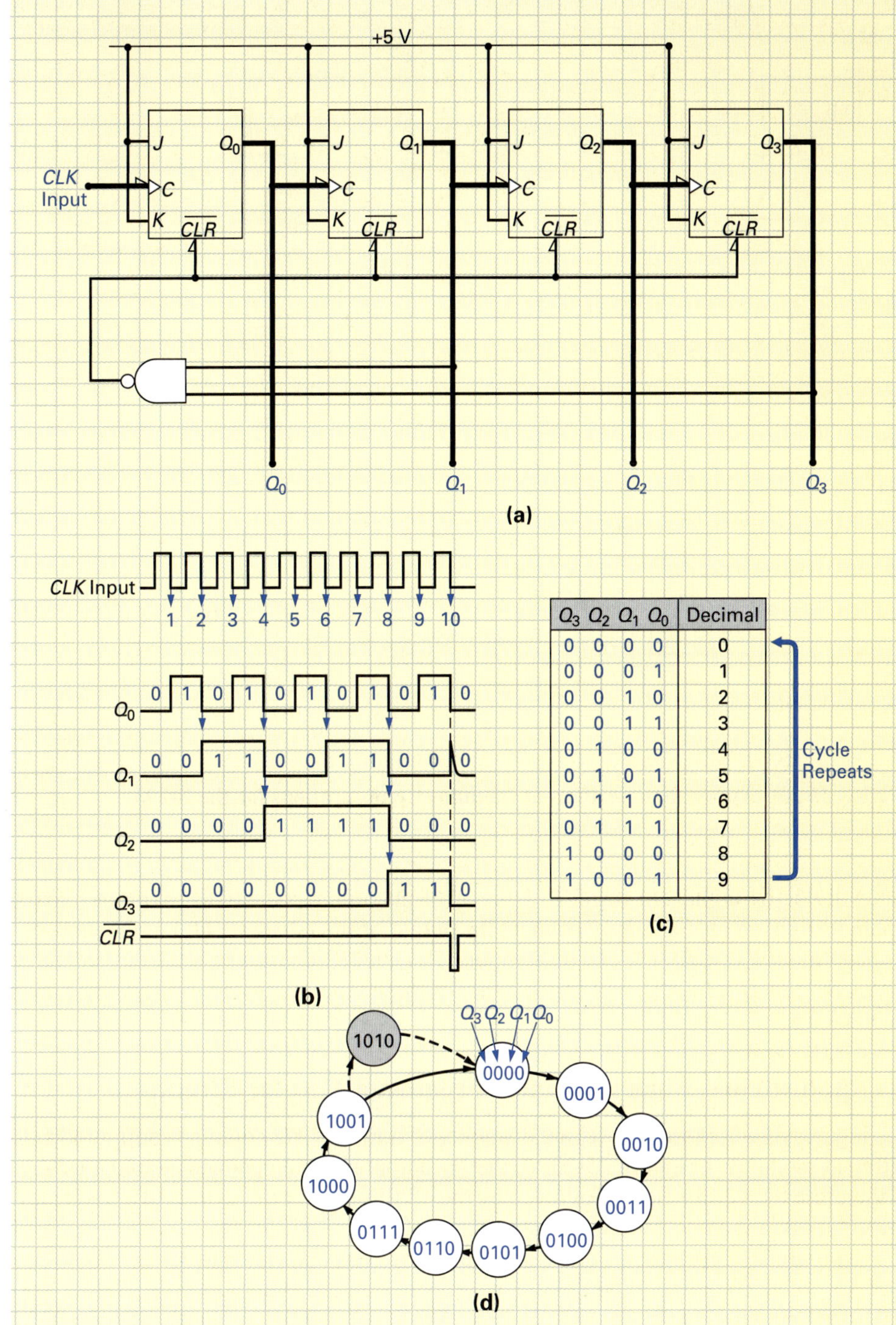

FIGURE 25-22 **An Asynchronous Decade Counter. (a) Logic Circuit. (b) Timing Waveforms. (c) Truth Table. (d) State Transition Diagram.**

In the counter applications section of this chapter, you will see how the decade counter is frequently chosen in applications that have a decimal display, such as digital clocks, digital voltmeters, and frequency counters. This is because the decade counter allows us to interface the binary information within a digital system to a decimal display that you and I understand.

Figure 25-23 shows the 74293 counter IC. This device contains four master-slave *J*-K flip-flops that for versatility are connected as a single divide-by-two (÷2) stage and a ripple

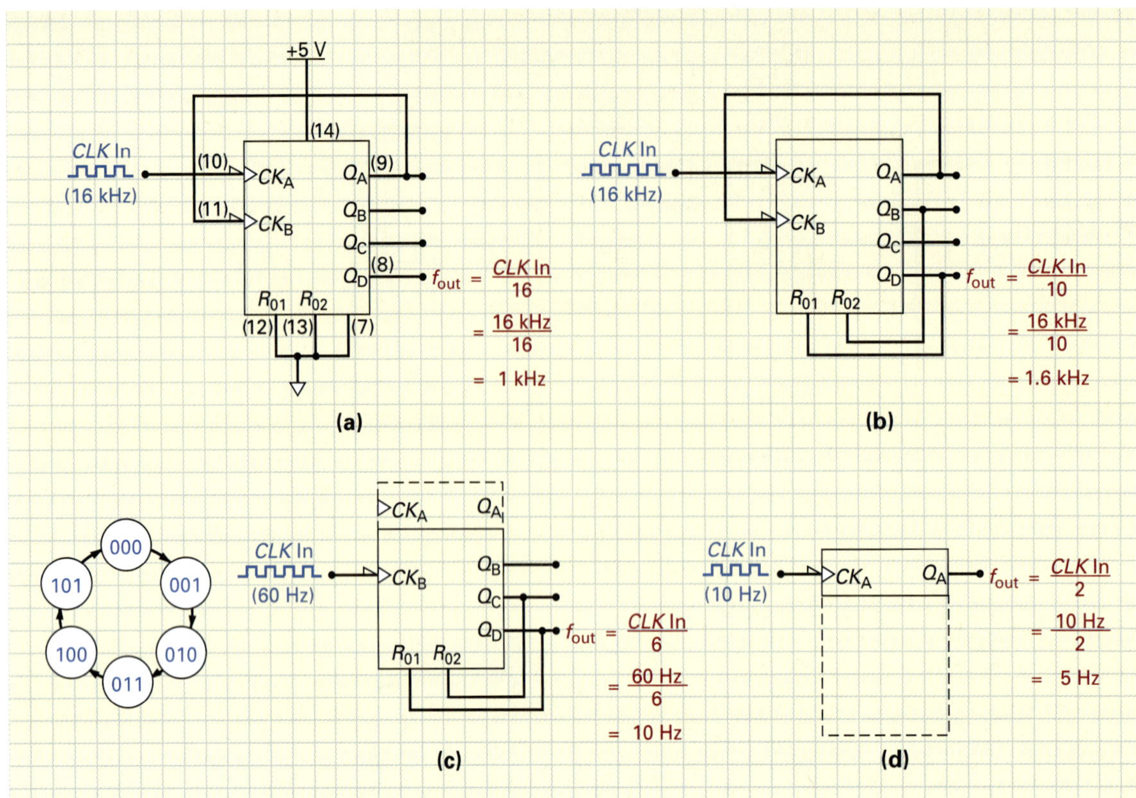

FIGURE 25-23 **Application—Different Modulus Configurations for the 74293.**

counter divide-by-eight (÷8) stage. To achieve the maximum count (0 to 15) or the maximum frequency division (÷16), simply connect the output of the ÷2 stage (Q_A) to the input of the ÷8 stage (input *B* or *CLK B*). The reset inputs ($R_{0(1)}$ and $R_{0(2)}$) must be HIGH in order to RESET the outputs Q_D, Q_C, Q_B, and Q_A. If either of these inputs is inactive (LOW), the counter will count up.

Application—Binary Counter/Frequency Divider Circuits

Figures 25-23(a) through (c) are examples of how a 74293 could be configured to have a different modulus. In Figure 25-23(a), the Q_A output has been connected to the CK_B input and therefore the ÷2 flip-flop is followed by the ÷8 flip-flops, producing a ÷16 frequency divider circuit or a mod-16 binary counter circuit (0000 through 1111).

In Figure 25-23(b), the Q_B and Q_D outputs have been connected to the reset inputs ($R_{0(1)}$ and $R_{0(2)}$), so the counter will be RESET when both of these *Q* outputs go HIGH (at a count of 1010). The circuit will therefore function as a mod-10 counter circuit (0000 through 1001) in which all four outputs will be used or as a ÷10 frequency divider circuit in which the Q_D output is one-tenth that of the clock input frequency applied to CK_A.

In Figure 25-23(c), only the ÷8 ripple counter section of the 74293 is being used to create a mod-6 counter or ÷6 frequency divider circuit. Since the Q_C and Q_D outputs have been connected to the reset inputs ($R_{0(1)}$ and $R_{0(2)}$), the counter will RESET the moment the counter reaches a count of 110 (decimal 6). The circuit will therefore function as a mod-6 counter circuit (000 through 101) in which outputs Q_B, Q_C, and Q_D will be used, or as a ÷6 frequency divider circuit in which the Q_D output is one-sixth that of the clock input frequency applied to CK_B.

In Figure 25-23(d), only the ÷2 section of the 74293 is being used to create a mod-2 counter or ÷2 frequency divider.

25-5-5 *Asynchronous Presettable Counters*

Figure 25-24(a) shows how four *J-K* flip-flops and several gates could be connected to construct an **asynchronous presettable counter.** As an example, this circuit is loading a preset data value of 1001 (decimal 9) into the counter directly following the maximum count of 1111 (decimal 15), as shown in the state transition diagram in Figure 25-24(b). Once the value of 1001 has been loaded, the counter will again count up to a maximum of 1111, reload the preset value of 1001 once again, and then repeat the cycle.

Asynchronous Presettable Counter

A counter that will count from its loaded preset value up to its maximum value. By changing the preset value, the modulus of the counter can be changed.

The preset value, which in this example is 1001, is applied to the data inputs D_0, D_1, D_2, and D_3. The logic levels applied to these inputs are normally prevented from reaching the *J-K* flip-flop's $\overline{PRE}$ and $\overline{CLR}$ inputs because all of the NAND gates are disabled by a LOW on the LOAD control line. The logic level on the LOAD control line is determined by a NOR, whose four inputs are connected to the counter's four outputs, Q_0, Q_1, Q_2, and Q_3.

To understand the operation of the circuit in Figure 25-24(a), let us begin by assuming that the counter is at its maximum count of 1111. On the next negative edge of the clock, the counter will increment to 0000, and only this output will cause the NOR gate, and therefore the LOAD control line, to go HIGH. With the LOAD control line HIGH, the data inputs and their complements are allowed to pass through the NAND gates to activate either the PRESET or CLEAR inputs of the flip-flops. In this example, the unshaded NAND gates will produce a HIGH output, while the shaded NAND gates will produce a LOW output. Since the flip-flop's $\overline{PRE}$ and $\overline{CLR}$ inputs are both active-LOW, Q_0 will be preset to 1, Q_1 will be cleared to 0, Q_2 will be cleared to 0, and Q_3 will be preset to 1. The 4-bit data value of 1001 is therefore asynchronously loaded into the counter (independent of the clock signal) the moment the counter goes to 0000. This 0000 state is only a temporary state, as shown in the state transition diagram in Figure 25-24(b), since it is almost immediately changed to the loaded state. Successive clock pulses at the input will cause the counter to count from 1001 to 1010, to 1011, to 1100, to 1101, to 1110, and then to the maximum count of 1111. The next clock pulse will reset the counter's count to 0000, enable the LOAD control line, load the data value, and then repeat the cycle.

One of the big advantages with a presettable counter is that it has a *programmable modulus.* For example, in Figure 25-24 we loaded a data value of 1001 (decimal 9) and then counted to 1111 (decimal 15), resulting in a total of seven different output combinations (1001, 1010, 1011, 1100, 1101, 1110, and 1111). As a result, this circuit will function as a mod-7 binary counter or ÷7 frequency divider.

By changing the data value loaded into a presettable counter, we can create a counter of any modulus. The simple formula for calculating the modulus of a presettable counter is

$$M_P = \text{mod} - D$$

M_P = modulus of the presettable counter
mod = natural modulus of the counter
D = preset data value

In our example circuit in Figure 25-24,

$$\begin{aligned} M_P &= \text{mod} - D \\ &= 16 - 9 \qquad \text{(counter skips states 0 to 8)} \\ &= 7 \end{aligned}$$

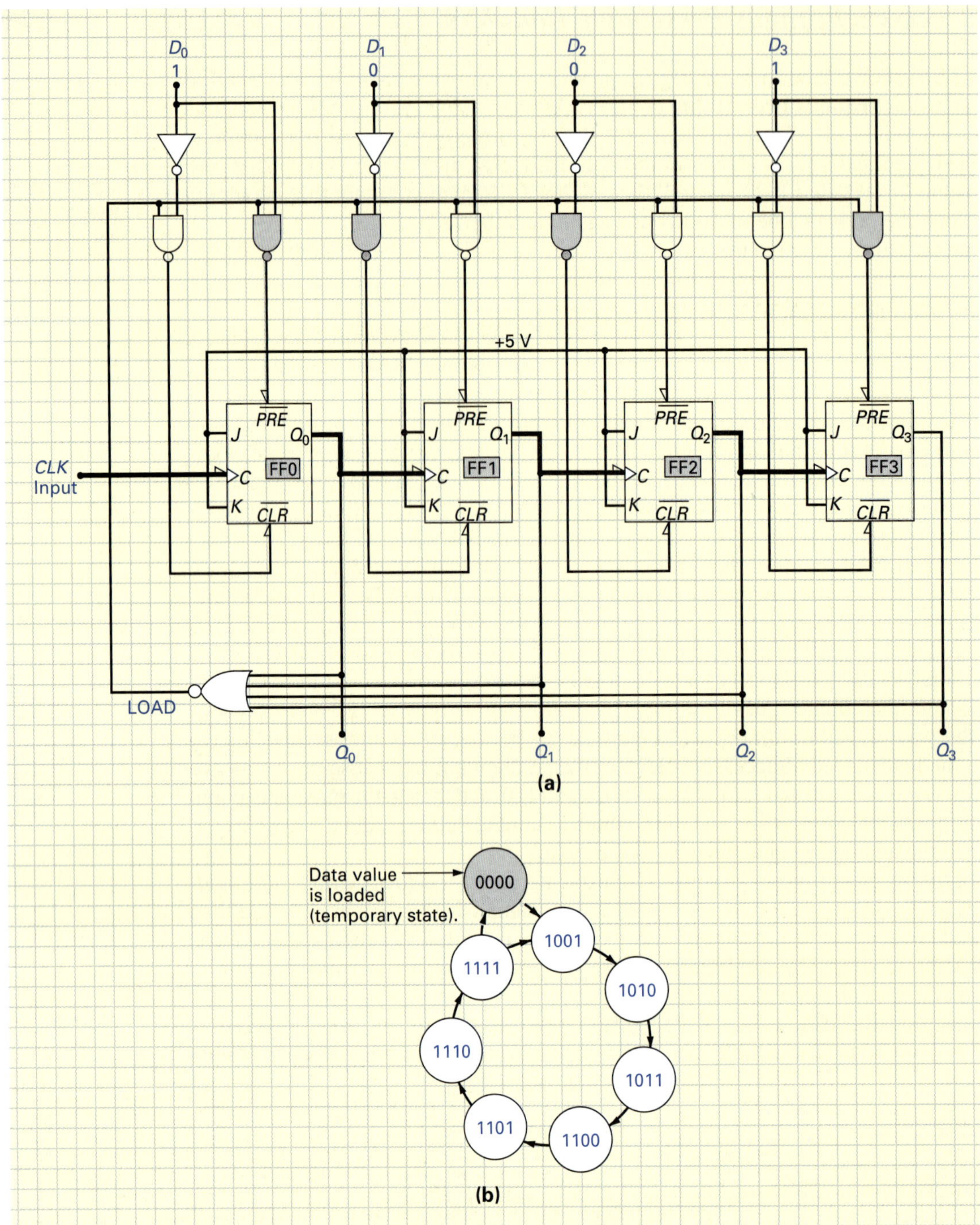

FIGURE 25-24 **An Asynchronous Presettable Counter. (a) Logic Counter. (b) State Transition Diagram for a Preset Value of 1001.**

EXAMPLE:

Assuming a 6-bit asynchronous presettable counter, calculate

a. its natural modulus

b. its preset modulus if 001100_2 is loaded

■ *Solution:*

a.
$$\text{mod} = 2^n$$
$$= 2^6 = 64$$

b.
$$M_P = \text{mod} - D$$
$$= 64 - 12 = 52$$

SELF-TEST EVALUATION POINT FOR SECTION 25-5

Use the following questions to test your understanding of Section 25-5.

1. Most of the flip-flops within an asynchronous counter ___________ (are/are not) connected to the clock input and therefore they ___________ (do/do not) change state in sync with the master clock signal.
2. What is the natural modulus and maximum count of an 8-bit counter?
3. What would be the upper frequency limit of an 8-bit counter if each flip-flop within an asynchronous counter had a propagation delay time of 30 ns?
4. What would be the natural modulus and preset modulus of an 8-bit counter if 0001 1110 was the load value?

25-6 SYNCHRONOUS COUNTERS

As mentioned previously, asynchronous counters are often referred to as ripple counters. The term *ripple* is used because the flip-flops making up the asynchronous counter are strung together with the output of one flip-flop driving the input of the next. As a result, the flip-flops do not change state simultaneously or in sync with the input clock pulses, since the new count has to ripple through and update all of the flip-flops. This ripple-through action leads to propagation delay times that limit the counter's counting speed. This limitation can be overcome with a **synchronous counter** in which all of the flip-flops within the counter are triggered simultaneously and therefore synchronized to the master timing clock signal.

Synchronous Counter
A counter in which all operations are controlled by the master clock signal.

25-6-1 *Synchronous Binary Up Counters*

Figure 25-25(a) shows how four *J-K* flip-flops and two AND gates can be connected to form a 4-bit synchronous mod-16 up counter. The clock input signal line has been drawn as a heavier line in this illustration to show how all of the flip-flops in a synchronous counter circuit are triggered simultaneously by the common clock input signal. This parallel connection makes the counter synchronous, since all of the flip-flops will now operate in step with the input clock signal.

The synchronous binary up counter shown in Figure 25-25(a) will operate as follows. The *J* and *K* inputs of FF0 are tied HIGH, and therefore the Q_0 output will continually toggle, like a single-stage asynchronous counter. This fact is verified by looking at the Q_0 column in Figure 25-25(b), which shows how the Q_0 output continually alternates HIGH and LOW. The *J* and *K* inputs of FF1 are controlled by the divide-by-two output of FF0. This means that when Q_0 is LOW, the Q_1 output of FF1 will not change; however, when Q_0 is HIGH, the Q_1 output of FF1 will toggle. This action can be confirmed by referring to the Q_1 column in Figure 25-25(b), which shows that when the Q_0 output goes HIGH, the Q_1 output will toggle or change state on the next active edge of the clock signal. The *J* and *K* inputs of FF2 are controlled by the ANDed outputs of Q_0 and Q_1. This means that only when Q_0 and Q_1 are both HIGH will the output of AND gate *A* be HIGH, and this HIGH will enable FF2 to toggle. Referring to the Q_2 column in Figure 25-25(b), you can see that only when Q_0 and

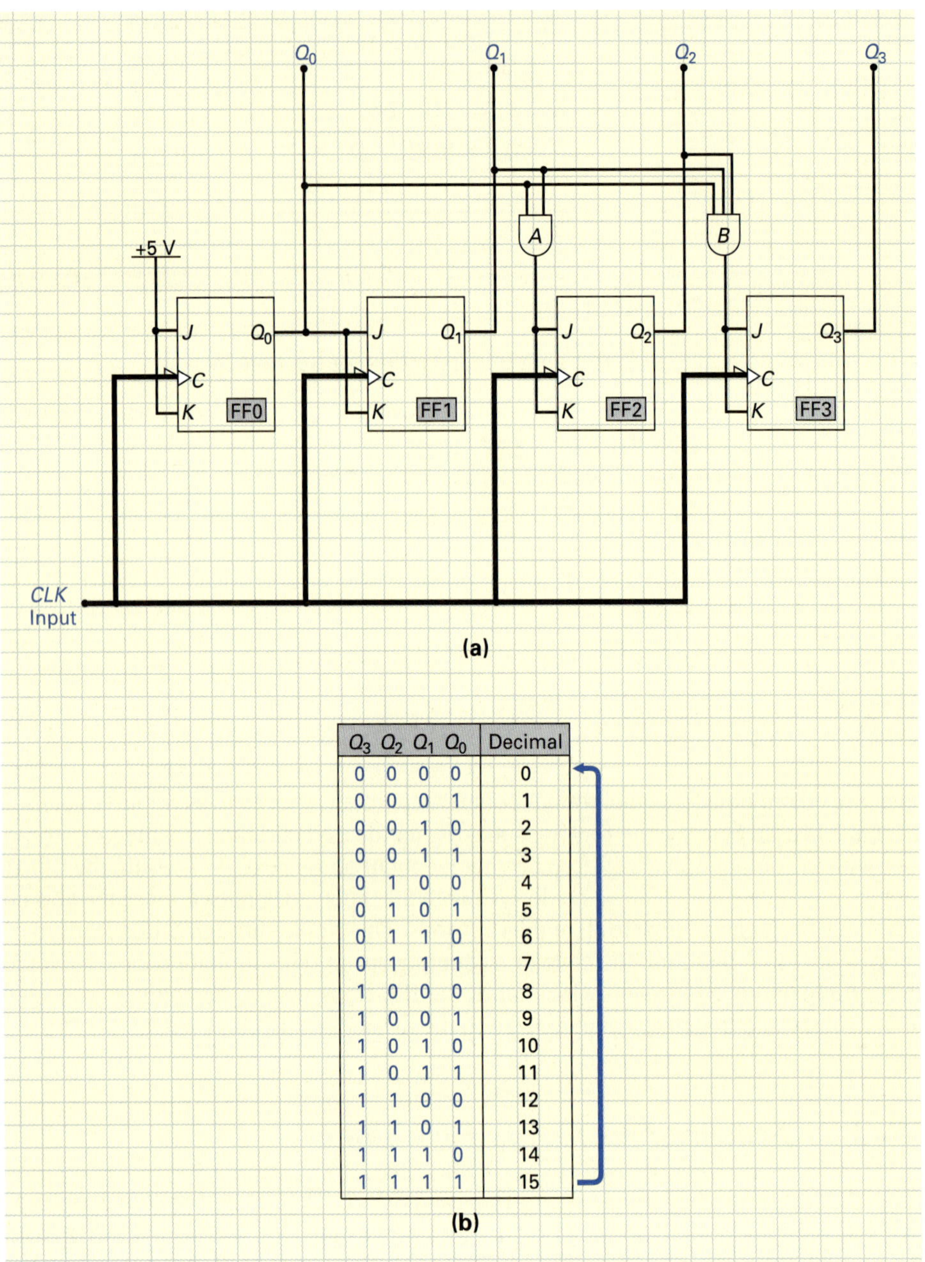

Q_3 Q_2 Q_1 Q_0	Decimal
0 0 0 0	0
0 0 0 1	1
0 0 1 0	2
0 0 1 1	3
0 1 0 0	4
0 1 0 1	5
0 1 1 0	6
0 1 1 1	7
1 0 0 0	8
1 0 0 1	9
1 0 1 0	10
1 0 1 1	11
1 1 0 0	12
1 1 0 1	13
1 1 1 0	14
1 1 1 1	15

FIGURE 25-25 A Synchronous Binary Up Counter. (a) Logic Circuit. (b) Truth Table.

Q_1 are HIGH will Q_2 toggle or change state on the next active edge of the clock. The J and K inputs of FF3 are controlled by the ANDed outputs of Q_0, Q_1, and Q_2. This means that only when Q_0, Q_1, and Q_2 are HIGH will the output of AND gate B be HIGH, enabling FF3 to toggle. If you look at the Q_3 column in Figure 25-25(b), you will notice that only when Q_0, Q_1, and Q_2 are all HIGH will Q_3 toggle or change state on the next active edge of the clock.

25-6-2 *Synchronous Counter Advantages*

The key advantage of the asynchronous or ripple counter is its circuit simplicity, as can be seen by comparing the synchronous binary up counter in Figure 25-25(a) to the asynchronous binary up counter in Figure 25-19(a). The key disadvantage of the asynchronous counter is its frequency of operation or counting speed limitation. Since the input is applied only to the first flip-flop, it takes a certain time for all of the outputs to be updated to reflect the new count. This propagation delay time of the counter is equal to the sum of all the flip-flop propagation delay times. This limitation means that we cannot trigger or clock the input of an asynchronous counter before all of the outputs have settled into their new state, and therefore the clock input frequency (or pulses to be counted) has a speed or frequency limitation. Asynchronous counters are sometimes constructed using ECL flip-flops, which have an extremely small propagation delay time, and therefore a high input clock frequency can be applied. These ECL flip-flops, however, are more expensive and have a higher power consumption.

Synchronous counters are a direct solution to the asynchronous counter limitations, since they have small propagation delay times and are less expensive and consume less power than ECL asynchronous counters. The key to the synchronous counter's small propagation delay time is that all of its flip-flops are triggered or clocked simultaneously by a common clock input signal. This means that all of the flip-flops will be changing their outputs at the same time, so the total counter propagation delay time is equal to the delay of only one flip-flop. In reality, we should also take into account the time it takes for a Q output to propagate through a single AND gate and reach the J-K inputs of the following stage. Taking these two factors into consideration, we arrive at the following formula for calculating the delay time for a synchronous counter:

$$t_p = \text{single flip-flop } t_p + \text{single AND gate } t_p$$

EXAMPLE:

Assuming a J-K flip-flop propagation delay time of 30 ns and an AND gate propagation delay time of 10 ns, calculate the counter propagation time for a four-stage asynchronous counter and four-stage synchronous counter.

Solution:

$$t_{p(asynchronous)} = \text{single flip-flop } t_p \times \text{number of flip-flops}$$
$$= 30 \text{ ns} \times 4 = 120 \text{ ns}$$
$$t_{p(synchronous)} = \text{single flip-flop } t_p + \text{single AND gate } t_p$$
$$= 30 \text{ ns} + 10 \text{ ns} = 40 \text{ ns}$$

In summary, the two advantages of a synchronous counter are that no transient output states occur due to the ripple through additive propagation delays and that synchronous counters can operate at much higher clock input frequencies.

25-6-3 *Synchronous Presettable Binary Counters*

Like the presettable asynchronous counter, the presettable synchronous counter has a programmable modulus. The 74LS163A, shown in Figure 25-26, is an example of a presettable synchronous up counter IC. A 4-bit binary value can be loaded into this counter to obtain any desired modulus by applying a LOW to the $\overline{\text{LOAD}}$ control input (pin 9). When the $\overline{\text{LOAD}}$ input is made LOW, the Q outputs (Q_A, Q_B, Q_C, and Q_D) will reflect the data inputs (A, B, C, and D) after the next positive edge of the clock input (pin 2). There are two active-HIGH enable inputs (pins 7 and 10), and because of the ANDing of these inputs, both must be HIGH for the counter to count. For example, let us assume that the counter is first loaded with a preset value of 1100 (decimal 12) and then counted up to its maximum count of 1111 (decimal 15). At this time, the ripple carry output (*RCO*, pin 15) generates an output pulse. This last-stage output is used to enable the input of another counter, if two

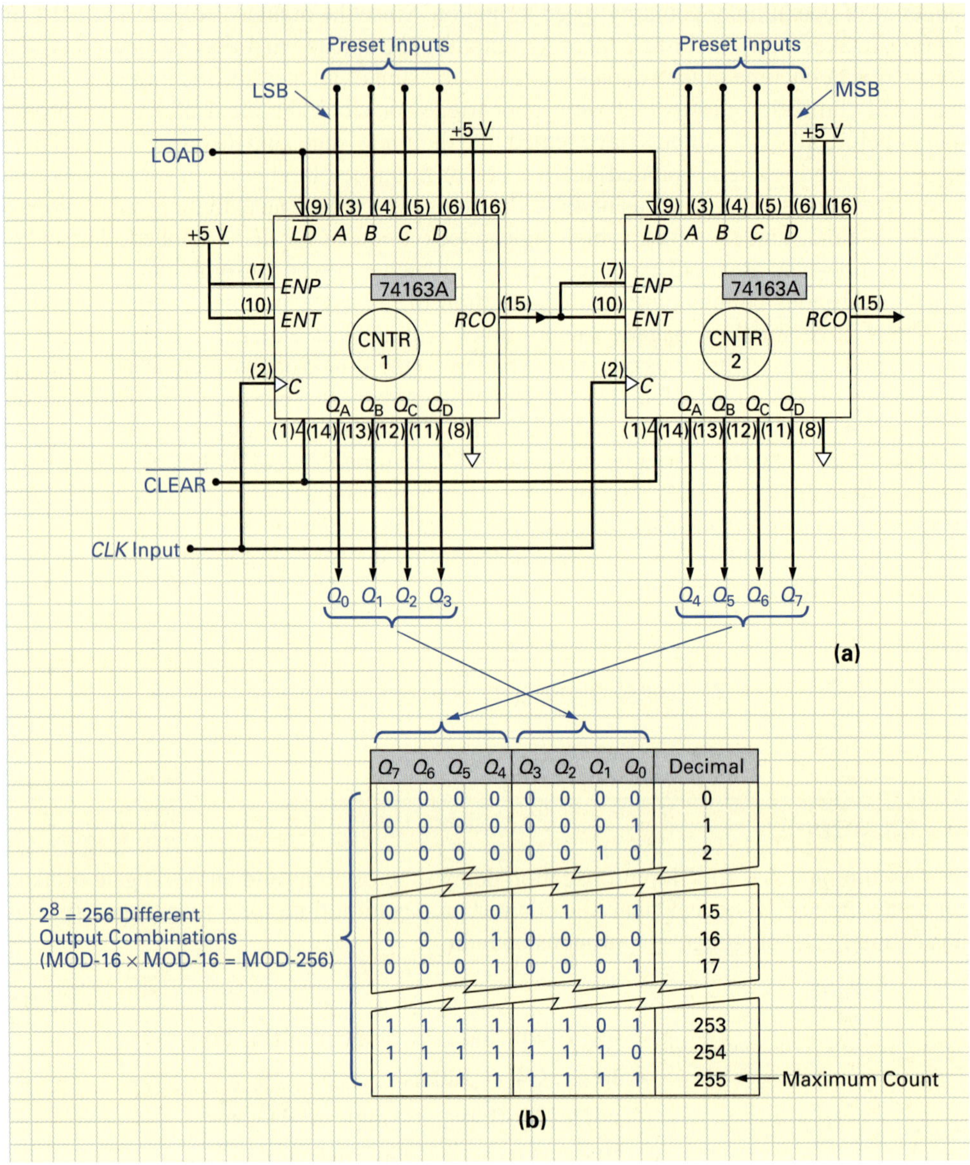

(a)

Q_7	Q_6	Q_5	Q_4	Q_3	Q_2	Q_1	Q_0	Decimal
0	0	0	0	0	0	0	0	0
0	0	0	0	0	0	0	1	1
0	0	0	0	0	0	1	0	2
0	0	0	0	1	1	1	1	15
0	0	0	1	0	0	0	0	16
0	0	0	1	0	0	0	1	17
1	1	1	1	1	1	0	1	253
1	1	1	1	1	1	1	0	254
1	1	1	1	1	1	1	1	255 ← Maximum Count

$2^8 = 256$ Different Output Combinations (MOD-16 × MOD-16 = MOD-256)

(b)

FIGURE 25-26 Application—Cascading Counters to Obtain a Higher Modulus.

counters were needed in a circuit application. For example, when a counter has reached a count of

Q_3	Q_2	Q_1	Q_0	
1	1	1	1	$= 15_{10}$

a reset-and-carry action occurs from right to left as follows:

Carry Bit	Q_3	Q_2	Q_1	Q_0	
①	0	0	0	0	$= 0_{10}$

If only a 4-bit counter is needed, all of the Q outputs are reset to 0, the cycle repeats, and the carry bit is lost. If, however, two 4-bit counters are needed, the *RCO* of the LSB counter will be connected to the input of the MSB counter so that the Q_3 output will carry into the Q_4 flip-flop. In this situation the *carry bit* is not lost and the 8-bit counter counts beyond 15 as follows:

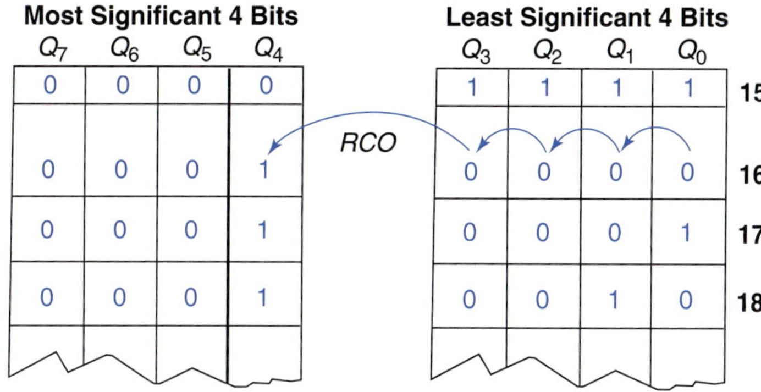

Application—Cascaded Counter Circuit

Figure 25-26(a) shows how two mod-16 74163A ICs can be *cascaded,* or strung together, to obtain a mod-256 counter. When connecting synchronous counters in cascade, the *RCO* (so called because it ripples out of the counter's last stage) is connected to the enable inputs of counter 2 (CNTR 2). As a result, when CNTR 1 reaches its maximum count (1111), the *RCO* from CNTR 1 will go HIGH and enable CNTR 2 so that on the next positive edge of the clock CNTR 2 will have a count of 0001 and CNTR 1 will have reset to 0000.

Since the counter in Figure 25-26(a) contains eight outputs (Q_0 through Q_7), it will produce a total of 2^8, or 256, different output combinations, with a maximum count of 255, as shown in Figure 25-26(b). If this circuit were used as a frequency divider, it would produce a single output pulse at the Q_7 output for every 256 input clock pulses (a divide-by-256 circuit).

If some other counter modulus or frequency division were required (anywhere between 2 and 256), the preset inputs could be used to load in any value. For example, if 0001 1000 (decimal 24) were loaded into the 8-bit counter, the counter would have a modulus of

$$M_p = \text{mod} - D$$
$$= 256 - 24 = 232$$

This counter would now count from 24 to 255 (a total of 232 different output combinations), or it would divide the input clock frequency by 232.

25-6-4 *Synchronous Decade (mod-10) Counters*

Like the asynchronous decade counter, a synchronous decade (mod-10) counter generates only the BCD outputs 0000 though 1001 (decimal 0 through 9). In the counter applications section of this chapter, you will see how decade counters are frequently chosen in applications that have a decimal display, such as digital clocks, digital voltmeters, and frequency counters. This is because the decade counter allows us to interface the binary information within a digital system to a decimal display that you and I understand.

The 74LS160A, shown in Figure 25-27, is a good example of a synchronous decade counter IC. The 74LS160A has exactly the same input and output pin assignments as the 74LS163A covered in Figure 25-26; however, this counter has only ten different output combinations.

Application—A 0-to-99 Counter and Display Circuit

Figure 25-27 shows how two 74LS160A synchronous decade counter ICs, two 7447 BCD-to-seven-segment decoder ICs, and two common-anode seven-segment LED displays could be connected to construct a 0-to-99 counter and display circuit. Up until this point, we have always followed the industry schematics standard of having inputs on the left side of the page and outputs on the right side of the page. A difficulty arises, however, when we are showing the various digits of a display, since the units digit is normally always on the right

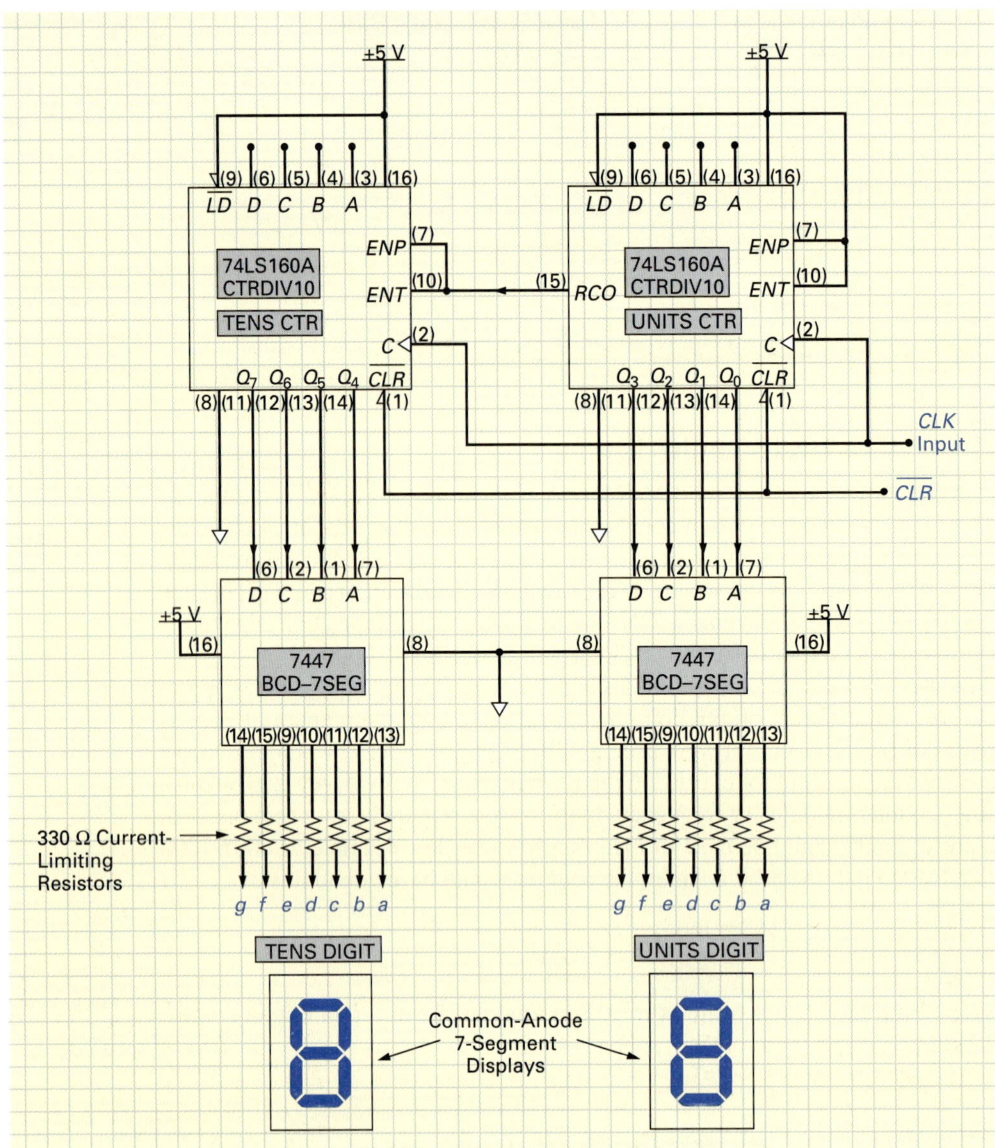

FIGURE 25-27 **Application—A 0-to-99 Counter.**

side and digits of increasing weight (such as tens, hundreds, and thousands) are to the left of the units digit. In this illustration, we will have the input clock signal applied at the right and propagate to the left so that the digital displays will be in their correct position.

The operation of this circuit is as follows. The units counter counts the input clock pulses, and after every ten clock pulses it activates the *RCO*. This enables the tens counter, which counts up by 1 on the next positive edge of the clock. The two decade counters will produce BCD outputs that range from 0000 0000_{BCD} (decimal 0) to 1001 1001_{BCD} (decimal 99). The 4-bit BCD output codes from the decade counters are decoded by the 7447s and displayed as their decimal equivalents on the seven-segment displays.

25-6-5 *Synchronous Up/Down Counters*

Bidirectional Counter
A counter that can be controlled to either count up or count down.

All up/down counters are sometimes referred to as **bidirectional counters** since they can count in one of two directions—up or down. A control line is generally used to determine whether the count is incremented (advanced upward) or decremented (advanced downward).

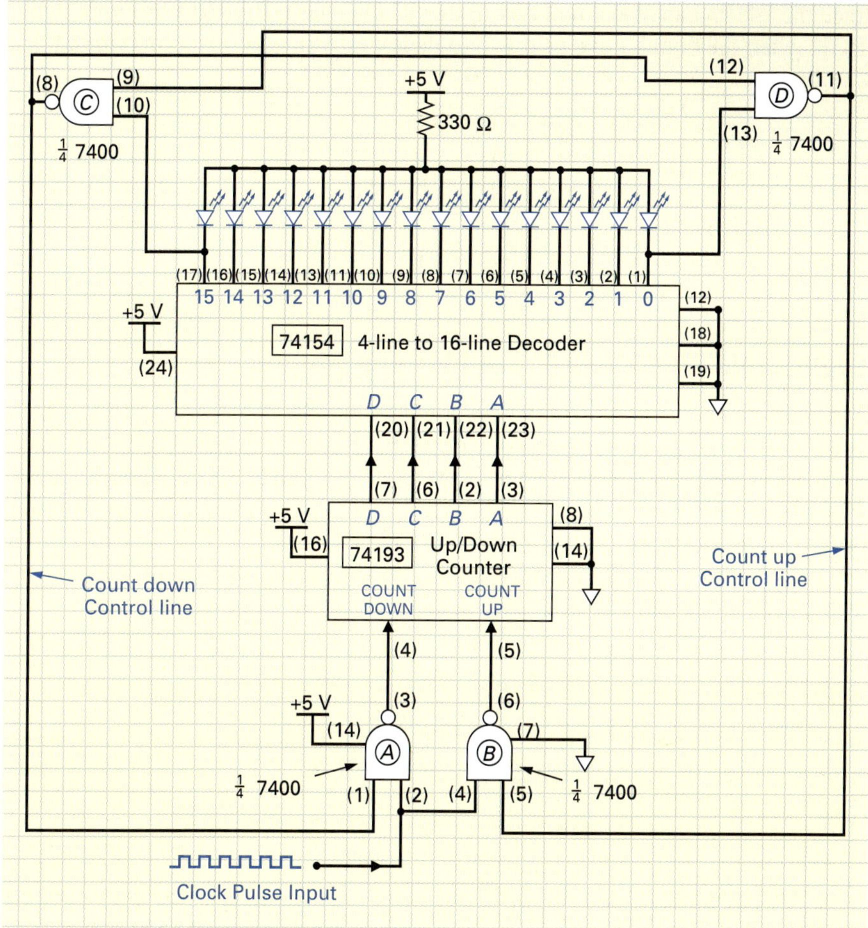

FIGURE 25-28 **Application—Up/Down Flashing Circuit.**

For example, if a 2-bit counter's DOWN/$\overline{\text{UP}}$ control line were made LOW, the counter would count up (0,1,2,3,0,1,2, . . .), whereas if the DOWN/$\overline{\text{UP}}$ control were made HIGH, the counter would count down (3,2,1,0,3,2,1, . . .).

The 74193 IC is described in data books as a presettable mod-16 up/down synchronous counter with asynchronous loading and asynchronous master reset. This IC is shown in Figure 25–28. This IC counter contains four *J-K* flip-flops and a variety of control gates. The counter can be used as a mod-*N* counter, or divider, by applying any preset value to the data inputs (*A, B, C,* and *D*) and activating the active-LOW load input (pin 11). An asynchronous CLEAR or master reset input (pin 14) is available and will force all outputs LOW when a HIGH level is applied. Synchronous operation is provided by having all flip-flops clocked simultaneously by a LOW-to-HIGH transition (positive edge-triggered). The direction of counting (up or down) is controlled by the two trigger inputs COUNT UP (pin 5) and COUNT DOWN (pin 4). To count up, simply apply the clock input to the COUNT UP input while the COUNT DOWN input is HIGH; whereas to count down, apply the clock input to the COUNT DOWN input while the COUNT UP input is HIGH.

The 74193 up/down counter was also designed to be cascaded, and therefore a CARRY output will pulse HIGH when the counter cycles beyond its maximum count, as follows:

		(8)	(4)	(2)	(1)		(10)	(1)
		1	1	1	0	=	1	4
Carry	1	1	1	1	1	=	1	5
	1	0	0	0	0	=	1	6
	1	0	0	0	1	=	1	7

Since the 74193 can also count down, this IC also has a BORROW output, which will pulse HIGH when the counter cycles down to its lowest value, resets to maximum, and then repeats, as follows:

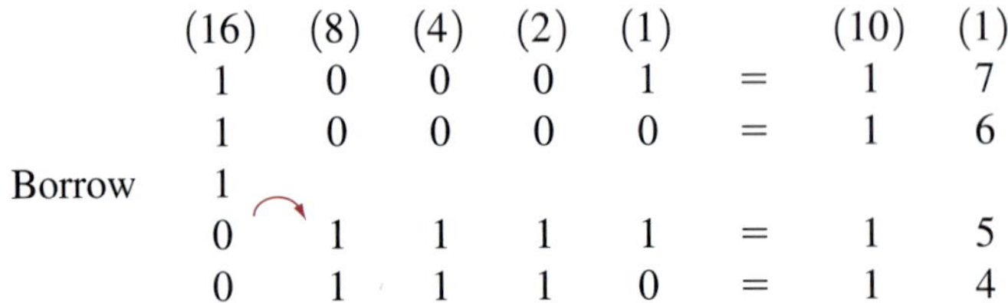

	(16)	(8)	(4)	(2)	(1)		(10)	(1)
	1	0	0	0	1	=	1	7
	1	0	0	0	0	=	1	6
Borrow	1							
	0	1	1	1	1	=	1	5
	0	1	1	1	0	=	1	4

Application—Back-and-Forth Flashing Circuit

Figure 25-28 shows how a 74193 up/down counter, a 74154 four-line to sixteen-line decoder, and a 7400 could be connected to form a back-and-forth flashing circuit. In this circuit, the 74193 is made to count up from 0000 to 1111 and is then switched to count down from 1111 to 0000, and then switched again to repeat the cycle. The NAND gates provide the count-up and count-down control, while the 74154 enables (makes LOW) its 0 through 15 outputs in response to the binary count from the 74193 counter. As a result, the sixteen LEDs in the display are sequenced ON and OFF in a continual back-and-forth action.

SELF-TEST EVALUATION POINT FOR SECTION 25-6

Use the following questions to test your understanding of Section 25-6.

1. Synchronous counters are also referred to as ripple counters. (True/False)
2. What are the main advantages of the asynchronous counter and the synchronous counter?
3. The ____________ output from an up/down counter will go active when the counter counts up to its maximum count and then resets, while the ____________ output will go active when the counter goes down to 0 and then recycles to its maximum count.
4. By cascading counters we can obtain a higher ____________.

25-7 COUNTER APPLICATIONS

The counter is one of the most versatile of all the digital circuits and is used in a wide variety of applications. In most cases, however, it is basically used to count or to divide the input signal frequency. As a counter, it generates a parallel output that represents the number of clock pulses that have occurred. As a frequency divider, it generates a serial output that is a fraction of the input frequency.

In the three circuit applications that follow we will see how counters are used as a counter and as a frequency divider.

25-7-1 *A Digital Clock*

Counters are probably most frequently used to form a digital clock circuit, like the one seen in Figure 25-29. This digital time-keeping circuit displays the hours, minutes, and seconds of the day.

Starting at the top of this circuit you can see that the power supply unit is converting the 120 V ac input into a +5 V dc output, which is used to power all of the digital logic circuits. Since the 120 V ac input from the power company alternates at an extremely accurate 60 cycles per second (60 Hz), we can make use of this timing signal to generate a one-cycle-

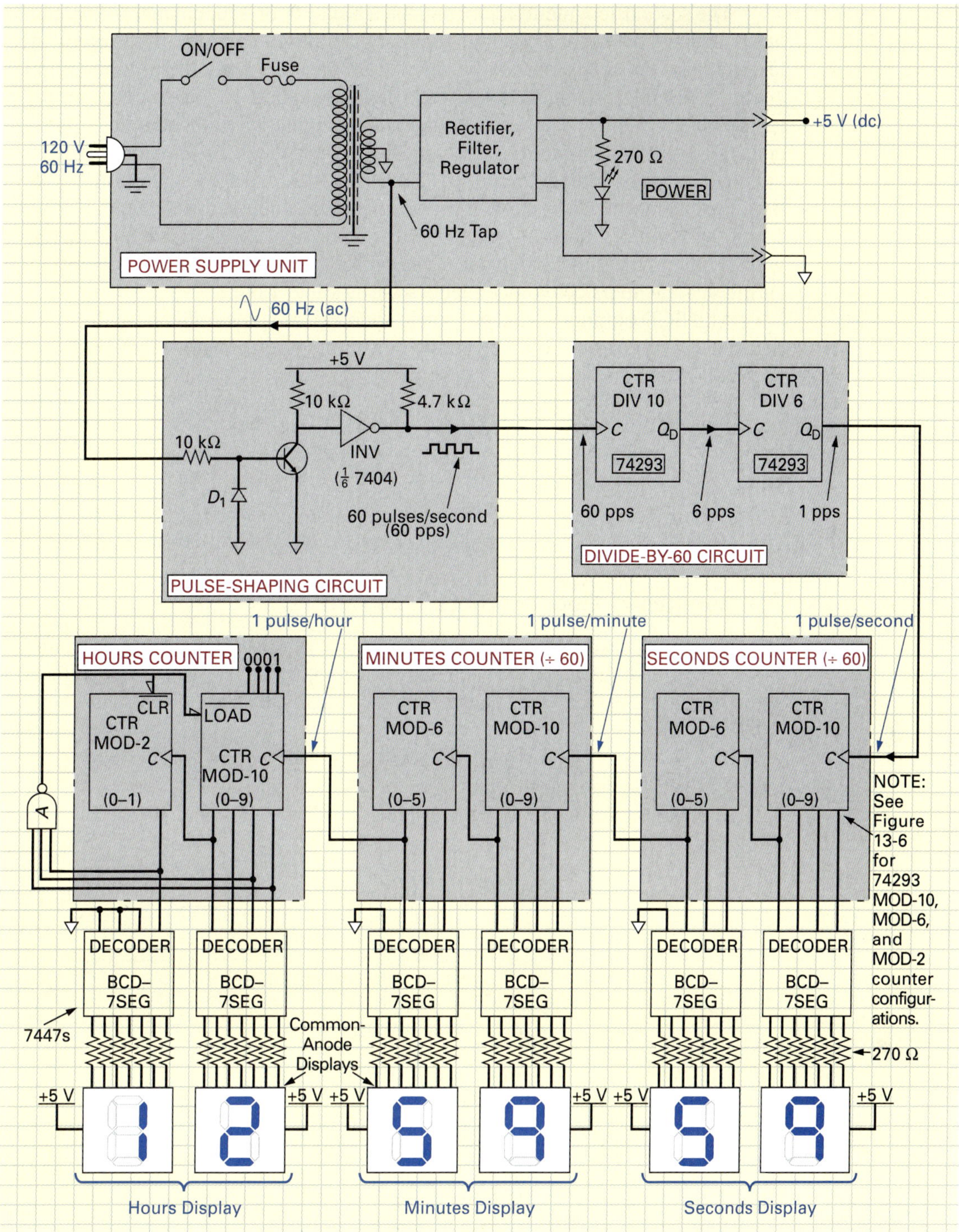

FIGURE 25-29 Application—A Digital Clock Circuit.

per-second (1 Hz) master timing, or clock, signal. This is achieved by tapping off the 60 Hz ac from the secondary of the power supply transformer and then feeding this signal into a pulse-shaping circuit that will convert the 60 Hz sine wave input into a sixty-pulses-per-second (60 pps) square wave output. The 60 pps output from the pulse-shaping circuit is then applied to a divide-by-ten counter and then to a divide-by-six counter that in combination will divide the 60 pps input by 60 to produce a one-pulse-per-second output from the divide-by-sixty circuit. This 1 pps clock signal from the divide-by-sixty circuit is the master timing signal for the seconds counter, minutes counter, and hours counter circuits.

The seconds counter circuit contains a mod-10, or ÷10, counter followed by a mod-6, or ÷6, counter. These two counters in combination will count the 1 pps clock input, and after every sixty seconds or sixty pulses will produce a 1 ppm output to the minutes counter circuit. The minutes counter circuit also contains a mod-10, or ÷10, counter followed by a mod-6, or ÷6, counter. These two counters in combination will count the 1 ppm clock input, and after every sixty minutes or sixty pulses will produce a l pph output to the hours counter circuit. The hours counter circuit contains a mod-10, or ÷10, counter followed by a mod-2, or ÷2, counter (which is a single flip-flop). These two counters in combination will count the 1 pph clock input. Once the count advances from 12 ($1\ 0010_{BCD}$) to 13 ($1\ 0011_{BCD}$), NAND gate *A* will almost immediately clear and preset load the hours counter to 1 ($0\ 0001_{BCD}$), ensuring that the hours count increments from 12:59 to 1:00.

The BCD outputs from the seconds counters, minutes counters, and hours counters are decoded by BCD-to-seven-segment 7447s, and these seven-segment codes drive the common-anode displays.

25-7-2 *A Frequency Counter*

Frequency Counter
A test instrument used to detect and display the frequency of a tested signal.

A **frequency counter** is a good example of a digital system that makes use of a clock oscillator, frequency dividers, counters, registers, decoders, and a seven-segment display.

Frequency Counter Block Diagram

Before we discuss the frequency counter circuit diagram, let us first examine the basic block diagram of the frequency counter, shown in Figure 25-30. To start at the far left of the timing diagram, you can see that the counter is cleared by a HIGH pulse just before the count

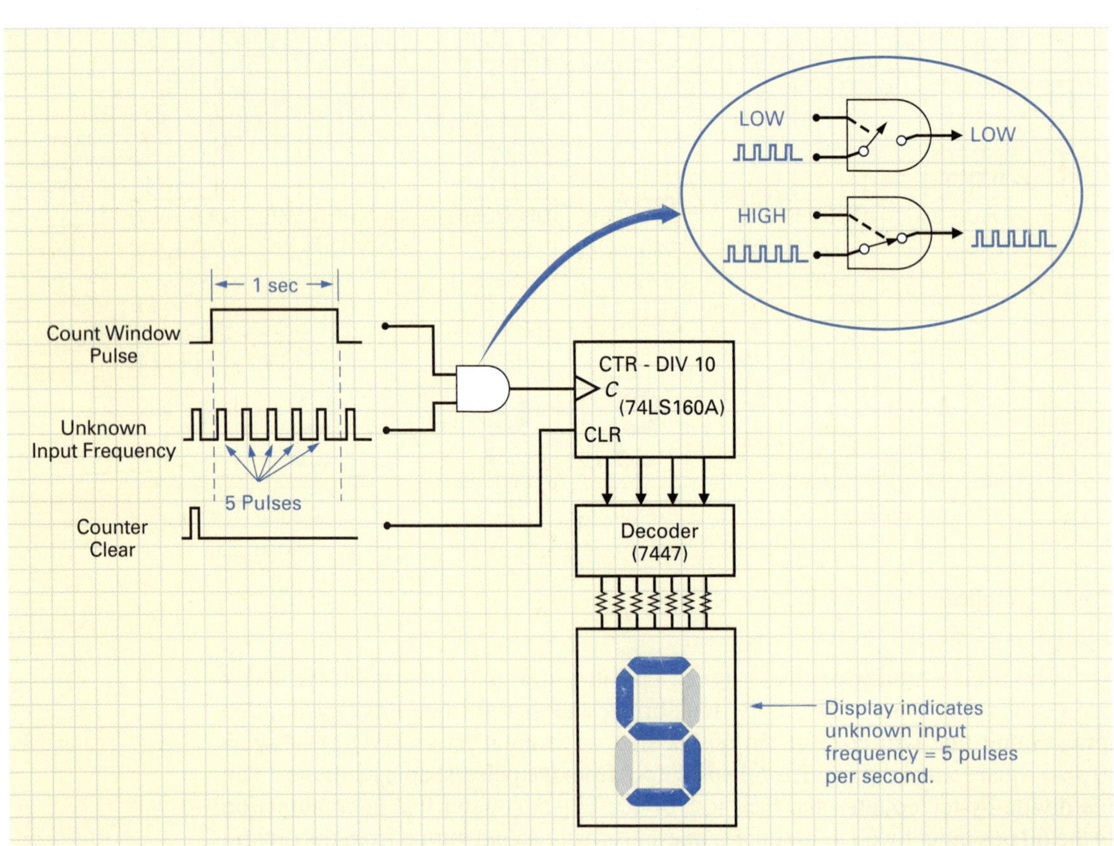

FIGURE 25-30 Simplified Frequency Counter Circuit.

window pulse begins. The count window pulse and the unknown input frequency are both applied to an AND gate, which acts as a controlled switch, as shown in the inset in Figure 25-30. When the count window pulse applies a LOW to the AND gate, it acts as an open switch since any LOW input to an AND gate produces a constant LOW out. On the other hand, when the count window pulse applies a HIGH to the AND gate, it acts as a closed switch connecting the unknown input frequency pulses through to the output. In the example shown in Figure 25-30, five pulses are switched through the AND gate to the clock input of the counter during the one-second count window pulse. These five pulses will trigger the decade counter five times, causing it to count to 5 (0101). This BCD count of 5 from the counter will be decoded by the BCD-to-seven-segment decoder and displayed on the seven-segment LED display. Since five pulses occurred in one second, the unknown input frequency is 5 pps.

The count window pulse is sometimes called the *sample pulse,* because the unknown input frequency is sampled during the time that this signal is active. The simple frequency counter circuit shown in Figure 25-30 is extremely limited, since it can only determine the unknown input frequency of inputs between 1 and 9 pps. To construct a more versatile frequency counter, we will have to include some additional digital ICs in the circuit.

Frequency Counter Circuit Diagram

Figure 25-31 shows a more complete frequency-counter circuit diagram. Although at first glance this appears to be a complex circuit, it can easily be understood by examining one block at a time.

The accuracy of a frequency counter is directly dependent on the accuracy of the count window pulse. To ensure this accuracy, a 100 kHz crystal oscillator is used to generate a very accurate square wave. The output from the crystal oscillator is used to clock a chain of divide-by-ten ICs (74LS160As) in the count window generator circuit block. The 100 kHz input at the right is successively divided by 10 to produce a 1 kHz, 100 Hz, 10 Hz, and 1 Hz output. Any one of these count window pulse frequencies can be selected by SW_1 and applied to a divide-by-two stage (1/2 7476) in the divide-by-two and monostable block. The waveforms shown in the inset in Figure 25-31 illustrate what occurs when SW_1 is placed in the 1 Hz position. The square wave input to the 7476 completes one cycle in one second; however, the positive half cycle of this waveform only lasts for half a second. By including the divide-by-two stage, the 1 Hz input is divided by 2 to produce an output frequency of 0.5 Hz, which means one complete cycle lasts for two seconds, and therefore the positive alternation, or count window pulse, lasts for one second. Including a ÷2 stage in this way means that when SW_1 selects

a. 1 Hz, the count window pulse out of the ÷2 stage will have a duration of one second.
b. 10 Hz, the count window pulse out of the ÷2 stage will have a duration of 100 ms (0.1 s).
c. 100 Hz, the count window pulse out of the ÷2 stage will have a duration of 10 ms (0.01 s).
d. 1 kHz, the count window pulse out of the ÷2 stage will have a duration of 1 ms (0.001 s).

As we will see later, these smaller count window pulses will allow us to measure higher input frequencies.

The count window pulse out of the ÷2 stage is applied to an AND gate along with the unknown input frequency. In the example timing waveforms shown in the inset in Figure 25-31, you can see that ten pulses are switched through the AND gate during the window of one second. These ten pulses are used to clock a four-stage counter in the counter, register, decoder, and display circuit block. The units counter on the right will count from 0 through 9, and if the count exceeds 9, it will reset to zero and carry a single tens pulse into the tens counter. In this example, this is exactly what will happen, and the final display will show decimal 10, meaning that the unknown input frequency is 10 pps. Similarly, the tens counter will reset-and-carry into the hundreds counter, and the hundreds counter will reset and carry into the thousands counter, as is expected when counting in decimal.

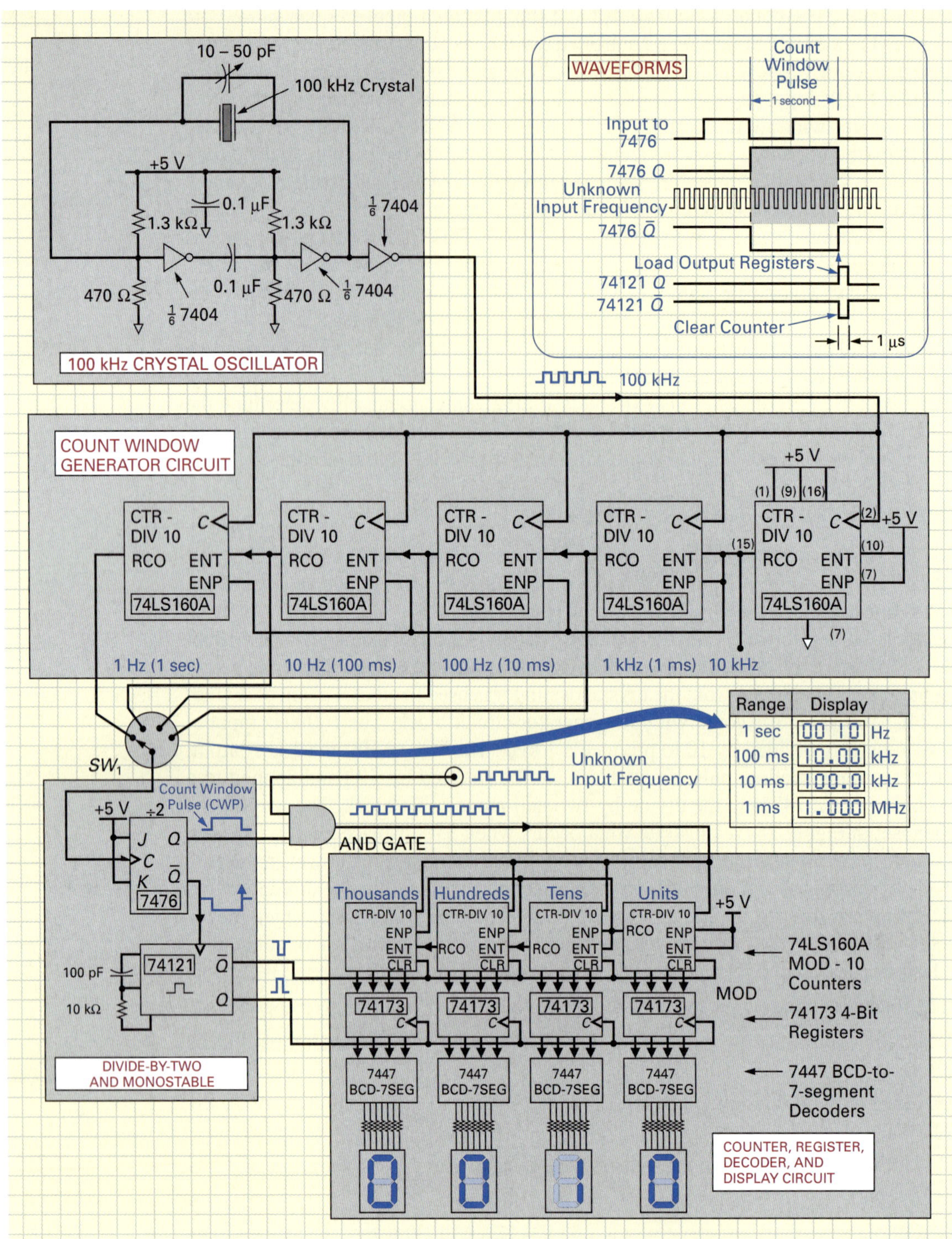

FIGURE 25-31 **Application—A Frequency Counter Circuit.**

The BCD outputs of the four-stage counter are applied to a set of registers, and the output of these registers drives a four-digit seven-segment display via BCD-to-seven-segment display decoders. Referring to the divide-by-two and monostable block, you can see that the 7476 $\overline{Q}$ output is used to trigger a 74121 one-shot, whose *RC* network has been chosen to generate a 1 μs pulse. Referring again to the timing waveforms in the inset in Figure 25-31, you can see that the positive edge of the 7476 $\overline{Q}$ output triggers the 74121, which generates a positive 1 μs pulse at its *Q* output and a negative 1 μs pulse at its $\overline{Q}$ output. The positive edge of the 74121 Q output clocks the four 74173 4-bit registers, which store the new frequency count. This stored value appears at the Q outputs of the 74173s, where it is decoded by the 7447s and displayed on the seven-segment displays. The 74173s are included to hold the display value constant, except for the few microseconds when the display is being updated to a new value. The negative 1 μs pulse at the 74121 $\overline{Q}$ output is used to clear the four-stage counter once the new frequency count has been loaded into the 74173 registers.

To explain this circuit in a little more detail, let us examine each of the frequency counter's different range selections (controlled by SW_1), which are shown in the table in Figure 25-31.

1 Hz or 1 sec Range. When SW_1 selects this frequency, a count window pulse of one second is generated. Using this window of time and a 4-digit display, we can measure any input frequency from 0001 Hz to 9999 Hz. If this maximum of 9999 Hz is exceeded, the counter will reset all four digits. For example, an input of 9999 Hz will be displayed as 9999, whereas an input of 10000 Hz will be displayed as 0000. To measure an input frequency in excess of 9999 Hz, therefore, we will have to use a smaller window so that fewer pulses are counted.

10 Hz or 100 ms Range. When SW_1 selects this frequency, a count window pulse of 1/l0 of a second is generated. This means that if the unknown input frequency is 10,000 Hz, only 1/l0 of these pulses will be switched through the AND gate to the counter (1/10 of 10,000 = 1000). This value of 1000 is smaller and can be displayed on a 4-digit display; however, a display of 1000 Hz is not accurate since the input frequency is 10,000 Hz. Since 10,000 Hz = 10 kHz, on this range we will add a decimal point in the mid position and add a prefix of kHz after the value so that the readout is correct, as shown in the table in Figure 25-31. An input of 10,000 Hz, therefore, will be displayed as 10.00 kHz.

100 Hz or 10 ms Range. When SW_1 selects this frequency, a count window pulse of 1/100 of a second is generated. This means that if the unknown input frequency is 100,000 Hz, only 1/100 of these pulses will be switched through the AND gate to the counter (1/100 of 100,000 = 1000). This value of 1000 is smaller and can be displayed on a 4-digit display; however, a display of 1000 Hz is not accurate since the input frequency is 100,000 Hz. Since 100,000 Hz = 100 kHz, on this range we will move the position of the decimal point and keep the prefix of kHz after the value so that the readout is correct, as shown in the table in Figure 25-31. An input of 100,000 Hz, therefore, will be displayed as 100.0 kHz.

1 kHz or 1 ms Range. When SW_1 selects this frequency, a count window pulse of 1/1000 of a second is generated. This means that if the unknown input frequency is 1,000,000 Hz, only 1/1000 of these pulses will be switched through the AND gate to the counter (1/1000 of 1,000,000 = 1000). This value of 1000 is smaller and can be displayed on a 4-digit display; however, a display of 1000 Hz is not accurate since the input frequency is 1,000,000 Hz. Since 1,000,000 Hz = 1 MHz, on this range we will again move the position of the decimal point and add a prefix of MHz after the value so that the readout is correct, as shown in the table in Figure 25-31. An input of 1,000,000 Hz, therefore, will be displayed as 1.000 MHz.

25-7-3 *The Multiplexed Display*

The display circuitry used for the previously described digital clock circuit and the frequency counter circuit has two basic disadvantages: It uses a large number of ICs and it uses a large amount of power. Both of these disadvantages can be solved by using a **multiplexed display circuit** similar to the one shown in Figure 25-32. This circuit needs only one 7447 decoder IC instead of the six that were needed for the digital clock circuit and the four that were needed for the frequency counter circuit. Second, since a multiplexed display has only one digit ON at a time, the total circuit current, and therefore power consumption, is dramatically less. For example, each LED in each seven-segment display draws approximately 20 mA, and therefore each seven-segment display can draw a maximum of 140 mA. If all six digits

Multiplexed Display Circuit
A display circuit in which the drive circuitry is time-shared.

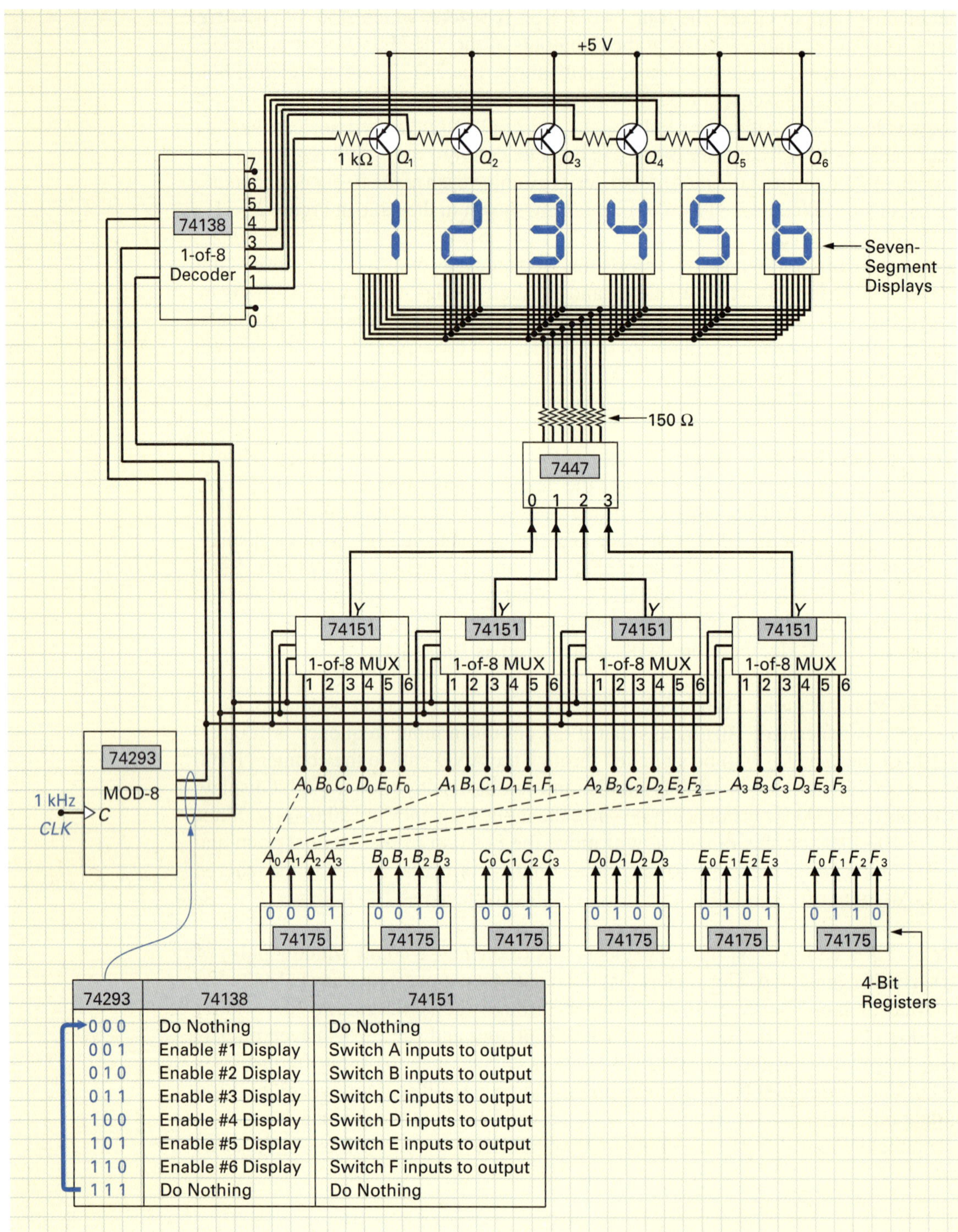

74293	74138	74151
000	Do Nothing	Do Nothing
001	Enable #1 Display	Switch A inputs to output
010	Enable #2 Display	Switch B inputs to output
011	Enable #3 Display	Switch C inputs to output
100	Enable #4 Display	Switch D inputs to output
101	Enable #5 Display	Switch E inputs to output
110	Enable #6 Display	Switch F inputs to output
111	Do Nothing	Do Nothing

FIGURE 25-32 **Application—A 6-Digit Multiplexed Display Circuit.**

in the digital clock circuit are at their maximum, a total current of 6 × 140 mA = 840 mA (0.84 A) will be needed. With a multiplexed display, only one seven-segment display is enabled at a time, and therefore the maximum display current at any one time is only 140 mA. If you combine this with the fact that each 7447 draws 260 mA, and only one 7447 (instead of four or six) is needed, you can see that the current savings are dramatic.

The multiplexed display circuit shown in Figure 25-32 operates in the following way. To *multiplex* is to simultaneously transmit two or more signals over a single channel. With the multiplexed display circuit in Figure 25-32, the 74293 mod-8 counter controls the switching operation by controlling both the 74138 1-of-8 decoder and the four 74151 1-of-8 multiplexers. Referring to the table below the circuit in Figure 25–32, you can see that when the 74293 reaches a count of 001 (decimal 1), the 74138 will make LOW its active-LOW 1 output, which will turn ON Q_1 and switch the +5 V supply to only the first seven-segment display. The 001 output from the 74293 is also applied to the four 74151 multiplexers, which will respond by switching all of their 1 inputs through to their Y outputs. This will connect the A_0, A_1, A_2, and A_3 4-bit word from the left-most 74175 register through to the 7447, where it will be decoded from BCD to a corresponding seven-segment code. This seven-segment code is applied to all six seven-segment displays; however, since only one of the displays is receiving a +5 V supply, the data will only be displayed on the left-most seven-segment display.

Referring again to the table in Figure 25-32, you can see that as the count of the 74293 advances, the displays are enabled in turn by the 74138, and the corresponding input words from the registers are switched through by the 74151 multiplexers. If the cycle is repeated at a fast enough rate, your eyes will not see the ON/OFF switching of each display, only a continuous display. **Time division multiplexing (TDM),** or the timesharing of several signals over one set of wires or a single channel, is a method frequently used in many electronic circuits and systems.

Time Division Multiplexing (TDM)
A timesharing technique in which several signals can be transmitted within a single channel.

SELF-TEST EVALUATION POINT FOR SECTION 25-7

Use the following questions to test your understanding of Section 25-7.

1. When a counter is used as a __________ it provides a parallel output that represents the number of input clock pulses applied. As a __________, a counter provides a serial output that is a fraction of the input frequency.
2. Do the counters in the seconds counter, minutes counter, and hours counter circuit in Figure 25-29 function as counter or frequency dividers?
3. Do the mod-10 counters in the count window generator circuit in Figure 25-31 function as counters or frequency dividers?
4. Do the mod-10 counters in the counter, register, decoder, and display circuit in Figure 25-31 function as counter or frequency dividers?

25-8 TROUBLESHOOTING COUNTER CIRCUITS

To review, the procedure for fixing a failure can be broken down into three steps—diagnose, isolate, and repair. Since this chapter has been devoted to counters, let us first examine the operation of a typical counter circuit and then apply our three-step troubleshooting process to this circuit.

25-8-1 *A Counter Circuit*

As an example, we will apply our three-step troubleshooting procedure to the frequency counter circuit previously discussed and shown in Figure 25-31.

Step 1: Diagnose

As before, it is extremely important that you first understand the operation of a circuit and how all of the devices within it are supposed to work so that you are able to determine whether or not a circuit malfunction really exists. If you were preparing to troubleshoot the circuit in Figure 25-31, your first step should be to read through the circuit description and review the operation of each integrated circuit until you feel completely confident with the correct operation of the entire circuit. Once you are fully familiar with the operation of the circuit, you will easily be able to diagnose the problem as either an *operator error* or a *circuit malfunction.*

Distinguishing an operator error from an actual circuit malfunction is an important first step, and a wrong diagnosis can waste much time and effort. For example, the following could be interpreted as circuit malfunctions, when in fact they are simply operator errors.

Example 1.
Symptom: SW_1 has selected the 1 ms range, input frequency = 100 Hz, display shows 0000.

Diagnosis: Operator Error—A 1 ms window will not sample any of the input since each cycle of the input lasts for 10 ms (0.01 s). To determine low input frequencies, the operator will need to use a wider window.

Example 2.
Symptom: SW_1 has selected the 100 ms range, input frequency = 12.3 kHz, display shows 1230.

Diagnosis: Operator Error—Since no circuitry is included to turn ON the decimal point in the correct position and display kHz, the display has to be interpreted based on the range selected. For example, 1230 on the 100 ms range = 12.30 kHz.

Once you have determined that the problem is not an operator error, but is in fact a circuit malfunction, proceed to step 2 and isolate the circuit failure.

Step 2: Isolate

No matter what circuit or system failure has occurred, you should always follow a logical and sequential troubleshooting procedure. Let us review some of the isolating techniques and then apply them to our example circuit in Figure 25-31.

a. Use a cause-and-effect troubleshooting process, which means study the effects you are getting from the faulty circuit and then logically reason out what could be the cause.

b. Check first for obvious errors before leaping into a detailed testing procedure. Is the power OFF or not connected to the circuit? Are there wiring errors? Are all of the ICs correctly oriented?

c. Using a logic probe or voltmeter, test that power and ground are connected to the circuit and are present at all points requiring power and ground. If the whole circuit, or a large section of the circuit, is not operating, the problem is normally power. Using a multimeter, check that all of the dc voltages for the circuit are present at all IC pins that should have power or a HIGH input and are within tolerance. Second, check that 0 V, or ground, is connected to all IC ground pins and to all inputs that should be tied LOW.

d. Use your senses to check for broken wires, loose connectors, overheating or smoking components, pins not making contact, and so on.

e. Test the clock signals using the logic probe or the oscilloscope. Although clock signals can be easily checked with a logic probe, the oscilloscope is best for checking the signal's amplitude, frequency, pulse width, and so on. The oscilloscope will also display timing problems and noise spikes (glitches), which could be false-triggering a circuit into operation at the wrong time. With the frequency counter circuit in Figure 25-31, you should first test the 100 kHz output from the crystal oscillator and then test in sequence the 10 kHz, 1 kHz, 100 Hz, 10 Hz, and 1 Hz outputs from the count window generator circuit. Problems in frequency divider chains are generally easily detected since an incorrect frequency output normally indicates that the previous stage has malfunctioned.

f. Perform a static test on the circuit. With our circuit example in Figure 25-31, you should disconnect the inputs to the counter, register, decoder, and display circuit and then use a logic pulser to clock the counter and a logic probe to test that the valid count is appearing at the outputs of each stage of the counter. For example, after nineteen clock pulse inputs from the logic pulser, the logic probe should indicate the following logic levels at the counter outputs.

0000	0000	1000	1001_{BCD}
0	0	1	9

By pulsing the register clock input, this value should appear at the output of the 74173s, be decoded by the 7447s, and be displayed. By pulsing the counter clear input, you should be able to clear all of the outputs of the counter. The divide-by-two and monostable circuit can also be tested in this way by disconnecting the inputs and using the logic pulser as a signal source and the logic probe as a signal detector.

g. With a dynamic test, the circuit is tested while it is operating at its normal clock frequency (which in Figure 25-31 is 100 kHz), and therefore all of the inputs and outputs are continually changing. Although a logic probe can detect a pulse waveform, the oscilloscope will display more signal detail, and since it can display more than one signal at a time, it is ideal for making signal comparisons and looking for timing errors. If the clock section is suspected, you can check the rest of the circuit by disconnecting the clock input to the circuit and then using a function generator to generate a 100 kHz clock.

In some instances, you will discover that some circuits will operate normally when undergoing a static test and yet fail a dynamic test. This effect usually points to a timing problem involving the clock signals and/or the propagation delay times of the ICs used within the circuit.

h. Noise due to electromagnetic interference (EMI) can false-trigger an IC, such as the counters in Figure 25-31. This problem can be overcome by not leaving any of the IC's inputs unconnected and, therefore, floating. Connect unneeded active-HIGH inputs to ground and active-LOW inputs to $+V_{CC}$.

i. Apply the half-split method of troubleshooting first to a circuit and then to a section of a circuit to help speed up the isolation process. With our circuit example in Figure 25-31, a good midpoint check would be to test the different frequency outputs from the count window generator circuit.

Also remember that a load problem can make a source appear at fault. If an output is incorrect, it would be safer to disconnect the source from the rest of the circuit and then recheck the output. If the output is still incorrect, the problem is definitely within the source; however, if the output is correct, the load is probably shorting the line either HIGH or LOW.

j. Substitution can be used to help speed up your troubleshooting process. Once the problem is localized to an area containing only a few ICs, substitute suspect ICs with known-working ICs (one at a time) to see if the problem can be quickly remedied.

k. Many electronic system manufacturers provide troubleshooting trees in the system technical manual. These charts are a graphical means to show the sequence of tests to be performed on a suspected circuit or system. Figure 25-33 shows a simple troubleshooting tree for our frequency counter circuit.

Step 3: Repair

Once the fault has been found, the final step is to repair the circuit and then perform a final test to see that the circuit and the system are now fully operational.

25-8-2 *Sample Problems*

Once you have constructed a circuit like the circuit in Figure 25-31, introduce a few errors to see what effects or symptoms they produce. Then logically reason out why a particular error or cause has a particular effect on the circuit. Never short any two points together unless you have carefully thought out the consequences, but generally it is safe to open a path and see the results. Here are some problems for our example circuit in Figure 25-31.

a. Remove the 100 kHz crystal.

b. Disconnect one of the *RCO* outputs in the count window generator circuit.

c. Disconnect the clock input to one of the 74173s.

d. Disconnect the $\overline{Q}$ output of the divide-by-two circuit.

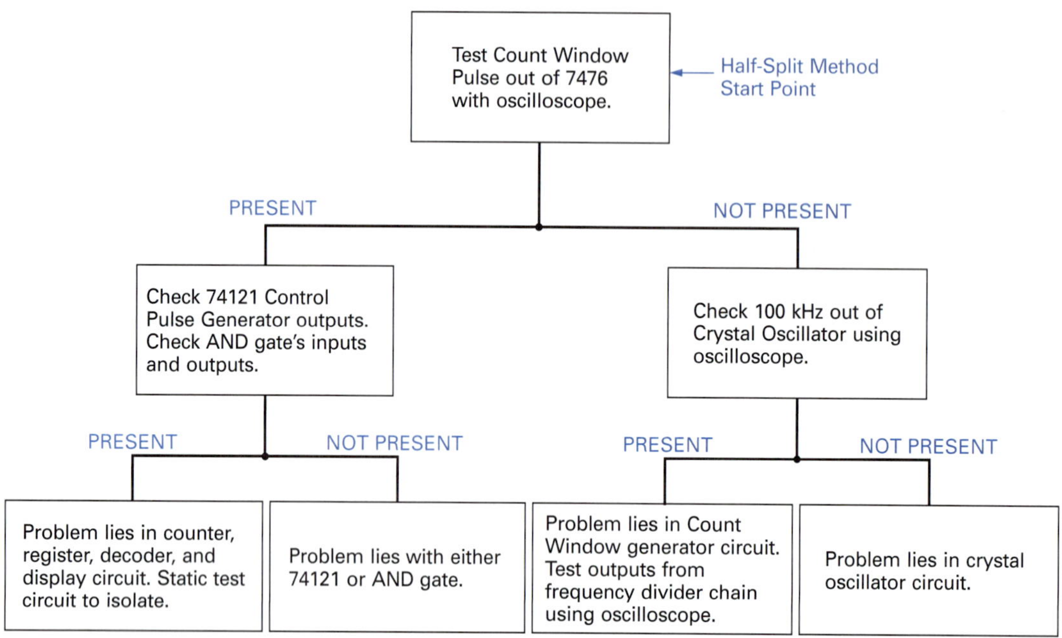

FIGURE 25-33 **A Troubleshooting Tree for the Frequency Counter Circuit.**

e. Disconnect the Q output of the 74121.

f. Disconnect the *ENT* and *ENP* inputs of the first 74LS160A counter from the +5 V and connect these lines to ground.

g. Disconnect one of the inputs to a 7447.

SELF-TEST EVALUATION POINT FOR SECTION 25-8

Use the following questions to test your understanding of Section 25-8.

1. What are the three basic troubleshooting steps?
2. Which of the 74LS160As in Figure 25-31 function as frequency dividers and which function as counters?
3. To test the count window generator circuit in Figure 25-31, the clock oscillator input is disconnected and a logic pulser is used to inject a 500 kHz input. Does a problem exist if outputs of 5 kHz, 500 Hz, 50 Hz, and 5 Hz are measured with the oscilloscope at each of the SW_1 positions?
4. What symptom would you get from the circuit in Figure 25-31 if you were to apply a permanent HIGH to the unknown frequency input?

REVIEW QUESTIONS

Multiple-Choice Questions

1. The ____________ of a counter describes the number of different output combinations generated by the counter.
   a. division factor
   b. modulus
   c. maximum count
   d. propagation delay time
2. The ____________ of a counter describes the time it will take for all of the outputs to be updated.
   a. division factor
   b. modulus
   c. maximum count
   d. propagation delay time
3. The ____________ of a counter describes how much lower the final flip-flop's output frequency will be compared to the input clock frequency.
   a. division factor
   b. modulus
   c. maximum count
   d. propagation delay time

4. A 4-bit SISO shift register will take ____________ clock pulse(s) to store a 4-bit word.
   a. one **c.** three
   b. two **d.** four

5. A 4-bit PISO shift register will take ____________ clock pulse(s) to store a 4-bit word.
   a. one **c.** three
   b. two **d.** four

6. Which of the following shift registers could be used for serial-to-parallel data conversion?
   a. SISO **d.** PIPO
   b. PISO **e.** All of the above
   c. SIPO

7. Which of the following shift registers could be used for parallel-to-serial data conversion?
   a. SISO **d.** PIPO
   b. PISO **e.** All of the above
   c. SIPO

8. Which of the following is considered a serial-in, serial-out shift register operation?
   a. Shift-right
   b. Shift-left
   c. Rotate-right
   d. Rotate-left
   e. All of the above

9. When an operation is in step with the clock signal, it is called a(an) ____________ operation.
   a. asynchronous
   b. UART
   c. sequenced
   d. synchronous

10. When an operation is not in step with the clock signal, it is called a(an) ____________ operation.
    a. asynchronous
    b. UART
    c. sequenced
    d. synchronous

11. A buffer register with three-state output control will generally have three control line inputs called ____________, ____________ and ____________.
    a. V_{CC}, clock, S_1
    b. clock, enable, load
    c. SRI, load, S_0
    d. shift-left, shift-right, inhibit

12. New data is ____________ into a memory register, while stored data is ____________ out of a memory register.
    a. read, written
    b. written, read

13. A universal asynchronous receiver/transmitter (UART) has a ____________ and a ____________ internally for interfacing between a bus and a single line.
    a. serial-to-parallel shift register, parallel-to-serial shift register
    b. memory register, arithmetic shift register
    c. ring counter, johnson counter
    d. buffer register, PIPO register

14. What will happen to the value of a binary number stored in a shift register if it is shifted to the left by three positions and zeros are shifted in through the LSB position?
    a. Value will be divided by 8.
    b. Value will be multiplied by 6.
    c. Value will be multiplied by 8.
    d. Value will be divided by 4.

15. With a ring counter, the ____________ output of the last flip-flop is connected back to the input of the first flip-flop, whereas with a johnson counter the ____________ output of the last flip-flop is connected back to the input of the first flip-flop.
    a. Q, CLEAR
    b. $\overline{Q}$, Q
    c. PRESET, $\overline{Q}$
    d. $Q, \overline{Q}$

16. The ____________ of a counter describes the highest binary value that will appear at the parallel outputs.
    a. division factor
    b. modulus
    c. maximum count
    d. propagation delay time

17. A mod-10 or decade counter will divide the input clock frequency by a factor of ____________.
    a. 16 **c.** 10
    b. 8 **d.** 100

18. A four-stage binary counter will have a natural modulus of ____________.
    a. 16 **c.** 10
    b. 8 **d.** 100

19. A(an) ____________ counter will start counting from a loaded value up to its natural modulus.
    a. synchronous
    b. asynchronous
    c. presettable
    d. cascaded

20. The maximum counting speed of a counter is limited by the counter's ____________.
    a. division factor
    b. modulus
    c. maximum count
    d. propagation delay time

21. What is the primary advantage of a synchronous counter?
    a. High counting speed
    b. Long propagation delay
    c. Circuit simplicity
    d. Ripple action

22. What is the primary advantage of an asynchronous counter?
    a. High counting speed
    b. Long propagation delay
    c. Circuit simplicity
    d. Ripple action

23. Cascaded up counters generate a ____________ output to the next significant counter.
    a. carry output **c.** *RCO*
    b. borrow output **d.** both a. and c. are true

24. Cascaded down counters generate a ____________ output to the next significant counter.
   a. carry output
   b. borrow output
   c. *RCO*
   d. both a. and c. are true

25. A state transition diagram shows a counter's ____________.
   a. output states **c.** propagation delay
   b. division factor **d.** modulus

Practice Problems

To practice your circuit recognition and operation, refer to Figure 25-34 and answer the following four questions.

26. Describe the 74175 IC in one sentence.

27. Which type of flip-flop is being used to form these two register circuits?

28. The circuit configuration in Figure 25-34(a) is a ____________.
   a. PIPO **c.** SIPO
   b. SISO **d.** PISO

29. The circuit configuration in Figure 25-34(b) is a ____________.
   a. PIPO **c.** SIPO
   b. SISO **d.** PISO

30. Calculate the frequency division factor of the following counters:
   a. mod-12 counter
   b. mod-4096 counter
   c. a counter with 8,192 different output states
   d. a counter with a maximum count of 15

31. Calculate the maximum count of a
   a. 4-bit counter.
   b. 6-bit counter.
   c. 24-bit counter.

32. Calculate the modulus of a
   a. 4-bit counter.
   b. 6-bit counter.
   c. 24-bit counter.

33. Assuming a flip-flop propagation delay of 25 ns, calculate the total asynchronous counter propagation delay time and the upper clock pulse frequency limit for the following:
   a. 4-bit counter
   b. 6-bit counter
   c. 24-bit counter

To practice your circuit recognition and operation skills, refer to Figure 25-35 and answer the following seven questions.

34. Is this circuit a synchronous or an asynchronous counter?

35. What is this counter's natural modulus?

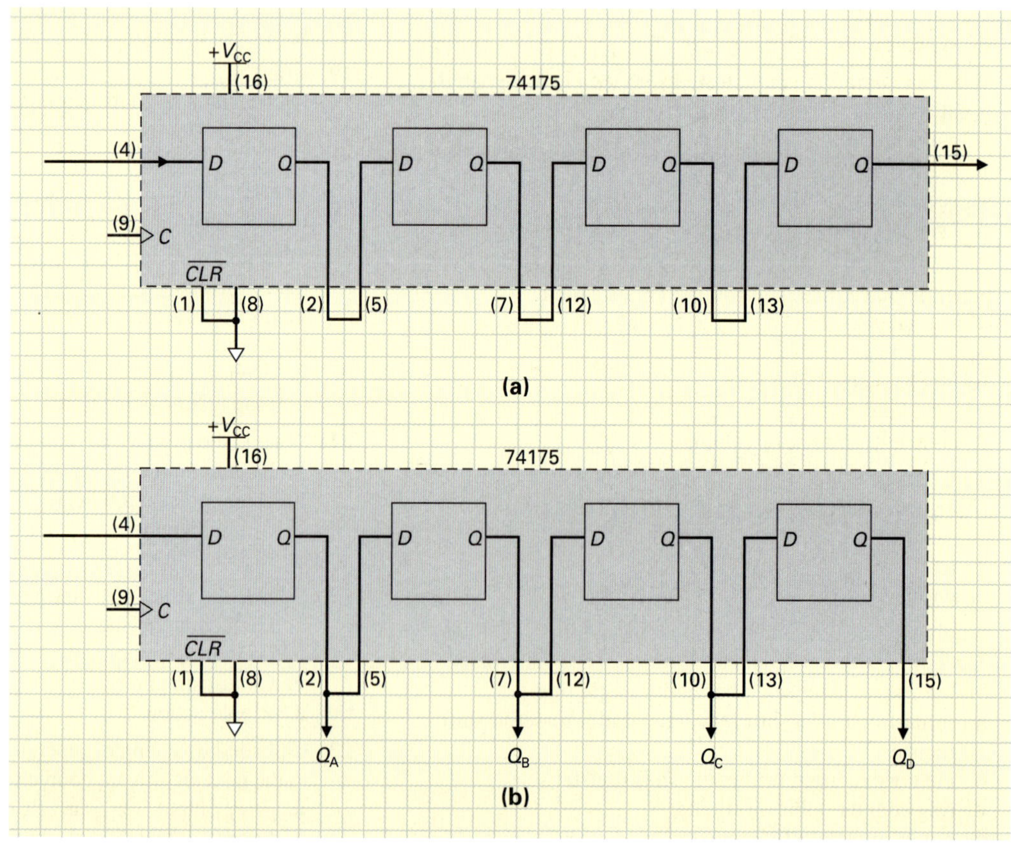

FIGURE 25-34 74175 Register Configurations.

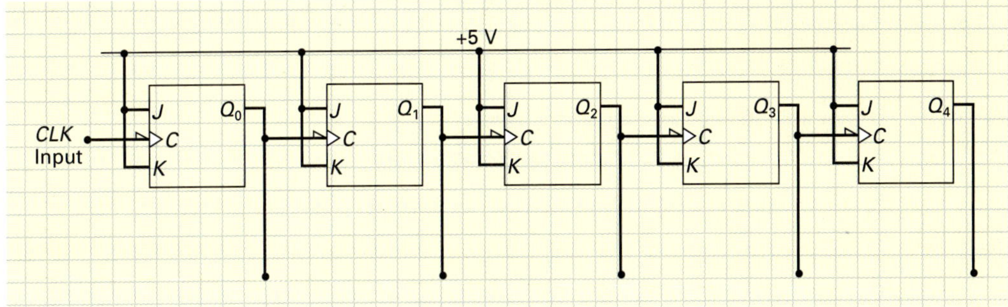

FIGURE 25-35 **Application—A 5-Stage Up Counter.**

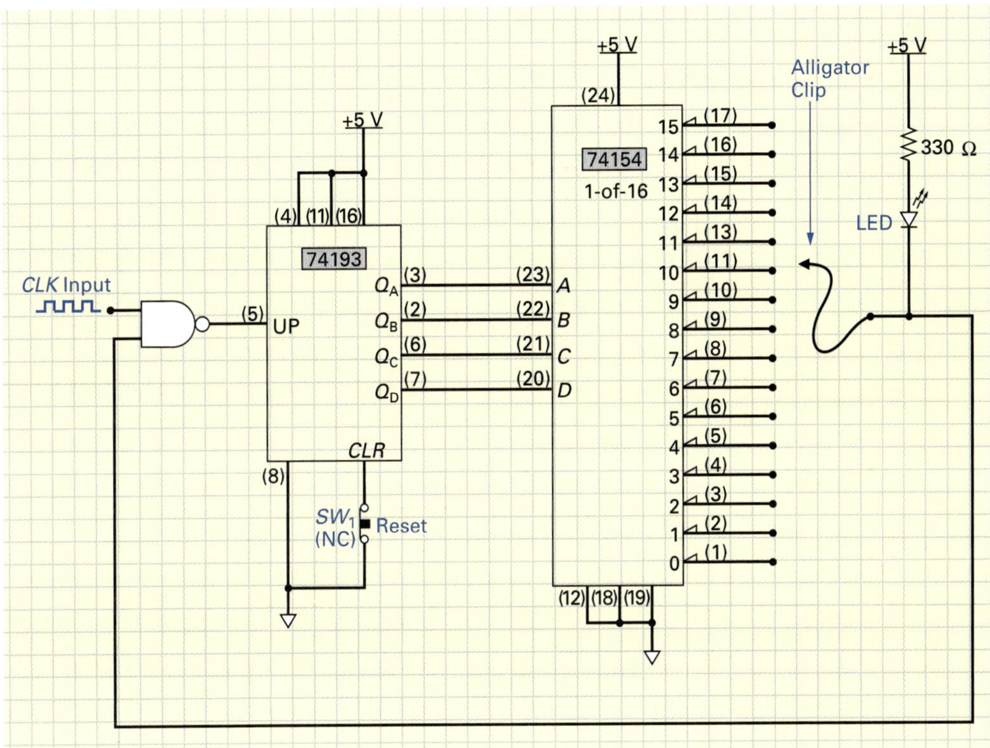

FIGURE 25-36 **Application—Count Up to a Selected Value and Stop Circuit.**

36. What is this counter's maximum count?

37. What is this counter's frequency division between the clock input and Q_4?

38. What would be the Q_4 output frequency if the clock input were 2 kHz?

39. What would be this counter's upper clock pulse frequency limit if each flip-flop had a propagation delay of 22 ns?

40. Assuming an 8-bit asynchronous presettable counter, calculate

a. its natural modulus.

b. its preset modulus if 0011 0100 is loaded.

To practice your circuit recognition and operation skills, refer to Figure 25-36 and answer the following question.

41. Briefly describe the operation of the

a. 74193 IC in this circuit including the NAND gate's function at the clock input and the function of SW_1.

b. 74154 IC in this circuit.

c. If the alligator clip is connected to the 74154's pin 14 output and SW_1 is pressed, how will this circuit operate?

Web Site Questions

Go to the web site http://www.prenhall.com/cook, select the textbook *Electronics: A Complete Course,* select this chapter, and then follow the instructions when answering the multiple choice practice problems.

Arithmetic Operations and Circuits

The Wizard of Menlo Park

Thomas Alva Edison was born to Samuel and Nancy Edison on February 11, 1847. As a young boy he had a keen and inquisitive mind, yet he did not do well at school, so his mother, a former schoolteacher, withdrew him from school and tutored him at home. In later life he said that his mother taught him to read well and instilled a love for books that lasted the rest of his life. In fact, the inventor's personal library of more than 10,000 volumes is preserved at the Edison Laboratory National Monument in West Orange, New Jersey.

At the age of twenty-nine, after several successful inventions, Edison put into effect what is probably his greatest idea—the first industrial research laboratory. Choosing Menlo Park in New Jersey, which was then a small rural village, Edison had a small building converted into a laboratory for his fifteen-member staff and a house built for his wife and two small daughters. When asked to explain the point of this lab, Edison boldly stated that it would produce "a minor invention every ten days and a big thing every six months or so." At the time, most of the scientific community viewed his prediction as preposterous; however, in the next ten years Edison would be granted 420 patents, including those for the electric light bulb, the motion picture, the phonograph, the universal electric motor, the fluorescent lamp, and the medical fluoroscope.

Over 1,000 patents were granted to Edison during his lifetime; and his achievements at what he called his "invention factory" earned him the nickname "the wizard of Menlo Park." When asked about his genius he said, "Genius is two percent inspiration and ninety-eight percent perspiration."

26

Introduction

At the heart of every digital electronic computer system is a circuit called a microprocessor unit (MPU). To perform its function, the MPU contains several internal circuits, and one of these circuits is called the arithmetic-logic unit, or ALU. This unit is the number-crunching circuit of every digital system and, as its name implies, it can be controlled to perform either arithmetic operations (such as addition and subtraction) or logic operations (such as AND and OR).

In order to understand the operation of an ALU, we must begin by discussing how binary numbers can represent any value, whether it is large, small, positive, or negative. We must also discuss how these numbers can be manipulated to perform arithmetic operations such as addition, subtraction, multiplication, and division. All of these topics will be covered in the first section of this chapter.

In the second section of this chapter, we will discuss how logic circuits can be connected to form arithmetic circuits, and how these circuits can perform arithmetic operations.

In the third section of this chapter, we will combine all of the topics covered in this chapter and see how an ALU can perform either arithmetic operations or logic operations.

In the fourth section of this chapter, we will practice circuit troubleshooting techniques by applying our three-step troubleshooting procedure to a typical arithmetic circuit.

26-1 ARITHMETIC OPERATIONS

Before we discuss the operation of arithmetic circuits, we must first understand the arithmetic operations these circuits perform. In the first section we will see how two binary numbers are added together, how one binary number can be subtracted from another, how one binary number is multiplied by another, and how one binary number is divided into another—in short, the principles of binary arithmetic.

26-1-1 *Binary Arithmetic*

The addition, subtraction, multiplication, and division of binary numbers is performed in exactly the same way as the addition, subtraction, multiplication, and division of decimal numbers. The only difference is that the decimal number system has ten digits, whereas the binary number system has two.

As we discuss each of these arithmetic operations, we will use decimal arithmetic as a guide to binary arithmetic so that we can compare the known to the unknown.

Binary Addition

Figure 26-1(a) illustrates how the decimal values 29,164 and 63,729 are added together. First the units 4 and 9 are added, producing a sum of 13, or 3 carry 1. The carry is then added to the 6 and 2 in the tens column, producing a sum of 9 with no carry. This process is then repeated for each column, moving right to left, until the total sum of the addend, augend, and their carries has been obtained.

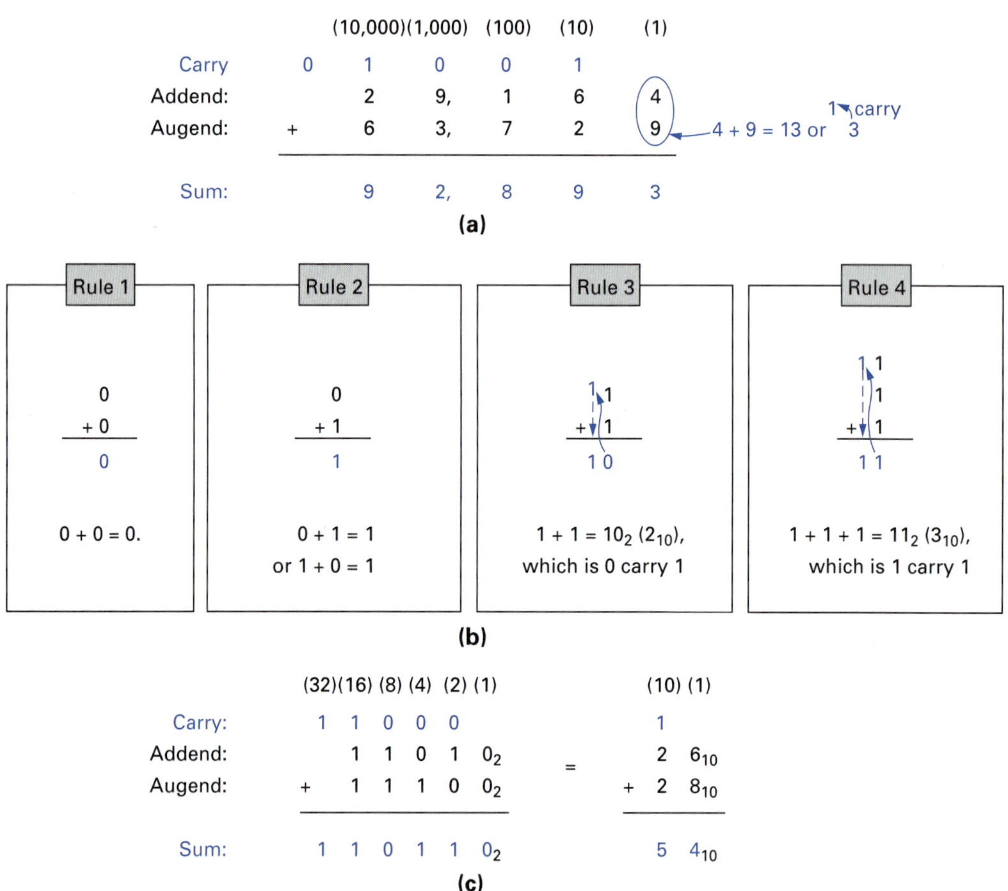

FIGURE 26-1 Binary Addition.

When adding two decimal values, many combinations are available since decimal has ten different digits. With binary, however, only two digits are available and therefore only four basic combinations are possible. These four rules of binary addition are shown in Figure 26-1(b). To help understand these rules, let us apply them to the example in Figure 26-1(c) in which we will add 11010 (decimal 26) to 11100 (decimal 28). Like decimal addition, we will begin at the right in the ones column and proceed to the left. In the ones column 0 + 0 = 0 with a carry of 0 (Rule 1). In the twos column 1 + 0 and the 0 carry from the ones column = 1 with no carry (Rule 2). In the fours column we have 0 + 1 plus a carry of 0 from the twos column. This results in a fours column total of 1, with a carry of 0 (Rule 2). In the eights column we have 1 + 1 plus a 0 carry from the fours column, producing a total of 10_2 (1 + 1 = decimal 2, which is 10 in binary). In the eights column this 10 (decimal 2) is written down as a total of 0 with a carry of 1 (Rule 3). In the sixteens column, we have 1 + 1 plus a carry of 1 from the eights column, producing a total of 11_2 (1 + 1 + 1 = decimal 3, which is 11 in binary). In the sixteens column, this 11 (decimal 3) is written down as a total of 1 with a carry of 1 (Rule 4). Since there is only a carry of 1 in the thirty-twos column, the total in this column will be 1.

Converting the binary addend and augend to their decimal equivalents as seen in Figure 26-1(c), you can see that 11010 (decimal 26) plus 11100 (decimal 28) will yield a sum or total of 110110 (decimal 54).

EXAMPLE:

Find the sum of the following binary numbers.

a. 1011 + 1101 = ? b. 1000110 + 1100111 = ?

Solution:

	Carry:		1	1	1	1			
	Augend			1	0	1	1_2		(11_{10})
a.	*Addend:*	+		1	1	0	1_2		(13_{10})
	Sum:		1	1	0	0	0_2		(24_{10})

	Carry:		1	0	0	0	1	1	0		
	Augend			1	0	0	0	1	1	0_2	(70_{10})
b.	*Addend:*	+		1	1	0	0	1	1	1_2	(103_{10})
	Sum:		1	0	1	0	1	1	0	1_2	(173_{10})

Binary Subtraction

Figure 26-2(a) reviews the decimal subtraction procedure by subtracting 4,615 from 7,003. Starting in the units column, you can see that since we cannot take 5 away from 3, we will have to go to a higher-order minuend unit and borrow. Since the minuend contains no tens or hundreds, we will have to go to the thousands column. From this point a chain of borrowing occurs as a thousand is borrowed and placed in the hundreds column (leaving 6 thousands), one of the hundreds is borrowed and placed in the tens column (leaving 9 hundreds), and one of the tens is borrowed and placed in the units column (leaving 9 tens). After borrowing 10, the minuend units digit has a value of 13, and if we now perform the subtraction, the result or difference of 13 − 5 = 8. Since all of the other minuend digits are now greater than their respective subtrahend digits, there will be no need for any further borrowing to obtain the difference.

When subtracting one decimal value from another, many combinations are available since decimal has ten different digits. With binary, however, only two digits are available and therefore only four basic combinations are possible. These four rules of binary subtraction are shown in Figure 26-2(b). To help understand these rules, let us apply them to the example in Figure 26-2(c), in which we will subtract 11100 (decimal 28) from 1010110 (decimal 86). Like

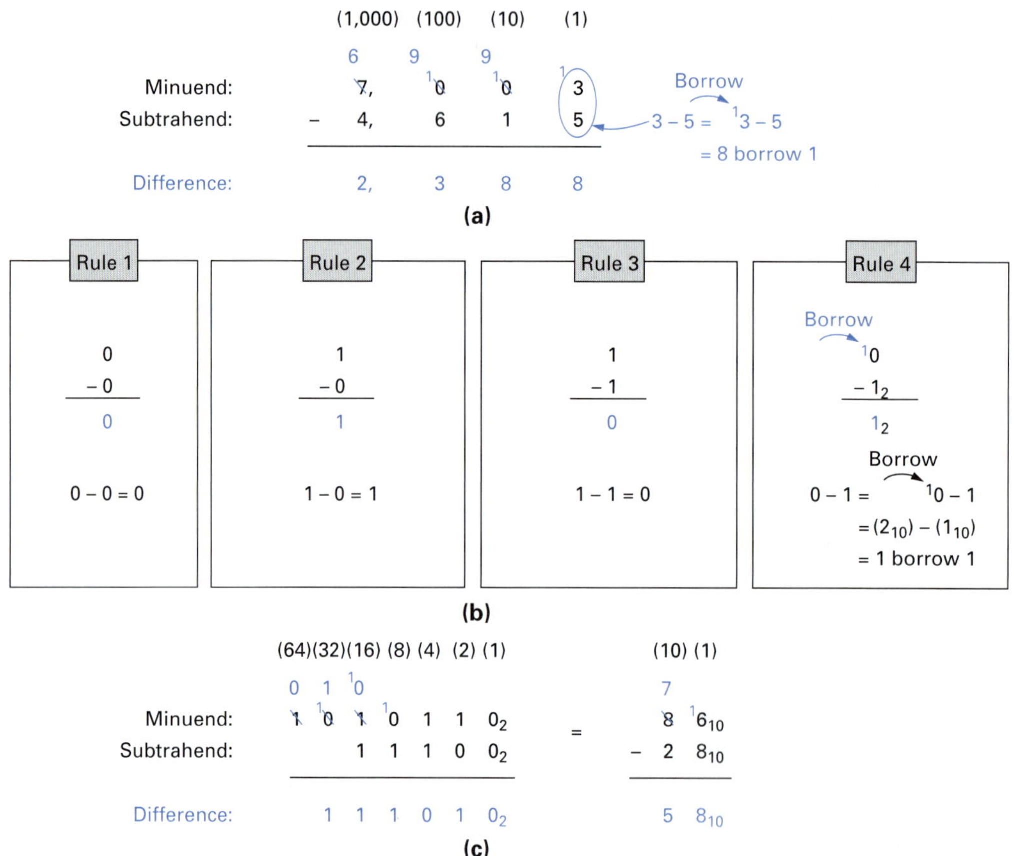

FIGURE 26-2 **Binary Subtraction.**

decimal subtraction, we will begin at the right in the ones column and proceed to the left. In the ones column $0 - 0 = 0$ (Rule 1). In the twos column $1 - 0 = 1$ (Rule 2), and in the fours column $1 - 1 = 0$ (Rule 3). In the eights column we cannot subtract 1 from nothing or 0, so therefore we must borrow 1 from the sixteens column, making the minuend in the eights column 10 (decimal 2). The difference can now be calculated in the eights column since $10 - 1 = 1$ (decimal $2 - 1 = 1$, Rule 4). Due to the previous borrow, the minuend in the sixteens column is now 0 and therefore we need to borrow a 1 from the thirty-twos column. Since the thirty-twos column is also 0, we will need to borrow from the sixty-fours column. Borrowing 1 from the sixty-fours column will leave a minuend of 0 and make the minuend in the thirty-twos column equal to 10 (decimal 2). Borrowing 1 from the 10 (borrowing decimal 1 from 2) in the thirty-twos column will leave a minuend of 1 and make the minuend in the sixteens column equal to 10 (decimal 2). We can now subtract the subtrahend of 1 from the minuend of 10 (decimal 2) in the sixteens column to obtain a difference of 1. Due to the previous borrow, the minuend in the thirty-twos column is equal to 1, and $1 - 0 = 1$ (Rule 2). Finally, since the minuend and subtrahend are both 0 in the sixty-fours column, the subtraction is complete.

Converting the binary minuend and subtrahend to their decimal equivalents, as seen in Figure 26-2(c), you can see that 1010110 (decimal 86) minus 11100 (decimal 28) will result in a difference of 111010 (decimal 58).

EXAMPLE:

Find the difference between the following binary numbers.

a. $100111 - 11101 = ?$

b. $101110111 - 1011100 = ?$

■ ***Solution:***

	Minuend:		$^{0}\not{1}$	$^{1}\not{0}$	$^{1}0$	1	1	1_2	(39_{10})		
a.	*Subtrahend:*	−		1	1	1	0	1_2	(29_{10})		
	Difference:		0	0	1	0	1	0_2	(10_{10})		

	Minuend:		1	0	1	$^{0}\not{1}$	$^{10}\not{1}$	$^{1}0$	1	1	1_2	(375_{10})
b.	*Subtrahend:*	−			1	0	1	1	1	0	0_2	(92_{10})
	Difference:		1	0	0	0	1	1	0	1	1_2	(283_{10})

Binary Multiplication

Figure 26-3(a) reviews the decimal multiplication procedure by multiplying 463 by 23. To perform this operation, you would begin by multiplying each digit of the multiplicand by the units digit of the multiplier to obtain the first partial product. Second, you would multiply each digit of the multiplicand by the tens digit of the multiplier to obtain the second partial product. Finally, all of the partial products are added to obtain the final product.

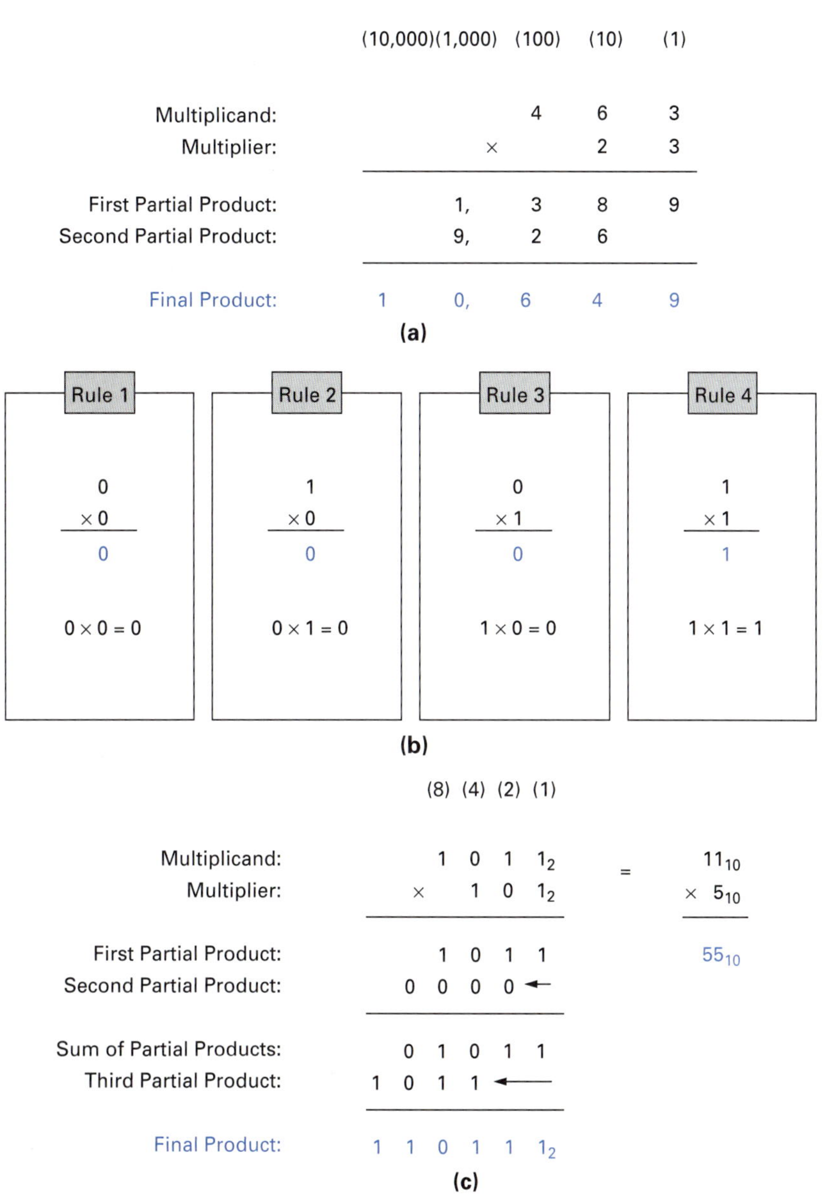

FIGURE 26-3 Binary Multiplication.

When multiplying one decimal value by another, many combinations are available since decimal has ten different digits. With binary multiplication, however, only two digits are available and therefore only four basic combinations are possible. These four rules of binary multiplication are shown in Figure 26-3(b). To help understand these rules, let us apply them to the example in Figure 26-3(c), in which we will multiply 1011 (decimal 11) by 101 (decimal 5). Like decimal multiplication, we begin by multiplying each bit of the multiplicand by the ones column multiplier bit to obtain the first partial product. Second, we multiply each bit of the multiplicand by the twos column multiplier bit to obtain the second partial product. Remember that the LSB of the partial product must always be directly under its respective multiplier bit, and therefore the LSB of the second partial product should be under the twos column multiplier bit. In the next step, we add together the first and second partial products to obtain a sum of the partial products. With decimal multiplication we usually calculate all of the partial products, and then add them together to obtain the final product. This procedure can also be used with binary multiplication; however, it is better to add only two partial products at a time since the many carries in binary become hard to keep track of and easily lead to errors. In the next step, we multiply each bit of the multiplicand by the fours column multiplier bit to obtain the third partial product. Remember that the LSB of this third partial product must be placed directly under its respective multiplier bit. The final step is to add the third partial product to the previous partial sum to obtain the final product or result.

If you study the partial products obtained in this example, you will notice that they were either exactly equal to the multiplicand (when the multiplier was 1) or all 0s (when the multiplier was 0). You can use this shortcut in future examples; however, be sure to always place the LSB of the partial products directly below their respective multiplier bits.

EXAMPLE:

Multiply 1101_2 by 1011_2.

Solution:

Multiplicand:					1	1	0	1_2	13_{10}
Multiplier:				×	1	0	1	1_2	$\times\ 11_{10}$
First Partial Product:					1	1	0	1	143_{10}
Second Partial Product:				1	1	0	1	←	
Sum of Partial Product:			1	0	0	1	1	1	
Third Partial Product:			0	0	0	0	←	←	
Sum of Partial Product:			1	0	0	1	1	1	
Fourth Partial Product:		1	1	0	1	←	←	←	
Final Product:	1	0	0	0	1	1	1	1_2	

Binary Division

Figure 26-4(a) reviews the decimal division procedure by dividing 830 by 23. This long-division procedure begins by determining how many times the divisor (23) can be subtracted from the first digit of the dividend (8). Since the dividend is smaller, the quotient is 0. Next, we see how many times the divisor (23) can be subtracted from the first two digits of the dividend (83). Since 23 can be subtracted three times from 83, a quotient of 3 results, and the product of 3×23, or 69, is subtracted from the first two digits of the dividend, giving a remainder of 14. Then the next digit of the dividend (0) is brought down to the remainder (14), and we see how many times the divisor (23) can be subtracted from the new remainder (140). In this example, 23 can be subtracted six times from 140, a quotient of 6 results, and the product of 6×23, or 138, is subtracted from the remainder, giving a new remainder of 2. Therefore, $830 \div 23 = 36$, remainder 2.

Figure 26-4(b) illustrates an example of binary division, which is generally much simpler than decimal division. To perform this operation, we must first determine how many times the divisor (101_2) can be subtracted from the first bit of the dividend (1_2). Since the dividend is smaller than the divisor, a quotient of 0_2 is placed above the first digit of the dividend. Next, we

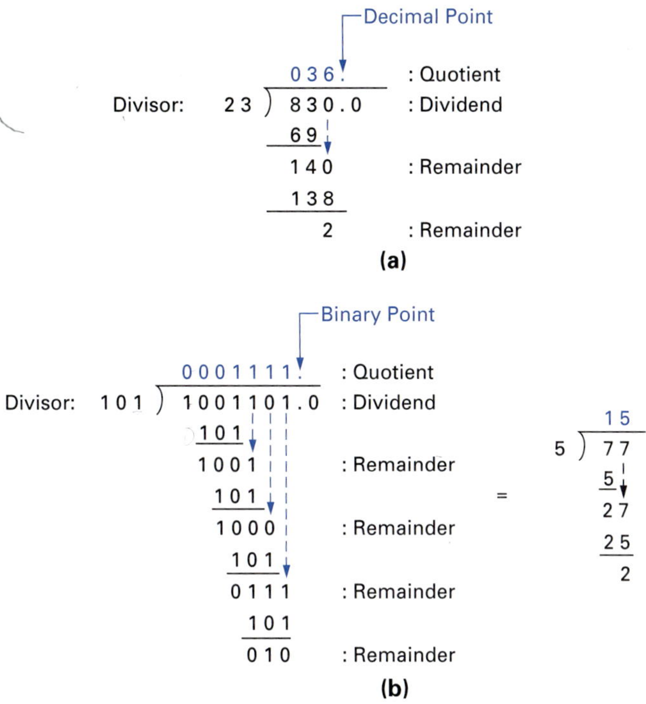

FIGURE 26-4 Binary Division.

see how many times the divisor (101_2) can be subtracted from the first two bits of the dividend (10_2), and this again results in a quotient of 0_2. Similarly, the divisor (101_2) cannot be subtracted from the first three bits of the dividend, so the quotient is once again 0_2. Finally, we find a dividend (1001_2) that is greater than the divisor (101_2), and therefore a 1_2 is placed in the quotient, and the divisor is subtracted from the first four bits of the dividend, resulting in a remainder of 100_2. Then the next bit of the dividend (1_2) is brought down to the remainder (100_2), and we see how many times the divisor (101_2) can be subtracted from the new remainder (1001_2). Since 101_2 is smaller than 1001_2, a quotient of 1_2 results. The divisor is subtracted from the remainder, resulting in a new remainder of 100_2. Then the next bit of the dividend (0_2) is brought down to the remainder (100_2), and we see how many times the divisor (101_2) can be subtracted from the new remainder (1000_2). Since 101_2 is smaller than 1000_2, a quotient of 1 results. The divisor is subtracted from the remainder, resulting in a new remainder of 11_2. Then the next bit of the dividend (1_2) is brought down to the remainder (11_2), and we see how many times the divisor (101_2) can be subtracted from the new remainder (111_2). Since 101_2 is smaller than 111_2, a quotient of 1 results. The divisor is subtracted from the remainder, resulting in a final remainder of 10. Therefore, $1001101_2 \div 101_2 = 1111_2$, remainder 10 ($77 \div 5 = 15$, remainder 2).

EXAMPLE:

Divide 1110101_2 by 110_2.

Solution:

	0	0	1	0	0	1	1_2	*:Quotient*	019_{10}
Divisor:	110) 1	1	1	0	1	0	1_2	*:Dividend*	6_{10}) 117_{10}
	1	1	0	↓	↓	↓	↓		
	0	0	1	0	1	0	↓	*:Remainder*	
				1	1	0	↓		6
				1	0	0	1	*:Remainder*	57
					1	1	0		54
				0	0	1	1_2	*:Remainder*	3_{10}

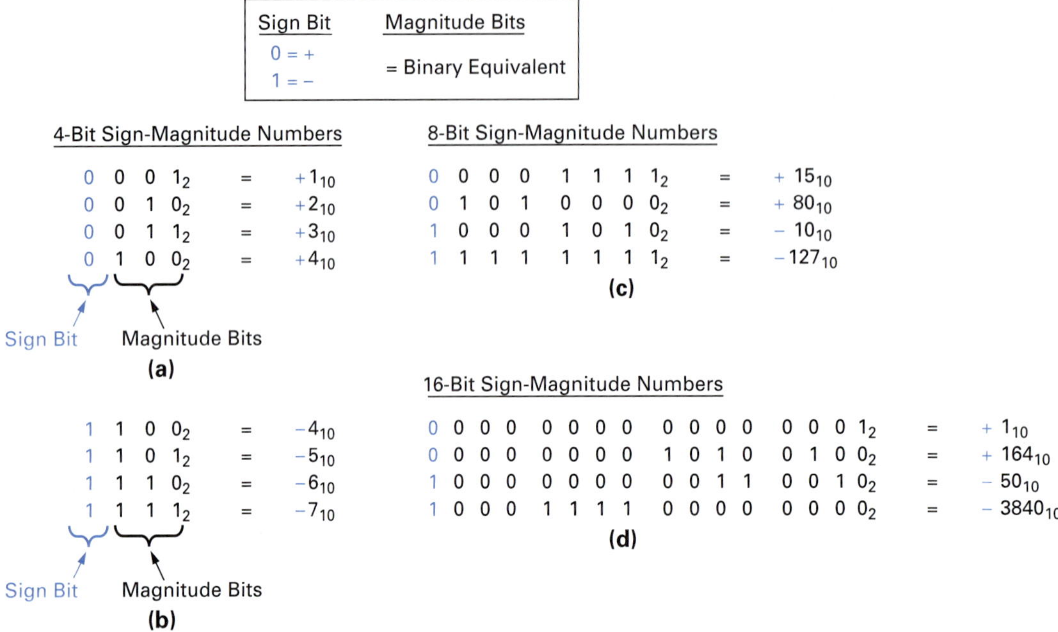

FIGURE 26-5 **The Sign-Magnitude Number System.**

26-1-2 *Representing Positive and Negative Numbers*

Many digital systems, such as calculators and computers, are used to perform mathematical functions on a wide range of numbers. Some of these numbers are positive and negative signed numbers; therefore, a binary code is needed to represent these numbers.

The Sign and Magnitude Number System

Sign and Magnitude Number System
A binary code used to represent positive and negative numbers in which the MSB signifies sign and the following bits indicate magnitude.

The **sign and magnitude number system** is a binary code system used to represent positive and negative numbers. A sign magnitude number contains a *sign bit* (0 for positive, 1 for negative) followed by the *magnitude bits.*

As an example, Figure 26-5(a) shows some 4-bit sign-magnitude positive numbers. All of these sign-magnitude binary numbers—0001, 0010, 0011, and 0100—have an MSB, or sign bit, of 0, and therefore they are all positive numbers. The remaining three bits in these 4-bit words indicate the magnitude of the number based on the standard 421 binary column weight ($001_2 = 1_{10}$, $010_2 = 2_{10}$, $011_2 = 3_{10}$, $100_2 = 4_{10}$).

As another example, Figure 26-5(b) shows some 4-bit sign-magnitude negative numbers. All of these sign-magnitude binary numbers—1100, 1101, 1110, and 1111—have an MSB, or sign bit, of 1, and therefore they are all negative numbers. The remaining three bits in these 4-bit words indicate the magnitude of the number based on the standard 421 binary column weight ($100_2 = 4_{10}$, $101_2 = 5_{10}$, $110_2 = 6_{10}$, $111_2 = 7_{10}$).

If you need to represent larger decimal numbers, you simply use more bits, as shown in Figure 26-5(c) and (d). The principle still remains the same in that the MSB indicates the sign of the number, and the remaining bits indicate the magnitude of the number.

The sign-magnitude number system is an ideal example of how binary numbers could be coded to represent positive and negative numbers. This code system, however, requires complex digital hardware for addition and subtraction and is therefore seldom used.

EXAMPLE:

What are the decimal equivalents of the following sign-magnitude number system codes?

a. 0111 b. 1010 c. 0001 1111 d. 1000 0001

Solution:

a. +7 b. −2 c. +31 d. −1

EXAMPLE:

What are the 8-bit sign-magnitude number system codes for the following decimal values?

a. +8 b. −65 c. +127 d. −127

Solution:

a. 0000 1000 b. 1100 0001 c. 0111 1111 d. 1111 1111

The One's Complement (1's Complement) Number System

The **one's complement number system** became popular in early digital systems. It also used an MSB sign bit that was either a 0 (indicating a positive number) or a 1 (indicating a negative number).

One's Complement Number System
A binary code used to represent positive and negative numbers in which negative values are determined by inverting all of the binary digits of an equivalent positive value.

Figure 26-6(a) shows some examples of 8-bit one's complement positive numbers. As you can see, one's complement positive numbers are no different from sign-magnitude positive numbers. The only difference between the one's complement number system and the sign-magnitude number system is the way in which negative numbers are represented. To understand the code used to represent negative numbers, we must first understand the meaning of the term *one's complement.* To one's complement a binary number means to invert or change all of the 1s in the binary word to 0s, and all of the 0s in the binary word to 1s. Performing the one's complement operation on a positive number will change the number from a one's complement positive number to a one's complement negative number. For example, by one's complementing the number 00000101, or +5, we will get 11111010, which is −5. In Figure 26-6(b), you will see the negative one's complement codes of the positive one's complement codes shown in Figure 26-6(a).

The next question you may have is, how can we determine the decimal value of a negative one's complement code? The answer is to do the complete opposite, or complement, to that of the positive one's complement code. This means count the 1s in a positive one's complement code, and count the 0s in a negative one's complement code. For example, with a positive one's complement code (MSB = 0), the value is determined by simply adding together all of the column weights that contain a 1 (00000110 has a 1 in the fours column and

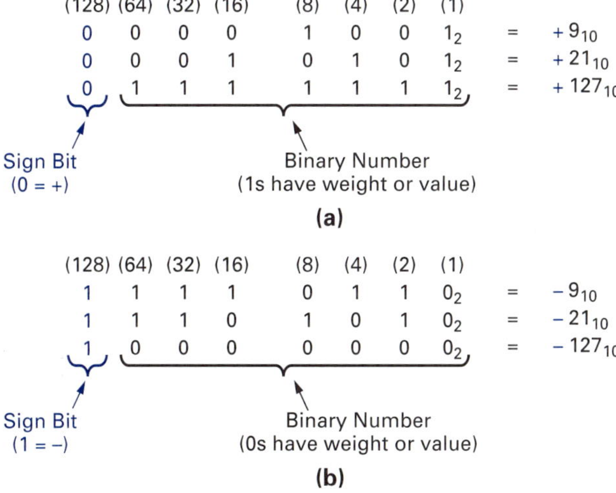

FIGURE 26-6 The One's Complement Number System.

twos column, so this number is 4 + 2, or +6). On the other hand, for a negative one's complement code (MSB = 1), the value is determined by doing the opposite and adding together all of the column weights that contain a 0 (11110101 has a 0 in the eights column and twos column, so this number is 8 + 2, or −10).

EXAMPLE:

What are the decimal equivalents of the following one's complement number system codes?

a. 0010 0001 b. 1101 1101

Solution:

a. MSB = 0 (number is positive, so add together all of the column weights that contain a 1). The number 0010 0001 has a 1 in the thirty-twos column and a one in the units column, so 0010 0001 = +33.

b. MSB = 1 (number is negative, so add together all of the column weights that contain a 0). The number 1101 1101 has a 0 in the thirty-twos column and a 0 in the twos column, so 1101 1101 = −34.

Since the term *one's complement* can be used to describe both an operation (change all of the 1s to 0s, and all of the 0s to 1s) and a positive and negative number system, some confusion can occur. To avoid this, remember that if you are asked to "one's complement a number," it means to change all of the 1s in the binary word to 0s and all of the 0s to 1s. If, on the other hand, you are asked to "find the decimal equivalent of a one's complement number," you should decode the sign bit and binary value to determine the decimal equivalent value.

EXAMPLE:

One's complement the following numbers:

a. 0101 1111 b. 1110 0101

Solution:

To one's complement a number means to change all of the 1s to 0s and all of the 0s to 1s.

a. 1010 0000 b. 0001 1010

EXAMPLE:

Find the decimal equivalent of the following one's complement numbers:

a. 0101 1111 b. 1110 0101

Solution:

a. MSB = 0 (number is positive, so add together all of the column weights that contain a 1). The number 0101 1111 = +95.

b. MSB = 1 (number is negative, so add together all of the column weights that contain a 0). The number 1110 0101 = −26.

Two's Complement Number System

A binary code used to represent positive and negative numbers in which negative values are determined by inverting all of the binary digits of an equivalent positive value and then adding 1.

The Two's Complement (2's Complement) Number System

The **two's complement number system** is used almost exclusively in digital systems to represent positive and negative numbers.

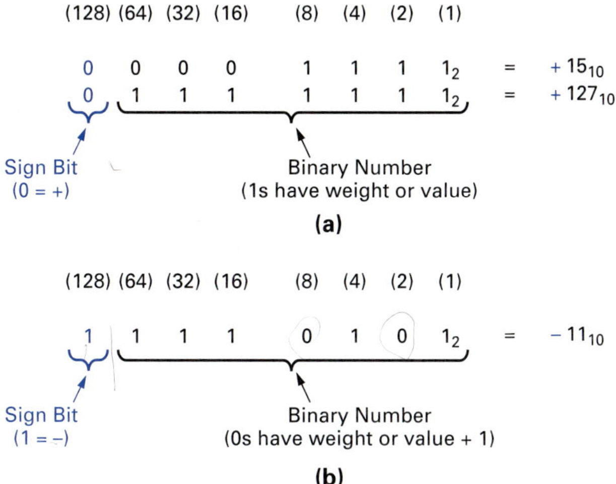

FIGURE 26-7 **The Two's Complement Number System.**

Positive two's complement numbers are no different from positive sign-magnitude and one's complement numbers in that an MSB sign bit of 0 indicates a positive number and the remaining bits of the word indicate the value of the number. Figure 26-7(a) shows two examples of positive two's complement numbers.

Negative two's complement numbers are determined by two's complementing a positive two's complement number code. To two's complement a number, first one's complement it (change all the 1s to 0s and 0s to 1s) and then add 1. For example, if we were to two's complement the number 0000 0101 (+5), we would obtain the two's complement number code for −5. The procedure to follow to two's complement 0000 0101 (+5), would be as follows:

0000 0101	Original Number (+5)
1111 1010	One's Complement (inverted)
+ 1	Plus 1
1111 1011	Two's Complement Code for −5

The new two's complement code obtained after this two's complement operation is the negative equivalent of the original number. In our example, therefore:

Original Number:	0000 0101 = +5
After Two's Complement Operation:	1111 1011 = −5

Any negative equivalent two's complement code therefore can be obtained by simply two's complementing the same positive two's complement number. For example, to find the negative two's complement code for −125, we would perform the following operation:

Two's	0111 1101	Original Number (+125)
Complement →	1 000 0010	One's Complement (inverted)
Operation	+ 1	Plus 1
	1000 0011	Two's Complement Code for −125

EXAMPLE:

Find the negative two's complement codes for

a. −26

b. −67

■ ***Solution:***

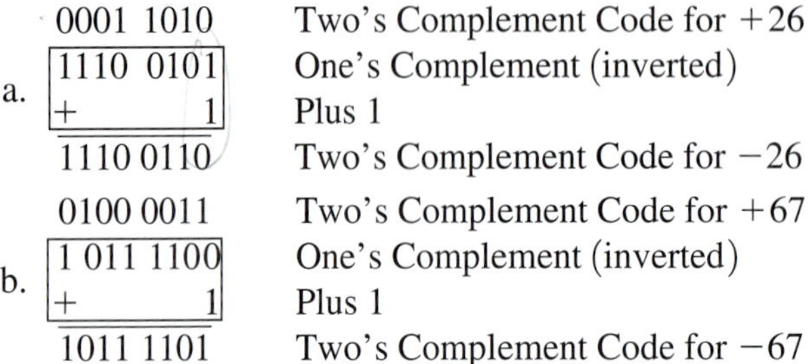

Performing the two's complement operation on a number twice will simply take the number back to its original value and polarity. For example, let's take the two's complement code for +6 (0000 0110) and two's complement it twice.

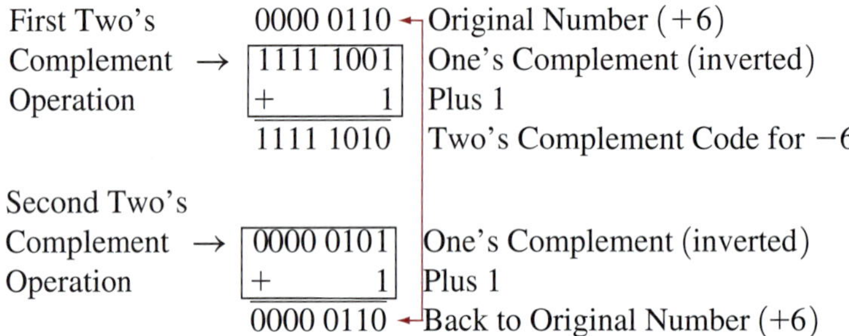

The decimal equivalent of a positive two's complement number (MSB = 0) is easy to determine, since the binary value following the sign bit applies the same as any binary number, as shown in Figure 26-7(a). On the other hand, the decimal equivalent of a negative two's complement number (MSB = 1) is not determined in the same way. To determine the value of a negative two's complement number, refer to the remaining bits following the sign bit, add the weight of all columns containing a 0, and then add 1 to the result. For example, the negative two's complement number in Figure 26-7(b) has 0s in the eights and twos columns, and therefore its value will equal the sum of 8 and 2, plus the additional 1 (due to two's complement). As a result,

$$1\,1\,1\,1\,\overset{(8)}{0}\,\overset{(4)}{1}\,\overset{(2)}{0}\,\overset{(1)}{1} = (8 + 2) + 1 = -11$$

■ **EXAMPLE:**

Determine the decimal equivalent of the following negative two's complement numbers:

a. 1110 0001 b. 1011 1011

c. 1000 0000 d. 1111 1111

■ ***Solution:***

a. $\overset{(\pm)}{1}\,\overset{(64)}{1}\,\overset{(32)}{1}\,\overset{(16)}{0}\,\overset{(8)}{0}\,\overset{(4)}{0}\,\overset{(2)}{0}\,\overset{(1)}{1}_2 = (16 + 8 + 4 + 2) + 1 = -31_{10}$

b. $\overset{(\pm)}{1}\,\overset{(64)}{0}\,\overset{(32)}{1}\,\overset{(16)}{1}\,\overset{(8)}{1}\,\overset{(4)}{0}\,\overset{(2)}{1}\,\overset{(1)}{1}_2 = (64 + 4) + 1 = -69_{10}$

c. $\overset{(\pm)}{1}\,\overset{(64)}{0}\,\overset{(32)}{0}\,\overset{(16)}{0}\,\overset{(8)}{0}\,\overset{(4)}{0}\,\overset{(2)}{0}\,\overset{(1)}{0}_2 = (64 + 32 + 16 + 8 + 4 + 2 + 1) + 1 = -128_{10}$

d. $\overset{(\pm)}{1}\,\overset{(64)}{1}\,\overset{(32)}{1}\,\overset{(16)}{1}\,\overset{(8)}{1}\,\overset{(4)}{1}\,\overset{(2)}{1}\,\overset{(1)}{1}_2 = (0) + 1 = -1_{10}$

4-Bit Word		Standard Binary Value	2's Complement Value	
	0 0 0 0	0	0	
	0 0 0 1	1	+1	
	0 0 1 0	2	+2	
	0 0 1 1	3	+3	8 Positive
	0 1 0 0	4	+4	Numbers
	0 1 0 1	5	+5	
	0 1 1 0	6	+6	
2^4 =	0 1 1 1	7	+7	
16 Words	1 0 0 0	8	−8	
	1 0 0 1	9	−7	
	1 0 1 0	10	−6	
	1 0 1 1	11	−5	8 Negative
	1 1 0 0	12	−4	Numbers
	1 1 0 1	13	−3	
	1 1 1 0	14	−2	
	1 1 1 1	15	−1	

FIGURE 26-8 4-Bit Two's Complement Codes.

As with the sign-magnitude and one's complement number systems, if you need to represent larger decimal numbers with a two's complement code, you will have to use a word containing more bits. For example, a total of 16 different combinations is available when we use a 4-bit word ($2^4 = 16$), as shown in Figure 26-8. With standard binary, these 16 different combinations give us a count from 0 to 15, as shown in the center column in Figure 26-8. As far as two's complement is concerned, we split these 16 different codes in half and ended up with 8 positive two's complement codes and 8 negative two's complement codes, as shown in the right column in Figure 26-8. Since decimal zero uses one of the positive two's complement codes, the 16 different combinations give us a range of -8_{10} to $+7_{10}$ (the maximum positive value, +7, is always one less than the maximum negative value, −8, since decimal zero uses one of the positive codes).

Using an 8-bit word will give us 256 different codes ($2^8 = 256$), as shown in Figure 26-9. Dividing this total in half will give us 128 positive two's complement codes and 128 negative two's complement codes. Since decimal zero uses one of the positive two's complement codes, the 256 different combinations gives us a range of -128_{10} to $+127_{10}$.

EXAMPLE:

What would be the two's complement range for a 16-bit word?

Solution:

$2^{16} = 65{,}536$; half of 65,536 = 32,768; therefore, the range will be −32,768 to +32,767.

26-1-3 *Two's Complement Arithmetic*

To perform its function, a computer's MPU contains several internal circuits, and one of these circuits is called the **arithmetic-logic unit,** or **ALU.** This unit is the number-crunching circuit of every digital system and, as its name implies, it can be controlled to perform either arithmetic operations (such as addition or subtraction) or logic operations (such as AND and OR). The arithmetic-logic unit has two parallel inputs, one parallel output, and a set of function-select control lines, as shown in Figure 26-10.

Arithmetic-Logic Unit (ALU)
The section of a computer that performs all arithmetic and logic operations.

	8-Bit Word	Standard Binary Value	2's Complement Value	
	0 0 0 0 0 0 0 0	0	0	128 Positive Numbers
	0 0 0 0 0 0 0 1	1	+1	
	0 0 0 0 0 0 1 0	2	+2	
	0 0 0 0 0 0 1 1	3	+3	
	0 0 0 0 0 1 0 0	4	+4	
	0 0 0 0 0 1 0 1	5	+5	
	0 0 0 0 0 1 1 0	6	+6	
	0 0 0 0 0 1 1 1	7	+7	
	⋮	⋮	⋮	
	0 1 1 1 1 0 1 1	123	+123	
	0 1 1 1 1 1 0 0	124	+124	
	0 1 1 1 1 1 0 1	125	+125	
	0 1 1 1 1 1 1 0	126	+126	
2^8 = 256 Words	0 1 1 1 1 1 1 1	127	+127	
	1 0 0 0 0 0 0 0	128	−128	128 Negative Numbers
	1 0 0 0 0 0 0 1	129	−127	
	1 0 0 0 0 0 1 0	130	−126	
	1 0 0 0 0 0 1 1	131	−125	
	1 0 0 0 0 1 0 0	132	−124	
	1 0 0 0 0 1 0 1	133	−123	
	⋮	⋮	⋮	
	1 1 1 1 1 0 1 1	251	−5	
	1 1 1 1 1 1 0 0	252	−4	
	1 1 1 1 1 1 0 1	253	−3	
	1 1 1 1 1 1 1 0	254	−2	
	1 1 1 1 1 1 1 1	255	−1	

FIGURE 26-9 8-Bit Two's Complement Codes.

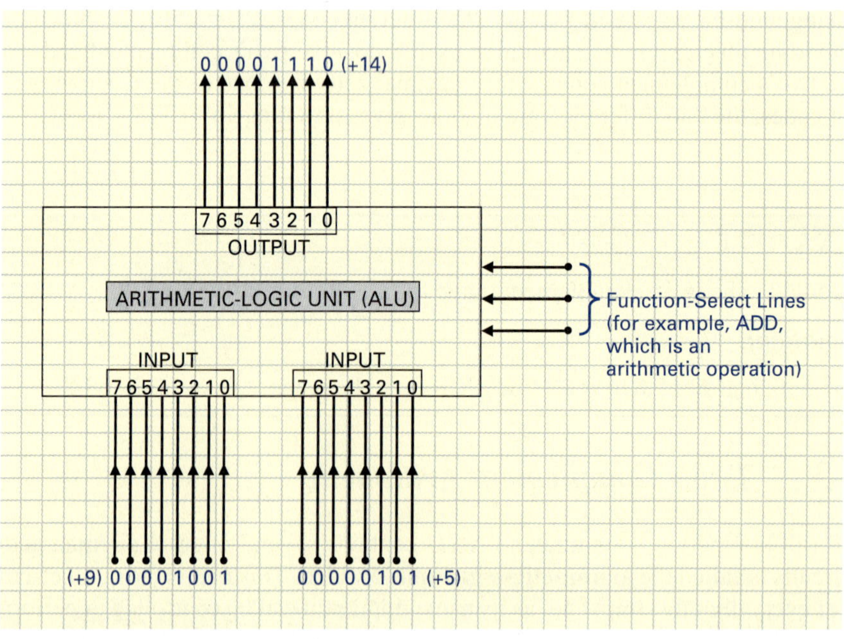

FIGURE 26-10 Block Diagram of the Arithmetic-Logic Unit (ALU).

As far as arithmetic operations are concerned, the ALU has to be able to perform the four basic operations, which, as you know, are addition, subtraction, multiplication, and division. One would expect, therefore, that an ALU would contain a separate circuit for each of these operations. This is not so, since all four operations can be performed using a binary-adder circuit. To explain this point, because the binary-adder circuit's function is to perform addition, and since multiplication is simply repeated addition, we can also use the binary-adder circuit to perform any multiplication operations. For example, 3×4 means that you are to take the number 4 and add it three times ($4 + 4 + 4 = 12$, therefore $3 \times 4 = 12$). But what about subtraction and division? The answer to this problem is the two's complement number system. As you will see in this section, by representing all values as two's complement numbers, we can perform a subtraction operation by slightly modifying the second number and then adding the two numbers together. This means that our binary-adder circuit can also be used to perform subtraction operations, and since division is simply repeated subtraction, we can also use the subtraction operation of the binary-adder circuit to perform any division operation. For example, $12 \div 4$ means that you are to take the number 12 and see how many times you can subtract 4 from it ($12 - 4 - 4 - 4 = 0$; therefore there are three 4s in 12, $12 \div 4 = 3$).

By using two's complement numbers, therefore, we can use a binary-adder circuit within an ALU to add, subtract, multiply, and divide. Having one circuit able to perform all four mathematical operations means that the circuit will be small in size, fast to operate, consume very little power, and be cheap to manufacture. In this section we will see how we can perform all four mathematical operations by always adding two's complement numbers.

Adding Positive Numbers

To begin with, let us prove that the addition of two positive two's complement numbers will render the correct result.

EXAMPLE:

Add +9 (0000 1001) and +5 (0000 0101).

Solution:

$$
\begin{array}{rr}
 & 0000\,1001 \\
 & +\ \underline{0000\,0101} \\
\text{MSB of } 0 = \text{Positive number} \rightarrow & 0000\,1110
\end{array}
$$

When both of the numbers to be added are positive two's complement numbers (both have an MSB of 0), it makes sense that the sum will have an MSB of 0, indicating a positive result. This is always true, unless the range of the two's complement word is exceeded. For example, the maximum positive number that can be represented by an 8-bit word is +127 (0111 1111). If the sum of the two positive two's complement numbers to be added exceeds the upper positive range limit, there will be a *two's complement overflow* into the sign bit. For instance:

$$
\begin{array}{rcr}
 & (\pm)(64)(32)(16)\ (8)\ (4)\ (2)\ \ (1) & \\
 & 0\ 1\ 1\ 0\ 0\ 1\ 0\ 0 & (+100) \\
 & +\ \underline{0\ 0\ 0\ 1\ 1\ 1\ 1\ 0} & +\ \underline{(+30)} \\
\text{MSB of } 1 = \text{negative number} \rightarrow & 1\ 0\ 0\ 0\ 0\ 0\ 1\ 0 & (-126)
\end{array}
$$

If two positive numbers are applied to an arithmetic circuit, and it is told to add the two, the result should also be a positive number. Most arithmetic circuits are able to detect this problem if it occurs by simply monitoring the MSB of the two input words and the MSB of the output word, as shown in Figure 26-11. If the two input sign bits are the same but the output sign bit is different, then a two's complement overflow has occurred.

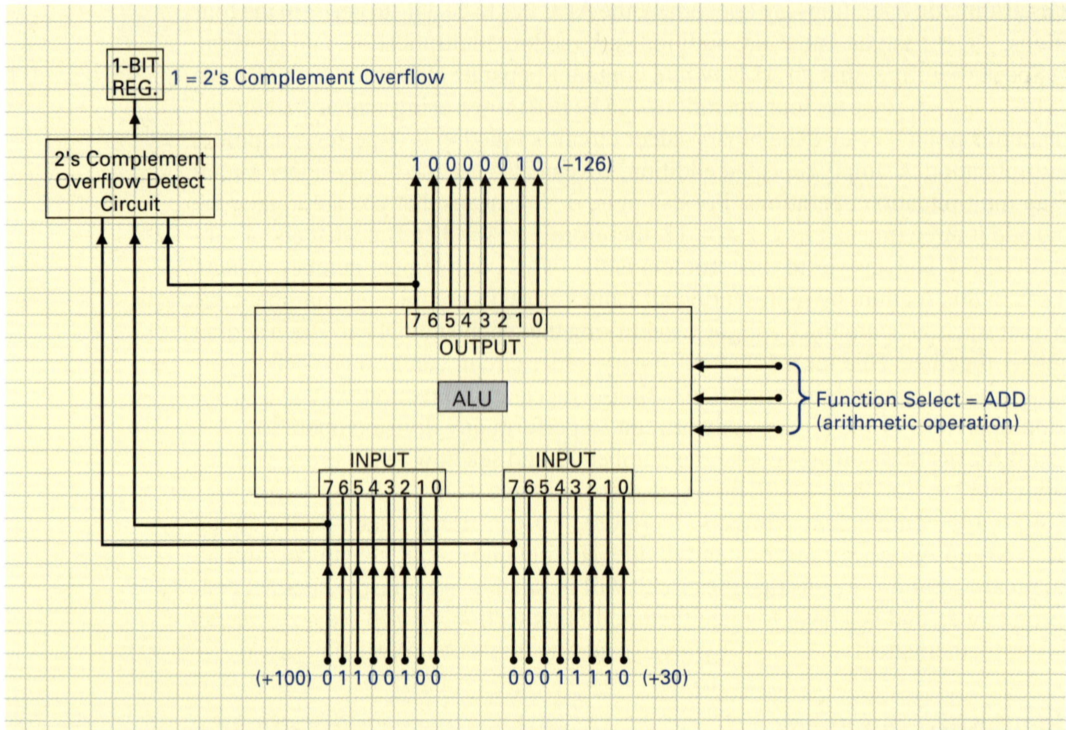

FIGURE 26-11 **Two's Complement Overflow Detection.**

Adding Positive and Negative Numbers

By using the two's complement number system, we can also add numbers with unlike signs and obtain the correct result. As an example, let us add +9 (0000 1001) and −5 (1111 1011).

	0000 1001	(+9)
	+1111 1011	+(−5)
Ignore the final carry → 1	0000 0100	(+4)

A final carry will always be generated whenever the sum is a positive number; however, if this carry is ignored because it is beyond the 8-bit two's complement word, the answer will be correct.

EXAMPLE:

Add +10 (0000 1010) and −12 (1111 0100).

Solution:

0000 1010	(+10)
+ 1111 0100	+ (−12)
1111 1110	(−2)

Adding Negative Numbers

As a final addition test for the two's complement number system, let us add two negative numbers to see if we can obtain the correct negative sum.

EXAMPLE:

Add −5 (1111 1011) and −4 (1111 1100).

■ ***Solution:***

	1111 1011	(−5)
	+ 1111 1100	+ (−4)
Ignore the final carry → 1	1111 0111	(−9)

The sum in this example is the correct two's complement code for −9.

Two's complement overflow can also occur in this condition if the sum of the two negative numbers exceeds the negative range of the word. For example, with an 8-bit two's complement word, the largest negative number is −128, and if we add −100 (1001 1100) and −30 (1110 0010), we will get an error, as follows:

	1001 1100	(−100)
	+ 1110 0010	+ (−30)
Ignore the final carry → 1	0111 1110	(+126)

Once again, the two's complement overflow detect circuit, shown in Figure 26-11, will indicate such an error since two negative numbers were applied to the input and a positive number appeared at the output.

In all of the addition examples mentioned so far (adding positive numbers, adding positive and negative numbers, and adding negative numbers) you can see that an ALU added the bits of the input bytes without any regard for whether they were signed or unsigned numbers. This brings up a very important point, which is often misunderstood. The ALU adds the bits of the input words using the four basic rules of binary addition, and it does not know if these input and output bit patterns are coded two's complement words or standard unsigned binary words. It is therefore up to us to maintain a consistency, which means that if we are dealing with coded two's complement words at the input of an ALU, we must interpret the output of an ALU as a two-coded two's complement word. On the other hand, if we are dealing with standard unsigned binary words at the input of an ALU, we must interpret the output of an ALU as a standard unsigned binary word.

Subtraction Through Addition

The key advantage of the two's complement number system is that it enables us to perform a subtraction operation using the adder circuit within the ALU. To explain how this is done, let us take an example and assume that we want to subtract 3 (0000 0011) from 8 (0000 1000), as follows:

Minuend:	0000 1000	(+8)
Subtrahend:	− 0000 0011	− (+3)
Difference:	0000 0101	(+5)

If you subtract the subtrahend of +3 from the minuend of +8, you will obtain the correct result of +5 (0000 0101). However, we must find the difference in this problem by using addition and not subtraction. The solution to this is achieved by *two's complementing the subtrahend and then adding the result to the minuend.* The following shows how this is done:

Minuend:	0000 1000	(+8)
Subtrahend:	− 0000 0011	− (+3)

0000 0011	2's Complement Code for +3
1111 1100	1's Complement (inverted)
+ 1	Plus 1
1111 1101	2's Complement Code for −3

Minuend:		0000 1000	(+8)
Two's Complement Subtrahend:	+	1111 1101	+ (−3)
Difference (ignore carry):	1	0000 0101	+5

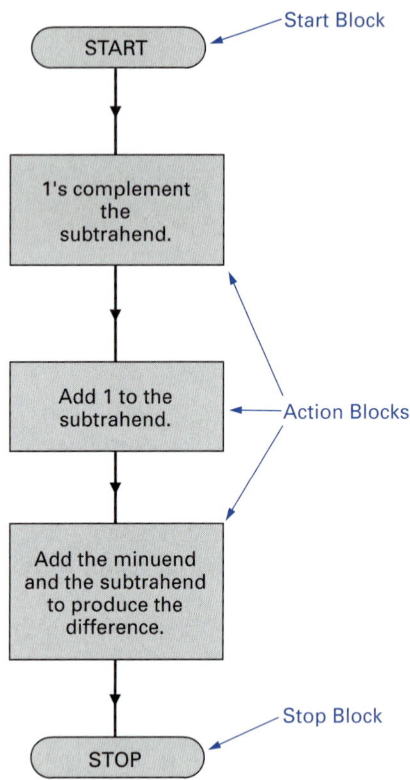

FIGURE 26-12 Flow Chart for Subtraction through Addition.

As you can see from this operation, adding the -3 to $+8$ achieves exactly the same result as subtracting $+3$ from $+8$. We can therefore perform a subtraction by two's complementing the subtrahend of a subtraction problem and then adding the result to the minuend.

Digital electronic calculators and computer systems perform the previously described steps automatically. That is, they first instruct the ALU to one's complement the subtrahend input (function-select control inputs = invert, which is a logic operation), then add 1 to the result (function-select control inputs = + 1, which is an arithmetic operation), and finally add the result of the two previous operations (which is the two's complemented subtrahend) to the minuend (function select = ADD, arithmetic operation). Although this step-by-step procedure or **algorithm** seems complicated, it is a lot simpler than having to include a separate digital electronic circuit to perform all subtraction operations. Having one circuit able to perform both addition and subtraction means that the circuit will be small in size, fast to operate, consume very little power, and be cheap to manufacture.

Algorithm
A set of rules for the solution of a problem.

Flow Chart
A graphic representation of a step-by-step procedure or program of instructions.

In Figure 26-12, you can see a **flow chart** summarizing the subtraction through addition procedure. Flow charts provide a graphic way of describing a step-by-step procedure (algorithm) or a program of instructions.

Multiplication Through Repeated Addition

Most digital electronic calculators and computers do not include a logic circuit that can multiply, since the adder circuit can also be used to perform this arithmetic operation. For example, to multiply 6×3, you simply add 6 three times.

Multiplicand:	6		6
Multiplier:	× 3	=	+ 6
Product:	18		+ 6
			18

As in subtraction through addition, an algorithm must be followed when a digital system needs to perform a multiplication operation. Figure 26-13 shows the flow chart for multipli-

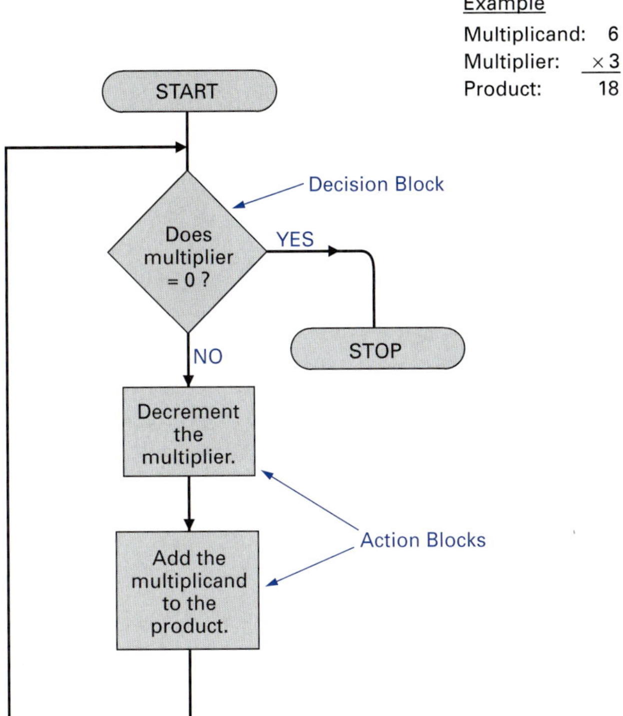

FIGURE 26-13 Flow Chart for Multiplication through Repeated Addition.

cation through repeated addition. Let us apply this procedure to our previous example of 6 × 3 and see how the multiplicand 6 is added three times to obtain a final product of 18. The steps would be as follows:

Start:	Multiplicand = 6,	Multiplier = 3,	Product = 0.
After First Pass:	Multiplicand = 6,	Multiplier = 2,	Product = 6.
After Second Pass:	Multiplicand = 6,	Multiplier = 1,	Product = 12.
After Third Pass:	Multiplicand = 6,	Multiplier = 0,	Product = 18.
Stop			

In this example we cycled through the flow chart three times (until the multiplier was 0) and then finally obtained a product of 18.

Although this algorithm seems complicated, it is a lot simpler than having to include a separate digital circuit for multiplication. Having one circuit able to perform addition, subtraction, and multiplication means that the circuit will be small in size, fast to operate, consume very little power, and be cheap to manufacture.

Division Through Repeated Subtraction

Up to this point, we have seen how we can use the two's complement number system and the adder circuit within an ALU for addition, subtraction, and multiplication. Just as multiplication can be achieved through repeated addition, division can be achieved through repeated subtraction. For example, to divide 31 by 9 you simply see how many times 9 can be subtracted from 31, as follows:

$$\begin{array}{rl} & 3 \text{ :Quotient} \\ \text{Divisor: } 9\overline{)31} & \text{:Dividend} \\ \underline{27} & \\ 4 & \text{:Remainder} \end{array} \quad = \quad \begin{array}{rl} 31 & \\ \underline{-9} & \text{:First Subtraction} \\ 22 & \\ \underline{-9} & \text{:Second Subtraction} \\ 13 & \\ \underline{-9} & \text{:Third Subtraction} \\ 4 & \end{array}$$

Since we were able to subtract 9 from 31 three times with a remainder of 4, then 31 ÷ 9 = 3, remainder 4.

As in multiplication, an algorithm must be followed when a digital system needs to perform a division. Figure 26-14 shows the flow chart for division through repeated subtraction. Let us apply this procedure to our previous example of 31 ÷ 3, and see how the quotient is incremented to 3, leaving a remainder of 4. The steps would be as follows:

Start:	Dividend = 31,	Divisor = 9,	Quotient = 0,	Remainder = 31.
After First Pass:	Dividend = 22,	Divisor = 9,	Quotient = 1,	Remainder = 22.
After Second Pass:	Dividend = 13,	Divisor = 9,	Quotient = 2,	Remainder = 13.
After Third Pass:	Dividend = 4,	Divisor = 9,	Quotient = 3,	Remainder = 4.
Stop				

When this program of instructions cycles or loops back for the third pass, the divisor of 9 is subtracted from the remaining dividend of 4, resulting in −5. When the question, Is the dividend negative? is asked, the answer will be yes, so the branch on the right is followed. This branch of the flow chart will add the divisor of 9 to the dividend of −5 to restore it to the correct remainder of 4 before the program stops. To summarize, these are the steps followed by this program:

$$\begin{array}{rl} (+31) & \\ \underline{-(+\ 9)} & \\ (+22) & \text{:Quotient} = 1 \\ \underline{-(+\ 9)} & \\ (+13) & \text{:Quotient} = 2 \\ \underline{-(+\ 9)} & \\ (+\ 4) & \text{:Quotient} = 3 \\ \underline{-(+\ 9)} & \\ (-\ 5) & \\ \underline{+(+\ 9)} & \text{:Add Divisor to Dividend} \\ (+\ 4) & \text{:Remainder} = 4 \end{array}$$

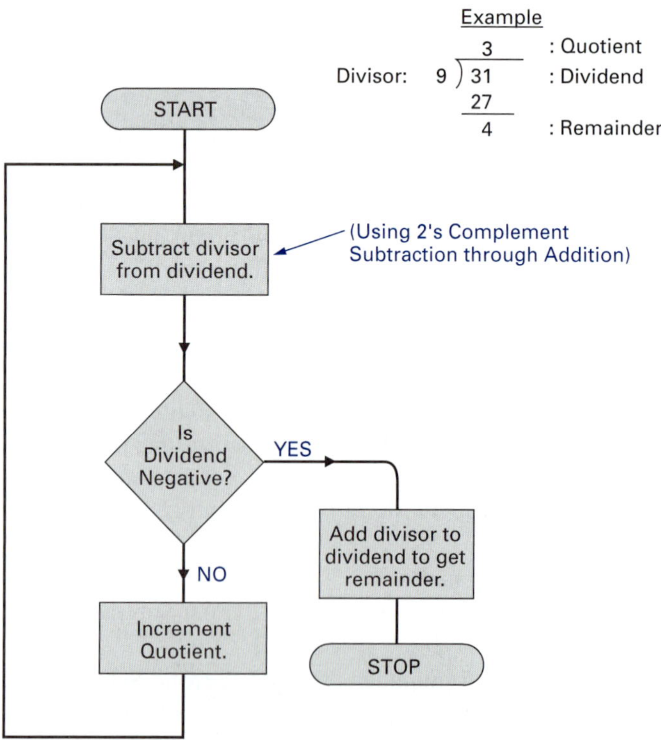

FIGURE 26-14 Flow Chart for Division through Repeated Subtraction.

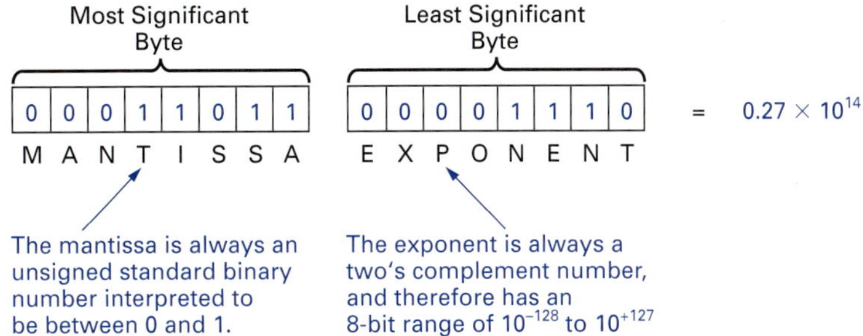

FIGURE 26-15 The Floating-Point Number System.

When a digital system is instructed to perform a division, it will use the ALU for all arithmetic and logic operations. Since division is achieved through repeated subtraction, and subtraction is achieved through addition, the ALU's adder circuit can also be used for division.

Although this algorithm seems complicated, it is a lot simpler than having to include a separate digital circuit for division. As previously mentioned, having one circuit able to perform addition, subtraction, multiplication, and division means that the circuit will be small in size, fast to operate, consume very little power, and be cheap to manufacture.

26-1-4 *Representing Large and Small Numbers*

While two's complementing enables us to represent positive and negative numbers, the range is limited to the number of bits in the word. For example, the range of an 8-bit two's complement word is from -128 to $+127$, while the range of a 16-bit two's complement word is from $-32{,}768$ to $+32{,}767$. To extend this range, we can keep increasing the number of bits; however, these large words become difficult to work with. Another problem arises when we wish to represent fractional numbers such as 1.57.

These problems are overcome by using the **floating-point number system,** which uses scientific notation to represent a wider range of values. Figure 26-15 explains the two-byte (16-bit) floating-point number system. The first byte is called the **mantissa,** and it is always an unsigned binary number that is interpreted as a value between 0 and 1. The second byte is called the **exponent,** and it is always a two's complement number and therefore represents a range from 10^{-128} to 10^{+127}. As an example, let us see how the floating-point number system can be used to represent a large-value number such as 27,000,000,000,000 and a small-value number such as 0.0000000015.

Floating-Point Number System
A number system in which a value is represented by a mantissa and an exponent.

Mantissa
The fractional part of a logarithm.

Exponent
A small number placed to the right and above a value to indicate the number of times that symbol is a factor.

EXAMPLE:

Give the two-byte floating-point binary values for 27,000,000,000,000.

Solution:

The value 27,000,000,000,000 would be written in scientific notation as 0.27×10^{14} and when coded as a floating-point number, would appear as follows:

$$27{,}000{,}000{,}000{,}000$$
$$= 0.27 \times 10^{14}$$

=	0001 1011	0000 1110
	Mantissa (unsigned binary) = 27, which is interpreted as 0.27.	Exponent (two's complement) = +14, which is interpreted as 10^{14}.

In this example, the mantissa byte (standard binary number) is equal to 27 (0001 1011) and is therefore interpreted as 0.27, and the exponent byte (two's complement number) is equal to +14 (0000 1110) and is therefore interpreted as 10^{14}.

EXAMPLE:

Give the two-byte floating-point binary values for 0.0000000015.

Solution:

The value 0.0000000015 would be written in scientific notation as 0.15×10^{-8} and when coded as a floating-point number, would appear as follows:

	0.0000000015	
=	0.15×10^{-8}	
=	0000 1111	1111 1000
	Mantissa (unsigned binary) = 15, which is interpreted as 0.15.	Exponent (two's complement) = −8, which is interpreted as 10^{-8}.

In this example, the mantissa byte (standard binary number) is equal to 15 (0000 1111) and is therefore interpreted as 0.15, and the exponent byte (two's complement number) is equal to −8 (1111 1000) and is therefore interpreted as 10^{-8}.

As with all binary number coding methods, we must know the number representation system being used if the bit pattern is to be accurately deciphered. For instance, the 16-bit word in Figure 26-15 would be very different if it were first interpreted as an unsigned binary number and then compared to a two's complement number and to a floating-point number.

SELF-TEST EVALUATION POINT FOR SECTION 26-1

Use the following questions to test your understanding of Section 26-1.

1. Perform the following binary arithmetic operations:
   a. 1011 + 11101 = ? b. 1010 − 1011 = ? c. 101 × 10 = ? d. 10111 ÷ 10 = ?
2. Two's complement the following binary numbers:
   a. 1011 b. 0110 1011 c. 0011 d. 1000 0111
3. What is the decimal equivalent of the following two's complement numbers?
   a. 1011 b. 0110 c. 1111 1011 d. 0000 0110
4. The two's complement number system makes it possible for us to perform addition, subtraction, multiplication, and division all with a/an ____________ circuit.

26-2 ARITHMETIC CIRCUITS

In the previous section, we discovered that there is no need to have a separate circuit to add, another circuit to subtract, another circuit to multiply, and yet another circuit to divide. If we use two's complement numbers, these four mathematical operations can be performed with a simple binary-adder circuit. In this section, we will examine the different types of binary-adder circuits and a typical binary-adder IC.

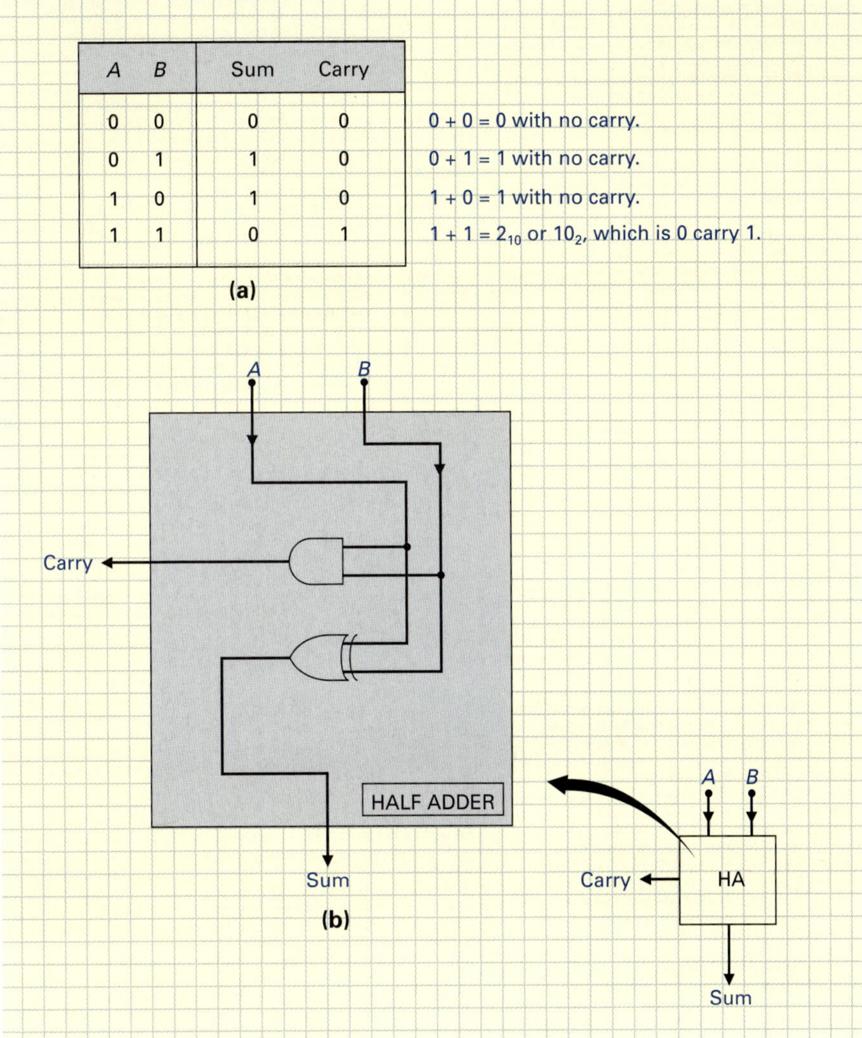

A	B	Sum	Carry
0	0	0	0
0	1	1	0
1	0	1	0
1	1	0	1

FIGURE 26-16 **The Half Adder (HA). (a) Truth Table. (b) Logic Circuit.**

26-2-1 *Half-Adder Circuit*

Figure 26-16(a) reviews the four rules of binary addition. Studying the sum column of the truth table, you may recognize that the output generated for each input combination is exactly the same as that generated by an XOR gate (odd number of ls at the input gives a 1 at the output). Referring now to the carry column in the truth table, you may recognize that in this case the output generated for each input combination is exactly the same as that generated by an AND gate (any 0 in gives a 0 out). Figure 26-16(b) shows how these two gates could be connected to sense and add an A and B input and generate both a sum and carry output that follows the truth table in Figure 26-16(a).

Half-Adder Circuit
A digital circuit that can sum a 2-bit binary input and generate a SUM and CARRY output.

The circuit in Figure 26-16(b) is called a **half-adder circuit,** and if we ever need to add only two input bits, this circuit would be ideal. In most binary addition operations, however, we will need to add three input bits at a time. To explain this, let us add 11_2 (3_{10}) and 01_2 (1_{10}).

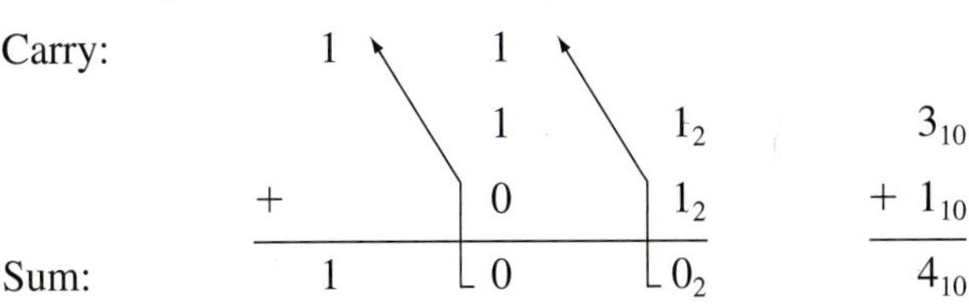

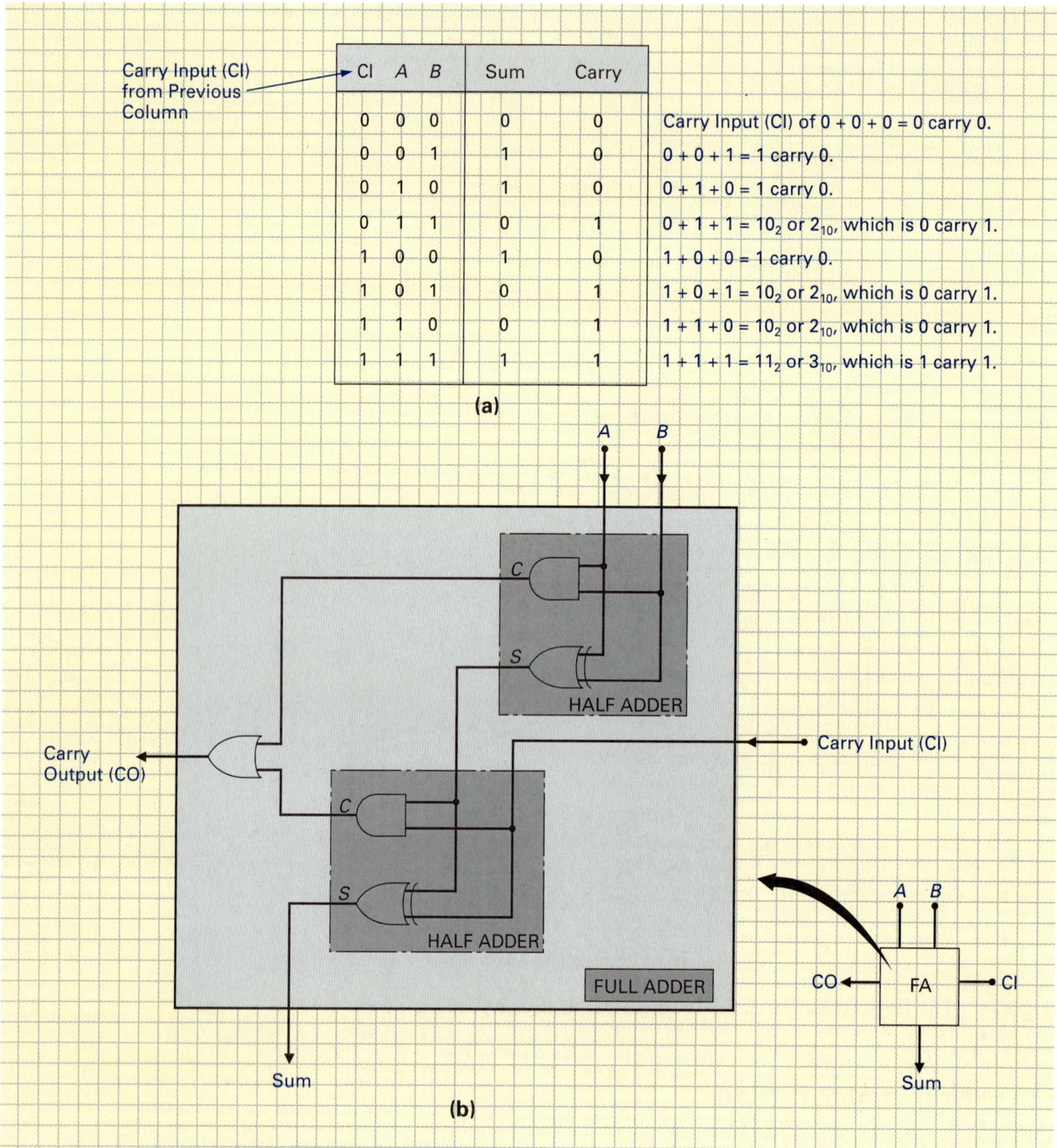

CI	A	B	Sum	Carry	
0	0	0	0	0	Carry Input (CI) of 0 + 0 + 0 = 0 carry 0.
0	0	1	1	0	0 + 0 + 1 = 1 carry 0.
0	1	0	1	0	0 + 1 + 0 = 1 carry 0.
0	1	1	0	1	0 + 1 + 1 = 10_2 or 2_{10}, which is 0 carry 1.
1	0	0	1	0	1 + 0 + 0 = 1 carry 0.
1	0	1	0	1	1 + 0 + 1 = 10_2 or 2_{10}, which is 0 carry 1.
1	1	0	0	1	1 + 1 + 0 = 10_2 or 2_{10}, which is 0 carry 1.
1	1	1	1	1	1 + 1 + 1 = 11_2 or 3_{10}, which is 1 carry 1.

FIGURE 26-17 The Full Adder (FA). (a) Truth Table. (b) Logic Circuit.

In the first column (units column) we only need to add two bits, 1 + 1, and therefore a half-adder circuit could be used to perform this addition. In the second column, however, we need to add three bits, 1 + 0 plus the 1 carry from the first column. A different logic circuit that can add a 3-bit input (*A*, *B* and carry) and then generate a sum and carry output is needed.

Full-Adder Circuit

A digital circuit that can sum a 3-bit binary input and generate a SUM and CARRY output.

26-2-2 *Full-Adder Circuit*

A full-adder circuit is able to add a 3-bit input and generate a sum and carry output. Figure 26-17(a) summarizes the addition operation that must be performed by a full-adder circuit. The full-adder circuit shown in Figure 26-17(b) has the three inputs *A*, *B*, and carry input (CI) and the two outputs sum and carry output (CO). The full-adder circuit contains two half adders, which in combination will generate the correct sum and carry output logic levels for any of the input combinations listed in Figure 26-17(a).

If we ever need to add only three input bits, the full-adder circuit would be ideal. In most binary addition operations, however, we will need to add several 3-bit columns. To explain this, let us add 0111_2 (7_{10}) and 0110_2 (6_{10}).

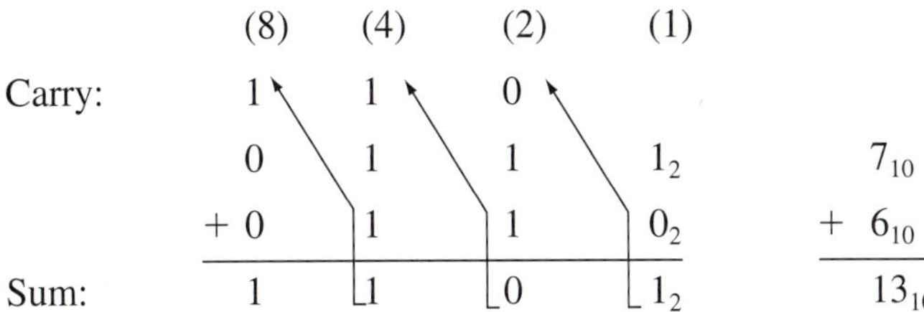

In the first column (units) we only need to add 2 bits; therefore, a half-adder circuit could be used to find the sum of this 2-bit column. In the remaining columns, however, we will always have a carry input from the previous column, and therefore a parallel-adder circuit will be needed to sum these 3-bit columns.

26-2-3 *Parallel-Adder Circuit*

Figure 26-18(a) shows a 4-bit **parallel-adder circuit.** The half adder (HA) on the right side will add A_1 and B_1 and produce at its output S_1 (sum 1 output) and C_1 (carry 1 output). All of the remaining adders will have to be full adders (FA) since these columns will always need

Parallel-Adder Circuit
A digital circuit that can simultaneously add the corresponding parts of two binary input words.

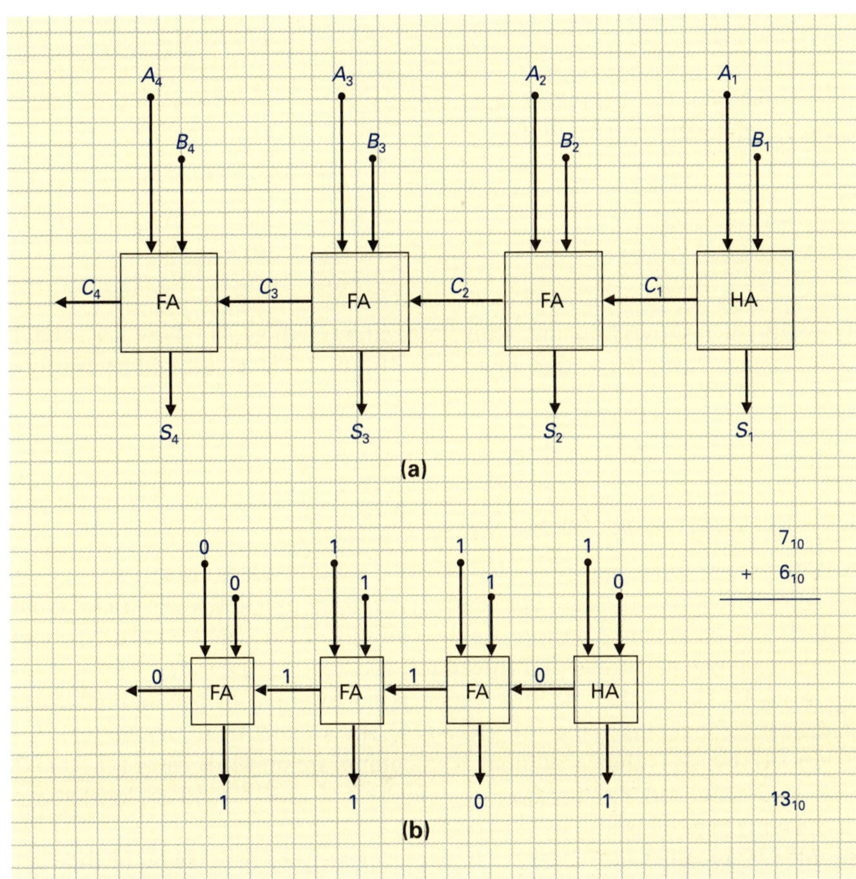

FIGURE 26-18 The Parallel Adder.

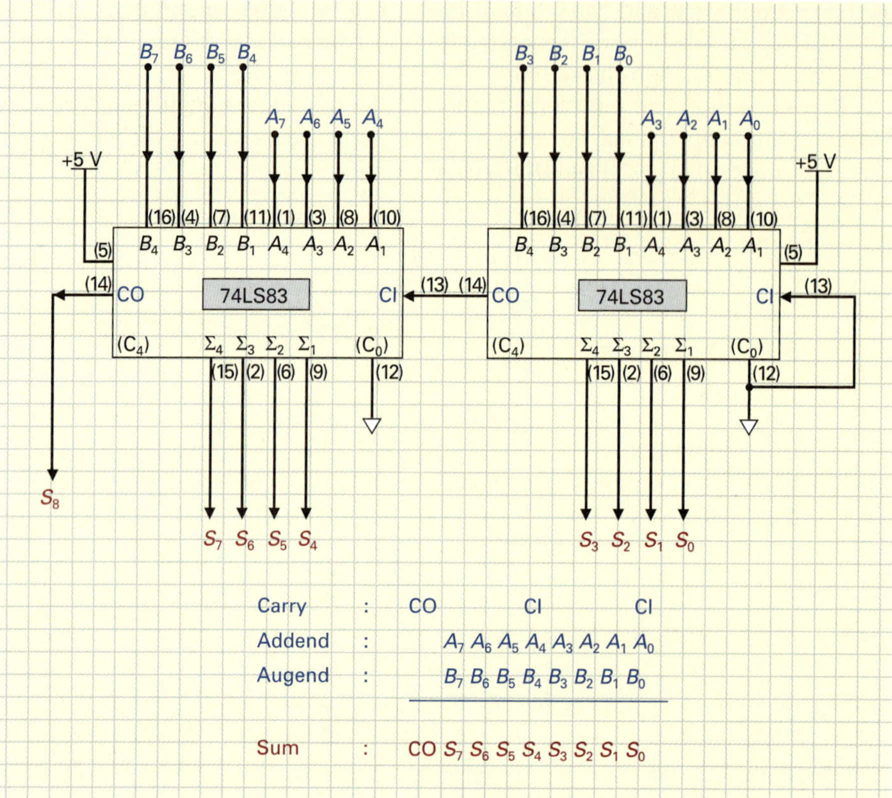

FIGURE 26-19 Application—An 8-Bit Parallel-Adder Circuit.

to add three input bits: the A input, B input, and the carry input from the previous column. This 4-bit parallel-adder circuit will therefore perform the following addition:

Carry:		C_4	C_3	C_2	C_1	
Addend:			A_4	A_3	A_2	A_1
Augend:	+		B_4	B_3	B_2	B_1
Sum:		C_4	S_4	S_3	S_2	S_1

Figure 26-18(b) shows how this 4-bit adder circuit will add an A_4, A_3, A_2, and A_1 input of 0111_2 (7_{10}) and a B_4, B_3, B_2, and B_1 input of 0110_2 (6_{10}). The result will be as follows:

Carry:		0 1 1 0
Addend:		0 1 1 1
Augend:	+	0 1 1 0
Sum:		0 1 1 0 1

By adding more full adders to the left side of a parallel-adder circuit, you can construct a parallel-adder circuit of any size. For example, an 8-bit parallel-adder circuit would contain one half adder (in the units position) and seven full adders, while a 16-bit parallel-adder circuit would need one half adder and fifteen full adders.

Figure 26-19 shows the 74LS83, which is an example of a commonly used 4-bit parallel-adder IC. The inputs to the 7483 IC are A_4, A_3, A_2, and A_1; B_4, B_3, B_2, and B_1; and an LSB C_0 (pin 13), which is the carry input. The outputs are an MSB carry (C_4) and the sum bits, which are symbolized with the Greek capital letter *sigma* (Σ) and are labeled Σ_4, Σ_3, Σ_2, and Σ_1. The sigma symbol is also used to denote an adder IC.

The 7483 IC is called a *fast carry* or *look-ahead carry* IC because of the time delay that occurs as the carry ripples from the LSB column to the second column, from the second to the third, and from the third to the fourth. For example, consider the following addition:

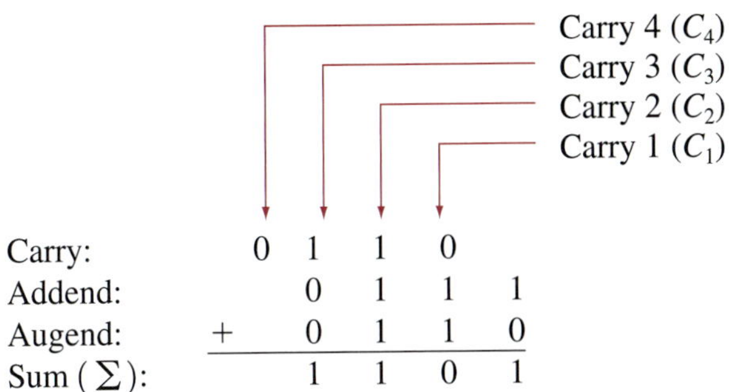

If each full adder within the 7483 had a propagation delay of 30 ns, it would take 30 ns for the first FA to generate C_1, another 30 ns for the second FA to generate C_2, another 30 ns for the third FA to generate C_3, and another 30 ns for the fourth FA to generate C_4. This means that the sum will not be available for 120 ns (4×30 ns) after the 4-bit A input and B input have been applied. This *carry propagation delay,* or *carry ripple delay,* becomes a real problem with 16-bit and 32-bit adder circuits. To overcome this problem, IC logic designers include several logic gates that look at the lower-order bits of the two inputs and generate higher-order carry bits. Most commercially available adder ICs feature full internal look-ahead across all 4-bits, generating the carry in typically 10 ns.

Application—An 8-Bit Parallel Circuit

Figure 26-19 shows how two 74LS83 4-bit adders could be cascaded to form an 8-bit parallel-adder circuit. The carry input of the first 74LS83 is connected LOW, while the carry output of the first 74LS83 is connected to the carry input of the second 74LS83. This circuit will add the two 8-bit A and B inputs and produce an 8-bit sum output. In the following section we will be using this circuit to form a complete two's complement adder/subtractor circuit.

SELF-TEST EVALUATION POINT FOR SECTION 26-2

Use the following questions to test your understanding of Section 26-2.

1. A half adder can be used to add ____________ bits, while a full adder can add ____________ bits.
2. Which of the following adder circuits could be used to add two binary words?
   a. Half adder b. Full adder c. Parallel adder d. Diamond head adder
3. How would you describe the 7483 IC?
4. The sum outputs of an adder circuit are normally symbolized with the Greek capital letter ____________.

26-3 ARITHMETIC CIRCUIT APPLICATIONS

In this section we will be examining some of the applications for arithmetic circuits.

26-3-1 *Basic Two's Complement Adder/Subtractor Circuit*

Figure 26-20 shows how two 7483 ICs could be connected to form an 8-bit two's complement adder/subtractor. In this circuit, there is also an 8-bit input switch circuit, an 8-bit output LED display circuit, and two 8-bit registers (A and B). This circuit operates as follows.

A set of eight input switches and two 74126 three-state buffers (controlled by the active-HIGH input enable, E_I) are used to connect any 8-bit word (set by the switches) onto

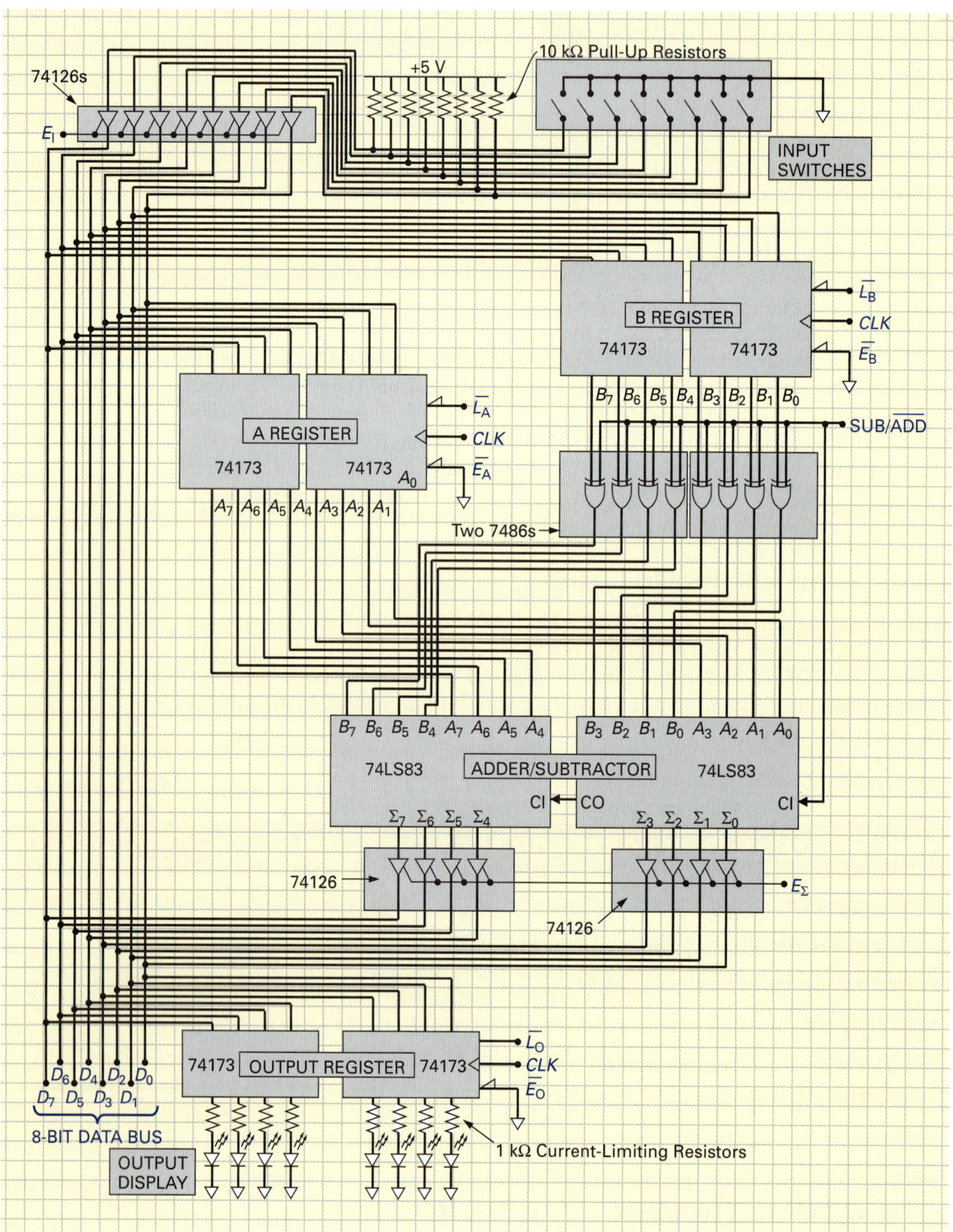

FIGURE 26-20 **Application—An 8-Bit Bus-Organized Two's Complement Adder/Subtractor Circuit.**

the 8-bit data bus (D_0 through D_7). The 8-bit input data can be loaded into the A register (by pulsing LOW, $\overline{L}_A$), the B register (by pulsing LOW, $\overline{L}_B$), or the output register (by pulsing LOW, $\overline{L}_O$).

Referring to the two 74LS83s (which form an 8-bit adder circuit), you can see that the A register supplies the 8-bit A input (A_7 through A_0), while the B register supplies the

8-bit B input (B_7 through B_0). The 8-bit sum output from the two 74LS83s (Σ_7 through Σ_0) can be switched onto the 8-bit data bus by pulsing the sum enable control line (E_Σ) HIGH.

The two 7486s connected to the B input of the adder/subtractor contain eight XOR gates which are included to function as controlled inverters under the control of the SUB/$\overline{\text{ADD}}$ input. When the SUB/$\overline{\text{ADD}}$ control line is LOW, inputs B_7 through B_0 will pass through the XOR gates without inversion, and therefore the sum output of the 74LS83s will be

$$\Sigma_7 \text{ through } \Sigma_0 = A_7 \text{ through } A_0 + B_7 \text{ through } B_0$$

The LSB carry input (CI) is also connected to the SUB/$\overline{\text{ADD}}$ control line, and since this input is LOW in this instance, it will have no effect on the adder circuit's result.

On the other hand, when the SUB/$\overline{\text{ADD}}$ control line is HIGH, the XOR gates will act as inverters and invert the output from the B register. The B input to the 74LS83s will therefore be the one's complement of the original word stored in the B register. Since the SUB/$\overline{\text{ADD}}$ control line is HIGH in this instance, the LSB carry input will also be HIGH, and therefore a 1 will be added to the LSB of the adder. As a result, when the SUB/$\overline{\text{ADD}}$ control line is HIGH, the B input word is first one's complemented and then a 1 is added to the LSB, so the overall effect is that the B input is two's complemented. Since the B input word is two's complemented and added to the A input word, the result will be a two's complement subtraction through addition operation or

$$\Sigma_7 \text{ through } \Sigma_0 = A_7 \text{ through } A_0 - B_7 \text{ through } B_0$$

By using the input switches, the operator is able to load an 8-bit two's complement word into the A register (E_I = HIGH, $\overline{L}_A$ = LOW) and then load a different 8-bit two's complement word into the B register (E_I = HIGH, $\overline{L}_B$ = LOW). Then, by controlling the logic level on the SUB/$\overline{\text{ADD}}$ control line (HIGH = subtract, LOW = add) you can control whether the sum or the difference appears at the output of the 74LS83s. To display the output of the 74LS83s on the 8-bit LED display, simply enable the 74126 three-state output buffers (E_Σ = HIGH) and then load the result from the data bus into the output register ($\overline{L}_O$ = LOW).

26-3-2 *An Arithmetic-Logic Unit (ALU) IC*

The 74181 IC is a good example of an arithmetic-logic unit (ALU) that can perform sixteen different arithmetic operations and sixteen different logic operations on two 4-bit input words.

Application—Single-Chip Arithmetic-Logic Unit (ALU Circuit)

Figure 26-21 shows some of the details for this IC. Referring to the Function Select Table, you can see that the mode control input M, pin 8, determines whether the ALU performs a logic operation (M = HIGH) or arithmetic operation (M = LOW). The four select inputs (S_3, S_2, S_1, and S_0) are used to select which one of the sixteen different logic functions or arithmetic operations are performed on the A input (A_3, A_2, A_1, or A_0) and B input (B_3, B_2, B_1, or B_0) words. For example, if M = HIGH (logic function) and S_3, S_2, S_1, and S_0 = HHHL ($F = A + B$, $F = A$ OR B), then the output word F (F_3, 210) will be the result of ORing input word A with input word B. As another example, if M = LOW (arithmetic operation), S_3, S_2, S_1, and S_0 = HLLH ($F = A$ plus B) when the active-LOW carry input is HIGH ($\overline{C}_n$ = HIGH, no carry), then the output word F will be the sum of input word A plus input word B. Figure 26-21 shows how a 74181 4-bit ALU IC could be connected so that its sixteen different logic functions and sixteen different arithmetic operations could be tested.

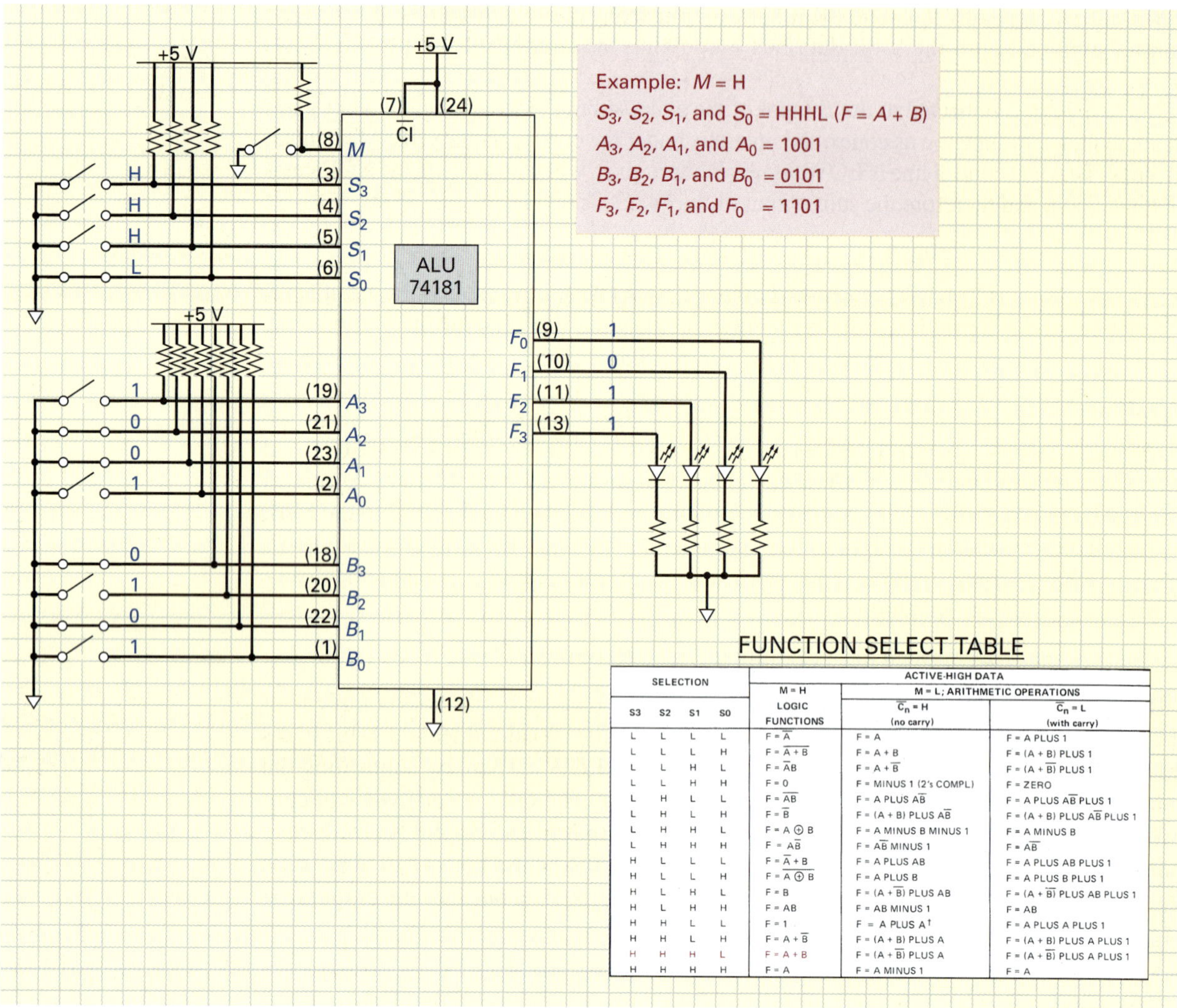

SELECTION				ACTIVE-HIGH DATA		
				M = H	M = L; ARITHMETIC OPERATIONS	
S3	S2	S1	S0	LOGIC FUNCTIONS	$\overline{C}_n$ = H (no carry)	$\overline{C}_n$ = L (with carry)
L	L	L	L	F = $\overline{A}$	F = A	F = A PLUS 1
L	L	L	H	F = $\overline{A + B}$	F = A + B	F = (A + B) PLUS 1
L	L	H	L	F = $\overline{A}$B	F = A + $\overline{B}$	F = (A + $\overline{B}$) PLUS 1
L	L	H	H	F = 0	F = MINUS 1 (2's COMPL)	F = ZERO
L	H	L	L	F = $\overline{AB}$	F = A PLUS A$\overline{B}$	F = A PLUS A$\overline{B}$ PLUS 1
L	H	L	H	F = $\overline{B}$	F = (A + B) PLUS A$\overline{B}$	F = (A + B) PLUS A$\overline{B}$ PLUS 1
L	H	H	L	F = A ⊕ B	F = A MINUS B MINUS 1	F = A MINUS B
L	H	H	H	F = A$\overline{B}$	F = A$\overline{B}$ MINUS 1	F = A$\overline{B}$
H	L	L	L	F = $\overline{A}$ + B	F = A PLUS AB	F = A PLUS AB PLUS 1
H	L	L	H	F = $\overline{A \oplus B}$	F = A PLUS B	F = A PLUS B PLUS 1
H	L	H	L	F = B	F = (A + $\overline{B}$) PLUS AB	F = (A + $\overline{B}$) PLUS AB PLUS 1
H	L	H	H	F = AB	F = AB MINUS 1	F = AB
H	H	L	L	F = 1	F = A PLUS A†	F = A PLUS A PLUS 1
H	H	L	H	F = A + $\overline{B}$	F = (A + B) PLUS A	F = (A + B) PLUS A PLUS 1
H	H	H	L	F = A + B	F = (A + $\overline{B}$) PLUS A	F = (A + $\overline{B}$) PLUS A PLUS 1
H	H	H	H	F = A	F = A MINUS 1	F = A

FIGURE 26-21 **Single-Chip 74181 ALU Test Circuit.**

SELF-TEST EVALUATION POINT FOR SECTION 26-3

Use the following questions to test your understanding of Section 26-3.

1. With the two's complement adder/subtractor circuit in Figure 26-20, which ICs are responsible for one's complementing during a subtraction operation?
2. With the two's complement adder/subtractor circuit in Figure 26-20, how is the additional 1 added so that we can two's complement the B input word?
3. What operation would a 74181 IC perform if its control inputs were
   a. M = H, S = LHLH?
   b. M = L, S = LHHL, $\overline{C}_n$ = L?
4. What would appear at the output of a 74181 if A = 0011, B = 0110, M = L, S = HLLH?

REVIEW QUESTIONS

Multiple-Choice Questions

1. The sum of 101_2 and 011_2 is
- **a.** 0111_2.
- **b.** 8_{10}.
- **c.** 1001_2.
- **d.** 6_{10}.

2. Which digital logic unit is used to perform arithmetic operations and logic functions?
- **a.** MPU
- **b.** ALU
- **c.** CPU
- **d.** ICU

3. With binary addition, $1 + 1 = ?$
- **a.** 10_2
- **b.** 0 carry 1
- **c.** 2_{10}
- **d.** all of the above

4. With binary subtraction, $0 - 1 = ?$
- **a.** 10_2
- **b.** 1 borrow 0
- **c.** 1 borrow 1
- **d.** all of the above

5. What would be the result of $1011101_2 \times 01_2$?
- **a.** 1011101_2
- **b.** 1110101_2
- **c.** 000111_2
- **d.** 1011100_2

6. The four rules of binary multiplication are $0 \times 0 =$ ____________, $1 \times 0 =$ ____________, $0 \times 1 =$ ____________, and $1 \times 1 =$ ____________.
- **a.** 0,1,0,1
- **b.** 0,0,1,1
- **c.** 0,0,0,0
- **d.** 0,0,0,1

7. What would be the result of $1001_2 \div 0011_2$?
- **a.** 3_{10}
- **b.** 3, remainder zero
- **c.** 0011_2
- **d.** All of the above

8. The one's complement of $0101\ 0001_2$ would be
- **a.** $0101\ 0010_2$.
- **b.** $1010\ 1110_2$.
- **c.** $1010\ 1111_2$.
- **d.** all of the above.

9. What would be the decimal equivalent of 1001_2 if it were a two's complement number?
- **a.** +9
- **b.** −9
- **c.** +7
- **d.** −7

10. Which number system makes it possible for us to add, subtract, multiply, and divide binary numbers all with an adder circuit?
- **a.** Sign-magnitude
- **b.** One's complement
- **c.** Two's complement
- **d.** Floating-point

11. Which binary code system enables us to represent extremely large numbers or small fractions with fewer bits?
- **a.** Sign-magnitude
- **b.** One's complement
- **c.** Two's complement
- **d.** Floating-point

12. With a half-adder circuit, the sum output of the 2-bit input is generated by a/an ____________ gate, while the carry output of the 2-bit input is generated by a/an ____________ gate.
- **a.** XOR, AND
- **b.** XNOR, NAND
- **c.** NOR, NAND
- **d.** AND, XOR

13. Which circuit is used to add a 2-bit input and a carry from the previous column?
- **a.** Half-adder
- **b.** Full-adder
- **c.** Parallel-adder
- **d.** All of the above

14. The 7483 IC is a ____________ -bit adder circuit.
- **a.** 2
- **b.** 6
- **c.** 4
- **d.** 3

15. The 74181 IC could be described as a/an ____________.
- **a.** 4-bit FA
- **b.** 8-bit FA
- **c.** 4-bit PA
- **d.** 8-bit PA

Practice Problems

16. Add the following 8-bit numbers:
- **a.** 0110 1000 and 0001 1111
- **b.** 0111 1111 and 0100 0011
- **c.** 0001 1010 and 0011 1010

17. Subtract the following 8-bit numbers:
- **a.** 0001 1010 from 1010 1010
- **b.** 0001 1111 from 0011 0000
- **c.** 0010 0000 from 1000 0000

18. Multiply the following binary numbers:
- **a.** 0101 1011 by 011
- **b.** 0001 1111 by 1011
- **c.** 1110 by 010

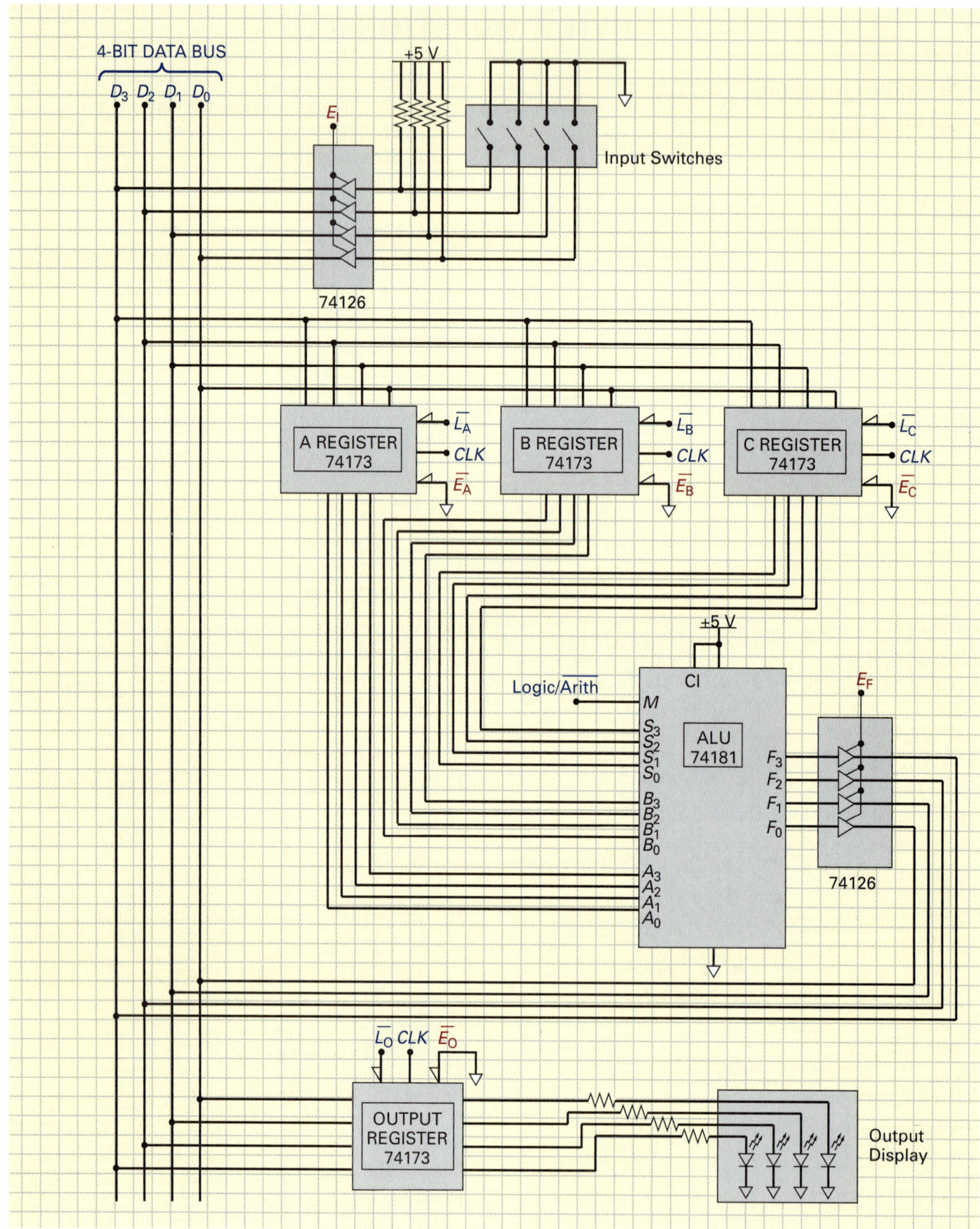

FIGURE 26-22 **Application—A 4-Bit Arithmetic-Logic Circuit.**

19. Divide the following binary numbers:
  a. 0110 into 0111 0110
  b. 101 into 1010 1010
  c. 1001 into 1011 1110

20. Convert each of the following decimal numbers to an 8-bit sign-magnitude number:
  a. +34
  b. −83
  c. +83

21. One's complement the following numbers:
  a. 0110 1001
  b. 1010 1010
  c. 1111 0000

22. Convert all of the following two's complement numbers to their decimal equivalents:
  a. 0001 0000
  b. 1011 1110
  c. 1111 0101

23. Convert each of the following decimal numbers to an 8-bit two's complement number:
   a. +119
   b. −119
   c. −34
24. Two's complement the following numbers:
   a. 0101 1010
   b. 1000 0001
   c. 1111 1101
25. What would be the two's complement range of a 6-bit binary number?

   To practice your circuit recognition and operation skills, refer to Figure 26-22 and answer the following five questions.
26. Give a one-line definition of each of the ICs used in this circuit.
27. What is the purpose of the C register in this circuit?
28. Describe the sequence of events that would have to be followed if we wanted to
   a. input a word into the A register.
   b. input a word into the B register.
   c. display the output of the ALU on the LED display.
   d. connect a 4-bit word from the input switches directly to the output display.
29. Describe the sequence of events that would have to be followed if we wanted to
   a. input and add two 4-bit words and display the result.
   b. input and OR two 4-bit words and display the result.
   c. input and AND two 4-bit words and display the result.
30. Describe the sequence of events that would have to be followed if we wanted to input and add two 4-bit words and then store the result in the A register.

Web Site Questions

Go to the web site http://www.prenhall.com/cook, select the textbook *Electronics: A Complete Course,* select this chapter, and then follow the instructions when answering the multiple choice practice problems.

Semiconductor Memory

Flying Too High

The term "software"—as opposed to "hardware" or physical components of the computer—was first used in the 1960s to describe the program of instructions that tells a computer what to do. At first, the binary words representing instructions and data were laboriously entered by setting up countless switches and dials. In modern computers, binary coded instructions and data are stored as magnetic signals on disks and then input into the computer's internal memory where the program will tell the computer what to do.

Computer software today is organized into several basic types. There are "translator" programs written to convert high-level computer languages, which are like human speech, into machine language, the ones and zeros understood by the machine or computer. There are also "utility" programs designed to sort and merge data files. Two of the biggest groups of software are "operating systems" and "application programs." An operating system coordinates the operation of the computer system controlling the transfer of data between the computer and its peripheral devices, such as the keyboard, monitor, printer, and disk drive. Application programs are the software that change the personality of a computer from a word processor generating a contract, to an accounting system controlling payroll, to an industrial controller directing a robotics assembly line.

In this vignette, let us examine some of the software-industry folklore relating to operating systems. Most of the 8-bit computers of the late 1970s used an operating system developed by Gary Kildall in 1974 called CP/M, standing for "Control Program for Microcomputers." Just before IBM officially entered into the personal computer market in August of 1981, two IBM executives went to a prearranged meeting with Kildall, who was at the time flying his private plane. After being contacted by radio, Kildall declined to come down and discuss an operating system for their proposed personal computer or "PC." After rescheduling the meeting, the executives were surprised to discover that he had flown off to the Caribbean for a vacation. When he came back, IBM had made a deal with Bill Gates of Microsoft for an operating system. Called MS-DOS, which stood for "Microsoft—Disk Operating System" (DOS means an operating system that is loaded into the computer from a disk), it became an instant success largely due to its blessing from IBM. Hundreds of computer manufacturers started producing computers that were compatible with the IBM PC and therefore used Gates's MS-DOS. Even in an expanding market, Kildall's CP/M could not hold on to any business and is today extinct.

Bill Gates's Microsoft is the largest computer software company in the world, and history is once again repeating itself. Not liking their dependency on Microsoft, IBM has developed their own operating system, which is now in its second version called "OS/2." Another battle begins.

27

Introduction

In Chapter 25 we saw how several flip-flops could be connected to form a register circuit, and how a register could be used to store a binary word. When a register is used to store binary data, it is called a memory register, and in digital systems, memory storage is a very important function.

In a digital electronic computer system, the microprocessor unit (MPU) decodes and executes binary instruction codes and directs the flow of binary data codes in order to perform its function. The instruction codes and the data codes within a digital computer system are stored in either a *primary memory* or a *secondary memory.* These two basic memory categories are shown in Figure 27-1.

The computer's primary memory is made up of two semiconductor memory IC types called ROM (read-only memory) and RAM (random-access memory). The MPU has the binary data that it needs to quickly access data stored in these high-speed ROM and RAM ICs. The ROM and RAM ICs, however, have a high cost per bit of storage, which means that their high speed of operation comes at a high cost. Secondary memory devices, such as floppy disks, hard disks, and optical disks, will store data at a cheaper cost per bit; therefore these devices are used to store the large volume of data that is not presently being used by the MPU.

In this chapter, we will be discussing in detail the semiconductor ROM and RAM primary memory devices.

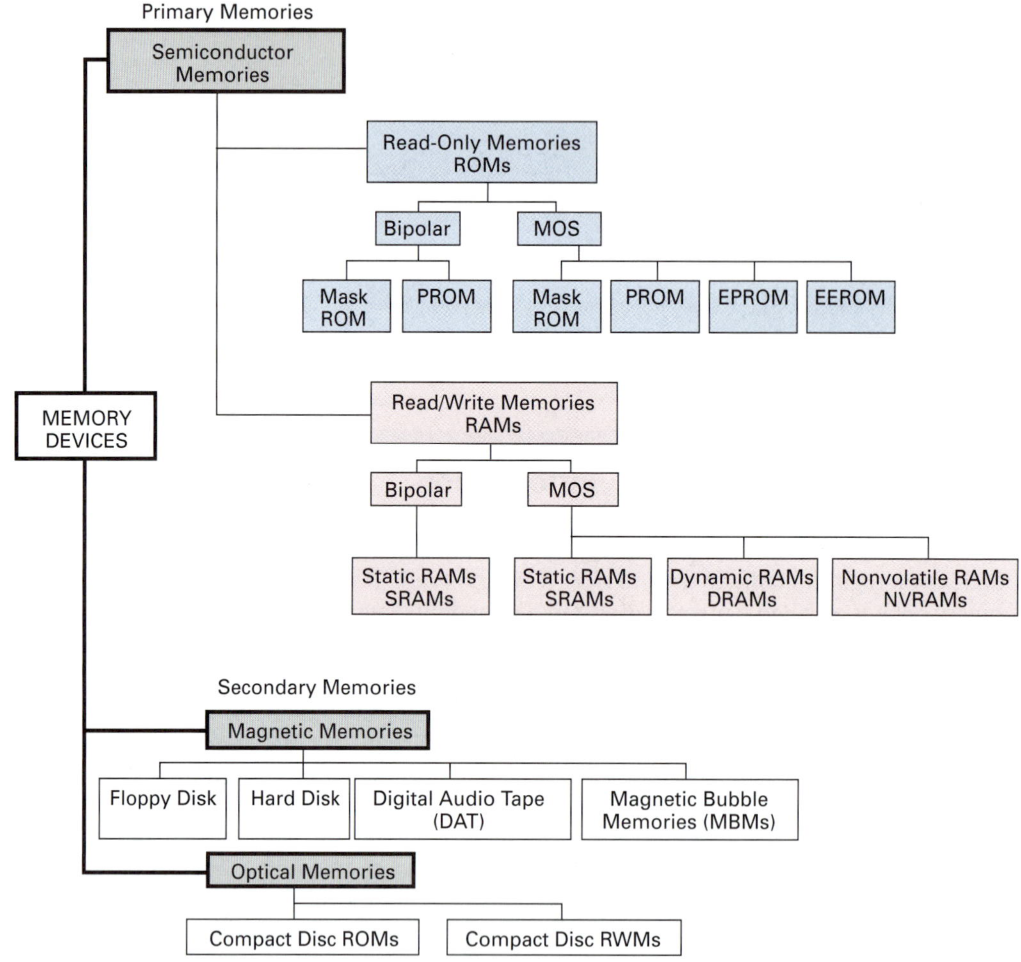

FIGURE 27-1 **Memory Devices.**

27-1 SEMICONDUCTOR READ-ONLY MEMORIES (ROMs)

Read-Only Memory (ROM)
A nonvolatile random-access memory in which the stored binary data is loaded into the device before it is installed into a system. Once installed into a system, data can only be read out of ROM.

A **read-only memory,** or **ROM,** is an electronic circuit that permanently stores binary data. In this section we will examine the operation, characteristics, applications, and testing of ROM ICs, the different types of which were shown in Figure 27-1.

27-1-1 *A Basic Diode ROM*

Figure 27-2(a) shows how a simple ROM circuit could be constructed using an 8-position rotary switch, a set of diodes, and four pull-up resistors. Each of the eight rows (rows 0 through 7) is equivalent to a 4-bit register, and when selected, the 4-bit binary word stored will be present at the 4-bit output (D_3, D_2, D_1, and D_0).

When the rotary switch is placed in position 0, all three diodes in row 0 will be turned ON, producing a LOW output at D_3, D_2, and D_1. The 4-bit word stored at memory location 0 is therefore 0001.

When the rotary switch is placed in position 1, a different pattern of diodes will be turned ON, producing a LOW output at D_3, D_1, and D_0. The 4-bit output word stored at memory location 1 is therefore 0100.

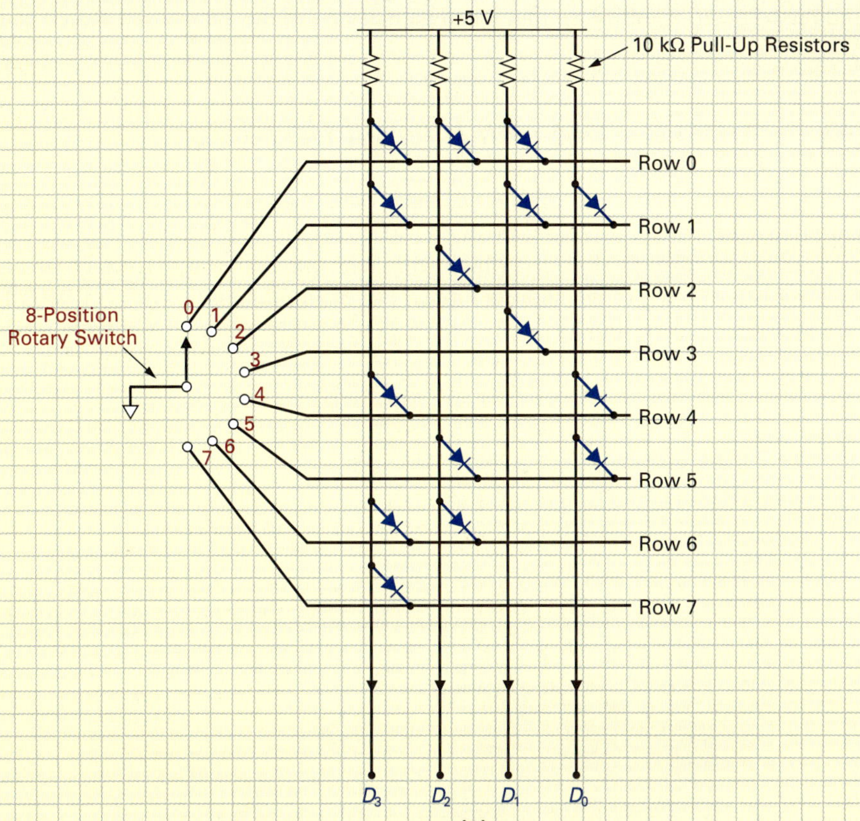

(a)

Switch Position	Output Word
0	0001
1	0100
2	1011
3	1101
4	0110
5	1010
6	0011
7	0111
Address	Data

(b)

FIGURE 27-2 A Basic ROM Circuit.

Figure 27-2(b) shows all of the 4-bit words stored in each of the eight memory locations. By adding or removing diodes from the matrix, you can change the 4-bit word stored in any of the eight memory locations.

Referring to the bottom of the table in Figure 27-2(b), you can see that the switch position is called the **address,** while the contents of each memory location are called the **data.** Each and every memory location is assigned its own unique address, so we can access the data in any one of the desired memory locations. This method is similar to the postal system, which gains access to us through our unique address. The address of a memory location and the data stored in that memory location are completely unrelated. For example, at address 2 in Figure 27-2, we have a stored data value of 1011 (decimal 11). Similarly, the postal system address and data at that address are also unrelated, in that address 10752 A Street can

Address
A binary code that designates a particular location in either a storage, input, or output device.

Data
A general term used for meaningful information.

have any number of letters in its mail box. In a digital electronic computer system, the MPU will send out a specific address so that it can gain access to the data stored at that address.

27-1-2 *A Diode ROM with Internal Decoding*

Figure 27-3(a) shows how a one-of-eight decoder could be used to select the desired memory location instead of using an eight-position rotary switch. The three address inputs A_2, A_1 and A_0 drive the one-of-eight decoder, which will make one of the eight rows LOW based on the binary address applied to the address inputs. For example, if an address of 010 (decimal 2) is applied to the address inputs A_2, A_1, and A_0, only NAND gate C will have a HIGH on

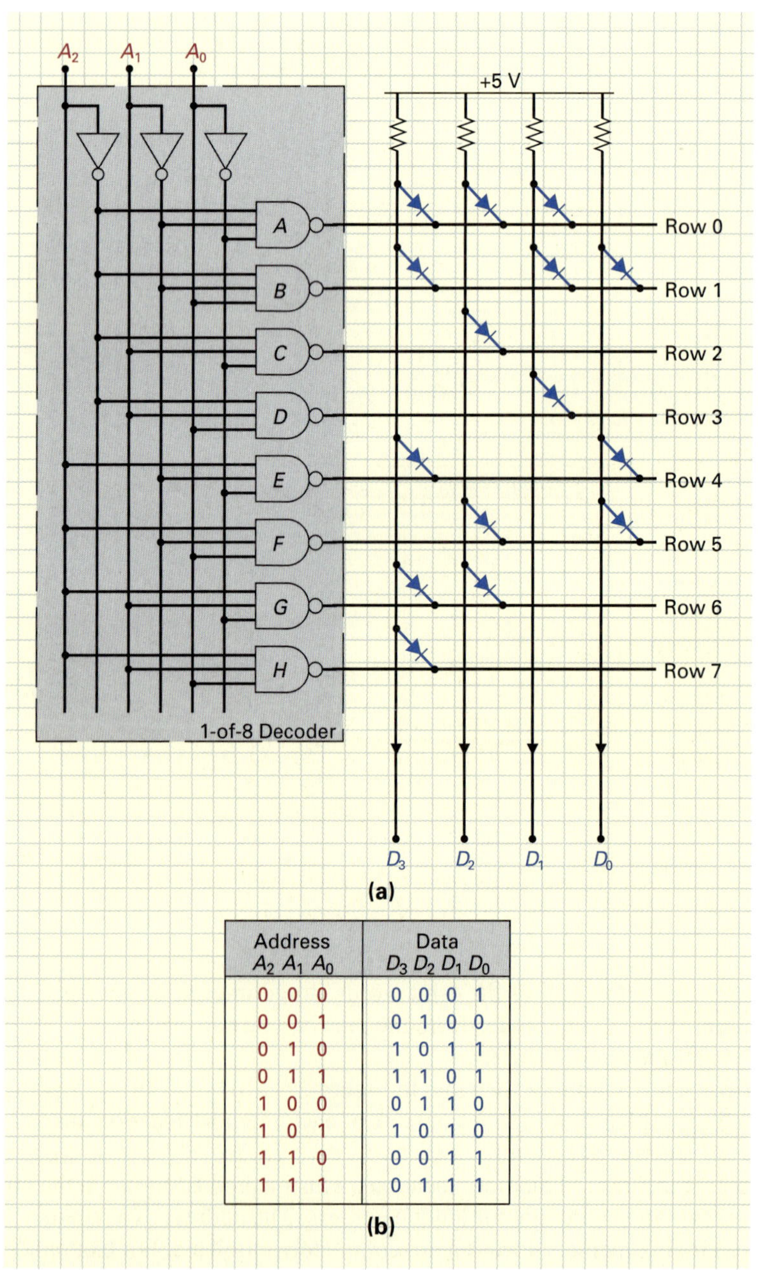

Address A_2 A_1 A_0	Data D_3 D_2 D_1 D_0
0 0 0	0 0 0 1
0 0 1	0 1 0 0
0 1 0	1 0 1 1
0 1 1	1 1 0 1
1 0 0	0 1 1 0
1 0 1	1 0 1 0
1 1 0	0 0 1 1
1 1 1	0 1 1 1

FIGURE 27-3 A Basic ROM Circuit with Internal Decoding.

all three of its inputs. This will generate a LOW out on row 2, producing a data output of 1011 at D_3, D_2, D_1, and D_0, respectively. The data stored at each address is listed in the table in Figure 27-3(b).

27-1-3 *Semiconductor ROM Characteristics*

In this section we will examine characteristics that are common to all semiconductor ROMs. These include number of address lines, memory size and organization, read-only operation, volatility, access time, and input/output timing.

Number of Address Lines

Although the ROM circuit in Figure 27-3(a) is more complex than the ROM circuit in Figure 27-2(a), the decoder enables the circuit to be electronically accessed using three address lines instead of manually accessed by eight switched input lines. When an on-chip decoder is included with the ROM memory circuit, the number of address lines (n) applied to the IC will indicate how many memory locations are available within the IC (2^n). For example, the three address lines applied to the ROM circuit in Figure 27-3(a) could address one of eight memory locations ($2^n = 2^3 = 8$ memory locations).

EXAMPLE:

Calculate the number of memory locations that can be accessed by a ROM memory IC, if it has

a. four address lines applied
b. eight address lines applied
c. sixteen address lines applied

Solution:

a. Four address lines can address $2^4 = 16$ memory locations.
b. Eight address lines can address $2^8 = 256$ memory locations.
c. Sixteen address lines can address $2^{16} = 65{,}536$ memory locations.

As you can see from these examples, there is a direct relationship between the number of address lines applied to a memory circuit and the number of memory locations available within that memory circuit. The left and center columns in Table 27-1 show the correlation between the two. The right column in Table 27-1 shows the abbreviation that is used to represent these large values. The label "k" indicates 1000 in decimal; however, here we are dealing with binary (base-2), where K indicates 1,024. A 2K memory, therefore, has 2,048 memory locations (2 × 1,024), and a 1M (one meg) memory has 1,048,576 memory locations (1,024 × 1,024).

Memory Size and Memory Configuration

The term **memory size** describes the number of bits of data stored in the memory. For example, the basic ROM in Figure 27-3 is a 32-bit memory since it can store thirty-two binary digits of data. The term **memory configuration** describes how the memory circuit is organized into groups of bits or words. For example, the thirty-two storage bits in the ROM circuit in Figure 27-3 are organized into eight 4-bit words (8 by 4, 8 × 4).

Memory Size
A term used to describe the total number of binary digits stored in a memory.

Memory Configuration
A term used to describe the organization of the storage bits within a memory.

TABLE 27-1 The Relationship Between Address Bits and Memory Locations

NUMBER OF ADDRESS BITS (n)	EXACT NUMBER OF MEMORY LOCATIONS (2^n)	ABBREVIATED NUMBER OF MEMORY LOCATIONS
8	256	256
9	512	512
10	1,024	1K
11	2,048	2K
12	4,096	4K
13	8,192	8K
14	16,384	16K
15	32,768	32K
16	65,536	64K
17	131,072	128K
18	262,144	256K
19	524,288	512K
20	1,048,576	1M
21	2,097,152	2M
22	3,194,304	4M
23	8,388,608	8M
24	16,777,216	16M
25	33,554,432	32M
26	67,108,864	64M

Storage Density
A relative term used to compare the memory storage ability of one device to another.

Stored
A term used to describe when data is placed in a memory location or output device.

Memory Write Operation
To store information into a location in memory or an output device.

Retrieved
A term used to describe when data is extracted from a memory location or input device.

Memory Read Operation
To extract information from a location in memory or an input device.

Programming
A process of preparing a sequence of operating instructions for a computer system and then loading these instructions in the computer's memory.

The term **storage density** is used to compare one memory to another. For example, if memory A IC can store thirty-two bits and memory B IC can store sixty-four bits, then memory B is said to contain more storage bits per chip.

EXAMPLE:

The 8355 ROM IC shown in Figure 27-4 has eleven address inputs and eight data outputs. What is the size and configuration of this read-only memory?

Solution:

The eleven address line inputs to this IC indicate that there are 2^{11}, or 2,048, different memory locations. The eight data lines to this IC indicate that an 8-bit word is stored at each of the memory locations. Therefore, the 8355 ROM IC has a memory size of 16,384 bits ($2{,}048 \times 8$) and is configured as 2,048 by 8.

The Read-Only Operation of a ROM

When a binary word is **stored** in a memory location, the process is called a **memory-write operation.** The term *write* is used, since writing a word into memory is similar to writing a word onto a page.

When a binary word is **retrieved** from a memory location, the process is called a **memory-read operation.** The term *read* is used, since reading a word out of memory is similar to reading a word from a page.

With a ROM, the binary information contained in each memory location is stored or written into the ROM only once. This process of entering data into the ROM is called **programming** or *burning in* the ROM. Once a ROM IC has been programmed, it is installed in a circuit within a digital system, and from that point on the MPU only reads or *fetches* data out of the ROM. Once it has been installed in a circuit, therefore, the ROM will only ever operate as a read-only memory.

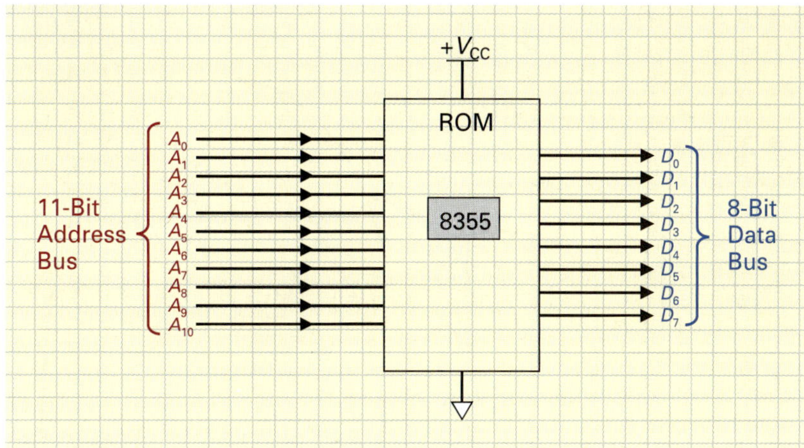

FIGURE 27-4 **A Read-Only Memory IC.**

Bipolar and MOS ROMs

Figure 27-5(a) shows the basic internal structure of a ROM IC. This **architecture** is essentially the same, irrespective of whether the device is a bipolar ROM or MOS ROM.

Architecture
The basic structure of a device, circuit, or system.

As far as operation, the *chip enable* $(\overline{CE})$ or *chip select* $(\overline{CS})$ is normally an active-LOW control input that either enables $(\overline{CE} = \text{LOW})$ or disables $(\overline{CE} = \text{HIGH})$ the ROM. It achieves this by disabling the address decoder's AND gates and the three-state output buffers.

When enabled, the one-of-eight decoder selects one of the eight rows depending on the address applied to the three address inputs A_2, A_1, and A_0, $(2^3 = 8)$. The decoder will give a HIGH output at only one of its AND gate outputs, and this HIGH will be applied to four blocks in the ROM matrix. Each of these blocks contains a transistor and a *programmable link,* as shown in the inset in Figure 27-5(a). It is the presence or absence of the programmable link that determines whether a 1 or 0 is stored in each of these blocks. If this were a bipolar ROM, the programmable link would connect the row line from the decoder to the base of an NPN bipolar transistor. If this were a MOS ROM, the programmable link would connect the row line from the decoder to the gate of a field-effect transistor. In both cases, the presence of a link will allow the HIGH on the row line from the decoder to turn ON the transistor and connect a HIGH onto the column line and to the data output line below. On the other hand, the absence of a link will keep the transistor OFF and therefore maintain the column line and the data output line below as LOW.

The logic symbol for this 32-bit ROM, which is organized as an 8 by 4 (eight 4-bit words), is shown in Figure 27-5(b).

Most semiconductor ROM ICs use MOSFETs because their relatively small storage cell size makes it possible for a greater number of storage bits per IC. MOS ROMs also cost less per bit and consume less power than bipolar ROMs. Bipolar TTL and ECL ROM ICs are available; however, their small storage capacity, high cost per bit, and high power consumption make them undesirable except for high-speed applications.

Access Time and Input/Output ROM Timing

The **access time** of a memory is the time interval between the memory receiving a new address input and the data stored at that address being available at the data outputs. To explain this in more detail, Figure 27-5(c) shows the typical timing waveforms for a ROM read operation. The upper waveform represents the address inputs, the center waveform represents the active-LOW chip-enable $(\overline{CE})$ input, and the lower waveform represents the data outputs. Since the address and data waveforms represent more than one line, they are shown as a block on the timing waveform since some of the lines are HIGH and some are LOW. Whenever these blocks criss-cross, as can be seen with the address waveform at time t_1, a change in logic levels has occurred on the lines.

Access Time
The time interval between a device being addressed and the data at that address being available at the device's output.

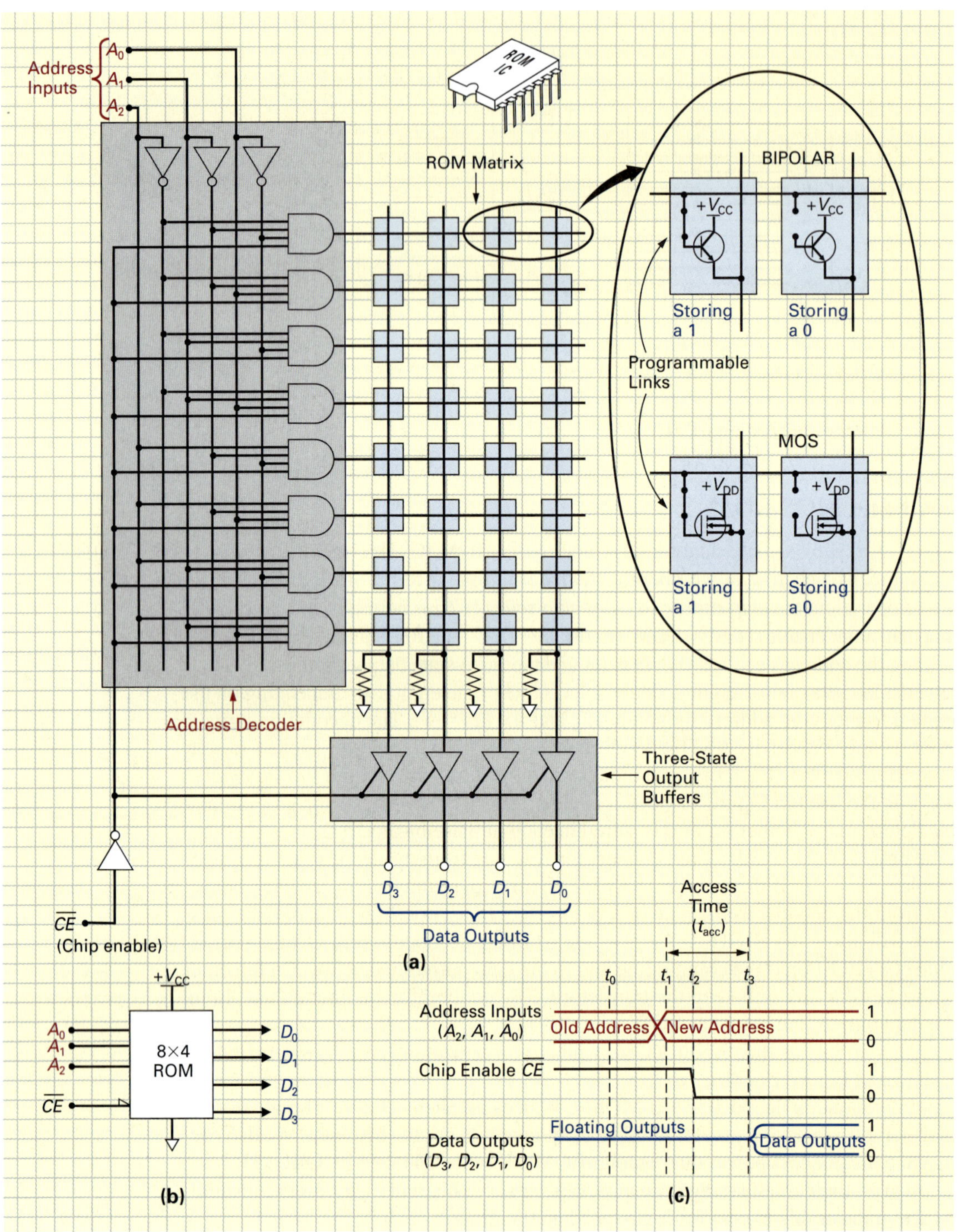

FIGURE 27-5 Typical ROM IC. (a) Basic Architecture. (b) Logic Symbol. (c) Timing Waveform.

Referring to the timing points shown in Figure 27-5(c), you can see that at time t_0, the chip-enable input is HIGH and therefore the ROM IC is disabled. At time t_1, a new address is placed on the 3-bit address bus (A_2, A_1, and A_0), and the ROM's internal address decoder will begin to decode the new address input to determine which memory location is to be enabled. At time t_2, the $\overline{CE}$ input is made active, enabling the final stage of the address decoder and the output three-state buffers. At time t_3, the data outputs (D_3, D_2, D_1, and D_0) change from a floating output state to a logic level that reflects the data stored in the addressed memory location.

Since the access time of a memory is the time interval between the memory receiving a new address input and the data being available at the data output, the access time (t_{acc}) is the time interval between t_1 and t_3 in Figure 27-5(c). Typically, bipolar ROMs have access times in the 10 ns to 50 ns range, while MOS ROMs will have access times from 35 ns to 500 ns.

27-1-4 *ROM Types*

There are three basic types of ROM—mask programmable ROMs (MROMs), programmable ROMs (PROMs), and erasable programmable ROMs (EPROMs, EEPROMs, and flash memories). Let us now examine each of these types and see how they differ.

Mask Programmable ROMs (MROMs)

Mask programmable ROMs, or **MROMs,** are permanently programmed by the manufacturer by simply adding or leaving out diodes or transistors, as previously shown. A customer sends the manufacturer a truth table stating what data should be stored at each address, and then the manufacturer generates a photographic negative called a *mask* that is used to produce the interconnections on the ROM matrix. These masks are normally expensive to develop and therefore this type of ROM only becomes economical if a large number of ROMs are required.

Mask Programmable ROM (MROM)
A read-only memory in which the manufacturer generates a photographic negative or mask designed to store the customer-specified data requirements.

The disadvantage of these mask programmed ROMs is that they cannot be reprogrammed with any design modifications, since programming is permanent.

As an example, the TMS47256 is available from Texas Instruments as an NMOS version (TMS47256) with an access time of 200 ns and a standby power dissipation of 82.5 mW, or as a CMOS version (TMS47C256) with an access time of 150 ns and a standby power dissipation of 2.8 mW.

Programmable ROMS (PROMS)

Although the MROM is extremely expensive, the cost per IC can become minimal when a large volume of ROMs is needed (thousands). On the other hand, if only a small number of ROMs is needed, one option is to use a fusible-link **programmable ROM,** or **PROM.** Unlike the MROM whose data is etched in stone, so to speak, by the manufacturer, the PROM comes with a clean slate and can be programmed by the user. Like the MROM, however, once the PROM has been programmed, the etched data cannot be erased.

Programmable ROM (PROM)
A read-only memory in which the customer can store any desired data codes by controlling fusible links.

Figure 27-6(a) shows how the architecture of a PROM is similar to that of the MROM shown in Figure 27-5(a). The difference with the PROM, however, is that each of its programmable links is fusible. All of these fusible links are initially intact when the PROM arrives from the manufacturer, and it is up to the user to either blow a fuse link (to store a 0) or leave it intact (to store a 1).

The programming or burning-in process is performed by a *PROM programmer unit,* similar to the one shown in Figure 27-6(b). By placing the PROM into the IC socket on the unit and entering the desired truth table using the keyboard, the PROM programmer unit will then automatically step through each address and blow or leave the fuse intact in each memory location. Once the data has been written into every memory location, the programmer can read back the stored data to verify that it matches the desired truth table.

Both bipolar (low-density, high-speed, high power dissipation) and MOS (high-density, low-speed, low power dissipation) PROMs are available. For example, the 74186 is a 512-bit bipolar PROM, which is organized as sixty-four words of 8 bits each, has an access time of 50 ns. As another example, the TMS27PC256 is a 256K-bit CMOS PROM which is organized as 32K words of eight bits each and has an access time of 120 ns.

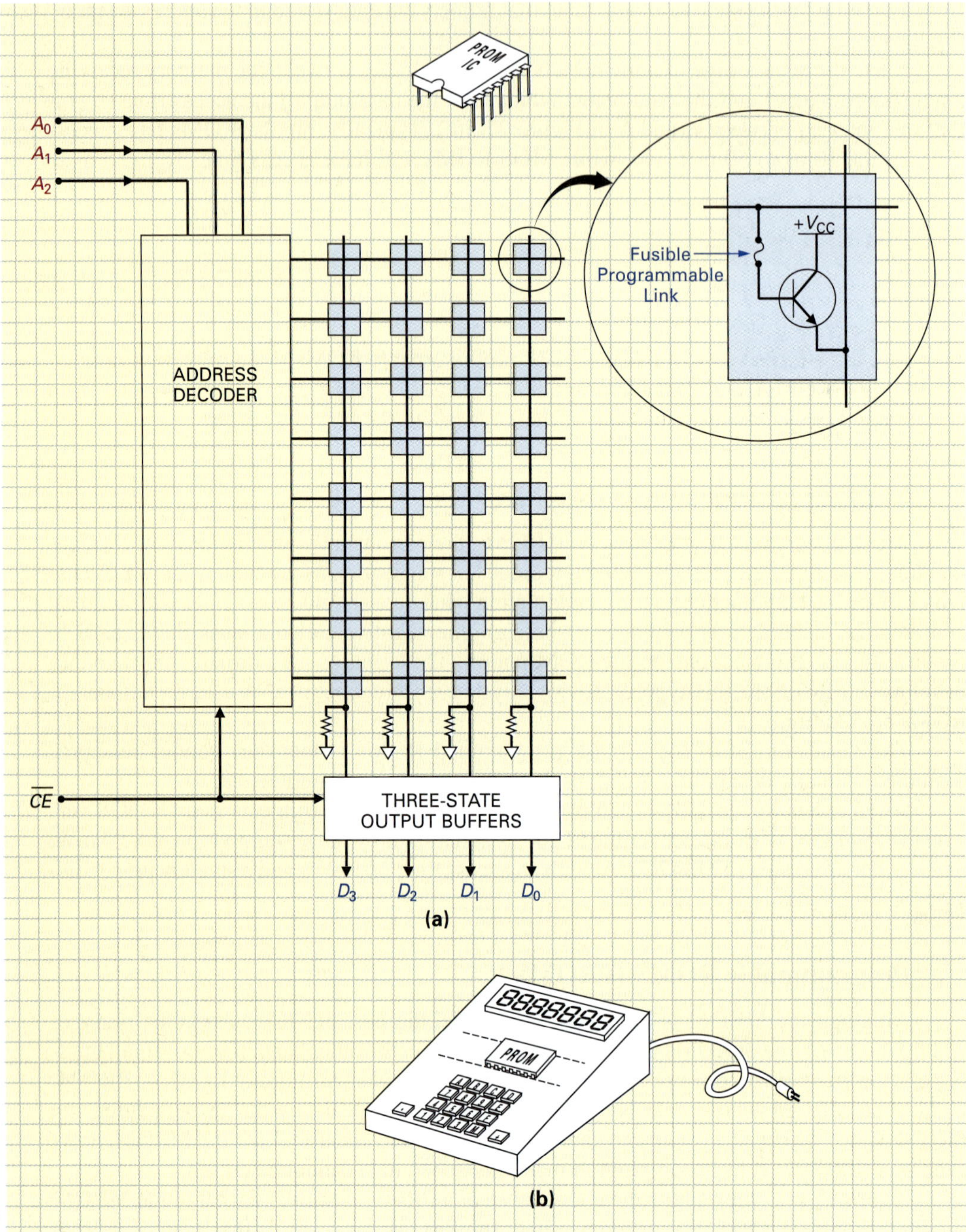

Erasable PROM (EPROM)
A PROM device that can have its contents erased using an ultraviolet light source and then be reprogrammed with new data.

Electrically Erasable PROM (EEPROM)
A PROM device that can have its contents erased using an electrical pulse, and then be reprogrammed with new data.

FIGURE 27-6 **Typical PROM IC. (a) Basic Architecture. (b) Programming.**

Erasable Programmable ROMs (EPROMs, EEPROMs, and Flash)

The ultraviolet-light **erasable PROM (EPROM),** the **electrically erasable PROM (EEPROM),** and *flash memories* are MOS circuits that enable the user to store data in the same way as the standard PROM, using the PROM programmer unit. The important difference with these devices is that if a modification to the stored data needs to be made at a later time, the contents of the IC can be erased and then reprogrammed with the new data.

The EPROM. With the ultraviolet erasable PROM, or EPROM, the data is erased by shining a high-intensity ultraviolet light through a quartz window on the top of the IC for about twenty minutes. Data in this IC type is stored as a charge or no charge, and the ultraviolet light causes these charges to leak off. The disadvantage with the EPROM is that the light erases the entire contents of the IC, and therefore if you only wish to change one byte or one bit, you will still have to erase everything and then reprogram the IC with the entire truth table.

The EEPROM. The electrically erasable PROM, or EEPROM, uses an electrical pulse to program data (like the EPROM); however, unlike the EPROM, it also makes use of an electrical pulse for erasing the stored data. This means that an ultraviolet light source is not needed for this type of PROM, since programming and erasing can be performed without removing the IC package from the circuit.

When programming an EEPROM, a 21 V pulse of typically 10 ms is needed to store a byte (an EPROM requires a 21 V pulse of typically 50 ms). When many thousands of bytes are involved, this makes the EEPROM programming and verifying process a lot less time-consuming. The key advantage of the EEPROM, however, is its ability to erase only a single byte of stored data, or to erase the entire device with a single 21 V/10 ms pulse. This means that a single byte modification can be performed without having to erase the entire device and then completely reprogram, as was the case with the EPROM. The erase time of 10 ms for the EEPROM is also a lot faster than that of the ultraviolet EPROM, which requires twenty minutes of UV exposure to erase its contents.

The EPROM is normally cheaper than the EEPROM, and both are generally used to develop a product. Once the design is finalized, however, a mask ROM, or MPROM, will be ordered in bulk for mass production.

Flash Memory
A reprogrammable ROM device that has extremely fast erase-data and write-data speeds.

The flash memory. To compare, EPROMs have fast access times, a high storage density, and a low cost per bit of storage. The EPROM's disadvantage, however, is that it has to be removed from its circuit to be erased, which takes twenty minutes. On the other hand, EEPROMs have fast access times and can be erased and reprogrammed without being removed from the circuit. The EEPROM's disadvantage, however, is that it has a low storage density and a much higher cost per bit of storage.

A **flash memory** has the advantages of both the EPROM and the EEPROM by combining the in-circuit erasability of the EEPROM with the high storage density and low cost of the EPROM. It is called a flash memory because of its high-speed erase and write data times. This new technology is rapidly improving, and at present the cost of a flash memory is much less than an EEPROM but not as low as an EPROM. As far as write time is concerned, a typical flash memory will have a write time of 10 μs per byte, compared to 100 μs per byte for a fast EPROM, and 5 ms for an EEPROM.

27-1-5 *ROM Applications*

As you study all of the ROM applications discussed in this section, you will notice that all the ROM is really doing is converting a coded input word at the address inputs into a desired output word at the data outputs.

Code Conversion

Figure 27-7 shows how a ROM could be used as a code converter. The input code is applied to the address input pins of the ROM. This code will be decoded by the ROM's internal decoder, which will select one of its internal memory locations and then apply the data content or desired output code to the data output pins.

Some standard ROM code converters are available. For example, there is a ROM code converter to convert the 7-bit alphanumeric ASCII (American Standard Code for Information Interchange) code into another code such as the 7-bit alphanumeric BCDIC (Binary Coded Decimal Interchange Code) code, which is used by some computers instead of ASCII. There is also a ROM code converter available that will achieve the opposite by converting BCDIC codes to ASCII codes.

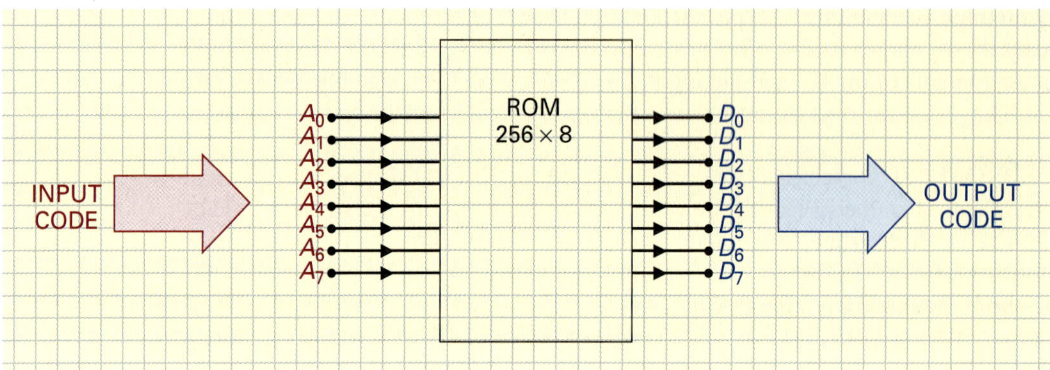

FIGURE 27-7 The ROM as a Code Converter.

Arithmetic Operations

As we discovered in the previous chapter, a simple adder circuit and the two's complement number system can be used to perform addition, subtraction, multiplication, and division. More complex mathematical functions, such as trigonometric and logarithmic operations, would require a more complex circuit and algorithm. In many of these instances, a ROM is used to implement the arithmetic operation. To use a simple example, Figure 27-8 shows how a ROM could be used to multiply two 2-bit words. In this circuit the 2-bit multiplicand and 2-bit multiplier are applied to the ROM's 4-bit address input. Since the correct product for every possible input combination is stored in the ROM, the correct result will always be present at the data outputs.

Look-Up Tables

A look-up table is not that different from a code converter. For example, the MM4232 ROM will convert a 9-bit binary input representing an angle between 0° and 90° to an 8-bit binary output that is equivalent to the sine of the input angle.

Logic Circuit Replacement

A ROM can also be used to replace a logic circuit. For example, an eight-input/four-output logic circuit could be replaced by an 8-bit address input/4-bit data output ROM, which could be programmed to give the desired output for any of the 8-bit input combinations.

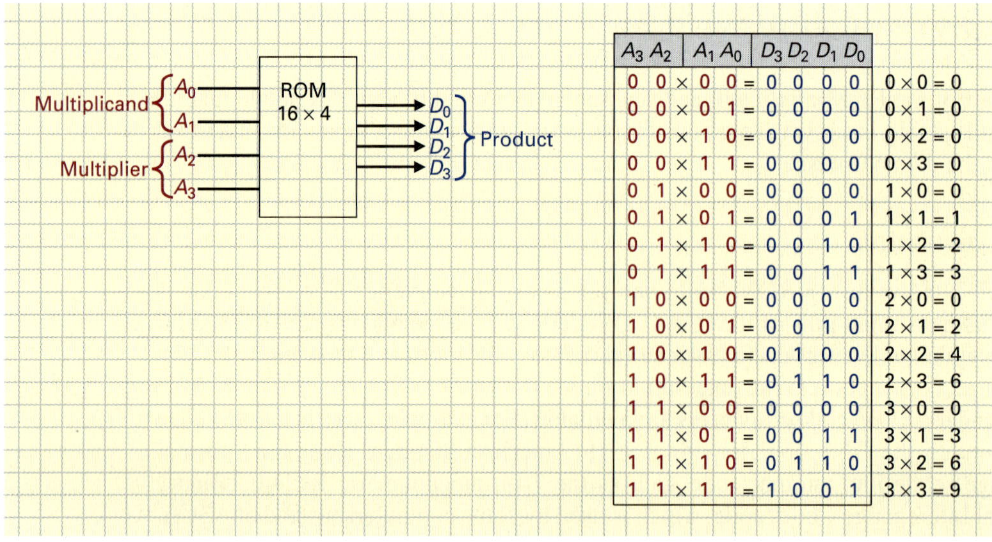

A_3 A_2		A_1 A_0		D_3 D_2 D_1 D_0	
0 0	×	0 0	=	0 0 0 0	0 × 0 = 0
0 0	×	0 1	=	0 0 0 0	0 × 1 = 0
0 0	×	1 0	=	0 0 0 0	0 × 2 = 0
0 0	×	1 1	=	0 0 0 0	0 × 3 = 0
0 1	×	0 0	=	0 0 0 0	1 × 0 = 0
0 1	×	0 1	=	0 0 0 1	1 × 1 = 1
0 1	×	1 0	=	0 0 1 0	1 × 2 = 2
0 1	×	1 1	=	0 0 1 1	1 × 3 = 3
1 0	×	0 0	=	0 0 0 0	2 × 0 = 0
1 0	×	0 1	=	0 0 1 0	2 × 1 = 2
1 0	×	1 0	=	0 1 0 0	2 × 2 = 4
1 0	×	1 1	=	0 1 1 0	2 × 3 = 6
1 1	×	0 0	=	0 0 0 0	3 × 0 = 0
1 1	×	0 1	=	0 0 1 1	3 × 1 = 3
1 1	×	1 0	=	0 1 1 0	3 × 2 = 6
1 1	×	1 1	=	1 0 0 1	3 × 3 = 9

FIGURE 27-8 The ROM as a Multiplier.

Character Generator

Another typical application for ROMs is as a character generator. For example, the Motorola MC6571 ROM is programmed with all of the 7 by 9 dot matrix codes used to display a number, letter, or symbol on a video display monitor screen.

Keyboard Encoder

Keyboard encoders are also not that different from code converters. For example, the MM5740 ROM is used within some computer keyboards to convert the keyboard switch matrix code generated when a key is pressed to a corresponding ASCII code that can be recognized by the computer.

Nonvolatile Computer Memory

The term **nonvolatile** describes a device that will retain its information in the absence of power. ROMs are mainly used in computer systems to store instruction codes and data codes that must not be lost when the system power is turned off. Several ROMs are often used to store information, such as system start-up instructions, keyboard encoding codes, character generator codes, and look-up tables.

Nonvolatile
A memory storage device that will retain its stored data in the absence of power.

27-1-6 *ROM Testing*

A **ROM listing** sheet lists the data values that should be stored in every memory location of a ROM. A ROM can be tested, therefore, by simply accessing or addressing each memory location, and then checking that the stored data in that memory location matches what is listed in the ROM listing.

ROM Listing
A table showing the binary codes stored in every memory location of a ROM.

Figure 27-9 shows a simple ROM test circuit. In this circuit, a push-button switch is used to clock a counter, causing it to advance its count. The counter's output is used to supply a

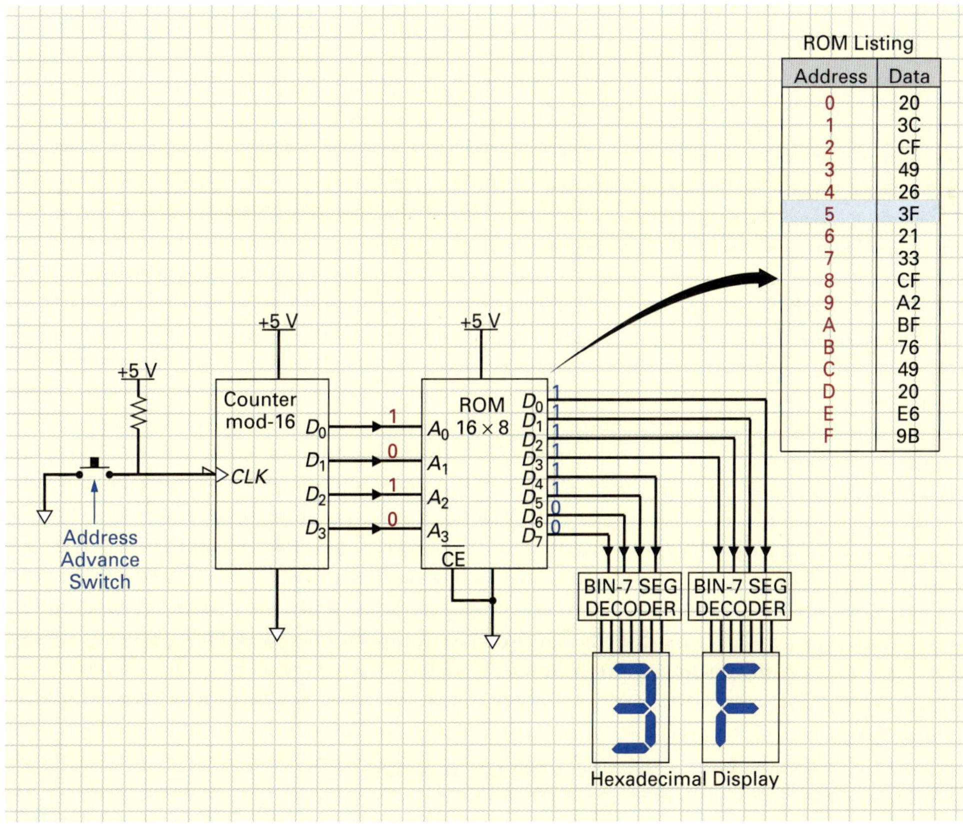

FIGURE 27-9 **Simple ROM Test Circuit.**

sequential address to the ROM being tested. The data output of the ROM is displayed on a seven-segment hexadecimal display. This circuit, therefore, will allow us to step through each address and check the stored data value against the value shown in the ROM listing. This check will verify the operation and contents of the ROM.

Self-Diagnostic Program
A test program stored within a computer to test itself.

Check-Sum Method
A ROM testing method in which the computer reads out and sums the entire contents of the memory and then compares the total to a known-good value.

The test circuit in Figure 27-9 would be ideal for testing a small 16 by 8 ROM; however, most digital system ROMs will be much larger, having thousands of memory locations, making a complete test with this circuit long and laborious. In these cases, a ROM tester can be used to check the memory's contents. This test instrument automatically reads the contents of the ROM being tested and compares it to the stored data within a known-good reference ROM to check for any false operation or storage errors.

Most ROMs within digital electronic computer systems are tested by a built-in **self-diagnostic program.** This test program will generally use the **check-sum method** to verify the ROM's contents. With the check-sum method, the computer reads out and adds all of the data stored in every one of the ROM's memory locations to obtain a sum (ignoring carries). The resultant sum is compared with a sum value that was stored in a specific address in the ROM when it was programmed. If a difference exists between the present check-sum and the previously stored check-sum, then there is definitely a fault with the programmed data.

SELF-TEST EVALUATION POINT FOR SECTION 27-1

Use the following questions to test your understanding of Section 27-1.

1. Is a ROM a volatile or nonvolatile device?
2. What is a memory's address input, and how is the number of address input lines related to the number of memory locations?
3. Define the following terms:
   a. storage density **b.** memory configuration **c.** burning in **d.** access time
4. Give the full names for the following abbreviations:
   a. MROM **b.** PROM **c.** EPROM **d.** EEPROM
5. How is a ROM used as a code converter?
6. What is a ROM listing, and how can it be used to test a ROM?

27-2 SEMICONDUCTOR READ/WRITE MEMORIES (RWMs)

Read/Write Memory
A memory that can have data written into or read out from any addressable storage location.

Random-Access Memory
A memory circuit characteristic in which any storage location can be directly accessed.

Sequential-Access Memory
A memory circuit characteristic in which any storage location can only be accessed in sequence or one after the other.

As its name implies, the previously discussed read-only memory, or ROM, can only read out or retrieve data from its memory locations once it is installed within a digital system. Data, such as ASCII coded text, numerical codes, or microprocessor instruction codes, are permanently stored in the ROM's memory locations before the device is installed in a digital circuit.

Read/write memories (RWMs) can perform either a read or a write operation after they are installed in a digital circuit. Unlike the ROM, which contains permanent storage locations, a RWM acts as a temporary storage location for binary data. When data is stored in a RWM, the data is said to be written into the memory, whereas when data is retrieved from an RWM, it is said that data is being read from memory. In this section, we will be concentrating on the semiconductor read/write memories shown previously in Figure 27-1. To begin with, let us discuss why read/write memories (RWMs) are more commonly referred to as **random-access memories (RAMs).**

27-2-1 *SAMs Versus RAMs*

The shift register discussed in Chapter 25 is an example of a **sequential-access memory (SAM),** or *serial memory.* To explain this, Figure 27-10 shows a 512-bit shift register which

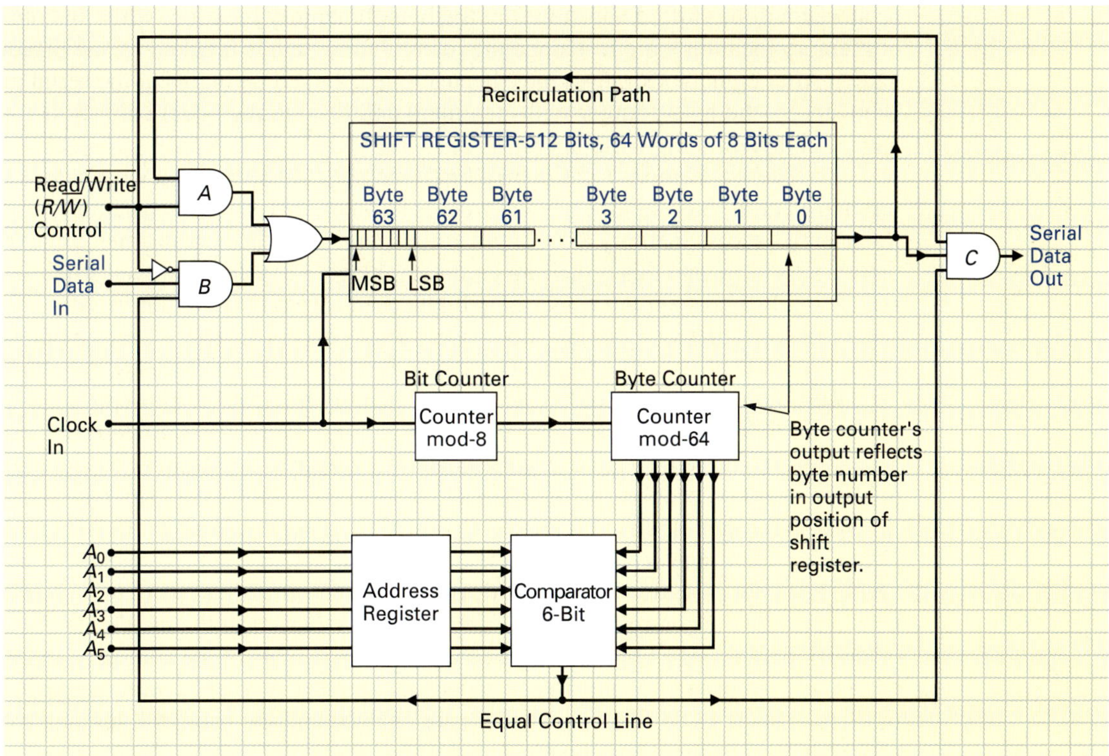

FIGURE 27-10 **A 64 × 8 Sequential-Access Read/Write Memory.**

can store sixty-four bytes (a 64 by 8 SAM). These sixty-four bytes of data are numbered byte 0 through byte 63 and are stored serially in 512 adjacent shift register storage elements. To store an 8-bit word in this serial memory, you must first apply a 6-bit address to specify the storage location for the data byte. This 6-bit address is loaded into the address register and then applied to the left input of the comparator.

For every clock pulse at the input, the data within the shift register is shifted one position to the right. The clock input is also applied to a mod-8 counter, which clocks the byte counter each time a complete word or eight bits have been shifted right. The byte counter, therefore, keeps track of the number of bytes shifted and which byte is presently being applied to the output.

To access memory location 9, for example, the address 001001 (decimal 9) is loaded into the address register and applied to the comparator. To read or retrieve the contents from memory location 9, the read/write control line is made HIGH, enabling AND gates *A* and *C*. The data bytes stored in the shift register are shifted right and recirculated back to the input via AND gate *A* until byte 9 is in the output position of the shift register. Since the byte counter's contents always reflect the number of the byte in the output position, the byte counter will also contain 9, and since both inputs to the comparator are equal, its output will go active or HIGH. This signal will enable AND gate *C,* allowing the 8-bit data contents in memory location 9 to be serially shifted out of the memory. After eight clock pulses have occurred and the entire contents of memory location 9 have been applied to the output, the byte counter will increment and the comparator's output will go inactive, completing the read-from-memory-9 operation.

To write or store an 8-bit data value into memory location 9, for example, we must once again load 9 into the address register and then shift and recirculate the contents of the shift register until byte 9 is in the output position. At this time, the byte counter will contain 9, the equal control line from the comparator will go HIGH, and this signal will be applied to AND gates *B* and *C.* Since the read/write control line is LOW in this instance, AND gates *A* and *C* will be disabled while AND gate *B* will be enabled. The 8-bit serial word to be stored will be shifted and stored in the byte 9 memory location, while the old contents of memory location 9 are lost since the recirculation path is blocked by the disabled AND gate *A*. After eight clock pulses have been applied, and all eight bits of the serial input have been stored, the byte counter will increment and disable the equal control line and AND gate *B*, completing the write operation.

To access a memory location in a serial memory, therefore, you must step or sequence through all of the bits until the desired word is located. This is why this memory type is called a sequential-access memory (SAM). As you can imagine, the access times for SAMs memory are very slow, and this is their major disadvantage. They are, however, still used today in serial data, low-cost applications where high-speed access is not required. Other sequential-access memory devices that are not widely used, except in specialized applications, are *charge-coupled devices (CCDs)* and *magnetic bubble memories (MBMs)*.

Unlike the serial memory, which has its memory locations end-to-end, a *parallel memory* has all of its memory locations stacked, as seen in Figure 27-11. This means that there is no need to sequence through every memory location, since the address decoder will enable only one memory location and therefore allow us to randomly access any byte. For example, the 512-bit register block in Figure 27-11 is organized as sixty-four words of eight bits each. These sixty-four bytes are numbered byte 0 through byte 63 and are stacked in parallel. To access a memory location in this RAM, first apply a 6-bit address word to the decoder, then take the read/write control line HIGH (read) or LOW (write). If the read/write control line is made HIGH, the output buffers are enabled and the contents of the addressed memory location is read out in parallel onto the 8-bit data output. On the other hand, if the read/write control line is made LOW, the input buffers are enabled and the parallel 8-bit data word being applied to the data input lines is stored in the memory location enabled by the address decoder.

The RAM is more widely used than the SAM, because its parallel format results in faster memory access times. Before we leave this topic of random access, it is important to point out that ROMs are random access in nature, even though the term RAM is now only associated with read/write memories.

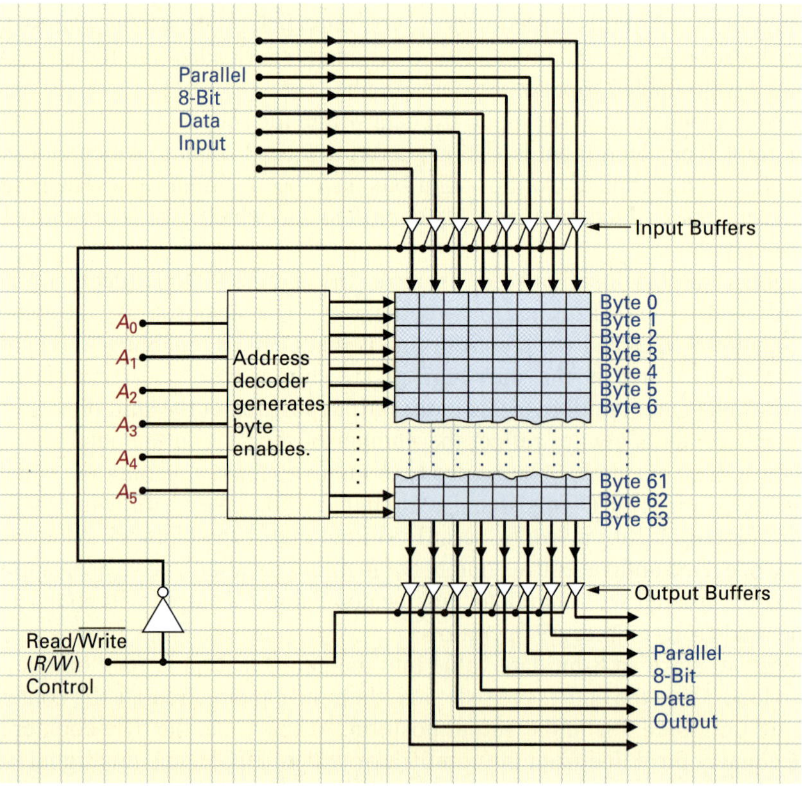

FIGURE 27-11 A 64 × 8 Random-Access Read/Write Memory.

27-2-2 *RAM Types*

There are two basic types of RAM—the static RAM (SRAM) and the dynamic RAM (DRAM). To begin with, the SRAM and DRAM are both **volatile,** which means that they will lose their data if power is interrupted. As far as storing data is concerned, both of these RAM types use a different technique. Static RAMs, for instance, latch bits of data in high-speed bistable flip-flops, whereas dynamic RAMs store each bit of data as a charge level in a capacitor. The DRAM capacitor circuit is smaller than the SRAM bistable flip-flop circuit; however, the DRAM needs additional refresh circuitry to replenish the charge lost from the capacitors due to capacitor leakage. Each RAM type, therefore, has its own advantage that makes it more suitable for certain applications. For example, the DRAM has a higher storage density but requires refresh circuitry, while the SRAM operates at a faster speed but is larger in size. Let us now examine these two types in more detail, beginning with the static RAM.

Volatile
A memory storage device that will not retain its stored data in the absence of power.

Static Random-Access Memories (SRAMs)

There are basically two types of **static RAMs** available. *Bipolar SRAMs* use TTL devices to construct bistable flip-flop memory circuits, which have a low storage density (small amount of storage per chip) but operate at a high speed. *MOS SRAMs* use either MOS or CMOS technology along with a capacitor to construct bistable flip-flop storage cells that have a high storage density but operate at a slower speed.

Static RAM
A RAM device in which data is stored in a bistable flip-flop circuit.

SRAM storage cell. Figure 27-12(a) shows the basic storage cell for a TTL static RAM. This storage cell, which could also be constructed using N-channel E-type MOSFETs, is capable of storing one single bit of data.

Write operation: To latch a binary 1 or binary 0 into the bistable flip-flop, the word-select control line must be pulsed HIGH to turn ON Q_3 and Q_4. If a HIGH is to be stored, it will appear on the data input line (Data Input = 1, $\overline{\text{Data Input}}$ = 0), and this positive voltage will be applied through Q_3 to the base of Q_2. This HIGH will force Q_2 to conduct and its dropping collector voltage will be applied to the base of Q_1, turning Q_1 OFF. With Q_1 OFF, its collector voltage will go HIGH, maintaining Q_2 ON. When the word-select control line goes inactive or LOW, the storage cell will remain in the same condition (Q_2 ON, Q_1 OFF) with a binary 1 latched in the flip-flop due to the cross coupling between Q_2 and Q_1. On the other hand, if we wanted to store or write a binary 0 into this storage cell, the data input line is made LOW (Data Input = 0, $\overline{\text{Data Input}}$ = 1), and then the word-select control line will again be pulsed HIGH.

When in the read mode, the data inputs are both floating and therefore no data is being applied to the flip-flop. To read or retrieve data out of the storage cell, the word-select line is pulsed HIGH in exactly the same way as in the write operation, turning ON Q_3 and Q_4 and switching the stored logic level to the output. For example, if the flip-flop has a logic 1 stored (Q_1 OFF, Q_2 ON), the left input of the sense amplifier (a comparator) will receive a HIGH voltage, while the right input of the sense amplifier will receive a LOW voltage. In this instance, the sense amplifier will produce a HIGH data output. On the other hand, if the flip-flop has a logic 0 stored (Q_1 ON, Q_2 OFF), the left input of the sense amplifier will receive a LOW voltage, while the right input of the sense amplifier will receive a HIGH voltage. In this instance, the sense amplifier will produce a LOW data output.

Figure 27-12(b) shows how sixty-four storage cells could be arranged to form a 16 by 4 (sixteen 4-bit words) static RAM. In this illustration, each of the squares represents one of the storage cells shown in Figure 27-12(a). For simplicity, only three of the sixteen 4-bit sets have been shown in Figure 27-12(b). Each of the data inputs is inverted to provide a *Data Input* and a $\overline{\textit{Data Input}}$. A sense amplifier is required for each of the four data output lines, so four sense comparator amplifiers are included at the bottom of the storage array.

SRAM storage array. Figure 27-12(c) shows all of the basic blocks in a 16 by 4 static RAM. The four-line to sixteen-line address decoder will decode the 4-bit address and make active one of the sixteen word-select control lines. The three-state input buffers are enabled by a HIGH signal on the write control line, while the three-state output buffers are enabled by a

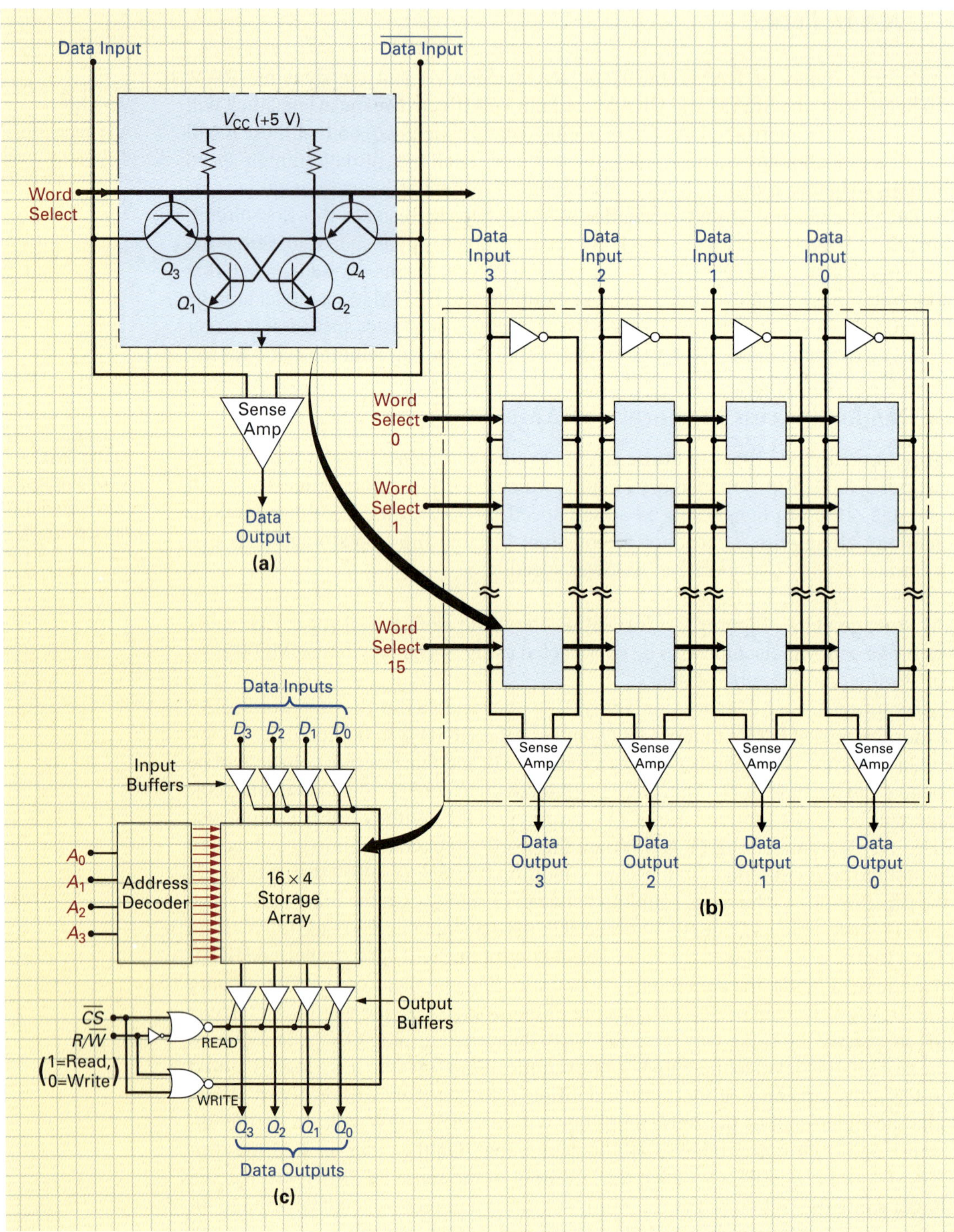

FIGURE 27-12 **Static RAM. (a) Storage Cell. (b) Storage Array. (c) Internal Architecture.**

HIGH on the read control line. These write and read control lines are controlled by the chip-select ($\overline{CS}$) and read/write ($R/\overline{W}$) inputs. When the $\overline{CS}$ and $R/\overline{W}$ control lines are both LOW, the input buffers are enabled and data is written into the addressed memory location. On the other hand, when the $\overline{CS}$ is LOW and the $R/\overline{W}$ control line is HIGH, the output buffers are enabled and the data at the addressed memory location is read out and applied to the Q outputs. Therefore, when the $\overline{CS}$ line is LOW (SRAM chip is selected or enabled), the logic level on the $R/\overline{W}$ control line will govern what operation is performed (1 = read, 0 = write). On the other hand, if the $\overline{CS}$ line is switched HIGH, both the input buffers and output buffers are disabled, and the SRAM chip is effectively disconnected from the data input and data output bus lines.

SRAM read and write timing. The logic symbol for the 16 by 4 SRAM is shown in Figure 27-13(a). Now that we have covered the operation of the SRAM cell and the operation of an SRAM's internal architecture, let us see how all of the control and input/output lines are used to perform a memory-read and a memory-write operation.

Figure 27-13(b) shows the typical timing waveforms for an SRAM read operation. Working from the top waveform to the bottom, you can see that the $R/\overline{W}$ control line input is first made HIGH (read operation). The desired address is then applied to the 4-bit address inputs, and then the chip-select control input is activated ($\overline{CS}$ = LOW). After a short delay,

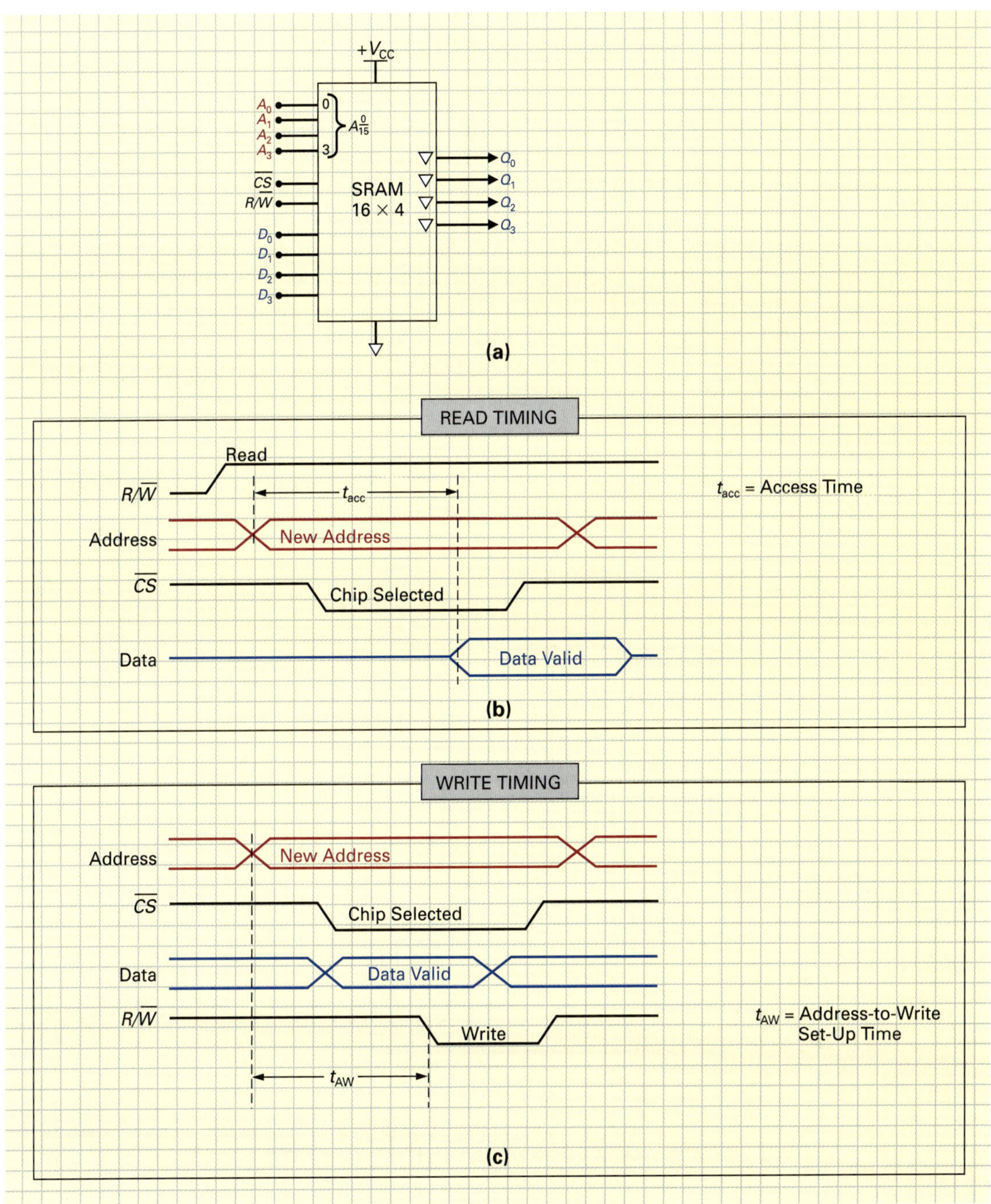

FIGURE 27-13 **Static RAM (SRAM) Logic Symbol and Read/Write Timing.**

the data stored at the addressed memory location will appear at the Q outputs. The *access time* (t_{acc}) of a memory IC is governed by the IC's internal address decoder Assuming no delays are introduced by the $R/\overline{W}$ or $\overline{CS}$ control lines, the read-access time of a memory IC is the time it takes for a new address to be decoded and the contents of the memory location to appear at the output.

Figure 27-13(c) shows the typical timing waveforms for an SRAM write operation. Working from the top waveform to the bottom, you can see that the desired address is first applied to the 4-bit address inputs, and then the chip-select control input is activated ($\overline{CS}$ = LOW). The data to be stored is then applied to the data inputs, and when the $R/\overline{W}$ control line input is brought LOW, the data is written into the addressed memory location. The *address-to-write set-up time* (t_{AW}) is a time delay that must be introduced between a new address being applied and the $R/\overline{W}$ control line going LOW. This time delay is needed so that the memory IC's internal decoder has enough time to decode and activate the new address before it is signaled to perform a write operation.

SRAM example. Figure 27-14 shows the details for the 2114A, which is a typical static RAM IC. The ten address input lines (A_0 through A_9) are used to select one of the 1,024 internal memory locations. The active-LOW chip-select ($\overline{CS}$) control line input is used to enable the IC, and the active-LOW write-enable ($\overline{WE}$) control line input operates in exactly the same way as the previously discussed $R/\overline{W}$ input. Each memory location in this IC can store a 4-bit word, and a set of four data lines is used for both input (write) and output (read) operations. During a read operation, these four data input/output (I/O) lines become output lines, and the 4-bit word from the addressed memory location appears on these four I/O lines. During a write operation, these four data input/output lines become input lines, and the 4-bit word applied to these four I/O inputs is written into the addressed memory location. It is the $\overline{WE}$ control line that determines whether these four I/O data lines are being used for input or output operations.

It is probably safe to say that a single RAM IC chip is not normally large enough for a digital system's memory requirements. In most applications, several RAM ICs will be interconnected to form a larger RAM memory circuit.

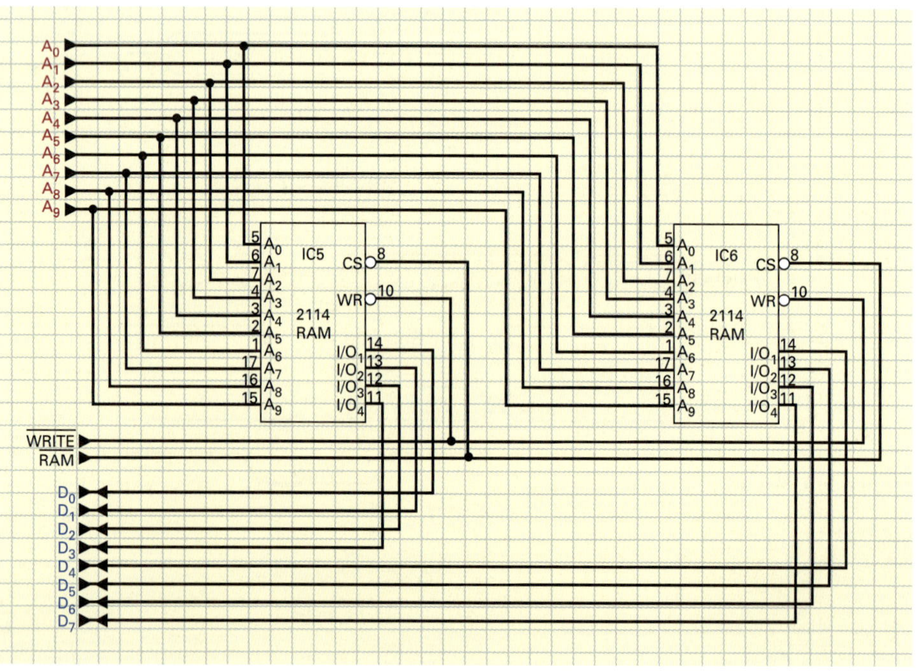

FIGURE 27-14 **1,024 by 4-Bit Bipolar Static RAM IC.**

Dynamic Random-Access Memories (DRAMs)

As mentioned previously, there are static RAMs and **dynamic RAMs.** As we saw in the previous section, memory systems using static RAMs appear to be ideal since they are fast, there are many types to choose from, and their circuitry is simple. The question, therefore, is, why bother having another type of RAM? There are two good reasons why dynamic RAMs are ideal in certain applications. The first is that you can package more storage cells into a dynamic RAM IC than you can into a static RAM IC. This can be easily seen by comparing the single static RAM cell shown in Figure 27-15(a), to the single dynamic RAM cell shown in Figure 27-15(b). A dynamic

Dynamic RAM
A RAM device in which data is stored as a charge which needs to be refreshed.

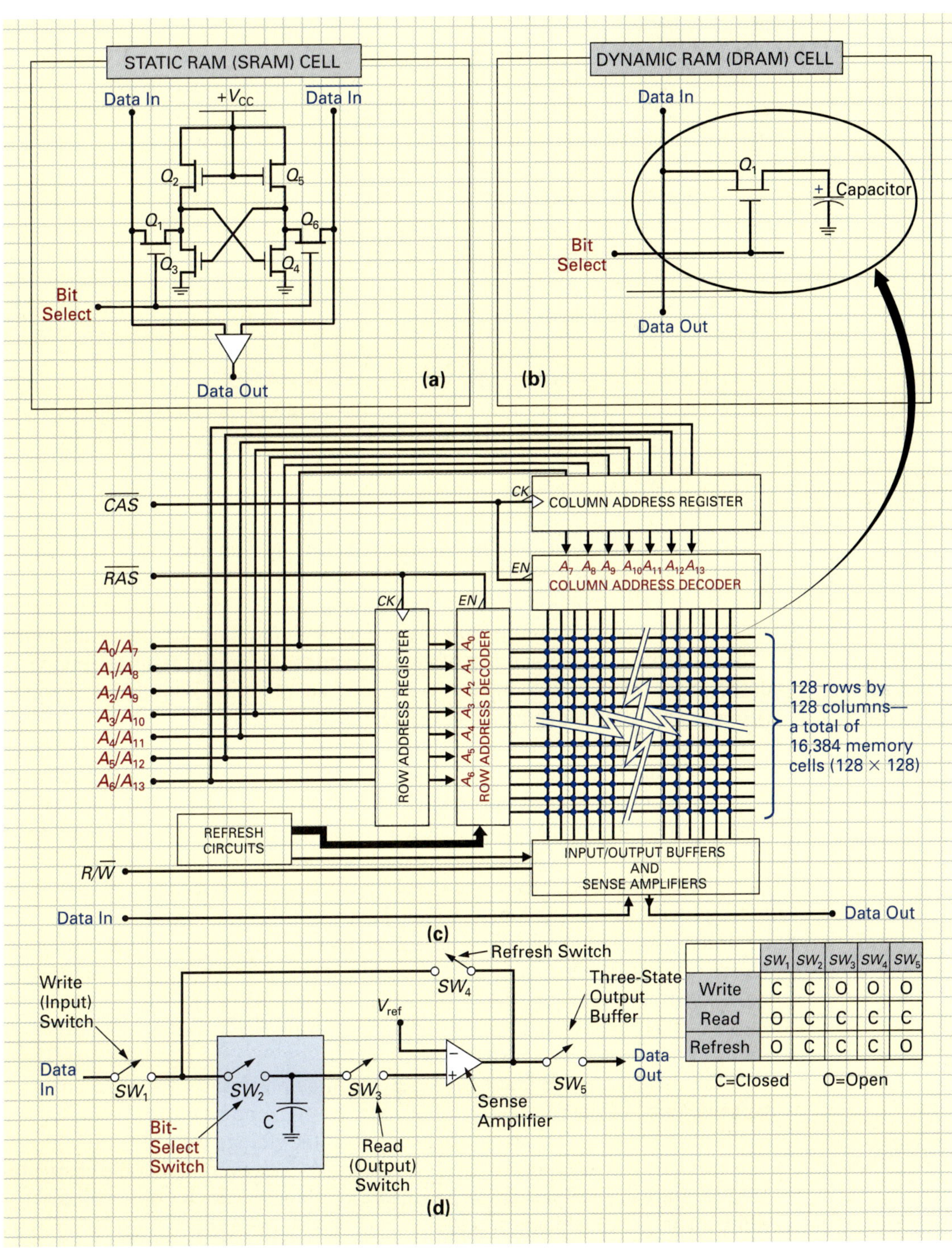

	SW_1	SW_2	SW_3	SW_4	SW_5
Write	C	C	O	O	O
Read	O	C	C	C	C
Refresh	O	C	C	C	O

C=Closed O=Open

FIGURE 27-15 Dynamic RAMs. (a) Static Cell. (b) Dynamic Cell. (c) Internal Architecture. (d) Simplified DRAM Cell Operation.

RAM storage cell has only one MOSFET and a capacitor compared to a static RAM storage cell, which normally contains six MOSFETs. Therefore, approximately three dynamic storage cells could fit into the same space as one static storage cell, giving the DRAM a large space-saving advantage. The second reason for using dynamic RAMs is power dissipation. Since power dissipation is dependent on the number of devices per storage cell, it is easy to imagine why DRAMs have a lower power dissipation since fewer MOSFETs are used for each storage cell.

DRAM storage cell. The DRAM storage cell shown in Figure 27-15(b) operates in the following way.

To write data into this storage cell, the data bit to be stored is first applied to the data-in line, and then the bit-select control line is pulsed HIGH. A HIGH on the bit-select control line will cause Q_1 to conduct (N-channel MOSFET), allowing either the HIGH or LOW data input to be applied directly to the capacitor, causing it to either charge (a 1 is stored) or discharge (a 0 is stored).

To read data out of this storage cell, the bit-select control line is switched HIGH to turn ON Q_1, allowing either the stored binary 1 (capacitor charged) or binary 0 (capacitor discharged) to pass through to the data-out line.

Now let us see how these dynamic storage cells are connected to form a complete dynamic RAM circuit.

DRAM storage array. Nearly all dynamic RAM ICs are organized as multiple 1-bit read/write memories. This means that they have multiple memory locations; however, each memory location has only a 1-bit word (for example, 16K by 1-bit, 64K by 1-bit, 256K by 1-bit, and so on).

Figure 27-15(c) shows the internal architecture of a typical dynamic RAM circuit. This dynamic RAM circuit has 16,384 1-bit memory locations, arranged into a 128 by 128 matrix. Each of the dynamic cells in this matrix has its own unique row and column position in the matrix. To address each of these 16,384 1-bit memory locations will require fourteen address inputs ($2^{14} = 16{,}384$). Looking at the dynamic RAM circuit in Figure 27-15(c), you will notice that this circuit only has seven address input pins. To reduce the number of pins on the IC package, manufacturers often use an **address multiplexing** technique, in which an address input pin is used to input two different address bits. With this technique, the 14-bit address is applied to the dynamic RAM address decoder in two 7-bit groups. First, the seven lower-order address bits (A_0 through A_6) are applied to the seven address inputs and are stored in the row address register and decoded by the row address decoder by activating the *row address strobe* ($\overline{\text{RAS}}$). Second, the seven higher-order address bits (A_7 through A_{13}) are applied to the seven address inputs and are stored in the column address register and decoded by the column address decoder by activating the *column address strobe* ($\overline{\text{CAS}}$). The row address strobe and column address strobe control-line inputs, therefore, allow us to input a 14-bit address on seven address input pins, which decreases the size of the IC package, decreasing the size of memory circuits and the final digital system. The outputs of the row and column address registers drive the row and column address decoders, and these decoders will bit-select one of the 16,384 dynamic RAM storage cells. Once activated, the read /write control-line input will determine whether a read operation (storage cell $\rightarrow$ data-out pin) or write operation (data-in pin $\rightarrow$ storage cell) is performed.

Address Multiplexing

A time-division technique in which two or more address codes are switched through a single input at different times.

The dynamic RAM, therefore, uses an on-chip capacitor for each storage element. A charge is stored on the capacitor to indicate a binary 1, and no charge on the capacitor is used to indicate a binary 0. This technique simplifies the storage cell, permitting denser memory chips that dissipate less power. There is a problem, however, in that the charge leaks off the capacitor due to capacitor leakage, and therefore a binary 1 can become a binary 0 in just a few milliseconds. When a binary 1 is stored in a DRAM, therefore, it must be *refreshed* continually so that the data stored within the DRAM is accurate. Most DRAM ICs today have on-chip refresh circuits, as shown in Figure 27-15(c), to automatically rewrite the same data back into each memory location. This refresh action will typically take place every two ms. After a refresh, all of the binary 1s are restored to full charge and all of the binary 0s are left as no charge.

DRAM write, read, and refresh. To explain the write, read, and refresh operations in more detail, Figure 27-15(d) shows a simplified DRAM cell and its input/output circuitry. In this diagram the circuit's three-state input buffer, three-state output buffer, and the internal cell MOSFETs have been represented as switches, which are under the control of the read/write input. The table associated with this illustration indicates which of the switches are open and closed for each of the three operations.

To write data to the cell, SW_1 and SW_2 are closed so that the data input can be applied directly to the capacitor. At the same time, SW_3, SW_4, and SW_5 are opened to disconnect the storage capacitor from the rest of the circuit.

To read data from the cell, SW_3 is closed to connect the stored capacitor voltage to the sense amplifier. The sense amplifier compares the stored capacitor voltage to a reference voltage in order to determine whether a logic 0 or logic 1 was stored, and then produces a valid 0 V or 5 V (binary 0 or 1) at its output. SW_5 switches the stored bit from the sense amplifier through to the data output. In this mode, SW_4 is also closed, along with SW_2, to refresh the capacitor voltage (recharge to reinforce the HIGH, or discharge to reinforce the LOW). During each read operation, therefore, a refresh action is performed.

To only refresh the cell, SW_2, SW_3, and SW_4 are closed, so that the logic level stored in the capacitor can be fed back to the capacitor to reinforce the stored HIGH or LOW.

DRAM read and write timing. The logic symbol for the previously discussed 16,384 by 1-bit DRAM is shown in Figure 27-16(a). Now that we have covered the operation of the DRAM cell and the operation of a DRAM's internal architecture, let us see how all of the control and input/output lines are used to perform a memory-read and a memory-write operation.

Figure 27-16(b) shows the typical timing waveforms for a DRAM read operation. Working from the top waveform to the bottom, you can see that first the seven lower-order address bits (row address) are applied to the seven address input pins. This row address input is strobed, or latched, into the row address register when the row address strobe control line ($\overline{\text{RAS}}$) is brought LOW. Then the seven higher-order address bits (column address) are applied to the seven address input pins. This column address input is strobed, or latched, into the column address register when the column address strobe control line ($\overline{\text{CAS}}$) is brought LOW. The $R/\overline{W}$ control line is used to control the input and output buffers and therefore control whether a read or write operation is performed. In this example, the $R/\overline{W}$ control line goes HIGH after the $\overline{\text{CAS}}$ goes LOW, to enable a read operation. The three-state output buffers will therefore be taken out of their high impedance state and the stored binary 1 or binary 0 will appear at the data-out pin.

Figure 27-16(c) shows the typical timing waveforms for a DRAM write operation. Working from the top waveform to the bottom, you can see that the row address and column address are latched into the DRAM in exactly the same way as in the read operation. Once the 14-bit address has been inputted, the $R/\overline{W}$ control line will control whether a read or write operation is performed. In this example, the $R/\overline{W}$ control line goes LOW after the $\overline{CAS}$ goes LOW to enable a write operation. The input switch will therefore be enabled, and the binary 1 or binary 0 input being applied to the data-in pin will be stored or written into the addressed memory location.

As mentioned previously, nearly all dynamic RAM ICs are organized as multiple 1-bit read/write memories. This means that they have multiple memory locations; however, each memory location has only a 1-bit word (for example, 16K by 1-bit, 64K by 1-bit, 256K by 1-bit, and so on). Figure 27-17 shows how eight 16K by 1-bit DRAM ICs can be connected to form a 16K by 8-bit DRAM read/write memory circuit. An address multiplexer is used to switch the seven lower-order address bits through to the DRAMs just before $\overline{RAS}$ goes active and then switch the seven higher-order address bits through to the DRAMs just before the $\overline{CAS}$ goes active. The $\overline{RAS}$, $\overline{CAS}$, $R/\overline{W}$, and $\overline{CS}$ control line inputs are connected to all eight DRAM ICs so that they are all activated simultaneously.

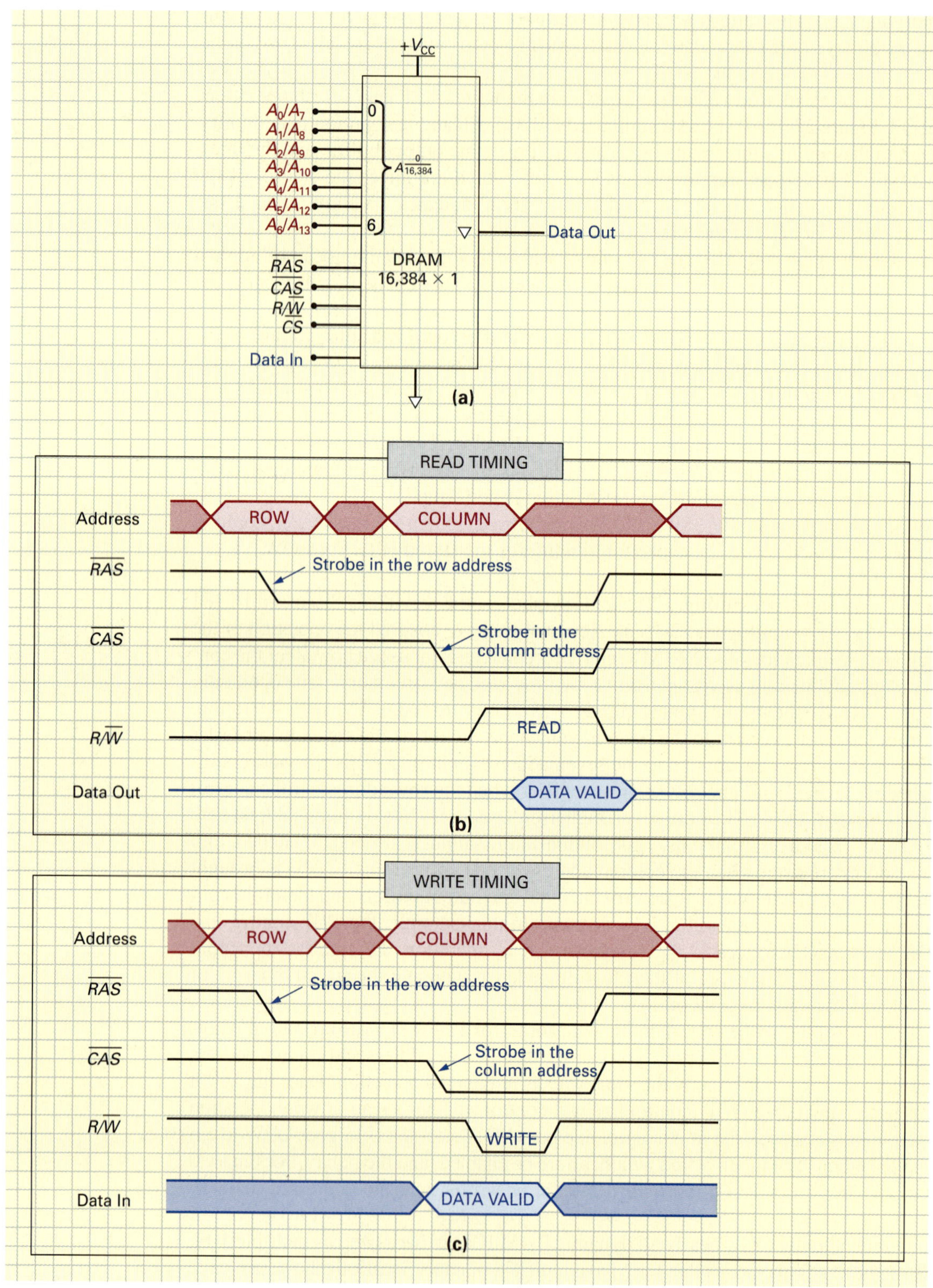

FIGURE 27-16 Dynamic RAM (DRAM) Logic Symbol and Read/Write Timing.

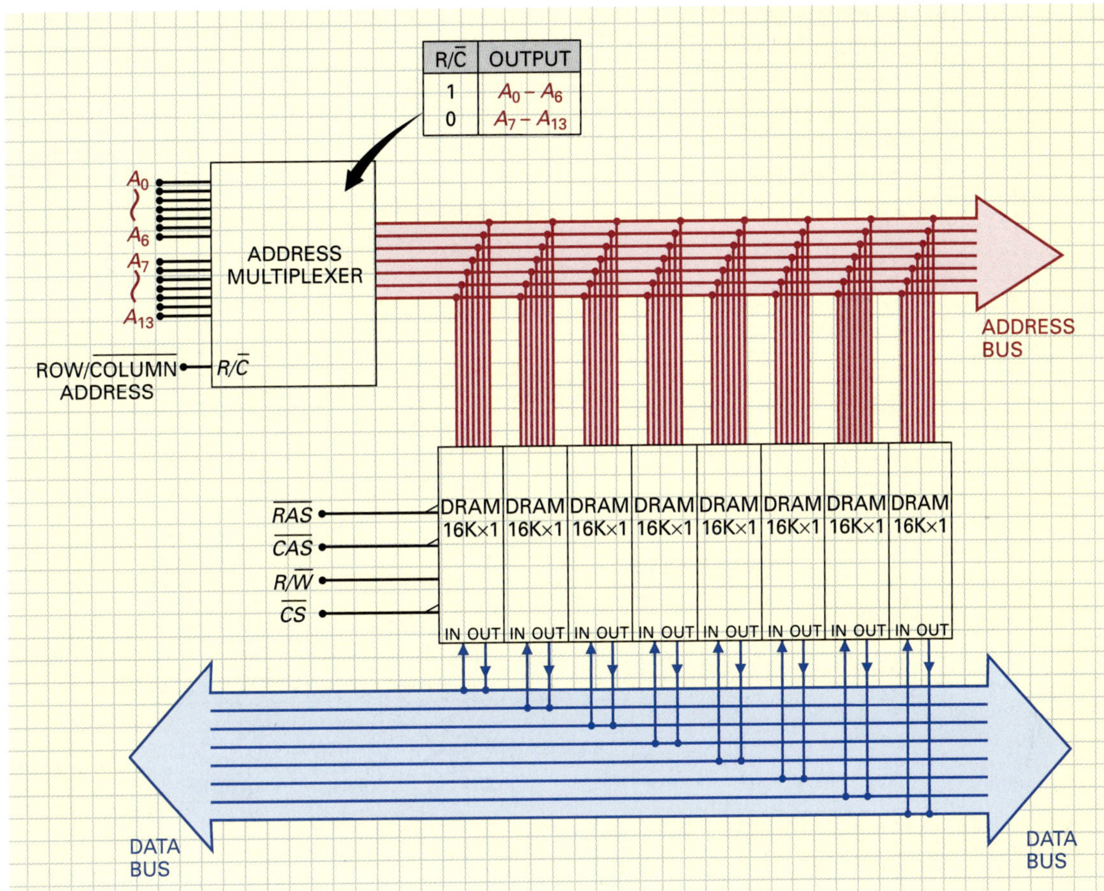

FIGURE 27-17 **A 16K by 8-Bit DRAM Memory Circuit.**

27-2-3 *RAM Applications*

As far as applications are concerned, dynamic RAMs are more widely used than static RAMs because they are smaller in size, lower in cost, and dissipate less power than an equivalent static RAM. Static RAMs, however, have faster access times and do not require refresh circuitry and the special timing requirements that go along with refreshing the stored data (you must refresh at least every 2 ms). Semiconductor memory manufacturers, however, are continually improving the DRAM, making its disadvantages easier to deal with.

SRAMs or DRAMs

In general, static RAMs are used when only a small amount of read/write memory is required (64K or less) or when high-speed access is needed. Dynamic RAMs, on the other hand, are used in applications where a large amount of read/write memory is needed (megabytes). Some systems employ a small amount of SRAM for high-speed read/write memory requirements and a large amount of DRAM for the bulk of the system's read/write memory needs.

A Typical RAM Memory Circuit

Whether the application is a personal computer or a digital oscilloscope, all digital computer systems use RAM. As discussed previously, a microprocessor unit (MPU) is at the heart of every digital computer system. The MPU is a complex integrated circuit that contains control logic circuits, instruction decoding circuits, temporary storage circuits, and arithmetic processing circuits. To be useful, the microprocessor IC must be able to exchange information with memory ICs (both ROM and RAM) input devices, such as a keyboard, and output devices, such as a display.

Figure 27-18(a) shows how a typical RAM memory circuit could be connected in a digital computer system. At this time, we will simply think of the MPU as a system controller that can select or address any memory location in RAM and then either read data from that memory location or write data to that memory location. It achieves this by controlling the ad-

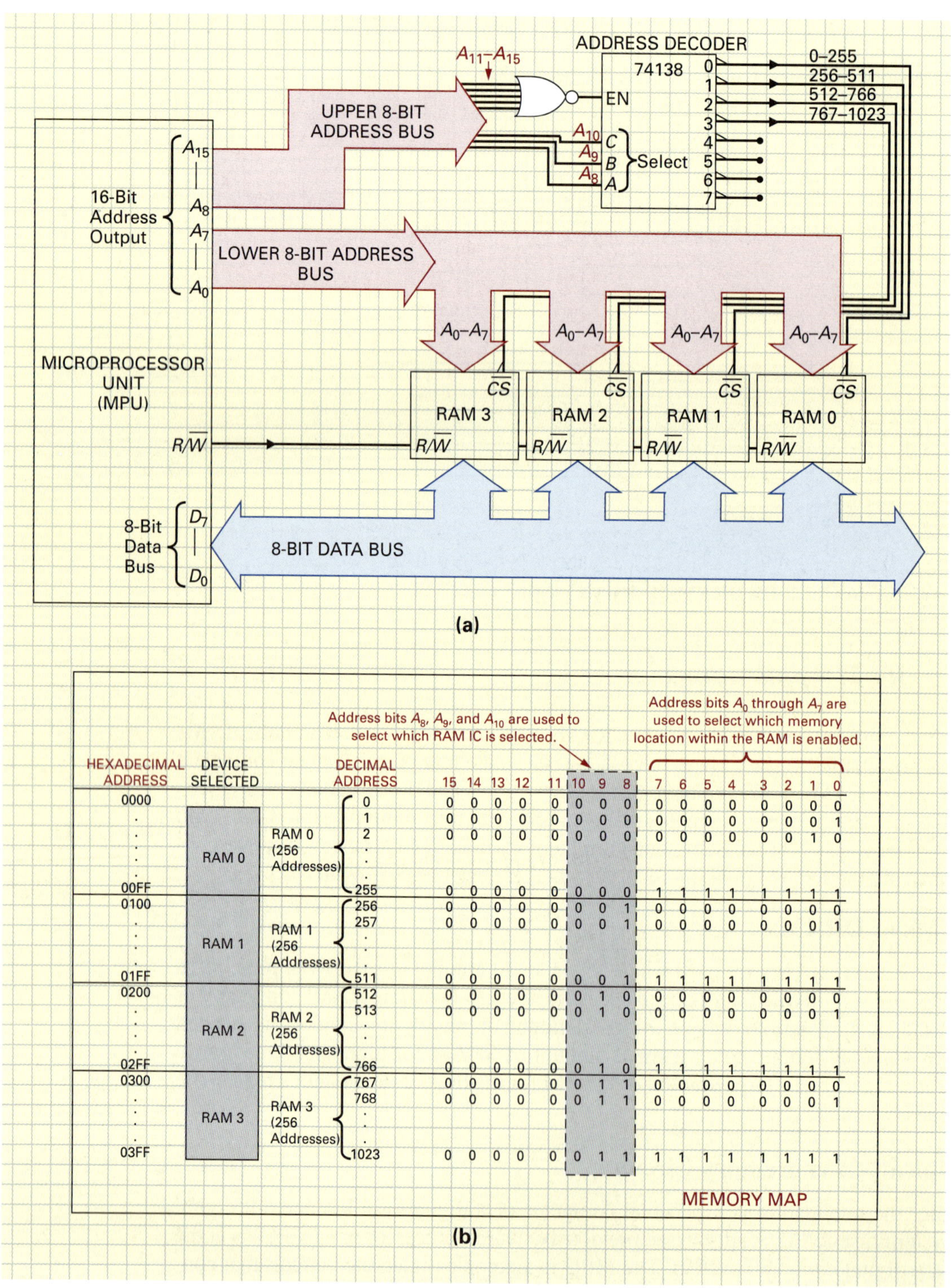

FIGURE 27-18 A Typical RAM Memory Circuit.

dress bus and the data bus and generating control signals. To review, the data bus is used for data transfers, and since all devices share this bus, the microprocessor will only activate one device at a time and then read data from that device via the data bus or write data to that device via the data bus (under the control of the $R/\overline{W}$ control line from the MPU). Before a data transfer can take place on the data bus, however, the microprocessor must output an address onto the address bus. This address specifies which memory location the microprocessor wishes to access.

In Figure 27-18(a), four 256 by 8-bit RAM ICs have been connected to form a 1,024 (4 × 256) by 8-bit RAM memory. Since most computer systems use more than one memory IC, the address from the MPU must indicate which memory IC should be selected and which word within that IC should be accessed. This is achieved by assigning a unique set of addresses to each memory IC.

The **memory map** shown in Figure 27-18(b) shows how each of the four RAM ICs has its own unique set of 256 addresses. For example, addresses 0_{10} through 255_{10} (0000_{16} through $00FF_{16}$) are assigned to RAM 0, addresses 256_{10} through 511_{10} (0100_{16} through $01FF_{16}$) are assigned to RAM 1, addresses 512_{10} through 766_{10} (0200_{16} through $02FF_{16}$) are assigned to RAM 2, and addresses 767_{10} through 1023_{10} (0300_{16} through $03FF_{16}$) are assigned to RAM 3.

Memory Map
An address-listing diagram showing the boundaries of the address space.

Referring to the 16-bit address bus from the MPU in Figure 27-18(a), you can see that the upper eight bits of the address bus are applied to an **address decoder,** and the lower eight bits of the address bus are applied to all four RAMs. By decoding address bits A_8 through A_{15}, the 74138 address decoder can determine which one of the four RAM ICs should be enabled. The eight lower-order address bits from the MPU (A_0 through A_7) will then be decoded by the selected RAM IC to determine which word within the RAM IC is accessed. In the circuit in Figure 27-18(a), you can see that address bits A_8, A_9, and A_{10} are applied to the *ABC* select inputs of the 74138.

Address Decoder
A circuit that decodes the applied addresses and activates certain outputs based on which zone the address is within.

In the memory map in Figure 27-18(b), you can see how address bits A_8, A_9, and A_{10} will advance by one for every 256 addresses. For example, address bits A_8, A_9, and A_{10} will remain at 000 (decimal 0) for the first 256 addresses, advance to 001 (decimal 1) for the next 256 addresses, advance to 010 (decimal 2) for the next 256 addresses, and so on. Since these three bits are applied to the 74138s *ABC* select inputs, output 0 will go active (LOW) for the first 256 addresses, output 1 will go active (LOW) for the next 256 addresses, output 2 will go active (LOW) for the next 256 addresses, and so on. Therefore, by connecting each of the 74138's active-LOW outputs to each of the active-LOW chip-select inputs of the RAM ICs, as shown in Figure 27-18(a), we will ensure that RAM 0 is only enabled whenever any address from 0000_{16} to $00FF_{16}$ is generated by the MPU, RAM 1 is only enabled whenever any address from 0100_{16} to $01FF_{16}$ is generated by the MPU, RAM 2 is only enabled whenever any address from 0200_{16} to $02FF_{16}$ is generated by the MPU, and RAM 3 is only enabled whenever any address from 0300_{16} to $03FF_{16}$ is generated by the MPU. The rest of the higher-order address bits (A_{11} through A_{15}) are used to enable (EN) the decoder only when they are all LOW.

To summarize, with the address decoding technique, the lower-order address bits are sent directly to the memory's address line, and the higher-order address bits are decoded to generate the chip selects, with only one IC selected at any given time. In other circuit designs, you may see that the number of bits fed directly to the memory devices may change (depending on the number of words that need to be addressed, 2^n), and this will change which higher-order address bits are decoded to generate chip selects. In the following troubleshooting memory circuits section, you will see another example of how an address decoder is used to create 1K (1,024) memory map blocks instead of the 256 memory map blocks shown in Figure 27-18(b).

27-2-4 *RAM Testing*

When testing a ROM IC, all we had to do was read out the contents of each memory location and compare it to the ROM listing to see if a memory cell error was present. With a RAM IC, we have to test both the read and the write ability of all the memory cells, making the testing of a RAM a little more lengthy.

Figure 27-19 shows a simple RAM test circuit. In this circuit, a 10-bit counter is triggered to sequence through the 1,024 addresses needed for a 2114 IC being tested. Four data switches can be used to generate any 4-bit word combination, which can then be applied to the 2114 and stored in any memory location (write operation). A 4-bit output register and LED display are included so that a 4-bit word retrieved from any memory location can be viewed (read operation). RAM ICs are normally tested by first storing 0s in every memory cell and then reading out all of these 0s and checking that no 1s are present. The second step is to store 1s in every memory cell and then read out all of these 1s and check that no 0s are present. These first two checks can be seen in the inset in Figure 27-19. These two checks, however, do not test for a memory cell short between two adjacent memory cells. For example, if one memory cell is shorted to an adjacent memory cell, storing all 1s or storing all 0s will not expose this fault since the short will cause both memory cells to be at the same logic level anyway.

Checkerboard Test Pattern

A RAM testing method in which the computer first stores 0s in every memory location, then stores 1s, and then alternates the storage of 1s and 0s.

To determine whether a memory cell short exists within a RAM IC, a **checkerboard test pattern** is normally written into the RAM IC, as shown in the inset in Figure 27-19. With

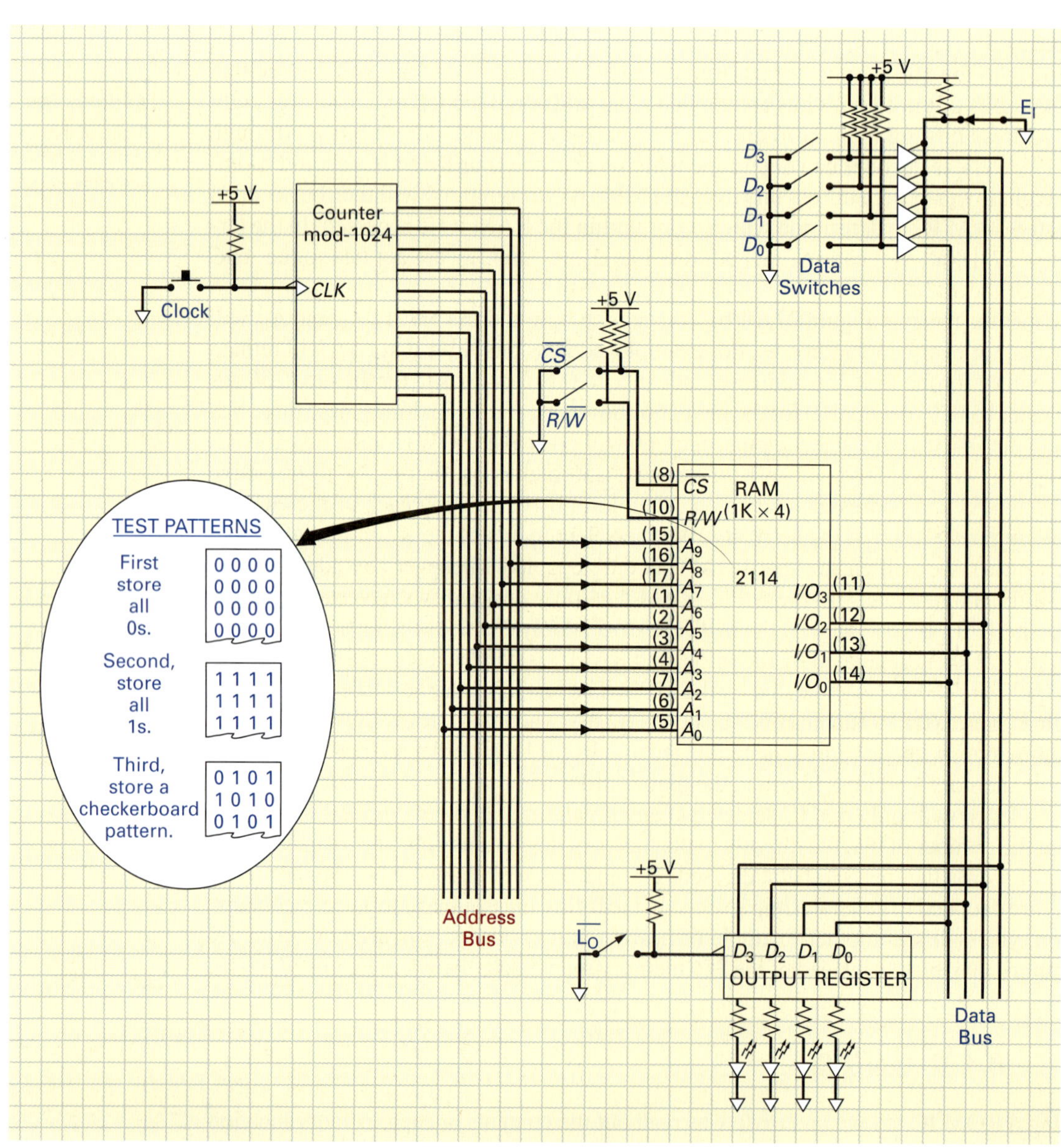

FIGURE 27-19 **A Simple RAM Test Circuit.**

the checkerboard test pattern, adjacent cells will have opposite bits stored, and if the logic levels do not alternate when the pattern is read out of the RAM, a memory cell problem exists. After the RAM is checked with this checkerboard pattern, the pattern is normally reversed to further test the RAM and then read out and checked again. Since there is no way to repair an internal RAM IC failure, the IC must be discarded and replaced if a problem is found.

The test circuit in Figure 27-19 would be ideal for testing a small 1K by 4 RAM; however, most digital system RAMs will be much larger, having possibly millions of memory locations, making a complete test with this circuit long and laborious. In these cases, a RAM tester can be used to check the RAM's read/write ability. This test instrument automatically writes several test patterns into the RAM and then reads back the data and compares it to check for any operation or storage error.

Most RAMs within digital electronic computer systems are tested by a built-in self-diagnostic program. This test program will generally use the checkerboard test pattern method to verify the RAM's operation.

SELF-TEST EVALUATION POINT FOR SECTION 27-2

Use the following questions to test your understanding of Section 27-2.

1. What is the difference between a read/write memory and a read-only memory?
2. What is the difference between a SAM and a RAM?
3. Are ROMs RAMs?
4. Is a RAM a volatile or nonvolatile device?
5. Describe the difference between a static RAM and a dynamic RAM.
6. What are the advantages and disadvantages of SRAMs and DRAMs?
7. What is a memory map?
8. How are RAMs usually tested?

REVIEW QUESTIONS

Multiple-Choice Questions

1. How many memory locations can be accessed by a ROM IC if it has ten address line inputs?
   a. 16
   b. 1,024
   c. 256
   d. 2,048
2. A memory is said to be ____________ if it loses its data whenever power is removed.
   a. nonvolatile
   b. accessible
   c. volatile
   d. sequential
3. If a RAM IC has twelve address inputs and eight data input/outputs, what is its size and configuration?
   a. 2K by 8
   b. 4K by 8
   c. $1{,}024 \times 8$
   d. 8,192 by 8
4. The ____________ of a memory is the time interval between the memory receiving a new address input and the data at that address being available at the data outputs.
   a. storage density
   b. programmable link
   c. access time
   d. memory configuration
5. The total storage capacity of an 8K by 8 RAM is ____________ bits.
   a. 16,384
   b. 65,536
   c. 64,000
   d. 49,152
6. Which of the following ROM types is the cheapest when ordered in large quantities?
   a. MROM
   b. PROM
   c. EPROM
   d. EEPROM

7. Which of the following ROM types can easily be erased and reprogrammed?
   a. MROM
   b. PROM
   c. EPROM
   d. EEPROM
8. Which of the following ROM types can be erased and is low in cost?
   a. MROM
   b. PROM
   c. EPROM
   d. EEPROM
9. A read/write memory is normally always referred to as a __________.
   a. ROM
   b. EEPROM
   c. RAM
   d. MBM
10. The storage cell in a static RAM is a __________.
    a. bistable flip-flop
    b. FET and capacitor
    c. monostable flip-flop
    d. Both a. and c. are true.
11. Which of the RAM types has a higher storage density but operates slower?
    a. SRAM
    b. DRAM
    c. SAM
    d. All of the above
12. Which of the RAM types has a lower storage density but operates faster?
    a. SRAM
    b. DRAM
    c. SAM
    d. All of the above
13. Which of the following RAM types has a lower power consumption?
    a. SRAM
    b. DRAM
    c. SAM
    d. All of the above
14. Static RAMs and dynamic RAMs are both nonvolatile.
    a. True
    b. False
15. Dynamic RAM ICs will normally use an address multiplexing technique to __________.
    a. refresh
    b. reduce the number of data inputs
    c. enable an address decoder
    d. reduce the number of address inputs
16. The notebook computer used to develop the manuscript for this textbook has 32 megabytes of RAM. Which RAM type do you think is being used in this application?
    a. SRAM
    b. DRAM
    c. SAM
    d. ERAM
17. If address bits 5, 6, and 7 are applied to the *ABC* select inputs of an address decoder and address bits 0 through 4 are applied directly to the memory, what is the size of the memory map blocks?
    a. 16
    b. 32
    c. 8
    d. 64
18. If address bits 8, 9, and 10 are applied to the *ABC* select inputs of an address decoder, and address bits 0 through 7 are applied directly to the memory, what is the size of the memory map blocks?
    a. 256
    b. 1,024
    c. 512
    d. 128
19. If address bits 13, 14, and 15 are applied to the *ABC* select inputs of an address decoder, and address bits 0 through 12 are applied directly to the memory, what is the size of the memory map blocks?
    a. 4K
    b. 32K
    c. 8K
    d. 16K
20. The memory-mapped input/output technique assigns a block of addresses to memory and input and output ports.
    a. True
    b. False

Practice Problems

To practice your circuit recognition and operation skills, refer to Figure 27-20 and answer the following four questions.

21. What is the purpose of the $\overline{ROM}$ control signal input?
22. What is the purpose of the $\overline{READ}$ control signal input?
23. How is data written into this IC?
24. How is data read from this IC?

To practice your circuit recognition and operation skills, refer to the SAM schematic diagram on pages 1009–1012 and answer the following five questions.

25. Where is the 2316 ROM circuit shown in Figure 27-20 on the SAM schematic diagram?
26. The NRD (not read) control line input to the 2316 ROM IC comes from what IC?
27. The NROM (not ROM) control line input to the 2316 ROM IC comes from what IC?
28. Which two ICs output the 11-bit address that is applied to the 2316 IC?
29. The 8-bit data output of the 2316 ROM IC should connect to the data bus and then back to the MPU IC. Which IC do you think is the MPU?

To practice your circuit recognition and operation skills, refer to Figure 27-21 and answer the following four questions.

30. What is the purpose of the $\overline{RAM}$ control signal input?
31. What is the purpose of the $\overline{WRITE}$ control signal input?

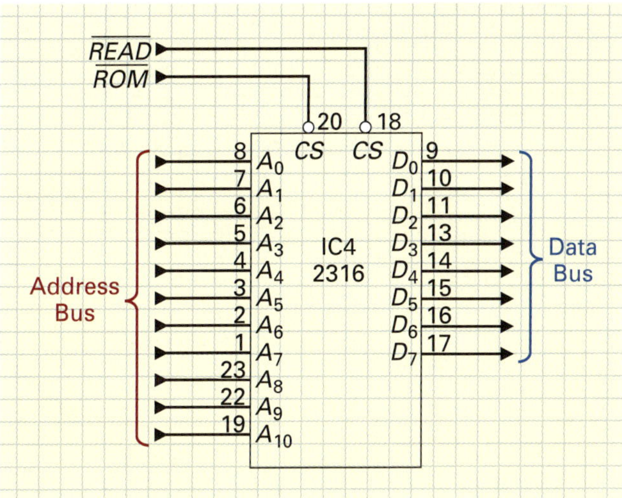

FIGURE 27-20 **Application—The ROM Circuit from SAM (Simple All-Purpose Microcomputer).**

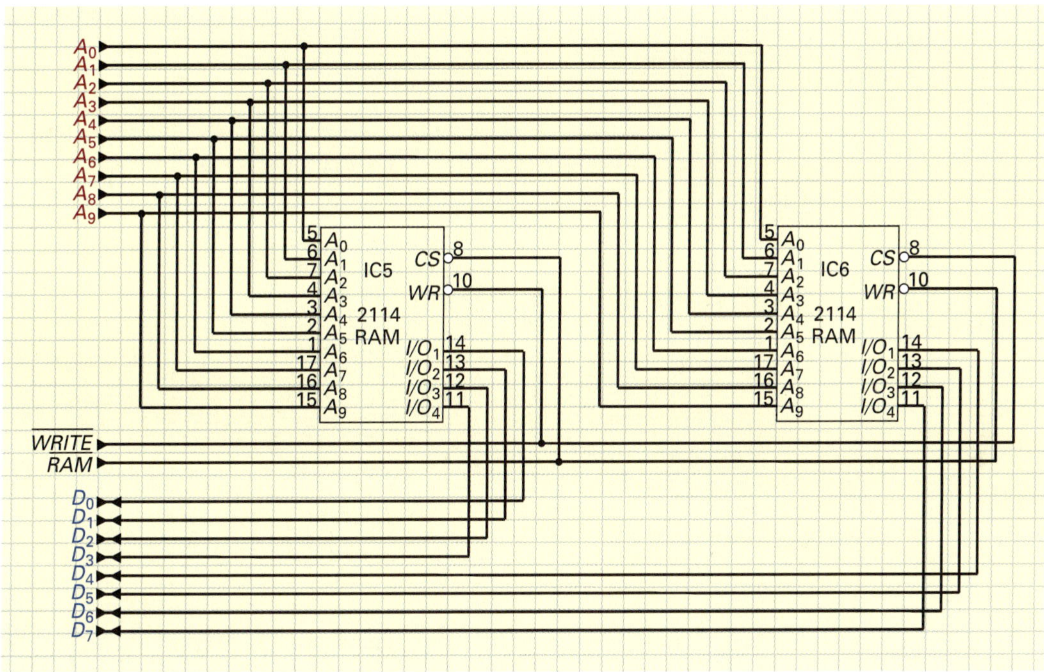

FIGURE 27-21 **Application—The RAM Circuit from SAM (Simple All-Purpose Microcomputer).**

32. How is data written into this IC?

33. How is data read from this IC?

To practice your circuit recognition and operation skills, refer again to the SAM schematic diagram on pages 1009–1012 and answer the following three questions.

34. Where is the 2114 RAM circuit shown in Figure 27-21 on the SAM schematic diagram?

35. The NWR (not write) control line input to the 2114 RAMs comes from what IC?

36. The NRAM (not RAM) control line input to the 2114 RAMs comes from what IC?

To practice your circuit recognition and operation skills, refer to Figure 27-22 and answer the following three questions.

37. Which address bits are applied to the address decoder's *ABC* select inputs?

38. What is the size of each block in the memory map?

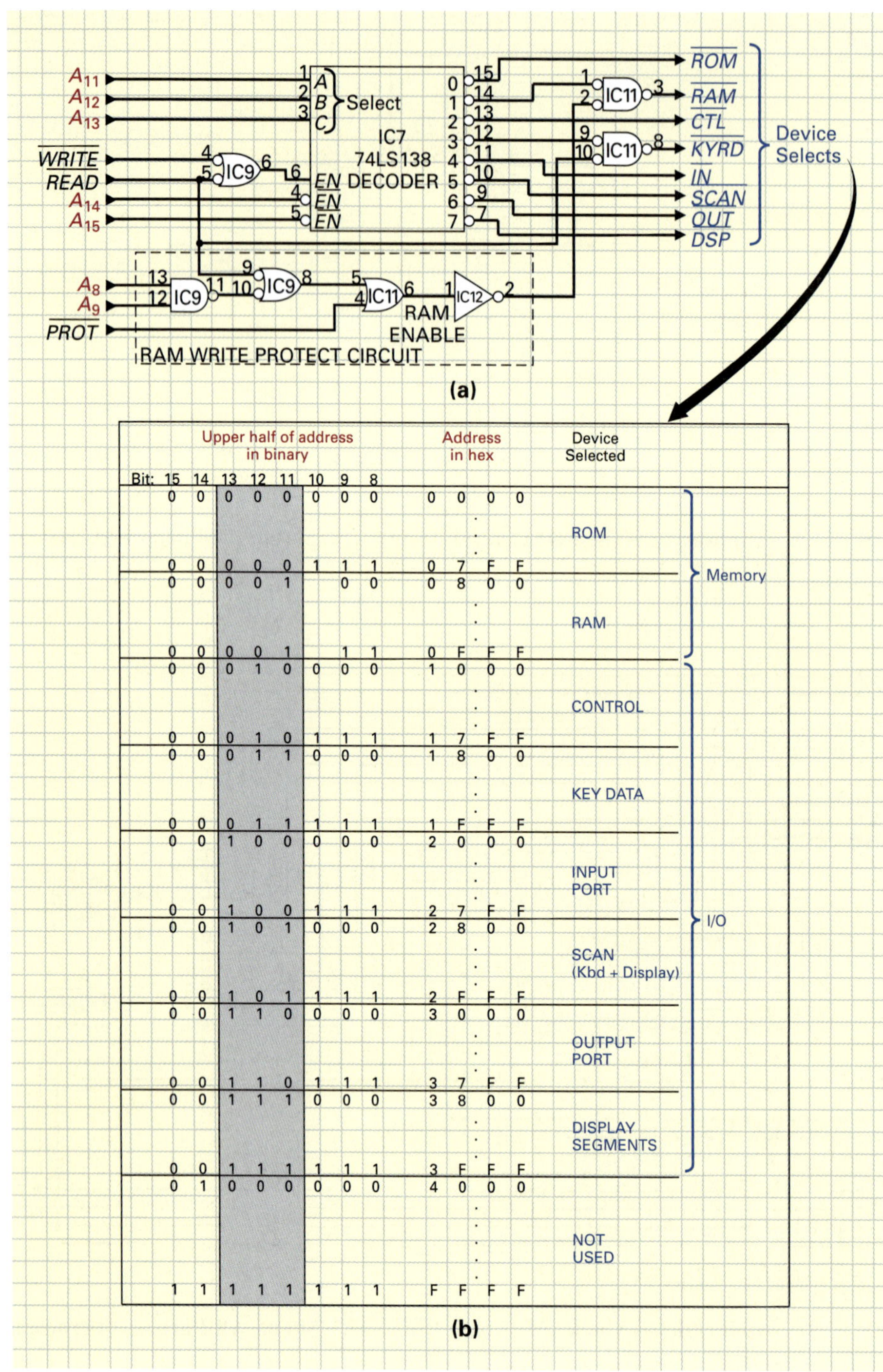

Bit: 15	14	13	12	11	10	9	8	Address in hex	Device Selected	
0	0	0	0	0	0	0	0	0 0 0 0	ROM	Memory
0	0	0	0	0	1	1	1	0 7 F F		
0	0	0	0	1		0	0	0 8 0 0	RAM	
0	0	0	0	1		1	1	0 F F F		
0	0	0	1	0	0	0	0	1 0 0 0	CONTROL	I/O
0	0	0	1	0	1	1	1	1 7 F F		
0	0	0	1	1	0	0	0	1 8 0 0	KEY DATA	
0	0	0	1	1	1	1	1	1 F F F		
0	0	1	0	0	0	0	0	2 0 0 0	INPUT PORT	
0	0	1	0	0	1	1	1	2 7 F F		
0	0	1	0	1	0	0	0	2 8 0 0	SCAN (Kbd + Display)	
0	0	1	0	1	1	1	1	2 F F F		
0	0	1	1	0	0	0	0	3 0 0 0	OUTPUT PORT	
0	0	1	1	0	1	1	1	3 7 F F		
0	0	1	1	1	0	0	0	3 8 0 0	DISPLAY SEGMENTS	
0	0	1	1	1	1	1	1	3 F F F		
0	1	0	0	0	0	0	0	4 0 0 0	NOT USED	
1	1	1	1	1	1	1	1	F F F F		

(Upper half of address in binary)

FIGURE 27-22 **Application—The Address Decoder and Memory Map For a SAM (Simple All-Purpose Microcomputer).**

39. What device is selected when A_{11}, A_{12}, and A_{13} equal:
   a. 000
   b. 001
   c. 010
   d. 011
   e. 100
   f. 101
   g. 110
   h. 111

To practice your circuit recognition and operation skills, refer again to the SAM schematic diagram on pages xxx-xxx and answer the following four questions.

40. Where is the address decoder shown in Figure 27-22 located on the SAM schematic diagram?
41. What is the purpose of the four-LED displays DS1, DS2, DS3, and DS4?
42. What is the purpose of the displays DS7 and DS8?
43. Is the SAM system using memory-mapped input/output?

Web Site Questions

Go to the web site http://www.prenhall.com/cook, select the textbook *Electronics: A Complete Course,* select this chapter, and then follow the instructions when answering the multiple choice practice problems.

Analog and Digital Signal Converters

Go and Do "Something More Useful"

Vladimir Zworykin was born in Mourom, Russia, on July 30, 1889. He graduated in 1912 from the St. Petersburg Institute of Technology and then did his postgraduate work at the College de France in Paris. After World War I, he immigrated to the United States and joined the research staff of Westinghouse Electric Corporation in Pittsburgh. Here he invented the iconoscope (electronic camera) and the kinescope (CRT picture tube). In 1923 he proudly demonstrated to his employers what he considered his biggest achievement—a cloudy image of boats on the river outside his lab on a small screen. Management was not at all impressed, and Zworykin later said that he was told "to spend my time on something more useful."

In 1929 David Sarnoff, the founder of the Radio Corporation of America (RCA), asked Zworykin what it would take to develop television for commercial use. Zworykin, who at that time was not at all sure, answered confidently, "Eighteen months and $100,000." Sarnoff liked what he heard and hired Zworykin, putting him in charge of RCA's electronic research laboratory at Camden, New Jersey. Years later Sarnoff loved pointing out the slight difference in development cost and time, which finally worked out to be twenty years and $50 million.

While at RCA, Zworykin also developed the electron microscope (in only three months), the electron multiplier tube, and the first infrared sniper scopes, which were used in World War II.

In 1967 Zworykin received the National Medal of Science for his contributions to the instruments of science, engineering, and television and for his stimulation of the application of engineering to medicine. Recognized as the "father of television," he lived long enough to see all of his inventions flourish before he died in 1982. When asked in a television interview in 1980 what he thought of television today, he answered, "The technique is wonderful. It is beyond my expectation. But the programs! I would never let my children even close to this thing."

28

Introduction

Electronic information exists as either analog data or as digital data. With analog data, the signal voltage change is an analogy of the physical information it is representing. For example, if a person were to speak into a microphone, the pitch and loudness of the sound wave would determine the frequency and amplitude of the voltage signal generated by the microphone. With a digital signal, information is first converted into a group of HIGH and LOW voltage pulses that represent the binary digits 1 and 0. For example, the binary digit code 1101001 is generated whenever the lower case "i" key on an ASCII computer keyboard is pressed. In this case, therefore, the information "i" is encoded into a digital code, or a code that makes use of binary digits.

Whenever you have two different forms, there is always a need to convert back and forth between these two different forms. For example, in previous chapters we have seen how we have needed circuits to convert between binary and decimal, bipolar logic levels and MOS logic levels, serial data and parallel data, and so on. In this chapter, we will examine data conversion techniques and the circuits that are used to perform digital-to-analog (D/A) data conversion and analog-to-digital (A/D) data conversion. We will also discuss the application, testing, and troubleshooting of these data conversion circuits.

28-1 ANALOG AND DIGITAL SIGNAL CONVERSION

Today, microcomputers are used in almost every electronic system because of their ability to quickly process and store large amounts of data, make systems more versatile, and perform more functions. Digital processing within electronic systems has therefore become more and more predominant. Many of the input signals applied to a microcomputer for storage or processing are analog in nature, and therefore a data conversion circuit is needed to interface these analog inputs to a digital system. Similarly, a data conversion circuit may also be needed to interface the digital microcomputer system to an analog output device.

28-1-1 *Connecting Analog and Digital Devices to a Computer*

Figure 28-1 shows a simplified block diagram of a microcomputer. No matter what the application, a microcomputer has three basic blocks—a microprocessor unit, a memory unit,

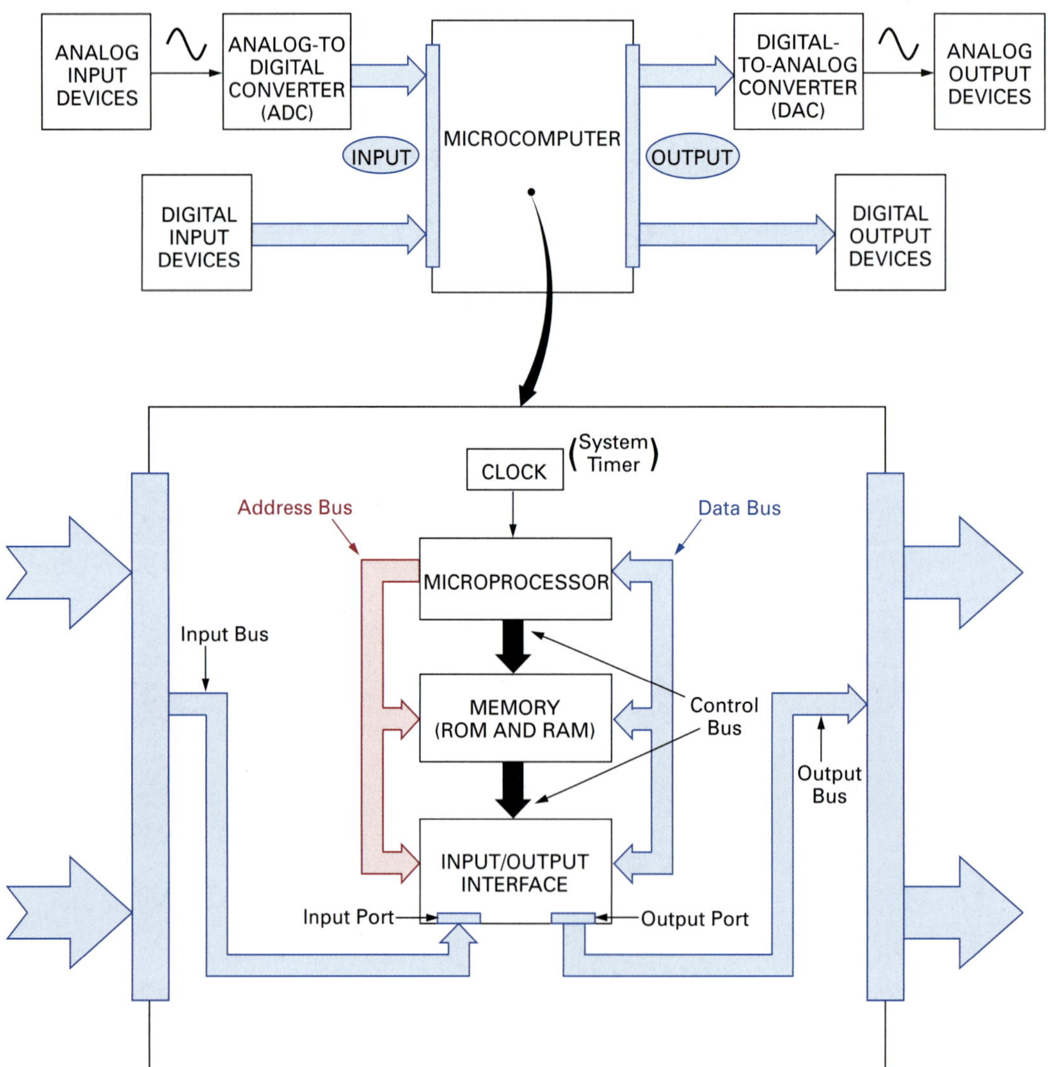

FIGURE 28-1 **Connecting Analog and Digital Devices to a Computer.**

and an input/output unit. Getting information into the computer is the job of input devices, and as you can see in Figure 28-1, there are two input paths into a computer—one path from analog input devices and the other path from digital input devices. Information in the form of light, sound, heat, pressure, or any other real-world quantity is analog in nature and has an infinite number of input levels. Sensors or transducers of this type, such as photoelectric cells for light or microphones for sound, will generate analog signals. Since the computer operates on information in only digital or binary form, a translation or conversion is needed if the computer is going to be able to understand or interpret an analog signal. This signal processing is achieved by an analog-to-digital converter (ADC), which transforms the varying analog input voltage into digital codes that the computer can understand. Once the analog information has been encoded into digital form, the information can then enter the computer through an electronic doorway called the **input port.** Some input devices, such as a computer keyboard, automatically generate a digital code for each key that is pressed (ASCII codes). The codes generated by these digital input devices can be connected directly to the computer without having to pass through a converter.

Input Port
An electronic doorway used for transmitting information into a computer.

Similarly, the digital output information from a microcomputer's **output port** can be used to drive either an analog device or a digital device. If an analog output is desired, as in the case of a voice signal for a loudspeaker, the digital computer's output will have to be converted into a corresponding analog signal by a digital-to-analog converter (DAC). If only a digital output is desired, as in the case of a printer, which translates the binary codes into printed characters, the digital computer output can be connected directly to the output device without the need for a converter.

Output Port
An electronic doorway used for transmitting information out of a computer.

28-1-2 *Converting Information Signals*

Figure 28-2(a) shows how analog data can be converted into digital data, and Figure 28-2(b) shows how digital data can be converted into analog data. To give this process purpose, let us consider how the music information stored on a compact disc (CD) is first recorded and then played back. Like all sound, music is made up of waves of compressed air. When these waves strike the diaphragm of a microphone, an analog voltage signal is generated. In the past, this analog data was recorded on magnetic tapes or as a grooved track on a record. These data storage devices were susceptible to wear and tear, temperature, noise, and age. Using digital recording techniques, binary codes are stored on a compact disc to achieve near-perfect fidelity to live sound.

Figure 28-2(a) shows how an **analog-to-digital converter (ADC)** is used during the recording process to convert the analog music input into a series of digital output codes. These digital codes are used to control the light beam of a recording laser so that it will engrave the binary 0s and 1s onto a compact disc (CD) in the form of pits or spaces. The ADC is triggered into operation by a sampling pulse that causes it to measure the input voltage of the analog signal at that particular time and generate an equivalent digital output code. For example, on the active edge of sampling pulse 1, the analog input voltage is at 2 V so the binary code 010 (decimal 2) is generated at the output of the ADC. On the active edge of sampling pulse 2, the analog voltage has risen to 4 V, so the binary code 100 (decimal 4) is generated at the output of the ADC. On the active edge of sampling pulses 3, 4, and 5, the binary codes 110 (decimal 6), 111 (decimal 7), and 110 (decimal 6) are generated at the output, representing the analog voltages of 6 V, 7 V, and 6 V, respectively, and so on.

Analog-to-Digital Converter (ADC)
A circuit that converts analog input signals into equivalent digital output signals.

Figure 28-2(b) shows how a **digital-to-analog converter (DAC)** is used during the music playback process to convert the digital codes stored on a CD to an analog output signal. Another laser is used during playback to read the pits and spaces on the compact disc as 0s and 1s. These codes are then applied to the DAC, which converts the digital input codes into discrete voltages. The DAC is triggered into operation by a strobe pulse that causes it to convert the code currently being applied at the input into an equivalent output voltage.

Digital-to-Analog Converter (DAC)
A circuit that converts digital input signals into equivalent analog output signals.

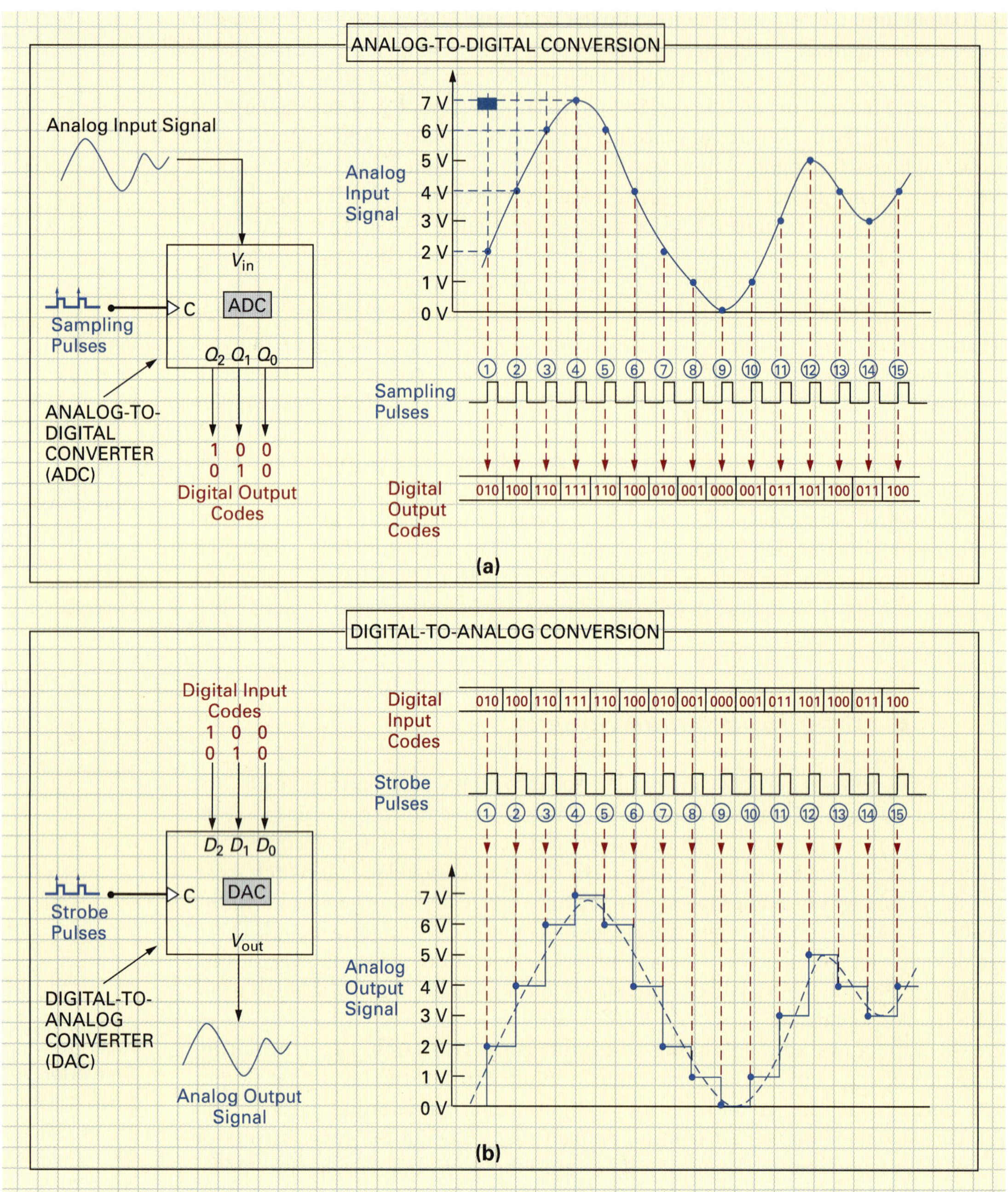

FIGURE 28-2 **Converting Information Signals.**

For example, on the active edge of strobe pulse 1, the digital code 010 (decimal 2) is being applied to the DAC, so it will generate 2 V at its output. On the active edge of strobe pulse 2, the digital code 100 (decimal 4) is being applied to the DAC, so it will generate 4 V at its output. On the active edge of strobe pulse 3, the digital code 110 (decimal 6) is being applied to the DAC, so it will generate 6 V at its output. On the active edge of strobe pulse 4, the digital code 111 (decimal 7) is being applied to the DAC, so it will generate 7 V at its output. On the active edge of strobe pulse 5, the digital code 110 (decimal 6) is being applied to the DAC, so it will generate 6 V at its output, and so on. If the output of a DAC is then applied to a lowpass filter, the discrete voltage steps can be blended into a smooth wave that closely approximates the original analog wave, as shown by the dashed line in Figure 28-2(b).

SELF-TEST EVALUATION POINT FOR SECTION 28-1

Use the following questions to test your understanding of Section 28-1.

1. Give the full names of the following abbreviations:
   a. ADC b. DAC
2. To interface a microcomputer with analog devices, a ____________ would be connected to the input port, and a ____________ would be connected to the output port.
3. An ADC will convert a/an ____________ input into an equivalent ____________ output.
4. A DAC will convert a/an ____________ input into an equivalent ____________ output.

28-2 DIGITAL-TO-ANALOG CONVERTERS (DACs)

A DAC (pronounced "dak") is a circuit that generates an analog output voltage that is proportional to the value of the binary input code. There are a variety of digital-to-analog converter ICs available with varying characteristics. In this section, we will be discussing digital-to-analog conversion techniques and characteristics, a typical DAC data sheet, and DAC applications and testing.

28-2-1 *Binary-Weighted Resistor DAC*

Binary-Weighted Resistor Circuit

A DAC circuit that uses a power-of-2 weighted resistor network to convert a digital signal input into an equivalent analog signal output.

One of the simplest digital-to-analog converter circuits is the **binary-weighted resistor DAC,** shown in Figure 28-3(a). In this circuit, the 4-bit digital input is used to control four HIGH-LOW input switches (SW_1 through SW_4). The setting of these input switches will determine the analog output voltage generated at the output, as listed in the table in Figure 28-3(b). A -10 V reference voltage is applied to a binary-weighted input resistor network, and an inverting op-amp is used to convert a current input into a positive analog output voltage.

Let us now examine the operation of this circuit in more detail. In the example in Figure 28-3(a), a digital input of 1011 (decimal 11) is being applied to the DAC circuit, so SW_1 is set HIGH, SW_2 is set HIGH, SW_3 is set LOW, and SW_4 is set HIGH. This means that input resistors $8R$, $4R$, and R will contribute to the final output, whereas input resistor $2R$ will not contribute to the final output. A negative reference voltage has been used because the op-amp is connected as an inverting amplifier, so a negative reference voltage input will generate positive analog output voltages, as listed in the table in Figure 28-3(b). The final analog output voltage from this circuit is determined by the reference voltage and the ratio of the feedback resistor R_f to the contributing input resistors. The formula is as follows:

$$V_{\text{out}} = -V_{\text{ref}}\left(\frac{R_f}{R} + \frac{R_f}{2R} + \frac{R_f}{4R} + \frac{R_f}{8R}\right)$$

For the 1101 (decimal 11) digital input example shown in Figure 28-3(a), the analog output voltage will be

$$\begin{aligned} V_{\text{out}} &= -V_{\text{ref}}\left(\frac{R_f}{R} + \frac{R_f}{2R} + \frac{R_f}{4R} + \frac{R_f}{8R}\right) \\ &= -(-10\text{ V})\left(\frac{8\text{ k}\Omega}{10\text{ k}\Omega} + 0 + \frac{8\text{ k}\Omega}{40\text{ k}\Omega} + \frac{8\text{ k}\Omega}{80\text{ k}\Omega}\right) \\ &= 10\text{ V}(0.8 + 0 + 0.2 + 0.1) \\ &= 10\text{ V} \times 1.1 = 11\text{ V} \end{aligned}$$

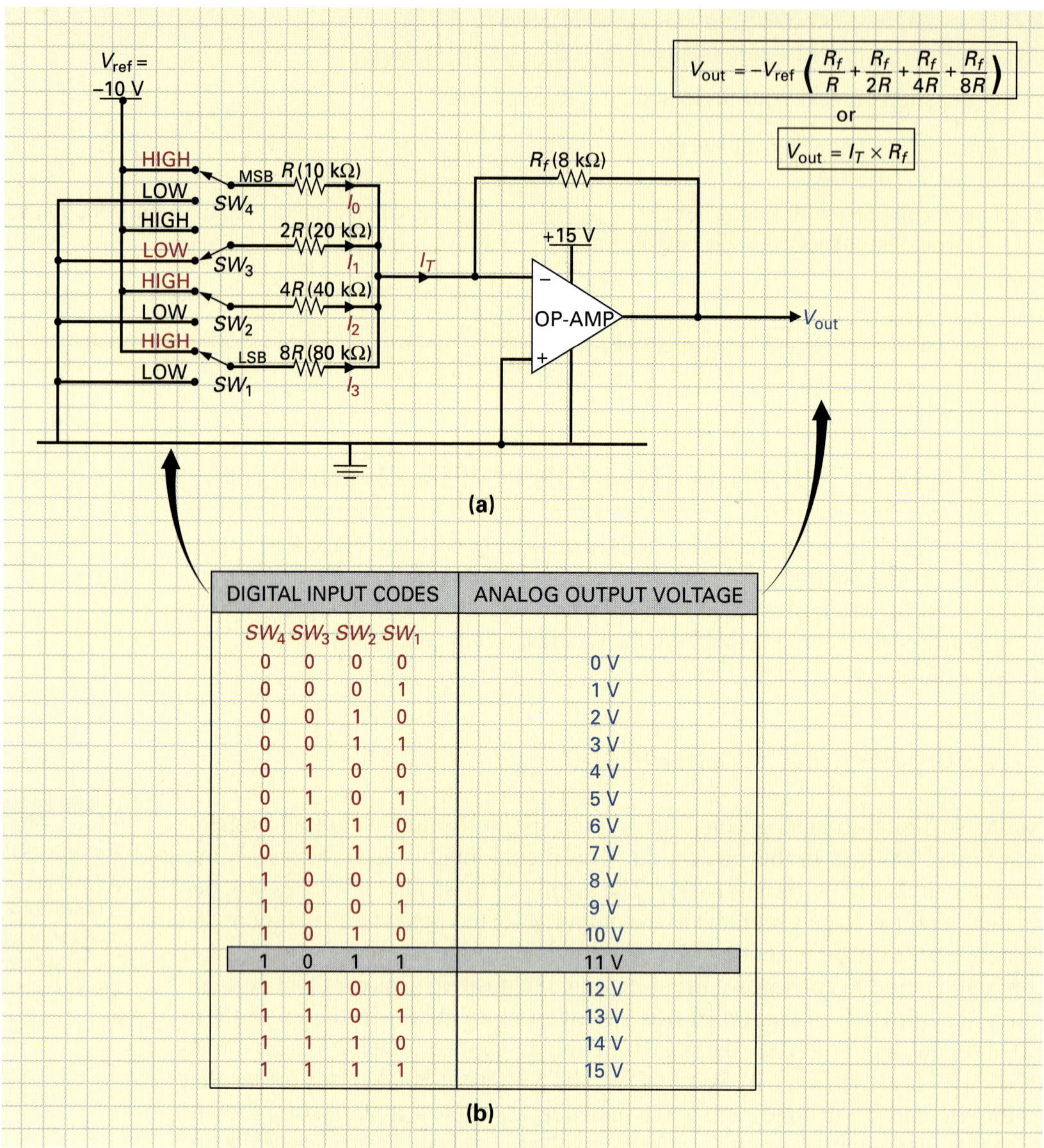

DIGITAL INPUT CODES SW_4	SW_3	SW_2	SW_1	ANALOG OUTPUT VOLTAGE
0	0	0	0	0 V
0	0	0	1	1 V
0	0	1	0	2 V
0	0	1	1	3 V
0	1	0	0	4 V
0	1	0	1	5 V
0	1	1	0	6 V
0	1	1	1	7 V
1	0	0	0	8 V
1	0	0	1	9 V
1	0	1	0	10 V
1	0	1	1	11 V
1	1	0	0	12 V
1	1	0	1	13 V
1	1	1	0	14 V
1	1	1	1	15 V

FIGURE 28-3 **Binary-Weighted Resistor Digital-to-Analog Converter (DAC).**

EXAMPLE:

What analog output voltage will be generated by the circuit in Figure 28-3(a) if a digital input of 1001 is applied?

Solution:

If a digital input of 1001 (decimal 9) is applied to the DAC circuit, SW_4 will be set HIGH, SW_3 will be set LOW, SW_2 will be set LOW, and SW_1 will be set HIGH. This means that input resistors $8R$ and R will contribute to the final output, whereas input resistors $4R$ and $2R$ will not contribute to the final output. The circuit will therefore generate the following analog output voltage:

$$V_{out} = -V_{ref}\left(\frac{R_f}{R} + \frac{R_f}{2R} + \frac{R_f}{4R} + \frac{R_f}{8R}\right)$$

$$= -(-10\text{ V})\left(\frac{8\text{ k}\Omega}{10\text{ k}\Omega} + 0 + 0 + \frac{8\text{ k}\Omega}{80\text{ k}\Omega}\right)$$

$$= 10\text{ V}(0.8 + 0 + 0 + 0.1)$$
$$= 10\text{ V} \times 0.9 = 9\text{ V}$$

The analog output voltage generated by the circuit in Figure 28-3(a) can also be calculated using the formula

$$V_{out} = I_T \times R_f$$

For a 1101 (decimal 11) input, therefore, the circuit in Figure 28-3(a) will generate the following analog output voltage:

$$V_{out} = I_T \times R_f$$
$$I_0 = \frac{-V_{ref}}{R} = \frac{-(-10\text{ V})}{10\text{ k}\Omega} = \frac{10\text{ V}}{10\text{ k}\Omega} = 1\text{ mA}$$
$$I_1 = 0\text{ mA}$$
$$I_2 = \frac{-V_{ref}}{4R} = \frac{-(-10\text{ V})}{40\text{ k}\Omega} = \frac{10\text{ V}}{40\text{ k}\Omega} = 0.25\text{ mA}$$
$$I_3 = \frac{-V_{ref}}{8R} = \frac{-(-10\text{ V})}{80\text{ k}\Omega} = \frac{10\text{ V}}{80\text{ k}\Omega} = 0.125\text{ mA}$$
$$I_T = I_1 + I_2 + I_3 + I_4 = 1\text{ mA} + 0\text{ mA} + 0.25\text{ mA} + 0.125\text{ mA}$$
$$= 1.375\text{ mA}$$
$$V_{out} = I_T \times R_f = 1.375\text{ mA} \times 8\text{ k}\Omega = 11\text{ V}$$

EXAMPLE:

Using the $V_{out} = I_T \times R_f$ formula, calculate the analog output voltage generated by the circuit in Figure 28-3(a) if a digital input of 1001 were applied.

Solution:

$$V_{out} = I_T \times R_f$$
$$I_0 = \frac{-V_{ref}}{R} = \frac{-(-10\text{ V})}{10\text{ k}\Omega} = \frac{10\text{ V}}{10\text{ k}\Omega} = 1\text{ mA}$$
$$I_1 = 0\text{ mA}$$
$$I_2 = 0\text{ mA}$$
$$I_3 = \frac{-V_{ref}}{8R} = \frac{-(-10\text{ V})}{80\text{ k}\Omega} = \frac{10\text{ V}}{80\text{ k}\Omega} = 0.125\text{ mA}$$
$$I_T = I_1 + I_2 + I_3 + I_4 = 1\text{ mA} + 0\text{ mA} + 0\text{ mA} + 0.125\text{ mA} = 1.125\text{ mA}$$
$$V_{out} = I_T \times R_f = 1.125\text{ mA} \times 8\text{ k}\Omega = 9\text{ V}$$

28-2-2 *R/2R Ladder DAC*

One of the key disadvantages of the binary-weighted resistor DAC is that the weighted resistors create difficulties as the number of digital input bits increases. For example, an 8-bit DAC will need eight weighted resistors of *R*, 2*R*, 4*R*, 8*R*, 16*R*, 32*R*, 64*R*, and 128*R*. A 16-bit DAC would need sixteen weighted resistors, with the largest resistance being 32,768 times larger than the smallest resistor. The large difference between the LSB resistor value and the MSB resistor value creates several problems. Another disadvantage is the extreme difference in current in each branch of the binary-weighted resistor network. For example, having a very large LSB resistor value will result in an extremely small current value, which is susceptible to noise. In addition, having an extremely small MSB resistor value will result in an extremely large current value, which will cause temperature and load changes.

***R/2R* Ladder Circuit**

A DAC circuit that uses a resistor network having only two different values—*R* and 2*R*.

The problems of the binary-weighted resistor DAC can be overcome by using the ***R/2R* ladder DAC,** shown in Figure 28-4(a). This *R*/2*R* resistor network does not have the resistance range problems of the weighted-resistor network because it only makes use of two resistance values, *R* and 2*R*. Like the binary-weighted resistor DAC, the *R*/2*R* ladder DAC circuit uses binary-controlled switches to switch binary-weighted currents through to the summing junction of the op-amp. The op-amp is included, as in the binary-weighted resistor DAC, to convert the input current into a proportional output voltage.

The *R*/2*R* ladder resistor network operates as a current divider. Figures 28-4(b) through (j) show how this circuit operates as a current divider. In Figure 28-4(b), you can see that below junction *D* we have a 2*R* resistor in parallel (‖) with a 2*R* resistor, which is equivalent to *R*, as redrawn in Figure 28-4(c). Moving on to the redrawn circuit in Figure 28-4(c), you can see that we now have *R* in series with *R* directly below junction *C*, which is equivalent to 2*R*, as redrawn in Figure 28-4(d). Moving on to the redrawn circuit in Figure 28-4(d), you can see that we now have 2*R* in parallel with 2*R* below junction *C*, which is equivalent to *R*, as redrawn in Figure 28-4(e). Moving on to the redrawn circuit in Figure 28-4(e), you can see that we now have *R* in series with *R* directly below junction *B*, which is equivalent to 2*R*, as redrawn in Figure 28-4(f). Moving on to the redrawn circuit in Figure 28-4(f), you can see that we now have 2*R* in parallel with 2*R* below junction *B*, which is equivalent to *R*, as redrawn in Figure 28-4(g). Moving on to the redrawn circuit in Figure 28-4(g), you can see that we now have *R* in series with *R* directly below junction *A*, which is equivalent to 2*R* as redrawn in Figure 28-4(h). Moving on to the redrawn circuit in Figure 28-4(h), you can see that we now have 2*R* in parallel with 2*R* below junction *A*, which is equivalent to *R*, as redrawn in Figure 28-4(i). The *R*/2*R* ladder resistor network is therefore equivalent to *R*, so the total current can be calculated with Ohm's law as follows:

$$I_T = \frac{V_{\text{ref}}}{R}$$

In the example circuit in Figure 28-4(a), $V_{\text{ref}} = -16$ V and $R = 1$ kΩ, so the total current will be

$$I_T = \frac{V_{\text{ref}}}{R} = \frac{-16\text{ V}}{1\text{ k}\Omega} = 16\text{ mA}$$

The *R*/2*R* ladder circuit will divide this total current into binary levels, as shown in Figure 28-4(j). The 16 mA arriving at junction *A* will see a parallel path made up of 2*R* and 2*R*, as was shown in Figure 28-4(h), so the 16 mA will be halved, with 8 mA flowing through each branch, as seen in Figure 28-4(j). The 8 mA arriving at junction *B* will see a parallel path made up of 2*R* and 2*R*, as was shown in Figure 28-4(f), so the 8 mA will be halved, with 4 mA flowing through each branch, as seen in Figure 28-4(j). The 4 mA arriving at junction *C* will see a parallel path made up of 2*R* and 2*R*, as was shown in Figure 28-4(d), so the 4 mA will be halved, with 2 mA flowing through each branch, as seen in Figure 28-4(j). The 2 mA arriving at junction *D* will see a parallel path made up of 2*R* and 2*R*, as was shown in Figure 28-4(b), so the 2 mA will be halved, with 1 mA flowing through each branch, as seen in Figure 28-4(j). These binary-weighted current values are switched through to the summing junction of the op-amp based on the condition of the binary-controlled electronic switches. For example, if a 1011 digital input were applied to this *R*/2*R* DAC circuit, SW_1 would be in its HIGH position, SW_2 would be set LOW, SW_3 would be set HIGH, and SW_4 would be set HIGH, as shown in Figure 28-4(a).

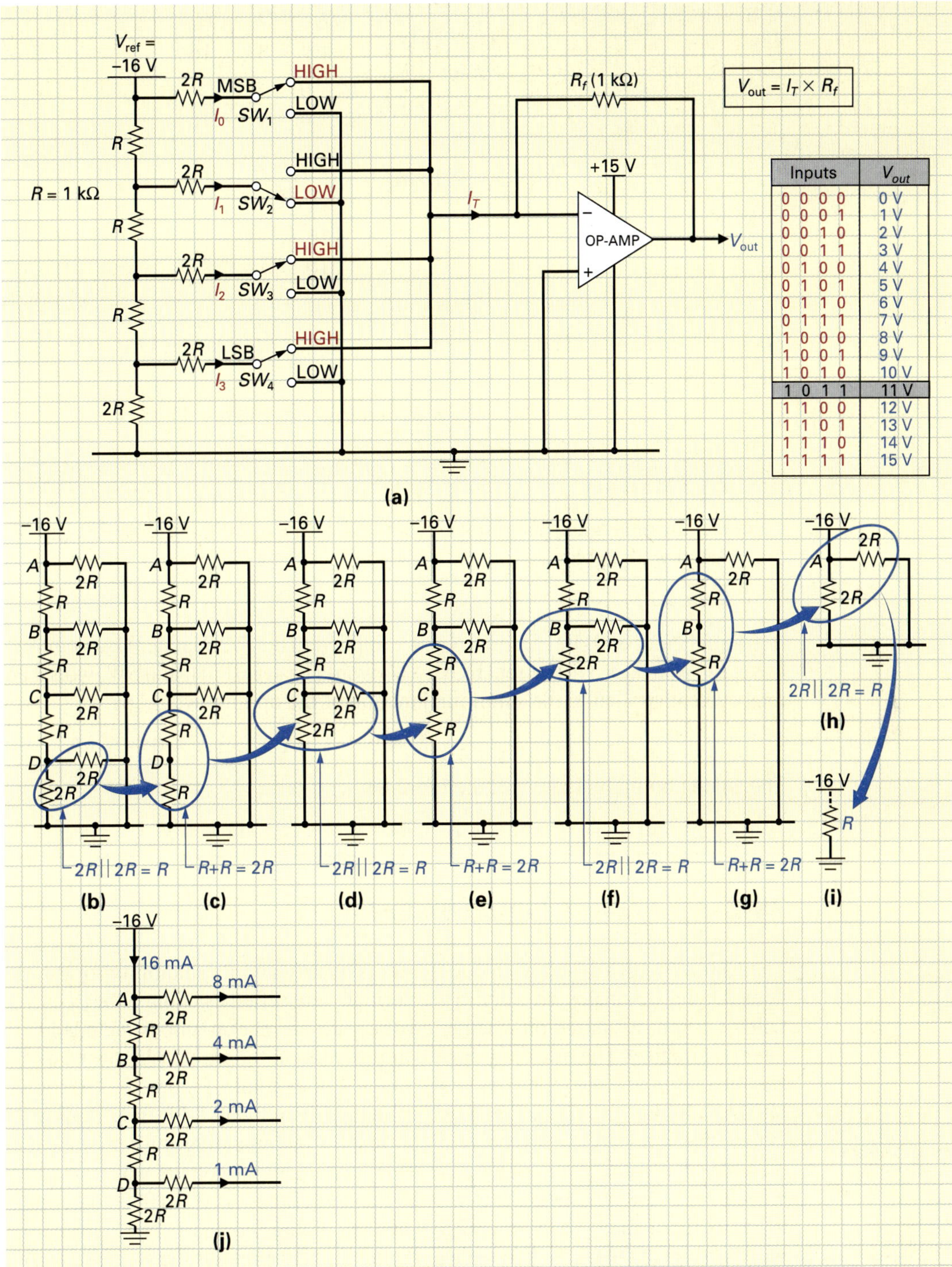

Inputs	V_{out}
0 0 0 0	0 V
0 0 0 1	1 V
0 0 1 0	2 V
0 0 1 1	3 V
0 1 0 0	4 V
0 1 0 1	5 V
0 1 1 0	6 V
0 1 1 1	7 V
1 0 0 0	8 V
1 0 0 1	9 V
1 0 1 0	10 V
1 0 1 1	11 V
1 1 0 0	12 V
1 1 0 1	13 V
1 1 1 0	14 V
1 1 1 1	15 V

FIGURE 28-4 ***R*/2*R* Digital-to-Analog Converter (DAC).**

This means that the 8 mA, 2 mA, and 1 mA current values will be switched through to the summing junction of the op-amp, so I_T will equal 11 mA (8 mA + 2 mA + 1 mA = 11 mA). Stated as a formula, I_T is equal to

$$I_{\text{out}} = I_{\text{ref}}\left(\frac{D_3}{2} + \frac{D_2}{4} + \frac{D_1}{8} + \frac{D_0}{16}\right)$$

$$I_{out} = I_{ref}\left(\frac{D_3}{2} + \frac{D_2}{4} + \frac{D_1}{8} + \frac{D_0}{16}\right)$$
$$= 16 \text{ mA}\left(\frac{1}{2} + \frac{0}{4} + \frac{1}{8} + \frac{1}{16}\right)$$
$$= 16 \text{ mA}\left(\frac{8 + 0 + 2 + 1}{16}\right)$$
$$= 16 \text{ mA} \times \left(\frac{11}{16}\right)$$
$$= 11 \text{ mA}$$

For a 1011 (decimal 11) digital input, therefore, this *R*/2*R* ladder DAC will generate an analog output voltage of

$$V_{out} = I_T \times R_f = 11 \text{ mA} \times 1 \text{ k}\Omega = 11 \text{ V}$$

EXAMPLE:

Calculate the analog output voltage generated by the circuit in Figure 28-4(a) if a digital input of 0110 were applied.

Solution:

If a digital input of 0110 (decimal 6) were applied to the *R*/2*R* ladder DAC circuit, SW_4 would be set LOW, SW_3 would be set HIGH, SW_2 would be set HIGH, and SW_1 would be LOW. The total current applied to the summing junction of the op-amp would be

$$I_T = I_0 + I_1 + I_2 + I_3 = 0 + 4 \text{ mA} + 2 \text{ mA} + 0 = 6 \text{ mA}$$

The analog output voltage will therefore be

$$V_{out} = I_T \times R_f = 6 \text{ mA} \times 1 \text{ k}\Omega = 6 \text{ V}$$

28-2-3 *DAC Characteristics*

Staircase Waveform

A waveform that continually steps up in value from zero to a maximum value and then repeats.

The characteristics of a device determine its quality and its suitability for different applications. Digital-to-analog converters are normally classified as being either a current output or voltage output IC, based on whether a voltage-to-current op-amp is included within the IC's internal circuitry. The other key specifications to look for are the number of digital input bits and the output voltage range or current range. For example, Figure 28-5(a) shows a 0–5 V 8-bit DAC. In this test circuit, the DAC is being driven by an 8-bit free-running counter. This means that the DAC will receive a continually incrementing digital count input and will therefore generate a continually increasing analog output voltage waveform. This **staircase waveform** output will begin at 0 V when the digital input is 0000 0000 (decimal 0), will rise to a maximum analog output voltage when the digital input is 1111 1111 (decimal 255), and will then repeat the cycle. The table in Figure 28-5(b) explains the staircase output waveform in more detail. As you know, each of the digital input bits is weighted, and therefore:

the LSB will have a weight of $^{5\text{ V}}/_{256} = 19.5$ mV,
the next bit will have a weight of $^{5\text{ V}}/_{128} = 39$ mV,
the next bit will have a weight of $^{5\text{ V}}/_{64} = 78$ mV,
the next bit will have a weight of $^{5\text{ V}}/_{32} = 156$ mV,
the next bit will have a weight of $^{5\text{ V}}/_{16} = 312$ mV,
the next bit will have a weight of $^{5\text{ V}}/_{8} = 625$ mV,
the next bit will have a weight of $^{5\text{ V}}/_{4} = 1.25$ V,
and the MSB will have a weight of $^{5\text{ V}}/_{2} = 2.5$ V.

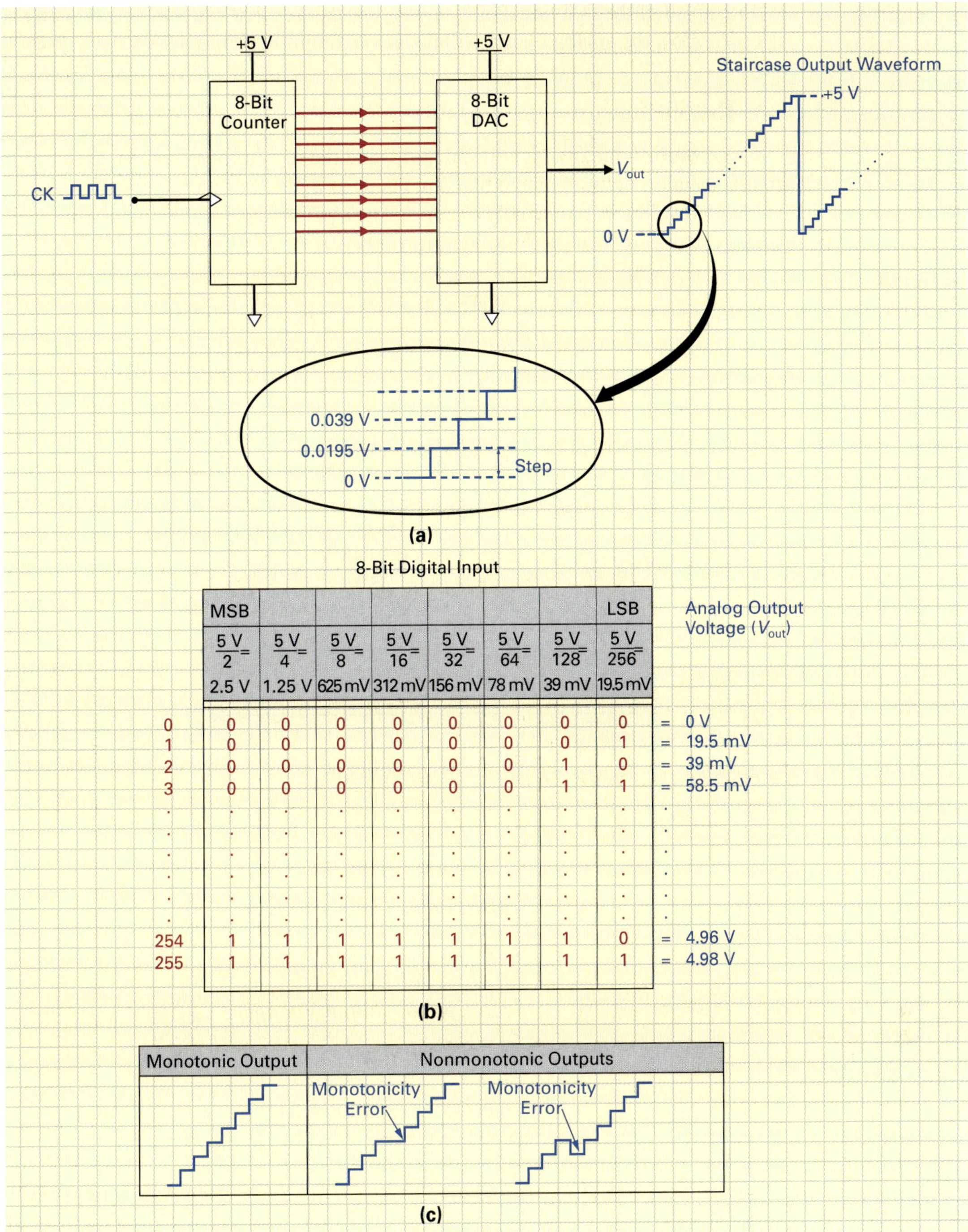

FIGURE 28-5 **DAC Characteristics.**

As the digital input count advances by 1, the analog output of the DAC will increase in 19.5 mV steps, as shown in the inset in Figure 28-5(a). The DAC's analog output, therefore, will generate 255 steps and have a **step size** of 19.5 mV.

$$\textit{Step Size} = \frac{\textit{Max. Rated Output Voltage}}{2^n}$$

$$\textit{Step Size} = \frac{\textit{Max. Rated Output Voltage}}{2^n} = \frac{5\text{ V}}{2^8} = \frac{5\text{ V}}{256} = 19.5\text{ mV}$$

Adding up all of the individual bit weights, you can see that when the digital input is 1111 1111 (decimal 255), the analog output will be its maximum, which is 4.98 V.

Resolution

A term used to describe the amount of detail present.

Resolution

Resolution means "smallest detail," and the resolution of a DAC is its step size, or the smallest analog output change that can occur as a result of an increment in the digital input. A DAC's resolution is normally expressed as a percentage of the maximum rated output and can be calculated with the following formula:

$$\textit{Percent Resolution} = \frac{\textit{Step Size}}{\textit{Max. Rated Output Voltage}} \times 100$$

To apply this formula to an example, the DAC in Figure 28-5(a) will have a resolution of

$$\begin{aligned}\textit{Percent Resolution} &= \frac{\textit{Step Size}}{\textit{Max. Rated Output Voltage}} \times 100\\ &= \frac{0.0195\text{ V}}{5\text{ V}} \times 100\\ &= 0.0039 \times 100 = 0.39\%\end{aligned}$$

EXAMPLE:

Calculate the resolution of a 0–5 V 10-bit DAC.

Solution:

First calculate the step size, and then calculate the DAC's resolution.

$$\textit{Step Size} = \frac{\textit{Max. Rated Output Voltage}}{2^n} = \frac{5\text{ V}}{2^{10}} = \frac{5\text{ V}}{1{,}024} = 4.9\text{ mV}$$

$$\begin{aligned}\textit{Percent Resolution} &= \frac{\textit{Step Size}}{\textit{Max. Rated Output Voltage}} \times 100\\ &= \frac{4.9\text{ mV}}{5\text{ V}} \times 100\\ &= 0.0009765 \times 100 = 0.098\%\end{aligned}$$

As you can see from this example, a 0-5 V 10-bit DAC has a much better resolution than a 0–5 V 8-bit DAC.

Monotonic

A term used to describe the condition of a DAC's output. A monotonic output from a DAC will have an increasing analog output for every increase in the digital input.

Monotonicity

When a free-running counter is connected to the input of a DAC, its output voltage should continually step up, producing a staircase waveform. A DAC is said to be **monotonic** if its output increases for each successive increase in the digital input, as shown in the left portion of Figure 28-5(c). If an increase in the digital input does not cause the output of the DAC to step up, or if the output steps backwards, as shown in the right portion of Figure 28-5(c), the DAC is said to be nonmonotonic. Monotonicity errors can be caused by circuit malfunctions within the DAC IC, and they can also be caused by circuit problems external to the IC. As always, you will need to isolate the device to determine whether the problem is source or load. We will be discussing the testing of DACs and the troubleshooting of data conversion circuits in a later section in this chapter.

Settling Time

The **settling time** of a DAC is the time it takes for the DAC's analog output to settle to 99.95% of its new value after a digital input has been applied. This DAC characteristic describes how fast a conversion can be made, and so it will determine the maximum frequency at which the DAC can operate. As an example, the MC1408 current DAC has a settling time of about 300 ns, whereas most voltage output DACs will have settling times ranging from 50 ns to 20 μs.

Settling Time
The time it takes for a DAC's analog output to settle to 99.95% of its new value.

Accuracy

The **relative accuracy** of a DAC describes how much the output level has deviated from its ideal theoretical output value. This accuracy specification is usually expressed as a percentage and describes the percentage of error rather than the percentage of accuracy. As an example, the MC1408 DAC has a ±0.19% error relative to the full-scale output.

Relative Acccuracy
The amount a DAC's output level has deviated relative to an ideal theoretical output value.

28-2-4 *A DAC Application Circuit*

Figure 28-6 shows the MC1408 8-bit current DAC. This IC contains an $R/2R$ ladder circuit, a reference current circuit, and eight binary current steering switches. In Figure 28-6 this IC is connected to function as a 0 to +5 V 8-bit digital-to-analog converter. A +5 V reference voltage ($+V_{ref}$), in conjunction with rheostat R_1, sets up a reference current of 2 mA for the DAC's internal ladder circuit. When all of the digital inputs are HIGH, therefore, the analog output current (I_{out}) will be

$$\begin{aligned} I_{out} &= I_{ref}\left(\frac{A_1}{2} + \frac{A_2}{4} + \frac{A_3}{8} + \frac{A_4}{16} + \frac{A_5}{32} + \frac{A_6}{64} + \frac{A_7}{128} + \frac{A_8}{256}\right) \\ &= 2\text{ mA}\left(\frac{1}{2} + \frac{1}{4} + \frac{1}{8} + \frac{1}{16} + \frac{1}{32} + \frac{1}{64} + \frac{1}{128} + \frac{1}{256}\right) \\ &= 2\text{ mA} \times \left(\frac{255}{256}\right) \\ &= 1.99\text{ mA} \end{aligned}$$

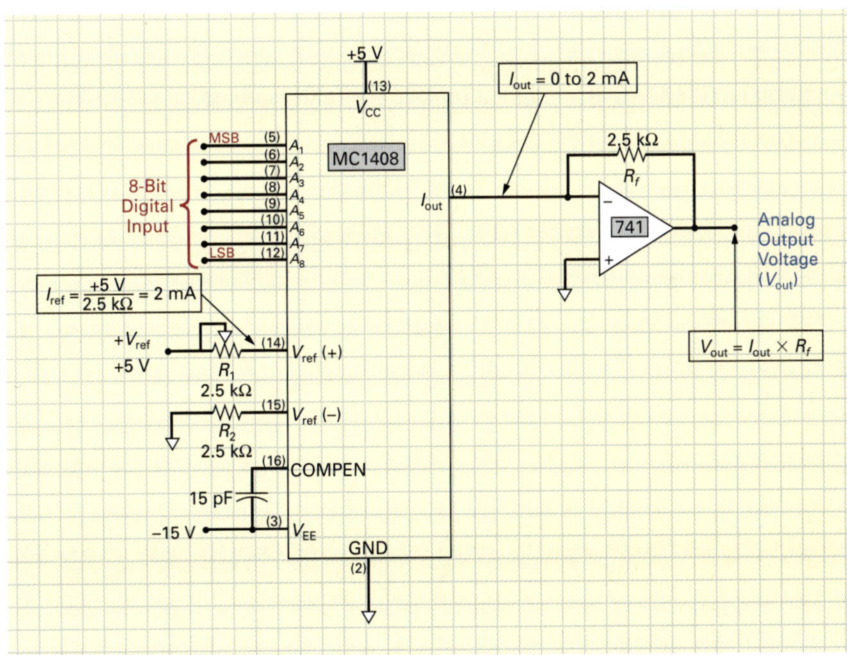

FIGURE 28-6 **Application—An 8-Bit Digital-to-Analog Converter Circuit.**

The analog output current is applied to a 741 op-amp, which will develop an output voltage that is equal to

$$V_{out} = I_T \times R_f$$

EXAMPLE:

What will be the analog output voltage from the circuit in Figure 28-6 if a digital input of 1100 0001 is applied?

Solution:

The first step should be to calculate I_{out} and then calculate V_{out}.

$$I_{out} = I_{ref}\left(\frac{A_1}{2} + \frac{A_2}{4} + \frac{A_3}{8} + \frac{A_4}{16} + \frac{A_5}{32} + \frac{A_6}{64} + \frac{A_7}{128} + \frac{A_8}{256}\right)$$

$$= 2\text{ mA}\left(\frac{1}{2} + \frac{1}{4} + \frac{0}{8} + \frac{0}{16} + \frac{0}{32} + \frac{0}{64} + \frac{0}{128} + \frac{1}{256}\right)$$

$$= 2\text{ mA} \times \left(\frac{128 + 64 + 1}{256}\right)$$

$$= 2\text{ mA} \times \left(\frac{193}{256}\right) = 1.51\text{ mA}$$

$$V_{out} = I_T \times R_f = 1.51\text{ mA} \times 2.5\text{ k}\Omega = 3.8\text{ V}$$

28-2-5 *Testing DACs*

One of the easiest ways to test a DAC is to have a free-running counter generate the digital input and have an oscilloscope monitor the analog output, as shown in Figure 28-7. Ideally, the output should be a straight-line monotonic staircase, as shown in Figure 28-7. If a fault exists, you may obtain an incorrect output similar to one of the waveform examples also shown in Figure 28-7. In this situation, you will need to isolate whether the problem is internal or external to the DAC IC and then replace the faulty device.

SELF-TEST EVALUATION POINT FOR SECTION 28-2

Use the following questions to test your understanding of Section 28-2:

1. What is the main disadvantage of the weighted-resistor DAC?

2. What is the purpose of the op-amp in a DAC circuit?

3. Define the following terms:
   a. staircase output
   b. resolution
   c. step size
   d. monotonicity
   e. settling time
   f. accuracy

4. How can a DAC be tested?

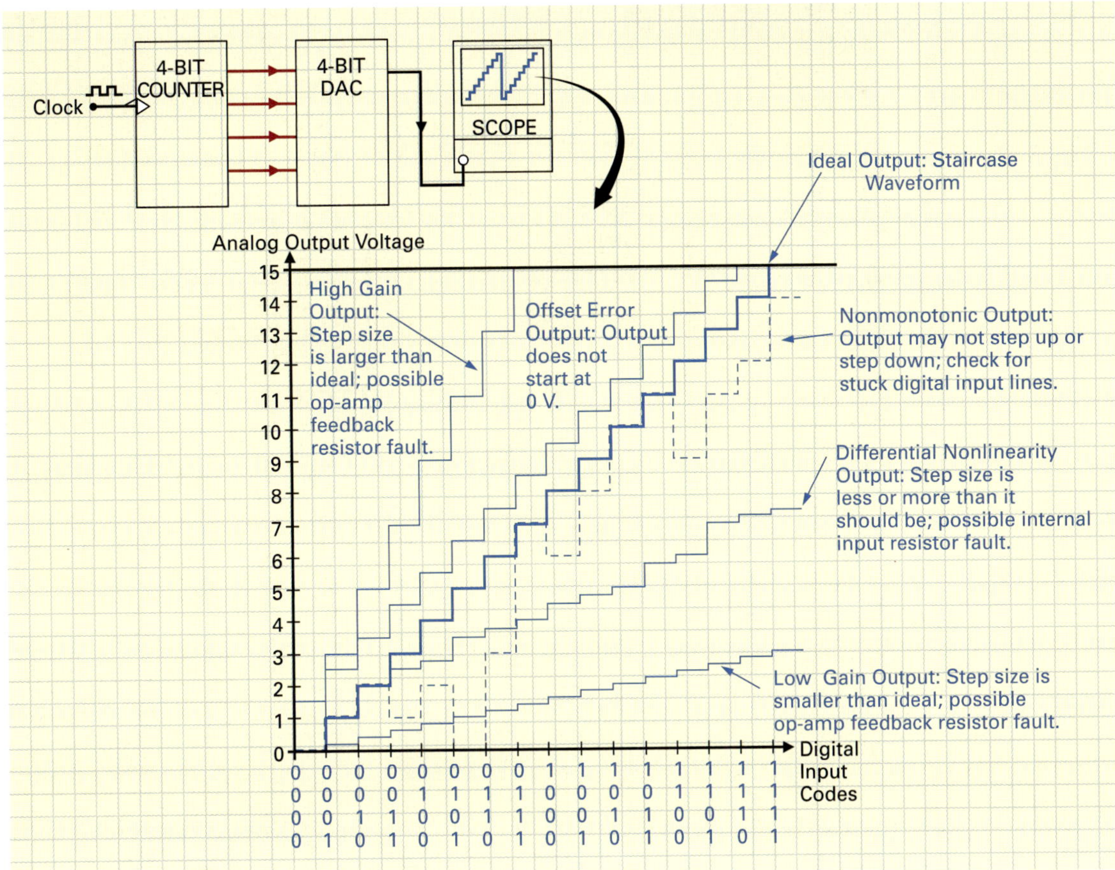

FIGURE 28-7 **Testing Digital-to-Analog Converters.**

28-3 ANALOG-TO-DIGITAL CONVERTERS (ADCs)

An ADC (pronounced "aye-dee-see") is a circuit that generates a binary output code that is proportional to an analog input voltage. A variety of analog-to-digital converter ICs are available with varying characteristics. In this section, we will be discussing analog-to-digital conversion techniques and characteristics, a typical ADC data sheet, and ADC applications and testing.

28-3-1 *Staircase ADC*

One of the simplest analog-to-digital converter circuits is the basic **staircase ADC,** shown in Figure 28-8(a). To convert an analog input into a digital output, this ADC circuit has a counter, comparator, AND gate, and a DAC. The analog input voltage is applied to the negative input of the comparator. The positive input of the comparator is a staircase voltage waveform from the DAC. The DAC generates a staircase voltage waveform because its digital input is connected to a counter incremented by a clock signal. This clock signal is either switched through the AND gate or blocked by the AND gate, based on the logic level output from the comparator.

Staircase ADC
An ADC circuit that uses a staircase waveform and comparator to determine the value of the analog input voltage and generate an equivalent digital output.

To explain the operation of the circuit in Figure 28-8(a), let us step through the timing waveforms shown in Figure 28-8(b), which show how an analog input voltage of 5.9 V is converted to the equivalent 4-bit digital code 0110 (decimal 6). To begin with, the active-LOW start conversion ($\overline{SC}$) input is used to RESET the counter to 0 V. This zero count will be

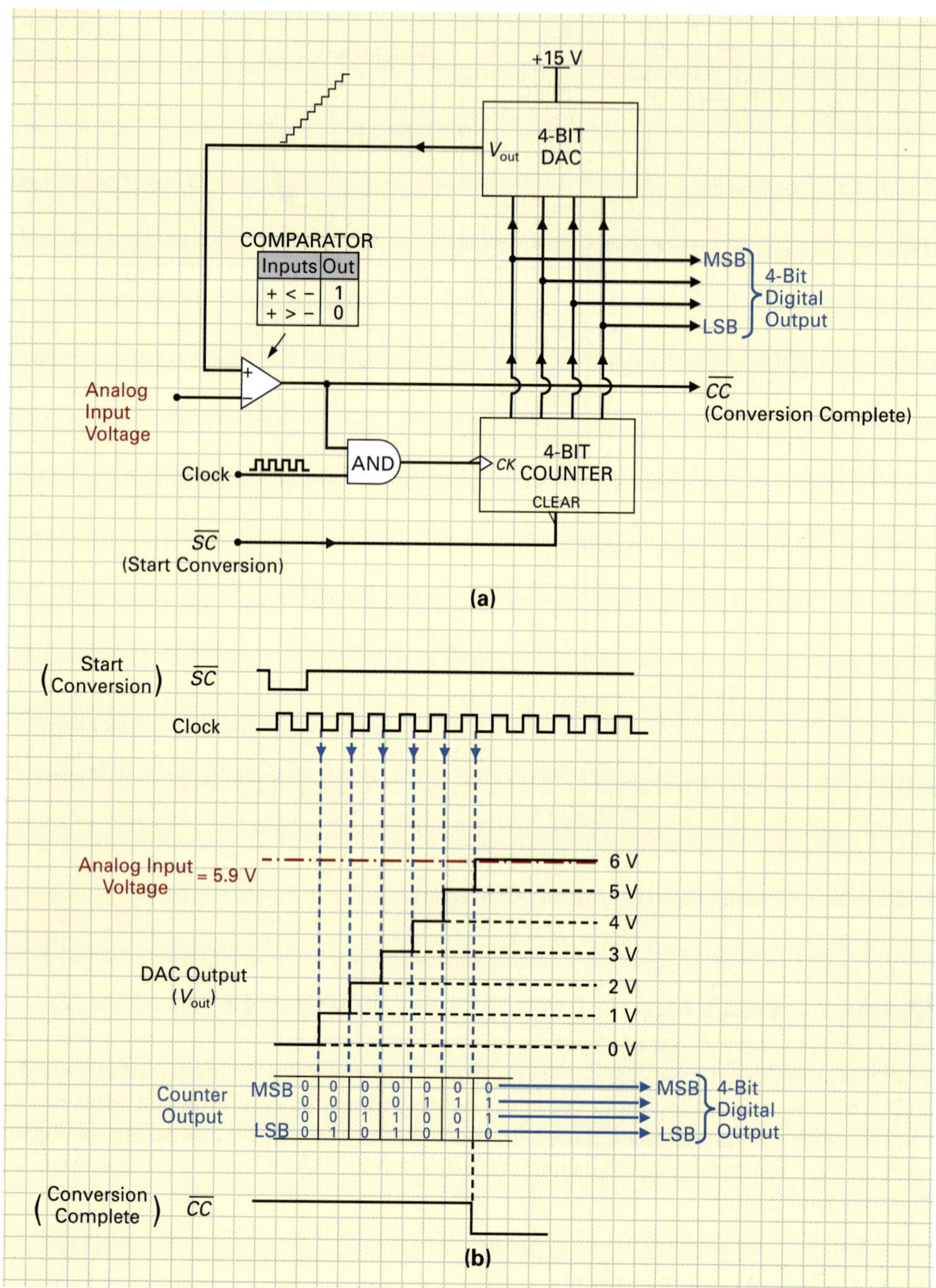

FIGURE 28-8 **A Basic Staircase Analog-to-Digital Converter (ADC).**

applied to the comparator's positive input. Since the comparator's positive input of 0 V is less than the negative analog input of 5.9 V at this time, the comparator will give a HIGH output.

The HIGH from the comparator will enable the AND gate, allowing the clock signal to pass through the AND gate and increment the counter to 0001 (decimal 1). The 0001 count will cause the DAC to generate a 1 V output, which is still less than the analog input voltage of 5.9 V, so the comparator will continue to enable the AND gate and allow the counter to be incremented further to 0010 (decimal 2). The 0010 count will cause the DAC to generate a 2 V output, which is still less than the analog input voltage of 5.9 V, so the comparator will continue to enable the AND gate and allow the counter to be incremented further to 0011 (decimal 3). This process will continue until the counter has been incremented to 0110 (decimal 6), and the DAC is therefore generating an output of 6 V. At this time, the comparator's positive input is greater than the analog input voltage of 5.9 V, so the comparator's output will go LOW.

The LOW output will disable the AND gate, preventing the clock signal from passing through to the counter, so the counter will hold its count of 0110. In addition to controlling the AND gate, the output of the comparator is also used to generate an active-LOW conversion complete ($\overline{CC}$) output. This $\overline{CC}$ output is used to indicate that the present 4-bit code appearing at the digital output is the binary equivalent of the analog input voltage.

EXAMPLE:

What digital output will be generated by the circuit in Figure 28-8(a) if an analog input voltage of 11.8 V were applied?

Solution:

If an analog input of 11.8 V were applied to the ADC circuit, the counter would be incremented until it reached a count of 1100 (decimal 12).

28-3-2 *Successive Approximation ADC*

One of the key disadvantages with the staircase ADC is that its conversion speed is dependent on the analog input voltage. For example, with the circuit in Figure 28-8, it would take fifteen clock pulses to convert an analog input voltage of 15 V to a digital output of 1111. For an 8-bit ADC, in the worst case, it would take 255 clock pulses to convert a maximum analog input voltage to 1111 1111. The worst-case conversion time, therefore, is equal to

$$t_{\text{conv(max)}} = (2^n - 1) \times t_{ck}$$

$t_{\text{conv(max)}}$ = maximum conversion time

n = number of bits

t_{ck} = clock pulse period

EXAMPLE:

Calculate the maximum conversion time of a 12-bit staircase ADC converter with a 20 kHz clock pulse applied.

Solution:

First, we will need to determine the clock pulse input period, and then we can use the maximum conversion time formula.

$$t_{ck} = \frac{1}{f} = \frac{1}{20\text{ kHz}} = 50\ \mu\text{s}$$

$$\begin{aligned} t_{\text{conv (max)}} &= (2^n - 1) \times t_{ck} \\ &= (2^{12} - 1) \times 50\ \mu\text{s} \\ &= (4{,}095) \times 50\ \mu\text{s} \\ &= 0.2\text{ s} \end{aligned}$$

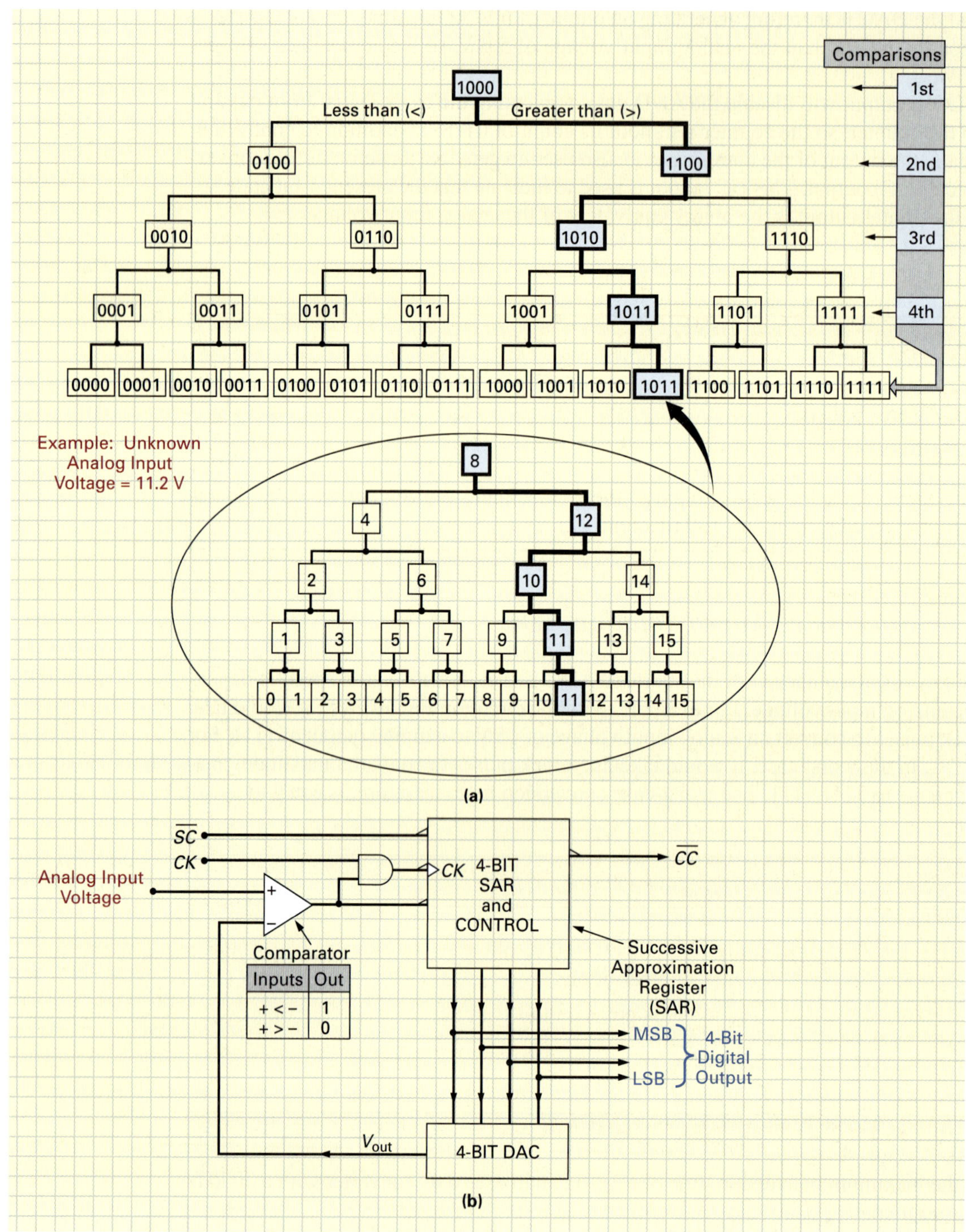

FIGURE 28-9 A Successive Approximation Analog-to-Digital Converter (ADC).

Successive Approximation ADC

An ADC circuit tests and sets each of the digital output word's bits in sequence to find the digital output code that is equivalent to the analog input.

The slow conversion time of the staircase ADC can be overcome by using the **successive approximation ADC,** shown in Figure 28-9(b). Before we discuss the operation of this circuit, let us first examine the successive approximation (SA) method. As an example, Figure 28-9(a) shows how the SA method can be used to determine the digital equivalent of the analog input by making only four comparisons. In the example shown highlighted in Figure 28-9(a), an analog input voltage of 11.2 V has been applied. Following the comparisons shown in binary (and decimal in the inset), you can see that the SA method first sets the MSB HIGH (1000, or decimal 8) and then tests to see whether this value is greater than or less than the analog input voltage. Since the analog input voltage of 11.2 V is greater than the first SA value of 1000, the MSB is kept HIGH, the next MSB is set HIGH (1100, or decimal 12), and another comparison is made. Since the analog input voltage of 11.2 V is less than the second SA value of 1100, the second MSB is reset LOW, the next MSB is set HIGH (1010, or decimal 10), and another comparison is made. Since the analog input voltage of 11.2 V is greater than the third SA value of 1010, the third MSB is kept HIGH, the LSB is set HIGH (1011, or decimal 11), and a final comparison is made. Since the analog input voltage of 11.2 V is greater than the fourth SA value of 1011, the LSB is kept HIGH, and an equivalent digital code has been found.

Referring to the successive approximation ADC block diagram in Figure 28-9(b), you can see that this circuit also contains a DAC and a comparator. This circuit, however, does not contain a counter since the *successive approximation register (SAR)* and control circuits will generate the special sequence of binary codes needed to converge on the digital equivalent of the analog input voltage. The start conversion ($\overline{SC}$) input is used to reset the SAR, and initiate the series of four comparisons. The SA values at the output of the SAR register are applied to the DAC, and as with the staircase ADC, the DAC output controls a comparator, which in turn controls an AND gate, which in turn controls the clock input. After each of the bits has been set and tested, the conversion complete output ($\overline{CC}$) is used to indicate that the present 4-bit code appearing at the digital output is the binary equivalent of the analog input voltage.

The key advantage of the successive approximation ADC is its conversion speed. Unlike the staircase ADC, the SA ADC does not need to sequence through every possible binary combination to find a match. An SA ADC will only need to set and test each of the digital output bits, so a 4-bit ADC will only need to make four comparisons, an 8-bit ADC will only need to make eight comparisons, and so on. This means that an SA ADC's conversion time will be the same for any input value. Having a fixed conversion time for any input value will make system timing a lot easier.

28-3-3 *Flash ADC*

One of the simplest ways to convert an analog input voltage into an equivalent digital output is to use a **flash ADC.** To explain the operation of this ADC type, Figure 28-10(a) shows a 3-bit flash converter circuit. Flash ADCs use comparators to compare the analog input voltage to successively smaller equally spaced reference voltages provided by a resistive voltage-divider network. A comparator will generate a HIGH output whenever the analog input voltage is greater than the applied reference voltage. The outputs from the comparators are applied to a 3-bit priority encoder. The 3-bit binary output code generated by the priority encoder will be based on which of the highest-order inputs is HIGH, as shown in Figure 28-10(b). For example, when an analog input of 4.1 V is applied, the priority encoder inputs 4, 3, 2, and 1 will be HIGH, so the binary output code generated will be 100 (decimal 4) since 4 is the highest-order input.

Flash ADC
An ADC circuit that uses a resistive voltage divider and comparators for a fast conversion time.

The advantage of this analog-to-digital conversion method is its fast conversion time, which is typically in the nanoseconds (hence the name "flash"). The disadvantage of this ADC type is the large number of comparators that are needed. In most cases, $2^n - 1$ comparators are needed, so an 8-bit flash ADC will need 255 comparators, a 16-bit flash ADC will need 65,535 comparators, and so on.

28-3-4 *An ADC Application Circuit*

Figure 28-11 shows an ADC0803 8-bit successive approximation ADC that has an on-chip clock circuit connected to function as an 8-bit analog-to-digital converter. In this example circuit, the 8-bit digital output from the ADC0803 (pins 11 through 18) have been applied to an LED display. In most digital system applications, the parallel digital output will be under three-state control, based on the active-LOW $\overline{RD}$ control input. The active-LOW $\overline{WR}$ control input

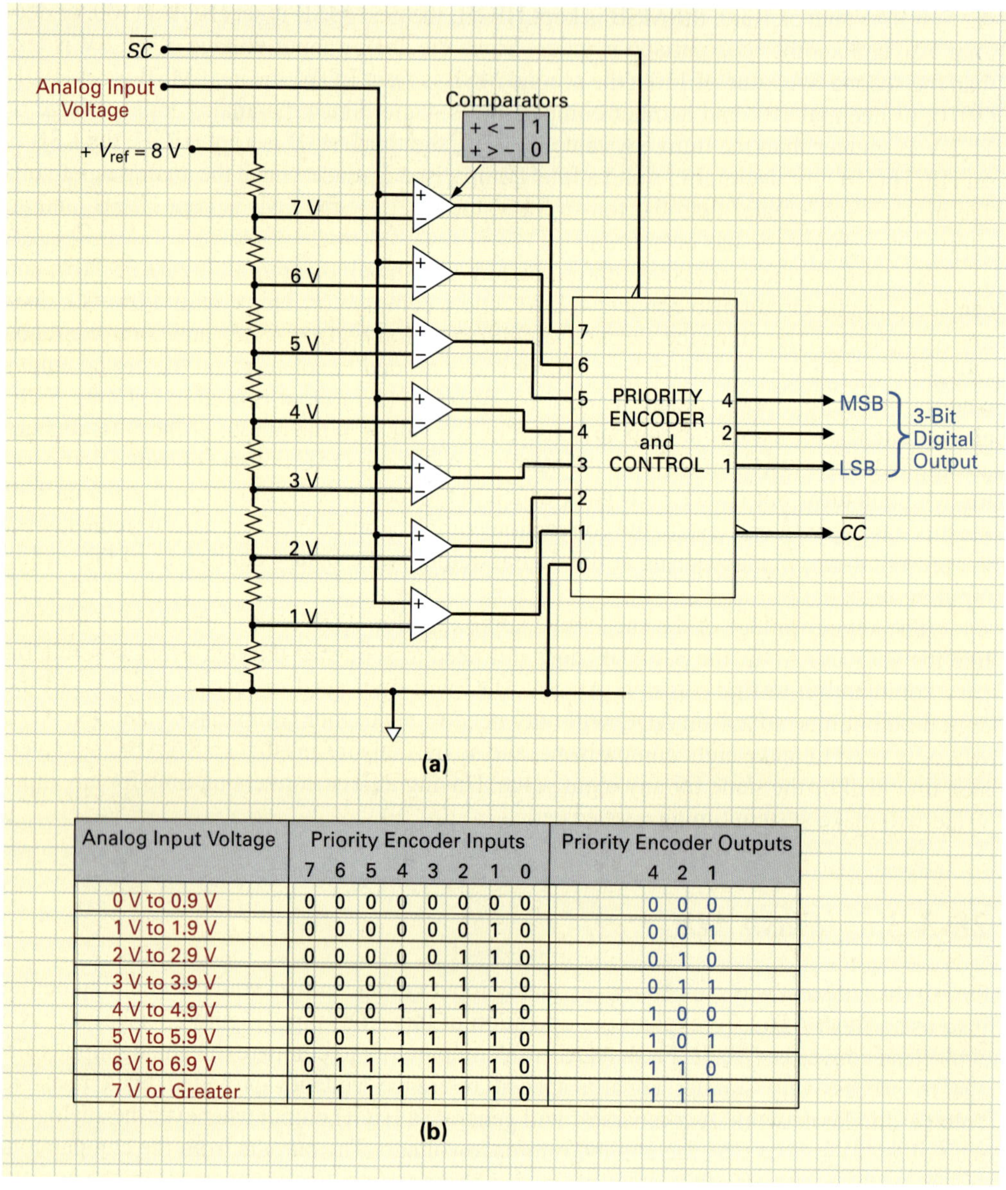

Analog Input Voltage	Priority Encoder Inputs 7 6 5 4 3 2 1 0	Priority Encoder Outputs 4 2 1
0 V to 0.9 V	0 0 0 0 0 0 0 0	0 0 0
1 V to 1.9 V	0 0 0 0 0 0 1 0	0 0 1
2 V to 2.9 V	0 0 0 0 0 1 1 0	0 1 0
3 V to 3.9 V	0 0 0 0 1 1 1 0	0 1 1
4 V to 4.9 V	0 0 0 1 1 1 1 0	1 0 0
5 V to 5.9 V	0 0 1 1 1 1 1 0	1 0 1
6 V to 6.9 V	0 1 1 1 1 1 1 0	1 1 0
7 V or Greater	1 1 1 1 1 1 1 0	1 1 1

(b)

FIGURE 28-10 A Flash Analog-to-Digital Converter.

is used to start a conversion, and the active-LOW $\overline{INTR}$ output is used to indicate when a conversion is complete. The $\overline{INTR}$ output is connected directly to the $\overline{WR}$ input to ensure that the conversion is continuous. The resistor, capacitor, and buffer circuit are included to ensure that the $\overline{WR}$ input is taken LOW when power is first applied to the circuit.

The *RC* circuit connected to the *CLK-R* and *CLK-IN* pins will determine the converter's internal clock circuit frequency based on the following formula:

$$f = \frac{1}{1.1\,RC}$$

In the example circuit in Figure 28–11, the converter clock frequency will be

$$f = \frac{1}{1.1\,RC} = \frac{1}{1.1 \times 10\text{ k}\Omega \times 150\text{ pF}} = 606\text{ kHz}$$

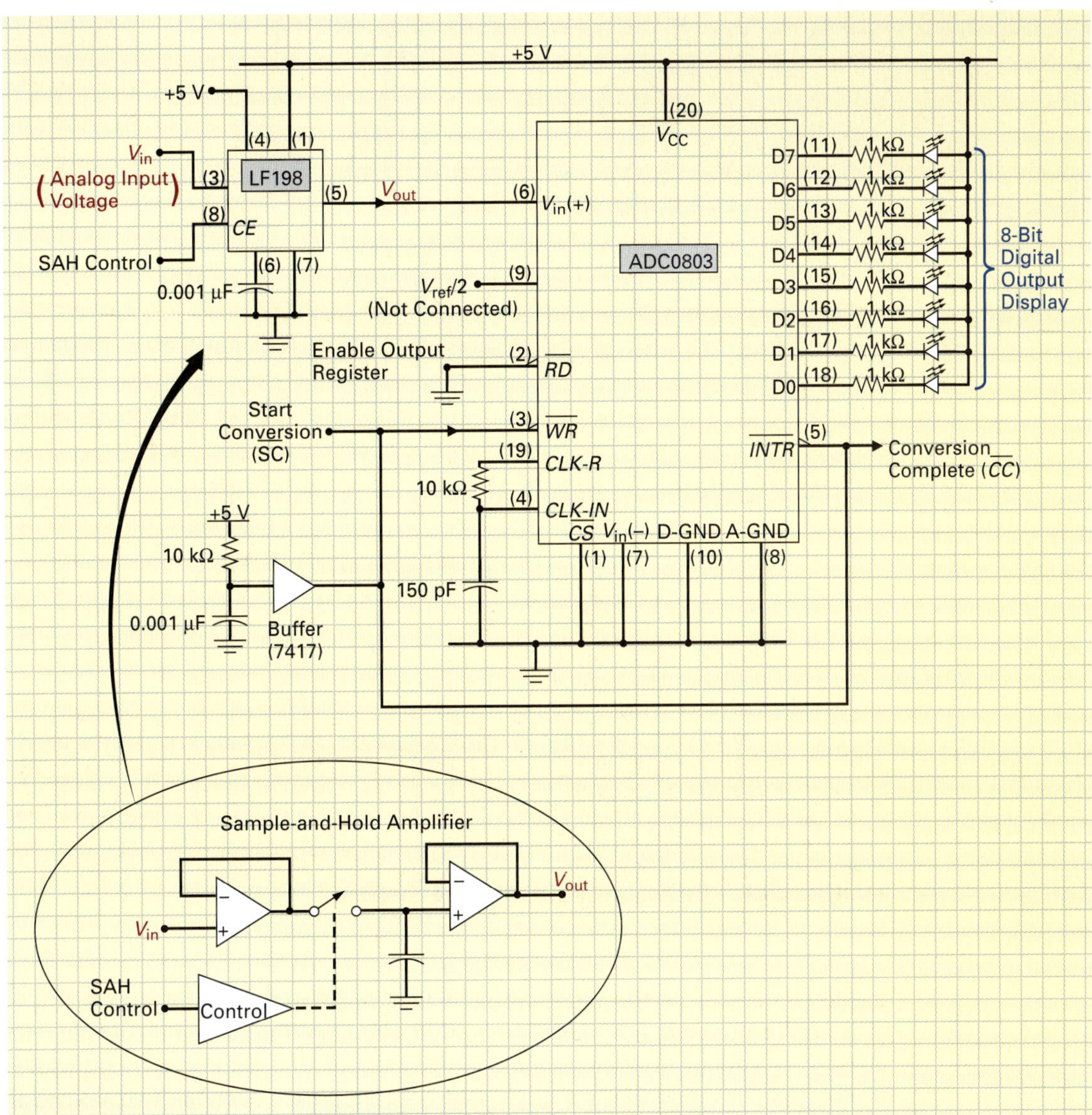

FIGURE 28-11 **Application—An 8-Bit Analog-to-Digital Converter Circuit.**

The ADC0803 has both a positive analog input (pin 6) and a negative analog input (pin 7). This versatility means that you can ground pin 7 and have a positive input, ground pin 6 and have a negative input, or have the ADC generate an output code based on the difference in voltage between pins 6 and 7 (differential input).

The +5 V supply voltage is applied to pin 20, and both the analog and digital ground pins must be grounded. The $V_{ref}/2$ input to the ADC0803 is used to determine which analog input voltage will generate the maximum digital output of 1111 1111 (decimal 255). If this input is not connected, the supply voltage of +5 V will mean that the analog input range will be from 0 V to +5 V. When a voltage is applied to the $V_{ref}/2$ input, a different analog input range can be obtained. For example, to generate a maximum digital output code of 1111 1111 when an analog input of +4 V is applied, connect +2 V to the $V_{ref}/2$ input (+4 V/2 = 2 V).

The application circuit in Figure 28–11 includes an LF198 **sample-and-hold amplifier** at the analog input. This circuit has been included to prevent any error due to analog input voltage changes during conversion. Up until this time, we have assumed that the analog input voltage remains constant while the converter is making the conversion. This does not always happen, and if a change in the analog input voltage were to occur while

Sample-and-Hold Amplifier
A circuit that will sample the value of a changing analog input signal when triggered and hold this value constant at its output.

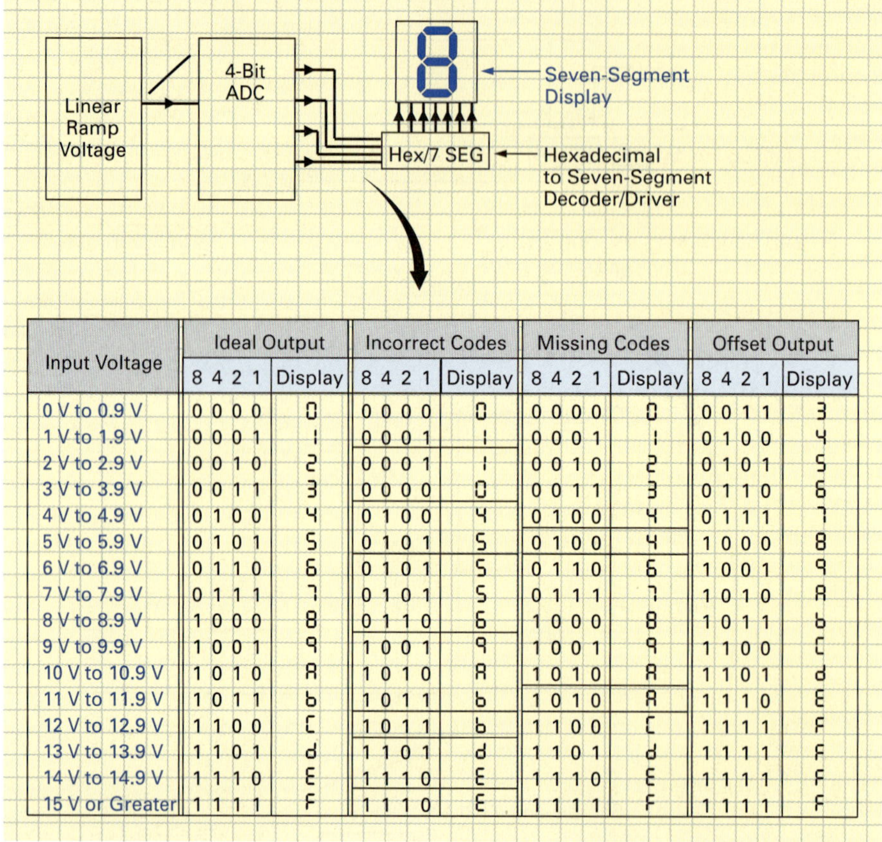

Input Voltage	Ideal Output		Incorrect Codes		Missing Codes		Offset Output	
	8 4 2 1	Display	8 4 2 1	Display	8 4 2 1	Display	8 4 2 1	Display
0 V to 0.9 V	0 0 0 0	0	0 0 0 0	0	0 0 0 0	0	0 0 1 1	3
1 V to 1.9 V	0 0 0 1	1	0 0 0 1	1	0 0 0 1	1	0 1 0 0	4
2 V to 2.9 V	0 0 1 0	2	0 0 0 1	1	0 0 1 0	2	0 1 0 1	5
3 V to 3.9 V	0 0 1 1	3	0 0 0 0	0	0 0 1 1	3	0 1 1 0	6
4 V to 4.9 V	0 1 0 0	4	0 1 0 0	4	0 1 0 0	4	0 1 1 1	7
5 V to 5.9 V	0 1 0 1	5	0 1 0 1	5	0 1 0 0	4	1 0 0 0	8
6 V to 6.9 V	0 1 1 0	6	0 1 0 1	5	0 1 1 0	6	1 0 0 1	9
7 V to 7.9 V	0 1 1 1	7	0 1 0 1	5	0 1 1 1	7	1 0 1 0	A
8 V to 8.9 V	1 0 0 0	8	0 1 1 0	6	1 0 0 0	8	1 0 1 1	b
9 V to 9.9 V	1 0 0 1	9	1 0 0 1	9	1 0 0 1	9	1 1 0 0	C
10 V to 10.9 V	1 0 1 0	A	1 0 1 0	A	1 0 1 0	A	1 1 0 1	d
11 V to 11.9 V	1 0 1 1	b	1 0 1 1	b	1 0 1 0	A	1 1 1 0	E
12 V to 12.9 V	1 1 0 0	C	1 0 1 1	b	1 1 0 0	C	1 1 1 1	F
13 V to 13.9 V	1 1 0 1	d	1 1 0 1	d	1 1 0 1	d	1 1 1 1	F
14 V to 14.9 V	1 1 1 0	E	1 1 1 0	E	1 1 1 0	E	1 1 1 1	F
15 V or Greater	1 1 1 1	F	1 1 1 0	E	1 1 1 1	F	1 1 1 1	F

FIGURE 28-12 Testing Analog-to-Digital Converters.

the converter was performing the conversion, the digital output would be inaccurate. To prevent this problem, a sample-and-hold (SAH) amplifier can be included at the input of an ADC to sample the analog input voltage and hold this sample constant while the ADC makes the conversion. The sample-and-hold amplifier contains two unity-gain noninverting amplifiers ($V_{out} = V_{in}$), a logic-controlled switch, and an external capacitor, as shown in the inset in Figure 28-11. When the SAH control input is made HIGH, the logic-controlled switch is closed, and the external capacitor charges quickly to equal the analog input voltage. When the logic-controlled switch is made LOW, the capacitor will retain the analog input voltage level, and this unchanging voltage will be applied to the ADC. Using a capacitor of 0.001 μF, the LF198 must have its switch closed for at least 4 μs in order to get an accurate sample (acquisition time), and when in the hold condition the output voltage will decrease due to leakage paths at a rate of 30 mV/s (droop rate).

28-3-5 *Testing ADCs*

One of the easiest ways to test an ADC is to apply a linear ramp to the analog input and have a display monitor the digital output, as shown in Figure 28-12. If operating correctly, the output should be an incrementing digital code, as listed in the table in Figure 28-12 in the ideal output column. If a fault exists, you may obtain incorrect output codes, missing codes, or an offset output, similar to the other columns listed in the table in Figure 28-12. In this situation, you will need to isolate whether the problem is internal or external to the ADC IC and then replace the faulty device.

SELF-TEST EVALUATION POINT FOR SECTION 28-3

Use the following questions to test your understanding of Section 28-3.

1. What is the main disadvantage of the staircase ADC?
2. What two types of ADCs make use of a DAC?
3. How does a successive approximation ADC operate?
4. What is the fastest type of ADC?
5. How can an ADC be tested?

REVIEW QUESTIONS

Multiple-Choice Questions

1. A ____________ converts an analog input voltage into a proportional digital output code.
   a. ROM **c.** ADC
   b. DAC **d.** RWM
2. A ____________ converts a digital input code into a proportional analog output voltage.
   a. ROM **c.** ADC
   b. DAC **d.** RWM
3. Which of the following signal converters is the most frequently used DAC?
   a. Binary-weighted **d.** Successive approximation
   b. Staircase **e.** Flash
   c. *R*/2*R* ladder
4. Which of the following signal converters is the fastest ADC?
   a. Binary-weighted **d.** Successive approximation
   b. Staircase **e.** Flash
   c. *R*/2*R* ladder
5. Most DACs give a/an ____________ output.
   a. parallel binary **c.** serial binary
   b. analog current **d.** analog voltage
6. Most ADCs give a/an ____________ output.
   a. parallel binary **c.** serial binary
   b. analog current **d.** analog voltage
7. The op-amp in a DAC circuit is used to convert a/an ____________ input into a ____________ output.
   a. parallel binary, current **c.** current, voltage
   b. serial binary, voltage **d.** analog voltage, current
8. Which of the following data converter types can be used to generate a staircase output waveform?
   a. Staircase **c.** *R*/2*R*
   b. Flash **d.** Successive approximation
9. The ____________ of a DAC describes the smallest analog output change.
   a. monotonicity **c.** accuracy
   b. settling time **d.** resolution
10. What circuit can be used to hold the input to an ADC constant during the conversion cycle?
    a. *R*/2*R* circuit **c.** *V*-to-*I* circuit
    b. Flash circuit **d.** Sample-and-hold circuit

Practice Problems

11. If a 4-bit DAC has a 12 V reference applied, calculate each of the following:
    a. step size
    b. percent resolution
    c. the weight of each binary column
    d. the output voltage when 0010 is applied
    e. the output voltage when 1011 is applied
12. If an 8-bit DAC has a 5 V reference applied, calculate each of the following:
    a. step size
    b. percent resolution
    c. the weight of each binary column
    d. the output voltage when 0000 0101 is applied
    e. the output voltage when 1011 0111 is applied
13. If an MC1408 has a 5 mA reference current applied, what will be the output current for the following digital inputs?
    a. 1010 0001 **c.** 1111 1111
    b. 0001 0101 **d.** 0001 0001
14. Calculate the maximum conversion time of a 16-bit staircase ADC converter with a 200 kHz clock.

To practice your circuit recognition and operation skills, refer to Figure 28-13 and answer the following five questions.

15. Would the circuit indicated by the ? in this circuit be a DAC or an ADC?
16. Sketch the output you would expect from the ? circuit.
17. Sketch the output you would expect from the lowpass filter circuit.
18. Why was the lowpass filter included in this circuit?
19. If the converter in this circuit had a 3 V reference applied, calculate each of the following:
    a. step size
    b. percent resolution
    c. the weight of each binary column
    d. the output voltage when 0000 0101 is applied
    e. the output voltage when 1011 0111 is applied

To practice your circuit recognition and operation skills, refer to Figure 28-14 and answer the following four questions.

20. What is the purpose of this circuit?

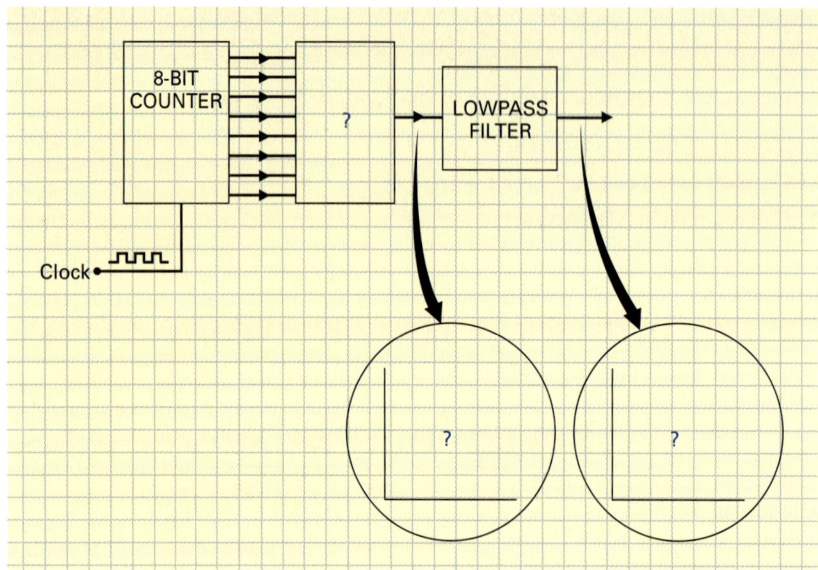

FIGURE 28-13 Application—A Ramp Generator Circuit.

21. Would the circuit indicated by the ? in this circuit be a DAC or an ADC?

22. What is the function of the PLD in this circuit?

23. Briefly describe the operation of this circuit.

To practice your circuit recognition and operation skills, refer to Figure 28-15 and answer the following four questions.

24. Is this circuit a DAC or an ADC?

25. What circuit name is given to this type of converter?

26. Indicate what logic level will appear at test points 1, 2, and 3 for each of the four input voltage ranges listed in the table in this figure.

27. Complete the table shown in this figure to show what logic levels will be present at the MSB and LSB outputs for the four input voltage ranges listed in the table.

To practice your circuit recognition and operation skills, refer to Figure 28-16 and answer the following three questions.

28. Which of the blocks in this system will perform the following conversions?
   a. Parallel digital input to a serial digital output
   b. Serial digital input to a parallel digital output
   c. Digital input to an analog output
   d. Analog input to a digital output

29. Sketch the output for the input shown in Figure 28-16(b).

30. Sketch the output for the input shown in Figure 28-16(c).

Web Site Questions

Go to the web site http://www.prenhall.com/cook, select the textbook *Electronics: A Complete Course,* select this chapter, and then follow the instructions when answering the multiple choice practice problems.

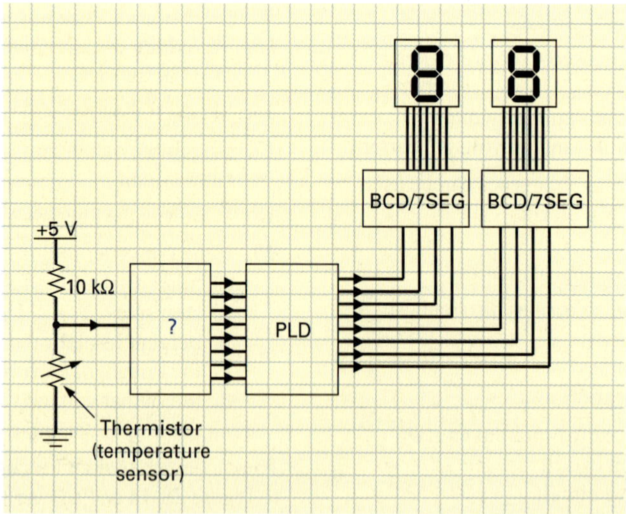

FIGURE 28-14 Application—A Digital Thermometer.

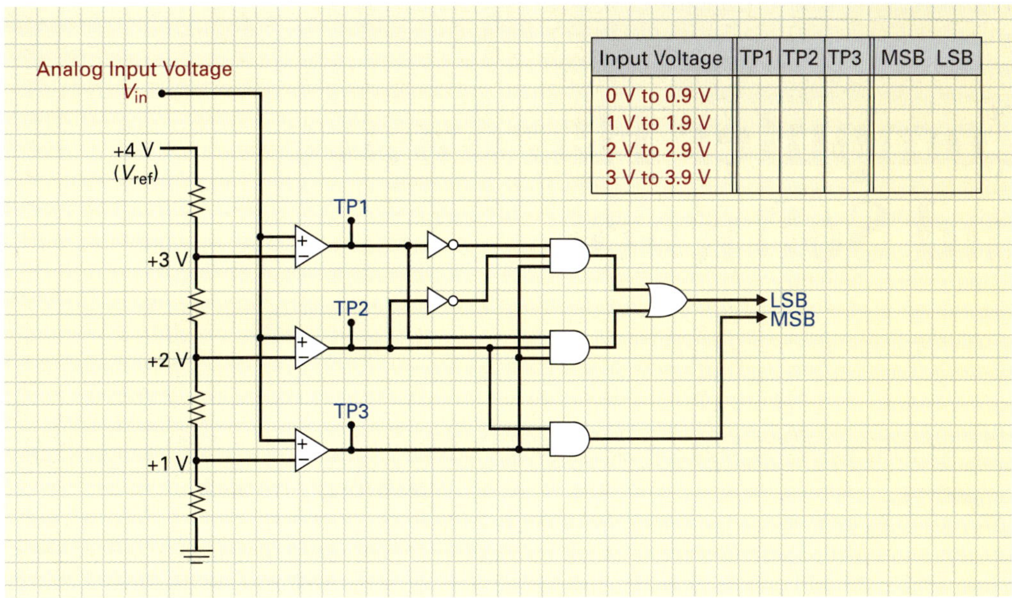

Input Voltage	TP1	TP2	TP3	MSB LSB
0 V to 0.9 V				
1 V to 1.9 V				
2 V to 2.9 V				
3 V to 3.9 V				

FIGURE 28-15 Application—A Converter.

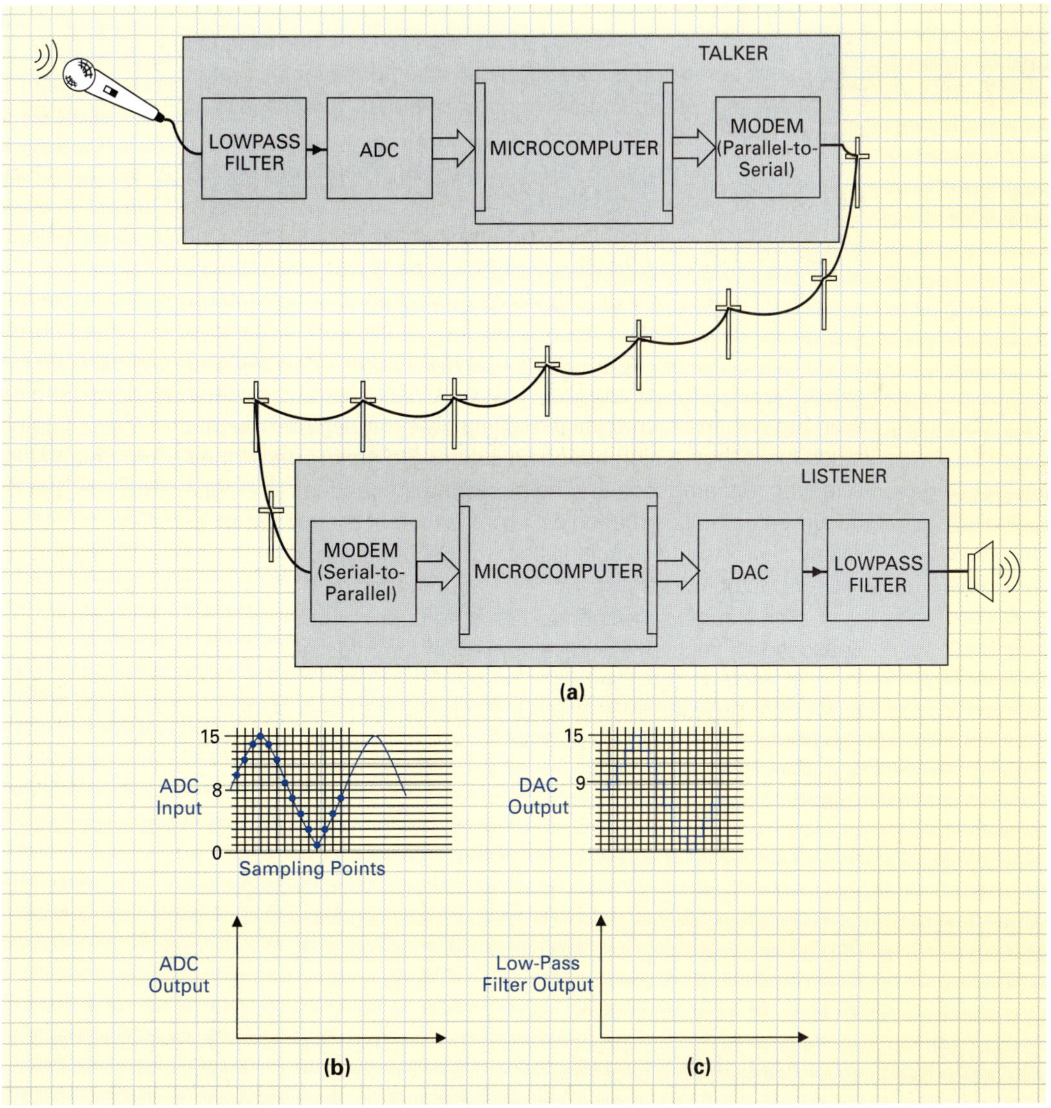

TALKER
LOWPASS FILTER
ADC
MICROCOMPUTER
MODEM (Parallel-to-Serial)

LISTENER
MODEM (Serial-to-Parallel)
MICROCOMPUTER
DAC
LOWPASS FILTER

(a)

15
8
ADC Input
0
Sampling Points

15
9
DAC Output

ADC Output

Low-Pass Filter Output

(b)

(c)

FIGURE 28-16 Application—Digital Voice Communication.

Computer Hardware and Software

Making an Impact

John Von Neuman, a mathematics professor at the Institute of Advanced Studies, delighted in amazing his students by performing complex computations in his head faster than they could with pencil, paper, and reference books. He possessed a photographic memory, and at his frequently held lavish parties in his home in Princeton, New Jersey, he gladly occupied center stage to recall from memory entire pages of books read years previously, the lineage of European royal families, and a store of controversial limericks. His memory, however, failed him in his search for basic items in a house he had lived in for seventeen years. On many occasions when traveling, he would become so completely absorbed in mathematics that he would have to call his office to find out where he was going and why.

Born in Hungary, he was quick to demonstrate his genius. At the age of six he would joke with his father in classical Greek. At the age of eight, he had mastered calculus, and in his mid-twenties he was teaching and making distinct contributions to the science of quantum mechanics, which is the cornerstone of nuclear physics.

Next to fine clothes, expensive restaurants, and automobiles, which he had to replace annually due to smash-ups, was his love of his work. His interest in computers began when he became involved in the top-secret Manhattan Project at Los Alamos, New Mexico, where he proved mathematically the implosive method of detonating an atom bomb. Working with the then-available computers, he became aware that they could become much more than a high-speed calculator. He believed that they could be an all-purpose scientific research tool, and he published these ideas in a paper. This was the first document to outline the logical organization of the electronic digital computer and was widely circulated to all scientists throughout the world. In fact, even to this day, scientists still refer to computers as "Von Neuman machines."

Von Neuman collaborated on a number of computers of advanced design for military applications, such as the development of the hydrogen bomb and ballistic missiles.

In 1957, at the age of 54, he lay in a hospital dying of bone cancer. Under the stress of excruciating pain, his brilliant mind began to break down. Since Von Neuman had been privy to so much highly classified information, the Pentagon had him surrounded with only medical orderlies specially cleared for security for fear he might, in pain or sleep, give out military secrets.

29

Introduction

The microprocessor is a large, complex integrated circuit containing all the computation and control circuitry for a small computer. The introduction of the microprocessor has caused a dramatic change in the design of digital electronic systems. Before microprocessors, random or hard-wired digital circuits were designed using individual logic blocks, such as gates, flip-flops, registers, counters, and so on. These building blocks were interconnected to achieve the desired end, as required by the application. Using random logic, each application required a unique design and there was little similarity between one system and another. This approach was very similar to analog circuit design, in that the structure of the circuit was governed by the function that needed to be performed. Once constructed, the function of the circuit was difficult to change.

The microprocessor, on the other hand, provides a general-purpose control system which can be adapted to a wide variety of applications with only slight circuit modification. The individuality of a microprocessor-based system is provided by a list of instructions (called the program) that controls the system's operation. A microprocessor-based system therefore has two main elements—the actual components, or hardware, and the programs, or software.

In the first section of this chapter, you will be introduced to the hardware that makes up a microprocessor system and the software that controls it. In the second section of this chapter, you will see how all of the circuits discussed in previous chapters come together to form a complete microprocessor-based system. The system is called "SAM," which is an acronym for "simple all-purpose microcomputer." We will be examining SAM's schematic diagram down to the component level to remove any mystery associated with microcomputers and to give you the confidence that comes from understanding a complete digital electronic microprocessor-based system. In addition, we will discuss general microprocessor system troubleshooting techniques and the detailed troubleshooting procedures for the SAM system.

FIGURE 29-1 **A Modern Microcomputer.**

29-1 MICROCOMPUTER BASICS

The earliest electronic computers were constructed using thousands of vacuum tubes. These machines were extremely large and unreliable and were mostly a laboratory curiosity. The *electronic numeric integrator and computer,* or ENIAC, was unveiled in 1946. The ENIAC weighed 38 tons, measured 18 feet wide and 88 feet long, and used 17,486 vacuum tubes. These vacuum tubes produced a great deal of heat and developed frequent faults requiring constant maintenance. The personal computer shown in Figure 29-1, on the other hand, makes use of semiconductor integrated circuits and is far more powerful, versatile, portable, and reliable than the ENIAC. Another advantage is cost: In 1946 the ENIAC calculator cost $400,000 to produce, whereas a present-day high-end personal computer can be purchased for about $1,200.

29-1-1 *Hardware*

Figure 29-2 shows the block diagram of a basic microcomputer system. The microprocessor (also called central processor unit, or CPU) is the "brains" of the system. It contains all of the logic circuitry needed to recognize and execute the program of instructions stored in

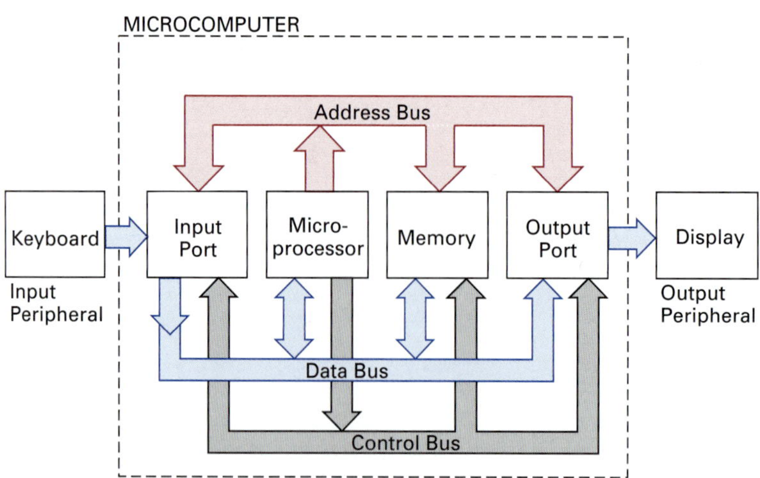

FIGURE 29-2 **Basic Microcomputer System.**

memory. The input port connects the processor to the keyboard so that data can be read from this input device. The output port connects the processor to the display so that we can write data to this output device. Combined in this way, the microprocessor unit, memory unit, and input/output unit form a **microcomputer.**

Microcomputer
Complete system, including CPU, memory, and I/O interfaces.

The blocks within the microcomputer are interconnected by three buses. The microprocessor uses the address bus to select which memory location, input port, or output port it wishes to put information into or take information out of. Once the microprocessor has selected the location using the address bus, data or information is transferred via the data bus. In most cases, data will travel either from the processor to memory, from memory to the processor, from the input port to the processor, or from the processor to the output port. The control bus is a group of control signal lines that are used by the processor to coordinate the transfer of data within the microcomputer.

A list of instructions is needed to direct a microcomputer system so that it will perform a desired task. For example, if we wanted the system in Figure 29-2 to display the number of any key pressed on the keyboard, the program would be as follows:

1. Read the data from the keyboard.
2. Write the data to the display.
3. Repeat (go to step 1).

For a microprocessor to perform a task from a list of instructions, the instructions must first be converted into codes that the microprocessor can understand. These instruction codes are stored, or programmed, into the system's memory. When the program is run, the microprocessor begins by reading the first coded instruction from memory, decoding its meaning, and then performing the indicated operation. The processor then reads the instruction from the next location in memory, decodes the meaning of this next instruction, and then performs the indicated operation. This process is then repeated, one memory location after another, until all of the instructions within the program have been fetched, decoded and then executed.

The input/output devices connected to the input and output ports (the keyboard and display, for example) are called the **peripherals.** Peripherals are the system's interface with the user, or the system's interface with other equipment such as printers or data storage devices. As an example, Figure 29-3 shows how a microcomputer could be made to operate as a microprocessor-based digital voltmeter. Its input peripherals are an analog-to-digital converter (ADC), and a range of function switches. The output peripheral is a digital display.

Peripheral
Any interface device connected to a computer. Also, a mass storage or communications device connected to a computer.

As you come across different microcomputer applications, you will see that the basic microcomputer system is always the same. The only differences between one system and another are in the program and peripherals. Whether the system is a personal computer or a digital voltmeter, the only differences are in the program of instructions that controls the microcomputer and the peripherals connected to the microcomputer.

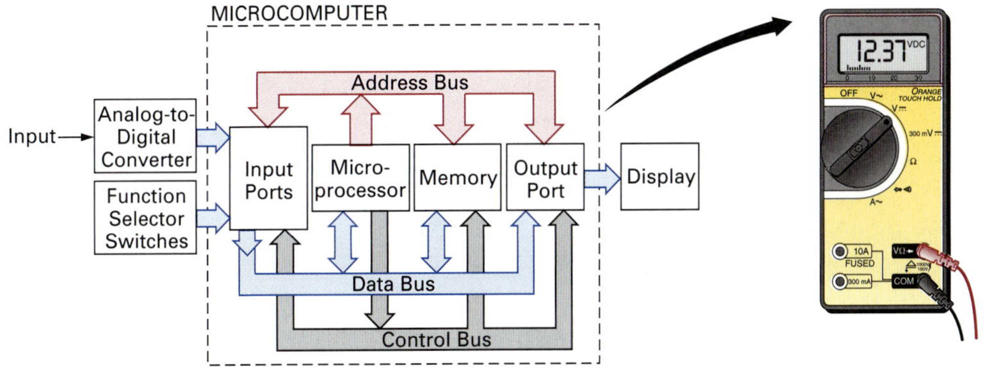

FIGURE 29-3 A Microprocessor-Based Digital Voltmeter.

29-1-2 *Software*

Microcomputer programs are first written in a way that is convenient for the person writing the program, or *programmer.* Once written, the program must then be converted and stored as a code that can be understood by the microprocessor. When writing a computer program, the programmer must tell the computer what to do down to the most minute detail. Computers can act with tremendous precision and speed for long periods of time, but they must be told exactly what to do. A computer can respond to a change in conditions, but only if it contains a program that says "if this condition occurs, do this."

The Microcomputer as a Logic Device

To understand how a program controls a microcomputer, let us see how we could make a microcomputer function as a simple two-input AND gate. As can be seen in Figure 29-4, the two AND gate inputs will be applied to the input port, and the single AND gate output will appear at the output port. To make the microcomputer operate as an AND gate, a program of instructions will first have to be loaded into the system's memory. This list of instructions will be as follows:

1. Read the input port.
2. Go to step 5 if all inputs are HIGH; otherwise, continue.
3. Set output LOW.
4. Go to step 1.
5. Set output HIGH.
6. Go to step 1.

Studying the steps in this program, you can see that first the input port is read. Then the inputs are examined to see if they are all HIGH, since that is the function of an AND gate. If the inputs are all HIGH, the output is set HIGH. If the inputs are not all HIGH, the output is set LOW. Once the procedure has been completed, the program jumps back to step 1 and repeats indefinitely, so that the output will continuously follow any changes at the input.

You may be wondering why we would use this complex system to perform a simple AND gate function. If this were the only function that we wanted the microcomputer to perform, then it would definitely be easier to use a simple AND logic gate; however, the microcomputer provides tremendous flexibility. It allows the function of the gate to be arbitrarily redefined just by changing the program. You could easily add more inputs, and the gate function could be extremely complex. For example, we could have the microcomputer function as an eight-input electronic lock. The output of this lock will only go HIGH if the inputs are turned ON in a specific order. Using traditional logic, this would require a complex circuit; however, this function could be easily implemented using the same microprocessor system

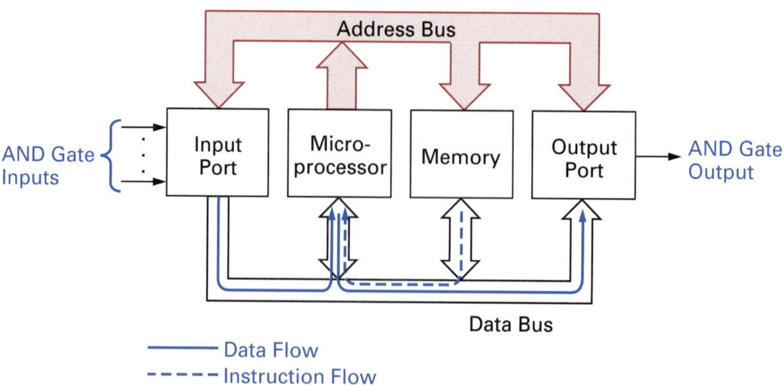

FIGURE 29-4 A Microprocessor-Based AND Gate.

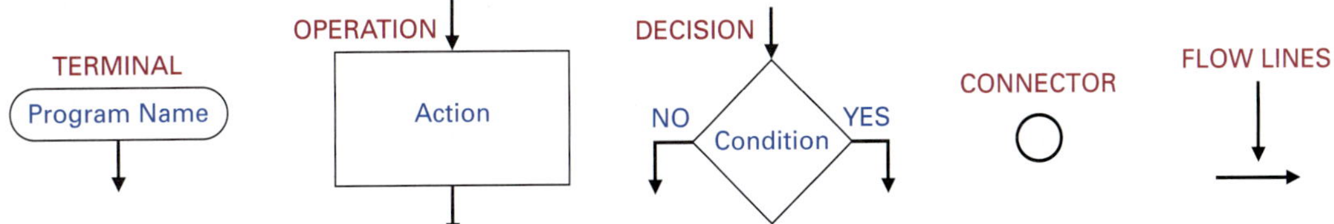

FIGURE 29-5 Flowchart Symbols.

used for the simple AND gate. Of course, a new program would be needed, more complex than the simple AND gate program, but the hardware will not change. Additionally, the combination of this electronic lock could easily be changed by simply modifying the program.

Flowcharts

To develop a program of instructions to solve a problem, a programmer would generally follow three steps:

1. Define the problem.
2. Flowchart the solution.
3. Write the program.

To understand these steps, let us begin by examining the elements of the flowchart in more detail.

Flowcharts provide a graphic way of describing the operation of a program. They are composed of different types of blocks interconnected with lines, as shown in Figure 29-5. A rectangular block describes each action that the program takes. A diamond-shaped block is used for each decision, such as testing the value of a variable. An oval block is used at the beginning of the flowchart with the name of the program placed inside it. This oval block can also be used to mark the end of the flowchart.

As an example, Figure 29-6 shows a flowchart for the AND gate program discussed previously. For each line of the program there is a block, except for the two "go to" instructions. These are represented as lines indicating how the program will flow from one block to

Flowchart or Flow Diagram
Graphical representation of program logic. Flowcharts enable the designer to visualize the procedure necessary for each item in the program. A complete flowchart leads directly to the final code.

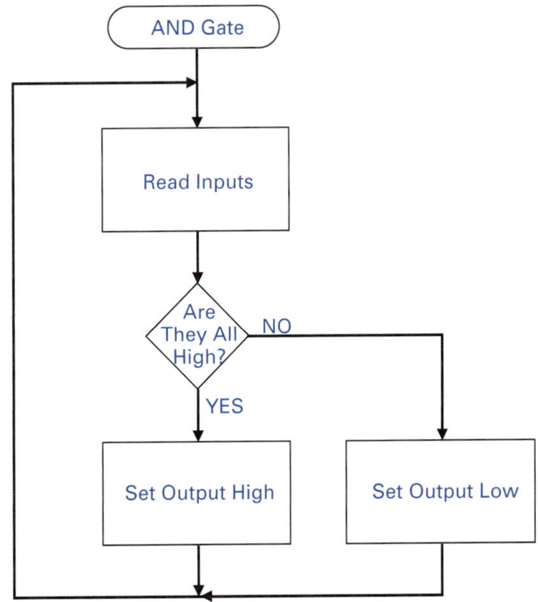

FIGURE 29-6 AND Gate Flowchart.

another. The flowchart contains the same information as the program list but in a graphic form. When you first set out to write a program, a flowchart is a good way to organize your thoughts and document what the program must do. By going through the flowchart manually, you can check the logic of the flowchart and then write the actual program from it. Flowcharts are also useful for understanding a program that has been written in the past.

Machine Language
Binary language (often represented in hex) that is directly understood by the processor. All other programming languages must be translated into binary code before they can be entered into the processor.

Mnemonic Code
To assist the human memory, the binary numbered codes are assigned groups of letters (or mnemonic symbols) that suggest the definition of the instruction.

Assembly Language
A program written as a series of statements using mnemonic symbols that suggest the definition of the instruction. It is then translated into machine language by an assembler program.

Instruction Set
Total group of instructions that can be executed by a given microprocessor.

High-Level Language
A language closer to the needs of the problem to be handled than to the language of the machine on which it is to be implemented.

Compiler
Translation program that converts high-level instructions into a set of binary instructions (machine code) for execution.

BASIC
An easy-to-learn and easy-to-use language available on most microcomputer systems.

Programming Languages

Writing programs in English is convenient, since it is the language most people understand. Unfortunately, English is meaningless to a microprocessor. The language understood by the microprocessor is called **machine language,** or machine code. Because microprocessors deal directly with digital signals, machine language instructions are binary codes, such as 00111100, and 11100111.

Machine language is not easy for people to use, since 00111100 has no obvious meaning. It can be made easier to work with by using the hexadecimal representation of 0011 1100, which is 3C; however, this still does not provide the user with any clue as to the meaning of this instruction. To counteract this problem, microprocessor manufacturers replace each instruction code with a short name called a **mnemonic,** or memory aid. As an example, Intel's 8085 microprocessor uses the mnemonic "INR A" for the code 3C, since this code instructs the 8085 microprocessor to "increment its internal A register." The mnemonics are much easier to remember than the machine codes. By assigning a mnemonic to each instruction code, you can write programs using the mnemonics instead of machine codes. Once the program is written, the mnemonics can easily be converted to their equivalent machine codes. Programs written using mnemonics are called **assembly language** programs.

The machine language is generally determined by the design of the microprocessor chip and cannot be modified. The assembly language mnemonics are made up by the microprocessor's manufacturer as a convenience for programmers and are not set by the microprocessor's design. For example, you could write INC A instead of INR A, as long as both were translated to the machine code 3C. A microprocessor is designed to recognize a specific list or group of codes called the **instruction set,** and each microprocessor type has its own set of instructions.

Although assembly language is a vast improvement over machine language, it is still difficult to use for complex programs. To make programming easier, **high-level languages** have been developed. These are similar to English and are generally independent of any particular microprocessor. For example, a typical instruction might be "LET COUNT = 10" or "PRINT COUNT." These instructions give a more complicated command than those that the microprocessor can understand. Therefore, microcomputers using high-level languages also contain long, complex programs (permanently stored in their memory) that translate the high-level language program into a machine language program. A single high-level instruction may translate into dozens of machine language instructions. Such translator programs are called **compilers.**

A Programming Example

To show the difference between machine language, assembly language, and a high-level language, let us use a simple programming example. Figure 29-7(a) shows the flowchart for a program that counts to ten. There are no input or output operations in this program, since all we will be doing is having the contents of a designated memory location count from zero to ten, and then repeat.

The translation from the flowchart to a high-level language is fairly simple, as seen in Figure 29-7(b). The high-level language used in this example is called **BASIC,** which stands for Beginner's All-purpose Symbolic Instruction Code and has the advantage of being simple and similar to English. Following the program listing shown in Figure 29-7(b), you can see that the first two lines of the program correspond exactly to the first two action blocks of the count-to-ten flowchart. In the first line, the memory location called COUNT is set to 0. In the second line, LET COUNT = COUNT + 1 is simply a way of saying "increment the count." Lines 3 and 4 perform the function of the decision block in the count-

Line No.	Instruction	Description
1	LET COUNT = 0	Set Count to 0
2	LET COUNT = COUNT + 1	Increment Count
3	IF COUNT = 10 THEN 1	Go to 1 if Count = 10
4	GO TO 2	Otherwise go to 2

(b)

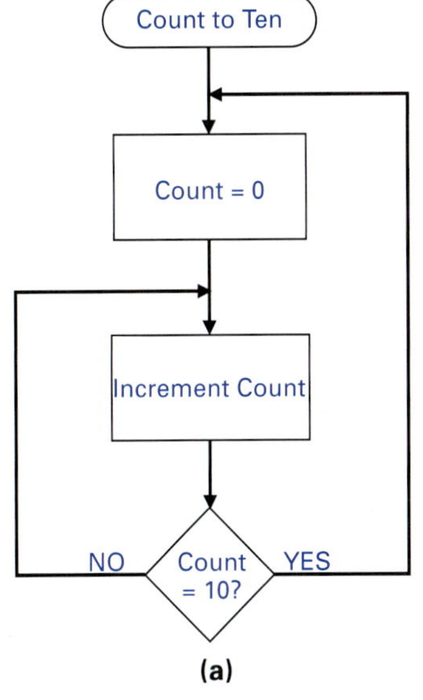

(a)

Label	Instruction		Comments
START:	MVI	A,0	;Set A register to 0
LOOP:	INR	A	;Increment A register
	CPI	10	;Compare A register to 10
	JZ	START	;Go to beginning if A = 10
	JMP	LOOP	;Repeat

(c)

Memory Address (Hex)	Memory Contents (Hex)	(Binary)
07F0	3E	00111110
07F1	00	00000000
07F2	3C	00111100
07F3	FE	11111110
07F4	0A	00001010
07F5	CA	11001010
07F6	F0	11110000
07F7	07	00000111
07F8	C3	11000011
07F9	F2	11110010
07FA	07	00000111

(d)

BASIC Language		8085 Assembly Language			8085 Machine Language		
Line No.	Instruction	Label	Instruction		Address	Contents	
1	LET COUNT = 0	START:	MVI	A,0	07F0	3E	*Opcode*
					07F1	00	*Data*
2	LET COUNT = COUNT + 1	LOOP:	INR	A	07F2	3C	*Opcode*
3	IF COUNT = 10 THEN 1		CPI	10_{10}	07F3	FE	*Opcode*
					07F4	0A	*Data*
			JZ	START	07F5	CA	*Opcode*
					07F6	F0	*Address*
					07F7	07	
4	GO TO 2		JMP	LOOP	07F8	C3	*Opcode*
					07F9	F2	*Address*
					07FA	07	

(e)

FIGURE 29-7 A Count-to-Ten Programming Example. (a) Flowchart. (b) High-Level Language Program. (c) Assembly Language Program. (d) Machine Language Program. (e) Summary of Program Languages.

to-ten flow chart. Line 3 specifies that if COUNT = 10, then the next instruction executed should be line 1. If the count is not equal to ten, the COUNT = 10 instruction has no effect, and the program continues with line 4, which says "go to line 2." To test whether this BASIC program will count to 10 and then repeat, try following the program step by step to see if it works, starting with a value of zero.

Assembly language is not one specific language but a class of languages. Each microprocessor has its own machine language and therefore its own assembly language, which is defined by the manufacturer. Figure 29-7(c) shows the assembly language listing for the count-to-10 program. This program is certainly more cryptic than the BASIC language program, but it performs exactly the same function.

The three columns in Figure 29-7(c) are for labels, instructions, and comments. The label provides the same function as the line number in the BASIC program. Instead of numbering every line, you simply make up a name (called a label) for each line to which you need to refer. A colon (:) is used to identify the label. A line needs a label only if there is another instruction in the program that refers to that line. The comments are an aid to understanding the program, and a semicolon (;) is used to identify the beginning of a comment. High-level language programs do not always need many comments because the instructions themselves are more descriptive. For assembly language programs, however, comments are an invaluable aid for people other than the programmer, or when the programmer returns to a program after some time.

Accumulator

One or more registers associated with the arithmetic and logic unit (ALU), which temporarily store sums and other arithmetical and logical results of the ALU.

The first instruction is MVI A, 0, which means "move immediately to the accumulator the data zero." The **accumulator** is also called the A register and is a storage location inside the microprocessor. This assembly language instruction is equivalent to the BASIC instruction LET COUNT = 0, except instead of making up a name for the variable (COUNT), we used a preassigned name (A) for a register inside the microprocessor. This MVI A, 0 instruction will therefore load the data 0 into the microprocessor's internal A register. The next instruction, INR A, means "increment the value in the accumulator," and this assembly language instruction is equivalent to the BASIC instruction LET COUNT = COUNT + 1. The next three instructions implement the decision function. The instruction CPI 10 means "compare the value in the accumulator with the value 10." The result of this comparison will determine whether a special flip-flop within the microprocessor called the *zero flag* is SET or RESET. If the value in the accumulator is not equal to 10, the zero flag is RESET LOW, whereas if the value in the accumulator is equal to 10, the zero flag is SET HIGH. The next instruction, JZ START, means "jump if the zero flag is SET to the line with the label START." This instruction tests the zero flag, and if it is SET (accumulator = 10), it will cause the program to jump to the line with the label START. Together, these two instructions (CPI 10 and JZ START) perform the function of the BASIC statement IF COUNT = 0 THEN 1. The last instruction, JMP LOOP, means "jump to the line with the label LOOP." This instruction simply causes the program to jump to the line with the label LOOP and is equivalent to the BASIC statement GO TO 2.

Opcode

The first part of a machine language instruction that specifies the operation to be performed. The other parts specify the data, address, or port. For the 8085, the first byte of each instruction is the opcode.

Figure 29-7(d) shows the machine language listing for the count-to-ten program. Although this language looks the most alien to us, its sequence of 1s and 0s is the only language that the microprocessor understands. In this example, each memory location holds eight bits of data. To program the microcomputer, we will have to store these 8-bit codes in the microcomputer's memory. Each instruction begins with an **opcode,** or operation code, that specifies the operation to be performed, and since all 8085 opcodes are eight bits, each opcode will occupy one memory location. An 8085 opcode may be followed by one byte of data, two bytes of data, or no bytes, depending upon the instruction type used.

Stepping through the machine language program shown in Figure 29-7(d), you can see that the first byte ($3E_{16}$) at address $07F0_{16}$ is the opcode for the instruction MVI A. The MVI A instruction is made up of two bytes. The first byte is the opcode specifying that you want to move some data into the accumulator, and the second byte stored in the very next memory location (address $07F1_{16}$) contains the data 00_{16} to be stored in the accumulator. The third memory location (address 07F2) contains the opcode for the second instruction, INR A. This opcode (3C) tells the microprocessor to increment the accumulator. The INR A instruction has no additional data, and therefore the instruction only occupies one memory location. The next code, FE, is the opcode for the compare instruction, CPI. Like the MVI A, 0 instruction, the memory location following the opcode contains the data required by the instruction. Since the machine language program is shown in hexadecimal notation, the data 10 (decimal) appears as 0A (hex). This instruction compares the accumulator with the value 10 and sets the zero flag if they are equal, as described earlier. The next instruction, JZ, has the opcode CA and appears at address 07F5. This opcode tells the microprocessor to jump if the zero flag is set. The next two memory locations contain the address to jump to. Because addresses in an 8085 system are sixteen bits long, it takes two memory locations (eight bits each) to store an address. The two parts of the address are stored in an order that is the reverse of what you might expect. The least significant half of the address is stored first and then the most significant half of the address is stored next. The address 07F0 is therefore stored as F0 07. The assembly language instruction JZ START means that the processor should jump to the instruction labeled START. The machine code must therefore use the actual address that corresponds to the label START, which in this case is address 07F0. The last instruction, JMP LOOP, is coded in the same way as the previous jump instruction. The only difference is that this jump instruction is independent of any flag condition, and therefore no flags will be tested to determine whether this jump should occur. When the program reaches the C3 jump opcode, it will always jump to the address specified in the two bytes following the opcode, which in this example is address 07F2. The machine language program, therefore, contains a series of bytes, some of which are opcodes, some are data, and some are addresses.

You must know the size and format of an instruction if you want to be able to interpret the operation of the program correctly. To compare the differences and show the equivalents, Figure 29-7(e) combines the high-level language, assembly language, and machine language listings for the count-to-ten program.

SELF-TEST EVALUATION POINT FOR SECTION 29-1

Use the following questions to test your understanding of Section 29-1.

1. Briefly describe the three following languages:
   a. machine language b. assembly language c. high-level language
2. What is a flowchart?
3. Define the following terms:
   a. instruction set b. peripheral c. programmer d. mnemonic e. compiler f. accumulator g. flag register h. opcode

29-2 A MICROCOMPUTER SYSTEM

In this section, you will see how all of the circuits discussed in the previous chapters come together to form a complete microprocessor-based system. The system is called "SAM," which is an acronym for "simple all-purpose microcomputer." We will be examining SAM's schematic diagram down to the component level to remove any mystery associated with microcomputers and give you the confidence that comes from understanding a complete digital electronic microprocessor-based system. In addition, we will discuss general microprocessor system troubleshooting techniques and the detailed troubleshooting procedures for the SAM system.

This section of the textbook will read very similar to a manufacturer's technical service manual. This has been done intentionally to introduce you to the format and style of a typical tech manual and so you can see how this resource will provide you with the detailed data that you need to have before attempting to troubleshoot any electronic system.

29-2-1 *Theory of Operation*

On pages 1009 through 1012 you will find the schematic diagram for the SAM system. You may wish to copy and combine this schematic, since we will be referring to it constantly throughout the circuit description. You may also want to color code your copy of the schematic to help you more easily see the circuit interconnections. For example, using highlighters, you could make all address lines yellow, all data lines blue, and all control signal lines green.

Introduction

The simple all-purpose microcomputer system was designed specifically for educational use. It uses Intel's 8085 microprocessor and has 2kB (kilobytes) of ROM and 1kB of RAM. The system contains two input ports that are connected to a keyboard from which you can enter programs, store data, and give commands to control the operation of the microcomputer, and a set of slide switches. The output port display allows you to view the contents of the memory and registers. In addition, there is an output port using LEDs and a speaker that is controlled by the processor. There are also LEDs on the address bus, the data bus, and the major control lines so you can monitor their activity. The ROM in the SAM contains programs to read the keyboard, execute keyboard commands, and send data to the display. The entire operation of the system is governed by a built-in software program called the *monitor.* Using the keyboard and the display, a program can be loaded into RAM, and then it can be run at full speed or executed one step at a time to allow you to follow the operation of the program in detail. Special switches are also provided to establish loop operations for test purposes, and moveable jumpers are included to insert faults in the circuits for training in troubleshooting microprocessor systems.

Block Diagram Description

A simplified block diagram of the SAM system is shown in Figure 29-8. The three main functional blocks of this microcomputer are the microprocessor, the memory, and the input/output ports. These blocks within the microcomputer are interconnected by three groups of signals—address, data, and control.

Unidirectional Bus

Wire or group of wires in which data flows in only one direction. Each device connected to a unidirectional bus is either a transmitter, or a receiver, but not both.

Address bus. The microprocessor is the only address bus "talker," which means that it is the only device that ever sends out addresses onto the 16-bit address bus. Every other device connected to the address bus, therefore, is a "listener" since these devices only receive addresses from the microprocessor. This one-direction flow of address information on this bus accounts for why the address bus is called a **unidirectional bus.** Before any data transfer can take place via the data bus, the microprocessor must first output an address onto the address bus to specify which memory location or input/output port the processor wishes to access. In this way, the processor can select any part of the system it wishes to communicate with.

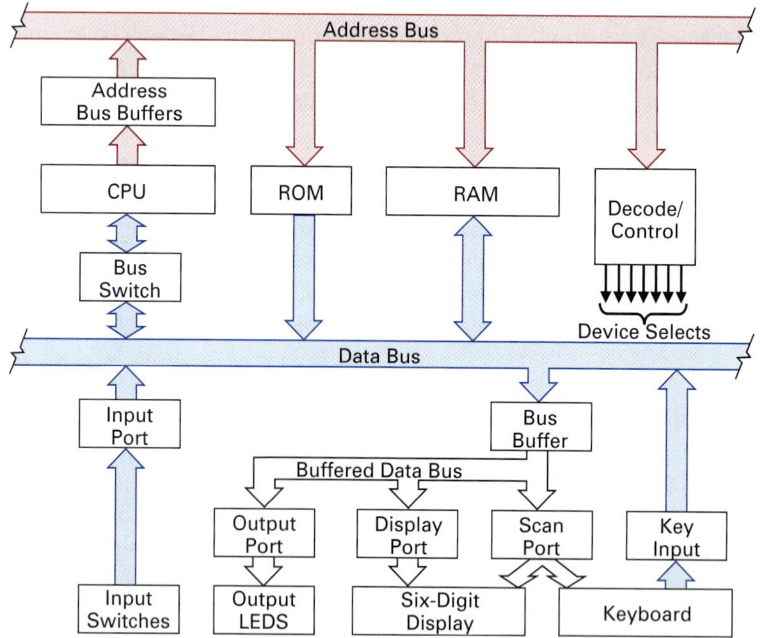

FIGURE 29-8 **The SAM System Block Diagram.**

Data bus. The 8-bit data bus is a **bidirectional bus** (two-direction), and therefore any device that talks on this bus must have eight driver outputs, and any device that listens must have eight data inputs. The microprocessor and the RAM are data bus talkers and listeners, while the ROM is only a data bus talker. The input ports are talkers since they input data from outside the system and put it on the data bus, whereas the output ports are listeners since they take data off the data bus and send it outside the system. The arrows on the data bus lines in Figure 29-8 summarize how each of the devices communicates with the data bus.

The microprocessor, RAM, ROM, and input ports contain three-state drivers on their data outputs. The select input to the RAM, ROM, and input ports enables the three-state drivers and causes the data from the selected device to appear on the data bus. The microprocessor is the controller of the system, so it will ensure that no more than one device is trying to use the data bus at any given time. For example, if the microprocessor wants to read data from the ROM, it would three-state its own data lines and generate the control signals needed to cause the ROM's select input to be true. The ROM's outputs would then appear on the data bus, and the microprocessor would read the data off the bus. Reading the RAM or an input port is done in a similar manner. On the other hand, if the microprocessor wanted to write data to the RAM or an output port, the microprocessor would first place the data to be written on its data lines. It would then generate the control signals needed to cause a write pulse to be sent to the appropriate device, which would then input the data from the bus.

Bidirectional Bus

Indicates that signal flow may be in either direction. Common bidirectional buses are three-state or open collector TTL.

Control bus. The address bus is used to select a particular memory location or input/output port. The data bus is then used to carry the data, and the control bus is used to control this process. The control bus consists of a number of control signals, most of which are generated by the microprocessor.

The three main control signals generated by the microprocessor are $\overline{RD}$ (not read, or *NRD*), $\overline{WR}$ (not write, or *NWR*), and $IO/\overline{M}$ (input/output/not memory, or *IO/NM*). If *NRD* is LOW (active), it indicates that a read is in progress and the microprocessor will be expecting the device that is being addressed to put data on the data bus. If *NWR* is LOW (active), a write is in progress and the microprocessor will be putting data on the data bus, expecting the addressed device to store this data. If *IO/NM* is LOW, the operation in progress (which may be a read or write) is a memory operation. If *IO/NM* is HIGH, the operation in progress (which may be an output or input) is an input/output port operation.

Inside the 8085

At the heart of the SAM system is an 8085 microprocessor (U3 on the SAM schematic). In most microcomputer circuit descriptions, the microprocessor is treated as a device with known characteristics but whose internal structure is of no concern. Since a knowledge of the microprocessor's internal operation will help you gain a clearer understanding of the operation of the SAM system, we will take a look at the internal structure and operation of the 8085.

Figure 29-9 shows the basic signals that connect to a microprocessor. There are sixteen address outputs that drive the address bus, and eight data lines connected to the data bus. The $\overline{RD}$ and $\overline{WR}$ output control signals coordinate the movement of data on the data bus, and the RESET input is used to initialize the microprocessor's internal circuitry. The two connections at the top of the microprocessor are for an external crystal, which is used as the frequency-determining device for the microprocessor's internal clock oscillator. The output from the microprocessor's internal clock oscillator is called the *system clock,* and it is used to synchronize all devices in the system and set the speed at which instructions are executed.

The inset in Figure 29-9 shows a simplified internal block diagram of the 8085 microprocessor. The accumulator, or A register, connects to the data bus and the arithmetic and logic unit (ALU). The ALU performs all data manipulation, such as incrementing a value or adding two numbers. The temporary register is automatically controlled by the microprocessor's control circuitry, and this register feeds the ALU's other input. The flags in the flag register are a collection of flip-flops that indicate certain characteristics about the result of the most recent operation performed by the ALU. For example, the zero flag is set if the

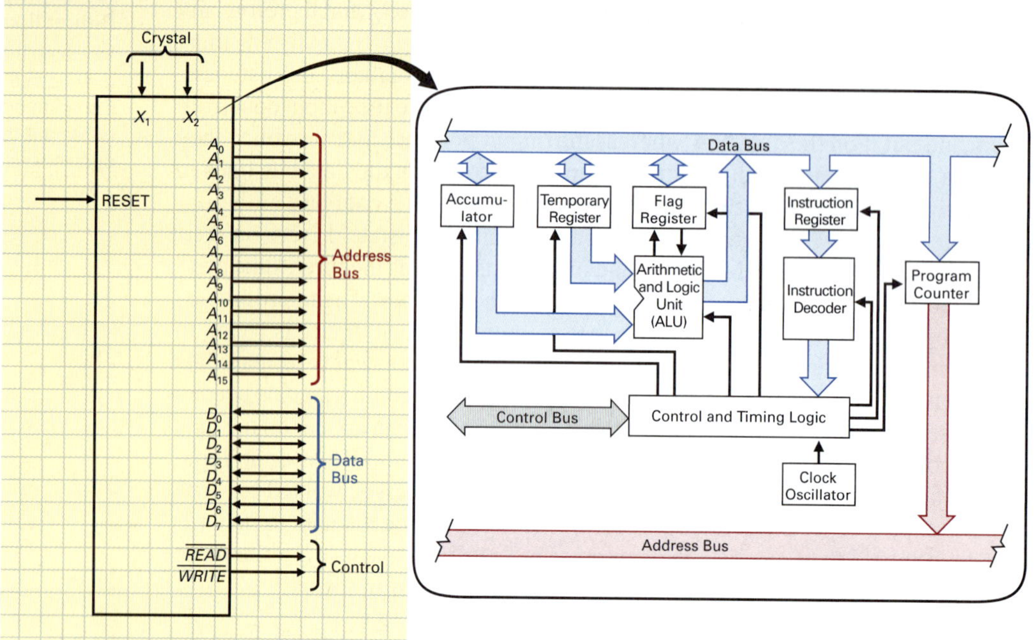

FIGURE 29-9 **The Microprocessor.**

result from an operation is 0, and as discussed earlier, this flag is tested by the JZ (jump if 0) instruction. The instruction register, instruction decoder, program counter, and control and timing logic are used for fetching instructions from memory and directing their execution.

Let us now take a closer look at the microprocessor's *fetch-execute cycle.* As an example, let us imagine that we need to read an MVI A instruction from ROM at address 0200. The first step is to retrieve the opcode for this instruction. This process is called the *instruction fetch* and is shown in Figure 29-10(a). As its name implies, the program counter is used to keep count by pointing to the current address in the program. In this example, the program counter will contain the address 0200, and this will be placed on the address bus and cause memory location 0200 to be selected. The ROM will then place the contents of memory location 0200 on the data bus, and the microprocessor will store this data, which is the opcode for MVI A (3E) in the instruction register.

Once the opcode is stored, the *instruction execute* process begins, as shown in Figure 29-10(b). The instruction register feeds the instruction decoder, which recognizes the opcode and provides control signals to the control and timing circuitry. For example, for an MVI A instruction, the control and timing logic circuit first reads the opcode 3E and then increments the address in the program counter. The instruction decoder determines that this opcode is always followed by a byte of data, so the contents of the memory location pointed to by the program counter is read, and the second byte of this instruction is placed into the accumulator. The program counter is again incremented, and the next byte of the main program (which will be the next instruction's opcode) is read into the instruction register, and then the execution of this instruction begins.

The real work is done in the execute phase of the instruction. Four basic types of operations can be performed by the 8085 microprocessor:

1. Read data from memory or an input port.
2. Write date to memory or an output port.
3. Perform an operation internal to the microprocessor.
4. Transfer control to another memory location.

The first two operation types are self-explanatory. The third involves manipulating the 8085's internal registers (such as the accumulator) without accessing the external memory or

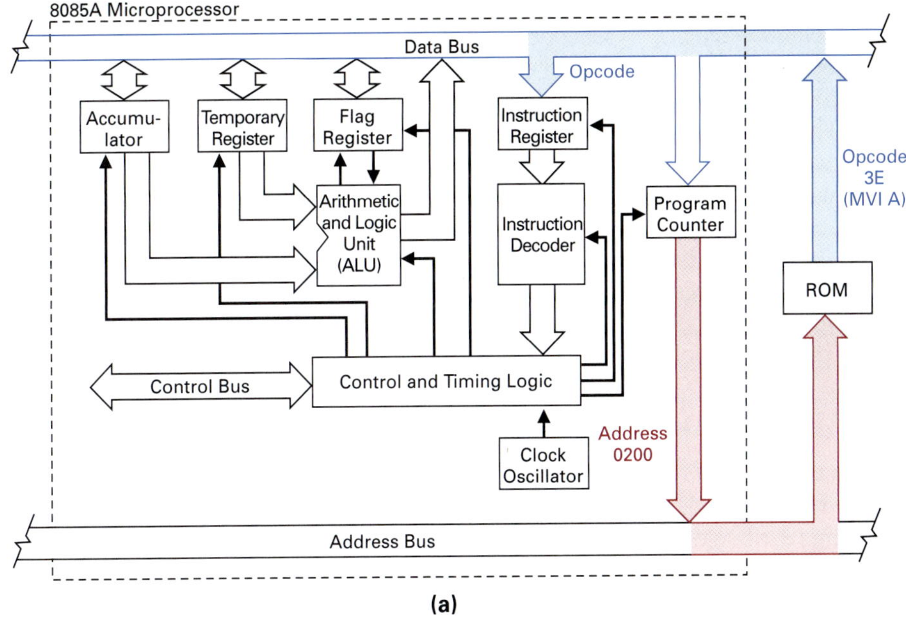

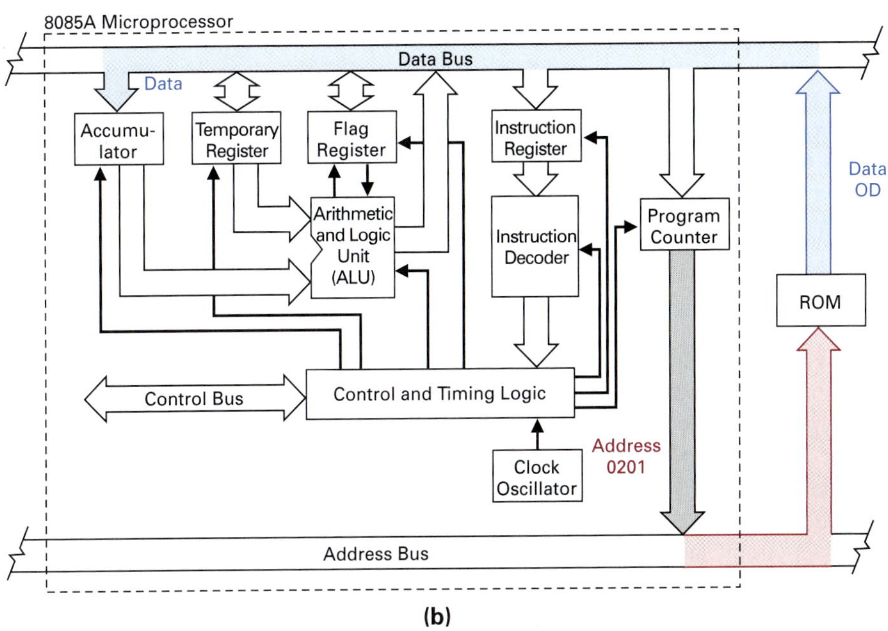

FIGURE 29-10 **The Fetch-Execute Cycle for an MVI A Instruction.**

input/output ports. For example, the contents of one register may be incremented or decremented. The fourth operation type includes instructions such as JMP, which causes the program counter and therefore the program to shift to point to a new memory location.

To summarize the operation of the microprocessor, we can say that the microprocessor keeps reading sequentially through the memory, one location after another, performing the indicated operations. Exceptions to this occur when a jump instruction is executed or at the occurrence of an **interrupt.** The 8085 has a few interrupt input pins, which are used to tell the microprocessor to stop its current task and perform another task immediately. Jump instructions or interrupts will cause the microprocessor to stop its sequential flow through a program and begin executing instructions from another address.

Remember that opcodes and data are intermixed in memory. One address might contain an opcode, the next two addresses a jump address, the next address an opcode, and the

Interrupt
Involves suspension of the normal program that the microprocessor is executing in order to handle a sudden request for service (interrupt). The processor then jumps from the program it was executing to the interrupt service routine. When the interrupt service routine is completed, control returns to the interrupted program.

next address a piece of data. It is the programmer's responsibility to be sure that the memory contains a valid sequence of opcodes and data. The microprocessor cannot distinguish among opcodes, jump addresses, and data since they are all simply bit patterns stored in the memory. All such information is read in exactly the same way and it travels over the same data bus. The microprocessor must always keep track of whether it is reading an opcode or data and treat each appropriately. The processor always assumes that the first location it reads contains an opcode and goes from there. If the opcode requires a second byte of data, the microprocessor is aware of this (because of the instruction decoder) and treats the next byte accordingly. It then assumes that the byte following the last byte in an instruction is the next opcode. The microprocessor will not know if a programmer has placed data in a position that should have an opcode. Such errors will cause the system to "crash" since the microprocessor would decode mispositioned data as opcodes.

SAM Circuit Description

As we proceed through this section, you should have your SAM schematic open, since we will be referring to devices on the schematic along with simplified SAM circuit diagrams shown in this textbook.

Decoding. The SAM system does not use I/O mapping for its I/O ports, so the IO/NM output from the microprocessor is not used for decoding. The SAM system uses memory-mapped input/output, which means that the I/O ports are treated as addressable devices similar to memory space.

Figure 29-11(a) shows the SAM address decoder (U7 on the SAM schematic), and Figure 29-11(b) shows the memory map or address map for the SAM system. Only the first quarter of the address space is used, so address bits 14 and 15 are always 0. The address space used in the SAM system is divided into eight equal sections, each of which has 2K, or 2,048, locations. The ROM occupies the first 2K addresses (0000 to 07FF hex), and the RAM has been assigned the next 2K addresses (0800 to 0FFF). The six address sections following the RAM are used for I/O ports. The control port is used by the monitor program (system control program in ROM) to provide some special functions, which will be described later. The key data, scan, and display segment ports control the keyboard and display, and the input and output ports are used for the switches and LEDs.

To simplify the address decoding hardware, 2,048 addresses have been assigned to each memory type and each of the input and output ports. The SAM system makes use of 16K addresses for 2K of ROM, 1K of RAM, and six I/O ports. The SAM system could have used only $2{,}048 + 1{,}024 + 6 = 3{,}078$ addresses, which would result in more unused address space and a much more complicated address decoding circuit. Referring to the simplified SAM address decoder circuit shown in Figure 29-11(a) and U7 on the SAM schematic, let us now see how this circuit operates. There are two things to note in this decoding arrangement:

1. Some of the read/write control is mixed with the decoding.
2. There is a special circuit for RAM write protect.

The binary addresses listed in the memory map in Figure 29-11(b) show that the A_{11}, A_{12}, and A_{13} lines specify which section is to be addressed. Therefore, these lines are used to provide the binary-select inputs to U7 (a 74138 binary to one-of-eight decoder IC). This provides eight separate active-LOW outputs, one for each of the 2K byte blocks in the SAM memory map. Address decoder U7 has three enable inputs, two that are active-LOW and one that is active-HIGH. All three enable inputs must be true to allow any of the outputs to be true. This is achieved by connecting the A_{14} and A_{15} lines to the two active-LOW enables. This will prevent any of the outputs from being true unless both A_{14} and A_{15} are LOW.

The read/write control gating is distributed throughout the circuitry, using the enables of the decoder, the memory devices, and the I/O ports to reduce the number of gates required. The decoder's third enable input is connected to a gate that generates the OR of *NRD* and *NWR*. This has the effect of allowing the device-select output of U7 to be true only when either a read or a write is in progress. This is necessary because the address bus will not contain meaningful in-

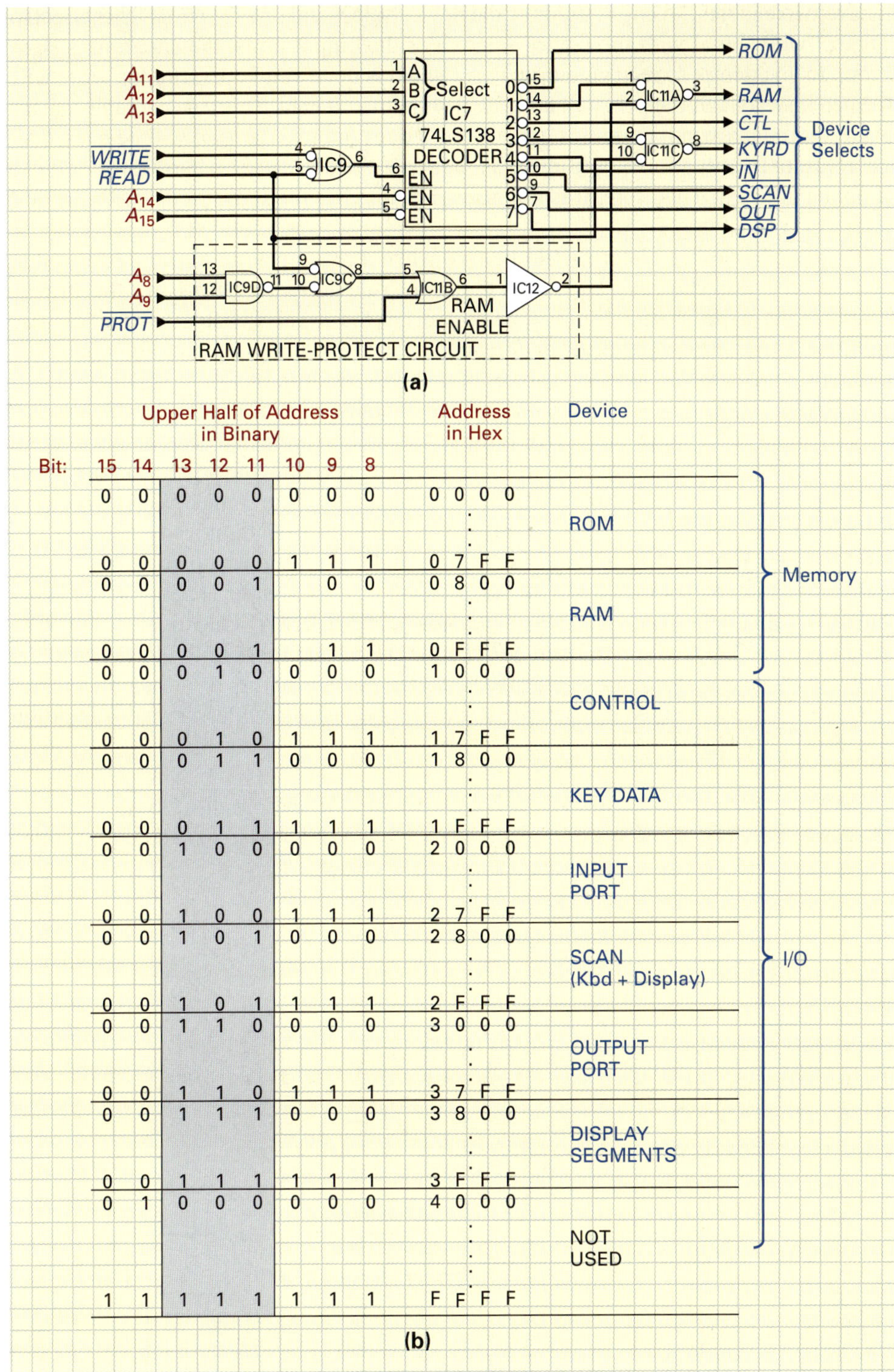

Bit: 15	14	13	12	11	10	9	8	Address in Hex	Device
0	0	0	0	0	0	0	0	0 0 0 0	ROM
0	0	0	0	0	1	1	1	0 7 F F	
0	0	0	0	1		0	0	0 8 0 0	RAM
0	0	0	0	1		1	1	0 F F F	
0	0	0	1	0	0	0	0	1 0 0 0	CONTROL
0	0	0	1	0	1	1	1	1 7 F F	
0	0	0	1	1	0	0	0	1 8 0 0	KEY DATA
0	0	0	1	1	1	1	1	1 F F F	
0	0	1	0	0	0	0	0	2 0 0 0	INPUT PORT
0	0	1	0	0	1	1	1	2 7 F F	
0	0	1	0	1	0	0	0	2 8 0 0	SCAN (Kbd + Display)
0	0	1	0	1	1	1	1	2 F F F	
0	0	1	1	0	0	0	0	3 0 0 0	OUTPUT PORT
0	0	1	1	0	1	1	1	3 7 F F	
0	0	1	1	1	0	0	0	3 8 0 0	DISPLAY SEGMENTS
0	0	1	1	1	1	1	1	3 F F F	
0	1	0	0	0	0	0	0	4 0 0 0	NOT USED
1	1	1	1	1	1	1	1	F F F F	

FIGURE 29-11 SAM's Address Decoder and Address Map.

formation if either *NRD* or *NWR* are true. In addition, the ROM and the input ports are to be selected only if a read is being performed. If they respond to both a read or a write, a bus conflict could occur. For example, if a write to the ROM is attempted, the microprocessor would put data on the data bus to be written to the ROM. If the ROM were allowed to be enabled by a write, then it would also put data on the data bus, which is an unacceptable situation.

To solve this problem, U11C ANDs the *NRD* signal with the device select. This is shown for the KYRD port, but not for the ROM and IN ports. The ROM and IN port chips each have two enables, so one is used for the device select and one for *NRD*. This effectively ANDs the *NRD* signal with the device select. For the output ports, the situation is slightly different. In an attempt to read an output port (which is not a valid operation), a write will be performed instead, and the port will be loaded with invalid data. This is acceptable, since the software should know not to do this and even if it does, no real damage will be caused. This is in contrast to the situation of writing to an input port, which causes a hardware conflict and must not be allowed. Therefore, it is not necessary to AND the *NWR* signal with the device select for the output port U15.

The RAM's device select should be true when a read or write to the RAM's address space is in progress. The gate (U11A) on the RAM's device-select line is for the write protect circuit. The write protect circuitry helps prevent the RAM's contents from being accidentally lost. A programming error may result in the microprocessor running wild (usually by interpreting data as an opcode), and often this will result in incorrect data being written into the RAM. To prevent this, the SAM system contains control latch U8, which provides an *NPROT* input. When this latch is SET, the RAM will be protected. The monitor program will set this protect latch whenever a program is running and reset the protect latch at all other times so that data can be stored in the RAM so SAM can be programmed. Since the program may want to use the RAM to store data during program execution, only the first three-quarters of the RAM is protected. Address lines A_8 and A_9 indicate which quarter of the RAM is being addressed, and if they are both HIGH, then the last quarter is being addressed and the memory will not be protected. To achieve this, A_8 and A_9 are ANDed together in U9D and the result is then ORed with *NRD* and *NPRT* in U11B. This produces the RAM enable signal, which will be true if A_8 and A_9 are HIGH if a read is in progress or if the protect latch is not set. If the protect latch is set, then the RAM will be disabled unless a read is in progress or if A_8 and A_9 are HIGH.

RAMs. Figure 29-12 shows the SAM system's RAM circuit (U5 and U6 on the SAM schematic). The SAM system uses two 1K by 4-bit SRAMs connected to form a 1K

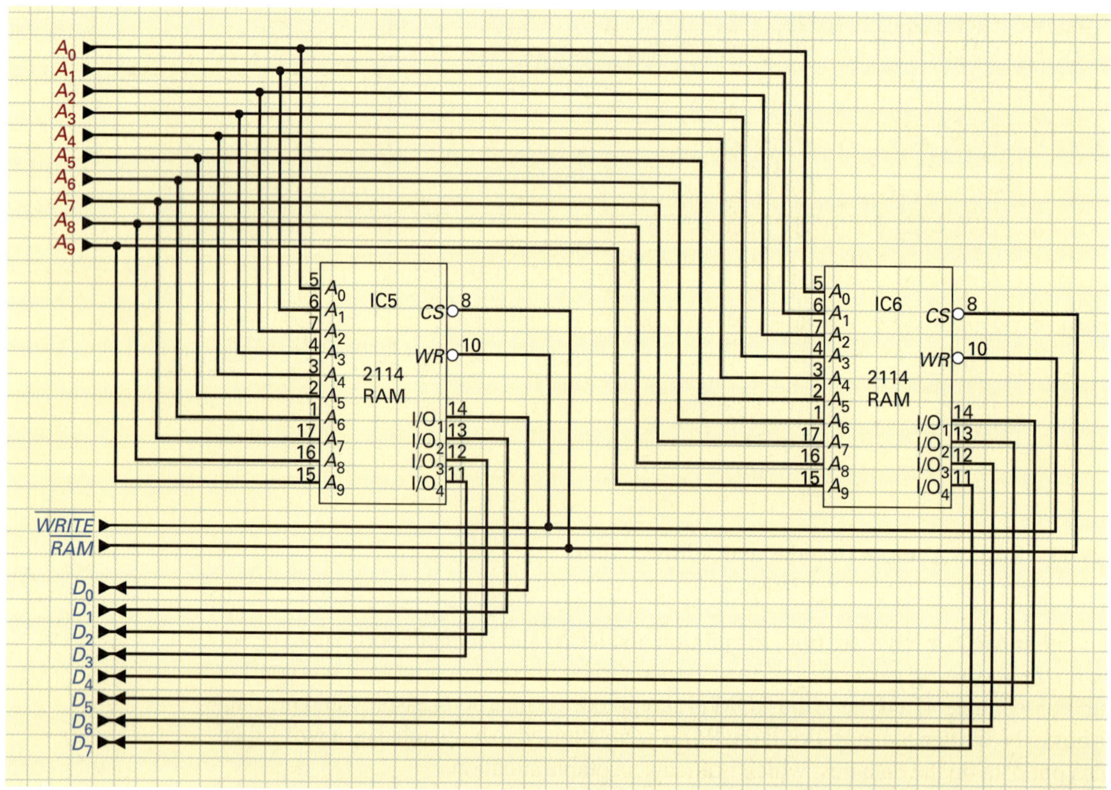

FIGURE 29-12 **SAM's RAM Circuit.**

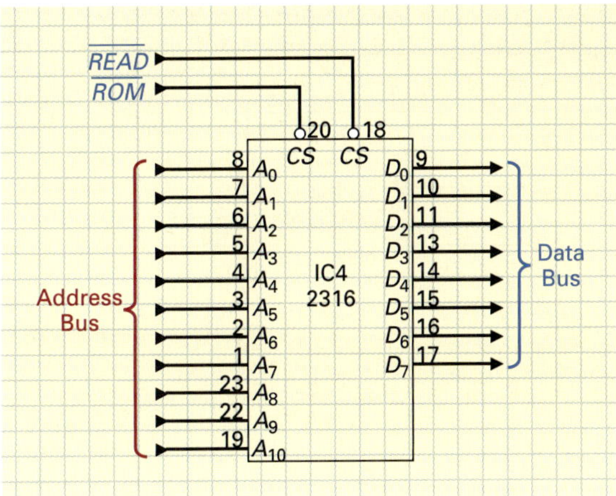

FIGURE 29-13 SAM's ROM Circuit.

by 8-bit RAM memory circuit. The address and control pins for both chips are parallel connected together, while U5 connects to data lines 0 through 3 and U6 connects to data lines 4 through 7.

ROM. Figure 29-13 shows the SAM system's ROM circuit (U4 on the SAM schematic). The eleven low-order address lines (A_0 through A_{10}) supply an address to this 2K by 8-bit masked-program ROM. The 8-bit data output from this IC is connected to the system's data bus. The two selects for this IC must be true for the ROM's three-state output drivers to be enabled. The ROM will therefore only drive the data bus if the ROM select is true and the operation is a read.

Speaker. Figure 29-14 shows the SAM system's serial output circuit, which will have to be traced carefully on the SAM schematic. The speaker is driven by the microprocessor's serial output, so this is in effect a 1-bit output port. The U3 (8085) SOD output is buffered by U18A and sent to the edge connector for use by external hardware, as desired. It is then buffered again before driving the speaker. The speaker draws so much current that the edge connector would not have valid logic levels if the speaker buffer were not included. A 100 Ω resistor in series with the speaker limits the current to levels that will not damage the buffer. The other connection to the speaker is tied to +5 V because the TTL buffer can sink more current than it can source (it can pull more current through the speaker than it can push through the speaker). All of the actual tone generation is controlled by the software. A beep program within the ROM turns the serial output ON and OFF several hundred times a second. This several-hundred-hertz square wave is applied to the speaker for a few seconds.

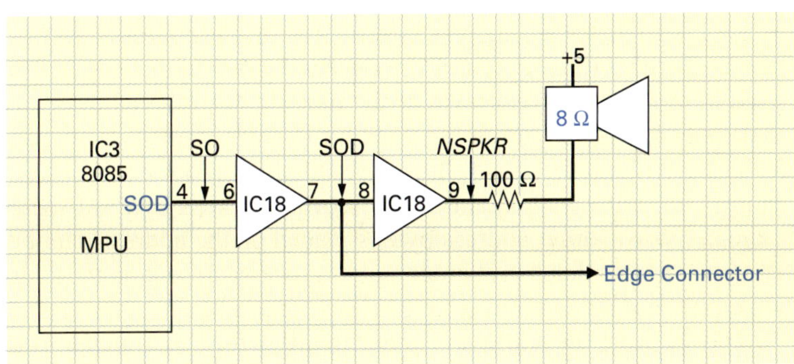

FIGURE 29-14 SAM's Serial Output Circuit.

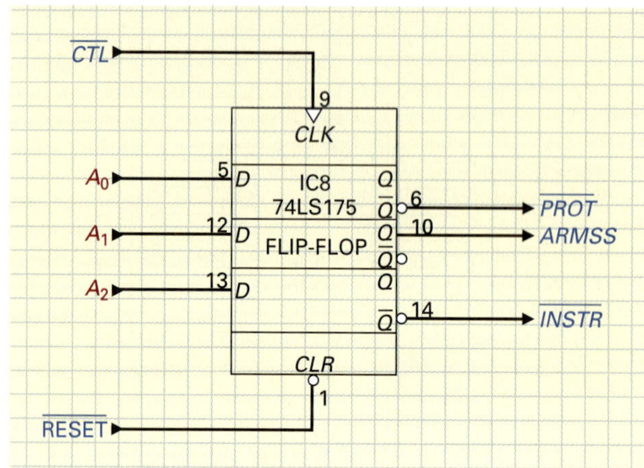

FIGURE 29-15 **SAM's Control Port Register.**

The control port. Figure 29-15 shows the SAM system's control port register (U8 on the SAM schematic). The control port is used by the microprocessor to send signals to special circuits. The control port IC (U8) is a 4-bit register clocked by the control port select signal, which is generated by address decoder U7. This is similar to the other output ports; however, it is unusual because the data inputs are connected to the address bus instead of the data bus. Therefore, the data written to the port is independent of the data on the data bus. The control port will be selected by any address from 1000 to 17FF. This allows the eleven low-order address lines to contain any value and still select the port. Note that A_0, A_1, and A_2 provide the data inputs to this register, and the data sent to the port is determined by the address used to select the port. For example, an address of 1000 would RESET or clear all of the bits, whereas an address of 1001 would SET the Q output and therefore RESET the $\overline{PROT}$ bit. As another example, an address of 1004 would SET the Q output and therefore RESET the $\overline{INSTR}$ bit. This technique simplifies the software. Since it does not matter what data is sent to the port (only the address matters), the software does not need to set up a value within the microprocessor before it writes the value to the port. The $\overline{PROT}$ bit of this port is used to control the memory protect circuit, and if active, the first three-fourths of the RAM will be write protected as described earlier. The other two control bits from this port are used to control the single-step circuitry.

The multiplexed bus. Figure 29-16(a) shows the SAM system demultiplexing circuit (U1 and U2 on the SAM circuit diagram). To keep the number of IC pins to a minimum, the 8085 microprocessor chip (U3) multiplexes the data bus with the lower half of the address bus. An 8-bit address bus from the 8085 carries the upper half of the address bus, and an 8-bit address/data bus carries the data and the lower half of the address. The address latch enable (*ALE*) signal is generated by the microprocessor to indicate when the address/data bus contains address information. This signal is used to latch the address/data bus contents into the address latch IC (U2) when the lower half of the address is being generated by the 8085. The 8-bit latch U2 latches the address information off the address/data bus on the negative edge of the *ALE* signal (the inverter U12 is needed to select this edge). The three-state buffer U1 is used to boost the current level of the upper half of the address bus for the address bus LED display and is not part of the demultiplexing circuit.

The waveforms in Figure 29-16(b) show the 8085 microprocessor's bus timing. The A_8 through A_{15} address lines always contain the high-order address byte. At the beginning of each memory cycle, the low-order address byte is placed on the address/data bus. The trailing edge of the *ALE* indicates that a low-order address byte is present on the address/data bus and causes the demultiplexing latch to store the low-order address byte. The address information is then removed from the address/data bus to allow for data transfer. If a read operation is in progress, the microprocessor will generate a read signal, and the addressed memory

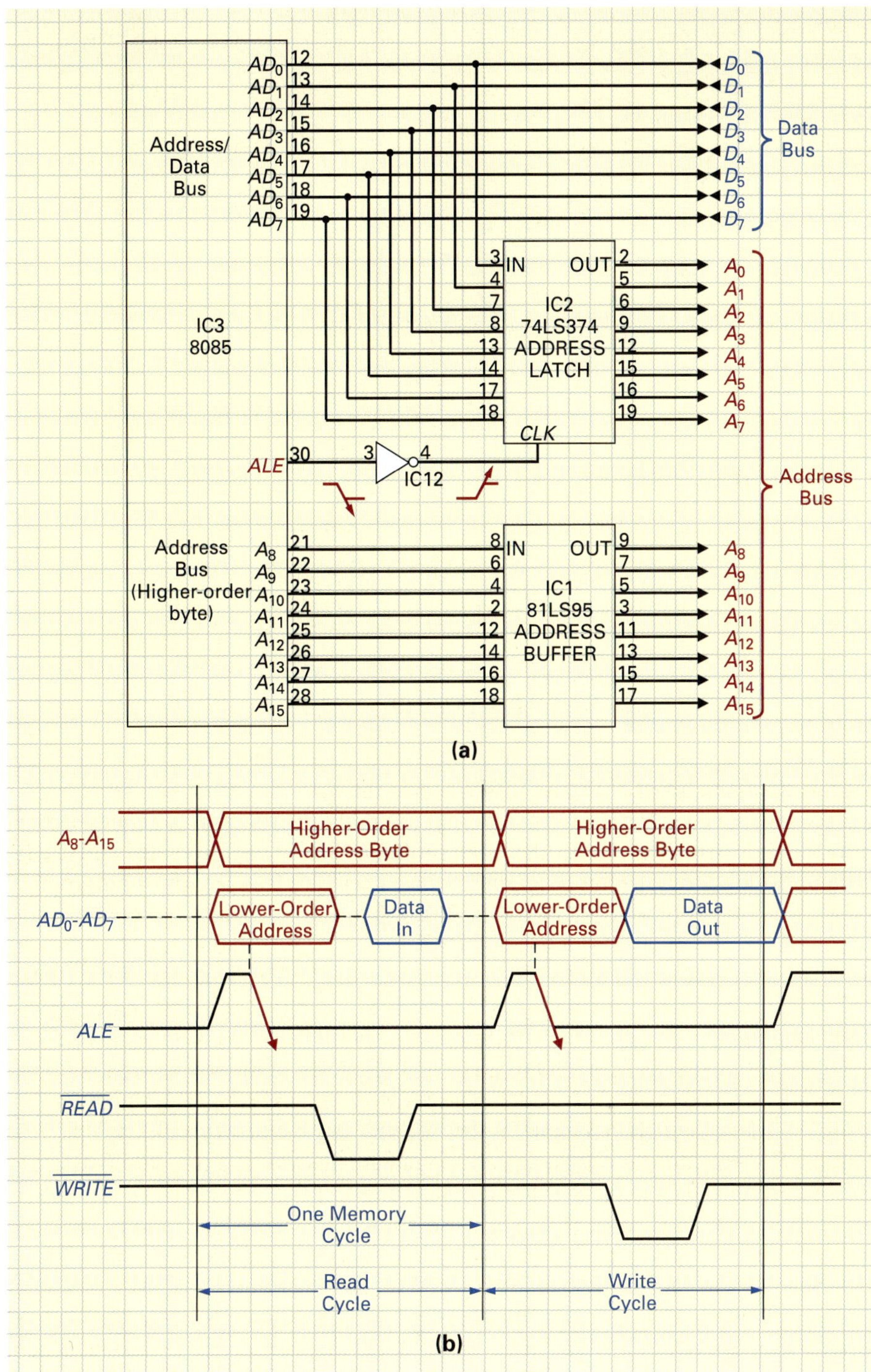

FIGURE 29-16 **SAM's Multiplexed Bus.**

or I/O device will place the data on the address/data bus. On the trailing edge of the *NRD* signal, the microprocessor will read the data off the bus.

The write cycle is similar, except that the direction of the data transfer is reversed. At the beginning of the cycle, the low-order address byte is placed on the address data bus and the *ALE* line is pulsed. The microprocessor then issues a write pulse and places the data on the address/data bus. On the trailing edge of the *NWR,* the addressed memory device or I/O device will store the data from the bus.

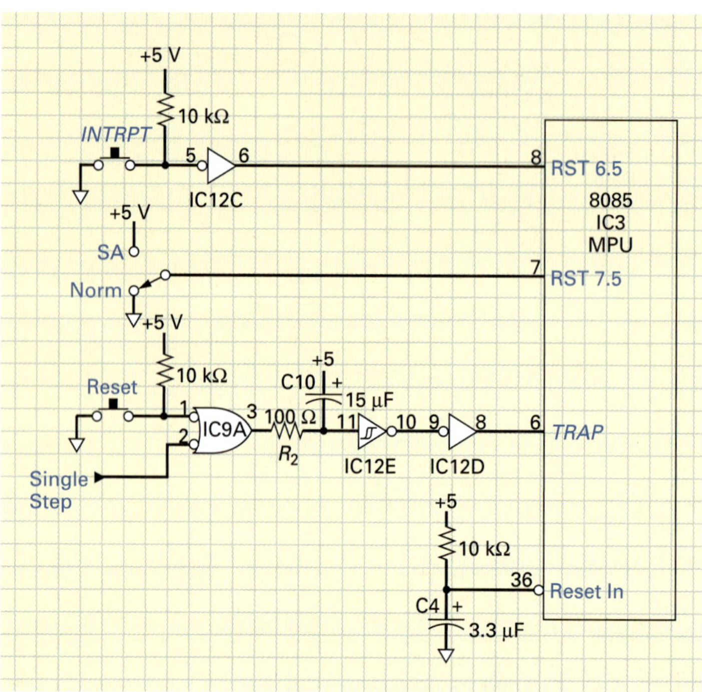

FIGURE 29-17 SAM's Interrupt and Reset Circuitry.

Interrupts and reset. Figure 29-17 shows the SAM interrupt and reset circuit. Interrupts provide a means for hardware external to the microprocessor to request immediate action by the processor. They allow the usual program flow to be interrupted and cause control to be transferred to a special program routine. There are two groups of interrupts available on the 8085 microprocessor: TRAP, RST 5.5, 6.5, and 7.5, which are controlled by the individual pins on the microprocessor; and RST 1, 2, 3, 4, 5, 6, and 7, which are controlled by INTR (interrupt request) and INTA (interrupt acknowledge). Only the TRAP, RST 7.5, and RST 6.5 interrupts are used in the SAM system. All that is required to initiate one of the interrupts is to apply a signal to the corresponding pin on the microprocessor. The RST 5.5 and 6.5 interrupt inputs will respond to a HIGH logic level, whereas the RST 7.5 interrupt input will only respond to a positive edge. The SAM system uses the *TRAP* input for the keyboard's RESET key, the RST 6.5 input for the INTRPT key on the keyboard, and the RST 7.5 input for the SA (signature analysis) switch. (Signature analysis is a troubleshooting method that will be described later in the SAM troubleshooting section.) The NAND gate U9A (which is equivalent to a bubbled-input OR) is used to allow the single-step circuitry to access the *TRAP* input. The 100 Ω resistor and the capacitor are used to debounce the RESET switch, which is necessary to cause only one interrupt. With the other interrupts, switch debouncing is not necessary because the software disables the interrupt as soon as it is acknowledged and so prevents a second interrupt. The *TRAP* input cannot be disabled, so it must be debounced using hardware.

Some allowances must be made for the fact that more than one interrupt may occur at the same time. To accommodate this, each interrupt is assigned a priority, and the interrupt with the highest priority will be acknowledged first. The 8085 *TRAP* interrupt has the highest priority, followed by the RST 7.5, 6.5, and 5.5, in that order, with INTR having the lowest priority.

The reset-in pin on U3 (pin 36) is used for power-up initialization. When a LOW level is applied to this pin, the microprocessor's internal circuitry is cleared. The program counter is set to 0000 (hex), so program execution begins from this address. This address is in ROM and contains a start-up program that initializes the complete SAM system. The slow-start circuit connected to the reset-in line will provide an automatic power-ON pulse.

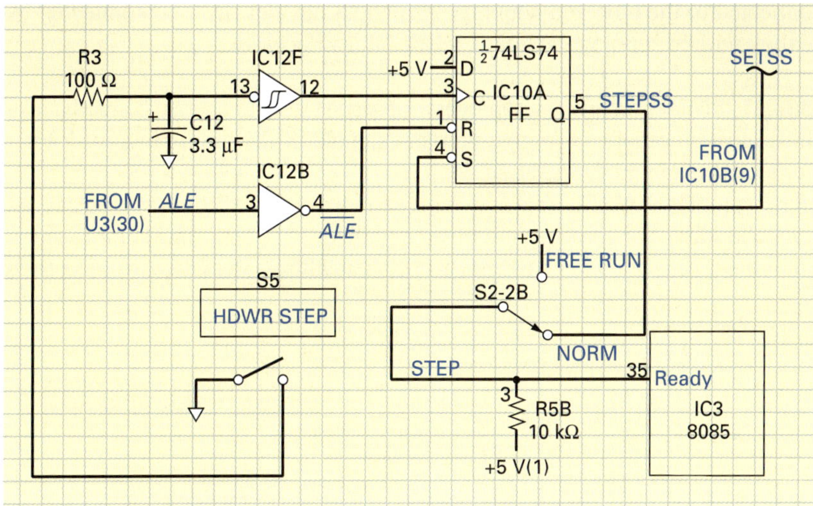

FIGURE 29-18 **SAM's Single-Step Circuit (Advances Microprocessor One Machine Cycle).**

Ready. Figure 29-18 shows the SAM system's single-step circuit and, as before, the IC numbers on this circuit correspond to the "U" numbers on the SAM schematic. When the ready input to the microprocessor (pin 35) is brought LOW, the microprocessor enters a wait state. The buses are not put in a three-state position, but remain at their current status until the ready input is brought HIGH again. This input is used for the hardware (HDWR) single-step mode, which allows the observation of the address bus, data bus, and status LEDs for each step of a program.

Data bus buffer. The 8085 microprocessor will not generate enough current at its address and data bus outputs to drive all of the devices connected to the buses plus a set of address bus monitoring LEDs (DS1 through DS4) and data bus monitoring LEDs (DS5 and DS6). Bus buffers are therefore included to boost current. Figure 29-19 shows how a data bus buffer IC14 (U14 on the SAM schematic) is included in this circuit. Since data flow can only travel in one direction (from the data bus to the buffered data bus), only output devices

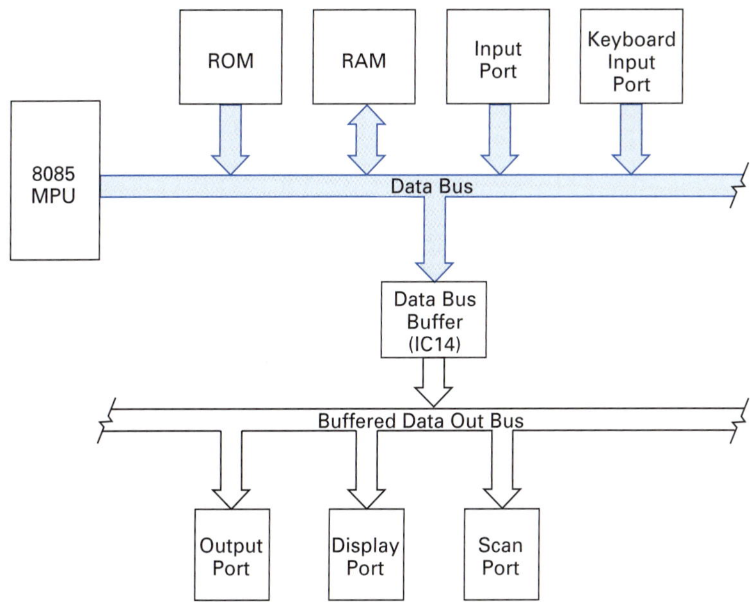

FIGURE 29-19 **SAM's Data Bus Buffer.**

can be connected to the buffered data bus. Any input devices must be connected to the data bus since they will be sending data back to the microprocessor.

Timing. In order for any microprocessor-based system to operate correctly, many timing relationships must be satisfied. To give you an example of the timing involved, Figure 29-20 shows a write-and-read timing operation.

Referring to the write timing waveforms shown in Figure 29-20(a), you can see that the address must be stable for some period of time (called the access time) before any operation can be performed. This allows the memory's internal address decoders to select the specified

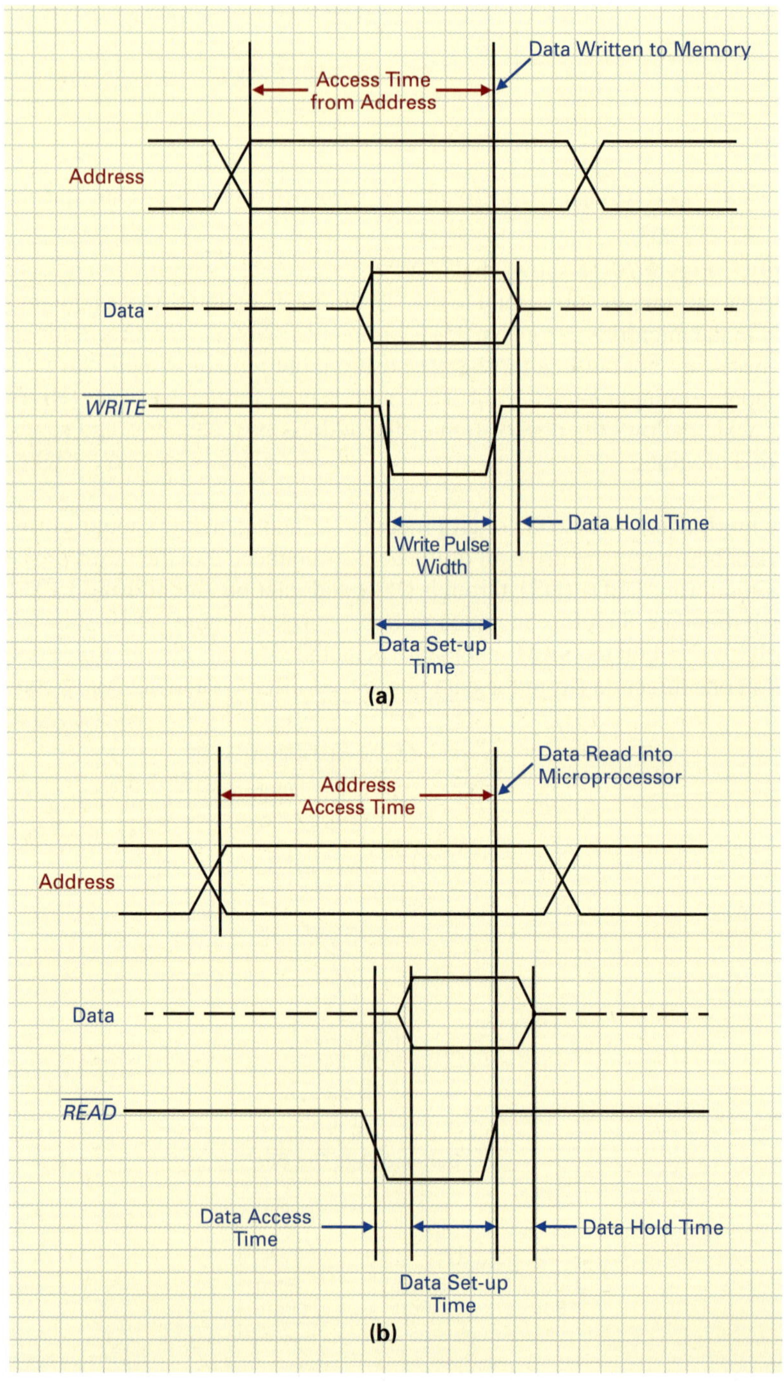

FIGURE 29-20 **Timing. (a) Write Timing. (b) Read Timing.**

memory location. The data must then be stable for some time before the write pulse (called set-up time) and after the write pulse (called hold time).

Referring to the read timing waveforms shown in Figure 29-20(b), you can see that, as with the write operation, the address must be stable for some time to allow the memory's internal decoders to settle. A read pulse is then generated, and after a short time (equal to the memory IC's access time), the accessed data will be placed on the data bus. This data must be allowed to stabilize (set-up time) before the data is read into the microprocessor on the rising edge of the *NRD* pulse, and the data must also be stable for some time after the read pulse (hold time).

Figure 29-21(a) shows the microprocessor timing for a typical instruction. In microprocessor-based systems, the basic unit of time is called the *state,* or timing state

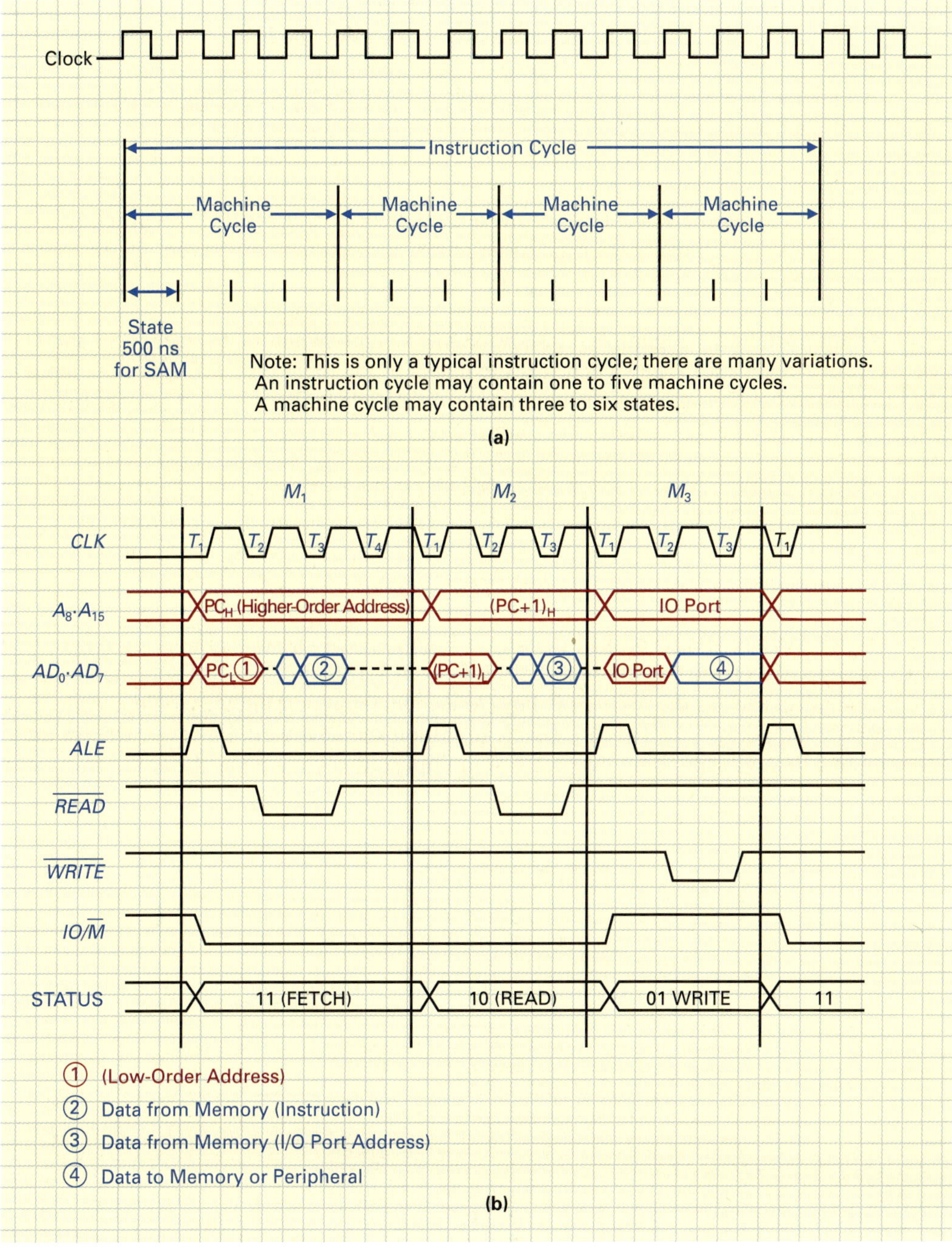

FIGURE 29-21 Instruction Cycle. (a) Summary of Microprocessor Timing. (b) Fetch and Execution of an OUT Instruction.

(T-state), and is equal to one clock period. The 8085 microprocessor in SAM generates a 2 MHz system clock signal, and so each T-state is equal to 500 ns (1/2 MHz = 500 ns). A **machine cycle** consists of between three and six T-states, and it is the time it takes the machine or microprocessor to complete an operation. An **instruction cycle** is the time it takes a microprocessor to fetch and execute a complete instruction. Most simple instructions have an instruction cycle that has only one machine cycle, whereas more complex instructions may consist of five machine cycles.

Machine Cycle
Basic period of time required to manipulate data in a system.

Instruction Cycle
All of the machine cycles necessary to fully execute an instruction.

As an example, Figure 29-21(b) shows the complete system timing for the two-byte OUT instruction. Timing states in this diagram use the prefix T (T_1, T_2, and so on), and machine cycles use the prefix M (M_1, M_2, and so on). The first byte of this two-byte instruction (the opcode) tells the microprocessor to send the contents of the accumulator to the output port specified in the second byte of the instruction. In the first machine cycle (M_1), the opcode is fetched from memory. In the second byte of the instruction (M_2), the output port address is read from memory. In the third machine cycle (M_3), the instruction is executed and the contents of the accumulator are written to the output port.

All instructions require one machine cycle to fetch the opcode. Single-byte simple instructions (such as the MOV A, B instruction, which transfers the data from one of the microprocessor's internal registers to another) can be performed in only one machine cycle. On the other hand, more complex multiple-byte instructions (such as STA 0837, which tells the microprocessor to store the contents of the accumulator at address 0837) will require four machine cycles since three bytes need to be retrieved from memory before the instruction can be executed.

Input switches. Figure 29-22 shows SAM's eight-switch input circuit (S3 and U13 on the SAM schematic). The switches at the inputs of the input port will cause the input to be LOW if the switch is closed. The input port's three-state buffer will only be enabled when both the *NRD* and the *NIN* control lines are active.

Output LEDs. Figure 29-23 shows SAM's eight-LED output circuit (DS12 through DS19—U15 on the SAM schematic). Since the LEDs form a common-anode display, a LOW stored in the output register will turn ON a LED. The output port's register is clocked by the *NOUT* control line.

Input keyboard and output display. Figure 29-24 shows a simplified diagram of SAM's keyboard circuit (S4 through S29, U17, and U18B on the SAM schematic). If we were to connect each of the twenty-six keys to its own input port bit, we would need to have four 8-bit input ports (4 × 8 = 32). Using a multiplexed keyboard, however, we can interface up to 256 keys using only two ports. A multiplexed keyboard has the keys arranged in

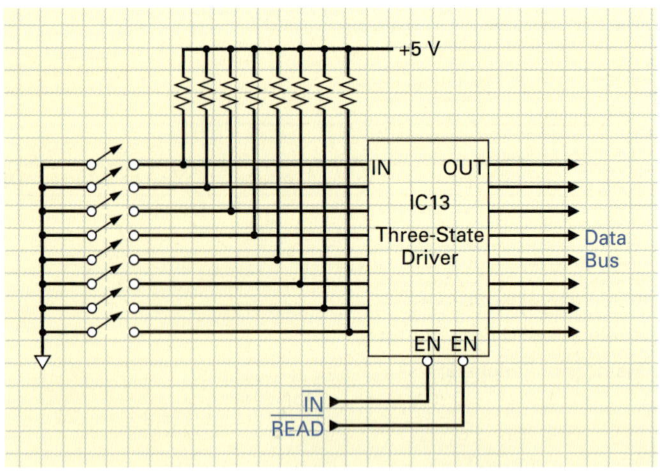

FIGURE 29-22 **SAM's Eight-Switch Input Port.**

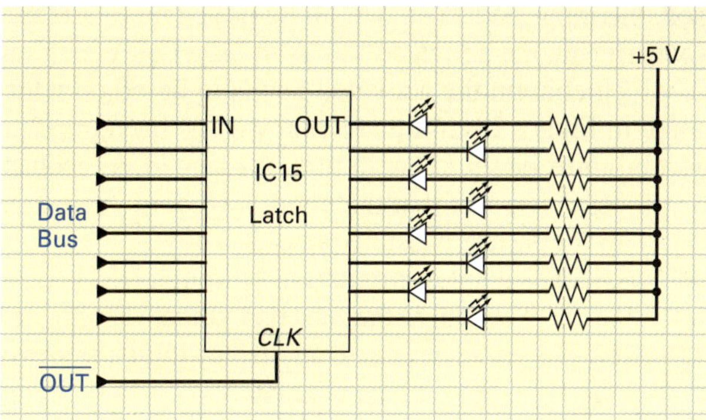

FIGURE 29-23 SAM's Eight-LED Output Port.

a matrix that has intersecting column lines and row lines. An output port (U17, address 2800) drives the rows, and an input port (U18B, address 1800) reads the columns. The SAM monitor program scans the keyboard by making LOW only one row at a time and then reading in the contents of the input port. If any of the keys are pressed in the row that is LOW, then the column line which that key is on will be forced LOW, and so that input port bit will be LOW.

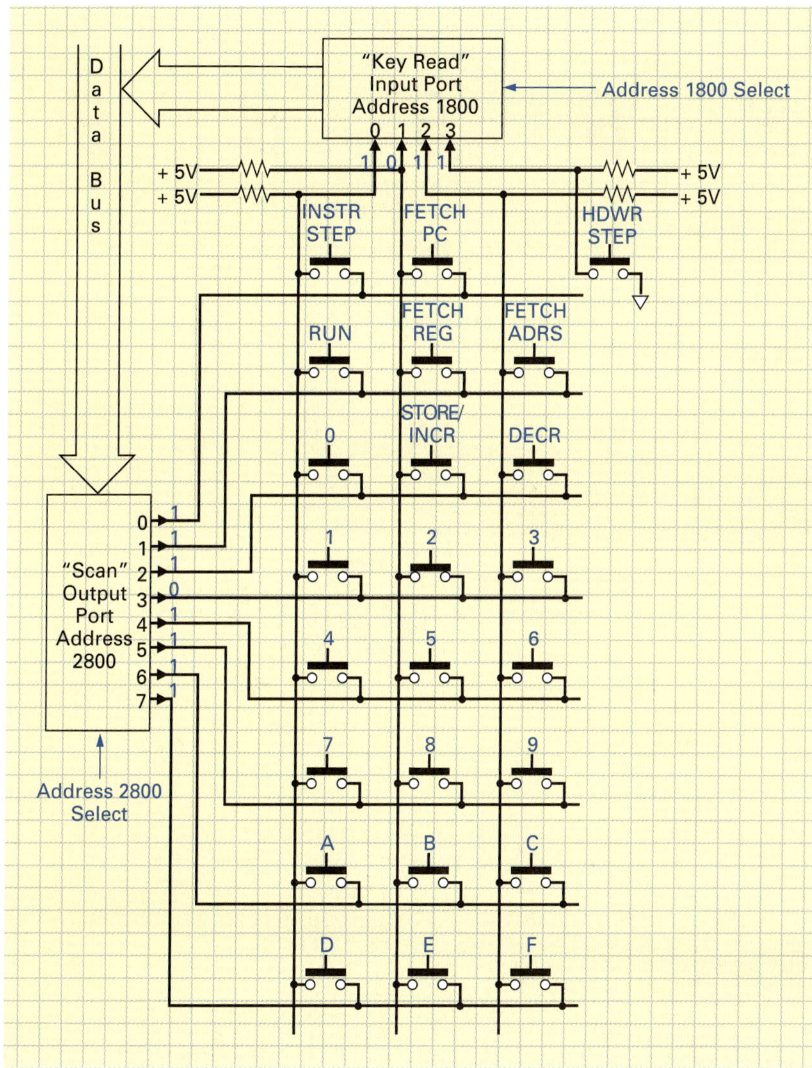

FIGURE 29-24 SAM's Keyboard Interface.

If no keys are pressed, the input port lines will all be pulled up to a HIGH level. The monitor program knows which row the activated key is in based on which output port bit is set LOW. It also knows which column the activated key is in based on which port bit is LOW (as seen in the example in Figure 29-24 in which key 2 has been pressed). To check all the keys, each output port bit must be set LOW in turn or scanned. At each step, only four keys are checked, and the process is so fast that the entire keyboard can be checked in less time than it takes a person to do the fastest possible press of a key.

Figure 29-25 shows SAM's display circuit (DS9 through DS11, U19, U17, and U20 on the SAM schematic). Like the keyboard, the seven-segment display also uses a scanning technique. The SAM system has six seven-segment display digits; however, only one display

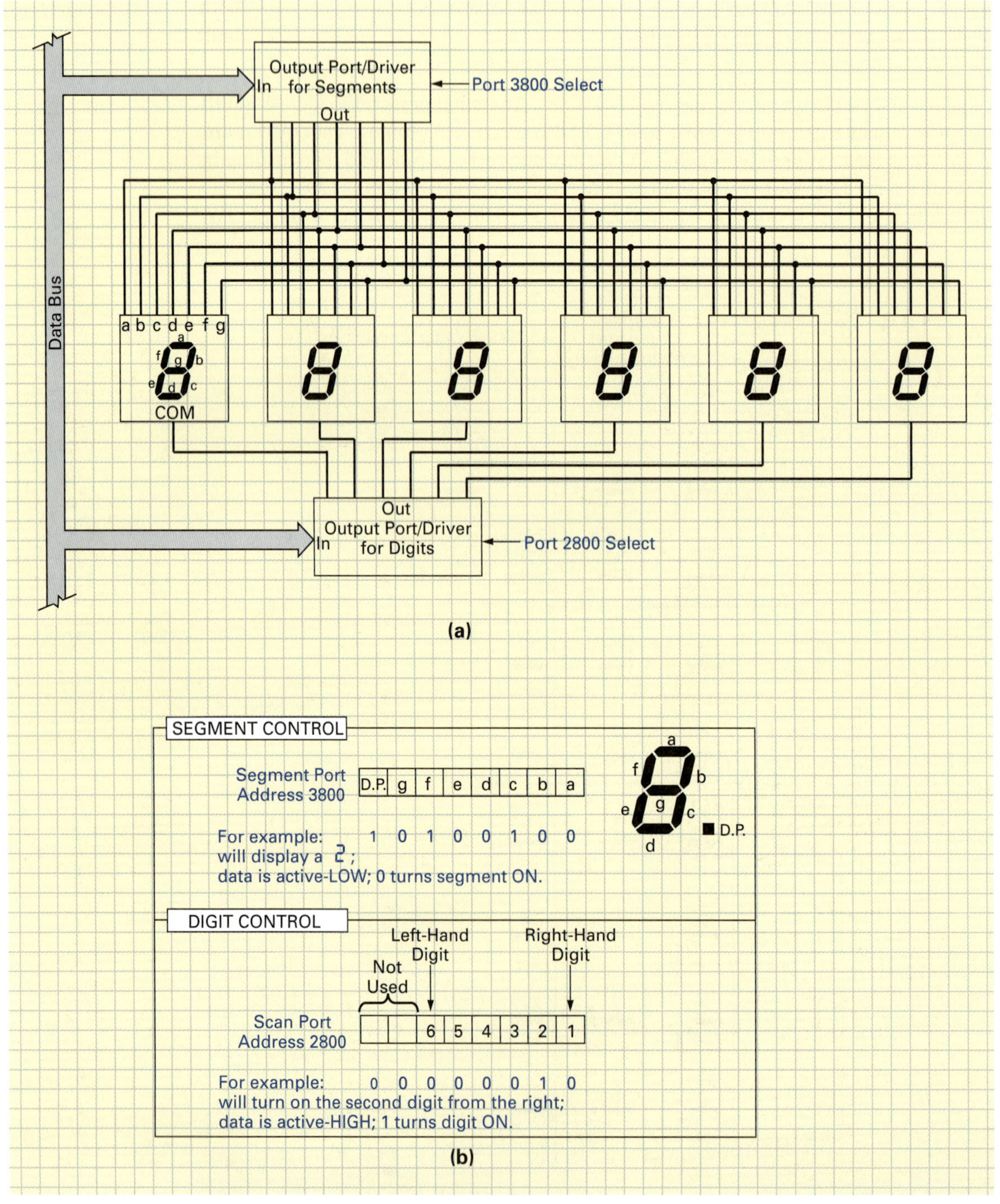

FIGURE 29-25 **SAM's Display Interface.**

is on at any instant. They are each turned on in sequence, but this happens so fast that they appear to all be on at the same time. A character is displayed by putting a HIGH level on the common connection and a LOW on the segment inputs based on which of the segments we wish to turn ON. In this multiplexed display, one 8-bit output port (U17 and U20, address 2800) is used to select one of the six digits, and another output port (U19, address 3800) supplies the segment information to the selected digit. A control program within the SAM monitor program can be run to control the display and display data stored in a set of six preselected RAM addresses. The control program operates in the following way. First, the segment information for digit 1 is sent to the segment port and the digit port is loaded with a value that will activate only digit 1. Then, the segment information for digit 2 is sent to the segment port and the digit port is loaded with a value that will activate only digit 2. Then, the segment information for digit 3 is sent to the segment port and the digit port is loaded with a value that will activate only digit 3. This continues until all six digits have been driven with segment data. The program then jumps back to its start point and the process is repeated.

SELF-TEST EVALUATION POINT FOR SECTION 29-2

Use the following questions to test your understanding of Section 29-2.

1. What is the difference between a unidirectional and bidirectional bus?
2. Define the function of the following 8085 microprocessor pins.
   a. NWR c. IO/NM e. READY g. HOLD
   b. NRD d. RESET f. ALE h. TRAP
3. Briefly describe a microprocessor's fetch-execute cycle.
4. What is a microprocessor
   a. interrupt? b. T-state? c. machine cycle? d. instruction cycle?

REVIEW QUESTIONS

Multiple-Choice Questions

1. Microprocessor-based systems are more flexible than hard-wired logic designs because ______________.
   a. they are faster
   b. they use LSI devices
   c. their operation is controlled by software
   d. the hardware is specialized
2. The peripherals of a microcomputer system are ______________.
   a. the memory devices **c.** the software
   b. the microprocessor **d.** the I/O devices
3. The personality of a microprocessor-based system is determined primarily by ______________.
   a. the microprocessor used
   b. the program and peripherals
   c. the number of data bus lines
   d. the type of memory ICs used
4. The language that the microprocessor understands directly is called ______________.
   a. assembly language
   b. high-level language
   c. English language
   d. machine language
5. The main purpose of a microprocessor's accumulator is ______________.
   a. keeping track of the next instruction to be executed
   b. temporary data storage
   c. selecting which interrupts should be enabled
   d. storing opcodes
6. The microprocessor's program counter is used for ______________.
   a. keeping track of the next instruction to be executed
   b. temporary data storage
   c. selecting which interrupts should be enabled
   d. storing opcodes
7. Interrupts are used primarily for ______________.
   a. breaking a program into modular segments
   b. responding quickly to unpredictable events
   c. speeding up program execution
   d. halting the system
8. When an interrupt occurs, the microprocessor will ______________.
   a. jump to the interrupt service routine
   b. halt until another request is made
   c. continue executing the main program
   d. complete the current program and then stop

9. The microprocessor knows which bytes to interpret as opcodes because _____________.
   a. every byte is an opcode
   b. every third byte is an opcode
   c. each opcode indicates the number of information bytes that follow
   d. the programmer has highlighted the opcodes
10. The purpose of the ALU in a microprocessor is to _____________.
   a. interpret the opcodes
   b. perform arithmetic and logic operations
   c. control the address bus
   d. calculate the number of machine cycles required
11. In a microprocessor system, the data bus is _____________.
   a. unidirectional and three-state
   b. bidirectional and three-state
   c. unidirectional and bidirectional
   d. All of the above are true.
12. In a system using several 1 kB memory devices, which address lines are connected to the memory chips?
   a. A0–A9
   b. A10–A15
   c. A0–A15
   d. A0–A7
13. In a system with a 16-bit address bus, what is the maximum number of 2 kByte memory devices it could contain?
   a. 16
   b. 32
   c. 64
   d. 128
14. Memory-mapped input/output _____________.
   a. treats I/O ports as memory locations
   b. tends to be more wasteful of address space than I/O-mapped decoding
   c. is used in the SAM system
   d. is all of the above
15. A bus conflict will occur when _____________.
   a. an output port is enabled by a *NRD* signal
   b. more than one device is reading from the data bus
   c. more than one device is writing to the data bus
   d. MOS and TTL devices are both connected to a data bus

Practice Problems

16. Describe the function of the SAM system program shown in Figure 29-26.

To practice your circuit recognition and operation skills, refer to U4 on the SAM schematic and answer the following three questions.

17. What is the size and organization of this IC?
18. How is this IC enabled, and where do the enables originate from?
19. Is this IC a data bus talker or listener or both?

To practice your circuit recognition and operation skills, refer to U5 and U6 on the SAM schematic and answer the following four questions.

20. What type of ICs are these?
21. What is the size and organization of these ICs?
22. How are these ICs enabled, and where do the enables originate from?
23. Are these ICs data bus talkers or listeners or both?

To practice your circuit recognition and operation skills, refer to U3 on the SAM schematic and answer the following seven questions.

24. Where does the signal input to pin 8 originate from?
25. What is connected between pins 1 and 2, and how is this related to the output on pin 37?
26. How are the address and data information separated if both appear on pins 12 through 19?
27. Where does the signal output on pin 4 go to?
28. How does the *RC* connected to pin 36 input operate?
29. What is the function of the signal outputs on pins 30, 31, and 32?
30. Are pins 21 through 28 multiplexed?

To practice your circuit recognition and operation skills, refer to the SAM schematic and answer the following ten questions.

31. What is the function of DS1 through DS4?
32. Describe the operation of U7 and how it governs the system memory map.
33. What is the function of DS5 and DS6?

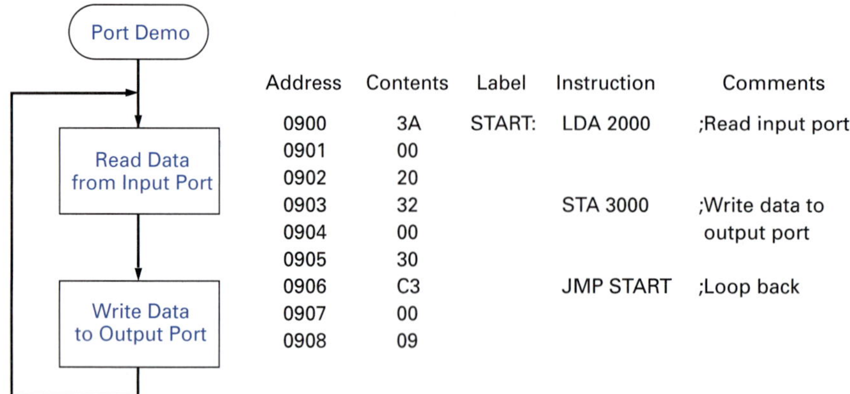

Address	Contents	Label	Instruction	Comments
0900	3A	START:	LDA 2000	;Read input port
0901	00			
0902	20			
0903	32		STA 3000	;Write data to output port
0904	00			
0905	30			
0906	C3		JMP START	;Loop back
0907	00			
0908	09			

FIGURE 29-26 An Echo Program for the SAM System.

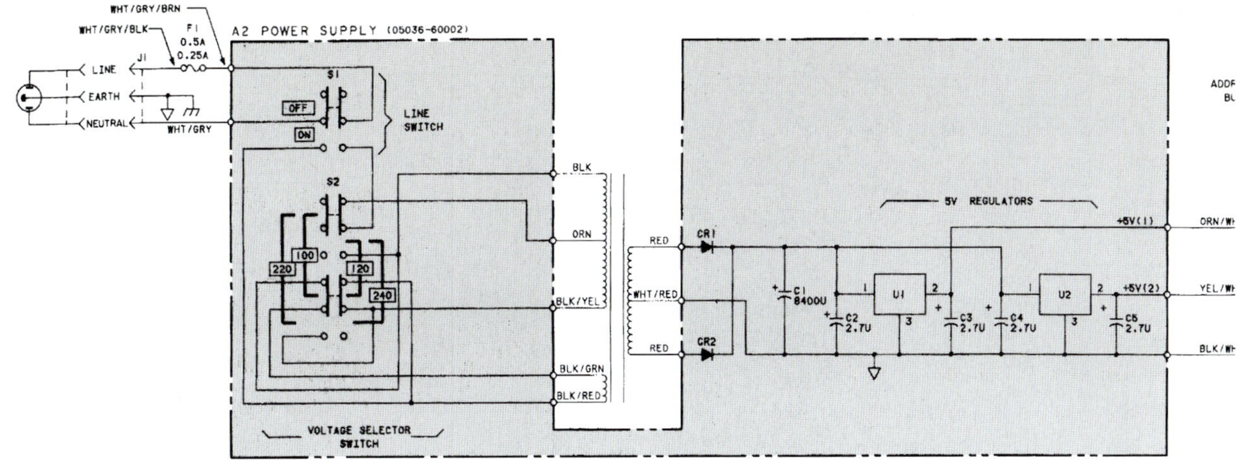

REFERENCE DESIGNATIONS

A1
C1-C16 CR1, CR2 DS1-DS19 LS1 R1-R8 S1-S29 U1-U20 W1-W14 XU4 Y1
A2
C1-C5 CR1, CR2 S1, S2 U1, U2

TABLE OF ACTIVE ELEMENTS

Reference Designation	HP Part Number	Mfg or Industry Part Number
A1		
CR1	1901-0518	1901-0518
CR2	1901-0731	1901-0731
DS1-DS6, DS8	1990-0652	HLMP-6220 1X4
DS7A, DS7B	1990-0685	1990-0685
DS9-DS11	1990-0667	1990-0667
DS12, DS16	1990-0673	5082-4690
DS13, DS15, DS17, DS19	1990-0675	5082-4590
DS14, DS18	1990-0674	5082-4990
U1, U13, U14	1820-1794	DM81LS95N
U2	1820-1997	SN74LS374PC
U3	1820-2074	P8085
U4	1818-0773	1818-0773
U5, U6	1818-0438	P2114
U7	1820-1216	SN74LS138N
U8	1820-1195	SN74LS175N
U9	1820-1197	SN74LS00N
U10	1820-1112	SN74LS74N
U11	1820-1208	SN74LS32N
U12	1820-1416	SN74LS14N
U15, U16, U17	1820-1730	SN74LS273N
U18	1820-1759	DM81LS97N
U19	1820-2138	DS8871N
U20	1820-1231	SN75492N
Y1	0410-1142	0410-1142
A2		
CR1, CR2	1901-0662	MR751
U1, U2	1826-0122	7805UC

NOTES

1. THE TRACE BETWEEN THE TERMINALS AT J1, J2, J3 AND J4 MUST BE CUT IF THESE LINES ARE TO BE USED WITH PERIPHERALS.
2. S3 SWITCHES ARE SHOWN IN CLOSED (LOGIC Ø) POSITION.
3. S1 SWITCHES ARE SHOWN IN CLOSED (NORM) POSITION. ALL SECTIONS OF S1 MUST BE OPEN FOR FREERUN MODE.
4. SIGNAL MNEMONICS PRECEDED BY THE LETTER 'N' INDICATE LOW POLARITY.

SAM Schematic Diagram (through p. 1012).

34. Describe the operation of the eight-switch input port.

35. Describe the operation of the eight-LED output port.

36. How are the scanning of the keyboard circuit and the scanning of the seven-segment display circuit related?

37. Describe the operation of the keyboard circuit.

38. Describe the operation of the seven-segment display circuit.

39. What is the function of DS7 and DS8?

40. Why does this circuit have a buffered data bus?

Web Site Questions

Go to the web site http://www.prenhall.com/cook, select the textbook *Electronics: A Complete Course,* select this chapter, and then follow the instructions when answering the multiple choice practice problems.

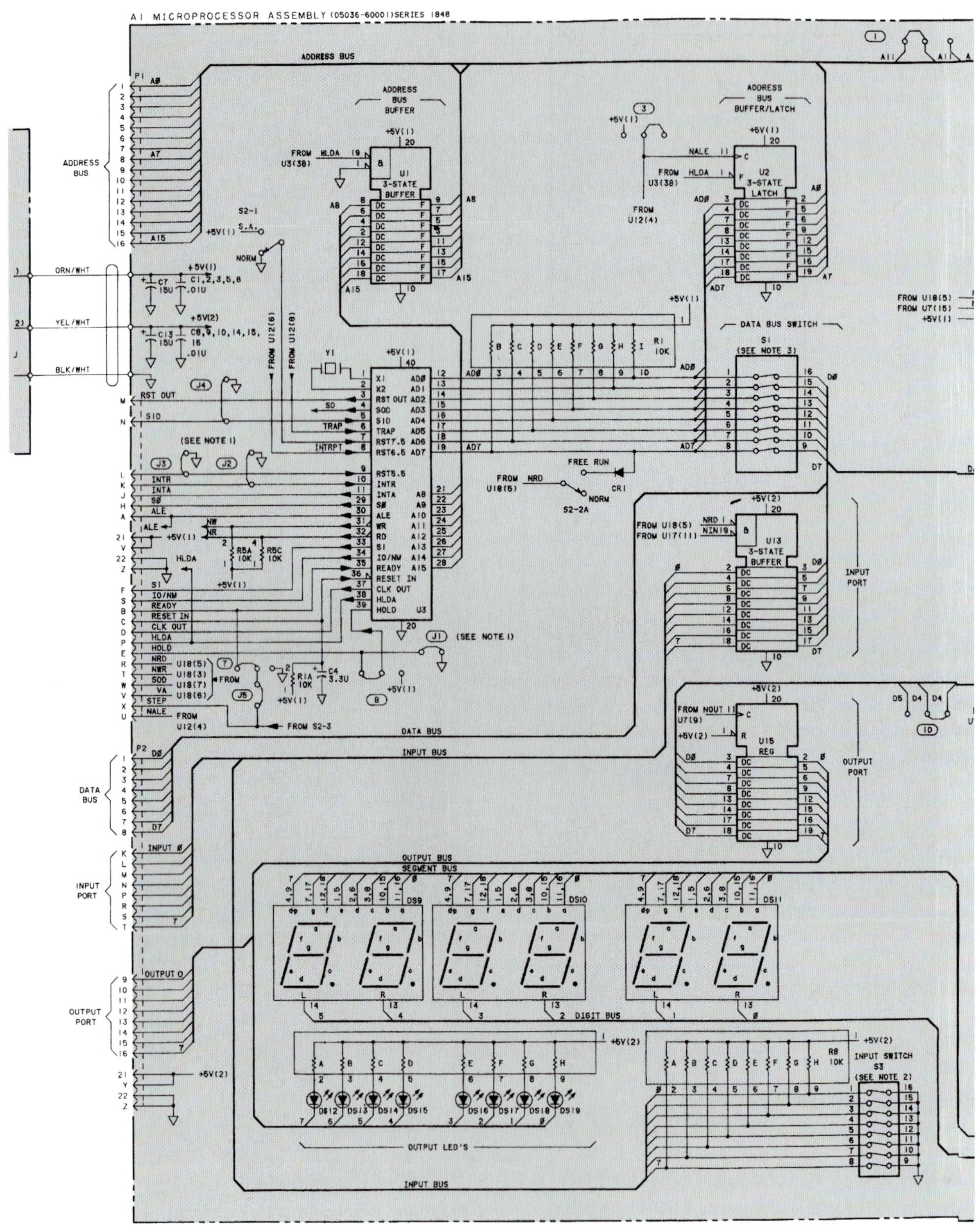
A1 MICROPROCESSOR ASSEMBLY (05036-60001) SERIES 1848
ADDRESS BUS
ADDRESS BUS BUFFER
ADDRESS BUS BUFFER/LATCH
DATA BUS SWITCH
INPUT PORT
OUTPUT PORT
DATA BUS
INPUT BUS
OUTPUT BUS
SEGMENT BUS
DIGIT BUS
OUTPUT LED'S
INPUT SWITCH S3 (SEE NOTE 2)
FREE RUN
NORM
(SEE NOTE 1)
(SEE NOTE 3)

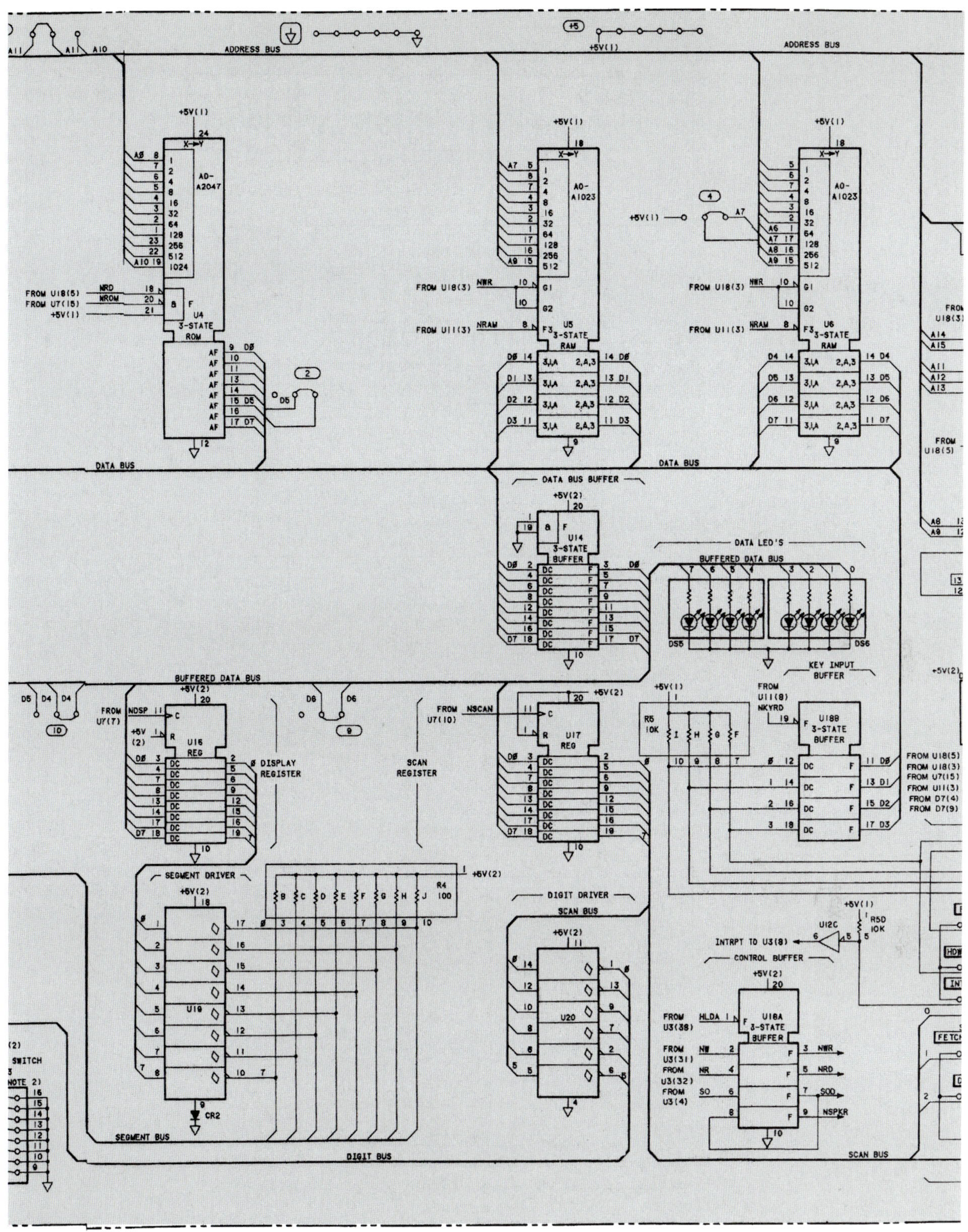

ADDRESS BUS
ADDRESS BUS
+5V(1)
AD-A2047
FROM U18(5)
FROM U7(15)
NRD
NROM
U4
3-STATE
ROM
A0-A1023
FROM U18(3)
FROM U11(3)
NWR
NRAM
U5
3-STATE
RAM
U6
3-STATE
RAM
DATA BUS
DATA BUS
DATA BUS BUFFER
+5V(2)
U14
3-STATE
BUFFER
DATA LED'S
BUFFERED DATA BUS
DS5
DS6
KEY INPUT
BUFFER
BUFFERED DATA BUS
FROM U7(7)
NDSP
U15
REG
DISPLAY
REGISTER
SCAN
REGISTER
FROM U7(10)
NSCAN
U17
REG
R5
10K
FROM U11(8)
NKYRD
U18B
3-STATE
BUFFER
FROM U18(5)
FROM U18(3)
FROM U7(15)
FROM U11(3)
FROM D7(4)
FROM D7(9)
SEGMENT DRIVER
R4
100
U19
CR2
DIGIT DRIVER
SCAN BUS
U20
U12C
R50
10K
INTRPT TO U3(8)
CONTROL BUFFER
FROM U3(38)
HLDA
U18A
3-STATE
BUFFER
FROM U3(31)
FROM U3(32)
FROM U3(4)
NW
NR
SO
NWR
NRD
SOD
NSPKR
SEGMENT BUS
DIGIT BUS
SCAN BUS
SWITCH
NOTE 2)

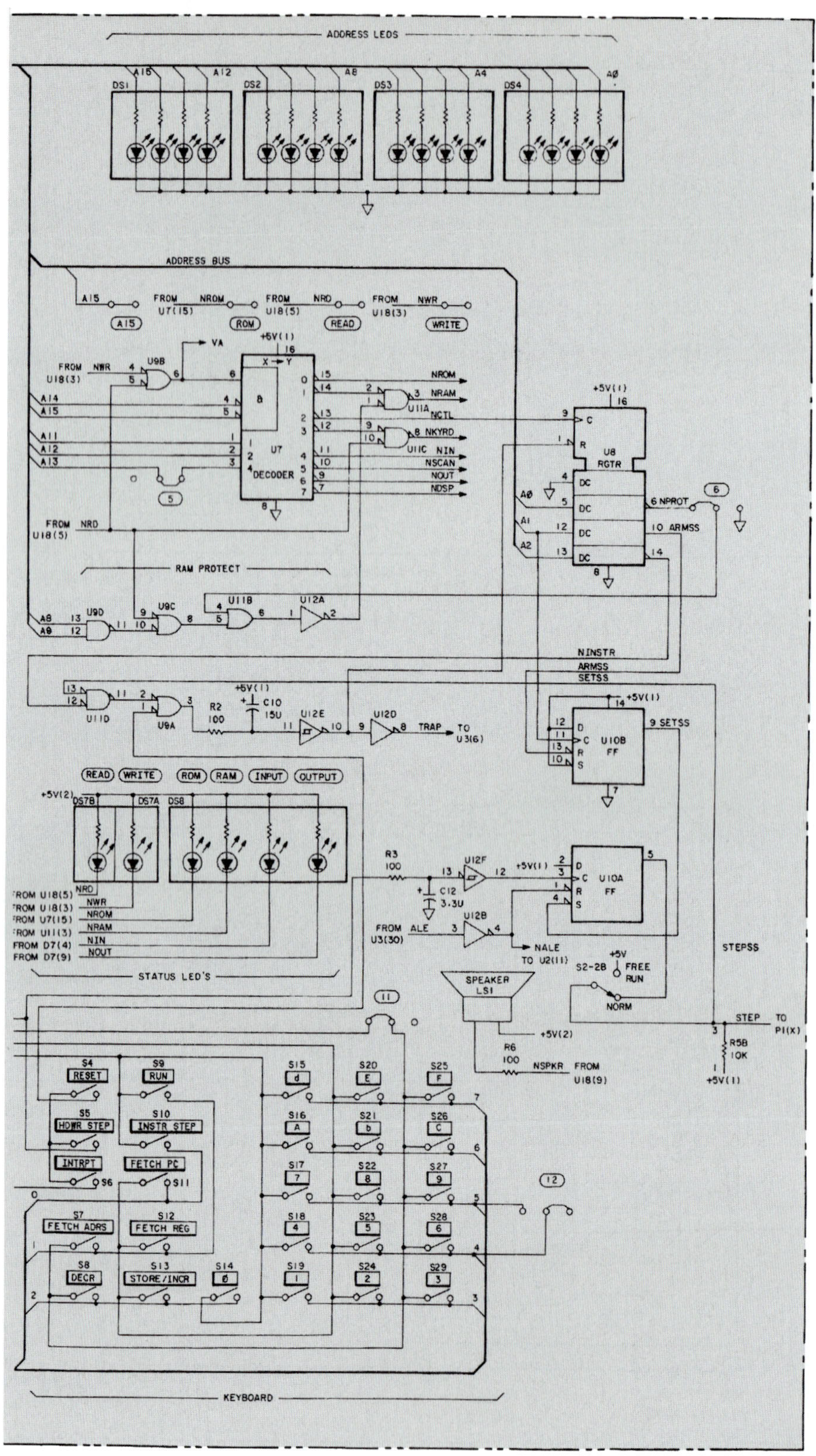

ADDRESS LEDS
ADDRESS BUS
RAM PROTECT
STATUS LED'S
KEYBOARD
U7 DECODER
U8 RGTR
U10B FF
U10A FF
SPEAKER LS1
NROM
NRAM
NCTL
NKYRD
NIN
NSCAN
NOUT
NDSP
NPROT
ARMSS
NINSTR
SETSS
TRAP TO U3(6)
NALE TO U2(11)
STEPSS
STEP TO P1(X)
NSPKR FROM U18(9)
RESET
RUN
HDWR STEP
INSTR STEP
INTRPT
FETCH PC
FETCH ADRS
FETCH REG
DECR
STORE/INCR
FREE RUN
NORM

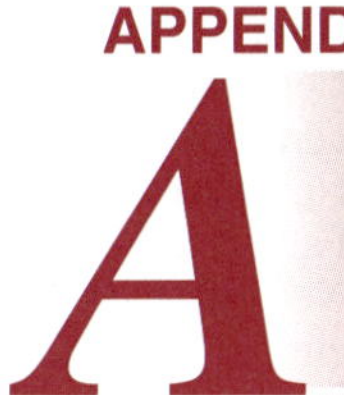

APPENDIX

Answers to Self-Test Evaluation Points

STEP 1-1
1. Elements are made up of similar atoms; compounds are made up of similar molecules.
2. Protons, neutrons, electrons
3. Copper
4. Like charges repel; unlike charges attract.

STEP 1-2
1. Amp.
2. Current = Q/t (number of coulombs divided by time in seconds)
3. The direction of flow is different: negative to positive is termed electron current flow; positive to negative is known as conventional current flow.
4. The ammeter

STEP 1-3
1. Volts
2. 3000 kV
3. The voltmeter

STEP 1-4
1. No
2. Yes
3. A short circuit provides an unintentional path for current to flow. A closed circuit has a complete path for current.
4. An open switch opens a path in a closed circuit. A closed switch closes a path in a closed circuit.
5. Negative plate, positive plate, electrolyte
6. A secondary cell can be recharged.
7. A power supply unit can have its output voltage adjusted and it will not run down.

STEP 2-1
1. A circuit is said to have a resistance of 1 ohm when 1 volt produces a current of 1 ampere.
2. $I = V/R = 24\ \text{V}/6\ \Omega = 4\ \text{A}$.
3. A memory aid to help remember Ohm's law:

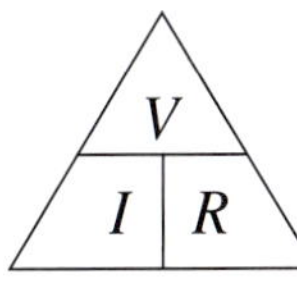

4. Current is proportional to voltage and inversely proportional to resistance.
5. $V = I \times R = 25\ \text{mA} \times 1\ \text{k}\Omega = (25 \times 10^{-3}) \times (1 \times 10^{3}) = 25\ \text{V}$.
6. $R = V/I = 12\ \text{V}/100\ \mu\text{A} = 12\ \text{V}/100 \times 10^{-6} = 120\ \text{k}\Omega$.
7. Ohmmeter
8. See Figure 2-10.

STEP 2-2
1. Light, heat, magnetic, chemical, electrical, mechanical
2. When energy is transformed, work is done. Power is the rate at which work is done or energy is transformed.
3. $W = Q \times V$, $P = V \times I$
4. When 1 kW of power is used in 1 hour

STEP 2-3
1. Type of conducting material used, cross-sectional area, length, temperature
2. The resistance of a conductor is proportional to its length and resistivity, and inversely proportional to its cross-sectional area.
3. True
4. True
5. A fuse is an equipment protection device that disconnects power from a circuit the moment current exceeds a safe level.

STEP 2-4
1. True
2. Mica
3. The voltage needed to cause current to flow through a material
4. Small

STEP 2-5
1. Carbon composition, carbon film, metal film, wirewound, metal oxide, thick film
2. SIPs have one row of connecting pins, while DIPs have two rows of connecting pins.
3. Rheostat has two terminals; potentiometer has three terminals.
4. Linear means that the resistance changes in direct proportion to the amount of change of the input, while a tapered potentiometer varies nonuniformly.

STEP 3-1 & 3-2
1. A circuit in which current has only one path
2. 8 A

STEP 3-3
1. $R_1 = R_1 + R_2 + R_3 + \cdots$
2. $R_T = R_1 + R_2 + R_3 = 2\ \text{k}\Omega + 3\ \text{k}\Omega + 4700 = 9.7\ \text{k}\Omega$

STEP 3-4
1. True
2. True
3. $R_T = R_1 + R_2 = 6 + 12 = 18\ \Omega$; $I_T = V_S/R_T = 18/18 = 1\ \text{A}$
   $V_{R1} = 1\ \text{A} \times 6\ \Omega = 6\ \text{V}$
   $V_{R2} = 1\ \text{A} \times 12\ \Omega = 12\ \text{V}$
4. $V_X = (R_X/R_T) \times V_S$
5. Potentiometer
6. No

STEP 3-5
1. $P = I \times V$ or $P = V^2/R$ or $P = I^2 \times R$
2. $P = V^2/R = 12^2/12 = 144/12 = 12\ \text{W}$
3. Wirewound, at least 12 W, ideally a 15 W
4. $P_T = P_1 + P_2 = 25\ \text{W} + 3800\ \text{mW} = 28.8\ \text{W}$

STEP 3-6

1. Component will open, component's value will change, and component will short.
2. No current will flow; source voltage dropped across it.
3. False
4. No voltage drop across component, yet current still flows in circuit; resistance equals zero for component.

STEP 4-1 & 4-2

1. When two or more components are connected across the same voltage source so that current can branch out over two or more paths
2. False
3. $V_{R_1} = V_S = 12\text{ V}$
4. No

STEP 4-3

1. The sum of all currents entering a junction is equal to sum of all currents leaving that junction.
2. $I_2 = I_T - I_1 = 4\text{ A} - 2.7\text{ A} = 1.3\text{ A}$
3. $I_X = (R_T/R_X) \times I_T$
4. $I_T = V_T/R_T = 12/1\text{ k}\Omega = 12\text{ mA}; I_1 = 1\text{ k}\Omega/2\text{ k}\Omega \times 12\text{ mA} = 6\text{ mA}$

STEP 4-4

1. $R_T = \dfrac{R_1 \times R_2}{R_1 + R_2}$
2. $R_T = \dfrac{1}{(1/R_1) + (1/R_2) + (1/R_3)} + \cdots$
3. $R_T = \dfrac{\text{common value of resistors } (R)}{\text{number of parallel resistors } (n)}$
4. $R_T = \dfrac{1}{(1/2.7\text{ k}\Omega) + (1/24\text{ k}\Omega) + (1/1\text{ M}\Omega)} = 2.421\text{ k}\Omega$

STEP 4-5

1. True
2. $P_1 = I_1 \times V = 2\text{ mA} \times 24\text{ V} = 48\text{ mW}$
3. $P_T = P_1 + P_2 = 22\text{ mW} + 6400\ \mu\text{W} = 28.4\text{ mW}$
4. Yes

STEP 4-6

1. No current will flow in the open branch; total current will decrease.
2. Maximum current is through shorted branch; total current will increase.
3. Will cause a corresponding opposite change in branch current and total current
4. False

STEP 5-1 & 5-2

1. By tracing current to see if it has one path (series connection) or more than one path (parallel connection)
2. $R_{1,2} = R_1 + R_2 = 12\text{ k}\Omega + 12\text{ k}\Omega = 24\text{ k}\Omega$
$$= R_{1,2,3} = \frac{R_{1,2} \times R_3}{R_{1,2} + R_3} = \frac{24\text{ k}\Omega \times 6\text{ k}\Omega}{24\text{ k}\Omega + 6\text{ k}\Omega}$$
$$= \frac{144\text{ k}\Omega}{30\text{ k}\Omega} = 4.8\text{ k}\Omega$$
3. Step A: Find equivalent resistances of series-connected resistors. Step B: Find equivalent resistances of parallel-connected combinations. STEP C: Find equivalent resistances of remaining series-connected resistances.
4. $R_{1,2} = R_1 + R_2 = 470 + 330 = 800\ \Omega$
$$R_{1,2,3} = \frac{R_{1,2} \times R_3}{R_{1,2} + R_3} = \frac{800 \times 270}{800 + 270} = \frac{216\text{ k}\Omega}{1.07\text{ k}\Omega} = 201.9\ \Omega$$

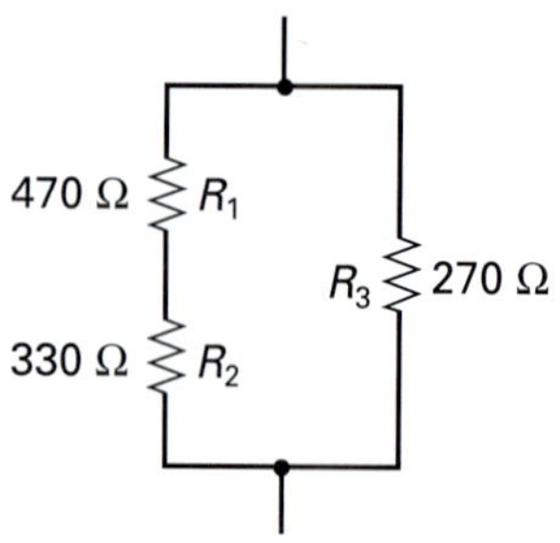

Figure 5-2, 4

STEP 5-3

1. Step 1: Find total resistance (Steps A, B and C). Step 2: Find total current. Step 3: Find voltage drop with $I_T \times R_X$.
2. The voltage drops previously calculated would not change since the ratio of the series resistor to the series equivalent resistors remains the same, and therefore the voltage division will remain the same.

STEP 5-4, 5-5, & 5-6

1. Find total resistance; find total current; find the voltage across each series and parallel combination resistors; find the current through each branch of parallel resistors; find the local and individual power dissipated.
2. This is a do-it-yourself question; each answer will vary.

STEP 5-7

1. When a load resistance changes the circuit and lowers output voltage

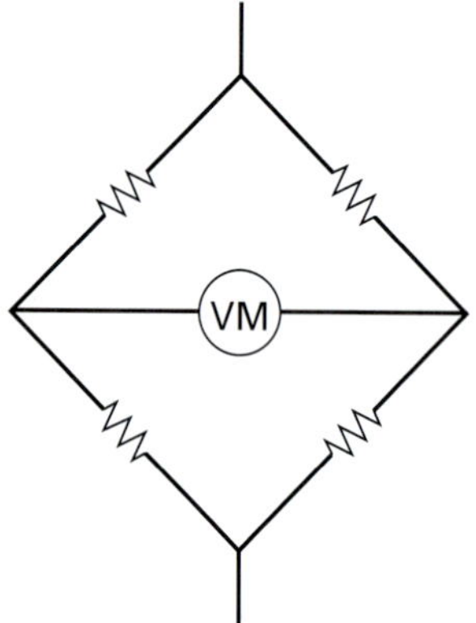

Figure 5-7, 2

2. Used to check for an unknown resistor's resistance.

STEP 5-8

1. If component is in series with circuit, no current will flow and source voltage will be across bad component. If component is parallel, no current will flow in that branch.
2. Total resistance will decrease and bad component will have 0 V dropped across it.
3. Will cause the circuit's behavior to vary

STEP 6-1

1. **a.** Alternating current
   b. Direct current
2. AC
3. DC

4. DC flows in one direction, whereas ac first flows in one direction and then in the opposite direction.

STEP 6-2-1

1. AC generators can be larger, less complex, and cheaper to run; transformers can be used with ac to step up/down, so low-current power lines can be used; easy to change ac to dc, but hard the other way around.
2. False
3. $P = I^2 \times R$
4. A device that can step up or down ac voltages
5. 120 V ac
6. AC, dc

STEP 6-2-2 & 6-2-3

1. Sound, electromagnetic, electrical
2. 1133 feet per second; 186,000 miles per second
3. Sound, electromagnetic, electrical
4. Electronic
5. Electrical

STEP 6.3

1.

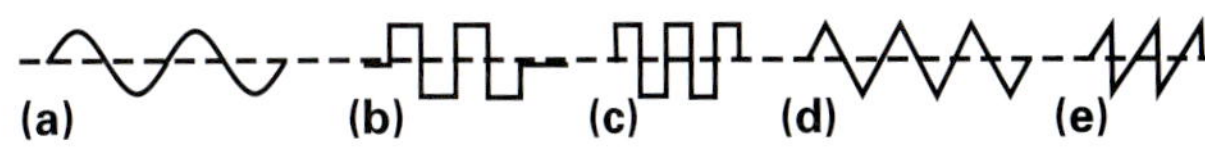

2. (Sound wave): $\lambda(\text{mm}) = \dfrac{344.4 \text{ m/s}}{f(\text{Hz})}$;

   (Electromagnetic wave): $\lambda(\text{m}) = \dfrac{3 \times 10^8 \text{ m/s}}{f(\text{Hz})}$;

   different because sound waves travel at a different speed than do electromagnetic waves.

STEP 6-4-1

1. Yes
2. A device or circuit that converts ac to pulsating dc

STEP 6-4-2

1. Cathode ray tube
2. Period = 80 μs
   Frequency = 12.5 kHz
3. Peak = 4 V
   Peak-to-Peak = 8 V
4. It allows us to make comparisons between the phase, amplitude, shape, and timing of two signals.

STEP 6-4-3

1. Sine, square, triangular, sawtooth
2. Analyze the frequency of any periodic wave

STEP 7-1

1. Plate area, distance between plates, type of dielectric used
2. $C = \dfrac{(8.85 \times 10^{-12}) \times K \times A}{d}$
3. Double
4. Double
5. 63.2%
6. 36.8%

STEP 7-2

1. $C_T = \dfrac{1}{(1/C_1) + (1/C_2) + (1/C_3)}$

   $= \dfrac{1}{(1/2\ \mu\text{F}) + (1/3\ \mu\text{F}) + (1/5\ \mu\text{F})}$

   $= 0.968\ \mu\text{F}$ or $1\ \mu\text{F}$
2. $C_T = C_1 + C_2 + C_3 = 7$ pF + 2 pF + 14 pF = 23 pF
3. $V_{CX} = (C_T/C_X) \times V_T$
4. True

STEP 7-3

1. 470 pF, 2% tolerance
2. 0.47 μF or 470 nF, 5% tolerance
3. **a.** Electrolytic
   b. ceramic

STEP 7-4

1. Opposition to current flow without the dissipation of energy
2. $X_C = 1/2\pi fC$
3. When frequency or capacitance goes up, there is more charge and discharge current, so X_C is lower.
4. $X_C = 1/2\pi fC = 1/2\pi \times 4$ kHz $\times 4$ μF = 9.95 Ω

STEP 7-5

1. Current leads voltage by some phase angle less than 90%.
2. Z = total opposition to current flow; $Z = \sqrt{R^2 + X_C{}^2}$

STEP 7-6

1. Resistor current is in phase with voltage; capacitor current is 90° out of phase (leading) with voltage.
2. No

STEP 7-7

1. 0.5 μF
2. Capacitance meter or analyzer

STEP 7-8

1. **a.** Filter
   b. Voltage divider
   c. Differentiator
2. Capacitor, resistor
3. Long
4. Differentiator

STEP 8-1

1. True
2. DC
3. AC
4. Magnetic-type circuit breaker, relays
5. An NO relay is open when off; NC is closed when off.

STEP 8-2

1. The voltage induced or produced in a coil as the magnetic lines of force link with the turns of a coil
2. Faraday's: When the magnetic flux linking a coil is changing, an emf is induced.
3. Sine wave

STEP 8-3

1. Number of turns, area of coil, length of coil, core material used
2. $L = \dfrac{N^2 \times A \times \mu}{l}$
3. Current in an inductive circuit builds up in the same way that voltage does in a capacitive circuit, but the capacitive time constant is proportional to resistance, where the inductive time constant is inversely proportional to resistance.

4. True
5. False
6. a. $L_T = L_1 + L_2 = 4\text{ mH} + 2\text{ mH} = 6\text{ mH}$
   b. $L_T = \frac{L_1 \times L_2}{L_1 + L_2}$ (using product over sum)
   $= \frac{4\text{ mH} \times 2\text{ mH}}{4\text{ mH} + 2\text{ mH}} = \frac{8\text{ mH}}{6\text{ mH}} = 1.33\text{ mH}$
7. False
8. $V_S = \sqrt{V_R^2 + V_L^2} = \sqrt{4^2 + 2^2} = \sqrt{16 + 4} = \sqrt{20} =$ 4.47 V
9. The total opposition to current flow offered by a circuit with both resistance and reactance: $Z = \sqrt{R^2 + X_L^2}$

STEP 8-4

1. True
2. False
3. Turns ratio $= N_s/N_p = 1608/402 = 4$; step up
4. $V_s = N_s/N_p \times V_p$
5. False
6. $I_s = N_p/N_s \times I_p$
7. Turns ratio $= \sqrt{Z_L/Z_S} = \sqrt{75\ \Omega/25\ \Omega} = \sqrt{3} = 1.732$
8. $V_s = (N_s/N_p) \times V_p = (200/112) \times 115 = 205.4$ V
9. 10 kVA is the apparent power rating, 200 V the maximum primary voltage, 100 V the maximum secondary voltage, at 60 cycles per second (Hz)
10. $R_L = V_s/I_s = 1\text{ kV}/8\text{ A} = 125\ \Omega$; $125 > 100$, so the transformer will overheat and possibly burn out.

STEP 9-1

1. Calculate the inductive and capacitive reactance (X_L and X_C), the circuit impedance (Z), the circuit current (I), the component voltage drops (V_R, V_L, and V_C), and the power distribution and power factor (PF).
2. a. $Z = \sqrt{R^2 + (X_L \sim X_C)^2}$
   b. $I = V_s/Z$
   c. Apparent power $= V_s \times I$ (volt-amperes)
   d. $V_S = \sqrt{V_R^2 + (V_L \sim V_C)^2}$
   e. True power $= I^2 \times R$(watts)
   f. $V_R = I_1 \times R$
   g. $V_L = I \times X_L$
   h. $V_C = I \times X_C$
   i. $\theta = \arctan \frac{V_L \sim V_C}{V_R}$
   j. PF $= \cos\theta$

STEP 9-2

1. a. $I_R = V/R$
   b. $I_T = \sqrt{I_R^2 + I_X^2}$
   c. $I_C = V/X_C$
   d. $I_L = V/X_L$
2. a. $P_R = I^2 \times R$
   b. $P_X = I^2 \times X_L$
   c. $P_A = V_S \times I_T$
   d. PF $= \cos\theta$

STEP 9-3-1

1. A circuit condition that occurs when the inductive reactance (X_L) and the capacitive reactance (X_C) have been balanced
2. A series *RLC* circuit that at resonance X_L equals X_C, so V_L and V_C will cancel, and $Z = R$
3. Voltage across L and C will measure 0; impedance only equals R; voltage drops across inductor or capacitor can be higher than source voltage
4. Q factor indicates the quality of the series resonant circuit, or is the ratio of the reactance to the resistance
5. Group or band of frequencies that causes the larger current flow
6. $\text{BW} = f_0/Q = 12\text{ kHz}/1000 = 12$ Hz

STEP 9-3-2

1. In a series resonant *RLC* circuit, source current is maximum; in a parallel resonant circuit, source current is minimum at resonance.
2. Oscillating effect with continual energy transfer between capacitor and inductor
3. Q equals $X_L/R = 50\ \Omega/25\ \Omega = 2$
4. Yes
5. Ability of a tuned circuit to respond to a desired frequency and ignore all others

STEP 9-4

1. a. High pass
   b. Low pass
   c. Band stop
   d. Band pass
2. Television, radio, and other communications equipment

STEP 10-1

1. Diodes, transistors and integrated circuits (ICs)
2. Current or voltage

STEP 10-2

1. They all have 4 valence electrons.
2. 1, 4
3. Covalent bond
4. Negative, decreases
5. Increases
6. Intrinsic, positive, negative

STEP 10-3

1. To increase their conductivity
2. Electrons, N
3. Holes, P
4. Electrons, holes

STEP 10-4

1. Depletion region
2. (a) 700 mV
3. True
4. Open, closed

STEP 11-1

1. 0.7 V
2. 0.7 V
3. One

STEP 11-2

1. True
2. Reverse
3. Zener has a "Z" shaped bar instead of a straight bar.

STEP 11-3

1. (b) 2 V
2. Series current limiting resistor

STEP 11-4

1. True
2. True

STEP 12-1
1. NPN and PNP
2. Emitter, base, and collector
3. As a switch, and as a variable-resistor
4. The two-state switching action is used in digital circuits, while the variable-resistor action is used in analog circuits.

STEP 12-2
1. Current
2. Forward, reverse
3. (b)
4. Open switch
5. Closed switch
6. **(a)** Common-base
   (b) Common-collector
   (c) Common-emitter
7. Voltage-divider bias
8. Base-biasing

STEP 13-1
1. Current, voltage
2. Reverse
3. True
4. False
5. Transconductance
6. Very high, reverse
7. R_S
8. Voltage divider
9. Common-source
10. Common-source
11. Common-drain
12. Its high input impedance
13. Because it responds well to the small signal voltages from the antenna, and because it is a low noise component

STEP 13-2-1
1. Depletion, enhancement
2. True
3. ON
4. MOSFET
5. (d)
6. Dual-gate D-MOSFET

STEP 13-2-2
1. False
2. OFF
3. (c) Drain feedback biased
4. True

STEP 14-1
1. Decrease
2. False
3. Bidirectional
4. DIAC
5. True
6. One
7. False
8. Gate

STEP 14-2
1. Resistor
2. Electrical, solar
3. Photoconductive and photovoltaic
4. Decrease
5. Electrical to light, electrical to light
6. True
7. False
8. PTC

STEP 15-1
1. Yes
2. An open-loop op-amp circuit has no feedback, whereas a closed-loop op-amp circuit does have a feedback path.
3. High gain, very high input impedance, and very low output impedance
4. Differential amplifier, darlington-pair voltage amplifier, emitter-follower output amplifier
5. (a) Comparator
6. Negative or degenerative feedback will lower the op-amp's gain to
   a. prevent output waveform distortion,
   b. prevent the amplifier from going into oscillation, and
   c. reduce the gain of the op-amp to a consistent value.
7. Voltage follower
8. Summing amplifier
9. The differentiator circuit has an input capacitor and a feedback resistor, while the integrator circuit has an input resistor and a feedback capacitor.
10. Differential or difference amplifier
11. See Figure 13-11(b)
12. An active filter circuit will filter and amplify the signal input, while a passive filter will only filter the signal input.

STEP 15-2
1. Op-amps, voltage regulators, function generators, phase-locked loops
2. Input voltage, load resistance
3. Variable resistor
4. Switch

STEP 15-3
1. Free-running
2. Mono-stable
3. Bistable
4. Because it can be latched in the set or reset condition
5. The three 5 kΩ voltage divider
6. Astable, monostable
7. Better frequency stability

STEP 15-4
1. Voltage controlled oscillator
2. Control voltage
3. Sine, square, triangular

STEP 15-5
1. Phase comparator, amplifier and low-pass filter, and voltage controlled oscillator
2. A phase-locked loop circuit consists of a phase comparator that compares the output frequency of a voltage controlled oscillator with an input frequency. The error voltage out of the phase comparator is then coupled via an amplifier and low-pass filter to the control input of the voltage controlled oscillator to keep it in phase, and therefore at exactly the same frequency as the input frequency.
3. A frequency divider circuit
4. Pin 7

STEP 16-1
1. Section 16-1-1
2. Section 16-1-1
3. Section 16-1-2
4. Circuit simplicity and accuracy

5. 2
6. Section 16-1-2

STEP 16-2

1. CPU, memory, I/O
2. Analog to Digital Converter
   Digital to Analog Converter
3. Analog, digital
4. Digital, analog

STEP 17-1

1. No difference
2. Base 10
3. 2 = 1000, 6 = 100, 3 = 10, 9 = 1
4. Reset and carry

STEP 17-2

1. 26_{10}
2. 10111_2
3. LSB = Least Significant Bit, MSB = Most Significant Bit
4. 1101110_2

STEP 17-3

1. (a) Base 10 (b) Base 2 (c) Base 16
2. 21_{10} 15_{16}
3. BF7A
4. Binary equivalent = 100001
Hexadecimal equivalent = 21

STEP 17-4

1. 8
2. 232_B
3. 4
4. 4267_B

STEP 18-1

1. OR, AND
2. Binary 1
3. AND gate
4. OR
5. Binary 0
6. AND

STEP 18-2

1. NOT gate
2. Binary 0
3. NAND
4. NOR
5. Binary 1
6. NAND

STEP 18-3

1. OR
2. XOR
3. XNOR
4. XNOR

STEP 18-4

1. IEEE/ANSI
2. Instead of using distinctive shapes to represent logic gates, it uses a special dependency notation system to indicate how the output is dependent on the input.

STEP 19-1

1. **a.** AND
   b. XOR
   c. OR
   d. NAND
   e. NOR
   f. NAND
2. $\overline{A} \cdot \overline{B} = \overline{A + B}$
3. **a.** $\overline{A} \cdot \overline{B}$ **b.** $(\overline{A} \cdot \overline{B}) \cdot C$
4. **a.** A, B, C → Y
   b. L, D → Y
   c. A, B → Y
   d. A, B, C → Y
   e. A, L, D → Y
   f. A, B, C → Y

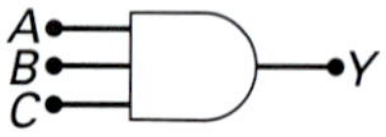

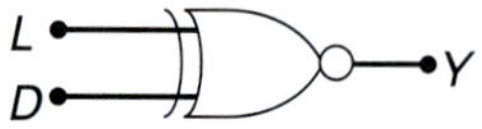

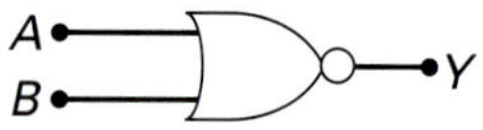

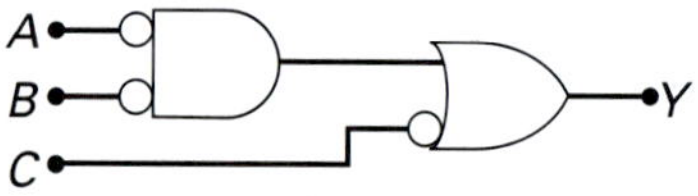

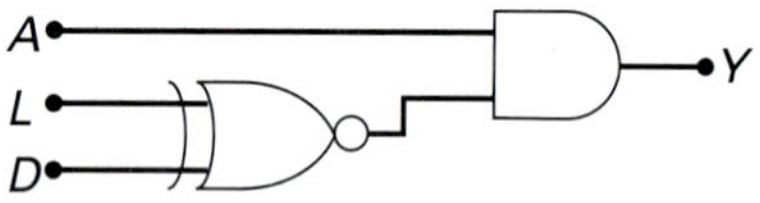

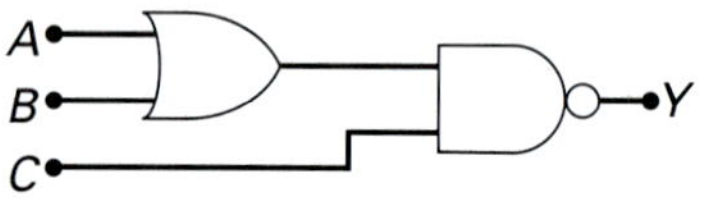

STEP 19-2

1. **a.** 1
   b. 1
   c. 0
   d. A
2. $A(B + C) = (AB) + (AC)$
3. $(AB)C$ or $(AC)B$
4. $A(B + C + D)$

STEP 19-3

1. Product, sum
2. **a.** $A\,\overline{B}\,C\,D$
   b. $\overline{A}\,B\,C\,\overline{D}$
3. SOP
4. Fundamental products are ORed and then the circuit is developed.

STEP 19-4

1. 4
2. $y = B(A + \overline{A}) = B(1) = B$
3. Larger
4. With 1s for larger grouping

STEP 20-1

1. Transistor to transistor logic
2. Ten
3. HIGH
4. LOW, HIGH, and high impedance
5. High-speed applications
6. Higher packing density

STEP 20-2

1. True
2. Voltage control, high input impedance, and low power dissipation

3. P channel and N channel
4. Low power

STEP 20-3

1. **(a)** Transistor outline
   (b) Dual-in-line package
   (c) Surface-mount technology
2. SMT
3. (a) SSI
4. E-MOSFET, MOS

STEP 20-4

1. Lower cost
2. CMOS
3. Signal voltage and current incompatibility
4. Provides interfacing by shifting signal voltage for compatibility
5. BiCMOS family
6. TTL series

STEP 21-1

1. Section 21-1, margin definitions
2. Section 21-1
3. Section 21-1
4. Section 21-1

STEP 21-2

1. Section 21-2, margin definitions
2. Section 21-2
3. Section 21-2
4. Section 21-2

STEP 22-1

1. Voltmeter
2. Oscilloscope
3. Pulser
4. Current tracer, logic probe

STEP 22-2

1. New chip substitution
2. Open, short
3. Midpoint start during isolation
4. 2-state timing signal
5. Logic pulser and current tracer, logic probe

STEP 22-3

1. **a.** Record orientation of IC
   b. Desolder using solder wick or vacuum bulb (grounded iron is required with MOS devices)
   c. Avoid overheating board or component
2. A final test is the application of power and clock to ensure all aspects of the system operate properly.

STEP 23-1

1. All HIGH except line 9 which would be LOW
2. **a.** 10
   b. 16
3. Output "*b*" would be the only HIGH
4. Matching low power consumption

STEP 23-2

1. Ten to four line
2. Inverter gates wired to each 74147 output
3. 0111 (7 has priority)
4. **a.** 1101
   b. 0001
   c. HLHH
   d. OFF, ON, OFF, ON

STEP 23-3

1. (b) One of several, one
2. 4
3. False
4. Parallel
5. Parallel
6. 74157

STEP 23-4

1. (a) One, one of several
2. Decoder
3. $C = 1, B = 0, A = 0$
4. True

STEP 23-5

1. XNOR
2. The three outputs are labeled $A < B$, $A > B$, and $A = B$
3. For additional chips to handle larger word size
4. $A < B = 1, A > B = 0, A = B$ is 0

STEP 23-6

1.

Decimal	Gray Code	Odd-Parity Bit
0	0000	1
1	0001	0
2	0011	1
3	0010	0
4	0110	1
5	0111	0

2. True
3. Both, as a parity generator and a parity checker
4. 8-bits (7-bit ASCII code plus 1-bit for parity)

STEP 23-7

1. Diagnose, isolate, and repair
2. A single pulse condition with no clock
3. Midpoint selection to begin isolation
4. Both displays would be OFF at the same time and ON at the same time. Displays would therefore show same data.

STEP 24-1

1. Latch
2. Level-triggered, edge-triggered, and pulse-triggered
3. No change, race
4. **a.** Edge triggering
   b. Pulse triggering
5. A latch to eliminate switch noise
6. Input delay is accomplished by capacitor charge time

STEP 24-2

1. Data
2. No
3. D flip-flop has a single bit input.
4. (d) All of the above

STEP 24-3

1. (d)
2. HIGH, toggle
3. (b) A 63 kHz square wave
4. (c) *J-K*, *D*-type, *S-R*

STEP 24-4

1. Monostable
2. Bistable
3. Pulse generation, pulse delay, and timing
4. An internal 5Ω voltage divider

STEP 25-1

1. (c) PIPO
2. Data storage

STEP 25-2

1. 8 pulses
2. SIPO, PISO
3. 1 pulse
4. Bidirectional

STEP 25-3

1. High impedance (floating)
2. BUS
3. Bus conflict occurs when more than one signal seeks control of the bus due to lack of signal isolation. It is avoided through the use of tristate devices.
4. HIGH, LOW, and control inputs

STEP 25-4

1. LOAD = bus data enters register. ENABLE = data read (output) is allowed
2. **a.** A bus is a group of parallel wires designed to carry binary data.
   b. A bus organized digital system occurs when one set of wires is connected to all devices—as long as those that transmit data onto the bus have three-state output control.
   c. Bus conflict occurs when more than one signal seeks control of the bus due to lack of signal isolation.
   d. A 16-bit bus utilizes 16 wires.
3. Universal Asynchronous Receiver and Transmitter. It can be used to interface the parallel data format within a digital electronics system to the serial data format of external equipment.
4. 64 times larger
5. The feedback path is twisted with the johnson counter.

STEP 25-5

1. Are not, do not
2. Modulus = 256, maximum count = 255
3. 30 ns × 8 = 240 ns, $1 \times 10^9 \div 240 = 4.2$ MHz
4. Natural modulus = 256, preset modulus = 226

STEP 25-6

1. False
2. Asynchronous has fewer gates, synchronous counter is faster
3. Carry, borrow
4. Count

STEP 25-7

1. Binary counter, frequency divider
2. They function as counters.
3. They function as frequency dividers.
4. They function as counters.

STEP 25-8

1. Diagnose, isolate, and repair
2. Count window 74LS160s function as frequency dividers. Mod-10 74LS160s function as counters.
3. No problem exists within dividers.
4. Display should read 0000 Hz.

STEP 26-1

1. **(a)** 1011 + 11101 = 101000
   (b) 1010 − 1011 = 1111 0001
   (c) 101 × 10 = 1010
   (d) 10111 ÷ 10 = 1011
2. **(a)** 0101
   (b) 1001 0101
   (c) 1101
   (d) 0111 1001
3. **(a)** −5
   (b) +6
   (c) −5
   (d) +6
4. Adder circuit

STEP 26-2

1. 2, 3 bits
2. (c) Parallel adder
3. The 7483 is an 8-bit, parallel, 2-word adder or a 4-bit binary full-adder with fast carry.
4. SIGMA

STEP 26-3

1. ICs are the two 7486 XOR chips.
2. A 1 is added via the carry-in input of the 7483.
3. **(a)** $F = \overline{B}$
   (b) $F = A$ minus B
4. 1001 ($F = A$ plus B)

STEP 27-1

1. Nonvolatile
2. A memory's address input is a gating system designed to access a certain cell. The number of address lines is related to these memory locations as function of 2^n.
3. **(a)** Storage density is a comparison of one memory content to another.
   (b) Memory configuration describes how the circuit is organized into groups of bits or words.
   (c) Burning-in is the process of entering data into a ROM.
   (d) Access time is the interval between a new address input and the data stored at that address being available at the output.
4. **(a)** MROM = Masked Read-Only Memory
   (b) PROM = Programmable Read-Only Memory
   (c) EPROM = Erasable Programmable Read-Only Memory
   (d) EEPROM = Electrically Erasable PROM
5. Input code is applied to the address inputs of a ROM where it is decoded internally to select that memory location, which, in turn, supplies the desired output code.
6. A ROM listing is a sheet that lists the data values that should be stored in each memory location within the ROM. ROM tests may be made by accessing each word location to check stored content with the listing.

STEP 27-2

1. Once programmed, the contents of a ROM cannot be changed.
2. With a SAM, one must sequence through all bits until the desired word is located. This is accomplished directly with a RAM due to its random-access capabilities.
3. No. Although access is the same, a ROM has no "write" capability.
4. A volatile device

5. A dynamic RAM MOS cell capacitor requires refreshing, whereas a static RAM has bipolar cells.
6. A SRAM has fast, simple circuitry, whereas the DRAM has more packing density and lower power consumption due to fewer MOSFETs per cell.
7. A memory map is an address-listing diagram showing which address blocks are assigned to each part of the system.
8. RAMs are tested by writing a test pattern into every memory location and then reading out and testing the stored contents. The stored patterns could be all 0s, then all 1s and then a checkerboard pattern.

STEP 28-1

1. (a) ADC = Analog-to-Digital Converter
   (b) DAC = Digital-to-Analog Converter
2. ADC, DAC
3. Analog, digital
4. Digital, analog

STEP 28-2

1. A weighted resistor DAC disadvantage is that as bits increase, large branch current differences may introduce noise or cause output load change.
2. The op-amp within a DAC is used to convert a current input into an equivalent voltage output.
3. (a) A staircase is a continually increasing analog output.
   (b) Resolution of a DAC is the smallest analog output change that may occur as a result of an increment in the digital input.
   (c) Step size is the DAC resolution.
   (d) Monotonicity describes correct staircase waveform behavior.
   (e) Settling time is the time it takes for a change in a DAC's analog output to settle to 99.95% of the new value.
   (f) The relative accuracy of a DAC describes how much the output level has deviated from its ideal theoretical output value.
4. DACs may be tested with an input counter to produce an output that may be viewed on an oscilloscope to determine high gain, offset, nonmonotonicity, and other problems.

STEP 28-3

1. The main disadvantage of the staircase ADC is that its conversion speed is dependent on the analog input voltage.
2. The staircase and successive approximation ADC circuits make use of a DAC.
3. A successive approximation ADC operates by first setting and then comparing bits to obtain an equivalent digital output.
4. The flash converter is the fastest type of ADC.
5. An ADC may be tested by applying a linear ramp to the analog input while monitoring the digital output with a display.

STEP 29-1

1. a. Machine language—the digital code understood by the MPU
   b. Assembly language—programs written using mnemonics, which are easier to remember
   c. High-level language—an independent language designed to make programming easier that later is compiled into machine language for the MPU
2. A flowchart is a graphic form that describes the operation of a program using oval, rectangular, and diamond-shaped blocks.
3. a. Instruction set—contains a list of the machine codes that are recognized by a specific type of microprocessor
   b. Peripherals—input and output devices connected to microprocessor system ports
   c. Programmer—a person who writes a program in any language that will be understood by the MPU
   d. Mnemonic—a short name, or memory aid, used in assembly language programming
   e. Compiler—a program that translates high-level language into machine code
   f. Accumulator—an MPU internal register (register A), used in conjunction with the ALU; a storage location within an MPU chip
   g. Flag register—an internal register within an MPU chip that is a collection of flip-flops used to indicate the results of certain instructions
   h. Op-code—the operation code or "do" portion of an instruction

STEP 29-2

1. A unidirectional bus receives data from only one transmitter chip, or "talker." Data transfers are therefore one-way only. A bidirectional bus has several "talkers"; therefore, data flow may be reversed. However, those transmitters must have three-state output control to avoid bus conflict.
2. 8085 pin-out signals are:
   a. *WR*—indicates data bus data is to be written into the selected memory or I/O location
   b. *RD*—indicates selected memory or I/O location is to be read. Data bus is ready for transfer.
   c. *IO/NM*—HIGH indicates I/O read/write is to I/O; LOW indicates read/write is to memory
   d. *RESET*—resets CPU. Once the program counter, interrupt enable, and HLDA flip-flops are reset, it allows a peripheral read.
   e. *READY*—HIGH allows completion of read/write cycle, while a LOW creates a wait state
   f. *ALE*—the address-latch enable signal clocks the lower-order address bits for storage
   g. *HOLD*—causes CPU to relinquish the use of buses at the end of the present machine cycle for use by another master, usually a peripheral
   h. *TRAP*—highest-priority signal of any interrupt request; recognized as INTR occurs
3. During FETCH, the op-code is read from memory and transferred to the instruction register. During EXECUTE, control/timing logic decodes the op-code, increments the program register, and retrieves the bytes necessary to complete the task.
4. a. Interrupt—provides a means for devices external to the microprocessor to request immediate action by the processor
   b. T-state—the basic unit of time for a microprocessor system equal to one clock period
   c. Machine cycle—the time it takes an operation to be completed; from three to six states
   d. Instruction cycle—the time it takes to fetch and execute a complete instruction; usually from one to five machine cycles.

APPENDIX

B

Answers to Odd-Numbered Problems

Chapter 1

1. b **3.** d **5.** a **7.** a **9.** b **11.** d **13.** d **15.** d

17. $n = Q \times (6.24 \times 10^{18}) = 6.5 \times (6.24 \times 10^{18}) = 4.056 \times 10^{19}$ electrons

19. **a.** 14 mA **b.** 1.374 kA **c.** 776 nA **d.** 910 μA

21. 0.945/1 = 0.945

23. 10 V/750 kV = 0.0000133 cm or 0.000133 mm

25. 35 × (2000 kV/cm ÷ 10) = 7000 kV

Chapter 2

1. a **3.** a **5.** d **7.** a **9.** d **11.** a **13.** c **15.** b

17. $V = I \times R = 8\text{ mA} \times 16\text{ k}\Omega = 128\text{ V}$

19. $P = I \times V$, 2.4 kW, 1.024 W, 1.2 kW

21. Total cost = power in (kW) × time × cost per hour
- **a.** 0.3 × 10 × 9 = 27 cents
- **b.** 0.1 × 10 × 9 = 9 cents
- **c.** 0.06 × 10 × 9 = 5 cents
- **d.** 0.025 × 10 × 9 = 2 cents

23. $R = \rho \times l/a = 10.7 \times 200/80^2 = 334\text{ m}\Omega$

25. $V = I \times R = 7.5\text{ A} \times 0.2485\ \Omega = 1.86375\text{ V}$

27. $W = Q \times V$, $V = W/Q = 1000\text{ J}/40\text{ C} = 25\text{ V}$

29. $R = V/I$
- **a.** 120 V/20 mA = 6 kΩ
- **b.** 12 V/2 A = 6 Ω
- **c.** 9 V/100 μA = 90 kΩ
- **d.** 1.5 V/4 mA = 375 Ω; so **c.** has the largest, **b.** the smallest

31. $P = W/t = 5000\text{ J}/25\text{ s} = 200\text{ W}$

33. $P = I^2 \times R$, $R = P/I^2 = 100\text{ W}/4\text{ A}^2 = 6.25\ \Omega$

35. $P = V \times I = 12\text{ V} \times 300\text{ mA} = 3.6\text{ W}$

37. **a.** $P = I^2 \times R = 50\text{ mA}^2 \times 10\text{ k}\Omega = 25\text{ W}$; **b.** no

39.
- **a.** 1.2 MΩ × 0.1 = ±120 kΩ = 1.08 MΩ to 1.32 MΩ
- **b.** 10 × 0.05 = ±0.5 = 9.5 Ω to 10.5 Ω
- **c.** 27 kΩ × 0.2 = ±5.4 kΩ = 21.6 kΩ to 32.4 kΩ
- **d.** 273 kΩ × 0.005 = ±1.365 kΩ = 271.635 kΩ to 274.365 kΩ
- **e.** 33 × 0.2 = ±6.6 = 26.4 Ω to 39.6 Ω
- **f.** 22.5 kΩ × 0.02 = ±450 = 22.05 kΩ to 22.95 kΩ
- **g.** 10 kΩ × 0.05 = ±500 = 9.5 kΩ to 10.5 kΩ
- **h.** 910 × 0.1 = ±91 = 819 Ω to 1001 Ω

41. See Section 2-5-4

43. In tolerance

Chapter 3

1. d
3. c
5. a
7. d
9. a
11. c
13. d
15. d

17. $I = \dfrac{V_S}{R_T}$, $R_T = R_1 + R_2 = 40 + 35 = 75\ \Omega$

$I = 24/75 = 320\text{ mA}$; 150 Ω (double 75 Ω) needed to half current

19. 40 Ω, 20 Ω, 60 Ω. (Any values can be used, as long as the ratio remains the same.)

21. $I_{R1} = I_T = 6.5\text{ mA}$

23. $P_T = P_1 + P_2 + P_3 = 120 + 60 + 200 = 380\text{ W}$;
$I_T = P_T/V_S = 380/120 = 3.17\text{ A}$
$V_1 = P_1/I_T = 120\text{ W}/3.17\text{ A} = 38\text{ V}$
$V_2 = P_2/I_T = 60\text{ W}/3.17\text{ A} = 18.9\text{ V}$
$V_3 = P_3/I_T = 200\text{ W}/3.17\text{ A} = 63.1\text{ V}$

25.
a. $R_T = R_1 + R_2 + R_3 = 22\text{ k}\Omega + 3.7\text{ k}\Omega + 18\text{ k}\Omega = 43.7\text{ k}\Omega$
$I = V/R = 12\text{ V}/43.7\text{ k}\Omega = 274.6\ \mu\text{A}$

b. $R_T = V/I = 12\text{ V}/10\text{ mA} = 1.2\text{ k}\Omega$
$P_T = V \times I = 12\text{ V} \times 10\text{ mA} = 120\text{ mW}$

c. $R_T = R_1 + R_2 + R_3 + R_4 = 5 + 10 + 6 + 4 = 25\ \Omega$
$V_S = I \times R_T = 100\text{ mA} \times 25\ \Omega = 2.5\text{ V}$
$V_{R_1} = I \times R_1 = 100\text{ mA} \times 5 = 500\text{ mV}$
$V_{R_2} = I \times R_2 = 100\text{ mA} \times 10 = 1\text{ V}$
$V_{R_3} = I \times R_3 = 100\text{ mA} \times 6 = 600\text{ mV}$
$V_{R_4} = I \times R_4 = 100\text{ mA} \times 4 = 400\text{ mV}$
$P_1 = I \times V_1 = 100\text{ mA} \times 500\text{ mV} = 50\text{ mW}$
$P_2 = I \times V_2 = 100\text{ mA} \times 1\text{ V} = 100\text{ mW}$.
$P_3 = I \times V_3 = 100\text{ mA} \times 600\text{ mV} = 60\text{ mW}$
$P_4 = I \times V_4 = 100\text{ mA} \times 400\text{ mV} = 40\text{ mW}$

d. $P_T = P_1 + P_2 + P_3 + P_4 = 12\text{ mW} + 7\text{ mW} + 16\text{ mW} + 3\text{ mW} = 38\text{ mW}$
$I = P_T/V_S = 38\text{ mW}/12.5\text{ V} = 3.04\text{ mA}$
$R_1 = P_1/I^2 = 12\text{ mW}/(3.04\text{ mA})^2 = 1.3\text{ k}\Omega$
$R_2 = P_2/I^2 = 7\text{ mW}/(3.04\text{ mA})^2 = 757.4\ \Omega$
$R_3 = P_3/I^2 = 16\text{ mW}/(3.04\text{ mA})^2 = 1.73\text{ k}\Omega$
$R_4 = P_4/I^2 = 3\text{ mW}/(3.04\text{ mA})^2 = 324.6\ \Omega$

27. Zero voltage drop across the shorted component, while there is also an increase in voltage across the others.

29.
- **a.** No current at all (zero)
- **b.** Go to infinity (∞)
- **c.** Measure source voltage
- **d.** No voltage across any other component

Chapter 4

1. b
3. d
5. c
7. b
9. a

11. $R_T = R/\text{no. of } R\text{'s} = 30\text{ k}\Omega/4 = 7.5\text{ k}\Omega$

13. $R_T = R/\text{no. of } R\text{'s} = 25/3 = 8.33\ \Omega$
$I_T = V_S\, R_T = 10/8.33 = 1.2\text{ A}$
$I_1 = I_2 = I_3 = R_T/R_x \times I_T = 8.33/25 \times 1.2 = I_T/\text{no. of } R\text{'s} = 1.2\text{ A}/3 = 400\text{ mA}$

15. $I_T = V_S/R_t = 14/700 = 20$ mA
$I_X = I_T/\text{no. of } R = 20 \text{ mA}/3 = 6.67$ mA

17. a. $R_T = \dfrac{R_1 \times R_2}{R_1 + R_3} = \dfrac{33 \times 22}{33 + 22} = \dfrac{726 \text{ k}\Omega}{55 \text{ k}\Omega} = 13.2 \text{ k}\Omega$
b. $I_T = V_S/R_T = 20/13.2 \text{ k}\Omega = 1.5$ mA
c. $I_1 = R_T/R_1 \times I_T = (13.2 \text{ k}\Omega/33 \text{ k}\Omega) \times 1.5 \text{ mA} = 600\ \mu\text{A}$
$I_2 = R_T/R_2 \times I_T = 13.2 \text{ k}\Omega/22 \text{ k}\Omega \times 1.5 \text{ mA} = 900\ \mu\text{A}$
d. $P_T = I_T \times V_S = 1.5 \text{ mA} \times 20 = 30$ mW
e. $P_1 = I_1 \times V_1, (V_1 = V_S = 20 \text{ V})$
$600\ \mu\text{A} \times 20 = 12$ mW
$P_2 = I_2 \times V_2, (V_2 = V_S = 20 \text{ V}), 900\ \mu\text{A} \times 20 = 18$ mW

19. a. $R_T = \dfrac{R_1 \times R_2}{R_1 + R_2} = \dfrac{22 \text{ k}\Omega \times 33 \text{ k}\Omega}{22 \text{ k}\Omega + 33 \text{ k}\Omega} = \dfrac{726 \text{ k}\Omega}{55 \text{ k}\Omega} = 13.2 \text{ k}\Omega$
$I_T = \dfrac{V_S}{R_T} = \dfrac{10}{13.2 \text{ k}\Omega} = 757.6\ \mu\text{A}$
$I_1 = \dfrac{R_T}{R_1} \times I_T = \dfrac{13.2 \text{ k}\Omega}{22 \text{ k}\Omega} \times 757.6\ \mu\text{A} = 454.56\ \mu\text{A}$
$I_2 = \dfrac{R_T}{R_1} \times I_T = \dfrac{13.2 \text{ k}\Omega}{33 \text{ k}\Omega} \times 757.6\ \mu\text{A} = 303.04\ \mu\text{A}$
b. $R_T = \dfrac{1}{(1/R_1) + (1/R_2) + (1/R_3)} = \dfrac{1}{(1/220\ \Omega) + (1/330\ \Omega) + (1/470\ \Omega)} = 103\ \Omega$
$I_T = V_S/R_T = 10/103 = 97 \text{ mA}, I_1 = R_T/R_1 \times I_T = 103/220 \times 97 \text{ mA} = 45.4$ mA
$I_2 = (R_T/R_2) \times I_T = (103/330) \times 97 \text{ mA} = 30.3$ mA
$I_3 = (R_T/R_2) \times I_T = (103/470) \times 97 \text{ mA} = 21.3$ mA

21. a. $G_T = \dfrac{1}{R_1} + \dfrac{1}{R_2} + \dfrac{1}{R_3} = \dfrac{1}{5} + \dfrac{1}{5} + \dfrac{1}{5} = 0.6$ S
$R_T = \dfrac{1}{G} = \dfrac{1}{0.6} = 1.67\ \Omega$
b. $G_T = \dfrac{1}{R_1} + \dfrac{1}{R_2} = \dfrac{1}{200} + \dfrac{1}{200} = 10$ mS
$R_T = \dfrac{1}{G} = \dfrac{1}{10 \text{ mS}} = 100\ \Omega$
c. $G_T = \dfrac{1}{R_1} + \dfrac{1}{R_2} + \dfrac{1}{R_3} = \dfrac{1}{1 \text{ M}\Omega} + \dfrac{1}{500 \text{ M}\Omega} + \dfrac{1}{3.3 \text{ M}\Omega} = 1.305\ \mu\text{S}, R_T = \dfrac{1}{G} = \dfrac{1}{1.305\ \mu\text{S}} = 766.3 \text{ k}\Omega$
d. $G_T = \dfrac{1}{R_1} + \dfrac{1}{R_2} + \dfrac{1}{R_3} = \dfrac{1}{5} + \dfrac{1}{3} + \dfrac{1}{2} = 1.033$ S
$R_T = \dfrac{1}{G} = \dfrac{1}{1.033} = 967.7 \text{ m}\Omega$

23. a. $R_T = \dfrac{R_1 \times R_2}{R_1 + R_2} = \dfrac{15 \times 7}{15 + 7} = \dfrac{105}{22} = 4.77\ \Omega$
b. $R_T = \dfrac{1}{(1/R_1) + (1/R_2) + (1/R_3)} = \dfrac{1}{(1/26\ \Omega) + (1/15\ \Omega) + (1/30\ \Omega)} = 7.22\ \Omega$
c. $R_T = \dfrac{R_1 \times R_2}{R_1 + R_2} = \dfrac{5.6 \text{ k}\Omega \times 2.2 \text{ k}\Omega}{5.6 \text{ k}\Omega + 2.2 \text{ k}\Omega} = \dfrac{12.32 \text{ M}\Omega}{7.8 \text{ k}\Omega} = 1.58 \text{ k}\Omega$
d. $R_T = \dfrac{1}{(1/R_1) + (1/R_2) + (1/R_3) + (1/R_4) + (1/R_5)} = \dfrac{1}{(1/1 \text{ M}\Omega) + (1/3 \text{ M}\Omega) + (1/4.7 \text{ M}\Omega) + (1/10 \text{ M}\Omega) + (1/33 \text{ M}\Omega)} = 596.5 \text{ k}\Omega$

25. a. $I_2 = I_T - I_1 - I_3 = 6 \text{ mA} - 2 \text{ mA} - 3.7 \text{ mA} = 300\ \mu\text{A}$
b. $I_T = I_1 + I_2 + I_3 = 6 \text{ A} + 4 \text{ A} + 3 \text{ A} = 13$ A
c. $R_T = \dfrac{R_1 \times R_2}{R_1 + R_2} = \dfrac{5.6 \text{ M} \times 3.3 \text{ M}}{5.6 \text{ M} + 3.3 \text{ M}} = \dfrac{18.48 \text{ M}\Omega}{8.9 \text{ M}\Omega} = 2.08 \text{ M}\Omega$
$V_S = I_T \times R_T = 100 \text{ mA} \times 2.08 \text{ M}\Omega = 208$ kV
$I_1 = \dfrac{R_T}{R_1} \times I_T = \dfrac{2.08 \text{ m}\Omega}{5.6 \text{ M}\Omega} \times 100 \text{ mA} = 63$ mA
$I_2 = \dfrac{R_T}{R_2} \times I_T = \dfrac{2.08 \text{ m}\Omega}{3.3 \text{ M}\Omega} \times 100 \text{ mA} = 63$ mA
d. $I_1 = V_{R_1}/R_1 = 2/200 \text{ k}\Omega = 10\ \mu\text{A}$
$I_2 = I_T - I_1 = 100 \text{ mA} - 10\ \mu\text{A} = 99.99$ mA
$R_2 = \dfrac{V_{R_2}}{I_2} = \dfrac{2}{99.99 \text{ mA}} = 20.002\ \Omega$
$P_T = I_T \times V_S = 100 \text{ mA} \times 2 \text{ V} = 200$ mW

27. a

29. Total current would increase, and the branch current with the shorted resistor would increase to $20 \text{ V}/1\ \Omega = 20$ A.

Chapter 5

1. c
3. b
5. d
7. d
9. c

11. a. $R_{1,2} = R_1 + R_2 = 2.5 \text{ k}\Omega + 10 \text{ k}\Omega = 12.5 \text{ k}\Omega$;
$R_{3,4} = R_3 + R_4 = 7.5 \text{ k} + 2.5 \text{ k} = 10 \text{ k}\Omega$
$R_{3,4,5} = \dfrac{R_{3,4} \times R_5}{R_{3,4} + R_5} = \dfrac{10 \text{ k}\Omega \times 2.5 \text{ M}\Omega}{10 \text{ k}\Omega + 2.5 \text{ M}\Omega} = \dfrac{256\ \Omega}{2.51 \text{ M}\Omega} = 9.96 \text{ k}\Omega$
$R_{1,2,3,4,5} = R_{1,2} + R_{3,4,5} = 12.5 \text{ k}\Omega + 9.96 \text{ k}\Omega = 22.46 \text{ k}\Omega$
b. $I_T = V_S/R_T = 100/22.46 \text{ k}\Omega = 4.45$ mA
c. $V_{R_1} = I_T \times R_1 = 4.45 \text{ mA} \times 2.5 \text{ k}\Omega = 11.125$ V
$V_{R_2} = I_T \times R_2 = 4.45 \text{ mA} \times 10 \text{ k}\Omega = 44.5$ V
$V_{R_{3,4,5}} = I_T \times R_{3,4,5} = 4.45 \text{ mA} \times 9.96 \text{ k}\Omega = 44.3$ V
d. $I_{R_1} = I_T = 4.45 \text{ mA}, I_{R_2} = I_T = 4.45$ mA
$I_{R_3} = I_{R_4} = V_{3,4,5}/R_{3,4} = \dfrac{44.3}{10 \text{ k}\Omega} = 4.43$ mA
$I_{R_5} = V_{3,4,5}/R_5 = 44.3/2.5 \text{ M}\Omega = 17.73\ \mu\text{A}$

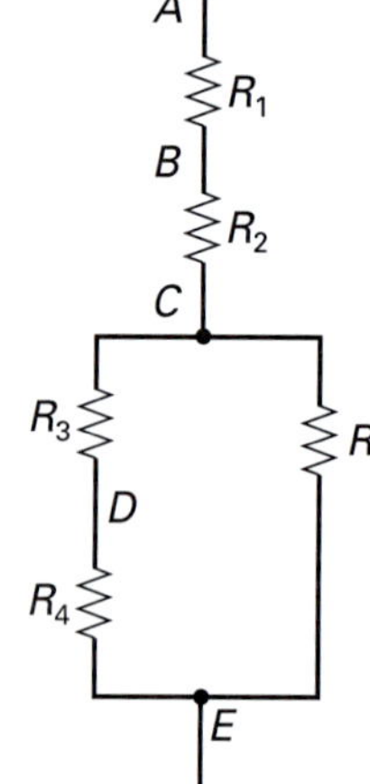

Figure Ch. 5

e. $P_T = I_T \times V_S = 4.45 \text{ mA} \times 100 \text{ V} = 445$ mW
$P_{R_1} = I_{R_1} \times V_{R_1} = 4.45 \text{ mA} \times 11.125 \text{ V} = 49.5$ mW
$P_{R_2} = I_{R_2} \times V_{R_2} = 4.45 \text{ mA} \times 44.5 \text{ V} = 198.025$ mW
$P_{R_3} = I_3^2 \times R_3 = (4.43 \text{ mA})^2 \times 7.5 \text{ k}\Omega = 147.2$ mW
$R_{R_4} = I_4^2 \times R_4 = (4.43 \text{ mA})^2 \times 2.5 \text{ k}\Omega = 49.1$ mW
$P_{R_5} = V_{R_{3,4,5}} \times I_{R_5} = 44.3 \text{ V} \times 17.73\ \mu\text{A} = 785\ \mu\text{W}$

13. $R_{3,4} = \frac{R_3 \times R_4}{R_3 + R_4} = \frac{200 \times 300}{200 + 300} = \frac{60\text{ k}\Omega}{500} = 120\ \Omega$

$R_{2,3,4} = R_2 + R_{3,4} = 100 + 120 = 220$

$R_{1,2,3,4} = \frac{R_1 \times R_{2,3,4}}{R_1 + R_{2,3,4}} = \frac{100 \times 220}{100 + 220} = \frac{22\text{ k}\Omega}{320} = 68.75 = R_T$

$I_T = V_S/R_T = 10/68.75 = 145.45\text{ mA}$

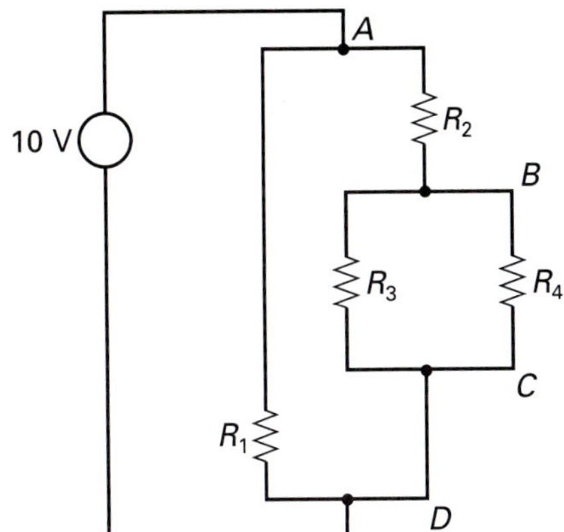

Figure Ch. 5

$V_{R_2} = I_{R_2} \times R_2$

$I_{R_2} = I_{R_{2,3,4}} = \frac{R_T}{R_{2,3,4}} \times I_T = \frac{68.76}{220} \times 145.45\text{ mA}$

$= 45.45\text{ mA} = I_2$

$V_{R_2} = 45.45\text{ mA} \times 100 = 4.545\text{ V}$

$V_{R_{3,4}} = I_{R_{3,4}} \times R_{3,4} = 45.45\text{ mA} \times 120 = 5.45\text{ V}$

$V_{R_1} = V_S = 10\text{ V}, I_{R_1} = V_S/R_1 = 10/100 = 100\text{ mA}$

$I_{R_2} = 45.45\text{ mA}$

$I_{R_3} = (R_{R_{3,4}}/R_3) \times I_{R_{3,4}} = (120/200) \times 45.45\text{ mA}$

$= 27.27\text{ mA}$

$I_{R_4} = I_{R_{3,4}} - I_{R_3} = 45.45\text{ mA} - 27.27\text{ mA} = 18.18\text{ mA}$

$P_T = V_S \times I_T = 10 \times 145.45\text{ mA} = 1.4545\text{ W}$

$P_{R_1} = I_{R_1} \times V_{R_1} = 100\text{ mA} \times 10\text{ V} = 1\text{ W}$

$P_{R_2} = I_{R_2} \times V_{R_2} = 45.45\text{ mA} \times 4.545\text{ V} = 207\text{ mW}$

$P_{R_3} = I_{R_3} \times V_{R_3} = 27.27\text{ mA} \times 5.45 = 149\text{ mW}$

$P_{R_4} = I_{R_4} \times V_{R_4} = 18.18\text{ mA} \times 5.45 = 99\text{ mW}$

$V_A = V_S = 10\text{ V}, V_R = V_A - V_{R_2} = 10 - 4.545 = 5.455$

$V_C = V_D = 0\text{ V}$

15. This answer will vary with each person

Chapter 6

1. a
3. b
5. a
7. d
9. b
11. c
13. b
15. d
17. Frequency = 1/time:
   - **a.** 1/16 ms = 62.5 Hz;
   - **b.** 1/1 s = 1 Hz;
   - **c.** 1/15 μs = 66.67 kHz;
   - **d.** 1/0.05 s = 20 Hz;
   - **e.** 1/200 μs = 5 kHz;
   - **f.** 1/350 ms = 2.86 Hz
19. **a.** Peak = 1.414 × rms = 1.414 × 40 mA = 56.56 mA
   b. Peak to peak = 2 × peak = 2 × 56.56 mA = 113.12 mA
   c. Average = 0.637 × peak = 0.637 × 56.56 mA = 36 mA
21. **a.** peak to peak = 5 cm × 0.5 V/cm = 2.5 $V_{p\text{–}p}$;
   t = 4 cm × 20 μs/cm = 80 μs,
   frequency = $1/t$ = 1/80 μs = 12.5 kHz;
   period = t = 80 μs; peak = p–p/2 =
   2.5/2 = 1.25 V_{pk};
   rms = 0.707 × peak = 0.707 × 1.25 V = 883.75 mV;
   average = 0.637 × peak = 0.637 × 1.25 V
   = 796.25 mV_{avg}
   b. peak to peak = 5 cm × 10 V/cm = 50 $V_{p\text{–}p}$;
   t = 4 cm × 10 ms/cm = 40 ms;
   frequency = $1/t$ = 1/40 ms = 25 Hz;
   period = t = 40 ms, peak = pk–pk/2 =
   50/2 = 25 $V_{pk,}$
   rms = 0.707 × peak = 0.707 × 25 = 17.675 V;
   average = 0.637 × peak = 0.637 × 25 = 15.925 V_{avg}
   c. peak to peak = 5 cm × 50 mV/cm = 250 $mV_{p\text{–}p}$,
   t = 4 cm × 0.2 μs/cm =
   800 ns; frequency = $1/t$ = 1/800 ns = 1.25 MHz;
   period = t = 800 ns: peak =
   pk–pk/2 = 250 mV_{pk}/2 = 125 mV_{pk};
   rms = 0.707 peak = 0.707 × 125 mV_{pk}
   = 88.375 mV; average = 0.637 × peak
   = 0.637 × 125 mV_{pk} = 79.625 mV
23. 3.5 cm × 10 V/cm = 35 V p–p
25. There would be no need to select a range setting since an autoset scopemeter would automatically adjust settings for best possible viewing.

Chapter 7

1. d
3. d
5. a
7. a
9. b
11. b
13. b
15. d
17. b
19. d
21. a
23. e
25. b
27. $V = Q/C = 125 \times 10^{-6}/0.006\ \mu\text{F} = 20.83\text{ kV}$
29. **a.** $C_T = C_1 + C_2 + C_3 + C_4 = 1.7\ \mu\text{F} + 2.6\ \mu\text{F} + 0.03\ \mu\text{F} + 1200\text{ pF} = 4.3312\ \mu\text{F}$
   b. $C_T = \frac{1}{(1/C_1) + (1/C_2) + (1/C_3)} = \frac{1}{(1/1.6\ \mu\text{F}) + (1/1.4\ \mu\text{F}) + (1/4\ \mu\text{F})} = 0.629\ \mu\text{F}$
31. **a.** 10 + 4 zeros = 100,000 pF or 0.1 μF
   b. 12 + 5 zeros = 1,200,000 pF or 1.2 μF
   c. 0.01 μF
   d. 220 pF
33. **a.** V_{ITC} = 63.2% of V_S = 0.632 × 10 V = 6.32 V;
   5TC = 5 × 84 ms = 420 ms
   b. V_{ITC} = 63.2% of V_S = 0.632 × 10 V = 6.32 V;
   5TC = 5 ×16.8 ms = 84 ms
   c. V_{ITC} = 63.2% of V_S = 0.632 × 10 V = 6.32 V;
   5TC = 5 × 4.08 ms = 20.4 ms
   d. V_{ITC} = 63.2% of V_S = 0.632 × 10 V = 6.32 V;
   5TC = 5 × 980 μs = 4.9 ms
35. $V_S = \sqrt{V_R^2 + V_C^2} = \sqrt{6^2 + 12^2} = \sqrt{36 + 144} = \sqrt{180}$
   = 13.4 V
37. **a.** $I^R = V/R$ = 12 V/4 MΩ = 3 μA
   b. $I_C = V/X_C$ = 12 V/1.3 kΩ = 9.23 mA
   c. $I_T = \sqrt{I_R^2 + I_C^2} = \sqrt{(3\ \mu\text{A})^2 + (9.23\text{ mA})^2} = 9.23\text{ mA}$
   d. $Z = V/I_T$ = 12/9.23 mA = 1.3 k
   e. θ = arctan (R/X_C) = arctan (4 MΩ/1.3 kΩ) = 89.98°
39. $C = \frac{1}{2\pi f X_C} = \frac{1}{2\pi(20\text{ kHz})\ 10\text{ k}\Omega} = 795.8\text{ pF}$
41.

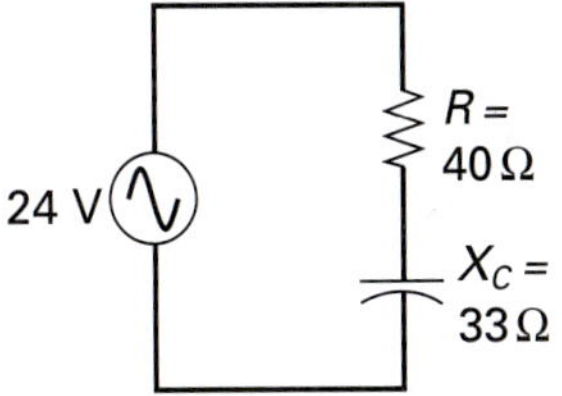

Figure Ch. 7, 41(a)

b. $Z = \sqrt{R^2 + X_C^2} = \sqrt{40^2 + 33^2} = 51.9\ \Omega$
$I = V/Z = 24\ \text{V}/51.9\ \Omega = 462.4\ \text{mA}$
$V_R = I \times R = 462.4\ \text{mA} \times 40\ \Omega = 18.496\ \text{V}$
$V_C = I \times X_C = 462.4\ \text{mA} \times 33\ \Omega = 15.2592\ \text{V}$
$I_R = I_C = I_T = 462.4\ \text{mA}$
$\theta = \arctan(X_C/R) = \arctan(33\ \Omega/40\ \Omega)\ 39.5°$

43. a. $Z = \sqrt{R^2 + X_C^2} = \sqrt{1\ \text{M}^2 + 2.5\ \text{M}^2} = 2.69\ \text{M}\Omega$
$I = V/Z = 12/2.69\ \text{M}\Omega = 4.5\ \mu\text{A}$
$V_R = I \times R = 4.5\ \mu\text{A} \times 1\ \text{M}\Omega = 4.5\ \text{V}$
$V_C = I \times X_C = 4.5\ \mu\text{A} \times 2.5\ \text{M}\Omega = 11.25\ \text{V}$

b. $Z = \sqrt{R^2 + X_C^2} = \sqrt{300^2 + 200^2} = 360.6\ \Omega$
$I = V/Z = 50\ \text{V}/360.6\ \Omega = 138.7\ \text{mA}$
$V_R = I \times R = 138.7\ \text{mA} \times 300\ \Omega = 41.61\ \text{V}$
$V_C = I \times R = 138.7\ \text{mA} \times 200\ \Omega = 27.74\ \text{V}$

45. Lag, 90

47. $X_C = \dfrac{1}{2\pi fc} = \dfrac{1}{2\pi(35\ \text{kHz})\ 10\ \mu\text{F}} = 455\ \text{m}\Omega$

$Z = \sqrt{R^2 + X_C^2} = \sqrt{100\ \text{k}\Omega^2 + 455\ \text{m}\Omega^2} = 100\ \text{k}\Omega$
$I = V/Z = 24\ \text{V}/100\ \text{k}\Omega = 240\ \mu\text{A}$
True power $= I^2 \times R = 240\ \mu\text{A}^2 \times 100\ \text{k}\Omega = 5.76\ \text{mW}$
Reactive power $= I^2 \times X_C = 240\ \mu\text{A}^2 \times 455\ \text{m}\Omega$
$= 26\ \text{nVAR}$
Apparent power $= \sqrt{P_R^2 + P_X^2} = 5.76\ \text{mVA}$
Power factor $= R/Z = 100\ \text{k}\Omega/100\ \text{k}\Omega = 1$

Chapter 8

1. e **3.** a **5.** c **7.** b **9.** b **11.** b **13.** b **15.** d **17.** a **19.** b **21.** d **23.** c **25.** a

27. a. 22 MΩ
b. 78.6 kΩ
c. 314.2 kΩ

29. a. $L_T = \dfrac{L_1 \times L_2}{L_1 + L_2} = \dfrac{12\ \text{mH} \times 8\ \text{mH}}{12\ \text{mH} + 8\ \text{mH}}$
$= \dfrac{96\ \mu\text{H}}{20\ \text{mH}} = 4.8\ \text{mH}$

b. $L_T = \dfrac{1}{(1/L_1) + (1/L_2) + (1/L_3)}$
$= \dfrac{1}{(1/75\ \mu\text{H}) + (1/34\ \mu\text{H}) + (1/27\ \mu\text{H})} = 12.53\ \mu\text{H}$

31. a. $V_S = \sqrt{V_R^2 + V_L^2} = \sqrt{12^2 + 6^2} = \sqrt{144 + 36}$
$= \sqrt{180} = 13.4\ \text{V}$
b. $I = V_S/Z = 13.4\ \text{V}/14\ \text{k}\Omega = 957.1\ \mu\text{A}$
c. $\angle = \arctan V_L/V_R = \arctan 2 = 63.4°$
d. $Q = V_L/V_R = 12/6 = 2$
e. $\text{PF} = \cos\theta = 0.448$

33. $\tau = \dfrac{L}{R} = \dfrac{400\ \text{mH}}{2\ \text{k}\Omega} = 200\ \mu\text{s}$

V_L will start at 12 V and then exponentially drop to 0 V

Time	Factor	V_S	V_L
0	1.0	12	12.
1 T_C	0.365	12	4.416
2 T_C	0.135	12	1.62
3 T_C	0.05	12	0.6
4 T_C	0.018	12	0.216
5 T_C	0.007	12	0.084

35. a. $R_T = R_1 + R_2 = 250 + 700 = 950\ \Omega$
b. $L_T = L_1 + L_2 = 800\ \mu\text{H} + 1200\ \mu\text{H} = 2\ \text{mH}$
c. $X_L = 2\pi fL = 2\pi(350\ \text{Hz})\ 2\ \text{mH} = 4.4\ \Omega$
d. $Z = \dfrac{R \times X_L}{\sqrt{R^2 + X_L^2}} = \dfrac{950 \times 4.4}{\sqrt{950^2 + 4.42^2}} = 4.39$
e. $V_{RT} = V_{LT} = V_S = 20\ \text{V}$
f. $I_{RT} = \dfrac{V_S}{R_T} = \dfrac{20\ \text{V}}{950\ \Omega} = 21\ \text{mA}$
$I_{LT} = \dfrac{V_{S1}}{X_{LT}} = \dfrac{20\ \text{V}}{4.4\ \Omega} = 4.5\ \text{A}$
g. $I_T = \sqrt{I_R^2 + I_L^2} = \sqrt{21\ \text{mA}^2 + 4.5\ \text{A}^2} = 4.5\ \text{A}$
h. $\theta = \arctan(R/X_L) = \arctan 950/4.4 = 89.7°$
i. $P_R = I^2 \times R = 21\ \text{mA}^2 \times 950 = 418.95\ \text{mW}$
$P_X = I^2 \times X_L = 4.5\ \text{A}^2 \times 4.4 = 89.1\ \text{VAR}$
$P_A = \sqrt{P_R^2 + P_X^2} = \sqrt{418.95\ \text{mW}^2 + 89.1\ \text{A}^2} = 89.1\ \text{VA}$
j. $\text{PF} = P_R/P_A = 418.95\ \text{mW}/89.1\ \text{W} = 0.0047$

37. Turns ratio $= \sqrt{Z_L/Z_S} = \sqrt{8\ \Omega/24\ \Omega} = \sqrt{1/3} = 0.58$

39. Follow polarity dots.

Chapter 9

1. b
3. b
5. d
7. c
9. c

11. a. $X_C = \dfrac{1}{2\pi fC} = \dfrac{1}{2\pi(60)0.02\ \mu\text{F}} = 132.6\ \text{k}\Omega$
b. $X_C = \dfrac{1}{2\pi fC} = \dfrac{1}{2\pi(60)18\ \mu\text{F}} = 147.4\ \Omega$
c. $X_C = \dfrac{1}{2\pi fC} = \dfrac{1}{2\pi(60)360\ \text{pF}} = 7.37\ \text{M}\Omega$
d. $X_C = \dfrac{1}{2\pi fC} = \dfrac{1}{2\pi(60)2700\ \text{nF}} = 982.4\ \Omega$
e. $X_L = 2\pi fL = 2(60)4\ \text{mH} = 1.5\ \Omega$
f. $X_L = 2\pi fL = 2(60)8.18\ \text{H} = 3.08\ \text{k}\Omega$
g. $X_L = 2\pi fL = 2(60)150\ \text{mH} = 56.5\ \Omega$
h. $X_L = 2\pi fL = 2(60)2\ \text{H} = 753.98\ \Omega$

13. a. $X_L = 2\pi fL = 2(60\ \text{Hz})150\ \text{mH} = 56.5\ \Omega$
b. $X_C = \dfrac{1}{2\pi fc} = \dfrac{1}{2\pi(60\ \text{Hz})20\ \mu\text{F}} = 132.6\ \Omega$
c. $I_R = V/R = 120\ \text{V}/270\ \Omega = 444.4\ \text{mA}$
d. $I_L = V/X_L = 120\ \text{V}/56.5\ \Omega = 2.12\ \text{A}$
e. $I_C = V/X_C = 120\ \text{V}/132.6\ \Omega = 905\ \text{mA}$
f. $I_T = \sqrt{I_R^2 + I_X^2} = \sqrt{(444.4\ \text{mA})^2 + (1.215)^2} = 1.29\ \text{A}$
g. $Z = V/I_T = 120\ \text{V}/1.29\ \text{A} = 93.02\ \Omega$
h. Resonant frequency
$= \dfrac{1}{2\pi\sqrt{LC}} = \dfrac{1}{2\pi\sqrt{150\ \text{mH} \times 20\ \mu\text{F}}} = 91.89\ \text{Hz}$
i. $X_L = 2\pi fL$
$= 6.28 \times 91.89\ \text{Hz} \times 150\ \text{mH}$
$= 86.54\ \Omega$
$Q = \dfrac{X_L}{R} = \dfrac{86.54\ \Omega}{270\ \Omega} = 0.5769$
j. $\text{BW} = \dfrac{f_0}{Q} = \dfrac{91.89\ \text{Hz}}{0.5769} = 159.28\ \text{Hz}$

15. Using a source voltage of 1 volt:
$Z = V/I_T,\ I_T = \sqrt{I_R^2 + I_X^2},\ I_R = V/R = 1/750 = 1.33\ \text{mA}$
$I_L = V/X_L = 1/25 = 40\ \text{mA}.$
$I_C = V/X_C = 1/160 = 6.25\ \text{mA}$
$I_X = I_L - I_C = 40\ \text{mA} - 6.25\ \text{mA} = 33.75\ \text{mA}$
$I_T = \sqrt{(1.33\ \text{mA})^2 + (33.75\ \text{mA})^2} = 33.78\ \text{mA}$
$Z = 1/33.78\ \text{mA} = 29.6$

Chapter 10

1. a **3.** d **5.** d **7.** d **9.** b **11.** b **13.** b **15.** c

17. a. P-N junction is forward biased. $I = V_S - V_{P\text{-}N}/R$ $= 5\text{ V} - 0.7\text{ V}/330\ \Omega = 13\text{ mA}$.
b. P-N junction is forward biased. $I = V_S - V_{P\text{-}N}/R$ $= 15\text{ V} - 0.3\text{ V}/15\ \Omega = 980\ \mu\text{A}$.
c. P-N junction is reverse biased. $I = 0$ A.

19. a. $V_{R_1} = V_S - V_{P-N} = 5\text{ V} - 0.7\text{ V} = 4.3\text{ V}$
b. $V_{R_1} = V_S - V_{P-N} = 15\text{ V} - 0.3\text{ V} = 14.7\text{ V}$
c. Since the P-N junction is open, or reverse biased, all of the applied voltage will appear across this series open. Therefore, the voltage across $R_1 = 0$ V.

Chapter 11

1. c **3.** d **5.** d **7.** a **9.** b **11.** a **13.** b **15.** c **17.** d **19.** d **21.** c **23.** a **25.** c **27.** d **29.** d **31.** a **33.** a **35.** c **37.** c **39.** c **41.** d **43.** b

45. a. reverse biased
b. reverse biased
c. reverse biased
d. forward biased
e. forward biased
f. reverse biased

47. a. $V_{\text{diode}} = 10$ V
b. $V_{\text{diode}} = 0.7$ V

49. a. Polarity correct (+ → cathode, − → anode)
b. Polarity incorrect (+ → anode, − → cathode)
c. Polarity incorrect
d. Polarity correct
e. Polarity correct
f. Polarity for both D_1 and D_2 are correct

51. a. $I_S = V_{in} - V_Z/R_S = 10\text{ V} - 6.8\text{ V}/200\ \Omega = 16\text{ mA}$
b. Since the zener diode is forward biased, we will assume a 0.7 V forward voltage drop. Therefore $I_S = V_{in} - V_Z/R_S = 20\text{ V} - 0.7\text{ V}/570\ \Omega = 33.9\text{ mA}$
c. Since the zener diode is forward biased, we will assume a 0.7 V forward voltage drop. Therefore $I_S = V_{in} - V_Z/R_S = 5\text{ V} - 0.7\text{ V}/400\ \Omega = 10.75\text{ mA}$
d. Since the input voltage is not large enough to send the zener into its reverse zener breakdown region, the circuit current will be equal to that of the reverse leakage current, which is almost zero.
e. $I_S = V_{in} - V_Z/R_S = 6\text{ V} - 4.7\text{ V}/200\ \Omega = 6.5\text{ mA}$
f. For zener D_1, $I_S = V_{in} - V_Z/R_S = 10\text{ V} - 6.8\text{ V}/220\ \Omega$ $= 14.5\text{ mA}$
For zener D_2 since the input voltage is not large enough to send the zener into its reverse zener breakdown region, the circuit current will be equal to that of the reverse leakage current, which is almost zero.

53. $P_D = I_{ZM} \times V_Z = 6.5\text{ mA} \times 4.7\text{ V} = 30.55\text{ mW}$
A 50 mW zener diode would be adequate in this application.

55. By reversing the input voltage polarity, and the zener diode's orientation. For example, if the polarity of the 10 V supply voltage was reversed so that the negative terminal connects to R_S, and the zener's connection was also reversed so that its anode was connected to point *X*, the output voltage will be 5.6 V. Similarly, if the polarity of the 20 V supply voltage was reversed so that the positive terminal connects to R_S, and the zener's connection was also reversed so that its cathode was connected to point *Y*, the output voltage will be +12 V.

57. $V_{in} = 12\text{ V}, V_Z = 6.8\text{ V}, V_{RL} = 6.8\text{ V}$
$V_{RS} = V_{in} - V_Z = 12\text{ V} - 6.8\text{ V} = 5.2\text{ V}$
$I_{RS} = V_{RS}/R_S = 5.2\text{ V}/120\ \Omega = 43.3\text{ mA}$
$R_L = 500\ \Omega$
$I_{RL} = V_{RL}/R_L = 6.8\text{ V}/500\ \Omega = 13.6\text{ mA}$
$I_Z = I_{RS} - I_{RL} = 43.3\text{ mA} - 13.6\text{ mA} = 29.7\text{ mA}$
$V_{in} = 15\text{ V}, V_Z = 6.8\text{ V}$
$V_{RS} = V_{in} - V_Z = 15\text{ V} - 6.8\text{ V} = 8.2\text{ V}$
$I_{RS} = V_{RS}/R_S = 8.2\text{ V}/120\ \Omega = 68.3\text{ mA}$
$R_L = 500\ \Omega$
$I_{RL} = V_{RL}/R_L = 6.8\text{ V}/500\ \Omega = 13.6\text{ mA}$
$I_Z = I_S - I_{RL} = 68.3\text{ mA} - 13.6\text{ mA} = 54.7\text{ mA}$

59. a. Forward biased (+ → anode − → cathode)
b. Forward biased
c. D_1 is reverse biased, D_2 is forward biased

61. Yes

63.

A	B	LED
0 V	0 V	OFF
0 V	+5 V	ON (D_2 turns ON and connects input power to D_3)
+5 V	0 V	ON (D_1 turns ON and connects input power to D_3)

65. Input = +10 V, Output = Green; Input = −10 V, Output = Red

67. Input = + 10 V
$I_{LED\text{-}GREEN} = V_S - V_{LED}/R_1 + R_2 = 10\text{ V} - 2.5\text{ V}/100\ \Omega + 200\ \Omega = 15\text{ mA}$
Input = −10 V
$I_{LED\text{-}GREEN} = V_{Source} - V_{D1} - V_{LED}/R_2 = 10\text{ V} - 0.7\text{ V} - 2.5\text{ V}/200\ \Omega = 34\text{ mA}$

69. Common cathode. The cathodes of all the LEDs have a common connection.

71. Switch closed (CL) = LED ON.
Switch open (OP) = LED OFF.

DIGIT	SW1	SW2	SW3	SW4	SW5	SW6	SW7
0	CL	CL	CL	CL	CL	CL	OP
1	OP	CL	CL	OP	OP	OP	OP
2	CL	CL	OP	CL	CL	OP	CL
3	CL	CL	CL	CL	OP	OP	CL
4	OP	CL	CL	OP	OP	CL	CL
5	CL	OP	CL	CL	OP	CL	CL
6	CL	OP	CL	CL	CL	CL	CL
7	CL	CL	CL	OP	OP	CL	OP
8	CL	CL	CL	CL	CL	CL	CL
9	CL	CL	CL	OP	OP	CL	CL

73. a. $I_{\text{DIODE}} = \dfrac{V_S - 0.7\text{ V}}{R} = \dfrac{5.1\text{ V} - 0.7\text{ V}}{250\ \Omega} = 17.6\text{ mA}$
b. Position I, since it brings into circuit five P-N junction encoder diodes and therefore current will equal 5 × 17.6 mA = 88 mA.
c. $I_{LED} = \dfrac{V_S - V_{LED}}{R} = \dfrac{5.1\text{ V} - 2\text{ V}}{250\ \Omega} = 12.4\text{ mA}.$
d. Maximum display current will occur when the display is showing 8, since this requires all seven segments ON. The value of current will be 7 × 12.4 mA = 86.8 mA.
e. An encoder diode, since it draws 17.6 mA compared to a display diode which draws 12.4 mA.
f. The load resistance changes as different digits are encoded and displayed since some circuit states require more current than others. Knowing load current and load

voltage we can use Ohm's law to calculate the changes in load resistance.

$$R_L = \frac{V_L}{I_L} = \frac{5.1\text{ V}}{86.8\text{ mA}} = 58.8\ \Omega$$

$$R_L = \frac{V_L}{I_L} = \frac{5.1\text{ V}}{112.8\text{ mA}} = 45.2\ \Omega$$

75. $V_{out} = 1/2\ V_{S\,pk} - 0.7\text{ V} = 13\text{ V} - 0.7\text{ V} = 12.3\text{ V}$

77. $V_{out} = V_{S\,pk} - 1.4\text{ V} = 37\text{ V} - 1.4\text{ V} = 35.6\text{ V}_{peak}$
$V_{avg} = 0.636 \times \text{V}_{peak} = 0.636 \times 35.6\text{ V} = 22.6\text{ V}$

79. Percent of Regulation $= V_{NL} - V_{FL}/V_{FL} \times 100$
$= 12\text{ V} - 10.6\text{ V}/12\text{ V} \times 100 = 11.67\%$

Chapter 12

1. d **11.** d
3. a **13.** d
5. b **15.** a
7. b **17.** c
9. a **19.** d

21. a. NPN, x = base, y = collector, z = emitter.
b. PNP, x = base, y = collector, z = emitter.

23. a. $I_E = 25$ mA, $I_C = 24.6$ mA, $I_B = I_E - I_C = 0.4$ mA or 400 μA
b. $I_B = 600$ μA, $I_C = 14$ mA, $I_E = I_B + I_C = 14.6$ mA
c. $I_E = 4.1$ mA, $I_B = 56.7$ μA, $I_C = I_E - I_B = 4.04$ mA

25. a. Common-collector circuit, since the input is applied to the base and the output is taken from the emitter (collector is common to both input and output).
b. Common-emitter circuit, since the input is applied to the base and the output is taken from the collector (emitter is common to both input and output).
c. Common-emitter circuit, since the input is applied to the base and the output is taken from the collector (emitter is common to both input and output).
d. Common-emitter circuit, since the input is applied to the base and the output is taken from the collector (emitter is common to both input and output).

27. a. $V_{R_B} = 20\text{ V} - 0.7\text{ V} = 19.3\text{ V}$
$I_B = V_{R_B}/R_B = 19.3\text{ V}/2.5\text{ M}\delta = 7.72\ \mu\text{A}$
b. $I_C = I_B \times B_{DC} = 7.72\ \mu\text{A} \times 150 = 1.16\text{ mA}$
c. $V_{R_C} = I_C \times R_C = 1.16\text{ mA} \times 10\text{ k}\Omega = 11.6\text{ V}$
$V_{CE} = V_{CC} - V_{R_C} = 20\text{ V} - 11.6\text{ V} = 8.4\text{ V}$

29. a. $V_B = R_2/R_1 + R_2 \times V_{CC} = 1\text{ k}\Omega/6.8\text{ k}\Omega + 1\text{ k}\Omega \times 20\text{ V} = 2.56\text{ V}$
$V_E = V_B - 0.7\text{ V} = 2.56\text{ V} - 0.7\text{ V} = 1.86\text{ V}$
b. $I_E = V_E/R_E = 1.86\text{ V}/1.1\text{ k}\Omega = 1.7\text{ mA}$
$I_E = I_C = 1.7\text{ mA}$
c. $V_{R_C} = I_C \times R_C = 1.7\text{ mA} \times 4.7\text{ k}\Omega = 8\text{ V}$
V_C or $V_{out} = V_{CC} - V_{R_C} = 20\text{ V} - 8\text{ V} = 12\text{ V}$
d. $V_{CE} = V_{CC} - (V_{R_C} - V_E) = 20\text{ V} - (8\text{ V} + 1.86\text{ V}) = 10.14\text{ V}$

31. Yes, since: $V_{R_C} + V_{CE} + V_{R_E} = V_{CC}$
$8\text{ V} + 10.14\text{ V} + 1.86\text{ V} = 20\text{ V}$

Chapter 13

1. b **11.** b
3. b **13.** a
5. b **15.** a
7. d **17.** a
9. b **19.** b

21. a. $\delta_m = \Delta I_D/\Delta V_{GS} = 8\text{ mA} - 2\text{ mA}/(-3.5\text{ V}) - (-0.5\text{ V})$
$= 6\text{ mA}/3\text{ V} = 2$ millisiemens
b. $A_V = \delta_m \times R_D = 2\text{ mS} \times 10\text{ k}\Omega = 20$

23. a. $V_S = I_S \times R_S = 2\text{ mA} \times 390\ \Omega = 0.78\text{ V}$
b. $V_{GS} = V_G - V_S = 0\text{ V} - 0.78\text{ V} = -0.78\text{ V}$
c. $V_{DS} = V_{DD} - (V_{RD} + V_{RS}) = 15\text{V} - [(I_D R_D) + 2\text{ V}]$
$= 15\text{ V} - [(2\text{ mA} \times 2\text{ k}\Omega) + 0.78\text{ V}] = 15\text{ V} - (4 + 0.78) = 15\text{ V} - 4.78\text{ V} - 10.22\text{ V}$

25. a. $I_{D(\text{Sat.})} = V_{DD}/R_D = 6\text{ V}/1\text{ k}\Omega = 6\text{ mA}$
b. $V_{DS(\text{OFF})} = V_{DD} = 6\text{ V}$

27. Figure 13–29 uses zero biasing
Figure 13–30 uses drain feedback biasing

Chapter 14

1. e
3. a
5. e
7. d
9. c

11. a. DIAC **e.** Photoresistor
b. SCR **f.** Solar cell
c. LED or ILD **g.** Phototransistor
d. Photodiode **h.** LASCR

13. A LOW input will turn OFF Q_1, causing its collector output to go HIGH, which will trigger the SCR ON, and connect power to the relay's energizing coil

15. When a trigger input is applied to the 555, it will generate a 1-second active-HIGH output, which will energize the relay coil, closing its normally open contacts, connecting ac power to the load under the control of the TRIAC.

17. a. LED, LASCR **b.** LED, Light activated TRIAC

19. Phototransistor

21. When light is present, phototransistor is ON, NPN transistor is OFF, relay is de-energized, normally closed contacts are closed, normally open contacts are open. When light is not present, phototransistor is OFF, NPN transistor is ON, relay is energized, normally closed contacts open, normally open contacts closed.

Chapter 15

1. b **11.** c
3. d **13.** b
5. a **15.** d
7. c **17.** c
9. e **19.** b

21. a. LM = National Semiconductor
318 = Op-Amp Type
C = Commercial 0 to 70° C
N = Plastic DIP with longer leads
b. NE = Signetics
101 = Op-Amp Type
C = Commercial 0 to 70° C
D = Plastic DIP

23. A single polarity supply (+5 V) inverting op-amp. The inverting amplifier would have to have a negative input because the output can only be between 0 V ($-V$ supply voltage) and +5 V ($+V$ supply voltage).

25. a. Differentiator
b. Integrator

27. a. Summing amplifier
$V_{\text{OUT}} = -(V_{\text{IN1}} + V_{\text{IN2}}) = -(3\text{ V} + 6\text{ V}) = -9\text{ V}$
b. Difference amplifier
$V_{\text{OUT}} = V_{\text{IN2}} - V_{\text{IN1}} = 10\text{ V} - 6.5\text{ V} = 3.5\text{ V}$

29. $f_C = 1/2\pi RC = 1/2\pi \times 10\text{ k}\Omega \times 0.1\ \mu F = 159.2\text{ Hz}$

31. a. A = +15 V
B = −15 V
C = +5 V
D = −5 V
b. 7815 and 7805 are fixed-positive output voltage regulators, while the 7915 and 7905 are fixed-negative output voltage regulators.
c. Positive and negative clipper circuit
d. The 555 Schmitt trigger will sharpen the rise and fall times of the clipped sine wave input producing a square wave output that is at the same frequency as the input (60 Hz).

33. a. A voltage controlled oscillator (VCO)
b. Pin 3 = square wave, pin 4 = triangular wave

c. R_1 adjusts the control voltage input to the VCO, which determines its frequency.

Chapter 16

1. b **3.** d **5.** c **7.** b

Chapter 17

1. b
3. c
5. d
7. a
9. d
11. 1001, Decimal 9
13. Decimal equivalents are:
a. 57 **b.** 39 **c.** 06
15. BCD equivalents are:
a. 0010 0011 0110 0101 **b.** 0010 0100
17. Codes in order are: BCD, Gray, Excess-3, and ASCII.
19. Stx practice makes perfect etx.

Chapter 18

1. a **11.** b
3. b **13.** d
5. b **15.** c
7. b **17.** b
9. b **19.** a
21. OR gate
23. Light, $Y = 1$; dark, $Y = 0$
25. OR
27. Truth table outputs (Y) should be:

a.	c.	e.
0 (AND)	0 (XOR)	1 (NOR)
0	1	0
0	1	0
1	0	0

b.	d.	f.
0 (OR)	1 (XNOR)	0 (NOT)
1	0	1
1	0	
1	1	

29. C = HIGH

Y_0	0
Y_1	1
Y_2	1
Y_3	0

Chapter 19

1. c **11.** c
3. c **13.** a
5. b **15.** b
7. c **17.** c
9. a **19.** c
21. **a.** $Y = (\overline{A} \cdot B) + (B \cdot C) + D$
b. $Y = \overline{(A \cdot B)} + \overline{(\overline{A} \cdot \overline{B})}$
c. $Y = \overline{\overline{(\overline{A} \cdot \overline{B} \cdot \overline{C})} + \overline{(A \cdot B \cdot \overline{C})}}$
d. Y = (LIT + NLT) × (SLW + FST)
23. **a.**

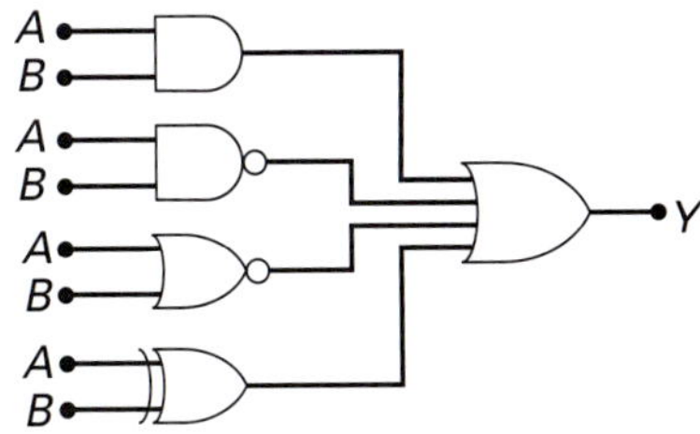

b.

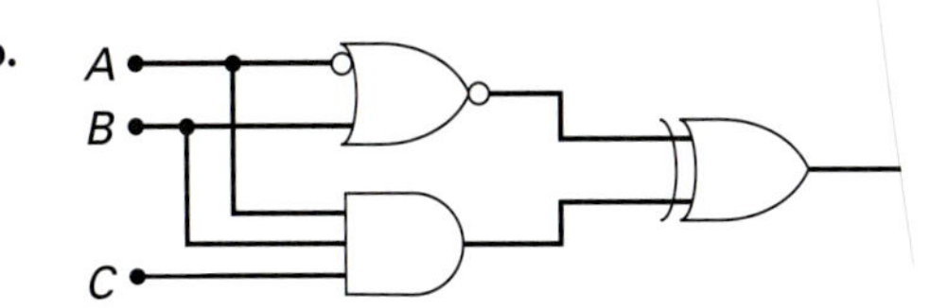

25.

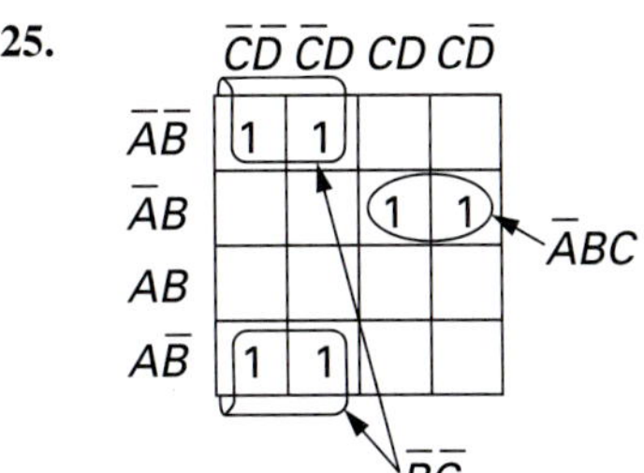

$Y_2 = \overline{B}\overline{C} + \overline{A}BC$

27. $Y_1 = \overline{A}\overline{B}\overline{C}\overline{D} + \overline{A}\overline{B}\overline{C}D + \overline{A}\overline{B}C\overline{D} + \overline{A}\overline{B}CD + A\overline{B}\overline{C}D + A\overline{B}CD + AB\overline{C}\overline{D} + AB\overline{C}D + ABC\overline{D} + ABCD$

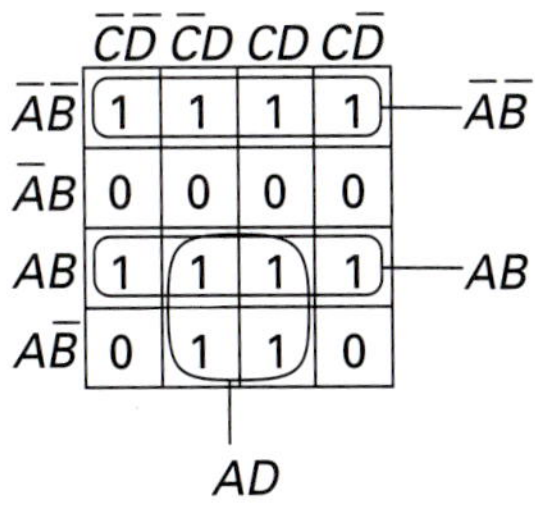

$Y_1 = \overline{A}\overline{B} + AB + AD$
(50 inputs to 9 inputs)

29.

(Question 25, Y_1)

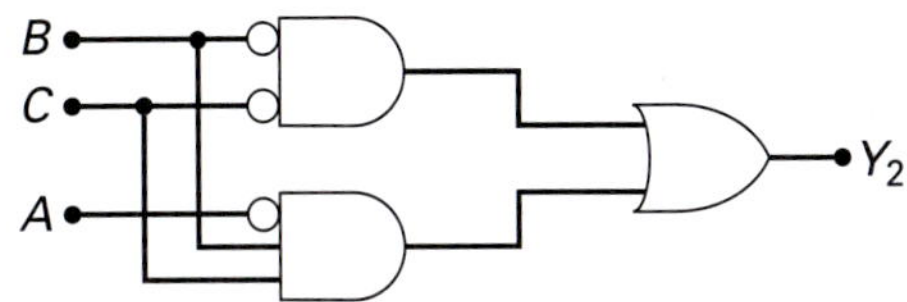

(Question 27, Y_2)

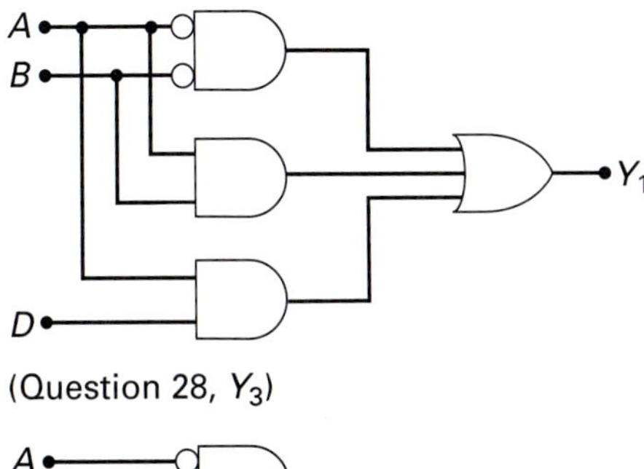

(Question 28, Y_3)

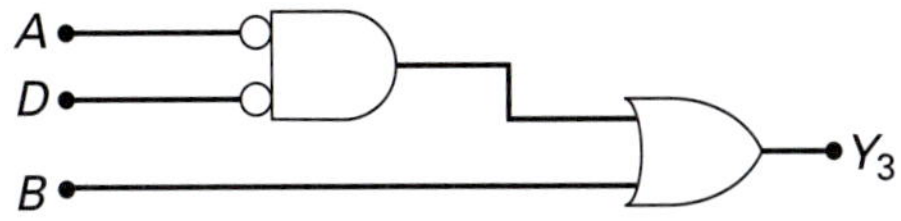

Chapter 20

1. c
3. c
5. a
7. a
9. d
11. a
13. d
15. a
17. NOR logic, any HIGH in = LOW out
19. Symbol means device will have open collector output
21. Yes, 24 mA needed, 33 mA is maximum
23. 74H05
25. Truth table output is:
    a. 0
    1
    Float
    Float
27. The 74LS125 has a LOW activated control input while the 72LS126 has a HIGH activated control input.
29. Circuit clips, or rectifies, an ac signal, which is then formed into a positive pulse waveshape within the trigger circuit.
31. N-type E-MOSFET
33. NOR Function

Chapter 21

1. c
3. c
5. a
7. a
9. e
11. c
13. b
15. d
17. b
19. PLA
21. Input/output function table
23. **a.** 0000110 **b.** 0100100 **c.** 0100100 **d.** 0001100
25. An intact fuse at the NOR inputs will generate a LOW output.

Chapter 22

1. b
3. a
5. b
7. b
9. b
11. Pulse or rectangular wave (positive going)
13. **a.** P_w = 50 ns
    b. Period = 500 μs
    c. f = 2 kHz
    d. Duty Cycle = 10%
15. No
17. Replace 7427
19. 7408 has open output

Chapter 23

1. c
3. c
5. a
7. d
9. a
11. b
13. b
15. a
17. d
19. d
21. 4-Line BCD to 10-Line Decimal Decoder
23. Common-Anode
25.

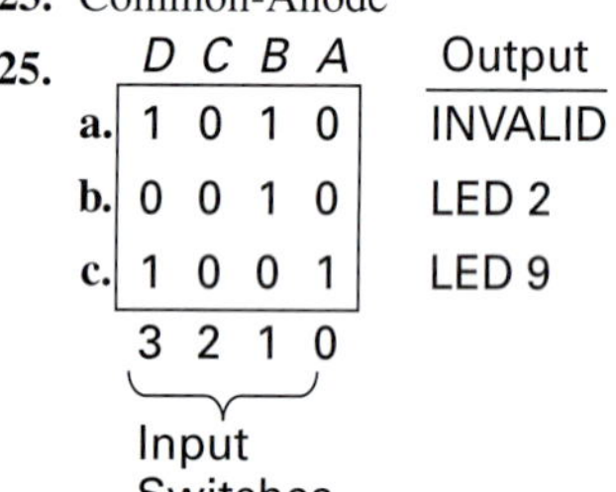

	D C B A	Output
a.	1 0 1 0	INVALID
b.	0 0 1 0	LED 2
c.	1 0 0 1	LED 9
	3 2 1 0 (Input Switches)	

27. BCD to seven-segment decoder/driver circuit
29. Active-HIGH
31. Continuous counting
33. Encoder
35.

	A	B	C
(1)	1	1	1
(2)	1	1	0
(3)	1	0	1
(4)	0	0	0
(5)	0	0	1
(6)	0	1	0
(7)	0	1	1

37. D_0 D_7
    1 0 1 1 0 0 1 1
39. 4(100), block, OFF-ON-ON

Chapter 24

1. c
3. d
5. c
7. b
9. d
11. b
13. c
15. b
17. All zeros
19. Negative edge triggering
21. Reset all Q outputs LOW when $\overline{CLR}$ is taken LOW.
23. OFF; $X = 0$ ON; $X = 1$
25. Positive edge-triggered D-type flip-flop
27.

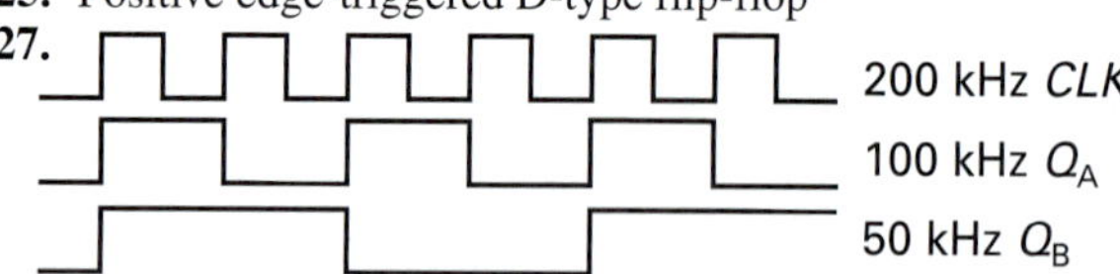

29. Positive edge triggering
31. When D input and clock are in phase, Q will go HIGH, and LED will turn OFF.
33. NAND S-R flip-flop
35. Edge triggered
39.

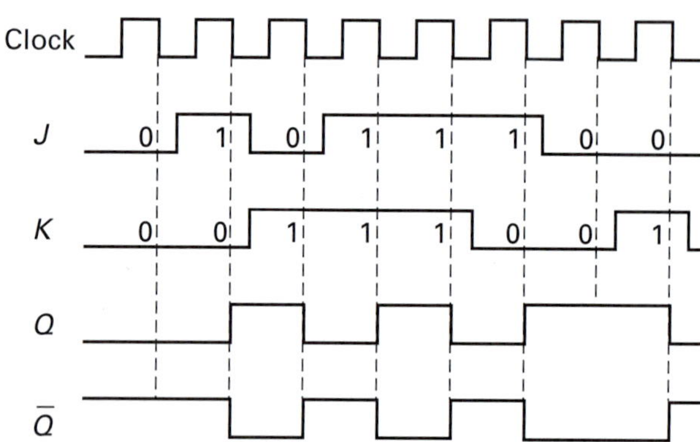

41. Astable
43. Non-retriggerable
45. Negative-edge triggered
47. 24–34(a)

Chapter 25

1. b
3. a
5. a
7. b
9. d
11. b
13. a
15. d
17. c
19. c
21. a
23. d
25. a

27. D-type
29. SIPO
31. **a.** 15
b. 63
c. 16,777,199
33. $f = \dfrac{1 \times 10^9}{n \times t_p}$
a. 100 ns, 10 MHz
b. 150 ns, 6.6 MHz
c. 600 ns, 1.7 MHz
35. 32
37. ÷32
39. 9.1 MHz
41. **a.** Depressing SW1 momentarily will FLOAT the CLR input HIGH, clearing the counter. The clock NAND gate is controlled by the selected output from 74154.
b. The 74154 decoder output lines produce a LOW decoded output.
c. After being cleared, the counter will count from 0000 to 1100. Line 12 (pin 14) will go LOW when the counter applies 1100 to the decoder's input, stopping the clock and turning the LED ON.

Chapter 26

1.	b	**9.**	d
3.	d	**11.**	d
5.	a	**13.**	b
7.	d	**15.**	c

17. **a.** 1001 0000
b. 0001 0001
c. 0110 0000
19. **a.** 1011 (R 110)
b. 0010 0010
c. 0001 0101 (R 001)
21. **a.** 1001 0110
b. 0101 0101
c. 0000 1111
23. **a.** +119 = 0111 0111
b. −119 = 1000 1001
c. −34 = 1101 1110
25. −32 to +31
27. To store and apply the control "s" inputs to the 74181.
29. **a.** Set input switches for A PLUS B "s" code (HLLH), pulse E_I HIGH, pulse $\overline{L_C}$ LOW. Set input switches for A word value, pulse E_I HIGH, $\overline{L_A}$ LOW. Set input switches for B word value, pulse E_I HIGH, pulse $\overline{L_B}$ LOW. Set Logic/$\overline{\text{Arith}}$ input LOW, pulse E_F HIGH, pulse $\overline{L_0}$ LOW.
b. Input words as in **a.**, except code for A or B will be LLLH (M = L) or HHHL (M = H). Set Logic/$\overline{\text{Arith}}$ input to match "s" control code, pulse E_F HIGH, pulse $\overline{L_0}$ LOW.
c. Input words as in **a.**, except code for A and B will be HLHH. Set Logic/$\overline{\text{Arith}}$ input HIGH, pulse E_F HIGH, pulse $\overline{L_0}$ LOW.

Chapter 27

1.	b	**11.**	b
3.	b	**13.**	b
5.	b	**15.**	d
7.	d	**17.**	a
9.	c	**19.**	a

21. Enable ROM IC (Chip Select)
23. When installed in a circuit, this MROM is a read-only device.
25. 2316 is designated as U4 on schematic.
27. U7, pin 15
29. U3, 8085
31. Control read/$\overline{\text{write}}$ control
33. $\overline{CS}$ should be active, $\overline{WE}$ should be HIGH
35. U18 pin 3
37. A_{11}, A_{12}, and A_{13}
39. **a.** ROM **e.** INPUT
b. RAM **f.** SCAN
c. CONTROL **g.** OUTPUT
d. KEY DATA **h.** DISPLAY
41. Display address bus logic levels
43. Yes

Chapter 28

1. c
3. c
5. b
7. b
9. d
11. **a.** Step size $= \dfrac{V_0}{2^n} = \dfrac{12\text{ V}}{16} = 0.75\text{ V}$
b. % Res. $= \dfrac{\text{s/size}}{V_0} = \dfrac{0.75\text{ V}}{12\text{ V}} = 6.25\%$
c. LSB = 12 V/16 = 0.75 V
1 = 12V/8 = 1.5 V
2 = 12V/4 = 3.0 V
MSB = 12V/2 = 6.0 V
d. 0010 = 1.5 V
e. 1011 = 8.25 V
13. **a.** $5\left(\dfrac{64 + 32 + 1}{256}\right) = 1.9\text{ mA}$
b. $5\left(\dfrac{16 + 5}{256}\right) = 0.4\text{ mA}$
c. $5\left(\dfrac{255}{256}\right) = 4.9\text{ mA}$
d. $5\left(\dfrac{16 + 1}{256}\right) = 0.33\text{ mA}$
15. DAC
17.

19. **a.** 0.012 mV
b. 0.0004%
c. LSB = 12 mV, 24 mV, 48 mV, 96 mV, 384 mV, 768 mV, MSB = 1.54 V
d. 60 mV
e. 2.2 V
21. ADC
23. Temperature change will vary dc input voltage to ADC. Equivalent digital output is decoded by PLD and displayed in decimal on 7-segment displays.
25. Flash converter
27.

MSB	LSB
0	0
0	1
1	0
1	1

29. 1010, 1100, 1110, 1111, 1110, 1100, 1001, 0111, 0101, 0011, 0001, 0011, 0101, 0111.

Chapter 29

1.	c	**9.**	c
3.	b	**11.**	b
5.	b	**13.**	b
7.	b	**15.**	c

17. U4 is a 2 k × 8 ROM.
19. A talker (read-only)
21. Together U5 and U6 form a 1 k × 8 RAM.
23. Both talkers (read) and listeners (write)
25. External crystal, determines clock signal frequency output (pin 37)
27. The serial output data goes to U18A (6), and then on to the speaker.
29. Pin 30 (address latch enable, ALE): when active it indicates an address is present at the microprocessor's address and address/data output pins (active-LOW).
    Pin 31 (write control output): indicates to external devices that data on data bus should be written into addressed location (active-LOW).
    Pin 32 (read control output): indicates to external devices that addressed device should place addressed data on data bus (active-LOW).
31. Displays contents of address bus.
33. Displays contents of buffered data bus
35. When U15 is enabled by NOUT, the contents on the data bus will be latched and displayed on the eight common-anode LED display.
37. See section 29-2-1
39. These LEDs show which device has been enabled by the address decoder.

Index